Fundamentals of Mathematical Statistics

S.C. Gupta
Associate Professor (Retd.)
Department of Statistics
Hindu College
University of Delhi, Delhi

V.K. Kapoor
Associate Professor (Retd.)
Department of Mathematics
Shri Ram College of Commerce
University of Delhi, Delhi

Dedicated to the Service of Education
SULTAN
1950

SULTAN CHAND & SONS

SULTAN CHAND & SONS®
23, Daryaganj, New Delhi-110002
Phones : 011-23281876, 23266105, 41625022 (*Showroom & Shop*)
011-23247051 , 40234454 (*Office*)
E-mail : sultanchand74@yahoo.com; info@sultanchandandsons.com
Fax : 011-23266357; Website : www.sultanchandandsons.com

First Edition : September 1970
Twelfth Thoroughly Revised Edition : July 2020; Reprint: 2021, 2022, 2023, 2024, 2025, 2026

ISBN: 978-93-5161-173-8 (TC-499)

Price: ₹ 895.00

Author's Acknowledgement: The writing of a Textbook always involves creation of a huge debt towards innumerable authors and publications. We owe our gratitude to all of them. We acknowledge our indebtedness in extensive footnotes throughout the book. If, for any reason, any acknowledgement has been left out, we beg to be excused. We assure to carry out correction in the subsequent edition as and when it is known.

Printed at: Taj Press, Noida (UP)

Preface to the Twelfth Edition

Knowledge updating is a never-ending process and so should be the revision of an effective textbook. The book originally written **fifty years ago** has, during the intervening period, been revised and reprinted several times. The authors have, however, been thinking, for the last few years that the book needed not only a thorough revision but rather a substantial rewriting. They now take great pleasure in presenting to the readers the **twelfth,** thoroughly revised and enlarged, **Golden Jubilee edition** of the book. The subject-matter in the entire book has been re-written in the light of numerous criticisms and suggestions received from the users of the earlier editions in India and abroad. The basis of this revision has been the emergence of new literature on the subject, the constructive feedback from students and teaching fraternity, as well as those changes that have been made in the syllabi and/or the pattern of examination papers of numerous universities.

Brief recall of some salient features of the present edition:

- The entire text has been restructured, rewritten and updated to cater to the revised syllabi of almost all universities, and various other professional examinations.
- During the course of rewriting, it has been specially borne in mind to retain all the basic features of the previous editions especially the simplicity of presentation, lucidity of style and analytical approach which have been appreciated by the readers throughout the country.
- A conscious effort has been made to make the presentation of all topics discussed very simple and to develop the subject-matter in such step-by-step manner that the book can serve as self-study text for the students.
- A number of typical problems have been added as solved examples in each chapter. These will enable the reader to have a better and thoughtful understanding of the basic concepts of the theory and its various applications.
- Several new topics have been added at appropriate places in various chapters to make the treatment of subject matter more exhaustive and up-to-date.

Some prominent additions are given below:

§ 6·5·1	:	Page 6·11	:	Variance of Degenerate Random Variable
§ 6·5·2	:	Page 6·12	:	Approximate Expression for Expectation and Variance
Remark 6	:	Page 6·40	:	A·M ≥ G.M., Inequality
§ 6·7·4	:	Page 6·42	:	Lyapounov's Inequality
§ 6·7·5	:	Page 6·42	:	C_r Inequality
§ 6·7·6	:	Page 6·44	:	Holder's Inequality
§ 6·7·7	:	Page 6·45	:	Minkowski's Inequality

§ 6·9·1	: Page 6·48	: Double Expectation Rule or Double-E Rule
§ 6·9·2	: Page 6·50	: Conditional Variance
§ 6·9·3	: Page 6·50	: Wald's Theorem on Mean and Variance of a Random Sum
§ 6·9·4	: Page 6·51	: Multistage Rule for Expectation (Multi-stage-E Rule)
§ 7·4	: Page 7·25 to Page 7·27	: Theorem 7·1 to Theorem 7·5
§ 7·5·2	: Page 7·38	: Generalisation of Chebychev's Inequality
§ 7·5·3	: Page 7·39	: Frechet Inequality
§ 7·5·4	: Page 7·39	: Bernstein's Inequality
§ 7·6	: Page 7·40	: Modes of Convergence of Sequence of Random Variables
§ 7·6·1	: Page 7·40 to Page 7·46	: Convergence in Probability Theorems 7·11 to Theorem 7·16
§ 7·6·2	: Page 7·46	: Almost Sure Convergence or Convergence With Probability One
	: Page 7·47	: Theorem 7·17
§ 7·6·3	: Page 7·48	: Convergence in rth-Mean Theorems 7·18 and 7·19
§ 7·6·4	: Page 7·50	: Convergence in Law or Convergence in Distribution Theorems 7·20, 7·20(a), 7·20(b), 7·21
§ 7·6·5	: Page 7·55	: Relationship Between Various Modes of Convergence
§ 7·8	: Page 7·72	: Strong Law of Large Numbers
§ 7·8·1	: Page 7·72	: Kolmogorov's Inequality
§ 7·8·2	: Page 7·75	: Strong Law of Large Numbers (S.L.L.N.)
§ 7·8·3	: Page 7·75	: Kolmogorov's First Theorem on S.L.L.N.
§ 7·8·4	: Page 7·76	: Kolmogorov's Second Theorem on S.L.L.N.
§ 7·10·2	: Page 7·85	: Factorial Moments
§ 7·10·3	: Page 7·85	: Factorial Moment Generating Function
§ 8·4·3 (i)	: Page 8·13	: Factorial Moment Generating Function (Binomial Distribution)
§ 8·6·3	: Page 8·54	: Additive Property of Negative Binomial Distribution
§ 8·9·2	: Page 8·70	: Applications of Multinomial Distribution
Remarks 1, 2, 3	: Page 9·67	: On Cauchy Distribution
§ 9·13	: Page 9·68	: Erlang Distribution
§ 9·13·1	: Page 9·69	: Moments
§ 9·13·2	: Page 9·69	: Uses
§ 9·13·3	: Page 9·70	: Erlang Family of Distributions

§ 9·14	:	Page 9·71	:	Pareto Distribution
§ 9·14·1	:	Page 9·71	:	Moments
§ 9·14·2	:	Page 9·72	:	Conditional Distribution
§ 9·14·3	:	Page 9·73	:	Characterisation of Pareto Distribution
§ 9·14·4	:	Page 9·73	:	Basic Pareto Distribution
§ 9·14·5	:	Page 9·74	:	Applications
§ 9·14·6	:	Page 9·74	:	Relation With Other Distributions
§ 9·15	:	Page 9·75	:	Raleigh Distribution
§ 9·15·1	:	Page 9·76	:	Relation With Other Distributions
§ 9·15·2	:	Page 9·78	:	Moments of Raleigh Distribution
§ 9·15·3	:	Page 9·78	:	Applications and Some Illustrations
§ 11·2·5	:	Page 11·8	:	Explained and Un-explained Variation
§ 12·2·6	:	Page 11·9	:	Coefficient of Determination
§ 12·5·2	:	Page 12·12	:	Coefficient of Multiple Determination
§ 12·6·1	:	Page 12·16	:	Coefficient of Partial Determination
§ 15·6·6	:	Page 15·41	:	Test for Equality of Several Population Proportions
§ 17·6·5	:	Page 17·46	:	Method of Minimum Chi-Square
§ 19·5·5	:	Page 19·43	:	Weddle's Rule
§ 20·2	:	Page 20·17	:	Multivariate Normal (MVN) Distribution
§ 20·2·1	:	Page 20·18	:	MVND p.d.f.
§ 20·2·2	:	Page 20·22	:	M.G.F. of M.V.N.D. N_p (μ, Σ)
§ 20·2·3	:	Page 20·24	:	Marginal Distributions
§ 20·2·4	:	Page 20·25	:	Conditional Distributions
§ 20·2·5	:	Page 20·27	:	Standard M.V.N. Distribution
§ 20·2·6	:	Page 20·29	:	Characterisation of M.V.N. Distribution
§ 21·4	:	Page 21·25	:	Hermite Polynomials
§ 21·4·1	:	Page 21·27	:	Properties of Hermite Polynomials
§ 21·4·2	:	Page 21·30	:	Physicist's Hermite Polynomials
§ 21·4·3	:	Page 21·31	:	Expected Value of Hermite Polynomials
§ 21·5	:	Page 21·32	:	Gram Charlier Type-A Series
§ 21·6	:	Page 21·33	:	Edgeworth Type-A Series
§ 21·7	:	Page 21·36	:	Gram Charlier Polynomials
§ 21·7·1	:	Page 21·38	:	Properties of Gram Charlier Polynomials
§ 21·7·2	:	Page 21·38	:	Gram Charlier Type-B Series

Problem sets at the end of each chapter are substantially reorganised. Many new problems are included at the end of each chapter. Repetition of problems of the same type (more than what is necessary) has been avoided. Further, the problems have been carefully arranged and properly graded according to the level of difficulty.

- Solved examples and unsolved problems have been drawn from the latest examination papers of various universities, professional examinations, etc.

- An attempt has been made to rectify the errors in the previous edition.
- The present edition incorporates modern viewpoints. In fact with the addition of new topics, rewriting and revision of many others and restructuring of problem sets, altogether a new book, covering the revised syllabi of almost all the universities, is being presented to the reader. It is earnestly hoped that, in the new form, the book will prove of much greater utility to the students as well as teachers of the subject and research scholars.

A list of all the individuals to whom we owe gratitude would exhaust readers' patience and intimidate our memory. A less personal but still sincere way is to extend thanks to the students who endured, to the fellow colleagues and associates who contributed, to an environment that inspired, to the authors that permitted, to the reviewers that critiqued, and to our family that waited. To these donors, we dedicate whatever is worthwhile in the book; the shortcomings and vagaries are, however, our responsibility.

We express our deep sense of gratitude to our Publishers, SULTAN CHAND & SONS for their untiring efforts, unfailing courtesy, and co-operation in bringing out the book, in such an elegant form.

Suggestions and constructive criticism for further improvement of the book as well as intimation of errors and misprints will be most gratefully acknowledged. And of course, given complexity of the subject-matter, we crave for readers' indulgence for any omission, error or discrepancy that might have crept in the text despite due care and caution on our part.

July 31, 2020 **S.C. GUPTA • V.K. KAPOOR**

Preface to the First Edition

Although there are a large number of books available covering various aspects in the field of Mathematical Statistics, there is no comprehensive book dealing with the various topics on Mathematical Statistics for the students. The present book is a modest though determined bid to meet the requirements of the students of Mathematical Statistics at Degree, Honours and Post-graduate levels. The book will also be found of use by the students preparing for various competitive examinations. While writing this book our goal has been to present a clear, interesting, systematic and thoroughly teachable treatment of Mathematical Statistics and to provide a textbook which should not only serve as an introduction to the study of Mathematical Statistics but also carry the student on to such a level that he can read with profit the numerous special monographs which are available on the subject. In any branch of Mathematics, it is certainly the teacher who holds the key to successful learning. Our aim in writing this book has been simply to assist the teacher in conveying to the students more effectively a thorough understanding of Mathematical Statistics.

The book contains sixteen chapters. The first chapter is devoted to a concise and logical development of the subject. The second and third chapters deal with the frequency distributions, and measures of average and "dispersion. Mathematical treatment has been given to the proofs of various articles included in these chapters in a very logical and simple manner. The theory of probability which has been developed by the application of the set theory has been discussed quite in detail. A large number of theorems have been deduced using the simple tools of set theory. The simple applications of probability are also given. The chapters on mathematical expectation and theoretical distributions (discrete as well as continuous) have been written keeping the latest ideas in mind. A new treatment has been given to the chapters on correlation, regression and bivariate normal distribution using the concepts of mathematical expectation. The thirteenth and fourteenth chapters deal mainly with the various sampling distributions and the various tests of significance which can be derived from them. In chapter 15, we have discussed concisely statistical inference (estimation and testing of hypothesis). Abundant material is given in the chapter on finite differences and numerical integration. The whole of the relevant theory is arranged in the form of serialised articles which are concise and to the point without being insufficient. The more difficult sections will, in general, be found towards the end of each chapter. We have tried our best to present the subject so as to be within the easy grasp of students with varying degrees of intellectual attainment.

Due care has been taken of the examination needs of the students and, wherever possible, indication of the year, when the articles and problems were set in the examination has been given. While writing this text, we have gone through the syllabi and examination papers of almost all Indian universities where the subject is taught so as to make it as comprehensive as possible. Each chapter contains a large number of carefully graded worked problems mostly drawn from university papers with a view to acquainting the student with the typical questions pertaining to each topic. Furthermore, to assist the student to gain proficiency in the subject, a large number of properly graded problems, mainly drawn from examination papers of various universities, are given at the end of each chapter. The questions and problems given at the end of each chapter, usually require for their solution, a thoughtful use of concepts. During the preparation of the text we have gone through a vast body of literature available on the subject, a list of which is given at the end of the book. It is expected that the bibliography given at the end of the book will considerably help those who want to make a detailed study of the subject.

The lucidity of style and simplicity of expression have been our twin objects to remove the awe which is usually associated with most mathematical and statistical textbooks.

While every effort has been made to avoid printing and other mistakes, we crave for the indulgence of the readers for the errors that might have inadvertently crept in. We shall consider our efforts amply rewarded if those for whom the book is intended are benefited by it. Suggestions for the improvement of the book will be highly appreciated and will be duly incorporated.

September 7, 1970 **S.C. GUPTA • V.K. KAPOOR**

We dedicate this book to

Our respected teacher Prof. Harish Chander Gupta who initiated the teaching of Mathematical Statistics at Post-graduate level at the University of Delhi in the year 1958.

... And with special thanks to our wives, who have cheerfully endured the hardships of an author's life and are always very supportive. We are proud of them.

... And most importantly, we must also put on record our gratitude to two great esteemed institutions, Hindu College and Shriram College of Commerce , (the constituent colleges of University of Delhi), for all that we learnt as a teacher for more than four decades from these premier institutions of South-east Asia

... Last but not the least, thanks to almighty God for giving us the talent to write and share the fruits of this wonderful gift with students.

Brief Contents

Contents

CHAPTER

1 Introduction

(Meaning and Scope)

LEARNING OBJECTIVES

Upon completion of this Chapter, you should be able to:

- Know a short historical development of the subject of Statistics.
- Discuss the two interpretations of the term Statistics, viz.,
  (a) Statistics as numerical data and
  (b) Statistics as statistical methods.
- Give overview of Statistics in various sectors and disciplines.
- Explain the applicability and use of Statistics in various diverse fields.
- Understand various limitations of Statistics.
- Understand the role of computers in solving Statistics problems.

CHAPTER OUTLINES

1·1. Origin and development of Statistics
1·2. Statistics defined
1·3. Importance and scope of Statistics
1·4. Limitations of Statistics
1·5. Distrust of Statistics
1·6. Role of computers in solving statistics problems

1·1. ORIGIN AND DEVELOPMENT OF STATISTICS

The subject of Statistics, as it seems, is not a new discipline but it is as old as the human society itself. Its origin can be traced to the old days when it was regarded as the 'Science of State-craft' and was the by-product of the administrative activity of the State. The word 'Statistics' seems to have been derived from the Latin word '*status*' or the Italian word '*statista*' or the German word '*statistik*' each of which means a 'political state'. In ancient times, the government used to collect the information regarding the population and 'property or wealth' of the country; the former enabling the government to have an idea of the man-power of the country (to safeguard itself against external aggression, if any), and the latter providing it a basis for introducing new taxes and levies.

In India, an efficient system of collecting official and administrative statistics existed even more than 2,000 years ago, in particular, during the reign of Chandra Gupta Maurya (324 – 300 B.C.). From Kautilya's *Arthshastra* it is known that even before 300 B.C. a very good system of collecting 'Vital Statistics' and registration of births and deaths was in vogue. During Akbar's reign (1556 – 1605 A.D.), Raja Todarmal, the then land and revenue minister, maintained good records of land and agricultural statistics. In *Aina-e-Akbari* written by Abul Fazl (in 1596 – 97), one of the nine gems of Akbar, we find detailed accounts of the administrative and statistical surveys conducted during Akbar's reign.

In Germany, the systematic collection of official statistics originated towards the end of the 18th century when, in order to have an idea of the relative strength of different German States, information regarding population and output – industrial and agricultural – was collected. In England, statistics were the outcome of Napoleonic wars. The wars necessitated the systematic collection of numerical data to enable the government to assess the revenues and expenditure with greater precision and then to levy new taxes in order to meet the cost of war.

Seventeenth century saw the origin of the 'Vital Statistics.' Captain John Grant of London (1620 – 1674), known as the 'father' of Vital Statistics, was the first man to study the statistics of births and deaths. Computation of mortality tables and the calculation of expectation of life at different ages by a number of persons, *viz.*, Casper Newman, Sir William Petty (1623 – 1687), James Dodson, Dr. Price, to mention only a few, led to the idea of 'life insurance' and the first life insurance institution was founded in London in 1698.

The theoretical development of the so-called modern statistics came during the mid-seventeenth century with the introduction of *'Theory of Probability'* and *'Theory of Games and Chance'*, the chief contributors being mathematicians and gamblers of France, Germany and England. The French mathematician Pascal (1623 – 1662), after lengthy correspondence with another French mathematician P. Fermat (1601 – 1665) solved the famous *'Problem of Points'* posed by the gambler Chevalier de-Mere. His study of the problem laid the foundation of the theory of probability which is the backbone of the modern theory of statistics. Pascal also investigated the properties of the co-efficients of binomial expansions and also invented mechanical computation machine. Other notable contributors in this field are: James Bernouli (1654 – 1705), who wrote the first treatise on the *'Theory of Probability'*; De-Moivre (1667 – 1754) who also worked on probabilities and annuities and published his important work *"The Doctrine of Chances"* in 1718, Laplace (1749 – 1827) who published in 1782 his monumental work on the theory of probability, and Gauss (1777 – 1855), perhaps the most original of all writers on statistical subjects, who gave the 'Principle of Least Squares' and the 'Normal Law of Errors'. Later on, most of the prominent mathematicians of 18th, 19th and 20th centuries, *viz.*, Euler, Lagrange, Bayes, A. Markoff, Khintchin, Kolmogorov, to mention only a few, have made very outstanding contributions to the modern theory of probability. Modern Stalwarts in the development of the subject of Statistics are Englishmen who did pioneering work in the application of Statistics to different disciplines. Francis Galton (1822 – 1921) pioneered the study of *'Regression Analysis'* in Biometry; Karl Pearson (1857 – 1936) who founded the greatest statistical laboratory in England pioneered the study of *'Correlation Analysis'*. His Chi-square Test (χ^2 – test) of Goodness of fit is the first and most important of the tests of significance in Statistics. W.S. Gosset with his t-test ushered in an era of exact (small) sample tests. Perhaps most of the work in the statistical theory during the past few decades can be attributed to a single person Sir Ronald A. Fisher (1890 – 1962) who applied statistics to a variety of diversified fields such as genetics, biometry, psychology and education, agriculture, etc., and who is rightly termed as the Father of Statistics. In addition to enhancing the existing statistical theory he is the pioneer in *Estimation Theory* (Point Estimation and Fiducial Inference); *Exact Sampling distributions, Analysis of Variance* and *Design of Experiments*. His contributions to the subject of Statistics are described in the following words:

'R.A. Fisher is the real giant in the development of the theory of Statistics'.

The varied and outstanding contributions of R.A. Fisher put the subject of Statistics on a very firm footing and earned for it the status of full-fledged science.

Indian statisticians have also made notable contributions to the development of Statistics in various diversified fields. The valuable contributions of P.C. Mahalanobis and P.V. Sukhatme (Sample Surveys); R.C. Bose, Panse, J.N. Srivastava (Design of Experiments in Agriculture); S.N. Roy (Multivariate Analysis); C.R. Rao (Statistical Inference); Parthasarthy (Theory of Probability), to mention only a few, have earned for India a high position in the world map of Statistics.

1·2. STATISTICS DEFINED

Statistics has been defined differently by different authors from time to time. The reasons for a variety of definitions are primarily two. *First,* in modern times the field of utility of Statistics has widened considerably. In ancient times Statistics was confined only to the affairs of State but now it embraces almost every sphere of human activity. Hence, a number of old definitions which were confined to a very narrow field of enquiry were replaced by new definitions which are much more comprehensive and exhaustive. *Secondly,* Statistics has been defined in two ways; some writers define it as *'statistical data'*, *i.e.*, numerical statement of facts, while others define it as *'statistical methods'*, *i.e.*, complete body of the principles and techniques used in collecting and analysing such data. Some of the important definitions used in plural sense are given below.

STATISTICS AS NUMERICAL DATA — SOME DEFINITIONS

1. *"Statistics are the classified facts representing the conditions of the people in a State...specially those facts which can be stated in number or in tables of numbers or in any tabular or classified arrangement".*
   —*Webster*
2. *"Statistics are numerical statement of facts in any department of enquity placed in relation to each other".*
   –*Bowley.*
3. *"By Statistics we mean quantitative data affected to a marked extent by multiplicity of causes".*
   —*Yule and Kendall*
4. *"Statistics are measurements, enumerations or estimates of natural phenomenon, usually systematically arranged, analysed and presented as to exhibit important interrelationships among them".* **—*A.M. Tuttle***
5. *"Statistics may be defined as the aggregate of facts affected to a marked extent by multiplicity of causes, numerically expressed, enumerated or estimated according to a reasonable standard of accuracy, collected in a systematic manner, for a predetermined purpose and placed in relation to each other".*
   —*Prof. Horace Secrist*

Secrist's definition seems to be the most comprehensive and exhaustive of all the five and highlights the following important characteristics of Statistics.

(*i*) Statistics are aggregates of facts.
(*ii*) Statistics are effected to a marked extent by multiplicity of causes.
(*iii*) Statistics are numerically expressed.
(*iv*) Statistics are enumerated or estimated according to reasonable standards of accuracy.
(*v*) Statistics are collected in a systematic manner for a pre-determined purpose.
(*vi*) Statistics are placed in relation to each other.

We give below the definitions of Statistics used in singular sense.

STATISTICS AS STATISTICAL METHODS — SOME DEFINITIONS

1. (i) *Statistics may be called the science of counting.*
(ii) *Statistics may rightly be called the science of averages.*
(iii) *Statistics is the science of the measurement of social organism, regarded as a whole in all its manifestations.* —**Bowley A.L.**

2. *"Statistics is the science of estimates and probabilities".* —**Boddington**

3. *"Statistics is the science and art of handling aggregate of facts – observing, enumeration, recording, classifying and otherwise systematically treating them".* —**Harlow**

4. *".....Statistical method include all those devices of analysis and synthesis by means of which statistics (data) are scientifically collected and used to explain or describe phenomena either in their individual or related capacities".* —**Secrist**

5. *"Statistics is the branch of scientific method which deals with the data obtained by counting or measuring the properties of population of natural phenomena".* —**Kendall**

6. *Statistics is the science which deals with the method of collecting, classifying, presenting, comparing and interpreting numerical data collected to throw some light on any sphere of inquiry".* —**Seligman**

7. *"Statistics refers to the body of technique or methodology which has been developed for the collection, presentation and analysis of quantitative data and for the use of such data in decision-making".* —**Netter & Wasserman**

8. *"Modern statistics refers to a body of methods and principles that have been developed to handle the collection, description, summarisation and analysis of numerical data. Its primary objective is to assist the researcher in making decisions or generalizations about the nature and characteristics of all the potential observations under consideration of which the collected data form only a small part".* —**Lincoln L. Chao**

9. *"Statistics is a method of decision making in the face of uncertainty on the basis of numerical data and calculated risks".* —**Prof. Ya-Lun-Chou**

10. *"The science of Statistics is the method of judging collective, natural or social phenomenon from the results obtained from the analysis or enumeration or collection of estimates".* —**King**

11. *"Statistics may be regarded as a body of methods for making wise decisions in the face of uncertainty".* —**Wallis and Roberts**

12. *"Statistics may be defined as the science of collection, presentation, analysis and interpretation of numerical data".* —**Croxton and Cowden**

From a logical analysis of the above definitions, it is clear that the definition of Statistics given by Croxton and Cowden is the most scientific and realistic one, and will form the subject-matter of our study.

1·3. IMPORTANCE AND SCOPE OF STATISTICS

In modern times, Statistics is viewed not as a mere device for collecting numerical data but as a means of developing sound techniques for their handling and analysis and drawing valid inferences from them. As such it is not confined to the affairs of the State but is intruding constantly into various diversified spheres of life – social as well as physical – such as biology, psychology, education, economics, business management, etc. It is hardly possible to enumerate even a single department of human activity where statistics does not creep in. It has rather become indispensable in all phases of human endeavour.

We now discuss briefly the importance of Statistics in some different sectors and disciplines.

1. *Statistics in Planning*

Statistics is indispensable to planning. In the modern age which is termed as *"the age of planning'*, almost all organisations in the Government or management of business are resorting to planning for efficient working and for formulating policy decisions. To achieve this end, the statistical data relating to production, consumption, prices, investment, income, expenditure, etc., and various advance statistical techniques for handling and analysing such complex data are of paramount importance.

2. *Statistics in Mathematics*

Statistics is intimately related to and essentially dependent upon mathematics. The modern theory of statistics has its foundations on the theory of probability which in turn is a particular branch of more advanced mathematical theory of Measure and Integration. Moreover, the developments of statistical techniques and theories for application to various sciences-social, physical and natural, are based on fitting different mathematical models to the observed data under certain assumptions and the whole process of such assumptions, analysis and testing is basically mathematical in character, making wide applications of mathematical tools of integration, differentiation, algebra, trigonometry, matrix theory and so on. The main stalwarts in the theory of modern Statistics namely Laplace, Bernoulli, Pascal, De-Moivre, Gauss, Chebychev, Kolmogorov, R.A. Fisher, S.N. Roy, R.C. Bose, H. Cramer to mention only a few of them were primarily talented and skilled mathematicians. Ever increasing role of mathematics in Statistics has resulted in the development of a new branch of Statistics called *'Mathematical Statistics.'* In the words of Connor:

'Statistics is a branch of applied mathematics which specialises in data.'

3. *Statistics in Economics*

Statistical data and techniques of statistical analysis have proved immensely useful in solving a variety of economic problems, such as wages, prices, consumption, production, distribution of income and wealth etc. Statistical tools likes Index Numbers, Time Series Analysis, Demand Analysis and Forecasting Techniques are extensively used for efficient planning and economic development of a country. [For a detailed discussion on these topics, the reader may refer to the book *'Fundamentals of Applied Statistics'* by the same authors.

Empirical studies based on sound statistical analysis have led to the formulation of many economic laws. For example :

(*i*) *'Engel's Law of Consumption'*, (1895) was based on detailed and systematic studies of the family budgets of a number of families.

(ii) *'Pereto's Law of Income Distribution'* is based on the empirical study of the income data of different countries of the world at different times.

(iii) Empirical studies based on the observation of the actual behaviour of the buyers in the market led to *'Revealed Preference Analysis'* of Prof. Samuelson.

The extensive use of mathematics and statistics in the study of economics have led to the development of new disciplines called *Economic Statistics* and *Econometrics.* These days, advanced statistical techniques are used to fit economic models for obtaining optimum results subject to a number of constraints on the resources like capital, labour, production capacity, technology precision, etc.

4. *Statistics in Business and Industry*

Statistics is an indispensable tool of production control also. Business executives are relying more and more on statistical techniques for studying the needs and the desires of the consumers and for many other purposes. The success of a businessman more or less depends upon the accuracy and precision of his statistical forecasting. Wrong expectations, which may be the result of faulty and inaccurate analysis of various causes affecting a particular phenomenon, might lead to his disaster. Suppose a businessman wants to manufacture readymade garments. Before starting with the production process he must have an overall idea as to 'how many garments are to be manufactured', 'how much raw material and labour is needed for that', and 'what is the quality, shape, colour, size, etc., of the garments to be manufactured'. Thus, the formulation of a production plan in advance is a must which cannot be done without having quantitative facts about the details mentioned above. As such most of the large industrial and commercial enterprises are employing trained and efficient statisticians.

Statistical techniques have also been used widely by business organisations in :

(i) Carrying out *Time and Motion* studies (which are a part of the scientific management).

(ii) *Marketing Decisions* (based on the statistical analysis of consumer preference studies — demand analysis).

(iii) *Investment* (based on sound study of individual shares and debentures).

(iv) *Personnel Administration* (for the study of statistical data relating to wages, cost of living, incentive plans, effect of labour dispute/unrest on the production, performance standards, etc.).

(v) *Credit policy.*

(vi) *Inventory Control* (for co-ordination between production and sales).

(vii) *Accounting* (for evaluation of the assets of the business concerns).

(viii) *Sales Control* (through the statistical data pertaining to market studies, consumer preference studies, trade channel studies and readership surveys, etc.) and so on.

Statistical tools of probability and expectation are extremely useful in *Life Insurance* which is one of the pioneer branches of Business and Commerce to use Statistics since the end of the seventeenth century.

In industry, Statistics is very widely used in 'Quality Control'. In production engineering, to find whether the production process is in control and the product is conforming to specifications or not, statistical tools, *viz.*, control charts, inspection plans, etc., are of extreme importance. In inspection plants we have to resort to some kind of sampling – a very important aspect of Statistics.

5. *Statistics and Biology, Astronomy and Medical Science*

The association between statistical methods and biological theories was first studied by Francis Galton in his work in 'Regression'. According to Prof. Karl Pearson, the whole 'theory of heredity' rests on statistical basis. He says, *"The whole problem of evolution is a problem of vital statistics, a problem of longevity, of fertility, of health, of disease and it is impossible for the Registrar General to discuss the national mortality without an enumeration of the population, a classification of deaths and knowledge of statistical theory"*.

In **astronomy**, the theory of Gaussian 'Normal Law of Errors' for the study of the movement of stars and planets is developed by using the 'Principle of Least Squares'.

In **medical science** also, the statistical tools for the collection, presentation and analysis of observed facts relating to the causes and incidence of diseases and the results obtained from the use of various drugs and medicines, are of great importance. Moreover, the efficacy of a manufactured drug or injection or medicine is tested by using the 'tests of significance' – (t-test).

6. *Statistics and Psychology and Education*

In education and psychology, too, Statistics has found wide applications, *e.g.*, to determine the reliability and validity of a test, 'Factor Analysis', etc., so much so that a new subject called 'Psychometry' has come into existence.

7. *Statistics in Research*

Statistics is of great use in the field of research. Most of the advancement in knowledge has taken place because of experiments conducted with the help of statistical methods. Statistical methods are now widely applied in market and production research. This includes advertising and promotional strategies, sales forecasting, decisions regarding introducing new products, etc. Also in scientific research, various advance statistical techniques are extensively used.

8. *Statistics and War*

In war, the theory of '*Decision functions*' can be of great assistance to military and technical personnel to plan 'maximum destruction with minimum effort'.

Thus, we see that the science of Statistics is associated with almost all the sciences – social as well as physical. Bowley has rightly said, *"A knowledge of Statistics is like a knowledge of foreign language or of algebra; it may prove of use at any time under any circumstance"*.

1·4. LIMITATIONS OF STATISTICS

Statistics, with its wide applications in almost every sphere of human activity, is not without limitations. The following are some of its important limitations:

1. *Statistics is not suited to the study of qualitative phenomenon*

Statistics, being a science dealing with a set of numerical data, is applicable to the study of only those subjects of enquiry which are capable of quantitative measurement. As such, qualitative phenomena like honesty, poverty, culture, etc., which cannot be expressed numerically, are not capable of direct statistical analysis. However, statistical techniques may be applied indirectly by first reducing the qualitative expressions to precise quantitative terms. For example, the intelligence of a group of candidates can be studied on the basis of their scores in a certain test.

2. *Statistics does not study individuals*

Statistics deals with an aggregate of objects and does not give any specific recognition to the individual items of a series. Individual items, taken separately, do not constitute statistical data and are meaningless for any statistical enquiry. For example, the individual figures of agricultural production, industrial output or national income of any country for a particular year are meaningless unless, to facilitate comparison, similar figures of other countries or of the same country for different years are given. Hence, statistical analysis is suited to only those problems where group characteristics are to be studied.

3. *Statistical laws are not exact*

Unlike the laws of physical and natural sciences, statistical laws are only approximations and not exact. On the basis of statistical analysis we can talk only in terms of probability and chance and not in terms of certainty. Statistical conclusions are not universally true — they are true only on an average. For example, let us consider the statement: "It has been found that 20% of a certain surgical operations by a particular doctor are successful". The statement does not imply that if the doctor is to operate on 5 persons on any day and four of the operations have proved fatal, the fifth must be a success. It may happen that fifth man also dies of the operation or it may also happen that of the five operations on any day, 2 or 3 or even more may be successful. By the statement we mean that as number of operations becomes larger and larger we should expect, on the average, 20% operations to be successful.

4. *Statistics is liable to be misused*

Perhaps the most important limitation of Statistics is that it must be used by experts. As the saying goes, "Statistical methods are the most dangerous tools in the hands of the inexpert. Statistics is one of those sciences whose adepts must exercise the self-restraint of an artist". The use of statistical tools by inexperienced and untrained persons might lead to very fallacious conclusions. One of the greatest shortcomings of Statistics is that they do not bear on their face the label of their quality and as such can be moulded and manipulated in any manner to support one's way of argument and reasoning.

As King says, "*Statistics are like clay of which one can make a god or devil as one pleases*". The requirement of experience and skill for judicious use of statistical methods restricts their use to experts only and limits the chances of the mass popularity of this useful and important science.

1·5. DISTRUST OF STATISTICS

We often hear the following interesting comments on Statistics :

(*i*) 'An ounce of truth will produce tons of Statistics'.

(*ii*) 'Statistics can prove anything'.

(*iii*) 'Figures do not lie. Liars figure'.

(*iv*) 'If figures say so it can't be otherwise'.

(*v*) 'There are three type of lies – lies, damned lies, and Statistics – wicked in the order of their naming, and so on.

Some of the reasons for the existence of such divergent views regarding the nature and function of Statistics are as follows:

(*i*) Figures are innocent, easily believable and more convincing. The facts supported by figures are psychologically more appealing.

(*ii*) Figures put forward for arguments may be inaccurate or incomplete and thus might lead to wrong inferences.

(*iii*) Figures, though accurate, might be moulded and manipulated by selfish persons to conceal the truth and present a distorted picture of facts to the public to meet their selfish motives. When the skilled talkers, writers or politicians through their forceful writings and speeches; or the business and commercial enterprises through advertisements in the press, mislead the public or belie their expectations by quoting wrong statistical statements or manipulating statistical data for personal motives, the public loses its faith and belief in the science of Statistics and starts condemning it. We cannot blame the layman for his distrust of Statistics, as he, unlike statistician, is not in a position to distinguish between valid and invalid conclusions from statistical statements and analysis.

It may be pointed out that Statistics neither proves anything nor disproves anything. It is only a tool which, if rightly used, may prove extremely useful and if misused, might be disastrous. According to Bowley, *"Statistics only furnishes a tool, necessary though imperfect, which is dangerous in the hands of those who do not know its use and its deficiencies"*. It is not the subject of Statistics that is to be blamed but those people who twist the numerical data and misuse them either due to ignorance or deliberately for personal selfish motives. As King points out, *"Science of Statistics is the most useful servant but only of great value to those who understand its proper use"*.

We discuss below a few interesting examples of misrepresentation of statistical data.

1. A statistical report: "The number of accidents taking place in the middle of the road is much less than the number of accidents taking place on its side. Hence, it is safer to walk in the middle of the road". This conclusion is obviously wrong since we are not given the proportion of the number of accidents to the number of persons walking in the two cases.
2. "The number of students taking up Mathematics Honours in a University has increased 5 times during the last 3 years. Thus, Mathematics is gaining popularity among the students of the university". Again, the conclusion is faulty since we are not given any such details about the other subjects and hence comparative study is not possible.
3. "75% of the people who drink alcohol die before attaining the age of 80 years. Hence drinking is harmful for longevity of life". This statement, too, is incorrect since nothing is mentioned about the number of persons who do not drink alcohol and die before attaining the age of 80 years.

Thus, statistical arguments based on incomplete data often lead to fallacious conclusions.

Remark. Incomplete data usually leads us to fallacious conclusions.

1·6. ROLE OF COMPUTERS IN SOLVING STATISTICS PROBLEMS

The advent of fast-speed computers have substantially changed the field of Statistics in the last few decades. Recent developments in the field of computer technology have enabled Statistics to integrate their models into information systems and thus make Statistics a part of decision-making procedures of many organisations.

Use of a digital computer has become an integral part of the statistical approach to decision-making. The computer may be required due to the complexity of the large volume of data required or the computations to be made. In other words, computer in today's scenario has become an indispensable tool for solving statistical problems. Many statistical techniques are available today in the form of 'canned' programmes.

The statistical problems are time – consuming and involve tedious computations. Even a simple problem sometimes takes a long time to solve manually and even by a hand calculator. For highly complex, real life problems, the task of performing computations, even with a hand calculator, is simply out of question. For this reason many of the techniques were not widely used until 60's. The advent of computers accelerated the wide use of statistical techniques for solving complex problems faced by administrators in business and government. Computers provide the much needed computational support for many of the techniques. The automation of computational algorithm allows decision-makers to concentrate on problem's formulation and the interpretation of the solutions. Major computer manufacturers and vendors have developed software packages for the various computer systems providing computational support for problems to be solved by the application of statistical techniques. Further, commercial software houses and academic departments in universities have also produced software packages for solving the various Statistical problems. Computer manufacturers like IBM, CDC, Honeywell, UNIVAC, ICL, etc., have invested substantial amounts in developing software programs for solving the design of experiments, forecasting, simulation problems etc.

During the past three decades, the field of Statistics has been changed dramatically by the development of computer softwares specially written for statistical analysis. During the 1980's statistical software experienced a vast technological revolution. Besides the usual improvements manifested in periodic updates, the recent technological advances in personal computers led to the development of new packages. In addition, rapid advances in computer hardware meant that larger amounts of computer memory were available at lower cost. This enabled package developers to include additional, more sophisticated statistical procedures in each subsequent version of their packages.

The use of computer softwares, such as SAS and SPSS, have brought about a major technological revolution. The increasing use of spreadsheet packages like Lotus 1-2-3 and Microsoft Excel have led to the incorporation of statistical features in these packages. These softwares have made the task of statistical analysis quite convenient and feasible.

Although statistical softwares have made even the most sophisticated analysis feasible, problem arises when statistically unsophisticated users, who do not understand the assumptions behind procedures or the limitations of the results obtained are misled by computer – generated statistical output.

CHAPTER CONCEPTS QUIZ

1. Comment on the following statements indicating whether they are TRUE or FALSE.

(*i*) Statistics stands merely for business data.

(*ii*) Statistics also stands for methods used in analysing data.

(*iii*) It is necessary that Statistics must be numerically expressed.

(*iv*) Figures cannot be manipulated in Statistics.

(*v*) Statistical techniques are helpful in conducting marketing research as well as for organising advertisement campaigns.

(*vi*) Statistics methods help in simplifying unwieldy mass of data.

(*vii*) Statistics helps in getting an accurate answer for any economic problem.
(*viii*) Analytical statistics deals with application of statistical techniques on a given data.
(*ix*) Statistics does not help in prediction about future.
(*x*) "Statistics is never a lie".
(*xi*) National income data is not statistics.
(*xii*) Statistical laws are exact.
(*xiii*) "Statistics is the science of counting".
(*xiv*) "Statistics conceals more than it reveals".
(*xv*) "All statistics are numerical statements but all numerical statements are net statistics".
(*xvi*) Two Indian statisticians who have made significant contribution in the development of statistics are P.C. Mahalanobis and C.R. Rao.

2. Fill in the blanks :
(*i*) The definition "Statistics is the science which deals with the collection, classification and tabulation of numerical facts as the basis for explanation, description and comparison of phenomena," was given by ....
(*ii*) Statistics can prove ... .
(*iii*) Use of statistical methods is most dangerous in the hands of ...
(*iv*) Statistical analysis helps in the ... of results.
(*v*) Statistics are numerical ... of facts, but all numerical statements are not ... .
(*vi*) Statistics is liable to be ... .
(*vii*) The statement, "Planning is the order of the day and without statistics planning is inconceivable" was given by ... ".
(*viii*) "Statistics is a body of methods for making wise decisions in the face of uncertainty" is the definition of Statistics given by ... and ...
(*ix*) Statistics is not a science, it is a ... .
(*x*) Statistics is not applicable to ... observation.

3. Interpret the following statements :
(*a*) Accidents to planes decreased as the time between overhauls of planes is increased; so the time between overhauls be increased.
(*b*) The estimates of German industrial output and capacity prepared by British statistician on the basis of the serial number of captured equipment were as good as those prepared by Germans themselves.
(*c*) It is three times as dangerous to be a pedestrian while intoxicated as to be a driver. This is shown by the fact that last year 13,943 intoxicated pedestrians were injured and only 4,399 intoxicated drivers.
(*d*) The frequency of divorce for couples with children is only about one-half of that for childless couples; so producing children is an effective check on divorce.
(*e*) The average school teacher in Delhi has become younger in 1978 than 1988.
(*f*) A foreign news journal reports that there is a drop of 400 per cent in farm labour during last two decades, *i.e.*, from 12·7 million in 1951 to 3 million in 1971.

DISCUSSION AND REVIEW QUESTIONS

1. Explain the meaning of the word 'Statistics' as used in different sense. Discuss the scope of the study of the science.
2. Briefly highlight a few major characteristics of Statistics.
3. "Statistics was orginally concerned with matters of State and was regarded as the science of statecraft". Explain the statement.
4. Elucidate the statement: "Statistics are measurements, enumerations or estimates of natural or social phenomena, systematically arranged so as to exhibit their inter-relation". (Connor).

5. (*a*) "Statistics are like clay of which you can make a God or Devil as you please". In the light of this statement discuss the uses and limitations of statistics.
   (*b*) "The science of statistics, then, is a most useful servant but only of great value to those who understand its proper use". (King).
   Comment on the above statement and discuss the limitations of statistics.
6. Define statistics. Discuss its functions, importance and limitations. Explain its use in commerce and economics.
7. "Statistics are straws out of which I, like every other economist, have to make bricks". (*Marshall*). Elucidate this statement and indicate the utility of statistics in economic planning of India.
8. "Statistics affects everybody and touches life at many points. It is both a science and art". Explain the statement with suitable examples.
9. "The term Statistics no longer simply refers to the collection and compilation of data; instead Statistics is often called the science of decision making in the face of uncertainty. It has to do with both the deductive and inductive process, that is, both mathematical and scientific procedures". Fully comment on the statement.
10. (*a*) Write a critical note on the limitations and distrust of statistics. Discuss the important causes of distrust.
    (*b*) "Statistical methods are most dangerous tools in the hands of the inexperts". Discuss and explain the limitations of statistics.
11. "Statistics rose from practical requirements of problems in various spheres and its importance is due to its use in treating such problems". Discuss the statement by giving suitable examples.
12. Critically examine the following statements :
    (*a*) "Statistics can prove anything".
    (*b*) "Statistics can prove nothing".
    (*c*) "There are three degrees of lies – lies, damned lies and Statistics".
    (*d*) "Figures would not lie but liars use figures to lie".
13. "Without adequate understanding of Statistics, the Statistician in social, physical and mathematical sciences may frequently be like the blind man groping in a dark closet for a black cat that is not there". Comment. Also give few examples of the use and abuse of Statistics in business and economics.
14. Briefly explain various 'Divisions of Statistics'.
    Hint:
    (*i*) **Descriptive Statistics.** It deals with the numerical data.
    (*ii*) **Inferential Statistics.** It includes statistical methods which facilitate estimating the characteristic of a population or making decision concerning a population on the basis of sample results. 'Sample' and 'population' are two relative terms. A population is treated as universe and a sample is a fraction (or segment) of that universe.
    (*iii*) **Inductive Statistics.** It covers those methods and techniques which are used to estimate a particular phenomenon on the basis of random observation.
    (*iv*) **Mathematical Statistics.** It is concerned with the application of various mathematical techniques such as differentiation and integration etc., to develop various statistical techniques.
    (*v*) **Applied Statistics.** It covers those methods and techniques which are generally applied to solve specific problems of real life situations.
15. Describe briefly the use of Computers in Staistics.

CHAPTER

2 Descriptive Measures

—Measures of Central Tendency
—Measures of Dispersion
—Measures of Skewness
—Measures of Kurtosis

LEARNING OBJECTIVES

Upon completion of this Chapter, you should be able to:

- Explain and illustrate the concept of a discrete and continuous frequency distribution.
- Describe central tendency.
- Enumerate the requisites of an ideal measure of central tendency or averages.
- Define and compute the various measures of central tendency, viz., arithmetic mean, median, mode, geometric mean and harmonic mean, for different types of data, i.e., (i) raw data, (ii) data arranged in the form of a frequency distribution, (iii) data arranged in the form of grouped or continuous frequency distribution.
- Define different kinds of partition values, viz., quartiles, deciles and percentiles, and compute them
  (i) algebraically, and (ii) graphically.
- Locate the value of mode graphically.
- Know the relationship between different measures of central tendency.
- Summarize the limitations and the uses of different measures of central tendency.
- Explain the concept of dispersion and the significance of measuring it.
- Define and compute different measures of dispersion, viz., range, quartile deviation, mean deviation and standard deviation, for different types of data and learn about their merits and demerits.
- Differentiate between absolute and relative measures of dispersion, and understand their use in comparing variability of different distributions.
- Decide about the use of an appropriate measure of dispersion under different situations.
- Explain the concept of skewness and the significance of its study.

- Define, compute and interpret different measures of skewness—absolute and relative (Karl Pearson's and Bowley's coefficients) — for different types of data.
- Decide about the appropriate measure of skewness in different situations.
- Define and compute raw moments (μ_r') and central moments (μ_r) for different types of data.
- Explain the concept of factorial moments and absolute moments.
- Apply Sheppard's correction (for grouping) for moments.
- Understand the use of moments in studying various characteristics of a distribution.
- Explain the concept of kurtosis.
- Define and compute Karl Pearson's Beta coefficients (β_1 and β_2) and Gamma coefficients (γ_1 and γ_2) and interpret their values regarding skewness and kurtosis of the distribution.

CHAPTER OUTLINES

2·1. INTRODUCTION

Quantitative data in a mass exhibit certain general characteristics or they differ from each other in the following ways :

1. They show a tendency to concentrate at certain values, usually somewhere in the centre of the distribution. Measures of this tendency are called *measures of central tendency* or *averages.*
2. The data vary about a measure of central tendency and these measures of deviation are called *measures of variation or dispersion.*
3. The data in a frequency distribution may fall into symmetrical or asymmetrical patterns. The measures of the direction and degree of asymmetry are called *measures of skewness.*
4. Polygons of frequency distributions exhibit flatness or peakedness of the frequency curves. The measures of peakedness or flatness of the frequency curves are called *measure of kurtosis.*

2·2. FREQUENCY DISTRIBUTION

When observations, discrete or continuous, are available on a single characteristic of a large number of individuals, often it becomes necessary to condense the data as far as possible without losing any information of interest. Let us consider the marks in Statistics obtained by 250 candidates selected at random from among those appearing in a certain examination.

TABLE 2·1 : MARKS IN STATISTICS OF 250 CANDIDATES

32	47	41	51	41	30	39	18	48	53
54	32	31	46	15	37	32	56	42	48
38	26	50	40	38	42	35	22	62	51
44	21	45	31	37	41	44	18	37	47
68	41	30	52	52	60	42	38	38	34
41	53	48	21	28	49	42	36	41	29
30	33	37	35	29	37	38	40	32	49
43	32	24	38	38	22	41	50	17	46
46	50	26	15	23	42	25	52	38	46
41	38	40	37	40	48	45	30	28	31
40	33	42	36	51	42	56	44	35	38
31	51	45	41	50	53	50	32	45	48
40	43	40	34	34	44	38	58	49	28
40	45	19	24	34	47	37	33	37	36
36	32	61	30	44	43	50	31	38	45
46	40	32	34	44	54	35	39	31	48
48	50	43	55	43	39	41	48	53	34
32	31	42	34	34	32	33	24	43	39
40	50	27	47	34	44	34	33	47	42
17	42	57	35	38	17	33	46	36	23
48	50	31	58	33	44	26	29	31	37
47	55	57	37	41	54	42	45	47	43
37	52	47	46	44	50	44	38	42	19
52	45	23	41	47	33	42	24	48	39
48	44	60	38	38	44	38	43	40	48

This representation of the data does not furnish any useful information and is rather confusing to mind. A better way may be to express the figures in an ascending or descending order of magnitude, commonly termed as *array*. But this does not reduce the bulk of the data. A much better representation is given in *Table 2·2* :

TABLE 2·2 : MARKS IN STATISTICS OF 250 CANDIDATES

Marks	*No. of Students — Tally Marks*	*Total Frequency*	*Marks*	*No. of Students — Tally Marks*	*Total Frequency*
15	II	= 2	40	𝍸 𝍸 I	= 11
17	III	= 3	41	𝍸 𝍸	= 10
18	II	= 2	42	𝍸 𝍸 III	= 13
19	II	= 2	43	𝍸 III	= 8
21	II	= 2	44	𝍸 𝍸 II	= 12
22	II	= 2	45	𝍸 II	= 7
23	III	= 3	46	𝍸 II	= 7
24	IIII	= 4	47	𝍸 III	= 8
25	I	= 1	48	𝍸 𝍸 II	= 12
26	III	= 3	49	III	= 3
27	I	= 1	50	𝍸 𝍸	= 10
28	III	= 3	51	IIII	= 4
29	II	= 2	52	𝍸	= 5
30	𝍸	= 5	53	IIII	= 4
31	𝍸 𝍸	= 10	54	III	= 3
32	𝍸 𝍸	= 10	55	II	= 2
33	𝍸 III	= 8	56	II	= 2
34	𝍸 𝍸 I	= 11	57	II	= 2
35	𝍸	= 5	58	II	= 2
36	𝍸	= 5	60	III	= 3
37	𝍸 𝍸 II	= 12	61	I	= 1
38	𝍸 𝍸 𝍸 II	= 17	62	I	= 1
39	𝍸 I	= 6	68	I	= 1

A bar (I) called *tally mark* is put against the number when it occurs. Having occurred four times, the fifth occurrence is represented by putting a cross tally (/) on the first four tallies. This technique facilitates the counting of the tally marks at the end.

The representation of the data as above is known as *frequency distribution*. Marks are called the *variable* (x) and the *'number of students'* against the marks is known as the *'frequency'* (f) of the variable. The word *'frequency'* is derived from 'how frequently' a variable occurs. For example, in the above case the frequency of 31 is 10 as there are ten students getting 31 marks. This representation, though better than an 'array', does not condense the data much and it is quite cumbersome to go through this huge mass of data.

If the identity of the individuals about whom a particular information is taken is not relevant, nor the order in which the observations arise, then the first real step of condensation is to divide the observed range of variable into a suitable number of class-intervals and to record the number of observations in each class. For example, in the above case, the data may be expressed as shown in *Table 2·3*.

TABLE 2·3 : FREQUENCY TABLE

Marks (x)	*No. of Students* (f)
15 — 19	9
20 — 24	11
25 — 29	10
30 — 34	44
35 — 39	45
40 — 44	54
45 — 49	37
50 — 54	26
55 — 59	8
60 — 64	5
65 — 69	1
Total	250

Such a table showing the distribution of the frequencies in the different classes is called a *frequency table* and the manner in which the class frequencies are distributed over the class intervals is called the *grouped frequency distribution* of the variable.

Remark. The classes of the type 15—19, 20—24, 25—29 etc., in which both the upper and lower limits are included are called *'inclusive classes'*. For example, the class 20—24, includes all the values from 20 to 24, both inclusive and the classification is termed as *inclusive type classification.*

Inspite of great importance of classification in statistical analysis, no hard and fast rules can be laid down for it. The following points may be kept in mind for classification :

1. The classes should be clearly defined and should not lead to any ambiguity.
2. The classes should be exhaustive, *i.e.*, each of the given values should be included in one of the classes.
3. The classes should be mutually exclusive and non-overlapping.
4. The classes should be of equal width. The principle, however, cannot be rigidly followed. If the classes are of varying width, the different class frequencies will not be comparable. Comparable figures can be obtained by dividing the value of the frequencies by the corresponding widths of the class intervals. The ratios thus obtained are called 'frequency densities'.
5. Indeterminate classes, *e.g.*, the open-end classes like less than *'a'* or greater than *'b'* should be avoided as far as possible since they create difficulty in analysis and interpretation.
6. The number of classes should neither be too large nor too small. It should preferably lie between 5 and 15. However, the number of classes may be more than 15 depending upon the total frequency and the details required, but it is desirable that it is not less than 5 since in that case the classification may not reveal the essential characteristics of the population. The following formula due to *Struges* may be used to determine an approximate number k of classes :

$$k = 1 + 3{\cdot}322 \log_{10} N, \text{ where } N \text{ is the total frequency.}$$

The Magnitude of the Class Interval. Having fixed the number of classes, divide the range (the difference between the greatest and the smallest observation) by it and the nearest integer to this value gives the magnitude of the class interval. Broad class intervals (i.e., less number of classes) will yield only rough estimates while for high degree of accuracy small class intervals (*i.e.*, large number of classes) are desirable.

Class Limits. The class limits should be chosen in such a way that the mid-value of the class interval and actual average of the observations in that class interval are as near to each other as possible. If this is not the case then the classification gives a distorted picture of the characteristics of the data. If possible, class limits should be located at the points which are multiple of 0, 2, 3, 10, ... etc., so that the midpoints of the classes are the common figures, *viz.*, 0, 2, 5, 10, ... etc., the figures capable of easy and simple analysis.

2·2·1. Continuous Frequency Distribution.

If we deal with a continuous variable, it is not possible to arrange the data in the class intervals of above type.

Let us consider the distribution of age in years. If class intervals are 15—19, 20—24, etc. then the persons with ages between 19 and 20 years are not taken into consideration. In such a case, we form the class intervals as shown in the adjoining *Table* 2·4.

TABLE 2·4

Age (in years)
Below 5
5 or more but less than 10
10 or more but less than 15
15 or more but less than 20
20 or more but less than 25
and so on.

Here all the persons with any fraction of age are included in one group or the other. For practical purposes we re-write the above classes as shown in the adjoining *Table* 2·4 (*a*).

TABLE 2·4 (*a*)

0 — 5
5 — 10
10 — 15
15 — 20
20 — 25

The form of the frequency distribution with such classes is known as *continuous frequency distribution*. It should be clearly understood that in the above classes, the upper limits of each class are excluded from the respective classes. Such classes in which the upper limits of each class are excluded from the respective classes and are included in the immediate next class are known as *'exclusive classes'* and the classification is termed as *'exclusive type classification'*.

2·3. GRAPHIC REPRESENTATION OF A FREQUENCY DISTRIBUTION

It is often useful to represent a frequency distribution by means of a diagram which makes the unwieldy data intelligible and conveys to the eye the general run of the observations. Diagrammatic representation also facilitates the comparison of two or more frequency distributions. We consider below some important types of graphic representation.

2·3·1. Histogram

In drawing the histogram of a given continuous frequency distribution we first mark off along the x-axis all the class intervals on a suitable scale. On each class interval erect rectangles with heights proportional to the frequency of the corresponding class interval so that the area of the rectangle is proportional to the frequency of the class. If, however, the classes are of unequal width then the height of the rectangle will be proportional to the ratio of the frequencies to the width of the classes. The diagram of continuous rectangles so obtained is called *histogram.*

Remarks:

1. To draw the histogram for an ungrouped frequency distribution of a variable we shall have to assume that the frequency corresponding to the variate value x is spread over the interval $x - \frac{h}{2}$ to $x + \frac{h}{2}$, where h is the jump from one value to the next.
2. If the grouped frequency distribution is not continuous, first it is to be converted into continuous distribution and then the histogram is drawn.
3. Although the height of each rectangle is proportional to the frequency of the corresponding class, the height of a fraction of the rectangle is not proportional to the frequency of the corresponding fraction of the class, so that histogram cannot be directly used to read frequency over a fraction of a class interval.
4. The histogram of the distribution of marks of 250 students in *Table 2·3* (page 2·6) is obtained in Fig. 2·1.

Since the grouped frequency distribution is not continuous, we first convert it into a continuous distribution with exclusive type classes as given in the following *Table 2·5* :

TABLE 2·5

Marks	*No. of Students*
14·5 — 19·5	9
19·5 — 24·5	11
24·5 — 29·5	10
29·5 — 34·5	44
34·5 — 39·5	45
39·5 — 44·5	54
44·5 — 49·5	37
49·5 — 54·5	26
54·5 — 59·5	8
59·5 — 64·5	5
64·5 — 69·5	1

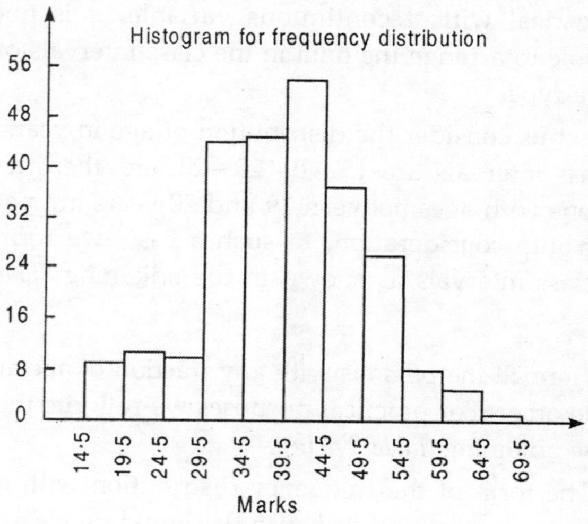

Fig. 2·1

Note. The upper and lower class limits of the *new exclusive type classes* are known as *class boundaries*.

If d is the gap between the upper limit of any class and the lower limit of the succeeding class, the class boundaries for any class are then given by :

Upper class boundary = Upper class limit + $(d/2)$; Lower class boundary = Lower class limit – $(d/2)$.

2·3·2. Frequency Polygon

For an ungrouped distribution, the frequency polygon is obtained by plotting points with abscissa as the variate values and the ordinate as the corresponding frequencies and joining the plotted points by means of straight lines. For a grouped frequency distribution, the abscissa of points are mid-values of the class intervals. For equal class intervals the frequency polygon can be obtained by joining the middle points of the upper sides of the adjacent rectangles of the histogram by means of straight lines. If the class intervals are of small width, the polygon can be obtained by drawing a smooth freehand curve through the vertices of the frequency polygon.

The frequency polygon so obtained should be extended to the base line (x-axis) at both the ends so that it meets the x-axis at the mid-points of two hypothetical classes, *viz.*, the class before the first class and the class after the last class, each assumed to have zero frequency.

2·4. AVERAGES (OR MEASURES OF CENTRAL TENDENCY)

According to Professor Bowley, averages are "*statistical constants which enable us to comprehend in a single effort the significance of the whole.*" They give us an idea about the concentration of the values in the central part of the distribution and hence the name measures of central tendency. Plainly speaking, an average of a statistical series is the value of the variable which is representative of the entire distribution. The following are the five measures of central tendency that are in common use :

(i) Arithmetic Mean or Simple Mean, *(ii) Median,* *(iii) Mode,*
(iv) Geometric Mean, and *(v) Harmonic Mean.*

2·4·1. Requisites for an Ideal Measure of Central Tendency

According to Professor Yule, the following are the characteristics to be satisfied by an ideal measure of central tendency :

(i) It should be rigidly defined.
(ii) It should be readily comprehensible and easy to calculate.
(iii) It should be based on all the observations.
(iv) It should be suitable for further mathematical treatment. By this we mean that if we are given the averages and sizes of a number of series, we should be able to calculate the average of the composite series obtained on combining the given series.
(v) It should be affected as little as possible by fluctuations of sampling.

In addition to the above criteria, we may add the following (which is not due to Prof. Yule).

(vi) It should not be affected much by extreme values.

2·5. ARITHMETIC MEAN

Arithmetic mean of a set of observations is their sum divided by the number of observations, *e.g.*, the arithmetic mean $\bar{x}$ of n observations $x_1, x_2, ..., x_n$ is given by :

$$\bar{x} = \frac{1}{n}(x_1 + x_2 + ... + x_n) = \frac{1}{n}\sum_{i=1}^{n} x_i \qquad ... (2{\cdot}1)$$

In case of the frequency distribution $x_i \mid f_i$, $i = 1, 2, ..., n$ where f_i is the frequency of the variable x_i,

$$\bar{x} = \frac{f_1x_1 + f_2x_2 + ... + f_nx_n}{f_1 + f_2 + ... + f_n} = \frac{\sum_{i=1}^{n} f_i x_i}{\sum_{i=1}^{n} f_i} = \frac{1}{N}\sum_{i=1}^{n} f_i x_i, \quad \left[\sum_{i=1}^{n} f_i = N\right] \qquad ... (2{\cdot}1a)$$

In case of grouped or continuous frequency distribution, x is taken as the mid-value of the corresponding class.

Remark. The symbol Σ is the letter capital sigma of the Greek alphabet and is used in mathematics to denote the sum of values.

Example 2·1. *(a) Find the arithmetic mean of the following frequency distribution :*

x :	1	2	3	4	5	6	7
f :	5	9	12	17	14	10	6

(*b*) *Calculate the arithmetic mean of the marks from the following table :*

Marks	:	*0—10*	*10—20*	*20—30*	*30—40*	*40—50*	*50—60*
No. of Students	:	*12*	*18*	*27*	*20*	*17*	*6*

Solution. (*a*)

TABLE 2·6 : COMPUTATION OF MEAN

x	f	fx
1	5	5
2	9	18
3	12	36
4	17	68
5	14	70
6	10	60
7	6	42
Total	73	299

$$\bar{x} = \frac{1}{N}\Sigma fx = \frac{299}{73} = 4.09$$

(*b*)

TABLE 2·7 : COMPUTATION OF MEAN

Marks	*No. of Students* (*f*)	*Mid-point* (x)	fx
0 — 10	12	5	60
10 — 20	18	15	270
20 — 30	27	25	675
30 — 40	20	35	700
40 — 50	17	45	765
50 — 60	6	55	330
Total	100		2,800

$$\text{Arithmetic mean } (\bar{x}) = \frac{1}{N}\Sigma fx = \frac{1}{100}\times 2,800 = 28$$

It may be noted that if the values of x or (and) f are large, the calculation of mean by formula (2·1*a*) is quite time-consuming and tedious. The arithmetic is reduced to a great extent by taking the deviations of the given values from any arbitrary point '*A*' as explained below :

Let $d_i = x_i - A$. Then $f_i d_i = f_i(x_i - A) = f_i x_i - A f_i$

Summing both sides over i from 1 to n, we get

$$\sum_{i=1}^{n} f_i d_i = \sum_{i=1}^{n} f_i x_i - A \sum_{i=1}^{n} f_i = \sum_{i=1}^{n} f_i x_i - A \, . \, N$$

$$\Rightarrow \quad \frac{1}{N}\sum_{i=1}^{n} f_i d_i = \frac{1}{N}\sum_{i=1}^{n} f_i x_i - A = \bar{x} - A,$$

where $\bar{x}$ is the arithmetic mean of the distribution.

$$\therefore \quad \bar{x} = A + \frac{1}{N}\sum_{i=1}^{n} f_i d_i \qquad \text{... (2·2)}$$

This formula is much more convenient to apply than formula (2·1*a*).

Any number can serve the purpose of arbitrary point '*A*' but, usually, the value of x corresponding to the middle part of the distribution will be much more convenient.

In case of grouped or continuous frequency distribution, the arithmetic is reduced to still greater extent by taking $d_i = (x_i - A)/h$, where A is an arbitrary point and h is the common magnitude of class interval. In this case, we have $hd_i = x_i - A$ and proceeding exactly similarly as above, we get

$$\bar{x} = A + \frac{h}{N}\sum_{i=1}^{n} f_i d_i \qquad \text{... (2·3)}$$

Example 2·2. *Calculate the mean for the following frequency distribution :*

Class interval :	0—8	8—16	16—24	24—32	32—40	40—48
Frequency :	8	7	16	24	15	7

Solution. Here we take $A = 28$ and $h = 8$.

TABLE 2·8 : COMPUTATION OF MEAN

Class interval	*Mid-value* (x)	*Frequency* (f)	$d = \frac{x - A}{h}$	fd
0—8	4	8	–3	–24
8—16	12	7	–2	–14
16—24	20	16	–1	–16
24—32	28	24	0	0
32—40	36	15	1	15
40—48	44	7	2	14
Total		77		–25

$$\bar{x} = A + \frac{h\,\Sigma fd}{N} = 28 + \frac{8 \times (-25)}{77} = 28 - \frac{200}{77} = 25.404$$

2·5·1. Properties of Arithmetic Mean

Property 1. *Algebraic sum of the deviations of a set of values from their arithmetic mean is zero. If* $x_i \mid f_i$; $i = 1, 2, ..., n$ *is the frequency distribution, then*

$$\sum_{i=1}^{n} f_i(x_i - \bar{x}) = 0, \ \bar{x} \text{ being the mean of distribution}$$

Proof.
$$\sum_i f_i(x_i - \bar{x}) = \sum_i f_i x_i - \bar{x}\sum_i f_i = \sum_i f_i x_i - \bar{x}\,.\,N \qquad \text{... (*)}$$

Also
$$\bar{x} = \frac{\sum_i f_i x_i}{N} \quad \Rightarrow \quad \sum_i f_i x_i = N\,\bar{x}$$

Substituting in (*), we get :
$$\sum_{i=1}^{n} f_i(x_i - \bar{x}) = N\,.\,\bar{x} - \bar{x}\,.\,N = 0$$

Property 2. *The sum of the squares of the deviations of a set of values is minimum when taken about mean.*

Proof. For frequency distribution $x_i \mid f_i$, $i = 1, 2, ..., n$, let $Z = \sum_{i=1}^{n} f_i(x_i - A)^2$, be the sum of the squares of the deviations of given values from any arbitrary point 'A'. We have to prove that Z is minimum when $A = \overline{x}$.

Applying the principles of maxima and minima from differential calculus, Z will be minimum for variations in A if $\frac{\partial Z}{\partial A} = 0$ and $\frac{\partial^2 Z}{\partial A^2} > 0$.

$$\frac{\partial Z}{\partial A} = -2 \sum_i f_i(x_i - A) = 0 \quad \Rightarrow \quad \sum_i f_i(x_i - A) = 0$$

$$\Rightarrow \quad \sum f_i x_i - A \sum_i f_i = 0 \quad \text{or} \quad A = \frac{\sum_i f_i x_i}{N} = \overline{x}$$

Again,
$$\frac{\partial^2 Z}{\partial A^2} = -2 \sum_i f_i(-1) = 2 \sum_i f_i = 2N > 0$$

Hence, Z is minimum at the point $A = \overline{x}$. This establishes the result.

Property 3. *(Mean of the composite series). If $\overline{x}_i$ $(i = 1, 2, ..., k)$ are the means of k component series of sizes n_i, $(i = 1, 2, ..., k)$ respectively, then the mean $\overline{x}$ of the composite series obtained on combining the component series is given by the formula :*

$$\overline{x} = \frac{n_1 \overline{x}_1 + n_2 \overline{x}_2 + ... + n_k \overline{x}_k}{n_1 + n_2 + ... + n_k} = \left[\sum_{i=1}^{k} n_i x_i \middle/ \sum_{i=1}^{k} n_i\right. \qquad ... (2·4)$$

Proof. Let $x_{11}, x_{12}, ..., x_{1n1}$ be n_1 members of the first series ; $x_{21}, x_{22}, ..., x_{2n2}$ be n_2 members of the second series, ..., $x_{k1}, x_{k2}, ..., x_{knk}$ be n_k members of the kth series.

Then, by *def.*,

$$\left.\begin{aligned} \overline{x}_1 &= \frac{1}{n_1}(x_{11} + x_{12} + ... + x_{1n_1}) \\ \overline{x}_2 &= \frac{1}{n_2}(x_{21} + x_{22} + ... + x_{2n_2}) \\ &\vdots \qquad \vdots \qquad \vdots \\ \overline{x}_k &= \frac{1}{n_k}(x_{k1} + x_{k2} + ... + x_{kn_k}) \end{aligned}\right\} \qquad ... (*)$$

The mean $\overline{x}$ of composite series of size $n_1 + n_2 + ... + n_k$ is given by

$$\overline{x} = \frac{(x_{11} + x_{12} + ... + x_{1n_1}) + (x_{21} + x_{22} + ... + x_{2n_2}) + (x_{k1} + x_{k2} + ... + x_{kn_k})}{n_1 + n_2 + ... + n_k}$$

$$= \frac{n_1 \overline{x}_1 + n_2 \overline{x}_2 + ... - n_k \overline{x}_k}{n_1 + n_2 + ... + n_k} = \sum_i n_i \overline{x}_i \,/\, (\sum_i n_i) \qquad \text{[From (*)]}$$

Example 2·3. *The average weekly salary of male employees in a firm was ₹ 5,200 and that of females was ₹ 4,200. The mean salary of all the employees was ₹ 5,000. Find the percentage of male and female employees.*

Solution. Let n_1 and n_2 denote respectively the number of male and female employees in the firm, and $\overline{x}_1$ and $\overline{x}_2$ denote respectively their average salary (in rupees). Let $\overline{x}$ denote the average salary of all the workers in the firm.

We are given that : $\overline{x}_1 = 5,200, \quad \overline{x}_2 = 4,200 \quad \overline{x} = 5,000.$

Also we know $\bar{x} = \dfrac{n_1 \bar{x}_1 + n_2 \bar{x}_2}{n_1 + n_2}$ $\Rightarrow$ $5{,}000(n_1 + n_2) = 5{,}200 n_1 + 4{,}200 n_2$

$\Rightarrow$ $(5{,}200 - 5{,}000)n_1 = (5{,}000 - 4{,}200)n_2$ $\Rightarrow$ $20 n_1 = 80 n_2$ or $\dfrac{n_1}{n_2} = \dfrac{4}{1}$.

$\therefore$ The percentage of male employees in the firm $= \dfrac{4}{4+1} \times 100 = 80$

and the percentage of female employees in the firm $= \dfrac{1}{4+1} \times 100 = 20$.

2·5·2. Merits and Demerits of Arithmetic Mean

Merits	*Demerits*
1. It is rigidly defined.	1. It cannot be determined by inspection nor it can be located graphically.
2. It is easy to understand and easy to calculate.	2. Arithmetic mean cannot be used if we are dealing with qualitative characteristics which cannot be measured quantitatively; such as, intelligence, honesty, beauty, etc.
3. It is based upon all the observations.	3. Arithmetic mean cannot be obtained if a single observation is missing or lost or is illegible unless we drop it out and compute the arithmetic mean of the remaining values.
4. It is amenable to algebraic treatment. The mean of the composite series in terms of the means and sizes of the component series is given by : $\bar{x} = \sum_{i=1}^{k} n_i \bar{x}_i / (\sum_{i=1}^{k} n_i)$	4. Arithmetic mean is affected very much by extreme values. In case of extreme items, arithmetic mean gives a distorted picture of the distribution and no longer remains representative of the distribution.
5. Of all the averages, arithmetic mean is affected least by fluctuations of sampling. This property is sometimes described by saying that arithmetic mean is a *stable* average. Thus, we see that arithmetic mean satisfies all the properties laid down by Prof. Yule for an ideal average.	5. Arithmetic mean may lead to wrong conclusions if the details of the data from which it is computed are not given. Let us consider the following marks obtained by two students *A* and *B* in three tests, *viz.*, terminal test, half-yearly examination and annual examination respectively. *Marks* — *I Test*, *II Test*, *III Test*, *Average* A — 50%, 60%, 70%, 60% B — 70%, 60%, 50%, 60% Thus average marks obtained by each of the two students at the end of the year are 60%. If we are given the average marks alone we conclude that the level of intelligence of both the students at the end of the year is same. This is a fallacious conclusion since we find from the data that student *A* has improved consistently while student *B* has deteriorated consistently.
	6. Arithmetic mean cannot be calculated if the extreme class is open, *e.g.*, below 10 or above 90.
	7. In extremely asymmetrical (skewed) distribution, usually arithmetic mean is not a suitable measure of location.

2·5·3. Weighted Mean

In calculating arithmetic mean we suppose that all the items in the distribution have equal importance. But in practice this may not be so. If some items in a distribution are more important than others, then this point must be borne in mind, in order that average computed is representative of the distribution. In such cases, proper weightage is to be given to various items; the weights attached to each item being proportional to the importance of the item in the distribution. For example if we want to have an idea of the change in cost of living of a certain group of people, then the simple mean of the prices of the commodities consumed by them will not do, since all the commodities are not equally important, *e.g.*, wheat, rice and pulses are more important than cigarettes, tea, confectionery, etc.

Let w_i be the weight attached to the item x_i, $i = 1, 2, ..., n$. Then we define :

$$\text{Weighted arithmetic mean (or weighted mean)} = \sum_i w_i x_i \Big/ \sum_i w_i \qquad \text{... (2·5)}$$

It may be observed that the formula for weighted mean is the same as the formula for simple mean with f_i, $(i = 1, 2, ..., n)$, *the frequencies replaced by* w_i, $(i = 1, 2, ..., n)$, *the weights.*

Weighted mean gives the result equal to the simple *mean* if the weights assigned to each of the variate values are equal. It results in higher value than the simple mean if smaller weights are given to smaller items and larger weights to larger items. If the weights attached to larger items are smaller and those attached to smaller items are larger, then the weighted mean results in smaller value than the simple mean.

Example 2·4. *Find the simple and weighted arithmetic mean of the first n natural numbers, the weights being the corresponding numbers.*

Solution. The first n natural numbers are : 1, 2, 3, ..., n. We know that

X	w	wX
1	1	1^2
2	2	2^2
3	3	3^2
.	.	.
.	.	.
.	.	.
n	n	n^2

$$1 + 2 + 3 + ... + n = \frac{n(n+1)}{2}$$

$$1^2 + 2^2 + 3^2 + ... + n^2 = \frac{n(n+1)(2n+1)}{6}$$

$$\text{Simple A.M. } (\bar{X}) = \frac{\Sigma X}{n} = \frac{1+2+3+...+n}{n} = \frac{n+1}{2}$$

$$\text{Weighted A.M. } (\bar{X}_w) = \frac{\Sigma\, w\, X}{\Sigma\, w} = \frac{1^2+2^2+...+n^2}{1+2+...+n} = \frac{n(n+1)(2n+1)}{6} \cdot \frac{2}{n(n+1)} = \frac{2n+1}{3}$$

2·6. MEDIAN

Median of a distribution is the value of the variable which divides it into two equal parts. It is the value which exceeds and is exceeded by the same number of observations, *i.e.*, it is the value such that the number of observations above it is equal to the number of observations below it. The median is thus a *positional* average.

In case of ungrouped data, if the number of observations is odd then median is the middle value after the values have been arranged in ascending or descending order of magnitude. In case of even number of observations, there are two middle terms and median is obtained by taking the arithmetic mean of the middle terms. For example, the median of the values 25, 20, 15, 35, 18, *i.e.*, 15, 18, 20, 25, 35 is 20 and the median of 8, 20, 50, 25, 15, 30, *i.e.*, of 8, 15, 20, 25, 30, 50 is $\frac{1}{2}(20 + 25) = 22·5$.

Remark. In case of even number of observations, in fact any value lying between the two middle values can be taken as median but conventionally we take it to be the mean of the middle terms.

In case of discrete frequency distribution median is obtained by considering the cumulative frequencies. The steps for calculating median are given below :

(*i*) Find $\frac{1}{2}N$, where $N = \sum_i f_i$.

(*ii*) See the (less than) cumulative frequency (*c.f.*) just greater than $\frac{N}{2}$.

(*iii*) The corresponding value of x is median.

Example 2·5. *Obtain the median for the following frequency distribution :*

x :	1	2	3	4	5	6	7	8	9
f :	8	10	11	16	20	25	15	9	6

Solution. TABLE 2·9 : COMPUTATION OF MEDIAN

x	*f*	*c.f.*
1	8	8
2	10	18
3	11	29
4	16	45
5	20	65
6	25	90
7	15	105
8	9	114
9	6	120
Total	N = 120	

Here $N = 120 \quad \Rightarrow \quad \frac{N}{2} = 60$

The cumulative frequency (*c.f.*) just greater than $\frac{1}{2}N$ is 65 and the value of x corresponding to 65 is 5. Therefore, median is 5.

2·6·1. Median for Continuous Frequency Distribution

In the case of continuous frequency distribution, the class corresponding to the *c.f.* just greater than (N/2) is called the *median class* and the value of median is obtained by the following formula :

$$\text{Median} = l + \frac{h}{f}\left(\frac{N}{2} - c\right) \qquad \text{... (2·6)}$$

where l is the lower limit of the median class,

f is the frequency of the median class,

h is the magnitude of the median class,

'c' is the *c.f.* of the class *preceding* the median class,

and $N = \Sigma f$.

Derivation of the Median Formula (2·6). Let us consider the following continuous frequency distribution, $(x_1 < x_2 < ... < x_{n+1})$:

Class interval :	$x_1 - x_2$	$x_2 - x_3$	...	$x_k - x_{k+1}$	...	$x_n - x_{n+1}$
Frequency :	f_1	f_2	...	f_k	...	f_n

The cumulative frequency distribution is given by :

Class interval :	$x_1 - x_2$	$x_2 - x_3$	...	$x_k - x_{k+1}$	...	$x_n - x_{n+1}$
Frequency :	F_1	F_2	...	F_k	...	F_n

where $F_i = f_1 + f_2 + ... + f_i$. The class $x_k - x_{k+1}$ is the median class if and only if $F_{k-1} < \frac{1}{2}N < F_k$.

If we assume that the variate values are uniformly distributed over the median-class which implies that the ogive is a straight line in the median-class, then we get from the *Fig.* 2·2.

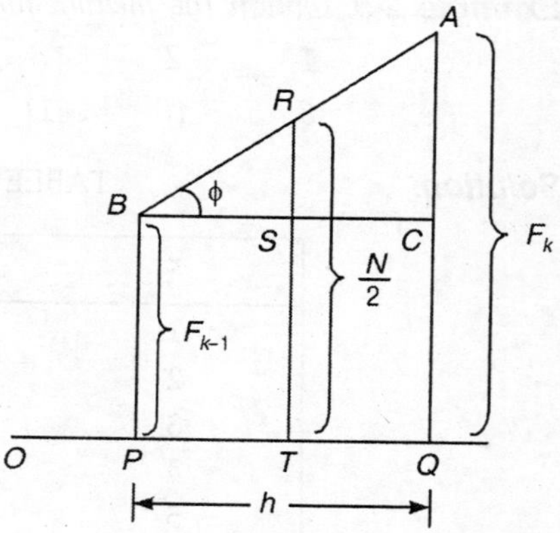

Fig. 2·2

$$\tan\phi = \frac{RS}{BS} = \frac{AC}{BC}$$

i.e.,
$$\frac{RT - TS}{BS} = \frac{AQ - CQ}{BC}$$

or
$$\frac{RT - BP}{BS} = \frac{AQ - BP}{PQ}$$

or
$$\frac{\frac{1}{2}N - F_{k-1}}{BS} = \frac{F_k - F_{k-1}}{PQ} = \frac{f_k}{h},$$

where f_k is the frequency and h the magnitude of the median class.

$$\therefore \quad BS = \frac{h}{f_k}\left(\frac{N}{2} - F_{k-1}\right)$$

Hence

$$\text{Median} = OT = OP + PT = OP + BS = l + \frac{h}{f_k}\left(\frac{N}{2} - F_{k-1}\right),$$ which is the required formula.

Remark. The median formula (2.6) can be used only for continuous classes without any gaps, *i.e.*, for *'exclusive type'* classification. If we are given a frequency distribution in which classes are of *'inclusive type'* with gaps, then it must be converted into a continuous *'exclusive type'* frequency distribution without any gaps before applying (2·6). This will affect the value of l in (2·6). As an illustration see *Example* 2·7.

Example 2·6. *Find the median wage of the following distribution :*

Wages (in ₹) :	2,000—3,000	3,000—4,000	4,000—5,000	5,000—6,000	6,000—7,000
No. of workers :	3	5	20	10	5

Solution.

TABLE 2·10 : COMPUTATION OF MEDIAN

Wages (in ₹)	*No. of employees*	*c.f.*
2,000 — 3,000	3	3
3,000 — 4,000	5	8
4,000 — 5,000	20	28
5,000 — 6,000	10	38
6,000 — 7,000	5	43

Here $\frac{1}{2}N = \frac{1}{2}(43) = 21{\cdot}5.$

Cumulative frequency just greater than 21.5 is 28 and the corresponding class is 4,000—5,000. Thus median class is 4,000—5,000.

Hence, using (2·6), Median $= 4{,}000 + \frac{1{,}000}{20}(21.5 - 8) = 4{,}000 + 675 = 4{,}675.$

Thus median wage is ₹ 4,675.

Example 2·7. *In a factory employing 3,000 persons, in a day 5 per cent work less than 3 hours, 580 work from 3·01 to 4·50 hours, 30 per cent work 4·51 to 6·00 hours, 500 work from 6·01 to 7·50 hours, 20 per cent work from 7·51 to 9·00 hours and the rest work 9·01 or more hours. What is the median hours of work ?*

Solution. The given information can be expressed in tabular from as follows :

TABLE 2·11 : CALCULATIONS FOR MEDIAN WAGES

Work hours	*No. of employees (f)*	*Less than c.f.*	*Class boundaries*
Less than 3	$\frac{5}{100} \times 3{,}000 = 150$	150	Below 3·005
3·01 — 4·50	580	730	3·005 — 4·505
4·51 — 6·00	$\frac{30}{100} \times 3{,}000 = 900$	1,630	4·505 — 6·005
6·01 — 7·50	500	2,130	6·005 — 7·505
7·51 — 9·00	$\frac{20}{100} \times 3{,}000 = 600$	2,730	7·505 — 9·005
9·01 and above	3,000 − 2,730 = 270	N = 3,000 = N	9·005 and above

Here $N = 3{,}000 \Rightarrow \frac{1}{2}N = 1{,}500.$

The *c.f.* just greater than 1500 is 1,630. The corresponding class 4·51—6·00, whose class boundaries are 4·505 — 6·005, is the median class.

Using the median formula, we get

$$\text{Median} = l + \frac{h}{f}\left(\frac{N}{2} - c\right) = 4{\cdot}505 + \frac{1{\cdot}5}{900}(1{,}500 - 730) = 4{\cdot}505 + 1{\cdot}283 \approx 5{\cdot}79$$

Hence, the median hours of work are 5·79.

Example 2·8. *An incomplete frequency distribution is given as follows :*

Variable	*Frequency*	*Variable*	*Frequency*
10—20	12	50—60	?
20—30	30	60—70	25
30—40	?	70—80	18
40—50	65	*Total*	229

Given that the median value is 46, determine the missing frequencies, using the median formula.

Solution. Let the frequency of the class 30—40 be f_1 and that of 50—60 be f_2.

Then $$f_1 + f_2 = 229 - (12 + 30 + 65 + 25 + 18) = 79.$$

Since median is given to be 46, the class 40—50 is the median class.

Hence, using median formula (2·6), we get

$$46 = 40 + \frac{114\cdot5 - (12 + 30 + f_1)}{65} \times 10$$

$$\Rightarrow \quad 46 - 40 = \frac{72\cdot5 - f_1}{65} \times 10 \quad \Rightarrow \quad 6 = \frac{72\cdot5 - f_1}{6\cdot5}$$

$$\Rightarrow \quad f_1 = 72\cdot5 - 39 = 33\cdot5 \approx 34 \qquad \text{(Since frequency is never fractional)}$$

$$\therefore \quad f_2 = 79 - 34 = 45 \qquad \text{(Since } f_1 + f_2 = 79\text{)}$$

2·6·2. Merits and Demerits of Median

Merits	*Demerits*
1. It is rigidly defined.	1. In case of even number of observations, median cannot be determined exactly. We merely estimate it by taking the mean of two middle terms.
2. It is easily understood and is easy to calculate. In some cases it can be located merely by inspection.	2. It is not based on all the observations. For example, the median of 10, 25, 50, 60 and 64 is 50. We can replace the observations 10 and 25 by any two values which are smaller than 50 and the observations 60 and 65 by any two values greater than 50, without affecting the value of median. This property is sometimes described by saying that median is *insensitive.*
3. It is not at all affected by extreme values.	3. It is not amenable to algebraic treatment.
4. It can be calculated for distributions with open-end classes.	4. As compared with mean, it is affected much by fluctuations of sampling.

Uses.

(*i*) Median is the only average to be used while dealing with qualitative data which cannot be measured quantitatively but still can be arranged in ascending or descending order of magnitude, *e.g.*, to find the average intelligence or average honesty among a group of people.

(*ii*) It is to be used for determining the typical value in problems concerning wages, distribution of wealth, etc.

2·7. MODE

Let us consider the following statements :

(*i*) The average height of an Indian (male) is 5′ - 6″.

(*ii*) The average size of the shoes sold in a shop is 7.

(*iii*) An average student in a hostel spends ₹ 750 per month.

In all the above cases, the average referred to is mode. *Mode is the value which occurs most frequently in a set of observations and around which the other items of the set cluster densely. In other words, mode is the value of the variable which is predominant in the series.* Thus, in the case of discrete frequency distribution, mode is the value of x corresponding to maximum frequency. For example, in the frequency distribution :

x :	1	2	3	4	5	6	7	8
f :	4	9	16	25	22	15	7	3

value of x corresponding to the maximum frequency, *viz.*, 25 is 4. Hence, mode is 4.

But in any one (or more) of the following cases :

(*i*) if the maximum frequency is repeated,

(*ii*) if the maximum frequency occurs in the very beginning or at the end of the distribution, and

(*iii*) if there are irregularities in the distribution,

the value of mode is determined by the *method of grouping*, which is illustrated below by an example.

Example 2·9. *Find the mode of the following frequency distribution :*

Size (x) :	*1*	*2*	*3*	*4*	*5*	*6*	*7*	*8*	*9*	*10*	*11*	*12*
Frequency (f) :	*3*	*8*	*15*	*23*	*35*	*40*	*32*	*28*	*20*	*45*	*14*	*6*

Solution. Here we see that the distribution is not regular since the frequencies are increasing steadily up to 40 and then decrease but the frequency 45 after 20 does not seem to be consistent with the distribution. Here we cannot say that since maximum frequency is 45, mode is 10. Here we shall locate mode by the method of grouping as explained below in *Table 2·12*.

TABLE 2·12

Size	*Frequency*					
(*x*)	(*i*)	(*ii*)	(*iii*)	(*iv*)	(*v*)	(*vi*)
1	3	11				
2	8		23	26		
3	15	38			46	
4	23		58			73
5	35	**75**		**98**		
6	40		**72**		**107**	
7	32	60				**100**
8	28		48	80		
9	20	65			93	
10	**45**		59			79
11	14	20		65		
12	6					

The frequencies in column (*i*) are the original frequencies. Column (*ii*) is obtained by combining the frequencies tow by two. If we leave the first frequency and combine the remaining frequencies two by two, we get column (*iii*). Combining the frequencies two by two after *leaving* the first two frequencies results in a repetition of column (*ii*). Hence, we proceed to combine the frequencies three by three, thus getting column (*iv*). The combination of frequencies three by three after leaving the first frequency results in column (*v*) and after leaving the first two frequencies results in column (*vi*). The maximum frequency in each column is given in black type.

To find mode we form the following *Table 2·13.*

TABLE 2·13 : ANALYSIS TABLE

Column Number of Table 2.12 (1)	*Maximum Frequency* (2)	*Value or combination of values of x giving max. frequency in (2)* (3)
(i)	45	10
(ii)	75	5, 6
(iii)	72	6, 7
(iv)	98	4, 5,, 6
(v)	107	5, 6, 7
(vi)	100	6, 7, 8

On examining the values in column (3) above, we find that the value 6 is repeated the maximum number of times and hence the value of mode is 6 and not 10 which is an irregular item.

2·7·1. Mode for Continuous Frequency Distribution

In case of continuous frequency distribution, mode is given by the formula :

$$\text{Mode} = l + \frac{h(f_1 - f_0)}{(f_1 - f_0) - (f_2 - f_1)} = l + \frac{h(f_1 - f_0)}{2f_1 - f_0 - f_2}, \qquad \text{... (2·7)}$$

where l is the lower limit, h the magnitude and f_1 the frequency of the modal class, and f_0 and f_2 are frequencies of the classes preceding and succeeding the modal class respectively.

Derivation of the Mode Formula (2·7). Let us consider the following continuous frequency distribution :

Class :	$x_1 - x_2$	$x_2 - x_3$	...	$x_k - x_{k+1}$	...	$x_n - x_{n+1}$
Frequency :	f_1	f_2	...	f_k	...	f_n.

where all the classes are of equal magnitude, say, h units.

If f_k is the maximum of all the frequencies, then the model class is $(x_k - x_{k+1})$.

Let us further consider a portion of the histogram, namely, the rectangles erected on the modal class and the two adjacent classes. The mode is the value of x for which the frequency curve has a maxima. Let the modal point be Q.

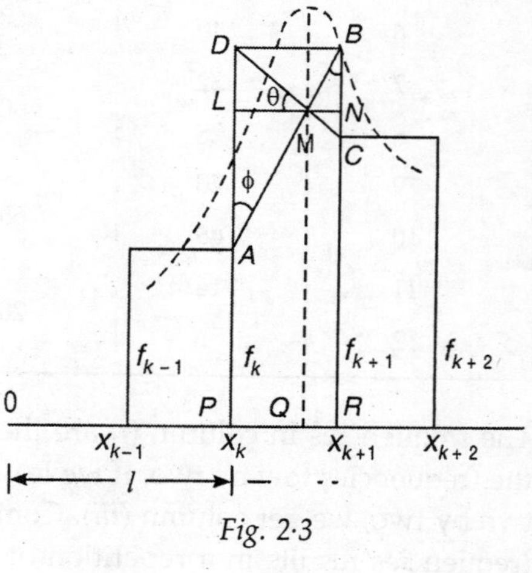

Fig. 2·3

From the adjoining figure,

$$\tan\theta = \frac{LD}{LM} = \frac{NC}{MN}$$

and
$$\tan\phi = \frac{LM}{AL} = \frac{MN}{NB}$$

$$\therefore \quad \frac{LM}{MN} = \frac{LD}{NC} = \frac{AL}{NB} = \frac{AL + LD}{NB + NC} = \frac{AD}{BC},$$

i.e.,
$$\frac{LM}{LN - LM} = \frac{PD - AP}{BR - CR}$$

or
$$\frac{LM}{h - LM} = \frac{f_k - f_{k-1}}{f_k - f_{k+1}},$$

where 'h' is the magnitude of the modal class. Thus, solving for LM, we get

$$LM = \frac{h(f_k - f_{k-1})}{(f_k - f_{k+1}) + (f_k - f_{k-1})} = \frac{h(f_k - f_{k-1})}{2f_k - f_{k-1} - f_{k+1}}.$$

Hence, $$\text{Mode} = OQ = OP + PQ = OP + LM = l + \frac{h(f_k - f_{k-1})}{2f_k - f_{k-1} - f_{k+1}}.$$

Example 2·10. *Find the mode for the following distribution :*

Class-Interval :	0 — 10	10 — 20	20 — 30	30 — 40	40 — 50	50 — 60	60 — 70	70 — 80
Frequency :	5	8	7	12	28	20	10	10

Solution. Here maximum frequency is 28. Thus the class 40 — 50 is the modal class. Using mode formula (2·7), the value of mode is given by :

$$\text{Mode} = 40 + \frac{10(28 - 12)}{(2 \times 28 - 12 - 20)} = 40 + 6{\cdot}666 = 46{\cdot}67 \text{ (approx).}$$

Remarks:

1. In case of irregularities in the distribution, or the maximum frequency being repeated or the maximum frequency occurring in the very beginning or at the end of the distribution, the modal class is determined by the method of grouping and the mode is obtained by using (2·7).
Sometimes, mode is estimated from the mean and the median. For a symmetrical distribution (See § 2·16), mean, median and mode coincide. If the distribution is *moderately asymmetrical*, the mean, median and mode obey the following empirical relationship (due to Karl Pearson) :

$$\text{Mean} - \text{Median} = \frac{1}{3}(\text{Mean} - \text{Mode}) \quad \Rightarrow \quad \text{Mode} = 3\text{ Median} - 2\text{ Mean} \qquad \text{... (2·7}a\text{)}$$

2. If the method of grouping gives the modal class which does not correspond to the maximum frequency, *i.e.*, the frequency of modal class is not the maximum frequency, then in some situations we may get $2f_k - f_{k-1} - f_{k+1} = 0$. In such cases, the value of mode can be obtained by the formula :

$$\text{Mode} = l + \frac{h(f_k - f_{k-1})}{|f_k - f_{k-1}| + |f_k - f_{k+1}|} \qquad \text{... (2·7}b\text{)}$$

2·7·2. Merits and Demerits of Mode

Merits	*Demerits*
1. Mode is readily comprehensible and easy to calculate. Like median, mode can be located in some cases merely by inspection.	1. Mode is ill-defined. It is not always possible to find a clearly defined mode. In some cases, we may come across distributions with two modes. Such distributions are called *bi-modal*. If a distribution has more than two modes, it is said to be *multimodal*.
2. Mode is not at all affected by extreme values.	2. It is not based upon all the observations.
3. Mode can be conveniently located even if the frequency distribution has class intervals of unequal magnitude provided the modal class and the classes preceding and succeeding it are of the same magnitude. Open-end classes also do not pose any problem in the location of mode.	3. It is not capable of further mathematical treatment. 4. As compared with mean, mode is affected to a greater extent, by fluctuations of sampling.

Uses. Mode is the average to be used to find the ideal size, *e.g.*, in business forecasting, in the manufacture of ready made garments, shoes, etc.

2·8. GEOMETRIC MEAN

Geometric mean of a set of n observations is the nth root of their product. Thus, the geometric mean G, of n observations x_i ; $i = 1, 2, ..., n$ is given by :

$$G = (x_1 \cdot x_2 \ldots x_n)^{1/n} \qquad \ldots (2\cdot8)$$

The computation is facilitated by use of logarithms. Taking logarithm of both sides, we have

$$\log G = \frac{1}{n}(\log x_1 + \log x_2 + \ldots + \log x_n) = \frac{1}{n}\sum_{i=1}^{n} \log x_i$$

$$\therefore \qquad G = \text{Antilog}\left(\frac{1}{n}\sum_{i=1}^{n} \log x_i\right) \qquad \ldots (2\cdot8a)$$

In case of frequency distribution $x_i \mid f_i$, $(i = 1, 2, ..., n)$ geometric mean, G is given by :

$$G = \left(x_1^{f_1} \cdot x_2^{f_2} \ldots x_n^{f_n}\right)^{1/N}, \text{ where } N = \sum_{i=1}^{n} f_i \qquad \ldots (2\cdot9)$$

Taking logarithms of both sides, we get

$$\log G = \frac{1}{N}(f_1 \log x_1 + f_2 \log x_2 + \ldots + f_n \log x_n) = \frac{1}{N}\sum_{i=1}^{n} f_i \log x_i \ldots (2\cdot9a)$$

Thus we find that logarithm of geometric mean is the arithmetic mean of the logarithms of the given values. From (2·9*a*), we get

$$G = \text{Antilog}\left(\frac{1}{N}\sum_{i=1}^{n} f_i \log x_i\right) \qquad \ldots (2\cdot9b)$$

In the case of grouped or continuous frequency distribution, x is taken to be the value corresponding to the mid-point of the class intervals.

2·8·1. Geometric Mean of the Combined Group

If n_1 and n_2 are the sizes, G_1 and G_2 the geometric means of two series respectively, the geometric mean G, of the combined series is given by :

$$\log G = \frac{n_1 \log G_1 + n_2 \log G_2}{n_1 + n_2} \qquad \ldots (2\cdot10)$$

Proof. Let x_{1i} $(i = 1, 2, ..., n_1)$ and x_{2j} $(j = 1, 2, ..., n_2)$ be n_1 and n_2 items of two series respectively. Then by *def.*,

$$G_1 = (x_{11} \cdot x_{12} \ldots x_{1n_1})^{1/n_1} \quad \Rightarrow \quad \log G_1 = \frac{1}{n_1}\sum_{i=1}^{n_1} \log x_{1i},$$

$$G_2 = (x_{21} \cdot x_{22} \ldots x_{2n_2})^{1/n_2} \quad \Rightarrow \quad \log G_2 = \frac{1}{n_2}\sum_{j=1}^{n_2} \log x_{2j}$$

The geometric mean G of the combined series is given by :

$$G = \left(x_{11} \cdot x_{12} \ldots x_{1n_1} \cdot x_{21} \cdot x_{22} \ldots x_{2n_2}\right)^{1/(n_1 + n_2)}$$

$$\Rightarrow \quad \log G = \frac{1}{n_1+n_2}\left(\sum_{i=1}^{n_1} \log x_{1i} + \sum_{j=1}^{n_2} \log x_{2j}\right) = \frac{1}{n_1+n_2}(n_1 \log G_1 + n_2 \log G_2)$$

$$\therefore \quad G = \text{Antilog}\left(\frac{n_1 \log G_1 + n_2 \log G_2}{n_1+n_2}\right) \quad \text{... (2·10}a\text{)}$$

In general, if $G_1, G_2, ..., G_k$ are the geometric means of k groups having $n_1, n_2, ..., n_k$ observations respectively, the geometric mean G of combined group consisting of $n_1 + n_2 + ... + n_k$ observations is given by :

$$G = \text{Antilog}\left(\frac{n_1 \log G_1 + n_2 \log G_2 + ... + n_k \log G_k}{n_1 + n_2 + ... + n_k}\right) \quad \text{... (2·10}b\text{)}$$

2·8·2. Merits and Demerits of Geometric Mean

Merits	*Demerits*
1. It is rigidly defined.	1. Because of its abstract mathematical character, geometric mean is not easy to understand and to calculate for a non-mathematics person.
2. It is based upon all the observations.	2. If any one of the observations is zero, geometric *mean* becomes zero and if any one of the observations is negative, geometric mean becomes imaginary regardless of the magnitude of the other items.
3. It is suitable for further mathematical treatment.	
4. It is not affected much by fluctuations of sampling.	
5. It gives comparatively more weight to small items.	

Uses. Geometric mean is used :

(*i*) To find the rate of population growth and the rate of interest.

(*ii*) In the construction of index numbers.

Example 2·11. *Show that in finding the arithmetic mean of a set of readings on a thermometer it does not matter whether we measure temperature in Centigrade or Fahrenheit, but that in finding the geometric mean it does matter which scale we use.*

Solution. Let $C_1, C_2, ..., C_n$ be the n readings on the Centigrade thermometer. Then their arithmetic mean is given by :

$$\bar{C} = \frac{1}{n}(C_1 + C_2 + ... + C_n)$$

If F and C be the readings in Fahrenheit and Centigrade respectively, then we have the relation:

$$\frac{F-32}{180} = \frac{C}{100} \quad \Rightarrow \quad F = 32 + \frac{9}{5}C.$$

Thus , the Fahrenheit equivalents of $C_1, C_2, ..., C_n$ are :

$$32 + \frac{9}{5}C_1, \quad 32 + \frac{9}{5}C_2, \quad ..., \quad 32 + \frac{9}{5}C_n, \text{ respectively.}$$

Hence, the arithmetic mean of the readings in Fahrenheit is

$$\overline{F} = \frac{1}{n}\left\{\left(32+\frac{9}{5}C_1\right)+\left(32+\frac{9}{5}C_2\right)+\ldots+\left(32+\frac{9}{5}C_n\right)\right\}$$

$$= \frac{1}{n}\left\{\left(32\,n+\frac{9}{5}(C_1+C_2+\ldots+C_n\right)\right\} = 32+\frac{9}{5}\left(\frac{C_1+C_2+\ldots+C_n}{n}\right)$$

$$= 32+\frac{9}{5}\overline{C}, \text{ which is the Fahrenheit equivalent of } \overline{C}.$$

Hence, in finding the arithmetic mean of a set of n readings on a thermometer, it is immaterial whether we measure temperature in Centigrade or Fahrenheit.

Geometric mean G, of n readings in Centigrade is : $G = (C_1 \cdot C_2 \ldots C_n)^{1/n}$

Geometric mean G_1, (say), of Fahrenheit equivalents of $C_1, C_2, \ldots, C_n$ is :

$$G_1 = \left\{\left(32+\frac{9}{5}C_1\right)\times\left(32+\frac{9}{5}C_2\right)\times\ldots\times\left(32+\frac{9}{5}C_n\right)\right\}^{1/n}$$

which is not equal to Fahrenheit equivalent of G, *viz.*, $\left\{\frac{9}{5}(C_1 \cdot C_2 \ldots C_n)^{1/n}+32\right\}$

Hence, in finding the geometric mean of the n readings on a thermometer, the scale (Centigrade or Fahrenheit) is important.

2·9. HARMONIC MEAN

Harmonic mean of a number of observations, none of which is zero, is the reciprocal of the arithmetic mean of the reciprocals of the given values. Thus, harmonic mean (H), of n observations x_i, $i = 1, 2, \ldots, n$ is given by :

$$H = \frac{1}{\frac{1}{n}\sum_{i=1}^{n}(1/x_i)} \qquad \ldots (2\cdot11)$$

In case of frequency distribution $x_i \mid f_i$, $(1, 2, \ldots, n)$,

$$H = \frac{1}{\frac{1}{N}\sum_{i=1}^{n}(f_i/x_i)}, \quad \left[N = \sum_{i=1}^{n} f_i\right] \qquad \ldots (2\cdot11a)$$

If $x_1, x_2, \ldots, x_n$ are n observations with weights $w_1, w_2, \ldots, w_n$ respectively, their weighted harmonic mean is defined as :

$$H = \frac{\Sigma w_i}{\Sigma(w_i / x_i)}. \qquad \ldots (2\cdot11b)$$

2·9·1. Merits and Demerits of Harmonic Mean

Merits	*Demerits*
Harmonic mean is rigidly defined, based upon all the observations and is suitable for further mathematical treatment. Like geometric mean, it is not affected much by fluctuations of sampling. It gives greater importance to small items and is useful only when small items have to be given a greater weightage.	Harmonic mean is not easily understood and is difficult to compute.

Example 2·12. *A cyclist pedals from his house to his college at a speed of 10 km.p.h. and back from the college to his house at 15 km.p.h. Find the average speed.*

Solution. Let the distance from the house to the college be x km. In going from house to college, the distance (x kilometre) is covered in $\frac{x}{10}$ hours, while in coming from college to house, the distance is covered in $\frac{x}{15}$ hours. Thus, a total distance of $2x$ km. is covered in $\left(\frac{x}{10}+\frac{x}{15}\right)$ hours.

$$\text{Hence,} \quad \text{Average speed} = \frac{\text{Total distance travelled}}{\text{Total time taken}} = \frac{2x}{\left(\frac{x}{10}+\frac{x}{15}\right)} = 12 \text{ km.p.h.}$$

In this case, the average speed is given by the harmonic mean of 10 and 15 and not by the arithmetic mean.

Remarks:

1. If equal distances are covered (travelled) per unit of time with speeds equal to $V_1, V_2, ..., V_n$, say, then the average speed is given by the harmonic mean of $V_1, V_2, ..., V_n$, *i.e.,*

$$\text{Average speed} = \frac{1}{\frac{1}{n}\left(\frac{1}{V_1}+\frac{1}{V_2}+...+\frac{1}{V_n}\right)} = \frac{n}{\sum_i \left(\frac{1}{V_i}\right)} \quad ... (2\cdot 11c)$$

2. Instead of fixed (constant) distance being travelled with varying speed, let us now suppose that different distances, say, $S_1, S_2, ..., S_n$ are travelled with different speeds, say, $V_1, V_2, ..., V_n$ respectively. In that case, the average speed is given by the weighted harmonic mean of the speeds, the weights being the corresponding distances travelled, *i.e.,*

$$\text{Average speed} = \frac{S_1 + S_2 + ... + S_n}{\left(\frac{S_1}{V_1}+\frac{S_2}{V_2}+...+\frac{S_n}{V_n}\right)} = \frac{\sum_i S_i}{\sum_i \left(\frac{S_i}{V_i}\right)} \quad ... (2\cdot 11d)$$

3. The harmonic mean, like arithmetic mean, is also used in averaging of ratios like price per unit, km. per hour, work done per hour, etc. under certain conditions. It may be noted here that a rate represents a ratio, e.g., $\text{price} = \frac{\text{money}}{\text{quantity}}$, $\text{speed} = \frac{\text{distance}}{\text{time}}$, $\text{work done per hour} = \frac{\text{work done}}{\text{time taken}}$, etc.

   The average of a rate, defined by the ratio p/q, is given by the arithmetic mean of its values in different situations if the conditions are given in terms of q and by the harmonic mean if the conditions are given in terms of p.

Example 2·13. *(a) Mineral Water is sold at the rates of 8, 10, 12 and 15 rupees per litre in four different months. Assuming that equal amounts are spent on water by a family in the four months, find the average price in rupees per month.*

(b) An individual purchases three qualities of pencils. The relevant data are given in adjoining Table. Calculate the average price per pencil.

Quality	*Price per pencil* (₹)	*Money spent* (₹)
A	1·00	50
B	1·50	30
C	2·00	20

Solution. (*a*) Since equal amounts of money are spent by the family for each of the four months, the average price of mineral water per month is given by the harmonic mean of 8, 10, 12 and 15.

$\therefore$ Average price of mineral water per month

$$= ₹\frac{4}{\frac{1}{8}+\frac{1}{10}+\frac{1}{12}+\frac{1}{15}} = ₹\frac{4\times 120}{15+12+10+8} = ₹\frac{4\times 120}{45} = ₹\ 10{\cdot}67$$

(*b*) Here ,we are given : Total expenditure = ₹ (50 + 30 + 20) = ₹ 100

$$\text{Total number of pencils purchased} = \frac{50}{1}+\frac{30}{1.50}+\frac{20}{2} = 80$$

$$\text{Average price per pencil} = \frac{\text{Total expenditure}}{\text{Total number of pencils}} = \frac{100}{80} = ₹\ 1{\cdot}25.$$

Remark. Average price of ₹ 1·25 can also be obtained by finding the weighted harmonic mean (H.M.) of 1, 1·5, and 2 with corresponding weights 50, 30 and 20, respectively.

Example 2·14. *You can take a trip which entails travelling 900 km. by train at an average speed of 60 km. per hour, 3,000 km. by boat at an average speed of 25 km. p.h., 400 km. by plane at 350 km. per hour and finally 15 km. by taxi at 25 km. per hour. What is your average speed for the entire distance ?*

Solution. Since different distances are covered with varying speeds, the required average speed for the entire distance is given by the weighted harmonic mean of the speeds (in km. p.h.), the weights being the corresponding distances covered (in km.).

TABLE 2·14 : COMPUTATION OF WEIGHTED H.M.

Speed (km./hr.) x	*Distance (in km.)* w	w/x
60	900	15·00
25	3,000	120·00
350	400	1·43
25	15	0·60
Total	4,315	137·03

$$\text{Average speed} = \frac{\Sigma w_i}{\Sigma (w_i / x_i)} = \frac{4315}{137{\cdot}03} = 31{\cdot}489 \text{ km.p.h.}$$

2·10. SELECTION OF AN AVERAGE

From the preceding discussion, it is evident that no single average is suitable for all practical purposes. Each one of the averages has its own merits and demerits and thus its own particular field of importance and utility. We cannot use the averages indiscriminately. A judicious selection of the average depending on the nature of the data and the purpose of the inquiry is essential for sound statistical analysis. Since arithmetic mean satisfies all the properties of an ideal average as laid down by Prof. Yule ; is familiar to a layman and further has wide applications in statistical theory at large, it may be regarded as the best of all the averages.

2·11. PARTITION VALUES

These are the values which divide the series into a number of equal parts.

The three points which divide the series into four equal parts are called *quartiles.* The first, second and third points are known as the first, second and third quartiles respectively. The first quartile, Q_1 is the value which exceed 25% of the observations and is exceeded by 75% of the observations. The second quartile, Q_2, coincides with median. The third quartile, Q_3, is the point which has 75% observations before it and 25% observations after it.

The nine points which divide the series into ten equal parts are called *deciles* whereas *percentiles* are the ninety-nine points which divide the series into hundred equal parts. For example, D_7, the seventh decile, has 70% observations before it and P_{47}, the forty-seventh percentile, is the point which exceeds 47% of the observations. The methods of computing the partition values are the same as those of locating the median in the case of both discrete and continuous distributions.

Example 2·15. *Eight coins were tossed together and the number of heads resulting was noted. The operation was repeated 256 times and the frequencies (f) that were obtained for different values of x, the number of heads, are shown in the following table. Calculate median, quartiles, 4th decile and 27th percentile.*

x :	0	1	2	3	4	5	6	7	8
f :	1	9	26	59	72	52	29	7	1

Solution.

x :	0	1	2	3	4	5	6	7	8
f :	1	9	26	59	72	52	29	7	1
c.f. :	1	10	36	95	167	219	248	255	256

Median : Here $\frac{1}{2}N = \frac{1}{2} \times 256 = 128$. Cumulative frequency (*c.f.*) just greater than 128 is 167. Thus, median = 4.

Q_1 : Here $\frac{1}{4}N = 64$ and *c.f.* just greater than 64 is 95. Hence, $Q_1 = 3$.

Q_3 : Here $\frac{3}{4}N = 192$ and *c.f.* just greater than 192 is 219. Thus, $Q_3 = 5$.

D_4 : $\frac{4}{10}N = 4 \times 25{\cdot}6 = 102{\cdot}4$ and *c.f.* just greater than 102·4 is 167. Hence, $D_4 = 4$.

P_{27} : $\frac{27}{100}N = 27 \times 2{\cdot}56 = 69{\cdot}12$ and *c.f.* just greater than 69·12 is 95. Hence, $P_{27} = 3$.

Example 2·16. *Following is the distribution of marks obtained by 500 candidates in Statistics paper of a civil services examination :*

Marks more than	:	0	10	20	30	40	50
Number of Candidates	:	500	460	400	200	100	30

Calculate the lower quartile marks. If 70% of the candidates pass in the paper, find the minimum marks obtained by a pass candidate.

Solution.

TABLE 2·15 : COMPUTATION OF LOWER QUARTILE AND THIRD DECILE

Marks more than	*Cumulative frequency*	*Class intervals*	*Frequency (f)*	*Less than c.f.*
0	500 = N	0—10	500 – 460 = 40	40
10	460	10—20	460 – 400 = 60	100
20	400	20—30	400 – 200 = 200	300
30	200	30—40	200 – 100 = 100	400
40	100	40—50	100 – 30 = 70	470
50	30	50 and above	30	500

$$\frac{1}{4}N = \frac{1}{4}(500) = 125.$$

The less than *c.f.* just greater than 125 is 300. Hence, the corresponding class 20—30 contains Q_1.

$$\therefore \qquad Q_1 = l + \frac{h}{f}\left(\frac{N}{4} - C\right) = 20 + \frac{10}{200}(125 - 100) = 21{\cdot}25$$

Since out of 500 students, 70% pass the test, *i.e.*, 30% fail in the test, the minimum marks obtained by a pass candidate are given by D_3 or P_{30}.

$\frac{3N}{10} = \frac{3 \times 500}{10} = 150$. The *c.f.* just greater than 150 is 300. Therefore, D_3 lies in the corresponding class 20—30, and is given by :

$$D_3 = P_{30} = l + \frac{h}{f}\left(\frac{3N}{100} - C\right) = 20 + \frac{10}{200}(150 - 100) = 22{\cdot}5$$

Hence, minimum marks obtained by the pass candidate are 22·5.

2·11·1. Graphical Location of the Partition Values

The partition values, *viz.*, quartiles, deciles and percentiles, can be conveniently located with the help of a curve called the '*cumulative frequency curve*' or '*Ogive*'. The procedure is illustrated below :

First, form the 'less than cumulative frequency' table. Take the class intervals (or the variate values) along the *x*-axis and plot the corresponding cumulative frequencies along the *y*-axis against the *upper limit* of the class interval (or against the variate value in the case of discrete frequency distribution). The curve obtained on joining the points so obtained by means of free hand drawing is called the *less than cumulative frequency curve* or *less than ogive.* Similarly, by plotting the more than *c.f.* against the lower limit of the corresponding class and joining the points so obtained by a smooth free hand curve, we obtain *'more than ogive'*. The graphical location of partition values from this curve is explained below by means of an example.

Example 2·17. *Draw the cumulative frequency curve for the following distribution showing the number of marks of 59 students in Statistics.*

Marks-group :	*0 — 10*	*10 — 20*	*20 — 30*	*30 — 40*	*40 — 50*	*50 — 60*	*60 — 70*
No. of Students :	4	8	11	15	12	6	3

Solution.

TABLE 2·16 : c.f. TABLE

Marks group	*No. of Students*	*Less than c.f.*	*More than c.f.*
0—10	4	4	59
10—20	8	12	55
20—30	11	23	47
30—40	15	38	36
40—50	12	50	21
50—60	6	56	9
60—70	3	59	3

Taking the marks-group along *x*-axis and cumulative frequencies along *y*-axis, we plot the less than cumulative frequencies, *viz.*, 4, 12, 23, ..., 59 against the upper limits of the corresponding classes, *viz.*, 10, 20, 30, ..., 70, respectively.

The smooth curve obtained on joining these points is called *'less than'* ogive.

If we plot the 'more than' cumulative frequencies, *viz.*, 59, 55, ..., 3 against the lower limits of the corresponding classes, *viz.*, 0, 10, ..., 60 and join the points by a smooth curve, we get 'more than cumulative frequency curve' or *'more than'* ogive.

To locate graphically the value of median, mark a point corresponding to $\frac{1}{2}N$ along y-axis. At this point draw a line parallel to x-axis meeting the ogive at the point 'A' (say). From 'A' draw a *line* perpendicular to x-axis meeting it in 'M' (say). Then abscissa of 'M' *viz.*, OM gives the value of median.

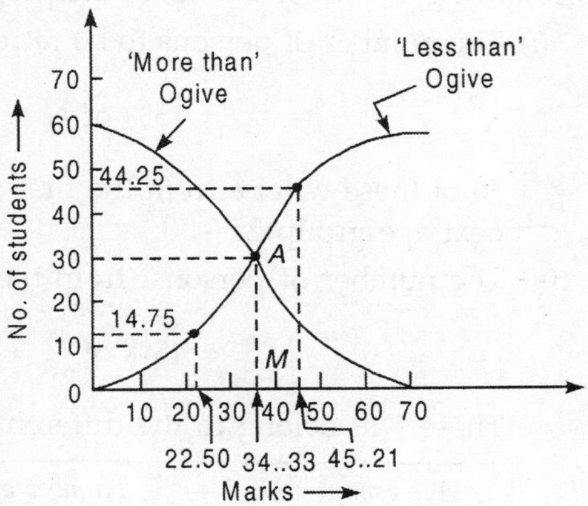

Fig. 2·4

To locate the values of Q_1 (or Q_3), we mark the points along y-axis corresponding to $\frac{1}{4}N$ (or $\frac{3}{4}N$) and proceed exactly similarly.

In the above example, we get

$$\text{Median} = 34{\cdot}33, \quad Q_1 = 22{\cdot}50 \quad \text{and} \quad Q_3 = 45{\cdot}21.$$

Remarks:

1. The median can also be located as follows :
   From the point of intersection of 'less than' ogive and 'more than' ogive, draw perpendicular to OX. The abscissa of the point so obtained gives median.
2. Other partition values, *viz.*, deciles and percentiles, can be similarly located from 'ogive'.

SUPPLEMENTARY EXAMPLES ON MEASURES OF CENTRAL TENDENCY

Example 2·18. *The following is the age distribution of 1,000 persons working in a large industrial house :*

Age - group	*No. of persons*	*Age - group*	*No. of persons*
20—25	30	45—50	105
25—30	160	50—55	70
30—35	210	55—60	60
35—40	180	60—65	40
40—45	145		

Due to continuous heavy losses the management decides to bring down the strength to 30% of the present number according to the following scheme :

(i) To retrench the first 15% from the lower group. *(ii) To absorb the next 45% in other branches.*

(iii) To make 10% from the highest age group retire permanently, if necessary.

Calculate the age limits of the persons retained and those to be transferred to other departments. Also find the average age of those retained.

Solution. Total number of persons in the industrial house is N = 1,000.

According to the conditions of the problem:

(*i*) The number of persons to be retrenched from the lower group

$$= 15\% \text{ of } N = \frac{15}{100} \times 1{,}000 = 150.$$

30 of these will be from the first group 20—25 and the remaining 150 – 30 = 120, from the next age group 25—30.

(*ii*) The number of persons (from the next groups) to be absorbed in other branches :

$$= 45\% \text{ of } N = \frac{45}{100} \times 1{,}000 = 450.$$

These will belong to the different age-groups as detailed below :

Age-group	*Number of persons*
25—30	160 – 120* = 40
30—35	210
35—40	180
40—45	450 – (40 + 210 + 180) = 20

[* Because 120 persons from this group have been retrenched in (*i*).]

(*iii*) The number of persons to retire (from the highest age-groups)

$$= 10\% \text{ of } N = \frac{10}{100} \times 1{,}000 = 100.$$

Age-group	*Number of persons*
55—60	60
60—65	40

These 100 persons are from the highest age groups as shown :

Hence, incorporating the steps in (*i*), (*ii*) and (*iii*), the frequency distribution of the number of persons retained in the industrial house is as shown in the adjoining table :

Age-group	*Number of persons*
40—45	145 – 20 = 125
45—50	105
50—55	70

TABLE 2·17 : CALCULATIONS FOR AVERAGE AGE OF PERSONS RETAINED

Age-group	*Mid-value (x)*	*f*	$d = \dfrac{x - 47.5}{5}$	*fd*
40—45	42·5	125	–1	–125
45—50	47·5	105	0	0
50—55	52·5	70	1	70
Total	N = 300			$\Sigma fd = -55$

The average age of persons retained is given by :

$$\overline{X} = A + \frac{h\,\Sigma fd}{N} = 47.5 + \frac{5 \times (-55)}{300}$$

$$= 46{\cdot}5833 \simeq 47$$

Hence, the average age of those retained in the industrial house is 47 years.

Example 2·19. *A number of particular articles has been classified according to their weights. After drying for two weeks the same articles have again been weighed and similarly classified. It is known that the median weight in the first weighing was 20·83 gm. while in the second weighing it was 17·35 gm. Some frequencies a and b in the first weighing and x & y in the second are missing. It is known that a = (1/3)x and b = (1/2)y. Find out the values of the missing frequencies.*

Class	Frequencies 1st weighing	Frequencies 2nd weighing	Class	Frequencies 1st weighing	Frequencies 2nd weighing
0—5	a	x	15—20	52	50
5—10	b	y	20—25	75	30
10—15	11	40	25—30	22	28

Solution. We are given : $a = \frac{1}{3}x$ and $b = \frac{1}{2}y$ $\Rightarrow$ $x = 3a$ and $y = 2b$.

First weighing

N_1 = Total frequency

$= a + b + 11 + 52 + 75 + 22$

$= 160 + a + b$

Since median is given to be 20·83, the class 20—25 is the median class.

Using the median formula, we get

$$Md = l + \frac{h}{f}\left(\frac{N_1}{2} - c\right)$$

$$\Rightarrow \quad 20.83 = 20 + \frac{5}{75}\left[\frac{N_1}{2} - (63 + a + b)\right]$$

$$\Rightarrow 15(20.83 - 20) = \frac{160 + a + b}{2} - (63 + a + b)$$

$$\Rightarrow \quad 12.45 = 17 - \frac{a + b}{2}$$

$$\therefore \quad a + b = 2(17 - 12·45)$$

$$= 9·10 \simeq 9 \qquad \ldots (*)$$

Second Weighing

N_2 = Total frequency

$= x + y + 40 + 50 + 30 + 28$

$= 148 + x + y = 148 + 3a + 2b$

Since median is given to be 17·35, the class 15—20 is the median class.

Using the median formula, we get

$$17.35 = 15 + \frac{5}{50}\left[\frac{148 + 3a + 2b}{2} - (40 + x + y)\right]$$

$$\Rightarrow 10 \times 2.35 = 74 + \frac{3a + 2b}{2} - 40 - 3a - 2b$$

$$\Rightarrow \quad \frac{3a + 2b}{2} = 34 - 23·5 = 10·5$$

$$\Rightarrow \quad 3a + 2b = 21 \qquad \ldots (**)$$

Multiplying (*) by 3, we get $3a + 3b = 27$ ... (***)

Subtracting (**) from (***), we get $b = 6$.

Substituting in (*), we have $a = 9 - 6 = 3$.

$\therefore$ $a = 3,\ b = 6;$ $x = 3a = 9,$ $y = 2b = 12.$

Example 2·20. *The median and mode of the wage distribution are known to be ₹ 3,350 and ₹ 3,400 respectively. Find the values of f_3, f_4 and f_5 :*

Wages (in ₹)	No. of Employees	Wages (in ₹)	No. of Employees
0—1,000	4	4,000—5,000	f_5
1,000—2,000	16	5,000—6,000	6
2,000—3,000	f_3	6,000—7,000	4
3,000—4,000	f_4	Total	230

Solution.

TABLE 2·18 : CALCULATIONS FOR MEDIAN AND MODE

Wages (in ₹)	*Frequency (f)*	*Less than c.f.*
0—1,000	4	4
1,000—2,000	16	20
2,000—3,000	f_3	$20 + f_3$
3,000—4,000	f_4	$20 + f_3 + f_4$
4,000—5,000	f_5	$20 + f_3 + f_4 + f_5$
5,000—6,000	6	$26 + f_3 + f_4 + f_5$
6,000—7,000	4	$N = 30 + f_3 + f_4 + f_5$
Total	N = 230	

From the above table,

$$N = \Sigma f = 30 + f_3 + f_4 + f_5 = 230$$

$$\Rightarrow \qquad f_3 + f_4 + f_5 = 230 - 30 = 200 \qquad \ldots (*)$$

Since median is 3,350, which lies in the class 3,000—4,000, 3,000—4,000 is the median class.

Using the median formula : $Md = l + \dfrac{h}{f}\left(\dfrac{N}{2} - c\right)$, we get

$$3{,}350 = 3{,}000 + \frac{1{,}000}{f_4}\{115 - (20 + f_3)\} \quad \text{or} \quad \frac{3{,}350 - 3{,}000}{1{,}000} = \frac{95 - f_3}{f_4}$$

$$\Rightarrow \quad 0{\cdot}35\, f_4 = 95 - f_3 \quad \Rightarrow \quad f_3 = 95 - 0{\cdot}35\, f_4 \qquad \ldots (**)$$

Mode being 3,400, the modal class is also 3,000—4,000. Using mode formula :

$$3{,}400 = 3{,}000 + \frac{1{,}000(f_4 - f_3)}{2f_4 - f_3 - f_5} \quad \Rightarrow \quad \frac{3{,}400 - 3{,}000}{1{,}000} = \frac{f_4 + 0{\cdot}35 f_4 - 95}{2f_4 - (200 - f_4)} \quad \text{[Using (*) and (**)]}$$

$$\Rightarrow \quad 0{\cdot}4 = \frac{1{\cdot}35 f_4 - 95}{3f_4 - 200} \quad \Rightarrow \quad f_4 = \frac{95 - 80}{1{\cdot}35 - 1{\cdot}20} = \frac{15}{0{\cdot}15} = 100$$

Substituting in (**), we have $f_3 = 95 - 0{\cdot}35 \times 100 = 60$

Substituting the values of f_3 and f_4 in (*), we get $f_5 = 200 - f_3 - f_4 = 40$

Hence $f_3 = 60, \quad f_4 = 100$ and $f_5 = 40.$

Example 2·21. *The geometric mean of 10 observations on a certain variable was calculated as 16·2. It was later discovered that one of the observations was wrongly recorded as 12·9; in fact it was 21·9. Apply appropriate correction and calculate the correct geometric mean.*

Solution. The geometric mean G of n observations $x_1, x_2, \ldots, x_n$ is given by :

$$G = (x_1 \cdot x_2 \ldots x_n)^{1/n} \qquad \ldots (*)$$

If x_1 is the observation copied wrongly instead of correct value x_1', then the *corrected* geometric mean G' is given by :

$$G' = (x_1' \, x_2 \, x_3 \ldots x_n)^{1/n} = \left(\frac{x_1'}{x_1} . x_1 \, x_2 \ldots x_n\right)^{1/n}$$

$$\Rightarrow \quad G' = (x_1 x_2 \ldots x_n)^{1/n} \cdot \left(\frac{x_1'}{x_1}\right)^{1/n} = G \cdot \left(\frac{x_1'}{x_1}\right)^{1/n} \quad \text{[From (*)]} \quad \ldots (**)$$

In the given problem, $G = 16{\cdot}2,\ n = 10,\ x_1 = 12{\cdot}9,\ x_1' = 21{\cdot}9$

$$\therefore \quad \text{Corrected } G.M. \ (G') = 16{\cdot}2 \times \left(\frac{21{\cdot}9}{12{\cdot}9}\right)^{1/10} \quad \text{[From (**)]}$$

$$\Rightarrow \quad \log_{10} G' = \log_{10} 16{\cdot}2 + \frac{1}{10}(\log_{10} 21{\cdot}9 - \log_{10} 12{\cdot}9)$$

$$= 1{\cdot}2095 + \frac{1}{10}(1{\cdot}3404 - 1{\cdot}1106) = 1{\cdot}23248$$

$$\therefore \quad G' = \text{Antilog } (1{\cdot}2325) = 17{\cdot}08.$$

Example 2·22. *In a frequency table, the upper boundary of each class interval has a constant ratio to the lower boundary. Show that the geometric mean G may be expressed by the formula :*

$$\log G = x_0 + \frac{c}{N} \sum_i f_i (i-1),$$

where x_0 is the logarithm of the mid-value of the first interval and c is the logarithm of the ratio between upper and lower boundaries.

Solution. Let the ith class be denoted by $l_i - l_{i+1}$, with the corresponding frequency f_i ; $i = 1, 2, 3, \ldots$

Let the constant ratio between upper and lower limits of each class be λ, so that :

$$\frac{l_2}{l_1} = \frac{l_3}{l_2} = \ldots = \frac{l_i}{l_{i-1}} = \ldots = \lambda, \text{ (say)} \quad \Rightarrow \quad l_2 = \lambda\, l_1,\ \ l_3 = \lambda\, l_2 = \lambda^2 l_1, \ldots,\ l_i = \lambda\, l_{i-1} = \lambda^{i-1} \cdot l_1 \quad \ldots (1)$$

If x_i is the mid-point of the ith class, then we have

$$x_1 = \frac{1}{2}(l_1 + l_2) = \frac{1}{2}(1+\lambda) l_1 \quad \text{[From (1)]} \quad \ldots (2)$$

and

$$x_i = \frac{1}{2}(l_i + l_{i+1}) = \frac{1}{2}(\lambda^{i-1} l_1 + \lambda^i l_1)$$

$$= \frac{1}{2} l_1 (1+\lambda) \lambda^{i-1} = x_1 \cdot \lambda^{i-1} \quad \text{[From (2)]} \quad \ldots (3)$$

Geometric mean G of the distribution is given by :

$$\log G = \frac{\Sigma f_i \log x_i}{\Sigma f_i} = \frac{1}{N} \sum_i f_i \log (x_1 \cdot \lambda^{i-1}) \quad \text{[From (3)]}$$

$$= \frac{1}{N} \sum_i f_i \log x_1 + \frac{1}{N} \sum_i f_i \cdot (i-1) \log \lambda$$

$$= \log x_1 \cdot \frac{1}{N} \sum_i f_i + \log \lambda \cdot \frac{1}{N} \sum_i (i-1) f_i$$

$$= x_0 + \frac{c}{N} \sum_i (i-1) f_i \quad \left[\because x_0 = \log x_1 \text{ and } c = \log \lambda \text{ (Given)}\right]$$

Example 2·23. *For a group of 5,000 shopkeepers the hourly earnings vary from ₹ 200 to ₹ 800. The earnings of 4 per cent of the shopkeepers are under ₹ 250 and those of 10 per cent are under ₹ 300 ; 15 percent of the shopkeepers earn ₹ 600 and over, and 5 per cent of them earn ₹ 700 and over. The quartile earnings are ₹ 400 and ₹ 540, and the sixth decile is ₹ 500. Put this information in the form of a frequency table.*

Solution. We are given $N = 5{,}000$.

(*i*) 4% of 5,000 = (4 × 5,000/100) = 200 shopkeepers earn under ₹ 250.

(*ii*) 10% of 5,000 = 500 shopkeepers earn under ₹ 300.

(*iii*) 15% of 5,000 = 750 shopkeepers earn ₹ 600 and over, *i.e.*, ≥ ₹ 600.

(*iv*) 5% of 5,000 = 250 shopkeepers earn ≥ ₹ 700.

(*v*) Q_1 = ₹ 400 ⇒ 25% of 5,000 = 1,250 shopkeepers earn less than ₹ 400.

(*vi*) Q_3 = ₹ 540 ⇒ 75% of 5,000 = 3,750 shopkeepers earn less than ₹ 540.

(*vii*) D_6 = ₹ 500 ⇒ 60% of 5,000 = 3,000 shopkeepers earn less than ₹ 500.

Since the earnings vary from ₹ 200 to ₹ 800, the first class will start with 200 and the last class will end with 800.

The above information can be expressed in the following *Table 2·19 (a)*.

TABLE 2·19(*a*) : FREQUENCY DISTRIBUTION OF EARNINGS OF SHOPKEEPERS

Value of the Variable (Hourly earnings in ₹)	*No. of Shopkeepers*	*Hourly Earnings (₹)*	*No. of Shopkeepers*
Less than 250	200	200—250	200
Less than 300	500	250—300	500 − 200 = 300
Less than 400	1,250	300—400	1,250 − 500 = 750
Less than 500	3,000	400—500	3,000 − 1,250 = 1,750
Less than 540	3,750	500—540	3,750 − 3,000 = 750
		540—600	5,000 − (3,750 + 750) = 500
More than 600	750	600—700	750 − 250 = 500
More than 700	250	700—800	250

In the above table, the various classes are of unequal widths. Rearranging and combining them to have classes with equal magnitude of 100 each, the final frequency distribution of earning of 5,000 shopkeepers is as shown in *Table 2·19 (b)*.

TABLE 2·19(*b*) : FREQUENCY DISTRIBUTION OF EARNINGS OF SHOPKEEPERS

Hourly Earnings (₹)	*No. of Shopkeepers (f)*
200 and under 300	200 + 300 = 500
300 and under 400	750
400 and under 500	1,750
500 and under 600	750 + 500 = 1,250
600 and under 700	500
700 and under 800	250

ASSORTED REVIEW PROBLEMS ON MEASURES OF CENTRAL TENDENCY

1. What are grouped and ungrouped frequency distributions ? What are their uses ? What are the considerations that one has to bear in mind while forming the frequency distribution ?

2. Explain the method of constructing Histogram and Frequency Polygon. Which, out of these two, is better representative of frequencies of (*i*) a particular group, and (*ii*) whole group.

3. What are the principles governing the choice of :

(*i*) Number of class intervals, (*ii*) The length of the class interval,

(*iii*) The mid-point of the class intervals.

4. Write short notes on :

(*i*) Frequency distribution, (*ii*) Histogram, frequency polygon and frequency curve, (*iii*) Ogive.

5. (*a*) What are the properties of a good average ? Examine these properties with reference to the Arithmetic Mean, the Geometric Mean and the Harmonic Mean, and give an example of situations in which each of them can be the appropriate measure for the average.

(*b*) Compare mean, median and mode as measures of location of a distribution.

(*c*) Explain the graphic method of locating the values of median, quartiles and mode.

(*d*) Explain how (*i*) the ogives, and (*ii*) histogram, can be drawn for any frequency distribution. Point out the method of finding out the values of median, mode, quartile, deciles and percentiles graphically. Also, write down the formula for the computation of each of them for any frequency distribution.

6. (*a*) "The mean is the most common measure of central tendency of the data. It satisfies almost all the requirements of a good average. The median is also an average, but it does not satisfy all the requirements of a good average. However, it carries certain merits and hence is useful in particular fields." Critically examine both the averages.

(*b*) Describe the different measures of central tendency of a frequency distribution, mentioning their merits and demerits.

7. (*a*) Define (*i*) arithmetic mean, (*ii*) geometric mean, and (*iii*) harmonic mean of grouped and ungrouped data. Compare and contrast their merits and demerits. Show that the geometric mean is capable of further mathematical treatment.

(*b*) When is an average a meaningful statistics ? What are the requisites of a satisfactory average ? In this light compare the relative merits and demerits of three well-known averages.

8. Show that (*i*) Sum of deviations about arithmetic mean is zero.

(*ii*) Sum of absolute deviations about median is least.

(*iii*) Sum of the squares of deviations about arithmetic mean is least.

9. The following numbers give the weights of 55 students of a class. Prepare a suitable frequency table :

42	74	40	60	82	115	41	61	75	83	63
53	110	76	84	50	67	65	78	77	56	95
68	69	104	80	79	79	54	73	59	81	100
66	49	77	90	84	76	42	64	69	70	80
72	50	79	52	103	96	51	86	78	94	71

(*i*) Draw the histogram and frequency polygon of the above data. From the histogram, obtain an approximate value of mode.

(*ii*) For the above weights, prepare a cumulative frequency table and draw the less than ogive. Hence, obtain an approximate value of median.

10. What are the points to be borne in mind in the formation of a frequency table ?

Choosing appropriate class intervals, form a frequency table for the following data :

10·2	0·5	5·2	6·1	3·1	6·7	8·9	7·2	8·9
5·4	3·6	9·2	6·1	7·3	2·0	1·3	6·4	8·0

4·3	4·7	12·4	8·6	13·1	3·2	9·5	7·6	4·0
5·1	8·1	1·1	11·5	3·1	6·8	7·0	8·2	2·0
3·1	6·5	11·2	12·0	5·1	10·9	11·2	8·5	2·3
3·4	5·2	10·7	4·9	6·2				

11. What are the considerations one has to bear in mind while forming a frequency distribution ?

A sample consists of 34 observations recorded correct to the nearest integer, ranging in value from 201 to 337. If it is decided to use seven classes of width 20 integers and to begin the first class at 199·5, find the class limits and class marks of the seven classes.

12. The class marks in a frequency table (of whole numbers) are given to be 5, 10, 15, 20, 25, 30, 35, 40, 45 and 50. Find out the following :

(*i*) the true classes, (*ii*) the true class limits, (*iii*) the true upper class limits.

13. (*a*) The following table shows the distribution of the number of students per teacher in 750 colleges :

Students	1	4	7	10	13	16	19	22	25	28
Frequency	7	46	165	195	189	89	28	19	9	3

Draw the histogram for the data and superimpose on it the frequency polygon.

(*b*) Draw the histogram and frequency curve for the following data :

Monthly wages in '000 ₹	11—13	13—15	15—17	17—19	19—21	21—23	23—25
No. of workers	6	53	85	56	21	16	8

14. (*a*) Three persons *A*, *B* and *C* were given the job of finding the average of 5,000 numbers. Each one did his own simplification. *A*'s method : Divide the sets into sets of 1,000 each, calculate the average in each set and then calculate the average of these averages. *B*'s method : Divide the set into 2,000 and 3,000 numbers, take average in each set and then take the average of the averages. *C*'s method : 500 numbers were unities. He averaged all other numbers and then added one. Are these methods correct ?

Ans. (*a*) Correct, not correct, not correct

(*b*) The total sale (in '000 rupees) of a particular item in shop, on 10 consecutive days, is reported by a clerk as, 35·00, 29·60, 38·00, 30·00, 40·00, 41·00, 42·00, 45·00, 3·60, 3·80. Calculate the average. Later it was found that there was a number 10·00 in the machine and the reports of 4th to 8th days were 10·00 more than the true values and in the last 2 days he put a decimal in the wrong place thus for example 3.60 was really 36·0. Calculate the true mean value.

15. (*a*) Given below is the distribution of 40 candidates obtaining marks X or higher in a certain examination (all marks are given in whole numbers) :

X	10	20	30	40	50	60	70	80	90	100
More than *c.f.*	140	133	118	100	75	45	25	9	2	0

Calculate the mean, median and mode of the distribution.

(*b*) The four parts of a distribution are as follows :

Part	*Frequency*	*Mean*
1	50	61
2	100	70
3	120	80
4	30	83

Find the mean of the distribution.

16. (*a*) Define a 'weighted mean'. If several sets of observations are combined into a single set, show that the mean of the combined set is the weighted mean of several sets.

(*b*) The weighted geometric mean of three numbers 229, 275 and 125 is 203. The weights for the first and second numbers are 2 and 4 respectively. Find the weight of third. **[Ans.** 3]

17. The following table shows some data collected for three regions of a country :

Region	*Number of inhabitants (million)*	*Percentage of literates*	*Average daily income per person* (₹)
A	10	52	850
B	5	68	620
C	18	39	730

Obtain the overall figures for the three regions taken together. Prove the formulae you use.

18. For the two frequency distributions given in the following table, the mean calculated from the first was 25·4 and that from the second was 32·5. Find the values of x and y.

Class	*Distribution I Frequency*	*Distribution II Frequency*
10—20	20	4
20—30	15	8
30—40	10	4
40—50	x	$2x$
50—60	y	y

Ans. $x = 3$, $y = 2$

19. Draw the ogives and hence estimate the median.

Class :	0—9	10—19	20—29	30—39	40—49	50—59	60—69	70—79
Frequency :	8	32	142	216	240	206	143	13

20. The following data relate to the ages of a group of workers in a factory. Draw the cumulative frequency curve and find from the graph the number of workers between the ages 28—48.

Ages	*No. of workers*	*Ages*	*No. of workers*
20—25	35	40—45	90
25—30	45	45—50	74
30—35	70	50—55	51
35—40	105	55—60	30

21. (*a*) The mean of marks obtained in an examination by a group of 100 students was found to be 49·96. The mean of the marks obtained in the same examination by another group of 200 students was 52·32. Find the mean of the marks obtained by both the groups of students taken together.

(*b*) A distribution consists of three components with frequencies 300, 200 and 600, having their means 16, 8 and 4 respectively. Find the mean of the combined distribution.

(*c*) The mean marks obtained by 300 students in the subject of Statistics are 45. The mean of the top 100 of them was found to be 70 and the mean of the last 100 was known to be 20. What is the mean of the remaining 100 students ?

(*d*) The mean weight of 150 students in a certain class is 60 kilograms. The mean weight of boys in the class is 70 kilograms and that of the girls is 55 kilograms. Find the number of boys and number of girls in the class. **[Ans.** (*a*) 51·33, (*b*) 8, (*c*) 45, (*d*) 50, 100.**]**

22. From the following table showing the wage distribution in a certain factory, determine :

(*a*) the mean wages, (*b*) the median wages, (*c*) the modal wages,

(*d*) the wage limits for the middle 50% of the wage earners,

(*e*) the percentages of workers who earned between ₹ 75 and ₹ 125, (*f*) the percentage of workers who earned more than ₹ 150 per hour, and

(*g*) the percentage of workers who earned less than ₹ 100 per hour.

Hourly wages (₹)	*No. of employees*	*Hourly wages* (₹)	*No. of employees*
20—40	8	120—140	35
40—60	12	140—160	18
60—80	20	160—180	7
80—100	30	180—200	5
100—120	40		

Ans. (*a*) $\overline{X}$ = 108·5, (*b*) Md. = 108·75, (*c*) Mode = 118·3, (*d*) 81·25, 129·3, (*e*) 48, (*f*) = 12, (*g*) 40

23. (*a*) The following table gives the frequency distribution of marks in a class of 65 students :

Marks	*No. of Students*	*Marks*	*No. of Students*
0—4	10	14—18	4
4—8	12	18—20	3
8—12	18	20—25	4
12—14	7	25 and over	6
Total			65

Calculate : (*i*) Upper and lower quartiles.

(*ii*) Number of students who secured marks more than 17.

(*iii*) Number of students who secured marks between 10 and 15.

(*b*) The following table shows the age distribution, (in years), of heads of families in a certain country during the year 2014. Find the median, the third quartile and the second decile of the distribution. Check your results by the graphical method.

Age of head of family :	Under 25	25—29	30—34	35—44	45—54	55—64	65—74	above 74
Number (million) :	2.3	4·1	5·3	10·6	9·7	6·8	4·4	1·8

Ans. Md = 45·2 years, Q_3 = 57·5 years, D_2 = 32·5 years.

24. The following data represent travel expenses (other than transportation) for 7 trips made during November by a salesman for a small firm :

Trip	*Days*	*Expenses* (₹)	*Expenses per day* (₹)
1	0·5	1,350	2,700
2	2·0	1,200	600
3	3·5	1,750	500
4	1·0	900	900
5	9·0	2,700	300
6	0·5	900	1,800
7	8·5	1,700	200
Total	25·0	10,500	7,000

An auditor criticised these expenses as excessive, asserting that the average expenses per day is ₹ 1,000 (₹ 7,000 divided by 7). The salesman replied that the average is only ₹ 420 (₹ 10,500 divided by 25) and that in any event the median is the appropriate measure and is only ₹ 300. The auditor rejoined that the arithmetic mean is the appropriate measure, but that the median is ₹ 600.

You are required to :

(i) Explain the proper interpretation of each of the four averages mentioned.

(ii) Which average seems appropriate to you ?

25. (a) An assessee depreciated the machinery of his factory by 10% each in the first two years and by 40% in the third year and thereby claimed 21% average depreciation relief from taxation department, but the I.T.O. objected and allowed only 20%. Show which of the two is right.

(b) A given machine is assumed to depreciate 40% in value in the first year, 25% in the second year and 10% per annum for the next three years, each percentage being calculated on the diminishing value. What is the average percentage depreciation, reckoned on the diminishing value for the five years ? **[Ans.** *20%***]**

(c) An economy grows at the rate of 2% in the first year, 2.5% in the second, 3% in the third, 4% in the fourth, ... and 10% in the tenth year. What is the average rate of growth of the economy ?

Hint. Average rate of growth of economy per annum

$$= [(100 + 2)\ (100 + 2{\cdot}5)\ (100 + 3)\ ...\ (100 + 10)]^{1/10} - 100.$$

(d) The export of a commodity increased by 30% in 2010, decreased by 22% in 2011 and then increased by 45% in the following year. The increase/ decrease in each year, being measured in comparison to its previous year. Calculate the average rate of change of the exports per annum.

26. (a) You take a tip which entails travelling 900 miles by train at an average speed of 60 m.p.h.; 300 miles by boat at an average of 25 m.p.h.; 400 miles by plane at 350 m.p.h. and finally 15 miles by taxi at 25 m.p.h. What is your speed for the entire distance ?

(b) A train runs 25 miles at a speed of 30 m.p.h.; another 50 miles at a speed of 40 m.p.h.; then due to repairs of the track, travels for 6 minutes at a speed of 100 m.p.h. and finally covers the remaining distance of 24 miles at a speed of 24 m.p.h. What is the average speed in m.p.h. ?

(c) A man motors from A to B. A large part of the distance is uphill and he gets a mileage of only 10 per gallon of gasoline. On the return trip he makes 15 miles per gallon. Find the harmonic mean of his mileage. Verify the fact this is the proper average to be used by assuming that the distance from A to B is 60 miles.

(d) Calculate the average speed of a car running at the rate of 15 km.p.h. during the first 30 km.; at 20 km.p.h. during the second 30 km. and at 25 km.p.h. during the third 30 km.

2·12. DISPERSION

Averages (or the measures of central tendency) give us an idea of the concentration of the observations about the central part of the distribution. If we know the average alone, we cannot form a complete idea about the distribution as will be clear from the following example.

Consider the series (i) 7, 8, 9, 10, 11, (ii) 3, 6, 9, 12, 15 and (iii) 1, 5, 9, 13, 17. In all these cases we see that n, the number of observations, is 5 and the mean is 9. If we are given that the mean of 5 observations is 9, we cannot form an idea as to whether it is the average of first series or second series or third series or of any other series of 5 observations whose sum is 45. Thus we see that the measures of central tendency are inadequate to give us a complete idea of the distribution. They must be supported and supplemented by some other measures. One such measure is *Dispersion*.

Literal meaning of dispersion is *'scatteredness'*. We study dispersion to have an idea about the homogeneity or heterogeneity of the distribution. In the above case we say that series (i) is more homogeneous (less dispersed) than the series (ii) or (iii) or we say that series (iii) is more heterogeneous (more scattered) than the series (i) or (ii).

Some important definitions of dispersion are given below :

(i) *"Dispersion is the measure of extent to which individual items vary."*
—L.R. Connor

(ii) *"The measure of the scatteredness of the mass of figures in a series about an average is called the measure of variation or dispersion."*
—Simpson and Kafka

(iii) *"The degree to which numerical data tend to spread about an average value is called variation or dispersion of the data."*
—Spiegel

2·12·1. Characteristics for an Ideal Measure of Dispersion

The desiderata for an ideal measure of dispersion are the same as those for an ideal measure of central tendency, *viz.*,

(i) It should be rigidly defined.
(ii) It should be easy to calculate and easy to understand.
(iii) It should be based on all the observations.
(iv) It should be amenable to further mathematical treatment.
(v) It should be affected as little as possible by fluctuations of sampling.

2·13. MEASURES OF DISPERSION

Various measures of dispersion can be classified into two broad categories :

(a) The measures which express the spread of observations in terms of distance between the values of selected observations. These are also termed as *distance measures, e.g.*, range and interquartile range (or quartile deviation).

(b) The measures which express the spread of observations in terms of the average of deviations of observations from some central value, *e.g.*, mean deviation and standard deviation.

2·13·1. Range

The range is the difference between two extreme observations of the distribution. If A and B are the greatest and smallest observations respectively in a distribution, then its range is given by :

$$\text{Range} = X_{max} - X_{min} = A - B \qquad \text{... (2·12)}$$

Range is the simplest but a crude measure of dispersion. Since it is based on two extreme observations which themselves are subject to chance fluctuations, it is not at all a reliable measure of dispersion.

2·13·2. Quartile Deviation

Quartile deviation or *semi-interquartile range* Q is given by :

$$Q = \frac{1}{2}(Q_3 - Q_1), \qquad \text{... (2·13)}$$

where Q_1 and Q_3 are the first and third quartiles of the distribution respectively.

Quartile deviation is definitely a better measure than the range as it makes use of 50% of the data. But since it ignores the other 50% of the data, it cannot be regarded as a reliable measure.

2·13·3. Mean Deviation

If $x_i \mid f_i$, $i = 1, 2, ..., n$ is the frequency distribution, then mean deviation from the average A (usually mean, median or mode) is given by :

$$\text{Mean deviation from average } A = \frac{1}{N}\sum_{i=1}^{n} f_i \mid x_i - A \mid, \quad \sum_i f_i = N \qquad \text{... (2·14)}$$

where $|x_i - A|$ represents modulus or the absolute value of the deviation $(x_i - A)$, where the negative sign is ignored.

Since mean deviation is based on all the observations, it is a better measure of dispersion than range or quartile deviation. But the step of ignoring the signs of the deviations $(x_i - A)$ creates artificiality and renders it useless for further mathematical treatment.

Remark: It may be pointed out here that mean deviation is least when taken from median. (The proof is given for continuous variable in Chapter 5.

Example 2·24. *Calculate : (i) Quartile deviation (Q.D.), and (ii) Mean Deviation (M.D.) from mean, for the following data :*

Marks. :	0—10	10—20	20—30	30—40	40—50	50—60	60—70
No. of Students :	6	5	8	15	7	6	3

Solution.

TABLE 2·20 : CALCULATIONS FOR Q.D. AND M.D. FROM MEAN

Marks	*Mid-value* (x)	*No. of Students* (f)	$d = \frac{x - 35}{10}$	*fd*	$\lvert x - \bar{x} \rvert = \lvert x - 33·4 \rvert$	$f \lvert x - \bar{x} \rvert$	*Less than c.f.*
0—10	5	6	–3	–18	28·4	170·4	6
10—20	15	5	–2	–10	18·4	92·0	11
20—30	25	8	–1	–8	8·4	67·2	19
30—40	35	15	0	0	1·6	24·0	34
40—50	45	7	1	7	11·6	81·2	41
50—60	55	6	2	12	21·6	129·6	47
60—70	65	3	3	9	31·6	94·8	50
Total		50		– 8		659·2	

(*i*) Here $\quad N = 50 ; \quad \frac{1}{4}N = 12·75 ; \quad \frac{3}{4}N = 37·25$

The *c.f.* just greater than 12·75 is 19. Hence, the corresponding class 20—30 contains Q_1.

$$\therefore \quad Q_1 = 20 + \frac{10}{8}(12·75 - 11) = 22·19$$

The *c.f.* just greater than 37·25 is 41. Hence, the corresponding class 40—50 contains Q_3.

$$\therefore \quad Q_3 = 40 + \frac{10}{7}(37·25 - 34) = 44·64$$

Hence, $$Q.D. = \frac{1}{2}(Q_3 - Q_1) = \frac{1}{2}(44·64 - 22·19) = 11·23$$

(*ii*) $$\text{Mean, } (\bar{x}) = A + \frac{h\,\Sigma fd}{N} = 35 + \frac{10 \times (-8)}{50} = 33·4 \text{ marks.}$$

$$\therefore \quad \text{M.D. (from mean)} = \frac{1}{N}\Sigma f \,|\, x - \bar{x} \,| = \frac{659·2}{50} = 13·184.$$

2·13·4. Standard Deviation and Root Mean Square Deviation

Standard deviation, usually denoted by the Greek letter small sigma (σ), is the positive square root of the arithmetic mean of the squares of the deviations of the given values from their arithmetic mean. For the frequency distribution $x_i \mid f_i$; $i = 1, 2, ..., n$,

$$\sigma = \sqrt{\frac{1}{N} \sum_i f_i (x_i - \bar{x})^2} \qquad \text{... (2·15)}$$

where $\bar{x}$ is the arithmetic mean of the distribution and $\sum_i f_i = N$.

The step of squaring the deviations $(x_i - \bar{x})$ overcomes the drawback of ignoring the signs in mean deviation. Standard deviation is also suitable for further mathematical treatment [*see equation* (2·21)]. Moreover, of all the measures, standard deviation is affected least by fluctuations of sampling.

Thus, we see that standard deviation satisfies almost all the properties laid down for an ideal measure of dispersion except for the general nature of extracting the square root which is not readily comprehensible for a non-mathematical person. It may also be pointed out that standard deviation gives greater weight to extreme values and as such has not found favour with economists or businessmen who are more interested in the results of the modal class. Taking into consideration the pros and cons and also the wide applications of standard deviation in statistical theory, we may regard *standard deviation as the best and the most powerful measure of dispersion.*

The square of standard deviation is called the *variance* and is given by :

$$\sigma^2 = \frac{1}{N} \sum_i f_i (x_i - \bar{x})^2 \qquad \text{... (2·15a)}$$

Root mean square deviation, denoted by 's', is given by :

$$s = \sqrt{\frac{1}{N} \sum_i f_i (x_i - A)^2} \qquad \text{... (2·16)}$$

where A is any arbitrary number. s^2 is called *mean square deviation.*

Remark. *Relation between σ and s.* By def., we have

$$s^2 = \frac{1}{N} \sum_i f_i (x_i - A)^2 = \frac{1}{N} \sum_i f_i (x_i - \bar{x} + \bar{x} - A)^2$$

$$= \frac{1}{N} \sum_i f_i \left\{ (x_i - \bar{x})^2 + (\bar{x} - A)^2 + 2(\bar{x} - A)(x_i - \bar{x}) \right\}$$

$$= \frac{1}{N} \sum_i f_i (x_i - \bar{x})^2 + (\bar{x} - A)^2 \frac{1}{N} \sum_i f_i + 2(\bar{x} - A) \frac{1}{N} \sum_i f_i (x_i - \bar{x}),$$

$(\bar{x} - A)$, being constant is taken outside the summation sign. But $\sum_i f_i (x_i - \bar{x}) = 0$, being the algebraic sum of the deviations of the given values from their mean. Thus,

$$s^2 = \sigma^2 + (\bar{x} - A)^2 = \sigma^2 + d^2, \text{ where } d = \bar{x} - A$$

Obviously, s^2 will be least when $d = 0$, *i.e.*, $\bar{x} = A$. Hence mean square deviation and consequently root mean square deviation is least when the deviations are taken from $A = \bar{x}$, *i.e., standard deviation is the least value of root mean square deviation.*

The same result could be obtained alternatively as follows :

Mean square deviation is given by : $s^2 = \frac{1}{N}\sum_i f_i (x_i - A)^2$

It has been shown in § 2·5·1 Property 2, that $\sum_i f_i (x_i - A)^2$ is minimum when $A = \bar{x}$. Thus, mean square deviation is minimum when $A = \bar{x}$ and its minimum value is:

$$(s^2)_{min} = \frac{1}{N}\sum_i f_i (x_i - \bar{x})^2 = \sigma^2 \qquad \text{... (2·16}b\text{)}$$

Hence, variance is the minimum value of mean square deviation or standard deviation is the minimum value of root mean square deviation.

Different Formulae for Calculating Variance. By *def.*, $\sigma^2 = \frac{1}{N}\sum_i f_i (x_i - \bar{x})^2$

More precisely we write it as σ_x^2, *i.e.*, variance of x. Thus,

$$\sigma_x^2 = \frac{1}{N}\sum_i f_i (x_i - \bar{x})^2 \qquad \text{... (2·17)}$$

If $\bar{x}$ is not a whole number but comes out to be in fractions, the calculation of σ_x^2 by using (2·17) is very cumbersome and time consuming. In order to overcome this difficulty, we shall develop different forms of the formula (2·17) which reduce the arithmetic to a great extent and are very useful for computational work. In the following discussion, the summation is extended over i from 1 to n.

$$\sigma_x^2 = \frac{1}{N}\sum_i f (x_i - \bar{x})^2 = \frac{1}{N}\sum_i f_i (x_i^2 + \bar{x}^2 - 2x_i\bar{x})$$

$$= \frac{1}{N}\sum_i f_i x_i^2 + \bar{x}^2 \frac{1}{N}\sum_i f_i - 2\bar{x} . \frac{1}{N}\sum_i f_i x_i$$

$$= \frac{1}{N}\sum_i f_i x_i^2 + \bar{x}^2 - 2\bar{x}^2 = \frac{1}{N}\sum_i f_i x_i^2 - \bar{x}^2 \qquad \text{... (2·18)}$$

$$\therefore \quad \sigma_x^2 = \frac{1}{N}\sum_i f_i x_i^2 - \left(\frac{1}{N}\sum_i f_i x_i\right)^2 \qquad \text{... (2·18}a\text{)}$$

If the values of x and f are large, the calculation of fx, fx^2 is quite tedious. In that case we take the deviations from any arbitrary point 'A'. Generally the point in the middle of the distribution is much convenient though the formula is true in general.

Let $d_i = x_i - A$ (*)

Multiplying by f_i, summing over i from 1 to n, and dividing by N, we get

$$\frac{1}{N}\sum_i f_i d_i = \frac{1}{N}\sum_i f_i x_i - A . \frac{1}{N}\sum f_i \quad \Rightarrow \quad \bar{d} = \bar{x} - A \qquad \text{... (**)}$$

Subtracting (**) from (*), we get $d_i - \bar{d} = x_i - \bar{x}$

$$\therefore \quad \sigma_x^2 = \frac{1}{N}\sum_{i=1}^{n} f_i (x_i - \bar{x})^2 = \frac{1}{N}\sum_{i=1}^{n} f_i (d_i - \bar{d})^2 = \sigma_d^2 \qquad \text{... (2·19)}$$

Hence, *variance and consequently standard deviation is independent of change of origin.*

If $\quad d_i = \dfrac{x_i - A}{h}$, then $x_i = A + hd_i$

$$\therefore \quad \bar{x} = A + h \cdot \frac{1}{N} \sum_i f_i d_i = A + h\bar{d} \quad \Rightarrow \quad x_i - \bar{x} = h(d_i - \bar{d}) \quad \text{... (***)}$$

and
$$\sigma_x^2 = \frac{1}{N} \sum_i f_i (x_i - \bar{x})^2 = h^2 \cdot \frac{1}{N} \sum_i f_i (d_i - \bar{d})^2 = h^2 \sigma_d^2 \qquad \text{[From (***)]} \quad \text{... (2·20)}$$

Hence, *variance is independent of change of origin but not of scale.*

Variance of the Combined Series. *If n_1, n_2 are the sizes, $\bar{x}_1, \bar{x}_2$ the means, and σ_1, σ_2 the standard deviations of two series, then the standard deviation σ of the combined series of size $n_1 + n_2$ is given by :*

$$\sigma^2 = \frac{1}{n_1 + n_2} \left[n_1(\sigma_1^2 + d_1^2) + n_2(\sigma_2^2 + d_2^2) \right], \quad \text{... (2·21)}$$

where $d_1 = \bar{x}_1 - \bar{x}$, $d_2 = \bar{x}_2 - \bar{x}$ *and* $\bar{x} = \dfrac{n_1 \bar{x}_1 + n_2 \bar{x}_2}{n_1 + n_2}$, *is the mean of the combined series.*

Proof. Let x_{1i} ; $i = 1, 2, ..., n_1$ and x_{2j} ; $j = 1, 2, ..., n_2$, be the two series, then

$$\left.\begin{aligned} \bar{x}_1 &= \frac{1}{n_1} \sum_{i=1}^{n_1} x_{1i} \\ \bar{x}_2 &= \frac{1}{n_2} \sum_{j=1}^{n_2} x_{2j} \end{aligned}\right\} \text{... (*)} \quad \text{and} \quad \left.\begin{aligned} \sigma_1^2 &= \frac{1}{n_1} \sum_{i=1}^{n_1} (x_{1i} - \bar{x}_1)^2 \\ \sigma_2^2 &= \frac{1}{n_2} \sum_{j=1}^{n_2} (x_{2j} - \bar{x}_2)^2 \end{aligned}\right\} \text{... (**)}$$

The mean $\bar{x}$ of the combined series is given by :

$$\bar{x} = \frac{1}{n_1 + n_2} \left(\sum_{i=1}^{n_1} x_{1i} + \sum_{j=1}^{n_1} x_{2j} \right) = \frac{n_1 \bar{x}_1 + n_2 \bar{x}_2}{n_1 + n_2} \qquad \text{[From (*)]}$$

The variance σ^2 of the combined series is given by :

$$\sigma^2 = \frac{1}{n_1 + n_2} \left[\sum_{i=1}^{n_1} (x_{1i} - \bar{x})^2 + \sum_{j=1}^{n_2} (x_{2j} - \bar{x})^2 \right] \quad \text{... (2·22)}$$

$$\sum_{i=1}^{n_1} (x_{1i} - \bar{x})^2 = \sum_{i=1}^{n_1} (x_{1i} - \bar{x}_1 + \bar{x}_1 - \bar{x})^2$$

$$= \sum_{i=1}^{n_1} (x_{1i} - \bar{x}_1)^2 + n_1(\bar{x}_1 - \bar{x})^2 + 2(\bar{x}_1 - \bar{x}) \sum_{i=1}^{n_1} (x_{1i} - \bar{x}_1) \quad \text{... (2·22a)}$$

But $\sum_{i=1}^{n_1} (x_{1i} - \bar{x}_1) = 0$, being the algebraic sum of the deviations of the values of first series from their mean. Hence from (2·22*a*), on using (**), we get

$$\sum_{i=1}^{n_1} (x_{1i} - \bar{x})^2 = n_1 \sigma_1^2 + n_1(\bar{x}_1 - \bar{x})^2 = n_1 \sigma_1^2 + n_1 d_1^2, \text{ where } d_1 = \bar{x}_1 - \bar{x}. \quad \text{... (2·22b)}$$

Similarly, we get

$$\sum_{j=1}^{n_2} (x_{2j} - \bar{x})^2 = \sum_{j=1}^{n_2} (x_{2j} - \bar{x}_2 + \bar{x}_2 - \bar{x})^2$$

$$= \sum_{j=1}^{n_2} (x_{2j} - \bar{x}_2)^2 + n_2(\bar{x}_2 - \bar{x})^2 = n_2\,\sigma_2^{\,2} + n_2\, d_2^{\,2}, \text{ where } d_2 = \bar{x}_2 - \bar{x}. \qquad \text{... (2·22c)}$$

Substituting from (2·22*b*) and (2·22*c*) in (2·22), we get the required formula

$$\sigma^2 = \frac{1}{n_1 + n_2}\left[n_1(\sigma_1^{\,2} + d_1^{\,2}) + n_2(\sigma_2^{\,2} + d_2^{\,2})\right] \qquad \text{... (2·22d)}$$

This formula can be simplified still further. We have

$$d_1 = \bar{x}_1 - \bar{x} = \bar{x}_1 - \frac{n_1\,\bar{x}_1 + n_2\,\bar{x}_2}{n_1 + n_2} = \frac{n_2(\bar{x}_1 - \bar{x}_2)}{n_1 + n_2}$$

and
$$d_2 = \bar{x}_2 - \bar{x} = \bar{x}_2 - \frac{n_1\,\bar{x}_1 + n_2\,\bar{x}_2}{n_1 + n_2} = \frac{n_1(\bar{x}_2 - \bar{x}_1)}{n_1 + n_2}$$

Hence,

$$\sigma^2 = \frac{1}{n_1 + n_2}\left[n_1\,\sigma_1^{\,2} + n_2\,\sigma_2^{\,2} + \left\{\frac{n_1\,n_2^{\,2}(\bar{x}_1 - \bar{x}_2)^2}{(n_1 + n_2)^2} + \frac{n_2\,n_1^{\,2}(\bar{x}_2 - \bar{x}_1)^2}{(n_1 + n_2)^2}\right\}\right]$$

$$= \frac{1}{n_1 + n_2}\left[n_1\,\sigma_1^{\,2} + n_2\,\sigma_2^{\,2} + \frac{n_1\,n_2}{n_1 + n_2}(\bar{x}_1 - \bar{x}_2)^2\right] \qquad \text{... (2·23)}$$

Remark. The formula (2·21) can be easily generalised to the case of more than two series. If n_i, $\bar{x}_i$ and σ_i ; $i = 1, 2, ..., k$ are the sizes, means and standard deviations, respectively, of *k*-component series, then the standard deviation σ of the combined series of size $\sum_{i=1}^{k} n_i$, is given by :

$$\sigma^2 = \frac{1}{n_1 + n_2 + ... + n_k}\left[n_1(\sigma_1^{\,2} + d_1^{\,2}) + n_2(\sigma_2^{\,2} + d_2^{\,2}) + ... + n_k(\sigma_k^{\,2} + d_k^{\,2})\right] \qquad \text{... (2·24)}$$

where $d_i = \bar{x}_i - \bar{x}$; $i = 1, 2, ..., k$ and $\bar{x} = \dfrac{n_1\bar{x}_1 + n_2\bar{x}_2 + ... + n_k\bar{x}_k}{n_1 + n_2 + ... + n_k}$ ... (2·24*a*)

Example 2·25. *Calculate the mean and standard deviation for the following table giving the age distribution of 542 members of a club :*

Age (in years) :	20 — 30	30 — 40	40 — 50	50 — 60	60 — 70	70 — 80	80 — 90
No. of members :	3	61	132	153	140	51	2

Solution. Here we take $d = \dfrac{x - A}{h} = \dfrac{x - 55}{10}$

TABLE 2·21 CALCULATIONS FOR MEAN AND S.D.

Age group	*Mid-value* x	*Frequency* f	$d = \frac{x-55}{10}$	*fd*	fd^2
20—30	25	3	−3	−9	27
30—40	35	61	−2	−122	244
40—50	45	132	−1	−132	132
50—60	55	153	0	0	0
60—70	65	140	1	140	140
70—80	75	51	2	102	204
80—90	85	2	3	6	18
Total		542		−15	765

$$\bar{x} = A + h\frac{\Sigma fd}{N}$$

$$= 55 + \frac{10 \times (-15)}{542} = 55 - 0{\cdot}28 = 54{\cdot}72 \text{ years}$$

$$\sigma^2 = h^2\left[\frac{1}{N}\Sigma fd^2 - \left(\frac{1}{N}\Sigma fd\right)^2\right] = 100\left[\frac{765}{542} - (0{\cdot}028)^2\right]$$

$$= 100 \times 1{\cdot}4107 = 141{\cdot}07$$

$\therefore$ σ (standard deviation) $= \sqrt{141{\cdot}07} = 11{\cdot}88$ years.

Example 2·26. *For a group of 200 candidates, the mean and standard deviation of scores were found to be 40 and 15 respectively. Later on it was discovered that the scores 43 and 35 were misread as 34 and 53 respectively. Find the corrected mean and standard deviation corresponding to the correct figures.*

Solution. Let x be the given variable. We are given $n = 200$, $\bar{x} = 40$ and $\sigma = 15$.

$$\bar{x} = \frac{1}{n}\sum_{i=1}^{n} x_i \quad \Rightarrow \quad \sum_i x_i = n\bar{x} = 200 \times 40 = 8{,}000$$

$$\sigma^2 = \frac{1}{n}\sum_i x_i^2 - \bar{x}^2 \quad \Rightarrow \quad \sum_i x_i^2 = n(\sigma^2 + \bar{x}^2) = 200(225 + 1{,}600) = 3{,}65{,}000$$

Corrected $\sum_i x_i = 8{,}000 - 34 - 53 + 43 + 35 = 7{,}991$; Corrected mean $= \frac{7{,}991}{200} = 39{\cdot}995$

and Corrected $\sum_i x_i^2 = 3{,}65{,}000 - (34)^2 - (53)^2 + (43)^2 + (35)^2 = 3{,}64{,}109$

$\Rightarrow$ Corrected $\sigma^2 = \frac{3{,}64{,}109}{200} - (39{\cdot}995)^2 = 1{,}820{\cdot}54 - 1{,}596{\cdot}40 = 224{\cdot}14$

$\therefore$ Corrected standard deviation $= \sqrt{224{\cdot}14} = 14{\cdot}97$.

Example 2·27. *After shift of origin and change of scale, a frequency distribution of a continuous variate takes the form as under :*

Step deviation :	−4	−3	−2	−1	0	1	2	3
Frequency :	2	5	7	13	21	16	8	3

If the mean and variance of the distribution are respectively 21.9 and 63.9725, find the original frequency distribution.

Solution. Let A be assumed mean and h be the magnitude of class interval & let $d = (x - A)/h$.

fd	−8	−15	−14	−13	0	16	16	9	$\Sigma fd = -9$
fd^2	32	45	28	13	0	16	32	27	$\Sigma fd^2 = 193$

We are given : $\bar{x} = 21{\cdot}9$ and $\sigma^2 = 63{\cdot}9725$. If $d = (x - A)/h$, then

$$\bar{x} = A + \frac{h \Sigma fd}{N} \quad \Rightarrow \quad 21{\cdot}9 = A + \frac{h(-9)}{75} \quad \text{... (1)}$$

and
$$\sigma^2 = h^2\left[\frac{\Sigma fd^2}{N} - \left(\frac{\Sigma fd}{N}\right)^2\right] \quad \Rightarrow \quad 63{\cdot}9725 = h^2\left[\frac{193}{75} - \left(\frac{-9}{75}\right)^2\right] \quad \text{... (2)}$$

(2) implies
$$h^2 = \frac{63{\cdot}9725}{2{\cdot}5733 - 0{\cdot}0144} = \frac{63{\cdot}9725}{2{\cdot}5589} = 25 \quad \Rightarrow \quad h = 5 \quad \text{... (3)}$$

Substituting in (1), we get
$$A = 21{\cdot}9 + \frac{9 \times 5}{75} = 22{\cdot}5 \quad \text{... (4)}$$

Hence, the mid-value of the class corresponding to $d = 0$ is $x = A = 22.5$. Let x_1, x_2 be the lower and upper limit of this class. Then

$$\text{Mid-value} = \frac{1}{2}(x_1 + x_2) = 22{\cdot}5 \quad \Rightarrow \quad x_1 + x_2 = 45 \quad \text{... (5)}$$

Magnitude of the class = $x_2 - x_1 = h = 5$ ... (6)

Solving (5) and (6), we obtain $x_1 = 20$ and $x_2 = 25$.

Hence, the class interval corresponding to $d = 0$ is x_1—x_2 = 20—25. Since the magnitude (h) of each class is 5, the class before this is 15—20 and the class after this is 25—30. Other classes can also be obtained similarly and are given in the following *Table 2.22*.

TABLE 2.22 : ORIGINAL FREQUENCY DISTRIBUTION

d	*Mid-value of class*	*Class interval*	*Frequency*
− 4	2·5	0—5	2
− 3	7·5	5—10	5
− 2	12·5	10—15	7
− 1	17·5	15—20	13
0	22·5	20—25	21
1	27·5	25—30	16
2	32·5	30—35	8
3	37·5	35—40	3

Example 2·28. *The first of the two samples has 100 items with mean 15 and standard deviation 3. If the whole group has 250 items with mean 15·6 and standard deviation $\sqrt{13{\cdot}44}$, find the standard deviation of the second group.*

Solution. In usual notations, we are given :

$$n_1 = 100, \ \bar{x}_1 = 15, \ \sigma_1 = 3, \ n = n_1 + n_2 = 250, \ \bar{x} = 15.6, \ \sigma = \sqrt{13{\cdot}44}. \text{ We want } \sigma_2.$$

$$n_2 = 250 - n_1 = 250 - 100 = 150.$$

We have $\bar{x} = \dfrac{n_1 \bar{x}_1 + n_2 \bar{x}_2}{n_2 + n_2} \Rightarrow 15·6 = \dfrac{100 \times 15 + 150 \times \bar{x}_2}{250}$

$\Rightarrow \quad 150\,\bar{x}_2 = 250 \times 15·6 - 1{,}500 = 2{,}400 \quad$ or $\quad \bar{x}_2 = \dfrac{2{,}400}{150} = 16$

Hence, $d_1 = \bar{x}_1 - \bar{x} = 15 - 15·6 = -0·6$ and $d_2 = \bar{x}_2 - \bar{x} = 16 - 15·6 = 0·4$

The variance σ^2 of the combined group is given by the formula :

$$(n_1 + n_2)\,\sigma^2 = n_1(\sigma_1^2 + d_1^2) + n_2(\sigma_2^2 + d_2^2)$$

$$\Rightarrow \quad 250 \times 13·44 = 100(9 + 0·36) + 150\,(\sigma_2^2 + 0·16)$$

$$\Rightarrow \quad 150\,\sigma_2^2 = 250 \times 13·44 - 100 \times 9·36 - 150 \times 0·16 = 2{,}400$$

$$\therefore \quad \sigma_2^2 = \frac{2{,}400}{150} = 16 \quad \Rightarrow \quad \sigma_2 = \sqrt{16} = 4$$

2·14. COEFFICIENTS OF DISPERSION

Whenever we want to compare the variability of the two series which differ widely in their averages or which are measured in different units, we do not merely calculate the measures of dispersion but we calculate the *coefficients of dispersion* which are pure numbers independent of the units of measurement. The coefficients of dispersion (C.D.) based on different measures of dispersion are as follows :

1. C.D based upon range $= \dfrac{A - B}{A + B} = \dfrac{X_{max} - X_{min}}{X_{max} + X_{min}}$

   where A and B are the greatest and the smallest items in the series.

2. Based upon quartile deviation : $\quad$ C.D. $= \dfrac{(Q_3 - Q_1)/2}{(Q_3 + Q_1)/2} = \dfrac{Q_3 - Q_1}{Q_3 + Q_1}$

3. Based upon mean deviation : C.D. $= \dfrac{\text{Mean deviation}}{\text{Average from which it is calculated}}$

4. Based upon standard deviation : $\quad$ C.D. $= \dfrac{\text{S.D.}}{\text{Mean}} = \dfrac{\sigma}{\bar{x}}$

2·14·1. Coefficient of Variation

100 times the coefficient of dispersion based upon standard deviation is called *coefficient of variation* (*C.V.*), *i.e.,*

$$\text{C.V.} = 100 \times \frac{\sigma}{\bar{x}} \qquad \ldots (2·25)$$

According to Professor Karl Pearson who suggested this measure, *C.V. is the percentage variation in the mean, standard deviation being considered as the total variation in the mean.*

For comparing the variability of two series, we calculate the coefficient of variations for each series. The series having greater C.V. is said to be more variable than the other and the series having lesser C.V. is said to be more consistent (or homogeneous) than the other.

Example 2·29. *An analysis of daily wages paid to the workers of two firms A and B belonging to the same industry gives the following results :*

	Firm A	*Firm B*
Number of workers	*500*	*600*
Average daily wage	*₹ 186·00*	*₹ 175·00*
Variance of distribution of wages	*81*	*100*

(*i*) *Which firm, A or B, has a larger wage bill ?*

(*ii*) *In which firm, A or B, is there greater variability in individual wages ?*

(*iii*) *Calculate (a) the average daily wage, and (b) the variance of the distribution of wages, of all the workers in the firms A and B taken together.*

Solution. (*i*) *Firm A* :

No. of wage-earners, (say), $n_1 = 500$, Average daily wage, (say), $\bar{x}_1$ = ₹ 186

$$\text{Average daily wage} = \frac{\text{Total wages paid}}{\text{No. of workers}}$$

Hence, Total wages paid to the workers = $n_1 \bar{x}_1 = 500 \times 186$ = ₹ 93,000

Firm B :

No. of wage-earners, (say), $n_2 = 600$; Average daily wage, (say), $\bar{x}_2$ = ₹ 175

∴ Total daily wages paid to the workers = $n_2 \bar{x}_2 = 600 \times 175$ = ₹ 1,05,000

Thus we see that the firm *B* has larger wage bill.

(*ii*) Variance of distribution of wages in firm *A*, (say), $\sigma_1^2 = 81$

Variance of distribution of wages in firm *B*, (say), $\sigma_2^2 = 100$

$$\text{C.V. of distribution of wages for firm } A = 100 \times \frac{\sigma_1}{\bar{x}_1} = \frac{100 \times 9}{186} = 4{\cdot}84$$

$$\text{C.V. of distribution of wages for firm } B = 100 \times \frac{\sigma_2}{\bar{x}_2} = \frac{100 \times 10}{175} = 5{\cdot}71$$

Since C.V. for firm *B* is greater than C.V. for firm *A*, firm *B* has greater variability in individual wages.

(*iii*) (*a*) The average daily wage (say) $\bar{x}$, of all the workers in the two firms *A* and *B* taken together is given by :

$$\bar{x} = \frac{n_1 \bar{x}_1 + n_2 \bar{x}_2}{n_1 + n_2} = \frac{500 \times 186 + 600 \times 175}{500 + 600} = \frac{1,98,000}{1,100} = ₹\ 180$$

(*b*) The combined variance σ^2 is given by the formula :

$$\sigma^2 = \frac{1}{n_1 + n_2}\left[n_1(\sigma_1^2 + d_1^2) + n_2(\sigma_2^2 + d_2^2)\right], \text{ where } d_1 = \bar{x}_1 - \bar{x} \text{ and } d_2 = \bar{x}_2 - \bar{x}.$$

Here, $d_1 = 186 - 180 = 6$ and $d_2 = 175 - 180 = -5$

$$\text{Hence, } \sigma^2 = \frac{500(81 + 36) + 600(100 + 25)}{500 + 600} = \frac{1,33,500}{1,100} = 121{\cdot}36.$$

ASSORTED REVIEW PROBLEMS ON MEASURES OF DISPERSION

1. Explain with suitable examples the term "dispersion". State the relative and absolute measures of dispersion and describe the merits and demerits of standard deviation.
2. (a) Explain the main difference between mean deviation and standard deviation. Show that standard deviation is independent of change of origin and scale.
   (b) Distinguish between absolute and relative measures of dispersion.
3. Compute quartile deviation graphically for the following data :

Marks	:	20—30	30—40	40—50	50—60	60—70	70 & over
Number of students	:	5	20	14	10	8	5

4. Compute a suitable measure of dispersion for the following frequency distribution, giving reasons :

Classes	*Frequency*
Less than 20	30
20—30	20
30—40	15
40—50	10
50—60	5

5. Age distribution of hundred life insurance policy holders is as follows :

Age as on nearest birthday	*Number*	*Age as on nearest birthday*	*Number*
17—19·5	9	41—50·5	14
20—25·5	16	51—55·5	12
26—35·5	12	56—60·5	6
36—40·5	26	61—70·5	5

   Calculate mean deviation from median age.

Ans. Median = 38·25, Mean deviation = 10·605

6. Calculate the mean and standard deviation of the following distribution :

x :	2·5—7·5	7·5—12·5	12·5—17·5	17·5—22·5
f :	12	28	65	121
x :	22·5—27·5	27·5—32·5	32·5—37·5	37·5—42·5
f :	175	198	176	120
x :	42·5—47·5	47·5—52·5	52·5—57·5	57·5—62·5
f :	66	27	9	3

Ans. Mean = 30·005, S.D. = 0·01

7. Explain clearly the ideas implied in using arbitrary working origin, and scale for the calculation of the arithmetic mean and standard deviation of a frequency distribution. The values of the arithmetic mean and standard deviation of the following frequency distribution of a continuous variable derived from the analysis in the above manner are 40·604 lb. and 7·92 lb. respectively.

x :	−3	−2	−1	0	1	2	3	4	Total
f :	3	15	45	57	50	36	25	9	240

   Determine the actual class intervals.
8. (a) The arithmetic mean and variance of a set of 10 figures are known to be 17 and 33 respectively. Of the 10 figures, one figure (*i.e.*, 26) was subsequently found inaccurate, and was weeded out. What is the resulting (a) arithmetic mean, and (b) standard deviation.

(*b*) The mean and standard deviation of 20 items is found to be 10 and 2 respectively. At the time of checking it was found that one item 8 was incorrect. Calculate the mean and standard deviation if (*i*) the wrong items is omitted, and (*ii*) it is replaced by 12.

9. For a frequency distribution of marks in Statistics of 200 candidates (grouped in intervals 0—5, 5—10, ..., etc.), the mean and standard deviation were found to be 40 and 15 respectively. Later is was discovered that the score 43 was misread as 53 in obtaining the frequency distribution. Find the corrected mean and standard deviation corresponding to the corrected frequency distribution.

[**Ans.** Mean = 39·95, S.D. = 14·974]

10. (*a*) Complete a table showing the frequencies with which words of different numbers of letters occur in the extract reproduced below (omitting punctuation marks) treating as the variable the number of letters in each word, and obtain the mean, median and coefficient of variation of the distribution :

"Her eyes were blue : blue as autumn distance - blue as the blue we see, between the retreating mouldings of hills and woody slopes on a sunny September morning : a misty and shady blue, that had no beginning or surface, and was looked into rather than at."

Ans. Mean = 4·35, Median = 4, σ = 2·23 and C.V. = 51·26

(*b*) Treating the number of letters in each word in the following passage as the variable x, prepare the frequency distribution table and obtain its mean, median, mode and variance.

"The reliability of data must always be examined before any attempt is made to base conclusions upon them. This is true of all data, but particularly so of numerical data, which do not carry their quality written large on them. It is a waste of time to apply the refined theoretical methods of Statistics to data which are suspect from the beginning."

Ans. Mean = 4·565, Median = 4, Mode = 3, S.D. = 2·673, Variance = 7·145

11. The mean of 5 observations is 4·4 and variance is 8·24. If three of the five observations are 1, 2 and 6, find the other two.

12. Scores of two golfers for 24 rounds were as follows :

Golfer *A* : 74, 75, 78, 72, 77, 79, 78, 81, 76, 72, 72, 77, 74, 70, 78, 79, 80, 81, 74, 80, 75, 71, 73.

Golfer *B* : 86, 84, 80, 88, 89, 85, 86, 82, 82, 79, 86, 80, 82, 76, 86, 89, 87, 83, 80, 88, 86, 81, 81, 87.

Find which golfer may be considered to be more consistent player ?

13. The sum and sum of squares corresponding to length X (in cms.) and weight Y (in gms.) of 50 tapioca tubers are given below :

$$\Sigma X = 212, \ \Sigma X^2 = 902{\cdot}8, \ \Sigma Y = 261, \ \Sigma Y^2 = 1457{\cdot}6.$$

Which is more varying, the length or weight ?

14. Lives of two models of refrigerators turned in for new models in a recent survey are given in the following table. What is the average life of each model of these refrigerators ? Which model shows more uniformity ?

Life (No. of years)	*Model A*	*Model B*
0—2	5	2
2—4	16	7
4—6	13	12
6—8	7	19
8—10	5	9
10—12	4	1

Ans. C.V. (Model *A*) = 54·9%, C.V. (Model *B*) = 36·2%

15. Goals scored by two teams A and B in a football season were as shown in the following table. Find out which team is more consistent ?

No. of goals scored in a match	No. of matches	
	Team A	Team B
0	27	17
1	9	9
2	8	6
3	5	5
4	4	3

Ans. C.V. (Team A) = 122, C.V. (Team B) = 108·3

16. The following data give the arithmetic averages and standard deviations of three sub-groups. Calculate the arithmetic average and standard deviation (s.d.) of the whole group :

Sub-group	No. of men	Average wages in (00 ₹)	s.d. wages (₹)
A	20	61·0	8·0
B	100	70·0	9·0
C	120	80·5	10·0

Ans. Combined mean = ₹ 7,300, combined s.d. = ₹ 1190

17. Find the missing information from the following data :

	Group I	Group II	Group III	Combined
Number	50	?	90	200
Standard Deviation	6	7	?	7·746
Mean	113	?	115	116

Ans. $n_2 = 60$, $\overline{x}_2 = 120$ and $\sigma_3 = 8$

18. A collar manufacturer is considering the production of a new style collar to attract young men. The following statistics of neck circumference are available based on the measurement of a typical group of students :

Mid-value (in inches)	:	12·5	13·0	13·5	14·0	14·5	15·0	15·5	16·0
No. of students	:	4	19	30	63	66	29	18	1

Compute the mean and standard deviation and use the criterion $\overline{x} \pm 3\sigma$, to obtain the largest and smallest size of collar he should make in order to meet needs of practically all his customers, bearing in mind that the collars are worn on an average 3/4 inch larger than neck size.

Ans. Mean = 14·232″, S.D. = 0·72″, Largest size = 17·14″, Smallest size 12·83″.

19. A frequency distribution is divided into two parts. The mean and standard deviation of the first part are m_1 and s_1 and those of the second part are m_2 and s_2 respectively. Obtain the mean and standard deviation for the combined distribution.

20. (*a*) The means of two samples of size 50 and 100 respectively are 54·1 and 50·3 and the standard deviations are 8 and 7. Obtain the mean and standard deviation of the sample of size 150 obtained by combining the two samples.

Ans. Combined mean = 51·57, Combined S.D. = 7·5

(*b*) A distribution consists of three components with frequencies 200, 250 and 300 having means 25, 10 and 15 and standard deviations 3, 4 and 5 respectively. Show that the mean of the combined group is 16 and its standard deviation is 7·2 approximately.

2·15. MOMENTS

The rth moment of a variable x about any point $x = A$, usually denoted by μ_r' is given by :

$$\mu_r' = \frac{1}{N}\sum_i f_i (x_i - A)^r, \quad \sum_i f_i = N \qquad \text{... (2·26)}$$

$$= \frac{1}{N}\sum_i f_i d_i^r, \text{ where } d_i = x_i - A. \qquad \text{... (2·26a)}$$

The rth moment of a variable x about the mean $\bar{x}$, usually denoted by μ_r is given by :

$$\mu_r = \frac{1}{N}\sum_i f_i (x_i - \bar{x})^r = \frac{1}{N}\sum_i f_i z_i^r, \text{ where } z_i = x_i - \bar{x}. \qquad \text{... (2·27)}$$

In particular, $\mu_0 = \frac{1}{N}\sum_i f_i (x_i - \bar{x})^0 = \frac{1}{N}\sum_i f_i = 1$ and $\mu_1 = \frac{1}{N}\sum_i f_i (x_i - \bar{x}) = 0$,

being the algebraic sum of deviations from the mean.

Also
$$\mu_2 = \frac{1}{N}\sum_i f_i (x_i - \bar{x})^2 = \sigma^2 \qquad \text{... (2·28)}$$

These results, *viz.*, $\mu_0 = 1$, $\mu_1 = 0$ and $\mu_2 = \sigma^2$, are of fundamental importance and should be committed to memory. We know that if $d_i = x_i - A$, then

$$\bar{x} = A + \frac{1}{N}\sum_i f_i d_i = A + \mu_1' \qquad \text{... (2·29)}$$

2·15·1. Relation between Moments about Mean in Terms of Moments about any Point and *Vice Versa*

We have

$$\mu_r = \frac{1}{N}\sum_i f_i(x_i - \bar{x})^r = \frac{1}{N}\sum_i f_i(x_i - A + A - \bar{x})^r = \frac{1}{N}\sum_i f_i(d_i + A - \bar{x})^r, \text{ where } d_i = x_i - A.$$

$$\mu_r = \frac{1}{N}\sum_i f_i(d_i - \mu_1')^r \qquad \text{[From (2·29)]}$$

$$= \frac{1}{N}\sum_i f_i\left[d_i^r - {}^rC_1\, d_i^{r-1}\mu_1' + {}^rC_2\, d_i^{r-2}\mu_1'^2 - {}^rC_3\, d_i^{r-3}\mu_1'^3 + \ldots + (-1)^r \mu_1'^r\right] \qquad \text{... (2·30)}$$

$$= \mu_r' - {}^rC_1\, \mu'_{r-1}\mu_1' + {}^rC_2\, \mu'_{r-2}\mu_1'^2 - \ldots + (-1)^r \mu_1'^r \qquad \text{[On using (2·26 a)]}$$

In particular, on putting $r = 2$, 3 and 4 in (2·30) and simplifying, we get

$$\left.\begin{aligned} \mu_2 &= \mu_2' - \mu_1'^2 \\ \mu_3 &= \mu_3' - 3\mu_2'\mu_1' + 2\mu_1'^3 \\ \mu_4 &= \mu_4' - 4\mu_3'\mu_1' + 6\mu_2'\mu_1'^2 - 3\mu_1'^4 \end{aligned}\right\} \qquad \text{... (2·31)}$$

$$\mu_r' = \frac{1}{N}\sum_i f_i (x_i - A)^r = \frac{1}{N}\sum_i f_i (x_i - \bar{x} + \bar{x} - A)^r = \frac{1}{N}\sum_i f_i (z_i + \mu_1')^r,$$

where $x_i - \bar{x} = z_i$ and $\bar{x} = A + \mu_1'$

Thus $\mu_r' = \frac{1}{N}\sum_i f_i \left(z_i^r + {}^rC_1\, z_i^{r-1}\, \mu_1' + {}^rC_2\, z_i^{r-2}\, \mu_1'^2 + \ldots + \mu_1'^r\right)$

$= \mu_r + {}^rC_1\, \mu_{r-1}\, \mu_1' + {}^rC_2\, \mu_{r-2}\, \mu_1'^2 + \ldots + \mu_1'^r.$ [From (2·27)]

In particular, putting $r = 2, 3, 4$ and noting that $\mu_1 = 0$, we get

$$\left.\begin{aligned} \mu_2' &= \mu_2 + \mu_1'^2 \\ \mu_3' &= \mu_3 + 3\mu_2\, \mu_1' + \mu_1'^3 \\ \mu_4' &= \mu_4 + 4\mu_3\, \mu_1' + 6\mu_2\, \mu_1'^2 + \mu_1'^4 \end{aligned}\right\} \quad \ldots (2·32)$$

These formulae enable us to find the moments about any point, once the mean and the moments about mean are known.

2·15·2. Effect of Change of Origin and Scale on Moments

Let $u = \frac{x - A}{h}$, so that $x = A + hu$, $\bar{x} = A + h\bar{u}$ and $x - \bar{x} = h(u - \bar{u})$.

Thus, rth moment of x about any point $x = A$ is given by :

$$\mu_r' = \frac{1}{N}\sum_i f_i\, (x_i - A)^r = \frac{1}{N}\sum_i f_i\, (hu_i)^r = h^r \frac{1}{N}.\sum_i f_i u_i^r$$

And the rth moment of x about mean is: $\mu_r = \frac{1}{N}\sum_i f_i\, (x_i - \bar{x})^r = \frac{1}{N}\sum_i f_i\, \{h(u_i - \bar{u})\}^r$

$$\Rightarrow \quad \mu_r\,(x) = h^r \frac{1}{N}\sum_i f_i\, (u_i - \bar{u})^r = h^r\, \mu_r(u) \quad \ldots (2·33)$$

Thus, the rth moment of the variable x about mean is h^r times the *rth* moment of the variable u about mean.

2·15·3. Sheppard's Correction for Moments

In case of grouped frequency distribution, while calculating moments we assume that the frequencies are concentrated at the middle point of the class intervals. If the distribution is symmetrical or slightly symmetrical and the class intervals are not greater than one-twentieth of the range, this assumption is very nearly true. But since this assumption is not in general true, some error, called the *'grouping error'*, creeps into the calculation of the moments. W. F. Sheppard proved that if

(*i*) the frequency distribution is continuous, and (*ii*) the frequency tapers off to zero in both directions, the effect due to grouping at the mid-point of the intervals can be corrected by the following formulae, known as *Sheppard's corrections* :

$$\left.\begin{aligned} \mu_2\ (\text{corrected}) &= \mu_2 - \frac{h^2}{12} \\ \mu_3\ (\text{corrected}) &= \mu_3 \\ \mu_4\ (\text{corrected}) &= \mu_4 - \frac{1}{2}h^2\mu_2 + \frac{7}{240}h^4 \end{aligned}\right\} \quad \ldots (2·34)$$

where h is the width of the class interval.

2·15·4. Charlier's Checks

The following identities :

$$\left.\begin{aligned}\Sigma f(x+1) &= \Sigma fx + N; \quad \Sigma f(x+1)^2 = \Sigma fx^2 + 2\Sigma fx + N\\ \Sigma f(x+1)^3 &= \Sigma fx^3 + 3\Sigma fx^2 + 3\Sigma fx + N\\ \Sigma f(x+1)^4 &= \Sigma fx^4 + 4\Sigma fx^3 + 6\Sigma fx^2 + 4\Sigma fx + N,\end{aligned}\right\} \quad \text{... (2·34a)}$$

are often used in checking the accuracy in the calculation of first four moments and are known as *'Charlier Checks'*.

2·15·5. Pearson's β and γ Coefficients

Karl Pearson defined the following four coefficients, based upon the first four moments about mean :

$$\beta_1 = \frac{\mu_3^2}{\mu_2^3}, \quad \gamma_1 = +\sqrt{\beta_1} \quad \text{and} \quad \beta_2 = \frac{\mu_4}{\mu_2^2}, \quad \gamma_2 = \beta_2 - 3 \quad \text{... (2·35)}$$

It may be pointed out that these co-efficients are pure numbers independent of units of measurement. The practical utility of these co-efficients is discussed in § 2·16 and § 2·17.

Remark. Sometimes, another coefficient based on moments, *viz.*, alpha (α) coefficient is used. Alpha coefficients are defined as :

$$\alpha_1 = \frac{\mu_1}{\sigma} = 0, \quad \alpha_2 = \frac{\mu_2}{\sigma^2} = 1, \quad \alpha_3 = \frac{\mu_3}{\sigma^3} = \sqrt{\beta_1} = \gamma_1, \quad \alpha_4 = \frac{\mu_4}{\sigma^4} = \beta_2$$

2·15·6. Factorial Moments

Factorial moment of order r about the origin of the distribution $x_i \mid f_i$, $(i = 1, 2, ..., n)$, is defined as follows :

$$\mu_{(r)}' = \frac{1}{N}\sum_{i=1}^{n} f_i\, x_i^{(r)}, \text{ where } x^{(r)} = x(x-1)(x-2)\ ...\ (x-r+1) \text{ and } N = \sum_{i=1}^{n} f_i \quad \text{... (2·36)}$$

Thus, the factorial moment of order r about any point $x = a$ is given by :

$$\mu_{(r)}' = \frac{1}{N}\sum_i f_i\,(x_i - a)^{(r)}, \text{ where } (x-a)^{(r)} = (x-a)(x-a-1)\ ...\ (x-a-r+1) \quad \text{... (2·37)}$$

In particular from (2·36), we have

$$\mu_{(1)}' = \frac{1}{N}\Sigma f_i\, x_i = \mu_1' \text{ (about origin)} = \text{Mean } (\overline{x})$$

$$\mu_{(2)}' = \frac{1}{N}\Sigma f_i\, x_i^{(2)} = \frac{1}{N}\sum_i f_i\, x_i(x_i - 1) = \frac{1}{N}\sum_i f_i\, x_i^{(2)} - \frac{1}{N}\sum_i f_i\, x_i = \mu_2' - \mu_1' \quad \text{... (2·38)}$$

$$\begin{aligned}\mu_{(3)}' &= \frac{1}{N}\sum_i f_i\, x_i^{(3)} = \frac{1}{N}\sum_i f_i\, x_i(x_i-1)(x_i-2) = \frac{1}{N}\sum_i f_i\, x_i^3 - 3\frac{1}{N}\sum_i f_i\, x_i^2 + 2\frac{1}{N}\sum_i f_i\, x_i\\ &= \mu_3' - 3\mu_2' + 2\mu_1' \end{aligned} \quad \text{... (2·38a)}$$

$$\begin{aligned}\mu_{(4)}' &= \frac{1}{N}\sum_i f_i\, x_i^{(4)} = \frac{1}{N}\sum_i f_i\, x_i(x_i-1)(x_i-2)(x_i-3)\\ &= \frac{1}{N}\sum_i f_i\, x_i(x_i^3 - 6x_i^2 + 11x_i - 6)\end{aligned}$$

$$= \frac{1}{N} \sum_i f_i x_i^4 - 6.\frac{1}{N} \Sigma f_i x_i^3 + 11.\frac{1}{N} \Sigma f_i x_i^2 - 6.\frac{1}{N} \Sigma f_i x_i$$

$$= \mu_4' - 6\mu_3' + 11\mu_2' - 6\mu_1' \quad \text{... (2·38b)}$$

Conversely, we will get

$$\left.\begin{aligned} \mu_1' &= \mu_{(1)}' \\ \mu_2' &= \mu_{(2)}' + \mu_{(1)}' \\ \mu_3' &= \mu_{(3)}' + 3\mu_{(2)}' + \mu_{(1)}' \\ \mu_4' &= \mu_{(4)}' + 6\mu_{(3)}' + 7\mu_{(2)}' + \mu_{(1)}' \end{aligned}\right\} \quad \text{... (2·38c)}$$

2·15·7. Absolute Moments

For the frequency distribution $x_i \mid f_i$; $i = 1, 2, ..., n$, the *rth absolute moment* of the variable x about the origin is given by :

$$\frac{1}{N} \sum_{i=1}^{n} f_i \left| x_i^r \right|, \; N = \Sigma f_i, \quad \text{... (2·39)}$$

where $\mid x_i^r \mid$ represents the absolute or modulus value of x_i^r.

The rth absolute moment of the variable about the mean $\bar{x}$ is given by :

$$\frac{1}{N} \sum_{i=1}^{n} f_i \left| x_i - \bar{x} \right|^r \quad \text{... (2·39a)}$$

Example 2·30. *The first four moments of a distribution about the value 4 of the variable are : −1·5, 17, −30 and 108. Find the moments about mean,* β_1 *and* β_2.
Find also the moments about (i) the origin, and (ii) the point $x = 2$.

Solution. In the usual notations, we are given

$A = 4, \quad \mu_1' = -1·5, \quad \mu_2' = 17, \quad \mu_3' = -30 \quad$ and $\quad \mu_4' = 108.$

Moments about mean :

$$\mu_2 = \mu_2' - \mu_1'^2 = 17 - (-1·5)^2 = 17 - 2·25 = 14·75$$

$$\mu_3 = \mu_3' - 3\mu_2' \mu_1' + 2\mu_1'^3 = -30 - 3 \times (17) \times (-1·5) + 2(-1·5)^3 = 39·75$$

$$\mu_4 = \mu_4' - 4\mu_3' \mu_1' + 6\mu_2' \mu_1'^2 - 3\mu_1'^4$$

$$= 108 - 4(-30)(-1·5) + 6(17)(-1·5)^2 - 3(-1·5)^4 = 142·3125$$

Hence, $\quad \beta_1 = \dfrac{\mu_3^2}{\mu_2^3} = \dfrac{(39·75)^2}{(14·75)^3} = 0·4926, \qquad \beta_2 = \dfrac{\mu_4}{\mu_2^2} = \dfrac{(142·3125)}{(14·75)^2} = 0·6543$

Also, $\quad$ Mean $(\bar{x}) = A + \mu_1' = 4 + (-1·5) = 2·5$

Moments about origin : We have

$\bar{x} = 2·5, \quad \mu_2 = 14·75, \quad \mu_3 = 39·75 \quad$ and $\quad \mu_4 = 142·31$ (approx).

We know : $\bar{x} = A + \mu_1'$, where μ_1' is the first moment about the point $x = A$.

Taking $A = 0$, we get the first moment about origin as μ_1' = mean, $\bar{x} = 2·5$.

Using (2·32), we get

$$\mu_2' = \mu_2 + \mu_1'^2 = 14·75 + (2·5)^2 = 14·75 + 6·25 = 21$$

$$\mu_3' = \mu_3 + 3\mu_2\,\mu_1' + \mu_1'^3 = 39{\cdot}75 + 3(14{\cdot}75)\,(2{\cdot}5) + (2{\cdot}5)^3 = 166$$
$$\mu_4' = \mu_4 + 4\mu_3\,\mu_1' + 6\mu_2\,\mu_1'^2 + \mu_1'^4$$
$$= 142{\cdot}3125 + 4(39{\cdot}75)\,(2{\cdot}5) + 6\,(14{\cdot}75)\,(2{\cdot}5)^2 + (2{\cdot}5)^4 = 1{,}132.$$

Moments about the point $x = 2$. We have $\bar{x} = A + \mu_1'$. Taking $A = 2$, the first moment about the point $x = 2$ is $\quad \mu_1' = \bar{x} - 2 = 2{\cdot}5 - 2 = 0{\cdot}5.$

Hence, $\quad \mu_2' = \mu_2 + \mu_1'^2 = 14{\cdot}75 + 0{\cdot}25 = 15$
$$\mu_3' = \mu_3 + 3\mu_2\,\mu_1' + \mu_1'^3 = 39{\cdot}75 + 3\,(14{\cdot}75)\,(0{\cdot}5) + (0{\cdot}5)^3 = 62$$
$$\mu_4' = \mu_4 + 4\mu_3\,\mu_1' + 6\mu_2\,\mu_1'^2 + \mu_1'^4$$
$$= 142{\cdot}3125 + 4(39{\cdot}75)\,(0{\cdot}5) + 6\,(14{\cdot}75)\,(0{\cdot}5)^2 + (0{\cdot}5)^4 = 244.$$

Example 2·31. *Calculate the first four moments of the following distribution about the mean and hence find* β_1 *and* β_2.

x :	0	1	2	3	4	5	6	7	8
f :	1	8	28	56	70	56	28	8	1

Solution.

TABLE 2·23 : CALCULATIONS FOR MOMENTS

x	f	$d = x - 4$	fd	fd^2	fd^3	fd^4
0	1	– 4	– 4	16	– 64	256
1	8	– 3	– 24	72	– 216	648
2	28	– 2	– 56	112	– 224	448
3	56	– 1	– 56	56	– 56	56
4	70	0	0	0	0	0
5	56	1	56	56	56	56
6	28	2	56	112	224	448
7	8	3	24	72	216	648
8	1	4	4	16	64	256
Total	256	0	0	512	0	2,816

Moments about the point $x = 4$ are

$$\mu_1' = \frac{1}{N}\Sigma fd = 0, \qquad \mu_2' = \frac{1}{N}\Sigma fd^2 = \frac{512}{256} = 2,$$

$$\mu_3' = \frac{1}{N}\Sigma fd^3 = 0, \quad \text{and} \quad \mu_4' = \frac{1}{N}\Sigma fd^4 = \frac{2816}{256} = 11$$

Moments about mean are :

$$\mu_1 = 0, \quad \mu_2 = \mu_2' - \mu_1'^2 = 2, \quad \mu_3 = \mu_3' - 3\mu_2'\,\mu_1' + 2\mu_1'^3 = 0$$
$$\mu_4 = \mu_4' - 4\mu_3\,\mu_1' + 6\mu_2\,\mu_1'^2 - 3\mu_1'^4 = 11$$
$$\beta_1 = \frac{\mu_3^2}{\mu_2^3} = 0, \qquad \beta_2 = \frac{\mu_4}{\mu_2^2} = \frac{11}{4} = 2{\cdot}75.$$

Example 2·32. *Show that for discrete distribution* $\beta_2 > 1$.

Solution. We have to show that $\mu_4 / \mu_2^2 > 1$, *i.e.*, $\mu_4 > \mu_2^2$. If $x_i \mid f_i$; $i = 1, 2, ..., n$, be the given discrete frequency distribution, then we have to prove that

$$\mu_4 > \mu_2^2, \qquad i.e., \qquad \frac{1}{N}\sum_i f_i\,(x_i - \bar{x})^4 > \left[\frac{1}{N}\sum_i f_i\,(x_i - \bar{x})^2\right]^2$$

Putting $(x_i - \bar{x})^2 = z_i$, we have to show that

$$\frac{1}{N}\sum_i f_i z^2 > \left(\frac{1}{N}\sum_i f_i z_i\right)^2 \Rightarrow \frac{1}{N}\sum_i f_i z_i^2 - \left(\frac{1}{N}\sum_i f_i z_i\right)^2 > 0 \quad \text{or} \quad \sigma_z^2 > 0,$$

which is always true, since variance is always positive.

Hence, $\beta_2 > 1$.

2·16. SKEWNESS

Literally, skewness means *'lack of symmetry'*. We study skewness to have an idea about the shape of the curve which we can draw with the help of the given data. A distribution is said to be skewed if —

(*i*) Mean, median and mode fall at different points, *i.e.*, Mean ≠ Median ≠ Mode;

(*ii*) Quartiles are not equidistant from median; and

(*iii*) The curve drawn with the help of the given data is not symmetrical but stretched more to one side than to the other.

2·16·1. Measures of Skewness

Various measures of skewness are :

(1) $S_k = M - M_d$, (2) $S_k = M - M_0$, (3) $S_k = (Q_3 - M_d) - (M_d - Q_1)$.

where M is the mean, M_d, the median and M_0, the mode of the distribution.

These are the *absolute measures* of skewness. As in dispersion, for comparing two series we do not calculate these absolute measures but we calculate the *relative measures* called the *coefficients of skewness* which are pure numbers independent of units of measurement. The following are the coefficients of skewness.

I. *Prof. Karl Pearson's Coefficient of Skewness.*

$$S_k = \frac{(M - M_0)}{\sigma}, \text{ where } \sigma \text{ is the standard deviation of the distribution.} \qquad \text{... (2·40)}$$

If mode is ill-defined, then using the empirical relation, $M_0 = 3M_d - 2M$, for a moderately asymmetrical distribution, we get

$$S_k = \frac{3(M - M_d)}{\sigma} \qquad \text{... (2·40}a\text{)}$$

From (2·40) and 2·40 (*a*), we observe that $S_k = 0$ if $M = M_0 = M_d$.

Hence, *for a symmetrical distribution, mean, median and mode coincide.*

Skewness is positive if $M > M_0$ or $M > M_d$ and negative if $M < M_0$ or $M < M_d$.

Remark. *Limits of Karl Pearson's Coefficient of Skewness* : $S_k = \dfrac{3(M - M_d)}{\sigma}$

$$|M - M_d| = \left|\frac{1}{n}\sum_{i=1}^{n} x_i - M_d\right| = \left|\frac{1}{n}\sum_{i=1}^{n}(x_i - M_d)\right| \le \frac{1}{n}\sum_{i=1}^{n}|x_i - M_d| \le \frac{1}{n}\sum_{i=1}^{n}|x_i - \bar{x}| \qquad \text{... (*)}$$

(∵ The sum of the absolute deviations is minimum when taken about median.)

$$|S_k|^2 = \left|\frac{3(M - M_d)}{\sigma}\right|^2 \le \frac{\left[3 \cdot \frac{1}{n}\sum_{i=1}^{n}|x_i - \bar{x}|\right]^2}{\frac{1}{n}\sum_{i=1}^{n}(x_i - \bar{x})^2} = \frac{\left[3\sum_{i=1}^{n}|x_i - \bar{x}|\right]^2}{n\sum_{i=1}^{n}(x_i - \bar{x})^2} \quad \text{[From (*)]} \qquad \text{... (**)}$$

∴ Using Cauchy-Schwartz inequality : {*c.f.* Chapter 6}

$$\left(\sum_{i=1}^{n} a_i\, b_i\right)^2 \le \left(\sum_{i=1}^{n} a_i^2\right)\left(\sum_{i=1}^{n} b_i^2\right), \quad \text{with } b_i = 1;\ i = 1, 2, ..., n, \text{ we get}$$

$$\left(\sum_{i=1}^{n} a_i\right)^2 \le \left(\sum_{i=1}^{n} a_i^2\right)\left(\sum_{i=1}^{n} 1\right) \Rightarrow \frac{\left(\sum_{i=1}^{n} a_i\right)^2}{n \sum_{i=1}^{n} a_i^2} \le 1 \qquad ...(***)$$

∴ $|S_k|^2 \le 3^2$ [From (**) and (***)] $\Rightarrow$ $|S_k| \le 3$ or $-3 \le S_k \le 3$.

Hence, the limits for Karl Pearson's coefficient of skewness are ± 3. However, in practice, these limits are rarely attained.

II. *Prof. Bowley's Coefficient of Skewness.* Based on quartiles

$$S_k = \frac{(Q_3 - M_d) - (M_d - Q_1)}{(Q_3 - M_d) + (M_d - Q_1)} = \frac{Q_3 + Q_1 - 2M_d}{Q_3 - Q_1} \qquad ...(2{\cdot}41)$$

Remarks:

1. Bowley's coefficient of skewness is also known as *Quartile Coefficient of Skewness* and is especially useful in situations where quartiles and median are used, *viz.*,
   (*i*) When the mode is ill-defined and extreme observations are present in the data.
   (*ii*) When the distribution has open end classes or unequal class intervals.
   In these situations Pearson's coefficient of skewness cannot be used.
2. From (2·41), we observe that : $S_k = 0$, if $Q_3 - M_d = M_d - Q_1$.
   This implies that for a symmetrical distribution ($S_k = 0$), median is equidistant from the upper and lower quartiles.
   Moreover, skewness is positive if : $Q_3 - M_d > M_d - Q_1 \Rightarrow Q_3 + Q_1 > 2M_d$
   and skewness is negative if $Q_3 - M_d < M_d - Q_1 \Rightarrow Q_3 + Q_1 < 2M_d$
3. *Limits for Bowley's Coefficient of Skewness.* We know that for two real positive number a and b (*i.e.*, $a > 0$ and $b > 0$), the modulus value of the difference $(a - b)$ is always less than or equal to the modulus value of the sum $(a + b)$, *i.e.*,

$$|a - b| \le |a + b| \Rightarrow \left|\frac{a-b}{a+b}\right| \le 1 \qquad ...(*)$$

   We also know that $(Q_3 - M_d)$ and $(M_d - Q_1)$ are both non-negative.
   Taking $a = Q_3 - M_d$ and $b = M_d - Q_1$ in (*), we get

$$\left|\frac{(Q_3 - M_d) - (M_d - Q_1)}{(Q_3 - M_d) + (M_d - Q_1)}\right| \le 1 \Rightarrow |S_k \text{ (Bowley)}| \le 1 \quad \text{or} \quad -1 \le S_k \text{ (Bowley)} \le 1.$$

   Thus, Bowley's coefficient of skewness ranges from –1 to 1.
   Further, we note from (2·41) that :

$$S_k = +1, \quad \text{if } M_d - Q_1 = 0, \quad \textit{i.e.,} \quad \text{if } M_d = Q_1$$
$$S_k = -1, \quad \text{if } Q_3 - M_d = 0, \quad \textit{i.e.,} \quad \text{if } Q_3 = M_d$$

4. It should be clearly understood that the values of the coefficients of skewness obtained by Bowley's formula and Pearson's formula are not comparable, although in each case, $S_k = 0$ implies the absence of skewness, *i.e.*, the distribution is symmetrical. It may even happen that one of them gives positive skewness while the other gives negative skewness.
5. In Bowley's coefficient of skewness, the disturbing factor of variations is eliminated by dividing the absolute measure of skewness, *viz.*, $(Q_3 - M_d) - (M_d - Q_1)$ by the measure of dispersion $(Q_3 - Q_1)$, *i.e.*, quartile range.

6. The only and perhaps quite serious limitations of this coefficient is that it is based only on the central 50% of the data and ignores the remaining 50% of the data towards the extremes.

III. *Based upon moments, coefficient of Skewness is :*

$$S_k = \frac{\sqrt{\beta_1}(\beta_2 + 3)}{2(5\beta_2 - 6\beta_1 - 9)} \quad \text{... (2·42)}$$

where symbols have their usual meaning. Thus $S_k = 0$ if either $\beta_1 = 0$ or $\beta_2 = -3$. But since $\beta_2 = \mu_4/\mu_2^2$, cannot be negative, $S_k = 0$ if and only if $\beta_1 = 0$. Thus, for a symmetrical distribution, $\beta_1 = 0$. In this respect β_1 is taken to be a measure of skewness. The coefficient in (2·42) is to be regarded as without sign.

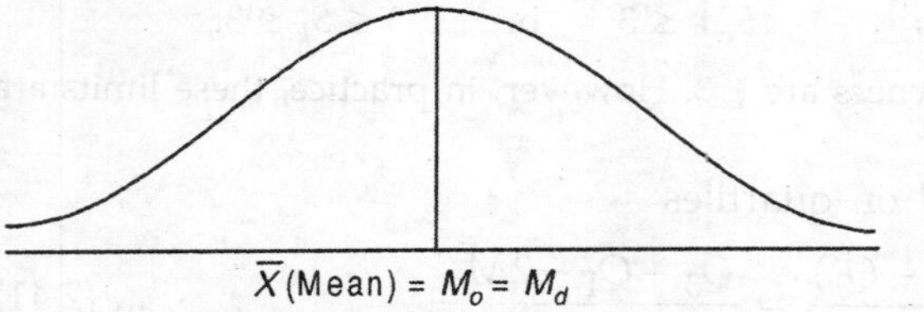

(*Symmetrical Distribution*)

Fig. 2·5

We observe in (2·40) and (2·41) that skewness can be positive as well as negative. The skewness is positive if the larger tail of the distribution lies towards the higher values of the variate (the right), *i.e.*, if the curve drawn with the help of the given data is stretched more to the right than to the left and is negative in the contrary case.

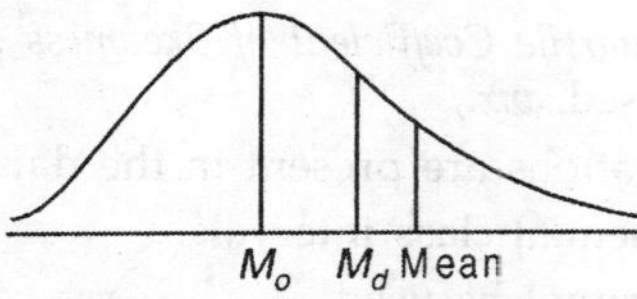

(*Positively Skewed Distribution*)

Fig. 2·6(a)

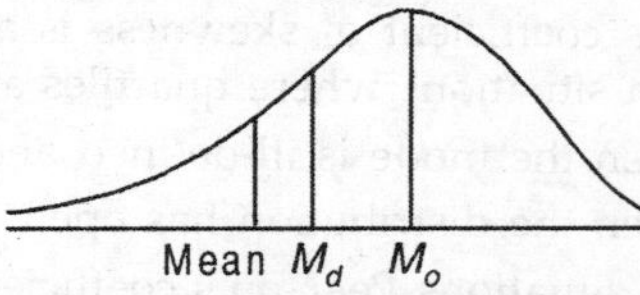

(*Negatively Skewed Distribution*)

Fig. 2·6(b)

2·17. KURTOSIS

If we know the measures of central tendency, dispersion and skewness, we still cannot form a complete idea about the distribution as will be clear from the following figure in which all the three curves *A*, *B* and *C* are symmetrical about the mean '*m*' and have the same range.

In addition to these measures, we should know one more measure which Prof. Karl Pearson calls as the '*Convexity of the Frequency Curve*' or *Kurtosis*. Kurtosis enables us to have an idea about the '*flatness or peakedness*' of the frequency curve. It is measured by the coefficient β_2 or its derivation γ_2 given by :

$$\beta_2 = \frac{\mu_4}{\mu_2^2}, \qquad \gamma_2 = \beta_2 - 3 \quad \text{... (2·43)}$$

Curve of the type '*A*' which is neither flat nor peaked is called the *normal curve* or *mesokurtic curve* and for such a curve $\beta_2 = 3$, *i.e.*, $\gamma_2 = 0$. Curve of the type '*B*' which is flatter than the normal curve is known as *platykurtic* and for such a curve $\beta_2 < 3$, *i.e.*, $\gamma_2 < 0$. Curve of the type '*C*' which is more peaked than the normal curve is called *leptokurtic* and for such a curve $\beta_2 > 3$, *i.e.*, $\gamma_2 > 0$.

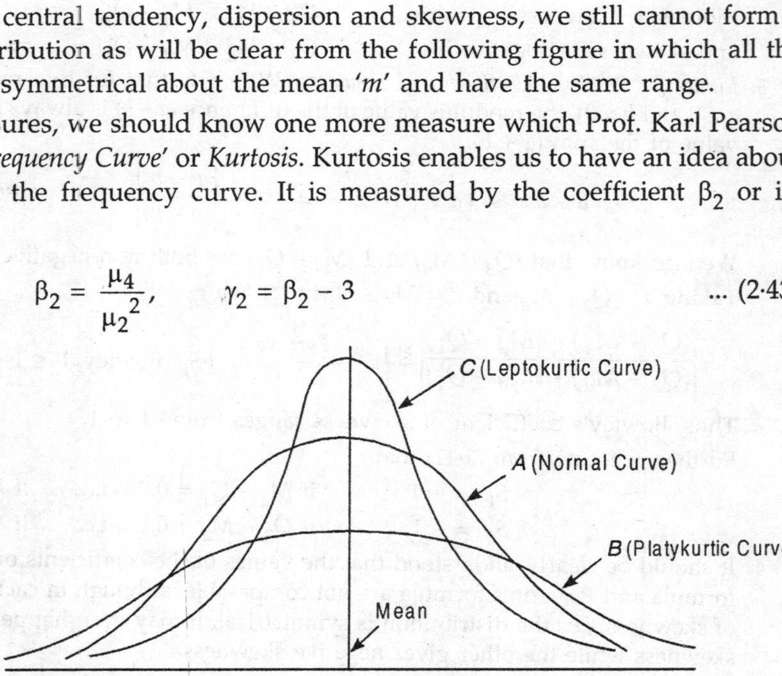

Fig. 2·7

Example 2·33. *For a distribution, the mean is 10, variance is 16, γ_1 is +1 and β_2 is 4. Obtain the first four moments about the origin, i.e., zero. Comment upon the nature of distribution.*

Solution. We are given : Mean = 10, $\mu_2 = 16 \Rightarrow \sigma^2 = 16$ or $\sigma = 4$, $\gamma_1 = +1$, $\beta_1 = 4$

First four moments about origin (μ_1', μ_2', μ_3', μ_4').

$$\mu_1' = \text{First moment about origin} = \text{Mean} = 10$$

$$\mu_2 = \mu_2' - \mu_1'^2 \quad \Rightarrow \quad \mu_2' = \mu_2 + \mu_1'^2 = 16 + 10^2 = 116$$

We have $$\gamma_1 = +1 \quad \Rightarrow \quad \frac{\mu_3}{\mu_2^{3/2}} = \frac{\mu_3}{\sigma^3} = 1 \quad \text{or} \quad \mu_3 = \sigma^3 = 4^3 = 64$$

$$\mu_3 = \mu_3' - 3\mu_2'\mu_1' + 2\mu_1'^3$$

$$\Rightarrow \quad \mu_3' = \mu_3 + 3\mu_2'\mu_1' - 2\mu_1'^3 = 64 + 3 \times 116 \times 10 - 2 \times 1{,}000 = 1{,}544$$

$$\beta_2 = \frac{\mu_4}{\mu_2^2} = 4 \quad \Rightarrow \quad \mu_4 = 4 \times 16^2 = 1024$$

and $$\mu_4 = \mu_4' - 4\mu_3'\mu_1' + 6\mu_2'\mu_1'^2 - 3\mu_1'^4$$

$$\Rightarrow \quad \mu_4' = 1{,}024 + 4 \times 1{,}544 \times 10 - 6 \times 116 \times 100 + 3 \times 10{,}000 = 23{,}184.$$

Comments on nature of the distribution. Since $\gamma_1 = +1$, the distribution is moderately positively skewed, *i.e.*, if we draw the curve of the given distribution, it will have longer tail towards the right. Further since $\beta_2 = 4 > 3$, the distribution is leptokurtic, *i.e.*, it will be slightly more peaked then the normal curve.

Example 2·34. *For the frequency distribution of scores in mathematics of 50 candidates selected at random from among those appearing at a certain examination, compute the first four moments about the mean of the distribution.*

Scores :	50—60	60—70	70—80	80—90	90—100	100—110	110—120	120—130
Frequency :	1	0	0	1	1	2	1	0
Scores :	130—140	140—150	150—160	160—170	170—180	180—190	190—200	200—210
Frequency :	4	4	2	5	10	11	4	1
Scores :	210—220	220—230						
Frequency :	1	2						

Find also the corrected values of the moments after Sheppard's corrections are applied. Also obtain moment coefficients of skewness and kurtosis and comment on the nature of the distribution.

Solution.

TABLE 2·24 : CALCULATIONS FOR MOMENTS

Mid-value (x)	*f*	$d = \frac{x - 135}{10}$	fd	fd^2	fd^3	fd^4
55	1	– 8	– 8	64	– 512	4096
65	0	– 7	0	0	0	0
75	0	– 6	0	0	0	0
85	1	– 5	– 5	25	– 125	625
95	1	– 4	– 4	16	– 64	256
105	2	– 3	– 6	18	– 54	162
115	1	– 2	– 2	4	– 8	16
125	0	– 1	0	0	0	0

135	4	0	0	0	0	0
145	4	1	4	4	4	4
155	2	2	4	8	16	32
165	5	3	15	45	135	405
175	10	4	40	160	640	2,560
185	11	5	55	275	1,375	6,875
195	4	6	24	144	864	5,184
205	1	7	7	49	343	2,401
215	1	8	8	64	512	4,096
225	2	9	18	162	1458	13,122
Total	50		150	1,038	4,584	39,834

The raw moments of variable *d* (about origin) are computed as follows :

$$\mu_1' = \frac{\Sigma fd}{\Sigma f} = \frac{150}{50} = 3{\cdot}00\ ; \qquad \mu_2' = \frac{\Sigma fd^2}{\Sigma f} = \frac{1{,}038}{50} = 20{\cdot}76\ ;$$

$$\mu_3' = \frac{\Sigma fd^3}{\Sigma f} = \frac{4{,}584}{50} = 91{\cdot}68\ ; \qquad \mu_4' = \frac{\Sigma fd^4}{\Sigma f} = \frac{39{,}834}{50} = 796{\cdot}68$$

The central moments of variable *X* are then computed as shown below :

$$\mu_2 = (\mu_2' - \mu_1'^2) \times h^2 = (20{\cdot}76 - 9{\cdot}00) \times 100 = 1{,}176$$

$$\mu_3 = (\mu_3' - 3\mu_2'\,\mu_1' + 2\mu_1'^3) \times h^3$$

$$= \{91{\cdot}68 - 3 \times 20{\cdot}76 \times 3 + 2\,(27)\} \times 1{,}000 = -\,41{,}160$$

$$\mu_4 = (\mu_4' - 4\mu_3'\,\mu_1' + 6\mu_2'\,\mu_1'^2 - 3\mu_1'^4) \times h^4$$

$$= [796{\cdot}68 - 4\,(91{\cdot}68)\,(3) + 6(20{\cdot}76)\,(9) - 3(81)] \times 10{,}000 = 57{,}45{,}600.$$

Sheppard's Corrections for Moments

$$\bar{\mu}_2 = \mu_2 - \frac{h^2}{12} = 1{,}176 - \frac{100}{12} = 1{,}167.67, \qquad \bar{\mu}_3 = \mu_3 = -\,41{,}160$$

$$\bar{\mu}_4 = \mu_4 - \frac{h^2}{2}\mu_2 + \frac{7}{240}h^4 = 57{,}45{,}600 - 58{,}800 + 291{\cdot}67 = 56{,}87{,}091{\cdot}67$$

Moment coefficient of skewness is given by :

$$\gamma_1 = \sqrt{\beta_1} = \frac{\mu_3}{(\mu_2)^{3/2}} = \frac{-\,41{,}160}{1{,}176\sqrt{1{,}176}} = -\,\frac{41{,}160}{40{,}328{\cdot}40} = -\,1.02$$

$$\text{Moment coefficient of kurtosis} = \beta_2 = \frac{\mu_4}{\mu_2^{\,2}} = \frac{57{,}45{,}600}{(1{,}176)^2} = 4{\cdot}15.$$

Interpretation. Since $\gamma_1 = -\,1{\cdot}02$, *i.e.*, (negative), the distribution is moderately negatively skewed, *i.e.*, the frequency curve of the given distribution has a longer tail towards the left.

Further, since $\beta_2 = 4{\cdot}15 > 3$, the given distribution is *leptokurtic, i.e.*, the frequecny curve is more peaked than the normal curve.

SUPPLEMENTARY EXAMPLES ON MEASURES OF DISPERSION, SKEWNESS AND KURTOSIS

Example 2·35. *Prove that the mean deviation about the mean $\bar{x}$ of the variable x, the frequency of whose ith size x_i is f_i is given by :*

$$\frac{2}{N}\left(\bar{x} \sum_{x_i < \bar{x}} f_i - \sum_{x_i < \bar{x}} f_i x_i\right), \quad N = \sum_i f_i$$

Solution. Mean deviation (M.D.) about mean $= \frac{1}{N} \sum_i f_i \mid x_i - \bar{x} \mid$

$$= \frac{1}{N}\left(\sum_{x_i < \bar{x}} f_i \mid x_i - \bar{x} \mid + \sum_{x_i > \bar{x}} f_i \mid x_i - \bar{x} \mid\right) = \frac{1}{N}\left[\sum_{x_i < \bar{x}} f_i (\bar{x} - x_i) + \sum_{x_i > \bar{x}} f_i (x_i - \bar{x})\right]$$

$$= \frac{1}{N}\left[-\sum_{x_i < \bar{x}} f_i (x_i - \bar{x}) + \sum_{x_i > \bar{x}} f_i (x_i - \bar{x})\right] \qquad \ldots (*)$$

Since, the algebraic sum of deviations about mean is zero, we have

$$\sum_i f_i (x_i - \bar{x}) = 0 \Rightarrow \sum_{x_i > \bar{x}} f_i (x_i - \bar{x}) + \sum_{x_i < \bar{x}} f_i (x_i - \bar{x}) = 0 \Rightarrow \sum_{x_i > \bar{x}} f_i (x_i - \bar{x}) = -\sum_{x_i < \bar{x}} f_i (x_i - \bar{x})$$

Substituting in (*), we get

$$\text{M.D.} = \frac{1}{N}\left[-2 \sum_{x_1 < \bar{x}} f_i (x_i - \bar{x})\right] = \frac{2}{N}\left[-\sum_{x_i < \bar{x}} f_i x_i + \bar{x} \sum_{x_i < \bar{x}} f_i\right] = \frac{2}{N}\left(\bar{x} \sum_{x_i < \bar{x}} f_i - \sum_{x_i < \bar{x}} f_i x_i\right)$$

Example 2·36. *Prove that for any discrete distribution, standard deviation is not less than mean deviation from mean.*

Solution. Let $x_i \mid f_i$; $i = 1, 2, 3, \ldots, n$ be any discrete frequency distribution. Then we have to prove that

S.D. $\nless$ Mean deviation from mean *i.e.,* $(\text{S.D.})^2 \nless (\text{Mean deviation from mean})^2$

$$\Rightarrow \quad (\text{S.D.})^2 \geq (\text{M.D. from mean})^2$$

$$\Rightarrow \quad \frac{1}{N} \sum_{i=1}^{n} f_i (x_i - \bar{x})^2 \geq \left(\frac{1}{N} \sum_{i=1}^{n} f_i \mid x_i - \bar{x} \mid\right)^2$$

If we put $\mid x_i - \bar{x} \mid = z_i$, then we have to prove that

$$\frac{1}{N} \sum_{i=1}^{n} f_i z_i^2 \geq \left(\frac{1}{N} \sum_{i=1}^{n} f_i z_i\right)^2 \quad \textit{i.e.,} \quad \frac{1}{N} \sum_{i=1}^{n} f_i z_i^2 - \left(\frac{1}{N} \sum_{i=1}^{n} f_i z_i\right)^2 \geq 0, \quad \textit{i.e.,} \quad \frac{1}{N} \sum_{i=1}^{n} f_i (z_i - \bar{z})^2 \geq 0$$

i.e., $\quad \text{Var}(Z) = \sigma_Z^2 \geq 0$, which is always true.

Hence, the result.

Example 2·37. *Find the mean deviation from the mean and standard deviation of A.P. series :*

a, a + d, a + 2d, ..., a + 2nd and verify that the latter is greater than the former.

Solution. We know that the mean of a series in A.P. is the mean of its first and last terms.

Hence, the mean of the given series is: $\bar{x} = \frac{1}{2}(a + a + 2nd) = a + nd$

TABLE 2·25: CALCULATIONS FOR M.D. AND S.D.

x	$\lvert x - \bar{x} \rvert$	$(x - \bar{x})^2$
a	nd	$n^2 d^2$
$a + d$	$(n - 1) d$	$(n - 1)^2 d^2$
$a + 2d$	$(n - 2) d$	$(n - 2)^2 d^2$
$\vdots$	$\vdots$	$\vdots$
$a + (n - 2) d$	$2d$	$2^2 \cdot d^2$
$a + (n - 1) d$	d	$1^2 \cdot d^2$
$a + nd$	0	0
$a + (n + 1) d$	d	$1^2 \cdot d^2$
$a + (n + 2) d$	$2d$	$2^2 \cdot d^2$
$\vdots$	$\vdots$	$\vdots$
$a + (2n - 2) d$	$(n - 2) d$	$(n - 2)^2 d^2$
$a + (2n - 1) d$	$(n - 1) d$	$(n - 1)^2 d^2$
$a + 2nd$	nd	$n^2 d^2$

$$\text{Mean deviation from mean} = \frac{1}{2n+1} \Sigma \lvert x_i - \bar{x} \rvert$$

$$= \frac{1}{2n+1}\left[2 \cdot d(1 + 2 + 3 + \dots + n)\right] = \frac{n(n+1)d}{(2n+1)}$$

$$\sigma^2 = \frac{1}{2n+1} \Sigma (x_i - \bar{x})^2 = \frac{1}{2n+1}\left[2 \cdot d^2 \, (1^2 + 2^2 + 3^2 + \dots + n^2)\right]$$

$$= \frac{1}{2n+1} \, 2d^2 \cdot \frac{n(n+1)(2n+1)}{6} = \frac{n(n+1)\, d^2}{3}$$

Hence, standard deviation, $\sigma = \sqrt{\dfrac{n(n+1)}{3}} \times d$

Verification. S.D. > M.D. from mean if $(\text{S.D.})^2 > (\text{M.D. from mean})^2$

i.e., if $\dfrac{n(n+1)\, d^2}{3} > \left\{\dfrac{n(n+1)\, d}{2n+1}\right\}^2$ or if $(2n + 1)^2 > 3n(n + 1)$, *i.e.*, if $n^2 + n + 1 > 0$

which is always true, since n is a positive integer.

Example 2·38. *Show that in a discrete series if deviations are small compared with mean M so that $(x/M)^3$ and higher powers of (x/M) are neglected, we have*

(i) $G = M\left(1 - \dfrac{1}{2} \cdot \dfrac{\sigma^2}{M^2}\right)$, *(ii)* $M^2 - G^2 = \sigma^2$, and *(iii)* $H = M\left(1 - \dfrac{\sigma^2}{M^2}\right)$

where M is the arithmetic mean, G is geometric mean, H is the harmonic mean and σ is the standard deviation of the distribution.

Solution. Let $X_i \mid f_i$; $i = 1, 2, 3, \dots, n$ be the given frequency distribution. Then we are given that $x_i = X_i - M$, *i.e.*, $X_i = x_i + M$, where M is the mean of the distribution.

We have $$\sum_i f_i \, x_i = \sum_i f_i \, (X_i - M) = 0 \qquad \dots (1)$$

being the algebraic sum of the deviations of the given values from their mean.

Also, $$\sum_i f_i x_i^2 = \sum_i f_i (X_i - M)^2 = N\sigma^2 \qquad \text{... (1a)}$$

(*i*) By def., $G = (X_1^{f_1} . X_2^{f_2} ... X_3^{f_n})^{1/N}$, where $N = \Sigma f_i$

$$\Rightarrow \quad \log G = \frac{1}{N}\sum_i f_i \log X_i = \frac{1}{N}\sum_i f_i \log (x_i + M)$$

$$= \frac{1}{N}\sum_i \left[f_i \log\left\{ M\left(1+\frac{x_i}{M}\right)\right\}\right] = \frac{1}{N}\sum_i \left[f_i \left\{\log M + \log\left(1+\frac{x_i}{M}\right)\right\}\right]$$

$$= \log M + \frac{1}{N}\sum_i f_i \log\left(1+\frac{x_i}{M}\right) = \log M + \frac{1}{N}\sum_i f_i \left\{\frac{x_i}{M} - \frac{1}{2}\frac{x_i^2}{M^2} + \frac{1}{3}\left(\frac{x_i}{M}\right)^3 + ...\right\},$$

the expansion of $\log\left(1+\frac{x_i}{M}\right)$ in ascending powers of $\frac{x_i}{M}$ being valid since $\left|\frac{x_i}{M}\right| < 1$. Neglecting $(x_i/M)^3$ and higher powers of (x_i/M), we get, on using (1) and (1*a*) :

$$\log G = \log M + \frac{1}{NM}\sum_i f_i x_i - \frac{1}{2M^2}\cdot\frac{1}{N}\sum_i f_i x_i^2 = \log M - \frac{\sigma^2}{2M^2} = \log\left(Me^{-\sigma^2/2M^2}\right)$$

$$\Rightarrow \quad G = Me^{-\sigma^2/2M^2} = M\left(1-\frac{\sigma^2}{2M^2}\right), \text{ neglecting higher powers.}$$

Hence, $$G = M\left(1-\frac{1}{2}\cdot\frac{\sigma^2}{M^2}\right) \qquad \text{... (2)}$$

(*ii*) Squaring both sides in (2), we get

$$G^2 = M^2\left(1-\frac{1}{2}\cdot\frac{\sigma^2}{M^2}\right) = M^2\left(1-\frac{\sigma^2}{M^2}\right) = M^2 - \sigma^2, \text{ neglecting } (\sigma/M)^4.$$

$$\therefore \quad M^2 - G^2 = \sigma^2 \qquad \text{... (3)}$$

(*iii*) By def., harmonic mean H is given by :

$$\frac{1}{H} = \frac{1}{N}\sum_i (f_i / X_i) = \frac{1}{N}\sum_i \{f_i / (x_i + M)\} = \frac{1}{MN}\sum_i \frac{f_i}{\{1+(x_i/M)\}} = \frac{1}{MN}\sum_i f_i\left(1+\frac{x_i}{M}\right)^{-1}$$

Since $\left|\frac{x_i}{M}\right| < 1$, the expansion of $\left(1+\frac{x_i}{M}\right)^{-1}$ in ascending powers of (x/M) is valid. Neglecting $(x_i/M)^3$ and higher powers of (x_i/M), we get

$$\frac{1}{H} = \frac{1}{MN}\sum_i f_i\left(1-\frac{x_i}{M}+\frac{x_i^2}{M^2}\right)$$

$$= \frac{1}{M}\left(\frac{1}{N}\sum_i f_i - \frac{1}{MN}\sum_i f_i x_i + \frac{1}{M^2}\cdot\frac{1}{N}\sum_i f_i x_i^2\right) = \frac{1}{M}\left(1+\frac{\sigma^2}{M^2}\right)$$

[From (1) and (1*a*)]

$$\therefore \qquad H = M\left(1+\frac{\sigma^2}{M^2}\right)^{-1} = M\left(1-\frac{\sigma^2}{M^2}\right), \text{ higher powers being neglected.}$$

Hence,
$$H = M\left(1-\frac{\sigma^2}{M^2}\right).$$

Example 2·39. *From a sample of n observations, the arithmetic mean and variance are calculated. It is then found that one of the values, x_1, is in error and should be replaced by x_1'. Show that the adjustment to the variance to correct this error is :*

$$\frac{1}{n}(x_1'-x_1)\left(x_1'+x_1-\frac{x_1'-x_1+2T}{n}\right), \text{ where } T \text{ is the total of the original results.}$$

Solution. Let $x_1, x_2, ..., x_n$ be the given sample of n observations. Their variance (σ^2) is given by :

$$\sigma^2 = \frac{1}{n}\sum_{i=1}^{n} x_i^2 - \bar{x}^2 = \frac{1}{n}(x_1^2+x_2^2+...+x_n^2) - \frac{T^2}{n^2},$$

where $T = x_1 + x_2 + ... + x_n$, (Given).

Let σ_1^2 be the corrected variance, if the wrong observation x_1, is replaced by the correct value x_1'. Then

$$\sigma_1^2 = \frac{1}{n}(x_1'^2 + x_2^2 + ... + x_n^2) - \left\{\frac{T-x_1+x_1'}{n}\right\}^2$$

Adjustment to the variance to correct the error is :

$$\sigma_1^2 - \sigma^2 = \frac{1}{n}\left\{x_1'^2 - x_1^2\right\} - \frac{1}{n^2}\left\{(T-x_1+x_1')^2 - T^2\right\}$$

$$= \frac{1}{n}(x_1'+x_1)(x_1'-x_1) - \frac{1}{n^2}\{(x_1'-x_1)\times(2T-x_1+x_1')\}$$

$$= \frac{1}{n}(x_1'-x_1)\left(x_1'+x_1-\frac{x_1'-x_1+2T}{n}\right).$$

Example 2·40. *If for a random variable x, the absolute moment β_k of order k about origin exists for ordinary k = 1, 2, ... n, then the following inequalities :*

$$(i)\ \beta_k^{2k} \le \beta_{k-1}^k \cdot \beta_{k+1}^k, \qquad (ii)\ \beta_k^{1/k} \le \beta_{k+1}{}^{1/(k+1)}$$

hold for k = 1, 2, ..., n − 1, where β_k is the kth absolute moment about the origin.

Solution. If $x_i \mid f_i$; $i = 1, 2, 3, ..., n$ is the given frequency distribution, then

$$\beta_k = \frac{1}{N}\Sigma f_i \mid x_i^k \mid \qquad ...(1)$$

Let u and v be arbitrary real numbers, then the expression :

$$\sum_{i=1}^{n} f_i\left[u \mid x_i^{(k-1)/2} \mid + v \mid x_i^{(k+1)/2} \mid\right]^2, \text{ is non-negative.}$$

$$\Rightarrow \qquad \sum_{i=1}^{n} f_i\left[u \mid x_i^{(k-1)/2} \mid + v \mid x_i^{(k+1)/2} \mid\right]^2 \ge 0$$

$$\Rightarrow \qquad u^2\,\Sigma f_i\,|\,x_i\,|^{k-1} + v^2\,\Sigma f_i\,|\,x_i\,|^{k+1} + 2uv\,\Sigma f_i\,|\,x_i\,|^{k} \geq 0.$$

Dividing throughout by N and using relation (1), we get

$$u^2\,\beta_{k-1} + v^2\,\beta_{k+1} + 2uv\beta_k \geq 0, \text{ i.e., } \quad u^2\,\beta_{k-1} + 2uv\beta_k + v^2\,\beta_{k+1} \geq 0 \qquad \text{... (2)}$$

We know that the condition for the expression $ax^2 + 2hxy + by^2$ to be non-negative for all values of x and y is that: $\begin{vmatrix} a & h \\ h & b \end{vmatrix} \geq 0.$

Using this result, we get from (2), $\begin{vmatrix} \beta_{k-1} & \beta_k \\ \beta_k & \beta_{k+1} \end{vmatrix} \geq 0 \quad \Rightarrow \quad \beta_{k-1} \cdot \beta_{k+1} - \beta_k^{\,2} \geq 0$ ... (3)

Raising both sides of (3) to power k, we get $\quad \beta_k^{2k} \leq \beta_{k-1}^k \cdot \beta_{k+1}^k$ ... (4)

Replacing k by 1, 2, ..., $k-1$ successively in (4), we get respectively

$$\begin{aligned}
\beta_1^{\,2} &\leq \beta_0\,\beta_2 \\
\beta_2^4 &\leq \beta_1^2\,\beta_3^2 \\
\beta_3^6 &\leq \beta_2^3\,\beta_4^3 \\
&\vdots \\
\beta_{k-1}^{2(k-1)} &\leq \beta_{k-2}^{k-1} \cdot \beta_k^{k-1} \\
\beta_k^{2k} &\leq \beta_{k-1}^{k} \cdot \beta_{k+1}^{k}
\end{aligned}$$

Multiplying these inequalities and noting that $\beta_0 = 1$, we get $\beta_k^{k+1} \leq \beta_{k+1}^{k}$, for $k = 1, 2, ..., n-1$.

Raising both sides of the inequality to the power $\dfrac{1}{k(k+1)}$, we get $\quad \beta_k^{1/k} \leq \beta_{k+1}^{1/(k+1)}$ ... (5)

Remark. Result (5) shows that $\beta_k^{\,1/k}$ is an increasing function of k.

Example 2·41. *For a random variable x moments of all order exist. Denoting by* μ_j, *the jth central moment, show that*

$$\mu^2_{2j+1} \leq \mu_{2j}\,\mu_{2j+2}.$$

Solution. If $x_i \mid f_i$; $i = 1, 2, ..., n$ is the given frequency distribution, then by *def.*,

$$\mu_j = \frac{1}{N} \sum_i f_i\,(x_i - \bar{x})^j \qquad \text{... (1)}$$

Let u and v be arbitrary positive numbers. Then the expression

$$\sum_i f_i \left\{u(x_i - \bar{x})^j + v(x_i - \bar{x})^{j+1}\right\}^2 \text{ is non-negative for all real values of } u \text{ and } v.$$

$$\Rightarrow \qquad \sum_i f_i \left\{u(x_i - \bar{x})^j + v(x_i - \bar{x})^{j+1}\right\}^2 \geq 0, \ \forall\, u \text{ and } v.$$

$$\Rightarrow \qquad u^2 \sum_i f_i\,(x_i - \bar{x})^{2j} + 2uv \sum_i f_i\,(x_i - \bar{x})^{2j+1} + v^2 \sum_i f_i\,(x_i - \bar{x})^{2j+2} \geq 0$$

Dividing throughout by $N = \Sigma f$ and using relation (1), we get

$$u^2 \cdot \mu_{2j} + 2uv \cdot \mu_{2j+1} + v^2 \cdot \mu_{2j+2} \geq 0 \ \ \forall\, u \text{ and } v. \qquad \text{... (2)}$$

We know that the quadratic form $ax^2 + 2hxy + by^2 \geq 0$, for all real x and y if :

$$a \geq 0 \quad \text{and} \quad \begin{vmatrix} a & h \\ h & b \end{vmatrix} \geq 0. \qquad \text{... (3)}$$

Hence, on using (3), we get from (2)

$$\begin{vmatrix} \mu_{2j} & \mu_{2j+1} \\ \mu_{2j+1} & \mu_{2j+2} \end{vmatrix} \geq 0 \quad \Rightarrow \quad \mu_{2j}\,\mu_{2j+2} - \mu_{2j+1}^2 \geq 0 \quad \Rightarrow \quad \mu_{2j+1}^2 \leq \mu_{2j}\,\mu_{2j+2}$$

Example 2·42. *For a distribution of 250 heights, calculations showed that the mean, standard deviation, β_1 and β_2 were 54 inches, 3 inches, 0 and 3 respectively. It was however, discovered on checking that the two items 64 and 50 in the original data were wrongly written in place of the correct values 62 and 52 inches respectively. Calculate the correct frequency constants.*

Solution. We are given : $N = 250,\ \bar{x} = 54,\ \sigma = 3,\ \beta_1 = 0$ and $\beta_2 = 3.$ ... (*)

Wrong observation used : 64 and 50, and Correct observations are : 62 and 52.

(i) $$\text{Uncorrected } (\Sigma x) = N.\bar{x} = 250 \times 54 = 13{,}500$$

$$\therefore \quad \text{Corrected } (\Sigma x) = 13{,}500 - (64 + 50) + (62 + 52) = 13{,}500$$

$$\Rightarrow \quad \text{Corrected Mean} = \frac{13{,}500}{250} = 54 = \text{Original mean.}$$

(ii) $$\text{Uncorrected } \left\{\Sigma (x - \bar{x})^2\right\} = N\sigma^2 = 250 \times 9 = 2{,}250$$

$$\therefore \quad \text{Corrected } \left\{\Sigma (x - \bar{x})^2\right\} = 2{,}250 - \{(64 - 54)^2 + (50 - 54)^2\} + \{(62 - 54)^2 + (52 - 54)^2\}$$
$$= 2{,}202$$

$$\Rightarrow \quad \text{Corrected } \sigma^2\ (\mu_2) = \frac{2{,}202}{250} = 8{\cdot}81 \quad \Rightarrow \quad \text{Corrected S.D} = \sqrt{8.81} = 2{\cdot}97$$

(iii) $$\beta_1 = \frac{\mu_3^2}{\mu_2^3} \quad \Rightarrow \quad \mu_3^2 = \beta_1 \,.\, \mu_2^3 = 0 \times 9 = 0 \quad \Rightarrow \quad \mu_3 = 0 \qquad \text{[From (*)]}$$

$$\Rightarrow \quad \text{Uncorrected } \left\{\Sigma (x - \bar{x})^3\right\} = N \,.\, \mu_3 = 0$$

$$\Rightarrow \quad \text{Corrected } \left\{\Sigma (x - \bar{x})^3\right\} = 0 - \left\{(64 - 54)^3 + (50 - 54)^3\right\} + \left\{(62 - 54)^3 + (52 - 54)^3\right\} = -544$$

$$\therefore \quad \text{Corrected } \mu_3 = \frac{-544}{250} = -\,2{\cdot}18$$

Hence, $$\text{Corrected } \beta_1 = \frac{(\text{Corrected } \mu_3)^2}{(\text{Corrected } \mu_2)^3} = \frac{(-2{\cdot}18)^2}{(8{\cdot}81)^3} = 0{\cdot}0070 \simeq 0$$

(iv) $$\beta_2 = \frac{\mu_4}{\mu_2^2} \quad \Rightarrow \quad \mu_4 = \beta_2 \,.\, \mu_2^2 = 3 \times 9^2 = 243 \qquad \text{[From (*)]}$$

$$\Rightarrow \quad \text{Uncorrected } \left\{\Sigma (x - \bar{x})^4\right\} = N\mu_4 = 250 \times 243 = 60{,}750$$

$$\therefore \quad \text{Corrected } \left\{\Sigma (x - \bar{x})^4\right\} = 60{,}750 - \left\{(64 - 54)^4 + (50 - 54)^4\right\} + \left\{(62 - 54)^4 + (52 - 54)^4\right\}$$
$$= 54{,}606$$

$$\Rightarrow \qquad \text{Corrected } \mu_4 = \frac{54,606}{250} = 218{\cdot}42$$

Hence,
$$\text{Corrected } \beta_2 = \frac{\text{Corrected } \mu_4}{\text{Corrected } \mu_2{}^2} = \frac{218{\cdot}42}{(8{\cdot}81)^2} = 2{\cdot}81.$$

Example 2·43. *Show that if a range of six times the standard deviation covers at least 18 class intervals, Sheppard's correction will make a difference of less than 0.5 per cent in the uncorrected value of the standard deviation.*

Solution. Let σ be the calculated standard deviation and σ_1 be the corrected standard deviation after applying Sheppard's correction for grouping. If h is the width of the class intervals, then we are given :

$$6\,\sigma \geq 18h \qquad \Rightarrow \qquad h \leq \frac{\sigma}{3} \qquad \ldots (*)$$

After applying Sheppard's correction for μ_2, we get

$$\sigma_1{}^2 = \sigma^2 - \frac{h^2}{12} \quad \Rightarrow \quad \sigma^2 = \sigma_1{}^2 + \frac{h^2}{12} \leq \sigma_1{}^2 + \frac{\sigma^2}{12 \times 9} \text{ [From (*)]} \quad \Rightarrow \quad \sigma^2\left(1 - \frac{1}{108}\right) \leq \sigma_1{}^2$$

$$\Rightarrow \qquad \sigma \leq \sigma_1\left(1 - \frac{1}{108}\right)^{-1/2} = \sigma_1\left\{1 + \left(-\frac{1}{2}\right)\left(-\frac{1}{108}\right) + \frac{(-1/2)(-3/2)}{2!}\left(-\frac{1}{108}\right)^2 + \ldots\right\}$$

$$\Rightarrow \qquad \sigma \leq \sigma_1\left(1 + \frac{1}{216}\right), \text{ approx.} \quad \Rightarrow \quad \sigma - \sigma_1 \leq \frac{\sigma_1}{216} < \frac{\sigma_1}{200} = 0{\cdot}005\,\sigma_1 = 0{\cdot}5\% \text{ of } \sigma_1.$$

Hence, the result.

ASSORTED REVIEW PROBLEMS ON MOMENTS & MEASURES OF SKEWNESS & KURTOSIS

1. (*a*) Define the raw and central moments of a frequency distribution. Obtain the relation between the central moments of order r in terms of the raw moments. What are Sheppard's corrections to the central moments ?
   (*b*) Define moments. Establish the relationship between the moments about mean, *i.e.*, Central Moments in terms of moments about any arbitrary point and *vice versa.*
2. What do you understand by skewness ? How is it measured ? Distinguish clearly, by giving figures, between positive and negative skewness. Also show the relative positions of mean, median and mode in the figures, for positively and negatively skewed distributions.
3. Explain the methods of measuring skewness and kurtosis of a frequency distribution.
4. (*a*) Why do we calculate in general only the first four moments about mean of a distribution and not the higher moments ?
   (*b*) "Measures of central tendency, dispersion, skewness and kurtosis are complementary to one another in understanding a frequency distribution." Explain.
5. (*a*) Define Pearsonian coefficients β_1 and β_2, and discuss their utility in Statistics.
   (*b*) What do you mean by skewness and kurtosis of a distribution ? Show that the Pearson's Beta coefficients satisfy the inequality $\beta_2 - \beta_1 - 1 \geq 0$. Also deduce that $\beta_2 \geq 1$.
   (*c*) Define Karl Pearson's coefficients γ_1 and γ_2 and discuss their utility in Statistics.

6. Obtain Karl Peason's measure of skewness for the following data :

Values	5—10	10—15	15—20	20—25
Frequency	6	8	17	21
Values	25—30	30—35	35—40	
Frequency	15	11	2	

7. (*a*) The first three moments of a distribution about the value 2 of the variable are 1, 16 and – 40. Show that the mean is 3, the variance is 15 and $\mu_3 = -86$. Also show that the first three moments about $x = 0$ are 3, 24 and 76.

(*b*) For a distribution the mean is 10, variance is 16, γ_1 is +1 and β_2 is 4. Find the first four moments about the origin. **[Ans.** $\mu_1' = 10$, $\mu_2' = 116$, $\mu_3' = 1544$, and $\mu_4' = 23184$.]

8. The first four moments of distribution about the value 5 of the variable are 2, 20, 40 and 50. Obtain, as far as possible, the various characteristics of the distribution on the basis of the information given.

Ans. Mean = 7, $\mu_2 = 16$, $\mu_3 = -64$, $\mu_4 = 162$, $\beta_1 = 1$ and $\beta_2 = 0·63$

9. (*a*) If the first four moments of a distribution about the value 5 are equal to – 4, 22, –117 and 560, determine the corresponding moments (*i*) about the mean, and (*ii*) about zero.

(*b*) The first four moments of a distribution about $x = 4$ are 1, 4, 10 and 45. Show that the mean is 5 and the variance is 3 and μ_3 and μ_4 are 0 and 26 respectively.

10. (*a*) (*i*) Karl Pearson's coefficient of skewness of a distribution is 0·32, its *s.d.* is 6·5 and mean is 29·6. Find the mode of the distribution.

(*ii*) If the mode of the above distribution 24·8, what will be the s.d. ?

(*b*) In a frequency distribution, the coefficient of skewness based upon the quartiles is 0.6. If the sum of the upper and lower quartiles is 100 and median is 38, find the value of the upper and lower quartiles.

Hint. We are given :

$$S_k = \frac{Q_3 + Q_1 - 2Md}{Q_3 - Q_1} = 0·6; \qquad Q_3 + Q_1 = 100 \text{ and Median} = 38 \qquad \dots (*)$$

$$\therefore \qquad \frac{100 - 2 \times 38}{Q_3 - Q_1} = 0·6 \qquad \Rightarrow \qquad Q_3 - Q_1 = 40 \qquad \dots (**)$$

Solving (*) and (**), we get $Q_1 = 30$, $Q_3 = 70$.

11. (*a*) A frequency distribution gave the following results :

(*i*) C.V. = 5, (*ii*) Karl Pearson's coefficient of skewness = 0·5, (*iii*) $\sigma = 2$.

Find the mean and mode of the distribution.

(*b*) Find the C.V. of a frequency distribution given that its mean is 120, mode is 123 and Karl Pearson's coefficient of skewness is – 0·3.

(*c*) The first three moments of distribution about the value 2 are 1, 16 and – 40 respectively. Examine the skewness of the distribution.

12. For the following data, calculate (*i*) Mean, (*ii*) Median, (*iii*) Quartile Deviation, (*iv*) Coefficient of variation, (*v*) Karl Pearson's coefficient of skewness, and (*vi*) β_1.

Wages in Rupees :	170—180	180—190	190—200	200—210	210—220	220—230	230—240	240—250
Number of Persons :	52	68	85	92	100	95	70	28

Ans. Mean = 209, Median = 209·8, Q.D. = 15·8, $\sigma = 19·7$, C.V. = 9·4, Sk (Karl Pearson) = – 0·1589, $\beta_1 = 0·003$

13. Find the second, third and fourth central moments of the frequency distribution given below. Hence find measure of skewness (γ_1) and measure of kurtosis (γ_2).

Class Limits	110·0—114·9	115·0—119·9	120·0—124·9	125·0—129·9	130·0—134·9	135·0—139·9	140·0—144·9
Frequency	5	15	20	35	10	10	5

Also apply Sheppard's corrections for moments.

Ans. $\mu_2 = 2{\cdot}16,\quad \mu_3 = 0{\cdot}804,\quad \mu_4 = 12{\cdot}5232,\quad \gamma_1 = \sqrt{\beta_1} = 0{\cdot}253,\quad \gamma_2 = \beta_2 - 3 = -0{\cdot}317$

14. The standard deviation of a symmetrical distribution is 5. What must be the value of the fourth moment about the mean in order that the distribution be

(i) leptokuric, (ii) mesokurtic, and (iii) platykurtic

Hint. $\mu_1 = \mu_3 = 0$ (∵ distribution is symmetrical)

$$\sigma = 5 \quad \Rightarrow \quad \sigma^2 = \mu_2 = 25\ ; \quad \beta_2 = \frac{\mu_4}{\mu_2^2} = \frac{\mu_4}{625}$$

(i) Distribution is leptokuric if $\beta_2 > 3$, *i.e.,* if $\frac{\mu_4}{625} > 3 \Rightarrow \mu_4 > 1{,}875$

(ii) Distribution is mesokurtic if $\beta_2 = 3, \Rightarrow$ if $\mu_4 = 1{,}875$

(iii) Distribution is platykurtic if $\beta_2 < 3, \Rightarrow$ if $\mu_4 < 1{,}875$

CHAPTER CONCEPTS QUIZ

1. Which measure of location will be suitable to compare :

(i) heights of students in two classes ; (ii) size of agricultural holdings ;
(iii) average sales for various years ; (iv) intelligence of students ;
(v) per capita income in several countries ; (vi) marks obtained 10, 8, 12, 4, 7, 11 and $X (X < 5)$.

2. Which of the following are true for all sets of data ?

(i) Arithmetic mean ≤ median ≤ mode, (ii) Arithmetic mean ≥ median ≥ mode,
(iii) Arithmetic mean = median = mode, (iv) None of these.

3. Which of the following are true in respect of any distribution ?

(i) The percentile points are in the ascending order.
(ii) The percentile points are equispaced.
(iii) A unique median value exists for each and every distribution.

4. Find out the missing figures :

(i) Mean = ? (3 Median – Mode). (ii) Mean – Mode = ? (Mean – Median).
(iii) Median = Mode + ? (Mean – Mode). (iv) Mode = Mean – ? (Mean – Median).

5. Fill in the blanks :

(i) Harmonic mean of a number of observations is ....
(ii) The geometric mean of 2, 4, 16 and 32 is ....
(iii) The strength of 7 colleges in a city are 385; 1,748; 1,343; 1,935; 786; 2,874 and 2,108. Then the median strength is ....
(iv) The geometric mean of a set of values lies between arithmetic mean and ....
(v) The mean and median of 100 items are 50 and 52 respectively. The value of the largest items is 100. It was later found that it is actually 110. Therefore, the true mean is .... and the true median is ....
(vi) The algebraic sum of the deviations of 20 observations measured from 30 is 2. Therefore, mean of these observations is ....
(vii) The relationship between A.M., G.M., and H.M. is ....
(viii) The mean of 20 observations is 15. On checking it was found that two observations were wrongly copied as 3 and 6. If wrong observations are replaced by correct values 8 and 4, then the correct mean is ....

(ix) Median = ... Quartile; (x) Median is the average suited for .... classes.
(xi) A distribution with two modes is called .... and with more than two modes is called ....
(xii) .... is not affected by extreme observations.

6. For the questions given below, give correct answer.
(i) The algebraic sum of deviations of a set of n values from their arithmetic mean is
(a) n, (b) 0, (c) 1, (d) none of these.
(ii) The most stable measure of central tendency is :
(a) the mean, (b) the median, (c) the mode, (d) none of these.
(iii) 10 is the mean of a set of 7 observations and 5 is the mean of a set of 3 observations. The mean of a combined set is given by
(a) 15, (b) 10, (c) 8·5, (d) 7·5.
(iv) The mean of the distribution, in which the values of x are 1, 2, ..., n, the frequency of each being unity is :
(a) $n(n + 1)/2$, (b) $n/2$, (c) $(n + 1)/2$, (d) none of these.
(v) The arithmetic mean of the numbers 1, 2, 3, ..., n is :
(a) $\frac{n(n+1)(2n+1)}{6}$, (b) $\frac{n(n+1)^2}{4}$, (c) $\frac{n(n+1)}{2}$, (d) none of these.
(vi) The point of intersection of 'less than' ogive and 'more than' ogive corresponds to
(a) the mean, (b) the median, (c) the geometric mean, (d) none of these.
(vii) When x_i and y_i $(i = 1, 2, ..., n)$ are the values of two variables X and Y with G.M.'s G_1 and G_2 respectively, then the geometric mean of X/Y is :
(a) G_1/G_2, (b) antilog (G_1/G_2),
(c) $n(\log G_1 - \log G_2)$, (d) Antilog $(\log G_1 - \log G_2/2n)$.

7. State which of the following statements are True and which are False. In case of false statements, give the correct statement.
(i) The harmonic mean of n number is the reciprocal of the arithmetic mean of the reciprocals of the numbers.
(ii) For the wholesale manufactures interested in the type which is usually in demand, median is the most suitable average.
(iii) The algebraic sum of the deviations of a series of individual observations from their mean is always zero.
(iv) Geometric mean is the appropriate average when emphasis is on the rate of change rather than the amount of change.
(v) Harmonic mean becomes zero when one of the items is zero.
(vi) Mean lies between median and mode.
(vii) Cumulative frequency is non-decreasing.
(viii) Geometric mean is the arithmetic mean of harmonic mean and arithmetic mean.
(ix) Mean, median mode have the same unit.
(x) One quintal of wheat was purchased at 0·8 kg. per rupee and another quintal at 1·2 kg. per rupee. The average rate per rupee is 1 kg.
(xi) One limitation of the median is that it cannot be calculated from a frequency distribution with open end classes.
(xii) The arithmetic mean of a frequency distribution is always located in the class which has the greatest number of frequencies.
(xiii) In a moderately asymmetrical distribution, the mean, median and mode are the same.
(xiv) It is really immaterial in which class an item falling at the boundary between two classes is listed.

(*xv*) The median is not affected by extreme items.

(*xvi*) The median is the point about which the sum of squared deviations is minimum.

(*xvii*) In construction of the frequency distribution, the selection of the class interval is arbitrary.

(*xviii*) Usual attendance of B.Sc. class is 35 per day. So for 100 working days, total attendance is 3,500.

(*xix*) A car travels 100 km. at a speed of 40 km. p.h. and another 400 km. at a speed of 30 km.p.h. So the average speed for the whole journey is either 35 km.p.h. or 33 km.p.h.

(*xx*) In calculating the mean for grouped data, the assumption is made that the mean fo the items in each class is equal to the mid-value of the class.

(*xxi*) The geometric mean of a group of numbers is less than the arithmetic mean in all cases, except in the special case in which the numbers are all the same.

(*xxii*) The median may be considered more typical than the mean because the median is not affected by the size of the extremes.

(*xxiii*) The harmonic mean of a series of fractions is the same as the reciprocal of the arithmetic mean fo the series.

(*xxiv*) In a frequency distribution, the true value of mode cannot be calculated exactly.

8. In each of the following cases, explain whether the description applies to mean, median or both.

(*i*) It can be calculated from a frequency distribution with open-end classes.

(*ii*) The values of all items are taken into consideration in the calculation.

(*iii*) The values of extreme items do not influence the average.

9. Be brief in your answer :

(*a*) The production in an industrial unit was 10,000 units during 2001 and in 2000 the production was 25,000 units. Hence the production has declined by 150 percent. Comment.

(*b*) A man travels by a car for 4 days. He travelled for 10 hours each day. He drove on the first day at the rate of 45 km. per hour, second day at the rate of 40 km. per hour, third day at the rate of 38 km. per hour and the fourth day at the rate of 37 km. per hour.

Which average, harmonic mean, arithmetic mean or median will give us his average speed ? Why ?

(*c*) It is seen from records that a country does not export more than 5% of its total production. Hence export trade is not vital to the economy of that country. Is the conclusion right ?

(*d*) A survey revealed that the children of engineers, doctors and lawyers have high intelligence quotients. It further revealed that the grandfathers of these children were also highly intelligent. Hence the inference is that intelligence is hereditary. Do you agree ?

10. Do you agree with the following interpretations made on the basis of the facts given. Explain briefly your answer.

(*a*) The number of deaths in military in the recent war was 10 out of 1,000 while the number of deaths in Hyderabad in the same period was 18 per 1,000. Hence it is safe to join military service than to live in the city of Hyderabad.

(*b*) The examination result in a college *A* was 70% in the year 2001. In the same year and at the same examination only 500 out of 750 students were successful in college *B*. Hence the teaching standard in college *A* was better.

(*c*) The average daily production in a small-scale factory in January 2001 was 4,000 candles and 3,800 candles in February 2001. So the workers were more efficient in January.

(*d*) The increase in the price of a commodity was 25%. Then the price decreased by 20% and again increased by 10%. So the resultant increase in the price was 25 – 20 + 10 = 15%.

(*e*) The rate of tomato in the first week of January was 2 kg. for a rupee and in the 2nd week was 4 kg. for a rupee. Hence the average price of tomato is $\frac{1}{2}(2 + 4) = 3$ kg. for a rupee.

11. (*a*) The mean mark of 100 students was given to be 40. It was found later that a mark 53 was read as 83. What is the corrected mean mark ?

(*b*) The mean salary paid to 1,0C0 employees of an establishment was found to be ₹ 108·40. Later on, after disbursement of salary it was discovered that the salary of two employees was wrongly entered as ₹ 297 and ₹ 165. Their correct salaries were ₹ 197 and ₹ 185. Find the correct arithmetic mean.

(*c*) Twelve persons gambled on a certain night. Seven of them lost at an average rate of ₹ 10·50 while the remaining five gained at an average of ₹ 13·00. Is the given information correct ? If not why ?

12. Which value of '*a*' gives the minimum ?

(*i*) Mean square deviation from '*a*', (*ii*) Mean deviation from '*a*'.

13. Mean of 100 observations is 50 and S.D. is 10. What will be the new mean and S.D. if

(*i*) 5 is added to each observation, (*ii*) each observation is multiplied by 3

(*iii*) 5 is subtracted from each observation and then it is divided by 4 ?

14. Fill in the blanks :

(*i*) (*a*) Absolute sum of deviation is minimum from ......

(*b*) Least value of root mean square deviation is ......

(*ii*) The sum of squares of deviations is least when measured from ......

(*iii*) The sum of 10 items is 12 and sum of their squares is 16.9. Variance = ...

(*iv*) In any distribution, the standard deviation is always ...... the mean deviation from mean.

(*v*) The relationship between root mean square deviation and standard deviation σ is ......

(*vi*) If 25% of the items are less than 10 and 25% are more than 40, the coefficient of quartile deviation is ......

(*vii*) The median and standard deviation of a distribution are 20 and 4 respectively. If each item is increased by 2, the new median will be ....,

(*viii*) In a symmetric distribution, the mean and the mode are ....,

(*ix*) In a symmetric distribution, the upper and the lower quartile are equidistant from ...

(*x*) If the mean, mode and standard deviation of a frequency distribution are 41, 45 and 8, respectively, then its Pearson's coefficient of skewness is ....,

(*xi*) For a symmetrical distribution β_1 = ..., ; (*xii*) If $\beta_2 > 3$, the distribution is said to be ....,

(*xiii*) For a symmetric distribution μ_{2n+1} = ...,

(*xiv*) If the mean and the mode of a given distribution are equal, then its coefficient of skewness is ....,

(*xv*) If the kurtosis of a distribution is 3, it is called ...... distribution.

(*xvi*) In a perfectly symmetrical distribution 50% of items are above 60 and 75% items are below 75. Therefore, the coefficient of quartile deviation is ...... and the coefficient of skewness is ....,

(*xvii*) The relation between β_1 and β_2 is given by ....,

15. For the following questions give correct answers :

(*i*) Sum of absolute deviations about median is

(*a*) least (*b*) greatest (*c*) zero (*d*) none of these

(*ii*) The sum of squares of deviations is least when measured from

(*a*) Median, (*b*) Mean, (*c*) Mode, (*d*) None of them

(*iii*) In any discrete series (when all the values are not same) the relationship between M.D. about mean and S.D. is

(*a*) M.D. = S.D., (*b*) M.D. ≥ S.D., (*c*) M.D. < S.D., (*d*) M.D. ≤ S.D.,

(*e*) None of these.

(*iv*) If each of a set of observations of a variable is multiplied by a constant (non-zero) value, the variance fo the resultant variable

(*a*) is unaltered, (*b*) increases, (*c*) decreases, (*d*) is unknown.

(*v*) The appropriate measure whenever the extreme items are to be disregarded and when the distribution contains indefinite classes at the end is

(*a*) Median (*b*) Mode (*c*) Quartile deviation (*d*) Standard Deviation

(*vi*) A.M., G.M., and H.M. in any series are equal when
(*a*) the distribution is symmetric, (*b*) all the values are same,
(*c*) the distribution is positively skewed, (*d*) the distribution is unimodal.

(*vii*) The limits for quartile coefficient of skewness are
(*a*) ± 3, (*b*) 0 and 3, (*c*) ± 1, (*d*) ± ∞.

(*viii*) The statement that the variance is equal to the second central moment is :
(*a*) always true, (*b*) sometimes true, (*c*) never true, (*d*) ambiguous.

(*ix*) The standard deviation of a distribution is 5. The value of the fourth central moment (μ_4), in order that the distribution be mesokurtic, should be
(*a*) equal to 3, (*b*) greater than 1,875, (*c*) equal to 1,875, (*d*) less than 1,875.

(*x*) In a frequency curve of scores, the mode was found to be higher then the mean. This shows that the distribution is
(*a*) symmetric, (*b*) negatively skewed, (*c*) positively skewed, (*d*) normal.

(*xi*) For any frequency distribution, the kurtosis is
(*a*) greater than 1, (*b*) less than 1, (*c*) equal to 1.

(*xii*) The measure of kurtosis is :
(*a*) $\beta_2 = 0$, (*b*) $\beta_2 = 3$, (*c*) $\beta_2 = 4$.

(*xiii*) For the distribution

X :	– 4	–3	–2	–1	0	1	2	3	4	Total
f :	2	4	5	7	10	7	5	4	2	46

(*a*) $\mu_4 = 0$ (*b*) Median = 0 (*c*) The distribution of x is symmetrical

16. State which of the following statements are *True* and which are *False*. In each of false statements give the correct statement.

(*i*) Mean, standard deviation and variance have the same unit.

(*ii*) Standard deviation of every distribution is unique and always exists.

(*iii*) Median is the value of the variable which divides the total frequency in two equal parts.

(*iv*) Mean – Mode = 3(mean – median), is often approximately satisfied.

(*v*) Mean deviation = $\frac{4}{5}$(standard deviation), is always satisfied.

(*vi*) $\beta_2 \geq 1$ is always satisfied.

(*vii*) $\beta_1 = 0$ is a conclusive test for a distribution to be symmetrical.

(*viii*) Two distributions having the same values of mean, *s.d.*, and skewness must have the same kurtosis.

(*ix*) For a symmetrical distribution $\mu_1 = \mu_3 = \mu_5 = ... = 0$.

(*x*) For platykurtic distribution, $\gamma_2 > 0$.

SUPPLEMENTARY REVIEW PROBLEMS FOR SELF-ASSESSMENT

2·1. The numbers 3·2, 5·8, 7·9 and 4·5 have frequencies x, $(x + 2)$, $(x - 3)$ and $(x + 6)$, respectively. If their arithmetic mean is 4·876, find the value of x.

2·2. If $M_{g.x}$ is the geometric mean of Nx's and $M_{g.y}$ is the geometric mean of Ny's, then the geometric mean M_g of the 2N values is given by $M_g^2 = M_{g.x}\, M_{g.y}$.

2·3. If $\bar{x}_1 = \frac{1}{n}\sum_{i=1}^{n} x_i$, $\bar{x}_2 = \frac{1}{n}\sum_{i=2}^{n+1} x_i$ and $\bar{x}_3 = \frac{1}{n}\sum_{i=3}^{n+2} x_i$, then show that

$$(a)\ \bar{x}_2 = \bar{x}_1 + \frac{1}{n}(x_{n+1} - x_1), \quad \text{and} \quad (b)\ \bar{x}_3 = \bar{x}_2 + \frac{1}{n}(x_{n+2} - x_2).$$

2·4. Let x_1 and x_2 be two distinct values of a variable x with respective weights w_1 and w_2. If $\bar{x}$ and $\bar{x}_W$ denote their simple and weighted arithmetic means, respectively, find the conditions under which :

(i) $\bar{x} = \bar{x}_W$; (ii) $\bar{x} > \bar{x}_W$ and (iii) $\bar{x} < \bar{x}_W$

Ans. (i) $\bar{x} = \bar{x}_W$ if $w_1 = w_2$ (ii) $\bar{x} > \bar{x}_W$ if greater weight is applied to the smaller value

(iii) $\bar{x} < \bar{x}_W$ if greater weight is applied to the greater value

2·5. If $\bar{x}_w$ is the weighted mean of x_i's with weights w_i, prove that

$$\left(\sum_{i=1}^{n} w_i\right)\left(\sum_{i=1}^{n} w_i(x_i - \bar{x}_w)^2\right) = \sum_{i=1}^{n}\sum_{j>1}^{n} w_i\, w_j,(x_i - x_j)^2 \text{ where } \sum_{i>1}^{n} w_i \neq 0.$$

2·6. A distribution $x_1, x_2, \ldots, x_n$ with frequencies $f_1, f_2, \ldots, f_n$ is transformed into the distribution $X_1, X_2, \ldots, X_n$ with the same corresponding frequencies by the relation $X_r = ax_r + b$, where a and b are constants. Show that the mean, median and mode of the new distribution are given in terms of those of the first distribution by the same transformation.

2·7. A variable x takes the positive value $a, ar, ar^2, \ldots, ar^{n-1}$, $r > 0$, each with frequencies 1. A, G and H are respectively the arithmetic, geometric and harmonic means, prove that :

$$A = \frac{a(1-r^n)}{n(1-r)}, \qquad G = ar^{(n-1)/2}, \qquad H = \frac{an\,(1-r)\,r^{n-1}}{1-r^n}$$

Also show that :

(i) $G^2 = AH$ and (ii) $A > G > H$ unless $n = 1$ when $A = G = H$.

2·8. Show that for a set of positive values of the variable X, the arithmetic, geometric and harmonic means are special cases of the pth root of the pth powers of the variable.

2·9. For two real numbers $a > 0$, $b > 0$, define Arithmetic means (A.M.), Geometric Mean (G.M), Harmonic Mean (H.M) and Root Mean Square (R.M.S.) and establish the following inequality between them :

$$\text{R.M.S.} \geq \text{A.M.} \geq \text{G.M.} \geq \text{H.M.}$$

State the conditions under which equality sign holds.

Hint. $\text{R.M.S.} = \sqrt{\frac{1}{2}(a^2 + b^2)}$

Equality sign holds. If and only if $a = b$.

2·10. (a) Find the mean and variance of the first n natural numbers.

(b) Prove that coefficient of variation (C.V.) of the first n-natural numbers is : $\text{C.V.} = \sqrt{\frac{(n-1)}{3(n+1)}}$

(c) Find the weighted mean and the variance of the first n-natural numbers, the weights, being the corresponding numbers.

Ans. (a) Mean $= (n + 1)/2$; Variance $= (n^2 - 1)/12$ (c) $\bar{x}_w = \frac{2n+1}{3}$; $\text{Var}(X) = \frac{(n+2)(n-1)}{18}$

2·11. If the mean and standard deviation of a variable x are m and σ respectively, obtain the mean and standard deviation of $u = (ax + b)/c$, where a, b and c are constants.

Ans. $\bar{u} = \frac{1}{c}(a\bar{x} + b)$; $\sigma_u = \left|\frac{a}{c}\right| \sigma_x$

2·12. In a series of measurements we obtain m_1 values of magnitude x_1, m_2 values of magnitude x_2, and so on. If $\bar{x}$ is the mean value of all the measurements, prove that the standard deviation is :

$$\sqrt{\frac{\Sigma m_r\,(k - x_r)^2}{\Sigma m_r} - \delta^2}, \quad \text{were } \bar{x} = k + \delta \text{ and } k \text{ is any constant.}$$

2·13. (*a*) Show that in a discrete series if deviations are small compared with mean M so that $(x/M)^3$ and higher powers of (x/M) are neglected, prove that

(*i*) $MH = G^2$, (*ii*) $M - 2G + H = 0$.

where G is geometric mean and H is harmonic mean.

(*b*) The mean and standard deviation of a variable x are m and σ, respectively. If the deviations are small compared with the value of the mean, show that

(*i*) $\text{Mean}\,(\sqrt{x}) = \sqrt{m}\left(1 - \dfrac{\sigma^2}{8m^2}\right)$ (*ii*) $\text{Mean}\left(\dfrac{1}{\sqrt{x}}\right) = \dfrac{1}{\sqrt{m}}\left(1 + \dfrac{3\sigma^2}{8m^2}\right)$ approximately.

(*c*) If the deviation $X_i = x_i - M$ is very small in comparison with mean M and (X_i/M) are neglected,

prove that $V = \sqrt{\dfrac{2(M-G)}{M}}$, where G is the geometric mean of the values $x_1, x_2, ..., x_n$ and V is the coefficient of variation (σ/M).

2·14. Show that, if the variable takes the value 0, 1, 2, ..., n with frequencies proportional to the binomial coefficients $^nC_0, {}^nC_1, {}^nC_2, ..., {}^nC_n$ respectively then the mean of the distribution is $(n/2)$, the mean square deviation about $x = 0$ is $n(n + 1)/4$ and the variance is $n/4$.

Hint. $N = \Sigma f = {}^nC_0 + {}^nC_1 + {}^nC_2 + ... + {}^nC_n = (1 + 1)^n = 2^n$

$$\Sigma fx = 0 \,.\, {}^nC_0 + 1.\, {}^nC_1 + 2.\, {}^nC_2 + ... + n.\, {}^nC_n = n\left\{1 + (n-1) + \frac{(n-1)(n-2)}{2!} + ... + 1\right\}$$

$$= n(1+1)^{n-1} = n \,.\, 2^{(n-1)}$$

$$\therefore \quad \text{Mean}\,(\bar{x}) = \frac{1}{N}\Sigma fx = \frac{n.2^{(n-1)}}{2^n} = \frac{n}{2}$$

The mean square deviation s^2, (say), about the point $x = 0$ is given by

$$s^2 = \frac{1}{N}\Sigma fx^2 = \frac{1}{2^n}\left[1^2 \,.\, {}^nC_1 + 2^2 \,.\, {}^nC_2 + 3^3 \,.\, {}^nC_3 + ... + n^2 \,.\, {}^nC_n\right]$$

$$= \frac{n}{2^n}\left[1 + 2(n-1) + \frac{3}{2}(n-1)(n-2) + ... + n\right]$$

$$= \frac{n}{2^n}\left[\left\{1 + (n-1) + \frac{(n-1)(n-2)}{2!} + ... + 1\right\} + \left\{(n-1) + (n-1)(n-2) + ... + (n-1)\right\}\right]$$

$$= \frac{n}{2^n}\left[\left({}^{n-1}C_0 + {}^{n-1}C_1 + {}^{n-1}C_2 + ... + {}^{n-1}C_{n-1}\right) + (n-1)\left({}^{n-2}C_0 + {}^{n-2}C_1 + ... + {}^{n-2}C_{n-2}\right)\}\right]$$

$$= \frac{n}{2^n}\left[(1+1)^{n-1} + (n-1)(1+1)^{n-2}\right] = \frac{n(n+1)}{4} \quad \Rightarrow \quad \sigma^2 = \frac{n(n+1)}{4} - \frac{n^2}{4} = \frac{n}{4}.$$

2·15. (*a*) Let r be the range and s be the standard deviation of a set of observations $x_1, x_2, ..., x_n$; then prove by general reasoning or otherwise that $s \le r$.

Hint. Since $x_i - \bar{x} \le r, \; i = 1, 2, ..., n$, we have

$$s^2 = \frac{1}{N}\sum_{i=1}^{n} f_i(x_i - \bar{x})^2 \le \frac{1}{N}\sum_{i=1}^{n} f_i\,(r^2) = r^2\,\frac{1}{N}\sum_{i=1}^{n} f_i = r^2 \quad \Rightarrow \quad s \le r$$

(*b*) Let r be the range and $S = \left[\dfrac{1}{n-1}\sum_{i=1}^{n}(x_i - \bar{x})^2\right]^{\frac{1}{2}}$, be the standard deviation of a set of observations $x_1, x_2, ..., x_n$, then prove that

$$S \le r\left(\frac{n}{n-1}\right)^{\frac{1}{2}}$$

2·16. Comment on the following :

(*i*) The variance of a distribution is invariant under the change of origin and scale.

(*ii*) If the mean and s.d. of n observations $(x_1, x_2, ..., x_n)$ are $\bar{x}$ and σ, respectively, then the mean and s.d. of $(-x_1, -x_2, ..., -x_n)$ are $(-\bar{x})$ and $(-\sigma)$ respectively.

(*iii*) Pearson's coefficients β_1 and β_2 are independent of change of origin and scale.

Ans. (*i*) Wrong. [Variance is independent of change of origin but not of scale]

(*ii*) Wrong. New mean $= -\bar{x}$; New s.d. $= \sigma$. (*iii*) True

2·17. Show that for any discrete distribution $\beta_2 \geq 1$, Explain, that the sign of equality holds if x takes only two values $(\bar{x} \pm \sigma)$, each with probability (1/2).

2·18. Show that for any frequency distribution :

(*i*) Kurtsies is greater than unity

(*ii*) Bowleys coefficient of skewness is less than unity numerically.

(*iii*) Karl Pearson's coefficient of skewness lies between –3 and 3.

2·19. (*a*) The scores in Economics of 250 candidates appearing at an examination have :

$$\text{Mean} = \bar{x} = 39{\cdot}72 \ ; \qquad \text{Variance} = \sigma^2 = 97{\cdot}80 \ ;$$

Third Central moment $= \mu_3 = -114{\cdot}18$; Fourth central moment $= \mu_4 = 28{,}396{\cdot}14$

It was later found on scrutiny that the score 61 of a candidate has been wrongly recorded as 51. Make necessary corrections in the given values of the mean and the central moments.

(*b*) In calculating the moments of a frequency distribution based on 100 observations, the following results are obtained :

$$\text{Mean} = 9, \qquad \text{Variance} = 19, \qquad \beta_1 = 0{\cdot}7(\mu_3 + \text{ive}), \qquad \beta_2 = 4$$

But later on it was found that one observation 12 was read as 21. Obtain the correct values of the first four central moments, and β_1 and β_2.

Ans. Corrected mean $= 8{\cdot}91$, $\mu_2 = 17{\cdot}64$, $\mu_3 = 57{\cdot}05$, $\mu_4 = 1257{\cdot}15$, $\beta_1 = 0{\cdot}59$ and $\beta_2 = 4{\cdot}04$.

2·20. The first three moments about the origin are given by :

$$\mu_1' = \frac{n+1}{2}, \qquad \mu_2' = \frac{(n+1)(2n+1)}{6} \qquad \text{and} \qquad \mu_3' = \frac{n(n+1)^2}{4}.$$

Examine the skewness of the data.

2·21. Show that, if the class interval of a grouped distribution is less than one third of the calculated standard deviation, Sheppard's adjustment makes a difference of less than $\frac{1}{2}$% in the estimate of the standard deviation.

2·22. Show that if a range of six times the standard deviation covers at least 18 class intervals, Sheppard's correction will make a difference of less than 0.5 per cent in the corrected value of the standard deviation.

2·23. If ∂_r is the rth absolute moment about zero, use the mean value of

$$\left\{u \mid x \mid^{(r-1)/2} + v \mid x \mid^{(r+1)/2}\right\}^2 \quad \text{to show that } \partial_r^{2r} \leq \partial_{r-1}^{r}\, \partial_{r+1}^{r}.$$

From this derive the following inequalities : (*i*) $\partial_r^{\,r+1} \leq \partial_{r+1}^{\,r}$, (*ii*) $\partial_r^{\,1/r} \leq \partial_{r+1}^{\,1/(r+1)}$.

2·24. For a random variable X moments of all orders exist. Denoting by μ_j and δ_j, the jth central moment and jth absolute moment respectively, show that

(*i*) $(\mu_{2j+1})^2 \leq \mu_{2j}\, \mu_{2j+2}$ (*ii*) $(\partial_j)^{1/j} \leq (\partial_{j+1})^{1/(j+1)}$.

CHAPTER

3 Theory of Probability – I

LEARNING OBJECTIVES

Upon completion of this Chapter, you should be able to:

- Learn about the different approaches to the theory of probability like the classical/mathematical probability, empirical probability and axiomatic probability, along with their limitations.
- Understand the addition theorem, the multiplication theorem and other important theorems on probability, and their use in solving problems in various diversified situations.
- Demonstrate the concepts of conditional probability, pairwise independence and mutual independence of events.
- Appreciate the use of probability theory in our day-to-day life and in the decision-making in the face of uncertainty.

CHAPTER OUTLINES

3·1. INTRODUCTION

If an experiment is repeated under essentially homogeneous and similar conditions, we generally come across two types of situations:

(i) The result or what is usually known as the *'outcome'* is unique or certain.

(ii) The result is not unique but may be one of the several possible outcomes.

The phenomena covered by (i) are known as *'deterministic'* or 'predictable' phenomena. By a deterministic phenomenon we mean one in which the result can be predicted with certainty. For example,

(a) For a perfect gas, $V \propto \frac{1}{P}$, *i.e.*, PV = constant, where V is the volume and P is the pressure of the gas, provided the temperature remains the same.

(b) Ohm's law, *viz.*, $C = \frac{E}{R}$, where C is the flow of current, E the potential difference between the two ends of the conductor and R the resistance, uniquely determines the value C as soon as E and R are given.

A deterministic model is defined as a model which stipulates that the conditions under which an experiment is performed determine the outcome of the experiment. For a number of situations the deterministic model suffices. However, there are phenomena [as covered by (ii) above] which do not lend themselves to deterministic approach and are known as *'unpredictable'* or *'probabilistic'* phenomena. Such phenomena are frequently observed in business, economics and social sciences or even in our day-to-day life.

For example:

(i) In a random toss of a uniform coin we are not sure of getting the head or tail.

(ii) A manufacturer cannot ascertain the future demand of his product with certainity.

(iii) A sales manager cannot predict with certainty about the sales target next year.

(iv) If an electric tube has lasted for one year, nothing can be predicted about its future life.

Probability is also used informally in day-to-day life. We daily come across the sentences like:

1. Possibly, it will rain to-night.
2. There is a high chance of my getting the job next month.
3. This year's demand for the product is likely to exceed that of the last year's.
4. The odds are 3 : 2 in favour of getting the contract applied for.

All the above sentences, with words like *'possibly'*, *'high chance'*, *'likely'* and *'odds'* are expressions indicating a degree of uncertainty about the happening of the event. A numerical measure of uncertainty is provided by a very important branch of Mathematics called *"Theory of Probability"*. Broadly, there are three possible states of expectation—*'certainty'*, *'impossibility'* and *'Probability'*. The probability theory describes certainty by 1. impossibility by 0 and the various grades of uncertainties by coefficients ranging between 0 and 1.

According to Ya-Lin Chou:

"Probability is the science of decision-making with calculated risks in the face of uncertainty."

3·2. SHORT HISTORY

Galileo (1564-1642), an Italian mathematician, was the first to attempt at a quantitative measure of probability while dealing with some problems related to the theory of dice in gambling. But the first foundation of the mathematical theory of probability was laid in the mid-seventeenth century by two French mathematicians B. Pascal (1623–62) and P. Fermat (1601–65), while solving a number of problems posed by French gambler and noble man Chevalier-De-Mere to Pascal. The famous *'problem of points'* posed by De-Mere to Pascal is : "Two persons play a game of chance." The person who first gains a certain number of points wins the stake. They stop playing before the game is completed. How is the stake to be decided on the basis of the number of points each has won?" The two mathematicians, after a lengthy correspondence between themselves ultimately solved this problem and this correspondence laid the first foundation of the science of probability. Next stalwart in this field was J. Bernoulli (1654–1705) whose *'Treatise on probability'* was published posthumously by his nephew N. Bernoulli in 1713. De-Moivre (1667–1754) also did considerable work in this field and published his famous *'Doctrine of Chances'* in 1718. Other main contributors are: T. Bayes (Inverse probability), P.S. Laplace (1749–1827) who after extensive research over a number of years finally published *'Theoric analytique des probabilities'* in 1812. In addition to these, other outstanding contributors are Levy, Mises and R.A. Fisher.

Russian mathematicians also have made very valuable contributions to the modern theory of probability. Chief contributors, to mention only a few of them are : Chebychev (1821–94) who founded the Russian School of Statisticians; A Markoff (1856–1922); Liapounoff (Central Limit Theorem); A Khintchine (Law of Large Numbers) and A. Kolmogorov, who axiomised the calculus of probability.

3·3. BASIC TERMINOLOGY

In this section we shall explain the various terms which are used in the definition of probability under different approaches.

1. *Random Experiment.* *If in each tiral of an experiment conducted under identical conditions, the outcome is not unique, but may be any one of the possible outcomes, then such an experiment is called a random experiment.*

Examples of random experiments are : tossing a coin, throwing a die, selecting a card from a pack of playing cards, selecting a family out of a given group of families, etc. In all these cases, there are a number of possible results which can occur but there is an uncertainty as to which one of them will actually occur.

Notes. (*i*) A *die* is a small cube used in gambling. On its six faces, dots are marked as

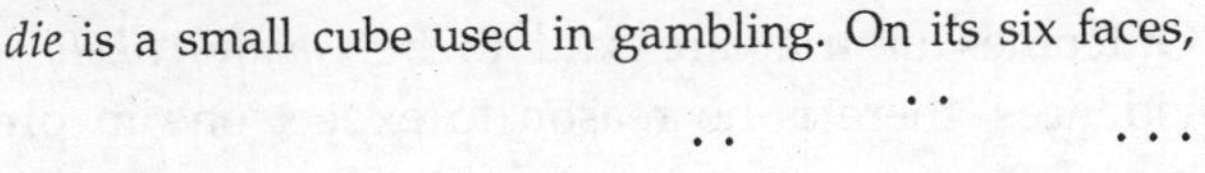

Plural of die is *dice*. The outcome of throwing a die is the number of dots on its uppermost face.

(*ii*) A *pack of cards* consists of four suits called *Spades, Hearts, Diamonds* and *Clubs*. Each suit consists of 13 cards, of which nine cards are numbered from 2 to 10, an ace, king, a queen and a jack (or knave). Spades and clubs are black-faced cards, while hearts and diamonds are red-faced cards.

2. *Outcome.* The result of a random experiment will be called an *outcome.*

3. *Trial and Event.* Any particular performance of a random experiment is called a *trial* and outcome or combination of outcomes are termed as *events.* For example,

(*i*) If a coin is tossed repeatedly, the result is not unique. We may get any of the two faces, head or tail. Thus tossing of a coin is a random experiment or trial and getting of a head or tail is an event.

(*ii*) In an experiment which consists of the throw of a six-faced die and observing the number of points that appear, the possible outcomes are 1, 2, 3, 4, 5, 6

In the same experiment, the possible events could also be stated as

'Odd number of points'; *'Even number of points'*; *'Getting a point greater than 4'*; and so on.

Remark. Event is called *simple* if it corresponds to a single possible outcome of the experiment otherwise it is known as a *compound or composite* event. Thus in tossing of a single die the event of getting '6' is a simple event but the event of getting an even number is a composite event.

4. *Exhaustive Events or Cases.* The total number of possible outcomes of a random experiment is known as the *exhaustive events or cases*. For example,

(*i*) In tossing of a coin, there are two exhaustive cases, *viz.*, head and tail (the possibility of the coin standing on an edge being ignored).

(*ii*) In throwing of a die, there are 6 exhaustive cases since any one of the 6 faces 1, 2, ..., 6 may come uppermost.

(*iii*) In drawing two cards from a pack of cards, the exhaustive number of cases is ${}^{52}C_2$, since 2 cards can be drawn out of 52 cards in ${}^{52}C_2$ ways.

(*iv*) In throwing of two dice, the exhaustive number of cases is $6^2 = 36$, since any of the numbers 1 to 6 on the first die can be associated with any of the 6 numbers on the other die. In general, in throwing of n dice, the exhaustive number of cases is 6^n.

5. *Favourable Events or Cases.* The number of cases favourable to an event in a trial is the number of outcomes which entail the happening of the event. For example,

(*i*) In drawing a card from a pack of cards the number of cases favourable to drawing of an ace is 4, for drawing a spade is 13 and for drawing a red card is 26.

(*ii*) In throwing of two dice, the number of cases favourable to getting the sum 5 is : (1, 4), (4, 1), (2, 3), (3, 2), *i.e.*, 4.

6. *Mutually Exclusive Events.* Events are said to be *mutually exclusive or incompatible* if the happening of any one of them precludes the happening of all the others, *i.e.*, if no two or more of them can happen simultaneously in the same trial. For example,

(*i*) In throwing a die all the 6 faces numbered 1 to 6 are mutually exclusive since if any one of these faces comes, the possibility of others, in the same trial, is ruled out.

(*ii*) Similarly in tossing a coin the events head and tail are mutually exclusive.

7. *Equally Likely Events.* Outcomes of trial are said to be *equally likely* if taking into consideration all the relevant evidences, there is no reason to expect one in preference to the others. For example,

(*i*) In a random toss of an unbiased or uniform coin, head and tail are equally likely events.

(*ii*) In throwing an unbiased die, all the six faces are equally likely to come.

8. *Independent Events.* Several events are said to be independent if the happening (or non-happening) of an event is not affected by the supplementary knowledge concerning the occurrence of any number of the remaining events. For example,

(*i*) In tossing an unbiased coin, the event of getting a head in the first toss is independent of getting a head in the second, third and subsequent throws.

(*ii*) When a die is thrown twice, the result of the first throw does not affect the result of the second throw.

(*iii*) If we draw a card from a pack of well-shuffled cards and replace it before drawing the second card, the result of the second draw is independent of the first draw. But, however, if the first card drawn is not replaced, then the second draw is dependent on the first draw.

3·4. MATHEMATICAL (OR CLASSICAL OR 'A PRIORI') PROBABILITY

Defintion. *If a random experiment or a trial results in 'n' exhaustive, mutually exclusive and equally likely outcomes (or cases), out of which m are favourable to the occurrence of an event E, then the probability 'p' of occurrence (or happening) of E, usually denoted by P(E), is given by :*

$$p = P(E) = \frac{\text{Number of favourable cases}}{\text{Total number of exhaustive cases}} = \frac{m}{n} \qquad ...(3\cdot 1)$$

This definition was given by James Bernoulli who was the first person to obtain a quantitative measure of uncertainty.

Remarks:

1. Since $m \geq 0$, $n > 0$ and $m \leq n$, we get from (3.1):
$$P(E) \geq 0 \quad \text{and} \quad P(E) \leq 1 \quad \Rightarrow \quad 0 \leq P(E) \leq 1$$
2. Sometimes we express (3·1) by saying that 'the odds' in favour of E are $m : (n - m)$ or the odds against E are $(n - m) : m$.
3. The non-happening of the event E is called the *complementary event of* E and is denoted by $\bar{E}$ or E^c. The number of cases favourable to $\bar{E}$, *i.e.*, non-happening of E is $(n - m)$. Then the probability q that E will not happen is given by :
$$q = P(\bar{E}) = \frac{n-m}{n} = 1 - \frac{m}{n} = 1 - p \quad \Rightarrow \quad p + q = 1 \qquad ...(3\cdot 1a)$$
$$\therefore \quad q = P(\bar{E}) = 1 - P(E) \quad \Rightarrow \quad P(E) = 1 - P(\bar{E}) \quad \text{or} \quad P(E) + P(\bar{E}) = 1 \qquad ...(3\cdot 1b)$$
If the event E represents the happening of at least one of the events $E_1, E_2, ..., E_n$, then its complementary event $\bar{E}$ represents the happening of none of the events E_1, E_2, E_n. Substituting in (3.1*b*), we get
$$P[\text{Happening of at least one of the events } E_1, E_2, \ldots E_n]$$
$$= 1 - P\,(\text{None of the events } E_1, E_2, ..., E_n \text{ happens}). \qquad ...(3\cdot 1c)$$
4. Probability 'p' of the happening of an event is also known as the probability of success and the probability 'q' of the non-happening of the event as the probability of failure, *i.e.*, $(p + q = 1)$.
5. If $P(E) = 1$, E is called a *certain event* and if $P(E) = 0$, E is called an *impossible event*.
6. We can compute the probability in (3·1) by logical reasoning, without conducting any experiment. Since the probability in (3·1) can be computed prior to obtaining any experimental data, it is also termed as *'a priori'* or *'mathematical probability.'*

3·4·1. Limitations of Classical Definition.

This definition of classical probability breaks down in the following cases:

(*i*) If the various outcomes of the random experiment are not equally likely or equally probable. For example,

(*a*) The probability that a candidate will pass in a certain test is not 50%, since the two possible outcomes, *viz.*, success and failure (excluding the possibility of a compartment) are not equally likely.

(*b*) The probability that a ceiling fan in a room will fall is not 1/2, since the events of the fan 'falling' and 'not falling' though mutually exclusive and exhaustive, are not equally likely. In fact, the probability of the fan falling will be almost zero.

(c) If a person jumps from a running train, then the probability of his survival will not be 50%, since in this case the events survival and death, though exhaustive and mutually exclusive, are not equally likely.

(ii) If the exhaustive number of outcomes of the random experiment is infinite or unknown. In such cases, we use Geometric Probability discussed in chapter 4, § 4·3.

3·5. STATISTICAL (OR EMPIRICAL) PROBABILITY

Definition. *(VON MISES). If an experiment is performed repeatedly under essentially homogenous and identical conditions, then the limiting value of the ratio of the number of times the event occurs to the number of trials, as the number of trials becomes indefinitely large, is called the probability of happening of the event, it being assumed that the limit is finite and unique.*

Symbolically, if in N trials an event E happens M times, then the probability of the happening of E, denoted by $P(E)$ is given by :

$$P(E) = \lim_{N \to \infty} \frac{M}{N} \qquad ...(3·2)$$

Remarks:

1. Since in the relative frequency approach, the probability is obtained objectively by repetitive empirical observations, it is also known as *'Empirical Probability'*.
2. An experiment is unique and non-repeating only in the case of *subjective probability*. In other cases, there are a large number of experiments or trials to establish the chance of occurrence of an event. This is particularly so in case of empirical probability. In classical probability also, repeated experiments may be made to verify whether a deduction on the basis of certain axioms or undisputed laws is justified. Only after repeated trials it can be established that the chance of head in a toss of a coin is $\frac{1}{2}$. J. E. Kerrich conducted coin tossing experiment with 10 sets of 1,000 tosses each during his confinement in World War II. The number of heads found by him were :

   502, 511, 497, 529, 504, 476, 507, 520, 504, 529.

   This gives the probability of getting a head in a toss of a coin as : $\frac{5,079}{10,000} = 0·5079 \simeq \frac{1}{2}$.

Thus, the empirical probability approaches the classical probability as the number of trials becomes indefinitely large.

3·5·1. Limitations of Empirical Probability.

(i) If an experiment is repeated a large number of times, the experimental conditions may not remain identical and homogenous.

(ii) The limit in (3·2) may not attain a unique value, however large N may be.

Example 3·1. *What is the chance that a leap year selected at random will contain 53 Sundays ?*

Solution. In a leap year (which consists of 366 days), there are 52 complete weeks and 2 days over. The following are the possible combinations for these two *'over'* days :

(i) Sunday and Monday, (ii) Monday and Tuesday,
(iii) Tuesday and Wednesday, (iv) Wednesday and Thursday,
(v) Thursday and Friday, (vi) Friday and Saturday,
(vii) Saturday and Sunday.

In order that a leap year selected at random should contain 53 Sundays, one of the two 'over' days must be Sunday. Since out of the above 7 possibilities, 2, *viz.*, (*i*) and (*vii*), are favourable to this event.

$$\therefore \qquad \text{Required probability} = \frac{2}{7}.$$

Example 3·2. *Two unbiased dice are thrown. Find the probability that :*

(*a*) *both the dice show the same number,*

(*b*) *the first die shows 6,*

(*c*) *the total of the numbers on the dice is 8,*

(*d*) *the total of the numbers on the dice is greater than 8,*

(*e*) *the total of the numbers on the dice is 13, and*

(*f*) *the total of the numbers on the dice is any number from 2 to 12, both inclusive.*

Solution. In a random throw of two dice, since each of the six faces of one die can be associated with each of six faces of the other die, the total number of cases is 6 × 6 = 36, as given below:

(1, 1)	(2, 1)	(3, 1)	(4, 1)	(5, 1)	(6, 1)
(1, 2)	(2, 2)	(3, 2)	(4, 2)	(5, 2)	(6, 2)
(1, 3)	(2, 3)	(3, 3)	(4, 3)	(5, 3)	(6, 3)
(1, 4)	(2, 4)	(3, 4)	(4, 4)	(5, 4)	(6, 4)
(1, 5)	(2, 5)	(3, 5)	(4, 5)	(5, 5)	(6, 5)
(1, 6)	(2, 6)	(3, 6)	(4, 6)	(5, 6)	(6, 6)

Here, the expression, say (i, j) means that the first die shows the number i and the second die shows the number j. Obviously, $(i, j) \neq (j, i)$ if $i \neq j$.

$$\therefore \qquad \text{Exhaustive number of cases } (n) = 36.$$

(*a*) The favourable cases that both the dice show the same number are :

(1, 1), (2, 2), (3, 3), (4, 4), (5, 5) and (6, 6), *i.e.*, $m = 6$.

$$\therefore \qquad \text{Probability that the two dice show the same number} = \frac{6}{36} = \frac{1}{6}.$$

(*b*) The favourable cases that the first die shows 6 are :

(6, 1), (6, 2) (6, 3), (6, 4), (6, 5) and (6, 6), *i.e.*, 6 in all.

$$\therefore \qquad \text{Probability that the first die shows '6'} = \frac{6}{36} = \frac{1}{6}.$$

(*c*) The cases favourable to getting a total of 8 on the two dice are :

(2, 6), (3, 5), (4, 4), (5, 3), (6, 2), *i.e.*, $m = 5$.

$$\therefore \qquad \text{Probability that total of numbers on two dice is '8'} = \frac{5}{36}.$$

(*d*) The cases favourable to getting a total of more than 8 are :

(3, 6), (6, 3), (4, 5), (5, 4), (4, 6), (6, 4), (5, 5), (5, 6), (6, 5), (6, 6), *i.e.*, $m = 10$.

$$\therefore \qquad \text{Probability that the total of numbers on two dice is greater than } 8 = \frac{10}{36} = \frac{5}{18}.$$

(*e*) This is an example of an *impossible event*, since the maximum total can be 6 + 6 = 12. Therefore, the required probability is 0.

(*f*) The probability is 1, as the total of the numbers on the two dice certainly ranges from 2 to 12. The given event is called a *certain event*.

Example 3·3. *(a) Among the digits 1, 2, 3, 4, 5 at first one is chosen and then a second selection is made among the remaining four digits. Assuming that all twenty possible outcomes have equal probabilities, find the probability that an odd digit will be selected.*

(i) the first time, *(ii) the second time, and* *(iii) both times.*

(b) From 25 tickets, marked with first 25 numerals, one is drawn at random. Find the chance that

(i) it is multiple of 5 or 7, and *(ii) it is a multiple of 3 or 7.*

Solution. (*a*) Total number of cases = 5 × 4 = 20.

(*i*) There are 12 cases in which the first digit drawn is odd, *viz.*, (1, 2), (1, 3), (1, 4), (1, 5), (3, 1), (3, 2), (3, 4), (3, 5), (5, 1), (5, 2), (5, 3) and (5, 4).

$$\therefore \qquad \text{The probability that the first digit drawn is odd} = \frac{12}{20} = \frac{3}{5}.$$

(*ii*) Also there are 12 cases in which the second digit drawn is odd, viz., (2, 1), (3, 1), (4, 1), (5, 1), (1, 3), (2, 3), (4, 3), (5, 3), (1, 5), (2, 5), (3, 5) and (4, 5).

$$\therefore \qquad \text{The probability that the second digit drawn is odd} = \frac{12}{20} = \frac{3}{5}.$$

(*iii*) There are six cases in which both the digits drawn are odd, *viz.*, (1, 3), (1, 5), (3, 1), (3, 5), (5, 1) and (5, 3).

$$\therefore \qquad \text{The probability that both the digits drawn are odd} = \frac{6}{20} = \frac{3}{10}.$$

(*b*) (*i*) Numbers (out of the first 25 numerals) which are multiples of 5 are 5, 10, 15, 20 and 25, *i.e.*, 5 in all and the numbers which are multiples of 7 are 7, 14 and 21, *i.e.*, 3 in all. Hence, required number of favourable cases are 5 + 3 = 8.

$$\therefore \qquad \text{Required probability} = \frac{8}{25}.$$

(*ii*) Numbers (among the first 25 numerals) which are multiples of 3 are 3, 6, 9, 12, 15, 18, 21, 24, *i.e.*, 8 in all; and the numbers, which are multiples of 7 are 7, 14, 21 *i.e.*, 3 in all. Since the number 21 is common in both the cases, required number of distinct favourable cases is 8 + 3 – 1 = 10.

$$\therefore \qquad \text{Required probability} = \frac{10}{25} = \frac{2}{5}.$$

Example 3·4. *(a) Four cards are drawn at random from a pack of 52 cards. Find the probability that*

(i) They are a king, a queen, a jack and an ace.;

(ii) Two are kings and two are queens;

(iii) Two are black and two are red;

(iv) There are two cards of hearts and two cards of diamonds.

(b) In shuffling a pack of cards, four are accidentally dropped. Find the chance that the missing cards should be one from each suit.

Solution. (*a*) Four cards can be drawn from a well-shuffled pack of 52 cards in ${}^{52}C_4$ ways, which gives the exhaustive number of cases.

(*i*) 1 king can be drawn out of the 4 kings in ${}^4C_1$ ways. Similarly, 1 queen, 1 jack and an ace can each be drawn in ${}^4C_1 = 4$ ways. Since any one of the ways of drawing a king can be associated with any one of the ways of drawing a queen, a jack and an ace, the favourable number of cases are ${}^4C_1 \times {}^4C_1 \times {}^4C_1 \times {}^4C_1$.

Hence, $$\text{Required probability} = \frac{{}^4C_1 \times {}^4C_1 \times {}^4C_1 \times {}^4C_1}{{}^{52}C_4} = \frac{256}{{}^{52}C_4}.$$

(*ii*) $$\text{Required probability} = \frac{{}^4C_2 \times {}^4C_2}{{}^{52}C_4}.$$

(*iii*) Since there are 26 black cards (of spades and clubs) and 26 red cards (of diamonds and hearts) in a pack of cards, the required probability $= \dfrac{{}^{26}C_2 \times {}^{26}C_2}{{}^{52}C_4}$.

(*iv*) $$\text{Required probability} = \frac{{}^{13}C_2 \times {}^{13}C_2}{{}^{52}C_4}.$$

(*b*) There are ${}^{52}C_4$ possible ways in which four cards can slip while shuffling a pack of cards. The favourable number of cases in which the four cards can be one from each suit is :

$${}^{13}C_1 \times {}^{13}C_1 \times {}^{13}C_1 \times {}^{13}C_1$$

$\therefore$ $$\text{Required probability} = \frac{{}^{13}C_1 \times {}^{13}C_1 \times {}^{13}C_1 \times {}^{13}C_1}{{}^{52}C_4} = \frac{2197}{20825}.$$

Example 3·5. *A man is dealt 4 spade cards from an ordinary pack of 52 cards. If he is given three more cards, find the probability p that at least one of the additional cards is also a spade.*

Solution. After a man has dealt 4 spade cards from an ordinary pack of 52 cards, there are $52 - 4 = 48$ cards left in the pack, out of which 9 are spade cards and 39 are non-spade cards.

Since 3 more cards can be dealt to the same man out of the 48 cards in ${}^{48}C_3$ ways, the exhaustive number of outcomes $= {}^{48}C_3$.

If none of these 3 additional cards is a spade card, then the additional cards must be drawn out of the 39 non-spade cards, which can be done in ${}^{39}C_3$ ways. The probability that none of the three additional cards dealt to the man is a spade card is given by ${}^{39}C_3/{}^{48}C_3$.

Hence, probability p that at least one of the three additional cards is also a spade is given by:

$$p = 1 - P\,[\text{None of the three additional cards is a spade.}]$$

$$= 1 - \frac{{}^{39}C_3}{{}^{48}C_3} = 1 - \frac{39 \times 38 \times 37}{3!} \times \frac{3!}{48 \times 47 \times 46} = 1 - \frac{13 \times 19 \times 37}{16 \times 47 \times 23} = 0{\cdot}4718.$$

Example 3·6. *A committee of 4 persons is to be appointed from 3 executives of the production department, 4 executives of the purchase department, 2 executives of the marketing department and 1 chartered accountant. Find the probability of forming the committee in the following manner:*

(*i*) *There must be one from each category.*

(*ii*) *It should have at least one from the purchase department.*

(*iii*) *The chartered accountant must be in the committee.*

Solution. There are 3 + 4 + 2 + 1 = 10 persons in all and a committee of 4 people can formed out of them in $^{10}C_4$ ways. Hence, exhaustive number of cases is:

$$^{10}C_4 = \frac{10 \times 9 \times 8 \times 7}{4!} = 210$$

(*i*) Favourable number of cases for the committee to consist of 4 members, one from each category, is : $^4C_1 \times {}^3C_1 \times {}^2C_1 \times 1 = 4 \times 3 \times 2 = 24$

$$\therefore \qquad \text{Required probability} = \frac{24}{210} = \frac{4}{35}.$$

(*ii*) P[Committee has at least one purchase executive]

$= 1 - P$ (Committee has no purchase executive)

In order that the committee has no purchase executive, all the 4 members are to be selected from amongst executives of production department, sales department and chartered accountant, *i.e.*, out of 3 + 2 + 1 = 6 members and this can be done in $^6C_4 = \dfrac{6 \times 5}{1 \times 2} = 15$ ways.

Hence,

$$P \text{ (Committee has no purchase executive)} = \frac{15}{210} = \frac{1}{14}$$

$$\therefore \qquad P \text{ (Committee has at least one purchase executive)} = 1 - \frac{1}{14} = \frac{13}{14}.$$

(*iii*) Favourable number of cases that the committee consists of a chartered accountant as a member and three others are : $1 \times {}^9C_3 = \dfrac{9 \times 8 \times 7}{1 \times 2 \times 3} = 84$ ways, since a chartered accountant can be selected out of one chartered accountant in only 1 way and the remaining 3 members can be selected out of the remaining 10 − 1 = 9 persons in $^9C_3$ ways.

$$\text{Hence,} \qquad \text{Required probability} = \frac{84}{210} = \frac{2}{5}.$$

Example 3·7. *An urn contains 6 white, 4 red and 9 black balls. If 3 balls are drawn at random, find the probability that : (i) two of the balls drawn are white, (ii) one is of each colour, (iii) none is red, (iv) at least one is white.*

Solution. Total number of balls in the urn is 6 + 4 + 9 = 19. Since 3 balls can be drawn out of 19 in $^{19}C_3$ ways, the exhaustive number of cases are $^{19}C_3$.

(*i*) If 2 balls of the 3 drawn balls are to be white, these two balls should be drawn out of 6 white balls which can be done in $^6C_2$ ways, and the third ball can be drawn out of the remaining 19 − 6 = 13 balls, which can be done in $^{13}C_1$ ways. Since any of the former ways can be associated with any one of the later ways, the number of favourable cases = $^6C_2 \times {}^{13}C_1$.

$$\text{Hence,} \qquad \text{Required probability} = \frac{^6C_2 \times {}^{13}C_1}{^{19}C_3}.$$

(*ii*) Since the number of favourable cases of getting one ball of each colour is

$${}^{6}C_1 \times {}^{4}C_1 \times {}^{9}C_1, \text{ the required probability } = \frac{{}^{6}C_1 \times {}^{4}C_1 \times {}^{9}C_1}{{}^{19}C_3}.$$

(*iii*) If none of the drawn balls is red, then all the 3 balls must be out of the whilte and black balls, *viz.*, out of 6 + 9 = 15 balls. Hence, the number of favourable cases for this event is ${}^{15}C_3$.

$$\therefore \qquad \text{Required probability } = \frac{{}^{15}C_3}{{}^{19}C_3}.$$

(*iv*) P (at least one ball is white) = 1 – P (none of the three balls is white) ...(*)

In order that none of the three balls is white, all the three balls must be drawn out of the red and black balls, *i.e.*, out of 4 + 9 = 13 balls and this can be done in ${}^{13}C_3$ ways.

Hence, $$P(\text{none of the three balls is white}) = \frac{{}^{13}C_3}{{}^{19}C_3}.$$

Substituting in (*), we obtain $$P(\text{at least one ball is white}) = 1 - \frac{{}^{13}C_3}{{}^{19}C_3}.$$

Example 3·8. *(a) If the letters of the word 'REGULATIONS' be arranged at random, what is the chance that there will be exactly 4 letters between R and E ?*

(b) What is the probability that four S's come consecutively in the word 'MISSISSIPPI'?

Solution. (*a*) The word 'REGULATIONS' consists of 11 letters. The two letters R and E can occupy ${}^{11}P_2$, *i.e.*, 11 × 10 = 110 positions.

The number of ways in which there will be exactly 4 letters between R and E are enumerated below:

(*i*) R is in the 1st place and E is in the 6th place.
(*ii*) R is in the 2nd place and E is in the 7th place.

... ...
... ...
... ...

(*vi*) R is in the 6th place and E is in the 11th place.

Since R and E can interchange their positions, required number of favourable cases is

$$2 \times 6 = 12.$$

$$\therefore \qquad \text{The required probability } = \frac{12}{110} = \frac{6}{55}.$$

(*b*) Total number of permutations of the 11 letters of the word '*MISSISSIPPI*' in which 4 are of one kind (*viz. S*), 4 of other kind (*viz. I*), 2 of third kind (*viz. P*) and 1 of fourth kind (*viz. M*) are

$$11\,!/4!\ 4!\ 2!\ 1!.$$

Following are the 8 possible combinations of 4 S's coming consecutively:

(*i*) S S S S
(*ii*) — S S S S
(*iii*) — — S S S S
⋮ ⋮ ⋮
(*viii*) — — — — — — — S S S S

Since in each of the above cases, the total number of arrangements of the remaining 7 letters, *viz.*, *MIIIPPI* of which 4 are of one kind, 2 of other kind and one of third kind are $\frac{7!}{4!2!1!}$, the required number of favourable cases $= \frac{8 \times 7!}{4!2!1!}$.

$$\therefore \quad \text{Required probability} = \frac{8 \times 7!}{4!2!1!} \div \frac{11!}{4!4!2!1!} = \frac{8 \times 7! \times 4!}{11!} = \frac{4}{165}.$$

Example 3·9. *Twenty-five books are placed at random in a shelf. Find the probability that a particular pair of books shall be : (i) Always together, and (ii) Never together.*

Solution. Since 25 books can be arranged among themselves in 25! ways, the exhaustive number of cases is 25!

(*i*) Let us now regard that the two particular books are tagged together so that we shall regard them as a single book. Thus, now we have (25 – 1) = 24 books which can be arranged among themselves in 24! ways. Hence, associating these two operations, the number of favourable cases for getting a particular pair of books always together is 24! × 2!.

$$\therefore \quad \text{Required probability} = \frac{24! \times 2!}{25!} = \frac{2}{25}.$$

(*ii*) Total number of arrangements of 25 books among themselves is 25! and the total number of arrangements that a particular pair of books will always be together is 24! × 2. Hence, the number of arrangements in which a particular pair of books is never together is :

$$25! - 2 \times 24! = (25 - 2) \times 24! = 23 \times 24!$$

$$\therefore \quad \text{Required probability} = \frac{23 \times 24!}{25!} = \frac{23}{25}.$$

Aliter. P[A particular pair of books shall never be together.]

$$= 1 - P[\text{A particular pair of books is always together.}]$$

$$= 1 - \frac{2}{25} = \frac{23}{25}.$$

Example 3·10. *n persons are seated on n chairs at a round table. Find the probability that two specified persons are sitting next to each other.*

Solution. Since n persons can be seated in n chairs at a round table in $(n-1)!$ ways, the exhaustive number of cases $= (n-1)!$.

Assuming the two specified persons A and B who sit together as one, we get $(n-1)$ persons in all, who can be seated at a round table in $(n-2)!$ ways. Further, since A and B can interchange their positions in 2! ways, total number of favourable cases of getting A and B together is $(n-2)! \times 2!$.

$$\therefore \quad \text{Required probability} = \frac{(n-2)! \times 2!}{(n-1)!} = \frac{2}{n-1}.$$

Example 3·11. *A five-figure number is formed by the digits 0, 1, 2, 3, 4 (without repetition). Find the probability that the number formed is divisible by 4.*

Solution. The total number of ways in which the five digits 0, 1, 2, 3, 4, can be arranged among themselves is 5!. Out of these, the number of arrangements which begin with 0 (and, therefore, will give only 4-digited numbers) is 4!. Hence the total number of five-digited numbers that can be formed from the digits 0, 1, 2, 3, 4 is:

$$5! - 4! = 120 - 24 = 96.$$

The number formed will be divisible by 4 if number formed by the two digits on extreme right (*i.e.*, the digits in the unit and tens places) is divisible by 4. Such numbers are:

04, 12, 20, 24, 32, and 40.

If the numbers end in 04, the remaining three digits, *viz.*, 1, 2, and 3 can be arranged among themselves in 3! ways. Similarly, the number of arrangements of the numbers ending with 20 and 40 is 3! in each case.

If the numbers end with 12, the remaining three digits 0, 3, 4 can be arranged in 3! ways. Out of these we shall reject those numbers which start with 0 (*i.e.*, have 0 as the first digit). There are $(3 - 1)! = 2!$ such cases. Hence, the number of five-digited numbers ending with 12 is $3! - 2! = 4$.

Similarly, the number of five digited numbers ending with 24 and 32 each is 4. Hence the total number of favourable cases is $3 \times 3! + 3 \times 4 = 18 + 12 = 30$.

$$\therefore \qquad \text{Required probability} = \frac{30}{96} = \frac{5}{16}.$$

Example 3·12. *(a) Twelve balls are distributed at random among three boxes. What is the probability that the first box will contain 3 balls?*

(b) If n biscuits he distributed among N persons, find the chance that a particular person receives r(< n) biscuits.

Solution. (*a*) Since each ball can go to any one of the three boxes, there are 3 ways in which a ball can go to any one of the three boxes. Hence there are 3^{12} ways in which 12 balls can be placed in the three boxes.

Number of ways in which 3 balls out of 12 can go to the first box is ${}^{12}C_3$. Now the remaining 9 balls are to be placed in remaining 2 boxes and this can be done in 2^9 ways. Hence, the total number of favourable cases $= {}^{12}C_3 \times 2^9$.

$$\therefore \qquad \text{Required probability} = \frac{{}^{12}C_3 \times 2^9}{3^{12}}.$$

(*b*) Take any one biscuit. This can be given to any one of the N beggars so that there are N ways of distributing any one biscuit. Hence, the total number of ways in which n biscuits can be distributed at random among N beggars $= N \,.\, N \ldots$ (n times) $= N^n$.

'r' biscuits can be given to any particular beggar in ${}^nC_r$ ways. Now we are left with $(n - r)$ biscuits which are to be distributed among the remaining $(N - 1)$ beggars and this can be done in $(N - 1)^{n-r}$ ways.

$$\therefore \qquad \text{Number of favourable cases} = {}^nC_r \cdot (N-1)^{n-r}$$

Hence,
$$\text{Required probability} = \frac{{}^nC_r\,(N-1)^{n-r}}{N^n}.$$

Example 3.13. *A car is parked among N cars in a row, not at either end. On his return the owner finds that exactly r of the N places are still occupied. What is the probability that both neighbouring places are empty?*

Solution. Since the owner finds on return that exactly r of the N places (including owner's car) are occupied, the exhaustive number of cases for such an arrangement is ${}^{N-1}C_{r-1}$.

[Since the remaining $r - 1$ cars are to be parked in the remaining $N - 1$ places and this can be done in ${}^{N-1}C_{r-1}$ ways.]

Let A denote the event that both the neighbouring places to owner's car are empty. This requires the remaining $(r - 1)$ cars to be parked in the remaining $N - 3$ places and hence, the number of cases favourable to A is ${}^{N-3}C_{r-1}$. Hence,

$$P(A) = \frac{{}^{N-3}C_{r-1}}{{}^{N-1}C_{r-1}} = \frac{(N-r)(N-r-1)}{(N-1)(N-2)}.$$

Example 3·14. *What is the probability that at least two out of n people have the same birthday? Assume 365 days in a year and that all days are equally likely.*

Solution. Since the birthday of any person can fall on any one of the 365 days, the exhaustive number of cases for the birthday of n persons is 365^n.

If the birthday of all n persons fall on different days, then the number of favourable cases is :

$$365\,(365-1)\,(365-2)\,\ldots\,[365-(n-1)],$$

because in this case the birthday of the first person can fall on any one of 365 days, the birthday of the second person can fall on any one of the remaining 364 days, and so on. Hence, the probability (p) that birthdays of all the n persons are different is given by :

$$p = \frac{365\,(365-1)\,(365-2)\ldots[365-(n-1)]}{365^n}$$

$$= \left(1-\frac{1}{365}\right)\left(1-\frac{2}{365}\right)\left(1-\frac{3}{365}\right)\ldots\left(1-\frac{n-1}{365}\right)$$

Hence, the required probability that at least two persons have same birthday is:

$$1 - p = 1 - \left(1-\frac{1}{365}\right)\left(1-\frac{2}{365}\right)\left(1-\frac{3}{365}\right)\ldots\left(1-\frac{n-1}{365}\right).$$

Example 3·15. *Compare the chances of throwing 4 with one die, 8 with two dice and 12 with three dice.*

Solution. (*i*) *Probability of throwing 4 with one die:* There are 6 possible ways in which the die can fall, and of these one is favourable to the required event.

$\therefore$ Required probability $(p_1) = \dfrac{1}{6}$.

(*ii*) *Probability of throwing 8 with two dice:* Exhaustive number of cases in single throw with two dice is $6^2 = 36$. Now the sum of '8' can be obtained on the two dice in the following ways: (2, 6), (3, 5), (4, 4), (5, 3), (6, 2), *i.e.*, 5 cases in all, where the first and second number in the brackets ( ) refer to the numbers on the 1st and 2nd die, respectively.

$\therefore$ Required probability $(p_2) = \dfrac{5}{36}$.

(*iii*) *Probability of throwing 12 with three dice:* The exhaustive number of ways in a single throw of three dice $= 6 \times 6 \times 6 = 216$.

To make a throw of 12, the three dice must show the faces either (6, 1, 5) or (6, 2, 4) or (6, 3, 3) or (5, 2, 5) or (5, 3, 4) or (4, 4, 4). The first two of these arrangements can occur in $3! = 6$ ways each, the second two (*i.e.*, third and fourth arrangement in $\dfrac{3!}{2!\,1!} = 3$ ways each, the fifth in $3! = 6$ ways and the last in one way only.

Thus, the total number of favourable cases = 6 + 6 + 3 + 3 + 6 + 1 = 25.

$\therefore$ Required probability $(p_3) = \dfrac{25}{216}$.

Hence, the chances of throwing 4 with one die, 8 with two dice, and 12 with three dice are :

$$p_1 : p_2 : p_3 :: \frac{1}{6} : \frac{5}{36} : \frac{25}{216} \quad \text{or} \quad 36 : 30 : 25.$$

Example 3·16. *A and B throw with three dice; if A throws 14, find B's chance of throwing a higher number.*

Solution. To throw higher number than A, B must throw either 15 or 16 or 17 or 18.

Now a throw amounting to 18 must be made up of (6, 6, 6), which can occur in one way; 17 can be made up of (6, 6, 5), which can occur in $\dfrac{3!}{2!1!} = 3$ ways; 16 may be made up of (6, 6, 4), and (6, 5, 5), each of which arrangements can occur in $\dfrac{3!}{2!1!} = 3$ ways; 15 can be made up of (6, 4, 5), or (6, 3, 6), or (5, 5, 5), which can occur in 3!, 3 and 1 way, respectively.

$\therefore$ The number of favourable cases = 1 + 3 + 3 + 3 + 6 + 3 + 1 = 20.

In a random throw of 3 dice, the exhaustive number of cases = $6^3 = 6 \times 6 \times 6 = 216$.

Hence, Required probability $= \dfrac{20}{216} = \dfrac{5}{54}$.

Example 3·17. *Each coefficient in the equation $ax^2 + bx + c = 0$, is determined by throwing an ordinary die. Find the probability that the equation will have real roots.*

Solution. The roots of the equation $ax^2 + bx + c = 0$ will be real if its discriminanat is non-negative, *i.e.*, if

$$b^2 - 4ac \geq 0 \quad \Rightarrow \quad b^2 \geq 4ac$$

Since each coefficient in equation $ax^2 + bx + c = 0$ is determined by throwing an ordinary die, each of the coefficients a, b and c can take the values from 1 to 6.

$\therefore$ Total number of possible outcomes (all being equally likely) = 6 × 6 × 6 = 216.

The number of favourable cases can be enumerated as follows:

ac		*a*	*c*	*4ac*	*b* (so that $b^2 \geq 4ac$)	*No. of cases*
1		1	1	4	2, 3, 4, 5, 6	1 × 5 = 5
2	(*i*)	1	2	8	3, 4, 5, 6	2 × 4 = 8
	(*ii*)	2	1			
3	(*i*)	1	3	12	4, 5, 6	2 × 3 = 6
	(*ii*)	3	1			
4	(*i*)	1	4			
	(*ii*)	4	1	16	4, 5, 6	3 × 3 = 9
	(*iii*)	2	2			

5	(i)	1	5	20	5, 6	2 × 2 = 4
	(ii)	5	1			
6	(i)	1	6			
	(ii)	6	1			
	(iii)	3	2	24	5, 6	4 × 2 = 8
	(iv)	2	3			
7	($ac = 7$ *is not possible*)					
8	(i)	2	4			
	(ii)	4	2	32	6	2 × 1 = 2
9		3	3	36	6	1
						Total = 43

Since $b^2 > 4ac$ and since the maximum value of b^2 is 36, ac = 10, 11, 12, ... etc. is not possible. Hence, total number of favourable cases = 43.

$$\therefore \quad \text{Required probability} = \frac{43}{216}.$$

Remark. The probability p that in the above case, the equation $ax^2 + bx + c = 0$, will have imaginary roots is given by:

$$p = 1 - P \text{ (The equation } ax^2 + bx + c = 0, \text{ has real roots.)}$$

$$= 1 - \frac{43}{216} = \frac{173}{216}.$$

Example 3·18. *The sum of two positive quantities is equal to 2n. Find the chance that their product is not less than 3/4 times their greatest product.*

Solution. Let $x > 0$ and $y > 0$, be the given quantities so that $x + y = 2n$. ...(*)

We know that the product of two positive quantities whose sum is constant (fixed) is greatest when the quantities are equal. Thus, the product x and y is maximum when $x = y = n$.

$$\therefore \quad P\left(xy \not< \frac{3}{4}n^2\right) = P\left(xy \geq \frac{3}{4}n^2\right) = P\left(x(2n - x) \geq \frac{3}{4}n^2\right) \quad \text{[From (*)]}$$

$$= P\left[(4x^2 - 8nx + 3n^2) \leq 0\right] = P\left[(2x - 3n)(2x - n) \leq 0\right]$$

$$= P\left(x \text{ lies between } \frac{n}{2} \text{ and } \frac{3n}{2}\right)$$

$$\therefore \quad \text{Favourable range} = \frac{3n}{2} - \frac{n}{2} = n, \quad \text{Total range} = 2n$$

Hence,
$$\text{Required probability} = \frac{\text{Favourable range}}{\text{Total range}} = \frac{n}{2n} = \frac{1}{2}.$$

Example 3·19. *Out of (2n + 1) tickets consecutively numbered, three are drawn at random. Find the chance that the numbers on them are in A.P.*

Solution. Since out of $(2n + 1)$ tickets, 3 tickets can be drawn in ${}^{2n+1}C_3$ ways,

$$\text{Exhaustive number of cases} = {}^{2n+1}C_3 = \frac{(2n+1)\,2n(2n-1)}{3!} = \frac{n(4n^2 - 1)}{3}$$

To find the favourable number of cases we are to enumerate all the cases in which the numbers on the drawn tickets are in A.P. with common difference, (say $d = 1, 2, 3, \ldots, n-1, n$).

If $d = 1$, the possible cases are as follows: If $d = 2$, the possible cases are as follows:

$$\left.\begin{array}{ccc} 1, & 2, & 3 \\ 2, & 3, & 4 \\ \vdots & \vdots & \vdots \\ 2n-1, & n & 2n+1 \end{array}\right\} \qquad \left.\begin{array}{ccc} 1, & 3, & 5 \\ 2, & 4, & 6 \\ \vdots & \vdots & \vdots \\ 2n-3, & 2n-1, & 2n+1 \end{array}\right\}$$

i.e., $(2n - 1)$ cases in all | *i.e.*, $(2n - 3)$ cases in all; and so on

If $d = n - 1$, the possible cases are:

$1, n, 2n - 1$; $\quad 2, n + 1, 2n$; $\quad 3, n + 2, 2n + 1$, $\quad$ *i.e.*, $\quad$ 3 cases in all.

If $d = n$, there is only one case, *viz.*, $(1, n + 1, 2n + 1)$.

Thus, total number of favourable cases $= (2n - 1) + (2n - 3) + \ldots + 5 + 3 + 1$
$= 1 + 3 + 5 + \ldots + (2n - 1)$,

which is a series in A.P. with $a = 1$, $d = 2$ and n terms.

$$\therefore \qquad \text{Number of favourable cases} = \frac{n}{2}[1 + (2n - 1)] = n^2.$$

$$\text{Hence,} \qquad \text{Required probability} = \frac{n^2}{n(4n^2 - 1)/3} = \frac{3n}{(4n^2 - 1)}$$

Example 3·20. *If 6n tickets numbered 0, 1, 2, …, 6n – 1 are placed in a bag and three are drawn out, show that the chance that the sum of the numbers on them is equal to 6n is {3n/(6n – 1) (6n – 2)}.*

Solution. The total number of ways of drawing 3 tickets out of $6n$ is given by:

$$^{6n}C_3 = n(6n - 1)(6n - 2).$$

Favourable cases for obtaining a sum of $6n$ on the three drawn tickets are given below:

$(0, 1, 6n - 1)$; $\quad (0, 2, 6n - 2)$; $\quad \ldots$; $\quad (0, 3n - 1, 3n + 1)$, $\quad$ *i.e.*, $(3n - 1)$ cases
$(1, 2, 6n - 3)$; $\quad (1, 3, 6n - 4)$; $\quad \ldots$; $\quad (1, 3n\ 1, 3n)$, $\quad$ *i.e.*, $(3n - 2)$ cases
$(2, 3, 6n - 5)$; $\quad (2, 4, 6n - 6)$; $\quad \ldots$; $\quad (2, 3n - 2, 3n)$, $\quad$ *i.e.*, $(3n - 4)$ cases
$(3, 4, 6n - 7)$; $\quad (3, 5, 6n - 8)$; $\quad \ldots$; $\quad (3, 3n - 2, 3n - 1)$, $\quad$ *i.e.*, $(3n - 5)$ cases
$\vdots \qquad \vdots \qquad \vdots \qquad \vdots$
$(2n - 2, 2n - 1, 2n + 3)$; $\quad (2n - 2; 2n; 2n + 2)$ $\quad$ *i.e.*, 2 cases
$(2n - 1, 2n, 2n + 1)$ $\quad$ *i.e.*, 1 case

$\therefore$ Total number of favourable cases

$$= [(3n - 1) + (3n - 4) + \ldots + 5 + 2] + [(3n - 2) + (3n - 5) + \ldots + 4 + 1] \quad \ldots(*)$$

The expression in each bracket { } of (*) is the sum of n terms of an Arithmetic Progression (A.P.) series.

$\therefore$ Total number of favourable cases

$$= [2 + 5 + \ldots + (3n - 4) + (3n - 1)] + [1 + 4 + \ldots + (3n - 5) + (3n - 2)]$$

$$= \frac{n}{2}\{2 + (3n - 1)\} + \frac{n}{2}\{1 + (3n - 2)\} = \frac{n}{2} \times [(3n + 1) + (3n - 1)] = 3n^2$$

$$\text{Hence,} \qquad \text{Required probability} = \frac{3n^2}{[n(6n - 1)(6n - 2)]} = \frac{3n}{(6n - 1)(6n - 2)}.$$

3·6. SUBJECTIVE PROBABILITY

Quite often we come across statements as follows:

(*i*) An old man says that the chance of rain on a particular given day is 60 per cent.

(*ii*) An expert of cricket says that the chance of India's winning the triangular one day cricket series against South Africa and Kenya are 40 per cent.

(*iii*) In the view of stocks and share expert, the chance that the price of the share of a particular company will double in the next month is 75%.

(*iv*) The odds that a particular horse will win in the race are 4 : 1.

In all the above cases, the probabilities (chances) of occurrence of the corresponding events are assigned by individuals and are based on their personal judgement, wisdom, intuition and expertise. These probabilities are called the *subjective probability* and represent the degree of belief and the confidence, one has in the occurrence of the respective event. Such probabilities can be calculated in all the cases, though the probabilities assigned by different persons to the same event may vary from person to person.

3·7. MATHEMATICAL TOOLS : PRELIMINARY NOTIONS OF SETS

The set theory was developed by the German mathematician, G. Cantor (1845-1918).

3·7·1. Sets and Elements of Sets.

A set is a well-defined collection or aggregate of all possible objects having given properties and specified according to a well-defined rule. The objects comprising a set are called elements, members or points of the set. Sets are often denoted by capital letters, *viz.*, A, B, C, etc. If x is an element of the set A, we write symbolically $x \in A$ (x belongs to A). If x is not a member of the set A, we write $x \notin A$(x does not belong to A). Sets are often described by describing the properties possessed by their members. Thus the set A of all non-negative rational numbers with square less than 2 will be written as:

$$A = \{x : x \text{ rational}, x > 0, x^2 < 2\}.$$

If every element of the set A belongs to the set B, i.e., if $x \in A \Rightarrow x \in B$ then we say that A is a subset of B and symbolically $A \subseteq B$ (A is contained in B) or $B \supseteq A$ (B contains A). Two sets A and B are said to be *equal* or *identical* if $A \subseteq B$ and $B \subseteq A$ and write $A = B$ or $B = A$.

A *null* or an *empty* set is one which does not contain any element at all and is denoted by ϕ.

Remarks:

1. Every set is a subset of itself.
2. An empty set is subset of every set.
3. A set containing only one element is conceptually distinct from the element itself, but will be represented by the same symbol for the sake of convenience.
4. As will be the case in all our applications of set theory, especially to probability theory, we shall have a fixed set S, (say), given in advance, and we shall be concerned only with subsets of this given set. The underlying set S may vary from one application to another, it will be referred to as *universal set* of each particular discourse.

3·7·2. Operations on Sets.

The *union* of two given sets A and B, denoted by $A \cup B$, is defined as a set consisting of all those points which belong to either A or B or both.

Thus symbolically : $A \cup B = \{x : x \in A \text{ or } x \in B\}$.

Similarly, $\bigcup_{i-1}^{n} A_i = \{x : x \in A_i \text{ for at least one } i = 1, 2, , \ldots, n\}$

The *intersection* of two sets *A* and *B*, *denoted by* $A \cap B$, is defined as a set consisting of those elements which belong to both *A* and *B*. Thus

$$A \cap B = \{x : x \in A \text{ and } x \in B\}$$

Similarly, $\bigcap_{i-1}^{n} A_i = \{x : x \in A_i \text{ for all } i = 1, 2, \ldots, n\}$.

For example, if $A = \{1, 2, 5, 8, 10\}$ and $B = \{2, 4, 8, 12\}$, then

$$A \cup B = \{1, 2, 4, 5, 8, 10, 12\} \quad \text{and} \quad A \cap B = \{2, 8\}$$

If *A* and *B* have no common point, *i.e.*, $A \cap B = \phi$, then the sets *A* and *B* are said to be disjoint, *mutually exclusive* or *non-overlapping*.

The *relative difference* of a set *A* from another set *B*, denoted by $A - B$ is defined as a set consisting of those elements of *A* which do not belong to *B*. Symbolically

$$A - B = \{x : x \in A \text{ and } x \notin B\}.$$

The *complement* or *negative* of any set *A*, denoted by $\overline{A}$, is a set containing all elements of the universal set *S*, (say), that are not elements of *A*, *i.e.*, $\overline{A} = S - A$.

3·7·3. Algebra of Sets

Now we state certain important properties concerning operations on sets. If *A*, B and *C* are the subsets of a universal set *S*, then the following laws hold:

Commutative Laws : $A \cup B = B \cup A, \qquad A \cap B = B \cap A$

Associative Law : $(A \cup B) \cup C = A \cup (B \cup C)$

$$(A \cap B) \cap C = A \cap (B \cap C)$$

Distributive Law : $A \cap (B \cup C) = (A \cap B) \cup (A \cap C)$

$$A \cup (B \cap C) = (A \cup B) \cap (A \cup C)$$

Complementary Law : $A \cup \overline{A} = S, \quad A \cap \overline{A} = \phi$

$$A \cup S = S, \text{ and } A \cap S = A \qquad (\because A \subset S),$$

$$A \cup \phi = A, \quad A \cap \phi = \phi.$$

Difference Law : $A - B = A \cap \overline{B}$

$$A - B = A - (A \cap B) = (A \cup B) - B$$

$$A - (B - C) = (A - B) \cup (A - C)$$

$$(A \cup B) - C = (A - C) \cup (B - C)$$

$$A - (B \cup C) = (A - B) \cap (A - C)$$

$$(A \cap B) \cup (A - B) = A, \; (A \cap B) \cap (A - B) = \phi$$

De-Morgan's Law : $\overline{(A \cup B)} = \overline{A} \cap \overline{B} \quad \text{and} \quad \overline{(A \cap B)} = \overline{A} \cup \overline{B}$

More generally,

$$\overline{\left(\bigcup_{i=1}^{n} A_i\right)} = \bigcap_{i=1}^{n} \overline{A}_i \quad \text{and} \quad \overline{\left(\bigcap_{i=1}^{n} A_i\right)} = \bigcup_{i=1}^{n} \overline{A}_i$$

Involution Law : $\overline{(\overline{A})} = A$

Idempotency Law : $A \cup A = A, \qquad A \cap A = A$

3·7·4. Limit of Sequence of Sets

Let $\{A_n\}$ be a sequence of sets in S. The *limit supremum* or *limit superior* of the the sequence, usually written as lim sup A_n, is the set of all those elements which belong to A_n for infinitely many n. Thus

$$\limsup_{n\to\infty} A_n = \{x : x \in A_n \text{ for infinitely many } n\} \qquad ...(3\cdot3)$$

The set of all those elements which belong to A_n for all but a finite number of n is called *limit infinimum or lime inferior* of the sequence and is denoted by lim inf A_n.

Thus
$$\liminf_{n\to\infty} A_n = \{x : x \in A_n \text{ for all but a finite number of } n\} \qquad ...(3\cdot3a)$$

The sequence $\{A_n\}$ is said to have a limit if and only if lim sup A_n = lim inf A_n and this common value gives the limit of the sequence

Theorem 3·1.
$$\textit{lim sup } A_n = \bigcap_{m=1}^{\infty}\left(\bigcup_{n=m}^{\infty} A_n\right)$$

and
$$\textit{lim inf } A_n = \bigcup_{m=1}^{\infty}\left(\bigcap_{n=m}^{\infty} A_n\right)$$

Definition. $\{A_n\}$ is a monotone (infinite) sequence of sets if

either (*i*) $A_n \subset A_{n+1} \quad \forall\ n$ or (*ii*) $A_n \supset A_{n+1}\ \forall\ n$.

In the former case the sequence $\{A_n\}$ is said to be *non-decreasing* (*or increasing*) *sequence* and is usually expressed as $A_n \uparrow$ and in the latter case it is said to be *non-increasing* (*or decreasing*) sequence and is expressed as $A_n \downarrow$.

For a monotone sequence (non-increasing or non-decreasing), the limit always exists and we have:

$$\lim_{n\to\infty} A_n = \begin{cases} \bigcup_{n=1}^{\infty} A_n \text{ in case } (i), \textit{i.e.}, A_n \uparrow \\ \bigcap_{n=1}^{\infty} A_n \text{ in case } (ii), \textit{i.e.}, A_n \downarrow \end{cases}$$

3·7·5. Classes of Sets

A group of sets will be termed as a class (of sets). Below we shall define some useful types of classes.

A ring **R** is a *non-empty* class of sets which is closed under the formation of finite 'unions' and 'difference'.

i.e., if $A \in \mathbf{R}$, $B \in \mathbf{R}$, then $A \cup B \in \mathbf{R}$ and $A - B \in \mathbf{R}$.

Obviously ϕ is a member of every ring.

A field **F** (or an *algebra*) is a non-empty class of sets which is closed under the formation of finite unions and under complementation. Thus

(*i*) $A \in \mathbf{F}, B \in \mathbf{F} \Rightarrow A \cup B \in \mathbf{F}$ *and* (*iii*) $A \in \mathbf{F} \Rightarrow \bar{A} \in \mathbf{F}$.

A σ-ring **C** is a non-empty class of sets which is closed under the formation of 'countable unions' and 'difference'. Thus

$$(i)\ A_i \in \mathbf{C},\ i = 1, 2, \dots \Rightarrow \bigcup_{i=1}^{\infty} A_i \in \mathbf{C} \qquad (ii)\ A \in \mathbf{C},\ B \in \mathbf{C} \Rightarrow A - B \in \mathbf{C}.$$

More precisely σ-ring is a ring which is closed under the formation of countable unions.

A σ-field (or σ-algebra) **B** is a non-empty class of sets that is closed under the formation of 'countable unions' and complementations, *i.e.,*

$$(i)\ A_i \in \mathbf{B},\ i = 1, 2, \dots \Rightarrow \bigcup_{i=1}^{\infty} A_i \in \mathbf{B}. \qquad (ii)\ A \in \mathbf{B} \Rightarrow \bar{A} \in \mathbf{B}.$$

σ-field may also be defined as a field which is closed under the formation of countable unions.

3·8. AXIOMATIC APPROACH TO PROBABILITY

The axiomatic approach to probability, which closely relates the theory of probability with the modern metric theory of functions and also set theory, was proposed by A.N. Kolmogorov, a Russian mathematician, in 1933. The axiomatic definition of probability includes 'both' the classical and the statistical definitions as particular cases and overcomes the deficiencies of each of them. On this basis, it is possible to construct a logically perfect structure of the modern theory of probability and at the same time to satisfy the enhanced requirements of modern natural science. The axiomatic development of mathematical theory of probability relies entirely upon the logic of deduction.

The diverse theorems of probability, as were available prior to 1933, were finally brought together into a unified axiomised system in 1933. It is important to remark that probability theory, in common with all axiomatic mathematical systems, is concerned solely with relations among undefined things.

The axioms thus provide a set of rules which define relationships between abstract entities. These rules can be used to deduce theorems, and the theorems can be brought together to deduce more complex theorems. These theorems have no empirical meaning although they can be given an interpretation in terms of empirical phenomenon. The important point, however, is that the mathematical development of probability theory is, in no way, conditional upon the interpretation given to the theory.

More precisely, under axiomatic approach, the probability can be deduced from mathematical concepts. To start with some concepts are laid down. Then some statements are made in respect of the properties possessed by these concepts. These properties, often termed as "*axioms*" or rules of the theory, are used to frame some theorems. These theorems are framed without any reference to the real world and are deductions from the axioms of the theory. It does not attempt to give any physical interpretation to the probability of an event.

3·8·1. Random Experiment, Sample Space and Elementary Events.

The theory of probability provides *mathematical models* for 'real-world phenomenon' involving games of chance such as tossing of coins and dice. We feel intuitively that statements such as

(*i*) "The probability of getting a 'head' in one toss of an unbiased coin is 1/2"

(*ii*) "The probability of getting a 'four' in a single toss of an unbiased die is 1/6", should hold. We also feel that the probability of obtaining *either* a "5" or a "6" in a single throw of a die, should be the sum of the probabilities of a "5" and a"6", *viz.*, $\frac{1}{6}+\frac{1}{6}=\frac{1}{3}$. In other words, probabilities should have some kind of *additive* property. Finally, we feel that in a large number of repetitions of experiment, for example, a coin tossing experiment, the proportion of heads should be approximately $\frac{1}{2}$, *i.e.*, the probability should have a *frequency interpretation.*

To deal with these properties sensibly, we need a *mathematical description* or *model* for the probabilistic situation we have. Any such probabilistic situation is referred to as a *random experiment,* denoted by **E**. **E** may be a coin or die throwing experiment, drawing of cards from a well-shuffled pack of cards, an agricultural experiment to determine the effect of fertilizers on yield of a commodity, and so on.

Each performance in a random experiment, is called a *trial*. That is, all the trials conducted under the same set of conditions form a *random experiment*. The result of a trial in a random experiment is called an *outcome,* an *elementary* event or *a sample point.* The totality of all possible outcomes (*i.e.*, sample points) of a random experiment constitutes the *sample space.*

Suppose $e_1, e_2, ..., e_N$ are possible outcomes of a random experiment **E** such that no two or more of them can occur simultaneously and exactly one of the outcomes $e_1, e_2, ..., e_N$ must occur. More specifically, with an experiment **E**, we associate a set $S = (e_1, e_2, ..., e_N)$ of possible outcomes with the following properties:

(*i*) each element of S denotes a possible outcome of the experiment, and

(*ii*) any trial results in an outcome that corresponds to one and only one element of the set S.

The set S associated with an experiment **E**, real or conceptual, satisfying the above two properties is called the *sample space* of the experiment.

The elements or the points of a sample space associated with a random experiment are called ***elementary*** ***events*** *of the experiment.*

In general, suppose we have a random experiment with m outcomes say $x_1, x_2, ..., x_m$ and another experiment with n outcomes $y_1, y_2, ..., y_n$; then the sample space S of the experiment which consists of carrying out the two experiments together has mn outcomes and can be described by the cross product of the two sets $\{x_1, x_2, ..., x_m\}$ and $\{y_1, y_2, ..., y_n\}$.

Hence, in this case,

$$\begin{aligned} S &= \{x_1, x_2, ..., x_m\} \times \{y_1, y_2, ..., y_n\} \\ &= \{(x_1\, y_1),\ (x_1\, y_2)\ ,\ ...,\ (x_1\, y_n), \\ &\quad\ (x_2\, y_1),\ (x_2\, y_2)\ ,\ ...,\ (x_2\, y_n), \\ &\quad\ \vdots \\ &\quad\ (x_m\, y_1),\ (x_m\, y_2)\ ,\ ...,\ (x_m\, y_n)\} \end{aligned}$$

The sample space of a random experiment is usually denoted by S or Ω and the outcomes of the random experiment are denoted by $e_1, e_2, e_3, ...$ or by $\omega_1, \omega_2, \omega_3, ...$

Definition. *The set of all possible outcomes of a given random experiment is called the sample space associated with the experiment. Each possible outcome or element in a sample space is called a sample point or an elementary event.*

The number of sample points in the sample space S are denoted by $n(S)$ and in the classical probability theory, represent the exhaustive number of cases associated with the random experiment **E**.

Remarks.

1. The sample space serves as universal set for all questions concerned with the experiment.
2. A sample space S is said to be finite (infinite) sample space if the number of elements in S is finite (infinite).
3. A sample space is called discrete if it contains only finitely or infinitely many points which can be arranged into a simple sequence $w_1, w_2, \ldots$, while a sample space containing non-denumerable number of points is called a continuous sample space. In this book, we shall restrict ourselves to discrete sample spaces only.

3·8·2. Event

Definition. *Every non-empty subset A of S, which is a disjoint union of single element subsets of the sample space S of a random experiment* **E** *is called an event.*

The notion of an event may also be defined as follows:

"Of all the possible outcomes in the sample space of an experiment, some outcomes satisfy a specified description, which we call an event."

Remarks.

1. As the empty set ϕ is a subset of S, ϕ is also an event, known as *impossible* event.
2. An event A, in particular, can be a single element subset of S, in which case it is known as *elementary* event.

Some Illustrations. We give below some illustrations to explain the concepts of the sample space and event.

1. Consider random tossing of a single coin. The possible outcomes for this experiment are (writing H for a "head" and T for a "tail") : H and T. Thus, the sample space S consists of two points (ω_1, ω_2), corresponding to each possible outcome or elementary event listed, *i.e.*, $S = \{H, T\} = \{\omega_1, \omega_2\}$ and $n(S) = 2$, where $n(S)$ is the total number of sample points in S.

It may again be remarked that no two *elementary events* can occur simultaneously and exactly one of these must occur in a single trial. Thus, in the toss of a coin we cannot obtain head and tail simultaneously and more over, we must obtain either head or tail.

2. (*a*) In random toss of two coins (or two tosses of a coin), the sample space S is given by:

$$S = \{H, T\} \times \{H, T\} = \{HH, HT, TH, TT\} = \{\omega_1, \omega_2, \omega_3, \omega_4\} \text{ and } n(S) = 4 = 2^2.$$

(*b*) Similarly, in a random toss of three coins (or three tosses of a coin), the sample space S is give by:

$$\begin{aligned} S &= \{H, T\} \times \{H, T\} \times \{H, T\} \\ &= \{H, T\} \times \{HH, HT, TH, TT\} \\ &= \{HHH, HHT, HTH, HTT, THH, THT, TTH, TTT\} \\ &= \{\omega_1, \omega_2, \omega_3, \omega_4, \omega_5, \omega_6, \omega_7, \omega_8\}, \text{ (say); and } n(S) = 8 = 2^3. \quad \ldots(*) \end{aligned}$$

Combination of these outcomes (elementary events) form the so-called events, *e.g.*, let us consider the following events associated with sample space in (*).

E_1 : Three heads ; E_2 : Three tails.
E_3 : Exactly one head ; E_4 : Exactly two heads
E_5 : At least one head ; E_6 : At least two heads.

Then these events are represented by the following subsets of S :

$$E_1 = \{HHH\} = \{\omega_1\}; \quad E_2 = \{TTT\} = (\omega_8); \quad E_3 = \{\omega_4, \omega_6, \omega_7\}$$

$$E_4 = \{\omega_2, \omega_3, \omega_5\}; \quad E_5 = E_1 \cup E_3 \cup E_4 = \{\omega_1, \omega_2, \omega_3, \omega_4, \omega_5, \omega_6, \omega_7\}$$

$$E_6 = E_1 \cup E_4 = \{\omega_1, \omega_2, \omega_3, \omega_5\}$$

(*c*) In general, in a random toss of N coins, $n(S) = 2^N$.

3. (*a*) Let us consider a random toss of a single dice. Since we can obtain any one of the six faces 1, 2, 3, ..., 6; the sample space S is given by:

$$S = \{1, 2, 3, 4, 5, 6\} = \{\omega_1, \omega_2, \omega_3, \omega_4, \omega_5, \omega_6\}, \quad \text{and} \quad n(S) = 6.$$

(*c*) In a random toss of two dice (or two tosses of a die), the sample space S is given by:

$$\left.\begin{aligned} S &= \{1, 2, 3, 4, 5, 6\} \times \{1, 2, 3, 4, 5, 6\} \\ &= \Big\{(1, 1), \quad (1, 2), \quad (1, 3), \ldots, \quad (1, 6), \\ &\quad\ (2, 1), \quad (2, 2), \quad (2, 3), \ldots, \quad (2, 6) \\ &\quad\ \vdots \qquad\qquad\qquad\quad \vdots \qquad\quad\ \vdots \\ &\quad\ (6, 1), \quad (6, 2), \quad (6, 3), \ldots, \quad (6, 6)\Big\} \end{aligned}\right\} \quad \ldots(**)$$

where, in the point (a, b), 'a' represents the number obtained on the first dice and 'b' represents the number on the second dice. Obviously, $(a, b) \neq (b, a)$, unless $a = b$.

$$\text{Exhaustive number of cases} = n(S) = 36 = 6^2.$$

Let us consider the following events associated with sample space S in (**).

E_1 : Sum of points on two dice is 5;

E_2 : Sum of points on two dice is 6

E_3 : Sum of points on the two dice is even;

E_4 : Sum of points on the two dice is odd

E_5 : Sum of the points on the two dice is greater than 12

E_6 : Sum of points on the two dice is divisible by 3

E_7 : Sum is greater than or equal to two and is less than or equal to 12

The events E_1, E_2, E_3 and E_4 are represented by the following subsets of S:

$$E_1 = \{(1, 4), (4, 1), (2, 3), (3, 2)\} \quad \text{and} \quad n(S) = 4$$

$$E_2 = \{(1, 5), (5, 1), (2, 4), (4, 2), (3, 3)\} \quad \text{and} \quad n(S) = 5$$

The sum of points on the two dice is even if

(*i*) the points obtained on each dice are even

or (*ii*) the points obtained on each dice are odd.

$\therefore$ $E_3 = \{2, 4, 6\} \times \{2, 4, 6\} \cup \{1, 3, 5\} \times \{1, 3, 5\}$

$$= \begin{Bmatrix} (2, 2), (2, 4), (2, 6), (4, 2), (4, 4) (4, 6), (6, 2), (6, 4), (6, 6) \\ (1, 1), (1, 3), (1, 5), (3, 1), (3, 3), (3, 5), (5, 1), (5, 3), (5, 5) \end{Bmatrix}$$

and $n(E_3) = 3 \times 3 + 3 \times 3 = 18$

Similarly,

$$E_4 = \{2, 4, 6\} \times \{1, 3, 5\} \cup \{1, 3, 5\} \times \{2, 4, 6\}.$$

$$= \begin{Bmatrix} (2,1),(2,3),(2,5),(4,1),(4,3),(4,5),(6,1),(6,3),(6,5) \\ (1,2),(1,4),(1,6),(3,2),(3,4),(3,6),(5,2),(5,4),(5,6) \end{Bmatrix}$$

and $n(E_4) = 3 \times 3 + 3 \times 3 = 18.$

$E_5 = \{\phi\}$, $n(E_5) = 0$; $E_7 = S$, $n(E_7) = n(S) = 36.$

The sum of the points on the two dice is divisible by 3 if their sum is 3, 6, 9 or 12.

$\therefore$ $E_6 = \{(1, 2), (2, 1), (1, 5), (5, 1), (2, 4), (4, 2), (3, 3), (3, 6), (6, 3), (4, 5), (5, 4), (6, 6)\}$ and $n(E_6) = 12$

Here E_5 is an 'impossible event' and E_7 is a 'certain event'.

(c) In general, in a random toss of N dice, exhaustive number of cases $= n(S) = 6^N$.

4. A coin and a dice are tossed together. For this experiment, the sample space consists of twelve points given by:

$$S = \{H, T\} \times \{1, 2, \ldots, 6\}$$
$$= \{(H, 1), (H, 2), \ldots, (H, 6), (T, 1), (T, 2), \ldots, (T, 6)\} = \{\omega_1, \omega_2, \omega_3, \ldots, \omega_{12}\}$$

where ω_i $(i = 1, 2, \ldots, 6)$ represents a head on the coin together with the appearance of ith number on the die and ω_i $(i = 7, 8, \ldots, 12)$ represents a tail on the coin together with the appearance of ith number on die. Obviously, $n(S) = 2 \times 6 = 12.$

Remark: In the above illustrations 1 to 4, if the coin and die are unbiased, we can see intuitively that in each of the above illustrations, the outcomes (sample points) are equally likely to occur.

5. From an urn containing 4 balls of different colours red (R), blue (B), yellow (Y) and white (W) draw two balls simultaneously.

The sample space for this experiment is: $S = \{RB, RY, RW, BY, BW, YW\}$.

where 'RB' stands for 'the balls drawn are red and blue', etc. Note that RB and BR represents the same outcome because we are drawing the balls simultaneously. Here, $n(S) = {}^4C_2 = 6$

6. In the above example, suppose that instead of drawing both the balls simultaneously, the balls are drawn in succession with replacement, *i.e.*, after the first draw, the colour is noted and the ball drawn is put again in the urn before drawing the second ball.

The sample space now consists of 16 elementary events, and is given by:

$$S = \{R, B, Y, W\} \times \{R, B, Y, W\}$$
$$= \{RR, RB, RY, RW, BR, BB, BY, BW, YR, YB, YY, YW, WR, WB, WY, WW\}$$

where 'RB' stands for 'the first ball is red and second blue', etc. and $n(S) = 4 \times 4 = 16.$

Remark. Illustrations 1 to 6 given above are all examples of finite sample space, since the sample points in each one of them are finite in number. We shall now give below illustrations of infinite sample spaces, having infinite number of sample points.

7. Consider a random experiment of throwing a coin until head appears. The sample space S in this experiment is given by:

$$S = \{H, TH, TTH, TTTH, \ldots \ldots\} = \{\omega_1, \omega_2, \omega_3, \omega_4, \ldots\}$$

where TTH represents tail in the first throw, tail in the second throw and head in the third throw, and so on. Obviously $n(S)$ is infinite.

3·8·3. Acceptable Assignment of Probabilities

Let $e_1, e_2, ..., e_N$ be mutually disjoint and exhaustive outcomes of a random experiment so that its sample space is:

$$S = \{e_1, e_2, ..., e_N\}.$$

To each elementary event e_i *belonging to S,* let us assign a *real number* called the *probability* of the elementary event e_i; and *denoted by* $P(e_i)$, *such that:*

(*i*) The probability of each elementary event is a non-negative real number, *i.e.,*

$$P(e_i) \geq 0, \text{ for } i = 1, 2, ..., N; \text{ and}$$

(*ii*) The sum of probabilities assigned to all elementary events of the sample space is 1, *i.e.,*

$$\sum_{i=1}^{N} P(e_i) = 1.$$

Such an assignment of real numbers to be elementary events of the sample space is called an *acceptable assignment of probabilities.*

Remark. There can be many, in fact infinite, assignments of probabilities to the elementary events of a sample space which are *acceptable* in the sense that they satisfy the two conditions, mentioned above, required for an acceptable assignment of probabilities.

The probability of no elementary event can be greater than 1. This follows immediately from the conditions (*i*) and (*ii*) given above. Thus, we have $0 \leq P(e_i) \leq 1$, for $i = 1, 2, ..., N$.

3·8·4. Natural Assignment of Probabilities

Definition. *An assignment of probabilities is called natural (or equiprobable) assignment if each elementary event is assigned the same probability.*

More precisely, if e_i and e_j are two elementary events of the sample space S, then in the case of natural assignment of probabilities,

$$P(e_i) = P(e_j), \forall\, i \neq j = 1, 2, ..., N \qquad ...(3·4)$$

Since, $$P(S) = \sum_{i=1}^{N} P(e_i) = 1, \text{ on using } (3 \cdot 4), \text{ we get}$$

$$P(e_i) = \frac{1}{N}; i = 1, 2, ..., N, \qquad ...(3·5)$$

in the case of natural assignment of probabilities.

For example, in the case of random toss of a coin, $S = (H, T)$ and the natural assignment of probabilities gives:

$$P(H) = P(T) = \frac{1}{2}.$$

Similarly, in a random throw of a die, $S = \{1, 2, ..., 6\} = \{e_1, e_2, ..., e_6\}$, (say) and the natural assignment of probabilities is :

$$P(e_i) = \frac{1}{6}; i = 1, 2, ..., 6.$$

In the problems relating to random throw(s) of coin(s) or die (dice), the natural assignment of probabilities is possible only if the coins/dice are fair, uniform or unbiased. Similarly, in drawing cards from a well-shuffled pack of cards, for natural assignment of probabilities, all the draws must be made at random.

In the following discussion, whenever we use the adjective such as: *'fair coin'*, *'fair* or *unbiased die'*, card or a number drawn at *'random'*, etc., we will assume that the natural assignment is understood. Also if no assignment of probabilities is given in any question, the natural assignment is to be taken.

3·8·5. Axiomatic Probability

A purely mathematical definition of probability cannot give us the actual value of $P(A)$, the probability of occurrence of the event A and this must be considered as a function defined on all events. Accordingly, as in the case of any function we need a *domain space* which is the σ-field **B** of the events, generated by S; a *range space* which is the closed interval [0, 1] on the real line, and a rule which assigns a value to every element of the domain space **B**.

The whole field of probability theory for finite sample spaces is based on three simple axioms or postulates, given below:

Definition. Probability Function. *P(A) is the probability function defined on a σ-field* **B** *of events if the following properties or axioms hold.*

1. *For each $A \in$ **B**, $P(A)$ is defined, is real and $P(A) \geq 0$.. (Axiom of non-negativity)*

2. $P(S) = 1$ *(Axiom of certainty)*

3. *If $\{A_n\}$ is any finite or infinite sequence of disjoint event in* **B**, *then* ...(3·6)

$$P\left(\bigcup_{i=1}^{n} A_i\right) = \sum_{i=1}^{n} P(A_i) \qquad \textit{(Axiom of additivity)}$$

An obvious advantage of this definition is that it is not concerned with assigning numerical values to the probabilities of events. The following points may be noted.

1. The set function **P** defined on σ-field **B**, taking its values in the real line and satisfying the above three axioms is called the *probability measure*.

2. The same definition of probability applies to *uncountable sample space* except that special restrictions must be placed on S and its subsets. It is important to realise that for a complete description of a probability measure, three things must be specified, *viz.*, the sample space S, the σ-field (σ-algebra) **B** formed from certain subsets of S and set function **P**. The triplet (S, **B**, **P**,) is often called the *probability space.* In most elementary applications, S is finite and the σ-algebra **B** is taken to be the collection of all subsets of S.

3. ***Classical Probability as a Special Case of Axiomatic Probability.*** Let us consider a random experiment whose sample space S has finite number of points, say N, given by:

$$S = \{e_1, e_2, \ldots, e_N\},$$

where obviously $e_1, e_2, \ldots, e_N$ are mutually disjoint events.

Let us now suppose that all the outcomes of the random experiment are equally likely so that the probability of each of the singleton events $\{e_i\}$ is $1/N$, *i.e.,*

$$P\{e_i\} = \frac{1}{N}\ ;\quad i = 1, 2, \ldots, n \qquad \ldots(*)$$

Further, let an event A consist of k distinct point of S, say $e_{i1}, e_{i2}, \ldots, e_{ik}$. Then by the axiom of additivity (Axiom 3), we get

$$P(A) = \sum_{j=1}^{k} P\{e_{ij}\} = \sum_{j=1}^{k} \frac{1}{N} = \frac{k}{N} \qquad \text{[Using (*)]}$$

$$= \frac{\text{Number of sample points in } A}{\text{Number of sample points in } S} = \frac{n(A)}{n(S)} \qquad ...(3\cdot7)$$

$$= \frac{\text{Number of cases favourable to the event } A}{\text{Exhaustive number of cases}},$$

which is the result obtained by the classical definition of probability.

Hence, *the classical probability may be regarded as a special case of axiomatic probability.*

4. It is interesting to see that there are some formal statements of the properties of events derived from the frequency approach.

From axiomatic definition of probability, we get $P(A) = \dfrac{n(A)}{n(S)} \geq 0$, as in Axiom 1.

Also, since $n(S) = N$, we have on using $(3\cdot7)$, $\quad P(S) = \dfrac{n(S)}{n(S)} = 1$, as in Axiom 2.

Further in the case of two mutually exclusive (or disjoint) events A and B defined by sample points $n(A)$ and $n(B)$, the number of sample points belonging to $A \cup B$ is $n(A) + n(B)$.

$$\therefore \qquad P(A \cup B) = \frac{n(A \cup B)}{n(S)} = \frac{n(A) + n(B)}{n(S)}$$

$$= \frac{n(A)}{n(S)} + \frac{n(B)}{n(S)} = P(A) + P(B), \text{ as in Axiom 3.}$$

Hence, the statistical or frequency approach of probability may be regarded as a special case of axiomatic probability.

3·8·6. Algebra of Events

For events A, B, C

(*i*) $A \cup B = \{\omega \in S : \omega \in A \text{ or } \omega \in B\}$ (*ii*) $A \cap B = \{\omega \in S : \omega \in A \text{ and } \omega \in B\}$

(*iii*) $\bar{A}\,(A \textit{ complement}) = \{\omega \in S : \omega \notin A\}$ (*iv*) $A - B = \{\omega \in S : \omega \in A \text{ but } \omega \notin B\}$

(*v*) Similar generalisation for $\bigcup_{i=1}^{n} A_i, \ \bigcap_{i=1}^{n} A_i$, etc.

(*vi*) $A \subset B \quad \Rightarrow \quad$ for every $\omega \in A$, $\omega \in B$,

(*vii*) $B \supset A \quad \Rightarrow \quad A \subset B$.

(*viii*) $A = B$ if and only if A and B have same elements, i.e., if $A \subset B$ and $B \subset A$.

(*ix*) A and B disjoint (mutually exclusive) $\Rightarrow \quad A \cap B = \phi$ (null set)

(x) $A \cup B$ can be denoted by $A + B$ if A and B are disjoint.

(*xi*) $A \,\Delta\, B$ denotes those ω belonging to exactly one of A and B, *i.e.*,

$$A \,\Delta\, B = \bar{A}\,B \cup A\,\bar{B} = \bar{A}\,B + A\,\bar{B} \text{ (disjoint events)}$$

Remark. Since the events are subsets of S, all the laws of set theory, *viz.*, commutative laws, associative laws, distributive laws, De-Morgan's law, etc., hold for algebra of events.

TABLE 3.1 GLOSSARY OF PROBABILITY TERMS

Statement	*Meaning in terms of set theory*
1. At least one of the events A or B occurs	$\omega \in A \cup B$
2. Both the events A and B occur.	$\omega \in A \cap B$
3. Neither A nor B occurs	$\omega \in \bar{A} \cap \bar{B}$
4. Event A occurs and B does not occur.	$\omega \in A \cap \bar{B}$
5. Exactly one of the events A or B occurs.	$\omega \in A \Delta B$
6. Not more than one of the events A or B occurs.	$\omega \in (A \cap \bar{B}) \cup (\bar{A} \cap B) \cup (\bar{A} \cap \bar{B})$
7. If event A occurs, so does B.	$A \subset B$
8. Events A and B are mutually exclusive.	$A \cap B = \phi$
9. Complementary event of A.	$\bar{A}$ or A^c
10. Sample space	Universal set S.

Example 3·21. *A, B and C are three arbitrary events. Find expression for the events noted below, in the context of A, B and C:*

(i) Only A occurs, *(ii) Both A and B, but not C, occur*
(iii) All three events occur, *(iv) At least one occurs,*
(v) At least two occur, *(vi) One and no more occurs,*
(vii) Two and no more occur, *(viii) None occurs.*

Solution.

(i) $A \cap \bar{B} \cap \bar{C}$, *(ii)* $A \cap B \cap \bar{C}$, *(iii)* $A \cap B \cap C$, *(iv)* $A \cup B \cup C$,

(v) $(A \cap B \cap \bar{C}) \cup (A \cap \bar{B} \cap C) \cup (\bar{A} \cap B \cap C) \cup (A \cap B \cap C)$

(vi) $(A \cap \bar{B} \cap \bar{C}) \cup (\bar{A} \cap B \cap \bar{C}) \cup (\bar{A} \cap \bar{B} \cap C)$

(vii) $(A \cap B \cap \bar{C}) \cup (\bar{A} \cap B \cap C) \cup (A \cap \bar{B} \cap C)$

(viii) $(\bar{A} \cap \bar{B} \cap \bar{C})$ or $\overline{A \cup B \cup C}$

3·9. SOME THEOREMS ON PROBABILITY

In this section, we shall prove a few simple theorems which help us evaluate the probabilites of some complicated events in a rather simple way. In proving these theorems, we shall follow the axiomatic approach, based on the three axioms, discussed in §3.8.5.

Theorem 3·1. *Probability of the impossible event is zero, i.e.,* $P(\phi) = 0$. ...(3·8)

Proof. Impossible event contains no sample point and hence the certain event S and the impossible event ϕ are mutually exclusive.

$$\therefore \quad S \cup \phi = S \quad \Rightarrow \quad P(S \cup \phi) = P(S)$$

Hence, using Axiom 3 of probability, *i.e.,* Axiom of Additivity, we get

$$P(S) + P(\phi) = P(S) \quad \Rightarrow \quad P(\phi) = 0$$

Remark. It may be noted $P(A) = 0$, does not imply that A is necessarily an empty set. In practice, probability '0' is assigned to the events which are so rare that they happen only once in a lifetime. For example, if a person who does not know typing is asked to type one page of the manuscript of a book, the probability of the event that he will type it correctly without any mistake is 0.

As another illustration, let us consider the random tossing of a coin. The event that the coin will stand erect on its edge, is assigned the probability 0.

The study of continuous random variable provides another illustration to the fact that $P(A) = 0$, does not imply $A = \phi$, because in case of continuous random variable X, the probability at a point is always zero, *i.e.*, $P(X = c) = 0$ [See Chapter 5].

Theorem 3·2. *Probability of the complementary event $\bar{A}$ of A is given by*

$$P(\bar{A}) = 1 - P(A) \qquad \text{...(3·9)}$$

Proof. A and $\bar{A}$ are mutually disjoint events, so that

$$A \cup \bar{A} = S \quad \Rightarrow \quad P(A \cup \bar{A}) = P(S)$$

Hence, from Axioms 2 and 3 of probability, we have

$$P(A) + P(\bar{A}) = P(S) = 1 \quad \Rightarrow \quad P(\bar{A}) = 1 - P(A)$$

Cor. 1. We have $P(A) = 1 - P(\bar{A}) \leq 1$ $\quad [\because P(\bar{A}) \geq 0$, by Axiom 1]

Further, since $P(A) \geq 0$ (Axiom 1)

$$\therefore \quad 0 \leq P(A) \leq 1. \qquad \text{...(3·9}a\text{)}$$

Cor. 2. $P(\phi) = 0$, since $\phi = \bar{S}$ and $P(\phi) = P(\bar{S}) = 1 - P(S) = 1 - 1 = 0.$

Theorem 3·3. *For any two events A and B, we have*

$$(i)\ P(\bar{A} \cap B) = P(B) - P(A \cap B) \qquad (ii)\ P(A \cap \bar{B}) = P(A) - P(A \cap B)$$

Proof. From the Venn diagram, we get $B = (A \cap B) \cup (\bar{A} \cap B)$,

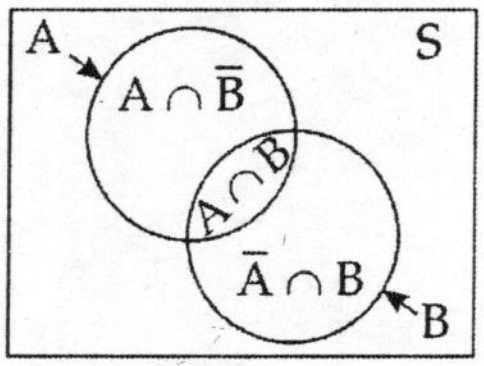

where $\bar{A} \cap B$ and $A \cap B$ are disjoint events.

Hence by Axiom (3), we get

$$P(B) = P(A \cap B) + P(\bar{A} \cap B)$$

$$\Rightarrow \quad P(\bar{A} \cap B) = P(B) - P(A \cap B) \qquad \text{...(3·10)}$$

(*ii*) Similarly, we have $A = (A \cap B) \cup (A \cap \bar{B})$

where $(A \cap B)$ and $A \cap \bar{B}$ are disjoint events. Hence, by Axiom 3, we get

$$P(A) = P(A \cap B) + P(A \cap \bar{B}) \Rightarrow P(A \cap \bar{B}) = P(A) - P(A \cap B) \qquad \text{...(3·11)}$$

Theorem 3·4. *If $B \subset A$, then*

$$(i)\ P(A \cap \bar{B}) = P(A) - P(B), \qquad (ii)\ P(B) \leq P(A)$$

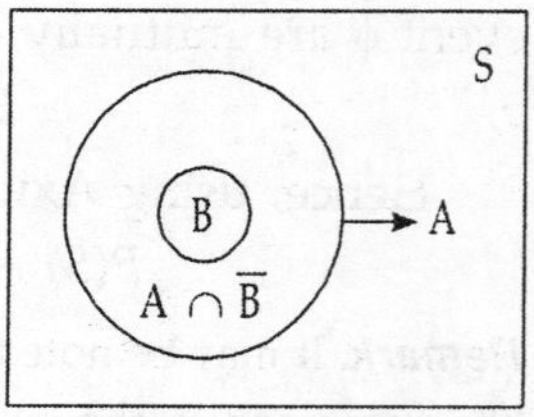

Proof. (*i*) When $B \subset A$, B and $A \cap \bar{B}$ are mutually exclusive events so that

$$A = B \cup (A \cap \bar{B})$$

$$\Rightarrow \quad P(A) = P[B \cup (A \cap \bar{B})]$$

$$= P(B) + P(A \cap \bar{B}) \quad \text{(By Axiom 3)}$$

$$\Rightarrow \quad P(A \cap \bar{B}) = P(A) - P(B)$$

(*ii*) $P(A \cap \bar{B}) \geq 0 \quad \Rightarrow \quad P(A) - P(B) \geq 0 \quad \Rightarrow \quad P(B) \leq P(A)$

Hence, $\quad B \subset A \quad \Rightarrow \quad P(B) \leq P(A) \qquad \text{...(3·12)}$

3·9·1. Addition Theorem of Probability

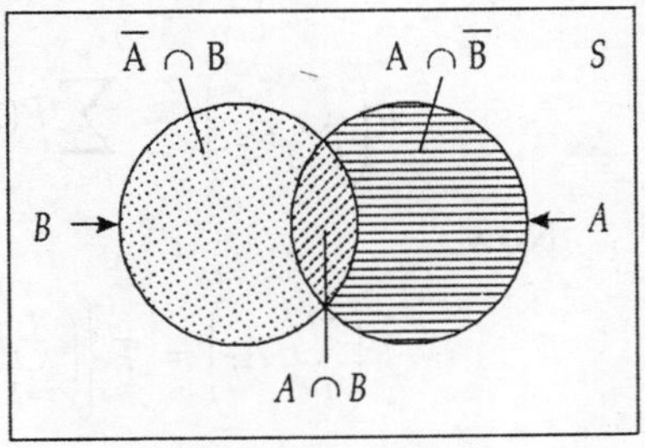

Theorem 3·5. *If A and B are any two events (subsets of sample space S) and are not disjoint, then*

$$P(A \cup B) = P(A) + P(B) - P(A \cap B) \quad ...(3·13)$$

Proof. From the Venn diagram, we have

$$A \cup B = A \cup (\bar{A} \cap B),$$

where A and $\bar{A} \cap B$ are mutually disjoint.

$$\therefore \quad P(A \cup B) = P[A \cup (\bar{A} \cap B)]$$

$$= P(A) + P(\bar{A} \cap B) \quad \text{[By Axiom 3]} \quad ...(*)$$

$$= P(A) + P(B) - (A \cap B) \quad \text{[From Theorem 3·3(}i\text{)]}$$

OR From (*) onwards

$$P(A \cup B) = P(A) + [P(\bar{A} \cap B) + P(A \cap B)] - P(A \cap B)$$

$$= P(A) + P[(\bar{A} \cap B) \cup (A \cap B)] - P(A \cap B)$$

$$[\because (\bar{A} \cap B) \text{ and } (A \cap B) \text{ are disjoint}]$$

$$\Rightarrow \quad P(A \cup B) = P(A) + P(B) - P(A \cap B)$$

Aliter.

$$P(A \cup B) = \frac{n(A \cup B)}{n(S)} = \frac{n(A) + n(B) - n(A \cap B)}{n(S)}$$

$$= \frac{n(A)}{n(S)} + \frac{n(B)}{n(S)} - \frac{n(A \cap B)}{n(S)} = P(A) + P(B) - P(A \cap B)$$

Cor. 1. If the events A and B are mutually disjoint, then

$$A \cap B = \phi \Rightarrow P(A \cap B) = P(\phi) = 0$$

$\therefore$ $P(A \cup B) = P(A) + P(B)$, which is Axiom 3 of probability.

Cor. 2. For three non mutually exclusive events A, B and C, we have

$$P(A \cup B \cup C) = P(A) + P(B) + P(C) - P(A \cap B) - P(B \cap C) - P(C \cap A) + P(A \cap B \cap C) \quad ...(3·13a)$$

Proof.

$$P(A \cup B \cup C) = P[A \cup (B \cup C)]$$

$$= P(A) + P(B \cup C) - P[A \cap (B \cup C)] \quad \text{[From (3·13)]}$$

$$= P(A) + [P(B) + P(C) - P(B \cap C)] - P[(A \cap B) \cup (A \cap C)]$$

$$= P(A) + P(B) + P(C) - P(B \cap C) - [P(A \cap B) + P(A \cap C) - P(A \cap B \cap C)]$$

$$= P(A) + P(B) + P(C) - P(A \cap B) - P(B \cap C) - P(C \cap A) + P(A \cap B \cap C).$$

3·9·2. Extension of Addition Theorem of Probability to *n* Events

Theorem 3·6. *For n events $A_1, A_2, \ldots, A_n$, we have*

$$P\left(\bigcup_{i=1}^{n} A_i\right) = \sum_{i=1}^{n} P(A_i) - \sum\sum_{1 \le i < j \le n} P(A_i \cap A_j) + \sum\sum\sum_{1 \le i < j < k \le n} P(A_i \cap A_j \cap A_k)$$

$$+ \ldots + (-1)^{n-1} P(A_1 \cap A_2 \cap \ldots \cap A_n) \quad ...(3·14)$$

Proof. For two events A_1 and A_2, we have

$$P(A_1 \cup A_2) = P(A_1) + P(A_2) - P(A_1 \cap A_2) \quad ...(*)$$

Hence, (3·14) is true for $n = 2$.

Let us now suppose that (3·14) is true for $n = r$, (say) so that

$$P\left(\bigcup_{i=1}^{r} A_i\right) = \sum_{i=1}^{r} P(A_i) - \sum\sum_{1 \le i < j \le r} P(A_i \cap A_j) + \ldots + (-1)^{r-1} P(A_1 \cap A_2 \cap \ldots \cap A_r) \quad \ldots(**)$$

Now

$$P\left(\bigcup_{i=1}^{r+1} A_i\right) = P\left\{\left(\bigcup_{i=1}^{r} A_i\right) \cup A_{r+1}\right\}$$

$$= P\left(\bigcup_{i=1}^{r} A_i\right) + P(A_{r+1}) - P\left\{\left(\bigcup_{i=1}^{r} A_i\right) \cap A_{r+1}\right\} \quad \text{[Using (*)]}$$

$$= P\left(\bigcup_{i=1}^{r} A_i\right) + P(A_{r+1}) - P\left\{\bigcup_{i=1}^{r} (A_i \cap A_{r+1})\right\} \quad \text{(By Distributive Law)}$$

$$= \sum_{i=1}^{r} P(A_i) - \sum\sum_{1 \le i < j \le r} P(A_i \cap A_j) + \ldots + (-1)^{r-1} P(A_1 \cap A_2 \cap \ldots \cap A_r)$$

$$+ P(A_{r+1}) - P\left\{\bigcup_{i=1}^{r} (A_i \cap A_{r+1})\right\} \quad \text{[From (**)]}$$

$$= \sum_{i=1}^{r+1} P(A_i) - \sum\sum_{1 \le i < j \le r} P(A_i \cap A_j) + \ldots + (-1)^{r-1} P(A_1 \cap A_2 \cap \ldots \cap A_r)$$

$$- \left\{\sum_{i=1}^{r} P(A_i \cap A_{r+1}) - \sum\sum_{1 \le i < j \le r} P(A_i \cap A_j \cap A_{r+1})\right.$$

$$\left. + \ldots + (-1)^{r-1} P(A_1 \cap A_2 \cap \ldots \cap A_r \cap A_{r+1})\right\} \quad \ldots \text{[From (**)]}$$

$$\Rightarrow \quad P\left(\bigcup_{i=1}^{r+1} A_i\right) = \sum_{i=1}^{r+1} P(A_i) - \left\{\sum\sum_{1 \le i < j \le r} P(A_i \cap A_j) + \sum_{i=1}^{r} P(A_i \cap A_{r+1})\right\}$$

$$+ \ldots + (-1)^r P\{(A_1 \cap A_2 \cap \ldots \cap A_{r+1})\}$$

$$= \sum_{i=1}^{r+1} P(A_i) - \sum_{1 \le i <} \sum_{j \le (r+1)} P(A_i \cap A_j) + \ldots + (-1)^r P(A_1 \cap A_2 \cap \ldots \cap A_{r+1})$$

Hence, if (3·14) is true for $n = r$, it is also true for $n = (r + 1)$. But we have proved in (*) that (3·14) is true for $n = 2$. Hence by the principle of mathematical induction, it follows that (3·14) is true for all positive integral values of n.

Remarks:

1. If we write $P(A_i) = p_i$, $P(A_i \cap A_j) = p_{ij}$, $P(A_i \cap A_j \cap A_k) = p_{ijk}$ and so on and

$$S_1 = \sum_{i=1}^{n} p_i = \sum_{i=1}^{n} P(A_i)$$

$$S_2 = \sum_{1 \le i < j \le n}\sum p_{ij} = \sum_{1 \le i < j \le n}\sum P(A_i \cap A_j)$$

$$S_3 = \sum_{1 \le i < j < k \le n}\sum\sum p_{ijk}$$ and so on, then from (3·14), we get:

$$P\left(\bigcup_{i=1}^{n} A_i\right) = S_1 - S_2 + S_3 - \ldots + (-1)^{n-1} S_n \qquad \ldots(3\cdot 14a)$$

2. If all the events A_i, $(i = 1, 2, \ldots, n)$ are mutually disjoint then (3·14) gives

$$P\left(\bigcup_{i=1}^{n} A_i\right) = \sum_{i=1}^{n} P(A_i)$$

3. *Practical use of (3·14).* From practical point of view the theorem can be restated in a slightly different form. Let us suppose that an event A can materialise in several mutually exclusive forms, *viz.*, $A_1, A_2, \ldots, A_n$ which may be regarded as that many mutually exclusive events. If A happens then any one of the events A_i, $(i = 1, 2, \ldots, n)$ must happen and conversely if any one of the events A_i, $(i = 1, 2, \ldots, n)$ happens, then A happens. Hence the probability of happening of A is same as the probability of happening of any one of its (unspecified) mutually exclusive forms. From this point of view, the total addition probability theorem can be restated as follows:

The probability of happening of an event A is the sum of the probabilities of happening of its mutually exclusive forms A_1 $A_2, \ldots, A_n$. Symbolically,

$$P(A) = P(A_1) + P(A_2) + \ldots + P(A_n) \qquad \ldots(3\cdot 14b)$$

The probabilities $P(A_1), P(A_2), \ldots, P(A_n)$ of the mutually exclusive forms of A are known as the *partial probabilities*. Since $P(A)$ is their sum, it may be called the *total probability* of A. Hence the name of the theorem.

3·9·3. Boole's Inequality

***Statement. For n events** $A_1, A_2, \ldots, A_n$, we have*

(a) $$P\left(\bigcap_{i=1}^{n} A_i\right) \ge \sum_{i=1}^{n} P(A_i) - (n-1) \qquad \ldots(3\cdot 15)$$

(b) $$P\left(\bigcup_{i=1}^{n} A_i\right) \le \sum_{i=1}^{n} P(A_i) \qquad \ldots(3\cdot 15a)$$

Proof. (a) $$P(A_1 \cup A_2) = P(A_1) + P(A_2) - P(A_1 \cap A_2) \le 1$$

$\Rightarrow$ $$P(A_1 \cap A_2) \ge P(A_1) + P(A_2) - 1 \qquad \ldots(*)$$

Hence, (3·15) is true for $n = 2$.

Let us now suppose that (3·15) is true for $n = r$ (say), such that

$$P\left(\bigcap_{i=1}^{r} A_i\right) \ge \sum_{i=1}^{r} P(A_i) - (r-1) \qquad \ldots(**)$$

Then, $$P\left(\bigcap_{i=1}^{r+1} A_i\right) = P\left(\bigcap_{i=1}^{r} A_i \cap A_{r+1}\right)$$

$$\geq P\left(\bigcap_{i=1}^{r} A_i\right) + P(A_{r+1}) - 1 \qquad \text{[From (*)]}$$

$$\geq \sum_{i=1}^{r} P(A_i) - (r-1) + P(A_{r+1}) - 1 \qquad \text{[From ((**)]}$$

$$\Rightarrow \qquad P\left(\bigcap_{i=1}^{r+1} A_i\right) \geq \sum_{i=1}^{r+1} P(A_i) - r \quad \Rightarrow \quad \text{(3·15) is true for } n = (r+1) \text{ also}$$

The result now follows by the principle of mathematical induction.

(*b*) Applying the inequality (3·15) to the event $\bar{A}_i, \bar{A}_2, \ldots, \bar{A}_n$, we get

$$P(\bar{A}_1 \cap \bar{A}_2 \cap \ldots \cap \bar{A}_n) \geq \left[P(\bar{A}_1) + P(\bar{A}_2) + \ldots + P(\bar{A}_n)\right] - (n-1)$$

$$= [1 - P(A_1)] + [1 - P(A_2)] + \ldots + [1 - P(A_n)] - (n-1)$$

$$= 1 - P(A_1) - P(A_2) - \ldots - P(A_n)$$

$$\Rightarrow \qquad P(A_1) + P(A_2) + \ldots + P(A_n) \geq 1 - P(\bar{A}_1 \cap \bar{A}_2 \cap \ldots \cap \bar{A}_n)$$

$$= 1 - P(\overline{A_1 \cup A_2 \cup \cdots \cup A_n})$$

$$= P(A_1 \cup A_2 \cup \ldots \cup A_n)$$

$$\Rightarrow \qquad P(A_1 \cup A_2 \cup \ldots \cup A_n) \leq P(A_1) + P(A_2) + \cdots + P(A_n),$$

as desired.

Aliter for (b), i.e., (3·15a). We have

$$P(A_1 \cup A_2) = P(A_1) + P(A_2) - P(A_1 \cap A_2) \leq P(A_1) + P(A_2) \quad [\because P(A_1 \cap A_2) \geq 0] \quad \ldots(***)$$

Hence, (3·15*a*) is true for $n = 2$.

Let us now suppose that (3·15*a*) is true for $n = r$ (say), so that

$$P\left(\bigcup_{i=1}^{r} A_i\right) \leq \sum_{i=1}^{r} P(A_i) \qquad \ldots(****)$$

$$\therefore \qquad P\left(\bigcup_{i=1}^{r+1} A_i\right) = P\left(\bigcup_{i=1}^{r} A_i \cup A_{r+1}\right)$$

$$\leq P\left(\bigcup_{i=1}^{r} A_i\right) + P(A_{r+1}) \qquad \text{[Using (***)]}$$

$$\leq \sum_{i=1}^{r} P(A_i) + P(A_{r+1}) \qquad \text{[Using (****)]}$$

$$\Rightarrow \qquad P\left(\bigcup_{i=1}^{r+1} A_i\right) \leq \sum_{i=1}^{r+1} P(A_i)$$

Hence, if (3·15*a*) is the true for $n = r$, then it is also true for $n = r + 1$. But we have proved in (***) that (3·15*a*) is true for $n = 2$. Hence by mathematical induction we conclude that (3·15*a*) is true for all positive integer values of n.

Theorem 3·7. *For n events $A_1, A_2, \ldots, A_n$,*

$$P\left(\bigcup_{i=1}^{n} A_i\right) \geq \sum_{i=1}^{n} P(A_i) - \sum_{1 \leq i < j \leq n} P(A_i \cap A_j) \qquad \text{...(3·16)}$$

Proof. We shall prove this theorem by the principle of mathematical induction.

We know that:

$$P(A_1 \cup A_2 \cup A_3) = P(A_1) + P(A_2) + P(A_3) - [P(A_1 \cap A_2) + P(A_2 \cap A_3) + P(A_3 \cap A_1)] + P(A_1 \cap A_2 \cap A_3)$$

$$\Rightarrow \quad P\left(\bigcup_{i=1}^{3} A_i\right) \geq \sum_{i=1}^{3} P(A_i) - \sum_{1 \leq i < j \leq 3} P(A_i \cap A_j) \qquad \text{...(*)}$$

Thus, the result is true for $n = 3$. Let us suppose that the result is true for $n = r$, (say), so that :

$$P\left(\bigcup_{i=1}^{r} A_i\right) \geq \sum_{i=1}^{r} P(A_i) - \sum\sum_{1 \leq i < j \leq r} P(A_i \cap A_j) \qquad \text{...(**)}$$

Now

$$P\left(\bigcup_{i=1}^{r+1} A_i\right) = P\left(\bigcup_{i=1}^{r} A_i \cup A_{r+1}\right)$$

$$= P\left(\bigcup_{i=1}^{r} A_i\right) + P(A_{r+1}) - P\left[\left(\bigcup_{i=1}^{r} A_i\right) \cap A_{r+1}\right]$$

$$= P\left(\bigcup_{i=1}^{r} A_i\right) + P(A_{r+1}) - P\left\{\bigcup_{i=1}^{r} (A_i \cap A_{r+1})\right\}$$

$$\geq \left\{\sum_{i=1}^{r} P(A_i) - \sum_{1 \leq i < j \leq r} \sum P(A_i \cap A_j)\right\}$$

$$+ P(A_{r+1}) - P\left\{\bigcup_{i=1}^{r} (A_i \cap A_{r+1})\right\} \qquad \text{[From (**)]} \qquad \text{...(***)}$$

From Boole's inequality, we get

$$P\left[\bigcup_{i=1}^{r} (A_i \cap A_{r+1})\right] \leq \sum_{i=1}^{r} P(A_i \cap A_{r+1}) \Rightarrow -P\left[\bigcup_{i=1}^{r} (A_i \cap A_{r+1})\right] \geq -\sum_{i=1}^{r} P(A_i \cap A_{r+1})$$

$\therefore$ From (***), we get

$$P\left(\bigcup_{i=1}^{r+1} A_i\right) \geq \sum_{i=1}^{r+1} P(A_i) - \sum\sum_{1 \leq i < j \leq r} P(A_i \cap A_j) - \sum_{i=1}^{r} P(A_i \cap A_{r+1})$$

$$\Rightarrow \quad P\left(\bigcup_{i=1}^{r+1} A_i\right) \geq \sum_{i=1}^{r+1} P(A_i) - \sum_{1 \leq i < j \leq r+1} \sum P(A_i \cap A_j)$$

Hence, if the theorem is true for $n = r$, it is also true for $n = r + 1$. But we have seen in (*) that the result is true for $n = 3$. Hence, by the principle of mathematical induction, the result is true for all positive integral values of n.

Example 3·22. *A letter of the English alphabet is chosen at random. Calculate the probability that the letter so chosen.*

(i) is a vowel, (ii) precedes m and is a vowel, (iii) follows m and is a vowel.

Solution. The sample space of the experiment is :

$$S = (a, b, c, d, \ldots, x, y, z), \qquad n(S) = 26$$

(*i*) Let E_1 be the event that the letter chosen is a vowel. Then,

$$E_1 = (a, e, i, o, u) \qquad n(E_1) = 5 \qquad \therefore \quad P(E_1) = \frac{n(E_1)}{n(S)} = \frac{5}{26}$$

(*ii*) Let E_2 be the event that the letter precedes m and is a vowel, Then,

$$E_2 = \{a, e, i\}; \qquad n(E_2) = 3 \qquad \therefore \quad P(E_2) = \frac{n(E_2)}{n(S)} = \frac{3}{26}$$

(*iii*) Let E_3 be the event that the letter follows m and is a vowel. Then,

$$E_3 = \{o, u\}; \qquad n(E_3) = 2 \qquad \therefore \quad P(E_3) = \frac{n(E_3)}{n(S)} = \frac{2}{26} = \frac{1}{13}$$

Example 3·23. *A, B and C are three mutually exclusive and exhaustive events associated with a random experiment. Find P(A) given that:*

$$P(B) = \frac{3}{2} P(A) \quad \text{and} \quad P(C) = \frac{1}{2} P(B)$$

Solution. Let $P(A) = p$, then $P(B) = \frac{3}{2} p$ and $P(C) = \frac{1}{2} \times \frac{3}{2} p = \frac{3}{4} p.$

Since A, B, C mutually exclusive and exhaustive events.

$$P(A) + P(B) + P(C) = 1 \;\Rightarrow\; p + \frac{3}{2} p + \frac{3}{4} p = 1 \;\Rightarrow\; \frac{13}{4} p = 1 \;\Rightarrow\; p = \frac{4}{13}.$$

Example 3·24. *If* $p_1 = P(A), \quad p_2 = P(B), \quad p_3 = P(A \cap B), \quad (p_1, p_2, p_3 > 0)$, *express the following in terms of* p_1, p_2, p_3:

(*a*) $P(\overline{A \cup B})$, (*b*) $P(\bar{A} \cup \bar{B})$, (*c*) $P(\bar{A} \cap B)$, (*d*) $P(\bar{A} \cup B)$, (*e*) $P(\bar{A} \cap \bar{B})$,

(*f*) $P(A \cap \bar{B})$, (g) $P(A \mid B)$, (*h*) $P(B \mid \bar{A})$, (*i*) $P[\bar{A} \cap (A \cup B)]$.

Solution.

(*a*) $P(\overline{A \cup B}) = 1 - P(A \cup B) = 1 - [P(A) + P(B) - P(A \cap B)] = 1 - p_1 - p_2 + p_3.$

(*b*) $P(\bar{A} \cup \bar{B}) = P(\overline{A \cap B}) = 1 - P(A \cap B) = 1 - p_3$

(*c*) $P(\bar{A} \cap B) = P(B) - P(A \cap B) = p_2 - p_3$

(*d*) $P(\bar{A} \cup B) = P(\bar{A}) + P(B) - P(\bar{A} \cap B) = 1 - p_1 + p_2 - (p_2 - p_3) = 1 - p_1 + p_3$ [Part (c)]

(e) $P(\bar{A} \cap \bar{B}) = P(\overline{A \cup B}) = 1 - p_1 - p_2 + p_3.$ [Part (a)]

(f) $P(A \cap \bar{B}) = P(A) - P(A \cap B) = p_1 - p_3$

(g) $P(A \mid B) = P(A \cap B)/P(B) = p_3/p_2$

(h) $P(B \mid \bar{A}) = P(\bar{A} \cap B)/P(\bar{A}) = (p_2 - p_3)/(1 - p_1)$ [Part (c)]

(i) $P[\bar{A} \cap (A \cup B)] = P[(\bar{A} \cap A) \cup (\bar{A} \cap B)] = P(\bar{A} \cap B) = p_2 - p_3$ $[\because A \cap \bar{A} = \phi]$

Note. For (g) and (h), see Theorem 3·8, § 3·11.

Example 3·25. *Let $P(A) = p$, $P(A \mid B) = q$, $P(B \mid A) = r$. Find relations between the number p, q, r for the following cases:*

(a) Events A and B are mutually exclusive.

(b) A and B are mutually exclusive and collectively exhaustive.

(c) A is sub-event of B; B is a sub-event of A.

(d) $\bar{A}$ and $\bar{B}$ are mutually exclusive.

Solution. From the given data: $P(A) = p$, $P(A \cap B) = P(A)\, P(B \mid A) = rp$

$$\therefore \quad P(B) = \frac{P(A \cap B)}{P(A \mid B)} = \frac{rp}{q} \quad \text{and} \quad P(A) + P(B) = p + \frac{rp}{q} = \frac{p(q+r)}{q}$$

(a) Since A and B are mutually exclusive

$$A \cap B = \phi \Rightarrow P(A \cap B) = P(\phi) = 0 \Rightarrow rp = 0.$$

(b) Since A and B are mutually exclusive and collectively exhaustive.

$$P(A \cap B) = 0 \quad \text{and} \quad P(A) + P(B) = 1$$

$$\Rightarrow \quad p(q + r) = q;\ rp = 0 \quad \text{or} \quad pq = q \quad \Rightarrow \quad p = 1 \quad \text{or} \quad q = 0$$

(c) $A \subseteq B \Rightarrow A \cap B = A$ or $P(A \cap B) = P(A) \Rightarrow rp = p$, *i.e.*, $r = 1$ or $p = 0$

$$B \subseteq A \Rightarrow A \cap B = B \quad \text{or} \quad P(A \cap B) = P(B)$$

$$\Rightarrow \quad rp = (rp/q) \quad \Rightarrow \quad rp(q - 1) = 0 \quad \Rightarrow \quad r = 0 \quad \text{or} \quad p = 0 \quad \text{or} \quad q = 1.$$

(d) Since $\bar{A}$ and $\bar{B}$ are mutually exclusive, $P(\bar{A} \cap \bar{B}) = 1 - P(A \cup B) = 0$

$$\Rightarrow \quad 1 - [P(A) + P(B) - P(A \cap B)] = 0 \quad \Rightarrow \quad P(A) + P(B) = 1 + P(A \cap B)$$

$$\Rightarrow \quad p[1 + (r/q)] = 1 + rp \quad \Rightarrow \quad p(q + r) = q(1 + pr).$$

Example 3·26. *A die is loaded in such a manner that for n = 1, 2, 3, 4, 5, 6, the probability of the face marked n, landing on top when the die is rolled is proportional to n. Find the probability that an odd number will appear on tossing the die.*

Solution. Hence, we are given:

$$P(n) \propto n \quad \Rightarrow \quad P(n) = kn \qquad \ldots(*),$$

where k is the constant of proportionality.

Also $\quad P(1) + P(2) + \ldots + P(6) = 1 \quad \Rightarrow \quad k(1 + 2 + 3 + 4 + 5 + 6) = 1 \Rightarrow k = \dfrac{1}{21}$

$$\text{Required Probability} = P(1) + P(3) + P(5) = \frac{1+3+5}{21} = \frac{3}{7} \qquad \text{[Using (*)]}$$

Example 3·27. *If two dice are thrown, what is the probability that the sum is*

(a) greater than 8, and (b) neither 7 nor 11?

Solution. (*a*) If S denotes the sum on the two dice, then we want $P(S > 8)$.

The required event can happen in the following mutually exclusive ways:

(*i*) $S = 9$ (*ii*) $S = 10$ (*iii*) $S = 11$ (*iv*) $S = 12$

Hence, by addition theorem of probability

$$P(S > 8) = P(S = 9) + P(S = 10) + P(S = 11) + P(S = 12) \quad ...(*)$$

In a throw of two dice, the sample space contains $6^2 = 36$ points. The number of favourable cases can be enumerated as follows:

$S = 9$: (3, 6), (6, 3), (4, 5), *i.e.*, 4 sample points $\therefore$ $P(S = 9) = \dfrac{4}{36}$

$S = 10$: (4, 6), (6, 4), (5, 5), *i.e.*, 3 sample points $\therefore$ $P(S = 10) = \dfrac{3}{36}$

$S = 11$: (5, 6), (6, 5), *i.e.*, 2 sample points $\therefore$ $P(S = 11) = \dfrac{2}{36}$

$S = 12$: (6, 6), *i.e.*, 1 sample point. $\therefore$ $P(S = 12) = \dfrac{1}{36}$

$$\therefore \quad P(S > 8) = \frac{4}{36} + \frac{3}{36} + \frac{2}{36} + \frac{1}{36} = \frac{10}{36} = \frac{5}{18} \quad \text{[From (*)]}$$

(*b*) Let A denote the event of getting the sum of 7 and B denote the event of getting the sum of 11 with a pair of dice.

$S = 7$: (1, 6), (6, 1), (2, 5), (5, 2), (3, 4), (4, 3) $\Rightarrow n(S = 7) = 6$ $\therefore$ $P(A) = P(S = 7) = \dfrac{6}{36} = \dfrac{1}{6}$

$S = 11$: (5, 6), (6, 5) $\Rightarrow$ $n(S = 11) = 2$; $\therefore$ $P(B) = P(S = 11) = \dfrac{2}{36} = \dfrac{1}{18}$

$\therefore$ Required probability $= P(\bar{A} \cap \bar{B}) = 1 - P(A \cup B)$

$$= 1 - [P(A) + P(B)] = 1 - \frac{1}{6} - \frac{1}{18} = \frac{7}{9}. \quad (\because \ A \text{ and } B \text{ are disjoint events.})$$

Example 3·28. *Two dice are tossed. Find the probability of getting 'an even number on the first die or a total of 8'.*

Solution. In a random toss of two dice, sample space S is given by:

$$S = \{1, 2, 3, 4, 5, 6\} \times \{1, 2, 3, 4, 5, 6\} \Rightarrow n(S) = 6 \times 6 = 36$$

Let us define the events:

A : Getting an even number on the first dice

B : The sum of the points obtained on the two dice is 8.

These events are represented by the following subsets of S.

$A = \{2, 4, 6\} \times \{1, 2, 3, 4, 5, 6\} \quad \Rightarrow \quad n(A) = 3 \times 6 = 18$

$B = \{(2, 6), (6, 2), (3, 5), (5, 3), (4, 4)\} \quad \Rightarrow \quad n(B) = 5$

Also $A \cap B = \{(2, 6), (6, 2), (4, 4)\} \quad \Rightarrow \quad n(A \cap B) = 3$

$$\therefore \quad P(A) = \frac{n(A)}{n(S)} = \frac{18}{36} = \frac{1}{2}, \quad P(B) = \frac{n(B)}{n(S)} = \frac{5}{36}, \quad \text{and} \quad P(A \cap B) = \frac{3}{36} = \frac{1}{12}$$

Hence, the required probability is given by:

$$P(A \cup B) = P(A) + P(B) - P(A \cap B) = \frac{18}{36} + \frac{5}{36} - \frac{3}{36} = \frac{20}{36} = \frac{5}{9}.$$

Example 3·29. *An integer is chosen at random from two hundred digits. What is the probability that the integer is divisible by 6 or 8?*

Solution. The sample space of the random experiment is:

$$S = \{1, 2, 3, \ldots, 199, 200\} \Rightarrow n(S) = 200$$

The event A : 'integer chosen is divisible by 6' has the sample points given by:

$$A = \{6, 12, 18, \ldots, 198\} \Rightarrow n(A) = \frac{198}{6} = 33. \quad \therefore \quad P(A) = \frac{n(A)}{n(S)} = \frac{33}{200}$$

Similarly, the event B: 'integer chosen is divisible by 8' has the sample points given by:

$$B = \{8, 16, 24, \ldots, 200\} \Rightarrow n(B) = \frac{200}{8} = 25 \quad \therefore \quad P(B) = \frac{n(B)}{n(S)} = \frac{25}{200}$$

The LCM of 6 and 8 is 24. Hence, a number is divisible by both 6 and 8, if it is divisible by 24.

$$\therefore \quad A \cap B = \{24, 48, 72, \ldots, 192\} \Rightarrow n(A \cap B) = \frac{192}{24} = 8 \Rightarrow P(A \cap B) = \frac{8}{200}$$

Hence, the required probability is :

$$P(A \cup B) = P(A) + P(B) - P(A \cap B) = \frac{33}{200} + \frac{25}{200} - \frac{8}{200} = \frac{1}{4}.$$

Example 3·30. *The probability that a student passes a Physics test is $\frac{2}{3}$ and the probability that he passes both a Physics test and an English test is $\frac{14}{45}$. The probability that he passes at least one test is $\frac{4}{5}$. What is the probability that he passes the English test?*

Solution. Let us define the following events:

A : The students passes a Physics test; B : The students passes an English test.

$$P(A) = \frac{2}{3}, \quad P(A \cap B) = \frac{14}{45}, \quad P(A \cup B) = \frac{4}{5} \text{ and we want, } P(B)$$

$$P(A \cup B) = P(A) + P(B) - P(A \cap B) \quad \Rightarrow \quad \frac{4}{5} = \frac{2}{3} + P(B) - \frac{14}{45}$$

$$\therefore \quad P(B) = \frac{4}{5} + \frac{14}{45} - \frac{2}{3} = \frac{36 + 14 - 30}{45} = \frac{4}{9}.$$

Example 3·31. *An investment consultant predicts that the odds against the price of a certain stock will go up during the next week are 2 : 1 and the odds in favour of the price remaining the same are 1 : 3. What is the probability that the price of the stock will go down during the next week ?*

Solution. Let A denote the event that 'stock price will go up', and B be the event 'stock price will remain same'. Then

$$P(A)=\frac{1}{2+1}=\frac{1}{3} \quad \text{and} \quad P(B)=\frac{1}{1+3}=\frac{1}{4}.$$

$\therefore$ P(stock price will either go up or remain same) $= P(A \cup B) = P(A)+P(B)=\frac{1}{3}+\frac{1}{4}=\frac{7}{12}$

Hence, probability that stock price will go down $= P(\bar{A}\cap\bar{B})=1-P(A\cup B)=1-\frac{7}{12}=\frac{5}{12}.$

Example 3·32. *A person with MCA degree applies for a job in two firms X and Y. The probability of his being selected in firm X is 0.7 and being rejected at Y is 0.5. The probability of at least one of his applications being rejected is 0.6. What is probability that he will be selected in one of the firms?*

Solution. Let A and B denote the events that the person is selected in firms X and Y, respectively. Then in the usual notations, we are given:

$$P(A) = 0{\cdot}7 \quad \Rightarrow \quad P(\bar{A})=1-0{\cdot}7=0{\cdot}3; \quad P(\bar{B})=0{\cdot}5 \quad \Rightarrow \quad P(B)=1-0{\cdot}5=0{\cdot}5 \qquad ...(*)$$

and $$P(\bar{A}\cup\bar{B}) = 0{\cdot}6=P(\bar{A})+P(\bar{B})-P(\bar{A}\cap\bar{B}) \qquad ...(**)$$

The probability that the persons will be selected in one of the two firms X or Y is given by:

$$P(A \cup B) = 1-P(\bar{A}\cap\bar{B})=1-\{P(\bar{A})+P(\bar{B})-P(\bar{A}\cup\bar{B})\} \qquad \text{[From (**)]}$$

$$= 1 - (0{\cdot}3 + 0{\cdot}5 - 0.6) = 0{\cdot}8 \qquad \text{[From (*)]}$$

Example 3·33. *Three newspapers A, B and C are published in a certain city. It is estimated from a survey that of the adult population: 20% read A, 16% read B, 14% read C, 8% read both A and B, 5% read both A and C, 4% read both B and C, 2% read all three. Find what percentage read at least one of the papers?*

Solution. Let E, F and G denote the events that the adult reads newspapers A, B and C, respectively. Then we are given:

$$P(E) = \frac{20}{100}, \quad P(F) = \frac{16}{100}, \quad P(G) = \frac{14}{100}, \quad P(E \cap F) = \frac{8}{100}$$

$$P(E \cap G) = \frac{5}{100}, \quad P(F \cap G) = \frac{4}{100}, \quad \text{and} \quad P(E \cap F \cap G) = \frac{2}{100}$$

The required probability that an adult reads at least one of the newspapers (by addition theorem) is given by:

$$P(E \cup F \cup G) = P(E) + P(F) + P(G) - P(E \cap F) - P(F \cap G) - P(E \cap G) + P(E \cap F \cap G)$$

$$= \frac{20}{100}+\frac{16}{100}+\frac{14}{100}-\frac{8}{100}-\frac{4}{100}-\frac{5}{100}+\frac{2}{100}=\frac{35}{100}=0{\cdot}35$$

Hence, 35% of the adult population reads at least one of the newspapers.

Example 3·34. *A card is drawn from a pack of 52 cards. Find the probability of getting a king or a heart or a red card.*

Solution. Let us define the following events:

A : the card drawn is a king; B: the card drawn is a heart;

C: the card drawn is a red card.

Then A, B and C are not mutually exclusive.

$A \cap B$	: the card drawn is the king of hearts	$\Rightarrow$	$n(A \cap B) = 1$
$B \cap C = B$	: the card drawn is a heart $(\because B \subset C)$	$\Rightarrow$	$n(B \cap C) = 13$
$C \cap A$	: the card drawn is a red king	$\Rightarrow$	$n(C \cap A) = 2$
$A \cap B \cap C = A \cap B$	: the card drawn is the king of hearts	$\Rightarrow$	$n(A \cap B \cap C) = 1$

$$\therefore \quad P(A) = \frac{n(A)}{n(S)} = \frac{4}{52}; \qquad P(B) = \frac{13}{52}; \qquad P(C) = \frac{26}{52}$$

$$P(A \cap B) = \frac{1}{52}; \quad P(B \cap C) = \frac{13}{52}; \quad P(C \cap A) = \frac{2}{52}; \quad P(A \cap B \cap C) = \frac{1}{52}$$

The required probability of getting a king or heart or a red card is given by:

$$P(A \cup B \cup C) = P(A) + P(B) + P(C) - P(A \cap B) - P(B \cap C) - P(C \cap A) + P(A \cap B \cap C)$$

$$= \frac{4}{52} + \frac{13}{52} + \frac{26}{52} - \frac{1}{52} - \frac{13}{52} - \frac{2}{52} + \frac{1}{52} = \frac{28}{52} = \frac{7}{13}.$$

3·10. CONDITIONAL PROBABILITY

As discussed earlier, the probability $P(A)$ of an event A represents the likelihood that a random experiment will result in an outcome in the set A relative to the sample space S of the random experiment. However, quite often, while evaluating some event probability, we already have some information stemming from the experiment. For example, if we have prior information that the outcome of the random experiment must be in a set B of S, then this information must be used to re-appraise the likelihood that the outcome will also be in B. This re-appraised probability is denoted by $P(A \mid B)$ and is read as the conditional probability of the event A, given that the event B has already happened.

We give below some illustrations to explain this concept.

***Illustrations* 1.** Let us consider a random experiment of drawing a card from a pack of cards. Then the probability of happening of the event A: "The card drawn is a king", is given by:

$$P(A) = \frac{4}{52} = \frac{1}{13}.$$

Now suppose that a card is drawn and we are informed that the drawn card is red. How does this information affect the likelihood of the event A ?

Obviously, if the event B: 'The card drawn is red', has happened, the event 'Black card' is not possible. hence the probability of the event A must be computed relative to the new sample space 'B' which consists of 26 sample points (red cards only), *i.e.*, $n(B) = 26$. Among these 26 red cards, there are two (red) kings so that $n(A \cap B) = 2$. Hence, the required probability is given by :

$$P(A \mid B) = \frac{n(A \cap B)}{n(B)} = \frac{2}{26} = \frac{1}{13}.$$

2. Consider a random experiment of tossing three fair coins. Then, as explained earlier, the sample space S is :

$$S = \{HHH, HHT, HTH, HTT, THH, THT, TTH, TTT\} = \{\omega_1, \omega_2, \omega_3, \omega_4, \omega_5, \omega_6, \omega_7, \omega_8\}, \text{(say)},$$

so that :
$$P(\{\omega_i\}) = \frac{1}{8}\,; \; i = 1, 2, \ldots 8.$$

Now suppose that the same experiment is performed by another person and nothing is known about its outcome. However, we have the information that he obtained 'at least two heads'. We are interested to find how this additional information affects the probabilities of the elementary outcomes.

This means that if the event A : 'At least two heads are obtained', has happened then the elementary outcomes ω_4, ω_6, ω_7, and ω_8 could not have happened. However, the remaining four outcomes ω_1, ω_2, ω_3 and ω_5 are still possible and we assign the probability $\frac{1}{4}$ to each one of them.

$$\therefore \quad P(\omega_1 \mid A) = P(\omega_2 \mid A) = P(\omega_3 \mid A) = P(\omega_5 \mid A) = \frac{1}{4}.$$

From the above illustrations we observe that some additional information may change the probability of the happening of some event. We now proceed to develop procedure to calculate the probabilities of events when we know some addition information.

Remark. When we know that a particular event B has occurred, instead of S, we concentrate our attention on B only and the conditional probability of A given B will be analogously the ratio of the probability of that part of A which is included in B (*i.e.*, $A \cap B$) to the probability of B. It, therefore, reflects the charge of viewpoint only, namely, instead of S we have to concentrate on B only.

3·11. MULTIPLICATION THEOREM OF PROBABILITY

Theorem 3.8. *For two events A and B,*

$$\left.\begin{aligned} P(A \cap B) &= P(A) \cdot P(B \mid A),\ P(A) > 0 \\ &= P(B) \cdot P(A \mid B),\ P(B) > 0 \end{aligned}\right\}, \qquad \ldots(3\cdot17)$$

where P(B | A) represents conditional probability of occurrence of B when the event A has already happened and P(A | B) is the conditional probability of happening of A, given that B has already happened.

Proof. In the usual notations, we have

$$P(A) = \frac{n(A)}{n(S)}; \quad P(B) = \frac{n(B)}{n(S)} \quad \text{and} \quad P(A \cap B) = \frac{n(A \cap B)}{n(S)} \qquad \ldots(*)$$

For the conditional event $A \mid B$, the favourable outcomes must be one of the sample points of B, *i.e.*, for the event $A \mid B$, the sample space is B and out of the $n(B)$ sample points, $n(A \cap B)$ pertain to be occurrence of the event A. Hence

$$P(A \mid B) = \frac{n(A \cap B)}{n(B)}$$

Rewriting (*), we get: $$P(A \cap B) = \frac{n(B)}{n(S)} \times \frac{n(A \cap B)}{n(B)} = P(B) \cdot P(A \mid B) \qquad \ldots(**)$$

Similarly, we get from (*) $$P(A \cap B) = \frac{n(A)}{n(S)} \times \frac{n(A \cap B)}{n(A)} = P(A) \cdot P(B \mid A) \qquad \ldots(***)$$

From (**) and (***), we get the result (3·17).

Thus, we have proved that *"the probability of the simultaneous occurrence of two events A and B is equal to the product of the probability of one of these events and the conditional probability of the other, given that the first one has occurred"*. Any of the events may be called the first event.

Remark. $P(B \mid A) = \dfrac{P(A \cap B)}{P(A)}$ and $P(A \mid B) = \dfrac{P(A \cap B)}{P(B)}$

Thus, the conditional probabilities $P(B \mid A)$ and $P(A \mid B)$ are defined if and only if $P(A) \neq 0$ and $P(B) \neq 0$, respectively.

3·12. INDEPENDENT EVENTS

Two or more events are said to be *independent* if the happening or non-happening of any one of them, does not, in any way, affect the happening of others.

Consider the experiment of throwing two dice, say die 1 and die 2. It is obvious that the occurrence of a certain number of dots on the die 1 has nothing to do with a similar event for the die 2. The two are quite independent of each other, so to say. But suppose, the two dice were connected with a piece of thread before being thrown. The situation changes. This time the two events are not independent in as much as that the uppermost face of one die will have something to do in causing a particular face of the other die to be uppermost; and the shorter the thread, the more is this influence or dependence.

Similarly, if we draw two cards from a pack of cards in succession, then the results of the two draws are independent if the cards are drawn with replacement (*i.e.*, if the first card drawn is placed back in the pack before drawing the second card) and the results of the two draws are not independent if the cards are drawn without replacement.

Definition. *An event A is said to be independent (or statistically independent) of another event B, if the conditional probability of A given B, i.e., P(A | B) is equal to the unconditional probability of B, i.e., if*

$$P(A \mid B) = P(A) \qquad ...(3·19)$$

It may be noted that the above definition is meaningful only when $P(A \mid B)$ is defined, *i.e.*, if $P(B) \neq 0$.

Similarly, an event B is said to be independent (or statistically independent) of event A, if

$$P(B \mid A) = P(B); \quad P(A) \neq 0 \qquad ...(3·20)$$

Theorem 3·9. *If the events A and B are such that $P(A) \neq 0$, $P(B) \neq 0$ and A is independent of B, then B is independent of A.*

Proof. Since the event A is independent of B, we have

$$P(A \mid B) = P(A) \Rightarrow \frac{P(A \cap B)}{P(B)} = P(A) \Rightarrow P(A \cap B) = P(A)\, P(B)$$

$$\therefore \quad \frac{P(B \cap A)}{P(A)} = P(B) \qquad [\because P(A) \neq 0 \text{ and } A \cap B = B \cap A]$$

$$P(B \mid A) = P(B) \Rightarrow B \text{ is independent of } A.$$

Remarks:

1. Thus, we see that if A is independent of B, then B is independent of A. Hence, instead of saying that 'A' is independent of B or '*B is independent of A*', we may say that A and B are independent events.
2. For any event A in S,
   (*a*) A and the null event ϕ are independent
   (*b*) A and S are independent.

Proof. (*a*) $P(A \cap \phi) = P(\phi) = 0 = P(A) \,.\, P(\phi) \Rightarrow A$ and ϕ are independent.

(*b*) $P(A \cap S) = P(A) = P(A) \,.\, 1 = P(A)\, P(S)$ $[\because\ A \subset S$ and $P(S) = 1]$

$\Rightarrow$ A and S are independent.

3·13. MULTIPLICATION THEOREM OF PROBABILITY FOR INDEPENDENT EVENTS

Theorem 3·10. *If A and B are two events with positive probabilities $P(A) \neq 0$, $P(B) \neq 0$, then A and B are independent if and only if* $P(A \cap B) = P(A) \,.\, P(B)$. ...(3.21)

Proof. We have:

$$P(A \cap B) = P(A)\, P(B \mid A) = P(B)\, P(A \mid B);\ P(A) \neq 0,\quad P(B) \neq 0 \qquad ...(*)$$

If A and B are indepedent, *i.e.*, A is independent of B and B is independent of A, then we have

$$P(A \mid B) = P(A) \quad \text{and} \quad P(B \mid A) = P(B) \qquad ...(**)$$

From (*) and (**), we get : $P(A \cap B) = P(A) \cdot P(B)$, as required.

Conversely, if (3·21) holds, then we get

$$\left.\begin{aligned} \frac{P(A \cap B)}{P(B)} &= P(A) \quad \Rightarrow \quad P(A \mid B) = P(A) \\ \text{and} \quad \frac{P(A \cap B)}{P(A)} &= P(B) \quad \Rightarrow \quad P(B \mid A) = P(B) \end{aligned}\right\} \qquad ...(***)$$

(***) implies that A and B are independent events.

Hence, for independent events A and B, probability that both of these occur simultaneously is the product of their respective probabilities.

This rule is known as the ***Multiplication Rule of Probability.***

3·14. EXTENSION OF MULTIPLICATION THEOREM OF PROBABILITY TO *n* –EVENTS

Theorem 3·11. *For n events $A_1, A_2, \ldots, A_n$, we have*

$$P(A_1 \cap A_2 \cap \ldots \cap A_n) = P(A_1)\, P(A_2 \mid A_1)\, P(A_3 \mid A_1 \cap A_2) \times \ldots \times P(A_n \mid A_1 \cap A_2 \cap \ldots \cap A_{n-1}) \qquad ...(3·22)$$

where $P(A_i \mid A_j \cap A_k \cap \ldots \cap A_l)$ represents the conditional probability of the event A_i given that the events $A_j, A_k, \ldots, A_l$ have already happened.

Proof. For two events A_1 and A_2, $\quad P(A_1 \cap A_2) = P(A_1)\, P(A_2 \mid A_1)$

We have for three events A_1, A_2, and A_3

$$\begin{aligned} P(A_1 \cap A_2 \cap A_3) &= P[A_1 \cap (A_2 \cap A_3)] \\ &= P(A_1)\, P[(A_2 \cap A_3) \mid A_1] && \text{[Using (3·17)]} \\ &= P(A_1)\, P[A_2 \mid A_1]\, P[(A_3 \mid (A_1 \cap A_2)] && \text{[Using (3·17)]} \end{aligned}$$

Thus, we find that (3·22) is true for $n = 2$ and $n = 3$. Let us suppose that (3·22) is true for $n = k$, so that

$$P(A_1 \cap A_2 \cap \dots \cap A_k) = P(A_1)\, P(A_2 \mid A_1)\, P(A_3 \mid A_1 \cap A_2) \dots P(A_k \mid A_1 \cap A_2 \cap \dots \cap A_{k-1}) \quad \dots(3·23)$$

We have:

$$P[(A_1 \cap A_2 \cap \dots \cap A_k) \cap A_{k+1}] = P(A_1 \cap A_2 \cap \dots \cap A_k) \times P(A_{k+1} \mid A_1 \cap A_2 \cap \dots \cap A_K)$$

[Using (3·17)]

$$= P(A_1)\, P(A_2 \mid A_1) \dots P(A_k \mid A_1 \cap A_2 \cap \dots \cap A_{k-1}) \times P(A_{k+1} \mid A_1 \cap A_2 \cap \dots \cap A_k)$$

[Using (3·23)]

Thus, (3·22) is true for $n = k + 1$ also. Since (3.22) is true for $n = 2$ and $n = 3$, by the principle of mathematical induction, it follows that (3·22) is true for all positive integer values of n.

3·14·1. Extension of Multiplication Theorem of Probability for *n*-Independent Events

Theorem 3·12. *Necessary and sufficient condition for independence of n events $A_1, A_2, \dots, A_n$ is that the probability of their simultaneous happening is equal to the product of their respective probabilities, i.e.,*

$$P(A_1 \cap A_2 \cap \dots \cap A_n) = P(A_1)\, P(A_2)\, P(A_3) \dots P(A_n) \quad \dots(3·24)$$

Proof. If $A_1, A_2, \dots, A_n$ are independent events then

$$P(A_2 \mid A_1) = P(A_2),\ P(A_3 \mid A_1 \cap A_2) = P(A_3),\ \dots,\ P(A_n \mid A_1 \cap A_2 \cap \dots \cap A_{n-1}) = P(A_n)$$

Hence, from (3.22), we get

$$P(A_1 \cap A_2 \cap \dots \cap A_n) = P(A_1)\, P(A_2) \dots P(A_n)$$

Conversely if (3.24) holds, then from (3.22) and (3.24), we get

$$P(A_2) = P(A_2 \mid A_1);\ P(A_3) = P(A_3 \mid A_1 \cap A_2),\ \dots,\ P(A_n) = P(A_n \mid A_1 \cap A_2 \cap \dots \cap A_{n-1})$$

$\Rightarrow$ $A_1, A_2, A_3, \dots A_n$ are independent events.

Hence the theorem.

Remark. **Mutually Exclusive (Disjoint) Events and Independent Events.** Let A and B be mutually exclusive (disjoint) events with positive probabilities $[P(A) > 0,\ P(B) > 0]$,

$$A \cap B = \phi \quad \Rightarrow \quad P(A \cap B) = P(\phi) = 0 \quad \dots(i)$$

Further, by compound probability theorem, we have

$$P(A \cap B) = P(A).\ P(B \mid A) = P(B) \cdot P(A \mid B) \quad \dots(ii)$$

Since $P(A) \neq 0$; $P(B) \neq 0$, from (*i*) and (*ii*), we get

$$P(A \mid B) = 0 \neq P(A), \qquad P(B \mid A) = 0 \neq P(B) \quad \dots(iii)$$

$\Rightarrow$ A and B are dependent events.

Hence, *two mutually disjoint events with positive probabilities are always dependent* (*not independent*) *events.* However, if A and B are independent events with $P(A) > 0$ and $P(B) > 0$, then

$$P(A \cap B) = P(A)\, P(B) \neq 0$$

$\Rightarrow$ A and B cannot be mutually exclusive.

Hence, *two independent events (both of which are possible events), cannot be mutually disjoint.*

Theorem 3·13. *For a fixed B with P(B) > 0, P(A | B) is probability function.*

Proof. (*i*) $P(A \mid B) = \dfrac{P(A \cap B)}{P(B)} \geq 0$

(*ii*) $P(S \mid B) = \dfrac{P(S \cap B)}{P(B)} = \dfrac{P(B)}{P(B)} = 1$

(*iii*) If $\{A_n\}$ is any finite or infinite sequence of disjoint events, then

$$P\left[\underset{n}{\cup} A_n \mid B\right] = \frac{P\left[\left(\underset{n}{\cup} A_n\right) \cap B\right]}{P(B)} = \frac{P\left[\left(\underset{n}{\cup} (A_n \cap B)\right)\right]}{P(B)}$$

$$= \frac{\sum_n P(A_n \cap B)}{P(B)} = \sum_n \left[\frac{P(A_n \cap B)}{P(B)}\right] = \sum_n P(A_n \mid B)$$

Hence, the theorem.

Remark. For given B satisfying $P(B) > 0$, the conditional probability $P[\cdot \mid B]$ also enjoys the same properties as the unconditional probability.

For example, in the usual notations, we have

(*i*) $P(\phi \mid B) = 0$

(*ii*) $P(\bar{A} \mid B) = 1 - P(A \mid B)$,

(*iii*) $P\left[\overset{n}{\underset{i=1}{\cup}} A_i \mid B\right] = \sum_{i=1}^{n} P(A_i \mid B)$, where $A_1, A_2, ..., A_n$ are mutually disjoint events.

(*iv*) $P(A_1 \cup A_2 \mid B) = P(A_1 \mid B) + P(A_2 \mid B) - P(A_1 \cup A_2 \mid B)$

(*v*) $P(A \cap \bar{B} \mid C) + P(A \cap B \mid C) = P(A \mid C)$

(*vi*) If $E \subset F$, then $P(E \mid B) \leq P(F \mid B)$

and so on.

The proofs of results (*iv*), (*v*) and (*vi*) are given in Theorems 3·14, 3·15 and 3·16, respectively. Others are left as exercises to the reader.

Theorem 3·14. *For any three events A, B and C,*

$$P[(A \cup B) \mid C] = P(A \mid C) + P(B \mid C) - P(A \cap B \mid C)$$

Proof. We have

$$P(A \cup B) = P(A) + P(B) - P(A \cap B)$$

$$\Rightarrow \quad P[(A \cap C) \cup (B \cap C)] = P(A \cap C) + P(B \cap C) - P(A \cap B \cap C)$$

Dividing both sides by $P(C)$, we get

$$\frac{P[(A \cap C) \cup (B \cap C)]}{P(C)} = \frac{P(A \cap C) + P(B \cap C) - P(A \cap B \cap C)}{P(C)}, \; P(C) > 0$$

$$= \frac{P(A \cap C)}{P(C)} + \frac{P(B \cap C)}{P(C)} - \frac{P(A \cap B \cap C)}{P(C)}$$

$$\Rightarrow \quad \frac{P[(A \cup B) \cap C]}{P(C)} = P(A \mid C) + P(B \mid C) - P(A \cap B \mid C)$$

$$\Rightarrow \quad P[(A \cup B) \mid C] = P(A \mid C) + P(B \mid C) - P(A \cap B \mid C)$$

Theorem 3·15. *For any three events A, B and C,*

$$P(A \cap \overline{B} \mid C) + P(A \cap B \mid C) = P(A \mid C)$$

Proof. $P(A \cap \overline{B} \mid C) + P(A \cap B \mid C)$

$$= \frac{P(A \cap \overline{B} \cap C)}{P(C)} + \frac{P(A \cap B \cap C)}{P(C)} = \frac{P(A \cap \overline{B} \cap C) + P(A \cap B \cap C)}{P(C)}$$

$$= \frac{P(A \cap C)}{P(C)} = P(A \mid C).$$

Theorem 3·16. *For any three events A, B and C defined on the sample space S such that $B \subset C$ and $P(A) > 0$,*

$$P(B \mid A) \leq P(C \mid A).$$

Proof.
$$P(C \mid A) = \frac{P(C \cap A)}{P(A)} = \frac{P[(B \cap C \cap A) \cup (\overline{B} \cap C \cap A)]}{P(A)}$$

$$= \frac{P[(B \cap C \cap A)]}{P(A)} + \frac{P(\overline{B} \cap C \cap A)}{P(A)} \qquad \text{(Using Axiom 3)}$$

$$= P(B \cap C \mid A) + P(\overline{B} \cap C \mid A)$$

$$= P(B \mid A) + P(\overline{B} \cap C \mid A) \qquad [\because B \subset C \Rightarrow B \cap C = B]$$

$$\Rightarrow \quad P(C \mid A) \geq P(B \mid A) \qquad [\because P(\overline{B} \cap C \mid A) \geq 0]$$

Theorem 3·17. *If A and B are independent events, then*

(i) A and $\overline{B}$ (ii) $\overline{A}$ and B (iii) $\overline{A}$ and $\overline{B}$, are also independent

Proof. Since A and B are independent, $P(A \cap B) = P(A)\, P(B)$ ...(*)

(a)
$$P(A \cap \overline{B}) = P(A) - P(A \cap B) = P(A) - P(A)\, P(B) \qquad \text{[From (*)]}$$

$$= P(A)[1 - P(B)] = P(A)\, P(\overline{B})$$

$\Rightarrow$ A and $\overline{B}$ are independent events

Aliter. $P(A \cap B) = P(A)\, P(B) = P(A)\, P(B \mid A) = P(B)\, P(A \mid B)$

i.e., $P(B \mid A) = P(B) \Rightarrow$ B is independent of A.

and $P(A \mid B) = P(A) \Rightarrow$ A is independent of B.

Also $P(B \mid A) + P(\overline{B} \mid A) = 1 \Rightarrow P(B) + P(\overline{B} \mid A) = 1$

$\therefore$ $P(\overline{B} \mid A) = 1 - P(B) = P(\overline{B})$

$\therefore$ $\overline{B}$ is independent of A and by symmetry we say that A is independent of $\overline{B}$.

Hence, A and $\bar{B}$ are independent events.

(*ii*)
$$P(\bar{A} \cap B) = P(B) - P(A \cap B) = P(B) - P(A)\, P(B) \qquad \text{[From (*)]}$$
$$= P(B)\,[1 - P(A)] = P(\bar{A})\, P(B)$$

$\Rightarrow$ $\bar{A}$ and B are independent events.

(*iii*)
$$P(\bar{A} \cap \bar{B}) = P(\overline{A \cup B}) = 1 - P(A \cup B) = 1 - [P(A) + P(B) - P(A \cap B)]$$
$$= 1 - P(A) - P(B) + P(A)\, P(B) \qquad \text{[From (*)]}$$
$$= [1 - P(B)] - P(A)\,[1 - P(B)]$$
$$= [1 - P(A)]\,[1 - P(B)] = P(\bar{A})\, P(\bar{B})$$

$\therefore$ $\bar{A}$ and $\bar{B}$ are independent events.

Aliter. We know that : $P(\bar{A} \mid \bar{B}) + P(A \mid \bar{B}) = 1$ ...(**)

Since A and B are independent, by Part (*i*) of the above theorem A and $\bar{B}$ are also independent.

$\therefore$ $P(A \mid \bar{B}) = P(A)$. Hence from (**), we get

$$P(\bar{A} \mid \bar{B}) + P(A) = 1 \quad \Rightarrow \quad P(\bar{A} \mid \bar{B}) = 1 - P(A) = P(\bar{A})$$

Hence, $\bar{A}$ and $\bar{B}$ are independent events.

3·15. PAIRWISE INDEPENDENT EVENTS

Consider n events $A_1, A_2, \ldots, A_n$ defined on the same sample space so that $P(A_i) > 0$; i = 1, 2, ... *n*. *These events are said to be pairwise independent if every pair of two events is independent in the sense of the definition given in* § *3.13.*

> **Definition.** *The events $A_1, A_2, \ldots, A_n$ are said to be pairwise independent if and only if :*
> $$P(A_i \cap A_j) = P(A_i)\, P(A_j),\ i \neq j = 1, 2, \ldots, n \qquad \ldots(3\cdot25)$$

In particular, three events A_1, A_2, A_3 are pairwise independent if and only if :

$$\left.\begin{aligned} P(A_1 \cap A_2) &= P(A_1)\, P(A_2) \\ P(A_1 \cap A_3) &= P(A_1)\, P(A_3) \\ P(A_2 \cap A_3) &= P(A_2)\, P(A_3) \end{aligned}\right\} \qquad \ldots(3\cdot26)$$

3·15·1. Mutually Independent Events

Let S denote the sample space for a number of events. *The events in S are said to be mutually independent if the probability of the simultaneous occurrence of (any) finite number of them is equal to the product of their separate probabilities.*

> **Definition.** *The n events $A_1, A_2, \ldots, A_n$ in a sample space S are said to be mutually independent if*
> $$P(A_{i1} \cap A_{i2} \cap \ldots \cap A_{ik}) = P(A_{i1})\, P(A_{i2}) \cdots P(A_{ik});\ k = 2, 3, \ldots, n \qquad \ldots(3\cdot27)$$

Hence, the events are mutually independent if they are independent by pairs, and by triplets, and by quadruples, and so on.

Conditions for mutual independence of n events. Mathematically, n events $A_1, A_2, \ldots, A_n$ are mutually independent if and only if the following conditions hold.

$$\left.\begin{array}{ll} (i) & P(A_i \cap A_j) = P(A_i)\, P(A_j),\ (i \neq j;\ i, j = 1, 2, \ldots, n) \\ (ii) & P(A_i \cap A_j \cap A_k) = P(A_i)\, P(A_j)\, P(A_k),\ (i \neq j \neq k;\ i, j, k = 1, 2, \ldots, n) \\ & \vdots \\ (n-1): & P(A_1 \cap A_2 \cap \ldots \cap A_n) = P(A_1)\, P(A_2) \ldots P(A_n) \end{array}\right\} \quad \ldots(3.28)$$

It is interesting to note that the above equations in (3·28) give, respectively, ${}^nC_2, {}^nC_3, \ldots, {}^nC_n$ conditions to be satisfied by $A_1, A_2, \ldots, A_n$.

Hence, the total number of conditions for mutual independence of $A_1, A_2, \ldots, A_n$ is :

$${}^nC_2 + {}^nC_3 + \ldots + {}^nC_n = {}^nC_0 + {}^nC_1 + {}^nC_2 + \ldots + {}^nC_n - ({}^nC_0 + {}^nC_1) = 2^n - 1 - n$$

In particular for three events A_1, A_2, and A_3, $(n = 3)$, we have the following, $2^3 - 1 - 3 = 4$, conditions for their mutual independence.

$$\left.\begin{array}{r} P(A_1 \cap A_2) = P(A_1)\, P(A_2) \\ P(A_2 \cap A_3) = P(A_2)\, P(A_3) \\ P(A_1 \cap A_3) = P(A_1)\, P(A_3) \\ P(A_1 \cap A_2 \cap A_3) = P(A_1)\, P(A_2\, P(A_3) \end{array}\right\} \quad \ldots(3·29)$$

Remarks:

1. It may be observed that pairwise or mutual independence of events : $A_1, A_2, \ldots, A_n$, is defined only when $P(A_i) \neq 0$, for $i = 1, 2, \ldots, n$.
2. From (3.26) and (3.29), it is obvious that mutual independence of events implies that they are pairwise independent. However, the converse is not true, *i.e.*, the events may be pairwise independent but not mutually independent. For illustrations, see Examples 3.59 and 3.60.

Theorem 3·18. *If A, B, C are mutually independent events then $A \cup B$ and C are also independent.*

Proof. We required to prove :

$$P[(A \cup B) \cap C] = P(A \cup B)\, P(C)$$

$$\begin{aligned} \text{L.H.S.} &= P[(A \cap C) \cup P(B \cap C)] && \text{[By Distributive Law]} \\ &= P(A \cap C) + P(B \cap C) - P(A \cap B \cap C) \\ &= P(A)\, P(C) + P(B)\, P(C) - P(A)\, P(B)\, P(C) \\ & && [\because A, B \text{ and } C \text{ mutually independent}] \\ &= P(C)\, [P(A) + P(B) - P(A \cap B)] = P(C)\, P(A \cup B) = \text{R.H.S.} \end{aligned}$$

Hence $(A \cup B)$ and C are independent.

Theorem 3·19. *If A, B and C are random events in a sample space and if A, B and C are pairwise independent and A is independent of $(B \cup C)$, then A, B and C are mutually independent.*

Proof. We are given

$$\left.\begin{array}{r} P(A \cap B) = P(A)\, P(B) \\ P(B \cap C) = P(B)\, P(C) \\ P(A \cap C) = P(A)\, P(C) \\ P[A \cap (B \cup C)] = P(A)\, P(B \cup C) \end{array}\right\} \quad \ldots(*)$$

But $P[A \cap (B \cup C)] = P[(A \cap B) \cup (A \cap C)]$

$= P(A \cap B) + P(A \cap C) - P[(A \cap B) \cap (A \cap C)]$ [From (*)]

$= P(A) \cdot P(B) + + P(A) \cdot P(C) - P(A \cap B \cap C)$ ..(**)

and $P(A)\, P(B \cup C) = P(A)\, [P(B) + P(C) - P(B \cap C)]$

$= P(A) \cdot P(B) + P(A) \cdot P(C) - P(A) \cdot P(B \cap C)$ ...(***)

From (**) and (***), on using (*), we get

$P(A \cap B \cap C) = P(A)\, P(B \cap C) = P(A)\, P(B)\, P(C)$ [From (*)]

Hence A, B, C are mutually independent.

3·15·2. Given *n* independent events A_i, ($i = 1, 2, \ldots, n$) with respective probabilities of occurrence p_i, to find the probability of occurrence of at least one of them

We have : $P(A_i) = p_i \Rightarrow P(\bar{A}_i) = 1 - p_i\,;\, i = 1, 2, \ldots, n$ ...(*)

Hence, the probability 'p' of happening of at least one of the events is given by:

$$\left.\begin{aligned} p &= P(A_1 \cup A_2 \cup \ldots \cup A_n) = 1 - P(\overline{A_1 \cup A_2 \cup \ldots \cup A_n}) \\ &= 1 - P(\bar{A}_1 \cap \bar{A}_2 \cap \ldots \cap \bar{A}_n) = 1 - P(\bar{A}_1)\, P(\bar{A}_2) \ldots P(\bar{A}_n) \end{aligned}\right\} \quad \ldots(**)$$

($\because A_1, A_2, \ldots, A_n$ are independent $\Rightarrow \bar{A}_1, \bar{A}_2, \ldots \bar{A}_n$ are also independent)

$$= 1 - [(1 - p_1)(1 - p_2) \ldots (1 - p_n)]$$

$$= \left[\sum_{i=1}^{n} p_i - \sum_{\substack{i,j=1 \\ i<j}}^{n} \sum (p_i\, p_j) + \sum_{\substack{i,j,k=1 \\ i<j<k}}^{n} \sum \sum (p_i\, p_j\, p_k) - \ldots + (-1)^{n-1} (p_1\, p_2 \ldots p_n)\right]$$

Remark. The results in (**) are very important and are used quite often in numerical problems. Result (**) stated in words gives:

P [happening of at least one of the events $A_1, A_2, \ldots, A_n$]

$= 1 - P$ (none of the events $A_1, A_2, \ldots, A_n$ happens) ...(3·30)

or equivalently,

P [none of the given events happens] $= 1 - P$ [at least one of them happens]. ...(3·30*a*)

Example 3·35. *If $A \cap B = \phi$, then show that $P(A) \le P(\bar{B})$.*

Solution. We have

$A = (A \cap B) \cup (A \cap \bar{B}) = \phi \cup (A \cap \bar{B}) = A \cap \bar{B}$ [$\because A \cap B = \phi$ (Given)]

$\therefore \quad A \subseteq \bar{B} \quad \Rightarrow \quad P(A) \le P(\bar{B})$, as desired.

Aliter. Since $A \cap B = \phi$, we have $A \subset \bar{B}$, which implies that $P(A) \le P(\bar{B})$.

Example 3·36. *Let A and B be two events such that $P(A) = \frac{3}{4}$ and $P(B) = \frac{5}{8}$, show that*

(a) $P(A \cup B) \ge \frac{3}{4}$, *and* *(b)* $\frac{3}{8} \le P(A \cap B) \le \frac{5}{8}$.

Solution. (*a*) We have

$$A \subset (A \cup B) \quad \Rightarrow \quad P(A) \le P(A \cup B) \quad \Rightarrow \quad \frac{3}{4} \le P(A \cup B) \quad \Rightarrow \quad P(A \cup B) \ge \frac{3}{4}$$

(*b*) $A \cap B \subseteq B \Rightarrow P(A \cap B) \le P(B) = \dfrac{5}{8}$ ...(*i*)

Also $\quad P(A \cup B) = P(A) + P(B) - P(A \cap B) \le 1 \Rightarrow \dfrac{3}{4} + \dfrac{5}{8} - 1 \le P(A \cap B)$

$\therefore \quad \dfrac{6+5-8}{8} \le P(A \cap B) \Rightarrow \dfrac{3}{8} \le P(A \cap B)$ ...(*ii*)

From (*i*) and (*ii*), we get $\dfrac{3}{8} \le P(A \cap B) \le \dfrac{5}{8}$.

Example 3·37. For any two events A and B,

$$P(A \cap B) \le P(A) \le P(A \cup B) \le P(A) + P(B)$$

Proof. We have: $A = (A \cap \bar{B}) \cup (A \cap B)$

Using axiom 3, we get:

$$P(A) = P[(A \cap \bar{B}) \cup (A \cap B)] = P(A \cap \bar{B}) + P(A \cap B)$$

$P[(A \cap \bar{B})] \ge 0 \Rightarrow P(A) \ge P(A \cap B)$ [From axiom 1] ...(*i*)

Similarly, $P(B) \ge P(A \cap B) \Rightarrow P(B) - P(A \cap B) \ge 0$ ...(*ii*)

Now $\quad P(A \cup B) = P(A) + [P(B) - P(A \cap B)]$ ...(*iii*)

$\therefore \quad P(A \cup B) \ge P(A) \Rightarrow P(A) \le P(A \cup B)$ [Using (*ii*)] ...(*iv*)

Also $\quad P(A \cup B) \le P(A) + P(B)$ [See (*) below] ...(*v*)

Hence, from (*i*), (*iv*) and (*v*), we get

$$P(A \cap B) \le P(A) \le P(A \cup B) \le P(A) + P(B)$$

Aliter. Since $A \cap B \subset A$, by Theorem 4.6 (*ii*), we get

$$P(A \cap B) \le P(A).$$

Also $\quad A \subset (A \cup B) \Rightarrow P(A) \le P(A \cup B)$

$$P(A \cup B) = P(A) + P(B) - P(A \cap B)$$

$\le P(A) + P(B) \quad [\because P(A \cap B) \ge 0]$ ... (*)

Combining the above results, we get

$$P(A \cap B) \le P(A) \le P(A \cup B) \le P(A) + P(B).$$

SUPPLEMENTARY EXAMPLES ON ADDITION AND MULTIPLICATION THEOREMS

Example 3·38. *The odds against Manager X settling the wage dispute with the workers are 8 : 6 and odds in favour of manager Y settling the same dispute are 14 : 16.*

(*i*) *What is the chance that neither settles the dispute, if they both try, independently of each other ?*

(*ii*) *What is the probability that the dispute will be settled?*

Solution.

(*i*) Let *A* be the event that the manager *X* will settle the dispute and *B* be the event that the Manager *Y* will settle the dispute. Then clearly

$$P(\bar{A}) = \frac{8}{8+6} = \frac{4}{7} \quad \Rightarrow \quad P(A) = 1 - P(\bar{A}) = \frac{6}{14} = \frac{3}{7}$$

$$P(B) = \frac{14}{14+16} = \frac{7}{15} \quad \Rightarrow \quad P(\bar{B}) = 1 - P(B) = \frac{16}{14+16} = \frac{8}{15}$$

The required probability that neither settles the dispute is given by :

$$P(\bar{A} \cap \bar{B}) = P(\bar{A}) \times P(\bar{B}) = \frac{4}{7} \times \frac{8}{15} = \frac{32}{105}$$

[Since A and B independent $\Rightarrow$ $\bar{A}$ and $\bar{B}$ are also independent]

(ii) The dispute will be settled if at least one of the managers X and Y settles the dispute. Hence, the required probability is given by :

$P(A \cup B)$ = P[At least one of X and Y settles the dispute.]

= 1 – P[None settles the dispute.] = $1 - P(\bar{A} \cap \bar{B}) = 1 - \frac{32}{105} = \frac{73}{105}$. [From Part (i)]

Example 3·39. *The odds that person X speaks the truth are 3 : 2 and the odds that person Y speaks the truth are 5 : 3. In what percentage of cases are they likely to contradict each other on an identical point.*

Solution. Let us define the events; A : X speaks the truth, B : Y speaks the truth

Then $\bar{A}$ and $\bar{B}$ represent the complementary events that X and Y tell a lie, respectively.

We are given:

$$P(A) = \frac{3}{3+2} = \frac{3}{5} \quad \Rightarrow \quad P(\bar{A}) = 1 - \frac{3}{5} = \frac{2}{5} \quad \text{and} \quad P(B) = \frac{5}{5+3} = \frac{5}{8} \quad \Rightarrow \quad P(\bar{B}) = 1 - \frac{5}{8} = \frac{3}{8}$$

The event E that X and Y contradict each other on an identical point can happen in the following mutually exclusive ways:

(i) X speaks the truth and Y tells a lie, *i.e.*, the event $A \cap \bar{B}$ happens,

(ii) X tells a lie and Y speaks the truth, *i.e.*, the event $\bar{A} \cap B$ happens.

Hence, by addition theorem of probability, the required probability is given by :

$$P(E) = P(i) + P(ii) = P(A \cap \bar{B}) + P(\bar{A} \cap B)$$

$$= P(A) \times P(\bar{B}) + P(\bar{A}) \times P(B) \qquad \text{[Since } A \text{ and } B \text{ are independent]}$$

$$= \frac{3}{5} \times \frac{3}{8} + \frac{2}{5} \times \frac{5}{8} = \frac{19}{40} = 0{\cdot}475$$

Hence, A and B are likely to contradict each other on an identical point in 47·5% of the cases.

Example 3·40. *One shot is fired from each of the three guns. E_1, E_2, E_3 denote the events that the target is hit by the first, second and third guns respectively,. If $P(E_1) = 0{\cdot}5$, $P(E_2) = 0{\cdot}6$ and $P(E_3) = 0{\cdot}8$ and E_1, E_2, E_3, are independent events, find the probability that*

(a) exactly one hit is registered and *(b) at least two hits are registered*

Solution. We are given : $P(E_1) = 0{\cdot}5$, $P(E_2) = 0{\cdot}6$ and $P(E_3) = 0{\cdot}8$

$\Rightarrow$ $P(\bar{E}_1) = 0{\cdot}5$, $P(\bar{E}_2) = 0{\cdot}4$ and $P(\bar{E}_3) = 0{\cdot}2$

(*a*) Exactly one hit can be registered in the following mutually exclusive ways :

(*i*) $E_1 \cap \bar{E}_2 \cap \bar{E}_3$ happens, (*ii*) $\bar{E}_1 \cap E_2 \cap \bar{E}_3$ happens, (*iii*) $\bar{E}_1 \cap \bar{E}_2 \cap E_3$ happens.

Hence, by addition probability theorem, the required probability '*p*' is given by :

$$p = P(E_1 \cap \bar{E}_2 \cap \bar{E}_3) + P(\bar{E}_1 \cap E_2 \cap \bar{E}_3) + P(\bar{E}_1 \cap \bar{E}_2 \cap E_3)$$

$$= P(E_1)\,P(\bar{E}_2)\,P(\bar{E}_3) + P(\bar{E}_1)\,P(E_2)\,P(\bar{E}_3) + P(\bar{E}_1)\,P(\bar{E}_2)\,P(E_3)$$

[Since E_1, E_2 and E_3 are independent.]

$$= 0{\cdot}5 \times 0{\cdot}4 \times 0{\cdot}2 + 0{\cdot}5 \times 0{\cdot}6 \times 0{\cdot}2 + 0{\cdot}5 \times 0{\cdot}4 \times 0{\cdot}8 = 0{\cdot}26$$

(*b*) At least two hits can be registered in the following mutually exclusive ways :

(*i*) $E_1 \cap E_2 \cap \bar{E}_3$ happens (*ii*) $E_1 \cap \bar{E}_2 \cap E_3$ happens

(*iii*) $\bar{E}_1 \cap E_2 \cap E_3$ happens (*iv*) $E_1 \cap E_2 \cap E_3$ happens

$$\therefore \text{ Required prob.} = P(E_1 \cap E_2 \cap \bar{E}_3) + P(E_1 \cap \bar{E}_2 \cap E_3) + P(\bar{E}_1 \cap E_2 \cap E_3) + P(E_1 \cap E_2 + E_3)$$

$$= 0{\cdot}5 \times 0{\cdot}6 \times 0{\cdot}2 + 0{\cdot}5 \times 0{\cdot}4 \times 0{\cdot}8 + 0{\cdot}5 \times 0{\cdot}6 \times 0{\cdot}8 + 0{\cdot}5 \times 0{\cdot}6 \times 0{\cdot}8$$

$$= 0{\cdot}06 + 0{\cdot}16 + 0{\cdot}24 + 0{\cdot}24 = 0{\cdot}70.$$

Example 3·41. *An urn contains 4 tickets numbered 1, 2, 3, 4 and another contains 6 tickets numbered 2, 4, 6, 7, 8, 9. If one of the two urns is chosen at random and a ticket is drawn at random from the chosen urn, find the probabilities that the ticket drawn bears the number (i) 2 or 4, (ii) 3, (iii) 1 or 9.*

Solution.

(*i*) Required event can happen in the following mutually exclusive ways :

(*I*) First urn is chosen and then a ticket is drawn.

(*II*) Second urn is chosen and then a ticket is drawn.

Since the probability of choosing any urn is $\frac{1}{2}$, required probability '*p*' is given by :

$$p = P(I) + P(II) = \frac{1}{2} \times \frac{2}{4} + \frac{1}{2} \times \frac{2}{6} = \frac{5}{12}$$

(*ii*) Required probability $= \frac{1}{2} \times \frac{1}{4} + \frac{1}{2} \times 0 = \frac{1}{8}$

($\because$ In the 2nd urn there is no ticket with number 3.)

(*iii*) Required probability $= \frac{1}{2} \times \frac{1}{4} + \frac{1}{2} \times \frac{1}{6} = \frac{5}{24}$.

Example 3·42. *A box contains 6 red, 4 white and 5 black balls. A person draws 4 balls from the box at random. Find the probability that among the balls drawn there is at least one ball of each colour.*

Solution. The required event *E* that 'in a draw of 4 balls from the box at random there is at least one ball of each colour', can materialise in the following mutually disjoint ways :

(*i*) 1 Red, 1 White, 2 Black balls; (*ii*) 2 Red, 1 White, 1 Black balls;

(*iii*) 1 Red, 2 White, 1 Black balls.

Hence, by addition theorem of probability, the required probability is given by :

$$P(E) = P(i) + P(ii) + P(iii) = \frac{{}^6C_1 \times {}^4C_1 \times {}^5C_2}{{}^{15}C_4} + \frac{{}^6C_2 \times {}^4C_1 \times {}^5C_1}{{}^{15}C_4} + \frac{{}^6C_1 \times {}^4C_2 \times {}^5C_1}{{}^{15}C_4}$$

$$= \frac{1}{{}^{15}C_4}\left[6 \times 4 \times 10 + 15 \times 4 \times 5 + 6 \times 6 \times 5\right] = \frac{4!}{15 \times 14 \times 13 \times 12}(240 + 300 + 180)$$

$$= \frac{24 \times 720}{15 \times 14 \times 13 \times 12} = 0{\cdot}5275.$$

Example 3·43. *Three groups of children contain, respectively 3 girls and 1 boy, 2 girls and 2 boys, and 1 girl and 3 boys. One child is selected at random from each group. Show that the chance that the three selected consist of 1 girl and 2 boys is 13/32.*

Solution. The required event of getting 1 girl and two boys among the three selected children can materialise in the following three mutually disjoint cases:

Group No. →	I	II	III
(*i*)	Girl	Boy	Boy
(*ii*)	Boy	Girl	Boy
(*iii*)	Boy	Boy	Girl

Hence, by addition theorem of probability, Required probability = $P(i) + P(ii) + P(iii)$ ...(*)

Since the probability of selecting a girl from the first group is 3/4, of selecting a boy from the second group is 2/4, and of selecting a boy from the third group is 3/4, and since these three events of selecting children from three groups are independent of each other, by compound probability theorem, we have

$$P(i) = \frac{3}{4} \times \frac{2}{4} \times \frac{3}{4} = \frac{9}{32}; \quad P(ii) = \frac{1}{4} \times \frac{2}{4} \times \frac{3}{4} = \frac{3}{32}; \quad \text{and} \quad P(iii) = \frac{1}{4} \times \frac{2}{4} \times \frac{1}{4} = \frac{1}{32}$$

Substituting in (*), we get : Required probability $= \frac{9}{32} + \frac{3}{32} + \frac{1}{32} = \frac{13}{32}.$

Example 3·44. *It is 8 : 5 against the wife who is 40 years old living till she is 70 and 4 : 3 against her husband now 50 living till he is 80. Find the probability that 30 years hence,*

(*i*) *Both will be alive,* (*ii*) *None will be alive,* (*iii*) *Only wife will be alive,*
(*iv*) *Only husband will be alive,* (*v*) *Only one will be alive,* (*vi*) *At least one will be alive.*

Solution. Let us define the events :

A : Wife will be alive, and B : Husband will be alive; 30 years hence.

Then, we are given :

$$P(A) = \frac{5}{8+5} = \frac{5}{13} \Rightarrow P(\bar{A}) = 1 - P(A) = \frac{8}{13}; \quad P(B) = \frac{3}{4+3} = \frac{3}{7} \Rightarrow P(\bar{B}) = 1 - P(B) = \frac{4}{7}$$

If we assume that A and B are independent so that A and $\bar{B}$, $\bar{A}$ and B, $\bar{A}$ and $\bar{B}$ are also independent, then the required probabilities are given by :

(*i*) $$P(A \cap B) = P(A)\,P(B) = \frac{5}{13} \times \frac{3}{7} = \frac{15}{91} \qquad (\because A \text{ and } B \text{ are independent.})$$

(ii) $P(\bar{A} \cap \bar{B}) = P(\bar{A})\, P(\bar{B}) = \dfrac{8}{13} \times \dfrac{4}{7} = \dfrac{32}{91}$ $(\because \bar{A}, \bar{B}$ are independent.)

(iii) $P(A \cap \bar{B}) = P(A) - P(A \cap B) = \dfrac{5}{13} - \dfrac{15}{91} = \dfrac{20}{91}$ [From Part (i)]

(iv) $P(\bar{A} \cap B) = P(B) - P(A \cap B) = \dfrac{3}{7} - \dfrac{15}{91} = \dfrac{24}{91}$ [From Part (i)]

(v) $P(A \cap \bar{B}) + P(\bar{A} \cap B) = \dfrac{20}{91} + \dfrac{24}{91} = \dfrac{44}{91}$ [From Parts (iii) and (iv)]

(vi) $P(A \cup B) = 1 - P(\bar{A} \cap \bar{B}) = 1 - \dfrac{32}{91} = \dfrac{59}{91}$ [From Part (ii)]

Example 3·45. *A problem in Statistics is given to three students A, B and C whose chances of solving it are 1/2, 3/4 and 1/4, respectively.*

What is the probability that the problem will be solved if all of them try independently ?

Solution. Let A, B, C denote the events that the problem is solved by the students A, B, C, respectively. Then

$$P(A) = \frac{1}{2}, \quad P(B) = \frac{3}{4}, \quad \text{and} \quad P(C) = \frac{1}{4}$$

The problem will be solved if at least one of them solves the problem. Thus we have to calculate the probability of occurrence of at least one of the three events A, B, C, *i.e.*, $P(A \cup B \cup C)$.

$$P(A \cup B \cup C) = P(A) + P(B) + P(C) - P(A \cap B) - P(A \cap C) - P(B \cap C) + P(A \cap B \cap C)$$

$$= P(A) + P(B) + P(C) - P(A)\,P(B) - P(A)\,P(C) - P(B)\,P(C) + P(A)\,P(B)\,P(C)$$

($\because$ A, B, C are mutually independent events.)

$$= \frac{1}{2} + \frac{3}{4} + \frac{1}{4} - \frac{1}{2}\cdot\frac{3}{4} - \frac{3}{4}\cdot\frac{1}{4} - \frac{1}{2}\cdot\frac{1}{4} + \frac{1}{2}\cdot\frac{3}{4}\cdot\frac{1}{4} = \frac{29}{32}.$$

Aliter. $P(A \cup B \cup C) = 1 - P(\overline{A \cup B \cup C}) = 1 - P(\bar{A} \cap \bar{B} \cap \bar{C}) = 1 - P(\bar{A})\,P(\bar{B})\,P(\bar{C})$

[$\because$ A, B, C are mutually independent $\Rightarrow \bar{A}, \bar{B}$, and $\bar{C}$ are mutually independent.]

$$= 1 - \left(1 - \frac{1}{2}\right)\left(1 - \frac{3}{4}\right)\left(1 - \frac{1}{4}\right) = 1 - \frac{3}{32} = \frac{29}{32}.$$

Example 3·46. *A manager has two assistants and he bases his decision on information supplied independently by each one of them. The probability that he makes a mistake in his thinking is 0·005. The probability that an assistant gives wrong information is 0·3. Assuming that the mistakes made by the manager are independent of the information given by the assistants, find the probability that he reaches a wrong decision.*

Solution. Let us define the following events :

A : The manager makes a mistake in his thinking.

B : The 1st assistant gives him wrong information.

C : The 2nd assistant gives him wrong information.

In usual notations, we are given:

$$P(A)=0{\cdot}005, \quad P(B)=0{\cdot}3=P(C) \quad \Rightarrow \quad P(\bar{A})=0{\cdot}995, \quad P(\bar{B})=P(\bar{C})=0{\cdot}7$$

Assuming that the mistakes made by the manager are independent of the information supplied independently by each of the two assistants, we conclude that A, B and C, and consequently $\bar{A}$, $\bar{B}$ and $\bar{C}$ are mutually independent.

$\therefore \quad p = P$[Manager reaches a wrong decision]

$= 1 - P$[Manager reaches a correct decision] $= 1-P(\bar{A}\cap\bar{B}\cap\bar{C})$;

because manager will reach a correct decision if he does not make a mistake in his thinking (*i.e.*, $\bar{A}$ happens) *and* both the assistants supply him correct information (*i.e.*, $\bar{B}\cap\bar{C}$ happens).

$\therefore \quad p = 1-P(\bar{A})\times P(\bar{B})\times P(\bar{C})$ [Since the events, $\bar{A}, \bar{B}$ and $\bar{C}$ are independent.]

$= 1 - 0{\cdot}995 \times 0{\cdot}7 \times 0{\cdot}7 = 1 - 0{\cdot}48755 = 0{\cdot}51245.$

Example 3·47. *The odds that a book on Statistics will be favourably reviewed by 3 independent critics are 3 to 2, 4 to 3 and 2 to 3 respectively. What is the probability that of the three reviews:*

(i) All will be favourable; *(ii) Majority of the reviews will be favourable;*

(iii) Exactly one review will be favourable; *(iv) Exactly two reviews will be favourable, and*

(v) At least one of the reviews will be favourable.

Solution. Let A, B and C denote respectively the events that the book a favourably reviewed by first, second and third critic respectively. Then we are given :

$$P(A)=\frac{3}{5},\ P(B)=\frac{4}{7} \text{ and } P(C)=\frac{2}{5} \quad \Rightarrow \quad P(\bar{A})=\frac{2}{5},\ P(\bar{B})=\frac{3}{7} \text{ and } P(\bar{C})=\frac{3}{5}$$

(*i*) The probability that all the three reviews will be favourable is :

$$P(A\cap B\cap C)=P(A)\times P(B)\times P(C)=\frac{3}{5}\times\frac{4}{7}\times\frac{2}{5}=\frac{24}{175}$$

($\because$ A, B and C are mutually independent)

(*ii*) The event that majority, *i.e.*, at least 2 reviews are favourable can materialise in the following mutually exclusive ways:

(*a*) $A\cap B\cap\bar{C}$ happens, (*b*) $A\cap\bar{B}\cap C$ happens,

(*c*) $\bar{A}\cap B\cap C$ happens, (*d*) $A\cap B\cap C$ happens.

Hence, the required probability is :

$$P(A\cap B\cap\bar{C})+P(A\cap\bar{B}\cap C)+P(\bar{A}\cap B\cap C)+P(A\cap B\cap C)$$

$$= P(A)\,P(B)\,P(\bar{C})+P(A)\,P(\bar{B})\,P(C)+P(\bar{A})\,P(B)\,P(C)+P(A)\,P(B)\,P(C)$$

$$= \frac{3}{5}\times\frac{4}{7}\times\frac{3}{5}+\frac{3}{5}\times\frac{3}{7}\times\frac{2}{5}+\frac{2}{5}\times\frac{4}{7}\times\frac{2}{5}+\frac{3}{5}\times\frac{4}{7}\times\frac{2}{5}=\frac{94}{175}$$

(*iii*) Arguing as in case (*ii*), the probability that exactly one review will be favourable is :

$$P(A\cap\bar{B}\cap\bar{C})+P(\bar{A}\cap B\cap\bar{C})+P(\bar{A}\cap\bar{B}\cap C)$$

$$= \frac{3}{5}\times\frac{3}{7}\times\frac{3}{5} + \frac{2}{5}\times\frac{4}{7}\times\frac{3}{5} + \frac{2}{5}\times\frac{3}{7}\times\frac{2}{5} = \frac{63}{175}$$

(*iv*) Similarly, the probability that exactly two reviews will be favourable is :

$$P(A\cap B\cap \bar{C}) + P(A\cap \bar{B}\cap C) + P(\bar{A}\cap B\cap C)$$

$$= P(A)\times P(B)\times P(\bar{C}) + P(A)\times P(\bar{B})\times P(C) + P(\bar{A})\times P(B)\times P(C)$$

$$= \frac{3}{5}\times\frac{4}{7}\times\frac{3}{5} + \frac{3}{5}\times\frac{3}{7}\times\frac{2}{5} + \frac{2}{5}\times\frac{4}{7}\times\frac{2}{5} = \frac{70}{175}$$

(*v*) The probability that at least one of the reviews will be favourable is :

$$P(A\cup B\cup C) = 1 - P(\bar{A}\cap\bar{B}\cap\bar{C}) = 1 - P(\bar{A})\times P(\bar{B})\times P(\bar{C}) = 1 - \frac{2}{5}\times\frac{3}{7}\times\frac{3}{5} = \frac{157}{175}$$

In (*ii*) to (*v*) we have used that A, B and C are mutually independent.

Example 3·48. *A and B alternately cut a pack of cards and the pack is shuffled after each cut. If A starts and the game is continued until one cuts a diamond, what are the respective chances of A and B first cutting a diamond ?*

Solution. Let A_i and B_i denote the events of A and B cutting a diamond, respectively, in the ith trial. Then, we are given :

$$P(A_i) = P(B_i) = \frac{13}{52} = \frac{1}{4} \quad\Rightarrow\quad P(\bar{A}_i) = P(\bar{B}_i) = \frac{3}{4};\ i = 1, 2, 3, \ldots$$

If A starts the game, he can first cut the diamond in the following mutually exclusive ways :

(*i*) A_1 happens, (*ii*) $\bar{A}_1\cap\bar{B}_2\cap A_3$ happens, (*iii*) $\bar{A}_1\cap\bar{B}_2\cap\bar{A}_3\cap\bar{B}_4\cap A_5$ happens, so on.

Hence , by addition theorem of probability, the probability 'p' that A first cuts a diamond is given by :

$$p = P(i) + P(ii) + P(iii) + \ldots\ldots$$

$$= P(A_1) + P(\bar{A}_1\cap\bar{B}_2\cap A_3) + P(\bar{A}_1\cap\bar{B}_2\cap\bar{A}_3\cap\bar{B}_4\cap A_5) + \cdots$$

$$= P(A_1) + P(\bar{A}_1)\,P(\bar{B}_2)\,P(A_3) + P(\bar{A}_1)\,P(\bar{B}_2)\,P(\bar{A}_3)\,P(\bar{B}_4)\,P(A_5) + \ldots$$

(By multiplication Law of Probability for independent events.)

$$= \frac{1}{4} + \frac{3}{4}\times\frac{3}{4}\times\frac{1}{4} + \frac{3}{4}\times\frac{3}{4}\times\frac{3}{4}\times\frac{3}{4}\times\frac{1}{4} + \ldots = \left[\frac{(1/4)}{1-(9/16)}\right] = \frac{4}{7}$$

$\therefore$ The probability that B first cuts a diamond $= 1 - p = 1 - \frac{4}{7} = \frac{3}{7}$.

Example 3·49. (*Huyghen's Problem*) *A and B throw alternatively with a pair of balanced dice. A wins if he throws a sum of six points before B throws a sum of seven points, while B wins if he throws a sum of seven points before A throws a sum of six points. If A begins the game, show that his probability of winning is 30/61.*

Solution. Let A_i denote the event of A's throwing '6' in the ith throw, $i = 1, 2, \ldots$, and B_i denote the events of B's throwing '7' in the ith throw, $i = 1, 2, \ldots$; with a pair of dice. Then $\bar{A}_i$ and $\bar{B}_i$ are the complementary events.

'6' can be obtained with two dice in the following ways:

(1, 5), (5, 1), (2, 4) , (4, 2), (3, 3), i.e., in 5 distinct ways.

$$\therefore \quad P(A_i) = \frac{5}{36} \quad \Rightarrow \quad P(\bar{A}_i) = 1 - \frac{5}{36} = \frac{31}{36};\ i = 1, 2, 3, \ldots \qquad \ldots(*)$$

'7' can be obtained with two dice as follows:

(1, 6), (6, 1), (2, 5), (5, 2), (3, 4), (4, 3),i.e., in 6 distinct ways.

$$\therefore \quad P(B_i) = \frac{6}{36} = \frac{1}{36} \quad \Rightarrow \quad P(\bar{B}_i) = 1 - \frac{1}{6} = \frac{5}{6};\ i = 1, 2, 3, \ldots \qquad \ldots(**)$$

If A starts the game, he will win in the following mutually exclusive ways:

(*i*) A_1 happens, (*ii*) $\bar{A}_1 \cap \bar{B}_2 \cap A_3$ happens

(*iii*) $\bar{A}_1 \cap \bar{B}_2 \cap \bar{A}_3 \cap \bar{B}_4 \cap A_5$ happens, and so on.

Hence, by addition theorem of probability, the required probability of A's wining , (say) $P(A)$ is given by :

$$P(A) = P(i) + P(ii) + P(iii) + \ldots$$

$$= P(A_1) + P(\bar{A}_1 \cap \bar{B}_2 \cap A_3) + P(\bar{A}_1 \cap \bar{B}_2 \cap \bar{A}_3 \cap \bar{B}_4 \cap A_5) + \ldots$$

$$= P(A_1) + P(\bar{A}_1)\, P(\bar{B}_2)\, P(A_3) + P(\bar{A}_1)\, P(\bar{B}_2)\, P(\bar{A}_3)\, P(\bar{B}_4)\, P(A_5) + \ldots$$

[$\because$ The events $A_i, B_i;\ i = 1, 2, 3, \ldots$ are mutually independent.]

$$= \frac{5}{36} + \frac{31}{36} \times \frac{5}{6} \times \frac{5}{36} + \frac{31}{36} \times \frac{5}{6} \times \frac{31}{36} \times \frac{5}{6} \times \frac{5}{36} + \ldots = \frac{\frac{5}{36}}{1 - \frac{31}{36} \times \frac{5}{6}} = \frac{30}{61}.$$

[Using (*) and (**)]

Example 3·50. *A, B and C play a game and the chance of their wining it in an attempt are* $\frac{2}{3}, \frac{1}{2}$ *and* $\frac{1}{4}$ *respectively. A has the first chance, followed by B and then by C. This cycle is repeated till one of them wins the game. Find their respective chances of winning the game.*

Solution. Let A_i, B_i and C_i denote the events that A, B and C, respectively, win the game in their ith ($i = 1, 2, 3, \ldots$) attempt. Then, we have:

$$P(A_i) = \frac{2}{3},\quad P(B_i) = \frac{1}{2},\quad P(C_i) = \frac{1}{4} \quad \Rightarrow \quad P(\bar{A}_i) = \frac{1}{3},\quad P(\bar{B}_i) = \frac{1}{2},\quad (\bar{C}_i) = \frac{3}{4}$$

Since A starts the game, he can win in the following mutually exclusive ways :

(*i*) He wins in the first attempt, *i.e.*, the event A_1 happens.

(*ii*) A does not win in the first attempt, B does not win in the first attempt, C does not win in first attempt and then A wins in the second attempt, *i.e.*, the event $\bar{A}_1 \cap \bar{B}_1 \cap \bar{C}_1 \cap A_2$, happens.

(*iii*) Arguing as in (*ii*), the event : $\bar{A}_1 \cap \bar{B}_1 \cap \bar{C}_1 \cap \bar{A}_2 \cap \bar{B}_2 \cap \bar{C}_2 \cap A_3$ happens, and so on.

The probability of *A*'s winning (by addition theorem) is given by :

$$P(A) = P(i) + P(ii) + P(iii) + \ldots$$

$$= P(A_1) + P(\bar{A}_1 \cap \bar{B}_1 \cap \bar{C}_1 \cap A_2) + P(\bar{A}_1 \cap \bar{B}_1 \cap \bar{C}_1 \cap \bar{A}_2 \cap \bar{B}_2 \cap \bar{C}_2 \cap A_3) + \ldots$$

$$= P(A_1) + P(\bar{A}_1)\, P(\bar{B}_1)\, P(\bar{C}_1)\, P(A_2) + P(\bar{A}_1)\, P(\bar{B}_1)\, P(\bar{C}_1)\, P(\bar{A}_2)\, P(\bar{B}_2)\, P(\bar{C}_2)\, P(A_3) + \ldots$$

[$\because$ The events A_i, B_i, C_i; $i = 1, 2, 3, \ldots$ are mutually independent.]

$$= \frac{2}{3} + \frac{1}{3} \times \frac{1}{2} \times \frac{3}{4} \times \frac{2}{3} + \frac{1}{3} \times \frac{1}{2} \times \frac{3}{4} \times \frac{1}{3} \times \frac{1}{2} \times \frac{3}{4} \times \frac{2}{3} + \ldots = \frac{2}{3} + \frac{1}{12} + \frac{1}{96} + \ldots$$

$$= \frac{(2/3)}{[1-(1/8)]} = \frac{2}{3} \times \frac{8}{7} = \frac{16}{21}.$$

If *A* starts the game, then *B* can win in the following mutually exclusive ways:

(*i*) $\bar{A}_1 \cap B_1$ happens; (*ii*) $\bar{A}_1 \cap \bar{B}_1 \cap \bar{C}_1 \cap \bar{A}_2 \cap B_2$ happens

(*iii*) $\bar{A}_1 \cap \bar{B}_1 \cap \bar{C}_1 \cap \bar{A}_2 \cap \bar{B}_2 \cap \bar{C}_2 \cap \bar{A}_3 \cap B_3$ happens, and so on.

The chance of B's winning is:

$$P(B) = P(\bar{A}_1 \cap B_1) + P(\bar{A}_1 \cap \bar{B}_1 \cap \bar{C}_1 \cap \bar{A}_2 \cap B_2)$$

$$+ P(\bar{A}_1 \cap \bar{B}_1 \cap \bar{C}_1 \cap \bar{A}_2 \cap \bar{B}_2 \cap \bar{C}_2 \cap \bar{A}_3 \cap B_3) + \cdots$$

$$= P(\bar{A}_1) \times P(B_1) + P(\bar{A}_1) \times P(\bar{B}_1) \times P(\bar{C}_1) \times P(\bar{A}_2) \times P(B_2)$$

$$+ P(\bar{A}_1) \times P(\bar{B}_1) \times P(\bar{C}_1) \times P(\bar{A}_2) \times P(\bar{B}_2) \times P(\bar{C}_2) \times P(\bar{A}_3) \times P(B_3) + \ldots$$

[$\because$ The events A_i, B_i, C_i; i = 1, 2, 3, ... are mutually independent.]

$$= \frac{1}{3} \times \frac{1}{2} + \frac{1}{3} \times \frac{1}{2} \times \frac{3}{4} \times \frac{1}{3} \times \frac{1}{2} + \frac{1}{3} \times \frac{1}{2} \times \frac{3}{4} \times \frac{1}{3} \times \frac{1}{2} \times \frac{3}{4} \times \frac{1}{3} \times \frac{1}{2} + \ldots = \frac{1/6}{[1-(1/8)]} = \frac{4}{21}$$

Hence, the chance of C's winning is : $P(C) = 1 - P(A) - P(B) = 1 - \frac{16}{21} - \frac{4}{21} = \frac{1}{21}.$

Example 3·51. *Data on the readership of a certain magazine show that the proportion of 'male readers under 35 is 0·40 and over 35 is 0·20. If the proportion of readers under 35 is 0·70, find the proportion of subscribers that are 'females over 35 years'. Also calculate the probability that a randomly selected male subscriber is under 35 years of age.*

Solution. Let us define the following events :

A : Reader of the magazine is a male; *B* : Reader of the magazine is over 35 years of age.

Then in usual notations, we are given:

$$P(A \cap B) = 0{\cdot}20,\ P(A \cap \bar{B}) = 0{\cdot}40 \quad \text{and} \quad P(\bar{B}) = 0{\cdot}70 \quad \Rightarrow \quad P(B) = 0{\cdot}30$$

(*i*) The proportion of subscribers that are 'females over 35 years' is :

$$P(\bar{A} \cap B) = P(B) - P(A \cap B) = 0{\cdot}30 - 0{\cdot}20 = 0{\cdot}10$$

(*ii*) The probability that a randomly selected male subscriber is under 35 years is :

$$P(\bar{B} \mid A) = \frac{P(A \cap \bar{B})}{P(A)} = \frac{0.40}{0.60} = \frac{2}{3}.$$

$$[\because P(A) = P(A \cap B) + P(A \cap \bar{B}) = 0{\cdot}20 + 0{\cdot}40 = 0{\cdot}60]$$

Example 3·52. *From a city population, the probability of selecting (i) a male or a smoker is 7/10, (ii) a male smoker is 2/5, and (iii) a male, if a smoker is already selected is 2/3. Find the probability of selecting (a) a non-smoker, (b) a male, and (c) a smoker, if a male is first selected.*

Solution. Define the events : A : a male is selected, B : a smoker is selected

We are given : $P(A \cup B) = \dfrac{7}{10}$, $P(A \cap B) = \dfrac{2}{5}$, $P(A \mid B) = \dfrac{2}{3}$

(*a*) The probability of selecting a non-smoker is :

$$P(\bar{B}) = 1 - P(B) = 1 - \frac{P(A \cap B)}{P(A \mid B)} \qquad \left[\because P(A \mid B) = \frac{P(A \cap B)}{P(B)}\right]$$

$$\therefore \quad P(\bar{B}) = 1 - \frac{2/5}{2/3} = 1 - \frac{3}{5} = \frac{2}{5} \Rightarrow P(B) = 1 - \frac{2}{5} = \frac{3}{5}$$

(*b*) The probability of selecting a male (by Addition theorem) is :

$$P(A) = P(A \cup B) + P(A \cap B) - P(B) = \frac{7}{10} + \frac{2}{5} - \frac{3}{5} = \frac{1}{2} \qquad \text{[Part } (a)]$$

(*c*) The probability of selecting a smoker if a male is first selected is :

$$P(B \mid A) = \frac{P(A \cap B)}{P(A)} = \frac{2/5}{1/2} = \frac{4}{5}. \qquad \text{[Part } (b)]$$

Example 3·53. *Two computers A and B are to be marketed. A salesman who is assigned the job of finding customers for them has 60% and 40% chances respectively of succeeding in case of computer A and B. The two computers can be sold independently. Given that he was able to sell at least one computer, what is the probability that computer A has been sold ?*

Solution. Let E denote the event that computer A is marketed and F denote the event that computer B is marketed. We are given :

$$P(E) = \frac{60}{100} = 0{\cdot}60 \Rightarrow P(\bar{E}) = 0{\cdot}40 \quad \text{and} \quad P(F) = \frac{40}{100} = 0{\cdot}40 \Rightarrow P(\bar{F}) = 0{\cdot}60$$

$$\text{Required probability} = P[E \mid (E \cup F)] = \frac{P[E \cap (E \cup F)]}{P(E \cup F)}$$

$$= \frac{P(E)}{1 - P(\bar{E} \cap \bar{F})} = \frac{P(E)}{1 - P(\bar{E})\,P(\bar{F})} = \frac{0{\cdot}6}{1 - 0{\cdot}4 \times 0{\cdot}6} = \frac{0{\cdot}60}{0{\cdot}76} = 0{\cdot}79.$$

Example 3·54. *A certain drug manufactured by a company is tested chemically for its toxic nature. Let the event 'the drug is toxic' be denoted by E and the event 'the chemical test reveals that the drug is toxic' be denoted by F. Let $P(E) = \theta$, $P(F \mid E) = P(\bar{F} \mid \bar{E}) = 1 - \theta$. Then show that probability that the drug is not toxic given that the chemical test reveals that it is toxic is free from θ.*

Solution. We are given $P(E) = \theta \Rightarrow P(\bar{E}) = 1 - \theta$ and $P(F \mid E) = P(\bar{F} \mid \bar{E}) = 1 - \theta$. ...(*)

We want $P(\bar{E} \mid F)$ and we have to show that it is independent of θ.

$$P(\bar{E} \mid F) = 1 - P(E \mid F) = 1 - \frac{P(E \cap F)}{P(F)} = 1 - \frac{P(E) \cdot P(F \mid E)}{P(F)} = 1 - \frac{\theta(1-\theta)}{P(F)} \quad \text{[From (*)]} \quad ...(**)$$

The event F can materialise in the following mutually disjoint and exhaustive cases : $E \cap F$ and $\bar{E} \cap F$. Hence, by addition theorem of probability, we have

$$\begin{aligned} P(F) &= P(E \cap F) + P(\bar{E} \cap F) = P(E)\,P(F \mid E) + P(\bar{E}) \cdot P(F \mid \bar{E}) \\ &= \theta\,(1-\theta) + (1-\theta)\,[1 - P(\bar{F} \mid \bar{E})] \quad \text{[Using (*)]} \\ &= \theta(1-\theta) + (1-\theta)\,[1 - (1-\theta)] \quad \text{[From (*)]} \\ &= \theta(1-\theta) + (1-\theta)\,\theta = 2\theta\,(1-\theta). \end{aligned}$$

Substituting in (**), we get : $$P(\bar{E} \mid F) = 1 - \frac{\theta\,(1-\theta)}{2\theta\,(1-\theta)} = 1 - \frac{1}{2} = \frac{1}{2}$$

which is independent of θ as desired.

Example 3·55. *A bag contains 17 counters marked with the numbers 1 to 17. A counter is drawn and replaced; a second drawing is then made. What is the probability that:*

(i) the first number drawn is even and the second odd?

(ii) the first number is odd and the second even ?

How will your results in (i) and (ii) be effected if the first counter drawn is not replaced?

Solution. (*i*) Let A denote the event of getting even numbered counter on the first draw and B denote the event of getting odd numbered counter on the second draw. Since the counter drawn is replaced, events A and B are independent.

From 1 to 17, the even numbers are 2, 4, 6, 8, 10, 12, 14, and 16, *i.e.*, 8 and odd numbers are 9.

$$\therefore \quad P(A) = \frac{8}{17} \quad \text{and} \quad P(B) = \frac{9}{17}$$

Using multiplication theorem of probability, the probability of getting even number on the first draw and odd number on the second draw is given by :

$$P(A \cap B) = P(A) \times P(B) = \frac{8}{17} \times \frac{9}{17} = \frac{72}{289}$$

However, if the first counter drawn is not replaced before the second counter is drawn, the events A and B are not independent. In this case,

$$P(A \cap B) = P(A) \times P(B \mid A) = \frac{8}{17} \times \frac{9}{16} = \frac{9}{34}$$

(*ii*) The probabilities of the first counter drawn being odd and the second counter drawn being even are:

$$\frac{9}{17} \times \frac{8}{17} = \frac{72}{289}, \quad \text{if replacement is made, and}$$

and $$\frac{9}{17} \times \frac{8}{16} = \frac{9}{34}, \quad \text{if the replacement is not made}$$

Example 3·56. *A bag contains 10 gold and 8 silver coins. Two successive drawings of 4 coins are made such that : (i) coins are replaced before the second trial, (ii) the coins are not replaced before the second trial. Find the probability that the first drawing will give 4 gold and the second 4 silver coins.*

Solution. Let A denote the event of drawing 4 gold coins in the first draw and B denote the event of drawing 4 silver coins in the second draw. Then we have to find the probability of $P(A \cap B)$.

(*i*) *Draws with replacement.* If the coins drawn in the first draw are replaced back in the bag before the second draw then the events A and B are independent and the required probability is given by the expression :

$$P(A \cap B) = P(A) \cdot P(B) \qquad ...(*)$$

1st draw. Four coins can be drawn out of 10 + 8 = 18 coins in ${}^{18}C_4$ ways, which gives the exhaustive number of cases. In order that all these coins are of gold, they must be drawn out of the 10 gold coins and this can be done in ${}^{10}C_4$ ways. Hence,

$$P(A) = \frac{{}^{10}C_4}{{}^{18}C_4}$$

2nd draw. When the coins drawn in the first draw are replaced before the 2nd draw, the bag contains 18 coins. The probability of drawing 4 silver coins in the 2nd draw is given by :

$$P(B) = \frac{{}^{8}C_4}{{}^{18}C_4}.$$

Substituting in (*), $$P(A \cap B) = \frac{{}^{10}C_4}{{}^{18}C_4} \times \frac{{}^{8}C_4}{{}^{18}C_4}$$

(*ii*) *Draws without replacement.* If the coins drawn are not replaced back before the second draw, then the events A and B are not independent and the required probability is given by :

$$P(A \cap B) = P(A) \cdot P(B \mid A) \qquad ...(**)$$

As discussed in part (*i*), $$P(A) = \frac{{}^{10}C_4}{{}^{18}C_4}$$

If the 4 gold coins which were drawn in the first draw are not replaced back, there are 18 – 4 = 14 coins left in the bag and $P(B \mid A)$ is the probability of drawing 4 silver coins from the bag containing 14 coins out of which 6 are gold coins and 8 are silver coins.

Hence, $P(B \mid A) = \dfrac{{}^{8}C_4}{{}^{14}C_4}$. Substituting in (**), we get $P(A \cap B) = \dfrac{{}^{10}C_4}{{}^{18}C_4} \times \dfrac{{}^{8}C_4}{{}^{14}C_4}$.

Example 3·57. *A consignment of 15 record players contains 4 defectives. The record players are selected at random, one by one, and examined. Those examined are not put back. What is the probability that the 9th one examined is the last defective?*

Solution. Let A be the event of getting exactly 3 defectives in examination of 8 record players and let B denote the event that the 9th piece examined is a defective one.

Since it is a problem of sampling without replacement and since there are 4 defectives (and 11 non-defectives) out of 15-record players, $P(A) = {}^{4}C_3 \times {}^{11}C_5 / {}^{15}C_8$.

$P(B \mid A) =$ [Probability that the 9th examined record player is defective given that there were 3 defectives in the first pieces examined] $= \dfrac{4-3}{15-8} = \dfrac{1}{7}$,

since there is only one defective piece left among the remaining 15 – 8 = 7, record players.

Hence, Required probability $= P(A \cap B) = P(A)\,P(B \mid A) = \dfrac{{}^4C_3 \times {}^{11}C_5}{{}^{15}C_8} \times \dfrac{1}{7} = \dfrac{8}{195}$.

Example 3·58. *Two dice, one green and the other red, are thrown. Let A be the event that the sum of the points on the faces is odd, and B be the event of at least one ace (number '1').*

(a) Describe the (i) complete sample space, (ii) events $A, B, \bar{B}, A \cap B, A \cup B$, and $A \cap \bar{B}$ and find their probabilities assuming that all the 36 sample points have equal probabilities.

(b) Find the probabilities of the events :

(i) $(\bar{A} \cup \bar{B})$ *(ii)* $(\bar{A} \cap \bar{B})$ *(iii)* $(A \cap \bar{B})$ *(iv)* $(\bar{A} \cap B)$

(v) $\overline{(A \cap B)}$ *(vi)* $(\bar{A} \cup B)$ *(vii)* $\overline{(A \cup B)}$ *(viii)* $\bar{A} \cap (A \cup B)$

(ix) $A \cup (\bar{A} \cap B)$ *(x)* $(A \mid B)$ *and* $(B \mid A)$, *and (xi)* $(\bar{A} \mid \bar{B})$ *and* $(\bar{B} \mid \bar{A})$.

Solution. (*a*) (*i*) The sample space S, consisting of the 36 elementary events is given by :

$$S = \{1, 2, 3, 4, 5, 6\} \times \{1, 2, 3, 4, 5, 6\}$$

$$= \begin{Bmatrix} (1,1); & (1,2); & (1,3); & (1,4); & (1,5); & (1,6), \\ (2,1); & (2,2); & (2,3); & (2,4); & (2,5); & (2,6), \\ (3,1); & (3,2); & (3,3); & (3,4); & (3,5); & (3,6); \\ (4,1); & (4,2); & (4,3); & (4,4); & (4,5); & (4,6); \\ (5,1); & (5,2); & (5,3); & (5,4); & (5,5); & (5,6); \\ (6,1); & (6,2); & (6,3); & (6,4); & (6,5); & (6,6); \end{Bmatrix}$$

where, for example, the ordered pair (4, 5) refers to the elementary event that the green die shows 4 and the red die shows 5.

$A =$ The event that the sum of the numbers shown by the two dice is odd.

$$= \{1, 3, 5\} \times \{2, 4, 6\} \cup \{2, 4, 6\} \times \{1, 3, 5\}$$

$$= \begin{Bmatrix} (1,2), & (2,1), & (1,4), & (2,3), & (3,2), & (4,1), & (1,6), & (2,5), & (3,4) \\ (4,3), & (5,2), & (6,1), & (3,6), & (4,5), & (5,4), & (6,3), & (5,6), & (6,5) \end{Bmatrix}$$

$$\therefore \quad P(A) = \frac{n(A)}{n(S)} = \frac{18}{36}$$

$B =$ The event that at least one face is 1,

$= \{(1, 1); (1, 2); (1, 3); (1, 4); (1, 5); (1, 6); (2, 1); (3, 1), (4, 1); (5, 1); (6, 1)\}$

$$\therefore \quad P(B) = \frac{n(B)}{n(S)} = \frac{11}{36}$$

$\bar{B}$ = The event that none of the faces obtained in an ace.
$= \{2, 3, 4, 5, 6\} \times \{2, 3, 4, 5, 6\}$
$= \{(2, 2); (2, 3); (2, 4); (2, 5); (2, 6), (3, 2), (3, 3); (3, 4);$
$(3, 5); (3, 6); (4, 2); (4, 3); (4, 4); (4, 5); (4, 6); (5, 2);$
$(5, 3); (5, 4); (5, 5); (5, 6); (6, 2); (6, 3); (6, 4); (6, 5); (6, 6)\}$

$$\therefore \quad P(\bar{B}) = \frac{n(\bar{B})}{n(S)} = \frac{25}{36} \quad \text{or} \quad P(\bar{B}) = 1 - P(B) = \frac{25}{36}$$

$A \cap B$ = The event that sum is odd and at least one face is an ace.
$= \{(1, 2); (2, 1); (1, 4); (4, 1); (1, 6); (6, 1)\}$

$$\therefore \quad P(A \cap B) = \frac{n(A \cap B)}{n(S)} = \frac{6}{36} = \frac{1}{6}$$

$A \cup B = \{(1, 2); (2, 1); (1, 4); (2, 3); (3, 2); (4, 1); (1, 6); (2, 5);$
$(3, 4); (4, 3); (5, 2); (6, 1); (3, 6); (4, 5); (5, 4); (6, 3);$
$(5, 6); (6, 5); (1, 1); (1, 3); (1, 5); (3, 1); (5, 1)\}$

$$\therefore \quad P(A \cup B) = \frac{n(A \cup B)}{n(S)} = \frac{23}{36}$$

$A \cap \bar{B} = \{(2, 3); (2, 5); (3, 2); (3, 4); (3, 6); (4, 3); (4, 5); (5, 2); (5, 4); (5, 6); (6, 3); (6, 5)\}$

$$P(A \cap \bar{B}) = \frac{n(A \cap \bar{B})}{n(S)} = \frac{12}{36} = \frac{1}{3}$$

(b) (i) $P(\bar{A} \cup \bar{B}) = P(\overline{A \cap B}) = 1 - P(A \cap B) = 1 - \frac{1}{6} = \frac{5}{6}$

(ii) $P(\bar{A} \cap \bar{B}) = P(\overline{A \cup B}) = 1 - P(A \cup B) = 1 - \frac{23}{36} = \frac{13}{36}$

(iii) $A(A \cap \bar{B}) = P(A) - P(A \cap B) = \frac{18}{36} - \frac{6}{36} = \frac{12}{36} = \frac{1}{3}$

(iv) $P(\bar{A} \cap B) = P(B) - P(A \cap B) = \frac{11}{36} - \frac{6}{36} = \frac{5}{36}$

(v) $P(\overline{A \cap B}) = 1 - P(A \cap B) = 1 - \frac{1}{6} = \frac{5}{6}$

(vi) $P(\bar{A} \cup B) = P(\bar{A}) + P(B) - P(\bar{A} \cap B) = \left(1 - \frac{18}{36}\right) + \frac{11}{36} - \frac{5}{36} = \frac{2}{3}$

(vii) $P(\overline{A \cup B}) = 1 - P(A \cup B) = 1 - \frac{23}{36} = \frac{13}{36}$

(viii) $P[\bar{A} \cap (A \cup B)] = P[(A \cap \bar{A}) \cup (\bar{A} \cap B)] = P(\bar{A} \cap B) = \frac{5}{36}$ $\quad [\because A \cap \bar{A} = \phi]$

(ix) $P[A \cup (\bar{A} \cap B)] = P(A) + P(\bar{A} \cap B) - P(A \cap \bar{A} \cap B) = \frac{18}{36} + \frac{5}{36} = \frac{23}{36}$

(x) $P(A \mid B) = \frac{P(A \cap B)}{P(B)} = \frac{6/36}{11/36} = \frac{6}{11}$; $P(B \mid A) = \frac{P(A \cap B)}{P(A)} = \frac{6/36}{18/36} = \frac{6}{18} = \frac{1}{3}$

(xi) $P(\bar{A} \mid \bar{B}) = \frac{P(\bar{A} \cap \bar{B})}{P(\bar{B})} = \frac{13/36}{25/36} = \frac{13}{25}$; $P(\bar{B} \mid \bar{A}) = \frac{P(\bar{A} \cap \bar{B})}{P(\bar{A})} = \frac{13/36}{18/36} = \frac{13}{18}$.

SOME MISCELLANEOUS EXAMPLES

Example 3·59. *An urn contains four tickets marked with numbers 112, 121, 211, 222 and one ticket is drawn at random. Let A_i, (i = 1, 2, 3) be the event that ith digit of the number of the ticket drawn is 1. Discuss the independence of the events A_1, A_2, and A_3.*

Solution. A_1 is the event that the first digit of the number of the ticket drawn is 1 and the favourable cases for this are 112 and 121, *i.e.*, two cases.

$\therefore \quad P(A_1) = \frac{2}{4} = \frac{1}{2}$. Similarly, we get $\quad P(A_2) = P(A_3) = \frac{2}{4} = \frac{1}{2}$

$A_1 \cap A_2$ is the event that the first two digits in the number which the selected ticket bears are each equal to unity and the only favourable case is ticket with number 112.

$$\therefore \quad P(A_1 \cap A_2) = \frac{1}{4} = \frac{1}{2} \cdot \frac{1}{2} = P(A_1)\,P(A_2)$$

Similarly, $\quad P(A_2 \cap A_3) = \frac{1}{4} = P(A_2)\,P(A_3)$ and $P(A_3 \cap A_1) = \frac{1}{4} = P(A_3)\,P(A_1)$

Thus, we conclude that the events A_1, A_2 and A_3 are pairwise independent. Now

$$P(A_1 \cap A_2 \cap A_3) = P[\text{all the three digits in the number selected are 1's}] = P(\phi) = 0$$
$$\neq P(A_1)\,P(A_2)\,P(A_3)$$

Hence, A_1, A_2 and A_3, though pairwise independent, are not mutually independent.

Example 3·60. *Two fair dice are thrown independently. Three events A, B and C are defined as follows:*

A : Odd face with first dice; B : Odd face with second dice ; C : Sum of points on two dice is odd.

Are the events A, B and C (i) Pairwise independent, (ii) Mutually independent ?

Solution. In a random toss of two dice

$S = \{1, 2, 3, 4, 5, 6\} \times \{1, 2, 3, 4, 5, 6\} \Rightarrow n(S) = 6 \times 6 = 36$

Event	*Favourable cases*	*No. of favourable cases*
A	{1, 3, 5} × {1, 2, 3, 4, 5, 6}	3 × 6 = 18
B	{1, 2, 3, 4, 5, 6} × {1, 3, 5}	6 × 3 = 18
*C	{1, 3, 5} × {2, 4, 6} ∪ {2, 4, 6} × (1, 3, 5)	3 × 3 + 3 × 3 = 18
$A \cap B$	{1, 3, 5} × {1, 3, 5}	3 × 3 = 9
**$A \cap C$	{1, 3, 5} × {2, 4, 6}	3 × 3 = 9
**$B \cap C$	{2, 4, 6} × {1, 3, 5}	3 × 3 = 9
$A \cap B \cap C$	ϕ	0

* The sum of points on two dice is odd if one shows odd number and the other shows even number.

** If one die shows odd number and the sum is also odd, then the other die must show even number.

$$\therefore \quad \left.\begin{aligned} P(A) &= \frac{18}{36} = \frac{1}{2} = P(B) = P(C) \\ P(A \cap B) &= \frac{9}{36} = \frac{1}{4} = P(A) \times P(B) \\ P(A \cap C) &= \frac{9}{36} = \frac{1}{4} = P(A) \times P(C) \\ P(B \cap C) &= \frac{9}{36} = \frac{1}{4} = P(B) \times P(C) \end{aligned}\right\} \quad ...(*)$$

and $$P(A \cap B \cap C) = P(\phi) = 0 \neq P(A)\, P(B)\, P(C) \quad ...(**)$$

Hence (*) implies that the events A, B and C are pairwise independent but (**) implies that they not mutually independent.

Example 3·61. *(a).* **(Chebychev's problem).** *What is the chance that two numbers chosen at random, will be prime to each other ?*

(b) Four positive integers are chosen at random from the set of positive integers. Find the probability that they have a common factor.

Solution. *(a)* If any number 'a' is divided by a prime number 'r', then the possible remainders are 0, 1, 2, ..., $(r - 1)$. Out of these r cases, only one *viz.*, '0' is favourable to the event that 'a' is divisible by 'r'.

$$\therefore \quad p_1 = P[\text{'}a\text{' is divisible by '}r\text{'}] = \frac{1}{r} \quad ...(i)$$

Similarly, the probability, (p_2) that any other number, choosen at random say, 'b' is divisible by 'r' is given by :

$$p_2 = (1/r).$$

Since, the numbers 'a' and 'b' are chosen at random, the probability 'p' that both of them are divisible by r is given by using the compound probability theorem as follows:

$$p = p_1 \times p_2 = \frac{1}{r^2}$$

$$\therefore \quad P[\text{None of } a \text{ and } b \text{ is divisible by } r] = 1 - \frac{1}{r^2}; \quad r = 2, 3, 5, 7, ...$$

Hence, the required probability P that the two numbers chosen at random are prime to each other is given by :

$$P = \prod_r \left(1 - \frac{1}{r^2}\right), \quad \text{where } r \text{ is a prime number.}$$

$$= \frac{6}{\pi^2} \qquad \text{[From Trigonometry]}$$

(b) Arguing as in Part *(a)*, the required probability P is given by :

$$P = 1 - Pr\,[\text{None of the four chosen numbers have a common factor}]$$
$$= 1 - Pr\,[\text{The four chosen numbers are prime to each other}]$$

$$= 1 - \prod_r \left(1 - \frac{1}{r^4}\right); \qquad [r = \text{Prime number} = 2, 3, 5, 7, \ldots]$$

$$= 1 - \frac{90}{\pi^4} \qquad [\text{From Trigonometry}]$$

Example 3·62. (de Mere's Problem) *Why does it pay to bet consistently on seeing 6 at least once in 4 throws of a die, but not on seeing a double six at least once in 24 throws with two dice ?*

Solution. The probability of getting a '6' in throw of die $= \frac{1}{6}$

$\therefore$ The probability of not getting a '6' in a throw of die $= 1 - \frac{1}{6} = \frac{5}{6}$

By compound probability theorem, the probability that in 4 throws of die, no '6' is obtained $= (5/6)^4$.

Hence, the probability of obtaining '6' at least once in 4 throws of a die $= 1 - (5/6)^4 = 0.516$

If a trial consists of throwing two dice at a time, then the probability of getting a 'double' of '6' in a trial $= \frac{1}{36}$.

Thus, the probability of not getting a 'double of 6' in a trial $= \frac{35}{36}$.

The probability that in 24 throws, with two dice each, no 'double of 6' is obtained $= (35/36)^{24}$

Hence the probability of getting a 'double of 6' at least once in 24 throws

$$= 1 - (35/36)^{24} = 0.491$$

Since the probability in the first case is greater than the probability in the second case, the result follows.

Example 3.63. *Let $A_1, A_2, \ldots A_n$ be independent events and $P\{A_k\} = p_k$. Further, let p be the probability that none of the events occurs; then show that:* $p \le \exp\left(-\sum_k p_k\right)$.

Solution. We have $p_i = P(A_i);\ i = 1, 2, \ldots, n$

$$p = P(\bar{A}_1 \cap \bar{A}_2 \cap \ldots \cap \bar{A}_n)$$

$$= \prod_{i=1}^n P(\bar{A}_i) = \prod_{i=1}^n [1 - P(A_i)] = \prod_{i=1}^n (1 - p_i) \qquad (\text{Since } A_i\text{'s are independent})$$

$$\le \prod_{i=1}^n \exp(-p_i) \qquad [\because 1 - x \le e^{-x} \text{ for } 0 \le x \le 1 \text{ and } 0 \le p_i \le 1]$$

$$\Rightarrow \qquad p \le \exp\left[-\sum_{i=1}^n p_i\right], \text{ as desired.}$$

Remark. We have $1 - x \le \exp(-x)$ for $0 \le x \le 1$. ...(*)

Proof. The inequality (*) is obvious for $x = 0$ and $x = 1$. Consider $0 < x < 1$. Then

$$\log (1-x)^{-1} = -\log(1-x) = \left(x + \frac{x^2}{2} + \frac{x^3}{3} + \frac{x^4}{4} + \ldots\right) \qquad ...(**)$$

the expansion being valid since $0 < x, < 1$. Further since $x > 0$, we get from (**)

$$\log (1-x)^{-1} > x \quad \Rightarrow \quad -\log(1-x) > x \quad \Rightarrow \quad \log(1-x) < -x, \ i.e., \ 1-x < e^{-x}.$$

Example 3·64. *p is the probability that a man aged x years will die in a year. Find the probability that out of n men $A_1, A_2, \ldots, A_n$ each aged x, A_1 will die in a year and will be the first to die.*

Solution. Let E_i, $(i = 1, 2, \ldots, n)$ denote the event that A_i dies in a year. Then

$$P(E_i) = p, (i = 1, 2, \ldots, n) \quad \text{and} \quad P(\bar{E}_i) = 1 - p.$$

The probability that none of n men $A_1, A_2, \ldots, A_n$ dies in a year

$$= P(\bar{E}_1 \cap \bar{E}_2 \cap \ldots \cap \bar{E}_n) = P(\bar{E}_1)\,P(\bar{E}_2) \ldots P(\bar{E}_n) = (1-p)^n$$

(By compound probability theorem)

$\therefore$ The probability that at least one of $A_1, A_2, \ldots, A_n$ dies in a year

$$= 1 - P(\bar{E}_1 \cap \bar{E}_2 \cap \ldots \cap \bar{E}_n) = 1 - (1-p)^n$$

The probability that among n men, A_1 is the first to die is $1/n$ and since this event is independent of the event that at least one man dies in a year, the required probability is :

$$\frac{1}{n}\left[1 - (1-p)^n\right].$$

Example 3·65. *Prove or disprove the following:*

$$(i)\ P(A \mid B) + P(A \mid \bar{B}) = 1, \qquad (ii)\ P(A \mid B) + P(\bar{A} \mid B) = 1.$$

Solution. (*i*) Consider the following assignment of probabilities :

$$P(A \cap B) = P(A \cap \bar{B}) = P(\bar{A} \cap B) = P(\bar{A} \cap \bar{B}) = \frac{1}{4} \qquad ...(*)$$

$$\text{Then, } P(A \mid B) + P(A \mid \bar{B}) = \frac{P(A \cap B)}{P(B)} + \frac{P(A \cap \bar{B})}{P(\bar{B})}$$

$$= \frac{(1/4)}{(1/2)} + \frac{(1/4)}{(1/2)} = 1 \qquad \text{[using (*)]} \qquad ...(1)$$

$$\left[\because P(B) = P(A \cap B) + P(\bar{A} \cap B) = \frac{1}{4} + \frac{1}{4} = \frac{1}{2} \quad \Rightarrow \quad P(\bar{B}) = \frac{1}{2}\right]$$

Again consider the following assignment of probabilities :

$$P(A \cap B) = P(A \cap \bar{B}) = P(\bar{A} \cap B) = \frac{1}{k}, \ P(\bar{A} \cap \bar{B}) = 1 - \frac{3}{k}; (k > 4)$$

$$\text{Then, } \quad P(B) = P(A \cap B) + P(\bar{A} \cap B) = \frac{1}{k} + \frac{1}{k} = \frac{2}{k} \quad \Rightarrow \quad P(\bar{B}) = 1 - \frac{2}{k}$$

$$\therefore \quad P(A \mid B) + P(A \mid \bar{B}) = \frac{P(A \cap B)}{P(B)} + \frac{P(A \cap \bar{B})}{P(\bar{B})} = \frac{(1/k)}{(2/k)} + \frac{(1/k)}{[1 - (2/k)]}$$

$$= \frac{1}{2} + \frac{1}{k-2} = \frac{k}{2(k-2)} \neq 1; \quad k > 4$$

From (1) and (2), we conclude that

$$P(A \mid B) + P(A \mid \bar{B}) = 1, \text{ is not true, in general.}$$

(*ii*)
$$P(A \mid B) + P(\bar{A} \mid B) = \frac{P(A \cap B)}{P(B)} + \frac{P(\bar{A} \cap B)}{P(B)}$$

$$= \frac{P(A \cap B) + P(\bar{A} \cap B)}{P(B)} = \frac{P(B)}{P(B)} = 1.$$

$\therefore$
$$P(A \mid B) + P(\bar{A} \mid B) = 1, \text{ is true.}$$

3·16. MATCHING PROBLEM

Let us have n letters corresponding to which there exist n envelopes bearing different addresses. Considering various letters being put in various envelopes at random, *a match is said to occur* if a letter goes into the right envelope. (Alternatively, if in a party there are n persons with n different hats, *a match is said to occur* if in the process of selecting hats *at random, the* ith person rightly gets the ith hat.)

A match at the kth position for k = 1, 2, ..., n. Let us first consider the event A_k when a match occurs at the kth place. For better understanding let us put the envelopes bearing numbers 1, 2, ..., n in ascending order. When A_k occurs, kth letter goes to the kth envelope but $(n - 1)$ letters can go to the remaining $(n - 1)$ envelopes in $(n - 1)!$ ways. Hence

$$P(A_k) = \frac{(n-1)!}{n!} = \frac{1}{n},$$

where $P(A_k)$ denotes the probability of the kth match. It is interesting to see that $P(A_k)$ *does not depend on* k.

Example 3·66. *(a) 'n' different objects 1, 2, ..., n are distributed at random in n places marked 1, 2 ... n, Find the probability that none of the objects occupies the place corresponding to its number.*

(b) If n letters are randomly placed in correctly addressed envelopes, prove that the probability that exactly r letters are placed in correct envelopes is given by :

$$\frac{1}{r!}\sum_{k=0}^{n-r}(-1)^k \frac{1}{k!} \quad ; \; r = 0, 1, 2, \ldots, n$$

***Solution.* (Probability of no match).** Let E_i, $(i = 1, 2, \ldots, n)$ denote the event that the ith object occupies the place corresponding to its number so that $\bar{E}_i$, is the complementary event. Then probability 'p' that none of the objects occupies the place corresponding to its number is given by

$$p = P(\bar{E}_1 \cap \bar{E}_2 \cap \bar{E}_3 \cap \ldots \cap \bar{E}_n)$$

$$= 1 - P(E_1 \cup E_2 \cup E_3 \cup \ldots \cup E_n)$$

$$= 1 - \left[\sum_{i=1}^{n} P(E_i) - \sum_{i,j=1}^{n}\sum_{i<j} P(E_i \cap E_j) + \sum_{i,j,k=1}^{n}\sum\sum_{i<j<k} P(E_i \cap E_j \cap E_k) - \ldots + (-1)^{n-1} P(E_1 \cap E_2 \cap \ldots \cap E_n)\right] \quad \ldots(*)$$

$$P(E_i) = \frac{1}{n}, \forall i$$

$$P(E_i \cap E_j) = P(E_i)\, P(E_j \mid E_i) = \frac{1}{n} \times \frac{1}{n-1}, \forall\, i, j\, (i < j)$$

$$P(E_i \cap E_j \cap E_k) = P(E_i)\, P(E_j \mid E_i)\, P(E_k \mid E_i \cap E_j)$$

$$= \frac{1}{n} \times \frac{1}{n-1} \times \frac{1}{n-2}, \forall i, j, k\, (i < j < k), \text{ and so on.}$$

Finally, $P(E_1 \cap E_2 \cap E_3 \cap \ldots \cap E_n) = \dfrac{1}{n} \times \dfrac{1}{n-1} \times \dfrac{1}{n-2} \ldots \dfrac{1}{2} \times 1 = \dfrac{1}{n!}$

Substituting in (*), we get

$$p = 1 - \left\{ {}^nC_1 \frac{1}{n} - {}^nC_2 \frac{1}{n(n-1)} + {}^nC_3 \frac{1}{n(n-1)(n-2)} - \ldots + (-1)^{n-1} \frac{1}{n!} \right\}$$

$$= 1 - \left\{ 1 - \frac{1}{2!} + \frac{1}{3!} - \ldots + (-1)^{n-1} \frac{1}{n!} \right\} = \frac{1}{2!} - \frac{1}{3!} + \frac{1}{4!} - \ldots + (-1)^n \frac{1}{n!} = \sum_{k=0}^{n} \frac{(-1)^k}{k!}. \qquad \ldots(**)$$

Remarks:

1. If n objects are distributed at random in n places, the expression in (**) gives the probability of 'no match'.
2. For large n, (*i.e.*, $n \to \infty$)

$$p = 1 - 1 + \frac{1}{2!} - \frac{1}{3!} + \frac{1}{4!} - \ldots = e^{-1} = 0{\cdot}36787$$

Hence, the probability of at least one match is given by :

$$1 - p = 1 - \frac{1}{2!} + \frac{1}{3!} - \ldots + \frac{(-1)^n}{n!} \simeq 1 - \frac{1}{e}, \text{ (for large } n).$$

(*b*) **[Probability of exactly *r* matches {r ≤ (n – 2)}]** Let A_i, (i = 1, 2, ..., n) denote the event that ith letter goes to the correct envelope. Then the probability that none of the n letters goes to the correct envelope is :

$$P(\bar{A}_1 \cap \bar{A}_2 \cap \ldots \cap \bar{A}_n) = \sum_{k=0}^{n} [(-1)^k / k!] \qquad [\text{(c.f. part } (a))] \qquad \ldots(***)$$

The probability that each of the 'r' letters is in the right envelope is $\dfrac{1}{n(n-1)(n-2)\ldots(n-r+1)}$, and the probability that none of the remaining $(n-r)$ letters goes in the correct envelope is obtained by replacing n by $(n-r)$ in (***) and is thus given by $\displaystyle\sum_{k=0}^{n-r} \frac{(-1)^k}{k!}$. Hence, by compound probability theorem, the probability that out of n letters, exactly r letters go to correct envelopes, (in a specified order), is :

$$\frac{1}{n(n-1)(n-2)\ldots(n-r+1)} \sum_{k=0}^{n-r} \frac{(-1)^k}{k!}\,;\; r \le n-2$$

Since r letters can go to n envelopes in ${}^nC_r$ mutually exclusive ways, the required probability of exactly r letters going to correct envelopes (in any order, whatsoever) is

$$= {}^nC_r \times \frac{1}{n(n-1)(n-2)\ldots(n-r+1)} \sum_{k=0}^{n-r} \frac{(-1)^k}{k!}$$

$$= \frac{1}{r!} \sum_{k=0}^{n-r} (-1)^k \frac{1}{k!}; \quad r = 1, 2, 3, \ldots, (n-2), \text{ as required}$$

Example 3·67. *A lady declares that by taking a cup of tea with milk she can discriminate whether the milk or tea infusion was first made in the cup. It is proposed to test this assertion by means of an experiment with 12 cups of tea, 6 made in one way and 6 in the other, and presenting them to the lady for judgment in a random order. Calculate the probability that on the null hypothesis (that the lady has no discriminating power) the lady would judge correctly all the twelve cups, it being known to her that 6 are of each kind.*

If, however, the 12 cups were presented to the lady in 6 pairs, each pair to consist of cups of each kind, in a random order, how would the probability of correctly judging with every cup on the null hypothesis be altered ?

Which of the two designs would you prefer and why ?

Solution. Case (i). *When all the cups are presented to the lady in random order.* We know the total number of permutations of n things, of which n_1 are of one kind and n_2 of the other kind, are $\frac{n!}{n_1!\,n_2!}$. Hence, the total number of ways in which 12 cups of tea, 6 made in one way and 6 in the other, can be presented to the lady are $\frac{12!}{6!\,6!} = 924$. Since there is only one way in which the lady can judge all the cups correctly.

$$\text{The required probability} = \frac{1}{924}.$$

Case (ii). When the 12 cups are presented to the lady in 6 pairs, each pair consisting of either kind of tea preparation. Since each pair consists of cups of each of the two kinds, the probability that the lady correctly judges each pair is $\frac{1}{2}$ and since the 6 presentations of the cups (in pairs) to the lady are independent of each other,

$$\text{The probability that the lady will correctly judge all the six pairs} = \left(\frac{1}{2}\right)^6 = \frac{1}{64}.$$

The first method of testing is preferable to the second because the probability of correctly judging all the cups is much less in the first case as compared with the corresponding probability in the second case.

Example 3·68. (a) *n a town consisting of (n + 1) inhabitants, a person narrates a rumour to a second person, who in turn narrates it to a third person, and so on. At each step the recipient of the rumour is chosen at random from the n available persons, excluding the narrator himself. Find the probability that the rumour will be told r times without :*

(i) returning to the originator and (ii) being narrated to any person more than once.

(b) Do the above problem when, at each step the rumour is told by one person to a gathering of N randomly chosen people.

Solution. (a) Since any person can narrate the rumour to any one of the n available persons in n ways, exhaustive number of cases that rumour will be told r times is n^r.

(*i*) Let A denote the event that the rumour will be told r times without returning to the originator. The originator can tell the rumour to any one of the remaining n persons in n ways, and each of the $(r-1)$ recipients of the rumour can tell it to any one of the remaining $(n-1)$ persons (without returning to the originator) in $(n-1)$ ways. Hence, the favourable number of cases for A is $n(n-1)^{r-1}$.

$$\therefore \qquad P(A) = \frac{n(n-1)^{r-1}}{n^r} = \left(\frac{n-1}{n}\right)^{r-1} = \left(1-\frac{1}{n}\right)^{r-1}$$

(*ii*) Let B denote the event that rumour is told r times wihtout being repeated to any person. In this case the first person (originator) can narrate the rumour to any one of the available n persons; 2nd person can tell it to any one of the remaining $(n-1)$ persons; ...; the rth person can tell it to any one of the remaining $(n-r+1)$ persons.

Hence, the favourable number of cases for B is $n(n-1)(n-2)\ldots(n-r+1)$.

$$\therefore \qquad P(B) = \frac{n(n-1)(n-2)\ldots(n-r+1)}{n^r}.$$

(*b*) When we consider the case that the rumour is told by a person to a gathering of N randomly chosen persons simultaneously, the exhaustive number of cases for rumour to be told r times is $({}^nC_N)^r$, since each person can tell the rumour in ${}^nC_N$ ways. Arguing exactly as in part (*a*), the number of cases favourable to A are : ${}^nC_N\,({}^{n-1}C_N)^{r-1}$

$$\therefore \qquad P(A) = \frac{{}^nC_N\,({}^{n-1}C_N)^{r-1}}{({}^nC_N)^r} = \left(\frac{{}^{n-1}C_N}{{}^nC_N}\right)^{r-1} = \left(\frac{n-N}{n}\right)^{r-1} = \left(1-\frac{N}{n}\right)^{r-1}$$

$$\left[\because {}^nC_N = \frac{n!}{N!(n-N)!} = \frac{n(n-1)!}{N!(n-N)(n-N-1)!} = \frac{n}{n-N}\times {}^{n-1}C_N\right]$$

Similarly, we shall get

$$P(B) = \frac{{}^nC_N \times {}^{n-N}C_N \times {}^{n-2N}C_N \times \ldots \times {}^{n-(r-1)N}C_N}{({}^nC_N)^r}$$

$$= \left[\frac{N!(n-N)!}{n!}\right]^r \times \left\{\frac{n!}{N!(n-N)!}\times\frac{(n-N)!}{N!(n-2N)!}\times\frac{(n-2N)!}{N!(n-3N)!}\times\ldots\times\frac{\{n-(r-1)N\}!}{N!(n-rN)!}\right\}$$

$$= \frac{n!}{(n-rN)!}\times\left[\frac{(n-N)!}{n!}\right]^r = \frac{{}^nC_{rN}}{({}^nC_N)^r}.$$

Remark. Part (*a*) is a particular case of part (*b*) with $N = 1$.

Example 3·69. *Two squares are selected at random from 64 squares in a chess-board. Find the probability that they have a common side.*

a	b	b	b	b	b	b	a
b	c	c	c	c	c	c	b
b	c	c	c	c	c	c	b
b	c	c	c	c	c	c	b
b	c	c	c	c	c	c	b
b	c	c	c	c	c	c	b
b	c	c	c	c	c	c	b
a	b	b	b	b	b	b	a

Chess Board

Solution. Two squares can be selected at random out of 64 squares in ${}^{64}C_2$ distinct equally likely ways.

$$\therefore \qquad \text{Exhaustive Number of cases} = n(S) = {}^{64}C_2 = \frac{64\times 63}{2} = 2016$$

Let us mark the four corner squares by the letter '*a*'; the remaining $4 \times 6 = 24$ squares, of the first and last rows and columns by letter '*b*' and the remaining $6 \times 6 = 36$ squares, by the letter '*c*' as shown in the above figure.

Favourable Cases. The number of cases favourable to the event :

E : The two squares selected at random have a common, side are enumerated below :

(*i*) If one of the two selected squares is marked '*a*', then for the other square there are two options as shown.

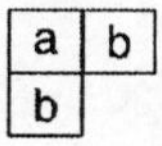

The number of such pairs $= {}^4C_1 \times {}^2C_1 = 4 \times 2 = 8$

(*ii*) If one of the two selected squares is marked '*b*', then for the second square we have 3 options as shown.

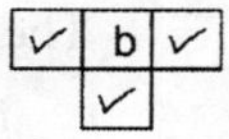

The number of such pairs $= {}^{24}C_1 \times {}^3C_1 = 24 \times 3 = 72$

(*iii*) If one of the selected squares is marked '*c*', then for the second square there are 4 options as shown.

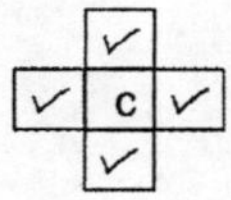

The number of such pair $= {}^{36}C_1 \times {}^4C_1 = 36 \times 4 = 144$

In all the cases enumerated above in (*i*), (*ii*) and (*iii*), each square is counted twice.

$\therefore$ Number of cases favourable to 'E' $= \dfrac{1}{2}(8 + 72 + 144) = \dfrac{1}{2} \times 224 = 112$

$\therefore$ Required probability $= P(E) = \dfrac{n(E)}{n(S)} = \dfrac{112}{2016} = \dfrac{1}{18}$

Example 3·70. *If four squares are chosen at random on a chess-board, find the chance that they should be in a diagonal line.*

Solution. In a chess-board there are $8 \times 8 = 64$ squares as shown in the following diagram.

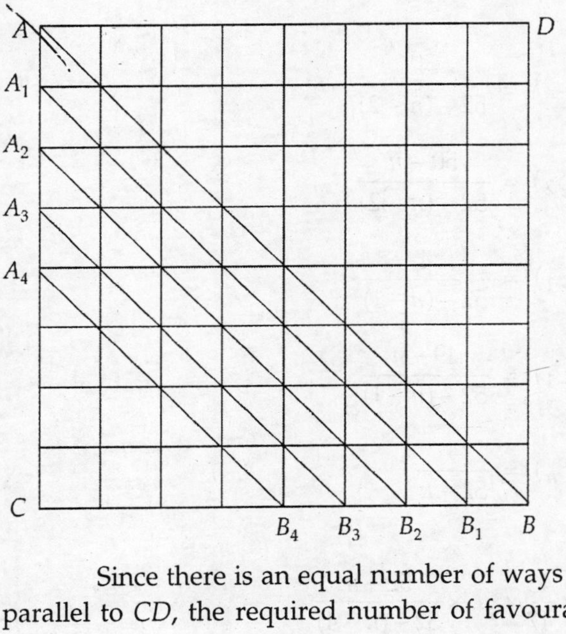

Let us consider the number of ways in which the 4 squares selected at random are in a diagonal line parallel to AB. Consider the $\Delta\, ABC$. Number of ways in which 4 selected squares are along the lines A_4B_4, A_3B_3, A_2B_2, A_1B_1 and AB are ${}^4C_4$, ${}^5C_4$, ${}^6C_4$, ${}^7C_4$ and ${}^8C_4$ respectively.

Similarly, in ΔABD there are an equal number of ways of selecting 4 squares in a diagonal line parallel to AB.

Hence, total number of ways in which the 4 selected squares are in a diagonal line parallel to AB are :

$$2({}^4C_4 + {}^5C_4 + {}^6C_4 + {}^7C_4) + {}^8C_4$$

Since there is an equal number of ways in which 4 selected squares are in a diagonal line parallel to CD, the required number of favourable cases is given by :

$$2[2({}^4C_4 + {}^5C_4 + {}^6C_4 + {}^7C_4) + {}^8C_4]$$

Since 4 squares can be selected out of 64 in ${}^{64}C_4$ ways, the required probability is :

$$= \frac{2[2({}^4C_4 + {}^5C_4 + {}^6C_4 + {}^7C_4) + {}^8C_4]}{{}^{64}C_4} = \frac{[4(1+5+15+35)+140] \times 4!}{64 \times 63 \times 62 \times 61} = \frac{91}{158844}.$$

Example 3·71. *Cards are dealt one by one from a well-shuffled pack until an ace appears. Show that the probability that exactly n cards are deatl before the first ace appears is :*

$$\frac{4(51-n)(50-n)(49-n)}{52\cdot 51\cdot 50\cdot 49}$$

Solution. Let E_i denote the event that an ace appears when the ith card is dealt.

Then the required probability 'p' is given by :

$$p = P(\text{Exactly } n \text{ cards are dealt before the first ace appears.})$$
$$= P[\text{The first ace appears at the } (n+1)\text{th dealing.}]$$
$$= P(\bar{E}_1 \cap \bar{E}_2 \cap \bar{E}_3 \cap \ldots \cap \bar{E}_{n-1} \cap \bar{E}_n \cap E_{n+1})$$
$$= P(\bar{E}_1)\, P(\bar{E}_2 \mid \bar{E}_1)\, P(\bar{E}_3 \mid \bar{E}_1 \cap \bar{E}_2) \ldots P(\bar{E}_n \mid \bar{E}_1 \cap \bar{E}_2 \cap \ldots \cap \bar{E}_{n-1})$$
$$\times P(E_{n+1} \mid \bar{E}_1 \cap \bar{E}_2 \cap \ldots \cap \bar{E}_n) \quad \ldots(*)$$

$$P(E_1) = \frac{4}{52} \quad \Rightarrow \quad P(\bar{E}_1) = \frac{48}{52}$$

$$P(E_2 \mid \bar{E}_1) = \frac{4}{51} \quad \Rightarrow \quad P(\bar{E}_2 \mid \bar{E}_1) = \frac{47}{51}$$

$$P(E_3 \mid \bar{E}_1 \cap \bar{E}_2) = \frac{4}{50} \quad \Rightarrow \quad P(\bar{E}_3 \mid \bar{E}_1 \cap \bar{E}_2) = \frac{46}{50}$$

$$\vdots \qquad\qquad\qquad \vdots$$

$$P(E_{n-1} \mid \bar{E}_1 \cap \bar{E}_2 \cap \ldots \cap \bar{E}_{n-2}) = \frac{4}{52-(n-2)}$$

$$\Rightarrow \quad P(\bar{E}_{n-1} \mid \bar{E}_1 \cap \bar{E}_2 \cap \ldots \cap \bar{E}_{n-2}) = \frac{50-n}{52-(n-2)}$$

$$P(E_n \mid \bar{E}_1 \cap \bar{E}_2 \cap \ldots \cap \bar{E}_{n-1}) = \frac{4}{52-(n-1)}$$

$$\Rightarrow \quad P(\bar{E}_n \mid \bar{E}_1 \cap \bar{E}_2 \cap \ldots \cap \bar{E}_{n-1}) = \frac{49-n}{52-(n-1)}$$

and

$$P(E_{n+1} \mid \bar{E}_1 \cap \bar{E}_2 \cap \ldots \cap \bar{E}_n) = \frac{4}{52-n}$$

Hence, from (*), we get

$$p = \left[\frac{48}{52} \times \frac{47}{51} \times \frac{46}{50} \times \frac{45}{49} \times \frac{44}{48} \times \frac{43}{47} \times \ldots \times \frac{52-n}{52-(n-4)}\right.$$
$$\left. \times \frac{51-n}{52-(n-3)} \times \frac{50-n}{52-(n-2)} \times \frac{49-n}{52-(n-1)} \times \frac{4}{52-n}\right]$$

$$= \frac{4(51-n)(50-n)(49-n)}{52 \times 51 \times 50 \times 49}$$

Aliter. A more elegant way of doing the problem is given below:

$$p = P[\text{First ace appears at the } (n+1)\text{th dealing}]$$

$$= P\left[\underbrace{\left\{\begin{matrix}n \text{ cards are dealt out}\\ \text{of 48 non-ace cards}\end{matrix}\right\}}_{A} \cap \underbrace{\left\{\begin{matrix}(n+1)\text{th card (ace) is drawn}\\ \text{out of remaining } (52-n) \text{ cards}\end{matrix}\right\}}_{B}\right]$$

$$= P(A \cap B) = P(A)\cdot P(B \mid A) = \frac{^{48}C_n}{^{52}C_n} \times \frac{4}{(52-n)}$$

$$= \frac{48!\; n!\; (52-n)!}{n!\; (48-n)!\; (52)!} \times \frac{4}{(52-n)} = \frac{4(51-n)(50-n)(49-9)}{52 \times 51 \times 50 \times 49},$$

Example 3·72. *A coin is tossed (m + n) times, (m > n). Show that*

$$P[\textit{At least m consecutive heads}] = \frac{m+2}{2^{m+1}}.$$

Solution. Since $m > n$, only one sequence of m consecutive heads is possible. This sequence may start either with the first toss or second toss or third toss, and so on, the last one will be starting with $(n + 1)$th toss.

Let E_1 denote the event that the sequence of m consecutive heads starts with ith toss. Then the required probability is :

$$P(E_1) + P(E_2) + \dots + P(E_{n+1}) \qquad \dots(*)$$

$$P(E_1) = P[\text{Consecutive heads in first } m \text{ tosses and head or tail in the rest}] = \left(\frac{1}{2}\right)^m.$$

$$P(E_2) = P\left[\begin{matrix}\text{Tail in the first toss, followed by } m \text{ consecutive}\\ \text{heads and head or tail in the next}\end{matrix}\right] = \frac{1}{2}\left(\frac{1}{2}\right)^m = \frac{1}{2^{m+1}}$$

In general,

$$P(E_r) = P\left[\begin{matrix}\text{Tail in the } (r-1)\text{th trial followed by } m \text{ consecutive}\\ \text{heads and head or tail in the next}\end{matrix}\right]$$

$$= \frac{1}{2}\left(\frac{1}{2}\right)^m = \frac{1}{2^{m+1}}, \forall\, r = 2, 3, \cdots, n+1.$$

Substituting in (*), $\quad$ Required probability $= \dfrac{1}{2^m} + \dfrac{n}{2^{m+1}} = \dfrac{2+n}{2^{m+1}}.$

Example 3·73. *A player tosses a coin and is to score one point for every head and two points for every tail turned up. He is to play on until his score reaches or passes n. If p_n is the chance of attaining exactly n score, show that*

$$(a)\ p_n = \frac{1}{2}(p_{n-1} + p_{n-2}); \qquad (b)\ p_n + \frac{1}{2}p_{n-1} = 1$$

and hence find the value of p_n.

Solution. (*a*) The score n can be reached in the following two mutually exclusive ways :

(*i*) By throwing a tail when score is $(n-2)$, and (*ii*) By throwing a head when score is $(n-1)$. Hence, by addition theorem of probability, we get

$$p_n = p(i) + P(ii) = \frac{1}{2}\cdot p_{n-2} + \frac{1}{2}\cdot p_{n-1} = \frac{1}{2}(p_{n-1} + p_{n-2}) \qquad \text{...(*)}$$

(*b*) To find p_n explicitly, (*) may be re-written as

$$p_n + \frac{1}{2}p_{n-1} = p_{n-1} - \frac{1}{2}p_{n-2} = p_{n-2} + \frac{1}{2}p_{n-3} = \dots = p_2 + \frac{1}{2}p_1 \qquad \text{...(**)}$$

Since the score 2 can be obtained in the following mutually disjoint ways :
(*i*) Head in first throw and head in second throw, (*ii*) Tail in the first throw,

we have $\quad p_2 = \frac{1}{2}\cdot\frac{1}{2} + \frac{1}{2} = \frac{1}{4} + \frac{1}{2} = \frac{3}{4}\quad$ and obviously $\quad p_1 = \frac{1}{2}.$

Hence, from (**), we get

$$p_n + \frac{1}{2}p_{n-1} = \frac{3}{4} + \frac{1}{2}\times\frac{1}{2} = 1. \qquad \text{...(***)}$$

Explicit Expression For p_n. Re-writing (***), we get

$$p_n + \frac{1}{2}p_{n-1} = 1 = \frac{2}{3} + \frac{1}{3} = \frac{2}{3} + \frac{1}{2}\cdot\frac{2}{3}$$

$$\therefore \qquad p_n - \frac{2}{3} = \left(-\frac{1}{2}\right)\left(p_{n-1} - \frac{2}{3}\right)$$

$$p_{n-1} - \frac{2}{3} = \left(-\frac{1}{2}\right)\left(p_{n-2} - \frac{2}{3}\right)$$

$$\vdots \qquad \vdots$$

$$p_2 - \frac{2}{3} = \left(-\frac{1}{2}\right)\left(p_1 - \frac{2}{3}\right)$$

Multiplying all the above equations, we get

$$p_n - \frac{2}{3} = \left(-\frac{1}{2}\right)^{n-1}\left(p_1 - \frac{2}{3}\right) = \left(-\frac{1}{2}\right)^{n-1}\left(\frac{1}{2} - \frac{2}{3}\right) = (-1)^n\cdot\frac{1}{2^n}\cdot\frac{1}{3}$$

$$\Rightarrow \qquad p_n = \frac{2}{3} + (-1)^n\frac{1}{2^n}\cdot\frac{1}{3} = \frac{1}{3}\left[2 + (-1)^n\frac{1}{2^n}\right]$$

Example 3·74. *Each of the n urns contains 'a' white balls and 'b' black balls. One ball is transferred from the first urn to the second, then one ball from the latter into the third, and so on. If p_k is the probability of drawing a white ball from the kth urn, show that*

$$p_{k+1} = \frac{a+1}{a+b+1}p_k + \frac{a}{a+b+1}(1-p_k)$$

Hence, for the last urn, prove that: $\qquad p_n = \frac{a}{a+b}.$

Solution. The event of drawing a white ball from the kth urn can materialise in the following two mutually disjoint ways :

(*i*) The ball transferred from the $(k-1)$th urn is white and then a white ball is drawn from the kth urn.

(*ii*) The ball transferred from the $(k-1)$th urn is black and then a white ball is drawn from the kth urn.

The probability of case (*i*) is : $\quad p_{k-1} \times \dfrac{a+1}{a+b+1}$,

since the probability of drawing a white ball from the $(k-1)$th urn is p_{k-1} and then the probability of drawing a white ball from the kth urn is $\dfrac{a+1}{a+b+1}$.

Since the probability of drawing a black ball from the $(k-1)$th urn is $[1-p_{k-1}]$ and then the probability of drawing a white ball from the kth urn is $\dfrac{a}{a+b+1}$, the probability of case (*ii*) is given by :

$$\frac{a}{a+b+1}[1-p_{k-1}].$$

Since cases (*i*) and (*ii*) are mutually exclusive, by addition theorem of probability

$$p_k = \frac{a+1}{a+b+1}p_{k-1} + \frac{a}{a+b+1}[1-p_{k-1}] \qquad \text{...(*)}$$

$$\Rightarrow \qquad p_k = \frac{1}{a+b+1}p_{k-1} + \frac{a}{a+b+1} \qquad \text{...(1)}$$

Replacing k by $k+1$ in (*), we get the required result.

Changing k to $k-1$, $k-2$, ... and so on in (1), we get

$$p_{k-1} = \frac{1}{a+b+1}p_{k-2} + \frac{a}{a+b+1} \qquad \text{...(2)}$$

$$p_{k-2} = \frac{1}{a+b+1}p_{k-3} + \frac{a}{a+b+1} \qquad \text{...(3)}$$

$$\vdots \qquad\qquad \vdots \qquad\qquad \vdots$$

$$p_2 = \frac{1}{a+b+1}p_1 + \frac{a}{a+b+1} \qquad \text{... }(k-1)$$

But $\qquad p_1 = $ Probabillity of drawing a white ball from the first urn $= \dfrac{a}{a+b}$

Multiplying (1) by 1, (2) by $\dfrac{1}{a+b+1}$, (3) by $\left(\dfrac{1}{a+b+1}\right)^2$, ..., and $(k-1)$th equation by $\left(\dfrac{1}{a+b+1}\right)^{k-2}$ and adding, we get

$$p_k = \left(\frac{1}{a+b+1}\right)^{k-1} p_1 + \frac{a}{a+b+1}\left[1+\frac{1}{a+b+1}+\frac{1}{(a+b+1)^2}+\ldots+\left(\frac{1}{a+b+1}\right)^{k-2}\right]$$

$$= \left(\frac{1}{a+b+1}\right)^{k-1} \times \frac{a}{(a+b)} + \frac{a}{a+b+1}\left[\frac{1-\left(\frac{1}{a+b+1}\right)^{k-1}}{\left(1-\frac{1}{a+b+1}\right)}\right]$$

$$= \frac{a}{a+b}\left(\frac{1}{a+b+1}\right)^{k-1} + \frac{a}{a+b}\left[1-\left(\frac{1}{a+b+1}\right)^{k-1}\right]$$

$$= \frac{a}{a+b}\left[\left(\frac{1}{a+b+1}\right)^{k-1} + \left\{1-\left(\frac{1}{a+b+1}\right)^{k-1}\right\}\right] = \frac{a}{a+b}, (k = 1, 2, \ldots, n)$$

Since the probability of drawing a white ball from the kth urn is independent of k, we have

$$p_n = \frac{a}{a+b}, \text{ as required.}$$

Example 3·75. *The following weather forecasting is used by an amateur forecaster. Each day is classified as 'dry' or 'wet' and the probability that any given day is same as the preceding one is assumed to be a constant p, (0 < p < 1). Based on past records, it is supposed that January 1 has a probability b of being dry. Letting.*

$$\beta_n = \text{Probability that nth day of the year is dry,}$$

Obtain an expression for β_n *in terms of* β *and* p. *Also evaluate* $\lim_{n\to\infty} \beta_n$.

Solution. The event 'that nth day of the year is dry' can materialise in the following mutually disjoint ways :

(*i*) $(n - 1)$th day is dry and nth day is dry, and the probability for this compount event is $\beta_{n-1} \times p$.

(*ii*) $(n - 1)$th day is wet and nth day is dry, and the probability for this compound event is $(1 - \beta_{n-1})(1 - p)$.

Hence, by addition theorem of probability, we get

$$\beta_n = P_r \text{ (nth day of the year is dry)} = P(i) + P(ii).$$

$$\Rightarrow \quad \beta_n = p \cdot \beta_{n-1} + (1-\beta_{n-1})(1-p)$$

$$= p \cdot \beta_{n-1} + 1 - p - (1-p)\beta_{n-1}$$

$$= (p - 1 + p)\beta_{n-1} + 1 - p$$

$$\Rightarrow \quad \beta_n = (2p - 1)\beta_{n-1} + (1 - p); \ (n = 2, 3, 4, \ldots) \qquad \ldots(*)$$

which is the difference equation in β_n.

Rewriting (*), we get

$$\beta_n - \frac{1}{2} = (2p-1)\beta_{n-1} + (1-p) - \frac{1}{2} = (2p-1)\beta_{n-1} - \frac{1}{2}(2p-1)$$

$$\Rightarrow \quad \beta_n - \frac{1}{2} = (2p-1)\left(\beta_{n-1} - \frac{1}{2}\right)$$

$$\Rightarrow \quad U_n = (2p-1)\,U_{n-1}, \text{ where } U_n = \beta_n - \frac{1}{2} \qquad ...(**)$$

Replacing n by $n-1, n-2, \ldots, 2$ successively in (**) and multiplying the resulting equations, we get

$$U_n = (2p-1)^{n-1}\,U_1 = (2p-1)^{n-1}\left(\beta_1 - \frac{1}{2}\right) \qquad \text{[From (**)]}$$

But $\beta_1 = P(\text{1st day of the year is dry}) = \beta$ (Given)

$$\therefore \quad U_n = (2p-1)^{n-1}\left(\beta - \frac{1}{2}\right) \Rightarrow \beta_n - \frac{1}{2} = (2p-1)^{n-1}\left(\beta - \frac{1}{2}\right)$$

$$\Rightarrow \quad \beta_n = (2p-1)^{n-1}\left(\beta - \frac{1}{2}\right) + \frac{1}{2}$$

Proceeding to the limit as $n \to \infty$, we get

$$\lim_{n\to\infty} \beta_n = \frac{1}{2} + \left(\beta - \frac{1}{2}\right)\lim_{n\to\infty}(2p-1)^{n-1} \qquad ...(***)$$

Since	$p < 1$	Also	$p > 0$
$\Rightarrow$	$2p < 2$	$\Rightarrow$	$2p > 0$
$\Rightarrow$	$2p - 1 < 1$	$\Rightarrow$	$2p - 1 > -1$
		$\Rightarrow$	$-1 < 2p - 1$

$$\therefore \quad -1 < 2p - 1 < 1 \Rightarrow |2p-1| < 1 \Rightarrow \lim_{n\to\infty}(2p-1)^{n-1} \to 0$$

Substituting in (***), we get : $\lim_{n\to\infty} \beta_n = \dfrac{1}{2}$

Example 3·76. *Two urns contain, respectively, 'a white and b black' and 'b white and a black' balls. A series of drawings is made according to the following rules :*

(i) Each time only one ball is drawn and immediately returned to the same urn it came from.

(ii) If the ball drawn is white, the next drawing is made from the first urn.

(iii) If it is black, the next drawing is made from the second urn.

(iv) The first ball drawn comes from the first urn.

What is the probability that nth ball drawn will be white ?

Solution. Let p_r be the probability of drawing a white ball at the rth draw.

We can draw white ball at the rth draw in two ways:

(*a*) a white ball at the $(r-1)$th draw and a white ball at the rth draw.

(*b*) a black ball at the $(r-1)$th draw and a white ball at the rth draw.

Probability of drawing a white ball at the $(r-1)$ th draw $= p_{(r-1)}$ and then the probability of drawings a white ball at the rth draw $= \dfrac{a}{(a+b)}$, since in this case, rth draw is made from the first urn.

Probability of drawing black ball at the $(r-1)$th draw $= 1 - p_{(r-1)}$ and then the probability of drawing a white ball at the rth draw $= \dfrac{b}{(a+b)}$, since in this case, rth draw is made from the second urn.

$$\therefore \quad p_r = \frac{a}{(a+b)} p_{(r-1)} + \frac{b}{(a+b)}\left[1 - p_{(r-1)}\right]$$

$$= \left[\frac{a}{(a+b)} - \frac{b}{(a+b)}\right] p_{(r-1)} + \frac{b}{(a+b)}$$

$$\Rightarrow \quad p_r = \frac{(a-b)}{(a+b)} p_{(r-1)} + \frac{b}{(a+b)}$$

Similarly,
$$p_{(r-1)} = \frac{(a-b)}{(a+b)} p_{(r-2)} + \frac{b}{(a+b)}$$

$$\vdots \qquad \vdots \qquad \vdots$$

$$p_2 = \frac{(a-b)}{(a+b)} p_1 + \frac{b}{(a+b)}$$

But p_1 = Probability of drawing a white ball at the first draw $= \dfrac{a}{a+b}$.

Multiplying the above equations successively by

$$1, \left(\frac{a-b}{a+b}\right), \left(\frac{a-b}{a+b}\right)^2, \ldots, \left(\frac{a-b}{a+b}\right)^{r-2},$$

respectively, and adding, we have

$$p_r = \left(\frac{a-b}{a+b}\right)^{r-1} p_1 + \frac{b}{(a+b)}\left[1 + \left(\frac{a-b}{a+b}\right) + \left(\frac{a-b}{a+b}\right)^2 + \ldots + \left(\frac{a-b}{a+b}\right)^{r-2}\right]$$

$$= \left(\frac{a-b}{a+b}\right)^{r-1} \cdot \frac{a}{(a+b)} + \frac{b}{(a+b)}\left[\frac{1-\left(\frac{a-b}{a+b}\right)^{r-1}}{1-\left(\frac{a-b}{a+b}\right)}\right]$$

$$= \left(\frac{a-b}{a+b}\right)^{r-1} \cdot \frac{a}{a+b} + \frac{1}{2}\left[1-\left(\frac{a-b}{a+b}\right)^{r-1}\right]$$

$$= \frac{1}{2} + \left(\frac{a-b}{a+b}\right)^{r-1} \cdot \left[\frac{a}{a+b} - \frac{1}{2}\right]$$

$$= \frac{1}{2} + \frac{1}{2}\left(\frac{a-b}{a+b}\right)^{r} ; r = 1, 2, \ldots, n$$

Substituting $r = n$, we have : $\quad p_n = \dfrac{1}{2} + \dfrac{1}{2}\left(\dfrac{a-b}{a+b}\right)^{n}.$

Example 3·77. *In the following Fig. (a) and (b) assume that the probability of a relay being closed is p and that a relay is open or closed independently of any other. In each case find the probability that current flows from L to R.*

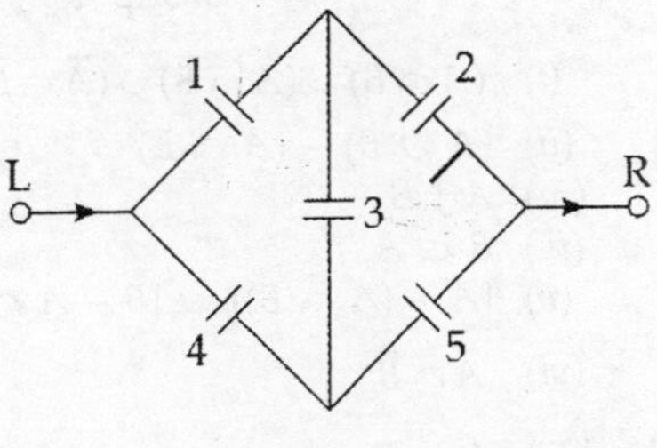

(a)

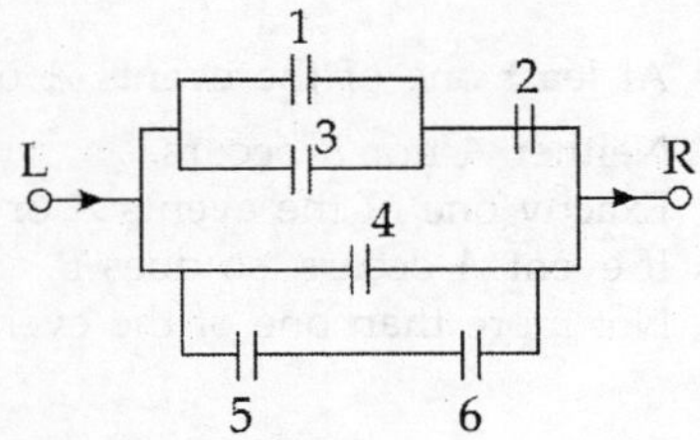

(b)

Solution. Let A_i denote the event that the relay i. (i = 1, 2, 3, ..., 6) is closed. Let E be the event that current flows from L to R.

In Fig. (a), the current will flow from L to R if at least one of the circuits from L to R is closed. Thus for the current to flow from L to R we have the following favourable cases :

(*i*) $A_1 \cap A_2 = B_1$, (*ii*) $A_4 \cap A_5 = B_2$, (*iii*) $A_1 \cap A_3 \cap A_5 = B_3$, (*iv*) $A_4 \cap A_3 \cap A_2 = B_4$.

The probability p_1 that current flows from L to R is given by :

$$p_1 = P(B_1 \cup B_2 \cup B_3 \cup B_4) = \sum_i P(B_i) - \sum_{i<j} P(B_i \cap B_j) + \sum_{i<j<k} P(B_i \cap B_j \cap B_k)$$
$$- P(B_1 \cap B_2 \cap B_3 \cap B_4) \quad ...(*)$$

Since the relays operate independently of each other, we have

$$P(B_1) = P(A_1 \cap A_2) = P(A_1) \cdot P(A_2) = p \cdot p = p^2$$
$$P(B_2) = P(A_4 \cap A_5) = P(A_4) \cdot P(A_5) = p \cdot p = p^2$$
$$P(B_3) = P(A_1)\, P(A_3)\, P(A_5) = p^3, \qquad P(B_4) = P(A_4)\, P(A_3)\, P(A_2) = p^3$$

Similarly,

$$P(B_1 \cap B_2) = P(A_1 \cap A_2 \cap A_4 \cap A_5) = P(A_1)\, P(A_2)\, P(A_4)\, P(A_5) = p^4$$
$$P(B_1 \cap B_2 \cap B_3) = P(A_1 \cap A_2 \cap A_3 \cap A_4 \cap A_5) = p^5, \text{ and so on.}$$

Finally, substituting in (*), we get

$$p_1 = (p^2 + p^2 + p^3 + p^3) - (p^4 + p^4 + p^4 + p^4 + p^4 + p^5) + (p^5 + p^5 + p^5 + p^5) - p^5$$
$$= 2p^2 + 2p^3 - 5p^4 + 2p^5$$

In Fig. (*b*), arguing as in the above case, the required probability p_2 that the current flows from L to R is given by :

$$p_2 = P(E_1 \cup E_2 \cup E_3 \cup E_4)$$

where $\quad E_1 = A_1 \cap A_2, \quad E_2 = A_3 \cap A_2, \quad E_3 = A_4, \quad E_4 = A_5 \cap A_6$

$$\therefore \quad p_2 = \sum_i P(E_i) - \sum_{i<j} P(E_i \cap E_j) + \sum_{i<j<k} P(E_i \cap E_j \cap E_k) - P(E_1 \cap E_2 \cap E_3 \cap E_4)$$
$$= (p^2 + p^2 + p + p^2) - (p^3 + p^3 + p^4 + p^3 + p^4 + p^3) + (p^4 + p^5 + p^5 + p^5) - p^6$$
$$= p + 3p^2 - 4p^3 - p^4 + 3p^5 - p^6.$$

CHAPTER CONCEPTS QUIZ

1. Find the correct answer from group Y for each item of group X.

Group X	Group Y
(*a*) At least one of the events A or B occurs.	(*i*) $(\bar{A} \cap B) \cup (A \cap \bar{B}) \cup (\bar{A} \cap \bar{B})$
(*b*) Neither A nor B occurs.	(*ii*) $(A \cup B) - (A \cap B)$
(*c*) Exactly one of the events A or B occurs.	(*iii*) A Ì B
(*d*) If event A occurs, so does B.	(*iv*) $B \subset A$
(*e*) Not more than one of the events A or B occurs.	(*v*) $[A - (A \cap B)] \cup [B - A \cap B)]$
	(*vi*) $\bar{A} \cap \bar{B}$
	(*vii*) $A \cup B$
	(*viii*) $S - (A \cup B)$

2. Match the correct expression of probabilities on the left:

(*a*) $P(\phi)$, where ϕ is null set	(*i*) $1 - P(A)$
(*b*) $P(A \mid B)\, P(B)$	(*ii*) $P(A \cap B)$
(*c*) $P(\bar{A})$	(*iii*) P(A) − P(A Ç B)
(*d*) $P(\bar{A} \cap \bar{B})$	(*iv*) 0
(*e*) $P(A \sim B)$	(*v*) $1 - P(A) - P(B) + P(A \cap B)$
(*f*) $P(A \cup B)$	(*vi*) $P(A) + P(B) - P(A \cap B)$

3. Given that A, B and C are mutually exclusive events, explain why the following are not permissible assignments of probabilities :

(*i*) $P(A) = 0{\cdot}24$, $\quad P(B) = 0{\cdot}4 \quad$ and $\quad P(A \cup C) = 0{\cdot}2$

(*ii*) $P(A) = 0{\cdot}4$, $\quad P(B) = 0{\cdot}61$

4. In each of the following, indicate whether events A and B are :

(*i*) Independent, (*ii*) mutually exclusive, (*iii*) dependent but not mutually exclusive.

(*a*) $P(A \cap B) = 0$

(*b*) $P(A \cap B) = 0{\cdot}3$, $\quad P(A) = 0{\cdot}45$

(*c*) $P(A \cup B) = 0{\cdot}85$, $\quad P(A) = 0{\cdot}3$, $\quad P(B) = 0{\cdot}6$

(*d*) $P(A \cup B) = 0{\cdot}70$, $\quad P(A) = 0{\cdot}5$, $\quad P(B) = 0{\cdot}4$

(*e*) $P(A \cup B) = 0{\cdot}90$, $\quad P(A \mid B) = 0{\cdot}8$, $\quad P(B) = 0{\cdot}5$.

5. Give the correct label as answer like *a* or *b* etc., for the following questions :

(*i*) The probability of drawing any one spade card from a pack of cards is

(*a*) $\dfrac{1}{52}$ $\qquad$ (*b*) $\dfrac{1}{13}$ $\qquad$ (*c*) $\dfrac{4}{13}$ $\qquad$ (*d*) $\dfrac{1}{4}$

(*ii*) The probability of drawing one white ball from a bag containing 6 red, 8 black, 10 yellow and 1 green balls is

(*a*) $\dfrac{1}{25}$ $\qquad$ (*b*) 0 $\qquad$ (*c*) 1 $\qquad$ (*d*) $\dfrac{24}{25}$ $\qquad$ (*e*) $\dfrac{15}{20}$

(*iii*) A coin is tossed three times in succession, the number of sample points in sample space is

(*a*) 6 $\qquad$ (*b*) 8 $\qquad$ (*c*) 3 $\qquad$ (*d*) 9

(*iv*) In the simultaneous tossing of two perfect coins, the probability of having at least one head is

(*a*) $\dfrac{1}{2}$ $\qquad$ (*b*) $\dfrac{1}{4}$ $\qquad$ (*c*) $\dfrac{3}{4}$ $\qquad$ (*d*) 1 (*v*) In the simultaneous tossing of two perfect dice, the probability of obtaining 4 as the sum of the resultant faces is

(*a*) $\frac{4}{12}$ (*b*) $\frac{1}{12}$ (*c*) $\frac{3}{12}$ (*d*) $\frac{2}{12}$

(*vi*) A single letter is selected at random from the word 'probability'. The probability that it is a vowel is

(*a*) $\frac{3}{11}$ (*b*) $\frac{2}{11}$ (*c*) $\frac{4}{11}$ (*d*) 0

(*vii*) An urn contains 9 balls, two of which are red, three blue and four black, Three balls are drawn at random. The chance that they are of the same colour is

(*a*) $\frac{5}{84}$ (*b*) $\frac{3}{9}$ (*c*) $\frac{3}{7}$ (*d*) $\frac{7}{17}$

(*viii*) A number is chosen at random among the first 120 natural numbers. The probability of the number chosen being a multiple of 5 or 15 is

(*a*) $\frac{1}{5}$ (*b*) $\frac{1}{8}$ (*c*) $\frac{1}{16}$ (*d*) $\frac{1}{9}$

(*ix*) If A and B are mutually exclusive events, then

(*a*) $P(A \cup B) = P(A) \cdot P(B)$, (*b*) $P(A \cup B) = P(A) + P(B)$, (*c*) $P(A \cup B) = 0$.

(*x*) If A and B are two independent events, the probability that both A and B occur is (1/8) and the probability that neither of them occurs is (3/8). If $P(A) < P(B)$, then the probability of the occurrence of A is :

(*a*) $\frac{1}{2}$, (*b*) $\frac{1}{3}$, (*c*) $\frac{1}{4}$, (*d*) $\frac{1}{5}$,

(*xi*) If A and B are two events, the probability that exactly one of them occurs is given by :

(*a*) $P(A) + P(B) - 2P(A \cap B)$, (*b*) $P(A) + P(B) - P(A \cap B)$

(*c*) $P(\bar{A}) + P(\bar{B}) - 2P(\bar{A} \cap \bar{B})$, (*d*) $P(A \cap \bar{B}) + P(\bar{A} \cap B)$

(*xii*) If $P(A \cap B) = \frac{1}{2}$, $P(\bar{A} \cap \bar{B}) = \frac{1}{3}$, and $P(A) = P(B) = p$, then the value of p is given by :

(*a*) $\frac{1}{2}$, (*b*) $\frac{7}{8}$, (*c*) $\frac{1}{3}$, (*d*) $\frac{7}{12}$.

(*xiii*) If $P(A \cap B) = \frac{1}{2}$, $P(\bar{A} \cap \bar{B}) = \frac{1}{2}$ and $2P(A) = P(B) = p$, then the value of p is given by :

(*a*) $\frac{1}{4}$, (*b*) $\frac{1}{2}$, (*c*) $\frac{1}{3}$, (*d*) $\frac{2}{3}$.

(*xiv*) A and B are two independent events such that $P(A \cap \bar{B}) = \frac{3}{25}$ and $P(\bar{A} \cap B) = \frac{8}{25}$. If $P(A) < P(B)$, then $P(A)$ is :

(*a*) $\frac{1}{5}$, (*b*) $\frac{2}{5}$, (*c*) $\frac{3}{5}$, (*d*) $\frac{4}{5}$.

(*xv*) A and B are two independent events such that $P(\bar{A}) = 0{\cdot}7$, $P(\bar{B}) = k$ and $P(A \cup B) = 0{\cdot}8$, then k is

(*a*) $\frac{5}{7}$, (*b*) 1, (*c*) $\frac{2}{7}$, (*d*) none of these.

(*xvi*) The probability that a 3-card hand drawn at random and without replacement from an ordinary deck consists entirely of black cards is :

(*a*) $\frac{1}{17}$, (*b*) $\frac{2}{17}$, (*c*) $\frac{1}{8}$, (*d*) $\frac{3}{17}$ (*e*) $\frac{4}{17}$.

(*xvii*) What is the probability that a bridge hand contains one card of each denomination (*i.e.*, one ace, one king, one queen, ..., one three, one two) ?

(*a*) $\frac{13!}{13^{13}}$, (*b*) $\frac{4^{13}}{^{52}C_{13}}$, (*c*) $\frac{^{52}C_4}{^{52}C_{13}}$ (*d*) $\left(\frac{1}{13}\right)^{13}$ (*e*) $\frac{13^4}{^{52}C_{13}}$

(*xviii*) If the events *S* and *T* have equal probability & are independent with $P(S \cap T) = p > 0$, then $P(S)$

(*a*) $\sqrt{p}$, (*b*) p^2, (*c*) $\frac{p}{2}$, (*d*) p, (*e*) $2p$

(*xix*) The probability that both *S* and *T* occur, the probability that *S* occurs and *T* does not, and the probability that *T* occurs and *S* does not are all equal to *p*. The probability that either *S* or *T* occurs is :

(*a*) p, (*b*) $2p$, (*c*) $3p$, (*d*) $3p^2$, (*e*) p^3

(*xx*) Events *S* and *T* are independent with $P(S) < P(T)$, $P(S \cap T) = 6/25$, and $P(S \mid T) + P(T \mid S) = 1$. Then $P(S)$ is

(*a*) $\frac{1}{25}$, (*b*) $\frac{1}{5}$, (*c*) $\frac{6}{25}$, (*d*) $\frac{2}{5}$, (*e*) $\frac{3}{5}$

(*xxi*) An unbiased die is thrown two independent times, Given that the first throw resulted in an even number, the probability that the sum obtained is 8 is :

(*a*) $\frac{5}{36}$, (*b*) $\frac{1}{6}$, (*c*) $\frac{4}{21}$, (*d*) $\frac{7}{36}$, (*e*) $\frac{1}{3}$

6. Fill in the blanks :

(*i*) Two events are said to be equally likely if ...

(*ii*) A set of events is said to be independent if ...

(*iii*) If $P(A) \cdot P(B) \cdot P(C) = P(A \cap B \cap C)$, then the events *A*, *B*, *C* are ...

(*iv*) Two events *A* and *B* are mutually exclusive if $P(A \cap B) = \ldots$ and are independent if $P(A \cap B) = \ldots$

(*v*) The probability of getting a multiple of 2 in a throw of a dice is 1/2 and of getting a multiple of 3 is 1/3. Hence probability of getting a multiple of 2 or 3 is ...

(*vi*) Let *A* and *B* be independent events and suppose the event *C* has probability 0 or 1. Then *A*, *B* and *C* are ... events.

(*vii*) If *A*, *B*, *C* are pairwise independent and *A* is independent of $B \cup C$, then, *A*, *B*, *C* are ... independent.

(*viii*) A man has tossed 2 fair dice. The conditional probability that he has tossed two sixes, given that he has tossed at least one six is ...

(*ix*) Let *A* and *B* be two events such that $P(A) = 0{\cdot}3$ and $P(A \cup B) = 0.8$. If *A* and *B* are independent events than $P(B) = \ldots$

(*x*) If $(1 + 3p)/3$, $(1 - p)/4$ and $(1 - 2p)/2$ are probabilities of three mutually exclusive events, then the set of all values of *p* is ...

7. Each of the following statements is either true or false. If it is true prove it, otherwise, give a counter example to show that it is false :

(*i*) The probability of occurrence of at least one of two events is the sum of the probability of each of the two events.

(ii) Mutually exclusive events are independent.

(iii) For any two events A and B, $P(A \cap B)$ cannot be less than either $P(A)$ or $P(B)$.

(iv) The conditional probability of A given B is always greater than $P(A)$.

(v) If the occurrence of an event A implies the occurrence of another event B then $P(A)$ cannot exceed $P(B)$.

(vi) For any two events A and B, $P(A \cup B)$ cannot be greater than either $P(A)$ or $P(B)$.

(vii) Mutually exclusive events are not independent.

(viii) Pairwise independence does not necessarily imply mutual independence.

(ix) Let A and B be events neither of which has probability zero. Then if A and B are disjoint, A and B are independent.

(x) The probability of any event is always a proper fraction.

(xi) If $0 < P(B) < 1$ so that $P(A \mid B)$ and $P(A \mid \bar{B})$ are both defined, then

$$P(A) = P(B)\, P(A \mid B) + P(\bar{B})\, P(A \mid \bar{B}).$$

(xii) For two events A and B if $P(A) = P(A \mid B) = 1/4$ and $P(A \mid \bar{B}) = 1/2$, then

(a) A and B are mutually exclusive. (b) A and B are independent.

(c) A is a sub-event of B. (d) $P(\bar{A} \mid B) = 3/4$.

(xiii) Two events can be independent and mutually exclusive simultaneously.

(xiv) Let A and B be events, neither of which has probability zero. Prove or disprove the following:

(a) If A and B are disjoint, A and B are independent.

(b) If A and B are independent, A and B are disjoint.

(xv) If $P(A) = 0$, then $A = \phi$.

(xvi) For two events A and B we have the following probabilities :

$$P(A) = P(A \mid B) = \frac{1}{4} \quad \text{and} \quad P(B \mid A) = \frac{1}{2}.$$

Check whether the following statements are true or false :

(a) A and B are mutually exclusive, (b) A and B are independent,

(b) A is a sub-event of B, (c) $P(\bar{A} \mid B) = \frac{3}{4}$.

DISCUSSION AND REVIEW QUESTIONS

1. (a) Give the classical and statistical definitions of probability. What are the objections raised in these definitions ?

(b) When are a number of cases said to be equally likely ? Give an example of the following :

(i) the equally likely cases,

(ii) four cases which are not equally likely, and

(iii) five cases in which one case is more likely than the other four.

(c) What is meant by mutually exclusive events ? Give an example of

(i) three mutually exclusive events,

(ii) three events which are not mutually exclusive.

(d) Can

(i) events be mutually exclusive and exhaustive ?

(*ii*) events be exhaustive and independent ?

(*iii*) events be mutually exclusive and independent ?

(*iv*) events be exhaustive, mutually exclusive and independent ?

2. (*i*) If *A*, *B* and *C* are any three events, write down the theoretical expressions for the following events :

(*a*) Only *A* occurs, (*b*) *A* and *B* occur but *C* does not,

(*c*) At least one occurs (*d*) *A*, *B*, and *C* all the three occur,

(*e*) At least two occur, (*f*) One does not occur,

(*g*) Two do not occur, (*h*) None occurs.

(*ii*) *A*, *B* and *C* are three events. Express the following events in appropriate symbols :

(*a*) Simultaneous occurrence of *A*, *B* and *C*;

(*b*) Occurrence of at least one of them;

(*c*) *A*, B and *C* are mutually exclusive events ;

(*d*) Every point of *A* is contained in *B*;

(*e*) The event *B* but not *A* occurs.

3. A sample space *S* contains four points x_1, x_2, x_3 and x_4 and the values of a set function *P*(*A*) are known for the following sets :

$$A_1 = (x_1, x_2) \text{ and } P(A_1) = \frac{4}{10}; \qquad A_2 = (x_3, x_4) \text{ and } P(A_2) = \frac{6}{10};$$

$$A_3 = (x_1, x_2, x_3) \text{ and } P(A_3) = \frac{4}{10}; \qquad A_4 = (x_2, x_3, x_4) \text{ and } P(A_4) = \frac{7}{10}$$

Show that :

(*i*) The total number of sets (including the "null" set of number points) of points of *x* is 16.

(*ii*) Although the set containing no sample point has zero probability, the converse is not always true, *i.e.*, set may have zero probability and yet it may be the set of a number of points.

4. Describe explicitly the sample spaces for each of the following experiments :

(*i*) The tossing of four coins.

(*ii*) The throwing of three dice.

(*iii*) The tossing of ten coins with the aim of observing the number of tails coming up.

(*iv*) Two cards are selected from a standard deck of cards.

(*v*) Four successive draws (*a*) with replacement, and (*b*) without replacement, from a bag containing fifty coloured balls out of which ten are white, twenty blue and twenty red.

(*vi*) A survey of families with two children is conducted and sec of the children (the older child first) is recorded.

(*vii*) A survey of families with three children is made and the sex of the children (in order of age, oldest child first) are recorded.

(*viii*) Three distinguishable objects distributed in three numbered cells.

(*ix*) A poker hand (five cards) is dealt from an ordinary deck of cards.

(*x*) Selecting *r* screws from the lot produced by a machine, a screw can be defective or non-defective.

5. In an experiment a coin is thrown five times. Write down the sample space. How many points are there in the sample space ?

6. Describe sample space appropriate in each of the following cases :

(*i*) *n*-tosses of a coin with head or tail as outcomes in each toss.

(ii) Successive tosses of a coin until a head turns up.

(iii) Two successive draws, (a) with replacement, and (b) without replacement, from a bag containing 4 coloured toys out of which one is white, one black and 2 red toys.

7. (a) An experiment consists of tossing an unbiased coin until the same result appears twice in succession for the first time. To every possible outcome requiring n tosses attribute probability $1/2^n$. Describe the sample space.

(b) A coin is tossed until there are either two consecutive heads or two consecutive tails or the number of tosses becomes five. Describe the sample space along with the probability associated with each sample point, if every sequence of n tosses has probability 2^{-n}.

8. Urn 1 contains two white, one red and 3 black balls. Urn 2 contains one white, 3 red and 2 black balls. An experiment consists of first selecting an urn and then drawing a ball from this urn. Define a suitable sample space for this experiment.

9. Suppose an experiment has n outcomes $A_1, A_2, \ldots, A_n$ and that it is repeated r times. Let $x_1, x_2, \ldots, x_n$ record the number of occurrences of $A_1, A_2, \ldots, A_n$. Describe the sample space. Show that the number of sample points is ${}^{n+r-1}C_{r-1}$.

10. A manufacturer buys parts from four different vendors numbered 1, 2, 3 and 4. Referring to orders placed on two successive days, (1, 4) denotes the event that on the first day, the order was given to vendor 1 and on the second day it was given to vendor 4. Letting A represent the event that vendor 1 gets at least one of these two orders, B the event that the same vendor gets both orders and C the event that vendors 1 and 3 do not get either order. List the elements of :

(a) Entire sample space, (b) A, (c) B', (d) C, (e) $\bar{A}$, (f) $\bar{B}$,

(g) $B \cup C$, (h) $A \cap B$ (i) $A \cap C$, (j) $\overline{A \cup B}$, and, (k) $A - B$

11. (a) With usual notations, prove that $P(A \cup B) = P(A) + P(B) - P(A \cap B)$.

Deduce a similar result for $P(A \cup B \cup C)$, where C is one more event.

(b) For any event : E_i, $P(E_i) = p_i$, $(i = 1, 2, 3)$; $P(E_i \cap E_j) = p_{ij}$, $(i, j = 1, 2, 3)$ and $P(E_1 \cap E_2 \cap E_3) = p_{123}$, find the probability that of the three events, (i) at least one, and (ii) exactly one happens.

(c) Discuss briefly the axiomatic approach to probability, illustrating with examples how it meets the deficiencies of the classical approach.

12. (a) What is a 'probability space' ? State (i) the 'law of total probability' and (ii) Boole's inequality, for events not necessarily mutually exclusive.

(b) Explain the following with examples :

(i) random experiment, (ii) an event, (iii) an event space.

State the axioms of probability and explain their frequency interpretations.

(c) Define conditional probability and give its frequency interpretation. Show that conditional probabilities satisfy the axioms of probability.

13. Prove the following laws, in each case assuming the conditional probabilities being defined.

(a) $P(E \mid E) = 1$; (b) $P(\phi \mid F) = 0$

(c) If $E_1 \subseteq E_2$, then $P(E_1 \mid F) < P(E_2 \mid F)$; (d) $P(\bar{E} \mid F) = 1 - P(E \mid F)$

(e) $P(E_1 \cup E_2 \mid F) = P(E_1 \mid F) + P(E_2 \mid F) - P(E_1 \cap E_2 \mid F)$ (f) If $P(F) = 1$ then $P(E \mid F) = P(E)$;

(g) $P(E - F) = P(E) - P(E \cap F)$

(h) If $P(F) > 0$, and E and F are mutually exclusive then $P(E \mid F) = 0$

(i) If $P(E \mid F) = P(E)$, then $P(E \mid \bar{F}) = P(E)$ and $P(\bar{E} \mid F) = P(\bar{E})$

14. (a) If $P(\bar{A}) = a$, $P(\bar{B}) = b$, then prove that $P(A \cap B) \geq 1 - a - b$.

(b) $P(a) = \alpha$, $P(B) = \beta$, then prove that $P(A \mid B) \geq (\alpha + \beta - 1)/\beta$.

(c) If $\bar{A}$ denotes the non-occurrence of event A, then prove that

$$P(\bar{A}_1 \cup \bar{A}_2 \cup \bar{A}_3) = 1 - P(A_1)\,P(A_2 \mid A_1)\,P(A_3 \mid A_1 \cap A_2)$$

(d) If A, B and C are three arbitrary events and

$$S_1 = P(A) + P(B) + P(C), \quad S_2 = P(A \cap B) + P(B \cap C) + P(C \cap A), \quad S_3 = P(A \cap B \cap C),$$

prove that the probability that exactly one of the three events occurs is given by :

$$S_1 - 2S_2 + 3S_3$$

15. (a) If $P(A) > 0$, $P(B) > 0$ and $P(A \mid B) = P(B \mid A)$, then $P(A) = P(B)$.

(b) Let A and B be two events, neither of which has probability zero. Then if A and B are disjoint. A and B are independent.

(c) Under what conditions does the following equality hold ?

$$P(A) = P(A \mid B) + P(A \mid \bar{B})$$

16. If A and B are two events and $P(B) \neq 1$, prove that

(i) $P(A \mid \bar{B}) = \dfrac{[P(A) - P(A \cap B)]}{[1 - P(B)]}$,

where $\bar{B}$ denotes the event complementary to B and hence deduce that

(ii) $P(A \cap B) \geq P(A) + P(B) - 1$

(iii) Also show that $P(A) >$ or $< P(A \mid B)$ according as $P(A \mid \bar{B}) >$ or $< P(A)$.

Hint. (i) and (ii) $P(A \mid \bar{B}) = \dfrac{P(A \cap \bar{B})}{P(\bar{B})} = \dfrac{[P(A) - P(A \cap B)]}{[1 - P(B)]} \leq 1$

$\Rightarrow \quad P(A) + P(B) - 1 \leq P(A \cap B)$

(iii) $\dfrac{P(A \mid \bar{B})}{P(A)} = \dfrac{P(\bar{B} \mid A)}{P(\bar{B})} = \dfrac{1 - P(B \mid A)}{1 - P(B)}$

Now $P(A \mid \bar{B}) > P(A)$ if $\{1 - P(B \mid A)\} > \{1 - P(B)\}$

i.e., if $P(B \mid A) < P(B)$, i.e., if $\dfrac{P(B \mid A)}{P(B)} < 1$ i.e., if $\dfrac{P(A \mid B)}{P(A)} < 1$, i.e., if $P(A) > P(A \mid B)$

17. (a) If A and B are two mutually exclusive events, show that : $P(A \mid \bar{B}) = P(A)/[1 - P(B)]$

(b) If A and B are two mutually exclusive events and $P(A \cup B) \neq 0$, then :

$$P(A \mid A \cup B) = \frac{P(A)}{P(A) + P(B)}$$

(c) Let A, B and C denote events. If $P(A \mid C) > P(B \mid C)$ and $P(A \mid \bar{C}) \geq P(B \mid \bar{C})$, then show that $P(A) \geq P(B)$.

18. (a) If A and B are independent events defined on a given probability space $\{\Omega, A, P(\cdot)\}$, then prove that (i) A and $\bar{B}$ are independent, (ii) $\bar{A}$ and $\bar{B}$ are independent.

(b) A, B and C are three events such that A and B are independent, $P(C) = 0$. Show that A, B and C are independent.

(c) An event A is known to be independent of the events B, $B \cup C$ and $B \cap C$. Show that it is also independent of C.

(d) Show that if an event C is independent of two mutually exclusive events A and B, then C is also independent of $A \cup B$.

19. Prove or disprove :

(a) (i) If $P(A \mid B) \geq P(A)$, then $P(B \mid A) \geq P(B)$ and (ii) If $P(A) = P(\bar{B})$, then $A = \bar{B}$.

(b) If $P(A) = 0$, then $A = \phi$.

(c) For possible events A, B, C :

(i) If $P(A) > P(B)$, then $P(A \mid C) > P(B \mid C)$.

(ii) If $P(A \mid C) \geq P(B \mid C)$ and $P(A \mid \bar{C}) \geq P(B \mid \bar{C})$, then $P(A) \geq P(B)$.

(d) If $P(A) = 0$, then $P(A \cap B) = 0$.

(e) (i) If $P(A) = P(B) = p$, then $P(A \cap B) \leq p^2$.

(ii) If $P(B \mid \bar{A}) = P(B \mid A)$, then A and B are independent.

20. (a) Prove that for any integer $m \geq 1$.

(i) $P\left(\bigcap\limits_{i=1}^{m} A_i\right) \leq P(A_i) \leq P\left(\bigcup\limits_{i=1}^{m} A_i\right) \leq \sum\limits_{i=1}^{m} P(A_i)$; (ii) $P\left(\bigcap\limits_{i=1}^{m} A_i\right) \geq 1 - \sum\limits_{i=1}^{m} P(\bar{A}_i)$

(b) For the events $A_1, A_2, \ldots, A_n$ assuming, $P\left(\bigcup\limits_{i=1}^{n} A_i\right) \leq \sum\limits_{i=1}^{n} P(A_i)$, prove that

(i) $P\left(\bigcap\limits_{i=1}^{n} A_i\right) \geq 1 - \sum\limits_{i=1}^{n} P(\bar{A}_i)$ and (ii) $P\left(\bigcap\limits_{i=1}^{n} A_i\right) \geq \sum\limits_{i=1}^{n} P(A_i) - (n-1)$

ASSORTED REVIEW PROBLEMS FOR SELF-ASSESSMENT

3.1. With reference to the pair of dice (red and green) throwing experiment, calculate the probabilities of the following events :

(i) The sum of the numbers is divisible by four.

(ii) The number on the green die is greater than the number on the red die.

(iii) The sum of the numbers is divisible by either 4 or 3 or both 4 and 3.

(iv) The sum of the numbers is divisible by 3 and 4 both.

(v) The sum of the numbers is odd and the first die shows an even number.

(vi) The sum of the numbers is less than five or the same number occurs on both the dice.

(vii) The sum of the numbers is 3 or 11.

Ans. (i) 7/36, (ii) 15/36, (iii) 1/2, (iv) 1/36, (v) 1/4, (vi) 1/6, (vii) 1/9.

3·2. There are 30 balls in an urn. The balls are numbered from 1 to 30, one number per ball. If one of the balls is drawn at random, what is the probability that the number on it is:

(i) divisible by 2 and 3 both; (ii) a perfect square;

(iii) a 2-digit number; (iv) a prime number, and

(v) a prime number that is 2 more than another prime number.

Ans. (i) 1/6, (ii) 1/6, (iii) 7/10, (iv) 1/3, (v) 2/15.

3·3. A letter of the English alphabet is chosen at random. Calculate the probability that the letter so chosen :

(i) is a consonant ; (ii) precedes r (in alphabetical order) ;

(iii) precedes x and is a vowel ; (iv) follows w and is a vowel ;

(v) follows p and precedes y and is a vowel

Ans. (i) 21/26, (ii) 17/26, (iii) 5/26, (iv) 0, (v) 1/26.

3·4. (*a*) In a single throw of 3 dice, find the probability of getting the same number on all the three dice.

(*b*) In a single throw of three dice, find the probability of getting a sum of (*i*) 5 and (*ii*) at least 5.

Ans. (*a*) 1/36, (*b*) (*i*) 1/36, (*ii*) 53/54.

3·5. (*a*) From a pack of playing cards, 2 cards are drawn at random. Find the probability that one is a king and the other is a queen.

(*b*) Two cards are drawn at random from a well-shuffled pack of 52 cards. Show that the chance of drawing two aces is (1/221).

3·6. (*a*) A bag contains 6 red, 5 white and 4 black balls. If two balls are drawn, find the probability that none of them is red.

(*b*) A bag contains 7 red balls and 5 white balls. 4 balls are drawn at random. What is the probability that :

(*i*) all of them are red; (*ii*) two of them would be red and two white ?

(*c*) An urn contains 9 balls, two of which are red, three blue and four black. Three balls are drawn from the urn at random. What is the probability that :

(*i*) the balls are of different colours,

(*ii*) two balls are of the same colour and third is of different colour,

(*iii*) the three balls are of the same colour.

(*d*) From a bag containing 4 white and 5 black balls, 3 balls are drawn at random. What are the odds against these being all black ?

Ans. (*a*) 12/35, (*b*) (*i*) 7/99, (*ii*) 14/33, (*c*) (*i*) 2/7, (*ii*) 55/84, (*d*) 13 : 5.

3·7. In a lottery of 50 tickets numbered 1 to 50, two tickets are drawn simultaneously. Find the probability when :

(*i*) both the tickets drawn have prime numbers,

(*ii*) none of the tickets drawn has prime number.

Ans. (*i*) 21/245, (*ii*) 119/245.

3·8. What is the probability of getting 9 cards of the same suit in one hand at a game of bridge ?

Ans. $\left({}^{13}C_9 \times {}^{39}C_4 \times 4\right) / {}^{52}C_{13}$.

3·9. (*a*) The number 1, 2, 3, 4, 5 and 6 are written on slips of paper and two of these slips are drawn. What is the probability

(*i*) That the sum of the numbers drawn is 9 ?

(*ii*) That the sum of the numbers drawn is 5 or less ?

(*iii*) That one of the numbers drawn is odd or greater than 3 ?

(*iv*) That the numbers drawn are both odd ?

Ans. (*a*) (*i*) 2/15, (*ii*) 4/15, (*iii*) 1, (*iv*) 1/5;

(*b*) A fair coin and a fair die are tossed and a card is randomly selected from a standard deck of 52 playing cards. Determine the probability that the coin, die and card, respectively show (*i*) a head, a5, and the king of diamonds, (*ii*) a head, a 5, and king, (*iii*) a tail, a 3 or 5, and a king, and (*iv*) a tail, an odd number, and a heart.

Ans. (*b*) (*i*) 1/624, (*ii*) 4/624, (*iii*) 8/624, (*iv*) 39/624.]

(*c*) Two different digits are chosen at random from the set 1, 2, 3, ..., 8. Show that the probability that the sum of the digits will be equal to 5 is same as the probability that their sum will exceed 13, each being 1/14. Also show that the chance of both digits exceeding 5 is 3/28.

(*d*) A die is loaded in such a manner that for $n = 1, 2, 3, 4, 5$ and 6 the probability of the face marked n, landing on top when the die is rolled is proportional to n. Find the probability that an odd number will appear on tossing the die.

Hint. $P(n) \propto n$ or $P(n) = kn$, where k is the constant of proportionality.

Also $P(1) + P(2) + \dots P(6) = 1 \Rightarrow k(1 + 2 + 3 + 4 + 5 + 6) = 1 \Rightarrow k = 1/21$

$$\text{Required probability} = P(1) + P(3 + P(5) = \frac{1+3+5}{21} = \frac{3}{7}.$$

3·10. A bag contains 50 tickets numbered 1, 2, 3, ..., 50 of which five are drawn at random and arranged in ascending order of the magnitude ($x_1 < x_2 < x_3 < x_4 < x_5$). What is the probability that $x_3 = 30$?

Ans. ${}^{29}C_2 \times {}^{20}C_2 / {}^{50}C_5$

3·11. In a party, n persons are seated at a round table. Find the odds against two specified individuals sitting next to each other.

Ans. (*a*) $\dfrac{n-3}{n-1} : \dfrac{2}{n-1} = (n-3):2$

3·12. In a random arrangement of the letter of the word 'MATHEMATICS', find the probability that all the vowels come together.

Hint. Total number of permutations of the letters of the word 'MATHEMATICS' are : $\dfrac{11!}{2!\,2!\,2!}$,
because it contains 11 letters, of which 2 are A's, 2M's, 2T's and remaining are all different.

The word 'MATHEMATICS' contains 4 vowels, *viz.*, *AEAI*, (2 A's being identical). To obtain the total number of arrangements in which these 4 vowels come together, we regard them as tied together, forming only one word so that total number of letters in MATHEMATICS may be taken as 11 – 3 = 8, out of which 2 are M's, 2 are T's and rest distinct and, therefore, their number of arrangements is given by 8!/2! 2!

Further, the four vowels *AEAI*, two of which are identical and rest distinct can be arranged among themselves in 4!/2! ways. Hence, the total number of arrangements favourable to getting all vowels together is : $\dfrac{8!}{2!\,2!} \times \dfrac{4!}{2!}$.

$$\therefore \quad \text{Required probability} = \frac{8!4!}{2!\,2!\,2!} \div \frac{11!}{2!2!2!} = \frac{4}{165}$$

3·13. In a random arrangement of the letter of the word 'COMMERCE', find the probability that all the vowels come together.

Ans. $\dfrac{6! \times 3!}{2!\,2!\,2!} \div \dfrac{8!}{2!\,2!\,2!} = \dfrac{3}{28}$.

3·14. (*a*) A five digit number is to be formed by using the numbers, 1, 2, 3, 4, 5. Find the probability that the number is greater than 23,000.

Hint. Exhaustive cases = 5! = 120. The favourable cases for a five digit number less than 23,000 are:

Number beginning with 1	1	•	•	•	•	Cases 4! = 24
Number beginning with 21	2	1	•	•	•	Cases 3! = 6

∴ Number of favourable cases for five digit number to be greater than 23,000 are 120 – 24 – 6 = 90.

Hence, $\quad \text{Required probability} = \dfrac{90}{120} = \dfrac{3}{4} = 0{\cdot}75.$

(*b*) A six digit number is to be formed using odd and even digits at odd and even places respectively. Find the probability that the number so formed is divisible by 4.

Hint. Exhaustive cases :

(i)	(ii)	(iii)	(iv)	(v)	(vi)
•	•	•	•	•	•

The digits at places (i), (iii) and (v) are to be filled with odd digits 1, 3, 5, 7, 9 which can be done in ${}^5P_3$ ways. Similarly, for even digits 0, 2, 4, 6, 8 to be placed in places (ii), (iv) and (vi), there are ${}^5P_3$ ways.

$\therefore$ Exhaustive number of ways $= {}^5P_3 \times {}^5P_3 = (5 \times 4 \times 3)^2 = 3600.$

The number is divisible by 4, if the last two digits (ten's and unit's digits) are divisible by 4 and these cases, keeping in mind that (i), (iii) and (v) places are to be filled with odd digits and (ii), (iv) and (vi) places are filled with even digits are :

12, 32, 52, 72, 92 ; 16, 36, 56, 76, 96 *i.e.*, 10 cases in all [For filling places (v) and (vi)] and ${}^4P_2 \times {}^4P_2$ cases for filling places (i) to (iv).

$\therefore$ Number of favourable cases $= 4P_2 \times 4P_2 \times 10 = 4 \times 3 \times 4 \times 3 \times 10 = 1440$

Hence, Required probability $= \dfrac{1440}{3600} = \dfrac{2}{5} = 0{\cdot}4.$

3·15. In a group of 25 persons, what is the probability that all 25 will have different birthday ? Assume a 365 – day year and that all days are equally likely.

Ans. $365 \times 364 \times \ldots \times 341/(365)^{25}$

3·16. Three numbers are selected at random one after another and without replacement from m numbers 1, 2, ..., m. Find the probability that :

(i) The first number drawn is the smallest and the second number the largest, and

(ii) The first number is smaller than the second number.

Ans. (i) 1/6, (ii) (3/6) = 1/2

3·17. (a) The chance of one event happening is the square of the chance of a second event but the odds against the first are the cubes of the odds against the second. Find the chances of these events.

Hint. Solve : $p_1 = p_2^2$ and $\left(\dfrac{1-p_1}{p_1}\right) = \left(\dfrac{1-p_2}{p_2}\right)^3$

Ans. $p_1 = 1/9,\ p_2 = 1/3.$

(b) A pack of cards is divided by cutting it into two unequal portions. If a card is drawn from portion 'A', the odds are two to one against it being a red card. A red card is now transferred from portion B to portion A. Now, the odds are 2 to 1 against a card drawn at random from portion B being black. How was the pack originally divided in portions A and B ?

Ans. Portion A : Red balls = 9 ; Black balls = 18 ; Total = 27

Portion B : Red balls = 17 ; Black balls = 8; Total = 25.

(c) A bag contains n white and n black balls. Pairs of balls are drawn without replacement until the bag is empty. Show that the probability that each pair consists of one white and one black ball is $\{2^n / {}^{2n}C_n\}$.

Hint. Exhaustive number of cases $= {}^{2n}C_2 \cdot {}^{2n-2}C_2 \cdot {}^{2n-4}C_2 \cdots {}^4C_2 \cdot {}^2C_2$

Favourable number of cases $= n^2 \cdot (n-1)^2 \ldots 2^2 \cdot 1^2.$

(d) From an urn containing 'a' white balls and 'b' black balls, a certain number of balls, 'k' is drawn and they are laid aside their colour unnoticed. Then one more ball is drawn. Find the probability that it is

(i) white and (ii) black.

Hint. (i) Required probability $= \dfrac{a \times {}^{a+b-1}P_k}{{}^{a+b}P_{k+1}} = \dfrac{a}{a+b}$ (ii) Required probability $= \dfrac{b}{a+b}$

(e) From an urn containing 'a' white balls and 'b' black balls, balls are drawn one by one until balls of one colour only are left ? What is the probability that only black balls are left ?

Hint. Let us suppose that after drawing 'k' balls from the urn containing $(a + b)$ balls, only black balls are left in the urn. Then the $(k + 1)$th ball, in that order is a black ball. Now the problem reduces to the problem 3·17(d).

3·18. (a) A person applies for the post of manager in two firms A and B. He estimates that the probability of his being selected in firm A is 0·75, being rejected in firm B is 0·45, and being rejected in at least one of the firms is 0·55. Find the probability that he will be selected in at least one of the firms.

(b) A factory needs two raw materials, say X and Y. The probability of not having an adequate supply of material X is 0·06, whereas the probability of not having an adequate supply of material Y is 0·04. A study shows that the probability of a shortage of both X and Y is 0·02. What is the probability of the factory being short of either material X or Y ?

Ans. (a) 0·85, (b) 0·08.

3·19. (a) A card is drawn from a deck of cards. Find the probability of getting a king or a heart or a red card.

(b) Suppose that if a person visits his dentist, the probability that he will have his teeth cleaned is 0·44, the probability that he will have a cavity filled is 0·24, the probability that he will have a tooth extracted is 0·21, the probability that he will have his teeth cleaned and a cavity filled is 0·08, the probability that he will have his teeth cleaned and a tooth extracted is 0·11, the probability that he will have a cavity filled and a tooth extracted is 0·07, and the probability that he will have his teeth cleaned, a cavity filled, and a tooth extracted is 0·03. What is the probability that a person visiting his dentist will have at least one of these things done to him ?

Ans. (a) 7/13, (b) 0·66.

3·20. (a) A die is thrown twice and the sum of the numbers appearing is observed to be 7. What is the conditional probability that the number 2 has appeared atleast once?

(b) A pair of fair dice is thrown. If the two numbers appearing are different, find the probability that the sum is 4 or less.

(c) A pair of fair dice is thrown. Find the probability that the sum is 10 or greater if 5 appears on the first die.

Ans. (a) 1/3, (b) 2/15, (c) 1/3.

3·21. A natural number is selected at random from amongst the first 100 natural numbers.

(i) What is the probability that the second digit of the number so selected is greater than 6, given that the first digit is 4 ?

(ii) What is the probability that, if the first digit of the number selected is 7, then the second digit is 4 ?

Ans. (i) 3/10, (ii) 1/10.

3·22. (a) A bag contains 5 white and 3 black balls. Two balls are drawn one after the other without replacement. Find the probability that second ball drawn is black given that the first ball drawn is white.

(b) A bag contains 6 white and 9 black balls. Four balls are drawn at a time. Find the probability for the first draw to give 4 white balls and the second to give 4 black balls when the balls are not replaced before the second draw.

(c) A box contains 5 white and 7 black balls. There balls are drawn in succession. What is the probability that all three are white if

(i) each ball is replaced before the next one is drawn,

(ii) none of the balls is replaced before the next draw ?

Ans. (a) 1/3, (b) 1066/5005, (c) (i) 125/1728 (ii) 1/22.

3·23. (a) The probability that automobile sales will increase next month is 0·40. The probability that sale of replacement parts will increase next month is 0·50. The probability that both the industries will experience an increase in sales is estimated to be 0·10. What is the probability that automobiles sales have increased during the month given that replacement part sales have increased ?

(*b*) The probability that a construction job will be finished on time is 3/5; the probability that there will be no strikes is 2/3, and the probability that the job will be finished on time given that there are no strikes is 4/5. (*i*) What is the probability that the job will be finished on time and there will be no strikes ? (*ii*) What is the probability that there will have been no strikes given that the job is finished on time ?

Ans. (*a*) 1/5, (*b*) (*i*) 8/15, (*ii*) 8/9.

3·24. (*a*) The H.R. department of a firm has records which show the following analysis of its 200 project managers:

Age	*Bachelor's degree only*	*Master's degree*	*Total*
Under 30	90	10	100
30 to 40	20	30	50
Over 40	40	10	50
Total	150	50	200

If one manager is selected at random from the company, find :

(*i*) The probability he has only a bachelor's degree.

(*ii*) The probability he has a master's degree given that he is over 40.

(*iii*) The probability he is under 30, given that he has only a bachelor's degree.

Ans. (*i*) 3/4, (*ii*) 1/5, (*iii*) 3/5.

(*b*) In a certain town, males and females form 50 percent of the population. It is known that 20 percent of the males and 5 percent of the females are unemployed. A research organization studying the employment situation selects an unemployed person at random. What is the probability that the person selected is a (*i*) male, (*ii*) female ? How these conditional probabilities change if the person selected was found to be employed?

Hint. The information can be displayed in the adjoining table :

Sex	Proportion of persons		Total
	Employed	Unemployed	
Male	0·400	0·100	0·5
Female	0·475	0·025	0·5
Total	0·875	0·125	1·0

Let *M*, *F* and *U* denote respectively the events that a male, a female and an unemployed person is chosen. Then the required probabilities are given by :

$$(i)\ P(M \mid U) = \frac{P(M \cap U)}{P(U)} = \frac{0{\cdot}100}{0{\cdot}125} = 0{\cdot}8 \quad \text{and} \quad (ii)\ P(F \mid U) = \frac{P(F \cap U)}{P(U)} = \frac{0{\cdot}025}{0{\cdot}125} = 0{\cdot}2$$

If the person selected happened to be an employed one, then the above probabilities would have been :

$$P(M \mid E) = \frac{P(M \cap E)}{P(E)} = \frac{0{\cdot}4}{0{\cdot}875} = 0{\cdot}457 \quad \text{and} \quad P(F \mid E) = \frac{P(F \cap E)}{P(E)} = \frac{0{\cdot}475}{0{\cdot}875} = 0{\cdot}543,$$

where *E* denotes the event of selecting an employed person.

3·25. (*a*) If $P(A \cap B) = \frac{5}{6}$, $P(A \cap B) = \frac{1}{3}$ and $P(\bar{A}) = \frac{1}{2}$, find $P(A)$ and $P(B)$.

Hence, show that *A* and *B* are independent.

Ans. $P(A) = 1/2$, $P(B) = 2/3$; $P(A \cap B) = P(A)\ P(B)$.

(*b*) Let $P(A) = p$, $P(A \mid B) = q$, $P(B \mid A) = r$. Find the relation between the numbers p, q and r such that $\bar{A}$ and $\bar{B}$ are mutually exclusive.

Hint. $P(A \cap B) = P(A)\,P(B \mid A) = P(B) \cdot P(A \mid B) \;\Rightarrow\; P(A \cap B) = pr = P(B) \cdot q \;\Rightarrow\; P(B) = pr/q$

If $\bar{A}$ and $\bar{B}$ are mutually disjoint, then $P(\bar{A} \cap \bar{B}) = 0. \;\Rightarrow\; 1 - P(A \cup B) = 0.$

3·26. (*a*) In terms of probabilities, $p_1 = P(A)$, $p_2 = P(B)$ and $p_3 = P(A \cap B)$, express :

(*i*) $P(A \cup B)$, (*ii*) $P(A \mid B)$ (*iii*) $(\bar{A} \cap \bar{B})$ under the condition that

(*i*) A and B are mutually exclusive, (*ii*) A and B are mutually independent.

(*b*) Let A and B be the possible outcomes of an experiment and suppose

$$P(A) = 0{\cdot}4, \quad P(A \cup B) = 0{\cdot}7 \quad \text{and} \quad P(B) = p$$

(*i*) For what choice of p are A and B mutually exclusive.

(*ii*) For what choice of p are A and B independent.

Ans. (*i*) 0·3, (*ii*) 0·5

(*c*) Let A, B, C be three mutually exclusive events such that :

$$P(A) = \{(1 - 3x)/2\}, \quad P(B) = \{(1 - 4x)/3\}, \quad P(C) = \{(1 + x)/6\}.$$

Show that events A, B and C are exhaustive and $(-1/4) \le x \le (1/3)$.

(*d*) A and B are independent events such that the probability that both occur simultaneously is 1/6 and neither of them occurs is 1/3. Is it possible to find $P(A)$ and $P(B)$? If so, find them.

Ans. Yes; $\{P(A) = 1/3,\ P(B) = 1/2\}$ or $\{P(A) = 1/2,\ P(B) = 1/3\}$.

3·27. (*a*) Prove that for n arbitrary independent events $A_1, A_2, \ldots, A_n$,

$$P(A_1 \cup A_2 \cup A_3 \cup \ldots \cup A_n) + P(\bar{A}_1)\,P(\bar{A}_2)\,P(\bar{A}_3) \ldots P(\bar{A}_n) = 1.$$

(*b*) $A_1, A_2, \ldots, A_n$ are n independent events with $P(A_i) = \dfrac{1}{\alpha^i}$; $i = 1, 2, \ldots, n$. Show that

$$P(A_1 \cup A_2 \ldots \cup A_n) = 1 - \alpha^{-n(n+1)/2}$$

(*c*) Suppose the events $A_1, A_2, \ldots, A_n$ are independent and that $P(A_i) = \dfrac{1}{i+1}$ for $1 \le i \le n$. Find the probability that none of the n events occurs, justifying each steps in your calculation.

Ans. $1/(n + 1)$.

(*d*) n dice are rolled. What is the probability that there is at least one 6, assuming that the dice are unbiased. Also obtain the probability if the dice are loaded, the probability of obtaining a '6' on ith dice being p_i ; $i > 1, 2, \ldots, n$.

Ans. $1 - \left(\dfrac{5}{6}\right)^n$; $1 - \prod_{i=1}^{n} (1 - p_i)$.

(*e*) When two dice are thrown n number of times, the probability of getting at least one double six is greater than 99%. What is the least numerical value of n ?

Hint. Given $1 - (35/36)^n > 0{\cdot}99$; $n = 164$.

3·28. (*a*) A group of $2N$ boys and $2N$ girls is divided into two equal groups. Find the probability 'p' that each group will be equally divided into boys and girls. Estimate 'p' using Stirling's formula for N ! for large N.

Ans. $p = \dfrac{{}^{2N}C_N \times {}^{2N}C_N}{{}^{4N}C_{2N}} = \dfrac{[(2N)!]^4}{(N!)^4\,(4N)!}$. Using $N! \simeq \sqrt{2\pi} \cdot e^{-N} N^{N+\frac{1}{2}}$, we get $p \simeq \sqrt{\dfrac{2}{\pi N}}$

(b) An urn contains r red and b black balls. Three random drawings are made such that the ball drawn is always replaced, and in addition, c balls of the colour drawn are added to the urn. Find the probability of a run of three black balls.

Hint. Let A_i (1 = 1, 2, 3) denote the event of drawing *a* black ball in the *i*th draw. Then

$$P(A_1)=\frac{b}{b+r};\quad P(A_2 \mid A_1)=\frac{b+c}{b+c+r};\quad P(A_3 \mid A_1 \cap A_2)=\frac{b+2c}{b+2c+r}$$

$$\text{Required probability} = P(A_1 \cap A_2 \cap A_3) = P(A_1)\,P(A_2 \mid A_1)\,P(A_3 \mid A_1 \cap A_2)$$

$$= \frac{b(b+c)(b+2c)}{(b+r)(b+c+r)(b+2c+r)}$$

3.29. Let A_1, A_2, A_3, A_4 be four independent events for which $P(A_1) = p$, $P(A_2) = q$, $P(A_3) = r$ and $P(A_4) = s$. Find the probability that

(i) at least one of the events occurs, (ii) exactly two of the events occur, and

(iii) at most three of the events occur.

3·30. A and B throw alternately a pair of unbiased dice, A beginning. A wins if he throws 7 before B throws 6, and B wins if he throws 6 before A throws 7. If A and B respectively denote the events that A wins and B wins the series, and a and b, respectively denote the events that it is A's and B's turn to throw the dice, show that

(i) $P(A \mid a)=\frac{1}{6}+\frac{5}{6}P(A \mid b)$, (ii) $P(A \mid b)=\frac{31}{36}P(A \mid a)$,

(iii) $P(B \mid a)=\frac{5}{6}P(B \mid b)$, and (iv) $P(B \mid b)=\frac{5}{36}+\frac{31}{36}P(B \mid a)$.

Hence, or otherwise, find $P(A \mid a)$ and $P(B \mid a)$. Also comment on the result :

$$P(A \mid a) + P(B \mid a) = 1.$$

3·31. (a) A bag contains an assortment of blue and red balls. if two balls are drawn at random, the probability of drawing two red balls is five times the probability of drawing two blue balls. Furthermore, the probability of drawing one ball of each colour is six times the probability of drawing two blue balls. How many red and blue balls are there in the bag ?

Hint. Let number of red and blue balls in the bag be r and b, respectively. Then

$$p_1 = \text{Probability of drawing two red balls} = \frac{r(r-1)}{(r+b)(r+b-1)}$$

$$p_2 = \text{Probability of drawing two blue balls} = \frac{b(b-1)}{(r+b)(r+b-1)}$$

$$p_3 = \text{Probability of drawing one red and one blue ball} = \left[\frac{2br}{(r+b)(r+b-1)}\right]$$

$$p_1 = 5p_2 \quad \text{and} \quad p_3 = 6p_2 \qquad \text{(Given)}$$

$$\therefore \quad r(r-1) = 5b\,(b-1) \quad \text{and} \quad 2br = 6b\,(b-1)$$

Ans. Red balls = 6, Blue balls = 3.

(b) Three players A, B and C alternately throw a die each in that order, the first player to throw a 6 being deemed the winner. A's die is fair, whereas B and C throw biased dice with probabilities p_1 and p_2 respectively of throwing a '6'. Find the probability of A's winning the game. Obtain the values of p_1 and p_2 so that the game is equiprobable to the three players.

Ans. $P(A)=\dfrac{1}{6-5q_1q_2}$, $P(B)=\dfrac{5p_1}{6-5q_1q_2}$, $P(C)=\dfrac{5q_1p_2}{6-5q_1q_2}$

Game will be equiprobable to the three players if $P(A) = P(B) = P(C) \Rightarrow p_1 = \dfrac{1}{5}, p_2 = \dfrac{1}{4}$.

3·32. (*a*) The probability that at least 2 of 3 people *A*, *B* and *C* will survive for 10 years is 247/315. The probability that *A* alone will survive for 10 years is 4/105 and the probability that *C* alone will die within 10 years is 2/21. Assuming that the events of the survival of *A*, *B* and *C* can be regarded as independent, calculate the probability of surviving 10 years for each person.

Ans. (*a*) 3/5, 5/7, 7/9.

(*b*) Three newspapers *A*, *B* and *C* are published in a certain city. It is estimated from a survey that 20% read *A*, 16% read *B*, 14% read *C*, 8% read *A* and *B*, 5% read *A* and *C*, 4% read *B* and *C*, and 2% read all the three newspapers. What is the probability that a randomly chosen person

(*i*) Does not read any paper, (*ii*) does not read C,

(*iii*) reads *A* but not *B*, (*iv*) reads only one of these papers, and

(*v*) reads only two of these papers.

Ans. (*b*) (*i*) 0·65, (*ii*) 0·86, (*iii*) 0·12, (*iv*) 0·22, (*v*) 0·11.

3·33. (*a*) Sixty per cent of the employees of the *XYZ* Corporation are college graduates. Of these, ten per cent are in sales. Of the employees who did not graduate from college, eighty per cent are in sales. What is the probability that

(*i*) an employee selected at random is in sales?

(*ii*) an employee selected at random is neither in sales nor a college graduate?

Ans. (*i*) 0·38, (*ii*) 0·08

(*b*) Plant I of *XYZ* manufacturing organization employs 5 production and 3 maintenance engineers, another plant II of same organization employs 4 production and 5 maintenance engineers. From any one of these plants, a single selection of two engineers is made. Find the probability that one of them would be production engineer and the other maintenance engineer.

Ans. $\dfrac{1}{2}\left[{}^5C_1 \times {}^3C_1/{}^8C_2\right] + \dfrac{1}{2}\left[{}^4C_1 \times {}^5C_1/{}^9C_2\right] = \dfrac{275}{504}$

(*c*) Urn A contains 3 white and 4 black balls and urn *B* contains 5 white and 3 black balls. One ball is chosen from urn *A* and transferred to urn *B*. Then two balls are chosen from *B* without replacement. Find the probability that both balls are of the same colour.

Ans. $\dfrac{3}{7}\left(\dfrac{{}^6C_2 + {}^3C_2}{{}^9C_2}\right) + \dfrac{4}{7}\left(\dfrac{{}^5C_2 + {}^4C_2}{{}^9C_2}\right) = \dfrac{59}{126}$

3·34. (*a*) An elevator starts with 7 passengers and stops at 10 floors. Show that

(*i*) P{No two passengers leave at the same floor} = $10 \cdot 9 \ldots 5 \cdot 4/(10)^7$

(*ii*) P{All the passengers are discharged on the 5th floor} = $(10)^{-7}$

(*iii*) P{All the passengers are discharged on the same unspecified floor} = $(10)^{-6}$

(*iv*) P{Exactly three of them are discharged on the 2nd floor} = ${}^7C_3\,(7)^4/(10)^7$

(*b*) An elevator starts with 6 passengers and stops at 100 floors. Assuming that any of the passengers may get down at any floor with equal probability, find the probability that no two or more of them get down at the same floor.

Ans. $(100 \times 99 \times 98 \times 97 \times 96 \times 95)/100^6 \approx 0{\cdot}85$

3·35. (*a*) An urn contains 4 tickets numbered 1, 2, 3, 4 and another contains 6 tickets numbered 2, 4, 6, 7, 8, 9. If one of the two urns is chosen at random and a ticket is drawn at random from the chosen urn, find the probability that ticket drawn bears the number

(*i*) 2 or 4, (*ii*) 3, (*iii*) 1 or 9

(*b*) One bag contains 5 white and 4 black balls. Another bag contains 7 white and 9 black balls. A ball is transferred from the first bag to the second bag and then a ball is drawn from the second. Find the probability that the ball is white.

(*c*) *A*, *B* and *C* are three urns which contain 2 white, 1 black ; 3 white, 2 black; and 2 white, 2 black balls respectively. One balls is drawn from urn *A* and put into the urn *B*. Then a ball is drawn from urn *B* and put into urn *C*. Then a ball is drawn from urn *C*. Find the chance that the ball drawn is white.

Ans. (*a*) (*i*) 5/12, (*ii*) 1/8, (*iii*) 5/24 (*b*) 4/9 (*c*) 47/90.

(*d*) A bag contains *n* balls, all possible proportions of white and black balls being equally likely. A ball is drawn from the bag and it turns out to be white. This ball is replaced back and then another ball is drawn which also turns out to be white. If this ball is replaced back, prove that the chance of a black ball in the next draw is :

$$\frac{1}{2}(n-1)(2n+1)^{-1}$$

3·36. (*a*) The sample space consists of (1, 2, 3, 4) and p_i is the probability assigned to i, i = 1, 2, 3, 4.

Given that : $p_1 = \frac{1}{\sqrt{2}} - \frac{1}{4}$, $p_2 = \frac{1}{4}$, $p_3 = \frac{3}{4} - \frac{1}{\sqrt{2}}$, $p_4 = \frac{1}{4}$.

examine the independence of the events $E_1 = (1, 3)$, E_2 (2, 3), $E_3 = (3, 4)$

(*b*) Eight tickets numbered 111, 121, 122, 122, 211, 212, 212, 221 are placed in a hat and stirred. One of them is then drawn at random. Show that the event *A* : "the first digit on the ticket drawn will be 1", *B* : "the second digit on the ticket drawn will be 1" and C; "the third digit on the ticket drawn will be 1", are not pairwise independent although

$$P(A \cap B \cap C) = P(A)\,P(B)\,P(C).$$

(*c*) Four identical marbles marked 1, 2, 3 and 123 respectively are put in an urn and one is drawn at random. Let A_i, (i = 1, 2, 3), denote the event that the number *i* appears on the drawn marble. Prove that the events A_1, A_2 and A_3 are pairwise independent but not mutually independent.

(*d*) A die is of the shape of a regular tetrahedron whose faces bear the numbers 111, 112, 121, 122. A_1, A_2, A_3 are respectively the events that the first two, the last two and the extreme two (first and last) digits are the same, when the die is tossed at random. Find whether or not the events A_1, A_2, A_3 are

(*i*) pairwise independent, and (*ii*) mutually (i.e., completely) independent.

(*e*) Two fair dice are thrown independently. Define the following events :

A : Even number on the first dice ; *B* : Even number on the second dice ;

C : Same number on both dice.

Prove that the events *A*, *B* and *C* are pairwise independent but not mutually independent.

3·37. (*a*) A town has two doctors *X* and *Y* operating independently. If the probability that doctor *X* is available is 0·9 and that for *Y* is 0·8, what is the probability that at least one doctor is available when needed ?

(*b*) The odds against a certain event are 5 to 2 and odds in favour of another (independent) event are 6 to 5. Find the chance that at least one of the events will happen.

(*c*) The odds that a book will be favourably reviewed by 3 independent critics are 5 to 2, 4 to 3 and 3 to 4 respectively. What is the probability that, of the three reviews, a majority will be favourable?

(*d*) *A*, *B* and *C* are independent witnesses of an event which is known to have occurred. *A* speaks the truth three times out of four, *B* four times out of five and *C* five times out of six.

What is the probability that the occurrence will be reported truthfully by majority of three witnesses ?

Ans. (*a*) 0·98, (*b*) 52/77, (*c*) 209/343, (*d*) 31/60

3·38. (*a*) A man seeks advice regarding one of two possible courses of action from three advisers who arrived at their recommendations independently. He follows the recommendation of the majority. The probability that the individual advisers are wrong are 0·1, 0·05 and 0·05, respectively. What is the probability that the man takes incorrect advice ?

Ans. *P*(Man takes at least two wrong advices) = 0·012.

(*b*) A person takes four tests in succession. The probability of his passing the first test is p, that of his passing each succeeding test is p or $p/2$ according as he passes or fails the preceding one. He qualifies provided he passes at least three tests. What is his chance of qualifying ?

(*c*) *A* is one of six horses entered for a race and is to be ridden by one of the two jockeys *B* and C. It is 2 to 1 that B rides *A*, in which case all the horses are equally likely to win; with rider *C*, *A*'s chance is trebled.

(*i*) Find the probability that *A* wins; (*ii*) What are odds against *A*'s winning ?

Hint. (*c*) E_1 : *B* rides '*A*' ; E_2 : *C* rides '*A*', *E* : *A* wins

$$P(E_1)=\frac{2}{3},\qquad P(E_2)=\frac{1}{3}.\qquad P(E\mid E_1)=\frac{1}{6},\qquad P(E\mid E_2)=\frac{3}{6}.$$

$$(i)\ P(A \text{ wins}) = P(E)=\sum_{i=1}^{2}P(E_i)\cdot P(E\mid E_i)=\frac{2}{18}+\frac{3}{18}=\frac{5}{18}$$

$$(ii)\ P(\bar{E})\ :\ P(E)=\frac{13}{18}\ :\ \frac{5}{18}=13\ :\ 5.$$

3·39. Defects are classified as *A*, *B* or *C*, and the following probabilities have been determined from available production data :

$P(A) = 0{\cdot}2$, $P(B) = 0{\cdot}16$, $P(C) = 0{\cdot}14$, $P(A \cap B) = 0{\cdot}08$, $P(A \cap C) = 0{\cdot}05$, $P(B \cap C) = 0{\cdot}04$, and $P(A \cap B \cap C) = 0{\cdot}02$

What is the probability that a randomly selected item of product will exhibit at least one type of defects ? What is the probability that it exhibits both *A* and *B* defects but is free from type *C* defect?

3·40. A supplier of a certain testing device claims that his device has high reliability in as much as $P(A\mid B)=P(\bar{A}\mid \bar{B})=0.95$, where *A* = {Device indicates component is faulty} and *B* = {Component is faulty}. Device is to be used for locating faulty components in a large batch of components of which 5% are faulty. Show that $P(B \mid A) = 0{\cdot}05$

3·41. An urn contains '*a*' white balls and '*b*' black balls. A sample of two balls is drawn without replacement. Under what conditions will the probability that both the drawn balls are of the same colour to be greater than 2.

Ans. $(a - b) > (a + b)$

3·42. (*a*) An urn contains ten balls of which three are black and seven are white. The following game is played : At each trial a ball is selected at random, its colour is noted, and is replaced along with two additional balls of the same colour. What is the probability that a black ball is selected in each of the first three trials ?

Ans. $\frac{3}{10}\times\frac{5}{12}\times\frac{7}{14}=\frac{1}{16}.$

(*b*) A bag contains *m* white and 3 black balls. Balls are drawn one by one without replacement till all the black balls are drawn. Show that the probability that this procedure for drawing balls will come to an end at the *r*th draw is $3(r-1)(r-2)/[(m+1)(m+2)(m+3)]$.

(*c*) An urn contains a white and b black balls and a series of drawings of one ball at a time is made, the ball removed being returned to the urn immediately after the next drawing is made. If p_n denotes the probability that the *n*th ball drawn is black, show that

$$p_n=(b-p_{n-1})/(a+b-1). \qquad \text{Hence find } p_n.$$

Hint. $p_n=p_{n-1}\times\left(\frac{b-1}{a+b-1}\right)+(1-p_{n-1})\times\left(\frac{b}{a+b-1}\right);\qquad p_n=\frac{b}{a+b}$

3·43. (*a*) What is the probability that (*i*) the birthdays of twelve people will fall in twelve different calendar months (assume equal probabilities for the twelve months), and (*ii*) the birthdays of six people will fall in exactly two calendar months ?

Hint. (*i*) The birthday of the first person, for instance, can fall in 12 different ways and so for the second, and so on.

∴ The total number of cases = 12^{12}.

Now there are 12 months in which the birthday of one person can fall and 11 months in which the birthday of the second person can fall and 10 months for another third person, and so on.

∴ The total number of favourable cases = 12·11·10 ... 3·2·1

$$\text{Required probability} = \frac{(12)!}{12^{12}}$$

(*ii*) The total number of ways in which the birthdays of 6 persons can fall in any of the month $= 12^6$.

$$\therefore \qquad \text{Required probability} = \frac{{}^{12}C_2(2^6-2)}{12^6}$$

(*b*) Of three independent events, the probability that the first only should happen is 1/4, the probability that the second only should happen is 1/8, and the probability that the third only should happen is 1/12. Obtain the unconditional probabilities of the three events.

Ans. 1/2, 1/3, 1/4

(*c*) A person is to be tested to see whether he can differentiate between the taste of two brands of cigarettes. If he cannot differentiate, it is assumed that the probability is one-half that he will identify a cigarette correctly. Under which of the following two procedures is there less chance that he will make all correct identifications when he actually cannot differentiate between the two brands ?

(*i*) The subject is given four pairs each containing both brands of cigarettes (this is known to the subject), he must identify for each pair which cigarette represents each brand.

(*ii*) The subject is given eight cigarettes and is told that the first four are of one brand and the last four of the other brand.

How do you explain the difference in results despite the fact that eight cigarettes are tested in each case ?

Ans. (*i*) 1/16, (*ii*) 1/2.

3·44. A total of n shells are fired at a target. the probability of the ith shell hitting the target is $p_i = 1, 2, 3, \ldots, n$. Assuming that the n firings are n mutually independent events, find the probability that at least two shells out of n hit the target.

Ans. $1 - q_1 q_2 \cdots q_n \{(p_1/q_1) + (p_2/q_2) + \ldots + (p_n/q_n)\}$

3·45. A sportsman's chance of shooting an animal at a distance $r(> a)$ is a^2/r^2. He fires when $r = 2a$, and if he misses he reloads and fires when $r = 3a, 4a \ldots$ If he misses at distance na, the animal escapes. Find the odds against the sportsman.

Hint. $P[\text{Sportsman shoots at a distance } ia] = \dfrac{a^2}{(ia)^2}$

$$\Rightarrow \quad P[\text{Sportsman misses the shot at a distance } ia] = 1 - \frac{1}{i^2}$$

$$\therefore \quad P[\text{Animal escapes}] = \prod_{i=2}^{n}\left(1-\frac{1}{i^2}\right) = \prod_{i=2}^{n}\left[\left(\frac{i-1}{i}\right)\left(\frac{i+1}{i}\right)\right] = \prod_{i=2}^{n}\left(\frac{i-1}{i}\right)\prod_{i=2}^{n}\left(\frac{i+1}{i}\right) = \frac{n+1}{2n}$$

$$\text{Required ratio} = \frac{n+1}{2n} : \left(1 - \frac{n+1}{2n}\right) = (n+1) : (n-1).$$

3·46. A carpenter has a tool chest with two compartments, each one having a lock. He has two keys for each lock, and he keeps all four keys in the same ring. His habitual procedure in opening a compartment is to select a key at random and try it. If it fails, he selects one of the remaining three and tries it and so on. Show that the probability that he succeeds on the first, second and third try is $\frac{1}{2}, \frac{1}{3}, \frac{1}{6}$, respectively.

3·47. (*a*) A closet contains n pairs of shoes. If $2r$ shoes are chosen at random (with $2r < n$), what is the probability that there will be :

(*i*) no complete pair, (*ii*) exactly one complete pair,

(*iii*) exactly two complete pairs, among them?

(*b*) Show that the probability of getting no right pair out of n, when the left foot shoes are paired randomly with the right foot shoes, is the sum of the first $(n + 1)$ terms in the expansion of e^{-1}.

(*c*) A drawer contains 8 pairs of socks. If 6 socks are taken at random and without replacement, show that the chance that there is at least one matching pair among these 6 socks is (111/143).

3·48. Out of 3n consecutive natural numbers, 3 natural numbers are chosen at random without replacement. Show that the probability that the sum of these numbers is divisible by 3, is :

$$(3n^2 - 3n + 2)/(3n - 1)\ (3n - 2).$$

3·49. A chess game between two grand masters A and B is won by whoever first wins a total of two games. A's chances of winning, drawing or losing a particular game are p, q, r, respectively. The games are independent and $p + q + r = 1$.

Find the probability that A wins the match. What is the probability that neither A not B wins the match ?

Ans. $P(A) = P(A \text{ wins}) = \dfrac{p^2\ (p + 3r)}{(p + r)^3}$; $P(B) = P(B \text{ wins}) = \dfrac{r^2\ (r + 3p)}{(p + r)^3}$;

$P(\text{Neither } A \text{ nor } B \text{ wins}) = 0$; $P(A) + P(B) = 1$.

3·50. A and B play a match, the winner being the one who first wins two games in succession, no games being drawn. their respective chances of winning a particular game are $p : q$.

Find (*i*) A's initial chance of winning. (*ii*) A's chance of winning after having won the first game.

Ans. (*i*) $\dfrac{p^2(1 + q)}{1 - qp}$; (*ii*) $\dfrac{p}{1 - qp}$

3·51. Three players A, B and C agree to play a series of games observing the following rules : two players participate in each game, while the third is idle, and the game is to be won by one of them. The loser in each game quits and his place in the next game is taken by the player who was idle. The player who succeeds in winning over both of his opponents without interruption, wins the whole series of games. Supposing the probabilities for each player to win a single game is 1/2, and that the first game is played by A and B, find the probability for A, B and C, respectively to win the whole series if the number of games is unlimited.

Ans. $P(A) = 5/14$; $P(B) = 5/14$; $P(C) = 4/14$

3·52. An astrologer claims that he can predict before birth the sex of a baby just to be born. Suppose that the astrologer has no real power but he tosses a coin just once before every birth and if the head turns up he predicts a boy for that birth and if the tail turns up he predicts a girl. Let p be the probability of the event that at a certain birth a male child is born, and p' the probability of a head turning up in a single toss with astrologer's coin. Find the probability of :

(*a*) a correct prediction, and (*b*) at least one correct prediction, in n predictions.

Ans. (*a*) $pp' + (1 - p)\ (1 - p') = 1 - p - p' + 2pp'$, (*b*) $1 - (p + p' - 2\ pp')^n$

3·53. A certain mathematician keeps two exactly similar match boxes with him (initially each containing n match-sticks). Whenever required he selects a box at random and takes a match stick. When he finds a box empty, the other box contains exactly r match-sticks.

Prove that the probability for such an occurrence is : ${}^{2n-r}C_n \div 2^{2n-r}$

3·54. There are k-varieties of objects, each variety consisting of the same number of objects. These objects are drawn one at a time and replaced before the next drawing. Show that the probability p_n that n and no less drawings will be required to produce objects of all varieties is given by :

$$k^{n-1}p_n = (k-1)^{n-1} - {}^{k-1}C_1 (k-2)^{n-1} + {}^{k-1}C_2 (k-3)^{n-1} - .$$

3·55. In a certain book of N pages, no page contains more than four errors, n_1 of them contain one error, n_2 contain two errors, n_3 contain three errors and n_4 contain four errors. Two copies of the book are opened at any two given pages. Show that the probability that the number of errors in these two pages shall not exceed five is :

$$1 - \frac{1}{N^2}\left(n_3^2 + n_4^2 + 2n_2 n_4 + 2n_3 n_4\right)$$

Hint. Let $E_i\text{I}$ = The event that a page of first book contains i errors.

$E_i\text{II}$ = The event that a page of second book contains i errors.

P(Number of errors in the two pages shall not exceed 5)

$$= 1 - P\left[E_2\text{I}\,E_4\text{II} + E_3\text{I}\,E_4\text{II} + E_4\text{I}\,E_4\text{II} + E_3\text{I}\,E_3\text{II} + E_4\text{I}E_3\text{II} + E_4\text{I}\,E_2\text{II}\right]$$

3·56. Of three independent events, the chance that the first only should happens is a, the chance of the second only is b and the chance of the third only is c. Show that the independent chances of the three events are, respectively:

$$\frac{a}{a+x}, \quad \frac{b}{b+x}, \quad \frac{c}{c+x},$$

where x is the root of the equation $(a + x)\ (b + x)\ (c + x) = x^2$

Hint. $P(E_1 \cap \bar{E}_2 \cap \bar{E}_3) = P(E_1)[1 - P(E_2)][1 - P(E_3)] = a$ ...(*)

$$P(\bar{E}_1 \cap E_2 \cap \bar{E}_3) = [1 - P(E_1)]\, P(E_2)\, [1 - P(E_3)] = b \qquad ...(**)$$

$$P(\bar{E}_1 \cap \bar{E}_2 \cap E_3) = [1 - P(E_1)][1 - P(E_2)]\, P(E_3) = c \qquad ...(***)$$

Multiplying (*), (**) and (***), we get :

$$P(E_1)\, P(E_2)\, P(E_3)\, x^2 = abc, \quad \text{where} \quad x = [1 - P(E_1)][1 - P(E_2)][1 - P(E_3)]$$

Multiplying (*) by $1 - P(E_1)$, we get : $P(E_1) = \dfrac{a}{a+x}$, and so on.

3·57. For three independent events A, B and C, the probability for A to occur is a, the probability that A, B and C will not occur is b, and the probability that at least one of the three events will not occur is c. If p denotes the probability that C occurs but neither A nor B occurs, prove that p satisfies the quadratic equation :

$$ap^2 + [ab - (1 - a)\ (a + c - 1)]\ p + b\ (1 - a)\ (1 - c) = 0$$

and hence deduce that : $\quad c > \dfrac{(1-a)^2 + ab}{(1-a)}.$

Hint. Let $P(A) = x$, $P(B) = y$ and $P(C) = z$.

Then $x = a$, $(1 - x)\ (1 - y)\ (1 - z) = b$, $1 - xyz = c$ and $p = z(1 - x)\ (1 - y)$.

Elimination of x, y and z gives quadratic equation in p.

3·58. Two players A and B agree to contest a match consisting of a series of games, the match to be won by the player who first wins three games, with the provision that if the players win two games each, the match is to continue until it is won by one player winning two games more than his opponent. The probability of A winning any given game is p, and the games cannot be drawn.

(i) Prove that $f(p)$, the initial probability of A winning the match is given by:

$$= p^3 (4 - 5p + 2p^2)/(1 - 2p + 2p^2)$$

(ii) Show that the equation $f(p) = p$ has five real roots, of which three are admissible values of p. Find these three roots and explain their significance

Ans. (ii) $p = 0, 1, \frac{1}{2}, \frac{1 \pm \sqrt{5}}{2}$ (Discarded)

3·59. A thief has a bunch of n keys, exactly one of which fits a lock. If the thief tries to open the lock by trying the keys at random, what is the probability that he requires exactly k attempts, if he rejects the keys already tried ? Find the same probability if he does not reject the keys already tried.

Ans. (i) $\frac{1}{n}$, (ii) $\left(1 - \frac{1}{n}\right)^{k-1} \frac{1}{n}$

3·60. A particular machine is prone to three similar types of faults A_1, A_2 and A_3. Past records on breakdowns of the machine show the following : the probability of a breakdown (i.e., at least one fault) is 0·1; for each i, the probability that fault A_i occurs and the others do not is 0·02; for each pair i, j the probability that A_i and A_j occur but the third fault does not is 0·012. Determine the probabilities of

(a) the fault of type A_1 occurring irrespective of whether the other faults occur or not,

(b) a fault of type A_1 given that A_2 has occurred, and

(c) faults of type A_1 and A_2 given that A_3 has occurred.

Ans. (a) 0·048, (b) 1/3, (c) 1/12

3·61. In each of a packet of cigarettes, there is one of set cards numbered 1 to n. If a number N of packets is bought at random (the population of packets is large and the numbers are equally frequent), show that the probability of getting a complete set of cards is :

$$1 - {}^nC_1\left(1 - \frac{1}{n}\right)^N + {}^nC_2\left(1 - \frac{2}{n}\right)^N - \ldots + (-1)^{n-1} \cdot {}^nC_{n-1}\left(\frac{1}{n}\right)^N$$

3·62. (a) The chance of success in each trial is p. If p_k is the probability that there are even number of successes in k independent trials, prove that

$$p_k = p + p_{k-1}(1 - 2p)$$

Deduce that : $$p_k = \frac{1}{2}\left[1 + (1 - 2p)^k\right]$$

(b) For a sequence of n Bernoulli trials in which the probability of success (S) is p and that of failure (F) is q, $p + q = 1$, show that p_n, the probability that the pattern SF does not occur in the entire sequence, satisfies the difference equation :

$$p_n - p_{n-1} + pq \cdot p_{n-2} = 0, \quad \text{for } n \geq 2.$$

Hence, or otherwise obtain an explict expression for p_n when (i) $p \neq q$ and (ii) $p = q$.

Ans. (i) $p_n = \frac{(p^{n+1} - q^{n+1})}{p - q}$; $p \neq q$ (ii) $p_n = (n+1)\,2^{-n}$; $p = q$.

3·63. If a day is dry, the conditional probability that the following day will also be dry is p; if a day is wet, the conditional probability that the following day will be dry is p'. If u_n is the probability that the nth day will be dry, prove that

$$u_n - (p - p')\,u_{n-1} - p = 0\ ;\ n \geq 2.$$

If the first day is dry, $p = 3/4$ and $p' = 1/4$, find u_n. **Ans.** $u_n = \frac{2^n + 1}{2^{n+1}}$

3·64. There are n similar biased dice such that the probability of obtaining a 6 with each one of them is the same and equal to p. If all the dice are rolled once, show that p_n, the probability that an odd number of 6's is obtained satisfies the difference equation:

$$p_n + (2p - 1)\, p_{n-1} = p \text{ and hence derive an explicit expression for } p_n.$$

Ans. $p_n = \frac{1}{2}[1 + (1-2p)^n]$.

3·65. Suppose that each day the weather can be uniquely classified as 'fine' or 'bad'. Suppose further that the probability of having fine weather on the last day of a certain year is P_0 and we have the probability p that the weather on an arbitrary day will be of the same kind as on the preceding day. Let the probability of having fine weather on the nth day of the following year be P_n.

Show that $$P_n = (2p - 1)\, P_{n-1} + (1 - p).$$

Deduce that : $$P_3 = (2p-1)^3\left(P_0 - \frac{1}{2}\right) + \frac{1}{2}.$$

3·66. Two players A and B start playing a series of games with ₹ a and b respectively. The stake is ₹ 1 on a game and no game can be drawn. If the probability of A winning any game is a constant p, find the initial probability of his exhausting the funds of B or his own. Also show that if the resources of B are unlimited, then

(*i*) A is certain to be ruined if $p = 1/2$, and

(*ii*) A has an even chance of escaping ruin if $p = 2^{1/a}/(1 + 2^{1/a})$

Hint. Let u_n be the probability of A,s final win when he has ₹ n.

Then, $$u_n = pu_{n+1} + (1-p)u_{n-1} \text{ where } u_0 = 0 \text{ and } u_{a+b} = 1$$

$$\therefore \quad u_{n+1} - u_n = \left(\frac{1-p}{n}\right)(u_n - u_{n-1})$$

Hence, $$u_{n+1} - u_n = \left(\frac{1-p}{p}\right)^n . u_1, \text{ by repeated application,}$$

so that, $$u_n = u_1\left[1 - \left(\frac{1-p}{p}\right)^n\right] \Big/ \left[1 - \left(\frac{1-p}{p}\right)\right]$$

Hence, using $u_{a+b} = 1$, $$u_n = \left[1 - \left(\frac{1-p}{p}\right)^n\right] \Big/ \left[1 - \left(\frac{1-p}{p}\right)^{a+b}\right]$$

$\therefore$ Initial probability of A's win is $$u_a = \frac{p^a - (1-p)^a}{p^{a+b} - (1-p)^{a+b}} \cdot p^b$$

Probability of A's ruin $= 1 - u_a$

For $p = \frac{1}{2}$, $u_a = \frac{a}{a+b} \to 0 \text{ as } b \to \infty$ and for $p \neq \frac{1}{2}, u_a = \frac{1}{2} \text{ if } p = 2^{1/a}(1 + 2^{1/a}), \& \, b \to \infty.$

3·67. If n letters are placed at random in n correctly addressed envelopes, find the probability that :

(*i*) none of the letters is placed in the correct envelope.

(*ii*) At least one letter goes to the correct envelope.

(*iii*) All letters go to the correct envelopes, and

(*iv*) Exactly r letters are placed in the correct envelopes.

Ans. $(i)\; p = \sum_{k=0}^{n} \frac{(-1)^k}{k!}, \quad (ii)\; 1 - p, \quad (iii)\; 1, \quad (iv)\; \frac{1}{r!}\sum_{k=0}^{n-r}(-1)^k \cdot \frac{1}{k!} \;;\; r = 0, 1, 2, \ldots, n$

3·68. In a lottery m tickets are drawn at a time out of the total number of n tickets, and returned before the next drawing is made. Show that the chance that in k drawings, each of the numbers 1, 2, 3, ..., n will appear at least once is given by :

$$p_k = 1 - {}^nC_1\left(1-\frac{m}{n}\right)^k + {}^nC_2\left(1-\frac{m}{n}\right)^k\left(1-\frac{m}{n-1}\right)^k - \ldots$$

3·69. An urn contains n white and m black balls, a second urn contains N white and M black balls. A ball is randomly transferred from the first to the second urn and then from the second to the first urn. If a ball is now selected randomly from the first urn, prove that the probability that it is white is:

$$\frac{n}{n+m} + \frac{mN - nM}{(n+m)^2\,(N+M+1)}$$

Hint. Let us define the following events :

B_i : Drawing of a black ball from the ith urn, i = 1, 2

W_i : Drawing of a white ball from the ith urn, i = 1, 2

The four distinct possibilities for the first two exchanges are $B_1\,W_2$, $B_1\,B_2$, $W_1\,B_2$, $W_1\,W_2$. Hence if E denotes the event of drawing a white ball from the first urn after the exchanges, then

$$P(E) = P(B_1 \cap W_2 \cap E) + P(B_1 \cap B_2 \cap E) + P(W_1 \cap B_2 \cap E) + P(W_1 \cap W_2 \cap E) \quad \ldots(*)$$

We have :

$$P(B_1 \cap W_2 \cap E) = P(B_1)\cdot P(W_2 \mid B_1)\,P(E \mid B_1\,W_2) = \frac{m}{m+n} \times \frac{N}{M+N+1} \times \frac{n+1}{m+n}$$

$$P(B_1 \cap B_2 \cap E) = P(B_1)\cdot P(B_2 \mid B_1)\cdot P(E \mid B_1\,B_2) = \frac{m}{m+n} \times \frac{M+1}{M+N+1} \times \frac{n}{m+n}$$

$$P(W_1 \cap B_2 \cap E) = P(W_1)\cdot P(B_2 \mid W_1)\cdot P(E \mid W_1\,B_2) = \frac{n}{m+n} \times \frac{M}{M+N+1} \times \frac{n-1}{m+n}$$

$$P(W_1 \cap W_2 \cap E) = P(W_1)\cdot P(W_2 \mid W_1)\cdot P(E \mid W_1\,W_2) = \frac{n}{m+n} \times \frac{N+1}{M+N+1} \times \frac{n}{m+n}$$

Substituting in (*) and simplifying we get the result.

3·70. (*Sampling with replacement*) A sample of size r is taken with replacement from a population of n people. Find the probability U_r that N given people will be included in the sample.

Show further that if $n \to \infty$ and $r \to \infty$ so that $(r/n) \to p$, a constant, then $(U_r \to 1 - e^{-p})^N$.

Ans. $U_r = \sum_{k=0}^{N} (-1)^k\,{}^NC_k \left(1-\frac{k}{n}\right)^r$.

3·71. Of three urns, the first contains a proportion x of white balls and a proportion $(1-x)$ of blue balls; the second contains blue and red balls in the proportions y and $(1-y)$ respectively, and the third contains red and white balls in the proportions z and $(1-z)$ respectively, none of the given proportions being zero. A ball is drawn randomly from the first urn and then replaced. If it was white, a ball is drawn from the second urn and replaced, but if the first ball drawn was blue, a ball is drawn from the third urn and replaced. The process continues, a ball being drawn each time from the first, second or third urn according as the previous ball was red, white or blue respectively. Every ball is replaced before the next is drawn.

If r_n, w_n and b_n denote respectively the probabilities at the outset that the nth ball drawn will be red, white or blue, prove that as $n \to \infty$, these probabilities tend to fixed limits, and find the limiting values.

Ans. As $n \to \infty$,
$$\left[\begin{array}{l} r_n \to (1-y+yz)/\{3-(x+y+z)+(xy+yz+zx)\} \\ w_n \to (1-z+zx)/\{3-(x+y+z)+(xy+yz+zx)\} \\ b_n \to (1-x+xy)/\{3-(x+y+z)+(xy+yz+zx)\} \end{array}\right]$$

CHAPTER 4

Theory of Probability – II

LEARNING OBJECTIVES

Upon completion of this Chapter, you should be able to:

- Explain the Extended Axiom of Addition and Axiom of Continuity and relationship between them.
- Understand the concept of prior and posterior probability.
- Establish the Bayes Theorem using prior and posterior probability.
- Demonstrate the use of Bayes Theorem in decision making.
- Examine the concept and use of Geometric Probability.

CHAPTER OUTLINES

4·1. EXTENDED AXIOM OF ADDITION AND AXIOM OF CONTINUITY

In Chapter 3, § 3·8, Axiomatic definition of probability, we discussed the three axioms *viz.*, the axiom of non-negativity, the axiom of certainty and the axiom of addition or union. In this section we shall have some further discussion on the last axiom *viz.*, the axiom of addition.

Extended Axiom of Addition. If an event A can materialise in the occurrence of any one of the pairwise disjoint events $A_1, A_2, A_3, \ldots$ so that

$$A = \bigcup_{i=1}^{\infty} A_i; \quad A_i \cap A_j = \phi, (i \neq j)$$

then

$$P(A) = P\left(\bigcup_{i=1}^{\infty} A_i\right) = \sum_{i=1}^{\infty} P(A_i) \qquad \ldots (4\cdot1)$$

Axiom of Continuity. If $B_1, B_2, \ldots, B_n, \ldots$ be a countable sequences of events such that

(*i*) $B_i \supset B_{i+1}$, $(i = 1, 2, 3, \ldots)$ and (*ii*) $\bigcap_{n=1}^{\infty} B_n = \phi$ *i.e.*, if each succeeding event implies the preceding event and if their simultaneous occurrence is an impossible event then

$$\lim_{n \to \infty} P(B_n) = 0 \qquad \ldots (4\cdot2)$$

We shall now prove that these two axioms, *viz.*, the extended axiom of addition and axiom of continuity are equivalent, *i.e.*, each implies the other. In other words., (4·1) $\Leftrightarrow$ (4·2).

Theorem 4·1. *Axiom of continuity follows from the extended axiom of addition and vice versa.*

Proof. (*a*) (4·1) $\Rightarrow$ (4·2). Let $\{B_n\}$ be a countable sequence of events such that

$$B_1 \supset B_2 \supset B_3 \supset \ldots, \supset B_n \supset B_{n+1} \supset \ldots$$

and let for any $n \geq 1$,

$$\bigcap_{k \geq n} B_k = \phi \qquad \ldots (*)$$

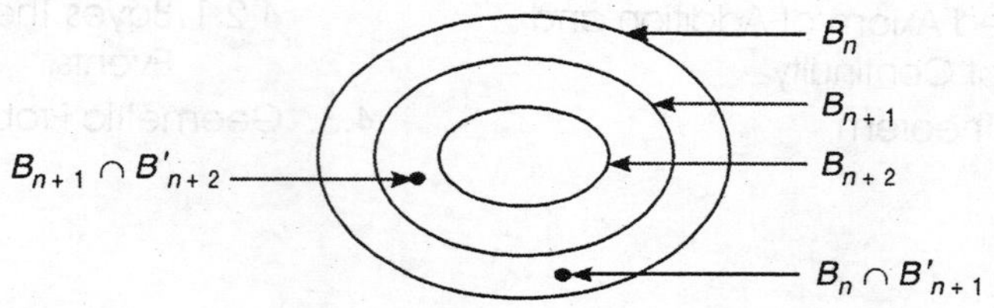

Fig. 4·1

Then, it is obvious from the diagram that

$$B_n = B_n B'_{n+1} \cup B_{n+1} B'_{n+2} \cup \ldots \cup \left(\bigcap_{k \geq n} B_k\right)$$

$$\Rightarrow \qquad B_n = \left(\bigcup_{k=n}^{\infty} B_k B'_{k+1}\right) \cup \left(\bigcap_{k \geq n} B_k\right),$$

where the events $B_k B'_{k+1}$, $(k = n, n+1, \ldots)$ are pairwise disjoint and each is disjoint with $\bigcap_{k \geq n} B_k$.

Thus, B_n has been expressed as the countable union of pairwise disjoint events and hence, by the extended axiom of addition, we get

$$P(B_n) = \sum_{k=n}^{\infty} P(B_k B'_{k+1}) + P\left(\bigcap_{k \geq n} B_k\right)$$

$$= \sum_{k=n}^{\infty} P(B_k B'_{k+1}), \qquad \text{... (**)}$$

since from (*), $P\left(\bigcap_{k \ge n} B_k\right) = P(\phi) = 0$

Further, from (**), since $\sum_{k=1}^{\infty} P(B_k B'_{k+1}) = P(B_1) \le 1,$

the right hand sum in (**), being the remainder after n terms of a convergent series tends to zero as $n \to \infty$.

Hence, $\lim_{n \to \infty} P(B_n) = \lim_{n \to \infty} \sum_{k=n}^{\infty} P(B_k B'_{k+1}) = 0$

Thus the extended axiom of addition implies the axiom of continuity.

(b) Conversely (2) ⇒ (1), *i.e., the extended axiom of addition follows from the axiom of continuity.*

Let $\{A_n\}$ be a countable sequence of pairwise disjoint events and let

$$A = \bigcup_{i=1}^{\infty} A_i = \left(\bigcup_{i=1}^{n} A_i\right) \cup \left(\bigcup_{i=n+1}^{\infty} A_i\right) \qquad \text{... (4·3)}$$

Let us define a countable sequence $\{B_n\}$ of events by $B_n = \bigcup_{i=n}^{\infty} A_i$ ... (4·4)

Obviously, B_n is a decreasing sequence of events, *i.e.*, $B_1 \supset B_2 \supset \ldots \supset B_n \supset B_{n+1} \supset \ldots$ ... (4·5)

Also, we have $A = \left(\bigcup_{i=1}^{n} A_i\right) \cup B_{n+1}$ ... (4·6)

Since A_i's are pairwise disjoint, $A_i \cap B_{n+1} = \phi, (i = 1, 2, \ldots, n)$ ... (4·6*a*)

From (4·4) we see that if the event B_n has occurred it implies the occurrence of any one of the events $A_{n+1}, A_{n+2}, \ldots$. Without loss of generality let us assume that this event is A_i $(i = n+1, n+2, \ldots)$. Further since A_i's are pairwise disjoint, the occurrence of A_i implies that events $A_{i+1}, A_{i+2}, \ldots$ do not occur leading to the conclusion that $B_{i+1}, B_{i+2}, \ldots$ will not occur, *i.e.*,

$$\bigcap_{i=n}^{\infty} B_i = \phi \qquad \text{... (4·7)}$$

From (4·5) and (4·7), we observe that both the conditions of axiom of continuity are satisfied and hence we get : $\lim_{n \to \infty} P(B_n) = 0$ ... (4·8)

From (4·6), we get

$$P(A) = P\left[\left(\bigcup_{i=1}^{n} A_i\right) \cup B_{n+1}\right] = \sum_{i=1}^{n} P(A_i) + P(B_{n+1}) \quad \text{(By axiom of Additivity)}$$

$$\Rightarrow \quad P\left(\bigcup_{i=1}^{\infty} A_i\right) = \lim_{n \to \infty} \sum_{i=1}^{n} P(A_i) + \lim_{n \to \infty} (B_{n+1}) = \sum_{i=1}^{\infty} P(A_i), \qquad \text{[From (4·8)]}$$

which is the extended axiom of addition.

4·2. BAYES' THEOREM

In the discussion of conditional probability we indicated that revising probability when new information is obtained is an important phase of probability analysis. Often, we begin our analysis with initial or *prior* probability estimates for specific events of interest. Then, from sources such as a sample, a special report, a product test, and so on we obtain some additional information about the events. Given this new information, we update the prior probability values by calculating revised probabilities, referred to as *posterior probabilities*. Bayes' theorem *which was given by Thomas Bayes, a British Mathematician,* in 1763, provides a means for making these probability calculations. The steps in this probability revision process are shown in the following diagram :

Prior Probabilities → New Information → Application of Bayes' Theorem → Posterior Probabilities

Theorem 4·2. *Bayes' Theorem. If $E_1, E_2, E_3 \ldots, E_n$ are mutually disjoint events with $P(E_i) \neq 0$, $(i = 1, 2, \ldots, n)$, then for any arbitrary event E which is a subset of $\bigcup_{i=1}^{n} E_i$ such that $P(E) > 0$, we have*

$$P(E_i \mid E) = \frac{P(E_i)\,P(E \mid E_i)}{\sum_{i=1}^{n} P(E_i)\,P(E \mid E_i)} = \frac{P(E_i)\,P(E \mid E_i)}{P(E)}\,;\; i = 1, 2, \ldots, n \qquad \ldots (4·9)$$

Proof. Since $E \subset \bigcup_{i=1}^{n} E_i$, we have, $E = E \cap \left(\bigcup_{i=1}^{n} E_i\right) = \bigcup_{i=1}^{n} (E \cap E_i)$ [By distributive law]

Since $(E \cap E_i) \subset E_i$, $(i = 1, 2, \ldots, n)$ are mutually disjoint events, we have by addition theorem of probability :

$$P(E) = P\left\{\bigcup_{i=1}^{n} (E \cap E_i)\right\} = \sum_{i=1}^{n} P(E \cap E_i) = \sum_{i=1}^{n} P(E_i)\,P(E \mid E_i), \qquad \ldots (4·10)$$

by multiplication theorem of probability.

Also we have $P(E \cap E_i) = P(E)\,P(E_i \mid E)$

$$\Rightarrow \quad P(E_i \mid E) = \frac{P(E \cap E_i)}{P(E)} = \frac{P(E_i)\,P(E \mid E_i)}{\sum_{i=1}^{n} P(E_i)\,P(E \mid E_i)} \qquad \text{[From (4·10)]}$$

Remarks:

1. The probabilities $P(E_1), P(E_2), \ldots, P(E_n)$ are termed as the *'a prior probabilities'* because they exist before we gain any information from the experiment itself.
2. The probabilities $P(E \mid E_i)$; $i = 1, 2, \ldots, n$ are called *'likelihoods'* because they indicate how likely the event E under consideration is to occur, given each and every *a prior* probability.
3. The probability $P(E_i \mid E)$; $i = 1, 2, \ldots, n$ are called *'posterior probabilities'* because they are determined after the results of the experiment are known.
4. From (4·10) we get the following important result :

   "If the events $E_1, E_2, \ldots, E_n$ constitute a disjoint partition of the sample space S and $P(E_i) \neq 0$; $i = 1, 2, \ldots, n$, then for any event E in S, we have

$$P(E) = \sum_{i=1}^{n} P(E_i)\,P(E \mid E_i) \qquad \ldots (4·11)$$

4·2·1. Bayes' Theorem for Future Events

The probability of the materialisation of another event C, given $P(C \mid E \cap E_1)$, $P(C \mid E \cap E_2)$, ..., $P(C \mid E \cap E_n)$, *is given by :*

$$P(C \mid E) = \frac{\sum_{i=1}^{n} P(E_i)\, P(E \mid E_i)\, P(C \mid E_i \cap E)}{\sum_{i=1}^{n} P(E_i)\, P(E \mid E_i)} \qquad \text{... (4·12)}$$

Proof. Since the occurrence of event E implies the occurrence of one and only one of the events $E_1, E_2, ..., E_n$, the event C (granted that E has occurred) can occur in the following mutually exclusive ways:

$$C \cap E_1,\ C \cap E_2,\ ...,\ C \cap E_n$$

i.e.,
$$C = (C \cap E_1) \cup (C \cap E_2) \cup ... \cup (C \cap E_n)$$

$$\Rightarrow \quad C \mid E = \{(C \cap E_1) \mid E\} \cup \{(C \cap E_2) \mid E\} \cup ... \cup \{(C \cap E_n) \mid E\}$$

$$\Rightarrow \quad P(C \mid E) = P\{(C \cap E_1) \mid E\} + P\{(C \cap E_2) \mid E\} + ... + P\{(C \cap E_n) \mid E\}$$

$$= \sum_{i=1}^{n} P\{(C \cap E_i) \mid E)\} = \sum_{i=1}^{n} P(E_i \mid E)\, P\{C \mid (E_i \cap E)\}$$

Substituting the value of $P(E_i \mid E)$ from (4·9), we get

$$P(C \mid E) = \frac{\sum_{i=1}^{n} P(E_i)\, P(E \mid E_i)\, P(C \mid E_i \cap E)}{\sum_{i=1}^{n} P(E_i)\, P(E \mid E_i)}$$

Remarks:

1. It may happen that materialisation of the event E_i makes C independent of E, then we have

$$P(C \mid E_i \cap E) = P(C \mid E_i),$$

and the above formula reduces to :
$$P(C \mid E) = \frac{\sum_{i=1}^{n} P(E_i)\, P(E \mid E_i)\, P(C \mid E_i)}{\sum_{i=1}^{n} P(E_i)\, P(E \mid E_i)} \qquad \text{... (4·13)}$$

The event C can be considered in regard to E as Future Event.

2. Bayes theorem is extensively used in statistical inference, and by business and management executives in arriving at valid decisions in the face of uncertainty.

The Tabular Approach. The posterior probabilities, using Bayes' Theorem can be obtained conveniently in a tabular form, which involves the following steps.

Steps 1 : Prepare the following three columns:

Column 1 — The list of all mutually exclusive events E_i, $i = 1, 2, ..., n$; that occur in the problem.

Column 2 — The prior probabilities $P(E_i)$; $i = 1, 2, ..., n$ for the events.

Column 3 — The conditional probabilities of the new information (E) given each event, *viz.*, $P(E \mid E_i)$, $i = 1, 2, ..., n$

2 : In column 4, compute joint probabilities for each event using the formula :

$$P(E_i \cap E) = P(E_i)\, P(E \mid E_i)$$

3 : Sum the joint probability column to find the probability of the new information *viz.*, $P(E)$.

4 : In column 5, compute the posterior probabilities using the basic relationship of conditional probability, *viz.*,

$$P(E_i \mid E) = \frac{P(E_i \cap E)}{P(E)}\ ;\ i = 1, 2, ..., n.$$

Example 4·1. *In a certain town there are equal number of male and female residents. It is known that 5% females and 20% males are unemployed. If any unemployed person is picked-up at random, what is the probability that:*

(a) it is male *and* *(b) it is female ?* *Use Bayes' law.*

Solution. Define the events:

E_1 : Person is a male ; E_2 : Person is a female ;

and E : Person selected at random is un-employed. Then, we are given:

$$P(E_1) = 50\% = \frac{1}{2} = P(E_2)\ ; \quad P(E \mid E_1) = 0{\cdot}20\ ; \quad P(E \mid E_2) = 0{\cdot}05.$$

$$\therefore \quad P(E) = \sum_{i=1}^{2} P(E \cap E_i) = P(E_1) \,.\, P(E \mid E_1) + P(E_2) \,.\, P(E \mid E_2)$$

$$= \frac{1}{2}(0{\cdot}20 + 0{\cdot}05) = \frac{0{\cdot}25}{2} = 0{\cdot}125$$

By Bayes' Rule, we have

(*a*) Required probability $= P(E_1 \mid E) = \dfrac{P(E_1 \cap E)}{P(E)} = \dfrac{P(E_1) \cdot P(E \mid E_1)}{P(E)}$

$$= \frac{\frac{1}{2} \times 0{\cdot}20}{0{\cdot}125} = \frac{0{\cdot}100}{0{\cdot}125} = 0{\cdot}8$$

(*b*) Required probability $= P(E_2 \mid E) = \dfrac{P(E_2 \cap E)}{P(E)} = \dfrac{P(E_2) \cdot P(E \mid E_2)}{P(E)}$

$$= \frac{\frac{1}{2} \times 0{\cdot}05}{0{\cdot}125} = \frac{0{\cdot}025}{0{\cdot}125} = 0{\cdot}2$$

Example 4·2. *The prior probabilities for events E_1 and E_2 are $P(E_1) = 0{\cdot}40$ and $P(E_2) = 0{\cdot}60$. It is also known that $P(E_1 \cap E_2) = 0$. Suppose $P(E \mid E_1) = 0{\cdot}20$ and $P(E \mid E_2) = 0{\cdot}05$.*

(i) Are E_1 and E_2 mutually exclusive ? Explain.

(ii) Compute $P(E_1 \cap E)$ and $P(E_2 \cap E)$.

(iii) Compute $P(E)$.

(iv) Apply Bayes' theorem to compute $P(E_1 \mid E)$ and $P(E_2 \mid E)$.

Solution. (*i*) Since, we are given $P(E_1 \cap E_2) = 0$, the events E_1 and E_2 are mutually exclusive.

(*ii*) $P(E_1 \cap E) = P(E_1) \,.\, P(E \mid E_1) = 0{\cdot}40 \times 0{\cdot}20 = 0{\cdot}08$

$P(E_2 \cap E) = P(E_2) \cdot P(E \mid E_2) = 0{\cdot}60 \times 0{\cdot}05 = 0{\cdot}03$

(*iii*) The event E can materialise in the following mutually disjoint ways: $E \cap E_1$, $E \cap E_2$,

i.e., $$E = (E \cap E_1) \cup (E \cap E_2)$$

$$\Rightarrow \quad P(E) = P(E \cap E_1) + P(E \cap E_2) = 0{\cdot}08 + 0{\cdot}03 = 0{\cdot}11 \qquad \text{[From par (ii)]}$$

(*iv*) By Bayes' theorem :

$$P(E_1 \mid E) = \frac{P(E \cap E_1)}{P(E)} = \frac{0{\cdot}08}{0{\cdot}11} = \frac{8}{11} = 0{\cdot}7273$$

and $$P(E_2 \mid E) = \frac{P(E \cap E_2)}{P(E)} = \frac{0{\cdot}03}{0{\cdot}11} = \frac{3}{11} = 0{\cdot}2727$$

Example 4·3. *Suppose that a product is produced in three factories X, Y and Z. It is known that factory X produces thrice as many items as factory Y, and that factories Y and Z produce the same number of items. Assume that it is known that 3 per cent of the items produced by each of the factories X and Z are defective while 5 per cent of those manufactured by factory Y are defective. All the items produced in the three factories are stocked, and an item of product is selected at random.*

(*i*) *What is the probability that this item is defective ?*

(*ii*) *If an item selected at random is found to be defective, what is the probability that it was produced by factory X, Y and Z respectively ?*

Solution. Let the number of items produced by each of the factories Y and Z be n. Then the number of items produced by the factory X is $3n$. Let E_1, E_2 and E_3 denote the events that the items are produced by factory X, Y and Z, respectively and let E be the event of the item being defective. Then we have

$$P(E_1) = \frac{3n}{3n+n+n} = 0{\cdot}6 \ ; \qquad P(E_2) = \frac{n}{5n} = 0{\cdot}2 \ ; \qquad P(E_3) = \frac{n}{5n} = 0{\cdot}2$$

Also, $$P(E \mid E_1) = P(E \mid E_3) = 0{\cdot}03 \quad \text{and} \quad P(E \mid E_2) = 0{\cdot}05 \text{ (Given)}$$

(*i*) The probability that an item selected at random from the stock is defective is given by :

$$P(E) = \sum_{i=1}^{3} P(E \cap E_i) = \sum_{i=1}^{3} P(E_i)\, P(E \mid E_i)$$

$$= P(E_1)\, P(E \mid E_1) + P(E_2)\, P(E \mid E_2) + P(E_3)\, P(E \mid E_3)$$

$$= 0{\cdot}6 \times 0{\cdot}03 + 0{\cdot}2 \times 0{\cdot}05 + 0{\cdot}2 \times 0{\cdot}03 = 0{\cdot}034$$

(*ii*) By Bayes' Rule, the required probabilities are given by :

$$P(E_1 \mid E) = \frac{P(E_1)\, P(E \mid E_1)}{P(E)} = \frac{0{\cdot}6 \times 0{\cdot}03}{0{\cdot}034} = \frac{0{\cdot}018}{0{\cdot}034} = \frac{9}{17}.$$

$$P(E_2 \mid E) = \frac{P(E_2)\, P(E \mid E_2)}{P(E)} = \frac{0{\cdot}2 \times 0{\cdot}05}{0{\cdot}034} = \frac{0{\cdot}010}{0{\cdot}034} = \frac{5}{17}.$$

$$P(E_3 \mid E) = \frac{P(E_3)\, P(E \mid E_3)}{P(E)} = \frac{0{\cdot}2 \times 0{\cdot}03}{0{\cdot}034} = \frac{0{\cdot}006}{0{\cdot}034} = \frac{3}{17}.$$

Aliter. $$P(E_3 \mid E) = 1 - \left[P(E_1 \mid E) + P(E_2 \mid E)\right] = 1 - \left(\frac{9}{17} + \frac{5}{17}\right) = \frac{3}{17}.$$

Example 4·4. *In 2021 there will be three candidates for the position of principal – Dr. Chatterji, Dr. Ayangar and Dr. Singh – whose chances of getting the appointment are in the proportion 4 : 2 : 3 respectively. The probability that Dr. Chatterji if selected would introduce co-education in the college is 0·3. The probabilities of Dr. Ayangar and Dr. Singh doing the same are respectively 0·5 and 0·8.*

(i) *What is probability that there will be co-education in the college in 2021 ?*

(ii) *If there is co-education in the college in 2021, what is the probability that Dr. Singh is the principal.*

Solution. Let us define the following events :

A : Introduction of co-education ; E_1 : Dr. Chatterji is selected as principal ;

E_2 : Dr. Ayangar is selected as principal ; E_3 : Dr. Singh is selected as principal

Then we are given:

$$P(E_1) = \frac{4}{9}, \quad P(E_2) = \frac{2}{9}, \quad \text{and} \quad P(E_3) = \frac{3}{9}$$

$$P(A \mid E_1) = \frac{3}{10}, \quad P(A \mid E_2) = \frac{5}{10}, \quad \text{and} \quad P(A \mid E_3) = \frac{8}{10}$$

(i) The required probability that there will be co-education in the college in 2021 is given by :

$$\begin{aligned} P(A) &= P[(A \cap E_1) \cup P(A \cap E_2) \cup P(A \cap E_3)] \\ &= P(A \cap E_1) + P(A \cap E_2) + P(A \cap E_3) \\ &= P(E_1)\, P(A \mid E_1) + P(E_2)\, P(A \mid E_2) + P(E_3)\, P(A \mid E_3) \\ &= \frac{4}{9}\cdot\frac{3}{10} + \frac{2}{9}\cdot\frac{5}{10} + \frac{3}{9}\cdot\frac{8}{10} = \frac{46}{90} = \frac{23}{45} \end{aligned}$$

(ii) The required probability is obtained by using Bayes' Theorem as follows :

$$P(E_3 \mid A) = \frac{P(E_3)\, P(A \mid E_3)}{P(A)} = \frac{\frac{3}{9} \times \frac{8}{10}}{\frac{46}{90}} = \frac{24}{46} = \frac{12}{23}.$$

Example 4·5. *The probabilities of A, B and C becoming managers of a company are* $\frac{4}{9}, \frac{2}{9}$ *and* $\frac{1}{3}$*, respectively. The probabilities that the Bonus Scheme will be introduced if A, B and C become managers are* $\frac{3}{10}, \frac{1}{2}$ *and* $\frac{4}{5}$*, respectively.*

(a) *What is the probability that Bonus Scheme will be introduced, and*

(b) *If the Bonus Scheme has been introduced, what is the probability that manager appointed was A ?*

Solution. Let E_1, E_2, E_3 denote the events that A, B and C become managers respectively and E denote the event that 'Bonus Scheme' is introduced. We are given :

$$P(E_1) = \frac{4}{9},\ P(E_2) = \frac{2}{9},\ P(E_3) = \frac{1}{3};\ P(E \mid E_1) = \frac{3}{10},\ P(E \mid E_2) = \frac{1}{2},\ P(E \mid E_3) = \frac{4}{5}$$

(a) The event E can materialise in the following mutually exclusive ways :

(i) A becomes manager and bonus scheme is introduced, *i.e.*, $E_1 \cap E$ happens,

(ii) $E_2 \cap E$ happens, (iii) $E_3 \cap E$ happens.

Thus, $$E = (E_1 \cap E) \cup (E_2 \cap E) \cup (E_3 \cap E),$$

where $E_1 \cap E$, $E_2 \cap E$ and $E_3 \cap E$ are disjoint.

Using Addition theorem of probability, the required probability that the bonus scheme is introduced is given by :

$$P(E) = P(E_1 \cap E) + P(E_2 \cap E) + P(E_3 \cap E)$$
$$= P(E_1)\, P(E \mid E_1) + P(E_2)\, P(E \mid E_2) + P(E_3)\, P(E \mid E_3)$$
$$= \frac{4}{9} \times \frac{3}{10} + \frac{2}{9} \times \frac{1}{2} + \frac{1}{3} \times \frac{4}{5} = \frac{23}{45}$$

(*b*) Using Bayes' Theorem, the required probability is :

$$P(E_1 \mid E) = \frac{P(E_1) \times P(E \mid E_1)}{P(E)} = \frac{12/90}{23/45} = \frac{6}{23}$$

Aliter. The above probabilities can be elegantly obtained in a tabular form as given below.

Events E_i	*Prior Probabilities* $P(E_i)$	*Conditional Probabilities* $P(E \mid E_i)$	*Joint Probabilities* $P(E_i \cap E)$	*Posterior Probabilities* $P(E_i \mid E)$
(1)	(2)	(3)	(4) = (2) × (3)	(5) = (4) ÷ P(E)
E_1	$\frac{4}{9}$	$\frac{3}{10}$	$\frac{12}{90}$	$\frac{6}{23}$
E_2	$\frac{2}{9}$	$\frac{1}{2}$	$\frac{1}{9} = \frac{10}{90}$	$\frac{5}{23}$
E_3	$\frac{1}{3}$	$\frac{4}{5}$	$\frac{4}{15} = \frac{24}{90}$	$\frac{12}{23}$
Total			$P(E) = \frac{46}{90}$	1

Example 4·6. *A company uses three different assembly lines — A_1, A_2 and A_3 to manufacture particular component. Of those manufactured by line A_1, there are 5% that need rework to remedy a defect, whereas 8% of A_2's components need rework and 10% of A_3's need rework. Suppose that 50% of all components are produced by line A_1, while 30% are produced by line A_2 and 20% come from line A_3. What is the probability that a randomly selected component :*

(i) came from line A_1 and needed rework ?
(ii) needed rework ?
(iii) came from line A_2, given that it needed rework ?
(iv) came from line A_3 and needed rework ?

Solution. Define the following events:

E_i : The component is manufactured by the assembly line A_i ; $i = 1, 2, 3$.
E : Component needs rework to remedy a defect.

Then, we are given :

$P(E_1) = 0{\cdot}50 \quad : \quad P(E \mid E_1) = 0{\cdot}05 \quad \Rightarrow \quad P(E_1) \cdot P(E \mid E_1) = 0{\cdot}025$
$P(E_2) = 0{\cdot}30 \quad : \quad P(E \mid E_2) = 0{\cdot}08 \quad \Rightarrow \quad P(E_2) \cdot P(E \mid E_2) = 0{\cdot}024$
$P(E_3) = 0{\cdot}20 \quad : \quad P(E \mid E_3) = 0{\cdot}10 \quad \Rightarrow \quad P(E_3) \cdot P(E \mid E_3) = 0{\cdot}020$

(*i*) Required Probability $= P(E_1 \cap E) = P(E_1)\, P(E \mid E_1) = 0{\cdot}025$

(*ii*) Required Probability $= P(E) = \sum_{i=1}^{3} P(E_i)\, P(E \mid E_i) = 0{\cdot}025 + 0{\cdot}024 + 0{\cdot}020 = 0{\cdot}069$

(iii) $P(E_2 \mid E) = \dfrac{P(E_2 \cap E)}{P(E)} = \dfrac{P(E_2) \cdot P(E \mid E_2)}{P(E)} = \dfrac{0{\cdot}024}{0{\cdot}069} = 0{\cdot}3478$

(iv) $P(E_3 \cap E) = P(E_3) \cdot P(E \mid E_3) = 0{\cdot}020$

Example 4·7. *A factory produces a certain type of outputs by three types of machine. The respective daily production figures are :*

Machine I : 3,000 Units ; Machine II : 2,500 Units ; Machine III : 4,500 Units

Past experience shows that 1 per cent of the output produced by Machine I is defective. The corresponding fraction of defectives for the other two machines are 1.2 per-cent and 2 per-cent, respectively. An item is drawn at random from the day's production run and is found to be defective. What is probability that it comes from the output of

(i) Machine I, (ii) Machine II, (iii) Machine III ?

Solution. Let E_1, E_2 and E_3 denote the events that the output is produced by machines I, II and III, respectively, and let E denote the event that the output is defective. Then we have :

$$P(E_1) = \frac{3000}{10{,}000} = 0{\cdot}30, \qquad P(E_2) = \frac{2500}{10{,}000} = 0{\cdot}25, \qquad P(E_3) = \frac{4500}{10{,}000} = 0{\cdot}45$$

$$P(E \mid E_1) = 1\% = 0{\cdot}01, \qquad P(E \mid E_2) = 1{\cdot}2\% = 0{\cdot}012, \qquad P(E \mid E_3) = 2\% = 0{\cdot}02$$

The probability that an item selected at random from day's production is defective is given by :

$$P(E) = \sum_{i=1}^{3} P(E \cap E_i) = \sum_{i=1}^{3} P(E_i) \cdot P(E \mid E_i)$$

$$= 0{\cdot}30 \times 0{\cdot}01 + 0{\cdot}25 \times 0{\cdot}012 + 0{\cdot}45 \times 0{\cdot}02 = 0{\cdot}015$$

By Bayes rule, the required probabilities are given by :

(i) $P(E_1 \mid E) = \dfrac{P(E_1) \cdot P(E \mid E_1)}{P(E)} = \dfrac{0{\cdot}003}{0{\cdot}015} = \dfrac{1}{5}$

(ii) $P(E_2 \mid E) = \dfrac{P(E_2) \cdot P(E \mid E_3)}{P(E)} = \dfrac{0{\cdot}003}{0{\cdot}015} = \dfrac{1}{5}$

(iii) $P(E_3 \mid E) = \dfrac{P(E_3) \cdot P(E \mid E_3)}{P(E)} = \dfrac{0{\cdot}009}{0{\cdot}015} = \dfrac{3}{5}$

The probabilities in (*i*), (*ii*) and (*iii*) are known as posterior probabilities of events E_1, E_2 and E_3, respectively.

Aliter. The posterior probabilities can be obtained elegantly in a tabular form as given below.

Events E_i	*Prior Probabilities* $P(E_i)$	*Conditional Probabilities* $P(E \mid E_i)$	*Joint Probabilities* $P(E_i \cap E)$	*Posterior Probabilities* $P(E_i \mid E)$
(1)	(2)	(3)	(4) = (2) × (3)	(5) = (4) ÷ $P(E)$
E_1	0·30	0·010	0·003	$\dfrac{0{\cdot}003}{0{\cdot}015} = \dfrac{1}{5}$
E_2	0·25	0·012	0·003	$\dfrac{0{\cdot}003}{0{\cdot}015} = \dfrac{1}{5}$
E_3	0·45	0·020	0·009	$\dfrac{0{\cdot}009}{0{\cdot}015} = \dfrac{3}{5}$
Total	1·00		$P(E) = 0{\cdot}015$	1

Example 4·8. *There are two bags A and B. A contains n white and 2 black balls and B contains 2 white and n black balls. One of the two bags is selected at random and two balls are drawn from it without replacement. If both the balls drawn are white and the probability that the bag A was used to draw the balls is* $\frac{6}{7}$, *find the value of n.*

Solution. Let E_1 denote the event that bag A is selected and E_2 denote the event that bag B is selected. Let E be the event that two balls drawn are white. We have

$$P(E_1) = P(E_2) = \frac{1}{2};\ P(E \mid E_1) = \frac{{}^nC_2}{{}^{n+2}C_2} = \frac{n(n-1)}{(n+2)(n+1)} \text{ and } P(E \mid E_2) = \frac{{}^2C_2}{{}^{n+2}C_2} = \frac{2}{(n+2)(n+1)}$$

Using Bayes' Theorem, the probability that the two white balls drawn are from the bag A, is given by :

$$P(E_1 \mid E) = \frac{P(E_1)\,P(E \mid E_1)}{P(E_1)\,P(E \mid E_1) + P(E_2)\,P(E \mid E_2)} = \frac{6}{7} \text{ (Given)}$$

$$\Rightarrow \quad \frac{\frac{1}{2}\cdot\frac{n(n-1)}{(n+2)(n+1)}}{\frac{1}{2}\cdot\frac{n(n-1)}{(n+2)(n+1)} + \frac{1}{2}\cdot\frac{2}{(n+2)(n+1)}} = \frac{6}{7} \quad \Rightarrow \quad \frac{n(n-1)}{n(n-1)+2} = \frac{6}{7}$$

$$\therefore \quad 7n(n-1) = 6n(n-1) + 12 \quad \Rightarrow \quad n^2 - n - 12 = 0 \quad \Rightarrow \quad n = 4, -3.$$

Since n cannot be negative, we get $n = 4$.

Example 4·9. *A letter is known to have come either from TATANAGAR or from CALCUTTA. On the envelop just two consecutive letters TA are visible. What is the probability that the letter came from CALCUTTA ?*

Solution. Let E_1 and E_2 denote the events the letter came from TATANAGAR and CALCUTTA, respectively. Let E denote the event that two consecutive visible letters on the envelop are TA.

Then $$P(E_1) = P(E_2) = \frac{1}{2}$$

Writing letters of the words TATANAGAR and CALCUTTA in consecutive pairs, we have

(*a*) TA, AT, TA, AN, NA, AG, GA, AR ; 8 pairs ; two pairs favourable to the event E

$$\Rightarrow \quad P(E \mid E_1) = \frac{2}{8}$$

(*b*) CA, AL, LC, CU, UT, TT, TA ; 7 pairs; one pair favourable to the event E

$$\Rightarrow \quad P(E \mid E_2) = \frac{1}{7}.$$

Using Bayes' theorem, we get

$$P(E_2 \mid E) = \frac{P(E_2)\,P(E \mid E_2)}{P(E_1)\,P(E \mid E_1) + P(E_2)\,P(E \mid E_2)} = \frac{\frac{1}{2}\cdot\frac{1}{7}}{\frac{1}{2}\cdot\frac{2}{8} + \frac{1}{2}\cdot\frac{1}{7}} = \frac{4}{11}.$$

Example 4·10. *The chances that doctor A will diagnose a disease X correctly is 60%. The chances that a patient will die by his treatment after correct diagnosis is 40% and the chance of death by wrong* *diagnosis is 70%. A patient of doctor A, who had disease X, died. What is the chance that his disease was diagnosed correctly ?*

Solution. Let us define the following events :

E_1 : Disease X is diagnosed correctly by doctor A.

E_2 : Disease X is not diagnosed correctly by doctor A.

E : A patient (of Dr A) who had disease X dies.

Then, we are given :

$$P(E_1) = 0{\cdot}6 \quad ; \quad P(E \mid E_1) = 0{\cdot}4$$

$$P(E_2) = P(\bar{E}_1) = 1 - P(E_1) = 0{\cdot}4 \quad ; \quad P(E \mid E_2) = 0{\cdot}7$$

$$\therefore \quad P(E) = \sum_{i=1}^{2} P(E_i)\, P(E \mid E_i) = 0{\cdot}6 \times 0{\cdot}4 + 0{\cdot}4 \times 0{\cdot}7 = 0{\cdot}52$$

Using Bayes theorem, the required probability is given by :

$$P(E_1 \mid E) = \frac{P(E_1)\, P(E \mid E_1)}{P(E)} = \frac{0{\cdot}6 \times 0{\cdot}4}{0{\cdot}52} = \frac{0{\cdot}24}{0{\cdot}52} = \frac{6}{13}.$$

Example 4·11. *The contents of urns, I, II and III are as follows:*
1 white, 2 black and 3 red balls ; 2 white, 1 black and 1 red balls, and 4 white, 5 black and 3 red balls.

One urn is chosen at random and two balls drawn from it. They happen to be white and red. What is the probability that they come from urns I, II or III ?

Solution. Let E_1, E_2 and E_3 denote the events that the urn, I, II and III is chosen, respectively, and let E be the event that the two balls taken from the selected urn are white and red.

Then $$P(E_1) = P(E_2) = P(E_3) = \frac{1}{3}$$

$$P(E \mid E_1) = \frac{1 \times 3}{{}^6C_2} = \frac{1}{5}, \quad P(E \mid E_2) = \frac{2 \times 1}{{}^4C_2} = \frac{1}{3}, \text{ and } P(E \mid E_3) = \frac{4 \times 3}{{}^{12}C_2} = \frac{2}{11}$$

$$\therefore \quad P(E_2 \mid E) = \frac{P(E_2)\, P(E \mid E_2)}{\sum_{i=1}^{3} P(E_i)\, P(E \mid E_i)} = \frac{\frac{1}{3} \times \frac{1}{3}}{\frac{1}{3} \times \frac{1}{5} + \frac{1}{3} \times \frac{1}{3} + \frac{1}{3} \times \frac{2}{11}} = \frac{55}{118}$$

Similarly, $$P(E_3 \mid E) = \frac{\frac{1}{3} \times \frac{2}{11}}{\frac{1}{3} \times \frac{1}{5} + \frac{1}{3} \times \frac{1}{3} + \frac{1}{3} \times \frac{2}{11}} = \frac{30}{118}$$

$$\therefore \quad P(E_1 \mid E) = 1 - \frac{55}{118} - \frac{30}{118} = \frac{33}{118}.$$

Example 4·12. *From a vessel containing 3 white and 5 black balls, 4 balls are transferred into an empty vessel. From this vessel a ball is drawn and is found to be white. What is the probability that out of four balls transferred 3 are white and 1 is black ?*

Solution. Let us define the following events :

E_1 : Transfer of 0 white and 4 black balls ; E_2 : Transfer of 1 white and 3 black balls

E_3 : Transfer of 2 white and 2 black balls ; E_4 : Transfer of 3 white and 1 black balls

(Since the urn contains 3 white balls, more than 3 white balls cannot be transferred from the vessel)

E : Drawing of a white ball from the second vessel.

$$\therefore\ P(E_1) = \frac{^5C_4}{^8C_4} = \frac{1}{14};\ P(E_2) = \frac{^3C_1 \times {^5C_3}}{^8C_4} = \frac{3}{7};\ P(E_3) = \frac{^3C_2 \times {^5C_2}}{^8C_4} = \frac{3}{7};\ P(E_4) = \frac{^3C_3 \times {^5C_1}}{^8C_4} = \frac{1}{14}$$

Also $\quad P(E \mid E_1) = 0,\ P(E \mid E_2) = \frac{1}{4},\ P(E \mid E_3) = \frac{2}{4}$ and $P(E \mid E_4) = \frac{3}{4}$

Hence, by Bayes Theorem, the probability that out of four balls transferred, 3 are white and 1 is black is :

$$P(E_4 \mid E) = \frac{\frac{1}{14} \times \frac{3}{4}}{\frac{1}{14} \times 0 + \frac{3}{7} \times \frac{1}{4} + \frac{3}{7} \times \frac{1}{2} + \frac{1}{14} \times \frac{3}{4}} = \frac{3}{6+12+3} = \frac{1}{7} = 0{\cdot}14.$$

Example 4·13. *A and B are two weak students of statistics and their chances of solving a problem in statistics correctly are* $\frac{1}{6}$ *and* $\frac{1}{8}$ *respectively. If the probability of their making a common error is* $\frac{1}{525}$ *and they obtain the same answer, find the probability that their answer is correct.*

Solution. Let us define the following events:

E_1 : Both A and B solve the problem correctly ;
E_2 : Exactly one of them solves the problem correctly.
E_3 : Neither of them solves the problem correctly ;
E : They get the same answer.

Then, according to the data given in the problem, assuming that A and B try the problem independently, we have:

$$P(E_1) = \frac{1}{6} \times \frac{1}{8} = \frac{1}{48} \quad ; \quad P(E \mid E_1) = 1$$

$$P(E_2) = \frac{1}{6} \times \frac{7}{8} + \frac{5}{6} \times \frac{1}{8} = \frac{12}{48} \quad ; \quad P(E \mid E_2) = 0$$

$$P(E_3) = \frac{5}{6} \times \frac{7}{8} = \frac{35}{48} \quad ; \quad P(E \mid E_3) = \frac{1}{525} \quad \text{(Given)}$$

Hence, by Bayes' Rule, the required probability is given by :

$$P(E_1 \mid E) = \frac{P(E_1)\, P(E \mid E_1)}{\sum_{i=1}^{3} P(E_i)\, P(E \mid E_i)}$$

$$= \frac{\frac{1}{48} \times 1}{\frac{1}{48} \times 1 + \frac{12}{48} \times 0 + \frac{35}{48} \times \frac{1}{525}} = \frac{\frac{1}{48}}{\frac{1}{48} + \frac{1}{48 \times 15}} = \frac{1}{48} \times \frac{48 \times 15}{(15+1)} = \frac{15}{16}.$$

Aliter. Let us define the events :

E_1 : A and B get the same correct answer
E_2 : A and B get the same wrong answer
E : A and B get the same answer

Then we have :

$$P(E_1) = \frac{1}{6} \times \frac{1}{8} = \frac{1}{48} \quad ; \quad P(E_2) = \frac{5}{6} \times \frac{7}{8} \times \frac{1}{525} = \frac{1}{720}$$

$$P(E) = P(E_1 \cup E_2) = P(E_1) + P(E_2) = \frac{1}{48} + \frac{1}{720} = \frac{15+1}{720} = \frac{1}{45}$$

The required probability that the answer is correct is given by :

$$P(E_1 \mid E) = \frac{P(E_1 \cap E)}{P(E)} = \frac{P(E_1)}{P(E)} = \frac{\frac{1}{48}}{\frac{1}{45}} = \frac{45}{48} = \frac{15}{16}. \qquad (\because E_1 \subset E)$$

SUPPLEMENTARY EXAMPLES ON BAYES THEOREM

Example 4.14. *In answering a question on a multiple choice test a student either knows the answer or he guesses. Let p be the probability that he knows the answer and 1 – p the probability that he guesses. Assume that a student who guesses at answer will be correct with probability 1/5, where 5 is the number of multiple-choice alternatives. What is the conditional probability that a student knew the answer to a question given that he answered it correctly ?*

Solution. Let us define the following events :

E_1 : The student knows the right answer ; E_2 : The student guesses the right answer.
E : The student gets the right answer.

Then we are given : $P(E_1) = p,\quad P(E_2) = 1 - p,\quad P(E \mid E_2) = \frac{1}{5}$

$P(E \mid E_1) = P$ (student gets right answer given that he knew the right answer) = 1

We want $P(E_1 \mid E)$. Using Bayes' Theorem, we get

$$P(E_1 \mid E) = \frac{P(E_1) \cdot P(E \mid E_1)}{P(E_1)\, P(E \mid E_1) + P(E_2) \cdot P(E \mid E_2)} = \frac{p \times 1}{p \times 1 + (1-p) \times \frac{1}{5}} = \frac{5p}{4p+1}.$$

Example 4·15. *(Bayes' Theorem for repeated trials). In n independent trials with constant probability p of success in each trial, m successes have been observed. p is not known, but it is known that it takes one of the values $p_1, p_2, ..., p_k$ and the probability with which p takes the value p_i is α_i $(i = 1, 2, ..., k)$. What is the probability that*

(a) $p = p_i$,

(b) p lies between the two given limits α and β, $(0 \le \alpha < \beta \le 1)$ or what is the probability of the inequality $\alpha \le p \le \beta$?

Solution. Let us define the events as follows :

$E = \{m$ successes in n trials$\}$; $E_i = \{p$ takes the value p_i, $(i = 1, 2, ..., k)\}$.

Then $\quad P(E_i) = \alpha_i \quad$ (Given)

and $\quad P(E \mid E_i) = {}^nC_m\, p_i^{\,m}\, (1 - p_i)^{n-m}. \quad$ (By Binomial Probability Law)

Using Bayes' Theorem, we get

$$P(E_i \mid E) = \frac{\alpha_i \cdot {}^nC_m\, p_i^{\,m}\, (1-p_i)^{n-m}}{\sum_{i=1}^{k} \alpha_i \cdot {}^nC_m\, p_i^{\,m}\, (1-p_i)^{n-m}} = \frac{\alpha_i\, p_i^{\,m}\, (1-p_i)^{n-m}}{\sum_{i=1}^{k} \alpha_i\, p_i^{\,m}\, (1-p_i)^{n-m}}$$

Now applying the Theorem of Total Probability, the probability of inequality $\alpha \le p \le \beta$ is given by :

$$P(\alpha \le p \le \beta \mid E) = \frac{\sum\limits_{p_i = \alpha}^{\beta} \alpha_i \, p_i^{\,m} (1 - p_i)^{n-m}}{\sum\limits_{i=1}^{k} \alpha_i \, p_i^{\,m} (1 - p_i)^{n-m}}$$

Example 4·16. *A printing machine can print n 'letters', say $\alpha_1, \alpha_2, ..., \alpha_n$. It is operated by electrical impulses, each letter being produced by a different impulse. Assume that p is the constant probability of printing the correct letter and the impulses are independent. One of the n impulses, chosen at random, was fed into the machine twice and both times the letter α_1 was printed. Compute the probability that the impulse chosen was meant to print α_1.*

Solution. Let us defined the following events :

E : Letter α_1 is printed both the times.

B_i : The impulse chosen is meant to print α_i, $(i = 1, 2, ..., n)$.

E_i : Letter α_i is printed.

Then we have: $\quad P(B_i) = \dfrac{1}{n}, \ (i = 1, 2, ..., n).$

$$P(E_i \mid B_i) = P[\text{A letter is printed correctly}] = p, \ (i = 1, 2, ..., n)$$

and $\quad P(E_i \mid B_j) = P[\text{A letter is printed wrongly}] = 1 - p, \ (i \ne j = 1, 2, ..., n)$

We want to compute $P(B_1 \mid E)$. By Bayes Rule, we have :

$$P(B_1 \mid E) = \frac{P(B_1)\,P(E \mid B_1)}{\sum\limits_{i=1}^{n} P(B_i)\,P(E \mid B_i)} = \frac{P(B_1)\,P(E \mid B_1)}{P(B_1)\,P(E \mid B_1) + \sum\limits_{i=2}^{n} P(B_i)\,P(E \mid B_i)} \quad \text{... (*)}$$

Since the impulses operate independently, we have

$$P(E \mid B_1) = p \cdot p = p^2$$

$$P(E \mid B_j) = P\Big[\alpha_1 \text{ is printed both the times when the impulse was meant to print } \alpha_j, \ (j = 2, 3, ..., n)\Big]$$

$$= \left(\frac{1-p}{n-1}\right)^2, \ (j = 2, 3, ..., n)$$

Finally, substituting in (*), we get

$$P(B_1 \mid E) = \frac{\frac{1}{n} \cdot p^2}{\frac{1}{n} \cdot p^2 + \sum\limits_{i=2}^{n} \frac{1}{n}\left(\frac{1-p}{n-1}\right)^2} = \frac{(p^2 / n)}{\frac{p^2}{n} + (n-1) \cdot \frac{1}{n}\left(\frac{1-p}{n-1}\right)^2}$$

$$= \frac{(n-1)\,p^2}{(n-1)\,p^2 + (1-p)^2} = \frac{(n-1)\,p^2}{(n-1)\,p^2 + 1 + p^2 - 2p}$$

$$= \frac{(n-1)\,p^2}{np^2 - 2p + 1}.$$

Example 4·17. *(i) Let the probability p_n that a family has exactly n children be αp^n when $n \ge 1$ and $p_0 = 1 - ap\,(1 - p + p^2 + ...)$. Suppose that all the sex distributions of n children have the same probability. Show that for $k \ge 1$, the probability that a family contains exactly k boys is $2\alpha \cdot p^k / (2-p)^{k+1}$.*

(ii) Given that a family includes at least one boy, show that the probability that there are two or more boys is $p / (2-p)$.

Solution. (*i*) We are given:

$$\left.\begin{aligned} p_n &= P(\text{that a family has exactly } n \text{ children}) = \alpha\, p^n,\ n \ge 1 \\ \text{and} \quad p_0 &= 1 - \alpha\, p\,(1 + p + p^2 + ...) \end{aligned}\right\} \qquad ...(*)$$

Let E_j be the event that the number of children in a family is j and let A be the event that a family contains exactly k boys. Then $P(E_j) = p_j;\ j = 0, 1, 2, ...$

Since each child can have any of the two sex distributions (either boy or girl), the total number of possible distributions for a family to have 'j' children is 2^j.

$$\therefore \qquad P(A \mid E_j) = \frac{{}^jC_k}{2^j},\ j \ge k \qquad ...(**)$$

and
$$P(A) = \sum_{j=k}^{\infty} P(E_j)\, P(A \mid E_j) = \sum_{j=k}^{\infty} p_j\, P(A \mid E_j)$$

$$= \sum_{j=k}^{\infty} \alpha\, p^j \left(\frac{{}^jC_k}{2^j}\right), (j \ge k \ge 1) = \alpha \sum_{j=k}^{\infty} \left(\frac{p}{2}\right)^j {}^jC_k \qquad \text{[From (*) and (**)]}$$

$$= \alpha \sum_{r=0}^{\infty} {}^{k+r}C_k \left(\frac{p}{2}\right)^{k+r} \qquad [\text{Put } j - k = r]$$

$$= \alpha \left(\frac{p}{2}\right)^k \sum_{r=0}^{\infty} {}^{k+r}C_r \left(\frac{p}{2}\right)^r \qquad [\because {}^nC_r = {}^nC_{n-r}]$$

We know that

$${}^{-n}C_r = (-1)^r \cdot {}^{n+r-1}C_r \Rightarrow (-1)^r \cdot {}^{-n}C_r = {}^{n+r-1}C_r \Rightarrow (-1)^r \cdot {}^{-(k+1)}C_r = {}^{k+r}C_r$$

$$\therefore \qquad P(A) = \alpha \left(\frac{p}{2}\right)^k \sum_{r=0}^{\infty} (-1)^r \cdot {}^{-(k+1)}C_r \left(\frac{p}{2}\right)^r$$

$$= \alpha \left(\frac{p}{2}\right)^k \sum_{r=0}^{\infty} {}^{-(k+1)}C_r \left(-\frac{p}{2}\right)^r = \alpha \left(\frac{p}{2}\right)^k \left(1 - \frac{p}{2}\right)^{-(k+1)}$$

$$= \alpha \left(\frac{p}{2}\right)^k \frac{2^{k+1}}{(2-p)^{k+1}} = \frac{2\alpha \cdot p^k}{(2-p)^{k+1}} \qquad ...(1)$$

(*ii*) Let B denote the event that a family includes at least one boy and C denote the event that a family has two or more boys. Then

$$P(B) = \sum_{k=1}^{\infty} P[\text{family has exactly } k \text{ boys}] = \sum_{k=1}^{\infty} \frac{2\alpha p^k}{(2-p)^{k+1}} = \frac{2\alpha}{2-p} \sum_{k=1}^{\infty} \left(\frac{p}{2-p}\right)^k$$

[From (1)]

$$= \frac{2\alpha}{2-p} \times \frac{p/(2-p)}{1 - [p/(2-p)]} = \frac{\alpha p}{(1-p)(2-p)} \qquad ...(2)$$

$$P(C) = \sum_{k=2}^{\infty} P[\text{family has exactly } k \text{ boys}] = \sum_{k=2}^{\infty} \frac{2\alpha p^k}{(2-p)^{k+1}} = \frac{2\alpha}{2-p} * \sum_{k=2}^{\infty} \left(\frac{p}{2-p}\right)^k$$

$$= \frac{2\alpha}{2-p} \cdot \frac{[p/(2-p)]^2}{1-[p/(2-p)]} = \frac{\alpha p^2}{(2-p)^2 (1-p)} \qquad \text{... (3)}$$

Since $C \subset B$ and $B \cap C = C$, $P(B \cap C) = P(C) \Rightarrow P(B)\, P(C \mid B) = P(C)$

Therefore, the required probability is given by :

$$P(C \mid B) = \frac{P(C)}{P(B)} = \frac{\alpha p^2}{(2-p)^2 (1-p)} \times \frac{(1-p)(2-p)}{\alpha p} = \frac{p}{2-p}. \qquad \text{[From (2) and (3)]}$$

Example 4·18. *A slip of paper is given to person A who marks it either with a plus sign or a minus sign; the probability of his wirting a plus sign is 1/3. A passes the slip to B, who may either leave it alone or change the sign before passing it to C. Next C passes the slip to D after perhaps changing the sign. Finally D passes it to a referee after perhaps changing the sign. The referee sees a plus sign on the slip. It is known that B, C and D each change the sign with probability 2/3. Find he probability that A originally wrote a plus.*

Solution. Let us define the following events :

E_1 : A wrote a plus sign ; E_2 : A wrote a minus sign

E : The referee observes a plus sign on the slip.

We are given: $P(E_1) = \dfrac{1}{3}$, $P(E_2) = 1 - \dfrac{1}{3} = \dfrac{2}{3}$

We want $P(E_1 \mid E)$, which by Bayes rule is given by :

$$P(E_1 \mid E) = \frac{P(E_1)\, P(E \mid E_1)}{P(E_1)\, P(E \mid E_1) + P(E_2)\, P(E \mid E_2)} \qquad \text{... (i)}$$

$P(E \mid E_1) = P$[Referee observes the plus sign given that 'A' wrote the plus sign on the slip]

$= P$[(Plus sign was not changed at all) $\cup$ (Plus sign was changed exactly twice in passing from 'A' to referee through B, C and D]

$= P(E_3 \cup E_4)$, (say). $= P(E_3) + P(E_4)$ ... (ii)

Let A_1, A_2 and A_3, respectively, denote the events that B, C and D change the sign on the slip. Then we are given :

$$P(A_1) = P(A_2) = P(A_3) = \frac{2}{3}; \qquad P(\bar{A}_1) = P(\bar{A}_2) = P(\bar{A}_3) = \frac{1}{3}$$

We have

$$P(E_3) = P(\bar{A}_1 \cap \bar{A}_2 \cap \bar{A}_3) = P(\bar{A}_1)\, P(\bar{A}_2)\, P(\bar{A}_3) = \left(\frac{1}{3}\right)^3 = \frac{1}{27}$$

$$P(E_4) = P[(A_1 A_2 \bar{A}_3) \cup (A_1 \bar{A}_2 A_3) \cup (\bar{A}_1 A_2 A_3)]$$

$$= P(A_1 A_2 \bar{A}_3) + P(A_1 \bar{A}_2 A_3) + P(\bar{A}_1 A_2 A_3)$$

$$= P(A_1)\, P(A_2)\, P(\bar{A}_3) + P(A_1)\, P(\bar{A}_2)\, P(A_3) + P(\bar{A}_1)\, P(A_2)\, P(A_3)$$

$$= \frac{2}{3} \cdot \frac{2}{3} \cdot \frac{1}{3} + \frac{2}{3} \cdot \frac{1}{3} \cdot \frac{2}{3} + \frac{1}{3} \cdot \frac{2}{3} \cdot \frac{2}{3} = \frac{4}{9}$$

Substituting in (*ii*), we get $\quad P(E \mid E_1) = \dfrac{1}{27} + \dfrac{4}{9} = \dfrac{13}{27}$ ... (*iii*)

Similarly,

$P(E \mid E_2) = P$[Referee observes plus sign given that '*A*' wrote the minus sign on the slip]

$= P$[(Minus sign was changed exactly once) $\cup$ (Minus sign was changed thrice)]

$= P(E_5 \cup E_6)$, (say). $= P(E_5) + P(E_6)$ ... (*iv*)

$$P(E_5) = p\left[(A_1 \bar{A}_2 \bar{A}_3) \cup (\bar{A}_1 A_2 \bar{A}_3) \cup (\bar{A}_1 \bar{A}_2 A_3)\right]$$

$$= P(A_1)\, P(\bar{A}_2)\, P(\bar{A}_3) + P(\bar{A}_1)\, P(A_2)\, P(\bar{A}_3) + P(\bar{A}_1)\, P(\bar{A}_2)\, P(A_3)$$

$$= \frac{2}{3}\cdot\frac{1}{3}\cdot\frac{1}{3} + \frac{1}{3}\cdot\frac{2}{3}\cdot\frac{1}{3} + \frac{1}{3}\cdot\frac{1}{3}\cdot\frac{2}{3} = \frac{2}{9}$$

$$P(E_6) = P(A_1 A_2 A_3) = P(A_1)\, P(A_2)\, P(A_3) = \frac{2}{3}\cdot\frac{2}{3}\cdot\frac{2}{3} = \frac{8}{27}$$

Substituting in (*iv*), we get : $\quad P(E \mid E_2) = \dfrac{2}{9} + \dfrac{8}{27} = \dfrac{14}{27}$. ... (*v*)

Substituting from (*iii*) and (*v*) in (*i*), we get

$$P(E_1 \mid E) = \frac{\frac{1}{3} \times \frac{13}{27}}{\frac{1}{3} \times \frac{13}{27} + \frac{2}{3} \times \frac{14}{27}} = \frac{13}{13 + 28} = \frac{13}{41}.$$

Example 4·19. *Three urns of the same appearance have the following proportion of balls.*

First urn : 2 black, 1 white ; Second urn : 1 black, 2 white ; Third urn : 2 black, 2 white

One of the urns is selected and one ball is drawn. It turns out to be white. What is the probability of drawing a white ball again, the first one not having been returned ?

Solution. Let us define the events :

E_i = The event of selection of *i*th urn, $(i = 1, 2, 3)$

and $\quad A$ = The event of drawing a white ball.

Then $\quad P(E_1) = P(E_2) = P(E_3) = \dfrac{1}{3}$

and $\quad P(A \mid E_1) = \dfrac{1}{3}, \quad P(A \mid E_2) = \dfrac{2}{3} \quad$ and $\quad P(A \mid E_3) = \dfrac{1}{2}$

Let *C* denote the future event of drawing another white ball from the urns. Then

$$P(C \mid E_1 \cap A) = 0, \qquad P(C \mid E_2 \cap A) = \frac{1}{2}, \qquad \text{and} \qquad P(C \mid E_3 \cap A) = \frac{1}{3}$$

$$\therefore \quad P(C \mid A) = \frac{\sum_{i=1}^{3} P(E_i)\, P(A \mid E_i)\, P(C \mid E_i \cap A)}{\sum_{i=1}^{3} P(E_i)\, P(A \mid E_i)} = \frac{\frac{1}{3}\cdot\frac{1}{3}\cdot 0 + \frac{1}{3}\cdot\frac{2}{3}\cdot\frac{1}{2} + \frac{1}{3}\cdot\frac{1}{2}\cdot\frac{1}{3}}{\frac{1}{3}\cdot\frac{1}{3} + \frac{1}{3}\cdot\frac{2}{3} + \frac{1}{3}\cdot\frac{1}{2}} = \frac{1}{3}.$$

Example 4·20. *An urn contains 5 white and 5 black balls, 4 balls are drawn from this urn and put into another urn. From this second urn a ball is drawn and is found to be white. What is the probability of drawing a white ball again at the next draw. (The first white ball drawn is not replaced.)*

Solution. The event of drawing 4 balls from the urn is associated with five mutually exclusive events as follows :

E_0 = The event of drawing 0 white and 4 black balls
E_1 = The event of drawing 1 white and 3 black balls
E_2 = The event of drawing 2 white and 2 black balls
E_3 = The event of drawing 3 white and 1 black ball
E_4 = The event of drawing 4 white and 0 black balls

Then,

$$P(E_0) = \frac{^5C_4}{^{10}C_4} = \frac{1}{42} \; ; \; P(E_1) = \frac{^5C_1 \times {^5C_3}}{^{10}C_4} = \frac{5}{21} \; ; \; P(E_2) = \frac{^5C_2 \times {^5C_2}}{^{10}C_4} = \frac{10}{21},$$

$$P(E_3) = \frac{^5C_3 \times {^5C_1}}{^{10}C_4} = \frac{5}{21} \quad \text{and} \quad P(E_4) = \frac{^5C_4}{^{10}C_4} = \frac{1}{42}.$$

Let A denote the event of drawing a white ball from the second urn in the first draw. Then,

$$P(A \mid E_0) = 0, \qquad P(A \mid E_1) = \frac{1}{4}, \qquad P(A \mid E_2) = \frac{2}{4},$$

$$P(A \mid E_3) = \frac{3}{4} \qquad \text{and} \qquad P(A \mid E_4) = \frac{4}{4} = 1.$$

Let C denote the future event, *i.e.*, the next ball drawn from the remaining 3 balls in the second urn is white.

$$P(C \mid A \cap E_0) = 0, \qquad P(C \mid A \cap E_1) = 0, \qquad P(C \mid A \cap E_2) = \frac{1}{3},$$

$$P(C \mid A \cap E_3) = \frac{2}{3} \qquad \text{and} \qquad P(C \mid A \cap E_4) = \frac{3}{3} = 1.$$

Hence,
$$P(C \mid A) = \frac{\sum_{i=0}^{4} P(E_i)\, P(A \mid E_i)\, P(C \mid A \cap E_i)}{\sum_{i=0}^{4} P(E_i)\, P(A \mid E_i)}$$

$$= \frac{\frac{1}{42} \times 0 \times 0 + \frac{5}{21} \times \frac{1}{4} \times 0 + \frac{10}{21} \times \frac{2}{4} \times \frac{1}{3} + \frac{5}{21} \times \frac{3}{4} \times \frac{2}{3} + \frac{1}{42} \times 1 \times 1}{\frac{1}{42} \times 0 + \frac{5}{21} \times \frac{1}{4} + \frac{10}{21} \times \frac{2}{4} + \frac{5}{21} \times \frac{3}{4} + \frac{1}{42} \times 1} = \frac{4}{9}.$$

Example 4·21. *There are (N + 1) identical urns marked 0, 1, 2, ..., N each of which contains N white and red balls. The kth urn contains k red and N – k white balls, (k = 0, 1, 2, ..., N). An urn is chosen at random and n random drawings of a ball are made from it, the ball drawn being replaced after each draw. If the balls drawn are all red, show that the probability that the next drawing will also yield a red ball is approximately (n + 1) / (n + 2) when N is large.*

Solution. Let E_k denote the event that the kth urn is chosen, (k = 0, 1, 2, ..., N). Then the events, $E_0, E_1, E_2, ..., E_n$ are pairwise mutually exclusive, one of which certainly occurs.

Then
$$P(E_0) = P(E_1) = P(E_2) = ... = P(E_n) = \frac{1}{N+1} \qquad \text{[From symmetry]} \qquad ... (i)$$

Let A denote the event of getting n red balls successively in n draws (with replacement from an urn chosen at random).

Then $$P(A \mid E_k) = \frac{k}{N} \cdot \frac{k}{N} \cdot \ldots \cdot \frac{k}{N} = \left(\frac{k}{N}\right)^n \qquad \ldots (ii)$$

and $$P(A) = \sum_{k=0}^{N} P(E_k)\, P(A \mid E_k) = \frac{1}{N+1} \sum_{k=0}^{N} \left(\frac{k}{N}\right)^n \qquad \ldots (iii)$$

Using Bayes' theorem, we get,

$$P(E_k \mid A) = \frac{P(E_k)\, P(A \mid E_k)}{\sum\limits_{k=0}^{N} P(E_k)\, P(A \mid E_k)} = \frac{\frac{1}{N+1} \cdot \left(\frac{k}{N}\right)^n}{\frac{1}{N+1} \sum\limits_{k=0}^{N} \left(\frac{k}{N}\right)^n}$$

Let C be the future event that $(n + 1)$th draw also yields a red ball. Then on the same lines, it can be shown that

$$P(A \cap C) = \frac{1}{N+1} \sum_{k=0}^{N} \left(\frac{k}{N}\right)^{n+1} \qquad \text{[Replacing } n \text{ by } (n + 1) \text{ in } (iii)]$$

Hence, the required probability is,

$$P(C \mid A) = \frac{P(A \cap C)}{P(A)} = \frac{\frac{1}{N+1} \sum\limits_{k=0}^{N} \left(\frac{k}{N}\right)^{n+1}}{\frac{1}{N+1} \sum\limits_{k=0}^{N} \left(\frac{k}{N}\right)^n} \qquad \ldots (iv)$$

If N is very large, then

$$P(A) \simeq \frac{1}{N} \sum_{k=0}^{N} \left(\frac{k}{N}\right)^n \simeq \int_0^1 x^n \, dx = \frac{1}{n+1}$$

and $$P(A \cap C) \simeq \frac{1}{N} \sum_{k=0}^{N} \left(\frac{k}{N}\right)^{n+1} \simeq \int_0^1 x^{n+1} \, dx = \frac{1}{n+2}$$

$$\therefore \quad P(C \mid A) = \frac{n+1}{n+2}. \qquad \text{[From } (iv)]$$

This is known is *Law of Succession* due to Laplace.

Example 4·22. *Die A has four red and two white faces whereas die B has two red and four white faces. A coin is flipped once. If it falls heads, the game continues by throwing die A and if it falls tails, die B is to be used.*

(i) Show that the probability of getting a red face at any throw is $\frac{1}{2}$.

(ii) If the first two throws resulted in red faces, what is the probability of red face at the 3rd throw ?

(iii) If red face turns up at the first n throws, what is the probability that die A is being used ?

Solution. Let us define the events E_1, E_2, E_3 and A as follows :

E_1 : Die A is used, $\qquad E_2$: Die B is used

E_3 : Getting a red face in any throw.

and A : Getting a red face in each of the first n throws.

(i) $P(E_3 \mid E_1) = \frac{4}{6} = \frac{2}{3}$, $\qquad P(E_3 \mid E_2) = \frac{2}{6} = \frac{1}{3}$,

Since $P(E_1) = P(E_2) = \frac{1}{2}$, the probability of getting a red face in any throw is given by :

$$P(E_3) = P(E_1) \cdot P(E_3 \mid E_1) + P(E_2) \cdot P(E_3 \mid E_2) = \frac{1}{2} \times \frac{2}{3} + \frac{1}{2} \times \frac{1}{3} = \frac{1}{2}$$

(*ii*) Clearly by the '*Law of Succession*', the probability of getting a red face at the third throw when the first two gave red faces is :

$$\frac{\frac{1}{2}+1}{\frac{1}{2}+2} = \frac{3}{5}.$$

Aliter. Let $R_3 \equiv$ Third throw results in red face.

$$P(R_3 \mid R_1 \cap R_2) = \frac{P(R_1 \cap R_2 \cap R_3)}{P(R_1 \cap R_2)} = \frac{\frac{1}{2}\left[\left(\frac{4}{6}\right)^3 + \left(\frac{2}{6}\right)^3\right]}{\frac{1}{2}\left[\left(\frac{4}{6}\right)^2 + \left(\frac{2}{6}\right)^2\right]} = \frac{72}{120} = \frac{3}{5}$$

(*iii*) $\quad P(A \mid E_1) = (2/3)^n, \qquad P(A \mid E_2) = (1/3)^n.$

Using Bayes' Theorem, the required probability is given by :

$$P(E_1 \mid A) = \frac{P(E_1) \cdot P(A \mid E_1)}{\sum_{i=1}^{2} P(E_i) \cdot P(A \mid E_i)} = \frac{\frac{1}{2} \cdot \left(\frac{2}{3}\right)^n}{\frac{1}{2} \cdot \left(\frac{2}{3}\right)^n + \frac{1}{2} \cdot \left(\frac{1}{3}\right)^n} = \frac{2^n}{2^n + 1}.$$

Example 4·23. *In n trials, m successes have been observed. Assuming that p takes any value* $\frac{i}{k}$ *(i = 1, 2, ..., k) with the priori probabilities* $\alpha_1 = \alpha_2 = \alpha_3 = \dots = \alpha_k = \frac{1}{k}$, *find the probability that in the next* n_1 *trials,* m_1 *successes will occur on the assumption that priori probability is uniformly distributed in* (0, 1).

Find the limiting case when $n \to \infty$.

Solution. Let us define the events as follows :

$A = m$ successes in n trials., $E_i = p_i = \frac{i}{k}$ $(i = 1, 2, ..., k)$, $C = m_1$ successes in further n_1 trials.

Then, $\quad P(E_i) = \alpha_i = \frac{1}{k}, \qquad P(A \mid E_i) = {}^nC_m \, p_i^{\,m} \, (1 - p_i)^{n-m}$

and $\quad P(C \mid E_i) = {}^{n_1}C_{m_1} \, p_i^{\,m_1} (1 - p_i)^{n_1 - m_1}$ (By Binomial probability law.)

By Bayes' Theorem, the required probability is given by

$$P(C \mid A) = \frac{\sum_{i=1}^{k} \alpha_i \cdot {}^nC_m \, p_i^{\,m} (1 - p_i)^{n-m} \cdot {}^{n_1}C_{m_1} \left(\frac{i}{k}\right)^{m_1} \left(1 - \frac{i}{k}\right)^{n_1 - m_1}}{\sum_{i=1}^{k} \frac{1}{k} \cdot {}^nC_m \left(\frac{i}{k}\right)^m \left(1 - \frac{i}{k}\right)^{n-m}}$$

$$= \frac{\sum_{i=1}^{k} \frac{1}{k} \cdot {}^{n}C_m \left(\frac{i}{k}\right)^m \left(1-\frac{i}{k}\right)^{n-m} \times {}^{n_1}C_{m_1} \left(\frac{i}{k}\right)^{m_1} \left(1-\frac{i}{k}\right)^{n_1-m_1}}{\sum_{i=1}^{k} \frac{1}{k} \cdot {}^{n}C_m \left(\frac{i}{k}\right)^m \left(1-\frac{i}{k}\right)^{n-m}}$$

$$\therefore \quad \lim_{k\to\infty} P(C \mid A) = \lim_{k\to\infty} \frac{{}^{n_1}C_{m_1} \sum_{i=1}^{k} \frac{1}{k} \left(\frac{i}{k}\right)^{m+m_1} \left(1-\frac{i}{k}\right)^{n+n_1-m-m_1}}{\sum_{i=1}^{k} \frac{1}{k} \left(\frac{i}{k}\right)^m \left(1-\frac{i}{k}\right)^{n-m}}$$

$$= \frac{{}^{n_1}C_{m_1} \int_0^1 x^{m+m_1} (1-x)^{n+n_1-m-m_1}\, dx}{\int_0^1 x^m (1-x)^{n-m}\, dx} \qquad \left[\text{Replace } \frac{i}{k} \text{ by } x \text{ and } \frac{1}{k} \text{ by } dx\right]$$

$$= {}^{n_1}C_{m_1} \cdot \frac{B(m+m_1+1, n-m+n_1-m_1+1)}{B(m+1, n-m+1)}$$

In particular, when $m_1 = n_1 = 1$, then

$$P(C \mid A) = \frac{B(m+2, n-m+1)}{B(m+1, n-m+1)} = \frac{\Gamma(m+2)\,\Gamma(n-m+1)\,\Gamma(n+2)}{\Gamma(n+3)\,\Gamma(m+1)\,\Gamma(n-m+1)} = \frac{m+1}{n+2}.$$

This is the *'Law of Succession'* established by Laplace.

4·3. GEOMETRIC PROBABILITY

In Chapter 3, we confined our discussion to the problems when the sample space is discrete *i.e.*, it has a finite or a countable number of points. However, in case of continuous sample space, where the total number of outcomes of a random experiment is infinite, we cannot define probability in terms of the number of points. Thus, for example, if we are interested in finding the probability that a point selected at random in a given region will lie in a specified part of it, the classical definition of probability is modified and extended to what is called *geometrical probability or probability in continuum*. In this case, the general expression for probability *'p'* is given by :

$$p = \frac{\textit{Measure of specified part of the region}}{\textit{Measure of the whole region}} \qquad \text{... (4·14)}$$

where 'measure' refers to the length, area or volume of the region if we are dealing with one, two or three dimensional space respectively.

Example 4·24. *The outcomes of an experiment are represented by the points in the square bounded by $x = 0$, $y = 0$, $x = a$, $y = a$ in the $x - y$ plane. If the probability is distributed uniformly, determine the probability that $x^2 + y^2 > a^2$.*

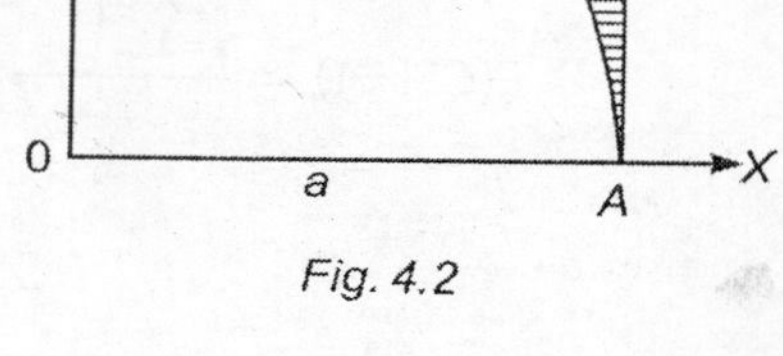

Fig. 4.2

Solution. Here, the sample space S is the square $OABC$ with side 'a' units. The region (R) favourable to the event $\{x^2 + y^2 > a^2\}$, is shown as the shaded region in the diagram. The required probability p is given by :

$$p = P(x^2 + y^2 > a^2) = \frac{\text{Area of region } R}{\text{Area of square } OABC} \qquad \text{... (*)}$$

Area of region R = (Area of square $OABC$) – [Area of quadrant of circle $C(O, a)$]

$$= a^2 - \frac{1}{4}.\pi a^2 = a^2\left(1 - \frac{\pi}{4}\right) \text{ sq. units.}$$

Substituting in (*), we obtain $$p = \frac{a^2\left(1-\frac{\pi}{4}\right)}{a^2} = 1 - \frac{\pi}{4}.$$

Example 4·25. *C is the mid-point of a line AB of length a units. Two points are selected at random on the lines AC and CB. Find the probability that the distance between them is less than λa, where $0 < \lambda < 1$.*

Solution. Let D and E be any two points selected at random on AC and CB respectively. Let $DC = x$ units and $CE = y$ units. Then the points (x, y) can lie anywhere inside a square of side $(a/2)$ units.

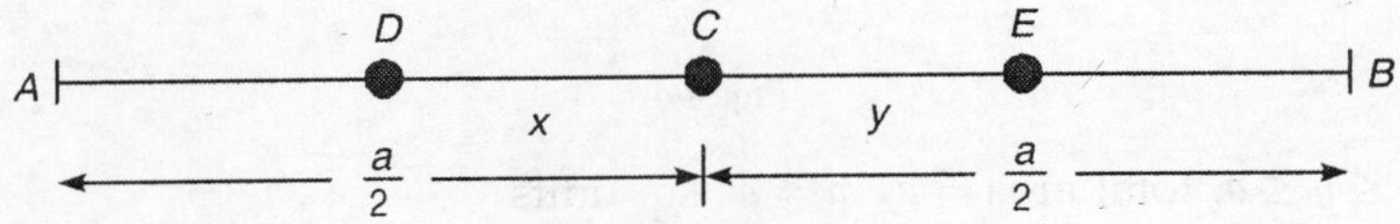

Fig. 4·3(a)

Hence, the sample space S is a square, say, $PQRS$ with side $(a/2)$ units and $P = P(0, 0)$.

We want: $P(DE < \lambda a) = P(x + y < \lambda a),\ 0 < \lambda < 1$

The region favourable to the event :

$x + y < \lambda a$, $(0 < \lambda < 1)$, is the shaded region R_1 in the Fig. 4·3(*b*).

The required probability 'p' is given by:

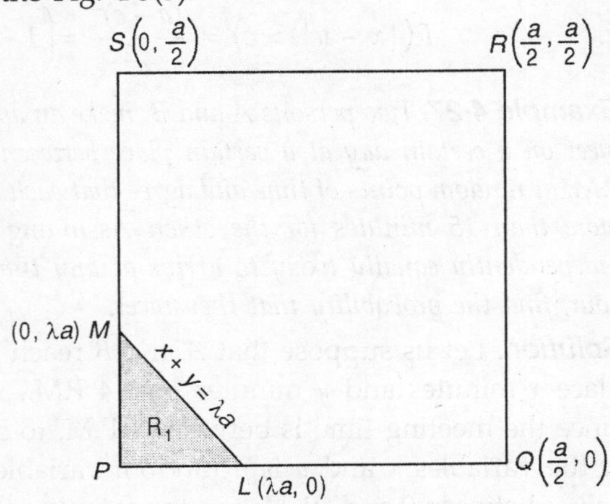

Fig. 4·3(b)

$$p = \frac{\text{Area of region } R_1}{\text{Area of square } PQRS}$$

$$= \frac{\text{Area of Right } \Delta MPL}{\frac{a}{2} \times \frac{a}{2} \text{ sq. units}}$$

$$= \frac{\frac{1}{2} \times \lambda a \times \lambda a}{\frac{a^2}{4}} = 2\lambda^2.$$

SUPPLEMENTARY EXAMPLES ON GEOMETRIC PROBABILITY

Example 4·26. *Two points are taken at random on the given straight line of length a. Prove that the probability of their distance exceeding a given length c (< a) is equal to* $\left(1 - \frac{c}{a}\right)^2$.

Solution. Let P and Q be any two points taken at random on the given straight line AB of length 'a'. Let $AP = x$ and $AQ = y$, $(0 \le x \le a,\ 0 \le y \le a)$.

Then we want $P\{|x - y| > c\}$.

The probability can be easily calculated geometrically. Plotting the lines $x - y = c$ and $y - x = c$ along the co-ordinates axes, we get the diagram, shown in Fig. 4.4.

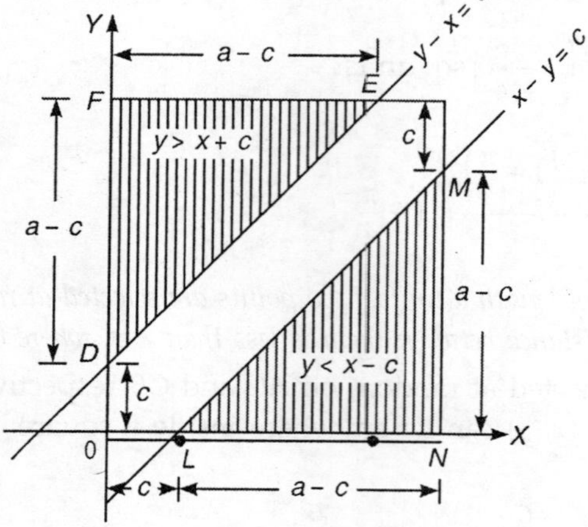

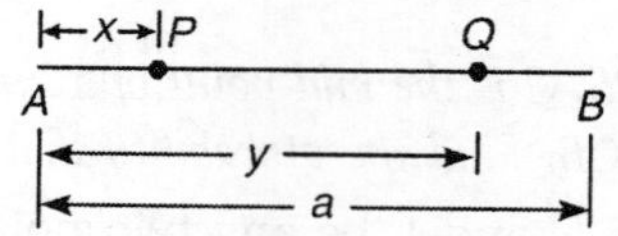

Fig. 4·4

Since $0 \le x \le a,\ 0 \le y \le a$, total area $= a \cdot a = a^2$ sq. units

Area favourable to the event $|x - y| > c$ is given by :

$$\text{Area } (\Delta\, LMN) + \text{area } (\Delta\, DEF) = \frac{1}{2} LN \cdot MN + \frac{1}{2} EF \cdot DF$$

$$= \frac{1}{2}(a-c)^2 + \frac{1}{2}(a-c)^2 = (a-c)^2 \text{ sq. units}$$

$$\therefore \qquad P(|x - y| > c) = \frac{(a-c)^2}{a^2} = \left(1 - \frac{c}{a}\right)^2 .$$

Example 4·27. *Two persons A and B, make an appointment to meet on a certain day at a certain place between 4 P.M. to 5 P.M. at random points of time and agree that each is to wait not more than 15 minutes for the other. Assuming that each is independently equally likely to arrive at any time during the hour, find the probability that they meet.*

Solution. Let us suppose that A and B reach the meeting place x minutes and y minutes past 4 P.M., respectively. Since the meeting time is between 4 P.M. to 5 P.M., each of the variables x and y is a random variable taking the values between 0 and 60. Hence, the sample space S, in this case, is a square, say, $OABC$ with side 60 units.

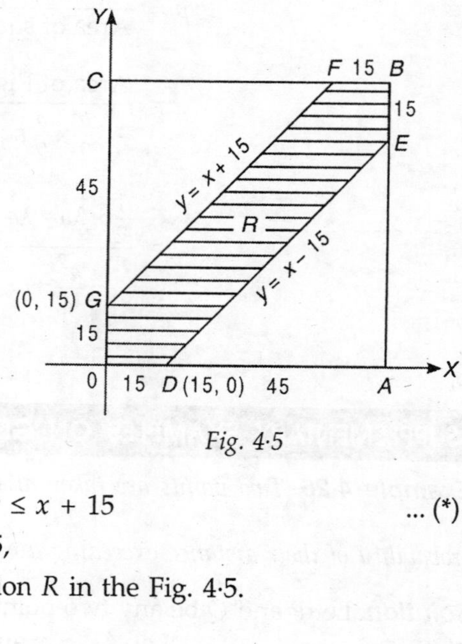

Fig. 4·5

As per the conditions of the problem, A and B will meet if and only if :

$$|x - y| \le 15 \quad \Rightarrow \quad x - 15 \le y \le x + 15 \qquad \ldots (*)$$

Plotting the lines $\quad y = x - 15 \quad$ and $\quad y = x + 15,$

the region favourable to the event in (*) is the shaded region R in the Fig. 4·5.

The required probability 'p' is given by :

$$p = P(|x - y| \le 15) = \frac{\text{Area of shaded region } R}{\text{Area of square } OABC} \qquad \ldots (**)$$

Area of square $OABC = 60 \times 60 = 3600$ sq. units.

Area of region R = Area of square $ABCD$ – 2 (Area of $\Delta\, DAE$)

$$= 60^2 - 2 \times \frac{1}{2} \times 45 \times 45 = 60^2 - 45^2 = 1575 \text{ sq. units.}$$

Substituting in (**), we obtain $p = \frac{1575}{3600} = \frac{7}{16}$.

Aliter. $P(|x - y| \leq 15) = 1 - P(|x - y| > 15) = 1 - \left(1 - \frac{15}{60}\right)^2 = \frac{7}{16}$.

[Taking $a = 60$ and $c = 15$ in *Example 4·26.*]

Example 4·28. *(BERTRAND'S PROBLEM). If a chord is taken at random in circle, what is the chance its length is not less than 'a', the radius of the circle ?*

Solution. Let the chord AB make an angle θ with the diameter AOA' of the circle with centre O and radius $OA = a$. Obviously θ lies between $-\pi/2$ and $\pi/2$. Let $AB = l$

Since all the positions of the chord AB and consequently all the values of θ are equally likely, θ may be regarded as a random variable which is uniformly distributed (*c.f.* Chapter 9) over $(-\pi/2, \pi/2)$ with probability density function.

$$f(\theta) = \frac{1}{\pi}; \; -\frac{\pi}{2} < \theta \leq \frac{\pi}{2}$$

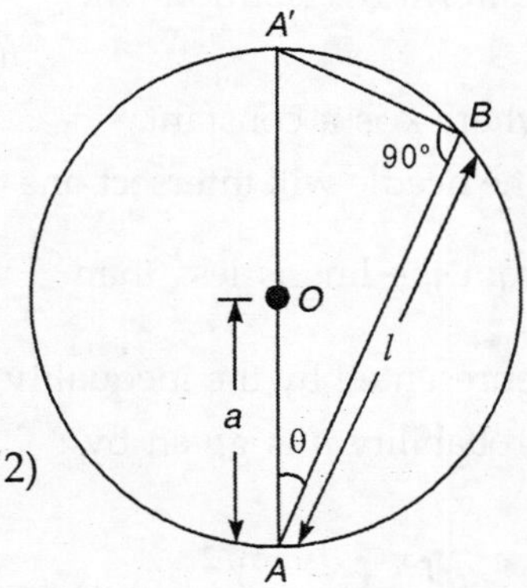

Fig. 4·6

$\angle ABA'$, being the angle in a semi-circle, is a right angle.

From $\Delta\, ABA'$ we have

$$\frac{AB}{AA'} = \cos\theta \quad \Rightarrow \quad l = 2a \cos\theta$$

The required probability 'p' is given by

$$p = P(l \geq a) = P(2a \cos\theta \geq a) = P(\cos\theta \geq 1/2)$$

$$= P(|\theta| \leq \pi/3)$$

$$= \int_{-\pi/3}^{\pi/3} f(\theta)\, d\theta = \frac{1}{\pi} \int_{-\pi/3}^{\pi/3} d\theta = \frac{2}{3}.$$

Example 4·29. *A rod of length 'a' is broken into three parts at random. What is the probability that a triangle can be formed from these parts ?*

Solution. Let the lengths of three parts of the rod be x, y and $a - (x + y)$. Obviously, we have

$$x > 0 \;\; ; \;\; y > 0 \quad \text{and} \quad x + y < a \quad \Rightarrow \quad y < a - x \qquad \text{... (*)}$$

In order that these three parts form the sides of a triangle, we should have

$$\left.\begin{aligned} x + y > a - (x + y) &\quad \Rightarrow \quad y > \frac{a}{2} - x \\ \text{and} \quad x + a - (x + y) > y &\quad \Rightarrow \quad y < \frac{a}{2} \\ y + a - (x + y) > x &\quad \Rightarrow \quad x < \frac{a}{2} \end{aligned}\right\} \qquad \text{... (**)}$$

since in a triangle, the sum of any two sides is greater than the third. Equivalently, (**) can be written as

$$\frac{a}{2} - x < y < \frac{a}{2} \wedge 0 < x < \frac{a}{2}. \qquad \text{... (***)}$$

Hence, on using (*) and (***), the required probability is given by :

$$\frac{\int_0^{a/2}\int_{(a/2)-x}^{a/2} dy\,dx}{\int_0^{a}\int_0^{a-x} dy\,dx} = \frac{\int_0^{a/2}\left[\frac{a}{2}-\left(\frac{a}{2}-x\right)\right]dx}{\int_0^{a}(a-x)\,dx} = \frac{\left|\frac{x^2}{2}\right|_0^{a/2}}{\left|\frac{-(a-x)^2}{2}\right|_0^{a}} = \frac{a^2/8}{a^2/2} = \frac{1}{4}.$$

Example 4·30. **(Buffon's Needle Problem).** *A vertical board is ruled with horizontal parallel lines at constant distance 'a' apart. A needle of length l (< a) is thrown at random on the table. Find the probability that it will intersect one of the lines.*

Solution. Let y denote the distance from the centre of the needle to the nearest parallel and ϕ be angle formed by the needle with this parallel. The quantities y and ϕ fully determine the position of the needle. Obviously y ranges from 0 to $a/2$ (since $l < a$) and ϕ from 0 to π.

Since the needle is dropped randomly, all possible values of y and ϕ may be regarded as equally likely and consequently the joint probability density function $f(y, \phi)$ of y and ϕ is given by the uniform distribution (c.f. § 9·3) by

$$f(y, \phi) = k \,;\, 0 \le \phi \le \pi,\ 0 \le y \le a/2, \qquad \text{... (*)}$$

where k is a constant.

The needle will intersect one of the lines if the distance of its centre from the line is less than $\frac{1}{2} l \sin\phi$, *i.e.*, the required event can be represented by the inequality $0 < y < \frac{1}{2} l \sin\phi$. Hence, the required probability p is given by :

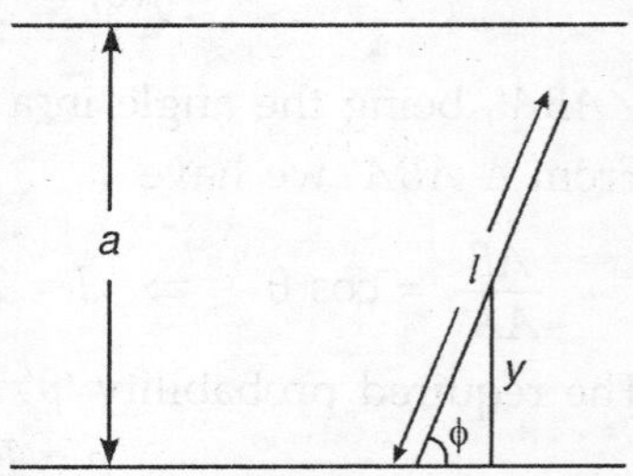

Fig. 4·7.

$$p = \frac{\int_0^{\pi}\int_0^{(l\sin\phi)/2} f(y,\phi)\,dy\,d\phi}{\int_0^{\pi}\int_0^{a/2} f(y,\phi)\,dy\,d\phi} = \frac{\frac{l}{2}\int_0^{\pi}\sin\phi\,d\phi}{(a/2)\cdot\pi} = \frac{l}{a\pi}\cdot\left|-\cos\phi\right|_0^{\pi} = \frac{2l}{a\pi}.$$

DISCUSSION AND REVIEW QUESTIONS

1. (*a*) State and prove Bayes' Theorem.

   (*b*) The set of events A_k, ($k = 1, 2, ..., n$) are (*i*) exhaustive and (*ii*) pairwise mutually exclusive. If for all, k the probabilities $P(A_k)$ are known, calculate $P(A_k \mid E)$, where E is an arbitrary event. Indicate where conditions (*i*) and (*ii*) are used.

2. The events $E_1, E_2, ..., E_n$ are mutually exclusive and $E = E_1 \cup E_2 \cup ... \cup E_n$. Show that if $P(A \mid E_i) = P(B \mid E_i)$; $i = 1, 2, ..., n$, then $P(A \mid E) = P(B \mid E)$. Is this conclusion true if the events E_i are not mutually exclusive ?

3. What are the criticisms against the use of Bayes theorem in probability theory.

4. Using the fundamental addition and multiplication rules of probability, show that

$$P(B \mid A) = \frac{P(B)\,P(A \mid B)}{P(B)\,P(A \mid B) + P(\bar{B})\,P(A \mid \bar{B})},$$

   where $\bar{B}$ is the event complementary to the event B.

ASSORTED REVIEW PROBLEMS FOR SELF-ASSESSMENT

PROBLEMS ON BAYES' THEOREM

4·1. (*a*) Two groups are competing for the positions on the Board of Directors of a corporation. The probabilities that the first and second groups will win are 0·6 and 0·4, respectively. Furthermore, if the first group wins the probability of introducing a new product is 0·8 and the corresponding probability if the second group wins is 0·3. What is the probability that the new product will be introduced ? **Ans.** 0·6

(*b*) A bag contains $(n + 1)$ coins. It is known that one of these coins has head on both the sides whereas the remaining coins are fair. One of these coins is selected at random and is tossed. If the probability that the toss results in a head is (5/9), find the value of n.

Hint. E_1 : Fair coin is chosen ; E_2 : Coin with head on both sides is chosen ;

E : Head is obtained

$$P(E) = \sum_{i=1}^{2} P(E_i)\, P(E \mid E_i) = \left[\frac{n}{n+1}\right] \times \frac{1}{2} + \left[\frac{1}{n+1}\right] \times 1 = \frac{5}{9} \text{ (Given)}$$ **[Ans.** $n = 8$.**]**

(*c*) The chances of X, Y, Z becoming managers of a certain company are 4 : 2 : 3. The probabilities that bonus scheme will be introduced, if X, Y, Z become managers are 0·3, 0·5 and 0·8, respectively. If the scheme has been introduced, what is the probability that X is appointed as the manager.

Ans. (6/23).

(*d*) A restaurant serves two special dishes, A and B to its customers consisting of 60% men and 40% women. 80% of men order dish A and the rest B. 70% of women order dish B and the rest A. In what ratio of A to B should the restaurant prepare the two dishes.

Ans. Required ratio = 0·6 : 0·4 = 3 : 2.

4·2. (*a*) In a factory machines A and B are producing springs of the same type. Of this production, machines A and B produce 5% and 10% defective springs, respectively. Machines A and B produce 40% and 60% of the total output of the factory. One spring is selected at random and it is found to be defective. What is the probability that this defective spring was produced by machine A ?

(*b*) If a machine is correctly set up it will produce 90% acceptable items. If it is incorrectly set up it will produce 30% acceptable items. Past experience shows that 80% of set ups are correctly done. If after a certain set up, first item produced is acceptable, what is the probability that machine is correctly set up ?

Ans. Required probability $= \dfrac{0.40 \times 0.80}{0.60 \times 0.25 + 0.40 \times 0.80} = \dfrac{32}{47} = 0.68$

4·3. The residents of a locality are examined for cancer. The examination results are classified as positive (+), if a malignancy is suspected, and as negative (–), if there is no indication of malignancy. If a person has cancer, the probability of a suspected malignancy is 0·95 and the probability of cancer where none existed is 0·12. If 8% of the residents have cancer, what is the probability of a person not having cancer if the examination is positive ?

Ans. Required probability $= \dfrac{0.92 \times 0.12}{0.08 \times 0.95 + 0.92 \times 0.12} = 0.5923$

4·4. An absent minded nurse is supposed to give the patient a pill each day. The probability that the nurse forgets to give the pill is 0·40. If the patient receives the pill, the probability that he will die is 0·25. If he does not get the pill, the probability that he will die is 0·80.

The patient died. What is the probability that the nurse forgot to give pill to the patient ?

4·5. In a group of 20,000 men and 10,000 women, 6 per cent of the men and 3 per cent of the women have certain affliction. What is the probability that an afflicted member of the group is a man ? **[Ans.** 0·8.**]**

4·6. (*a*) Suppose that there is a chance for a newly constructed building to collapse, whether the design is faulty or not. The chance that the design is faulty is 10%. The chance that the building collapses is 95% if the design is faulty and otherwise it is 45%. It is seen that the building collapsed. What is the probability that it is due to faulty design ? **[Ans.** 0·1919.]

(*b*) Two cannons No. 1 and 2 fire at the same target. Cannon No. 1 gives on an average 9 shots in the time in which cannon No. 2 fires 10 projectiles. But on an average 8 out of 10 projectiles from cannon No. 1 and 7 out of 10 from cannon No. 2 strike the target. In the course of shooting, the target is struck by one projectile. What is the probability that the projectile which has struck the target belongs to cannon No. 2 ? **[Ans.** 0·493]

Hint. $\dfrac{P(E_1)}{P(E_2)}=\dfrac{9}{10}$, where E_i is the event that the projectile from cannon i, (i = 1, 2), strikes the target.

(*c*) State and prove Bayes rule and explain why, in spite of its easy deductibility from the postulates of probability, it has been the subject of such extensive controversy.

In the chest X-ray test, it is found that the probability of detection when a person has actually T.B. is 0·95 and probability of diagnosing incorrectly as having T.B. is 0·002. In a certain city 0·1% of the adult population is suspected to be suffering from T.B. If an adult is selected at random and is diagnosed as having T.B. on the basis of the X-ray test, what is the probability of his actually having a T.B. ?

4·7. (*a*) Assume that an urn contains white and black balls marked X and Y as follows.

Marked X : 6 balls, including 3 white and 3 black

Marked Y : 6 balls, including 4 white and 2 black

Total : 12 balls, including 7 white and 5 black

A white ball is drawn randomly from the bag. Determine the probability that the white ball is marked X and Y. **[Ans.** (3/7, 4/7).]

(*b*) There are two urns with parts of the same type. The first contains '*a*' good parts and '*b*' defective parts and the second contains '*c*' good parts and '*d*' defective parts. An urn is selected at random and one part is drawn from it and is found to be good. What is the probability that :

(*i*) Urn 1 is selected and (*ii*) Urn II is selected ?

Ans. (*i*) $\dfrac{p_1}{p_1+p_2}$, (*ii*) $\dfrac{p_2}{p_1+p_2}$, where $p_1=\dfrac{a}{a+b}$ and $p_2=\dfrac{c}{c+d}$.

(*c*) In factory 2 machines M_1 and M_2 are used for manufacturing screws which may be uniquely classified as good or bad. M_1 produces per day n_1 boxes of screws, of which on the average, p_1 % are bad while the corresponding numbers for M_2 and n_2 and p_2. From the total production of both M_1 and M_2 for a certain day, a box is chosen at random, a screw taken out of it and it is found to be bad. Find the chance that selected box is manufactured by (*i*) M_1, (*ii*) M_2.

Ans. (*i*) $n_1 p_1/(n_1 p_1 + n_2 p_2)$, (*ii*) $n_2 p_2/(n_1 p_1 + n_2 p_2)$.

4·8. (*a*) Urn A contains 2 white, 1 black and 3 red balls; urn B contain 3 white, 2 black and 4 red balls, and urn C contains 4 white, 3 black and 2 red balls. One urn is chosen at random and 2 balls are drawn. They happen to be red and black. What is the probability that both balls came from urn '*B*' ? **[Ans.** (20/53)]

(*b*) There are three urns having the following compositions of black and white balls.

Urn 1 : 7 white, 3 black balls ; Urn 2 : 4 white, 6 black balls ; Urn 3 : 2 white, 8 black balls.

One of these urns is chosen at random with probabilities 0.20, 0.60 and 0.20, respectively. From the chosen urn two balls are drawn at random without replacement. Calculate the probability that both these balls are white. **[Ans.** (8/45).]

4·9. (*a*) In a bolt factory machines A, B and C manufacture respectively 25%, 35% and 40% of the total. Of their output 5, 4, 2 per cent, respectively are defective bolts. A bolt is drawn at random from the product and is found to be defective. What are the probabilities that it was manufactured by machines (*i*) A, (*ii*) B, and (*iii*) C ? **[Ans.** (*i*) {(25/69), (*ii*) (28/69), (*iii*) (16/69)}.]

(*b*) There are three machines producing 10,000 ; 20,000 and 30,000 bullets per hour, respectively. These machines are known to produce 5%, 4% and 2% defective bullets respectively. One bullet is taken at random from an hour's production of the three machines. What is the probability that it is defective ? If the drawn bullet is defective, what is the probability that this was produced by the second machine ?

(*c*) Three machines *X*, *Y*, *Z* with capacities proportional to 2 : 3 : 4 are producting bullets. The probabilities that the machines produce defective bullet are 0·1, 0·2 and 0·1 respectively. A bullet is taken from a day's production and found to be defective. What is the probability that it came from machine *X* ?

4·10. (*a*) A company produces certain type of sophisticated item by three machines. The respective daily production figures are : Machine *A*, 300 units ; Machine *B*, 450 units and Machine *C*, 250 units. Past experience shows that the percentage of defectives in three machines are 0·1, 0·2 and 0·7 respectively for the machines *A*, *B* and *C*. An item is drawn at random from a day's production and is found to be defective. What is the probability that it is not produced by machine *C* ?

Ans. (24/59).

(*b*) A manufacturing firm produces steel pipes in three plants with daily production volumes of 500, 1,000 and 2,000 units respectively. According to past experience it is known that the fraction of defective outputs produced by the three plants are respectively 0·005, 0·008 and 0·010. If a pipe is selected at random from a day's total production and found to be defective, from which plant is that pipe likely to have come ?

Ans. $\left(\frac{5}{61}, \frac{16}{61}, \frac{40}{61}\right)$; Third plant.

4·11. A company has four production sections *viz.*, S_1, S_2, S_3 and S_4 which contribute 30%, 20%, 28% and 22% respectively to the total output. It was observed that these sections respectively produced 1%, 2%, 3% and 4% defective units. If a unit is selected at random and found to be defective, what is the probability that the unit so selected has come from either section one or section four ?

Ans. $\frac{0{\cdot}0030}{0{\cdot}0242} + \frac{0{\cdot}0088}{0{\cdot}0242} = 0{\cdot}4876.$

4·12. (*a*) *A* and *B* are two very weak students of Statistics and their chances of solving a problem correctly are 1/8 and 1/12 respectively. If the probability of their making a common mistake is 1/1001 and they obtain the same answer, find chance that their answer is correct.

Ans. $$\frac{\frac{1}{8} \times \frac{1}{12}}{\frac{1}{8} \times \frac{1}{12} + \left(1 - \frac{1}{8}\right) \times \left(1 - \frac{1}{12}\right) \times \frac{1}{1001}} = \frac{13}{14}$$

(*b*) The chance that *A* solves a problem correctly is 0·1 and *B* solves it correctly is 0·2. The probability of getting an identical wrong answer is 0·001. If both get the same answer for the problem, what is the chance that the answer is correct?

(*c*) *A* and *B* are two independent witnesses (*i.e.*, there is no collusion between them) in a case. The probability that *A* will speak the truth is p_1 and for *B* it is p_2. *A* and *B* agree in a certain statement. Show that the probability that this statement is true is

$$[\, p_1 p_2 / \{1 - p_1 - p_2 + 2p_1 p_2\}]$$

Hint. E_1 (E_2) : Event that *A* and *B* agree in asserting that statement is true (false).
E : Event that *A* and *B* agree in a statement ; Required probability = $P(E_1 \mid E)$

4·13. Three boxes practically indistinguishable in appearance, have two drawers each. Box I contains a gold coin in one and a silver coin in the other drawer; box II contains a gold coin in each drawer, and box III contains a silver coin in each drawer. One box is chosen at random and one of its drawers is opened at random and a gold coin found. What is the probability that the other drawer contains a coin of silver ?

Hint. E_i : ith box is chosen ; $i = 1, 2, 3, ...$

E : The opened drawer of the selected box contains a gold coin

Required probability $= P(E_1 \mid E) = (1/3)$ [$\because$ Only first box contains gold and silver coin]

4·14. (*a*) A doctor is to visit the patient and from past experience it is known that the probabilities that he will come by train, bus or scooter are respectively 3/10, 1/5 and 1/10; the probability that he will use some other means of transport being, therefore, 2/5. If he comes by train, the probability that he will be late is 1/4; if by bus 1/3; and if by scooter 1/12; if he uses some other means of transport it can be assumed that he will not be late. When he arrives he is late. What is the probability that (*i*) he comes by train (*ii*) he is not late ? **Ans.** (*i*) $\frac{1}{2}$, (*ii*) $\frac{17}{20}$.

(*b*) The police plans to enforce speed limits by using radar traps at 4 different locations within the city limits. The radar traps at each of the locations L_1, L_2, L_3 and L_4 are operated 40%, 30%, 20% and 30% of the time. If a person who is speeding on his way to work has probabilities of 0·2, 0·1, 0·5 and 0·2 respectively, of passing through these locations, what is the probability that he will receive a speeding ticket ?

4·15. The sample space consists of integers from 1 to $2n$ which are assigned probabilities proportional to their logarithms. Find the probabilities and show that the conditional probability of the integer 2, given that an even integer occurs, is :

$$\frac{\log 2}{[n \log 2 + \log (n!)]}$$

Hint. Let E_i : the event that the integer $2i$ is drawn, $(i = 1, 2, 3, ..., n)$.

E : the event of drawing an even integer.

$$\Rightarrow \quad E = E_1 \cup E_2 \cup ... \cup E_n \quad \Rightarrow \quad P(E) = \sum_{i=1}^{n} P(E_i)$$

But $\quad P(E_i) = k \log (2i)$ (Given)

$$\therefore \quad P(E) = k \sum_{i=1}^{n} \log (2i) = k \log \prod_{i=1}^{n} (2i) = k[n \log 2 + \log (n!)]$$

$$\therefore \quad P(E_i \mid E) = \frac{P(E_i \cap E)}{P(E)} = \frac{P(E_i)}{P(E)} = \frac{\log (2i)}{[n \log 2 + \log (n!)]} \qquad (\because E_i \subset E)$$

4·16. In a test, an examinee either guesses or copies or knows the answer to a multiple choice question with four choices. The probability that he makes a guess is 1/3 and the probability that he copies the answer is 1/6. The probability that his answer is correct, given that he copied it is 1/8. Find the probability that he knew the answer to the question, given that he correctly answered it.

Hint. $P(\text{Examinee knows the answer}) = \left(1 - \frac{1}{3} - \frac{1}{6}\right) = \frac{1}{2}$;

and $P(\text{Examinee gets the correct answer given that he knew the answer}) = 1$. **[Ans. (24/29)]**

4·17. In answering a question on a multiple choice test, an examinee either knows the answer (with probability p), or he guesses (with probability $1 - p$). Assume that the probability of answering a question correctly is unity for an examinee who knows the answer and $1/m$ for the examinee who guesses, where m is the number of multiple choice alternatives. Supposing an examinee answers a question correctly, what is the probability p_m that he really knows the answer ?

Show further that $p_m > p$ and that p_m is a strictly increasing sequence for fixed p, $(0 < p < 1, m \geq 2)$.

Hint. Show that $p_{m+1} - p_m > 0$ for $0 < p < 1$ and $m \geq 2$.

Ans. $p_m = \dfrac{mp}{\{(m-1)p + 1\}}$; $\quad p_m - p = \dfrac{(m-1)p(1-p)}{\{(m-1)p + 1\}} > 0$, for $m \geq 2$, $0 < p < 1$.

4·18. (*a*) Three urns are given, each containing red and white balls as indicated :

Urn 1 : 6 red and 4 white ; Urn 2 : 2 red and 6 white ; Urn 3 : 1 red and 8 white

(*i*) An urn is chosen at random and a ball is drawn from this urn. The ball is red. Find the probability that the urn chosen was urn I.

(*ii*) An urn is chosen at random and two balls are drawn without replacement from this urn. If both balls are red, find the probability that urn I was chosen. Under these conditions, what is the probability that urn III was chosen?

(*b*) Three urns are marked *A*, *B* and *C*. *A* contains 3 blue and 2 white balls, *B* contains 4 blue and 3 white balls, and *C* contains 3 blue and 4 white balls. A ball is drawn randomly from urn *A* in the first instance. If its colour is white, the second draw is made from urn *B* and if its colour is blue the second draw is made from *C*. If the colour of the ball in the second draw is white, what is the probability that the urn *B* was selected ?

Hint. E_1 : First ball drawn from urn *A* is white ; E_2 : First ball drawn from urn *A* is blue

E : Second ball drawn is white ; Required probability $= P(E_1 \mid E)$ **[Ans.** 1/3**]**

4·19. There are ten urns of which each of three contains 1 white and 9 black balls; each of other three contains 9 white and 1 black ball, and of the remaining four each contains 5 white and 5 black balls. One of the urns is selected at random and a ball taken blindly from it turns out to be white. What is the probability that an urn containing 1 white and 9 black balls was selected ?

Hint. Let E_i : The event of drawing urn of type *i*, (i = 1, 2, 3). Then we have :

$$P(E_1) = \frac{3}{10}, \quad P(E_2) = \frac{3}{10} \quad \text{and} \quad P(E_3) = \frac{4}{10}.$$

Let *E* be the event of drawing a white ball.

$$P(E) = \frac{3}{10}\times\frac{1}{10}+\frac{3}{10}\times\frac{9}{10}+\frac{4}{10}\times\frac{5}{10}=\frac{1}{2} \quad ; \quad P(E \mid E_1) = \frac{1}{10} \text{ and } P(E_1 \mid E) = \frac{3}{50}$$

4·20. (*a*) It is known that an urn containing altogether 10 balls was filled in the following manner. A coin was tossed 10 times, and according as it showed heads or tails, one white or one black ball was put into the urn. Balls are drawn from this urn one at a time, 10 times in succession (with replacement) and every one turns out to be white. Find the chance that the urn contains nothing but white balls. **[Ans.** 0·0702**]**

(*b*) From a vessel containing 3 white and 5 black balls, 4 balls are transferred into an empty vessel. From this vessel a ball is drawn and is found to be white. What is the probability that out of four balls transferred 3 are white and 1 black ?

Hint. Let the five mutually exclusive events for the four balls transferred be E_0, E_1, E_2, E_3 and E_4 where E_i denotes the event that *i* white balls are transferred and let *E* be the event of drawing a white ball from the new vessel. Then

$$P(E_0) = \frac{^5C_4}{^8C_4}, \quad P(E_1) = \frac{^3C_1 \times {^5C_3}}{^8C_4}, \quad P(E_2) = \frac{^3C_2 \times {^5C_2}}{^8C_4}, \quad P(E_3) = \frac{^3C_3 \times {^5C_1}}{^8C_4} \text{ and } P(E_4) = 0$$

Also $P(E \mid E_0) = 0, \quad P(E \mid E_1) = \frac{1}{4}, \quad P(E \mid E_2) = \frac{2}{4}, \quad P(E \mid E_3) = \frac{3}{4}$, and $P(E \mid E_4) = 1$.

Hence, $$P(E_3 \mid E) = \frac{1}{7}.$$

4·21. The contents of the urns 1 and 2 are as follows :

Urn 1 : 4 white and 5 black balls ; Urn 2 : 3 white and 6 black balls.

One urn is chosen at random and a ball is drawn and its colour noted and replaced back to the urn. Again a ball is drawn from the same urn, colour noted and replaced. The process is repeated 4 times and as a result one ball of white colour and three balls of black colour are obtained. What is the probability that the urn chosen was the urn 1 ?

Hint. $P(E_1) = P(E_2) = 1/2,$

$P(E \mid E_1) = 4/9, \quad 1 - P(E \mid E_1) = 5/9\,; \quad P(E \mid E_2) = 1/3, \; 1 - P(E \mid E_2) = 2/3$

The probability that the urn chosen was the urn 1 = $\dfrac{\frac{1}{2}\cdot\frac{4}{9}\cdot\left(\frac{5}{9}\right)^3}{\frac{1}{2}\cdot\frac{4}{9}\cdot\left(\frac{5}{9}\right)^3 + \frac{1}{2}\cdot\frac{1}{3}\cdot\left(\frac{2}{3}\right)^3}$

4·22. There are five urns numbered 1 to 5. Each urn contains 10 balls. The *i*th urn has *i* defective balls and $(10 - i)$ non-defective balls; $i = 1, 2, ..., 5$. An urn is chosen at random and then a ball is selected at random from that urn.

(*i*) What is the probability that a defective ball is selected ?

(*ii*) If the selected ball is defective, find the probability that it came from urn *i*, $(i = 1, 2, ..., 5)$.

Hint. Define the following events :

E_i = *i*th urn is selected at random, $(i = 1, 2, 3, 4, 5)$; $\quad E$: Defective ball is selected.

$$P(E_i) = 1/5\,;\; i = 1, 2, ..., 5.$$

$$P(E \mid E_i) = P(\text{Defective ball from } i^{th} \text{ urn}) = i/10,\; (i = 1, 2, ..., 5)$$

$$P(E_i)\cdot P(E \mid E_i) = \frac{1}{5}\times\frac{i}{10} = \frac{i}{50},\quad (i = 1, 2, ..., 5)$$

(*i*) $$P(E) = \sum_{i=1}^{5} P(E_i)\,P(E \mid E_i) = \sum_{i=1}^{5}\left(\frac{i}{50}\right) = \frac{1+2+3+4+5}{50} = \frac{3}{10}$$

(*ii*) $$P(E_i \mid E) = \frac{P(E_i)\,P(E \mid E_i)}{\sum P(E_i)\,P(E \mid E_i)} = \frac{i/50}{3/10} = \frac{i}{15};\; i = 1, 2, ..., 5.$$

For example, the probability that the defective ball came from 5th urn = $\dfrac{5}{15} = \dfrac{1}{3}$.

4·23. An urn contains *n* balls, each of different colour, of which one is white. Two independent observers, each with probability 0.1 of telling the truth, assert that a ball drawn at random is white. Prove that the probability that the ball is, in fact, white is $(n - 1)/(n + 80)$. Also show that if $n < 20$, this probability is less than the probability that at least one of the observers is telling the truth.

4·24. A piece of mechanism consists of 11 components, 5 of type *A*, 3 of type *B*, 2 of type *C* and 1 of type *D*. The probability that any particular component will function for a period of 24 hours from the commencement of operations without breaking down is independent of whether or not any other component breaks down during that period and can be obtained from the following table :

Component Type :	*A*	*B*	*C*	*D*
Probability :	0·6	0·7	0·3	0·2

(*i*) Calculate the probability that 2 components chosen at random from the 11 components will both function for a period of 24 hours from the commencement of operations without breaking down.

(*ii*) If at the end of 24 hours of operations neither of the 2 components chosen in (*i*) has broken down, what is the probability that they are both type C components.?

Hint. (*i*) Required probability is :

$$\frac{1}{{}^{11}C_2}\left[\begin{array}{l} {}^5C_2\times(0{\cdot}6)^2 + {}^3C_2\,(0{\cdot}7)^2 + {}^2C_2\,(0{\cdot}3)^2 + {}^5C_1\times{}^3C_1\times 0{\cdot}6\times 0{\cdot}7 \\ +\,{}^5C_1\times{}^2C_1\times(0{\cdot}6)\times(0{\cdot}3) + {}^5C_1\times{}^1C_1\times(0{\cdot}6)\times(0{\cdot}2) + {}^3C_1\times{}^2C_1\times 0{\cdot}7\times 0{\cdot}3 \\ +\,{}^3C_1\times{}^1C_1\times 0{\cdot}7\times 0{\cdot}2 + {}^2C_1\times{}^1C_1\times 0{\cdot}3\times 0{\cdot}2 \end{array}\right] = p \text{ (say)}$$

(*ii*) Required probability (By Bayes theorem) = $\dfrac{{}^2C_2\times(0{\cdot}3)^2}{p} = \dfrac{0{\cdot}09}{p}$

4·25. There is a series of n urns. In the ith urn there are i white and $(n - i)$ black balls, $i = 1, 2, 3, ..., n$. One urn is chosen at random and 2 balls are drawn from it. Both turn out to be white. What is the probability that the jth urn was chosen, where j is a particular number between 3 and n ?

Hint. Let E_j denote the event of selection of jth urn, $j = 3, 4, ..., n$ and E denote the event of drawing of 2 white balls, then

$$P(E \mid E_j) = \left(\frac{j}{n}\right)\left(\frac{j-1}{n-1}\right), \qquad P(E_j) = \frac{1}{n}, \qquad P(E) = \sum_{i=1}^{n} \frac{1}{n}\left(\frac{i}{n}\right)\left(\frac{i-1}{n-1}\right)$$

$$\therefore \qquad P(E_j \mid E) = \frac{\frac{1}{n}\left(\frac{j}{n}\right)\left(\frac{j-1}{n-1}\right)}{\sum_{i=1}^{n}\left(\frac{1}{n}\right)\left(\frac{i}{n}\right)\left(\frac{i-1}{n-1}\right)}$$

4·26. Three prisoners A, B and C with apparently equally good records have applied for parole. The parole board has decided to release two of the three and the prisoners know this but not which two. A warder friend of prisoner A knows who are to be released. Prisoner A realises that it would be unethical to ask the warder if he, (A), is to be released, but thinks of asking for the name of one prisoner other than himself who is to be released. He thinks that before he asks, his chances of release are (2/3). He thinks that if the warder says, "B will be released" his own chances have now gone down to (1/2) because either A and B or B and C are to be released. So he decides not to reduce his chance by asking. Are his calculations justified ?

PROBLEMS ON GEOMETRIC PROBABILITY

4·27. Two points are selected at random in a line AC of length 'a' so as to lie on the opposite sides of its mid-point O. Find the probability that the distance between them is less than $a/3$.

4·28. Two points are selected at random on a line of length a. What is the probability that none of three sections in which the line is thus divided is less than $a/4$. **[Ans.** $\frac{1}{16}$**]**

4·29. A rectilinear segment AB is divided by a point C into two parts $AC = a$, $CB = b$. Points X and Y are taken at random on AC and CB respectively. What is the probability that AX, XY and BY can form a triangle ? **[Ans.** $b/(2a)$**]**

4·30. ABG is straight line such that AB is 6 inches and BG is 5 inches. A point Y is chosen at random on the BG part of the line. If C lies between B and G in such a way that $AC = t$ inches, find

(*i*) the probability that Y will lie in BC. ; (*ii*) the probability that Y will lie in CG.

What can you say about the sum of these probabilities ?

4·31. The sides of a rectangle are taken at random each less than a and all lengths are equally likely. Find the chance that the diagonal is less than a.

4·32. (*a*) Three points are taken at random on the circumference of a circle. Find the chance that they lie on the same semi-circle. **[Ans. (3/4)]**

(*b*) A chord is drawn at random in a given circle. What is the probability that it is greater than the side of an equilateral triangle inscribed in that circle ?

(*c*) Show that the probability of choosing two points randomly from a line segment of length 2 inches and their being at a distance of at least 1 inch from each other is 1/4.

4·33. (*a*) Two points are selected at random on a line segment of length 'a' units. Prove that the probability that the distance between them is less than 'b' units is $[b(2a - b)/a^2]$, $b < a$.

(*b*) One takes at random two points P and Q on a segment AB of length a.

(*i*) What is the probability for the distance PQ is less than b ($< a$) ?

(*ii*) Find the chance that the distance between them is greater than a given length 'b'.

Hint. See Example 4·26.

4·34. (*a*) A point is selected at random inside a circle. Find the probability that the point is closer to the centre of the circle than to its circumference.

(*b*) n points are chosen at random and independently of one another inside a sphere of radius R units. Find the probability that the distance from the centre of the sphere to the nearest point is not less than r units, $(r < R)$. **[Ans.** $\{1 - (r^3/R^3)\}^n$**]**

4·35. (*a*) **(The Encounter Problem).** Two persons A and B agree to meet on a particular day at a specified place between 6 P.M. and 7 P.M. Whosoever comes first, waits for 20 minutes and then leaves. If each one of them comes to the place during the specified time at random and independent of each other, show that the probability that they meet is 5/9.

(*b*) Two persons A and B, make an appointment to meet on a certain day at a certain place, but without fixing the time further than that is to be between 2 p.m. and 3 p.m. and that each is to wait not longer than ten minutes for the other.

(*i*) Assuming that each is independently equally likely to arrive at any time during the hour, find the probability that they meet.

(*ii*) Thir-d person C, is to be at the same place from 2·10 p.m. until 2·40 p.m. on the same day. Find the probabilities of C being present when A and B are there together.

(A) When A and B remain after they meet,

(B) When A and B leave as soon as they meet.

4·36. The outcomes of an experiment are represented by points in the square bounded by $x = 0$, $x = 2$ and $y = 2$ in the xy - plane. If the probability is distributed uniformly, determine the probability that $x^2 + y^2 > 1$. **Ans.** $[1 - (\pi/16)]$

4·37. POQ is a fixed diameter of a circle of radius r units. A point is taken at random within the circle. Show that the probability that the point lies within an inscribed isosceles triangle with POQ as one of the three sides is $(2/\pi)$.

4·38. A floor is paved with rectangular bricks, each of length a units and breadth b units. A circular disc of diameter c units is thrown on the floor. Show that the probability that it falls entirely on one brick is $(a - c)(b - c)/ab$, $(c < a, c < b)$.

4·39. A floor is paved with tiles, each tile being a parallelogram such that the distance between pairs of opposite sides are a and b respectively, the length of the diagonal being l. A stick of length c falls on the floor parallel to the diagonal. Show that the probability that it will lie entirely on one tile is :

$$\left(1-\frac{c}{l}\right)^2$$

If a circle of diameter d is thrown on the floor, show that the probability that it will lie on one tile is :

$$\left(1-\frac{d}{a}\right)\left(1-\frac{d}{b}\right).$$

4·40. Circular discs of radius r are thrown at random on a plane circular table of radius R which is surrounded by a border of uniform width r lying in the same plane as the table. If the discs are thrown independently and at random, and N stay on the table, show that the probability that a fixed point on the table but not on the border, will be covered is :

$$\left\{1-\left(1-\frac{r^2}{(R+r)^2}\right)^N\right\}.$$

CHAPTER

5 Random Variables and Distribution Functions

LEARNING OBJECTIVES

Upon completion of this Chapter, you should be able to:

- Explain and illustrate the concept of a random variable and its probability distribution.
- Elaborately illustrate the concept of distribution function.
- Distinguish between a discrete and a continuous random variable.
- Understand the notion of probability mass function and probability density function.
- Demonstrate the concept of joint, marginal and conditional probability distributions involving two random variables.
- Discuss transformation of one-dimensional, two-dimensional and n-dimensional random variables and obtain the probability distributions of the transformed variables.

CHAPTER OUTLINES

5·1. INTRODUCTION

In the previous chapters, we have discussed that the concept of the random experiment leads to the notation of a sample space. The assignment and computation of probabilities of events were studied in detail. In many experiments, we are interested not in knowing which of the outcomes has occurred, but in the numbers associated with them. For example, when n coins are tossed, one may be interested in knowing the number of heads obtained. When a pair of dice are tossed, one may seek information about the sum of points. Thus, we associate a real number with each outcome of an experiment. In other words, we are considering a function whose domain is the set of possible outcomes, and whose range is a subset of the set of reals. Such a function is called a *random variable.*

Intuitively by a *random variable (r.v.)* we mean a real number X connected with the outcome of a random experiment E. For example, if E consists of two tosses of a coin, the random variable which is the number of heads takes the values (0, 1 or 2).

Outcome :	*HH*	*HT*	*TH*	*TT*
Value of X :	2	1	1	0

Thus, to each outcome ω, there corresponds a real number $X(\omega)$. Since the points of the sample space S correspond to outcomes, this means that a real number, which we denote by $X(\omega)$, is defined for each $\omega \in S$. From this standpoint, we define random variable to be a real function on S as follows :

"Let S be the sample space associated with a given random experiment. A real-valued function defined on S and taking values in R $(-\infty, \infty)$ is called a one-dimensional random variable. If the function values are ordered pairs of real numbers (i.e., vectors in two-space), the function is said to be a two-dimensional random variable. More generally, an n-dimensional random variable is simply a function whose domain is S and whose range is a collection of n-tuples of real numbers (vectors in n-space)."

For a mathematical and rigorous definition of the random variable, let us consider the probability space, the triplet (**S, B, P**), where **S** is the sample space, *viz*, space of outcomes, **B** is the σ-field of subsets in S, and **P** is a probability function on **B**.

Def. *A random variable (r.v) is a function $X(\omega)$ with domain S and range $(-\infty, \infty)$ such that for every real number a, the event $[\omega : X(\omega) \le a] \in$* **B**.

Remarks:

1. The refinement above is the same as saying that the function $X(\omega)$ is *measurable* real function on **(S, B)**.
2. We shall need to make probability statements about a random variable X such as $P\{X \le a\}$. For the simple example given above we should write $P\{X \le 1\} = P\{HH, HT, TH\} = \frac{3}{4}$ *i.e.,* $P(X \le a)$ is simply the probability of the set of outcome ω for which $X(\omega) \le a$ or
$$P(X \le a) = P\{\omega : X(\omega) \le a\}$$
Since **P** is a measure on (**S, B**) *i.e.,* **P** is defined on subsets of **B**, the above probability will be defined *only* if $\{\omega : X(\omega) \le a\} \in$ **B**, which implies that $X(\omega)$ is a measurable function on **(S, B)**.
3. One-dimensional random variables will be denoted by capital letters, X, Y, Z, ..., etc. A typical outcome of the experiment (*i.e.,* a typical element of the sample space) will be denoted by ω or e. Thus $X(\omega)$ represents the real number which the random variable X associates with the outcome ω. The values which X, Y, Z, ..., etc., can assume are denoted by lower case letters, *viz., x, y, z,* ...., etc.
4. *Notations.* If *x is a real number,* the set of all ω in **S** such that $X(\omega) = x$ is, denoted briefly by writing $X = x$. Thus $P(X = x) = P\{\omega : X(\omega) = x\}$

Similarly $P(X \le a) = P\{\omega : X(\omega) \in (-\infty, a]\}$ and $P(a < X \le b) = P\{\omega : X(\omega) \in (a, b]\}$

Analogous meanings are given to $P(X = a \text{ or } X = b) = P\{(X = a) \cup (X = b)\}$,

$P(X = a \text{ and } X = b) = P\{(X = a) \cap (X = b)\}$, etc.

Illustrations:

1. If a coin is tossed, then

$$S = \{\omega_1, \omega_2\}, \text{ where } \omega_1 = H, \omega_2 = T \; ; \; X(\omega) = \begin{cases} 1, \text{ if } \omega = H \\ 0, \text{ if } \omega = T \end{cases}$$

$X(\omega)$ is a Bernoulli random variable. Here $X(\omega)$ takes only two values.

2. An experiment consists of rolling a die and reading the number of points on the upturned face. The most natural *r.v.* X to consider is : $X(\omega) = \omega$; $\omega = 1, 2, ..., 6$.

If we are interested in knowing whether the number of points is even or odd, we consider a random variable Y defined as follows : $Y(\omega) = \begin{cases} 0, \text{ if } \omega = \text{is even} \\ 1, \text{ if } \omega = \text{is odd} \end{cases}$

3. If a dart is thrown at a circular target, the sample space S is the set of all points ω on the target. By imagining a coordinate system placed on the target with the origin at the centre, we can assign various random variables to this experiment. A natural one is the two-dimensional random variable which assigns to the point ω, its rectangular coordinates (x, y). Another is that which assigns ω its polar coordinates (r, θ). A one-dimensional random variable assigns to each ω only one of the coordinates x or y (for cartesian system), r or θ (for polar system). The event E, "that the dart will land in the first quadrant" can be described by a random variable which assigns to each point ω its polar coordinate θ so that $X(\omega) = \theta$ and then $E = \{\omega : 0 \le X(\omega) \le \pi/2\}$.

4. If a pair of fair dice is tossed then $S = \{1, 2, 3, 4, 5, 6\} \times \{1, 2, 3, 4, 5, 6\}$ and $n(S) = 36$. Let X be a random variable with image set $X(S) = \{1, 2, 3, 4, 5, 6\}$.

$$P(X = 1) = P\{1, 1\} = \frac{1}{36}, \qquad P(X = 2) = P\{(2, 1), (2, 2), (1, 2)\} = \frac{3}{36}$$

$$P(X = 3) = P\{(3, 1), (3, 2), (3, 3), (2, 3), (1, 3)\} = \frac{5}{36}$$

$$P(X = 4) = P\{(4, 1), (4, 2), (4, 3), (4, 4), (3, 4), (2, 4), (1, 4)\} = \frac{7}{36}$$

Similarly, $P(X = 5) = \dfrac{9}{36}$ and $P(X = 6) = \dfrac{11}{36}$

Some theorems on Random Variables. Here we shall state (without proof) some of the fundamental results and theorems on random variables.

Consider the probability space $(S, \mathbf{B}, P)$.

1. A function $X(\omega)$ from S to $R(-\infty, \infty)$ is a *r.v.* if and only if for real a,

$$\{\omega : X(\omega) < a\} \in \mathbf{B}.$$

2. If X_1 and X_2 are random variables and C is a constant then CX_1, $X_1 + X_2$, $X_1 X_2$ are also random variables.

Remark. It will follow that $C_1X_1 + C_2X_2$ is a *r.v.* for constants C_1 and C_2. In particular, $X_1 - X_2$ is a *r.v.*

3. If X is a random variable then

(*i*) $\frac{1}{X}$, where $\left(\frac{1}{X}\right)(\omega) = \infty$ if X (ω) = 0, (*ii*) $X_+(\omega) = \max\{0, X(\omega)\}$,

(*iii*) $X_-(\omega) = -\min\{0, X(\omega)\}$, and (*iv*) $|X|$, are random variables.

4. If X_1 and X_2 are random variables, then (*i*) max $[X_1, X_2]$, and (*ii*) min $[X_1, X_2]$ are also random variables.
5. If X is a *r.v.* and *f* (.) is a continuous function, then *f* (X) is a *r.v.*
6. If X is a *r.v.* and *f* (.) is an increasing function, then *f* (X) is a *r.v.*

Corollary. If *f* is a function of bounded variations on every finite interval [**a**, **b**], and X is a *r.v.* then *f*(X) is a *r.v.*

(*Proofs of the above results or theorems are beyond the scope of this book.*)

5·2. DISTRIBUTION FUNCTION

Definition. *Let X be a random variable. The function F defined for all real x by*

$$F(x) = P(X \le x) = P[\omega : X(\omega) \le x], -\infty < x < \infty, \quad \text{... (5·1)}$$

is called the distribution function (d.f.) of the r.v. (X).

Remark. A distribution function is also called the cumulative distribution function. Sometimes, the notation $F_X(x)$ is used to emphasise the fact that the distribution function is associated with the particular random variable X. Clearly, the domain of the distribution function is (− ∞, ∞) and its range is [0, 1].

5·2·1. Properties of Distribution Function

We now proceed to derive a number of properties common to all distribution functions.

1. *If F is the d.f. of the r.v. X and if a < b, then P (a < X ≤ b) = F (b) − F (a).*

Proof. The events '$a < X \le b$' and '$X \le a$' are disjoint and their union is the event '$X \le b$'., Hence, by addition theorem of probability :

$$P(a < X \le b) + P(X \le a) = P(X \le b)$$

$$\Rightarrow \quad P(a < X \le b) = P(X \le b) - P(X \le a) = F(b) - F(a) \quad \text{... (5·2)}$$

Cor. 1. $P(a \le X \le b) = P\{(X = a) \cup (a < X \le b)\} = P(X = a) + P(a < X \le b)\}$

$$\{(X = a) \cup (a < X \le b)\} = P(X = a) + P(a < X \le b)\}$$

$$= P(X = a) + [F(b) - F(a)] \quad \text{... (5·2a)}$$

Similarly, we get

$$P(a < X < b) = P(a < X \le b) - P(X = b) = F(b) - F(a) - P(X = b) \quad \text{... (5·2b)}$$

$$P(a \le X < b) = P(a < X < b) + P(X = a)$$

$$= F(b) - F(a) - P(X = b) + P(X = a) \quad \text{... (5·2c)}$$

Remark. When $P(X = a) = 0$ and $P(X = b) = 0$, all four events $a \le X \le b$, $a < X < b$, $a \le X < b$ and $a < X \le b$ have the same probability $F(b) - F(a)$.

2. *If F is d.f. of one-dimensional r.v. X, then (i) $0 \le F(x) \le 1$, (ii) $F(x) \le F(y)$ if $x < y$.*

In other words, all distribution functions are monotonically non-decreasing and lie between 0 and 1.

3. *If F is d.f. of one-dimensional r.v. X, then*

$$F(-\infty) = \lim_{x \to -\infty} F(x) = 0 \quad \text{and} \quad F(\infty) = \lim_{x \to \infty} F(x) = 1$$

Proof. Let us express the whole sample space S as a countable union of disjoint events as follows :

$$S = \left\{ \bigcup_{n=1}^{\infty} (-n < X \le -n+1) \right\} \cup \left\{ \bigcup_{n=0}^{\infty} (n < X \le n+1) \right\}$$

$$\Rightarrow \quad P(S) = \sum_{n=1}^{\infty} P(-n < X \le -n+1) + \sum_{n=0}^{\infty} P(n < X \le n+1)$$

$$\Rightarrow \quad 1 = \lim_{a \to \infty} \sum_{n=1}^{a} \{F(-n+1) - F(-n)\} + \lim_{b \to \infty} \sum_{n=0}^{b} \{F(n+1) - F(n)\}$$

$$= \lim_{a \to \infty} \{F(0) - F(-a)\} + \lim_{b \to \infty} \{F(b+1) - F(0)\}$$

$$= \{F(0) - F(-\infty)\} + \{F(\infty) - F(0)\}$$

$$\therefore \quad 1 = F(\infty) - F(-\infty) \quad \text{... (*)}$$

Since $-\infty < \infty$, $F(-\infty) \le F(\infty)$. Also $F(-\infty) \ge 0$ and $F(\infty) \le 1$

$$\therefore \quad 0 \le F(-\infty) \le F(\infty) \le 1 \quad \text{... (**)}$$

From (*) and (**), we get $F(-\infty) = 0$ and $F(\infty) = 1$.

Remarks:

1. Discontinuities of $F(x)$ are at most countable.
2. $F(a) - F(a-0) = \lim_{h \to 0} P(a - h \le X \le a),\ h < 0$

   and $F(a+0) - F(a) = \lim_{h \to 0} P(a \le X \le a + h),\ h > 0$

5·3. DISCRETE RANDOM VARIABLE

A variable which can assume only a countable number of real values and for which the value which the variable takes depends on chance, is called a *discrete random variable* (or discrete stochastic variable or discrete chance variable). In other words, *a real valued function defined on a discrete sample space is called a discrete random variable.*

Examples of discrete random variables are marks obtained in a test, number of accidents per month, number of telephone calls per unit time, number of successes in n trials, and so on.

5·3·1. Probability Mass Function

If X is a one-dimensional discrete random variable taking at most a countably infinite number of values $x_1, x_2, \dots$ then its probabilistic behaviour at each real point is described by a function called the probability mass function (or discrete density function) which is defined below :

Definition. *If X is a discrete random variable with distinct values $x_1, x_2, \dots, x_n, \dots$, then the function p(x) defined as :*

$$p_X(x) = \begin{cases} P(X = x_i) = p_i, \text{if } x = x_i \\ 0, \text{ if } x \ne x_i \ ; i = 1, 2 \dots \end{cases}$$

is called the probability mass function of r.v. X.

The set of ordered pairs $\{x_i, p(x_i); i = 1, 2, ..., n, ...\}$ or $\{(x_1, p_1), (x_2, p_2), ..., (x_n, p_n), ...\}$, specifies the *probability distribution* of the *r.v.* X .

Remarks:

1. The numbers $p(x_i)$; i = 1, 2, ... must satisfy the following conditions :

$$(i)\ p(x_i) \geq 0\ \forall\ i, \qquad \text{and} \qquad (ii)\ \sum_{i=1}^{\infty} p(x_i) = 1$$

2. The set of values which X takes is called the *spectrum* of the random variable.

3. For discrete *r.v.*, a knowledge of the probability mass function enables us to compute probabilities of arbitrary events. In fact, if E is a set of real numbers,

$$P(X \in E) = \sum_{x \in E \cap S} p(x), \text{ where } S \text{ is the sample space.}$$

Illustration:

In toss of a coin, S = {*H*, *T*}. Let X be the random variable defined by :

X (*H*) = 1, *i.e.*, X = 1, if *'Head'* occurs.

X (*T*) = 0, *i.e.*, X = 0, if *'Tail'* occurs.

If the coin is 'fair', the probability function is : $P(\{H\}) = P(\{T\}) = \frac{1}{2}$, and we can speak of the probability distribution of the *r.v.* X as :

$$P(X = 1) = P(\{H\}) = \frac{1}{2}, \qquad \text{and} \qquad P(X = 0) = P(\{T\}) = \frac{1}{2}.$$

5·3·2. Discrete Distribution Function

In this case there are a countable number of points $x_1, x_2, x_3, ...$ and number

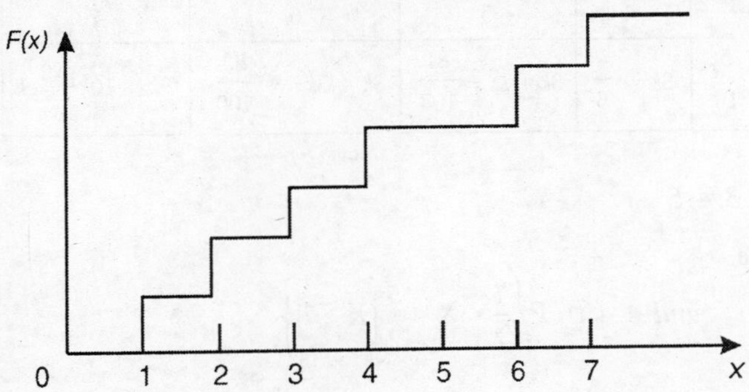

$$p_i \geq 0, \sum_{i}^{\infty} p_i = 1, \text{ such that}$$

$$F(x) = \sum_{i: x_i \leq x} p_i .$$

For example, if x_i is just the integer i, so that $P(X = i) = p_i$; i = 1, 2, 3, ..., then $F(x)$ is a *"step function"* having jump p_i at i, and being constant between each pair of integers.

Theorem 5·1. $p(x_j) = P(X = x_j) = F(x_j) - F(x_{j-1})$, *where F is the d.f. of X.*

Proof. Let $x_1 < x_2 < ...$ We have

$$F(x_j) = P(X \leq x_j) = \sum_{i=1}^{j} P(X = x_i) = \sum_{i=1}^{j} p(x_i) \text{ and } F(x_{j-1}) = P(X \leq x_{j-1}) = \sum_{i=1}^{j-1} p(x_i)$$

$$\therefore \qquad F(x_j) - F(x_{j-1}) = p(x_j) \qquad ...(5·5)$$

Thus, given the distribution function of discrete random variable, we can compute its probability mass function.

Example 5·1. *A random variable X has the following probability function :*

Values of X, x :	0	1	2	3	4	5	6	7
p(x) :	0	k	2k	2k	3k	k^2	$2k^2$	$7k^2 + k$

(i) Find k, (ii) Evaluate $P(X < 6)$, $P(X \geq 6)$, and $P(0 < X < 5)$,

(iii) If $P(X \leq a) > \frac{1}{2}$, find the minimum value of a, and

(iv) Determine the distribution function of X.

Solution.

(i) Since $\sum_{x=0}^{7} p(x) = 1$, $\quad k + 2k + 2k + 3k + k^2 + 2k^2 + 7k^2 + k = 1$

$\Rightarrow \quad 10k^2 + 9k - 1 = 0 \quad \Rightarrow \quad (10k - 1)(k + 1) = 0 \quad \Rightarrow \quad k = \frac{1}{10}$ or $k = -1$.

But since *p(x) cannot be negative,* $k = -1$, *is rejected.* Hence, $k = \frac{1}{10}$.

(ii) $P(X < 6) = P(X = 0) + P(X = 1) + ... + P(X = 5) = \frac{1}{10} + \frac{2}{10} + \frac{2}{10} + \frac{3}{10} + \frac{1}{100} = \frac{81}{100}$

$P(X \geq 6) = 1 - P(X < 6) = 1 - \frac{81}{100} = \frac{19}{100}$

$P(0 < X < 5) = P(X = 1) + P(X = 2) + P(X = 3) + P(X = 4) = 8k = \frac{4}{5}$.

(iii) $P(X \leq a) > \frac{1}{2}$. By trial, we get $a = 4$.

(iv) The distribution function $F_X(x) = P(X \leq x)$ of X is given in the following Table.

x	0	1	2	3	4	5	6	7
$F_X(x)$	0	$k = \frac{1}{10}$	$3k = \frac{3}{10}$	$5k = \frac{5}{10}$	$8k = \frac{4}{5}$	$8k + k^2 = \frac{81}{100}$	$8k + 3k^2 = \frac{83}{100}$	$9k + 10k^2 = 1$

Example 5·2. If, $\quad p(x) = \begin{cases} \frac{x}{15}\,; x = 1, 2, 3, 4, 5 \\ 0, \quad elsewhere \end{cases}$

Find (i) $P\{X = 1 \text{ or } 2\}$ and (ii) $P\left\{\frac{1}{2} < X < \frac{5}{2} \middle| X > 1\right\}$.

Solution.

(i) $P(X = 1 \text{ or } 2) = P(X = 1) + P(X = 2) = \frac{1}{15} + \frac{2}{15} = \frac{1}{5}$.

$$\text{(ii)}\ P\left(\frac{1}{2} < X < \frac{5}{2} \middle| X > 1\right) = \frac{P\left\{\left(\frac{1}{2} < X < \frac{5}{2}\right) \cap (X > 1)\right\}}{P(X > 1)} = \frac{P\{(X = 1 \text{ or } 2) \cap (X > 1)\}}{P(X > 1)}$$

$$= \frac{P(X = 2)}{1 - P(X = 1)} = \frac{2/15}{1 - (1/15)} = \frac{1}{7}.$$

Example 5·3. *Two dice are rolled. Let X denote the random variable which counts the total number of points on the upturned faces, Construct a table giving the non-zero values of the probability mass function and draw the probability chart. Also find the distribution function of X.*

Solution. If both dice are unbiased and the two rolls are independent, then each sample point of sample space S has probability $\frac{1}{36}$. Then

$$p(2) = P(X = 2) = P\{1, 1)\} = \frac{1}{36}; \qquad p(3) = P(X = 3) = P\{(1, 2), (2, 1)\} = \frac{2}{36}$$

$$p(4) = P(X = 4) = P\{(1, 3), (2, 2), (3, 1)\} = \frac{3}{36}$$

$$p(5) = P(X = 5) = P\{(1, 4), (2, 3)\ (3, 2), (4, 1)\} = \frac{4}{36}$$

$$\vdots \qquad \vdots$$

$$p(11) = P(X = 11) = P\{(6, 5), (5, 6)\} = \frac{2}{36}; \qquad p(12) = P(X = 12) = P\{(6, 6)\} = \frac{1}{36}$$

These values are summarized in the following *probability table* :

x :	2	3	4	5	6	7	8	9	10	11	12
$p(x)$:	$\frac{1}{36}$	$\frac{2}{36}$	$\frac{3}{36}$	$\frac{4}{36}$	$\frac{5}{36}$	$\frac{6}{36}$	$\frac{5}{36}$	$\frac{4}{36}$	$\frac{3}{36}$	$\frac{2}{36}$	$\frac{1}{36}$

The chart of the probability distribution is given below :

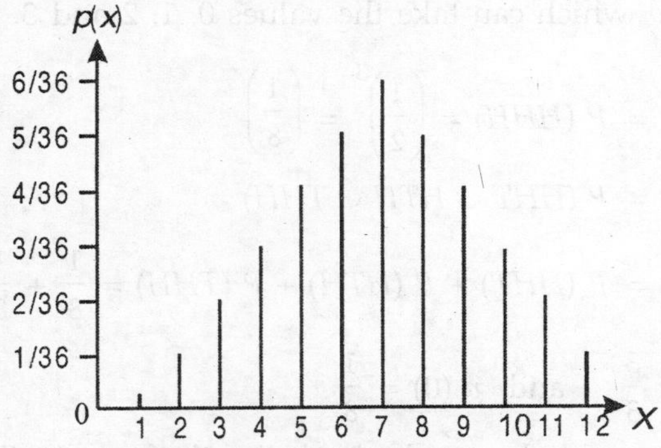

Fig. 5·1. ***Probability Function of X***

$$F(1) = P(X \le 1) = 0,$$

$$F(2) = P(X \le 2) = \frac{1}{36}$$

$$F(3) = P(X \le 3) = P(X = 2) + P(X = 3)$$

$$= p(2) + p(3) = \frac{1}{36} + \frac{2}{36} = \frac{3}{36}$$

$$F(4) = P(X \le 4) = P(X = 2) + P(X = 3) + P(X = 4)$$

$$= \frac{1}{36} + \frac{2}{36} + \frac{3}{36} = \frac{6}{36}, \text{ and so on}$$

The distribution function of X is shown in the adjoining Table.

$$F(x) = \begin{cases} 0, & \text{for } x < 2 \\ \frac{1}{36}, & \text{for } 2 \le x < 3 \\ \frac{3}{36}, & \text{for } 3 \le x < 4 \\ \frac{6}{36}, & \text{for } 4 \le x < 5 \\ \vdots & \vdots \\ \frac{35}{36}, & \text{for } 11 \le x < 12 \\ 1, & \text{for } x \ge 12 \end{cases}$$

Example 5·4. *An experiment consists of three independent tosses of a fair coin. Let :*

X = The number of heads, Y = the number of head runs, Z = The length of head runs, a head run being defined as consecutive occurrence of at least two heads, its length then being the number of heads occurring together in three tosses of the coin.

Find the probability function of (i) X, (ii) Y, (iii) Z, (iv) X + Y, and (v) XY and construct probability tables and draw their probability charts.

Solution.

TABLE 5·1

S. No.	Elementary event	Random Variables X	Y	Z	X + Y	XY
1	HHH	3	1	3	4	3
2	HHT	2	1	2	3	2
3	HTH	2	0	0	2	0
4	HTT	1	0	0	1	0
5	THH	2	1	2	3	2
6	THT	1	0	0	1	0
7	TTH	1	0	0	1	0
8	TTT	0	0	0	0	0

Here, sample space is : $S = \{HHH, HHT, HTH, HTT, THH, THT, TTH, TTT\}$

(*i*) Obviously X is a *r.v.* which can take the values 0, 1, 2 and 3.

$$p(3) = P(HHH) = \left(\frac{1}{2}\right)^3 = \left(\frac{1}{8}\right)$$

$$p(2) = P(HHT \cup HTH \cup THH)$$

$$= P(HHT) + P(HTH) + P(THH) = \frac{1}{8} + \frac{1}{8} + \frac{1}{8} = \frac{3}{8}.$$

Similarly, $p(1) = \frac{3}{8}$ and $p(0) = \frac{1}{8}$.

These probabilities could also be obtained directly from the Table 5·1.

TABLE 5·2 : PROBABILITY DISTRIBUTION OF X

Values of X, (x)	0	1	2	3
p(x)	$\frac{1}{8}$	$\frac{3}{8}$	$\frac{3}{8}$	$\frac{1}{8}$

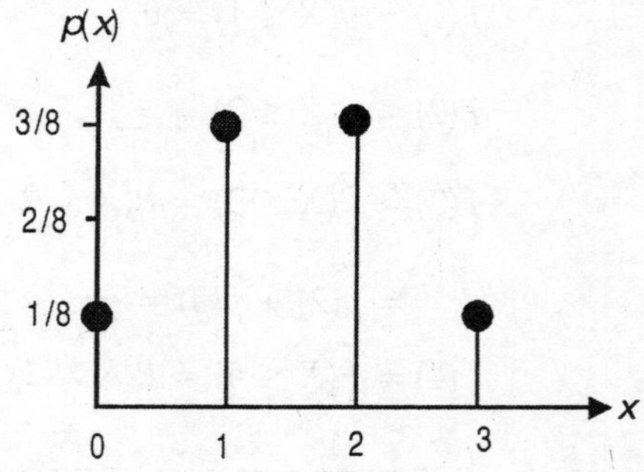

Fig. 5·2 : Probability chart of X

(*ii*)

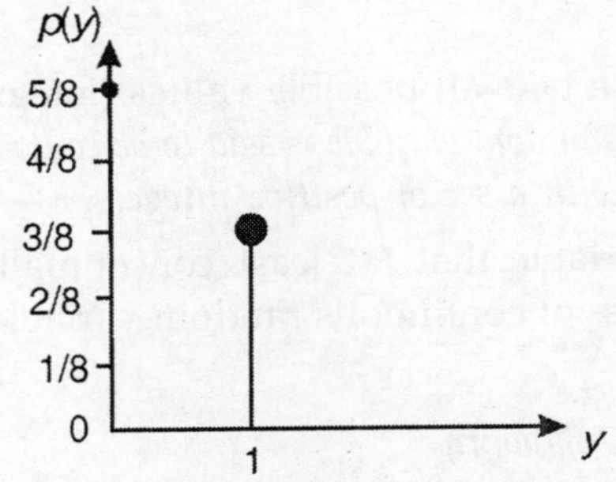

Fig. 5·3 : Probability chart of Y

TABLE 5·3 : PROBABILITY DISTRIBUTION OF Y

Values of Y, (y)	0	1
$p(y)$	$\frac{5}{8}$	$\frac{3}{8}$

(*iii*) From Table 5·1, we have

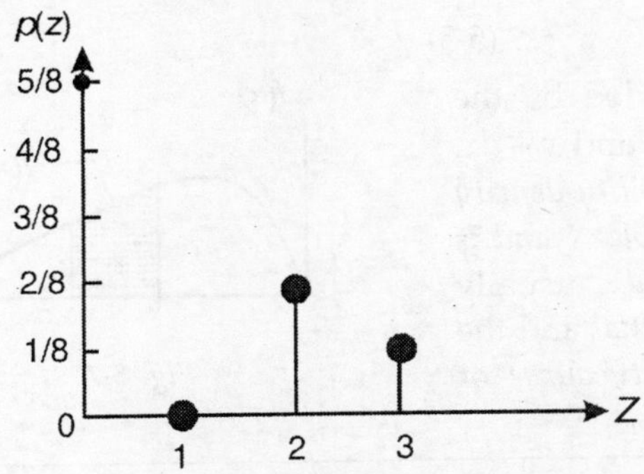

Fig. 5·4 : Probability chart of Z

TABLE 5·4 : PROBABILITY DISTRIBUTION OF Z

Values of Z, (z)	0	1	2	3
$p(z)$	$\frac{5}{8}$	0	$\frac{2}{8}$	$\frac{1}{8}$

(*iv*) Let $U = X + Y$. From Table 5·1, we get

TABLE 5·5 : PROBABILITY DISTRIBUTION OF U

Values of U, (u)	0	1	2	3	4
$p(u)$	$\frac{1}{8}$	$\frac{3}{8}$	$\frac{1}{8}$	$\frac{2}{8}$	$\frac{1}{8}$

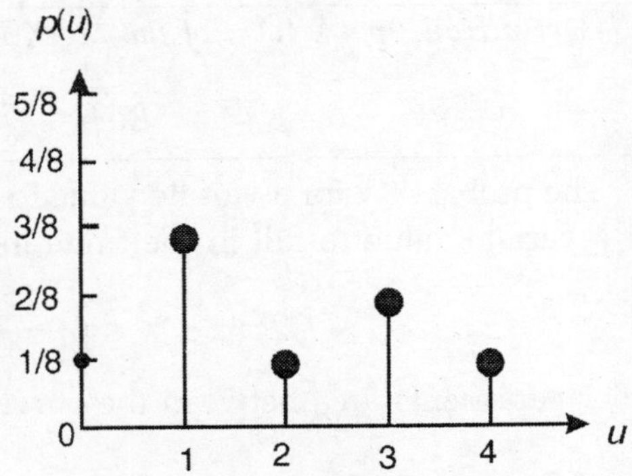

Fig. 5·5 : Probability chart of U = X+Y

(*v*) Let $V = XY$

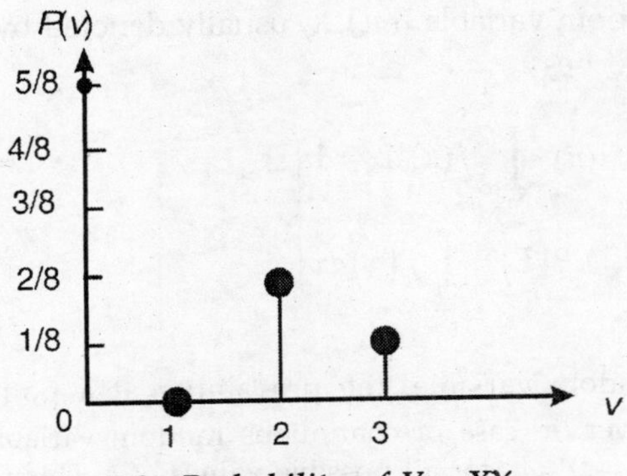

Fig. 5·6 : Probability of V = XY

TABLE 5·6 : PROBABILITY DISTRIBUTION OF V

Values of V, (v)	0	1	2	3
$p(v)$	$\frac{5}{8}$	0	$\frac{2}{8}$	$\frac{1}{8}$

5·4. CONTINUOUS RANDOM VARIABLE

A random variable X is said to be continuous if it can take all possible values (integral as well as fractional) between certain limits. In other words, *a random variable is said to be continuous when its different values cannot be put in 1-1 correspondence with a set of positive integers.*

A continuous random variable is a random variable that (at least conceptually) can be measured to any desired degree of accuracy. Examples of continuous random variables are age, height, weight, etc.

5·4·1. Probability Density Function (*Concept and Definition*)

Consider the small interval $(x, x + dx)$ of length dx round the point x. Let $f(x)$ be any continuous function of x so that $f(x)dx$ represents the probability that X falls in the infinitesimal interval $(x, x + dx)$. Symbolically,

$$P(x \le X \le x + dx) = f_X(x)\, dx \qquad \dots (5\cdot5)$$

In the figure, $f(x)\, dx$ represents the area bounded by the curve $y = f(x)$, x-axis and ordinates at the points x and $x + dx$. The function $f_X(x)$ so defined is known as *probability density function or simply density function of random variable X and is usually abbreviated as p.d.f.* The expression, $f(x)dx$, usually written as $dF(x)$, is known as the *probability differential* and the curve $y = f(x)$ is known as the *probability density curve* or simply *probability curve.*

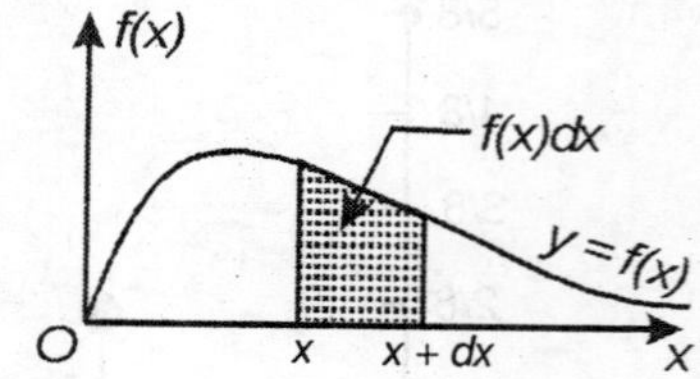

Fig. 5·7

Definition. *p.d.f.* $f_X(x)$ *of the r.v. X is defined as :*

$$f_X(x) = \lim_{\delta x \to 0} \frac{P(x \le X \le x + \delta x)}{\delta x} \qquad \dots (5\cdot5\ a)$$

The probability for a variate value to lie in the interval dx is $f(x)\, dx$ and hence the probability for a variate value to fall in the finite interval $[\alpha, \beta]$ is :

$$P(\alpha \le X \le \beta) = \int_\alpha^\beta f(x)\, dx, \qquad \dots (5\cdot5\ b)$$

which represents area between the curve $y = f(x)$, x-axis and the ordinates at $x = \alpha$ and $x = \beta$.

Further, since total probability is unity, we have $\int_a^b f(x)dx = 1$, where $[a, b]$ is the range of the random variable X. The range of the variable may be finite or infinite.

The probability density function (*p.d.f.*) of a random variable (*r.v.*) X, usually denoted by $f_X(x)$ or simply by $f(x)$ has the following obvious properties :

$$\left.\begin{array}{ll} (i)\ f(x) \ge 0, & (ii)\ \int_{-\infty}^{\infty} f(x)\, dx = 1 \\ (iii)\ \text{The probability } P(E) \text{ given by} : & P(E) = \int_E f(x)dx \end{array}\right\} \qquad \dots (5\cdot5\ c)$$

is well defined for any event E.

Important Remark. In case of discrete random variable, the probability at a point, *i.e.*, $P(x = c)$ is not zero for some fixed c. However, in case of continuous random variables the probability at a point is always zero, *i.e.*, $P(x = c) = 0$, for all possible values of c. This follows directly from (5·5*b*) by taking $\alpha = \beta = c$. This also agrees with our discussion earlier that $P(E) = 0$ does not imply that the event E is null or impossible event. This property of continuous *r.v.*, *viz.*,

$$P(X = c) = 0, \ \forall \ c, \qquad \text{... (5·5 } d)$$

leads us to the following important result :

$$P(\alpha \le X \le \beta) = P(\alpha \le X < \beta) = P(\alpha < X \le \beta) = P(\alpha < X < \beta) \qquad \text{... (5·5 } e)$$

i.e., in case of continuous *r.v.*, it does not matter whether we include the end points of the interval from α to β.

However, this result is, in general, not true for discrete random variables.

5·4·2. Various Measures of Central Tendency, Dispersion, Skewness and Kurtosis for Continuous Probability Distribution

The formulae for these measures in case of discrete frequency distribution can be easily extended to the case of continuous probability distribution by simply replacing $p_i = f_i/N$ by $f(x)\,dx$, x_i by x and the summation over '*i*' by integration over the specified range of the variable X.

Let $f_X(x)$ or $f(x)$ be the *p.d.f.* of a *r.v.* X, where X is defined from a to b. Then

(*i*) *Arithmetic Mean* $= \int_a^b x\, f(x)\, dx$ ... (5·6)

(*ii*) *Harmonic Mean.* Harmonic mean H is given by : $\dfrac{1}{H} = \int_a^b \dfrac{1}{x} \cdot f(x)\, dx$ ... (5·6 *a*)

(*iii*) *Geometric Mean.* Geometric mean G is given by : $\log G = \int_a^b \log x \,.\, f(x)\, dx$ ... (5·6 *b*)

(*iv*) μ_r' (about origin) $= \int_a^b x^r \,.\, f(x)\, dx$ ... (5·7)

μ_r' (about the point $x = A$) $= \int_a^b (x - A)^r \,.\, f(x)\, dx$ ... (5·7 *a*)

and μ_r (about mean) $= \int_a^b (x - \text{mean})^r \,.\, f(x) dx$ ... (5·7 *b*)

In particular, from (5·6) and (5·7), we have

$$\mu_1' \text{ (about origin)} = \text{Mean} = \int_a^b x\, f(x)\, dx \quad \text{and} \quad \mu_2' = \int_a^b x^2\, f(x)\, dx$$

Hence,

$$\mu_2 = \mu_2' - \mu_1'^2 = \int_a^b x^2\, f(x)\, dx - \left(\int_a^b x\, f(x) dx \right)^2 \qquad \text{... (5·7 } c)$$

From (5·7), on putting $r = 3$ and 4 respectively, we get the values of μ_3' and μ_4' and consequently the moments about mean can be obtained by using the relations :

$$\left.\begin{aligned} \mu_3 &= \mu_3' - 3\mu_2'\,\mu_1' + 2\mu_1'^3 \\ \text{and} \qquad \mu_4 &= \mu_4' - 4\mu_3'\,\mu_1' + 6\mu_2'\,\mu_1'^2 - 3\mu_1'^4 \end{aligned}\right\} \qquad \text{... (5·7 } d)$$

and hence β_1 and β_2 can be computed.

(*v*) *Median.* Median is the point which divides the entire distribution in two equal parts. In case of continuous distribution, median is the point which divides the total area into two equal parts. Thus if M is the median, then

$$\int_a^M f(x)\,dx = \int_M^b f(x)\,dx = \frac{1}{2} \qquad \text{... (5·8)}$$

Thus, solving $\int_a^M f(x)\,dx = \frac{1}{2}$ or $\int_M^b f(x)\,dx = \frac{1}{2}$ ... (5·8 *a*)

for M, we get the value of median.

(*vi*) *Mean Deviation.* Mean deviation about the mean μ_1' is given by :

$$M.D. = \int_a^b |x - \text{mean}|\, f(x)\,dx \qquad \text{... (5·9)}$$

In general, mean diviation about an average 'A' is given by :

$$M.D. \text{ about } 'A' = \int_a^b |x - A|\, f(x)\,dx \qquad \text{... (5·9 a)}$$

(*vii*) *Quartiles and Deciles.* Q_1 and Q_3 are given by the equations :

$$\int_a^{Q_1} f(x)\,dx = \frac{1}{4} \quad \text{and} \quad \int_a^{Q_3} f(x)\,dx = \frac{3}{4} \qquad \text{... (5·10)}$$

D_i, *i*th decile is given by : $\int_a^{D_i} f(x)\,dx = \frac{i}{10}$; $i = 1, 2, ..., 9$ ... (5·10 *a*)

(*viii*) *Mode.* Mode is the value of x for which $f(x)$ is maximum. Mode is thus the solution of $f'(x) = 0$ and $f''(x) < 0$, provided it lies in $[a, b]$. ... (5·11)

Remark. Various Measures of Central Tendency, Dispersion, Skewness and Kurtosis for Discrete Random Variable. Let X be a discrete *r.v.* with probability mass function (*p.m.f.*) $f_X(x)$ or $f(x)$. Then the various measures of central tendency, dispersion, skewness and kurtosis can be obtained similarly, on replacing integration ($\int$) with summation (Σ), over the given range of the variable X, in the formulae (5·6) to (5·10*a*).

Example 5·5. *The diameter of an electric cable, say X, is assumed to be a continuous random variable with p.d.f. :*

$$f(x) = 6x(1 - x),\ 0 \le x \le 1.$$

(*i*) *Check that f(x) is p.d.f.,* and (*ii*) *Determine a number b such that* $P(X < b) = P(X > b)$.

Solution.

(*i*) Obviously, for $0 \le x \le 1$, $f(x) \ge 0$.

$$\int_0^1 f(x)\,dx = 6\int_0^1 x(1-x)\,dx = 6\int_0^1 (x - x^2)\,dx = 6\left|\frac{x^2}{2} - \frac{x^3}{3}\right|_0^1 = 1$$

Hence, $f(x)$ is the *p.d.f.* of *r.v.* X.

(*ii*) $P(X < b) = P(X > b)$

$$\Rightarrow \int_0^b f(x)\,dx = \int_b^1 f(x)\,dx \Rightarrow 6\int_0^b x(1-x)\,dx = 6\int_b^1 x(1-x)\,dx$$

$$\Rightarrow \left|\frac{x^2}{2} - \frac{x^3}{3}\right|_0^b = \left|\frac{x^2}{2} - \frac{x^3}{3}\right|_b^1 \Rightarrow \left(\frac{b^2}{2} - \frac{b^3}{3}\right) = \left[\left(\frac{1}{2} - \frac{1}{3}\right) - \left(\frac{b^2}{2} - \frac{b^3}{3}\right)\right]$$

$$\Rightarrow 3b^2 - 2b^3 = (1 - 3b^2 + 2b^3) \Rightarrow 4b^3 - 6b^2 + 1 = 0 \quad \Rightarrow (2b - 1)(2b^2 - 2b - 1) = 0$$

$$\therefore \quad 2b - 1 = 0 \Rightarrow b = \frac{1}{2} \quad \text{or} \quad 2b^2 - 2b - 1 = 0 \Rightarrow b = \frac{2 \pm \sqrt{4+8}}{4} = \frac{1 \pm \sqrt{3}}{2}.$$

Hence, $b = \frac{1}{2}$, is the only real value lying between 0 and 1 and satisfying (*).

Example 5·6. *A continuous random variable X has a p.d.f. : $f(x) = 3x^2,\ 0 \le x \le 1$.*

Find a and b such that (i) $P(X \le a) = P(X > a)$, and (ii) $P(X > b) = 0{\cdot}05$.

Solution.

(*i*) Since $P(X \le a) = P(X > a)$, each must be equal to $\frac{1}{2}$, because total probability is always unity.

$$\therefore \quad P(X \le a) = \frac{1}{2} \Rightarrow \int_0^a f(x)\,dx = \frac{1}{2}$$

$$\Rightarrow \quad 3\int_0^a x^2 dx = \frac{1}{2} \Rightarrow 3\left|\frac{x^3}{3}\right|_0^a = \frac{1}{2} \Rightarrow a = \left(\frac{1}{2}\right)^{\frac{1}{3}}$$

$$(ii)\ P(X > b) = 0{\cdot}05 \Rightarrow \int_b^1 f(x)\,dx = 0{\cdot}05$$

$$\Rightarrow \quad 3\left|\frac{x^3}{3}\right|_b^1 = \frac{1}{20} \Rightarrow 1 - b^3 = \frac{1}{20} \Rightarrow b = \left(\frac{19}{20}\right)^{\frac{1}{3}}$$

Example 5·7. *A continuous r.v. X has the p.d.f.,*

$$f(x) = \begin{cases} 2x & ; \ 0 < x < \frac{1}{2} \\ \frac{4}{3} - \frac{2}{3}x & ; \ \frac{1}{2} < x < 2 \end{cases} ; \quad f(x) = 0, \text{otherwise}$$

Find Median.

Solution. Median 'm' is the solution of

$$\int_0^m f(x)\,dx = \int_m^2 f(x)\,dx = \frac{1}{2} \qquad \ldots (*)$$

We have : $\int_0^{1/2} f(x)\,dx = \int_0^{1/2} 2x\,dx = \left|x^2\right|_0^{1/2} = \left|\frac{1}{4}\right| < \frac{1}{2}.$

Hence, median $m > \frac{1}{2}$ *i.e.*, m lies in the interval $\frac{1}{2} \le x < 2$, and is given by : $\left[m > \frac{1}{2}\right]$

$$\int_m^2 f(x)\,dx = \frac{1}{2} \Rightarrow \int_m^2 \left(\frac{4}{3} - \frac{2}{3}x\right)dx = \frac{1}{2} \Rightarrow \left|\frac{4}{3}x - \frac{x^2}{3}\right|_m^2 = \frac{1}{2}$$

$$\Rightarrow \quad \left(\frac{8}{3} - \frac{4}{3}\right) - \left(\frac{4m}{3} - \frac{m^2}{3}\right) = \frac{1}{2} \Rightarrow 2(4 - 4m + m^2) = 3$$

$$\therefore \quad 2m^2 - 8m + 5 = 0 \quad \Rightarrow \quad m = \frac{8 \pm \sqrt{64-40}}{4} = 2 \pm \sqrt{3/2}$$

Since $\frac{1}{2} < m < 2$, the only admissible value of $m = 2 - \sqrt{3/2}$.

Example 5·8. *Prove that the geometric mean G of the distribution :*

$$dF = 6\,(2-x)\,(x-1)\,dx,\ 1 \le x \le 2$$

is given by $\qquad 6 \log (16G) = 19.$

Solution. By def., we have

$$\log G = \int_1^2 \log x\, f(x)\, dx = 6 \int_1^2 (2-x)\,(x-1) \log x\, dx = -6 \int_1^2 (x^2 - 3x + 2) \log x\, dx$$

$$\Rightarrow \log G = -6 \left\{ \left| \left(\frac{x^3}{3} - \frac{3x^2}{2} + 2x \right) \log x \right|_1^2 - \int_1^2 \left(\frac{x^3}{3} - \frac{3x^3}{2} + 2x \right) \frac{1}{x} dx \right\}. \quad \text{(Integrating by parts)}$$

$$= -4 \log 2 + 6 \times \frac{19}{36} \qquad \text{(On simplification)}$$

$$\therefore \ \log G + 4 \log 2 = \frac{19}{6} \Rightarrow \log (G \times 2^4) = \frac{19}{6} \Rightarrow \log (16\,G) = \frac{19}{6} \Rightarrow 6 \log (16\,G) = 19.$$

Example 5·9. *The probability distribution of a r.v. X is :* $f(x) = k \sin \left(\frac{1}{5} \pi x \right),\ 0 \le x \le 5.$

Determine the constant k and obtain the median and quartiles of the distribution.

Solution. Since $k \int_0^5 \sin \left(\frac{1}{5} \pi\, x\, dx \right) = 1 \Rightarrow \frac{5k}{\pi} \int_0^{\pi} \sin y\, dy = 1$, where $\frac{1}{5} \pi\, x = y$

$$\therefore \qquad k = \frac{\pi}{10} \qquad \text{... (*)}$$

The median M, is given by :

$$k \int_0^M \sin \frac{1}{5} \pi\, x\, dx = \frac{1}{2} \quad \Rightarrow \quad \frac{5}{\pi} k \int_0^{\frac{1}{5}\pi M} \sin y\, dy = \frac{1}{2}, \qquad \text{where } \frac{1}{5} \pi\, x = y$$

$$\Rightarrow \quad \frac{5k}{\pi} \left| -\cos y \right|_0^{\frac{1}{5}\pi M} = \frac{1}{2} \quad \Rightarrow \quad \frac{1}{2} \left(1 - \cos \frac{1}{5} \pi\, M \right) = \frac{1}{2} \quad \Rightarrow \quad M = \frac{5}{2} \qquad \text{[From (*)]}$$

The first quartile Q_1 is given by :

$$k \int_0^{Q_1} \sin \left(\frac{1}{5} \pi x \right) dx = \frac{1}{4} \quad \Rightarrow \quad \frac{5k}{\pi} \int_0^{\frac{1}{5}\pi Q_1} \sin y\, dy = \frac{1}{4} \Rightarrow \frac{1}{2} \left(1 - \cos \frac{1}{5} \pi Q_1 \right) = \frac{1}{4} \qquad \text{[From (*)]}$$

$$\therefore \qquad \cos \frac{(\pi Q_1)}{5} = \frac{1}{2} = \cos \left(\frac{\pi}{3} \right) \quad \Rightarrow \quad Q_1 = \frac{\pi}{3} \times \frac{5}{\pi} = \frac{5}{3}.$$

The third quartile Q_3 is given by :

$$k\int_0^{Q_3} \sin\left(\frac{1}{5}\pi x\right) dx = \frac{3}{4} \Rightarrow \frac{5k}{\pi}\int_0^{\frac{1}{5}\pi Q_3} \sin y\, dy = \frac{3}{4} \Rightarrow \frac{1}{2}\left(1-\cos\frac{1}{5}\pi Q_3\right) = \frac{3}{4}$$

$$\therefore \cos\left(\frac{\pi Q_3}{5}\right) = -\frac{1}{2} = \cos\left(\pi - \frac{\pi}{3}\right) = \cos\left(\frac{2\pi}{3}\right) \Rightarrow Q_3 = \frac{2\pi}{3} \times \frac{5}{\pi} = \frac{10}{3}.$$

Example 5·10. *A variable X is distributed at random between the values 0 and 4 and its probability density function is given by :* $f(x) = kx^3 (4 - x)^2$.

Find the value of k, the mean and standard deviation of the distribution.

Solution. Since $\int_{-\infty}^{\infty} f(x)\, dx = 1,\quad k\int_0^4 x^3(4-x^2)\, dx = 1 \Rightarrow k = \frac{15}{1024}.$

$$\text{Mean} = \mu_1' = \int_{-\infty}^{\infty} x f(x)\, dx = k\int_0^4 x^4 (4-x)^2\, dx = \frac{16}{7}$$

$$\mu_2' = \int_{-\infty}^{\infty} x^2 f(x)\, dx = k\int_0^4 x^5 (4-x)^2\, dx = \frac{40}{7}$$

$$\text{Variance} = \mu_2 = \mu_2' - \mu_1'^2 = \left\{\frac{40}{7} - \left(\frac{16}{7}\right)^2\right\} = \frac{24}{49}$$

$$\therefore \quad \text{Standard deviation} = \frac{2\sqrt{6}}{7}.$$

Example 5·11. *In a continuous distribution whose relative frequency density is given by :*

$$f(x) = y_0 .\ x\ (2 - x),\ 0 \le x \le 2,$$

find mean, variance, β_1, and β_2 and hence show that the distribution is symmetrical. Also (i) find mean deviation about mean, and (ii) show that for this distribution $\mu_{2n+1} = 0$, and (iii) find the mode, harmonic mean and median.

Solution. Since total probability is unity, we have

$$\int_0^2 f(x)\, dx = 1 \Rightarrow y_0 \int_0^2 x(2-x)\, dx = 1 \Rightarrow y_0 = \frac{3}{4}$$

$$\therefore \quad f(x) = \frac{3}{4} x\ (2 - x),\ 0 \le x \le 2$$

$$\mu_r' = \int_0^2 x^r f(x)\, dx = \frac{3}{4}\int_0^2 x^{r+1}(2-x)\, dx = \frac{3\cdot 2^{r+1}}{(r+2)(r+3)}$$

In particular

$$\text{Mean} = \mu_1' = \frac{3\cdot 2^2}{3\cdot 4} = 1,\quad \mu_2' = \frac{3\cdot 2^3}{4\cdot 5} = \frac{6}{5},\quad \mu_3' = \frac{3\cdot 2^4}{5\cdot 6} = \frac{8}{5},\quad \text{and} \quad \mu_4' = \frac{3\cdot 2^5}{6\cdot 7} = \frac{16}{7}$$

Hence, $\quad$ variance $= \mu_2 = \mu_2' - \mu_1'^2 = \dfrac{6}{5} - 1 = \dfrac{1}{5}$

$$\mu_3 = \mu_3' - 3\mu_2'\,\mu_1' + 2\mu_1'^3 = \frac{8}{5} - 3\cdot\frac{6}{5}\,1 + 2 = 0$$

$$\mu_4 = \mu_4' - 4\,\mu_3'\,\mu_1' + 6\mu_2'\,\mu_1'^2 - 3\mu_1'^4 = \frac{16}{7} - 4\cdot\frac{8}{5}\cdot 1 + 6\cdot\frac{6}{5}\cdot 1 - 3\cdot 1 = \frac{3}{35}$$

$$\therefore \quad \beta_1 = \frac{\mu_3^2}{\mu_2^3} = 0 \quad \text{and} \quad \beta_2 = \frac{\mu_4}{\mu_2^2} = \frac{3/35}{(1/5)^2} = \frac{15}{7}$$

Since $\beta_1 = 0$, the distribution is *symmetrical*. ... (*)

Mean deviation about mean

$$= \int_0^2 |x-1|\, f(x)\,dx = \int_0^1 |x-1|\, f(x)\,dx + \int_1^2 |x-1|\, f(x)\,dx$$

$$= \frac{3}{4}\left[\int_0^1 (1-x)\,x(2-x)\,dx + \int_1^2 (x-1)\,x(2-x)\,dx\right]$$

$$= \frac{3}{4}\left[\int_0^1 (2x-3x^2+x^3)\,dx + \int_1^2 (3x^2-x^3-2x)\,dx\right]$$

$$= \frac{3}{4}\left[\left|x^2 - \frac{3\cdot x^3}{3} + \frac{x^4}{4}\right|_0^1 + \left|3\cdot\frac{x^3}{3} - \frac{x^4}{4} - \frac{2x^2}{2}\right|_1^2\right] = \frac{3}{8}$$

$$\mu_{2n+1} = \int_0^2 (x-\text{mean})^{2n+1} f(x) = \frac{3}{4}\int_0^2 (x-1)^{2n+1}\,x(2-x)\,dx$$

$$= \frac{3}{4}\int_{-1}^1 t^{2n+1}(t+1)(1-t)\,dt = \frac{3}{4}\int_{-t}^1 t^{2n+1}(1-t^2)\,dt, \text{ where } \quad t = x-1$$

Since t^{2n+1} is an odd function of t and $(1-t^2)$ is an even function of t, the integrand $t^{2n+1}(1-t^2)$ is an odd function of t.

Hence, $\quad \mu_{2n+1} = 0.$

$$f'(x) = \frac{3}{4}(2-2x) = 0 \quad \Rightarrow \quad x = 1 \quad \text{and} \quad f''(x) = \frac{3}{4}(-2) = -\frac{3}{2} < 0.$$

Hence, $\quad$ Mode $= 1$

Harmonic mean H is given by :

$$\frac{1}{H} = \int_0^2 \frac{1}{x} f(x)\,dx = \frac{3}{4}\int_0^2 (2-x)\,dx = \frac{3}{2} \quad \Rightarrow \quad H = \frac{2}{3}.$$

If M is the median, $\displaystyle\int_0^M f(x)\,dx = \frac{1}{2} \Rightarrow \frac{3}{4}\int_0^M x(2-x)\,dx = \frac{1}{2} \Rightarrow \left|x^2 - \frac{x^3}{3}\right|_0^M = \frac{2}{3}$

$$\Rightarrow \quad 3M^2 - M^3 = 2 \quad \Rightarrow \quad M^3 - 3M^2 + 2 = 0 \quad \Rightarrow \quad (M-1)(M^2 - 2M - 2) = 0$$

$$\therefore \quad M = 1 \quad \text{or} \quad M = \frac{2 \pm \sqrt{4+8}}{2} = 1 \pm \sqrt{3}$$

The only value of M lying in [0, 2] is $M = 1$. Hence median is 1.

Aliter. Since, the distribution is symmetrical, [From (*)] Mode = Median = Mean = 1.

Example 5·12. *Find the mean, variance and the coefficients* β_1, β_2 *of the distribution :*

$$dF = kx^2 e^{-x}\, dx,\ 0 < x < \infty .$$

Solution. Since $\int_0^\infty f(x)\, dx = 1$,

$$\int_0^\infty kx^2 e^{-x}\, dx = 1 \quad \Rightarrow \quad k\,\Gamma(3) = 1 \quad \Rightarrow \quad k = \frac{1}{2} \quad \left[\because \int_0^\infty e^{-x} x^{n-1}\, dx = \Gamma(n)\right]$$

$$\mu_r' = \int_0^\infty x^r f(x)\, dx = \frac{1}{2}\int_0^\infty e^{-x} x^{r+2}\, dx = \frac{1}{2}\Gamma(r+3) = \frac{1}{2}(r+2)!$$

$$\therefore \quad \mu_1' = \frac{1}{2}(3!) = 3, \quad \mu_2' = \frac{1}{2}(4!) = 12, \quad \mu_3' = \frac{1}{2}(5!) = 60, \quad \mu_4' = \frac{1}{2}(6!) = 360,$$

$$\mu_2 = \mu_2' - \mu_1'^2 = 12 - (3)^2 = 3$$

$$\mu_3 = \mu_3' - 3\mu_2'\,\mu_1' + 2\mu_1'^3 = 60 - 3\,(12)\,(3) + 2\,(3)^3 = 6$$

$$\mu_4 = \mu_4' - 4\mu_3'\,\mu_1' + 6\,\mu_2'\,\mu_1'^2 - 3\mu_1'^4 = 45$$

$$\text{Mean} = \mu_1' = 3, \text{ Variance} = \mu_2 = 3,\ \beta_1 = \frac{\mu_3^2}{\mu_2^3} = \frac{(6)^2}{(3)^3} = \frac{4}{3} \text{ and } \beta_2 = \frac{\mu_4}{\mu_2^2} = \frac{45}{(3)^2} = 5.$$

Example 5·13. *Show that for the symmetrical distribution :* $f(x) = \frac{2a}{\pi}\left(\frac{1}{a^2+x^2}\right),\ -a \le x \le a;$

$$\mu_2 = \frac{a^2(4-\pi)}{\pi} \quad \textit{and} \quad \mu_4 = a^4\left(1 - \frac{8}{3\pi}\right).$$

Solution. Since $\int_{-a}^{a} f(x)\, dx = \frac{2a}{\pi}\int_{-a}^{a}\frac{1}{a^2+x^2}\, dx = \frac{2a}{\pi}\left|\frac{1}{a}\tan^{-1}\frac{x}{a}\right|_{-a}^{a} = 1,$

$f(x)$ represents a probability density function.

$$\mu_1' \text{ (about origin)} = \frac{2a}{\pi}\int_{-a}^{a} x\left(\frac{1}{a^2+x^2}\right) dx = \frac{a}{\pi}\left|\log(a^2+x^2)\right|_{-a}^{a} = 0$$

$$\mu_2 = \mu_2' - \mu_1'^2 = \mu_2' = \frac{2a}{\pi}\int_{-a}^{a}\frac{x^2}{a^2+x^2}\, dx = \frac{4a}{\pi}\int_0^a \frac{x^2+a^2-a^2}{a^2+x^2}\, dx$$

$$= \frac{4a}{\pi}\left[x - a\tan^{-1}\frac{x}{a}\right]_0^a = \frac{4a}{\pi}\left(a - a\cdot\frac{\pi}{4}\right) = \frac{a^2(4-\pi)}{\pi}$$

$$\mu_4 = \mu_4' = \frac{2a}{\pi}\int_{-a}^{a} x^4\left(\frac{1}{a^2+x^2}\right) dx = \frac{4a}{\pi}\int_0^a\left(x^2 - a^2 + \frac{a^4}{a^2+x^2}\right) dx$$

$$= \frac{4a}{\pi}\left|\frac{1}{3}x^3 - a^2x + a^3\tan^{-1}\frac{x}{a}\right|_0^a = a^4\left(1-\frac{8}{3\pi}\right).$$

Example 5·14. *Suppose that the life in hours of a certain part of radio tube is a continuous random variable X with p.d.f. given by :*

$$f(x) = \begin{cases} \dfrac{100}{x^2}, & \text{when } x \geq 100 \\ 0, & \text{elsewhere} \end{cases} \qquad \text{... (*)}$$

(*i*) *What is the probability that all of three such tubes in a given radio set will have to be replaced during the first 150 hours of operation ?*

(*ii*) *What is the probability that none of three of the original tubes will have to be replaced during that first 150 hours of operation ?*

(*iii*) *What is the probability that a tube will last less than 200 hours if it is known that the tube is still functioning after 150 hours of service ?*

(*iv*) *What is the maximum number of tubes that may be inserted into a set so that there is a probability of 0·5 that after 150 hours of service all of them are still functioning ?*

Solution.

(*i*) A tube in a radio set will have to be replaced during the first 150 hours, if its life is ≤ 150 hours. Hence, the required probability 'p' that a tube is replaced during the first 150 hours, on using (*) is given by :

$$p = P(X \leq 150) = \int_{100}^{150} f(x)\, dx = \int_{100}^{150} \frac{100}{x^2}\, dx = \frac{1}{3}$$

∴ By compound probability theorem, the probability that all three of the original tubes will have to be replaced during the first 150 hours $= p^3 = \left(\frac{1}{3}\right)^3 = \frac{1}{27}$.

(*ii*) The proability that a tube is not replaced during the first 150 hours of operation is given by :

$$P(X > 150) = 1 - P(X \leq 150) = 1 - p = 1 - \frac{1}{3} = \frac{2}{3}.$$

Hence, by compound probability theorem, the probability that none of the three tubes will be replaced during the first 150 hours of operation $= \left(\frac{2}{3}\right)^3 = \frac{8}{27}$.

(*iii*) Probability that a tube will last less than 200 hours given that the tube is still functioning after 150 hours is :

$$P\left(X < 200 \mid X > 150\right) = \frac{P(150 < X < 200)}{P(X > 150)} = \frac{\int_{150}^{200} \frac{100}{x^2}\, dx}{\int_{150}^{\infty} \frac{100}{x^2}\, dx} = \frac{1}{6} \times \frac{3}{2} = 0{\cdot}25.$$

(*iv*) Let n be the maximum number of tubes that may be inserted into a set so that there is a probability of 0·5 that after 150 hours of service all of them are still functioning. Then we have to find n so that

$$0{\cdot}5 = \left(\frac{2}{3}\right)^n \Rightarrow n \log(0{\cdot}6667) = \log 0{\cdot}5 \Rightarrow n = \frac{\log 0{\cdot}5}{\log 0{\cdot}6667} = \frac{-0{\cdot}3010}{-0{\cdot}1760} = 1{\cdot}7.$$

Hence, the maximum number of tubes is 2.

Example 5·15. *The time one has to wait for a bus at a downtown bus stop is observed to be random phenomenon (X) with the following probability density function :*

$$f_X(x) = \begin{cases} 0, & \text{for } x < 0 \\ \frac{1}{9}(x+1), & \text{for } 0 \le x < 1 \\ \frac{4}{9}\left(x - \frac{1}{2}\right), & \text{for } 1 \le x < \frac{3}{2} \\ \frac{4}{9}\left(\frac{5}{2} - x\right), & \text{for } \frac{3}{2} \le x < 2 \\ \frac{1}{9}(4 - x), & \text{for } 2 \le x < 3 \\ \frac{1}{9}, & \text{for } 3 \le x < 6 \\ 0, & \text{for } x \ge 6 \end{cases}$$

Let the events A and B be defined as follows :

A : One waits between 0 to 2 minutes inclusive.

B : One waits between 1 to 3 minutes inclusive.

(i) Draw the graph of probability density function.

(ii) Show that (a) $P(B \mid A) = \frac{2}{3}$, *(b)* $P(\bar{A} \cap \bar{B}) = \frac{1}{3}$.

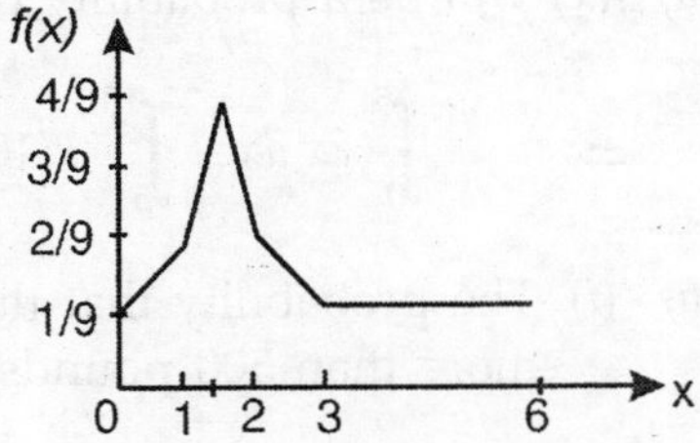

Fig. 5·8.

Solution.

(i) The graph of *p.d.f.* is given in Fig. 5·8. :

(ii) (a) $P(A) = P(X \le 2) = \int_0^2 f(x)\, dx$

$$= \int_0^1 \frac{1}{9}(x+1)\, dx + \int_1^{3/2} \frac{4}{9}\left(x - \frac{1}{2}\right) dx + \int_{3/2}^2 \frac{4}{9}\left(\frac{5}{2} - x\right) dx$$

$= \frac{1}{2}$. (On simplification).

$$P(A \cap B) = P[(0 \le X \le 2) \cap (1 \le X \le 3)] = P(1 \le X \le 2) = \int_1^2 f(x)\, dx$$

$$= \int_1^{3/2} \frac{4}{9}\left(x - \frac{1}{2}\right) dx + \int_{3/2}^2 \frac{4}{9}\left(\frac{5}{2} - x\right) dx$$

$$= \frac{4}{9}\left|\frac{x^2}{2} - \frac{x}{2}\right|_1^{3/2} + \frac{4}{9}\left|\frac{5}{2}x - \frac{x^2}{2}\right|_{3/2}^2 = \frac{1}{3} \quad \text{(On simplification)}$$

$$\therefore \quad P(B \mid A) = \frac{P(A \cap B)}{P(A)} = \frac{1/3}{1/2} = \frac{2}{3}.$$

(b) $\bar{A} \cap \bar{B}$ means that waiting time is more than 3 minutes.

$$\therefore \ P(\bar{A} \cap \bar{B}) = P(X > 3) = \int_3^{\infty} f(x) = \int_3^6 f(x)\,dx + \int_6^{\infty} f(x)\,dx = \int_3^6 \frac{1}{9}\,dx = \frac{1}{9} \Big| x \Big|_3^6 = \frac{1}{3}.$$

Example 5·16 *The amount of bread (in hundreds of pounds) x that a certain bakery is able to sell in a day is found to be a numerical valued random phenomenon, with a probability function specified by the p.d.f. f(x), given by :*

$$f(x) = \begin{cases} k \cdot x, & \text{for } 0 \le x < 5 \\ k \cdot (10 - x), & \text{for } 5 \le x < 10 \\ 0, & \text{otherwise} \end{cases}$$

(a) Find the value of k such that f(x) is a probability density function.

(b) What is the probability that the number of pounds of bread that will be sold tomorrow is : (i) more than 500 pounds, (ii) less than 500 pounds, and (iii) between 250 and 750 pounds?

(c) Denoting by A, B and C the events that the pounds of bread sold are as in b (i), b (ii) and b (iii) respectively, find P (A | B), P (A | C). Are (i) A and B independent events ? (ii) Are A and C independent events ?

Solution.

(a) $f(x)$ will be a probability density function if $\int_{-\infty}^{\infty} f(x)\,dx = 1$

$$\Rightarrow \quad \int_0^5 kx\,dx + \int_5^{10} k(10 - x)\,dx = 1 \quad \text{or} \quad k = \frac{1}{25}. \quad \text{(On simplification)}$$

(b) (i) The probability that the number of pounds of bread that will be sold tomorrow is more than 500 pounds is given by :

$$P(5 \le X \le 10) = \int_5^{10} \frac{1}{25}(10 - x)\,dx = \frac{1}{25}\left| 10x - \frac{x^2}{2} \right|_5^{10} = \frac{1}{25}\left(\frac{25}{2}\right) = 0{\cdot}5$$

(ii) The probability that the number of pounds of bread that will be sold tomorrow is less than 500 pounds, is given by :

$$P(0 \le X < 5) = \int_0^5 \frac{1}{25} \cdot x\,dx \text{ c} = \frac{1}{25}\left| \frac{x^2}{2} \right|_0^5 = \frac{1}{2} = 0{\cdot}5$$

OR $\quad P(X < 5) = 1 - P(X \ge 5) = 1 - 0{\cdot}5 = 0{\cdot}5$ [From part (i)]

(iii) The required probability is given by :

$$P(2{\cdot}5 \le X \le 7{\cdot}5) = \int_{2{\cdot}5}^5 \frac{1}{25} x\,dx + \int_5^{7{\cdot}5} \frac{1}{25}(10 - x)\,dx = \frac{3}{4}. \quad \text{(On simplification)}$$

(c) The events A, B and C are given by

$$A : 5 < X \le 10 \quad ; \quad B : 0 \le X \le 5 \quad ; \quad C : 2{\cdot}5 \le X \le 7{\cdot}5.$$

Then from parts b (i), (ii) and (iii), $P(A) = 0{\cdot}5$, $P(B) = 0{\cdot}5$, $P(C) = \frac{3}{4}$.

The events $A \cap B$ and $A \cap C$ are given by : $A \cap B = \phi$ and $A \cap C : 5 < X < 7{\cdot}5$.

$$\therefore \quad P(A \cap B) = P(\phi) = 0$$

and
$$P(A \cap C) = \int_5^{7\cdot5} f(x)\,dx = \frac{1}{25}\int_5^{7\cdot5}(10 - x)\,dx = \frac{1}{25}\times\frac{75}{8} = \frac{3}{8}$$

$$P(A)\cdot P(C) = \frac{1}{2}\times\frac{3}{4} = \frac{3}{8} = P(A \cap C) \quad \Rightarrow \text{A and C are independent.}$$

Again
$$P(A)\cdot P(B) = \frac{1}{4} \neq P(A \cap B) \quad \Rightarrow \text{A and B are not independent.}$$

$$P(A \mid B) = \frac{P(A \cap B)}{P(B)} = 0 \quad \text{and} \quad P(A \mid C) = \frac{P(A \cap C)}{P(C)} = \frac{3/8}{3/4} = \frac{1}{2}.$$

Example 5·17. *A random variable X has the probability law :*

$$dF(x) = \frac{x}{b^2}\cdot e^{-x^2/2b^2}\,dx, 0 \le x < \infty$$

Find the distance between the quartiles and show that the ratio of this distance to the standard deviation of X is independent of the parameter 'b'.

Solution. If Q_1 and Q_3 are the first and third quartiles, respectively, we have

$$\int_0^{Q_1} f(x)\,dx = \frac{1}{4} \Rightarrow \frac{1}{b^2}\int_0^{Q_1} xe^{-x^2/2b^2}\,dx = \frac{1}{4}. \text{ Put } y = \frac{x^2}{2b^2} \text{ so that } dy = \frac{x}{b^2}dx.$$

$$\therefore \int_0^{Q_1^2/2b^2} e^{-y}\,dy = \frac{1}{4} \Rightarrow \left|\frac{e^{-y}}{-1}\right|_0^{Q_1^2/2b^2} = \frac{1}{4} \Rightarrow 1 - e^{-Q_1^2/2b^2} = \frac{1}{4} \Rightarrow e^{-Q_1^2/2b^2} = \frac{3}{4}$$

Thus
$$-\frac{Q_1^2}{2b^2} = \log\frac{3}{4} \Rightarrow \frac{Q_1^2}{2b^2} = \log\frac{4}{3} \Rightarrow Q_1 = b\sqrt{2}\sqrt{\log(4/3)}$$

Again we have $\int_0^{Q_3} f(x)\,dx = \frac{3}{4}$ which, on proceeding similarly, will give

$$1 - e^{-Q_3^2/2b^2} = \frac{3}{4} \Rightarrow e^{-Q_3^2/2b^2} = \frac{1}{4} \Rightarrow Q_3 = b\sqrt{2}\sqrt{\log 4}$$

The distance between the quartiles is : $Q_3 - Q_1 = b\sqrt{2}\left\{\sqrt{\log 4} - \sqrt{\log(4/3)}\right\}$

$$\mu_1' = \int_0^\infty x\,f(x)\,dx = \int_0^\infty x\frac{x}{b^2}e^{-x^2/2b^2}\,dx = \int_0^\infty b\sqrt{2}y^{1/2}e^{-y}\,dy$$

$$= b\sqrt{2}\int_0^\infty e^{-y}y^{\frac{3}{2}-1}\,dy = b\sqrt{2}\,\Gamma(3/2) = b\sqrt{2}\,\frac{1}{2}\Gamma(1/2) = b\sqrt{\pi/2}$$

$$\mu_2' = \int_0^\infty x^2 f(x)\,dx = \int_0^\infty x^2 \frac{x}{b^2} e^{-x^2/2b^2}\,dx$$

$$= 2b^2 \int_0^\infty y\,e^{-y}\,dy, (y = x^2/2b^2) = 2b^2\,\Gamma(2) = 2b^2.\ 1\,! = 2b^2$$

$$\therefore \quad \sigma^2 = \mu_2 = \mu_2' - \mu_1'^2 = 2b^2 - b^2\,\frac{\pi}{2} = b^2\left(2 - \frac{\pi}{2}\right) \quad \Rightarrow \quad \sigma = b\sqrt{\{2-(\pi/2)\}}$$

Hence, $\dfrac{Q_3 - Q_1}{\sigma} = \dfrac{\sqrt{2}\left[\sqrt{\log 4} - \sqrt{\log(4/3)}\right]}{\sqrt{\{2-(\pi/2)\}}}$, which is independent of parameter 'b'.

Example 5·18. *For the following probability distribution :*

$$dF = y_0 \cdot\ e^{-|x|}\,dx, \ -\infty < x < \infty$$

show that : $\quad y_0 = \dfrac{1}{2} \qquad \mu_1' = 0, \qquad \sigma = \sqrt{2}$ *and mean deviation about mean* = 1.

Solution. Since $\displaystyle\int_{-\infty}^{\infty} f(x)\,dx = 1, \quad y_0 \int_{-\infty}^{\infty} e^{-|x|}\,dx = 1$

$$\Rightarrow 2y_0 \int_0^\infty e^{-|x|}\,dx = 1, \qquad (\text{since } e^{-|x|} \text{ is an even function of } x)$$

$$\Rightarrow \ 2y_0 \int_0^\infty e^{-x}\,dx = 1 \quad \Rightarrow \quad 2y_0 \left|\frac{e^{-x}}{-1}\right|_0^\infty = 1 \quad \Rightarrow \quad 2y_0 = 1 \quad \Rightarrow \quad y_0 = \frac{1}{2}.$$

$$\mu_1' \text{ (about origin)} = \int_{-\infty}^{\infty} x\,f(x)\,dx = \frac{1}{2}\int_{-\infty}^{\infty} x\,e^{-|x|}\,dx = 0$$

[since the integrand $x\,e^{-|x|}$ is an odd function of x]

$$\mu_2' = \int_{-\infty}^{\infty} x^2 f(x)\,dx = \frac{1}{2}\int_{-\infty}^{\infty} x^2 e^{-|x|}dx = \frac{1}{2}\cdot 2\int_0^\infty x^2 e^{-|x|}dx$$

[since the integrand $x^2\,e^{-|x|}$ is an even function of x]

$$\therefore \qquad \mu_2' = \int_0^\infty x^2 e^{-x} dx = \Gamma\,(3) = 2\,! = 2. \qquad \text{[On using Gamma Integral]}$$

Hence, $\qquad \sigma^2 = \mu_2 = \mu_2' - \mu_1'^2 = 2$

$$\text{M.D. (about mean)} = \int_{-\infty}^{\infty} \left|x - \text{mean}\right| f(x)\,dx = \frac{1}{2}\int_{-\infty}^{\infty} |x|\,e^{-|x|}dx \qquad [\because \text{ Mean} = \mu_1' = 0]$$

$$= \frac{1}{2}\cdot 2\int_0^\infty |\,x\,|\,e^{-|x|}\,dx = \int_0^\infty x\,e^{-x}\,dx = \Gamma\,(2) = 1.$$

Example 5·19. *The elementary probability law of a continuous r.v. is :*

$$f(x) = y_0\,e^{-b(x-a)}, \ a \le x < \infty, \ b > 0, \text{ where } a, b \text{ and } y_0 \text{ are constants.}$$

Show that $y_o = b = 1/\sigma$ *and* $a = m - \sigma$, *where* m *and* σ *are respectively the mean and standard deviation of the distribution. Show also that* $\beta_1 = 4$ *and* $\beta_2 = 9$.

Solution. Since total probability is unity,

$$\int_a^\infty f(x)\,dx = 1 \Rightarrow y_0 \int_a^\infty e^{-b(x-a)}\,dx = 1 \Rightarrow y_0 \left| \frac{e^{-b(x-a)}}{-b} \right|_a^\infty = 1 \Rightarrow y_0\left(\frac{1}{b}\right) = 1 \Rightarrow y_0 = b.$$

$$\mu_r' = (r\text{th moment about the point } `x = a\text{'}) = \int_a^\infty (x-a)^r f(x)\,dx$$

$$= b\int_a^\infty (x-a)^r e^{-b(x-a)}\,dx = b\int_a^\infty t^r e^{-bt}\,dt \qquad [\text{On putting } x - a = t]$$

$$= b\,\frac{\Gamma(r+1)}{b^{r+1}} = \frac{r!}{b^r} \qquad [\text{Using Gamma Integral}]$$

In particular $\mu_1' = \dfrac{1}{b}$, $\mu_2' = \dfrac{2}{b^2}$, $\mu_3' = \dfrac{6}{b^3}$, $\mu_4' = \dfrac{24}{b^4}$

$$m = \text{Mean} = a + \mu_1' = a + \frac{1}{b} \quad \text{and} \quad \sigma^2 = \mu_2 = \mu_2' - \mu_1'^2 = \frac{1}{b^2}$$

$$\therefore \quad \sigma = \frac{1}{b} \quad \text{and} \quad m = a + \frac{1}{b} = a + \sigma\,.$$

Hence, $y_0 = b = \dfrac{1}{\sigma}$ and $a = m - \dfrac{1}{b} = m - \sigma.$

Also, $\mu_3 = \mu_3' - 3\mu_2'\,\mu_1' + 2\mu_1'^3$

$$= \frac{1}{b^3}(6 - 3.2 + 2) = \frac{2}{b^3} = 2\sigma^3$$

and $\mu_4 = \mu_4' - 4\,\mu_3'\,\mu_1' + 6\,\mu_2'\,\mu_1'^2 - 3\,\mu_1'^4$

$$= \frac{1}{b^4}(24 - 4.6.1 + 6.2.1 - 3) = \frac{9}{b^4} = 9\,\sigma^4$$

Hence, $\beta_1 = \dfrac{\mu_3^2}{\mu_2^3} = \dfrac{4\sigma^6}{\sigma^6} = 4$ and $\beta_2 = \dfrac{\mu_4}{\mu_2^2} = \dfrac{9\sigma^4}{\sigma^4} = 9.$

Example 5·20. *Show that for the distribution :*

$$dF = y_0\left(1 - \frac{x^2}{a^2}\right)^{-p} dx\ ;\ -a \le x \le a,\ 0 < p < 1.$$

$$\mu_r = \frac{(r-1)\,a^2}{r+1-2p}\,\mu_{r-2}. \quad \textit{Express `a' and `p' in terms of } \sigma \textit{ and } \beta_2.$$

Solution. Mean $= \mu_1' = y_0 \displaystyle\int_{-a}^{a} x\left(1 - \frac{x^2}{a^2}\right)^{-p} dx = 0$

($\because$ Integrand is an odd function of x.)

$$\therefore \quad \mu_r = \mu_r' = y_0 \int_{-a}^{a} x^r \left(1 - \frac{x^2}{a^2}\right)^{-p} dx$$

$$= y_0\, a^2 \int_{-a}^{a} x^{r-2} \left(\frac{x^2}{a^2} - 1 + 1\right)\left(1 - \frac{x^2}{a^2}\right)^{-p} dx$$

$$= a^2\, y_0 \int_{-a}^{a} x^{r-2} \left(1 - \frac{x^2}{a^2}\right)^{-p} dx - a^2 y_0 \int_{-a}^{a} x^{r-2} \left(1 - \frac{x^2}{a^2}\right)^{1-p} dx$$

$$= a^2\, \mu_{r-2} - a^2 y_0 \left\{ \left| \left(1 - \frac{x^2}{a^2}\right)^{1-p} \frac{x^{r-1}}{r-1} \right|_{-a}^{a} + \frac{2(1-p)}{a^2} \int_{-a}^{a} \frac{x^r}{r-1} \left(1 - \frac{x^2}{a^2}\right)^{-p} dx \right\}$$

$$= a^2\, \mu_{r-2} - \frac{2(1-p)}{r-1} \mu_r$$

$$\Rightarrow \quad \mu_r + \frac{2(1-p)}{r-1}\, \mu_r = a^2\, \mu_{r-2} \quad \Rightarrow \quad \mu_r = \frac{(r-1)\, a^2}{r+1-2p}\, \mu_{r-2} \qquad \dots (*)$$

Putting $r = 2$ and 4 in (*), we get, respectively :

$$\mu_2 = \sigma^2 = \frac{a^2}{3-2p}\, \mu_0 = \frac{a^2}{3-2p} \quad \Rightarrow \quad a^2 = (3 - 2p)\, \sigma^2 \qquad \dots (**)$$

$$\mu_4 = \frac{3a^2}{5-2p}\, \mu_2 = \frac{3a^4}{(5-2p)(3-2p)} \qquad \text{[From (**)]} \qquad \dots (***)$$

$$\therefore \quad \beta_2 = \frac{\mu_4}{\mu_2^{\,2}} = \frac{3a^4}{(5-2p)(3-2p)} \cdot \frac{(3-2p)^2}{a^4} = \frac{3(3-2p)}{5-2p} \qquad \text{[From (**) and (***)]}$$

$$\Rightarrow \quad (5 - 2p)\, \beta_2 = 9 - 6p \quad \Rightarrow \quad 2p\, (\beta_2 - 3) = 5\beta_2 - 9 \quad \Rightarrow \quad p = \frac{5\beta_2 - 9}{2(\beta_2 - 3)}$$

5·4·3. Continuous Distribution Function

Definition. *If X is a continuous random variable with the p.d.f. f(x), then the function :*

$$F_X(x) = P\,(X \le x) = \int_{-\infty}^{x} f(t)\, dt, \; -\infty < x < \infty \qquad \dots (5\cdot 12)$$

is called the distribution function (d.f.) or sometimes the cumulative distribution function (c.d.f.) of the random variable X.

Properties of Distribution Function

1. $0 \le F(x) \le 1, \; -\infty < x < \infty$.
2. From analysis (Riemann integral), we know that

$$F'(x) = \frac{d}{dx} F(x) = f\,(x) \ge 0 \qquad [\because f(x) \text{ is } p.d.f.]$$

$\Rightarrow$ $F(x)$ is non-decreasing function of x.

3. $$F(-\infty) = \lim_{x \to -\infty} F(x) = \lim_{x \to -\infty} \int_{-\infty}^{x} f(x)\,dx = \int_{-\infty}^{-\infty} f(x)\,dx = 0$$

and $$F(+\infty) = \lim_{x \to \infty} F(x) = \lim_{x \to \infty} \int_{-\infty}^{x} f(x)\,dx = \int_{-\infty}^{\infty} f(x)\,dx = 1$$

$\Rightarrow$ $$0 \le F(x) \le 1.$$

4. $F(x)$ is a continuous function of x on the right.
5. The discontinuities of $F(x)$ are at the most countable.
6. It may be noted that

$$P(a \le X \le b) = \int_a^b f(x)\,dx = \int_{-\infty}^{b} f(x)\,dx - \int_{-\infty}^{a} f(x)\,dx$$
$$= P(X \le b) - P(X \le a) = F(b) - F(a) \quad \text{...(5·12a)}$$

Similarly, $$P(a < X < b) = P(a < X \le b) = P(a \le X < b) = \int_a^b f(t)\,dt \quad \text{...(5·12b)}$$

7. $$F'(x) = \frac{d}{dx}F(x) = f(x) \quad \Rightarrow \quad dF(x) = f(x)dx$$

$dF(x)$ is known as probability differential of X.

Remarks:

1. It may be pointed out that the properties (2), (3) and (4) above uniquely characterise the distribution functions. This means that any function $F(x)$ satisfying (2) to (4) is the distribution function of some random variable, and any function $F(x)$ violating any one or more of these three properties cannot be the distribution function of any random variable.
2. Often, one can obtain a *p.d.f.* from a distribution function $F(x)$ by differentiating $F(x)$, provided the derivative exists, *e.g.*, consider :

$$F_X(x) = \begin{cases} 0, & \text{for } x < 0 \\ x, & \text{for } 0 \le x \le 1 \\ 1, & \text{for } x > 1 \end{cases}$$

The graph of $F(x)$ is given in Fig. (5·9) by bold lines. Obviously we see that $F(x)$ is continuous from right as stipulated in (4) and we also see that $F(x)$ is not continuous at $x = 0$ and $x = 1$ and hence is not derivable at $x = 0$ and $x = 1$.

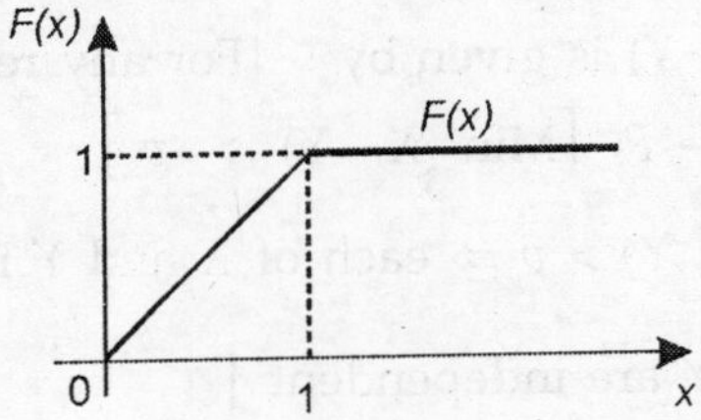

Fig. 5·9

Differentiating $F(x)$ *w.r.* to x, we get

$$\frac{d}{dx}F(x) = \begin{cases} 1, & 0 < x < 1 \\ 0, & \text{otherwise} \end{cases}$$

[Note the strict inequality in $0 < x < 1$, since $F(x)$ is not derivable at $x = 0$ and $x = 1$.]

Let us define : $$f(x) = \begin{cases} 1, & 0 < x < 1 \\ 0, & \text{otherwise} \end{cases}$$

Then, $f(x)$ is a *p.d.f.* for F.

Example 5·21. *Verify that the following is a distribution function :*

$$F(x) = \begin{cases} 0 & , \quad x < -a \\ \dfrac{1}{2}\left(\dfrac{x}{a}+1\right) & , -a \le x \le a \\ 1 & , \quad x > a \end{cases}$$

Solution. Obviously the properties (*i*), (*ii*), (*iii*) and (*iv*) are satisfied. Also we observe that $F(x)$ is continuous at $x = a$ and $x = -a$ as well.

$$\frac{d}{dx}F(x) = \begin{cases} \dfrac{1}{2a}, & -a \le x \le a \\ 0, & \text{otherwise} \end{cases} = f(x), \text{ say}$$

In order that $F(x)$ is a distribution function, $f(x)$ must be a *p.d.f.* Thus we have to show that

$$\int_{-\infty}^{\infty} f(x)\, dx = 1.$$

$$\text{Now} \quad \int_{-\infty}^{\infty} f(x)\, dx = \int_{-a}^{a} f(x)\, dx = \frac{1}{2a}\int_{-a}^{a} 1 \cdot dx = 1.$$

Hence $F(x)$ is a distribution function.

Example 5·22. *If X and Y are independent random variables with the distribution functions F_X and F_Y respectively, obtain the distribution functions of :*

(a) U = Max (X, Y) and *(b) V = Min (X, Y).*

Hence, obtain their p.d.f.'s.

Solution.

(*a*) The distribution function $G_U(\bullet)$ of $U = \text{Max}(X, Y)$ is given by : [For any real U]

$$G_U(u) = P(U \le u) = P\left[\text{Max}(X, Y) \le u\right] = P\left[(X \le u) \cap (Y \le u)\right]$$

$$\left[\because \text{Max}(X, Y) \le u \Rightarrow \text{each of X and Y is } \le u\right]$$

$$= P(X \le u) \cdot P(Y \le u) \qquad \left[\because X \text{ and } Y \text{ are independent}\right]$$

$$\therefore \qquad G_U(u) = F_X(u) \cdot F_Y(u) \qquad \text{... (1)}$$

(*b*) The distribution function $H_V(\bullet)$ of $V = \text{Min}(X, Y)$ is given by : [For any real v]

$$H_V(v) = P(V \le v) = P\left[\text{Min}(X, Y) \le v\right] = 1 - P\left[\text{Min}(X, Y) > v\right]$$

$$= 1 - P\left[(X > v) \cap (Y > v)\right] \quad \left[\because \text{Min}(X, Y) > v \Rightarrow \text{each of } X \text{ and } Y \text{ is } > v\right]$$

$$= 1 - P(X > v) \cdot P(Y > v) \quad \left[\because X \text{ and } Y \text{ are independent}\right]$$

$$= 1 - \left[1 - P(X \le v)\right] \cdot \left[1 - P(Y \le v)\right]$$

$$\therefore \qquad H_V(v) = 1 - \left[1 - F_X(v)\right]\left[1 - F_Y(v)\right] \qquad \text{... (2)}$$

If F_X and F_Y have p.d.f.'s f_X and f_Y respectively, then differentiating (1) and (2), the p.d.f.'s of U and V are given by :

$$g_U(u) = f_X(u) \cdot F_Y(u) + F_X(u) \cdot f_Y(u)$$

and $$h_V(v) = f_X(v)\,[1 - F_Y(v)] + f_Y(v)\,[1 - F_X(v)]$$

Example 5·23. *The diameter, say X, of an electric cable, is assumed to be a continuous random variable with p.d.f. :* $f(x) = 6x(1-x),\ 0 \le x \le 1$ ... (*)

(i) Check that f (x) is a p.d.f., ; *(ii) Obtain an expression for the c.d.f. of X,*

(iii) Compute $P\left(X \le \frac{1}{2} \,\middle|\, \frac{1}{3} \le X \le \frac{2}{3}\right)$, *and*

(iv) Determine the number k such that P(X < k) = P (X > k).

Solution.

(*i*) Since $\int_0^1 f(x)\,dx = \int_0^1 6x(1-x)\,dx = 6\left|\frac{x^2}{2} - \frac{x^3}{3}\right|_0^1 = 1$, $f(x)$ is a *p.d.f.*

$$(ii)\ F(x) = \begin{cases} 0, & \text{if } x \le 0 \\ \int_0^x 6t(1-t)\,dt = (3x^2 - 2x^3), & 0 < x \le 1 \\ 1, & \text{if } x > 1 \end{cases}$$

$$(iii)\ P\left(X \le \frac{1}{2} \,\middle|\, \frac{1}{3} \le X \le \frac{2}{3}\right) = \frac{P\left(\frac{1}{3} \le X \le \frac{1}{2}\right)}{P\left(\frac{1}{3} \le X \le \frac{2}{3}\right)} = \frac{\int_{1/3}^{1/2} 6x(1-x)\,dx}{\int_{1/3}^{2/3} 6x(1-x)\,dx} = \frac{11/54}{13/27} = \frac{11}{26}$$

(*iv*) We have $P(X < k) = P(X > k)$

$$\Rightarrow \int_0^k 6x(1-x)\,dx = \int_k^1 6x(1-x)\,dx \quad \text{or} \quad 3k^2 - 2k^3 = 3(1-k^2) - 2(1-k^3)$$

$$\Rightarrow 4k^3 - 6k^2 + 1 = 0 \quad \Rightarrow \quad k = \frac{1}{2}, \frac{1 \pm \sqrt{3}}{2}$$

The only admissible value of k in the given range is $\frac{1}{2}$. Hence the value of k is $\frac{1}{2}$.

Note 1. $k = \frac{1}{2}$, gives the value of the median of the distribution.

2. In fact, $f(x)$ is symmetric about $x = \frac{1}{2}$ because $f\left(\frac{1}{2} - x\right) = f\left(\frac{1}{2} + x\right)$

Example 5·24. *Let X be a continuous random variable with p.d.f. given by :*

$$f(x) = \begin{cases} kx, & 0 \le x < 1 \\ k, & 1 \le x < 2 \\ -kx + 3k, & 2 \le x < 3 \\ 0, & \text{elsewhere} \end{cases}$$

(i) Determine the constant k, *(ii) Determine F(x), the c.d.f., and*

(iii) If x_1, x_2 and x_3 are three independent observations from X, what is the probability that exactly one of these three numbers is larger than 1·5 ?

Solution.

(i) Since $f(x)$ in the *p.d.f.* of X, we have :

$$\int_0^3 f(x)\,dx = \int_0^1 f(x)\,dx + \int_1^2 f(x)\,dx + \int_2^3 f(x)\,dx = 1$$

$$\Rightarrow \qquad \int_0^1 k\,x\,dx + \int_1^2 k\,dx + \int_2^3 (-kx + 3k) = 1$$

$$\therefore \qquad \left| k\frac{x^2}{2}\right|_0^1 + \left| kx \right|_1^2 + \left| -k\cdot\frac{x^2}{2} + 3kx\right|_2^3 = 1 \quad \Rightarrow \quad k = \frac{1}{2}.$$

(ii) For any x such that $-\infty < x < 0$;

$$F(x) = \int_{-\infty}^x 0\cdot dt = 0$$

For any x, where $0 \le x < 1$;

$$F(x) = \int_{-\infty}^0 0\cdot dt + \int_0^x \frac{t}{2}\,dt = \frac{x^2}{4}$$

For x, where $1 \le x < 2$,

$$F(x) = \int_{-\infty}^0 0\cdot dt + \int_0^1 \frac{t}{2}\,dt + \int_1^x \frac{1}{2}\,dt = \frac{2x-1}{4}$$

For any x, where $2 \le x < 3$,

$$F(x) = \int_{-\infty}^0 0\cdot dt + \int_0^1 \frac{t}{2}\,dt + \int_1^2 \frac{1}{2}\,dt + \int_2^x \left(-\frac{t}{2} + \frac{3}{2}\right)dt$$

$$= \frac{1}{4} + \left(1 - \frac{1}{2}\right) + \left(-\frac{x^2}{4} + \frac{3x}{2} - 2\right) = -\frac{x^2}{4} + \frac{3x}{2} - \frac{5}{4}$$

For any x, where $3 \le x < \infty$

$$F(x) = \int_{-\infty}^0 0\cdot dt + \int_0^1 \frac{t}{2}\,dt + \int_1^2 \frac{1}{2}\,dt + \int_2^3 \left(-\frac{t}{2} + \frac{3}{2}\right)dt + \int_3^x 0\cdot dt$$

$$= \frac{1}{4} + \left(1 - \frac{1}{2}\right) + \left(-\frac{9}{4} + \frac{9}{2} + 1 - 3\right) = 1$$

Hence, the distribution function $F(x)$ is given by :

$$F(x) = \begin{cases} 0, & \text{for } -\infty \le x < 0 \\ \dfrac{x^2}{4}, & \text{for } 0 \le x < 1 \\ \dfrac{2x-1}{4}, & \text{for } 1 \le x < 2 \\ -\dfrac{x^2}{4} + \dfrac{3x}{2} - \dfrac{5}{4}, & \text{for } 2 \le x < 3 \\ 1, & \text{for } 3 \le x < \infty \end{cases}$$

(*iii*) The probability that X is larger than 1·5 is given by :

$$P\,(X > 1\cdot5) = 1 - P\,(X \le 1\cdot5) = 1 - F\,(1\cdot5) = \left(1 - \frac{3-1}{4}\right) = \frac{1}{2}.$$

$\therefore$ The probability that X is not larger than 1·5 = $P\,(X \le 1\cdot5) = 1 - \dfrac{1}{2} = \dfrac{1}{2}.$

Hence, out of three numbers x_1, x_2 and x_3, the probability that exactly one is larger than 1·5 is : $\quad 3\left(\frac{1}{2}\right)\left(\frac{1}{2}\right)\left(\frac{1}{2}\right) = \frac{3}{8}$

Example 5·25. *A bombing plane carrying three bombs flies directly above a railroad track. If a bomb falls within 40 metres of track, the track will be sufficiently damaged to disrupt the traffic. With a certain bomb site the points of impact of a bomb have the probability density function :*

$$f(x) = \begin{cases} \dfrac{100+x}{10{,}000}, & \text{when, } -100 \le x < 0 \\ \dfrac{100-x}{10{,}000}, & \text{when, } 0 \le x < 100 \\ 0, & \text{elsewhere} \end{cases}$$

where x represents the vertical deviation (in metres) from the aiming point, which is the track in this case. Find the distribution function. If all the three bombs are used, what is the probability that the track will be damaged ?

Solution. If $x < -100$, $\quad F\,(x) = 0$

If, $-100 \le x < 0$, $F(x) = \displaystyle\int_{-100}^{x} \frac{100+t}{10{,}000}\,dt = \frac{1}{10^4}\left(100x + \frac{x^2}{2} + \frac{10^4}{2}\right)$

If $0 \le x < 100$,

$$F(x) = \int_{-\infty}^{-100} 0\,dt + \int_{-100}^{0} \frac{100+t}{10{,}000}\,dt + \int_{0}^{x} \frac{100-t}{10{,}000}\,dt$$

$$= \frac{1}{10{,}000}\left[\left|100\,t + \frac{t^2}{2}\right|_{-100}^{0}\right] + \frac{1}{10{,}000}\left[\left|100\,t - \frac{t^2}{2}\right|_{0}^{x}\right]$$

$$= \frac{1}{10{,}000}\left(100x - \frac{x^2}{2} + \frac{10^4}{2}\right)$$

and if $x > 100$,

$$F(x) = \int_{-\infty}^{-100} 0\,dt + \int_{-100}^{0} \frac{100+t}{10{,}000}\,dt + \int_{0}^{100} \frac{100-t}{10{,}000}\,dt + \int_{100}^{x} 0\,dt = 1$$

Probability that the track will be damaged by a bomb is :

$$P\left(|X| < 40\right) = P\left(-40 < X < 40\right) - \int_{-40}^{0} \frac{100+x}{10,000}\, dx + \int_{0}^{40} \frac{100-x}{10,000}\, dx = \frac{16}{25}$$

$\therefore$ The probability that a bomb will not damage the track $= 1 - \dfrac{16}{25} = \dfrac{9}{25}$.

If three bombs are dropped, the required probability p that the track will be damaged is given by :

$$\begin{aligned} p &= P \text{ [At least one of the three bombs damages the track.]} \\ &= 1 - P \text{ [None of the three bombs damages the track.]} \\ &= 1 - \left(\frac{9}{25}\right)^3 = 1 - 0{\cdot}0467 = 0{\cdot}9533. \end{aligned}$$

Example 5·26. *Suppose that the time in minutes that a person has to wait at a certain bus stop for a bus is found to be a random phenomenon, with a probability function specified by the distribution function:*

$$F(x) = \begin{cases} 0, & x < 0 \\ \dfrac{x}{8}, & 0 \le x < 2 \\ \dfrac{x^2}{16}, & 2 \le x < 4 \\ 1, & x \ge 4 \end{cases} \qquad \dots (*)$$

(*i*) *Is the distribution function continuous ? If so, give the formula for its probability density function.*

(*ii*) *What is the probability that a person will have to wait*

(*a*) *more than 2 minutes,* (*b*) *less than 2 minutes, and* (*c*) *between 1 and 2 minutes ?*

(*iii*) *What is the conditional probability that the person will have to wait for a bus for*

(*a*) *more than 2 minutes, given that it is more than 1 minute,*

(*b*) *less than 2 minutes given that it is more than 1 minute ?*

Solution.

(*i*) Since the value of the distribution function is the same at the points $x = 0$, $x = 1$, $x = 2$ and $x = 4$ given by the two forms of $F(x)$ for $x < 0$ and $0 \le x < 2$; $0 \le x < 2$ and $2 \le x < 4$; $2 \le x < 4$ and $x \ge 4$, the distribution function is continuous.

$$\text{Probability density function} = f(x) = \frac{d}{dx} F(x)$$

$$\therefore \qquad f(x) = \begin{cases} 0, & \text{for } x < 0 \\ \dfrac{1}{8}, & \text{for } 0 \le x < 2 \\ \dfrac{x}{8}, & \text{for } 2 \le x < 4 \\ 0, & \text{for } x \ge 4 \end{cases}$$

(*ii*) Let the random variable X represent the waiting time in minutes, then

(*a*) Required probability $= P(X > 2) = 1 - P(X \le 2) = 1 - F(2) = 1 - \frac{2^2}{16} = \frac{3}{4}$. [From(*)]

(*b*) Required probability $= P(X < 2) = P(X \le 2) - P(X = 2) = F(2) = \frac{1}{4}$.

(Since, the probability that a continuous variable takes a fixed value is zero.)

(*c*) Required probability $= P(1 < X < 2) = F(2) - F(1) = \frac{1}{4} - \frac{1}{8} = \frac{1}{8}$. [From (*)]

(*iii*) Let A denote the event that a person has to wait for more than 2 minutes and B the event that he has to wait for more than 1 minute. Then $P(A) = P(X > 2) = \frac{3}{4}$.

$$P(B) = P(X > 1) = 1 - P(X \le 1) = 1 - F(1) = 1 - \frac{1}{8} = \frac{7}{8}. \quad \text{[From (*)]}$$

$$P(A \cap B) = P(X > 2 \cap X > 1) = P(X > 2) = \frac{3}{4}$$

(*a*) Required probability $= P(A \mid B) = \dfrac{P(A \cap B)}{P(B)} = \dfrac{3/4}{7/8} = \dfrac{6}{7}$

(*b*) Required probability $= P(\bar{A} \mid B) = \dfrac{P(\bar{A} \cap B)}{P(B)}$

$$P(\bar{A} \cap B) = P(X \le 2 \cap X > 1) = P(1 < X \le 2) = F(2) - F(1) = \frac{1}{4} - \frac{1}{8} = \frac{1}{8} \quad \text{[From (*)]}$$

OR $\quad P(\bar{A} \cap B) = P(B) - P(A \cap B) = \dfrac{7}{8} - \dfrac{3}{4} = \dfrac{1}{8}$

$$\therefore \quad P(\bar{A} \mid B) = \frac{1/8}{7/8} = \frac{1}{7}$$

Example 5·27. *For any continuous distribution, show that the mean deviation is least when measured from the median.*

Solution. If $f(x)$ is probability density function of a random variable X, $a \le x \le b$, then mean deviation $M(A)$, say, about the point $x = A$ is given by :

$$M(A) = \int_a^b |x - A|\, f(x)\, dx$$

$$= \int_a^A |x - A|\, f(x)\, dx + \int_A^b |x - A|\, f(x)\, dx$$

$$= \int_a^A (A - x)\, f(x)\, dx + \int_A^b (x - A)\, f(x)\, dx \qquad \text{... } (i)$$

We want to find the value of 'A' so that $M(A)$ is minimum. From the principle of maximum and minimum in differential calculus, $M(A)$ will be minimum for variations in A if

$$\frac{\partial M(A)}{\partial A} = 0 \quad \text{and} \quad \frac{\partial^2 M(A)}{\partial A^2} > 0 \qquad \text{... } (ii)$$

Differentiating (*i*) *w.r.* to '*A*' under the integral sign, since the functions $(A - x) f(x)$ and $(x - A) f(x)$ vanish at the point $x = A$, we get

$$\frac{\partial M(A)}{\partial A} = \int_a^A f(x)\,dx - \int_A^b f(x)\,dx \qquad \text{... } (iii)$$

[$\because$ If $f(x, \theta)$ is a continuous function of both variables x and θ, possessing continuous partial derivatives $\frac{\partial^2 f}{\partial x\,\partial\theta}, \frac{\partial^2 f}{\partial\theta\,\partial x}$ and a and b are differentiable functions of θ, then,

$$\frac{\partial}{\partial\theta}\left\{\int_a^b f(x, \theta)\,dx\right\} = \int_a^b \frac{\partial f}{\partial\theta}\,dx + f(b, \theta)\frac{db}{d\theta} - f(a, \theta)\frac{da}{d\theta}\Bigg]$$

Also $$\frac{\partial M(A)}{\partial A} = \int_a^A f(x)\,dx - \left(1 - \int_a^A f(x)\,dx\right), \qquad \left[\because \int_a^b f(x)\,dx = 1\right]$$

$$= 2\int_a^A f(x)\,dx - 1 = 2F(A) - 1,$$

where $F(.)$ is the distribution function of X. Differentiating again *w.r.* to A, we get

$$\frac{\partial^2}{\partial A^2} M(A) = 2 f(A) \qquad \text{... } (iv)$$

$\frac{\partial M(A)}{\partial A} = 0$, on using (*iii*) gives $\int_a^A f(x)\,dx = \int_A^b f(x)\,dx$, *i.e.*, A is the median value.

Also from (*iv*), we find that $\frac{\partial^2 M(A)}{\partial A^2} > 0$, assuming that $f(x)$ does not vanish at the median value. Thus mean deviation is least when taken from median.

5·5. TWO-DIMENSIONAL RANDOM VARIABLES

So far we have defined only one random variable on a sample space. It is also possible to define more than one random variable on the same sample space. For example, we may be interested in recording the height and weight of every person in a certain educational institution. To describe such experiments mathematically we introduce the study of two random variables.

Definition. *Let X and Y be two random variables defined on the same sample space S, then the function (X, Y) that assigns a point in* R^2 *(= R × R), is called a two-dimensional random variable.*

Let (X, Y) be a two-dimensional random variable, defined on the sample space S and $\omega \in S$. The value of (X, Y) at ω is given by the pair of real numbers $\{X(\omega), Y(\omega)\}$. The notation $\{X \le a, Y \le b\}$ denotes the event of all elements $\omega \in S$, such that $X(\omega) \le a$ and $Y(\omega) \le b$. The probability of the event $\{X \le a, Y \le b\}$ will be denoted by $P(X \le a, Y \le b)$.

Let $A = \{a < X \le b\}$, $B = \{c < Y \le d)$ be two events. Then the event

$$\{a < X \le b, c < Y \le d\} = \{a < X \le b\} \cap \{c < Y \le d\} = A \cap B$$

$$\therefore \quad P(a < X \le b, c < Y \le d) = P(A \cap B)$$

Remarks:

1. A two-dimensional random variable is said to be discrete if it takes at most a countable number of points in R^2.

2. Two random variables X and Y are said to be jointly distributed if they are defined on the same probability space. The sample points consist of 2-tuples. If the joint probability function is denoted by $P_{XY}(x, y)$ then the probability of a certain event E is given by :

$$P_{XY}(x, y) = P[(X, Y) \in E] \quad \text{... (5·13)}$$

5·5·1. Two-dimensional or Joint Probability Mass Function

Let X and Y be random variables on a sample space S with respective image sets

$X(S) = \{x_1, x_2, ... x_n\}$ and $Y(S) = \{y_1, y_2, ... y_m\}$. We make the product set

$$X(S) \times Y(S) = \{x_1, x_2, ... x_n\} \times \{y_1, y_2, ... y_m\}$$

into a probability space by defining the probability of the ordered pair (x_i, y_i) to be $P(X = x_i, Y = y_i)$ which we write $p(x_i, y_i)$. The function p on $X(S) \times Y(S)$ defined by :

$P_{ij} = P(X = x_i \cap Y = y_i) = p(x_i, y_i)$ is called the *joint probability function* of X and Y and is usually represented in the form of the following table :

X \ Y	y_1	y_2	y_3	...	y_j	...	y_m	*Total*
x_1	p_{11}	p_{12}	p_{13}	...	p_{1j}	...	p_{1m}	$p_{1.}$
x_2	p_{21}	p_{22}	p_{23}	...	p_{2j}	...	p_{2m}	$p_{2.}$
x_3	p_{31}	p_{32}	p_{33}	...	p_{3j}	...	p_{3m}	$p_{3.}$
⋮	⋮	⋮	⋮	⋮	⋮	⋮	⋮	⋮
x_i	p_{i1}	p_{i2}	p_{i3}	...	p_{ij}	...	p_{im}	$p_{i.}$
⋮	⋮	⋮	⋮	⋮	⋮	⋮	⋮	⋮
x_n	p_{n1}	p_{n2}	p_{n3}	...	p_{nj}	...	p_{nm}	$p_{n.}$
Total	$p_{.1}$	$p_{.2}$	$p_{.3}$	...	$p_{.j}$	...	$p_{.m}$	1

Definition. *If (X, Y) is a two-dimensional discrete random variable, then the joint discrete function of X, Y, also called the joint probability mass function of X, Y, denoted by $p_{X, Y}$ is defined as :*

$$\left.\begin{aligned} p_{XY}(x_i, y_j) &= P(X = x_i, Y = y_j) \text{ for a value } (x_i, y_j) \text{ of } (X, Y) \\ \text{and} \quad p_{XY}(x_i, y_j) &= 0, \text{ otherwise.} \end{aligned}\right\} \quad \text{... (5·14)}$$

Remark. It may be noted that $\sum\sum p_{XY}(x_i, y_j) = 1$, where the summation is taken over all possible values of (X, Y).

Marginal Probability Function. Let (X, Y) be a discrete two-dimensional *r.v.* which takes up countable number of values (x_i, y_i). Then the probability distribution of X is determined as follows :

$$\begin{aligned} p_X(x_i) &= P(X = x_i) \\ &= P(X = x_i \cap Y = y_1) + P(X = x_i \cap Y = y_2) + \text{......} + P(X = x_i \cap Y = y_m) \end{aligned}$$

$$= p_{i1} + p_{i2} + ... + p_{ij} + ... + p_{im} = \sum_{j=1}^{m} p_{ij} = \sum_{j=1}^{m} p\left(x_i, y_j\right) = p_{i}. \quad ... (5·14\ a)$$

and is known as *marginal probability mass function or discrete marginal density function of X.*

Also, $$\sum_{i=1}^{n} p_{i}. = p_{1}. + p_{2}. + ... + p_{n}. = \sum_{i=1}^{n}\sum_{j=1}^{m} p_{ij} = 1$$

Similarly, we can prove that

$$p_Y\left(y_j\right) = P\left(Y = y_j\right) = \sum_{i=1}^{n} p_{ij} = \sum_{i=1}^{n} p\left(x_i, y_j\right) = p._j \quad ... (5·14\ b)$$

which is the *marginal probability mass function of Y.*

Conditional Probability Function

> **Definition.** *Let (X, Y) be a discrete two-dimensional random variable. Then the conditional discrete density function or the conditional probability mass function of X, given Y = y, denoted by $p_{X|Y}(x|y)$, is defined as :*
>
> $$p_{X|Y}(x|y) = \frac{P(X = x, Y = y)}{P(Y = y)}, \text{ provided } P(Y = y) \neq 0 \quad ... (5·14\ c)$$

Since for a fixed y,

$$\sum_i \frac{P\left(X=x_i, Y=y\right)}{P(Y=y)} = \frac{1}{P(Y=y)} \sum_i P\left(X = x_i, Y = y\right) = \frac{1}{P(Y=y)} P(Y=y) = 1,$$

it follows that the conditional discrete density function $f_{X|Y}(x|y)$ is a discrete density function, when considered as a function of the values of X.

The conditional probability mass function $p_{Y|X}(y|x)$ is similarly defined, *i.e.,*

$$p_{Y|X}(y|x) = \frac{P(X = x, Y = y)}{P(X = x)}$$

A necessary and sufficient condition for the discrete random variables X and Y to be *independent* is :

$$P\left(X = x_i, Y = y_j\right) = P\left(X = x_i\right) P\left(Y = y_j\right) \text{ for all values } \left(x_i, = y_j\right) \text{ of } (X, Y) \quad ... (5·14\ d)$$

5·5·2. Two-Dimensional Distribution Function

> **Definition.** *The distribution function of the two-dimensional random variable (X, Y) is a real valued function F defined for all real x and y by the relation :*
>
> $$F_{XY}(x, y) = P(X \leq x, Y \leq y). \quad ... (5·15)$$

Properties of Joint Distribution Function

1. (*i*) For the real numbers a_1, b_1, a_2 and b_2; ($a_1 < b_1$; $a_2 < b_2$),

$$P\left(a_1 < X \leq b_1, a_2 < Y \leq b_2\right) = F_{XY}\left(b_1, b_2\right) + F_{XY}\left(a_1, a_2\right) - F_{XY}\left(a_1, b_2\right) - F_{XY}\left(b_1, a_2\right)$$

[For proof see Example 5·49]

(*ii*) From the adjoining diagram for $a_1 < b_1, a_2 < b_2$; we have

$$P(X \le a_1, Y \le a_2) + P(a_1 \le X \le b_1, Y \le a_2) = P(X \le b_1, Y \le a_2)$$

$$\Rightarrow \quad P(X \le b_1, Y \le a_2) - P(X \le a_1, Y \le a_2) = P[a_1 \le X \le b_1, Y \le a_2] \ge 0$$

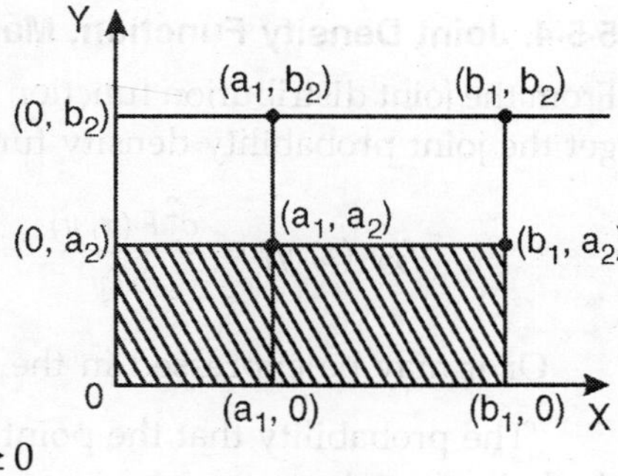

Fig. 5·9a

$$\Rightarrow \quad F(b_1, a_2) - F(a_1, a_2) \ge 0$$

$$\Rightarrow \quad F(b_1, a_2) \ge F(a_1, a_2) \qquad \ldots (*)$$

Similarly it follows that :

$$F(a_1, b_2) - F(a_1, a_2) = P(X \le a_1, a_2 \le Y \le b_2) \ge 0$$

$$\Rightarrow \quad F(a_1, b_2) \ge F(a_1, a_2) \qquad \ldots (**)$$

(*) and (**) show that $F(x, y)$ is monotonic non-decreasing function in both variables x and y.

2. $F(-\infty, y) = 0 = F(x, -\infty), \quad F(+\infty, +\infty) = 1.$

3. If the density function $f(x, y)$ is continuous at (x, y), $\dfrac{\partial^2 F}{\partial x \, \partial y} = f(x, y)$.

5·5·3. Marginal Distribution Functions

From the knowledge of joint distribution function $F_{XY}(x, y)$, it is possible to obtain the individual distribution functions, $F_X(x)$ and $F_Y(y)$ which are termed as marginal distribution function of X and Y, respectively, with respect to the joint distribution function $F_{XY}(x, y)$.

$$F_X(x) = P(X \le x) = P(X \le x, Y < \infty) = \lim_{y \to \infty} F_{XY}(x, y) = F_{XY}(x, \infty) \qquad \ldots (5·16)$$

Similarly, $$F_Y(y) = P(Y \le y) = P(X < \infty, Y \le y) = \lim_{x \to \infty} F_{XY}(x, y) = F_{XY}(\infty, y) \qquad \ldots 5·16(a)$$

$F_X(x)$ is termed as the marginal distribution function of X corresponding to the joint distribution function $F_{XY}(x, y)$ and similarly $F_Y(y)$ is called marginal distribution function of the random variable Y corresponding to the joint distribution function $F_{XY}(x, y)$.

In the case of jointly discrete random variables, the marginal distribution functions are given as :

$$F_X(x) = \sum_y P(X \le x, Y = y), \quad \text{and} \quad F_Y(y) = \sum_x P(X = x, Y \le y)$$

Similarly, in the case of jointly continuous random variables, the marginal distribution functions are given as :

$$F_X(x) = \int_{-\infty}^{x} \left\{ \int_{-\infty}^{\infty} f_{XY}(x, y)\, dy \right\} dx, \quad F_Y(y) = \int_{-\infty}^{y} \left\{ \int_{-\infty}^{\infty} f_{XY}(x, y)\, dx \right\} dy$$

5·5·4. Joint Density Function, Marginal Density Function

From the joint distribution function $F_{XY}(x, y)$ of two-dimensional continuous random variable, we get the joint probability density function by differentiation as follows :

$$f_{XY}(x, y) = \frac{\partial^2 F(x, y)}{\partial x \partial y} = \lim_{\delta x \to 0,\, \delta y \to 0} \frac{P\left(x \le X \le x + \delta x,\, y \le Y \le y + \delta y\right)}{\delta x\, \delta y} \qquad \text{... (5·17)}$$

Or it may be expressed in the following way also :

"The probability that the point (x, y) will lie in the infinitesimal rectangular region, of area $dx\, dy$ is given by :

$$P\left(x - \frac{1}{2}dx \le X \le x + \frac{1}{2}dx,\quad y - \frac{1}{2}dy \le Y \le y + \frac{1}{2}dy\right) = dF_{XY}(x, y) \qquad \text{... (5·17}a\text{)}$$

and is denoted by $f_{XY}(x, y)\, dx\, dy$, where the function $f_{XY}(x, y)$ is called the joint probability density function of X and Y.

The marginal probability functions of X and Y are given respectively as follows :

$$f_X(x) = \begin{cases} \displaystyle\sum_y p_{XY}(x, y) & \text{(for discrete variables)} \\ \displaystyle\int_{-\infty}^{\infty} f_{XY}(x, y)\, dy, & \text{(for continuous variables)} \end{cases} \qquad \text{... (5·17 } b\text{)}$$

and

$$f_Y(y) = \begin{cases} \displaystyle\sum_x p_{XY}(x, y) & \text{(for discrete variables)} \\ \displaystyle\int_{-\infty}^{\infty} f_{XY}(x, y)\, dx, & \text{(for continuous variables)} \end{cases} \qquad \text{... (5·17 } c\text{)}$$

The marginal density functions of X and Y can be obtained in the following manner also :

$$\left.\begin{aligned} f_X(x) &= \frac{dF_X(x)}{dx} = \int_{-\infty}^{\infty} f_{XY}(x, y)\, dy \\ \text{and} \quad f_Y(y) &= \frac{dF_Y(y)}{dy} = \int_{-\infty}^{\infty} f_{XY}(x, y)\, dx \end{aligned}\right\} \qquad \text{... (5·17 } d\text{)}$$

Important Remark. If we know the joint *p.d.f.* (*p.m.f.*) $f_{XY}(x, y)$ of two random variables X and Y, we can obtain the individual distributions of X and Y in the form of their marginal *p.d.f.'s* (*p.m.f.'s*) $f_X(x)$ and $f_Y(y)$ by using (5·17 *b*) and (5·17 *c*). However, the converse is not true, *i.e.*, *from the marginal distributions of two jointly distributed random variables, we cannot determine the joint distributions of these two random variables.*

To verify this, it will suffice to show that two different joint *p.m.f.'s* (*p.d.f.'s*) have the same marginal distribution for X and the same marginal distribution for Y. We give below two joint discrete probability distributions which have the same marginal distributions.

JOINT DISTRIBUTIONS HAVING SAME MARGINALS

Probability Distribution I

Y \ X	0	1	$f_Y(y)$
0	0·28	0·37	0·65
1	0·22	0·13	0·35
$f_X(x)$	0·50	0·50	1·00

Probability Distribution II

Y \ X	0	1	$f_Y(y)$
0	0·35	0·30	0·65
1	0·15	0·20	0·35
$f_X(x)$	0·50	0·50	1·00

As an illustration for continuous *r.v.'s*, let (X, Y) be continuous *r.v.* with joint *p.d.f.* :

$$f_{XY}(x, y) = x + y\ ;\ 0 \le (x, y) \le 1 \qquad \text{... (5·17 e)}$$

The marginal *p.d.f.'s* of X and Y are given by :

$$f_X(x) = \int_0^1 f(x, y)\, dy = \int_0^1 (x+y)\, dy = \left| xy + \frac{y^2}{2} \right|_0^1$$

$$\Rightarrow \qquad f_X(x) = x + \frac{1}{2}\ ;\ 0 \le x \le 1$$

$$\text{Similarly,} \qquad f_Y(y) = \int_0^1 f(x, y)\, dx = y + \frac{1}{2}\ ;\ 0 \le y \le 1 \qquad \text{... (5·17 f)}$$

Consider another continuous joint *p.d.f.* :

$$g(x, y) = \left(x + \frac{1}{2}\right)\left(y + \frac{1}{2}\right); 0 \le (x, y) \le 1 \qquad \text{... (5·17 g)}$$

Then marginal *p.d.f.'s* of X and Y are given by :

$$g_1(x) = \int_0^1 g(x, y)\, dy = \left(x + \frac{1}{2}\right)\int_0^1 \left(y + \frac{1}{2}\right) dy = \left(x + \frac{1}{2}\right)\left| \frac{y^2}{2} + \frac{1}{2} y \right|_0^1$$

$$\Rightarrow \qquad g_1(x) = x + \frac{1}{2}\ ;\quad 0 \le x \le 1$$

$$\text{Similarly,} \qquad g_2(y) = y + \frac{1}{2}\ ;\quad 0 \le y \le 1 \qquad \text{... (5·17 h)}$$

(5·17 *f*) and (5·17 *h*) imply that the two joint *p.d.f.'s* in (5·17 *e*) and (5·17 *g*) have the same marginal *p.d.f.'s* (5·17 *f*) or (5·17 *h*).

5·5·5. The Conditional Distribution Function and Conditional Probability Density Function

For two-diamensional random variable (X, Y), the joint distribution function $F_{XY}(x, y)$ for any real numbers *x* and *y* is given by :

$$F_{XY}(x, y) = P(X \le x, Y \le y)$$

Now let *A* be the event $(Y \le y)$ such that the event *A* is said to occur when *Y* assumes values upto and inclusive of *y*. Using conditional probabilities we may now write

$$F_{XY}(x, y) = \int_{-\infty}^{x} P(A \mid X = x)\, dF_X(x) \qquad \text{... (5·18)}$$

The *conditional distribution function* $F_{Y|X}(y \mid x)$ denotes the distribution function of *Y* when X has already assumed the particular value *x*. Hence,

$$F_{Y|X}(y \mid x) = P(Y \le y \mid X = x) = P(A \mid X = x)$$

Using this expression, the joint distribution function $F_{XY}(x, y)$ may be expressed in terms of the conditional distribution function as follows :

$$F_{XY}(x, y) = \int_{-\infty}^{x} F_{Y|X}(y \mid x)\, dF_X(x) \qquad \text{... (5·18 a)}$$

Similarly,
$$F_{XY}(x, y) = \int_{-\infty}^{y} F_{X|Y}(x \mid y)\, dF_Y(y) \qquad \text{... (5·18 b)}$$

The *conditional probability density function* of Y given X for two random variables X and Y which are jointly continuously distributed is defined as follows, for two real numbers x and y :

$$f_{Y|X}(y \mid x) = \frac{\partial}{\partial y} F(y \mid x) \qquad \text{... (5·19)}$$

Remarks:

1. If $f_X(x) > 0$, then $f_{Y|X}(y \mid x) = \dfrac{f_{XY}(x,y)}{f_X(x)}$.

Proof. We have

$$F_{XY}(x, y) = \int_{-\infty}^{x} f_{Y|X}(y \mid x)\, dF_X(x) = \int_{-\infty}^{x} f_{Y|X}(y \mid x)\, f_X(x)\, dx$$

Differentiating *w.r.* to x,
$$\frac{\partial}{\partial x} F_{XY}(x, y) = F_{Y|X}(y \mid x)\, f_X(x)$$

Differentiating *w.r.* to y, we get
$$\frac{\partial}{\partial y}\left\{\frac{\partial}{\partial x} F_{XY}(x, y)\right\} = f_{Y|X}(y \mid x)\, f_X(x)$$

$\Rightarrow$
$$f_{XY}(x, y) = f_{Y|X}(y \mid x)\, f_X(x)$$

$\therefore$
$$f_{Y|X}(y \mid x) = \frac{f_{XY}(x, y)}{f_X(x)}$$

2. If $f_Y(y) > 0$, then
$$f_{X|Y}(x \mid y) = \frac{f_{XY}(x, y)}{f_Y(y)}$$

3. In terms of the differentials, we have

$$P\left(x < X \le x + dx \mid y < Y \le y + dy\right) = \frac{P\left(x < X \le x + dx,\, y < Y \le y + dy\right)}{P\left(y < Y \le y + dy\right)} = \frac{f_{XY}(x,y)\, dx\, dy}{f_Y(y)\, dy}$$

$$= f_{X|Y}(x \mid y)\, dx$$

Hence, $f_{X|Y}(x \mid y)$ may be interpreted as the conditional density function of X on the assumption $Y = y$.

Note. The conditional probability $P(a \le X \le b \mid Y = y)$, as such is meaningless because for a continuous r.v. X, the probability at a point is zero, *i.e.*, $P(Y = y) = 0$, *i.e.*, the event $\{Y = y\}$ is an impossible event. However this difficulty is avoided, by considering '$Y = y$' as a limiting value :

$$(Y = y) = \lim_{\partial y \to 0}\left[y \le Y \le y + \delta y\right].$$

5·5·6. Stochastic Independence

Let us consider two random variables X and Y (of discrete or continuous type) with joint *p.d.f.* (*p.m.f.*) $f_{XY}(x, y)$ and marginal *p.d.f.*'s (*p.m.f.'s*) $f_X(x)$ and $g_Y(y)$, respectively. Then by the compound probability theorem :

$$f_{XY}(x, y) = f_X(x)\, g_{Y|X}(y \mid x)$$

where $g_{Y|X}(y \mid x)$ is the conditional *p.d.f.* of Y for given value of $X = x$.

If we assume that $g(y \mid x)$ does not depend on x, then by the definition of marginal *p.d.f.'s*, we get for continuous *r.v.'s* :

$$g(y) = \int_{-\infty}^{\infty} f(x, y)\, dx = \int_{-\infty}^{\infty} f_X(x)\, g(y \mid x)\, dx$$

$$= g(y \mid x) \int_{-\infty}^{\infty} f_X(x)\, dx \qquad \text{[since } g(y \mid x) \text{ does not depend on } x]$$

$$= g(y \mid x) \qquad [\because f(.) \text{ is } p.d.f. \text{ of } X]$$

Hence $\quad g(y) = g(y \mid x)$ and $f_{XY}(x, y) = f_X(x) \cdot g_Y(y)$ ... (*)

provided $g(y \mid x)$ does not depend on x. This motivates the following definition of independent random variables.

> ***Independent Random Variables.*** *Two r.v.'s X and Y with joint p.d.f. (p.m.f.) $f_{XY}(x,y)$ and marginal p.d.f.'s (p.m.f's) $f_X(x)$ and $g_Y(y)$, respectively, are said to be stochastically independent if and only if*
>
> $$f_{XY}(x, y) = f_X(x) \cdot g_Y(y) \qquad \text{... (5·20)}$$

Remarks:

1. In terms of the distribution function, we have the following definition :
Two jointly distributed random variables X and Y are stochastically independent if and only if their joint distribution function $F_{XY}(.,.)$ is the product of their marginal distribution function $F_X(.)$ and $G_Y(.)$, *i.e.*, if for real (x, y)

$$F_{X,Y}(x, y) = F_X(x) \cdot G_Y(y) \qquad \text{... (5·20 a)}$$

2. The variables which are not stochastically independent are said to be stochastically dependent.

Theorem 5·1. *Two random variables X and Y with joint p.d.f. f(x, y) are stochastically independent if and only if $f_{XY}(x, y)$ can be expressed as the product of a non-negative function of x alone and a non-negative function of y alone, i.e., if*

$$f_{XY}(x, y) = h_X(x) . k_Y(y), \text{ where } h(.) \geq 0 \text{ and } k(.) \geq 0. \qquad \text{... (5·20 b)}$$

Proof. If X and Y are independent, then by *def.*, $\quad f_{X,Y}(x, y) = f_X(x) . g_Y(y)$,

where $f(x)$ and $g(y)$ are marginal *p.d.f.'s* of X and Y respectively. Thus condition (5·20*b*) is satisfied.

Conversely if (5·20*b*) holds, then we have to prove that X and Y are independent. For continuous random variables X and Y, the marginal *p.d.f.'s* are given by :

$$f_X(x) = \int_{-\infty}^{\infty} f(x, y)\, dy = \int_{-\infty}^{\infty} h(x)\, k(y)\, dy = h(x) \int_{-\infty}^{\infty} k(y)\, dy = c_1\, h(x), \text{ say ... (*)}$$

$$\text{and} \quad g_Y(y) = \int_{-\infty}^{\infty} f(x, y)\, dx = \int_{-\infty}^{\infty} h(x)\, k(y)\, dx = k(y) \int_{-\infty}^{\infty} h(x)\, dx = c_2\, k(y), \text{ say ... (**)}$$

where c_1 and c_2 are constants independent of x and y. Moreover,

$$\int_{-\infty}^{\infty}\int_{-\infty}^{\infty} f(x, y)\, dx\, dy = 1 \quad \Rightarrow \quad \int_{-\infty}^{\infty}\int_{-\infty}^{\infty} h(x)\, k(y)\, dx\, dy = 1$$

$$\Rightarrow \left(\int_{-\infty}^{\infty} h(x)\, dx\right)\left(\int_{-\infty}^{\infty} k(y)\, dy\right) = 1 \quad \Rightarrow \quad c_2\, c_1 = 1 \qquad \text{[From (**) and (*)]} \quad \ldots (***)$$

Finally, we get

$$f_{XY}(x, y) = h_X(x)\, k_Y(y) = c_1 c_2\, h_X(x)\, k_Y(y) \qquad \text{[Using (***)]}$$
$$= \{c_1\, h_X(x)\}\, \{c_2\, k_Y(y)\}$$
$$= f_X(x)\, g_Y(y) \qquad \text{[From (*) and (**)]}$$

$\Rightarrow$ X and Y are stochastically independent.

Theorem 5·2. *If the random variables X and Y are stochastically independent, then for all possible selections of the corresponding pairs of real numbers (a_1, b_1), (a_2, b_2) where $a_i \le b_i$ for all $i = 1, 2$ and where the values $\pm\infty$ are allowed, the events $(a_1 < X \le b_1)$ and $(a_2 < Y \le b_2)$ are independent, i.e.,*

$$P\{(a_1 < X \le b_1) \cap (a_2 < Y \le b_2)\} = P(a_1 < X \le b_1) \cdot P(a_2 < Y \le b_2)$$

Proof. Since X and Y are stochastically independent, in the usual notations :

$$f_{X,Y}(x, y) = f_X(x)\, g_Y(y) \qquad \ldots (*)$$

In case of continuous *r.v.*'s, we have

$$P[(a_1 < X \le b_1) \cap (a_2 < Y \le b_2)] = \int_{a_1}^{b_1}\int_{a_2}^{b_2} f(x, y)\, dx\, dy$$

$$= \left(\int_{a_1}^{b_1} f_X(x)\, dx\right)\left(\int_{a_2}^{b_2} g_Y(y)\, dy\right) \qquad \text{[From (*)]}$$

$$= P(a_1 < X \le b_1)\, P(a_2 < Y \le b_2), \text{ as desired.}$$

$\Rightarrow$ The events : $(a_1 < X \le b_1)$ and $(a_2 < Y \le b_2)$ are indepndent.

Remark. In case of discrete *r.v.'s* theorems 5·2 and 5·3 can be proved on replacing integration by summation over the given range of the variables.

5·5·7. Generalisation to *n*-Dimensional Random Variable

The concept of two-dimensional random variables and their joint and marginal distributions in § 5·5 to § 5·5·6 can be easily generalised to the case of n-dimensional random variable.

Joint and Marginal Probability Mass Function

Let $(X_1, X_2, \ldots, X_n)$ be a discrete n-dimensional *r.v.*, assuming discrete values, in some region, say, R^n of the n-dimensional space. Then the joint *p.m.f.* of $(X_1, X_2, \ldots, X_n)$ is defined as :

$$p_{X_1, X_2, \ldots, X_n}(x_1, x_2, \ldots, x_n) = P[X_1 = x_1, X_2 = x_2, \ldots, X_n = x_n]$$

$$= P\left[\bigcap_{i=1}^{n}(X_i = x_i)\right] \qquad \ldots (5\cdot21)$$

where,

(i) $p(x_1, x_2, ..., x_n) \geq 0, \forall (x_1, x_2, ..., x_n) \in R^n$, and

(ii) $\sum_{x_1, x_2, ..., x_n} p(x_1, x_2, ..., x_n) = 1$

The marginal *p.m.f.* of any *r.v.*, say, X_i, is obtained on summing $p(x_1, x_2, ..., x_n)$, over the values of all other variables except X_i. Thus,

$$P_{X_i}(x_i) = \sum_{\substack{x_1, x_2, ..., x_n \\ \text{except } x_i}} p(x_1, x_1, ..., x_n) \quad ... (5·21a)$$

In particular, if $p(x_1, x_2, x_3)$ is the joint *p.m.f.* of three *r.v.'s* X_1, X_2 and X_3, then the marginal *p.m.f.* of, say, X_1 is given by :

$$P_{X_1}(x_1) = \sum_{x_2, x_3} p(x_1, x_2, x_3), \quad ... (5·21b)$$

and so on.

As, in the case of two random variables, *the r.v.'s $X_1, X_2, ..., X_n$ are independent if and only if their joint p.m.f. is equal to the product of their marginal p.m.f.'s, i.e., iff :*

$$p_{X_1, X_2, ..., X_n}(x_1, x_2, ..., x_n) = p_{X_1}(x_1) \cdot p_{X_2}(x_2) \; ... \; p_{Xn}(x_n) \quad ... (5·21c)$$

Joint and Marginal Probability Density Function

Let $(X_1, X_2, ..., X_n)$ be *n*-dimensional continuous *r.v.* assuming all the values in some region, say, $R_1{}^n$ of the *n*-dimensional space. Then the joint *p.d.f.* of $(X_1, X_2, ..., X_n)$ is given by :

$$f_{X_1, X_2, ..., X_n}(x_1, x_2, ..., x_n) = \lim_{dx_1 \to 0, dx_2 \to 0, ..., dx_n \to 0} \frac{p\left[\bigcap_{i=1}^{n} (x_i < X_i < x_i + dx_i)\right]}{dx_1 \cdot dx_2, ..., dx_n} \quad ... (5·21d)$$

where :

(i) $f(x_1, x_2, ..., x_n) \geq 0, \forall \; (x_1, x_2, ..., x_n) \in R_1{}^n$, and

(ii) $\int_{-\infty}^{\infty} \int_{-\infty}^{\infty} \cdots \int_{-\infty}^{\infty} f(x_1, x_2, ..., x_n) \, dx_1 \, dx_2 \, ... \, dx_n = 1$

The marginal *p.d.f.* of any variable, say, X_i, is obtained on integrating the joint *p.d.f.* over the range of all the variables except X_i. Thus,

$$f_{X_i}(x_i) = \int_{-\infty}^{\infty} \int_{-\infty}^{\infty} \cdots \int_{-\infty}^{\infty} f(x_1, x_2, ..., x_n) \, dx_1 \ldots dx_{i-1} dx_{i+1} \ldots dx_n \quad ... (5·21e)$$

In particular, for three *r.v.'s* X_1, X_2, X_3 with joint *p.d.f.* $f(x_1, x_2, x_3)$, the marginal *p.d.f.* of, say, X_2 is given by :

$$f_{X_2}(x_2) = \int_{-\infty}^{\infty}\int_{-\infty}^{\infty} f(x_1, x_2, x_3)\, dx_1\, dx_3 \qquad \text{... (5·21f)}$$

and so on.

The necessary and sufficient condition for the independence of r.v.'s $X_1, X_2, ..., X_n$ is that their joint p.d.f. is the product of their marginal p.d.f.'s i.e.,

$$f_{X_1, X_2, \ldots, X_n}(x_1, x_2, \ldots, x_n) = \prod_{i=1}^{n} f_{X_i}(x_i) \qquad \text{... (5·21g)}$$

Example 5·28. *X_1 and X_2 are independent random variables with common probability function :*

$$p(x) = \frac{x}{6}\ ; \quad x = 1, 2, 3 \qquad \text{... (*)}$$

Find the probability distribution of $X_1 + X_2$.

Solution. Since X_1 and X_2 each takes values 1, 2, 3 ; $Y = X_1 + X_2$, takes the values 2, 3, 4, 5 and 6. Further, since X_1 and X_2 are independent, their joint probability function is given by :

$$p(x_1, x_2) = P(X_1 = x_1, X_2 = x_2) = P(X_1 = x_1) \cdot P(X_2 = x_2) \cdot$$

$$= \frac{x_1 x_2}{36}\ ; \quad x_1 = 1, 2, 3\ ; \quad x_2 = 1, 2, 3 \qquad \text{[From (*)]} \qquad \text{... (**)}$$

Probability Function of $Y = X_1 + X_2$:

$Y = X_1 + X_2$	Favourable cases (X_1, X_2)	Probability [Using (**)]
2	(1, 1)	$p(1, 1) = \frac{1\times 1}{36} = \frac{1}{36}$
3	(1, 2), (2, 1)	$p(1, 2) + p(2, 1) = \frac{1\times 2}{36} + \frac{2\times 1}{36} = \frac{4}{36}$
4	(1, 3), (3, 1), (2, 2)	$\frac{1\times 3}{36} + \frac{3\times 1}{36} + \frac{2\times 2}{36} = \frac{10}{36}$
5	(2, 3), (3, 2)	$\frac{2\times 3}{36} + \frac{3\times 2}{36} = \frac{12}{36}$
6	(3, 3)	$\frac{3\times 3}{36} = \frac{9}{36}$

Hence, the probability distribution of $Y = X_1 + X_2$ is :

$y = x_1 + x_2$	2	3	4	5	6	Total
$g(y)$	$\frac{1}{36}$	$\frac{4}{36}$	$\frac{10}{36}$	$\frac{12}{36}$	$\frac{9}{36}$	1

Example 5·29. *A random observation on a bivariate population (X, Y) can yield one of the adjoining pairs of values with probabilities noted against them :*

For each observation pair	*Probability*
(1,1) ; (2,1) ; (3,3) ; (4,3)	$\frac{1}{20}$
(3,1) ; (4,1) ; (1,2) ; (2,2) ; (3,2) ; (4,2) ; (1,3) ; (2,3)	$\frac{1}{10}$

Find the probability that Y = 2 given that X = 4. Also find the probability that Y = 2. Examine if the two events X = 4 and Y = 2 are independent.

Solution. $P(Y = 2) = P\{(1\ 2) \cup (2, 2) \cup (3, 2) \cup (4, 2)\} = \frac{1}{10}+\frac{1}{10}+\frac{1}{10}+\frac{1}{10} = \frac{2}{5}$.

$$P(X = 4) = P\{(4, 1) \cup (4, 2) \cup (4, 3)\} = \frac{1}{10}+\frac{1}{10}+\frac{1}{20} = \frac{1}{4}$$

$$P(X = 4, Y = 2) = P\{(4, 2)\} = \frac{1}{10}$$

$$P(Y = 2 \mid X = 4) = \frac{P(X = 4 \cap Y = 2)}{P(X = 4)} = \frac{1/10}{1/4} = \frac{2}{5}$$

$$P(X = 4).\ P(Y = 2) = \frac{1}{4}.\frac{2}{5} = \frac{1}{10} = P(X = 4 \cap Y = 2)$$

Hence, the events X = 4 and Y = 2 are independent.

Example 5·30. *The joint probability distribution of two random variables X and Y is given by :*

$$P(X = 0, Y = 1) = \frac{1}{3}, \qquad P(X = 1, Y = -1) = \frac{1}{3}, \qquad \text{and} \qquad P(X = 1, Y = 1) = \frac{1}{3}.$$

Find (i) Marginal distributions of X and Y, and

(ii) The conditional probability distribution of X given Y = 1.

Solution. $P(X = -1)$

$$= \sum_y P(X = -1, Y = y)$$

$$= P(X = -1, Y = -1) + P(X = -1, Y = 0) + P(X = -1, Y = 1) = 0$$

Similarly $P(X = 0) = \frac{1}{3}$

and $P(X = 1) = \frac{2}{3}$

Y \ X	−1	0	1	Marginal (Y)
−1	0	0	$\frac{1}{3}$	$\frac{1}{3}$
0	0	0	0	0
1	0	$\frac{1}{3}$	$\frac{1}{3}$	$\frac{2}{3}$
Marginal (X)	0	$\frac{1}{3}$	$\frac{2}{3}$	1

Marginal distribution of X is :

Values of X, x :	−1	0	1
$P(X = x)$:	0	$\frac{1}{3}$	$\frac{2}{3}$

Marginal distribution of Y is :

Values of Y, y :	−1	0	1
$P(Y = y)$:	$\frac{1}{3}$	0	$\frac{2}{3}$

(*ii*) The conditional probability distribution of X given Y is :

$$P(X = x \mid Y = y) = \frac{P(X = x, Y = y)}{P(Y = y)}.$$

$$P(X = -1 \mid Y = 1) = \frac{P(X = -1, Y = 1)}{P(Y = 1)} = 0,$$

$$P(X = 0 \mid Y = 1) = \frac{P(X = 0, Y = 1)}{P(Y = 1)} = \frac{1/3}{2/3} = \frac{1}{2}$$

$$P(X = 1 \mid Y = 1) = \frac{P(X = 1, Y = 1)}{P(Y = 1)} = \frac{1/3}{2/3} = \frac{1}{2}$$

Thus, the conditional distribution of X given $Y = 1$ is :

Values of $X = x$	-1	0	1
$P(X = x \mid Y = 1)$	0	$\frac{1}{2}$	$\frac{1}{2}$

Example 5·31. *For the adjoining bivariate probability distribution of X and Y, find :*

(*i*) $P(X \le 1, Y = 2)$,

(*ii*) $P(X \le 1)$,

(*iii*) $P(Y \le 3)$, *and*

(*iv*) $P(X < 3, Y \le 4)$

X \ Y	1	2	3	4	5	6
0	0	0	$\frac{1}{32}$	$\frac{2}{32}$	$\frac{2}{32}$	$\frac{3}{32}$
1	$\frac{1}{16}$	$\frac{1}{16}$	$\frac{1}{8}$	$\frac{1}{8}$	$\frac{1}{8}$	$\frac{1}{8}$
2	$\frac{1}{32}$	$\frac{1}{32}$	$\frac{1}{64}$	$\frac{1}{64}$	0	$\frac{2}{64}$

Solution. The marginal distributions are given below :

X \ Y	1	2	3	4	5	6	$p_X(x)$
0	0	0	$\frac{1}{32}$	$\frac{2}{32}$	$\frac{2}{32}$	$\frac{3}{32}$	$\frac{8}{32}$
1	$\frac{1}{16}$	$\frac{1}{16}$	$\frac{1}{8}$	$\frac{1}{8}$	$\frac{1}{8}$	$\frac{1}{8}$	$\frac{10}{16}$
2	$\frac{1}{32}$	$\frac{1}{32}$	$\frac{1}{64}$	$\frac{1}{64}$	0	$\frac{2}{64}$	$\frac{8}{64}$
$p_Y(y)$	$\frac{3}{32}$	$\frac{3}{32}$	$\frac{11}{64}$	$\frac{13}{64}$	$\frac{6}{32}$	$\frac{16}{64}$	$\Sigma p(x) = 1$ $\Sigma p(y) = 1$

(*i*) $P(X \le 1, Y = 2) = P(X = 0, Y = 2) + P(X = 1, Y = 2) = 0 + \frac{1}{16} = \frac{1}{16}$

(ii) $P(X \le 1) = P(X = 0) + P(X = 1) = \dfrac{8}{32} + \dfrac{10}{16} = \dfrac{7}{8}$

(iii) $P(Y \le 3) = P(Y = 1) + P(Y = 2) + P(Y = 3) = \dfrac{3}{32} + \dfrac{3}{32} + \dfrac{11}{64} = \dfrac{23}{64}$

(iv) $P(X < 3, Y \le 4) = P(X = 0, Y \le 4) + P(X = 1, Y \le 4) + P(X = 2, Y \le 4)$

$$= \left(\frac{1}{32}+\frac{2}{32}\right) + \left(\frac{1}{16}+\frac{1}{16}+\frac{1}{8}+\frac{1}{8}\right) + \left(\frac{1}{32}+\frac{1}{32}+\frac{1}{64}+\frac{1}{64}\right) = \frac{9}{16}.$$

Example 5·32. *For the joint probability distribution of two random variables X and Y given below :*

X \ Y	1	2	3	4	*Total*
1	$\frac{4}{36}$	$\frac{3}{36}$	$\frac{2}{36}$	$\frac{1}{36}$	$\frac{10}{36}$
2	$\frac{1}{36}$	$\frac{3}{36}$	$\frac{3}{36}$	$\frac{2}{36}$	$\frac{9}{36}$
3	$\frac{5}{36}$	$\frac{1}{36}$	$\frac{1}{36}$	$\frac{1}{36}$	$\frac{8}{36}$
4	$\frac{1}{36}$	$\frac{2}{36}$	$\frac{1}{36}$	$\frac{5}{36}$	$\frac{9}{36}$
Total	$\frac{11}{36}$	$\frac{9}{36}$	$\frac{7}{36}$	$\frac{9}{36}$	1

Find :

(i) the marginal distributions of X and Y, and

(ii) conditional distribution of X given the value of Y = 1 and that of Y given the value of X = 2.

Solution. The marginal distribution of X is defined as : $P(X = x) = \sum_y P(X = x, Y = y)$

$$\therefore \quad P(X = 1) = \sum_y P(X = 1, Y = y)$$

$$= P(X = 1, Y = 1) + P(X = 1, Y = 2) + P(X = 1, Y = 3) + P(X = 1, Y = 4)$$

$$= \frac{4}{36} + \frac{3}{36} + \frac{2}{36} + \frac{1}{36} = \frac{10}{36}.$$

Similarly $\quad P(X = 2) = \sum_y P(X = 2, Y = y) = \dfrac{9}{36};$

$$P(X = 3) = \sum_y P(X = 3, Y = y) = \frac{8}{36};$$

and $\quad P(X = 4) = \sum_y P(X = 4, Y = y) = \dfrac{9}{36}.$

Similarly, we can obtain the marginal distribution of Y.

MARGINAL DISTRIBUTION OF X

Values of X, x	1	2	3	4
$P(X = x)$	$\frac{10}{36}$	$\frac{9}{36}$	$\frac{8}{36}$	$\frac{9}{36}$

MARGINAL DISTRIBUTION OF Y

Values of Y, y	1	2	3	4
$P(Y = y)$	$\frac{11}{36}$	$\frac{9}{36}$	$\frac{7}{36}$	$\frac{9}{36}$

(*ii*) The conditional probability function of X given Y is defined as follow:

$$P(X=x \mid Y=y) = \frac{P(X=x, Y=y)}{P(Y=y)}.$$

$$\therefore \quad P(X=-1 \mid Y=1) = \frac{P(X=1, Y=1)}{P(Y=1)} = \frac{4/36}{11/36} = \frac{4}{11}$$

$$P(X=2 \mid Y=1) = \frac{P(X=2, Y=1)}{P(Y=1)} = \frac{1/36}{11/36} = \frac{1}{11}$$

$$P(X=3 \mid Y=1) = \frac{P(X=3, Y=1)}{P(Y=1)} = \frac{4/36}{11/36} = \frac{5}{11}$$

$$P(X=4 \mid Y=1) = \frac{P(X=4, Y=1)}{P(Y=1)} = \frac{1/36}{11/36} = \frac{1}{11}$$

Hence, the conditional distribution of X given $Y = 1$ is :

x	1	2	3	4
$P(X=x \mid Y=1)$	$\frac{4}{11}$	$\frac{1}{11}$	$\frac{5}{11}$	$\frac{1}{11}$

Similarly, we can obtain the conditional distribution of Y for $X = 2$ as given below :

y	1	2	3	4
$P(Y=y \mid X=2)$	$\frac{1}{9}$	$\frac{1}{3}$	$\frac{1}{3}$	$\frac{2}{9}$

Example 5·33.

(*a*) *A two-dimensional r.v. (X, Y) have a bivariate distribution given by :*

$$P(X=x, Y=y) = \frac{x^2+y}{32}, \text{ for } x = 0, 1, 2, 3 \text{ and } y = 0, 1.$$

Find the marginal distributions of X and Y.

(*b*) *A two-dimensional r.v. (X, Y) have a joint probability mass function :*

$$p(x, y) = \frac{1}{27}(2x+y), \text{ where } x \text{ and } y \text{ can assume only the integer values } 0, 1 \text{ and } 2.$$

Find the conditional distribution of Y for X = x.

Solution.

(*a*) We have

Y \ X	0	1	2	3	Marginal distribution of Y, $P(Y=y)$
0	0	$\frac{1}{32}$	$\frac{4}{32}$	$\frac{9}{32}$	$\frac{14}{32}$
1	$\frac{1}{32}$	$\frac{2}{32}$	$\frac{5}{32}$	$\frac{10}{32}$	$\frac{18}{32}$
Marginal distribution of X, $P(X=x)$	$\frac{1}{32}$	$\frac{3}{32}$	$\frac{9}{32}$	$\frac{19}{32}$	1

The marginal probability distribution of X is given by :

$$P(X = x) = \sum_y P(X = x, Y = y)$$

and is tabulated in last row of above table.

The marginal probability distribu-tion of Y is given by :

$$P(Y = y) = \sum_y P(X = x, Y = y) \text{ and is tabulated in last column of above table.}$$

(*b*) The joint probability function :

$$p_{XY}(x, y) = \frac{1}{27}(2x + y) \; ; x = 0, 1, 2 \; ; \; y = 0, 1, 2$$

gives the following table of joint probability distribution of X and Y.

JOINT PROBABILITY DISTRIBUTION $p(x, y)$ OF X AND Y

X \ Y	0	1	2	$f_X(x)$
0	0	$\frac{1}{27}$	$\frac{2}{27}$	$\frac{3}{27}$
1	$\frac{2}{27}$	$\frac{3}{27}$	$\frac{4}{27}$	$\frac{9}{27}$
2	$\frac{4}{27}$	$\frac{5}{27}$	$\frac{6}{27}$	$\frac{15}{27}$

For example, $p(0, 0) = \frac{1}{27}(0 + 2 \times 0) = 0$, $\quad p(1, 0) = \frac{1}{27}(0 + 2 \times 1) = \frac{2}{27}$;

$p(2, 0) = \frac{1}{27}(0 + 2 \times 2) = \frac{4}{27}$;

and so on,

The conditional distribution of Y for $X = x$ is given by :

$$P_{Y|X}(Y = y \mid X = x) = \frac{p_{XY}(x, y)}{p_X(x)}$$

and is obtained in the adjoining table .

CONDITIONAL DISTRIBUTION OF Y FOR $X = x$

Y \ X	0	1	2
0	0	$\frac{1}{3}$	$\frac{2}{3}$
1	$\frac{2}{9}$	$\frac{3}{9}$	$\frac{4}{9}$
2	$\frac{4}{15}$	$\frac{5}{15}$	$\frac{6}{15}$

Example 5·34. *In the random placement of three balls in three cells, describe the possible outcomes of the experiment. Let X_i denote the number of balls in cell i; i = 1, 2, 3; and N, the number of cells occupied. Obtain the joint distribution of : (a) (X_1, N)* and *(b) (X_1, X_2).*

Solution.

(*a*) Let the three balls be denoted by *a*, *b* and *c*. Then the possible outcomes of placing the three balls in three cells are as follows :

1. {*a* | *b* | *c*}
2. {*a* | *c* | *b*}
3. {*b* | *a* | *c*}
8. {*ab* | – | *c*}
9. {– | *ab* | *c*}
10. {*ac* | *b* | –}
15. {– | *bc* | *a*}
16. {*a* | *bc* | –}
17. {*a* | – | *bc*}
22. {*c* | *ab* | –}
23. {*c* | – | *ab*}
24. {– | *c* | *ab*}

4. $\{b \mid c \mid a\}$	11. $\{ac \mid - \mid b\}$	18. $\{ - \mid a \mid bc\}$	25. $\{abc \mid - \mid - \}$
5. $\{c \mid a \mid b\}$	12. $\{- \mid ac \mid b\}$	19. $\{b \mid ca \mid -\}$	26. $\{- \mid abc \mid - \}$
6. $\{c \mid b \mid a\}$	13. $\{bc \mid a \mid -\}$	20. $\{b \mid - \mid ca\}$	27. $\{- \mid - \mid abc\}$
7. $\{ab \mid c \mid -\}$	14. $\{bc \mid - \mid a\}$	21. $\{- \mid b \mid ca\}$	

Each of these arrangements represents a sample event *i.e.*, a sample point. The sample space contains 27 poins.

Let N denote the number of occupied cells. The favourable cases for $N = 1$ are at numbers 25, 26, and 27, *i.e.*, 3; for $N = 2$ are at numbers 7 to 24, i.e., 18; and for $N = 3$ are at numbers 1 to 6, *i.e.*, 6. Accordingly, the probability distribution of N is :

$$P(N=1)=\frac{3}{27}, \quad P(N=2)=\frac{18}{27}, \quad P(N=3)=\frac{6}{27}.$$

Let X_1 denote the number of balls placed in the first cell. Then from the above table of sample points, we get

$$P(X_1=0)=\frac{8}{27}, \quad P(X_1=1)=\frac{12}{27}, \quad P(X_1=2)=\frac{6}{27} \quad \text{and} \quad P(X_1=3)=\frac{1}{27}$$

The joint distribution of N and X_1 can be obtained as follows :

$$P(N=1, X_1=0)=\frac{2}{27}, \quad P(N=1, X_1=1)=0, \quad P(N=1, X_1=2)=0,$$

$$P(N=1, X_1=3)=\frac{1}{27}, \quad P(N=2, X_1=0)=\frac{6}{27}, \quad P(N=2, X_1=1)=\frac{6}{27},$$

$$P(N=2, X_1=2)=\frac{6}{27}, \quad P(N=2, X_1=3)=0, \quad P(N=3, X_1=0)=0,$$

$$P(N=3, X_1=1)=\frac{6}{27}, \text{ and } P(N=3, X_1=2)=0, \quad P(N=3, X_1=3)=0.$$

JOINT PROBABILITY DISTRIBUTION OF N AND X_1

X_1 \ N	1	2	3	*Distribution of X_1*
0	$\frac{2}{27}$	$\frac{6}{27}$	0	$\frac{8}{27}$
1	0	$\frac{6}{27}$	$\frac{6}{27}$	$\frac{12}{27}$
2	0	$\frac{6}{27}$	0	$\frac{6}{27}$
3	$\frac{1}{27}$	0	0	$\frac{1}{27}$
Distribution of N	$\frac{3}{27}$	$\frac{18}{27}$	$\frac{6}{27}$	1

(b) Proceeding on the same lines, the joint distribution of X_1 and X_2 can be obtained as shown in the adjoining table :

JOINT DISTRIBUTION OF X_1 and X_2

X_1 \ X_2	0	1	2	3	Distribution of X_1
0	$\frac{1}{27}$	$\frac{3}{27}$	$\frac{3}{27}$	$\frac{1}{27}$	$\frac{8}{27}$
1	$\frac{3}{27}$	$\frac{6}{27}$	$\frac{3}{27}$	0	$\frac{12}{27}$
2	$\frac{3}{27}$	$\frac{3}{27}$	0	0	$\frac{6}{27}$
3	$\frac{1}{27}$	0	0	0	$\frac{1}{27}$
Distribution of X_2	$\frac{8}{27}$	$\frac{12}{27}$	$\frac{6}{27}$	$\frac{1}{27}$	1

SUPPLMENTARY EXAMPLES

Example 5·35. *Two discrete random variables X and Y have the joint probability density function :*

$$p_{XY}(x,y) = \frac{\lambda^x e^{-\lambda} p^y (1-p)^{x-y}}{y!(x-y)!}, \quad y = 0, 1, 2, ..., x \; ; \; x = 0, 1, 2, ...,$$

where λ, p are constants with $\lambda > 0$ and $0 < p < 1$.

Find (i) The marginal probability density functions of X and Y.

(ii) The conditional distribution of Y for a given X and of X for a given Y.

Solution.

(i) The marginal *p.m.f.* of X is given by :

$$p_X(x) = \sum_{y=0}^{x} p(x, y) = \sum_{y=0}^{x} \frac{\lambda^x e^{-\lambda} p^y (1-p)^{x-y}}{y!(x-y)!} = \frac{\lambda^x e^{-\lambda}}{x!} \sum_{y=0}^{x} \frac{x!\, p^y (1-p)^{x-y}}{y!(x-y)!}$$

$$= \frac{\lambda^x e^{-\lambda}}{x!} \sum_{y=0}^{x} {}^xC_y\, p^y (1-p)^{x-y} = \frac{\lambda^x e^{-\lambda}}{x!} [p + (1-p)]^x = \frac{\lambda^x e^{-\lambda}}{x!}, x = 0, 1, 2, ...$$

which is the probability function of a Poisson distribution with parameter λ.

$$p_Y(y) = \sum_{x=y}^{\infty} p(x, y) = \sum_{x=y}^{\infty} \frac{\lambda^x e^{-\lambda} p^y (1-p)^{x-y}}{y!\,(x-y)!} = \frac{(\lambda p)^y e^{-\lambda}}{y!} \sum_{x=y}^{\infty} \frac{[\lambda(1-p)]^{x-y}}{(x-y)!}$$

$$[\because y = 0, 1, 2, ..., x \Rightarrow y \le x \Rightarrow x \ge y]$$

$$= \frac{(\lambda p)^y e^{-\lambda}}{y!} e^{\lambda(1-p)} = \frac{e^{-\lambda p} (\lambda p)^y}{y!}; \quad y = 0, 1, 2, ...$$

which is the probability function of a Poisson distribution with parameter λp.

(*ii*) The conditional distribution of Y for given X is :

$$p_{Y|X}(y \mid x) = \frac{p_{XY}(x, y)}{p_X(x)} = \frac{\lambda^x e^{-\lambda} p^y (1-p)^{x-y} x!}{y!(x-y)!\lambda^x e^{-\lambda}} = \frac{x!}{y!(x-y)!} p^y (1-p)^{x-y}$$

$$= {}^xC_y \, p^y (1-p)^{x-y}, \quad x \geq y \text{ i.e., } y = 0, 1, 2, \ldots x.$$

Example 5·36. *If* $f_1(x) = qp^{x-1} \; ; x = 1, 2, 3, \ldots \; ;$

$$f_2(y \mid x) = xy^{x-1} \; ; 0 \leq y \leq 1, \qquad \ldots (1)$$

determine the unconditional density function of Y and the conditional density function of X given Y.

Also evaluate : (*i*) $P\left(Y > \frac{1}{2} \,\middle|\, X = k\right)$ *and* (*ii*) $P\left(X = k \,\middle|\, Y < \frac{1}{2}\right)$.

Solution. The joint p.d.f. of X and Y is given by :

$$f(x, y) = f_1(x) \cdot f_2(y \mid x) = q\,x\,(py)^{x-1} \; ; x = 1, 2, 3, \ldots \; ; 0 \leq y \leq 1 \qquad \ldots (2)$$

The marginal density of Y is given by :

$$g(y) = \sum_{x=1}^{\infty} f(x, y) = q \sum_{x=1}^{\infty} \left[x\,(py)^{x-1} \right]$$

$$\Rightarrow \quad g(y) = \frac{q}{(1-py)^2} \; ; \quad 0 \leq y \leq 1 \qquad \ldots (3)$$

$$\left[\because \sum_{x=1}^{\infty} x\,a^{x-1} = 1 + 2a + 3a^2 + \ldots = (1-a)^{-2} = \frac{1}{(1-a)^2} \right]$$

The conditional density of X given Y is :

$$f_3(x \mid y) = \frac{f(x, y)}{g(y)} = \frac{q\,x\,(py)^{x-1}\,(1-py)^2}{q}$$

$$= x\,(py)^{x-1}\,(1-py)^2 \; ; 0 \leq y \leq 1 \; ; x = 1, 2, 3, \ldots \qquad \ldots (4)$$

$$(i) \; P\left(Y > \frac{1}{2} \,\middle|\, X = k\right) = \int_{1/2}^{1} f_2(y \mid x = k)\,dy = \int_{1/2}^{1} k\,y^{k-1}\,dy \qquad \left[\text{From (1)}\right]$$

$$= k \left| y^k / k \right|_{1/2}^{1} = \left\{ 1 - (1/2)^k \right\}$$

$$(ii) \quad P\left(X = k \,\middle|\, y < \frac{1}{2}\right) = \frac{P\left(x = k \cap Y < \frac{1}{2}\right)}{P\left(Y < \frac{1}{2}\right)} \qquad \ldots (*)$$

$$P\left(Y < \frac{1}{2}\right) = \int_0^{1/2} g(y)\,dy = q \int_0^{1/2} (1-py)^{-2}\,dy$$

$$= \left| \frac{q}{p(1-py)} \right|_0^{1/2} = \frac{q}{p} \left[\frac{1}{1 - \frac{1}{2}p} - 1 \right] = \frac{q}{2-p}$$

$$P\left(x = k \cap Y < \frac{1}{2}\right) = \int_{y<1/2} \left[f(x, y) \right]_{x=k} dy$$

$$= qk \int_0^{1/2} (py)^{k-1} dy \qquad \text{[From (2)]}$$

$$= q\, p^{k-1}\, k \left| \frac{y^k}{k} \right|_0^{1/2} = q\, p^{k-1} \left(\frac{1}{2}\right)^k$$

Substituting in (*), we get

$$P\left(X = k \,\middle|\, Y < \frac{1}{2}\right) = \frac{q\, p^{k-1}\, (1/2)^k}{q/(2-p)} = (2-p)\, p^{k-1} \left(\frac{1}{2}\right)^k.$$

Example 5·37. *If X, Y, Z are i.i.d. continuous random variables, then prove that ;*

(*a*) $P(X < Y) = \frac{1}{2}$, (*b*) $P(X < Y < Z) = \frac{1}{6}$,

(*c*) $P(X > Y \cap X > Z) = \frac{1}{3}$ (*d*) $P[(X > Y) \cap (Y < Z)] = \frac{1}{3}$.

Solution. Let X, Y, Z be *i.i.d.* random variables with common p.d.f. $f(\bullet)$ and distribution function $F(\bullet)$. Let $g(x, y)$ and $h(x, y, z)$ be the joint p.d.f.'s of (X, Y) and (X, Y, Z) respectively so that :

$$g(x, y) = f(x) \cdot f(y) \quad \text{and} \quad h(x, y, z) = f(x) \cdot f(y) \cdot f(z) \qquad [\because \text{X, Y, Z are } i.i.d.]$$

(*a*)
$$P(X < Y) = \int_{-\infty}^{\infty} \left(\int_{-\infty}^{y} g(x, y)\, dx \right) dy = \int_{-\infty}^{\infty} \left(\int_{-\infty}^{y} f(x)\, dx \right) f(y)\, dy$$

$$= \int_{-\infty}^{\infty} f(y) \cdot F(y)\, dy = \int_0^1 t\, dt = \frac{1}{2} \left| t^2 \right|_0^1 = \frac{1}{2}$$

$$[F(y) = t \;;\quad F(-\infty) = 0 ,\quad F(\infty) = 1 \;;\quad f(y)\, dy = dt]$$

(*b*)
$$P(X < Y < Z) = \int_{-\infty}^{\infty} \left[\int_{-\infty}^{z} \left\{ \int_{x}^{z} h(x, y, z)\, dy \right\} dx \right] dz \qquad [\because x < y < z \Rightarrow x < z]$$

$$= \int_{-\infty}^{\infty} \left[\int_{-\infty}^{z} \left\{ \int_{x}^{z} f(y)\, dy \right\} f(x)\, dx \right] f(z)\, dz$$

$$= \int_{-\infty}^{\infty} \left[\int_{-\infty}^{z} \{ F(z) - F(x) \} f(x)\, dx \right] f(z)\, dz$$

$$= \int_{-\infty}^{\infty} \left[F(z) \int_{-\infty}^{z} f(x)\, dx - \int_{-\infty}^{z} F(x)\, f(x)\, dx \right] f(z)\, dz$$

$$= \int_{-\infty}^{\infty} \left[F(z) \cdot F(z) - \int_0^{F(z)} t.dt \right] f(z)\, dz$$

$$[F(x) = t \Rightarrow f(x)\, dx = dt \;;\quad x = -\infty \Rightarrow F(-\infty) = 0, \quad \text{and} \quad x = z \Rightarrow t = F(z)]$$

$$= \int_{-\infty}^{\infty} \left[\{F(z)\}^2 - \frac{1}{2}\{F(z)\}^2 \right] f(z)\, dz$$

$$= \frac{1}{2}\int_{-\infty}^{\infty} \{F(z)\}^2 \cdot f(z)\, dz = \frac{1}{2}\int_0^1 u^2\, du = \frac{1}{2}\left| \frac{u^3}{3} \right|_0^1 = \frac{1}{6} = \frac{1}{3!}$$

$$[\, F(z) = u \Rightarrow f(z)\, dz = du\ ;\ F(-\infty) = 0,\ F(\infty) = 1\,]$$

Note on Part (a) and (b). In general, if $X_1, X_2, X_3, \ldots, X_n$ are *i.i.d.* continuous random variables, then

$$P\left(X_1 < X_2 < X_3 < \ldots < X_n\right) = \frac{1}{n!}, \quad \text{where } n \geq 2 \text{ is a positive integer.}$$

(c) $$P(X > Y \cap X > Z) = P[X > \text{Max}(Y, Z)] = \int_{-\infty}^{\infty} P[x > \text{Max}(Y, Z)]\, f(x)\, dx$$

$$P[X > \text{Max}(Y, Z)] = P[(Y < x) \cap (Z < x)] = P(Y < x) \cdot P(Z < x) = F(x) \cdot F(x)$$

$$[\because\ Y \text{ and } Z \text{ are independent and identically distributed}]$$

$$\therefore\ P(X > Y \cap X > Z) = \int_{-\infty}^{\infty} \{F(x)\}^2 f(x)\, dx = \int_0^1 v^2 dv = \left| \frac{v^3}{3} \right|_0^1 = \frac{1}{3}, \quad [\text{where } F(x) = v]$$

(d) $$P(X > Y \cap Y < Z) = P(Y < X \cap Y < Z) = P[Y < \text{Min}(X, Z)]$$

$$= \int_{-\infty}^{\infty} P[y < \text{Min}(X, Z)]\, f(y)\, dy \quad \ldots (*)$$

$$P[y < \text{Min}(X, Z)] = P[\text{Min}(X, Z) > y] = P[(X > y) \cap (Z > y)]$$

$$= P(X > y) \cdot P(Z > y) = [1 - P(X \leq y)] \cdot [1 - P(Z \leq y)]$$

$$= [(1 - F(y)) \cdot (1 - F(y))] \quad (\because X \text{ and } Z \text{ are } i.i.d.) \quad \ldots (**)$$

$$\therefore\ P(X > Y \cap Y < Z) = \int_{-\infty}^{\infty} [1 - F(y)]^2 f(y)\, dy \quad [\text{From } (*) \text{ and } (**)]$$

$$= \int_1^0 w^2(-dw) = \int_0^1 w^2 dw = \frac{1}{3}; \quad [1 - F(y) = w]$$

Example 5·38. *If X and Y are two random variables having joint density function :*

$$f(x, y) = \begin{cases} \frac{1}{8}(6 - x - y)\ ; 0 \leq x < 2, 2 \leq y < 4 \\ 0, \quad \text{otherwise} \end{cases}$$

Find (i) $P(X < 1 \cap Y < 3)$, (ii) $P(X + Y < 3)$ and (iii) $P(X < 1 \mid Y < 3)$.

Solution. We have

$$(i)\ P(X < 1 \cap Y < 3) = \int_{-\infty}^{1} \int_{-\infty}^{3} f(x, y)\, dx\, dy$$

$$= \int_0^1 \int_2^3 \frac{1}{8}(6-x-y)\,dx\,dy = \frac{3}{8}$$

(ii) $$P(X+Y<3) = \int_0^1 \int_2^{3-x} \frac{1}{8}(6-x-y)\,dx\,dy = \frac{5}{24}$$

(iii) $$P(X<1 \mid Y<3) = \frac{P(X<1 \cap Y<3)}{P(Y<3)} = \frac{3/8}{5/8} = \frac{3}{5}$$

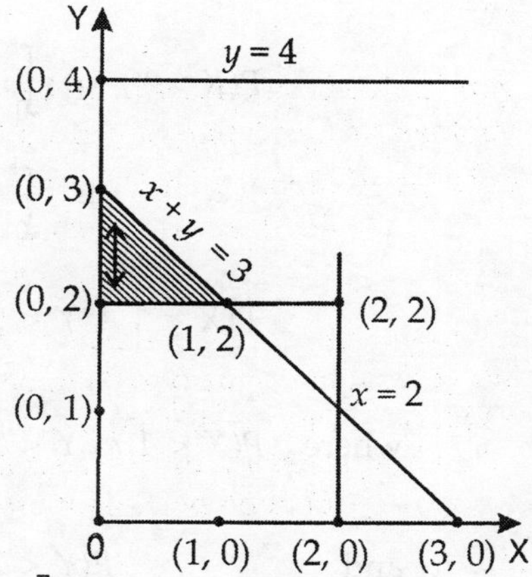

Fig. 5·10

$$\left[\text{From part } (i) \text{ and } P(Y<3) = \int_0^2 \int_2^3 \frac{1}{8}(6-x-y)\,dx\,dy = \frac{5}{8}\right]$$

Example 5·39. *Suppose that two-dimensional continuous random variable (X, Y) has joint p.d.f. given by :*

$$f(x, y) = \begin{cases} 6x^2 y, & 0<x<1,\ 0<y<1 \\ 0, & \text{otherwise} \end{cases}$$

(i) Verify that $\int_0^1 \int_0^1 f(x, y)\,dx\,dy = 1.$

(ii) Find $P\left(0<X<\frac{3}{4}, \frac{1}{3}<Y<2\right)$, $P(X+Y<1)$, $P(X>Y)$ *and* $P(X<1 \mid Y<2)$.

Solution.

(i) $$\int_0^1 \int_0^1 f(x, y)\,dx\,dy = \int_0^1 \int_0^1 6x^2 y\,dx\,dy = \int_0^1 6x^2 \left|\frac{y^2}{2}\right|_0^1 dx = \int_0^1 3x^2\,dx = \left|x^3\right|_0^1 = 1$$

(ii) $$P\left(0<X<\frac{3}{4}, \frac{1}{3}<Y<2\right) = \int_0^{3/4} \int_{1/3}^1 6x^2 y\,dx\,dy + \int_0^{3/4} \int_1^2 0.\,dx\,dy$$

$$= \int_0^{3/4} 6x^2 \left|\frac{y^2}{2}\right|_{1/3}^1 dx = \frac{8}{9} \int_0^{3/4} 3x^2\,dx = \frac{8}{9}\left|x^3\right|_0^{3/4} = \frac{3}{8}.$$

$$P(X+Y<1) = \int_0^1 \int_0^{1-x} 6x^2 y\,dx\,dy$$

(See Fig. 5·11)

$$= \int_0^1 6x^2 \left|\frac{y^2}{2}\right|_0^{1-x} dx$$

$$= \int_0^1 3x^2 (1-x)^2\,dx = \frac{1}{10}$$

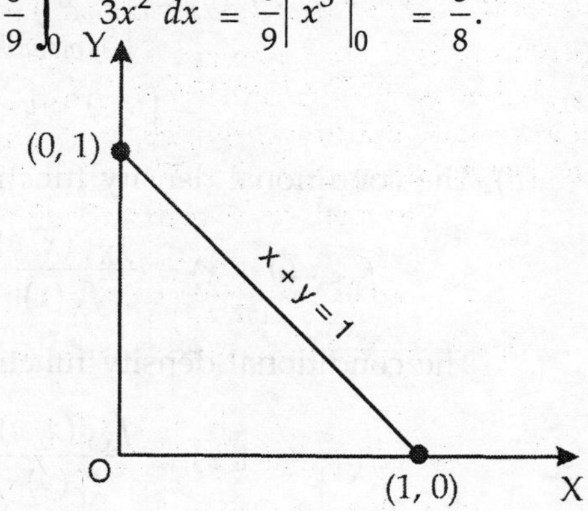

Fig. 5·11

$$P(X > Y) = \int_0^1 \int_0^x 6x^2 y \, dx \, dy \ ; \quad (x > y \ \Rightarrow \ y < x)$$

$$= \int_0^1 3x^2 \left| y^2 \right|_0^x dx = \int_0^1 3x^4 \, dx = \frac{3}{5}$$

$$P(X < 1 \mid Y < 2) = \frac{P(X < 1 \cap Y < 2)}{P(Y < 2)}$$

where $P(X < 1 \cap Y < 2) = \int_0^1 \int_0^1 6x^2 \, y \, dx \, dy + \int_0^1 \int_1^2 0 \cdot dx \, dy = 1$

and $P(Y < 2) = \int_0^1 \int_0^2 f(x, y) \, dx \, dy = \int_0^1 \int_0^1 6x^2 \, y \, dx \, dy + \int_0^1 \int_1^2 0 \cdot dx \, dy = 1$

$$\therefore \quad P(X < 1 \mid Y < 2) = \frac{P(X < 1 \cap Y < 2)}{P(Y < 2)} = 1.$$

Example 5·40. *The joint probability density function of a two-dimensional random variable (X, Y) is given by:*

$$f(x, y) = \begin{cases} 2 \ ; 0 < x < 1, 0 < y < x \ ; \\ 0, \quad elsewhere \end{cases}$$

(i) Find the marginal density functions of X and Y.

(ii) Find the conditional density function of Y given X = x and conditional density function of X given Y = y.

(iii) Check for independence of X and Y.

Solution. Evidently $f(x,y) \geq 0$ and $\int_0^1 \int_0^x 2 \, dx \, dy = 2 \int_0^1 x \, dx = 1.$

(*i*) The marginal *p.d.f.'s* of X and Y are given by :

$$f_X(x) = \begin{cases} \int_{-\infty}^{\infty} f_{XY}(x, y) \, dy \\ 0, \ elsewhere \end{cases} = \int_0^x 2dy = 2x, \ 0 < x < 1$$

$$f_Y(y) = \begin{cases} \int_{-\infty}^{\infty} f_{XY}(x, y) \, dx \\ 0, \ elsewhere \end{cases} = \int_y^1 2dx = 2(1 - y), \ 0 < y < 1$$

$$[\because 0 < y < x \quad \Rightarrow \quad x > y \text{ and } 0 < x < 1 \quad \Rightarrow \quad y < x < 1]$$

(*ii*) The conditional density function of Y given X , $(0 < x < 1)$ is :

$$f_{Y \mid X}(y \mid x) = \frac{f_{XY}(x, y)}{f_X(x)} = \frac{2}{2x} = \frac{1}{x}, \quad 0 < y < x.$$

The conditional density function of X given Y, $(0 < y < 1)$ is :

$$f_{X \mid Y}(x \mid y) = \frac{f_{XY}(x, y)}{f_Y(y)} = \frac{2}{2(1-y)} = \frac{1}{(1-y)}, \quad y < x < 1$$

(*iii*) Since $f_X(x) \, f_Y(y) = 2(2x)(1 - y) \neq f_{XY}(x, y)$, X and Y are not independent.

Example 5·41. *The joint p.d.f. of two random variables X and Y is given by :*

$$f(x, y) = \frac{9(1+x+y)}{2(1+x)^4\,(1+y)^4}\;;\; 0 \le x < \infty, 0 \le y < \infty$$

Find the marginal distributions of X and Y, and the conditional distribution of Y for X = x.

Solution. Marginal *p.d.f.* of X is given by :

$$f_X(x) = \int_0^\infty f(x, y)\, dy = \frac{9}{2(1+x)^4}\int_0^\infty \frac{(1+y)+x}{(1+y)^4}\, dy$$

$$= \frac{9}{2(1+x)^4}\int_0^\infty \left\{(1+y)^{-3} + x(1+y)^{-4}\right\} dy$$

$$= \frac{9}{2(1+x)^4}\left(\left|\frac{-1}{2(1+y)^2}\right|_0^\infty + x\left|\frac{-1}{3(1+y)^3}\right|_0^\infty\right)$$

$$= \frac{9}{2(1+x)^4}\left(\frac{1}{2}+\frac{x}{3}\right) = \frac{3}{4}\cdot\left(\frac{3+2x}{(1+x)^4}\right)\;;\; 0 < x < \infty$$

Since $f(x, y)$ is symmetric in x and y, the marginal *p.d.f.* of Y is given by :

$$f_Y(y) = \int_0^\infty f(x, y)\, dx = \frac{3}{4}\cdot\left(\frac{3+2y}{(1+y)^4}\right)\;;\; 0 < y < \infty$$

The conditional distribution of Y for X = x is given by :

$$f_{Y\,|\,X}(Y = y \mid X = x) = \frac{f_{XY}(x, y)}{f_X(x)} = \frac{9(1+x+y)}{2(1+x)^4(1+y)^4}\times\frac{4(1+x)^4}{3(3+2x)} = \frac{6(1+x+y)}{(1+y)^4(3+2x)}\;;\; 0 < y < \infty$$

Example 5·42. *The joint p.d.f. of two random variables X and Y is given by :*

$$f(x, y) = \begin{cases} k\;; & x^2 \le y \le x,\ 0 < x < 1 \\ 0\;; & \text{otherwise} \end{cases}$$

Find : (a) k ,

(b) The marginal p.d.f.'s of X and Y.

(c) The conditional p.d.f. of X given Y = y and of Y given X = x.

Solution.

(*a*) Since total probability is 1 ;

$$\int_{x=0}^{1}\int_{y=x^2}^{x} f(x, y)\, dx\, dy = 1$$

$$\therefore \quad k\int_{x=0}^{1}\left(\int_{y=x^2}^{x} 1\cdot dy\right) dx = 1 \quad\Rightarrow\quad k\int_{x=0}^{1}(x - x^2)\, dx = 1$$

$$\Rightarrow \quad k\left|\frac{x^2}{2} - \frac{x^3}{3}\right|_0^1 = 1 \quad\Rightarrow\quad k\left(\frac{1}{2} - \frac{1}{3}\right) = 1 \Rightarrow k = 6$$

(b) The region ($x^2 \le y \le x,\ 0 < x < 1$ is bounded by the curves : $x^2 = y,\ \ y = x,\ \ x = 0$ and $x = 1$, and is shown as shaded region in the following Fig. 5·12.

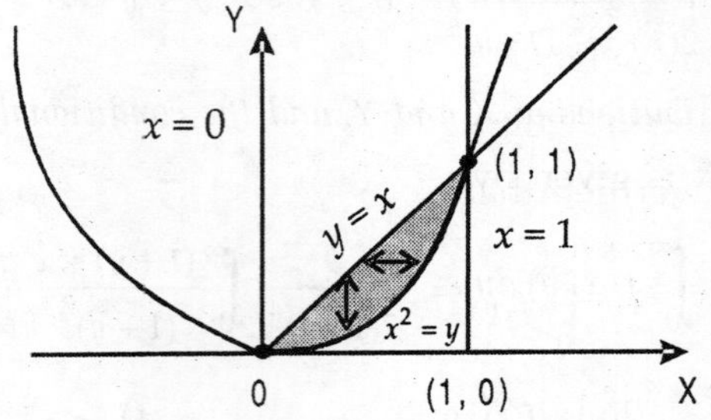

Fig. 5·12

The marginal *p.d.f.'s* of X and Y are given by :

$$f_X(x) = \int_{x^2}^{x} f(x, y)\, dy = 6\int_{x^2}^{x} 1 \cdot dy = 6\left(x - x^2\right)\ ;\ 0 < x < 1$$

$$f_Y(y) = \int_{y}^{\sqrt{y}} f(x, y)\, dx = 6\int_{y}^{\sqrt{y}} 1 \cdot dx = 6\left(\sqrt{y} - y\right)\ ;\ 0 < y < 1$$

(c) Conditional *p.d.f.'s* are given by :

$$f_{X|Y}(x \mid y) = \frac{f(x, y)}{f_Y(y)} = \frac{6}{6\left(\sqrt{y} - y\right)} = \frac{1}{\sqrt{y} - y}\ ;\ y < x < \sqrt{y}\ ,\ 0 < y < 1$$

$$f_{Y|X}(y \mid x) = \frac{f(x, y)}{f_X(x)} = \frac{6}{6\left(x - x^2\right)} = \frac{1}{x - x^2}\ ;\ x^2 \le y \le x\ ,\ 0 < x < 1$$

Example 5·43. *Joint distribution of X and Y is given by :* $f(x, y) = 4xy\, e^{-\left(x^2 + y^2\right)}$; $x \ge 0,\ y \ge 0$.

Test whether X and Y are independent. For the above joint distribution, find the conditional density of X given Y = y.

Solution. Joint *p.d.f.* of X and Y is : $f_{XY}(x, y) = 4xy\, e^{-(x^2 + y^2)}\ ;\ x \ge 0,\ y \ge 0.$

Marginal density of X is given by :

$$f_X(x) = \int_0^\infty f_{XY}(x, y)\, dy = \int_0^\infty 4xy\, e^{-\left(x^2 + y^2\right)}\, dy = 4x\, e^{-x^2}\int_0^\infty y e^{-y^2}\, dy$$

$$= 4x\, e^{-x^2} \cdot \int_0^\infty e^{-t} \cdot \frac{dt}{2} = 2x \cdot e^{-x^2}\left|-e^{-t}\right|_0^\infty$$

$\therefore$ $$f_X(x) = 2x\, e^{-x^2}\ ;\ x \ge 0$$

Similarly, the marginal *p.d.f.* of Y is given by :

$$f_Y(y) = \int_0^\infty f_{XY}(x, y)\, dx = 2y\, e^{-y^2}\ \ ;\ y \ge 0$$

Aliter. *For Independence of X and Y.*

$$f(x, y) = 4\,xy\,e^{-(x^2+y^2)} = \left(2x\,e^{-x^2}\right)\left(2y\,e^{-y^2}\right) = f_1(x)\cdot f_2(y), \quad \text{(say).}$$

$\Rightarrow$ X and Y are independent.

Note. It can be easily verified that $f_1(x)$ and $f_2(y)$ represent *p.d.f.'s* of X and Y, *i.e.,*

$$\int_0^\infty f_1(x)\,dx = \int_0^\infty f_2(y)\,dy = 1$$

Since $f_{XY}(x, y) = f_X(x)\cdot f_Y(y)$, X and Y are independtly distributed. The conditional distribution of X given Y is given by :

$$f_{X\mid Y}(x \mid y) = \frac{f(x, y)}{f_Y(y)} = 2x\,e^{-x^2}\ ;\ x \ge 0.$$

Example 5·44. *Let X and Y be jointly distributed with p.d.f. :*

$$f_{XY}(x, y) = \begin{cases} \frac{1}{4}(1+xy), & |x|<1,\ |y|<1 \\ 0, & \text{otherwise} \end{cases}$$

Show that X and Y are not independent but X^2 and Y^2 are independent.

Solution. $$f_X(x) = \int_{-1}^{1} f(x, y)\,dy = \frac{1}{4}\left|y + \frac{xy^2}{2}\right|_{-1}^{1} = \frac{1}{2},\quad -1 < x < 1\ ;$$

Similarly, $$f_Y(y) = \int_{-1}^{1} f(x, y)\,dx = \frac{1}{2},\quad -1 < y < 1$$

Since $f_{XY}(x, y) \ne f_X(x)\,f_Y(y)$, X and Y are not independent. However,

$$P\left(X^2 \le x\right) = P\left(|X| \le \sqrt{x}\right) = \int_{-\sqrt{x}}^{\sqrt{x}} f_X(x)\,dx = \sqrt{x} \qquad \text{... (*)}$$

$$P\left(X^2 \le x \cap Y^2 \le y\right) = P\left(|X| \le \sqrt{x}\ \cap\ |Y| \le \sqrt{y}\right)$$

$$= \int_{-\sqrt{x}}^{\sqrt{x}}\left[\int_{-\sqrt{y}}^{\sqrt{y}} f(u, v)\,dv\right]du = \sqrt{x}\,\sqrt{y}$$

$$= P\,(X^2 \le x)\,.\,P\,(Y^2 \le y)$$

Hence, X^2 and Y^2 are independent.

Example 5·45. *The joint p.d.f of X and Y is :*

$$f(x, y) = k\,(x^2 + y^2),\ \text{for}\ \ x^2 + y^2 < 1.$$

Find k and the marginal and conditional distributions of X and Y.

Solution. Since total probability is 1, we have

$$1 = \iint_{x^2+y^2<1} k(x^2+y^2)\,dx\,dy = k\int_{r=0}^{1}\int_{\theta=0}^{2\pi} r^2 . r\,dr\,d\theta$$

[Transforming to Polar coordinates : $x = r \cos\theta,\ y = r \sin\theta \Rightarrow J = r$

and $x^2 + y^2 < 1$ transforms to : $0 < r < 1$ and $0 < \theta < 2\pi$]

$$\Rightarrow k \cdot 2\pi \cdot \int_0^1 r^3\, dr = 1 \quad \Rightarrow \quad k \cdot 2\pi \cdot \left| \frac{r^4}{4} \right|_0^1 = 1 \quad \Rightarrow \quad k = \frac{2}{\pi}$$

$$\therefore \quad f(x, y) = \frac{2}{\pi}\left(x^2 + y^2\right) \ ; \text{ for } x^2 + y^2 < 1.$$

The marginal *p.d.f.* of X is given by :

$$f_X(x) = \int_{-\sqrt{1-x^2}}^{\sqrt{1-x^2}} f(x, y)\, dy = \frac{2}{\pi} \int_{-\sqrt{1-x^2}}^{\sqrt{1-x^2}} \left(x^2 + y^2\right) dy = \frac{2}{\pi} \left| yx^2 + \frac{y^3}{3} \right|_{-\sqrt{1-x^2}}^{\sqrt{1-x^2}}$$

$$= \frac{2}{\pi} \times 2 \left[x^2 \sqrt{1 - x^2} + \frac{(1 - x^2)^{3/2}}{3} \right] = \frac{4}{\pi} \sqrt{1 - x^2} \left[x^2 + \frac{1 - x^2}{3} \right]$$

$$= \frac{4}{3\pi} \sqrt{1 - x^2} \cdot (2x^2 + 1) \ ; \ -1 < x < 1$$

Similary, proceeding the same way or by symmetry of $f(x, y)$ in x and y, we will get

$$f_Y(y) = \frac{4}{3\pi} \sqrt{1 - y^2} \cdot \left(2y^2 + 1\right) \ ; \ -1 < y < 1$$

The conditional distributions are given by :

$$f_{Y \mid X}(y \mid x) = \frac{f(x, y)}{f_X(x)} = \frac{\frac{2}{\pi}\left(x^2 + y^2\right)}{\frac{4}{3\pi}\sqrt{1 - x^2} \cdot \left(2x^2 + 1\right)}$$

$$\Rightarrow \quad f_{Y \mid X}(y \mid x) = \frac{3}{2} \cdot \frac{x^2 + y^2}{\sqrt{1 - x^2} \cdot \left(2x^2 + 1\right)} \ ; -1 < y < 1, -1 < x < 1$$

Similary, proceeding the same way (or by symmetry), we get

$$f_{X \mid Y}(x \mid y) = \frac{f(x, y)}{f_Y(y)} = \frac{3}{2} \cdot \frac{\left(x^2 + y^2\right)}{\sqrt{1 - y^2}\left(2y^2 + 1\right)} ; -1 < y < 1, -1 < x < 1$$

Example 5·46. *A gun is aimed at a certain point (origin of the coordinate system). Because of the random factors, the actual hit point can be any point (X, Y) in a circle of radius R about the origin. Assume that the joint density of X and Y is constant in this circle given by :*

$$f_{XY}(x, y) = \begin{cases} k, & \text{for } x^2 + y^2 \le R^2 \\ 0, & \text{otherwise} \end{cases}$$

(i) Compute k, *(ii) show that*
$$f_X(x) = \begin{cases} \dfrac{2}{\pi R}\left\{1 - \left(\dfrac{x}{R}\right)^2\right\}^{1/2}, & \text{for } -R \le x \le R \\ 0, & \text{otherwise} \end{cases}$$

Solution.

(*i*) The constant k is computed from the consideration that the total probability is 1, *i.e.*,

$$\int_{-\infty}^{\infty}\int_{-\infty}^{\infty} f(x, y)\,dx\,dy = 1 \quad\Rightarrow\quad \iint_{x^2+y^2\le R^2} k\,dx\,dy = 1 \quad\Rightarrow\quad 4\iint_I k\,dx\,dy = 1$$

where, region I is the first quadrant of the circle $x^2 + y^2 = R^2$.

$$\Rightarrow \quad 4k\int_0^R\left(\int_0^{\sqrt{R^2-x^2}} 1\cdot dy\right)dx = 1 \quad\Rightarrow\quad 4k\int_0^R \sqrt{R^2-x^2}\,dx = 1$$

$$\Rightarrow\ 4k\left| x\sqrt{R^2-x^2} + \frac{R^2}{2}\sin^{-1}\left(\frac{x}{R}\right)\right|_0^R = 1 \quad\Rightarrow\quad 4k\left(\frac{R^2}{2}\cdot\frac{\pi}{2}\right) = 1 \quad\Rightarrow\quad k = \frac{1}{\pi R^2}.$$

$$\therefore \quad f_{XY}(x, y) = \begin{cases} 1/(\pi R^2)\,; & x^2+y^2 \le R^2 \\ 0, & \textit{otherwise} \end{cases}$$

(*ii*)
$$f_X(x) = \int_{-\infty}^{\infty} f(x, y)\,dy = \frac{1}{\pi R^2}\int_{-\sqrt{R^2-x^2}}^{\sqrt{R^2-x^2}} 1\cdot dy$$

$$\left[\because\ x^2 + y^2 \le R^2 \ \Rightarrow\ -(R^2 - x^2)^{1/2} \le y \le (R^2 - x^2)^{1/2}\right]$$

$$= \frac{2}{\pi R^2}\int_0^{\sqrt{R^2-x^2}} 1\cdot dy = \frac{2}{\pi R^2}(R^2-x^2)^{1/2} = \frac{2}{\pi R}\left\{1-\left(\frac{x}{R}\right)^2\right\}^{1/2}, -R \le x \le R$$

Example 5·47. *Given :* $f(x, y) = e^{-(x+y)}\ I_{(0,\ \infty)}(x)\,.\,I_{(0,\ \infty)}(y)$. *Are X and Y independent ?*

Find (*i*) $P(X > 1)$, (*ii*) $P(X < Y \mid X < 2Y)$, (*iii*) $P(1 < X + Y < 2)$.

Solution. We are given :

$$f(x, y) = e^{-(x+y)}\ ;\ 0 \le x < \infty,\ 0 \le y < \infty$$

$$= (e^{-x})(e^{-y}) = f_X(x)\cdot f_Y(y)\ ;\ 0 \le x < \infty\ ,\ 0 \le y < \infty \qquad \ldots (1)$$

$\Rightarrow$ X and Y are independent and $f_X(x) = e^{-x}\ ;\ x \ge 0$ and $f_Y(y) = e^{-y}\ ;\ y \ge 0$ ... (2)

(*i*)
$$P(X > 1) = \int_1^{\infty} f_X(x)\,dx$$

$$= \int_1^{\infty} e^{-x}\,dx = \left|\frac{e^{-x}}{-1}\right|_1^{\infty} = \frac{1}{e}$$

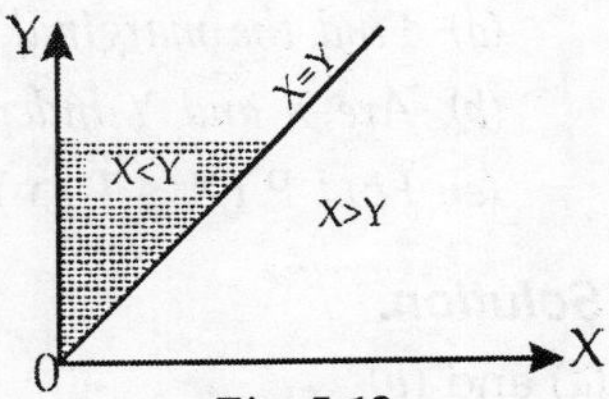

Fig. 5·13

(*ii*)
$$P(X < Y \mid X < 2Y) = \frac{P(X<Y \cap X<2Y)}{P(X<2Y)} = \frac{P(X<Y)}{P(X<2Y)} \qquad \ldots (3)$$

$$P(X < Y) = \int_0^{\infty}\left\{\int_0^{y} f(x, y)\,dx\right\}dy = \int_0^{\infty}\left\{e^{-y}\left|\frac{e^{-x}}{-1}\right|_0^{y}\right\}dy$$

$$= -\int_0^{\infty} e^{-y}(e^{-y}-1)\,dy = -\left|\frac{e^{-2y}}{-2}+e^{-y}\right|_0^{\infty} = 1-\frac{1}{2} = \frac{1}{2}$$

$$P(X < 2Y) = \int_0^{\infty} \left\{ \int_0^{2y} f(x, y)\, dx \right\} dy = -\int_0^{\infty} e^{-y}(e^{-2y} - 1)\, dy$$

$$= -\left| \frac{e^{-3y}}{-3} + e^{-y} \right|_0^{\infty} = 1 - \frac{1}{3} = \frac{2}{3}.$$

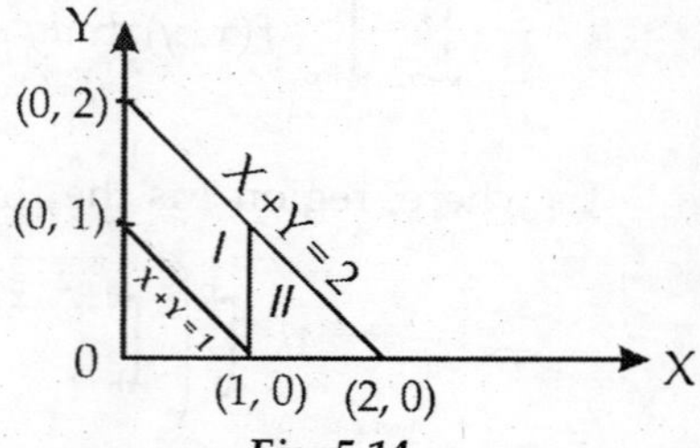

Fig. 5·14

Substituting in (3),

$$P(X < Y \mid X < 2Y) = \frac{1/2}{2/3} = \frac{3}{4}.$$

(*iii*) $$P(1 < X + Y < 2) = \iint_I f(x, y)\, dx\, dy + \iint_{II} f(x, y)\, dx\, dy$$

$$= \int_0^1 \left(\int_{1-x}^{2-x} f(x, y)\, dy \right) dx + \int_1^2 \left(\int_0^{2-x} f(x, y)\, dy \right) dx$$

$$= \int_0^1 \left(e^{-x} \int_{1-x}^{2-x} e^{-y}\, dy \right) dx + \int_1^2 \left(e^{-x} \int_0^{2-x} e^{-y} dy \right) dx$$

$$= \int_0^1 \frac{e^{-x}}{-1} \left(e^{x-2} - e^{x-1} \right) dx + \int_1^2 \frac{e^{-x}}{-1} \left(e^{x-2} - 1 \right) dx$$

$$= -\left(e^{-2} - e^{-1} \right) \int_0^1 1 \cdot dx - \int_1^2 \left(e^{-2} - e^{-x} \right) dx$$

$$= -\left(e^{-2} - e^{-1} \right) \left[\left| x \right|_0^1 \right] - \left[\left| e^{-2}.x + e^{-x} \right|_1^2 \right] = \frac{2}{e} - \frac{3}{e^2}$$

Example 5·48. *If the joint distribution function of X and Y is given by :*

$$F_{XY}(x,y) = \begin{cases} 1 - e^{-x} - e^{-y} + e^{-(x+y)} \ ; & x > 0, y > 0 \\ 0 \ ; & \textit{elsewhere} \end{cases}$$

(*a*) *Find the marginal densities of X and Y.*

(*b*) *Are X and Y independent ?*

(*c*) *Find* $P(X \le 1 \cap Y \le 1)$ *and* $P(X + Y \le 1)$.

Solution.

(*a*) and (*b*)

The joint *p.d.f.* of the *r.v.*'s (*X*, *Y*) is given by :

$$f_{XY}(x, y) = \frac{\partial^2 F_{XY}(x, y)}{\partial x \partial y} = \frac{\partial}{\partial x}\left[e^{-y} - e^{-(x+y)} \right] = \begin{cases} e^{-(x+y)} \ ; & x \ge 0, y \ge 0 \\ 0 \ ; & \text{otherwise} \end{cases} \quad \text{... (i)}$$

We have $f_{XY}(x, y) = e^{-x}.\ e^{-y} = f_X(x)\, f_Y(y)$ ... (*ii*)

where $f_X(x) = e^{-x}\ ;\ x \ge 0\ ;\quad f_Y(y) = e^{-y}\ ;\ y \ge 0$ ... (*iii*)

(*ii*) ⇒ *X* and *Y* are independent, and (*iii*) gives the marginal *p.d.f.*'s of *X* and *Y*.

(c) $P(X \le 1 \cap Y \le 1) = \int_0^1 \int_0^1 f(x, y)\, dx\, dy$

$$= \left(\int_0^1 e^{-x} dx \right) \left(\int_0^1 e^{-y}\, dy \right) = \left(1 - e^{-1}\right)^2$$

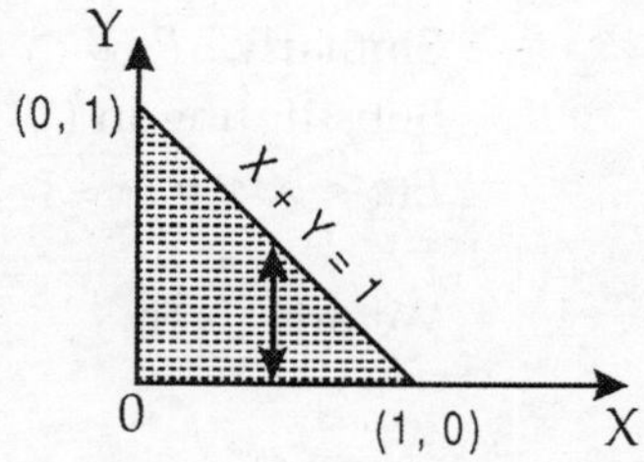

Fig. 5·15

$$P(X + Y \le 1) = \iint\limits_{X+Y \le 1} f(x, y) = \int_0^1 \left\{ \int_0^{1-x} f(x, y)\, dy \right\} dx$$

$$= \int_0^1 \left[e^{-x} \int_0^{1-x} e^{-y}\, dy \right] dx$$

$$= \int_0^1 e^{-x} \left\{ 1 - e^{-(1-x)} \right\} dx = \int_0^1 \left(e^{-x} - e^{-1} \right) dx = 1 - 2\, e^{-1}.$$

Example 5·49.

(i) Let F (x, y) be the distribution function of X and Y. Show that

$$P(a < X \le b,\ c < Y \le d) = F(b, d) - F(b, c) - F(a, d) + F(a, c),$$

where a, b, c and d are real constants a < b ; c < d.

$$\text{Deduce that if : } F(x, y) = \begin{cases} 1, & \text{for } x + 2y \ge 1 \\ 0, & \text{for } x + 2y < 1, \end{cases}$$

then F (x, y) cannot be joint distribution function of variables X and Y.

(ii) Show that, with usual notations : for all x, y,

$$F_X(x) + F_Y(y) - 1 \le F_{XY}(x, y) \le \sqrt{F_X(x)\, F_Y(y)}\,.$$

Solution.

(i) Let us define the events :

$A : \{X \le a\}$; $B : \{X \le b\}$; $C = \{Y \le c\}$; $D = \{Y \le d\}$; for $a < b$; $c < d$.

$P(a < X \le b \cap c < Y \le d)$

$= P\left[(B - A) \cap (D - C)\right]$

$= P\left[B \cap (D - C) - A \cap (D - C)\right]$ ... (*)

(By distributive property of sets)

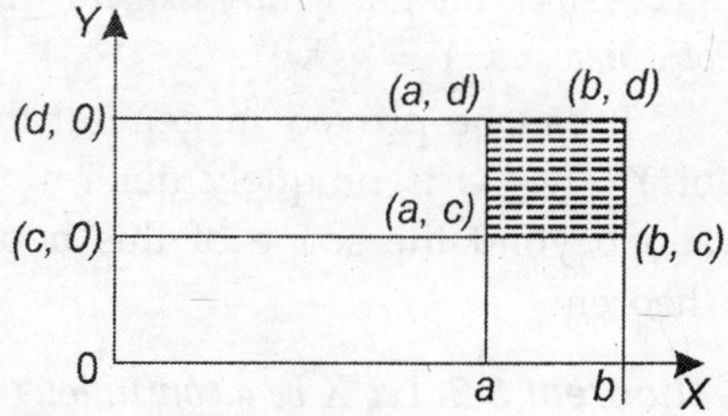

Fig. 5.16

We know that if $E \subset F \Rightarrow E \cap F = E$, then

$P(F - E) = P(\bar{E} \cap F) = P(F) - P(E \cap F) = P(F) - P(E)$ ... (**)

Obviously $A \subset B \Rightarrow [A \cap (D - C)] \subset [B \cap (D - C)]$

Hence, using (**), we get from (*)

$$P(a < X \le b \cap c < Y \le d) = P[B \cap (D - C)] - P[A \cap (D - C)]$$
$$= P[(B \cap D) - (B \cap C)] - P[(A \cap D) - (A \cap C)]$$
$$= P(B \cap D) - P(B \cap C) - P(A \cap D) + P(A \cap C) \quad \text{... (***)}$$

[On using (**), since $C \subset D \Rightarrow (B \cap C) \subset (B \cap D)$ and $(A \cap C) \subset (A \cap D)$]

We have : $P(B \cap D) = P[X \le b \cap Y \le d] = F(b, d)$.

Similarly, $P(B \cap C) = F(b, c)$; $P(A \cap D) = F(a, d)$ and $P(A \cap C) = F(a, c)$

Substituting in (***), we get

$$P(a < X \leq b \cap c < Y \leq d) = F(b, d) - F(b, c) - F(a, d) + F(a, c) \quad \text{... (1)}$$

We are given
$$F(x, y) = \begin{cases} 1, & \text{for } x+2y \geq 1 \\ 0, & \text{for } x+2y < 1 \end{cases} \quad \text{... (2)}$$

In (1) let us take $a = 0, b = \frac{1}{2}; c = \frac{1}{4}, d = \frac{3}{4}$ s.t. $a < b$ and $c < d$. Then using (2), we get

$F(b, d) = 1$; $F(b, c) = 1$; $F(a, d) = 1$; $F(a, c) = 0$.

Substituting in (1), $P(a < X \leq b \cap c < Y \leq d) = 1 - 1 - 1 + 0 = -1$;

which is not possible since $P(.) \geq 0$.

Hence $F(x, y)$ defined in (2) cannot be the distribution function of *r.v.*'s X and Y.

(*ii*) Let us define the events : $A \{X \leq x\}$; $B = \{Y \leq y\}$

Then : $P(A) = P(X \leq x) = F_X(x)$; $P(B) = P(Y \leq y) = F_Y(y)$ ... (3)

and $P(A \cap B) = P(X \leq x \cap Y \leq y) = F_{XY}(x, y)$

$(A \cap B) \subset A \Rightarrow P(A \cap B) \leq P(A) \Rightarrow F_{XY}(x, y) \leq F_X(x)$

$(A \cap B) \subset B \Rightarrow P(A \cap B) \leq P(B) \Rightarrow F_{XY}(x, y) \leq F_Y(y)$

Multiplying these inequalities, we get

$$F_{XY}^2(x, y) \leq F_X(x) F_Y(y) \Rightarrow F_{XY}(x, y) \leq \sqrt{F_X(x) F_Y(y)} \quad \text{... (4)}$$

Also $P(A \cup B) \leq 1 \Rightarrow P(A) + P(B) - P(A \cap B) \leq 1 \Rightarrow P(A) + P(B) - 1 \leq P(A \cap B)$

$$\therefore \quad F_X(x) + F_Y(y) - 1 \leq F_{XY}(x, y) \quad \text{... (5)}$$

From (4) and (5), $F_X(x) + F_Y(y) - 1 \leq F_{XY}(x, y) \leq \sqrt{F_X(x) F_Y(y)}$, as required.

5·6. TRANSFORMATION OF ONE-DIMENSIONAL RANDOM VARIABLE

Let X be a *r.v.* defined on the event space S and let $g(.)$ be a function such that $Y = g(X)$ is also a *r.v.* defined on S. In this section we shall deal with the following problem :

"Given the probability density function of a r.v. X, to determine the probability density function of a new r.v. Y = g(X)."

It can be proved in general that, if $g(.)$ is any continuous function, then the distribution of $Y = g(X)$ is uniquely determined by that of X. The proof of this result is rather difficult and beyond the scope of this book. Here, we shall consider the following, relatively simple theorem.

Theorem 5·3. *Let X be a continuous r.v. with p.d.f. $f_X(x)$. Let $y = g(x)$ be strictly monotonic (increasing or decreasing) function of x. Assume that g(x) is differentiable (and hence continuous) for all x. Then the p.d.f. h(.) of the r.v. Y is given by :*

$$h_Y(y) = f_X(x) \left| \frac{dx}{dy} \right|, \quad \text{... (5·22)}$$

where x is expressed in terms of y, and the range of Y is determined from the given range of the variable X, on using the transformation y = g(x).

Proof.

Case (i) $y = g(x)$ is strictly increasing function of x $\left(i.e., \frac{dy}{dx} > 0\right)$.

The *d.f.* of Y is given by : $H_Y(y) = P(Y \le y) = P\{g(X) \le y\} = P\{X \le g^{-1}(y)\}$,

the inverse exists and is unique, since $g(.)$ is strictly increasing.

$$\therefore \quad H_Y(y) = F_X\left[g^{-1}(y)\right], \text{ where } F \text{ is the } d.f. \text{ of } X$$

$$= F_X(x) \qquad [\because y = g(x) \Rightarrow g^{-1}(y) = x]$$

Differentiating *w.r.* to y, we get

$$h_Y(y) = \frac{d}{dy}\{F_X(x)\} = \frac{d}{dx}\{F_X(x)\}\frac{dx}{dy} = f_X(x) \cdot \frac{dx}{dy} \qquad \dots (*)$$

Case (ii) *$y = g(x)$ is strictly monotonic decreasing function of x.*

$$H_Y(y) = P(Y \le y) = P\{g(X) \le y\} = P\{X \ge g^{-1}(y)\}$$

$$= 1 - P\{X \le g^{-1}(y)\} = 1 - F_X\{g^{-1}(y)\} = 1 - F_X(x),$$

where $x = g^{-1}(y)$, the inverse exists and is unique. Differentiating *w.r.* to y, we get

$$h_Y(y) = \frac{d}{dx}[1 - F_X(x)]\frac{dx}{dy} = -f_X(x). \frac{dx}{dy} = f_X(x) \cdot \left(\frac{-dx}{dy}\right) \qquad \dots (**)$$

It may be noted that the algebraic sign (–ve) obtained in (**) is correct, since y is a decreasing function of x $\Rightarrow$ x is a decreasing function of y or $\frac{dx}{dy} < 0$.

The results in (*) and (**) can be combined to give $h_Y(y) = f_X(x)\left|\frac{dx}{dy}\right|$.

Example 5·50. *If the cumulative distribution function of a continuous r.v. X is F(x), find the cumulative distribution function of :*

(i) $Y = X + a$, *(ii)* $Y = X - b$, *(iii)* $Y = aX$, *(iv)* $Y = X^3$, *and*
(v) $Y = X^2$.

What are the corresponding probability density functions ?

Solution. Let $G(.)$ be the *c.d.f.* of Y. Then

(i) $G(x) = P(Y \le x) = P(X + a \le x) = P(X \le x - a) = F(x - a)$

(ii) $G(x) = P(Y \le x) = P(X - b \le x) = P(X \le x + b) = F(x + b)$

(iii) $G(x) = P(aX \le x) = P\{X \le (x/a)\} = F(x/a)$, if $a > 0$ and

$G(x) = P\{X \ge (x/a)\} = 1 - P\{X \le (x/a)\} = 1 - F(x/a)$, if $a < 0$.

(iv) $G(x) = P(Y \le x) = P(X^3 \le x) = P(X \le x^{1/3}) = F(x^{1/3})$

(v) $G(x) = (X^2 \le x) = P(-x^{1/2} \le X \le x^{1/2}) = P(X \le x^{1/2}) - P(X \le -x^{1/2})$

$$= \begin{cases} 0, & \text{if } x < 0 \\ F\left(\sqrt{x}\right) - F\left(-\sqrt{x} - 0\right), & \text{if } x > 0 \end{cases}$$

Variable	*Distribution function*	*p.d.f.*
X	$F(x)$	$f(x)$
X − a	$F(x+a)$	$f(x+a)$
aX	$\left.\begin{array}{l} F(x/a), a>0 \\ 1-F(x/a), a<0 \end{array}\right\}$	$\left.\begin{array}{l} \frac{1}{a} f(x/a),\ a>0 \\ -\frac{1}{a} f(x/a),\ a<0 \end{array}\right\}$
X^2	$\left.\begin{array}{ll} F(\sqrt{x})-F(-\sqrt{x}-0), & \text{for } x>0 \\ 0, & \text{otherwise} \end{array}\right\}$	$\left.\begin{array}{ll} \frac{1}{2(\sqrt{x})}\left[f(\sqrt{x})+f(-\sqrt{x})\right], & \text{for } x>0 \\ 0, & \text{for } x \le 0 \end{array}\right\}$
X^3	$F(x^{1/3})$	$\frac{1}{3} f\left(x^{1/3}\right) \cdot \frac{1}{x^{2/3}}\ ;\ x \ne 0.$

Example 5.51. *If the p.d.f. of r.v. X is :*

$$f(x) = \begin{cases} \frac{2}{9}(x+1) & ; \ -1<x<2 \\ 0 & ; \ \text{otherwise} \end{cases}$$

Obtain the p.d.f. of $Y = X^2$.

Solution. We split the range of X in two parts $-1 < x < 1$ and $1 \le x < 2$. Let $Y = X^2$

$$-1 < x < 1 \quad \Rightarrow \quad 0 < y < 1 \quad \text{and} \quad x = \pm \sqrt{y}$$

and
$$1 \le x < 2 \quad \Rightarrow \quad 1 \le y < 4 \quad \text{and} \quad 1 < x < \sqrt{y}$$

The distribution function G (·) of $Y = X^2$, is given by :

$$G_Y(y) = P(Y \le y) = P\left(X^2 \le y\right)$$

$$= \begin{cases} P\left(-\sqrt{y} < X < \sqrt{y}\right) = \int_{-\sqrt{y}}^{\sqrt{y}} \frac{2}{9}(x+1)\,dx\ ;\ 0<y<1 \\ P\left(1 < X < \sqrt{y}\right) = \int_{1}^{\sqrt{y}} \frac{2}{9}(x+1)\,dx\ ;\ 1 \le y < 4 \end{cases}$$

$$= \begin{cases} \frac{2}{9}\left|\frac{x^2}{2}+x\right|_{-\sqrt{y}}^{\sqrt{y}} = \frac{4}{9}\sqrt{y}\ ;\ 0<y<1 \\ \frac{2}{9}\left|\frac{x^2}{2}+x\right|_{1}^{\sqrt{y}} = \frac{2}{9}\left(\frac{y}{2}+\sqrt{y}-\frac{3}{2}\right); 1 \le y \le 4 \end{cases}$$

Differentiating *w.r.t. y*, the *p.d.f.* $g(\cdot)$ of Y is given by :

$$g(y) = \begin{cases} \dfrac{2}{9} \cdot \dfrac{1}{\sqrt{y}} \ ; & 0 < y < 1 \\[2ex] \dfrac{1}{9}\left(1 + \dfrac{1}{\sqrt{y}}\right) ; & 1 \le y < 4 \end{cases}$$

5·7. TRANSFORMATION OF TWO-DIMENSIONAL RANDOM VARIABLE

In this section we shall consider the problem of change of variables in the two-dimensional case. Let *r.v.'s* X and Y be transformed to the *r.v's* U and V by the transformation $u = u\ (x, y\ ), v = v\ (x, y)$, where u and v are continuously differentiable functions, for which Jacobian of transformation :

$$J = \frac{\partial(x, y)}{\partial(u, v)} = \begin{vmatrix} \dfrac{\partial x}{\partial u} & \dfrac{\partial y}{\partial u} \\[2ex] \dfrac{\partial x}{\partial v} & \dfrac{\partial y}{\partial v} \end{vmatrix}, \qquad \text{... (5·23)}$$

is either > 0 or < 0 throughout the (x, y) plane so that the inverse transformation is uniquely given by $x = x\ (u, v), y = y\ (u, v)$.

Theorem 5·4. *The joint p.d.f.* $g_{UV}\ (u, v)$ *of the transformed variables U and V is:*

$$g_{UV}\ (u, v) = f_{XY}(x, y)\ |J|, \qquad \text{... (5·24)}$$

where $|J|$ *is the modulus value of the Jacobian of transformation and f(x, y) is expressed in terms of u and v.*

Proof. $P\ (x < X \le x + dx, y < Y \le y + dy) = P\ (u < U \le u + du, v < V \le v + dv)$

$$\Rightarrow \qquad f_{XY}(x, y)\ dx\ dy = g_{UV}(u, v)\ du\ dv$$

$$\Rightarrow \qquad g_{UV}(u, v)\ du\ dv = f_{XY}\ (x, y) \left|\frac{\partial(x, y)}{\partial(u, v)}\right| du\ dv$$

$$\therefore \qquad g_{UV}(u, v) = f_{XY}\ (x, y) \left|\frac{\partial(x, y)}{\partial(u, v)}\right| = f_{XY}\ (x, y)\ |J|$$

Theorem 5·5. *(a) If X & Y are independent continuous r.v.'s, then the p.d.f. of U = X +Y is given by :*

$$h\ (u) = \int_{-\infty}^{\infty} f_X(v)\, f_Y(u-v)\, dv \qquad \text{... (5·25)}$$

Proof. Let $f_{XY}\ (x, y)$ be the joint *p.d.f.* of independent continuous *r.v.*'s X and Y and let us make the transformation : $u = x + y, \quad v = x \quad \Rightarrow \quad x = v, \quad y = u - v.$

$$J = \frac{\partial(x, y)}{\partial(u, v)} = \begin{vmatrix} \dfrac{\partial x}{\partial u} & \dfrac{\partial y}{\partial u} \\[2ex] \dfrac{\partial x}{\partial v} & \dfrac{\partial y}{\partial v} \end{vmatrix} = \begin{vmatrix} 0 & 1 \\ 1 & -1 \end{vmatrix} = -1$$

Thus the joint *p.d.f.* of *r.v.'s* U and V is given by :

$$g_{UV}\ (u, v) = f_{XY}\ (x, y\ )\ |J| = f_X\ (x)\ f_Y\ (y)\ |J| \qquad \text{(Since X and Y are independent)}$$

$$= f_X\ (v)\ f_Y\ (u - v)$$

The marginal density of U is given by :

$$h(u) = \int_{-\infty}^{\infty} g_{UV}(u, v)\, dv = \int_{-\infty}^{\infty} f_X(v)\, f_Y(u - v)\, dv$$

Remark. The function h (.) in (5·25) is given a special name and is said to be the *convolution of* f_X(.) and f_Y(.) and we write

$$h(.) = f_X(.) * f_Y(.).$$

Alter. The distribution function G(.) of $U = X + Y$, is given by :

$$G_U(u) = P(U \le u) = \iint_A f(x, y)\, dx\, dy,$$

where $A = \{(x, y) : x + y \le u\}$. Hence on $A : y \le u - x$.

$$\therefore \quad G_U(u) = \int_{-\infty}^{\infty} \left(\int_{-\infty}^{u-x} f(x, y)\, dy \right) dx$$

$$= \int_{-\infty}^{\infty} \left(\int_{-\infty}^{u} f(x, z - x)\, dz \right) dx \qquad [x + y = z \Rightarrow dy = dz]$$

$$= \int_{-\infty}^{u} \left(\int_{-\infty}^{\infty} f(x, z - x)\, dx \right) dz$$

(By Fubini's Theorem for non-negative functions)

Differentiating *w.r.* to u, the *p.d.f.* of $U = X + Y$, is given by :

$$g(u) = \int_{-\infty}^{\infty} f(x, u - x)\, dx \qquad \text{... (5·26)}$$

If X and Y are independent, then

$$g(u) = \int_{-\infty}^{\infty} f_X(x) \cdot f_Y(u - x)\, dx, \qquad \text{... (5·26}a\text{)}$$

an expression obtained in (5·25).

Theorem 5·5 (b) Distribution of the Difference of Two Random Variables. *If X and Y are independent continuous random variables, then the p.d.f. of U = X – Y, is given by :*

$$h(u) = \int_{-\infty}^{\infty} f(x, x - u)\, dx = \int_{-\infty}^{\infty} f_X(x)\, f_Y(x - u)\, dx \qquad \text{... (5·27)}$$

Proof. The distribution function G(.) of $U = X - Y$, is given by :

$$G_U(u) = P(U \le u) = \iint_{(x,y) : x - y \le u} f(x, y)\, dx\, dy$$

$$= \int_{-\infty}^{\infty} \left(\int_{x-u}^{\infty} f(x, y)\, dy \right) dx, \qquad [\because x - y \le u \Rightarrow y \ge x - u]$$

Put $x - y = z \Rightarrow x - z = y \Rightarrow dy = -dz$; when $y = x - u$, $z = x - (x - u) = u$ ... (*)

$$\therefore \quad G_U(u) = \int_{-\infty}^{\infty} \left(\int_{-\infty}^{u} f(x, x - z)\, dz \right) dx$$

$$= \int_{-\infty}^{u} \left(\int_{-\infty}^{\infty} f(x, x - z)\, dx \right) dz$$

Differentiating *w.r. to u*, the *p.d.f.* of $U = X - Y$, is given by :

$$g(u) = \int_{-\infty}^{\infty} f(x, x-u)\, dx \qquad \text{... (5·28)}$$

$$= \int_{-\infty}^{\infty} f(u+v, v)\, dv \qquad \text{... (5·28a)}$$

If X and Y are independent random variables, then the *p.d.f.* of $U = X - Y$, is given by :

$$g(u) = \int_{-\infty}^{\infty} f_X(x) \cdot f_Y(x-u)\, dx = \int_{-\infty}^{\infty} f_X(u+v)\, f_Y(v)\, dv$$

The range of U is obtained from the given range of X and Y, on using the transformation (*). ... (5·29)

Similarly, proceeding, the *p.d.f.* of $W = Y - X$, is given by :

$$g_1(w) = \int_{-\infty}^{\infty} f(v, w+v)\, dv \qquad \text{... (5·30)}$$

$$= \int_{-\infty}^{\infty} f_X(v)\, f_Y(w+v)\, dv, \text{ if } X \text{ and } Y \text{ are independent.} \qquad \text{...(5·30a)}$$

Aliter. Let us transform (X, Y) to (U, V) by the transformation :

$$x - y = u \quad \text{and} \quad y = v \quad \Rightarrow \quad x = u + v, \ y = v.$$

The Jacobian of transformation is given by :

$$J = \begin{vmatrix} \frac{\partial x}{\partial u} & \frac{\partial x}{\partial v} \\ \frac{\partial y}{\partial u} & \frac{\partial y}{\partial v} \end{vmatrix} = \begin{vmatrix} 1 & 1 \\ 0 & 1 \end{vmatrix} = 1$$

Hence, the joint *p.d.f.* of U and V becomes :

$$g_{UV}(u, v) = f_{XY}(x, y)\ |J| = f(u + v, v).\ 1$$

Integrating *w.r.* to v, the *p.d.f.* of U is given by :

$$g_1(u) = \int_{-\infty}^{\infty} f(u+v, v)\, dv = \int_{-\infty}^{\infty} f(z, z-u)\, dz$$

an expression obtained in (5·28*a*) or (5·28).

Theorem 5·5 (c) Distribution of the Product of two Random Variables. *Let X and Y be two independent continuous r.v.'s. Then the p.d.f. of their product U = XY, is given by :*

$$g(u) = \int_{-\infty}^{\infty} f_{XY}\left(x, \frac{u}{x}\right) \frac{dx}{|x|} \qquad \text{... (5·31)}$$

Proof. Let us transform (X, Y) to (U, V) by the transformation : $U = XY$ and $V = X$ so that :

$$u = xy, \quad v = x \quad \Rightarrow \quad x = v, \quad y = \frac{u}{v} \qquad \text{... (*)}$$

Jacobian of transformation is given by :

$$J = \begin{vmatrix} \dfrac{\partial x}{\partial u} & \dfrac{\partial x}{\partial v} \\ \dfrac{\partial y}{\partial u} & \dfrac{\partial y}{\partial v} \end{vmatrix} = \begin{vmatrix} 0 & 1 \\ \dfrac{1}{v} & \dfrac{-u}{v^2} \end{vmatrix} = \dfrac{-1}{v}$$

Hence, the joint *p.d.f.* of U and V is given by :

$$g(u, v) = f(x, y) \; |J| = f\left(v, \frac{u}{v}\right) \cdot \left|\frac{1}{v}\right|$$

Integrating *w.r.* to v, the *p.d.f.* of $U = XY$, is given by :

$$g(u) = \int_{-\infty}^{\infty} f\left(v, \frac{u}{v}\right)\left(\frac{dv}{|v|}\right) = \int_{-\infty}^{\infty} f\left(x, \frac{u}{x}\right)\frac{dx}{|x|} \qquad \text{... (5·32)}$$

$$= \int_{-\infty}^{\infty} f_X(x) \cdot f_Y\left(\frac{u}{x}\right) \cdot \frac{dx}{|x|}, \text{ if } X \text{ and } Y \text{ are independent.} \qquad \text{... (5·32}a\text{)}$$

The range of U is determined from the given range of X and Y on using the transformation (*).

Theorem 5·5 (d) Distribution of the Quotient of Two Random Variables. *If X and Y are independent continuous random variables, then the p.d.f. of the quotient Z = X/Y, is given by :*

$$g_1(z) = \int_{-\infty}^{\infty} f(vz, v) \, |v| \, dv = \int_{-\infty}^{\infty} f_X(vz) \, f_Y(v) \, |v| \, dv \qquad \text{... (5·33)}$$

Proof. Let us transform (X, Y) to (Z, V) by the transformation $Z = X/Y$ and $V = Y$, so that :

$$z = \frac{x}{y}, \; v = y \quad \Rightarrow \quad x = vz, \quad y = v \qquad \text{... (*)}$$

$$J = \begin{vmatrix} \dfrac{\partial x}{\partial z} & \dfrac{\partial y}{\partial z} \\ \dfrac{\partial x}{\partial v} & \dfrac{\partial y}{\partial v} \end{vmatrix} = \begin{vmatrix} v & 0 \\ z & 1 \end{vmatrix} = v$$

Hence, the joint *p.d.f.* of Z and V is given by :

$$g(z, v) = f(x, y) \; |J| = f(vz, v). \; |v|$$

Integrating *w.r.* to v, the *p.d.f.* of $Z = X/Y$, is given by :

$$g_1(z) = \int_{-\infty}^{\infty} f(vz, v) \, |v| \, dv \qquad \text{... (5·34)}$$

$$= \int_{-\infty}^{\infty} f_X(vz) \, f_Y(v) \, |v| \, dv, \text{ if } X \text{ and } Y \text{ are independent.} \qquad \text{... (5·35)}$$

The new range of the variable Z is determined from the range of the variables X and Y, on using the transformation (*).

Example 5·52. *Let (X, Y) be a two-dimensional non-negative continuous r.v. having the joint density :*

$$f(x, y) = \begin{cases} 4xy\, e^{-(x^2+y^2)}; & x \ge 0, y \ge 0 \\ 0 \quad ; & elsewhere \end{cases}$$

Prove that the density function of $U = \sqrt{X^2 + Y^2}$ *is :*

$$h(u) = \begin{cases} 2u^3 e^{-u^2} ; & 0 \le u < \infty \\ 0 \quad ; & elsewhere \end{cases}$$

Solution. Let us make the transformation : $u = \sqrt{x^2 + y^2}$ and $v = x$

$\Rightarrow \quad v \ge 0, \quad u \ge 0$ and $u \ge v$ or $u \ge 0$ and $0 \le v \le u$.

The Jacobian of transformation J is given by :

$$\frac{1}{J} = \frac{\partial(u, v)}{\partial(x, y)} = \begin{vmatrix} \frac{\partial u}{\partial x} & \frac{\partial v}{\partial x} \\ \frac{\partial u}{\partial y} & \frac{\partial v}{\partial y} \end{vmatrix} = \begin{vmatrix} \frac{x}{\sqrt{x^2+y^2}} & 1 \\ \frac{y}{\sqrt{x^2+y^2}} & 0 \end{vmatrix} = -\frac{y}{\sqrt{x^2+y^2}}$$

The joint *p.d.f.* of U and V is given by :

$$g(u, v) = f(x, y)\ |J| = 4xy\ e^{-(x^2+y^2)} \left| -\frac{\sqrt{x^2+y^2}}{y} \right| = 4x\sqrt{x^2+y^2}\ e^{-(x^2+y^2)}$$

$$= \begin{cases} 4vu \cdot e^{-u^2} ; & u \ge 0,\ 0 \le v \le u \\ 0 \quad ; & \text{otherwise} \end{cases}$$

Hence, the marginal density function of $U = \sqrt{X^2 + Y^2}$ is :

$$h(u) = \int_0^u g(u, v)\, dv$$

$$= 4ue^{-u^2} \int_0^u v\, dv = \begin{cases} 2u^3\, e^{-u^2}, & u \ge 0 \\ 0, & elsewhere \end{cases}$$

Example 5·53. *Let the p.d.f. of the random variable (x, y) be :*

$$f(x, y) = \begin{cases} \alpha^{-2} \exp\left[-\left(\frac{x+y}{\alpha}\right)\right]; & x, y > 0, \alpha > 0 \\ 0 \quad ; & elsewhere \end{cases}$$

Find the distribution of $\frac{1}{2}(X - Y)$.

Solution. Consider the transformation :

$$u = \frac{1}{2}(x-y) \quad \text{and} \quad v = y \quad \Rightarrow \quad x = 2u + v \quad \text{and} \quad y = v$$

The Jacobian of the transformation is :

$$J = \begin{vmatrix} \frac{\partial x}{\partial u} & \frac{\partial x}{\partial v} \\ \frac{\partial y}{\partial u} & \frac{\partial y}{\partial v} \end{vmatrix} = \begin{vmatrix} 2 & 1 \\ 0 & 1 \end{vmatrix} = 2$$

Thus, the joint *p.d.f.* of the random variables (U, V) is given by :

$$g(u, v) = \begin{cases} \frac{2}{\alpha^2} e^{-(2/\alpha)(u+v)} & ; -\infty < u < \infty, \ v > -2u, \text{ if } u < 0; \\ & ; v > 0 \text{ if } u \geq 0 \text{ and } \alpha > 0 \\ 0 & , \text{elsewhere} \end{cases}$$

The marginal *p.d.f.* of U is given by :

$$g_U(u) = \begin{cases} \int_{-2u}^{\infty} \frac{2}{\alpha^2} \exp\{-(2/\alpha)(u+v)\} \, dv = \frac{1}{\alpha} e^{2u/\alpha}, & u < 0 \\ \int_{0}^{\infty} \frac{2}{\alpha^2} \exp\{-(2/\alpha)(u+v)\} \, dv = \frac{1}{\alpha} e^{-2u/\alpha}, & u \geq 0 \end{cases}$$

Hence, $$g_U(u) = \frac{1}{\alpha} \cdot \exp\left\{-\frac{2}{\alpha} | u |\right\}, \quad -\infty < u < \infty.$$

Example 5·54. *Given the joint density function of X and Y as :*

$$f(x, y) = \begin{cases} \frac{1}{2} x e^{-y} ; & 0 < x < 2, y > 0 \\ 0 ; & \textit{elsewhere} \end{cases}$$

find the distribution of X + Y.

Solution. Consider the transformation :

$u = x + y$ and $v = y \quad \Rightarrow \quad y = v, \quad x = u - v$

The Jacobian of transformation $J = \frac{\partial(x, y)}{\partial(u, v)} = 1$

and the region $0 < x < 2$ and $y > 0$ transforms to $0 < u - v < 2$ and $v > 0$ as shown in the adjoining Fig. 5·17. The joint density function of U and V is given by :

$$g(u, v) = \frac{1}{2}(u - v) \cdot e^{-v} ; 0 < v < u, u > 0$$

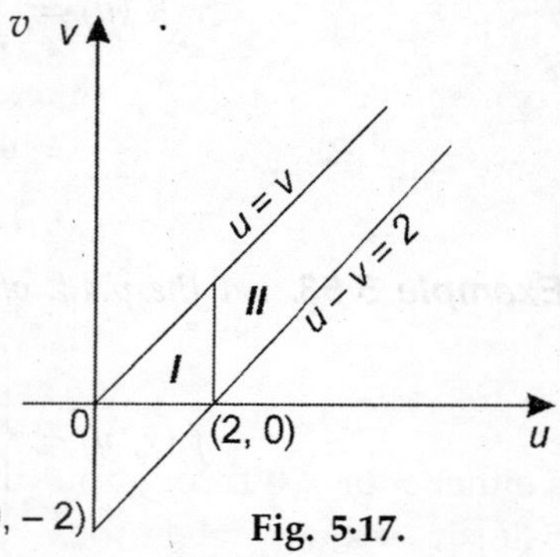

Fig. 5·17.

To find the density of $U = X + Y$, we split the range of U into two parts,

(*i*) $0 < u \leq 2$ (region *I*), and

(*ii*) $u > 2$ (region *II*) (which is suggested by the diagram).

For $0 < u \le 2$, (Region I) :

$$\begin{cases} h(u) = \int_0^u g(u, v)\, dv = \frac{1}{2}\int_0^u (u-v)\, e^{-v}\, dv \\ \quad = \frac{1}{2}\left| -e^{-v}(u-v) + e^{-v} \right|_{v=0}^{v=u} = \frac{1}{2}\left(e^{-u} + u - 1\right) \end{cases}$$

For $2 < u < \infty$, (Region II) :

$$\begin{cases} h(u) = \frac{1}{2}\int_{u-2}^{u} (u-v)\, e^{-v}\, dv \\ \quad = \frac{1}{2}\left| e^{-v}(1+v-u) \right|_{v=u-2}^{v=u} = \frac{1}{2} e^{-u}\left(1+e^2\right) \end{cases}$$

(On simplification)

Hence,
$$g(u) = \begin{cases} \frac{1}{2}(e^{-u} + u - 1), & 0 < u \le 2 \\ \frac{1}{2}e^{-u}(1+e^2), & 2 < u < \infty \\ 0, & \text{elsewhere} \end{cases}$$

5·8. TRANSFORMATION OF *n* - DIMENSIONAL RANDOM VARIABLE

The concepts in § 5·6 and § 5·7 can be similarly generalised to the case of transformation of n-dimensional ($n > 2$), random variables.

Let the n-dimensional continuous random variable $(X_1, X_2, \ldots, X_n)$ be transformed to the n-dimensional *r.v.* $(U_1, U_2, \ldots, U_n)$ by the transformation :

$$U_i = U_i\ (x_1, x_2, \ldots, x_n)\ ;\ i = 1, 2, \ldots, n \qquad \ldots (5·36)$$

where $U_1, U_2, \ldots, U_n$ are continuous differentiable functions for which the Jacobian of transformation,

$$J = \frac{\partial(x_1, x_2, \ldots, x_n)}{\partial(u_1, u_2, \ldots, u_n)} = \begin{vmatrix} \frac{\partial x_1}{\partial u_1} & \frac{\partial x_2}{\partial u_1} & \cdots & \frac{\partial x_n}{\partial u_1} \\ \frac{\partial x_1}{\partial u_2} & \frac{\partial x_2}{\partial u_2} & \cdots & \frac{\partial x_n}{\partial u_2} \\ \vdots & \vdots & & \vdots \\ \frac{\partial x_1}{\partial u_n} & \frac{\partial x_2}{\partial u_n} & \cdots & \frac{\partial x_n}{\partial u_n} \end{vmatrix} \qquad \ldots (5·36\ a)$$

is either > or < 0 throughout the $(x_1, x_2, \ldots, x_n)$ hyper-plane so that the inverse transformation of (5·36) is uniquely given by :

$$x_i = x_i\ (u_1, u_2, \ldots, u_n)\ ;\ i = 1, 2, \ldots, n \qquad \ldots (5·36\ b)$$

Then the joint *p.d.f.* $g\ (.)$ of $U_1, U_2, \ldots, U_n$ is given by :

$$g_{U_1, U_2, \ldots, U_n}(u_1, u_2, \ldots, u_n) = f_{X_1, X_2, \ldots, X_n}(x_1, x_2, \ldots, x_n) \times |J| \qquad \ldots (5·37)$$

where $|J|$ is the modulus value of the Jacobian of transformation J in (5·36*a*) and the right-hand side in (5·37) is expressed in term of $u_1, u_2, ..., u_n$ by using the inverse transformation (5·36*b*). The range of the transformed variables $U_1, U_2, ..., U_n$ is determined from the given range of the variables $(X_1, X_2, ..., X_n)$ on using the transformation (5·36).

CHAPTER CONCEPT QUIZ

1. Which of the following statements are TRUE or FALSE ?

(*i*) Given a continuous random variable X with probability density function $f(x)$, then $f(x)$ cannot exceed unity.

(*ii*) A random variable X has the probability density function : $f(x) = \begin{cases} x, & 0 < x < 1 \\ 0, & \text{elsewhere} \end{cases}$

(*iii*) The function defined as : $f(x) = \begin{cases} |x|, & -1 < x < 1 \\ 0, & \text{elsewhere} \end{cases}$, is a possible probability density function.

(*iv*) The following represents joint probability distribution.

Y \ X	1	2	3
– 1	1/9	1/18	1/18
0	1/18	2/9	3/9
1	1/18	1/18	1/18

2. Are any of the following probability mass or density functions ?
Prove your answer in each case.

(*a*) $f(x) = x \; ; \; x = \dfrac{1}{16}, \dfrac{3}{16}, \dfrac{1}{4}, \dfrac{1}{2}$

(*b*) $f(x) = \lambda\, e^{-\lambda x};\; x \geq 0 \; ; \; \lambda > 0,$

(*c*) $f(x) = \begin{cases} 2x & ; 0 < x < 1 \\ 4 - 2x & ; 1 \leq x < 2 \\ 0 & ; \text{elsewhere} \end{cases}$

3. If f_1 and f_2 are *p.d.f.*'s and $\theta_1 + \theta_2 = 1$, check if $g(x) = \theta_1 f_1(x) + \theta_2 f_2(x)$, is a p.d.f.

Ans. $g(x)$ is a p.d.f. if $\theta_1 > 0,\; \theta_2 > 0$ so that $\theta_1 + \theta_2 = 1$.

4. Fill in the blanks :

(*i*) If $p_1(x)$ and $p_2(y)$ be the marginal probability function of two independent discrete random variables X and Y, then their joint probability function $p(x, y) = ...$

(*ii*) The function $f(x)$ defined as

$$f(x) = \begin{cases} |x|, & -1 < x < 1 \\ 0, & \text{elsewhere} \end{cases}$$, is possible......

5. Find out the correct alternative :

(*i*) Let X be a random variable. Then for $f(x) = \begin{cases} k\,e^{-2x}, & x \geq 0 \\ 0, & \text{otherwise} \end{cases}$

to be density function, k must be equal to : (*a*) 2, (*b*) 1/2, (*c*) 0, (*d*) 1.

(*ii*) Let X be a random variable. Then $F(x) = \begin{cases} k(1-e^{-x})^2, & x \geq 0 \\ 0, & \text{otherwise} \end{cases}$

is a distribution function, provided k is : (*a*) – 1, (*b*) 1/2, (*c*) 1, (*d*) 2.

(*iii*) Let X and Y be two random variables. Then for

$$f(x, y) = \begin{cases} c(2x+y), & 0<x<1,\ 0<y<2 \\ 0, & \text{otherwise} \end{cases}$$

to be a joint density function, c must be equal to : (*a*) 1/3, (*b*) 1/4, (*c*) 1/5, (*d*) 1.

(*iv*) Let X and Y be two random variables. Then for

$$f(x, y) = \begin{cases} kxy, & \text{for } 0<x<4 \text{ and } 1<y<5 \\ 0, & \text{otherwise} \end{cases}$$

to be a joint density function, k must be equal to :

(*a*) 1/100, (*b*) 1/96, (*c*) 1/48, (*d*) none of these.

(*v*) Let the distribution function of a random variable X be : $F(x) = \begin{cases} 1-e^{-2x}, & \text{for } x \geq 0 \\ 0, & \text{otherwise} \end{cases}$

Then the density function is :

(*a*) $\begin{cases} 1-2e^{-x}, & x>0 \\ 0, & \text{otherwise} \end{cases}$ (*b*) $\begin{cases} 2e^{-2x}, & x>0 \\ 0, & \text{otherwise} \end{cases}$ (*c*) $\begin{cases} 1-2e^{-2x}, & x \geq 0 \\ 0, & \text{otherwise} \end{cases}$ (*d*) $\begin{cases} e^{-2x}, & x \geq 0 \\ 0, & \text{otherwise} \end{cases}$

(*vi*) If the joint *p.d.f.* of a two-dimensional random variable (X, Y) is given by :

$$f(x, y) = \begin{cases} 2, & \text{for } 0<x<1 \text{ and } 0<y<x \\ 0, & \text{otherwise} \end{cases}$$

then the conditional density function of X given Y is :

(*a*) $\begin{cases} 2y \text{ for } 0<y<1 \\ 0, \text{ otherwise} \end{cases}$ (*b*) $\begin{cases} 1-2y \text{ for } 0<y<1 \\ 0, \text{ otherwise} \end{cases}$ (*c*) $\begin{cases} 1-y \text{ for } 0<y<1 \\ 0, \text{ otherwise} \end{cases}$ (*d*) none of these.

(*vii*) Let $g(x) = \int_0^2 \left(x^2 + \frac{1}{3}xy\right) dy$ and $h(y) = \int_0^1 \left(x^2 + \frac{1}{3}xy\right) dx$,

Then the conditional density function of X given Y is :

(*a*) $(6x^2 + 2xy)/(2 + y)$ for $0 \leq x \leq 1$ and $0 \leq y \leq 2$

(*b*) $(3x + y)/(6x + 2)$ for $0 \leq x \leq 1$ and $0 \leq y \leq 2$

(*c*) $(6x^2 + 2xy)/(2 + y)$ for $0 \leq x \leq 2$ and $0 \leq y \leq 1$

(*d*) $(3x + y)/(6x + 2)$ for $0 \leq x \leq 2$ and $0 \leq y \leq 1$

(*viii*) Let the joint density function of a two-dimensional random variable (X, Y) be :

$$f(x, y) = \begin{cases} x+y \text{ for } 0 \leq x \leq 1 \text{ and } 0 \leq y \leq 1 \\ 0, \text{ otherwise} \end{cases}$$

Then the conditional density function of Y given X is

(*a*) $\begin{cases} \dfrac{(x+y)}{\left(x+\frac{1}{2}\right)}, & \text{for } 0 \le x \le 1 \text{ and } 0 \le y \le 1 \\ 0, & \text{otherwise} \end{cases}$ (*b*) $\begin{cases} \dfrac{(x+y)}{\left(y+\frac{1}{2}\right)}, & \text{for } 0 \le x \le 1 \text{ and } 0 \le y \le 1 \\ 0 & \text{otherwise} \end{cases}$

(*c*) $\begin{cases} \dfrac{(x+y)}{\left(x+\frac{1}{2}\right)}, & \text{for } 0 \le x \le 1 \text{ and } 0 \le y \le 1 \\ 0, & 0 \le x \le 1 \text{ and other } y \end{cases}$ (*d*) $\begin{cases} \dfrac{(x+y)}{\left(x+\frac{1}{2}\right)}, & \text{for } 0 \le x \le 1 \text{ and } 0 \le y \le 1 \\ 0, & \text{other } x \text{ and } 0 \le y \le 1 \end{cases}$

6. (*a*) Let (*X*, *Y*) be a pair of discrete random variables each taking three values 1, 2 and 3 with the following joint distribution :

Y \ X	1	2	3
1	$\frac{5}{27}$	$\frac{4}{27}$	$\frac{2}{27}$
2	$\frac{1}{27}$	$\frac{3}{27}$	$\frac{3}{27}$
3	$\frac{3}{27}$	$\frac{4}{27}$	$\frac{2}{27}$

Below are given a number of statements which are either true or false. Indicate in each case whether the statement is true or false, adding a remark in support of your answer.

(*i*) The marginal distributions of *X* and *Y* are the same.

(*ii*) The conditional distribution of *X* given that the value of *Y* is 2 is

x :	1	2	3
$P(X = x \mid Y = 2)$:	5/9	1/9	3/9

(*iii*) *X* and *Y* are dependent variables.

(*b*) Consider two events *A* and *B* such that $P(A) = \frac{1}{4}$, $P(B \mid A) = \frac{1}{2}$ and $P(A \mid B) = \frac{1}{4}$. Let the random variables *X* and *Y* be defined as : *X* = 1 or 0 depending on whether the event *A* has or has not occurred, and *Y* = 1 or 0 depending on whether the event *B* has or has not occurred. State whether each of the following statements is true or false :

(*i*) The random variables *X* and *Y* are independent,

(*ii*) $P(X^2 + Y^2 = 1) = \frac{1}{4}$,

(*iii*) $P(XY = X^2Y^2) = 1$, and

(*iv*) The random variables *X* and *Y* are identically distributed.

Ans. (*i*) True (*ii*) False (*iii*) True (*iv*) False.

DISCUSSION AND REVIEW QUESTIONS

1. Explain the concepts : (*i*) random variable, (*ii*) independence of random variables, and (*iii*) marginal and conditional probability distributions.

2. Define the "distribution function" (or cumulative distribution function) of a random variable and state its important properties.

3. Explain the terms (*i*) probability differential, (*ii*) probability density function, and (*iii*) distribution function.

4. Explain what is meant by a random variable. Distinguish between a discrete and a continuous random variable. Define distribution function of a random variable and show that it is monotonic non-decreasing everywhere and continuous on the right at every point.

5. Show that the distribution function $F(x)$ of a random variable *X* is a non-decreasing function of x. Determine the jump of $F(x)$ at a point x_0 of its discontinuity in terms of the probability that the random variable has the value x_0.

6. (a) Explain the notion of the joint distribution of two random variables. If $F(x, y)$ be the joint distribution function of X & Y, what will be the distribution functions for *the marginal distribution of X and Y ?*

(b) What is meant by the *conditional distribution* of Y under the condition that X = x ? Consider separately the cases where (i) X and Y are both discrete, and (ii) X & Y are both continuous.

ASSORTED REVIEW PROBLEMS FOR SELE-ASSESSMENT

5·1. (a) A random variable X can take all non-negative integral values, and the probability that X takes the value r is proportional to a^r, $(0 < a < 1)$. Find $P(X = 0)$.

(b) Suppose that the random variable X has possible values 1, 2, 3, ... and $P(X = j) = \frac{1}{2^j}$; j = 1, 2, 3, ... Compute (i) P (X is even), (ii) $P(X \ge 5)$, and (iii) P (X is divisible by 3).

Ans. (a) $P(X = r) = A.a^r$; r = 0, 1, 2, ... ; $A = 1 - a$; $P(X = 0) = A = 1 - a$

(b) (i) $\frac{1}{3}$, (ii) $\frac{1}{16}$, and (iii) $\frac{1}{7}$.

5·2. (a) Let X be a random variable such that :

$P(X = -2) = P(X = -1)$, $P(X = 2) = P(X = 1)$ and $P(X > 0) = P(X < 0) = P(X = 0)$.

Obtain the probability mass function of X and its distribution function.

(b) A random variable X assumes the values – 3, – 2, – 1, 0, 1, 2, 3 such that :

$$P(X = -3) = P(X = -2) = P(X = -1),\quad P(X = 1) = P(X = 2) = P(X = 3)$$

and $P(X = 0) = P(X > 0) = P(X < 0)$. Obtain the probability mass function of X and its distribution function, and find the probability mass function of $Y = 2X^2 + 3X + 4$.

Ans. (a)

x :	–2	–1	0	1	2
p (x) :	$\frac{1}{6}$	$\frac{1}{6}$	$\frac{2}{6}$	$\frac{1}{6}$	$\frac{1}{6}$
F(x) :	$\frac{1}{6}$	$\frac{2}{6}$	$\frac{4}{6}$	$\frac{5}{6}$	1

(b)

x :	–3	–2	–1	0	1	2	3
p(x) :	$\frac{1}{9}$	$\frac{1}{9}$	$\frac{1}{9}$	$\frac{1}{3}$	$\frac{1}{9}$	$\frac{1}{9}$	$\frac{1}{9}$
y :	13	6	3	4	9	18	31
p (y) :	$\frac{1}{9}$	$\frac{1}{9}$	$\frac{1}{9}$	$\frac{1}{3}$	$\frac{1}{9}$	$\frac{1}{9}$	$\frac{1}{9}$

5·3. (a) A random variable X has the following probability function :

Values of X, x :	– 2	– 1	0	1	2	3
p(x) :	0·1	*k*	0·2	2*k*	0·3	*k*

(i) Find the value of *k*, and calculate mean and variance.

(ii) Construct the *c.d.f.* F(x) and draw its graph.

(b) Given the probability function :

x	0	1	2	3
p (x)	0·1	0·3	0·5	0·1

let $Y = X^2 + 2X$. Find (i) the probability function of Y, (ii) mean and variance of Y.

Ans. (i)

y	0	3	8	15
p (y)	0·1	0·3	0·5	0·1

(ii) 6·4, 16·24

5·4. A random variable X has the following probability distribution :

Values of X, x	0	1	2	3	4	5	6	7	8
$p(x)$	k	$3k$	$5k$	$7k$	$9k$	$11k$	$13k$	$15k$	$17k$

(*i*) Determine the value of k.

(*ii*) Find $P(X < 3)$, $P(X \geq 3)$, $P(0 < X < 5)$.

(*iii*) Find the distribution function of X ?

(*iv*) What is the smallest value of x for which $P(X \leq x) > 0{\cdot}5$?

Ans. (*i*) $k = \dfrac{1}{81}$ (*ii*) $\dfrac{9}{81}, \dfrac{72}{81}, \dfrac{24}{81}$, (*iv*) 6

(*iii*)

x	0	1	2	3	4	5	6	7	8
$F(x)$	k	$4k$	$9k$	$16k$	$25k$	$36k$	$49k$	$64k$	$81k$

$; k = \dfrac{1}{81}$.

5·5. (*a*) Let $p(x)$ be the probability function of a discrete random variable X which assumes the values x_1, x_2, x_3, x_4, such that $2p(x_1) = 3p(x_2) = p(x_3) = 5p(x_4)$. Find probability distribution and cumulative probability distribution of X.

Ans.

x	x_1	x_2	x_3	x_4
$p(x)$	$\dfrac{15}{61}$	$\dfrac{10}{61}$	$\dfrac{30}{61}$	$\dfrac{6}{61}$

(*b*) The following is the distribution function of a discrete random variable X :

x	– 3	– 1	0	1	2	3	5	8
$F(x)$	0·10	0·30	0·45	0·65	0·75	0·90	0·95	1·00

(*i*) Find the probability distribution of X, (*ii*) Find P (*X is even*) and $P\ (1 \leq X \leq 8)$, and (*iii*) Find $P\ (X = -3 \mid X < 0)$ and $P(X \geq 3 \mid X > 0)$.

Ans. (*ii*) 0·30, 0·55, (*iii*) $\dfrac{1}{3}, \dfrac{5}{11}$

5·6. The probability mass function of a random variable X is zero except at the points $x = 0, 1, 2$. At these points it has the values $p\ (0) = 3c^3$, $p\ (1) = 4c - 10c^2$ and $p(2) = 5c - 1$ for some $c > 0$.

(*i*) Determine value of c.

(*ii*) Compute the probabilities, $P(X < 2)$ and $P\ (1 < X \leq 2)$.

(*iii*) Describe the distribution function and draw its graph.

(*iv*) Find the largest x such that $F(x) < \dfrac{1}{2}$.

(*v*) Find the smallest x such that $F(x) \geq \dfrac{1}{3}$.

Ans. (*i*) $\dfrac{1}{3}$, (*ii*) $\dfrac{1}{3}, \dfrac{2}{3}$, (*iv*) 1, (*v*) 1.

5·7. Suppose that the random variable X assumes three values 0, 1 and 2 with probabilities 1/3, 1/6 and 1/2 respectively. Obtain the distribution function of X.

5·8. Given that $f(x) = k\,(1/2)^x$, is a probability distribution for a random variable which can take on the values $x = 0, 1, 2, 3, 4, 5, 6$, find k and find an expression for the corresponding cumulative probability function $F(x)$.

5·9. If a die is thrown n times, find the probability that (*i*) the greatest, (*ii*) the least, number obtained will be a given value k.

Ans. (*i*) $\frac{1}{6^n}\left\{k^n-(k-1)^n\right\}$; (*ii*) $\frac{1}{6^n}\left\{(7-k)^n-(6-k)^n\right\}$

5·10. (*a*) A continuous *r.v.* X follows the probability law : $f(x) = Ax^2, 0 \le x \le 1$

Determine *A* and find the probability that

(*i*) X lies between 0·2 and 0·5, (*ii*) X is less than 0·3,

(*iii*) $\frac{1}{4} < X < \frac{1}{2}$ and (*iv*) $X > \frac{3}{4}$ given $X > \frac{1}{2}$.

Ans. *A* = 0·3, (*i*) 0·117, (*ii*) 0·027, (*iii*) $\frac{15}{256}$ and (*iv*) $\frac{27}{56}$.

5·10. (*b*) A *r.v.* X has the density function : $f(x) = \begin{cases} \frac{1}{4}, -2<x<2 \\ 0 \text{ , elsewhere} \end{cases}$

Obtain (*i*) $P(X < 1)$, (*ii*) $P(|X| > 1)$ (*iii*) $P(2X + 3 > 5)$

Hint. (*ii*) $P(|X| > 1) = P(X > 1 \text{ or } X < -1) = \int_{-2}^{-1} f(x)\,dx + \int_1^2 f(x)\,dx$

or $P(|X| > 1) = 1 - P(|X| \le 1) = 1 - P(-1 \le X \le 1)$

Ans. (*i*) 3/4, (*ii*) 1/2 (*iii*) 1/4.

5·10. (*c*) A random process gives measurements X between 0 and 1 with a *p.d.f.* :

$$f(x) = \begin{cases} 12x^3 - 21x^2 + 10x, & 0 \le x \le 1 \\ 0 & \text{, elsewhere} \end{cases}$$

(*i*) Find $P\left(X \le \frac{1}{2}\right)$ and $P\left(X > \frac{1}{2}\right)$, (*ii*) Find a number *k* such that $P(X \le k) = \frac{1}{2}$.

Ans. (*i*) $\frac{9}{16}, \frac{7}{16}$ (*ii*) $k = 0·452$.

5·11. (*a*) A continuous random variable X has the probability density function :

$$f(x) = A + Bx, \quad 0 \le x \le 1.$$

If the mean of the distribution is $\frac{1}{2}$, find *A* and *B*.

Hint. Solve $\int_0^1 f(x)\,dx = 1$ and $\int_0^1 x\,f(x)\,dx = \frac{1}{2}$ for *A* and *B*.

Ans. *A* = 1, *B* = 0.

5·12. (*a*) For the probability density function :

$$f(x) = cx^2(1-x), \; 0 < x < 1,$$

find (*i*) the constant *c*, and (*ii*) mean.

Ans. (*i*) c = 12, (*ii*) Mean = 3/5.

5·12. (*b*) The distribution of a continuous random variable X in range (– 3, 3) is given by p.d.f. :

$$f(x) = \begin{cases} \frac{1}{16}(3+x)^2 & , -3 \le x \le -1 \\ \frac{1}{16}\left(6-2x^2\right) & , -1 \le x \le 1 \\ \frac{1}{16}(3-x)^2 & , 1 \le x \le 3 \end{cases}$$

(i) Verify that the area under the curve is unity.

(ii) Find the mean and variance of the above distribution.

Hint. $\int_{-3}^{3} f(x)\,dx = \int_{-3}^{-1} f(x)\,dx + \int_{-1}^{1} f(x)\,dx + \int_{1}^{3} f(x)\,dx$

Ans. (ii) Mean = 0, variance = 1.

5·13. For the continuous distribution : $dF = y_0\left(x - x^2\right)dx\,;\ 0 \le x \le 1, y_0$ being a constant, find

(i) arithmetic mean, median, mode,

(ii) harmonic mean, geometric mean and

(iii) rth moment about mean. Hence find β_1 and β_2 and show that the distribution is symmetrical.

(iv) Also find mean deviation about mean.

Ans. $y_0 = 6$;

(i) Mean = Median = Mode = $\frac{1}{2}$,

(ii) H.M. = $\frac{1}{3}$; G.M. = exp (– 5/6)

(iii) $\mu_{2n+1} = 0,\ \mu_{2n} = \frac{3}{2^{2n}(2n+1)(2n+3)};\ n = 0, 1, 2, \ldots\ ;\quad \beta_1 = 0\ ;\quad \beta_2 = (15/7) = 2{\cdot}14$

(iv) M.D. about mean = (3/16).

5.14. The distribution of r.v. *X* in the range (0, 2) is defined by :

$$f(x) = \begin{cases} x^3 & , \ 0 < x \le 1 \\ 3(2-x)^3 & , \ 1 < x \le 2 \end{cases}$$

Calculate the mean, standard deviation and the mean deviation about mean of the distribution.

Ans. Mean = 1·1 ; S.D. = $\sqrt{0{\cdot}5667}$ = 0·238 ; M.D. about mean = $\frac{3}{10}\cdot\left(\frac{9}{10}\right)^5$

5·15. (a) A continuous random variable X has a *p.d.f.* given by : $f(x) = \begin{cases} kx\,e^{-\lambda x} & ; \ x \ge 0, \lambda > 0 \\ 0 & ; \ \text{otherwise} \end{cases}$

Determine the constant *k* and obtain the mean and variance of X.

(b) For the *p.d.f.* of *r.v.* *X* given by : $f(x) = \begin{cases} \dfrac{2(b+x)}{b(a+b)} & , \ -b \le x < 0 \\ \dfrac{2(a-x)}{a(a+b)} & , \ 0 \le x \le a \end{cases}$

find mean, median and variance.

Ans. (b) Mean = $\frac{1}{3}(a-b)$, median = $a - \sqrt{\frac{1}{2}a(a+b)}$ and variance = $\frac{1}{18}(a^2 + b^2 + ab)$. .

5·16. (a) Show that for the exponential distribution : $dP = y_o \cdot \exp\left(-x/\sigma\right)dx\,;0 \le x < \infty, \sigma > 0$, the mean and *S.D.* are both equal to σ and that the interquartile range is $\sigma \log_e 3$. Also find μ_r' and show that $\beta_1 = 4,\ \beta_2 = 9$.

Ans. $\mu_r' = \sigma^r \cdot (r\,!)$; $r = 1, 2, 3, \ldots$

(b) Express the constants y_0, a and m of the distribution :

$$dF = y_0\left(1-\frac{x^2}{a^2}\right)^m dx \ ; \quad -a \le x \le a \text{ , in terms of } \mu_2 \text{ and } \beta_2.$$

Ans. $a = \sqrt{(2m+3)\ \mu_2}$; $m = \dfrac{9-5\beta_2}{2\beta_2-6}$; $y_0 = \dfrac{1}{a\sqrt{\pi}} \cdot \dfrac{\Gamma(m+3/2)}{\Gamma(m+1)}$

(c) A variable X can assume values only between 0 and 5 and the equation of its frequency curve is :

$y = A \sin\left(\frac{1}{5}\pi x\right)$, $0 \le x \le 5$, where A is a constant such that the area under the curve is unity.

Determine the value of A and obtain the median and quartiles of the distribution.

Show also that the variance of the distribution is $50\left(\dfrac{1}{8}-\dfrac{1}{\pi^2}\right)$.

Ans. $A = \dfrac{1}{10}$, Median = 2·5, $Q_1 = \dfrac{4}{3}$, $Q_3 = \dfrac{10}{3}$.

5·17. Show that the following is a density function and find its median :

$$f(x) = \frac{\alpha^2(\alpha+2x)}{x^2(\alpha+x)^2}\, I_{(\alpha,\infty)}(x) + \frac{x(2\alpha+x)}{\alpha(\alpha+x)^2}\, I_{(0,\alpha)}(x), \text{ for } \alpha > 0$$

Hint.

$$f(x) = \begin{cases} \dfrac{\alpha(\alpha+2x)}{x^2(\alpha+x)^2} = \dfrac{\alpha}{x^2} - \dfrac{\alpha}{(\alpha+x)^2}, & \text{if } \alpha < x < \infty \\[2ex] \dfrac{x(2\alpha+x)}{\alpha(\alpha+x)^2} = \dfrac{1}{\alpha} - \dfrac{\alpha}{(\alpha+x)^2}, & \text{if } 0 < x \le \alpha \end{cases}$$

Since
$$\int_0^\infty f(x)\,dx = \int_0^\alpha \left\{\frac{1}{\alpha} - \frac{\alpha}{(\alpha+x)^2}\right\} dx + \int_\alpha^\infty \left\{\frac{\alpha}{x^2} - \frac{\alpha}{(\alpha+x)^2}\right\} dx$$

$$= \alpha\cdot\frac{1}{\alpha} + \alpha\left|\frac{1}{\alpha+x}\right|_0^\alpha - \left|\frac{\alpha}{x}\right|_\alpha^\infty + \left|\frac{\alpha}{\alpha+x}\right|_\alpha^\infty = 1,$$

$\therefore$ $f(x)$ is a *p.d.f.*

Also
$$\int_0^\alpha f(x)\,dx = \int_0^\alpha \left\{\frac{1}{\alpha} - \frac{\alpha}{(\alpha+x)^2}\right\} dx = 1+\alpha\left(\frac{1}{2\alpha}-\frac{1}{\alpha}\right) = \frac{1}{2} \quad ; \text{ Median} = \alpha$$

5·18. Pareto distribution with parameters r and A is given by the *p.d.f.* :

$$f(x) = \begin{cases} rA^r \dfrac{1}{x^{r+1}}, & \text{for } x \ge A, \quad r > 0 \\ 0, & \text{elsewhere} \end{cases}$$

Show that it has a finite nth moment if and only if $n < r$. Find the mean, median and variance of the distribution.

Ans. $\mu_n' = \dfrac{rA^n}{r-n}$; $n < r$; Mean $= \dfrac{rA}{r-1}$, Variance $= \dfrac{rA^2}{(r-2)(r-1)^2}$; Median $= A.(2^{1/r})$

5·19. For a continuous random variable X, defined in the range $(0 \le x < \infty)$, the probability distribution is such that $P(X \le x) = 1 - \exp(-\beta x^2)$, where $\beta > 0$.

Find the median of the distribution. Also if m, m_0 and σ denote the mean, mode and standard deviation respectively of the distribution, prove that $2m_0^2 - m^2 = \sigma^2$ and $m_0 = m.\sqrt{2/\pi}$.

What is the sign of skewness of the distribution ?

Ans. Mean $(m) = \dfrac{1}{2}\cdot\dfrac{\sqrt{\pi}}{\sqrt{\beta}}$; Mode $(m_0) = \dfrac{1}{\sqrt{2\beta}}$; $\sigma^2 = \dfrac{4-\pi}{4\beta}$; Median $= \sqrt{\dfrac{\log_e 2}{\beta}}$

Since, Mean – Mode $= \dfrac{\sqrt{\pi} - \sqrt{2}}{2\sqrt{\beta}} > 0$, skewness is positive.

5·20. A man leaves his house at the same time every morning and the time taken for journey to work has the following probability density function : less than 30 minutes, zero; between 30 minutes and 60 minutes, uniform with density k; between 60 minutes and 70 minutes, uniform with density $2k$; and more than 70 minutes, zero. What is the probability that on one particular day he arrives at work later than on the previous day but not more than 5 minutes later.

5·21. A batch of small calibre ammunition is accepted as satisfactory if none of a sample of five shots falls more than 2 metres from the centre of the target at a given range. If X, the distance from the centre of the target to a given impact point, actually has the *p.d.f.* :

$$f(x) = k.\ 2x \exp(-x^2),\ 0 < x < 3,$$

where k is a number which makes it probability density function, what is the value of k and what is the probability that the batch will be accepted ?

Hint. $\displaystyle\int_0^3 f(x)\,dx = 1 \quad \Rightarrow \quad k = \frac{1}{1-\exp(-9)}.$

Required probability = P [Each of a sample of 5 shots falls within a distance of 2 ft. from the centre]

$$= [P(0 < X < 2)]^5 = \left[\int_0^2 f(x)\,dx\right]^5 = \left[\frac{1-e^{-4}}{1-e^{-9}}\right]^5$$

5·22. A *r.v.* X has the *p.d.f.* : $f(x) = \begin{cases} 2x, & 0 < x < 1 \\ 0, & \text{otherwise} \end{cases}$

Find (*i*) $P\left(X < \dfrac{1}{2}\right)$, (*ii*) $P\left(\dfrac{1}{4} < X < \dfrac{1}{2}\right)$, (*iii*) $P\left(X > \dfrac{3}{4} \,\middle|\, X > \dfrac{1}{2}\right)$, and (*iv*) $P\left(X < \dfrac{3}{4} \,\middle|\, X > \dfrac{1}{2}\right)$.

Ans. (*i*) $\dfrac{1}{4}$, (*ii*) $\dfrac{3}{16}$, (*iii*) $\dfrac{P(X > 3/4)}{P(X > 1/2)} = \dfrac{7/16}{3/4} = \dfrac{7}{12}$; (*iv*) $\dfrac{P(1/2 < X < 3/4)}{P(X > 1/2)}$.

5·23. The length (in hours) X of a certain type of light bulb may be supposed to be a continuous random variable with *p.d.f.* : $f(x) = \begin{cases} \dfrac{a}{x^3}, & 1{,}500 < x < 2{,}500 \\ 0, & \text{elsewhere} \end{cases}$

Determine the constant a, the distribution function of X, and compute the probability of the event $1{,}700 \le X \le 1{,}900$.

Hint. $a = 7031250$; $F(x) = \frac{a}{2}\left(\frac{1}{22{,}50{,}000} - \frac{1}{x^2}\right)$ and

$$P\,(1{,}700 < X < 1{,}900) = F(1{,}900) - F(1{,}700) = \frac{a}{2}\left(\frac{1}{28{,}90{,}000} - \frac{1}{36{,}10{,}000}\right)$$

5·24. *(a)* The kms X in thousands of kms, which car owners get with a certain kind of tyre is a random variable having probability density function :

$$f(x) = \begin{cases} \frac{1}{2}\cdot \exp\left(-\frac{x}{2}\right), & \text{for } x > 0 \\ 0, & \text{for } x \le 0 \end{cases}$$

Find the probability that one of these tyres will last :
(i) at most 10,000 kms., *(ii)* anywhere from 16,000 to 24,000 kms, *(iii)* at least 30,000 kms.

Ans. *(i)* 0·3935, *(ii)* 0·1481, *(iii)* 0·2231.

5·24. *(b)* A petrol pump is supplied with petrol once a day. If its daily volume of sales (X) in thousands of litres is distributed by :

$$f\,(x) = 5\,(1 - x)^4\ ,\ 0 \le x \le 1.$$

What must be the capacity of its tank in order that the probability that its supply will be exhausted in a given day shall be 0·01 ?

Hint. Let the capacity of the tank (in '000 of litres) be *'a'* such that :

$$\text{P}\,(X \ge a) = \int_a^1 f(x)\,dx = 0.01 \quad \Rightarrow \quad a = \left\{1 - (0.01)^{1/5}\right\} = 0{\cdot}6019$$

Ans. 601·9 liters.

5·25. Calculate the standard deviation and mean deviation from mean if probability function $f(x)$ has the form:

$$f\,(x) = \begin{cases} \frac{3+2x}{18}, & 2 \le x \le 4 \\ 0, & \text{otherwise} \end{cases}$$

Ans. Mean $= \frac{83}{27}$; S.D. $= \sqrt{\frac{239}{729}} = 0{\cdot}57$; M.D. from mean = 0·49.

5·26. A random variable X is distributed at random between the value 0 and 1 so that its probability density function : $f\,(x) = kx^2\,(1 - x^3)$, where k is a constant. Find the value of k. Using this value of k, find its mean and variance.

Ans. $k = 6$, Mean = (9/14), Variance = (9/245).

5·27. For the p.d.f. : $f_X\,(x) = k\,e^{-ax}\,(1 - e^{-ax})\,I_{0,\,\infty}\,(x)$,
find the normalising constant k, $F_X\,(x)$ and evalute $P\,(X > 1)$.

Ans. $k = 2a$, $F\,(x) = 1 - 2e^{-ax} + e^{-2ax}$; $P(X > 1) = 2\,e^{-a} - e^{-2a}$.

5·28. A random variable X has the density function :

$$f(x) = \begin{cases} k\cdot\frac{1}{1+x^2}, & \text{if } -\infty < x < \infty \\ 0, & \text{otherwise} \end{cases}$$

Determine k and the distribution function. Evaluate the probability $P(X \geq 0)$. Also find the mean and variance, mode and quartile range of X.

Ans. $k = \frac{1}{\pi}$, $F(x) = \frac{1}{\pi}\left(\tan^{-1} x + \frac{\pi}{2}\right)$, $P(X \geq 0) = \frac{1}{2}$, Mean $= \frac{1}{\pi}\int_{-\infty}^{\infty} \frac{x}{1+x^2}\, dx = 0$

$$\text{Variance} = \mu_2 = \mu'_2 = \int_{-\infty}^{\infty} x^2 f(x)\, dx$$

$$= \frac{1}{\pi}\int_{-\infty}^{\infty} \frac{x^2}{1+x^2}\, dx = \frac{2}{\pi}\int_{0}^{\infty} \frac{x^2}{1+x^2}\, dx = \frac{2}{\pi}\int_{0}^{\infty} \left(1 - \frac{1}{1+x^2}\right) dx = \infty$$

$\therefore$ Mean = 0 and variance does not exist ; Mode = 0, Quartile Range = 2

5·29 A continuous random variable X has the distribution function :

$$F(x) = \begin{cases} 0 & , \quad \text{if } x \leq 1 \\ k(x-1)^4 & , \quad \text{if } 1 < x \leq 3 \\ 1 & , \quad \text{if } x > 3 \end{cases}$$

Find (*i*) k, (*ii*) the probability density function $f(x)$, and (*iii*) the mean and the median of X.

Ans. (*i*) $k = \frac{1}{16}$, (*ii*) $f(x) = \frac{1}{4}(x-1)^3$, $1 \leq x \leq 3$.

5·30. The length of time (in minutes) that a certain lady speaks on the telephone is found to be random phenomenon, with a probability function specified by the probability density function $f(x)$ as :

$$f(x) = \begin{cases} A\, e^{-x/5}, & \text{for } x \geq 0 \\ 0 & , \text{otherwise} \end{cases}$$

(*a*) Find the value of A that makes $f(x)$ a *p.d.f.*

(*b*) What is the probability that the number of minutes that she will talk over the phone is : (*i*) more than 10 minutes, (*ii*) less than 5 minutes, and (*iii*) between 5 and 10 minutes.

Ans. $A = \frac{1}{5}$, (*i*) $\frac{1}{e^2}$, (*ii*) $\frac{e-1}{e}$, (*iii*) $\frac{e-1}{e^2}$.

5·31. Suppose that the time in minutes that a person has to wait at a certain station for a train is found to be a random phenomenon with a probability function specified by the following distribution function :

$$F(x) = \begin{cases} 0, \text{ for } x \leq 0 \\ \frac{x}{2}, \text{ for } 0 \leq x < 1 \\ \frac{1}{2}, \text{ for } 1 \leq x < 2 \\ \frac{x}{4}, \text{ for } 2 \leq x < 4 \\ 1, \text{ for } x \geq 4 \end{cases}$$

(*a*) Is the distribution function continuous ? If so, give its p.d.f.

(*b*) What is the probability that a person will have to wait :
(*i*) more than 3 minutes, (*ii*) less than 3 minutes, and (*iii*) between 1 and 3 minutes ?

(c) What is the conditional probability that the person will have to wait for a train for
(i) more than 3 minutes, given that it is more than 1 minute,
(ii) less than 3 minutes, given that it is more than 1 minute ?

Ans. (a) $f(x) = \begin{cases} 0, & \text{for } x < 0 \\ \frac{1}{2}, & \text{for } 0 \le x < 1 \\ 0, & \text{for } 1 \le x < 2 \\ \frac{1}{4}, & \text{for } 2 \le x < 4 \\ 0, & \text{for } x \ge 4 \end{cases}$ (b) (i) $\frac{1}{4}$, (ii) $\frac{3}{4}$, (iii) $\frac{1}{4}$ (c) (i) $\frac{1}{2}$, (ii) $\frac{1}{2}$.

5·32. (a) Two fair dice are tossed simultaneously. Let X denote the number on the first die and Y denote the number on the second die,
(i) Write down the sample space of this experiment.
(ii) Find the following probabilities :
(1) $P(X + Y = 8)$, (2) $P(X + Y \ge 8)$, (3) $P(X = Y)$, (4) $P(X + Y = 6 \mid Y = 4)$, (5) $P(X - Y = 2)$.

Ans. (i) $S = \{1, 2, 3, 4, 5, 6\} \times \{1, 2, 3, 4, 5, 6\} \Rightarrow n(S) = 36$

(ii) (1) $\frac{5}{36}$, (2) $\frac{15}{36}$, (3) $\frac{1}{6}$, (4) $\frac{1}{6}$, (5) $\frac{1}{9}$.

(b) Two dice are rolled, $S = \{(a, b) \mid a, b = 1, 2 \ldots, 6\}$. Let X denote the sum of the two faces and Y the absolute value of their difference, *i.e.*, X is distributed over the integers 2, 3, ... , 12 and Y over 0, 1, 2, ..., 5. Assuming the dice are fair, find the probabilities :
(i) $X = 5 \cap Y = 1$, (ii) $X = 7 \cap Y \ge 3$, (iii) $X = Y$, and (iv) $(X + Y) = 4 \cap (X - Y) = 2$.

Ans. (i) $\frac{1}{18}$, (ii) $\frac{1}{9}$, (iii) 0, and (iv) $\frac{1}{18}$.

5·33. (a) The joint probability distribution of a pair of random variables is given by the following table :

Y \ X	1	2	3
1	0·1	0·1	0·2
2	0·2	0·3	0·1

Find :
(i) The marginal distributions.
(ii) The conditional distribution of X given $Y = 1$.
(iii) $P\{(X + Y) < 4\}$.

Ans. (i)

x	1	2	3
$p_1(x)$	0·3	0·4	0·3

;

y	1	2
$p_2(y)$	0·4	0·6

(ii)

x	1	2	3
$p(x \mid Y = 1)$	0·25	0·25	0·50

(iii) 0·4

(b) What do you mean by marginal and conditional distributions ? The following table represents the joint probability distribution of the discrete random variable (X, Y) :

Y \ X	1	2	3
1	$\frac{1}{12}$	$\frac{1}{6}$	0
2	0	$\frac{1}{9}$	$\frac{1}{5}$
3	$\frac{1}{18}$	$\frac{1}{4}$	$\frac{2}{15}$

(i) Evaluate marginal distribution of X.
(ii) Evaluate conditional distribution of Y given X = 2.

Ans. (*i*)

x	1	2	3
$p_1(x)$	5/36	19/36	5/15

, (*ii*)

y	1	2	3
$p(y \mid X=2)$	6/19	4/19	9/19

5·34. Let the joint *p.m.f.* of X_1 and X_2 be : $p(x_1, x_2) = \begin{cases} \dfrac{x_1+x_2}{21} & ; x_1 = 1, 2, 3 ; x_2 = 1, 2 \\ 0 & , \text{otherwise} \end{cases}$

Show that marginal *p.m.f.*'s of X_1 and X_2 are :

$$p_1(x_1) = \frac{2x_1+3}{21} \ ; x_1 = 1, 2, 3; \qquad p_2(x_2) = \frac{6+3x_2}{21} \ ; x_2 = 1, 2$$

5·35. One bag contains three identical cards marked 1, 2, 3 and another contains two cards marked 1, 2. Two cards are randomly chosen one from each bag and the numbers observed. Denote the minimum of the two by X and the sum of the two by Y. Find the joint probability distribution of the random variable pair (X, Y). Find also the two marginal distributions and the conditional distribution of X when $Y \le 4$.

Ans. JOINT & MARGINAL DISTRIBUTION OF X & Y

X \ Y	2	3	4	5	$p_1(x)$
1	$\frac{1}{6}$	$\frac{2}{6}$	$\frac{1}{6}$	–	$\frac{4}{6}$
2	–	–	$\frac{1}{6}$	$\frac{1}{6}$	$\frac{2}{6}$
$p_2(y)$	$\frac{1}{6}$	$\frac{2}{6}$	$\frac{2}{6}$	$\frac{1}{6}$	1

CONDITIONAL DISTRIBUTION OF $(X \mid Y \le 4)$

x	1	2
$p(x \mid y \le 4)$	4/5	1/5

5·36. If $f(x, y) = \begin{cases} e^{-(x+y)} & ; x \ge 0, y \ge 0 \\ 0 & , \text{elsewhere} \end{cases}$

is the joint probability density function of random variables X and Y, find

(*i*) $P(X < 1)$, (*ii*) $P(X > Y)$, and (*iii*) $P(X + Y < 1)$.

Ans. (*i*) $1 - \frac{1}{e}$, (*ii*) $\frac{1}{2}$, (*iii*) $1 - \frac{2}{e}$.

5·37. The joint probability density function of the two-dimensional variable (X, Y) is of the form :

$$f(x, y) = \begin{cases} k\, e^{-(x+y)}, & 0 \le y < x < \infty \\ 0 & , \text{elsewhere} \end{cases}$$

(*i*) Determine the constant *k*.

(*ii*) Find the marginal density function $f_1(x)$ of X and $f_2(y)$ of Y.

(*iii*) Examine if X and Y are independent.

(*iv*) Find the conditional p.d.f.'s $f_3(x \mid y)$, $f_4(y \mid x)$.

(*v*) Compute $P(Y \ge 3)$.

Ans. (*i*) $k = 2$,

(*ii*) $f_1(x) = 2e^{-x}(1 - e^{-x})$, $0 \le x < \infty$; $\quad f_2(y) = 2e^{-2y}$, $0 \le y < \infty$

(*iii*) $f(x, y) \ne f_1(x) f_2(y) \quad \Rightarrow \quad$ X and Y are not independent.

(iv) $f_3(x \mid y) = e^{-(x-y)}$; $y < x < \infty$, $f_4(y \mid x) = \dfrac{e^{-y}}{1-e^{-x}}$; $0 < y < x$, (v) e^{-6}.

5·38. Suppose that the random variables X and Y have the joint *p.d.f.*

$$f(x, y) = \begin{cases} kx(x-y), & 0 < x < 2,\ -x < y < x \\ 0 & \text{, elsewhere} \end{cases}$$

(i) Evaluate the constant k.

(ii) Find the marginal probability density functions of the random variables.

Ans. (i) $k = \dfrac{1}{8}$; (ii) $f_X(x) = \dfrac{1}{4}x^3$; $0 < x < 2$

$$f_Y(y) = \begin{cases} \dfrac{1}{3} + \dfrac{1}{8}\left(\dfrac{5}{6}y^3 - 2y\right) & ;\ -2 \le y < 0 \\ \dfrac{1}{3} + \dfrac{1}{8}\left(\dfrac{1}{6}y^3 - 2y\right) & ;\ 0 \le y \le 2 \end{cases}$$

5·39. Two-dimensional random variable (X, Y) have the joint density :

$$f(x, y) = \begin{cases} 8xy, & 0 < x < y < 1 \\ 0, & \text{otherwise} \end{cases}$$

(i) Find the marginal and conditional distributions.

(ii) Are X and Y independent ? Give reasons for your answer .

(iii) Find $P\left(X < \dfrac{1}{2} \cap Y < \dfrac{1}{4}\right)$. (iv) Find $P\left(X \le \dfrac{1}{4} \,\middle|\, \dfrac{1}{2} \le Y \le 1\right)$.

Ans. (i) $f_1(x) = \begin{cases} 4x(1-x^2), & 0 < x < 1 \\ 0 & \text{, otherwise} \end{cases}$ $f_1(x \mid y) = 2x/y^2$; $0 < x < y$, $0 < y < 1$

$f_2(y) = 4y^3$, $0 < y < 1$ $f_2(y \mid x) = 2y/(1-x^2)$; $x < y < 1, 0 < x < 1$

(ii) $f(x, y) \ne f_1(x) \cdot f_2(y)$ $\Rightarrow$ X and Y are not independent.

(iii) $\dfrac{1}{256}$, (iv) $\dfrac{1}{10}$.

5·40. The random variables X and Y have the joint density function :

$$f(x, y) = \begin{cases} 2, & \text{if } x + y \le 1, x \ge 0 \text{ and } y \ge 0 \\ 0 & \text{, otherwise} \end{cases}$$

Find the conditional distribution of (i) Y, given $X = x$, and (ii) X given $Y = y$.

Ans. $f_{X \mid Y}(x \mid y) = \dfrac{1}{1-x}$; $0 < y < 1 - x$; $0 < x < 1$

$f_{Y \mid X}(y \mid x) = \dfrac{1}{1-y}$; $0 < x < 1 - y$; $0 < y < 1$

5·41. The random variables X and Y have the joint distribution given by the *p.d.f.* :

$$f(x, y) = \begin{cases} 6(1-x-y), & \text{for } x > 0,\ y > 0,\ x + y < 1 \\ 0 & \text{, elsewhere} \end{cases}$$

Find the marginal distributions of X and Y. Hence, examine if X and Y are independent.

Ans. $f_X(x) = 3(1-x)^2,\ 0 < x < 1\ ;\quad f_Y(y) = 3(1-y)^2,\ 0 < y < 1.$

Since $f_{XY}(x, y) \neq f_X(x)\, f_Y(y)$, X and Y are not independent.

5·42. The joint probability density function of the two-dimensional random variable (X, Y) is given by :

$$f(x, y) = \begin{cases} k\,xy, & 1 \le x \le y \le 2 \\ 0, & \text{elsewhere} \end{cases}$$

(i) Find k (ii) Find the marginal density functions of X and Y.

(iii) Find the conditional density functions of Y given X = x, and of X given Y = y.

Ans. (i) $k = \dfrac{8}{9}$, (ii) $f_X(x) = \displaystyle\int_x^2 f(x,y)\,dy = \begin{cases} \dfrac{4}{9}x(4-x^2), & 1 \le x \le 2 \\ 0, & \text{otherwise} \end{cases}$

$$f_Y(y) = \int_1^y f(x, y)\,dx = \begin{cases} \dfrac{4}{9}y(y^2-1), & 1 \le y \le 2 \\ 0, & \text{otherwise} \end{cases}$$

(iii) $f_{X|Y}(x \mid y) = \dfrac{2x}{y^2-1};\ \left.\begin{matrix} 1 \le x \le y \\ 1 \le y \le 2 \end{matrix}\right\}$ and $f_{Y|X}(y \mid x) = \dfrac{f(x,y)}{f_X(x)} = \dfrac{2y}{4-x^2};\ \left.\begin{matrix} x \le y \le 2 \\ 1 < x < 2 \end{matrix}\right\}$

5·43. Let $f(x_1, x_2) = \begin{cases} A\left(x_1 x_2 + e^{x_1}\right); & 0 < (x_1, x_2) < 1 \\ 0, & \text{elsewhere} \end{cases}$

(i) Determine A. (ii) Examine whether X_1 and X_2 are stochastically independent.

Hint. (i) $A = \dfrac{4}{4e-3}$, (ii) $g_1(x_1) = A\left(\dfrac{1}{2}x_1 + e^{x_1}\right)$, $g_2(x_2) = A\left(\dfrac{1}{2}x_2 + e - 1\right)$, $[0 < (x_1, x_2) < 1]$

Since $g_1(x_1) \cdot g_2(x_2) \neq f(x_1, x_2)$, X_1 and X_2 are not statistically independent.

5·44. The two random variables X and Y have the joint probability density function :

$$f(x, y) = \frac{1}{2x^2 y},\ \text{for } 1 \le x < \infty \text{ and } \frac{1}{x} < y < x$$

Derive the marginal distributions of X and Y. Further obtain the conditional distribution of Y for X = x and also that of X given Y = y.

Hint. $f_X(x) = \displaystyle\int_y f(x, y)\,dy = \int_{1/x}^{x} f(x, y)\,dy = \frac{\log x}{x^2};\ 1 \le x < \infty$

$$f_Y(y) = \begin{cases} \displaystyle\int_{1/y}^{\infty} f(x, y)\,dx\ ; & 0 \le y \le 1 \\ \displaystyle\int_y^{\infty} f(x, y)\,dx\ ; & 1 \le y < \infty \end{cases} = \begin{cases} \dfrac{1}{2}, & 0 < y \le 1 \\ \dfrac{1}{2y^2}, & 1 < y < \infty \end{cases}$$

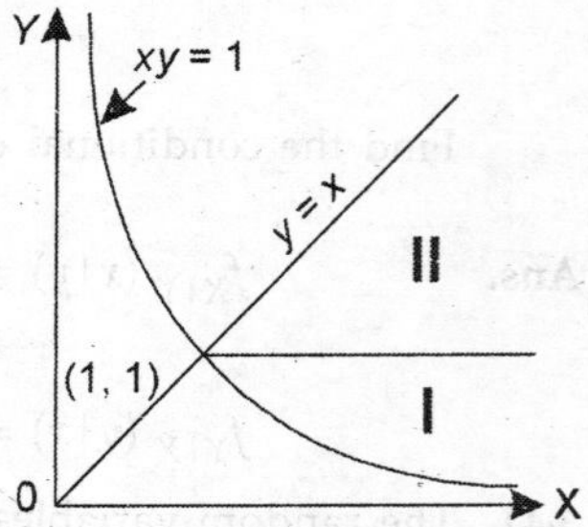

Fig. 5·18

$$f_{X|Y}(x \mid y) = \begin{cases} \dfrac{1}{x^2 y}; & \dfrac{1}{y} < x < \infty, & (0 < y \le 1) \\ \dfrac{y}{x^2}; & y < x < \infty, & (1 < y < \infty) \end{cases};\quad f_{Y|X}(y \mid x) = \frac{1}{2y \log x};\ \frac{1}{x} \le y < x;\ 1 \le x < \infty$$

5·45. If the joint distribution function of X and Y is given by :

$$F(x, y) = \begin{cases} (1-e^{-x})(1-e^{-y}) & ; \text{ for } x>0, y>0 \\ 0 & \text{elsewhere} \end{cases}$$

Find, $P(1 < X < 3 \cap 1 < Y < 2)$.

Hint. $f(x, y) = \dfrac{\partial^2 F}{\partial x\, \partial y} = e^{-(x+y)}$

Ans. $P(1 < X < 3 \cap 1 < Y < 2) = \dfrac{(e+1)(e-1)^2}{e^5}$

5·46. Let $f(x, y) = \begin{cases} {}^yC_x\, p^x (1-p)^{y-x} \left(\dfrac{e^{-\lambda}\lambda^y}{y!}\right) & ; x=0,1,2,\ldots \;;\; y=0,1,2,\ldots; \text{ with } y \ge x \\ 0 & , \text{elsewhere} \end{cases}$

Find the marginal density function of X and the marginal density function of Y. Also determine whether the random variables X and Y are independent.

Ans. $f_1(x) = \dfrac{e^{-\lambda p}(\lambda p)^x}{x!}$; $x = 0, 1, 2, \ldots$; $f_2(y) = \dfrac{e^{-\lambda}\lambda^y}{y!}$; $y = 0, 1, 2, \ldots$;

X and Y are not independent.

5·47. Consider the function : $f(x \mid y) = \begin{cases} \dfrac{y^x e^{-y}}{x!} & ; x = 0, 1, 2, \ldots \\ 0 & , \text{otherwise} \end{cases}$

(*i*) Show that $f(x \mid y)$ is the conditional probability function of X given Y ; $y \ge 0$.

(*ii*) If the marginal *p.d.f.* of Y is : $f_Y(y) = \begin{cases} \lambda e^{-\lambda y} & ; y>0, \lambda>0 \\ 0 & , y \le 0 \end{cases}$

what is the joint *p.d.f.* of X and Y ?

(*iii*) Obtain the marginal probability function of X.

Ans. (*ii*) $f(x, y) = \dfrac{\lambda}{x!}\cdot \exp\left[-(1+\lambda)\, y\right] y^x$, $y>0$; $x=0,1,2,\ldots$

(*iii*) $f_X(x) = pq^x$; $x = 0, 1, 2, \ldots$; $\left[p = \dfrac{\lambda}{1+\lambda}, q = 1-p\right]$

5·48. The joint probability density function of two dimensional random variable (X, Y) is given by :

$$f(x, y) = \begin{cases} x^3y^3/16, & 0 \le x \le 2, \; 0 \le y \le 2 \\ 0 & , \text{elsewhere} \end{cases}$$

Find the marginal densities of X and Y. Also find the cumulative distribution functions for X and Y.

Ans. $f_X(x) = \dfrac{x^3}{4}$; $0 \le x < 2$; $f_Y(y) = \dfrac{y^3}{4}$; $0 \le y \le 2$

$$F_X(x) = \begin{cases} 0 & ; x<0 \\ x^4/16 & ; 0 \le x \le 2 \\ 1 & ; x>2 \end{cases} \qquad F_Y(y) = \begin{cases} 0 & ; y<0 \\ y^4/16 & ; 0 \le y \le 2 \\ 1 & ; y>2 \end{cases}$$

5·49. (*a*) Let $f(x) = \begin{cases} \frac{1}{2}, & -1 < x < 1 \\ 0, & \text{elsewhere} \end{cases}$

be the *p.d.f.* of the *r.v.* X. Find the distribution function and the *p.d.f.* of $Y = X^2$.

Hint. $F(x) = P(X \le x) = \int_{-1}^{x} f(u)\,du = \frac{1}{2}(x+1)$ ... (*)

Distribution function $G(\bullet)$ of $Y = X^2$ is given by :

$$G_Y(y) = F(\sqrt{y}) - F(-\sqrt{y})\ ;\ y > 0 = \frac{1}{2}(\sqrt{y}+1) - \frac{1}{2}(-\sqrt{y}+1) = \sqrt{y}\ ;\ 0 < y < 1 \qquad \text{[From (*)]}$$

(As $-1 < x < 1$, $Y = X^2$ lies between 0 and 1)

p.d.f. of $Y = X^2$ is : $g(y) = G_Y'(y) = \dfrac{1}{2\sqrt{y}}$; $0 < y < 1$.

(*b*) Let X be a continuous random variable with *p.d.f.* $f(x)$. Let $Y = X^2$. Show that the random variable Y has *p.d.f.* given by : $g(y) = \begin{cases} \frac{1}{2\sqrt{y}}\left\{f(\sqrt{y}) + f(-\sqrt{y})\right\}, & y > 0 \\ 0, & y \le 0 \end{cases}$

5·50. (*a*) Find the distribution and density functions for (*i*) $Y = aX + b$, $a \ne 0$, b real (*ii*) $Y = e^X$, assuming that $F(x)$ and $f(x)$, the distribution and the density functions respectively of X are known.

Ans. (*i*) $G(y) = \begin{cases} F[(y-b)/a], & \text{if } a > 0 \\ 1 - F[(y-b)/a], & \text{if } a < 0 \end{cases}$ and $g_Y(y) = \dfrac{1}{|a|} f\left(\dfrac{y-b}{a}\right)$

(*ii*) $G(y) = \begin{cases} F(\log_e y) & , y > 0 \\ 0 & , y \le 0 \end{cases}$ and $g(y) = \begin{cases} \frac{1}{y} f(\log_e y) & , y > 0 \\ 0 & , y \le 0 \end{cases}$

(*b*) Suppose that X has *p.d.f.* : $f(x) = \begin{cases} 2x & , 0 < x < 1 \\ 0 & , \text{elsewhere} \end{cases}$, find the *p.d.f.* of $Y = 3X + 1$.

Ans. $g(y) = \frac{2}{9}(y-1), 1 < y < 4$.

5·51. (*a*) Let the p.d.f. of X be : $f(x) = \begin{cases} \frac{1}{6}, & -3 \le x \le 3 \\ 0, & \text{elsewhere} \end{cases}$. Find the p.d.f. of $Y = 2X^2 - 3$.

Ans. $g_Y(y) = \dfrac{1}{6\sqrt{2}}(3+y)^{-1/2}$; $-3 \le y \le 15$.

(*b*) A *r.v.* X has *p.d.f.* $f(x) = \frac{1}{3}$, $-1 \le x \le 2$. Find the distribution function and *p.d.f* of $Y = X^2$.

Ans. $g(y) = \begin{cases} \frac{1}{3} y^{-1/2} & ; \ 0 \le y < 1 \\ \frac{1}{6} y^{-1/2} & ; \ 1 \le y < 4 \\ 0 & , \ \text{otherwise} \end{cases}$

5·52. Let X be a *r.v.* with the distribution function : $F_X(x) = \begin{cases} 0, & x < 0 \\ x, & 0 \le x \le 1 \\ 1, & x > 1 \end{cases}$

Determine the distribution function $G_Y(y)$ of the random variable $Y = \sqrt{X}$ and hence compute mean of Y.

Ans. $G_Y(y) = \begin{cases} 0 \;; & y < 0 \\ y^2 \;; & 0 \le y \le 1 \\ 1 \;; & y > 1 \end{cases} \Rightarrow g_Y(y) = \begin{cases} 2y \;, & 0 \le y \le 1 \\ 0 \;, & \text{elsewhere} \end{cases}$; Mean $= \frac{2}{3}$.

5·53. The probability density function of (x_1, x_2) is given as :

$$f(x_1, x_2) = \begin{cases} \theta_1 \theta_2 \exp(-\theta_1 x_1 - \theta_2 x_2), & \text{if } x_1, x_2 > 0 \\ 0 & , \text{ otherwise} \end{cases}$$

Find the density function of (Y_1, Y_2), where $y_1 = \frac{2x_1}{x_2} + 1$, $y_2 = 3x_1 + x_2$, almost everywhere.

Hence, obtain the marginal *p.d.f.'s* of Y_1 and Y_2.

Ans. $g(y_1, y_2) = \theta_1 \theta_2 . \exp\left[-\frac{y_2}{(3y_1 - 1)} \{\theta_1(y_1 - 1) + 2\theta_2\} \right] \times \frac{2\,y_2}{(3y_1 - 1)^2}$; $y_1 > 0,\; y_2 > 0$

$$g_1(y_1) = \frac{2\,\theta_1\,\theta_2}{[\theta_1(y_1 - 1) + 2\,\theta_2]^2} \;; y_1 \ge 1 \;;$$

$$g_2(y_2) = \left(\frac{\theta_1\,\theta_2}{\theta_1 - 3\theta_2} \right) \left[\exp(-\theta_2\, y_2) - \exp\left(-\frac{\theta_1}{3} y_2 \right) \right]; \quad y_2 \ge 0$$

5·54. Let X_1, X_2 be a random sample of size 2 from a distribution with *p.d.f.*, :

$$f(x) = \begin{cases} \exp(-x), 0 < x < \infty \\ 0 \quad , \text{elsewhere} \end{cases}$$

Show that $Y_1 = X_1 + X_2$ and $Y_2 = [X_1 / (X_1 + X_2)]$, are independent.

5·55. If the probability density functions of the random variables X and $Y | X$ are given by :

$$f_X(x) = \begin{cases} e^{-x}, x \ge 0 \\ 0 \;, \text{elsewhere} \end{cases} \quad \text{and} \quad f_{Y|X}(y | x) = \begin{cases} \dfrac{e^{-x} x^y}{y!} & ; y = 0, 1, 2, 3, ..., \\ 0, \text{elsewhere} \end{cases}$$

respectively, find the probability density function of the *r.v.* Y and identify the distribution..

Ans. $f_Y(y) = \frac{1}{2}\left(\frac{1}{2}\right)^y$; $y = 0, 1, 2, ...$

$\Rightarrow$ Y has geometric distribution with parameter $p = 1/2 (\Rightarrow q = 1/2)$. [See Chapter 8].

5·56. If X, Y, Z are *i.i.d.* continuous variates, prove that :

(i) $P(X > Y \mid X > Z) = \frac{2}{3}$ (ii) $P(X > Y \mid X < Z) = \frac{1}{3}$

(iii) $P(X > Y \mid Y > Z) = \frac{1}{3}$ (iv) $P(X > Y \mid Y < Z) = \frac{2}{3}$

Ans. (i) $\dfrac{P[X > \max.(Y, Z)]}{P(X > Z)} = \dfrac{(1/3)}{(1/2)} = \dfrac{2}{3}$ (ii) $\dfrac{(1/3!)}{(1/2)} = \dfrac{1}{3}$

(iii) $\dfrac{(1/3!)}{(1/2)} = \dfrac{1}{3}$ (iv) $\dfrac{P[Y < \text{Min}(X,Z)]}{P(Y < Z)} = \dfrac{(1/3)}{(1/2)} = \dfrac{2}{3}$

5·57. The joint *pd.f.* of two random variables X and Y is :

$$f(x, y) = \begin{cases} k\,e^{-y} & ;\ y > |x|, -\infty < x < \infty \\ 0 & ;\ \text{otherwise} \end{cases}$$

Find k, and the marginal and conditional distributions of X and Y.

Ans. $k = \frac{1}{2}$; $f_X(x) = \frac{1}{2} e^{-|x|}, \ -\infty < x < \infty$; $f_Y(y) = y\,e^{-y}$; $0 < y < \infty$

$$f_{X|Y}(x \mid y) = \frac{1}{2y};\ -y < x < y,\ 0 < y < \infty\ ;$$

$$f_{Y|X}(y \mid x) = e^{-(y - |x|)},\ y > |x|,\ -\infty < x < \infty.$$

5·58. The joint p.d.f. (X, Y) is :

$$f(x, y) = \begin{cases} 1/2 & ,\ \text{for } |x| + |y| \le 1 \\ 0 & ;\ \text{otherwise} \end{cases}$$

Obtain the marginal and conditional p.d.f.'s of X.

Ans. $f_X(x) = \begin{cases} 1 - |x| & ;\ |x| \le 1 \\ 0 & ;\ \text{otherwise} \end{cases}$; $f_Y(y) = \begin{cases} 1 - |y| & ;\ |y| \le 1 \\ 0 & ;\ \text{otherwise} \end{cases}$

$$f_{X|Y}(x \mid y) = \begin{cases} \dfrac{1}{2(1-|y|)} & ;\ |x| \le 1 - |y|\ ;\ |y| \le 1 \\ 0 & ;\ |x| > 1 - |y|\ ;\ |y| \le 1 \end{cases}$$

5·59. The joint p.d.f. of r.v.'s X, Y, Z is given by :

$$f(x, y, z) = \frac{6}{(1 + x + y + z)^4}\ ;\ 0 < (x, y, z) < \infty$$

Prove that the p.d.f. of $U = X + Y + Z$ is :

$$g(u) = \frac{3u^2}{(1+u)^4}\ ; 0 < u < \infty$$

Hint. Transform (X, Y, Z) to (U, V, W) : $U = X + Y + Z$; $V = X + Y$, $W = X$.

Then $v = w + y > w \Rightarrow 0 < w < v$;

$u = v + z > v \Rightarrow 0 < v < u$; and $u \ge 0$

5·60. The joint p.d.f. of X and Y is given by :

$$f(x,y) = \begin{cases} x+y & ; \quad 0<x<1 \quad , \quad 0<y<1 \\ 0 & ; \quad \text{Otherwise} \end{cases}$$

Obtain the *p.d.f.* of $U = XY$ (*a*) By transformation technique
(*b*) By finding its distribution function.

Hint. and **Ans.** (*a*) $U = XY, \ V = X; \qquad J = -\dfrac{1}{v};$

$0 < u < 1$ and $u < v < 1$ (Note it)

p.d.f. of $U = XY$: $g(u) = 2\ (1 - u)\ ; \quad 0 < u < 1$

(*b*) $G_U(u) = P(XY \le u) = \displaystyle\int_{y=0}^{1}\int_{x=0}^{u} f(x,y)\,dx\,dy + \int_{x=u}^{1}\int_{y=0}^{u/x} f(x,y)\,dx\,dy$

5·61. If the joint p.d.f of r.v.'s X and Y is :

$$f(x,y) = \begin{cases} x+y & ; \quad 0<x<1, \quad 0<y<1 \\ 0 & ; \quad \text{otherwise} \end{cases}$$

show that the p.d.f. of U = X + Y is : $p(u) = \begin{cases} u^2 & ; \quad 0<u\le 1 \\ u(2-u) & ; \quad 1<u<2 \end{cases}$

Hint. $u = x + y \ ; \ v = x \quad \Rightarrow \quad x = v, \quad y = u - v.$

Range : $(0 < x < 1, \ 0 < y < 1)$ transforms to : $(0 < v < 1, \ 0 < u - v < 1)$.

which is bounded by : $v = 0, \ v = 1, \ u - v = 0$ and $u - v = 1$.

5·62. Joint p.d.f. of *r.v.'s* (X, Y) is : $f(x, y) = x\,e^{-x(1+y)}\ ; \ x > 0\ ; \ y > 0.$

Find the p.d.f. of : (*i*) $U = XY$ (*ii*) $V = \text{Max}\ (X, Y)$.

Hint. Distribution function of $U = XY$ is given by :

$$G_U(u) = P(U \le u) = P(XY \le u) = \iint_{xy \le u} f(x,y)\,dx\,dy$$

$$= \int_0^{\infty}\left(x\int_0^{u/x} e^{-x(1+y)}\,dy\right)dx = 1 - e^{-u}\ ; u > 0$$

$\Rightarrow \qquad g_U(u) = G'(u) = e^{-u}\ ; \ u > 0$

(*ii*) $\qquad H_V(v) = P(V \le v) = P\,[\text{Max}\ (X, Y) \le v] = P\ (X \le v \cap Y \le v)$

$$= \int_0^{v}\left(x\int_0^{v} e^{-x(1+y)}\,dy\right)dx = 1 - e^{-v} - \left[\frac{1-e^{-v(1+v)}}{1+v}\right];\ v > 0$$

$$\Rightarrow \qquad h(v) = H'(v) = e^{-v} + \frac{1-e^{-v(1+v)}}{(1+v)^2} - \frac{(1+2v)\,e^{-v(1+v)}}{1+v}\ ;\ v > 0$$

5·63. If $f(x)$ is an odd function of x of period $\dfrac{1}{2}$, show that :

$$\int_0^{\infty} x^r \cdot x^{-\log x} f(\log x)\, dx = 0, \text{ for all integral values of } r.$$

Hence, show that : $dF = x^{-\log x}\left[1 - \lambda \sin(4\pi \cdot \log x\right] dx \;\; ; 0 < x < \infty, 0 \le \lambda \le 1,$

have the same moments, whatever be the value of λ.

5·64. The random variables X and Y have a joint *p.d.f.* $f(x, y)$ given by :

$$f(x, y) = \begin{cases} g(x+y, & x > 0, y > 0 \\ 0 \quad , & \text{otherwise} \end{cases}$$

Obtain the distribution function $H(z)$ of $Z = X + Y$ and hence show that its *p.d.f.* is :

$$h(z) = \begin{cases} z\, g(z), & z > 0 \\ 0 & , z \le 0 \end{cases}$$

Ans. $H_Z(z) = \int_0^z u\, g(u)\, du \quad \Rightarrow \quad h_Z(z) = \begin{cases} z\, g(z) \;; & z > 0 \\ 0 \;; & \text{otherwise} \end{cases}$

5·65. The joint density function of two *r.v.'s* is given by : $f(x, y) = e^{-(x+y)}$; $x > 0, y > 0$

Show that the *p.d.f.* of $U = \dfrac{X+Y}{2}$ is $g(u) = 4u\, e^{-2u}$, $u > 0$.

5·66. The time X taken by the garage to repair a car is a continuous random variable with probability density function :

$$f_1(x) = \begin{cases} \frac{3}{4} x(2-x) \;, & 0 \le x \le 2 \\ 0 \quad , & \text{elsewhere} \end{cases}$$

If, on leaving his car, a motorist goes to keep an engagement lasting for a time Y, where Y is a continuous random variable, independent of X, with probability density function :

$$f_2(y) = \begin{cases} \frac{1}{2} y, & 0 \le y \le 2 \\ 0 \;, & \text{elsewhere} \end{cases}$$

determine the probability that the car will not be ready on his return. **[Ans. 3/10.]**

5·67. If X and Y are two independent random variables such that

$$f(x) = e^{-x}\;, x \ge 0 \quad \text{and} \quad g(y) = 3e^{-3y}\;, y \ge 0\;;$$

find the probability distribution of $Z = X/Y$.

Ans. $g(z) = \dfrac{3}{(z+3)^2}$; $z \ge 0$.

5·68. The random variables X and Y are independent and their probability density functions are, respectively, given by :

$$f(x) = \frac{1}{\pi} \cdot \frac{1}{\sqrt{1-x^2}}, \; |x| < 1 \qquad \text{and} \qquad g(y) = y \exp\left(-\frac{1}{2} y^2\right), \; y > 0$$

Find the joint probability density function of Z and W, where $Z = XY$ and $W = X$. Deduce the probability density function of Z.

Ans. $g(z, w) = \dfrac{1}{\pi}\left[\dfrac{z}{w\sqrt{1-w^2}}\right] \cdot \exp\left[-\dfrac{z^2}{2w^2}\right] \cdot \dfrac{1}{|w|}$;

Take $0 < w < 1$, otherwise p.d.f. is negative.

$$g_Z(z) = \frac{1}{\sqrt{2\pi}} \cdot \exp\left[-\frac{1}{2} z^2\right]; \; -\infty < z < \infty$$

CHAPTER

6 Mathematical Expectation

LEARNING OBJECTIVES

Upon completion of this Chapter, you should be able to:

- Understand and interpret the expected value and variance of a random variable.
- State with proof the various properties of expectation and variance.
- Explain the concept of covariance.
- Discuss some inequality relationship involving expectation, like:
  - (i) Cauchy – Schwartz Inequality
  - (ii) Jenson's Inequality
  - (iii) Gurland's Inequality
  - (iv) Lyapounov's Inequality
  - (v) C_r – Inequality
  - (vi) Holder's Inequality
  - (vii) Minkowski's Inequality
- Obtain moments of bivariate probability distributions.
- Understand the concepts of conditional expectation and conditional variance, and its use in:
  - (a) Double Expectation Rule or Double - E Rule
  - (b) Multistage Rule of Expectation (Multistage- E Rule)
  - (c) Wald's Theorem on mean and variance of a random sum.

CHAPTER OUTLINES

6·1. INTRODUCTION

Many frequently used *r.v.*'s can be both characterized and dealt with effectively for practical purposes by consideration of quantities called their *expectation.* For example, a gambler might be interested in his average winnings at a game, a businessman in his average profits on a product, a physicist in the average charge of a particle, and so on. The *'average'* value of a random phenomenon is also termed as its *mathematical expectation* or *expected value.* In this chapter we will define and study this concept in detail, which will be used extensively in subsequent chapters.

6·2. MATHEMATICAL EXPECTATION OR EXPECTED VALUE OF A RANDOM VARIABLE

Once we have constructed the probability distribution for a random variable, we often want to compute the mean or expected value of the random variable. The *expected value* of a discrete random variable is a weighted average of all possible values of the random variable, where the weights are the probabilities associated with the corresponding values. The mathematical expression for computing the expected value of a discrete random variable X with probability mass function (*p.m.f.*) $f(x)$ is given below :

$$E(X) = \sum_x x.f(x), \text{ (for discrete r.v.)} \quad \text{... (6·1)}$$

The mathematical expression for computing the expected value of a continuous random variable X with probability density function (*p.d.f.*) $f(x)$ is, however, as follows :

$$E(X) = \int_{-\infty}^{\infty} x\, f(x)\, dx, \text{ (for continuous r.v.)} \quad \text{... (6·1a)}$$

provided the right hand integral in (6·1*a*) or series in (6·1) is absolutely convergent, *i.e.,* provided

$$\int_{-\infty}^{\infty} |x f(x)|\, dx = \int_{-\infty}^{\infty} |x|\, f(x)\, dx < \infty \quad \text{... (6·2)}$$

or

$$\sum_x |x.f(x)| = \sum_x |x|.f(x) < \infty \quad \text{... (6·2a)}$$

Remarks :

1. Since absolute convergence implies ordinary convergence, if (6·2) or (6·2 *a*) holds then the series or integral in (6·1) and (6·1*a*) also exists, *i.e.,* has a finite value and in that case we define $E(X)$ by (6·1) or (6·1*a*). It should be clearly understood that although X has an expectation only if L.H.S. in (6·2) or (6·2*a*) exists, *i.e.,* converges to a finite limit, its value is given by (6·1) or (6·1*a*).
2. $E(X)$ exists if $E|X|$ exists.
3. *Expected Value and Variance of an Indicator Variable.* Consider the indicator variable :

$$X = I_A = \begin{cases} 1, \text{ if } A \text{ happens} \\ 0, \text{ if } \bar{A} \text{ happens} \end{cases}$$

$$E(X) = 1.\, P(X = 1) + 0.\, P(X = 0) \quad \Rightarrow \quad E(I_A) = 1.\, P(I_A = 1) + 0.\, P(I_A = 0)$$

$$\therefore \quad E(I_A) = P(A)$$

This gives us a very useful tool to find $P(A)$, rather than to evaluate $E(X)$.

Thus
$$P(A) = E(I_A) \quad \text{... (6·2b)}$$

$$E(X^2) = 1^2 \cdot P(X = 1) + 0^2 .\, P(X = 0) = P(I_A = 1) = P(A)$$

$$\therefore \quad \text{Var } X = E(X^2) - [E(X)]^2 = P(A) - [P(A)]^2 = P(A)\,[1 - P(A)] = P(A) \cdot P(\bar{A}) \quad \text{... (6·2c)}$$

$$\Rightarrow \quad \text{Var } (I_A) = P(A) \cdot P(\bar{A}) \quad \text{... (6·2d)}$$

4. If the *r.v.* X takes the values 0 !, 1 !, 2 !, ...with probability law:

$$P(X = x!) = \frac{e^{-1}}{x!}\ ;\ x = 0, 1, 2, ..., \text{ then } \sum_{x=0}^{\infty} x!\ P(X = x!) = e^{-1}\sum_{x=0}^{\infty} 1,$$

which is a divergent series. In this case $E(X)$ does not exist.

More rigorously, let us consider a random variable X which takes the values

$x_i = (-1)^{i+1}(i+1)\ ;\ i = 1, 2, 3, ...$ with the probability : $p_i = P(X = x_i) = \dfrac{1}{i(i+1)}\ ;\ i = 1, 2, 3, ...$

Here, $$\sum_{i=1}^{\infty} x_i\, P(X = x_i) = \sum_{i=1}^{\infty}(-1)^{i+1}(1/i) = 1 - \frac{1}{2} + \frac{1}{3} - \frac{1}{4} + ...$$

Using Leibnitz test for alternating series, the series on right-hand side is conditionally convergent, since the terms alternate in sign, are monotonically decreasing and coverge to zero. By conditional convergence we mean that although $\sum_{i=1}^{\infty} p_i\, x_i$ converges, $\sum_{i=1}^{\infty} |p_i\, x_i|$ does not converge. So, rigorously speaking, in the above example $E(X)$ does not exist, although $\sum_{i=1}^{\infty} p_i\, x_i$ is finite, *viz.*, $\log_e 2$.

As another example, let us consider the *r.v.* X which takes the values $x_k = \dfrac{(-1)^k . 2^k}{k}$, $(k = 1, 2, 3, ...)$, with probabilities $p_k = 2^{-k}$. Here also we get

$$\sum_{k=1}^{\infty} x_k\, p_k = \sum_{k=1}^{\infty} \frac{(-1)^k}{k} = -\left[1 - \frac{1}{2} + \frac{1}{3} - \frac{1}{4} + ...\right] = -\log_e 2 \quad \text{and} \quad \sum_{k=1}^{\infty} |x_k|\, p_k = \sum_{k=1}^{\infty} \frac{1}{k},$$

which is a divergent series. Hence in this case also expectation does not exist.

As an illustration of a continuous *r.v.*, let us consider the *r.v.* X with *p.d.f.* :

$$f(x) = \frac{1}{\pi} \cdot \frac{1}{1+x^2}\ ;\ -\infty < x < \infty$$

which is *p.d.f.* of standard Cauchy distribution. [*c.f.* Chapter 9].

$$\int_{-\infty}^{\infty} |x|\, f(x)\, dx = \frac{1}{\pi}\int_{-\infty}^{\infty} \frac{|x|}{1+x^2}\, dx = \frac{2}{\pi}\int_0^{\infty} \frac{x}{1+x^2}\, dx = \frac{1}{\pi}\left|\log\left(1+x^2\right)\right|_0^{\infty} \to \infty$$

($\because$ Integrand is an even function of x.)

Since this integral does not converge to a finite limit, $E(X)$ does not exist.

6·3. EXPECTED VALUE OF FUNCTION OF A RANDOM VARIABLE

Consider a *r.v.* X with *p.d.f.* (*p.m.f.*) $f(x)$ and distribution function $F(x)$. If $g(\cdot)$ is a function such that $g(X)$ is a *r.v.* and $E[g(X)]$ exists (*i.e.*, is defined), then

$$E[g(X)] = \int_{-\infty}^{\infty} g(x)\, f(x)\, dx \quad \text{(For continuous r.v.)} \qquad ...\ (6·3)$$

$$E[g(X)] = \sum_x g(x)\, f(x) \quad \text{(For discrete r.v.)} \qquad ...\ (6·3a)$$

By definition, the expectation of $Y = g(X)$ is :

$$E[g(X)] = E(Y) = \int_y y \, dH_Y(y) = \int_y y \, h(y) \, dy \quad \text{...(6·4)} \quad \text{or} \quad E(Y) = \sum_y y \, h(y), \text{... (6·4a)}$$

where $H_Y(y)$ is the distribution function of Y and $h(y)$ is *p.d.f.* of Y.

[The proof of equivalence of (6·3) and (6·4) is beyond the scope of the book.]

This result extends into higher dimensions. If X and Y have a joint *p.d.f.*, $f(x, y)$ and $Z = h(x, y)$ is a random variable for some function h and if $E(Z)$ exists, then

$$E(Z) = \int_{-\infty}^{\infty} \int_{-\infty}^{\infty} h(x, y) \, f(x, y) \, dx \, dy \quad \text{... (6.5)}$$

or

$$E(Z) = \sum_x \sum_y h(x, y) \, f(x, y) \quad \text{... (6·5a)}$$

Particular Cases

1. If we take $g(X) = X^r$, r being a positive integer, in (6·3),

$$E(X^r) = \int_{-\infty}^{\infty} x^r f(x) \, dx, \quad \text{... (6·5b)}$$

which is defined as μ_r', the rth moment (about origin) of the probability distribution.

Thus μ_r' *(about origin)* $= E(X^r)$. In particular

μ_1' *(about origin)* $= E(X)$ and μ_2' *(about origin)* $= E(X^2)$

Hence, Mean $= \bar{x} = \mu_1'$ *(about origin)* $= E(X)$ ... (6·6)

and $\mu_2 = \mu_2' - \mu_1'^2 = E(X^2) - \{E(X)\}^2$ ... (6·6a)

2. If $g(X) = [X - E(X)]^r = (X - \bar{x})^r$, then from (6·3), we obtain

$$E[X - E(X)]^r = \int_{-\infty}^{\infty} [x - E(x)]^r f(x) \, dx = \int_{-\infty}^{\infty} (x - \bar{x})^r f(x) \, dx, \quad \text{... (6·7)}$$

which is μ_r, the rth moment about mean.

In particular, if $r = 2$, we get

$$\mu_2 = E[X - E(X)]^2 = \int_{-\infty}^{\infty} (x - \bar{x})^2 f(x) \, dx, \quad \text{... (6·8)}$$

Formulae (6·6a) and (6·8) give the variance of the probability distribution of a continuous *r.v.* X in terms of expectation.

3. Taking $g(x) = constant = c$, say in (6·3), we get

$$E(c) = \int_{-\infty}^{\infty} c \, f(x) \, dx = c \int_{-\infty}^{\infty} f(x) \, dx = c \quad \Rightarrow \quad E(c) = c \quad \text{... (6·9)}$$

Remark. The corresponding results for a discrete *r.v.* X can be obtained on replacing integration by summation (Σ) over the given range of the variable X in the formulae (6·5) to (6·9).

In the following sections, we shall establish some more results on 'Expectation' in the form of Theorems, for continuous *r.v.'s*. The corresponding results for discrete *r.v.'s* can be obtained similarly on replacing integration by summation (Σ) over the given range of the variable X and are left as an exercise to the reader.

6·4. PROPERTIES OF EXPECTATION

Property 1. Addition Theorem of Expectation.

If X and Y are random variables, then :

$$E(X + Y) = E(X) + E(Y), \qquad \text{... (6·10)}$$

provided all the expectations exist.

Proof. Let X and Y be continuous *r.v.*'s with joint *p.d.f.* $f_{XY}(x, y)$ and marginal *p.d.f.*'s $f_X(x)$ and $f_Y(y)$ respectively. Then by definition,

$$E(X) = \int_{-\infty}^{\infty} x\, f_X(x)\, dx \qquad \text{... (6·11)}$$

and
$$E(Y) = \int_{-\infty}^{\infty} y\, f_Y(y)\, dy \qquad \text{... (6·12)}$$

$$E(X + Y) = \int_{-\infty}^{\infty}\int_{-\infty}^{\infty} (x+y)\, f_{XY}(x,y)\, dx\, dy$$

$$= \int_{-\infty}^{\infty}\int_{-\infty}^{\infty} x f_{XY}(x,y)\, dx\, dy + \int_{-\infty}^{\infty}\int_{-\infty}^{\infty} y\, f_{XY}(x,y)\, dx\, dy$$

$$= \int_{-\infty}^{\infty} x\left[\int_{-\infty}^{\infty} f_{XY}(x,y)\, dy\right] dx + \int_{-\infty}^{\infty} y\left[\int_{-\infty}^{\infty} f_{XY}(x,y)\, dx\right] dy$$

$$= \int_{-\infty}^{\infty} x\, f_X(x) dx + \int_{-\infty}^{\infty} y\, f_Y(y)\, dy$$

$$= E(X) + E(Y) \qquad \text{[On using (6·11) and (6·12)]}$$

The result in (6·10) can be extended to *n* variables as given below.

Generalisation*. The mathematical expectation of the sum of n random variables is equal to the sum of their expectations, provided all the expectations exist. Symbolically, if* $X_1, X_2, ..., X_n$ *are random variables then*

$$E(X_1 + X_2 + ... + X_n) = E(X_1) + E(X_2) + ... + E(X_n) \qquad \text{... (6·13)}$$

or
$$E\left(\sum_{i=1}^{n} X_i\right) = \sum_{i=1}^{n} E(X_i), \text{ if all the expectations exist.} \qquad \text{... (6·13a)}$$

Proof. Using (6·10), for two *r.v.*'s X_1 and X_2, we get

$$E(X_1 + X_2) = E(X_1) + E(X_2) \quad \Rightarrow \quad \text{(6·13) is true for } n = 2. \qquad \text{... (*)}$$

Let us now suppose that (6·13) is true for $n = r$, (say), so that

$$E\left(\sum_{i=1}^{r} X_i\right) = \sum_{i=1}^{r} E(X_i) \qquad \text{... (6·14)}$$

$$E\left(\sum_{i=1}^{r+1} X_i\right) = E\left[\sum_{i=1}^{r} X_i + X_{r+1}\right] = E\left(\sum_{i=1}^{r} X_i\right) + E(X_{r+1}) \qquad \text{[Using (6·10)]}$$

$$= \sum_{i=1}^{r} E(X_i) + E(X_{r+1}) \qquad \text{[Using (6·14)]}$$

$$= \sum_{i=1}^{r+1} E(X_i)$$

Hence, if (6·13) is true for $n = r$, it is also true for $n = r + 1$. But we have proved in (*) above that (6·13) is true for $n = 2$. Hence it is true for $n = 2 + 1 = 3$; $n = 3 + 1 = 4$; ...and so on. Hence by the principle of mathematical induction, (6·13) is true for all positive integral values of n.

Property 2. Multiplication Theorem of Expectation.

If X and Y are independent random variables, then :

$$E(XY) = E(X) \cdot E(Y) \qquad \text{...(6·15)}$$

provided all the expectations exist.

Proof. Proceeding as in property 1, we have

$$E(XY) = \int_{-\infty}^{\infty} \int_{-\infty}^{\infty} xy\, f_{XY}(x, y)\, dx\, dy$$

$$= \int_{-\infty}^{\infty} \int_{-\infty}^{\infty} xy\, f_X(x)\, f_Y(y)\, dx\, dy \qquad \text{[Since X and Y are independent]}$$

$$= \int_{-\infty}^{\infty} x\, f_X(x)\, dx \int_{-\infty}^{\infty} y\, f_Y(y)\, dy$$

$= E(X)\, E(Y)$, provided X and Y are independent [Using (6·11) and (6·12)]

Generalisation. *The mathematical expectation of the product of a number of independent random variables is equal to the product of their expectations. Symbolically, if $X_1, X_2, ..., X_n$ are n independent r.v.'s, then*

$$E(X_1, X_2, \dots X_n) = E(X_1)\, E(X_2) \dots E(X_n)$$

i.e.,

$$E\left(\prod_{i=1}^{n} X_i\right) = \prod_{i=1}^{n} E(X_i), \qquad \text{...(6·16)}$$

provided all the expectations exist.

Proof. Using (6·15), for two *independent r.v.'s* X_1 and X_2, we get

$$E(X_1 X_2) = E(X_1)\, E(X_2) \quad \Rightarrow \quad \text{(6·16) is true for } n = 2. \qquad \text{... (*)}$$

Let us now suppose that (6·16) is true for $n = r$, (say), so that

$$E\left(\prod_{i=1}^{r} (X_i)\right) = \prod_{i=1}^{r} E(X_i) \qquad \text{... (6·17)}$$

Thus,

$$E\left(\prod_{i=1}^{r+1} X_i\right) = E\left(\prod_{i=1}^{r} X_i \cdot X_{r+1}\right) = E\left(\prod_{i=1}^{r} X_i\right) . E(X_{r+1}) \qquad \text{[Using (6·15)]}$$

$$= \left[\prod_{i=1}^{r} E(X_i)\right] E(X_{r+1}) \qquad \text{[Using (6·17)]}$$

$$= \prod_{i=1}^{r+1} E(X_i)$$

Hence, if (6·16) is true for $n = r$, it is also true for $n = r + 1$. Hence using (*), by the principle of mathematical induction we conclude that (6·16) is true for all positive integral values of n.

Property 3. *If X is a random variable and 'a' is constant, then*

(i) $$E[a\,\Psi(X)] = a\,E[\Psi(X)] \qquad \text{... (6·18)}$$

(ii) $$E[\Psi(X) + a] = E[\Psi(X)] + a, \qquad \text{... (6·19)}$$

where $\Psi(X)$, *a function of X, is a r.v. and all the expectations exist.*

Proof.

(i) $$E[a\,\Psi(X)] = \int_{-\infty}^{\infty} a\psi(x).f(x)\,dx = a\cdot\int_{-\infty}^{\infty}\psi(x).f(x)\,dx = a\,E[\Psi(X)]$$

(ii) $$E[\Psi(X) + a] = \int_{-\infty}^{\infty}\{\psi(x) + a\}f(x)\,dx = \int_{-\infty}^{\infty}\psi(x)f(x)\,dx + a\cdot\int_{-\infty}^{\infty}f(x)\,dx$$

$$= E[\Psi(X)] + a \qquad \left(\because \int_{-\infty}^{\infty} f(x)\,dx = 1\right)$$

Corollary

(i) If $\Psi(X) = X$, in (6·18) then

$$E(aX) = a\,E(X) \quad \text{and} \quad E(X + a) = E(X) + a \qquad \text{... (6·20)}$$

(ii) If $\Psi(X) = 1$ in (6·18) then $E(a) = a$. ... (6·21)

Property 4. *If X is a random variable and a and b are constants, then*

$$E(aX + b) = a\,E(X) + b, \qquad \text{... (6·22)}$$

provided all 6the expectations exist.

Proof. By def., we have

$$E(aX + b) = \int_{-\infty}^{\infty}(ax + b)\,f(x)\,dx = a\int_{-\infty}^{\infty}xf(x)\,dx + b\int_{-\infty}^{\infty}f(x)\,dx = a\,E(X) + b$$

Cor. 1. If $b = 0$, then we get $\quad E(aX) = a\,.\,E(X)$ ... (6·22*a*)

Cor. 2. Taking $a = 1$, $b = -\bar{X} = -E(X)$, we get $\quad E(X - \bar{X}) = 0$

Remark.

If we write $\quad g(X) = aX + b$ ... (6·23)

then $\quad g[E(X)] = a\,E(X) + b$ ... (6·23*a*)

Hence, from (6·22) and (6·23*a*), $\quad E[g(X)] = g[E(X)]$ ... (6·24)

Now (6·23) and (6·24) imply *that expectation of a linear function is the same linear function of the expectation.* The result, however, is not true if $g(\cdot)$ is not linear. For instance,

$$E(1/X) \neq \{1/E(X)\}\,; \qquad E[(X^{1/2})] \neq [E(X)]^{1/2}$$

$$E[\log(X)] \neq \log[E(X)]\,; \qquad E(X^2) \neq [E(X)]^2,$$

since all the functions stated above are non-linear. As an illustration, let us consider a random variable X which assumes only two values +1 and –1, each with equal probability $\frac{1}{2}$. Then

$$E(X) = 1 \times \frac{1}{2} + (-1) \times \frac{1}{2} = 0 \text{ and } E(X^2) = 1^2 \times \frac{1}{2} + (-1)^2 \times \frac{1}{2} = 1.$$

Thus $\quad E(X^2) \neq [E(X)]^2$

For a non-linear function $g(X)$, it is difficult to obtain expressions for $E[g(X)]$ in terms of $g[E(X)]$, say, for $E[\log(X)]$ or $E(X^2)$ in terms of $\log[E(X)]$ or $[E(X)]^2$. However, some results in the form of inequalities between $E[g(X)]$ and $g[E(X)]$ are available, as discussed in later part of the chapter.

Property 5. Expectation of a Linear Combination of Random Variables :

Let $X_1, X_2, ..., X_n$ be any n random variables and if $a_1, a_2, ..., a_n$ are any n constants, then

$$E\left(\sum_{i=1}^{n} a_i X_i\right) = \sum_{i=1}^{n} a_i E(X_i) \quad \text{... (6·25)}$$

provided all the expectations exist.

The result is obvious from (6·13) and (6·20).

Property 6. *If $X \geq 0$ then $E(X) \geq 0$.*

Proof. If X is a continuous random variable *s.t.* $X \geq 0$, then

$$E(X) = \int_{-\infty}^{\infty} x \cdot f(x)\, dx = \int_{0}^{\infty} x \cdot f(x)\, dx > 0, \qquad [\because \text{ If } X \geq 0,\ p(x) = 0 \text{ for } x < 0]$$

provided the expectation exists.

Property 7. *If X and Y are two random variables such that $Y \leq X$, then $E(Y) \leq E(X)$, provided all the expectations exist.*

Proof. Since $Y \leq X$, we have the *r.v.* $Y - X \leq 0 \quad \Rightarrow \quad X - Y \geq 0$.

Hence $\quad E(X - Y) \geq 0 \quad \Rightarrow \quad E(X) - E(Y) \geq 0$

$\Rightarrow \quad E(X) \geq E(Y) \quad \Rightarrow \quad E(Y) \leq E(X)$, as desired.

Property 8 . $|E(X)| \leq E|X|$, *provided the expectations exist.* ... (6·26)

Proof. Since $\quad X \leq |X|$, we have by Property 7, $E(X) \leq E|X|$ ... (*)

Again since $-X \leq |X|$, we have by Property 7, $E(-X) \leq E|X|$

$\therefore \quad -E(X) \leq E|X|$ ... (**)

From (*) and (**), we get the desired result $|E(X)| \leq E|X|$.

Property 9. *If μ_r' exists, then μ_s' exists for all $1 \leq s \leq r$.*

Mathematically, if $E(X^r)$ exists, then $E(X^s)$ exists for all $1 \leq s \leq r$, i.e.,

$$E(X^r) < \infty \quad \Rightarrow \quad E(X^s) < \infty, \text{ for all } 1 \leq s \leq r \quad \text{... (6·27)}$$

Proof. $$\int_{-\infty}^{\infty} |x|\, dF(x) = \int_{-1}^{1} |x|^s\, dF(x) + \int_{|x|>1} |x|^s\, dF(x)$$

If $s < r$, then $\quad |x|^s < |x|^r$, for $|x| > 1$.

$$\therefore \quad \int_{-\infty}^{\infty} |x|^s\, dF(x) \leq \int_{-1}^{1} |x|^s\, dF(x) + \int_{|x|>1} |x|^r\, dF(x),$$

$$\leq \int_{-1}^{1} dF(x) + \int_{|x|>1} |x|^r \, dF(x),$$

since for $\quad -1 < x < 1, \quad |x|^s < 1.$

$$\therefore \quad \int_{-\infty}^{\infty} |x|^s \, dF(x) \leq 1 + E\,|X|^r < \infty \Rightarrow E(X^s) \text{ exists}, \ \forall\ 1 \leq s \leq r \quad [\because E(X^r) \text{ exists}]$$

Remark.
The above result states that if the moments of a specified order exist, then all the lower order moments automatically exist. However, the converse is not true, *i.e.*, we may have distributions for which all the moments of a specified order exist but no higher order moment exist, *e.g.*, for the *r.v.* with *p.d.f.*:

$$f(x) = \begin{cases} 2/x^3 & ; \ x \geq 1 \\ 0 & ; \ x < 1 \end{cases}$$

$$E(X) = \int_1^{\infty} x\, f(x)\, dx = 2\int_1^{\infty} x^{-2}\, dx = \left|\left(\frac{-2}{x}\right)\right|_1^{\infty} = 2\ ; \ E(X^2) = \int_1^{\infty} x^2 f(x)\, dx = 2\int_1^{\infty} \frac{1}{x} \cdot dx = \infty$$

Thus for the above distribution, 1st order moment (mean) exists but 2nd order moment (variance) does not exist.

As another illustration, consider a *r.v.* X with *p.d.f.*: $\quad f(x) = \dfrac{(r+1)a^{r+1}}{(x+a)^{r+2}}\ ; \quad x \geq 0,\ a > 0$

$$\mu_r' = E(X^r) = (r+1)\, a^{r+1} \int_0^{\infty} \frac{x^r}{(x+a)^{r+2}}\, dx$$

Put $x = ay$ and using Beta integral: $\displaystyle\int_0^{\infty} \frac{x^{m-1}}{(1+x)^{m+n}} = \beta(m, n)$, we shall get, on simplification:

$$\mu_r' = (r+1)\, a^r \,.\, \beta(r+1, 1) = a^r$$

However, $\quad \mu_{r+1}' = E(X^{r+1}) = (r+1)\, a^{r+1} \displaystyle\int_0^{\infty} \frac{x^{r+1}}{(x+a)^{r+2}}\, dx \to \infty,$

as the integral is not convergent. Hence in this case only the moments up to rth order exist and higher order moments do not exist.

Property 10. *If X and Y are independent random variables, then*

$$E[h(X).\, k(Y)] = E[h(X)]\, E[k(Y)] \qquad \text{... (6·28)}$$

where $h(\cdot)$ is a function of X alone and $k(\cdot)$ is a function of Y alone, provided expectations on both sides exist.

Proof. Let $f_X(x)$ and $g_Y(y)$ be the marginal *p.d.f.*'s of X and Y respectively. Since X and Y are independent, their joint *p.d.f.* $f_{XY}(x, y)$ is given by :

$$f_{XY}(x, y) = f_X(x)\, f_Y(y) \qquad \text{... (*)}$$

By def., for continuous *r.v.'s*

$$E[h(X) \,.\, k(Y)] = \int_{-\infty}^{\infty}\int_{-\infty}^{\infty} h(x)\, k(y)\, f(x, y)\, dx\, dy = \int_{-\infty}^{\infty}\int_{-\infty}^{\infty} h(x)\, k(y)\, f(x)\, g(y)\, dx\, dy \ \text{[From (*)]}$$

Since $E[h(X).\, k(Y)]$ exists, the integral on the right-hand side is absolutely convergent and hence by Fuibini's theorem for integrable functions, we can change the order of integration to get

$$E[h(X)\, k(Y)] = \left[\int_{-\infty}^{\infty} h(x)\, f(x)\, dx\right]\left[\int_{-\infty}^{\infty} k(y)\, g(y)\, dy\right] = E[h(X)] \,.\, E[k(Y)],$$

as desired.

Remark: The result can be proved for discrete random variables X and Y on replacing integration by summation over the given range of X and Y.

6·5. PROPERTIES OF VARIANCE

If X is a random variable, then $V(aX + b) = a^2 . V(X)$, *where a and b are constants.* ... (6·29)

Proof. Let $Y = aX + b.$

Then $E(Y) = a\, E(X) + b$

$\therefore \quad Y - E(Y) = a\,[X - E(X)]$

Squaring and taking expectation of both sides, we get

$$E\,[Y - E(Y)]^2 = a^2\, E\,[X - E(X)]^2$$

$\Rightarrow \quad V(Y) = a^2\, V(X) \quad$ or $\quad V(aX + b) = a^2\, V(X)$,

where $V(X)$ is written for variance of X.

Corollary

(*i*) If $b = 0$, then $V(aX) = a^2\, V(X) \Rightarrow$ Variance is not independent of change of scale. ... (6·29*a*)

(*ii*) If $a = 0$, then $V(b) = 0 \Rightarrow$ Variance of a constant is zero. ... (6·29*b*)

(*iii*) If $a = 1$, then $V(X + b) = V(X) \Rightarrow$ Variance is independent of change of origin.... (6·29*c*)

Hence, Nariance is independent of change of origin but not of scale.

6·5·1. Variance of Degenerate Random Variable.

(*a*) *If X is a discrete r.v., then Var (X) = 0, if and only if X is degenerate.*

(*b*) *If X is a continuous r.v. then Var (X)* $\neq 0$.

Proof.

(*a*) Let X be a *discrete r.v.* with *p.m.f.* $P(X = x_i) = p_i$ and $E(X) = \sum_i x_i\, P(X = x_i) = \mu$

Then, $\quad \text{Var}(X) = E(X - \mu)^2 = \sum_i (x_i - \mu)^2 \cdot p_i$

$\therefore \quad \text{Var}(X) = 0, \quad$ iff $\quad (x_i - \mu) = 0\ \forall\ i \quad \Rightarrow \quad x_i = \mu\ \forall\ i$

$\Rightarrow \quad P(X = \mu) = 1.$

Hence, Var (X) = 0 iff X is a degenerate *r.v.* at $X = \mu$.

Note. If $X = c$ (constant), then $E(X) = c$ and $E(X^2) = c^2$

$\Rightarrow \quad \text{Var}(X) = E(X^2) - [E(X)]^2 = c^2 - c^2 = 0$

(*b*) If X is a continuous *r.v.* with *p.d.f.* $f(x)$, then

$$\text{Var}(X) = E(X - \mu)^2 = \int_{-\infty}^{\infty} (x - \mu)^2 . f(x)\, dx = 0,$$

iff $f(x) = 0$, whenever $x - \mu \neq 0$ *i.e.,* $x \neq \mu$, which is not possible for any *p.d.f.* defined on the real line $R^1 :]-\infty, \infty[$.

Hence, if X is a continuous *r.v.*, then Var (X) $\neq 0$.

6·5·2. Approximate Expression for Expectation and Variance

Theorem. *Let X be a r.v. with* $E(X) = \mu$ *and* $Var(X) = \sigma^2$. *Then, if* $Y = G(X)$, *then*

$$E(Y) \simeq G(\mu) + \frac{1}{2} G''(\mu) . \sigma^2 \qquad \text{... (6·30)}$$

and

$$Var(Y) \simeq [G'(\mu)]^2 \sigma^2, \qquad \text{... (6·30a)}$$

provided G is twice differentiable at $X = \mu$.

Proof. We have : $Y = G(X) = G[\mu + (X - \mu)]$

Expanding $G(X)$ as a Taylor's series (upto two terms) about $X = \mu$, we get

$$Y = G(X) = G(\mu) + (X - \mu) G'(\mu) + \frac{(X - \mu)^2}{2!} G''(\mu) + R_1 \qquad \text{... (*)}$$

where R_1 is the *Remainder.* If we discard the remainder term, then on taking expectation of both sides in (*), we get

$$E(Y) \simeq G(\mu) + G'(\mu) E(X - \mu) + \frac{1}{2} G''(\mu) . E(X - \mu)^2$$

$$\Rightarrow \qquad E(Y) \simeq G(\mu) + \frac{1}{2} G''(\mu) . \sigma^2$$

To obtain Var (Y), we expand $G(X)$ as a Taylor's series (*upto one term*) about $X = \mu$ to get :

$$Y = G(X) = G[\mu + (X - \mu)]$$
$$= G(\mu) + (X - \mu) . G'(\mu) + R_2$$

where R_2 is the *Remainder.* Discarding R_2, we get

$$Var(Y) \simeq Var[G(\mu) + (X - \mu) G'(\mu)]$$
$$= Var[G(\mu)] + [G'(\mu)]^2 . Var(X - \mu) \; ; \quad [\because Var(aX) = a^2 Var(X)]$$
$$= 0 + [G'(\mu)]^2 . Var(X) \; ; \; [\because Var(C) = 0 \text{ and } Var(X \pm a) = Var X]$$

$$\Rightarrow \qquad Var(Y) \simeq [G'(\mu)]^2 . \sigma^2.$$

The above result can be extended to two dimensional *r.v.*, as given in the following Theorem.

Theorem. *Let (X, Y) be two dimensional r.v. with :*

$$E(X) = \mu_X, \; Var(X) = \sigma_X^2, \; E(Y) = \mu_Y \quad \text{and} \quad Var(Y) = \sigma_Y^2, \text{ and let } Z = G(X, Y).$$

If X and Y are independent, then

$$E(Z) \simeq G(\mu_X, \mu_Y) + \frac{1}{2}\left[\frac{\partial^2 G}{\partial x^2} \cdot \sigma_X^2 + \frac{\partial^2 G}{\partial y^2} \cdot \sigma_Y^2\right] \qquad \text{... (6·31)}$$

and

$$Var(Z) \simeq \left(\frac{\partial G}{\partial x}\right)^2 \cdot \sigma_X^2 + \left(\frac{\partial G}{\partial y}\right)^2 \sigma_Y^2, \qquad \text{... (6·31a)}$$

where all the partial derivatives are evaluated at the point (μ_X, μ_Y)

Proof. The proof of this theorem is based on Taylor series expansion of the function $f(x, y)$ of two variables about the point (a, b), stated below.

$$f(x, y) = f[a + (x - a), b + (y - b)] = f(a + h, b + k) \; ; \quad (h = x - a, k = y - b)$$

$$= f(a, b) + \left[h.\frac{\partial f}{\partial x} + k.\frac{\partial f}{\partial y}\right] + \frac{1}{2!}\left[h^2.\frac{\partial^2 f}{\partial x^2} + 2hk\frac{\partial^2 f}{\partial x \partial y} + k^2.\frac{\partial^2 f}{\partial y^2}\right] + \ldots,$$

where $h = x - a$, $k = y - b$ and $f \equiv f(a, b)$.

Using Taylor's series expansion of $G(X, Y)$ about (μ_X, μ_Y), we get :

$$\therefore \quad Z = G(X, Y) = G[\mu_X + (X - \mu_X), \mu_Y + (Y - \mu_Y)]$$

$$= G(\mu_X, \mu_Y) + \left[(X - \mu_X)\frac{\partial G}{\partial x} + (Y - \mu_Y)\left(\frac{\partial G}{\partial y}\right)\right]$$

$$+ \frac{1}{2!}\left[(X - \mu_X)^2.\frac{\partial^2 G}{\partial x^2} + 2(X - \mu_X)(Y - \mu_Y)\frac{\partial^2 G}{\partial x \partial y} + (Y - \mu_Y)^2 \cdot \frac{\partial^2 G}{\partial y^2}\right] + R_1 \quad \ldots (*)$$

where all partial derivatives are evaluated at (μ_X, μ_Y) *i.e.*,

$$\frac{\partial^r G}{\partial x^r} = \frac{\partial^r}{\partial x^r}[G(\mu_X, \mu_Y)] \quad ; \quad r = 1, 2, \text{ and so on, and } R_1 \text{ is the } \textit{remainder term.}$$

Discarding R_1 and taking expectation of both sides in (*), we get

$$E(Z) \simeq G(\mu_X, \mu_Y) + \frac{1}{2!}\left[\frac{\partial^2 G}{\partial x^2}E(X - \mu_X)^2 + \frac{\partial^2 G}{\partial y^2}E(Y - \mu_Y)^2\right]$$

$[\because \quad E(X - \mu_X) = E(Y - \mu_Y) = 0$

and $\quad E[(X - \mu_X)(Y - \mu_Y)] = \text{Cov}(X, Y) = 0$, because X and Y are independent$]$

$$\Rightarrow \quad E(Z) = G(\mu_X, \mu_Y) + \frac{1}{2!}\left[\frac{\partial^2 G}{\partial x^2}.\sigma_X^2 + \frac{\partial^2 G}{\partial y^2}\sigma_Y^2\right],$$

where partial derivatives are evaluated at the point (μ_X, μ_Y).

To obtain the expression for Var (Z), writing Taylor's series expansion(*) upto one term only, we get

$$Z = G(\mu_X, \mu_Y) + \left[(X - \mu_X)\frac{\partial G}{\partial x} + (Y - \mu_Y)\frac{\partial G}{\partial Y}\right] + R_2 \quad \ldots (**)$$

where R_2 is the *remainder* term.

Discarding R_2 and taking variance of both sides in (**), we get

$$\text{Var}(Z) = V[G(\mu_X, \mu_Y)] + \left(\frac{\partial G}{\partial x}\right)^2.\text{Var}(X - \mu_X) + \left(\frac{\partial G}{\partial y}\right)^2.\text{Var}(Y - \mu_Y),$$

the covariance term vanishes because X and Y are independent.

$$\Rightarrow \quad \text{Var}(Z) = \left(\frac{\partial G}{\partial x}\right)^2.\text{Var}(X) + \left(\frac{\partial G}{\partial y}\right)^2.\text{Var}(Y)$$

$[\because \text{Var}(C) = 0, \ \text{Var}(X \pm C) = \text{Var } X \ ; \ \text{Var}(Y \pm k) = \text{Var } Y]$

$$\Rightarrow \quad \text{Var}(Z) \approx \left(\frac{\partial G}{\partial x}\right)^2.\sigma_X^2 + \left(\frac{\partial G}{\partial y}\right)^2.\sigma_Y^2,$$

where the partial derivatives are evaluated at the point (μ_X, μ_Y).

Note. The above result may be extended to a function of n *independent r.v.'s*.

If $Z = G(X_1, X_2, ..., X_n)$ with $E(X_i) = \mu_i$, $\text{Var}(X_i) = \sigma_i^2$; $i = 1, 2, ..., n$, then assuming that all the partial derivatives exist, we have the following approximation

$$E(Z) = G(\mu_1, \mu_2, ..., \mu_n) + \frac{1}{2}\sum_{i=1}^{n}\left[\frac{\partial^2 G}{\partial x_i^2}.\sigma_i^2\right] \quad ...(6·32)$$

and

$$\text{Var}(Z) = \sum_{i=1}^{n}\left[\left(\frac{\partial G}{\partial x_i}\right)^2.\sigma_i^2\right] \quad ...(6·32a)$$

where all the partial derivatives are evaluated at the point $(\mu_1, \mu_2, ..., \mu_n)$.

6·6. COVARIANCE

If X and Y are two random variables, then covariance between them is defined as

$$Cov(X,Y) = E[\{X - E(X)\}\{Y - E(Y)\}] \quad ...(6·33)$$

$$= E[XY - X E(Y) - Y E(X) + E(X) E(Y)]$$

$$= E(XY) - E(Y) E(X) - E(X) E(Y) + E(X) E(Y)$$

$$= E(XY) - E(X) E(Y) \quad ...(6·33a)$$

If X and Y are independent then $E(XY) = E(X) E(Y)$ and hence in this case

$$Cov(X, Y) = E(X) E(Y) - E(X) E(Y) = 0 \quad ...(6·33b)$$

Remarks :

1. $$Cov(aX, bY) = E\left[\{aX - E(aX)\} \cdot \{bY - E(bY)\}\right]$$
$$= E\left[a\{X - E(X)\} \cdot b\{Y - E(Y)\}\right]$$
$$= ab\, E\left[\{X - E(X)\} \cdot \{Y - E(Y)\}\right]$$
$$= ab\, \text{Cov}(X, Y) \quad ...(6·34)$$

2. $$\text{Cov}(X + a, Y + b) = \text{Cov}(X, Y) \quad ...(6·34a)$$

3. $$\text{Cov}\left(\frac{X - \bar{X}}{\sigma_X}, \frac{Y - \bar{Y}}{\sigma_Y}\right) = \frac{1}{\sigma_X \sigma_Y}\text{Cov}(X, Y) \quad ...(6·34b)$$

4. Similarly, we shall get :

$$\text{Cov}(aX + b, cY + d) = ac\, Cov(X, Y) \quad ...(6·34c)$$

$$\text{Cov}(X + Y, Z) = \text{Cov}(X, Z) + \text{Cov}(Y, Z) \quad ...(6·34d)$$

$$\text{Cov}(aX + bY, cX + dY) = a\,c\,\sigma_X^2 + b\,d\,\sigma_Y^2 + (a\,d + b\,c)\,\text{Cov}(X, Y) \quad ...(6·34e)$$

5. If X and Y are independent, $\text{Cov}(X, Y) = 0$. [c.f. (6·33b)]. However, the converse is not true. For illustrations see Chapter 10 on Correlation.

6·6·1. Variance of a Linear Combination of Random Variables

Let $X_1, X_2, ..., X_n$ be n random variables, then

$$V\left(\sum_{i=1}^{n} a_i X_i\right) = \sum_{i=1}^{n} a_i^2 V(X_i) + 2\sum_{i=1}^{n}\sum_{\substack{j=1 \\ i<j}}^{n} a_i a_j\, Cov(X_i, X_j) \quad ...(6·35)$$

Proof. Let $U = a_1 X_1 + a_2 X_2 + ... + a_n X_n$

so that $E(U) = a_1 E(X_1) + a_2 E(X_2) + ... + a_n E(X_n)$

$\therefore \quad U - E(U) = a_1 [X_1 - E(X_1)] + a_2 [X_2 - E(X_2)] + ... + a_n [X_n - E(X_n)]$

Squaring and taking expectation of both sides, we get

$$E[U - E(U)]^2 = a_1^2 E[X_1 - E(X_1)]^2 + a_2^2 E[X_2 - E(X_2)]^2 + ... + a_n^2 E[X_n - E(X_n)]^2$$

$$+ 2 \sum_{i=1}^{n} \sum_{\substack{j=1 \\ i<j}}^{n} a_i a_j E\Big[\{X_i - E(X_i)\}\{X_j - E(X_j)\}\Big]$$

$$\Rightarrow \quad V(U) = a_1^2 V(X_1) + a_2^2 V(X_2) + ... + a_n^2 V(X_n) + 2 \sum_{i=1}^{n} \sum_{\substack{j=1 \\ i<j}}^{n} a_i a_j \text{Cov}(X_i, X_j)$$

$$\Rightarrow \quad V\left(\sum_{i=1}^{n} a_i X_i\right) = \sum_{i=1}^{n} a_i^2 V(X_i) + 2 \sum_{i=1}^{n} \sum_{\substack{j=1 \\ i<j}}^{n} a_i a_j \text{Cov}(X_i, X_j)$$

Remarks :

1. If $a_i = 1$; $i = 1, 2, ..., n$, then

$$V(X_1 + X_2 + ... + X_n) = V(X_1) + V(X_2) + ... + V(X_n) + 2 \sum_{i=1}^{n} \sum_{\substack{j=1 \\ i<j}}^{n} \text{Cov}(X_i, X_j) \quad ...(6·35a)$$

2. If $X_1, X_2, ..., X_n$ are independent (pairwise), then $\text{Cov}(X_i, X_j) = 0, (i \neq j)$.

Thus, from (6·35) and (6·35*a*), we get

$$\left.\begin{aligned} V(a_1X_1 + a_2X_2 + ... + a_n X_n) &= a_1^2 V(X_1) + a_2^2 V(X_2) + ... + a_n^2 V(X_n) \\ \text{and} \quad V(X_1 + X_2 + ... + X_n) &= V(X_1) + V(X_2) + ... + V(X_n), \end{aligned}\right\} \quad ...(6·35b)$$

provided $X_1, X_2, ..., X_n$ are independent.

3. If $a_1 = 1 = a_2$ and $a_3 = a_4 = ... = a_n = 0$, then from (6·35), we get

$$V(X_1 + X_2) = V(X_1) + V(X_2) + 2 \text{Cov}(X_1, X_2)$$

Again if $a_1 = 1, a_2 = -1$ and $a_3 = a_4 = ... = a_n = 0$, then

$$V(X_1 - X_2) = V(X_1) + V(X_2) - 2 \text{Cov}(X_1, X_2)$$

Thus we have :

$$V(X_1 \pm X_2) = V(X_1) + V(X_2) \pm 2 \text{Cov}(X_1, X_2) \quad ...(6·35c)$$

If X_1 and X_2 are independent, then $\text{Cov}(X_1, X_2) = 0$ and we get

$$V(X_1 \pm X_2) = V(X_1) + V(X_2) \quad ...(6·35d)$$

Example 6·1. *Let X be a random variable with the following probability distribution :*

x	:	-3	6	9
$P(X = x)$	:	1/6	1/2	1/3

Find E (X) and E (X²) and using the laws of expectation, evaluate E (2X + 1)².

Solution. $E(X) = \sum x\, p(x) = (-3) \times \frac{1}{6} + 6 \times \frac{1}{2} + 9 \times \frac{1}{3} = \frac{11}{2}$

$$E(X^2) = \sum x^2\, p(x) = 9 \times \frac{1}{6} + 36 \times \frac{1}{2} + 81 \times \frac{1}{3} = \frac{93}{2}$$

$$\therefore \quad E(2X+1)^2 = E(4X^2 + 4X + 1) = 4E(X^2) + 4E(X) + 1 = 4 \times \frac{93}{2} + 4 \times \frac{11}{2} + 1 = 209.$$

Example 6·2.

(a) Find the expectation of the number on a dice when thrown.

(b) Two unbiased dice are thrown. Find the expected values of the sum of numbers of points on them.

Solution.

(*a*) Let X be the random variable representing the number on a dice when thrown. Then, X can take any one of the values 1, 2, 3, ..., 6, each with equal probability $\frac{1}{6}$. Hence,

$$E(X) = \frac{1}{6} \times 1 + \frac{1}{6} \times 2 + \frac{1}{6} \times 3 + \ldots + \frac{1}{6} \times 6 = \frac{1}{6}(1 + 2 + 3 + \ldots + 6) = \frac{1}{6} \times \frac{6 \times 7}{2} = \frac{7}{2} \ldots (*)$$

Remark. This does not mean that in a random throw of a dice, the player will get the number $\frac{7}{2} = 3{\cdot}5$. In fact, one can never get this (fractional) number in a throw of a dice. Rather, this implies that if the player tosses the dice for a "long" period, then on the average toss he will get $\frac{7}{2} = 3{\cdot}5$.

(*b*) The probability function of X (the sum of numbers obtained on two dice), is

Value of X : x	2	3	4	5	6	7	....	11	12
Probability p (x)	1/36	2/36	3/36	4/36	5/36	6/36	....	2/36	1/36

$$E(X) = \sum_x x . p(x)$$

$$= 2 \times \frac{1}{36} + 3 \times \frac{2}{36} + 4 \times \frac{3}{36} + 5 \times \frac{4}{36} + 6 \times \frac{5}{36} + 7 \times \frac{6}{36} + 8 \times \frac{5}{36} + 9 \times \frac{4}{36}$$
$$+ 10 \times \frac{3}{36} + 11 \times \frac{2}{36} + 12 \times \frac{1}{36}$$

$$= \frac{1}{36}(2 + 6 + 12 + 20 + 30 + 42 + 40 + 36 + 30 + 22 + 12) = \frac{1}{36} \times 252 = 7.$$

Aliter. Let X_i be the number obtained on the *i*th dice ($i = 1, 2$) when thrown. Then the sum of the number of points on two dice is given by :

$$S = X_1 + X_2 \quad \Rightarrow \quad E(S) = E(X_1) + E(X_2) = \frac{7}{2} + \frac{7}{2} = 7 \qquad \text{[On using (*)]}$$

Remark. This result can be generalised to the sum of points obtained in a random throw of *n* dice. Then,

$$E(S) = \sum_{i=1}^{n} E(X_i) = \sum_{i=1}^{n} \frac{7}{2} = \frac{7n}{2}.$$

Example 6·3. *In four tosses of a coin, let X be the number of heads. Tabulate the 16 possible outcomes with the corresponding values of X. By simple counting, derive the probability distribution of X and hence calculate the expected value and variance of X.*

Solution. Let H represent a head, T a tail and X, the *r.v.* denoting the number of heads.

S.No.	Outcomes	No. of Heads (X)	S.No.	Outcomes	No. of Heads (X)
1	H H H H	4	9	H T H T	2
2	H H H T	3	10	T H T H	2
3	H H T H	3	11	T H H T	2
4	H T H H	3	12	H T T T	1
5	T H H H	3	13	T H T T	1
6	H H T T	2	14	T T H T	1
7	H T T H	2	15	T T T H	1
8	T T H H	2	16	T T T T	0

The random variable X takes the values 0, 1, 2, 3 and 4. Since, from the above table, we find that the number of cases favourable to the coming of 0, 1, 2, 3 and 4 heads are 1, 4, 6, 4 and 1 respectively, we have

$$P(X=0)=\frac{1}{16},\qquad P(X=1)=\frac{4}{16}=\frac{1}{4},\qquad P(X=2)=\frac{6}{16}=\frac{3}{8},$$

$$P(X=3)=\frac{4}{16}=\frac{1}{4},\qquad P(X=4)=\frac{1}{16}.$$

The probability distribution of X can be summarized as follows :

x	:	0	1	2	3	4
$p(x)$	:	$\frac{1}{16}$	$\frac{1}{4}$	$\frac{3}{8}$	$\frac{1}{4}$	$\frac{1}{16}$

$$\therefore\qquad E(X)=\sum_{x=0}^{4} x.p(x)=1\cdot\frac{1}{4}+2\cdot\frac{3}{8}+3\cdot\frac{1}{4}+4\cdot\frac{1}{16}=2.$$

$$E(X^2)=\sum_{x=0}^{4} x^2.p(x)=1^2\cdot\frac{1}{4}+2^2\cdot\frac{3}{8}+3^2\cdot\frac{1}{4}+4^2\cdot\frac{1}{16}=\frac{1+6+9+4}{4}=5$$

$$\therefore\qquad \text{Var}(X)=E(X^2)-[E(X)]^2=5-2^2=1.$$

Aliter. We can use Binomial distribution (Chapter 8).

If $X \sim B\left(n=4, p=\frac{1}{2}\right)$, then $E(X)=np=4\times\frac{1}{2}=2$ and $\text{Var}(X)=np(1-p)=4\times\frac{1}{2}\times\frac{1}{2}=1.$

Example 6·4. *A gamester has a disc with a freely revolving needle. The disc is divided into 20 equal sectors by thin lines and the sectors are marked 0, 1, 2, ..., 19. The gamester treats 5 or any multiple of 5 as lucky numbers and zero as a special lucky number. He allows a player to whirl the needle on a charge of 10 monetary units. When the needle stops at the lucky number the gamester pays back the player twice the sum charged and at the special lucky number the gamester pays to the player 5 times of the sum charged. Is the game fair? What is the expectation of the player ?*

Solution.

Event	*Favourable*	*p (x)*	*Player's Gain (x)*
Lucky number	5, 10, 15	3/20	20 – 10 = 10
Special lucky No.	0	1/20	50 – 10 = 40
Other numbers	1, 2, 3, 4, 6, 7, 8, 9, 11, 12, 13, 14, 16, 17, 18, 19	16/20	– 10

$\therefore \quad E(X) = \frac{3}{20} \times 10 + \frac{1}{20} \times 40 - \frac{16}{20} \times 10 = -\frac{9}{2} \neq 0$, *i.e.*, the game is *not fair*.

Example 6·5. *A box contains 2^n tickets among which ${}^nC_i$ tickets bear the number i ; i = 0, 1, 2, ..., n. A group of m tickets is drawn. What is the expectation of the sum of their numbers ?*

Solution. Let X_i ; $i = 1, 2, ..., m$ be the variable representing the number on the *i*th ticket drawn. Then the sum '*S*' of the numbers on the tickets drawn is given by :

$$S = X_1 + X_2 + ... + X_m = \sum_{i=1}^{m} X_i \text{, so that } E(S) = \sum_{i=1}^{m} E(X_i)$$

X_i is a random variable which can take any one of the possible values 0, 1, 2, ..., *n* with respective probabilities : ${}^nC_0 / 2^n, {}^nC_1 / 2^n, {}^nC_2 / 2^n, ..., {}^nC_n / 2^n$.

$$\therefore \quad E(X_i) = \frac{1}{2^n}\left(1 \cdot {}^nC_1 + 2 \cdot {}^nC_2 + 3 \cdot {}^nC_3 + ... + n \cdot {}^nC_n\right)$$

$$= \frac{1}{2^n}\left[1 \cdot n + 2 \cdot \frac{n(n-1)}{2!} + 3 \cdot \frac{n(n-1)(n-2)}{3!} + ... + n \cdot 1\right]$$

$$= \frac{n}{2^n}\left\{1 + (n-1) + \frac{(n-1)(n-2)}{2!} + ... + 1\right\}$$

$$= \frac{n}{2^n}\left\{{}^{n-1}C_0 + {}^{n-1}C_1 + {}^{n-1}C_2 + ... + {}^{n-1}C_{n-1}\right\} = \frac{n}{2^n} \cdot (1+1)^{n-1} = \frac{n}{2}$$

$$\therefore \quad E(S) = \sum_{i=1}^{m} E(X_i) = \sum_{i=1}^{m} (n/2) = \frac{mn}{2}.$$

Example 6·6. *Prove that the events $E_1, E_2, ..., E_n$ are independent iff their corresponding indicator variates $I_1, I_2, ..., I_n$ are independent.*

Solution. By definition : $I_i(w) = \begin{Bmatrix} 1, & \text{if } w \in E_i \\ 0, & \text{if } w \notin E_i \end{Bmatrix}$ ... (*)

If Part. Let the events E_i, $i = 1, 2, ..., n$ be independent. By definition in (*), $I_i(w)$ is a function of E_i. Hence, if E_i's are independent, their corresponding indicator functions $I_i(w)$ are also independent.

Only if Part. Let us assume that the indicator functions $I_1, I_2, ..., I_n$ are independent. Then, we have.

$$E[I_1 \cdot I_2 ... I_n] = E(I_1) \cdot E(I_2) ... E(I_n) \quad ... (**)$$

But $E[I_1 \cdot I_2 ... I_n] = E\left[I_{E_1 \cap E_2 \cap ... \cap E_n}\right] = P[E_1 \cap E_2 \cap ... \cap E_n]$

and $E(I_i) = P(E_i)$; $i = 1, 2, ... n$

Substituting in (**), we get

$$P(E_1 \cap E_2 \cap ... \cap E_n) = P(E_1) \cdot P(E_2) ... P(E_n) \Rightarrow E_1, E_2, ..., E_n \text{ are independent.}$$

Example 6·7. *A coin is tossed until a head appears. What is the expectation of the number of tosses required?*

Solution. Let X denote the number of tosses required to get the first head. Then X can materialise in the following ways :

Event	x	*Probability, p(x)*
H	1	$\frac{1}{2}$
TH	2	$\frac{1}{2}\times\frac{1}{2}=\frac{1}{4}$
TTH	3	$\frac{1}{2}\times\frac{1}{2}\times\frac{1}{2}=\frac{1}{8}$
⋮	⋮	⋮

$$\therefore \qquad E(X) = \sum_{x=1}^{\infty} x.p(x) = 1\times\frac{1}{2}+2\times\frac{1}{4}+3\times\frac{1}{8}+4\times\frac{1}{16}+\ldots \qquad \ldots (*)$$

This is an arithmetic-geometric series with ratio of GP being $r = \frac{1}{2}$.

$$\text{Let} \qquad S = 1\cdot\frac{1}{2}+2\cdot\frac{1}{4}+3\cdot\frac{1}{8}+4\cdot\frac{1}{16}+\ldots$$

$$\text{Then} \qquad \frac{1}{2}S = \qquad \frac{1}{4}+2\cdot\frac{1}{8}+3\cdot\frac{1}{16}+\ldots$$

$$\therefore \qquad \left(1-\frac{1}{2}\right)S = \frac{1}{2}+\frac{1}{4}+\frac{1}{8}+\frac{1}{16}+\ldots \quad \Rightarrow \quad \frac{1}{2}S = \frac{\frac{1}{2}}{1-\frac{1}{2}} = 1 \quad \Rightarrow \quad S = 2.$$

[Since the sum of an infinite G.P. with first term a and common ratio r (< 1) is $\frac{a}{1-r}$.]

Hence, substituting in (*), we have : $E(X) = 2$.

Example 6·8. *What is the expectation of the number of failures preceding the first success in an infinite series of independent trials with constant probability p of success in each trial ?*

Solution. Let the random variable X denote the number of failures preceding the first success. Then X can take the values 0, 1, 2, ..., ∞. We have

$$P(X = x) = p(x) = P(x \text{ failures precede the first success}) = q^x p,$$

where $q = 1 - p$, is the probability of failure in a trial. Then by *def.*,

$$E(X) = \sum_{x=0}^{\infty} x\, p(x) = \sum_{x=0}^{\infty} x\, q^x p = p\,q \sum_{x=0}^{\infty} x\, q^{x-1} = pq\,(1+2q+3q^2+4q^3+\ldots) \ \ldots (*)$$

$1 + 2q + 3q^2 + 4q^3 + \ldots$ is an infinite arithmetic-geometric series.

$$\text{Let} \qquad S = 1 + 2q + 3q^2 + 4q^3 + \ldots$$

$$qS = \qquad q + 2q^2 + 3q^3 + \ldots$$

$$(1-q)\,S = 1 + q + q^2 + q^3 + \ldots = \frac{1}{1-q} \quad \Rightarrow \quad S = \frac{1}{(1-q)^2}$$

$$\therefore \quad 1 + 2q + 3q^2 + 4q^3 \ldots = \frac{1}{(1-q)^2}. \quad \text{Hence, } E(X) = \frac{pq}{(1-q)^2} = \frac{pq}{p^2} = \frac{p}{q} \quad \text{[From (*)]}$$

Example 6·9. *A box contains 'a' white and 'b' black balls. 'c' balls are drawn at random. Find the expected value of the number of white balls drawn.*

Solution. Let the variable X_i, associated with ith draw, be defined as follows :

$$X_i = \begin{cases} 1, & \text{if } i\text{th ball drawn is white} \\ 0, & \text{if } i\text{th ball drawn is black} \end{cases} ; \quad i = 1, 2, \ldots, c.$$

Then the number 'S' of the white balls among 'c' balls drawn is given by :

$$S = X_1 + X_2 + \ldots + X_c = \sum_{i=1}^{c} X_i \quad \Rightarrow \quad E(S) = \sum_{i=1}^{c} E(X_i) \qquad \ldots (*)$$

$$P(X_1 = 1) = P(\text{First ball drawn is white}) = \frac{a}{a+b}$$

$$P(X_2 = 1) = P(X_1 = 1) \cdot P(X_2 = 1 \mid X_1 = 1) + P(X_1 = 0) \cdot P(X_2 = 1 \mid X_1 = 0)$$

$$= \frac{a}{(a+b)} \cdot \frac{a-1}{(a+b-1)} + \frac{b}{(a+b)} \cdot \frac{a}{(a+b-1)}$$

$$= \frac{a}{(a+b)(a+b-1)} \left[(a-1)+b\right] = \frac{a}{a+b}; \quad \text{and so on.}$$

In general, $P(X_i = 1) = P$ (of drawing a white ball at the ith draw) $= \dfrac{a}{a+b}$

$\Rightarrow \quad P(X_i = 0) = P$ (of drawing a black ball at the ith draw) $= \dfrac{b}{a+b}$

$$\therefore \quad E(X_i) = 1 \,.\, P(X_i = 1) + 0 \,.\, P(X_i = 0) = \frac{a}{a+b}$$

$$\text{Hence,} \quad E(S) = \sum_{i=1}^{c} \left(\frac{a}{a+b}\right) = \frac{c\,a}{a+b}. \qquad \text{[From (*)]}$$

SUPPLEMENTARY EXAMPLES ON EXPECTATION

Example 6·10. *Let the r.v. X have the distribution :*

$$P(X = 0) = P(X = 2) = p\,; \qquad P(X = 1) = 1 - 2p, \text{ for } 0 \le p \le \frac{1}{2}.$$

For what value of p is the Var (X) a maximum ?

Solution. Here the *r.v.* X takes the values 0, 1 and 2 with respective probabilities p, $1 - 2p$ and p, $0 \le p \le \frac{1}{2}$. Thus,

$$E(X) = 0 \times p + 1 \times (1 - 2p) + 2 \times p = 1,$$
$$E(X^2) = 0 \times p + 1^2 \times (1 - 2p) + 2^2 \times p = 1 + 2p$$
$$\therefore \quad \text{Var}(X) = E(X^2) - \{E(X)\}^2 = 2p\ ;\ 0 \le p \le \frac{1}{2}$$

Obviously, for $0 \le p \le \frac{1}{2}$, Var (X) is maximum when $p = \frac{1}{2}$, and $[\text{Var}(X)]_{max} = 2 \times \frac{1}{2} = 1$.

Example 6·11. *(Random Walk Problem). Starting from the origin, unit steps are taken to the right with probability p and to the left with probability q (= 1 – p). Assuming independent movements, find the mean and variance of the distance moved from origin after n steps .*

Solution. Let us associate a variable X_i with the *i*th step defined as follows :

$$X_i = \begin{cases} +1, & \text{if the } i\text{th step is towards the right,} \\ -1, & \text{if the } i\text{th step is towards the left.} \end{cases}$$

Then, $$S_n = X_1 + X_2 + \dots + X_n = \sum X_i,$$

represents the random distance moved from origin after *n* steps.

$$E(X_i) = 1 \times p + (-1) \times q = p - q$$

and $$E(X_i^2) = 1^2 \times p + (-1)^2 \times q = p + q = 1$$

$$\therefore \quad \text{Var}(X_i) = E(X_i^2) - \{E(X_i)\}^2 = (q + p)^2 - (p - q)^2 = 4pq$$

Hence, $$E(S_n) = \sum_{i=1}^{n} E(X_i) = n(p - q) \quad \text{and} \quad V(S_n) = \sum_{i=1}^{n} V(X_i) = 4npq$$

[∵ Movements of steps are independent.]

Example 6·12. *Let X be a r.v. with mean μ and variance σ^2. Show that $E(X - b)^2$, as a function of b, is minimised when b = μ.*

Solution. $$E(X - b)^2 = E[(X - \mu) + (\mu - b)]^2$$
$$= E(X - \mu)^2 + (\mu - b)^2 + 2(\mu - b)E(X - \mu) = \text{Var}(X) + (\mu - b)^2$$
$$[\because E(X - \mu) = 0]$$

$$\Rightarrow \quad E(X - b)^2 \ge Var(X), \qquad (*)$$

since $(\mu - b)^2$, being the square of a real quantity is always non-negative.

The sign of equality holds in (*) iff $(\mu - b)^2 = 0 \Rightarrow \mu = b$.

Hence, $E(X - b)^2$ is minimised when $\mu = b$ and its minimum value is :

$$E(X - \mu)^2 = \sigma_X^2.$$

Remark. This result states that the sum of squares of deviations is minimum when taken about mean.

Example 6·13. *In a sequence of Bernoulli trials, let X be the length of the run of either successes or failures starting with the first trial. Find E (X) and V (X).*

Solution. Let '*p*' denote the probability of success. Then $q = 1 - p$, is the probability of failure. X = 1 means that we can have any of the possibilities *SF* and *FS* with respective probabilities *pq* and *qp*.

$\therefore \quad P(X = 1) = P(SF) + P(FS) = pq + qp = 2pq$

Similarly, $P(X = 2) = P(SSF) + P(FFS) = p^2q + q^2p$

In general $P(X = r) = P(SSS...SF) + P(FFF...FS) = p^r.q + q^r.p$

$$\therefore \quad E(X) = \sum_{r=1}^{\infty} r\,P(X = r) = \sum_{r=1}^{\infty} r\left(p^r.q + q^r.p\right) = pq\left(\sum_{r=1}^{\infty} r.p^{r-1} + \sum_{r=1}^{\infty} r.q^{r-1}\right)$$

$$= pq\left[(1 + 2p + 3p^2 + ...) + (1 + 2q + 3q^2 + ...)\right] = pq\left[(1-p)^{-2} + (1-q)^{-2}\right]$$

$$= pq\,(q^{-2} + p^{-2}) = pq\left(\frac{1}{q^2} + \frac{1}{p^2}\right) = \frac{p}{q} + \frac{q}{p}$$

(See Remark 1 to Example 6·15)

$$V(X) = E(X^2) - [E(X)]^2 = E[X(X-1)] + E(X) - [E(X)]^2 \qquad ...(*)$$

$$E[X(X-1)] = \sum_{r=2}^{\infty} r(r-1)\,P(X = r) = \sum_{r=2}^{\infty} r(r-1)\,(p^r q + q^r p)$$

$$= \sum_{r=2}^{\infty} r(r-1)\,p^r q + \sum_{r=2}^{\infty} r(r-1)\,q^r p$$

$$= p^2 q \sum_{r=2}^{\infty} r(r-1)\,p^{r-2} + q^2 p \sum_{r=2}^{\infty} r(r-1)\,q^{r-2}$$

$$= 2p^2 q \sum_{r=2}^{\infty} \frac{r(r-1)}{2}\,p^{r-2} + 2q^2 p \sum_{r=2}^{\infty} \frac{r(r-1)}{2}\,q^{r-2}$$

$$= 2p^2 q(1-p)^{-3} + 2q^2 p(1-q)^{-3} = 2\left(\frac{p^2}{q^2} + \frac{q^2}{p^2}\right)$$

$$\therefore \quad V(X) = 2\left(\frac{p^2}{q^2} + \frac{q^2}{p^2}\right) + \left(\frac{p}{q} + \frac{q}{p}\right) - \left(\frac{p}{q} + \frac{q}{p}\right)^2 = \left(\frac{p}{q} - \frac{q}{p}\right)^2 + \left(\frac{p}{q} + \frac{q}{p}\right) \qquad \text{[From (*)]}$$

Example 6·14. (MATCHING PROBLEM) *A deck of n numbered cards is thoroughly shuffled and the cards are inserted into n numbered cells one by one. If the card number 'i' falls in the cell 'i', we count it as a match, otherwise not. Find the mean and variance of total number of such matches.*

Solution. Let us associate a *r.v.*, X_i with the ith draw defined as follows :

$$X_i = \begin{cases} 1, & \text{if the } i\text{th card falls in the } i\text{th cell} \\ 0, & \text{otherwise} \end{cases}$$

The total number of matches 'S' is given by

$$S = X_1 + X_2 + ... + X_n = \sum_{i=1}^{n} X_i \quad \Rightarrow \quad E(S) = \sum_{i=1}^{n} E(X_i)$$

$$E(X_i) = 1.P(X_i = 1) + 0.P(X_i = 0) = P(X_i = 1) = \frac{1}{n}$$

Hence, $$E(S) = \sum_{i=1}^{n} \left(\frac{1}{n}\right) = n \cdot \frac{1}{n} = 1$$

$$V(S) = V(X_1 + X_2 + ... + X_n) = \sum_{i=1}^{n} V(X_i) + 2 \sum_{\substack{i=1 \\ i<j}}^{n} \sum_{j=1}^{n} \text{Cov}(X_i, X_j) \quad ...(1)$$

$$V(X_i) = E(X_i^2) - [E(X_i)]^2 = 1^2 \cdot P(X_i = 1) + 0^2 \cdot P(X_i = 0) - \left(\frac{1}{n}\right)^2 = \frac{1}{n} - \frac{1}{n^2} = \frac{n-1}{n^2} \quad ...(2)$$

$$\text{Cov}(X_i, X_j) = E(X_i X_j) - E(X_i) E(X_j) \quad ...(3)$$

$$E(X_i X_j) = 1 \cdot P(X_i X_j = 1) + 0 \cdot P(X_i X_j = 0) = \frac{(n-2)!}{n!} = \frac{1}{n(n-1)},$$

since $X_i X_j = 1$ if and only if both card numbers i and j are in their respective matching places and there are $(n-2)!$ arrangements of the remaining cards that correspond to this event. Substituting in (3), we get

$$\text{Cov}(X_i, X_j) = \frac{1}{n(n-1)} - \frac{1}{n} \cdot \frac{1}{n} = \frac{1}{n^2(n-1)} \quad ...(4)$$

Substituting from (2) and (4) in (1), we have

$$V(S) = \sum_{i=1}^{n} \left(\frac{n-1}{n^2}\right) + 2 \sum_{\substack{i=1 \\ i<j}}^{n} \sum_{j=1}^{n} \left\{\frac{1}{n^2(n-1)}\right\}$$

$$= n\left(\frac{n-1}{n^2}\right) + 2 \cdot {}^nC_2 \frac{1}{n^2(n-1)} = \frac{n-1}{n} + \frac{1}{n} = 1.$$

***Example 6·15**. If t is any positive real number, show that the function defined by*

$$p(x) = e^{-t}(1 - e^{-t})^{x-1} \quad ...(1)$$

can represent a probability function of a random variable X assuming the values 1, 2, 3, ...Find E (X) and Var (X) of the distribution.

Solution. We have $e^t > 1, \forall t > 0 \Rightarrow e^{-t} < 1$ or $1 - e^{-t} > 0, \forall t > 0$

Also $$e^{-t} = \frac{1}{e^t} > 0, \forall t > 0$$

Hence, $$p(x) = e^{-t}(1 - e^{-t})^{x-1} \geq 0 \ \forall t > 0, x = 1, 2, 3, ...$$

Also $$\sum_{x=1}^{\infty} p(x) = e^{-t} \sum_{x=1}^{\infty} (1 - e^{-t})^{x-1} = e^{-t} \sum_{x=1}^{\infty} a^{x-1}, \quad (a = 1 - e^{-t})$$

$$= e^t (1 + a + a^2 + a^3 + ...) = e^{-t} \cdot \frac{1}{(1-a)}$$

$$= e^{-t}[1 - (1 - e^{-t})]^{-1} = e^{-t} \cdot e^t = 1$$

Hence, $p(x)$ defined in (1) represents the probability function of a *r.v.* X.

$$E(X) = \sum_x x \cdot p(x) = e^{-t} \sum_{x=1}^{\infty} x\left(1-e^{-t}\right)^{x-1} = e^{-t} \sum_{x=1}^{\infty} x \cdot a^{x-1}, \quad (a = 1-e^{-t})$$

$$= e^{-t}(1 + 2a + 3a^2 + 4a^3 + ...) = e^{-t}(1-a)^{-2} \qquad \text{(See Remark 1)}$$

$$= e^{-t}(e^{-t})^{-2} = e^{t}$$

$$E(X^2) = \sum x^2 p(x) = e^{-t} \sum_{x=1}^{\infty} x^2 \cdot a^{x-1} = e^{-t}(1 + 4a + 9a^2 + 16a^3 + ...)$$

$$= e^{-t}(1+a)(1-a)^{-3} = e^{-t}(2-e^{-t})\,e^{3t} \qquad \text{(See Remark 2)}$$

Hence, $\quad \text{Var}(X) = E(X^2) - [E(X)]^2 = e^{-t}(2-e^{-t})\,e^{3t} - e^{2t}$

$$= e^{2t}\left[(2-e^{-t})-1\right] = e^{2t}(1-e^{-t}) = e^{t}(e^{t}-1).$$

Remarks :

1. Consider

$$S = 1 + 2a + 3a^2 + 4a^3 + ... \qquad \text{(Arithmetic-geometric series)}$$

$$\Rightarrow \quad aS = a + 2a^2 + 3a^3 + ...$$

$$\Rightarrow \quad (1-a)S = 1 + a + a^2 + a^3 + ... = \frac{1}{(1-a)} \quad \text{or} \quad S = (1-a)^{-2}$$

$$\therefore \quad \sum_{x=1}^{\infty} x\,a^{x-1} = 1 + 2a + 3a^2 + 4a^3 + ... = (1-a)^{-2} \qquad ...(*)$$

2. Consider

$$S = 1 + 2^2 . a + 3^2 . a^2 + 4^2 . a^3 + 5^2 . a^4 + ...$$

$$\Rightarrow \quad S = 1 + 4a + 9a^2 + 16a^3 + 25a^4 + ...$$

$$-3a\,S = -3a - 12a^2 - 27a^3 - 48a^4 - ...$$

$$+3a^2 S = +3a^2 + 12a^3 + 27a^4 + ...$$

$$-a^3 S = -a^3 - 4a^4 - ...$$

Adding the above equations, we get

$$(1-a)^3 S = 1 + a \quad \Rightarrow \quad S = (1+a)(1-a)^{-3} \qquad ...(**)$$

$$\therefore \quad \sum_{x=1}^{\infty} x^2 a^{x-1} = 1 + 4a + 9a^2 + 16a^3 + ... = (1+a)(1-a)^{-3}$$

The results (*) and (**) are quite useful for numerical problems and should be committed to memory.

Example 6·16. *A man with n keys wants to open his door and tries the keys independently and at random. Find the mean and variance of the number of trials required to open the door,*

(i) if unsuccessful keys are not eliminated from further selection, and

(ii) if they are.

Solution.

(*i*) Suppose the man gets the first success at *x*th trial, *i.e.*, he is unable to open the door in the first $(x-1)$ trials. If unsuccessful keys are not eliminated then X is a random variable which can take the values 1, 2, 3, ..., ∞.

Probability of success at the first trial $= \dfrac{1}{n}$

$\therefore$ Probability of failure at the first trial $= 1 - \dfrac{1}{n}$

If unsuccessful keys are not eliminated then the probability of success and consequently of failure is constant for each trial.

Hence, $p(x)$ = Probability of 1st success at the x th trial $= \left(1-\dfrac{1}{n}\right)^{x-1} \cdot \dfrac{1}{n}$.

$$\therefore \quad E(X) = \sum_{x=1}^{\infty} x\,p(x) = \sum_{x=1}^{\infty} x\left(1-\frac{1}{n}\right)^{x-1} \frac{1}{n} = \frac{1}{n}\sum_{x=1}^{\infty} x\,A^{x-1}, \text{ where } A = 1-\frac{1}{n}.$$

$$E(X) = \frac{1}{n}(1 + 2A + 3A^2 + 4A^3 + \ldots) = \frac{1}{n}(1-A)^{-2} \text{[See (*), Example (6·15)]}$$

$$= \frac{1}{n}\left\{1-\left(1-\frac{1}{n}\right)\right\}^{-2} = n$$

$$E(X^2) = \sum_{x=1}^{\infty} x^2 . p(x) = \sum_{x=1}^{\infty} x^2\left(1-\frac{1}{n}\right)^{x-1} . \frac{1}{n}$$

$$= \frac{1}{n}\sum_{x=1}^{\infty} x^2 A^{x-1} = \frac{1}{n}\left(1+2^2 \cdot A+3^2 \cdot A^2+4^2 \cdot A^3+\ldots\right)$$

$$= \frac{1}{n}(1 + A)(1 - A)^{-3} \qquad \text{[See (**), Example (6·15)]}$$

$$= \frac{1}{n}\left\{1+\left(1-\frac{1}{n}\right)\right\}\left\{1-\left(1-\frac{1}{n}\right)\right\}^{-3} = (2n-1)n$$

Hence, $V(X) = E(X^2) - [E(X)]^2 = (2n-1)\,n - n^2 = n^2 - n = n\,(n-1)$.

(*ii*) If unsuccessful keys are eliminated from further selection, then the random variable X will take the values from 1 to n. In this case, we have

Probability of success at the first trial $= \dfrac{1}{n}$

Probability of success at the 2nd trial $= \dfrac{1}{n-1}$

Probability of success at the 3rd trial $= \dfrac{1}{n-2}$, and so on.

Hence, probability of 1st success at the 2nd trial $= \left(1-\dfrac{1}{n}\right)\dfrac{1}{n-1} = \dfrac{1}{n}$

Probability of first success at the third trial $= \left(1-\dfrac{1}{n}\right)\left(1-\dfrac{1}{n-1}\right).\dfrac{1}{n-2} = \dfrac{1}{n}$,

and so on. In general, we have

$$p(x) = \text{Probability of first success at the } x\text{th trial} = \frac{1}{n}$$

$$\therefore \quad E(X) = \sum_{x=1}^{n} x\, p(x) = \frac{1}{n} \sum_{x=1}^{n} x = \frac{n+1}{2}$$

$$E(X^2) = \sum_{x=1}^{n} x^2 p(x) = \frac{1}{n} \sum_{x=1}^{n} x^2 = \frac{(n+1)(2n+1)}{6}$$

Hence, $$V(X) = E(X^2) - [E(X)]^2 = \frac{(n+1)(2n+1)}{6} - \left(\frac{n+1}{2}\right)^2$$

$$= \frac{n+1}{12}\,[2(2n+1) - 3(n+1)] = \frac{n^2-1}{12}.$$

Example 6·17. *In a lottery m tickets are drawn at a time out of n tickets numbered 1 to n. Find the expectation and the variance of the sum S of the numbers in the tickets drawn.*

Solution. Let X_i denote the score on the ith ticket drawn.

Then $$S = X_1 + X_2 + \ldots + X_m = \sum_{i=1}^{m} X_i,$$ is the total score on the m tickets drawn.

$$\therefore \quad E(S) = \sum_{i=1}^{m} E(X_i)$$

Now each X_i is a random variable which assumes the values 1, 2, 3, ... , n each with equal probability $1/n$.

$$\therefore \quad E(X_i) = \frac{1}{n}(1 + 2 + 3 + \ldots + n) = \frac{(n+1)}{2}.$$

Hence, $$E(S) = \sum_{i=1}^{m} \left(\frac{n+1}{2}\right) = \frac{m(n+1)}{2}$$

$$V(S) = V(X_1 + X_2 + \ldots + X_m) = \sum_{i=1}^{m} V(X_i) + 2 \sum_{\substack{i,j \\ i<j}} \sum \text{Cov}(X_i, X_j)$$

$$E(X_i^2) = \frac{1}{n}(1^2 + 2^2 + 3^2 + \ldots + n^2) = \frac{1}{n} \cdot \frac{n(n+1)(2n+1)}{6} = \frac{(n+1)(2n+1)}{6}$$

$$\therefore \quad V(X_i) = E(X_i^2) - [E(X_i)]^2 = \frac{(n+1)(2n+1)}{6} - \left(\frac{n+1}{2}\right)^2 = \frac{n^2-1}{12}$$

Also $$\text{Cov}(X_i, X_j) = E(X_i X_j) - E(X_i)E(X_j)$$

To find $E(X_i X_j)$, we note that the variables X_i and X_j can take the values as shown below :

X_i	X_j
1	2, 3, ..., n
2	1, 3, ..., n
⋮	⋮
n	1, 2, ..., $(n-1)$

In other words, the variable $X_i X_j$ can take $n(n-1)$ possible values and

$$P(X_i = l \cap X_j = k) = \frac{1}{n(n-1)}, k \neq l.$$

Hence, $$E(X_i X_j) = \frac{1}{n(n-1)} \begin{bmatrix} 1.2 + 1.3 + \text{.........} + \text{.........} 1.n \\ + 2.1 + 2.3 + \text{.........} + \text{.........} 2.n \\ + \text{.................................} \\ \vdots \quad \vdots \quad \vdots \\ + n.1 + n.2 + \text{.........} + n.(n-1) \end{bmatrix}$$

$$= \frac{1}{n(n-1)} \begin{bmatrix} 1(1+2+3+\text{......}+n) - 1^2 \\ + 2(1+2+3+\text{......}+n) - 2^2 \\ + \text{.................................} \\ \text{.................................} \\ + n(1+2+\ldots+\overline{n-1}+n) - n^2 \end{bmatrix}$$

$$= \frac{1}{n(n-1)} \left\{ (1+2+3+\ldots n)^2 - (1^2+2^2+\ldots+n^2) \right\}$$

$$= \frac{1}{n(n-1)} \left[\left\{ \frac{n(n+1)}{2} \right\}^2 - \frac{n(n+1)(2n+1)}{6} \right] = \frac{(n+1)(3n^2-n-2)}{12(n-1)}$$

$$\therefore \text{Cov}(X_i, X_j) = \frac{(n+1)(3n^2-n-2)}{12(n-1)} - \left(\frac{n+1}{2} \right)^2 = \frac{(n+1)}{12(n-1)} \left[3n^2 - n - 2 - 3(n^2-1) \right] = -\frac{(n+1)}{12}$$

Hence, $$V(S) = \sum_{i=1}^{m} \left(\frac{n^2-1}{12} \right) + 2 \sum_{i<j=1}^{m} \sum \left\{ -\frac{(n+1)}{12} \right\} = \frac{m(n^2-1)}{12} + 2.\frac{m(m-1)}{2!} \left\{ \frac{-(n+1)}{12} \right\},$$

[Since there are ${}^mC_2$ covariance terms in Cov (X_i, X_j)]

$$\therefore \quad V(S) = \frac{m(n+1)}{12} \left[(n-1) - (m-1) \right] = \frac{m(n+1)(n-m)}{12}.$$

Example 6·18. *A dice is thrown $(n + 2)$ times. After each throw a '+' is recorded for 4, 5, or 6 and '–' for 1, 2 or 3, the signs forming an ordered sequence. To each, except the first and the last sign, is attached a characteristic random variable which takes the value 1 if both the neighbouring signs differ from the one between them and 0 otherwise. If $X_1, X_2, \ldots, X_n$ are characteristic random variables, find the mean and variance of $X = \sum_{i=1}^{n} X_i$.*

Solution. $$X = \sum_{i=1}^{n} X_i \Rightarrow E(X) = \sum_{i=1}^{n} E(X_i) \quad \text{... (i)}$$

$$E(X_i) = 1.P(X_i = 1) + 0.\,P(X_i = 0) = P(X_i = 1) \quad \text{... (ii)}$$

For $X_i = 1$, there are the following two mutually exclusive possibilities :

$$(i) \quad - \quad + \quad - \quad ; \qquad (ii) \quad + \quad - \quad +$$

and since the probability of each sign is $\frac{1}{2}$, we have by addition probability theorem :

$$P(X_i = 1) = P(i) + P(ii) = \left(\frac{1}{2}\right)^3 + \left(\frac{1}{2}\right)^3 = \frac{1}{4} \qquad ...(iii)$$

$$\therefore \quad E(X_i) = \frac{1}{4} \quad \text{[From } (ii)\text{]}. \qquad \text{Hence,} \quad E(X) = \sum_{i=1}^{n} \left(\frac{1}{4}\right) = \frac{n}{4}, \quad \text{[From } (i)\text{]}$$

$$V(X) = V(X_1 + X_2 + ... + X_n) = \sum_{i=1}^{n} V(X_i) + 2 \sum_{\substack{i=1 \\ i<j}}^{n} \sum_{j=1}^{n} \text{Cov}(X_i, X_j) \qquad ...(*)$$

$$E(X_i^2) = 1^2 . P(X_i = 1) + 0^2 . P(X_i = 0) = P(X_i = 1) = \frac{1}{4} \qquad \text{[From } (iii)\text{]}$$

$$\therefore \quad V(X_i) = E(X_i^2) - \{E(X_i)\}^2 = \frac{1}{4} - \frac{1}{16} = \frac{3}{16}$$

$$E(X_i X_j) = 1.P(X_i = 1 \cap X_j = 1) + 0.P(X_i = 0 \cap X_j = 0)$$
$$+ 0.P(X_i = 1 \cap X_j = 0) + 0.P(X_i = 0 \cap X_j = 1)$$
$$= P(X_i = 1 \cap X_j = 1)$$

Since there are following two mutually exclusive possibilities for the event : $(X_i = 1 \cap X_j = 1)$,

$(i) \; - \; + \; - \; + \qquad (ii) \; + \; - \; + \; -$, we have

$$P(X_i = 1 \cap X_j = 1) = P(i) + P(ii) = \left(\frac{1}{2}\right)^4 + \left(\frac{1}{2}\right)^4 = \frac{1}{8}$$

$$\therefore \quad \text{Cov}(X_i, X_j) = E(X_i X_j) - E(X_i) E(X_j) = \frac{1}{8} - \frac{1}{4} \times \frac{1}{4} = \frac{1}{16}$$

$$\text{Hence,} \quad V(X) = \sum_{i=1}^{n} \frac{3}{16} + 2 \sum_{i<j=1}^{n} \sum \text{Cov}(X_i, X_j) \qquad \text{[From (*)]}$$

$$= \frac{3n}{16} + 2 \left[\text{Cov}(X_1, X_2) + \text{Cov}(X_2, X_3) + ... + \text{Cov}(X_{n-1}, X_n)\right]$$

$$= \frac{3n}{16} + 2(n-1) . \frac{1}{16} = \frac{5n-2}{16}.$$

Example 6·19. *From a point on the circumference of a circle of radius 'a', a chord is drawn in a random direction, (all directions being equally likely). Show that the expected value of the length of the chord is $(4a / \pi)$ and that the variance of the length is $2a^2 [1 - (8 / \pi^2)]$. Also show that the chance is 1/3 that the length of the chord will exceed the length of the side of an equilateral triangle inscribed in the circle.*

Solution. Let P be any point on the circumference of a circle of radius 'a' and centre 'O'. Let PQ be any chord drawn at random and let $\angle OPQ = \theta$. Obviously, θ ranges from $-\pi/2$ to $\pi/2$.

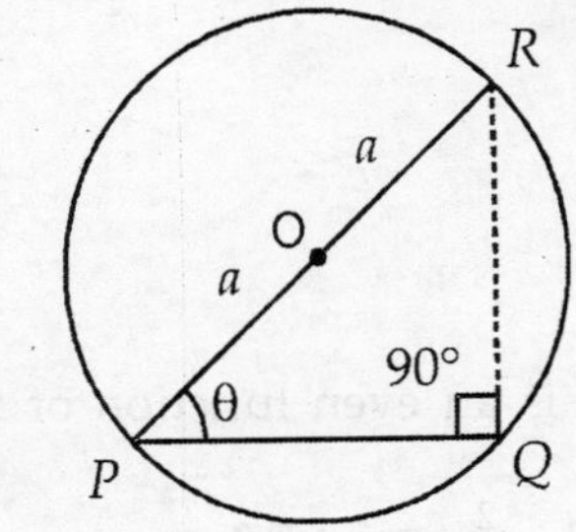

Fig 6·1

Since all the directions are equally likely, the probability differential of θ is given by the rectangular distribution (*c.f. Chapter 9*) :

$$dF(\theta) = f(\theta)\, d\theta = \frac{d\theta}{\{\pi/2-(-\pi/2)\}} = \frac{d\theta}{\pi}, \quad -\frac{\pi}{2} \le \theta \le \frac{\pi}{2}$$

Since $\angle PQR$ is a right angle, (angle in a semi-circle), we have

$$\frac{PQ}{PR} = \cos\theta \quad \Rightarrow \quad PQ = PR\cos\theta = 2a\cos\theta$$

$$E(PQ) = \int_{-\pi/2}^{\pi/2} (PQ)\, f(\theta)\, d\theta = \frac{2a}{\pi}\int_{-\pi/2}^{\pi/2} \cos\theta\, d\theta = \frac{2a}{\pi}\Big|\sin\theta\Big|_{-\pi/2}^{\pi/2} = \frac{4a}{\pi}.$$

$$E\{(PQ)^2\} = \int_{-\pi/2}^{\pi/2} (PQ)^2 f(\theta)\, d\theta = \frac{4a^2}{\pi}\int_{-\pi/2}^{\pi/2} \cos^2\theta\, d\theta = \frac{4a^2}{\pi}\int_0^{\pi/2} 2\cos^2\theta\, d\theta$$

(Since $\cos^2\theta$ is an even function of θ)

$$= \frac{4a^2}{\pi}\int_0^{\pi/2} (1+\cos 2\theta)\, d\theta = \frac{4a^2}{\pi}\left|\theta + \frac{\sin 2\theta}{2}\right|_0^{\pi/2} = \frac{4a^2}{\pi}\cdot\frac{\pi}{2} = 2a^2$$

$$\therefore \quad V(PQ) = E[(PQ)^2] - \{E(PQ)\}^2 = 2a^2 - \frac{16a^2}{\pi^2} = 2a^2\left(1-\frac{8}{\pi^2}\right)$$

We know that the length of the side of an equilateral triangle inscribed in a circle of radius '*a*' is $(a\sqrt{3})$. Hence,

$$P\left(PQ > a\sqrt{3}\right) = P\left(2a\cos\theta > a\sqrt{3}\right) = P\left(\cos\theta > \frac{\sqrt{3}}{2}\right)$$

$$= P\left(|\theta| < \frac{\pi}{6}\right) = P\left(-\frac{\pi}{6} < \theta < \frac{\pi}{6}\right)$$

$$= \int_{-\pi/6}^{\pi/6} f(\theta)\, d\theta = \frac{1}{\pi}\int_{-\pi/6}^{\pi/6} 1\cdot d\theta = \frac{1}{\pi}\cdot\frac{\pi}{3} = \frac{1}{3}.$$

Example 6·20. *A chord of a circle of radius 'a' is drawn parallel to a given straight line, all distance from the centre of the circle being equally likely. Show that the expected value of the length of the chord is $\pi a/2$ and that the variance of the length is $a^2(32-3\pi^2)/12$. Also show that the chance is 1/2 that the length of the chord will exceed the length of the side of an equilateral triangle inscribed in the circle.*

Solution. Let PQ be the chord of a circle with centre O and radius '*a*' drawn at random parallel to the given straight line AB. Draw $OM \perp PQ$. Let $OM = x$.

Obviously x ranges from $-a$ to a. Since all distances from the centre are equally likely, the probability that a random value of x will lie in the small interval 'dx' is given by the rectangular distribution [*c.f. Chapter 9*] :

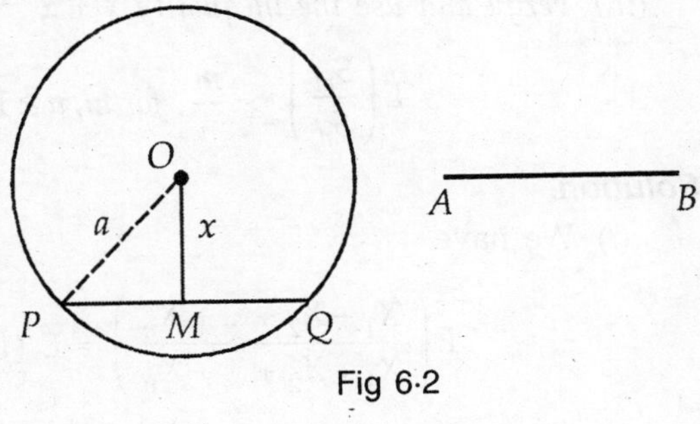

Fig 6·2

$$dF(x) = f(x)dx = \frac{dx}{a-(-a)} = \frac{dx}{2a}, \quad -a \le x \le a$$

Length of the chord is :

$$PQ = 2PM = 2\sqrt{a^2 - x^2}$$

Hence, $E(PQ) = \int_{-a}^{a} PQ\, dF(x) = \frac{2}{2a}\int_{-a}^{a} \sqrt{a^2 - x^2}\, dx$

$$= \frac{2}{a}\int_0^a \sqrt{a^2 - x^2}\, dx, \quad \text{(Since integrand is an even function of } x\text{).}$$

$$= \frac{2}{a}\left[\left|\frac{1}{2}x\sqrt{a^2 - x^2} + \frac{1}{2}a^2 \sin^{-1}\left(\frac{x}{a}\right)\right|_0^a\right] = \frac{2}{a}\cdot\frac{a^2}{2}\cdot\frac{\pi}{2} = \frac{\pi a}{2}$$

$$E\{(PQ)^2\} = \int_{-a}^{a} (PQ)^2\, dF(x) = \frac{4}{2a}\int_{-a}^{a} (a^2 - x^2)\, dx$$

$$= \frac{4}{a}\int_0^a (a^2 - x^2)\, dx = \frac{4}{a}\left[\left|a^2 x - \frac{x^3}{3}\right|_0^a\right] = \frac{4}{a}\cdot\frac{2a^3}{3} = \frac{8a^2}{3}$$

Hence, $\text{Var(length of chord)} = E\{(PQ)^2\} - \{E(PQ)\}^2 = \frac{8a^2}{3} - \frac{\pi^2 a^2}{4} = \frac{a^2}{12}(32 - 3\pi^2)$

The length of the chord is greater than the side of the equilateral triangle inscribed in the circle of radius 'a' if

$$2\sqrt{a^2 - x^2} > a\sqrt{3} \Rightarrow 4(a^2 - x^2) > 3a^2, \quad i.e., \quad x^2 < \frac{1}{4}a^2 \quad \text{or} \quad |x| < \frac{1}{2}a$$

Hence, the required probability is :

$$P\left(|X| < \frac{a}{2}\right) = P\left(-\frac{a}{2} < X < \frac{a}{2}\right) = \int_{-a/2}^{a/2} dF(x) = \frac{1}{2a}\int_{-a/2}^{a/2} 1 \cdot dx = \frac{1}{2}.$$

Example 6·21. *Let $X_1, X_2, \ldots, X_n$ be a sequence of mutually independent random variables with common distribution. Suppose X_k assumes only positive integral values and $E(X_k) = a$, exists; $k = 1, 2, \ldots, n$. Let $S_n = X_1 + X_2 + \ldots + X_n$.*

(i) Show that $E\left(\frac{S_m}{S_n}\right) = \frac{m}{n}$, *for* $1 \le m \le n$

(ii) Show that $E(S_n^{-1})$ exists and

$$E\left(\frac{S_m}{S_n}\right) = 1 + (m - n)\, a\, E(S_n^{-1}), \text{ for } 1 \le n \le m$$

(iii) Verify and use the inequality $x + x^{-1} \ge 2$, $(x > 0)$ to show that

$$E\left(\frac{S_m}{S_n}\right) \ge \frac{m}{n}, \text{ for } m, n \ge 1$$

Solution.

(*i*) We have

$$E\left(\frac{X_1 + X_2 + \ldots + X_n}{X_1 + X_2 + \ldots + X_n}\right) = E(1) \quad \Rightarrow \quad E\left(\frac{X_1 + X_2 + \ldots + X_n}{S_n}\right) = 1$$

$$\therefore \quad E\left(\frac{X_1}{S_n}\right)+E\left(\frac{X_2}{S_n}\right)+\ldots+E\left(\frac{X_n}{S_n}\right) = 1$$

Since X_i's, $(i = 1, 2, \ldots, n)$ are identically distributed random variables, (X_i / S_n), $(i = 1, 2, \ldots, n)$ are also identically distributed random variables.

$$\therefore \quad n\,E\left(\frac{X_i}{S_n}\right) = 1 \quad \Rightarrow \quad E\left(\frac{X_i}{S_n}\right) = \frac{1}{n}\,; \quad i = 1, 2, \ldots, n \qquad \ldots (*)$$

$$\therefore \quad E\left(\frac{S_m}{S_n}\right) = E\left(\frac{X_1+X_2+\ldots+X_m}{S_n}\right) = E\left(\frac{X_1}{S_n}+\frac{X_2}{S_n}+\ldots+\frac{X_m}{S_n}\right)$$

$$= E\left(\frac{X_1}{S_n}\right)+E\left(\frac{X_2}{S_n}\right)+\ldots+E\left(\frac{X_m}{S_n}\right) = \frac{1}{n}+\frac{1}{n}+\ldots+\frac{1}{n}\left[(m \text{ times})\right]$$

[Using (*)]

$$= \frac{m}{n}, \; (m < n)$$

(*ii*) Since X_i's assume only positive integral values, we have

$$n \le X_1 + X_2 + \ldots + X_n < \infty \;\Rightarrow\; \frac{1}{n} \ge \frac{1}{S_n} > 0 \;\Rightarrow\; 0 < S_n^{-1} \le \frac{1}{n}$$

Since S_n^{-1} lies between two finite quantities 0 and $\frac{1}{n}$, $\quad 0 < E\,(S_n^{-1}) \le \frac{1}{n}$

Hence, $E\,(S_n^{-1})$ exists.

$$E\left(\frac{S_m}{S_n}\right) = E\left(\frac{X_1+X_2+\ldots+X_n+X_{n+1}+\ldots+X_m}{S_n}\right), \text{ for } 1 \le n \le m.$$

$$= E\left(1+\frac{X_{n+1}}{S_n}+\ldots+\frac{X_m}{S_n}\right) = 1+E\left(\frac{X_{n+1}}{S_n}\right)+\ldots+E\left(\frac{X_m}{S_n}\right)$$

Since $X_{n+1}, X_{n+2}, \ldots, X_m$ are independent of $S_n = X_1 + X_2 + \ldots + X_n$, they are independent of S_n^{-1} also.

$$\therefore \quad E\,(S_m/S_n) = 1 + E\,(X_{n+1})\,E\,(S_n^{-1}) + \ldots + E\,(X_m)\,E\,(S_n^{-1})$$
$$= 1 + \left[E\,(X_{n+1}) + \ldots + E\,(X_m)\right] E\,(S_n^{-1})$$
$$= 1 + (m-n)\,a\,E\,(S_n^{-1}),\; 1 \le n \le m \qquad [\because\; E\,(X_i) = a \;\forall\, i]$$

(*iii*) Verification of $x + \frac{1}{x} \ge 2, \; (x > 0)$. ... (**)

$$x + \frac{1}{x} \ge 2 \;\Rightarrow\; x^2 + 1 \ge 2x \qquad \text{(multiplication valid only if } x > 0)$$

$$\Rightarrow \quad (x-1)^2 \ge 0, \qquad \text{which is always true for } x > 0.$$

If $1 \le m \le n$, result follows from (*).

If $1 \le n \le m$, then using (**), we have to prove that

$$1 + (m-n)\, a\, E(S_n^{-1}) \ge \frac{m}{n} \Rightarrow (m-n)\, a\, E(S_n^{-1}) \ge \frac{m-n}{n} \Rightarrow E(S_n^{-1}) \ge \frac{1}{a\,n} \quad \text{... (***)}$$

In (**), taking $x = \dfrac{S_n}{a\,n} > 0$, we get

$$E(S_n/an) + E(S_n/an)^{-1} \ge 2 \quad \text{or} \quad \frac{1}{an} E(S_n) + an\, E(S_n^{-1}) \ge 2$$

$$\Rightarrow \quad \frac{1}{an}\, an + an\, E(S_n^{-1}) \ge 2 \quad \text{or} \quad an\, E(S_n^{-1}) \ge 1$$

$$\therefore \quad E(S_n^{-1}) \ge \frac{1}{a\,n}, \text{ which was to be proved in (***).}$$

Example 6·22. *Let X be a r.v. for which β_1 and β_2 exist. Then for any real k, prove that :*
$\beta_2 \ge \beta_1 - (2k + k^2)$. Deduce that (i) $\beta_2 \ge \beta_1$, (ii) $\beta_2 \ge 1$. When is $\beta_2 = 1$?

Solution. Without any loss of generality we can take $E(X) = 0$.
[If $E(X) \ne 0$, then we may start with the random variable $Y = X - E(X)$ so that $E(Y) = 0$.]

Consider the real valued function of the real variable t defined by :

$$Z(t) = E\left(X^2 + tX + k\,\mu_2\right)^2 \ge 0, \ \forall\ t,$$

where $\mu_r = E(X^r)$, is the rth moment of X about mean. ... (i)

$$\therefore \quad Z(t) = E\left(X^4 + t^2 X^2 + k^2 \mu_2^2 + 2t X^3 + 2k\,\mu_2 X^2 + 2k\,\mu_2\, tX\right)$$
$$= \mu_4 + t^2 \mu_2 + k^2 \mu_2^2 + 2t\,\mu_3 + 2k\,\mu_2^2 \qquad [\text{Using } (i) \text{ and } E(X) = 0]$$
$$= t^2 \mu_2 + 2t\,\mu_3 + \mu_4 + k^2 \mu_2^2 + 2k\,\mu_2^2 \ge 0, \ \text{ for all } t. \quad \text{... (ii)}$$

Since $Z(t)$ is a quadratic form in t, $Z(t) \ge 0$ for all t iff its discriminant is ≤ 0, *i.e.*, iff

$$(2\,\mu_3)^2 - 4\mu_2\,(\mu_4 + k^2 \mu_2^2 + 2k\,\mu_2^2) \le 0 \quad \Rightarrow \quad \frac{\mu_3^2}{\mu_2^3} - \left(\frac{\mu_4}{\mu_2^2} + k^2 + 2k\right) \le 0$$

[Dividing by $4\mu_2^3 > 0$]

$$\Rightarrow \quad \beta_1 - \beta_2 - (2k + k^2) \le 0 \quad \Rightarrow \quad \beta_2 \ge \beta_1 - (2k + k^2) \quad \text{... (*)}$$

Deductions.

(*i*) Taking $k = 0$ in (*), we get $\beta_2 \ge \beta_1$ ... (**)

(*ii*) Taking $k = -1$ in (*), we get $\beta_2 \ge \beta_1 + 1$ ... (***)

(*iii*) Since $\beta_1 = \mu_3^2/\mu_2^3$ is always non-negative, we get from (***), $\beta_2 \ge 1$. ... (****)

Remark. The sign of equality holds in (****), *i.e.*, $\beta_2 = 1$ iff $\beta_2 = \dfrac{\mu_4}{\mu_2^2} = 1 \Rightarrow \mu_4 = \mu_2^2$

$$\Rightarrow \quad E[X - E(X)]^4 = \left[E[X - E(X)]^2\right]^2 \quad \text{or} \quad E(Y^2) - [E(Y)]^2 = 0, \qquad \left[Y = [X - E(X)]^2\right]$$

$$\Rightarrow \quad \text{Var}(Y) = 0$$

$$\Rightarrow \quad P\{Y = E(Y)\} = 1$$

$$\Rightarrow \quad P\left\{\left[\{X - E(X)\}\right]^2 = E\left[X - E(X)]^2\right]\right\} = 1$$

$$\Rightarrow \quad P\left[(X-\mu)^2 = \sigma^2\right] = 1 \quad \Rightarrow \quad P\left\{(X - \mu) = \pm\,\sigma\right\} = 1$$

Hence, $P(X = \mu \pm \sigma) = 1$

Thus, X takes only two values $\mu + \sigma$ and $\mu - \sigma$ with respective probabilities p and q, (say).

$\therefore \quad E(X) = p(\mu + \sigma) + q(\mu - \sigma) = \mu \quad \Rightarrow \quad p + q = 1$ and $(p - q)\sigma = 0$

But since $\sigma \neq 0$, (since in this case β_2 is defined), we have

$$p + q = 1 \text{ and } p - q = 0 \quad \Rightarrow \quad p = q = \frac{1}{2}.$$

Hence, $\beta_2 = 1$ iff the r.v. X assumes only two values, each with equal probability $\frac{1}{2}$.

Example 6·23. *Let X and Y be two r.v.'s having finite means. Prove or disprove the following :*

(a) $E[\text{Min}(X, Y)] \leq \text{Min}[E(X), E(Y)]$,

(b) $E[\text{Max}(X, Y)] \geq \text{Max}[E(X), E(Y)]$.

(c) $E[\text{Min}(X,Y) + \text{Max}(X, Y)] = E(X) + E(Y)$.

Solution. We know that

$$\text{Min}(X, Y) = \frac{1}{2}(X + Y) - |X - Y| \quad \text{... (1)}$$

and

$$\text{Max}(X, Y) = \frac{1}{2}(X + Y) + |X - Y| \quad \text{... (2)}$$

(*a*) $E[\text{Min}(X, Y)] = \frac{1}{2} E|X + Y| - E|X - Y|$ [From (1)] ...(3)

We have $|E(X - Y)| \leq E|X - Y| \quad \Rightarrow \quad -|E(X - Y)| \geq -E|X - Y|$

$\therefore \quad -E|X - Y| \leq -|E(X - Y)| = -|E(X) - E(Y)|$ ... (*)

Substituting in (3), we get

$$E[\text{Min}(X, Y)] \leq \frac{1}{2}[E(X) + E(Y)] - |E(X) - E(Y)| \quad \text{... [From (*)]}$$

$\Rightarrow \quad E[\text{Min}(X, Y)] \leq \text{Min}[E(X), E(Y)]$

(*b*) Similarly. from (2), we get

$$E[\text{Max}(X, Y)] = \frac{1}{2} E(X+Y) + E|X + Y|$$

$$\geq \frac{1}{2} E(X+Y) + |E(X + Y)| \qquad [\because |E(X + Y)| \leq E|X + Y|]$$

$$= \frac{1}{2}[E(X) + E(Y)] + |E(X) + E(Y)| = \text{Max}[E(X), E(Y)]$$

i.e., $E[\text{Max}(X, Y)] \geq \text{Max}[E(X), E(Y)]$

(*c*) $[\text{Min}(X, Y) + \text{Max}(X, Y)] = (X + Y)$

$\Rightarrow E[\text{Min}(X, Y) + \text{Max}(X, Y)] = E(X + Y) = E(X) + E(Y)$, as required.

Hence, all the results in (*a*), (*b*) and (*c*) are true.

Example 6·24. *Use the relation* $E(AX^a + BX^b + CX^c)^2 \geq 0$, *X being a random variable with* $E(X) = 0$, *E denoting the mathematical expectation, to show that*

$$\begin{vmatrix} \mu_{2a} & \mu_{a+b} & \mu_{a+c} \\ \mu_{a+b} & \mu_{2b} & \mu_{b+c} \\ \mu_{a+c} & \mu_{b+c} & \mu_{2c} \end{vmatrix} \geq 0, \quad \mu_n \text{ denoting the nth moment about mean.} \quad \text{... (*)}$$

Hence, or otherwise show that Pearson Beta-coefficients satisfy the inequality $\beta_2 - \beta_1 - 1 \geq 0$. *Also deduce that* $\beta_2 \geq 1$.

Solution. Since $E(X) = 0$, we have $\quad E(X^r) = \mu_r$ ... (**)

We are given that: $\quad E(AX^a + BX^b + CX^c)^2 \geq 0$

$\Rightarrow E(A^2 X^{2a} + B^2 X^{2b} + C^2 X^{2c} + 2ABX^{a+b} + 2ACX^{a+c} + 2BCX^{b+c}) \geq 0$

$\Rightarrow \quad A^2\mu_{2a} + B^2\mu_{2b} + C^2\mu_{2c} + 2AB\mu_{a+b} + 2AC\mu_{a+c} + 2BC\mu_{b+c} \geq 0$, [From (**)] ... (***)

for all values of A, B, C.

We know from matrix theory that the conditions for the quadratic form

$$a'x^2 + b'y^2 + c'z^2 + 2f'yz + 2g'zx + 2h'xy,$$

to be non-negative for all values of x, y and z are :

(i) $a' \geq 0$, $\qquad$ (ii) $\begin{vmatrix} a' & h' \\ h' & b' \end{vmatrix} \geq 0$, $\qquad$ and $\qquad$ (iii) $\begin{vmatrix} a' & h' & g' \\ h' & b' & f' \\ g' & f' & c' \end{vmatrix} \geq 0$

Comparing with (***), we have

$$a' = \mu_{2a}, \quad b' = \mu_{2b}, \quad c' = \mu_{2c}, \quad f' = \mu_{b+c}, \quad g' = \mu_{a+c}, \quad h' = \mu_{a+b}$$

Substituting these values in condition (iii) above, we get the required result.

Taking $a = 0$, $b = 1$ and $c = 2$ in (*) and noting that $\mu_0 = 1$ and $\mu_1 = 0$, we get

$$\begin{vmatrix} 1 & 0 & \mu_2 \\ 0 & \mu_2 & \mu_3 \\ \mu_2 & \mu_3 & \mu_4 \end{vmatrix} \geq 0 \quad \Rightarrow \quad \mu_2\,\mu_4 - \mu_3^2 + \mu_2\,(-\mu_2^2) \geq 0$$

Dividing throughout by μ_2^3 (assuming that $\mu_2 > 0$, for otherwise β_2 will become infinite), we get

$$\frac{\mu_4}{\mu_2^2} - \frac{\mu_3^2}{\mu_2^3} - 1 \geq 0 \quad \Rightarrow \quad \beta_2 - \beta_1 - 1 \geq 0 \quad \Rightarrow \quad \beta_2 \geq \beta_1 + 1$$

Further, since $\beta_1 \geq 0$, we get : $\qquad \beta_2 \geq 1$.

Example 6·25. *Let X be a non-negative random variable with distribution function F. Show that*

$$E(X) = \int_0^\infty [1 - F(x)]\,dx. \qquad \text{... (i)}$$

Conjecture a corresponding expression for $E(X^2)$.

Solution. Since $X \geq 0$, we have

$$\text{R.H.S.} = \int_0^\infty [1 - P(X \leq x)]\,dx = \int_0^\infty \left\{1 - \int_0^x f(u)\,du\right\} dx,$$

where $f(.)$ is the *p.d.f.* of *r.v.* X.

$$\therefore \qquad \text{R.H.S.} = \int_0^\infty \left\{\int_x^\infty f(u)\,du\right\} dx \qquad \text{... (ii)}$$

From the integral in bracket in (ii), we have, $u \geq x$ and since x ranges from 0 to ∞, u also range from 0 to ∞. Further $u \geq x \Rightarrow x \leq u$ and since x is non-negative, we have $0 \leq x \leq u$. [See Region R_1 in Fig. 6·3 on page 6·35.] Hence, changing the order of integration in (ii), (by Fubini's theorem for non-negative functions), we get

$$\text{R.H.S.} = \int_0^\infty \left\{\int_0^u 1 \cdot dx\right\} f(u)\,du = \int_u^\infty u\,f(u)\,du = E(X) \qquad [\text{Since } f(.) \text{ is } p.d.f. \text{ of } X]$$

Conjecture for $E(X^2)$. Consider the integral :

$$\int_0^\infty 2x\{1-F(x)\}\,dx = \int_0^\infty 2x\left(\int_x^\infty f(u)\,du\right)dx = \int_0^\infty\left(\int_0^u 2x\,dx\right)f(u)\,du,$$

(By Fubini's theorem for non-negative functions)

$$= \int_0^\infty u^2 f(u)\,du = E\left(X^2\right)$$

Remarks :

1. If X is a non-negative *r.v.* then

$$\text{Var}(X) = E(X^2) - [E(X)]^2 = \int_0^\infty 2x[1-F(x)]\,dx - \mu_X^2 \quad \text{... (iii)}$$

2. If we do not restrict ourselves to non-negative random variables only, we have the following more generalised result.

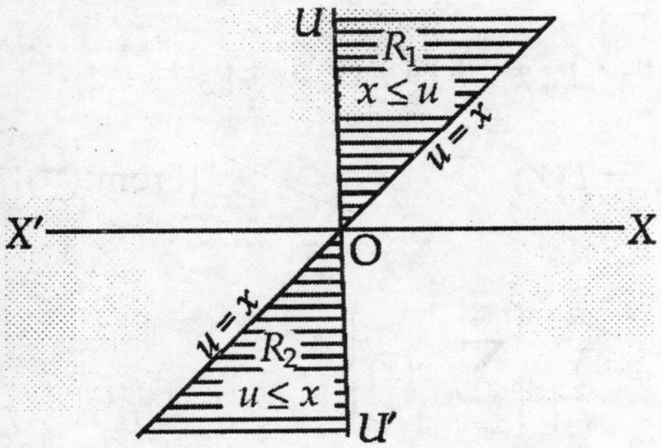

Fig 6.3

If F denotes the distribution function of the random variable X, then :

$$E(X) = \int_0^\infty \{1-F(x)\}\,dx - \int_{-\infty}^0 F(x)\,dx, \quad \text{... (iv)}$$

provided the integrals exist finitely.

Proof of (iv). The first integral has already been evaluated in the above example, *i.e.*,

$$\int_0^\infty \{1-F(x)\}\,dx = \int_0^\infty u\,f(u)\,du \quad \text{... (v)}$$

Consider:

$$\int_{-\infty}^0 F(x)\,dx = \int_{-\infty}^0 P(X \le x)\,dx = \int_{-\infty}^0\left(\int_{-\infty}^x f(u)\,du\right)dx = \int_{-\infty}^0\left(\int_u^0 1\cdot dx\right)f(u)\,du$$

[Changing the order of integration in the Region R_2 in Fig. 6·3 where $u \le x$].

$$= -\int_{-\infty}^0 u\,f(u)\,du \quad \text{... (vi)}$$

Subtracting (*vi*) from (*v*), we get

$$\int_0^\infty\{1-F(x)\}\,dx - \int_{-\infty}^0 F(x)\,dx = \int_0^\infty u\,f(u)\,du + \int_{-\infty}^0 u\,f(u)\,du = \int_{-\infty}^\infty u\,f(u)\,du = E(X),$$

[Since $f(.)$ is *p.d.f.* of X]

as desired.

In this generalised case,

$$\text{Var}(X) = \int_0^\infty 2x\{1-F_X(x)+F_X(-x)\}\,dx - \{E(X)\}^2$$

3. The corresponding analogue of the above result for discrete random variable is given in the next Example 6·26.

Example 6·26. *If the possible values of a variate X are 0, 1, 2, 3, ... then prove that*

(i) $$E(X) = \sum_{n=0}^{\infty} P(X > n)$$

(ii) $$\sum_{n=0}^{\infty} n \cdot P(X > n) = \frac{1}{2}\left[E(X^2) - E(X)\right]$$

Solution. Let $P(X = n) = p_n$, $n = 0, 1, 2, 3, ...$ ... (*)

If $E(X)$ exists, then by def., $$E(X) = \sum_{n=0}^{\infty} n\,P(X = n) = \sum_{n=1}^{\infty} n\,p_n \quad \text{... (**)}$$

Consider

$$\sum_{n=0}^{\infty} P(X > n) = P(X > 0) + P(X > 1) + P(X > 2) + ...$$
$$= P(X \geq 1) + P(X \geq 2) + P(X \geq 3) + ...$$
$$= (p_1 + p_2 + p_3 + p_4 + ...) + (p_2 + p_3 + p_4 + ...) + (p_3 + p_4 + p_5 + ...) + ...$$
$$= p_1 + 2p_2 + 3p_3 + ... = \sum_{n=1}^{\infty} n\,p_n = E(X) \quad \text{[From (**)]}$$

Aliter.

(i) $$\text{R.H.S.} = \sum_{n=0}^{\infty} P(X > n) = \sum_{n=1}^{\infty} P(X \geq n) = \sum_{n=1}^{\infty}\left\{\sum_{x=n}^{\infty} p(x)\right\},$$

Since the series is convergent and $p(x) \geq 0 \ \forall\ x$, by Fubini's theorem, changing the order of summation, we get

$$\text{R.H.S.} = \sum_{x=1}^{\infty}\left\{\sum_{n=1}^{x} p(x)\right\} = \sum_{x=1}^{\infty}\left\{p(x)\sum_{n=1}^{x} 1\right\},$$

since $x \geq n \Rightarrow n \leq x$ and x assumes only positive integral values.

$$\therefore \quad \text{R.H.S.} = \sum_{x=1}^{\infty} x\,p(x) = \sum_{x=0}^{\infty} x\,p(x) = E(X)$$

(ii) $$\sum_{n=0}^{\infty} n\,P(X > n) = \sum_{n=1}^{\infty} n\,P(X > n) = \sum_{n=1}^{\infty} n\,.\,P(X \geq n + 1)$$

$$= \sum_{n=1}^{\infty}\left[n \sum_{j=n+1}^{\infty} P(X = j)\right] = \sum_{j=1}^{\infty}\left[P(X = j)\left\{\sum_{n=0}^{j-1} n\right\}\right], \quad \text{... (***)}$$

$j \geq n + 1 \Rightarrow n \leq j - 1$ and $j \geq 1$. Since the R.H.S. series is convergent series of positive terms, changing the order of summation by Fubini's Theorem, we get the result in (***)]

$$\therefore \quad \sum_{n=0}^{\infty} n\,P(X > n) = \sum_{j=1}^{\infty}\left[P(X = j)\left\{1 + 2 + 3 + ... + (j - 1)\right\}\right]$$

$$= \sum_{j=1}^{\infty} P(X=j) \cdot \left\{ \frac{(j-1)\cdot j}{2} \right\} = \frac{1}{2}\left[\sum_{j=1}^{\infty} j^2 P(X=j) - \sum_{j=1}^{\infty} j P(X=j) \right]$$

$$= \frac{1}{2}\left[E(X^2) - E(X) \right]$$

Example 6·27. *Explain by means of an example that a probability distribution is not uniquely determined by its moments.*

Solution. Consider a *r.v.* X with *p.d.f.* [*c.f.* Log-Normal distribution, Chapter 9]

$$\left. \begin{aligned} f(x) &= \frac{1}{\sqrt{2\pi}\, x} \cdot \exp\left\{ -\frac{1}{2}(\log x)^2 \right\};\ x > 0 \\ &= 0, \quad \text{otherwise} \end{aligned} \right\} \quad \ldots (*)$$

Consider, another *r.v.* Y with *p.d.f.*,

$$g(y) = \left[1 + a \sin(2\pi \log y) \right] f(y) = g_a(y),\ \text{(say)},\ y > 0, \quad \ldots (**)$$

which, for $-1 \le a \le 1$, represents a family of probability distributions.

$$E(Y^r) = \int_0^{\infty} y^r\, g(y)\, dy = \int_0^{\infty} y^r \{1 + a \sin(2\pi \log y)\}\, f(y)\, dy$$

$$= \int_0^{\infty} y^r f(y)\, dy + a \int_0^{\infty} y^r \sin(2\pi \log y)\, f(y)\, dy$$

$$= E(X^r) + a \frac{1}{\sqrt{2\pi}} \int_0^{\infty} y^r \sin(2\pi \log y) \cdot \frac{1}{y} \exp\left\{ -\frac{1}{2}(\log y)^2 \right\} dy$$

$$= E(X^r) + \frac{a}{\sqrt{2\pi}} \int_{-\infty}^{\infty} e^{rz - z^2/2} \sin(2\pi z)\, dz \qquad [\log y = z \ \Rightarrow\ y = e^z]$$

$$= E(X^r) + \frac{a \cdot e^{r^2/2}}{\sqrt{2\pi}} \int_{-\infty}^{\infty} e^{-(z-r)^2/2} \sin(2\pi z)\, dz$$

$$= E(X^r) + \frac{a \cdot e^{r^2/2}}{\sqrt{2\pi}} \int_{-\infty}^{\infty} e^{-t^2/2} . \sin(2\pi t)\, dt = E(X^r),$$

$\left[\because\ z - r = t \ \Rightarrow\ \sin(2\pi z) = \sin(2\pi r + 2\pi t) = \sin 2\pi t,\ r \text{ being a positive integer.} \right]$

the value of the integral being zero, since the integrand is an odd function of t.

$\therefore \qquad E(Y^r)$ is independent of 'a' in (**).

Hence, $\{g(y) = g_a(y);\ -1 \le a \le 1\}$, represents a family of distributions, each different from the other, but having the same moments. This explains that the moments may not determine a distribution uniquely.

6·7. SOME INEQUALITIES INVOLVING EXPECTATION

6·7·1. Cauchy-Schwartz Inequality

If X and Y are random variables taking real values, then

$$[E(XY)]^2 \le E(X^2).\ E(Y^2) \quad \text{... (6·36)}$$

Proof. Let us consider a real valued function of the real variable t, defined by

$$Z(t) = E(X + tY)^2$$

which is always non-negative, since $(X + tY)^2 \ge 0$, for all real X, Y and t.

Thus, $\quad Z(t) = E(X + tY)^2 \ge 0 \ \forall\ t.$

$\Rightarrow \quad Z(t) = E[X^2 + 2tXY + t^2 Y^2]$

$\qquad\qquad = E(X^2) + 2t\, E(XY) + t^2\, E(Y^2) \ge 0$, for all t. $\quad$ (*)

Obviously, $Z(t)$ is a quadratic expression in 't'.

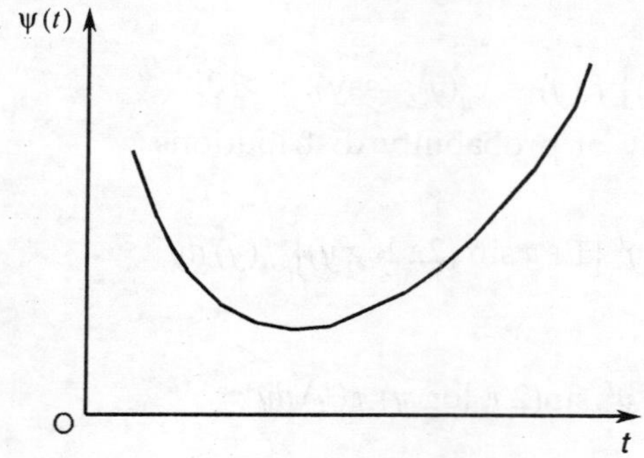

Fig 6·4 (a)

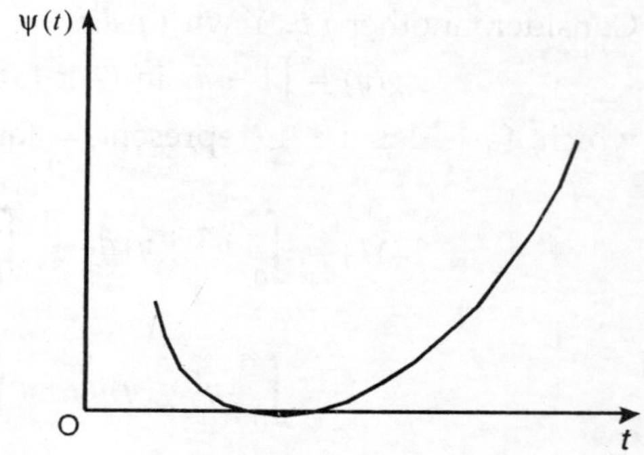

Fig 6·4 (b)

We know that the quadratic expression of the form :

$\psi(t) = At^2 + Bt + C \ge 0$ for all t, implies that the graph of the function $\psi(t)$ either touches the t-axis at only one point or not at all, as exhibited in the above diagrams.

This is equivalent to saying that the discriminant of the function $\psi(t)$, *viz.*, $B^2 - 4AC \le 0$, since the condition $B^2 - 4AC > 0$ implies that the function $\psi(t)$ has two distinct real roots which is a contradiction to the fact that $\psi(t)$ meets the t – axis either at only one point or not at all. Using this result, we get from (*),

$$4.\ \{E(XY)\}^2 - 4.\ E(X^2) \cdot E(Y^2) \le 0 \quad \Rightarrow \quad \{E(XY)\}^2 \le E(X^2).\ E(Y^2)$$

Remarks :

1. The sign of equality holds in (6·36) or (*) iff

$$E(X + tY)^2 = 0\ \forall\ t \quad \Rightarrow \quad P[(X + tY)^2 = 0] = 1,$$

$$\Rightarrow \quad P[(X + tY) = 0] = 1 \quad \text{or} \quad P\left(Y = -\frac{X}{t}\right) = 1$$

$$\therefore \quad P(Y = AX) = 1\ ;\ (A = -1/t) \quad \text{... (6·36}a\text{)}$$

2. If the *r.v.* X takes the real values $x_1, x_2, \ldots, x_n$ and the *r.v.* Y takes the real values $y_1, y_2, \ldots, y_n$ then Cauchy-Schwartz inequality implies :

$$\left(\frac{1}{n}\sum_{i=1}^{n} x_i\, y_i\right)^2 \le \left(\frac{1}{n}\sum_{i=1}^{n} x_i^2\right) \cdot \left(\frac{1}{n}\sum_{i=1}^{n} y_i^2\right) \quad \Rightarrow \quad \left(\sum_{i=1}^{n} x_i\, y_i\right)^2 \le \left(\sum_{i=1}^{n} x_i^2\right) \cdot \left(\sum_{i=1}^{n} y_i^2\right)$$

the sign of equality holding if and only if :

$\dfrac{x_i}{y_i}$ = *constant* = k , *(say)*, for *all* $i = 1, 2, ..., n,$ *i.e., iff* $\dfrac{x_1}{y_1} = \dfrac{x_2}{y_2} = \dots = \dfrac{x_n}{y_n} = k$, *(say)*,

3. Replacing X by $|X - E(X)| = |X - \mu_x|$. and taking $Y = 1$ in (6·36), we get

$$[E\ |X - \mu_x|]^2 \le E\ |X - \mu_x|^2 . E\ (1)$$

$\Rightarrow$ *(Mean Deviation about mean)*^{2} $\le$ *Variance* (X)

$\therefore$ *M.D.* $\le$ *S.D.* $\Rightarrow$ *S.D.* $\ge$ *M.D.* ... (6·36 *b*)

6·7·2. Jenson's Inequality

Continuous Convex Function (*Definition*). *A continuous function $g(x)$ on the interval I is convex if for every x_1 and x_2, $(x_1 + x_2)/2 \in I$, we have*

$$g\left(\frac{x_1 + x_2}{2}\right) \le \frac{1}{2}\, g(x_1) + \frac{1}{2}\, g(x_2) \qquad \text{... (6·37)}$$

Remarks :

1. If $x_1, x_2 \in I$, then $(x_1 + x_2)/2 \in I$.
2. Sometimes (6·37) is replaced by the stronger condition :

   For $x_1, x_2 \in I$, $\quad g[\lambda x_1 + (1 - \lambda)\, x_2] \le \lambda\, g\,(x_1) + (1 - \lambda)\, g\,(x_2)\ ;\ 0 \le \lambda \le 1$ ... (6·38)

   (6·37) and (6·38) agree at $\lambda = \dfrac{1}{2}$.
3. If we do not assume the continuity of $g(x)$, then (6·38) is required to define convexity. There are certain *'non-measurable'* non-constant functions g (.) satisfying (6·37) but not (6·38). If g (.) is measurable, then (6·37) and (6·38) are equivalent.
4. A function satisfying (6·38) is continuous except possibly at the end points of the interval I (if it has end points).
5. If g is twice differentiable, *i.e.*, $g''(x)$ exists for $x \in$ [interior of I], and $g''(x) \ge 0$ for such x, then g is convex on the interior points.
6. For any point x_o interior to I, there exists a straight line $y = ax + b$, which passes through $[x_o, g\,(x_o)]$ and satisfies $g\,(x) \ge ax + b$, for all $x \in I$.

Jenson's Inequality. *If g is continuous and convex function on the interval I, and X is a random variable whose values are in I with probability 1, then*

$$E\ [g\ (X)] \ge g\ [E\ (X)], \text{ provided the expectations exist.} \qquad \text{... (6·39)}$$

Proof. First of all, we shall show that $E\ (X) \in I$.

The various possible cases for I are :

$$I = (-\infty, \infty)\ ;\quad I = (a, \infty);\quad I = (-\infty, b]\ ;\quad I = (a, b) \text{ and variations of this.}$$

If E (X) exists, then $-\infty < E\ (X) < \infty$.

If $X \ge a$ almost surely (*a. s.*), *i.e.*, with probability 1, then $E\ (X) \ge a$.

If $X \le b$, a.s. then $E\ (X) \le b$.

Thus, $E\ (X) \in I$. Now E (X) can be either a left end or a right end point (if end points exist) of I or an interior point of I.

Suppose I has a left end point *'a'*, *i.e.*, $X \ge a$ and $E\ (X) = a$. Then,

$$X - a \ge 0 \text{ a.s. and } E\ (X - a) = 0.$$

Thus, $\quad P\ (X = a) = 1 \quad$ or $\quad P\ [(X - a) = 0)] = 1.$

$\therefore \quad E[g(X)] = E[g(a)] \quad [\because g(x) = g(a) \text{ a.s.}]$

$= g(a) \quad [\because g(a) \text{ is a constant}]$

$= g\{E(X)\}.$

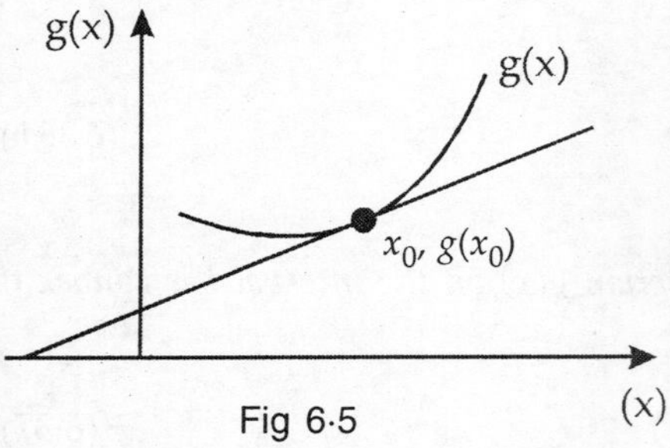

Fig 6·5

The result can be established similarly if I has a right end point 'b' and $E(X) = b$.

Thus we are now required to establish (6·39) when $E(X) = x_o$, is an interior point of I.

Let $ax + b$ pass through the point $\{x_o, g(x_o)\}$ and let it be below g. [c.f. Remark 6 on page 6·39].

$$\therefore \quad E\{g(X)\} \geq E(aX + b) = aE(X) + b$$

$$= ax_o + b = g(x_o) = g\{E(X)\}$$

Hence, $E\{g(X)\} \geq g\{E(X)\}$.

Continuous Concave Function (*Definition*).

A continuous function g is concave on an interval I if $(-g)$ is convex.

Corollary to Jenson's Inequality. *If g is a continuous and concave function on the interval I and X is a r.v. whose values are in I with probability 1, then*

$$E[g(X)] \leq g[E(X)] \quad \text{... (6·40)}$$

provided the expectations exist.

Remarks:

1. Equality holds in (6·39) and corollary (6·40) if the only if

$$P[g(X) = aX + b] = 1, \text{ for some } a \text{ and } b.$$

2. Jenson's inequality extends to random vectors. If I is a convex set in n dimensional Euclidean space, *i.e.*, the interval I in (6·39) is transferred to convex set, g is continuous on I, (6·37) holds whenever X_i and X_j are any arbitrary vectors in I. The condition $g''(x) \geq 0$ for x interior to I implies

$\left(\dfrac{\partial^2 g(x)}{\partial x_i \, \partial x_j}\right) = M(x)$, (say), is non-negative definite for all x interior to I.

3. If $E(X^2)$ exists, then $\quad E(X^2) \geq \{E(X)\}^2, \quad$ ... (6·41)

since $g(X) = X^2$, is convex function of X as $g''(X) = 2 > 0$.

4. If $X > 0$, *i.e.*, X assumes only positive values and $E(X)$ and $E(1/X)$ exist then

$$E\left(\frac{1}{X}\right) \geq \frac{1}{E(X)} \quad \text{... (6·41a)}$$

because $g(X) = \dfrac{1}{X}$, is a convex function of X since $g''(X) = \dfrac{2}{X^3} > 0$, for $X > 0$.

5. If $X > 0$, *a.s.* then $\quad E(X^{1/2}) \leq \{E(X)\}^{1/2} \quad$ ... (6·41b)

since $g(X) = X^{1/2}$, $X > 0$ is a concave function as $g''(X) = -\dfrac{1}{4} X^{-3/2} < 0$, for $X > 0$.

6. AM $\geq$ G.M. Inequality. If $X > 0$, then $f(x) = \log x \Rightarrow f'(x) = \dfrac{1}{x} \Rightarrow f''(x) = -\dfrac{1}{x^2} < 0$, is a concave function of x. Hence, by Jenson's inequality :

$$E(\log X) \leq \log[E(X)] \Rightarrow \sum_{i=1}^{n} p_i \log x_i \leq \log\left(\sum_{i=1}^{n} p_i x_i\right) \Rightarrow \log \prod_{i=1}^{n}\left(x_i^{p_i}\right) \leq \log\left(\sum_{i=1}^{n} p_i x_i\right) \quad \text{... (*)}$$

$$\therefore \quad \prod_{i=1}^{n}\left(x_i^{p_i}\right) \leq \sum_{i=1}^{n} p_i x_i \quad \text{... (*)}$$

If we assume that all the values of x are equiprobable so that :

$$P(X = x_i) = p_i = \frac{1}{n} \; ; \; i = 1, 2, ..., n \quad \text{then, from (*) , we get}$$

$$\prod_{i=1}^{n} (x_i)^{1/n} \le \sum_{i=1}^{n} \left(\frac{1}{n} x_i\right) \Rightarrow (x_1 \, x_2 \ldots x_n)^{1/n} \le \frac{1}{n} \sum_{i=1}^{n} x_i \Rightarrow G.M. \le A.M. \Rightarrow A.M. \ge G.M. \quad \ldots (6{\cdot}41c)$$

6·7·3. Gurland's Inequality

Let f and g be monotone functions on some subset of the real line and X be a r.v. whose range is in the subset almost surely (a.s.). If the expectations exist, then

$$E[f(X)\, g(X)] \ge E[f(X)].\, E[g(X)] \quad \ldots (6{\cdot}42)$$

or
$$E[f(X)\, g(X)] \le E[f(X)].\, E[g(X)] \quad \ldots (6{\cdot}42a)$$

according as f and g are monotone in the same or in the opposite directions.

Proof. Let us consider the case when both the functions f and g are monotone in the same direction *i.e.*, both are non-decreasing or both are non-increasing. Let x and y lie in the domain of f and g respectively.

If f and g are both monotonically increasing, then

$$y \ge x \Rightarrow f(y) \ge f(x) \quad \text{and} \quad g(y) \ge g(x)$$

$$\Rightarrow \quad f(y) - f(x) \ge 0 \quad \text{and} \quad g(y) - g(x) \ge 0$$

$$\Rightarrow \quad [f(y) - f(x)].\, [g(y) - g(x)] \ge 0 \quad \ldots (*)$$

If f and g are both monotonically decreasing then for $y \ge x$, we have

$$f(y) \le f(x) \quad \text{and} \quad g(y) \le g(x)$$

$$\Rightarrow \quad f(y) - f(x) \le 0 \quad \text{and} \quad g(y) - g(x) \le 0$$

$$\therefore \quad [f(y) - f(x)] \,.\, [g(y) - g(x)] \ge 0 \quad \ldots (**)$$

Hence, if f and g are both monotonic in the same direction, then from (*) and (**), we get the same result, *viz.*, $[f(y) - f(x)] \cdot [g(y) - g(x)] \ge 0$.

Let us now consider independently & identically distributed (*i.i.d.*) random variables X and Y. Then from above, we get

$$E[\{f(Y) - f(X)\}\, \{g(Y) - g(X)\}] \ge 0.$$

$$\Rightarrow \quad E[f(Y) \,.\, g(Y)] - E[f(Y).\, g(X)] - E[f(X)\, g(Y)] + E[f(X).\, g(X)] \ge 0 \quad \ldots (6{\cdot}43)$$

Since X and Y are *i.i.d. r.v.*'s, we have

$$E[f(Y)\, g(Y)] = E[f(X)\, g(X)],$$

$$E[f(Y)\, g(X)] = E[f(Y)]\, E[g(X)] = E[f(X)]\, E[g(X)]$$

($\because$ X and Y are independent) ($\because$ X and Y are identical)

and $\quad E[f(X)\, g(Y)] = E[f(X)] \,.\, E[g(Y)] = E[f(X)] \,.\, E[g(X)]$

Substituting in (6·43), we get

$$2E[f(X).\, g(X)] - 2E[f(X)] \,.\, E[g(X)] \ge 0 \quad \Rightarrow \quad E[f(X) \,.\, g(X)] \ge E[f(X)].\, E[g(X)],$$

which establishes the result in (6·42).

Similarly, (6·42*a*) can be established, if f and g are monotonic in opposite directions, *i.e.*, if f is monotonically increasing (decreasing) and g is monotonically decreasing (increasing). *The proof is left as an exercise to the reader.*

Remarks :

1. If X is a r. v. which takes only non-negative values, *i.e.*, if $X \geq 0$ a.s. then for $\alpha > 0$, $\beta > 0$, $f(X) = X^\alpha$ and $g(X) = X^\beta$ are monotonic in the same direction. Hence, if the expectations exist,

$$E(X^\alpha \cdot X^\beta) \geq E(X^\alpha) \cdot E(X^\beta)$$

$$\Rightarrow \quad E(X^{\alpha+\beta}) \geq E(X^\alpha)\, E(X^\beta)\ ;\ \alpha > 0,\ \beta > 0 \qquad \text{...(6·44)}$$

In particular, taking $\alpha = \beta = 1$, $E(X^2) \geq [E(X)]^2$, a result already obtained in (6·41).

2. If $X \geq 0$, a.s. and $E(X^\alpha)$ and $E(X^{-1})$ exist, then for $\alpha > 0$, we get from (6·42*a*)

$$E(X^\alpha \cdot X^{-1}) \leq E(X^\alpha) \cdot E(X^{-1})$$

$$\Rightarrow \quad E(X^\alpha)\, E(1/X) \geq E(X^{\alpha-1})\ ;\ \alpha > 0. \qquad \text{...(6·45)}$$

In particular with $\alpha = 1$, $E(X)\, E(1/X) \geq 1$, a result already obtained in (6·41*a*).

Taking $\alpha = 2$ in (6·45), $E(X^2) \cdot E(1/X) \geq E(X)$

$$\Rightarrow \quad E(X^2) \geq \frac{E(X)}{E(1/X)} \geq \frac{1}{\{E(1/X)\}^2}, \text{ on using (6·41}a\text{).} \qquad \text{... (6·46)}$$

Taking $\alpha = 2$ and $\beta = -2$ in $f(X) = X^\alpha$, $g(X) = X^\beta$ and using (6·42*a*), we get

$$E(X^2) \cdot E(X^{-2}) \geq E(X^2 \cdot X^{-2}) = 1 \quad \Rightarrow \quad E(X^2) \geq \frac{1}{E(X^{-2})} \qquad \text{... (6·46}a\text{)}$$

which is a weaker inequality than (6·46).

6·7·4. Lyapounov's Inequality

If $E(|X|^r) < \infty$ *i.e., if the rth order absolute moment of X about origin exists, then*

$$[E(|X|^a)]^{1/a} \leq [E(|X|^b)]^{1/b}, \qquad \text{... (6·47)}$$

for all $a > 0$, $b > 0$ and $a < b$.

Proof. Consider the function : $\phi(z) = z^\lambda$; $z > 0$, $\lambda > 1$.

$$\phi'(z) = \lambda z^{\lambda - 1}, \quad \phi''(z) = \lambda(\lambda - 1) z^{\lambda - 1} > 0 \text{ for } \lambda > 1.$$

Hence, $\phi(z) = z^\lambda$, is a convex function of z for $\lambda > 1$. By Jensen's inequality,

$$E[\phi(Z)] \geq \phi\{E(Z)\} \Rightarrow E(Z^\lambda) \geq [E(Z)]^\lambda\ ;\ \lambda > 1 \qquad \text{... (*)}$$

Let us takeZ $= |X|^a$ and $\lambda = (b/a) > 1 \Rightarrow b > a$ in (*), so that

$$E[(|X|^a)^{b/a}] \geq [E|X|^a]^{b/a} \Rightarrow E(|X|^b) \geq (E|X|^a)^{b/a}$$

$$\therefore \quad [E(|X|^b]^{1/b} \leq [E|X|^a]^{1/a} \Rightarrow [E|X|^a]^{1/a} \leq [E|X|^b]^{1/b}\ ;\ a < b$$

Corollary. Taking $a = k - 1$ and $b = k$, $(a < b)$, in (6·47), we get

$$[E|X|^{k-1}]^{1/(k-1)} \leq [E|X|^k]^{1/k} \qquad \text{...(6·48)}$$

If $\beta_k = E|X|^k$, is the kth absolute moment of X about origin, then we get from (6·48) :

$$(\beta_{k-1})^{1/(k-1)} \leq (\beta_k)^{1/k}\ ;\ k = 2, 3, ..., n.$$

$$\Rightarrow \quad \beta_1 \leq (\beta_2)^{1/2} \leq (\beta_3)^{1/3} \leq (\beta_4)^{1/4} \leq ...$$

$$\Rightarrow \quad (\beta_k)^{1/k} = [E|X|^k]^{1/k}, \text{ is a decreasing function of } k = 1, 2, 3, ...$$

6·7·5. C_r – Inequality

If X and Y are two random variables and r is a positive integer, then

$$E|X + Y|^r \leq C_r [E|X|^r + E|Y|^r] \qquad \text{... (6·49)}$$

where $$C_r = \begin{cases} 1, & r \le 1 \\ 2^{r-1}, & r \ge 2 \end{cases} \quad \text{... (6·50)}$$

In other words C_r Inequality can be stated as follows :

If rth absolute moments of r.v.'s X and Y exist and are finite, then the rth absolute moment of their sum (X + Y) also exists.

Proof. In proving this inequality, we need a result which we prove below in the form of a Lemma.

Lemma. *If r is any non-negative integer, then for any two real numbers a and b, a ≥ 0 and b ≥ 0,*

$$(a + b)^r \le C_r\ (a^r + b^r)\ , \quad \text{... (6·51)}$$

where $$C_r = \begin{cases} 1, & r \le 1 \\ 2^{r-1}, & r > 1 \end{cases} \quad \text{... (6·52)}$$

Proof. The inequality (6·51) is trivial for $r = 0$ and 1. We prove it for $r = 2$.

We have : $(a - b)^2 + (a + b)^2 = 2\ (a^2 + b^2)$

$\Rightarrow$ $$(a - b)^2 = 2\ (a^2 + b)^2\ - (a + b)^2 \ge 0 \quad \text{[Square of real no.} \ge 0]$$

$\Rightarrow$ $$(a + b)^2 \le 2^{2-1}\ (a^2 + b^2) \quad \text{... (6·53)}$$

$\Rightarrow$ (6·51) is true for $r = 2$.

Let (6·51) be true for $r = k$ so that

$$(a + b)^k \le C_k\ (a^k + b^k) = 2^{k-1}\ (a^k + b^k) \quad \text{... (6·54)}$$

For $r = k + 1$, we have

$$(a + b)^{k+1} = (a + b)\ (a + b)^k$$
$$\le (a + b)\ .\ 2^{k-1}\ (a^k + b^k) \quad \text{[From (6·54)]}$$

Subtracting $2^k\ (a^{k+1} + b^{k+1})$ from both sides, we get

$$(a+b)^{k+1} - 2^k\ (a^{k+1} + b^{k+1}) \le (a + b)\ 2^{k-1}\ (a^k + b^k) - 2^k\ (a^{k+1} + b^{k+1})$$
$$= 2^{(k-1)}\ [a^{k+1} + ab^k + ba^k + b^{k+1} - 2\ (a^{k+1} + b^{k+1})]$$
$$= -2^{k-1}\ [a^{k+1} + b^{k+1} - ab^k - ba^k] = -2^{k-1}\ [a^k\ (a - b) - b^k\ (a - b)]$$
$$= -2^{k-1}\ [(a - b)\ (a^k - b^k)] \quad \text{... (6·55)}$$

Since real numbers $a \ge 0$ and $b \ge 0$, for any positive integer k, both $(a - b)$ and $(a^k - b^k)$ have the same sign so that the product $(a - b)\ (a^k - b^k) \ge 0$.

Hence, the R.H.S. in (6·55) is ≤ 0.

$\Rightarrow$ $$(a + b)^{k+1} - 2^k\ (a^{k+1} + b^{k+1}) \le 0$$

$\Rightarrow$ $$(a + b)^{k+1} \le 2^k\ (a^{k+1} + b^{k+1}) = c_{k+1}\ (a^{k+1} + b^{k+1})$$

$\Rightarrow$ (6·51) is true for $r = (k + 1)$ also.

But we have proved in (6·53) that (6·51) is true for $r = 2$. Hence, (6·51) is true for $r = 2 + 1 = 3$, $3 + 1 = 4$, $4 + 1 = 5$, ... and so on. Hence, by the principle of mathematical induction, (6·51) is true for all positive integer values of r.

Proof of Main Inequality (6·49.) For *r.v.*'s X and Y, we have

$$|\ X + Y\ | \le |\ X\ | + |\ Y\ | \quad \text{(By Triangular Inequality)} \quad \text{... (6·56)}$$

$\Rightarrow$ $$E\ |\ X + Y\ | \le 2^{1-1}\ [E\ |\ X\ | + E\ |\ Y\ |] \quad \text{... (6·57)}$$

$\Rightarrow$ (6·49) is true for $r = 1$.

For $r \geq 2$, using the result in the above Lemma with $a = |X|$ & $b = |Y|$, for any $k \geq 2$, we get

$$(|X| + |Y|)^k \leq 2^{k-1}(|X|^k + |Y|^k)$$

$$\Rightarrow \quad |X|^k + |Y|^k \geq 2^{-(k-1)}(|X| + |Y|)^k$$

$$\Rightarrow \quad |X|^k + |Y|^k \geq 2^{-(k-1)} \cdot |X+Y|^k \qquad \text{[Using Triangular Inequality in (6·56)]}$$

$$\Rightarrow \quad |X+Y|^k \leq 2^{k-1}(|X|^k + |Y|^k)$$

$$\Rightarrow \quad |X+Y|^k \leq C_k(|X|^k + |Y|^k)$$

Taking expectation of both sides, we get

$$E(|X+Y|^k) \leq C_k(E|X|^k + E|Y|^k) \quad ; \quad k \geq 2$$

$$\Rightarrow \quad E(|X+Y|^r) \leq C_r(E|X|^r + E|Y|^r) \quad ; \quad r \geq 2$$

6·7·6. Holder's Inequality

If random variables X and Y assume positive values, then for $p > 1$, $q > 1$ so that $\frac{1}{p} + \frac{1}{q} = 1,$

$$E(XY) \leq [E(X^p)]^{1/p} \cdot [E(Y^q)]^{1/q} \qquad \text{... (6·58)}$$

Proof. Consider the function :

$$g(x) = -\log x \quad ; \quad x > 0 \qquad \text{... (*)}$$

$$\Rightarrow \quad g'(x) = -\frac{1}{x} \quad \text{and} \quad g''(x) = \frac{1}{x^2} > 0.$$

$\Rightarrow$ $g(x)$ is a convex function of x.

Hence, for $0 < \lambda < 1$, $x > 0$, $y > 0$, we have

$$g[\lambda x + (1-\lambda)y] \leq \lambda g(x) + (1-\lambda) g(y)$$

$$\Rightarrow \quad -\log[\lambda x + (1-\lambda)y] \leq -[\lambda \log x + (1-\lambda)\log y] \quad ; \qquad \text{[From (*)]}$$

$$\Rightarrow \quad \log[\lambda x + (1-\lambda)y] \geq \lambda \log x + (1-\lambda)\log y$$

$$\Rightarrow \quad \log[\lambda x + (1-\lambda)y] \geq \log(x^\lambda \cdot y^{1-\lambda})$$

Taking antilog on both sides, we get

$$[\lambda x + (1-\lambda)y] \geq x^\lambda y^{1-\lambda} \quad \text{... (**)} \qquad [\because \log(.) \text{ is } \uparrow \text{ function}]$$

Let us now take : $\lambda = \frac{1}{p}$ and $1 - \lambda = \frac{1}{q}$ $\Rightarrow$ $\frac{1}{p} + \frac{1}{q} = 1$ ... (***)

$$\left[p = \frac{1}{\lambda} > 0 \; ; \; q = \frac{1}{1-\lambda} > 0 \; ; \; 0 < \lambda < 1\right]$$

and $x = \dfrac{X^p}{E(X^p)}$ and $y = \dfrac{Y^q}{E(Y^q)}$ in (**), so that

$$\frac{1}{p}\left(\frac{X^p}{E(X^p)}\right) + \frac{1}{q}\left(\frac{Y^q}{E(Y^q)}\right) \geq \left(\frac{X^p}{E(X^p)}\right)^{1/p} \cdot \left(\frac{Y^q}{E(Y^q)}\right)^{1/q}$$

$$\Rightarrow \quad \frac{1}{p}\left(\frac{X^p}{E(X^p)}\right) + \frac{1}{q}\left(\frac{Y^q}{E(Y^q)}\right) \geq \frac{(XY)}{[E(X^p)]^{1/p} \cdot [E(Y^q)]^{1/q}}$$

Taking expectation on both sides and noting that $E(X^p)$ and $E(Y^q)$ are constants, we get

$$\frac{1}{p}+\frac{1}{q} \geq \frac{E(XY)}{\left[E(X^p)\right]^{1/p}\cdot\left[E(Y^q)\right]^{1/q}}$$

Cross multiplying and using (***), we get

$$[E(X^p)]^{1/p}\,.\,[E(Y^q)]^{1/q} \geq E(XY) \quad \Rightarrow \quad E(XY) \leq [E(X^p)]^{1/p}\,.\,[E(Y^q)]^{1/q}$$

Remark. Taking $p = q = 2$ in Holder's inequality, we get

$$E(XY) \leq [E(X^2)]^{1/2}\cdot[E(Y^2)]^{1/2} \qquad \text{... (6·58a)}$$

$$\Rightarrow \quad [E(XY)]^2 \leq E(X^2)\cdot E(Y^2),$$

which is *Cauchy-Schwartz* Inequality.

6·7·7. Minkowski's Inequality

If X and Y are two random variables with $E(X^r) < \infty$ and $E(Y^r) < \infty$, for every positive integer $r \geq 1$, then

$$\left[E\,|\,X+Y\,|^r\right]^{1/r} \leq \left[E\,|\,X\,|^r\right]^{1/r}.\left[E\,|\,Y\,|^r\right]^{1/r} \qquad \text{... (6·59)}$$

Proof. For any integer $r > 1$, we have

$$|\,X+Y\,|^r = |\,X+Y\,|^{r-1}\,.\,|\,X+Y\,| \leq |\,X+Y\,|^{r-1}\left[\,|\,X\,|+|\,Y\,|\,\right]$$

(By Triangular Inequality)

$$\Rightarrow \quad |\,X+Y\,|^r \leq |\,X+Y\,|^{r-1}\,.\,|\,X\,| + |\,X+Y\,|^{r-1}\,.\,|\,Y\,|$$

Taking expectation on both sides, we get

$$E\left[\,|\,X+Y\,|^r\right] \leq E\left[\,|\,X+Y\,|^{r-1}\,.\,|\,X\,|\,\right] + E\left[\,|\,X+Y\,|^{r-1}\,|\,Y\,|\,\right] \qquad \text{...(*)}$$

Let us choose another constant $s \geq 2$, so that $\frac{1}{r}+\frac{1}{s}=1$ and use Holder's inequality on each term on the R.H.S. in (*), we get

$$E\left[\,|\,X\,|\;|\,X+Y\,|^{r-1}\right] \leq \left[E\,|\,X^r\,|\right]^{1/r}.\left[E\,|\,X+Y\,|^{s(r-1)}\right]^{1/s}$$

and

$$E\left[\,|\,Y\,|\;|\,X+Y\,|^{r-1}\right] \leq \left[E\,|\,Y\,|^r\right]^{1/r}.\left[E\,|\,X+Y\,|^{s(r-1)}\right]^{1/s}$$

Substituting in (*), we get

$$\left[E\,|\,X+Y\,|^r\right] \leq \left[E\,|\,X\,|^r\right]^{1/r}.\left[E\,|\,X+Y\,|^r\right]^{1/s} + \left[E\,|\,Y\,|^r\right]^{1/r}\left[E\,|\,X+Y\,|^r\right]^{1/s}$$

$$\left[\because \frac{1}{r}+\frac{1}{s}=1 \quad\Rightarrow\quad r+s=rs \quad\Rightarrow\quad r=s(r-1)\right]$$

$$\Rightarrow \quad \left[E\,|\,X+Y\,|^r\right]^{1-\frac{1}{s}} \leq \left[E\,|\,X\,|^r\right]^{1/r} + \left[E\,|\,Y\,|^r\right]^{1/r}$$

$$\Rightarrow \quad \left[E\,|\,X+Y\,|^r\right]^{1/r} \leq \left[E\,|\,X\,|^r\right]^{1/r} + \left[E\,|\,Y\,|^r\right]^{1/r}; \qquad \left(\because \frac{1}{r}+\frac{1}{s}=1\right)$$

6·8. MOMENTS OF BIVARIATE PROBABILITY DISTRIBUTIONS

The mathematical expectation of a function $g(x, y)$ of two-dimensional random variable (X, Y) with p.d.f. $f(x, y)$ is given by :

$$E[g(X, Y)] = \int_{-\infty}^{\infty}\int_{-\infty}^{\infty} g(x,y)\,f(x,y)\,dx\,dy, \text{ (If X and Y are continuous variables.)} \qquad \text{... (6·60)}$$

$$= \sum_i \sum_j g(x_i, y_j)\,P(X = x_i \cap Y = y_j), \text{ (If X and Y are discrete variables.)} \qquad \text{... (6·60a)}$$

provided the expectation exists.

In particular, the rth and sth product moment about origin of the random variables X and Y respectively is defined as :

$$\mu_{rs}' = E(X^r Y^s) = \int_{-\infty}^{\infty}\int_{-\infty}^{\infty} x^r y^s f(x,y)\,dx\,dy$$

or
$$\mu_{rs}' = \sum_i \sum_j x_i^r y_j^s P(X = x_i \cap Y = y_j) \quad \text{... (6·61)}$$

The joint rth central moment of X and sth central moment of Y is given by :

$$\mu_{rs} = E[\{X - E(X)\}^r \{Y - E(Y)\}^s]$$
$$= E[(X - \mu_X)^r (Y - \mu_Y)^s], \quad [E(X) = \mu_X, E(Y) = \mu_Y] \quad \text{... (6·62)}$$

In particular,

$$\mu_{00}' = 1 = \mu_{00}, \quad \mu_{10} = 0 = \mu_{01}, \quad \mu_{10}' = E(X), \quad \mu_{01}' = E(Y) \quad \text{... (6·62a)}$$
$$\mu_{20} = \sigma_X^2, \quad \mu_{02} = \sigma_Y^2 \quad \text{and} \quad \mu_{11} = \text{Cov}(X, Y).$$

Example 6·28. *Two random variables X and Y have the following joint probability density function :*

$$f(x, y) = \begin{cases} 2 - x - y; & 0 \le x \le 1,\ 0 \le y \le 1 \\ 0, & otherwise \end{cases}$$

Find

(i) Marginal probability density functions of X and Y ;
(ii) Conditional density functions, *(iii) Var (X) and Var (Y) ; and*
(iv) Covariance between X and Y.

Solution.

(i)
$$f_X(x) = \int_{-\infty}^{\infty} f(x,y)\,dy = \int_0^1 (2 - x - y)\,dy = \frac{3}{2} - x$$

∴
$$f_X(x) = \frac{3}{2} - x,\ 0 < x < 1 \quad ; \quad f_X(x) = 0, \text{ otherwise}$$

Similarly,
$$f_Y(y) = \frac{3}{2} - y,\ 0 < y < 1 \quad ; \quad f_Y(y) = 0, \text{ otherwise}$$

(ii)
$$f_{X|Y}(x\,|\,y) = \frac{f_{XY}(x,y)}{f_Y(y)} = \frac{(2 - x - y)}{\left(\frac{3}{2} - y\right)}, \quad 0 < (x, y) < 1$$

$$f_{Y|X}(y\,|\,x) = \frac{f_{XY}(x,y)}{f_X(x)} = \frac{(2 - x - y)}{\left(\frac{3}{2} - x\right)}, \quad 0 < (x, y) < 1$$

(iii)
$$E(X) = \int_0^1 x\, f_X(x)\,dx = \int_0^1 x\left(\frac{3}{2} - x\right)dx = \frac{5}{12}$$

$$E(Y) = \int_0^1 y\, f_Y(y)\,dy = \int_0^1 y\left(\frac{3}{2} - y\right)dy = \frac{5}{12}$$

$$E(X^2) = \int_0^1 x^2\left(\frac{3}{2} - x\right)dx = \left|\frac{3}{6}x^3 - \frac{x^4}{4}\right|_0^1 = \frac{1}{4}$$

$$V(X) = E(X^2) - \{E(X)\}^2 = \frac{1}{4} - \frac{25}{144} = \frac{11}{144}$$

Similarly, by symmetry, $V(Y) = \dfrac{11}{144}$

(iv) $$E(XY) = \int_0^1 \int_0^1 xy(2-x-y)\,dx\,dy = \int_0^1 \left| 2\frac{x^2y}{2} - \frac{x^3y}{3} - \frac{x^2y^2}{2} \right|_{x=0}^{x=1} dy$$

$$= \int_0^1 \left(\frac{2}{3}y - \frac{1}{2}y^2\right) dy = \left| \frac{y^2}{3} - \frac{y^3}{6} \right|_0^1 = \frac{1}{6}$$

$$\therefore \quad \text{Cov}(X, Y) = E(XY) - E(X)E(Y) = \frac{1}{6} - \frac{5}{12} \cdot \frac{5}{12} = \frac{-1}{144}.$$

Example 6·29. *Two ideal dice are thrown. Let X_1 be the score on the first dice and X_2 the score on the second dice. Let Y denote the maximum of X_1 and X_2, i.e., $Y = \max(X_1, X_2)$.*

(i) *Write down the joint distribution of Y and X_1, and*

(ii) *find the mean and variance of Y, and covariance (Y, X_1).*

Solution.

(i) Each of the random variables X_1 and X_2 can take six values 1, 2, 3, 4, 5, 6, each with probability $\frac{1}{6}$, *i.e.*,

$$P(X_1 = i) = P(X_2 = i) = \frac{1}{6}\ ; \quad i = 1, 2, 3, 4, 5, 6 \qquad \text{... (i)}$$

$$Y = \text{Max}(X_1, X_2).$$

Obviously, $P(X_1 = i, Y = j) = 0$, if $j < i = 1, 2, ..., 6$.

$$P(X_1 = i, Y = i) = P(X_1 = i \cap X_2 \le i) = \sum_{j=1}^{i} P(X_1 = i \cap X_2 = j)$$

$$= \sum_{j=1}^{i} P(X_1 = i) \cdot P(X_2 = j) \quad (\because X_1, X_2 \text{ are independent.})$$

$$= \sum_{j=1}^{i} \left(\frac{1}{36}\right) = \frac{i}{36}\ ; \ i = 1, 2, ... 6$$

$$P(X_1 = i, Y = j) = P(X_1 = i, X_2 = j) = P(X_1 = i)\,P(X_2 = j) = \frac{1}{36}\ ; \ j > i = 1, 2, ... 6.$$

The joint probability table of X_1 and Y is given as follows :

X_1 \ Y	1	2	3	4	5	6	*Marginal Totals*
1	1/36	1/36	1/36	1/36	1/36	1/36	6/36
2	0	2/36	1/36	1/36	1/36	1/36	6/36
3	0	0	3/36	1/36	1/36	1/36	6/36
4	0	0	0	4/36	1/36	1/36	6/36
5	0	0	0	0	5/36	1/36	6/36
6	0	0	0	0	0	6/36	6/36
Marginal Totals	1/36	3/36	5/36	7/36	9/36	11/36	1

Aliter. For an alternate solution, proceed as in Example 6·34 .

$$(ii)\ E(Y) = 1\cdot\frac{1}{36} + 2\cdot\frac{3}{36} + 3\cdot\frac{5}{36} + 4\cdot\frac{7}{36} + 5\cdot\frac{9}{36} + 6\cdot\frac{11}{36} = \frac{161}{36}$$

$$E(Y^2) = 1^2\cdot\frac{1}{36} + 2^2\cdot\frac{3}{36} + 3^2\cdot\frac{5}{36} + 4^2\cdot\frac{7}{36} + 5^2\cdot\frac{9}{36} + 6^2\cdot\frac{11}{36} = \frac{791}{36}$$

$$V(Y) = E(Y^2) - \{E(Y)\}^2 = \frac{791}{36} - \left(\frac{161}{36}\right)^2 = \frac{2555}{1296}$$

$$E(X_1) = \frac{6}{36}(1+2+3+4+5+6) = \frac{126}{36} = \frac{21}{6}$$

$$E(X_1 Y) = 1\cdot\frac{1}{36} + 2\cdot\frac{1}{36} + 3\cdot\frac{1}{36} + 4\cdot\frac{1}{36} + 5\cdot\frac{1}{36} + 6\cdot\frac{1}{36}$$
$$+ \left(4\cdot\frac{2}{36}+6\cdot\frac{1}{36}+8\cdot\frac{1}{36}+10\cdot\frac{1}{36}+12\cdot\frac{1}{36}\right)$$
$$+ \left(9\cdot\frac{3}{36}+12\cdot\frac{1}{36}+15\cdot\frac{1}{36}+18\cdot\frac{1}{36}\right) + \left(16\cdot\frac{4}{36}+20\cdot\frac{1}{36}+24\cdot\frac{1}{36}\right)$$
$$+ \left(25\cdot\frac{5}{36}+30\cdot\frac{1}{36}\right)+36\cdot\frac{6}{36}$$
$$= \frac{1}{36}(21+44+72+108+155+216) = \frac{1}{36}\times 616$$

$$\text{Cov}(X_1, Y) = E(X_1, Y) - E(X_1)\,E(Y) = \frac{616}{36} - \frac{21}{6}\cdot\frac{161}{36} = \frac{3696-3381}{216} = \frac{315}{216} = \frac{35}{24}.$$

6·9. CONDITIONAL EXPECTATION AND CONDITIONAL VARIANCE

Definition. Let (X, Y) be a two dimensional random variable and $g(.,.)$ be a function of these two variables. The conditional expectation of $g(X, Y)$ given $X = x$, denoted by $E[g(X, Y) \mid X = x]$ or more precisely $E_Y[g(X, Y) \mid X]$ is defined as :

$$E[g(X, Y) \mid X = x] = \int_{-\infty}^{\infty} g(x,y)\,.\,f_{Y\mid X}(y \mid x)\,dy, \text{ if } (X, Y) \text{ are jointly continuous} \quad \dots (6{\cdot}63)$$

$$\text{and } E[g(X, Y) \mid X = x] = \sum_y g(x,y)\, f_{Y\mid X}(y \mid x), \text{ if } (X, Y) \text{ are jointly discrete.} \quad \dots (6{\cdot}64)$$

where summation is taken over all possible values of y.

It may be noted that in (6·63) or (6·64), x is a constant.

6·9·1. Double Expectation Rule or Double – E Rule

Theorem 6·1. *Let (X, Y) be two dimensional random variable. Then*

$$E[g(Y)] = E[E\{g(Y) \mid X\}] = E_X[E_Y\{g(Y) \mid X\}] \quad \dots (6{\cdot}65)$$

In particular, $E(Y) = E[E(Y \mid X)] = E_X[E_Y(Y \mid X)]$ ... (6·66)

Proof. We will prove the result for jointly continuous *r.v.*'s X and Y with joint *p.d.f.* $f(x, y)$ and marginal *p.d.f.'s* $f_X(x)$ and $f_Y(y)$.

[The proof for jointly discrete *r.v.*'s X and Y is similar and is obtained on replacing integration with summation over the range of the variables.]

We note that $E[g(Y) \mid X]$ is some function of X, say, $h(X)$.

$$\therefore\ E[E\{g(Y) \mid X\}] = E[h(X)] = \int_{-\infty}^{\infty} h(x)\, f_X(x)\, dx$$

$$= \int_{-\infty}^{\infty} E\{g(Y) \mid X\}\, f_X(x)\, dx$$

$$= \int_{-\infty}^{\infty} \left[\left\{\int_{-\infty}^{\infty} g(y) \cdot f_{Y \mid X}(y \mid x)\, dy\right\} f_X(x)\, dx\right]$$

$$= \int_{-\infty}^{\infty} \int_{-\infty}^{\infty} g(y) \,.\, f_{Y \mid X}(y \mid x) \,.\, f_X(x)\, dx\, dy$$

$$= \int_{-\infty}^{\infty} \int_{-\infty}^{\infty} g(y) \,.\, f_{XY}(x, y)\, dx\, dy \qquad \left[\because f_{Y|X}(y \mid x) = \frac{f_{XY}(x,y)}{f_X(x)}\right]$$

$$= E[g(Y)]$$

$$\Rightarrow \qquad E[g(Y)] = E[E\{g(Y) \mid X\}] = E_X[E_Y\{g(Y) \mid X\}]$$

In particular, if we take $g(Y) = Y$, then

$$E(Y) = E[E(Y \mid X)] = E_X[E_Y(Y \mid X)]$$

Note. Similarly proceeding, we can prove that

$$E[g(X)] = E[E\{g(X) \mid Y\}] = E_Y[E_X\{g(X) \mid Y\}] \qquad \text{... (6·67)}$$

and

$$E(X) = E[E(X \mid Y)] = E_Y[E_X(X \mid Y)] \qquad \text{... (6·68)}$$

***Theorem 6·2.* (Double–E Rule of Expectation).** *For two dimensional r.v.* (X, Y),

(*i*) $$E(X) = E[E(X \mid Y)] = E_Y[E_X(X \mid Y)] \qquad \text{... (6·69)}$$

(*ii*) $$E(Y) = E[E(Y \mid X)] = E_X[E_Y(Y \mid X)] \qquad \text{... (6·70)}$$

Proof. As in the above theorem, we will prove the result for continuous *r.v.* (X, Y) with joint *p.d.f.* $f_{XY}(x, y)$ and marginal *p.d.f.'s* $f_X(x)$ and $f_Y(y)$.

(*i*) $$E[E(X \mid Y)] = E\left[\int_{-\infty}^{\infty} x \cdot f_{X \mid Y}(x \mid y)\, dx\right]$$

It may be noted that the expression in bracket [ ], is a function of *y*.

$$\therefore\ E[E(X \mid Y)] = \int_{-\infty}^{\infty} \left(\int_{-\infty}^{\infty} x\, f_{X \mid Y}(x \mid y)\, dx\right) \cdot f_Y(y)\, dy$$

$$= \int_{-\infty}^{\infty} \int_{-\infty}^{\infty} x \cdot f_{XY}(x, y)\, dx\, dy \qquad \left[\because f_{X \mid Y}(x \mid y) = \frac{f_{XY}(x, y)}{f_Y(y)}\right]$$

$$= \int_{-\infty}^{\infty} \left[x \left(\int_{-\infty}^{\infty} f_{XY}(x, y)\, dy\right) dx\right]$$

$$= \int_{-\infty}^{\infty} x \cdot f_X(x)\, dx = E(X)$$

Similarly, we can prove :

(ii) $$E(Y) = E[E(Y \mid X)] = E_X[E_Y(Y \mid X)]$$

Note. The proof for joint discrete *r.v.* (X, Y) is exactly similar and is obtained on replacing integration with summation over the range of the variables.

6·9·2. Conditional Variance

For *r.v.* U, we have

$$\text{Var}(U) = E[U - E(U)]^2 = E(U^2) - [E(U)]^2$$

Replacing U by the conditioned variable (X | Y), we get

$$\text{Var}(X \mid Y) = E[\{X - E(X \mid Y)\}^2 \mid Y] = E(X^2 \mid Y) - [E(X \mid Y)]^2 \quad \text{... (6·71)}$$

Similarly,

$$\text{Var}(Y \mid X) = E[\{Y - E(Y \mid X)\}^2 \mid X] = E(Y^2 \mid X) - [E(Y \mid X)]^2 \quad \text{... (6·72)}$$

Theorem 6·3. *The variance of X can be regarded as consisting of two parts, the expectation of the conditional variance and the variance of the conditional expectation. Symbolically, we have*

$$Var(X) = E[Var(X \mid Y)] + Var[E(X \mid Y)] \quad \text{... (6·73)}$$

Proof. By definition, we have

$$\begin{aligned} E[\text{Var}(X \mid Y)] &= E[E(X^2 \mid Y) - \{E(X \mid Y)\}^2] \\ &= E[E(X^2 \mid Y)] - E\{E(X \mid Y)\}^2 \\ &= E(X^2) - E[E(X \mid Y)]^2 \quad \text{(By Double - E Rule)} \quad \text{... (*)} \\ \text{Var}[E(X \mid Y)] &= E[E(X \mid Y)]^2 - [E\{E(X \mid Y)\}]^2 \\ &= E[E(X \mid Y)]^2 - \{E(X)\}^2 \quad \text{(By Double - } E \text{ Rule)} \quad \text{... (**)} \end{aligned}$$

Adding (*) and (**), we get

$$E[\text{Var}(X \mid Y)] + \text{Var}[E(X \mid Y)] = E(X^2) - \{E(X)\}^2 = \text{Var}(X), \text{ as required.}$$

Note. Similarly, it can be proved that

$$\text{Var}(Y) = E[\text{Var}(Y \mid X)] + \text{Var}[E(Y \mid X)] \quad \text{... (6·74)}$$

6·9·3. Wald's Theorem on Mean and Variance of a Random Sum.

Let $\{X_i\}$, $i = 1, 2, 3...$ *be a sequence of i.i.d. random variables and*

$$S_N = X_1 + X_2 + X_3 + \ldots + X_N$$

where N is any other integer valued random variable independent of X_i*'s, then :*

(a) $$E(S_N) = E(X_1) \,.\, E(N) \quad \text{... (6·75)}$$

(b) $$Var(S_N) = Var(X_1) \,.\, E(N) + [E(X_1)]^2 \,.\, Var\, N. \quad \text{... (6·76)}$$

Solution.

(a) $$\begin{aligned} E(S_N) &= E(X_1 + X_2 + \ldots + X_N) = E_N[E(X_1 + X_2 + \ldots + X_N \mid N)] \\ &= E_N[N \,.\, E(X_1)] \quad (\because X_i\text{'s are i.i.d.}) \quad \text{... (i)} \\ &= E(X_1)\, E(N) \end{aligned}$$

(b) $$\text{Var}(S_N) = E[\text{Var}(S_N \mid N)] + \text{Var}[E(S_N \mid N)] \quad \text{... (ii)}$$

$$\text{Var}(S_N \mid N) = \text{Var}[(X_1 + X_2 + \ldots + X_N) \mid N]$$

$$= \sum_{i=1}^{N} Var(X_i) = N.\text{Var}(X_1) \qquad [\because X_i\text{'s are } i.i.d.\ r.v.\text{'s}]$$

$$\therefore\ E\,[\text{Var}(S_N \mid N)] = E\,[N \text{ Var}(X_1)] = \text{Var}(X_1) \,.\, E(N) \qquad \ldots (iii)$$

$$E(S_N \mid N) = E\,[(X_1 + X_2 + \ldots + X_N) \mid N] = N \,.\, E(X_1)$$

$$\Rightarrow \text{Var}\,[E(S_N \mid N)] = \text{Var}\,[N \,.\, E(X_1)] = [E(X_1)]^2 \,.\, \text{Var}(N) \qquad \ldots (iv)$$

Substituting from *(iii)* and *(iv)* in *(ii)*, we get

$$\text{Var}(S_N) = \text{Var}(X_1) \,.\, E(N) + [E(X_1)]^2 \,.\, \text{Var}(N).$$

6·9·4. Multistage Rule for Expectation (Multistage – E Rule)

Theorem 6·4. *For two events A and B with* $P(A \cup B) > 0$,

$$E(X \mid A \cup B) = \frac{P(A) \,.\, E(X \mid A) + P(B) \cdot E(X \mid B) - P(A \cap B) \cdot E(X \mid AB)}{P(A \cup B)} \qquad \ldots (6\cdot77)$$

where by def., $\quad E(X \mid A) = \dfrac{1}{P(A)} \displaystyle\sum_{x_i \in A} x_i\, P(X = x_i). \qquad \ldots (*)$

Proof. $\quad E(X \mid A \cup B) = \dfrac{1}{P(A \cup B)} \displaystyle\sum_{x_i \in A \cup B} x_i \cdot P(X = x_i) \qquad \ldots (**)$

We have :

$$\sum_{x_i \in A \cup B} x_i\, P(X = x_i) = \sum_{x_i \in A} x_i\, P(X = x_i) + \sum_{x_i \in B} x_i\, P(X = x_i) - \sum_{x_i \in (A \cap B)} x_i \cdot P(X = x_i)$$

$$\therefore\ E(X \mid A \cup B) = \frac{1}{P(A \cup B)} \left[P(A)\, E(X \mid A) + P(B)\, E(X \mid B) - P(AB) \cdot E(X \mid AB) \right]$$

[From (**) and (*)]

Cor 1. If the events A & B are disjoint then $A \cap B = \phi$ & $P(A \cap B) = 0$. Hence, (6·77) becomes :
For two mutually disjoint events A and B with $P(A \cap B) > 0$,

$$E(X \mid A \cup B) = \frac{P(A) \cdot E(X \mid A) + P(B)\, E(X \mid B)}{P(A \cup B)} \qquad \ldots (6\cdot77a)$$

Cor 2. **Multistage – E rule.** For a random variable X associated with the occurrence or non-occurrence of event A, taking $B = \bar{A} = A^c$ in (6·77*a*), we get

$$E(X) = P(A)\, E(X \mid A) + P(\bar{A}) \,.\, E(X \mid \bar{A})\,; \qquad \ldots (6\cdot77b)$$

because $P(A \cup \bar{A}) = P(S) = 1$ and $E(X \mid A \cup \bar{A}) = E(X \mid S) = \dfrac{1}{P(S)} \cdot \displaystyle\sum_{x \in S} x \cdot P(X = x) = E(X)$

The result in (6·77*b*) can be extended to n events.

If the *r.v.* X is associated with the occurrence of the events $A_1, A_2, ..., A_n$ where $\bigcup_{i=1}^{n} A_i = S$, then

$$E(X) = P(A_1)\,E(X \mid A_1) + P(A_2)\,.\,E(X \mid E_2) + ... + P(A_n)\,E(X \mid E_n)$$

$$\Rightarrow \quad E(X) = \sum_{i=1}^{n} P(A_i)\,E(X \mid A_i) \qquad ... (6·77c)$$

Example 6·30. *A number X is chosen at random from the numbers 1, 2, 3, ..., n. Then a number Y is chosen at random between 1 and X. Find E (Y).*

Solution. Define the events :

$A_i : (X = i)$ *i.e.,* the number '*i*' is chosen from the numbers 1, 2, ..., *n*

Thus, $$P(A_i) = \frac{1}{n}\,; \quad i = 1, 2, ..., n \qquad ... (1)$$

Then, by the Multistage *E*–rule,

$$E(Y) = \sum_{i=1}^{n} P(A_i) \cdot E(Y \mid A_i), \qquad ... (2)$$

where $(Y \mid A_i)$ represents the event that the number *Y* is chosen at random from amongst the numbers 1, 2, 3, ..., *i* so that :

$$P(Y \mid A_i) = \frac{1}{i}$$

$$\therefore \quad E(Y \mid A_i) = \frac{1}{i}(1 + 2 + 3 + ... + i) = \frac{(i+1)}{2} \qquad ... (3)$$

[Here *Y* is a *r.v.* which can take any one of the values 1, 2, 3, ..., *i* each with probability (1/*i*)].

Substituting from (1) and (3) in (2), we get

$$E(Y) = \sum_{i=1}^{n} \left[\frac{1}{n}\left(\frac{i+1}{2}\right)\right] = \frac{1}{2n}\left[\sum_{i=1}^{n} i + n\right]$$

$$= \frac{1}{2n}\left[\frac{n(n+1)}{2} + n\right] = \frac{1}{2}\left[\frac{n+1}{2} + 1\right] = \frac{1}{4}(n+3)$$

Example 6·31. *A prisoner is trapped in a cell containing three doors. The first door leads to a tunnel which takes him to freedom after 2 hours travel. The second door leads to a tunnel which returns him to the cell after 3 hours travel. The third door leads to a tunnel which returns him to his cell after 4 hours. Assuming that the prisoner is at all times equally likely to choose any one of the doors, what is the expected length of time until the prisoner becomes free?*

Solution. The diagrammatic representation of the prisoner's route of escape is given in the figure on page 6·53, where D_i, $(i = 1, 2, 3)$ denotes the event that the prisoner chooses the *i*th door in the cell to escape.

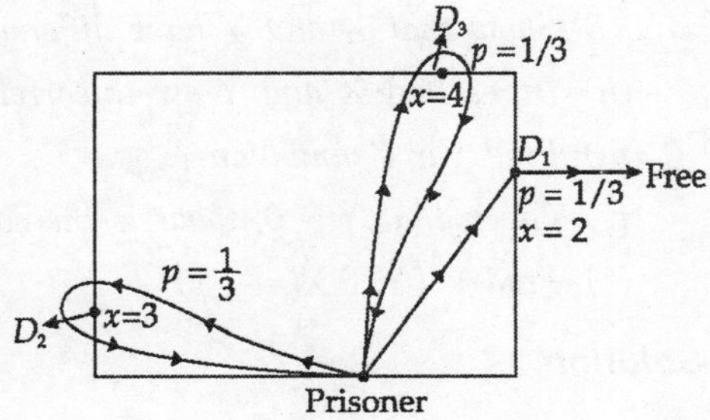

Fig 6.6

Since the prisoner is at all times equally likely to choose any one of the three doors, we have

$$P(D_1) = P(D_2) = P(D_3) = \frac{1}{3} \quad \text{... (1)}$$

Let the r.v. X denote the time in hours in the cell before the prisoner becomes free and let us suppose that the expected time to freedom is :

$$E(X) = \mu \quad \text{... (2)}$$

Suppose that the prisoner chooses the door number 3 for escape. As per conditions of the question, after travelling in the tunnel for 4 hours, he returns to his initial position in the cell, from where he has to start afresh for some other door and his *expected additional time* till he becomes free is just $E(X) = \mu$.

$$\therefore \quad E(X \mid D_3) = 4 + \mu \quad \text{... (3)}$$

By similar arguments, we have

$$E(X \mid D_2) = 3 + \mu \quad \text{and} \quad E(X \mid D_1) = 2 \quad \text{... (3a)}$$

By the multistage E-rule, we have

$$E(X) = P(D_1) \cdot E(X \mid D_1) + P(D_2) \cdot E(X \mid D_2) + P(D_3) \cdot E(X \mid D_3)$$

$$\Rightarrow \quad \mu = \frac{1}{3}\left[2 + (3 + \mu) + (4 + \mu)\right]; \quad \text{[From (1), (3) and (3a)]}$$

$$\Rightarrow \quad 3\mu = 9 + 2\mu \quad \Rightarrow \quad \mu = 9$$

Hence, the expected time till the prisoner becomes free is 9 hours.

Example 6·32. *If $X_1, X_2, ..., X_k$ are independent and identically distributed random variables, then prove that:*

$$E\left[X_1 \mid (X_1 + X_2 + ... + X_k = \lambda)\right] = \frac{\lambda}{k}$$

Solution. We have

$$E\left[(X_1 + X_2 + ... + X_k) \mid (X_1 + X_2 + ... + X_k = \lambda)\right] = E(\lambda) = \lambda$$

$$\Rightarrow \quad E\left[X_1 \mid (X_1 + X_2 + ... + X_k = \lambda)\right] + E\left[X_2 \mid (X_1 + X_2 + ... + X_k = \lambda)\right] + ... + E\left[X_k \mid (X_1 + X_2 + ... + X_k = \lambda)\right] = \lambda$$

$$\Rightarrow \quad k \cdot E\left[X_1 \mid (X_1 + X_2 + ... + X_k = \lambda)\right] = \lambda. \quad \left[\because X_1, X_2, ..., X_k \text{ are identically distributed}\right]$$

$$\text{Hence,} \quad E\left[X_1 \mid (X_1 + X_2 + ... + X_k = \lambda)\right] = \frac{\lambda}{k}.$$

Example 6·33. *Let X and Y be two random variables each taking three values –1, 0 and 1, and having the joint probability distribution*

Y \ X	–1	0	1	Total
–1	0	·1	·1	·2
0	·2	·2	·2	·6
1	0	·1	·1	·2
Total	·2	·4	·4	1·0

(i) *Show that X and Y have different expectations.*

(ii) *Prove that X and Y are uncorrelated.*

(iii) *Find Var X and Var Y.*

(iv) *Given that Y = 0, what is the conditional probability distribution of X?*

(v) *Find* $V(Y \mid X = -1)$.

Solution.

(i) $$E(Y) = \sum p_i y_i = -1(\cdot 2) + 0(\cdot 6) + 1(\cdot 2) = 0$$

$$E(X) = \sum p_i x_i = -1(\cdot 2) + 0(\cdot 4) + 1(\cdot 4) = \cdot 2$$

$$\therefore \quad E(X) \neq E(Y)$$

(ii) $$E(XY) = \sum p_{ij} x_i y_j = (-1)(-1)(0) + 0(-1)(\cdot 1) + 1(-1)(\cdot 1) + 0(-1)(\cdot 2) + 0(0)(\cdot 2) + 0(1)(\cdot 2) + 1(-1)(0) + 1(0)(\cdot 1) + 1(1)(\cdot 1) = -0\cdot 1 + 0\cdot 1 = 0$$

$$\therefore \quad \text{Cov}(X, Y) = E(XY) - E(X)E(Y) = 0$$

$\Rightarrow$ X and Y are uncorrelated. (*c.f.* Chapter 10)

(iii) $$E(Y^2) = (-1)^2(0\cdot 2) + 0(0\cdot 6) + 1^2(0\cdot 2) = 0\cdot 4$$

$$\therefore \quad V(Y) = E(Y^2) - [E(Y)]^2 = 0\cdot 4$$

$$E(X^2) = (-1)^2(0\cdot 2) + 0(0\cdot 4) + 1^2(0\cdot 4) = 0\cdot 2 + 0\cdot 4 = 0\cdot 6\ ;$$

$$V(X) = E(X^2) - \{E(X)\}^2 = 0\cdot 6 - 0\cdot 04 = 0\cdot 56$$

(iv) $$P(X = -1 \mid Y = 0) = \frac{P(X = -1 \cap Y = 0)}{P(Y = 0)} = \frac{0\cdot 2}{0\cdot 6} = \frac{1}{3}\ ;$$

$$P(X = 0 \mid Y = 0) = \frac{P(X = 0 \cap Y = 0)}{P(Y = 0)} = \frac{0\cdot 2}{0\cdot 6} = \frac{1}{3}$$

$$P(X = 1 \mid Y = 0) = \frac{P(X = 1 \cap Y = 0)}{P(Y = 0)} = \frac{0\cdot 2}{0\cdot 6} = \frac{1}{3}$$

(v) $$V(Y \mid X = -1) = E(Y \mid X = -1)^2 - [E(Y \mid X = -1)]^2 \quad \text{... (*)}$$

$$E(Y \mid X = -1) = \sum_y y\, P(Y = y \mid X = -1) = (-1)0 + 0(\cdot 2) + 1(0) = 0$$

$$E(Y \mid X = -1)^2 = \sum_y y^2\, P(Y = y \mid X = -1) = 1(0) + 0(\cdot 2) + 1(0) = 0$$

$$\therefore \quad V(Y \mid X = -1) = 0. \quad \text{[From (*)]}$$

Example 6·34. *Two tetrahedra with sides numbered 1 to 4 are tossed. Let X denote the number on the downturned face of the first tetrahedron and Y denote the larger of the downturned numbers. Investigate the following :*

(a) *Joint density function X, Y and marginals* f_X *and* f_Y,

(b) $P(X \le 2, Y \le 3)$,

(c) $\text{Cov}(X, Y)$,

(*d*) $E(Y \mid X = 2)$,

(*e*) *Construct joint density different from that in part (a) but processing same marginals f_X and f_Y.*

Solution.

(*a*) The sample space is $S = \{1, 2, 3, 4\} \times \{1, 2, 3, 4\}$ & each of the 16 sample points (outcomes) has probability $p = \frac{1}{16}$, of occurrence.

Let X : Number on the first tetrahedron and Y : Larger of the numbers on the two tetrahedron. Then the above 16 sample points, in that order, give the following distribution of X and Y.

Sample Point	:	(1, 1)	(1, 2)	(1, 3)	(1, 4)	(2, 1)	(2, 2)	(2, 3)	(2, 4)
X	:	1	1	1	1	2	2	2	2
Y	:	1	2	3	4	2	2	3	4
Sample Point	:	(3, 1)	(3, 2)	(3, 3)	(3, 4)	(4, 1)	(4, 2)	(4, 3)	(4, 4)
X	:	3	3	3	3	4	4	4	4
Y	:	3	3	3	4	4	4	4	4

Since each sample point has probability $p = \frac{1}{16}$, the joint density functions of X and Y and the marginal densities f_X and f_Y are given below :

(*i*)

		X				
		1	2	3	4	*Total* (f_Y)
	1	p	0	0	0	p
	2	p	$2p$	0	0	$3p$
Y	3	p	p	$3p$	0	$5p$
	4	p	p	p	$4p$	$7p$
Total (f_X)		$4p$	$4p$	$4p$	$4p$	$16\,p = 1$

(*ii*)

		X				
		1	2	3	4	*Total* (f_Y)
	1	p	0	0	0	p
	2	p	$2p$	0	0	$3p$
Y	3	p	$p + \varepsilon$	$3p - \varepsilon$	0	$5p$
	4	p	$p - \varepsilon$	$p + \varepsilon$	$4p$	$7p$
Total (f_X)		$4p$	$4p$	$4p$	$4p$	1

(*b*) $$P(X \le 2, Y \le 3) = p + p + 2p + p + p = 6p = \frac{3}{8}. \qquad \left(\because p = \frac{1}{16}\right)$$

(*c*) $$\text{Var}(X) = E(X^2) - \{E(X)\}^2 = \frac{15}{2} - \frac{25}{4} = \frac{5}{4} \qquad \text{(Try it)}$$

$$\text{Var}(Y) = E(Y^2) - \{E(Y)\}^2 = \frac{85}{8} - \left(\frac{25}{8}\right)^2 = \frac{55}{64} \qquad \text{(Try it)}$$

$$\text{Cov}(X, Y) = E(XY) - E(X)\,E(Y) = \frac{135}{16} - \frac{5}{2} \times \frac{25}{8} = \frac{5}{8} \qquad \text{(Try it)}$$

(*d*) $$E(Y \mid X = 2) = \sum y \cdot f(y \mid x = 2) = \sum \left\{ y \cdot \frac{f(x = 2 \cap y)}{f(x = 2)} \right\}$$

$$= 4 \sum y\, f(2, y) = 4\,(0 + 4p + 3p + 4p) = 44p = \frac{11}{4}$$

(*e*) Let $0 < \varepsilon < p$. The joint density of X and Y given in Table in (*ii*) on page 6·55 is different from that in Table in (*i*), but has the same marginals as in (*i*).

Example 6·35.

(*a*) *Given two variates X_1 and X_2 with joint density function $f(x_1, x_2)$, prove that conditional mean of X_2 (given X_1) coincides with (unconditional) mean only if X_1 and X_2 are independent (stochastically).*

(*b*) *Let $f(x_1, x_2) = 21x_1^2 x_2^3$, $0 < x_1 < x_2 < 1$, and zero elsewhere, be the joint p.d.f. of X_1 and X_2. Find the conditional mean and variance of X_1 given $X_2 = x_2$, $0 < x_2 < 1$.*

Solution.

(*a*) Conditional mean of X_2 given X_1 is given by :

$$E(X_2 \mid X_1 = x_1) = \int_{x_2} x_2\, f(x_2 \mid x_1)\, dx_2, \qquad \text{... (*)}$$

where $f(x_2 \mid x_1)$ is conditional *p.d.f.* of X_2 given $X_1 = x_1$.

The joint *p.d.f.* of X_1 and X_2 is given by :

$$f(x_1, x_2) = f_1(x_1) \cdot f(x_2 \mid x_1) \quad \Rightarrow \quad f(x_2 \mid x_1) = \frac{f(x_1, x_2)}{f_1(x_1)}$$

where $f_1(.)$ is marginal *p.d.f.* of X_1.

Substituting in (*),

$$E(X_2 \mid X_1 = x_1) = \int_{x_2} \left\{ \frac{x_2\, f(x_1, x_2)}{f_1(x_1)} \right\} dx_2, \qquad \text{... (**)}$$

Unconditional mean of X_2 is given by :

$$E(X_2) = \int_{x_2} x_2\, f_2(x_2)\, dx_2, \text{ where } f_2(.) \text{ is marginal } p.d.f. \text{ of } X_2. \qquad \text{... (***)}$$

From (**) and (***), we conclude that the conditional mean of X_2 (given X_1) will coincide with unconditional mean of X_2 if only if

$$\frac{f(x_1, x_2)}{f_1(x_1)} = f_2(x_2) \quad \Rightarrow \quad f(x_1, x_2) = f_1(x_1) \cdot f_2(x_2)$$

i.e., if X_1 and X_2 are (stochastically) independent.

(b) $$f(x_1, x_2) = \begin{cases} 21\, x_1^2\, x_2^3\ ; 0 < x_1 < x_2 < 1 \\ 0, \text{otherwise} \end{cases}$$

Marginal *p.d.f.* of X_2 is given by :

$$f_2(x_2) = \int_0^{x_2} f(x_1, x_2)\, dx_1 = 21 x_2^3 \int_0^{x_2} x_1^2\, dx_1 = 21 x_2^3 \left| \frac{x_1^3}{3} \right|_0^{x_2} = 7x_2^6\ ; 0 < x_2 < 1$$

∴ Conditional *p.d.f.* of X_1 (given X_2) is given by :

$$f_1(x_1 \mid x_2) = \frac{f(x_1, x_2)}{f_2(x_2)} = 3.\frac{x_1^2}{x_2^3}\ ;\ 0 < x_1 < x_2\ ;\ 0 < x_2 < 1$$

Conditional mean of X_1 is :

$$E(X_1 \mid X_2 = x_2) = \int_0^{x_2} x_1 f_1(x_1 \mid x_2)\, dx_1 = \frac{3}{x_2^3} \int_0^{x_2} x_1^3\, dx_1 = \frac{3}{x_2^3} \left| \frac{x_1^4}{4} \right|_0^{x_2} = \frac{3x_2}{4}$$

$$E(X_1^2 \mid X_2 = x_2) = \int_0^{x_2} x_1^2 f_1(x_1 \mid x_2)\, dx_1 = \frac{3}{x_2^3} \int_0^{x_2} x_1^4\, dx_1 = \frac{3}{x_2^3} \cdot \frac{x_2^5}{5} = \frac{3}{5} x_2^2$$

$$\therefore \quad \text{Var}(X_1 \mid X_2 = x_2) = E(X_1^2 \mid X_2 = x_2) - [E(X_1 \mid X_2 = x_2)]^2$$

$$= \frac{3}{5} x_2^2 - \frac{9}{16} x_2^2 = \frac{3}{80} x_2^2\ ;\ 0 < x_2 < 1.$$

Example 6·36. *X and Y have joint p.d.f.*

$$f(x, y) = \begin{cases} 2\ , & 0 < x < y < 1 \\ 0\ , & \textit{otherwise} \end{cases}$$

(a) *Show that* : $E(X \mid Y = y) = \frac{1}{2} y$ *and* $Var(X \mid Y = y) = \frac{1}{12} y^2$

(b) *Verify that* : $E[E(X \mid Y)] = E(X)$

Solution. $0 < x < y < 1 \quad \Rightarrow \quad \{0 < x < 1,\ \ x < y < 1 \text{ or } 0 < y < 1\ ,\ \ 0 < x < y\}$

The marginal *p.d.f.*'s of X and Y are given by :

$$f_X(x) = \int_x^1 f(x, y)\, dy = \int_x^1 2\, dy = 2\,|\,y\,|_x^1 = 2(1 - x)\ ;\ 0 < x < 1.$$

$$f_Y(y) = \int_0^y f(x, y)\, dx = \int_0^y 2 \cdot dx = 2\,|\,x\,|_0^y = 2y\ ;\ 0 < y < 1.$$

$$\therefore \quad f_{X|Y}(x \mid y) = \frac{f(x, y)}{f_Y(y)} = \frac{2}{2y} = \frac{1}{y}\ ;\ 0 < x < y.$$

(a) $$E(X \mid Y = y) = \int_{x=0}^{y} x\, f(x \mid y)\, dx = \int_{x=0}^{y} x \cdot \frac{1}{y}\, dx = \frac{1}{y} \cdot \left| \frac{x^2}{2} \right|_0^y = \frac{1}{2} y \quad \text{... (*)}$$

and $E(X^2 \mid Y = y) = \int_{x=0}^{y} x^2 f(x \mid y)\, dx = \int_{x=0}^{y} \frac{x^2}{y}\, dx = \frac{1}{y} \cdot \left| \frac{x^3}{3} \right|_0^y = \frac{y^2}{3}$

$$\therefore \quad \text{Var}(X \mid Y = y) = E(X^2 \mid Y = y) - [E(X \mid Y = y)]^2 = \frac{y^2}{3} - \left(\frac{y}{2}\right)^2 = \frac{1}{12} y^2$$

(b) $E[E(X \mid Y)] = E_Y[E_X(X \mid Y)] = E_Y\left(\frac{1}{2} Y\right),$ [From (*)]

$$= \int_0^1 \frac{1}{2} y \cdot f_Y(y)\, dy = \int_0^1 \frac{1}{2} y \cdot 2y\, dy = \left| \frac{y^3}{3} \right|_0^1 = \frac{1}{3}$$

$$E(X) = \int_0^1 x\, f_X(x)\, dx = \int_0^1 x \cdot 2(1-x)\, dx = 2 \left| \frac{x^2}{2} - \frac{x^3}{3} \right|_0^1 = 2 \cdot \left(\frac{1}{2} - \frac{1}{3}\right) = \frac{1}{3}.$$

$$\therefore \quad E[E(X \mid Y)] = E(X) = \frac{1}{3}.$$

SUPPLEMENTARY EXAMPLES ON CONDITIONAL EXPECTATION

Example 6·37. *Let* $f(x, y) = 8xy,\ 0 < x < y < 1\ ;\ f(x, y) = 0$ *elsewhere.*

Find

(a) $E(Y \mid X = x)$,

(b) $E(XY \mid X = x)$,

(c) $\text{Var}(Y \mid X = x)$.

Solution.

$$f_X(x) = \int_{-\infty}^{\infty} f(x, y)\, dy = 8x \int_x^1 y\, dy = 4x(1 - x^2), 0 < x < 1$$

$$f_Y(y) = \int_{-\infty}^{\infty} f(x, y)\, dx = 8y \int_0^y x\, dx = 4y^3,\ 0 < y < 1$$

$$f_{X \mid Y}(x \mid y) = \frac{f(x, y)}{f_Y(y)} = \frac{2x}{y^2}; \quad f_{Y \mid X}(y \mid x) = \frac{2y}{1 - x^2};\ 0 < x < y < 1.$$

(a) $$E(Y \mid X = x) = \int_x^1 y \left(\frac{2y}{1 - x^2}\right) dy = \frac{2}{3}\left(\frac{1 - x^3}{1 - x^2}\right) = \frac{2}{3}\left(\frac{1 + x + x^2}{1 + x}\right)$$

(b) $$E(XY \mid X = x) = x\, E(Y \mid X = x) = \frac{2}{3} \times \frac{x(1 + x + x^2)}{(1 + x)}$$

(c) $$E(Y^2 \mid X = x) = \int_x^1 y^2 \left(\frac{2y}{1 - x^2}\right) dy = \frac{1}{2}\left(\frac{1 - x^4}{1 - x^2}\right) = \frac{1 + x^2}{2}$$

$$\text{Var}(Y \mid X = x) = E(Y^2 \mid X = x) - \{E(Y \mid X = x)\}^2 = \frac{1 + x^2}{2} - \frac{4}{9} \cdot \frac{(1 + x + x^2)^2}{(1 + x)^2}$$

Example 6·38. *Let* $f(x, y) = (x + y)\, I_{(0,\,1)}(x)\, I_{(0,\,1)}(y)$. *Find*

(i) $f_{Y|X}(y \mid x)$,

(ii) $E[Y \mid X = x]$,

(iii) $Var[Y \mid X = x]$.

Solution. Here, the joint *p.d.f.* of X and Y is :

$$f(x, y) = \begin{cases} x + y \; ; & 0 < x < 1,\; 0 < y < 1 \\ 0 \; ; & \text{otherwise.} \end{cases}$$

(i) The marginal distribution of X is :

$$f_X(x) = \int_{-\infty}^{\infty} f(x, y)\, dy = \int_0^1 (x + y)\, dy = \left| xy + \frac{1}{2} y^2 \right|_{y=0}^{1} = x + \frac{1}{2}$$

$$\Rightarrow \qquad f_X(x) = x + \frac{1}{2},\; 0 < x < 1$$

Similarly, by symmetry,

$$f_Y(y) = y + \frac{1}{2},\; 0 < y < 1$$

The conditional distribution of Y given X = x, (0 < x < 1) is :

$$f_{Y \mid X}(y \mid x) = \frac{f(x, y)}{f_X(x)} = \frac{x + y}{x + \frac{1}{2}} \; ; \; 0 < y < 1.$$

(ii) For 0 < x < 1, we have

$$E[Y \mid X = x] = \int_0^1 y\, f_{Y|X}(y \mid x)\, dy = \int_0^1 y \left(\frac{x + y}{x + \frac{1}{2}} \right) dy = \frac{1}{\left(x + \frac{1}{2} \right)} \left| x . \frac{1}{2} y^2 + \frac{1}{3} y^3 \right|_{y=0}^{1}$$

$$= \frac{2}{2x + 1} \left(\frac{1}{2} x + \frac{1}{3} \right) = \frac{3x + 2}{3(2x + 1)} ; 0 < x < 1.$$

$$E[Y^2 \mid X = x] = \int_0^1 y^2 f_{Y|X}(y \mid x)\, dy = \int_0^1 y^2 \left(\frac{x + y}{x + \frac{1}{2}} \right) dy$$

$$= \frac{1}{\left(x + \frac{1}{2} \right)} \left(\frac{1}{3} x + \frac{1}{4} \right) = \frac{4x + 3}{6(2x + 1)} ; \; 0 < x < 1.$$

(iii) $$\text{Var}[Y \mid X = x] = E[Y^2 \mid X = x] - [E(Y \mid X) = x]^2 = \frac{4x + 3}{6(2x + 1)} - \frac{(3x + 2)^2}{9(2x + 1)^2}$$

$$= \frac{3(4x + 3)(2x + 1) - 2(3x + 2)^2}{18(2x + 1)^2} = \frac{6x^2 + 6x + 1}{18(2x + 1)^2} ; \; 0 < x < 1.$$

Example 6·39. *Let the two-dimensional random variable (X, Y) have the joint density :*

$$f(x, y) = \frac{1}{8}(6 - x - y)\, I_{(0,\,2)}(x) \cdot I_{(2,\,4)}(y)$$
$$= 0, \text{ otherwise}$$

(a) Find $E[Y \mid X = x]$ *(b) Find* $E[Y^2 \mid X = x]$ *(c) Find* $Var[Y \mid X = x]$
(d) Show that $E[Y] = E[E(Y \mid X)]$ *(e) Find* $E[XY \mid X = x]$.

Solution. Here $f(x, y) = \frac{1}{8}(6 - x - y)$, for $0 < x < 2,\ 2 < y < 4$
$= 0$, otherwise.

The marginal distribution of X is given by :

$$f_X(x) = \int_{-\infty}^{\infty} f(x, y)\, dy = \int_{y=2}^{4} \frac{1}{8}(6 - x - y)\, dy = \frac{1}{8}\left|\left(6y - xy - \frac{y^2}{2}\right)\right|_{y=0}^{4}$$

$$\therefore \quad f_X(x) = \frac{1}{8}(24 - 4x - 8 - 12 + 2x + 2) = \frac{1}{4}(3 - x),\ 0 < x < 2$$

Similarly, $$f_Y(y) = \int_{x=0}^{2} \frac{1}{8}(6 - x - y)\, dx = \frac{1}{4}(5 - y),\ 2 < y < 4$$

The conditional distribution of Y for $X = x$ is given by :

$$f_{Y \mid X}(y \mid x) = \frac{f(x, y)}{f_X(x)} = \frac{\frac{1}{8}(6 - x - y)}{\frac{1}{4}(3 - x)} = \frac{6 - x - y}{2(3 - x)}\ ;\ 2 < y < 4,\ 0 < x < 2$$

(*a*) $$E(Y \mid X = x) = \int_2^4 y\, f_{Y \mid X}(y \mid x)\, dy = \frac{1}{2(3 - x)} \int_2^4 y(6 - x - y)\, dy$$

$$= \frac{1}{2(3 - x)}\left[3y^2 - \frac{xy^2}{2} - \frac{y^3}{3}\right]_{y=2}^{4} = \frac{26 - 9x}{3(3 - x)},\ \text{for } 0 < x < 2$$

(*b*) $$E(Y^2 \mid X = x) = \int_2^4 y^2 f_{Y \mid X}(y \mid x)\, dy = \frac{1}{2(3 - x)} \int_2^4 y^2 (6 - x - y)\, dy$$

$$= \frac{78 - 28x}{3(3 - x)},\ \text{for } 0 < x < 2$$

(*c*) $$Var[Y \mid X = x] = E[Y^2 \mid X = x] - [E(Y \mid X = x)]^2$$

$$= \frac{78 - 28x}{3(3 - x)} - \frac{(26 - 9x)^2}{9(3 - x)^2} = \frac{\left[6(39 - 14x)(3 - x) - (26 - 9x)^2\right]}{9(3 - x)^2}$$

$$\therefore \text{Var}[Y \mid X = x] = \frac{3x^2 - 18x + 26}{9(3 - x)^2},\ 0 < x < 2$$

(*d*) We have

$$E[E(Y \mid X)] = \int_{x=0}^{2} E\left[Y \mid X = x\right] f_X(x)\, dx = \int_0^2 \frac{26 - 9x}{3(3 - x)} \cdot \frac{1}{4}(3 - x)\, dx$$

$$= \frac{1}{12}\int_0^2 (26 - 9x)\, dx = \frac{1}{12}\left(26x - \frac{9}{2}x^2\right)\Big|_0^2 = \frac{1}{12}(52 - 18) = \frac{17}{6}.$$

Also $$E[Y] = \int_2^4 y\, f_Y(y)\, dy = \int_2^4 y \cdot \frac{1}{4}(5-y)\, dy = \frac{1}{4}\left[\left|\frac{5}{2}y^2 - \frac{y^3}{3}\right|_2^4\right]$$

$$= \frac{1}{4}\left[40 - \frac{64}{3} - 10 + \frac{8}{3}\right] = \frac{17}{6}$$

Hence, $$E[Y] = E[E(Y \mid X)]$$

(e) $$E[XY \mid X = x] = x\, E[Y \mid X = x] = x \cdot \frac{(26-9x)}{3(3-x)},\ 0 < x < 2.$$

CHAPTER CONCEPTS QUIZ

1. Fill in the blanks :

(i) Expected value of a random variable X exists if —

(ii) If $E(X^r)$ exists then $E(X^s)$ also exists for ....

(iii) If X is a random variable, expectation of $(X - \text{constant})^2$ is minimum when the constant is ...

(iv) $E \mid X - A \mid$ is minimum when A is ...

(v) Var (c) = ..., where c is a constant.

(vi) Var (X + c) = ..., where c is a constant.

(vii) Var (aX + b) = ..., where a and b are constants.

(viii) If X is a r.v. with mean m and variance σ^2, then $E\left(\frac{X-\mu}{\sigma}\right)$ = ..., $\text{Var}\left(\frac{X-\mu}{\sigma}\right)$ = ...

(ix) $\{E(XY)\}^2$ ... $E(X^2) \cdot E(Y^2)$. (Inequality relationship)

(x) $V(aX \pm bY)$ = ..., where a and b are constants.

(xi) If $f(x, y) = 4xy$, for $0 < x < 1$, $0 < y < 1$, then

(a) $E(Y \mid x)$ = ...   (b) $V(Y \mid x)$ = ...

(xii) $P(X > Y) = 1$ implies that $E(X)$ ... $E(Y)$.

(xiii) If X, Y and Z are three random variables, the covariance between X and Y for a given value of Z, i.e., cov $(X, Y \mid Z)$ = ...

2. Mark the correct answer in the following :

(i) For two random variables X and Y, the relation $E(XY) = E(X)\, E(Y)$, holds good

(a) if X and Y are statistically independent,

(b) for all X and Y,

(c) if X and Y are identical.

(ii) If Var (X) = 1, then Var (2X ± 3) is

(a) 5   (b) 13   (c) 4

(iii) $E(X - k)^2$ is minimum when

(a) $k < E(X)$,   (b) $k > E(X)$,   (c) $k = E(X)$.

3. Comment on the following :

If X and Y are mutually independent variables, then

(i) $E(XY + Y + 1) - E(X + 1)\, E(Y) = 0$

(ii) X and Y are independent if and only if Cov $(X, Y) = 0$

(iii) For every univariable distribution :

(a) $V(cX) = c^2 V(X)$ (b) $E(c / X) = c / E(X)$

(iv) Expected value of a *r.v.* always exists.

(v) If X and Y are two random variables such that their expectations exist and $P(X \geq Y) = 1$, then

(a) $E(X) \leq E(Y)$, (b) $E(X) \geq E(Y)$

(c) $E(X) = E(Y)$, (d) None of the above

4. Mark true or false with reasons for your answers :

(i) Cov $(X, Y) = 0 \Rightarrow X$ and Y are dependent.

(ii) If Var $(X) >$ Var (Y), then $X + Y$ and $X - Y$ are independent.

(iii) If Var $(X) =$ Var (Y) and if $2X + Y$ and $X - Y$ are independent, then X and Y are independent.

(iv) If Cov $(aX + bY, bX + aY) \neq ab$ Var $(X + Y)$, then X and Y are dependent.

DISCUSSION AND REVIEW QUESTIONS

1. (a) Define a random variable and its mathematical expectation.

(b) Show that the mathematical expectation of the sum of two random variables is the sum of their individual expectations and if two variables are independent, the mathematical expectation of their product is the product of their expectations.

(c) If X is a random variable, prove that $| E(X) | \leq E(| X |)$.

(d) If X and Y are two random variables such that $X \leq Y$, prove that $E(X) \leq E(Y)$.

(e) Prove that $E[(X - c)^2] = [\text{var}(X)] + [E(X) - c]^2$, where c is a constant.

2. Prove that

(a) $E(aX + bY) = a E(X) + b E(Y)$, where a and b are any constants.

(b) $E(a) = a$, a being a constant.

(c) $E[a\, g(X)] = a\, E[g(X)]$

(d) $E[g_1(X) + g_2(X) + ... + g_n(X)] = E\{g_1(X)\} + E\{g_2(X)\} + ... + E\{g_n(X)\}$

(e) $| E\{g(X)\} | \leq E\{| g(X) |\}$.

(f) If $g(X) \geq 0$ everywhere, then $E[g(X)] \geq 0$.

(g) If $g(X) \geq 0$ everywhere and $E[g(X)] = 0$, then $g(X) = 0$, *i.e.*, the random variable $g(X)$ has a one point distribution at $X = 0$.

3. Show that if X is non-negative random variable such that both $E(X)$ and $E(1/X)$ exist, then

$$E(1/X) \geq 1/E(X).$$

4. (a) If X and Y are two independent random variables, show that

$$\text{Var}(aX + bY) = a^2 \text{Var}(X) + b^2 \text{Var}(Y).$$

(b) With usual notations, show that

$$\text{Cov}(aX + bY, cX + dY) = a\,c\, \text{Var}(X) + b\,d\, \text{Var}(Y) + (a\,d + b\,c)\, \text{Cov}(X, Y)$$

(c) Show that

$$\text{Cov}\left(\sum_{i=1}^{n} a_i X_i, \sum_{j=1}^{n} b_j X_j\right) = \sum_{j=1}^{n} \sum_{i=1}^{n} a_i b_j \text{Cov}(X_i, X_j)$$

5. (a) Define the indicator function $I_A(X)$ and show that $E[I_A(X)] = P(A)$.

(b) Prove that the probability function $P(X \in A)$ for set A and the distribution function $F_X(x)$, $(-\infty < x < \infty)$, can be regarded as expectations of some random variable.

ASSORTED REVIEW PROBLEMS FOR SELE-ASSESSMENT

6·1. Given the following table :

x	– 3	– 2	– 1	0	1	2	3
$p(x)$	0·05	0·10	0·30	0	0·30	0·15	0·10

Compute : (*i*) $E(X)$, (*ii*) $E(2X \pm 3)$, (*iii*) $E(4X + 5)$, (*iv*) $E(X^2)$, (*v*) $V(X)$, and (*vi*) $V(2X \pm 3)$.

6·2. (*a*) *A* and *B* throw with one dice for a stake of ₹ 44,000 which is to be won by the player who first throws a 6. If *A* has the first throw, what are their respective expectations ?

Ans. (*a*) ₹ 24,000; ₹ 20,000.

(*b*) A contractor has to choose between two jobs. The first promises a profit of ₹ 1,20,000 with a probability of $\frac{3}{4}$ or a loss of ₹ 30,000 due to delays with a probability of $\frac{1}{4}$; the second promises a profit of ₹ 1,80,000 with a probability of $\frac{1}{2}$ or a loss of ₹ 45,000 with a probability of $\frac{1}{2}$. Which job should the contractor choose so as to maximise his expected profit ?

(*c*) The monthly demand for transistors is known to have the following probability distribution :

Demand :	1	2	3	4	5	6
Probability :	0·10	0·15	0·20	0·25	0·18	0·12

Determine the expected demand for transistors. Also obtain the variance. Suppose that the cost (C) of producing (*n*) transistors is given by the rule C = 10,000 + 500 *n*. Determine the expected cost.

Hint. $E(C) = E\{10,000 + 500n\} = 10,000 + 500\,E(n)$

(*d*) The ₹ 100 fully paid up share of *XYZ* Co. is currently selling at ₹ 720. Possible dividend payments and the corresponding market prices at the end of the year are shown below :

Dividend per share (₹)	*Probability*	*Market price ex-dividend at the end of the year* (₹)
10	0·1	760
20	0·5	840
30	0·3	900
40	0·1	1000

Compute the expected value of the possible returns (dividend appreciation in share value.)

Hint. $E(X) = (760 - 720 + 10)\ 0{\cdot}1 + 140 \times 0{\cdot}5 + 210 \times 0{\cdot}3 + 320 \times 0{\cdot}1.$

6·3. (*a*) A company, which operates a chain of lunch canteens, plans to install a new unit in either of two locations. The company feels that the probability of a unit being successful in location *X* is 3/4 and that, if it is successful it will make annual profit of ₹ 4,00,000. If it is not successful, the company will lose ₹ 1,00,000 per year. The probability of a unit making success in location *Y* is only 1/2, but if it does succeed, the annual profit will be ₹ 6,00,000. If it does not succeed in location *Y*, the annual loss will be ₹ 1,20,000.

Where should the company locate the new unit so as to maximize its expected gain ?

(*b*) A building contractor is considering bidding on one of two contracts for new downtown buildings *A* and *B*. It has been estimated that a profit of ₹ 2,00,000 would be made on building *A*. Bidding costs for the contractor on building *A* would be ₹ 10,000. On building *B*, the estimated profit is ₹ 5,00,000 and bidding costs would be ₹ 20,000. The probability of being awarded the contract is $\frac{2}{5}$ on building *A* and $\frac{1}{5}$ on building *B*. (Assume bidding costs are incurred only in case the contract is not obtained.)

(*i*) What is the contractor's expectation for building *A* ? Building B ?

(*ii*) For which job should the contractor bid ?

Hint. (*i*) Contractor's expectation for building *A* : $E(X) = ₹\left\{2,00,000 \times \frac{2}{5} + (-10,000) \times \frac{3}{5}\right\}$

Contractor's expectation for building *B* : $E(Y) = ₹\left\{5,00,000 \times \frac{1}{5} + (-20,000) \times \frac{4}{5}\right\}$

6·4. (*a*) Three tickets are chosen at random without replacement from 100 tickets numbered 1, 2, 3,...., 100. Find the mathematical expectation of the sum of the numbers on the tickets drawn.

(*b*) Three urns contain respectively 3 green and 2 white balls, 5 green and 6 white balls and 2 green and 4 white balls. One ball is drawn from each urn. Find the expected number of white balls drawn out.

Hint. Let $X_i = \begin{cases} 1 \text{ , if the ball drawn from } i\text{th urn is white} \\ 0 \text{ , otherwise} \end{cases}$

Then, the number of white balls drawn is : $S = X_1 + X_2 + X_3$.

$$E(S) = E(X_1) + E(X_2) + E(X_3) = 1 \times \frac{2}{5} + 1 \times \frac{6}{11} + 1 \times \frac{4}{6} = \frac{266}{165}.$$

(*c*) An urn contains 7 white and 3 red balls. Two balls are drawn together at random from this urn. Compute the probability that neither of them is white. Find also the probability of getting one white and one red ball. Hence, compute the expected number of white balls drawn.

Hint. X : Number of white balls drawn.

$$P(X = 0) = \frac{{}^3C_2}{{}^{10}C_2} = \frac{1}{15}\ ;\ P(X=1) = \frac{{}^7C_1 \times {}^3C_1}{{}^{10}C_2} = \frac{7}{15}\ ;\ P(X=2) = \frac{{}^7C_2}{{}^{10}C_2} = \frac{7}{15}$$

Ans. $$E(X) = 0 \times \frac{1}{15} + 1 \times \frac{7}{15} + 2 \times \frac{7}{15} = \frac{21}{15} = \frac{7}{5}$$

6·5. (*a*) A random variable X can assume any positive integral value *n* with a probability proportional to $1/3^n$. Find the expectation of X.

(*b*) A coin is tossed until a tail appears. What is the expectation of the number of tosses ?

(*c*) Find the expectation of (*i*) the sum, and (*ii*) the product, of number of points on *n* dice when thrown.

Ans. (*a*) (3/2) ; (*b*) 2 ; (*c*) (*i*) $(7n/2)$, (*ii*) $(7/2)^n$.

6·6 (*a*) A person draws cards one by one from a pack of cards until he draws all the aces. Find the expected value of the number of cards drawn.

Hint. X : No. of cards drawn to obtain all aces.

$$P(X = r) = \left[\frac{{}^4C_3 \times {}^{48}C_{r-4}}{{}^{52}C_{r-1}}\right] \cdot \frac{1}{52 - (r - 1)}$$

$$E(X) = \sum_{r=4}^{52} r\,P(X = r) = \frac{4}{49 \times 50 \times 51 \times 52}\left[\sum_{r=4}^{52} r^4 - 6\sum_{r=4}^{52} r^3 + 11\sum_{r=4}^{52} r^2 - 6\sum_{r=4}^{52} r\right]$$

(*b*) An urn contains n cards marked from 1 to n. Two cards are drawn at a time. Find the mathematical expectation of the product of the numbers on the cards.

Ans. $(n + 1)(3n + 2)/12$

(*c*) In a lottery, m tickets are drawn out of n tickets numbered from 1 to n. What is the expectation of the sum of the squares of numbers drawn ?

Ans. $m(n + 1)(2n + 1)/6$

(*d*) A bag contains n white and 2 black balls. Balls are drawn one by one without replacement until a black ball is drawn. If 0, 1, 2, 3, ..., white balls are drawn before the first black ball, a man is to receive $0^2, 1^2, 2^2, 3^2$, ..., rupees respectively. Find his expectation.

Ans. $[n(n + 1)/6]$

(*e*) Find the expectation and the variance of the number of successes in a series of independent trials, the probability of success in the ith trial being p_i, $(i = 1, 2, ..., n)$.

Hint. Define : $X_i = \begin{cases} 1, \text{ if the } i\text{th trial results in a success} \\ 0, \text{ if the } i\text{th trial results in a failure} \end{cases}$

Then, $$S = \sum_{i=1}^{n} X_i, \text{ gives the number of successes in } n \text{ trials.}$$

$$P(X_i = 1) = p_i \ ; \qquad P(X_i = 0) = 1 - p_i = q_i \ ; \qquad i = 1, 2, ..., n$$

Ans. $$E(S) = \sum_{i=1}^{n} E(X_i) = \sum_{i=1}^{n} p_i \ ; \qquad \text{Var}(S) = \sum_{i=1}^{n} V(X_i) = \sum_{i=1}^{n} p_i q_i$$

Note. If the probability of success is constant (p) for each trial, then the mean and variance of number of successes in n trials is np and $npq = np(1 - p)$, respectively.

(*f*) In a multiple choice question (MCQ) test paper containing n questions, a student gets 4 marks for correct answer and (–1) mark for wrong answer. What should be the least number of options (choices) given for each question so that student's expected score is minimum provided he answers the questions at random.

Hint. Let each question have 'r' choices. Define the *r.v.* X_i so that :

$$X_i = \begin{cases} 1, \text{ if } i\text{th question is answered correctly} \\ 0, \text{ if } i\text{th question is answered wrongly} \end{cases}$$

Then, if student answers questions at random,

$$E(X_i) = 1 \times \frac{1}{r} + 0 \times \left(1 - \frac{1}{r}\right) = \frac{1}{r}$$

$S = \sum_{i=1}^{n} X_i$, is the number of questions answered correctly.

Student's total score 'T' is given by :

$$T = 4 \times S + (-1)(n - S) = 5S - n = 5\sum_{i=1}^{n} X_i - n$$

$$\therefore \qquad E(T) = 5\sum_{i=1}^{n} E(X_i) - n = 5\sum_{i=1}^{n}\left(\frac{1}{r}\right) - n = n\left(\frac{5}{r} - 1\right)$$

We want r if $E(T)$ is minimum.

$$E(T) = 0 \quad \Rightarrow \quad \frac{5}{r} - 1 = 0 \quad \Rightarrow \quad 5 = r.$$

$\therefore$ Required number of choices for each question is 5.

6·7. Balls are taken one by one out of an urn containing w white and b black balls until the first white ball is drawn. Prove that the expectation of number of black balls preceding the first white ball is $b/(w+1)$.

6·8. (*a*) X and Y are independent variables with means 10 and 20, and variances 2 and 3 respectively. Find the variance of $3X + 4Y$.

(*b*) Obtain the variance of $Y = 2X_1 + 3X_2 + 4X_3$, where X_1, X_2 and X_3 are three random variables with means given by 3, 4, 5 respectively; variances by 10, 20, 30 respectively, and co-variances by $\sigma_{X_1 X_2} = 0$, $\sigma_{X_2 X_3} = 0$, $\sigma_{X_1 X_3} = 5$, where $\sigma_{X_r X_s}$ stands for the co-variance of X_r and X_s.

(*c*) Suppose that X is a random variable for which $E(X) = 10$ and $\text{Var}(X) = 25$. Find the positive values of a and b such that $Y = aX - b$, has expectation 0 and variance 1.

(*d*) Let X_1 and X_2 be two stochastic random variables having variances k and 2 respectively. If the variance of $Y = 3X_2 - X_1$ is 25, find k.

Ans. (*a*) 66, (*b*) 710, (*c*) $a = 1/5$, $b = 2$, (*d*) $k = 7$.

6·9. Show that the variance of the weighted mean $\left[\sum_{i=1}^{n} w_i x_i \Big/ \sum_{i=1}^{n} w_i\right]$ of n independent *r.v.*'s x_i ; $i = 1, 2, \ldots, n$ is minimum when the weights are inversely proportional to the corresponding variances. Also show that the minimum variance is (H/n), where H is the harmonic mean of the variances $\sigma_i^2 = \text{Var}(x_i)$, $i = 1, 2, \ldots, n$.

6·10. For any constants a and b, show that the variates X and Y are independent iff,

$$E[u(a - X) \,.\, u(b - Y)] = E[u(a - X)] \,.\, E[u(b - Y)]$$

where $u(t)$ is a unit step function defined by :

$$u(t) = \begin{cases} 1, & \text{if } t \geq 0 \\ 0, & \text{if } t < 0. \end{cases}$$

Hint. $u(a - X) = 1$ if $a - X \geq 0 \quad \Rightarrow \quad X \leq a.$

$$\therefore \quad E[u(a - X)] = \int_{-\infty}^{\infty} u(a - x) f(x)\, dx = \int_{-\infty}^{a} f(x)\, dx = F_X(a)$$

Similarly, $E[u(b - Y)] = F_Y(b)$ and $E[u(a - X) \,.\, u(b - Y)] = F_{XY}(a, b)$.

6·11. A bag contains $2n$ counters, of which half are marked with odd numbers and half with even numbers, the sum of all the numbers being S. A man is to draw two counters. If the sum of the numbers drawn is odd, he is to receive that number of rupees; if even he is to pay that number of rupees. Show that his expectation is $S/\{n(2n - 1)\}$ rupees.

6·12. A jar has n chips numbered 1, 2, ..., n. A person draws a chip, returns it, draws another, returns it, and so on, until a chip is drawn that has been drawn before. Let X be the number of drawings. Find $E(X)$.

Hint. Obviously $P(X > 1) = 1$, because we must have at least two draws to get the chip which has been drawn before.

$$P(X > r) = P[\text{Distinct number on } i\text{th draw} ;\ i = 1, 2, \ldots, r]$$

$$= \frac{n}{n} \times \frac{n-1}{n} \times \frac{n-2}{n} \times \ldots \times \frac{n-(r-1)}{n}$$

$$\Rightarrow \quad P(X > r) = \left(1-\frac{1}{n}\right)\left(1-\frac{2}{n}\right) \cdots \left(1-\frac{r-1}{n}\right); \ r = 1, 2, 3, \ldots \qquad \ldots(*)$$

Hence,
$$E(X) = \sum_{r=1}^{\infty} P(X>r) = P(X>1)+P(X>2)+P(X>3)+\ldots$$

$$= 1+\left(1-\frac{1}{n}\right)+\left(1-\frac{1}{n}\right)\left(1-\frac{2}{n}\right)+\ldots+\left(1-\frac{1}{n}\right)\left(1-\frac{2}{n}\right)\ldots\left(1-\frac{r-1}{n}\right)+\ldots \quad \text{[Using (*)]}$$

6·13. A coin is tossed four times. Let X denote the number of times a head is followed immediately by a tail. Find the distribution, mean and variance of X.

Hint.
$$S = \{H, T\} \times \{H, T\} \times \{H, T\} \times \{H, T\}$$
$$= [HHHH, \ HHHT, \ HHTH, \ HTHH, \ HTHT, \ \ldots\ldots, \ TTTT]$$
$$X: \quad 0, \quad 1, \quad 1, \quad 1, \quad 2, \quad \ldots \quad 0$$

x :	0	1	2
$p(x)$:	5/16	10/16	1/16

$$E(X) = \frac{3}{4}, \quad E(X^2) = \frac{7}{8}$$
$$\text{Var}(X) = \frac{7}{8} - \frac{9}{16} = \frac{5}{16}.$$

6·14. An urn contains balls numbered 1, 2, 3. First a ball is drawn from the urn and then a fair coin is tossed the number of times shown on the drawn ball. Find the expected number of heads.

Hint. B_j : Event of drawing the ball numbered j, $\quad P(B_j) = \frac{1}{3}$; $j = 1, 2, 3.$

X : Number of heads shown, $\quad X$ is a *r.v.* taking the values 0, 1, 2, and 3.

$$P(X = x) = \sum_{j=1}^{3} P(B_j) \cdot P(X = x \mid B_j) = \frac{1}{3}\sum_{j=1}^{3} P(X = x \mid B_j)$$

$$\therefore \quad P(X=0) = \frac{1}{3}\{P(X=0 \mid B_1)+P(X=0 \mid B_2)+P(X=0 \mid B_3)\} = \frac{1}{3}\left(\frac{1}{2}+\frac{1}{4}+\frac{1}{8}\right) = \frac{7}{24}$$

$$P(X=1) = \frac{1}{3}\{P(X=1 \mid B_1)+P(X=1 \mid B_2)+P(X=1 \mid B_3)\} = \frac{1}{3}\left(\frac{1}{2}+\frac{2}{4}+\frac{3}{8}\right) = \frac{11}{24}$$

e.g.,
$$P(X = 0 \mid B_2) = P(\text{No head when two coins are tossed}) = \frac{1}{4}$$
$$P(X = 1 \mid B_3) = P(\text{1 head when three coins are tossed}) = \frac{3}{8}$$

Similarly,
$$P(X=2) = \frac{1}{3}\left(0+\frac{1}{4}+\frac{3}{8}\right) = \frac{5}{24}; \quad P(X=3) = \frac{1}{3}\left(0+0+\frac{1}{8}\right) = \frac{1}{24}$$

$$\therefore \quad E(X) = \sum_{x=0}^{3} x\,P(X=x) = \frac{11}{24}+\frac{10}{24}+\frac{3}{24} = 1.$$

6·15. (*a*) An urn contains pN white and qN black balls, the total number of balls being N, $p + q = 1$. Balls are drawn one by one (without being returned to the urn) until a certain number n of balls is reached.

Let
$$X_i = \begin{cases} 1, \text{ if the } i\text{th ball drawn is white.} \\ 0, \text{ if the } i\text{th ball drawn is black.} \end{cases}$$

(i) Show that $E(X_i) = p$, $\text{Var}(X_i) = pq$.

(ii) Show that the covariance between X_j and X_k is $-\dfrac{pq}{n-1}$, $(j \neq k)$.

(iii) From (i) and (ii), obtain the variance of $S_n = X_1 + X_2 + \ldots + X_n$.

Ans. (iii) $\text{Var}(S_n) = npq\left(\dfrac{N-n}{N-1}\right)$

(b) In a lottery containing n numbers $(1, 2, 3, \ldots, n)$, m numbers are drawn at a time. Let X_i represent the frequency of a specified number i in N drawings. Prove that

$$E(X_i) = Np, \quad E(X_i - Np)^2 = Npq \quad ; \quad E(X_i - Np)(X_j - Np) = Np(p' - p),\ i \neq j$$

where $p = \dfrac{m}{n}$, $q = 1 - p$ and $p' = \dfrac{m-1}{n-1}$.

6·16. Two similar decks of n distinct cards each are put into random order and are matched against each other. Prove that the probability of having exactly r matches is given by :

$$\frac{1}{r!}\sum_{k=0}^{n-r}\frac{(-1)^k}{k!},\ r = 0, 1, 2, \ldots, n$$

Prove further that the expected number of matches and its variance are equal and are independent of n.

Ans. S = Number of matches; $E(S) = \text{Var}(S) = 1$.

6·17. (a) If X and Y are two independent random variables, such that $E(X) = \lambda_1$, $V(X) = \sigma_1^2$, and $E(Y) = \lambda_2$, $V(Y) = \sigma_2^2$, then prove that $V(XY) = \sigma_1^2\sigma_2^2 + \lambda_1^2\sigma_2^2 + \lambda_2^2\sigma_1^2$.

(b) If X and Y are two independent random variables, show that

$$\frac{V(XY)}{[E(X)]^2[E(Y)]^2} = C_X^2 C_Y^2 + C_X^2 + C_Y^2, \text{ where } C_X = \frac{\sqrt{V(X)}}{E(X)}, \quad C_Y = \frac{\sqrt{V(Y)}}{E(Y)}$$

are the so-called coefficients of variation of X and Y.

6·18. (a) If X, Y are independent random variables with

$E(X) = \alpha$, $E(X^2) = \beta$, and $E(Y^k) = a_k$; $k = 1, 2, 3, 4$, find $E(XY + Y^2)^2$.

(b) Show that : $E(X - \mu)^3 = E(X^3) - 3\mu\sigma^2 - \mu^3$,

where μ and σ^2 are the mean and variance of X respectively.

6·19. Let $a_1, a_2, \ldots, a_n$ be arbitrary real numbers and $A_1, A_2, \ldots, A_n$ be the events. Prove that :

$$\sum_{i=1}^{n}\sum_{j=1}^{n} a_i a_j P(A_i A_j) \geq 0.$$

Hint. Define the indicator function I_j of A_j so that $E(I_j) = P(A_j)$ and $E(I_i I_j) = P(A_i A_j)$.

$$\text{Now use : } E\left(\sum_{i=1}^{n} a_i I_i\right)^2 = E\left[\left(\sum_{i=1}^{n} a_i I_i\right)\left(\sum_{j=1}^{n} a_j I_j\right)\right] \geq 0 \Rightarrow E\left[\sum_{i=1}^{n}\sum_{j=1}^{n} a_i a_j I_i I_j\right] \geq 0.$$

6·20. Two players A and B alternately roll a pair of fair dice. A wins if he gets six points before B gets seven points and B wins if he gets seven points before A gets six points. If A takes the first turn, find the probability that B wins and the expected number of trials for A to win.

6·21. A box contains 2^n tickets among which ${}^nC_i$ tickets bear the numbers i $(i = 0, 1, 2, \ldots, n)$. A group of m tickets is drawn. Let S denote the sum of their numbers. Find the expectation and variance of S.

Ans. $\frac{1}{2}mn, \quad \frac{mn}{4}\left[1-\frac{m-1}{2^n-1}\right]$

6·22. A newspaper boy buys N newspapers at a price of ₹ 'a' each and sells them at a price of ₹ 'b' each ($b > a$). The unsold newspapers are returned at a price ₹ 'c' each ($c > a$). The probability of the boy selling n newspapers is p_n ($n = 0, 1, 2, \ldots$). Find his expected profit. How many newspapers he should buy to maximize his expected profit ?

6·23. In a certain society, an individual pays income tax only if his income 'x' is greater than 'a', the amount of tax being $c\,(x - a)$, where $0 < c < 1$. The distribution of the incomes of individuals liable to tax has density function :

$$f(x) = \begin{cases} \dfrac{k}{x^{\theta+1}} & ; \ x > a,\ \theta > 1 \\ 0 & ; \ \text{otherwise.} \end{cases}$$

(*i*) Show that the average tax paid is $ca/(\theta - 1)$.

(*ii*) It is proposed to increase the tax on each individual whose income x is greater than 'b', ($b > a$) by an amount $c\,(x - b)$. Show that this will increase the average tax paid by 20%, provided 'b' is chosen so that $(b/a)^{\theta-1} = 5$.

Hint. $\displaystyle\int_{-\infty}^{\infty} f(x)\,dx = 1 \quad \Rightarrow \quad k\int_a^{\infty} x^{-\theta-1}\,dx = 1 \quad \Rightarrow \quad k = \theta a^{\theta}$

$$\therefore \quad f(x) = \begin{cases} (\theta a^{\theta} / x^{\theta+1}) & ; \ x > a,\ \theta > 1 \\ 0 & , \ \text{otherwise} \end{cases}$$

Let the *r.v.* $Y = c\,(X - a)$, denote the amount of tax paid. Then average tax paid is :

$$E(Y) = \int_a^{\infty} y \cdot f(x)\,dx = \theta\, a^{\theta} \int_a^{\infty} \frac{c\,(x-a)}{x^{\theta+1}}\,dx = \frac{ca}{\theta - 1}$$

Let the *r.v.* $Z = c\,(X - b)\ ;\ X > b$ denote the increase in tax. Then

$$E(Z) = \int_b^{\infty} z\, f(x)\,dx = \theta\, a^{\theta} \int_b^{\infty} c\,(x-b) \cdot x^{-\theta-1}\,dx = \frac{ca^{\theta}}{(\theta-1)\, b^{\theta-1}}$$

Percentage increase in tax paid is given by :

$$\text{P.I.} = 100 \times \frac{E(Z)}{E(Y)} = 100 \times \frac{c\,a^{\theta}}{(\theta-1)\, b^{\theta-1}} \times \frac{\theta - 1}{c\,a} = 100\left(\frac{a}{b}\right)^{\theta-1}$$

The average increase in tax paid will be 20% if

$$\text{P.I.} = 100\left(\frac{a}{b}\right)^{\theta-1} = 20 \quad \Rightarrow \quad \left(\frac{b}{a}\right)^{\theta-1} = 5.$$

6·24. In the simplest type of weather forecasting —"rain" or "no rain" in the next 24 hours — suppose the probability of raining is p ($> 1/2$), and that a forecaster scores a point if his forecast proves correct and zero otherwise. In making n independent forecasts of this type, a forecaster, who has no genuine ability, predicts "rain" with probability λ and "no rain" with probability $(1 - \lambda)$. Prove that probability of the forecast being correct for any one day is $[1 - p + (2p - 1)\,\lambda]$.

Hence, derive the expectation of the total score (S_n) of the forecaster for n days, and show that this attains its maximum value for $\lambda = 1$. Also, prove that :

$$\text{Var}(S_n) = n\,[\,p - (2p - 1)\,\lambda]\,[1 - p + (2p - 1)\,\lambda]\,,$$

and thereby deduce that, for fixed n, this variance is maximum for $\lambda = (1/2)$.

Hint. P [Raining on any day] $= p\ ;\ (p > 1/2)$

P [Forecaster's forecast is true]

$= P$ [He forecast rain and it rains] + P [He forecast no-rain and it does not rain]

$= \lambda \cdot p + (1 - \lambda)(1 - p)$ ... (*)

Let X_i be a *r.v.* defined as follows :

$$X_i = 1, \text{ if forecast is correct for } i\text{th day}$$
$$= 0, \text{ if forecast is incorrect for } i\text{th day}, \ i = 1, 2, ..., n$$

$\therefore \quad E(X_i) = 1 \cdot P(X_i = 1) + 0 \cdot P(X_i = 0) = P(X_i = 1)$

$= P$ [Forecast is correct for ith day]

$= \lambda p + (1 - \lambda)(1 - p)$ [From (*)]

$= 1 - p - \lambda(1 - p) + \lambda p$

$= 1 - p + (2p - 1)\lambda = \mu$ (say). ... (**)

Forecaster's total score (S_n) for n days is given by :

$$S_n = \sum_{i=1}^{n} X_i \quad \Rightarrow \quad E(S_n) = \sum_{i=1}^{n} E(X_i) = n[1 - p + (2p - 1)\lambda] \qquad \text{... (***)}$$

Since $p > \frac{1}{2}$, $2p - 1 > 0$. Hence $E(S_n)$ is maximum when λ is maximum $\Rightarrow \quad \lambda = 1$

$$\text{Var}(S_n) = \text{Var}\left(\sum_{i=1}^{n} X_i\right) = \sum_{i=1}^{n} \text{Var}(X_i) \quad (\because X_1, X_2, ..., X_n \text{ are independent}) \qquad \text{... (****)}$$

$$\text{Var}(X_i) = E(X_i^2) - [E(X_i)]^2 = [1^2 \cdot P(X_i = 1) + 0^2 \cdot P(X_i = 0)] - \mu^2 = \mu - \mu^2 = \mu(1 - \mu)$$

Substituting in (****), we get

$$\text{Var}(S_n) = n\mu(1 - \mu) = n[1 - p + (2p - 1)\lambda][p - (2p - 1)\lambda]$$
$$= n(a + b\lambda - b\lambda^2) = f(\lambda), \text{ (say); where } b = (2p - 1) > 0, \quad a = p(1 - p).$$

$f'(\lambda) = 0 \quad \Rightarrow \quad \lambda = 1/2 \quad$ and $\quad f''(\lambda) = -2b < 0$

$\Rightarrow$ Var (S_n) is maximum when $\lambda = 1/2$.

6·25. Consider a sequence of Bernoulli trials with a constant probability p of success in a single trial. Let X_k denote the number of failures following the $(k - 1)$th and preceding the kth success and let

$$S_r = \sum_{k=1}^{r} X_k .$$

Derive the probability distribution of X_k. Hence, derive the probability distribution of S_r. Find $E(S_r)$ and Var (S_r).

6·26. Let r tickets be drawn successively with replacements from an urn containing n tickets numbered 1, 2, 3, ..., n. If X denotes the greatest number drawn,

(*i*) Find the probability distribution of X *i.e.*, $P(X = i)$; $i = 1, 2, ..., n$

(*ii*) Find $E(X)$. Also obtain its limiting value for large n.

Ans. (*i*) $\quad P(X = i) = n^{-r}[i^r - (i - 1)^r]$; $\quad i = 1, 2, ..., n.$

(*ii*)
$$E(X) = n - n^{-r} \sum_{i=1}^{n} (i - 1)^r$$
$$\simeq n - n\int_0^1 x^r \, dx = n\left(1 - \frac{1}{r+1}\right) = \frac{nr}{r+1}, \text{ for large } n.$$

6·27. A *r.v.* X assumes possible values $x_1, x_2, ..., x_n$ where $x_i < x_{i+1}$, $1 \le i \le n-1$. Prove that :

$$E(X) = \sum_{i=1}^{n} (x_i - x_{i-1}) \cdot P(X \ge x_i) \; ; \quad x_0 \equiv 0.$$

Another variate Y assumes the same set of values as X and also satisfies the condition :

$$P(X \ge x) \ge P(Y \ge x) \text{ for all } x.$$

Show that $\quad E(X) \ge E(Y).$

Hint. $E(X) = \sum_{i=1}^{n} x_i P(X = x_i) = \sum_{i=1}^{n} x_i \{P(X \ge x_i) - P(X \ge x_{i+1})\}$ (Suggested by the answer).

6·28. (*a*) An urn contains *k* varieties of objects in equal numbers. The objects are drawn one at a time and replaced before the next drawing. Show that the probability that *n* and no less drawings will be required to produce objects of all varieties is $\sum_{r=0}^{k-1} (-1)^r \cdot {}^{k-1}C_r \left(\frac{k-1-r}{k}\right)^{n-1}$.

Hence or otherwise, find the expected number of drawings in a simple form.

(*b*) A box contains *k* varieties of objects, the number of objects of each variety being the same. These objects are drawn one at time and put back before the next drawing. Denoting by *n* the smallest number of drawings which produce objects of all varieties, find *E* (*n*) and *V* (*n*).

Ans. $E(n) = k\left[1 + \frac{1}{2} + \frac{1}{3} + ... + \frac{1}{k}\right]$

$$Var(n) = k^2\left[1 + \frac{1}{2^2} + \frac{1}{3^2} + ... + \frac{1}{k^2}\right] - k\left[1 + \frac{1}{2} + \frac{1}{3} + ... + \frac{1}{n}\right]$$

6·29. The probability of person suffering from diabetes is *p*. Blood sample from *N*[= *nk*] persons are pooled into *k* groups of blood from *n persons* in each. If the pooled sample is negative, no further test is carried out. If the pooled sample is positive, each of the persons is tested separately. Show that the expected number of tests is :

$$N\left(1 - q^n + \frac{1}{n}\right) ; \quad q = 1 - p.$$

Hint. Let X_i denote the number of tests required for the *i*th group (*i* = 1, 2, ..., *k*), of *n* persons each. Then

$$X = X_1 + X_2 + ... + X_k = \sum_{i=1}^{k} X_i, \text{ gives total number of tests to be carried out.}$$

$$X_i = \begin{cases} 1 & , \text{ if the pooled test is negative} \\ (n+1) & , \text{ if the pooled test is positive} \end{cases} ; \quad i = 1, 2, 3, ..., k.$$

$$P(X_i = 1) = P\,[\text{Each of the } n \text{ individuals in the group tests negative}] = (1-p)^n = q^n$$

$$P(X_i = n+1) = P\,[\text{At least one of the } n \text{ individuals in the group tests positive}]$$

$$= 1 - P\,[\text{Each of } n \text{ individuals in the group tests negative}] = 1 - q^n$$

$$\therefore \quad E(X_i) = 1 \cdot q^n + (n+1)(1-q^n) = 1 + n(1-q^n).$$

$$E(X) = \sum_{i=1}^{k} E(X_i) = k\left[1 + n(1-q^n)\right] = \frac{nk}{n} + nk(1-q^n) = N\left[1 - q^n + \frac{1}{n}\right]$$

6·30. An urn contain a white and b black balls. After a ball is drawn, it is to be returned to the urn if it is white, but if it is black, it is to be replaced by a white ball from another urn. Show that the probability of drawing a white ball after the foregoing operation has been repeated x times is :

$$1-\frac{b}{a+b}\left(1-\frac{1}{a+b}\right)^x$$

6·31. There is a lot of N objects from which objects are taken at random one by one with replacement. Prove that the expected value and variance of the least number of drawings needed to get n different objects are respectively given by :

$$N\left(\frac{1}{N}+\frac{1}{N-1}+\ldots+\frac{1}{N-n+1}\right) \text{ and } N\left\{\frac{1}{(N-1)^2}+\frac{2}{(N-2)^2}+\ldots+\frac{n-1}{(N-n+1)^2}\right\}$$

6·32. A large population consists of equal number of individuals of c different types. Individuals are drawn at random one by one until at least one individual of each type has been found, whereupon, sampling ceases. Show that the mean number of individuals in the sample is :

$$c\left(1+\frac{1}{2}+\frac{1}{3}+\ldots+\frac{1}{c}\right)$$

and the variance of the number is : $c^2\left(1+\frac{1}{2^2}+\frac{1}{3^2}+\ldots+\frac{1}{c^2}\right)-c\left(1+\frac{1}{2}+\frac{1}{3}+\ldots+\frac{1}{c}\right).$

6·33. A bag contains a coin of value M and a number of other coins whose aggregate value is m. A person draws one coin at a time till he gets the coin of value M. Prove that the value of his expectation is

$$\left(M+\frac{m}{2}\right).$$

6·34. (*a*) Under a newly proposed motor insurance policy, the premium is ₹ α in the first year. If no claim is made in the first year, the premium is ₹ $\lambda\alpha$ in the second year, where $(0<\lambda<1)$ and λ is fixed. If no claim is made in the first or second years, the premium is ₹ $\lambda^2\cdot\alpha$ in the third year; and, in general, if no claim is made in any of the first r years $(r \geq 1)$, the premium is ₹ $\lambda^r\alpha$ in the $(r+1)$th year.

If in any year a claim is made, the premium in that year is unaffected, but the next year's premium reverts to ₹ α, and this year is then treated as if it were the first year of the insurance for the purpose of calculating further reductions. Assuming that the probability that no claim will arise in any year is constant and equal to q, prove that in the nth year $(n \geq 2)$ of the policy, the probabilities that the premium paid is

₹ $\lambda^{n-1}\alpha$ or ₹ $\lambda^{n-j-1}\alpha$, $(1 \leq j \leq n-1)$, are q^{n-1} and $(1-q)\,q^{n-j-1}$ respectively.

Hence, calculate the expected amount of the premium payable in the nth year and show that if the limiting value of this mean as $n \to \infty$ must always exceed $k\alpha$, $(k>0)$, then: $\lambda > [(k+q-1)/kq]$.

(*b*) In a game of skill a player has probability 1/3, 5/12 and 1/4 of scoring 0, 1 and 2 points respectively at each trial, the game terminating on first realisation of a zero score at a trial. Assuming that the trials are independent, prove that the probability u_n of the player obtaining a total score of n points is :

$$u_n=\frac{3}{13}\left(\frac{3}{4}\right)^n+\frac{4}{39}\left(-\frac{1}{3}\right)^n.$$

Also, prove that the expectation of his total score is (11/4).

Hint. E (Total score) $= \sum_{n=0}^{\infty} n\,u_n$. Now use $\sum_{n=0}^{\infty} n\,a^n = a(1-a)^{-2}$.

6·35. (*Banach's Match-box-Problem*). A certain mathematician always carries two match boxes (initially containing N match-sticks). Each time he wants a match-stick, he selects a box at random. Inevitably a moment comes when he finds a box empty. Show that the probability that there are exactly r match-sticks in one box when the other box is found empty is :

$$^{2N-r}C_N \times \frac{1}{2^{2N-r}}.\text{ Prove also that the number of matches is } {}^{2N}C_N \times \frac{2N+1}{2^{2N}} - 1.$$

6·36. n couples procreate independently with no limits on family size. Births are single and independent and for the ith couple, the probability of a baby is p_i. The sex ratio S is defined as :

$$S = \frac{\text{Mean number of all boys}}{\text{Mean number of all children}}$$

Show that if all couples,

(*i*) Stop procreating on the birth of a boy, then $S = n \Big/ \sum_{i=1}^{n} \frac{1}{p_i}$.

(*ii*) Stop procreating on birth of a girl, then

$$S = 1 - \left(n \Big/ \sum_{i=1}^{n} \frac{1}{q_i} \right), \text{ where } q_i = 1 - p_i.$$

(*iii*) Stop procreating when they have children of both sexes, then

$$S = \left(\sum_{i=1}^{n} \frac{1}{q_i} - \sum_{i=1}^{n} p_i \right) \Bigg/ \left(\sum_{i=1}^{n} \frac{1}{p_i\, q_i} - n \right)$$

6·37. The probability of a shot hitting at a point distant r from the centre c of the target follows the law :

$$p(r) = \frac{2}{\pi(1+r^2)}$$

The target is divided into four regions as follows :

S_1 : Within the circle with centre c and radius $1/\sqrt{3}$.
S_2 : Without S_1 but within the concentric circle of radius 1.
S_3 : Without S_2 but within the concentric circle of radius $\sqrt{3}$.
S_4 : Without S_3.

The scores for hits in the regions S_1, S_2, S_3 and S_4 are 4, 3, 2 and 0 respectively. Let X_1, X_2, X_3 and X_4 be the number of hits registered in the region S_1, S_2, S_3 and S_4 respectively, out of a total of N shots. Find

(*i*) the probability distribution of X_1, X_2, X_3, X_4
(*ii*) the expected score after 5 shots, and
(*iii*) the probability that the score exceeds 15 in 5 shots.

Hint.

(*i*) Let X : No. of hits ; Y : Score

X	Y	*Probability*
X_1	4	1/3
X_2	3	1/6
X_3	2	1/6
X_4	0	1/3

(*ii*) $$E(Y) = \frac{4}{3} + \frac{3}{6} + \frac{2}{6} = \frac{13}{6}$$

Expected score after 5 shots

$$= \sum_{i=1}^{5} E(Y_i) = 5 \times \frac{13}{6} = \frac{65}{6}$$

where Y_i is score of ith hit.

(iii) $P(Y > 15) = P(\Sigma Y_i = 16) + P(\Sigma Y_i = 17) + P(\Sigma Y_i = 18) + P(\Sigma Y_i = 19) + P(\Sigma Y_i = 20)$.

($\because$ In 5 shots maximum score can be 20)

For example $\Sigma Y_i = 17$, if scores in 5 shots are : 4, 4, 4, 3, 2

$$\therefore \quad P(\Sigma Y_i = 17) = \frac{5!}{3!1!1!}\left(\frac{1}{3}\right)^3 \cdot \frac{1}{6} \cdot \frac{1}{6} \text{ and so on.}$$

6·38. (a) A point is chosen at random on a circle of radius a. Show that the expectation of its distance from another fixed point also on the circle is $4a/\pi$.

(b) Two points P and Q are selected at random in a square of side a. Prove that

$$E(|PQ|^2) = \frac{a^2}{3}$$

6·39. (a) State end prove Cauchy's Inequality on Expectation.

(b) For two random variables X and Y, prove that :

$$[E(X+Y)^2]^{1/2} \leq [E(X^2)]^{1/2} + [E(Y^2)]^{1/2}$$

Hint. Squaring both sides and simplifying, we get

$$[E(XY)]^2 \leq E(X^2) \cdot E(Y^2), \text{ which is Cauchy's inequality.}$$

6·40. Show that if X is a random variable such that $P(a \leq X \leq b) = 1$, then $E(X)$ and Var (X) exist, and

$$a \leq E(X) \leq b \quad \text{and} \quad \text{Var}(X) \leq (b-a)^2 / 4.$$

6·41. Let X be a continuous random variable with median m. Minimise $E[|X - b|]$, as a function of b.

Ans. $E|X - b|$ is minimum when b = Median $(X) = m$.

6·42. Let X be a $r.v.$ with distribution function F. Show that

$$E(X) = \int_0^{\infty} [1 - F_X(x)]\,dx - \int_{-\infty}^{0} F_X(x)\,dx.$$

in the sense that, if either side exists, so does the other and the two are equal.

6·43. (a) Show that if Y and Z are independent random values of a variable X, the expected value of $(Y - Z)^2$ is twice the variance of the distribution of X.

Hint.
$$E(Y) = E(Z) = E(X) = \mu, \text{ (say)} \;;\; \sigma_y^2 = \sigma_z^2 = \sigma_x^2 = \sigma^2, \text{ (say)}. \quad \ldots (*)$$
$$E(Y-Z)^2 = E(Y^2) + E(Z^2) - 2E(Y)E(Z) \quad (\because Y, Z \text{ are independent})$$
$$= (\sigma_y^2 + \mu_y^2) + (\sigma_z^2 + \mu_z^2) - 2\mu^2 = 2\sigma_x^2 \quad \text{[From (*)]}$$

(b) If X and Y are symmetric, show that

$$E\left(\frac{X}{X+Y}\right) = \frac{1}{2}$$

Hint.
$$1 = E\left(\frac{X+Y}{X+Y}\right) = E\left(\frac{X}{X+Y}\right) + E\left(\frac{Y}{X+Y}\right)$$
$$\Rightarrow \quad 1 = 2E\left(\frac{X}{X+Y}\right) \quad (\because X \text{ and } Y \text{ are symmetric})$$

(c) If a $r.v.$ X has a symmetric density about the point 'a' and if $E(X)$ exists, then

$$\text{Mean}(X) = \text{Median}(X) = a$$

Hint. Given $f(a - x) = f(a + x)$ where $f(x)$ is $p.d.f.$ of X. ... (*)

$$E(X - a) = \int_{-\infty}^{\infty} (x-a)\,f(x)\,dx = \int_{-\infty}^{a} (x-a)\,f(x)\,dx + \int_a^{\infty} (x-a)\,f(x)\,dx$$

$$= \int_{-\infty}^{0} y\, f(a+y)\, dy + \int_{0}^{-\infty} (-y)\, f(a-y)(-dy) = 0 \qquad \text{[From (*)]}$$

$$\Rightarrow \qquad E(X) = a = \text{Mean}$$

Also $\int_{-\infty}^{a} f(x)\, dx = \int_{-\infty}^{0} f(a+y)\, dy$ and $\int_{a}^{\infty} f(x)\, dx = \int_{0}^{-\infty} f(a-y)(-dy) = \int_{-\infty}^{0} f(a-y)\, dy$

$$\therefore \quad \int_{-\infty}^{a} f(x)\, dx = \int_{a}^{\infty} f(x)\, dx \quad \text{[From (*)]} \quad \Rightarrow \quad \text{Median } (X) = a.$$

(*d*) If X, Y, Z are *i.i.d r.v.'s*, then prove that :

$$E\left[X \mid (X+Y+Z=k)\right] = \frac{k}{3}.$$

6·44. (*a*) Variates X and Y take the values 1, 2, 3, along with the probabilities shown in the adjoining table. Find *k*, E(X + Y), Var (X) and Cov (X, Y) :

Y \ X	1	3	9
1	*k*	*k*	*k*
2	*k*	2*k*	*k*
3	*k*	*k*	*k*

(*b*) The joint probability distribution of X and Y is given in the adjoining table :

(*i*) Find the marginal probability distribution of Y.

(*ii*) Find the conditional distribution of Y given that X = 2.

(*iii*) Find the covariance between X and Y.

(*iv*) Are X and Y independent ?

X \ Y	1	3	9
2	$\frac{1}{8}$	$\frac{1}{24}$	$\frac{1}{12}$
4	$\frac{1}{4}$	$\frac{1}{4}$	0
6	$\frac{1}{8}$	$\frac{1}{24}$	$\frac{1}{12}$

6·45. A fair coin is tossed four times. Let X denote the number of heads occurring and let Y denote the longest string of heads occurring.

(*i*) Determine the joint distribution of X and Y, and (*ii*) Find Cov (X, Y).

Hint.

X \ Y	0	1	2	3	4	Total
0	1/16	0	0	0	0	1/16
1	0	4/16	0	0	0	4/16
2	0	3/16	3/16	0	0	6/16
3	0	0	2/16	2/16	0	4/16
4	0	0	0	0	1/16	1/16
Total	1/16	7/16	5/16	2/16	1/16	1

6·46. A process for producing an industrial chemical yields a product containing two types of impurities. For a specified sample from this process, let X_1 denote the proportion of impurities in the sample and let X_2 denote the proportion of type 1 impurities among all impurities found. Suppose that the joint distribution of X_1 and X_2 can be modeled by the following probability density function :

$$f(x_1, x_2) = \begin{cases} 2(1-x_1) & ; \; 0 \le x_1 \le 1, \; 0 \le x_2 \le 1 \\ 0 & ; \; \text{elsewhere} \end{cases}$$

Find the expected value of the proportion of type 1 impurities in the sample.

Hint. Find $E(X_1X_2)$.

Ans. $E(X_1X_2) = \frac{1}{6} \Rightarrow$ (1/6)th of the sample is made up of type 1 impurities.

6·47. Let the joint probability density function of the random variables X and Y be :

$$f(x, y) = \begin{cases} 2(x + y - 3xy^2) \; ; \; 0 < x < 1, \; 0 < y < 1 \\ 0, \text{ otherwise} \end{cases}$$

(*i*) Find the marginal distributions of X and Y.

(*ii*) Is $E(XY) = E(X) \cdot E(Y)$?

(*iii*) Find $E(X + Y)$ and $E(X - Y)$.

6·48. An electronic system has one each of two different types of components in joint operation. Let X_1 and X_2 denote the random lengths of life of the components of type 1 and type 2, respectively. The joint density function is given by :

$$f(x_1, x_2) = \begin{cases} \frac{1}{8} x_1 \cdot \exp\left[-\frac{(x_1 + x_2)}{2}\right] & ; \; x_1 > 0, x_2 > 0 \\ 0 & ; \; \text{elsewhere} \end{cases}$$

(Measurements are in hundreds of hours.)

(*i*) Are X_1 and X_2 independent ?

(*ii*) One way to measure the relative efficiency of the two components is to compute the ratio (X_2/X_1). Find $E(X_2/X_1)$.

(*iii*) The cost C of replacing the two components depends upon their length of life at failure and is given by $C = 50 + 2X_1 + 4X_2$. Find $E(C)$ and $V(C)$.

Ans. $f_1(x_1) = \frac{1}{4} x_1 \exp\left(-\frac{1}{2} x_1\right) \; ; \; x_1 > 0 \; ; \quad f_2(x_2) = \frac{1}{2} \exp\left(-\frac{1}{2} x_2\right) \; ; \; x_2 > 0.$

(*i*) $f(x_1, x_2) = f_1(x_1) \cdot f_2(x_2) \Rightarrow X_1$ and X_2 are independent.

(*ii*) $E(X_2/X_1) = 1$

(*iii*) $E(X_1) = 4, \quad \text{Var}(X_1) = 8 \; ; \quad E(X_2) = 2, \quad \text{Var}(X_2) = 4 \; ;$

$E(C) = 50 + 2E(X_1) + 4E(X_2) = 66 \; ; \quad \text{Var}(C) = 4 \,\text{Var}(X_1) + 16 \,\text{Var}(X_2) = 96.$

6·49. The joint *p.d.f.* of X and Y is given by :

$$f(x, y) = 2x(x - y) \; ; \; -x < y < x \; ; \; 0 < x < 1.$$

Find the marginals and show that :

$$E(X^m Y^n) = \frac{2}{m + n + 4}\left[\frac{1 + (-1)^m}{m + 1} - \frac{1 + (-1)^{m+1}}{m + 2}\right]$$

6·50. X_1 and X_2 have a bivariate distribution given by

$$P(X_1 = x_1 \cap X_2 = x_2) = \frac{x_1 + 3x_2}{24}, \text{ where } (x_1, x_2) = (1, 1), (1, 2), (2, 1), (2, 2).$$

Find the conditional mean and variance of X_1 given $X_2 = 2$.

6·51. The joint *p.d.f.* of (X, Y) is given by :

$$f(x, y) = n(n - 1)(y - x)^{n-2} \; ; \; 0 \le x \le y \le 1$$
$$= 0, \text{ otherwise}$$

(*a*) Find $E(Y \mid X = x)$ and hence find $E(Y)$.

(*b*) Find $E(X \mid Y = y)$ and hence find $E(X)$.

Ans. $E(Y \mid X) = \frac{n - 1 + X}{n} \; ; \quad E(X) = \frac{1}{n + 1}$

$$E(Y) = E[E(Y \mid X)] = E\left[\frac{n-1+X}{n}\right] = \frac{n}{n+1}$$

$$E(X \mid Y) = \frac{Y}{n} \; ; \qquad E(X) = E[E(X \mid Y)] = E(Y/n) = \frac{1}{n+1}$$

6·52. The joint *p.d.f.* of two *r.v.'s* X and Y is :

$$f(x, y) = \begin{cases} 2, & 0<x<y<1 \\ 0, & \text{otherwise} \end{cases}$$

Find : (*i*) $E(X)$ and $E(Y)$, (*ii*) Var (X) and Var (Y), (*iii*) $E(Y \mid X)$

Hence, verify that :

(*a*) $E(Y) = E[E(Y \mid X)]$,

(*b*) $\text{Var}(Y) = E[\text{Var}(Y \mid X)] + \text{Var}[E(Y \mid X)]$

Hint. $0 < x < y < 1 \;\Rightarrow\; (0 < x < 1,\; x < y < 1)$ or $(0 < y < 1,\; 0 < x < y)$

$$f_X(x) = \int_x^1 f(x, y)\, dy = 2(1-x), 0<x<1 \; ; \quad f_Y(y) = \int_0^y f(x, y)\, dx = 2y, \; 0<y<1$$

(*i*) $$E(X) = \frac{1}{3}, \quad E(X^2) = \frac{1}{6} \; ; \; E(Y) = \frac{2}{3}, \quad E(Y^2) = \frac{1}{2} \; ;$$

(*ii*) $$\text{Var}(X) = \frac{1}{18}, \quad \text{Var}(Y) = \frac{1}{18}.$$

(*iii*) $$f_{Y \mid X = x} = \frac{f(x, y)}{f_X(x)} = \frac{1}{1-x} \; ; \; x<y<1.$$

$$E_Y(Y \mid X = x) = \int_x^1 y \cdot \frac{1}{(1-x)}\, dy = \frac{1+x}{2} \; ; \; 0<x<1$$

(*a*) $$E_X[E_Y(Y \mid X)] = E_X\left[\frac{1+X}{2}\right] = \frac{1}{2}[1+E(X)] = \frac{2}{3} = E(Y)$$

(*b*) $$\text{Var}(Y \mid X) = E(Y^2 \mid X) - [E(Y \mid X)]^2 = \int_x^1 y^2 \cdot \frac{1}{(1-x)}\, dy - \left(\frac{1+x}{2}\right)^2$$

$$= \frac{1+x+x^2}{3} - \frac{1+x^2+2x}{4} = \frac{(1-x)^2}{12} \qquad [\because 0 < x < 1]$$

$$E[\text{Var}(Y \mid X)] = E\left[\frac{(1-X)^2}{12}\right] = \frac{1}{12} E[1+X^2-2X]$$

$$= \frac{1}{12}[1+E(X^2)-2E(X)] = \frac{1+\frac{1}{6}-\frac{2}{3}}{12} = \frac{1}{24}$$

Or $$E[\text{Var}(Y \mid X)] = \frac{1}{12}\int_0^1 (1-x)^2 f_X(x)\, dx = \frac{1}{24}.$$

$$\text{Var}[E(Y \mid X)] = \text{Var}\left[\frac{1+X}{2}\right] = \frac{1}{4}\text{Var}(1+X) = \frac{1}{4}\text{Var}(X) = \frac{1}{4} \times \frac{1}{18} = \frac{1}{72}.$$

$$\therefore E[\text{Var}(Y \mid X)] + \text{Var}[E(Y \mid X)] = \frac{1}{24} + \frac{1}{72} = \frac{4}{72} = \frac{1}{18} = \text{Var}(Y)$$

6·53. The joint *p.d.f.* of two random variables X and Y is :

$$f(x, y) = \begin{cases} 6 \; ; & x^2<y<x, \; 0<x<1 \\ 0 \; ; & \text{otherwise} \end{cases}$$

Prove that :

$$(a)\ E\,(X \mid Y = y) = \frac{1}{2}\left(\sqrt{y} + y\right) \ ; \ 0 < y < 1 \ ;$$

$$(b)\ E\,(Y \mid X = x) = \frac{1}{2}(x + x^2) \ ; \ y < x < \sqrt{y} \ ;$$

$$(c)\ E\,(XY) = \frac{1}{4}.$$

Hint. $x^2 < y < x\ ,\ 0 < x < 1 \ \Rightarrow \ (0 < x < 1 \text{ and } x^2 < y < x)$ or $(0 < y < 1 \text{ and } y \le x \le \sqrt{y}\,)$

6·54. If $f(x, y) = \frac{2}{3}(x + 2y),\ 0 < (x, y) < 1$, find the marginal densities, conditional density of X for a given Y, and the conditional mean and variance of X when $Y = 1/2$.

Ans. $f_X(x) = \frac{2}{3}(x+1),\ 0 < x < 1 \ ; \quad f_Y(y) = \frac{1}{3}(1 + 4y),\ 0 < y < 1$

$$f_{X \mid Y}(x \mid y) = \frac{2(x+2y)}{1+4y};\ 0<x<1,\ 0<y<1\ ; \quad E\left(X \,\middle|\, Y = \frac{1}{2}\right) = \frac{5}{9}; \quad \text{Var}\left(X \,\middle|\, Y = \frac{1}{2}\right) = \frac{13}{162}$$

6·55. (a) Let Y have a *p.d.f.* $g(y) = \frac{3}{4}y(2-y)\,;0 \le y \le 2$. Given $Y = y,\ 0 \le y \le 1$, the conditional distribution of a variable X is uniform over [0, 1] and given $Y = y,\ 1 < y < 2$, the conditional distribution is given by :

$$P\,(X = 1 \mid Y = y) = \frac{1}{2} \quad \text{and} \quad P\,(X = y \mid Y = y) = \frac{1}{2}.$$

Using the result $E\,[E\,(X \mid Y)] = E\,(X)$ or otherwise, find $E\,(X)$.

Hint. For $0 \le y \le 1,\quad E\,(X \mid Y = y) = \frac{1}{2};\quad$ For $1 < y \le 2,\ \ E\,(X \mid Y = y) = \frac{1}{2}(1+y)$

$$E\,(X) = E_Y\,[E_X\,(X \mid Y = y)] = \int_0^1 \frac{1}{2}\cdot g(y)\,dy + \int_1^2 \frac{1}{2}(1+y)\,g(y)\,dy = \frac{27}{32}.$$

(b) The joint *p.d.f.* of *r.v.* (X, Y) is given by :

$$f\,(x, y) = \frac{1}{\sqrt{2\pi y}}.\exp\left[-\frac{(x-y)^2}{2y}\right];\ 0<y<1,\ -\infty<x<\infty$$

Find $E(X)$, Var (X), Var (Y), and Cov (X, Y).

Ans. $E\,(X) = \frac{1}{2},\quad \text{Var}\,(X) = \frac{7}{12},\quad \text{Var}\,(Y) = \frac{1}{12},\quad \text{Cov}\,(X, Y) = \frac{1}{12}$

6·56. Give an example to show that $E\,(Y)$ may not exist though $E\,(XY)$ and $E\,(Y \mid X)$ may both exist ?

Hint. Consider the joint *p.d.f.* :

$$f\,(x, y) = \begin{cases} x\,e^{-x(1+y)}\,; & x \ge 0,\quad y \ge 0. \\ 0, \text{ otherwise} \end{cases}$$

Then, we shall get

$$f_X(x) = \int_0^\infty f\,(x, y)\,dy = e^{-x}\,;\ x \ge 0 \quad \text{and} \quad f_Y(y) = \int_0^\infty f(x, y)\,dx = \frac{1}{(1+y)^2};\quad y \ge 0$$

$$f(y \mid x) = \frac{f(x, y)}{f_X(x)} = x \cdot \exp(-xy); \; y \geq 0$$

$$\therefore \quad E(Y) = \int_0^\infty y\, f(y) = \infty \quad \Rightarrow \quad E(Y) \text{ does not exist.}$$

$$E(XY) = \int_0^\infty \int_0^\infty xy\, f(x, y)\, dx\, dy = \int_0^\infty \left(\int_0^\infty y e^{-xy}\, dy \right) x^2 e^{-x}\, dx$$

$$= \int_0^\infty \frac{\Gamma(2)}{x^2} \cdot x^2 e^{-x}\, dx = \left[\left| \frac{e^{-x}}{-1} \right|_0^\infty \right] = 1$$

$$E(Y \mid X = x) = \int_0^\infty y \cdot f(y \mid x)\, dy = x \int_0^\infty y\, e^{-xy}\, dy = x \cdot \frac{\Gamma(2)}{x^2} = \frac{1}{x}$$

$\Rightarrow$ Both $E(XY)$ and $E(Y \mid X = x)$ exist, though $E(Y)$ does not exist.

6·57. An urn contains 3 red and 2 white balls. A random sample of two balls is drawn with replacement. Let $X = 0$ if the first ball drawn is white and $X = 1$ if the first ball drawn is red. Let $Y = 0$ if the second ball drawn is white and $Y = 1$ if the second ball drawn is red. Find the conditional expectation $E(X \mid Y)$.

6·58. Three coins are tossed. Let X denote the number of heads on the first two coins, Y denote the number of tails on the last two and Z denote the number of heads on the last two.

(*i*) Find the joint distribution of

(*a*) X and Y, (*b*) X and Z.

(*ii*) Find the conditional distribution of Y given X = 1.

(*iii*) Find $E(Z \mid X = 1)$. (*iv*) Find Cov (X, Y) and Cov (X, Z).

(*v*) Give a joint distribution that is not the joint distribution of X and Z in (*i*) (*b*) and yet has the same marginals as $f(x, z)$ has in part (*i*) (*b*).

Hint. The sample space is $S = \{H, T\} \times \{H, T\} \times \{H, T\} = \{H, T\} \times \{HH, HT, TH, TT\}$ and each of the 8 sample points (outcomes) has the probability $p = \frac{1}{8}$ of occurrence.

X : Number of heads on the 1st two coins.

Y : Number of tails on the last two coins.

Z : Number of heads on the last two coins.

Then the distribution of X, Y and Z is given below :

Sample Point	*HHH*	*HHT*	*HTH*	*HTT*	*THH*	*THT*	*TTH*	*TTT*
Probability	*p*	*p*	*p*	*p*	*p*	*p*	*p*	*p*
X	2	2	1	1	1	1	0	0
Y	0	1	1	2	0	1	1	2
Z	2	1	1	0	2	1	1	0

(*i*) (*a*) JOINT DISTRIBUTION OF X AND Y

		y		Total
	0 *f*	1	2	$f_X(x)$
x 0	0	1/8	1/8	1/4
1	1/8	2/8	1/8	1/2
2	1/8	1/8	0	1/4
Total $f_Y(y)$	1/4	1/2	1/4	1

(*b*) JOINT DISTRIBUTION OF X AND Z

		z		Total
	0	1	2	$f_X(x)$
x 0	1/8	1/8	0	1/4
1	1/8	2/8	1/8	1/2
2	0	1/8	1/8	1/4
Total $f_Z(z)$				1

(*ii*) $$P(Y=0 \mid X=1) = \frac{P(Y=0, X=1)}{P(X=1)} = \frac{1/8}{1/2} = \frac{1}{4}$$

Similarly, $$P(Y=1 \mid X=1) = \frac{2/8}{1/2} = \frac{1}{2}\ ;\quad P(Y=2 \mid X=1) = \frac{1/8}{1/2} = \frac{1}{4}$$

(*iii*) $$E(Z \mid X=1) = \sum z.P(Z \mid X=1) = 0\times\frac{1/8}{1/2}+1\times\frac{2/8}{1/2}+2\times\frac{1/8}{1/2} = 1$$

(*iv*) Cov $(X, Y) = -1/4$ and Cov $(X, Z) = 1/4$.

(*v*) Let $0 \le \varepsilon \le 1/8$. The joint probability distribution of (X, Z) given below has the same marginal as in part (*i*) (*b*).

			Z		Total
		0	1	2	$f_X(x)$
X	0	1/8	1/8	0	1/4
	1	1/8	(2/8) + ε	(1/8) − ε	1/2
	2	0	(1/8) − ε	(1/8) + ε	1/4
Total $f_Z(z)$		1/4	1/2	1/4	1

6·59. Let $$f_{XY}(x, y) = e^{-(x+y)}\ ;\ 0 < x < \infty,\ 0 < y < \infty$$

Find :

(*i*) $P(X > 1)$

(*ii*) $P[1 < (X+Y) < 2]$

(*iii*) $P(X < Y \mid X < 2Y)$

(*iv*) m so that $P[(X+Y) < m] = \frac{1}{2}$

(*v*) $P(0 < X < 1 \mid Y = 2)$

(*vi*) Cov (X, Y)

Hint. $f_X(x) = e^{-x}\ ;\ x \ge 0\ ;\ f_Y(y) = e^{-y}\ ;\ y \ge 0.$

(*i*) $\frac{1}{e}$ (*ii*) $(X + Y)$ is a Gamma variable with parameter $n = 2$. (See chapter 9); $\left(\frac{2}{e}-\frac{3}{e^2}\right)$

(*iii*) $$P(X < Y \mid X < 2Y) = \frac{P(X<Y \cap X<2Y)}{P(X<2Y)} = \frac{P(X<Y)}{P(X<2Y)} = \frac{1/2}{2/3} = \frac{3}{4}$$

(*iv*) Use hint in Part (*ii*) ; $e^{-m}(1+m) = 1/2$; (*v*) $(e-1)/e$

(*vi*) $f_{XY}(x, y) = f_X(x)\, f_Y(y)$ ⇒ X and Y are independent ⇒ Cov $(X, Y) = 0$.

6·60. The joint *p.d.f.* of X and Y is given by :

$f(x, y) = 3(x + y)\ ;\ 0 \le x \le 1,\ 0 \le y \le 1;\ 0 \le x + y \le 1$

Find (*i*) Marginal density of X; (*ii*) $P(X + Y) < \frac{1}{2}$; (*iii*) $E(Y \mid X = x)$, and (*iv*) Cov (X, Y).

Ans. (*i*) $f_X(x) = \frac{3}{2}(1-x^2)\ ;\ 0 < x < 1$ (*ii*) $\frac{1}{8}$ (*iii*) $\frac{(1-x)(x+2)}{3(1+x)}$ (*iv*) $\frac{-13}{320}$

CHAPTER

7 Generating Functions and Law of Large Numbers

LEARNING OBJECTIVES

Upon completion of this Chapter, you should be able to:

- Understand various types of generating functions like Moment Generating Function (m.g.f.) Characteristic Function (c.f.), Probability Generating Function (p.g.f.).
- Differentiate between moments and cumulants and establish the relationship between them.
- Explain the difference between moment generating function and characteristic function and their utility in Statistics.
- Understand about the various 'Modes of Convergence' in Probability Theory and inter-relationships between them.
- Develop the various forms of Chebychev's inequality and understand their use in Statistics.
- Discuss in detail the concepts relating to various forms of Week Law of Large Numbers (W.L.L.N.).
- Learn about
  - (a) Kolmogorov's Inequality
  - (b) Kolmogorov's Theorems on Strong Law of Large Numbers (S.L.L.N.).
- Define Probability Generating Function (p.g.f.) and
  - (a) Establish the relationship between p.g.f. and m.g.f.
  - (b) Discuss its utility in Statistics.

CHAPTER OUTLINES

7·1. MOMENT GENERATING FUNCTION

The moment generating function (*m.g.f.*) of a random variable X (about origin) having the probability function $f(x)$ is given by:

$$M_X(t) = E(e^{tX}) = \begin{cases} \int e^{tx} f(x)\,dx, \text{(for continuous probability distribution)} \\ \sum_x e^{tx} f(x), \text{(for discrete probability distribution)} \end{cases} \quad \dots (7·1)$$

the integration or summation being extended to the entire range of x, t being the real parameter and it is being assumed that the right-hand side of (7·1) is absolutely convergent for some positive number h such that $-h < t < h$. Thus

$$M_X(t) = E(e^{tX}) = E\left(1 + tX + \frac{t^2X^2}{2!} + \dots + \frac{t^rX^r}{r!} + \dots\right)$$

$$= 1 + t\,E(X) + \frac{t^2}{2!}E(X^2) + \dots + \frac{t^r}{r!}E(X^r) + \dots$$

$$= 1 + t\mu_1' + \frac{t^2}{2!}\mu_2' + \dots + \frac{t^r}{r!}\mu_r' + \dots \quad \dots (7·2)$$

$$= \sum_{r=0}^{\infty} \frac{t^r}{r!}\mu_r' \quad \dots (7·2a)$$

where $$\mu_r' = E(X^r) = \begin{cases} \int_{-\infty}^{\infty} x^r f(x)\,dx, \text{ for continuous probability distribution,} \\ \sum_x x^r f(x), \text{ for discrete distribution.} \end{cases}$$

is the rth moment of X about origin. Thus we see that the coefficient of $\frac{t^r}{r!}$ in $M_X(t)$ gives μ_r' (about origin). Since $M_X(t)$ generates moments, it is known as *moment generating function.*

Differentiating (7·2) *w.r.t.* t, r times and then putting $t = 0$, we get

$$\left|\frac{d^r}{dt^r}\{M_X(t)\}\right|_{t=0} = \left|\frac{\mu_r'}{r!}\cdot r! + \mu'_{r+1}\cdot t + \mu'_{r+2}\cdot\frac{t^2}{2!} + \dots\right|_{t=0}$$

$$\Rightarrow \quad \mu_r' = \left|\frac{d^r}{dt^r}\{M_X(t)\}\right|_{t=0} \quad \dots (7·3)$$

In general, the moment generating function X about the point X = a is defined as :

$$M_{X-a}(t) = E\left[e^{t(X-a)}\right]$$

$$= E\left\{1 + t(X-a) + \frac{t^2}{2!}(X-a)^2 + \dots + \frac{t^r}{r!}(X-a)^r + \dots\right\}$$

$$= 1 + t\mu_1' + \frac{t^2}{2!}\mu_2' + \dots + \frac{t^r}{r!}\mu_r' + \dots \quad \dots (7·4)$$

where $\mu_r' = E\{(X-a)^r\}$ is the rth moment about the point X = a.

7·1·1. Some Limitations of Moment Generating Function

Moment generating function suffers from drawbacks which have restricted its use in Statistics. We give below some of the deficiencies of *m.g.f.* with illustrative examples.

1. *A random variable X may have no moments although its m.g.f. exists.* For example, let us consider a discrete *r.v.* with probability function :

$$f(x) = \begin{cases} \dfrac{1}{x(x+1)} \ ; & x = 1, 2, \ldots \\ 0, \text{otherwise} \end{cases}$$

Here, $$E(X) = \sum_{x=1}^{\infty} x\, f(x) = \sum_{x=1}^{\infty} \frac{1}{(x+1)} = \frac{1}{2} + \frac{1}{3} + \frac{1}{4} + \ldots = \left(\sum_{x=1}^{\infty} \frac{1}{x} \right) - 1$$

Since, $\sum\limits_{x=1}^{\infty} \dfrac{1}{x}$ is a divergent series, $E(X)$ does not exist and consequently no moment of X exists. However, the *m.g.f.* of X is given by :

$$M_X(t) = \sum_{x=1}^{\infty} e^{tx} f(x) = \sum_{x=1}^{\infty} \frac{e^{tx}}{x(x+1)} = \sum_{x=1}^{\infty} \frac{z^x}{x(x+1)}, \ (z = e^t) \qquad \ldots (*)$$

$$= \frac{z}{1.2} + \frac{z^2}{2.3} + \frac{z^3}{3.4} + \frac{z^4}{4.5} + \ldots$$

$$= z\left(1 - \frac{1}{2}\right) + z^2\left(\frac{1}{2} - \frac{1}{3}\right) + z^3\left(\frac{1}{3} - \frac{1}{4}\right) + z^4\left(\frac{1}{4} - \frac{1}{5}\right) + \ldots$$

$$= \left(z + \frac{z^2}{2} + \frac{z^3}{3} + \frac{z^4}{4} + \ldots\right) - \frac{z}{2} - \frac{z^2}{3} - \frac{z^3}{4} - \frac{z^4}{5} - \ldots$$

$$= -\log(1 - z) - \frac{1}{z}\left(\frac{z^2}{2} + \frac{z^3}{3} + \frac{z^4}{4} + \ldots\right)$$

$$= -\log(1 - z) + 1 + \frac{1}{z}\log(1 - z), \ |z| < 1$$

$$= 1 + \left(\frac{1}{z} - 1\right)\log(1 - z), \ |z| < 1$$

$$= 1 + (e^{-t} - 1)\log(1 - e^t), \ t < 0 \text{ [from (*)]} \qquad [\because |z| < 1 \Rightarrow |e^t| < 1 \Rightarrow t < 0]$$

Also $M_X(t) = 1$ for $t = 0$.

Hence, $M_X(t)$ exists for $t \leq 0$.

2. *A random variable X can have m.g.f. and some (or all) moments, yet the m.g.f. does not generate the moments. For example, consider a discrete r.v. with probability function :*

$$P(X = 2^x) = \frac{e^{-1}}{x!} \; ; x = 0, 1, 2, ...$$

$$E(X^r) = \sum_{x=0}^{\infty} (2^x)^r \, P(X = 2^x) = e^{-1} \sum_{x=0}^{\infty} \frac{(2^r)^x}{x!} = e^{-1} . \exp(2^r) = \exp(2^r - 1)$$

Hence, all the moments of X exist. The *m.g.f.* of X, if it exists, is given by :

$$M_x(t) = \sum_{x=0}^{\infty} \exp(t . 2^x) \left(\frac{e^{-1}}{x!}\right) = e^{-1} \sum_{x=0}^{\infty} \exp(t . 2^x) \frac{1}{x!}$$

By D' Alembert's ratio test, the series on R.H.S. converges for $t \leq 0$ and diverges for $t \geq 0$. Hence $M_X(t)$ cannot be differentiated at $t = 0$ and has no Maclaurin's expansion and consequently it does not generate moments.

3. *A r.v. X can have all or some moments, but m.g.f. does not exist except perhaps at one point.*

For example, let X be a r.v. with probability function :

$$P(X = \pm\, 2^x) = \begin{cases} \dfrac{e^{-1}}{2\, x!} \; ; \; x = 0, 1, 2, ... \\ 0, \text{otherwise} \end{cases}$$

Since the distribution is symmetric about the line X = 0, all moments of odd order about origin vanish, *i.e.*, $E(X^{2r+1}) = 0 \;\Rightarrow\; \mu_{2r+1} = 0$.

$$E(X^{2r}) = \sum_{x=0}^{\infty} (\pm 2^x)^{2r} \left(\frac{1}{2e\, x!}\right) = \frac{1}{e} \sum_{x=0}^{\infty} \frac{(2^{2r})^x}{x!} = \frac{1}{e} \exp(2^{2r}) = \exp(2^{2r} - 1)$$

Thus all the moments of X exist. The *m.g.f.* of X, if it exists, is given by

$$M_X(t) = \sum_{x=0}^{\infty} \left\{ \left(e^{t \cdot 2^x} + e^{-t \cdot 2^x}\right) \frac{1}{2e\, x!} \right\} = e^{-1} \sum_{x=0}^{\infty} \left\{ \frac{\cosh\left(t\, 2^x\right)}{x!} \right\},$$

which converges only for $t = 0$. [By D – Alembert's Ratio Test]

As an illustration of a continuous probability distribution, consider Pareto distribution with *p.d.f.* :

$$p(x) = \frac{\theta\, a^{\theta}}{x^{\theta + 1}} ; x \geq a, \theta > 1.$$

$$E(X^r) = \theta\, a^{\theta} \int_a^{\infty} x^{r - \theta - 1}\, dx = \theta\, a^{\theta} \left| \frac{x^{r - \theta}}{r - \theta} \right|_a^{\infty},$$

which is finite iff $r - \theta < 0 \;\Rightarrow\; \theta > r$

and then $E(X^r) = \theta\, a^{\theta} \left[0 - \dfrac{a^{r - \theta}}{r - \theta} \right] = \dfrac{\theta . a^r}{\theta - r} \; ; \; \theta > r$

However, the *m.g.f.* is given by :

$M_X(t) = \theta\, a^{\theta} \displaystyle\int_a^{\infty} \frac{e^{tx}}{x^{\theta + 1}}\, dx$, which does not exist, since e^{tx} dominates $x^{\theta + 1}$ and $\left(e^{tx} / x^{\theta + 1}\right) \to \infty$ as $x \to \infty$ and hence the integral is not convergent.

For more illustrations see Student's *t*-distribution and Snedecor's *F*-distribution, for which *m.g.f.'s* do not exist, though moments of all orders exist. [*c.f.* Chapter 16].

Remark. The reason that *m.g.f.* is a poor tool in comparison with characteristic function (*c.f.* § 7×3) is that the domain of the dummy parameter '*t*' of the *m.g.f.* depends on the distribution of the *r.v.* under consideration, while characteristic function exists for all real *t*, $(-\infty < t < \infty)$. If *m.g.f.* is valid for *t* lying in an interval containing zero, then *m.g.f.* can be expanded with perhaps some additional restrictions.

7·1·2. Properties of Moment Generating Function

Property 1. $M_{cX}(t) = M_X(ct)$, *c being a constant.* ... (7·5)

Proof. By *def.*, L.H.S. $= M_{cX}(t) = E(e^{tcX})$, R.H.S. $= M_X(ct) = E(e^{ctX}) =$ L.H.S.

Property 2. *The moment generating function of the sum of a number of independent random variables is equal to the product of their respective moment generating functions.*

Symbolically, if $X_1, X_2, \ldots, X_n$ are independent random variables, then the moment generating function of their sum $X_1 + X_2 + \ldots + X_n$ is given by :

$$M_{X_1+X_2+\ldots+X_n}(t) = M_{X_1}(t), M_{X_2}(t), \ldots, M_{X_n}(t) \quad \ldots (7·6)$$

Proof. By def.,

$$M_{X_1+X_2+\ldots+X_n}(t) = E\left[e^{t(X_1+X_2+\ldots+X_n)}\right] = E\left[e^{tX_1}.e^{tX_2}\ldots e^{tX_n}\right]$$

$$= E\left(e^{tX_1}\right).E\left(e^{tX_2}\right)\ldots E\left(e^{tX_n}\right) \qquad (\because X_1, X_2, \ldots, X_n \text{ are independent})$$

$$= M_{X_1}(t)\, M_{X_2}(t) \ldots M_{X_n}(t).$$

Property 3. *Effect of change of origin and scale on M.G.F.* Let us transform X to the new variable *U* by changing both the origin and scale in X as follows :

$$U = \frac{X-a}{h}, \text{ where } a \text{ and } h \text{ are constants.}$$

M.G.F. of *U* (about origin) is given by :

$$M_U(t) = E\left(e^{tU}\right) = E\left[\exp\{t(X-a)/h\}\right] = E\left[e^{tX/h}\, e^{-at/h}\right] = e^{-at/h}\, E\left(e^{tX/h}\right)$$

$$= e^{-at/h}\, E\left(e^{Xt/h}\right) = e^{-at/h}\, M_X(t/h), \quad \ldots (7·7)$$

where, $M_X(t)$ is the *m.g.f.* of X about origin.

In particular, if we take $a = E(X) = \mu$, (say), and $h = \sigma_X = \sigma$, (say), then

$$U = \frac{X - E(X)}{\sigma_X} = \frac{X-\mu}{\sigma} = Z \text{ (say)},$$

is known as a *standard variate.* Thus, the *m.g.f.* of a standard variate *Z* is given by

$$M_Z(t) = e^{-\mu t/\sigma} M_X(t/\sigma) \quad \ldots (7·8)$$

Remark. $E(Z) = E\left(\frac{X-\mu}{\sigma}\right) = \frac{1}{\sigma}E(X-\mu) = \frac{1}{\sigma}\{E(X)-\mu\} = \frac{1}{\sigma}(\mu-\mu) = 0$

and $V(Z) = V\left(\frac{X-\mu}{\sigma}\right) = \frac{1}{\sigma^2}V(X-\mu) = \frac{1}{\sigma^2}V(X) = \frac{1}{\sigma}.\sigma^2 = 1.$

$\therefore$ $E(Z) = 0$ and $V(Z) = 1$,

i.e., mean and variance of a standard variate are 0 and 1, respectively.

7·1·3. Uniqueness Theorem of Moment Generating Function

The moment generating function of a distribution, if it exists, uniquely determines the distribution. This implies that corresponding to a given probability distribution, there is only one *m.g.f.* (provided it exists) and corresponding to a given *m.g.f.*, there is only one probability distribution. Hence $M_X(t) = M_Y(t) \Rightarrow X$ and Y are identically distributed. [For detailed discussion, see Uniqueness Theorem of Characteristic Functions. Theorem 7·7. Page 7·31]

7·2. CUMULANTS

Cumulants generating function $K(t)$ is defined as : $K_X(t) = \log_e M_X(t)$, provided the right-hand side can be expanded as a convergent series in powers of t. Thus

$$K_X(t) = \kappa_1 t + \kappa_2 \frac{t^2}{2!} + \ldots + \kappa_r \frac{t^r}{r!} + \ldots = \log M_X(t)$$

$$= \log\left(1 + \mu_1' t + \mu_2' \frac{t^2}{2!} + \mu_3' \frac{t^3}{3!} + \ldots + \mu_r' \frac{t^r}{r!} + \ldots\right), \quad \ldots (7·9a)$$

where, κ_r = coefficient of $\frac{t^r}{r!}$ in $K_X(t)$ is called the rth cumulant. Hence

$$\kappa_1 t + \kappa_2 \frac{t^2}{2!} + \kappa_3 \frac{t^3}{3!} + \kappa_4 \frac{t^4}{4!} + \ldots$$

$$= \left(\mu_1' t + \mu_2' \frac{t^2}{2!} + \mu_3' \frac{t^3}{3!} + \mu_4' \frac{t^4}{4!} + \ldots\right) - \frac{1}{2}\left(\mu_1' t + \mu_2' \frac{t^2}{2!} + \mu_3' \frac{t^3}{3!} + \ldots\right)^2$$

$$+ \frac{1}{3}\left(\mu_1' t + \mu_2' \frac{t^2}{2!} + \ldots\right)^3 - \frac{1}{4}\left(\mu_1' t + \mu_2' \frac{t^2}{2!} + \ldots\right)^4 + \ldots$$

Comparing the coefficients of like powers of 't' on both sides, we get the relationship between the moments and cumulants. Hence, we have

$$\kappa_1 = \mu_1' = \text{Mean}, \qquad \frac{\kappa_2}{2!} = \frac{\mu_2'}{2!} - \frac{\mu_1'^2}{2!} \quad \Rightarrow \quad \kappa_2 = \mu_2' - \mu_1'^2 = \mu_2$$

$$\frac{\kappa_3}{3!} = \frac{\mu_3'}{3!} - \frac{1}{2} \cdot \frac{2\mu_1' \mu_2'}{2!} + \frac{\mu_1'^3}{3} \quad \Rightarrow \quad \kappa_3 = \mu_3' - 3\mu_2' \mu_1' + 2\mu_1'^3 = \mu_3$$

Also,
$$\frac{\kappa_4}{4!} = \frac{\mu_4'}{4} - \frac{1}{2}\left(\frac{\mu_2'^2}{4} + \frac{2\mu_1' \mu_3'}{3!}\right) + \frac{1}{3} \frac{3\mu_1'^2 \mu_2'}{2} - \frac{\mu_1'^4}{4}$$

$$\Rightarrow \quad \kappa_4 = \mu_4' - 3\mu_2'^2 - 4\mu_1' \mu_3' + 12\mu_1'^2 \mu_2' - 6\mu_1'^4$$

$$= \left(\mu_4' - 4\mu_3' \mu_1' + 6\mu_2' \mu_1'^2 - 3\mu_1'^4\right) - 3\left(\mu_2'^2 - 2\mu_2' \mu_1'^2 + \mu_1'^4\right)$$

$$= \mu_4 - 3\left(\mu_2' - \mu_1'^2\right)^2 = \mu_4 - 3\mu_2^2 = \mu_4 - 3\kappa_2^2 \qquad (\because \mu_2 = \kappa_2)$$

$\therefore \quad \mu_4 = \kappa_4 + 3\kappa_2^2$

Hence, we have obtained :

$$\left.\begin{aligned} \text{Mean} &= \kappa_1 \\ \mu_2 &= \kappa_2 = \text{Variance} \\ \mu_3 &= \kappa_3 \\ \mu_4 &= \kappa_4 + 3\kappa_2^2 \end{aligned}\right\} \quad \dots (7{\cdot}9b)$$

Remarks:

1. These results are of fundamental importance and should be committed to memory.
2. If we differentiate both sides of (7·9*a*) *w. r.* to *t*, '*r*' times and then put $t = 0$, we get :

$$\kappa_r = \left| \frac{d^r}{dt^r} K_X(t) \right|_{t=0} \quad \dots (7{\cdot}9c)$$

7·2·1. Properties of Cumulants

Property 1. ***Additive Property of Cumulants.*** *The rth cumulants of the sum of the independent random variables is equal to the sum of the rth cumulants of the individual variables. Symbolically,*

$$\kappa_r (X_1 + X_2 + \dots + X_n) = \kappa_r(X_1) + \kappa_r(X_2) + \dots + \kappa_r(X_n), \quad \dots (7{\cdot}10)$$

where X_i *;* $i = 1, 2, \dots, n$ *are independent random variables.*

Proof. Since X_i's are independent,

$$M_{X_1+X_2+\dots+X_n}(t) = M_{X_1}(t)\, M_{X_2}(t) \dots M_{X_n}(t)$$

Taking logarithm of each side,

$$K_{X_1+X_2+\dots+X_n}(t) = K_{X_1}(t) + K_{X_2}(t) + \dots + K_{X_n}(t)$$

Differentiating both sides *w.r.* to *t*, '*r*' times and putting $t = 0$, we get

$$\left[\frac{d^r}{dt^r} K_{X_1+X_2+\dots+X_n}(t)\right]_{t=0} = \left[\frac{d^r}{dt^r} K_{X_1}(t)\right]_{t=0} + \left[\frac{d^r}{dt^r} K_{X_2}(t)\right]_{t=0} + \dots + \left[\frac{d^r}{dt^r} K_{X_n}(t)\right]_{t=0}$$

$\therefore \quad \kappa_r (X_1 + X_2 + \dots + X_n) = \kappa_r (X_1) + \kappa_r (X_2) + \dots + \kappa_r (X_n),$

which established the result.

Property 2. ***Effect of Change of Origin and Scale on Cumulants.*** If we take

$$U = \frac{X - a}{h}, \text{ then } M_U(t) = \exp(-at/h)\, M_X(t/h).$$

$$\therefore \quad K_U(t) = \log M_U(t) = -\frac{at}{h} + K_X(t/h)$$

$$\kappa_1' t + \kappa_2' \frac{t^2}{2!} + \dots + \kappa_r' \frac{t^r}{r!} + \dots = -\frac{at}{h} + \kappa_1 (t/h) + \kappa_2 \frac{(t/h)^2}{2!} + \dots + \kappa_r \frac{(t/h)^r}{r!} + \dots$$

where κ_r' and κ_r are the *r*th cumulants of *U* and *X* , respectively.

Comparing coefficients, we get

$$\kappa_1' = \frac{\kappa_1 - a}{h} \quad \text{and} \quad \kappa_r' = \frac{\kappa_r}{h^r}\ ;\ r = 2, 3, \ldots \qquad \ldots (7·10a)$$

Thus we see that except the first cumulant, all cumulants are independent of change of origin. But the cumulants are not invariant of change of scale as the rth cumulants of U is $(1/h^r)$ times the rth cumulant of the distribution of X. Therefore, the cumulants are sometimes called *semi-invariants*. The cumulants were originally defined and studied by Thiele (1903).

Example 7·1. *Let the r.v. X assume the value 'r' with the probability law :*
$P(X = r) = q^{r-1}\, p\ ;\ r = 1, 2, 3.$ Find the m.g.f. of X and hence its mean and variance.

Solution.
$$M_X(t) = E(e^{tX}) = \sum_{r=1}^{\infty} e^{tr}\, q^{r-1}\, p = \frac{p}{q}\sum_{r=1}^{\infty} (q e^t)^r$$

$$= \frac{p}{q}\, q e^t \sum_{r=1}^{\infty} (q e^t)^{r-1} = p e^t \left\{1 + q e^t + (q e^t)^2 + \ldots\right\} = \left(\frac{p e^t}{1 - q e^t}\right)$$

If dash (′) denotes the differentiation *w. r.* to t, then we have

$$M_X'(t) = \frac{p e^t}{(1 - q e^t)^2}, \qquad M_X''(t) = p e^t\, \frac{(1 + q e^t)}{(1 - q e^t)^3}$$

$$\therefore \quad \mu_1' \text{ (about origin)} = M_X'(0) = \frac{p}{(1-q)^2} = \frac{1}{p}$$

$$\mu_2' \text{ (about origin)} = M_X''(0) = \frac{p(1+q)}{(1-q)^3} = \frac{1+q}{p^2}.$$

Hence, mean μ_1' (about origin) $= \dfrac{1}{p}$ and variance $= \mu_2 = \mu_2' - \mu_1'^2 = \dfrac{1+q}{p^2} - \dfrac{1}{p^2} = \dfrac{q}{p^2}$.

Example 7·2. *The probability density function of the random variable X follows the probability law :*

$$p(x) = \frac{1}{2\theta} \exp\left(-\frac{|x-\theta|}{\theta}\right), -\infty < x < \infty$$

Find M.G.F. of X. Hence or otherwise find E(X) and V(X).

Solution. The moment generating function of X is :

$$M_X(t) = E(e^{tX}) = \int_{-\infty}^{\infty} \frac{1}{2\theta} \exp\left(-\frac{|x-\theta|}{\theta}\right) e^{tx}\, dx$$

$$= \int_{-\infty}^{\theta} \frac{1}{2\theta} \exp\left(-\frac{\theta - x}{\theta}\right) e^{tx} dx + \int_{\theta}^{\infty} \frac{1}{2\theta} \exp\left(-\frac{x-\theta}{\theta}\right) e^{tx} dx$$

[For $x \in (-\infty, \theta)$, $x - \theta < 0 \quad \Rightarrow \quad \theta - x > 0$

$\therefore \quad |x - \theta| = \theta - x \ \forall\ x \in (-\infty, \theta).$

Similarly, $\quad |x - \theta| = x - \theta \ \forall\ x \in (\theta, \infty)$]

$$\therefore \quad M_X(t) = \frac{e^{-1}}{2\theta}\int_{-\infty}^{\theta} \exp\left\{x\left(t+\frac{1}{\theta}\right)\right\}dx + \frac{e}{2\theta}\int_{\theta}^{\infty} \exp\left\{-x\left(\frac{1}{\theta}-t\right)\right\}dx$$

$$= \frac{e^{-1}}{2\theta}\cdot\frac{1}{\left(t+\frac{1}{\theta}\right)}\cdot\exp\left\{\theta\left(t+\frac{1}{\theta}\right)\right\} + \frac{e}{2\theta}\cdot\frac{1}{\left(\frac{1}{\theta}-t\right)}\cdot\exp\left\{-\theta\left(\frac{1}{\theta}-t\right)\right\}$$

$$= \frac{e^{\theta t}}{2(\theta t+1)} + \frac{e^{\theta t}}{2(1-\theta t)} = \frac{e^{\theta t}}{1-\theta^2 t^2} = e^{\theta t}\left(1-\theta^2 t^2\right)^{-1}$$

$$= \left(1+\theta t+\frac{\theta^2 t^2}{2!}+\ldots\right)\left(1+\theta^2 t^2+\theta^4 t^4+\ldots\right) = 1+\theta t+\frac{3\theta^2 t^2}{2!}+\ldots$$

$$\therefore \quad E(X) = \mu_1' = \text{Coeff. of } t \text{ in } M_X(t) = \theta, \qquad \mu_2' = \text{Coeff. of } \frac{t^2}{2!} \text{ in } M_X(t) = 3\theta^2$$

Hence, $\quad \text{Var}(X) = \mu_2' - \mu_1'^2 = 3\theta^2 - \theta^2 = 2\theta^2.$

Example 7·3.

(a) If the moments of variable X are defined by $E(X^r) = 0·6$; $r = 1, 2, 3, \ldots$, show that

$$P(X = 0) = 0·4, \qquad P(X = 1) = 0·6, \qquad P(X \geq 2) = 0.$$

(b) X_1 and X_2 are independent random variables with common probability function:

$$p(x) = \frac{x}{6}\ ;\ x = 1, 2, 3$$

Obtain the probability distribution of $Y = X_1 + X_2$, using the moment generating function technique.

Solution.

(a) The *m.g.f.* of variate X is :

$$M_X(t) = \sum_{r=0}^{\infty}\frac{t^r}{r!}\mu_r' = 1 + \sum_{r=1}^{\infty}\frac{t^r}{r!}(0·6) = 0·4 + 0·6\sum_{r=0}^{\infty}\frac{t^r}{r!} = 0·4 + 0·6\,e^t \quad \ldots (*)$$

But $$M_X(t) = E(e^{tX}) = \sum_{x=0}^{\infty} e^{tx}\, P(X = x)$$

$$= P(X = 0) + e^t \cdot P(X = 1) + \sum_{x=2}^{\infty} e^{tx}\cdot P(X = x) \quad \ldots (**)$$

From (*) and (**), we get

$$P(X = 0) = 0·4\ ; \qquad P(X = 1) = 0·6\ ; \qquad P(X \geq 2) = 0.$$

(b) $$M_Y(t) = M_{X_1+X_2}(t) = M_{X_1}(t).\,M_{X_2}(t)$$

$$= [M_{X_1}(t)]^2 \qquad (\because X_1 \text{ and } X_2 \text{ are } i.i.d.)$$

$$M_{X_1}(t) = E(e^{tX_1}) = \sum_{x_1=1}^{3} e^{tx_1}\, P(X = x_1)$$

$$= \frac{1}{6}e^t + \frac{2}{6}e^{2t} + \frac{3}{6}e^{3t}$$

$$\therefore \quad M_Y(t) = \left[\frac{e^t}{6}\left(1+2e^t+3e^{2t}\right)\right]^2$$

$$= \frac{e^{2t}}{36}\left[1+4e^{2t}+9e^{4t}+4e^t+6e^{2t}+12e^{3t}\right]$$

$$= \frac{e^{2t}}{36}\left[1+4e^t+10e^{2t}+12e^{3t}+9e^{4t}\right]$$

$$\Rightarrow \quad M_Y(t) = \frac{1}{36}e^{2t} + \frac{4}{36}e^{3t} + \frac{10}{36}e^{4t} + \frac{12}{36}e^{5t} + \frac{9}{36}e^{6t} \qquad \text{... (*)}$$

Also $$M_Y(t) = \sum_{y=2}^{6} e^{ty}\, P(Y=y) \qquad \text{... (**)}$$

From (*) and (**), the probability distribution of Y is :

y	2	3	4	5	6	Total
$p(y)$	$\frac{1}{36}$	$\frac{4}{36}$	$\frac{10}{36}$	$\frac{12}{36}$	$\frac{9}{36}$	1

Example 7·4. *Find the moment generating function of the random variable whose moments are :*

$$\mu_r' = (r+1)\,!\,2^r$$

Solution. The *m.g.f.* is given by :

$$M_X(t) = \sum_{r=0}^{\infty} \frac{t^r}{r!}\mu_r' = \sum_{r=0}^{\infty} \frac{t^r}{r!}(r+1)!\,2^r = \sum_{r=0}^{\infty}(r+1)(2t)^r$$

$$= 1 + 2(2t) + 3(2t)^2 + 4(2t)^3 + \ldots = (1-2t)^{-2}.$$

Aliter. The R.H.S. is arithmetic–geometric series with ratio $r = (2t)$.

Let $$S = 1 + 2r + 3r^2 + 4r^3 + \ldots \ldots$$

Then $$r\,S = r + 2r^2 + 3r^3 + \ldots \ldots$$

$$\therefore \quad (1-r)\,S = 1 + r + r^2 + \ldots = \frac{1}{(1-r)}$$

$$\Rightarrow \quad S = \frac{1}{(1-r)^2} = (1-2t)^{-2}$$

Remark. This is *m.g.f.* of Chi-square (χ^2) distribution with parameter (degrees of freedom) $n = 2$ [*c.f.* Chapter 15].

Example 7·5. *If μ_r' is the rth moment about origin, prove that*

$$\mu_r' = \sum_{j=1}^{r}\binom{r-1}{j-1}\mu'_{r-j}\,k_j, \text{ where } k_j \text{ is the jth cumulant.}$$

Solution. Differentiating both sides of (7·9*a*) in page 7·7, *w.r. to t,* we get

$$\kappa_1 + \kappa_2\, t + \kappa_3 \frac{t^2}{2\,!} + \ldots + \kappa_r \frac{t^{r-1}}{(r-1)!} + \ldots$$

$$= \frac{\mu_1' + \mu_2' t + \mu_3' \dfrac{t^2}{2\,!} + \ldots + \mu_r' \dfrac{t^{r-1}}{(r-1)\,!} + \ldots}{1 + \mu_1' t + \mu_2' \dfrac{t^2}{2\,!} + \ldots + \mu_r' \dfrac{t^r}{r\,!} + \ldots}$$

$$\Rightarrow \left\{\kappa_1 + \kappa_2 t + \kappa_3 \frac{t^2}{2\,!} + \ldots + \kappa_r \frac{t^{r-1}}{(r-1)\,!} + \ldots\right\} \times \left\{1 + \mu_1' t + \mu_2' \frac{t^2}{2\,!} + \ldots \mu_r' \frac{t^r}{r\,!} + \ldots\right\}$$

$$= \mu_1' + \mu_2' t + \mu_3' \frac{t^2}{2\,!} + \ldots + \mu_r' \frac{t^{r-1}}{(r-1)\,!} + \ldots$$

Comparing the coefficient of $\dfrac{t^{r-1}}{(r-1)\,!}$ on both sides, we get

$$\mu_r' = \kappa_1 \cdot \mu'_{r-1} + (r-1)\, \kappa_2\, \mu'_{r-2} + \binom{r-1}{2} \kappa_3\, \mu'_{r-3} + \ldots + \kappa_r$$

$$= \binom{r-1}{0} \mu'_{r-1}\, \kappa_1 + \binom{r-1}{1} \mu'_{r-2}\, \kappa_2 + \binom{r-1}{2} \mu'_{r-3}\, \kappa_3 + \ldots + \binom{r-1}{r-1} \mu'_0\, \kappa_r$$

$$= \sum_{j=1}^{r} \binom{r-1}{j-1} \mu'_{r-j}\, \kappa_j, \text{ which is the required result.}$$

7·3. CHARACTERISTIC FUNCTION

In some cases *m.g.f.* does not exist, since the integral $\int_{-\infty}^{\infty} e^{tx} f(x)\, dx$ or the series $\sum_x e^{tx} f(x)$ does not converge absolutely for real values of t for some distributions. For example, for the continuous probability distribution :

$$dF(x) = C \frac{1}{\left(1 + x^2\right)^m} dx \;\; ; m > 1, -\infty < x < \infty$$

the *m.g.f.* does not exist, since the integral : $M_X(t) = C \int_{-\infty}^{\infty} e^{tx} \dfrac{1}{(1 + x^2)^m} dx$ does not converge absolutely for finite positive values of m because the function e^{tx} dominates the function x^{2m} so that $(e^{ax}/x^{2m}) \rightarrow \infty$ as $x \rightarrow \infty$.

Again, for the discrete probability distribution :

$$f(x) = \begin{cases} \dfrac{6}{\pi^2 x^2} \; ; x = 1, 2, 3, \ldots \\ 0, \text{elsewhere} \end{cases}$$

$$M_X(t) = \sum_x e^{tx} f(x) = \frac{6}{\pi^2} \sum_{x=1}^{\infty} \left(\frac{e^{tx}}{x^2}\right)$$

The series is not convergent (by D′ Alembert's Ratio Test) for $t > 0$. Thus, there does not exist a positive number h such that $M_X(t)$ exist for $-h < t < h$. Hence, $M_X(t)$ does not exist in this case also.

A more serviceable function than the *m.g.f.* is what is known as *characteristic function* and is defined as

$$\phi_X(t) = E\left(e^{itX}\right) = \begin{cases} \displaystyle\int_{-\infty}^{\infty} e^{itx} f(x)\, dx & (\textit{for continuous probability distributions}) \\ \displaystyle\sum_x e^{itx} f(x) & (\textit{for discrete probability distributions}) \end{cases} \quad \text{... (7·11)}$$

If $F_X(x)$ is the distribution function of a continuous random variable X, then

$$\phi_X(t) = \int_{-\infty}^{\infty} e^{itx}\, dF(x) \quad \text{... (7·11}a\text{)}$$

Obviously, $\phi(t)$ is a complex valued function of real variable t. It may be noted that

$$|\phi(t)| = \left|\int_{-\infty}^{\infty} e^{itx} f(x) dx\right| \le \int_{\infty}^{\infty} \left|e^{itx}\right| f(x)\, dx = \int_{-\infty}^{\infty} f(x)\, dx = 1 \quad \text{... (7·11}b\text{)}$$

$$\left[\because \left|e^{itx}\right| = |\cos tx + i \sin tx| = \left(\cos^2 tx + \sin^2 tx\right)^{1/2} = 1\right]$$

Since $|\phi(t)| \le 1$, characteristic function $\phi_X(t)$ always exists.

Remarks:

1.

$$\phi_X(t) = E(e^{itX}) = E\left[1 + itX + \frac{(it)^2}{2!}X^2 + \ldots + \frac{(it)^r}{r!}X^r + \ldots\right]$$

$$= 1 + it\, E(X) + \frac{(it)^2}{2!}E(X^2) + \ldots + \frac{(it)^r}{r!}E(X^r) + \ldots$$

$$= 1 + it\, \mu_1' + \frac{(it)^2}{2!}\mu_2' + \ldots + \frac{(it)^r}{r!}\mu_r' + \ldots \quad \text{... (7·11}c\text{)}$$

where $\mu'_r = E(X^r)$, is the rth moment of X about origin.

$\therefore$ μ'_r = Coefficient of $\dfrac{(it)^r}{r!}$ in $\phi_X(t)$ ... (7·11*d*)

Hence, like *m.g.f.*, the characteristic function $\phi(t)$, also generates moments. Re-writing (7·11c), we get

$$\phi_X(t) = \sum_{r=0}^{\infty} \frac{(it)^r}{r!}\mu'_r \quad \text{... (7·11}e\text{)}$$

2. Cumulant generating function $K_X(t)$ in terms of $\phi_X(t)$ is given by :

$$K(t) = \log \phi(t) \equiv \kappa_1 \cdot (it) + \kappa_2 \cdot \frac{(it)^2}{2!} + \ldots + \kappa_r \cdot \frac{(it)^r}{r!} + \ldots = \sum_{r=1}^{\infty} \frac{(it)^r}{r!}\kappa_r \quad \text{... (7·12)}$$

were, κ_r = Coefficient of $\dfrac{(it)^r}{r!}$ in $K_X(t)$, is the rth cumulant.

3. Yet another advantage of characteristic function lies in the fact that it uniquely determines the distribution function, *i.e.*, if the characteristic function of a distribution is given, the distribution can be uniquely determined by the theorem, known as the *Uniqueness Theorem of Characteristic Functions* discussed later in the chapter [*Theorem 7·7*, Page 7·31].

7·3·1. Properties of Characteristic Function

Property 1. *For all real t, we have*

(*i*) $\phi(0) = \int_{-\infty}^{\infty} dF(x) = 1$ ... (7·13*a*)

(*ii*) $|\phi(t)| \le 1 = \phi(0)$ [From (7·11*b*)] ... (7·13*b*)

Property 2. *$\phi(t)$ is continuous everywhere, i.e., $\phi(t)$ is continuous function of t in $(-\infty, \infty)$. Rather $\phi(t)$ is uniformly continuous in 't'.*

Proof. For $h \neq 0$,

$$|\phi_X(t+h) - \phi_X(t)| = \left|\int_{-\infty}^{\infty} \left[e^{i(t+h)x} - e^{itx}\right] dF(x)\right|$$

$$\le \int_{-\infty}^{\infty} \left|e^{itx}(e^{ihx} - 1)\right| dF(x) = \int_{-\infty}^{\infty} \left|e^{ihx} - 1\right| dF(x) \quad \text{... (*)}$$

The last integral does not depend on *t*. If it tends to zero as $h \to 0$ then $\phi_X(t)$ uniformly continuous in *'t'*.

Now $$\left|e^{ihx} - 1\right| \le \left|e^{ihx}\right| + 1 \le 1 + 1 = 2$$

$$\therefore \quad \int_{-\infty}^{\infty} \left|e^{ihx} - 1\right| dF(x) \le 2\int_{-\infty}^{\infty} dF(x) = 2$$

Hence, by Dominated Convergence Theorem (D.C.T.), taking the limit inside the integral sign in (*), we get

$$\lim_{h \to 0} |\phi_X(t+h) - \phi_X(t)| \le \int_{-\infty}^{\infty} \lim_{h \to 0} \left|e^{ihx} - 1\right| dF(x) = 0$$

$$\Rightarrow \quad \lim_{h \to 0} \phi_X(t+h) = \phi_X(t), \forall\, t$$

Hence, $\phi_X(t)$ is uniformly continuous in *'t'* for all $t \in$ real line R' $(-\infty, \infty)$.

Property 3. *$\phi_X(-t)$ and $\phi_X(t)$ are conjugate functions, i.e. $\phi_X(-t) = \overline{\phi_X(t)}$,*

where $\bar{a}$ is the complex conjugate of a. .... (7·13*c*)

Proof. $$\phi_X(t) = E(e^{itX}) = E[\cos tX + i \sin tX]$$

$$\therefore \quad \overline{\phi_X(t)} = E(\cos tX - i \sin tX) = E[\cos(-tX) + i \sin(-tX)]$$

$$= E(e^{-itX}) = \phi_X(-t)$$

Property 4. *If the distribution function of a r.v. X is symmetrical about zero, i.e., if*

$$1 - F(x) = F(-x) \quad \Rightarrow \quad f(-x) = f(x),$$

then $\phi_X(t)$ is real valued and even function of t.

Proof. By *def.*, we have

$$\therefore \quad \phi_X(t) = \int_{-\infty}^{\infty} e^{itx} f(x)\, dx = \int_{-\infty}^{\infty} e^{-ity} f(-y)\, dy \qquad (x = -y)$$

$$= \int_{-\infty}^{\infty} e^{-ity} f(y)\, dy \qquad [\because f(-y) = f(y)]$$

$$= \phi_X(-t) \qquad \text{... (7·13}d)$$

$\Rightarrow \quad \phi_X(t)$ is an even function of t.

From (7·13*c*) and (7·13*d*), we get

$$\phi_X(t) = \phi_X(-t) = \overline{\phi_X(t)}$$

Hence, $\phi_X(t)$ is a real valued and even function of t.

Property 5. *If X is some r.v. with characteristic function $\phi_X(t)$, and if $\mu_r' = E(X^r)$ exists,*

then
$$\mu_r' = (-i)^r \left| \frac{\partial^r}{\partial t^r} \phi(t) \right|_{t=0} \qquad \text{... (7·14)}$$

Proof.
$$\phi(t) = \int_{-\infty}^{\infty} e^{itx} f(x)\, dx$$

Differentiating (under the integral sign) '*r*' times *w.r.* to *t*, we get

$$\frac{\partial^r}{\partial t^r} \phi(t) = \int_{-\infty}^{\infty} (ix)^r e^{itx} f(x)\, dx = (i)^r \int_{-\infty}^{\infty} x^r e^{itx} f(x)\, dx$$

$$\therefore \quad \left| \frac{\partial^r}{\partial t^r} \phi(t) \right|_{t=0} = (i)^r \left| \int_{-\infty}^{\infty} x^r e^{itx} f(x)\, dx \right|_{t=0}$$

$$= (i)^r \int_{-\infty}^{\infty} x^r f(x)\, dx = i^r E(X^r) = i^r \mu_r'$$

Hence,
$$\mu_r' = \left(\frac{1}{i}\right)^r \left| \frac{\partial^r}{\partial t^r} \phi(t) \right|_{t=0} = (-i)^r \left| \frac{\partial^r}{\partial t^r} \phi(t) \right|_{t=0}$$

Aliter. We have: [*c.f.* equation (7·11*b*)]

$$\phi_X(t) = 1 + it\, \mu_1' + \frac{(it)^2}{2!} \mu_2' + \ldots + \frac{(it)^r}{r!} \mu_r' + \frac{(it)^{r+1}}{(r+1)!} \mu_{r+1}' + \ldots$$

Differentiating r times w.r. to t and then putting $t = 0$, we get

$$\left| \frac{\partial^r}{\partial t^r} \phi_X(t) \right|_{t=0} = i^r \mu_r' \quad \Rightarrow \quad \mu_r' = (-i)^r \left| \frac{\partial^r}{\partial t^r} \phi_X(t) \right|_{t=0} \qquad \left(\because i = -\frac{1}{i}\right)$$

Property 6. $\phi_{cX}(t) = \phi_X(ct)$, *c being a constant.*

Property 7. *If X_1 and X_2 are independent random variables, then*

$$\phi_{X_1+X_2}(t) = \phi_{X_1}(t)\cdot\phi_{X_2}(t) \qquad \text{... (7·14a)}$$

More generally, for independent random variables X_i ; $i = 1, 2, ..., n$, we have

$$\phi_{X_1+X_2+...X_n}(t) = \phi_{X_1}(t)\cdot\ \phi_{X_2}(t)\ \dots\ \phi_{X_n}(t) \qquad \text{...(7·14b)}$$

Important Remarks. Converse of (*) is not true, *i.e.*

$$\phi_{X_1+X_2}(t) = \phi_{X_1}(t)\cdot\phi_{X_2}(t),$$

does not imply that X_1 and X_2 are independent.

For example, let X_1 be a standard Cauchy variate with *p.d.f.* :

$$f(x) = \frac{1}{\pi(1+x^2)},\ -\infty < x < \infty$$

Then $\phi_{X_1}(t) = e^{-|t|}$ [*c.f.* Chapter 9] ... (*)

Let $X_2 \equiv X_1$, almost surely (*a.s.*), *i.e.*, $P(X_1 = X_2) = 1$. ... (**)

Then $\phi_{X_2}(t) = e^{-|t|}$

Now $\phi_{X_1+X_2}(t) = \phi_{2X_1}(t) = \phi_{X_1}(2t) = e^{-2|t|} = \phi_{X_1}(t)\cdot\phi_{X_2}(t)$ [From (**) and (*)]

i.e., (7·14*a*) is satisfied but obviously X_1 and X_2 are not independent, because of (**).

In fact, (7·14*a*) will hold even if we take $X_1 = aX$ and $X_2 = bX$, a and b being real numbers so that X_1 and X_2 are connected by the releation :

$$\frac{X_1}{a} = \frac{X_2}{b} \quad\Rightarrow\quad aX_2 = bX_1$$

As another example let us consider the joint *p.d.f.* of two *r.v.'s* X and Y given by :

$$f(x, y) = \begin{cases}\frac{1}{4a^2}\left\{1+xy(x^2-y^2)\right\}; & |x|\le a,\ |y|\le a,\ a>0\\ 0, \text{ elsewhere}\end{cases}$$

Then the marginal *p.d.f.'s* of X and Y are given by :

$$g(x) = \int_{-a}^{a} f(x, y)\,dy = \frac{1}{2a};\ |x|\le a \qquad \text{(On simplitication)}$$

$$h(y) = \int_{-a}^{a} f(x, y)\,dx = \frac{1}{2a};\ |y|\le a$$

Then $$\phi_X(t) = \int_{-\infty}^{\infty} e^{itx} g(x)\,dx = \frac{1}{2a}\int_{-a}^{a} e^{itx}\,dx = \frac{e^{iat}-e^{-iat}}{2ait} = \frac{\sin at}{at}$$

Similarly, $$\phi_Y(t) = \frac{\sin at}{at}$$

$$\therefore \qquad \phi_X(t)\ \phi_Y(t) = \left(\frac{\sin at}{at}\right)^2 \qquad \ldots (*)$$

The *p.d.f.* $k(z)$ of the *r.v.* $Z = X + Y$ is given by the convolution of *p.d.f.'s* of X and Y, *viz.*,

$$k(z) = \int f(u, z-u)\, du = \frac{1}{4a^2}\int \left[1 + u(z-u)\left\{u^2 - (z-u)^2\right\}\right] du$$

$$= \frac{1}{4a^2}\int \left(1 + 3z^2u^2 - 2z\,u^3 - z^3u\right) du$$

the limits of integration for u being in terms of z and are given by :

$$-a \le u \le z + a\ ;\ u \le 0 \qquad \text{and} \qquad z - a \le u \le a\ ;\ u > 0$$

(*Left as an exercise to the reader.*)

Thus

$$k(z) = \begin{cases} \dfrac{1}{4a^2}\displaystyle\int_{-a}^{z+a} \left(1 + 3z^2u^2 - 2zu^3 - z^3u\right) du = \dfrac{2a+z}{4a^2}\ ; & -2a \le z \le 0 \\[2ex] \dfrac{1}{4a^2}\displaystyle\int_{z-a}^{a} \left(1 + 3z^2u^2 - 2zu^3 - z^3u\right) du = \dfrac{2a-z}{4a^2}\ ; & 0 \le z \le 2a \end{cases}$$

Now

$$\phi_{X+Y}(t) = \phi_Z(t) = \int_{-2a}^{2a} e^{itz} k(z)\, dz$$

$$= \int_{-2a}^{0} \left(\frac{2a+z}{4a^2}\right) e^{itz}\, dz + \int_{0}^{2a} \left(\frac{2a-z}{4a^2}\right) e^{itz}\, dz$$

$$= \int_0^{2a} \left(e^{-itz} + e^{itz}\right)\left(\frac{2a-z}{4a^2}\right) dz \text{ [Changing } z \text{ to } -z \text{ in the first integral]}$$

$$= \frac{1}{2a^2}\int_0^{2a} (2a-z)\cos tz\, dz = \frac{2 - 2\cos 2at}{4a^2t^2} = \frac{1-\cos 2at}{2a^2t^2} = \left(\frac{\sin at}{at}\right)^2$$

$$= \phi_X(t)\cdot\phi_Y(t) \qquad \text{[From (*)]}$$

But $\quad g(x).\ h(y) \ne f(x, y) \quad \Rightarrow \quad X$ and Y are not independent.

However, the joint characteristic function :

$$\phi_{X_1, X_2}(t_1, t_2) = E\left(e^{it_1X_1 + it_2X_2}\right) = \phi_{X_1}(t_1)\cdot\phi_{X_2}(t_2),$$

$\Rightarrow$ X_1 and X_2 are independent. [For details, see *Theorem 7·8, page 7·31*]

Property 8. Effect of Change of Origin and Scale on Characteristic Function. If

$$U = \frac{X-a}{h}, \textit{ a and h being constants, then: } \quad \phi_U(t) = e^{-iat/h}\,\phi_X(t/h).$$

In particular if we take $a = E(X) = \mu$ (say) and $h = \sigma_X = \sigma$, then the characteristic function of the standard variate

$$Z = \frac{X - E(X)}{\sigma_X} = \frac{X-\mu}{\sigma}, \text{ is given by :}$$

$$\phi_Z(t) = e^{-i\mu t/\sigma} \cdot \phi_X(t/\sigma) \qquad \text{... (7·15)}$$

Proofs of properties 6, 7 and 8 are similar to the properties 1, 2 and 3 of moment generating functions and are left as an exercise to the reader.

Definition. *A random variable X is said to be a Lattice variable or be lattice distributed, if for some $h > 0$,*

$$P[(X/h) \text{ is an integer}] = 1 \text{ ; } h \text{ is called a mesh.}$$

Property 9. *If $|\phi_X(s)| = 1$ for some $s \neq 0$, then for some real a, $X - a$ is a lattice variable with mesh,* $h = 2\pi / |s|$.

Proof. Consider any fixed t. We can write

$$\phi_X(t) = |\phi_X(t)|\, e^{iat}, \text{ ('a' dependent on } t\text{)},$$

since any complex number z can be written as : $z = |z|\, e^{i\theta}$.

$$\therefore \quad |\phi_X(t)| = e^{-iat}\,\phi_X(t) = \phi_{X-a}(t)$$

$$= E\{\cos t(X-a) + i \sin t(X-a)\} = E\{\cos t(X-a)\}$$

since left-hand side being real, we must have $E\{\sin t(X-a)\} = 0$.

$$\therefore \quad 1 - |\phi_X(t)| = E\{1 - \cos t(X-a)\} \qquad \text{... (*)}$$

If $|\phi_X(s)| = 1$, for some $s \neq 0$, then for some a dependent on s, we have from (*)

$$E\{1 - \cos s(X-a)\} = 0 \qquad \text{... (**)}$$

But since $1 - \cos s(X-a)$ is a non-negative random variable, (**) implies that

$$P\{1 - \cos s(X-a) = 0\} = 1 \Rightarrow P\{\cos s(X-a) = 1\} = 1 \Rightarrow P\{s(X-a) = 2n\pi\} = 1$$

$$\therefore \quad P\left[(X-a) = \frac{2n\pi}{|s|}\right] = 1, \text{ for some } n = 0, 1, 2, \ldots$$

Thus, $(X-a)$ is a Lattice variable with mesh, $h = \dfrac{2\pi}{|s|}$.

7·3·2 Necessary and Sufficient Conditions for a Function $\phi(t)$ to be Characteristic Function

Properties 1 to 3 in § 7·3·1 are merely the necessary conditions for a function $\phi(t)$ to be the characteristic function of a *r.v.* X. Thus, if a function $\phi(t)$ does not satisfy any one of these conditions, it cannot be the characteristic function of any *r.v.* X, *e.g.* the function $\phi(t) = \log(1+t)$ cannot be the *c.f.* of *r.v.* X, since $\phi(0) = \log 1 = 0 \neq 1$.

These conditions are, however, not sufficient. It has been shown (*c.f.* Methods of Mathematical Statistics by H. Cramer) that if $\phi(t)$ is near $t = 0$ of the form,

$$\phi(t) = 1 + o\left(t^{2+\delta}\right), \delta > 0 \qquad \text{... (7·15a)}$$

where $o(t^r)$ divided by t^r tends to zero as $t \to 0$, then $\phi(t)$ cannot be the characteristic function unless it is identically equal to one. Thus, the functions

$$(i)\ \phi(t) = \exp\left(-t^4\right) = 1 + o(t^4), \quad \text{and} \quad (ii)\ \phi(t) = \frac{1}{1+t^4} = 1 + o(t^4),$$

being of the form (7·15*a*) are not characteristic functions, though both satisfy all the necessary conditions.

We give below, without proof, another important result on characteristic functions.

Marcinkiewies Theorem. *If $\phi_X(t)$ is of the form $\exp[P(t)]$, where P (t) is a polynomial in t, then P (t) must be of the form :*

$$P(t) = i\,\alpha\, t - h^2 t^2, \text{ (}\alpha \text{ and } h \text{ being real).} \quad \text{... (7·15b)}$$

From this result, it follows that $\exp\left[-t^{\beta}\right]$ is a characteristic function iff $\beta = 2$.

Polya's Sufficient (But not Necessary) Conditions. *We give below a set of sufficient but not necessary conditions, due to Polya, for a function $\phi(t)$ to be the characteristic function :*

$\phi(t)$ is a characteristic function if

1. $\phi(0) = 0$,
2. $\phi(t) = \phi(-t)$,
3. $\phi(t)$ is continuous,
4. $\phi(t)$ is convex for $t > 0$, *i.e.*, for $t_1, t_2 > 0$, $2\phi\left[\frac{1}{2}(t_1 + t_2)\right] \le \phi(t_1) + \phi(t_2)$
5. $\lim_{t \to \infty} \phi(t) = 0$.

Hence, by Polya's conditions, the functions $e^{-|t|}$ and $\left[1 + |t|\right]^{-1}$ are characteristic functions. However, Polya's conditions are only sufficient and not necessary for a characteristic function.

For example if $X \sim N(\mu, \sigma^2)$, *i.e.*, X has normal distribution with mean μ and variance σ^2, then :

$$\phi(t) = \exp\left[it\mu - (t^2\sigma^2)/2\right] \qquad \text{and} \qquad \phi(-t) \ne \phi(t)$$

Various necessary and sufficient conditions are known, some of which are given below, without proof.

Cramer's Necessary and Sufficient Conditions. *In order that a given, bounded and continuous function $\phi(t)$ should be the characteristic function of a distribution, it is necessary and sufficient that $\phi(0) = 1$ and that the function :*

$$\phi(x, A) = \int_0^A \int_0^A \phi(t-u)\exp\left[(t-u)ix\right] dt\, du$$

is real and non-negative for all x and all A > 0.

Bochner's Criterion. *The necessary and sufficient conditions for a bounded and uniformly continuous function $\phi(t)$ to be the characteristic function of some r.v., X are :*

(i) $\phi(0) = 1$ *and*

(ii) $\phi(t)$ *is non-negative definite* $\Rightarrow \displaystyle\sum_{i=1}^{n}\sum_{j=1}^{n} \phi(t_i - t_j)\, h(t_i) \cdot \overline{h(t_j)} \ge 0$,

for all integral values of n and set of real numbers $t_1, t_2, ..., t_n$ and complex valued function h(t) of t.

7·3·3. Multivariate Characteristic Function

Let $X = \begin{bmatrix} X_1 \\ X_2 \\ \vdots \\ X_n \end{bmatrix}$ and $t = \begin{bmatrix} t_1 \\ t_2 \\ \vdots \\ t_n \end{bmatrix}$, *t real,* be $n \times 1$ column vectors.

Then characteristic function of X is defined as :

$$\phi_X(t) = E\left[e^{it'X}\right] = E\left[e^{i(t_1X_1 + t_2X_2 + \dots + t_nX_n)}\right] \qquad \dots (7\cdot16)$$

We may also write it as $\phi_{X_1, X_2, \dots X_n}(t_1, t_2, \dots t_n)$ or $\phi_X(t_1, t_2, \dots, t_n)$.

Some Properties

(i) $\phi_X(0, 0, \dots, 0) = 1$,

(ii) $\phi_{-X}(t) = \overline{\phi_X(t)}$,

(iii) $|\phi_X(t)| \le 1$,

(iv) $\phi_X(t)$ is uniformly continuous in n-dimensional Euclidian space,

(v) If $f_X(.)$ is $p.d.f.$ of X,

$$\phi_X(t) = \int_{-\infty}^{\infty} \dots \int_{-\infty}^{\infty} f_X(x_1, x_2, \dots, x_n) \exp\left\{i\sum_j t_j x_j\right\} dx_1\, dx_2 \dots dx_n$$

(vi) $\phi_X(t) = \phi_Y(t)$ for all t, then X and Y have the same distribution.

(vii) **Multivariate Inversion Theorem.** If $\int_{-\infty}^{\infty} \dots \int_{-\infty}^{\infty} |\phi_X(t_1, t_2, \dots, t_n)|\, dt_1\, dt_2 \dots dt_n < \infty$ then X is absolutely continuous and has a uniformly continuous $p.d.f.$

$$f_X(x_1, x_2, \dots, x_n) = \frac{1}{(2\pi)^n} \int_{-\infty}^{\infty} \dots \int_{-\infty}^{\infty} \exp\left(-i\sum_j t_j x_j\right) \phi_X(t_1, t_2, \dots, t_n)\, dt_1\, dt_2 \dots dt_n$$

Thus, for two random variables (X, Y) with joint characteristic function :

$\phi(t_1, t_2) = \phi_{X,Y}(t_1, t_2)$, the joint $p.d.f.$ $f_{X,Y}(x, y)$ is given by :

$$f_{X,Y}(x, y) = \frac{1}{(2\pi)^2} \int_{-\infty}^{\infty} \int_{-\infty}^{\infty} \exp\left[-i(t_1x + t_2y)\right] \cdot \phi(t_1, t_2)\, dt_1 dt_2,$$

provided $\int_{-\infty}^{\infty} \int_{-\infty}^{\infty} |\phi(t_1, t_2)|\, dt_1\, dt_2 < \infty$.

(viii) The random variables $X_1, X_2, \dots, X_n$ are (mutually) independent if and only if

$$\phi_{X_1, X_2, \dots X_n}(t_1, t_2, \dots, t_n) = \phi_{X_1}(t_1)\, \phi_{X_2}(t_2) \dots \phi_{X_n}(t_n)$$

7·3·4 Multivariate Moment Generating Function

The $m.g.f.$ of vector $X = (X_1, X_2, \dots, X_n)$, is given by :

$$M_X(t) = E\left(e^{t'X}\right) = E\left[\exp\left(t_1X_1 + t_2X_2 + \dots + t_nX_n\right)\right]$$

Also $$M_X(t) = M_{X_1, X_2, \dots, X_n}(t_1, t_2, \dots, t_n) = E\left[\exp\left(t_1X_1 + t_2X_2 + \dots + t_nX_n\right)\right] \qquad \dots (7\cdot17)$$

In particular, for two variates X_1 and X_2,

$$M_X(t) = M_{X_1, X_2}(t_1, t_2) = E\left(e^{t_1X_1 + t_2X_2}\right) = \sum_{r=0}^{\infty} \sum_{s=0}^{\infty} \frac{t_1^r t_2^s}{r!s!} E\left(X_1^r X_2^s\right), \qquad \dots (7\cdot18)$$

provided it exists for $-h_1 < t_1 < h_1$ and $-h_2 < t_2 < h_2$, where h_1 and h_2 are positive.

$$M_{X_1,X_2}(t_1, 0) = E(e^{t_1 X_1}) = M_{X_1}(t_1) \quad \text{... (7·18a)}$$

$$M_{X_1,X_2}(0, t_2) = E(e^{t_2 X_2}) = M_{X_2}(t_2) \quad \text{... (7·18b)}$$

If $M(t_1, t_2)$ exists, the moments of all orders of X and Y exist and are given by :

$$E\left(X_1^r\right) = \left[\frac{\partial^r M(t_1, t_2)}{\partial t_1^r}\right]_{t_1 = t_2 = 0} = \frac{\partial^r M(0,0)}{\partial t_1^r} \quad \text{... (7·19)}$$

$$E\left(X_2^r\right) = \left[\frac{\partial^r M(t_1, t_2)}{\partial t_2^r}\right]_{t_1 = t_2 = 0} = \frac{\partial^r M(0,0)}{\partial t_2^r} \quad \text{... (7·19a)}$$

$$E\left(X_1^r X_2^s\right) = \left[\frac{\partial^{r+s} M(t_1, t_2)}{\partial t_1^r \, \partial t_2^s}\right]_{t_1 = t_2 = 0} = \frac{\partial^{r+s} M(0,0)}{\partial t_1^r \, \partial t_2^s} \quad \text{... (7·19b)}$$

Cumulant generating function of $\mathbf{X} = (X_1, X_2)'$ is given by :

$$K_{X_1, X_2}(t_1, t_2) = \log M_{X_1, X_2}(t_1, t_2).$$

Example 7·6. *For a distribution, the cumulants are given by :*

$$\kappa_r = n\,\{(r-1)!\},\ n > 0. \text{ Find the characteristic function.}$$

Solution. The cumulant generating function $K(t)$, if it exists, is given by :

$$K(t) = \sum_{r=1}^{\infty} \frac{(it)^r}{r!} \kappa_r = \sum_{r=1}^{\infty} \frac{(it)^r}{r!} n\{(r-1)!\} = n \sum_{r=1}^{\infty} \frac{(it)^r}{r}$$

$$= n\left\{ it + \frac{(it)^2}{2} + \frac{(it)^3}{3} + \dots \right\} = n\,\{-\log(1-it)\}$$

$$= -n \log(1-it) = \log(1-it)^{-n}$$

Also, we have $\quad K(t) = \log \phi(t) = \log(1-it)^{-n} \quad \Rightarrow \quad \phi(t) = (1-it)^{-n}.$

Note : This is the characteristic function of Gamma distribution with parameter n *(Chapter 9).*

Example 7·7. *The moments about origin of a distribution are given by :*

$$\mu_r' = \frac{\Gamma(v+r)}{\Gamma(v)}. \text{ Find the characteristic function.}$$

Solution. We have

$$\phi(t) = \sum_{r=0}^{\infty} \frac{(it)^r}{r!} \mu_r' = \sum_{r=0}^{\infty} \frac{(it)^r}{r!} \cdot \frac{\Gamma(v+r)}{\Gamma(v)} = \sum_{r=0}^{\infty} \frac{(it)^r}{r!} \cdot \frac{(v+r-1)!}{(v-1)!}$$

$$= \sum_{r=0}^{\infty} (it)^r \cdot {}^{v+r-1}C_r = \sum_{r=0}^{\infty} (-1)^r \cdot {}^{-v}C_r \, (it)^r$$

$$\left[\because\ {}^{-v}C_r = (-1)^r \cdot {}^{v+r-1}C_r \quad \Rightarrow \quad (-1)^r \cdot {}^{-v}C_r = {}^{v+r-1}C_r \right]$$

$$\therefore \qquad \phi(t) = \sum_{r=0}^{\infty} {}^{-v}C_r(-it)^r = (1-it)^{-v}$$

Example 7·8. *Examine if the following functions represent the characteristic function of some r.v. X. Justify your answer.*

(*i*) $\frac{2}{5}e^{it} + 1$, (*ii*) $\exp\left[iat + t^2\right]$, (*iii*) $\frac{1}{1-t}$, (*iv*) $1 - it$

(*v*) e^{iat}, (*vi*) $\cos at$, (*vii*) $\frac{1}{2}\left(1 + e^{it}\right)$,

Solution. We will first verify if the necessary conditions for $\phi(t)$ to be *c.f.* are satisfied or not.

(*i*) $\phi(t) = \frac{2}{5}e^{it} + 1$; $\quad |\phi(t)| = \frac{2}{5}\left|e^{it}\right| + 1 = \frac{2}{5} + 1 > 1$

$$\left[\because \left|e^{it}\right| = |\cos t + i \sin t| = \sqrt{\cos^2 t + \sin^2 t} = 1\right]$$

or $\qquad \phi(0) = \frac{2}{5}e^0 + 1 = \frac{2}{5} + 1 \neq 1.$

Hence, $\phi(t) = \frac{2}{5}e^{it} + 1$, is not a *c.f.*

(*ii*) $\qquad \phi(t) = e^{iat+t^2} \Rightarrow |\phi(t)| = \left|e^{iat}\right| . \left|e^{t^2}\right| = 1 . e^{t^2} > 1 \left[\because e^{t^2} = 1 + t^2 + \frac{(t^2)^2}{2!} + ... > 1\right]$

$\Rightarrow \qquad e^{iat + t^2}$ cannot be a *c.f.*

(*iii*) $\qquad \phi(t) = \frac{1}{1-t} \Rightarrow \phi(-t) = \frac{1}{1+t}$ and $\overline{\phi(t)} = \frac{1}{1-t} \neq \phi(-t)$

$\Rightarrow \qquad \phi(t)$ cannot be a *c.f.*

(*iv*) $\qquad \phi(t) = 1 - it \Rightarrow |\phi(t)| = \sqrt{\{1^2 + (-t)^2\}} = \sqrt{1+t^2} > 1$

$\Rightarrow \qquad \phi(t)$ cannot be a *c.f.*

(*v*) $\qquad \phi(t) = e^{iat}$. $\qquad$ Here $\phi(0) = 1$, $|\phi(t)| = 1$ and $\overline{\phi(t)} = \phi(-t)$

The necessary conditions are satisfied. Thus, $\phi(t)$ may be the *c.f.* of some *r.v.* X. Consider a *r.v.* X degenerate at the point '$X = a$' so that $P(X = a) = 1$. Then

$$\phi_X(t) = E(e^{itX}) = e^{iat} . P(X = a) = e^{iat} . 1 = e^{iat}$$

Hence, $\phi_X(t) = e^{iat}$, is the *c.f.* of degenerate random variable at $X = a$.

(*vi*) $\phi(t) = \cos at$, satisfies the necessary conditions because:

$$\phi(0) = \cos 0 = 1,$$

$$|\phi(t)| = |\cos at| \leq 1 \text{ and } \overline{\phi(t)} = \cos at = \cos(-at) = \phi(-t)$$

Hence, $\phi(t)$ may be *c.f.* of some *r.v.* X. Consider the *r.v.* X. which assumes only two values a and $(-a)$, each with equal probability $\left(\frac{1}{2}\right)$.

$$\phi(t) = E(e^{itX}) = e^{iat}\cdot\left(\frac{1}{2}\right) + e^{-iat}\cdot\left(\frac{1}{2}\right) = \frac{1}{2}\left(e^{iat} + e^{-iat}\right) = \cos at$$

Hence, $\phi(t) = \cos at$ is the *c.f.* of *r.v.* X with probability distribution:

x	a	$-a$
$p(x)$	$1/2$	$1/2$

(*vii*) $$\phi(t) = \frac{1}{2}\left(1 + e^{it}\right) = \left(\frac{1}{2} + \frac{1}{2}e^{it}\right)$$

Consider the Bernoulli variate X which takes only two values 1 and 0 each with equal probability $\left(1/2\right)$, *i.e.*,

$$P(X = 1) = \frac{1}{2} \text{ and } P(X = 0) = \frac{1}{2}$$

$$\phi_X(t) = E(e^{itX}) = e^{it.1}P(X = 1) + e^{it.0}P(X = 0) = \frac{1}{2}\left(e^{it} + 1\right)$$

Hence, $\phi(t) = \frac{1}{2}\left(1 + e^{it}\right)$ is the *c.f.* of Bernoulli variate with *p.m.f.*

x	1	0
$p(x)$	$1/2$	$1/2$

Example 7·9. *If $\phi(t)$ is the characteristic function of a r.v. X so that* $\phi(t) = \phi_R(t) + i\,\phi_I(t)$ ...(*)
where $\phi_R(t)$ is the real part and $\phi_I(t)$ is the imaginary part of $\phi(t)$, then

(*a*) $1 - \phi_R(2t) \le 4\{1 - \phi_R(t)\}$

(*b*) $|\phi_I(t)| \le \sqrt{\{1 - \phi_R(2t)\}}$

Solution. By definition :

$$\begin{aligned}\phi(t) = \phi_X(t) = E(e^{itX}) &= E[\cos tX + i\sin tX]\\ &= E(\cos tX) + i\,E(\sin tX)\\ &= \phi_R(t) + i\,\phi_I(t), \quad \text{(say)}\end{aligned} \qquad \text{... } (i)$$

so that :

$$\left.\begin{aligned}\phi_R(t) &= \text{Real part of } \phi(t) = E(\cos tX)\\ \phi_I(t) &= \text{Imaginary part of } \phi(t) = E(\sin tX)\end{aligned}\right\} \qquad \text{... } (ii)$$

(*a*)
$$\begin{aligned}1 - \phi_R(2t) &= 1 - E(\cos 2tX) \qquad \text{[from } (ii)\text{]}\\ &= E\left[1 - \cos 2tX\right] = E\left[2\sin^2 tX\right] = 2\cdot E\left(\sin^2 tX\right) = 2\cdot E\left[1 - \cos^2 tX\right]\\ &= 2E\left[\left(1 + \cos tX\right)\left(1 - \cos tX\right)\right]\\ &\le 2 \times 2\,E\,(1 - \cos tX) \qquad [\because (1 + \cos tX) \le 2]\\ &= 4\left[1 - E\left(\cos tX\right)\right]\end{aligned}$$

$$\therefore \quad 1 - \phi_R(2t) \le 4[1 - \phi_R(t)] \qquad \text{[from } (ii)\text{]}$$

(*b*)
$$|\phi_I(t)|^2 = \left|E\left(\sin tX\right)\right|^2 \le E\left(\sin^2 tX\right)$$

$$\left[\because \sigma_X^2 = E\left(X^2\right) - \{E(X)\}^2 \ge 0 \Rightarrow \{E(X)\}^2 \le E\left(X^2\right)\right]$$

$$\Rightarrow \quad |\phi_I(t)|^2 \leq E\left[\frac{1}{2}(1-\cos 2tX)\right] = \frac{1}{2}\left[1-E(\cos 2tX)\right]$$

$$\Rightarrow \quad |\phi_I(t)|^2 \leq \frac{1}{2}\left[1-\phi_R(2t)\right] \leq 1-\phi_R(2t)$$

$$\therefore \quad |\phi_I(t)| \leq \sqrt{\{1-\phi_R(2t)\}}$$

Example 7·10. *If $\phi(t)$ is the real valued characteristic function of a r.v. X, then prove that :*

(*i*) $1-\phi(2\theta) \leq 4\left[1-\phi(\theta)\right]$; (*ii*) $1+\phi(2\theta) \geq 2\left[\phi(\theta)\right]^2$

Solution. $\phi(t) = \phi_X(t) = E(e^{itX}) = E(\cos tX) + iE(\sin tX)$

Since $\phi(t)$ is given to be real valued, we have $E(\sin tX) = 0$.

$$\therefore \quad \phi(t) = E[\cos(tX)] \qquad \text{... (*)}$$

(*i*) We have : $[\sin\theta X]^2 = \left[2\sin\left(\frac{\theta X}{2}\right)\cos\left(\frac{\theta X}{2}\right)\right]^2 = 4\sin^2\left(\frac{\theta X}{2}\right)\cos^2\left(\frac{\theta X}{2}\right)$

$$\Rightarrow \quad \frac{1}{2}(1-\cos 2\theta X) \leq 4\sin^2\left(\frac{\theta X}{2}\right) \qquad \left[\because \cos\left(\frac{\theta X}{2}\right) \leq 1\right]$$

$$\Rightarrow \quad 1-\cos 2\theta X \leq 4(1-\cos\theta X)$$

Taking expectation on both sides and using (*), we get

$$1-\phi(2\theta) \leq 4[1-\phi(\theta)]$$

(*ii*) We have :

$$1+\cos 2\theta X = 2\cos^2\theta X$$

Taking expectation on both sides and using (*), we get

$$1+\phi(2\theta) = 2E(\cos^2\theta X) \geq 2[E(\cos\theta X)]^2$$

$$[\because E(X^2) \geq \{E(X)\}^2, \text{ by Cauchy-Schwartz Inequality}]$$

$$\Rightarrow \quad 1+\phi(2\theta) > 2[\phi(\theta)]^2$$

Example 7·11. *Show that*

$$e^{itx} = 1+(e^{it}-1)x^{(1)}+(e^{it}-1)^2 \cdot \frac{x^{(2)}}{2!}+\ldots+(e^{it}-1)^r \cdot \frac{x^{(r)}}{r!}+\ldots, \qquad \text{... (*)}$$

where $x^{(r)} = x(x-1)(x-2)\ldots(x-r+1)$. *Hence show that*

$$\mu'_{(r)} = \left[D^r\phi(t)\right]_{t=0}, \text{ where } D = \frac{d}{d(e^{it})} \text{ and } \mu'_{(r)} \text{ is the } r^{th} \text{ factorial moment.}$$

Solution. We have

$$\text{R.H.S.} = 1+(e^{it}-1)x^{(1)}+(e^{it}-1)^2 \cdot \frac{x^{(2)}}{2!}+\ldots+(e^{it}-1)^r \cdot \frac{x^{(r)}}{r!}+\ldots$$

$$= 1+(e^{it}-1)({}^xC_1)+(e^{it}-1)^2({}^xC_2)+(e^{it}-1)^3({}^xC_3)+\ldots+(e^{it}-1)^r({}^xC_r)$$

$$= [1+(e^{it}-1)]^x = e^{itx} = \text{L.H.S.}$$

By def, $\phi(t) = \int_{-\infty}^{\infty} e^{itx} f(x)\,dx$

$$= \int_{-\infty}^{\infty} \left\{ 1 + (e^{it} - 1)x^{(1)} + (e^{it} - 1)^2 . \frac{x^{(2)}}{2!} + ... + (e^{it} - 1)^r . \frac{x^{(r)}}{r!} + ... \right\} f(x) dx \qquad \text{[from(*)]}$$

$$= 1 + (e^{it} - 1) \int_{-\infty}^{\infty} x^1 f(x)\, dx + \frac{1 + (e^{it} - 1)^2}{2!} \int_{-\infty}^{\infty} x^{(2)} f(x)\, dx + ... + \frac{(e^{it} - 1)^r}{r!} \int_{-\infty}^{\infty} x^{(r)} f(x)\, dx + ...$$

$$\therefore \qquad \left[D^r \phi(t)_{t=0} \right] = \left[\frac{d^r \phi(t)}{d(e^{it})^r} \right]_{t=0} = \int_{-\infty}^{\infty} x^{(r)} f(x) dx = \mu'_{(r)}$$

where, $\mu'_{(r)}$ is the rth factorial moment.

7·4. SOME IMPORTANT THEOREMS ON CHARACTERISTIC FUNCTIONS

Theorem 7·1. *If $\phi_X(t)$ is the characteristic function of a r.v. X, then its reciprocal $\{1/\phi_X(t)\}$ is also a characteristic function if and only if X belongs to degenerate family of distributions.*

Proof. **If part (Condition is Sufficient)**

Let X be a *r.v.* degenerate at the point $X = a$, so that : $\quad P(X = a) = 1 \qquad$ ...(*)

$$\phi_X(t) = E\left(e^{itX}\right) = e^{iat} \cdot P(X = a) = e^{iat} \qquad \text{[From(*)]}$$

$$\Rightarrow \qquad [\phi_X(t)]^{-1} = e^{-iat}$$

$\Rightarrow \qquad [\phi_X(t)]^{-1}$ is the *c.f.* of a degenerate *r.v.* at the point $X = -a$.

Only if Part (Condition is Necessary).

If both $\phi_X(t)$ and $[\phi_X(t)]^{-1}$ are the *c.f.'s* then :

$$|\phi_X(t)| \le 1 \qquad ... (i)$$

and $\quad \left|[\phi_X(t)]^{-1}\right| \le 1 \quad \Rightarrow \quad |\phi_X(t)|^{-1} \le 1 \quad \Rightarrow \quad |\phi_X(t)| \ge 1 \qquad$ ... (*ii*)

From (*i*) and (*ii*), we get

$$|\phi_X(t)| = 1 \quad \Rightarrow \quad \phi_X(t) = e^{iat} \text{ for same } a. \qquad ... (iii)$$

But (*iii*) is the *c.f.* of a *r.v.* degenerate at the point $(X = a)$. Hence, $\phi_X(t)$ and $[\phi_X(t)]^{-1}$ are both the *c.f.'s* if the *r.v.* X belongs to degenerate family of distributions.

Theorem 7·2. (Convolution Theorem). *If $f_1(.)$ and $f(.)$ are the p.d.f.'s of two distributions, then*

$$f(x) = \int_{-\infty}^{\infty} f_1(y)\, f_2(x - y)\, dy, \qquad ... (*)$$

is called their convolution. [See Theorem 5·5(a)].

Let $\phi_1(t)$, $\phi_2(t)$ *and* $\phi(t)$ *be the characteristic functions of the r. v.'s with p.d.f.'s* $f_1(\cdot)$, $f_2(\cdot)$ *and* $f(\cdot)$ *respectively. Then* $\phi(t) = \phi_1(t)\cdot\phi_2(t)$.

Proof. The characteristic function $\phi(t)$ of the convolution of the *p.d.f.'s* $f_1(\cdot)$ and $f_2(\cdot)$ is given by :

$$\phi(t) = E\left(e^{itX}\right) = \int_{-\infty}^{\infty} e^{itx} f(x)\,dx$$

$$= \int_{-\infty}^{\infty}\int_{-\infty}^{\infty} e^{itx} f_1(y)\cdot f_2(x-y)\,dy\,dx$$

Since $|\phi(t)| \le 1$ *i.e.,* the integrand in the R.H.S. is bounded by an integrable function, we can change the order of integration by Fubini's theorem.

$$\therefore \quad \phi(t) = \int_{-\infty}^{\infty}\left[e^{ity}\cdot f_1(y)\int_{-\infty}^{\infty} e^{it(x-y)}\cdot f_2(x-y)\,dx\right]dy$$

$$= \int_{-\infty}^{\infty}\left[e^{ity} f_1(y)\left(\int_{-\infty}^{\infty} e^{itu} f_2(u)\,du\right)\right]dy\,; \quad u = x-y$$

$$= \int_{-\infty}^{\infty} e^{ity} f_1(y)\,.\,\phi_2(t)\,dy = \phi_2(t)\int_{-\infty}^{\infty} e^{ity} f_1(y)\,dy$$

$$\Rightarrow \quad \phi(t) = \phi_2(t)\,.\,\phi_1(t) = \phi_1(t)\,.\,\phi_2(t).$$

Theorem 7·3. *If* $\phi(t)$ *is the characteristic function of some random variable X, then :*

(*i*) $\overline{\phi(t)}$ (*ii*) $\phi(-t)$,

(*iii*) $|\phi(t)|^2$ (*iv*) *Real Part of* $\phi(t)$, *are all characteristic functions.*

Proof. Since $\phi_X(t)$ is the *c.f.* of *r.v.* X, it satisfies the necessary conditions :

(*a*) $\phi(0) = 1$ (*b*) $|\phi(t)| \le 1$ (*c*) $\overline{\phi_X(t)} = \phi_X(-t)$.

(*i*) and (*ii*). Let $\psi(t) = \overline{\phi_X(t)} = \overline{E\left(e^{itX}\right)} = E\left(e^{-itX}\right) = \phi_X(-t) = \phi_{-X}(t)$

Hence, $\overline{\phi_X(t)}$ and $\phi_X(-t)$, both are *c.f.* of the *r.v.* (–X).

(*iii*) $$|\phi(t)|^2 = \phi_X(t)\cdot\overline{\phi_X(t)} = \phi_X(t)\cdot\phi_X(-t) = \phi_X(t)\cdot\phi_{-X}(t)$$

which is the *c.f.* of the convolution of X and –X.

$\Rightarrow$ $|\phi(t)|^2$ is a *c.f.*

(*iv*) Let $\phi(t) = \phi_1(t) + i\phi_2(t) \Rightarrow \text{Real}\left[\phi(t)\right] = \phi_1(t)$

Since $\phi(t)$ is a characteristic function, $\phi(0) = \phi_1(0) + i\phi_2(0) = 1$

Comparing real and imaginary parts on both sides, we get : $\phi_1(0) = 1$ ... (*)

Since $\phi(t)$ and $\overline{\phi(t)}$ are both characteristic functions, they are both non-negative definite

$$\Rightarrow \quad \left[\frac{\phi(t) + \overline{\phi(t)}}{2}\right] = \text{Real}\,[\phi(t)] = \phi_1(t) \qquad \text{... (**)}$$

is also non-negative definite.

Moreover, since $\phi(t)$ and $\overline{\phi(t)}$ are *c.f.ns* they are continuous and from (**), $\phi_1(t)$, being a linear function of continuous functions is also continuous.

Hence, by Bochner's Criteria, $\phi_1(t) = \text{Real}\,[\phi(t)]$ is a characteristic function.

Theorem 7·4. *The product of characteristic functions is also a characteristic function.*

If $\phi_i(t) = \phi_{X_i}(t)\,;\, i = 1, 2, ..., n,$ *then* $\prod_{i=1}^{n} \phi_i(t) = \prod_{i=1}^{n} \phi_{X_i}(t)$, *is also a characteristic function.*

Proof. If $X_1, X_2, ..., X_n$ are independent *r.v.'s* then

$$\phi_{\sum_{i=1}^{n} X_i}(t) = \prod_{i=1}^{n} \phi_{X_i}(t)$$

Hence, the product of *c.f.ns viz,* $\prod_{i=1}^{n} \phi_i(t)$ is the *c.f.* of the sum of independent random variables $X_1, X_2, ..., X_n$.

Remark. If $\phi_1(t), \phi_2(t)$ are *c.f.'s* of two *r.v.'s.* X_1 and X_2, then $\phi_1(t) \cdot \phi_2(t)$ is the *c.f.* of sum of independent random variables X_1 and X_2.

Moreover, $\phi_1(t) \cdot \phi_2(t)$ is also the *c.f.* of the convolution of X and *Y*. (*c.f.* Theorem. 7·2)

Theorem 7·5. *If* $\phi_1(t)$ *and* $\phi_2(t)$ *are c.f.' ns then their linear combination :*

$$\psi(t) = (a\,\phi_1 + b\,\phi_2), \text{ is a c.f. if } a \geq 0,\ b \geq 0 \text{ and } a + b = 1.$$

Proof. Since ϕ_1 and ϕ_2 are *c.f.'s,* $\phi_1(0) = \phi_2(0) = 1$

$$\psi(t) = a\,\phi_1 + b\,\phi_2 \quad \Rightarrow \quad \psi(0) = a\,\phi_1(0) + b\,\phi_2(0) = 1 \quad \Rightarrow \quad a + b = 1.$$

Also, ϕ_1 *and* ϕ_2 are uniformly continuous, $\psi(t)$, being their linear combination is also uniformly continuous.

Further, since ϕ_1 and ϕ_2 are *c.f.'s*, by Bochner's criteria, they are non-negative definite, *i.e.,*

$$\sum_j \sum_{k=1}^{n} \phi_1(t_j - t_k) w_j \,.\, \overline{w_k} \geq 0 \quad \text{and} \quad \sum_j \sum_k \phi_2\,(t_j - t_k)\, w_j\, \overline{w_k} \geq 0 \qquad \text{... (*)}$$

$$\Rightarrow \quad \sum_j \sum_{k=1}^{n} \psi(t_j - t_k) \,.\, w_j \,.\, \overline{w_k} = a \sum_j \sum_k \phi_1(t_j - t_k)\, w_j\, \overline{w_k} + b \sum_j \sum_k \phi_2(t_j - t_k) w_j\, \overline{w_k} \geq 0$$

[From (*)]

$\therefore$ $\psi(t)$ satisfies all the conditions of Bochner's criteria.

Hence, $\psi(t) = a\phi_1(t) + b\phi_2(t)$, is a characteristic function.

Generalisation. The above result can be generalised to n *c.f.'s.*

If $\phi_i(t), i = 1, 2, ..., n$ are n c.f.'s, then their proper linear combination,

$$\sum_{i=1}^{n} \left[a_i\, \phi_i(t)\right] \text{ is a c.f.} \quad \text{if } a_i \geq 0 \text{ for } i = 1, 2, ..., n \quad \text{and} \quad \sum_{i=1}^{n} a_i = 1$$

Theorem 7·6. **Inversion Theorem. (Levy Theorem)**

Let $F(x)$ and $\phi(t)$ denote, respectively, the distribution function and the characteristic function of a r.v. X. If $(a-h,\ a+h)$ is the continuity interval of the distribution function $F(x)$, then

$$F(a+h)-F(a-h)=\lim_{T\to\infty}\frac{1}{\pi}\int_{-T}^{T}\frac{\sin ht}{t}\,.\,e^{-ita}\phi(t)\,dt \qquad \text{... (7·20)}$$

Proof. We establish this theorem for continuous random variable. In proving this theorem we need the following important result from real analysis :

If
$$S(h,T)=\frac{2}{\pi}\int_{0}^{T}\frac{\sin ht}{t}\,dt, \qquad \text{... }(i)$$

where, h is real and $T>0$, then

$$\lim_{T\to\infty}S(h,T)=\begin{cases}1, & \text{for } h>0\\ 0 & \text{for } h=0\\ -1, & \text{for } h<0\end{cases} \qquad \text{... }(ii)$$

To prove the theorem, let us consider

$$J=\frac{1}{\pi}\int_{-T}^{T}\frac{\sin ht}{t}\,e^{-ita}\,\phi(t)\,dt \qquad \text{... }(iii)$$

$$=\frac{1}{\pi}\int_{-T}^{T}\left\{\frac{\sin ht}{t}\,e^{-ita}\int_{-\infty}^{\infty}e^{itx}f(x)\,dx\right\}dt$$

where, $dF(x)=f(x)\,dx$, $f(x)$ being the *p.d.f.* of X.

$$\therefore\qquad J=\frac{1}{\pi}\int_{-T}^{T}\left\{\int_{-\infty}^{\infty}\frac{\sin ht}{t}\,e^{it(x-a)}\,f(x)\,dx\right\}dt \qquad \text{... }(iv)$$

Since the limits of integral over t are finite and the integral in the bracket $\{\ \}$ is absolutely convergent, *i.e.*,

$$\int_{-\infty}^{\infty}\left|\frac{\sin ht}{t}\cdot e^{it(x-a)}f(x)\right|dx\le\int_{-\infty}^{\infty}\left|\frac{\sin ht}{t}\right|f(x)\,dx \qquad (\because\ |e^{itx}|=1)$$

$$=\int_{-\infty}^{\infty}\left|\frac{\sin ht}{ht}\cdot h\right|f(x)\,dx\le|h|\int_{-\infty}^{\infty}f(x)\,dx=|h|,$$

By Fubini's theorem we can change the order of integration in (iv).

$$\therefore\qquad J=\frac{1}{\pi}\int_{-\infty}^{\infty}\left\{\int_{-T}^{T}\frac{\sin ht}{t}\,.\,e^{it(x-a)}\,f(x)\,dt\right\}dx$$

$$=\frac{1}{\pi}\left[\int_{-\infty}^{\infty}\left\{\int_{-T}^{T}\frac{\sin ht}{t}\,.\cos t(x-a)\,dt\right\}f(x)\,dx\right]$$

$$+\frac{i}{\pi}\left[\int_{-\infty}^{\infty}\left\{\int_{-T}^{T}\frac{\sin ht}{t}\,.\sin t(x-a)\,dt\right\}f(x)\,dx\right]$$

Since $\dfrac{\sin ht}{t}\cdot\cos t\,(x-a)$ is an even function of t and $\dfrac{\sin ht}{t}\cdot\sin t\,(x-a)$ is an odd function of t, the second integral vanishes and we get

$$J = \frac{2}{\pi} \int_{-\infty}^{\infty} \left\{ \int_0^T \frac{\sin ht}{t} . \cos t(x-a)\, dt \right\} f(x)\, dx$$

$$= \int_{-\infty}^{\infty} g(x, T)\, f(x)\, dx,$$

where,
$$g(x, T) = \frac{2}{\pi} \int_0^T \frac{\sin ht}{t} . \cos t(x-a)\, dt$$

$$= \frac{1}{\pi} \int_0^T \frac{1}{t} \{ 2 \cos t(x-a) . \sin ht \}\, dt$$

$$= \frac{1}{\pi} \int_0^T \frac{1}{t} \{ \sin(x-a+h)t - \sin(x-a-h)\, t \}\, dt$$

$$= \frac{1}{2} \int_0^T \frac{2}{\pi} . \frac{\sin(x-a+h)t}{t} \cdot dt - \frac{1}{2} \int_0^T \frac{2}{\pi} \cdot \frac{\sin(x-a-h)}{t} dt$$

$$= \frac{1}{2} . S(x-a+h, T) - \frac{1}{2} . S(x-a-h, T) \qquad \text{[Using } (i)\text{]}$$

$$= \frac{1}{2} S\{x-(a-h), T\} - \frac{1}{2} S\{x-(a+h), T\},$$

$$= I_1 - I_2, \text{ (say)} \qquad \dots (vi)$$

Since $S(h, T)$ is bounded, the right hand side in (*vi*) is bounded and consequently $g(x, T)$ is bounded, *i.e.*,

$$| g(x, T) | < | k |, \text{ say}, \forall T. \qquad \dots (vii)$$

Hence, by Dominated Convergence Theorem (DCT), we get from (*v*) :

$$\lim_{T \to \infty} J = \lim_{T \to \infty} \int_{-\infty}^{\infty} g(x, T)\, f(x)\, dx$$

$$= \int_{-\infty}^{\infty} \lim_{T \to \infty} g(x, T) f(x)\, dx \qquad \dots (viii)$$

We consider the values of I_1 and I_2 in (*vi*), on using (*ii*), for different ranges of x in the following table.

$x \rightarrow$	$x < a - h$	$x = a - h$	$a - h < x < a + h$	$x = a + h$	$x > a + h$	
I_1	–1/2	0	1/2	1/2	1/2	as $T \to \infty$
I_2	–1/2	–1/2	–1/2	0	1/2	as $T \to \infty$
$I_1 - I_2$	0	1/2	1	1/2	0	

$$\therefore \quad \lim_{T \to \infty} g(x, T) = \begin{cases} 0\,, & x < a-h \\ \frac{1}{2}, & x = a-h \\ 1\,, & a-h < x < a+h \\ \frac{1}{2}, & x = a+h \\ 0\,, & x > a+h \end{cases} \qquad \dots (ix)$$

Substituting these values in (*viii*), we get

$$\lim_{T\to\infty} J = \int_{-\infty}^{a-h} \lim_{T\to\infty} g(x,T)f(x)\,dx + \int_{a-h}^{a+h} \lim_{T\to\infty} g(x,T)f(x)\,dx + \int_{a+h}^{\infty} \lim_{T\to\infty} g(x,T)\,f(x)\,dx$$

$$= 0 + \int_{a-h}^{a+h} f(x)\,dx + 0 = \int_{a-h}^{a+h} f(x)\,dx = P(a-h \le X \le a+h)$$

$$= P(X \le a+h) - P(X \le a-h)$$

$$= F(a+h) - F(a-h) \qquad \ldots (x)$$

Substituting from (*x*) in (*iii*), we get

$$F(a+h) - F(a-h) = \lim_{T\to\infty} \frac{1}{\pi} \int_{-T}^{T} \frac{\sin ht}{t} e^{-ita} \phi(t)\,dt$$

Corollary. If $\phi(t)$ is absolutely integrable over R^1, *i.e.*, if $\int_{-\infty}^{\infty} |\phi(t)|\,dt < \infty$, then the derivative of $F(x)$ exists, which is bounded, continuous on R^1 and is given by

$$f(x) = F'(x) = \frac{1}{2\pi}\int_{-\infty}^{\infty} e^{-itx}\phi(t)\,dt, \qquad \ldots (7{\cdot}21)$$

for every $x \in R^1$.

Proof. In the above theorem replacing a by x and on dividing by $2h$, we have

$$\frac{F(x+h)-F(x-h)}{2h} = \frac{1}{2\pi}\cdot \lim_{T\to\infty} \int_{-T}^{T} \frac{\sin ht}{ht} e^{-itx} \phi(t)\,dt$$

$$= \frac{1}{2\pi} \int_{-\infty}^{\infty} \frac{\sin ht}{ht} e^{-itx} \phi(t)\,dt$$

$$\therefore \quad \lim_{h\to 0} \frac{F(x+h)-F(x-h)}{2h} = \frac{1}{2\pi} \lim_{h\to 0} \int_{-\infty}^{\infty} \frac{\sin ht}{ht} e^{-itx} \phi(t)\,dt$$

Since $\int_{-\infty}^{\infty} |\phi(t)|\,dt < \infty$, the integrand on the right-hand side is bounded by an integrable function and hence by Dominated Convergence Theorem, we get

$$\lim_{h\to 0} \frac{F(x+h)-F(x-h)}{2h} = \frac{2}{2\pi} \int_{-\infty}^{\infty} \lim_{h\to 0}\left(\frac{\sin ht}{ht}\right) e^{-itx} \phi(t)\,dt$$

By mean value theorem of differential calculus, we have

$$\lim_{h\to 0} \frac{F(x+h)-F(x-h)}{2h} = F'(x) = f(x)$$

where, $f(.)$ is the *p.d.f.* corresponding to $\phi(t)$. Thus,

$$f(x) = F'(x) = \frac{1}{2\pi} \int_{-\infty}^{\infty} e^{-itx} \phi(t)\,dt, \text{ as desired}$$

Remark. Consider the function F_c defined by $\quad F_c = \int_{-c}^{c} e^{-itx}\phi(t)\,dt$

If $F'(x) = f(x)$ exists, then

$$\lim_{c\to\infty}\frac{F_c}{2c} = \lim_{c\to\infty}\frac{1}{2}\int_{-c}^{c}\phi(t)\,e^{-itx}dt = \lim_{c\to\infty}\left\{\frac{1}{2c}\cdot 2\pi f(x)\right\} = 0 \qquad \text{[From (7·21)]}$$

Hence, $\dfrac{F_c}{2c}\to 0$ at all points where $F(x)$ is continuous. In other words, if the probability distribution is continuous, $\dfrac{F_c}{2c}\to 0$ as $c\to\infty$.

If, however, the frequency function is discontinuous, *i.e.*, distribution is discrete, consider one point of discontinuity, say, the frequency f_j at $x = x_j$. Then the contribution of x_j to $\phi(t)$ is $f_i\, e^{itx_j}$ and hence its contributions to F_c will be $\int_{-c}^{c} e^{itx_j}e^{-itx}\,dt$. Therefore,

$$\lim_{c\to\infty}\frac{F_c}{2c} = \lim_{c\to\infty}\frac{1}{2c}\, f_j\int_{-c}^{c} e^{it(x_j-x)}\,dt = \lim_{c\to\infty}\frac{1}{2c}\, f_j\left[\frac{e^{it(x_j-x)}}{i(x_j-x)}\right]_{-c}^{c} = \begin{cases}0, & \text{for } x\neq x_j\\ f_j, & \text{for } x = x_j\end{cases}$$

Hence, if $F_c/2c\to 0$ at a point, there is no discontinuity in the distribution function at that point, but if it tends to a positive number f_j, the distribution is discontinuous at that point and the frequency is f_j. This gives us a criterion whether a given characteristic function represents a continuous distribution or not.

Theorem 7·7. Uniqueness Theorem of Characteristic Functions. *Characteristic function uniquely determines the distribution, i.e., a necessary and sufficient condition for two distributions with p.d.f 's $f_1(.)$ and $f_2(.)$ to be identical is that their characteristic functions $\phi_1(t)$ and $\phi_2(t)$ are identical.*

Proof. If $f_1(.) = f_2(.)$, then from the definition of characteristic function, we get

$$\phi_1(t) = \int_{-\infty}^{\infty} e^{itx} f_1(x)\,dx = \int_{-\infty}^{\infty} e^{itx} f_2(x)\,dx = \phi_2(t)$$

Conversely if $\phi_1(t) = \phi_2(t)$ then from corollary to *Theorem 7·6,* we get

$$f_1(x) = \frac{1}{2\pi}\int_{-\infty}^{\infty} e^{-itx}\phi_1(t)\,dt = \frac{1}{2\pi}\int_{-\infty}^{\infty} e^{-itx}\phi_2(t)\,dt = f_2(x)$$

Remark. This is one of the most fundamental theorems in the distribution theory. It implies that corresponding to a given distribution, there is only one characteristic function and corresponding to a given characteristic function, there is only one distribution. This *one-to-one correspondence between characteristic functions and the p.d.f.'s enables us to identify the form of the p.d.f. from that of characteristic function.*

Thus, if the c.f. (or the m.g.f. if it exists) of one random variable approaches that of another random variable, then the distribution (p.d.f. or p.m.f.) of the first random variable approaches that of the second random variable under the same limiting conditions.

Theorem 7·8. *Necessary and sufficient condition for the random variables X_1 and X_2 to be independent is that their joint characteristic function is equal to the product of their individual characteristic function, i.e.,*

$$\phi_{X_1,X_2}(t_1,t_2) = \phi_{X_1}(t_1)\cdot\phi_{X_2}(t_2) \qquad \text{... (*)}$$

Proof.

(i) *Condition is necessary.* If X_1 and X_2 are independent then we have to show that (*) holds. By def.,

$$\phi_{X_1,X_2}(t_1,t_2) = E\left(e^{it_1X_1+it_2X_2}\right) = E\left(e^{it_1X_1}\cdot e^{it_2X_2}\right)$$

$$= E\left(e^{it_1X_1}\right) E\left(e^{it_2X_2}\right) \qquad (\because X_1, X_2 \text{ are independent.})$$

$$= \phi_{X_1}(t_1)\cdot\phi_{X_2}(t_2), \text{ as required,}$$

(ii) *Condition is sufficient.* We have to show that if (*) holds, then X_1 and X_2 are independent.

Let $f_{X_1,X_2}(x_1,x_2)$ be the joint *p.d.f.* of X_1 and X_2 and $f_1(x_1)$ and $f_2(x_2)$ be the marginal *p.d.f's* X_1 and X_2 respectively. Then by definition (for continuous *r.v.'s*), we get

$$\phi_{X_1}(t_1) = \int_{-\infty}^{\infty} e^{it_1x_1} f_1(x_1)\,dx_1 \quad \text{and} \quad \phi_{X_2}(t_2) = \int_{-\infty}^{\infty} e^{it_2x_2} f_2(x_2)\,dx_2$$

$$\therefore \quad \phi_{X_1}(t_1)\,\phi_{X_2}(t_2) = \left\{\int_{-\infty}^{\infty} e^{it_1x_1} f_1(x_1)\,dx_1\right\}\left\{\int_{-\infty}^{\infty} e^{it_2x_2} f_2(x_2)\,dx_2\right\}$$

$$= \int_{-\infty}^{\infty}\int_{-\infty}^{\infty} e^{i(t_1x_1+t_2x_2)} f_1(x_1) f_2(x_2)\,dx_1\,dx_2 \qquad \dots (**)$$

by Fubini's Theorem, since the integrand is bounded by an integrable function.

Also by *def.*, $$\phi_{X_1,X_2}(t_1,t_2) = \int_{-\infty}^{\infty}\int_{-\infty}^{\infty} e^{i(t_1x_1+t_2x_2)} f(x_1,x_2)\,dx_1\,dx_2 \qquad \dots (***)$$

If (*) holds, we get from (**) and (***)

$$\phi_{X_1,X_2}(t_1,t_2) = \int_{-\infty}^{\infty}\int_{-\infty}^{\infty} e^{i(t_1x_1+t_2x_2)} f_1(x_1) f_2(x_2)\,dx_1\,dx_2$$

Hence, by uniqueness theorem of characteristic functions, $f(x_1, x_2) = f_1(x_1) f_2(x_2)$, which implies that X_1 and X_2 are independent.

Remarks:

1. For discrete *r.v.'s* the result is established by using summation instead of integration over the given range of the variables.

2. The result can be generalized to the case of more than two variables.

Necessary and sufficient condition for the mutual independence of random variables X_i $(i = 1, 2, 3, \dots\dots, n)$ is that :

$$\phi_{X_1,X_2,\dots,X_n}(t_1,t_2,\dots,t_n) = \phi_{X_1}(t_1)\,\phi_{X_2}(t_2)\dots\phi_{X_n}(t_n) \qquad \dots (7{\cdot}22a)$$

3. In terms of moment generation functions, the necessary and sufficient condition for the *r.v.'s* X_1, X_2, ..., X_n to be mutually independent is that

$$M_{X_1,X_2,\dots,X_n}(t_1,t_2,\dots,t_n) = M_{X_1}(t_1)\,M_{X_2}(t_2)\dots M_{X_n}(t_n), \qquad \dots (7{\cdot}22b)$$

provided all the *m.g.f.'s* exist.

Theorem 7·9. **Hally-Bray Theorem.** *If the sequence of distribution functions* $\{F_n(x)\}$ *converges to the distribution function* $F(x)$ *at all the points of continuity of the latter and* $g(x)$ *is bounded continuous function over the line* $R^1(-\infty, \infty)$ *then*

$$\lim_{n \to \infty} \int_{-\infty}^{\infty} g(x)\, dF_n(x) = \int_{-\infty}^{\infty} g(x)\, dF(x) \quad \text{... (7·23)}$$

Corollary. If $F_n(x) \to F(x)$, then the corresponding sequence of characteristic functions $\phi_n(t)$ of $F_n(x)$ converges to the characteristic function $\phi(t)$ of F at every point 't'.

Proof of Corollary. cos tx and sin tx are continuous and bounded functions of x for all t and, hence, from (7·23), we get

$$\lim_{n \to \infty} \int_{-\infty}^{\infty} \cos tx\, dF_n(x) = \int_{-\infty}^{\infty} \cos tx\, dF(x)$$

and

$$\lim_{n \to \infty} \int_{-\infty}^{\infty} \sin tx\, dF_n(x) = \int_{-\infty}^{\infty} \sin tx\, dF(x)$$

$$\therefore \quad \lim_{n \to \infty} \int_{-\infty}^{\infty} (\cos tx + i \sin tx)\, dF_n(x) = \int_{-\infty}^{\infty} (\cos tx + i \sin tx)\, dF(x)$$

$$\Rightarrow \quad \lim_{n \to \infty} \int_{-\infty}^{\infty} e^{itx}\, dF_n(x) = \int_{-\infty}^{\infty} e^{itx} dF(x).$$ Hence $\phi_n(t) \to \phi(t)$ as $n \to \infty$.

Theorem 7·10. Levy's Continuity Theorem for Characteristic Functions. *For a sequence of distribution functions $\{F_n(x)\}$ with the corresponding sequence of characteristic functions $\{\phi_n(t)\}$, a necessary and sufficient condition that $F_n(x) \to F(x)$ at all points of continuity of F is that for every real t, $\phi_n(t) \to \phi(t)$, which is continuous at $t = 0$ and $\phi(t)$ is the characteristic function corresponding to F.*

Example 7·12. *Let $F_n(x)$ be the distribution function defined by :*

$$F_n(x) = \begin{cases} 0, & \text{for } x \le -n \\ \dfrac{x+n}{2n}, & \text{for } -n < x < n \\ 1, & \text{for } x \ge n \end{cases}$$

Is the limit $F_n(x)$ a distribution function ? If not, why ?

Solution.

$$\phi_n(t) = \int_{-n}^{n} e^{itx} \frac{1}{2n} dx = \frac{1}{2n}\left[\frac{e^{itn} - e^{-itn}}{it}\right] = \frac{\sin nt}{nt} \qquad [\because f_n(x) = F_n'(x)]$$

$$\phi(t) = \lim_{n \to \infty} \phi_n(t) = \lim_{n \to \infty} \frac{\sin nt}{nt} = \begin{cases} 1, & \text{if } t = 0 \\ 0, & \text{if } t \ne 0 \end{cases}$$

i.e., $\phi(t)$ is discontinuous at $t = 0$ and $\displaystyle\lim_{n \to \infty} F_n(x) = \frac{1}{2}$.

Hence, $F(x)$ is not a distribution function.

Example 7·13. *Find the distribution for which characteristic function is :*

(*a*) $\phi(t) = \left(q + pe^{it}\right)^n$, (*b*) $\phi(t) = e^{-t^2 \sigma^2 / 2}$

Solution.

(a) $$\phi(t) = \left(q + pe^{it}\right)^n = \sum_{j=0}^{n} {}^nC_j p^j q^{n-j} e^{itj}$$

We have $$F_c = \int_{-c}^{c} e^{-itx}\,\phi(t)\,dt = \int_{-c}^{c} \left\{ e^{-itx} \sum_{j=0}^{n} {}^nC_j\, p^j\, q^{n-j} e^{itj} \right\} dt$$

$$= \sum_{j=0}^{n} \left\{ {}^nC_j\, p^j\, q^{n-j} \int_{-c}^{c} e^{-it(x-j)}\, dt \right\} \qquad \ldots (*)$$

(i) If $x \neq j$,

$$F_c = \sum_{j=0}^{n} \left\{ {}^nC_j p^j q^{n-j} \left| \frac{e^{-it(x-j)}}{-i(x-j)} \right|_{t=-c}^{c} \right\} = \sum_{j=0}^{n} \left[{}^nC_j p^j q^{n-j} \left\{ \frac{e^{ic(x-j)} - e^{-ic(x-j)}}{i(x-j)} \right\} \right]$$

$$= \sum_{j=0}^{n} \left\{ {}^nC_j p^j q^{n-j} . \frac{2i \sin\{c(x-j)\}}{i\,(x-j)} \right\}$$

$$\therefore \quad \lim_{c\to\infty} \frac{F_c}{2c} \to 0 \;\forall\; x$$

Hence, there is no discontinuity in the distribution function when $x \neq j$.

(ii) If $x = j$,

$$F_c = \sum_{j=0}^{n} \left\{ {}^nC_j\, p^j q^{n-j} \int_{-c}^{c} dt \right\} = 2c \sum_{j=0}^{n} {}^nC_j p^j q^{n-j} = 2c\,(q+p)^n = 2c \qquad [\text{From}(*)]$$

Since $\frac{F_c}{2c} \to 1$ at $x = j$, the distribution function is discontinuous and its frequency is ${}^nC_j\; p^j\; q^{n-j}$.

(b) Let $$F_c = \int_{-c}^{c} e^{-itx}\,\phi(t)\,dt = \int_{-c}^{c} \exp\left\{ -itx - \frac{1}{2}t^2\sigma^2 \right\} dx$$

$$\therefore \quad |F_c| \leq \int_{-c}^{c} \left| \exp\left\{ -itx - \frac{1}{2}t^2\sigma^2 \right\} \right| dt \leq \int_{-c}^{c} e^{-t^2\sigma^2/2} dt \leq \int_{-\infty}^{\infty} e^{-t^2\sigma^2/2}\, dt = \frac{\sqrt{2\pi}}{\sigma}$$

$$\left(\because \int_{-\infty}^{\infty} e^{-a^2x^2}\, dx = \frac{\sqrt{\pi}}{a} \right)$$

Thus $\lim_{c\to\infty} \frac{F_c}{2c} = 0 \;\forall\; x.$

Hence, the distribution function is continuous for all x.

$$f(x) = \frac{1}{2\pi} \int_{-\infty}^{\infty} e^{-itx} e^{-t^2\sigma^2/2}\, dt = \frac{1}{2\pi} e^{-x^2/2\sigma^2} \int_{-\infty}^{\infty} \exp\left\{ -\frac{1}{2}\left(t\,\sigma + \frac{ix}{\sigma} \right)^2 \right\} dt$$

Put $\quad t\sigma + \frac{ix}{\sigma} = \xi, \quad i.e., \quad \sigma dt = d\xi.$

$$\therefore \quad f(x) = \frac{1}{2\pi} \exp\left(-\frac{x^2}{2\sigma^2}\right) \int_{-\infty}^{\infty} e^{-\xi^2/2} \cdot \frac{d\xi}{\sigma}$$

Hence, $$f(x) = \frac{1}{2\pi} \exp\left(-\frac{x^2}{2\sigma^2}\right) \frac{\sqrt{2\pi}}{\sigma} = \frac{1}{\sigma\sqrt{2\pi}} \exp\left(-\frac{x^2}{2\sigma^2}\right), -\infty < x < \infty$$

which is the *p.d.f.* of normal distribution. [See Chapter 9.]

Example 7·14. *Find the density function f(x) corresponding to the characteristic function defined as follows :*

$$\phi(t) = \begin{cases} 1-|t|, & |t| \le 1 \\ 0, & |t| > 1 \end{cases}$$

Solution. By Inversion Theorem, the *p.d.f.* of X is given by :

$$f(x) = \frac{1}{2\pi} \int_{-\infty}^{\infty} e^{-itx} \phi(t)\, dt = \frac{1}{2\pi} \int_{-1}^{0} e^{-itx} (1+t)\, dt + \frac{1}{2\pi} \int_{0}^{1} e^{-itx} (1-t)\, dt$$

$\{\because$ for $-1 < t < 0$, $|t| = -t$ and for $0 < t < 1$, $|t| = +t\}$

$$\int_{-1}^{0} e^{-itx} (1+t)\, dt = \left[\frac{e^{-itx}}{-ix}(1+t)\right]_{-1}^{0} + \frac{1}{ix} \int_{-1}^{0} e^{-itx}\, dt = -\frac{1}{ix} + \left[\frac{e^{-itx}}{-ix}\right]_{-1}^{0}$$

$$= -\frac{1}{ix} + \frac{1}{(ix)^2}\left(e^{ix} - 1\right)$$

Similarly, $$\int_{0}^{1} e^{-itx} (1-t)\, dt = \frac{1}{ix} + \frac{1}{(ix)^2}(e^{-ix} - 1)$$

$$\therefore \quad f(x) = \frac{1}{2\pi}\left[\frac{1}{(ix)^2}\left\{e^{ix} - 1 + e^{-ix} - 1\right\}\right] = \frac{1}{\pi x^2}\left(1 - \frac{e^{ix} + e^{-ix}}{2}\right)$$

$$= \frac{1}{\pi} \cdot \frac{1 - \cos x}{x^2}, -\infty < x < \infty.$$

7.5. CHEBYCHEV'S INEQUALITY

The role of standard deviation as a parameter to characterise variance is precisely interpreted by means of the well- known Chebychev's inequality. The theorem discovered in 1853 was later on discussed in 1856 by Bienayme.

Definition. *If X is a random variable with mean μ and variance σ^2, then for any positive number k, we have*

$$\left.\begin{aligned} &P\{|X-\mu| \ge k\sigma\} \le \frac{1}{k^2} \\ \text{or} \quad &P\{|X-\mu| < k\sigma\} \ge 1 - \frac{1}{k^2} \end{aligned}\right\} \quad \ldots (7{\cdot}24)$$

Proof.

Case (i) *X is a continuous r.v. By def.*

$$\sigma^2 = \sigma_X{}^2 = E\{X - E(X)\}^2 = E(X-\mu)^2$$

$$= \int_{-\infty}^{\infty} (x-\mu)^2 f(x)\, dx, \quad \text{where} \quad f(x) \text{ is } p.d.f. \text{ of X.}$$

$$= \int_{-\infty}^{\mu-k\sigma} (x-\mu)^2 f(x)\, dx + \int_{\mu-k\sigma}^{\mu+k\sigma} (x-\mu)^2 f(x)\, dx + \int_{\mu+k\sigma}^{\infty} (x-\mu)^2 f(x)\, dx$$

$$\geq \int_{-\infty}^{\mu-k\sigma} (x-\mu)^2 f(x)\, dx + \int_{\mu-k\sigma}^{\infty} (x-\mu)^2 f(x)\, dx \qquad \text{... (*)}$$

We know that :

$$x \leq \mu - k\sigma \quad \text{and} \quad x \geq \mu + k\sigma \quad \Leftrightarrow \quad |x-\mu| \geq k\sigma \quad \Rightarrow \quad (x-\mu)^2 \geq k^2\sigma^2$$

Substituting in (*), we get

$$\therefore \qquad \sigma^2 \geq k^2\sigma^2 \left[\int_{-\infty}^{\mu-k\sigma} f(x)\, dx + \int_{\mu+k\sigma}^{\infty} f(x)\, dx \right] \qquad \text{[From (**)]}$$

$$= k^2\sigma^2 \left[P(X \leq \mu - k\sigma) + P(X \geq \mu + k\sigma) \right]$$

$$= k^2\sigma^2 \cdot P\left[|X-\mu| \geq k\sigma \right]$$

$$\Rightarrow \qquad P\left[|X-\mu| \geq k\sigma \right] \leq 1/k^2 \qquad \text{... (***)}$$

Also, since

$$P\{|X-\mu| \geq k\sigma\} + P\{|X-\mu| < k\sigma\} = 1, \text{ we get}$$

$$P\{|X-\mu| < k\sigma\} = 1 - P\{|X-\mu| \geq k\sigma\} \geq 1 - \{1/k^2\} \qquad \text{[From (***)]}$$

which establishes (7·23).

Case (ii) In case of discrete random variable, the proof follows exactly similarly on replacing integration by summation.

Remark. In particular, if we take $k\sigma = c > 0$, then (*) and (**) give respectively

$$P\{|X-\mu| \geq c\} \leq \frac{\sigma^2}{c^2} \quad \text{and} \quad P\{|X-\mu| < c\} \geq 1 - \frac{\sigma^2}{c^2}$$

$$\Rightarrow \quad P\{|X-E(X)| \geq c\} \leq \frac{\text{Var}(X)}{c^2} \quad \text{and} \quad P\{|X-E(X)| < c\} \geq 1 - \frac{\text{Var}(X)}{c^2} \qquad \text{... (7·24a)}$$

7·5·1. Bienayme – Chebychev's Inequality

Let $g(X)$ be non-negative function of a random variable X. Then for every $k > 0$, we have

$$P\{g(X) \geq k\} \leq \frac{E\{g(X)\}}{k} \qquad \text{... (7·25)}$$

Proof. Here, we shall prove the theorem for continuous random variable. The proof can be adapted to the case of discrete random variable on replacing integration by summation over the given range of the variable.

Let S be the set of all X, where $g(X) \geq k$, *i.e.*, $S = \{x : g(x) \geq k\}$ then

$$\int_S dF(x) = P(X \in S) = P\{g(X) \geq k\}, \quad \text{...} (*)$$

where, $F(x)$ is the distribution function of X.

$$E\{g(X)\} = \int_{-\infty}^{\infty} g(x)\, dF(x) \geq \int_S g(x)\, dF(x)$$

$$\geq k \cdot P\{g(X) \geq k\} \qquad [\because \text{ on } S, g(x) \geq k \text{ and using } (*)]$$

$$\Rightarrow \qquad P\{g(X) \geq k\} \leq \frac{E[g(X)]}{k}$$

Remarks:

1. If we take $g(X) = \{X - E(X)\}^2 = \{X - \mu\}^2$ and replace k by $k^2\sigma^2$ in (7·25), we get

$$P\left[(X-\mu)^2 \geq k^2\sigma^2\right] \leq \frac{E(X-\mu)^2}{k^2\sigma^2} = \frac{\sigma^2}{k^2\sigma^2} = \frac{1}{k^2} \quad \Rightarrow \quad P\left[|X-\mu| \geq k\sigma\right] \leq \frac{1}{k^2} \quad \text{...} (7\cdot25a)$$

which is Chebychev's inequality.

2. *Markov's Inequality.* Taking $g(X) = |X|$ in (7·25) we get, for any $k > 0$

$$P\{|X| \geq k\} \leq \frac{E|X|}{k}, \quad \text{...} (7\cdot26)$$

which is Markov's inequality.

Rather, taking $g(X) = |X|^r$ and replacing k by k^r in (7·25), we get a more generalised form of Markov's inequality, *viz.*,

$$P\left(|X|^r \geq k^r\right) \leq \frac{E|X|^r}{k^r} \quad \text{...} (7\cdot26a)$$

3. If we assume the existence of only second-order moments of X, then we cannot do better than Chebychev's inequality (7·24). However, we can sometimes improve upon the results of Chebychev's inequality if we assume the existence of higher order moments. We give below (without proof) one such inequality which assumes the existence of moments of fourth order.

If $\qquad E|X|^4 < \infty, \qquad E(X) = 0 \qquad$ and $\qquad E(X^2) = \sigma^2$, then

$$P\{|X| > k\sigma\} \geq \frac{\mu_4 - \sigma^4}{\mu_4 + \sigma^4 k^4 - 2k^2\sigma^4} \quad \text{...} (7\cdot27)$$

If $X \sim U[0, 1]$, [*c.f. chapter* 9], with *p.d.f.* $p(x) = 1,\ 0 < x < 1$ and $= 0$, otherwise, then

$$E(X^r) = \int_0^1 x^r\, dx = \frac{1}{r+1}; \quad (r = 1, 2, 3, 4).$$

$$E(X) = \frac{1}{2},\ E(X^2) = \frac{1}{3},\ E(X^3) = \frac{1}{4},\ E(X^4) = \frac{1}{5};\quad \text{Var}(X) = E(X^2) - [E(X)]^2 = \frac{1}{12} \quad \text{...} (*)$$

$$\mu_4 = E(X-\mu)^4 = E\left(X - \frac{1}{2}\right)^4 = \frac{1}{80}. \qquad [\text{On using binomial expansion with } (*)]$$

Chebychev's inequality (7·24*a*) with $k = 2$ gives : $P\left(\left|X-\frac{1}{2}\right|<2\frac{1}{\sqrt{12}}\right) \geq 1-\frac{1}{4} = 0{\cdot}75$

With $k = 2$, (7·27) gives :

$$P\left(\left|X-\frac{1}{2}\right|>2\frac{1}{\sqrt{12}}\right) \leq \frac{(1/80)-(1/144)}{(1/80)+(1/9)-(1/18)} = \frac{4}{49}$$

$$\Rightarrow \quad P\left(\left|X-\frac{1}{2}\right|\leq 2\frac{1}{\sqrt{12}}\right) \geq 1-\frac{4}{49} = \frac{45}{49} = 0{\cdot}92.$$

which is a much better lower bound than the lower bound given by Chebychev's inequality.

7.5.2. Generalisation of Chebychev's Inequality. if X is a *r.v.* with mean zero and variance σ^2, then for $\varepsilon > 0$

$$(i)\ P(x \geq \varepsilon) \leq \frac{\sigma^2}{\varepsilon^2} \quad \text{and} \quad (ii)\ P(x \geq \varepsilon) \leq \frac{\sigma^2}{\sigma^2+\varepsilon^2} \qquad \ldots (7{\cdot}28)$$

Proof. Consider the set $M = [\varepsilon, \infty)$. Its indicator function is given by :

$$I_M(x) = \begin{cases} 1; & x \geq \varepsilon \\ 0; & x < \varepsilon \end{cases}$$

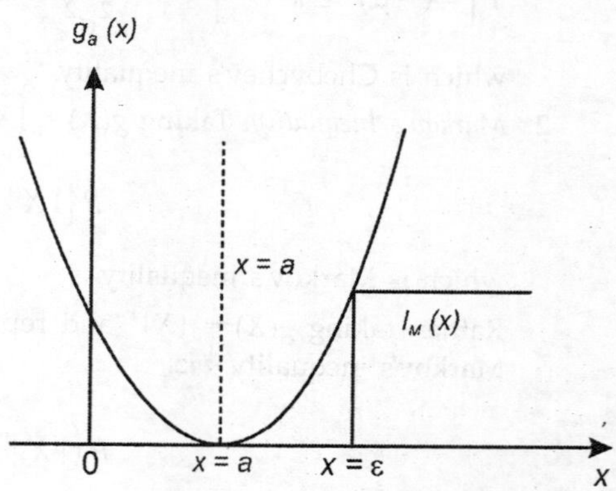

Again, consider the function :

$$g_a(x) = \frac{(x-a)^2}{(\varepsilon-a)^2}, -\infty < x < \infty$$

[This is the equation of a parabolic curve of the type $X^2 = 4AY$. It meets the x-axis at x = a and is symmetric about the line $x = a$]

We have :

$$g_a(x) \geq I_M(x) \Rightarrow I_M(x) \leq g_a(x), \text{ for all } a < \varepsilon$$

$$\Rightarrow E\left[I_M(x)\right] \leq E\left[g_a(x)\right] = E\left[\frac{(X-a)^2}{(\varepsilon-a)^2}\right] = \frac{\sigma^2+a^2}{(\varepsilon-a)^2} \qquad [\because E(x) = 0]$$

$$\Rightarrow \quad P(M) \leq \frac{\sigma^2+a^2}{(\varepsilon-a)^2} \Rightarrow P(X \geq \varepsilon) \leq \frac{\sigma^2+a^2}{(\varepsilon-a)^2} \qquad \ldots (*)$$

This result is true for all $a < \varepsilon$.

In particular, if we take $a = 0$, then $P(X \geq \varepsilon) \leq \frac{\sigma^2}{\varepsilon^2}$, a result already established.

Let $h(a) = \frac{\sigma^2+a^2}{(\varepsilon-a)^2}$. Differentiating twice *w.r.t.* '*a*' we shall get

$$h'(a) = 0 \Rightarrow a = (-\sigma^2/\varepsilon) = a_0, \text{ (say)} \quad \text{and} \quad [\,h''(a)\,]_{a=a_0} = \left(\varepsilon+\frac{\sigma^2}{\varepsilon}\right)^2 > 0$$

$\Rightarrow$ $h(a)$ is smallest at $a = (-\sigma^2/\varepsilon)$.

Hence, from (*), we get :

$$P(X \geq \varepsilon) \leq \frac{\sigma^2 + (\sigma^4/\varepsilon^2)}{\left[\varepsilon + (\sigma^2/\varepsilon)\right]^2} = \frac{\sigma^2}{\sigma^2 + \varepsilon^2}.$$

7·5·3. Frechet Inequality

If X is a r.v. with E(X) = μ and var(X) = σ², then for real a > 0 and b > 0,

$$p = P\left[\mu - a\sigma < X < \mu + b\sigma\right] \geq 1 - \left\{\frac{\left\{1 + \left(\frac{b-a}{2}\right)^2\right\}}{\left(\frac{b+a}{2}\right)^2}\right\} \quad \ldots (7\cdot29)$$

Proof. Let $Z = \dfrac{X - E(X)}{\sigma_x} = \dfrac{X - \mu}{\sigma}$, so that $E(Z) = 0$ and $\text{Var}(Z) = 1 = E(Z)^2$ ...(*)

$$\therefore \quad p = P[\mu - a\sigma < X < \mu + b\sigma] = P[-a < Z < b]$$

$$= P\left[-a - \frac{b-a}{2} < Z - \frac{b-a}{2} < b - \frac{b-a}{2}\right]$$

[Subtracting $(b - a)/2$ from each number]

$$= P\left[-\left(\frac{a+b}{2}\right) < Z - \frac{b-a}{2} < \frac{b+a}{2}\right]$$

$$= P\left[\left|Z - \frac{b-a}{2}\right| < \frac{a+b}{2}\right] \geq 1 - \frac{E\left[Z - \frac{b-a}{2}\right]^2}{\left(\frac{a+b}{2}\right)^2} \quad \ldots (**)$$

[Using Bienayme Chebychev's Inequality]

$$E\left(Z - \frac{b-a}{2}\right)^2 = E(Z^2) + \left(\frac{b-a}{2}\right)^2 - (b-a)\,E(Z) = 1 + \left(\frac{b-a}{2}\right)^2 \quad \text{[Using (*)]}$$

Substituting in (**),

$$p \geq 1 - \left[\frac{\left\{1 + \left[\frac{(b-a)}{2}\right]^2\right\}}{\frac{(b+a)^2}{2}}\right].$$

7·5·4. Bernstein's Inequality

Let X be a r.v. with m.g.f. $M_X(t)$. Then for any ε > 0, we have :

$$P(X \geq \varepsilon) \geq e^{-\varepsilon t} M_X(t) \quad \ldots (7\cdot30)$$

Proof. Taking $g(x) = e^{tx} \geq 0$ and $k = e^{\epsilon t} > 0$, $(\varepsilon > 0)$ in Bienayme-Chebychev's inequality (7·25), we get

$$P\left[e^{tX} \geq e^{\varepsilon t}\right] \leq \frac{E\left(e^{tX}\right)}{e^{\varepsilon t}} = e^{-\varepsilon t} M_X(t) \qquad \text{... (*)}$$

Also $\quad X \geq \varepsilon \quad \Rightarrow \quad e^X \geq e^{\epsilon} \quad \Rightarrow \quad e^{tX} \geq e^{\epsilon t}$

$$\Rightarrow \quad P(X \geq \varepsilon) = P\left(e^{tX} \geq e^{\epsilon t}\right) \qquad \text{... (*)}$$

From (*) and (**), we get

$$P(X \geq \varepsilon) \leq e^{-\varepsilon t} M_X(t)$$

Example 7·15. *When k ≤ 1, Chebychev's inequality is useless''. Comment.*

Solution. Chebychve's Inequality gives :

$$P\left(|X-\mu| \leq k\,\sigma\right) \geq 1 - \frac{1}{k^2}; \quad k > 0$$

i.e., it provides the lower bound to the probability that the *r.v.* X lies in the interval $(\mu \pm k\sigma)$. Since, this probability cannot be negative, we must have

$$1 - \frac{1}{k^2} \geq 0 \quad \Rightarrow \quad 1 \geq \frac{1}{k^2} \quad \Rightarrow \quad k^2 \geq 1 \quad \Rightarrow \quad |k| \geq 1.$$

For $k \leq 1$, this lower bound to the probability is negative or 0, which is meaningless. Hence, for $k \leq 1$, Chebychev's inequality is useless.

Hence, for the validity of Chebychev's inequality we must have $k > 1$. (∵ k is a positive constant).

7·6. MODES OF CONVERGENCE OF A SEQUENCE OF RANDOM VARIABLES

The convergence of sequence of random variables to some limit random variable is an important concept in probability theory and its applications to Statistics and Stochastic Processes.

In the following sections, we shall consider a sequence $\{X_n,\ n \geq 1\}$ random variables defined on some probability space, say, (Ω, F, P) and another random variable X defined on the same sample space.

In probability theory, there are several different notions of convergence of random variables as discussed below.

7·6·1. Convergence in Probability

A sequence $\{X_n;\ n \geq 1\}$ of random variables defined on some probability space (Ω, F, P) is said to converge in probability to the *r.v.* X defined on the same sample space, if for $\varepsilon > 0$,

$$\left.\begin{aligned} \lim_{n\to\infty} P\left[\left|X_n - X\right| > \varepsilon\right] = 0 \\ \lim_{n\to\infty} P\left[\left|X_n - X\right| \leq \varepsilon\right] = 1 \end{aligned}\right\} \qquad \text{... (7·31)}$$

More precisely the sequence of random variables $\{X_n ; n \geq 1\}$ is said to converge to the limit *r.v.* X in probability, if for every $\varepsilon > 0$ and for every $\delta > 0$ there is an integer N such that :

$$\left.\begin{aligned} &P\{\omega \in \Omega : |X_n(\omega) - X(\omega)| > \varepsilon\} < \delta, \text{ for all } n > N \\ \text{or} \qquad &P\left[|X_n - X| > \varepsilon\right] = 0 \text{ as } n \to \infty \end{aligned}\right\} \quad \ldots (7\cdot32)$$

Notation: $X_n \xrightarrow{p} X$, (for convergence in probability).

Remarks:

1. **Weak Law of Large Numbers.**

   Under appropriate assumptions, [See § 7.7] [X_i's are *i.i.d.* $(0, \sigma^2)$], we have
   $\lim_{n\to\infty} P(\bar{X}_n > \varepsilon) = 0$; for all $\varepsilon > 0$.

   Also if X_i's are *i.i.d* $N(\mu, \sigma^2)$, then $\lim_{n\to\infty} P\left[|\bar{X}_n - \mu| > \varepsilon\right] = 0.$

   Thus, weak law of large numbers is a special case of convergence in probability.
2. Convergence in probability is also known as *stochastic convergence.*
3. The concept of convergence in probability is used very often in Statistics. For, example, in 'Estimation Theory' an estimator is called consistent if it converges in probability to the quantity being estimated *i.e.,* the parameter.

Theorem 7·11. *If a sequence of r.v.'s $\{X_n\}$ converges to a r.v. X in probability, then it is unique. In other words,* *if* $X_n \xrightarrow{p} X$ and $X_n \xrightarrow{p} Y$ *then* $X = Y$ *a.s. i.e.,* $P(X = Y) = 1$.

Proof. Let $\varepsilon > 0$, be any positive number then

$$\left.\begin{aligned} &X_n \xrightarrow{p} X \;\Rightarrow\; \lim_{n\to\infty} P\left[|X_n - X| \geq \frac{\varepsilon}{2}\right] = 0 \\ \text{and} \quad &X_n \xrightarrow{p} Y \;\Rightarrow\; \lim_{n\to\infty} P\left[|X_n - Y| \geq \frac{\varepsilon}{2}\right] = 0 \end{aligned}\right\} \quad \ldots (*)$$

$$|X - Y| \geq \varepsilon \quad \Rightarrow \quad \varepsilon \leq |X - Y| = |(X - X_n) + (X_n - y)| \leq |X - X_n| + |X_n - Y|$$

$$\therefore \quad P\left[|X - Y| \geq \varepsilon\right] \leq P\left[\{|X - X_n| + |X_n - Y|\} \geq \varepsilon\right] \quad \ldots (**)$$

[$\because A \Rightarrow B$ means $A \subset B \quad \Rightarrow \quad P(A) \leq P(B)$]

We have :

$$\{|X - X_n| + |X_n - Y|\} \geq \epsilon \subset \left[|X - X_n| \geq \frac{\epsilon}{2}\right] \cup \left[|X_n - Y| \geq \frac{\epsilon}{2}\right]$$

$$\Rightarrow \quad P\left[\{|X - X_n| + |X_n - Y|\} \geq \varepsilon\right] \leq P\left[|X - X_n| \geq \frac{\varepsilon}{2}\right] + P\left[|X_n - Y| \geq \frac{\varepsilon}{2}\right] \quad \ldots (***)$$

From (**) and (***), we get

$$P\left[|X - Y| \geq \varepsilon\right] \leq P\left[|X - X_n| \geq \frac{\varepsilon}{2}\right] + P\left[|X_n - Y| \geq \frac{\varepsilon}{2}\right]$$

Taking limit on both sides as $n \to \infty$ and using (*), we get

$$\lim_{n\to\infty} P\left[|X-Y| \geq \varepsilon\right] \leq 0 \;\Rightarrow\; \lim_{n\to\infty} P\left[|X-Y| \geq \varepsilon\right] = 0 \left[\because P(.) \text{ can't be negative}\right]$$

$$\Rightarrow \qquad P(X = Y) = 1 \text{ as } n \to \infty \;\Rightarrow\; X = Y \; a.s.$$

Remarks:

1. From the definition of convergence in probability, we can easily deduce that:

$$X_n \xrightarrow{p} X \quad \longleftrightarrow \quad (X_n - X) \xrightarrow{p} 0.$$

2. If the limiting *r.v.* X is a degenerate *r.v.* say at $X = c$, with $P(X = c) = 1$, then

$$X_n \xrightarrow{p} X \text{ is also expressed as } X_n \xrightarrow{p} c.$$

3. **Sufficient Condition for Convergence in Probability** The following theorem gives the sufficient condition for the convergence of the sequence $\{X_n\}$ to $X(\equiv 0)$ in probability.

Theorem 7·12. *If* $E|X_n|^r \longrightarrow 0,\ r \geq 1$; *then* $X_n \xrightarrow{p} 0.$

Proof. For $\varepsilon > 0$, by Markov's Theorem, we get

$$P\left[|X_n| > \varepsilon\right] \leq \frac{E|X_n|^r}{\varepsilon^r};\ r \geq 1$$

If $E|X_n|^r \longrightarrow 0;\ r \geq 1$, as $n \longrightarrow \infty$,

then $$P\left[|X_n| > \varepsilon\right] \leq 0 \;\Rightarrow\; P\left[|X_n| > \varepsilon\right] = 0 \qquad \left[\because P(.) \text{ can't be negative}\right]$$

$$\therefore \qquad E|X_n|^r = 0;\ r \geq 1 \;\Rightarrow\; X_n \xrightarrow{p} 0, \text{ as } n \longrightarrow \infty.$$

Corollary. If $\quad X_n \xrightarrow{p} c \;\Rightarrow\; X_n - c \xrightarrow{p} 0,$

then by the above theorem :

$$X_n - c \xrightarrow{p} 0 \quad \text{if} \quad E|X_n - c|^r = 0, \quad r \geq 1 \text{ as } n \longrightarrow \infty \qquad \ldots (7\cdot32)$$

4. **Necessary and Sufficient condition for Convergence of $X_n \longrightarrow 0$ in Probability.** [Only statement]

Theorem 7·13. $X_n \xrightarrow{p} 0$, *if and only if:*

$$E\left[\frac{|X_n|}{1+|X_n|}\right] \longrightarrow 0 \qquad \ldots (7\cdot33)$$

Theorem 7·14. **Continuous Mapping Theorem**

If $X_n \xrightarrow{p} X$ *and f(.) is a real valued continuous functions defined on R : [−∞, ∞] then*

$$f(X_n) \xrightarrow{p} f(X).$$

Proof. $X_n \xrightarrow{p} X \Rightarrow$ for $\delta > 0$, we have

$$\lim_{n\to\infty} p\left[\left|X_n - X\right| \geq \delta\right] = 0 \qquad \ldots (*)$$

Since $f(x)$ is continuous on R, it is continuous in every interval, say, $I = (-a, a)$, $a > 0$. Hence, by definition of continuity, for every $\varepsilon > 0$, there exists a number $\delta > 0$ so that

$$\left|f(X_n - f(X))\right| < \varepsilon \quad \text{whenever } \left|X_n - X\right| < \delta \text{ and } X \; \varepsilon \, I \qquad \ldots (**)$$

Define the events:

$$A = \left\{\left|X_n - X\right| < \delta\right\}, \quad B = \left\{X \; \varepsilon \; I\right\}, \quad C = \left\{\left|f(X_n) - f(X)\right| < \varepsilon\right\} \qquad \ldots (***)$$

$$(**) \Rightarrow \qquad A \cap B \subset C \Rightarrow P(A \cap B) \leq P(C) \qquad \ldots (****)$$

Writing A = (Complementary event of A) = A^c and using De-Morgan's law, we get from $(****)$,

$$P\left[\overline{(A \cap B)}\right] \leq P(C) \Rightarrow P\left[\left(\bar{A} \cup \bar{B}\right)\right] \leq P(C) \Rightarrow 1 - P\left(\bar{A} \cup \bar{B}\right) \leq P(C)$$

$$\Rightarrow \qquad 1 - P(C) \leq P(\bar{A} \cup \bar{B}) \qquad \Rightarrow \qquad P(\bar{C}) \leq P(\bar{A}) + P(\bar{B}) \quad \text{(By Boole's inequality)}$$

$$\Rightarrow \qquad P\left[\left|f(X_n) - f(X) \geq \varepsilon\right|\right] \leq P\left[\left|X_n - X\right| \geq \delta\right] + P\left(X \notin I\right) \qquad \left[\text{Using } (***)\right]$$

Taking limit as $n \to \infty$ on both sides, and using $(*)$ we get :

$$\lim_{n\to\infty} P\left[\left|f(X_n) - f(X)\right| \geq \varepsilon\right] \leq \lim_{n\to\infty} P\left(X \notin I\right) \to 0, \text{ for sufficiently large } n$$

$$\Rightarrow \qquad f(X_n) \xrightarrow{p} f(X)$$

Remark. This theorem illustrates an important property, known as the **'Invariance Property'** of the convergence in probability.

This theorem is useful in asserting that if a statistic $t_n \xrightarrow{p} \theta$ (parameter), then $t_n^2 \xrightarrow{p} \theta^2$ or $a\, t_n + b \xrightarrow{p} a\,\theta + b$, which asserts the **consistency** of a continuous function of the consistent estimator, in the Theory of Estimation in Statistics.

Theorem 7·15. *If $X_n \xrightarrow{p} X$ and $Y_n \xrightarrow{p} Y$, then for real number a,*

$$(A)\ aX_n \xrightarrow{p} aX \qquad (B)\ X_n + Y_n \xrightarrow{p} X + Y$$

Proof.

(A) Consider real number $a \neq 0$. For $\varepsilon > 0$,

$$P\left[\left|aX_n - aX\right| \geq \varepsilon\right] = P\left[\left|a\right|\left|X_n - X\right| \geq \varepsilon\right] = P\left[\left|X_n - X\right| \geq \frac{\varepsilon}{|a|}\right] \quad \text{...}(*)$$

Since $X_n \xrightarrow{p} X$ (Given), for $\varepsilon > 0$ and $a \neq 0$, ...(**)

$$\lim_{n\to\infty} P\left[\left|X_n - X\right| \geq \frac{\varepsilon}{|a|}\right] = 0$$

Taking limit in (*) as $n \to \infty$ and using (**), we get :

$$\lim_{n\to\infty} P\left[\left|aX_n - aX\right| \geq \varepsilon\right] = 0 \quad \Rightarrow \quad aX_n \xrightarrow{p} aX$$

When $a = 0$, $aX_n = 0 = aX$, so that $aX_n \xrightarrow{p} aX$ trivially.

Hence, $aX_n \xrightarrow{p} aX$ as $n \to \infty$, for all real $a \neq 0$.

(B) Since $X_n \xrightarrow{p} X$ and $Y_n \xrightarrow{p} Y$ as $n \to \infty$, for $\varepsilon > 0$

$$\lim_{n\to\infty} P\left[\left|X_n - X\right| > \frac{\varepsilon}{2}\right] = 0 \quad \text{and} \quad \lim_{n\to\infty} P\left[\left|Y_n - Y\right| > \frac{\varepsilon}{2}\right] = 0 \quad \text{...}(i)$$

Define the sets :

$$A = \left\{\left|\left(X_n + Y_n\right) - \left(X + Y\right)\right| > \varepsilon\right\}$$

$$B = \left\{\left|X_n - X\right| > \frac{\varepsilon}{2}\right\} \quad \text{and} \quad C = \left\{\left|Y_n - Y\right| > \frac{\varepsilon}{2}\right\}$$

$$\left|X_n + Y_n - (X + Y)\right| = \left|\left(X_n - X\right) + \left(Y_n - Y\right)\right| \leq \left|X_n - X\right| + \left|Y_n - Y\right|$$

$$\Rightarrow \quad A \subset B \cup C \quad \Rightarrow \quad P(A) \leq P(B) + P(C)$$

$$\Rightarrow \quad P\left[\left|\left(X_n + Y_n\right) - (X + Y)\right| > \varepsilon\right] \leq P\left[\left|X_n - X\right| > \frac{\varepsilon}{2}\right] + P\left[\left|Y_n - Y\right| > \frac{\varepsilon}{2}\right]$$

Taking limit on both sides as $n \to \infty$, and using (*i*), we get

$$\lim_{n\to\infty} P\left[\left|(X_n + Y_n) - (X + Y)\right| > \varepsilon\right] \leq 0 \quad \Rightarrow \quad \lim_{n\to\infty} P\left[\left|X_n + Y_n) - (X + Y)\right| > \varepsilon\right] = 0$$

$$\left[\because P(.) \text{ can not be negative}\right]$$

$$\Rightarrow \quad (X_n + Y_n) \xrightarrow{p} (X + Y)$$

Remark. Combining the results in (A) and (B), we get

If $X_n \xrightarrow{p} X$ and $Y_n \xrightarrow{p} Y$, then for real numbers a and b,

$$aX_n + bY_n \xrightarrow{p} aX + bY, \text{ as } n \to \infty$$

Theorem 7·16. *If* $X_n \xrightarrow{p} X$ *and* $Y_n \xrightarrow{p} Y$, *then* $X_n Y_n \xrightarrow{p} XY$.

Proof. To prove this theorem, we shall first prove the two Lemmas which we then use in the proof of the theorem.

Lemma 1. Let Z be a *r.v.* which is identically equal to 0.

If $$X_n \xrightarrow{p} Z \text{ and } Y_n \xrightarrow{p} Z, \text{ then } X_n Y_n \xrightarrow{p} Z$$

Proof. Since $X_n \xrightarrow{p} Z\,(\equiv 0)$ and $Y_n \xrightarrow{p} Z\,(\equiv 0)$, for $\varepsilon > 0$, we have

$$\lim_{n\to\infty} P\left[|X_n| > \sqrt{\epsilon}\right] = 0 \quad \text{and} \quad \lim_{n\to\infty} P\left[|Y_n| > \sqrt{\epsilon}\right] = 0 \qquad \text{... (*)}$$

Consider the three events:

$$A : \{|X_n Y_n| > \varepsilon\} \;;\; A_1 = \{|X_n| > \sqrt{\varepsilon}\} \text{ and } A_2 = \{|Y_n| > \sqrt{\varepsilon}\}$$

Since $$|X_n Y_n| = |X_n| \cdot |Y_n|, \quad A \subset A_1 \cup A_2$$

$$\Rightarrow \quad P(A) \le P(A_1) + P(A_2) \qquad \text{[By Boole's Inequality]}$$

$$\therefore \quad \lim_{n\to\infty} P(A) \le \lim_{n\to\infty} P(A_1) + \lim_{n\to\infty} P(A_2)$$

$$\lim_{n\to\infty} P\left[|X_n Y_n| > \varepsilon\right] \le \lim_{n\to\infty} P\left[|X_n| > \sqrt{\varepsilon}\right] + \lim_{n\to\infty} P\left[|Y_n| > \sqrt{\varepsilon}\right]$$

$$\Rightarrow \quad \lim_{n\to\infty} P\left[|X_n Y_n| > \varepsilon\right] \le 0 \qquad \text{[Using (*)]}$$

$$\Rightarrow \quad \lim_{n\to\infty} P\left[|X_n Y_n| > \varepsilon\right] = 0 \qquad \left[\because P(.) \text{ cannot be negative}\right]$$

$$\Rightarrow \quad \lim_{n\to\infty} P\left[|X_n Y_n| > \varepsilon\right] = 0 \quad \Rightarrow \quad X_n Y_n \xrightarrow{p} Z\,(\equiv 0)$$

Lemma 2. If $X_n \xrightarrow{p} Z$ and Y is a *r.v.*, then $X_n Y \xrightarrow{p} Z$,

where, Z is a *r.v.*, identically equal to 0.

Proof. Since $X_n \xrightarrow{p} Z(\equiv 0)$, for $\varepsilon > 0$, we have

$$\lim_{n\to\infty} P\left[|X_n| > \epsilon\right] = 0 \qquad \text{... (**)}$$

Since Y is a *r.v.*, therefore we can make a statement :

$$\lim_{n\to\infty} P\left[\,|Y| \ge k\right] = 0,$$

so that given $\delta > 0$, there is a constant k_0 say, such that :

$$P[\,|Y| > k_o] < \frac{\delta}{2} \qquad \text{... (***)}$$

Define the events: $A = \{|X_n Y| > \varepsilon\}$; $B_1 = \left\{|X_n| > \frac{\varepsilon}{k_o}\right\}$, $B_2 = \{|Y| > k_o\}$

We have $A \subset B_1 \cup B_2 \Rightarrow P(A) \le P(B_1) + P(B_2)$

Taking limit on both sides as $n \to \infty$ and using (**) and (***), we get

$$\lim_{n\to\infty} P\left[|X_n Y| > \varepsilon\right] \le 0 \Rightarrow \lim_{n\to\infty} P\left[|X_n Y| > \varepsilon\right] = 0$$

$$\Rightarrow \qquad X_n Y \xrightarrow{p} Z \ (\equiv 0).$$

Proof of Main Theorem.

Since $X_n \xrightarrow{p} X \Rightarrow X_n - X \xrightarrow{p} 0 \equiv Z$ and $Y_n \xrightarrow{p} Y \Rightarrow Y_n - Y \xrightarrow{p} 0 \equiv Z$,
where, Z is a *r.v.* taking value equal to zero identically.

We have :

$$X_n Y_n - XY = (X_n - X)(Y_n - Y) + Y(X_n - X) + X(Y_n - Y) \qquad \text{... (****)}$$

$$(X_n - X)(Y_n - Y) \xrightarrow{p} Z \equiv 0 \qquad \text{(By Lemma 1)}$$

$$\left.\begin{array}{l} \text{and } Y(X_n - X) \xrightarrow{p} Z \equiv 0 \\ \text{and } X(Y_n - Y) \xrightarrow{p} Z \equiv 0 \end{array}\right\} \qquad \text{(By Lemma 2)}$$

Hence, taking limit on both sides of (****) as $n \to \infty$ and using above results, we get

$$\text{R.H.S. of (****)} \xrightarrow{p} Z \equiv 0$$

$$\therefore \qquad \text{L.H.S.} = X_n Y_n - XY \xrightarrow{p} 0 \Rightarrow X_n Y_n \xrightarrow{p} XY$$

7·6·2. Almost Sure Convergence or Convergence with Probability One

Definition 1. A sequence $\{X_n ; n \ge 1\}$ of *r.v.'s* is said to converge to a *r.v.* X with probability 1 (or equivalent almost surely), if

$$P\left[\lim_{n\to\infty} X_n = X\right] = 1 \qquad \text{... (7·32)}$$

i.e., for $\varepsilon > 0$, we have

$$P\left[\lim_{n\to\infty} |X_n - X| < \varepsilon\right] = 1 \qquad \text{... (7·32a)}$$

or

$$P\left[\lim_{n\to\infty} |X_n - X| \ge \varepsilon\right] = 0 \qquad \text{... (7·32b)}$$

Definition 2. A sequence $\{X_n ; n \ge 1\}$ of *r.v.'s* is said to converge to a *r.v.* X with probability 1, if for $\varepsilon > 0$,

$$\left.\begin{aligned}&\lim_{k\to\infty} P\left[\sup_{n\geq k}|X_n - X| < \varepsilon\right] = 1\\ \text{or} \quad &\lim_{k\to\infty} P\left[\sup_{n\geq k}|X_n - X| \geq \varepsilon\right] = 0\end{aligned}\right\} \quad \ldots (7{\cdot}33)$$

or equivalently :

$$\left.\begin{aligned}&\lim_{k\to\infty} P\left[\bigcup_{n\geq k}|X_n - X| \geq \varepsilon\right] = 0\\ \text{or} \quad &\lim_{k\to\infty} P\left[\bigcap_{n\geq k}|X_n - X| < \varepsilon\right] = 0\end{aligned}\right\} \quad \ldots (7{\cdot}33a)$$

Note. The two definitions of almost sure convergence are *equivalent.*

Notation. $X_n \xrightarrow{a.s.} X$ or $X_n \to X$ *a.s.*, for convergence almost surely or with probability 1.

Remark. The strong law of large numbers (*S.L.L.N.*; § 7.8.2) is a special case of the almost sure convergence.

Theorem 7·17. *If* $X_n \xrightarrow{a.s.} X$, then $X_n \xrightarrow{P} X$. *i.e. Almost sure convergence* ⇒ *Convergence in probability.*

Proof. Since $X_n \xrightarrow{a.s.} X$, for every ε > 0,

$$\lim_{N\to\infty} P\left[\bigcup_{n\geq N}\left(|X_n - X| \geq \varepsilon\right)\right] \Delta ABC = 0. \quad \ldots (*)$$

$$\{|X_N - X| \geq \varepsilon\} \Rightarrow \left\{\bigcup_{n\geq N}\left(|X_n - X| \geq \varepsilon\right)\right\}$$

$$\Rightarrow \quad \{|X_N - X| \geq \varepsilon\} \subset \left\{\bigcup_{n\geq N}\left(|X_n - X| \geq \varepsilon\right)\right\}$$

$$\Rightarrow \quad P\left[|X_N - X| \geq \varepsilon\right] \leq P\left[\bigcup_{n\geq N}\left(|X_n - X| \geq \varepsilon\right)\right]$$

$$\Rightarrow \quad \lim_{N\to\infty} P\left[|X_N - X| \geq \varepsilon\right] \leq \lim_{N\to\infty}\left[\bigcup_{n\geq N}\left(|X_n - X| \geq \varepsilon\right)\right]$$

$$\Rightarrow \quad \lim_{N\to\infty} P\left[|X_N - X| \geq \varepsilon\right] = 0 \quad \Rightarrow \quad X_n \xrightarrow{P} X \text{ as } n \to \infty. \quad \text{[From (*)]}$$

Hence, $$X_n \xrightarrow{a.s.} X \Rightarrow X_n \xrightarrow{P} X$$

Remark. The converse is not true *i.e.,* $X_n \xrightarrow{p} X \not\Rightarrow X_n \xrightarrow{a.s.} X$.

7·6·3. Convergence in *r*-th Mean

Definition. *A sequence of random variables* $\{X_n\,;\, n \ge 1\}$ *is said to converge to a r.v. X in r-th mean for some* $r \ge 1$, *if* $E|X|^r$ *and* $E|X_n|^r$ *exist (are finite) for all n and then*

$$\lim_{n\to\infty} E|X_n - X|^r = 0\,;\ r \ge 1 \qquad \text{... (7·34)}$$

Notation. $X_n \xrightarrow{r} X$: convergence in *r*-th mean.

Remarks:

1. If $E|X|^r$ and $E|X_n|^r$ exist, then $E|X_n - X|^r$ exists for $r \ge 1$.
2. If $r = 1$, we say that $\{X_n\}$ *converges in mean to* X and we write $X_n \xrightarrow{m} X$.
3. **Convergence in Mean Square or Quadratic Mean.** A sequence of *r.v.'s* $\{X_n,\ n \ge 1\}$ is said to converge in *mean square,* or *quadratic mean* if $E(X^2)$ and $E(X_n^2)$ exist for all *n* and

$$\lim_{n\to\infty} E(X_n - X)^2 = 0$$

We write: $X_n \xrightarrow{m.s} X$ or $X_n \xrightarrow{q.m.} X$.

Theorem 7·18. *If sequence* $\{X_n\}$ *of r.v.'s converges in mean square to zero in such a way that* $\sum_{n=1}^{\infty} E(X_n^2) < \infty$, *then* $\{X_n\}$ *converges to zero with probability 1, i.e.,* $X_n \xrightarrow{a.s} 0$.

Proof. Since $X_n \xrightarrow{m.s.} 0$, $\lim_{n\to\infty} E|X_n - 0|^2 = 0$

$$\Rightarrow \quad \lim_{n\to\infty} E\left(X_n^2\right) = 0 \Rightarrow \lim_{n\to\infty} X_n = 0 \qquad \text{... (*)}$$

We are given : $\sum_{n=1}^{\infty} E\left(X_n^2\right) < \infty \Rightarrow E\left(\sum_{n=1}^{\infty} X_n^2\right) < \infty$

This implies that $\left(\sum_{n=1}^{\infty} X_n^2\right)$ has a finite mean, *i.e.,* the series $\sum_{n=1}^{\infty} X_n^2$ is convergent.

In other words, X_n^2 assumes only finite values with probability 1, *i.e.,*

$$P(0 < X_n^2 < \infty) = 1 \Rightarrow P\left[\lim_{n\to\infty} X_n^2 = 0\right] = 1$$

[Here, we have used the result that if infinite series $\sum_n u_n$ is convergent then $\lim_{n\to\infty} u_n = 0$.]

$$\therefore \quad P\left[\lim_{n\to\infty} X_n = 0\right] = 1 \Rightarrow X_n \xrightarrow{a.s.} 0.$$

Theorem 7·19. *If* $X_n \xrightarrow{r} X$, *for some* $r > 0$, *then* $X_n \xrightarrow{p} X$

i.e., *Convergence in rth mean implies convergence in probability.*

Proof. Let $\in > 0$, be any real number. Then by Markov's Inequality, we get

$$P\left[\,|X_n - X| > \varepsilon\,\right] \le \frac{E|X_n - X|^r}{\varepsilon^r} \qquad \text{... (*)}$$

If $X_n \xrightarrow{r} X$, then for $\varepsilon > 0$,

$$\lim_{n\to\infty} E|X_n - X|^r = 0 \qquad \text{... (**)}$$

Taking limit in (*) as $n \to \infty$ and using (**), we get

$$\lim_{n\to\infty} P\left(|X_n - X| > \varepsilon\right) \le 0$$

$$\Rightarrow \quad \lim_{n\to\infty} P\left(|X_n - X| > \varepsilon\right) = 0 \quad \Rightarrow \quad X_n \xrightarrow{p} X$$

Hence, $\qquad X_n \xrightarrow{r} X \;\Rightarrow\; X_n \xrightarrow{p} X$

Remarks:

1. In particular, if we take $r = 2$, we get

$$X_n \xrightarrow{m.s.} X \quad \Rightarrow \quad X_n \xrightarrow{p} X$$

i.e., convergence in mean square or quadratic mean implies convergence in probability.

2. The converse of the above theorem is false, *i.e.,*

$$X_n \xrightarrow{p} X \quad \not\Rightarrow \quad X_n \xrightarrow{r} X \text{ for any } r.$$

Counter Example. Let $Z \sim U\,[0, 1]$ and

$$X_n = \begin{cases} 0, \text{ as } z \in \left[0, \dfrac{n-1}{n}\right] \\ n^{1/r}, \text{ otherwise } i.e., \text{ as } z \in \left[\dfrac{n-1}{n}, 1\right] \end{cases}$$

$$\therefore \quad P(X_n = 0) = 1 - \frac{1}{n}; \quad P\left(X_n = n^{1/r}\right) = P(X_n \ne 0) = \frac{1}{n}; \qquad [\because Z \sim U\,[0, 1]]$$

Let $X \equiv 0$. Then, for $\varepsilon > 0$

$$\lim_{n\to\infty} P\left[\,|X_n - X| > \varepsilon\,\right] = \lim_{n\to\infty} P\left[\,|X_n| > \varepsilon\,\right] \qquad (\because X \equiv 0)$$

$$= \lim_{n\to\infty} P\left(X_n \ne 0\right) = \lim_{n\to\infty}\left(\frac{1}{n}\right) = 0$$

$$\Rightarrow \quad X_n \xrightarrow{p} X \equiv 0. \qquad \text{... (*)}$$

But $\quad E\left[\left|X_n - X\right|^r\right] = E\left(\left|X_n\right|^r\right) = 0^r \cdot P\left(X_n = 0\right) + \left(n^{1/r}\right)^r \cdot P\left(X_n = n^{1/r}\right) \qquad (X \equiv 0)$

$$= n \cdot \frac{1}{n} = 1, \text{ for all } n$$

$$\therefore \lim_{n\to\infty} E\left[\left|X_n - X\right|^r\right] = 1 \quad \Rightarrow \quad X_n \overset{r}{\not\longrightarrow} X \equiv 0$$

i.e., X_n does not converge to $X \equiv 0$ in r^{th} mean.

There is a partial converse to the above Theorem which we state below without proof.

Theorem 7·19. (a) *If X_n's are almost surely bounded, i.e., if : $|X_n| = |X_n(w)| <$ some constant M for all n and all $w \in \Omega$, then*

$$X_n \xrightarrow{p} X \quad \Rightarrow \quad X_n \xrightarrow{r} X.$$

7·6·4. Convergence in Law or Convergence in Distribution

Definition. *A sequence $\{X_n, n \geq 1\}$ of r.v.'s having corresponding sequence $\left\{F_{X_n} = F_n \,;\, n \geq 1\right\}$ of distribution functions is said to converge to a r.v. X (whose distribution functions is F_X) in Law or in Distribution if*

$$\lim_{n\to\infty} F_{X_n}(x) = \lim_{n\to\infty} F_n(x) = F_X(x) \qquad \text{... (7·35)}$$

at all continuity points of $F_X(x)$.

Notation : Convergence in distribution : $X_n \xrightarrow{d} X$ or $X_n \xrightarrow{D} X$

Convergence in Law : $X_n \xrightarrow{L} X$ or $L(X_n) \rightarrow L(X)$

Note that both convergence in distribution and convergence in law refer to the same concept *i.e.* Law $\Leftrightarrow$ Distribution.

Remarks:

1. It is possible for $F_n(x)$ to converge to a continuous limit which is not a distribution function.

Illustration. Let X_n be a degenerate *r.v.* at $x = n$ *i.e.,* $P[X_n = n] = 1 \,;\, n \in N$.

Then $\quad F_n(x) = \text{Distribution function of } X_n = \begin{cases} 0 \,;\, x < n \\ 1 \,;\, x \geq n \end{cases}$

$$\Rightarrow \quad \lim_{n\to\infty} F_n(x) = 0, \ \forall\ x < \infty$$

Let $\quad F(x) = 0, \ \forall\ x < \infty$

Then $\quad \lim_{n\to\infty} F_n(x) = F(x) \ \Rightarrow\ F_n(x) \xrightarrow{L} F(x)$

But $F(x)$ is not the distribution function of any random variable.

2. It is unreasonable to require $F_{X_n}(x) \rightarrow F_X(x)$ at points of discontinuity in F_X, in the definition of convergence in distribution.

Illustration. Suppose $X_n \equiv a_n$, a real number for each n and $a_n \rightarrow a \equiv X$. Consider.

$$F_{X_n}(x) = F_n(x) = \begin{cases} 1 \text{ for } x \geq a_n \\ 0 \text{ for } x < a_n \end{cases} \quad \text{and} \quad F_X(x) = \begin{cases} 1 \text{ for } x \geq a \\ 0 \text{ for } x < a \end{cases}$$

If $x < a$, $\lim_{n\to\infty} F_n(x) = 0 = F_X(x)$

If $x > a$, $\lim_{n\to\infty} F_n(x) = 1 = F_X(x)$

Thus, $\lim_{n\to\infty} F_n(x) = F(x)$, for all points of continuity of $F(x)$.

For $X \equiv a$ where $F_X(x)$ *is discontinuous*, $F_X(a) = 1$.

$$\lim_{n\to\infty} F_n(a) = 1 \text{ iff } a_n > a \text{ for only finitely many } n.$$

$$\lim_{n\to\infty} F_n(a) = 0 \neq F_X(a) \text{ iff } a_n > a \text{ for all but finitely many } n.$$

Thus, if $a_n \uparrow a$, $F_n(x) \uparrow F_X(a)$ and if $a_n \downarrow a$, $F_n(x) \not\to F_X(a)$.

This unpleasant behaviour is typical of what happens at discontinuity of F_X. It may be recalled that the number of points, of discontinuity of $F_X(x)$ is atmost countable.

Another illustration, for the requirement that only continuity points of F should be considered, is given below.

Let X_n be uniformly distributed on $\left(0, \frac{1}{n}\right)$ so that its *p.d.f.* is :

$$f_n(x) = \frac{1}{n}\ ; 0 \leq x \leq \frac{1}{n}.$$

Distribution function $F_n(x)$ is given by :

$$F_n(x) = \int_{u=0}^{x} f_n(u)\, du = \frac{1}{n} \Big|\, u \Big|_{u=0}^{x} = \frac{x}{n}\ ;\ 0 \leq x \leq \frac{1}{n}$$

$$\therefore \qquad F_n(x) = \begin{cases} 0\ ; x \leq 0 \ \forall\ n \\ 1\ ; x \geq \frac{1}{n}\ ; n > 0 \end{cases}$$

As $n \to \infty$; $X_n \to 0 \equiv X$.

Let X be degenerate *r.v.* at the point '0' *i.e.*, $P(X = 0) = 1 \Rightarrow F_X(x) = 1;\ x = 0$

Thus, for the limiting *r.v.*, $X = 0$ (as $n \to \infty$), $F(0) = 1$.

But $F_n(0) = 0,\ \forall\ n.$

i.e. $\lim_{n\to\infty} F_n(0) \neq F(0).$

Thus, the convergence of $X_n \to X \equiv 0$ fails at the point $x = 0$, where $F(x)$ is discontinuous.

3. The points of possible non-convergence are determined by the discontinuities of the limiting distribution function $F_X(x)$ and have nothing to do with any discontinuity, the distribution function $F_n(x)$ might have.

4. We give below, (without proof), some important results on convergence of a sequence of *r.v.'s* to a *r.v.* in distribution or law.

Results

1. **Theorem 7.20 (Continuous Mapping Theorem).** *If* $X_n \xrightarrow{L} X$ *as* $n \to \infty$ *and* $g(\cdot)$ *is a continuous function, then* $$g(X_n) \xrightarrow{L} g(x)$$

2. $X_n \xrightarrow{L} X$ and $Y_n \xrightarrow{L} Y$, as $n \to \infty$, **does not imply that**

$$(a)\ X_n + Y_n \xrightarrow{L} X + Y$$

$$(b)\ X_n Y_n \xrightarrow{L} XY$$

Theorem 7·20 (a). **Levy's Continuity Theorem**

The sequence $\{X_n, n \geq 1\}$ *of r.v.'s converge in distribution to the r.v X if and only if the sequence* $\{\phi_n;\ n \geq 1\}$ *of the corresponding characteristic functions of* X_n*'s converges point-wise to the characteristic function* ϕ *of X.*

Theorem 7·20 (b). *Convergence in probability of a sequence of r.v.'s* $\{X_n\}$ *implies that* $\{X_n\}$ *has a limiting distribution, i.e.,*

$$X_n \xrightarrow{p} X \quad \Rightarrow \quad X_n \xrightarrow{L} X \qquad \text{... (7·36)}$$

Proof. Since $X_n \xrightarrow{p} X$, for $\varepsilon > 0$, we have

$$\lim_{n\to\infty} P\left(\left|X_n - X\right| > \varepsilon\right) = 0 \qquad \text{... (*)}$$

Let x be a point of continuity of the distribution function of X.

Define : $A = \{w \mid X(w) \leq x - \varepsilon\}$ $\quad B = \{w \mid X_n(w) \leq x\}$

We have : $A = (A \cap B) \cup (A \cap B') \quad \subset \quad B \cup (A \cap B')$

$$\Rightarrow \quad P(A) \leq P(B) + P(A \cap B')$$

$$\Rightarrow \quad P(X \leq x - \varepsilon) \leq P\left(X_n \leq x\right) + P\left[X \leq x - \varepsilon \cap X_n > x\right]$$

$$\Rightarrow \quad F_X(x - \varepsilon) \leq F_n(x) + P\left[X \leq x - \varepsilon \cap X_n > x\right]$$

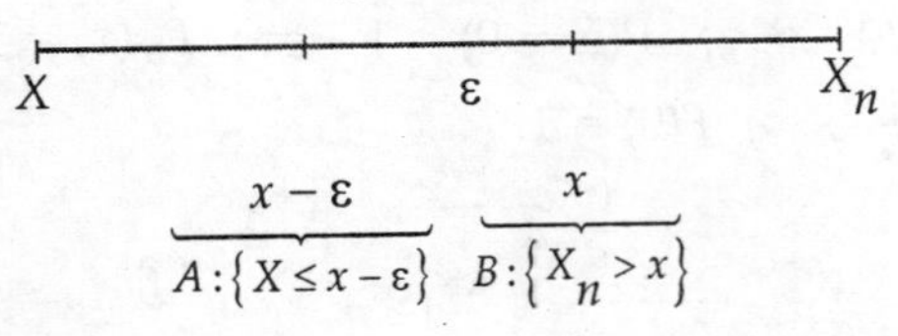

$$\Rightarrow \quad \left|X_n - X\right| > \varepsilon$$

$$\therefore \quad F_X(x - \varepsilon) \leq F_n(x) + P\left[\left|X_n - X\right| > \varepsilon\right]$$

Taking limit infinimum on both sides as $n \to \infty$, we get

$$F_X(x - \varepsilon) \leq \lim_{n\to\infty} \inf F_n(x) + \lim_{n\to\infty} \inf P\left[\left|X_n - X\right| > \varepsilon\right]$$

$$\Rightarrow \quad F_X(x - \varepsilon) \leq \lim_{n\to\infty} \inf F_n(x) \qquad \text{[From (*)]} \qquad \text{... (**)}$$

Define : $B = \{w \mid X_n(w) \le x\}$; $\quad C = \{w \mid X(w) \le x + \varepsilon\}$

Then $\quad B = (B \cap C) \cup (B \cap C') \quad \subset \quad C \cup (B \cap C')$

$$\Rightarrow \quad P(B) \le P(C) + P(B \cap C')$$

$$\Rightarrow \quad P(X_n \le x) \le P(X \le x + \varepsilon) + P[X_n \le x \cap (X > x + \varepsilon)]$$

$\Rightarrow$

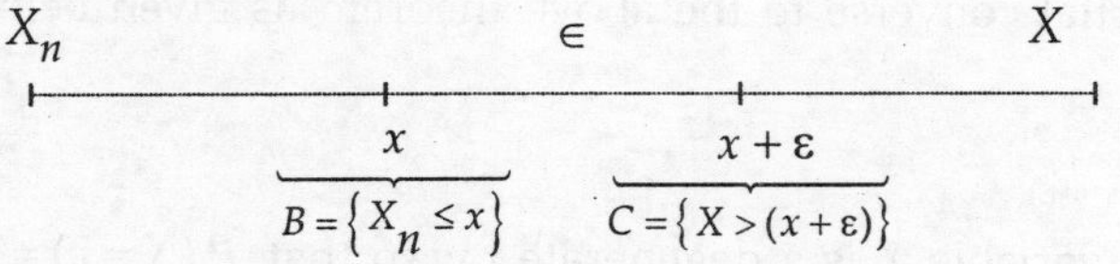

$$\Rightarrow \quad |X - X_n| = |X_n - X| > \varepsilon$$

$$\therefore \quad P(X_n \le x) \le P(X \le x+\varepsilon) + P[|X_n - X| > \varepsilon]$$

$$\Rightarrow \quad F_n(x) \le F(x+\varepsilon) + P[|X_n - X| > \varepsilon]$$

On taking limit superemum of both sides as $n \to \infty$

$$\lim_{n\to\infty} \sup F_n(x) \le F(x+\varepsilon) + \lim_{n\to\infty} \sup P[|X_n - X| > \varepsilon]$$

$$\Rightarrow \quad \lim_{n\to\infty} \sup F_n(x) \le F_X(x+\varepsilon) \qquad \text{[From (*)]} \qquad \ldots (***)$$

Also, $\lim\inf (F_n(x)) \le \lim_{n\to\infty} F_n(x) \le \lim\sup (F_n(x))$, we have from (**) and (***)

$$F(x-\varepsilon) \le \lim\inf F_n(x) \le \lim_{n\to\infty} F_n(x) \le \lim\sup (F_n(x)) \le F_X(x+\varepsilon) \qquad \ldots (****)$$

Futher, since x is a continuity point of $F_X(x)$, we have

$$\lim_{\varepsilon\to 0} F(x+\varepsilon) = \lim_{\varepsilon\to 0} F(x-\varepsilon) = F(x) \text{ and we finally get :}$$

$$\lim_{n\to\infty} F_n(x) = F_X(x) \quad \Rightarrow \quad \{X_n\} \xrightarrow{L} X$$

Remark. The converse of this theorem is false *i.e.*, $X_n \xrightarrow{L} X$ does not imply $X_n \xrightarrow{p} X$.

Counter Example. Let $X_1, X_2, \ldots, X_n$ be *i.i.d.* $N(0, 1)$.

Since $X_n \xrightarrow{L} X$, X is also $N(0, 1)$

$$\therefore \quad F_{X_n}(x) = F_X(x) = \Phi(x) = \int_{-\infty}^{x} \frac{1}{\sqrt{2\pi}} e^{-u^2/2} du, \ \forall\, x,'$$

where, $\Phi(.)$ is distribution funtion of S.N.V.

$$\therefore \quad V = X_n - X \sim N(0, 2), \text{ for all } n.$$

$$[\because X \sim N(0, 1) \text{ and } X_n \sim N(0, 1) \text{ are independent.}]$$

$\therefore$ For $\varepsilon > 0$,

$$P[|X_n - X| > \varepsilon] = P[|V| > \varepsilon] = P\left[|Z| > \frac{\varepsilon}{\sqrt{2}}\right]; \qquad Z \sim N(0,1)$$

$$= 2P\left[Z < -\frac{\varepsilon}{\sqrt{2}}\right] = 2\Phi\left(-\frac{\varepsilon}{\sqrt{2}}\right) > 0; \text{ for all } n$$

where, Φ is the distribution function of standard normal variate.

Since $\lim_{n\to\infty} P\left[|X_n - X| > \varepsilon\right] \not\longrightarrow 0, \qquad X_n \xrightarrow{p}\!\!\!\!\!/\;\; X$

However, there is an important partial converse to the above theorem as given below.

Partial Converse to Theorem *7·20 (b)*

Let $X_n \xrightarrow{L} X$. If the lionting random variable X is a degenerate r.v. so that $P(X = c) = 1$, then $X_n \xrightarrow{P} X$.

In other words : $X_n \xrightarrow{L} X\,(\equiv c) \quad \Rightarrow \quad X_n \xrightarrow{p} c$

Theorem 7·21. *$X_n \xrightarrow{p} c$ if and only if $F_n(x) \to 0$ or 1 according as $x < c$ or $x \geq c$; i.e.,*

$$\lim_{n\to\infty} F_n(x) = \begin{cases} 0, & x < c \\ 1, & x \geq c \end{cases}$$

Proof. *If part :*

$$X_n \xrightarrow{p} c \Rightarrow \lim_{n\to\infty} \left[|X_n - c| > \varepsilon\right] = 0 \,;\; \varepsilon > 0$$

$$\Rightarrow \quad \lim_{n\to\infty} P\left[(X_n - c) > \varepsilon \text{ or } (X_n - c) < -\varepsilon\right] = 0$$

$$\Rightarrow \quad \lim_{n\to\infty} P\left[(X_n - c) > \varepsilon\right] + \lim_{n\to\infty} P\left[(X_n - c) < -\varepsilon\right] = 0$$

$$\Rightarrow \quad \lim_{n\to\infty} P\left[(X_n - c) > \varepsilon\right] = 0 \quad \text{and} \quad \lim_{n\to\infty} P\left[(X_n - c) < -\varepsilon\right] = 0$$

$[\because P(.) \geq 0$ and if $a \geq 0$, $b \geq 0$, then $a + b = 0 \quad \Rightarrow \quad a = 0$ and $b = 0]$

$$\Rightarrow \quad \lim_{n\to\infty} P\left(X_n > c + \varepsilon\right) = 0 \quad \text{and} \quad \lim_{n\to\infty} P\left(X_n < c - \varepsilon\right) = 0$$

$$\Rightarrow \quad \lim_{n\to\infty} P\left(X_n \leq c + \varepsilon\right) = 1 \quad \text{and} \quad \lim_{n\to\infty} F_n\left(c - \varepsilon\right) = 0$$

$$\Rightarrow \quad \lim_{n\to\infty} F_n\left(c + \varepsilon\right) = 1 \quad \text{and} \quad \lim_{n\to\infty} F_n\left(c - \varepsilon\right) = 0$$

$$\Rightarrow \quad \lim_{n\to\infty} F_n\left(x\right) = 1,\ x > c \quad \text{and} \quad \lim_{n\to\infty} F_n\left(x\right) = 0,\ x < c$$

$$\therefore \quad \lim_{n\to\infty} F_n\left(x\right) = \begin{cases} 0; & x < c \\ 1; & x \geq c \end{cases} \qquad \ldots (*)$$

Only if part : Given that (*) is true, we have to prove that $X_n \xrightarrow{p} c$.

For any $\varepsilon > 0$,

$$P\left[|X_n - c| \leq \varepsilon\right] = P\left[c - \varepsilon \leq X_n \leq c + \varepsilon\right]$$

$$= P[X_n \le c + \varepsilon] - P[X_n \le c - \varepsilon]$$
$$= F_n(c + \varepsilon) - F_n(c - \varepsilon) \; ; \quad \varepsilon > 0$$
$$= 1 - 0 = 1 \qquad \text{[From(*)]}$$

$\Rightarrow$ $$X_n \xrightarrow{p} c.$$

Note. *This Theorem states that if the limit r.v. X is a degenerate random variable, then*

$$X_n \xrightarrow{L} X \quad \Leftrightarrow \quad X_n \xrightarrow{p} X$$

7·6·5. Relationship Between Various Modes of Convergence

In this section, we will summarise the important results concerning the different modes of convergence of sequence of random variables $\{X_n \; ; \; n \ge 1\}$ to a limit random variable X, as discussed in the earlier sections.

Notations

1. $X_n \xrightarrow{p} X$: Convergence in probability.
2. $X_n \xrightarrow{a.s.} X$: Almost sure convergence or convergence with probability 1.
3. $X_n \xrightarrow{r} X$: Convergence in r^{th} mean.
4. $X_n \xrightarrow{d}$ or $X_n \xrightarrow{L} X$: Convergence in distribution or in Law.

Result 1.

(*a*) If $X_n \xrightarrow{p} X$ and $X_n \xrightarrow{p} Y$, then $X = Y$ *a.s.*

(*b*) If $X_n \xrightarrow{a.s.} X$ and $X_n \xrightarrow{a.s.} Y$, then $X = Y$ *a.s.*

(*c*) If $X_n \xrightarrow{r} X$ and $X_n \xrightarrow{r} Y$, then $X = Y$ *a.s.*

Result 2.

(*a*) If $X_n \xrightarrow{p} X$ and $Y_n \xrightarrow{p} Y$, then

(*i*) $aX_n + bY_n \xrightarrow{p} aX + bY$, for real numbers *a* and *b*.

(*ii*) $X_n Y_n \xrightarrow{p} XY$.

(*b*) If $X_n \xrightarrow{a.s.} X$ and $Y_n \xrightarrow{a.s.} Y$, then :

(*i*) $aX_n + bY_n \xrightarrow{a.s.} aX + bY$, for real numbers *a* and *b*.

(*ii*) $X_n Y_n \xrightarrow{a.s.} XY$.

(*c*) If $X_n \xrightarrow{r} X$ and $Y_n \xrightarrow{r} Y$, then $aX_n + bY_n \xrightarrow{r} aX + bY$, for real numbers *a* and *b*.

None of the above results in 2(*a*), (*b*), (*c*) are true for convergence in distribution.

Inter-relationships Between Different Modes of Convergence.

(A) (2) $\Rightarrow$ (1) *i.e.,* Almost sure convergence inplies convergence in probability.

(B) (1) $\not\Rightarrow$ (2) *i.e.,* converse of (*A*) is not true.

(C) (3) $\Rightarrow$ (1) *i.e.,* convergence in rth mean implies convergence in probability (Theorem 7·19)

(D) (1) $\not\Rightarrow$ (3) *i.e.,* converse of (C) is not true [See Remark 2 to Theorem 7·19] However, there is a partial converse to (C). [See Theorem 7·19(a)]

(E) Convergence in *r*th mean does not imply and is not implied by convergence with probability 1, *i.e.,* $X_n \xrightarrow{r} X \not\Leftrightarrow X_n \xrightarrow{a.s.} X$

(F) (1) $\Rightarrow$ (4), *i.e.,* $X_n \xrightarrow{p} X \Rightarrow X_n \xrightarrow{L} X.$

Converse however is not ture, *i.e.,* $X_n \xrightarrow{L} X \not\Rightarrow X_n \xrightarrow{P} X.$

However, if X is a degenerate *r.v.* at '*c*', then $X_n \xrightarrow{p} X\,(\equiv c)$ *iff* $X_n \xrightarrow{L} X.$

The relationships between different modes of convergence of a sequence $\{X_n;\ n \geq 1\}$ of *r.v.'s* to the *r.v.* X are elegantly displayed in the following diagram.

Relationship Between Various Modes of Convergence

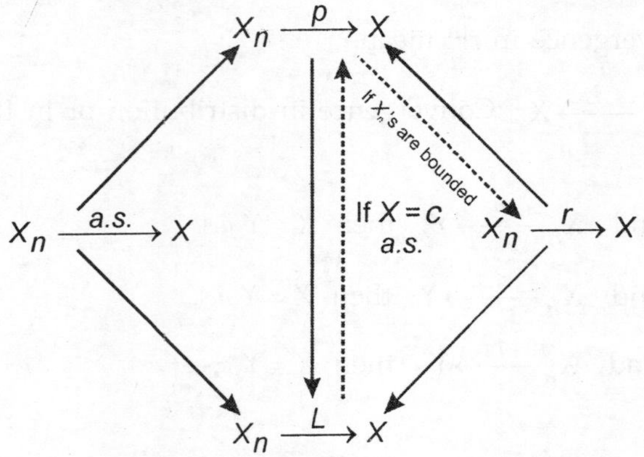

7·6·6. Chebychev's Theorem

As an immediate consequence of Chebychev's inequality, we have the following theorem on convergence in probability.

"If $X_1, X_2, \ldots, X_n$ is a sequence of random variables and if mean μ_n and standard deviation σ_n of X_n exists for all n and if $\sigma_n \to 0$ as $n \to \infty$, then $X_n - \mu_n \xrightarrow{p} 0$ as $n \to \infty$.

Proof. By Chebychev's inequality, for $\varepsilon > 0$,

$$P\{|X_n - \mu_n| \geq \varepsilon\} \leq \frac{\sigma_n^2}{\varepsilon^2} \longrightarrow 0 \text{ as } n \to \infty. \qquad (\because \sigma_n \to 0 \text{ as } n \to \infty)$$

Hence, $(X_n - \mu_n) \xrightarrow{p} 0$ as $n \to \infty$ provided $\sigma_n \to 0$ as $n \to \infty$.

7.7. WEAK LAW OF LARGE NUMBERS (W.L.L.N)

Statement. Let $X_1, X_2, ..., X_n$ *be a sequence of random variables and* $\mu_1, \mu_2, ..., \mu_n$ *be their respective expectations and let*

$$B_n = \text{Var}(X_1 + X_2 + ... + X_n) < \infty \quad ...(7·37)$$

Then, $$P\left\{\left|\frac{X_1+X_2+...+X_n}{n} - \frac{\mu_1+\mu_2+...+\mu_n}{n}\right| < \varepsilon\right\} \geq 1-\eta,$$

for all $n > n_0$, *where* ε *and* η *are arbitrary small positive numbers, provided*

$$\lim_{n\to\infty} \frac{B_n}{n^2} \to 0. \quad ...(7·37a)$$

Proof. Using Chebychev's Inequality (7·24), to the *r.v.* $(X_1 + X_2 + ... + X_n) / n$, we get for any $\varepsilon > 0$,

$$P\left\{\left|\left(\frac{X_1+X_2+...+X_n}{n}\right) - E\left(\frac{X_1+X_2+...+X_n}{n}\right)\right| < \varepsilon\right\} \geq 1-\frac{B_n}{n^2\varepsilon^2}$$

$$\left[\text{since Var}\left(\frac{X_1+X_2+...+X_n}{n}\right) = \frac{1}{n^2}\text{Var}\left(X_1+X_2+...+X_n\right) = \frac{B_n}{n^2}\right]$$

$$\Rightarrow \quad P\left\{\left|\frac{X_1+X_2+...+X_n}{n} - \frac{\mu_1+\mu_2+...+\mu_n}{n}\right| < \varepsilon\right\} \geq 1-\frac{B_n}{n^2\varepsilon^2}$$

So far, nothing is assumed about the behaviour of B_n for indefinitely increasing values of n.

Since ε is arbitrary, we assume $\frac{B_n}{n^2\varepsilon^2} \to 0$, as n becomes indefinitely large.

Thus, having chosen two arbitrary small positive numbers ε and η, number n_0 can be found so that the inequality $\frac{B_n}{n^2\varepsilon^2} < \eta$ will hold for $n > n_0$. Consequently, we shall have

$$P\left\{\left|\frac{X_1+X_2+...+X_n}{n} - \frac{\mu_1+\mu_2+...+\mu_n}{n}\right| < \varepsilon\right\} \geq 1-\eta, \text{ for all } n > n_0(\varepsilon, \eta).$$

This conclusion leads to the following important result, known as the (*Weak*) Law of Large Numbers:

"With the probability approaching unity or certainty as near as we please, we may expect that the arithmetic mean of values actually assumed by n random variables will differ from the arithmetic mean of their expectations by less than any given number, however small, provided the number of variables can be taken sufficiently large and provided the condition :

$$\frac{B_n}{n^2} \to 0 \text{ as } n \to \infty \textit{ is fulfilled."}$$

Remarks:

1. Weak law of large numbers can also be stated as follows:

$$\overline{X}_n \xrightarrow{p} \overline{\mu}_n, \text{ provided } \frac{B_n}{n^2} \to 0 \text{ as } n \to \infty, \text{ symbols having their usual meanings.} \quad ...(7·37b)$$

2. For the existence of the law we assume the following conditions :

(*i*) $E(X_i)$ exists for all i, (*ii*) B_n = Var $(X_1 + X_2 + ... + X_n)$ exists, and (*iii*) $B_n/n^2 \to 0$ as $n \to \infty$. Condition (*i*) is necessary, without it the law itself cannot be stated. But the conditions (*ii*) and (*iii*) are not necessary ; Condition (*iii*) is however a sufficient condition.

3. WLLN for *i.i.d.* random variables. If the variables $X_1, X_2, ..., X_n$ are independent and identically distributed, *i.e.*, if $E(X_i) = \mu$, (say), and Var $(X_i) = \sigma^2$,(say), for all $i = 1, 2, ..., n$, then

$$B_n = \text{Var}(X_1 + X_2 + ... + X_n) = \sum_{i=1}^{n} \text{Var}(X_i) = n\sigma^2 \qquad ... (*)$$

the covariance terms vanish, since the variables are independent.

Hence,
$$\lim_{n\to\infty} \frac{B_n}{n^2} = \lim_{n\to\infty} \frac{\sigma^2}{n} = 0$$

Thus, the weak law of large number holds for the sequence $\{X_n\}$ of *i.i.d. r.v.'s* and we get

$$P\left\{\left|\frac{X_1 + X_2 + ... + X_n}{n} - \mu\right| < \varepsilon\right\} > 1 - \eta, \forall\, n > n_0$$

i.e.,
$$P\left\{\left|\overline{X}_n - \mu\right| < \varepsilon\right\} \to 1 \text{ as } n \to \infty \;\Rightarrow\; P\left\{\left|\overline{X}_n - \mu\right| \geq \varepsilon\right\} \to 0 \text{ as } n \to \infty$$

where, $\overline{X}_n$ is the mean of n random variables $X_1, X_2, ... X_n$. This result implies that $\overline{X}_n$ converges in *probability* to μ, *i.e.*, $\overline{X}_n \xrightarrow{p} \mu$.

Note. If $\overline{X}_n$ is the mean of n *i.i.d. r.v.'s* $X_1, X_2, ... X_n$ with $E(X_i) = \mu$; Var $(X_i) = \sigma^2$, then

$$E(\overline{X}_n) = \mu \quad \text{and} \quad \text{Var}(\overline{X}_n) = \text{Var}\left(\sum_{i=1}^{n} x_i/n\right) = \frac{\sigma^2}{n} \qquad \text{[On using (*)]}$$

Theorem 7·22. *If the variables are uniformly bounded then the condition,*

$$\lim_{n\to\infty} \frac{B_n}{n^2} = 0,$$

is necessary as well as sufficient for WLLN to hold.

Proof. Let $\xi_i = X_i - a_i$, where $E(X_i) = a_i$; then $E(\xi_i) = 0$, $(i = 1, 2, ..., n)$

Since X_i's are uniformly bounded, there exists a positive number $c < \infty$ such that $|\xi_i| < c$.

If $\quad p = P\left\{\left|\xi_1 + \xi_2 + ... + \xi_n\right| \leq n\,\varepsilon\right\}$, $\quad$ then $\quad 1 - p = P\left[\left|\xi_1 + \xi_2 + ... + \xi_n\right| > n\,\varepsilon\right]$

Let $U_n = \xi_1 + \xi_2 + ... + \xi_n$, then $E(U_n) = \sum_{i=1}^{n} E(\xi_i) = 0$ and Var $(U_n) = E\left(U_n^2\right) = B_n$ (say).

Also, $$|U_n| = \left|\sum_{i=1}^{n} \xi_i\right| \le \sum_{i=1}^{n} |\xi_i| < \left.\sum_{i=1}^{n} c = nc\right\} \quad \text{... (7·40)}$$

and $P(|U_n| \le n\varepsilon) = p$ and $P(|U_n| \le n\varepsilon) = 1 - p.$ ... (7·40a)

$$\Rightarrow \quad B_n = \int_0^{\infty} U_n^2 \, dF(U_n), \text{ where } F(U_n) \text{ is } d.f. \text{ of } U_n$$

$$= \int_{U_n^2 \le n^2\varepsilon^2} U_n^2 \, dF + \int_{U_n^2 > n^2\varepsilon^2} U_n^2 \, dF$$

(where $F(U_n)$ is the distribution function of U_n)

$$\Rightarrow \quad B_n \le n^2\varepsilon^2 \int_{|\mu_n| \le n\varepsilon} dF + n^2c^2 \int_{|\mu_n| > n\varepsilon} dF \le n^2\varepsilon^2 p + n^2c^2(1-p)$$

[From (7·40) and (7·40a)]

$$\therefore \quad \frac{B_n}{n^2} \le \varepsilon^2 p + c^2(1-p)$$

If the law of large numbers holds, $1 - p = P[|\xi_1 + \xi_2 + \dots + \xi_n| > n\varepsilon] \to 0$ as $n \to \infty$.

Hence, as $n \to \infty$, $(1-p) \to 0$, and $\dfrac{B_n}{n^2} < \varepsilon^2 p + c^2\delta$, ε and δ being arbitrarily small +ve numbers.

Hence, $$\frac{B_n}{n^2} \to 0 \quad \text{as } n \to \infty.$$

7·7·1. Bernoulli's Law of Large Numbers

Let there be n trials of an event, each trial resulting in a success or failure. If X is the number of successes in n trials with constant probability p of success for each trial, then $E(X) = np$ and $\text{Var}(X) = npq$, $q = 1 - p$. [See Binomial Distribution, Chapter 8] The variable X/n represents the proportion of successes or the relative frequency of successes, and

$$E\left(\frac{X}{n}\right) = p, \quad \text{and} \quad \text{Var}\left(\frac{X}{n}\right) = \frac{1}{n^2}\text{Var}(X) = \frac{pq}{n}.$$

Then, $$P\left\{\left|\frac{X}{n} - p\right| < \varepsilon\right\} \to 1 \text{ as } n \to \infty \quad \Rightarrow \quad P\left\{\left|\frac{X}{n} - p\right| \ge \varepsilon\right\} \to 0 \text{ as } n \to \infty \quad \text{... (7·41)}$$

for any assigned $\varepsilon > 0$. This implies that (X/n) converges in probability to p as $n \to \infty$.

Proof. Applying Chebychev's inequality (7·24) to the variable X/n, we get for any $\varepsilon > 0$.

$$P\left\{\left|\frac{X}{n} - E\left(\frac{X}{n}\right)\right| \ge \varepsilon\right\} \le \frac{\text{Var}(X/n)}{\varepsilon^2} = \frac{pq}{n\varepsilon^2} \le \frac{1}{4n\varepsilon^2},$$

since the maximum value of pq is at $p = q = \dfrac{1}{2}$, *i.e.*, $\max(pq) = \dfrac{1}{4}$, *i.e.*, $pq \le \dfrac{1}{4}$.

Since ε is arbitrary, we get $P\left\{\left|\frac{X}{n}-p\right|\geq\varepsilon\right\}\to 0$ as $n\to\infty \Rightarrow P\left\{\left|\frac{X}{n}-p\right|<\varepsilon\right\}\to 1$ as $n\to\infty$.

7·7·2. Markov's Theorem

Theorem 7·23. *The weak law of large numbers holds if for some $\delta > 0$, all the mathematical expectations*

$$E\left(|X_i|^{1+\delta}\right);\quad i=1,2,\ldots \text{ exist and are bounded.} \qquad \ldots (7\cdot 42)$$

Markov's theorem provides only a necessary condition for the weak law of large numbers to hold good. This means that if, for $\delta > 0$, $E|X_i|^{1+\delta}$, is unbounded then W.L.L.N. cannot hold for the sequence of *r.v.'s* $\{X_n\}$.

7·7·3. Khinchin's Theorem (W.L.L.N.)

Theorem 7·24. *If X_i's; $i = 1, 2, \ldots, n$ are identically and independently distributed random variables, the only condition for the W.L.L.N. to hold is that $E(X_i) = \mu$; $i = 1, 2, \ldots, n$ should exist.*

OR

If X_i's, $(i = 1, 2, \ldots, n)$ are i.i.d. random variables, with $E(X_i) = \mu < \infty$, then $\overline{X}_n \xrightarrow{p} \mu$. ... (7·43)

Proof. Let X_i; $i = 1, 2, \ldots, n$ be *i.i.d. r.v.'s* with $E(X_i) = \mu < \infty$ and $\overline{X}_n = \frac{1}{n}\sum_{i=1}^{n} X_i$

The characteristic function of $\overline{X}_n$ is given by :

$$\phi_{\overline{X}_n}(t) = \phi_{\frac{1}{n}\sum_{j=1}^{n} X_j}(t) = \phi_{\sum_{j=1}^{n} X_j}(t/n) = [\phi_X(t/n)]^n, \qquad \ldots (*)$$

where, $\phi_X(t)$ is the common *c.f.* of each X_i.

Thus,
$$\phi_X(t) = 1 + it\,E(X) + \frac{(it)^2}{2}E(X^2) + \ldots$$
$$= 1 + it\mu + O(t^2),$$

where, $\left[O(t^r)/t^r\right] \to 0$ as $t \to 0$.

Substituting in (*), we get

$$\phi_{\overline{X}_n}(t) = \left[1 + \frac{it}{n}\mu + O\left(\frac{t}{n}\right)\right]^n \to \exp(i\,t\,\mu), \text{ as } n \to \infty \qquad \ldots (**)$$

Hence, by the Uniqueness Theorem of characteristic functions, the distribution of $\overline{X}_n$ tends to that of *a r.v.* whose characteristic function is $\exp(i\,t\,\mu)$.

Let us consider a *r.v.* Y defined as follows :

$$Y = \mu \text{ with probability 1 and } Y \neq \mu \text{ with probability 0.}$$

Then, the *c.f.* of Y is given by :

$$\phi_Y(t) = E(e^{itY}) = e^{it\mu}.\, 1 = e^{i\mu t} \qquad \text{... (****)}$$

which is the *c.f.* of a degenerate *r.v.*, degenerate at μ.

Hence, by Uniqueness Theorem of characteristic functions and Theorem 7·21,

$$\overline{X}_n \xrightarrow{distt.} \mu \text{ or } \overline{X}_n \xrightarrow{L} \mu = \text{costant} \quad \Rightarrow \quad \overline{X}_n \xrightarrow{P} \mu$$

Theorem 7·25. *Necessary and Sufficient Condition for W.L.L.N. to Hold.*

Let $\{X_n\}$ be any sequence of r.v.'s. Write :

$$Y_n = \{S_n - E(S_n)\}/n, \text{ where } S_n = X_1 + X_2 + \ldots + X_n.$$

A necessary and sufficient condition for sequence $\{X_n\}$ to satisfy the W.L.L.N. is:

$$E\left\{\frac{Y_n^2}{1+Y_n^2}\right\} \to 0 \text{ as } n \to \infty \qquad \text{... (7·44)}$$

Proof. *If Part* : Let us assume that (7·44) holds. We shall prove $\{X_n\}$ satisfies W.L.L.N.
For real numbers a, b ; $a \geq b > 0$, we have $\quad a \geq b \;\Rightarrow\; a + ab \geq b + ab$ ... (*)

Let us define the event $A = \{|Y_n| \geq \varepsilon\}$.

$$\omega \in A \;\Rightarrow\; |Y_n| \geq \varepsilon \;\Rightarrow\; |Y_n|^2 \geq \varepsilon^2 > 0$$

∴ Taking $a = Y_n^2$ and $b = \varepsilon^2$ in (*), we define another event B as follows :

$$B = \left\{\frac{Y_n^2(1+\varepsilon^2)}{\varepsilon^2(1+Y_n^2)} \geq 1\right\} = \left\{\frac{Y_n^2}{1+Y_n^2} \geq \frac{\varepsilon^2}{1+\varepsilon^2}\right\}$$

Since $\quad \omega \in A \;\Rightarrow\; \omega \in B, \quad A \subseteq B \;\Rightarrow\; P(A) \leq P(B).$

$$\therefore \quad P(|Y_n| \geq \varepsilon) \leq P\left(\frac{Y_n^2}{1+Y_n^2} \cdot \frac{1+\varepsilon^2}{\varepsilon^2} \geq 1\right)$$

$$\leq E\left\{\frac{Y_n^2/(1+Y_n^2)}{\varepsilon^2/(1+\varepsilon^2)}\right\} \qquad \text{[By Markov's Inequality]}$$

$$\to 0 \text{ as } n \to \infty \qquad \text{[By assumption (7·44)]}$$

$$\therefore \quad P(|Y_n| \geq \varepsilon) \to 0 \text{ as } n \to \infty \;\Rightarrow\; \lim_{n\to\infty} P\left\{\left|\frac{S_n - E(S_n)}{n}\right| \geq \varepsilon\right\} \to 0$$

Thus, W.L.L.N. holds for the sequence $\{X_n\}$ of *r.v.'s*.

Conversely, if $\{X_n\}$ satisfies W.L.L.N. we shall establish (7·44). Let us assume that X_i's are continuous and let Y_n have *p.d.f.* $f_n(y)$. Then

$$E\left(\frac{Y_n^2}{1+Y_n^2}\right) = \int_{-\infty}^{\infty} \frac{y^2}{1+y^2} f_n(y)\,dy = \left(\int_A + \int_{A^c}\right)\frac{y^2}{1+y^2} f_n(y)\,dy,$$

where, $A = \{|Y| \geq \varepsilon\}$ and $A^c = \{|Y| \leq \varepsilon\}$

$$E\left(\frac{Y_n^2}{1+Y_n^2}\right) \leq \int_A 1\cdot f_n(y)\,dy + \int_{A^c} y^2 \cdot f_n(y)\,dy \quad \left[\because \frac{y^2}{1+y^2} < 1 \text{ and } \frac{y^2}{1+y^2} < y^2\right]$$

$$\leq P(A) + \varepsilon^2 \int_{A^c} f_n(y)\,dy \qquad (\because \text{ On } A^c : |y| < \varepsilon)$$

$$= P(A) + \varepsilon^2 . P(A^c) \leq P(A) + \varepsilon^2 \qquad [\because P(A^c) < 1]$$

$$\Rightarrow \qquad E\left\{\frac{Y_n^2}{1+Y_n^2}\right\} \leq P\{|Y_n| \geq \varepsilon\} + \varepsilon^2 \qquad ...(**)$$

But since $\{X_n\}$ satisfies W.L.L.N., we have $\lim_{n\to\infty} P\left[|Y_n| \geq \varepsilon\right] \to 0$, and since ε^2 is arbitrarily small positive number, we get on taking limits in (**),

$$\lim_{n\to\infty}\left(\frac{Y_n^2}{1+Y_n^2}\right) \to 0$$

Cor. Let $X_1, X_2, ..., X_n$ be a sequence of independent *r.v.'s* such that Var $(X_i) < \infty$

for $i = 1, 2, ...$ and $\dfrac{B_n}{n^2} = \dfrac{\text{Var}\left(\sum_{i=1}^{n} X_i\right)}{n^2} = \dfrac{\text{Var}\,(S_n)}{n^2} \to 0$ as $n \to \infty$. Then W.L.L.N. holds.

Proof. We have

$$\frac{Y_n^2}{1+Y_n^2} \leq Y_n^2 = \left\{\frac{S_n - E(S_n)}{n}\right\}^2$$

$$\Rightarrow \qquad E\left(\frac{Y_n^2}{1+Y_n^2}\right) \leq \frac{1}{n^2} E\{S_n - E(S_n)\}^2 = \frac{\text{Var}\, S_n}{n^2} = \frac{B_n}{n^2}$$

$$\therefore \qquad \lim_{n\to\infty}\left(\frac{Y_n^2}{1+Y_n^2}\right) \leq \lim_{n\to\infty} \frac{B_n}{n^2} \to 0 \qquad \text{(By assumption)}$$

Hence, by the above theorem, W.L.L.N. holds for the sequence $\{X_n\}$ of *r.v.'s*.

Remark. The result of Theorem 7·25 holds even if $E(X_i)$ does not exist. In this case we simply define $Y_n = \{S_n / n\}$ rather than $\{S_n - E(S_n)\}/n$.

Example 7·16. *For geometric distribution* $p(x) = 2^{-x}$; $x = 1, 2, 3, \ldots,$

prove that Chebychev's inequality gives $P\{|X-2| \le 2\} > \frac{1}{2}$, *while the actual probablity is* $\frac{15}{16}$.

Solution.
$$E(X) = \sum_{x=1}^{\infty} \frac{x}{2^x} = \frac{1}{2} + \frac{2}{2^2} + \frac{3}{2^3} + \frac{4}{2^4} + \ldots$$
$$= \frac{1}{2}\left(1 + 2A + 3A^2 + 4A^3 + \ldots\right)$$
$$= \frac{1}{2}(1-A)^{-2} = 2, \left(A = \frac{1}{2}\right)$$
$$E(X^2) = \sum_{x=1}^{\infty} \frac{x^2}{2^x} = \frac{1}{2} + \frac{4}{2^2} + \frac{9}{2^3} + \ldots = \frac{1}{2}\left[1 + 4A + 9A^2 + \ldots\right];\ \left(A = \frac{1}{2}\right)$$
$$= \frac{1}{2}(1+A)(1-A)^{-3} = 6, \qquad \text{[See Remark to Example 6·15]}$$
$$\therefore \quad \text{Var}(X) = E(X^2) - \{E(X)\}^2 = 6 - 4 = 2 \quad \Rightarrow \quad \sigma = \sqrt{2}$$

Using Chebychev's inequality, we get $P\{|X - E(X) \le k\sigma|\} > 1 - \frac{1}{k^2}$

With $k = \sqrt{2}$, we get $P\{|X-2| \le \sqrt{2} \cdot \sqrt{2}\} > 1 - \frac{1}{2} = \frac{1}{2} \quad \Rightarrow \quad P[|X-2| \le 2] > \frac{1}{2}$

And the actual probability is given by

$$P\{|X-2| \le 2\} = P\{0 \le X \le 4\} = P\{X = 1, 2, 3 \text{ or } 4\} = \frac{1}{2} + \left(\frac{1}{2}\right)^2 + \left(\frac{1}{2}\right)^3 + \left(\frac{1}{2}\right)^4 = \frac{15}{16}.$$

Example 7·17. *Does there exist a variate X for which* $P[\mu_X - 2\sigma \le X \le \mu_X - 2\sigma] = 0.6$? ... (*)

Solution. We have

$$P(\mu_X - 2\sigma \le X \le \mu_X + 2\sigma) = P(|X - \mu_X| \le 2\sigma) \ge 1 - \frac{1}{4} = 0·75$$

(Using Chebychev's Inequality)

Since lower bound for the probability is 0·75, there does not exist a *r.v.* X for which (*) holds.

Example 7·18.

(*a*) *For the discrete variate with density :*

$$f(x) = \frac{1}{8} I_{(-1)}(x) + \frac{6}{8} I_{(0)}(x) + \frac{1}{8} I_{(1)}(x),$$

evaluate $P\{|X - \mu_X| \ge 2\sigma_X\}$.

(*b*) *Compare this result with that obtained on using Chebychev's inequality.*

Solution.

(*a*) Here, X has the probability distribution:

x:	-1	0	1
$p(x)$	$1/8$	$6/8$	$1/8$

$$\therefore \quad E(X) = -1 \times \frac{1}{8} + 1 \times \frac{1}{8} = 0$$

$$E(X^2) = 1 \times \frac{1}{8} + 1 \times \frac{1}{8} = \frac{1}{4}$$

$$\therefore \qquad \text{Var}(X) = E(X^2) - \{E(X)\}^2 = \frac{1}{4} \quad \Rightarrow \quad \sigma_X = \frac{1}{2}$$

$$P\{|X - \mu_X| \geq 2\sigma_X\} = P[|X| \geq 1] = 1 - P\{|X| < 1\}$$

$$= 1 - P(-1 < X < 1) = 1 - P(X = 0) = \frac{1}{4}.$$

(*b*) $$P\{|X - \mu_X| \geq 2\,\sigma_X\} \leq \frac{1}{4}$$ (By Chebychev's inequality)

In this case, both results are same.

Note. This example shows that, in general, Chebychev's inequality cannot be improved.

Example 7·19. *Two unbiased dice are thrown. If X is the sum of the numbers showing up, prove that* $P\{|X - 7| \geq 3\} \leq \frac{35}{54}$. *Compare this with the actual probability.*

Solution. The probability distribution of the *r.v.* X (the sum of the numbers on the two dice) is as given in the following table :

X	*Favourable cases (distinct)*	*Probability (p)*
2	(1, 1)	1/36
3	(1, 2), (2, 1)	2/36
4	(1, 3), (3, 1), (2, 2)	3/36
5	(1, 4), (4, 1), (2, 3), (3, 2)	4/36
6	(1, 5), (5, 1), (2, 4), (4, 2), (3, 3)	5/36
7	(1, 6), (6, 1), (2, 5), (5, 2), (3, 4), (4, 3)	6/36
8	(2, 6), (6, 2), (3, 5), (5, 3), (4, 4)	5/36
9	(3, 6), (6, 3), (4, 5), (5, 4)	4/36
10	(4, 6), (6, 4), (5, 5)	3/36
11	(5, 6), (6, 5)	2/36
12	(6, 6)	1/36

$$E(X) = \sum_x p \cdot x = \frac{1}{36}(2 + 6 + 12 + 20 + 30 + 42 + 40 + 36 + 30 + 22 + 12) = 7$$

$$E(X^2) = \sum_x p \cdot x^2 = \frac{1}{36}(4 + 18 + 48 + 100 + 180 + 294 + 320 + 324 + 300 + 242 + 144)$$

$$= \frac{1}{36}(1974) = \frac{1974}{36} = \frac{329}{6}$$

$$\therefore \qquad \text{Var}(X) = E(X^2) - \{E(X)\}^2 = \frac{329}{6} - (7)^2 = \frac{35}{6}$$

By Chebychev's inequality, for $k > 0$, we have

$$P(|X - \mu| \geq k) < \frac{\text{Var } X}{k^2} \quad \Rightarrow \quad P(|X - 7| \geq 3) \leq \frac{35/6}{9} = \frac{35}{54}, \quad \text{(Taking } k = 3\text{)}.$$

Actual Probability :

$$P(|X - 7| \geq 3) = 1 - P(|X - 7| < 3) = 1 - P(4 < X < 10)$$

$$= 1 - [P(X = 5) + P(X = 6) + P(X = 7) + P(X = 8) + P(X = 9)]$$

$$= 1 - \frac{1}{36}(4 + 5 + 6 + 5 + 4) = 1 - \frac{24}{36} = \frac{1}{3}.$$

Example 7·20. *If X is the number scored in a throw of a fair die, show that the Chebychev's ineqality gives* $P\{|X - \mu| > 2\cdot5\} < 0\cdot47$, *where μ is the mean of X, while the actual probability is zero.*

Solution. Here X is a random variable which takes the values 1, 2, ..., 6, each with probability $\frac{1}{6}$. Hence,

$$E(X) = \frac{1}{6}(1 + 2 + ... + 6) = \frac{1}{6} \times \frac{6 \times 7}{2} = \frac{7}{2}$$

$$E(X^2) = \frac{1}{6}(1^2 + 2^2 + ... + 6^2) = \frac{1}{6} \times \frac{6 \times 7 \times 13}{6} = \frac{91}{6}$$

$$\therefore \qquad \text{Var}(X) = E(X^2) - \{E(X)\}^2 = E(X^2) - \{E(X)\}^2 = \frac{91}{6} - \frac{49}{4} = \frac{35}{12} = 2\cdot9167$$

For $k > 0$, Chebychev's inequality gives $\quad P\{|X - E(X)| > k\} < \dfrac{\text{Var } X}{k^2}$.

Choosing $k = 2\cdot5$, we get

$$P\{|X - \mu| > 2\cdot5\} < \frac{2\cdot9167}{6\cdot25} = 0\cdot47$$

The actual probability p is given by :

$$p = P\{|X - 3\cdot5| > 2\cdot5\}$$

$$= P\{X \text{ lies outside the limits } (3\cdot5 - 2\cdot5, 3\cdot5 + 2\cdot5), \text{ i.e., } (1, 6)\}$$

But since X is the number on the dice when thrown, it cannot lie outside the limits of 1 and 6. In other words, $\quad p = P(\varphi) = 0$.

Example 7·21. *How large a sample must be taken in order that the probability will be at least 0·95 that $\overline{X}_n$ will be within 0·5 of μ. (μ is unknown and $\sigma = 1$.)*

Solution. We have

$$E(\overline{X}_n) = \mu \quad \text{and} \quad \text{Var}(\overline{X}_n) = \sigma^2 / n$$

Applying Chebychev's inequality to the *r.v.* $\overline{X}_n$, we get, for any $c > 0$

$$P\{|\overline{X}_n - E(\overline{X}_n)| < c\} \geq 1 - \frac{\text{Var}(\overline{X}_n)}{c^2}$$

$$\Rightarrow \qquad P\{|\overline{X}_n - \mu| < c\} \geq 1 - \frac{\sigma^2}{nc^2} \qquad \text{... (*)}$$

We want n so that $P\{|\overline{X}_n - \mu| < 0\cdot5\} \geq 0\cdot95$ ... (**)

Comparing (*) and (**), we get

$$c = 0\cdot5 = \frac{1}{2} \text{ and } 1 - \frac{\sigma^2}{nc^2} = 0\cdot95 \text{ and } \sigma = 1 \qquad \text{(Given).}$$

$$\therefore \qquad 1 - \frac{4}{n} = 0\cdot95 \quad \Rightarrow \quad \frac{4}{n} = 0\cdot05 = \frac{1}{20} \quad \Rightarrow \quad n = 80.$$

Hence, $\quad n \geq 80.$

Example 7·22. *A symmetric die is thrown 600 times. Find the lower bound for the probability of getting 80 to 120 sixes.*

Solution. Let S be total number of successes (sixes). Then

$$E(S) = np = 600 \times \frac{1}{6} = 100, \quad \text{and} \quad V(S) = npq = 600 \times \frac{1}{6} \times \frac{5}{6} = \frac{500}{6}$$

Using Chebychev's inequality, we get

$$P\{|S - E(S)| < k\sigma\} \geq 1 - \frac{1}{k^2} \quad \Rightarrow \quad P\{|S - 100| < k\sqrt{500/6}\} \geq 1 - \frac{1}{k^2}$$

$$\therefore \qquad P\{100 - k\sqrt{500/6} < S < 100 + k\sqrt{500/6}\} > 1 - \frac{1}{k^2}$$

Taking $k = \dfrac{20}{\sqrt{500/6}}$, we get $P(80 < S < 120) > 1 - \dfrac{1}{400 \times (6/500)} = \dfrac{19}{24}.$

Example 7·23. *Use Chebychev's inequality to determine how many times a fair coin must be tossed in order that the probability will be at least 0·90 that the ratio of the observed number of heads to the number of tosses will lie between 0·4 and 0·6.*

Solution. As in the proof of Bernoulli's Law of Large Numbers, we get for any $\varepsilon > 0$,

$$P\left\{\left|\frac{X}{n} - p\right| \geq \varepsilon\right\} \leq \frac{1}{4n\varepsilon^2} \quad \Rightarrow \quad P\left\{\left|\frac{X}{n} - p\right| < \varepsilon\right\} \geq 1 - \frac{1}{4n\varepsilon^2}$$

Since $p = 0\cdot5$ (as the coin is unbaised) and we want the proportion of successes X/n to lie between 0·4 and 0·6, we have $|(X/n) - p| < 0\cdot1.$

Thus choosing $\varepsilon = 0{\cdot}1$, we have $P\left\{\left|\frac{X}{n} - p\right| < 0{\cdot}1\right\} \geq 1 - \frac{1}{4n\,(0{\cdot}1)^2} = 1 - \frac{1}{0{\cdot}04n}$

Since we want this probability to be 0·9, we fix :

$$1 - \frac{1}{0\cdot 04n} = 0{\cdot}90 \quad \Rightarrow \quad 0{\cdot}10 = \frac{1}{0{\cdot}04n} \quad \Rightarrow \quad n = \frac{1}{0{\cdot}10 \times 0{\cdot}04} = 250$$

Hence, the required number of tosses is 250.

Example 7·24. *If you wish to estimate the proportion of engineers and scientists who have studied probability theory and you wish your estimate to be correct within 2% with probability 0·95 or more, how large a sample would you take :*

(*i*) *If you have no idea what the true proportion is,*

(*ii*) *If you are confident that the true proportion is less than 0·2 ?*

Solution.

(*i*) Proceeding as in example 7·23 and writing $(X/n) = f_n$, here ε = 2% or 0·02

so that $$P\{|f_n - p| < 0\cdot 02\} \geq 1 - \frac{1}{4n\,(0{\cdot}02)^2} = 0{\cdot}95$$

$$\Rightarrow \qquad 0{\cdot}05 = \frac{1}{4n\,(0{\cdot}02)^2} \quad \Rightarrow \quad n = 12{,}500$$

(*ii*) By Bernoull's W.L.L.N. $$P\{|f_n - p| \leq \varepsilon\} \geq 1 - \frac{p\,(1-p)}{n\,\varepsilon^2} \qquad \ldots (*)$$

Here, $p < 0{\cdot}2 \Rightarrow p(1-p) < 0{\cdot}16$. Taking ε = 2% = 1/50, in (*), we have to find n so that :

$$P\{|f_n - p| \leq \varepsilon\} > 1 - \frac{0{\cdot}16}{n\,\varepsilon^2} = 0{\cdot}95 \quad \Rightarrow n = 50 \times 50 \times 20 \times 0{\cdot}16 = 8{,}000.$$

Example 7·25. *If the variable* X_p *assumes values* $2^{p-2\log p}$ *with probability* 2^{-p}; $p = 1, 2, 3, \ldots$; $i = 1, 2, 3, \ldots$, *examine if the law of large numbers holds for the sequence* $\{ x_i \}$ *of independent r.v.'s.*

Solution. Putting $p = 1, 2, 3, \ldots$, we get $2^{1-2\log 1}$, $2^{2-2\log 2}$, $2^{3-2\log 3}$, ... as the values of the variables $X_1, X_2, X_3, \ldots$, respectively, and $\frac{1}{2}, \frac{1}{2^2}, \frac{1}{2^3}$ ... their corresponding probabilities.

$$\therefore \qquad E(X_k) = 2^{1-2\log 1} \cdot \frac{1}{2} + 2^{2-2\log 2} \cdot \frac{1}{2^2} + \ldots = \sum_{p=1}^{\infty} 2^{p-2\log p} \cdot 2^{-p} = \sum_{p=1}^{\infty} \frac{1}{2^{2\log p}}$$

Let $U = 2^{2\log p}$, then

$$\log U = 2 \log p \log 2 = \log p \cdot \log 2^2 = \log 4 \cdot \log p = \log p^{\log 4}$$

$$\Rightarrow \qquad U = p^{\log 4}$$

$$\therefore \qquad E(X_k) = \sum_{p=1}^{\infty} \frac{1}{p^{\log 4}} = 1 + \frac{1}{2^{\log 4}} + \frac{1}{3^{\log 4}} + \ldots,$$

which is a convergent series. $\left[\because \sum_{n=1}^{\infty} \frac{1}{n^p} \text{ is convergent if and only if } p > 1\right]$

Therefore, the mathematical expectation of the variables $X_1, X_2, \ldots,$ exists. Thus, by Khinchin's theorem, the (weak) law of large numbers holds in this case.

Example 7·26. *Let $X_1, X_2, \ldots, X_n$ be i.i.d. variables with mean μ and variance σ^2 and as $n \to \infty$, $\left(X_1^2 + X_2^2 + \ldots + X_n^2\right)/n \xrightarrow{p} c$, for some constant c ; $(0 \le c \le \infty)$. Find c.*

Solution. We are given :

$$E(X_i) = \mu, \quad \text{Var}(X_i) = \sigma^2 ; i = 1, 2, \ldots, n.$$

$$\therefore \quad E(X_i^2) = \text{Var}(X_i) + \{E(X_i)\}^2 = \sigma^2 + \mu^2 \text{ (finite) } ; i = 1, 2, \ldots, n.$$

Since $E(X_i^2)$ is finite ; by Khinchine's Theorem W.L.L.N. holds for the sequence X_i^2 of *i.i.d. r.v.*'s so that

$$\left(X_1^2 + X_2^2 + \ldots + X_n^2\right)/n \xrightarrow{p} E\left(X_i^2\right), \text{ as } n \to \infty$$

$$\Rightarrow \quad \left(X_1^2 + X_2^2 + \ldots + X_n^2\right)/n \xrightarrow{p} \sigma^2 + \mu^2 = c, \text{ as } n \to \infty$$

Hence, $\quad c = \sigma^2 + \mu^2$.

Example 7·27.

(a) *Let X_i assume that values i and – i with equal probabilities. Show that the law of large numbers cannot be applied to the independent variables $X_1, X_2, \ldots$, i.e., X_i's.*

(b) *If X_i can have only two values with equal probabilities i^α and $-i^\alpha$, show that law of large numbers can be applied to the independent variables $X_1, X_2, \ldots$, if $\alpha < \frac{1}{2}$.*

Solution.

(a) We have

$$P(X_i = i) = \frac{1}{2}, \quad P(X_i = -i) = \frac{1}{2}$$

$$\therefore \quad E(X_i) = \frac{1}{2}(i) + \frac{1}{2}(-i) = 0 ; i = 1, 2, 3, \ldots$$

$$V(X_i) = E(X_i^2) = \frac{i^2}{2} + \frac{i^2}{2} = i^2 \qquad [\because E(X_i) = 0] \qquad \ldots (*)$$

Since $X_1, X_2, \ldots, X_n$ are independent variables,

$$B_n = V(X_1 + X_2 + \ldots + X_n) = V(X_1) + V(X_2) + \ldots + V(X_n)$$

$$= (1^2 + 2^2 + \ldots + n^2) = \frac{n(n+1)(2n+1)}{6} \qquad \text{[From (*)]}$$

$\therefore \frac{B_n}{n^2} \to \infty$ as $n \to \infty$. Hence we cannot draw any conclusion whether W.L.L.N. holds or not. Here, we need to apply further tests.

$$E|X_i|^{1+\delta} = \left|i^{1+\delta}\right| \times \frac{1}{2} + \left|(-i)^{1+\delta}\right| \times \frac{1}{2} = i^{1+\delta}, \text{ which for } \delta > 0, \text{ is unbounded.}$$

Hence, by Markov's theorem, the W.L.L.N. cannot hold for the sequence $\{X_n\}$ of independent *r.v.*'s.

(*b*) $$E(X_i) = \frac{i^\alpha}{2} + \left(\frac{-i^\alpha}{2}\right) = 0, \qquad E(X_i^2) = \frac{(i^\alpha)^2}{2} + \frac{(-i^\alpha)^2}{2} = i^{2\alpha}$$

$$\therefore \quad V(X_i) = E(X_i^2) - \{E(X_i)\}^2 = i^{2\alpha}$$

$$B_n = V(X_1 + X_2 + \ldots + X_n) = \sum_{i=1}^{n} V(X_i) = \sum_{i=1}^{n} i^{2\alpha}$$

($\because X_1, X_2, \ldots, X_n$ are independent variables)

$$= 1^{2\alpha} + 2^{2\alpha} + \ldots + n^{2\alpha} \simeq \int_0^n x^{2\alpha}\, dx \text{ [From Euler Maclaurin's Formula]}$$

$$= \left| \frac{x^{2\alpha+1}}{2\alpha+1} \right|_0^n = \frac{n^{2\alpha+1}}{2\alpha+1}$$

$$\therefore \quad \frac{B_n}{n^2} = \frac{n^{2\alpha-1}}{2\alpha+1} \to 0 \text{ if } 2\alpha - 1 < 0 \quad \Rightarrow \quad \alpha < \frac{1}{2}.$$

Hence, the result.

Example 7·28. *Examine whether the weak law of large numbers holds for the sequence $\{X_k\}$ of independent random variables defined as follows:*

$$P(X_k = \pm 2^k) = 2^{-(2k+1)}, \qquad P(X_k = 0) = 1 - 2^{-2k}.$$

Solution. We have

$$E(X_k) = 2^k\, 2^{-(2k+1)} + (-2^k)\cdot 2^{-(2k+1)} + 0 \times (1 - 2^{-2k}) = 2^{-(2k+1)}\,(2^k - 2^k) = 0$$

$$E(X_k^2) = (2^k)^2 \cdot 2^{-(2k+1)} + (-2^k)^2 \cdot 2^{-(2k+1)} + 0^2 \times (1 - 2^{-2k})$$

$$= 2^{2k} \cdot 2^{-(2k+1)} + 2^{2k} \cdot 2^{-(2k+1)} = \frac{1}{2} + \frac{1}{2} = 1$$

$$\therefore \quad \text{Var}(X_k) = E(X_k^2) - \{E(X_k)\}^2 = 1 - 0 = 1$$

$$B_n = \text{Var}\left(\sum_{i=1}^{n} X_i\right) = \sum_{i=1}^{n} \text{Var}(X_i) = \sum_{i=1}^{n} (1) = n$$

[$\because X_i$'s, ($i = 1, 2, \ldots n$) are independent]

$$\therefore \quad \lim_{n\to\infty} \frac{B_n}{n^2} = \lim_{n\to\infty} \frac{1}{n} \to 0$$

Hence, (Weak) Law of large numbers, holds for sequence of independent *r.v.*'s $\{X_k\}$.

Example 7·29.

(a) *Examine if W.L.L.N. holds for the sequence $\{X_m\}$ of i.i.d. r.v.'s with*

$$P\{X_i = (-1)^{k-1} \cdot k\} = \frac{6}{\pi^2 k^2} \; ; k = 1, 2, 3, ..., ; i = 1, 2, 3, ...$$

(b) *The r.v.'s $X_1, X_2, ..., X_n$ have equal expectations and finite variation. Is the weak law of large numbers applicable to this sequence if all the co-variances σ_{ij} are negative ?*

Solution.

(a)
$$E\{X_i\} = \sum_{k=1}^{\infty} (-1)^{k-1} \cdot k \cdot \frac{6}{\pi^2 k^2} = \frac{6}{\pi^2} \sum_{k=1}^{\infty} (-1)^{k-1} \cdot (1/k)$$

$$= \frac{6}{\pi^2}\left(1 - \frac{1}{2} + \frac{1}{3} - \frac{1}{4} + \frac{1}{5} - ...\right) = \frac{6}{\pi^2} \cdot \log_e 2 \text{ (finite)}$$

[∵ The series in bracket is convergent by Leibnitz test for alternating series].

Hence by Khinchin's theorem, W.L.L.N. holds for the sequence $\{X_i\}$ of *i.i.d. r.v.'s.*

(b)
$$\frac{B_n}{n^2} = \frac{\text{Var}(X_1 + X_2 + ... + X_n)}{n^2} = \frac{1}{n^2}\left\{\sum_{i=1}^{n} \sigma_i^2 + 2 \sum_{i<j=1}^{n} \sum \sigma_{ij}\right\}$$

$$< \frac{1}{n^2}\left(\sum_{i=1}^{n} \sigma_i^2\right), \text{ if all the convariances } \sigma_{ij} \text{ are negative.}$$

$$\therefore \quad \lim_{n\to\infty} \frac{B_n}{n^2} < \lim_{n\to\infty} \frac{1}{n^2}\left(\sum_{i=1}^{n} \sigma_i^2\right) = 0 \quad (\because \sigma_i^2 \; ; i = 1, 2, ..., n \text{ are finite})$$

Hence, W.L.L.N. holds.

Example 7·30. *Let $\{X_k\}$ be mutually independent and identically distributed random variables with mean, μ and finite variance. If $S_n = X_1 + X_2 + ... + X_n$, prove that the law of large numbers does not hold for sequence $\{S_n\}$.*

Solution. The variables now are $S_1, S_2, ... S_n$.

$$B_n = V(S_1 + S_2 + ... + S_n)$$

$$= V\{X_1 + (X_1 + X_2) + (X_1 + X_2 + X_3) + ... + (X_1 + X_2 + ... + X_n)\}$$

$$= V\{n X_1 + (n-1) X_2 + ... + 2 X_{n-1} + X_n\}$$

$$= n^2 V(X_1) + (n-1)^2 V(X_2) + ... + 2^2 V(X_{n-1}) + 1^2 V(X_n)$$

(Covariance terms vanish since variable are independent.)

Let $V(X_i) = \sigma^2$ for all i (Since the variables are identically distributed.)

$$\therefore \quad B_n = (1^2 + 2^2 + \ldots + n^2)\,\sigma^2 = \frac{n(n+1)(2n+1)}{6}\sigma^2$$

Hence, $$\frac{B_n}{n^2} = \frac{(n+1)(2n+1)}{6n}\sigma^2 \to \infty \text{ as } n \to \infty$$

Hence, we cannot draw any conclusion whether W.L.L.N. holds or not. Here, we need to apply further tests.

Example 7·31. *Let $X_1, X_2, \ldots, X_n$ be jointly normal with $E(X_i) = 0$ and $E(X_1^2) = 1$ for all i and $Cov(X_i, X_j) = \rho$ if $|j - i| = 1$ and $= 0$, otherwise. Examine if W.L.L.N. holds for the sequence $\{X_n\}$.*

Solution. We have $E(X_i) = 0, \quad E(X_i^2) = 1, \forall\, i$

$$\text{Var}(X_i) = E(X_i^2) - \{E(X_i)\}^2 = 1, (i = 1, 2, \ldots n). \quad \ldots(*)$$

$$E(S_n) = E\left(\sum_{i=1}^{n} X_i\right) = \sum_{i=1}^{n} E(X_i) = 0$$

$$\text{Var}(S_n) = \text{Var}\left(\sum_{i=1}^{n} X_i\right) = \sum_{i=1}^{n} \text{Var}\, X_i + 2\sum_{i<j=1}^{n} \text{Cov}(X_i, X_j) = n + 2.(n-1)\,\rho$$

[On using (*) and Cov $(X_i, X_j) = \rho$, if $|j - i| = 1$ and zero otherwise]

Since $X_1, X_2, \ldots, X_n$ are jointly normal, (See Chapter 9)

$$S_n = \sum_{i=1}^{n} X_i \sim N(0, \sigma^2), \text{ where } \sigma^2 = [n + 2(n-1)\rho] > 0. \quad \ldots(**)$$

Taking $$Y_n = \frac{1}{n}\sum_{i=1}^{n} X_i = \frac{S_n}{n}, \text{ we have}$$

$$E\left(\frac{Y_n^2}{1+Y_n^2}\right) = E\left(\frac{S_n^2}{n^2 + S_n^2}\right) = \int_{-\infty}^{\infty}\left(\frac{x^2}{n^2 + x^2}\right)\frac{1}{\sqrt{2\pi}\,\sigma} e^{-x^2/2\sigma^2}\,dx$$

$$= \frac{2}{\sqrt{2\pi}}\int_0^{\infty} \frac{\sigma^2 y^2}{n^2 + \sigma^2 y^2} e^{-y^2/2}\,dy, \quad \left(\frac{x}{\sigma} = y\right)$$

$$= \frac{2}{\sqrt{2\pi}}\{n + 2(n-1)\rho\}\int_0^{\infty} \frac{y^2}{n^2 + y^2\{n + 2(n-1)\rho\}} \cdot e^{-y^2/2}\,dy \quad \text{[From (**)]}$$

$$\leq \frac{n + 2(n-1)\rho}{n^2} \cdot \frac{2}{\sqrt{2\pi}}\int_0^{\infty} y^2 \cdot e^{-y^2/2}\,dy \to 0 \text{ as } n \to \infty.$$

$$\therefore \quad \lim_{n\to\infty} E\left(\frac{Y_n^2}{1+Y_n^2}\right) \to 0$$

Hence, by *Theorem* 7·25, W.L.L.N. holds for $\{X_n\}$ *i.e.,* $\dfrac{S_n}{n} \xrightarrow{p} 0$ as $n \to \infty$.

Example 7·32. *Let $X_1, X_2, X_3, \ldots$ be a sequence of independent and identically distributed r.v.'s each uniform on [0, 1]. For the geometric mean $G_n = (X_1 X_2 \ldots X_n)^{1/n}$, show that $G_n \xrightarrow{p} c$ for some finite number c. Find c.*

Solution. If $X \sim U\,[0, 1]$, its *p.d.f.* is $p(x) = 1,\ 0 < x < 1$ [See Chapter 9]

Let $Y = -\log X$ *i.e.,* $y = -\log x \Rightarrow x = e^{-y}$. ... (*i*)

$x = 0 \Rightarrow y = \infty$ and $x = 1 \Rightarrow y = 0$. Hence, $0 \le y \le \infty$.

The *p.d.f.* $g(.)$ of Y is given by :

$$g(y) = p(x).\left|\frac{dx}{dy}\right| = 1 \cdot \left|-e^{-y}\right| = e^{-y},\quad y \ge 0$$

$$\therefore \quad E(Y) = \int_0^\infty y\, g(y)\, dy = \int_0^\infty y \cdot e^{-y}\, dy = \Gamma 2 = 1.$$

Hence, if $X_i \sim U\,[0, 1]$, $(i = 1, 2, \ldots, n)$; then $Y_i = -\log X_i$, $(i = 1, 2, \ldots, n)$ are *i.i.d.* random variables with $E(Y_i) = 1$; $i = 1, 2, 3, \ldots$ ... (*ii*)

Hence, by Khinchines Theorem,

$$\frac{1}{n}\sum_{i=1}^{n} Y_i \xrightarrow{p} E(Y_i) = 1 \qquad \text{[From (}ii\text{)]} \qquad \ldots (iii)$$

We are given : $G_n = (X_1 X_2 \ldots X_n)^{1/n}$ [From (*i*)]

$$\log G_n = \frac{1}{n}\sum_{i=1}^{n} \log X_i = -\frac{1}{n}\sum_{i=1}^{n} Y_i \qquad \ldots (iv)$$

From (*iii*) and (*iv*), we get: $-\log G_n \xrightarrow{p} 1 \Rightarrow G_n \xrightarrow{p} e^{-1} = c.$

7·8. STRONG LAW OF LARGE NUMBERS

To obtain the strong law of large numbers we make use of an inequality – Kolmogorov's Inequality, which has its own significance.

7·8·1. Kolmogorov's Inequality

Let $X_1, X_2, \ldots, X_n$ be independent random variables with zero means and finite variances so that :

$$E(X_i) = 0 \text{ and } \mathrm{Var}(X_i) = E(X_i^2) = \sigma_i^2 < \infty\ ;\ i = 1, 2, \ldots, n$$

Write : $S_k = X_1 + X_2 + \ldots + X_k$; $k = 1, 2, \ldots, n$

and
$$X = \max_{1\le k\le n} |S_k| = \max_{1\le k\le n}\left\{\left|\sum_{i=1}^{k} X_i\right|\right\}$$

Then, for every $\varepsilon > 0$,

$$P\left[\max_{1 \le k \le n} |S_k| > \varepsilon\right] \le \frac{1}{\varepsilon^2} \sum_{i=1}^{n} \sigma_i^2 \qquad \dots (7{\cdot}45)$$

Proof of Kolmogorov's Inequality. Let us write :

$$E = \{w : X(w) > \varepsilon\}$$

and $$E_k = \{w : |S_k(w)| \ge \varepsilon\} \cap \{w : |S_i(w)| < \varepsilon \,; i = 1, 2, \dots, k-1\}$$

i.e., k is the first integar for which $|S_R| \ge \varepsilon$.

Thus, E_k's, ($k = 1, 2, \dots, n$) are disjoint

and $$E = \bigcup_{k=1}^{n} E_k \quad \Rightarrow \quad I_E = \sum_{k=1}^{n} I_{E_k} \qquad \dots (*)$$

Obviously, I_{E_k} and $(S_n - S_k)$, $k = 1, 2, \dots, (n-1)$ are independent since the former depends on $X_1, X_2, \dots, X_k$ and the latter on $X_{k+1}, X_{k+2}, \dots, X_n$.

Of course $S_n \cdot I_{E_n}$ and $S_n - S_n = 0$, are independent.

$$0 < \sum_{k=1}^{n} I_{E_K} = I_E \le 1$$

We have : $$\sum_{k=1}^{n} \sigma_k^2 = \text{Var}(X_1 + X_2 + \dots + X_n) = \text{Var}(S_n)$$

$$= E(S_n^2) - [E(S_n)]^2 = E(S_n^2) \qquad \left[\because E(S_n) = \sum_{i=1}^{n} E(X_i) = 0\right]$$

$$\ge E\left[S_n^2 \sum_{k=1}^{n} I_{E_k}\right]$$

$$= \sum_{k=1}^{n} \left[E\left\{(S_n - S_k)^2 + 2S_k(S_n - S_k) + S_k^2\right\} I_{E_k}\right]$$

$$\ge 2 \sum_{k=1}^{n} E\left[(S_n - S_k) S_k I_{E_k}\right] + \sum_{k=1}^{n} E\left(S_k^2 I_{E_k}\right)$$

Since $(S_n - S_k)$ and $(S_k I_{E_k})$ are independent,

$$E\left[(S_n - S_k) S_k I_{E_k}\right] = E(S_n - S_k) \cdot E(S_k \cdot I_{E_k}) = 0 \cdot E\left[S_k I_{E_k}\right] = 0$$

$$\sum_{k=1}^{n} \sigma_k^2 \geq \sum_{k=1}^{n} E\left(S_k^2 I_{E_k}\right) \geq \epsilon^2 \sum_{k=1}^{n} E\left(I_{E_k}\right) \qquad [\because \text{ On } E_k ;\ |S_k| > \varepsilon]$$

$$\Rightarrow \quad \sum_{k=1}^{n} \sigma_k^2 \geq \varepsilon^2 \sum_{k=1}^{n} P\left(E_k\right) = \varepsilon^2 P(E) = \varepsilon^2 P[X > \varepsilon]$$

$$\left[\because E = \bigcup_{k=1}^{n} E_k \text{ and } E_k's \text{ are disjoint}\right]$$

$$\Rightarrow \quad \sum_{k=1}^{n} \sigma_k^2 \geq \varepsilon^2 P\left[\max_{1 \leq k \leq n} |S_k| > \varepsilon\right]$$

$$\Rightarrow \quad P\left[\max_{1 \leq k \leq n} |S_k| > \varepsilon\right] \leq \frac{1}{\varepsilon^2} \sum_{k=1}^{n} \sigma_k^2,$$

which is the **Kolmogorov's Inequality.**

Note. If X_i's are *i.i.d. r.v.'s* with $E\left(X_i\right) = 0$ and $\text{Var}\left(X_i\right) = 0 = \sigma^2 < \infty$; $i = 1, 2, ..., n$; then Kolmogorov's Inequality gives :

$$P\left[\max_{1 \leq k \leq n} |S_k| > \varepsilon\right] \leq \frac{n\sigma^2}{\epsilon^2}$$

i.e.,
$$P\left[|S_k| > \varepsilon, \text{ for some } k = 1, 2, ..., n\right] \leq \left(\frac{n\sigma^2}{\varepsilon^2}\right) \qquad ...\ (7\cdot45a)$$

Remarks:

1. Chebychev's inequality is a particular case of Kolmogorov's inequality with $n = 1$. Taking $n = 1$ in (7·45), we get :

$$P\left[|X_1| > \epsilon\right] < \frac{\sigma_1^2}{\epsilon^2} ; \epsilon > 0, \qquad ...\ (7\cdot45b)$$

a result obtained in Chebychev's inequality.

2. Kolmogorov's inequality is a much stronger result than Chebychev's inequality. We have, for $\epsilon > 0$:

Chebychev's Inequality : $P\left[|S_n| > \varepsilon\right] \leq \dfrac{\text{Var}\,(S_n)}{\varepsilon^2} = \displaystyle\sum_{k=1}^{n} \frac{\sigma_k^2}{\varepsilon^2}$

Kolmogorov's Inequality : $P\left[\displaystyle\bigcup_{k=1}^{n} \{|S_k| > \varepsilon\}\right] \leq \displaystyle\sum_{k=1}^{n} \frac{\sigma_k^2}{\varepsilon^2}$

Obviously, the set $\bigcup_{k=1}^{n} \{|S_k| > \epsilon\}$, is much larger than the set $\{|S_n| > \epsilon\}$.

7·8·2. Strong Law of Large Numbers (S.L.L.N.)

A sequence $\{X_n ; n \geq 1\}$ of random variables is said to obey the Strong Law of Large Numbers (S.L.L.N.) if

$$\left[\frac{S_n - E(S_n)}{n}\right] \xrightarrow{a.s.} 0, \text{ as } n \to \infty \quad \text{... (7·46)}$$

i.e.,

$$P\left[\lim_{n\to\infty}\left\{\frac{S_n - E(S_n)}{n}\right\} = 0\right] = 1, \text{ where } S_n = \frac{1}{n}\sum_{i=1}^{n} X_i \quad \text{... (7·46a)}$$

If we write : $\overline{X}_n = \frac{1}{n}\sum_{i=1}^{n} X_i = \frac{S_n}{n}$ and $\overline{\mu}_n = E(\overline{X}_n) = E\left[\frac{S_n}{n}\right]$,

then, S.L.L.N. can be restated as follows :

$$\frac{S_n}{n} - E\left(\frac{S_n}{n}\right) = \overline{X}_n - \overline{\mu}_n \xrightarrow{a.s.} 0, \text{ as } n \to \infty \quad \text{... (7·47)}$$

or $\overline{X}_n \xrightarrow{a.s.} \overline{\mu}_n$ as $n \to \infty$ *i.e.,* $P\left[\lim_{n\to\infty}\left(\overline{X}_n - \overline{\mu}_n\right) = 0\right] = 1$ ... (7·47a)

Remarks:

1. The above law is termed as Strong Law because it is concerned with the convergence of the random variables themselves rather than their distributions.
2. We have proved in Theorem 7·17 that :
   Almost sure converagence implies convergence in probability *i.e.,* $X_n \xrightarrow{a.s.} X \Rightarrow X_n \xrightarrow{p} X$
   Hence, if S.L.L.N. holds for a sequence $\{X_n , n \geq 1\}$ of *r.v.'s,* then W.L.L.N. also holds for the sequence.
   We state below, (without proof), two Theorems due to Kolmogorov on S.L.L.N. The proof of these theorems is however, beyond the scope of the book.

7·8·3. Kolmogorov's First Theorem on S.L.L.N.

Theorem 7·26. *Let* $\{X_k, k \geq 1\}$, *be a sequence of independent r.v.'s with:*

$$E(X_k) = \mu_k \text{ and } Var(X_k) = \sigma_k^2 < \infty ; k = 1, 2, 3, ...$$

If

$$\sum_{k=1}^{\infty}\left[\sigma_k^2 / k^2\right] < \infty \quad \text{... (7·48)}$$

then S.L.L.N. holds for the sequence $\{X_k ; k \geq 1\}$.

The condition (7·48) is only *a sufficient condition* for the S.L.L.N. to hold *i.e.,* if the condition (7·48) is not satisfied *i.e.,* if $\sum_{k=1}^{\infty}\frac{\sigma_k^2}{k^2}$, is *not* convergent, then the S.L.L.N. may or may not hold.

Deduction 1. Let $\{X_k, k = 1\}$, be a sequence of *independent r.v.'s* with common finite variance, say, $Var(X_k) = \sigma^2 < \infty$; $k = 1, 2, \ldots$ Then

$$\sum_{k=1}^{\infty} \frac{\sigma_k^2}{k^2} = \sigma^2 \sum_{k=1}^{\infty} \left(\frac{1}{k^2}\right) < \infty$$

because, $\sum_{k=1}^{\infty} \frac{1}{k^p}$ is convergent if and only if $p > 1$.

$$\left[\text{In fact, we have } \sum_{k=1}^{\infty} \frac{1}{k^2} = \frac{\pi^2}{6}\right]$$

Since, the sufficient condition (7·48) is satisfied, *the S.L.L.N. always holds for a sequence of independent r.v.'s with finite common variance.*

Deduction 2. **Borel's S.L.L.N.**

S.L.L.N. holds for a sequence of Bernoulli variates with constant probability of success 'p'.

Let $\{X_k ; k \geq 1\}$ be a sequence of Bernoulli variates with constant probability of success 'p' so that

$$P(X_k = 1) = p \quad \text{and} \quad P(X_k = 0) = q = 1 - p \,;\, k = 1, 2, 3, \ldots$$

Then :
$$E(X_k) = 1 \times p + 0 \times q = p \,;$$

$$E(X_k^2) = 1^2 \times p + 0^2 \times q = p$$

$$\therefore \quad \text{Var}(X_k) = \sigma_k^2 = p - p^2 = p(1 - p) = pq \leq \frac{1}{4}$$

$\left[\because \phi(p) = pq = p - p^2 \text{ is maximum if } p = \frac{1}{2} = q \,;\, \text{solution of } \phi'(p) = 0 \text{ and } \phi''(p) < 0\right]$

$$\therefore \quad \sum_{k=1}^{\infty} \frac{\sigma_k^2}{k^2} \leq \frac{1}{4} \sum_{k=1}^{\infty} \frac{1}{k^2} = \frac{\pi^2}{24} < \infty$$

Since, Kolmogorov's sufficient condition (7·48) is satisfied, S.L.L.N. holds.

Note : We have : $E(\overline{X}_n) = \frac{1}{n} \sum_{k=1}^{n} E(X_k) = \frac{1}{n} . np = p$

Hence, $\overline{X}_n \xrightarrow{a.s.} E(\overline{X}_n) = p$.

7·8·4. Kolmogorov's Second Theorem on S.L.L.N.

This theorem, (**Theorem 7·27**) gives the necessary and sufficient condition for a sequence $\{X_n ; n \geq 1\}$ of *i.i.d. r.v.'s* to obey the S.L.L.N.:

Theorem 7·27. *The necessary and sufficient condition that a sequence $\{X_k ; k \geq 1\}$ of i.i.d. r.v.'s obeys the S.L.L.N. is that*

$$E\{X_k\} = \mu \text{ exists and is finite for } k = 1, 2, 3, \ldots$$

In other words, S.L.L.N. holds for a sequence $\{X_k ; k \geq 1\}$ of i.i.d. r.v.'s if and only if

$$E\{X_k\} < \infty ; k = 1, 2, 3, \ldots$$

Example 7·33. *Examine if the S.L.L.N. holds for the sequence of independent r.v.'s with the following distributions :*

(i) $P(X_k = \pm 2^k) = 2^{-(2k+1)}$; $P(X_k = 0) = 1 - 2^{-2k}$ *(ii)* $P(X_k = \pm \sqrt{k}) = \frac{1}{2}$.

Solution.

(i) We have :

$$E(X_k) = 2^k \left[2^{-(2k+1)}\right] + (-2^k)\left[2^{-(2k+1)}\right] + 0 \times \left[1 - 2^{-2k}\right] = 0$$

and $$E(X_k^2) = 2^{2k} \cdot 2^{-(2k+1)} + 2^{2k} \cdot 2^{-(2k+1)} = 2^{-1} + 2^{-1} = 1$$

$$\sigma_k^2 = \text{Var}(X_k) = E(X_k^2) - \left[E(X_k)\right]^2 = 1 - 0 = 1$$

$$\therefore \quad \sum_{k=1}^{\infty} \frac{\sigma_k^2}{k^2} = \sum_{k=1}^{\infty} \frac{1}{k^2} = \frac{\pi^2}{6} < \infty$$

$\Rightarrow$ Kolmogorov's sufficient condition is satisfied.

Hence, S.L.L.N. holds in this case.

(ii) $$E(X_k) = \frac{1}{2} \times \sqrt{k} + \frac{1}{2} \times (-\sqrt{k}) = 0 \quad ; \quad E(X_k^2) = \frac{1}{2} \times k + \frac{1}{2} \times k = k$$

$$\text{Var}(X_k) = \sigma_k^2 = E(X_k^2) - \{E(X_k)\}^2 = k - 0 = k$$

$$\therefore \quad \sum_{k=1}^{\infty} \frac{\sigma_k^2}{k^2} = \sum_{k=1}^{\infty} \frac{k}{k^2} = \sum_{k=1}^{\infty} \frac{1}{k},$$

which is a divergent series.

$$\left[\because \sum_{k=1}^{\infty} \frac{1}{k^p}, \text{ is convergent if and only if } p > 1\right]$$

Since Kolmogorov's sufficient condition is not satisfied, S.L.L.N. may or may not hold in this case.

Example 7·34. *Discuss the validity of S.L.L.N. for the following sequence of i.i.d. random variables:*

$$P\left[X_i = (-1)^{k-1} \cdot k\right] = \frac{6}{\pi^2 k^2}\ ;\ i, k = 1, 2, 3, \ldots$$

Solution. $$\sum_{i=1}^{\infty} |x_i| \cdot P(X = x_i) = \sum_{i=1}^{\infty}\left[k \cdot \frac{6}{\pi^2 k^2}\right] = \frac{6}{\pi^2} \sum_{k=1}^{\infty} \frac{1}{k}$$

Since $\sum_{k=1}^{\infty} \left(1/k \right)$ is a divergent series, $E(X_i)$ does not exist.

Hence, by Kolmogorov's Second Theorem on S.L.L.N., S.L.L.N. does not hold in this case.

7·9. BOREL-CANTELLI LEMMA (ZERO-ONE LAW)

Let $A_1, A_2, \ldots$ be a sequence of events. Let A be the event "that an infinite number of A_n occur, *i.e.*, $\omega \in A$ if $\omega \in A_n$ for an infinite number of values of n (but not necessarily every n). But the set of such ω is precisely lim sup A_n, *i.e.*, $\overline{\lim}\, A_n$. Thus the event A, that an infinte number of A_n occur is just $\overline{\lim}\, A_n$. Sometimes we are interested in the probability that an infinite number of the events A_n occur. Often this question is answered by means of the Borel-Cantelli lemma or its converse.

Theorem 7·28. *(Borel-Cantelli Lemma) Let $A_1, A_2, \ldots$ be a sequence of events on the probability space (S, B, P) and let $A = \overline{\lim}\, A_n$.* *If $\sum_{n=1}^{\infty} P(A_n) < \infty$, then $P(A) = 0$.* ... (7·49)

In other words, this states that if $\Sigma\, P(A_n)$ converges then with probability one, only a finite number of $A_1, A_2, \ldots$ can occur.

Proof. Since $A = \overline{\lim}\, A_n = \bigcap_{n=1}^{\infty} \bigcup_{m=n}^{\infty} A_n$, we have $A \subset \bigcup_{m=n}^{\infty} A_m$, for every n.

Thus for each n, $P(A) \le \sum_{m=n}^{\infty} P(A_m)$

Since $\sum_{n=1}^{\infty} P(A_n)$ is convergent (given), $\sum_{m=n}^{\infty} P(A_m)$, being the remainder term of a convergent series, tends to zero as $n \to \infty$.

$$\therefore \qquad P(A) \le \sum_{m=n}^{\infty} P(A_m) \to 0 \text{ as } n \to \infty.$$

Thus $P(A) = 0$, as required.

The result just proved does not require events $A_1, A_2, \ldots$ considered to be independent. For the converse result it is necessary to make this further assumption.

Borel-Cantelli Lemma (Converse). *Let $A_1, A_2, \ldots$ be independent events on (S, B, P) and A equal to*

$\overline{\lim}\, A_n$. *If* $\sum_{n=1}^{\infty} P(A_n) = \infty$, *then* $P(A) = 1$. ... (7·50)

Proof. Writing in usual notation, $\overline{A}_n$ for the complement $(S - A_n)$ of A_n, we have for any $m, n\ (m > n)$.

$$\bigcap_{k=n}^{\infty} \overline{A_k} \subset \bigcap_{k=n}^{m} \overline{A_k} \quad \Rightarrow \quad P\left(\bigcap_{k=n}^{\infty} \overline{A}_k\right) \le P\left(\bigcap_{k=n}^{m} \overline{A}_k\right) = \prod_{k=n}^{m} P(\overline{A}_k),$$

because of the fact that if $(A_n, A_{n+1}, \ldots, A_m)$ are independent events, so are $(\overline{A}_n, \overline{A}_{n+1}, \ldots, \overline{A}_m)$.

Hence, $$P\left(\bigcap_{k=n}^{\infty} \overline{A_k}\right) \le \prod_{k=n}^{m}\{1 - P(A_k)\} \le \prod_{k=n}^{m} e^{-P(A_k)} \qquad [\because 1 - x \le e^{-x}, \text{ for } x \ge 0]$$

$$= \exp\left\{-\sum_{k=n}^{m} P(A_k)\right\}, \forall m$$

Since $\sum_{k=1}^{\infty} P(A_k) = \infty$, $\sum_{k=n}^{m} P(A_k) \to \infty$ as $m \to \infty$. Hence, $\exp\left\{-\sum_{k=n}^{m} P(A_k)\right\} \to 0$ as $m \to \infty$.

$$\therefore \qquad P\left(\bigcap_{k=n}^{\infty} \overline{A_k}\right) = 0 \qquad \ldots (*)$$

But $$A = \bigcap_{n=1}^{\infty} \bigcup_{k=n}^{\infty} A_k$$

$$\therefore \qquad \overline{A} = \bigcup_{n=1}^{\infty} \bigcap_{k=n}^{\infty} \overline{A}_k \qquad \text{(De-Morgan's Law)}$$

$$\Rightarrow \qquad P(\overline{A}) \le \sum_{n=1}^{\infty} P\left(\bigcap_{k=n}^{\infty} \overline{A_k}\right) = 0 \quad \Rightarrow \quad P(\overline{A}) = 0 \quad \text{[From (*)]}$$

Hence, $$P(A) = 1 - P(\overline{A}) = 1, \text{ as required.}$$

If $A_1, A_2, \ldots$ are independent events, it follows from Theorem 7·28 that the probability that an infinite number of them occur is either zero $\left(\text{when } \sum_{n=1}^{\infty} P(A_n) < \infty\right)$ or one $\left[\text{when } \sum_{n=1}^{\infty} P(A_n) = \infty\right]$. This statement is a special case of so-called *"Zero-one law"* which we now state.

Theorem 7·29. (Zero-One Law) *If $A_1, A_2, \ldots$ are independent and if E belongs to the σ-field generated by the class $(A_n, A_{n+1}, \ldots)$ for every n, then P(E) is zero or one.*

Example 7·35. *What is the probability that in a sequence of Bernoulli trials with probability of success p for each trial, the pattern SFS appears infinitely often?*

Solution. Let A_k be the event that the trial number k, $k + 1$, $k + 2$ produce the sequence SFS, $(k = 0, 1, 2, ...)$. The events A_k are not mutually independent but the sequence $A_1, A_4, A_7, A_{10}, ...$ contains only mutually independent events (since no two depend on the outcome of the same trials).

$$p_k = P\left(A_k\right) = P\,(SFS) = p^2 q, \ (q = 1 - p)$$

is independent of k, and hence, the series $p_1 + p_4 + p_7 + ...,$ diverges.

Hence, by Borel-Cantelli Lemma (Converse), the pattern SFS appears infinitely often with probability one.

Example 7·36. *A bag contains one black ball and m white balls. A ball is drawn at random. If a white ball is drawn, it is returned to the bag together with an additional white ball. If the black ball is drawn, it alone is returned to the bag.*

Let A_n denote the event that the black ball is not drawn in the first n trials. Discuss the converse to Borel-Cantelli Lemma with referance to events A_n, $n = 1, 2, ...$

Solution. A_n = The event that black ball is not drawn in the first n trials. ... (*)

= The event that each of the first n trials resulted in the draw of a white ball.

$$\Rightarrow \quad P\left(A_n\right) = P\left(E_1 \cap E_2 \cap ... \cap E_n\right),$$

where, E_i is the event of drawing a white ball in the ith trial.

$$\therefore \quad P\left(A_n\right) = P\left(E_1\right) P\left(E_2 \mid E_1\right) P\left(E_3 \mid E_1 \cap E_2\right) ... P\left(E_n \mid E_1 \cap E_2 ... \cap E_{n-1}\right)$$

$$= \frac{m}{m+1} \times \frac{m+1}{m+2} \times ... \times \frac{m+n-1}{m+n} = \frac{m}{m+n} \quad ... (**)$$

[Since if first ball drawn is white (W) it is returned together with an additional white ball, *i.e.*, for the second draw the box contains $1B$, $(m + 1)\, W$ balls and $P\left(E_2 \mid E_1\right) = \dfrac{m+1}{m+2}$, and so on.]

$$\therefore \quad \sum_{n=1}^{\infty} P\left(A_n\right) = \sum_{n=1}^{\infty} \frac{m}{m+n} = m \sum_{n=1}^{\infty} \frac{1}{m+n} = m\left(\frac{1}{m+1} + \frac{1}{m+2} + \frac{1}{m+3} + ...\right)$$

$$= m\left\{\sum_{n=1}^{\infty} \frac{1}{n} - \left(\sum_{n=1}^{m} \frac{1}{n}\right)\right\} \quad ... (***)$$

But $\displaystyle\sum_{n=1}^{\infty} \frac{1}{n}$ is divergent, $\left(\because \displaystyle\sum_{n=1}^{\infty} \frac{1}{n^p} \text{ is convergent, iff } p > 1\right)$ and $\displaystyle\sum_{n=1}^{m} \frac{1}{n}$ is finite,

$\therefore$ R.H.S. of (***) is divergent. Hence $\displaystyle\sum_{n=1}^{m} P\left(A_n\right) = \infty$

From the definition of A_n in (*) it is obvious that $A_n \downarrow$.

$$\therefore \quad A = \overline{\lim_n} = \lim_n \sup A_n = \varphi \quad \Rightarrow \quad P(A) = P(\varphi) = 0$$

This result is inconsistent with the converse of Barel-Cantelli Lemma, the reason being that the events A_n $(n = 1, 2, ...)$ considered here are not independent,

$$\because P(A_i \cap A_j) = P(A_i) = \frac{m}{m+i} \neq P(A_i)P(A_j),$$

since for $(i > j)$, $A_i \subset A_j$ as $A_n \downarrow$ [From (**)]

7·10. PROBABILTY GENERATING FUNCTION (p.g.f.)

Definition. If a_0, a_1, a_2 ... is a sequence of real numbers and if

$$A(s) = a_0 + a_1 s + a_2 s^2 + ... = \sum_{i=1}^{\infty} a_i s^i$$

converges in some interval $-s_0 < s < s_0$, when the sequence is infinite then the function $A(s)$ is known as the generating function of the sequence $\{a_i\}$.

The variable s has no significance of its own and is introduced to identify a_i as the coefficient of s^i in the expansion of $A(s)$. If the sequence (a_i) is bounded, then the comparison with the geometric series shows that $A(s)$ converges at least for $|s| < 1$.

In the particular case when a_i is the probability that an integral valued discrete variable X takes the value i, *i.e.*, $a_i = p_i = P(X = i)$; $i = 0, 1, 2, ...$ with $\Sigma p_i = 1$, then the probability generating function, abbreviated as *p.g.f.*, of *r.v.* X is defined as :

$$P(s) = E(s^X) = \sum_{x=0}^{\infty} s^x \cdot p_x \qquad ... (7·51)$$

Remarks:

1. Obviously, we have $P(1) = \Sigma p_x = 1$. Thus function $P(s)$ defined in (7·51) is a *p.g.f.* iff $p \geq 0 \; \forall x$ and $\Sigma p_x = 1$.

2. There is one to one correspondence between the *p.g.f.* and *p.m.f.* of a non-negative integer valued random variable.

Proof. We have :

$$P(s) = \sum_{k=0}^{\infty} s^k \cdot p_k \qquad ... (*)$$

$$\Rightarrow \quad |P(s)| = \left|\sum_{k=0}^{\infty} s^k \cdot p_k\right| \leq \sum_{k=0}^{\infty} s^k \; ; \quad [\because |p_k| \leq 1]$$

R.H.S., being an infinite series in G.P. with common ratio s is convergent if and only if $|s| < 1$.

Further,

$$P(1) = \sum_{k=0}^{\infty} p_k = 1.$$

Hence, $P(s)$ is convergent for all $s \leq 1$, and so term by term differentiation of the power-series (*) is valid. Differentiating (*) successfully *w.r.t.* s, we get

$$P'(s) = \sum_{k=1}^{\infty} k s^{k-1} \cdot p_k \,; \quad P''(s) = \sum_{k=z}^{\infty} k^{(2)} s^{k-2} \cdot p_k \,; \quad \dots; \quad P^r(s) = \sum_{k=r}^{\infty} k^{(r)} \cdot s^{k-r} \cdot p_k \qquad \dots (**)$$

where, $\quad k^{(r)} = k\,(k-1)\,(k-2) \dots (k-r+1) \qquad \dots (***)$

Putting, $s = 0$ in (**), we get

$$P^r(0) = \left[\frac{d^r}{ds^r} P(s)\right]_{s=0} = r^{(r)} \cdot p_r = (r!) \cdot p_r$$

$$\Rightarrow \qquad p_r = P(X = r) = \frac{P^r(0)}{r!} \qquad \dots (7·52)$$

Substituting in (*), we get

$$P(s) = \sum_{r=0}^{\infty} \left[P(X=r) \cdot s^r \right] = \sum_{r=0}^{\infty} p_r s^r = \sum_{r=0}^{\infty} \left[\frac{P^r(0)}{r!} \cdot s^r \right] \qquad \dots (7·53)$$

$\Rightarrow$ [*p.m.f.* uniquely determines its *p.g.f.* and conversely, from given *p.g.f.*, we can uniquely determine its *p.m.f.* using (7·52).]

3. ***Relation between p.g.f. and m.g.f.*** Taking $s = e^t$ in (7·51), we get

$P(e^t) = E(e^{tX}) = M_X(t)$, *i.e.,* from *p.g.f.* we can obtain *m.g.f.* on replacing s by e^t. ... (7·54)

4. ***Bivariate Probability Generating Function.*** The joint *p.g.f.* of two random variables X_1 and X_2 is a function of two variables s_1 and s_2 defined by :

$$P_{X_1, X_2}(s_1, s_2) = E\left(s_1^{X_1} \cdot s_2^{X_2}\right) = \sum_{x_1} \sum_{x_2} s_1^{x_1} \, s_2^{x_2} \, p(x_1, x_2) \qquad \dots (7·55)$$

Marginal *p.g.f.'s* can be obtained from (7·55) as given below :

$$P_{X_1}(s_1) = E\left(s_1^{X_1}\right) = P_{X_1, X_2}(s_1, 1); \qquad P_{X_2}(s_2) = E\left(s_2^{X_2}\right) = P_{X_1, X_2}(1, s_2) \qquad \dots (7·56)$$

5. ***Two r.v.'s X_1 and X_2 are independent if and only if :***

$$P_{X_1, X_2}(s_1, s_2) = P_{X_1}(s_1)\, P_{X_2}(s_2). \qquad \dots (7·57)$$

The above concepts can be generalised to n random variables.

Theorem 7·30. *If X is a random variable which assumes only integral values with probability distribution:*

$P(X = k) = p_k \,;\, k = 0, 1, 2, \dots$ *and* $P(X > k) = q_k,\ k \geq 0$ *so that*

$$q_k = p_{k+1} + p_{k+2} + \dots = 1 - \sum_{i=0}^{k} p_i \text{, and two generating functions are :}$$

$$P(s) = p_0 + p_1 s + p_2 s^2 + \dots \quad \text{and} \quad Q(s) = q_0 + q_1 s + q_2 s^2 + \dots$$

then for $-1 < s < 1$, $\qquad Q(s) = \dfrac{1 - P(s)}{1 - s} \qquad \dots (7·58)$

Proof. We have $\qquad q_{k-1} - q_k = p_k,\ k \geq 1$

$$\therefore \qquad \sum_{k=1}^{\infty} q_{k-1} s^k - \sum_{k=1}^{\infty} q_k s^k = \sum_{k=1}^{\infty} p_k s^k$$

$$\Rightarrow \qquad s\,Q(s) - Q(s) + q_0 = P(s) - p_0$$

$$\Rightarrow \qquad Q(s) = \frac{(p_0 + q_0) - P(s)}{(1-s)}$$

But $\qquad p_0 + q_0 = p_0 + p_1 + p_2 \ldots = 1$. Hence the theorem.

Theorem 7·31. *For a random variable X, which assumes only integral values, the expectation E(X) can be calculated either from probability distribution* $P(X = i) = p_i$ *or in terms of* $q_k = p_{k+1} + p_{k+2} + \ldots$

Thus, $$E(X) = \sum_{i=1}^{\infty} i\,p_i = \sum_{k=0}^{\infty} q_k$$

In terms of the generating functions : $$E(X) = P'(1) = Q(1) \qquad \ldots (7·59)$$

Proof. $$P(s) = \sum_{k=1}^{\infty} p_k s^k. \qquad \text{If } E(X) \text{ exists, then } E(X) = \sum_{k=1}^{\infty} k\,p_k \qquad \ldots (*)$$

$$P'(s) = \sum_{k=0}^{\infty} k p_k s^{k-1} \Rightarrow P'(1) = \sum_{k=1}^{\infty} k\,p_k = E(X) \qquad \text{[From (*)]}$$

We know that $\qquad Q(s)(1-s) = 1 - P(s)$

Differentiating both sides *w.r.* to s, we get

$$Q'(s)(1-s) - Q(s) = -P'(s) \Rightarrow Q(1) = P'(1). \qquad \ldots (**)$$

Hence, from (*) and (**) we get :

$$E(X) = P'(1) = Q(1).$$

Theorem 7·32. *If* $E(X^2) = \Sigma k^2 p_k$ *exists, then*

$$E(X^2) = P''(1) + P'(1) = 2Q'(1) + Q(1)$$

and hence $$V(X) = 2Q'(1) + Q(1) - \{Q(1)\}^2 = P''(1) + P'(1) - \{P'(1)\}^2 \qquad \ldots (7·60)$$

Proof. $$P(s) = \sum_{k=0}^{\infty} p_k s^k, \qquad P'(s) = \Sigma\, k\, p_k\, s^{k-1} \Rightarrow s\,P'(s) = \Sigma\, k\, p_k\, s^k.$$

Differentiating again, *w.r.* to s,

$$P'(s) P'(s) + sP''(s) = \Sigma\, k^2 p_k s^{k-1}$$

$$\therefore \qquad P'(1) + P''(1) = \Sigma k^2 p_k = E(X^2) \qquad \ldots (***)$$

Also, $\qquad Q''(s)(1-s) - 2Q'(s) = -P''(s) \qquad$ [Differentiating (**) again *w.r.* to s.]

Putting $s = 1$, we have $\quad 2Q'(1) = P''(1)$

Substituting in (***), we get, on using (7·59)

$$E\left(X^2\right) = P'(s)\,P'(1) + P''(1) = Q(1) + 2Q'(1) \qquad \text{... (7·61)}$$

$$\therefore \text{Var}(X) = E\left(X^2\right) - \{E(X)\}^2 = P''(1) + P'(1) - \{P'(1)\}^2 = 2Q'(1) + Q(1) - \{Q(1)\}^2$$

7·10·1. Probability Generating Function for the Sum of Independent Variables (Convolutions)

If X and Y are non-negative independent, integral valued discrete random variables with respective probability generating functions,

$$P(s) = \sum_{k=0}^{\infty} p_k s^k, \quad p_k = P(X = k)$$

and

$$R(s) = \sum_{k=0}^{\infty} r_k s^k, \quad r_k = P(Y = k),$$

it is possible to deduce probability generating function for the variable $Z = X + Y$, which is also clearly integral valued, in terms of $P(s)$ and $R(s)$.

Let w_k denote $P(Z = k)$. The event $Z = k$ is the union of the following mutually exclusive events,

$$(X = 0 \cap Y = k),\ (X = 1 \cap Y = k-1),\ (X = 2 \cap Y = k-2),\ \ldots,\ (X = k \cap Y = 0)$$

and since the variables X and Y are independent, each joint probability is the product of the appropriate individual probabilities. Therefore, the probability $w_k = P(Z = k)$ is given by :

$$w_k = p_0 r_k + p_1 r_{k-1} + p_2 r_{k-2} + \ldots + p_k r_0, \text{ for all integral } k \geq 0. \qquad \text{... (7·62)}$$

The new sequence of probabilities $\{w_k\}$ defined in terms of the sequences $\{p_k\}$ and $\{r_k\}$ is called the convolution of these sequences and is denoted by :

$$\{w_k\} = \{p_k\} * \{r_k\}. \qquad \text{... (7·63)}$$

Theorem 7·33. *If $\{p_k\}$ and $\{r_k\}$ are the sequences with probability generating functions $P(s)$ and $R(s)$ respectively, and $\{w_k\}$ is their convolution, then $W(s) = P(s), R(s)$ where $W(s) = \Sigma\, w_k s^k$ is the probability generating function of the sum $Z = X + Y$.*

Proof. Since the co-efficient of s^k in the product $P(s)\,R(s)$ is :

$$p_0 r_k + p_1 r_{k-1} + \ldots + p_{k-1} r_1 + p_k r_0 = w_k,$$

it follows that the probablity generating function for Z, namely,

$$W(S) = \sum_{k=0}^{\infty} w_k s^k, \text{ is equal to } P(s)\,R(s). \qquad \text{... (7·64)}$$

Corollary. If $X_1, X_2, \ldots, X_n$ are independent integral-valued discrete variables with respective probability generating functions $P_1(s), P_2(s), \ldots, P_n(s)$, and if $Z = X_1 + X_2 + \ldots + X_n$, the probability generating function for Z is given by :

$$P_Z(s) = \prod_{i=1}^{n} P_i(s)$$

In particular, when $X_1, X_2, \ldots, X_n$ all have a common distribution and hence common probability generating function $P(s)$, we have : $$P_Z(s) = \{P(s)\}^n. \quad \ldots (7\cdot64a)$$

7·10·2. Factorial Moments

Theorem 7·34. $$\mu'_{(r)} = E\{X(X-1)\ldots(X-r+1)\} = \left[\frac{\partial^r}{\partial s^r}P(s)\right]_{s=1} \quad \ldots (7\cdot65)$$

where, $\mu'_{(r)}$ is the rth factorial moment of X, about origin.

Proof. We have $$P(s) = \sum_x s^x \cdot p_x$$

Differentiating partially r times *w.r.* to s, we get

$$\frac{\partial^r P(s)}{\partial s^r} = \sum_x x(x-1)(x-2)\ldots(x-r+1)\, s^{x-r} p_x$$

$$\therefore \quad \left[\frac{\partial^r P(s)}{\partial s^r}\right]_{s=1} = \sum_x x(x-1)(x-2)\ldots(x-r+1)\, p_x = \mu'_{(r)}$$

7·10·3. Factorial Moment Generating Function

The factorial moment generating function (*f.m.g.f.*) of a discrete probability distribution of X about the origin is defined as :

$$f.m.g.f.\ (X) = E\left[(1+t)^X\right] \quad \ldots (7\cdot66)$$

$$= E\left[e^{X\log_e(1+t)}\right] = M_X\left[\log_e(1+t)\right] \quad \ldots (7\cdot66a)$$

where, $M_X(t)$ is the *m.g.f.* of X about origin.

We have : $$E\left[(1+t)^X\right] = E\left[\sum_{r=0}^{\infty}\left\{{}^XC_r\, t^r\right\}\right] = E\left[\sum_{r=0}^{\infty}\frac{X^{(r)}}{r!}t^r\right],$$

where, $X^{(r)} = X(X-1)(X-2)\ldots(X-r+1)$.

$$\therefore \quad E\left[(1+t)^X\right] = \sum_{r=0}^{\infty}\left[\frac{t^r}{r!}\cdot E\left\{X^{(r)}\right\}\right] = \sum_{r=0}^{\infty}\frac{t^r}{r!}\mu'_{(r)}$$

$\Rightarrow$ $\mu'_{(r)}$ = rth factorial moment of X about origin.

= Coefficient of $(t^r/r\,!)$ in the expansion of $E\left[(1+t)^X\right]$ ... (7·66b)

Also, we have :

$$\mu'_{(r)} = \left[\frac{d^r}{dt^r}(f.m.g.f.\text{ of X})\right]_{t=0} \quad \ldots (7\cdot67)$$

Example 7·37. *Can $P(s) = 2/(1 + s)$ be the p.g.f. of a r.v. X ? Give reasons.*

Solution. We have $P(1) = 2/2 = 1$.

Also, $$P(s) = \sum_{r=0}^{\infty} p_r s^r = 2(1+s)^{-1} = 2(1 - s + s^2 - s^3 + ...) = 2\sum_{r=0}^{\infty} (-1)^r . s^r$$

$\therefore$ $$p_r = 2(-1)^r \qquad ... (*)$$

Hence, $p_{2k+1} < 0$, *i.e.*, $p_1, p_3, p_5, ...$ are negative.

Since some coefficients in (*) are negative, $P(s)$ cannot be the *p.g.f.* of *r.v.* X.

Example 7·38. *If P(s) is the probability generating function for the r.v. X, find the probability generating function for (X – a)/b.*

Solution. $$P(s) = E\left(s^X\right)$$

$$\text{P.G.F. for } \frac{X-a}{b} = E\left\{s^{(X-a)/b}\right\} = s^{-a/b} \cdot E\left(s^{X/b}\right) = s^{-a/b} \cdot E\left[\left(s^{1/b}\right)^X\right] = s^{-a/b}\ P\left(s^{1/b}\right)$$

Example 7·39. *Let X be a random variable with probability generating function P(s). Find the probability generating function of* (a) X + 1, (b) 2X.

Solution.

(*a*) $$P(s) = \sum_{k=0}^{\infty} p_k s^k = E\left(s^X\right)$$

$$\text{P.G.F. of } (X + 1) = E\left(s^{X+1}\right) = sE\left(s^X\right) = sP(s)$$

(*b*) $$\text{P.G.F. of } 2X = E\left(s^{2X}\right) = E\left\{\left(s^2\right)^X\right\} = P\left(s^2\right).$$

Example 7·40. *Find the probabililty generating function of :*

(*a*) $P(X \le n)$, (*b*) $P(X < n)$, *and* (*c*) $P(X = 2n)$.

Solution.

(*a*) Let X be an integral valued random variable with the probability distribution:

$$P(X = n) = p_n \quad \text{and} \quad P(X \le n) = q_n, \text{ so that}$$
$$q_n = p_0 + p_1 + p_2 + ... + p_n \ ; n = 0, 1, 2, ...$$

$\therefore$ $$q_n - q_{n-1} = p_n, \ n \ge 1$$

$$\Rightarrow \sum_{n=1}^{\infty} q_n s^n - \sum_{n=1}^{\infty} q_{n-1} s^n = \sum_{n=1}^{\infty} p_n s^n \quad \Rightarrow \quad Q(s) - q_0 - sQ(s) = P(s) - p_0$$

Hence, $$Q(s) = \frac{P(s) + q_0 - p_0}{1-s} = \frac{P(s)}{1-s} \qquad [\because q_0 = p_0]$$

(*b*) Let $\quad q_n = P(X < n) = p_0 + p_1 + ... + p_{n-1}, \quad \Rightarrow \quad q_n - q_{n-1} = p_{n-1},\ n \geq 2$

$$\Rightarrow \sum_{n=2}^{\infty} q_n s^n - \sum_{n=2}^{\infty} q_{n-1}\, s^n = \sum_{n=2}^{\infty} p_{n-1}\, s^n = s \sum_{n=1}^{\infty} p_n s^n$$

$$\Rightarrow \quad Q(s) - q_1 s - sQ(s) = sP(s) - sp_0 \qquad [\because q_0 = 0]$$

$$\Rightarrow \quad \{1 - s\}Q(s) = sP(s) - sp_0 + q_1 s$$

$$\therefore \quad Q(s) = \frac{sP(s)}{1-s} \qquad [\because q_1 = p_0]$$

(*c*) Let $\quad P(X = 2n) = p_{2n}.\quad$ Its *p.g.f.* $Q(s)$ is given by :

$$Q(s) = \sum_{n=0}^{\infty} p_{2n} s^n = p_0 + p_2\, s + p_4\, s^2 + ...$$

$$\Rightarrow \quad 2Q(s) = 2p_0 + 2p_2 s + 2p_4 s^2 + ...$$

$$= (p_0 + p_1 s^{1/2} + p_2 s + p_3 s^{3/2} + p_4 s^2 + ...) + (p_0 - p_1 s^{1/2} + p_2 s - p_3 s^{3/2} + ...)$$

$$= \sum_{k=0}^{\infty} p_k (s^{1/2})^k + \sum_{k=0}^{\infty} p_k (-s^{1/2})^k = P(s^{1/2}) + P(-s^{1/2})$$

$$\therefore \quad Q(s) = \frac{P(s^{1/2}) + P(-s^{1/2})}{2}$$

Example 7·41. *Four tickets are drawn, one at a time with replacement, from a set of ten tickets numbered respectively, 1, 2, 3, ..., 10, in such a way that at each draw, each ticket is equally likely to be selected. What is the probability that the total of the numbers on the four drawn tickets is 20 ?*

Solution. If X_i denote the number on the *i*th ticket drawn then, for $i = 1, 2, 3, 4$, we observe that X_i is an integral-valued variate with possible values 1, 2, 3, ..., 10, each having associated probability $\frac{1}{10}$. Here each X_i has the :

$$p.g.f. = \frac{1}{10}s + \frac{1}{10}s^2 + ... + \frac{1}{10}s^{10} = \frac{s}{10}(1 - s^{10})(1 - s)^{-1}$$

Since X_i's are independent, it follows that the total of the numbers on the drawn tickets, *viz*, $Z = X_1 + X_2 + X_3 + X_4$, has probability generating function :

$$\left\{\frac{1}{10}s\left(1 - s^{10}\right)(1 - s)^{-1}\right\}^4 = \frac{1}{10^4}s^4\left(1 - s^{10}\right)^4(1 - s)^{-4}$$

The required probability is the co-efficient of s^{16} in $\frac{1}{10^4}(1 - s^{10})^4(1 - s)^{-4} = \frac{63}{10{,}000}$.

Example 7·42. *The six faces of an ordinary cubical die are numbered from 1 to 6. If two such unbiased dice are rolled once, find the probability distribution of the r.v. X. denoting the sum of points obtained.*

Also find an appropriate numbering of the twelve faces of two unbiased dice which would ensure that the probability $P(X = r)$ is the same for all r in the range $1 < r < 12$, and show that for such a pair of dice, the probability generating function of such a numbering scheme is :

$$G(\theta) = \frac{\theta\left(1 - \theta^{12}\right)}{12(1 - \theta)}.$$

Solution. The probability distribution of the sum of the points (X) obtained on two dice is :

x	2	3	4	5	6	7	8	9	10	11	12
$p(x)$	1/36	2/36	3/36	4/36	5/36	6/36	5/36	4/36	3/36	2/36	1/36

[See Example 7·19]

This can be expressed as :

$$p(x) = \frac{x-1}{36} \; ; \; x = 2, 3, 4, 5, 6, 7$$

$$= \frac{12-(x-1)}{36} \; ; \; x = 8, 9, 10, 11, 12.$$

2nd part. The numbering on the two dice should be serially given as 1, 2, 3, ..., 10, 11, 12, so that the probability of any number is 1/12 *i.e.,* all the numbers are equi-probable. The *p.g.f.* $G(\theta)$ for such a numbering scheme is :

$$G(\theta) = \sum_{i=1}^{12} p_i \theta^i = \frac{1}{12}\left[\theta + \theta^2 + ... + \theta^{12}\right] = \frac{\theta\left(1-\theta^{12}\right)}{12\,(1-\theta)}.$$

[The sum of *G.P.* series of 12 terms with $a = \theta$ and $r = \theta$].

Example 7·43.

(*a*) *From an urn containing (2n + 1) tickets numbered serially, three tickets are drawn at random without replacement. Prove that the probability that the numbers on them are in arithmetical progression is*

$$\left\{3n / \left(4n^2 - 1\right)\right\}.$$

(*b*) *Further, by considering the sample space corresponding to the possible realisation of the arithmetical progressions, show that the common difference of the arithmetical progression can be regarded as a discerete r.v. X. with the probability distribution defined by :*

$$P(X = r) = \frac{(2n+1) - 2r}{n^2} \; ; r = 1, 2, 3, ... \, n.$$

Hence, deduce that the probability-generating function of X is :

$$G(\theta) = \frac{4\theta}{n(1-\theta)^2}\left[1 - \frac{(2n+1-\theta^n)(1+\theta)}{4n}\right].$$

Solution.

(*a*) For the first part, see Example 3·19.

(*b*) If X denotes the common difference of the arithmetic progression, then from part (*a*), we conclude that X is a *r.v.* with the following distribution.

X :	1	2	3	...	r	...	$n-1$	n
Number of favourable cases :	$2n-1$	$2n-3$	$2n-5$	...	$2n+1-2r$	...	3	1
Probability* :	$\frac{2n-1}{n^2}$	$\frac{2n-3}{n^2}$	$\frac{2n-5}{n^2}$	...	$\frac{2n+1-2r}{n^2}$	...	$\frac{3}{n^2}$	$\frac{1}{n^2}$

*[Total No. of favourable cases = 1 + 3 + ... + (2n − 1) = n^2

$$\text{Probability} = \frac{\text{No. of favourable cases}}{\text{Total number of farourable cases}}.]$$

$$\therefore \qquad P(X = r) = \frac{2n+1-2r}{n^2}\ ;\ r = 1, 2, 3, ..., n.$$

The probability generating function $G(\theta)$ of X is given by :

$$G(\theta) = \sum_{r=1}^{n} P(X=r)\cdot\theta^r = \sum_{r=1}^{n}\left[\frac{(2n+1)-2r}{n^2}\cdot\theta^r\right]$$

$$= \frac{1}{n^2}\left[(2n+1)\sum_{r=1}^{n}\theta^r - 2\sum_{r=1}^{n} r\cdot\theta^r\right]$$

$$= \frac{1}{n^2}\left[(2n+1)\left(\theta+\theta^2+\theta^3+...+\theta^n\right)-2\left(\theta+2\theta^2+3\theta^3+...+n\theta^n\right)\right] \qquad ...(*)$$

Let $\qquad S = \theta + 2\theta^2 + 3\theta^3 + ... + (n-1)\theta^{n-1} + n\theta^n$

$\Rightarrow \qquad \theta S = \theta^2 + 2\theta^3 + ... + \ ...\ .... + (n-1)\theta^n + n\theta^{n+1}$

(Arithmetic geometric series)

$\Rightarrow \qquad (1-\theta)S = \theta + \theta^2 + ... + \theta^n - n\theta^{n+1}$

$$\Rightarrow \qquad (1-\theta)S = \frac{\theta(1-\theta^n)}{1-\theta} - n\theta^{n+1} \qquad \Rightarrow \qquad S = \frac{\theta(1-\theta^n)}{(1-\theta)^2} - \frac{n\theta^{n+1}}{(1-\theta)}$$

Substituting in (*), we get

$$G(\theta) = \frac{1}{n^2}\left[(2n+1)\cdot\frac{\theta(1-\theta^n)}{1-\theta} - 2\left\{\frac{\theta(1-\theta^n)}{(1-\theta)^2} - \frac{n\theta^{n+1}}{1-\theta}\right\}\right]$$

$$= \frac{\theta}{n^2(1-\theta)^2}\left[(2n+1)\left(1-\theta^n\right)(1-\theta) - 2\left\{\left(1-\theta^n\right) - n\theta^n(1-\theta)\right\}\right]$$

$$= \frac{\theta}{n^2(1-\theta)^2}\left[2n\left(1-\theta-\theta^n+\theta^{n+1}\right) + \left(1-\theta-\theta^n+\theta^{n+1}\right) + \left(-2+2n\theta^n-2n\theta^{n+1}+2\theta^n\right)\right]$$

$$\Rightarrow \qquad G(\theta) = \frac{\theta}{n^2(1-\theta)^2}\left[2n(1-\theta)-(1+\theta)+\theta^n(1+\theta)\right] \qquad ...(1)$$

Let us conside the given expression for $G(\theta)$ *viz.*,

$$\frac{4\theta}{n(1-\theta)^2}\left[1 - \frac{\left(2n+1-\theta^n\right)(1+\theta)}{4n}\right] = \frac{\theta}{n^2(1-\theta)^2}\left[4n-\left(2n+1-\theta^n\right)(1+\theta)\right]$$

$$= \frac{\theta}{n^2(1-\theta)^2}\left[4n-\left\{2n+1+2n\theta+\theta-\theta^n(1+\theta)\right\}\right]$$

$$= \frac{\theta}{n^2(1-\theta)^2}\left[2n-1-2n\theta-\theta+\theta^n(1+\theta)\right]$$

$$= \frac{\theta}{n^2(1-\theta)^2}\left[2n(1-\theta)-(1+\theta)+\theta^n(1+\theta)\right] \quad \text{... (2)}$$

which is same as (1).

From (1) and (2), we get :

$$G(\theta) = \frac{4\theta}{n(1-\theta)^2}\left[1-\frac{(2n+1-\theta^n)(1+\theta)}{4n}\right]$$

Example 7·44. Let $\{X_k\}$ *be mutually independent, each assuming the values* 0, 1, 2, 3, ..., $a - 1$ *with probability* $\frac{1}{a}$. *Let* $S_n = X_1 + X_2 + ... + X_n$. *Show that the probability generating function of* S_n *is :*

$$P(S) = \left\{\frac{1-s^a}{a(1-s)}\right\}^n \quad \textit{and} \quad \textit{hence } P(S_n = j) = \frac{1}{a^n}\sum_{v=0}^{\infty}(-1)^{v+j+av}\binom{n}{v}\binom{-n}{j-av}$$

Solution. As $\{X_k\}$ are mutually independent random variables and each X_i assumes the same values 0, 1, 2, ... $a - 1$ with the probability $\frac{1}{a}$, therefore, each will have the same generating function and the probability generating function of S_n will be the nth convolution of generating function of X_1. Now

$$P_{X_1}(s) = \sum_{k=0}^{a-1} p_k s^k = \frac{1}{a}\left(s^0+s^1+...+s^{a-1}\right) = \frac{1-s^a}{a(1-s)}$$

$$\therefore \quad P_{S_n}(s) = \left\{\frac{1-s^a}{a(1-s)}\right\}^n = \frac{1}{a^n}\left(1-s^a\right)^n(1-s)^{-n}$$

The probability $S_n = j$ is the co-efficient of s^j in $P_{S_n}(s)$.

If we take $(v + 1)$th term from $(1 - s^a)^n$, then it will have the power of s equivalent to s^{av} and hence to get the power of s as j, we must take the term from $(1 - s)^{-n}$ having the power of s as $j - av$.

$$\therefore \quad \text{Required probability} = \frac{1}{a^n}\sum_{v=0}^{\infty}\binom{n}{v}(-1)^v\binom{-n}{j-av}(-1)^{j-av}$$

$$= \frac{1}{a^n}\sum_{v=0}^{\infty}(-1)^{j+v-av}\binom{n}{v}\binom{-n}{j-av}$$

$$= \frac{1}{a^n}\sum_{v=0}^{\infty}(-1)^{j+v+av}\binom{n}{v}\binom{-n}{j-av}$$

$$\left[\because (-1)^{2av} = 1 \quad \Rightarrow \quad (-1)^{av} = (-1)^{-av}\right]$$

Example 7·45. *A random variable X assumes the value $\lambda_1, \lambda_2, \ldots$ with probabilities $u_1, u_2, \ldots$, respectively. Show that*

$$P_k = \frac{1}{k!}\sum_{j=0}^{\infty} u_j e^{-\lambda_j}(\lambda_j)^k \; ; \; \lambda_j > 0, \sum u_j = 1$$

is a probability distribution. Find its generating function and prove that its mean equals E(X) and variance equals V(X) + E(X).

Solution.

$$\sum_{k=0}^{\infty} p_k = \sum_{k=0}^{\infty}\left\{\frac{1}{k!}\sum_{j=0}^{\infty} u_j \exp(-\lambda_j)(\lambda_j^k)\right\}$$

$$= \sum_{j=0}^{\infty}\left\{u_j \exp(-\lambda_j)\sum_{k=0}^{\infty}(\lambda_j)^k / k!\right\} = \sum_{j=0}^{\infty}\left\{u_j \exp(-\lambda_j)\exp(\lambda_j)\right\} = \sum_{j=0}^{\infty} u_j = 1$$

Hence, p_k represents a probability distribution.

Let $P(s)$ be the probability generating function of p_k, then

$$P(s) = \sum_{k=0}^{\infty} p_k s^k = \sum_{k=0}^{\infty}\left\{\frac{1}{k!}\sum_{j=0}^{\infty} u_j e^{-\lambda_j}(\lambda_j)^k s^k\right\}$$

$$= \sum_{j=0}^{\infty}\left\{u_j e^{-\lambda j}\sum_{k=0}^{\infty}(s\lambda_j)^k / k!\right\} = \sum_{j=0}^{\infty} u_j \cdot e^{-\lambda_j} e^{s\lambda_j} = \sum_{j=0}^{\infty} u_j \cdot e^{\lambda_j(s-1)}$$

(By Fubini's Theorem)

Thus,

$$P(S) = \sum_{j=0}^{\infty} u_j\left\{1 + \lambda_j(s-1) + \frac{\lambda_j^2(s-1)^2}{2!} + \ldots\right\} \Rightarrow P'(1) = \sum_{j=0}^{\infty} u_j \lambda_j = E(X)$$

$$P''(s) = \sum_{j=0}^{\infty} u_j\left\{\lambda_j^2 + \lambda_j^3(s-1) + \ldots\right\} \Rightarrow P''(1) = \sum_{j=0}^{\infty} \lambda_j^2 u_j = E\left(X^2\right)$$

Hence, the variance V of probability distribution p_k is given by :

$$V = P''(1) + P'(1) - \{P'(1)\}^2 = E\left(X^2\right) + E(X) - \{E(X)\}^2 = E(X) + V(X).$$

CHAPTER CONCEPTS QUIZ

1. Comment on the following, giving examples, if possible :

(*i*) *M.g.f.* of a *r.v.* always exists.

(*ii*) Characteristic function of a *r.v.* always exists.

(*iii*) *M.g.f.* is not affected by change of origin or/and scale.

(*iv*) $\phi_{X+Y}(t) = \phi_X(t) . \phi_Y(t)$, implies X and Y are independent.

(v) $\phi(0) = 1$ and $|\phi(t)| \leq 1$.

(vi) Variance of a *r.v.* is 5 and its mean does not exist.

(vii) It is possible to find a *r.v.* whose first k moments exist but $(k + 1)$th moment does not exist.

(viii) If a *r.v.* X has a symmetrical distribution about origin then

(a) $\phi_X(t)$ is even valued function of t. (b) $\phi_X(t)$ is complex valued function of t.

2. (a) Can following be the characteristic function of any distribution ? Given reasons.

(i) $\log(1 + t)$, (ii) $\exp(-t^4)$ (iii) $1/(1+t^4)$.

(b) Prove that $\phi(t) = \exp(-t^{\alpha})$, cannot be characteristic function unless $\alpha = 2$.

3. State the relations, if any, between the following:

(i) $E(X^r)$ and $\phi_X(t)$,

(ii) $M_X(t)$ and $M_{X-a}(t)$, a being constant.

(iii) $M_X(t)$ and $M_{(X-a)/h}(t)$, a and h being constants ;

(iv) $\phi_X(t)$ and *p.d.f.* of X.

(v) μ_r and μ'_r.

(vi) First four cumulants in terms of first four moments about mean.

4. Let $X_1, X_2, ..., X_n$ be n *i.i.d.* (independent and identically distributed) *r.v.'s* with *m.g.f.* $M(t)$.

Then prove that $$M_{\overline{X}}(t) = \{M(t/n)\}^n, \text{ where } \overline{X} = \sum_{i=1}^{n} X_i / n$$

5. If $X_1, X_2, ..., X_n$ are independent *r.v.'s* then prove that *m.g.f.* of $\sum_{i=1}^{n} c_i X_i$ is $:= \prod_{i=1}^{n} M_{X_i}(c_i t)$

6. Fill in the blanks:

(i) If $\int_{-\infty}^{\infty} |\phi_X(t)|\, dt < \infty$, then *p.d.f.* of X is given by ... ...

(ii) X_1 and X_2 are independent if and only if ... ...

(iii) If X_1 and X_2 are independent, then $\phi_{X_1-X_2}(t)$ ... ...

(iv) $\phi(t)$ is defined and is .... for all t in $(-\infty, \infty)$.

7. Examine critically the following statements :

(a) Two distributions having the same set of moments are identical.

(b) The characteristic function of a certain non-degenerate distribution is $\exp(-t^3)$.

8. Mention the correct alternative:

(i) If X is a continuous random variable with mean μ and variance σ^2, then for any positive number k, $P\{|X - \mu| > k\sigma\} \leq 1/k^2$ is known as :

(a) Lyapunov's inequality. (b) Chebychev's inequality.

(c) Bienayme-Chebychev's inequality. (d) *Khintchine*'s-inequality.

(ii) According to Chebychev's inequality, the probability that :

(a) X differing from its mean by more than 2 standard deviations is less than or equal to 0·75.

(b) X will lie within 2 standard deviations of its mean, is greater than or equal to 0·75.

(c) X will lie within 2 standard deviations of its mean, is greater than or equal to 0·25.

(d) X will lie within 2 standard deviations of its mean, is greater than or equal to 0·95.

(iii) To facilitate the process of determining prbabilities and moments of a random variable, usually we take the help of some functions. For instance, for a random variable X with probability distribution $p_i = P(X = x_i)$; $i = 0, 1, 2, \ldots$, we often use a function $\sum p(x_i)\, s^i$. The function, thus defined, subject to the convergence, is known as :

(a) moment generating function, (b) probability generating function,
(c) probability distribution function, (d) characteristic function.

(iv) If X is a *r.v.* with probability distribution :

$p_i = P(X = x_i)$; $i = 0, 1, 2, \ldots$, then the function $\sum p(x_i)\, s^i$ is known as probability generating function if :

(a) $s < 1$, (b) $s < \infty$, (c) $s \le 1$, (d) $|s| \le 1$.

(v) If X is a random variable and $f(x)$ be the probability function, then subject to the convergence, the function, $\sum_x e^{tx} f(x)$ is known as :

(a) moment generating function, (b) probability generating function,
(c) probability distribution function, (d) characteristic function.

(vi) Let $M_X(t)$ be the moment generating function of a *r.v.* X, then subject to the convergence of the expansion of $\log_e M_X(t)$ in the powers of t, the function $K_X(t) = \log_e M_X(t)$ is known as :

(a) moment generating function, (b) probability generating function,
(c) cumulant generating function, (d) Maclaurin's expansion function.

DISCUSSION AND REVIEW QUESTIONS

1. Define *m.g.f.* of a *r.v.* Hence or otherwise find the *m.g.f.* of (i) $Y = aX + b$, (ii) $Y = \dfrac{X - m}{\sigma}$

2. Show that if $\overline{X}$ is mean of n *i.i.d.* random variables : $M_{\overline{X}}(t) = [M_X(t/n)]^n$.

3. Define moments and moment generating function (*m.g.f.*) of a *r.v.* X. If $M(t)$ is the *m.g.f.* of a *r.v.* X about the origin, show that the moment μ_r' is given by : $\mu_r' = \left[\dfrac{d^r M(t)}{dt^r}\right]_{t=0}$

4. If μ_r' is moment of rth order about the origin and κ_j is the cumulant of jth order, prove that :

$$\frac{\partial \mu_r'}{\partial k_j} = \binom{r-1}{j-1} \mu'_{r-j}$$

5. If μ_r' is the rth moment about the origin of a variable X and if $\mu_r' = r\,!$, find the *m.g.f.* of X.

6. Define cumulants and obtain the first four cumulants in terms of central moments.

7. If X is a variable with zero mean and cumulants κ_r, show that the first two cumulants l_1 and l_2 of X^2 are given by $l_1 = \kappa_2$ and $l_2 = 2k_2^2 + \kappa_4$.

8. If X is random variable with cumulants κ_r ; $r = 1, 2, \ldots$ Find the cumulants of

(i) cX, (ii) $c + X$, where c is a constant.

9. Define the characteristic function of a random variable. Show that the characteristic function of the sum of two independent variables is equal to the product of their characteristic functions.

10. State and prove Chebychev's inequality.

11. Explain the concept of 'convergence in probability'.

12. Derive Chebychev's inequality and show how it leads to the weak law of large numbers. Mention some important particular cases wherein the weak law of large numbers holds good.

13. State and prove the weak law of large numbers. Deduce as a corollary Bernoulli Theorem and comment on its applications.

14. State Kolmogorov's Inequality

15. State and prove strong law of large numbers.

16. Define the probability generating function (*p.g.f.*) of a random variable.

ASSORTED REVIEW PROBLEMS FOR SELF-ASSESSMENT

7·1. (*a*) The random variable X takes the value n with probability $1/2^n$; $n = 1, 2, 3, ...$ Find the moment generating function of X and hence find the mean and variance of X.

(*b*) A random variable 'X' has probability function :

$$p(x) = \frac{1}{2^x} ; x = 1, 2, 3 ... \text{ Find its M.G.F., mean and variance.}$$

Ans. $M_X(t) = \left[e^t / (2 - e^t)\right]$; Mean = 2, Variance = 6.

(*c*) Show that the *m.g.f.* of *r.v.* X having the *p.d.f.*

$$f(x) = \begin{cases} \frac{1}{3}, -1 < x < 2 \\ 0, \text{elsewhere} \end{cases} \quad \text{is } M(t) = \begin{cases} \frac{\exp(2t) - \exp(-t)}{3t}, t \neq 0 \\ 1 \quad , t = 0 \end{cases}$$

(*d*) A random variable 'X' has the density function :

$$f(x) = \begin{cases} \frac{1}{2\sqrt{x}}, 0 < x < 1 \\ 0, \text{ elsewhere} \end{cases}$$

Obtain the *m.g.f.* and hence the mean and variance.

(*e*) X is random variable and $p(x) = ab^x$, where a and b are positive, $a + b = 1$ and x taking the values 0, 1, 2, ... Find the moment generating function of X. Hence show that :

$m_2 = m_1 (2m_1 + 1)$, m_1 and m_2 being first two moments about origin.

Ans. $M_X(t) = \alpha (1 - b e^t)^{-1}$; $|b e^t| < 1$; $m_1 = \frac{b}{a}$; $m_2 = b (a + 2b) / a^2$

7·2. Obtain the *m.g.f.* of the random variable X having *p.d.f.* :

$$f(x) = \begin{cases} x \quad , \text{for } 0 \le x < 1 \\ 2 - x, \text{for } 1 \le x < 2 \\ 0 \quad , \text{elsewhere} \end{cases}$$

Determine Mean, μ_2, μ_3 and μ_4.

Ans. $M_X(t) = \left(\frac{e^t - 1}{t}\right)^2 \quad \Rightarrow \quad k_X(t) = 2\left[\log(e^t - 1) - \log t\right]$

Mean = 1 ; $\mu_2 = \frac{1}{6}$; $\mu_3 = 0$; $\mu_4 = \frac{53}{120}$.

7·3. The experiment is to toss two balls into five boxes in such a way that each ball is equally likely to fall in any box. Let X denote the number of balls in the first box.

(*i*) What is the *p.d.f.* of X ? (*ii*) Give the *m.g.f.* of X. (*iii*) Find the mean and variance of X.

7·4. Show that the rth cumulant for the distribution :

$$f(x) = c\, e^{-cx},\ c > 0 \text{ and } 0 \le x < \infty, \text{ is } (r-1)\,!\,/\,c^r.$$

7·5. If X is a *r.v.* having cumulants κ_r ; $r = 1, 2, \ldots$ given by :

$$\kappa_r = (r-1)\,!\; pa^{-r}\ ;\ p > 0,\ a > 0,$$

find the characteristic function of X.

Ans. $\phi_X(t) = \left(1 - \dfrac{it}{a}\right)^{-p}$

7·6. (*a*) Find the characteristic function of the following distributions and their variances :

(*i*) $dF(x) = a\, e^{-ax} dx,\ (a > 0,\ x > 0)$, (*ii*) $dF(x) = \dfrac{1}{2} e^{-|x|} dx, (-\infty < x < \infty)$

(*iii*) $P(X = j) = {}^nC_j\, p^j\, q^{n-j},\ (0 < p < 1,\ q = 1 - p,\ j = 0, 1, 2, \ldots\, n)$.

(*b*) Define $\phi(t)$, the characteristic function of a random variable.

Find the characteristic function of a *r.v.* X defined as follows:

$$f(x) = \begin{cases} 0, & x < 0 \\ 1, & 0 \le x \le 1 \\ 0, & x > 1 \end{cases}$$

7·7. For the joint distribution of two-dimensional random variable (X, Y) given by :

$$f(x, y) = \begin{cases} \dfrac{1}{4a^2}\left\{1 + xy(x^2 - y^2)\right\}, & |x| \le a;\ \ |y| \le a, a > 0 \\ 0, & \text{elsewhere,} \end{cases}$$

show that the characteristic function of X + Y is equal to the product of the characteristic functions of X and Y. Comment on the result.

7·8. (*a*) Let $K(t_1, t_2) = \log_e M(t_1, t_2)$, where $M(t_1, t_2)$ is the joint *m.g.f.* of X and Y. Show that :

$$\frac{\partial K(0,0)}{\partial t_1} = E(X)\,;\qquad \frac{\partial^2 K(0,0)}{\partial t_1^2} = \text{Var}\,X\,;\qquad \frac{\partial^2 K(0,0)}{\partial t_1\, \partial t_2} = \text{Cov}\,(X, Y)$$

(*b*) (X, Y) is a two-dimensional discrete random variable with the possible values 0 and 1 for X, and also 0 and 1 for Y, and with the joint probabilities given in the adjoining Table.

Y \ X	0	1
0	p_{00}	p_{10}
1	p_{01}	p_{11}

Find the characteristic functions $\psi_1(t_1)$, $\psi_2(t_2)$ and $\psi(t_1, t_2)$ for X, Y and (X, Y) respectiviely and show that

$$\psi(t_1, t_2) = \psi_1(t_1).\ \psi_2(t_2), \quad \text{when} \quad p_{00}\, p_{11} = p_{01}\, p_{10} \qquad \ldots(*)$$

Also interpret (*).

7·9. (*a*) A random variable X has a mean value of 5 and variance of 3.

(*i*) What is the least value of Prob $\{|X - 5| < 3\}$?

(*ii*) What value of h guarantees that Prob $\{|X - 5| < h\} \ge 0{\cdot}99$?

(*iii*) What is the least value of Prob $(|X - 5| < 7{\cdot}5)$?

(*b*) A random variable X takes the values –1, 1, 3, 5 with associated probabilities 1/6, 1/6, 1/6 and 1/2. Find by direct computation $P(|X - 3| \ge 1)$. Find an upper bound to this probability by applying Chebychev's inequality.

(*c*) If X denotes the sum of the numbers obtained when two dice are thrown, use Chebychev's inequality to obtain an upper bound for $P(|X-7| \geq 4)$. Compare this with the actual probability.

Ans. Upper bound $= \dfrac{35}{96}$; Actual probability $= \dfrac{1}{18}$.

(*d*) Let S_n be the total score in n tosses of a symmetric die. Using Chebychev's inequality, obtain an upperbound for the probability :

$$P\left[\left|\frac{S_n}{n} - 3{\cdot}5\right| > \varepsilon\right], \varepsilon > 0$$

Ans. $\{35/12\, n \epsilon^2\}$.

(*e*) In Chebychev's inequality, find the least value of k for which the probability that a *r.v.* X takes the value between $\mu \pm k\sigma$ is : (*i*) at least 0·95, (*ii*) at least 0·99.

Ans. (*i*) $\sqrt{20}$, (*ii*) 10.

7·10. (*a*) An unbiased coin is tossed 100 times. Show that the probability that the number of heads will be between 30 and 70 is greater than 0·93.

(*b*) Within what limits will the number of heads lie, with 95% probability, in 1000 tosses of a coin which is practically unbiased ?

(*c*) A symmetric die is thrown 720 times. Use Chebychev's inequality to find the lower bound for the probability of getting 100 to 140 sixes.

Ans. 0·75

(*d*) Use Chebychev's inequality to determine how many times a fair coin must be tossed in order that the probability will be at least 0·95 that the ratio of the number of heads to the number of tosses will be between 0·45 and 0·55.

Ans. 2000

(*e*) An unbiased coin is tossed 100 times. Using Chebychev's inequality, obtain upper bound to the chance that the difference between the number of heads and tails obtained is 20 or more.

Hint. X: Number of heads obtained in 100 tosses of coin.

$X \sim B(n = 100, p = 1/2)$. We want $P\left[|X - \mu_X| \geq 10\right]$.

Ans. 0·25.

7·11. (*a*) If X is a *r.v.* such that $E(X) = 3$ and $E(X^2) = 13$, use Chebychev's inequality to determine a lower bound for $P(-2 < X < 8)$.

(*b*) State and prove Chebychev's inequality. Use it to prove that in 2,000 throws with a coin the probability that the number of heads lies between 900 and 1,100 is at least 19/20.

(*c*) A random variable X has the density function e^{-x} for $x \geq 0$. Show that Chebychev's inequality gives, $P(|X-1| > 2) < \dfrac{1}{4}$ and show that the actual probability is e^{-3}.

(*d*) Let X have the *p.d.f.* :

$$f(x) = \begin{cases} \dfrac{1}{2\sqrt{3}}, -\sqrt{3} < x < \sqrt{3} \\ 0 \quad , \text{elsewhere} \end{cases}$$

Find the actual probability $P\left(|X-\mu| \geq \frac{3}{2}\sigma\right)$ and compare it with the upper bound obtained by Chebychev's inequality.

Ans. Upper bound by Chebychev's inequality = 0·444; Actual probabilty = 0·134.

7·12. (*a*) If $X_1, X_2, \ldots X_n$ be *r.v.'s.* with means $\mu_1, \mu_2, \ldots, \mu_n$ and standard deviations $\sigma_1, \sigma_2, \ldots, \sigma_n$ respectively and $\left[Var\left(X_1 + X_2 + \ldots + X_n\right)\right] / n^2 \to 0$ *as* $n \to \infty$, show that $\left(\bar{X}_n - \bar{\mu}_n\right)$ converges to zero in probability.

Hence show that if m is the number of successes in n independent trials, the probability of success at the ith trial being p_i then m/n converges in probability to $(p_1 + p_2 + \ldots + p_n)/n$.

(*b*) If X_n takes the values 1 and 0 with corresponding probabilities p_n and $1 - p_n$, examine whether the weak law of large numbers can be applied to the sequence $\{X_n\}$ where the variables X_n, $n = 1, 2, 3, \ldots$ are independent.

Hint. $\text{Var}(X_n) = p_n q_n \leq \frac{1}{4}$; yes, W.L.L.N. holds good.

7·13. (*a*) $\{X_k\}$, $k = 1, 2, \ldots$ is a sequence of independent random variables each taking the values : $-1, 0, 1$. Given that $P(X_k = 1) = \frac{1}{k} = P(X_k = -1)$, $P(X_k = 0) = 1 - \frac{2}{k}$, examine if the law of large numbers holds for this sequence.

Ans. WLLN holds.

(*b*) Examine whether the weak law of large numbers holds good for the sequence X_n of independent random variables, where $P\left(X_n = \frac{1}{\sqrt{n}}\right) = \frac{2}{3}$, $P\left(X_n = -\frac{1}{\sqrt{n}}\right) = \frac{1}{3}$.

Ans. WLLN holds.

7·14. (*a*) $\{X_n\}$ is a sequence of independent random variables such that

$$P\left(X_n = \frac{1}{\sqrt{n}}\right) = p_n, \quad P\left(X_n = 1 + \frac{1}{\sqrt{n}}\right) = 1 - p_n$$

Examine whether the weak law of large numbers is applicable to the sequence $\{X_n\}$.

Hint. $\text{Var}(X_n) = p_n q_n \leq \frac{1}{4}$. Yes.

(*b*) If X is a *r.v.* and $E(X^2) < \infty$, then prove that $P\{|X| \geq a\} \leq \frac{1}{a^2} E\left(X^2\right)$, for all $a > 0$.

Use Chebychev's inequality to show that for $n > 36$, the probability that in n throws of a fair die, the number of sixes lies between $\frac{1}{6}n - \sqrt{n}$ and $\frac{1}{6}n + \sqrt{n}$ is at least $\frac{31}{36}$.

Hint. Chebychev's Inequality gives :

$$P\left[\left|X - \frac{n}{6}\right| < k\right] \geq 1 - \frac{5n}{36k^2}; \quad \text{Choose } k = \sqrt{n} \;\Rightarrow\; P\left[\frac{n}{6} - \sqrt{n} < X < \frac{n}{6} + \sqrt{n}\right] \geq \frac{31}{36}$$

Since $X \geq 0$, $\quad \frac{n}{6} - \sqrt{n} > 0 \;\Rightarrow\; n > 36.$

7·15. (*a*) Let $\{X_n\}$ be a sequence of mutually independent random variables such that :

$$X_n = \pm 1 \text{ with probability } \frac{1-2^{-n}}{2} \text{ and } X^n = \pm 2^{-n} \text{ with probability } 2^{-n-1}.$$

Examine whether the weak law of large numbers can be applied to the sequence $\{X_n\}$.

Ans. WLLN holds for the sequence $\{X_n\ ; n = 1, 2, ...\}$

(*b*) Examine whether the law of large numbers holds for the sequence $\{X_k\}$ of independent random variables defined by $P\left(X_k = \pm k^{-1/2}\right) = \dfrac{1}{2}$. **[Ans. Yes]**

(*c*) State and prove Weak Law of Large Numbers.

7·16. (*a*) Let $\{X_n\}$ be any sequence of *r.v.'s*. Write $Y_n = \dfrac{1}{n}\left[\sum_{i=1}^{n} X_i - E\left(\sum_{i=1}^{n} Y_n\right)\right]$.

Prove that a necessary and sufficient condition for the sequence $\{X_n\}$ to satisfy the weak law of large numbers is that $E\left(\dfrac{Y_n^2}{1+Y_n^2}\right) \to 0$ as $n \to \infty$.

(*b*) Let $X_1, X_2, X_3, ...$ be *i.i.d.* standard Cauchy variates. Show that the W.L.L.N. does not hold good for the sequence $\{X_n\}$.

7·17. (*a*) X is a positive integral valued variable, such that $P(X = n) = p_n$, $n = 0, 1, 2, ...,$

Define the probability generating function $G(S)$ and the moment generating function $M(t)$ for X and show that, $M(t)$ is equal to $G(e^t)$. Hence or otherwise prove that

$$E(X) = G'(1), \qquad \text{Var}(X) = G''(1) + G'(1) - \{G'(1)\}^2$$

(*b*) If X and Y are non-negative integral valued independent *r.v.'s* with $P(s)$ and $Q(s)$ as their probability generating functions, show that their sum $X + Y$ has the *p.g.f.* $P(s)\ Q(s)$.

7·18. A test of the strength of a wire consists of bending and unbending until it breaks. Considering bending and unbending as two operations, let X denote the random variable corresponding to the number of operations necessary to break the wire. If $P(X = r) = (1 - p)\ p^{r-1}$; $r = 1, 2, 3, ...$ and $0 < p < 1$, find the probability generating function of X.

7·19. Define the generating function $A(s)$ of the sequence $\{a_j\}$. Let a_j be the number of ways in which the score j can be obtained by throwing a die any number of times. Show that the generating function, *g.f.* of $\{a_j\}$ is $\left\{\left(1 - s - s^2 - s^3 - s^3 - s^4 - s^5 - s^6\right)^{-1} - 1\right\}$.

7·20. (*a*) Find the generating function of (*i*) $P(X \geq n)$, (*ii*) $P(X > n + 1)$.

(*b*) Obtain the generating function of q_n, the probability that in n tosses of an ideal coin, no run of three heads occurs.

(*c*) Let X be a non-negative integral-valued random variable with probability generating function $P(s) = \sum_{n=0}^{\infty} p_n\, s^n$. After observing X, conduct X binomial trials with probability p of success and let Y denote the corresponding resulting number of successes.

Determine (*i*) the probability generating function of Y, and
(*ii*) probability generating function of X given that $Y = X$.

7·21. In a sequence of Bernoulli trials, let U_n be the probability that the first combination *SF* occurs at trials number $(n - 1)$ and n.

Find the probability generating function, mean and variance.

Hint. $P(s) = \dfrac{pq}{q-p}\left[\dfrac{s}{1-qs} - \dfrac{s}{1-ps}\right]$; Mean $= \dfrac{1}{pq}$; Variance $= \dfrac{1-3pq}{p^2q^2}$

7·22. In a sequence of Bernoulli trials, let U_n be the probabilty of an even number of successes. Prove the recursion formula : $U_n = qU_{n-1} + (1 - U_{n-1})\,p.$

From this derive the probability generating function and hence the explicit formula for U_n.

Ans. Generating Function : $G(t) = \sum\limits_{n=0}^{\infty} U_n t^n = \dfrac{1}{2}\left[\dfrac{1}{1-t} + \dfrac{1}{1-ct}\right]; c = 1-2p$; $U_n = \dfrac{1}{2}\left[1+(1-2p)^n\right]$

7·23. A series of independent Bernoulli trials with probability '*p*' of success for each trial is performed until an uninterrupted run of *r* successes is obtained for the first time where *r* is a given positive integer. Assuming that the probability of a success in any trial is $p = 1 - q$, show that the probability generating function of the number of trials is : $F(S) = \dfrac{p^r s^r(1-ps)}{1-s+qp^r s^{r+1}}$.

7·24. A population of *N* distinct elements in equal proportion is sampled with replacement. If S_r denotes the sample size necessary for the aequisition of *r* distinct varieties, and $p_n(r) = P(S_r = n)$, prove the recursion formula :

$$p_{n+1}(r) = \frac{r-1}{N} \cdot p_n(r) + \frac{N-r+1}{N} \cdot p_n(r-1)$$

Hence, or otherwise, derive the generating function $P_r(s)$ of $p_n(r)$.

Ans. $$P_r(S) = s^r\left[\frac{(N-1)(N-2)\dots(N-r+1)}{(N-s)(N-2s)\dots\{N-(r-1)s\}}\right]$$

7·25. There is a complete set of values (events) $x_0, x_1, x_2, \dots x_k, \dots$ with respective probabilities $p_0, p_1, p_2, \dots p_k, \dots$ which vary with time. These are related by the differential equation :

$$\frac{d}{dt}(p_k) = \lambda\,(p_{k-1} - p_k) \qquad \dots(*)$$

Find the value of p_k.

Hint. Post multiplying (*) with s^k and summing over *k* from 0 to ∞, we get

$$\frac{d}{dt}[G(s)] = \lambda\,[s\,G(s) - G(s)] = \lambda\,(s-1)\,G(s)$$

Integrate both sides *w.r.t.* '*t*' and use $G(1) = 1$ to determine the constant of integration.

Ans. $$p_k = \frac{e^{-\lambda t}\cdot(\lambda t)^k}{k!}; k = 0, 1, 2, \dots$$

7·26. In a sequence of Bernoulli trials, let A_n be the event that a run of *n* consecutive successes occurs between the 2^nth and 2^{n+1}th trials. Show that if $p \geq \frac{1}{2}$, there is probability one that infinitely many A_n occur; if $p < \frac{1}{2}$, then with probability one only finitely many A_n occur.

7·27. For a given sequence $\{X_n\}$ *of r.v.'s*, $\varphi_n(t, X_n) = (\sin nt)\,/\,nt$.

Determine the distribution function of X_n. Hence show that even though the sequence of characteristic function $\varphi_n(t)$ converges to a limit $\varphi(t)$, the sequence of distribution functions does not converge to a distribution function. What is the condition that is violated here ?

7·28. If $\Upsilon_2 = \dfrac{\mu_4}{\mu_2^2} - 3$, is known and finite, show that it is possible to improve Chebychev's inequality in the form :

$$P\left[\,|X-\mu| \geq k\,\sigma\,\right] \leq \frac{\Upsilon_2 + 2}{\Upsilon_2 + 2 + (k^2-1)^2},$$

where X is a random variable with mean μ and variance σ^2, and k is any positive integer greater than 1.

7·29. $\{X_n,\ n \geq 1\}$ is a sequence of random variables such that :

$$\text{Var}(X_i) = 1;\ i = 1, 2, 3, \ldots \text{ and } \text{cov}(X_i, X_j) = \begin{cases} \rho\,; & |i-j| = 1,2 \\ 0\,; & \text{otherwise} \end{cases}$$

Show that W.L.L.N. holds for the sequence $\{X_n\}$.

7·30. Let $\{X_n, n = 1, 2 \ldots\}$ be a sequence of standardised variates and

Corr $(X_m, X_n) = \exp\{-|m-n|\,\alpha\}, \alpha > 0$ and $m \neq n$. Show that W.L.L.N. holds for this sequence.

7·31. Two continuous variates X and Y have a joint *p.d.f.* with a joint characteristic function $\varphi(t_1, t_2)$. If $g_X(x)$ is the marginal density, show that $\mu'_r(x)$, the rth simple moment for the conditional distribution of Y given $X = x$, satisfies the equation :

$$\mu'_r(x).\ g(x) = \mu'_r(x) \cdot g(x) = \frac{1}{2\pi} \int_{-\infty}^{\infty} \frac{\partial^r \phi(t_1, 0)}{i^r\, \partial t_2^r} \cdot \exp(-i\,t\,x)\ dt_1$$

7·32. Prove that the real part of a characteristic function is again a characteristic function. Prove further that if $\psi_1(t) = a_1(t) + ib_1(t)$ and $\psi_2(t) = a_2(t) + ib_2(t)$, are characteristic function, then $a_1(t)\,a_2(t) - b_1(t)\,b_2(t)$ is a characteristic function.

7·33. Show that for any distribution : $\displaystyle\int_{-\infty}^{\infty} \left(1 - \frac{x^2}{t^2}\right) dF(x) \leq \int_{-t}^{t} dF(x)$

and hence deduce $P\{|X - E(X)| > k\,\sigma\} \leq \dfrac{1}{k^2}$, where $k > 0$ and $\text{Var}(X) = \sigma^2$.

7·34. Let X be a *r.v.* with moment generating function $M(t)$, $-h < t < h$. Prove that

$$P(X \geq a) \leq e^{-at}\, M(t),\ 0 < t < h,$$

and that

$$P(X \leq a) \leq e^{-at}\, M(t),\ -h < t < 0.$$

7·35. Let

$$f(x, y) = \begin{cases} x \exp\{-x(y+1)\};\ x > 0, y > 0 \\ 0, \text{ elsewhere} \end{cases}$$

Find moment generating function of $Z = XY$.

7·36. If X has a *p.d.f.* : $f(x) = 1 - |x|;\ x \leq 1$, show that $M_X(t) = \left\{\left(e^{-t} + e^{t} - 2\right)/t^2\right\}$.

Hence, deduce that :

$$\mu_{2n} = \frac{1}{(n+1)(2n+1)}\ ;\quad \mu_{2n+1} = 0\ ;\quad n = 0, 1, 2, \ldots$$

7·37. For the distribution :

$$dF(x) = k\left(1+\frac{x}{a}\right)^{m-1} \cdot \exp\left[-\frac{mx}{a}\right] \; ; \; -a \le x \le a,$$

Find: (*i*) the constant *k*, (*ii*) *M.g.f.* $M_X(t)$.

Hence, obtain the first four central moments and derive the linear relation between β_1 and β_2.

Ans. $k = \dfrac{m^m}{\Gamma m} \cdot \dfrac{e^{-m}}{a}$; $M_X(t) = \left(1 - \dfrac{at}{m}\right)^{-m}$; $t < \dfrac{m}{a}$

$$\mu_1 = 0, \quad \mu_2 = \frac{a^2}{m}, \quad \mu_3 = \frac{2a^3}{m^2}, \quad \mu_4 = \frac{3a^4(2+m)}{m^3},$$

Linear relationship between β_1 and β_2 : $2\beta_2 - 3\beta_1 = 6$.

7·38. If *n* letters are placed at random in *n* correctly addressed envelopes, prove that the probability that exactly '*r*' letters are placed in correct envelopes is given by :

$$\frac{1}{r!}\sum_{k=0}^{r-1}\left[\frac{(-1)^k}{k!}\right] \; ; \; r = 0, 1, 2, \ldots, n.$$

If the *r.v.* X denotes the number of letters going to the correct envelopes then prove that *m.g.f.* of X is given by :

$$M_X(t) = \sum_{y=0}^{n}\left[\frac{1}{y!}\left(e^t - 1\right)^y\right]$$

Hence, obtain the mean and variance of number of matches.

Hint. Use : $\mu_1' = \left(\dfrac{dM}{dt}\right)_{t=0}$ and $\mu_2' = \left[\dfrac{d^2M}{dt^2}\right]_{t=0}$

Ans. Mean = 1, Variance = 1.

7·39. Suppose $\phi(\theta)$ is the real valued characteristic function of a *r.v.* X. Prove that :

(*i*) $1 - \phi(2\theta) \le 4\left[1 - \phi(\theta)\right]$, (*ii*) $1 + \phi(2\theta) \ge 2\left[\phi(\theta)\right]^2$

7·40. If $Z = X_1 + X_2 + \ldots + X_n$, where X_i's are *i.i.d.* variates with characteristic function $\phi_1(t)$ and *n* is a variate taking on values 1, 2, 3, ... with characteristic function $\phi_2(t)$, show that :

$$\phi(t, Z) = \phi_2\left[-i \log \phi_1(t)\right].$$

Hint. Use : $\phi_Z(t) = E_n\left[E_Z\left(e^{itZ} \mid n\right)\right]$.

7·41. A random variable X follows normal distribution $N\left(0, \dfrac{1}{n}\right)$ with probability $\left(1 - \dfrac{1}{n}\right)$ and $N\left(1, \dfrac{1}{n}\right)$ with probability $(1/n)$. Find the characteristic function (*c.f.*) of X and hence find its mean and variance.

Hint. *p.d.f.* of X : $f(x) = \left(1 - \dfrac{1}{n}\right) f_1(x) + \dfrac{1}{n} f_2(x)$;

where, $f_1(x)$ is *p.d.f.* of $N\left(0, \dfrac{1}{n}\right)$ and $f_2(x)$ is *p.d.f.* of $N\left(1, \dfrac{1}{n}\right)$.

Ans. $\phi_X(t) = e^{-t^2/2n}\left[1 - \frac{1}{n} + \frac{1}{n}e^{it}\right]$; Mean $= \frac{1}{n}$, Variance $= \frac{(2n-1)}{n^2}$.

7·42. The probability of obtaining a 6 with a biased die is p, where $(0 < p < 1)$. Three players A, B and C roll this die in order, A starting. The first one to throw a 6 wins. Find the probability of winning for A, B and C.

If X is a random variable which takes the value r if the game finishes at the rth throw, determine the probability generating function of X and hence, or otherwise, evaluate $E(X)$ and Var (X).

7·43. From the probability generating function (*p.g.f.*) of two random variables X and Y given by :

$$P(s, t) = \exp\left(-\lambda - \mu - b + \lambda s + \mu t + bst\right)$$

(*i*) Obtain the marginal *p.g.f.'s* and identify them,

(*ii*) Obtain the *p.g.f.* of X + Y and $P\{X + Y = 0\}$, and

(*iii*) Interpret the case $b = 0$.

Ans. (*i*) $P_X(s) = \exp\left[(\lambda + b)(s-1)\right] \Rightarrow X \sim P(\lambda + b)$

$P_Y(t) = \exp\left[(\mu + b)(s-1)\right] \Rightarrow Y \sim P(\mu + b)$

(*ii*) $$P_{X+Y}(t) = \sum_{k=0}^{\infty} p_k t^k = \exp\left[-\lambda - \mu - b + (\lambda + \mu)t + bt^2\right], \quad ...(*)$$

where $p_k = P(X + Y = k)$.

$\therefore \quad P(X + Y = 0) = p_0 =$ constant term in (*) $= \exp[-\lambda - \mu - b]$.

7·44. Define convergence in probability. Let $X_1, X_2, X_3, \ldots$ be *i.i.d.* variates with :

$$f(x) = e^{-(x-1)}, \ x \geq 1$$

Show that $Y_n \to 1$ in probability, where $Y_n = \text{Min}(X_k);\ 1 \leq k \leq n$.

Hint. Prove that, for $\varepsilon > 0$,

$$P\left[\left|Y_n - 1\right| < \varepsilon\right] \to 1 \text{ as } n \to \infty.$$

CHAPTER

8 Special Discrete Probability Distributions

LEARNING OBJECTIVES

Upon completion of this Chapter, you should be able to:

- Explain and illustrate the concept of a discrete probability distribution.
- Enumerate the basic assumptions of several important discrete probability distributions.
- Write down the probability mass functions (p.m.f's) of various important discrete probability distributions.
- Derive moments, cumulants, moment generating functions (m.g.f's) and various constants of discrete probability distributions.
- Apply probability distributions to a variety of problems in various diversified fields.
- Appreciate the importance of various discrete probability distributions in statistical theory.

CHAPTER OUTLINES

8·1. INTRODUCTION

In this Chapter we shall study some of the probability distributions that figure most prominently in statistical theory and application. We shall also study their parameters, *i.e.*, the quantities that are constants for particular distributions but that can take on different values for different members of families of distributions of the same kind. We shall introduce number of discrete probability distributions that have been successfully applied in a wide variety of decision situations. The purpose of this chapter is to show the types of situations in which these distributions can be applied.

It may be mentioned that a theoretical probability distribution gives us a law according to which different values of the random variable are distributed with specified probabilities according to some definite law which can be expressed mathematically. It is possible to formulate such laws either on the basis of given conditions (a prior considerations) or on the basis of the results (a posterior inferences) of an experiment. The present study will also enable us to fit a mathematical model or a function of the form $y = p(x)$ to the observed data.

This chapter is devoted to the study of univariate (except for the multinomial) distributions like Rectangular, Binomial, Poisson, Negative Binomial, Geometric, Hypergeometric, Multinomial and Power-series distributions. We have already defined distribution function, mathematical expectation, *m.g.f.*, characteristic function and moments in earlier chapters. This prepares us for a study of discrete theoretical probability distributions.

8·2. DISCRETE UNIFORM DISTRIBUTION

Definition. *A r.v. X is said to have a discrete uniform distribution over the range [1, n] if its p.m.f. is expressed as follows :*

$$P(X = x) = \begin{cases} \dfrac{1}{n}, & \text{for } x = 1, 2, \ldots, n. \\ 0, & \text{otherwise} \end{cases} \qquad \ldots (8\cdot1)$$

Here n is known as the parameter of the distribution and lies in the set of all positive integers. Equation (8·1) is also called a discrete rectangular distribution.

Such distributions can be conceived in practice if under the given experimental conditions, the different values of the random variable become equally likely. Thus, for a die experiment, and for an experiment with a deck of cards, such distribution is appropriate.

8·2·1. Moments

$$E(X) = \frac{1}{n}\sum_{i=1}^{n} i = \frac{n+1}{2}, \quad E(X^2) = \frac{1}{n}\sum_{i=1}^{n} i^2 = \frac{(n+1)(2n+1)}{6}$$

$$V(X) = E(X^2) - [E(X)]^2 = \frac{(n+1)(n-1)}{12} = \frac{n^2-1}{12}.$$

The *m.g.f.* of X is : $$M_X(t) = E(e^{tX}) = \frac{1}{n}\sum_{x=1}^{n} e^{tx} = \frac{e^t(1-e^{nt})}{n(1-e^t)}$$

8·3. BERNOULLI DISTRIBUTION

Definition. *A r.v. X is said to have a Bernoulli distribution with parameter p if its p.m.f. is given by :*

$$P(X = x) = \begin{cases} p^x(1-p)^{1-x}, & \text{for } x = 0, 1 \\ 0, & \text{otherwise} \end{cases} \qquad \ldots (8\cdot2)$$

The parameter p satisfies $0 \le p \le 1$. Often $(1 - p)$ is denoted as q.

A random experiment whose outcomes are of two types, success S and failure F, occurring with probabilities p and q respectively, is called a *Bernoulli trial.* If for this experiment, a *r.v.* X is defined such that it takes value 1 when S occurs and 0 if F occurs, then X follows a Bernoulli distribution.

8·3·1. Moments of Bernoulli Distribution

The rth moment about origin is :

$$\mu_r' = E(X^r) = 0^r \cdot q + 1^r \cdot p = p; \; r = 1, 2, ... \qquad ...(8\cdot2a)$$

$$\mu_1' = E(X) = p, \; \mu_2' = E(X^2) = p, \text{ so that } \mu_2 = \text{Var}(X) = \mu_2' - \mu_1'^2 = p - p^2 = pq$$

The *m.g.f.* of Bernoulli variate is :

$$M_X(t) = e^{0 \times t} P(X = 0) + e^{1 \times t} P(X = 1) = q + pe^t \qquad ...(8\cdot2b)$$

Remark. *Degenerate Random Variable.* Sometimes we may come across a variate X which is degenerate at a point 'c', say, so that $P(X = c) = 1$ and $= 0$ otherwise, *i.e.,* the whole mass of the variable is concentrated at a single point 'c'.

Since $P(X = c) = 1$, Var $(X) = 0$. Hence, a degenerate *r.v.* X is characterised by : Var $(X) = 0$.

m.g.f. of degenerate *r.v.* is : $\quad M_X(t) = E(e^{tX}) = e^{tc} \, P(X = c) = e^{ct}$.

8·4. BINOMIAL DISTRIBUTION

Binomial distribution was discovered by Swiss Mathematician James Bernoulli (1654 - 1705) in the year 1700 and was first published posthumously in 1713, eight years after his death. Let a random experiment be performed repeatedly, each repetition being called a trial and let the occurrence of an event in a trial be called a success and its non-occurrence a failure. Consider a set of n independent Bernoullian trials (n being finite) in which the probability 'p' of success in any trial is constant for each trial, then $q = 1 - p$, is the probability of failure in any trial.

The probability of x successes and consequently $(n - x)$ failures in n independent trials, in a specified order (say) *SSFSFFFS...FSF* (where S represents success and F represents failure) is given by the compound probability theorem by the expression :

$$\begin{aligned} P(SSFSFFFS...FSF) &= P(S)\,P(S)\,P(F)\,P(S)\,P(F)\,P(F)\,P(F)\,P(S) \times ... \times P(F)\,P(S)\,P(F) \\ &= p \cdot p \cdot q \cdot p \cdot q \cdot q \cdot q \cdot p \cdots q \cdot p \cdot q \\ &= \underbrace{p \cdot p \cdot p \ldots p}_{\{x \text{ factors}\}} \cdot \underbrace{q \cdot q \cdot q \ldots\ldots q}_{\{(n-x) \text{ factors}\}} = p^x q^{n-x} \end{aligned}$$

But x successes in n trials can occur in $\binom{n}{x}$ ways and the probability for each of these ways is same, *viz.,* $p^x q^{n-x}$. Hence, the probability of x successes in n trials in *any order* is given by the *addition theorem* of probability by the expression $\binom{n}{x} p^x q^{n-x}$.

The probability distribution of the number of successes, so obtained is called the *Binomial probability distribution,* for the obvious reason that the probabilities of 0, 1, 2, ..., n successes, *viz.,* $q^n, \binom{n}{1} q^{n-1} p, \binom{n}{2} q^{n-2} p^2, ..., p^n$, are the successive terms of the binomial expansion $(q + p)^n$.

Definition. *A random variable X is said to follow binomial distribution if it assumes only non-negative values and its probability mass function is given by :*

$$P(X = x) = p(x) = \begin{cases} \binom{n}{x} p^x q^{n-x}; \; x = 0, 1, 2, ..., n; \; q = 1 - p \\ 0, \qquad \textit{otherwise} \end{cases} \qquad ...(8\cdot3)$$

The two independent constants n and p in the distribution are known as the *parameters* of the distribution. 'n' is also sometimes, known as the degree of the binomial distribution.

Binomial distribution is a discrete distribution as X can take only the integral values, *viz.*, 0, 1, 2, ..., n. Any random variable which follows binomial distribution is known as *binomial variate.*

We shall use the notation $X \sim B(n, p)$ to denote that the random variable X follows binomial distribution with parameters n and p.

The probability $p(x)$ in (8·3) is also sometimes denoted by $b(x, n, p)$.

Remarks :

1. The assignment of probabilities in (8·3) is permissible because

$$\sum_{x=0}^{n} p(x) = \sum_{x=0}^{n} \binom{n}{x} p^x q^{n-x} = (q+p)^n = 1$$

2. Let us suppose that n trials constitute an experiment. Then, if this experiment is repeated N times, the *frequency function* of the binomial distribution is given by :

$$f(x) = N \times p(x) = N \times \binom{n}{x} p^x q^{n-x} \; ; \; x = 0, 1, 2, ..., n \qquad (8\cdot3a)$$

and the expected frequencies of 0, 1, 2, ..., n successes are the successive terms of the binomial expansion, $N(q + p)^n$, $q + p = 1$.

3. Physical conditions for Binomial Distribution. We get the binomial distribution under the following experimental conditions

(*i*) The random experiment is performed under the same conditions for a fixed and finite (also discrete) number of trials, say n.

(*ii*) The outcomes of each trial in the process are complementary and one of them is referred to as a *success* and the other as a *failure.*

(*iii*) The probability of outcomes remains constant through the experiment. Thus, the probability of an outcome being classified as success (or failure) is constant regardless of the number of trials.

(*iv*) The outcome (*i.e.*, success or failure) of any trial is independent of the outcome of any other trial. The trials satisfying the conditions (*ii*), (*iii*) and (*iv*) are also called *Bernoulli trials.*

4. Binomial distribution is important not only because of its wide applicability, but because it gives rise to many other probability distributions. Tables for $p(x)$ are available for various values of n and p.

Example 8·1. *Ten coins are thrown simultaneously. Find the probability of getting at least seven heads.*

Solution. p = Probability of getting a head = $\frac{1}{2}$

q = Probability of not getting a head = $\frac{1}{2}$

The probability of getting x heads in a random throw of 10 coins is :

$$p(x) = \binom{10}{x} \left(\frac{1}{2}\right)^x \left(\frac{1}{2}\right)^{10-x} = \binom{10}{x} \left(\frac{1}{2}\right)^{10} ; \; x = 0, 1, 2, ..., 10$$

∴ Probability of getting at least seven heads is given by :

$$P(X \geq 7) = p(7) + p(8) + p(9) + p(10)$$

$$= \left(\frac{1}{2}\right)^{10} \left\{\binom{10}{7} + \binom{10}{8} + \binom{10}{9} + \binom{10}{10}\right\} = \frac{120 + 45 + 10 + 1}{1024} = \frac{176}{1024}.$$

Example 8·2. *A and B play a game in which their chances of winning are in the ratio 3 : 2. Find A's chance of winning at least three games out of the five games played.*

Solution. Let p be the probability that 'A' wins the game. Then we are given :

$$n = 5,\ p = \frac{3}{5} \quad \Rightarrow \quad q = 1 - p = \frac{2}{5}.$$

Hence, by Binomial Probability Law, the probability that out of 5 games played, A wins 'x' games is given by :

$$P(X = x) = p(x) = \binom{5}{x}\left(\frac{3}{5}\right)^x \left(\frac{2}{5}\right)^{5-x} \ ; \ x = 0, 1, 2, ..., 5$$

The required probability that 'A' wins at least three games is given by :

$$P(X \geq 3) = \sum_{r=3}^{5} \binom{5}{r} \frac{3^r \cdot 2^{5-r}}{5^5} = \frac{3^3}{5^5}\left[\binom{5}{3} 2^2 + \binom{5}{4} . 3 \times 2 + 1.3^2 \times 1\right]$$

$$= \frac{27 \times (40 + 30 + 9)}{3125} = 0{\cdot}68.$$

Example 8·3. *A coffee connoisseur claims that he can distinguish between a cup of instant coffee and a cup of percolator coffee 75% of the time. It is agreed that his claim will be accepted if he correctly identifies at least 5 of the 6 cups. Find his chances of having the claim*

(i) accepted, (ii) rejected, when he does have the ability he claims.

Solution. If p denotes the probability of a correct distinction between a cup of instant coffee and a cup of percolator coffee, then we are given :

$$p = \frac{75}{100} = \frac{3}{4} \quad \Rightarrow \quad q = 1 - p = \frac{1}{4}, \quad \text{and} \quad n = 6$$

If the random variable X denotes the number of correct distinctions, then by the Binomial Probability Law, the probability of x correct identifications out of 6 cups is given by :

$$P(X = x) = p(x) = \binom{6}{x}\left(\frac{3}{4}\right)^x \left(\frac{1}{4}\right)^{6-x} \ ; \ x = 0, 1, 2, ..., 6$$

(*i*) The probability of the claim being accepted is :

$$P(X \geq 5) = p(5) + p(6) = \binom{6}{5}\left(\frac{3}{4}\right)^5 \left(\frac{1}{4}\right)^{6-5} + \binom{6}{6}\left(\frac{3}{4}\right)^6 = \frac{1458}{4096} + \frac{729}{4096} = 0{\cdot}534.$$

(*ii*) The probability of the claim being rejected is :

$$P(X \leq 4) = 1 - P(X \geq 5) = 1 - 0{\cdot}534 = 0{\cdot}466.$$

Example 8·4. *A multiple-choice test consists of 8 questions with 3 answers to each question (of which only one is correct). A student answers each question by rolling a balanced die and checking the first answer if he gets 1 or 2, the second answer if he gets 3 or 4 and the third answer if he gets 5 or 6. To get a distinction, the student must secure at least 75% correct answers. If there is no negative marking, what is the probability that the student secures a distinction ?*

Solution. Since there are three answers to each question, out of which only one is correct, the probability of getting an answer to a question correctly is given by :

$$p = \frac{1}{3}, \quad \text{so that} \quad q = 1 - p = \frac{2}{3}$$

By Binomial Probability Law, the probability of getting r correct answers in a 8-question test is given by :

$$P(X = x) = p(x) = \binom{8}{x}\left(\frac{1}{3}\right)^x \left(\frac{2}{3}\right)^{8-x} ; \; x = 0, 1, 2, ..., 8$$

Hence, the required probability of securing a distinction (*i.e.*, of getting correct answers to at least 6 out of the 8 questions) is given by :

$$p(6) + p(7) + p(8) = \binom{8}{6}\left(\frac{1}{3}\right)^6 \left(\frac{2}{3}\right)^{8-6} + \binom{8}{7}\left(\frac{1}{3}\right)^7 \left(\frac{2}{3}\right)^{8-7} + \binom{8}{8}\left(\frac{1}{3}\right)^8 \left(\frac{2}{3}\right)^{8-8}$$

$$= \frac{1}{3^6}\left[28 \times \frac{4}{9} + 8 \times \frac{1}{3} \times \frac{2}{3} + \frac{1}{9}\right] = \frac{129}{729 \times 9} = 0{\cdot}0197.$$

Example 8·5. *An irregular six-faced dice is thrown and the expectation that in 10 throws it will give five even numbers is twice the expectation that it will give four even numbers. How many items in 10,000 sets of 10 throws each, would you expect it to give no even number.*

Solution. Let p be the probability of getting an even number in a throw of a die. Then the probability of getting x even numbers in ten throws of a die is given by :

$$P(X = x) = \binom{10}{x} p^x q^{10-x} ; \; x = 0, 1, 2, ..., 10 \qquad ... (*)$$

We are given that : $P(X = 5) = 2P(X = 4) \Rightarrow \binom{10}{5} p^5 q^5 = 2\binom{10}{4} p^4 q^6$

$$\Rightarrow \frac{10!\,p}{5!\,5!} = 2\frac{10!\,q}{4!\,6!} \Rightarrow \frac{p}{5} = \frac{2q}{6} = \frac{q}{3} \Rightarrow 3p = 5(1 - p) \Rightarrow p = \frac{5}{8} \text{ and } q = \frac{3}{8}.$$

Thus, $P(X = x) = \binom{10}{x}\left(\frac{5}{8}\right)^x \left(\frac{3}{8}\right)^{10-x}$ ... [From (*)]

Hence, the required number of times that in 10,000 sets of 10 throws each, we get no even number

$$= 10{,}000 \times P(X = 0) = 10{,}000 \times (3/8)^{10} = 1 \text{ (approx.)}.$$

Example 8·6. *The probability of a man hitting a target is 1/4.*

(i) If he fires 7 times, what is the probability of his hitting the target at least twice ?

(ii) How many times must he fire so that the probability of his hitting the target at least once is greater than 2/3 ?

Solution. p = Probability of the man hitting the target $= \frac{1}{4} \Rightarrow q = 1 - p = \frac{3}{4}.$

$$p(x) = \text{Probability of getting } x \text{ hits in 7 shots} = \binom{7}{x}\left(\frac{1}{4}\right)^x \left(\frac{3}{4}\right)^{7-x} ; \; x = 0, 1, ..., 7$$

(i) Probability of at least two hits

$$= 1 - \{p(0) + p(1)\} = 1 - \left\{\binom{7}{0}\left(\frac{1}{4}\right)^0 \left(\frac{3}{4}\right)^{7-0} + \binom{7}{1}\left(\frac{1}{4}\right)^1 \left(\frac{3}{4}\right)^{7-1}\right\} = \frac{4547}{8192}.$$

(ii) Probability of at least one hit in n shots $= 1 - p(0) = 1 - \left(\frac{3}{4}\right)^n.$

It is required to find n, so that $1-\left(\frac{3}{4}\right)^n > \frac{2}{3} \Rightarrow \frac{1}{3} > \left(\frac{3}{4}\right)^n$.

Taking logarithms of each side, $\log\frac{1}{3} > n\log\frac{3}{4} \Rightarrow \log 1 - \log 3 > n(\log 3 - \log 4)$

$$\Rightarrow \quad 0 - 0{\cdot}4771 > n(0{\cdot}4771 - 0{\cdot}6021) \quad \Rightarrow \quad 0{\cdot}4771 < 0{\cdot}1250\, n$$

$$\therefore \quad n > \frac{0{\cdot}4771}{0{\cdot}1250} = 3{\cdot}8$$

Since n cannot be fractional, the required number of shots is 4.

Example 8·7. *In a binomial distribution consisting of 5 independent trials, probabilities of 1 and 2 successes are found to be 0·4096 and 0·2048 respectively. Find the parameter 'p' of the distribution.*

Solution. Let $X \sim B(n, p)$. In usual notations, we are given :

$$n = 5, \quad p(1) = 0{\cdot}4096 \quad \text{and} \quad p(2) = 0{\cdot}2048.$$

According to Binomial probability distribution :

$$P(X = x) = p(x) = \binom{5}{x} p^x (1-p)^{5-x}; \; x = 0, 1, 2, ..., 5$$

$$p(1) = \binom{5}{1} p(1-p)^4 = 0{\cdot}4096 \quad ... (*) \qquad \text{and} \qquad p(2) = \binom{5}{2} p^2 (1-p)^3 = 0{\cdot}2048 \quad ... (**)$$

Dividing (*) by (**), we get

$$\frac{\binom{5}{1} p(1-p)^4}{\binom{5}{2} p^2 (1-p)^3} = \frac{0{\cdot}4096}{0{\cdot}2048} \quad \Rightarrow \quad \frac{5(1-p)}{10p} = 2 \quad \Rightarrow \quad p = \frac{1}{5} = 0{\cdot}2.$$

Example 8·8. *With the usual notations, find the probability of success 'p' for a binomial distribution, if* $n = 6$ *and* $9\,P(X = 4) = P(X = 2)$.

Solution. For the binomial random variable X with parameters $n = 6$ and p, the probability function is :

$$P(X = r) = \binom{6}{r} p^r q^{6-r}; \; r = 0, 1, 2, ..., 6$$

We are given: $9P(X = 4) = P(X = 2) \Rightarrow 9 \times \binom{6}{4} p^4 q^2 = \binom{6}{2} p^2 q^4$

$$\Rightarrow \quad 9p^2 = q^2 \quad \Rightarrow \quad 9p^2 = (1-p)^2 = 1 + p^2 - 2p$$

$$\Rightarrow \quad 8p^2 + 2p - 1 = 0 \quad \Rightarrow \quad p = \frac{-2 \pm \sqrt{4+32}}{2 \times 8} = \frac{-2 \pm 6}{16} = -\frac{1}{2}, \frac{1}{4}$$

Since probability cannot be negative, $p = -\frac{1}{2}$ is rejected. Hence, $p = \frac{1}{4}$.

8·4·1. Moments of Binomial Distribution

The first four moments about origin of binomial distribution are obtained as follows :

$$\mu_1' = E(X) = \sum_{x=0}^{n} x\binom{n}{x}p^x q^{n-x} = np\sum_{x=1}^{n}\binom{n-1}{x-1}p^{x-1}q^{n-x} = np\,(q+p)^{n-1} = np$$

$$\left[\because \binom{n}{x} = \frac{n}{x}\binom{n-1}{x-1} = \frac{n}{x}\cdot\frac{n-1}{x-1}\cdot\binom{n-2}{x-2} = \frac{n}{x}\cdot\frac{n-1}{x-1}\cdot\frac{n-2}{x-2}\binom{n-3}{x-3}, \text{ and so on}\right]$$

Thus, the mean of the binomial distribution is np.

$$\mu_2' = E(X^2) = \sum_{x=0}^{n} x^2\binom{n}{x}p^x q^{n-x}$$

$$= \sum_{x=0}^{n}\{x(x-1)+x\}\frac{n(n-1)}{x(x-1)}\cdot\binom{n-2}{x-2}p^x q^{n-x}$$

$$= n(n-1)p^2\left\{\sum_{x=2}^{n}\binom{n-2}{x-2}p^{x-2}q^{n-x}\right\} + np$$

$$= n(n-1)\,p^2\,(q+p)^{n-2} + np = n(n-1)\,p^2 + np$$

$$\mu_3' = E(X^3) = \sum_{x=0}^{n} x^3 p(x) = \sum_{x=0}^{n}\{x(x-1)(x-2) + 3x(x-1) + x\}\binom{n}{x}p^x q^{n-x}$$

$$= n(n-1)(n-2)p^3\sum_{x=3}^{n}\binom{n-3}{x-3}p^{x-3}q^{n-x} + 3n(n-1)\,p^2\sum_{x=2}^{n}\binom{n-2}{x-2}p^{x-2}q^{n-x} + np$$

$$= n(n-1)\,(n-2)\,p^3(q+p)^{n-3} + 3n(n-1)p^2(q+p)^{n-2} + np$$

$$= n(n-1)\,(n-2)\,p^3 + 3n(n-1)\,p^2 + np$$

Similarly

$$x^4 = x(x-1)\,(x-2)\,(x-3) + 6x\,(x-1)\,(x-2) + 7x(x-1) + x$$

Let $x^4 = Ax(x-1)\,(x-2)\,(x-3) + Bx\,(x-1)\,(x-2) + Cx(x-1) + x$

(By giving to x the values 1, 2 and 3, we find the values of arbitrary constants A, B and C.)

$$\therefore \qquad \mu_4' = E(X^4) = \sum_{x=0}^{n} x^4\binom{n}{x}p^x q^{n-x}$$

$$= n(n-1)\,(n-2)\,(n-3)\,p^4 + 6n(n-1)\,(n-2)\,p^3 + 7n\,(n-1)\,p^2 + np$$

[On simplification]

Central Moments of Binomial Distribution :

$$\mu_2 = \mu_2' - \mu_1'^2 = n^2p^2 - np^2 + np - n^2p^2 = np(1-p) = npq$$

$$\mu_3 = \mu_3' - 3\mu_2'\,\mu_1' + 2\mu_1'^3$$

$$= \{n(n-1)\,(n-2)\,p^3 + 3n\,(n-1)\,p^2 + np\} - 3\,\{n\,(n-1)\,p^2 + np\}\,np + 2\,(np)^3$$

$$= np\,\{-\,3np^2 + 3np + 2p^2 - 3p + 1 - 3npq\}$$
$$= np\,\{3np\,(1-p) + 2p^2 - 3p + 1 - 3npq\}$$
$$= np\,(2p^2 - 3p + 1) = np\,(2p^2 - 2p + q) = npq\,(1-2p)$$
$$= npq\,\{-\,q + p - 2p\} = npq\,(q-p)$$
$$\mu_4 = \mu_4' - 4\mu_3'\,\mu_1' + 6\mu_2'\,\mu_1'^2 - 3\mu_1'^4 = npq\,\{1 + 3(n-2)\,pq\} \qquad \text{[On simplification]}$$

Hence,

$$\beta_1 = \frac{\mu_3^2}{\mu_2^3} = \frac{n^2p^2q^2(q-p)^2}{n^3p^3q^3} = \frac{(q-p)^2}{npq} = \frac{(1-2p)^2}{npq} \qquad \text{... (8·5)}$$

$$\beta_2 = \frac{\mu_4}{\mu_2^2} = \frac{npq\,[1+3(n-2)pq]}{n^2p^2q^2} = \frac{1+3(n-2)pq}{npq} = 3 + \frac{1-6pq}{npq} \qquad \text{... (8·6)}$$

$$\gamma_1 = \sqrt{\beta_1} = \frac{q-p}{\sqrt{npq}} = \frac{1-2p}{\sqrt{npq}}, \qquad \gamma_2 = \beta_2 - 3 = \frac{1-6pq}{npq} \qquad \text{... (8·6a)}$$

Remarks:

1. If $X \sim B(n, p)$, then mean $= np$ and variance $= npq$
   $\mu_3 = npq\,(q-p)$ and $\mu_4 = npq\,[1 + 3(n-2)pq]$ ... (8·6b)
2. Variance $= npq < np =$ Mean $\quad (\because 0 < q < 1)$ ... (8·6c)

Hence, *for the binomial distribution, variance is less than mean.*

Example 8·9. *Comment on the following :*

(*a*) *The mean of a binomial distribution is 3 and variance is 4.*

(*b*) *The mean and variance of binomial distribution are 4 and $\frac{4}{3}$ respectively. Find $P(X \geq 1)$.*

Solution. (*a*) If the given binomial distribution has parameters n and p, then we are given :

$$\text{Mean} = np = 3 \quad \text{... (*)} \qquad \text{and} \qquad \text{Variance} = npq = 4 \quad \text{... (**)}$$

Dividing (**) by (*), $q = \frac{4}{3}$, which is impossible, since probability cannot exceed unity.

Hence, the given statement is wrong.

Aliter. Since for a binomial distribution variance is always less than mean, the given statement is wrong.

(*b*) Let $X \sim B\,(n, p)$. Then we are given :

$$\text{Mean} = np = 4 \quad \text{... }(i) \qquad \text{and} \qquad \text{Var}\,(X) = npq = \frac{4}{3}. \qquad \text{... }(ii)$$

Dividing (*ii*) by (*i*), we get : $q = \frac{1}{3} \Rightarrow p = \frac{2}{3}$. Substituting in (*i*), we obtain $n = \frac{4}{p} = \frac{4 \times 3}{2} = 6.$

$$\therefore\ P(X \geq 1) = 1 - P(X = 0) = 1 - q^n = 1 - (1/3)^6 = 1 - (1/729) = 0{\cdot}99863.$$

Example 8·10. *If $X \sim B\,(n, p)$, show that :*

$$E\left(\frac{X}{n} - p\right)^2 = \frac{pq}{n} \qquad Cov\left(\frac{X}{n}, \frac{n-X}{n}\right) = -\,\frac{pq}{n}$$

Solution. Since $X \sim B\,(n, p)$, $\quad E(X) = np \quad$ and $\quad \text{Var}(X) = npq$

$$\therefore \quad E\left(\frac{X}{n}\right) = \frac{1}{n}E(X) = p; \qquad \text{Var}\left(\frac{X}{n}\right) = \frac{1}{n^2}\,.\,\text{Var}\,(X) = \frac{pq}{n} \qquad \text{... (*)}$$

$$(i) \quad E\left(\frac{X}{n} - p\right)^2 = E\left[\frac{X}{n} - E\left(\frac{X}{n}\right)\right]^2 = \text{Var}\left(\frac{X}{n}\right) = \frac{pq}{n} \qquad \text{[From (*)]}$$

$$(ii) \quad \text{Cov}\left(\frac{X}{n}, \frac{n-X}{n}\right) = E\left[\left\{\frac{X}{n} - E\left(\frac{X}{n}\right)\right\}\left\{\frac{n-X}{n} - E\left(\frac{n-X}{n}\right)\right\}\right]$$

$$= E\left[\left(\frac{X}{n} - p\right)\left\{\left(1 - \frac{X}{n}\right) - (1-p)\right\}\right] = E\left[\left(\frac{X}{n} - p\right)\left\{-\left(\frac{X}{n} - p\right)\right\}\right]$$

$$= -E\left(\frac{X}{n} - p\right)^2 = -\,\text{Var}\left(\frac{X}{n}\right) = -\frac{pq}{n}.$$

8·4·2. Recurrence Relation for the moments of Binomial Distribution (*Renovsky Formula*).

By *def.*,
$$\mu_r = E[X - E(X)]^r = \sum_{x=0}^{n} (x - np)^r \binom{n}{x} p^x q^{n-x}$$

Differentiating *w.r.* to p, we get

$$\frac{d\mu_r}{dp} = \sum_{x=0}^{n} \binom{n}{x}\left[-nr\,(x-np)^{r-1} p^x q^{n-x} + (x-np)^r \left\{x p^{x-1} q^{n-x} - (n-x)\,p^x q^{n-x-1}\right\}\right]$$

$$= -nr \sum_{x=0}^{n} \binom{n}{x}(x-np)^{r-1} p^x q^{n-x} + \sum_{x=0}^{n}\binom{n}{x}(x-np)^r p^x q^{n-x}\left(\frac{x}{p} - \frac{n-x}{q}\right)$$

$$= -nr \sum_{x=0}^{n} (x-np)^{r-1} p(x) + \sum_{x=0}^{n} (x-np)^r p(x)\,\frac{(x-np)}{pq}$$

$$= -nr \sum_{x=0}^{n} (x-np)^{r-1} p(x) + \frac{1}{pq}\sum_{x=0}^{n} (x-np)^{r+1} p(x)$$

$$= -\,nr\,\mu_{r-1} + \frac{1}{pq}\,\mu_{r+1}$$

$$\Rightarrow \quad \mu_{r+1} = pq\left(nr\,\mu_{r-1} + \frac{d\mu_r}{dp}\right) \qquad \textbf{(Renovsky Formula)} \quad \text{... (8·7)}$$

Putting $r = 1$, 2 and 3 successively in (8·7), we get

$$\mu_2 = pq\left(n\,\mu_0 + \frac{d\mu_1}{dp}\right) = npq \qquad (\because\ \mu_0 = 1 \text{ and } \mu_1 = 0)$$

$$\mu_3 = pq\left[2n\,\mu_1 + \frac{d\mu_2}{dp}\right] = pq\,.\,\frac{d(npq)}{dp} = npq\,\frac{d}{dp}\{p(1-p)\}$$

$$= npq\,\frac{d}{dp}(p - p^2) = npq\,(1 - 2p) = npq\,(q - p)$$

and $$\mu_4 = pq\left[3n\mu_2 + \frac{d\mu_3}{dp}\right] = pq\left[3n\,.\,npq + \frac{d}{dp}\{npq\,(q-p)\}\right]$$

$$= pq\left[3n^2\,pq + n\frac{d}{dp}\{p\,(1-p)\,(1-2p)\}\right]$$

$$= pq\left[3n^2\,pq + n\frac{d}{dp}(p - 3p^2 + 2p^3)\right]$$

$$= pq\,[3n^2\,pq + n(1 - 6p + 6p^2)] = pq\,\{3n^2\,pq + n(1 - 6pq)\}$$

$$= npq\,[3npq + 1 - 6pq] = npq\,[1 + 3pq\,(n-2)].$$

Example 8·11. *Show that the rth moment μ_r' about the origin of the binomial distribution of degree n is given by :*

$$\mu_r' = \left(p\frac{\partial}{\partial p}\right)^r (q+p)^n \qquad \text{... (*)}$$

Solution. We shall prove this result by using the principle of mathematical induction. We have

$$(q+p)^n = \sum_{x=0}^{n}\binom{n}{x}p^x\,q^{n-x} \quad\Rightarrow\quad \frac{\partial}{\partial p}(q+p)^n = \sum_{x=0}^{n}\binom{n}{x}q^{n-x}\,xp^{x-1}$$

$$\therefore \quad p\frac{\partial}{\partial p}(q+p)^n = p\sum_{x=0}^{n}\binom{n}{x}q^{n-x}\,xp^{x-1} = \sum_{x=0}^{n}\binom{n}{x}p^x\,q^{n-x}\,x = \mu_1' \qquad \text{... }(i)$$

Thus, the result (*) is true for $r = 1$.
Let us now assume that the result (*) is true for $r = k$, say, so that

$$\left(p\frac{\partial}{\partial p}\right)^k (q+p)^n = \mu_k' = \sum_{x=0}^{n}\binom{n}{x}p^x\,q^{n-x}\,x^k \qquad \text{... }(ii)$$

Differentiating (*ii*) partially *w.r.* to p and then multiplying both sides by p, we get

$$p\left(\frac{\partial}{\partial p}\right)\left[\left(p\frac{\partial}{\partial p}\right)^k (q+p)^n\right] = \sum_{x=0}^{n}\binom{n}{x}p^x\,q^{n-x}\,x^{k+1} = E\,(X^{k+1})$$

$$\Rightarrow \qquad \left(p\frac{\partial}{\partial p}\right)^{k+1} (q+p)^n = \mu'_{k+1}$$

Hence if the result (*) is true for $r = k$, it is also true for $r = k + 1$. But we have proved in (*i*) that (*) is true for $r = 1$. Hence it is true for $r = 1 + 1 = 2$; $2 + 1 = 3$; and so on. Hence by the principle of mathematical induction, (*) is true for all positive integral values of r.

8·4·3. Factorial Moments of Binomial Distribution.

The rth factorial moment of the Binomial distribution is :

$$\mu_{(r)}' = E\{(X)^{(r)}\} = \sum_{x=0}^{n} x^{(r)}\,p(x) = \sum_{x=0}^{n} x^{(r)}\frac{n!}{x!\,(n-x)!}p^x\,q^{n-x}$$

$$= n^{(r)}\,p^r\sum_{x=r}^{n}\frac{(n-r)!}{(x-r)!\,(n-x)!}p^{x-r}\,q^{n-x} = n^{(r)}\,p^r\,(q+p)^{n-r} = n^{(r)}\,p^r \qquad \text{... (8·8)}$$

$$\therefore \quad \mu_{(1)}' = E\{(X)^{(1)}\} = np = \text{Mean}, \quad \mu_{(2)}' = E\{(X)^{(2)}\} = n^{(2)} p^2 = n(n-1)p^2$$

$$\mu_{(3)}' = E[(X)^{(3)}] = n^{(3)} p^3 = n(n-1)(n-2)p^3$$

$$\therefore \quad \mu_{(2)} = \mu_{(2)}' - \mu_{(1)}'^2 + \mu_{(1)}' = n^2p^2 - np^2 - n^2p^2 + np = npq$$

$$\mu_{(3)} = \mu_{(3)}' - 3\mu_{(2)}'\mu_{(1)}' + 2\mu_{(1)}'^3 - 2\mu_{(1)}'$$

$$= n(n-1)(n-2)p^3 - 3n(n-1)p^2 \cdot np + 2n^3p^3 - 2np$$

$$= -2npq(1+p) \qquad \text{[On simplification]}$$

8·4·3(1). Factorial Moment Generating Function.

A more elegant way of finding the factorial moments is through the use of Factorial Moment Generating Function, which is given below :

$$\omega_{(t)} = \text{Factorial m.g.f.} = E[(1+t)^X] = \sum_{x=0}^{n} (1+t)^x p(x)$$

$$= \sum_{x=0}^{n} \left[{}^nC_x \{p(1+t)\}^x q^{n-x} \right]$$

$$= [q + p(1+t)]^n = (q + p + pt)^n = (1 + pt)^n$$

$$= 1 + {}^nC_1(pt) + {}^nC_2(pt)^2 + \dots + {}^nC_r(pt)^r + \dots \qquad \dots (8·8a)$$

$$\mu'_{(r)} = r\text{th factorial moment} = \text{Coefficient of } \frac{t^r}{r!} \text{ in } (8·8a)$$

$$\Rightarrow \quad \mu'_{(r)} = {}^nC_r \cdot p^r \cdot r! = \frac{n!}{(n-r)!} p^r; \; r = 1, 2, 3, \dots$$

$$\therefore \quad \mu'_{(1)} = \frac{n!\,p}{(n-1)!} = np = \text{Mean} \quad ; \quad \mu'_{(2)} = \frac{n!\,p^2}{(n-2)!} = n(n-1)p^2,$$

$$\mu'_{(3)} = n(n-1)(n-2)p^3 \quad \text{and} \quad \mu'_{(4)} = n(n-1)(n-2)(n-3)p^4$$

8·4·4. Mean Deviation about Mean of Binomial Distribution.

The mean deviation η about the mean np of the binomial distribution is given by :

$$\eta = \sum_{x=0}^{n} |x - np|\, p(x) = \sum_{x=0}^{n} |x - np| \binom{n}{x} p^x q^{n-x}, \quad (x \text{ being an integer})$$

$$= \sum_{x=0}^{np} -(x-np)\binom{n}{x} p^x q^{n-x} + \sum_{x=np}^{n} (x-np)\binom{n}{x} p^x q^{n-x}$$

$$= 2\sum_{x=np}^{n} (x-np)\binom{n}{x} p^x q^{n-x}$$

$$\left[\because \sum_{x=0}^{n} x\binom{n}{x} p^x q^{n-x} = np \;\Rightarrow\; \sum_{x=0}^{n} (x-np)\binom{n}{x} p^x q^{n-x} = 0 \right]$$

$$= 2\sum_{\mu}^{n} (x-np)\binom{n}{x} p^x q^{n-x}, \qquad \text{where } \mu \text{ is greatest integer contained in } (np+1).$$

$$= 2\sum_{\mu}^{n}\left[\{xq-(n-x)p\}\binom{n}{x}p^x\, q^{n-x}\right]$$

$$= 2\sum_{\mu}^{n}\left[\frac{n!}{(x-1)!\,(n-x)!}p^x\, q^{n-x+1} - \frac{n!}{x!\,(n-x-1)!}p^{x+1}\, q^{n-x}\right]$$

$$= 2\sum_{x=\mu}^{n}(t_{x-1}-t_x), \qquad \text{where} \quad t_x = \frac{n!}{x!\,(n-x-1)!}p^{x+1}\, q^{n-x}$$

$$= 2\,(t_{\mu-1}-t_n) = 2\,t_{\mu-1}$$

This is obtained by summing over x and using $t_n = 0$.

$$\therefore \quad \eta = 2\,t_{\mu-1} = 2\frac{n!}{(\mu-1)!\,(n-\mu)!}\cdot p^{\mu}\, q^{n-\mu+1} = 2npq\binom{n-1}{\mu-1}p^{\mu-1}\, q^{n-\mu}. \qquad \text{... (8·9)}$$

8·4·5. Mode of Binomial Distribution.

We have

$$\frac{p(x)}{p(x-1)} = \binom{n}{x}p^x\, q^{n-x}\Big/\binom{n}{x-1}p^{x-1}\, q^{n-x+1}$$

$$= \frac{n!}{(n-x)!\;x!}p^x\, q^{n-x}\Big/\frac{n!}{(x-1)!\;(n-x+1)!}p^{x-1}\, q^{n-x+1}$$

$$= \frac{(n-x+1)p}{xq} = \frac{xq+(n-x+1)p-xq}{xq} = 1+\frac{(n+1)p-x(p+q)}{xq}$$

$$= 1+\frac{(n+1)p-x}{xq} \qquad \text{... (8·10)}$$

Mode is the value of x for which $p(x)$ is maximum.

We discuss the following two cases :

Case I. *When $(n+1)p$ is not an integer.* Let $(n+1)p = m+f$, where m is an integer and f is fractional such that $0 < f < 1$. Substituting in (8.10), we get

$$\frac{p(x)}{p(x-1)} = 1+\frac{(m+f)-x}{xq} \qquad \text{... (*)}$$

From (*), it is obvious that

$$\frac{p(x)}{p(x-1)} > 1 \text{ for } x = 0, 1, 2, \dots m \quad \text{and} \quad \frac{p(x)}{p(x-1)} < 1 \text{ for } x = m+1, m+2, \dots, n$$

$$\Rightarrow \quad \frac{p(1)}{p(0)}>1,\ \frac{p(2)}{p(1)}>1, \dots,\ \frac{p(m)}{p(m-1)}>1, \text{ and } \frac{p(m+1)}{p(m)}<1,\ \frac{p(m+2)}{p(m+1)}<1, \dots,\ \frac{p(n)}{p(n-1)}<1$$

$\therefore \quad p(0) < p(1) < p(2) < \dots < p(m-1) < p(m) > p(m+1) > p(m+2) > \dots > p(n).$

$\Rightarrow \quad p(x)$ is maximum at $x = m$.

Thus, in this case there exists unique modal value for binomial distribution and it is m, the integral part of $(n+1)p$.

Case II. *When* $(n + 1)p$ *is an integer.* Let $(n + 1)p = m$ *(an integer).*

Substituting in (8.10), we get $$\frac{p(x)}{p(x-1)} = 1 + \frac{m-x}{xq} \quad \text{... (**)}$$

From (**), it is obvious that : $$\frac{p(x)}{p(x-1)} \begin{cases} > 1 \text{ for } x = 1, 2, ..., m-1 \\ = 1 \text{ for } x = m \\ < 1 \text{ for } x = m+1, m+2, ..., n \end{cases}$$

Now proceeding as in case 1, we have

$$p(0) < p(1) < ... < p(m-1) = p(m) > p(m+1) > p(m+2) > ... > p(n)$$

Thus, in this case the binomial distribution is *bimodal* and two modal values are m and $m - 1$.

Example 8·12. *Determine the binomial distribution for which the mean is 4 and variance 3 and find its mode.*

Solution. Let $X \sim B(n, p)$, then we are given that

$$E(X) = np = 4 \quad \text{... (*)} \quad \text{and} \quad \text{Var}(X) = npq = 3 \quad \text{... (**)}$$

Dividing (**) by (*), we get $q = \frac{3}{4} \Rightarrow p = 1 - q = \frac{1}{4}$

Hence from (*), we obtain $n = \frac{4}{p} = 16$

Thus the given binomial distribution has parameters $n = 16$ and $p = \frac{1}{4}$.

Mode. We have $(n + 1)p = 4{\cdot}25$, which is not an integer. Hence the unique mode of the binomial distribution is 4, the integral part of $(n + 1)p$.

Example 8·13. *Show that for* $p = 0{\cdot}5$*, the binomial distribution has maximum probability at* $X = \frac{1}{2}n$*, if n is even, and at* $X = \frac{1}{2}(n-1)$ *as well as* $X = \frac{1}{2}(n+1)$*, if n is odd.*

Solution. Here we have to find the mode of the binomial distribution.

(*i*) Let n be even $= 2m$, (say), $m = 1, 2, ...$

$\therefore$ If $p = 0.5$, then $(n + 1)p = (2m+1) \times \frac{1}{2} = m + 0.5$. Hence in this case, the distribution is unimodal, the unique mode being at $X = m = \frac{n}{2}$.

(*ii*) Let n be odd $= (2m + 1)$ say. Then

$$(n+1)p = (2m+2) \times \frac{1}{2} = m + 1 \text{ (Integer)} = \frac{n-1}{2} + 1 = \frac{n+1}{2}$$

Since $(n + 1)p$ is an integer, the distribution is bimodal, the two modes being

$$\frac{1}{2}(n+1) \text{ and } \frac{1}{2}(n+1) - 1 = \frac{1}{2}(n-1).$$

8·4·6. Moment Generating Function of Binomial Distribution.

Let $X \sim B(n, p)$, then

$$M_X(t) = E(e^{tX}) = \sum_{x=0}^{n} e^{tx} \binom{n}{x} p^x q^{n-x} = \sum_{x=0}^{n} \binom{n}{x} (pe^t)^x q^{n-x} = (q + pe^t)^n \quad \text{... (8·11)}$$

***m.g.f.* about Mean of Binomial Distribution :**

$$E\{e^{t(X-np)}\} = e^{-tnp} \cdot E(e^{tX}) = e^{-tnp} \cdot M_X(t) = e^{-tnp}(q+pe^t)^n = (qe^{-pt}+pe^{tq})^n$$

$$= \left[q\left\{1 - pt + \frac{p^2t^2}{2!} - \frac{p^3t^3}{3!} + \frac{p^4t^4}{4!} - \ldots\right\} + p\left\{1 + tq + \frac{t^2q^2}{2!} + \frac{t^3q^3}{3!} - \ldots\right\}\right]^n$$

$$= \left[(q+p) + \frac{t^2}{2!}pq(q+p) + \frac{t^3}{3!}pq(q^2-p^2) + \frac{t^4}{4!}pq(q^3+p^3) + \ldots\right]^n$$

$$= \left[1 + \left\{\frac{t^2}{2!}\cdot pq + \frac{t^3}{3!}\cdot pq(q-p) + \frac{t^4}{4!}qp(1-3pq) + \ldots\right\}\right]^n$$

$$= \left[1 + \binom{n}{1}\left\{\frac{t^2}{2!}\cdot pq + \frac{t^3}{3!}\cdot pq(q-p) + \frac{t^4}{4!}qp(1-3pq) + \ldots\right\}\right.$$

$$\left. + \binom{n}{2}\left\{\frac{t^2}{2!}pq + \frac{t^3}{3!}pq(q-p) + \ldots\right\}^2 + \ldots\right]$$

$\therefore \quad \mu_2 = \text{Coefficient of } \dfrac{t^2}{2!} = npq, \qquad \mu_3 = \text{Coefficient of } \dfrac{t^3}{3!} = npq(q-p)$

$\mu_4 = \text{Coefficient of } \dfrac{t^4}{4!} = npq(1-3pq) + 3n(n-1)p^2q^2$

$= 3n^2p^2q^2 + npq(1-6pq)$.

Example 8·14. *X is binomially distributed with parameters n & p. What is distribution of Y = n – X?*

Solution. $X \sim B(n, p)$ represents the number of successes in n independent trails with constant probability p of success for each trail.

$\therefore Y = n - X$, represents the number of failures in n independent trails with constant probability 'q' of failure of each trial. Hence, $Y = (n-X) \sim B(n, q)$

Aliter. Since $X \sim B(n, p)$, $\qquad M_X(t) = E(e^{tX}) = (q+pe^t)^n$

$$\therefore \quad M_Y(t) = E(e^{tY}) = E[e^{t(n-X)}] = e^{nt} \cdot E(e^{-tX}) = e^{nt} M_X(-t)$$

$$= e^{nt}\cdot(q+pe^{-t})^n = [e^t(q+pe^{-t})]^n = (p+qe^t)^n$$

Hence, by uniqueness theorem of *m.g.f.*, $\qquad Y = (n-X) \sim B(n, q)$.

Example 8·15. *The m.g.f. of a r.v. X is* $\left(\dfrac{2}{3}+\dfrac{1}{3}e^t\right)^9$.

Show that : $\qquad P(\mu - 2\sigma < X < \mu + 2\sigma) = \sum_{x=1}^{5}\binom{9}{x}\left(\frac{1}{3}\right)^x\left(\frac{2}{3}\right)^{9-x}$.

Solution. Since $M_X(t) = \left(\dfrac{2}{3}+\dfrac{1}{3}e^t\right)^9 = (q+pe^t)^n$, by uniqueness theorem of *m.g.f.* —

$X \sim B\left(n = 9, p = \frac{1}{3}\right)$. Hence, $E(X) = \mu_X = np = 3$; $\sigma_X^2 = npq = 9 \times \frac{1}{3} \times \frac{2}{3} = 2.$

$$\mu \pm 2\sigma = 3 \pm 2 \times \sqrt{2} = 3 \pm 2 \times 1{\cdot}4 = (0{\cdot}2, 5{\cdot}8)$$

$$\therefore \quad P(\mu - 2\sigma < X < \mu + 2\sigma) = P(0{\cdot}2 < X < 5{\cdot}8) = P(1 \le X \le 5)$$

$$= \sum_{x=1}^{5} p(x) = \sum_{x=1}^{5} \binom{n}{x} p^x q^{n-x} = \sum_{x=1}^{5} \binom{9}{x} \left(\frac{1}{3}\right)^x \left(\frac{2}{3}\right)^{9-x}$$

8·4·7. Additive Property of Binomial Distribution.

Let $X \sim B(n_1, p_1)$ and $Y \sim B(n_2, p_2)$ be independent random variables. Then

$$M_X(t) = (q_1 + p_1 e^t)^{n_1}, \qquad M_Y(t) = (q_2 + p_2 e^t)^{n_2} \qquad \text{... (*)}$$

What is the distribution of $X + Y$?

We have $M_{X+Y}(t) = M_X(t) \,.\, M_Y(t)$ [$\because$ X and Y are independent]

$$= (q_1 + p_1 e^t)^{n_1} \,.\, (q_2 + p_2 e^t)^{n_2} \qquad \text{... (**)}$$

Since (**) cannot be expressed in the form $(q + p\, e^t)^n$, from uniqueness theorem of *m.g.f's* it follows that $X + Y$ is not a binomial variate. Hence, *in general the sum of two independent binomial variates is not a binomial variate. In other words, binomial distribution does not possess the additive or reproductive property.*

However, if we take $p_1 = p_2 = p$, (say), then from (**) $M_{X+Y}(t) = (q + pe^t)^{n_1 + n_2}$, which is the *m.g.f.* of a binomial variate with parameters $(n_1 + n_2, p)$. Hence by uniqueness theorem of *m.g.f.'s*

$$X + Y \sim B(n_1 + n_2, p).$$

Thus, the binomial distribution possesses the additive or reproductive property if $p_1 = p_2$.

Generalisation. *If* X_i, *(i = 1, 2, ..., k) are independent binomial variates with parameters* (n_i, p), *(i = 1, 2, ..., k), then their sum* $\sum_{i=1}^{k} X_i \sim B\left(\sum_{i=1}^{k} n_i, p\right)$.

The proof is left as an exercise to the reader.

Example 8·16. *If the independent random variables X, Y are binomially distributed, respectively with parameters (n = 3, p = 1/3), and (n = 5, p = 1/3), write down the probability that* $X + Y \ge 1$.

Solution. We are given : $X \sim B\left(3, \frac{1}{3}\right)$ and $Y \sim B\left(5, \frac{1}{3}\right)$.

Since X and Y are independent binomial random variables, with $p_1 = p_2 = \frac{1}{3}$, by the additive property of binomial distribution, we get

$$X + Y \sim B\left(3 + 5, \frac{1}{3}\right), \quad i.e., \quad X + Y \sim B\left(8, \frac{1}{3}\right)$$

$$\therefore \quad P(X + Y = r) = \binom{8}{r} \left(\frac{1}{3}\right)^r \left(\frac{2}{3}\right)^{8-r} ; r = 0, 1, 2, ..., 8 \qquad \text{... (*)}$$

Hence, $P(X + Y \ge 1) = 1 - P(X + Y < 1) = 1 - P(X + Y = 0) = 1 - \left(\frac{2}{3}\right)^8$. [From (*)]

8·4·8. Characteristic Function of Binomial Distribution.

$$\varphi_X(t) = E(e^{itX}) = \sum_{x=0}^{n} e^{itx}\, p(x) = \sum_{x=0}^{n} e^{itx} \binom{n}{x} p^x q^{n-x} = \sum_{x=0}^{n} \binom{n}{x} (pe^{it})^x q^{n-x}$$

$$= (q + pe^{it})^n \qquad \text{... (8·13)}$$

8·4·9. Cumulants of Binomial Distribution.

Cumulant generating function of binomial distribution is given by :

$$K_X(t) = \log M_X(t) = \log (q + pe^t)^n = n \log (q + p.e^t)$$

$$= n \log \left[q + p\left(1 + t + \frac{t^2}{2!} + \frac{t^3}{3!} + \frac{t^4}{4!} + \dots\right)\right]$$

$$= n \log \left[1 + p\left(t + \frac{t^2}{2!} + \frac{t^3}{3!} + \frac{t^4}{4!} + \dots\right)\right]$$

$$= n \left[p\left(t + \frac{t^2}{2!} + \frac{t^3}{3!} + \frac{t^4}{4!} + \dots\right) - \frac{p^2}{2}\left(t + \frac{t^2}{2!} + \frac{t^3}{3!} + \dots\right)^2 \right.$$

$$\left. + \frac{p^3}{3}\left(t + \frac{t^2}{2!} + \frac{t^3}{3!} + \dots\right)^3 - \frac{p^4}{4}\left(t + \frac{t^2}{2!} + \frac{t^3}{3!} + \dots\right)^4 + \dots \right]$$

Mean $= \kappa_1 =$ Coefficient of t in $K_X(t) = np$

$$\mu_2 = \kappa_2 = \text{Coefficient of } \frac{t^2}{2!} \text{ in } K_X(t) = n(p - p^2) = np(1 - p) = npq$$

The coefficient of t^3 in $K_X(t) = n\left[\frac{p}{3!} - \frac{p^2}{2!} \times 2 \times \frac{1}{2!} + \frac{p^3}{3}\right] = \frac{np}{3!}(1 - 3p + 2p^2)$

$$\therefore \quad \kappa_3 = \text{Coefficient of } \frac{t^3}{3!} \text{ in } K_X(t) = np\,(1 - 3p + 2p^2)$$

$$= np\,(1 - p)\,(1 - 2p) = npq\,(1 - p - p) = npq\,(q - p)$$

$$\Rightarrow \quad \mu_3 = \kappa_3 = npq\,(q - p)$$

The coefficient of t^4 in $K_X(t) = n\left[\frac{p}{4!} - \frac{p^2}{2!}\left(\frac{2}{3!} + \frac{1}{4}\right) + \frac{p^3}{3} \cdot \frac{3}{2!} - \frac{p^4}{4}\right] = \frac{np}{4!}(1 - 7p + 12p^2 - 6p^3)$

$$\therefore \quad \kappa_4 = \text{Coefficient of } \frac{t^4}{4!} \text{ in } K_X(t) = np\,(1 - p)\,(1 - 6p + 6p^2)$$

$$= npq\,[1 - 6p\,(1 - p)] = npq\,(1 - 6pq)$$

$$\therefore \quad \mu_4 = \kappa_4 + 3\kappa_2^{\,2} = npq\,(1 - 6pq) + 3n^2 p^2 q^2 = npq\,(1 - 6pq + 3npq)$$

$$= npq\,[1 + 3pq\,(n - 2)].$$

8·4·10. Recurrence Relation for Cumulants of Binomial Distribution.

By *def.*,

$$\kappa_r = \left[\frac{d^r}{dt^r}\log M_X(t)\right]_{t=0} = n\left[\frac{d^r}{dt^r}\log(q+pe^t)\right]_{t=0}$$

$$\therefore \quad \frac{d\kappa_r}{dp} = n\left[\frac{d^r}{dt^r}\cdot\frac{d}{dp}\log(q+pe^t)\right]_{t=0} = n\left[\frac{d^r}{dt^r}\cdot\frac{(-1+e^t)}{q+pe^t}\right]_{t=0}$$

$$\kappa_{r+1} = n\left[\frac{d^{r+1}}{dt^{r+1}}\log(q+pe^t)\right]_{t=0} = n\left[\frac{d^r}{dt^r}\cdot\frac{d}{dt}\log(q+pe^t)\right]_{t=0} = n\left[\frac{d^r}{dt^r}\left(\frac{pe^t}{q+pe^t}\right)\right]_{t=0}$$

$$= n\left[\frac{d^r}{dt^r}\left(1-\frac{q}{q+pe^t}\right)\right]_{t=0} = -nq\left[\frac{d^r}{dt^r}\left(\frac{1}{q+pe^t}\right)\right]_{t=0}$$

Hence,

$$\kappa_{r+1} - pq\frac{d\kappa_r}{dp} = -nq\left[\frac{d^r}{dt^r}\left(\frac{1}{q+pe^t}\right)\right]_{t=0} - npq\left[\frac{d^r}{dt^r}\left(\frac{e^t-1}{q+pe^t}\right)\right]_{t=0} = -nq\left[\frac{d^r}{dt^r}\left(\frac{1+pe^t-p}{q+pe^t}\right)\right]_{t=0}$$

$$= -nq\left[\frac{d^r}{dt^r}\left\{\frac{q+pe^t}{q+pe^t}\right\}\right]_{t=0} = -nq\left[\frac{d^r}{dt^r}(1)\right]_{t=0} = 0$$

$$\therefore \quad \kappa_{r+1} = pq\frac{d\kappa_r}{dp} \quad \text{... (8·14)}$$

In particular,

$$\kappa_2 = pq\cdot\frac{d\kappa_1}{dp} = pq\cdot\frac{d}{dp}(np) = npq \qquad (\because\ \kappa_1 = \text{mean} = np)$$

$$\kappa_3 = pq\cdot\frac{d\kappa_2}{dp} = pq\cdot\frac{d\,(npq)}{dp} = npq\,(q-p)$$

$$\kappa_4 = pq\cdot\frac{d\kappa_3}{dp} = pq\cdot\frac{d}{dp}\{npq\,(q-p)\} = npq\,\frac{d}{dp}\{p\,(1-p)\,(1-2p)\}$$

$$= npq\cdot\frac{d}{dp}(p-3p^2+2p^3) = npq\,(1-6p+6p^2) = npq\,[1-6p\,(1-p)]$$

$$= npq\,(1-6pq).$$

8·4·11. Probability Generating Function of Binomial Distribution.

$$P(s) = \sum_{k=0}^{n} P(X=k)\,s^k = \sum_{k=0}^{n}\binom{n}{k}(ps)^k q^{n-k} = (ps+q)^n \quad \text{... (8·14a)}$$

The fact that this generating function is nth power of $(q + ps)$ shows that $p(x) = \{b\,(x\,;\,n,\,p\}$ is the distribution of the sum $S_n = X_1 + X_2 + \ldots + X_n$, of n independent random variables with the common generating function $(q + ps)$. Each variable X, assumes the value 0 with probability q and 1 with probability p.

Thus $$\{b\,(k\,;\,n,\,p)\} = \{b\,(k\,;\,1,\,p)\}^n \quad \text{... (8·14b)}$$

Let X and Y be two independent *r.v.*'s having $b\,(k\,;\,m,\,p\}$ and $b\,(k\,;\,n,\,p)$ as their distributions, then

$$P_X(s) = (q + ps)^m \quad \text{and} \quad P_Y(s) = (q + ps)^n$$

$\therefore$ $$P_{X+Y}(s) = P_X(s) \cdot P_Y(s) = (q + ps)^m (q + ps)^n = (q + ps)^{m+n}$$

$\Rightarrow$ $$b(k ; m, p) * b(k ; n, p) = \{b(k ; m + n, p)\} \quad \text{... (8·14c)}$$

Also $$\mu_{(1)}' = \left[n (q + ps)^{n-1} p\right]_{s=1} = np$$

$$\mu_{(2)}' = \left[n (n-1) (q + ps)^{n-2} p^2\right]_{s=1} = n (n-1) p^2, \text{ and so on.}$$

$$\mu_{(r)}' = \left[n (n-1) (n-2) \ldots (n-r+1) (q + ps)^{n-r} p^r\right]_{s=1} = n (n-1) (n-2) \ldots (n-r+1) p^r.$$

Example 8·17. *Show that :* $E\left(\dfrac{1}{X+a}\right) = \displaystyle\int_0^1 t^{a-1} G(t)\, dt, a > 0$ ... (*)

where G(t) is the probability generating function of X. Find it when $X \sim B(n, p)$*, and* $a = 1$.

Solution. R.H.S. $= \displaystyle\int_0^1 t^{a-1} \cdot G(t)\, dt = \int_0^1 t^{a-1} E(t^X)\, dt$

$$= \int_0^1 \left\{t^{a-1} \left(\sum_x p(x) t^x\right)\right\} dt = \sum_x \left[p(x) \int_0^1 t^{x+a-1}\, dt\right] = \sum_x p(x) \cdot \frac{1}{(x+a)} = E\left(\frac{1}{X+a}\right) \quad \text{... (**)}$$

If $X \sim B(n, p)$, then $$G(t) = \sum_{x=0}^{n} t^x p(x) = (q + pt)^n$$

Hence, taking $a = 1$ in (*) and using (**), we get

$$E\left[\frac{1}{(X+1)}\right] = \int_0^1 (q + pt)^n\, dt = \left|\frac{(q+pt)^{n+1}}{(n+1)p}\right|_0^1 = \frac{1 - q^{n+1}}{(n+1)p}.$$

8·4·12. Recurrence Relation for the Probabilities of Binomial Distribution.

(Fitting of Binomial Distribution).

We have $$\frac{p(x+1)}{p(x)} = \frac{\dbinom{n}{x+1} p^{x+1} q^{n-x-1}}{\dbinom{n}{x} p^x q^{n-x}} = \frac{n-x}{x+1} \cdot \frac{p}{q} \quad \text{(On simplification)}$$

$\therefore$ $$p(x + 1) = \left\{\frac{n-x}{x+1} \cdot \frac{p}{q}\right\} p(x), \quad \text{... (8·15)}$$

which is the required recurrence formula for the probabilities of binomial distribution.

This formula provides us a very convenient method of graduating the given data by a binomial distribution. The only probability we need to calculate is $p(0)$ which is given by :

$p(0) = q^n$, where q is estimated from the given data by equating the mean $\bar{x}$ of the distribution to np, the mean of the binomial distribution. Thus $\hat{p} = \bar{x}/n$.

The remaining probabilities, *viz.*, $p(1)$, $p(2)$, ... can now be easily obtained from (8·15) as explained below :

$$p(1) = [p(x+1)]_{x=0} = \left(\frac{n-x}{x+1} \cdot \frac{p}{q}\right)_{x=0} \times p(0), \quad p(2) = [p(x+1)]_{x=1} = \left(\frac{n-x}{x+1} \cdot \frac{p}{q}\right)_{x=1} \times p(1)$$

$$p(3) = [p(x+1)]_{x=2} = \left(\frac{n-x}{x+1} \cdot \frac{p}{q}\right)_{x=2} \times p(2), \text{ and so on.} \quad \text{... (8·15a)}$$

Example 8·18. *The following data due to Weldon shows the results of throwing 12 fair dice 4,096 times; a throw of 4, 5 or 6 being called success.*

Success	*Frequency*	*Success*	*Frequency*
0	—	7	847
1	7	8	536
2	60	9	257
3	198	10	71
4	430	11	11
5	731	12	—
6	948		

Fit a binomial distribution and find the expected frequencies.

Solution. In the usual notations, we are given : $n = 12$, $N = 4,096$

$$p = \text{Probability of success } i.e., \text{ throw of 4, 5 or 6} = \frac{1}{2} \quad \Rightarrow \quad q = 1 - p = \frac{1}{2}$$

Thus, by the Binomial probability model, the probability of x successes in a throw of 12 dice is given by :

$$P(X = x) = p(x) = \binom{12}{x}\left(\frac{1}{2}\right)^x\left(\frac{1}{2}\right)^{12-x} = \binom{12}{x}\left(\frac{1}{2}\right)^{12} = \frac{1}{4096}\cdot\binom{12}{x}$$

Hence, in 4,096 throws of 12 dice, the frequencies of x successes are given by :

$$f(x) = N \,.\, p(x) = 4096 \cdot \frac{1}{4096}\binom{12}{x} = \binom{12}{x}.$$

Thus, the expected frequencies are given in following Table :

COMPUTATION OF EXPECTED FREQUENCIES

Success (x)	*Expected Frequency* $f(x) = N\,p(x)$	*Success* (x)	*Expected Frequency* $f(x) = N\,p(x)$
0	$\binom{12}{0} = 1$	7	$\binom{12}{7} = 792$
1	$\binom{12}{1} = 12$	8	$\binom{12}{8} = 405$
2	$\binom{12}{2} = 66$	9	$\binom{12}{9} = 220$
3	$\binom{12}{3} = 220$	10	$\binom{12}{10} = 66$
4	$\binom{12}{4} = 495$	11	$\binom{12}{11} = 12$
5	$\binom{12}{5} = 792$	12	$\binom{12}{12} = 1$
6	$\binom{12}{6} = 924$	Total	4,096

Example 8·19. *Seven coins are tossed and number of heads noted. The experiment is repeated 128 times and the following distribution is obtained :*

No. of heads	0	1	2	3	4	5	6	7	*Total*
Frequencies	7	6	19	35	30	23	7	1	128

Fit a binomial distribution assuming, (i) The coin is unbaised. (ii) The nature of the coin is not known. (iii) Probability of a head for four coins is 0·5 and for the remaining three coins is 0·45.

Solution. In fitting binomial distribution, first of all the mean and variance of the data are equated to np and npq respectively. Then the expected frequencies are calculated from these values of n and p. Here $n = 7$ and $N = 128$.

Case I. *When the coin is unbiased :* $p = q = \frac{1}{2}, \Rightarrow \frac{p}{q} = 1$

$$p(0) = q^n = \left(\frac{1}{2}\right)^7 = \frac{1}{128} \quad \text{so that} \quad f(0) = Nq^n = 128\left(\frac{1}{2}\right)^7 = 1$$

Using the recurrence formula (8·15), the various probabilities, *viz.*, $p(1)$, $p(2)$, ... can be easily calculated as shown below :

COMPUTATION OF EXPECTED BINOMIAL FREQUENCIES

x	f	fx	$\frac{n-x}{x+1}$	$\frac{n-x}{x+1}\cdot\frac{p}{q}$	*Expected frequency* $f(x) = N\,p(x)$
0	7	0	7	7	$f(0) = Np\,(0) = 1$
1	6	6	3	3	$f(1) = 1 \times 7 = 7$
2	19	38	$\frac{5}{3}$	$\frac{5}{3}$	$f(2) = 7 \times 3 = 21$
3	35	105	1	1	$f(3) = 21 \times \frac{5}{3} = 35$
4	30	120	$\frac{3}{5}$	$\frac{3}{5}$	$f(4) = 35 \times 1 = 35$
5	23	115	$\frac{1}{3}$	$\frac{1}{3}$	$f(5) = 35 \times \frac{3}{5} = 21$
6	7	42	$\frac{1}{7}$	$\frac{1}{7}$	$f(6) = 21 \times \frac{1}{3} = 7$
7	1	7			$f(7) = 7 \times \frac{1}{7} = 1$
Total	128	433			

Case II. *When the nature of the coin is not known, then*

$$\text{Mean} = np = \bar{x} = \frac{1}{N}\sum_{i=1}^{n} f_i x_i = \frac{433}{128} = 3{\cdot}3828\,; \qquad n = 7$$

$$\therefore \qquad p = \frac{3{\cdot}3828}{7} = 0{\cdot}48326 \quad \text{and} \quad q = 1 - p = 0{\cdot}51674 \quad \Rightarrow \quad \frac{p}{q} = 0{\cdot}93521$$

$$f(0) = Nq^7 = 128\,(0{\cdot}5167)^7 = 1{\cdot}2593 \text{ (using logarithms)}$$

COMPUTATION OF EXPECTED BINOMIAL FREQUENCIES

x	$\frac{n-x}{x+1}$	$\frac{n-x}{x+1}\cdot\frac{p}{q}$	Expected frequency : $f(x) = N\,p(x)$
0	7	6·54647	$f(0) = N\,p(0) = 1·2593 \simeq 1$
1	3	2·80563	$f(1) = 1·2593 \times 6·54647 = 8·2438 \simeq 8$
2	$\frac{5}{3}$	1·55868	$f(2) = 2·80563 \times 8·2438 = 23·129 \simeq 23$
3	1	0·93521	$f(3) = 1·55868 \times 23·129 = 36·05 \simeq 34$
4	$\frac{3}{5}$	0·56113	$f(4) = 0·93521 \times 36·05 = 33·715 \simeq 34$
5	$\frac{1}{3}$	0·31174	$f(5) = 0·56113 \times 33·715 = 18·918 \simeq 19$
6	$\frac{1}{7}$	0·13360	$f(6) = 0·31174 \times 18·918 = 5·897 \simeq 6$
7			$f(7) = 0·13360 \times 5·897 = 0·788 \simeq 1$

Case III. The probability generating functions (*p.g.f*), say $P_X(s)$ for the 4 coins and $P_Y(s)$ for the remaining 3 coins are given by :

$$P_X(s) = (0·50 + 0·50s)^4,\ P_Y(s) = (0·55 + 0·45s)^3 \qquad [c.f.\ 8·14\ (a)]$$

Since all the throws are independent, the *p.g.f.* $P_{X+Y}(s)$ for the whole experiment is given by :

$$P_{X+Y}(s) = P_X(s)\,P_Y(s) \qquad [c.f.\ 8·14\ (b)]$$

$$= (0·50 + 0·50\,s)^4 \times (0·55 + 0·45\,s)^3$$

$$= \left(0·0625 + 0·25s + 0·375s^2 + 0·25s^3 + 0·0625s^4\right) \times \left(0·166375 + 0·408375s + 0·334125s^2 + 0·091125s^3\right)$$

$$f(x) = N \times \text{Coefficient of } s^x \text{ in } P_{X+Y}(s)$$

$\therefore$ $f(0) = 128 \times ·0625 \times ·16637 = 1·13310$

$f(1) = 128\,(·25 + ·166375 + ·408375 \times ·0625) = 8·5910$

$f(2) = 128\,(·28396) = 36·3470$ $\qquad f(5) = 128\,(·14602) = 18·6934$

$f(3) = 128\,(·184117) = 23·5669$ $\qquad f(6) = 128\,(·04366) = 5·5889$

$f(4) = 128\,(·26057) = 33·3529$ $\qquad f(7) = 128\,(·005695) = 0·72896$

These frequencies, rounded to the nearest integer, keeping in mind that total frequency is $N = 128$, are given below :

x	0	1	2	3	4	5	6	7	Total
f	1	9	36	23	33	19	6	1	N = 128

Example 8·20. *Find the m.g.f. of standard binomial variate $(X - np)/\sqrt{npq}$ and obtain its limiting form as $n \to \infty$. Also interpret the result.*

Solution. We know that if $X \sim B\,(n, p)$ then $M_X(t) = (q + pe^t)^n$.

The *m.g.f.* of standard binomial variate : $\qquad Z = \dfrac{X - np}{\sqrt{npq}} = \dfrac{X - \mu}{\sigma}$, (say),

where $\qquad \mu = np$ and $\sigma^2 = npq$, is given by :

$$M_Z(t) = e^{-\mu t/\sigma}\, M_X(t/\sigma)$$

$$= e^{-npt/\sqrt{npq}} \cdot \left(q + p\, e^{t/\sqrt{npq}}\right)^n = \left[q e^{-pt/\sqrt{npq}} + p e^{qt/\sqrt{npq}}\right]^n$$

$$= \left[q\left\{1 - \frac{pt}{\sqrt{npq}} + \frac{p^2 t^2}{2npq} + O'(n^{-3/2})\right\} + p\left\{1 + \frac{qt}{\sqrt{npq}} + \frac{q^2 t^2}{2npq} + O''(n^{-3/2})\right\}\right]^n$$

where $O'(n^{-3/2})$ and $O''(n^{-3/2})$ involve terms containing $n^{3/2}$ and higher powers of n in the denominator. Thus

$$M_Z(t) = \left[(q+p) + \frac{t^2\, pq}{2npq}(p+q) + O(n^{-3/2})\right]^n = \left[1 + \frac{t^2}{2n} + O(n^{-3/2})\right]^n,$$

where $O(n^{-3/2})$ involves terms with $n^{3/2}$ and higher powers of n in the denominator.

$$\therefore \qquad \log M_Z(t) = n \log\left[1 + \frac{t^2}{2n} + O(n^{-3/2})\right]$$

$$= n\left[\left\{\frac{t^2}{2n} + O(n^{-3/2})\right\} - \frac{1}{2}\left\{\frac{t^2}{2n} + O(n^{-3/2})\right\}^2 + \ldots\right] = \frac{t^2}{2} + O'''(n^{-1/2}),$$

where $O'''(n^{-1/2})$ involves terms with $n^{1/2}$ and higher powers of n in the denominator. Proceeding to the limit as $n \to \infty$, we get

$$\lim_{n\to\infty} \log M_Z(t) = \frac{t^2}{2} \quad \Rightarrow \quad \lim_{n\to\infty} M_Z(t) = \exp(t^2/2) \qquad \ldots (**)$$

Interpretation. (**) is the *m.g.f.* of standard normal variate [*c.f.* Remark to 9·2·5]. Hence, by uniqueness theorem of moment generating functions, standard binomial variate tends to standard normal variate as $n \to \infty$. In other words, *binomial distribution tends to normal distribution as* $n \to \infty$.

Example 8·21. *A drunk person performs a random walk over positions 0, ± 1, ± 2, ..., as follows. He starts at 0. He takes successive one unit steps, going to the right with probability p and to the left with probability (1 – p). His steps are independent. Let X denote his position after n steps. Find the distribution of* $\frac{1}{2}(X+n)$ *and find E(X).*

Solution. With the ith step of the drunkard, let us associate a variable X_i defined as follows :

$$X_i = \begin{cases} 1, & \text{if he takes the step to the right} \\ -1, & \text{if he takes the step to the left} \end{cases}$$

Then, $X = X_1 + X_2 + \ldots + X_n$, gives the position of the drunkard after n steps.

Define : $Y_i = \frac{1}{2}(X_i + 1)$. Then $Y_i = \begin{cases} \frac{1}{2}(1+1) = 1, & \text{with probability } p \\ \frac{1}{2}(-1+1) = 0, & \text{with probability } 1-p = q, \text{ (say).} \end{cases}$

Since the n steps of drunkard are independent, Y_i's, $(i = 1, 2, \ldots n)$ are *i.i.d.* Bernoulli variates with parameter p. Hence,

$$\sum_{i=1}^{n} Y_i \sim B(n, p) \quad \Rightarrow \quad \sum_{i=1}^{n} Y_i = \sum_{i=1}^{n}\left(\frac{X_i + 1}{2}\right) = \frac{1}{2}\left(\sum_{i=1}^{n} X_i + n\right) = \frac{X+n}{2} \sim B(n, p),$$

where $X = \sum_{i=1}^{n} X_i$, is the position of the drunkard after n steps.

Since $\frac{1}{2}(X+n) \sim B(n,p), \quad E\left\{\frac{1}{2}(X+n)\right\} = np \quad \Rightarrow \quad E(X+n) = 2np$

$\Rightarrow \quad E(X) + n = 2np \quad \Rightarrow \quad E(X) = n(2p-1)$

Example 8·22. *Show that :*

(a) $b(n, p; k) = b(n, 1-p; n-k)$, *(b)* $\sum_{k=r}^{n} b(n, p; k) = 1 - \sum_{k=n-r+1}^{n} b(n, 1-p; k)$

(c) $b(n+1, p; k) = p.b(n, p; k-1) + q.b(n, p; k)$.

Solution. *(a)* $b(n, 1-p; n-k) = \binom{n}{n-k}(1-p)^{n-k}[1-(1-p)]^{n-(n-k)} = \binom{n}{k}p^k(1-p)^{n-k} = b(n,p;k)$

(b) $\sum_{k=r}^{n} b(n,p;k) = \sum_{k=r}^{n} b(n, 1-p; n-k)$ [From Part (a)]

$$= \sum_{j=0}^{n-r} b(n, 1-p; j) = 1 - \sum_{k=n-r+1}^{n} b(n, 1-p; k)$$

(c) R.H.S. $= p \cdot \binom{n}{k-1} p^{k-1} q^{n-k+1} + q \cdot \binom{n}{k} p^k q^{n-k}$

$$= \left[\binom{n}{k} + \binom{n}{k-1}\right] p^k q^{n-k+1} = \binom{n+1}{k} p^k q^{n-k+1} = b(n+1, p; k) = \text{L.H.S.}$$

Example 8·23. *If* $b(r; n, p) = \binom{n}{r} p^r q^{n-r}$, *is the binomial probability in the usual notation and if*

$$B(k; n, p) = P(X \le k) = \sum_{r=0}^{k} b(r; n, p), \text{ then prove that}$$

$$B(k; n, p) = (n-k)\binom{n}{k} \int_0^q t^{n-k-1}(I-t)^k\, dt, \; q = 1-p$$

Solution. $B(k; n, p) = \sum_{r=0}^{k} b(r; n, p) = \sum_{r=0}^{k} \binom{n}{r} p^r q^{n-r}$

Differentiating *w.r.* to q and noting that $q = 1-p \Rightarrow p = 1-q \Rightarrow \frac{dp}{dq} = -1$, we get

$$\frac{d}{dq} \cdot B(k; n, p) = \sum_{r=0}^{k} \left[\binom{n}{r}\left\{rp^{r-1}(-1) \cdot q^{n-r} + p^r \cdot (n-r)\, q^{n-r-1}\right\}\right]$$

$$= \sum_{r=0}^{k} \left[\frac{n!(-r)}{r!(n-r)!} p^{r-1} q^{n-r} + \frac{n!(n-r)}{r!(n-r)!} p^r q^{n-r-1}\right]$$

$$= \sum_{r=0}^{k} \left[-\frac{n(n-1)!}{(r-1)!\,(n-r)!} p^{r-1} q^{n-r} + \frac{n(n-1)!}{r!(n-r-1)!} p^r q^{n-r-1} \right]$$

$$= \sum_{r=0}^{k} \left[n \cdot \binom{n-1}{r} p^r q^{n-r-1} - n \binom{n-1}{r-1} p^{r-1} q^{n-r} \right]$$

$$= \sum_{r=0}^{k} \left[n\{t_r - t_{r-1}\} \right], \text{ where } t_r = \binom{n-1}{r} p^r q^{n-r-1}$$

$$= n[(t_0 - t_{-1}) + (t_1 - t_0) + (t_2 - t_1) + \ldots + (t_k - t_{k-1})] = nt_k, \; (t_{-1=0})$$

$$\therefore \quad \frac{d}{dq} B(k; n, p) = n \binom{n-1}{k} p^k q^{n-k-1}, \; p = 1 - q$$

On integration, we get $B(k; n, p) = n \cdot \binom{n-1}{k} \int_0^q (1-u)^k u^{n-k-1}\, du$

But $$n\binom{n-1}{k} = \frac{n(n-1)!}{k!\,(n-1-k)!} = \frac{n!(n-k)}{k!\,(n-k)!} = (n-k)\binom{n}{k}$$

$$\therefore \quad B(k; n, p) = (n-k)\binom{n}{k} \int_0^q (1-u)^k . u^{n-k-1}\, du, \text{ as desired.}$$

Remarks :

1. We further get :

$$B(k+1, n-k) = \frac{\Gamma(k+1)\,\Gamma(n-k)}{\Gamma(n+1)} = \frac{k!\,(n-k-1)!}{n!} \Rightarrow \frac{1}{\beta(k+1, n-k)} = \frac{n!}{k!(n-k-1)!} = (n-k)\binom{n}{k}$$

Hence, the result may be written as :

$$\therefore \quad B(k; n, p) = P(X \le k) = \frac{1}{\beta(k+1, n-k)} \int_0^q (1-u)^k\, u^{n-k-1}\, du$$

This result is of grat practical utility. It enables us to represent the cumulative binomial probabilities (which are generally quite tedious and time consuming to compute) in terms of *Incomplete Beta Functions,* which are tabulatd in Karl Pearson's Tables of the Incomplete Beta Functions.

2. Let us now work out the probability : $P(X \ge k) = \sum_{r=k}^{n} \binom{n}{r} p^r\, q^{n-r}$

Differentiating *w.r. to p*, and proceeding similarly, we shall get

$$\frac{d}{dp} P(X \ge k) = -\, n \sum_{r=k}^{n} (T_r - T_{r-1}), \text{ where } T_r = \binom{n-1}{r} p^r\, q^{n-r-1}, \; (T_n \equiv 0)$$

$$\therefore \quad \frac{d}{dp} P(X \ge k) = n\, T_{k-1} = n\binom{n-1}{k-1} p^{k-1} (1-p)^{n-k} \qquad (\because q = 1-p)$$

On integration, we shall get

$$P(X \ge k) = n\binom{n-1}{k-1} \int_0^p u^{k-1} (1-u)^{n-k} du = \frac{1}{\beta(k, n-k+1)} \int_0^p u^{k-1} (1-u)^{n-k}\, du$$

This is quite an important result and should be committed to memory. We shall use it in '*Order Statistics*'. This result can be stated as follows :

If $X \sim B(n, p)$ and Y has Beta disstribution with parameters k and $n - k + 1$ (*c.f.* Chapter 9), then

$$P(Y \leq p) = P(X \geq k) = 1 - P(X \leq k - 1) \quad \Rightarrow \quad F_Y(p) = 1 - F_X(k-1)$$

SUPPLEMENTARY EXAMPLES ON BINOMIAL DISTRIBUTION

Example 8·24. *A department in a works has 10 machines which may need adjustment from time to time during the day. Three of these machines are old, each having a probability of 1/11 of needing adjustment during the day, and 7 are new, having corresponding probabilities of 1/21.*

Assuming that no machine needs adjustment twice on the same day, determine the probabilities that on a particular day

(*i*) *Just 2 old and no new machines need adjustment.*

(*ii*) *If just 2 machines need adjustment, they are of the same type.*

Solution. Let p_1 = Probability that an old machine needs adjustment $= \dfrac{1}{11} \quad \Rightarrow \quad q_1 = \dfrac{10}{11}$.

and p_2 = Probability that a new machine needs adjustment $= \dfrac{1}{21} \quad \Rightarrow \quad q_2 = \dfrac{20}{21}$.

Then $P_1(x)$ = Probability that 'x' old machines need adjustment

$$= \binom{3}{x} p_1^{\,x} q_1^{\,3-x} = \binom{3}{x}\left(\frac{1}{11}\right)^x \left(\frac{10}{11}\right)^{3-x} \quad ; \; x = 0, 1, 2, 3$$

and $P_2(x)$ = Probability that 'x' new machines need adjustment

$$= \binom{7}{x} p_2^{\,x} q_2^{\,7-x} = \binom{7}{x}\left(\frac{1}{21}\right)^x \left(\frac{20}{21}\right)^{7-x} \quad ; \; x = 0, 1, 2, \ldots, 7$$

(*i*) The probability that just two old machines and no new machine need adjustment is given (by the compound probability theorem) by the expression :

$$P_1(2) \cdot P_2(0) = \binom{3}{2}\left(\frac{1}{11}\right)^2 \left(\frac{10}{11}\right)\left(\frac{20}{21}\right)^7 = 0{\cdot}016 \qquad \ldots (1)$$

(*ii*) Similarly, the probability that just 2 new machines and no old machine need adjustment is:

$$P_1(0) \cdot P_2(2) = \left(\frac{10}{11}\right)^3 \times \binom{7}{2}\left(\frac{1}{21}\right)^2 \left(\frac{20}{21}\right)^5 = 0{\cdot}028 \qquad \ldots (2)$$

The probability that 'if just two machines need adjustment, they are of the same type' is the same as the probability that 'either just 2 old and no new or just 2 new and no old machines need adjustment'.

∴ Required probability $= P_1(2) \,.\, P_2(0) + P_1(0) \,.\, P_2(2) = 0{\cdot}016 + 0{\cdot}028 = 0{\cdot}044$.

Example 8·25. *In a precision bombing attack there is a 50% chance that any one bomb will strike the target. Two direct hits are required to destroy the target completely. How many bombs must be dropped to give a 99% chance or better of completely destroying the target ?*

Solution. We have : p = Probability that the bomb strikes the target = 50% $= \dfrac{1}{2}$.

Let n be the number of bombs which should be dropped to ensure 99% chance or better of completely destroying the target. This implies that 'probability that out of n bombs, at least two strike the target, is greater than 0·99.

Let X be a *r.v.* representing the number of bombs striking the target. Then

$$X \sim B\left(n, p=\frac{1}{2}\right) \text{ with } P(X=x)=p(x)=\binom{n}{x}\left(\frac{1}{2}\right)^x\left(\frac{1}{2}\right)^{n-x}=\binom{n}{x}\left(\frac{1}{2}\right)^n ;\ x=0,1,2,\ldots,n$$

We should have: $P(X \ge 2) \ge 0{\cdot}99 \quad \Rightarrow \quad [1 - p(X \le 1)] \ge 0{\cdot}99$

$$\Rightarrow \quad [1 - \{p(0) + p(1)\}] \ge 0{\cdot}99 \quad \Rightarrow \quad 1 - \left\{\binom{n}{0} + \binom{n}{1}\right\}\left(\frac{1}{2}\right)^n \ge 0{\cdot}99$$

$$\Rightarrow \quad 0{\cdot}01 \ge \frac{1+n}{2^n} \quad \Rightarrow \quad 2^n \times (0{\cdot}01) \ge 1+n \quad \Rightarrow \quad 2^n \ge 100 + 100n \qquad \ldots (*)$$

By trial method, we find that the inequality (*) is satisfied by $n = 11$. Hence the minimum number of bombs needed to destroy the target completely is 11.

Example 8·26. *Let X and Y be independent binomial variates, each with parameters n and p. Find* $P(X - Y = k)$.

Solution. Since each of the variables X and Y takes the values 0, 1, 2, ..., n,

$Z = X - Y$, takes on the values $-n, -(n-1), \ldots, -1, 0, 1, \ldots, n$.

$$P(Z=k) = P(X-Y=k) = \sum_{r=0}^{n} P(X=k+r \cap Y=r)$$

$$= \sum_{r=0}^{n} P(X=k+r) \,.\, P(Y=r) \qquad (\because X \text{ and } Y \text{ are independent}).$$

$$= \sum_{r=0}^{n} \binom{n}{k+r} p^{k+r} q^{n-k-r} \binom{n}{r} p^r q^{n-r} = \sum_{r=0}^{n} \binom{n}{k+r}\binom{n}{r} p^{2r+k} q^{2n-2r-k}, \qquad \ldots (*)$$

where, $k = -n, -(n-1), \ldots, -2, -1, 0, 1, 2, \ldots, n$; and $q = 1 - p$.

In particular, we have $\quad P(Z=0) = \sum_{r=0}^{n} \left\{\binom{n}{r}\right\}^2 . p^{2r} q^{2n-2r}$

$P(Z=-n) = \sum_{r=0}^{n} \binom{n}{-n+r}\binom{n}{r} p^{2r-n} q^{3n-2r} = p^n q^n$, because we get the result when $r = n$ and for other values of $r < n$, $\binom{n}{-n+r}$ is not defined and hence taken as 0.

Example 8·27. *If a fair coins is tossed an even number 2n times, show that the probability of obtaining more heads than tails is :* $\frac{1}{2}\left\{1 - \binom{2n}{n}\left(\frac{1}{2}\right)^{2n}\right\}$.

Solution. Let X = Number of heads ; Y = Number of tails ; Number of trials = $2n$

$$P(X > Y) + P(X < Y) + P(X = Y) = 1 \quad \Rightarrow \quad 2P(X > Y) = 1 - P(X = Y) \qquad \ldots (*)$$

$$[\because \text{By symmetry, } p = q = \frac{1}{2} \quad \Rightarrow \quad P(X > Y) = P(X < Y)]$$

In $2n$ trials, $\quad X = Y \quad \Rightarrow \quad$ Number of heads = Number of tails = n

$$\therefore \quad P(X > Y) = \frac{1}{2}\left\{1 - \binom{2n}{n} p^n q^n\right\} = \frac{1}{2}\left\{1 - \binom{2n}{n}\left(\frac{1}{2}\right)^{2n}\right\} \qquad \left[\text{From } (*) \text{ and } p = q = \frac{1}{2}\right]$$

Example 8·28. *(a) If K(t) is the cumulant generating function about the origin of the binomial distribution of size n, show that*

$$\frac{d}{dt} K(t) = n\left[1 + e^{-(z+t)}\right]^{-1}, \text{ where } z = \log_e (p/q).$$

(b) By expanding the R.H.S. in powers of t by Taylor's Theorem, show that

$$\kappa_r = n \frac{d^{r-1} p}{dz^{r-1}}, \text{ where } \kappa_r \text{ is the rth cumulant.}$$

(c) Hence, (or otherwise) obtain the recurrence relation : $\kappa_{r+1} = pq \dfrac{d\kappa_r}{dp}, \; r > 1$

(d) Prove that : $\kappa_{r+1} = \dfrac{d\kappa_r}{dz}$, *where* $z = \log_e \left(\dfrac{p}{q}\right)$.

Solution. For binomial distribution with parameters n and p, we have

$$K(t) = \log M(t) = n \log (q + pe^t)$$

(a)
$$\frac{d}{dt} K(t) = \frac{npe^t}{q + pe^t} = n\left(1 + \frac{q}{p} e^{-t}\right)^{-1}$$

If $z = \log_e (p/q) \quad \Rightarrow \quad (p/q) = e^z \quad \Rightarrow \quad (q/p) = e^{-z}$,

$$\therefore \quad \frac{d}{dt} K(t) = n\left[1 + e^{-(z+t)}\right]^{-1}$$

(b)
$$\kappa_r = \left[\frac{d^r}{dt^r} K(t)\right]_{t=0} = \left[\frac{d^{r-1}}{dt^{r-1}} \frac{d}{dt} K(t)\right]_{t=0}$$

$$= n\left[\frac{d^{r-1}}{dt^{r-1}}\left[1 + e^{-(z+t)}\right]^{-1}\right]_{t=0} = n\left[\frac{d^{r-1}}{dt^{r-1}}\left(\frac{e^{z+t}}{1 + e^{z+t}}\right)\right]_{t=0} \qquad \text{[From Part (a)]} \quad \ldots (**)$$

By symmetry of the function $\left[e^{z+t}/(1 + e^{z+t})\right]$ in t and z, we have

$$\frac{d}{dt}\left(\frac{e^{z+t}}{1 + e^{z+t}}\right) = \frac{d}{dz}\left(\frac{e^{z+t}}{1 + e^{z+t}}\right) \Rightarrow \frac{d^{r-1}}{dt^{r-1}}\left(\frac{e^{z+t}}{1 + e^{z+t}}\right) = \frac{d^{r-1}}{dz^{r-1}}\left(\frac{e^{z+t}}{1 + e^{z+t}}\right)$$

Substituting in (**), we get

$$\kappa_r = n\left[\frac{d^{r-1}}{dz^{r-1}}\left(\frac{e^{z+t}}{1 + e^{z+t}}\right)\right]_{t=0} = n \frac{d^{r-1}}{dz^{r-1}}\left(\frac{e^z}{1 + e^z}\right)$$

$$= n \frac{d^{r-1}}{dz^{r-1}}\left(1 + e^{-z}\right)^{-1} = n \frac{d^{r-1}}{dz^{r-1}}\left(1 + \frac{q}{p}\right)^{-1} = n \frac{d^{r-1} p}{dz^{r-1}} \qquad \ldots (***)$$

(c)
$$\frac{d\kappa_r}{dp} = n \frac{d}{dp}\left(\frac{d^{r-1} p}{dz^{r-1}}\right) = n \frac{d}{dz}\left(\frac{d^{r-1} p}{dz^{r-1}}\right)\frac{dz}{dp} = n \frac{d^r p}{dz^r} \cdot \frac{1}{pq} \qquad [\because z = \log_e (p/q)]$$

$$= \frac{1}{pq} \cdot \kappa_{r+1} \qquad \text{[From (***)]}$$

(d) $$\frac{d\kappa_r}{dz} = \frac{d\kappa_r}{dp} \cdot \frac{dp}{dz} = \left(\frac{d\kappa_r}{dp} \Big/ \frac{dz}{dp}\right) = \left(\frac{d\kappa_r}{dp} \Big/ \frac{1}{pq}\right) = pq \cdot \frac{d\kappa_r}{dp}$$

$$\therefore \quad \frac{d\kappa_r}{dz} = \kappa_{r+1}. \qquad \text{[From Part (c)]}$$

8·5. POISSON DISTRIBUTION

Poisson distribution was discovered by the French mathematician and physicist Simeon Denis Poisson (1781 – 1840) who published it in 1837. Poisson distribution is a limiting case of the binomial distribution under the following conditions :

(i) n, the number of trials is indefinitely large, *i.e.*, $n \to \infty$.

(ii) p, the constant probability of success for each trial is indefinitely small, *i.e.*, $p \to 0$.

(iii) $np = \lambda$, (say) is finite.

Thus, $p = \lambda/n$, $q = 1 - (\lambda/n)$, where λ is a positive real number.

The probability of x successes in a series of n independent trials is :

$$b(x\,;\, n, p) = \binom{n}{x} p^x q^{n-x}\,;\ x = 0, 1, 2, \dots, n \qquad \dots (*)$$

We want the limiting form of (*) under the above conditions. Hence

$$\lim_{n\to\infty} b(x; n, p) = \lim_{n\to\infty} \frac{n!}{x!(n-x)!}\left(\frac{\lambda}{n}\right)^x \cdot \left(1 - \frac{\lambda}{n}\right)^{n-x}$$

Using Stirling's approximation for n ! as $n \to \infty$, *viz.*,

$$\lim_{n\to\infty} n! \approx \sqrt{2\pi}\, e^{-n}\, n^{n+(1/2)}, \text{ we get}$$

$$\lim_{n\to\infty} b(x; n, p) = \lim_{n\to\infty} \left\{\frac{\sqrt{2\pi}\, e^{-n} \,.\, n^{n+(1/2)}}{x!\sqrt{2\pi}\, e^{-(n-x)} \,.\, (n-x)^{n-x+(1/2)}}\right\}\left(\frac{\lambda}{n}\right)^x \left(1 - \frac{\lambda}{n}\right)^{n-x}$$

$$= \frac{\lambda^x}{e^x \,.\, x!} \,.\, \lim_{n\to\infty} \frac{n^{n-x+(1/2)}}{(n-x)^{n-x+(1/2)}} \cdot \left(1 - \frac{\lambda}{n}\right)^{n-x} = \frac{\lambda^x}{e^x\, x!} \,.\, \lim_{n\to\infty} \frac{\left(1 - \frac{\lambda}{n}\right)^{n-x}}{\left(1 - \frac{x}{n}\right)^{n-x+(1/2)}}$$

$$= \frac{\lambda^x}{e^x\, x!} \cdot \frac{\lim\limits_{n\to\infty}\left(1 - \frac{\lambda}{n}\right)^n \,.\, \lim\limits_{n\to\infty}\left(1 - \frac{\lambda}{n}\right)^{-x}}{\lim\limits_{n\to\infty}\left(1 - \frac{x}{n}\right)^n \ \lim\limits_{n\to\infty}\left(1 - \frac{x}{n}\right)^{-x+(1/2)}}$$

But we know that $$\left\{\begin{array}{l} \lim\limits_{n\to\infty}\left(1 - \dfrac{\lambda}{n}\right)^n = e^{-\lambda}, \\ \lim\limits_{n\to\infty}\left(1 - \dfrac{\lambda}{n}\right)^\alpha = 1, \alpha \text{ is not a function of } n \end{array}\right\} \qquad \dots (**)$$

$$\therefore \quad \lim_{n\to\infty} b(x; n, p) = \frac{\lambda^x}{e^x\, x!} \times \frac{e^{-\lambda} \,.\, 1}{e^{-x} \,.\, 1} = \frac{e^{-\lambda} \,.\, \lambda^x}{x!}\,;\ x = 0, 1, 2, \dots, \infty, \qquad \text{[Using (**)]}$$

which is the required probability function of the Poisson distribution. 'λ' *is known as the parameter of Poisson distribution.*

Aliter. Poisson distribution can also be derived without using Stirling's approximation as follows:

$$b(x\,;\,n,p) = \binom{n}{x} p^x (1-p)^{n-x} = \binom{n}{x}\left(\frac{p}{1-p}\right)^x (1-p)^n$$

$$= \frac{n(n-1)(n-2)\dots(n-x+1)}{x!}\cdot\frac{\left(\frac{\lambda}{n}\right)^x}{\left(1-\frac{\lambda}{n}\right)^x}\cdot\left(1-\frac{\lambda}{n}\right)^n.$$

$$= \frac{\left(1-\frac{1}{n}\right)\left(1-\frac{2}{n}\right)\dots\left(1-\frac{x-1}{n}\right)}{x!\left(1-\frac{\lambda}{n}\right)^x}\cdot\lambda^x\left(1-\frac{\lambda}{n}\right)^n$$

$$\therefore \quad \lim_{n\to\infty} b(x;n,p) = \frac{e^{-\lambda}\lambda^x}{x!};\ x = 0, 1, 2, \dots \qquad \text{[From (**)]}$$

Definition. *A random variable X is said to follow a Poisson distribution if it assumes only non-negative values and its probability mass function is given by :*

$$P(X = x) = p(x,\lambda) = \begin{cases} \dfrac{e^{-\lambda}\lambda^x}{x!};\ x = 0, 1, 2, \dots;\ \lambda > 0 \\ 0,\ otherwise \end{cases} \qquad \dots(8\cdot15)$$

Here, λ *is known as the parameter of the distribution.* We shall use the notation $X \sim P(\lambda)$, to denote that X is a Poisson variate with parameter λ.

Remarks:

1. It should be noted that $\sum_{x=0}^{\infty} P(X=x) = e^{-\lambda}\sum_{x=0}^{\infty}(\lambda^x / x!) = e^{-\lambda}e^{\lambda} = 1$

2. The corresponding distribution function is :

$$F(x) = P(X \le x) = \sum_{r=0}^{x} p(r) = e^{-\lambda}\sum_{r=0}^{x}\lambda^r / r!\,;\ x = 0, 1, 2, \dots$$

3. Poisson distribution occurs when there are events which do not occur as outcomes of a definite number of trials (unlike that in binomial distribution) of an experiment but which occur at random points of time and space wherein our interest lies only in the number of occurrences of the event, not in its non-occurrences.

4. Following are some instances where Poisson distribution may be successfully employed :
   - (*i*) Number of deaths from a disease (not in the form of an epidemic) such as heart attack or cancer or due to snake bite.
   - (*ii*) Number of suicides reported in a particular city.
   - (*iii*) The number of defective material in a packing manufactured by a good concern.
   - (*iv*) Number of faulty blades in a packet of 100.
   - (*v*) Number of air accidents in some unit of time.
   - (*vi*) Number of printing mistakes at each page of the book.
   - (*vii*) Number of telephone calls received at a particular telephone exchange in some unit of time or connections to wrong numbers in a telephone exchange.
   - (*viii*) Number of cars passing a crossing per minute during the busy hours of a day.
   - (*ix*) The number of fragments received by a surface area '*A*' from a fragment atom bomb.
   - (*x*) The emission of radioactive (alpha) particles.

8·5·1. The Poisson Process.

The Poisson distribution may also be obtained independently (*i.e.*, without considering it as a limiting form of the binomial distribution) as follows :

Let X_t be the number of telephone calls received in time interval 't' on a telephone switch board. Consider the following experimental conditions :

1. The probability of getting a call in small time interval $(t, t + dt)$ is $\lambda\, dt$, where λ is a positive constant and dt denotes a small increment in time 't'.
2. The probability of getting more than one call in this time interval is very small, *i.e.*, is of the order of $(dt)^2$, *i.e.*, $O\,[(dt)^2]$ such that $\lim\limits_{dt \to 0} \dfrac{O\,(dt)^2}{dt} = 0.$
3. The probability of any particular call in the time interval $(t, t + dt)$ is independent of the actual time t and also of all previous calls.

Under these conditions it can be shown that the probability of getting x calls in time 't', say, $P_x(t)$ is given by :

$$P_x(t) = \frac{e^{-\lambda t}\,(\lambda t)^x}{x\,!}\,;\ x = 0, 1, 2, \ldots \infty$$

which is a Poisson distribution with parameter λt.

Proof. Let $P_x(t) = P$ {*of getting x calls in a time interval of length 't'*}.

Also $\quad P$ {At least one call during $(t, t + dt)$} $= \lambda\, dt + O\,[(dt)^2]$

and $\quad P$ {More than one call during $(t, t + dt)$} $= O\,[(dt)^2]$.

The event of getting exactly x calls in time $t + dt$ can materialise in the following two mutually exclusive ways :

(*i*) x calls in $(0, t)$, and none during $(t, t + dt)$ and the probability of this event is :

$$P_x(t)\,\left[1 - \{(\lambda\, dt + O\,(dt)^2\}\right].$$

(*ii*) Exactly $(x - 1)$ calls during $(0, t)$ & one call in $(t, t + dt)$ and the probability of this event is : $\quad P_{x-1}(t) \cdot (\lambda\, dt)$

Hence, by the addition theorem of probability, we get

$$P_x(t + dt) = P_x(t)\,[1 - \lambda\, dt - O\,(dt)^2] + P_{x-1}(t)\,.\,\lambda\, dt.$$

$$= P_x(t)\,(1 - \lambda\, dt) + P_{x-1}(t)\,\lambda\, dt - O\,(dt)^2\, P_x(t) \qquad \ldots (1)$$

$$\Rightarrow \quad \frac{P_x(t+dt) - P_x(t)}{dt} = -\,\lambda\, P_x(t) + \lambda P_{x-1}(t) - \frac{O\,(dt)^2}{dt}\, P_x(t)$$

Proceeding to the limit as $dt \to 0$, we get

$$\lim_{dt \to 0} \frac{P_x(t+dt) - P_x(t)}{dt} = -\,\lambda\, P_x(t) + \lambda P_{x-1}(t)$$

$$\Rightarrow \quad P_x'(t) = -\,\lambda\, P_x(t) + \lambda P_{x-1}(t),\ x \geq 1 \qquad \ldots (2)$$

where (′) denotes differentiation *w.r.* to 't'.

For $x = 0$, $\quad P_{x-1}(t) = P_{-1}(t) = P$ {(–1) calls in time 't'} $\equiv 0$

Hence from (1), we get $\quad P_0(t + dt) = P_0(t)\,\{1 - \lambda\, dt\} - O\,(dt)^2$

which on taking the limit $dt \to 0$, gives, $\quad P_0'(t) = -\,\lambda\, P_0(t) \quad \Rightarrow \quad \dfrac{P_0'(t)}{P_0(t)} = -\,\lambda.$

Integrating *w.r.* to 't', $\log P_0(t) = -\lambda t + C,$
where C is an arbitrary constant to be determined from the condition $P_0(0) = 1$.

Hence, $C = \log 1 = 0 \quad \therefore \quad \log P_0(t) = -\lambda t \quad \Rightarrow \quad P_0(t) = e^{-\lambda t}$

Substituting this value of $P_0(t)$ in (2), we get, with $x = 1$,

$$P_1'(t) = -\lambda P_1(t) + \lambda e^{-\lambda t} \quad \Rightarrow \quad P_1'(t) + \lambda P_1(t) = \lambda e^{-\lambda t}$$

This is an ordinary linear differential equation whose integrating factor is $e^{\lambda t}$. Hence, its solution is :

$$e^{\lambda t} P_1(t) = \lambda \int e^{\lambda t} e^{-\lambda t} dt + C_1 = \lambda t + C_1,$$

where C_1 is an arbitrary constant to be determined from $P_1(0) = 0$, which gives $C_1 = 0$.

$$\therefore \quad P_1(t) = e^{-\lambda t} \lambda t$$

Again substituting this in (2) with $x = 2$, $P_2'(t) + \lambda P_2(t) = \lambda e^{-\lambda t} \lambda t.$

Integrating factor of this equation is $e^{\lambda t}$ and its solution is :

$$P_2(t) e^{\lambda t} = \lambda^2 \int t e^{-\lambda t} e^{\lambda t} dt + C_2 = \frac{\lambda^2 t^2}{2} + C_2,$$

where C_2 is an arbitrary constant to be determined from $P_2(0) = 0$, which gives $C_2 = 0$.

Hence, $P_2(t) = e^{-\lambda t} \dfrac{(\lambda t)^2}{2}$. Proceeding similarly step by step, we shall get

$$P_x(t) = \frac{e^{-\lambda t} (\lambda t)^x}{x!}; \; x = 0, 1, 2, \ldots, \infty \qquad \ldots (8{\cdot}15a)$$

which is the *p.m.f.* of Poisson distribution with parameter λt.

8·5·2. Moments of the Poisson Distribution.

$$\mu_1' = E(X) = \sum_{x=0}^{\infty} x\, p(x, \lambda) = \sum_{x=0}^{\infty} x \cdot \frac{e^{-\lambda} \lambda^x}{x!} = \lambda e^{-\lambda} \left\{ \sum_{x=1}^{\infty} \frac{\lambda^{x-1}}{(x-1)!} \right\}$$

$$= \lambda e^{-\lambda} \left(1 + \lambda + \frac{\lambda^2}{2!} + \frac{\lambda^3}{3!} + \ldots \right) = \lambda e^{-\lambda} \cdot e^{\lambda} = \lambda$$

Hence, *the mean of the Poisson distribution is* λ.

$$\mu_2' = E(X^2) = \sum_{x=0}^{\infty} x^2 p(x, \lambda) = \sum_{x=0}^{\infty} [x(x-1) + x] \frac{e^{-\lambda} \lambda^x}{x!}$$

$$= e^{-\lambda} \sum_{x=0}^{\infty} x(x-1) \frac{\lambda^x}{x!} + \sum_{x=0}^{\infty} x \frac{e^{-\lambda} \lambda^x}{x!} = \lambda^2 e^{-\lambda} \left[\sum_{x=2}^{\infty} \frac{\lambda^{x-2}}{(x-2)!} \right] + \lambda$$

$$= \lambda^2 e^{-\lambda} e^{\lambda} + \lambda = \lambda^2 + \lambda$$

$$\mu_3' = E(X^3) = \sum_{x=0}^{\infty} x^3 p(x, \lambda) = \sum_{x=0}^{\infty} [x(x-1)(x-2) + 3x(x-1) + x] \frac{e^{-\lambda} \lambda^x}{x!}$$

$$= \sum_{x=0}^{\infty} x(x-1)(x-2) \frac{e^{-\lambda} \lambda^x}{x!} + 3 \sum_{x=0}^{\infty} x(x-1) \frac{e^{-\lambda} \lambda^x}{x!} + \sum_{x=0}^{\infty} x \frac{e^{-\lambda} \lambda^x}{x!}$$

$$= e^{-\lambda}\lambda^3 \left\{\sum_{x=3}^{\infty} \frac{\lambda^{x-3}}{(x-3)!}\right\} + 3e^{-\lambda}\lambda^2 \left\{\sum_{x=2}^{\infty} \frac{\lambda^{x-2}}{(x-2)!}\right\} + \lambda$$

$$= e^{-\lambda}\lambda^3 e^{\lambda} + 3e^{-\lambda}\lambda^2 e^{\lambda} + \lambda = \lambda^3 + 3\lambda^2 + \lambda$$

$$\mu_4' = E(X^4) = \sum_{x=0}^{\infty} x^4 \cdot p(x, \lambda)$$

$$= \sum_{x=0}^{\infty} \left[x(x-1)(x-2)(x-3) + 6x(x-1)(x-2) + 7x(x-1) + x\right] \frac{e^{-\lambda}\lambda^x}{x!}$$

$$= e^{-\lambda}\lambda^4 \left\{\sum_{x=4}^{\infty} \frac{\lambda^{x-4}}{(x-4)!}\right\} + 6e^{-\lambda}\lambda^3 \left\{\sum_{x=3}^{\infty} \frac{\lambda^{x-3}}{(x-3)!}\right\} + 7e^{-\lambda}\lambda^2 \left\{\sum_{x=2}^{\infty} \frac{\lambda^{x-2}}{(x-2)!}\right\} + \lambda$$

$$= \lambda^4 (e^{-\lambda} e^{\lambda}) + 6\lambda^3 (e^{-\lambda} e^{\lambda}) + 7\lambda^2 (e^{-\lambda} e^{\lambda}) + \lambda$$

$$= \lambda^4 + 6\lambda^3 + 7\lambda^2 + \lambda$$

The four central moments are now obtained as follows :

$$\mu_2 = \mu_2' - \mu_1'^2 = (\lambda^2 + \lambda) - \lambda^2 = \lambda$$

Thus the mean and the variance of the Poisson distribution are each equal to λ.

$$\mu_3 = \mu_3' - 3\,\mu_2'\,\mu_1' + 2\mu_1'^3 = (\lambda^3 + 3\lambda^2 + \lambda) - 3\lambda(\lambda^2 + \lambda) + 2\lambda^3 = \lambda$$

$$\mu_4 = \mu_4' - 4\,\mu_3'\,\mu_1' + 6\mu_2'\,\mu_1'^2 - 3\mu_1'^4$$

$$= (\lambda^4 + 6\lambda^3 + 7\lambda^2 + \lambda) - 4\lambda(\lambda^3 + 3\lambda^2 + \lambda) + 6\lambda^2(\lambda^2 + \lambda) - 3\lambda^4 = 3\lambda^2 + \lambda$$

Co-efficients of skewness and kurtosis are given by :

$$\left.\begin{aligned} \beta_1 &= \frac{\mu_3^2}{\mu_2^3} = \frac{\lambda^2}{\lambda^3} = \frac{1}{\lambda} \quad \text{and} \quad \gamma_1 = \sqrt{\beta_1} = \frac{1}{\sqrt{\lambda}} \\ \text{Also,} \quad \beta_2 &= \frac{\mu_4}{\mu_2^2} = 3 + \frac{1}{\lambda} \quad \text{and} \quad \gamma_2 = \beta_2 - 3 = \frac{1}{\lambda} \end{aligned}\right\} \quad \text{... (8·16)}$$

Hence, the Poisson distribution is always a skewed distribution.

Proceeding to the limit as $\lambda \to \infty$, $\beta_1 = 0$ and $\beta_2 = 3$.

8·5·3. Mode of the Poisson Distribution.

$$\frac{p(x)}{p(x-1)} = \frac{\dfrac{e^{-\lambda}\lambda^x}{x!}}{\dfrac{e^{-\lambda}\lambda^{x-1}}{(x-1)!}} = \frac{\lambda}{x}\ ;\ x = 1, 2, 3, \ldots \quad \text{... (8·17)}$$

We discuss the following cases :

Case I. *When λ is not an integer.* Let us suppose that S is the integral part of λ, so that $\lambda = S + f,\ 0 < f < 1$. Hence from (8.17), we get

$$\frac{p(x)}{p(x-1)} = \frac{S+f}{x} = \begin{cases} > 1, \text{ if } x = 1, 2, \ldots, S \\ < 1, \text{ if } x = S+1, S+2, \ldots \end{cases}$$

$$\frac{p(1)}{p(0)} > 1, \quad \frac{p(2)}{p(1)} > 1, \quad ..., \quad \frac{p(S-1)}{p(S-2)} > 1, \quad \frac{p(S)}{p(S-1)} > 1,$$

and $$\frac{p(S+1)}{p(S)} < 1, \quad \frac{p(S+2)}{p(S+1)} < 1, \quad ...$$

Combining the above expressions into a single expression, we get

$$p(0) < p(1) < p(2) < ... < p(S-2) < p(S-1) < p(S) > p(S+1) > p(S+2) > ...,$$

which shows that $p(S)$ is the maximum value. Hence, *in this case, the distribution is unimodal and the integral part of λ is the unique modal value.*

Case II. *When $\lambda = k$ (say) is an integer.* Here, as in case I, we have

$$\frac{p(1)}{p(0)} > 1, \quad \frac{p(2)}{p(1)} > 1, \quad ..., \quad \frac{p(k-1)}{p(k-2)} > 1$$

and $$\frac{p(k)}{p(k-1)} = 1, \quad \frac{p(k+1)}{p(k)} < 1, \quad \frac{p(k+2)}{p(k+1)} < 1, \quad ...$$

$$\therefore \quad p(0) < p(1) < p(2) < ... < p(k-2) < p(k-1) = p(k) > p(k+1) > p(k+2) ...$$

In this case we have *two maximum values, viz., $p(k-1)$ and $p(k)$ and thus the distribution is bimodal and two modes are at $(k-1)$ and k, i.e., at $(\lambda - 1)$ and λ, (since $k = \lambda$).*

8·5·4. Recurrence Relation for Moments of the Poisson Distribution.

By *def.,*

$$\mu_r = E\,[X - E(X)]^r = \sum_{x=0}^{\infty} (x-\lambda)^r\, p(x, \lambda) = \sum_{x=0}^{\infty} (x-\lambda)^r \cdot \frac{e^{-\lambda}\lambda^x}{x!}$$

Differentiating *w.r.t.* to λ, we get

$$\frac{d\mu_r}{d\lambda} = \sum_{x=0}^{\infty} r(x-\lambda)^{r-1}(-1)\frac{e^{-\lambda}\lambda^x}{x!} + \sum_{x=0}^{\infty}\frac{(x-\lambda)^r}{x!}\left[x\lambda^{x-1}e^{-\lambda} - \lambda^x e^{-\lambda}\right]$$

$$= -r\sum_{x=0}^{\infty}(x-\lambda)^{r-1}\cdot\frac{e^{-\lambda}\lambda^x}{x!} + \sum_{x=0}^{\infty}\frac{(x-\lambda)^r}{x!}\left[\lambda^{x-1}e^{-\lambda}(x-\lambda)\right]$$

$$= -r\sum_{x=0}^{\infty}(x-\lambda)^{r-1}\cdot\frac{e^{-\lambda}\lambda^x}{x!} + \frac{1}{\lambda}\sum_{x=0}^{\infty}(x-\lambda)^{r+1}\cdot\frac{e^{-\lambda}\lambda^x}{x!}$$

$$= -r\,\mu_{r-1} + \frac{1}{\lambda}\mu_{r+1}$$

$$\Rightarrow \quad \mu_{r+1} = r\lambda\,\mu_{r-1} + \lambda\frac{d\mu_r}{d\lambda} \quad ...(8·18)$$

Putting $r = 1$, 2 and 3 successively, we get

$$\mu_2 = \lambda\,\mu_0 + \lambda\frac{d\mu_1}{d\lambda} = \lambda, \quad \mu_3 = 2\lambda\,\mu_1 + \lambda\frac{d\mu_2}{d\lambda} = \lambda, \quad \mu_4 = 3\lambda\,\mu_2 + \lambda\frac{d\mu_3}{d\lambda} = 3\lambda^2 + \lambda.$$

8·5·5. Moment Generating Function of the Poisson Distribution.

$$M_X(t) = \sum_{x=0}^{\infty} e^{tx} \cdot \frac{e^{-\lambda}\lambda^x}{x!} = \sum_{x=0}^{\infty} \frac{e^{-\lambda}(\lambda e^t)^x}{x!}$$

$$= e^{-\lambda}\left\{1 + \lambda e^t + \frac{(\lambda e^t)^2}{2!} + \dots\right\} = e^{-\lambda} \cdot e^{\lambda e^t} = e^{\lambda(e^t - 1)} \quad \dots (8\cdot19)$$

8·5·5(a). Poisson Distribution As a Limiting Case of Binomial Distribution.

Let $X \sim B(n, p)$, then

$$M_X(t) = \left(q + pe^t\right)^n = \left[1 + p(e^t - 1)\right]^n ; \qquad (q = 1 - p)$$

If $n \to \infty$ and $np = \lambda$ (finite) $\quad \Rightarrow \quad p = (\lambda/n);\ \lambda > 0$, then

$$\lim_{n\to\infty} M_X(t) = \lim_{n\to\infty}\left[1 + \frac{\lambda(e^t - 1)}{n}\right]^n = e^{\lambda(e^t - 1)}$$

which is the *m.g.f.* of Poisson distribution with parameter λ.

Hence, by uniqueness theorem of *m.g.f.*, the limiting form of binomial distribution as $n \to \infty$ and $np \to \lambda > 0$ (finite) is Poisson with parameter λ.

Note. If $\lim_{n\to\infty} \frac{\chi(n)}{n} = 0$, then $\lim_{n\to\infty}\left[1 + \frac{b}{n} + \frac{\chi(n)}{n}\right]^{cn} = e^{bc}$

8·5·6. Characteristic Function of the Poisson Distribution.

$$\phi_X(t) = \sum_{x=0}^{\infty} e^{itx} \cdot p(x, \lambda) = \sum_{x=0}^{\infty} e^{itx} \frac{e^{-\lambda}\lambda^x}{x!} = e^{-\lambda}\sum_{x=0}^{\infty}\frac{(\lambda e^{it})^x}{x!} = e^{-\lambda}\, e^{\lambda e^{it}} = e^{\lambda(e^{it} - 1)} \quad \dots (8\cdot20)$$

8·5·7. Cumulants of the Poisson Distribution.

$$K_X(t) = \log M_X(t) = \log\left[e^{\lambda(e^t - 1)}\right] = \lambda(e^t - 1)$$

$$= \lambda\left[\left(1 + t + \frac{t^2}{2!} + \frac{t^3}{3!} + \dots + \frac{t^r}{r!} + \dots\right) - 1\right] = \lambda\left[t + \frac{t^2}{2!} + \frac{t^3}{3!} + \dots + \frac{t^r}{r!} + \dots\right]$$

$$\kappa_r = r\text{th Cumulant} = \text{Coefficient of } \frac{t^r}{r!} \text{ in } K_X(t) = \lambda$$

$$\Rightarrow \qquad \kappa_r = \lambda;\ r = 1, 2, 3, \dots \quad \dots (8\cdot20a)$$

Hence, *all cumulants of the Poisson distribution are equal, each being equal to* λ.

In particular, we have

Mean $= \kappa_1 = \lambda, \quad \mu_2 = \kappa_2 = \lambda, \quad \mu_3 = \kappa_3 = \lambda \quad$ and $\quad \mu_4 = \kappa_4 + 3\kappa_2^2 = \lambda + 3\lambda^2$

$$\beta_1 = \frac{\mu_3^2}{\mu_2^3} = \frac{\lambda^2}{\lambda^3} = \frac{1}{\lambda} \quad \text{and} \quad \beta_2 = \frac{\mu_4}{\mu_2^2} = \frac{\lambda + 3\lambda^2}{\lambda^2} = \frac{1}{\lambda} + 3$$

Remark. If m is the mean and σ is the s.d. of Poisson distribution with parameter λ, then

$$m\,\sigma\,\gamma_1\,\gamma_2 = \lambda \cdot \sqrt{\lambda} \cdot \sqrt{\beta_1}\,(\beta_2 - 3) = \lambda\sqrt{\lambda} \cdot \frac{1}{\sqrt{\lambda}} \cdot \frac{1}{\lambda} = 1.$$

8·5·8. Additive or Reproductive Property of Independent Poisson Variates.

Sum of independent Poisson variates is also a Poisson variate. More elaborately, if X_i, ($i = 1, 2, ..., n$) are independent Poisson variates with parameters λ_i ; $i = 1, 2, ..., n$ respectively, then $\sum_{i=1}^{n} X_i$ is also a Poisson variate with parameter $\sum_{i=1}^{n} \lambda_i$.

Proof. $M_{X_i}(t) = e^{\lambda_i (e^t - 1)} ; i = 1, 2, ..., n$ $\quad [\because X_i \sim P(\lambda_i)]$

$$M_{X_1 + X_2 + ... + X_n}(t) = M_{X_1}(t)\, M_{X_2}(t) ... M_{X_n}(t), \quad \text{[Since } X_i ; i = 1, 2, ..., n \text{ are independent.]}$$

$$= e^{\lambda_1 (e^t - 1)}\, e^{\lambda_2 (e^t - 1)} ... e^{\lambda_n (e^t - 1)} = e^{(\lambda_1 + \lambda_2 + ... + \lambda_n)(e^t - 1)}$$

which is the *m.g.f.* of a Poisson variate with parameter $\lambda_1 + \lambda_2 + ... + \lambda_n$. Hence, by uniqueness theorem of *m.g.f.*'s, $\sum_{i=1}^{n} X_i$ is also a Poisson variate with parameter $\sum_{i=1}^{n} \lambda_i$

Remarks:

1. **In fact, the converse of the above result is also true,** *i.e., if $X_1, X_2, ..., X_n$ are independent and $\sum_{i=1}^{n} X_i$ has a Poisson distribution, then each of the random variables $X_1, X_2, ..., X_n$ has a Poisson distribution.*

Let X_1 and X_2 be independent *r.v.*'s so that $X_1 \sim P(\lambda_1)$ and $X_1 + X_2 \sim P(\lambda_1 + \lambda_2)$. Then we want to prove that $X_2 \sim P(\lambda_2)$.

Proof. Since X_1 and X_2 are independent, we have

$$M_{X_1 + X_2}(t) = M_{X_1}(t)\, M_{X_2}(t) \quad \Rightarrow \quad e^{(\lambda_1 + \lambda_2)(e^t - 1)} = e^{\lambda_1 (e^t - 1)}\, M_{X_2}(t)$$

$$\Rightarrow \quad M_{X_2}(t) = e^{\lambda_2 (e^t - 1)} \quad \Rightarrow \quad X_2 \sim P(\lambda_2), \text{ by uniqueness theorem of } m.g.f.$$

2. The difference of two independent Poisson variates is not a Poisson variate.

$$M_{X_1 - X_2}(t) = M_{X_1 + (-X_2)}(t) = M_{X_1}(t) . M_{(-X_2)}(t), \quad \text{(since } X_1 \text{ and } X_2 \text{ are independent)}$$

$$\therefore \quad M_{X_1 - X_2}(t) = M_{X_1}(t) . M_{X_2}(-t) = e^{\lambda_1 (e^t - 1)} . e^{\lambda_2 (e^{-t} - 1)} \quad [\because M_{cX}(t) = M_X(ct)]$$

which cannot be put in the form $e^{\lambda (e^t - 1)}$. Hence $(X_1 - X_2)$ is not a Poisson variate.

Moreover, the difference $(X_1 - X_2)$ cannot be a Poisson variate is evident from the fact that it may have positive as well as negative values, while a Poisson variate is always non-negative.

8·5·9. Probability Generating Function of Poisson Distribution.

$$\text{P.G.F. of } X = \sum_{k=0}^{\infty} \frac{e^{-\lambda} \lambda^k}{k!} . s^k = \sum_{k=0}^{\infty} e^{-\lambda} \frac{(\lambda s)^k}{k!} = e^{-\lambda}\, e^{\lambda s} = e^{\lambda (s - 1)} \quad ... (8·21)$$

Example 8·29. *A manufacturer of cotter pins knows that 5% of his product is defective. If he sells cotter pins in boxes of 100 and guarantees that not more than 10 pins will be defective, what is the approximate probability that a box will fail to meet the guaranteed quality ?*

Solution. In the usual notations, we are given: $n = 100$,

and p = Probability of a defective pin = 5% = 0.05

(Since 'p' is small, we may use Poisson distribution.)

$\therefore$ λ = Mean number of defective pins = $np = 100 \times 0{\cdot}05 = 5$

Let the random variable X denote the number of defective pins in a box of 100. Then by Poisson probability law, the probability of x defective pins in a box is :

$$P(X = x) = \frac{e^{-\lambda}\,\lambda^x}{x\,!} = \frac{e^{-5}\,5^x}{x\,!};\ x = 0, 1, 2, \ldots$$

Probability that a box will fail to meet the guaranteed quality is :

$$P(X > 10) = 1 - P(X \le 10) = 1 - \sum_{x=0}^{10} \frac{e^{-5}\,5^x}{x\,!} = 1 - e^{-5} \sum_{x=0}^{10} \frac{5^x}{x\,!}$$

Example 8·30. *A car hire firm has two cars, which it hires out day by day. The number of demands for a car on each day is distributed as a Poisson distribution with mean 1·5. Calculate the proportion of days on which (i) neither car is used, and (ii) the proportion of days on which some demand is refused.*

Solution. Here the random variable X, which denotes the number of demands for a car on any day follows Poisson distribution with mean $\lambda = 1{\cdot}5$. The proportion of days on which there are x demands for a car is given by :

$$P(X = x) = \frac{e^{-1\cdot5}\,(1{\cdot}5)^x}{x\,!};\ x = 0, 1, 2, \ldots$$

(*i*) Proportion of days on which neither car is used is given by :

$$P(X = 0) = e^{-1\cdot5} = \left\{1 - 1{\cdot}5 + \frac{(1{\cdot}5)^2}{2\,!} - \frac{(1{\cdot}5)^3}{3\,!} + \frac{(1{\cdot}5)^4}{4\,!} - \ldots\right\} = 0{\cdot}2231$$

(*ii*) Proportion of days on which some demand is refused is :

$$P(X > 2) = 1 - \mathrm{P}\,(X \le 2) = 1 - [P(X = 0) + P(X = 1) + P(X = 2)]$$

$$= 1 - e^{-1\cdot5}\left\{1 + 1{\cdot}5 + \frac{(1{\cdot}5)^2}{2\,!}\right\} = 1 - 0{\cdot}2231 \times 3{\cdot}625 = 0{\cdot}19126.$$

Example 8·31. *An insurance company insures 4,000 people against loss of both eyes in a car accident. Based on previous data, the rates were computed on the assumption that on the average 10 persons in 1,00,000 will have car accident each year that result in this type of injury. What is the probability that more than 3 of the insured will collect on their policy in a given year ?*

Solution. In usual notations, we are given : $n = 4{,}000$, and

$$p = \text{Probability of loss of both eyes in a car accident} = \frac{10}{1{,}00{,}000} = 0{\cdot}0001.$$

Since p is very small and n is large, we may approximate the given distribution by Poisson distribution. Thus the parameter λ of the Poisson distribution is :

$$\lambda = np = 4{,}000 \times 0{\cdot}0001 = 0{\cdot}4$$

Let the random variable X denote number of car accidents in the batch of 4,000 people. Then by Poisson probability law :

$$P(X = x) = \frac{e^{-\lambda} \cdot \lambda^x}{x!} = \frac{e^{-0\cdot4}(0\cdot4)^x}{x!}; x = 0, 1, 2 \ldots \qquad \ldots (*)$$

Hence, the required probability that more than 3 of the insured will collect on their policy is given by :

$$P(X > 3) = 1 - [P(X = 0) + P(X = 1) + P(X = 2) + P(X = 3)]$$

$$= 1 - e^{-0\cdot4}\left\{(0\cdot4)^0 + (0\cdot4) + \frac{(0\cdot4)^2}{2!} + \frac{(0\cdot4)^3}{3!}\right\}$$

$$= 1 - 0\cdot6703\ (1 + 0\cdot4 + 0\cdot08 + 0\cdot0107) = 1 - 0\cdot6703 \times 1\cdot4907 = 0\cdot0008.$$

Example 8·32. *A manufacturer, who produces medicine bottles, finds that 0.1% of the bottles are defective. The bottles are packed in boxes, each containing 500 bottles. A drug manufacturer buys 100 boxes from the producer of bottles. Using Poisson distribution, find how many boxes will contain :*

(i) no defective, (ii) at most 3 defective, and (iii) at least two defective bottles.

[Given $e^{-0\cdot5} = 0\cdot6065$]

Solution. In the usual notations, we are given :

$N = 100$, $n = 500$, p = Probability of a defective bottle = 0·001, and $\lambda = np = 500 \times 0\cdot001 = 0\cdot5$

Let the random variable X denote the number of defective bottles in a box of 500. Then by Poisson probability law, the probability of x defective bottles in a box is given by :

$$P(X = x) = \frac{e^{-0\cdot5}(0\cdot5)^x}{x!} = \frac{0\cdot6065 \times (0\cdot5)^x}{x!}; x = 0, 1, 2, \ldots$$

Hence, in a consignment of 100 boxes, the frequency (number) of boxes containing x defective bottles is :

$$f(x) = N.P\ (X = x) = \frac{100 \times 0\cdot6065 \times (0\cdot5)^x}{x!};\ 0, 1, 2, \ldots, 500.$$

(*i*) Number of boxes containing no defective bottle

$$= 100 \times P(X = 0) = 100 \times 0\cdot6065 \simeq 61$$

(*ii*) Number of boxes containing at most 3 defective bottles

$$= 100\ [P(X \le 3)] = 100\ [P(X = 0) + P(X = 1) + P(X = 2) + P(X = 3)]$$

$$= 100 \times 0\cdot6065\ [1 + 0\cdot5 + 0\cdot125 + 0\cdot0208]$$

$$= 100 \times 0\cdot6065 \times 1\cdot6458 = 100 \times 0\cdot9981 \simeq 100.$$

(*iii*) Number of boxes containing at least two defective bottles

$$= 100\ [P\ (X \ge 2)] = 100\ [1 - P(X = 0) - P(X = 1)]$$

$$= 100\ [1 - 0\cdot6065 - 0\cdot6065 \times 0\cdot5] = 100 \times 0\cdot09025 \simeq 9.$$

Example 8·33. *Six coins are tossed 6,400 times. Using the Poisson distribution, find the approximate probability of getting six heads r times.*

Solution. The probability of obtaining six heads in one throw of six coins (*a* single trial), is $p = (1/2)^6$, assuming that head and tail are equally probable.

$$\therefore \qquad \lambda = np = 6{,}400 \times (1/2)^6 = 100$$

Hence, using Poisson probability distribution, the required probability of getting 6 heads r times is given by :

$$P(X = r) = \frac{e^{-\lambda} \cdot \lambda^r}{r\,!} = \frac{e^{-100} \cdot (100)^r}{r\,!} \, ; \; r = 0, 1, 2, \ldots$$

Example 8·34. *In a book of 520 pages, 390 typographical errors occur. Assuming Poisson law for the number of errors per page, find the probability that a random sample of 5 pages will contain no error.*

Solution. The average number of typographical errors per page in the book is given by:

$$\lambda = (390/520) = 0{\cdot}75$$

Hence, using Poisson probability law, the probability of x errors per page is given by :

$$P(X = x) = \frac{e^{-\lambda} \cdot \lambda^x}{x\,!} = \frac{e^{-0{\cdot}75}(0{\cdot}75)^x}{x\,!} \, ; \; x = 0, 1, 2, \ldots$$

The required probability that a random sample of 5 pages will contain no error is given by :

$$[\, P(X = 0) \,]^5 = (\, e^{-0{\cdot}75} \,)^5 = e^{-3{\cdot}75}$$

Example 8·35. *In a Poisson frequency distribution, frequency corresponding to 3 successes is 2/3 times frequency corresponding to 4 successes. Find the mean and standard deviation of the distribution.*

Solution. Let X be a random variable following Poisson distribution with parameter λ and total frequency N. Then the frequency function is given by :

$$f(x) = N \,.\, p(x) = NP(X = x) = N \cdot \frac{e^{-\lambda} \cdot \lambda^x}{x\,!} \, ; \; x = 0, 1, 2, \ldots \qquad \ldots (*)$$

Putting $x = 3$ and 4 in (*), $\quad f(3) = N \cdot \dfrac{e^{-\lambda}\,\lambda^3}{3\,!} \quad$ and $\quad f(4) = N \cdot \dfrac{e^{-\lambda}\,\lambda^4}{4\,!}$

We are given : $\quad f(3) = \dfrac{2}{3} f(4) \quad \Rightarrow \quad N \cdot \dfrac{e^{-\lambda}\,\lambda^3}{3\,!} = \dfrac{2}{3} N \cdot \dfrac{e^{-\lambda}\,\lambda^4}{4\,!}$

$$\Rightarrow \quad \frac{1}{3\,!} = \frac{2}{3}\,\frac{\lambda}{4\,!} \qquad \Rightarrow \qquad \lambda = \frac{1}{3\,!} \times \frac{3}{2} \times 4\,! = 6.$$

Mean of the Poisson distribution $= \lambda = 6$ and *s.d.* of the distribution $= \sqrt{\lambda} = \sqrt{6}$.

Example 8·36. *Suppose that the number of telephone calls coming into a telephone exchange between 10 A.M. and 11 A.M. say, X_1 is a random variable with Poisson distribution with parameter 2. Similarly the number of calls arriving between 11 A.M. and 12 noon, say, X_2 has a Poisson distribution with parameter 6. If X_1 and X_2 are independent, what is the probability that more than 5 calls come in-between 10 A.M. and 12 noon ?*

Solution. We are given : $X_1 \sim P(2)$ and $X_2 \sim P(6)$. Let $X = X_1 + X_2$. By the additive property of Poisson distribution, X is also a Poisson variate with parameter (say) $\lambda = 2 + 6 = 8$.

Hence, the probability of x calls in-between 10 A.M. and 12 noon is given by :

$$P(X = x) = \frac{e^{-\lambda}\,\lambda^x}{x\,!} = \frac{e^{-8}\,8^x}{x\,!} \, ; \; x = 0, 1, 2, \ldots$$

Probability that more than 5 calls come in-between 10 A.M. and 12 noon is :

$$P(X > 5) = 1 - P(X \le 5) = 1 - \sum_{x=0}^{5} \frac{e^{-8}\,8^x}{x\,!} = 1 - 0{\cdot}1912 = 0{\cdot}8088.$$

Example 8·37. *A Poisson distribution has a double mode at x = 1 and x = 2. What is the probability that x will have one or the other of these two values ?*

Solution. We know that if the Poisson distribution is bimodal, then the two modes are at the points $x = \lambda - 1$ and $x = \lambda$, where λ is the parameter of the Poisson distribution. Therefore, since we are given that the two modes are at the points $x = 1$ and $x = 2$, we find that $\lambda = 2$.

$$\therefore \quad P(X = x) = \frac{e^{-\lambda}\lambda^x}{x!} = \frac{e^{-2}\, 2^x}{x!}; \; x = 0, 1, 2, \ldots$$

$$\Rightarrow \quad P(X = 1) = e^{-2}\, 2 \quad \text{and} \quad P(X = 2) = \frac{e^{-2}\, 2^2}{2!} = e^{-2} \cdot 2.$$

$$\text{Required probability} = P(X = 1) + P(X = 2) = 2e^{-2} + 2e^{-2} = 0{\cdot}542.$$

Example 8·38. *If X and Y are independent Poisson variates such that*

$$P(X = 1) = P(X = 2) \text{ and } P(Y = 2) = P(Y = 3) \qquad \ldots (*)$$

Find the variance of X – 2Y.

Solution. Let $X \sim P(\lambda)$ and $Y \sim P(\mu)$. Then we have

$$P(X = x) = \frac{e^{-\lambda}.\lambda^x}{x!},\; x = 0, 1, 2, \ldots ;\; \lambda > 0 \text{ and } P(Y = y) = \frac{e^{-\mu}.\mu^y}{y!},\; y = 0, 1, 2, \ldots ;\; \mu > 0$$

$$\text{Using } (*), \quad \lambda\, e^{-\lambda} = \frac{\lambda^2\, e^{-\lambda}}{2!} \quad \text{and} \quad \frac{\mu^2\, e^{-\mu}}{2} = \frac{\mu^3\, e^{-\mu}}{3!} \qquad \ldots (**)$$

Solving (**), we obtain

$$\lambda e^{-\lambda}(\lambda - 2) = 0 \text{ and } \mu^2 e^{-\mu}(\mu - 3) = 0 \quad \Rightarrow \quad \lambda = 2 \text{ and } \mu = 3, \text{ since } \lambda > 0, \mu > 0.$$

$$\text{Var}(X) = \lambda = 2, \quad \text{and} \quad \text{Var}(Y) = \mu = 3 \qquad \ldots (***)$$

$$\therefore \quad \text{Var}(X - 2Y) = 1^2 . \text{Var}(X) + (-2)^2 . \text{Var } Y,$$

covariance term vanishes since X and Y are independent.

Hence, on using (***), we get : $\quad \text{Var}(X - 2Y) = 2 + 4 \times 3 = 14.$

Example 8·39. *If X and Y are independent Poisson variates with means λ_1 and λ_2 respectively, find the probability that :*

(i) X + Y = k, *and* *(ii) X = Y.*

Solution. We have : $X \sim P(\lambda_1)$ and $Y \sim P(\lambda_2)$ are independent. Hence,

$$P(X = x) = \frac{e^{-\lambda_1}\lambda_1^{\,x}}{x!},\; x = 0, 1, 2, 3, \ldots;\; \lambda_1 > 0 \text{ and } P(Y = y) = \frac{e^{-\lambda_2}\lambda_2^{\,y}}{y!},\; y = 0, 1, 2, 3, \ldots;\; \lambda_2 > 0$$

$$(i)\; P(X + Y = k) = \sum_{r=0}^{k} P(X = r \cap Y = k - r)$$

$$= \sum_{r=0}^{k} P(X = r) \cdot P(Y = k - r) \qquad [\because X \text{ and } Y \text{ are independent}]$$

$$= \sum_{r=0}^{k} \frac{e^{-\lambda_1}\lambda_1^{\,r}}{r!} \cdot \frac{e^{-\lambda_2}\lambda_2^{\,k-r}}{(k-r)!} = e^{-(\lambda_1 + \lambda_2)} \sum_{r=0}^{k} \frac{\lambda_1^{\,r}\lambda_2^{\,k-r}}{r!\,(k-r)!}$$

$$= e^{-(\lambda_1+\lambda_2)}\left[\frac{\lambda_2^{\,k}}{k!}+\frac{\lambda_1\lambda_2^{\,k-1}}{1!(k-1)!}+\frac{\lambda_1^{\,2}\lambda_2^{\,k-2}}{2!(k-2)!}+\ldots+\frac{\lambda_1^{\,k}}{k!}\right]$$

$$= \frac{e^{-(\lambda_1+\lambda_2)}}{k!}\left[\lambda_2^{\,k}+{}^kC_1\,\lambda_2^{\,k-1}.\lambda_1+{}^kC_2\,\lambda_2^{\,k-2}.\lambda_1^{\,2}+\ldots+\lambda_1^{\,k}\right]$$

$$= \frac{e^{-(\lambda_1+\lambda_2)}}{k!}\times(\lambda_1+\lambda_2)^k;\ k=0,1,2,\ldots$$

which is the probability function of Poisson distribution with parameter $\lambda_1 + \lambda_2$.

Aliter. Since $X \sim P(\lambda_1)$ and $Y \sim P(\lambda_2)$ are independent, by the additive property of Poisson distribution,

$$X + Y \sim P(\lambda_1 + \lambda_2).$$

Hence,

$$P(X+Y=k)=\frac{e^{-(\lambda_1+\lambda_2)}\times(\lambda_1+\lambda_2)^k}{k!};\ k=0,1,2,\ldots$$

(*ii*) $$P(X=Y)=\sum_{r=0}^{\infty}P(X=r\cap Y=r)=\sum_{r=0}^{\infty}P(X=r)\,P(Y=r)\qquad[\because X \text{ and } Y \text{ are independent}]$$

$$=e^{-(\lambda_1+\lambda_2)}\sum_{r=0}^{\infty}\frac{(\lambda_1\lambda_2)^r}{(r!)^2}$$

Example 8·40. *Show that in Poisson distribution with unit mean, mean deviation about mean is (2/e) times the standard deviation.*

Solution. Let $X \sim P(\lambda)$. We are given : Mean $= \lambda = 1$.

$$\therefore\qquad P(X=x)=\frac{e^{-\lambda}\lambda^x}{x!}=\frac{e^{-1}.1}{x!}=\frac{e^{-1}}{x!};\ x=0,1,2,\ldots$$

Mean deviation about mean 1 is :

$$E(|X-1|)=\sum_{x=0}^{\infty}|x-1|P(X=x)=e^{-1}\sum_{x=0}^{\infty}\frac{|x-1|}{x!}=e^{-1}\left(1+\frac{1}{2!}+\frac{2}{3!}+\frac{3}{4!}+\ldots\right)$$

We have $$\frac{n}{(n+1)!}=\frac{(n+1)-1}{(n+1)!}=\frac{1}{n!}-\frac{1}{(n+1)!}$$

$\therefore$ Mean deviation about mean

$$=e^{-1}\left\{1+\left(1-\frac{1}{2!}\right)+\left(\frac{1}{2!}-\frac{1}{3!}\right)+\left(\frac{1}{3!}-\frac{1}{4!}\right)+\ldots\right\}$$

$$=e^{-1}(1+1)=\frac{2}{e}\times 1=\frac{2}{e}\times\text{standard deviation,}$$

since for the Poisson distribution, variance = mean = 1 (given) $\Rightarrow$ *s.d.* = 1.

Example 8·41. (*a*) *Let $X_1, X_2, \ldots, X_n$ be identically and independently distributed Bin (1, p) variates.*

Let $S_n = \sum_{j=1}^{n} X_j$, *be a binomial (n, p) variate and $M_n(t)$ be the m.g.f. of S_n.*

Find, $\lim_{n \to \infty} M_n(t)$, *using* $np = \lambda$ *(constant).*

(b) *If* $X \sim P(\lambda)$, *find* $E(\cos tX)$ *and* $E(\sin tX)$.

Solution. (*a*) Since X_i, $i = 1, 2, ..., n$ are *i.i.d.* binomial variates $B(1, p)$,

$$S_n = \sum_{j=1}^{n} X_j, \text{ is a binomial } B(n, p) \text{ variate.}$$

$$\therefore \qquad M_n(t) = m.g.f. \text{ of } S_n = (q + pe^t)^n = \left[1 + (e^t - 1)p\right]^n$$

If we take $np = \lambda \Rightarrow p = \lambda/n$ and let $n \to \infty$, we get

$$\lim_{n \to \infty} M_n(t) = \lim_{n \to \infty} \left[1 + \frac{(e^t - 1)\lambda}{n}\right]^n = \exp[\lambda(e^t - 1)],$$

which is the *m.g.f.* of Poisson distribution with parameter λ.

Hence, by uniqueness theorem of *m.g.f.*,

$$S_n = \sum_{j=1}^{n} X_j \to P(\lambda), \text{ as } n \to \infty, \text{ with } np = \lambda \text{ (fixed).}$$

(*b*) If $X \sim P(\lambda)$, its characteristic function $\Phi_X(t)$ is given by :

$$\Phi_X(t) = E(e^{itX}) = e^{\lambda(e^{it} - 1)}$$

$$\Rightarrow \qquad E[\cos tX + i \sin tX] = e^{\lambda(\cos t + i \sin t - 1)} = e^{\lambda(\cos t - 1)} \times e^{i\lambda \sin t}$$

$$\Rightarrow E\left[(\cos tX + i\, E(\sin tX)\right] = e^{\lambda(\cos t - 1)} \times \left[\cos(\lambda \sin t) + i \sin(\lambda \sin t)\right]$$

Equating real and magnify parts on both sides, we get :

$$E(\cos tX) = e^{\lambda(\cos t - 1)} \,.\, \cos(\lambda \sin t)$$

and

$$E(\sin tX) = e^{\lambda(\cos t - 1)} \,.\, \sin(\lambda \sin t)$$

Example 8·42. *If X and Y are independent Poisson variates with means* m_1 *and* m_2 *respectively, prove that the probability that* $X - Y$ *has the value 'r' is the coefficient of* t^r *in :* $\exp(m_1 t + m_2 t^{-1} - m_1 - m_2)$.

Solution. Since X and Y are independent Poisson variates with means m_1 and m_2 respectively,

$$P(X = x) = \frac{e^{-m_1} m_1^x}{x!};\; x = 0, 1, 2, ... \infty, \text{ and } \quad P(Y = y) = \frac{e^{-m_2} m_2^y}{y!};\; y = 0, 1, 2, ... \infty \qquad ...(1)$$

$$P(X - Y = r) = \sum_{s=0}^{\infty} P(X = r + s \cap Y = s) = \sum_{s=0}^{\infty} P(X = r + s) \cdot P(Y = s)$$

[Since X and Y are independent.]

$$= \sum_{s=0}^{\infty} \frac{e^{-m_1} m_1^{r+s}}{(r+s)!} \times \frac{e^{-m_2} m_2^s}{s!} = e^{-m_1 - m_2} \sum_{s=0}^{\infty} \frac{m_1^{r+s}\, m_2^s}{(r+s)!\, s!} \qquad ...(2)$$

We have : $\exp(m_1 t + m_2 t^{-1}) = \exp(m_1 t) \,.\, \exp(m_2 t^{-1})$

$$= \left\{1 + m_1 + \frac{(m_1 t)^2}{2!} + ... + \frac{(m_1 t)^{r+s}}{(r+s)!} + \cdots\right\} \times \left\{1 + m_2 t^{-1} + \frac{(m_2 t^{-1})^2}{2!} + ... + \frac{(m_2 t^{-1})^{r+s}}{s!} + \cdots\right\}$$

$\therefore$ Co-efficient of t^r in $\exp\,(m_1\, t + m_2\, t^{-1}) = \displaystyle\sum_{s=0}^{\infty} \frac{m_1^{r+s}\, m_2^{s}}{(r+s)!\, s!}$ ... (3)

Hence, from (2) and (3), we get

$$P(X - Y = r) = e^{-m_1 - m_2} \times \text{Coefficient of } t^r \text{ in } e^{m_1 t + m_2 t^{-1}}$$

$$= \text{Coefficient of } t^r \text{ in } \exp\,(m_1 t + m_2\, t^{-1} - m_1 - m_2),$$

which is the required result.

Example 8·43. *If X is a Poisson variate with mean m, show that* $\dfrac{X-m}{\sqrt{m}}$ *is a variable with mean zero and variance unity. Find the M.G.F. for this variable and show that it approaches* $\exp\,(t^2/2)$ *as* $m \to \infty$. *Also interpret the result.*

Solution. Let $Y = \dfrac{X-m}{\sqrt{m}}$ so that $E(Y) = \dfrac{1}{\sqrt{m}} E(X-m) = 0$

$$\therefore \quad V(Y) = E\left(\frac{X-m}{\sqrt{m}}\right)^2 = \frac{1}{m} E(X-m)^2 = \frac{1}{m}\,.\ \mu_2 = \frac{1}{m}\,.\, m = 1$$

Hence, $Y = (X-m)/\sqrt{m}$ has zero mean and unit variance.

$$\text{M.G.F. of } Y = M_Y(t) = E\,(e^{tY}) = E\left[e^{t(X-m)/\sqrt{m}}\right]$$

$$= e^{-t\sqrt{m}}\left[E(e^{tX/\sqrt{m}})\right] = e^{-t\sqrt{m}} \sum_{x=0}^{\infty} \frac{e^{-m}\, m^x}{x!}\,.\, e^{tx/\sqrt{m}} = e^{-t\sqrt{m}}\,.\, e^{-m} \sum_{x=0}^{\infty} \frac{(me^{t/\sqrt{m}})^x}{x!}$$

$$= e^{-m - t\sqrt{m}}\left[1 + \frac{me^{t/\sqrt{m}}}{1!} + \frac{(me^{t/\sqrt{m}})^2}{2!} + \ldots\right]$$

$$= e^{-m - t\sqrt{m}}\,.\exp\,(me^{t/\sqrt{m}}) = \exp\left[-m - t\sqrt{m} + me^{t/\sqrt{m}}\right]$$

$$= \exp\left\{-m - t\sqrt{m} + m\left(1 + \frac{t}{\sqrt{m}} + \frac{t^2}{2!\,m} + \frac{t^3}{3!\,m^{3/2}} + \ldots\right)\right\}$$

$$= \exp\left[\frac{1}{2}t^2 + \frac{1}{3!}\frac{t^2}{\sqrt{m}} + \ldots\right]$$

Proceeding to limit as $m \to \infty$, $\displaystyle\lim_{m\to\infty} M_Y(t) = \exp\,(t^2/2).$ ... (*)

Interpretation. (*) is the *m.g.f.* of Standard Normal Variate [*c.f.*, Remark to § 9·2·5]. Hence, by uniqueness theorem of *m.g.f.*'s, standard Poisson variate tends to standard normal variate as $m \to \infty$. Hence, *Poisson distribution tends to Normal distribution for large values of parameter m.*

Example 8·44. *Deduce the first four moments about the mean of the Poisson distribution from those of the Binomial distribution.*

Solution. The first four central moments of the binomial distribution are :

$$\left.\begin{aligned} &\mu_1 = 0, && \text{Mean} = np \\ &\mu_2 = npq, && \mu_3 = npq(q-p) \\ &\mu_4 = npq(1-6pq) + 3n^2 p^2 q^2 \end{aligned}\right\} \quad \ldots (*)$$

Poisson distribution is a limiting form of the binomial distribution under the following conditions :

(*i*) $n \to \infty$, (*ii*) $p \to 0$, *i.e.*, $q \to 1$, and (*iii*) $np = \lambda$ (say) is finite.

Using these conditions, we get from (*), the moments of Poisson distribution as :

$$\mu_1 = 0, \quad \text{Mean} = \lim (np) = \lambda\ ; \quad \mu_2 = \lim (npq) = \lim (np) \,.\, \lim (q) = \lambda \,.\, 1 = \lambda,$$

$$\mu_3 = \lim \left[npq\,(q - p)\right] = \lambda \,.\, 1 \,.\, (1 - 0) = \lambda\ ;$$

$$\mu_4 = \lim \left[npq\,(1 - 6pq) + 3\,(np)^2 q^2\right] = \left[\lambda \,.\, 1\,(1 - 6 \times 0 \times 1) + 3\lambda^2 \times 1\right] = \lambda + 3\lambda^2.$$

Example 8·45. *If X is a Poisson variate with parameter m and Y is another discrete variable whose conditional distribution for a given X is given by :*

$$P(Y = r \mid X = x) = \binom{x}{r} p^r (1 - p)^{x - r}; \quad 0 < p < 1,\ r = 0, 1, 2, ..., x\ ;$$

then show that the unconditional distribution of Y is a Poisson distribution with parameter mp.

Solution. We are given : $\quad P(X = x) = \dfrac{e^{-m}\, m^x}{x\,!};\ x = 0, 1, 2, ...$

and $\quad P(Y = r \mid X = x) = \dbinom{x}{r} p^r (1 - p)^{x - r};\ r \le x.$

$$\therefore\ P(X = x \cap Y = r) = P(X = x)\, P(Y = r \mid X = x)$$

$$= \frac{e^{-m}\, m^x}{x\,!} \binom{x}{r} p^r (1 - p)^{x - r};\ r \le x = 0, 1, 2, ...$$

$$P(Y = r) = \text{The unconditional distribution of } Y$$

$$= \sum_{x = r}^{\infty} \left\{ \frac{e^{-m}\, m^x}{x\,!} \binom{x}{r} p^r (1 - p)^{x - r} \right\} = e^{-m} \left\{ \sum_{x = r}^{\infty} \binom{x}{r} \frac{p^r m^x (1 - p)^{x - r}}{x\,!} \right\}$$

$$= e^{-m} \left\{ \sum_{x = r}^{\infty} \frac{m^x}{x\,!} \cdot \frac{x\,!}{r\,!\,(x - r)\,!} p^r (1 - p)^{x - r} \right\} = \frac{e^{-m}}{r\,!} \left\{ \sum_{x = r}^{\infty} \frac{m^x}{(x - r)\,!} p^r (1 - p)^{x - r} \right\}$$

$$= \frac{e^{-m}\, (mp)^r}{r\,!} \left\{ \sum_{x = r}^{\infty} \frac{m^{x - r} (1 - p)^{x - r}}{(x - r)\,!} \right\} = \frac{e^{-m}\, (mp)^r}{r\,!} \left[\sum_{x = r}^{\infty} \frac{\{m(1 - p)\}^{x - r}}{(x - r)\,!} \right]$$

$$= \frac{e^{-m}\, (mp)^r}{r\,!} e^{m(1 - p)} = \frac{e^{-mp}\, (mp)^r}{r\,!};\ r = 0, 1, 2, ...$$

Hence, Y is a Poisson variate with parameter mp.

Example 8·46. *If X and Y are independent Poisson variates, show that the conditional distribution of X given X + Y, is binomial.*

Solution. Let X and Y be independent Poisson variates with parameters λ and μ respectively. Then X + Y is also a Poisson variate with parameter $\lambda + \mu$.

$$P\left[(X = r \mid (X + Y) = n)\right] = \frac{P(X = r \cap X + Y = n)}{P(X + Y = n)} = \frac{P(X = r \cap Y = n - r)}{P(X + Y = n)}$$

$$= \frac{P(X=r)\,P(Y=n-r)}{P(X+Y=n)} \qquad \text{[Since X and Y are independent.]}$$

$$= \frac{\dfrac{e^{-\lambda}\,\lambda^r}{r\,!}\cdot\dfrac{e^{-\mu}\,\mu^{n-r}}{(n-r)\,!}}{\dfrac{e^{-(\lambda+\mu)}\,(\lambda+\mu)^n}{n\,!}} = \frac{n\,!}{r\,!\,(n-r)\,!}\left(\frac{\lambda}{\lambda+\mu}\right)^r\left(\frac{\mu}{\lambda+\mu}\right)^{n-r}$$

$$= \binom{n}{r} p^r\, q^{n-r}, \text{ where } p = \frac{\lambda}{\lambda+\mu},\; q = 1-p$$

Hence, the conditional distribution of X given X + Y = n, is a binomial distribution with parameters

$$n \text{ and } p = \frac{\lambda}{\lambda+\mu}.$$

Example 8·47. *If X has a Poisson distribution with parameter λ, show that the distribution function of X is given by :*

$$F(x) = \frac{1}{\Gamma(x+1)}\int_\lambda^\infty e^{-t}\,t^x\,dt;\; x = 0, 1, 2, \ldots$$

Solution. If $X \sim P(\lambda)$ is a Poisson variate, $\quad P(X=x) = \dfrac{e^{-\lambda}\,\lambda^x}{x\,!};\; x = 0, 1, 2, \ldots \quad \ldots (*)$

Consider the incomplete gamma integral ;

$$I_x = \frac{1}{x\,!}\int_\lambda^\infty e^{-t}\,t^x\,dt;\; (x \text{ is positive integer}) = \left|-\frac{e^{-t}t^x}{x\,!}\right|_\lambda^\infty + \frac{1}{(x-1)\,!}\int_\lambda^\infty e^{-t}\,t^{x-1}\,dt$$

$$= \frac{e^{-\lambda}\,\lambda^x}{x\,!} + I_{x-1} \qquad \ldots (**)$$

which is a reduction formula for I_x.

Repeated application of (**) gives : $\quad I_x = \dfrac{e^{-\lambda}\,\lambda^x}{x\,!} + \dfrac{e^{-\lambda}\,\lambda^{x-1}}{(x-1)\,!} + \ldots + \dfrac{e^{-\lambda}\,\lambda}{1\,!} + I_0 \qquad \ldots (***)$

But $\quad I_0 = \displaystyle\int_\lambda^\infty e^{-t}\,dt = \left|-e^{-t}\right|_\lambda^\infty = e^{-\lambda}$

$$\therefore \quad I_x = e^{-\lambda} + \lambda e^{-\lambda} + \frac{\lambda^2 e^{-\lambda}}{2\,!} + \ldots + \frac{\lambda^x e^{-\lambda}}{x\,!} = P(X=0) + P(X=1) + \ldots + P(X=x)$$

$$= P(X \le x) = F(x), \qquad \text{[From (***) and (*)]}$$

where $F(\cdot)$ is the distribution function of the *r.v.* X.

$$\Rightarrow \quad F(x) = \frac{1}{x\,!}\int_\lambda^\infty e^{-t}\,t^x\,dt = \frac{1}{\Gamma(x+1)}\int_\lambda^\infty e^{-t}\,t^x\,dt \;\left[\because \Gamma(x+1) = x\,!, \text{ since } x \text{ is a positive integer}\right]$$

Remark. This result is of great practical utility. It enables us to represent the cumulative Poisson probabilities (which are generally tedious to compute numerically) in terms of incomplete gamma integral, the values of which are tabulated for different values of λ by Karl Pearson in his Tables of Incomplete Γ-Functions.

8·5·10. Recurrence Formula for the Probabilities of Poisson Distribution.

(Fitting of Poisson Distribution).

For a Poisson distribution with parameter λ, we have

$$p(x) = \frac{e^{-\lambda}.\lambda^x}{x\,!};\; x = 0, 1, 2, \ldots, \infty \quad \text{and} \quad p(x+1) = \frac{e^{-\lambda}\,\lambda^{x+1}}{(x+1)\,!};\quad x = 0, 1, 2, \ldots, \infty$$

$$\therefore \quad \frac{p(x+1)}{p(x)} = \frac{\lambda}{(x+1)} \quad \Rightarrow \quad p(x+1) = \frac{\lambda}{x+1}\, p(x) \ ; \quad x = 0, 1, 2, 3 \ldots$$

which is the required recurrence formula. ... (8·21)

This formula provides us a very convenient method of graduating the given frequency distribution by a Poisson distribution. The only probability we need to calculate is $p(0)$ which is given by $p(0) = e^{-\lambda}$, where λ is estimated from the given frequency distribution by equating the mean of the distribution to λ. The other probabilities, *viz.*, $p(1)$, $p(2)$ ... can now be easily obtained as explained below :

$$p(1) = [p(x+1)]_{x=0} = \left[\frac{\lambda}{x+1}\right]_{x=0} \times p(0)\ ;$$

$$p(2) = [p(x+1)]_{x=1} = \left[\frac{\lambda}{x+1}\right]_{x=1} \times p(1),$$

$$p(3) = [p(x+1)]_{x=2} = \left[\frac{\lambda}{x+1}\right]_{x=2} \times p(2), \quad \text{and so on.}$$

Example 8·48. *Fit a Poisson distribution to the following data :*

Number of mistakes per page :	0	1	2	3	4	*Total*
Number of pages :	109	65	22	3	1	200

Solution. If the above distribution is approximated by a Poisson distribution, then the parameter (λ) of Poisson distribution is given by :

$$\lambda = \bar{x} = \frac{0 \times 109 + 1 \times 65 + 2 \times 22 + 3 \times 3 + 4 \times 1}{200} = \frac{122}{200} = 0{\cdot}61.$$

By Poisson probability distribution, the expected frequency (number) of pages containing x mistakes is given by :

$$f(x) = N \,.\, p(x) = 200 \times \frac{e^{-0\cdot61}\,(0{\cdot}61)^x}{x\,!}\ ; \ x = 0, 1, 2, \ldots$$

Putting $x = 0, 1, 2, \ldots$, we get the expected frequencies of Poisson distribution. Also

$$p(0) = e^{-0\cdot61} = (2{\cdot}71828)^{-0\cdot61} = \frac{1}{\text{Antilog}\,(0{\cdot}61 \times \log 2{\cdot}71828)}$$

$$= \frac{1}{\text{Antilog}\,(0{\cdot}61 \times \log 0{\cdot}4343)} = \frac{1}{\text{Antilog}\,(0{\cdot}26492)} = \frac{1}{1{\cdot}841} = 0{\cdot}5432.$$

CALCULATIONS FOR EXPECTED POISSON FREQUENCIES

No. of Mistakes	*Expected Frequency*
0	$f(0) = 200 \times e^{-0\cdot61} = 200 \times 0{\cdot}5432 = 108{\cdot}64 \simeq 108$
1	$f(1) = 200 \times e^{-0\cdot61} \times 0{\cdot}61 = 108{\cdot}64 \times 0.61 = 66{\cdot}27 \simeq 66$
2	$f(2) = 200 \times e^{-0\cdot61} \times \frac{(0\cdot61)^2}{2\,!} = 66{\cdot}27 \times \frac{0\cdot61}{2} = 20{\cdot}21 \simeq 20$
3	$f(3) = 200 \times e^{-0\cdot61} \times \frac{(0\cdot61)^3}{3\,!} = 20{\cdot}21 \times \frac{0\cdot61}{3} = 4{\cdot}11 \simeq 4$
4	$f(4) = 200 \times e^{-0\cdot61} \times \frac{(0\cdot61)^4}{4\,!} = 4{\cdot}11 \times \frac{0\cdot61}{4} = 0{\cdot}63 \simeq 1$

Example 8·49. *After correcting 50 pages of the proof of a book, the proof reader finds that there are, on the average, 2 errors per 5 pages. How many pages would one expect to find with 0, 1, 2, 3 and 4 errors, in 1,000 pages of the first print of the book ? (Given that $e^{-0.4} = 0.6703$)*

Solution. Let the random variable X denote the number of errors per page. Then the mean number of errors per page is given by : $\lambda = \frac{2}{5} = 0{\cdot}4.$

Using Poisson probability law, probability of x errors per page is given by:

$$P(X = x) = p(x) = \frac{e^{-\lambda}\,\lambda^x}{x\,!} = \frac{e^{-0\cdot4}\,(0{\cdot}4)^x}{x\,!};\ x = 0, 1, 2, ...$$

Expected number of pages with x errors per page in a book of $N = 1{,}000$ pages are :

$$f(x) = N \times P(X = x) = 1{,}000 \times \frac{e^{-0\cdot4}\,(0{\cdot}4)^x}{x\,!};\ x = 0, 1, 2, ...$$

Using the recurrence formula (8·21), various probabilities can be easily calculated as shown in the following table :

CALCULATIONS FOR EXPECTED POISSON FREQUENCIES

No. of errors per page (x)	*Probability p(x)*	*Expected number of pages, f(x) = Np(x) = 1000 . p(x)*
0	$p(0) = e^{-0\cdot4} = 0{\cdot}6703$	670·3 ≃ 670
1	$p(1) = \frac{0\cdot4}{0+1}p(0) = 0{\cdot}26812$	268·12 ≃ 268
2	$p(2) = \frac{0\cdot4}{1+1}p(1) = 0{\cdot}053624$	53·624 ≃ 54
3	$p(3) = \frac{0\cdot4}{2+1}p(2) = 0{\cdot}0071298$	7·1298 ≃ 7
4	$p(4) = \frac{0\cdot4}{2+1}p(2) = 0{\cdot}0071298$	0·71298 ≃ 1

Example 8·50. *Fit a Poisson distribution to the following data which gives the number of doddens in a sample of clover seeds.*

No. of doddens (x) :	0	1	2	3	4	5	6	7	8
Observed frequency (f) :	56	156	132	92	37	22	4	0	1

Solution. Mean $= \frac{1}{N}\sum fx = \frac{986}{500} = 1{\cdot}972$. Taking the mean of the given distribution as the mean of the Poisson distribution we want to fit, we get $\lambda = 1{\cdot}972,$

and $p(x) = \dfrac{e^{-\lambda}\,.\,\lambda^x}{x\,!};\ x = 0, 1, 2, ..., \infty \quad \Rightarrow \quad p(0) = e^{-\lambda} = e^{-1\cdot972}$, so that

$$\log_{10} p(0) = -1{\cdot}972 \log_{10} e = -\ 1{\cdot}972 \times 0{\cdot}43429 = -\ 0{\cdot}856419 = \bar{1}{\cdot}143581$$

$$\therefore \quad p(0) = \text{Antilog}\,(\bar{1}{\cdot}1436) = 0{\cdot}1392$$

Using the recurrence formula (8·21), the various Probabilities, *viz.*, $p(1)$, $p(2)$, ..., can be easily calculated as shown in the following Table :

CALCULATIONS FOR EXPECTED POISSON FREQUENCIES

x	$\frac{\lambda}{x+1}$	$p(x)$	*Expected frequency* $N.p(x) = 500.\ p(x)$
0	1·972	0·13920	69·6000 ≃ 70
1	0·986	0·27455	137·2512 ≃ 137
2	0·657	0·27006	135·3296 ≃ 135
3	0·493	0·17793	88·9566 ≃ 89
4	0·394	0·10964	43·8556 ≃ 44
5	0·328	0·03459	17·2966 ≃ 17
6	0·281	0·01137	5·6846 ≃ 6
7	0·247	0·00320	1·6013 ≃ 2
8	0·219	0·00078	0·3942 ≃ 0

Remark. In rounding the figures to the nearest integer it has to be kept in mind that the total of the observed and the expected frequencies should be same.

SUPPLEMENTARY EXAMPLES ON POISSON DISTRIBUTION

Example 8·51. *A manager accepts the work submitted by his typist only when there is no mistake in the work. The typist has to type on an average 20 letters per day of about 200 words each. Find the chance of her making a mistake :*

(*i*) *if less than 1% of the letters submitted by her are rejected.*

(*ii*) *if on 90% days all the letters submitted by her are accepted.*

[As the probability of making a mistake is small, you may use Poisson Distribution. Take e = 2·72]

Solution.

(*i*) Assuming that the number of mistakes per page (X) follows Poisson distribution with parameter λ, probability of no mistake in a page is $p(0) = e^{-\lambda}$.

If less than 1 % of the letters are rejected, then more than 99% of the letters are accepted, *i.e.*, probability of making no mistakes in a page is at least 0·99.

Hence, $e^{-\lambda} \geq 0{\cdot}99$

But $\lambda = np = 200p$, where p is the probability of making mistake in typing a word.

$$\therefore \quad e^{-200p} \geq 0{\cdot}99 \quad \Rightarrow \quad -200p \cdot (\log 2{\cdot}72) \geq \log 0{\cdot}99$$

$$\Rightarrow \quad -p(200 \times 0{\cdot}4346) \geq -0{\cdot}0044 \quad \Rightarrow \quad p \leq \frac{0{\cdot}0044}{86{\cdot}92} = 0{\cdot}0000506.$$

(*ii*) The day's work of 20 letters of 200 words each is accepted, when there is no mistake in any of the $n = 20 \times 200 = 4{,}000$ words. Assuming Poisson distribution, probability of no mistake in the day's work $= e^{-\lambda}$, where $\lambda = np = 4{,}000p$. We want to find p such that :

$$\therefore \quad e^{-4000p} = 0{\cdot}90 \quad \Rightarrow \quad -4000p\ (\log 2{\cdot}72) = \log 0{\cdot}90$$

$$\Rightarrow \quad -p(4{,}000 \times 0{\cdot}4346) = -0{\cdot}0458 \quad \Rightarrow \quad p = \frac{0{\cdot}0458}{1738{\cdot}4} = 0{\cdot}0000263.$$

Example 8·52. *If a random variable X follows a Poisson distribution such that*

$$P(X = 2) = 9P(X = 4) + 90P(X = 6) \qquad ...\ (*)$$

Find (*i*) λ, (*ii*) *the mean and variance of X,* (*iii*) β_1, *the coefficient of skewness.*

Solution. If X is a Poisson variate with parameter λ, then

$$P(X = x) = \frac{e^{-\lambda} . \lambda^x}{x!}; \; x = 0, 1, 2, ..., \lambda > 0$$

Hence, (*) gives

$$\frac{e^{-\lambda} . \lambda^2}{2!} = e^{-\lambda}\left(9\frac{\lambda^4}{4!} + 90\frac{\lambda^6}{6!}\right) = \frac{e^{-\lambda}\lambda^2}{8}(3\lambda^2 + \lambda^4) \quad \Rightarrow \quad \lambda^4 + 3\lambda^2 - 4 = 0.$$

Solving as a quadratic in λ^2, $\qquad \lambda^2 = \dfrac{-3 \pm \sqrt{9+16}}{2} = \dfrac{-3 \pm 5}{2} = (-4, 1).$

(*i*) Since $\lambda > 0$, $\qquad \lambda^2 = 1 \quad \Rightarrow \quad \lambda = 1.$

(*ii*) Since for a Poisson variate, mean = variance = λ ; $E(X) = \text{Var}(X) = \lambda = 1.$

(*iii*) β_1 = Coefficient of skewness = $(1/\lambda) = 1.$

Example 8·53. *(a) If X is a Poison variate with mean m, show that the expectation of e^{-kX} is : $\exp\left[-m(1 - e^{-k})\right]$. Hence show that if $\overline{X}$ is the arithmetic mean of n independent random variables X_1, X_2, ..., X_n, each having Poisson distribution with parameter m, then exp $(-\overline{X})$ as an estimate of e^{-m} is biased, although $\overline{X}$ is an unbiased estimate of m.*

(b) If X is a Poisson variate with mean m, what would be the expectation of [exp(–kX)]. kX, k being a constant.

Solution. (*a*) $\quad E(e^{-kX}) = \sum_{x=0}^{\infty} e^{-kx}\, p(x) = \sum_{x=0}^{\infty} e^{-kx} . \dfrac{e^{-m}\, m^x}{x!}$

$$= e^{-m}\sum_{x=0}^{\infty}\frac{(me^{-k})^x}{x!} = e^{-m}\left\{1 + me^{-k} + \frac{(me^{-k})^2}{2!} + ...\right\}$$

$$= \exp(-m) \cdot \exp(m\, e^{-k}) = \exp\left[-m(1 - e^{-k})\right] \qquad ... (*)$$

We have $\qquad E(\overline{X}) = E\left(\dfrac{1}{n}\sum_{i=1}^{n} X_i\right) = \dfrac{1}{n}\sum_{i=1}^{n} E(X_i).$

Since X_i; $i = 1, 2, ..., n$ are independent Poisson variates with parameter m, $E(X_i) = m.$

$\therefore \qquad E(\overline{X}) = \dfrac{1}{n}\sum_{i=1}^{n} m = \dfrac{1}{n}\, nm = m.$

Hence, $\overline{X}$ is an unbiased estimate of m.

$$E\left[\exp(-\overline{X})\right] = E\left[\exp\left(\frac{-1}{n}\sum_{i=1}^{n} X_i\right)\right] = E\left(e^{-X_1/n} . e^{-X_2/n} ... e^{-X_n/n}\right)$$

$$= E(e^{-X_1/n})\, E(e^{-X_2/n}) ... E(e^{-X_n/n}), \text{ (since } X_1, X_2, ..., X_n \text{ are independent)} \qquad ... (**)$$

$\therefore \qquad E\left[\exp(-\overline{X})\right] = \prod_{i=1}^{n} E(e^{-X_i/n})$

Using (*) with $k = 1/n$, we get

$$E(e^{-X_i/n}) = \exp\left[-m(1-e^{-1/n})\right] \qquad \text{(Since } X_i \text{ is a Poisson variate with parameter } m.)$$

$$\therefore \quad E\left[\exp(-\bar{X})\right] = \prod_{i=1}^{n}\left[\exp\{-m(1-e^{-1/n})\}\right] = \left[\exp\{-m(1-e^{-1/n})\}\right]^n$$

$$= \exp\left[-mn(1-e^{-1/n})\right] \neq e^{-m}$$

Hence, $e^{-\bar{X}}$ is not an unbiased estimated of e^{-m}, though $\bar{X}$ is an unbiased estimate of m.

(b) $$E(e^{-kX}\, kX) = \sum_{x=0}^{\infty} e^{-kx} kx \, . \, p(x, m) = k\sum_{x=1}^{\infty} e^{-kx}\, x \frac{e^{-m}\, m^x}{x\,!}$$

$$= ke^{-m}\sum_{x=1}^{\infty}\frac{(me^{-k})^x}{(x-1)\,!} = ke^{-m}\, me^{-k}\sum_{x=1}^{\infty}\frac{(me^{-k})^{x-1}}{(x-1)\,!}$$

$$= mke^{-m-k}\left\{1 + me^{-k} + \frac{(me^{-k})^2}{2\,!} + \dots\right\}$$

$$= mk\, e^{-m-k}\, . \, e^{me^{-k}} = mk \cdot \exp\left[\{m(e^{-k}-1)\} - k\right].$$

Example 8·54. *If X is a Poisson variate with parameter m & μ_r is the rth central moment, prove that*

$$m\left({}^rC_1\, \mu_{r-1} + {}^rC_2\, \mu_{r-2} + \dots + {}^rC_r\, \mu_0\right) = \mu_{r+1}.$$

Solution. Since X ~ $P(m)$, its probability function is given by :

$$p(x) = \frac{e^{-m}\, . \, m^x}{x\,!}, \quad x = 0, 1, 2, \dots;\ m > 0$$

By *def.*, $$\mu_{r+1} = E\left[X - E(X)\right]^{r+1} = E(X-m)^{r+1} = \sum_{x=0}^{\infty}(x-m)^{r+1}\, p(x)$$

$$= \sum_{x=0}^{\infty}(x-m)^r\, (x-m)\frac{e^{-m}\, . \, m^x}{x\,!}$$

$$= \sum_{x=0}^{\infty}\frac{x(x-m)^r\, e^{-m}\, m^x}{x\,!} - m\sum_{x=0}^{\infty}(x-m)^r\, . \, \frac{e^{-m}\, m^x}{x\,!}$$

$$= \sum_{x=1}^{\infty}\frac{(x-m)^r\, e^{-m}\, m^x}{(x-1)\,!} - m\mu_r = \sum_{y=0}^{\infty}\frac{(y-m+1)^r\, . \, e^{-m}\, m^{y+1}}{y\,!} - m\mu_r, \qquad (x-1=y)$$

$$= m\sum_{y=0}^{\infty}(y-m+1)^r\, . \, p(y) - m\mu_r$$

$$= m\sum_{y=0}^{\infty}\left[(y-m)^r + {}^rC_1(y-m)^{r-1} + \dots + {}^rC_{r-1}(y-m) + 1\right] p(y) - m\mu_r$$

$$= m(\mu_r + {}^rC_1\, \mu_{r-1} + {}^rC_2\, \mu_{r-2} + \dots + {}^rC_r\, \mu_0) - m\, \mu_r$$
$$= m({}^rC_1\, \mu_{r-1} + {}^rC_2\, \mu_{r-2} + \dots + {}^rC_r\, \mu_0).$$

8·6. NEGATIVE BINOMIAL DISTRIBUTION

The equality of the mean and variance is an important characteristic of the Poisson distribution, whereas for the binomial distribution the mean is always greater than the variance. Occasionally, however, observable phenomena give rise to empirical discrete distributions which show a variance larger than the mean. Some of the commonest examples of such behaviour are the frequency distributions of plant density obtained by quadrant sampling when the clustering of plants makes the simple Poisson model inapplicable. It has been shown by different investigators that in such cases the negative binomial distribution provides an excellent model because this distribution has a variance larger than the mean. Bacterial clustering (or contagion), *e.g.*, deaths of insects, number of insect bites leads to the negative binomial distribution, and the distribution also arises in inverse sampling from a binomial population or as a weighted average of Poisson distribution. This important probability distribution is sometimes also referred to as the Pascal distribution named after the French mathematician Blaise Pascal (1623-62), but there seems to be no historical justification. The negative binomial distribution can be derived from empirical considerations in many ways. Here we consider the Binomial probability situation with some modifications.

Suppose we have a succession of n Bernoulli trials. We assume that (*i*) the trials are independent, (*ii*) the probability of success 'p' in a trial remains constant from trial to trial.

Let $f(x; r, p)$ denote the probability that there are x failures preceding the rth success in $x + r$ trials. Now, the last trial must be a success, whose probability is p. In the remaining $(x + r - 1)$ trials we must have $(r - 1)$ successes whose probability is given by binomial probability distribution by the expression : $\binom{x+r-1}{r-1} p^{r-1} q^x$.

Therefore, by compound probability theorem, $f(x, r, p)$ is given by the product of these two probabilities, i.e., $f(x, r, p) = \binom{x+r-1}{r-1} p^{r-1} q^x . p = \binom{x+r-1}{r-1} p^r q^x$

Definition. *A random variable X is said to follow a negative binomial distribution with parameters r and p if its probability mass function is given by :*

$$P(X = x) = p(x) = \begin{cases} \binom{x+r-1}{r-1} p^r q^x \,;\, x = 0, 1, 2, \dots \\ 0, \text{ otherwise} \end{cases} \qquad \dots (8{\cdot}22)$$

Also $\binom{x+r-1}{r-1} = \binom{x+r-1}{x}$ $\left[\because \binom{n}{r} = \binom{n}{n-r}\right]$

$$= \frac{(x+r-1)(x+r-2)\dots(r+1)\,r}{x\,!}$$

$$= \frac{(-1)^x (-r)(-r-1)\dots(-r-x+2)(-r-x+1)}{x\,!} = (-1)^x \binom{-r}{x} \qquad \dots (*)$$

$$\therefore \quad p(x) = \begin{cases} \binom{-r}{x} p^r (-q)^x \,;\, x = 0, 1, 2, \dots \\ 0, \text{ otherwise} \end{cases} \qquad \dots (8{\cdot}22a)$$

which is the $(x + 1)$th term in the expansion of $p^r (1 - q)^{-r}$, a binomial expansion with a negative index. Hence, the distribution is known as negative binomial distribution. Also

$$\sum_{x=0}^{\infty} p(x) = p^r \sum_{x=0}^{\infty} \binom{-r}{x} (-q)^x = p^r \cdot (1-q)^{-r} = 1$$

Therefore, $p(x)$ represents the probability function and the discrete variable which follows this probability function is called the *negative binomial variate.*

If $p = \frac{1}{Q}$ and $q = \frac{P}{Q}$ so that $Q - P = 1$, ($\because p + q = 1$), then

$$p(x) = \begin{cases} \binom{-r}{x} Q^{-r} \left(-\frac{P}{Q}\right)^x & ; x = 0, 1, 2, \ldots \\ 0, & \text{otherwise} \end{cases} \qquad \ldots (8\cdot 22b)$$

This is the general term in the negative binomial expansion $(Q - P)^{-r}$.

Remarks:

1. $p(x)$ in (8·22) or (8·22*a*) or (8·22*b*) is also sometimes written as $f(x; r, p)$.
2. **Some Important Deductions:**

(*a*) ***Geometric Distribution.*** If we take $r = 1$ in (8·22), we have

$$p(x) = q^x\, p\ ;\ x = 0, 1, 2, \ldots,$$

which is the probability function of geometric distribution (*c.f.* § 8·7).
Hence, negative binomial distribution may be regarded as the generalisation of geometric distribution.

(*b*) ***Pascal's Distribution.*** The negative binomial distribution (8·22*a*) when regarded as one having two parameters p and r is known as *Pascal's distribution.*

(*c*) ***Polya's Distribution.*** If we take $r = \frac{1}{\beta}, p = \frac{1}{1+\beta\mu}, q = 1 - p = \frac{\beta\mu}{1+\beta\mu}$ in (8·22) and use (*), we get

$$p(x) = \frac{r(r+1)(r+2)\ldots(r+x-1)}{x!}\, p^r q^x$$

$$= \frac{(1+\beta)(1+2\beta)\ldots\{1+\beta(x-1)\}}{x!} \left(\frac{1}{1+\beta\mu}\right)^{1/\beta} \left(\frac{\mu}{1+\beta\mu}\right)^x ; x = 0, 1, 2, \ldots \qquad \ldots (8\cdot 22c)$$

which is known as *Polya's distribution* with two parameters, β and μ.

(*d*) ***Second Form of Geometric Distribution.*** Taking $\beta = 1$ in Polya's distribution (8·22*c*), we get

$$p(x) = \left(\frac{1}{1+\mu}\right)\left(\frac{\mu}{1+\mu}\right)^x ; x = 0, 1, 2, \ldots \qquad \ldots (8\cdot 22d)$$

which is geometric distribution (*c.f.* § 8·7) with $p = \frac{1}{1+\mu}$, $q = 1 - p = \frac{\mu}{1+\mu}$.

8·6·1. Moment Generating Function of Negative Binomial Distribution.

$$M_X(t) = E(e^{tx}) = \sum_{x=0}^{\infty} e^{tx}\, p(x) = \sum_{x=0}^{\infty} \binom{-r}{x} Q^{-r} \left(-\frac{Pe^t}{Q}\right)^x = (Q - Pe^t)^{-r} \qquad \ldots (8\cdot 23)$$

$$\mu_1' = \left[\frac{d}{dt} M(t)\right]_{t=0} = \left[-r(-Pe^t)(Q - Pe^t)^{-r-1}\right]_{t=0} = rP$$

$\therefore$ Mean of the negative binomial distribution is rP. ... (8·23*a*)

$$\mu_2' = \left[\frac{d^2}{dt^2} M(t)\right]_{t=0}$$

$$= \left[rPe^t (Q - Pe^t)^{-r-1} + (-r-1)\, rPe^t (Q - Pe^t)^{-r-2} (-Pe^t)\right]_{t=0}$$

$$= rP + r(r+1)\, P^2$$

$$\therefore \quad \mu_2 = \mu_2' - \mu_1'^2 = r(r+1)\, P^2 + rP - r^2P^2 = rPQ \qquad \ldots (8{\cdot}23b)$$

As $Q > 1$, $rP < rPQ$, *i.e., Mean < Variance, which is a distinguishing feature of the negative binomial distribution.*

Remark. Putting $p = 1/Q$ and $q = P/Q \quad \Rightarrow \quad P = q\,Q$, in (8·23), we get

$$M_X(t) = (Q - q\,Qe^t)^{-r} = \left[Q(1 - qe^t)\right]^{-r} = \left[\frac{1 - qe^t}{p}\right]^{-r}$$

$$\Rightarrow \quad M_X(t) = \left(\frac{p}{1 - qe^t}\right)^r \qquad \ldots (8{\cdot}23c)$$

8·6·2. Cumulants of Negative Binomial Distribution.

$$K_X(t) = \log M_X(t) = -r \log (Q - Pe^t)$$

$$= -r \log\left[Q - P\left(1 + t + \frac{t^2}{2!} + \frac{t^3}{3!} + \frac{t^4}{4!} + \ldots\right)\right]$$

$$= -r \log\left[1 - P\left(t + \frac{t^2}{2!} + \frac{t^3}{3!} + \frac{t^4}{4!} + \ldots\right)\right] \qquad (\because Q - P = 1)$$

Proceeding as in § 8·3·9, we will get (on replacing n with $-r$ and p with $-P$)

$$\text{Mean} = \kappa_1 = rP, \qquad \mu_2 = rP(1 + P) = rPQ \qquad \ldots (8{\cdot}24)$$

$$\mu_3 = \kappa_3 = rP(1 + 3P + 2P^2) = rP(1 + P)(1 + 2P) = rPQ\,(Q + P)$$

$$\kappa_4 = rP(1 + P)(1 + 6P + 6P^2) = rPQ(1 + 6PQ)$$

$$\therefore \quad \mu_4 = \kappa_4 + 3\kappa_2^2 = rPQ[1 + 3PQ(r + 2)]$$

Since $Q = 1/p$, $\quad P = qQ = q/p$, $\quad$ we have in terms of p and q,

$$\text{Mean} = \frac{rq}{p}, \quad \text{Variance} = \mu_2 = \frac{rq}{p^2}, \quad \mu_3 = \frac{rq(1+q)}{p^3}, \quad \mu_4 = \frac{rq[p^2 + 3q(r+2)]}{p^4}$$

$$\therefore \quad \beta_1 = \frac{\mu_3^2}{\mu_2^3} = \frac{(1+q)^2}{rq}, \qquad \beta_2 = \frac{\mu_4}{\mu_2^2} = \frac{p^2 + 3q(r+2)}{rq} \qquad \ldots (8{\cdot}24a)$$

Also $\quad \gamma_1 = \sqrt{\beta_1} = (1+q)/\sqrt{rq} \quad$ and $\quad \gamma_2 = \beta_2 - 3 = (p^2 + 6q)/rq$

8·6·3. Additive Property of Negative Binomial Distribution.

If X_i, $(i = 1, 2, \ldots, k)$ are independent Negative Binomial variates with parameters (r_i, p), $i = 1, 2, \ldots, k$ respectively, then $\displaystyle\sum_{i=1}^{k} X_i \sim \text{Negative Binomial}\left(\sum_{i=1}^{k} r_i, p\right)$.

Proof. $X_i \sim$ Negative Binomial $(r_i, p) \Rightarrow M_{X_i}(t) = \left(\dfrac{p}{1-qe^t}\right)^{r_i}$ ... (*)

The *m.g.f.* of the sum $\sum_{i=1}^{k} X_i$ is given by :

$$M_{\sum_{i=1}^{k} X_i}(t) = \prod_{i=1}^{k} M_{X_i}(t) \qquad (\because X_i\text{'s are independent})$$

$$= \prod_{i=1}^{k}\left(\frac{p}{1-qe^t}\right)^{r_i} = \left(\frac{p}{1-qe^t}\right)^{\sum_{i=1}^{k} r_i}$$

which is the m.g.f. of Negative Binomial Distribution with parameters $\left(\sum_{i=1}^{k} r_i, p\right)$.

Hence, by uniqueness theorem of m.g.f.'s, $\sum_{i=1}^{k} X_i \sim$ Negative Binomial $\left(\sum_{i=1}^{k} r_i, p\right)$, which establishes the additive or reproductive property of Negative Binomial Distribution.

8·6·4. Poisson Distribution as a Limiting Case of the Negative Binomial Distribution.

Negative binomial distribution tends to Poisson distribution as $P \to 0$, $r \to \infty$ such that $rP = \lambda$ (finite). Proceeding to the limits, we get

$$\lim p(x) = \lim \binom{x+r-1}{r-1} p^r q^x = \lim \binom{x+r-1}{x} Q^{-r}\left(\frac{P}{Q}\right)^x$$

$$= \lim_{r\to\infty} \frac{(x+r-1)(x+r-2)\ldots(r+1)\,r}{x\,!}\cdot(1+P)^{-r}\left(\frac{P}{1+P}\right)^x$$

$$= \lim_{r\to\infty}\left\{\frac{1}{x\,!}\left(1+\frac{x-1}{r}\right)\left(1+\frac{x-2}{r}\right)\ldots\left(1+\frac{1}{r}\right).1.r^x\,(1+P)^{-r}\left(\frac{P}{1+P}\right)^x\right\}$$

$$= \frac{1}{x\,!}\lim_{r\to\infty}\left\{(1+P)^{-r}\left(\frac{rP}{1+P}\right)^x\right\} = \frac{\lambda^x}{x\,!}\lim_{r\to\infty}\left[\left(1+\frac{\lambda}{r}\right)\right]^{-r}\lim_{r\to\infty}\left(1+\frac{\lambda}{r}\right)^{-x} \qquad [\because rP = \lambda]$$

$$= \frac{\lambda^x}{x\,!}.e^{-\lambda}.1 = \frac{e^{-\lambda}\lambda^x}{x\,!},$$

which is the probability function of the Poisson distribution with parameter 'λ'.

Aliter. **Using m.g.f.** $\quad NBD \to P.D.\quad$ as $P \to 0$, $r \to \infty$ such that $rP = \lambda$ (finite).

$$M_X(t) = (Q - Pe^t)^{-r} = (1 + P - Pe^t)^{-r}\ ;\ (Q = 1 - P)$$

$$\lim_{r\to\infty} M_X(t) = \lim_{r\to\infty}\left\{\left[1+\frac{\lambda(1-e^t)}{r}\right]^r\right\}^{-1} = \left[\exp\{\lambda(1-e^t)\}\right]^{-1} = \exp[\lambda(e^t-1)]$$

which is the m.g.f. of Poisson distribution with parameter λ. Hence, by uniqueness theorem of m.g.f., the result follows.

8·6·5. Probability Generating Function of Negative Binomial Distribution.

Let X be a random variable following negative binomial distribution, then

$$P_X(s) = E(s^X) = \sum_{x=0}^{\infty} s^x\, p(x) = \sum_{x=0}^{\infty} \binom{-r}{x} p^r (-qs)^x \qquad \text{[Using (8·22a)]}$$

$$= p^r(1-qs)^{-r} = [p/(1-qs)]^r \qquad \text{... (8·25)}$$

Example 8·55. *An item is produced in large numbers. The machine is known to produce 5% defectives. A quality control inspector is examining the items by taking them at random. What is the probability that at least 4 items are to be examined in order to get 2 defectives ?*

Solution. If 2 defectives are to be obtained then it can happen in 2 or more trials. The probability of success is 0·05 for every trial. It is a negative binomial situation with $r = 2$, $p = 0{\cdot}05$ and x replaced by $(x - 2)$ and the required probability is given by :

$$P(X = 4) + P(X = 5) + \ldots = \sum_{x=4}^{\infty} \binom{x-1}{2-1} (0{\cdot}05)^2 (0{\cdot}95)^{x-2}$$

$$= 1 - \sum_{2}^{3} \binom{x-1}{2-1} (0{\cdot}05)^2 (0{\cdot}95)^{x-2} = 1 - [(0{\cdot}05)^2 + 2(0{\cdot}05)^2\,(0{\cdot}95)] = 0{\cdot}9928.$$

Example 8·56. *If $X \sim B(n, p)$ and Y has negative binomial distribution with parameters r and p, prove that :* $F_X(r-1) = 1 - F_Y(n-r)$.

Solution. $1 - F_Y(n-r) = 1 - P(Y \le n-r) = P(Y > n-r)$

$$= \sum_{y=n-r+1}^{\infty} \binom{y+r-1}{r-1} p^r q^y = p^r q^{n-r+1} \cdot \sum_{z=0}^{\infty} \binom{z+n}{r-1} q^z; [z = y-(n-r+1)]$$

$$= p^r q^{n-r+1} \sum_{z=0}^{\infty} \left\{ \sum_{k=0}^{r-1} \binom{n}{k} \binom{z}{r-1-k} \right\} q^z \qquad \left[\because \sum_{k=0}^{r} \binom{a}{k}\binom{b}{r-k} = \binom{a+b}{r} \right]$$

$$= p^r q^{n-r+1} \sum_{k=0}^{r-1} \left\{ \binom{n}{k} \sum_{z=r-1-k}^{\infty} \binom{z}{r-1-k} q^z \right\}$$

[Changing the order of summation and noting that $\binom{n}{r} = 0;\ n < r$]

$$= p^r q^{n-r+1} \sum_{k=0}^{r-1} \left[\binom{n}{k} \sum_{t=0}^{\infty} \binom{t+r-1-k}{r-1-k} q^{t+r-1-k} \right], [t = z - (r-1-k)]$$

$$= p^r q^n \sum_{k=0}^{r-1} \left\{ \binom{n}{k} q^{-k} \cdot (1-q)^{-(r-k)} \right\} = \sum_{k=0}^{r-1} \binom{n}{k} p^k q^{n-k}$$

$$= P[X \le (r-1)] = F_X(r-1)$$

Example 8·57. ***(Banach's Match-box Problem).*** *A certain mathematician always carries two match boxes (initially containing N match sticks). Each time he wants a match-stick he selects a box at random. Inevitably a moment comes when he finds a box empty. Show that the probability that there are exactly r match-sticks in one box when the other box is found empty is :* $\binom{2N-r}{N} \times \left(\frac{1}{2}\right)^{2N-r}$

Solution. Let the two match boxes be numbered 1 and 2. Let the choice of the 1st box be regarded as failure and that of second box be regarded as a success.

Since the mathematician selects the match box at random,

$$p = \text{Probability of selecting second match box} = \frac{1}{2} \quad \Rightarrow \quad q = 1 - p = \frac{1}{2}$$

The second box will be found empty if it is selected for the $(N + 1)$st time. At this stage, the first box will contain exactly r matches if $(N - r)$ matches have already been drawn from it. Hence the second box will be found empty at the stage when the first box contains exactly r matches if and only if $(N - r)$ failures precede the $(N + 1)$st success. Thus in a total of $N + 1 + (N - r) = 2N - r + 1$, trials the last one must be success and out of the remaining $(2N - r)$ trials we should have $(N - r)$ failures and N successes.

∴ Probability that second box is found empty when there are exactly r matches in first box is :

$$p_1 = \binom{2N-r}{N}\left(\frac{1}{2}\right)^N\left(\frac{1}{2}\right)^{N-r}\frac{1}{2} = \binom{2N-r}{N}\left(\frac{1}{2}\right)^{2N-r+1}$$

Similarly, the probability that first box is found empty, when the second box contains exactly r matches is given by :

$$p_2 = \binom{2N-r}{N}\left(\frac{1}{2}\right)^{2N-r+1}$$

Hence, the required probability that one match box is found empty when the other contains exactly r matches is :

$$p_1 + p_2 = 2 \times \binom{2N-r}{N}\left(\frac{1}{2}\right)^{2N-r+1} = \binom{2N-r}{N}\left(\frac{1}{2}\right)^{2N-r}$$

Remark. The statement that '*he finds the box empty*' implies that when he used the last match in this box, he did not throw it away, but instead put it back in his pocket. Thus, there is a difference between 'the box is empty' and 'the box is found empty'.

The box becomes empty when the Nth match stick was taken from it but it is found to be empty only when it is selected for the $(N + 1)$st time.

Example 8·58. *X is negative binomial variate with p.m.f.,*

$$f(x) = \begin{cases} \binom{k+x-1}{x} q^x p^k, & x = 0, 1, 2, \ldots \\ 0, & \textit{otherwise} \end{cases}$$

Show that the moment recurrence formula is : $\quad \mu_{r+1} = q\left(\dfrac{d\mu_r}{dq} + \dfrac{r\,k}{p^2}\mu_{r-1}\right).$

State how the moments of negative binomial variate can be written from the corresponding formulae for binomial variate.

Solution. For Negative Binomial Distribution with parameter k and p,

$$\text{Mean} = (k\,.\,q/p) = \mu, \text{ (say).}$$

$$\therefore \quad \mu_r = \sum_{x=0}^{\infty}(x-\mu)^r f(x) = \sum_{x=0}^{\infty}\left[\left(x - \frac{kq}{p}\right)^r \cdot \binom{k+x-1}{x} q^x \cdot p^k\right]$$

Differentiating *w.r.* to q, we get

$$\frac{d\mu_r}{dq} = \sum_x \left[r\left(x - \frac{kq}{p}\right)^{r-1} \times \left\{ \frac{d}{dq}\left(x - \frac{kq}{p}\right) \right\} \binom{k+x-1}{x} q^x p^k \right]$$

$$+ \sum_x \left[\left(x - \frac{kq}{p}\right)^r \binom{k+x-1}{x} \left\{ x\, q^{x-1} p^k + q^x . kp^{k-1} . \frac{dp}{dq} \right\} \right]$$

But $\quad \dfrac{dp}{dq} = \dfrac{d}{dq}(1-q) = -1$ and $\dfrac{d}{dq}\left(x - \dfrac{kq}{p}\right) = \dfrac{d}{dq}\left[x - k\left(\dfrac{1}{p} - 1\right)\right] = \dfrac{k}{p^2} . \dfrac{dp}{dq} = -\dfrac{k}{p^2}$

$$\therefore \quad \frac{d\mu_r}{dq} = -\frac{rk}{p^2} \sum_x \left(x - \frac{kq}{p}\right)^{r-1} f(x) + \sum_x \left(x - \frac{kq}{p}\right)^r \binom{k+x-1}{x} q^{x-1} p^k \left(x - \frac{kq}{p}\right)$$

$$= -\frac{rk}{p^2} \mu_{r-1} + \frac{1}{q} \sum_x \left(x - \frac{kq}{p}\right)^{r+1} f(x) = -\frac{rk}{p^2} \mu_{r-1} + \frac{1}{q} \mu_{r+1}$$

$$\therefore \quad \mu_{r+1} = q\left(\frac{d\mu_r}{dq} + \frac{rk}{p^2} \mu_{r-1}\right) \quad ; \quad r = 1, 2, 3, \ldots$$

For the second part, see § 8·6·6 below.

8·6·6. Deduction of Moments of Negative Binomial Distribution from those of Binomial Distribution. If we write $p = 1/Q$, $q = P/Q$ such that $Q - P = 1$, then the *m.g.f.* of negative binomial variate X is given by [*c.f.* § 8·6·1] :

$$M_X(t) = (Q - Pe^t)^{-r} \qquad \ldots (*)$$

This is analogous to *m.g.f.* of binomial variate Y with parameters n and p', *viz.*,

$$M_Y(t) = (q' + p'\, e^t)^n ;\ q' = 1 - p' \qquad \ldots (**)$$

Comparing (*) and **), we get $\qquad q' = Q,\ p' = -P \quad$ and $\quad n = -r \qquad \ldots (***)$

Using the formulae for moments of binomial distribution, the moments of negative binomial distribution are given by :

$$\text{Mean} = np' = (-r)\,(-P) = rP$$

$$\text{Variance} = np'\, q' = (-r)\,(-P)\, Q = rPQ$$

$$\mu_3 = np'\, q'\, (q' - p') = (-r)\,(-P)\, Q\, (Q + P) = rPQ\, (Q + P)$$

$$\mu_4 = np'\, q' \left[1 + 3p'\, q'\, (n - 2)\right]$$

$$= (-r)\,(-P)\, Q \left[1 + 3(-P)\, Q\, (-r - 2)\right] = rPQ \left[1 + 3PQ\, (r + 2)\right].$$

Example 8·59. *Prove that the recurrence formula for the probabilities of negative binomial distribution is :*

$$f(x + 1; r, p) = \frac{x + r}{x + 1} q \cdot f(x; r, p).$$

Solution. We have

$$f(x\,;\, r, p) = \binom{x + r - 1}{r - 1} p^r q^x, \qquad f(x + 1\,;\, r, p) = \binom{x + r}{r - 1} p^r q^{x+1}$$

$$\therefore \quad \frac{f(x + 1\,;\, r, p)}{f(x\,;\, r, p)} = \frac{(x + r)!\,(r - 1)!\, x!}{(r - 1)!\,(x + 1)!\,(x + r - 1)!} q = \frac{x + r}{x + 1} q$$

$$\Rightarrow \qquad f(x+1\,;r,p) = \frac{x+r}{x+1}\,q\cdot f(x\,;r,p) \qquad \text{... (8·25a)}$$

This recurrence relation is useful for fitting of the negative binomial distribution to the given data as discussed in the following example :

Example 8·60. *Given the hypothetical distribution :*

No. of cells (x) :	0	1	2	3	4	5	*Total*
Frequency (f) :	213	128	37	18	3	1	400

Fit a negative binomial distribution and calculate the expected frequencies.

Solution. Let X be negative binomial variate with parameters r and p.

$$\mu_1' = \text{Mean} = \frac{\Sigma fx}{\Sigma f} = \frac{273}{400} = 0{\cdot}6825 = \frac{rq}{p}\,; \quad (q = 1-p) \qquad \text{... (*)}$$

$$\mu_2' = \frac{\Sigma fx^2}{\Sigma f} = \frac{511}{400} = 1{\cdot}2775$$

$$\mu_2 = \mu_2' - \mu_1'^2 = 1{\cdot}2775 - (0{\cdot}6825)^2 = 0{\cdot}8117$$

$$\therefore \qquad \text{Variance} = 0{\cdot}8117 = \frac{rq}{p^2} \qquad \text{... (**)}$$

Dividing (*) by (**), we get : $\quad p = \dfrac{0{\cdot}6825}{0{\cdot}8117} = 0{\cdot}8408, \qquad q = 1 - p = 0{\cdot}1592$

$$\therefore \qquad r = \frac{p \times 0{\cdot}6825}{q} = \frac{0{\cdot}5738}{0{\cdot}1592} = 3{\cdot}6043 \simeq 4\,, \qquad \text{[From (*)]} \qquad \text{... (***)}$$

since, r being the number of successes cannot be fractional.

Using the recurrence relation (8·25*a*), the negative-binomial probabilities are given by :

$$f_0 = p^r = (\cdot 8408)^4 = 0{\cdot}4978 \simeq 0{\cdot}5$$

$$f_1 = \frac{r+0}{0+1}\,q\,f_0 = rq\,f_0 = 0{\cdot}5738 \times 0{\cdot}5 = 0{\cdot}2869$$

$$[\because\ rq = p \times 0{\cdot}6825 = 0{\cdot}5738\ ] \qquad \text{[From (***)]}$$

$$f_2 = \frac{r+1}{1+1}\cdot q \cdot f_1 = \frac{5}{2} \times 0{\cdot}1592 \times 0{\cdot}2869 = 0{\cdot}1142$$

$$f_3 = \frac{r+2}{2+1}\cdot q \cdot f_2 = \frac{6}{3} \times 0{\cdot}1592 \times 0{\cdot}1142 = 0{\cdot}0364$$

$$f_4 = \frac{r+3}{3+1}\cdot q \cdot f_3 = \frac{7}{4} \times 0{\cdot}1592 \times 0{\cdot}0364 = 0{\cdot}0101$$

$$f_5 = \frac{r+4}{4+1}\cdot q \cdot f_4 = \frac{8}{5} \times 0{\cdot}1592 \times 0{\cdot}0101 = 0{\cdot}0026$$

$\therefore$ Expected frequencies are : ($N = 400$)

	Nf_0	Nf_1	Nf_2	Nf_3	Nf_4	Nf_5
	200	114·76	45·68	14·56	4·04	1·04
$\therefore$ *Observed Frequency* :	213	128	37	18	3	1
Expected Frequency :	200	115	46	14	4	1

Note. Expected frequencies are rounded so that sum of observed and expected frequencies is same viz., $N = 400$.

8·7. GEOMETRIC DISTRIBUTION

Suppose we have a series of independent trials or repetitions and in each trial the probability of success 'p' remains the same. Then the probability that there are x failures preceding the first success is given by $q^x\, p,\ q = 1 - p$.

Definition. *A random variable X is said to have a geometric distribution if it assumes only non-negative integer values and its probability mass function is given by :*

$$P(X = x) = \begin{cases} q^x\, p\,;\ x = 0, 1, 2, \ldots,\,; & 0 < p \le 1\,; \quad q = 1 - p \\ 0, \text{ otherwise} \end{cases} \qquad \ldots (8{\cdot}26)$$

Remarks:

1. Since the various probabilities for $x = 0, 1, 2, \ldots,$ are the terms of geometric progression series, hence the name geometric distribution.
2. Clearly, assignment of probabilities in (8.26) is permissible, since

$$\sum_{x=0}^{\infty} P(X = x) = \sum_{x=0}^{\infty} q^x p = p(1 + q + q^2 + \ldots) = \frac{p}{1-q} = 1.$$

8·7·1. Lack of Memory.

The geometric distribution is said to lack memory in a certain sense. Suppose an event E can occur at one of the times $t = 0, 1, 2, \ldots$ and the occurrence (waiting) time X has a geometric distribution with parameter p.

Thus, $\qquad P(X = t) = q^t \cdot p; \quad t = 0, 1, 2, \ldots$

Suppose we know that the event E has not occurred before k, *i.e.*, $X \ge k$. Let $Y = X - k$. Thus, Y is the amount of additional time needed for E to occur. We can show that

$$P\,(Y = t \mid X \ge k) = P(X = t) = pq^t \qquad \ldots (8{\cdot}27)$$

which implies that the additional time to wait has the same distribution as initial time to wait.

Since the distribution does not depend upon k, it, in a sense, *'lacks memory'* of how much we shifted the time origin. If 'B' were waiting for the event E and is relieved by 'C' immediately before time k, then the waiting time distribution of 'C' is the same as that of 'B'.

Proof of (8·27). We have

$$P(X \ge r) = \sum_{s=r}^{\infty} pq^s = p(q^r + q^{r+1} + q^{r+2} + \ldots) = \frac{pq^r}{(1-q)} = q^r \qquad \ldots (*)$$

$$P(Y \ge t \mid X \ge k) = \frac{P(Y \ge t \cap X \ge k)}{P(X \ge k)} = \frac{P(X - k \ge t \cap X \ge k)}{P(X \ge k)} \qquad (\because\ Y = X - k)$$

$$= \frac{P(X \ge k + t)}{P(X \ge k)} = \frac{q^{t+k}}{q^k} = q^t \qquad \text{[From (*)]} \qquad \ldots (**)$$

$$\therefore\ P(Y = t \mid X \ge k) = P(Y \ge t \mid X \ge k) - P(Y \ge t + 1 \mid X \ge k)$$

$$= q^t - q^{t+1} = q^t(1 - q) = pq^t = P(X = t) \qquad \text{[From (**)]}$$

8·7·2. Moments of Geometric Distribution.

$$\text{Mean} = \mu_1' = \sum_{x=0}^{\infty} x \,.\, P(X = x) = \sum_{x=1}^{\infty} x \,.\, pq^x = pq \sum_{x=1}^{\infty} xq^{x-1} = pq(1-q)^{-2} = \frac{q}{p}$$

$$V(X) = E(X^2) - [E(X)]^2 = E[X(X-1)] + E(X) - [E(X)]^2 \qquad ...\ (*)$$

$$E[X(X-1)] = \sum_{x=0}^{\infty} x(x-1)\, P(X = x) = \sum_{x=2}^{\infty} x(x-1)\, pq^x$$

$$= 2pq^2 \sum_{x=2}^{\infty} \left[\frac{x(x-1)}{2 \times 1} q^{x-2}\right] = 2pq^2\, (1-q)^{-3} = \frac{2q^2}{p^2}$$

$$\therefore \qquad V(X) = \mu_2 = \frac{2q^2}{p^2} + \frac{q}{p} - \frac{q^2}{p^2} = \frac{q^2}{p^2} + \frac{q}{p} = \frac{q}{p^2}. \qquad \text{[From (*)]}$$

Hence, for the geometric distribution with parameter p,

$$\text{Mean} = E(X) = \frac{q}{p} \quad \text{and} \quad \text{Var}(X) = \frac{q}{p^2}; \qquad q = 1 - p. \qquad ...\ (8·28)$$

8·7·3. Moment Generating Function of Geometric Distribution.

$$M_X(t) = E(e^{tX}) = \sum_{x=0}^{\infty} e^{tx} \,.\, q^x\, p = \sum_{x=0}^{\infty} p\, (e^t q)^x$$

$$= p(1 - qe^t)^{-1} = \frac{p}{1 - qe^t} \qquad ...\ (8·28a)$$

$$\mu_1' = \left[\frac{d}{dt} M(t)\right]_{t=0} = \left[\frac{d}{dt} p(1 - qe^t)^{-1}\right]_{t=0} = pq\,(1-q)^{-2} = \frac{q}{p}$$

$$\mu_2' = \left[\frac{d^2}{dt^2} M(t)\right]_{t=0} = \frac{q}{p} + \frac{2q^2}{p^2} \qquad \text{(On simplification)}$$

$$\mu_2 = \mu_2' - \mu_1'^2 = \frac{q}{p} + \frac{2q^2}{p^2} - \frac{q^2}{p^2} = \frac{q^2 + pq}{p^2} = \frac{q}{p^2}$$

Hence, *the mean and variance of the geometric distribution are* $\frac{q}{p}$ and $\frac{q}{p^2}$ *respectively.*

Remarks:

1. Variance $= \dfrac{q}{p^2} = \dfrac{q}{p} \cdot \dfrac{1}{p} = \dfrac{\text{Mean}}{p} > \text{Mean} \qquad (\because 0 < p < 1)$

   Hence, *for the geometric distribution, variance is greater than the mean*

2. The *p.g.f.* of the geometric distribution is obtained on replacing e^t by s in (8·28a) and is given by :

$$P_X(s) = p/(1 - qs) \qquad ...\ (8·28b)$$

Example 8·61. *Let* X_1, X_2 *be independent r.v.'s each having geometric distribution* $q^k\, p$; $k = 0, 1, 2, ...$ *Show that the conditional distribution of* X_1 *given* $X_1 + X_2$ *is uniform.*

Solution.

$$P[X_1 = r \mid (X_1 + X_2 = n)] = \frac{P(X_1 = r \cap X_1 + X_2 = n)}{P(X_1 + X_2 = n)}$$

$$= \frac{P(X_1 = r \cap P(X_2 = n - r)}{P(X_1 + X_2 = n)} = \frac{P(X_1 = r \cap X_2 = n - r)}{\sum_{s=0}^{n} P[(X_1 = s \cap X_2 = n - s)]}$$

$$= \frac{P(X_1 = r) . P(X_2 = n - r)}{\sum_{s=0}^{n} P[(X_1 = s) . P(X_2 = n - s)]} \quad \text{[Since } X_1 \text{ and } X_2 \text{ are independent]}$$

$$\therefore \; P[X_1 = r \mid (X_1 + X_2 = n)] = \frac{pq^r . pq^{n-r}}{\sum_{s=0}^{n} [pq^s . pq^{n-s}]} = \frac{p^2 q^n}{\sum_{s=0}^{n} (p^2 q^n)} = \frac{p^2 q^n}{(n+1)p^2 q^n}$$

$$= \frac{1}{n+1} ; r = 0, 1, 2, ..., n$$

Hence, the result.

Example 8·62. *Suppose X is a non-negative integral valued random variable. Show that the distribution of X is geometric if it 'lacks memory', i.e.,*

if for each $k \geq 0$ and $Y = X - k$, one has $P(Y = t \mid X \geq k) = P(X = t)$, for $t \geq 0$.

Solution. Let us suppose $P(X = r) = p_r ; r = 0, 1, 2, ...$

Define $q_k = P(X \geq k) = p_k + p_{k+1} + p_{k+2} + ...$ ... (*)

We are given : $P(Y = t \mid X \geq k) = P(X = t) = p_t$ ... (**)

$$\therefore \quad P(Y = t \mid X \geq k) = \frac{P(Y = t \cap X \geq k)}{P(X \geq k)} = \frac{P(X - k = t \cap X \geq k)}{P(X \geq k)} = \frac{P(X = k + t)}{P(X \geq k)} = \frac{p_{k+t}}{q_k}$$

$$\Rightarrow \quad p_t = \frac{p_{k+t}}{q_k}, \text{ for every } t \geq 0 \text{ and all } k \geq 0. \quad \text{[From (**)]}$$

In particular, taking $k = 1$, we get

$$p_{t+1} = q_1 \cdot p_t = (p_1 + p_2 + ...)p_t = (1 - p_0)p_t \quad \text{[From (*)]}$$

$$\Rightarrow \quad p_t = (1 - p_0)\, p_{t-1} = (1 - p_0)^2\, p_{t-2} = ... = (1 - p_0)^t\, p_0$$

Hence, $p_t = P(X = t) = p_0(1 - p_0)^t ; t = 0, 1, 2, ...$

$\Rightarrow$ X has a geometric distribution.

Example 8·63. *In an infinite sequence of independent Bernoulli trials with constant probability 'p' of success for each trial, find the mean and variance of the number of trials required to get the first success.*

Solution. Let the *r.v. Y.* denote the number of trials required to get the first success. Then Y can take the values 1, 2, 3, ... with probability function :

$$P(Y = y) = p(y) = P[\text{First success at the } y\text{th trial}]$$

$$= q^{y-1} p ; y = 1, 2, 3, ... \quad ... (*)$$

where $q = 1 - p$, is the probability of failure in a trial.

$$E(Y) = \sum_{y=1}^{\infty} y\, p(y) = \sum_{y=1}^{\infty} y\, q^{y-1} . p = p[1 + 2q + 3q^2 + 4q^3 + ...] = p(1-q)^{-2} = \frac{1}{p}$$

OR Let $S = 1 + 2q + 3q^2 + 4q^3 + ...$ (Arithmetico-geometric series)

$\Rightarrow \quad qS = \quad q + 2q^2 + 3q^3 + ...$

Subtracting, $(1 - q)S = 1 + q + q^2 + ... = \dfrac{1}{1-q} \quad \Rightarrow \quad S = \dfrac{1}{(1-q)^2} = \dfrac{1}{p^2}$

$$E[Y(Y-1)] = \sum_{y=1}^{\infty} y(y-1)\, p(y) = p \cdot \sum_{y=1}^{\infty} y(y-1)\, q^{y-1}$$

$$= p\left[0 + 2 . 1 . q + 3 . 2 . q^2 + 4 . 3 . q^3 + ...\right] = 2pq\left[1 + 3q + 6q^2 + ...\right]$$

$$= 2pq . (1-q)^{-3} = 2pq . \frac{1}{p^3} = \frac{2q}{p^2} \qquad ... (ii)$$

$\therefore \quad E(Y^2) = E[Y(Y-1) + Y] = E[Y(Y-1)] + E(Y) = \dfrac{2q}{p^2} + \dfrac{1}{p}$ [From (*i*) and (*ii*)]

$\therefore \quad \text{Var}(Y) = E(Y^2) - [E(Y)]^2 = \dfrac{2q}{p^2} + \dfrac{1}{p} - \dfrac{1}{p^2} = \dfrac{2q + p - 1}{p^2} = \dfrac{2q - q}{p^2} = \dfrac{q}{p^2}$

Hence, $E(Y) = \dfrac{1}{p}$ and $\text{Var}(Y) = \dfrac{q}{p^2}$ ... (8·28*c*)

Note. Some authors define the 'Geometric Distribution' with p.m.*f* in (*) viz ;

$p(y) = q^{y-1} . p \,;\; y = 1, 2, 3, ...$ ... (8·28*d*)

Here the underlying variable is

Y : "The number of trials required for the first success", rather than

X : "The number of failures preceding the first success", as in (8·26)

Note that X and Y are connected by the relation : $Y = X + 1$.

$\therefore \quad E(Y) = E(X) + 1 = \dfrac{q}{p} + 1 = \dfrac{q+p}{p} = \dfrac{1}{p}$ and $\text{Var}(Y) = \text{Var}(X+1) = \text{Var}(X) = \dfrac{q}{p^2}$ [From 8·28]

8·8. HYPERGEOMETRIC DISTRIBUTION

When the population is finite and the sampling is done without replacement, so that the events are stochastically dependent, although random, we obtain hypergeometric distribution. Consider an urn with N balls, M of which are white and $N - M$ are red. Suppose that we draw a sample of n balls at random (*without replacement*) from the urn. Then the probability of getting k white balls out of n, $(k < n)$ is : $\dbinom{M}{k}\dbinom{N-M}{n-k} \div \dbinom{N}{n}$

[Since k white balls can be drawn from 'M' white balls in $\dbinom{M}{k}$ ways and out of the remaining $(N - M)$ red balls, $(n - k)$ balls can be chosen in $\dbinom{N-M}{n-k}$ ways, the total number of favourable cases is $\dbinom{M}{k} \times \dbinom{N-M}{n-k}$.]

Definition. *A discrete random variable X is said to follow the hypergeometric distribution with parameters N, M and n if it assumes only non-negative values and its probability mass function is given by :*

$$P(X = k) = h(k\ ; N, M, n) = \begin{cases} \dfrac{\binom{M}{k}\binom{N-M}{n-k}}{\binom{N}{n}} ; & k = 0, 1, 2, ..., \min(n, M). \\ 0, & \textit{otherwise} \end{cases} \qquad ...(8·29)$$

where N is a positive integer, M is a positive integer not exceeding N and n is a positive integer that is at most N.

Remarks:

1. As it can be shown that $\sum_{k=0}^{n} \binom{M}{k}\binom{N-M}{n-k} \div \binom{N}{n} = 1$, this assignment of probabilities is permissible.

2. The formula in (8·29) is valid for all integer values of x such that, for each symbol of the form $\binom{a}{b}$ in (8·29), $a \geq b$. This means that $0 \leq k \leq M$ and $0 \leq (n - k) \leq N - M$, *i.e.*, $\max(0, n - N + M) \leq k \leq \min(M, n)$.

3. The quantities in the R.H.S. of (8·29) are successive terms in the expansion of :

$$\frac{(N-n)!(N-M)!}{N!(N-M-n)!} . F(-n-M, N-M-n+1, 1),$$

where $F(\alpha, \beta, \gamma, t) = 1 + \dfrac{\alpha\beta}{\gamma} . \dfrac{t}{1!} + \dfrac{\alpha(\alpha+1)\beta(\beta+1)}{\gamma(\gamma+1)} . \dfrac{t^2}{2!} + ...,$ is a hypergeometric function.

That is why the distribution is called a hypergeometric distribution.

8·8·1. Mean and Variance of Hypergeometric Distribution

$$E(X) = \sum_{k=0}^{n} k\, P(X = k) = \sum_{k=0}^{n} k \left\{ \binom{M}{k}\binom{N-M}{n-k} \div \binom{N}{n} \right\}$$

$$= \frac{M}{\binom{N}{n}} \sum_{k=1}^{n} \left\{ \binom{M-1}{k-1}\binom{N-M}{n-k} \right\} = \frac{M}{\binom{N}{n}} \sum_{x=0}^{m} \binom{A}{x}\binom{N-A-1}{m-x},$$

where $x = k - 1, \qquad m = n - 1, \qquad M - 1 = A.$

$$= \frac{M}{\binom{N}{n}} . \binom{N-1}{m} = \frac{M}{\binom{N}{n}} \binom{N-1}{n-1} = \frac{nM}{N} \qquad ...(8·29a)$$

$$E\{X(X-1)\} = \sum_{k=0}^{n} k(k-1) \left\{ \binom{M}{k}\binom{N-M}{n-k} \div \binom{N}{n} \right\}$$

$$= \frac{M(M-1)}{\binom{N}{n}} \sum_{k=2}^{n} \left\{ \binom{M-2}{k-2}\binom{N-M}{n-k} \right\}$$

$$= \frac{M(M-1)}{\binom{N}{n}} . \binom{N-2}{n-2} = \frac{M(M-1)\, n(n-1)}{N(N-1)}$$

$$\therefore \quad E(X^2) = E[X(X-1)] + E(X) = \frac{M(M-1)\,n(n-1)}{N(N-1)} + \frac{nM}{N} \qquad \ldots (8{\cdot}29b)$$

Hence,

$$V(X) = \frac{M(M-1)\,n(n-1)}{N(N-1)} + \frac{nM}{N} - \left(\frac{nM}{N}\right)^2$$

$$= \frac{n\,M(N-M)(N-n)}{N^2(N-1)} \qquad \text{(On simplification)}$$

8·8·2. Factorial Moments of Hypergeometric Distribution.

The rth factorial moment is :

$$E[X^{(r)}] = \sum_{k=r}^{n} k^{(r)}\, P(X=k) = \sum_{k=r}^{n} k^{(r)} \left\{\binom{M}{k}\binom{N-M}{n-k} \div \binom{N}{n}\right\}$$

$$= \sum_{k=r}^{n} M^{(r)} \left\{\binom{M-r}{k-r}\binom{N-M}{n-k} \div \binom{N}{n}\right\}$$

$$\left\{\because k^{(r)}\binom{M}{k} = \frac{k(k-1)(k-2)\ldots(k-r+1)\,M!}{k!\,(M-k)!} = \frac{M(M-1)(M-2)\ldots(M-r+1)(M-r)!}{(k-r)!\,(M-k)!} = M^{(r)}\binom{M-r}{k-r}\right\}$$

$$\therefore \quad E\left[X^{(r)}\right] = M^{(r)} \sum_{j=0}^{n-r} \left\{\binom{M-r}{j}\binom{(N-r)-(M-r)}{(n-r)-j} \div \binom{N}{n}\right\}, \text{ where } j = k-r$$

$$= \frac{M^{(r)}\,n^{(r)}}{N^{(r)}} \sum_{j=0}^{n-r} \left\{\binom{M-r}{j}\binom{(N-r)-(M-r)}{(n-r)-j} \div \binom{N-r}{n-r}\right\}$$

$$\left\{\because n^{(r)}\binom{N}{n} = N^{(r)}\binom{N-r}{n-r}\right\}$$

$$= \frac{M^{(r)}\,n^{(r)}}{N^{(r)}} \sum_{j=0}^{n-r} h(j\,;N-r, M-r, n-r) = \frac{M^{(r)}\,n^{(r)}}{N^{(r)}}\,.\,1$$

$$\therefore \quad E[X^{(r)}] = \frac{M^{(r)}\,n^{(r)}}{N^{(r)}} \qquad \ldots (8{\cdot}29c)$$

$$\Rightarrow \quad \mu_X = E(X) = \frac{nM}{N} \quad \text{and} \quad E[X^{(2)}] = \frac{M(M-1)\,n(n-1)}{N(N-1)}$$

$$\sigma_X{}^2 = E[X^{(2)}] + E(X) - [E(X)]^2 = n\,.\,\frac{M}{N}\,.\,\frac{N-M}{N}\,.\,\frac{N-n}{N-1} \qquad \text{(On simplification)} \qquad \ldots (8{\cdot}29d)$$

Remark. If we sample the n balls with replacement and denote by Y the number of white balls in the sample, then Y is a binomial variate with parameters n and p, where

$$p = \frac{M}{N}, \qquad q = 1 - p = \frac{N-M}{N}$$

$$\therefore \quad E(Y) = np = \frac{nM}{N} = E(X), \qquad \sigma_Y{}^2 = npq = n\,.\,\frac{M}{N}\,.\,\frac{N-M}{N} \geq \sigma_X{}^2, \qquad \text{[From (8·29}d\text{)]}$$

equality holding only if $n = 1$.

8·8·3. Approximation to Binomial Distribution.

Hypergeometric distribution tends to binomial distribution as $N \to \infty$ and $\dfrac{M}{N} \to p$.

$$h(k\,;N,M,n) = \binom{M}{k}\binom{N-M}{n-k} \div \binom{N}{n} = \frac{M!}{k!(M-k)!} \cdot \frac{(N-M)!}{(n-k)!(N-M-n+k)!} \cdot \frac{n!(N-n)!}{N!}$$

$$= \frac{M(M-1)(M-2)\ldots(M-k+1)}{k!} \times \frac{(N-M)(N-M-1)\ldots(N-M-n+k+1)}{(n-k)!}$$

$$\times \frac{n!}{N(N-1)(N-2)\ldots(N-n+1)}$$

$$= \frac{n!}{k!(n-k)!} \cdot \frac{M}{N}\left(\frac{M}{N}-\frac{1}{N}\right)\left(\frac{M}{N}-\frac{2}{N}\right)\ldots\left(\frac{M}{N}-\frac{k-1}{N}\right)$$

$$\times \frac{\left(1-\frac{M}{N}\right)\left(1-\frac{M}{N}-\frac{1}{N}\right)\ldots\left(1-\frac{M}{N}-\frac{(n-k-1)}{N}\right)}{\left(1-\frac{1}{N}\right)\left(1-\frac{2}{N}\right)\ldots\left(1-\frac{n-1}{N}\right)}.$$

Proceeding to the limit as $N \to \infty$ and putting $\dfrac{M}{N} = p$, we get

$$\lim_{N \to \infty} h(k\,;N,M,n) = \binom{n}{k} \underbrace{p\,.\,p \ldots p}_{k \text{ times}} \times \underbrace{(1-p)(1-p)\ldots(1-p)}_{(n-k) \text{ times}}$$

$$= \binom{n}{k} p^k (1-p)^{n-k} = b(k\,;p,1-p).$$

8·8·4. Recurrence Relation for the Probabilities of Hypergeometric Distribution.

We have

$$h(k\,;N,M,n) = \binom{M}{k}\binom{N-M}{n-k} \div \binom{N}{n}, \qquad h(k+1\,;N,M,n) = \binom{M}{k+1}\binom{N-M}{n-k-1} \div \binom{N}{n};$$

$$\therefore \quad \frac{h(k+1\,;N,M,n)}{h(k\,;N,M,n)} = \frac{(n-k)(M-k)}{(k+1)(N-M-n+k+1)}, \qquad \text{(On simplification)}$$

which is the required recurrence relation between the probabilities of hypergeometric distribution.

Example 8·64. *Explain how you will use hypergeometric model to estimate the number of fish in a lake.*

Solution. Let us suppose that in a lake there are N fish, N unknown. The problem is to estimate N. A catch of 'r' fish (all at the same time) is made and these fish are returned alive into the lake after marking each with a red spot. After a reasonable period of time, during which these 'marked' fish are assumed to have distributed themselves 'at random' in the lake, another catch of 's' fish (again, all at once) is made. Here r and s are regarded as fixed pre-determined constants. Among these s fish caught, there will be (say) X marked fish where X is a random variable following discrete probability function given by hypergeometric model :

$$f_X(x \mid N) = \binom{r}{x}\binom{N-r}{s-x} \div \binom{N}{s} = p(N), \text{ say} \qquad \ldots (*)$$

where x is an integer such that $\max(0, s - N + r) \le x \le \min(r, s)$ and $f_X(x \mid N) = 0$, otherwise. The value of N is estimated by the principle of Maximum Likelihood (*c.f.* Chapter 17), *i.e.*, we find the value $\hat{N} = \hat{N}(x)$ of N which maximises $p(N)$. Since N is a discrete *r.v.*, the principle of maxima and minima in calculus cannot be used here. Here we proceed as follows :

$$\lambda(N) = \frac{p(N)}{p(N-1)} = \frac{(N-r)(N-s)}{N(N-r-s+x)}. \qquad \text{(On simplification)}$$

$$\therefore \quad \lambda(N) > 1 \quad \text{iff} \quad N < \frac{rs}{x} \quad \Rightarrow \quad p(N) > P(N-1) \quad \text{iff} \quad N < \frac{rs}{x} \qquad \text{... } (i)$$

$$\text{and} \quad \lambda(N) < 1 \quad \text{iff} \quad N > \frac{rs}{x} \quad \Rightarrow \quad p(N) < P(N-1) \quad \text{iff} \quad N > \frac{rs}{x} \qquad \text{... } (ii)$$

From (*i*) and (*ii*) we see that $p(N) = f_X(x \mid N)$ reaches the maximum value (as a function of N) when N is approximately equal to rs/x. Hence, maximum likelihood estimate of N is given by :

$$\hat{N} = \frac{rs}{x} \quad \Rightarrow \quad \hat{N}(X) = \frac{rs}{X}.$$

8·9. MULTINOMIAL DISTRIBUTION

This distribution can be regarded as a generalisation of Binomial distribution.

When there are more than two mutually exclusive outcomes of a trial, the observations lead to multinomial distribution. Suppose $E_1, E_2, ..., E_k$ are k mutually exclusive and exhaustive outcomes of a trial with respective probabilities $p_1, p_2, ..., p_k$.

The probability that E_1 occurs x_1 times, E_2, occurs x_2 times ... and E_k, occurs x_k times in n independent observations, is given by $p(x_1, x_2, ..., x_k) = c\, p_1^{x_1} p_2^{x_2} ... p_k^{x_k}$, where $\Sigma x_i = n$ and c is the number of permutation of the events $E_1, E_2, ..., E_k$.

To determine c, we have to find the number of permutations of n objects of which x_1 are of one kind, x_2 of another kind, ..., x_k of the kth kind, which is given by :

$$c = \frac{n!}{x_1!\, x_2! ... x_k!}.$$

$$\text{Hence,} \quad p(x_1, x_2, ..., x_k) = \frac{n!}{x_1!\, x_2! ... x_k!} \cdot p_1^{x_1}\, p_2^{x_1} ... p_k^{x_k}, \quad 0 \le x_i \le n$$

$$= \frac{n!}{\prod_{i=1}^{k} x_i!} \cdot \prod_{i=1}^{k} p_i^{x_i}, \quad \sum_{i=1}^{k} x_i = n \qquad \text{... (8·30)}$$

which is the required probability function of the multinomial distribution. It is so called since (8·30) is the general term in the multinomial expansion :

$$(p_1 + p_2 + ... + p_k)^n, \quad \sum_{i=1}^{k} p_i = 1$$

Since, total probability is 1, we have

$$\sum_x p(x) = \sum_x \left[\frac{n!}{x_1!\, x_2! ... x_k!} p_1^{x_1}\, p_2^{x_2} ... p_k^{x_k} \right] = (p_1 + p_2 + ... + p_k)^n = 1 \qquad \text{... (8·30}a)$$

8·9·1. Moments of Multinomial Distribution.

The moment generating function is given by :

$$M_X(t) = M_{X_1, X_2, \ldots, X_k}(t_1, t_2, \ldots, t_k) = E\left[\exp\left\{\sum_{i=1}^{k} t_i X_i\right\}\right]$$

$$= \sum_x \left[\frac{n!}{x_1!\, x_2! \ldots x_k!} p_1^{x_1} p_2^{x_2} \ldots p_k^{x_k} \exp\left(\sum_{i=1}^{k} t_i x_i\right)\right]$$

$$= \sum_x \left[\frac{n!}{x_1!\, x_2! \ldots x_k!} (p_1 e^{t_1})^{x_1} (p_2 e^{t_2})^{x_2} \ldots (p_k e^{t_k})^{x_k}\right]$$

$$= (p_1 e^{t_1} + p_2 e^{t_2} + \ldots p_k e^{t_k})^n \qquad \text{[On using 8·30}a\text{]} \qquad \ldots (8·31)$$

where $X = (X_1, X_2, \ldots, X_k)$.

$$M_{X_1}(t_1) = M_X(t_1, 0, 0, \ldots, 0) = (p_1 e^{t_1} + p_2 + p_3 + \ldots p_k)^n$$

$$= \left[(1 - p_1) + p_1 e^{t_1}\right]^n$$

$\Rightarrow$ $X_1 \sim B(n, p_1)$ [By uniqueness theorem of *m.g.f.*]

Similarly, we shall get: $X_i \sim B(n, p_i)$; $i = 1, 2, \ldots, k.$ (8.31a)

$\Rightarrow$ $E(X_i) = np_i$ and $\text{Var}(X_i) = np_i(1 - p_i)$; $i = 1, 2, \ldots, k$

$$E(X_i X_j) = \left[\frac{\partial^2 M}{\partial t_i\, \partial t_j}\right]_{t=0}, i \neq j; \text{ where } t = (t_1, t_2, \ldots, t_k)$$

$$= \left[np_i e^{t_i}(n-1)(p_1 e^{t_1} + \ldots + p_k e^{t_k})^{n-2} p_j e^{t_j}\right]_{t=0}$$

$$= n(n-1)\, p_i p_j \qquad (\because p_1 + p_2 + \ldots + p_k = 1)$$

$$\text{Cov}(X_i, X_j) = E(X_i X_j) - E(X_i)\, E(X_j) = n(n-1)\, p_i p_j - n^2 p_i p_j = -np_i p_j$$

$$\therefore \quad \rho(X_i, X_j) = \frac{\text{Cov}(X_i, X_j)}{\sigma_{X_i} \sigma_{X_j}} = \frac{-np_i p_j}{\sqrt{np_i(1-p_i)\, np_j(1-p_j)}} = -\left\{\frac{p_i p_j}{(1-p_i)(1-p_j)}\right\}^{1/2}.$$

Note. Since cov $(X_i, X_j) < 0$, the correlation coefficients are negative.

Example 8·65. *The trinomial distribution of two r.v.'s X and Y is given by :*

$$f_{X,Y}(x, y) = \frac{n!}{x!\, y!\, (n-x-y)!} p^x q^y (1-p-q)^{n-x-y}$$

for $x, y = 0, 1, 2, \ldots, n$ and $x + y \le n$, where $0 \le p$, $0 \le q$ and $p + q \le 1$.

(i) Find the marginal distributions of X and Y.

(ii) Find the conditional distributions of X and Y and obtain

(a) $E(Y \mid X = x)$, and (b) $E(X \mid Y = y)$

Solution. (*i*) The joint *m.g.f.* of X and Y is given by :

$$M_{X,Y}(t_1, t_2) = E(e^{t_1 X + t_2 Y})$$

$$= \sum_{x=0}^{n} \sum_{y=0}^{n-x} \left[\frac{n!(pe^{t_1})^x (qe^{t_2})^y (1-p-q)^{n-x-y}}{x!\, y!\, (n-x-y)!}\right]$$

$$= \left[pe^{t_1} + qe^{t_2} + (1-p-q)\right]^n$$

$\therefore \quad M_X(t_1) = M(t_1, 0) = \left\{(1-p) + pe^{t_1}\right\}^n \quad \Rightarrow \quad X \sim B(n, p)$

and $\quad M_Y(t_2) = M(0, t_2) = \left\{(1-q) + qe^{t_2}\right\}^n \quad \Rightarrow \quad Y \sim B(n, q)$

Observe that $M(t_1, t_2) \neq M(t_1, 0) \times M(0, t_2) \quad \Rightarrow \quad X$ and Y are not independent.

(*ii*) The conditional distribution of X given $Y = y$ is given by :

$$f(X \mid Y = y) = \frac{f_{XY}(x, y)}{f_Y(y)} = \frac{f_{XY}(x, y)}{{}^nC_y\, q^y\, (1-q)^{n-y}} \qquad [\because Y \sim B(n, q)]$$

$$= \frac{(n-y)!}{x!\,(n-y-x)!}\left(\frac{p}{1-q}\right)^x \left(\frac{1-p-q}{1-q}\right)^{n-y-x}$$

$$= \binom{n-y}{x}\left(\frac{p}{1-q}\right)^x \left(1 - \frac{p}{1-q}\right)^{n-x-y} \quad ; x = 0, 1, \ldots, (n-y)$$

$\therefore \quad X \mid (Y = y) \sim B\{(n-y, p/(1-q)\} \quad \Rightarrow \quad E[X \mid Y = y] = (n-y)\,.\,[p/(1-q)]$

Similarly, we shall get

$$f(Y \mid X = x) = \frac{f_{XY}(x, y)}{f_X(x)} = \frac{f(x, y)}{{}^nC_x\, p^x\, (1-p)^{n-x}} \qquad [\because X \sim B(n, p)]$$

$$= \binom{n-x}{y}\left(\frac{q}{1-p}\right)^y \left(1 - \frac{q}{1-p}\right)^{n-x-y} \quad ; y = 0, 1, \ldots, (n-x)$$

$\therefore \quad Y \mid (X = x) \sim B[(n-x, q/(1-p)] \quad \Rightarrow \quad E[Y \mid X = x] = (n-x)\,.\,[q/(1-p)]$

Example 8·66. *If $X_1, X_2, \ldots, X_k$ are k independent Poisson variates with parameters $\lambda_1, \lambda_2, \ldots, \lambda_k$ respectively, prove that the conditional distribution :*

$P(X_1 \cap X_2 \cap \ldots \cap X_k \mid X)$, where $X = X_1 + X_2 + \ldots + X_k$ is fixed, is multinomial.

Solution. $P[X_1 \cap X_2 \cap \ldots \cap X_k \mid X = n]$

$$= P[X_1 = r_1 \cap X_2 = r_2 \cap \ldots \cap X_k = r_k \mid X = n]$$

$$= \frac{P[X_1 = r_1 \cap X_2 = r_2 \cap \ldots \cap X_k = r_k \cap X = n]}{P(X = n)}$$

$$= \frac{P[X_1 = r_1 \cap \ldots \cap X_{k-1} = r_{k-1} \cap X_k = n - r_1 - r_2 \ldots - r_{k-1}]}{P(X = n)}$$

$$= \frac{P(X_1 = r_1)\, P(X_2 = r_2) \ldots P(X_{k-1} = r_{k-1})\, P(X_k = n - r_1 - \ldots - r_{k-1})}{P(X = n)}$$

($\because X_1, X_2, \ldots, X_k$ are independent.)

Further, since X_i, $(i = 1, 2, \ldots, k)$ are independent Poisson variates with parameters λ_i respectively, $X = X_1 + X_2 + \ldots + X_k$ is also a Poisson variate with parameter $\lambda_1 + \lambda_2 + \ldots + \lambda_k = \lambda$ (say).

Hence, $P\left[X_1 \cap X_2 \cap \dots \cap X_k \mid X = n\right] = \dfrac{\dfrac{e^{-\lambda_1}\lambda_1^{r_1}}{r_1!} \cdots \dfrac{e^{-\lambda_{k-1}}\lambda_{k-1}^{r_{k-1}}}{r_{k-1}!} \cdot \dfrac{e^{-\lambda_k}\lambda_k^{\,n-r_1-\dots-r_{k-1}}}{(n-r_1-\dots-r_{k-1})!}}{\dfrac{e^{-\lambda}\lambda^n}{n!}}$

$$= \left\{\frac{n!}{r_1!\,r_2!\dots r_{k-1}!\,(n-r_1-\dots-r_{k-1})!}\right\} \times \left\{\left(\frac{\lambda_1}{\lambda}\right)^{r_1} \cdots \left(\frac{\lambda_{k-1}}{\lambda}\right)^{r_{k-1}} \left(\frac{\lambda_k}{\lambda}\right)^{n-r_1-\dots-r_{k-1}}\right\}$$

$$= \frac{n!}{r_1!\,r_2!\dots r_k!}\, p_1^{r_1} p_2^{r_2} \dots p_k^{r_k}, \text{ where } \sum_{i=1}^{k} r_i = n \text{ and } \sum_{i=1}^{k} p_i = \sum_{i=1}^{k}\left(\frac{\lambda_i}{\lambda}\right) = \frac{1}{\lambda}\sum_{i=1}^{k}\lambda_i = 1$$

Thus, the *conditional distribution* $P(X_1 \cap X_2 \cap \dots \cap X_k \mid X = n)$ *is multinomial with probabilities* $p_i = (\lambda_i/\lambda);\ i = 1, 2, \dots, k$ *in k classes.*

Remark. If X_i's are identically distributed independent Poisson variates with parameter m (say,) then

$$\lambda_i = m\ ;\ i = 1, 2, \dots, k \text{ and } \lambda = \sum_{i=1}^{k} \lambda_i = km. \qquad \therefore\ p_i = \frac{\lambda_i}{\lambda} = \frac{1}{k}.$$

Hence, in this case the conditional distribution of $X_1, X_2, \dots, X_k$, given that their sum $X_1 + X_2 + \dots + X_k = n$, is a multinomial distribution with index n and the probability in each class being equal to $1/k$.

8·9·2. Application of Multinomial Distribution.

An important application of Multinomial Distribution is in finding the probability of the sum of the scores obtained in a random toss of n fair dice, each of which has six faces marked 1 to 6.

Since any one of the six faces (marked 1 to 6) on any dice can be associated with any one of the six faces on each of the other $(n - 1)$ dice,

The exhaustive number of ways in which n dice can fall $= 6^n$.

Since the total number of permutations in which six numbers, *viz.*, 1, 2, ..., 6 taken 'n' at a time can add to s is the coefficient of x^s in the multinomial expansion of $(x + x^2 + \dots + x^6)^n$, the required number of favourable cases for getting a sum 's' on a dice is the co-efficient of x^s in the expansion of $(x + x^2 + \dots + x^6)^n$.

$$\therefore \text{ Required Probability } = \frac{1}{6^n}\left[\text{Coefficient of } x^s \text{ in } (x + x^2 + \dots + x^6)^n\right].$$

Identically, we have : $\quad x + x^2 + \dots + x^6 = x(1 + x + \dots + x^5) = x\left(\dfrac{1-x^6}{1-x}\right)$

and by binomial expansion : $\quad x^n(1 - x^6)^n = \displaystyle\sum_{k=0}^{n} (-1)^k\, {}^nC_k\, x^{n+6k}$

and $\quad (1-x)^{-n} = \displaystyle\sum_{r=0}^{\infty} {}^{-n}C_r\,(-x)^r = \sum_{r=0}^{\infty} {}^{n+r-1}C_r \,.\, x^r = \sum_{r=0}^{\infty} {}^{n+r-1}C_{n-1} \,.\, x^r$

$$\therefore \left[x\left(\frac{1-x^6}{1-x}\right)\right]^n = \sum_{k=0}^{n}\sum_{r=0}^{\infty} (-1)^k \,.\, {}^nC_k \,.\, {}^{n+r-1}C_{n-1} \,.\, x^{n+6k+r}$$

To find the co-efficient of x^s, we put $n + 6k + r = s$, *i.e.*, $n + r = s - 6k$

Thus, co-efficient of x^s in expansion of $(x + x^2 + \dots + x^6)^n = \sum_{k=0}^{(s-n)/6} (-1)^k \cdot {}^nC_k \cdot {}^{s-6k-1}C_{n-1}$,

the summation being extended over the integral values of k not exceeding $(s - n)/6$.

Remarks:

1. The probability of getting a sum 's' with a throw of n dice, each having 'f' faces marked 1 to f is the

$$\text{co-efficient of } x^s \text{ in } \frac{1}{f^n}\left[(x + x^2 + \dots + x^f)^n\right]$$

2. If n dice have faces $f_1, f_2, \dots f_n$ respectively, then the required probability of getting a sum 's' is the co-efficient of x^s in

$$\frac{1}{f_1 f_2 \cdots f_n}\left[(x + x^2 + \dots + x^{f_1})(x + x^2 + \dots + x^{f_2}) \quad \dots \quad (x + x^2 + \dots + x^{f_n})\right]$$

Example 8·67. *What is the probability of obtaining 15 points in a random throw of 5 dice together ?*

Solution. Here $n = 5$.

The number of exhaustive cases in throwing of 5 die is 6^5. The number of ways in which the 5 dice thrown will give 15 points is the co-efficient of x^{15} in the expansion of $(x^1 + x^2 + x^3 + \dots + x^6)^5$.

$$\therefore \quad \text{Favourable No. of cases} = \text{Co-efficient of } x^{15} \text{ in } (x + x^2 + \dots + x^6)^5$$

$$= \text{Co-efficient of } x^{10} \text{ in } (1 - x^6)^5 (1 - x)^{-5}$$

$$(1 - x^6)^5 = \left(1 - {}^5C_1\, x^6 + {}^5C_2\, x^{12} - \dots - x^{30}\right) = \left(1 - 5x^6 + 10x^{12} - \dots - x^{30}\right)$$

$$(1 - x)^{-5} = 1 + 5x + \frac{5 \times 6}{2!}x^2 + \frac{5 \times 6 \times 7}{3!}x^3 + \frac{5 \times 6 \times 7 \times 8}{4!}x^4 + \dots + \frac{5 \times 6 \times 7 \times \dots \times 14}{10!}x^{10} + \dots$$

$$= (1 + 5x + 15x^2 + 35x^3 + 70x^4 + \dots + 1001x^{10} + \dots)$$

$\therefore$ Favourable number of cases

$= \text{Coefficient of } x^{10} \text{ in } \left[(1 - 5x^6 + 10x^{12} - \dots - x^{30}) \times (1 + 5x + \dots + 70x^4 + \dots + 1001x^{10} + \dots)\right]$

$= (1001 - 5 \times 70) = 651$

Hence,
$$\text{Required probability} = \frac{651}{6^5} = \frac{651}{7776}.$$

8·10. POWER SERIES DISTRIBUTION (P.S.D.).

Definition. *A discrete r.v. X is said to follow a generalised power series distribution (G.P.S.D.), if its probability mass function is given by :*

$$P(X = x) = \begin{cases} \dfrac{a_x\, \theta^x}{f(\theta)}; & x = 0, 1, 2, \dots\ ;\ a_x \geq 0 \\ 0, & \textit{elsewhere} \end{cases} \qquad \dots (8·32)$$

where $f(\theta)$ is a generating function, i.e.,

$$f(\theta) = \sum_{x \in s} a_x\, \theta^x, \theta \geq 0 \qquad \dots (8·32a)$$

so that $f(\theta)$ is positive, finite and differentiable and S is a non-empty countable sub-set of non–negative integers.

Remarks:

1. By taking proper choice of S and $f(\theta)$, the *G.P.S.D.* can be reduced to binomial, Poisson and logarithmic series distribution and their truncated forms.
2. An inflated power series distribution (*P.S.D.*), inflated at zero is given by :

$$P(X = x) = \begin{cases} 1 - \alpha + \dfrac{\alpha\, a_0}{f(\theta)}, & x = 0 \\ \alpha\, \dfrac{a_x\, \theta^x}{f(\theta)}; & x = 1, 2, \ldots \end{cases} \qquad \ldots (8{\cdot}33)$$

where α $(0 < \alpha \le 1)$, is inflation parameter.

3. The truncated *P.S.D.* is given by :

$$P(X = x \mid S) = \begin{cases} \dfrac{a_x\, \theta^x}{f(\theta)} \div f(S), & x \in S \\ 0, & \text{otherwise} \end{cases}$$

$$\Rightarrow \qquad P(X = x \mid S) = \begin{cases} \dfrac{a_x\, \theta^x}{f_1(\theta)}; & x \in S, \text{ where } f_1(\theta) = \sum\limits_{x \in S} a_x\, \theta^x \\ 0, & \text{otherwise} \end{cases} \qquad \ldots (8{\cdot}34)$$

8·10·1. Moment Generating Function of P.S.D.

$$M_X(t) = \sum_{x=0}^{\infty} e^{tx}\, P(X = x) = \sum_{x=0}^{\infty} e^{tx} [a_x\, \theta^x / f(\theta)]$$

$$= \frac{1}{f(\theta)} \sum_{x=0}^{\infty} a_x\, (\theta\, e^t)^x = \frac{f(\theta\, e^t)}{f(\theta)} \qquad \ldots (8{\cdot}35)$$

8·10·2. Recurrence Relation for Cumulants of P.S.D.

The cumulant generating function is given by :

$$K_X(t) = \log M_X(t) = \log \left[\frac{f(\theta\, e^t)}{f(\theta)} \right] \quad \Rightarrow \quad \sum_{r=1}^{\infty} \kappa_r\, \frac{t^r}{r!} = \log f(\theta e^t) - \log f(\theta) \qquad \ldots (*)$$

Differentiating (*) partially w.r. to θ and r respectively, we get

$$\sum_{r=1}^{\infty} \frac{\partial}{\partial \theta} \kappa_r\, \frac{t^r}{r!} = \frac{e^t\, f'(\theta\, e^t)}{f(\theta\, e^t)} - \frac{f'(\theta)}{f(\theta)} \qquad \ldots (**)$$

and
$$\sum_{r=1}^{\infty} \kappa_r\, \frac{r\, t^{r-1}}{r!} = \frac{\theta e^t\, f'(\theta\, e^t)}{f(\theta\, e^t)} \qquad \ldots (***)$$

Subtracting (***) from θ times (**), we get

$$\theta \sum_{r=1}^{\infty} \frac{\partial}{\partial \theta} \kappa_r\, \frac{t^r}{r!} = \sum_{r=1}^{\infty} \kappa_r \cdot \frac{t^{r-1}}{(r-1)!} - \frac{\theta f'(\theta)}{f(\theta)}$$

Comparing like powers of t on both sides, we get

$$0 = \kappa_1 - \frac{\theta f'(\theta)}{f(\theta)} \quad \Rightarrow \quad \kappa_1 = \frac{\theta f'(\theta)}{f(\theta)} \qquad \ldots (8{\cdot}36)$$

and $\kappa_{r+1} = \theta \cdot \frac{d}{d\theta} \kappa_r \; ; \; r = 1, 2, ...$ (Comparing co-efficient of $t^r / r!$) ... (8·36a)

Remark. We have : $\text{Mean} = \kappa_1 = \frac{\theta f'(\theta)}{f(\theta)}$ ... (8·36b)

Alternatively : $\text{Mean} = \sum_{x=0}^{\infty} x \{a_x \theta^x / f(\theta)\} = \frac{\theta}{f(\theta)} \sum_{x=0}^{\infty} x\, a_x \theta^{x-1} = \frac{\theta f'(\theta)}{f(\theta)}$

8·10·3. Particular Cases of G.P.S.D.

1. Binomial Distribution

Let us take $\theta = p/(1-p)$, $f(\theta) = (1+\theta)^n$ and $S = \{0, 1, 2, ..., n\}$, a set of $(n+1)$ non-negative integers, then

$$f(\theta) = \sum_{x \in S} a_x \theta^x \quad \Rightarrow \quad (1+\theta)^n = \sum_{x=0}^{n} a_x \theta^x \quad \Rightarrow \quad a_x = \binom{n}{x}$$

$$\therefore \quad P(X = x) = \frac{\binom{n}{x}\left\{\frac{p}{(1-p)}\right\}^x}{\left\{1 + \frac{p}{(1-p)}\right\}^n} = \begin{cases} \binom{n}{x} p^x (1-p)^{n-x} \; ; x = 0, 1, ..., n \\ 0, \quad \text{otherwise} \end{cases} \qquad \text{[From (8·32)]}$$

which is the probability mass function of the binomial distribution with parameters n and p.

2. Negative Binomial Distribution

Let us take $\theta = p/(1+p)$, $f(\theta) = (1-\theta)^{-n}$ and S = [0, 1, 2, ... ad infinity], $0 \le \theta < 1$, $n > 0$.

$$\therefore \quad f(\theta) = \sum_{x \in S} a_x \theta^x \quad \Rightarrow \quad (1-\theta)^{-n} = \sum_{x=0}^{\infty} a_x \theta^x$$

$$\Rightarrow \quad a_x = (-1)^x \binom{-n}{x} = (-1)^x \cdot (-1)^x \binom{n+x-1}{x} = \binom{n+x-1}{x}$$

$$\therefore \quad P(X = x) = \binom{n+x-1}{x} \frac{[(p/(1+p)]^x}{[1 - \{p/(1+p)\}]^{-n}}$$

$$= \binom{n+x-1}{x} p^x (1+p)^{-(n+x)} ; \; x = 0, 1, 2, ...$$

$$= \binom{-n}{x} (1+p)^{-(n+x)} (-p)^x ; \; x = 0, 1, 2, ...$$

which is the probability mass function of the negative binomial distribution.

3. Logarithmic Series Distribution

Let $f(\theta) = -\log(1-\theta)$ and S = [1, 2, 3, ...].

Then, $f(\theta) = \sum_{x \in S} a_x \theta^x \quad \Rightarrow \quad -\log(1-\theta) = \sum_{x=1}^{\infty} a_x \theta^x, \quad \Rightarrow \quad a_x = \frac{1}{x}$

$$\therefore \quad P(X = x) = \begin{cases} \dfrac{a_x\, \theta^x}{f(\theta)} = \dfrac{\theta^x}{x[-\log(1-\theta)]} & ; x = 1, 2, 3, \ldots \\ 0, & \text{otherwise} \end{cases}$$

which is the *p.m.f.* of logarithmic series distribution with parameter θ.

4. Poisson Distribution

Let $f(\theta) = e^{\theta}$, $S = \{0, 1, 2, \ldots\}$. Then

$$f(\theta) = \sum_{x \in S} a_x\, \theta^x \quad \Rightarrow \quad e^{\theta} = \sum_{x=0}^{\infty} a_x\, \theta^x \quad \Rightarrow \quad a_x = \frac{1}{x\,!}$$

$$\therefore \quad P(X = x) = \frac{a_x\, \theta^x}{f(\theta)} = \frac{\theta^x}{x\,!\, e^{\theta}} = \frac{e^{-\theta}\, \theta^x}{x\,!} \quad ; \; x = 0, 1, 2, \ldots$$

which is the probability mass function of the Poisson distribution with parameter θ.

CHAPTER CONCEPTS QUIZ

1. Match the correct parts to make a valid statement :

(*a*) Binomial distribution applies to	1. rare events
(*b*) Poisson distribution applies to	2. repeated two alternatives.
(*c*) *The mean of hypergeometric distribution* is	3. $\dfrac{1-6pq}{npq}$
(*d*) The moment generating function of negative binomial distribution is	4. $n \cdot \dfrac{M}{n}\left(1 - \dfrac{M}{n}\right)\left(\dfrac{N-n}{N-1}\right)$
(*e*) The co-efficient of kurtosis of a binomial distribution is	5. $(Q - Pe^{t})^{-r}$
(*f*) The variance of geometric distribution is	6. $\dfrac{nM}{N}$
(*g*) Variance of Hypergeometric distribution is	7. $\dfrac{q}{p^2}$

2. Under what conditions binomial distribution tends to

   (*i*) Poisson distribution, (*ii*) Normal distribution, (*iii*) Geometric distribution.

   Give practical examples (one each) where you would expect binomial, Poisson, negative binomial and geometric distribution.

3. State the relationship between :

   (*i*) Mean and variance of Poisson distribution.

   (*ii*) Mean and variance of negative binomial distribution.

   (*iii*) Mean and variance of geometric distribution.

   (*iv*) Poisson distribution and binomial distribution.

   (*v*) Hypergeometric distribution and binomial distribution.

4. Name the discrete distribution for which

   (*i*) Mean and variance have the same value.

   (*ii*) Mean is greater than the variance.

5. State which of the following statement are True and which are False. In case of the false statement, give the correct statement :

(i) Mean of binomial distribution is 3 and variance is 5.

(ii) Mean of Poisson distribution is 2 and variance is 3.

(iii) The sum of two independent Poisson variates is also a Poisson variate. The result holds for the difference also.

(iv) For binomial distribution, Mean = Mode = Median

(v) The Poisson distribution is a limiting case of binomial distribution when $n \to \infty$, $p \to 0$, $np \to m$ (finite).

(vi) Nearly all the distributions are particular cases of Poisson distribution.

(vii) The sum of two binomial variates is a binomial variate if the variables are independent and have the different probabilities of success.

(viii) Negative binomial distribution may be regarded as the generalisation of geometric distribution.

6. Fill in the blanks :

(i) The variance of binomial distribution is ....

(ii) The β-co-efficient of kewness of the binomial distribution is ....

(iii) The moment generating function of Poisson distribution is ....

(iv) The characteristic function of negative binomial distribution is ....

(v) The co-efficient of skewness of Poisson distribution is ....

(vi) Poisson distribution is a limiting case of binomial distribution under the conditions ....

(vii) For Poisson distribution, all cumulants ....

(viii) Mean > variance for .... distribution.

(ix) For the Poisson distribution, the variance and the third central moment are ....

(x) Mean < variance for .... distribution.

7. Give the correct answer to each of the following :

(i) The skewness in a binomial distribution will be zero, if

(a) $p<\frac{1}{2}$, (b) $p=\frac{1}{2}$, (c) $p>\frac{1}{2}$, (d) $p<q$.

(ii) The mean and variance of negative binomial distribution

(a) are same, (b) cannot be same,

(c) are sometimes equal in limiting case, as $n \to \infty$.

(iii) The characteristic function of Poisson distribution is :

(a) $e^{m(it-1)}$, (b) $e^{m(e^{it}-1)}$, (c) e^{mit}, (d) none of these.

(iv) The coefficient of variation of Poisson distribution with mean 4 is :

(a) $\frac{1}{4}$, (b) $\frac{2}{4}$, (c) 4, (d) 2

(v) The coefficient of kurtosis of Poisson distribution with mean m is :

(a) $1/m$, (b) $-1/m$, (c) m, (d) $3 + (1/m)$

(vi) The mean of Hypergeometric distribution is

(a) $\frac{N(M-1)}{N(N-1)}$, (b) $\frac{M(M-1)}{N(N-1)}$, (c) $\frac{NM(M-1)}{N(N-1)}$, (d) None of these.

(vii) In a Poisson distribution, the second moment about the origin is 12. Then its third moment about mean is

(a) 2, (b) 3, (c) 5, (d) 10.

(*viii*) The mean of the binomial distribution $\binom{10}{x}\left(\frac{2}{5}\right)^x\left(\frac{3}{5}\right)^{10-x}$; $x = 0, 1, 2, ..., 10$ is

(*a*) 4, (*b*) 6, (*c*) 5, (*d*) 0.

(*ix*) The mean of Poisson variate is :

(*a*) greater than, (*b*) less than, (*c*) equal to, (*d*) twice, its variance.

8. By using the uniqueness property of *m.g.f.*'s, determine the distribution if the M.G.F. is as follows :

(*a*) $M(t) = \left(\frac{1}{2}+\frac{1}{2}e^t\right)^6$, (*b*) $M(t) = \frac{(1+e^t)^5}{32}$ (*c*) $M(t) = \frac{(1+2e^t)^3}{27}$,

(*d*) $M(t) = e^{3(e^t-1)}$ (*e*) $M(t) = e^{(e^t-1)/4}$, (*f*) $M(t) = \frac{1}{3}e^{-t}\left(e^{-t}-\frac{2}{3}\right)^{-1}$

(*g*) $M(t) = 4(3e^{-t}-1)^{-2}$ (*h*) $M(t) = (3e^{-t}-2)^{-3}$

DISCUSSION AND REVIEW QUESTIONS

1. (*a*) Describe the probability model from which the binomial distribution can be generated. Hence find the first four central moments.

(*b*) If p is the probability of 'success' at a single trial, obtain the probability of r 'successes' out of n independent trials. Determine the mode of the resulting distribution.

2. (*a*) Define the binomial distribution with parameters p and n, and give a situation in real life where the distribution is likely to be realized. Obtain the moment generating function of the binomial distribution and hence or otherwise obtain the mean, variance, skewness and kurtosis of the distribution.

(*b*) Using the Moment Generating Function of the Binomial Distribution, derive the result that the sum of two binomial variates is also a binomial variate if the variates are independent and have the same probability of success.

3. (*a*) By differentiating the following identity *w.r.* to p and then multiplying by p,

$$\sum_{x=0}^{n}\binom{n}{x}p^x q^{n-x} = (q+p)^n,\ q = 1-p;\ \text{prove that : } \mu_1' = np \text{ and } \mu_2 = npq.$$

(*b*) Let $X \sim b(x\,;n,p)$ and r be a non-negative integer. If the rth moment about the origin is denoted by $\mu_r' = E(X^r)$, prove that $$\mu'_{r+1} = np\,\mu'_r + p(1-p)\frac{d\mu'_r}{dp}.$$

(*c*) If $X \sim B(n, p)$, obtain the recurrence relation for its central moments and hence find values of β_1 and β_2.

4. (*a*) Show that the necessary and sufficient conditions for two given numbers a, b to be respectively the mean and the variance of some binomial distribution are that $a > b > 0$ and $\frac{a^2}{a-b}$ is an integer. Show further that when these conditions are satisfied, the binomial distribution is uniquely determined.

(*b*) Let X be a random variable whose probability function is $b(x\,;n,p)$. Let $Y = X/n$ be a new random variable. Show that the expected value of Y is p and the variance of Y is pq/n. If $p(y)$ is the probability function for Y, show that $p(y) = b(ny\,;n,p)$. What are the possible values that y can take on ?

5. (a) $x = r$ is unique mode of binomial distribution having mean np and variance $np(1 - p)$. Show that:

$$(n + 1)p - 1 < r < (n + 1)p.$$

Find the mode of the binomial distribution with $p = \frac{1}{2}$ and $n = 7$.

(b) Show that if np be a whole number, the mean of the binomial distribution coincides with the greatest term.

6. (a) Show that if two symmetrical binomial distributions $\left(p = q = \frac{1}{2}\right)$ of degree n (and of the same number of observations) are so superimposed that the rth term of one coincides with the $(r + 1)$th term of the other, the distribution formed by adding superimposed terms is a symmetrical binomial of degree $(n + 1)$.

(b) Let X denote a binomially distributed random variable. Show that

$$E\left\{ (X - np) / \sqrt{npq} \right\} = 0, \qquad E\left\{ (X - np) / \sqrt{npq} \right\}^2 = 1$$

and $$E\left[\exp\left\{t(X - np) / \sqrt{npq}\right\}\right] = \left[(1 - p)\exp\left\{-t\sqrt{p / nq}\right\} + p\exp\left\{t\sqrt{q / np}\right\}\right]^n$$

(c) Obtain the characteristic function of the standard binomial variate $(X - np)/\sqrt{npq}$, where X is the number of successes obtained in n independent trials, each with constant probability p of success, $q = 1 - p$. Obtain the limit of this function as $n \to \infty$. Also interpret the result.

(d) If $X \sim B(n, p)$, prove that : $\kappa_{r+1} = pq \cdot \frac{d}{dp}(\kappa_r)$, where κ_r is the rth cumulant.

Hence deduce the values of κ_2, κ_3 and κ_4.

7. (a) If X and Y are two independent identically distributed binomial variates, obtain the probability that the absolute difference $|X - Y|$ equals a given value, say r.

(b) If $X \sim B(n_1, p_1)$ and $Y \sim B(n_2, p_2)$ are independent binomial variates, find :

$$P[X + Y = r].$$

If $p_1 = p_2 = p$, (say), what is the distribution of $X + Y$?

Hint. (a) $$P\left[|X - Y| = r\right] = P\left[\{(X - Y) = r\} \cup \{Y - X = r\}\right]$$
$$= 2\,P\left[(X - Y) = r\right] \qquad (\because X \text{ and } Y \text{ are identically distributed})$$

Ans. $$2\sum_{k=0}^{n}\left[\binom{n}{k+r}\cdot\binom{n}{k}p^{2k+r}\,q^{2n-2k-r}\right]\ ; \ q = 1 - p.$$

(b) $$P[(X + y) = r] = \sum_{k=0}^{n}\left[\binom{n_1}{k}\cdot\binom{n_2}{r-k}p_1^k\,q_1^{n_1-k}\cdot p_2^{r-k}\,q_2^{n_2-r+k}\right]$$

If $p_1 = p_2 = p$, then $X + Y \sim B(n_1 + n_2, p)$.

8. Obtain the *m.g.f.* of the binomial distribution with parameters n and p. Hence or otherwise obtain the mean, variance and skewness of the distribution.

9. (a) Derive Poisson distribution as a limiting form of a binomial distribution. Hence, find β_1 and β_2 of the distribution.

Give some examples of the occurrence of Poisson distribution in different fields.

(b) Prove that under certain conditions to be stated by you, the number of telephone calls on a trunkline in a given interval of time has a Poisson distribution.

(c) State and prove the reproductive property of the Poisson distribution. Show that the mean and variance of the Poisson distribution are equal.

Find the mode of the Poisson distribution with mean value 5.

(d) Show that for a Poisson distribution, the coefficient of variation is the reciprocal of the standard deviation.

10. (a) If two independent variables X_1 and X_2 have Poisson distribution with means λ_1 and λ_2 respectively, then show that their sum $X_1 + X_2$ is a Poisson variate with mean $\lambda_1 + \lambda_2$.

Does the difference of two independent Poisson variates follow a Poisson distribution? Give reasons.

(b) Prove that the sum of two independent Poisson variates is a Poisson variate. Is the result true for the difference also ? Give reasons.

11. (a) Prove the recurrence relation between the moments of Poisson distribution :

$$\mu_{r+1} = \lambda\left(r\,\mu_{r-1} + \frac{d\mu_r}{d\lambda}\right), \text{ where } \mu_r \text{ is the } r\text{th moment about the mean } \lambda.$$

Hence obtain the skewness and kurtosis of Poisson distribution.

(b) Let X have a Poisson distribution with parameter $\lambda > 0$. If r is a non-negative integer and if

$$\mu_r' = E(X^r), \text{ prove that } \quad \mu'_{r+1} = \lambda\left(\mu'_r + \frac{d\mu'_r}{d\lambda}\right).$$

12. (a) What do you understand by (i) cumulants, (ii) cumulative function. Obtain the cumulant generating function of a Poisson distribution with parameter λ. Hence or otherwise show that for a Poisson distribution with parameter λ, all the cumulants are equal to λ.

(b) For Poisson distribution with parameter λ, show that the rth factorial moment is $\mu'_{(r)} = \lambda^r$.

Show further that $\quad \mu_{(2)} = \lambda, \quad \mu_{(3)} = -2\lambda \quad$ and $\quad \mu_{(4)} = 3\lambda(\lambda + 2)$.

13. (a) If X and Y are independent r.v.'s so that $X \sim P(\lambda)$ and $X + Y \sim P(\lambda + \mu)$, find the distribution of Y.

(b) If $X \sim P(\lambda)$ find (i) Karl Pearson's coefficient of skewness. (ii) Moment measure of skewness.

14. If $X \sim P(\lambda)$ and $(Y \mid X = x) \sim B(x, p)$, then prove that $Y - P(\lambda p)$.

15. Show that the sum of any finite number of Poisson variates is itself a Poisson variate with mean equal to the sum of separate means.

16. X is a Poisson variate with mean λ. Show that $\quad E(X^2) = \lambda \,.\, E(X + 1)$. If $\lambda = 1$, show that

$$E\,|X - 1| = (2/e).$$

17. Show that mean deviation about mean for Poisson distribution $p(x) = (e^{-m}\, m^x / x\,!)\,;\, x = 0, 1, 2, \ldots$ is

$$2\mu . \frac{e^{-m} \,.\, m^{\mu}}{\mu\,!},$$ where μ is the greatest integer contained in $(m + 1)$.

18. Show that the function which generates the central moments of the Poisson distribution with parameter λ is:

$$M(t) = \exp\,\{\lambda(e^t - 1 - t\}.$$

Show that it satisfies the equation : $$\frac{dM(t)}{dt} = \lambda\, t\, M(t) + \lambda \frac{dM\,(t)}{d\lambda}.$$

19. (a) Define negative binomial distribution. Give an example in which it occurs. Obtain its moment generating function. Hence or otherwise obtain its mean, variance and third central moment.

(b) If X denotes the number of failures preceding the rth success in an infinite series of independent trials with constant probability p of success for each trial, then identify the distribution of X and obtain $E(X)$. What is the distribution when $r = 1$?

20. In a series of independent trials with constant probability p of success in each trial, show that the number of successes in a fixed number n of independent trials follows a binomial distribution. Show further that the number of trials required for a specified number r of successes follows a negative binomial distribution. Obtain the mean and the variance of this distribution.

21. (a) Obtain the Poisson distribution as a limiting case of the negative binomial distribution.

(b) Show how the moments of negative binomial variate can be written from the corresponding formulae for the binomial variate.

(c) Consider a sequence of Bernoulli trials with constant probability p of success in a single trial. Find $P(x, k)$, the probability that exactly $x + k$ trials are required to get k successes, $x = 0, 1, 2, ...$ Show that $P(x, k)$ defines the probability function of the discrete random variable X. Find the moment generating function of X. Hence find $E(X)$ and $V(X)$.

22. (a) Derive moment generating function of negative binomial distribution and hence show that mean < variance.

(b) Derive negative binomial distribution in the following form :

$$f(x) = \binom{-k}{x}(-P)^x Q^{-k-x};\ x = 0, 1, 2, ..., \text{ and } Q = 1 + P.$$

Obtain (i) moment generating function, mean and variance of this distribution.

(ii) Co-efficient of skewness β_1. (iii) Give an example of its occurrence.

(c) Obtain the characteristic function of the negative binomial distribution given in the form :

$$f(x\ ;\ \alpha, \lambda) = \binom{-\lambda}{x}\left(\frac{\alpha}{1+\alpha}\right)^{\lambda}\left(\frac{-1}{1+\alpha}\right)^{x};\ x = 0, 1, 2, \text{ ... and hence evaluate its first two moments.}$$

23. (a) Show that for the negative binomial distribution $(Q - P)^{-r}$, where $Q - P = 1$, cumulant generating function:

$$K(t) = -r \log\,[1 - P(e^t - 1)].$$

Hence deduce that $\kappa_1 = r\,P$, $\kappa_2 = r\,PQ$. Also obtain κ_3 and κ_4.

(b) Show that the mean deviation about mean for the negative binomial distribution is :

$$2(\mu + 1)\binom{n+\mu}{\mu+1}p^{\mu+1}\,q^{-(n+\mu)},$$

where μ is the greatest integer contained in $np + 1$.

24. If X has negative binomial distribution with parameters (n, P), prove that : $M_X(t) = (Q - Pe^t)^{-n}$. Hence, find *m.g.f.* of $Z = (X - nP)/\sqrt{nPQ}$ and deduce that Z is asymptotically normal as $n \to \infty$.

25. Let Y have the negative binomial distribution. Let X_j be the number of failures between the $(j - 1)$th and jth success. Then $\sum_{j=1}^{r} X_j = Y$. Find $E(Y)$, and σ_Y^2 by obtaining the means and variances of the X_j's.

26. (a) Let X be a discrete random variable having geometric distribution with parameter p. Obtain its mean and variance. Also, show that for any two positive integers s and t.

$$P[X > s + t \mid X > s] = P[X > t]$$

(b) Find (i) the probability generating function, (ii) the moment generating function, and (iii) the cumulant generating function, for discrete random variable X following the geometric distribution:

$$P(X = r) = (1 - p)\,p^{r-1};\ r = 1, 2, ...$$

27. (a) What is a hypergeometric distribution ? Find the mean and variance of this distribution. How is this distribution related to the binomial ?

(b) Obtain binomial distribution as a limiting case of hypergeometric distribution.

(c) For the hypergeometric distribution :

$$h(N\ ;\ n, p, x) = \frac{\binom{Np}{x}\binom{Nq}{n-x}}{\binom{N}{n}};\ x = 0, 1, 2, ..., \text{ prove that: } \mu_1' = np \text{ and } \mu_2 = \frac{n(N-n)pq}{N-1}.$$

(*d*) A box contains N items of which 'a' items are defective and 'b' are non-defective, $(a + b = N)$. A sample of n items is drawn at random. Let X be number of defective items in the sample. Obtain the probability distribution of X and obtain the mean of the distribution.

(*e*) Show how the hypergeometric distribution arises, by giving an example. Obtain the frequency function of a random variable X following the above law. Derive $E(X)$ and $V(X)$. Show that under certain conditions to be stated, the Binomial and Poisson distributions are special cases of the hypergeometric distribution.

(*f*) Find the factorial moments of the hypergeometric distribution.

(*g*) X is a random variable distributed according to hypergeometric law :

$$P(X = x) = h(x\ ;\ n, a, b) = \frac{\binom{a}{x}\binom{b}{n-x}}{\binom{a+b}{n}};\ x = 0, 1, 2, \ldots \text{ Obtain the recurrence formula:}$$

$$h(x + 1\ ;\ n, a, b) = \frac{(n-x)(a-x)}{(x+1)(b-n+x+1)} \cdot h(x\ ;\ n, a, b)$$

28. (*a*) Suppose that from a population of N elements of which M are defective and $(N - M)$ are non-defective, a sample of size n is drawn without replacement. What is the probability that the sample contains exactly x defectives ? Name this probability distribution.

(*b*) Show that, for the distribution derived in (*a*),

(*i*) $E(X) = \dfrac{nM}{N}$ and (*ii*) $V(X) = \dfrac{nM}{N}\left(1 - \dfrac{M}{N}\right)\left(1 - \dfrac{n-1}{N-1}\right)$

(*c*) Show that, under certain conditions to be stated, the binomial distribution may be looked upon as a limiting form of the probability distribution as derived in (*a*).

(*d*) Explain how you will use hypergeometric model to estimate the number of wild animals in a dense forest.

29. If $X_1, X_2, \ldots, X_k$ have a multinomial distribution with the parameters n and p_i , $(i = 1, 2, \ldots, k)$ with $\Sigma p_i = 1$, obtain the joint probability $P(X_1 = x_1 \cap X_2 = x_2 \cap \ldots \cap X_k = x_k)$.
Obtain the corresponding moment generating function. Hence, or otherwise show that

$E(X_i) = np_i$, $V(X_i) = np_i\,(1 - p_i)$ and $\text{Cov}\,(X_i, X_j) = -\,np_i\,p_j$, $(i \neq j)$.

30. (*a*) Discuss the marginal and conditional distributions, associated with the multinomial distribution. If $(n_1, n_2, \ldots, n_k)$ have a multinomial distribution with parameters $(n, p_1, p_2, \ldots, p_k)$ and if c_i, d_i ; $i = 1, 2, \ldots, k$ are constants, find :

(*i*) the variance of $\sum_{i=1}^{k} c_i\, n_i$ and (*ii*) covariance between $\sum_{i=1}^{k} c_i\, n_i$ *and* $\sum_{i=1}^{k} d_i\, n_i$.

(*b*) If the random variables $X_1, X_2, \ldots, X_k$ have a multinomial distribution, show that the marginal distribution of X_i is a binomial distribution with parameters n and p_i, $i = 1, 2, \ldots, k$.

(*c*) Show that the mode of the multinomial distribution is given by $x_1, x_2, \ldots, x_k$, satisfying

$$np_i - 1 < x_i \le (n + k - 1)p_i;\ i = 1, 2, \ldots, k$$

[In order to establish this, show that $p_i\, x_j \le p_j(x_i + 1)$ for $1 \le i, j \le k$.]

ASSORTED REVIEW PROBLEMS FOR SELF-ASSESSMENT

PROBLEMS ON BINOMIAL DISTRIBUTION

8·1. (*a*) The mean and variance of a binomial variate X with parameters n and p are 16 and 8. Find
(*i*) $P(X = 0)$, (*ii*) $P(X = 1)$, (*iii*) $P(X \geq 2)$.

(*b*) For a Binomial distribution, the mean is 6 and the standard deviation is $\sqrt{2}$. Write out all the terms of the distribution.

8·2. (*a*) A perfect cube is thrown a large number of times in sets of 8. The occurrence of a 2 or 4 is called a success. In what proportion of the sets would you expect 3 successes.

(*b*) In a game of taking a chance, a contestant has to give correct answers to 4 out of 5 questions to win the contest. Questions are given with 3 answers each, out of which one is a correct answer. If a contestant answers the questions by selecting the answers at random, what is the probability that he will win the contest ?

(*c*) In eight throws of a die, 5 or 6 is considered a success. Find the mean number of successes and the standard deviation.

(*d*) A man tosses a fair coin 10 times. Find the probability that he will have :
(*i*) heads on the first five tosses and tails on the next five tosses,
(*ii*) heads on tosses 1, 3, 5, 7, 9 and tails on 2, 4, 6, 8, 10, tosses,
(*iii*) 5 heads and 5 tails (*iv*) at least 5 heads (*v*) not more than 5 heads.

(*e*) In 100 sets of ten tosses of an unbiased coin, in how many cases should we expect :
(*i*) Seven heads and three tails, (*ii*) At least seven heads ?

Ans. [(*a*) 27·31%, (*b*) $10/3^5$, (*c*) 2·66, 1·33 (*d*) (*i*) $\left(\frac{1}{2}\right)^{10}$, (*ii*) $\left(\frac{1}{2}\right)^{10}$, (*iii*) $\binom{10}{5}\left(\frac{1}{2}\right)^{10}$,
(*iv*) $\sum_{x=5}^{10}\binom{10}{x}\left(\frac{1}{2}\right)^{10}$, (*v*) $\sum_{x=0}^{5} {}^{10}C_x\left(\frac{1}{2}\right)^{10}$ (*e*) (*i*) 12, (*ii*) 17.]

8·3. (*a*) A student takes a true-false examination consisting of 10 questions. He is completely unprepared, so he plans to guess each answer. The guesses are to be made at random. For example, he may toss a fair coin and use the outcome to determine his guess.
(*i*) Compute the probability that he guesses correctly at least five times.
(*ii*) Compute the probability that he guesses correctly at least 9 times.
(*iii*) What is the smallest n that the probability of guessing at least n correct answers is less than (1/2).

(*b*) A multiple choice test consists of 8 questions and 3 answers to each question, of which only one is correct. If a student answers each question by rolling a balanced die and checking the first answer if he gets 1 or 2, the second answer if he gets 3 or 4, and the third answer if he gets 5 or 6, find the probability of getting :
(*i*) exactly 3 correct answers, (*ii*) no correct answer, and
(*iii*) at least 6 correct answers.

Ans. (*a*) (*i*) 319/512, (*ii*) 11/1024, (*iii*) 6

8·4. (*a*) A machine produces an average of 20% defective bolts. A batch is accepted if a sample of 5 bolts taken from that batch contains no defective and rejected if the sample contains 3 or more defectives. In other case, a second sample is taken. What is the probability that the second sample is required ?

Hint. Let the random variable X denote the number of defective bolts. Since $n = 5$, $p = 1/5$, $q = 4/5$, using the binomial distribution, we have

$$P(X = x) = {}^5C_x\,(1/5)^x\,(4/5)^{5-x}\ ;\ x = 0, 1, 2, ..., 5$$

Under the given conditions, if a sample of 5 bolts is taken, then a second sample will be required if the number of defectives in the sample is more than 0 and less than 3, *i.e.*, the number of defectives is either 1 or 2. Hence, the required probability is given by :

$$P(X = 1) + P(X = 2) = {}^5C_1\ (1/5)\ (4/5)^4 + {}^5C_2\ (1/5)^2\ (4/5)^3$$

$$= \frac{1}{3125}\left(5 \times 1 \times 256 + \frac{5 \times 4}{2} \times 64\right) = \frac{1920}{3125} = 0{\cdot}6144$$

(*b*) If $X \sim B\left(n, \frac{1}{3}\right)$, find the smallest value of n so that $P(X \geq 1) > 0{\cdot}80$.

Hint. $p(x) = P(X = x) = {}^nC_x\ (1/3)^x\ (2/3)^{n-x}\ ;\ x = 1, 2, ..., n$

$P(X \geq 1) > 0{\cdot}80 \quad \Rightarrow \quad 1 - P(X = 0) > 0{\cdot}80 \quad \Rightarrow \quad p(0) < 0{\cdot}2 \quad \Rightarrow \quad (2/3)^n < 0{\cdot}2$ ... (*)

n	1	2	3	4
$(2/3)^n$	0·667	$\frac{4}{9} = 0{\cdot}444$	$\frac{8}{27} = 0{\cdot}296$	$\frac{16}{81} = 0{\cdot}1975 < 0{\cdot}2$

Hence, the minimum value of n for which (*) holds is $n = 4$.

8·5. (*a*) In market region half of the households is known to use a particular brand of detergent powder. In a households survey, sample of 10 households are allotted to each investigator and 2,048 investigators are appointed for the survey. How many investigators are likely to report :

(*i*) 3 users, (*ii*) not more than 3 users, and (*iii*) at least 4 users.

(*b*) 20 men aged 45, who are members of the same occupational grouping, apply for a 15-year term life insurance policy. In order to know how much to charge for the policy, the issuing company needs to be able to estimate how many of the applicants will die before age 60, in which case the company will have to pay off on the policy. Actuarial studies indicate that, of all 45 year-old men in the applicants occupational grouping, 5% will die before they reach age 60 :

(*i*) Find the parameters of the random variable which denotes the number of applicants who die before they reach age 60.

(*ii*) What is the probability that none of the applicants will die before age 60, making it a very profitable deal for the insurance company ?

(*iii*) Find the probability that exactly one of the 20 will die during the 15-year term.

(*iv*) What are the chances that more than 5 of the 20 applicants will die before age 60 ?

(*v*) What is the probability that fewer than 10 of the applicants will die during the specified period ?

Ans. (*a*) (*i*) 240, (*ii*) 352, (*iii*) 1696; (*b*) (*i*) $n = 20$, $p = 0{\cdot}05$, (*ii*) 0·3585, (*iii*) 0·3773, (*iv*) 0·0003, (*v*) 1·0

8·6. (*a*) The probability of a man hitting a target is (1/3). How many times must he fire so that the probability of hitting the target at least once is more than 90%.

(*b*) A card is drawn and replaced in an ordinary deck of 52 cards. How many times must a card be drawn so that

(*i*) there is at least an even chance of drawing a heart,

(*ii*) the probability of drawing a heart is greater than (3/4) ?

(*c*) A communication system consists of n components, each of which will independently function with probability p. The total system will be able to operate effectively if at least one-half of its components function. For what values of p is a 5-component system more likely to operate effectively than *a* 3-component system ?

Ans. (*a*) 6, (*b*) (*i*) 3 (*ii*) 5, (*c*) $p \geq (1/2)$.

8·7. (*a*) In a binomial distribution consisting of 5 independents trials, probabilities of 1 and 2 successes are 0.4096 and 0.2048 respectively. Find the parameter p of the distribution.

(*b*) With the usual notations, find p for a binomial *r.v.* if $n = 6$ and if $4P(X = 4) = P(X = 2)$.

(*c*) X is a random variable following binomial distribution with mean 2·4 and variance 1·44.

Find : $P(X \geq 5)$, $P(1 < X \leq 4)$.

(*d*) An irregular 6 faced dice is such that the probability that it gives 3 even number in 5 throws is twice the probability that it gives 2 even numbers in 5 throws. How many sets of exactly 5 trials can be expected to give no even number out of 2500 sets ?

Ans. (*a*) $p = 1/5$, (*b*) $p = 1/3$ (*c*) $X \sim B(n = 6, p = 0\cdot4)$ (*d*) $2500 \times (1/3)^5 \simeq 10$.

8·8. The following data due to Weldon shows the results of throwing 12 dice 4,096 times, a throw of 4, 5 or 6 being called a success (x).

x :	0	1	2	3	4	5	6	7	8	9	10	11	12	Total
f :	—	7	60	198	430	731	948	847	536	257	71	11	—	4,096

Find the binomial distribution and calculate the expected frequencies. Compare the actual mean and S.D. with those of the expected ones for the distribution.

Ans. Expected frequencies: 1 12 66 220 495 792 924 792 495 220 66 12 0

8·9. (*a*) The following results were obtained when 100 batches of seeds were allowed to germinate on damp filter paper in a laboratory : $\beta_1 = \dfrac{1}{15}$ and $\beta_2 = \dfrac{89}{30}$.

Determine the binomial distribution and calculate the frequency for $X = 8$, considering $p > q$.

(*b*) If X and Y are two independent binomial variates with parameters ($n_1 = 3$, $p = 0.4$) and ($n_2 = 4$, $p = 0.4$) respectively, find :

(*i*) $P(X = Y)$, (*ii*) $P(X + Y \leq 2)$, (*iii*) $P(X = 3 \mid X + Y = 4)$.

Hint. (*a*) $\beta_1 = \dfrac{(q-p)^2}{npq} = \dfrac{1}{15}$ ... (1) and $\beta_2 = 3 + \dfrac{1-6pq}{npq} = \dfrac{89}{30}$ ... (2)

From (1) and (2), find the value of n, p and q.

Hint. (*b*) $X + Y \sim B(3 + 4, 0\cdot4) = B(7, 0\cdot4)$

(*i*) $P(X = Y) = \sum_{r=0}^{3} P(X = r \cap Y = r) = \sum_{r=0}^{3} P(X = r) \cap P(Y = r) = 0\cdot2871$

(*ii*) $P(X + Y \leq 2) = \sum_{r=0}^{2} \binom{7}{r} (0\cdot4)^r (0\cdot6)^{7-r} = 0\cdot420$

(*iii*) $P(X = 3 \mid X + Y = 4) = \dfrac{P(X = 3 \cap X + Y = 4)}{P(X + Y = 4)} = \dfrac{P(X = 3 \cap Y = 1)}{P(X + Y = 4)} = 0\cdot1141.$

8·10. (*a*) Obtain moment generating function of binomial distribution with $n = 7$ and $p = 0.6$. Find the first three moments of the distribution.

Ans. $(0\cdot4 + 0\cdot6e^t)^7$, mean $= 4\cdot2$, $\mu_2 = 1\cdot68$, $\mu_3 = -0\cdot336$

(*b*) Suppose that *m.g.f.* of a random variable X is of the form : $M_X(t) = (0\cdot4e^t + 0\cdot6)^8$.

What is the *m.g.f.* of a random variable $Y = 3X + 2$? Evaluate $E(X)$.

Ans. $E(X) = 3\cdot2$, $M_Y(t) = e^{2t} (0\cdot6 + 0\cdot4e^{3t})^8$.

(*c*) A random variable X has probability function :

$$p(x) = {}^7C_x (1/2)^x (1/2)^{7-x} \; ; x = 0, 1, 2, ..., 7.$$

Obtain : (*i*) m.g.f. $M_X(t)$; (*ii*) Mean and variance (*iii*) $P(0 \leq X \leq 1)$ and $P(X = 5)$

Ans. (*i*) $X \sim B(n = 7, p = 1/2)$; $M_X(t) = (q + pe^t)^n = \left(\dfrac{1}{2} + \dfrac{1}{2}e^t\right)^7$

(*ii*) Mean $= 7/2$, Var (X) $= 7/4$, (*iii*) (*a*) $p(0) + p(1) = 1/16$, (*b*) $p(5) = \dfrac{21}{128}$.

(*d*) If *m.g.f.* of a random variable X is $\left(\frac{1}{3}+\frac{2}{3}e^t\right)^5$, find $P(X = 2 \text{ or } 3)$.

Ans. $X \sim B(n = 5, p = 2/3)$, $p(2) + p(3) = \dfrac{40}{81}$

8·11. (*a*) If X has *B.D*, with parameters n and p, then prove that, $P[X \text{ is even}] = \frac{1}{2}\left[1+(q-p)^n\right]$.

(*b*) If the probability of hitting a target is 1/5 and if 10 shots are fired, what is the conditional probability of the target being hit at least twice assuming that at least one hit is already scored ?

(*c*) Let X be a $B(2, p)$ and Y be a $B(4, p)$. If $P(X \geq 1) = 5/9$, find $P(Y \geq 1)$.

Hint. (*b*) Let X denote the number of times a target is hit when 10 shots are fired. Then $X \sim B(10, 0{\cdot}2)$. The required probability is :

$$P(X \geq 2 \mid X \geq 1) = \frac{P[(X \geq 2) \cap (X \geq 1)]}{P(X \geq 1)} = \frac{P(X \geq 2)}{P(X \geq 1)} = \frac{1-[P(X=0)+P(X=1)]}{1-[P(X=0)]} = \frac{0{\cdot}625}{0{\cdot}893} = 0{\cdot}6999$$

(*c*) $P(X \geq 1) = 1 - P(X=0) = 1 - q^2 = \frac{5}{9} \Rightarrow q = \frac{2}{3},\ p = \frac{1}{3};$ $\quad P(Y \geq 1) = 1 - P(Y=0) = 1 - q^4 = \frac{65}{81}$

8·12. (*a*) A person is asked to throw a fair die a certain number of times. A prize is promised if he gets 7 successes. (A success is the occurrence of a 5 or 6). The catch is that at the outset, the person is to choose the number of throws he will take. How many throws he should choose to maximise the probability of his winning a prize.

Ans. $n = 21, 22$

Hint. $X \sim B\left(n, p = \frac{1}{3}\right)$. Find n so that distribution has unique mode at $x = 7$.

(*b*) A man takes a step forward with probability 0·6 and a step backward with probability 0.4. Find the probability that at the end of 15 steps, he is :

(*i*) 3 steps forward from the starting point.

(*ii*) 1 step backward from the starting point.

Ans. (*i*) $^{15}C_9\ (0{\cdot}6)^9\ (0{\cdot}4)^6$; (*ii*) $^{15}C_7\ (0{\cdot}6)^7\ (0{\cdot}4)^8$

8·13. A language class has only three students A, B and C and they independently attend the class. The probabilities of attendance of A, B and C on any given day are $\frac{1}{2}, \frac{2}{3}$ and $\frac{3}{4}$, respectively. Find the probability that the total number of attendances in two consecutive days is exactly three.

Hint. Let the *r.v.'s* X, Y, Z denote the number of days on which A, B and C attend the classes respectively. Then

$$X \sim B(n_1 = 2, p_1 = 1/2)\ ;\ Y \sim B(n_2 = 2, p_2 = 2/3)\ ;\ Z \sim B(n_3 = 2, p_3 = 3/4)$$

are independent binomial variates. We want $P(X + Y + Z = 3)$.

Try all possible combinations of (x, y, z), each taking values 0, 1, 2 so that: $x + y + z = 3$ and use independence of X, Y, Z.

Ans. $(1/4) = 0{\cdot}25$

8·14. If $X \sim B(n, p)$ and $Y \sim B(n, q)$, be independent r.v.'s $(p + q = 1)$, show that :

$X + Y$ and $\sum_{i=1}^{n} X_i$, where X_i's are *i.i.d.* variates with the probability distribution :

$$P(X_i = 0) = pq, \quad P(X_i = 1) = p^2 + q^2 \quad \text{and} \quad P(X_i = 2) = pq,$$

have the same distribution.

Hint. Prove that $M_{X+Y}(t) = M_{\sum_{i=1}^{n} X_i}(t)$

8·15. Let X_1, X_2 be independent binomial variates $B\left(n_i, \frac{1}{2}\right)$; $i = 1, 2$. Obtain the p.m.f. of $(X_1 - X_2 + n_2)$. Hence, find $P[(X_1 - X_2 + n_2) = k]$.

Ans. $(X_1 - X_2 + n_2) \sim B\left[(n_1 + n_2), \frac{1}{2}\right]$; $P[(X_1 - X_2 + n_2) = k] = \binom{n_1 + n_2}{k}\left(\frac{1}{2}\right)^{n_1 + n_2}$; $k = 0, 1, 2, \ldots, n_1 + n_2$

Hint. If $X \sim B(n, p)$ then $(n - X) \sim B(n, 1 - p)$ and now use Additive property of binomial distribution.

8·16. A and B play a game which must be either won or lost. If the probability that A wins any game is p, find the probability that A wins m games before B wins n games; $(m, n \geq 1)$.

Ans. $\sum_{r=m}^{m+n-1} \binom{m+n-1}{r} p^r (1-p)^{m+n-1-r}$

8·17. If $X \sim B(n, p)$ show that X is symmetrically distributed about 'c' if and only if $p = 1/2$ and $c = n/2$.

8·18. An unbiased coin is tossed $2n$ times. Find the probability $p(2n)$, that there will be an equal number of heads and tails. Prove that $p(2n)$ is a decreasing function of n.

Ans. $p(2n) = \binom{2n}{n}\left(\frac{1}{2}\right)^n$; Prove that $\dfrac{p(2n)}{p(2n-2)} < 1$.

8·19. Calculate the value of p if the ratio of the probability of an event happening exactly r times in n trials to the probability of the event happening exactly $(n - r)$ times is independent of n, $(0 < p < 1)$.

Ans. $p = 1/2$

8·20. X and Y have joint p.d.f.

$$P(X = x, Y = y) = \left[\frac{(1-\theta)\,\theta'}{\theta + \theta'}\right]^{(1-x)(1-y)} \cdot \left[\frac{\theta\,\theta'}{\theta + \theta'}\right]^{x(1-y)+y(1-x)} \cdot \left[\frac{\theta\,(1-\theta')}{\theta + \theta'}\right]^{xy}, \quad \ldots (*)$$

where $x = 0, 1$; $y = 0, 1$; $0 \leq \theta \leq 1$, $0 \leq \theta' \leq 1$,

Prove that the marginal distributions of X and Y are Bernoulli, each with parameter p. Also prove that
$$\rho(X, Y) = 1 - \theta - \theta'.$$

Hint. Find marginal p.m.f.'s of X and Y.

$$f(x) = \sum_{y=0}^{1} P(X = x \cap Y = y) \text{ and find } f(1) \text{ and } f(0) \text{ ; etc.}$$

8·21. (*a*) If A has $(n + 1)$ and B has n fair coins, which they flip, show that the probability that A gets more heads than B is $\frac{1}{2}$.

(*b*) If a coin is tossed repeatedly, show that the probability of getting m heads before n tails is :

$$\frac{1}{2^{m+n-1}} \sum_{i=m}^{m+n-1} {}^{m+n-1}C_i.$$

8·22. (*a*) For a binomial distribution, let $F_n(y) = \sum_{x=0}^{y} \binom{n}{x} p^x q^{n-x}$, where $q = 1 - p$, prove that

(*i*) $F_{n+1}(y) = p\,F_n(y-1) + q\,F_n(y)$, (*ii*) Cov $(X, n - X) = -\,npq$.

(*b*) The discrete density of X is given by : $f_X(x) = \frac{x}{3}$, for $x = 1, 2$ and $f_{Y|X}(y \mid x)$ is binomial with parameters x and $\frac{1}{2}$, *i.e.*,

$$f_{Y|X}(y \mid x) = P(Y = y \mid X = x) = \binom{x}{y} \cdot \left(\frac{1}{2}\right)^x ; \text{ for } y = 0, 1, ..., x \text{ and } x = 1, 2.$$

Find (*i*) E(X) and Var (X) (*ii*) E(Y), and (*iii*) the joint distribution of X and Y

Ans. (*i*) $E(X) = \frac{5}{3}, V(X) = \frac{2}{9}$, (*ii*) $E(Y) = \frac{5}{6}$, (*iii*) $f(x, y) = \binom{x}{y} \cdot \frac{x}{3} \cdot \left(\frac{1}{2}\right)^x ; x = 1, 2 ; y = 0, 1, ..., x$

(*c*) Let $X \sim B(n, p)$. For what value of 'p' is Var(x) maximized if we assume n is constant ? Also obtain the value of maximum variance. [**Ans.** $p = \frac{1}{2}$; $(\text{Var } x)_{\text{Max}} = n/4$]

8·23. (*a*) If m things are distributed among 'a' men and 'b' women, show that the probability that the number of things received by men is odd, is : $\frac{1}{2}\{(b+a)^m - (b-a)^m\} / (b+a)^m$.

(*b*) A and B have equal chances of winning a single game, A wants n games and B, $n + 1$ games to win a match. Show that the odds in favour of A are $1 + P$ to $1 - P$,

where $P = \frac{(2n)!}{n!\,n!\,2^{2n}}$.

Hint. (*a*) p = Probability that a thing is received by man = $\frac{a}{a+b}$, then $q = \frac{b}{a+b}$.

The probability that out of m things exactly x are received by men and the rest by women, is given by :

$$p(x) = {}^mC_x\, p^x q^{m-x} ; x = 0, 1, 2, ..., m$$

The probability P that the number of things received by men is odd is given by :

$$P = p(1) + p(3) + p(5) + ... = {}^mC_1\, q^{m-1} p + {}^mC_3\, q^{m-3} p^3 + {}^mC_5\, q^{m-5} p^5 + ...$$

Now, $$(q + p)^m = q^m + {}^mC_1\, q^{m-1} p + {}^mC_2\, q^{m-2} p^2 + {}^mC_3\, q^{m-3} p^3 + {}^mC_4\, q^{m-4} p^4 + ...$$

and $$(q - p)^m = q^m - {}^mC_1\, q^{m-1} p + {}^mC_2\, q^{m-2} p^2 - {}^mC_3\, q^{m-3} p^3 + {}^mC_4\, q^{m-4} p^4 - ...$$

$\therefore$ $$(q + p)^m - (q - p)^m = 2\left[{}^mC_1\, q^{m-1} p + {}^mC_3\, q^{m-3} p^3 + ...\right] = 2P$$

But $q + p = 1$ and $q - p = \frac{b-a}{b+a}$ $\quad \therefore 1 - \left(\frac{b-a}{b+a}\right)^m = 2P$.

8·24. (*a*) If $X \sim B(n, p)$, show that $P(X \leq 2) = P[X \geq (n - 2)]$, if and only if $p = \frac{1}{2}$

(*b*) Let $X_1 \sim B(n, p_1)$ and $X_2 \sim B(n, p_2)$. If $p_1 < p_2$, prove that :

$$P(X_1 \leq k) \geq P(X_2 \leq k), \text{ for } k = 0, 1, ..., n.$$

(*c*) A spider and a fly are situated at the corners (0, 0) and (n, n) of a rectangular grid. The spider walks only north or east, the fly only south or west. They take their steps simultaneously to an adjacent vertex of the grid. Show that, if the successive steps are independent and equally likely to go in each of the two possible directions, the probability that they will meet is $\binom{2n}{n}\left(\frac{1}{2}\right)^{2n}$

8·25. The chance of success in each Bernoulli trial is p. If p_k is the probability that there are even number of successes in k trials, prove that

$$p_k = p + p_{k-1}(1 - 2p) \qquad \ldots (*)$$

Deduce that

$$p_k = \frac{1}{2}\left[1 + (1 - 2p)^k\right]$$

(*i*) Also obtain the probability generating function of (*) and hence obtain an explicit expression for p_k.

(*ii*) Obtain an expression for p_k directly without using (*i*) or (*ii*).

Ans. (*i*) p.g.f. = $\sum_{k=0}^{\infty} \frac{1}{2}\left\{1 + (1 - 2p)^k\right\} . t^k$

8·26. If $b(r\ ;\ n,\ p) = {}^nC_r\ p^r\ q^{n-r}$, is the binomial probability in the usual notation and if $B(k\ ;\ n,\ p) = \sum_{r=0}^{k} b(r\ ; n, p)$, prove the following results for the "tails" of the binomial distribution.

(*i*) $1 - B(k - 1\ ;\ n,\ p) \quad \leq \quad \dfrac{n}{k - np}\, b(k\ ; n, p), \qquad k > np + 1$

(*ii*) $B(k\ ;\ n,\ p) \quad \leq \quad \dfrac{n}{np - k}\, b(k\ ; n, p), \qquad k < np$

(*iii*) $1 - B(k\ ;\ n,\ p) \quad = \quad n\dbinom{n-1}{k}\displaystyle\int_0^p t^k\,(1 - t)^{n-k-1}\,dt$

8·27. (*a*) If $X \sim B(n,\ p)$, show that

$$P(X \leq k) = \lambda \int_{p/q}^{\infty} \frac{y^k}{(1 + y)^{n+1}}\,dy \text{ where } \lambda^{-1} = \int_0^{\infty} \frac{y^k}{(1 + y)^{n+1}}\,dy = B(k + 1, n - k)$$

(*b*) For the binomial distribution, show that the probability that the number of successes in n trials should not exceed x is given by :

$$\frac{\displaystyle\int_{p/q}^{\infty} \frac{y^x}{(1 + y)^{n+1}}\,dy}{\displaystyle\int_0^{\infty} \frac{y^x}{(1 + y)^{n+1}}}, \quad p + q = 1$$

where p is the constant probability of success in a trial.

(*c*) Prove the identity :

$$p^n + \binom{n}{1}p^{n-1}q + \binom{n}{2}p^{n-2}q^2 + \ldots + \binom{n}{k}p^{n-k}q^k = \sum_{x=0}^{k}\binom{n}{x}p^{n-x}q^x = \frac{\displaystyle\int_0^p x^{n-k-1}(1 - x)^k\,dx}{\displaystyle\int_0^1 x^{n-k-1}(1 - x)^k\,dx}$$

8·28. (*a*) Prove the following identity :

$$\sum_{i=k}^{n}\binom{n}{i}p^i\,q^{n-i} = \frac{\displaystyle\int_0^p x^{k-1}(1 - x)^{n-k}\,dx}{\displaystyle\int_0^1 x^{k-1}(1 - x)^{n-k}\,dx},$$

where L.H.S. in the probability of at least k successes in n independent Bernoulli trials.

(*b*) If $X \sim B(n, p)$, prove that

(*i*) $P(X \geq k) = \dfrac{1}{B(k, n-k+1)} \cdot \displaystyle\int_0^p x^{k-1} (1-x)^{n-k}\, dx$

(*ii*) $P(X \geq k) = k\dbinom{n}{k} \displaystyle\int_0^p x^{k-1} (1-x)^{n-k}\, dx = n\dbinom{n-1}{k-1} \int_0^p x^{k-1} (1-x)^{n-k}\, dx$

(*c*) Deduce that : $P(X \geq k) \leq \dbinom{n}{k} p^k$

Hint. Use [*b* (*ii*)] with $(1 - x) \leq 1$, for $x \geq 0 \Rightarrow (1-x)^{n-k} \leq 1$

8·29. Suppose that the *r.v.* X is uniformly distributed on (0, 1), *i.e.*,

$$f_X(x) = 1, \quad 0 \leq x \leq 1 \qquad \text{... (*)}$$

Assume that the conditional distribution $Y \mid X = x$ has a binomial distribution with parameters n and $p = x$, *i.e.*,

$$P(Y = y \mid X = x) = \binom{n}{y} x^y (1-x)^{n-y};\ y = 0, 1, 2, ..., n \qquad \text{... (**)}$$

Find (*i*) $E(Y)$ and (*ii*) the distribution of Y.

Hint. (*i*) Since conditional distribution of $[Y \mid X = x] \sim B(n, x)$,

$$E(Y \mid X = x) = nx \qquad \text{... (1)}$$

$$\therefore \quad E(Y) = E[E(Y \mid X)] = E(nX) = nE(X) \qquad \text{[From (1)]} \qquad \text{... (2)}$$

$$E(X) = \int_0^1 x\, f(x)\, dx = \int_0^1 x\, dx = \frac{1}{2}$$ Substituting in (2), we get : $E(Y) = \dfrac{1}{2}n.$

(*ii*) We have $f_{XY}(xy) = f_X(x) \cdot f_{Y \mid X}(y \mid x)$

Since X has (continuous) uniform distribution on (0, 1), marginal distribution of Y is given by :

$$f_Y(y) = \int_{-\infty}^{\infty} f(x, y)\, dx = \int_0^1 f_{Y \mid X}(y \mid x) \cdot f_X(x)\, dx$$

$$= \int_0^1 {}^nC_y\, x^y (1-x)^{n-y} \cdot 1\, dx = {}^nC_y \int_0^1 x^y (1-x)^{n-y}\, dx \qquad \text{[From (*) and (**)]}$$

$$= {}^nC_y\, B(y+1, n-y+1) = \frac{n!}{y!\,(n-y)!} \cdot \frac{\Gamma(y+1)\,\Gamma(n-y+1)}{\Gamma(n+2)}$$

$$= \frac{n!}{y!\,(n-y)!} \times \frac{y!\,(n-y)!}{(n+1)!} = \frac{1}{n+1};\ y = 0, 1, 2, ..., n$$

Since Y takes the values 0, 1, 2, ..., n each with equal probability $\dfrac{1}{n+1}$, Y has discrete uniform distribution on $[0, n]$.

8·30. (*a*) $X \sim B(n = 100, p = 0.1)$. Find $P(X \leq \mu_x - 3\sigma_x)$

Ans. $\mu = 10$, $\sigma = 3$, $P(X \leq \mu_x - 3\sigma_x) = P(X \leq 1) = 10{\cdot}9 \times (0{\cdot}9)^{99}$

(*b*) Random variable X follows binomial distribution with parameters $n = 40$ and $p = \dfrac{1}{4}$.

Use Chebychev's inequality to find bounds for :

(*i*) $P[\,|X - 10| < 8\,]$;

(*ii*) $P[\,|X - 10| > 10\,]$. Compare these values with the actual values.

Ans. (*i*) 113/128 (lower bound), (*ii*) 0·075 (upper bound)

(c) X follows binomial distribution with parameters $n = 40, p = \frac{1}{2}$. Use Chebychev's inequality to

(i) find k such that $P\{|X - 20| > 10k\} \le 0{\cdot}25$, and

(ii) obtain a lower limit for $P\{|X - 20| \le 5\}$.

Ans. (i) $2/\sqrt{10}$, (ii) 3/5.

PROBLEMS ON POISSON DISTRIBUTION

8·31. (a) Records show that the probability is 0.00002 that a car will have a flat tyre while driving over a certain bridge. Use the Poisson probability distribution to determine the probability that among 20,000 car drivers over this bridge, not more than one will have a flat tyre. (Given $e^{-0{\cdot}4} = 0{\cdot}670$)

(b) A local electrical appliances shop has found from experience that the demand for tube lights is distributed as Poisson with a mean of 4 tube lights per day. If the shop keeps 6 tubes during a particular day, what is the probability that the demand will exceed supply during that day ? (Given $e^{-4} = 0{\cdot}0183$).

(c) An insurance company insures 5,000 people against loss of limbs in a car accident. Based on previous data, the rates were computed on the assumption that on the average 40 persons in 2,00,000 will have car accident each year that result in this type of injury. What is the probability that more than 5 of the insured will collect on their policy in a given year ?

Hint. (a) Let the random variable X denote the number of flat tyres, then X will follow Poisson distribution with parameter $\lambda = 20{,}000 \times 0{\cdot}00002 = 0{\cdot}4$ and its p.m.f. is given by :

$$P(X = x) = \frac{e^{-0{\cdot}4}(0{\cdot}4)^x}{x!};\ x = 0, 1, 2, ...$$

Required probability $= P(X \le 1) = P(X = 0) + P(X = 1) = e^{-0{\cdot}4}(1 + 0{\cdot}4) = 0{\cdot}670 \times 1{\cdot}4 = 0{\cdot}938$.

(b) Let the random variable X represent the number of tube lights to be demanded each day. Then X will follow Poisson distribution with parameter $\lambda = 4$ and its *p.m.f.* is given by :

$$P(X = x) = \frac{e^{-4}\,4^x}{x!};\ x = 0, 1, 2, ...$$

$$\text{Required probability} = P(X > 6) = 1 - P(X \le 6) = 1 - \sum_{x=0}^{6} e^{-4}\,4^x / x!$$

$$= 1 - 0{\cdot}0183[1 + 4 + 8 + (32/3) + (256/24) + (1024/120) + (4096/7720)] = 0{\cdot}1114.$$

(c) In usual notations : $n = 5000$ and $p = (40/2{,}00{,}000) = 0{\cdot}0002$.

Since p is very small and n is large, the given distribution may be approximated by Poisson distribution with parameter $\lambda = 5000 \times 0{\cdot}0002 = 1$. Let the random variable X denote number of car accidents in the batch of 5,000 people, then by Poisson probability law :

$$P(X = x) = \frac{e^{-1}(1)^x}{x!} = \frac{e^{-1}}{x!}\ ;\ x = 0, 1, 2, ...$$

Required probability $= P(X > 5) = 1 - P(X \le 5)$

8·32. (a) It is stated that 2% of mobile phones supplied by a manufacturer are defective. A random sample of 200 mobile phones is drawn from a lot. Find the probability that :

(i) 3 or more are defective, and (ii) 4 or less are defective.

Ans. (a) (i) $1 - e^{-4}\sum_{x=0}^{2}(4^r / r!) = 0{\cdot}7621$, (ii) $e^{-4}\sum_{x=0}^{4}(4^r / r!) = 0{\cdot}6283$.

(*b*) In a mass production process, where 0·1% of the articles are defective, they are packed in boxes of 1,000. Find what proportion would contain 2 or more defectives.

(*c*) An automatic machine makes paper clips from coils of wire. On the average, 1 in 400 paper clips is defective. If the paper clips are packed in boxes of 100, what is the probability that any given box of clips will contain :

(*i*) no defective, (*ii*) one or more defective, and (*iii*) less than two defectives ?

(*d*) A certain airline finds that 4% of the persons making reservations on certain flight will not show up for the flight. Consequently, their policy is to sell to 75 persons reserved seats on a plane that has exactly 73 seats. What is the probability that for every person who shows up for the flight, there will be a seat available ?

8·33. (*a*) On the average, there are 1·8 breakdowns every five working days in major factory equipment at the main plant of *XYZ* Fabricators, according to the plant's maintenance department :

(*i*) Find the probability that there will be fewer than 2 breakdowns during the next 5 working days.

(*ii*) The rate of breakdowns namely 1·8 per 5 working days, is equivalent to a rate of 3·6 per 10 working days. Find the probability that fewer than 4 breakdowns occur in the next 10 working days.

(*iii*) The rate of breakdowns is also equivalent to a rate of 0·36 breakdown per working day. What are the chances that there will be one or more breakdowns next Tuesday ?

(*iv*) What is the probability that there will be one or more breakdowns during the next three-working-day period ?

(*v*) Find the probability that fewer than two breakdowns will occur during the next six working days.

Ans. (*i*) 0·4628, (*ii*) 0·5152, (*iii*) 0·3023, (*iv*) 0·6604, (*v*) 0·3644.

(*b*) It is known that the probability that an item produced by a certain machine will be defective is 0·01. By applying Poisson's approximations, show that the probability that random sample of 100 items selected from the total output will contain no more than one defective item is $2/e$.

8·34. (*a*) Patients arrive randomly and independently at a doctor's O.P.D. clinic from 8·0 A.M. at an average rate of one in five minutes. The waiting room holds 20 persons. What is the probability that the room will be full when the doctor arrives at 9·0 A.M. (Estimate the probability to an accuracy of 5 per cent.)

(*b*) In a certain factory turning out optical lenses, there is a small chance 1/500 for any lens to be defective. The lenses are supplied in a packet of 10. Use Poisson distribution to calculate the approximate number of packets containing no defective, one defective, two defective and three defective lenses in a consignment of 20,000 packets.

(*c*) After correcting the proofs of the first 50 pages or so of a book, it is found that on the average there are 3 errors per 5 pages. Use Poisson probabilities and estimate the number of pages with 0, 1, 2 and 3 errors in the whole book of 1,000 pages (Given that $e^{-0·6} = 0·5488$.)

(*d*) A distributor of bean seeds determines from extensive tests that 5% of large batch of seeds will not germinate. He sells the seeds in packets of 200 and guarantees 90% germination. Determine the probability that a particular packet will violate guarantee.

(*e*) In an automatic telephone exchange the probability that any one call is wrongly connected is 0·001. What is the minimum number of independent calls required to ensure a probability of 0·90, that at least one call is wrongly connected ?

(*f*) Suppose that the number of telephone calls that an operator receives from 9·00 to 9·05 hours in a day follows a Poisson distribution with mean 3. Find the probability that

(*i*) the operator will receive no calls in that time interval tomorrow,

(*ii*) In the next three days the operator will receive a total of 1 call in that time interval.

(g) A large number of observations on a given solution which contained bacteria were made taking samples of 1 ml. each, noting down the number of bacteria present in each sample. Assuming the Poisson distribution, and given that 10% samples contained no bacteria, find the average number of bacteria per ml.

Ans. (b) 19604, 392, 4 and 0 packets, (c) 549, 329, 99 and 20, (d) $1 - \sum_{r=0}^{20} \left(e^{-10} \cdot 10^r / r!\right)$

(f) (i) e^{-3}, (ii) $(e^{-3})^2$, $(1 - e^{-3})$ (g) $\log_e 10$

8·35. (a) The first proof of 400 pages of a book containing 560 pages revealed the following distribution of the number of printing errors :

No. of errors in a page :	0	1	2	3	4	5	Total
No. of pages :	142	156	69	27	5	1	400

Fit a Poisson distribution corresponding to these data and calculate the theoretical frequencies.

(b) The following data give frequency of aircraft accidents experienced by 2,546 pilots during a four-year period :

No. of accidents :	0	1	2	3	4	5
Frequencies :	2,036	422	71	13	3	1

Calculate mean number of accidents and fit a Poisson distribution.

(c) Fit a Poisson distribution to the following data

x	0	1	2	3	4	5	6
f	143	90	42	12	9	3	1

Ans. (a) $\lambda = 1$; [147, 147, 74, 25, 6, 1], (c) [123, 110, 49, 14, 3, 1, 0].

8·36. (a) If X is the number of occurrences of the Poisson variate with mean λ ; show that

$$P(X \geq n) - P(X \geq n + 1) = P(X = n).$$

(b) Suppose that X has a Poisson distribution. If $P(X = 2) = \frac{2}{3} P(X = 1)$, evaluate

(i) $P(X = 0)$ and (ii) $P(X = 3)$.

(c) If X has a Poisson distribution such that $P(X = 1) = P(X = 2)$, find $P(X = 4)$.

(d) If X is a Poisson variate such that $P(X = 1) = 2P(X = 2)$, find

(i) $P(X = 0)$, (ii) mean, and (iii) variance.

(e) If for a Poisson variate X, $E(X^2) = 6$, what is $E(X)$?

(f) If X and Y are independent Poisson variates having means 1 and 3 respectively, find the variance of $3X + Y$.

(g) Show that for a Poisson distribution :

(i) $M \sigma \gamma_1 \gamma_2 = 1$, (ii) $\beta_1^{1/2} (\beta_2 - 3) \mu_1' \sigma = 1$.

Ans. (b) (i) $e^{-4/3}$, (ii) $\dfrac{e^{-4/3} (4/3)^3}{3!}$; (c) $\dfrac{e^{-2} \cdot 2^4}{4!}$; (d) (i) e^{-1}, (ii) 1, (iii) 1, (e) 2,

(f) 12.

8·37. (a) If $X \sim P(m)$, show that $E|X - 1| = m - 1 + 2\exp(-m)$.

Hint. $E|X-1| = \sum_{x=0}^{\infty}\left[|x-1|\dfrac{e^{-m}\cdot m^x}{x!}\right] = e^{-m} + \sum_{x=2}^{\infty}(x-1)\dfrac{e^{-m}\cdot m^x}{x!}$

$$= e^{-m} + e^{-m}\sum_{x=2}^{\infty}\left[m^x\left\{\frac{1}{(x-1)!}-\frac{1}{x!}\right\}\right]$$

(*b*) Let X, Y be independent Poisson variates. The variance of $X+Y$ is 9 and

$P(X=3 \mid X+Y=6) = \dfrac{5}{54}$. Find the mean of X.

Ans. (*b*) $\dfrac{1}{2}\left[9 \pm 3\sqrt{3}\right] = 1{\cdot}902$ or $7{\cdot}098$

8·38. The probability function $p(x)$ of a *r.v.* X is defined for non-negative integers x and is given by recurrence relation :

$$\left.\begin{aligned} p(x) &= \frac{4}{x}\,p(x-1)\ ;\ x=1,2,3,\ldots \\ &= 0 \qquad ;\ \text{elsewhere} \end{aligned}\right\} \qquad \ldots (*)$$

Find p(x) and identify the distribution.

Hint. $\sum_{x=0}^{\infty} p(x) = 1 \quad \Rightarrow \quad p(0)+p(1)+p(2)+p(3)+\ldots = 1$

$$\Rightarrow \quad p(0)\left[1+4+\frac{4^2}{2!}+\frac{4^3}{3!}+\frac{4^4}{4!}+\ldots\right]=1 \qquad \text{[From (*)]}$$

$$\Rightarrow \quad p(0)\,.\,e^4 = 1 \quad \Rightarrow \quad p(0) = e^{-4}.$$

$$\therefore \quad p(1) = 4\,e^{-4}\ ; \quad p(2) = \frac{4^2\,e^{-4}}{2!}\ ; \quad p(3) = \frac{4^3\,e^{-4}}{3!},\ldots \qquad \text{[From (*)]}$$

$$\Rightarrow \quad p(x) = \frac{e^{-4}\,.\,4^x}{x!}\ ; \quad x = 0, 1, 2, 3, \ldots$$
$$= 0 \quad ; \quad \text{otherwise}$$

Aliter. $p(x) = \dfrac{4}{x}\cdot p(x-1) \quad \Rightarrow \quad p(x+1) = \dfrac{4}{(x+1)}\cdot p(x)$ ... (*i*)

We know, if $X \sim P(\lambda)$, $\quad p(x+1) = \dfrac{\lambda}{x+1}\cdot p(x)$ ... (*ii*)

From (*i*) and (*ii*), we conclude that $X \sim P(\lambda = 4)$.

8·39. (*a*) If $X \sim P(\lambda)$ and $Y \sim P(\mu)$ are correlated Poisson variates, prove that $(X+Y)$ can not be a Poisson variate.

Hint. Prove that $E(X+Y) \neq \text{Var}\,(X+Y)$.

(*b*) If $X \sim P(\lambda)$ and $Y \sim P(\mu)$ are independent random variables, prove that the conditional distribution of $[X \mid (X+Y=n)]$ is binomial. Hence deduce that

$$E\left[X \mid (X+Y)=n\right] = \frac{n\lambda}{\lambda+\mu}$$

8·40. If $X \sim P(\lambda)$ and $Y \sim P(\mu)$ are independent Poisson variates, find :

(*i*) $P(X+Y=r)$, (*ii*) $P(X=Y)$, (*iii*) $P(X>Y)$.

8·41. If X and Y are i.i.d. Poisson variates, each with parameter λ, find $P(Z = k)$, where $Z = \text{Min}(X, Y)$. Hence deduce $P(Z = 1)$, when $\lambda = 1$.

Hint. $P(Z = k) = P(Z \geq k) - P(Z \geq k + 1)$; $\left[P(Z \geq k) = P(X \geq k) \cdot P(Y \geq k) = [P(X \geq k)]^2\right]$

$$= \left[1 - e^{-\lambda} \sum_{r=0}^{k-1} \frac{\lambda^r}{r!}\right]^2 - \left[1 - e^{-\lambda} \sum_{r=0}^{k} \frac{\lambda^r}{r!}\right]^2$$

If $\lambda = 1$, $P(Z = 1) = (2e - 3)/e^2$.

8·42. A *r.v.* Y has a distribution with mean μ and variance σ^2. By expressing Y in the form

$$Y = \mu\left[1 + \frac{Y - \mu}{\mu}\right], \qquad \ldots (*)$$

(*a*) Show that : $E\sqrt{Y} \simeq \sqrt{\mu}\left(1 - \frac{\sigma^2}{8\mu^2}\right)$,

if the variations in Y are sufficiently small compared with its mean value.

(*b*) If Y has a Poisson distribution with a large mean, prove that as a first approximation,

$$\text{Var}\left(\sqrt{Y}\right) = 1/4.$$

Hint. In part (*a*), use $\left(\frac{Y - \mu}{\mu}\right)^r \to 0$, $r \geq 3$.

In part (*b*) use $\left(1/\mu^{3/2}\right) \to 0$ as $\mu \to \infty$.

8·43. (*a*) The random variable X has *p.m.f.* : $p(x) = \begin{cases} \dfrac{\exp(-\theta) \cdot \theta^x}{x!} & ; x = 0, 1, 2, \ldots \\ 0, & \text{elsewhere} \end{cases}$

Find the *m.g.f.* of $Y = 2X - 1$ and Var (Y).

(*b*) Identity the distributions with the following *m.g.f.*'s :

$$M_X(t) = (0{\cdot}3 + 0{\cdot}7\, e^t)^{10}, \qquad M_Y(t) = \exp\{3(e^t - 1)\}$$

(*c*) The *m.g.f.* of a *r.v.* X is $\exp\left(4(e^t - 1)\right)$. Show that $P(\mu - 2\sigma < X < \mu + 2\sigma) = 0{\cdot}931$.

(*d*) Identify the variates X and Y where :

$$M_X(t) = \frac{1}{27}(1 + 2e^t)^3 . \exp\{3(e^t - 1)\}, \qquad M_Y(t) = \frac{1}{32}(1 + e^t)^5 . \exp\{-2(1 - e^t)\}.$$

Ans. $X = U + V$, where $U \sim B\left(n = 3, p = \frac{2}{3}\right)$ and $V \sim P(\lambda = 3)$ are independent.

$Y = W + Z$, where $W \sim B\left(n = 5, p = \frac{1}{2}\right)$ and $Z \sim P(\mu = 2)$ are independent.

(*e*) A random variable X is distributed in the Poisson's form with parameter λ. If λ is a random variable with a characteristic function $\phi(t)$, show that the characteristic function of X is $\phi\left(\frac{e^{it} - 1}{i}\right)$.

Hint. c.f. of $X = E_X(e^{itX}) = E_\lambda\left(E_X\, e^{itX} \mid \lambda \text{ is constant}\right)$.

8·44. (*a*) Let X, Y, Z be independent Poisson variates with parameters a, b and c respectively. Obtain :

(*i*) *m.g.f.* of $X + 2Y + 3Z$, (*ii*) Conditional expectation of X given $X + Y + Z = n$.

(b) The joint density of r.v.'s X and Y is :

$$f(x, y) = e^{-2}/[x\,!\,(y-x)\,!]\;;\; y = 0, 1, 2, \ldots\;;\; x = 0, 1, 2, \ldots, y.$$

Find the m.g.f. $M(t_1, t_2)$ of (X, Y) and covariance between X and Y. Show that the marginal distributions of X and Y are Poisson.

Hint. (a) (i) $M_{X+2Y+3Z}(t) = M_X(t) \cdot M_Y(2t) \cdot M_Z(3t) = \exp\left[a(e^t - 1) + b(e^{2t} - 1) + c(e^{3t} - 1)\right]$

(ii) $$P(X = x \mid X + Y + Z = n) = \frac{P(X = x \cap X + Y + Z = n)}{P(X + Y + Z = n)}$$

$$= \frac{P(X = x)\,P(Y + Z = n - x)}{P(X + Y + Z = n)} \qquad (\because X, Y, Z \text{ are independent})$$

$$= \frac{e^{-a} \cdot a^x}{x\,!} \times \frac{e^{-(b+c)} \cdot (b+c)^{n-x}}{(n-x)\,!} \times \left[\frac{n\,!}{e^{-(a+b+c)} \cdot (a+b+c)^n}\right]$$

$$= \frac{n\,!}{x\,!\,(n-x)\,!}\left(\frac{a}{a+b+c}\right)^x \cdot \left(\frac{b+c}{a+b+c}\right)^{n-x}$$

$$\therefore\; X \mid (X + Y + Z = n) \sim B\{n, p = a/(a+b+c)\} \Rightarrow E\left[X \mid X + Y + Z = n\right] = np = \frac{na}{a+b+c}$$

(b) $$M(t_1, t_2) = \sum_{y=0}^{\infty}\sum_{x=0}^{y} e^{t_1 x + t_2 y} \times \left[\frac{e^{-2}}{x\,!\,(y-x)\,!}\right] = e^{-2}\sum_{y=0}^{\infty}\left[\frac{e^{t_2 y}}{y\,!}\left\{\sum_{x=0}^{y} {}^yC_x \cdot (e^{t_1})^x\right\}\right]$$

$$= e^{-2}\sum_{y=0}^{\infty}\left\{\left[e^{t_2}(1 + e^{t_1})\right]^y / y\,!\right\} = e^{-2} \cdot \exp\left\{e^{t_2}(1 + e^{t_1})\right\}$$

$M(t_1, 0) = \exp[(e^{t_1} - 1)] \Rightarrow X \sim P(\lambda = 1)$ and $M(0, t_2) = \exp[2(e^{t_2} - 1)] \Rightarrow Y \sim P(\mu = 2)$

Observe $M(t_1, t_2) \neq M(t_1, 0) \times M(0, t_2) \Rightarrow$ X and Y are not independent

$E(X) = 1, \quad \text{Var}(X) = 1; \quad E(Y) = 2 = \text{Var}\, Y, \quad E(XY) = \left|\dfrac{\partial^2 M(t_1, t_2)}{\partial t_1\, \partial t_2}\right|_{t_1 = t_2 = 0} = 3$

$\therefore \quad \text{Cov}(X, Y) = E(XY) - E(X)\,E(Y) = 3 - 1 \times 2 = 1.$

8·45. Suppose (X, Y) have joint p.m.f. :

$$f(x, y) = \frac{e^{-(a+b)}\, a^x\, b^{y-x}}{x\,!\,(y-x)\,!}, \quad x = 0, 1, 2, \ldots\;;\; y = x, x+1, \ldots$$

Show that Cov (X, Y) = a and $r(X, Y) = \left(\dfrac{a}{a+b}\right)^{1/2}$

Also obtain the distribution of Y – X.

Hint. $$M_{X,Y}(t_1, t_2) = \sum_{x=0}^{\infty}\sum_{y=x}^{\infty}\left[\exp(t_1 x + t_2 y)\frac{e^{-(a+b)}\, a^x\, b^{y-x}}{x\,!\,(y-x)\,!}\right]$$

$$= e^{-(a+b)}\left[\sum_{x=0}^{\infty}\frac{(a\,e^{t_1}e^{t_2})^x}{x\,!}\sum_{z=0}^{\infty}\frac{(b\,e^{t_2})^z}{z\,!}\right]; \qquad (y - x = z)$$

$$= \exp\left[a\, e^{t_1} e^{t_2} + b\, e^{t_2} - a - b\right] \qquad \ldots (*)$$

$$M_X(t_1) = M(t_1, 0) = \exp\{a\,(e^{t_1} - 1)\} \quad \Rightarrow \quad X \sim P(a)$$

$$M_Y(t_2) = M(0, t_2) = \exp[(a+b)\{e^{t_2} - 1\}] \quad \Rightarrow \quad Y \sim P(a + b)$$

$$E(XY) = \left|\frac{\partial^2 M(t_1, t_2)}{\partial t_1\, \partial t_2}\right|_{t_1 = t_2 = 0} = a^2 + ab + a$$

$\therefore$ Cov $(X, Y) = E(XY) - E(X)\, E(Y) = a^2 + ab + a - a(a + b) = a$

Distribution of $Y - X$. Taking $t_1 + t_2 = 0 \Rightarrow t_1 = -t_2$ in (*),

$$M(-t_2, t_2) = E\left(e^{-t_2 X + t_2 Y}\right) = E\left[e^{(Y-X)t_2}\right] = \exp\{b(e^{t_2} - 1)\} \quad \Rightarrow \quad Y - X \sim P(b).$$

PROBLEMS ON NEGATIVE BINOMIAL DISTRIBUTION

8·46. The number of accidents among 414 machine operators was investigated for three successive months. The following table gives the distribution of the operators according to the number k, of accidents which happened to the same operator. Fit the distribution of the type

$$P(X = k) = (-1)^k \binom{-v}{k} p^v q^k \,;\; k = 0, 1, 2, \ldots \,;\; v > 0, \quad q = 1 - p, \quad 0 < p < 1$$

k	0	1	2	3	4	5	6	7	8
Observed frequency	296	74	26	8	4	4	1	0	1

8·47. Assume that the mutually independent random variables X_i, each has the negative binomial distribution with parameters r_i $(i = 1, 2, \ldots, n)$, where r_i are all positive integers, *i.e.*,

$$P(X_i = x) = \binom{r_i + x - 1}{x} p^{r_i} q^x \,;\; x = 0, 1, 2, \ldots$$

Show that the probability function of $\sum_{i=1}^{n} X_i$ is the negative binomial distribution with $r = \sum_{i=1}^{n} r_i$, *i.e.*, the negative binomial distribution (with fixed p) is reproductive with respect to r.

8·48. (*a*) If X is Negative binomial variate with parameters k and Q^{-1}, prove that

$$P(X \geq m) = \frac{1}{B(m, k)} \int_0^P \frac{x^{m-1}\, dx}{(x+1)^{k+m}} \,;\; Q - P = 1$$

(*b*) If X is Negative binomial variate with parameters k and p, show that

$$P(X \geq m) = \frac{1}{B(m, k)} \int_p^1 y^{k-1} (1 - y)^{m-1}\, dy$$

Hint. (*a*) $\quad P(X \geq m) = \sum_{r=m}^{\infty} \binom{-k}{r} Q^{-k-r} (-P)^r = \sum_{r=m}^{\infty} \binom{k+r-1}{r} Q^{-k-r} P^r$

$$\frac{d}{dP}\left[P(X \geq m)\right] = \sum_{r=m}^{\infty} (T_r - T_{r+1}), \quad T_r = r \cdot \binom{k+r-1}{r} P^{r-1} Q^{-k-r}$$

$$= T_m = \frac{1}{B(m, k)} \cdot P^{m-1} Q^{-k-m}$$

Integrating, We get : $$P(X \geq m) = \frac{1}{B(m,k)} \int_0^P \frac{x^{m-1}}{(1+x)^{k+m}}\,dx \qquad (\because Q - P = 1)$$

8·49. In a sequence of independent trials, the probability of a success on each trial is '*p*'. By considering the outcome of the first trial, show that $G_r(t)$, the *p.g.f.* of the number of trials required to achieve the *r*th success, satisfies : $G_r(t) = pt\, G_{r-1}(t) + qt\, G_r(t)$,

and hence obtain an explicit expression for $G_r(t)$. $\left[\textbf{Ans. } G_r(t) = \{pt/(1-qt)\}^r\right]$

8·50. (*a*) If the probability that a target is destroyed on any one shot is 0.5, what is the probability that it would be destroyed on 6th attempt ?

(*b*) A couple decides to have children until they have a male child. What is the probability distribution of the number of children they would have ? If the probability of a male child in their community is 1/3, how many children are they expected to have before the first male child is born ?

(*c*) A die is cast until 6 appears. What is the probability that it must be cast more than five times ?

Ans. (*a*) $(0{\cdot}5)^6$ (*c*) $P(X > 5) = 1 - P(X \leq 5) = 1 - \sum_{x=1}^{5} (5/6)^{x-1} \cdot (1/6)$

PROBLEMS ON GEOMETRIC AND HYPERGEOMETRIC DISTRIBUTION

8·51. (*a*) X_1 and X_2 are independent random variables with the same distribution $q^k p$; $k = 0, 1, \ldots$. Let Y be defined as the largest of X_1 and X_2, *i.e.*, $Y = \max(X_1, X_2)$. Obtain the joint distribution of Y and X_1 and the distribution of Y.

(*b*) Let X and Y be independent random variables such that :
$$P(X = r) = P(Y = r) = q^r p \;;\; r = 0, 1, 2, \ldots$$
where p and q are positive numbers such that $p + q = 1$.

Find (*i*) the distribution of $X + Y$, and

(*ii*) the conditional distribution of X given $X + Y = 3$.

8·52. For the geometric distribution with *p.m.f.* : $f(x) = 2^{-x}$; $x = 1, 2, 3, \ldots$;

show that Chebychev's inequality gives $P(|X - 2| \leq 2) > \frac{1}{2}$, while the actual probability is 15/16.

8·53. The conditional distribution of random variable X given $Y = y$ is $\frac{e^{-y} y^x}{x!}$ and the marginal probability density of Y is e^{-y}, where X is a discrete variable, *i.e.*, $x = 0, 1, 2, \ldots$ and Y is continuous, $y \geq 0$. Show that the marginal distribution of X is geometric.

8·54. If X and Y are two independent random variables, each representing the number of failures preceding the first success in a sequence of Bernoulli trials with p as probability of success in a single trial and q as probability of failure, show that: $$P(X = Y) = \frac{p}{1+q}.$$

Hint. We have $P(X = r) = P(Y = r) = q^r \cdot p$; $(r = 0, 1, 2, \ldots)$

$$P(X = Y) = \sum_{r=0}^{\infty} P(X = r \cap Y = r) = \sum_{r=0}^{\infty} P(X = r) \cdot P(Y = r) \qquad [\because X \text{ and } Y \text{ are independent } r.v.\text{'s}]$$

$$= p^2 \sum_{r=0}^{\infty} q^{2r} = p^2 (1 + q^2 + q^4 + \ldots) = \frac{p^2}{1-q^2} = \frac{p}{1+q}.$$

8·55. If $X \sim$ geom (p) and $Y \sim$ geom (p') are independent r.v.'s, obtain $P(X + Y = r)$.

Ans. $p\,p' \left\{ \frac{(q')^{r+1} - q^{r+1}}{q' - q} \right\}$

8·56. Let X and Y be independent geometric variates representing the number of trials required for the first success with constant probability 'p' of success for each trial. Prove that:

$$P[\max (X, Y) = r] = pq^{r-1}\left[2 - (1 + q)\, q^{r-1}\right].$$

8·57. Let X denote the number of trials required for the first success in an infinite series of independent Bernoulli trials with constant probability 'p' of success for each trial. Prove that :

$$P(X = x) = q^{x-1} \cdot p\ ;\ q = 1 - p,\ x = 1, 2, 3, \ldots$$

(*a*) Obtain the mean and variance of the distribution

(*b*) Find $M_X(t)$ and hence deduce that mean is $(1/p)$ and variance is (q/p^2).

Ans. $M_X(t) = pe^t\,(1 - qe^t)^{-1}$

8·58. Show that the necessary and sufficient condition for independent r.v.'s $X_1, X_2, \ldots, X_n$ to be geometrically distributed with parameter 'p' is that

$$Y = \min(X_1, X_2, \ldots, X_n) \text{ has geometric distribution with parameter } (1 - q^n);\ q = 1 - p.$$

8·59. Let X and Y be independent random variables with the same (geometric) distribution given by

$P(X = k) = p\,q^k\ ;\ k = 0, 1, 2, \ldots$. Let $Z = \max(X, Y)$.

(*i*) Find the probability distribution of Z.

(*ii*) Find the joint probability distribution of X and Z.

(*iii*) Find the conditional probability $P(X = k \mid Z = l)$ for all $k, l = 0, 1, 2, \ldots$.

(*iv*) Find the conditional probability $P(Z = l \mid X = k)$ for all $k, l = 0, 1, 2, \ldots$.

Hint. (*i*) $P(Z = l) = p\,q^l\,(2 - q^l - q^{l+1})\ ;\ l = 0, 1, 2, \ldots$

$$(ii)\ P(Z = l \cap X = k) = \begin{cases} 0, \text{ if } l < k \\ p\,q^k\,(1 - q^{k+1}), \text{ if } l = k = 0, 1, 2, \ldots \\ p^2\,q^{k+l}, \text{ if } l > k = 0, 1, 2, \ldots \end{cases}$$

$$(iii)\ P(X = k \mid Z = l) = \begin{cases} 0, \text{ if } l < k \\ (1 - q^{k+1}) / (2 - q^k - q^{k+1}), \text{ if } l = k = 0, 1, 2, \ldots \\ pq^k / (2 - q^l - q^{l+1}), \text{ if } l > k = 0, 1, 2, \ldots, \end{cases}$$

$$(iv)\ P(Z = l \mid X = k) = \begin{cases} 0, \text{ if } l < k \\ 1 - q^{k+1} \text{ if } l = k = 0, 1, 2, \ldots \\ pq^l \text{ if } l > k = 0, 1, 2, \ldots, \end{cases}$$

8·60. 200 students of the F.Y.B.Sc. class in a certain college are divided at random into 2 batches of 10 each for the annual practical examination in Statistics. Suppose the class consists of 40 resident students and 160 non-resident students; and let R denote the number of resident students in the first batch. Use the binomial approximation to find the probability that $R \geq 3$.

Hint. The probability distribution of R is hypergeometric with parameters : $N = 200$, $n = 10$, $M = 40$. Since N (= 200) is large, the hypergeometric distribution can be approximated by binomial distribution

with parameters : $n = 10,\ p = \dfrac{M}{N} = \dfrac{40}{200} = 0{\cdot}2.$

$$\therefore\ P(R = r) = \binom{10}{r}(0{\cdot}2)^r\,(0{\cdot}8)^{10-r}\ ;\ r = 0, 1, \ldots, 10, \text{ and required probability is :}$$

$$P(R \geq 3) = 1 - \left[P(R = 0) + P(R = 1) + P(R = 2)\right] = 0{\cdot}323$$

8·61. If X and Y are independent binomial variates with parameters (n_1, p) and (n_2, p) respectively, show

that $$P[X = r \mid (x + y) = n] = \frac{\binom{n_1}{r}\binom{n_2}{n-r}}{\binom{n_1+n_2}{n}}.$$

8·62. (*a*) From a finite population of N animals in a given region, W are caught, marked and then released again. The animals are caught again one by one until m (preassigned) marked animals are caught. The total number of animals caught is a random variable X. Find

$$P(X = n), \text{ for } m \le n \le N - W + m.$$

(*b*) An urn contains M balls numbered 1 to M, where the first K balls are defective and the remaining $(M - K)$ are non-defective. A sample of n balls is drawn from the urn. Let A_k be the event that the sample of n balls contains exactly k defectives.

Find $P(A_k)$ when the sample is drawn (*i*) with replacement, and (*ii*) without replacement.

Hint. (*a*) $P(X = n) = P\big[\text{Catching } (n-1) \text{ animals of whom } (m-1) \text{ are marked}\big]$
$\times P\big[\text{Catching the marked animal from the remaining } N-(n-1) \text{ animals}\big].$

$$= \frac{\binom{W}{m-1}\binom{N-W}{n-m}}{\binom{N}{n-1}} \times \frac{\{W-(m-1)\}}{\{N-(n-1)\}} = \binom{N-n}{W-m}\binom{n-1}{m-1} \div \binom{N}{W}$$

(*b*) If sampling is done without replacement, we get hypergeometric probability model.

$$P(A_k) = \binom{K}{k}\binom{M-K}{n-k} \div \binom{M}{n}$$

If sampling is done with replacement, then $X \sim B\big(n, p = K/M\big)$

$$\therefore \quad P(A_k) = \binom{n}{k}\left(\frac{K}{M}\right)^k \cdot \left(1 - \frac{K}{M}\right)^{n-k} = \binom{n}{k}\frac{K^k (M-K)^{n-k}}{M^n}$$

8·63. (*a*) Suppose that $X_1, X_2, \ldots, X_n$, are mutually independent indicator random variables, with $P(X_i = 1) = p,\ 0 < p < 1$. Show that for $1 \le M \le N$, $P\left(\sum_{i=1}^{M} X_i = k \,\middle|\, \sum_{i=1}^{N} X_i = n\right) = \dfrac{\binom{M}{k}\binom{N-M}{n-k}}{\binom{N}{n}}$.

(*b*) Let $X_1, X_2, \ldots, X_n$ be independent Bernoulli variates with common parameter $p = P(X_1 = 1)$. Let $S_j = X_1 + X_2 + \ldots + X_j$ for $1 \le j \le n$. Show that $P\big(S_j = r \mid S_n = s\big)$ does not depend on p, $(0 < p < 1)$ and takes the form of a hypergeometric probability for $1 \le j \le n$, $0 \le r \le s \le n$.

PROBLEMS ON MULTINOMIAL DISTRIBUTION

8·64. The joint *p.m.f.* of two discrete *r.v.*'s X_1 and X_2 is :

$$p(x_1, x_2) = \binom{n_1}{x_1}\binom{n_2}{x_2 - x_1} p^{x_2} (1-p)^{n_1 + n_2 - x_2} \text{ with } x_1 \le x_2 \le n_2 + x_1;\ \ 0 \le x_1 \le n_1.$$

Find the marginal distributions of X_1 and X_2.

Ans. $X_1 \sim B(n, p)$ and $X_2 \sim B(n_1 + n_2, p)$

8·65. For the trinomial distribution of two *r.v.*'s X and Y given by :

$$f(x, y) = \frac{n!}{x!\,y!\,(n-x-y)!} p_1^x p_2^y p_3^{n-x-y},$$ where x and y are non-negative integers with $x + y \le n$ and p_1, p_2, p_3 are proper positive fractions with $p_1 + p_2 + p_3 = 1$, and $f(x, y) = 0$ otherwise.

Show that (*i*) $X \sim B(n, p_1)$ and $Y \sim B(n, p_2)$

(*ii*) $X \mid (Y = y) \sim B[n - y, p_1/(1 - p_2)]$ and $Y \mid (X = x) \sim B\big[n - x, p_2/(1 - p_1)\big]$

8·66. Two discrete random variables X and Y have the joint probability distribution :

$$p(x, y) = \frac{9!}{x!\,y!\,(9-x-y)!}\left(\frac{1}{3}\right)^9, \text{ where, } 0 \le x \le 9,\ 0 \le y \le 9 \text{ and } 0 \le (x + y) \le 9.$$

(i) Show that the marginal distribution of X is binomial with parameters $n = 9$ and $p = 1 / 3$.

(ii) Show that the conditional distribution of Y given $X = 3$, is also binomial with parameters $n = 6$ and $p = 1 / 2$.

8·67. (a) Show that the variance-covariance of matrix (Σ) of a multinomial distribution with parameters $(n, p_1, p_2, ..., p_k)$ is singular with Rank (Σ) = $k - 1$.

(b) Find the chance of throwing a sum of 3 or 5 or 11 in a random throw of 2 dice.

(c) Obtain the probability of getting a total of 15 points in a random throw of 3 dice.

Hint. (b) Using multinomial distribution,

Favourable cases = Sum of the Coefficients of x^3, x^5 and x^{11} in the expansion of $\left[\dfrac{x(1-x^6)}{1-x}\right]^2$

$= 2 + 4 + 2 = 8$

Required Probability = $(8/6^2) = 2/9$

(c) Favourable cases = Coefficients of x^{15} in $\left[\dfrac{x(1-x^6)}{1-x}\right]^3 = 10$

Required Probability = $\dfrac{10}{6^3}$

8·68. A Polya process is defined by the quantities :

$$P_k(t) = \left[\frac{\lambda t}{1+b\lambda t}\right]^k \frac{1(1+b)\dots[1+(k-1)b]}{k!} P_0(t), \text{ where } P_0(t) = (1 + b\lambda t)^{-1/b}$$

and λ, b are parameters, t is a continuous variable and k may take zero or positive integral values only. Verify that the distribution satisfies the requirements for a probability distribution in K and find the expectation of K and its variance.

Hint. Let K be a random variable with the distribution,

$P(K = k) = P_k(t);\ k = 0, 1, 2, ..., \infty.$

$$\sum_{k=0}^{\infty} P_k(t) = (1+b\lambda t)^{-1/b} \sum_{k=0}^{\infty} \left[\frac{\lambda t}{1+b\lambda t}\right]^k \times \frac{b^k\left(\frac{1}{b}\right)\left(\frac{1}{b}+1\right)\dots\left(\frac{1}{b}+k-1\right)}{k!}$$

$$= (1+b\lambda t)^{-1/b} \sum_{k=0}^{\infty} \left(\frac{b\lambda t}{1+b\lambda t}\right)^k \binom{(1/b)+k-1}{k}$$

$$= (1+b\lambda t)^{-1/b} \frac{1}{\left(1-\dfrac{b\lambda t}{1+b\lambda t}\right)^{1/b}} = 1 \qquad \left[\because (1-x)^{-n} = \sum_{k=0}^{\infty}\binom{-n}{k}(-x)^k = \sum_{k=0}^{\infty}\binom{n+k-1}{k}x^k\right]$$

$\therefore$ $P_k(t)$ represents a probability distribution for every fixed b, λ and t.

$$\text{M.G.F. of } K = E(e^{Ku}) = \sum_{k=0}^{\infty} e^{uk} P_k(t)$$

$$= (1+b\lambda t)^{-1/b} \sum_{k=0}^{\infty}\left[e^{uk} \frac{(\lambda t)^k}{(1+b\lambda t)^k} \frac{b^k\left(\frac{1}{b}\right)\left(\frac{1}{b}+1\right)\dots\left(\frac{1}{b}+k-1\right)}{k!}\right]$$

$$= (1+b\lambda t)^{-1/b} \sum_{k=0}^{\infty} \left(\frac{e^u\, b\lambda t}{1+b\lambda t}\right)^k \binom{\frac{1}{b}+k-1}{k}$$

$$= (1+b\lambda t)^{-1/b} \frac{1}{\left(1-\dfrac{e^u\, b\lambda t}{1+b\lambda t}\right)^{1/b}} = \left[1+b\lambda t - e^u\, b\lambda t\right]^{-1/b} = g(u), \text{(say)}.$$

$$\therefore \qquad g'(u) = \frac{1}{b}\left[1+b\lambda t - e^u\, b\lambda t\right]^{-(1/b)-1} (e^u\, b\lambda t)$$

$$E(K) = \mu_1' \text{ (about origin)} = \text{Mean} = \left[g'(u)\right]_{u=0} = \left(\frac{1}{b}\right)\left[1+b\lambda t - b\lambda t\right]^{-(1/b)-1} (b\lambda t) = \lambda t$$

Similarly $\mu_2' = [g''(u)]_{u=0} = (b+1)\lambda^2 t^2 + \lambda t$

$$\therefore \qquad \text{Variance} = \mu_2' - \mu_1'^2 = \lambda t(1+b\lambda t).$$

8·69. Let X and Y be independent random variables each having uniform distribution on (1, 2, 3, ..., N). Find

(*i*) $P(X = Y)$ (*ii*) $P(X \geq Y)$ (*iii*) $P[\min(X, Y) = k]$ (*iv*) $P[\max(X, Y) = k]$

Ans. (*i*) $\dfrac{1}{N}$, (*ii*) $\dfrac{N+1}{2N}$, (*iii*) $\dfrac{1}{N^2}(2N-2k+1)$, (*iv*) $\dfrac{2k-1}{N^2}$

CHAPTER

9 Special Continuous Probability Distributions

LEARNING OBJECTIVES

Upon completion of this Chapter, you should be able to:

- Explain and illustrate the concept of a continuous probability distribution.
- Enumerate the basic assumptions of several important continuous probability distributions.
- Write down the p.d.f. of various important continuous probability distributions.
- Derive moments, cumulants, m.g.f. and various constants of continuous probability distributions.
- Apply various continuous probability distributions in various diversified fields and day-to-day real-life problems.
- Know about various types of 'Central Limit Theorems' (CLTs) and their use in Statistics.
- Understand the concept of 'Compound Distributions' and derive the probability functions of compound binomial distribution and compound Poisson distribution.
- Understand the concept of 'Order Statistics' in sampling from a given population with p.d.f. f (x) and their use in deriving the p.d.f's of the smallest and largest sample observations, the sample median, the sample range, etc.
- Know about the concept of 'Truncated Distributions', and derive the p.m.f. / p.d.f.'s of some standard truncated distributions.

CHAPTER OUTLINES

9·1. INTRODUCTION

In the present Chapter we propose to discuss some univariate continuous distributions. The main continuous distributions like uniform distribution, normal distribution, gamma, beta, exponential, Laplace, Weibul, Logistic, Cauchy, Erlang, Pareto and Raleigh distributions will be discussed in detail in the subsequent sections. We shall define these distributions in terms of the *p.d.f.*.

9·2. NORMAL DISTRIBUTION

The normal distribution was first discovered in 1733 by English mathematician De-Moivre, who obtained this continuous distribution as a limiting case of the binomial distribution and applied it to problems arising in the game of chance. It was also known to Laplace, no later than 1774 but through a historical error it was credited to Gauss, who first made reference to it in the beginning of 19th century (1809), as the distribution of errors in Astronomy. Gauss used the normal curve to describe the theory of accidental errors of measurements involved in the calculation of orbits of heavenly bodies. Throughout the eighteenth and nineteenth centuries, various efforts were made to establish the normal model as the underlying law ruling all continuous random variables. Thus, the name "*normal*". These efforts, however, failed because of false premises. The normal model has, nevertheless, become the most important probability model in statistical analysis.

Definition. *A r.v. X is said to have a normal distribution with parameters μ (called 'mean') and* σ^2 *(called 'variance') if its p.d.f. is given by the probability law :*

$$f(x;\ \mu, \sigma) = \frac{1}{\sigma\sqrt{2\pi}} \cdot exp \left\{ -\frac{1}{2}\left(\frac{x-\mu}{\sigma}\right)^2 \right\}$$

or

$$f(x;\ \mu, \sigma) = \frac{1}{\sigma\sqrt{2\pi}}\ e^{-(x-\mu)^2/2\sigma^2}; \quad -\infty < x < \infty, -\infty < \mu < \infty, \sigma > 0 \qquad \text{... (9·1)}$$

π and e are, respectively, the mathematical constants 3·14159 ... and 2·71828.

Remarks

1. When a *r.v.* is normally distributed with mean μ and standard deviation σ, it is customary to write X is distributed as $N(\mu, \sigma^2)$ and is expressed by $X \sim N(\mu, \sigma^2)$.

2. If $X \sim N(\mu, \sigma^2)$, then $Z = \dfrac{X-\mu}{\sigma}$, is a standard normal variate with $E(Z) = 0$ and $\text{Var}(Z) = 1$ and we write $Z \sim N(0, 1)$.

3. The *p.d.f.* of standard normal variate Z is given by :

$$\varphi(z) = \frac{1}{\sqrt{2\pi}} \cdot e^{-z^2/2}, \quad -\infty < z < \infty \qquad \text{... (9·1a)}$$

and the corresponding distribution function, denoted by $\Phi(z)$ is given by :

$$\Phi(z) = P(Z \le z) = \int_{-\infty}^{z} \varphi(u)\, du = \frac{1}{\sqrt{2\pi}} \int_{-\infty}^{z} e^{-u^2/2}\, du \qquad \text{... (9·1b)}$$

We shall prove below two important results on the distribution function $\Phi(.)$ of standard normal variate.

Result 1. $\Phi(-z) = 1 - \Phi(z), \quad z > 0$ ... (9·1c)

Proof. $\Phi(-z) = P(Z \le -z) = P(Z \ge z) = 1 - P(Z \le z) = 1 - \Phi(z)$

Result 2. $P(a \le X \le b) = \Phi\left(\dfrac{b-\mu}{\sigma}\right) - \Phi\left(\dfrac{a-\mu}{\sigma}\right)$, where $X \sim N(\mu, \sigma^2)$ ... (9·1d)

Proof. $$P(a \le X \le b) = P\left(\frac{a-\mu}{\sigma} \le Z \le \frac{b-\mu}{\sigma}\right), \quad \left(Z = \frac{X-\mu}{\sigma}\right)$$

$$= P\left(Z \le \frac{b-\mu}{\sigma}\right) - P\left(Z \le \frac{a-\mu}{\sigma}\right) = \Phi\left(\frac{b-\mu}{\sigma}\right) - \Phi\left(\frac{a-\mu}{\sigma}\right)$$

4. The graph of $f(x)$ is famous *'bell-shaped'* curve. The top of the bell is directly above the mean μ. For large values of σ, the curve tends to flatten out and for small values of σ, it has a sharp peak.

The parameters μ and σ, which define a normal distribution, are referred to as the *location* and *shape* parameters. The parameter m determines the location of the normal distribution along a number line, the parameter σ determines the relative width and height (or the spread) of the distribution. The graph of three normal distributions with the same standard deviation but different means is shown in *Fig. 9·1* and the graph of three normal distributions with the same mean but different standard deviations is shown in *Fig. 9·2*. The larger the value of the standard deviation σ, the wider and flatter is the normal curve, thus showing more variability in the data, as shown in *Fig. 9·2*. Thus, standard deviation σ determines the range of values that any random variable is likely to assume.

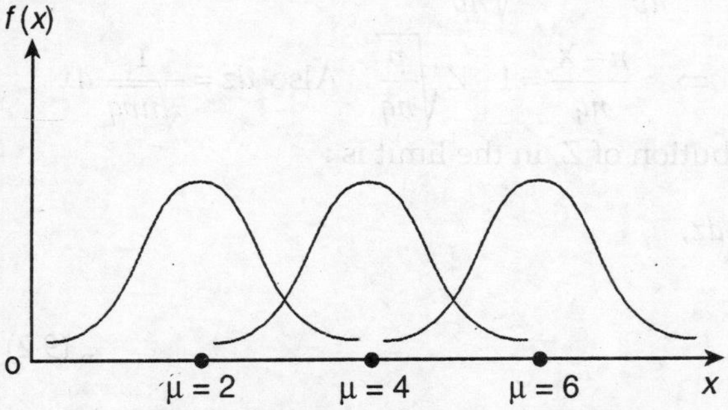

Fig. 9·1 : Normal distributions with μ = 2, 4, 6 and same standard deviation.

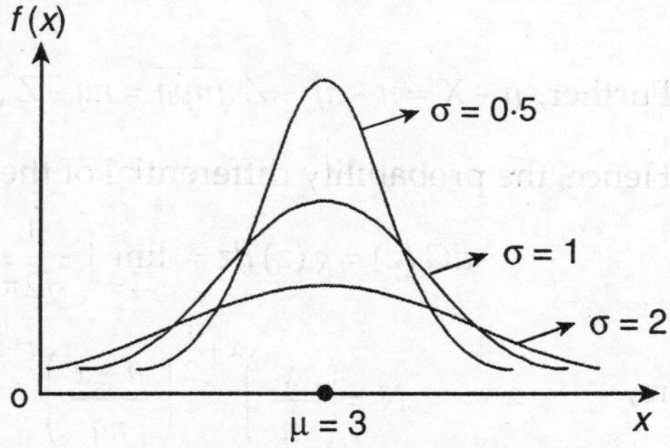

Fig. 9·2 : Normal distribution with μ = 3 and σ = 0·5, 1, 2.

9·2·1. Normal Distribution as a Limiting form of Binomial Distribution

Normal distribution is another limiting form of the binomial distribution under the following conditions:

(*i*) n, the number of trials is indefinitely large, *i.e.*, $n \to \infty$; and

(*ii*) neither p nor q is very small.

The *p.m.f.* of the binomial distribution with parameters n and p is given by:

$$p(x) = \binom{n}{x} p^x q^{n-x} = \frac{n!}{x!\,(n-x)!} p^x q^{n-x}; \quad x = 0, 1, 2, \ldots, n \qquad \ldots (*)$$

Let us now consider the standard binomial variate :

$$Z = \frac{X - E(X)}{\sqrt{V(X)}} = \frac{X - np}{\sqrt{npq}}; \quad X = 0, 1, 2, \ldots, n \qquad \ldots (**)$$

$$\text{when } X = 0,\ Z = \frac{-np}{\sqrt{npq}} = -\sqrt{\frac{np}{q}} \quad \text{and} \quad \text{when } X = n,\ Z = \frac{n - np}{\sqrt{npq}} = \sqrt{\frac{nq}{p}}$$

Thus, in the limit as $n \to \infty$, Z takes the values from $-\infty$ to ∞. Hence, the distribution of X will be a continuous distribution over the range $-\infty$ to ∞.

We want the limiting form of (*) under the above two conditions. Using Stirling's approximation to $r\,!$ for large r, *viz.*,
$$\lim_{r \to \infty} r! \simeq \sqrt{2\pi}\, e^{-r} r^{r + (1/2)},$$

We have in the limit as $n \to \infty$ and consequently $x \to \infty$,

$$\lim p(x) = \lim \left[\frac{\sqrt{2\pi}\, e^{-n}\, n^{n+\frac{1}{2}}\, p^x\, q^{n-x}}{\sqrt{2\pi}\, e^{-x}\, x^{x+\frac{1}{2}} \sqrt{2\pi}\, e^{-(n-x)}\, (n-x)^{n-x+\frac{1}{2}}} \right]$$

$$= \lim \left[\frac{1}{\sqrt{2\pi}} \cdot \frac{1}{\sqrt{npq}} \cdot \frac{(np)^{x+\frac{1}{2}}\, (nq)^{n-x+\frac{1}{2}}}{x^{x+\frac{1}{2}}\, (n-x)^{n-x+\frac{1}{2}}} \right]$$

$$= \lim \left[\frac{1}{\sqrt{2\pi}\sqrt{npq}} \left(\frac{np}{x} \right)^{x+\frac{1}{2}} \left(\frac{nq}{n-x} \right)^{n-x+\frac{1}{2}} \right] \qquad \text{... (***)}$$

From (**), we get $\quad X = np + Z\sqrt{npq} \quad \Rightarrow \quad \dfrac{X}{np} = 1 + Z\sqrt{\dfrac{q}{np}}$

Further, $n - X = n - np - Z\sqrt{npq} = nq - Z\sqrt{npq} \quad \Rightarrow \quad \dfrac{n-X}{nq} = 1 - Z\sqrt{\dfrac{p}{nq}}$. Also $dz = \dfrac{1}{\sqrt{npq}}\, dx$

Hence, the probability differential of the distribution of Z, in the limit is :

$$dG(z) = g(z)\, dz = \lim_{n \to \infty} \left(\frac{1}{\sqrt{2\pi}} \times \frac{1}{N} \right) dz,$$

where,
$$N = \left(\frac{x}{np} \right)^{x+\frac{1}{2}} \left(\frac{n-x}{nq} \right)^{n-x+\frac{1}{2}} \qquad \text{... (9·2)}$$

$$\Rightarrow \quad \log N = \left(x + \frac{1}{2} \right) \log (x/np) + \left(n - x + \frac{1}{2} \right) \log \{(n-x)/nq\}$$

$$= \left(np + z\sqrt{npq} + \frac{1}{2} \right) \log \left\{ 1 + z\sqrt{q/np} \right\} + \left(nq - z\sqrt{npq} + \frac{1}{2} \right) \log \left\{ 1 - z\sqrt{p/nq} \right\}$$

$$= \left(np + z\sqrt{npq} + \frac{1}{2} \right) \left\{ z\sqrt{q/np} - \frac{1}{2} z^2 (q/np) + \frac{1}{3} z^3 (q/np)^{3/2} - \dots \right\}$$

$$+ \left(nq - z\sqrt{npq} + \frac{1}{2} \right) \left\{ -z\sqrt{p/nq} - \frac{1}{2} z^2 (p/nq) - \frac{1}{3} z^3 (p/nq)^{3/2} - \dots \right\}$$

$$= \left[\left(z\sqrt{npq} - \frac{1}{2} z^2 q + \frac{1}{3} z^3 \frac{q^{3/2}}{\sqrt{np}} + z^2 q - \frac{1}{2} z^3 \frac{q^{3/2}}{\sqrt{np}} + \frac{1}{2} z \sqrt{\frac{q}{np}} - \frac{1}{4} z^2 \frac{q}{np} + \dots \right) \right.$$

$$\left. + \left(-z\sqrt{npq} - \frac{1}{2} z^2 p - \frac{1}{3} z^3 \frac{p^{3/2}}{\sqrt{nq}} + z^2 p + \frac{1}{2} z^3 \frac{p^{3/2}}{\sqrt{nq}} - \frac{1}{2} z \sqrt{\frac{p}{nq}} - \frac{1}{4} z^2 \frac{p}{nq} + \dots \right) \right]$$

$$= \left[-\frac{1}{2} z^2 (p+q) + z^2 (p+q) + \frac{z}{2\sqrt{n}} \left(\sqrt{\frac{q}{p}} - \sqrt{\frac{p}{q}} \right) + O(n^{-1/2}) \right]$$

$$= \frac{z^2}{2} + O'(n^{-1/2}) \quad \to \quad \frac{z^2}{2} \quad \text{as } n \to \infty \to 1$$

where, $O'(n^{-1/2})$ contains terms involving $n^{1/2}$ and higher powers of n in the denominator.

$$\therefore \qquad \lim_{n \to \infty} \log N = \frac{z^2}{2} \quad \Rightarrow \quad \lim_{n \to \infty} N = e^{z^2/2}$$

Substituting in (9·2), we get

$$dG(z) = g(z)\,dz = \frac{1}{\sqrt{2\pi}}\, e^{-z^2/2}\, dz, \qquad -\infty < z < \infty \qquad \text{... (9·2a)}$$

Hence, the probability function of Z is :

$$g(z) = \frac{1}{\sqrt{2\pi}}\, e^{-z^2/2}, \quad -\infty < z < \infty \qquad \text{... (9·2b)}$$

This is the probability density function of the *normal distribution* with mean 0 and unit variance.

If X is normal variate with mean μ and *s.d.* σ, then $Z = (X - \mu)/\sigma$, is standard normal variate. Jacobian of transformation is $1/\sigma$. Hence, substituting in {9·2(*b*)}, the *p.d.f.* of a normal variate X with $E(X) = \mu$, $\text{Var}(X) = \sigma^2$ is given by :

$$f_X(x) = \begin{cases} \dfrac{1}{\sigma\sqrt{2\pi}} \exp\left[-\dfrac{(x-\mu)^2}{2\sigma^2}\right], & -\infty < x < \infty \\ 0, \text{ otherwise} \end{cases}$$

Remark. Normal distribution can also be obtained as a limiting case of Poisson distribution with the parameter $\lambda \to \infty$.

9·2·2. Chief Characteristics of the Normal Distribution and Normal Probability Curve

The normal probability curve with mean μ and standard deviation σ is given by the equation :

$$f(x) = \frac{1}{\sigma\sqrt{2\pi}} \cdot e^{-(x-\mu)^2/2\sigma^2}, \qquad -\infty < x < \infty$$

and has the following properties :

(*i*) The normal curve is bell-shaped and is symmetrical around a vertical line erected at the mean μ with respect to the area under it, *i.e.*, 50% of the area of the curve lies on both sides of the mean. In other words, $P(X \le \mu) = P(X \ge \mu) = 0 \cdot 5.$

(*ii*) Mean, median and mode of the distribution coincide.

(*iii*) As x increases numerically, $f(x)$ decreases rapidly, the maximum probability occurring at the point $x = \mu$, and is given by : $[f(x)]_{\max} = \dfrac{1}{\sigma\sqrt{2\pi}}.$

(*iv*) $\beta_1 = 0$ and $\beta_2 = 3$.

(*v*) $\mu_{2r+1} = 0,\ (r = 0, 1, 2, \ldots)$, and $\mu_{2r} = 1.3.5 \ldots (2r-1)\,\sigma^{2r},\ (r = 0, 1, 2, \ldots)$.

(*vi*) Since $f(x)$ being the probability, can never be negative, no portion of the curve lies below the x-axis.

(*vii*) Linear combination of independent normal variates is also a normal variate.

(*viii*) x-axis is an asymptote to the curve.

(*ix*) The points of inflexion of the curve are :

$$x = \mu \pm \sigma, \quad f(x) = \frac{1}{\sigma\sqrt{2\pi}}\, e^{-1/2}.$$

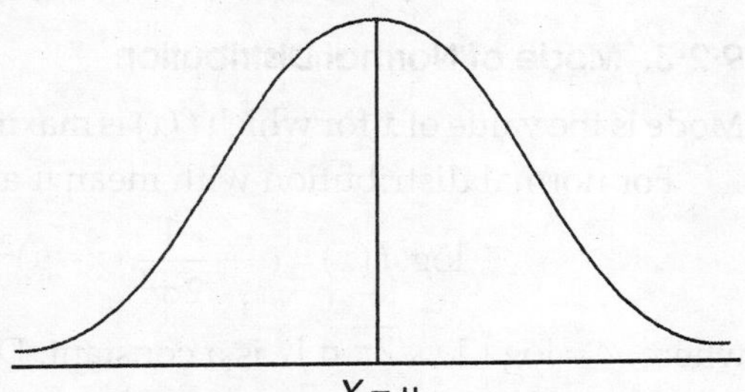

Fig. 9·3 : Normal Probability Curve

(*x*) Mean deviation about mean $= \sqrt{\dfrac{2}{\pi}}\,\sigma \simeq \dfrac{4}{5}\,\sigma$ (approx.)

(*xi*) Quartiles are given by : $Q_1 = \mu - 0 \cdot 6745\,\sigma; \qquad Q_3 = \mu + 0 \cdot 6745\,\sigma.$

(*xii*) $Q.D. = \dfrac{Q_3 - Q_1}{2} \simeq \dfrac{2}{3}\sigma$. We have (approximately)

$$Q.D.: M.D.: S.D. :: \frac{2}{3}\sigma : \frac{4}{5}\sigma : \sigma :: \frac{2}{3} : \frac{4}{5} : 1 \quad \Rightarrow \quad Q.D.: M.D.: S.D. :: 10:12:15$$

(*xiii*) Area Property : $P(\mu - \sigma < X < \mu + \sigma) = 0 \cdot 6826$; $P(\mu - 2\sigma < X < \mu + 2\sigma) = 0 \cdot 9544$,

and $P(\mu - 3\sigma < X < \mu + 3\sigma) = 0 \cdot 9973$.

The *percentage distribution* of area under standard normal curve for some important values of standard normal variate Z is broadly shown in the *Fig. 9·4*.

Distances from the mean ordinates in terms of $\pm\sigma$	*Area under the curve*
$Z = \pm 0{\cdot}6745$	50% = 0·50
$Z = \pm 1{\cdot}00$	68·26% = 0·6826
$Z = \pm 1{\cdot}96$	95% = 0·95
$Z = \pm 2{\cdot}0$	95·44% = 0·9544
$Z = \pm 2{\cdot}58$	99% = 0·99
$Z = \pm 3{\cdot}0$	99·73% = 0·9973

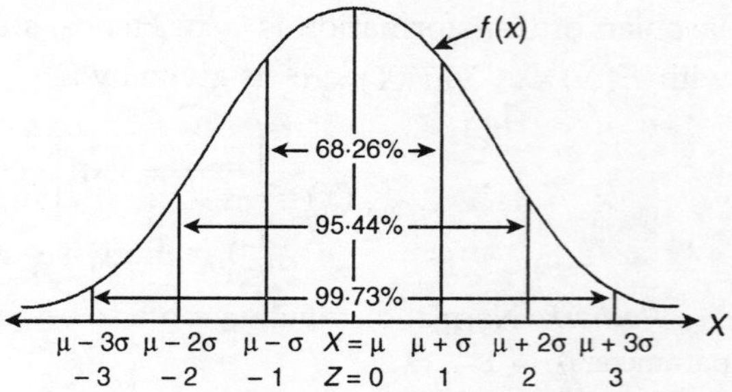

Fig. 9·4 : Normal Probability Curve

Remark. The above concept is discussed in detail in § 9·2·11.

(*xiv*) If X and Y are independent standard normal variates, then it can be proved that $U = X + Y$ and $V = X - Y$ are independently distributed, $U \sim N(0, 2)$ and $V \sim N(0, 2)$.

We state (without proof) the converse of this result which is due to *D.* Bernstein.

Bernstein's Theorem. *If X and Y are independent and identically distributed random variables with finite variances and if U = X + Y and V = X − Y are independent, then all r.v.'s X, Y, U and V are normally distributed.*

(*xv*) We state below another result which characterises the normal distribution.

If $X_1, X_2, \ldots, X_n$ are *i.i.d. r.v.*'s with finite variance, then the common distribution is normal if and only if :

$$\overline{X} = \frac{1}{n}\sum_{i=1}^{n} X_i \text{ and } \frac{1}{n}\sum_{i=1}^{n}\left(X_i - \overline{X}\right)^2 \quad \text{or} \quad \sum_{i=1}^{n} X_i \text{ and } \sum_{i=1}^{n}\left(X_i - \overline{X}\right)^2 \quad \text{are independent.}$$

In the following sequences, we shall establish some of these properties.

9·2·3. Mode of Normal Distribution

Mode is the value of x for which $f(x)$ is maximum, *i.e.*, mode is the solution of $f'(x) = 0$ and $f''(x) < 0$.

For normal distribution with mean μ and standard deviation σ,

$$\log f(x) = c - \frac{1}{2\sigma^2}(x - \mu)^2,$$

where, $c = \log\left(1/\sqrt{2\pi}\,\sigma\right)$, is a constant. Differentiating *w.r.* to x, we get

$$\frac{1}{f(x)} \cdot f'(x) = -\frac{1}{\sigma^2}(x - \mu) \quad \Rightarrow \quad f'(x) = -\frac{1}{\sigma^2}(x - \mu)\, f(x)$$

and $$f''(x) = -\frac{1}{\sigma^2}\left[1 \cdot f(x) + (x - \mu) f'(x)\right] = -\frac{f(x)}{\sigma^2}\left[1 - \frac{(x - \mu)^2}{\sigma^2}\right] \qquad \ldots (9{\cdot}3)$$

$$f'(x) = 0 \quad \Rightarrow \quad x - \mu = 0, \quad \Rightarrow \quad x = \mu.$$ At the point $x = \mu$, we have from (9·3) :

$$f''(x) = -\frac{1}{\sigma^2}[f(x)]_{x=\mu} = -\frac{1}{\sigma^2}\cdot\frac{1}{\sigma\sqrt{2\pi}} < 0$$

Hence, $x = \mu$, is the mode of the normal distribution.

9·2·4. Median of Normal Distribution

If M is the median of the normal distribution, we have

$$\int_{-\infty}^{M} f(x)\,dx = \frac{1}{2} \quad \Rightarrow \quad \frac{1}{\sigma\sqrt{2\pi}}\int_{-\infty}^{M} \exp\{-(x-\mu)^2/2\sigma^2\}\,dx = \frac{1}{2}$$

$$\Rightarrow \quad \frac{1}{\sigma\sqrt{2\pi}}\int_{-\infty}^{\mu} \exp\{-(x-\mu)^2/2\sigma^2\}\,dx + \frac{1}{\sigma\sqrt{2\pi}}\int_{\mu}^{M} \exp\{-(x-\mu)^2/2\sigma^2\}\,dx = \frac{1}{2} \quad \text{... (9·4)}$$

But $$\frac{1}{\sigma\sqrt{2\pi}}\int_{-\infty}^{\mu} \exp\{-(x-\mu)^2/2\sigma^2\}\,dx = \frac{1}{\sqrt{2\pi}}\int_{-\infty}^{0} \exp(-z^2/2)\,dz = \frac{1}{2}$$

$\therefore$ From (9·4), we have : $$\frac{1}{2} + \frac{1}{\sigma\sqrt{2\pi}}\int_{\mu}^{M} \exp\{-(x-\mu)^2/2\sigma^2\}\,dx = \frac{1}{2}$$

$$\Rightarrow \quad \frac{1}{\sigma\sqrt{2\pi}}\int_{\mu}^{M} \exp\{-(x-\mu)^2/2\sigma^2\}\,dx = 0, \qquad i.e., \quad \mu = M.$$

Hence, for the normal distribution, Mean = Median = μ.

Remark. From § 9·2·3. and § 9·2·4, we find that for the normal distribution mean, median and mode coincide. Hence, the distribution is *symmetrical*.

9·2·5. M.G.F. of Normal Distribution

The *m.g.f.* (about origin) is given by :

$$M_X(t) = \int_{-\infty}^{\infty} e^{tx} f(x)\,dx = \frac{1}{\sigma\sqrt{2\pi}}\int_{-\infty}^{\infty} e^{tx} \exp\{-(x-\mu)^2/2\sigma^2\}\,dx$$

$$= \frac{1}{\sqrt{2\pi}}\int_{-\infty}^{\infty} \exp\{t(\mu+\sigma z)\}\exp(-z^2/2)\,dz, \qquad \left(z = \frac{x-\mu}{\sigma}\right)$$

$$= e^{\mu t}\frac{1}{\sqrt{2\pi}}\int_{-\infty}^{\infty} \exp\left\{-\frac{1}{2}\left(z^2 - 2t\sigma z\right)\right\}dz$$

$$= e^{\mu t}\frac{1}{\sqrt{2\pi}}\int_{-\infty}^{\infty} \exp\left[-\frac{1}{2}\left\{(z-\sigma t)^2 - \sigma^2 t^2\right\}\right]dz$$

$$= e^{\mu t + t^2\sigma^2/2} \times \frac{1}{\sqrt{2\pi}}\int_{-\infty}^{\infty} \exp\left\{-\frac{1}{2}(z-\sigma t)^2\right\}dz$$

$$= e^{\mu t + t^2\sigma^2/2} \times \frac{1}{\sqrt{2\pi}}\int_{-\infty}^{\infty} \exp(-u^2/2)\,du$$

Hence, $$M_X(t) = \exp\left[\mu t + \frac{t^2\sigma^2}{2}\right] \quad \text{... (9·5)}$$

Remark. *M.G.F. of Standard Normal Variate.* If $X \sim N(\mu, \sigma^2)$, then standard normal variate is given by : $Z = (X - \mu)/\sigma$.

$$M_Z(t) = e^{-\mu t/\sigma} M_X(t/\sigma) = \exp(-\mu t/\sigma) \cdot \exp\left\{(\mu t/\sigma) + (t^2/\sigma^2)(\sigma^2/2)\right\} = \exp(t^2/2) \quad \text{... (9·5a)}$$

Aliter. $Z \sim N(0, 1)$. Hence, taking $\mu = 0$ and $\sigma^2 = 1$ in (9·5), we get : $M_Z(t) = \exp(t^2/2)$.

9·2·6. Cumulant Generating Function (c.g.f.) of Normal Distribution

The *c.g.f.* of normal distribution is given by :

$$K_X(t) = \log_e M_X(t) = \log_e \left[\exp\left(\mu t + \frac{t^2 \sigma^2}{2}\right)\right] = \mu t + \frac{t^2 \sigma^2}{2}$$

$\therefore$ Mean $= \kappa_1 =$ Coefficient of t in $K_X(t) = \mu$

Variance $= \kappa_2 =$ Coefficient of $\dfrac{t^2}{2!}$ in $K_X(t) = \sigma^2$

and $\kappa_r =$ Coefficient of $\dfrac{t^r}{r!}$ in $K_X(t) = 0,\ r \geq 3$.

Thus, $\mu_3 = \kappa_3 = 0$ and $\mu_4 = \kappa_4 + 3\kappa_2^2 = 3\sigma^4$

Hence,
$$\beta_1 = \frac{\mu_3^2}{\mu_2^3} = 0 \quad \text{and} \quad \beta_2 = \frac{\mu_4}{\mu_2^2} = 3 \quad \text{... (9·6)}$$

9·2·7. Moments of Normal Distribution

Odd order moments about mean are given by :

$$\mu_{2n+1} = \int_{-\infty}^{\infty} (x-\mu)^{2n+1} f(x)\, dx = \frac{1}{\sigma\sqrt{2\pi}} \int_{-\infty}^{\infty} (x-\mu)^{2n+1} \exp\left\{-(x-\mu)^2/2\sigma^2\right\} dx$$

$$\therefore \quad \mu_{2n+1} = \frac{1}{\sqrt{2\pi}} \int_{-\infty}^{\infty} (\sigma z)^{2n+1} \exp(-z^2/2)\, dz, \qquad \left(z = \frac{x-\mu}{\sigma}\right)$$

$$= \frac{\sigma^{2n+1}}{\sqrt{2\pi}} \int_{-\infty}^{\infty} z^{2n+1} \exp(-z^2/2)\, dz = 0, \quad \text{... (9·7)}$$

since the integrand $z^{2n+1} e^{-z^2/2}$ is an odd function of z.

Even order moments about mean are given by :

$$\mu_{2n} = \int_{-\infty}^{\infty} (x-\mu)^{2n} f(x)\, dx = \frac{1}{\sqrt{2\pi}} \int_{-\infty}^{\infty} (\sigma z)^{2n} \exp(-z^2/2)\, dz$$

$$= \frac{\sigma^{2n}}{\sqrt{2\pi}} \int_{-\infty}^{\infty} z^{2n} \exp(-z^2/2)\, dz = \frac{\sigma^{2n}}{\sqrt{2\pi}} \cdot 2 \int_0^{\infty} z^{2n} \exp(-z^2/2)\, dz$$

(Since integrand is an even function of z.)

$$= \frac{2\sigma^{2n}}{\sqrt{2\pi}} \cdot \int_0^{\infty} (2t)^n e^{-t} \frac{dt}{\sqrt{2t}}, \qquad \left(t = \frac{z^2}{2}\right)$$

$$\therefore \quad \mu_{2n} = \frac{2^n \sigma^{2n}}{\sqrt{\pi}} \int_0^{\infty} e^{-t}\, t^{\left(n+\frac{1}{2}\right)-1} dt \quad \Rightarrow \quad \mu_{2n} = \frac{2^n \sigma^{2n}}{\sqrt{\pi}} \cdot \Gamma\left(n + \frac{1}{2}\right)$$

Changing n to $(n-1)$, we get : $\mu_{2n-2} = \dfrac{2^{n-1}\cdot\sigma^{2n-2}}{\sqrt{\pi}}\cdot\Gamma\left(n-\dfrac{1}{2}\right)$

$$\therefore \quad \frac{\mu_{2n}}{\mu_{2n-2}} = 2\sigma^2\cdot\frac{\Gamma\left(n+\frac{1}{2}\right)}{\Gamma\left(n-\frac{1}{2}\right)} = 2\sigma^2\left(n-\frac{1}{2}\right) \qquad [\because \Gamma(r)=(r-1)\Gamma(r-1)]$$

$$\Rightarrow \quad \mu_{2n} = \sigma^2(2n-1)\,\mu_{2n-2} \qquad \text{... (9·8)}$$

which gives the *recurrence relation* for the moments of normal distribution.

From (9·8), we have

$$\begin{aligned}\mu_{2n} &= \left[(2n-1)\sigma^2\right]\left[(2n-3)\sigma^2\right]\mu_{2n-4}\\ &= \left[(2n-1)\sigma^2\right]\left[(2n-3)\sigma^2\right]\left[(2n-5)\sigma^2\right]\mu_{2n-6}\\ &= \left[(2n-1)\sigma^2\right]\left[(2n-3)\sigma^2\right]\left[(2n-5)\sigma^2\right]\ldots\left(3\sigma^2\right)\left(1\sigma^2\right)\cdot\mu_0\\ &= 1.3.5\ldots(2n-1)\sigma^{2n} \qquad \text{... (9·9)}\end{aligned}$$

From (9·7) and (9·9), we conclude that *for the normal distribution all odd order moments about mean vanish and even order moments about mean are given by (9·9).*

Aliter. The above result can also be obtained quite conveniently as follows :

The *m.g.f.* (about mean) is given by : $E\left[e^{t(X-\mu)}\right] = e^{-\mu t}E\left(e^{tX}\right) = e^{-\mu t}M_X(t)$,

where $M_X(t)$ is the *m.g.f.* (about origin).

$$\therefore \quad m.g.f.\text{ (about mean)} = e^{-\mu t}\,e^{\mu t + t^2\sigma^2/2} = e^{t^2\sigma^2/2}$$

$$= \left[1 + \left(t^2\sigma^2/2\right) + \frac{\left(t^2\sigma^2/2\right)^2}{2!} + \frac{\left(t^2\sigma^2/2\right)^3}{3!} + \cdots + \frac{\left(t^2\sigma^2/2\right)^n}{n!} + \cdots\right] \qquad \text{... (9·10)}$$

The coefficient of $\dfrac{t^r}{r!}$ in (9·10) gives μ_r, the rth moment about mean. Since there is no term with odd powers of t in (9·10), all moments of odd order about mean vanish, *i.e.*,

$$\mu_{2n+1} = 0; \quad n = 0, 1, 2, \cdots$$

and

$$\begin{aligned}\mu_{2n} &= \text{Coefficient of } \frac{t^{2n}}{(2n)!} \text{ in } (9\cdot 10) = \frac{\sigma^{2n}\times(2n)!}{2^n\, n!}\\ &= \frac{\sigma^{2n}}{2^n\, n!}\left[(2n(2n-1)(2n-2)(2n-3)\ldots 5\cdot 4\cdot 3\cdot 2\cdot 1\right]\\ &= \frac{\sigma^{2n}}{2^n\cdot n!}\left[1\cdot 3\cdot 5\ldots(2n-1)\right]\left[2\cdot 4\cdot 6\ldots(2n-2)\cdot 2n\right]\\ &= \frac{\sigma^{2n}}{2^n\cdot n!}\left[1\cdot 3\cdot 5\ldots(2n-1)\right]2^n\left[1\cdot 2\cdot 3\ldots n\right]\\ &= 1\cdot 3\cdot 5\ldots(2n-1)\,\sigma^{2n}\end{aligned}$$

Remark. In particular, from (9·7) and (9·9), $\mu_3 = 0$ and $\mu_2 = \sigma^2$, $\mu_4 = 1\cdot 3\sigma^4 = 3\sigma^4$

Hence, $\beta_1 = \dfrac{\mu_3^2}{\mu_2^3} = 0$ and $\beta_2 = \dfrac{\mu_4}{\mu_2^2} = \dfrac{3\sigma^4}{\sigma^4} = 3$, the results which have already been obtained in (9·6).

9·2·8. A Linear Combination of Independent Normal Variates is also a Normal Variate

Let $X_i, (i = 1, 2, 3, ..., n)$ be n independent normal variates with mean μ_i and variance σ_i^2 respectively.

Then $$M_{X_i}(t) = \exp\left\{\mu_i t + \left(t^2 \sigma_i^2 / 2\right)\right\} \quad ... (9·11)$$

The *m.g.f.* of their linear combination $\sum_{i=1}^{n} a_i X_i$, where $a_1, a_2, ..., a_n$ are constants, is given by :

$$M_{\sum a_i X_i}(t) = \prod_{i=1}^{n} M_{a_i X_i}(t) \qquad (\because X_i\text{'s are independent})$$

$$= M_{X_1}(a_1 t) \cdot M_{X_2}(a_2 t) \dots M_{X_n}(a_n t) \qquad [\because M_{cX}(t) = M_X(ct)] \quad ... (9·12)$$

From (9·11), we have $$M_{X_i}(a_i t) = \exp\left[\mu_i a_i t + t^2 a_i^2 \sigma_i^2 / 2\right]$$

$$\therefore \quad M_{\sum_i a_i X_i}(t) = \prod_{i=1}^{n} \exp\left\{\mu_i a_i t + t^2 a_i^2 \sigma_i^2 / 2\right\} \qquad [\text{From } (9·12)]$$

$$= \exp\left[\left(\sum_{i=1}^{n} a_i \mu_i\right) t + t^2 \left(\sum_{i=1}^{n} a_i^2 \sigma_i^2\right) / 2\right],$$

which is the *m.g.f.* of a normal variate with mean $\sum_{i=1}^{n} a_i \mu_i$ and variance $\sum_{i=1}^{n} a_i^2 \sigma_i^2$.

Hence, by uniqueness theorem of *m.g.f.*,

$$\sum_{i=1}^{n} a_i X_i \sim N\left[\sum_{i=1}^{n} a_i \mu_i, \ \sum_{i=1}^{n} a_i^2 \sigma_i^2\right] \quad ... (9·12a)$$

Remarks :

1. If we take $a_1 = a_2 = 1, a_3 = a_4 = ... = 0$, then $X_1 + X_2 \sim N(\mu_1 + \mu_2, \sigma_1^2 + \sigma_2^2)$

If we take $a_1 = 1, a_2 = -1, a_3 = a_4 = ... = 0$, then $X_1 - X_2 \sim N(\mu_1 - \mu_2, \sigma_1^2 + \sigma_2^2)$

Thus, we see that *the sum as well as the difference of two independent normal variates is also a normal variate.* This result provides a sharp contrast to the Poisson distribution, in which case though the sum of two independent Poisson variates is a Poisson variate, the difference is not a Poisson variate.

2. If we take $a_1 = a_2 = ... = a_n = 1$, then we get $$\sum_{i=1}^{n} X_i \sim N\left[\sum_{i=1}^{n} \mu_i, \ \sum_{i=1}^{n} \sigma_i^2\right] \quad ... (9·12b)$$

i.e., the sum of independent normal variates is also a normal variate, which establishes the *additive property* of the normal distribution.

3. If $X_i; i = 1, 2, ..., n$ are identically and independently distributed as $N(\mu, \sigma^2)$ and if we take $a_1 = a_2 = ... = a_n = 1/n$, then

$$\frac{1}{n}\sum_{i=1}^{n} X_i \sim N\left(\frac{1}{n}\sum_{i=1}^{n} \mu, \ \frac{1}{n^2}\sum_{i=1}^{n} \sigma^2\right) \Rightarrow \bar{X} \sim N\left(\mu, \sigma^2/n\right), \text{ where } \bar{X} = \frac{1}{n}\sum_{i=1}^{n} X_i$$

This leads to the following important conclusion :

If $X_i, (i = 1, 2, ..., n)$, are identically and independently distributed normal variates with mean μ and variance σ^2, then their mean $\bar{X}$ is also $N(\mu, \sigma^2/n)$.

9·2·9. Points of Inflexion of Normal Curve

At the point of inflexion of the normal curve, we should have : $f''(x)=0$, and $f'''(x)\neq 0$.

For normal curve, we have from (9·3), $f''(x)=-\dfrac{f(x)}{\sigma^2}\left\{1-\dfrac{(x-\mu)^2}{\sigma^2}\right\}$

$$\therefore \quad f''(x)=0 \quad\Rightarrow\quad 1-\frac{(x-\mu)^2}{\sigma^2}=0 \quad\Rightarrow\quad x=\mu\pm\sigma.$$

It can be easily verified that at the points $x=\mu\pm\sigma$, $f'''(x)\neq 0$. Hence, the points of inflexion of the normal curve are given by $\left[x=\mu\pm\sigma \text{ and } f(x)=\dfrac{1}{\sigma\sqrt{2\pi}}e^{-1/2}\right]$, *i.e.*, they are equi-distant (at a distance σ) from the mean.

9·2·10. Mean Deviation About the Mean for Normal Distribution

$$\text{M.D. (about mean)}=\int_{-\infty}^{\infty}|x-\mu|\,f(x)\,dx=\frac{1}{\sigma\sqrt{2\pi}}\int_{-\infty}^{\infty}|x-\mu|\,e^{-(x-\mu)^2/2\sigma^2}\,dx$$

$$=\frac{\sigma}{\sqrt{2\pi}}\int_{-\infty}^{\infty}|z|\,e^{-z^2/2}\,dz, \qquad \left(\frac{x-\mu}{\sigma}=z\right)$$

$$=\frac{2\sigma}{\sqrt{2\pi}}\cdot\int_{0}^{\infty}|z|\,e^{-z^2/2}\,dz,$$

($\because$ The integrand $|z|\,e^{-z^2/2}$ is an even function of z.)

Since in $[0,\infty]$, $|z|=z$, we have

$$\text{M.D. (about mean)}=\sqrt{2/\pi}\cdot\sigma\int_0^{\infty}z\,e^{-z^2/2}\,dz=\sqrt{2/\pi}\cdot\sigma\int_0^{\infty}e^{-t}\,dt, \qquad (z^2/2=t)$$

$$=\sqrt{2/\pi}\cdot\sigma\left|\frac{e^{-t}}{-1}\right|_0^{\infty}=\sqrt{2/\pi}\cdot\sigma\simeq\frac{4}{5}\sigma \text{ (approx.)}$$

9·2·11. Area Property (Normal Probability Integral)

If $X\sim N(\mu,\sigma^2)$, then the probability that random value of X will lie between $X=\mu$ and $X=x_1$ $(>\mu)$ is given by :

$$P(\mu<X<x_1)=\int_{\mu}^{x_1}f(x)\,dx=\frac{1}{\sigma\sqrt{2\pi}}\int_{\mu}^{x_1}e^{-(x-\mu)^2/2\sigma^2}\,dx\,; \quad (x_1>\mu)$$

Put $\dfrac{X-\mu}{\sigma}=Z \quad\Rightarrow\quad X-\mu=\sigma Z$

When $X=\mu$, $Z=0$ and when $X=x_1$, $Z=\dfrac{x_1-\mu}{\sigma}=z_1$, (say).

$$\therefore \quad P(\mu<X<x_1)=P(0<Z<z_1)=\frac{1}{\sqrt{2\pi}}\int_0^{z_1}e^{-z^2/2}\,dz=\int_0^{z_1}\varphi(z)\,dz\,; \quad z_1>0,$$

where, $\varphi(z)=\dfrac{1}{\sqrt{2\pi}}e^{-z^2/2}$, is the probability function of standard normal variate. The definite integral $\int_0^{z_1} \varphi(z)\, dz$ is known as *normal probability integral* and gives the area under standard normal curve between the ordinates at $Z = 0$ and $Z = z_1$. These areas have been tabulated for different values of z_1, at intervals of 0·01 in Table VI given in the Appendix at the end of the book.

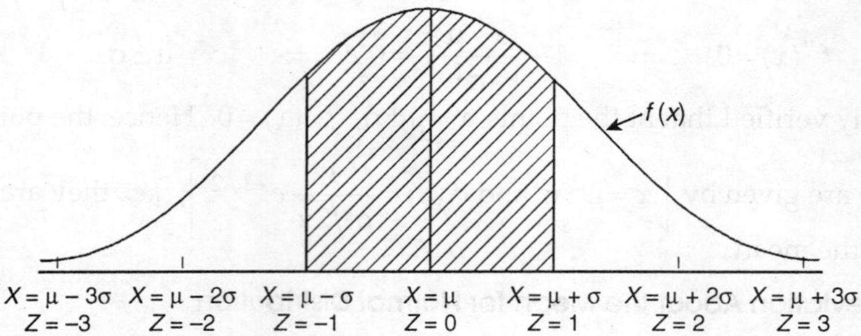

Fig. 9·5

In particular, the probability that a random value of X lies in the interval $(\mu - \sigma, \mu + \sigma)$ is given by :

$$P(\mu - \sigma < X < \mu + \sigma) = \int_{\mu-\sigma}^{\mu+\sigma} f(x)\, dx \qquad \text{[See Fig. 9·5]}$$

$$\Rightarrow \qquad P(-1 < Z < 1) = \int_{-1}^{1} \varphi(z)\, dz, \qquad \left[z = \frac{x-\mu}{\sigma}\right]$$

$$= 2\int_0^1 \varphi(z)\, dz \qquad \text{(By symmetry)}$$

$$= 2 \times 0{\cdot}3413 = 0{\cdot}6826 \qquad \text{(From Tables)} \qquad \ldots (9{\cdot}14)$$

Similarly, $$P(\mu - 2\sigma < X < \mu + 2\sigma) = P(-2 < Z < 2) = \int_{-2}^{2} \varphi(z)\, dz$$

$$= 2\int_0^2 \varphi(z)\, dz = 2 \times 0{\cdot}4772 = 0{\cdot}9544 \qquad \ldots (9{\cdot}15)$$

and $$P(\mu - 3\sigma < X < \mu + 3\sigma) = P(-3 < Z < 3) = \int_{-3}^{3} \varphi(z)\, dz$$

$$= 2\int_0^3 \varphi(z)\, dz = 2 \times 0{\cdot}49865 = 0{\cdot}9973 \qquad \ldots (9{\cdot}16)$$

Thus the probability that a normal variate X lies outside the range $\mu \pm 3\sigma$ is given by :

$$P(|X - \mu| > 3\sigma) = P(|Z| > 3) = 1 - P(-3 \le Z \le 3) = 0{\cdot}0027$$

Thus, in all probability, we should expect a normal variate to lie within the range $\mu \pm 3\sigma$, though theoretically, it may range from $-\infty$ to ∞.

Remarks :

1. The total area under normal probability curve is unity, *i.e.,*

$$\int_{-\infty}^{\infty} f(x)\, dx = \int_{-\infty}^{\infty} \varphi(z)\, dz = 1.$$

2. Since in the normal probability tables, we are given the areas under standard normal curve, in numerical problems we shall deal with the standard normal variate Z rather than the variable X itself.

3. If we want to find area under normal curve, we will somehow or other try to convert the given area to the form $P(0 < Z < z_1)$, since the areas have been given in this form in the Tables.

4. Error Function. If $X \sim N(0, \sigma^2)$, then $f(x) = \dfrac{1}{\sigma\sqrt{2\pi}} e^{-x^2/2\sigma^2}, \quad -\infty < x < \infty$

If we take $h^2 = \dfrac{1}{2\sigma^2}$, then $f(x) = \dfrac{h}{\sqrt{\pi}} e^{-h^2x^2}$

The probability P that a random value of the variate lies in the range $\pm x$ is :

$$P = \int_{-x}^{x} f(x)\,dx = \frac{h}{\sqrt{\pi}} \int_{-x}^{x} e^{-h^2x^2}\,dx = \frac{2h}{\sqrt{\pi}} \int_0^x e^{-h^2x^2}\,dx = \frac{2}{\sqrt{\pi}} \int_0^x e^{-h^2x^2}\,(hdx) \qquad \text{... (*)}$$

Taking $\psi(y) = \dfrac{2}{\sqrt{\pi}} \displaystyle\int_0^y e^{-y^2}\,dy$, (*) may be re-written as :

$$P = \psi(hx) = \frac{2}{\sqrt{\pi}} \int_0^x e^{-h^2x^2}\,(hdx) \qquad \text{... (**)}$$

The function $\psi(y)$, known as the *error function*, is of fundamental importance in the *theory of errors* in Astronomy.

9·2·12. Importance of Normal Distribution

Normal distribution plays a very important role in statistical theory because of the following reasons :

(*i*) Most of the discrete distributions occurring in practice, *e.g.*, Binomial, Poisson, Hypergeometric distributions, etc., can be approximated by normal distribution . Moreover, many of the sampling distributions, *e.g.*, Student's t, Snedecor's F, Chi-square distributions, etc., tend to normality for large samples.

(*ii*) Even if a variable is not normally distributed, it can sometimes be brought to normal form by simple transformation of variable. For example, if the distribution of X is skewed, the distribution of $\sqrt{X}$ might come out to be normal [*c.f.* Variate Transformations, § 20·3 in Chapter 20].

(*iii*) $X \sim N(\mu, \sigma^2)$, then $P(\mu - 3\sigma \le X < \mu + 3\sigma) = P(-3 < Z < 3) = 0{\cdot}9973$

$$\therefore \qquad P(|Z| > 3) = 1 - P(|Z| \le 3) = 0{\cdot}0027$$

This property of the normal distribution forms the basis of entire *Large Sample theory.*

(*iv*) Many of the distributions of sample statistics (*e.g.*, the distributions of sample mean, sample variance, etc.) tend to normality for large samples and as such they can best be studied with the help of the normal curves.

(*v*) The entire theory of small sample tests, *viz.*, t, F, χ^2 tests, etc., is based on the fundamental assumption that the parent populations from which the samples have been drawn follow normal distribution.

(*vi*) Normal distribution finds large applications in Statistical Quality Control in industry for setting control limits.

The following quotation due to Lipman rightly reveals the popularity and importance of normal distribution :

"Every body believes in the law of errors (the normal curve), the experimenters because they think it is a mathematical theorem, the mathematicians because they think it is experimental fact."

W.J. Youden of the National Bureau of Standards describes the importance of the Normal distribution artistically in the following words :

THE NORMAL
LAW OF ERRORS
STANDS OUT IN THE
EXPERIENCE OF MANKIND
AS ONE OF THE BROADEST
GENERALISATIONS OF NATURAL
PHILOSOPHY. IT SERVES AS THE
GUIDING INSTRUMENT IN RESEARCHES,
IN THE PHYSICAL AND SOCIAL SCIENCES
AND IN MEDICINE, AGRICULTURE AND
ENGINEERING. IT IS AN INDISPENSABLE TOOL FOR
THE ANALYSIS AND THE INTERPRETATION OF THE
BASIC DATA OBTAINED BY OBSERVATION AND EXPERIMENT.

9·2·13. Fitting of Normal Distribution

In order to fit normal distribution to the given data we first calculate the mean μ, (say), and standard deviation σ (say) from the given data. Then the normal curve fitted to the given data is given by :

$$f(x)=\frac{1}{\sigma\sqrt{2\pi}}\exp\left\{-(x-\mu)^2/2\sigma^2\right\},\quad -\infty<x<\infty$$

To calculate the expected normal frequencies we first find the standard normal variates corresponding to the lower limits of each of the class intervals, *i.e.*, we compute $z_i=(x_i'-\mu)/\sigma$, where x_i' is the lower limit of the ith class interval. Then the areas under the normal curve to the left of the ordinate at $z=z_i$, say, $\Phi(z_i)=P(Z\le z_i)$ are computed from the tables. Finally, the areas for the successive class intervals are obtained by subtraction, *viz.*, $\Phi(z_i+1)-\Phi(z_i)$, $(i=1, 2, ...)$ and on multiplying these areas by N, we get the expected normal frequencies.

Example 9·1. *Obtain the equation of the normal curve that may be fitted to the following data :*

Class :	*60—65*	*65—70*	*70—75*	*75—80*	*80—85*	*85—90*	*90—95*	*95—100*
Frequency :	*3*	*21*	*150*	*335*	*326*	*135*	*26*	*4*

Also obtain the expected normal frequencies.

Solution. For the given data, $N=1000$, $\mu=79{\cdot}945$ and $\sigma=5{\cdot}545$ (Try it.)

Hence, the equation of the normal curve fitted to the given data is :

$$f(x)=\frac{1000}{\sqrt{2\pi}\times 5{\cdot}545}\cdot\exp\left\{-\frac{1}{2}\left(\frac{x-79{\cdot}945}{5{\cdot}545}\right)^2\right\};\quad -\infty<x<\infty$$

COMPUTATION OF THEORETICAL NORMAL FREQUENCIES

Class	*Lower class boundary (X')*	$z=\frac{X'-\mu}{\sigma}$	$\Phi(z)=\frac{1}{\sqrt{2\pi}}\times\int_{-\infty}^{z}e^{-u^2/2}\,du$	$\Delta\Phi(z)=\Phi(z+1)-\Phi(z)$	*Expected frequency* $=N\Delta\Phi(z)$
(1)	*(2)*	*(3)*	*(4)*	*(5)*	*(6)*
Below 60	$-\infty$	$-\infty$	0	0·000112	0·12 ≅ 0
60—65	60	–3·663	0·000112	0·002914	2·914 ≅ 3

(1)	(2)	(3)	(4)	(5)	(6)
65—70	65	−2·745	0·003026	0·031044	31·044 ≅ 31
70—75	70	−1·826	0·034070	0·147870	147·870 ≅ 148
75—80	75	−0·908	0·181940	0·322050	322·050 ≅ 322
80—85	80	0·010	0·503990	0·919300	319·300 ≅ 319
85—90	85	0·928	0·823290	0·144072	144·072 ≅ 144
90—95	90	1·487	0·967362	0·029792	29·792 ≅ 30
95—100	95	2·675	0·997154	0·002733	2·733 ≅ 3
100 and over	100	3·683	0·999887		
Total					1,000

Example 9·2. *For a certain normal distribution, the first moment about 10 is 40 and the fourth moment about 50 is 48. What is the arithmetic mean and standard deviation of the distribution ?*

Solution. We know that if μ_1' is the first moment about the point $X = A$, then arithmetic mean is given by : Mean $= A + \mu_1'$

We are given : μ_1' (about the point $X = 10$) $= 40$ $\Rightarrow$ Mean $= 10 + 40 = 50$

Also μ_4' (about the point $X = 50$) $= 48$, *i.e.,* $\mu_4 = 48$ ($\because$ Mean = 50)

But for a normal distribution with standard deviation σ,

$$\mu_4 = 3\sigma^2 \quad \Rightarrow \quad 3\sigma^2 = 48 \quad \Rightarrow \quad \sigma = 2.$$

Example 9·3. *X is normally distributed and the mean of X is 12 and S.D. is 4. Find out the probability of the following :*

(a) (i) $X \geq 20$, *(ii)* $X \leq 20$, *and (iii)* $0 \leq X \leq 12$ *(b) Find* x', *when* $P(X > x') = 0{\cdot}24$.

(c) Find x_0' *and* x_1', *when* $P(x_0' < X < x_1') = 0{\cdot}50$ *and* $P(X > x_1') = 0{\cdot}25$.

Solution. (*a*) We have $\mu = 12,\ \sigma = 4$, *i.e.,* $X \sim N(12, 16)$.

(*i*) $P(X \geq 20)$?

When $X = 20$, $Z = \dfrac{20-12}{4} = 2$

$$\therefore \quad P(X \geq 20) = P(Z \geq 2) = 0{\cdot}5 - P(0 \leq Z \leq 2)$$
$$= 0{\cdot}5 - 0{\cdot}4772 = 0{\cdot}0228 \quad [\text{Fig. } 9{\cdot}6(a)]$$

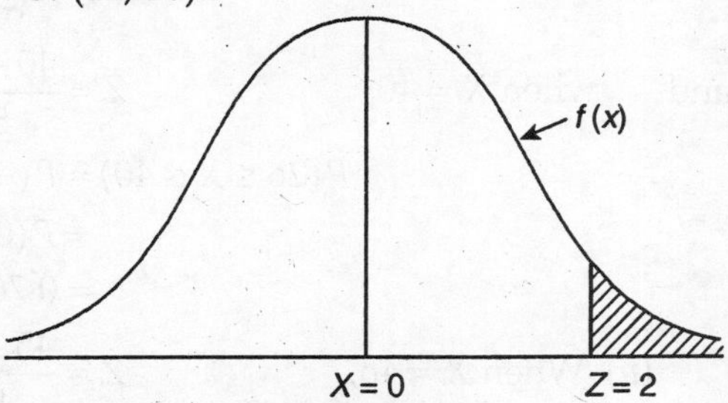

Fig. 9·6(a)

(*ii*) $P(X \leq 20) = 1 - P(X \geq 20)$
$$= 1 - 0{\cdot}0228 = 0{\cdot}9722 \quad [\text{From part } (i)]$$

(*iii*) $P(0 \leq X \leq 12) = P(-3 \leq Z \leq 0)$, $\left(Z = \dfrac{X-12}{4}\right)$
$$= P(0 \leq Z \leq 3) \quad (\text{From symmetry})$$
$$= 0{\cdot}49865$$

(*b*) When $X = x'$, $Z = \dfrac{x'-12}{4} = z_1$, (say).

Then, we are given :

$$P(X > x') = 0{\cdot}24 \quad [\text{Fig. } 9{\cdot}6(b)]$$
$$\Rightarrow \quad P(Z > z_1) = 0{\cdot}24$$
$$\Rightarrow \quad P(0 < Z < z_1) = 0{\cdot}26$$

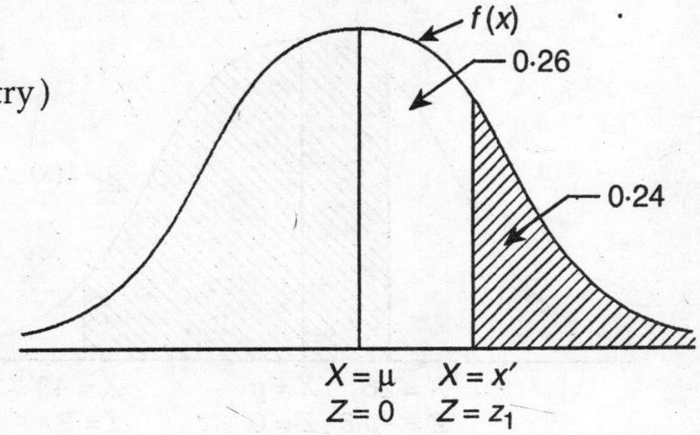

Fig. 9·6(b)

$\therefore$ From Normal Tables, $z_1 = 0{\cdot}71$ (approx.)

Hence, $$\frac{x' - 12}{4} = 0{\cdot}71$$

$$\Rightarrow \quad x' = 12 + 4 \times 0{\cdot}71 = 14{\cdot}84.$$

(c) We are given : $P(x_0' < X < x_1') = 0{\cdot}50$ and $P(X > x_1') = 0{\cdot}25$ ... (*)

From (*), obviously the points x_0' and x_1' are located as shown in *Fig. 9·6(c)*.

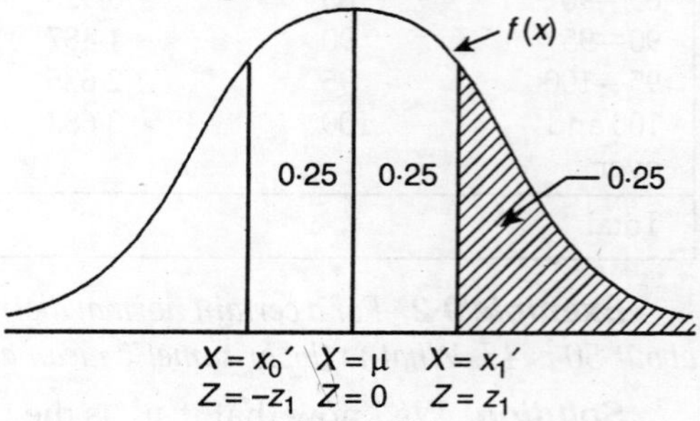

Fig. 9·6(c)

when $X = x_1'$, $$Z = \frac{x_1' - 12}{4} = z_1, \text{ (say)},$$

and when $X = x_0'$, $$Z = \frac{x_0' - 12}{4} = -z_1$$

(It is obvious from the figure.)

We have $P(Z > z_1) = 0{\cdot}25 \Rightarrow P(0 < Z < z_1) = 0{\cdot}25 \therefore z_1 = 0{\cdot}67$ (From Tables)

Hence, $$\frac{x_1' - 12}{4} = 0{\cdot}67 \Rightarrow x_1' = 12 + 4 \times 0{\cdot}67 = 14{\cdot}68$$

and $$\frac{x_0' - 12}{4} = -0{\cdot}67 \Rightarrow x_0' = 12 - 4 \times 0{\cdot}67 = 9{\cdot}32.$$

Example 9·4. *X is a normal variate with mean 30 and S.D. 5. Find the probabilities that*

(i) $26 \le X \le 40$, *(ii)* $X \ge 45$, *and* *(iii)* $|X - 30| > 5$.

Solution. Here $\mu = 30$ and $\sigma = 5$.

(i) When $X = 26$, $$Z = \frac{X - \mu}{\sigma} = \frac{26 - 30}{5} = -0{\cdot}8$$ [Fig. 9·7(a)]

and when $X = 40$, $$Z = \frac{40 - 30}{5} = 2$$

$$\therefore \quad P(26 \le X \le 40) = P(-0{\cdot}8 \le Z \le 2) = P(-0{\cdot}8 \le Z \le 0) + P(0 \le Z \le 2)$$
$$= P(0 \le Z \le 0{\cdot}8) + (0 \le Z \le 2) = 0{\cdot}2881 + 0{\cdot}4772 \quad \text{[By symmetry]}$$
$$= 0{\cdot}7653 \quad \text{(From Normal Probability Tables)}$$

(ii) When $X = 45$, $$Z = \frac{45 - 30}{5} = 3.$$

$$\therefore \quad P(X \ge 45) = P(Z \ge 3) = 0{\cdot}5 - P(0 \le X \le 3) = 0{\cdot}5 - 0{\cdot}49865 = 0{\cdot}00135$$

[See Fig. 9·7(b)]

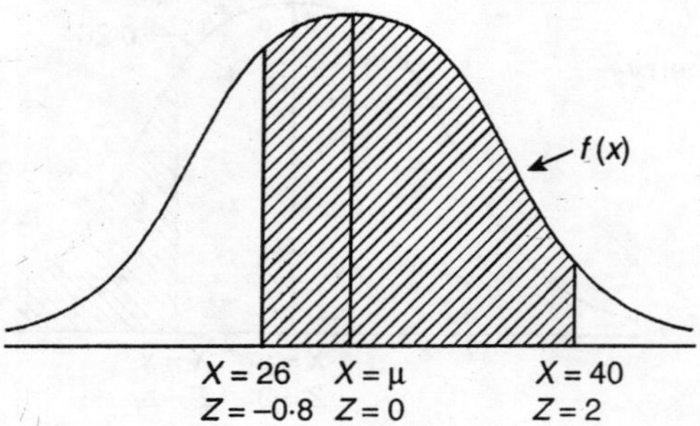

Fig. 9·7(a)

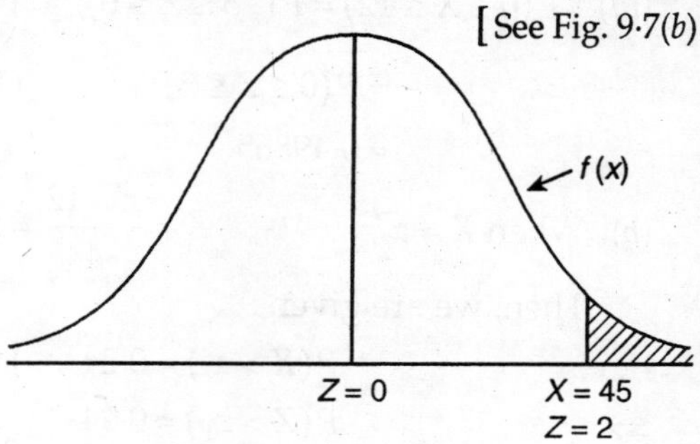

Fig. 9·7(b)

(*iii*) $$P(|X-30|\le 5) = P(25 \le X \le 35) = P(-1 \le Z \le 1)$$
$$= 2P(0 \le Z \le 1) = 2 \times 0{\cdot}3413 = 0{\cdot}6826 \qquad \text{[By symmetry]}$$
$$\therefore \quad P(|X-30| > 5) = 1 - P(|X-30| \le 5) = 1 - 0{\cdot}6826 = 0{\cdot}3174.$$

Example 9·5. *The mean yield for one-acre plot is 662 kilos with a s.d. 32 kilos. Assuming normal distribution, how many one-acre plots in a batch of 1,000 plots would you expect to have yield (i) over 700 kilos, (ii) below 650 kilos, and (iii) what is the lowest yield of the best 100 plots ?*

Solution. If the *r.v.* X denotes the yield (in kilos) for one-acre plot, then we are given that $X \sim N(\mu, \sigma^2)$, where $\mu = 662$ and $\sigma = 32$.

(*i*) The probability that a plot has a yield over 700 kilos is given by :

$$P(X > 700) = P(Z > 1{\cdot}19), \qquad \left(Z = \frac{700-662}{32}\right)$$
$$= 0{\cdot}5 - P(0 \le Z \le 1{\cdot}19) = 0{\cdot}5 - 0{\cdot}3830 = 0{\cdot}1170$$

Hence, in a batch of 1,000 plots, the expected number of plots with yield over 700 kilos is :
$$= 1{,}000 \times 0{\cdot}117 = 117.$$

(*ii*) Required number of plots with yield below 650 kilos is given by :

$$1000 \times P(X < 650) = 1{,}000 \times P(Z < -0{\cdot}38) \qquad \left(Z = \frac{650-662}{32}\right)$$
$$= 1{,}000 \times P(Z > 0{\cdot}38), \qquad \text{(By symmetry)}$$
$$= 1{,}000 \times [0{\cdot}5 - P(0 \le Z \le 0{\cdot}38)]$$
$$= 1{,}000 \times (0{\cdot}5 - 0{\cdot}1480) = 1{,}000 \times 0{\cdot}352 = 352$$

(*iii*) The lowest yield, say, x_1 of best 100 plots is given by : $P(X > x_1) = \dfrac{100}{1000} = 0{\cdot}1$

When $X = x_1$, $$Z = \frac{x_1 - \mu}{\sigma} = \frac{x_1 - 662}{32} = z_1, \text{ (say), such that} \qquad \ldots (*)$$

$$P(Z > z_1) = 0{\cdot}1 \Rightarrow P(0 \le Z \le z_1) = 0{\cdot}4 \Rightarrow z_1 = 1{\cdot}28 \text{ (approx.)} \qquad \text{[From Normal Tables]}$$

Substituting in (*), $$x_1 = 662 + 32z_1 = 662 + 32 \times 1{\cdot}28 = 662 + 40{\cdot}96 = 702{\cdot}96$$

Hence, the best 100 plots have yield over 702·96 kilos.

Example 9·6. *The local authorities in a certain city instal 10,000 electric lamps in the streets of the city. If these lamps have an average life of 1,000 burning hours with a standard deviation of 200 hours, assuming normality, what number of lamps might be expected to fail (i) in the first 800 burning hours ? (ii) between 800 and 1,200 burning hours ? After what period of burning hours would you expect that (a) 10% of the lamps would fail ? (b) 10% of the lamps would be still burning ?*

[In a normal curve, the area between the ordinates corresponding to $\{(X-\mu)/\sigma\} = 0$ *and* $\{(X-\mu)/\sigma\} = 1$ *is 0·34134 and 80% of the area lies between the ordinates corresponding to* $\{(X-\mu)/\sigma\} = \pm 1{\cdot}28$.*]*

Solution. If the variable X denotes the life of a bulb in burning hours, then we are given that $X \sim N(\mu, \sigma^2)$, where $\mu = 1{,}000$ and $\sigma = 200$.

(*i*) The probability '*p*' that bulb fails in the first 800 burning hours is given by :

$$p = P(X < 800) = P(Z < -1) = P(Z > 1), \qquad \left(Z = \frac{800-1000}{200}\right)$$
$$= 0{\cdot}5 - P(0 < Z < 1) = 0{\cdot}5 - 0{\cdot}3413 = 0{\cdot}1587$$

$\therefore$ Out of 10,000 bulbs, number of bulbs which fail in the first 800 hours

$$= 10{,}000 \times 0{\cdot}1587 = 1{,}587.$$

(*ii*) Required probability

$$= P(800 < X < 1{,}200) = P(-1 < Z < 1) = 2P(0 < Z < 1) = 2 \times 0{\cdot}3413 = 0{\cdot}6826$$

[By symmetry]

Hence, the expected number of bulbs with life between 800 and 1,200 hours of burning life is :

$$10{,}000 \times 0{\cdot}6826 = 6{,}826.$$

(*a*) Let 10% of the bulbs fail after x_1 hours of burning life. Then we have to find x_1 such that

$$P(X < x_1) = 0{\cdot}10$$

When $X = x_1$, $\quad Z = \dfrac{x_1 - 1000}{200} = -z_1$ (say), $\quad$ [Fig. 9·8(*a*)] $\quad$ ... (*)

$\therefore \quad P(Z < -z_1) = 0{\cdot}10 \quad \Rightarrow \quad P(Z > z_1) = 0{\cdot}10 \quad \Rightarrow \quad P(0 < Z < z_1) = 0{\cdot}40$ $\quad$ ... (**)

Also $\quad P(-1{\cdot}28 < z < 1{\cdot}28) = 0{\cdot}80$ (Given)

$\Rightarrow \quad 2P(0 < Z < 1{\cdot}28) = 0{\cdot}80$

$\Rightarrow \quad P(0 < Z < 1{\cdot}28) = 0{\cdot}40$ ... (***)

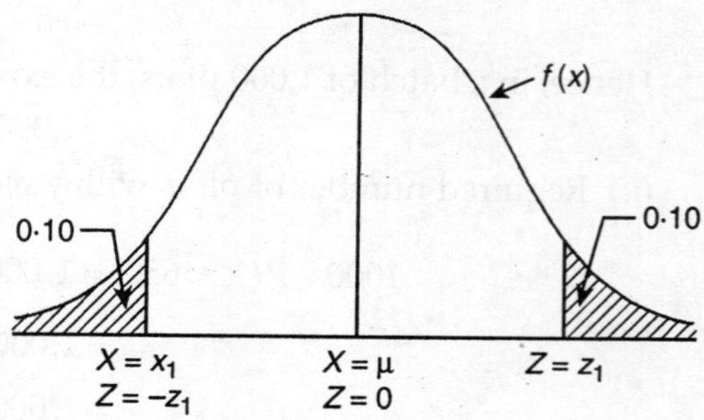

Fig. 9·8(a)

From (**) and (***), we get $z_1 = 1{\cdot}28$

$\therefore \quad \dfrac{x_1 - 1000}{200} = -1{\cdot}28$ $\quad$ [From (*)]

$\Rightarrow \quad x_1 = 1{,}000 - 256 = 744$

Thus, after 744 hours of burning life, 10% of the bulbs will fail.

(*b*) Let 10% of the bulbs be still burning after (say) x_2 hours of burning life. Then we have

$$P(X > x_2) = 0{\cdot}10 \qquad [\text{Fig. } 9{\cdot}8(b)]$$

$\Rightarrow \quad P(Z > z_2) = 0{\cdot}10, \quad \left(z_2 = \dfrac{x_2 - 1000}{200}\right)$

$\Rightarrow \quad P(0 < Z < z_2) = 0{\cdot}40$

$z_2 = 1{\cdot}28$, [From (***)]

i.e., $\quad \dfrac{x_2 - 1000}{200} = 1{\cdot}28$

$\Rightarrow \quad x_2 = 1{,}000 + 256 = 1{,}256$

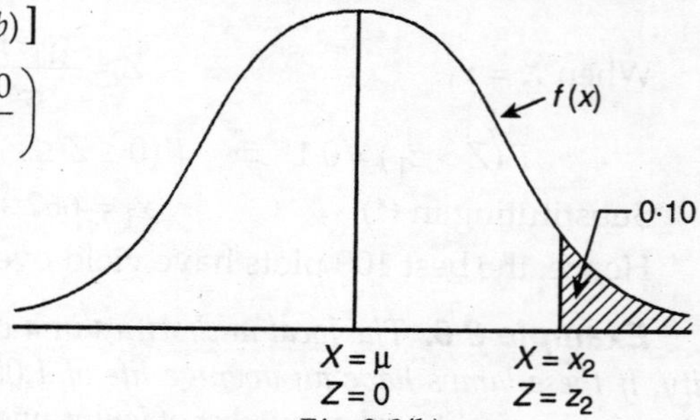

Fig. 9·8(b)

Hence, after 1,256 hours of burning life, 10% of the blubs will be still burning.

Example 9·7. *Prove that for the normal distribution, the quartile deviation, the mean deviation and standard deviation are approximately 10 : 12 : 15.*

Solution. Let $X \sim N(\mu, \sigma^2)$. If Q_1 and Q_3 are the first and third quartiles respectively, then by definition :

$$P(X < Q_1) = 0{\cdot}25 \text{ and } P(X > Q_3) = 0{\cdot}25$$

The points Q_1 and Q_3 are located as shown in the Fig. 9·9.

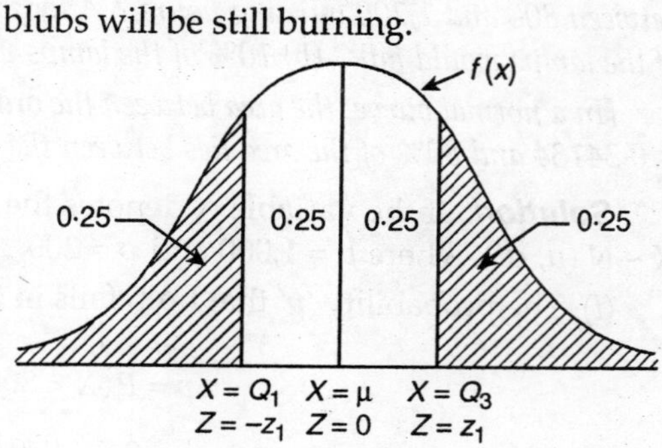

Fig. 9·9

When $X = Q_3$, $Z = \dfrac{Q_3 - \mu}{\sigma} = z_1$, (say) ... (*)

and when $X = Q_1$, $Z = \dfrac{Q_1 - \mu}{\sigma} = -z_1$ ... (**) (This is obvious form the figure.)

Subtracting, (**) from (*), we get $\dfrac{Q_3 - Q_1}{\sigma} = 2z_1$

The quartile deviation is given by : $Q.D. = \dfrac{Q_3 - Q_1}{2} = \sigma z_1$ ... (***)

From the figure, obviously $P(0 < Z < z_1) = 0{\cdot}25 \Rightarrow z_1 = 0{\cdot}67$ (From Normal Tables)

$\therefore$ $Q.D. = \sigma z_1 = 0{\cdot}67\,\sigma = \dfrac{2}{3}\sigma$ [From (***)]

For normal distribution, mean deviation about mean (*c.f.* § 9·2·10, page 9·13) is given by :

$$M.D. = \sqrt{\frac{2}{\pi}}\,\sigma \approx \frac{4}{5}\sigma$$

Hence, $Q.D. : M.D. : S.D. :: \dfrac{2}{3}\sigma : \dfrac{4}{5}\sigma : \sigma :: \dfrac{2}{3} : \dfrac{4}{5} : 1 :: 10 : 12 : 15.$

Remark. From (*) and (**), we get

$$Q_1 = \mu - \sigma z_1 = \mu - 0{\cdot}67\,\sigma \quad \text{and} \quad Q_3 = \mu + \sigma z_1 = \mu + 0{\cdot}67\sigma.$$

Example 9·8. *In a distribution exactly normal, 10·03% of the items are under 25 kilogram weight and 89·97% of the items are under 70 kilogram weight. What are the mean and standard deviation of the distribution ?*

Solution. Let X denote the weight (in kilograms) of the items. If $X \sim N(\mu, \sigma^2)$, then we are given :

$$P(X < 25) = 0{\cdot}1003 \quad \text{and} \quad P(X < 70) = 0{\cdot}8997$$

The points X = 25 and X = 70 are located as shown in Fig. 9·10 :

Since the value X = 25 is located to the left of the ordinate at X = μ, the corresponding value of Z is negative.

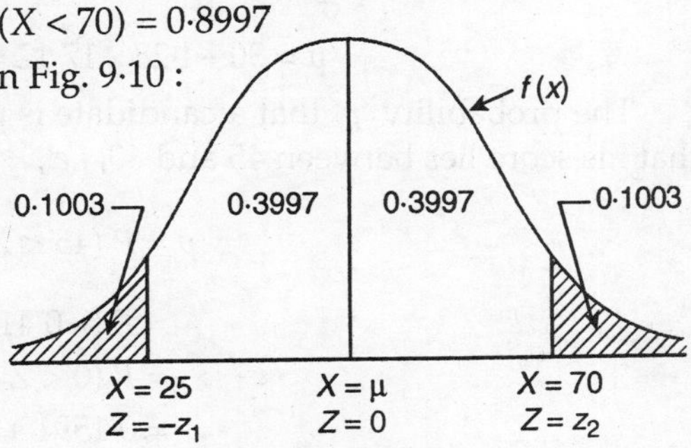

Fig. 9·10

When X = 25, $Z = \dfrac{25 - \mu}{\sigma} = -z_1$, (say) ... (1)

and when X = 70, $Z = \dfrac{70 - \mu}{\sigma} = z_2$ (say), ... (2)

From the diagram, it is obvious that

$$P(Z < -z_1) = 0{\cdot}1003 \quad \text{and} \quad P(Z < z_2) = 0{\cdot}8997$$

Now $P(0 < Z < z_2) = 0{\cdot}3997 \Rightarrow z_2 = 1{\cdot}28$ (From Normal Tables)

$P(Z < -z_1) = 0{\cdot}1003 \Rightarrow P(Z > z_1) = 0{\cdot}1003$ (By Symmetry)

$\therefore$ $P(0 < Z < z_1) = 0{\cdot}5 - 0{\cdot}1003 = 0{\cdot}3997 \Rightarrow z_1 = 1{\cdot}28$

Substituting the value of z_1 and z_2 in (1) and (2), we get

$\dfrac{25 - \mu}{\sigma} = -1{\cdot}28 \Rightarrow 25 - \mu = -1{\cdot}28\sigma$ ... (3)

and $\dfrac{70 - \mu}{\sigma} = 1{\cdot}28 \Rightarrow 70 - \mu = 1{\cdot}28\sigma$ ... (4)

Subtracting (3) from (4), we get $45 = 2{\cdot}56\sigma \Rightarrow \sigma = 17{\cdot}578.$

Substituting the value of σ in (3), we obtain $\mu = 25 + 17{\cdot}578 \times 1{\cdot}28 = 47{\cdot}5$

Hence, the mean is 47·5 kilogram and standard deviation is 17·578 kilogram.

Example 9·9. *In an examination it is laid down that a student passes if he secures 30 per cent or more marks. He is placed in the first, second or third division according as he secures 60% or more marks, between 45% and 60% marks and marks between 30% and 45% respectively. He gets distinction in case he secures 80% or more marks. It is noticed from the result that 10% of the students failed in the examination, whereas 5% of them obtained distinction. Calculate the percentage of students placed in the second division. (Assume normal distribution of marks.)*

Solution. Let variable X denote the marks (out of 100) in the examination and let $X \sim N(\mu, \sigma^2)$, then we are given $P(X < 30) = 0{\cdot}10$ and $P(X \geq 80) = 0{\cdot}05$. [See Fig. 9·11]

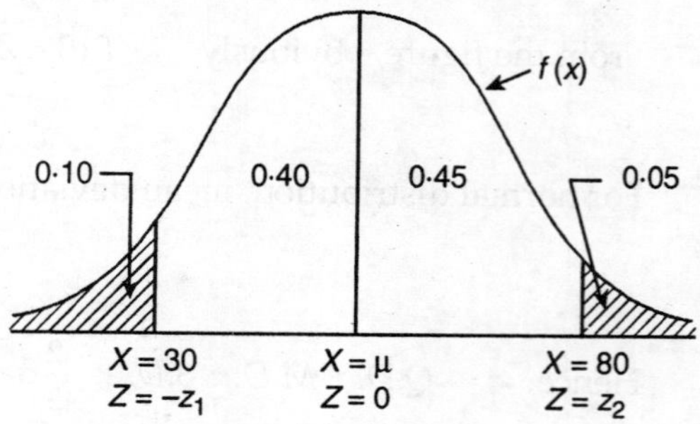

Fig. 9·11

$$\left.\begin{aligned} &\text{When } X = 30, \quad Z = \frac{30-\mu}{\sigma} = -z_1 \text{ (say)},\\ &\text{When } X = 80, \quad Z = \frac{80-\mu}{\sigma} = z_2 \text{ (say)}.\end{aligned}\right\} \quad \ldots (*)$$

$\therefore \quad P(0 < Z < z_2) = 0{\cdot}5 - 0{\cdot}05 = 0{\cdot}45$

and $\quad P(0 < Z < z_1) = P(-z_1 < Z < 0) = 0{\cdot}50 - 0{\cdot}10 = 0{\cdot}40$ (By symmetry)

$\therefore$ From Normal probability Tables : $z_1 = 1{\cdot}28$ and $z_2 = 1{\cdot}64$

Hence, $\dfrac{30-\mu}{\sigma} = -1{\cdot}28 \Rightarrow \dfrac{\mu - 30}{\sigma} = 1{\cdot}28$ and $\dfrac{80-\mu}{\sigma} = 1{\cdot}64$ ... [From(*)]

Adding, we get $\dfrac{50}{\sigma} = 2{\cdot}92 \Rightarrow \sigma = \dfrac{50}{2{\cdot}92} = 17{\cdot}12$

$\therefore \quad \mu = 30 + 1{\cdot}28 \times 17{\cdot}12 = 30 + 21{\cdot}9136 = 51{\cdot}9136 \simeq 52.$

The probability 'p' that a candidate is placed in the second division is equal to the probability that his score lies between 45 and 60, *i.e.*,

$$\begin{aligned} p = P(45 < X < 60) &= P(-0{\cdot}41 < Z < 0{\cdot}47), \qquad \left(Z = \frac{X-52}{17{\cdot}12}\right)\\ &= P(-0{\cdot}41 < Z < 0) + P(0 < Z < 0{\cdot}47)\\ &= P(0 < Z < 0{\cdot}41) + P(0 < Z < 0{\cdot}47) \qquad \text{(By symmetry)}\\ &= 0{\cdot}1591 + 0{\cdot}1808 = 0{\cdot}3399 = 0{\cdot}34 \text{ (approx.)}\end{aligned}$$

Hence, 34% candidates got second division in the examination.

Example 9·10. *If X, Y are independent normal variates with means 6, 7 and variances 9, 16 respectively, determine λ such that :* $P(2X + Y \leq \lambda) = P(4X - 3Y \geq 4\lambda)$.

Solution. Since X and Y are independent normal variates, using § 9·2·8

$$U = 2X + Y \sim N(2 \times 6 + 7, 4 \times 9 + 16), \quad \textit{i.e.,} \quad U \sim N(19, 52)$$

$$V = 4X - 3Y \sim N(4 \times 6 - 3 \times 7, 16 \times 9 + 9 \times 16), \quad \textit{i.e.,} \quad V \sim N(3, 288)$$

$$\therefore \quad P(2X + Y \leq \lambda) = P(U \leq \lambda) = P\left(Z \leq \frac{\lambda - 19}{\sqrt{52}}\right), \quad \text{where } Z \sim N(0, 1)$$

and $$P(4X - 3Y \geq 4\lambda) = P(V \geq 4\lambda) = P\left(Z \geq \frac{4\lambda - 3}{12\sqrt{2}}\right), \quad \text{where } Z \sim N(0, 1)$$

$$P(2X+Y \le \lambda) = P\{(4X-3Y) \ge 4\lambda\} \qquad \text{(Given)}$$

$$\Rightarrow \quad P\left(Z \le \frac{\lambda-19}{\sqrt{52}}\right) = P\left(Z \ge \frac{4\lambda-3}{12\sqrt{2}}\right) \quad \text{or} \quad \frac{\lambda-19}{\sqrt{52}} = -\left(\frac{4\lambda-3}{12\sqrt{2}}\right)$$

[$\because P(Z \le a) = P(Z \ge b) \Rightarrow a = -b$, since normal probability curve is symmetric about $Z = 0$.]

$$\Rightarrow \quad \frac{\lambda-19}{\sqrt{13}} = \frac{3-4\lambda}{6\sqrt{2}} \Rightarrow (6\sqrt{2}+4\sqrt{13})\lambda = 114\sqrt{2}+3\sqrt{13} \Rightarrow \lambda = \frac{114\sqrt{2}+3\sqrt{13}}{6\sqrt{2}+4\sqrt{13}}.$$

Example 9·11. *If two normal universes A and B have the same total frequency but the standard deviation of universe A is k times that of the universe B, show that maximum frequency of universe A is (1/k) times that of universe B.*

Solution. Let N be the same total frequency for each of the two universes A and B. If σ is the standard deviation of universe B, then the standard deviation of universe A is $k\sigma$. Let μ_1 and μ_2 be the means of the universes A and B, respectively.

The frequency function of universe A is : $f_A(x) = \dfrac{N}{k\sigma\sqrt{2\pi}} \exp\{-(x-\mu_1)^2/2k^2\sigma^2\}$

and the frequency function of universe B is : $f_B(x) = \dfrac{N}{\sigma\sqrt{2\pi}} \exp\{-(x-\mu_2)^2/2\sigma^2\}$

Since, for a normal distribution, the maximum frequency occurs at the point x = mean, we have

$$[f_A(x)]_{\max} = \text{Maximum frequency of universe } A$$

$$= [f_A(x)]_{x=\mu_1} = \left[\frac{N}{k\sigma\sqrt{2\pi}} \exp\{-(x-\mu_1)^2/2k^2\sigma^2\}\right]_{x=\mu_1} = \frac{N}{k\sigma\sqrt{2\pi}}$$

Similarly, $$[f_B(x)]_{\max} = [f_B(x)]_{x=\mu_2} = \left[\frac{N}{\sigma\sqrt{2\pi}} \exp\{-(x-\mu_2)^2/2\sigma^2\}\right]_{x=\mu_2} = \frac{N}{\sigma\sqrt{2\pi}}$$

$$\therefore \quad \frac{[f_A(x)]_{\max}}{[f_B(x)]_{\max}} = \frac{1}{k} \quad \Rightarrow \quad [f_A(x)]_{\max} = \frac{1}{k}\cdot[f_B(x)]_{\max}$$

Example 9·12. *Let X be a random variable following normal distribution with mean m and variance σ^2 and let r be a non-negative integer :*

If $\mu'_r = E(X^r)$ *and if* $\mu_{2r} = [E(X-m)^{2r}]$, *prove that*

(i) $\mu'_{r+2} = 2\mu\mu'_{r+1} + (\sigma^2-\mu^2)\mu'_r + \sigma^3 \dfrac{d\mu'_r}{d\sigma}$ *(ii)* $\mu_{2r+2} = \sigma^2\mu_{2r} + \sigma^3\dfrac{d\mu_{2r}}{d\sigma}$.

Solution. (*i*) $$\mu'_r = \int_{-\infty}^{\infty} x^r f(x)\,dx = \int_{-\infty}^{\infty} x^r \frac{1}{\sigma\sqrt{2\pi}} \exp\left[-\frac{(x-\mu)^2}{2\sigma^2}\right] dx$$

$$\therefore \quad \frac{d\mu'_r}{d\sigma} = \int_{-\infty}^{\infty} \frac{x^r}{\sqrt{2\pi}}\left(-\frac{1}{\sigma^2}\right) \exp\left[-\frac{(x-\mu)^2}{2\sigma^2}\right] dx$$

$$+ \int_{-\infty}^{\infty} \frac{x^r}{\sigma\sqrt{2\pi}}\left\{-\frac{(x-\mu)^2}{2}\times\left(\frac{-2}{\sigma^3}\right)\right\} \exp\left[-\frac{(x-\mu)^2}{2\sigma^2}\right] dx$$

$$= -\frac{1}{\sigma}\int_{-\infty}^{\infty} x^r \cdot f(x)\,dx + \frac{1}{\sigma^3}\int_{-\infty}^{\infty} x^r (x-\mu)^2 \cdot f(x)\,dx$$

$$= -\frac{1}{\sigma}\cdot \mu_r' + \frac{1}{\sigma^3}\int_{-\infty}^{\infty}\left(x^{r+2} - 2\mu x^{r+1} + \mu^2 x^r\right) f(x)\,dx$$

$$= -\frac{1}{\sigma}\cdot \mu_r' + \frac{1}{\sigma^3}\left(\mu_{r+2}' - 2\mu\mu_{r+1}' + \mu^2\mu_r'\right)$$

$$\Rightarrow \quad \sigma^3\frac{d\mu_r'}{d\sigma} = -\sigma^2\mu_r' + \mu_{r+2}' - 2\mu\mu_{r+1}' + \mu^2\mu_r'$$

$$\Rightarrow \quad \mu_{r+2}' = 2\mu\mu_{r+1}' + (\sigma^2 - \mu^2)\mu_r' + \sigma^3\cdot\frac{d\mu_r'}{d\sigma}.$$

(*ii*)
$$\mu_{2r} = \int_{-\infty}^{\infty}(x-\mu)^{2r}\cdot f(x)\,dx = \int_{-\infty}^{\infty}(x-\mu)^{2r}\frac{1}{\sigma\sqrt{2\pi}}\exp\left[-\frac{(x-\mu)^2}{2\sigma^2}\right]dx$$

$$\therefore \quad \frac{d\mu_{2r}}{d\sigma} = \int_{-\infty}^{\infty}\frac{(x-\mu)^{2r}}{\sqrt{2\pi}}\left(-\frac{1}{\sigma^2}\right)\cdot\exp\left[-\frac{(x-\mu)^2}{2\sigma^2}\right]dx$$

$$+\int_{-\infty}^{\infty}\frac{(x-\mu)^{2r}}{\sigma\sqrt{2\pi}}\cdot\exp\left[-\frac{(x-\mu)^2}{2\sigma^2}\right].\left\{\frac{(x-\mu)^2}{2}\left(\frac{-2}{\sigma^3}\right)\right\}dx$$

$$= -\frac{1}{\sigma}\int_{-\infty}^{\infty}(x-\mu)^{2r}f(x)\,dx + \frac{1}{\sigma^3}\int_{-\infty}^{\infty}(x-\mu)^{2r+2}\cdot f(x)\,dx$$

$$= -\frac{1}{\sigma}\mu_{2r} + \frac{1}{\sigma^3}\mu_{2r+2}$$

Hence,
$$\mu_{2r+2} = \sigma^2\mu_{2r} + \sigma^3\frac{d\mu_{2r}}{d\sigma}.$$

SUPPLEMENTARY EXAMPLES ON NORMAL DISTRIBUTION

Example 9·13. *The scores of an entrance examination are known to be normally distributed with a mean of 118 and standard deviation as 12.*

(i) If a candidate obtaining a score of at least 100 is declared successful, what is the probability that a candidate chosen at random will be successful ?

(ii) Obtain the lowest score of the top 25% of the candidates.

(iii) If it is decided to pass 70% of those taking the examination, what should be the lowest passing scores ?

Solution. If the *r.v.* X denotes the score of an entrance examination, then we are given that $X \sim N(\mu, \sigma^2)$, where $\mu = 118$ and $\sigma = 12$.

(*i*) The probability that a candidate will be successful is given by :

$$P(X \geq 100) = P(Z \geq -1{\cdot}5), \qquad \left[Z = \frac{100-118}{12} = -1{\cdot}5\right]$$

$$= P(-1{\cdot}5 \leq Z \leq 0) + 0{\cdot}5$$

$$= P(0 \leq Z \leq 1{\cdot}5) + 0{\cdot}5 \qquad \text{(By symmetry)}$$

$$= 0{\cdot}4332 + 0{\cdot}5 = 0{\cdot}9332.$$

(*ii*) Let x_1 be the lowest score of the top 25% of the candidates. Then we have to find x_1 such that

$$P(X \geq x_1) = 0{\cdot}25.$$

$$\Rightarrow \quad P(Z \geq z_1) = 0{\cdot}25, \qquad \left[Z = \frac{x_1 - 118}{12} = z_1, \text{ (say)} \right]$$

$$\Rightarrow \quad P(0 \leq Z \leq z_1) = 0{\cdot}5 - 0{\cdot}25 = 0{\cdot}25$$

$$\Rightarrow \quad z_1 = 0{\cdot}675 \qquad \text{(From normal Tables)}$$

$$\Rightarrow \quad \frac{x_1 - 118}{12} = 0{\cdot}675 \quad \Rightarrow \quad x_1 = 118 + 12 \times 0{\cdot}675 = 126{\cdot}1 \simeq 126$$

Thus, 126 is the lowest score of the top 25% of the candidates.

(*iii*) Let x_2 be the lowest passing scores, if it is decided to pass 70% of the candidates. Then we have to find x_2 such that :

$$P(X \geq x_2) = 0{\cdot}70 \quad \Rightarrow \quad P(Z \geq -z_2) = 0{\cdot}70, \qquad \left[Z = \frac{x_2 - 118}{12} = -z_2, \text{ (say)} \right]$$

$$\Rightarrow \quad P(-z_2 \leq Z \leq 0) + 0{\cdot}5 = 0{\cdot}70 \quad \Rightarrow \quad P(0 \leq Z \leq z_2) = 0{\cdot}20$$

$$\Rightarrow \quad z_2 = 0{\cdot}525 \qquad \text{(Using Normal probability Tables)}$$

$$\Rightarrow \quad \frac{x_2 - 118}{12} = -0{\cdot}525 \quad \Rightarrow \quad x_2 = 118 - 12 \times 0{\cdot}525 = 118 - 6{\cdot}3 = 111{\cdot}7 \simeq 112.$$

Example 9·14. *There are 600 students in the post-graduate department of a university, and the probability for any student to need a copy of a particular textbook from the university library on any day is 0·05. How many copies of the book should be kept in the university library so that the probability may be greater than 0·90 that none of the students needing a copy from the library has to come back disappointed. (Use normal approximation to the binomial probability distribution).*

Solution. Let n be the number of students and p the probability for any student to need a copy of particular textbook from the university library. We are given : $n = 600$, $p = 0{\cdot}05$,

Mean, $\mu = np = 600 \times 0{\cdot}05 = 30$ and *s.d.*, $\sigma = \sqrt{npq} = \sqrt{600 \times 0{\cdot}05 \times 0{\cdot}95} = \sqrt{28{\cdot}5} \simeq 3{\cdot}34.$

Let x_1 represent the number of copies of textbook required on any day. It is required to find x_1 such that

$$P(X < x_1) > 0{\cdot}90 \quad \Rightarrow \quad P(Z < z_1) > 0{\cdot}90 \qquad \left[z_1 = \frac{x_1 - 30}{5{\cdot}34} \right]$$

$$\Rightarrow \quad P(0 < Z < z_1) > 0{\cdot}40, \quad i.e., \quad z_1 > 1{\cdot}28 \qquad \text{[From Normal probability Tables]}$$

$$\Rightarrow \quad \frac{x_1 - 30}{5{\cdot}34} > 1{\cdot}28 \quad \Rightarrow \quad x_1 - 30 > 6{\cdot}835 \quad \Rightarrow \quad x_1 > 36{\cdot}835$$

Hence, the university library should keep at least 37 copies of the desired particular textbook.

Example 9·15. *The marks obtained by a number of students for a certain subject are assumed to be approximately normally distributed with mean value 65 and with a standard deviation of 5. If 3 students are taken at random from this set, what is the probability that exactly 2 of them will have marks over 70 ?*

Solution. Let the *r.v.* X denote the marks obtained by the given set of students in the given subject. Then we are given that $X \sim N(\mu, \sigma^2)$, where $\mu = 65$ and $\sigma = 5$. The probability 'p' that a randomly selected student from the given set gets marks over 70 is given by $p = P(X > 70)$.

When $X = 70$, $$Z = \frac{X - \mu}{\sigma} = \frac{70 - 65}{5} = 1.$$

$\therefore$ $$p = P(X > 70) = P(Z > 1) = 0{\cdot}5 - P(0 \le Z \le 1)$$
$$= 0{\cdot}5 - 0{\cdot}3413 = 0{\cdot}1587 \quad \text{[From Normal probability Tables]}$$

Since this probability is same for each student of the set, the required probability that 'out of 3 students selected at random from the set, exactly 2 will have marks over 70, is given by the binomial probability distribution :
$$^3C_2\, p^2 (1-p) = 3 \times (0{\cdot}1587)^2 \times (0{\cdot}8413) = 0{\cdot}06357.$$

Example 9·16. *(a) If $\log_{10} X$ is normally distributed with mean 4 and variance 4, find the probability of $1{\cdot}202 < X < 83180000$.*

(Given $\log_{10} 1202 = 3{\cdot}08$, $\log_{10} 8318 = 3{\cdot}92$).

(b) $\log_{10} X$ is normally distributed with mean 7 and variance 3, $\log_{10} Y$ is normally distributed with mean 3 and variance unity. If the distributions of X and Y are independent, find the probability of $1{\cdot}202 < (X/Y) < 83180000$.

[Given $\log_{10}(1202) = 3{\cdot}08$, $\log_{10}(8318) = 3{\cdot}92$]

Solution. (*a*) Since log X is a non-decreasing function of X, we have
$$P(1{\cdot}202 < X < 83180000) = P(\log_{10} 1{\cdot}202 < \log_{10} X < 7{\cdot}92) = P(0{\cdot}08 < Y < 7{\cdot}92),$$
where, $$Y = \log_{10} X \sim N(4, 4) \quad \text{(Given)}$$

When $Y = 0{\cdot}08$, $Z = \dfrac{0{\cdot}08 - 4}{2} = -1{\cdot}96$ and when $Y = 7{\cdot}92$, $Z = \dfrac{7{\cdot}92 - 4}{2} = 1{\cdot}96$

$\therefore$ Required probability $= P(0{\cdot}08 < Y < 7{\cdot}92)$
$$= P(-1{\cdot}96 < Z < 1{\cdot}96) = 2P(0 < Z < 1{\cdot}96) \quad \text{(By symmetry)}$$
$$= 2 \times 0{\cdot}4750 = 0{\cdot}9500.$$

(*b*) $$P[1{\cdot}202 < (X/Y) < 83180000]$$
$$= P[\log_{10} 1{\cdot}202 < \log_{10}(X/Y) < \log_{10} 83180000] = (0{\cdot}08 < U < 7{\cdot}92),$$
where, $$U = \log_{10}(X/Y) = \log_{10} X - \log_{10} Y.$$

Since $\log_{10} X \sim N(7, 3)$ and $\log_{10} Y \sim N(3, 1)$, are independent (Given)

$\therefore$ $$U = \log_{10} X - \log_{10} Y \sim N(7-3,\ 3+1) \quad \text{(c.f. Remark 1, § 9·2·8, page 9·12)}$$

$\Rightarrow$ $$U = (\log_{10} X - \log_{10} Y) \sim N(4, 4)$$

$\therefore$ Required probability is given by :
$$p = P(0{\cdot}08 < U < 7{\cdot}92), \text{ where } U \sim N(4, 4)$$
$$= 0{\cdot}95. \quad \text{[See part (a)]}$$

Example 9·17. *Two independent random variates X and Y are both normally distributed with means 1 and 2 and standard deviations 3 and 4 respectively. If $Z = X - Y$, write the probability density function of Z. Also state the median, s.d. and mean of the distribution of Z. Find Pro. $(Z + 1 \le 0)$.*

Solution. Since $X \sim N(1, 9)$ and $Y \sim N(2, 16)$ are independent,
$$Z = X - Y \sim N(1-2,\ 9+16), \quad i.e., \quad Z = X - Y \sim N(-1, 25).$$

Hence, *p.d.f.* of Z is given by : $$p(z) = \frac{1}{5\sqrt{2\pi}} \exp\left[-\frac{1}{2}\left(\frac{z+1}{5}\right)^2\right], \quad -\infty < z < \infty.$$

For the distribution of Z, Median = Mean = − 1 and *s.d.* $= \sqrt{25} = 5$.

and $$P(Z+1<0) = P(Z \le -1) = P(U \le 0) = 0{\cdot}5. \qquad \left[U = \frac{Z+1}{5} \sim N(0,1) \right]$$

Example 9·18. *If the skulls are classified as A, B and C according as the length-breadth index is under 75, between 75 and 80, or over 80, find approximately (assuming that the distribution is normal) the mean and standard deviation of a series in which A are 58%, B are 38% and C are 4%, being given that if*

$$f(t) = \frac{1}{\sqrt{2\pi}} \cdot \int_0^t \exp(-x^2/2)\, dx, \quad \textit{then} \quad f(0{\cdot}20) = 0{\cdot}08 \quad \textit{and} \quad f(1{\cdot}75) = 0{\cdot}46.$$

Solution. Let the length-breadth index be denoted by the variable X, then

$$P(X < 75) = 0{\cdot}58 \quad \text{and} \quad P(X > 80) = 0{\cdot}04 \quad \text{(Given)} \qquad \ldots (1)$$

Since $P(X < 75)$ represents the total area to the left of the ordinate at the point X = 75 and $P(X > 80)$ represents the total area to the right of the ordinate at the point X = 80, it is obvious from (1) that the points X = 75 and X = 80 are located at the positions shown in Fig. 9·12.

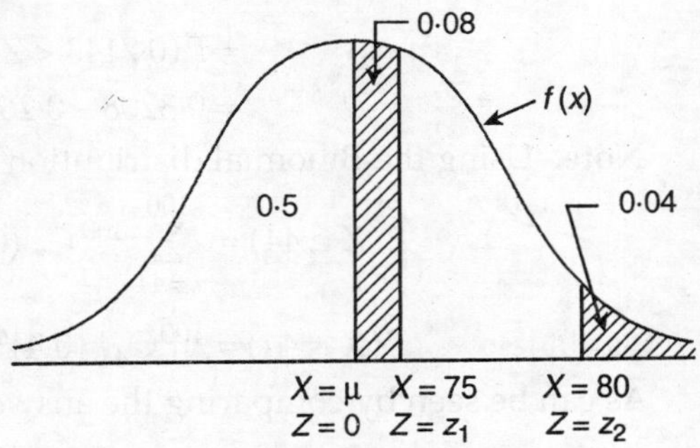

Fig. 9·12

$\frac{1}{\sqrt{2\pi}} \int_0^t \exp(-x^2/2)\, dx$ represents the area under standard normal curve between the ordinates at Z = 0 and Z = t, Z being N (0, 1) variate.

$$\therefore \quad f(t) = \frac{1}{\sqrt{2\pi}} \int_0^t \exp(-x^2/2)\, dx = P(0 < Z < t)$$

Hence, $$f(0{\cdot}20) = P(0 < Z < 0{\cdot}20) = 0{\cdot}08 \quad \text{and} \quad f(1{\cdot}75) = P(0 < Z < 1{\cdot}75) = 0{\cdot}46 \qquad \ldots (2)$$

Let μ and σ be the mean and *s.d.* of the distribution, respectively. Then $X \sim N(\mu, \sigma^2)$.

When X = 75, $$Z = \frac{75-\mu}{\sigma} = z_1 \text{ (say)}, \quad \text{and} \quad \text{when } X = 80,\ Z = \frac{80-\mu}{\sigma} = z_2 \text{ (say)}.$$

Then from the figure, it is obvious that : $P(X < 75) = 0{\cdot}58 \Rightarrow P(0 < Z < z_1) = 0{\cdot}08$

$\therefore$ Using (2), we have $$z_1 = \frac{75-\mu}{\sigma} = 0{\cdot}20 \qquad \ldots (3)$$

Also $$P(X > 80) = 0{\cdot}40 \quad \Rightarrow \quad P(0 < Z < z_2) = 0{\cdot}46$$

$\therefore$ From (2), we have $$z_2 = \frac{80-\mu}{\sigma} = 1{\cdot}75 \qquad \ldots (4)$$

Solving the equations (3) and (4), we obtain : $\mu = 74{\cdot}35$ (approx.) and $\sigma = 3{\cdot}23$ (approx.).

Example 9·19. *A sample of 100 items is taken at random from a batch known to contain 40% defectives. What is the probability that the sample contains :*

(i) at least 44 defectives, (ii) exactly 44 defectives ?

Solution. Since $n = 100$, is large, we may use the normal distribution as an approximation to the Binomial distribution.

Let X denote number of defectives with parameters,

$$\mu = np = 100(0{\cdot}4) = 40 \quad \text{and} \quad \sigma = \sqrt{npq} = \sqrt{100(0{\cdot}4)(0{\cdot}6)} = 4{\cdot}9$$

(*i*) It should be noted that the continuous normal distribution is approximating the discrete Binomial distribution so that the continuity correction has to be taken into account in determining the various probabilities. So finding the probability of at least 44 defectives in a sample of 100 items requires finding the area under the normal curve from 43·5 to 100·5.

∴ The probability of at least 44 defectives is :

$$P(43{\cdot}5 < X < 100{\cdot}5) = P\left(\frac{43{\cdot}5-40}{4{\cdot}9} < Z < \frac{100{\cdot}5-40}{4{\cdot}9}\right) = P(0{\cdot}7143 < Z < 12{\cdot}347)$$

$$= P(0 < Z < 12{\cdot}347) - P(0 < Z < 0{\cdot}7143) = 0{\cdot}5 - 0{\cdot}2624 = 0{\cdot}2376.$$

(*ii*) The probability of exactly 44 defectives is :

$$P(X = 44) = P(43{\cdot}5 < X < 44{\cdot}5) = P\left(\frac{43{\cdot}5-40}{4{\cdot}9} < Z < \frac{44{\cdot}5-40}{4{\cdot}9}\right)$$

$$= P(0{\cdot}7143 < Z < 0{\cdot}9184) = P(0 < Z < 0{\cdot}9184) - P(0 < Z < 0{\cdot}7143)$$

$$= 0{\cdot}3208 - 0{\cdot}2624 = 0{\cdot}0584.$$

Note. Using the Binomial distribution :

$$P(X \geq 44) = \sum_{r=44}^{100} {}^{100}C_r\, (0{\cdot}4)^r\, (0{\cdot}6)^{100-r} = 0{\cdot}2365 \text{ (using tables)}$$

$$P(X = 44) = {}^{100}C_{44}\, (0{\cdot}4)^{44}\, (0{\cdot}6)^{56} = 0{\cdot}0576.$$

As can be seen by comparing the answers, both sets of answers are remarkably close.

Example 9·20. *If X and Y are independent normal variates possessing a common mean μ such that*

$$P(2X + 4Y \leq 10) + P(3X + Y \leq 9) = 1 \quad \text{and} \quad P(2X - 4Y \leq 6) + P(Y - 3X \geq 1) = 1,$$

determine the values of μ and the ratio of the variances of X and Y.

Solution. Let $\text{Var}(X_1) = \sigma_1^2$ and $\text{Var}(Y) = \sigma_2^2$.

Since $E(X) = E(Y) = \mu$, (Given) and X & Y are independent normal variates, by § 9·2·8, we have

$$2X + 4Y \sim N(2\mu + 4\mu,\ 4\sigma_1^2 + 16\sigma_2^2), \quad \textit{i.e.,} \quad N(6\mu,\ 4\sigma_1^2 + 16\sigma_2^2)$$

$$3X + Y \sim N(3\mu + \mu,\ 9\sigma_1^2 + \sigma_2^2), \quad \textit{i.e.,} \quad N(4\mu,\ 9\sigma_1^2 + \sigma_2^2)$$

$$2X - 4Y \sim N(2\mu - 4\mu,\ 4\sigma_1^2 + 16\sigma_2^2), \quad \textit{i.e.,} \quad N(-2\mu,\ 4\sigma_1^2 + 16\,\sigma_2^2)$$

$$Y - 3X \sim N(\mu - 3\mu,\ \sigma_2^2 + 9\sigma_1^2), \quad \textit{i.e.,} \quad N(-2\mu,\ 9\sigma_1^2 + \sigma_2^2)$$

Let us further write : $4\sigma_1^2 + 16\sigma_2^2 = \alpha^2$ and $9\sigma_1^2 + \sigma_2^2 = \beta^2$ ... (1)

If Z denotes the standard normal variate, *i.e.*, if Z ~ (0, 1), we get

$$P(2X + 4Y \leq 10) + P(3X + Y \leq 9) = 1 \;\Rightarrow\; P\left(Z \leq \frac{10-6\mu}{\alpha}\right) + P\left(Z \leq \frac{9-4\mu}{\beta}\right) = 1$$

$$\Rightarrow \quad P\left(Z \leq \frac{10-6\mu}{\alpha}\right) = 1 - P\left(Z \leq \frac{9-4\mu}{\beta}\right) = P\left(Z \geq \frac{9-4\mu}{\beta}\right)$$

$$\therefore \quad \frac{10-6\mu}{\alpha} = -\left(\frac{9-4\mu}{\beta}\right)$$ (Since normal distribution is symmetric about Z = 0.) ... (2)

Similarly,

$$P(2X-4Y\le 6)+P(Y-3X\ge 1)=1 \;\Rightarrow\; P\left(Z\le\frac{6+2\mu}{\alpha}\right)+P\left(Z\ge\frac{1+2\mu}{\beta}\right)=1$$

$$\Rightarrow\quad P\left(Z\le\frac{6+2\mu}{\alpha}\right)=1-P\left(Z\ge\frac{1+2\mu}{\beta}\right)=P\left(Z\le\frac{1+2\mu}{\beta}\right)\;\Rightarrow\;\frac{6+2\mu}{\alpha}=\frac{1+2\mu}{\beta} \quad \text{... (3)}$$

Solving (2) and (3), we get $\dfrac{\alpha}{\beta}=\dfrac{6+2\mu}{1+2\mu}=\dfrac{10-6\mu}{4\mu-9}$ ... (4)

$$\Rightarrow\quad (6+2\mu)(4\mu-9)=(10-6\mu)(1+2\mu) \;\Rightarrow\; 5\mu^2-2\mu-16=0$$

$$\therefore\quad \mu=\frac{2\pm\sqrt{4+320}}{10}=\frac{2\pm 18}{10}=2 \;\text{ or }\; -1{\cdot}6$$

Substituting $\mu = 2$ in (4), $\dfrac{\alpha}{\beta}=\dfrac{10}{5}=2 \;\Rightarrow\; 4=\dfrac{\alpha^2}{\beta^2}$

From (1), we get $4=\dfrac{4\sigma_1^2+16\sigma_2^2}{9\sigma_1^2+\sigma_2^2}=\dfrac{4+16\lambda}{9+\lambda}$ $\left[\text{Taking } \lambda=\dfrac{\sigma_2^2}{\sigma_1^2}\right]$

$$\Rightarrow\quad 4(9+\lambda)=4+16\lambda \;\Rightarrow\; \lambda=\frac{32}{12}=\frac{8}{3}$$

Putting $\mu = -1{\cdot}6$ in (4), we obtain $\left(\dfrac{14}{11}\right)^2=\dfrac{\alpha^2}{\beta^2}=\dfrac{4+16\lambda}{9+\lambda} \;\Rightarrow\; \lambda=\dfrac{1280}{1740}=\dfrac{64}{87}.$

9·2·14. Log-normal Distribution

The positive *r.v.* X is said to have a log-normal distribution if $\log_e X$ is normally distributed.

Let $Y=\log_e X \sim N(\mu, \sigma^2)$.

For $x > 0$, $F_X(x)=P(X\le x)=P(\log_e X\le \log_e x)=P(Y\le \log_e x)$

(Since log X is monotonic increasing function.)

$$=\frac{1}{\sigma\sqrt{2\pi}}\int_{-\infty}^{\log x}\exp\{-(y-\mu)^2/2\sigma^2\}\,dy \qquad [\text{Since } Y\sim N(\mu,\sigma^2)]$$

$$=\frac{1}{\sigma\sqrt{2\pi}}\int_{0}^{x}\exp\{-(\log u-\mu)^2/2\sigma^2\}\,\frac{du}{u}, \qquad (y=\log u)$$

For $x \le 0$, $F_X(x) = P(X \le x) = 0$, because X is a positive *r.v.*

Let us define

$$f_X(u)=\begin{cases}\dfrac{1}{u\sigma\sqrt{2\pi}}\cdot\exp\{-(\log u-\mu)^2/2\sigma^2\},\ u>0\\ 0,\ u\le 0\end{cases} \quad \text{... (9·17)}$$

Then, $F_X(x)=\displaystyle\int_{-\infty}^{x} f_X(u)\,du$, for every x and hence $f(x)$ defined in (9·17) is a *p.d.f.* of X.

Remark. If $X \sim N(\mu, \sigma^2)$, then $Y = e^X$, is called a log-normal *r.v.*, since its logarithm $\log Y = X$, is a normal *r.v.*

Moments. The rth moment about origin is given by :

$$\mu'_r = E(X^r) = E(e^{rY}) \qquad \left[\because Y = \log X \Rightarrow X = e^Y\right]$$

$$= M_Y(r) \qquad \textit{(m.g.f. of Y, r being the parameter)}$$

$$= \exp\left(\mu r + \frac{1}{2} r^2 \sigma^2\right) \qquad \left[\because Y \sim N(\mu, \sigma^2)\right] \qquad \text{... (9·18)}$$

Remarks :

1. In particular if we take $\mu = \log \alpha, \alpha > 0,$ *i.e.,* $\log X \sim N(\log \alpha, \sigma^2)$, then

$$\mu'_r = E(X^r) = \exp\left\{r \cdot \log \alpha + \frac{1}{2} r^2 \sigma^2\right\} = \alpha^r \cdot \exp\left\{r^2 \sigma^2 / 2\right\} \qquad \text{... (9·18a)}$$

$$\therefore \qquad \text{Mean} = \mu'_1 = \alpha e^{\sigma^2/2} \quad \text{and} \quad \mu_2 = \mu'_2 - \mu_1'^2 = \alpha^2 e^{\sigma^2}\left(e^{\sigma^2} - 1\right)$$

2. Log normal distribution arises in problems of economics, biology, geology, and reliability theory. In particular, it arises in the study of dimensions of particles under pulverisation.

3. If $X_1, X_2, ..., X_n$ is a set of independently identically distributed random variables such that mean of each $\log X_i$ is μ and its variance is σ^2, then the product $X_1 \cdot X_2 \cdot ... \cdot X_n$ is asymptotically distributed according to logarithmic normal distribution with mean μ and variance $n\sigma^2$.

9·3. RECTANGULAR (OR UNIFORM) DISTRIBUTION

Definition. *A random variable X is said to have a continuous rectangular (uniform) distribution over an interval (a, b), i.e., (− ∞ < a < b < ∞), if its p.d.f. is given by :*

$$f(x; a, b) = \begin{cases} \dfrac{1}{b-a}, & \text{if } a < x < b \\ 0, & \text{otherwise} \end{cases} \qquad \text{... (9·19)}$$

Remarks :

1. a and b, $(a < b)$ are the two parameters of the distribution. The distributin is called uniform distribution on (a, b) since it assumes a constant (uniform) value for all x in (a, b).

2. The distribution is also known as rectangular distribution, since the curve $y = f(x)$ describes a rectangle over the x-axis and between the ordinates at $x = a$ and $x = b$.

3. A uniform or rectangular variate X on the interval (a, b) is written as : $X \sim U[a, b]$ or $X \sim R[a, b]$.

4. The cumulative distribution function $F(x)$ is given by :

$$F(x) = \begin{cases} 0 & , \quad x \le a \\ \dfrac{x-a}{b-a} & , \quad a < x < b \\ 1 & , \quad x \ge b \end{cases} \qquad \text{... (9·19a)}$$

Since $F(x)$ is not continuous at $x = a$ and $x = b$, it is not differentiable at these points. Thus,

$$\frac{d}{dx} F(x) = f(x) = \frac{1}{b-a} \ne 0, \text{ exists everywhere except at the points } x = a \text{ and } x = b,$$

and consequently *p.d.f.* $f(x)$ is given by (9·19).

5. The graphs of uniform *p.d.f.* $f(x)$ and the corresponding distribution function $F(x)$ are given in Fig. 9·13(*a*) and Fig. 9·13(*b*) on page 9·31.

6. For a rectangular or uniform variate X in $(-a, a)$, the *p.d.f.* is given by :

$$f(x) = \begin{cases} \dfrac{1}{2a}, & -a < x < a \\ 0, & \text{otherwise} \end{cases} \qquad \text{... (9·19b)}$$

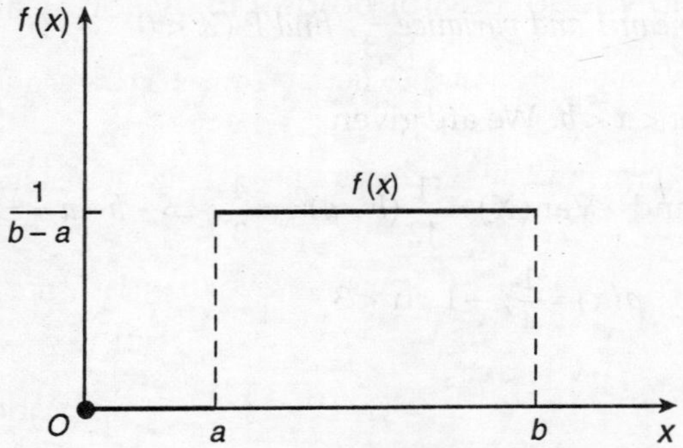

Fig. 9·13(a)

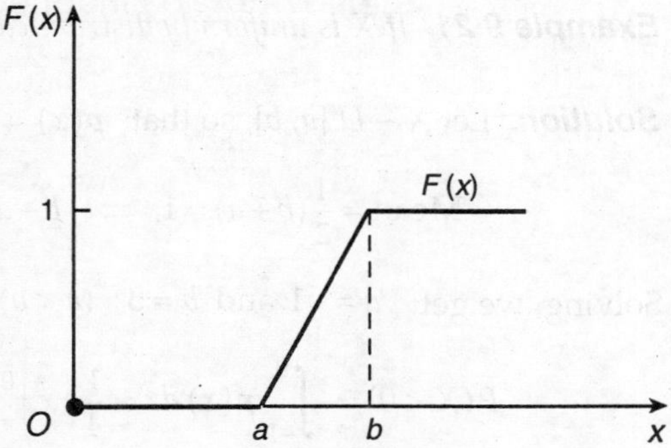

Fig. 9·13(b)

9·3·1. Moments of Rectangular Distribution

Let $X \sim U[a, b]$.

$$\mu_r' = \int_a^b x^r f(x)\,dx = \frac{1}{b-a}\int_a^b x^r\,dx = \frac{1}{b-a}\left(\frac{b^{r+1}-a^{r+1}}{r+1}\right) \qquad \ldots (9\cdot20)$$

In particular, $\text{Mean} = \mu_1' = \dfrac{1}{b-a}\left(\dfrac{b^2-a^2}{2}\right) = \dfrac{b+a}{2}$ ... (9·20*a*)

and $\mu_2' = \dfrac{1}{b-a}\left(\dfrac{b^3-a^3}{3}\right) = \dfrac{1}{3}(b^2+ab+a^2)$

$$\therefore \quad \text{Variance} = \mu_2' - \mu_1'^2 = \frac{1}{3}(b^2+ab+a^2) - \left\{\frac{1}{2}(b+a)\right\}^2 = \frac{1}{12}(b-a)^2 \qquad \ldots (9\cdot20b)$$

9·3·2. M.G.F. of Rectangular Distribution

M.G.F. of Rectangular Distribution is given by :

$$M_X(t) = \int_a^b e^{tx} f(x)\,dx = \int_a^b \frac{e^{tx}}{b-a}\,dx = \frac{e^{bt}-e^{at}}{t(b-a)},\ t \neq 0 \qquad \ldots (9\cdot20c)$$

9·3·3. Characteristic Function of Rectangular Distribution

Characteristic Function of Rectangular Distribution is given by :

$$\phi_X(t) = \int_a^b e^{itx}\,dx = \frac{e^{ibt}-e^{iat}}{it(b-a)},\ t \neq 0. \qquad \ldots (9\cdot20d)$$

9·3·4. Mean Deviation about Mean

M.D. about mean (η) of Rectangular Distribution is given by :

$$\eta = E\,|\,X - \text{Mean}\,| = \int_a^b |\,x - \text{Mean}\,|\,f(x)\,dx$$

$$= \frac{1}{b-a}\int_a^b \left|\,x - \frac{a+b}{2}\,\right| dx = \frac{1}{b-a}\int_{-(b-a)/2}^{(b-a)/2} |\,t\,|\,dt, \text{ where } t = x - \frac{a+b}{2}.$$

$$= \frac{1}{b-a}\cdot 2\int_0^{(b-a)/2} t\,dt = \frac{b-a}{4} \qquad \ldots (9\cdot20e)$$

Example 9·21. *If X is uniformly distributed with mean 1 and variance $\frac{4}{3}$, find P (X < 0).*

Solution. Let $X \sim U[a, b]$, so that $p(x) = \dfrac{1}{b-a},\ a < x < b$. We are given :

$$\text{Mean} = \frac{1}{2}(b+a) = 1 \Rightarrow b+a = 2 \quad \text{and} \quad \text{Var}(X) = \frac{1}{12}(b-a)^2 = \frac{4}{3} \Rightarrow b-a = \pm 4.$$

Solving, we get $a = -1$ and $b = 3$; $(a < b)$. $\therefore\ p(x) = \frac{1}{4};\ -1 < x < 3$

$$P(X < 0) = \int_{-1}^{0} p(x)\,dx = \frac{1}{4}\left| x \right|_{-1}^{0} = \frac{1}{4}.$$

Example 9·22. *Subway trains on a certain line run every half hour between mid-night and six in the morning. What is the probability that a man entering the station at a random time during this period will have to wait at least twenty minutes ?*

Solution. Let the *r.v.* X denote the waiting time (in minutes) for the next train. Under the assumption that a man arrives at the station at random, X is distributed uniformly on (0, 30), with *p.d.f.*,

$$f(x) = \begin{cases} \dfrac{1}{30}, & 0 < x < 30 \\ 0, & \text{otherwise} \end{cases}$$

The probability that he has to wait at least 20 minutes is given by :

$$P(X \geq 20) = \int_{20}^{30} f(x)\,dx = \frac{1}{30}\int_{20}^{30} 1 \cdot dx = \frac{1}{30}(30-20) = \frac{1}{3}.$$

Example 9·23. *Show that for rectangular distribution : $f(x) = \dfrac{1}{2a},\ -a < x < a$, m.g.f. about origin is $\dfrac{\sinh(at)}{at}$. Also show that moments of even order are given by $\mu_{2n} = \dfrac{a^{2n}}{(2n+1)}$.*

Solution. M.G.F. about origin is given by :

$$M_X(t) = E(e^{tX}) = \int_{-a}^{a} e^{tx} f(x)\,dx = \frac{1}{2a}\int_{-a}^{a} e^{tx}\,dx = \frac{1}{2a}\left|\frac{e^{tx}}{t}\right|_{-a}^{a} = \frac{1}{2at}(e^{at} - e^{-at}) = \frac{\sinh(at)}{at}$$

$$= \frac{1}{at}\left\{ at + \frac{(at)^3}{3!} + \frac{(at)^5}{5!} + \ldots + \frac{(at)^{2n+1}}{(2n+1)!} + \ldots \right\} = 1 + \frac{a^2t^2}{3!} + \frac{a^4t^4}{5!} + \ldots + \frac{a^{2n}t^{2n}}{(2n+1)!} + \ldots$$

Since there are no terms with odd powers of t in $M(t)$, all moments of odd order about origin vanish, *i.e.*, μ'_{2n+1} (about origin) = 0.

In particular, μ'_1 (about origin) = 0, *i.e.*, mean = 0

Thus μ'_r (about origin) = μ_r (Since mean is origin.)

Hence, $\mu_{2n+1} = 0;\ n = 0, 1, 2, \ldots$, *i.e.*, all moments of odd order about mean vanish. The moments of even order are given by :

$$\mu_{2n} = \text{coefficient of } \frac{t^{2n}}{(2n)!} \text{ in } M(t) = \frac{a^{2n}}{(2n+1)}.$$

Example 9·24. *If X is a random variable with a continuous distribution function F, then prove that F (X) has a uniform distribution on [0, 1].*

Hence, prove that : $P[a \le F(x) \le b] = b - a,\ 0 \le (a, b) \le 1.$

Solution. Since F is a distribution function, it is non-decreasing. Let $Y = F(X)$. Then, the distribution function G of Y is given by :

$$G_Y(y) = P(Y \le y) = P[F(X) \le y] = P[X \le F^{-1}(y)],$$

the inverse exists, since F is non-decreasing and given to be continuous.

$\therefore$ $G_Y(y) = F[F^{-1}(y)]$, since F is the distribution function of X.

Thus, $G_Y(y) = y$

Therefore, the *p.d.f.* of $Y = F(X)$ is given by : $g_Y(y) = \dfrac{d}{dy}[G_Y(y)] = 1$

Since F is a distribution function, $Y = F(X)$ takes the values in the range [0, 1].

Hence, $g_Y(y) = 1,\ 0 \le y \le 1 \quad \Rightarrow \quad Y$ is a uniform variate on [0, 1].

Since $Y = F(X) \sim U[0, 1]$,

$$P[a \le F(X) \le b] = P[a \le Y \le b] = \int_a^b g(y)\, dy$$

$$= \int_a^b 1 \cdot dy = \left| y \right|_a^b = b - a.$$

Remark. Suppose X is a random variable with *p.d.f.*,

$$f_X(x) = \begin{cases} e^{-x}, & x \ge 0 \\ 0, & \text{otherwise} \end{cases}, \quad \text{then } F(x) = \begin{cases} 0, & \text{if } x < 0 \\ 1 - e^{-x}, & \text{if } x \ge 0 \end{cases}$$

Then, by above result, $F(X) = 1 - e^{-X}$, is uniformly distributed on [0, 1].

Example 9·25. *If p and q are independent variates uniformly distributed over the interval (–1, 1), find the probability that the equation* $x^2 + 2px + q = 0$, *has real roots.*

Solution. Since $p \sim U[-1, 1]$ and $q \sim U[-1, 1]$ are independent, their joint *p.d.f.* is :

$$f(p, q) = f_1(p) \cdot f_2(q) = \frac{1}{2} \times \frac{1}{2} = \frac{1}{4};\quad 0 < (p, q) < 1. \qquad \text{... (*)}$$

The equation, $x^2 + 2px + q = 0$, has real roots if its discriminant (D) is non-negative, *i.e.*, if

$$D = b^2 - 4ac = (2p)^2 - 4 \cdot 1 \cdot q = 4(p^2 - q) \ge 0, \quad i.e., \ \text{if } p^2 \ge q.$$

The required probability (P) is given by :

$$P = P_r(p^2 \ge q) = P_r(q \le p^2) = \int_{p=-1}^{1} \left(\int_{q=-1}^{p^2} f(p, q)\, dq \right) dp$$

$$= \frac{1}{4} \int_{p=-1}^{1} \left(\int_{q=-1}^{p^2} 1 \cdot dq \right) dp = \frac{1}{4} \int_{p=-1}^{1} (p^2 + 1)\, dp = \frac{1}{4} \left| \frac{p^3}{3} + p \right|_{-1}^{1}$$

$$= \frac{2}{4}\left(\frac{1}{3} + 1\right) = \frac{2}{3}.$$

Example 9·26. *If X and Y are independent rectangular variates for the range – a to a each, then show that the sum X + Y = U, has the probability density function :*

$$\varphi(u) = \begin{cases} \dfrac{2a+u}{4a^2}, & -2a \le u \le 0 \\ \dfrac{2a-u}{4a^2}, & 0 \le u \le 2a \end{cases}$$

Solution. Since X & Y are independent rectangular variates, each in the interval $(-a, a)$, we have

$$f_1(x) = \begin{cases} \dfrac{1}{2a}, & -a < x < a \\ 0, & \text{elsewhere} \end{cases} \quad \text{and} \quad f_2(y) = \begin{cases} \dfrac{1}{2a}, & -a < y < a \\ 0, & \text{elsewhere} \end{cases}$$

Hence, by compound probability theorem, the joint probability differential of X & Y is given by :

$$dP(x, y) = f_1(x) \cdot f_2(y)\, dx\, dy = \frac{1}{4a^2}\, dx\, dy, \quad -a < (x, y) < a$$

Let us define new variables U and V as follows :

$$u = x + y, \quad v = x - y \quad \Rightarrow \quad x = \frac{u+v}{2} \quad \text{and} \quad y = \frac{u-v}{2}$$

Jacobian of the transformation J is given by :

$$J = \frac{\partial(x, y)}{\partial(u, v)} = \begin{vmatrix} \dfrac{\partial x}{\partial u} & \dfrac{\partial x}{\partial v} \\ \dfrac{\partial y}{\partial u} & \dfrac{\partial y}{\partial v} \end{vmatrix} = \begin{vmatrix} \dfrac{1}{2} & \dfrac{1}{2} \\ \dfrac{1}{2} & -\dfrac{1}{2} \end{vmatrix} = -\frac{1}{2}$$

Thus the probability differential of U and V becomes :

$$dG(u, v) = \frac{1}{4a^2} |J|\, du\, dv = \frac{1}{8a^2}\, du\, dv \qquad \text{... (*)}$$

Integrating *w.r.t.* to v over specified range, we can find the distribution of U.

Let us consider the region to the left of v–axis, *i.e.*, to the left of the line AC in Fig. 9·14. In this region, the values of v are bounded by the lines $x = -a$ and $y = -a$.

For fixed values of u,

$$x = -a \;\Rightarrow\; \frac{u+v}{2} = -a \;\Rightarrow\; v = -(u+2a)$$

and

$$y = -a \;\Rightarrow\; \frac{u-v}{2} = -a \;\Rightarrow\; v = (v+2a)$$

Thus integrating (*) *w.r.t.* to v between the limits $-(u + 2a)$ and $(u + 2a)$, the distribution of U becomes :

$$\varphi(u)\, du = \int_{-(u+2a)}^{u+2a} \frac{1}{8a^2}\, du\, dv$$

$$= \frac{1}{8a^2} \Big|\, v \,\Big|_{-(u+2a)}^{u+2a} du = \frac{u+2a}{4a^2}\, du$$

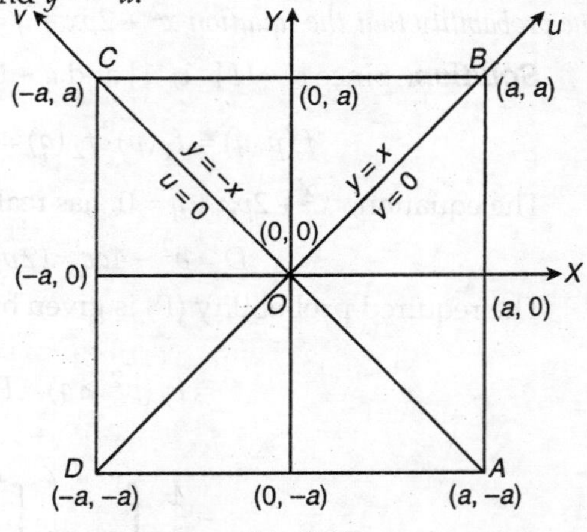

Fig. 9·14

In region to the left of v-axis, *i.e.*, below the line AC, u varies from the points $(x = -a, y = -a)$ to the point $(x = 0, y = 0)$ and since $u = x + y$, in this region u lies between $(-a - a)$ and $(0 + 0)$, *i.e.*, between $-2a$ to 0.

$$\therefore \qquad \varphi(u) = \frac{u+2a}{4a^2}, \; -2a \le u \le 0 \qquad \text{... (**)}$$

In region to the right of v-axis, *i.e.*, above the line AC, the values of v are bounded by the lines $x = a$ and $y = a$ and for fixed values of u,

$$x = a \quad \Rightarrow \quad \frac{u+v}{2} = a \quad \Rightarrow \quad v = 2a - u$$

$$y = a \quad \Rightarrow \quad \frac{u-v}{2} = a \quad \Rightarrow \quad v = -(2a-u)$$

In this region u varies from the point $(x = 0, y = 0)$ to the point $(x = a, y = a)$, *i.e.*, $u = x + y$ varies from 0 to $2a$. Thus integrating (*) *w.r.t.* to v between the limits $-(2a-u)$ to $(2a-u)$, we get the distribution of U as :

$$\varphi(u)\, du = \int_{-(2a-u)}^{2a-u} \frac{1}{8a^2}\, du\, dv = \frac{1}{8a^2} \Big| \; v \; \Big|_{-(2a-u)}^{2a-u} du = \frac{2a-u}{4a^2}\, du, \; 0 \le u \le 2a \qquad \text{... (***)}$$

Combining the results in (**) and (***), we get the required result.

Aliter. Example 9·27

9·4. TRIANGULAR DISTRIBUTION

Definition. *A random variable X is said to have a triangular distribution in the interval (a, b), if its p.d.f. is given by :*

$$f(x) = \begin{cases} \dfrac{2(x-a)}{(b-a)(c-a)}, & a < x \le c \\[2ex] \dfrac{2(b-x)}{(b-a)(b-c)}, & c < x < b \end{cases} \qquad \text{... (9·21)}$$

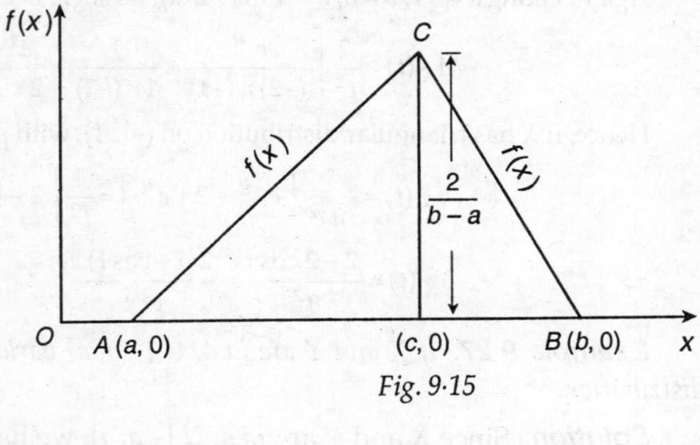

Fig. 9·15

Remarks :

1. We write $X \sim$ Trg. (a, b), with peak at $x = c$. The graph of the *p.d.f.* is shown in Fig. 9·15.

2. The distribution is so called because the graph of its *p.d.f.* is a triangle with peak at $x = c$.

3. The *m.g.f.* of Trg. (a, b) variate, with peak at $x = c$ is given by :

$$M_X(t) = \int_a^b e^{tx} f(x)\, dx = \left(\int_a^c + \int_c^b \right) e^{tx} f(x)\, dx$$

$$= \frac{2}{(b-a)(c-a)} \int_a^c e^{tx}(x-a)\, dx + \frac{2}{(b-a)(b-c)} \int_c^b e^{tx}(b-x)\, dx$$

$$= \frac{2}{t^2} \left\{ \frac{e^{at}}{(a-b)(a-c)} + \frac{e^{ct}}{(c-a)(c-b)} + \frac{e^{bt}}{(b-a)(b-c)} \right\}, \; a < c < b \qquad \text{(On integration by parts)} \qquad \text{... (9·21a)}$$

4. In particular, taking $a = 0$, $c = 1$ and $b = 2$, in (9·21), the *p.d.f.* of the Trg. (0, 2) variate with peak at $x = 1$ is given by :

$$f(x) = \begin{cases} x; & 0 < x \le 1 \\ 2 - x; & 1 \le x \le 2 \\ 0, & \text{otherwise} \end{cases} \qquad \text{... (9·21}b\text{)}$$

and its *m.g.f.* is : $M_X(t) = (e^t - 1)^2 / t^2$, [From (9·21*a*)] ... (9·21*c*)

which is left as an exercise to the reader.

5. In particular, replacing a by $-2a$, b by $2a$ and c by 0, the *p.d.f.* of triangular distribution on the interval $(-2a, 2a)$ with peak at $x = 0$ is given by :

$$f(x) = \begin{cases} (2a + x)/4a^2; & -2a < x \le 0 \\ (2a - x)/4a^2; & 0 < x < 2a \end{cases} \qquad \text{... (9·21}d\text{)}$$

The *m.g.f.* of (9·21*d*) is given by :

$$M_X(t) = \int_{-2a}^{2a} e^{tx} f(x)\, dx = \frac{1}{4a^2}\left\{ \int_{-2a}^{0} e^{tx}(2a + x)\, dx + \int_{0}^{2a} e^{tx}(2a - x)\, dx \right\}$$

$$= \frac{1}{4a^2}\left[e^{tx}\left\{ \frac{2a + x}{t} - \frac{1}{t^2} \right\} \right]_{-2a}^{0} + \frac{1}{4a^2}\left[e^{tx}\left\{ \frac{2a - x}{t} - \frac{1}{t^2} \right\} \right]_{0}^{2a} \quad \text{[On integrating by parts]}$$

$$= \frac{1}{4a^2}\left[-\frac{2}{t^2} + \frac{1}{t^2}\left\{ e^{2at} + e^{-2at} \right\} \right] = \frac{1}{4a^2 t^2}\left\{ e^{2at} + e^{-2at} - 2 \right\} = \left\{ \frac{1}{2at}\left(e^{at} - e^{-at} \right) \right\}^2$$

6. Triangular Distribution on (–1, 1). Taking $a = -1$, $c = 0$ and $b = 1$ in (9·21), the *p.d.f.* of triangular distribution on (–1, 1) with peak at $x = 0$ is given by :

$$f(x) = \begin{cases} (x + 1); & -1 < x \le 0 \\ (1 - x); & 0 < x < 1 \end{cases} \quad \Rightarrow \quad f(x) = \begin{cases} 1 - |x|, & -1 < x < 1 \\ 0, & \text{otherwise} \end{cases} \qquad \text{... (9·21}e\text{)}$$

Again, taking $a = -1$, $c = 0$, $b = 1$ in (9·21*a*), its *m.g.f.* is given by :

$$M_X(t) = \frac{2}{t^2}\left[\frac{e^{-t}}{(-2) \times (-1)} + \frac{1}{1 \times (-1)} + \frac{e^t}{2 \times 1} \right] \quad \Rightarrow \quad M_X(t) = \frac{1}{t^2}\left[e^{-t} - 2 + e^t \right] \qquad \text{... (9·21}f\text{)}$$

Hence, if X has triangular distribution on (–1, 1), with peak at $x = 0$, its characteristic function is given by :

$$\phi_X(t) = \frac{1}{(it)^2}\left[e^{-it} - 2 + e^{it} \right] = \frac{1}{t^2}\left[2 - \left(e^{it} + e^{-it} \right) \right]$$

$$\Rightarrow \qquad \phi_X(t) = \frac{2 - 2\cos t}{t^2} = \frac{2(1 - \cos t)}{t^2} \qquad \text{... (9·21}g\text{)}$$

Example 9·27. *If X and Y are i.i.d. U [– a, a] variates, find the p.d.f. of Z = X + Y and identify the distribution.*

Solution. Since X and Y are *i.i.d.* $U[-a. a]$, we have : [*c.f.* § 9·3·2],

$$M_X(t) = M_Y(t) = \left(e^{at} - e^{-at} \right) / (2at) \qquad \text{... (*)}$$

$$M_{X+Y}(t) = M_X(t)\, M_Y(t) = \left\{ \frac{1}{2at}\left(e^{at} - e^{-at} \right) \right\}^2, \text{ since } X \text{ and } Y \text{ are independent.}$$

But, this is the *m.g.f.* of Trg $(-2a, 2a)$ variate with peak at $x = 0$. [*c.f.* Remark 5]

Hence, by uniqueness theorem of *m.g.f.*, $Z = X + Y \sim$ Trg $(-2a, 2a)$ with peak at $x = 0$ and *p.d.f.* as given in (9·21*d*), Remark 5.

Aliter. $$M_{X+Y}(t) = \frac{1}{4a^2t^2}\left(e^{2at} - 2 + e^{-2at}\right) \qquad \text{[From (**)]}$$

$$= \frac{2}{t^2}\left[\frac{e^{-2at}}{(-2a-0)(-2a-2a)} + \frac{e^{0\cdot t}}{(0+2a)(0-2a)} + \frac{e^{2at}}{(2a-0)(2a+2a)}\right]$$

which is of the form (9·21*a*), [*c.f.* Remark 3], with *a* replaced by – 2*a* and *b* replaced by 2*a* and *c* by 0. Hence $X + Y \sim \text{Trg}(-2a, 2a)$ with peak at $x = 0$ and *p.d.f.* $p(x)$ given in (9·21*d*).

Remarks :

1. The distribution of X + Y has also been obtained in *Example 9·26*.
2. Similarly, we can find the distribution of X – Y.

$$M_{X-Y}(t) = M_X(t)\cdot M_Y(-t) = \left\{\frac{1}{2at}\left(e^{at} - e^{-at}\right)\right\}^2 \qquad \text{[From (*)]}$$

$$\Rightarrow \qquad X - Y \sim \text{Trg}(-2a, 2a), \text{ with peak at } x = 0.$$

Example 9·28. If X and Y are *i.i.d.* *U* [0, 1] *r.v.'s*, then obtain the *p.d.f.* of X – Y.

Solution. Since X and Y are identically distributed *U* [0, 1], *r.v.'s*,

$$\phi_X(t) = \phi_Y(t) = \left(\frac{e^{it} - 1}{it}\right) \qquad \text{... (*)}$$

$$\left[\phi_X(t) = \int_{-\infty}^{\infty} e^{itx}\cdot f(x)\,dx = \int_0^1 e^{itx}\,dx = \left|\frac{e^{itx}}{it}\right|_0^1 = \left(\frac{e^{it} - 1}{it}\right)\right]$$

The characteristic function of $U = (X - Y)$ is given by :

$$\phi_U(t) = \phi_{X-Y}(t) = \phi_{X+(-Y)}(t) = \phi_X(t)\cdot\phi_{(-Y)}(t) \qquad [\because \text{X and Y are independent}]$$

$$= \phi_X(t)\cdot\phi_Y(-t) \qquad [\because \phi_{cX}(t) = \phi_X(ct)]$$

$$= \left(\frac{e^{it} - 1}{it}\right)\cdot\left(\frac{e^{-it} - 1}{it}\right) \qquad \text{[From (*)]}$$

$$= \frac{1}{t^2}\left[1 - (e^{it} + e^{-it}) + 1\right] = \frac{2(1 - \cos t)}{t^2},$$

which is the *c.f.* of triangular distribution on (–1, 1) with peak at $x = 0$. Hence, by uniqueness theorem of *c.f.*'s, $U = (X - Y)$, has triangular distribution with *p.d.f.* [See equation (9·21*e*)]

$$f_U(u) = \begin{cases} 1 - |u|\,; & -1 < u < 1 \\ 0\,; & \text{otherwise} \end{cases}$$

9·5. GAMMA DISTRIBUTION

Definition. *A r.v. X is said to have a gamma distribution with parameter $\lambda > 0$, if its p.d.f. is given by :*

$$f(x) = \begin{cases} \dfrac{e^{-x}\,x^{\lambda-1}}{\Gamma(\lambda)}\,; & \lambda > 0,\ 0 < x < \infty \\ 0\,; & \textit{otherwise} \end{cases} \qquad \text{... (9·22)}$$

Remarks :

1. X is known as a Gamma variate with parameter λ and it is referred to as a $\gamma(\lambda)$ variate.

2. The function $f(x)$ defined in (9·22) represents a probability function, since

$$\int_0^\infty f(x)\,dx = \frac{1}{\Gamma(\lambda)}\int_0^\infty e^{-x}x^{\lambda-1}\,dx = \frac{1}{\Gamma(\lambda)}\cdot\Gamma(\lambda) = 1$$

3. The *cumulative distribution function*, called incomplete gamma function is defined as :

$$F_X(x) = \begin{cases} \displaystyle\int_0^x f(u)\,du = \frac{1}{\Gamma(\lambda)}\int_0^x e^{-u}u^{\lambda-1}\,du\,; & x>0,\ \lambda>0. \\ 0, & \text{otherwise} \end{cases} \qquad \dots (9\cdot 22a)$$

4. Relation Between Gamma and Normal Distributions.

(*i*) If $X \sim N(0,1)$, then $\left(\dfrac{X^2}{2}\right) \sim \gamma\left(\dfrac{1}{2}\right)$.

(*ii*) If $X \sim N(\mu, \sigma^2)$, then $\dfrac{1}{2}\left(\dfrac{X-\mu}{\sigma}\right)^2 \sim \gamma\left(\dfrac{1}{2}\right)$. [See Example 9·30]

9·5·1. M.G.F. of Gamma Distribution

M.G.F. about origin is given by :

$$M_X(t) = E(e^{tX}) = \int_0^\infty e^{tx} f(x)\,dx = \frac{1}{\Gamma(\lambda)}\int_0^\infty e^{tx}e^{-x}x^{\lambda-1}\,dx$$

$$= \frac{1}{\Gamma(\lambda)}\int_0^\infty e^{-(1-t)x}x^{\lambda-1}\,dx = \frac{1}{\Gamma(\lambda)}\cdot\frac{\Gamma(\lambda)}{(1-t)^\lambda},\ |t|<1$$

$$\therefore \quad M_X(t) = (1-t)^{-\lambda},\ |t|<1. \qquad \dots (9\cdot 23)$$

9·5·2. Cumulant Generating Function of Gamma Distribution

The cumulant generating function $K_X(t)$ is given by :

$$K_X(t) = \log M_X(t) = \log(1-t)^{-\lambda} = -\lambda\log(1-t)\,;\ |t|<1$$

$$= \lambda\left(t + \frac{t^2}{2} + \frac{t^3}{3} + \frac{t^4}{4} + \dots\right)$$

$\therefore$ Mean $= \kappa_1 =$ Coefficient of t in $K_X(t) = \lambda$

Variance $= \mu_2 = \kappa_2 =$ Coefficient of $\dfrac{t^2}{2!}$ in $K_X(t) = \lambda$

Hence, if $X \sim \gamma(\lambda)$, Mean = Variance = λ. ... (9·23*a*)

$\mu_3 = \kappa_3 =$ Coefficient of $\dfrac{t^3}{3!}$ in $K_X(t) = 2\lambda$

$\kappa_4 =$ Coefficient of $\dfrac{t^4}{4!}$ in $K_X(t) = 6\lambda \quad \Rightarrow \quad \mu_4 = \kappa_4 + 3\kappa_2^2 = 6\lambda + 3\lambda^2.$

In general, we have :

$$\kappa_r = \text{Coefficient of } \frac{t^r}{r!} \text{ in } K_X(t) = \lambda(r-1)!\,;\ r \ge 2. \qquad \dots (9\cdot 23b)$$

Hence, $$\beta_1 = \frac{\mu_3^2}{\mu_2^3} = \frac{4\lambda^2}{\lambda^3} = \frac{4}{\lambda} \quad \text{and} \quad \beta_2 = \frac{\mu_4}{\mu_2^2} = 3 + \frac{6}{\lambda}. \qquad \text{... (9·23c)}$$

Remarks :

1. Like Poisson distribution, the mean and variance of the Gamma distribution are also equal. However, Poisson distribution is discrete while Gamma distribution is continuous.

2. Limiting form of Gamma distribution as $\lambda \to \infty$. We know that if $X \sim \gamma(\lambda)$, then $E(X) = \lambda = \mu$, (say) and Var $(X) = \lambda = \sigma^2$, (say). Then standard gamma variate is given by : $Z = \dfrac{X-\mu}{\sigma} = \dfrac{X-\lambda}{\sqrt{\lambda}}$

$$M_Z(t) = \exp(-\mu t/\sigma)\, M_X(t/\sigma) = \exp(-\mu t/\sigma)\left(1 - \frac{t}{\sigma}\right)^{-\lambda} = e^{-t\lambda/\sqrt{\lambda}}\left(1 - \frac{t}{\sqrt{\lambda}}\right)^{-\lambda}$$

$$\Rightarrow \quad K_Z(t) = -\sqrt{\lambda}\cdot t - \lambda \log\left(1 - \frac{t}{\sqrt{\lambda}}\right) = -\sqrt{\lambda}\, t + \lambda\left(\frac{t}{\sqrt{\lambda}} + \frac{t^2}{2\lambda} + \frac{t^3}{3\lambda^{3/2}} + \ldots\right)$$

$$= -\sqrt{\lambda}\, t + \sqrt{\lambda}\, t + \frac{t^2}{2} + 0(\lambda^{-1/2}),$$

where, $0(\lambda^{-1/2})$ are terms containing $\lambda^{1/2}$ and higher powers of λ in the denominator.

$$\therefore \quad \lim_{\lambda \to \infty} K_Z(t) = \frac{t^2}{2} \quad \Rightarrow \quad \lim_{\lambda \to \infty} M_Z(t) = \exp(t^2/2)$$

which is the *m.g.f.* of a Standard Normal Variate. Hence, by uniqueness theorem of *m.g.f.*, Standard Gamma Variate tends to Standard Normal Variate as $\lambda \to \infty$. In other words, *Gamma distribution tends to Normal distribution for large value of parameter* λ.

3. Two Parameters Gamma Distribution. A continuous random variable X having the following *p.d.f.* is said to have a gamma distribution with two parameters *a* and λ.

$$f(x) = \begin{cases} \dfrac{a^\lambda}{\Gamma(\lambda)} e^{-ax} x^{\lambda-1}\,; & a > 0, \lambda > 0, 0 < x < \infty \\ 0, & \text{otherwise} \end{cases} \qquad \text{... (9·23d)}$$

Here, $X \sim \gamma(a, \lambda)$. Taking $a = 1$ in (9·23*d*), we get (9·22). Hence, we may write $X \sim \gamma(\lambda) = \gamma(1, \lambda)$.

4. For the two parameter gamma distribution (9·23*d*), we have

$$M_X(t) = \left(1 - \frac{t}{a}\right)^{-\lambda}; \quad t < a. \qquad \text{... (9·23e)}$$

Proof is left as an excercise to the reader.

Also $$K_X(t) = -\lambda \log\left(1 - \frac{t}{a}\right) = \lambda\left\{\frac{t}{a} + \frac{1}{2}\left(\frac{t}{a}\right)^2 + \frac{1}{3}\left(\frac{t}{a}\right)^3 + \ldots\right\}; \quad t < a$$

$$\therefore \quad \text{Mean} = \kappa_1 = \lambda/a \quad \text{and} \quad \text{Variance} = \kappa_2 = \lambda/a^2 = \text{Mean}/a \qquad \text{... (9·23f)}$$

Hence, Variance > Mean if $a < 1$; Variance = Mean if $a = 1$; and Variance < Mean if $a > 1$.

9·5·3. Additive Property of Gamma Distribution

The sum of independent Gamma variates is also a Gamma variate. More precisely, if $X_1, X_2, \ldots, X_k$ are independent Gamma variates with parameters $\lambda_1, \lambda_2, \ldots, \lambda_k$ respectively then $X_1 + X_2 + \ldots + X_k$ is also a Gamma variate with parameter $\lambda_1 + \lambda_2 + \ldots + \lambda_k$.

Proof. Since X_i is a $\gamma(\lambda_i)$ variate, $M_{X_i}(t) = (1-t)^{-\lambda_i}$

The *m.g.f.* of the sum $X_1 + X_2 + \ldots + X_k$ is given by :

$$M_{X_1+X_2+\ldots+X_k}(t) = M_{X_1}(t)\, M_{X_2}(t) \ldots M_{X_k}(t), \qquad (\because X_1, X_2, , X_k \text{ are independent})$$

$$= (1-t)^{-\lambda_1}(1-t)^{-\lambda_2}\ldots(1-t)^{-\lambda_k} = (1-t)^{-(\lambda_1+\lambda_2+\ldots+\lambda_k)}$$

which is the *m.g.f.* of a Gamma variate with parameter $\lambda_1 + \lambda_2 + \ldots + \lambda_k$. Hence, the result follows by the uniqueness theorem of *m.g.f.*'s.

Remark. In general, if $X_i \sim \gamma(a, \lambda_i)$; $i = 1, 2, \ldots\ldots, n$ are independent *r.v.*' s, then

$$\sum_{i=1}^{n} X_i \sim \gamma\left(a, \sum_{i=1}^{n} \lambda_i\right).$$

Example 9·29. *The daily consumption of milk in a city, in excess of 20,000 litres, is approximately distributed as a Gamma variate with parameters* $a = \dfrac{1}{10{,}000}$ *and* $\lambda = 2$. *The city has a daily stock of 30,000 litres. What is the probability that the stock is insufficient on a particular day ?*

Solution. If the *r.v.* X denotes the daily consumption of milk (in litres) in a city, then the *r.v.* $Y = (X - 20{,}000)$ has a gamma distribution with *p.d.f.* ;

$$g(y) = \frac{1}{(10{,}000)^2\, \Gamma(2)}\, y^{2-1}\, e^{-y/10{,}000} = \frac{y\, e^{-y/10{,}000}}{(10{,}000)^2}; \quad 0 < y < \infty$$

Since, the daily stock of the city is 30,000 litres, the required probability '*p*' that the stock is insufficient on a particular day is given by :

$$p = P(X > 30{,}000) = P(Y > 10{,}000) = \int_{10{,}000}^{\infty} g(y)\, dy = \int_{10{,}000}^{\infty} \frac{y\, e^{-y/10{,}000}}{(10{,}000)^2}\, dy$$

$$= \int_1^{\infty} z\, e^{-z}\, dz \qquad [\text{Taking } z = y/10{,}000]$$

Integrating by parts, $p = \left| -z e^{-z} \right|_1^{\infty} + \int_1^{\infty} e^{-z}\, dz = e^{-1} - \left| e^{-z} \right|_1^{\infty} = e^{-1} + e^{-1} = \dfrac{2}{e}$

Remark. Since $\lambda = 2$, the integration is easily done. However, for general values of a and λ, the integral is evaluated by using tables of Incomplete Gamma Integral [see Tables of Incomplete Gamma Functions, K. Pearson; Cambridge University Press] of the form :

$$\int_0^{\alpha} \frac{e^{-x}\, x^{n-1}}{\Gamma n}\, dx,$$ which have been tabulated for different values of α and n.

Example 9·30. *If* $X \sim N(\mu, \sigma^2)$, *obtain the p.d.f. of* $U = \frac{1}{2}\left(\dfrac{X-\mu}{\sigma}\right)^2$.

Solution. Since $X \sim N(\mu, \sigma^2) \quad \Rightarrow \quad Z = \dfrac{X-\mu}{\sigma} \sim N(0, 1)$ with *p.d.f.*:

$$\phi(z) = \frac{1}{\sqrt{2\pi}} \exp\left(-\frac{1}{2}z^2\right); \quad -\infty < z < \infty. \qquad \ldots (*)$$

The distribution function $G(\cdot)$ of $U = \frac{1}{2}\left(\dfrac{X-\mu}{\sigma}\right)^2$ is given by :

$$G_U(u) = P(U \le u) = P\left[\frac{1}{2}\left(\frac{X-\mu}{\sigma}\right)^2 \le u\right]$$

$$= P(Z^2 \le 2u) = P\left(-\sqrt{2u} \le Z \le \sqrt{2u}\right), \text{ where } Z \sim N(0,1)$$

$$= P\left(Z \le \sqrt{2u}\right) - P\left(Z \le -\sqrt{2u}\right)$$

$$= \Phi\left(\sqrt{2u}\right) - \Phi\left(-\sqrt{2u}\right) \qquad \text{... (**)}$$

where, $\Phi(\cdot)$ is the distribution function of standard normal variate (SNV) Z.

Differentiating *w.r.* to u, the *p.d.f.* $g(\cdot)$ of U is given by :

$$g(u) = \phi\left(\sqrt{2u}\right)\cdot\frac{d}{du}\left(\sqrt{2u}\right) - \phi\left(-\sqrt{2u}\right)\cdot\frac{d}{du}\left(-\sqrt{2u}\right) \quad [\phi(\cdot) \text{ is } p.d.f. \text{ of SNV } Z]$$

$$= \frac{1}{\sqrt{2u}}\left[\phi\left(\sqrt{2u}\right) + \phi\left(-\sqrt{2u}\right)\right]$$

$$= \frac{1}{\sqrt{2u}}\cdot 2\phi\left(\sqrt{2u}\right) \qquad [\because \phi(u) \text{ is even function of } u]$$

$$= \sqrt{\frac{2}{u}}\cdot\frac{1}{\sqrt{2\pi}}\exp\left(-\frac{1}{2}\cdot 2u\right) \qquad \text{[From (*)]}$$

$$= \frac{1}{\Gamma(1/2)}e^{-u}u^{(1/2)-1}, \; u \ge 0 \qquad \left[\because \sqrt{\pi} = \Gamma(1/2) \text{ and } u = \frac{1}{2}\left(\frac{x-u}{\sigma}\right)^2 \ge 0\right]$$

which is the *p.d.f.* of gamma distribution with parameter $\frac{1}{2}$.

Hence, $\quad U = \frac{1}{2}\left(\frac{X-\mu}{\sigma}\right)^2$ is a $\gamma(1/2)$ variate.

Example 9·31. *Show that the mean value of positive square root of a $\gamma(\mu)$ variate is $\Gamma\left(\mu+\frac{1}{2}\right)/\Gamma(\mu)$. Hence, prove that the mean deviation of a normal variate from its mean is $\sigma\cdot\sqrt{2/\pi}$, where σ is the standard deviation of the distribution.*

Solution. Let X be a $\gamma(\mu)$ variate. Then $\quad f(x) = \dfrac{e^{-x}x^{\mu-1}}{\Gamma(u)}; \quad \mu > 0, \; 0 < x < \infty$

$$\therefore \quad E\left(\sqrt{X}\right) = \int_0^\infty x^{1/2} f(x)\,dx = \frac{1}{\Gamma(\mu)}\int_0^\infty e^{-x}x^{\mu+(1/2)-1}\,dx = \frac{\Gamma\left(\mu+\frac{1}{2}\right)}{\Gamma(\mu)} \qquad \text{... (*)}$$

If $X \sim N(\mu, \sigma^2)$, then $U = \frac{1}{2}\left(\frac{X-\mu}{\sigma}\right)^2$ is a $\gamma\left(\frac{1}{2}\right)$ variate. [See Example 9·30]

$$\therefore \quad |X-\mu| = \sqrt{2}\,\sigma\,U^{1/2}, \text{ where } U \text{ is a } \gamma\left(\frac{1}{2}\right) \text{ variate.}$$

Hence, mean deviation of X about mean is given by :

$$E|X-\mu| = E\left(\sqrt{2}\,\sigma\,U^{1/2}\right) = \sqrt{2}\,\sigma\,E\left(U^{1/2}\right)$$

$$= \sqrt{2}\,\sigma\,\frac{\Gamma\left(\frac{1}{2}+\frac{1}{2}\right)}{\Gamma\left(\frac{1}{2}\right)} = \frac{\sqrt{2}\,\sigma}{\sqrt{\pi}} = \sigma\sqrt{2/\pi}. \qquad \left[\text{Using (*) with } \mu = \frac{1}{2}\right]$$

9·6. BETA DISTRIBUTION OF FIRST KIND

Definition. *A r.v. X is said to have a beta distribution of first kind with parameters μ and ν (μ > 0, ν > 0) if its p.d.f. is given by :*

$$f(x) = \begin{cases} \dfrac{1}{B(\mu, \nu)} \cdot x^{\mu-1}(1-x)^{\nu-1}\,; & (\mu, \nu) > 0,\ 0 < x < 1 \\ 0, \text{ otherwise} \end{cases}$$

[where, B (μ, ν) is the Beta function]. ... (9·24)

The *r.v.* X is known as a *Beta variate of the first kind with parameters μ and ν* and is referred to as $\beta_1(\mu, \nu)$ variate.

Remarks :

1. The cumulative distribution function, often called the *Incomplete Beta Function*, is given by :

$$F(x) = \begin{cases} 0, & x < 0 \\ \displaystyle\int_0^x \frac{1}{B(\mu, \nu)}\, u^{\mu-1}(1-u)^{\nu-1}\, du; & 0 < x < 1, (\mu, \nu) > 0 \\ 1, & x > 1 \end{cases} \qquad \text{... (9·24a)}$$

2. In particular, if we take μ = 1, ν = 1 in (9·24), we get : $f(x) = \dfrac{1}{\beta(1, 1)} = 1,\ 0 < x < 1$ ... (9·24b)

which is the *p.d.f.* of uniform distribution on [0, 1].

3. If $X \sim \beta_1(\mu, \nu)$, then it can be easily verified that $(1 - X) \sim \beta_1(\nu, \mu)$.

9·6·1. Constants of Beta Distribution of First Kind

$$\mu_r' = \int_0^1 x^r f(x)\, dx = \frac{1}{B(\mu, \nu)} \int_0^1 x^{\mu+r-1}(1-x)^{\nu-1}\, dx = \frac{1}{B(\mu, \nu)} \cdot B(\mu+r, \nu)$$

$$= \frac{\Gamma(\mu+r)\,\Gamma(\nu)}{\Gamma(\mu+r+\nu)} \cdot \frac{\Gamma(\mu+\nu)}{\Gamma(\mu)\,\Gamma(\nu)} = \frac{\Gamma(\mu+r)\,\Gamma(\mu+\nu)}{\Gamma(\mu+r+\nu)\,\Gamma(\mu)} \qquad \text{... (9·24c)}$$

In particular,

$$\text{Mean} = \mu_1' = \frac{\Gamma(\mu+1)}{\Gamma(\mu+1+\nu)} \cdot \frac{\Gamma(\mu+\nu)}{\Gamma(\mu)} = \frac{\mu\,\Gamma(\mu)\,\Gamma(\mu+\nu)}{(\mu+\nu)\,\Gamma(\mu+\nu)\,\Gamma(\mu)} = \frac{\mu}{\mu+\nu} \qquad \text{... (9·24d)}$$

$$[\because \Gamma(k) = (k-1)\,\Gamma(k-1)]$$

$$\mu_2' = \frac{\Gamma(\mu+2)\cdot\Gamma(\mu+\nu)}{\Gamma(\mu+\nu+2)\,\Gamma(\mu)} = \frac{(\mu+1)\,\mu\,\Gamma(\mu)\,\Gamma(\mu+\nu)}{(\mu+\nu+1)(\mu+\nu)\,\Gamma(\mu+\nu)\,\Gamma(\mu)} = \frac{\mu(1+\mu)}{(\mu+\nu)(\mu+\nu+1)}$$

Hence,

$$\mu_2 = \mu_2' - \mu_1'^2 = \frac{\mu(1+\mu)}{(\mu+\nu)(\mu+\nu+1)} - \left(\frac{\mu}{\mu+\nu}\right)^2$$

$$= \frac{\mu}{(\mu+\nu)^2(\mu+\nu+1)}\Big[(\mu+\nu)(\mu+1) - \mu(\mu+\nu+1)\Big] = \frac{\mu\nu}{(\mu+\nu)^2(\mu+\nu+1)}$$

Similarly, we have

$$\mu_3 = \mu_3' - 3\mu_2'\mu_1' + 2\mu_1'^3 = \frac{2\mu\nu(\nu-\mu)}{(\mu+\nu)^3(\mu+\nu+1)(\mu+\nu+2)}$$

and

$$\mu_4 = \mu_4' - 4\mu_3'\mu_1' + 6\mu_2'\mu_1'^2 - 3\mu_1'^4$$

$$= \frac{3\mu\nu\{\mu\nu(\mu+\nu-6)+2(\mu+\nu)^2\}}{(\mu+\nu)^4(\mu+\nu+1)(\mu+\nu+2)(\mu+\nu+3)}$$

$$\therefore \quad \beta_1 = \frac{\mu_3^2}{\mu_2^3} = \frac{4(\nu-\mu)^2(\mu+\nu+1)}{\mu\nu(\mu+\nu+2)^2} \quad \text{and} \quad \beta_2 = \frac{\mu_4}{\mu_2^2} = \frac{3(\mu+\nu+1)\mu\nu(\mu+\nu-6)+2(\mu+\nu)^2}{\mu\nu(\mu+\nu+2)(\mu+\nu+3)}$$

The harmonic mean H is given by :

$$\frac{1}{H} = \int_0^1 \frac{1}{x} f(x)\,dx = \frac{1}{B(\mu,\nu)} \int_0^1 x^{\mu-2}(1-x)^{\nu-1}\,dx$$

$$= \frac{1}{B(\mu,\nu)} B(\mu-1,\nu) = \frac{\Gamma(\mu-1)\Gamma(\nu)}{\Gamma(\mu+\nu-1)} \times \frac{\Gamma(\mu+\nu)}{\Gamma(\mu)\Gamma(\nu)}$$

$$= \frac{\Gamma(\mu-1)(\mu+\nu-1)\Gamma(\mu+\nu-1)}{\Gamma(\mu+\nu-1)(\mu-1)\Gamma(\mu-1)} = \frac{\mu+\nu-1}{\mu-1}$$

$$\therefore \quad H = \frac{\mu-1}{\mu+\nu-1} \qquad \text{... (9·24}d\text{)}$$

Mode. Mode is the solution of $f'(x) = 0$ and $f''(x) < 0$.

$$f(x) = c \cdot x^{\mu-1}(1-x)^{\nu-1}; \quad \text{where } c = \frac{1}{B(\mu,\nu)} \text{ is a constant.}$$

$$\Rightarrow \quad \log f(x) = \log c + (\mu-1)\log x + (\nu-1)\log(1-x)$$

Differentiating *w.r.t.* x : $\quad f'(x) = f(x)\left[\frac{\mu-1}{x} - \frac{\nu-1}{1-x}\right].$

$$f'(x) = 0 \quad \Rightarrow \quad (1-x)(\mu-1) - (\nu-1)x = 0$$

$$\Rightarrow \quad (\mu-1) - x[\mu-1+\nu-1] = 0 \quad \Rightarrow \quad x = \frac{\mu-1}{(\mu+\nu-2)}$$

It can be easily verified that at $x = (\mu-1)/(\mu+\nu-2)$, $f''(x) < 0$.

Hence, the unique mode of $\beta_1(\mu,\nu)$ distribution is at $x = \dfrac{\mu-1}{\mu+\nu-2}$; $\mu > 1$, $\nu > 1$. ... (9·24*e*)

9·6·2. Moment Generating Function of Beta Distribution of First Kind

If $X \sim \beta_1(\mu,\nu)$, its m.g.f. is given by :

$$M_X(t) = E(e^{tX}) = \frac{1}{B(\mu,\nu)} \int_0^1 e^{tx} \cdot x^{\mu-1}(1-x)^{\nu-1}\,dx$$

We cannot evaluate this integral by usual methods. Expanding e^{tx} as a series in powers of x and taking summation outside the integral sign, we get :

$$M_X(t) = \frac{1}{B(\mu, \nu)} \int_0^1 \left[\left\{ \sum_{r=0}^{\infty} \frac{(tx)^r}{r!} \right\} \cdot x^{\mu-1} (1-x)^{\nu-1} \right] dx$$

$$= \frac{1}{B(\mu, \nu)} \sum_{r=0}^{\infty} \left[\int_0^1 \frac{t^r}{r!} x^{\mu+r-1} (1-x)^{\nu-1} \, dx \right]$$

$$= \frac{1}{B(\mu, \nu)} \sum_{r=0}^{\infty} \left[\frac{t^r}{r!} \cdot B(\mu+r, \nu) \right]$$

$$= \sum_{r=0}^{\infty} \left[\frac{\Gamma(\mu+r)\,\Gamma(\mu+\nu)}{\Gamma(\mu+\nu+r)\,\Gamma(\mu)} \cdot \frac{t^r}{r!} \right] \quad \text{...} (*)$$

$$= \sum_{r=0}^{\infty} \left[\frac{(\mu+r-1)(\mu+r-2) \ldots \mu}{(\mu+\nu+r-1)(\mu+\nu+r-2) \ldots (\mu+\nu)} \times \frac{t^r}{r!} \right] \quad \text{... (9·24}f\text{)}$$

which is a slight variant of a hypergeometric series.

$$\therefore \quad \mu_r' = E(X^r) = \text{Coefficient of } \frac{t^r}{r!} \text{ in } (*) = \frac{\Gamma(\mu+r)\,\Gamma(\mu+\nu)}{\Gamma(\mu+\nu+r)\cdot\Gamma(\mu)}, \quad \text{[From ... (*)]}$$

an expression already obtained in (9·24c)

Example 9·32. *If X and Y are independent Gamma variates with parameters μ and ν ,respectively, show that the variables $U = X + Y$, $Z = \frac{X}{X+Y}$ are independent and that U is a $\gamma(\mu + \nu)$ variate and Z is a $\beta_1(\mu, \nu)$ variate.*

Solution. Since X is a $\gamma(\mu)$ variate and Y is a $\gamma(\nu)$ variate, we have

$$f_1(x)\,dx = \frac{1}{\Gamma(\mu)} e^{-x} x^{\mu-1} \,dx\,; \quad 0 < x < \infty, \mu > 0$$

$$f_2(y)\,dy = \frac{1}{\Gamma(\nu)} e^{-y} y^{\nu-1} \,dy\,; \quad 0 < y < \infty, \nu > 0$$

Since X and Y are independently distributed, their joint probability differential is given by the compound probability theorem as shown below :

$$dF(x, y) = f_1(x)\, f_2(y)\, dx\, dy = \frac{1}{\Gamma\mu\,\Gamma\nu} e^{-(x+y)} x^{\mu-1} y^{\nu-1} \,dx\, dy$$

Now, $\quad u = x + y, \quad z = \frac{x}{x+y}, \quad$ so that $\quad x = uz, \quad y = u - x = u(1-z)$

Jacobian of transformation J is given by :

$$J = \frac{\partial(x, y)}{\partial(u, z)} = \begin{vmatrix} \frac{\partial x}{\partial u} & \frac{\partial y}{\partial u} \\ \frac{\partial x}{\partial z} & \frac{\partial y}{\partial z} \end{vmatrix} = \begin{vmatrix} z & 1-z \\ u & -u \end{vmatrix} = -u$$

As X and Y range from 0 to ∞, u ranges from 0 to ∞ and z from 0 to 1. $\left(\because \ \frac{x}{x+y} \le 1 \right)$.

Hence, the joint distribution of U and Z is given by :

$$dG(u, z) = g(u, z)\, du\, dz = \frac{1}{\Gamma(\mu)\,\Gamma(\nu)} e^{-u} (uz)^{\mu-1} [u(1-z)]^{\nu-1} \mid J \mid du\, dz$$

$$= \frac{1}{\Gamma(\mu)\,\Gamma(\nu)} \cdot e^{-u}\, u^{\mu+\nu-1}\, z^{\mu-1} (1-z)^{\nu-1}\, du\, dz$$

$$= \left\{ \frac{e^{-u}\, u^{\mu+\nu-1}}{\Gamma(\mu+\nu)}\, du \right\} \left\{ \frac{1}{B(\mu, \nu)}\, z^{\mu-1} (1-z)^{\nu-1}\, dz \right\}.$$

$$= [g_1(u)\, du][g_2(z)\, dz], \text{ (say)}, \qquad \ldots (*)$$

where, $g_1(u) = \dfrac{1}{\Gamma(\mu+\nu)} e^{-u}\, u^{\mu+\nu-1},\ 0 < u < \infty$

and $g_2(z) = \dfrac{1}{B(\mu, \nu)} z^{\mu-1} (1-z)^{\nu-1},\ 0 < z < 1$ $\qquad \ldots (**)$

From (*) and (**), we conclude that U and Z are independently distributed, U as a $\gamma(\mu+\nu)$ variate and Z as a $\beta_1(\mu, \nu)$ variate.

Example 9·33. *Let $X \sim \beta_1(\mu, \nu)$ and $Y \sim \gamma(\lambda, \mu+\nu)$ be independent random variables ($\mu, \nu, \lambda > 0$). Find a p.d.f. for XY and identify its distribution.*

Solution. Since X and Y are independently distributed, their joint *p.d.f.* is :

$$f(x, y) = \frac{1}{B(\mu, \nu)} \cdot x^{\mu-1} (1-x)^{\nu-1} \times \frac{\lambda^{\mu+\nu}}{\Gamma(\mu+\nu)} e^{-\lambda y}\, y^{\mu+\nu-1};\ 0 < x < 1,\ 0 < y < \infty$$

Let us transform to the new variables U and Z by the transformation :

$$xy = u, \qquad x = z, \qquad i.e., \qquad x = z \quad \text{and} \quad y = u/z$$

Jacobian of transformation J is given by : $J = \dfrac{\partial(x, y)}{\partial(u, z)} = \begin{vmatrix} 0 & 1 \\ \dfrac{1}{z} & \dfrac{-u}{z^2} \end{vmatrix} = -\dfrac{1}{z}$

Thus, the *p.d.f.* of U and Z becomes :

$$g(u, z) = \frac{\lambda^{\mu+\nu}}{B(\mu, \nu)\Gamma(\mu+\nu)} (z)^{\mu-1} (1-z)^{\nu-1} e^{-\lambda u/z} \left(\frac{u}{z}\right)^{\mu+\nu-1} \mid J \mid;\ 0 < u < \infty, 0 < z < 1$$

Integrating *w.r.* to z in the range $0 < z < 1$, the marginal *p.d.f.* of U is given by :

$$g_1(u) = \frac{\lambda^{\mu+\nu}\, u^{\mu+\nu-1}}{\Gamma(\mu)\,\Gamma(\nu)} \int_0^1 \frac{(1-z)^{\nu-1}\, e^{-\lambda u/z}}{z^{\nu+1}}\, dz$$

$$= \frac{\lambda^{\mu+\nu}\, u^{\mu+\nu-1}}{\Gamma(\mu)\,\Gamma(\nu)} \int_0^1 \frac{1}{z^2} \left(\frac{1}{z} - 1\right)^{\nu-1} e^{-\lambda u/z}\, dz$$

$$= \frac{\lambda^{\mu+\nu}\, u^{\mu+\nu-1}}{\Gamma(\mu)\,\Gamma(\nu)} \int_0^\infty t^{\nu-1}\, e^{-\lambda u(1+t)}\, dt, \qquad \left(\frac{1}{z} - 1 = t\right)$$

$$= \frac{\lambda^{\mu+\nu}\, u^{\mu+\nu-1}\, e^{-\lambda u}}{\Gamma(\mu)\,\Gamma(\nu)} \int_0^\infty e^{-\lambda u t}\, t^{\nu-1}\, dt = \frac{\lambda^{\mu+\nu}\, u^{\mu+\nu-1}\, e^{-\lambda u}}{\Gamma(\mu)\,\Gamma(\nu)} \cdot \frac{\Gamma(\nu)}{(\lambda u)^\nu}$$

$$= \frac{\lambda^\mu}{\Gamma(\mu)} \cdot e^{-\lambda u} u^{\mu-1}, \quad 0 < u < \infty$$

Hence, $U = XY$, is distributed as a gamma variate with parameters λ and μ, *i.e.*, $XY \sim \gamma(\lambda, \mu)$.

9·7. BETA DISTRIBUTION OF SECOND KIND

Definition. *A r.v. X is said to have a beta distribution of the second kind with parameters μ and ν, ($\mu > 0$, $\nu > 0$), if its p.d.f. is given by :*

$$f(x) = \begin{cases} \dfrac{1}{B(\mu, \nu)} \cdot \dfrac{x^{\mu-1}}{(1+x)^{\mu+\nu}}, (\mu, \nu) > 0, \ 0 < x < \infty \\ 0, \ otherwise \end{cases} \quad \text{... (9·25)}$$

Remarks :

1. The *r.v.* X is known as a *Beta variate of second kind with parameters μ and ν* and is denoted as $\beta_2(\mu, \nu)$ variate.

2. Beta distribution of second kind is transformed to Beta distribution of first kind by the transformation :

$$1 + x = \frac{1}{y} \quad \Rightarrow \quad y = \frac{1}{1+x} \quad \text{... (*)}$$

Thus, if $X \sim \beta_2(\mu, \nu)$, then Y defined in (*) is a $\beta_1(\mu, \nu)$.

The proof is left as an exercise to the reader.

9·7·1. Constants of Beta Distribution of Second Kind.

$$\mu_r' = \int_0^\infty x^r f(x)\, dx = \frac{1}{B(\mu, \nu)} \int_0^\infty \frac{x^{\mu+r-1}}{(1+x)^{\mu+\nu}}\, dx$$

$$= \frac{1}{B(\mu, \nu)} \int_0^\infty \frac{x^{(\mu+r)-1}}{(1+x)^{\mu+r+\nu-r}}\, dx = \frac{1}{B(\mu, \nu)} \cdot B(\mu+r, \nu-r)$$

$$= \frac{\Gamma(\mu+r)\,\Gamma(\nu-r)}{\Gamma(\mu+\nu)} \cdot \frac{\Gamma(\mu+\nu)}{\Gamma(\mu)\,\Gamma(\nu)} = \frac{\Gamma(\mu+r)\,\Gamma(\nu-r)}{\Gamma(\mu)\,\Gamma(\nu)}; \quad \nu > r$$

In particular,

$$\mu_1' = \frac{\Gamma(\mu+1)\,\Gamma(\nu-1)}{\Gamma(\mu)\,\Gamma(\nu)} = \frac{\mu\,\Gamma(\mu)\,\Gamma(\nu-1)}{\Gamma(\mu)(\nu-1)\,\Gamma(\nu-1)} = \frac{\mu}{\nu-1}, \quad \nu > 1$$

$$\mu_2' = \frac{\Gamma(\mu+2)\,\Gamma(\nu-2)}{\Gamma(\mu)\,\Gamma(\nu)} = \frac{(\mu+1)\mu\,\Gamma(\mu)\,\Gamma(\nu-2)}{\Gamma(\mu)(\nu-1)(\nu-2)\,\Gamma(\nu-2)} = \frac{\mu(\mu+1)}{(\nu-1)(\nu-2)}, \quad \nu > 2$$

$$\therefore \quad \mu_2 = \mu_2' - \mu_1'^2 = \frac{\mu(\mu+1)}{(\nu-1)(\nu-2)} - \left(\frac{\mu}{\nu-1}\right)^2$$

$$= \frac{\mu}{\nu-1}\left[\frac{(\nu-1)(\mu+1) - \mu(\nu-2)}{(\nu-1)\,(\nu-2)}\right] = \frac{\mu(\mu+\nu-1)}{(\nu-1)^2\,(\nu-2)}$$

The harmonic mean H is given by :

$$\frac{1}{H} = E\left(\frac{1}{X}\right) = \int_0^\infty \frac{1}{x} \cdot f(x)\, dx = \frac{1}{B(\mu, \nu)} \int_0^\infty \frac{x^{\mu-2}}{(1+x)^{\mu+\nu}}\, dx$$

$$= \frac{1}{B(\mu, \nu)} \int_0^{\infty} \frac{x^{\mu-1-1}}{(1+x)^{\mu-1+\nu+1}} dx = \frac{1}{B(\mu, \nu)} \cdot B(\mu-1, \nu+1), \quad \mu > 1$$

$$= \frac{\Gamma(\mu-1)\,\Gamma(\nu+1)}{\Gamma(\mu+\nu)} \cdot \frac{\Gamma(\mu+\nu)}{\Gamma(\mu)\,\Gamma(\nu)} = \frac{\Gamma(\mu-1)\,\nu\,\Gamma(\nu)}{(\mu-1)\,\Gamma(\mu-1)\,\Gamma(\nu)} = \frac{\nu}{\mu-1}$$

Hence, $H = \dfrac{\mu-1}{\nu}; \quad \mu > 1.$

Example 9·34. *If X and Y are independent Gamma variates with parameters μ and ν, respectively, show that* $U = X + Y,\ Z = \dfrac{X}{Y}$ *are independent and that U is a γ(μ + ν) variate and Z is a* $\beta_2(\mu, \nu)$ *variate.*

Solution. Proceeding on the same lines as *Example 9·32*, we have

$$dF(x, y) = \frac{1}{\Gamma(\mu)\,\Gamma(\nu)} e^{-(x+y)} x^{\mu-1} y^{\nu-1}\, dx\, dy, \quad 0 < (x, y) < \infty$$

Since $u = x + y$ and $z = \dfrac{x}{y}$, $1 + z = 1 + \dfrac{x}{y} = \dfrac{u}{y} \Rightarrow y = \dfrac{u}{1+z}$ and $x = \dfrac{uz}{1+z} = u\left(1 - \dfrac{1}{1+z}\right)$

$$J = \frac{\partial(x, y)}{\partial(u, z)} = \frac{-u}{(1+z)^2}$$

As x and y range from 0 to ∞, both u and z range from 0 to ∞. Hence, the joint probability differential of random variables U and Z becomes :

$$dG(u, z) = \frac{1}{\Gamma(\mu)\,\Gamma(\nu)} e^{-u} \left(\frac{uz}{1+z}\right)^{\mu-1} \left(\frac{u}{1+z}\right)^{\nu-1} |\,J\,|\, du\, dx$$

$$= \left[\frac{e^{-u} u^{\mu+\nu-1}}{\Gamma(\mu+\nu)} du\right] \left[\frac{1}{B(\mu, \nu)} \cdot \frac{z^{\mu-1}}{(1+z)^{\mu+\nu}} dz\right]; \quad 0 < u < \infty,\ 0 < z < \infty$$

Hence, U and Z are independenly distributed, U as a $\gamma(\mu + \nu)$ variate and Z as a $\beta_2(\mu, \nu)$ variate.

Remark. The above example lead to the following important results :

If X is a γ (μ) variate and Y is an independent γ (ν) variate, then

(*i*) $X + Y$ is a γ (μ + ν) variate, *i.e.*, the sum of two independent Gamma variates is also a Gamma variate.

(*ii*) (X/Y) is a $\beta_2(\mu, \nu)$ variate, *i.e.*, the ratio of two independent Gamma variates is β_2 – variate.

SUPPLEMENTARY EXAMPLES ON UNIFORM, GAMMA AND BETA DISTRIBUTIONS

Example 9·35. *On the x-axis (n + 1) points are taken independently between the origin and x = 1, all positions being equally likely. Show that the (k + 1)th of these points, counted from origin, lies in the interval* $\left(x - \frac{1}{2}dx\right)$ *to* $\left(x + \frac{1}{2}dx\right)$ *is :*

$$\binom{n}{k} (n+1) x^k (1-x)^{n-k}\, dx$$

Verify that integral of this expression from x = 0 to x = 1 is unity.

Solution. Here X is given to be a *r.v.* uniformly distributed on [0, 1].

$$\Rightarrow \qquad f_X(x) = 1, \quad 0 \le x \le 1$$

$$\therefore \qquad P(0 < X < x) = \int_0^x f(x)\,dx = \int_0^x 1 \cdot dx = x \qquad \text{... (*)}$$

$$\Rightarrow \qquad P(X > x) = 1 - P(X \le x) = 1 - x \qquad \text{[From (*)]} \qquad \text{... (**)}$$

Also $$P\left(x - \frac{dx}{2} < X < x + \frac{dx}{2}\right) = \int_{x-\frac{dx}{2}}^{x+\frac{dx}{2}} f(x)\,dx = dx \qquad \text{... (***)}$$

Required probability 'p' is given by :

$p = P$ [out of $(n+1)$ points, k points lie in closed interval $\left(0, x - \dfrac{dx}{2}\right)$ and out of the remaining $(n+1-k)$ points, $(n-k)$ points lie in $\left(x + \dfrac{dx}{2}, 1\right)$ and one point lies in $\left(x - \dfrac{dx}{2}, x + \dfrac{dx}{2}\right)$]

$$= \left[\binom{n+1}{k} x^k\right] \times \left[\binom{n+1-k}{n-k}(1-x)^{n-k}\right] dx, \text{ on using (*), (**) and (***) respectively,}$$

$$\therefore \qquad p = \frac{(n+1)!}{k!(n+1-k)!} \cdot x^k \cdot \frac{(n+1-k)!}{(n-k)!} \cdot (1-x)^{n-k}\,dx = \binom{n}{k}(n+1)\,x^k(1-x)^{n-k}\,dx$$

Varification. $$I = \int_0^1 \binom{n}{k}(n+1)x^k(1-x)^{n-k}\,dx = \binom{n}{k}(n+1) \cdot B(k+1, n-k+1)$$

$$\left[\text{On using beta-Integral: } \int_0^1 x^{m-1}(1-x)^{n-1}\,dx = B(m, n) = \frac{\Gamma m\,\Gamma n}{\Gamma(m+n)};\; m > 0, n > 0.\right]$$

$$\therefore \qquad I = \frac{n!(n+1)}{k!(n-k)!} \cdot \frac{k!(n-k)!}{(n+1)!} = 1.$$

Example 9·36. *Let X and Y have joint p.d.f. :*

$$g(x, y) = \begin{cases} \dfrac{e^{-(x+y)}\,x^3\,y^4}{\Gamma 4\,\Gamma 5}\,; & x > 0,\ y > 0 \\ 0, \text{ elsewhere} \end{cases}$$

Find (i) p.d.f. of $U = \dfrac{X}{X+Y}$, *(ii)* $E(U)$, *and (iii)* $E[U - E(U)]^2$.

Solution. Let $u = \dfrac{x}{x+y}$ and $v = x + y \quad \Rightarrow \quad x = uv, \quad y = v - uv = v(1-u).$

Jacobian of transformation is : $$J = \frac{\partial(x, y)}{\partial(u, v)} = \begin{vmatrix} \dfrac{\partial x}{\partial u} & \dfrac{\partial x}{\partial v} \\ \dfrac{\partial y}{\partial u} & \dfrac{\partial y}{\partial v} \end{vmatrix} = \begin{vmatrix} v & u \\ -v & 1-u \end{vmatrix} = v$$

Hence, joint *p.d.f.* of U and V becomes :

$$p(u, v) = g(x, y) \cdot |J| = \frac{1}{\Gamma 4 \Gamma 5} e^{-v} \cdot (uv)^3 [v(1-u)]^4 \times v$$

$$= \frac{1}{\Gamma 4 \Gamma 5} e^{-v} \cdot v^8 \cdot u^3 (1-u)^4 ; \quad 0 \le u \le 1, \ v > 0$$

$$\left(\because u = \frac{x}{x+y} < 1 \text{ and since } x > 0, y > 0, \text{ we have } 0 < u < 1 \text{ and } v = x + y \ge 0 \right)$$

$$= \left(\frac{1}{\Gamma 9} e^{-v} v^8 \right) \left\{ \frac{\Gamma 9}{\Gamma 4 \Gamma 5} u^3 (1-u)^4 \right\}; \quad 0 < u < 1, \ v > 0$$

$\Rightarrow$
$$p(u, v) = p_1(v) \cdot p_2(u) \qquad \ldots (*)$$

where,
$$p_1(v) = \frac{1}{\Gamma 9} e^{-v} v^8 ; \ v > 0 \quad \text{and} \quad p_2(u) = \frac{\Gamma 9}{\Gamma 4 \Gamma 5} u^3 (1-u)^4 ; \ 0 < u < 1 \qquad \ldots (**)$$

From (*), we conclude that U & V are independently distributed & from (**), we conclude that :

$$U = \frac{X}{X+Y} \sim \beta_1(4, 5),$$ *i.e.*, U is a Beta variate of first kind with parameters (4, 5).

Aliter. We have

$$g(x, y) = \frac{1}{\Gamma 4 \Gamma 5} e^{-(x+y)} x^3 y^4 = \left[\frac{1}{\Gamma 4} e^{-x} x^3 \right] \left[\frac{1}{\Gamma 5} e^{-y} y^4 \right]$$

$$= g_1(x) \, g_2(y); \ x > 0, \ y > 0$$

$\Rightarrow$ X and Y are independently distributed and $X \sim \gamma(4)$ and $Y \sim \gamma(5)$.

Hence,
$$U = \frac{X}{X+Y} \sim \beta_1(4, 5)$$ [See Example 9·32]

$\therefore$
$$E(U) = \int_0^1 u \cdot p_2(u) \, du = \frac{1}{B(4, 5)} \cdot \int_0^1 u^4 (1-u)^4 \, du$$

$$= \frac{1}{B(4, 5)} \cdot B(5, 5)$$ [Using Beta integral]

$$= \frac{\Gamma 9}{\Gamma 4 \Gamma 5} \times \frac{\Gamma 5 \Gamma 5}{\Gamma 10} = \frac{\Gamma 9 \cdot 4 \Gamma 4}{\Gamma 4 \cdot 9 \Gamma 9} = \frac{4}{9}$$

$$E(U^2) = \frac{1}{B(4, 5)} \int_0^1 u^2 \cdot u^3 (1-u)^4 \, du$$

$$= \frac{1}{B(4, 5)} \times B(6, 5) = \frac{\Gamma 9}{\Gamma 4 \Gamma 5} \times \frac{\Gamma 6 \Gamma 5}{\Gamma 11} = \frac{5 \times 4}{10 \times 9} = \frac{2}{9}$$

$$E[U - E(U)]^2 = E(U^2) - [E(U)]^2 = \frac{2}{9} - \frac{16}{81} = \frac{2}{81}.$$

Example 9·37. *A random sample of size n is taken from a population with distribution :*

$$dP(x) = \frac{1}{\Gamma(\lambda)} e^{-x/a} \left(\frac{x}{a} \right)^{\lambda - 1} \frac{dx}{a} ; \quad 0 < x < \infty, \ a > 0, \ \lambda > 0$$

Find the distribution of the mean $\overline{X}$.

Solution.

$$M_X(t) = E(e^{tX}) = \int_0^\infty e^{tx} f(x)\,dx = \frac{1}{\Gamma(\lambda)}\int_0^\infty e^{tx} e^{-x/a}\left(\frac{x}{a}\right)^{\lambda-1}\frac{dx}{a}$$

$$= \frac{1}{\Gamma(\lambda)a^\lambda}\int_0^\infty \exp\left\{-\left(\frac{1}{a}-t\right)x\right\}\cdot x^{\lambda-1}\,dx = \frac{1}{\Gamma(\lambda)a^\lambda}\cdot\frac{\Gamma(\lambda)}{\left(\frac{1}{a}-t\right)^\lambda} = (1-at)^{-\lambda} \quad \text{... (*)}$$

$$\therefore \quad M_{\bar{X}}(t) = M_{(X_1+X_2+\ldots+X_n)/n}(t) = M_{(X_1+X_2+\ldots+X_n)}(t/n) \qquad [\because M_{cX}(t) = M_X(ct)]$$

$$= M_{X_1}(t/n)\,M_{x_2}(t/n)\ldots M_{X_n}(t/n), \qquad \text{(Since } X_1, X_2, \ldots, X_n \text{ are independent.)}$$

Hence, on using (*), we get $\quad M_{\bar{X}}(t) = \left[\left(1-\frac{at}{n}\right)^{-\lambda}\right]^n = \left(1-\frac{t}{n/a}\right)^{-n\lambda}$

which is the *m.g.f.* of a Gamma distribution (*c.f.* Remark 3, § 9·5·2). Hence, by uniqueness theorem of *m.g.f.*, $\bar{X} \sim \gamma(n/a, n\lambda)$ with *p.d.f.*:

$$g(\bar{x}) = \frac{(n/a)^{\lambda n}}{\Gamma(\lambda n)} e^{-n\bar{x}/a}(\bar{x})^{n\lambda-1}, \quad 0 < \bar{x} < \infty.$$

Example 9·38. *A sample of n values is drawn from a population with p.d.f.* ae^{-ax}, $(x \ge 0, a > 0)$. *If* $\bar{X}$ *is mean of the sample, show that* $na\bar{X}$ *is a* $\gamma(n)$ *variate and prove that*

$$E(\bar{X}) = \frac{1}{a} \quad \text{and} \quad S.E. \text{ of } (\bar{X}) = \frac{1}{a\sqrt{n}}.$$

Solution. $\quad f(x) = a\,e^{-ax}\,;\ 0 \le x < \infty,\ a > 0$

$$\therefore \quad M_X(t) = \int_0^\infty e^{tx} f(x)\,dx = a\int_0^\infty e^{-(a-t)x}\,dx = a\left|\frac{e^{-(a-t)x}}{-(a-t)}\right|_0^\infty = \frac{a}{a-t},\ (a > t) \quad \text{... (*)}$$

Also, $\quad an\bar{X} = an\left(\frac{X_1+X_2+\ldots+X_n}{n}\right) = a(X_1+X_2+\ldots+X_n)$

$$\Rightarrow \quad M_{an\bar{X}}(t) = M_{a(X_1+X_2+\ldots+X_n)}(t)$$

$$= M_{X_1}(at)\cdot M_{X_2}(at)\ldots M_{X_n}(at), \quad \text{(Since the sample values are independent.)}$$

$$\Rightarrow \quad M_{an\bar{X}}(t) = \prod_{i=1}^n M_{X_i}(at) = \left[M_{X_i}(at)\right]^n, \quad \text{(Since } X_1, X_2, \ldots, X_n \text{ are identically distributed.)}$$

$$\therefore \quad M_{an\bar{X}}(t) = \left(\frac{1}{1-t}\right)^n = (1-t)^{-n}, \qquad \text{[From (*)]}$$

which is the *m.g.f.* of a $\gamma(n)$ variate.

Hence, by uniqueness theorem of *m.g.f.*, $an\bar{X}$ is a $\gamma(n)$ variate.

Since the mean and variance of a $\gamma(n)$ variate are equal, each being equal to n,

$$E(an\bar{X}) = n \quad \Rightarrow \quad an\,E(\bar{X}) = n \quad \Rightarrow \quad E(\bar{X}) = \frac{1}{a}$$

and $\quad V(an\bar{X}) = n \quad \Rightarrow \quad a^2 n^2 V(\bar{X}) = n \quad \Rightarrow \quad V(\bar{X}) = \dfrac{1}{na^2}$

Hence, Standard error (*S.E.*) of $\bar{X} = \sqrt{V(\bar{X})} = \dfrac{1}{a\sqrt{n}}$.

Note. The concept of Standard Error (*S.E.*) is discussed in detail in Chapter Fourteen.

Example 9·39. *The joint probability function of two r.v.'s is given by :*

$$f(x, y) = e^{-(x+y)}\,;\ x > 0, y > 0.$$

Find the m.g.f. of $U = (X + Y)/2$ and identify its distribution. Hence, show that its p.d.f. is :

$$g(u) = 4u\,e^{-2u}\,;\ u > 0.$$

Solution. $\quad f(x, y) = e^{-(x+y)} = e^{-x} \cdot e^{-y}\,;\ x > 0, y > 0.$

$\Rightarrow \quad f(x, y) = f_1(x) \cdot f_2(y),$ (say) $\quad$ ... (*)

where, $\quad f_1(x) = e^{-x},\ x > 0 \quad$ and $\quad f_2(y) = e^{-y},\ y > 0 \quad$ ... (**)

$\Rightarrow$ (*i*) X and Y are independent and (*ii*) $X \sim \gamma(1)$ and $Y \sim \gamma(1)$.

Thus, X and Y are *i.i.d.* $\gamma(1)$ variates.

Hence, by additive property of Gamma distribution, $(X + Y) \sim \gamma(1 + 1) = \gamma(2)$.

$\therefore \quad M_{X+Y}(t) = (1-t)^{-2} \quad$ [If $X \sim \gamma(n)$, $M_X(t) = (1-t)^{-n}$] ... (***)

The *m.g.f.* of $U = \frac{1}{2}(X+Y)$ is given by:

$$M_U(t) = M_{\frac{1}{2}(X+Y)}(t) = M_{X+Y}(t/2) = \left(1 - \frac{t}{2}\right)^{-2} \qquad \text{[From (***)]}$$

which is *m.g.f.* of two parameter Gamma distribution with parameters $a = 2$ and $n = 2$.

Hence, by uniqueness theorem of *m.g.f.* :

$$U = \frac{1}{2}(X+Y) \sim \gamma(a = 2, n = 2), \text{ with } p.d.f.$$

$$g(u) = \frac{a^n}{\Gamma n}\, e^{-au}\, u^{n-1} = \frac{2^2}{\Gamma 2}\, e^{-2u}\, u\,;\ u > 0$$

$$\Rightarrow \quad g(u) = 4u\,e^{-2u}\,;\ u > 0.$$

Example 9·40. *Let $p \sim \beta_1(a, b)$ where a and b are positive integers. After one observes p, one secures a coin for which the probability of head is p. This coin is flipped n times. Let X denote the number of heads which result. Find $P(X = k)$ for $k = 0, 1, 2, ..., n$. Express the answer in terms of binomial coefficients.*

Solution. Since $p \sim \beta_1(a, b)$, its *p.d.f.* is given by :

$$f(p) = \frac{1}{B(a, b)} \cdot p^{a-1}(1-p)^{b-1},\ 0 < p < 1$$

$$P(X = k \mid \text{the probability of success in a single trial is } p) = \binom{n}{k} p^k q^{n-k},\ q = 1 - p$$

$$\therefore \qquad P(X=k) = \int_0^1 f(p)\, P(X=k \mid p)\, dp$$

$$= \int_0^1 \frac{1}{B(a,b)} \cdot p^{a-1}(1-p)^{b-1} \cdot \binom{n}{k} p^k (1-p)^{n-k}\, dp$$

$$= \frac{\binom{n}{k}}{B(a,b)} \int_0^1 p^{a+k-1}(1-p)^{n+b-k-1}\, dp = \frac{\binom{n}{k} B(a+k, n+b-k)}{B(a,b)}$$

We have $\dfrac{1}{B(m,n)} = \dfrac{\Gamma(m+n)}{\Gamma(m)\,\Gamma(n)} = \dfrac{(m+n-1)!}{(m-1)!\,(n-1)!} = \dfrac{mn}{m+n}\dbinom{m+n}{m}$ ... (*)

$$\therefore \qquad P(X=k) = \frac{\binom{n}{k} \dfrac{ab}{a+b} \binom{a+b}{a}}{\dfrac{(a+k)(n+b-k)}{(n+a+b)} \binom{n+a+b}{a+k}} = \frac{\binom{n}{k}\binom{a+b}{a}}{\binom{n+a+b}{a+k}} \cdot \frac{ab(n+a+b)}{(a+b)(a+k)(n+b-k)}.$$

Example 9·41. *Given the Incomplete Beta Function,*

$$B_x(l,m) = \int_0^x x^{l-1}(l-x)^{m-1}\, dx \qquad \textit{and} \qquad I_x(l,m) = B_x(l,m)/B(l,m),$$

show that $I_x(l,m) = 1 - I_{1-x}(m,l)$.

Solution. We have

$$I_x(l,m)\, B(l,m) = B_x(l,m) = \int_0^x x^{l-1}(1-x)^{m-1}\, dx, \; 0 \le x \le 1. \qquad \text{... (*)}$$

$$= \int_0^1 x^{l-1}(1-x)^{m-1}\, dx - \int_x^1 x^{l-1}(a-x)^{m-1}\, dx$$

$$= B(l,m) - \int_x^1 x^{l-1}(1-x)^{m-1}\, dx$$

In the integral, put $1-x=y$, then

$$I_x(l,m) \cdot B(l,m) = B(l,m) - \int_{1-x}^0 (1-y)^{l-1}\, y^{m-1}(-dy)$$

$$= B(l,m) - \int_0^{1-x} y^{m-1}(1-y)^{l-1}\, dy$$

$$= B(l,m) - B_{1-x}(m,l) = B(l,m) - I_{1-x}(m,l)\, B(m,l) \qquad \text{[From (*)]}$$

Since $B(l,m) = B(m,l)$, we get on dividing throughout by $B(l,m)$,

$I_x(l,m) = 1 - I_{1-x}(m,l)$.

Note. This result gives the incomplete Beta function in terms of incomplete Gamma function.

9·8. EXPONENTIAL DISTRIBUTION

Definition. *A r.v. X is said to have an exponential distribution with parameter $\theta > 0$, if its p.d.f. is given by :*

$$f(x, \theta) = \begin{cases} \theta e^{-\theta x}, & x \geq 0 \\ 0, & \text{otherwise} \end{cases} \qquad \text{... (9·26)}$$

and we write $X \sim expo\,(\theta)$.

The cumulative distribution function $F(x)$ is given by :

$$F(x) = \int_0^x f(u)\,du = \theta \int_0^x \exp(-\theta u)\,du$$

$$F(x) = \begin{cases} 1 - \exp(-\theta x), & x \geq 0 \\ 0, & \text{otherwise} \end{cases} \qquad \text{... (9·26a)}$$

Remark. Relation between Exponential Distribution and Gamma Distribution.
In the two parameter Gamma distribution $\gamma(a, \lambda)$, with *p.d.f.*

$$f(x) = \frac{a^\lambda}{\Gamma\lambda} e^{-ax} x^{\lambda-1}\,;\ x > 0,\ \lambda > 0,\ a > 0,$$

if we take $\lambda = 1$, we get

$$f(x) = a e^{-ax}\,;\ x > 0,\ a > 0$$

which is the *p.d.f.* of exponential distribution with parameter '*a*'.

Hence, the exponential distribution (9·26) may be regarded as a special case of Gamma distribution $\gamma(a = \theta, \lambda = 1)$. In other words :

$$X \sim \text{expo}\,(\theta) \quad \Rightarrow \quad X \sim \gamma(\theta, 1).$$

9·8·1. Moment Generating Function of Exponential Distribution

$$M_X(t) = E(e^{tX}) = \theta \int_0^\infty e^{tx} e^{-\theta x}\,dx = \theta \int_0^\infty \exp\{-(\theta - t)x\}\,dx$$

$$= \frac{\theta}{(\theta - t)} = \left(1 - \frac{t}{\theta}\right)^{-1} = \sum_{r=0}^{\infty} \left(\frac{t}{\theta}\right)^r,\ \theta > t \qquad \text{... (9·26b)}$$

$$\therefore \qquad \mu_r' = E(X^r) = \text{Coefficient of } \frac{t^r}{r!} \text{ in } M_X(t) = \frac{r!}{\theta^r}\,;\ r = 1, 2, \ldots$$

$$\Rightarrow \qquad \text{Mean} = \mu_1' = \frac{1}{\theta} \text{ and Variance} = \mu_2 = \mu_2' - \mu_1'^2 = \frac{2}{\theta^2} - \frac{1}{\theta^2} = \frac{1}{\theta^2}$$

Hence, if $X \sim$ expo (θ), then Mean $= \dfrac{1}{\theta}$ and Variance $= \dfrac{1}{\theta^2}$. ... (9·26c)

Remarks :

1. $\text{Variance} = \dfrac{1}{\theta^2} = \dfrac{1}{\theta} \cdot \dfrac{1}{\theta} = \dfrac{\text{Mean}}{\theta}$

$\therefore$ Variance > Mean, if $0 < \theta < 1$; Variance = Mean, if $\theta = 1$; and Variance < Mean, if $\theta > 1$.

Hence, for the exponential distribution,

Variance >, =, or < mean, for different values of the parameter θ. ... (9·26d)

2. Does Exponential Distribution Possess Additive Property ?

Let X_i ($i = 1, 2, ..., n$) be *i.i.d.* exponential variates each with parameter θ. Then,

$$M_{X_i}(t) = \left(1 - \frac{t}{\theta}\right)^{-1} \quad ; \; i = 1, 2, \ldots, n \qquad \ldots (9\cdot26e)$$

The *m.g.f.* of sum is given by :

$$M_{\sum_{i=1}^{n} X_i}(t) = \prod_{i=1}^{n} M_{X_i}(t) = \left[M_{X_i}(t)\right]^n \qquad [\because X_1, X_2, ..., X_n \text{ are } i.i.d.]$$

$$= \left[\left(1 - \frac{t}{\theta}\right)^{-1}\right]^n = \left(1 - \frac{t}{\theta}\right)^{-n}, \qquad \text{[From (9·26e)]}$$

which is the *m.g.f.* of two parameter Gamma distribution with parameters ($a = \theta$ and n). Hence, by uniqueness theorem of *m.g.f.*'s,

$$X = \sum_{i=1}^{n} X_i \sim \gamma(a = \theta, n) \text{ with } p.d.f.$$

$$g(x) = \frac{a^n}{\Gamma n} e^{-ax} x^{n-1} = \frac{\theta^n}{\Gamma n} e^{-\theta x} \cdot x^{n-1}; \quad x > 0, \; \theta > 0, \; n > 0 \qquad \ldots (9\cdot26f)$$

Hence, we have the following result.

If X_i, ($i = 1, 2, \ldots, n$) are *i.i.d.* expo (θ), then $\sum_{i=1}^{n} X_i \sim \gamma(\theta, n)$.

Thus, exponential distribution does not possess the additive property.

Theorem. *If $X_1, X_2, ..., X_n$ are independent r.v.'s, X_i having an exponential distribution with parameter θ_i ; $i = 1, 2, ..., n$; then $Z = \min(X_1, X_2, ..., X_n)$ has exponential distribution with parameter $\sum_{i=1}^{n} \theta_i$.*

Proof. $G_Z(z) = P(Z \le z) = 1 - P(Z > z) = 1 - P[\min(X_1, X_2, \ldots, X_n) > z]$

$$= 1 - P(X_i > z; i = 1, 2, \ldots, n) = 1 - \prod_{i=1}^{n} P(X_i > z) \quad (\because X_1, X_2, ..., X_n \text{ are independent})$$

$$= 1 - \prod_{i=1}^{n} [1 - P(X_i \le z)] = 1 - \prod_{i=1}^{n} \left[1 - F_{X_i}(z)\right]$$

(where F is the distribution function of X_i)

$$= 1 - \prod_{i=1}^{n} \left[1 - \left(1 - e^{-\theta_i z}\right)\right] = \begin{cases} 1 - \exp\left\{\left(-\sum_{i=1}^{n} \theta_i\right) z\right\}, & z > 0 \\ 0, & \text{otherwise} \end{cases}$$

$$\therefore \quad g_Z(z) = \frac{d}{dz}[G(z)] = \begin{cases} \left(\sum_{i=1}^{n} \theta_i\right) \exp\left\{\left(-\sum_{i=1}^{n} \theta_i\right) z\right\}, & z > 0 \\ 0, & \text{otherwise} \end{cases}$$

$\Rightarrow$ $Z = \min(X_1, X_2, ..., X_n)$ is an exponential variate with parameter $\sum_{i=1}^{n} \theta_i$.

Cor. If X_i, $i = 1, 2, ..., n$ are identically distributed, following exponential distribution with parameter θ, then $Z = \min(X_1, X_2, ..., X_n)$ is also exponentially distributed with parameter $(n\,\theta)$.

Example 9·42. *If X and Y are i.i.d. random variables with common p.d.f.*

$$f(x) = e^{-x}\,;\; x > 0,$$

find the p.d.f. of $U = (X/Y)$.

Solution. The distribution function $G(\cdot)$ of $U = (X/Y)$ is given by :

$$G(u) = P(U \le u) = P\left(\frac{X}{Y} \le u\right) = \iint_{\left(\frac{x}{y} \le u\right)} f(x, y)\, dx\, dy$$

$$= \iint_{\left(\frac{x}{y} \le u\right)} f_X(x)\, f_Y(y)\, dx\, dy = \int_{\frac{x}{y} \le u} e^{-(x+y)}\, dx\, dy$$

$$= \int_{y=0}^{\infty} e^{-y} \left(\int_{x=0}^{uy} e^{-x}\, dx \right) dy = \int_{y=0}^{\infty} e^{-y} \left| \frac{e^{-x}}{-1} \right|_{x=0}^{uy} dy$$

$$= \int_{y=0}^{\infty} e^{-y} \left(1 - e^{-uy}\right) dy = \left| \frac{e^{-y}}{-1} + \frac{e^{-(1+u)y}}{1+u} \right|_{y=0}^{\infty}$$

$$G(u) = 1 - \frac{1}{1+u}\,;\; u > 0.$$

Differentiating *w.r.* to u, the *p.d.f.* of $U = (X/Y)$ is given by :

$$g(u) = G'(u) = \frac{1}{(1+u)^2}\,;\; u > 0.$$

Aliter. Transform (X, Y) to $\left(U = \dfrac{X}{Y},\; V = Y\right)$ and proceed in the usual manner.

Example 9·43. *Show that the exponential distribution 'lacks memory', i.e., if X has an exponential distribution, then for every constant* $a \ge 0$, *one has*

$$P\left(Y \le x \mid X \ge a\right) = P(X \le x) \text{ for all } x, \text{ where } Y = X - a.$$

Solution. The *p.d.f.* of the exponential distribution with parameter θ is :

$$f(x) = \theta \exp(-\theta x)\,;\; \theta > 0, 0 < x < \infty$$

We have,

$$P(Y \le x \cap X \ge a) = P(X - a \le x \cap X \ge a) \qquad (\because\; Y = X - a)$$

$$= P(X \le a + x \cap X \ge a) = P(a \le X \le a + x)$$

$$= \theta \int_a^{a+x} e^{-\theta x}\, dx = e^{-a\theta}\left(1 - e^{-\theta x}\right)$$

and

$$P(X \ge a) = \theta \int_a^{\infty} e^{-\theta x}\, dx = e^{-a\theta}$$

$$\therefore \quad P(Y \le x \mid X \ge a) = \frac{P(Y \le x \cap X \ge a)}{P(X \ge a)} = 1 - e^{-\theta x} \quad \text{... (*)}$$

Also $$P(X \le x) = \theta \int_0^x e^{-\theta x}\, dx = 1 - e^{-\theta x} \quad \text{... (**)}$$

From (*) and (**), we get $P(Y \le x \mid X \ge a) = P(X \le x)$

i.e., *exponential distribution lacks memory.*

9·8·2. Characterisation of Exponential Distribution

If $X_1, X_2, ..., X_n$ are i.i.d. random variables with continuous p.d.f., then $Y = \min(X_1, X_2, ..., X_n)$ has exponential distribution with parameter $n\theta$ if and only if each $X_i \sim Expo(\theta)$.

Proof. See Example 9·51, page 9·92.

Example 9·44. *X and Y are independent with a common p.d.f. (exponential) :*

$$f(x) = \begin{cases} e^{-x}, & x \ge 0 \\ 0, & x < 0 \end{cases}$$

Find a p.d.f. for X – Y and identify the distribution.

Solution. Since X and Y are independent and identically distributed (*i.i.d.*), their joint *p.d.f.* is given by :

$$f_{XY}(x, y) = f_X(x) \cdot f_Y(y) = \begin{cases} e^{-(x+y)}; & x > 0, y > 0 \\ 0, & \text{otherwise} \end{cases}$$

Let $$\begin{cases} u = x - y \\ v = y \end{cases} \Rightarrow \begin{cases} x = u + v \\ y = v \end{cases} \quad \text{... (*)} \qquad J = \frac{\partial(x, y)}{\partial(u, v)} = \begin{vmatrix} 1 & 1 \\ 0 & 1 \end{vmatrix} = 1$$

Thus, the joint *p.d.f.* of *U* and *V* becomes :

$$g(u, v) = e^{-(u+2v)}; \; v > 0, -\infty < u < \infty$$

$$(*) \Rightarrow \quad u = x - v \quad \Rightarrow \quad v = x - u \quad \text{Thus,} \quad \left.\begin{matrix} v > -u, \text{ if } -\infty < u < 0 \\ v > 0, \text{ if } u > 0 \end{matrix}\right\} \quad \text{[From (*)]}$$

For $-\infty < u < 0$, $$g(u) = \int_{-u}^{\infty} g(u, v)\, du = \int_{-u}^{\infty} e^{-(u+2v)}\, dv = e^{-u} \left| \frac{e^{-2v}}{-2} \right|_{-u}^{\infty} = \frac{1}{2} e^{u}$$

and for $u > 0$, $$g(u) = \int_0^{\infty} g(u, v)\, dv = e^{-u} \left| \frac{e^{-2v}}{-2} \right|_0^{\infty} = \frac{1}{2} e^{-u}$$

Hence, the *p.d.f.* of $U = X - Y$ if given by : $$g(u) = \begin{cases} \frac{1}{2} e^{u}, & -\infty < u < 0 \\ \frac{1}{2} e^{-u}, & u > 0 \end{cases}$$

These results can be cambined to give : $$g(u) = \frac{1}{2} e^{-|u|}, -\infty < u < \infty$$

which is the *p.d.f.* of standard Laplace distribution (*c.f.* § 9·9).

Aliter.

$$M_X(t) = \int_0^{\infty} e^{tx} f(x)\,dx = \int_0^{\infty} e^{-(1-t)x}\,dx = \left| \frac{e^{-(1-t)x}}{-(1-t)} \right|_0^{\infty} = \frac{1}{1-t},\; t<1$$

∴ The characteristic function of X is :

$$\varphi_X(t) = \frac{1}{1-it} = \varphi_Y(t), \qquad \text{(Since } X \text{ and } Y \text{ are identically distributed)}$$

Hence, $\varphi_{X-Y}(t) = \varphi_{X+(-Y)}(t) = \varphi_X(t)\cdot\varphi_{-Y}(t)$ (∵ X, Y are independent.)

$$= \varphi_X(t)\cdot\varphi_Y(-t) = \frac{1}{(1-it)(1+it)} = \frac{1}{1+t^2},$$

which is the characteristic function of the Standard Laplace distribution (*c.f.* § 9·9·1) with *p.d.f.*

$$g(u) = \frac{1}{2} e^{-|u|},\; -\infty < u < \infty \qquad \ldots (*)$$

Hence, by the uniqueness theorem of characteristic functions, $U = X - Y$ has Standard Laplace distribution with the *p.d.f.* given in (*).

9·9. STANDARD LAPLACE (DOUBLE EXPONENTIAL) DISTRIBUTION

> ***Definition.*** *A continuous r.v. X is said to have standard Laplace (double exponential) distribution if its p.d.f. is given by :*
>
> $$f(x) = \frac{1}{2}\exp(-|x|);\; -\infty < x < \infty. \qquad \ldots (9{\cdot}27)$$

9·9·1. Characteristic Function of Standard Distribution

$$\varphi_X(t) = \int_{-\infty}^{\infty} e^{itx} f(x)\,dx = \frac{1}{2}\int_{-\infty}^{\infty} e^{itx} e^{-|x|}\,dx$$

$$= \frac{1}{2}\left[\int_{-\infty}^{\infty} \cos tx\cdot e^{-|x|}\,dx + i\int_{-\infty}^{\infty} \sin tx\, e^{-|x|}\,dx\right]$$

$$= \frac{1}{2}\cdot 2\int_0^{\infty} \cos tx\cdot e^{-|x|}\,dx,$$

since the integrands in the first and second integrals are even and odd function of x, respectively.

$$\therefore \quad \varphi_X(t) = \int_0^{\infty} e^{-x}\cos tx\,dx = 1 - t^2\int_0^{\infty} e^{-x}\cos tx\,dx \qquad \text{(On integration by parts)}$$

$$= 1 - t^2\,\varphi_X(t)$$

$$\Rightarrow \quad \varphi_X(t) = \frac{1}{1+t^2} = (1+t^2)^{-1}$$

$$\therefore \quad \kappa_X(t) = \log\varphi_X(t) = -\log(1+t^2)$$

$$= -\left[t^2 - \frac{(t^2)^2}{2} + \frac{(t^2)^3}{3} - \ldots\right] = \left[(it)^2 + \frac{(it)^4}{2} + \frac{(it)^6}{3} - \ldots\right] \qquad \ldots (*)$$

Since $\kappa_X(t)$ involves only even powers of t, all cumulants of odd order are zero, *i.e.*,

$$\kappa_1 = \kappa_3 = \kappa_5 = \ldots = 0 \quad \Rightarrow \quad \kappa_{2r+1} = 0\,;\; r = 0, 1, 2, \ldots \qquad \ldots (9\cdot27b)$$

Even order cumulants are given by :

$$\kappa_2 = \text{Coefficient of } \frac{(it)^2}{2!} \text{ in } (*) = 2 \quad ; \quad \kappa_4 = \text{Coefficient of } \frac{(it)^4}{4!} \text{ in } (*) = 4 \times 3 = 12$$

$$\therefore \quad \text{Mean} = \kappa_1 = 0 \quad ; \quad \text{Variance} = \mu_2 = \kappa_2 = 2 \qquad \ldots (9\cdot27c)$$

$$\mu_3 = \kappa_3 = 0 \quad ; \quad \mu_4 = \kappa_4 + 3\kappa_2^2 = 12 + 12 = 24$$

$$\therefore \quad \beta_1 = \frac{\mu_3^2}{\mu_2^3} = 0 \quad \text{and} \quad \beta_2 = \frac{\mu_4}{\mu_2^2} = \frac{24}{4} = 6 \qquad \ldots (9\cdot27d)$$

Remark. Mean deviation about mean for Standard Laplace distribution is 1. (Try it).

9·9·2. Two Parameter Laplace Distribution

Definition. *A continuous r.v. X is said to have a double exponential (Laplace) distribution with two parameters λ and μ if its p.d.f. is given by :*

$$f(x; \mu, \sigma) = \frac{1}{2\lambda} \exp\left(-|x-\mu|/\lambda\right); \; -\infty < x < \infty, \; \lambda > 0 \qquad \ldots (9\cdot27)$$

We write $X \sim \text{Lap}(\lambda, \mu)$.

Let
$$Y = \frac{X-\mu}{\lambda} \quad \Rightarrow \quad X = \mu + \lambda Y \qquad \ldots (*)$$

The *p.d.f.* $g(\cdot)$ of Y is given by :

$$g(y) = f(x)\left|\frac{dx}{dy}\right| = \frac{1}{2\lambda} \exp\left(-|y|\right) \cdot \lambda \qquad \text{[From (*)]}$$

$$= \frac{1}{2} \exp\left(-|y|\right); \; -\infty < y < \infty, \qquad \ldots (9\cdot27a)$$

which is the *p.d.f.* of standard Laplace distribution

Hence, if $X \sim \text{Lap}(\lambda, \mu)$, then $Y = \dfrac{X-\mu}{\lambda} \sim \text{Lap}(1, 0)$

i.e., Y has standard Laplace distribution with *p.d.f.* $g(y)$ given in (9·27*a*).

9·9·3. Characteristic Function of Two Parameter Laplace Distribution

If $X \sim \text{Lap}(\lambda, \mu)$, then

$$\Phi_X(t) = E\left(e^{itX}\right) = E\left[e^{it(\mu+\lambda Y)}\right], \quad \text{where } Y = \frac{X-\mu}{\lambda} \sim \text{Lap}(1, 0)$$

$$= e^{it\mu} \cdot E\left(e^{i \cdot \lambda t \cdot Y}\right) = e^{it\mu} \cdot \Phi_Y(\lambda t)$$

$$= \frac{e^{it\mu}}{1 + \lambda^2 t^2} \qquad \left[\because Y \text{ is standard Laplace variate with } \Phi(t) = \frac{1}{1+t^2}\right] \qquad \ldots (9\cdot27b)$$

9·9·4. Moments of Two Parameter Laplace Distribution

If $X \sim \text{Lap}(\lambda, \mu)$, the rth moment about origin is given by :

$$\mu_r' = E(X^r) = \frac{1}{2\lambda} \int_{-\infty}^{\infty} x^r \exp\left(\frac{-|x-\mu|}{\lambda}\right) dx$$

$$= \frac{1}{2}\int_{-\infty}^{\infty}(z\lambda+\mu)^r \exp(-|z|)\,dz, \qquad \left(z=\frac{x-\mu}{\lambda}\right)$$

$$= \frac{1}{2}\int_{-\infty}^{\infty}\left[\sum_{k=0}^{r}\binom{r}{k}(z\lambda)^k\mu^{r-k}\right]\exp(-|z|)\,dz$$

$$= \frac{1}{2}\sum_{k=0}^{r}\left[\binom{r}{k}\lambda^k\mu^{r-k}\int_{-\infty}^{\infty}z^k\exp(-|z|)\,dz\right]$$

$$= \frac{1}{2}\sum_{k=0}^{r}\left[\binom{r}{k}\lambda^k\mu^{r-k}\left\{\int_{-\infty}^{0}z^k e^{-|z|}\,dz+\int_{0}^{\infty}z^k e^{-|z|}\,dz\right\}\right]$$

$$= \frac{1}{2}\sum_{k=0}^{r}\left[\binom{r}{k}\lambda^k\mu^{r-k}\left\{(-1)^k\int_{0}^{\infty}e^{-z}z^k\,dz+\int_{0}^{\infty}e^{-z}z^k\,dz\right\}\right]$$

$$= \frac{1}{2}\sum_{k=0}^{r}\left[\binom{r}{k}\lambda^k\mu^{r-k}\,\Gamma(k+1)\{(-1)^k+1\}\right]$$

$$\Rightarrow \qquad \mu_r' = \frac{1}{2}\sum_{k=0}^{r}\left[\binom{r}{k}\lambda^k\mu^{r-k}\,k!\{1+(-1)^k\}\right] \qquad \ldots(9\cdot27c)$$

$\therefore$ Mean $=\mu_1'=\mu$, $\mu_2'=\mu^2+2\lambda^2$ and Variance $=\mu_2=\mu_2'-\mu_1'^2=2\lambda^2$.

Similarly, we can obtain higher order moments from (9·27*c*) and hence the values of β_1 and β_2 can be obtained.

Note. We can also obtain moments and β_1 and β_2 by using cumulant generating function :

$$\kappa_X(t)=\log\Phi_X(t)=i\mu t-\log(1+\lambda^2t^2), \qquad \text{[From (9·27}b\text{)]}$$

and proceeding as usual.

9·10. WEIBUL DISTRIBUTION

A random variable X has a Weibul distribution with three parameters $c\ (>0)$, $\alpha\ (>0)$ and μ if the *r.v.*

$$Y=\left(\frac{X-\mu}{\alpha}\right)^c \qquad \ldots(9\cdot28)$$

has the exponential distribution with *p.d.f.*

$$p_Y(y)=e^{-y},\ y>0 \qquad \ldots(9\cdot28a)$$

Definition. *A continuous r.v. X has a Weibul distribution with parameters c (> 0), α (> 0), and μ if its p.d.f. is :*

$$f(x;c,\alpha,\mu)=\frac{c}{\alpha}\left(\frac{x-\mu}{\alpha}\right)^{c-1}\exp\left\{-\left(\frac{x-\mu}{\alpha}\right)^c\right\};\quad x>\mu,\ c>0 \qquad \ldots(9\cdot28b)$$

The standard Weibul distribution is obtained on taking $\alpha=1$ and $\mu=0$, so that the *p.d.f.* of *standard Weibul distribution* which depends only on a single parameter *c* is given by :

$$p_X(x)=c\,x^{c-1}\cdot\exp(-x^c);\quad x>0,\ c>0 \qquad \ldots(9\cdot28c)$$

9·10·1. Moments of Standard Weibul Distribution

For standard Weibul distribution, $(\alpha = 1, \mu = 0)$, from (9·28), we get $Y = X^c$ which has the exponential distribution (9·28*a*). We have

$$\mu_r' = E(X^r) = E(Y^{1/c})^r = E(Y^{r/c}) = \int_0^\infty e^{-y} y^{r/c}\, dy \quad \Rightarrow \quad \mu_r' = \Gamma\left(\frac{r}{c} + 1\right)$$

$$\therefore \qquad \text{Mean} = E(X) = \Gamma\left(\frac{1}{c} + 1\right)$$

and
$$\text{Var}(X) = E(X^2) - [E(X)]^2 = \Gamma\left(\frac{2}{c} + 1\right) - \left[\Gamma\left(\frac{1}{c} + 1\right)\right]^2$$

Similarly, we can obtain expressions for higher order moments and hence for β_1 and β_2. For large c, the mean is approximated by :

$$E(X) \simeq 1 - \frac{\gamma}{c} + \frac{1}{2c^2}\left(\frac{\pi^2}{6} + \gamma^2\right)$$

$$= 1 - 0{\cdot}57722\, c^{-1} + 0{\cdot}98905\, c^{-2},$$

where $\gamma = 0{\cdot}57722$ is Euler's constant.

The distribution is named after Waloddi Weibul, a Swedish physicist, who used it in 1939 to represent the distribution of the breaking strength of materials. Kao, J.H.K. (1958-59) advocated the use of this distribution in reliability studies and quality control work. It is also used as a tolerance distribution in the analysis of quantum response data.

9·10·2. Characterisation of Weibul Distribution

Dubey, S.D. (1968) has obtained the following result :

"Let X_i (i = 1, 2, ..., n) be i.i.d. random variables. Then min $(X_1, X_2, ..., X_n)$ has a Weibul distribution if and only if the common distribution of X_i's is a Weibul Distribution."

Proof. Let X_i, $(i = 1, 2, ..., n)$ be *i.i.d. r.v.*'s each with Weibul distribution (9·28*b*) and let $Y = \min(X_1, X_2, ..., X_n)$. Then

$$P(Y > y) = P\left[\min(X_1, X_2, \ldots, X_n) > y\right] = P\left[\bigcap_{i=1}^{n} X_i > y\right]$$

$$= \prod_{i=i}^{n} P(X_i > y) = \left[P(X_i > y)\right]^n \qquad [\because X_i\text{'s are } i.i.d.\ r.v.\text{'s}] \qquad \ldots (*)$$

$$P(X_i > y) = \int_y^\infty c\,\alpha^{-1}\left(\frac{x-\mu}{\alpha}\right)^{c-1} \exp\left\{-\left(\frac{x-\mu}{\alpha}\right)^c\right\} dx$$

$$= \int_{[(y-\mu)/\alpha]^c}^\infty e^{-t}\, dt, \qquad t = \left(\frac{x-\mu}{\alpha}\right)^c$$

$$= \exp\left\{-\left(\frac{y-\mu}{\alpha}\right)^c\right\}$$

Substituting in (*), we get

$$P(Y > y) = \left[\exp\left\{ -\left(\frac{y-\mu}{\alpha}\right)^c \right\} \right]^n = \exp\left\{ -n\left(\frac{y-\mu}{\alpha}\right)^c \right\}$$

$$= \exp\left[-\left\{ \frac{n^{1/c}(y-\mu)}{\alpha} \right\}^c \right]$$

This implies that Y has the same Weibul distribution as X_i's with the difference that the parameter α is replaced by $\alpha n^{-1/c}$.

9·11. LOGISTIC DISTRIBUTION

Definition. *A continuous r.v. X is said to have a Logistic distribution with parameters α and β, if its distribution function is of the form :*

$$F_X(x) = \left[1 + exp\left\{-(x-\alpha)/\beta\right\}\right]^{-1}, \ \beta > 0 \qquad \text{... (9·29)}$$

$$= \frac{1}{2}\left[1 + \tanh\left\{\frac{1}{2}(x-\alpha)/\beta\right\}\right]; \ \beta > 0 \qquad \text{[See Remark 1 below]... (9·29a)}$$

The *p.d.f.* of Logistic distribution with parameters α and β (> 0) is given by :

$$f(x) = \frac{d}{dx}[F(x)]$$

$$= \frac{1}{\beta}[1 + \exp\{-(x-\alpha)/\beta]^{-2}\,[\exp\{-(x-\alpha)/\beta\}] \qquad \text{... (9·29b)}$$

$$= \frac{1}{4\beta}\operatorname{sech}^2\left\{\frac{1}{2}(x-\alpha)/\beta\right\} \qquad \text{... (9·29c)}$$

The *p.d.f.* of standard Logistic variate $Y = (X - \alpha)/\beta$, is given by :

$$g_Y(y) = f(x)\cdot\left|\frac{dx}{dy}\right| = e^{-y}(1+e^{-y})^{-2}; -\infty < y < \infty \qquad \text{... (9·29d)}$$

$$= \frac{1}{4}\operatorname{sech}^2\left(\frac{1}{2}y\right); \ -\infty < y < \infty \qquad \text{... (9·29e)}$$

The distribution function of Y is : $\quad G_Y(y) = (1+e^{-y})^{-1}; \ -\infty < y < \infty$ ... (9·29f)

Logistic distribution is extensively used as growth function in population and demographic studies and in time series analysis. Theoretically, Logistic distribution can be obtained as :

(*i*) The limiting distribution (as $n \to \infty$) of the standardised mid-range, (average of the smallest and the largest sample observations), in random samples of size n.

(*ii*) A mixture of extreme value distributions.

9·11·1. Moment Generating Function of Logistic Distribution

The *m.g.f.* of standard Logistic variate Y is given by :

$$\mu_Y(t) = E(e^{tY}) = \int_{-\infty}^{\infty} e^{ty}\cdot g(y)\,dy = \int_{-\infty}^{\infty} e^{ty}\cdot e^{-y}\left(1+e^{-y}\right)^{-2} dy$$

$$= \int_{-\infty}^{\infty} e^{ty} e^{-y} \left(\frac{1+e^y}{e^y} \right)^{-2} dy = \int_{-\infty}^{\infty} e^{ty} \cdot e^y (1+e^y)^{-2} \, dy$$

Put $\quad z = (1+e^y)^{-1} \quad \Rightarrow \quad e^y = \dfrac{1}{z} - 1 = \dfrac{1-z}{z}$

$$\therefore \quad M_Y(t) = \int_1^0 \left(\frac{1-z}{z} \right)^t \cdot (-dz) = \int_0^1 z^{-t} (1-z)^t \, dz = B(1-t, 1+t), \quad 1-t > 0$$

$$= \Gamma(1-t)\,\Gamma(1+t)/\Gamma(2) = \Gamma(1-t)\,\Gamma(1+t)$$

$$= \pi t \operatorname{cosec} \pi t; \quad t < 1$$

$$= 1 + \frac{\pi^2 t^2}{6} + \frac{7}{360} \pi^4 t^4 + \dots \qquad \text{(See Remark 2 below)} \quad \dots (9{\cdot}29g)$$

$\therefore \quad E(Y) = \text{Coefficient of } t \text{ in } (9{\cdot}29g) = 0 \quad \Rightarrow \quad \text{Mean} = 0$

$\therefore \quad \mu_2 = E(Y^2) = \text{Coefficient of } \dfrac{t^2}{2!} \text{ in } (9{\cdot}29g) = \dfrac{\pi^2}{3}, \quad \mu_3 = E(Y^3) = 0$

$$\mu_4 = E(Y^4) = \text{Coefficient of } \frac{t^4}{4!} \text{ in } (9{\cdot}29g) = \frac{7}{15}\pi^4$$

Hence, for standard Logistic distribution :

$$\text{Mean} = 0, \qquad \text{Variance} = \mu_2 = \frac{\pi^2}{3}, \qquad \beta_1 = \frac{\mu_3^2}{\mu_2^3} = 0, \qquad \beta_2 = \frac{\mu_4}{\mu_2^2} = \frac{7 \times 9}{15} = 4{\cdot}2.$$

Remarks :

1. We have

$$\tanh x = \frac{\sinh x}{\cosh x} = \frac{e^x - e^{-x}}{e^x + e^{-x}} = \frac{1 - e^{-2x}}{1 + e^{-2x}}$$

$$\Rightarrow \quad 1 + \tanh x = \frac{2}{1+e^{-2x}} \quad \Rightarrow \quad \frac{1}{2}(1 + \tanh x) = (1 + e^{-2x})^{-1}$$

2. $\quad x \operatorname{cosec} x = 1 + \dfrac{x^2}{6} + \dfrac{7}{360} x^4 + \dots$

Proof.

$$x \operatorname{cosec} x = \frac{x}{\sin x} = \frac{x}{\left(x - \dfrac{x^3}{3!} + \dfrac{x^5}{5!} - \dfrac{x^7}{7!} + \dots \right)} = \left[1 - \left(\frac{x^2}{6} - \frac{x^4}{120} + \frac{x^6}{7!} + \dots \right) \right]^{-1}$$

$$= 1 + \left(\frac{x^2}{6} - \frac{x^4}{120} + \dots \right) + \left(\frac{x^2}{6} - \frac{x^4}{120} + \dots \right)^2 + \dots$$

$$= 1 + \frac{x^2}{6} + x^4 \left(\frac{1}{36} - \frac{1}{120} \right) + \dots = 1 + \frac{x^2}{6} + \frac{7}{360} x^4 + \dots$$

3. We have : $\quad g(y) = e^{-y} \left(1 + \dfrac{1}{e^y} \right)^{-2} = e^y (1 + e^y)^{-2} = g(-y) \qquad \dots (*)$

$\Rightarrow \quad$ The probability curve of g is symmetric about the line $y = 0$.

Since *p.d.f.* $g(y)$ is symmetric about the origin $(y=0)$, all odd order moments about origin are zero, *i.e.*,

$$\mu'_{2r+1} = E(Y^{2r+1}) = 0,\ r = 0, 1, 2, \ldots \qquad \ldots (**)$$

In particular : Mean $= \mu'_1 = 0$

$\therefore \quad \mu'_r = r$th moment about origin $= r$th moment about mean $= \mu_r$

$\Rightarrow \quad \mu_{2r+1} = \mu'_{2r+1} = 0,$ [From (**)]

i.e., all odd order moments about mean of the standard logistic distribution are zero.

In particular, $\mu_3 = 0 \quad \Rightarrow \quad \beta_1 = 0.$

4. The mean and variance of the logistic Variable (X) with parameters α and β are given by :

$$E(X) = E(\alpha + \beta Y) = \alpha + \beta E(Y) = \alpha, \qquad \left(\because Y = \frac{X-\alpha}{\beta}\right)$$

$$\text{Var}(X) = \text{Var}(\alpha + \beta Y) = \beta^2\ \text{Var}(Y) = \beta^2 \pi^2/3.$$

5. We have : $G(y) = (1+e^{-y})^{-1} = \left(\dfrac{1+e^y}{e^y}\right)^{-1} = \dfrac{e^y}{1+e^y} \Rightarrow 1 - G(y) = 1 - \dfrac{e^y}{1+e^y} = \dfrac{1}{1+e^y}$

$$\therefore \quad G(y)\cdot[1-G(y)] = \frac{e^y}{(1+e^y)^2} = g(y) \qquad [\textit{c.f.}\ (*)\ \text{in Remark 3}] \qquad \ldots (9\cdot29h)$$

$$\text{Also,} \quad \frac{G(y)}{1-G(y)} = e^y \quad \Rightarrow \quad y = \log_e\left[\frac{G(y)}{1-G(y)}\right] \qquad \ldots (9\cdot29i)$$

6. Mean deviation for the standard Logistic distribution is :

$$2\left(1 - \frac{1}{2} + \frac{1}{3} - \frac{1}{4} + \ldots\right) = 2\sum_{i=1}^{\infty}\left[\frac{(-1)^{i-1}}{i}\right] = 2\log_e 2$$

Proof is left as an exercise to the reader.

9·12. CAUCHY DISTRIBUTION

Let us consider a roulette wheel in which the probability of the pointer stopping at any part of the circumference is constant. In other words, the probability that any value of θ lies in the interval $[-\pi/2, \pi/2]$ is constant and consequently θ is a rectangular variate in the range $[-\pi/2, \pi/2]$ with probability differential given by [See *Fig. 9·16*] :

$$dP(\theta) = \begin{cases} (1/\pi)\, d\theta, & -\pi/2 \le \theta \le \pi/2 \\ 0, & \text{otherwise} \end{cases} \qquad \ldots (9\cdot29j)$$

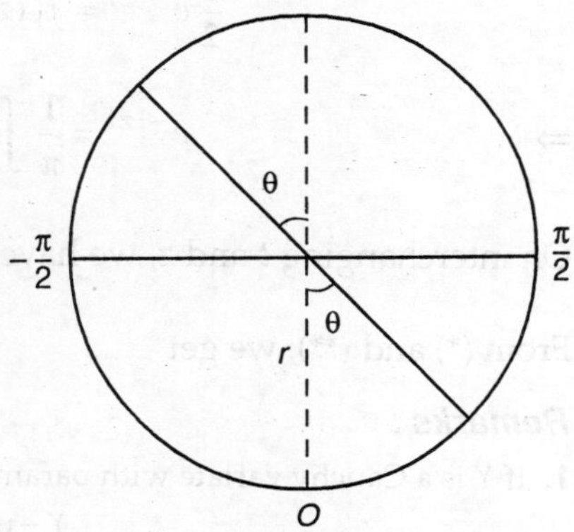

Fig. 9·16

Let us now transform to variable X by the substitution : $x = r\tan\theta \Rightarrow dx = r\sec^2\theta\, d\theta$

Since, $-\pi/2 \le \theta \le \pi/2$, the range for X is from $-\infty$ to ∞. Thus, the probability differential of X becomes :

$$dF(x) = \frac{1}{\pi}\cdot\frac{dx}{r\sec^2\theta} = \frac{1}{\pi}\cdot\frac{dx}{\left[r\{1+(x^2/r^2)\}\right]} = \frac{r}{\pi}\cdot\frac{dx}{r^2+x^2}; \quad -\infty < x < \infty$$

In particular if we take $r = 1$, we get : $f(x) = \dfrac{1}{\pi}\cdot\dfrac{1}{1+x^2}, \quad -\infty < x < \infty$

This is the *p.d.f.* of a standard Cauchy variate and we write $X \sim C(1, 0)$.

Definition. *A random variable X is said to have a standard Cauchy distribution if its p.d.f. is given by :*

$$f_X(x) = \frac{1}{\pi(1+x^2)}, \quad -\infty < x < \infty \qquad \text{... (9·30)}$$

and X is termed as standard Cauchy variate and we write $X \sim C(1, 0)$.
More generally, Cauchy distribution with two parameters λ and μ has the p.d.f.,

$$g_Y(y) = \frac{\lambda}{\pi\left[\lambda^2 + (y-\mu)^2\right]}, \quad -\infty < y < \infty;\ \lambda > 0 \qquad \text{... (9·30a)}$$

and we write $Y \sim C(\lambda, \mu)$.

But putting $X = (Y - \mu)/\lambda$ in (9·30*a*), we get (9·30).
Hence, if $Y \sim C(\lambda, \mu)$, then $X = (Y-\mu)/\lambda \sim C(1, 0)$ ... (9·30*b*)

9·12·1. Characteristics Function of (Standard) Cauchy Distribution

If *X* is a standard Cauchy variate then

$$\varphi_X(t) = \frac{1}{\pi}\int_{-\infty}^{\infty} \frac{e^{itx}}{1+x^2}\,dx \qquad \text{... (*)}$$

To evaluate (*) consider standard Laplace distribution $f_1(z) = \frac{1}{2}e^{-|z|},\ -\infty < z < \infty.$

Then $\varphi_1(t) = \phi_Z(t) = E(e^{itz}) = \dfrac{1}{1+t^2}.$

Since $\varphi_1(t)$ is absolutely integrable in $(-\infty, \infty)$, we have by Inversion theorem

$$\frac{1}{2}e^{-|z|} = f_1(z) = \frac{1}{2\pi}\int_{-\infty}^{\infty} e^{-itz}\,\varphi_1(t)\,dt = \frac{1}{2\pi}\int_{-\infty}^{\infty}\frac{e^{-itz}}{1+t^2}\,dt$$

$$\Rightarrow \quad e^{-|z|} = \frac{1}{\pi}\int_{-\infty}^{\infty}\frac{e^{-itz}}{1+t^2}\,dt = \frac{1}{\pi}\int_{-\infty}^{\infty}\frac{e^{itz}}{1+t^2}\,dt \qquad [\text{Changing } t \text{ to } -t]$$

On interchanging *t* and *z*, we have $e^{-|t|} = \dfrac{1}{\pi}\displaystyle\int_{-\infty}^{\infty}\frac{e^{itz}}{1+z^2}\,dz$ ... (**)

From (*) and (**), we get $\varphi_X(t) = e^{-|t|}$ ... (9·31)

Remarks :

1. If *Y* is a Cauchy variate with parameters λ and μ, then

$$X = \frac{Y-\mu}{\lambda} \sim C(1, 0), \quad \Rightarrow \quad Y = \mu + \lambda X$$

$$\therefore \quad \varphi_Y(t) = E\left(e^{itY}\right) = e^{i\mu t}\,E\left(e^{it\lambda X}\right) = e^{i\mu t}\,\varphi_X(t\lambda)$$

$$= e^{i\mu t - \lambda|t|}, \quad \lambda > 0 \qquad [\text{Using (9·31)}] \qquad \text{... (9·31a)}$$

2. ***Additive Property of Cauchy Distribution.*** *If* X_1 *and* X_2 *are independent Cauchy variates with parameters* (λ_1, μ_1) *and* (λ_2, μ_2) *respectively, then* $X_1 + X_2$ *is a Cauchy variate with parameters* $(\lambda_1 + \lambda_2, \mu_1 + \mu_2)$.

Proof. $\varphi_{X_j}(t) = \exp\left\{i\mu_j t - \lambda_j|t|\right\},\ (j = 1, 2)$ [From (9·31*a*)]

$$\therefore \quad \varphi_{X_1+X_2}(t) = \varphi_{X_1}(t)\,\varphi_{X_2}(t) \qquad (\text{Since } X_1, X_2 \text{ are independent})$$

$$= \exp\left[it(\mu_1+\mu_2) - (\lambda_1+\lambda_2)|t|\right]$$

and the result follows by uniqueness theorem of characteristic functions.

3. Since $\varphi'_X(t)$ in (9·31) [where (′) denotes differentiation *w.r.* to *t*] does not exist at $t = 0$, the mean of the Cauchy distribution does not exist.

4. Let $X_1, X_2, ..., X_n$ be a sample of *n* independent observations from a standard Cauchy distribution and define $\bar{X} = \frac{1}{n}\sum_{i=1}^{n} X_i$. Then

$$\varphi_{\bar{X}}(t) = \varphi_{\Sigma X_j}(t/n) = \prod_{j=1}^{n}\left[\varphi_{X_j}(t/n)\right] = \left[\varphi_{X_j}(t/n)\right]^n \qquad \text{[Since } X_j\text{'s are } i.i.d.]$$

$$= \left[e^{-|t/n|}\right]^n = e^{-|t|} = \varphi_X(t)$$

Hence, by uniqueness theorem of characteristic functions, we have :

"The arithmetic mean $\bar{X}$ of sample $X_1, X_2, ..., X_n$ of independent observations from a standard Cauchy distribution is also a standard Cauchy variate. In other words, the arithmetic mean of a random sample of any size yields exactly as much information as a single determination of X."

This implies that the sample mean $\bar{X}_n$ of random sample of size *n*, as an estimate of population mean does not improve with increasing *n*, which contradicts the Weak Law of Large Numbers (WLLN).

9·12·2. Moments of Cauchy Distribution

$$E(Y) = \int_{-\infty}^{\infty} y\, f(y)\, dy = \frac{\lambda}{\pi}\int_{-\infty}^{\infty} \frac{y}{\lambda^2 + (y-\mu)^2}\, dy = \frac{\lambda}{\pi}\int_{-\infty}^{\infty} \frac{(y-\mu)+\mu}{\lambda^2 + (y-\mu)^2}\, dy$$

$$= \mu \cdot \frac{\lambda}{\pi}\int_{-\infty}^{\infty} \frac{dy}{\lambda^2 + (y-\mu)^2} + \frac{\lambda}{\pi}\int_{-\infty}^{\infty} \frac{(y-\mu)}{\lambda^2 + (y-\mu)^2}\, dy$$

$$= \mu \cdot 1 + \frac{1}{\pi}\int_{-\infty}^{\infty} \frac{z}{\lambda^2 + z^2}\, dz,$$

Although the integral $\int_{-\infty}^{\infty} \frac{z}{\lambda^2 + z^2}\, dz$, is not completely convergent, *i.e.*, $\lim_{\substack{n\to\infty \\ n'\to\infty}} \int_{-n}^{n'} \frac{z}{\lambda^2 + z^2}\, dz$, does not exist, its principal value, *viz.*, $\lim_{n\to\infty} \int_{-n}^{n} \frac{z}{\lambda^2 + z^2}\, dz$, exists and is equal to zero. Thus, in the general sense, the mean of Cauchy distribution does not exist. But, if we conventionally agree to assume that the mean of Cauchy distribution exists (by taking the principal value), then it is located at $x = \mu$. Also, obviously, the probability curve is symmetrical about the point $y = \mu$. Hence, for this distribution, the mean, median and mode coincide at the point $y = \mu$.

$$\mu_2 = E(Y-\mu)^2 = \int_{-\infty}^{\infty} (y-\mu)^2 f(y)\, dy = \frac{\lambda}{\pi}\int_{-\infty}^{\infty} \frac{(y-\mu)^2}{\lambda^2 + (y-\mu)^2}\, dy,$$

which does not exist since the integral is not convergent. Thus, in general, for the Cauchy's distribution the moments $\mu_r, (r \geq 2)$ do not exist.

Remarks :

1. The role of Cauchy distribution in statistical theory often lies in providing counter examples, *e.g.*, it is often quoted as a distribution for which moments do not exist. It also provides an example to show that $\varphi_{X+Y}(t) = \varphi_X(t)\,\varphi_Y(t)$ does *not* imply that X and Y are independent.

Let $X_1, X_2, ..., X_n$ be a random sample of size *n* from a standard Cauchy distribution. Let $\bar{X} = \sum_{i=1}^{n} X_i / n$.

Since $E(X_i)$ does not exist ($\because$ mean of a Cauchy distribution does not exist), $E(\overline{X})$ does not exist either and the definition of an unbiased estimate does not apply to $\overline{X}$.

Cauchy distribution also contradicts the WLLN [see Remark 4, § 9·12·1]

2. Some Important Results on Cauchy Distribution :

(i) If X has standard Cauchy distribution, then

$$X^2 \sim \beta_2\left(\frac{1}{2}, \frac{1}{2}\right). \qquad \text{[See Example 9·45]}$$

(ii) If X and Y are independent standard normal variates, then

(X/Y) has Standard Cauchy distribution. [See Example 9·46]

(iii) If X and Y are *i.i.d.* standard Cauchy variates, then the *p.d.f.* of $U = XY$, is :

$$f(u) = \frac{2}{\pi^2}\left[\frac{\log|u|}{u^2-1}\right]; \ -\infty < u < \infty. \qquad \text{[See Example 9·47]}$$

Example 9·45. *Let X have a (standard) Cauchy distribution. Find a p.d.f. for X^2 and identify its distribution.*

Solution. Since X has a standard Cauchy distribution, its *p.d.f.* is :

$$f(x) = \frac{1}{\pi}\cdot\frac{1}{1+x^2}, \ -\infty < x < \infty$$

The distribution function $G(\cdot)$ of $Y = X^2$ is :

$$G_Y(y) = P(Y \le y) = P(X^2 \le y) = P(-\sqrt{y} \le X \le \sqrt{y})$$

$$= \int_{-\sqrt{y}}^{\sqrt{y}} f(x)\,dx = \frac{2}{\pi}\int_0^{\sqrt{y}} \frac{dx}{1+x^2} = \frac{2}{\pi}\tan^{-1}(\sqrt{y}), \ 0 < y < \infty$$

The *p.d.f.* $g_Y(y)$ of Y is given by :

$$g_Y(y) = \frac{d}{dy}[G_Y(y)] = \frac{2}{\pi}\cdot\frac{1}{(1+y)}\cdot\frac{1}{2\sqrt{y}} = \frac{1}{\pi}\cdot\frac{y^{-1/2}}{1+y} = \frac{1}{B\left(\frac{1}{2}, \frac{1}{2}\right)}\cdot\frac{y^{\frac{1}{2}-1}}{(1+y)^{\frac{1}{2}+\frac{1}{2}}}, \ y > 0$$

This is the *p.d.f.* of Beta distribution of second kind with parameters $\left(\frac{1}{2}, \frac{1}{2}\right)$, *i.e.*,

$$X^2 \sim \beta_2\left(\frac{1}{2}, \frac{1}{2}\right).$$

Example 9·46. *Let $X \sim N(0, 1)$ and $Y \sim N(0, 1)$ be independent random variables. Find the distribution of X/Y and identify it.*

Solution. Since X and Y are independent $N(0, 1)$, their joint *p.d.f.* is given by :

$$f_{XY}(x, y) = f_X(x)\cdot f_Y(y) = \frac{1}{2\pi}\cdot e^{-(x^2+y^2)/2}; \ -\infty < (x, y) < \infty$$

Let us make the following transformation of variables :

$u = x/y, \ v = y$ so that $x = uv, \ y = v$. Jacobian of transformation $J = v$.

Hence, the joint *p.d.f.*, of U and V becomes :

$$g_{UV}(u, v) = \frac{1}{2\pi}\cdot\exp\{-(u^2v^2+v^2)/2\}\,|J|$$

$$= \frac{1}{2\pi} \exp\{-(1+u^2)v^2/2\} \mid v \mid, \quad -\infty < (u, v) < \infty$$

The marginal *p.d.f.* of U is :

$$g_U(u) = \frac{1}{2\pi} \int_{-\infty}^{\infty} \exp\{-(1+u^2)v^2/2\} \mid v \mid dv$$

$$= \frac{1}{2\pi} \cdot 2 \int_0^{\infty} \exp\left[-\frac{1}{2}(1+u^2)v^2\right] \mid v \mid dv \quad [\because \text{ Integrand is an even function of } v]$$

$$= \frac{1}{\pi} \int_0^{\infty} \exp\left[-\frac{1}{2}(1+u^2)v^2\right] v\, dv; \quad [\because \text{ For } v \geq 0, \mid v \mid = v]$$

$$= \frac{1}{\pi} \int_0^{\infty} e^{-t} \cdot \frac{dt}{(1+u^2)}; \quad \left[\frac{1}{2}(1+u^2)v^2 = t\right]$$

$$= \frac{1}{\pi(1+u^2)} \cdot \left| -e^{-t} \right|_0^{\infty} = \frac{1}{\pi(1+u^2)}, \quad -\infty < u < \infty,$$

which is the *p.d.f.* of standard Cauchy distribution.

Thus, the ratio of two independent standard normal variates is a standard Cauchy variate.

Remarks :

1. The converse of the above result is not true, *i.e.,*

"If X and Y are i.i.d. random variables and (X/Y) has standard Cauchy distribution, then X and Y need not be i.i.d. N (0, 1)."

For example, if X and Y are *i.i.d. r.v.'s* with common *p.d.f.*

$$p(x) = \sqrt{\frac{2}{\pi}} \left(\frac{1}{1+x^4}\right); \ -\infty < x < \infty,$$

then (proceeding similarly as in the above example), it can be proved that $U = (X/Y)$ has standard Cauchy distribution. (Try it).

However, in this case X and Y are *not i.i.d.* $N(0, 1)$.

2. Distribution of $(X / \mid Y \mid)$.

If $Y \sim N(0, 1)$, then the *p.d.f.* of $\mid Y \mid$ is :

$$g_{\mid Y \mid}(y) = f(y) + f(-y) = 2f(y) \qquad [\text{By symmetry of } f(\cdot)]$$

$$= \frac{\sqrt{2}}{\sqrt{\pi}} e^{-y^2/2}; \ y \geq 0. \qquad \ldots (*)$$

Proceeding as in the Example, and taking $U = (x / \mid y \mid)$, $v = \mid y \mid$, and using (*), it can be proved that $U = (X / \mid Y \mid)$ also has standard Cauchy distribution.

3. The p.d.f. of $\mid U \mid = \mid X \mid / \mid Y \mid$.

If $f(\cdot)$ is the *p.d.f.* of $U = (X/Y) \sim C(1, 0)$ and $g(\cdot)$ is the *p.d.f.* of $\mid U \mid$, then

$$g(x) = f(x) + f(-x) = 2f(x) \qquad [\text{By symmetry of } f(\cdot)]$$

$$\Rightarrow \qquad g(x) = \frac{2}{\pi} \cdot \frac{1}{(1+x^2)}; \ x \geq 0.$$

Example 9·47. *Let X and Y be i.i.d. standard Cauchy variates. Prove that the p.d.f. of U = XY is :*

$$\frac{2}{\pi^2} \left\{\frac{\log \mid u \mid}{u^2 - 1}\right\}; \ -\infty < u < \infty$$

Solution. Since X & Y are independent standard Cauchy variates, their joint *p.d.f.* is given by :

$$f(x, y) = \frac{1}{\pi^2} \cdot \frac{1}{(1+x^2)(1+y^2)}\ ; \ -\infty < (x, y) < \infty.$$

Let $u = xy$ and $v = y$. Then Jacobian of transformation is given by :

$$J = \frac{\partial(x, y)}{\partial(u, v)} = \begin{vmatrix} \frac{1}{v} & -\frac{u}{v^2} \\ 0 & 1 \end{vmatrix} = \frac{1}{v} \qquad \left(\because y = v, x = \frac{u}{v}\right)$$

Thus the joint *p.d.f.* of U and V is given by :

$$g(u, v) = \frac{1}{\pi^2} \cdot \frac{1}{\left(1+\frac{u^2}{v^2}\right)(1+v^2)} \cdot \frac{1}{|v|} = \frac{1}{\pi^2} \cdot \frac{|v|}{(u^2+v^2)(1+v^2)}, \ -\infty < (u, v) < \infty$$

Integrating *w.r.* to v over the range $-\infty$ to ∞, marginal *p.d.f.* of U is given by :

$$g_1(u) = \int_{-\infty}^{\infty} g(u, v)\, dv = \frac{1}{\pi^2} \int_{-\infty}^{\infty} \frac{|v|}{(u^2+v^2)(1+v^2)}\, dv = \frac{2}{\pi^2} \int_0^{\infty} \frac{|v|}{(u^2+v^2)(1+v^2)}\, dv$$

(Since the integrand is an even function of v.)

$$\therefore \quad g_1(u) = \frac{2}{\pi^2} \int_0^{\infty} \frac{v}{(u^2+v^2)(1+v^2)}\, dv = \frac{1}{\pi^2} \int_0^{\infty} \frac{2v}{(u^2-1)} \left(\frac{1}{1+v^2} - \frac{1}{u^2+v^2}\right) dv$$

$$= \frac{1}{\pi^2 (u^2-1)} \Big| \log(1+v^2) - \log(u^2+v^2) \Big|_{v=0}^{\infty} = \frac{1}{\pi^2 (u^2-1)} \left| \log\left(\frac{1+v^2}{u^2+v^2}\right) \right|_{v=0}^{\infty}$$

$$= \frac{1}{\pi^2 (u^2-1)} \left\{ \left[\log \left(\frac{\frac{1}{v^2}+1}{\frac{u^2}{v^2}+1} \right) \right]_{v=\infty} - \log\left(\frac{1}{u^2}\right) \right\}$$

$$= \frac{1}{\pi^2 (u^2-1)} \left[\log 1 + 2 \log |u|\right] = \frac{2 \log |u|}{\pi^2 (u^2-1)}, \quad -\infty < u < \infty$$

9·13. ERLANG DISTRIBUTION

Erlang distribution was developed by A.K. Erlang to examine the number of telephone calls which might be made at the same time to the operators of the switching systems. The *p.d.f.* of Erlang distribution with parameters (λ, k) is given by :

$$f(x, \lambda, k) = \frac{\lambda^k}{(k-1)!} \cdot e^{-\lambda x}\, x^{k-1};\ x \geq 0,\ \lambda > 0, \qquad \text{... (9·32)}$$

where, k is a positive integer.

'k' is called the *shape parameter* and λ, the *rate parameter*. $\mu = (1/\lambda)$, the reciprocal of the rate parameter is called the *scale parameter*.

Note that :

(*i*) Erlang distribution is a special case of two parameter Gamma distribution with parameters (λ, k), with restriction that k is a positive integer.

(*ii*) For $k = 1$, we get $f(x) = \lambda e^{-\lambda x}$; $x > 0, \lambda > 0$, which is the *p.d.f.* of Exponential distribution with mean $(1/\lambda)$. Hence, *if* $k = 1$, Erlang (λ, k) = Erlang $(\lambda, 1) \sim$ Expo (λ).

Distribution Function. Unlike Gamma distribution, Erlang distribution has a cumulative distribution function given by :

$$F(x) = P(X \le x) = \frac{\lambda^k}{\Gamma k}\int_0^x e^{-\lambda u}\, u^{k-1}\, du$$

$$= 1 - \sum_{i=0}^{k-1} \frac{e^{-\lambda x}\cdot(\lambda x)^i}{i!} \qquad \text{... (9·33)}$$

9·13·1. Moments

The *m.g.f.* of Erlang (λ, k) distribution is :

$$M_X(t) = E(e^{tX}) = \int_0^\infty \frac{\lambda^k}{\Gamma k}\cdot e^{-(\lambda - t)x}\, x^{k-1}\, dx$$

$$= \frac{\lambda^k}{\Gamma k}\cdot\frac{\Gamma k}{(\lambda - t)^k} = \left(1 - \frac{t}{\lambda}\right)^{-k} ; \; t < \lambda \qquad \text{... (9·34)}$$

$$K_X(t) = \log M_X(t) = -k \log\left(1 - \frac{t}{\lambda}\right)$$

$$= k\left[\frac{t}{\lambda} + \frac{1}{2}\left(\frac{t}{\lambda}\right)^2 + \frac{1}{3}\left(\frac{t}{\lambda}\right)^3 + \ldots\right]; \; t < \lambda \qquad \text{... (9·34a)}$$

$\therefore$ Mean $= \kappa_1$ = Coefficient of $t = \dfrac{k}{\lambda}$

Variance $= \mu_2 = \kappa_2$ = Coefficient of $\dfrac{t^2}{2!} = \dfrac{k}{\lambda^2} = \dfrac{\text{Mean}}{\lambda}$

$\Rightarrow$ Variance $>, \; = \;$ or $\; <$ Mean, according as $\lambda < 1, \; = 1, \; > 1$, respectively.

From (9·34*a*), we can compute higher moments and hence β_1 and β_2.

9·13·2. Uses

A.K. Erlang worked a lot on traffic modelling — specially telephone systems. His work on traffic engineering has been expanded to predict the waiting times in queueing systems, etc. when a Poisson process is in operation. This distribution is now used in the fields of stochastic processes and of biomathematics.

Remark. *The distribution of the sum of i.i.d. exponential variates is a two parameter Erlang distribution.*

If X_i, *(i = 1, 2, ..., n) are i.i.d. Expo* (λ) *variates, then* $\sum_{i=1}^{n} X_i \sim$ *Erlang* (n, λ).

Proof. Since $X_i \sim$ Expo (λ), $i = 1, 2, ..., n$,

$$M_{X_i}(t) = \left(1 - \frac{t}{\lambda}\right)^{-1} ; \; t < \lambda \qquad \text{[From (9·26b)]} \qquad \text{... (*)}$$

$$\therefore \quad M_{\sum_{i=1}^{n} X_i}(t) = \prod_{i=1}^{n} M_{X_i}(t) = \left(1 - \frac{t}{\lambda}\right)^{-n} ; \; t < \lambda, \qquad \text{... (**)}$$

where n is a *positive integer*.

(**) is the *m.g.f.* of Erlang (n, λ) distribution. Hence, by uniqueness theorem of *m.g.f.*'s,

$$\sum_{i=1}^{n} X_i \sim \text{Erlang}(n, \lambda)$$

Example 9·48. *If $X \sim$ Expo (λ) and $Y \sim$ Erlang (n, λ) are independent r.v.'s prove that, (X/Y) has Pareto distribution.*

Solution. Since X and Y are independent, their joint *p.d.f.* is given by :

$$f(x, y) = f_X(x) \cdot f_Y(y) = (\lambda e^{-\lambda x}) \cdot \frac{\lambda^n}{(n-1)!} e^{-\lambda y} y^{n-1}$$

$$= \frac{\lambda^{n+1}}{(n-1)!} e^{-\lambda(x+y)} \cdot y^{n-1}; \quad x > 0, \ y > 0 \qquad \text{... (*)}$$

Transform $(X, Y) \to (U, V)$ by : $\quad U = \dfrac{X}{Y}, \quad Y = V \quad \Rightarrow \quad y = v, \ x = uv$

$$J = \frac{\partial(x, y)}{\partial(u, v)} = \begin{vmatrix} v & u \\ 0 & 1 \end{vmatrix} = v.$$

$$x \geq 0, \ y \geq 0 \quad \Rightarrow \quad u \geq 0, \ v \geq 0.$$

The joint *p.d.f.* $g(u, v)$ of U and Y becomes :

$$g(u, v) = f(x, y) | J | = \frac{\lambda^{n+1}}{\Gamma n} e^{-\lambda v(1+u)} \cdot v^{n-1} \cdot v$$

Integrating *w.r.* to v over $[0, \infty]$, the marginal *p.d.f.* of U is given by :

$$g_1(u) = \int_{v=0}^{\infty} g(u, v)\, dv = \frac{\lambda^{n+1}}{\Gamma n} \cdot \int_{v=0}^{\infty} e^{-\lambda(1+u)v} \cdot v^n \, dv$$

$$= \frac{\lambda^{n+1}}{\Gamma n} \cdot \frac{\Gamma(n+1)}{[\lambda(1+u)]^{n+1}} = \frac{1}{\Gamma n} \cdot \frac{n \Gamma n}{(1+u)^{n+1}}$$

$$\Rightarrow \qquad g_1(u) = \frac{n}{(1+u)^{n+1}}; \quad u \geq 0$$

which is the *p.d.f.* of Pareto distribution with parameters $(1, n)$. [See § 9·14·4]

Hence, $\quad U = (X/Y) \sim$ Pareto $(1, n)$

9·13·3. Erlang Family of Distributions

The generalised *Erlang family* of distributions with parameters $(k\lambda, k)$ is given by *p.d.f.* :

$[X \sim \text{Erlang}(k\lambda, k)]$.

$$f(x) = \frac{(k\lambda)^k}{(k-1)!} e^{-k\lambda x} \cdot x^{k-1}; \quad x \geq 0, \ \lambda \geq 0 \qquad \text{... (9·35)}$$

and k is a positive integer.

(9·35) gives the p.d.f. of the k th member of the family.

The *m.g.f.* of Erlang $(k\lambda, k)$ is given by :

$$M_X(t) = \left(1 - \frac{t}{k\lambda}\right)^{-k}; \quad t < k\lambda \qquad [\text{Changing } \lambda \text{ to } k\lambda \text{ in (9·34)}] \qquad \text{... (9·35a)}$$

$$\Rightarrow \qquad K_X(t) = -k\log\left(1-\frac{t}{k\lambda}\right) = k\left[\frac{t}{k\lambda}+\frac{1}{2}\left(\frac{t}{k\lambda}\right)^2+\ldots\right] \qquad \ldots (9\cdot 25b)$$

$$\therefore \qquad \text{Mean} = \text{Coefficient of } t \text{ in } K_X(t) = \frac{1}{\lambda}; \qquad \text{(Independent of } k) \quad \ldots (9\cdot 35c)$$

$\Rightarrow$ *Each member of the family has same mean $(1/\lambda)$.*

$$\text{Variance} = \mu_2 = \text{Coefficient of } \frac{t^2}{2!} \text{ in } K_X(t) = \frac{1}{k\lambda^2} \qquad \ldots (9\cdot 35d)$$

Hence, *the variance of the kth member of the family* is $1/(k\lambda^2)$, *i.e.*, inversely proportional to k, and decreases as k increases and tends to zero for large values of k.

Remarks :

1. Obviously for $k=1$, $f(x)=\lambda e^{-\lambda x}$; $x \ge 0, \lambda > 0$, *i.e., Exponential* (λ) distribution is the *first member of Erlang* $(k\lambda, k)$ *family of distributions.*

In fact, we have the following relationship between Exponential distribution and Erlang family of distributions.

If $X_1, X_2, \ldots, X_k$ are *i.i.d.* Expo $(k\lambda)$ variates, then the sum $\sum_{i=1}^{k} X_i \sim$ Erlang $(k\lambda, k)$, *i.e.*, the sum of *i.i.d.* Expo $(k\lambda)$ variates follows the kth Erlang distribution. ... (9·35)

Proof is left as an Exercise to the reader.

2. *If $X \sim$ Erlang $(k\lambda, k)$, then $k\lambda X$ has a gamma distribution with parameter k.*

Proof. If $X \sim$ Erlang $(k\lambda, k)$, $M_X(t) = \left(1-\dfrac{t}{k\lambda}\right)^{-k}$ [From (9·35a)]

$$\therefore \qquad M_{k\lambda X}(t) = M_X(k\lambda t) = (1-t)^{-k}$$

$\Rightarrow$ $k\lambda X$ has gamma distribution with parameter 'k'.

9·14. PARETO DISTRIBUTION

Pareto distribution, named after Italian civil engineer and economist, Vilfredo Pareto, is characterised by two parameters — a *scale parameter* x_m and a *shape parameter* 'α', which is also known as the tail index. When the distribution is used to model the distribution of wealth, then the parameter 'α' is called the *Pareto index.*

The *p.d.f.* of Pareto distribution with parameters 'x_m and α' is given by :

$$p(x) = \begin{cases} \dfrac{\alpha \cdot x_m^{\alpha}}{x^{\alpha+1}} & ; \ x \ge x_m, \ (x_m \le x < \infty) \\ 0 & ; \ x < x_m \end{cases} \qquad \ldots (9\cdot 36)$$

We write $\qquad X \sim \text{Pareto}\,(x_m, \alpha)$

9·14·1. Moments

$$\mu_r' = E(X^r) = \alpha \cdot x_m^{\alpha} \int_{x_m}^{\infty} x^{r-\alpha-1}\,dx = \alpha \cdot x_m^{\alpha} \cdot \left|\frac{x^{r-\alpha}}{r-\alpha}\right|_{x_m}^{\infty}$$

The limiting value of $[x^{r-\alpha}]_{x=\infty}$ exists if and only if $r-\alpha<0 \quad \Rightarrow \quad \alpha > r$.

$$\therefore \qquad \mu_r' = \alpha \cdot x_m^{\alpha}\left[0 - \frac{x_m^{r-\alpha}}{r-\alpha}\right] = \frac{\alpha \cdot x_m^{r}}{\alpha - r}\ ; \quad \alpha > r. \qquad \text{... (9·37)}$$

In particular, taking $r = 1, 2$, we get respectively :

$$\text{Mean} = E(X) = \mu_1' = \frac{\alpha\, x_m}{\alpha - 1}\ ; \quad \alpha > 1 \qquad \text{... (9·38)}$$

$$\text{Var}(X) = \mu_2' - \mu_1'^2 = \frac{\alpha\, x_m^2}{\alpha - 2} - \left(\frac{\alpha\, x_m}{\alpha - 1}\right)^2$$

$$= \left(\frac{x_m}{\alpha - 1}\right)^2 \cdot \alpha \left[\frac{(\alpha-1)^2 - \alpha(\alpha - 2)}{\alpha - 2}\right]$$

$$\text{Var}(X) = \left(\frac{x_m}{\alpha - 1}\right)^2 \cdot \left(\frac{\alpha}{\alpha - 2}\right); \quad \alpha > 2 \qquad \text{... (9·38a)}$$

Similarly, we can obtain the values of μ_3' and μ_4', and hence β_1 and β_2.

The Geometric Mean (G) and the Harmonic Mean (H) for Pareto distribution are given by : [Try it] :

$$\text{G} = \text{G.M.} = x_m \cdot e^{1/\alpha}\ ; \qquad \text{H} = \text{H.M.} = x_m\left(1 + \frac{1}{\alpha}\right) \qquad \text{... (9·39)}$$

9·14·2. Conditional Distribution

If the *r.v.* X has Pareto distribution (9·36) with parameters x_m and α, then the conditional probability distribution of X, given that it is greater than or equal to a particular number $x_1 > x_m$, is a Pareto distribution with the same Pareto index α but with min (x_1) instead of x_m.

Proof. If $X \sim$ Pareto (x_m, α), then its distribution function $F(\cdot)$ is given by :

$$F(x) = P(X \le x) = \int_{x_m}^{x} f(u)\, du$$

$$= \alpha \cdot x_m^{\alpha} \cdot \int_{x_m}^{x} u^{-\alpha - 1}\, du = \alpha \cdot x_m^{\alpha} \left| \frac{u^{-\alpha}}{-\alpha} \right|_{x_m}^{x}$$

$$= x_m^{\alpha} \cdot \left(x_m^{-\alpha} - x^{-\alpha}\right) = 1 - \left(\frac{x_m}{x}\right)^{\alpha}$$

$$\therefore \qquad F_X(x) = P(X \le x) = \begin{cases} 1 - \left(\dfrac{x_m}{x}\right)^{\alpha} & ; \quad x \ge x_m \\ 0 & ; \quad x < x_m \end{cases} \qquad \text{... (9·40)}$$

$$\Rightarrow \qquad P(X > x) = 1 - P(X \le x) = \begin{cases} \left(\dfrac{x_m}{x}\right)^{\alpha} & ; \quad x \ge x_m \\ 0 & ; \quad x < x_m \end{cases} \qquad \text{... (9·41)}$$

where, $x_m > 0$, is the minimum value of x. The function defined in (9·41) is called the *Survival Function* or the *Tail Function.*

We want the conditional probability :

$$P\left(X > x \mid X \geq x_1\right); \text{ where } x_1 > x_m \text{ and } x_m \leq x < \infty.$$

$$P\left(X > x \mid X \geq x_1\right) = \frac{P\left(X > x \cap X \geq x_1\right)}{P\left(X \geq x_1\right)} = \frac{P(X > x)}{P\left(X \geq x_1\right)} = \frac{(x_m / x)^{\alpha}}{(x_m / x_1)^{\alpha}} \qquad \text{[Using (9·41)]}$$

$$= \left(\frac{x_1}{x}\right)^{\alpha} \qquad \text{... (9·42)}$$

which is the Survival function for Pareto distribution with scale parameter x_1 (instead of x_m) and shape parameter 'α'.

Note on Survival Function. When plotted on linear axes, the graph of the Survival function is a familiar *J*-shaped curve, the two asymptotes being the *x*-axis $[y = p(x) = 0]$ and the line $x = x_m$; where x_m is the lowest value of x at which the curve begins. When plotted on a log-log plot, the graph is represented by a straight line.

9·14·3. Characterisation of Pareto Distribution

Suppose X_i ; $i = 1, 2, 3, ..., n$ are independent, indentically distributed *r.v.*'s whose *p.d.f.* is supported on the interval $[x_m, \infty]$, for some $x_m > 0$. Suppose that for all n, the two *r.v.*'s :

$$\min\left(X_1, X_2, \ldots, X_n\right) \quad \text{and} \quad \frac{X_1 + X_2 + \ldots + X_n}{\min\left(X_1, X_2, \ldots, X_n\right)},$$

are independent. Then, the common distribution of X_i's is a Pareto distribution.

9·14·4. Basic Pareto Distribution

In particular, if we take $x_m = 1$ in (9·36), then we obtain the *Basic Pareto distribution*, with the single *shape parameter* 'α' with *p.d.f.*

$$p(x) = \frac{\alpha}{x^{\alpha+1}}; 1 \leq x < \infty \qquad \text{... (9·43)}$$

We write $X \sim \text{Pareto}(\alpha)$.

Properties of (9·43)

(*i*) $p(x)$ is decreasing function of x.

(*ii*) Its mode occurs at $x = 1$.

(*iii*) $p(x) \to 0$ as $x \to \infty$, *i.e.*, *x*-axis is an asymptote to the curve.

Pareto distribution (9·43) is a skewed, heavily tailed distribution because the convergence in (*iii*) is at a power rate rather than an exponential rate.

(*iv*) Taking $x_m = 1$, in (9·37), (9·38) and (9·38*a*), we get, respectively :

$$\mu_r' = \frac{\alpha}{\alpha - r}; \quad \alpha > r \qquad \text{... (9·44)}$$

$$\left.\begin{aligned} \text{Mean} = \mu_1' &= \frac{\alpha}{\alpha - 1}; \quad \alpha > 1 \\ \text{Variance} = \mu_2 &= \frac{\alpha}{(\alpha-1)^2(\alpha-2)}; \quad \alpha > 2 \end{aligned}\right\} \qquad \text{... (9·45)}$$

Similarly, proceeding, we can obtain μ_3' and μ_4' from (9·44), and μ_3 and μ_4 in the usual manner. The coefficients of skewness (β_1) and Kurtosis (β_2), are then given by : [*Try it*]

$$\left.\begin{aligned}\text{Skewness } (\beta_1) &= \frac{2(1+\alpha)}{\alpha-3}\cdot\sqrt{1-\frac{2}{\alpha}}\ ;\ \alpha>3\\ \text{Kurtosis } (\beta_2) &= \frac{3(\alpha-2)(3\alpha^2+\alpha+2)}{\alpha(\alpha-3)(\alpha-4)}\ ;\ \alpha>4\end{aligned}\right\} \quad \ldots (9\cdot46)$$

Distribution is positively skewed, and $\beta_1 \to 2$ as $\alpha \to \infty$, and $\beta_2 \to 9$ as $\alpha \to \infty$.

(*v*) **Quartiles.** The *i*th quartile Q_i; $i = 1, 2, 3$; is given by :

$$\int_1^{Q_i} p(x)\,dx = \alpha\cdot\int_1^{Q_i} x^{-\alpha-1}\,dx = \frac{i}{4} \quad\Rightarrow\quad \alpha\cdot\left|\frac{x^{-\alpha}}{-\alpha}\right|_1^{Q_i} = \frac{i}{4}$$

$$1-Q_i^{-\alpha} = \frac{i}{4} \quad\Rightarrow\quad Q_i = \left(1-\frac{i}{4}\right)^{-1/\alpha}\ ;\ i=1,2,3,\ldots \quad \ldots (9\cdot47)$$

$$\therefore \quad Q_1 = \left(\frac{4}{3}\right)^{1/\alpha}\ ;\ Q_2 = \text{Median} = 2^{1/\alpha} \text{ and } Q_3 = 4^{1/\alpha} \quad \ldots (9\cdot48)$$

9·14·5. Applications

Vilfredo Pareto made an extensive study of the problem of distribution of income and wealth among the citizens of a State. Pareto propourded his famous law of income/wealth distribution in the 19th-20th century, primarily basing his arguments and reasoning on the empirical study of the income/wealth data of various countries of the world at different times. Pareto used this distribution (9·36) to model the distribution of the income/wealth among individuals, as it shows rather well the way a larger of portion of income/wealth of any society is owned by smaller percentage of people in that society. **Pareto's principle** or commonly known as **"80-20 Rule"**, says that **"20% of the population controls 80% of the wealth".**

This is well established by the graph of *p* (*x*) which shows that the probability of a person owning a small amount of wealth per person is rather high and then decreases steadily as wealth increases.

It may be pointed out that Pareto's **"80-20 Rule"** holds precisely when the Pareto index,

$$\alpha = \log_4 5 = \frac{\log 5}{\log 4} \simeq 1{\cdot}61 \quad \ldots (9\cdot49)$$

Income is distributed according to Pareto distribution if $\alpha > 1$.

Initially, Pareto used this distribution to model income and wealth of a State. But today, this distribution is used to describe social, scientific, geo-physical, actuarial, financial and many other types of observable phenomenon.

9·14·6. Relation With Other Distributions

(*a*) **Relation With Exponential Distribution.** *If* $X \sim$ *Pareto* (x_m, α), *then* $Y = \log\left(\frac{X}{x_m}\right)$ *has exponential distribution with parameter* α.

Note. If we take $x_m = 1$, then *X* has basic Pareto distribution with shape parameter 'α', *i.e.*, $X \sim$ Pareto (α). In this case we have the following result :

If $X \sim$ *Pareto* (α), *then* $Y = \log_e X \sim$ *Expo* (α).

Conversely, if $Y \sim$ *Expo* (α), *then* $X = e^Y$ *has basic Pareto distribution with shape Parameter* α.

(*b*) **Relation With Uniform Distribution.** *If* $X \sim U[0, 1]$, *then* $Y = \left(\frac{x_m}{X^{1/\alpha}}\right) \sim$ *Pareto* (x_m, α).

Note. In particular, if $x_m = 1$, then, we have the following result :

If $$X \sim U[0,1], \text{ then } Y = \frac{1}{X^{1/\alpha}} \sim \text{Pareto}(\alpha)$$

and *conversely, if* $$Y = \frac{1}{X^{1/\alpha}} \sim \text{Pareto}(\alpha), \text{ then } X = \frac{1}{Y^{\alpha}} \sim U[0,1].$$

(c) Relation With Beta-Distribution. *If* $X \sim$ *Pareto* (α), *then* $Y = \dfrac{1}{X}$, *has Beta distribution of first kind with parameters* $(\alpha, 1)$, *i.e.,* $Y \sim \beta_1(\alpha, 1)$.

Note. Converse of this result is also true.

If $$Y \sim \beta_1(\alpha, 1), \text{ then } X = \frac{1}{Y} \sim \text{Pareto}(\alpha).$$

Proofs of all the above results are left as an exercise for the reader.

9·15. RALEIGH DISTRIBUTION

Raleigh distribution was given by British physicist Lord Raleigh (or by his less glamorous name John William Strut).

The *p.d.f.* of Raleigh distribution with parameter σ is given by :

$$f(x, \sigma) = \frac{1}{\sigma^2} \cdot x \cdot \exp\left(-\frac{x^2}{2\sigma^2}\right); \quad x \geq 0 \qquad \text{... (9·50)}$$

$\sigma > 0$, is the scale parameter of the distribution. We write $X \sim R(\sigma)$.

Relation to Random Vector Lengths. Consider a two dimensional vector $Y = (U, V)$ whose components U and V are independently and normally distributed with zero mean and common variance σ^2, *i.e.,* $U \sim N(0, \sigma^2)$ and $V \sim N(0, \sigma^2)$ are independent. The magnitude (length) $X = |Y| = \sqrt{U^2 + V^2}$ of the vector Y has Raleigh distribution with parameter σ. Distribution Function of X is given by :

$$F(x) = P(X \leq x) = \int_A f_1(u) f_2(v)\, du\, dv$$

where, $$A : \{(u, v) \in R^2 \; ; \; u^2 + v^2 \leq x^2\}$$

$$\therefore \quad F(x) = \int_A \frac{1}{\sqrt{2\pi}\,\sigma} \exp\left(-\frac{u^2}{2\sigma^2}\right) \cdot \frac{1}{\sqrt{2\pi}\,.\,\sigma} \exp\left(-\frac{v^2}{2\sigma^2}\right) du\, dv$$

Transform to Polar coordinates : $(u, v) \to (r, \theta)$ by :

$$u = r\cos\theta, \quad v = r\sin\theta\,; \quad u^2 + v^2 = r^2\,; \quad J = \frac{\partial(u, v)}{\partial(r, \theta)} = r$$

The region A transforms to $0 \leq \theta \leq 2\pi$ and $0 \leq r \leq x$.

$$\therefore \quad F(x) = \frac{1}{2\pi\sigma^2} \cdot \int_0^{2\pi} 1 \cdot d\theta \cdot \int_0^x r \cdot \exp\left(-\frac{r^2}{2\sigma^2}\right) dr$$

$$= \frac{1}{2\pi\sigma^2} \cdot 2\pi \cdot \int_0^{x^2/2\sigma^2} \cdot e^{-t} \cdot \sigma^2\, dt\,; \quad \left(\frac{r^2}{2\sigma^2} = t\right)$$

$$F(x) = \left| -e^{-t} \right|_0^{x^2/2\sigma^2} = 1 - e^{-x^2/2\sigma^2} \; ; \; x > 0 \qquad \text{... (9·51)}$$

Differentiating *w.r.* to x, the *p.d.f.* of X is

$$f(x) = \frac{x}{\sigma^2} e^{-x^2/2\sigma^2} \; ; \; x > 0, \sigma > 0 \qquad \text{... (9·51a)}$$

which is same as given in (9·50).

Taking $\sigma = 1$, we obtain *the p.d.f. of standard Raleigh distribution*

$$f(x) = x\, e^{-x^2/2} ; \; x \geq 0 \qquad \text{... (9·52)}$$

and we write $X \sim R\,(1)$.

We have : $$f'(x) = \frac{1}{\sigma^2} \cdot e^{-x^2/2\sigma^2} \left(1 - \frac{x^2}{\sigma^2} \right) \quad \Rightarrow \quad f'(x) = 0 \textit{ if } x = \sigma \qquad (\because x > 0)$$

and $$f''(x) = -\frac{x}{\sigma^4} e^{-x^2/2\sigma^2} \left(3 - \frac{x^2}{\sigma^2} \right) < 0, \text{ when } x = \sigma.$$

Hence, the mode of $R\,(\sigma)$ distribution is at $x = \sigma$.

$$f'(x) > 0 \text{ if } x < \sigma \quad \text{and} \quad f'(x) < 0 \text{ if } x > \sigma.$$

Hence, $f(x)$ is an increasing function of x in $[0, \sigma]$ and a decreasing function of x in $[\sigma, \infty]$.

The point of inflexion is the solution of $f''(x) = 0$ and $f'''(x) \neq 0$ and is given by :

$$3 - \frac{x^2}{\sigma^2} = 0 \quad \Rightarrow \quad x = \sqrt{3} \cdot \sigma \qquad (\because x > 0)$$

Hence $f(x)$ is concave downwards on $\left[0, \sqrt{3}\,\sigma\right]$ and concave upward on $\left[\sqrt{3}\,\sigma, \infty\right]$.

Remark. If $U_1 \sim N\,(0, 1)$ and $U_2 \sim N\,(0, 1)$ are independent, then the *r.v.* $X = \sqrt{U_1^2 + U_2^2}$ has standard Raleigh distribution, *i.e.*, $X \sim R\,(1)$ with *p.d.f.*

$$g(x) = x \cdot \exp\left[-\frac{x^2}{2} \right] ; \; x \geq 0$$

and distribution function :

$$G(x) = P(X \leq x) = 1 - \exp\left[-\frac{x^2}{2} \right] ; \; x \geq 0.$$

9·15·1. Relation With Other Distributions

1. *Raleigh distribution is a special case of Weibul distribution [c.f. § 9·10] with* $\mu = 0$, $\alpha = \sigma\sqrt{2}$ *and* $c = 2$.

The *p.d.f.* of Weibul distribution with parameters $c > 0$, $\alpha > 0$ and μ is given by :

$$f(x, c, \alpha, \mu) = \frac{c}{\alpha} \left(\frac{x - \mu}{\alpha} \right)^{c-1} \cdot \exp\left[-\left(\frac{x - \mu}{\alpha} \right)^c \right]$$

Taking $\mu = 0$, $c = 2$ and $\alpha = \sigma\sqrt{2}$, we get

$$f\left(x, 2, \sigma\sqrt{2}, 0\right) = \frac{2}{\sigma\sqrt{2}} \cdot \frac{x}{\sigma\sqrt{2}} \cdot \exp\left[-\left(\frac{x}{\sigma\sqrt{2}} \right)^2 \right] = \frac{x}{\sigma^2} \cdot \exp\left(-\frac{x^2}{2\sigma^2} \right) ; \; x > 0,$$

which is the *p.d.f.* of Raleigh distribution with parameter σ.

2. Relation Between Raleigh Distribution and Exponential Distribution.

(a) If $X \sim Expo(\lambda)$, then $Y = \sqrt{2\sigma^2 \lambda X} \sim R(\sigma)$.

Proof. The *p.d.f.* of X is : $f(x) = \lambda e^{-\lambda x}$; $x \ge 0, \lambda > 0$

$$Y = \sqrt{2\sigma^2 \lambda X} \Rightarrow y^2 = 2\sigma^2 \lambda x \Rightarrow \frac{dx}{dy} = \frac{y}{\sigma^2 \lambda} ; \quad x \ge 0 \Rightarrow y \ge 0$$

The *p.d.f.* $g(\cdot)$ of Y is given by :

$$g(y) = f(x) \cdot \left| \frac{dx}{dy} \right| = \lambda e^{-y^2/2\sigma^2} \frac{y}{\sigma^2 \lambda} \Rightarrow g(y) = \frac{y}{\sigma^2} e^{-y^2/2\sigma^2} ; \quad y > 0, \sigma > 0$$

$\Rightarrow$ $Y \sim R(\sigma)$.

(b) If $X \sim R(\sigma)$, then $X^2 \sim Expo\left(\lambda = \frac{1}{2\sigma^2}\right)$.

Proof. $f(x) = \frac{x}{\sigma^2} e^{-x^2/2\sigma^2}$; $x \ge 0, \sigma > 0$.

Let $Y = X^2 \Rightarrow y = x^2 \Rightarrow dy = 2x\,dx$; $x \ge 0 \Rightarrow y \ge 0$.

Note that $y = x^2$ is strictly increasing, because $x > 0$.

The *p.d.f.* $g(\cdot)$ of $Y = X^2$ is given by :

$$g(y) = f(x) \cdot \left| \frac{dx}{dy} \right| = \frac{x}{\sigma^2} \cdot e^{-y/2\sigma^2} \cdot \frac{1}{2x} = \frac{1}{2\sigma^2} e^{-y/2\sigma^2} ; \quad y \ge 0$$

$\Rightarrow$ $Y = X^2 \sim \text{Expo}\left(\lambda = \frac{1}{2\sigma^2}\right)$.

3. Relation Between Raleigh Distribution and Chi-Square Distribution.

If $X \sim R(\sigma = 1)$, then $X^2 \sim \chi^2_{2 d.f.}$

Proof. *p.d.f.* of X with $\sigma = 1$ is : $f(x) = x e^{-x^2/2}$; $x > 0$

Let $Y = X^2 \Rightarrow y = x^2 \Rightarrow dy = 2x\,dx$ [$Y \uparrow$ because $X > 0$]

The *p.d.f.* $g(\cdot)$ of Y is :

$$g(y) = f(x) \left| \frac{dx}{dy} \right| = x e^{-y/2} \cdot \frac{1}{2x} = \frac{1}{2} e^{-y/2} ; \quad y > 0$$

which is *p.d.f.* of Chi-square distribution with 2 *d.f.*

Hence, $Y = X^2 \sim \chi^2_{2 d.f.}$

[The *p.d.f.* of Chi-square distribution with *n d.f.* is :

$$f(x) = \frac{1}{2^{n/2}\,\Gamma(n/2)} e^{-x/2} \cdot x^{(n/2)-1} ; \quad x \ge 0$$] (See Chapter 15)

9·15·2. Moments of Raleigh Distribution

$$\mu_r' = E(X^r) = \int_0^\infty x^r\, f(x)\, dx = \frac{1}{\sigma^2} \cdot \int_0^\infty x^{r+1} \cdot e^{-x^2/2\sigma^2}\, dx$$

$$= \frac{1}{2\sigma^2} \int_0^\infty x^r\, e^{-z/2\sigma^2}\, dz \qquad (x^2 = z)$$

$$= \frac{1}{2\sigma^2} \int_0^\infty z^{r/2} \cdot e^{-z/2\sigma^2}\, dz = \frac{1}{2\sigma^2} \cdot \frac{\Gamma\left(\frac{r}{2}+1\right)}{\left(\frac{1}{2\sigma^2}\right)^{\frac{r}{2}+1}}$$

$$\Rightarrow \qquad \mu_r' = 2^{r/2} \cdot \sigma^r\, \Gamma\left(1 + \frac{r}{2}\right); \quad r = 1, 2, \ldots \qquad \ldots (9\cdot53)$$

$$\therefore \qquad \text{Mean} = \mu_1' = 2^{1/2} \cdot \sigma \cdot \Gamma\left(\frac{3}{2}\right) = 2^{1/2} \cdot \sigma \cdot \frac{1}{2}\Gamma\left(\frac{1}{2}\right) = \sigma \cdot \sqrt{\frac{\pi}{2}} \simeq 1\cdot253\,\sigma$$

$$\mu_2' = 2\sigma^2\, \Gamma 2 = 2\sigma^2$$

$$\Rightarrow \qquad \text{Variance} = \mu_2 = \mu_2' - \mu_1'^2 = 2\sigma^2 - \frac{\pi\sigma^2}{2} = \left(\frac{4-\pi}{2}\right)\sigma^2 \simeq 0\cdot429\,\sigma^2 \qquad \ldots (9\cdot54)$$

Mode of the distribution is the solution of $f'(x) = 0$ and $f''(x) < 0$. $\Rightarrow$ Mode $= \sigma$. [*Try it*]

$$\therefore \qquad \text{Maximum probability} = [f(x, \sigma)]_{x=\sigma} = \frac{1}{\sigma} e^{-1/2}.$$

From (9·53), we can find μ_3' and μ_4' and hence μ_3 and μ_4. (*Try these*).

The coefficient of Skewness is : $\gamma_1 = \sqrt{\beta_1} = \dfrac{2\sqrt{\pi}\,(\pi - 3)}{(4-\pi)^{3/2}}$ ... (9·55)

Excess of Kurtosis is given by : $\gamma_2 = \beta_2 - 3 = \dfrac{6\pi^2 - 24\pi + 16}{(4-\pi)^2}$ ... (9·56)

9·15·3. Applications and Some Illustrations

1. Raleigh distribution is often observed when the overall magnitude of a vector is related to its directional components. Some illustrations are given below :

(*i*) If the component velocities of a particle in the x and y directions are *i.i.d.* normal variates with zero mean and variance σ^2, then the distance the particle travels per unit time has a Raleigh $R(\sigma)$ distribution.

(*ii*) If the wind velocity is analyzed into its orthogonal two dimensional vector components and the magnitudes of each component are *i.i.d.* $N(0, \sigma^2)$ variates, then the overall wind speed (vector magnitude) follows Raleigh distribution with parameter σ.

(*iii*) Consider a random complex number $Z = X + iY$, whose real and imaginary components X and Y are *i.i.d.* $N(0, \sigma^2)$, then the absolute value of the complex number, *viz.*, $|Z| = \sqrt{X^2 + Y^2}$ follows $R(\sigma)$ distribution.

2. Raleigh distribution is used to model :

(*a*) Wave heights in oceanography.

(*b*) The frequency of different wind speeds over a year at wind turbine sites.

3. In communication theory, Raleigh distribution is used :

(*a*) To model scattered signals, specially dense scatters that reach a receiver by multiple paths.

(*b*) To describe hourly median and instataneous peak power of received radio signals.

4. As pointed out earlier, Raleigh distribution is a special case of Weibul distribution (with $\mu = 0$, $\alpha = \sigma\sqrt{2}$ and $c = 2$) and as such is suitable for modelling the life time of a device that has a linearly increasing instantaneous failure rate $Z(x) = (x/\sigma^2)$.

9·16. CENTRAL LIMIT THEOREM (C.L.T.)

The central limit theorem in the mathematical theory of probability may be expressed as follows :

"If X_i, $(i = 1, 2, ..., n)$ be independent random variables such that $E(X_i) = \mu_i$ and $V(X_i) = \sigma_i^2$, then under certain very general conditions, the random variable $S_n = X_1 + X_2 + ... + X_n$, is asymptotically normal with mean μ and standard deviation σ, where

$$\mu = \sum_{i=1}^{n} \mu_i \quad \text{and} \quad \sigma^2 = \sum_{i=1}^{n} \sigma_i^2 \qquad \text{... (9·57)}$$

This theorem was first stated by Laplace in 1812 and a regorous proof under fairly general conditions was given by Liapounoff in 1901. Below we shall consider some particular cases of this general central limit theorem.

9·16·1. De Moivre - Laplace Theorem (1733)

A particular case of central limit theorem is De-Moivre's theorem which states as follows :

"If $$X_i = \begin{cases} 1, \text{ with probability } p \\ 0, \text{ with probability } q \end{cases}$$

then the distribution of the random variable $S_n = X_1 + X_2 + ... + X_n$, where X_i's are independent, is asymptotically normal as $n \to \infty$."

Proof. *m.g.f.* of X_i is given by : $\quad M_{X_i}(t) = E(e^{tX_i}) = e^{t\cdot 1}p + e^{t\cdot 0}q = (q + pe^t)$ ... (*)

m.g.f. of the sum $S_n = X_1 + X_2 + ... + X_n$ is given by :

$$M_{S_n}(t) = M_{X_1 + X_2 + ... + X_n}(t) = M_{X_1}(t)\cdot M_{X_2}(t) \ldots M_{X_n}(t) = [M_{X_i}(t)]^n$$

(since X_i's are independently and identically distributed.)

$$= (q + pe^t)^n, \qquad \text{[From (*)]} \qquad \text{... (**)}$$

which is the *m.g.f.* of a binomial variate with parameters n and p.

Hence by uniqueness theorem of *m.g.f.*'s, $\quad S_n \sim B(n, p)$.

$\therefore \quad E(S_n) = np = \mu$ (say), and $V(S_n) = npq = \sigma^2$, (say).

Let $$Z = \frac{S_n - E(S_n)}{\sqrt{V(S_n)}} = \frac{S_n - \mu}{\sigma}$$

$$M_Z(t) = e^{-\mu t/\sigma} M_{S_n}(t/\sigma) = e^{-npt/\sqrt{npq}} \left[q + pe^{t/\sqrt{npq}} \right]^n \qquad \text{[From (**)]}$$

$$= \left[1 + \frac{t^2}{2n} + O(n^{-3/2}) \right]^n,$$

where, $O(n^{-3/2})$ represents terms involving $n^{3/2}$ and higher powers of n in the denominator.

Proceeding to the limits as $n \to \infty$, we get

$$\lim_{n\to\infty} M_Z(t) = \lim_{n\to\infty}\left[1+\frac{t^2}{2n}+O(n^{-3/2})\right]^n = \lim_{n\to\infty}\left[1+\frac{t^2}{2n}\right]^n = e^{t^2/2},$$

which is the *m.g.f.* of a standard normal variate. Hence by the uniqueness theorem of *m.g.f.*'s,

$$Z = \frac{S_n - \mu}{\sigma} \text{ is asymptotically } N(0, 1).$$

Hence, $S_n + X_1 + X_2 + \ldots + X_n$ is asymptotically $N(\mu, \sigma^2)$ as $n \to \infty$.

Remarks :

1. From this theorem it follows that standard binomial variate tends to standard normal variate as $n \to \infty$. In other words, *binomial distribution tends to normal distribution as* $n \to \infty$.

2. *Convergence in Distribution or law.* Let $\{X_n\}$ be a sequence of *r.v.*'s and $\{F_n\}$ be the corresponding sequence of distribution functions. We say that X_n converges in distribution (or law) to X if there exists a *r.v.* X with distribution function F such that as $n \to \infty$, $F_n(x) \to F(x)$ at every point x at which F is continuous.

We write $X_n \xrightarrow{L} X$ or $X_n \xrightarrow{d} X$.

3. It may be remarked that convergence in probability discussed in Chapter 7 implies convergence in distribution (or law), *i.e.*, $X_n \xrightarrow{P} X \Rightarrow X_n \xrightarrow{L} X$. ... (*)

The converse is not true, *i.e.*, $X_n \xrightarrow{P} X$, in general, does not imply $X_n \xrightarrow{P} X$.

However, we have the following result.

Let k be a constant. Then, $X_n \xrightarrow{L} k \Rightarrow X_n \xrightarrow{P} k$. ...(**)

Combining (*) and (**), we get the following results.

Let k be a constant. Then $X_n \xrightarrow{L} k \Leftrightarrow X_n \xrightarrow{P} k$. ... (***)

9·16·2. Lindeberg-Levy Theorem

The following case of central limit theorem for independent and identically distributed (*i.i.d.*) random variables, was first proved by Lindeberg and Levy.

> *"If $X_1, X_2, \ldots X_n$ are independently and identically distributed random variables with*
>
> $$\left.\begin{aligned} E(X_i) &= \mu_1 \\ V(X_i) &= \sigma_1^2 \end{aligned}\right\},\ i = 1, 2, \ldots, n \qquad \ldots (9{\cdot}58)$$
>
> *then the sum $S_n = X_1 + X_2 + \ldots + X_n$ is asymptotically normal with mean $\mu = n\mu_1$ and variance $\sigma^2 = n\sigma_1^2$."*

Here we make the following assumptions :

(*i*) The variables are independent and identically distributed.

(*ii*) $E(X_i^2)$ exists for all $i = 1, 2, 3, \ldots$

Proof. Let $M_1(t)$ denote the *m.g.f.* of each of the deviation $(X_i - \mu_1)$ and $M(t)$ denote the *m.g.f.* of the standard variate $Z = (S_n - \mu)/\sigma$.

Since μ_1' and μ_2', (about origin) of the deviation $(X_i - \mu_1)$ are given by :

$$\mu_1' = E(X_i - \mu_1) = 0, \qquad \mu_2' = E(X_i - \mu_1)^2 = \sigma_1^2$$

We have $$M_1(t) = \left(1 + \mu_1' t + \mu_2' \frac{t^2}{2!} + \mu_3' \frac{t^3}{3!} + \ldots\right) = \left\{1 + \frac{t^2}{2!}\sigma_1^2 + O(t^3)\right\}, \qquad \ldots (*)$$

where, $O(t^3)$ contains terms with t^3 and higher powers of t.

We have $$Z = \frac{S_n - \mu}{\sigma} = \frac{(X_1 + X_2 + \ldots + X_n) - n\mu_1}{\sigma} = \sum_{i=1}^{n} \left(\frac{X_i - \mu_1}{\sigma} \right)$$

and since X_i's are independent, we get

$$M_Z(t) = M_{\sum_{i=1}^{n} (X_i - \mu_1)/\sigma}(t) = M_{\sum_{i=1}^{n} (X_i - \mu_1)}(t/\sigma)$$

$$= \prod_{i=1}^{n} \{ M_{(X_i - \mu_1)}(t/\sigma) \} = [M_1(t/\sigma)]^n = \left[M_1(t/\sqrt{n}\,\sigma_1) \right]^n$$

$$= \left[1 + \frac{t^2}{2n} + O(n^{-3/2}) \right]^n \qquad \text{[From (*)]}$$

For every fixed 't', the terms $O(n^{-3/2}) \to 0$ as $n \to \infty$. Therefore, as $n \to \infty$, we get

$$\lim_{n \to \infty} M_Z(t) = \lim_{n \to \infty} \left[1 + \frac{t^2}{2n} + O(n^{-3/2}) \right]^n = \exp(t^2/2),$$

which is the *m.g.f.* of standard normal variate. Hence by uniqueness theorem of *m.g.f.*'s : $Z = (S_n - \mu)/\sigma$ is asymptotically $N(0, 1)$, or $S_n = X_1 + X_2 + \ldots + X_n$ is asymptotically $N(\mu, \sigma^2)$, where

$$\mu = n\mu_1 \quad \text{and} \quad \sigma^2 = n\,\sigma_1^2.$$

Note. C.L.T. can be stated in another from as follows :

(*i*) If $X_1, X_2, \ldots, X_n$ are *i.i.d.* with mean μ_1 and variance σ_1^2 (finite) and $S_n = X_1 + X_2 + \ldots + X_n$, then for $-\infty < a < b < \infty$;

$$\lim_{n \to \infty} P\left[a \le \frac{S_n - \mu_1}{\sigma_1 \sqrt{n}} \le b \right] = \Phi(b) - \Phi(a) = \int_a^b \frac{1}{\sqrt{2\pi}} e^{-x^2/2}\, dx \qquad \ldots (9{\cdot}59)$$

where $\Phi(\cdot)$ is the distribution function of standard normal variate.

or (*ii*)
$$\lim_{n \to \infty} P\left[a \le \frac{S_n - E(S_n)}{\sqrt{\text{Var}(S_n)}} \le b \right] = \Phi(b) - \Phi(a) \qquad \ldots (9{\cdot}59a)$$

or, still another form :

(*iii*)
$$\lim_{n \to \infty} P\left[a \le \frac{\bar{X}_n - E(\bar{X}_n)}{\sqrt{\text{Var}(\bar{X}_n)}} \le b \right] = \Phi(b) - \Phi(a)$$

i.e.,
$$\lim_{n \to \infty} P\left[a \le \frac{\bar{X}_n - \mu_1}{\sigma_1 \sqrt{n}} \le b \right] = \Phi(b) - \Phi(a) \qquad \ldots (9{\cdot}59b)$$

Remarks :

1. We wrote the *C.L.T.* using non-strict inequalities $P\left(a \le (\cdot) \le b\right]$

It makes no difference whether one or both are changed to a strict inequality. The reason is that the limit distribution function (*d.f.*) $\Phi(\cdot)$ is a continuous *d.f.*

2. In the binomial case, *C.L.T.* gives good approximation if p is nearly 1/2. For p near about 0 or 1, *C.L.T.* approximation still holds but in that case n has to be sufficiently greater than in the case $p = 1/2$ approximately.

9·16·3. Applications of Central Limit Theorem. (*a*) If $X_1, X_2, \ldots$ are *i.i.d.* $B(r, p)$ and $S_n = X_1 + X_2 + \ldots + X_n$, then
$$E(S_n) = \sum_{i=1}^{n} E(X_i) = \sum_{i=1}^{n} (rp) = nrp$$

and
$$V(S_n) = V\left(\sum_{i=1}^{n} X_i\right) = \sum_{i=1}^{n} V(X_i) = \sum_{i=1}^{n} (rpq) = nrpq, \qquad (q = 1 - p)$$

Hence, (9·59*a*)
$$\Rightarrow \qquad \lim_{n\to\infty} P\left[a \le \frac{S_n - nrp}{\sqrt{nrp(1-p)}} \le b\right] = \Phi(b) - \Phi(a),\ 0 < p < 1$$

(*b*) If Y_n is binomial variate with parameters n and p, then
$$\lim_{n\to\infty} P\left[a \le \frac{Y_n - np}{\sqrt{np(1-p)}} \le b\right] = \Phi(b) - \Phi(a),\ 0 < p < 1$$

Proof. Let $X_1, X_2, \ldots$ be *i.i.d.* Bernoulli variates, *i.e.*, $B(1, p)$, then
$$S_n = X_1 + X_2 + \ldots + X_n = B(n, p). \text{ But } Y_n = B(n, p)$$

Hence, using Y_n instead of S_n in (9·59*a*), we get
$$\lim_{n\to\infty} P\left[a \le \frac{Y_n - E(Y_n)}{\sqrt{\text{Var } Y_n}} \le b\right] = \Phi(b) - \Phi(a)$$

i.e.,
$$\lim_{n\to\infty} P\left[a \le \frac{Y_n - np}{\sqrt{npq}} \le b\right] = \Phi(b) - \Phi(a),\ q = 1 - p$$

(*c*) If Y_n is distributed as $P(n)$, then $\lim_{n\to\infty} P\left[a \le \frac{Y_n - n}{\sqrt{n}} \le b\right] = \Phi(b) - \Phi(a)$

Thus, for instance $\lim_{n\to\infty} P(Y_n \le n) = \frac{1}{2}$, *i.e.*, $\sum_{k=0}^{n} \frac{e^{-n} n^k}{k!} = \frac{1}{2}$, as $n \to \infty$

Proof. Let $X_1, X_2, \ldots$ be *i.i.d.* $P(1)$. Then $S_n = X_1 + X_2 + \ldots + X_n \sim P(n) \Rightarrow Y_n = S_n$
$$\therefore \qquad \lim_{n\to\infty} P\left[a \le \frac{Y_n - n}{\sqrt{n}} \le b\right] = P\left[a \le \frac{S_n - n}{\sqrt{n}} \le b\right] \to \Phi(b) - \Phi(a) \text{ as } n \to \infty$$

In particular, let us take $a = -\infty$ and $b = 0$, then
$$P\left(a \le \frac{Y_n - n}{\sqrt{n}} \le b\right) = P\left(\frac{Y_n - n}{\sqrt{n}} \le 0\right) = P(Y_n \le n) \qquad \ldots (*)$$

Also, $$\Phi(b) - \Phi(a) = \Phi(0) - \Phi(-\infty) = \frac{1}{2} \qquad \ldots (**)$$

From (*) and (**), we get $P(Y_n \le n) \to \frac{1}{2}$ as $n \to \infty$.

Remark. This result could be generalised. In fact, on taking $a = -\infty$ and $b = 0$ in (9·59*a*), we get
$$P\left[\frac{S_n - E(S_n)}{\sqrt{\text{Var}(S_n)}} \le 0\right] \to \frac{1}{2} \quad \Rightarrow \quad P[S_n \le E(S_n)] \to \frac{1}{2}, \text{ as } n \to \infty$$

9·16·4. Liapounoff's Central Limit Theorem

Below we shall give (without proof) the central limit theorem for the generalised case when the variables are not identically distributed and where, in addition to the existence of the second moment for the variables X_i, we impose some further conditions.

Let $X_1, X_2, ..., X_n$ be independent random variables such that

$$\left.\begin{aligned} E(X_i) &= \mu_i \\ V(X_i) &= \sigma_i^2 \end{aligned}\right\} ; \; i = 1, 2, ..., n. \qquad ...(9{\cdot}60)$$

Let us suppose that third absolute moment, say ρ_i^3 of X_i about its mean exists, i.e.,

$$\rho_i^3 = E\{|X_i - \mu_i|^3\}; \; i = 1, 2, ..., n \text{ is finite.} \qquad ...(9{\cdot}60a)$$

Further let $$\rho^3 = \sum_{i=1}^{n} \rho_i^3 .$$

If $\lim\limits_{n\to\infty} \dfrac{\rho}{\sigma} = 0$, *then the sum* $X = X_1 + X_2 + ... + X_n$ *is asymptotically* $N(\mu, \sigma^2)$, *where*

$$\mu = \sum_{i=1}^{n} \mu_i \quad \text{and} \quad \sigma^2 = \sum_{i=1}^{n} \sigma_i^2 \qquad ...(9{\cdot}60b)$$

***Remarks* on Central Limit Theorems :**

1. (*About Liapounoff's theorem*). If the variables X_i ; $i = 1, 2, ..., n$ are identical, then

$$\rho^3 = \sum_{i=1}^{n} \rho_i^3 = n\rho_1^3 \text{ and } \sigma^2 = \sum_{i=1}^{n} \sigma_i^2 = n\sigma_1^2 \qquad \therefore \frac{\rho}{\sigma} = \frac{n^{1/3}\rho_1}{n^{1/2}\cdot\sigma_1} = \frac{\rho_1}{\sigma_1}\cdot\frac{1}{n^{1/6}} \to 0, \; n \to \infty.$$

Thus, for identical variables, the condition of Liapounoff's theorem is satisfied.

It may be pointed out here that Lindeberg – Levy theorem proved in § 9·16·2, should not be inferred as a particular case of Liapounoff's theorem, since the former does not assume the existence of the third moment.

2. Central limit theorem can be expected in the following cases :

(*i*) If a certain random variable X arises as cumulative effect of several independent causes each of which can be considered as a continuous random variable, then X obeys central limit theorem under certain regularity conditions.

(*ii*) If $\varphi(X_1, X_2, ..., X_n)$ is a function of X_i's having first and second continuous derivatives about the point $(\mu_1, \mu_2, ..., \mu_n)$, then under certain regularity conditions, $\varphi(X_1, X_2, ..., X_n)$ is asymptotically normal with mean $\varphi(\mu_1, \mu_2, ..., \mu_n)$.

(*iii*) Under certain conditions, the central limit theorem holds for variables which are not independent.

3. *Relation between CLT and WLLN* :

(*a*) Both the central limit theorem (CLT) and the weak law of large numbers (WLLN) hold for a sequence $\{X_n\}$ of *i.i.d.* random variables with finite mean μ and variance σ^2.

However, in this case, the CLT is a stronger result than the WLLN in the sense that the former provides an estimate of the $P\{|S_n - n\mu)/n| \geq \varepsilon\}$, as given below :

$$P\left[\left|\frac{S_n - n\mu}{n}\right| \geq \varepsilon\right] = P\left[\left|\bar{X}_n - \mu\right| \geq \varepsilon\right] = \left[\left|\frac{\bar{X}_n - \mu}{\sigma/\sqrt{n}}\right| \geq \frac{\varepsilon}{\sigma/\sqrt{n}}\right]$$

$$= P\left[|Z| \geq \varepsilon\sqrt{n}/\sigma\right]; \quad Z \sim N(0, 1)$$

$$= 1 - P\left[|Z| \leq \varepsilon\sqrt{n}/\sigma\right] = 1 - \left[\Phi\left(\frac{\varepsilon\sqrt{n}}{\sigma}\right) - \Phi\left(-\frac{\varepsilon\sqrt{n}}{\sigma}\right)\right]$$

where, $\Phi(\cdot)$ is the distribution function of standard normal variate.

However, WLLN does not require the existence of variance (*c.f. Khinchine's Theorem*).

(*b*) For the sequence $\{X_n\}$ of independent and uniformly bounded *r.v.*'s WLLN holds and CLT holds in this case provided $B_n = \text{Var}(X_1 + X_2 + ... + X_n) = \sigma_1^2 + \sigma_2^2 + ... + \sigma_n^2 \to \infty$ as $n \to \infty$.

(*c*) For the sequence $\{X_n\}$ of independent *r.v.*'s CLT may hold but the WLLN may not hold.

9·16·5. Cramer's Theorem

We state below (without proof), a useful result on the convergence of sequences of *r.v.*'s.

Cramer's Theorem. Let $\{X_n\}$ and $\{Y_n\}$ be sequences of *r.v.*'s such that

$$X_n \xrightarrow{L} X \quad \text{and} \quad Y_n \xrightarrow{P} c \text{ (constant), then } \quad \frac{X_n}{Y_n} \xrightarrow{L} \frac{X}{c} \text{ if } c \neq 0 \qquad ...(9·61)$$

Example 9·49. *Let $X_1, X_2, ...$ be i.i.d. Poisson variates with parameter λ. Use CLT to estimate :*

$$P(120 \le S_n \le 160), \text{ where } S_n = X_1 + X_2 + ... + X_n ; \ \lambda = 2 \text{ and } n = 75.$$

Solution. Since X_i's are *i.i.d.* $P(\lambda)$, $E(X_i) = \lambda$ and $\text{Var}(X_i) = \lambda$; $i = 1, 2, ..., n$

$$\therefore \quad E(S_n) = \sum_{i=1}^{n} E(X_i) = n\lambda \quad \text{and} \quad \text{Var}(S_n) = \text{Var}(X_1 + X_2 + ... + X_n) = \sum_{i=1}^{n} \text{Var } X_i = n\lambda$$

Hence, by Lindeberg-Levy CLT, (for large n)

$$S_n \sim N(n\lambda, n\lambda) = N(\mu = 150, \sigma^2 = 150) ; \ (n = 75 ; \ \lambda = 2)$$

$$\therefore \quad P(120 \le S_n \le 160) = P\left(\frac{120 - 150}{\sqrt{150}} \le Z \le \frac{160 - 150}{\sqrt{150}}\right) = P(-2·45 \le Z \le 0·82) ; \ Z \sim N(0, 1)$$

$$= P(-2·45 \le Z \le 0) + P(0 \le Z \le 0·82)$$

$$= P(0 \le Z \le 2·45) + P(0 \le Z \le 0·82) \qquad \text{[By symmetry]}$$

$$= 0·4929 + 0·2939 = 0·7868 \qquad \text{[From Normal Probability Tables]}$$

Example 9·50. *Let $X_1, X_2, ..., X_{2n}$ be i.i.d. standardised variates with $E(X_i^4) < \infty$. Find the limiting distribution of :*

$$Z_n = \sqrt{n}\left(X_1 X_2 + X_3 X_4 + ... + X_{2n-1} X_{2n}\right) \div \left(X_1^2 + X_2^2 + ... + X_{2n}^2\right)$$

Solution. Since X_i's are *i.i.d.* standardised variates, we have

$$E(X_i) = 0 ; \quad \text{Var}(X_i) = E(X_i^2) = 1, \ i = 1, 2, ..., 2n \qquad ...(*)$$

Let $Y_i = X_{2i-1} X_{2i} ; \ i = 1, 2, ..., n \quad \Rightarrow \quad E(Y_i) = E(X_{2i-1}) E(X_{2i}) = 0$ [From (*)]

($\because$ X_i's are independent.)

$$\therefore \quad \text{Var}(Y_i) = E(Y_i^2) = E(X_{2i-1}^2 X_{2i}^2) = E(X_{2i-1}^2) E(X_{2i}^2) = 1 \qquad \text{[From (*)]}$$

Hence, Y_i, $i = 1, 2, ..., n$ are also *i.i.d.* standardised variates. Hence by CLT for *i.i.d. r.v.*'s, $\left(S_n = \sum_{i=1}^{n} Y_i\right)$, we get

$$U_n = \frac{S_n - E(S_n)}{\sqrt{\text{Var}(S_n)}} = \frac{X_1 X_2 + X_3 X_4 + ... + X_{2n-1} X_{2n}}{\sqrt{n}} \xrightarrow{L} Z \sim N(0, 1) \text{ as } n \to \infty.$$

Also, $E(X_i^2) = 1$ (finite), $i = 1, 2, ..., n$.

Hence, by Khinchine's theorem, WLLN applied to sequence $\{X_i^2\}$, $i = 1, 2, ..., 2n$; so that

$$V_n = \frac{X_1^2 + X_2^2 + \ldots + X_{2n}^2}{2n} \xrightarrow{P} E(X_i^2) = 1, \text{ as } n \to \infty.$$

Hence, by Cramer's theorem :

$$\lim_{n\to\infty} \frac{U_n}{V_n} = \frac{2\sqrt{n}\,(X_1 X_2 + X_3 X_4 + \ldots + X_{2n-1} X_{2n})}{X_1^2 + X_2^2 + \ldots + X_{2n}^2} \xrightarrow{L} \frac{Z}{1} = Z \sim N(0, 1)$$

$$\Rightarrow \quad \lim_{n\to\infty} Z_n = \lim_{n\to\infty} \frac{\sqrt{n}\,(X_1 X_2 + X_3 X_4 + \ldots + X_{2n-1} X_{2n})}{X_1^2 + X_2^2 + \ldots + X_{2n}^2} \xrightarrow{L} \frac{Z}{2} \sim N\left(0, \frac{1}{4}\right)$$

$$[\because Z \sim N(0,1) \quad \Rightarrow \quad CZ \sim N(0, C^2)]$$

9·17. COMPOUND DISTRIBUTIONS

Consider a random variable X whose distribution depends on a single parameter θ, which instead of being regarded as a fixed constant, is also a random variable following a particular distribution. In this case, we say that the random variable X has a *compound or composed distribution.*

9·17·1. Compound Binomial Distribution

Let us suppose that $X_1, X_2, X_3, \ldots$ are identically and independently distributed Bernoulli variates with

$$P(X_i = 1) = p \quad \text{and} \quad P(X_i = 0) = q = 1 - p$$

For a fixed n, the random variable $X = X_1 + X_2 + \ldots + X_n$ is a Binomial variate with parameters n and p and probability function :

$$P(X = r) = \binom{n}{r} p^r q^{n-r} \; ; \; r = 0, 1, 2, \ldots, n \qquad \ldots (*)$$

which gives the probability of r successes in n independent trials with constant probability 'p' of success for each trial.

Now suppose that n, instead of being regarded as a fixed constant, is also a random variable following Poisson distribution with parameter λ. Then

$$P(n = k) = \frac{e^{-\lambda}\lambda^k}{k!} \; ; \; k = 0, 1, 2, \ldots \qquad \ldots (**)$$

In such a case X is said to have compound binomial distribution. The joint probability function of X and n is given by :

$$P(X = r \cap n = k) = P(n = k)\, P(X = r \mid n = k) = \frac{e^{-\lambda}\lambda^k}{k!} \binom{k}{r} p^r q^{k-r}, \qquad \text{[From (*) and (**)]}$$

since $P(X = r \mid n = k)$ is probability of r successes in k trials. Obviously, $r \le k \Rightarrow k \ge r$.

The marginal distribution of X is given by :

$$P(X = r) = \sum_{k=r}^{\infty} P(X = r \cap n = k)$$

$$= e^{-\lambda} p^r \sum_{k=r}^{\infty} \binom{k}{r} \frac{\lambda^k q^{k-r}}{k!} = \frac{e^{-\lambda}(\lambda p)^r}{r!} \sum_{k=r}^{\infty} \frac{(\lambda q)^{k-r}}{(k-r)!}$$

$$= \frac{e^{-\lambda}(\lambda p)^r}{r!} \sum_{j=0}^{\infty} \frac{(\lambda q)^j}{j!}, \quad (j = k - r)$$

$$= \frac{e^{-\lambda}(\lambda p)^r}{r!} \cdot e^{\lambda q} = \frac{e^{-\lambda p}(\lambda p)^r}{r!},$$

which is the probability function of a Poisson variate with parameter λp.

Hence, $\qquad E(X) = \lambda p \qquad$ and $\qquad \text{Var}(x) = \lambda p$

We give below some of the practical situations where we would come across compound Binomial distribution.

1. Suppose that the probability of an insect laying n eggs is given by the Poisson distribution $e^{-\lambda}\lambda^n / n!$ and the probability of an egg developing is p. Assuming natural independence of eggs, the probability of a total of k survivors is given by the Poisson distribution with parameter λp.

2. The probability that a radioactive substance gives off n Beta-particles in a unit of time is $P(\lambda)$, $(n = 0, 1, 2, \ldots)$. The probability that a given particle will strike a counter and be registered is p. Then, the probability of registering n Beta-particles in a unit of time is also $P(\lambda p)$.

3. If the probability of number of hits by lightning during any time interval t is $P(\lambda t)$ and if the probability of its hitting and damaging an individual is p, then (assuming stochastic independence) the total damage during time 't' is $P(\lambda t p)$.

9·17·2. Compound Poisson Distribution

Let X be a $P(\lambda)$ so that

$$P(X = r) = \frac{e^{-\lambda}\lambda^r}{r!}; \; r = 0, 1, 2, \ldots$$

where, λ itself is a continuous random variable with generalised gamma density

$$g(\lambda) = \begin{cases} \dfrac{a^v}{\Gamma(v)} e^{-a\lambda} \lambda^{v-1}; \; \lambda > 0, a > 0, v > 0 \\ 0, \; \lambda \le 0 \end{cases}$$

Let us consider the two dimensional random vector (X, λ) in which one variable is discrete and other is continuous. For a constant $h > 0$ and $\lambda_1 > 0$, the joint density of X and λ is given by :

$$P(X = r \cap \lambda_1 \le \lambda \le \lambda_1 + h) = P(\lambda_1 \le \lambda \le \lambda_1 + h)\, P(X = r \mid \lambda_1 \le \lambda \le \lambda_1 + h)$$

Dividing both sides by h and proceeding to the limits as $h \to 0$, we get

$$\lim_{h \to 0} \frac{P(X = r \cap \lambda_1 \le \lambda \le \lambda_1 + h)}{h} = \lim_{h \to 0} P(X = r \mid \lambda_1 \le \lambda \le \lambda_1 + h) \times \lim_{h \to 0} \frac{P(\lambda_1 \le \lambda \le \lambda_1 + h)}{h}$$

But $$\lim_{h \to 0} \frac{P(\lambda_1 \le \lambda \le \lambda_1 + h)}{h} = \lim_{h \to 0} \frac{G(\lambda_1 + h) - G(\lambda_1)}{h} = G'(\lambda_1) = g(\lambda_1),$$

where, $G(\cdot)$ is the distribution function and $g(\cdot)$ is *p.d.f.* of λ.

$$\therefore \quad \lim_{h \to 0} \frac{P(X = r \cap \lambda_1 \le \lambda \le \lambda_1 + h)}{h} = \frac{e^{-\lambda_1}\lambda_1^r}{r!} \cdot \frac{a^v}{\Gamma(v)} \lambda_1^{v-1} e^{-a\lambda_1}$$

Integrating *w.r.* to λ_1 over 0 to ∞ and using gamma integral, the marginal probability function of X is given by :

$$P(x=r)=\frac{a^v}{\Gamma(v)\,r!}\int_0^\infty e^{-(1+a)\lambda}\,\lambda^{r+v-1}\,d\lambda=\frac{a^v}{\Gamma(v)\,r!}\cdot\frac{\Gamma(r+v)}{(1+a)^{r+v}}$$

$$=\left(\frac{a}{1+a}\right)^v\frac{v(v+1)(v+2)\ldots(v+r-1)}{(1+a)^r\,r!}$$

$$=\left(\frac{a}{1+a}\right)^v(-1)^r\binom{-v}{r}\left(\frac{1}{1+a}\right)^r=\binom{-v}{r}p^v(-q)^r\;;\;r=0,1,2,\ldots$$

where, $p=a/(1+a), q=1-p=1/(1+a)$.

Thus, the marginal distribution of X is negative binomial with parameters (v, p).

9·18. ORDER STATISTICS

Let $X_1, X_2, \ldots, X_n$ be n independent and identically distributed variates, each with cumulative distribution function $F(x)$ and *p.d.f.* $f(x)$. If these variables are arranged in ascending order of magnitude and then written as $X_{(1)} \le X_{(2)} \le \ldots \le X_{(n)}$, we call $X_{(r)}$ as the rth order statistic, $r = 1, 2, \ldots, n$. The $X_{(r)}$'s because of the inequality relations among them are necessarily dependent.

Remark. If we write these ordered values as

$Y_1 \le Y_2 \le \ldots \le Y_n$, then

$Y_r = X_{(r)} =$ rth smallest of $X_1, X_2, \ldots, X_n$

$Y_1 = X_{(1)} =$ The smallest of $X_1, X_2, \ldots, X_n$

$Y_n = X_{(n)} =$ The largest of $X_1, X_2, \ldots, X_n$

9·18·1. Cumulative Distribution Function of a Single Order Statistics

Let $F_r(x), r = 1, 2, \ldots, n$ denote the *c.d.f.* of the rth order statistic $X_{(r)}$. Then the *c.d.f.* $F_n(x)$ of the largest order statistic $X_{(n)}$ is given by :

$$F_n(x)=P\left[X_{(n)}\le x\right]=P(X_i\le x;\;i=1,2,\ldots,n)$$

$$=P(X_1\le x\cap X_2\le x\cap\ldots\cap X_n\le x)$$

$$=P(X_1\le x)\cdot P(X_2\le x)\ldots P(X_n\le x)\qquad(\because\ X_i\text{'s are independent})$$

$$=[F(x)]^n, \qquad \ldots (9·62)$$

since $X_1, X_2, \ldots, X_n$ are identically distributed.

The *p.d.f.* $f_n(x)$ of $X_{(n)}$ is given by :

$$f_n(x)=\frac{d}{dx}[F_n(x)]=n[F(x)]^{n-1}\cdot f(x) \qquad \ldots (9·62a)$$

The *c.d.f.* $F_1(x)$ of the smallest order statistic $X_{(1)}$ is given by :

$$F_1(x)=P\left[X_{(1)}\le x\right]=1-P\left[X_{(1)}>x\right]=1-P[X_i>x;\;i=1,2,\ldots,n]$$

$$=1-\prod_{i=1}^n P(X_i>x)=1-\prod_{i=1}^n[1-P(X_i\le x)]$$

$$=1-[1-F(x)]^n, \qquad \ldots (9·63)$$

since $X_1, X_2, \ldots, X_n$ are *i.i.d. r.v.'s*.

The *p.d.f.* $f_1(x)$ of $X_{(1)}$ is given by :

$$f_1(x) = \frac{d}{dx}[F_1(x)] = n[1-F(x)]^{n-1} \cdot f(x) \qquad \text{... (9·63a)}$$

In general, the *c.d.f.* $F_r(x)$ of the *r*th order statistic $X_{(r)}$ is given by :

$$F_r(x) = P[X_{(r)} \le x] = P[\text{At least } r \text{ of the } X_i\text{'s are} \le x]$$

$$= \sum_{j=r}^{n} P[\text{Exatcly } j \text{ of the } n, X_i\text{'s are} \le x]$$

$$= \sum_{j=r}^{n} \binom{n}{j} F^j(x)[1-F(x)]^{n-j}, \qquad \text{... (9·64)}$$

by using Binomial probability model.

Remarks :

1. (9·64) can also be written as :

$$F_r(x) = I_{F(x)}(r, n-r+1), \qquad \text{... (9·65)}$$

where,

$$I_p(a, b) = \frac{1}{B(a,b)} \int_0^p t^{a-1}(1-t)^{b-1}\, dt \qquad \text{... (9·65a)}$$

is the '*incomplete Beta Function*' and has been tabulated in Biometrika tables by Pearson and Hartley.

(9·65) and (9·65*a*) show that the probability points of an order statistic can be obtained with the help of incomplete beta function.

2. Taking $r = 1$ and $r = n$ in (9·64), we get, respectively

$$F_1(x) = \sum_{j=1}^{n} \binom{n}{j} F^j(x)[1-F(x)]^{n-j} = 1 - \left\{ \binom{n}{j} F^j(x)[1-F(x)]^{n-j} \right\}_{j=0} = 1-[1-F(x)]^n \qquad \text{... (9·65b)}$$

and

$$F_n(x) = [F(x)]^n \qquad \text{... (9·65c)}$$

the results which have already been obtained in (9·63) and (9·62) respectively.

9·18·2. Probability Density Function (p.d.f.) of a Single Order Statistic

The results in (9·62) to (9·64) are valid for both discrete and continuous *r.v.*'s. We shall now assume that X_i's are *i.i.d.* continuous *r.v.*'s with *p.d.f.* $f(x) = F'(x)$. If $f_r(x)$ denotes the *p.d.f.* of $X_{(r)}$ then from (9·64) or (9·65), we get

$$f_r(x) = \frac{d}{dx}[F_r(x)] = \frac{d}{dx}[I_{F(x)}(r, n-r+1)]$$

$$= \frac{d}{dx}\left[\frac{1}{B(r, n-r+1)} \int_0^{F(x)} t^{r-1}(1-t)^{n-r}\, dt\right] \qquad \text{... (9·66)}$$

Let us write

$$g(t) = \int t^{r-1}(1-t)^{n-r}\, dt \quad \Rightarrow \quad g'(t) = t^{r-1}(1-t)^{n-r} \qquad \text{... (*)}$$

$$\Rightarrow \quad \int_0^{F(x)} t^{r-1}(1-t)^{n-r}\, dt = \Big| g(t) \Big|_0^{F(x)} = g[F(x)] - g(0)$$

$$\Rightarrow \quad \frac{d}{dx} \int_0^{F(x)} t^{r-1}(1-t)^{n-r}\, dt = g'[F(x)] \cdot f(x) \qquad [\because g(0) \text{ is constant}]$$

$$= [F(x)]^{r-1}[1-F(x)]^{n-r} f(x) \qquad [\text{Using } (*)]$$

Substituting in (9·66), we get

$$f_r(x) = \frac{1}{B(r, n-r+1)} \cdot F^{r-1}(x)[1-F(x)]^{n-r} \cdot f(x) \qquad \ldots (9{\cdot}67)$$

Aliter. By definition of a *p.d.f.*, we get

$$f_r(x) = \lim_{\delta x \to 0} \frac{P[x \le X_{(r)} \le x + \delta x]}{\delta x} \qquad \ldots (9{\cdot}68)$$

The event $E : x < X_{(r)} \le x + \delta x$ can materialise as follows :

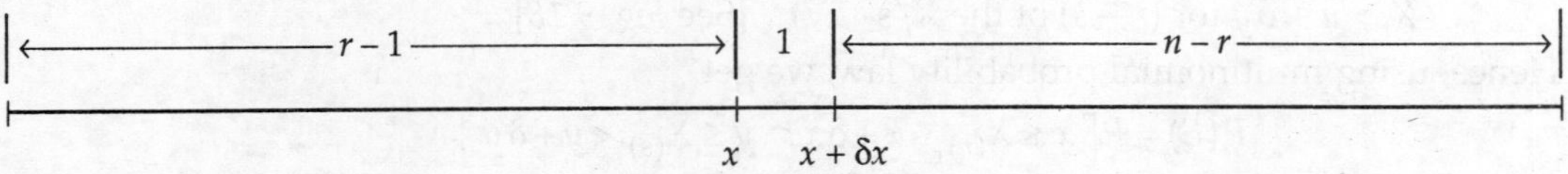

Fig. 9·17

$X_i \le x$ for $(r-1)$ of the X_i's

and $x < X_i \le x + \delta x$ for one X_i

and $X_i \ge x + \delta x$ for the remaining $(n-r)$ of the X_i's [See Fig. 9·17]

Hence, by the multinomial probability law, we have

$$P\left(x < X_{(r)} \le x + \delta x\right) = \frac{n!}{(r-1)!\,1!\,(n-r)!} p_1^{r-1} \cdot p_2 \cdot p_3^{n-r} \qquad \ldots (9{\cdot}69)$$

where,
$$p_1 = P(X_i \le x) = F(x)$$
$$p_2 = P(x < X_i \le x + \delta x) = F(x+\delta x) - F(x)$$
and
$$p_3 = P(X_i \ge x + \delta x) = 1 - P(X_i \le x + \delta x) = 1 - F(x + \delta x)$$

Substituting in (9·69) and using (9·68), we get

$$f_r(x) = \frac{1}{B(r, n-r+1)} \times F^{r-1}(x) \times \lim_{\delta x \to 0}\left[\frac{F(x+\delta x) - F(x)}{\delta x}\right] \times \lim_{\delta x \to 0}[1 - F(x+\delta x)]^{n-r}$$

$$= \frac{1}{B(r, n-r+1)} \cdot F^{r-1}(x) \cdot f(x) \cdot [1 - F(x)]^{n-r}$$

as in (9·67).

Deductions. In particular, taking $r = 1$ and $r = n$ in (9·67), we get respectively :

$$f_1(x) = \frac{1}{B(1,n)} \cdot f(x)[1-F(x)]^{n-1} = n[1-F(x)]^{n-1} f(x)$$

and
$$f_n(x) = \frac{1}{B(n,1)} \cdot [F(x)]^{n-1} \cdot f(x) = n[F(x)]^{n-1} \cdot f(x)$$

$$\left[\because \frac{1}{B(1,n)} = \frac{1}{B(n,1)} = \frac{\Gamma(n+1)}{\Gamma 1 \cdot \Gamma n} = \frac{n\,\Gamma n}{\Gamma n} = n\right]$$

9·18·3. Joint p.d.f. of two Order Statistics

Let us denote the joint *p.d.f.* of $X_{(r)}$ and $X_{(s)}$, where $1 \le r < s \le n$ by $f_{rs}(x, y)$. Then,

$$f_{rs}(x, y) = \lim_{\substack{\delta x \to 0 \\ \delta y \to 0}} \frac{P[x \le X_{(r)} \le x + \delta x \cap y \le X_{(s)} \le y + \delta y]}{\delta x\, \delta y} \qquad \ldots (9{\cdot}70)$$

The event $E=\{x \le X_{(r)} \le x+\delta x \cap y \le X_{(s)} \le y+\delta y\}$ can materialise as follows :

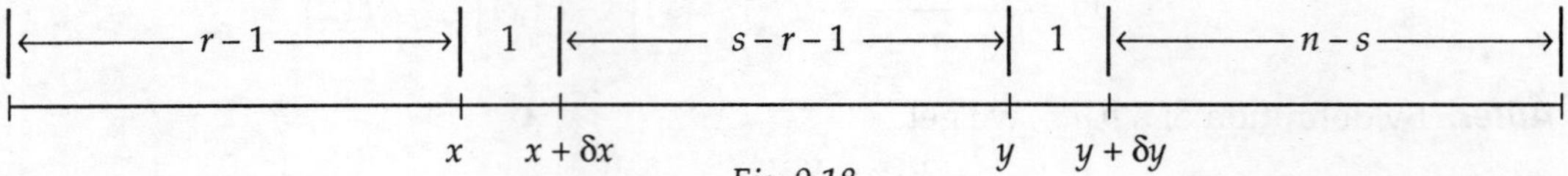

Fig. 9·18

$X_i \le x$ for $(r-1)$ of the X_i's ; $x < X_i \le x+\delta x$ for one X_i

$x+\delta x < X_i \le y$ for $(s-r-1)$ of X_i's ; $y < X_i \le y+\delta y$ for one X_i

and $X_i > y+\delta y$ for $(n-s)$ of the X_i's [See *Fig. 9·18*]

Hence, using multinomial probability law, we get

$$P(E)=P\left[x \le X_{(r)} \le x+\delta x \cap y \le X_{(s)} < y+\delta y\right]$$

$$=\frac{n!}{(r-1)!\,1!\,(s-r-1)!\,1!\,(n-s)!}p_1^{r-1}\cdot p_2\,p_3^{s-r-1}\cdot p_4\,p_5^{n-s} \quad \text{... (9·71)}$$

where, $p_1=P(X_i \le x)=F(x)$, $\quad p_2=P(x<X_i \le x+\delta x)=F(x+\delta x)-F(x)$

$p_3=P(x+\delta x<X_i \le y)=F(y)-F(x+\delta x)$, $\quad p_4=P(y<X_i \le y+\delta y)=F(y+\delta y)-F(y)$,

$p_5=P(X_i>y+\delta y)=1-P(X_i \le y+\delta y)=1-F(y+\delta y)$

Substituting in (9·71) and using (9·70), we get

$$f_{rs}(x,y)=\lim_{\substack{\delta x \to 0\\ \delta y \to 0}}\frac{P(E)}{\delta x\,\delta y}$$

$$=\frac{n!}{(r-1)!\,(s-r-1)!\,(n-s)!}\times F^{r-1}(x)\times\lim_{\delta x\to 0}\frac{[F(x+\delta x)-F(x)]}{\delta x}\times\lim_{\delta x\to 0}[F(y)-F(x+\delta x)]^{s-r-1}$$

$$\times\lim_{\delta y\to 0}\left[\frac{F(y+\delta y)-F(y)}{\delta y}\right]\times\lim_{\delta y\to 0}[1-F(y+\delta y)]^{n-s}$$

$$=\frac{n!}{(r-1)!\,(s-r-1)!\,(n-s)!}\cdot F^{r-1}(x)\cdot f(x)\cdot[F(y)-F(x)]^{s-r-1}f(y)\cdot[1-F(y)]^{n-s} \quad \text{... (9·72)}$$

9·18·4. Joint p.d.f. of *k*-Order Statistics

The joint *p.d.f.* of *k*-order statistics $X_{(r_1)}, X_{(r_2)}, \ldots, X_{(r_k)}$ where $1 \le r_1 < r_2 < \ldots < r_k \le n$ and $1 \le k \le n$ is for $x_1 \le x_2 \le .. \le x_k$ given by [on using the following configuration [*Fig. 9·19*] and the multinomial probability law] :

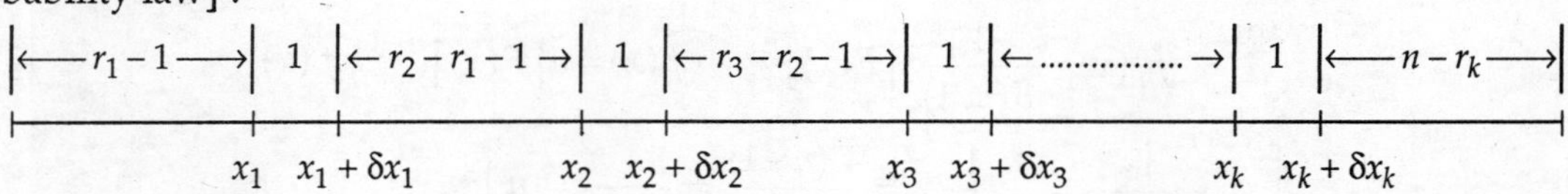

Fig. 9·19

$$f_{r_1,r_2,\ldots,r_k}(x_1,x_2,\ldots,x_k)=\frac{n!}{(r_1-1)!\,(r_2-r_1-1)!\,\ldots\,(r_k-r_{k-1}-1)!\,(n-r_k)!}$$

$$\times F^{r_1-1}(x_1)\times f(x_1)\times[F(x_2)-F(x_1)]^{r_2-r_1-1}\times f(x_2)$$

$$\times[F(x_3)-F(x_2)]^{r_3-r_2-1}\times f(x_3)\times\ldots\times f(x_k)[1-F(x_k)]^{n-r_k} \quad \text{... (9·73)}$$

9·18·5. Joint p.d.f. of all *n* - Order Statistics

In particular, the joint *p.d.f.* of all the *n* order statistics is obtained on taking $k = n$ in (9·73). This implies that $r_i = i$ for $i = 1, 2, ..., n$. Hence joint *p.d.f.* of $X_{(1)}, X_{(2)}, ..., X_{(n)}$ is given by :

$$f_{1, 2, ..., n}(x_1, x_2, ..., x_n) = n!\, f(x_1)\, f(x_2) \ldots f(x_n) \qquad \ldots (9·74)$$

Aliter. We can easily obtain (9·74) by using the following configuration [*Fig.* 9·20] and the multinomial probability law :

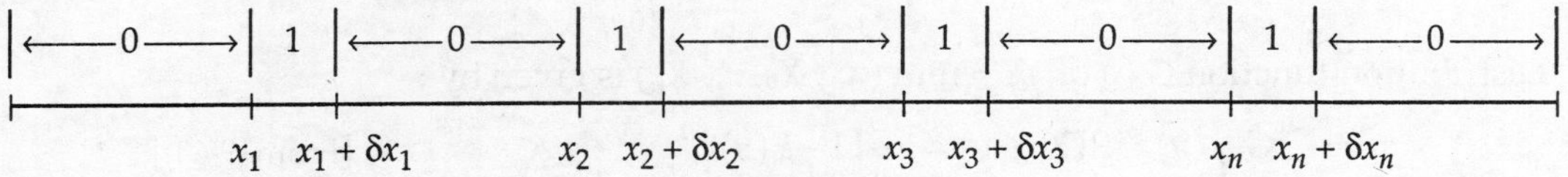

Fig. 9·20

9·18·6. Distribution of Range and Other Systematic Statistics

Let us obtain the *p.d.f.* of the statistics $W_{rs} = X_{(s)} - X_{(r)}$; $r < s$. We start with the joint *p.d.f.* of $X_{(r)}$ and $X_{(s)}$ given in (9·72) and transform $[X_{(r)}, X_{(s)}]$ to the new variables W_{rs} and $X_{(r)}$ such that

$$w_{rs} = y - x \ ; \quad x = x \qquad \text{such that} \qquad y = x + w_{rs} \quad \text{and} \quad x = x$$

$$\therefore \qquad J = \frac{\partial(x, y)}{\partial(x, w_{rs})} = \begin{vmatrix} 1 & 1 \\ 0 & 1 \end{vmatrix} = 1 \quad \Rightarrow \quad |J| = 1$$

The joint *p.d.f.* $f_{rs}(x, y)$ in (9·72) transforms to the joint *p.d.f.* of $X_{(r)}$ and W_{rs} as given below ·

$$g(x, w_{rs}) = c_{rs} \cdot F^{r-1}(x) \cdot f(x) [F(x + w_{rs}) - F(x)]^{s-r-1}$$
$$\times f(x + w_{rs}) \times [1 - F(x + w_{rs})]^{n-s} \qquad \ldots (9·75)$$

where
$$c_{rs} = \frac{n!}{(r-1)!\,(s-r-1)!\,(n-s)!} \qquad \ldots (9·76)$$

Integrating (9·75) *w.r.* to *x* from $-\infty$ to ∞, we obtain the *p.d.f.* of W_{rs} as :

$$g(w_{rs}) = c_{rs} \int_{-\infty}^{\infty} \Big[F^{r-1}(x)\, f(x) \big\{ F(x + w_{rs}) - F(x) \big\}^{s-r-1} \cdot f(x + w_{rs})$$
$$\times \big\{ 1 - F(x + w_{rs}) \big\}^{n-s} \Big] dx \qquad \ldots (9·77)$$

***Remark.* Distribution of Range $W = X_{(n)} - X_{(1)}$.** Taking $r = 1$ and $s = n$ in (9·77), we obtain the *p.d.f.* of the range $W = X_{(n)} - X_{(1)}$ as :

$$g(w) = n(n-1) \int_{-\infty}^{\infty} f(x) [F(x + w) - F(x)]^{n-2} \cdot f(x + w)\, dx \ ; \ w \geq 0 \qquad \ldots (9·78)$$

The *c.d.f.* of *W* is rather simple as given below :

$$G(w) = P(W \leq w) = \int_0^w g(u)\, du$$
$$= \int_0^w \left\{ n(n-1) \int_{-\infty}^{\infty} f(x) [F(x + u) - F(x)]^{n-2} f(x + u)\, dx \right\} du$$
$$= n \int_{-\infty}^{\infty} \left[f(x) \left\{ \int_0^w (n-1) f(x + u) [F(x + u) - F(x)]^{n-2} du \right\} \right] dx$$
$$= n \int_{-\infty}^{\infty} f(x) [F(x + w) - F(x)]^{n-1} dx \qquad \ldots (9·78a)$$

Example 9·51. *Let $X_1, X_2, ..., X_n$ be a random sample from a population with continuous density. Show that $Y_1 = \min(X_1, X_2, ..., X_n)$ is exponential with parameter $n\lambda$ if and only if each X_i is exponential with parameter λ.*

Solution. Let X_i be *i.i.d.* exponential variates with parameter λ and *p.d.f.*

$$f(x) = \lambda e^{-\lambda x}\ ;\ x \geq 0, \lambda > 0 \qquad \text{... (*)}$$

$$F(x) = P(X \leq x) = \int_0^x f(u)\, du = \lambda \int_0^x e^{-\lambda u}\, du = 1 - e^{-\lambda x} \qquad \text{... (**)}$$

Distribution function $G(\cdot)$ of $Y_1 = \min(X_1, X_2, ..., X_n)$ is given by :

$$G_{Y_1}(y) = P(Y_1 \leq y) = 1 - [1 - F(y)]^n \qquad \text{[From (9·63)]}$$

$$= 1 - \left[1 - (1 - e^{-\lambda y})\right]^n = 1 - e^{-n\lambda y}, \qquad \text{... (***)}$$

which is the distribution function of exponential distribution with parameter $n\lambda$. Hence $Y_1 = \min(X_1, X_2, ..., X_n)$ has exponential distribution with parameter $n\lambda$.

Conversely. Let $Y_1 = \min(X_1, X_2, ..., X_n) \sim \text{Exp}(n\lambda)$ so that

$$P(Y_1 \leq y) = 1 - e^{-n\lambda y} \quad \Rightarrow \quad P(Y_1 \geq y) = e^{-n\lambda y}, \qquad \text{... (****)}$$

$$\Rightarrow \quad P[\min(X_1, X_2, ..., X_n) \geq y] = e^{-n\lambda y},$$

$$\Rightarrow \quad P[(X_1 \geq y) \cap (X_2 \geq y) \cap ... (X_n \geq y)] = e^{-n\lambda y}$$

$$\Rightarrow \quad \prod_{i=1}^{n} P(X_i \geq y) = e^{-n\lambda y} \quad \Rightarrow \quad [P(X_i \geq y)]^n = e^{-n\lambda y} \qquad [\because X_i\text{'s are } i.i.d.]$$

$$\therefore \quad P(X_i \geq y) = e^{-\lambda y} \quad \Rightarrow \quad P(X_i \leq y) = 1 - e^{-\lambda y}$$

which is distribution function of Expo (λ) distribution. Hence, X_i's are *i.i.d.* Expo (λ).

Example 9·52. *For the exponential distribution $f(x) = e^{-x}, x \geq 0$; show that the cumulative distribution function (c.d.f.) of $X_{(n)}$ in a random sample of size n is : $F_n(x) = (1 - e^{-x})^n$. Hence, prove that as $n \to \infty$, the c.d.f. of $X_{(n)} - \log n$ tends to the limiting form $\exp\left[-\{\exp(-x)\}\right]$, $-\infty < x < \infty$.*

Solution. Here $f(x) = e^{-x},\ x \geq 0 \quad \Rightarrow \quad F(x) = P(X \leq x) = 1 - e^{-x}$ ... (*)

The *c.d.f.* of $X_{(n)}$ is given by [from (9·62)]

$$F_n(x) = P\left[X_{(n)} \leq x\right] = [F(x)]^n = (1 - e^{-x})^n \qquad \text{[From (*)]} \qquad \text{... (**)}$$

The *c.d.f.* $G_n(\cdot)$ of $X_{(n)} - \log n$ is given by :

$$G_n(x) = P\left[X_{(n)} - \log n \leq x\right] = P\left[X_{(n)} \leq x + \log n\right],$$

$$= [F_n(x + \log n)]^n = \left\{1 - e^{-x - \log n}\right\}^n \qquad \text{[From (**)]}$$

$$= \left(1 - \frac{e^{-x}}{n}\right)^n \qquad \left(\because e^{-\log n} = e^{\log n^{-1}} = \frac{1}{n}\right)$$

$$\therefore \quad \lim_{n \to \infty} G_n(x) = \lim_{n \to \infty}\left(1 - \frac{e^{-x}}{n}\right)^n = \exp\left(-e^{-x}\right), \qquad \left[\because \lim_{n \to \infty}\left(1 + \frac{m}{n}\right)^n = e^m\right]$$

Example 9·53. *Show that in odd samples of size n from U [0, 1] population, the mean and variance of the distribution of median are 1/2 and 1/[4(n + 2)], respectively.*

Solution. We have $f(x) = 1 \,;\; 0 \le x \le 1$

$$F(x) = P(X \le x) = \int_0^x f(u)\,du = \int_0^x 1 \cdot du = x$$

Let $n = 2m + 1$ (odd), where m is a positive integer ≥ 1. Then, median observation is $X_{(m+1)}$. Taking $r = (m + 1)$ in (9·67), the *p.d.f.* of median $X_{(m+1)}$ is given by :

$$f_{m+1}(x) = \frac{1}{\beta(m+1, m+1)} \cdot x^m (1-x)^m$$

$$\therefore \quad E\left[X_{(m+1)}\right] = \frac{1}{\beta(m+1, m+1)} \int_0^1 x\, x^m (1-x)^m\, dx = \frac{\beta(m+2, m+1)}{\beta(m+1, m+1)}$$

$$= \frac{\Gamma(m+2)\,\Gamma(m+1)}{\Gamma(2m+3)} \times \frac{\Gamma(2m+2)}{\Gamma(m+1)\,\Gamma(m+1)} = \frac{m+1}{2m+2} = \frac{1}{2}$$

$$E\left[X^2_{(m+1)}\right] = \int_0^1 x^2 f_{m+1}(x)\,dx = \frac{1}{\beta(m+1, m+1)} \cdot \int_0^1 x^{m+2} (1-x)^m\, dx$$

$$= \frac{\beta(m+3, m+1)}{\beta(m+1, m+1)} = \frac{m+2}{2(2m+3)}$$

$$\therefore \quad \text{Var}\left[X_{(m+1)}\right] = E\left[X^2_{(m+1)}\right] - \left[E\left\{X_{(m+1)}\right\}\right]^2 = \frac{m+2}{2(2m+3)} - \frac{1}{4} = \frac{1}{4(2m+3)} = \frac{1}{4(n+2)}.$$

Example 9·54. (a) *Let $X_1, X_2, \ldots, X_n$ be i.i.d. non-negative random variables of the continuous type with p.d.f. $f(\cdot)$ and distribution function $F(\cdot)$.*

If $E|X| < \infty$, *show that* $E|X_{(r)}| < \infty$.

(b) *Write $M_n = X_{(n)} = \max(X_1, X_2, \ldots, X_n)$. Show that*

$$E(M_n) = E(M_{n-1}) + \int_0^\infty F^{n-1}(x)\left[1 - F(x)\right] dx \,;\; n = 2, 3, \ldots$$

Hence, evaluate $E(M_n)$ if $X_1, X_2, \ldots, X_n$ have common distribution function :

$$F(x) = x \,;\; 0 < x < 1.$$

Solution. (a) $\quad E|X_{(r)}| = \int_0^\infty |x|\, f_r(x)\,dx \qquad$ ($\because$ X is non-negative continuous *r.v.*)

$$= \int_0^\infty |x| \frac{n!}{(r-1)!\,(n-r)!} f(x) \cdot F^{r-1}(x)\left[1 - F(x)\right]^{n-r} dx$$

$$\le n \binom{n-1}{r-1} \int_0^\infty |x|\, f(x)\,dx \le n \binom{n-1}{r-1} E|X|$$

Hence, $\quad E|X_{(r)}| < \infty$ if $E|X| < \infty$.

(b) The *p.d.f.* $f_n(x)$, of $M_n = X_{(n)}$ is given by : $\quad f_n(x) = n\left[F(x)\right]^{n-1} f(x)$

and distribution function $F_n(x)$ of M_n is given by : $F_n(x) = \left[F(x)\right]^n$.

$$E(M_n) = \int_0^\infty [1 - F_n(x)]\, dx \qquad \left[\because \text{ If } X \ge 0 \text{ a.s. then } E(X) = \int_0^\infty [1 - F(x)]\, dx\right] \quad \text{[See Example 6·25]}$$

$$\therefore \qquad E(M_n) = \int_0^\infty \left[1 - F^n(x)\right] dx \qquad \ldots (*)$$

$$= \int_0^\infty \left[1 - F^{n-1}(x) \cdot F(x)\right] dx = \int_0^\infty \left[1 - F^{n-1}(x)\{1 - (1 - F(x))\}\right] dx$$

$$= \int_0^\infty \left[1 - F^{n-1}(x)\right] dx + \int_0^\infty F^{n-1}(x)[1 - F(x)]\, dx$$

$$= E(M_{n-1}) + \int_0^\infty F^{n-1}(x)[1 - F(x)]\, dx \qquad \text{[From (*)]} \qquad \ldots (**)$$

If $X \sim U[0, 1]$, then $f(x) = 1 \,;\, 0 < x < 1$ and $F(x) = x \,;\, 0 < x < 1$

Substituting in (**), we get

$$E(M_n) - E(M_{n-1}) = \int_0^1 x^{n-1}(1 - x)\, dx = \frac{1}{n} - \frac{1}{n+1} \qquad \ldots (***)$$

Changing n to $n - 1, n - 2, \ldots, 1$ in (***), we get, respectively

$$E(M_{n-1}) - E(M_{n-2}) = \frac{1}{n-1} - \frac{1}{n}$$

$$\vdots \qquad \vdots$$

$$E(M_2) - E(M_1) = \frac{1}{2} - \frac{1}{3}$$

$$E(M_1) - E(M_0) = 1 - \frac{1}{2}$$

Adding (***) and the above equations and noting that $E(M_0) = 0$, we get

$$E(M_n) = 1 - \frac{1}{n+1} = \frac{n}{n+1}.$$

Example 9·55. *(a) Find the p.d.f. of $X_{(r)}$ in a random sample of size n from the exponential distribution :*

$$f(x) = \alpha e^{-\alpha x}, \ \alpha > 0, x \ge 0.$$

(b) Show that $X_{(r)}$ and $W_{rs} = X_{(s)} - X_{(r)}$, $r < s$, are independently distributed.

(c) What is the distribution of $W_1 = X_{(r+1)} - X_{(r)}$?

Solution. Here, $F(x) = P(X \le x) = \int_0^x \alpha \cdot e^{-\alpha u} du = 1 - e^{-\alpha x}$ ... (*)

The *p.d.f.* of $X_{(r)}$ is given by :

$$f_r(x) = \frac{1}{B(r, n-r+1)} \cdot [F(x)]^{r-1} \cdot [1 - F(x)]^{n-r} \cdot f(x)$$

$$= \frac{1}{B(r, n-r+1)} \cdot \left(1 - e^{-\alpha x}\right)^{r-1} \cdot e^{-\alpha x(n-r)} \cdot \alpha e^{-\alpha x}$$

$$= \frac{1}{B(r, n-r+1)} \cdot \alpha \cdot e^{-\alpha x(n-r+1)} \cdot \left[1 - e^{-\alpha x}\right]^{r-1}; \quad x > 0$$

(b) The joint *p.d.f.* of $X_{(r)}$ and $W_{rs} = X_{(s)} - X_{(r)}$ is given by : [From (9·75)]

$$g(x, w_{rs}) = c_{rs} \cdot F^{r-1}(x)\, f(x) \left[F(x+w_{rs}) - F(x)\right]^{s-r-1} \times f(x+w_{rs}) \left[1 - F(x+w_{rs})\right]^{n-s}$$

$$= \frac{n!}{(r-1)!\,(n-r)!} \times \frac{(n-r)!}{(s-r-1)!\,(n-s)!} \times \left[1 - e^{-\alpha x}\right]^{r-1} \cdot \alpha e^{-\alpha x}$$

$$\times \left[e^{-\alpha x} - e^{-\alpha(x+w_{rs})}\right]^{s-r-1} \times \alpha e^{-\alpha(x+w_{rs})} \times \left[e^{-\alpha(x+w_{rs})}\right]^{n-s}$$

$$= \left[\frac{1}{B(r, n-r+1)} \cdot \alpha e^{-\alpha x(n-r+1)} \left(1 - e^{-\alpha x}\right)^{r-1}\right]$$

$$\times \left[\frac{1}{B(s-r, n-s+1)} \cdot \alpha \cdot e^{-(n-s+1)\alpha w_{rs}} \left(1 - e^{-\alpha w_{rs}}\right)^{s-r-1}\right] \quad \text{... (**)}$$

$\Rightarrow$ $X_{(r)}$ and W_{rs} are independently distributed.

(c) Taking $s = r + 1$ in (**), the *p.d.f.* of $W_1 = X_{(r+1)} - X_{(r)}$ becomes :

$$g(w_1) = \frac{1}{B(1, n-r)} \cdot \alpha e^{-\alpha(n-r) w_1}$$

$$= (n-r)\, \alpha e^{-(n-r)\alpha w_1}\ ;\ w_1 \geq 0$$

which shows that W_1 has an exponential distribution with parameter $(n - r)\,\alpha$.

9·19. TRUNCATED DISTRIBUTIONS

Let X be a random variable with *p.d.f.* (or *p.m.f.*) $f(x)$. The distribution of X is said to be truncated at the point $X = a$ if all the values of $X \leq a$ are discarded. Hence, the *p.d.f.* (or *p.m.f.*) $g(\cdot)$ of the distribution, truncated at $X = a$ is given by :

$$g(x) = \frac{f(x)}{P(X > a)}\ ;\ x > a \quad \text{... (9·79)}$$

$$\Rightarrow \quad g(x) = \begin{cases} \dfrac{f(x)}{\sum\limits_{x>a} f(x)}\ ;\ x > a \text{ (For distcrete } r.v.) & \text{... (9·79}a) \\[2ex] \dfrac{f(x)}{\int_a^\infty f(x)\,dx}\ ;\ x > a \text{ (For continuous } r.v.) & \text{... (9·79}b) \end{cases}$$

For the continuous *r.v.* X, the *r*th moment (about origin) for the truncated distribution is given by :

$$\mu_r' = E(X^r) = \int_a^\infty x^r\, g(x)\, dx = \frac{\int_a^\infty x^r f(x)\, dx}{\int_a^\infty f(x)\, dx} \quad \text{... (9·80)}$$

Example 9·56. *Let* $X \sim B(n, p)$. *Find the mean and variance of the binomial distribution truncated at* $X = 0$.

Solution. Let $f(x)$ be the *p.m.f.* of $X \sim B(n, p)$ variate. Then, the *p.m.f.* $g(x)$ of the Binomial distribution truncated at $X = 0$, is given by :

$$g(x) = \frac{f(x)}{P(X > 0)} = \frac{f(x)}{1 - P(X = 0)} = \frac{f(x)}{1 - f(0)} = \frac{1}{1 - q^n} \cdot {}^nC_x\, p^x q^{n-x}\ ;\ x = 1, 2, \ldots, n \quad \text{... (*)}$$

$$E(X)=\sum_{x=1}^{n} x\,g(x)=\frac{1}{1-q^n}\sum_{x=1}^{n} x\cdot {}^nC_x\, p^x q^{n-x}$$

$$=\frac{1}{1-q^n}\left(\sum_{x=0}^{n} x\cdot {}^nC_x\, p^x q^{n-x}-0\right)=\frac{np}{1-q^n}\qquad [\because X\sim B(n,p);\ E(X)=np]\quad ...(**)$$

$$E(X^2)=\frac{1}{1-q^n}\sum_{x=1}^{n} x^2\cdot {}^nC_x\, p^x q^{n-x}=\frac{1}{1-q^n}\left(\sum_{x=0}^{n} x^2\cdot {}^nC_x\, p^x q^{n-x}-0\right)$$

$$=\frac{1}{1-q^n}(npq+n^2p^2)\qquad [\because X\sim B(n,p);\ E(X^2)=\text{Var } x+\{E(X)\}^2=npq+n^2p^2\,]$$

$$\therefore\quad \text{Var }(X)=E(X^2)-[E(X)]^2=\frac{1}{1-q^n}\left(npq+n^2p^2-\frac{n^2p^2}{(1-q^n)^2}\right)$$

Example 9·57. *Obtain the mean and variance of a standard Cauchy distribution truncated at both ends, with relevant range of variation as* $(-\beta,\beta)$.

Solution. Let $f(x)$ be the *p.d.f.* of standard Cauchy distribution. Then the *p.d.f.* $g(x)$ of truncated distribution with relevant range of variation as $(-\beta,\beta)$ is given by :

$$g(x)=\frac{f(x)}{P(-\beta\le X\le\beta)}=\frac{f(x)}{\int_{-\beta}^{\beta} f(x)\,dx}=\frac{\frac{1}{\pi}\cdot\frac{1}{1+x^2}}{\frac{1}{\pi}\int_{-\beta}^{\beta}\frac{dx}{1+x^2}}=\frac{1}{1+x^2}\cdot\frac{1}{\left|\tan^{-1}x\right|_{-\beta}^{\beta}}$$

$$=\frac{1}{2\tan^{-1}\beta}\cdot\frac{1}{(1+x^2)}\ ;\ -\beta\le x\le\beta\qquad ...(*)$$

$$\text{Mean}=\mu_1'=\int_{-\beta}^{\beta} x\,g(x)=\frac{1}{2\tan^{-1}\beta}\int_{-\beta}^{\beta}\frac{x}{1+x^2}\,dx=0$$

[$\because$ Integrand is an odd function of x.]

$$\text{Variance}=\mu_2'-\mu_1'^2=\mu_2'=\int_{-\beta}^{\beta} x^2 g(x)\,dx=\frac{2}{2\tan^{-1}\beta}\int_0^{\beta}\frac{x^2}{1+x^2}\,dx$$

$$=\frac{1}{\tan^{-1}\beta}\int_0^{\beta}\left(1-\frac{1}{1+x^2}\right)dx=\frac{1}{\tan^{-1}\beta}\left|\,x-\tan^{-1}x\,\right|_0^{\beta}$$

$$=\frac{1}{\tan^{-1}\beta}(\beta-\tan^{-1}\beta)=\frac{\beta}{\tan^{-1}\beta}-1.$$

Example 9·58. *Consider a truncated standard normal distribution truncated at both ends with relevant range of variation as (A, B). Obtain the p.d.f. mean, mode and variance of the truncated distribution.*

Solution. Let $Z\sim N(0,1)$ with *p.d.f.* $\varphi(z)$ and *c.d.f.* $\Phi(z)=P(Z\le z)$.

Then, the *p.d.f.* $g(\cdot)$ of the truncated normal distribution is given by :

$$g(z)=\frac{\varphi(z)}{P(A\le Z\le B)}=\frac{\varphi(z)}{\Phi(B)-\Phi(A)}=\frac{1}{k}\varphi(z)\ ;\ A\le z\le B\qquad ...(*)$$

where, $\quad k=\Phi(B)-\Phi(A).$

$$\text{Mean} = \int_A^B z\, g(z)\, dx = \frac{1}{k}\int_A^B z\,\varphi(z)\, dx \qquad \ldots (**)$$

We have $\varphi(z) = \dfrac{1}{\sqrt{2\pi}} e^{-z^2/2} \quad \Rightarrow \quad \dfrac{d}{dz}\varphi(z) = \dfrac{1}{\sqrt{2\pi}} e^{-z^2/2}(-z) = -z\,\varphi(z)$

$$\therefore \qquad \int z\,\varphi(z)\, dz = -\varphi(z) \qquad \ldots (*)$$

Substituting in (**), we get

$$\text{Mean} = \frac{1}{k}\left| -\varphi(z) \right|_A^B = \frac{\varphi(A)-\varphi(B)}{\Phi(B)-\Phi(A)} = \mu, \text{ (say)} \qquad \ldots (***)$$

$$\text{Mode} = \begin{cases} B, & \text{if } A<0,\ B<0 \\ 0, & \text{if } A<0,\ B>0 \\ A, & \text{if } A>0,\ B>0 \end{cases}$$

$$\text{Variance} = \int_A^B z^2\, g(z)\, dz - \mu^2 = \frac{1}{k}\left[\left| z\{-\varphi(z)\} \right|_A^B + \int_A^B \varphi(z)\, dz \right] - \mu^2$$

$$= \frac{-1}{k}\left[B\,\varphi(B) - A\,\varphi(A) - \{\Phi(B)-\Phi(A)\} \right] - \mu^2$$

$$= 1 + \frac{A\,\varphi(A) - B\,\varphi(B)}{\Phi(B)-\Phi(A)} - \mu^2, \text{ where } \mu \text{ is given in } (***).$$

CHAPTER CONCEPTS QUIZ

1. Choose the correct answer from *B* and match it with corresponding item in *A* :

A	*B*
(*a*) β_2 for a Normal distribution	(1) $3\sigma^4$
(*b*) β_1 for a Normal distribution	(2) 0
(*c*) μ_3 for a Normal distribution	(3) 3
(*d*) μ_4 for a Normal distribution	(4) $\exp\left(it\mu - \frac{1}{2}t^2\sigma^2\right)$
(*e*) Characteristic function for Normal distribution	(5) $\frac{4}{5}\sigma$
(*f*) Moment generating function of Normal distribution	(6) 0
(*g*) Mode of Normal distribution	(7) $\exp\left(t\mu + \frac{1}{2}t^2\sigma^2\right)$
(*h*) Mean deviation from mean for Normal distribution	(8) μ

2. Match the distributions :

(*a*) Uniform distribution	(1) $f(x) = \dfrac{\lambda}{\pi\{(x-\mu)^2+\lambda^2\}}, \ -\infty < x < \infty\ ;\ \lambda > 0$
(*b*) Normal distribution	(2) $f(x) = \dfrac{1}{\sqrt{2\pi}} e^{-x^2/2}, \ -\infty < x < \infty$
(*c*) Exponential distribution	(3) $f(x) = \dfrac{1}{b-a}, \ a \le x \le b$

(*d*) Beta distribution (4) $f(x) = \frac{1}{B(m, n)} x^{m-1}(1-x)^{n-1},\ 0 \le x \le 1$

(*e*) Cauchy distribution (5) $f(x) = \frac{1}{a} e^{-x/a},\ x \ge 0, \alpha > 0$

3. If X_i ($i = 1, 2, 3, ..., n$) are independent $N(0, 1)$, write (without proof), the distribution of :

(*i*) $X_1 - 2X_2 + X_3$, (*ii*) $\frac{X_2}{X_3}$, (*iii*) $\sum_{i=1}^{n} X_i$, (*iv*) $\frac{X_1^2}{X_1^2 + X_2^2}$,

(*v*) $\frac{X_1^2}{X_2^2 + X_3^2}$, (*vi*) $\frac{X_i}{X_j}, (i \ne j)$, (*vii*) $\sum_{i=1}^{n} X_i^2$, (*viii*) $\frac{X_i^2}{X_j^2}, i \ne j$

4. In each case, specify the distribution for which :

(*i*) Moments do not exist (*ii*) Mean = variance (*iii*) Mean < variance

(*iv*) Mean > variance (*v*) $\varphi(t) = e^{it - t^2}$ (*vi*) $\varphi(t) = e^{-|t|}$

5. State the conditions under which

(*i*) Binomial distribution, (*ii*) Poisson distribution, tend to Normal distribution.

6. (*i*) Give two examples of variates which you expect to be distributed normally.

(*ii*) Give two examples of variates which you expect to be distributed exponentially.

7. State which of the following statements are TRUE and which are FALSE. In case of false statements, give the correct statement.

(*i*) For normal distribution, mean deviation about mean is greater than quartile deviation.

(*ii*) X is a random variable following Cauchy distribution, for which mean does not exist but variance exists.

(*iii*) In case of normal distribution $\beta_1 = 3, \beta_2 = 0$.

(*iv*) If X and Y are two independent normal variates, then $X - Y$ is also a normal variate.

(*v*) Binomial distribution tends to normal distribution as $n \to \infty$.

(*vi*) For normal distribution, mean = mode = median.

(*vii*) It is possible to reduce every normal distribution to the standard normal distribution by a transformation.

(*viii*) In uniform distribution, the percentile points are equi-spaced.

(*ix*) Normal distribution can be obtained as limiting case of Poisson distribution with the parameter $\lambda \to \infty$.

8. Give the correct answer to each of the following :

(*i*) The mean and variance of Normal distribution

(*a*) are same, (*b*) cannot be same, (*c*) are sometimes equal,

(*d*) are equal in the limiting case, as $n \to \infty$.

(*ii*) The mean and variance of Gamma distribution

(*a*) are same, (*b*) cannot be same, (*c*) are sometimes equal,

(*d*) are equal in the limiting case, as $n \to \infty$.

(*iii*) X is normally distributed with zero mean and unit variance. The variance of X^2 is :

(*a*) 0, (*b*) 1, (*c*) 2, (*d*) 4.

(*iv*) The points of inflexion of Normal $N(m, \sigma^2)$ curve are :

(*a*) $m \pm \sigma$, (*b*) $m \pm 2\sigma$, (*c*) $m \pm 3\sigma$, (*d*) $m \pm \frac{2}{3}\sigma$

(*v*) The moment generating function of gamma distribution is :

(*a*) $(1+t)^{\lambda}$, (*b*) $(1-t)^{\lambda}$, (*c*) $(1-t)^{-\lambda}$, (*d*) $(1+t)^{-\lambda}$

(vi) The characteristic function of standard Cauchy distribution is :

(a) e^{-t}, (b) $e^{-|t|}$, (c) e^{t}, (d) $e^{|t|}$

(vii) Area to the right of the point x_1 is 0·6 and to the left of the point x_2 is 0·7. Which is the correct :

(a) $x_1 > x_2$, (b) $x_1 < x_2$, (c) $x_1 = x_2$?

(viii) The standard normal distribution is represented by :

(a) N (0, 0), (b) N (1, 1), (c) N (0, 1), (d) N (1, 0)

(ix) For a normal distribution, quartile deviation, mean deviation, standard deviation are in the ratio :

(a) $\frac{4}{5}:\frac{2}{3}:1$, (b) $\frac{2}{3}:\frac{4}{5}:1$, (c) $1:\frac{4}{5}:\frac{2}{3}$, (d) $\frac{1}{2}:1:\frac{4}{5}$

(x) The normal distribution is a limiting form of binomial distribution if :

(a) $a \to \infty, p \to 0$, (b) $n \to 0, p \to q$, (c) $n \to \infty, p \to n$,

(d) $n \to \infty$ and neither p nor q is small.

(xi) Normal curve is :

(a) Very flat, (b) bell-shaped symmetrical about mean, (c) very peaked,

(d) smooth.

(xii) The normal distribution is a limiting case of Poisson distribution when parameter :

(a) $\lambda \to 0$, (b) $\lambda \to \sigma$, (c) $\lambda \to \infty$, (d) $\lambda < \sigma$

(xiii) In a normal $N(m, \sigma^2)$ curve 95% of observations are included in the range :

(a) $m \pm 3\sigma$, (b) $m \pm 1{\cdot}96\sigma$, (c) $m \pm 2\sigma$, (d) $m \pm 0{\cdot}67\sigma$

(xiv) If X is a standard normal variate, then $\frac{1}{2}X^2$ is :

(a) Gamma variate with parameter $\frac{1}{2}$, (b) Normal variate, (c) Poisson variate.

(xv) The range of the beta variate of first kind is :

(a) $(0, \infty)$, (b) $(-\infty, \infty)$, (c) $(0, 1)$, (d) $(-1, +1)$

9. Fill in the blanks :

(i) The mean deviation of normal distribution is ...

(ii) The *p.d.f.* of Gamma distribution is ...

(iii) The relationship between Beta distribution of first and second kind is ...

(iv) The normal distribution is a limiting form of binomial distribution if ...

(v) Mean = Variance for ... distribution (continuous).

(vi) For the normal distribution :

$\beta_1 = ...$, $\beta_2 = ...$, Mean deviation = ..., Quartile deviation = ...

(vii) The characteristic function of a Gamma distribution is ...

(viii) The points of inflexion for a normal curve are ...

(ix) For the normal distribution with variance σ^2,

$$\mu_{2n+1} = \cdots ; \quad \mu_{2n} = \cdots ; \quad n = 0, 1, 2, \cdots$$

(x) A normal distribution is completely specified by the parameters ...

(xi) For normal distribution

S.D. : M.D. : Q.D. :: ... : ...

(xii) If $X \sim N(\mu, \sigma^2)$ then

$P(\mu - \sigma < X < \mu + \sigma) = ...$; $P(\mu - 2\sigma < X < \mu + 2\sigma) = ...$; $P(\mu - 3\sigma < X < \mu + 3\sigma) = ...$

(xiii) If X is a random variable with distribution function F then F (X) has ... distribution.

(xiv) If $X \sim N(0, 1)$, then $X^2/2$ has ... distribution with parameter ...

10. (*a*) Random variables X_i are independent and all of them have, the same distribution defined by :

$$f(x) = \frac{1}{\sqrt{8\pi}} \exp\left[-(x-1)^2/8\right]; -\infty < x < \infty$$

Find the distribution of : (*i*) $\sum_{i=1}^{10} X_i/10$; and (*ii*) $X_1 - 2X_2 + X_3$.

(*b*) The random variables X_i, $i = 1, 2, \ldots$ are independent and all of them have the same distribution defined by :

$$f(x) = \frac{1}{\sqrt{18\pi}} \exp\left[-\frac{(x-1)^2}{18}\right]; -\infty < x < \infty.$$

Find the distribution of : (*a*) $\frac{1}{5}\sum_{i=1}^{5} X_i$, and (*b*) $3X_1 - X_2 + 2X_3$.

(*c*) Find the mean and *s.d.* of a probability distribution whose *p.d.f.* is :

$$f(x) = C\exp\left[-\frac{1}{24}(x^2 = 6x + 9)\right]; -\infty < x < \infty$$

where C is a constant. Also find C.

(*d*) Identify the distribution for which characteristic function is : $\phi_X(t) = \exp\left[2it - 8t^2\right]$.

Ans. $X \sim N(\mu = 2, \sigma^2 = 16)$.

11. (*a*) If $f(x) = ke^{-3x^2+6x}$, $-\infty < x < \infty$, obtain the values of k, μ and σ^2.

(*b*) If X is a normal variate with mean μ and variance σ^2, show that the distribution of $Y = aX + b$ is $N(a\mu + b, a^2\sigma^2)$.

(*c*) If X is distributed normally with mean μ and standard deviation σ, write down the distribution of $U = 2X - 3$ and find the mean and variance of U.

(*d*) If the *m.g.f.* of a *r.v.* X is : $M_X(t) = \exp\left[2t + 32t^2\right]$, obtain the *p.d.f.* $f(x)$.

Ans.
$$X \sim N\left(\mu = 2, \frac{\sigma^2}{2} = 32 \Rightarrow \sigma = 8\right);$$

$$f(x) = \frac{1}{8\sqrt{2\pi}} \exp\left[-\frac{(x-2)^2}{128}\right]; -\infty < x < \infty.$$

(*e*) If $\exp\left[3t + 8t^2\right]$ is the *m.g.f.* of a *r.v.* X, find $P(-1 < X < 9)$.

Hint.
$$X \sim N\left(\mu = 3, \frac{\sigma^2}{2} = 8 \Rightarrow \sigma = 4\right)$$

$$P(-1 < X < 9) = P(-1 < Z < 1{\cdot}5) = 0{\cdot}7745.$$

12. (*a*) If X_1 is normally distributed with mean 10 and variance 16 and X_2 is normally distributed with mean 10 and variance 15 and if $W = X_1 + X_2$, what will be the values of the two parameters of the distribution of the variate W ? (Assume that X_1 and X_2 are independent).

(*b*) Let X and Y be independent and normally distributed as $N(\mu_1, \sigma_1^2)$ and $N(\mu_2, \sigma_2^2)$, respectively. Find mean and variance of $\frac{1}{2}(X + Y)$.

13. (*a*) If X_i, $(i = 1, 2, 3, \ldots, n)$ are *i.i.d.* standard Cauchy variates, write the distribution of $\bar{X} = \frac{1}{n}\sum_{i=1}^{n} X_i$.

(*b*) Is the sum of two independent Cauchy variates a Cauchy variate ? If X_i, $(i = 1, 2, 3, 4)$ are independent standard normal variates, what is the distribution of :

$$\frac{X_1}{X_2} + \frac{X_3}{X_4}\ ?$$

14. If $X_i, (i = 1, 2, 3, 4)$ are *i.i.d.* $N(0, 1)$, write the distribution of :

(*i*) $X_1 - X_2$ (*ii*) $X_1 + X_2$ (*iii*) $\dfrac{X_1}{X_2}$ (*iv*) $\dfrac{X_1}{X_2} + \dfrac{X_3}{X_4}$ (*v*) $X_1 - X_2 + X_3 - X_4$

(*vi*) $\frac{1}{2} X_1^2$ (*vii*) $\frac{1}{2}\left(X_2^2 + X_3^2\right)$ (*viii*) $\dfrac{X_1^2}{X_1^2 + X_2^2}$ (*ix*) $\dfrac{X_3^2}{X_4^2}$ (*x*) $\dfrac{X_1^2}{X_2^2 + X_3^2}$.

Ans. (*i*) $N(0, 2)$; (*ii*) $N(0, 2)$; (*iii*) Standard Cauchy ; (*iv*) Cauchy ; (*v*) $N(0, 4)$;

(*vi*) $\gamma\left(\frac{1}{2}\right)$; (*viii*) $\gamma(1)$; (*viii*) $\beta_1\left(\frac{1}{2}, \frac{1}{2}\right)$; (*ix*) $\beta_2\left(\frac{1}{2}, \frac{1}{2}\right)$; (*x*) $\beta_2\left(\frac{1}{2}, 1\right)$.

DISCUSSION AND REVIEW QUESTIONS

NORMAL DISTRIBUTION

1. "If the Poisson and the Normal distributions are limiting cases of Binomial distribution, then there must be a limiting relation between the Poisson and the Normal distributions." Investigate the relation.

2. (*a*) Derive the mathematical form and properties of normal distribution. Discuss the importance of normal distribution in Statistics.

(*b*) Mention the chief characteristics of Normal distribution and Normal probability curve.

3. (*a*) Explain, under what conditions and how the binomial distribution can be approximated to the normal distribution ?

(*b*) For a normal distribution with mean 'μ' and *s.d.* σ, show that the mean deviation from the mean 'μ' is equal $\sigma\sqrt{(2/\pi)}$. What will be the mean deviation from median ?

(*c*) Define Normal probability distribution. If the mean of a Normal population is μ and its variance σ^2, what are its (*i*) mode, (*ii*) median, (*iii*) β_1 and β_2 ?

(*d*) For a normal distribution $N(\mu, \sigma^2)$:

(*i*) Show that the mean, the median and the mode coincide.

(*ii*) Find the recurrence relation between μ_{2n} and μ_{2n-2}.

(*iii*) State and prove additive property of normal variates.

(*iv*) Obtain the points of inflexion for the normal distribution $N(\mu, \sigma^2)$.

(*v*) Obtain mean deviation about mean.

(*e*) Show that any linear combination of *n* independent normal variates is also a normal variate.

(*f*) Show that for the normal curve :

(*i*) The maximum probability occurs at the mean of the distribution, and

(*ii*) The points of inflexion lie at a distance of ± σ from the mean, where σ is the standard deviation.

(*g*) Describe the steps involved in fitting a normal distribution to the given data and computing the expected frequencies.

(*h*) Explain how the normal probability integral $\int_0^{z_1} \varphi(z)\, dz$, is used in computing normal probabilities.

4. Write a note on the salient features of a normal distribution. $N(\mu, \sigma^2)$ denotes the normal distribution of each of the random variables $X_1, X_2, X_3, ..., X_n$, where μ is the mean and σ^2 the variance. Prove the following :

(*i*) If $X_1, X_2, ..., X_n$ are independent, then $X_1 + X_2 + ... + X_n$ has the distribution $N(n\mu, n\sigma^2)$.

(*ii*) kX, where *k* is a constant has the distribution $N(k\mu, k^2\sigma^2)$.

(*iii*) $X + a$, where *a* is a constant has the distribution $N(\mu + a, \sigma^2)$.

(*iv*) In (*i*) if $\bar{X} = \dfrac{X_1 + X_2 + ... + X_n}{n}$, then $\sqrt{n}\,\dfrac{(\bar{X} - \mu)}{\sigma}$ has the distribution $N(0, 1)$.

5. (*a*) Show that for a normal distribution with mean μ and variance σ^2, the central moments satisfy the relation :
$$\mu_{2n} = (2n-1)\,\mu_{2n-2}\,\sigma^2\ ; \quad \mu_{2n+1} = 0.$$

Hence, show that $\mu_{2n} = \dfrac{(2n)!}{n!}\left(\dfrac{1}{2}\sigma^2\right)^n$ and $\mu_{2n+1} = 0\ ;\ n = 1, 2, \ldots$

(*b*) State the mathematical equation of a normal curve. Discuss its chief features.

(*c*) Find the moment generating function of the normal distribution $N(\mu, \sigma^2)$, and deduce that : $\mu_{2n+1} = 0$, and $\mu_{2n} = 1.3.5 \ldots (2n-1)\,\sigma^{2n}$, where μ_n denotes the *n*th central moment.

(*d*) Show that all central moments of a normal distribution can be expressed in terms of the standard deviation and obtain the expression in the general case.

6. (*a*) If X_1 and X_2 are two independent normal variates with means μ_1 and μ_2 and variances σ_1^2 and σ_2^2, respectively, show that the variables *U* and *V*, where $U = X_1 + X_2$ and $V = X_1 - X_2$, are independent normal variates. Find the means and variances of *U* and *V*.

(*b*) If X_1 and X_2 are independent standard normal variates, obtain the *p.d.f.* of $(X_1 - X_2)/\sqrt{2}$.

LOG-NORMAL DISTRIBUTION

7. (*a*) Let *X* be a non-negative random variable such that $\log X = Y$, (say), is normally distributed with mean μ and variance σ^2.

(*i*) Write down the *p.d.f.* of *X*. Find *E* (*X*) and Var (*X*).

(*ii*) Find the median and the mode of the distribution of *X*.

(*b*) If *X* has a uniform distribution in [0, 1], find the distribution (*p.d.f.*) of $-2\log X$. Identify the distribution also.

Hint. Let $Y = -2\log X$. Then the distribution function *G* of *Y* is given by :
$$G_Y(y) = P(Y \le y) = P(-2\log X \le y) = P(\log X \ge -y/2) = P(X \ge e^{-y/2})$$
$$= 1 - P(X \le e^{-y/2}) = 1 - \int_0^{e^{-y/2}} f(x)\,dx = 1 - \int_0^{e^{-y/2}} 1 \cdot dx = 1 - e^{-y/2}$$
$$\therefore \quad g_Y(y) = \frac{d}{dy}G(y) = \frac{1}{2}e^{-y/2},\ 0 < y < \infty \quad \Rightarrow \quad Y \sim \text{Expo}\left(\theta = \frac{1}{2}\right).$$

8. A random variable *X* has the probability density function :
$$f(x) = \frac{1}{\beta x\sqrt{2\pi}} \cdot \exp\left[-\frac{1}{2\beta^2}(\log x - \alpha)^2\right],\ x > 0. \text{ Find } E(X) \text{ and Var}(X).$$

9. (*a*) A random variate *X* has the *p.d.f.*,
$$f(x) = \begin{cases} \dfrac{1}{x\sqrt{2\pi}}\,e^{-(\log x)^2/2}, & x > 0 \\ 0, & \text{elsewhere} \end{cases}$$

Calculate the mean, mode, standard deviation and coefficient of skewness.

(*b*) Show that log-normal distribution is positively skewed, *i.e.*, mean > median > mode.

10. The *r.v.* *X* has mean *m* and standard deviation *s*. If $Y = \log X$ is normally distributed with mean *M* and standard deviation *S*, prove that
$$(i)\ m = \exp\left(M + \frac{1}{2}S^2\right), \qquad (ii)\ 1 + \frac{s^2}{m^2} = \exp(S^2).$$

11. Given that X_i are independent logarithmic normal variates with parameters μ_i and $\sigma_i\ ;\ i = 2, \ldots, n$, find the *s*th raw moment of the variable $Y = \prod (a_i X_i)\ ;\ i = 2, \ldots, n$.

GAMMA AND BETA DISTRIBUTION

12. (*a*) Suppose the probability function of a random variable X is given by :

$$f(x)=\begin{cases}\dfrac{x^k e^{-x}}{k!}, & \text{for } x>0\\ 0, & \text{otherwise}\end{cases}$$

where, k is non-negative integer.

(*i*) Find the moment generating function of this distribution.

(*ii*) Determine the mean and variance of this distribution, using moment generating function.

(*b*) If X is a Gamma variate with parameter λ, obtain its *m.g.f.* Hence deduce that the *m.g.f.* of standard gamma variate tends to $\exp\left(\frac{1}{2}t^2\right)$ as $\lambda \to \infty$. Also interpret the result.

(*c*) X and Y are two independent gamma variates, with parameters l and m. Prove that $(X + Y)$ is a gamma variate with parameter $(l + m)$.

(*d*) If $X_1, X_2, ..., X_n$ are independent and identically distributed gamma random variables, what is the distribution of $X_1 + X_2 + ... + X_n$?

13. (*a*) Define the Beta variate of first kind. Obtain its mean and variance. Also define the Beta variate of second kind and state its relation with Gamma variate.

(*b*) Write down the Beta probability functions of the first kind and the second kind with parameters μ and v. Show that a Beta variate of the first kind can be obtained by a transformation of a Beta variate of the second kind.

14. Let $X \sim \gamma(\lambda, a)$ and $Y \sim \gamma(\lambda, b)$, be independent random variables. Show that

$$E\left(\frac{X}{X+Y}\right)=\frac{E(X)}{E(X+Y)}$$

Hint. Since $X \sim \gamma(\lambda, a)$ and $Y \sim \gamma(\lambda, b)$ are independent, $U = X/(X + Y)$ and $V = X + Y$ are independent.

Hence, $\quad E(UV) = E(U)\,E(V) \quad \Rightarrow \quad E(X) = E\left(\dfrac{X}{X+Y}\right)\cdot E(X+Y).$

15. (*a*) If $X \sim \gamma(\lambda, \mu)$ and $Y \sim \gamma(\lambda, \nu)$ are independent *r.v.*'s, show that $\dfrac{X}{Y} \sim \beta_2(\mu, \nu)$.

(*b*) If X and Y are independent Gamma variates, find the distribution of $X/(X + Y)$.

(*c*) If X is random variable having as its rth moment $\mu_r' = \dfrac{(k+r)!}{k!}$, k being a positive integer, show that its probability density function is :

$$f(x)=\begin{cases}\dfrac{x^k}{k!}e^{-x}, & x>0\\ 0, & x<0.\end{cases}$$

(*d*) If the *r.v.* X is such that $E(X^n) = (n + k)!\,k^n/k!$; $n = 1, 2, 3, ...$ k being a positive integer, find the *p.d.f.* of X.

16. If X and Y are independent Gamma variates with parameters μ and ν respectively, show that the variables :

$$U = X + Y \quad \text{and} \quad V=\frac{X-Y}{X+Y} \quad \text{are independent variables.}$$

17. If X and Y are independent Gamma variates with parameters λ and μ respectively, show that the variables :

(*a*) $U + X + Y$ and $V = \dfrac{X}{X+Y}$ are independently distributed and identify their distributions.

(*b*) $U + X + Y$ and $V = X/Y$ are independently distributed, $U \sim \gamma(\lambda + \mu)$ and $V \sim \beta_2(\lambda, \mu)$.

18. A simple sample of n values $x_1, x_2, ..., x_n$ is drawn from the population :

$$dP(x) = \frac{1}{\Gamma(n)} e^{-x} x^{n-1} dx, \ 0 \le x < \infty$$

If $\bar{x}$ is the mean of the sample, find the distribution of $n\bar{x}$. Hence, find the mean and variance of the distribution.

19. (a) Show that for a $\gamma(\lambda)$ distribution, $\dfrac{\text{Mean} - \text{Mode}}{\sigma} = \dfrac{1}{\sqrt{\lambda}} = \dfrac{1}{2} \cdot \dfrac{\mu_3}{\sigma^3}$.

Show that the excess of kurtosis of the distribution is $6/\lambda$.

(b) Show that mean value of the positive square root of a $\gamma(\lambda, n)$ variate is :

$$\Gamma\left(n + \frac{1}{2}\right) \Big/ \left[\sqrt{\lambda}\,\Gamma(n)\right].$$

Hence, prove that the mean deviation of a $N(\mu, \sigma^2)$ variate from its mean is $\sigma\sqrt{2/\pi}$.

20. Show that the mean value of the positive square root of $\beta(\mu, \nu)$ variate is :

$$\frac{\Gamma\left(\mu + \frac{1}{2}\right)\Gamma(\mu + \nu)}{\Gamma(\mu)\,\Gamma\left(\mu + \nu + \frac{1}{2}\right)}.$$

21. (a) For the distribution : $dP(x) = \dfrac{1}{B(\mu, \nu)} \cdot \dfrac{x^{\mu-1}}{(1+x)^{\mu+\nu}}$; $0 < x < \infty$, $\nu > 2$

Show that variance is : $\dfrac{\mu(\mu + \nu - 1)}{(\nu - 1)^2 (\nu - 2)}$

Find also the mode and μ_r' (about origin). Also show that harmonic mean is $(\mu - 1)/\nu$.

(b) Find the arithmetic mean, harmonic mean and variance of a Beta distribution of first kind with parameter μ and ν. Verify that A.M. > H.M. Also prove that if G is the geometric mean, then

$$\log G = \frac{1}{B(\mu, \nu)} \cdot \frac{\partial}{\partial \nu} B(\mu, \nu) = \frac{\partial}{\partial \nu}\left[\log \Gamma v - \log \Gamma(\mu + \nu)\right]$$

22. Given the Beta distribution in the following form :

$$p(x) = \frac{1}{B(\alpha + 1, \lambda + 1)} \cdot x^{\alpha}(1 - x)^{\lambda}; \ \alpha > -1, \ \lambda > -1, \ 0 \le x < 1,$$

find its variance. Also find the distribution of : (i) $1/X$, (ii) $(1 - X)/X$.

EXPONENTIAL DISTRIBUTION, LAPLACE DISTRIBUTION AND WEIBUL DISTRIBUTION

23. (a) Show that for the exponential distribution : $p(x) = y_0 \cdot e^{-x/\sigma}$, $0 \le x < \infty$; $\sigma > 0$,

mean and variance are equal. Also obtain the interquartile range of the distribution.

(b) If $X_1, X_2, ..., X_n$ are independent random variables having exponential distribution with parameter λ, obtain the distribution of $Y = \sum_{i=1}^{n} X_i$.

(c) Obtain the moment generating function and the cumulant generating function of the distribution with *p.d.f.* :

$$f(x) = \frac{1}{\sigma} e^{-x/\sigma}; \ 0 < x < \infty, \ \sigma > 0.$$

Hence, or otherwise obtain the values of the constants $\beta_1, \beta_2, \gamma_1$ and γ_2.

24. Find the characteristic function of standard Laplace distribution and hence find its mean and standard deviation.

25. (a) Find the first four cumulants of the Laplace distribution defined by :

$$f(x) = \frac{1}{2\lambda}\left[\exp\{-|x-\mu|/\lambda\}\right]; -\infty < x < \infty, \lambda > 0$$

and hence find the values of mean, *s.d.*, β_1 and β_2. Calculate also the semi-interquartile range.

(b) The *p.d.f.* of a *r.v.*, X follows the following probability law :

$$p(x) = \frac{1}{2\theta}\exp\left(-\frac{|x-\theta|}{\theta}\right), -\infty < x < \infty.$$

Find *m.g.f.* of X. Hence or otherwise, find $E(X)$ and $\text{Var}(X)$.

26. X_i, $i = 1, 2, ..., n$ are *i.i.d. r.v.*'s having Weibul distribution with three parameters. Show that the variable $Y = \min(X_1, X_2, ..., X_n)$, also has Weibul distribution and identify the parameters.

LOGISTIC DISTRIBUTION

27. Obtain the moment generating function of Logistic distribution and hence find its mean and variance.

28. (a) Obtain the *p.d.f.* $g(y)$ and distribution function $G(y)$ of the standard logistic variate and prove that :

(i) $g(y)$ is symmetric about origin. (ii) $g(y) = G(y)[1 - G(y)]$, (iii) $y = \log_e\left[\dfrac{G(y)}{1-G(y)}\right]$.

(b) Obtain the *m.g.f.* of standard logistic variate and hence prove that :

$$\text{Mean} = 0, \quad \text{Variance} = \pi^2/3, \quad \beta_1 = 0, \quad \beta_2 = \frac{21}{5}; \quad \mu_{2r+1} = 0$$

and mean deviation about mean = $2\log_3 2$.

CAUCHY DISTRIBUTION, C.L.T. AND COMPOUND DISTRIBUTIONS

29. (a) Show that a function : $f(x; \mu, \lambda) = \dfrac{k}{\lambda^2 + (x-\mu)^2}; -\infty < x < \infty$

represents a frequency function of a distribution for a suitable value of k. Determine k and obtain median and quartiles of the distribution. Hence interpret the parameters λ and μ of the distribution.

(b) If X is a Cauchy variate with parameters λ and μ, find the characteristic function $\varphi_X(t)$. Discuss briefly the role of Cauchy distribution in Statistics.

(c) "The role of Cauchy distribution often lies in providing counter examples." Justify.

(d) Discuss briefly the role of Cauchy distribution in Statistics.

If $X_1, X_2, ..., X_n$ are independent standard Cauchy variates, show that the mean :

$\bar{X} = (X_1 + X_2 + ... + X_n)/n$, is also a Cauchy variate.

30. (a) If X and Y are independent random variables following Cauchy distribution with parameters (λ_1, μ_1) and (λ_2, μ_2) respectively, show that X + Y follows Cauchy distribution with parameters $\lambda_1 + \lambda_2$ and $\mu_1 + \mu_2$.

(b) Obtain the characteristic function of Cauchy distribution : $dF(x) = \dfrac{dx}{\pi(1+x^2)}, -\infty < x < \infty$

If $X_1, X_2, ..., X_n$ are independent Cauchy variates, show that the mean $\bar{X} = \dfrac{1}{n}\sum_i X$, is also a Cauchy variate.

(c) If $X_1 = X_2 = X$, where X has the standard Cauchy *p.d.f.* : $f(x) = \frac{1}{\pi} \cdot \frac{1}{1+x^2}$; $-\infty < x < \infty$,

prove that : $$\phi_{X_1+X_2}(t) = \phi_{X_1}(t) \cdot \phi_{X_2}(t),$$

even though X_1 and X_2 are not independent.

(d) If $\frac{X_1}{a} = X = \frac{X_2}{b}$; $a > 0$, $b > 0$ and X has standard Cauchy *p.d.f.*, prove that :

$$\phi_{X_1+X_2}(t) = \phi_{X_1}(t) \cdot \phi_{X_2}(t). \quad \text{... (*)}$$

What does this result signify ?

Ans. (d) (*) does not imply that X_1 and X_2 are independent.

31. Let X and Y be independent standard normal variates. Show that the *p.d.f.* $U = X/|Y|$ is :

$$f(u) = \frac{1}{\pi} \cdot \frac{1}{1+u^2} \ ; \ -\infty < u < \infty.$$

32. State and prove the central limit theorem for the sum of n independently and identically distributed random variables with positive finite variance under conditions to be stated.

33. State Lindeberg's sufficient condition for the central limit theorem to hold for a sequence $\{X_k\}$ of independent random variables. Show that every uniformly bounded sequence $\{X_k\}$ of mutually independent random variables obeys the central limit theorem.

Comment on the case when the random variables do not possess expectations.

34. (a) What do you mean by a compound distribution ? Obtain the probability function of compound Poisson distribution and identify it.

(b) What is the Compound Binomial distribution ? Obtain its probability function and identity the distribution.

ORDER STATISTICS

35. Obtain the distribution function and hence the *p.d.f.* of the rth order statistic $X_{(r)}$ in a random sample of size n from a population with continuous distribution function $F(\cdot)$. Deduce the *p.d.f.*'s of the smallest and the largest sample observations.

36. Let $X_1, X_2, ..., X_n$ be a random sample of size n from a population having continuous distribution function $F(x)$. Define the rth order statistic $X_{(r)}$ and obtain its distribution function and hence its *p.d.f.*

37. Define rth order statistic $X_{(r)}$. Obtain the joint *p.d.f.* of $X_{(r)}$ and $X_{(s)}$, $r < s$, in a random sample of size n from a population with continuous distribution function $F(\cdot)$. Hence deduce the *p.d.f.* of sample range $W = X_{(n)} - X_{(1)}$.

38. Obtain the distribution function and hence the *p.d.f.* of the smallest sample observation $X_{(1)}$ in a random sample of size n from a population with a continuous distribution function $F(x)$. Show that for random sample of size 2 from normal population $N(0, \sigma^2)$, $E[X_{(1)}] = -\sigma/\sqrt{\pi}$.

ASSORTED REVIEW PROBLEMS FOR SELF-ASSESSMENT

PROBLEMS ON NORMAL DISTRIBUTION

9·1. (a) Let X be normally distributed with mean 8 and *s.d.* 4. Find

(i) $P(5 \le X \le 10)$, (ii) $P(10 \le X \le 15)$, (iii) $P(X \ge 15)$, (iv) $P(X \le 5)$,

(b) If X is normally distributed with mean 2 and variance 1, find $P(|X-2|<1)$.

(c) If $X \sim N(\mu = 2, \sigma^2 = 2)$, find $P(|X-1| \le 2)$ in terms of distribution function of standard normal variate.

(*d*) If $X - N(30, 5^2)$ and $Y - N(15, 10^2)$, show that $P(26 \le X \le 40) = P(7 \le Y \le 35)$.

Ans. (*a*) (*i*) 0·4649, (*ii*) 0·2684, (*iii*) 0·0401, (*iv*) 0·2266, (*b*) 0·6826 or $[\Phi(1) - \Phi(-1)]$

(*c*) Probability $= P(-1 \le X \le 3) = \Phi(1/\sqrt{2}) - \Phi(-3/\sqrt{2})$

Hint. (*d*) Each probability $= P(-0{\cdot}8 \le Z \le 2)$, where $Z \sim N(0, 1)$.

9·2. If $X \sim N(30, 5^2)$, find the probabilities of :

(*i*) $26 \le X \le 40$, (*ii*) $|X - 30| > 5$, (*iii*) $X \ge 42$, (*iv*) $X < 28$.

Ans. (*i*) 0·7653, (*ii*) 0·3174, (*iii*) 0·0082, (*iv*) 0·3446.

9·3. (*a*) In a normal population with mean 15·00 and standard deviation 3·5, it is known that 647 observations exceed 16·25. What is the total number of observations in the population ?

Hint. Let $X \sim N(\mu, \sigma^2)$, where $\mu = 15$ and $\sigma = 3{\cdot}5$.

If N is the total number of observations in the population, then we have to find N such that :

$$N \times P(X > 16{\cdot}25) = 647$$

(*b*) Assume the mean heights of soldiers to be 68·22 inches with a variance of 10·8 (in.)^2. How many soldiers in a regiment of 1,000 would you expect to be over 6 feet tall ? (Given that the area under the standard normal curve between X = 0 and X = 0·35 is 0·1368 and between X = 0 and X = 1·15 is 0·3746.)

Ans. 125.

9·4. (*a*) In an intelligence test administered to 1,000 examinees, the average score is 42 and standard deviation 24.

(*i*) Find the number of examinees exceeding the score 60, and

(*ii*) Find the number of examinees with score lying between 20 and 40. (Assume normal distribution.)

(*b*) The mean I.Q. (intelligence quotient) of a large number of children of age 14 was 100 and the standard deviation 16. Assuming that the distribution was normal, find

(*i*) What % of the children had I.Q. under 80 ?

(*ii*) Between what limits the I.Q.'s of the middle 40% of the children lay ?

(*iii*) What % of the children had I.Q.'s within the range $\mu \pm 1{\cdot}96\,\sigma$?

Ans. (*a*) (*i*) 227, (*ii*) 289; (*b*) (*i*) 10·56%, (*ii*) 91·6, 108·4, (*iii*) 0·95.

9·5. The life-time in hours of a certain electrical equipment has the normal distribution with mean = 80 and standard deviation = 16. What is the probability that the equipment lasts at least 100 hours ? If the equipment has already lasted 88 hours, what is the conditional probability that it will last at least another 12 hours ?

9·6. (*a*) In a university examination of a particular year, 60% of the students failed when mean of the marks was 50% and *s.d.* 5%. University decided to relax the conditions of passing by lowering the pass marks, to show its result 70%. Find the minimum marks for a student to pass, supposing the marks to be normally distributed and no change in the performance of students takes place.

(*b*) The width of a slot on a forging is normally distributed with mean 0·900 inch and standard deviation 0·004 inch. The specification are 0·900 ± 0·005 inch. What percentage of forgings will be defective ?

Ans. (*a*) 47·375 ; (*b*) Let X denote the width (in inches) of the slot. We want

$$100 \times P(X \text{ lies outside specification limits}) = 100\,[1 - P(0{\cdot}895 < X < 0{\cdot}905)].$$

9·7. (*a*) The daily income of a group of 10,000 persons were found to be normally distributed with mean ₹ 750 and *s.d.* of ₹ 50. Show that of this group, about 95% had income exceeding ₹ 668 and only 5% had income exceeding ₹ 832. What was the lowest income among the richest 100 ?

(*b*) Given that X is normally distributed with mean 10 and $P(X > 12) = 0{\cdot}1587$, what is the probability that X will fall in the interval (9, 11) ?

Take $\Phi(1) = 0{\cdot}8413$ and $\Phi\left(-\frac{1}{2}\right) = 0{\cdot}3085$, where $\Phi(x) = \frac{1}{\sqrt{2\pi}} \int_{-\infty}^{x} \exp(-u^2/2)\, du$.

Ans. (*a*) ₹ 866·30, (*b*) 0·3830.

9·8. A normal distribution has mean 25 and variance 25. Find

(*i*) the limits which include the middle 50% of the area under the curve, and

(*ii*) the values of x corresponding to the points of inflexion of the curve.

Ans. (*i*) Limits which include the middle 50% of the area under the curve are :

$$Q_1 = \mu - 0{\cdot}6745\,\sigma = 21{\cdot}7275, \quad Q_3 = \mu + 0{\cdot}6734\,\sigma = 38{\cdot}2725\,; \quad (ii)\ (20, 30).$$

9·9. (*a*) In a distribution exactly normal, 7% of the items are under 35 and 89% are under 63. What are the mean and standard deviation of the distribution ?

(*b*) In a normal distribution, 31% of the items are under 45 and 8% are over 64. Find the mean and variance of the distribution.

Ans. (*a*) $\mu = 50{\cdot}3, \ \sigma = 10{\cdot}33$; (*b*) $\mu = 50, \ \sigma^2 = 100$.

9·10. A minimum height is to be prescribed for eligibility to government services such that 60% of the young men will have a fair chance of coming up to that standard. The heights of youngmen are normally distributed with mean 60·6″ and *s.d.* 2·55″. Determine the minimum specification.

Hint. We want x_1 such that $P(X > x_1) = 0{\cdot}6$.

When $\quad X = x_1, \ Z = \dfrac{x_1 - 60{\cdot}6}{2{\cdot}55} = -z_1$, [Note the negative sign] ... (*)

Obviously $\quad P(0 < Z < z_1) = 0{\cdot}10 \quad \Rightarrow \quad z_1 = 0{\cdot}254$

Substituting in (*), $\quad x_1 = 60{\cdot}6 - 2{\cdot}55 \times 0{\cdot}254 = 60{\cdot}6 - 0{\cdot}65 = 59{\cdot}95''$.

9·11. Marks secured by students in sections I and II of a class are independently normally distributed with means 50 and 60, respectively, and variances 10 and 6, respectively. What is the probability that a randomly chosen student from section II scores more marks than randomly chosen student from section I ? What percentage of students are expected to secure first division (*i.e.*, 60 marks or more) in section I ? Write down your results in terms of the standard normal distribution function.

Hint. We want : $\quad p = P(X_1 > X_2) = P(X_2 - X_1 > 0)$

$$U = (X_2 - X_1) \sim N\left(\mu = 60 - 50 = 10, \ \sigma^2 = 6 + 10 = 16\right)$$

$$\therefore \quad p = P(U > 0) = P\left[Z > \left(\frac{0 - 10}{4}\right)\right] = P[Z > -2{\cdot}5] = P[Z < 2{\cdot}5] = \Phi(2{\cdot}5).$$

9·12. (*a*) In an examination marks obtained by the students in Mathematics, Physics and Chemistry are distributed normally about the means 50, 52 and 48 with S.D. 15, 12, 16 respectively. Find the probability of securing total marks of (*i*) 180 or above, (*ii*) 90 or below.

(*b*) In a certain examination the percentage of passes and distinctions were 46 and 9, respectively. Estimate the average marks obtained by the candidates, the minimum pass and distinction marks being 40 and 75 respectively. (Assume the distribution of marks to be normal.)

Also determine what would have been the minimum qualifying marks for admission to a re-examination of the failed candidates, had it been desired that the best 25% of them should be given another opportunity of being examined.

Ans. (*a*) 0·1942, 0·0224 ; (*b*) 29.

9·13. (*a*) The quartiles of a normal distribution are 8 and 14 respectively. Estimate the mean and standard deviation.

(*b*) The third decile and the upper quartile of a normal distribution are 56 and 63 respectively. Find the mean and variance of the distribution.

Ans. (*a*) $\mu = 11, \sigma = 4{\cdot}4$; (*b*) $\mu = 59{\cdot}1, \sigma = 5{\cdot}8$.

9·14. 5,000 variates are normally distributed with mean 50 and probable error (semi-interquartile range) 13·49. Without using tables, find the values of the quartiles, median, mode, standard deviation and mean deviation. Find also the value of the variate for which cumulative frequency is 1,250.

Ans. (*a*) $Q_1 = 36{\cdot}51, \ Q_3 = 63{\cdot}49$, Mean = Median = Mode = 50, $\sigma = 20$, M.D. = 16, $x_1 = 36{\cdot}51$.

9·15. Steel rods are manufactured to be 3 inches in diameter but they are acceptable if they are inside the limits 2·99 inches and 3·01 inches. It is observed that 5% are rejected as oversize and 5% are rejected as undersize. Assuming that the diameters are normally distributed, find the standard deviation of the distribution. Hence calculate, what would be the proportion of rejects if the permissible limits were widened to 2·985 inches and 3·015 inches.

Hint. Let X denote the diameter of the rods in inches and let $X \sim N(\mu, \sigma^2)$.

Then $P(X > 3{\cdot}01) = 0{\cdot}05$ and $P(X < 2{\cdot}99) = 0{\cdot}05 \Rightarrow \dfrac{3{\cdot}01-\mu}{\sigma} = 1{\cdot}65$ and $\dfrac{2{\cdot}99-\mu}{\sigma} = -1{\cdot}65$

Solving, we get $\mu = 3$ and $\sigma = \dfrac{1}{165}$.

The probability that a random value of X lies within the rejection limits is :

$P(2{\cdot}985 < X < 3{\cdot}015) = P(-2{\cdot}475 < Z < 2{\cdot}475) = 2 \times P(0 < Z < 2{\cdot}475) = 0{\cdot}9866$

$\therefore$ The probability that X lies outside the rejection limits = $1 - 0{\cdot}9866 = 0{\cdot}0134$.

Hence, the proportion of the rejects outside the revised limits is 0·0134, *i.e.*, 1·34%.

9·16. The following table gives baseball throws for a distance by 303 first year high school girls :

(*i*) Fit a normal distribution and find the theoretical frequencies for the classes of this frequency distribution.

(*ii*) Find the expected number of girls throwing baseballs at a distance exceeding 105 feet on the basis that the data fit a normal distribution.

Distance in feet	*Number of girls*	*Distance in feet*	*Number of girls*
15 and under 25	1	85 and under 95	44
25 and under 35	2	95 and under 105	31
35 and under 45	7	105 and under 115	27
45 and under 55	25	115 and under 125	11
55 and under 65	33	125 and under 135	4
65 and under 75	53	135 and under 145	1
75 and under 85	64		

9·17. (*a*) Determine the constant *k* so that $k \exp(-2x^2 + x)\ -\infty < x < \infty$, is a density function. If the random variable X has the resulting density function, then find :

(*i*) the mean of X, (*ii*) variance of X, and (*iii*) $P\left(X \ge \dfrac{1}{4}\right)$.

(*b*) If $f(x) = k \exp\{-(9x^2 - 12x + 13)\}$, is *p.d.f.* of a normal distribution (*k*, being a constant), find the mean and *s.d.* of the distribution.

(*c*) If X is a normal variate with *p.d.f.* $f(x) = 0{\cdot}03989 \exp(-0{\cdot}005x^2 + 0{\cdot}5x - 12{\cdot}5)$, express $f(x)$ in standard form and hence find the mean and variance of X.

9·18. If X is a normal variate with mean 50 and *s.d.* 10, find $P(Y \le 3137)$, where $Y = X^2 + 1$.

Hint. Required probability = $P(X^2 + 1 \le 3137) = P(-56 \le X \le 56)$.

9·19. If X is a normal variate with mean 1 and *s.d.* 0·6, obtain

$P(X > 0)$, $P(|X - 1| \ge 0{\cdot}6)$ and $P(-1{\cdot}8 < X < 2{\cdot}0)$. What is the distribution of 4X + 5 ?

9·20. (*a*) If $X_i, (i = 1, 2, ..., n)$ are independent $N(\mu_i, \sigma_i^2)$ variates, obtain the distribution of $\sum_{i=1}^{n} a_i X_1$, where $a_i, (i = 1, 2, ..., n)$ are constants. Hence deduce the distribution of :

(*i*) $X_1 + X_2$, (*ii*) $X_1 - X_2$, (*iii*) $\bar{X} = \dfrac{1}{n}\sum_{i=1}^{n} X_i$; if X_i's are *i.i.d.* $N(\mu, \sigma^2)$.

(*b*) How do the results in (*i*) and (*ii*) compare with those in Poisson distribution, and result in (*iii*) compare with Cauchy distribution ?

9·21. If X is normal with mean 2 and standard deviation 3, describe the distribution of $Y = \dfrac{1}{2}X - 1$. Explain how you would find $P\left(Y \ge \dfrac{3}{2}\right)$ from the tables.

Hint. We are given that $X \sim N(\mu, \sigma^2)$, where $\mu = 2, \sigma = 3$. The distribution of the new variable : $Y = aX + b$ is also normal with

$$E(Y) = E(aX + b) = aE(X) + b = a\mu + b \quad \text{and} \quad \text{Var}(Y) = \text{Var}(aX + b) = a^2 \, \text{Var}(X) = a^2 \sigma^2 \quad \ldots (*)$$

Hence, $Y = \frac{1}{2} X - 1 \sim N(\mu_1, \sigma_1^2)$, where μ_1 and σ_1^2 are given by (*) with $a = (1/2)$ and $b = -1$,

i.e., $\mu_1 = (1/2) \cdot 2 - 1 = 0$; $\quad \sigma_1^2 = (1/2)^2 \cdot 9 = \frac{9}{4}$.

Thus, $Y \sim N(\mu_1, \sigma_1^2)$, where $\mu_1 = 0$, $\sigma_1 = (3/2)$

$$P[Y \ge (3/2)] = P(Z \ge 1) = 0{\cdot}5 - P(0 < Z < 1) = 0{\cdot}1587.$$

9·22. Prove that if the independent random variables X and Y have the probability densities,

$$\frac{h}{\sqrt{\pi}} e^{-h^2 x^2} \text{ and } \frac{k}{\sqrt{\pi}} e^{-k^2 y^2}, \; -\infty < (x, y) < \infty, \quad \text{then the } r.v. \; U = X + Y \text{ has the } p.d.f.,$$

$$\frac{l}{\sqrt{\pi}} \cdot e^{-l^2 u^2}, \; -\infty < u < \infty \quad \text{where} \quad \frac{1}{l^2} = \frac{1}{h^2} + \frac{1}{k^2}.$$

9·23. If $\left(\sum_{i=1}^{n} c_i \mu_i \right)^2 = 9 \sum_{i=1}^{n} c_i^2 \sigma_i^2$, find $P\left(0 \le Y \le 2 \sum_{i=1}^{n} c_i \mu_i \right)$, where $Y = \sum_{i=1}^{n} c_i X_i$, X_i being a normal variate with mean mean μ_i and variance σ_i^2 and X_i's are independent.

Hint. We know $Y = \sum_{i=1}^{n} c_i X_i \sim N(\mu, \sigma^2)$, where $\mu = \sum_{i=1}^{n} c_i \mu_i$ and $\sigma^2 = \sum_{i=1}^{n} c_i^2 \sigma_i^2$

Since $\left(\sum_i c_i \mu_i \right)^2 = 9 \left(\sum_i c_i^2 \sigma_i^2 \right)$, we have $\mu^2 = 9\sigma^2 \Rightarrow \frac{\mu}{\sigma} = 3. \quad \ldots (*)$

If we write $Z = \frac{Y - \mu}{\sigma}$, then $Z \sim N(0, 1)$.

$$P\left(0 \le Y \le 2 \sum_{i=1}^{n} c_i \mu_i \right) = P(0 \le Y \le 2\mu) = P(-3 \le Z \le 3) = 0{\cdot}9973. \quad \text{[From (*)]}$$

9·24. If $f(x)$ is density function for the normal distribution with mean zero and standard deviation σ, then show that $\int_{-\infty}^{\infty} \{f(x)\}^2 \, dx = \frac{1}{2\sigma\sqrt{\pi}}$.

Hence show that if the normal distribution is grouped in intervals with total frequency N_1, and N_2 is the sum of the squares of the frequencies, an estimate of σ is $N_1^2 / 2N_2 \sqrt{\pi}$.

Hint. $$\int_{-\infty}^{\infty} [f(x)]^2 \, dx = \int_{-\infty}^{\infty} \left\{ \frac{1}{\sqrt{2\pi}\,\sigma} \exp(-x^2/2\sigma^2) \right\}^2 dx$$

$$= \frac{1}{2\pi\sigma^2} \int_{-\infty}^{\infty} e^{-x^2/\sigma^2} \, dx = \frac{1}{2\pi\sigma^2} \cdot \frac{\sqrt{\pi}}{(1/\sigma)} = \frac{1}{2\sigma\sqrt{\pi}} \quad \left[\because \int_{-\infty}^{\infty} e^{-a^2 x^2} \, dx = \frac{\sqrt{\pi}}{a} \right]$$

$$\therefore \quad N_2 = \int_{-\infty}^{\infty} \{N_1 f(x)\}^2 \, dx = \frac{N_1^2}{2\sqrt{\pi}\,\sigma} \quad \Rightarrow \quad \sigma = \frac{N_1^2}{2N_2 \sqrt{\pi}}.$$

9·25. If $X \sim N(\mu, 9^2)$ and $Y \sim N(\mu, 12^2)$ are independent, and if

$$P(X + 2Y \le 3) = P(2X - Y \ge 4), \text{ determine } \mu.$$

9·26. If $X \sim N(0, 1)$ and $Y \sim N(0, 1)$ are independent *r.v.*'s, prove that

(*i*) $\text{Var}(\sin X) > \text{Var}(\cos X)$ $\qquad$ (*ii*) $E|X - Y| \le \sqrt{(8/\pi)}$

Hint. (*i*) $X \sim N(0,1) \quad \Rightarrow \quad \varphi_X(t) = E(\cos tX + i \sin tX) = e^{-t^2/2}$

$\Rightarrow \quad E(\cos tX) = e^{-t^2/2}$ and $E(\sin tX) = 0.$ Taking $t = 1$ and 2, we get

$E(\cos X) = e^{-1/2}\,; \quad E(\cos 2X) = e^{-2}\,; \quad E(\sin X) = E(\sin 2X) = 0.$

$$\text{Var}(\cos X) = E(\cos^2 X) - (E\cos X)^2 = E\left(\frac{1+\cos 2X}{2}\right) - (E\cos X)^2 = \frac{1}{2}(1-e^{-1})^2 \simeq 0{\cdot}1998$$

Similarly $\text{Var}(\sin X) = E\left(\dfrac{1-\cos 2X}{2}\right) - (E\sin X)^2 = \dfrac{1}{2}(1-e^{-2}) \simeq 0{\cdot}43.$

(*ii*) Use $\quad |X-Y| \le |X| + |Y|$ and $E|X| = E|Y| = \sqrt{2/\pi}$

or $\quad (X \sim Y) \sim N(0, \sigma^2 = 2)\,;\; E|X-Y| = \sqrt{2/\pi}\cdot\sigma = \sqrt{4/\pi} < \sqrt{(8/\pi)}.$

9·27. (*a*) If $X - N(\mu, \sigma^2)$, obtain the distribution of e^X. Find out the mean of the distribution.

(*b*) If $X \sim N(\mu, \sigma^2)$, find the *p.d.f.* of $Y = e^X$, using the result that $E(e^{tX}) = \exp\left[t\mu + t^2\sigma^2/2\right].$ Find the coefficient of variation of Y.

9·28. (*a*) If random variable X has a standard normal distribution and $Y = X^2$,

(*i*) Show that $F_Y(t) = \dfrac{2}{\sqrt{2\pi}} \displaystyle\int_0^{\sqrt{t}} \exp\left[-\frac{1}{2}u^2\right] du,\; t \ge 0.$

(*ii*) Determine $F_Y(t)$ when $t < 0$ and describe the density function $f_Y(t)$.

(*b*) Let $\Phi(x)$ be standard normal distribution function and let $\varphi(x)$ be the *p.d.f.* of standard normal distribution. Show that

(*i*) $\left(\dfrac{1}{x} - \dfrac{1}{x^3}\right)\varphi(x) \le 1 - \Phi(x) \le \dfrac{1}{x}\varphi(x),\; x > 0,$ (*ii*) $\displaystyle\lim_{x\to\infty} \frac{x[1-\Phi(x)]}{\varphi(x)} = 1.$

9·29. If X_1 and X_2 are independent normal variates with means μ_1 and μ_2 and variances σ_1^2 and σ_2^2 respectively, find the relation between α, β, γ and δ so that

$$P(c_1X_1 + c_2X_2 < \alpha) = \gamma \quad \text{and} \quad P(c_1X_1 + c_1X_2 < \beta) = \delta.$$

SUPPLEMENTARY PROBLEMS ON NORMAL DISTRIBUTION

9·30. (*a*) If $X \sim N(0,1)$ and $Y \sim N(0,1)$ are independent random variables, show that $(X+Y)$ is independent of $(X-Y)$.

(*b*) Suppose $X_1 \sim N(0,1)$ and $X_2 \sim N(0,1)$ are independent *r.v.*'s.

(*i*) Find the joint distribution of $(X_1+X_2)/\sqrt{2}$ and $(X_1-X_2)/\sqrt{2}.$

(*ii*) Argue that $2X_1X_2$ and $\left(X_2^2 - X_1^2\right)$ have the same distribution.

Ans. (*b*) (*i*) $U = (X_1+X_2)/\sqrt{2}$ and $V = (X_1-X_2)/\sqrt{2}$ are independent $N(0,1)$ variates.

Hint. (*ii*) $$X_2^2 - X_1^2 = 2\left(\frac{X_2+X_1}{\sqrt{2}}\right)\left(\frac{X_2-X_1}{\sqrt{2}}\right) = 2(UV)$$

= 2 × [Product of two independent SNV's]

$2X_1X_2$ = 2 × [Product of two independent SNV's]

Hence, the result.

9·31. If $X \sim N(0,1)$, prove that the *p.d.f.* of $|X|$ is :

$$h(x) = \begin{cases} \sqrt{2/\pi}\cdot\exp(-x^2/2)\,; & x \ge 0 \\ 0\,; & \text{otherwise} \end{cases}$$

Hence, or otherwise evaluate $E|X|$.

Hint. $E|X| = \text{M.D. of S.N.V.} = \sqrt{2/\pi}.$

9·32. If X and Y are independent standard normal variates and if $Z = aX + bY + c$, where a, b and c are constants, what will be the distribution of Z ? What is the mean, median and standard deviation of the distribution of Z ? Find $P(Z \le 0{\cdot}1)$, if $a = 1, b = -1$ and $c = 0$.

Hint. $Z \sim N(c, a^2 + b^2)$; Mean = Median = c; $s.d. = \sqrt{a^2 + b^2}$

If $a = 1, b = -1, c = 0$ then $Z = X - Y \sim N(0, 2)$

$$\therefore \quad P(Z \le 0{\cdot}1) = P\left(U \le \frac{1}{14{\cdot}142}\right), \quad U = \frac{Z - 0}{\sqrt{2}} \sim N(0, 1)$$

9·33. If $X \sim N(\mu, \sigma^2)$, prove that

$$P(a \le X \le b) = \Phi\left(\frac{b - \mu}{\sigma}\right) - \Phi\left(\frac{a - \mu}{\sigma}\right) \text{ and } P(|X - \mu| > a\sigma) = 2[1 - \Phi(a)];\ a > 0$$

where, $\Phi(\cdot)$ is the distribution function of standard normal distribution.

9·34. Find the normal distribution in which the mean, variance and mean deviation are all equal.

Ans. $N\left(\mu = \frac{16}{25}, \sigma^2 = \frac{16}{25}\right)$

Hint. $\text{M.D.} = \sigma^2 \Rightarrow \frac{4}{5}\sigma = \sigma^2 \Rightarrow \sigma = \frac{4}{5};\ \mu = \sigma^2 = \frac{16}{25}$.

9·35. If $X \sim N(0, 1)$, prove that : *(i)* $E(X^2) = 1$; *(ii)* $E(e^{aX}) = \exp[a^2/2]$.

9·36. Let X be normally distributed with mean μ and variance σ^2. Suppose σ^2 is some function of μ, say, $\sigma^2 = h(\mu)$. Pick $h(\cdot)$ so that $P(X \le 0)$ does not depend on μ for $\mu > 0$.

Hint. $P(X \le 0) = P\left[Z \le -\mu/\sqrt{h(\mu)}\right] = P(Z \le -1)$, independent of μ if we take $h(\mu) = \mu^2$.

9·37. $X \sim N(\mu, \sigma^2)$. Show that the mean of the conditional distribution of X given $a < X < b$ is :

$$\mu + \frac{f(a) - f(b)}{F(b) - F(a)},$$

where $f(x)$ is the *p.d.f.* of X and $F(x) = P(X \le x)$ is the distribution function of X.

9·38. Let X and Y be independent $N(0, 1)$ variates. Prove that $a_1X + b_1Y$ and $a_2X + b_2Y$ are independent if and only if $a_1a_2 + b_1b_2 = 0$.

Hint. Find joint *m.g.f.* of $U = a_1X + b_1Y$ and $V = a_2X + b_2Y$. U and V are independent if and only if

$$M_{U,V}(t_1, t_2) = M_U(t_1) \cdot M_V(t_2).$$

9·39. If X and Y are independent standard normal variates, find the mean value of $\max(|X|, |Y|)$.

Ans. $2/\sqrt{\pi}$.

9·40. The frequency function $f(x)$ is obtained from the normal distribution $N(t) = \frac{1}{\sqrt{2\pi}} e^{-t^2/2};\ -\infty < t < \infty$

by means of the equations :

$$\int_1^x f(u)\,du = \int_{-\infty}^t N(t)\,dt \quad ...(i) \quad \text{and} \quad t = a\log(x - 1) \quad ...(ii)$$

If $\exp(1/a^2) = 4$, find the mean, median and mode for $f(x)$.

Ans. $\mu_X = 3$, Median $(X) = 2$, Mode $(X) = 1{\cdot}25$.

9·41. Let X_1, X_2, X_3 be three independent normal variates with mean μ and variance σ^2. Let :

$$Y_1 = \frac{X_1 - X_3}{\sqrt{2}}, \quad Y_2 = \frac{X_1 - 2X_2 + X_3}{\sqrt{6}} \quad \text{and} \quad Y_3 = \frac{X_1 + X_2 + X_3}{\sqrt{3}}$$

Show that Y_1, Y_2 and Y_3 are independent normal variates. Show further that :

$$Y_1^2 + Y_2^2 = \sum_{i=1}^{3}(X_i - \bar{X})^2, \text{ where } \bar{X} = \frac{X_1 + X_2 + X_3}{3}$$

9·42. Let X_1, X_2, X_3 denote a random sample from $N(0, 1)$ distribution. Let the *r.v.*'s Y_1, Y_2, Y_3 be defined by :

$$X_1 = Y_1 \cos Y_2 \sin Y_3\,; \quad X_2 = Y_1 \sin Y_2 \sin Y_3\,; \quad X_3 = Y_1 \cos Y_3,$$

where, $0 \le Y_1 \le \infty,\ 0 \le Y_2 \le 2\pi,\ 0 \le Y_3 \le \pi$. Show that Y_1, Y_2, Y_3 are mutually stochastically independent.

Ans. $|J| = y_1^2 \sin y_3\,;\ x_1^2 + x_2^2 + x_3^2 = y_1^2.$

$$g(y_1, y_2, y_3) = \frac{1}{(2\pi)^{3/2}} \exp\left[-\frac{1}{2}\left(x_1^2 + x_2^2 + x_3^2\right)\right] \cdot |J|$$

$$= \frac{1}{(2\pi)^{3/2}} \cdot \exp\left(-\frac{1}{2} y_1^2\right) \cdot y_1^2 \cdot \sin y_3$$

$$= g_1(y_1) \cdot g_2(y_2) \cdot g_3(y_3)$$

$\Rightarrow Y_1, Y_2, Y_3$ are stochastically independent.

9·43. If X and Y are two independent log-normal variates, show that XY and X/Y are also log-normal variates.

Hint. Let $\log X \sim N(\mu_1, \sigma_1^2)$; $\log Y \sim N(\mu_2, \sigma_2^2)$; $U = XY$ and $V = (X/Y)$.

$$\left.\begin{aligned} \log U &= \log X + \log Y \sim N(\mu_1 + \mu_2,\ \sigma_1^2 + \sigma_2^2) \\ \log V &= \log X - \log Y \sim N(\mu_1 - \mu_2,\ \sigma_1^2 + \sigma_2^2) \end{aligned}\right\} \quad (\because X \text{ and } Y \text{ are independent.})$$

9·44. Let X be log-normal variate with *p.d.f.* :

$$f(x, \mu, \sigma) = \frac{1}{x\sigma\sqrt{2\pi}} \exp\left\{-(\log x - \mu)^2 / 2\sigma^2\right\},\ x > 0,\ \sigma > 0,\ |\mu| < \infty$$

Show that $e^a X^b$ has a log normal *p.d.f.* $f(x, a + b\mu, b\sigma)$. If $X_1, X_2, \ldots, X_n$ are n independent observations on X, then show that $G = (X_1 X_2 \ldots X_n)^{1/n}$ also has a log-normal *p.d.f.* $f\left(x, \mu, \sigma/\sqrt{n}\right)$.

9·45. A random variable X has *p.d.f.* $\quad f(x) = C\varphi(x),\ x \ge k,$

where, k is a given number, C is a constant chosen to ensure that $f(x)$ is a probability density function and

$$\varphi(x) = \frac{1}{\sqrt{2\pi}} \cdot \exp\left(-\frac{1}{2}x^2\right)$$

If $g(k) = \int_k^\infty \varphi(x)\,dx$, show that the arithmetic mean and the variance of X are

$$\mu = \frac{\varphi(k)}{g(k)} \quad \text{and} \quad \sigma^2 = 1 + \frac{\varphi(k)}{g(k)}\left\{k - \frac{\varphi(k)}{g(k)}\right\} = 1 + \mu(k - \mu), \text{ respectively.}$$

PROBLEMS ON UNIFORM DISTRIBUTION

9·46. A random variable X has a uniform distribution over $(-3, 3)$. Compute

(*i*) $P(X = 2),\ P(X < 2),\ P(|X| < 2)$ and $P(|X - 2| < 2)$.

(*ii*) Find k for which $P(X > k) = \frac{1}{3}$.

9·47. Suppose that X is uniformly distributed over $(-\alpha, \alpha)$, where $\alpha > 0$. Determine α so that

(*i*) $P(X > 1) = \frac{1}{3}$, (*ii*) $P\left(X < \frac{1}{2}\right) = 0{\cdot}3$ and (*iii*) $P(|X| < 1) = P(|X| > 1)$.

Ans. (*i*) $\alpha = 3$, (*ii*) $\alpha = 5/6$, (*iii*) $\alpha = 2$.

9·48. Calculate the coefficient of variation for the rectangular distribution in $(0, b)$ given that the probability law of the distribution is $P(X \le t) = \frac{t}{b}$.

9·49. If random variable X has the density function $f(x)$, prove that

$$Y = \int_{-\infty}^{x} f(x)\,dx, \text{ has a rectangular distribution over } (0,1).$$

If
$$f(x) = \begin{cases} \frac{1}{2}(x-1), & 1 \le x \le 3 \\ 0, & \text{otherwise} \end{cases}$$

determine what interval for Y will correspond to the interval $1{\cdot}1 \le X \le 2{\cdot}9$.

Ans. $y = F(x) = (x-1)^2/4\ ;\ 1 \le x \le 3\ ;\ 0{\cdot}0025 \le y \le 0{\cdot}9025$.

9·50. (*a*) Show that, whatever be the distribution function $F(x)$ of a continuous *r.v.* X,
$$P[a \le F(x) \le b] = b - a\ ;\ 0 \le (a, b) \le 1.$$

(*b*) What is the relation between X and Y if X has *p.d.f.* $f(x) = e^{-x},\ x \ge 0$ and Y has uniform distribution in the interval [0, 1] ?

(*c*) A variable X has uniform distribution over the unit interval. What function Y of X has the distribution :
$$dP = e^{-x}\,dx,\ x \ge 0.$$

Ans. (*b*) $Y = F(X) = 1 - e^{-X}$, (*c*) $Y = \log_e\left(\frac{1}{1-X}\right)$.

9·51. Two numbers are independently chosen at random in the interval (0, 1). Find the probability that :

(*a*) The distance between them is less than a fixed number k, $(0 < k < 1)$.

(*b*) Their product is less than a constant k, $(0 < k < 1)$.

Ans. (*a*) $k(2-k)$; (*b*) $k(1 - \log k)$.

9·52. X and Y are uniform variates each uniformly distributed over the interval (0, 1). Find the probability that the greater of X, Y is less than a fixed number k, $(0 < k < 1)$.

Ans. k^2

9·53. Let X and Y be independent variates which are uniformly distributed over the unit interval (0, 1). Find the distribution function and the *p.d.f.* of random variable $Z = X + Y$. Is Z a uniformly distributed variable ? Give reasons.

9·54. Let X_1 and X_2 be independent random variable uniformly distributed over the interval (0, 1). Find

(*i*) $P(X_1 + X_2 < 0{\cdot}5)$, (*ii*) $P(X_1 - X_2 < 0{\cdot}5)$, (*iii*) $P\left(X_1^2 + X_2^2 < 0{\cdot}5\right)$,

(*iv*) $P\left(e^{-X_1} < 0{\cdot}5\right)$, and (*v*) $P(\cos \pi X_2 < 0{\cdot}5)$.

Ans. (*i*) 0·125, (*ii*) 0·875, (*iii*) 0·393, (*iv*) $1 - \log 2$, and (*v*) 2/3.

9·55. (*a*) A random variable X is uniformly distributed over (0, 1), find the probability density functions of :

(*i*) $Y = X^2 + 1$, and (*ii*) $Z = 1/(X+1)$.

(*b*) If the random variable X is uniformly distributed over $\left(0, \frac{1}{2}\pi\right)$, compute the expectation of the function $\sin X$. Also find the distribution of $Y = \sin X$, and show that the mean of this distribution is the same as the above expectation.

Ans. (*b*) $(2/\pi)$, $f_Y(y) = 2\Big/\left\{\pi\sqrt{1-y^2}\right\},\ 0 < y < 1$.

9·56. Let $X_1, X_2, \ldots X_n$ be independent random variables having a common rectangular distribution over the interval $[a, b]$. Obtain the distribution of $Y = \max(X_1, X_2, \ldots, X_n)$.

9·57. If $X_1, X_2, \ldots, X_n$ are mutually independent rectangular variates on [0, 1], prove that the *p.d.f.* of the product $X_1 X_2 \ldots X_n$ is :
$$f(x) = \begin{cases} \dfrac{(-\log x)^{n-1}}{(n-1)!}, & 0 < x \le 1 \\ 0, & \text{otherwise} \end{cases}$$

9·58. Let X_1 & X_2 be independent *r.v.*'s, each uniform on [0, 1]. Show that $Y_1 = \sqrt{-2 \log X_1} \cdot \sin(2\pi X_2)$ and $Y_2 = \sqrt{-2 \log X_1} \cdot \cos(2\pi X_2)$ are independent *r.v.*'s and that each is $N(0, 1)$. [This is known as *Box and Muller transformation.*]

9·59. (*a*) The variates a and b are independently and uniformly distributed in the intervals [0, 6] and [0, 9] respectively. Find the probability that $x^2 - ax + b = 0$, has two real roots.

(*b*) Find the probability that the roots of the equation $x^2 + 2bx + x = 0$, should be real, given that $b \sim U[-\alpha, \alpha]$ and $c \sim U[-\beta, \beta]$ are independent.

(*c*) If a, b, c are randomly chosen between 0 and 1, find the probability that the quadratic equation $ax^2 + bx + c = 0$, has real roots.

Ans. (*a*) $$P\left(b \le \frac{a^2}{4}\right) = \int_{a=0}^{6} \int_{b=0}^{a^2/4} \left(\frac{1}{6 \times 9}\right) da\, db = \frac{1}{3}.$$

(*b*) $$\text{Probability} = P(b^2 > c) = P(c < b^2)$$
$$= \int_{b=-\alpha}^{\alpha} \left[\int_{c=-\beta}^{b^2} \frac{1}{2\alpha \cdot 2\beta}\, dc \right] db = \frac{1}{2\beta}\left(\frac{\alpha^2}{3} + \beta\right).$$

(*c*) $$P(b^2 \ge 4ac) = 1 - P(b^2 \le 4ac) = 1 - P\left(b \le \sqrt{4ac}\right); \quad (\because b > 0) = \frac{1}{9}.$$

9·60. A boy and a girl agree to meet at a certain park between 4 and 5 P.M. They agree that the one arriving first will wait t hours, $0 \le t \le 1$, for the other to arrive. Assuming that the arrival times are independent and uniformly distributed, find the probability that they meet.

Hence, obtain the probability that they meet if $t = 10$ minutes.

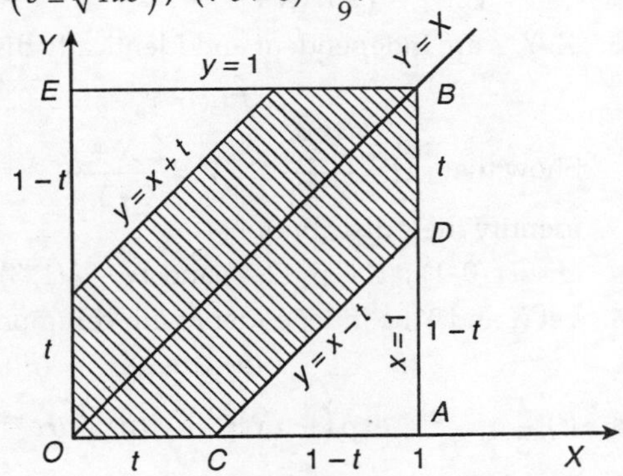

Fig. 9·21

Hint. $$p = P(|X - Y| \le t) = \frac{\text{Shaded Area}}{\text{Total Area}}$$ [See *Fig. 9·21*]

$$= 2\,[\text{Area } OAB - \text{Area } CAD]/1 \times 1$$
$$= 1 - (1 - t)^2 = 2t - t^2$$

Ans. $t = 10$ minutes, Probability = 11/36.

9·61. Let X_1 and X_2 be independent random variables with uniform distribution over the interval $\left[\theta - \frac{1}{2}, \theta + \frac{1}{2}\right]$. Show that $(X_1 - X_2)$ has a distribution independent of θ and obtain its *p.d.f.*

Hint. Use *m.g.f.* and triangular distribution on the interval (–1, 1), with peak at $c = 0$.

Ans. $$U = X_1 - X_2\ ; \qquad f(u) = \begin{cases} u + 1; & -1 < u \le 0 \\ 1 - u; & 0 < u < 1 \end{cases}$$

9·62. The random variables X and Y are independent and both have the uniform distribution on [0, 1]. Let $Z = |X - Y|$. Prove that for real θ, the characteristic function of Z is given by :

$$\varphi(Z, \theta) = 2\,[\exp(i\theta) - 1 - i\theta] / (i\theta)^2$$

Hence, deduce the general expression for $E(Z^n)$.

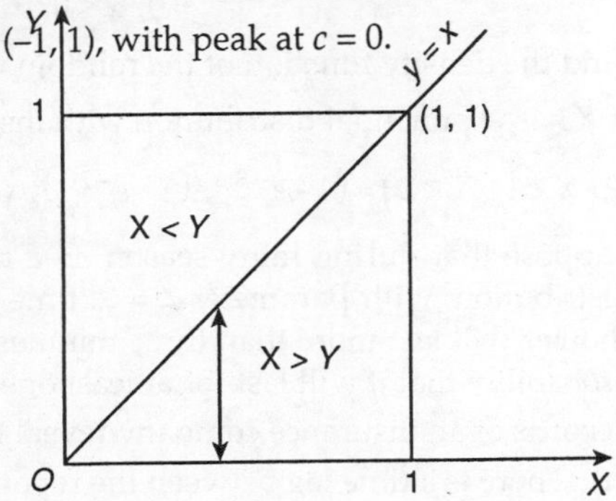

Fig. 9·22

Hint. $$\varphi(\theta\,; |X - Y|) = \int_0^1 \int_0^1 e^{i\theta|x-y|} f(x, y)\, dx\, dy$$
$$= 2 \int_0^1 \left(\int_0^x e^{i\theta(x-y)}\, dy \right) dx \qquad \text{[See } Fig.\ 9\cdot22]$$

Ans. $2/[(n + 1)(n + 2)]$.

9·63. Let X be a random number selected from (0, 1) and let :

$$Y=1+\left[\frac{\log X}{\log q}\right];\ q=1-p,\ 0<p<1$$

where, $[a]$ denotes the greatest integer contained in a. Prove that Y is geometric variate with parameter p.

PROBLEMS ON GAMMA, BETA AND EXPONENTIAL DISTRIBUTIONS

9·64. Let $X \sim \gamma(a, \lambda)$ and $Y \sim \gamma(b, \lambda)$ be independent Gamma variates.

Let $U=\dfrac{X}{Y},\quad V=\dfrac{X}{X+Y},\quad W=\dfrac{X-Y}{X+Y}$ and $Z=X+Y$. Prove that

(i) Z is independent of each of U, V and W.

(ii) $U \sim \beta_2(a, b)$ and $V \sim \beta_1(a, b)$.

(iii) The *p.d.f.* of W is : $g(w)=\dfrac{1}{2B(a,b)}\left(\dfrac{1+w}{2}\right)^{a-1}\left(\dfrac{1-w}{2}\right)^{b-1};\ -1\le w\le 1.$

(iv) $E\left[\dfrac{X}{X+Y}\right]=\dfrac{E(X)}{E(X+Y)}$ and $E\left[\dfrac{X-Y}{X+Y}\right]=\dfrac{E(X)-E(Y)}{E(X)+E(Y)}.$

9·65. X, Y, Z are independent and identically distributed random variables with common *p.d.f.*

$$f(x)=e^{-x};\ x>0.$$

Show that $U=\dfrac{X}{X+Y},\quad V=\dfrac{X+Y}{X+Y+Z}$ and $Z=(X+Y+Z)$ are independendly distributed and identify their distributions.

Ans. $U \sim U(0,1)$; $V \sim \beta_1(2, 1)$ and $W \sim \gamma(1, 3)$.

9·66. Let X and Y be independent with common *p.d.f.* $\alpha e^{-\alpha x},\ 0<x<\infty,\ \alpha>0$. Find the *p.d.f.* of :

(i) X^3, (ii) $3+2X$, (iii) $X-Y$, and (iv) $|X-Y|$

Ans. (i) $\frac{1}{3}\alpha x^{-2/3}\exp\left(-\alpha x^{1/3}\right)$, (ii) $\frac{1}{2}\alpha e^{-\alpha(x-3)/2},\ x>3$, (iii) $\frac{1}{2}\alpha e^{-\alpha|x|}, \forall x$, (iv) $\alpha e^{-\alpha x}, x>0.$

9·67. X and Y are independent random variables each exponentially distributed with the same parameter θ. Find *p.d.f.* for $\dfrac{X}{X+Y}$ and identify its distribution.

9·68. The density functions of the independent random variables X and Y are :

$$f_X(x)=\lambda e^{-\lambda x};\ x>0, \lambda>0 \qquad f_Y(y)=\lambda e^{-\lambda y};\ y>0, \lambda>0$$
$$=0,\ x\le 0 \qquad\qquad =0, \text{ otherwise}$$

Find the density function of the random variable $Z = X/Y$.

9·69. If X has exponential distribution with mean 2, find $P(X<1 \mid X<2)$.

Ans. $P(X<1 \mid X<2)=(1-e^{-\theta})/(1-e^{-2\theta})$, where $\theta=\frac{1}{2}$.

9·70. Suppose that during rainy season on a tropical island the length of the shower has an exponential distribution, with parameter λ = 2, time being measured in minutes. What is the probability that a shower will last more than three minutes ? If a shower has already lasted for 2 minutes, what is the probability that it will last for at least one more minute ?

9·71. Records of an insurance company reveal that :

(i) There is a time lag between the reporting of a claim and its settlement.

(ii) The frequency density function for the number of claims getting settled, out of an initial number N of claims, at a point of time in the vicinity of time t (in years) is $N k \cdot \exp(-kt),\ k>0$.

(*iii*) The amount paid out as claim increases with the time lag between the reporting and settlement of a claim, the average amount paid out in respect of claims settled at time t being $\exp(ct)$, $c > 0$.

(*a*) Show that the proportion of claims outstanding at time t is e^{-kt}.

(*b*) Find the average amount paid in respect of all claims.

(*c*) Show that the average amount paid in respect of the claims settled within the first year of its being reported is $k(e^k - e^c)/(k-c)(e^k - 1)$. How does this compare with the result of (*b*) ?

9·72. A continuous *r.v.* X has the *p.d.f.* $f(x)$ given by : $f(x) = \begin{cases} Ae^{-x/5}, & x > 0, \\ 0, & \text{otherwise} \end{cases}$

Find the value of A and show that for any two positive numbers s and t,

$$P[X > s + t \mid X > s] = P[X > t].$$

9·73. If $X_1, X_2, \ldots, X_n$ are independent *r.v.'s* X_i having an exponential distribution with parameter θ_i, $(i = 1, 2, \ldots, n)$, then prove that $Z = \min(X_1, X_2, \ldots, X_n)$ has an exponential distribution with parameter $\sum_{i=1}^{n} \theta_i$.

9·74. (*a*) If X has exponential distribution with mean 2, find $P[(X < 1) \mid (X < 2)]$.

(*b*) If $X \sim \text{Expo}(\lambda)$, show that $Y = -\frac{1}{\lambda} \cdot \log F(x) \sim \text{Expo}(\lambda)$.

Ans. (*a*) $P(X < 1 \mid X < 2) = (1 - e^{-\theta})/(1 - e^{-2\theta})$, where $\theta = \frac{1}{2}$.

Hint. (*b*) $M_Y(t) = E(e^{tY}) = E \exp\left\{-\frac{t}{\lambda} \log F(X)\right\} = E\left\{F(X)^{-t/\lambda}\right\}$

$$= E\left(Z^{-t/\lambda}\right), \text{ where } Z = F(X) \sim U[0, 1]$$

9·75. X is a positive random variable with distribution function $F(x)$ and *p.d.f.* $f(x)$. Show that $P(x \le X \le x + dx \mid X \le x) = g(x)\,dx$ where $g(x) = f(x)/[1 - F(x)]$.

Prove also that $\quad F(x) = 1 - \exp\left[-\int_0^x g(y)\,dy\right]$.

Hence, deduce that the negative expnential is the only absolutely continuous distribution with constant $g(x)$.

Hint. $$\int_0^x g(y)\,dy = \int_0^x \frac{f(y)}{1 - F(y)}\,dy = \left| -\log[1 - F(y)] \right|_{y=0}^{x}$$

$$g(x) = c \quad \Rightarrow \quad F(x) = 1 - \exp[-cx] \quad \Rightarrow \quad X \sim \text{Expo}(\theta = c).$$

9·76. If X and Y are *i.i.d.* exponential variates with *p.d.f.*

$$f(x) = \theta e^{-\theta x};\ x > 0,\ \theta > 0,$$

Obtain the joint distribution of $U = (X/Y)$ and $V = X + Y$. Are U and V independent ? Identify their distributions.

Ans. U and V are independent.

$$g_1(u) = \frac{1}{(1+u)^2};\ u > 0; \quad g_2(v) = \theta^2 \cdot ve^{-\theta v};\ \theta > 0,\ v > 0.$$

9·77. Let X and Y be independent variates each with *p.d.f.*

$$f(x) = \theta e^{-\theta x};\ x > 0,\ \theta > 0$$

Find the *p.d.f.* for (X/Y). Hence deduce that the variate $Y/(X + Y)$ is uniformly distributed over $(0, 1)$.

Hint. $$W = \frac{Y}{X+Y} = \frac{1}{1+U} \quad \Rightarrow \quad g_2(w) = g_1(u) \cdot \left|\frac{du}{dw}\right|; \quad U = \left(\frac{X}{Y}\right).$$

Ans. $U = (X/Y)$; $\quad g_1(u) = (1+u)^{-2}$; $u > 0$.

9·78. X_1 and X_2 are two independent observations of a *r.v.* X for which the *p.d.f.* is :

$$f(x, \theta) = \frac{1}{\theta} \exp\left(-\frac{x}{\theta}\right); \; 0 < x < \infty, \; \theta > 0.$$

(*a*) Derive the sampling distribution of the statistics $U = X_1 + X_2$ and $V = [X_1/(X_1 + X_2)]$, and hence prove that they are independently distributed.

(*b*) Find the mean and variance of V.

(*c*) Derive the distribution of $W = (X_2/X_1)$ and show that W and U are independently distributed.

Ans. (*a*) $U \sim \gamma\left(\frac{1}{\theta}, 2\right)$, $V \sim U[0, 1]$ are independent.

(*b*) Mean $(U) = \frac{1}{2}$; Variance $(V) = \frac{1}{12}$.

(*c*) Let $U = X_1 + X_2$, $W = \left(\frac{X_2}{X_1}\right)$; U and W are independent ;

p.d.f. of W: $\quad h(w) = \frac{1}{(1+w)^2}; \; 0 < w < \infty.$

9·79. $X_1, X_2, ..., X_n$ are independent *r.v.*'s, each having the same exponential distribution. Prove that :

$$X_1 + \frac{1}{2} X_2 + \frac{1}{3} X_3 + \ldots + \frac{1}{n} X_n \quad \text{and} \quad \max(X_1, X_2, X_3, \ldots, X_n),$$

have the same distribution.

Hint. Let $\quad Y = X_1 + \frac{1}{2} X_2 + \frac{1}{3} X_3 + \ldots + \frac{1}{n} X_n \quad$ and $\quad Z = \max(X_1, X_2, \ldots, X_n).$

Now prove that : $\quad M_Z(t) = M_Y(t).$

PROBLEMS ON LAPLACE AND CAUCHY DISTRIBUTIONS

9·80. The joint *p.d.f.* of *r.v.*'s X and Y is :

$$f(x, y) = \begin{cases} \alpha^{-2} \cdot \exp\left(-\frac{x+y}{\alpha}\right); \; x > 0, \; y > 0 \; ; \; \alpha > 0. \\ 0, \text{ otherwise} \end{cases}$$

Obtain the distribution of $U = \frac{1}{2}(X - Y)$. For what value of α, U follows standard Laplace distribution ?

Hint. X and Y are *i.i.d.* exponential with mean α.

$$\Phi_X(t) = \Phi_Y(t) = (1 - i\alpha t)^{-1}.$$

$$\Phi_U(t) = \Phi_X(t/2) \cdot \Phi_Y(-t/2) = \frac{1}{1 + (\alpha/2)^2 \cdot t^2}$$

$$\Rightarrow \quad U = \frac{1}{2}(X - Y) \sim L(\mu = 0, \; \lambda = \alpha/2), \text{ with } p.d.f.$$

$$g(u) = \frac{\exp\left[-\frac{|u|}{\alpha/2}\right]}{2 \cdot (\alpha/2)} = \frac{1}{\alpha} \exp\left[-\frac{2|u|}{\alpha}\right]; \; -\infty < u < \infty$$

U has standard Laplace distribution for $\alpha = 2$.

9·81. If X_1, X_2, X_3 and X_4 are *i.i.d.* $N(0, 1)$ variates, show that $Y = X_1 X_2 - X_3 X_4$ has *p.d.f.* :

$$f(y) = \frac{1}{2} \exp(-|y|), \; -\infty < y < \infty.$$

Hint. Show that $\varphi_Y(t) = 1/(1 + t^2) \quad \Rightarrow \quad Y$ has standard Laplace distribution.

Use : $$\int_{-\infty}^{\infty} \int_{-\infty}^{\infty} \exp\left[-(ax^2 + 2hxy + by^2)\right] dx\, dy = \frac{\pi}{\sqrt{ab - h^2}}.$$

9·82. A needle spins about the point $(0, b)$ of the $x-y$ plane with $b > 0$ and comes to a stop thereby making an angle φ with Y-axis. The direction of the needle then intersects the x-axis at a point $(X, 0)$. Assuming φ is a *r.v.* with uniform probability distribution on $(-\pi/2, \pi/2)$, what is the distribution function and hence *p.d.f.* of X? Also identify the distribution.

Ans. $F_X(x) = \dfrac{1}{\pi}\left(\tan^{-1}\dfrac{x}{b} + \dfrac{\pi}{2}\right)$, $\quad f_X(x) = \dfrac{1}{\pi}\cdot\dfrac{b}{x^2 + b^2}$, $-\infty < x < \infty$.

9·83. If X_1, X_2, X_3, X_4 are independent standard normal variates, find the distribution of $(X_1/X_2) + (X_3/X_4)$.

Ans. $C(2, 0)$.

9·84. $\bar{X}_n$ is the mean of n independent variables distributed like X, and X has a symmetric distribution. if $\bar{X}_n$ has exactly the same distribution of X for all n, then prove that the characteristic function of X is : $\Phi_X(t) = \exp(-c\,|\,t\,|)$, for some real constant $c > 0$. Identify this distribution.

Ans. $X \sim C(c, 0)$.

9·85. If $X \sim N(\mu_1, \sigma_1^2)$ and $Y \sim N(\mu_2, \sigma_2^2)$ are independent random variables, obtain the *p.d.f.* of :

$$U = (X - \mu_1) / (Y - \mu_2).$$

9·86. X and Y are *i.i.d.* random variables with common *p.d.f.* : $f(x) = \dfrac{c}{1+x^4}$; $-\infty < x < \infty$, where c is a constant. Prove that with proper choice of c, (X/Y) follows Standard Cauchy distribution.

Hint. Let $\quad U = (X/Y), \ V = Y.$

$$g(u, v) = \frac{c^2\,|\,v\,|}{(1+u^4v^4)(1+v^4)}\ ; \ -\infty < (u, v) < \infty.$$

Let
$$\frac{1}{(1+u^4v^4)(1+v^4)} \equiv \frac{1}{(1+kz)(1+z)}; \ k = u^4, \ z = v^4.$$

Resolve into partial fractions and find $g_1(u) = \int_{-\infty}^{\infty} g(u, v)\, dv.$

Ans. *p.d.f.* of $U = (X/Y)$ is : $\quad g_1(u) = \dfrac{\pi c^2}{2(1+u^2)}$; $-\infty < u < \infty$; $\ c = \dfrac{\sqrt{2}}{\pi}$.

PROBLEMS ON CENTRAL LIMIT THEOREM

9·87. Let $\bar{X}_n$ be the sample mean of a random sample of size n from Rectangular distribution on $[0, 1]$.

Let
$$U_n = \sqrt{n}\left(\bar{X}_n - \frac{1}{2}\right).$$

Show that $\quad F(u) = \lim\limits_{n\to\infty} P(U_n < u)$ exists and determine it.

Ans. $\Phi\left(\sqrt{12}\,u\right)$ where $\Phi(u) = \dfrac{1}{\sqrt{2\pi}}\displaystyle\int_{-\infty}^{u}\exp\left(-\frac{1}{2}x^2\right)dx.$

9·88. Let $X_1, X_2, \ldots$ be a sequence of independent, identically distributed non-negative random variables such that $E(\log X_1)^2$ is finite. $Z_n = (X_1 X_2 \ldots X_n)^{1/n}$. Show that the positive constant c can be so chosen that the random variable $(cZ_n)^{\sqrt{n}}$ has a non-degenerate limit distribution function $F(\cdot)$ and determine $F(\cdot)$.

Ans. $c = e^{-\mu}$, $\quad F(x) = \Phi(\log x/\sigma)$, $\quad \mu = E(\log X_1)$ and $\quad \sigma^2 = V(\log X_1)$.

9·89. $\{X_n\}$ is a sequence of *i.i.d.* random variables. If n is a perfect square, then X_n is Cauchy variate with density : $\dfrac{1}{\pi}\cdot\dfrac{1}{1+x^2}$, $-\infty < x < \infty$. Otherwise, X_n has a distribution function $F(x)$ with mean zero and finite variance σ^2. Discuss the asymptotic distribution of $(X_1 + X_2 + \ldots + X_n)/\sqrt{n}$.

9·90. Let $\{X_k\}$, $k \geq 1$ be a sequence of *i.i.d.* variates with $f(x) = \frac{1}{2} e^{-|x|}$, $-\infty < x < \infty$.

Find the constant a_n and b_n such that $\{|X_1| + |X_2| + \ldots + |X_n| - a_n\}/b_n \xrightarrow{L} N(0, 1)$.

9·91. Obtain $M_X(t)$; when $X \sim P(\lambda)$. Find the limit as $\lambda \to \infty$ of *m.g.f.* of $(X - \lambda)/\sqrt{\lambda}$ and interpret the result in the context of *C.L.T.* Also prove that :

$$\lim_{\lambda \to \infty} \sum_{k=\alpha}^{\beta} \left[\frac{e^{-\lambda} \cdot \lambda^k}{k!} \right] = \frac{1}{\sqrt{2\pi}} \int_a^b \exp\left(-\frac{1}{2} u^2 \right) du, \quad \alpha = \lambda + a\sqrt{\lambda}, \ \beta = \lambda + b\sqrt{\lambda}.$$

Show that $2X$ is not a Poisson variate. Give a set of conditions under which $X + Y$ too is a Poisson variate.

9·92. (*a*) Decide whether the central limit theorem holds for the sequence of independent random variables X_r with distribution defined as $P(X_r = 1) = p$, and $P(X_r = 0) = 1 - p$.

(*b*) Show that the central limit theorem applies if (*i*) $P\left(X_k = \pm k^{\alpha}\right) = \frac{1}{2}$, (*ii*) $P\left(X_k = \pm \sqrt{2k-1}\right) = \frac{1}{2}$,

and (*iii*) $P(X_k = 0) = 1 - k^{1-2\alpha}$, $P(X_k = \pm k\alpha) = \frac{1}{2} k^{-2\alpha}$, where $\alpha < \frac{1}{2}$.

9·93. Using C.L.T. show that : $\displaystyle \lim_{n \to \infty} \left(e^{-n} \sum_{k=0}^{n} \frac{n^k}{k!} \right) = \frac{1}{2} = \lim_{n \to \infty} \int_0^n \frac{e^{-t} \cdot t^{n-1}}{(n-1)!} dt.$

9·94. Let $\{X_n, n = 1, 2, \ldots\}$ be a sequence of independent Bernoulli variates such that :

$$P(X_n = 1) = p_n = 1 - P(X_n = 0), \ n = 1, 2, \ldots, (q_n = 1 - p_n).$$

Show that if $\sum p_n q_n = \infty$, $(n = 1, 2, \ldots, \infty)$, then the C.L.T. holds for the sequence $\{X_n\}$.

What happens if $\sum p_n q_n < \infty$.

9·95. Let $X_1, X_2, \ldots, X_n$ be independent identically distributed *r.v.*'s with $E(X_i) = \mu$; $\text{Var}(X_i) = \sigma^2$; $(0 < \sigma^2 < \infty)$; $i = 1, 2, \ldots, n$ and $E(X_i - \mu)^4 = \sigma^4 + 1$.

(*a*) If $\displaystyle P\left[\frac{1}{n}(X_1^2 + X_2^2 + \ldots + X_n^2) - c \right] \to 0$ as $n \to \infty$, find c.

(*b*) Find $\displaystyle \lim_{n \to \infty} P\left[\sigma^2 - \frac{1}{\sqrt{n}} \leq \frac{(X_1 - \mu)^2 + \ldots + (X_n - \mu)^2}{n} \leq \sigma^2 + \frac{1}{\sqrt{n}} \right].$

Hint. (*a*) By Khinchine's theorem, $c = E(X_i^2) = \sigma^2 + \mu^2$ (finite)

(*b*)
$$p_n = P\left[-\frac{1}{\sqrt{n}} \leq \frac{\Sigma (X_i - \mu)^2}{n} - \sigma^2 \leq \frac{1}{\sqrt{n}} \right] = P\left[-\sqrt{n} \leq \sum_i (X_i - \mu)^2 - n\sigma^2 \leq \sqrt{n} \right]$$

$$= P\left[-1 \leq \left\{ \sum_{i=1}^{n} \{(X_i - \mu)^2 - \sigma^2\} / \sqrt{n} \right\} \leq 1 \right] = P\left(-1 \leq S_n / \sqrt{n} \leq 1 \right) \quad \ldots (*)$$

where
$$S_n = \sum_{i=1}^{n} \{(X_i - \mu)^2 - \sigma^2\} = \sum_{i=1}^{n} U_i$$

and
$$U_i = (X_i - \mu)^2 - \sigma^2, \ i = 1, 2, \ldots, n \text{ ; are } i.i.d. \ r.v.\text{'s.}$$

$$E(U_i) = E(X_i - \mu)^2 - \sigma^2 = \sigma^2 - \sigma^2 = 0$$

and $\text{Var } U_i = \text{Var}\left[(X_i - \mu)^2 - \sigma^2 \right] = \text{Var}\left[(X_i - \mu)^2 \right] = E(X_i - \mu)^4 - \left[E(X_i - \mu)^2 \right]^2 = \sigma^4 + 1 - \sigma^4 = 1$

$\left[\because E(X_i - \mu)^4 = \sigma^4 + 1 \text{ (Given)} \right]$

$$\therefore \quad E(S_n) = \sum_{i=1}^{n} E(U_i) = 0\,; \quad \text{Var}(S_n) = \text{Var}\left(\sum_{i=1}^{n} U_i\right) = \sum_{i=1}^{n} \text{Var}(U_i) = n \qquad (\because U_i\text{'s are } i.i.d.)$$

Hence, by C.L.T. $\dfrac{S_n - E(S_n)}{\sqrt{\text{Var}(S_n)}} = \dfrac{S_n}{\sqrt{n}} \xrightarrow{L} N(0, 1)$ as $n \to \infty$ ... (**)

From (*) and (**), we get

$$\lim_{n \to \infty} p_n = \lim_{n \to \infty} \left(-1 \le S_n / \sqrt{n} \le 1\right) = P(-1 \le Z \le 1), \text{ where } Z \sim N(0, 1)$$

$$= 2 \times P(0 \le Z \le 1) = 2 \times 0{\cdot}3413 = 0{\cdot}6826.$$

9·96. (*a*) If $X_1, X_2, ..., X_n$ are independent random variables, having the probability law :

$$p(x_i) = \frac{e^{-\lambda_i} \lambda_i^{x_i}}{x_i\,!}\,;\; i = 1, 2, \cdots, n\,;\; x_i = 0, 1, 2, \cdots, \infty \quad \text{and if } X = \sum_{i=1}^{n} X_i \text{ and } \lambda = \sum_{i=1}^{n} \lambda_i\,,$$

then under certain conditions to be specified, prove that :

$$P\left\{\frac{X - \lambda}{\sqrt{\lambda}} < \alpha\right\} \to \frac{1}{\sqrt{2\pi}} \int_{-\infty}^{\alpha} \exp\left(-\frac{1}{2} t^2\right) dt \quad \text{as } n \to \infty.$$

9·97. (*a*) Let $X_1, X_2, ..., X_n$ be *i.i.d.* $N(0, 1)$ variates. Show that the limiting distribution of :

$$\sqrt{n}\,(X_1 + X_2 + ... + X_n)/(X_1^2 + X_2^2 + ... + X_n^2) \sim N(0, 1) \text{ as } n \to \infty.$$

(*b*) Let $X_1, X_2, ..., X_{2n}$ be *i.i.d.* $N(0, 1)$ variates. Find the limiting distribution of $Z_n = U_n / V_n$,

where, $$U_n = \left(\frac{X_1}{X_2} + \frac{X_3}{X_4} + ... + \frac{X_{2n-1}}{X_{2n}}\right), \quad V_n = X_1^2 + X_2^2 + ... + X_n^2.$$

(*c*) Let $\{X_n\}$ be sequence of *i.i.d. r.v.*'s with mean α and variance σ^2 and let $\{Y_n\}$ be another sequence of *i.i.d. r.v.*'s with mean β ($\neq 0$) and variance σ_1^2. Find the limiting distribution of :

$$Z_n = \sqrt{n}\left(\bar{X}_n - \alpha\right) / \bar{Y}_n\,, \quad \text{where} \quad \bar{X}_n = \frac{1}{n}\sum_{i=1}^{n} X_i\,, \quad \bar{Y}_n = \frac{1}{n}\sum_{i=1}^{n} Y_i\,.$$

Hint. (*a*) Use Cramer's Theorem.

(*b*) X_i are *i.i.d.* $N(0, 1)$; $i = 1, 2, ..., 2n$: $\quad E(X_1^2) = \text{Var}\, X_i = 1$

$\Rightarrow$ (X_{2i-1}/X_{2i}) are *i.i.d.* standard Cauchy variates ; $i = 1, 2, ..., n$

$$\Rightarrow \quad \lim_{n \to \infty} \frac{U_n}{n} \xrightarrow{L} \text{Standard Cauchy variate} = C(0, 1),$$

(Being the mean of i.i.d. standard Cauchy variates)

$$\lim_{n \to \infty} \frac{V_n}{n} = \frac{X_1^2 + X_2^2 + ... + X_n^2}{n} \xrightarrow{P} E(X_1^2) = 1, \qquad \textit{(By Khinchine's WLLN)}$$

$$\therefore \quad \frac{U_n}{V_n} = \left(\frac{U_n}{n}\right) \Big/ \frac{V_n}{n} \xrightarrow{L} C(0, 1) \qquad \textit{(By Cramer's Theorem)}$$

(*c*) $$U_n = \frac{\bar{X}_n - E\left(\bar{X}_n\right)}{\sigma / \sqrt{n}} \xrightarrow{L} Z \sim N(0, 1) \qquad \textit{(By CLT)}$$

$$V_n = \bar{Y}_n \xrightarrow{P} E(Y_n) = \beta \qquad \textit{(By WLLN)}$$

By Cramer's Theorem : $\lim\limits_{n \to \infty} \dfrac{U_n}{V_n} = \dfrac{\sqrt{n}\,(\bar{X}_n - \alpha)}{\sigma \bar{Y}_n} \xrightarrow{L} \dfrac{Z}{\beta}$, where $Z \sim N(0, 1)$

$$\Rightarrow \qquad \lim_{n \to \infty} \frac{\sqrt{n}\,(\bar{X}_n - \alpha)}{\bar{Y}_n} \xrightarrow{L} \frac{\sigma}{\beta} Z \sim N\left(0, \frac{\sigma^2}{\beta^2}\right).$$

9·98. Let n numbers $X_1, X_2, ..., X_n$ in decimal form, be each approximated by the closest integer. If X_i is the ith true number and Y_i is the nearest integer, then $U_i = X_i - Y_i$ is the error made by the rounding process. Suppose that $U_1, U_2, ..., U_n$ are independent and each is uniform on $(-0·5, 0·5)$.

(*i*) What is the probability that the true sum is within 'a' units of the approximated sum ?

(*ii*) If $n = 300$ terms are added, find 'a' so that we are 95% sure that the approximation is within 'a' units of the true sum.

Hint. Required Probability $\quad 'p' = P\left[\left|\sum\limits_{i=1}^{n}(X_i - Y_i)\right| \le a\right] = P\left[-a \le \sum\limits_{i=1}^{n} U_i \le a\right]$

Now use Lindberg-Levy C.L.T. for *i.i.d. r.v.*'s $U_i \sim U[-0·5, 0·5]$ with $E(U_i) = 0$, $\text{Var}(U_i) = \dfrac{1}{12}$.

Ans. (*i*) $p = 2\,\Phi\left(a\sqrt{12/n}\right) - 1$; (*ii*) $p = 0·95 \Rightarrow \Phi\left(a\sqrt{12/300}\right) = 0·975 \Rightarrow \dfrac{a}{5} = 1·96$ or $a = 9·8$.

PROBLEMS ON COMPOUND DISTRIBUTIONS

9·99. If X is a random variable with *p.d.f.*, $\quad f(x) = \dfrac{1}{\Gamma(n+1)} e^{-x} x^n, \; x \ge 0$

where n is a positive integer, and the discrete random variable Y has a Poisson distribution with parameter λ, show that $P(X \ge \lambda) = P(Y \le n)$.

Hint. $$P(X \ge \lambda) = 1 - \frac{1}{\Gamma(n+1)} \int_0^{\lambda} e^{-x} x^n \, dx.$$

Integrating by parts successively, we get the result.

9·100. If X has Poisson distribution : $\quad P(X = r) = \dfrac{e^{-\lambda} \lambda^r}{r!}$; $r = 0, 1, 2, \ldots$

where the parameter λ is a random variable of the continuous type with the density function :

$$f(\lambda) = \frac{a^v}{\Gamma(v)} \cdot e^{-a\lambda} \lambda^{(v-1)} \; ; \; \lambda \ge 0, \; v > 0, \; a > 0, \text{ derive the distribution of X.}$$

Show that the characteristic function of X is given by :

$$\Phi_X(t) = E(e^{itX}) = q^v (1 - pe^{it})^{-v}, \text{ where } p = 1/(1+a), \; q = 1 - p.$$

9·101. The conditional probability that the random variable X should lie within the range dx for a given σ is given by : $\quad \dfrac{1}{\sigma\sqrt{2\pi}} \exp\left\{-\dfrac{1}{2}(x-\mu)^2/\sigma^2\right\} dx, \; -\infty < x < \infty$

while the probability of σ itself within the range $d\sigma$ is :

$$\frac{1}{\sigma_0^2} \exp\left\{-\frac{1}{2}\sigma^2/\sigma_0^2\right\} \sigma \, d\sigma, \; 0 < \sigma < \infty;$$

where σ_0 is a constant. Show that the unconditional (*i.e.*, marginal) distribution of X has the probability function : $\quad \dfrac{1}{2\sigma_0} \exp\left\{-\dfrac{1}{\sigma_0}|x - \mu|\right\}, \; -\infty < x < \infty.$

9·102. The probability that a radio-active substance gives off n β-particles in a unit of time is Poisson with mean λ, $(n = 0, 1, 2, ...)$. The probability that a given particle will strike a counter and be registered is p. Show that the probability of registering n β-particles $(n = 0, 1, 2, ...)$, in the unit of time is also Poisson and find its mean.

Ans. Poisson with parameter $= \lambda p$; Mean $= \lambda p$.

9·103. Let X and Y be two random variables, each having spectrum $(-\infty, \infty)$. If the conditional density of X given $Y = y$, is $\frac{1}{\sqrt{\pi}} |y| \cdot \exp(-x^2 y^2)$, and the p.d.f. of Y is . exp $\frac{1}{\sqrt{\pi}} \cdot \lambda \exp(-\lambda^2 y^2)$, prove that the density. function of X is $\frac{\lambda}{\pi(\lambda^2 + x^2)}$.

Also identify the distribution.

Let X and Y be two random variables, each having rpectrum $(-\infty, \infty)$. It the conditional density of X given Y = y is $\frac{1}{\pi}$ $|y|$. $\exp(-x^2 y^2)$ and the p.d.f. of Y is $\frac{1}{\sqrt{\pi}}$. $\lambda \exp(-\lambda^2 y^2)$, prove the at the density function of x is $\frac{\lambda}{\pi(-\lambda^2 x^2)}$

Ans. $X \sim C(\lambda, 0)$.

9·104. The conditional distribution of a continuous random variable X for a discrete random variable Y, assuming a value n is : $dF = n(1-x)^{n-1} dx,\ 0 \le x \le 1.$

The distribution of Y is : $P(Y = n) = \left(\frac{1}{2}\right)^n$; $n = 1, 2, 3, ...$ Find the marginal distribution of X.

Ans. $f_1(x) = \frac{2}{(1+x)^2}$; $0 \le x \le 1$.

9·105. The conditional distribution of the *r.v.* X for given $\lambda > 0$ is Expo (λ) with *p.d.f.*

$$f(x \mid \lambda) = \lambda e^{-\lambda x} \; ; \; x > 0, \; \lambda > 0$$

If λ is a *r.v.* with Gamma *p.d.f.*

$$g(\lambda) = \frac{a^k}{\Gamma k} e^{-a\lambda} \lambda^{k-1} \; ; \; \lambda \ge 0, \; a > 0,$$

prove that the marginal distribution of X is Pareto distribution with parameters 'a' and 'k'.

Hint. Pareto distribution with parameters 'a' and 'k' has *p.d.f.*

$$f(x) = \frac{k \cdot a^k}{(x+a)^{k+1}} \; ; \; x > 0, \; k > 0.$$

9·106. Let $X \sim U[0, 1]$ and $[Y \mid (X = x)] \sim B(n, x)$, *i.e.*

$$P(Y = y \mid X = x) = \binom{n}{y} x^y (1-x)^{n-y} \; ; \; y = 0, 1, ..., n$$

Find the distribution of Y. Also find $E(Y)$.

Ans. Y has discrete uniform distribution on $(0, 1, 2, ..., n)$, $E(Y) = (n/2)$.

9·107. Prove that : $\dfrac{1}{n!}\displaystyle\int_{\lambda}^{\infty} e^{-x} x^n \, dx = \sum_{r=0}^{n} \frac{e^{-\lambda}\lambda^r}{r!}$; $n = 0, 1, 2, \ldots$

Using the above, write down the relation connecting the distribution functions of a Poisson and a Gamma variate.

9·108. A two-dimensional random variable (X, Y) has the joint *p.d.f.*,

$$f(x, y) = \frac{1}{\Gamma(\mu)\Gamma(\nu)} x^{\mu-1} (y-x)^{\nu-1} e^{-y}$$

on the $x-y$ plane, where $0 < x < y < \infty$ and zero elsewhere. Show that the marginal distribution of X and Y are gamma distributions.

PROBLEMS ON ORDER STATISTICS

9·109. Let $X_1, X_2, \ldots, X_n$ be a random sample with common *p.d.f.* : $f(x) = \begin{cases} 1, & 0 < x < 1 \\ 0, & \text{otherwise} \end{cases}$

(*i*) Find the *p.d.f.*, mean and variance of $X_{(1)}$.

(*ii*) Find the *p.d.f.*, mean and variance of $X_{(n)}$.

(*iii*) Find Corr. $\{X_{(1)}, X_{(n)}\}$.

Ans. (*i*) $f_1(x) = n(1-x)^{n-1},\ 0 \le x \le 1;\quad E[X_{(1)}] = 1/(n+1)$

$\text{Var}[X_{(1)}] = n/[(n+2)(n+1)^2]$

(*ii*) $f_n(x) = n x^{n-1};\ 0 \le x \le 1;\quad E[X_{(n)}] = n/(n+1)$

$\text{Var}[X_{(n)}] = n/[(n+2)(n+1)^2]$

(*iii*) **Hint.** $E[X_{(1)} \cdot X_{(n)}] = \displaystyle\int_0^y \int_0^1 xy\, f_{1n}(x, y)\, dx\, dy = n(n-1) \int_0^y \int_0^1 xy\,(y-x)^{n-2}\, dx\, dy$

$$\left[\because f_{1n}(x, y) = n(n-1) f(x) [F(y) - F(x)]^{n-2} f(y);\ 0 \le x < y \le 1\right]$$

$$\therefore \quad E[X_{(1)} \cdot X_{(n)}] = n(n-1) \int_0^y \int_0^1 x y^{n-1} \left(1 - \frac{x}{y}\right)^{n-2} dx\, dy$$

$$= n(n-1) \int_0^1 \int_0^1 y^{n+1}\, t(1-t)^{n-2}\, dt\, dy\ ; \left(\frac{x}{y} = t\right)$$

$$= 1/(n+2) \qquad \text{[On simplification]}$$

$\text{Cov}[X_{(1)}, X_{(n)}] = 1/\{(n+1)^2 \cdot (n+2)\}$ and $\text{Corr}[X_{(1)}, X_{(n)}] = 1/n.$

9·110. Let X_i $(i = 1, 2, \ldots, n)$, be *i.i.d.* non-negative *r.v.*'s of continuous type.
If $M_n = X_{(n)} = \max(X_1, X_2, \ldots, X_n)$ and $E(|X|) < \infty$, then prove that

$$E(M_n) = E(M_{n-1}) + \int_0^{\infty} F^{n-1}(x) [1 - F(x)]\, dx.$$

Hence find $E(M_n)$ if X_i's are *i.i.d.* exponential variates with parameter λ.

Ans. $E(M_n) = \dfrac{1}{\lambda}\left(1 + \dfrac{1}{2} + \dfrac{1}{3} + \ldots + \dfrac{1}{n}\right).$

9·111. Let $X_1, X_2, ..., X_n$ be a random sample of size n from a population with *p.d.f.*

$$f(x) = \begin{cases} 1, & 0 < x < 1 \\ 0, & \text{otherwise} \end{cases}$$

Show that $Y_1 = X_{(1)}/X_{(2)}$, $Y_2 = X_{(2)}/X_{(3)}$, ..., $Y_{n-1} = X_{(n-1)}/X_{(n)}$ and $Y_n = X_{(n)}$ are independently distributed and identify their distributions.

9·112. A random sample of size n is drawn from an exponential population :

$$p(x) = \frac{1}{\theta} e^{-x/\theta} \; ; \; \theta > 0, x \geq 0.$$

(*i*) Obtain the *p.d.f.* of $X_{(r)}$.

(*ii*) Show that $X_{(r)}$ and $W_{rs} = X_{(s)} - X_{(r)}$, $r < s$ are independently distributed.

(*iii*) Identify the distribution of $W_1 = X_{(r+1)} - X_{(r)}$.

9·113. Let $X_1, X_2, ..., X_n$ be a random sample from continuous population with *p.d.f.*

$$f(x) = \begin{cases} \beta e^{-x\beta} \; ; & x \geq 0, \beta > 0 \\ 0, & \text{otherwise} \end{cases}$$

(*i*) Show that $X_{(r)}$ and $X_{(s)} - X_{(r)}$ are independent for any $s > r$.

(*ii*) Find the *p.d.f.* of $X_{(r+1)} - X_{(r)}$.

(*iii*) Let $Z_1 = n X_{(1)}$, $Z_2 = (n-1)(X_{(2)} - X_{(1)})$, $Z_3 = (n-2)(X_{(3)} - X_{(2)})$, ..., $Z_n = (X_{(n)} - X_{(n-1)})$ ... (*)
Show that $(Z_1, Z_2, ..., Z_n)$ and $(X_1, X_2, ..., X_n)$ are identically distributed.

Hint. (*iii*) The joint *p.d.f.* of $X_{(1)}, X_{(2)}, ..., X_{(n)}$ is :

$$f(x_1, x_2, ..., x_n) = n! \, f(x_1) f(x_2) \ldots f(x_n) = n! \, \beta^n \cdot e^{-\beta x_1} \ldots e^{-\beta x_n} \quad \ldots (**)$$

Transformation (*) in the question gives :

$$X_{(1)} = \frac{Z_1}{n} \; ; \; X_{(2)} = \frac{Z_1}{n} + \frac{Z_2}{n-1} \; ; \; X_{(3)} = \frac{Z_1}{n} + \frac{Z_2}{n-1} + \frac{Z_3}{n-2} \; ; \ldots , \; X_{(n)} = \frac{Z_1}{n} + \frac{Z_2}{n-1} + \ldots + \frac{Z_{n-1}}{2} + \frac{Z_n}{1} \quad \ldots (***)$$

$$J = \frac{\partial(X_{(1)}, X_{(2)}, ..., X_{(n)})}{\partial(Z_1, Z_2, ..., Z_n)} = \frac{1}{n!}$$

$$0 < X_{(1)} < X_{(2)} < \ldots < X_{(n)} < \infty \quad \Rightarrow \quad 0 < Z_i < \infty \; ; \; i = 1, 2, ..., n.$$

Using (***) and $|J|$, (**) transforms to :

$$g(z_1, z_2, ..., z_n) = (\beta e^{-\beta z_1})(\beta e^{-\beta z_2}) \ldots (\beta e^{-\beta z_n}) \; ; \; 0 < z_i < \infty$$

$\Rightarrow$ $Z_1, Z_2, ..., Z_n$ are *i.i.d.* exponential variates with parameter β.

9·114. Let $X_1, X_2, ..., X_n$ be n independent variates, X_i having a geometric distribution with parameter p_i, *i.e.*,

$$P(X_i = x_i) = q_i^{x_i - 1} \cdot p_i \; ; \; q_i = 1 - p_i, \; x_i = 1, 2, 3, \ldots$$

Show that $X_{(1)}$ is distributed geometrically with parameter $(1 - q_1 q_2 \ldots q_n)$.

9·115. For a random sample of size n from a continuous population whose *p.d.f.* $p(x)$ is symmetrical at $x = \mu$, show that $f_r(\mu + x) = f_{n-r+1}(\mu - x)$, where $f_r(\cdot)$ is the *p.d.f.* of $X_{(r)}$.

Hint.

$$f(\mu + x) = f(\mu - x)$$

$$F(\mu + x) = P[X \leq \mu + x] = P[X \geq \mu - x] \qquad \text{(By symmetry)}$$

$$= 1 - P[X \leq \mu - x] = 1 - F(\mu - x).$$

9·116. Let $X_1, X_2, ..., X_n$ be a random sample of size n from a population having continuous distribution function $F(x)$. Define the order statistic of rank k, $1 \le k \le n$. Find its distribution function. Show that for the rectangular distribution : $f(x) = \dfrac{1}{\theta_2}$, $\theta_1 - \dfrac{1}{2}\theta_2 \le x \le \theta_1 + \dfrac{1}{2}\theta_2$,

$$E\left(\frac{X_{(r)} - \theta_1}{\theta_2}\right) = \frac{r}{n+1} - \frac{1}{2}.$$

9·117. Show that the *c.d.f.* of the mid-point (or mid-range) $M = \frac{1}{2}\left[X_{(1)} + X_{(n)}\right]$, in random sample of size n from a continuous population with *c.d.f.* $F(x)$ is :

$$F(m) = P(M \le m) = n \int_{-\infty}^{m} \left[F(2m - x) - F(x)\right]^{n-1} \cdot f(x)\, dx$$

9·118. Show by means of an example that there may exist *a r.v.* X for which $E(X)$ does not exist but $E(X_{(r)})$ exists for some r ;

Hint. Let $X_1, X_2, ..., X_n$ be a random sample from a population with *p.d.f.*

$$f(x) = \frac{1}{x^2}, \; 1 < x < \infty ; \quad F(x) = 1 - \frac{1}{x}$$

$E(X)$ does not exist but $E[X_{(r)}]$ exists for any $r < n$.

9·119. Let $X_1, X_2, ..., X_{2n+1}$ be an odd-size random sample from a $N(\mu, \sigma^2)$ population. Find *p.d.f.* of the sample median and show that it is symmetric about μ, and hence has the mean μ.

9·120. (*a*) Let $X_1, X_2, ..., X_n$ be a random sample of size n from a uniform population with *p.d.f.*

$$f(x) = \begin{cases} 1, & \text{if } 0 \le x \le 1 \\ 0, & \text{otherwise} \end{cases}$$

Show that : (*i*) $X_{(r)}$ is a $\beta_1(r, n - r + 1)$ variate.

(*ii*) $W_{rs} = X_{(s)} - X_{(r)}$, also has a Beta distribution which depends only on $s - r$ and not on s and r individually.

(*b*) In a random sample of size n from uniform $U[0, 1]$ population, obtain the *p.d.f.* of $W_{rs} = X_{(s)} - X_{(r)}$ and identify its distribution.

Ans. (*a*) (*ii*) $f(w_{rs}) = \dfrac{1}{B(s - r, n - s + r + 1)} \cdot w_{rs}^{s-r-1}(1 - w_{rs})^{n-s+r}$; $0 \le w_{rs} \le 1$

9·121. Let $X_1, X_2, ..., X_n$ be *i.i.d.* with *p.d.f.* : $f(x) = \begin{cases} \dfrac{1}{\sigma}\exp\left\{-\left(\dfrac{x-\theta}{\sigma}\right)\right\}; & x > \theta, \sigma > 0 \\ 0, & \text{otherwise} \end{cases}$

Show that $X_{(1)}$, $X_{(2)} - X_{(1)}$, $X_{(3)} - X_{(2)}$, ..., $X_{(n)} - X_{(n-1)}$ are independent.

Hint. Let $Z_1 = X_{(1)}$; $Z_2 = X_{(2)} - X_{(1)}$, ..., $Z_n = X_{(n)} - X_{(n-1)}$

$$\Rightarrow X_{(1)} = Z_1, \; X_{(2)} = Z_1 + Z_2, \; ..., \; X_{(n)} = Z_1 + Z_2 + ... + Z_n \quad \therefore \; J = \frac{\partial(X_{(1)}, X_{(2)}, ..., X_{(n)})}{\partial(Z_1, Z_2, ..., Z_n)} = 1$$

$$g(z_1, z_2, ..., z_n) = n! \prod_{i=1}^{n} f(x_i)\,|J| = \left(\frac{n}{\sigma}e^{-n(z_1-\theta)/\sigma}\right)\left(\frac{(n-1)}{\sigma}e^{-(n-1)z_2/\sigma}\right)...\left(\frac{2}{\sigma}e^{-2z_{n-1}/\sigma}\right)\left(\frac{1}{\sigma}e^{-z_n/\sigma}\right)$$

$\Rightarrow$ $Z_1, Z_2, ..., Z_n$ are independently distributed.

9·122. For the exponential distribution $f(x) = e^{-x}$, $x \ge 0$, find the *p.d.f.* of the range W in a random sample of size n and show that $E(W) = 1 + \dfrac{1}{2} + \dfrac{1}{3} + ... + \dfrac{1}{n-1}$.

Ans. $g(w) = (n-1)e^{-w}(1 - e^{-w})^{n-2}$; $w \ge 0$.

9·123. Let $X_1, X_2, ..., X_n$ be a random sample of size n from $U[a, b]$ population. Obtain the *p.d.f.*'s of

(*i*) $X_{(1)}$, (*ii*) $X_{(n)}$, and (*iii*) joint *p.d.f.* of $X_{(1)}$ and $X_{(n)}$.

9·124. Let X_1, X_2 be *i.i.d. r.v.*'s with *p.m.f.* $P(X_i = x) = \dfrac{e^{-\lambda}\lambda^x}{x!}$, $x = 0, 1, 2, ...; i = 1, 2,$ where $\lambda > 0$.

Let $M = \max(X_1, X_2)$ and $N = \min(X_1, X_2)$.

Find the marginal *p.m.f.*'s of M and N.

9·125. The numbers $X_1, X_2, ..., X_n$ are independently chosen at random in the interval (a, b). Prove that the probability density function of the *r.v.* $X = \min(X_1, X_2, ..., X_n)$ is given by :

$$f(x) = \frac{n(b-x)^{n-1}}{(b-a)^n},\ a < x < b.$$

9·126. Let $X_{(n)}^{(1)}, X_{(n)}^{(2)}, ..., X_{(n)}^{(k)}$ be the maxima in k independent samples each of size n drawn from $U[0, 1]$ population. Show that the *p.d.f.* of the product $V = X_{(n)}^{(1)} \cdot X_{(n)}^{(2)} ... X_{(n)}^{(k)}$, is given by :

$$g(v) = \frac{n^k}{\Gamma(k)} v^{n-1} (-\log v)^{k-1},\ 0 \le v \le 1.$$

PROBLEMS ON TRUNCATED DISTRIBUTIONS

9·127. (*a*) Let X be normally distributed with mean μ and variance σ^2. Truncate the density of X on the left at a and on the right at b, and then calculate the mean of the truncated distribution. [Note that if $a = \mu - c$ and $b = \mu + c$, then the mean of the truncated distribution should equal μ.]

(*b*) If X is normally distributed with mean μ and variance σ^2, find the mean of the conditional distribution of X given $a \le X \le b$.

Hint. In fact Problem in Part (*a*) is same as in Part (*b*), stated differently.

$$f(x) = \frac{1}{\sqrt{2\pi}\,\sigma} \exp\left[-\frac{1}{2}\left(\frac{x-\mu}{\sigma}\right)^2\right];\ -\infty < x < \infty$$

Mean of truncated distribution is :

$$\mu' = \frac{\int_a^b x f(x)\,dx}{\int_a^b f(x)\,dx} = \frac{\int_a^b (x-\mu+\mu) f(x)\,dx}{\int_a^b f(x)\,dx} = \mu + \frac{\int_a^b (x-\mu) f(x)\,dx}{\int_a^b f(x)\,dx} = \mu + \sigma^2 \left[\frac{f(a) - f(b)}{F(b) - F(a)}\right]$$

where $F(x) = P(X \le x)$, is the distribution function of X.

9·128. A truncated Poisson distribution is given by the mass function :

$$f(x) = \frac{1}{1-e^{-\lambda}} \cdot \frac{e^{-\lambda}\lambda^x}{x!};\ x = 1, 2, 3, ...;\ \lambda > 0$$

Find the *m.g.f.* and hence mean and variance of the distribution.

9·129. Consider the *p.d.f.* : $g(x) = \dfrac{f(x)}{1 - F(x_0)},\ x > x_0$ ... (*)

where $f(x) = \left(\sqrt{2\pi}\cdot\sigma\right)^{-1} \cdot \exp\{-(x-\mu)^2/2\sigma^2\},\ -\infty < \mu < \infty,\ \sigma > 0$

and $F(x_0) = \int_{-\infty}^{x_0} f(u)\,du.$

[(*) is the *p.d.f.* of $N(\mu, \sigma^2)$ distribution, truncated at the point $x = x_0$.]

Show that the first two raw moments can be expressed as :

$$\mu_1' = \mu + \lambda\sigma\ ;\ \mu_2' = \mu^2 + \lambda\sigma(x_0 + \mu) + \sigma^2, \text{ where } \lambda[1 - F(x_0)] = f[(x_0 - \mu)/\sigma].$$

9·130. Let $X \sim \gamma(\alpha)$. Obtain the *p.d.f.* of the truncated distribution, truncated at the point x_0 and prove that $\mu_r' = E(X^r)$ for the truncated gamma distribution is given by :

$$\mu_r' = \left[E(X^r) \text{ for the untruncated } \gamma(\alpha) \text{ distribution}\right] \times \left[\{1 - I_{x_0}(\alpha + \gamma)\}/\{1 - I_{x_0}(\alpha)\}\right]$$

where $$I_{x_0}(\alpha) = \int_0^{x_0} \frac{e^{-x}\, x^{\alpha-1}}{\Gamma(\alpha)}\, dx \qquad \text{(Incomplete Gamma Integral)}$$

PROBLEMS ON RALEIGH DISTRIBUTION

9·131. The *p.d.f.* of a *r.v.* X with parameter α is given by :

$$f(x) = \begin{cases} 2\alpha x e^{-\alpha x^2}\ ;\ x > 0\ ;\ \alpha > 0 \\ 0, \text{ otherwise} \end{cases}$$

Show that for this distribution :

$$(a)\ \mu_x = \frac{1}{2} \cdot \sqrt{\frac{\pi}{\alpha}} \quad \text{and} \quad (b)\ \sigma_x^2 = \frac{1}{\alpha}\left(1 - \frac{\pi}{4}\right).$$

[Note that, this is usual Raleigh's distribution with parameter $\alpha = (1/2\sigma^2)$.]

9·132. If X and Y are independent $N(0, \sigma^2)$ variates, obtain the joint *p.d.f.* of $U = \sqrt{X^2 + Y^2}$ and $V = (X/Y)$. Are U and V independent ? Also identify their distributions.

Ans. Yes, U and V are independent. $U \sim$ Raleigh (σ) and $V \sim$ Cauchy $C(1, 0)$.

CHAPTER

10 Correlation

LEARNING OBJECTIVES

Upon completion of this Chapter, you should be able to:

- Understand the meaning of the term correlation between two variables and significance of its study.
- Compute and interpret Karl Pearson's correlation coefficient r.
- Derive various important properties of correlation coefficient.
- Understand the meaning and the significance of probable error of correlation coefficient and interpret its value.
- Distinguish between Karl Pearson's correlation coefficient and Spearman's rank correlation coefficient.

CHAPTER OUTLINES

10·1. INTRODUCTION

So far we have confined ourselves to univariate distributions, *i.e.*, the distributions involving only one variable. Often we come across situations in which our focus is simultaneously on two or more variables and invariably, we observe that movements in one variable are accompanied by movements in other variable. For example, scores on an I.Q. test move with scores in university examination. Similarly, studies in income and expenditure on households or price and demand of commodities, exhibit accompaning movements of two variables. Notwithstanding the cases of spurious or nonsensical relations which we may enjoy through such funny combinations of variables as suggesting that the number of runs scored by a batsman increases with an increase in the consumption of fertilizer in the local market or the number of flights space is increasing with a decrease in the population of tigers, the study of variables indicating accompaning behaviour is of great interest in Statistics.

However, the above statements are not precise enough to be of use to decision makers. We are therefore on the look out for a *quantitative measure* of the relationship between the two variables, and also for an appropriate mathematical or statistical form of the relationship. While the second questions will be discussed in the next chapter, we will take up the first question in the present chapter.

10·2. MEANING OF CORRELATION

In a bivariate distribution we may be interested to find out if there is any correlation or covariation between the two variables under study. If the change in one variable affects a change in the other variable, the variables are said to be correlated. If the two variables deviate in the same direction, *i.e.*, if the increase (or decrease) in one results in a corresponding increase (or decrease) in the other, correlation is said to be *direct* or *positive*. But if they constantly deviate in the opposite directions, *i.e.*, if increase (or decrease) in one results in corresponding decrease (or increase) in the other, correlation is said to be *diverse* or *negative*. For example, the correlation between (*i*) the heights and weights of a group of persons, and (*ii*) the income and expenditure; is positive and the correlation between (*i*) price and demand of a commodity and (*ii*) the volume and pressure of a perfect gas; is negative. Correlation is said to be *perfect* if the deviation in one variable is followed by a corresponding and proportional deviation in the other.

10·3. SCATTER DIAGRAM

It is the simplest way of the diagrammatic representation of bivariate data. Thus, for the bivariate distribution (x_i, y_i); $i = 1, 2, \ldots, n$, if the values of the variables X and Y are plotted along the x-axis and y-axis respectively in the x-y plane, the diagram of dots so obtained is known as *scatter diagram*. From the scatter diagram, we can form a fairly good, though vague, idea whether the variables are correlated or not, *e.g.*, if the points are very dense, *i.e.*, very close to each other, we should expect a fairly good amount of correlation between the variables and if the points are widely scattered, a poor correlation is expected. This method, however, is not suitable if the number of observations is fairly large.

10·4. KARL PEARSON'S COEFFICIENT OF CORRELATION

As a measure of intensity or degree of linear relationship between two variables, Karl Pearson (1867-1936), *a British Biometrician,* developed a formula called *Correlation Coefficient.*

Correlation coefficient between two random variables X and Y, usually denoted by $r(X, Y)$ or simply r_{XY}, is a numerical measure of *linear relationship* between them and is defined as :

$$r(X, Y) = \frac{\text{Cov}(X, Y)}{\sigma_X \sigma_Y} \qquad \text{... (10·1)}$$

If (x_i, y_i); $i = 1, 2, \ldots, n$ is the bivariate distribution, then

$$\text{Cov}(X, Y) = E\left[\{X - E(X)\} \cdot \{Y - E(Y)\}\right]$$

$$= \frac{1}{n} \sum (x_i - \bar{x})(y_i - \bar{y}) = \mu_{11}$$

$$\sigma_X^2 = E[X - E(X)]^2 = \frac{1}{n} \sum (x_i - \bar{x})^2$$

$$\sigma_Y^2 = E[Y - E(Y)]^2 = \frac{1}{n} \sum (y_i - \bar{y})^2$$

the summation extending over i from 1 to n.

Another convenient form of the formula (10·2) for computational work is as follows:

$$\text{Cov}(X, Y) = \frac{1}{n} \sum (x_i - \bar{x})(y_i - \bar{y}) = \frac{1}{n} \sum (x_i y_i - x_i \bar{y} - \bar{x} y_i + \bar{x}\,\bar{y})$$

$$= \frac{1}{n} \sum x_i y_i - \bar{y} \frac{1}{n} \sum x_i - \bar{x} \frac{1}{n} \sum y_i + \bar{x}\,\bar{y}$$

$$\therefore \quad \text{Cov}(X, Y) = \frac{1}{n} \sum x_i y_i - \bar{x}\,\bar{y}, \quad \sigma_X^2 = \frac{1}{n} \sum x_i^2 - \bar{x}^2 \text{ and } \sigma_Y^2 = \frac{1}{n} \sum y_i^2 - \bar{y}^2 \qquad \text{...(10·2}a\text{)}$$

Remarks:

1. Following are the figures of the standard data for $r > 0$, < 0, $= 0$, and $r = \pm 1$:

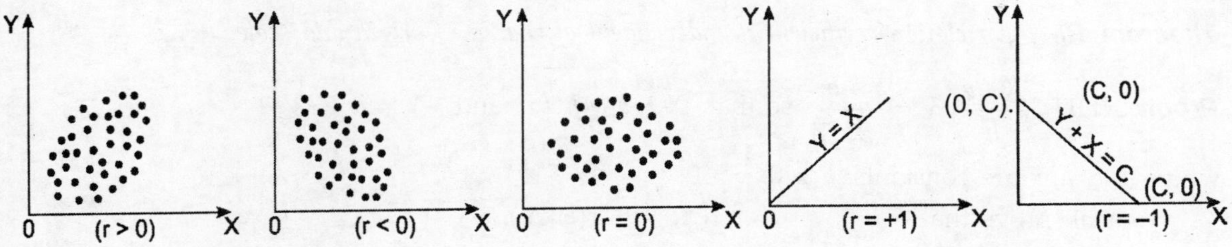

2. It may be noted that $r(X, Y)$ provides a measure of *linear relationship* between X and Y. For non-linear relationship, however, it is not very suitable.

3. Sometimes, it may be convenient to write : $\text{Cov}(X, Y) = \sigma_{XY}$.

4. Karl Pearson's correlation coefficient is also called *product-moment correlation coefficient*, since

$$\text{Cov}(X, Y) = E\left[\{X - E(X)\}\,\{Y - E(Y)\}\right] = \mu_{11}.$$

10·4·1. Limits for Correlation Coefficient

We have

$$r(X, Y) = \frac{\text{Cov}(X, Y)}{\sigma_X \sigma_Y} = \frac{\frac{1}{n} \sum (x_i - \bar{x})(y_i - \bar{y})}{\left[\frac{1}{n} \sum (x_i - \bar{x})^2 \cdot \frac{1}{n} \sum (y_i - \bar{y})^2\right]^{1/2}},$$

$$\therefore \qquad r^2(X, Y) = \frac{\left(\sum_i a_i b_i\right)^2}{\left(\sum_i a_i^2\right)\left(\sum_i b_i^2\right)}, \qquad \text{where } \begin{pmatrix} a_i = x_i - \bar{x} \\ b_i = y_i - \bar{y} \end{pmatrix} \qquad \text{...(*)}$$

We have the Schwartz inequality which states that if a_i, b_i; $i = 1, 2, \ldots, n$ are real quantities then

$$\left(\sum_{i=1}^{n} a_i b_i\right)^2 \le \left(\sum_{i=1}^{n} a_i^2\right)\left(\sum_{i=1}^{n} b_i^2\right),$$

the sign of equality holding if and only if $\quad \dfrac{a_1}{b_1} = \dfrac{a_2}{b_2} = \cdots = \dfrac{a_n}{b_n}$

Using Schwartz inequality, we get from (*) :

$$r^2(X, Y) \le 1 \text{ i.e., } |r(X, Y)| \le 1 \quad \Rightarrow \quad -1 \le r(X, Y) \le 1$$

Hence, correlation coefficient cannot exceed unity numerically. It always lies between – 1 & +1. *If r = +1, the correlation is perfect and positive and if r = – 1, correlation is perfect and negative.*

Aliter. If we write $E(X) = \mu_X$ and $E(Y) = \mu_Y$, $\quad$ then $E\left[\left(\dfrac{X - \mu_X}{\sigma_X}\right) \pm \left(\dfrac{Y - \mu_Y}{\sigma_Y}\right)\right]^2 \ge 0$

$$\Rightarrow \quad E\left(\frac{X - \mu_X}{\sigma_X}\right)^2 + E\left(\frac{Y - \mu_Y}{\sigma_Y}\right)^2 \pm 2\frac{E[(X - \mu_X)(Y - \mu_Y)]}{\sigma_X \sigma_Y} \ge 0 \quad \Rightarrow \quad 1 + 1 \pm 2r(X, Y) \ge 0$$

$$\therefore \qquad -1 \le r(X, Y) \le 1.$$

Theorem 10·1. *Correlation coefficient is independent of change of origin and scale.*

Proof. Let $U = \dfrac{X - a}{h}$, $V = \dfrac{Y - b}{k}$, so that $X = a + hU$ and $Y = b + kV$,

where a, b, h, k are constants; $h > 0, k > 0$.

We shall prove that $\quad r(X, Y) = r(U, V)$.

Since $\quad X = a + hU \quad$ and $\quad Y = b + kV$, on taking expectations, we get

$$E(X) = a + h\,E(U) \quad \text{and} \quad E(Y) = b + k\,E(V)$$

$$\therefore \quad X - E(X) = h[U - E(U)] \text{ and } Y - E(Y) = k[V - E(V)]$$

$$\Rightarrow \text{Cov}(X, Y) = E[\{X - E(X)\}\{Y - E(Y)\}] = E[h\{U - E(U)\}[k\{V - E(V)\}]$$

$$= hk\,E[\{U - E(U)\}\{V - E(V)\}] = h\,k\,\text{Cov}(U, V) \qquad \text{...(10·4)}$$

$$\sigma_X^2 = E[\{X - E(X)\}^2] = E[h^2\{U - E(U)\}^2] = h^2\sigma_U^2$$

$$\Rightarrow \qquad \sigma_X = h\,\sigma_U, \; (h > 0) \qquad \text{...(10·4a)}$$

$$\text{and} \qquad \sigma_Y^2 = E[\{Y - E(Y)\}^2] = E[k^2\{V - E(V)\}^2] = k^2\sigma_V^2$$

$$\Rightarrow \qquad \sigma_Y = k\cdot\sigma_V, \; (k > 0) \qquad \text{...(10·4b)}$$

Substituting from (10·4), (10·4*a*) and (10·4*b*) *in* (10·1), we get

$$r(X, Y) = \frac{Cov(X, Y)}{\sigma_X \sigma_Y} = \frac{hk\,Cov(U, V)}{hk\,\sigma_U \sigma_V} = \frac{Cov(U, V)}{\sigma_U \sigma_V} = r(U, V)$$

Remark. This theorem is of fundamental importance in the numerical computation of the correlation coefficient.

Corollary. *If X and Y are random variables and a, b, c d are any numbers provided only that* $a \neq 0$, $c \neq 0$, *then*

$$r(aX + b, cY + d) = \frac{ac}{|ac|} r(X, Y). \quad ...(10·4c)$$

Proof. With usual notations, we have

$$\text{Var}(aX + b) = a^2\sigma_X^2 ; \quad \text{Var}(cY + d) = c^2\sigma_Y^2 ; \text{ and}$$

$$\text{Cov}(aX + b, cY + d) = a\, c\, \sigma_{XY}$$

$$\therefore \quad r(aX + b, cY + d) = \frac{\text{Cov}(aX + b, cY + d)}{[Var(aX + b)\, Var(cY + d)]^{1/2}} = \frac{ac\,\sigma_{XY}}{|a||c|\,\sigma_X\sigma_Y} = \frac{ac}{|ac|} \cdot r(X, Y)$$

If $ac > 0$, *i.e.*, if a and c are of same signs, then $ac/|ac| = +1$.
If $ac < 0$, *i.e.*, if a and c are of opposite signs, then $ac/|ac| = -1$.

Remark. In particular, if we take $b = 0 = d$, then from (10·4c), we get

$$r(aX, cY) = \frac{ac}{|ac|} \cdot r(X, Y). \quad ...(10·4d)$$

Theorem 10·2. *Two independent variables are uncorrelated.*

Proof. If X and Y are independent variables, then

$$\text{Cov}(X, Y) = 0 \quad \Rightarrow \quad r(X, Y) = \frac{\text{Cov}(X, Y)}{\sigma_X\sigma_Y} = 0$$

Hence, two independent variables are uncorrelated.

But the *converse* of the theorem is not true, *i.e.*, two uncorrelated variables may not be independent as the following example illustrates :

							Total
X	– 3	– 2	– 1	1	2	3	ΣX = 0
Y	9	4	1	1	4	9	ΣY = 28
XY	– 27	– 8	– 1	1	8	27	ΣXY = 0

$$\bar{X} = \frac{1}{n}\sum X = 0 \qquad \text{Cov}(X, Y) = \frac{1}{n}\sum XY - \bar{X}\bar{Y} = 0$$

$$\therefore \quad r(X, Y) = \frac{Cov(X, Y)}{\sigma_X\sigma_Y} = 0$$

Thus, in the above example, the variables X and Y are uncorrelated. But on careful examinaion we find that X & Y are not independent but they are connected by relation $Y = X^2$. Hence, two uncorrelated variables need not necessarily be independent. A simple reasoning for this strange conclusion is that $r(X, Y) = 0$, *merely implies the absence of any linear relationship between the variables X and Y. There may, however, exist some other form of relationship between them, e.g., quadratic, cubic or trigonometric.*

***Remarks*:**

1. Following are some more examples where two variables are uncorrelated but *not* independent.

(*i*) $X \sim N(0, 1)$ and $Y = X^2$. Since $X \sim N(0, 1)$, $E(X) = 0 = E(X^3)$

$$\therefore \quad \text{Cov}(X, Y) = E(XY) - E(X)\,E(Y) = E(X^3) - E(X)\,E(Y) = 0 \qquad (\because Y = X^2)$$

$$\Rightarrow \quad r(X, Y) = \frac{\text{Cov}(X, Y)}{\sigma_X \sigma_Y} = 0.$$

Hence, X and Y are uncorrelated but not independent, ($\because Y = X^2$).

(*ii*) Let X be a *r.v.* with *p.d.f.*, $f(x) = \frac{1}{2}, -1 \le x \le 1$, and let $Y = X^2$

Here we shall get

$$E(X) = 0 \text{ and } E(XY) = E(X^3) = 0, \text{ so that } r(X, Y) = 0 \qquad \text{[See Remark 1]}$$

2. However, the converse of the theorem holds in the following cases :

(*a*) If X and Y are jointly normally distributed with $\rho = \rho(X, Y) = 0$, then they are independent. If $\rho = 0$, then [*c.f. Bivariate Normal Distribution, Chapter 20*]

$$f(x, y) = \frac{1}{\sigma_X\sqrt{2\pi}} \exp\left[-\frac{1}{2}\left(\frac{X-\mu_X}{\sigma_X}\right)^2\right] \times \frac{1}{\sigma_Y\sqrt{2\pi}} \exp\left[-\frac{1}{2}\left(\frac{Y-\mu_Y}{\sigma_Y}\right)^2\right] = f_1(x)\,f_2(y)$$

Hence, X and Y are independent.

(*b*) *If each of the two variables X and Y takes two values, 0, 1 with positive probabilities, then*

$$r(X, Y) = 0 \quad \Rightarrow \quad X \text{ and } Y \text{ are independent.}$$

Proof. Let X take the values 1 and 0 with positive probabilities p_1 and q_1 respectively and let Y take the values 1 and 0 with positive probabilities p_2 and q_2 respectively. Then

$$r(X, Y) = 0 \Rightarrow \text{Cov}(X, Y) = 0$$

$$\Rightarrow \quad 0 = E(X\,Y) - E(X)\,E(Y) = 1 \cdot P(X = 1 \cap Y = 1) - [1 \cdot P(X = 1) \times 1 \cdot P(Y = 1)]$$

$$= P(X = 1 \cap Y = 1) - p_1 p_2$$

$$\Rightarrow \quad P(X = 1 \cap Y = 1) = p_1 p_2 = P(X = 1) \cdot P(Y = 1).$$ Hence, X and Y are independent.

10·4·2. Assumptions Underlying Karl Pearson's Correlation Coefficient.

Pearsonian correlation coefficient r is based on the following assumptions:

(*i*) *The variables X and Y under study are linearly related.* In other words, the scatter diagram of the data will give a straight line curve.

(*ii*) *Each of the variables (series) is being affected by a large number of independent contributory causes of such a nature as to produce normal distribution.* For example, the variables (series) relating to ages, heights, weights, supply, price, etc. conform to this assumption.

In the words of Karl Pearson :

"The sizes of the complex of organs (something measurable) are determined by a great variety of independent contributory causes, for example, climate, nourishment, physical training and innumerable other causes which cannot be individually observed or their effects measured." Karl Pearson further observes, *"The variations in intensity of the contributory causes are small as compared with their absolute intensity and these variations follow the normal law of distribution."*

(iii) The forces so operating on each of the variable series are not independent of each other but are related in a causal fashion. In other words, cause and effect relationship exists between different forces operating on the items of the two variable series. These forces must be common to both the series. If the operating forces are entirely independent of each other and not related in any fashion, then there cannot be any correlation between the variables under study.

For example, the correlation coefficient between,

(*a*) the series of heights and incomes of individuals over a period of time,

(*b*) the series of marriage rate and the rate of agricultural production in a country over a period of time, and

(*c*) the series relating to the size of the shoe and intelligence of a group of individuals, should be zero, since the forces affecting the two variable series in each of the above cases are entirely independent of each other.

However, if in any of the above cases the value of r for a given set of data is not zero, then such correlation is termed as *chance correlation or spurious or nonsense correlation.*

Example 10·1. *Calculate the coefficient of correlation between X and Y from the following data :*

X :	65	66	67	67	68	69	70	72
Y :	67	68	65	68	72	72	69	71

Solution.

CALCULATIONS FOR CORRELATION COEFFICIENT

X	Y	X^2	Y^2	XY
65	67	4225	4489	4355
66	68	4356	4624	4488
67	65	4489	4225	4355
67	68	4489	4624	4556
68	72	4624	5184	4896
69	72	4761	5184	4968
70	69	4900	4761	4830
72	71	5184	5041	5112
Total 544	552	37028	38132	37560

$$\bar{X} = \frac{1}{n}\sum X = \frac{544}{8} = 68, \qquad \bar{Y} = \frac{1}{n}\sum Y = \frac{1}{8} \times 552 = 69$$

$$r(X, Y) = \frac{\text{Cov}(X, Y)}{\sigma_X \sigma_Y} = \frac{\frac{1}{n}\sum XY - \bar{X}\bar{Y}}{\sqrt{\left(\frac{1}{n}\sum X^2 - \bar{X}^2\right)\left(\frac{1}{n}\sum Y^2 - \bar{Y}^2\right)}}$$

$$= \frac{\frac{1}{8} \times 37560 - 68 \times 69}{\sqrt{\left\{\frac{37028}{8} - (68)^2\right\}\left\{\frac{38132}{8} - (69)^2\right\}}}$$

$$= \frac{4695 - 4692}{\sqrt{(4628{\cdot}5 - 4624)\,(4766{\cdot}5 - 4761)}} = \frac{3}{\sqrt{4{\cdot}5 \times 5{\cdot}5}} = 0{\cdot}603.$$

Aliter. (SHORT-CUT METHOD)

X	Y	U = X − 68	V = Y − 69	U^2	V^2	UV
65	67	− 3	− 2	9	4	6
66	68	− 2	− 1	4	1	2
67	65	− 1	− 4	1	16	4
67	68	− 1	− 1	1	1	1
68	72	0	3	0	9	0
69	72	1	3	1	9	3
70	69	2	0	4	0	0
72	71	4	2	16	4	8
Total		0	0	36	44	24

$$\bar{U} = \frac{1}{n}\sum U = 0, \quad \bar{V} = \frac{1}{n}\sum V = 0, \quad \text{Cov}(U, V) = \frac{1}{n}\sum UV - \bar{U}\bar{V} = \frac{1}{8} \times 24 = 3$$

$$\sigma_U^2 = \frac{1}{n}\sum U^2 - (\bar{U})^2 = \frac{1}{8} \times 36 = 4{\cdot}5, \qquad \sigma_V^2 = \frac{1}{n}\sum V^2 - (\bar{V})^2 = \frac{1}{8} \times 44 = 5{\cdot}5$$

$$\therefore \qquad r(U, V) = \frac{\text{Cov}(U, V)}{\sigma_U \sigma_V} = \frac{3}{\sqrt{4{\cdot}5 \times 5{\cdot}5}} = 0{\cdot}603 = r(X, Y)$$

Remark. The reader is advised to calculate the correlation coefficient by arbitrary origin method rather than by the direct method, since the later leads to much simpler arithmetical calculations.

Example 10·2. *A computer while calculating correlation coefficient between two variables X and Y from 25 pairs of observations obtained the following results :*

$$n = 25, \quad \sum X = 125, \quad \sum X^2 = 650, \quad \sum Y = 100, \quad \Sigma Y^2 = 460, \quad \sum XY = 508$$

If was, however, later discovered at the time of checking that he had copied down two pairs as

X	Y
6	14
8	6

while the correct values were

X	Y
8	12
6	8

. Obtain the correct value of correlation coefficient.

Solution.

$$\text{Corrected } \Sigma X = 125 - 6 - 8 + 8 + 6 = 125$$

$$\text{Corrected } \Sigma Y = 100 - 14 - 6 + 12 + 8 = 100$$

$$\text{Corrected } \Sigma X^2 = 650 - 6^2 - 8^2 + 8^2 + 6^2 = 650$$

$$\text{Corrected } \Sigma Y^2 = 460 - 14^2 - 6^2 + 12^2 + 8^2 = 436$$

$$\text{Corrected } \Sigma XY = 508 - 6 \times 14 - 8 \times 6 + 8 \times 12 + 6 \times 8 = 520$$

$$\bar{X} = \frac{1}{n}\sum X = \frac{1}{25}\times 125 = 5, \qquad \bar{Y} = \frac{1}{n}\sum Y = \frac{1}{25}\times 100 = 4$$

$$\text{Cov}(X, Y) = \frac{1}{n}\sum XY - \bar{X}\bar{Y} = \frac{1}{25}\times 520 - 5\times 4 = \frac{4}{5}$$

$$\sigma_X^2 = \frac{1}{n}\sum X^2 - \bar{X}^2 = \frac{1}{25}\times 650 - (5)^2 = 1 \quad \text{and} \quad \sigma_Y^2 = \frac{1}{n}\sum Y^2 - \bar{Y}^2 = \frac{1}{25}\times 436 - 16 = \frac{36}{25}$$

$$\therefore \qquad \text{Corrected } r(X, Y) = \frac{\text{Cov}(X, Y)}{\sigma_X\sigma_Y} = \frac{(4/5)}{[1\times(6/5)]} = \frac{2}{3} = 0{\cdot}67.$$

Example 10·3. *Show that if X', Y' are the deviations of the random variables X and Y from their respective means, then*

$$(i)\ r = 1 - \frac{1}{2N}\sum_i\left(\frac{X_i'}{\sigma_X} - \frac{Y_i'}{\sigma_Y}\right)^2 \quad \text{and} \quad (ii)\ r = -1 + \frac{1}{2N}\sum_i\left(\frac{X_i'}{\sigma_X} + \frac{Y_i'}{\sigma_Y}\right)^2.$$

Deduce that: $\quad -1 \le r \le +1.$

Solution. (*i*) Here, $X_i' = (X_i - \bar{X})$ and $Y_i' = (Y_i - \bar{Y})$

$$\text{R.H.S.} = 1 - \frac{1}{2N}\sum_i\left(\frac{X_i'}{\sigma_X} - \frac{Y_i'}{\sigma_Y}\right)^2 = 1 - \frac{1}{2N}\sum_i\left[\frac{X_i'^2}{\sigma_X^2} + \frac{Y_i'^2}{\sigma_Y^2} - \frac{2X_i'Y_i'}{\sigma_X\sigma_Y}\right]$$

$$= 1 - \frac{1}{2N}\left[\frac{1}{\sigma_X^2}\sum_i X_i'^2 + \frac{1}{\sigma_Y^2}\sum_i Y_i'^2 - \frac{2}{\sigma_X\sigma_Y}\sum X_i'Y_i'\right]$$

$$= 1 - \frac{1}{2N}\left[\frac{1}{\sigma_X^2}\sum_i(X_i - \bar{X})^2 + \frac{1}{\sigma_Y^2}\sum_i(Y_i - \bar{Y})^2 - \frac{2}{\sigma_X\sigma_Y}\sum_i(X_i - \bar{X})(Y_i - \bar{Y})\right]$$

$$= 1 - \frac{1}{2}\left(\frac{1}{\sigma_X^2}\cdot\sigma_X^2 + \frac{1}{\sigma_Y^2}\cdot\sigma_Y^2 - \frac{2}{\sigma_X\sigma_Y}\cdot r\sigma_X\sigma_Y\right) = 1 - \frac{1}{2}(1 + 1 - 2r) = r$$

(*ii*) Proceeding similarly, we will get: $\quad \text{R.H.S.} = -1 + \frac{1}{2}(1 + 1 + 2r) = r$

Deduction. Since $\left(\dfrac{X_i'}{\sigma_X} \pm \dfrac{Y_i'}{\sigma_Y}\right)^2$, being the square of a real quantity is always non-negative, $\sum_i\left(\dfrac{X_i'}{\sigma_X} \mp \dfrac{Y_i'}{\sigma_Y}\right)^2$ is also non-negative. From part (*i*), we get

$$r = 1 - (\text{some non-negative quantity}) \quad \Rightarrow \quad r \le 1 \qquad \text{...(*)}$$

Also, from part (*ii*), we get

$$r = -1 + (\text{some non-negative quantity}) \quad \Rightarrow \quad r \ge -1 \quad \Rightarrow \quad -1 \le r \qquad \text{...(**)}$$

The sign of equality in (*) and (**) holds if and only if

$$\frac{X_i'}{\sigma_X} - \frac{Y_i'}{\sigma_Y} = 0 \quad \text{and} \quad \frac{X_i'}{\sigma_X} + \frac{Y_i'}{\sigma_Y} = 0,\ (\forall\, i = 1, 2, \ldots, n), \text{ respectively,}$$

From (*) and (**), we conclude that : $\quad -1 \le r \le 1.$

Example 10·4. *The variables X and Y are connected by the equation $aX + bY + c = 0$. Show that the correlation between them is -1 if the signs of a and b are alike and $+1$ if they are different.*

Solution. $$aX + bY + c = 0 \quad \Rightarrow \quad aE(X) + bE(Y) + c = 0$$

$$\therefore \quad a[X - E(X)] + b[Y - E(Y)] = 0 \quad \Rightarrow \quad [X - E(X)] = -\frac{b}{a}[Y - E(Y)]$$

$$\therefore \quad \text{Cov}(X, Y) = E[\{X - E(X)\}Y - E(Y)] = -\frac{b}{a}E[Y - E(Y)^2] = -\frac{b}{a}\cdot\sigma_Y^2$$

and $$\sigma_X^2 = E[X - E(X)]^2 = \frac{b^2}{a^2}E[\{Y - E(Y)\}^2] = \frac{b^2}{a^2}\cdot\sigma_Y^2$$

$$\therefore \quad r = \frac{\text{Cov}(X, Y)}{\sigma_X\sigma_Y} = \frac{-\frac{b}{a}\cdot\sigma_Y^2}{\sqrt{\sigma_Y^2}\sqrt{\frac{b^2}{a^2}\sigma_Y^2}} = \frac{-\frac{b}{a}\sigma_Y^2}{\left|\frac{b}{a}\right|\sigma_Y^2}$$

$$= \begin{cases} +1, \text{ if } b \text{ and } a \text{ are of opposite signs.} \\ -1, \text{ if } b \text{ and } a \text{ are of same sign.} \end{cases}$$

Example 10·5. *(a) If $Z = aX + bY$ and r is the correlation coefficient between X and Y, show that :*

$$\sigma_Z^2 = a^2\sigma_X^2 + b^2\sigma_Y^2 + 2abr\,\sigma_X\,\sigma_Y.$$

(b) Show that the correlation coefficient r between two random variables X and Y is given by :

$$r = \left(\sigma_X^2 + \sigma_Y^2 - \sigma_{X-Y}^2\right)/2\sigma_X\sigma_Y,$$

where, σ_X, σ_Y and σ_{X-Y} are the standard deviations of X, Y and X – Y, respectively.

Solution. (a) Taking expectation of both sides of $Z = aX + bY$, we get

$$E(Z) = aE(X) + bE(Y)$$

$$\therefore \quad Z - E(Z) = a[X - E(X)] + b[Y - E(Y)]$$

Squaring and taking expectation of both sides, we get

$$\sigma_Z^2 = a^2\sigma_X^2 + b^2\sigma_Y^2 + 2ab\,\text{Cov}(X, Y) = a^2\sigma_X^2 + b^2\sigma_Y^2 + 2abr\sigma_X\sigma_Y$$

(b) Taking a = 1, $b = -1$ in the above case, $Z = X - Y$ and $\sigma_{X-Y}^2 = \sigma_X^2 + \sigma_Y^2 - 2r\sigma_X\sigma_Y$

$$\therefore \quad r = \frac{\sigma_X^2 + \sigma_Y^2 - \sigma_{X-Y}^2}{2\sigma_X\sigma_Y}.$$

Remark. In the above example, we have obtained

$$V(aX + bY) = a^2V(X) + b^2\,V(Y) + 2ab\text{ Cov }(X, Y)$$

Similarly, we could obtain the result $V(aX - bY) = a^2V(X) + b^2\,V(Y) - 2ab\text{ Cov }(X, Y)$

The above results are useful in solving theoretical problems.

Example 10·6. *X and Y are two random variables with variances σ_X^2 and σ_Y^2, respectively and r is the coefficient of correlation between them. If $U = X + kY$ and $V = X + (\sigma_X/\sigma_Y)Y$, find the value of k so that U and V are uncorrelated.*

Solution. Taking expectations of $U = X + kY$ and $V = X + \frac{\sigma_X}{\sigma_Y}Y$,

$$E(U) = E(X) + kE(Y) \text{ and } E(V) = E(X) + \frac{\sigma_X}{\sigma_Y}E(Y)$$

$$\therefore \quad U - E(U) = [X - E(X)] + k[Y - E(Y)]$$

$$\text{and} \quad V - E(V) = [X - E(X)] + \frac{\sigma_X}{\sigma_Y}[Y - E(Y)]$$

$$\text{Cov}(U, V) = E[\{U - E(U)\}\{V - E(V)\}]$$

$$= E\left[\left[\{X - E(X)\} + k\{Y - E(Y)\}\right] \times \left[\{X - E(X)\} + \frac{\sigma_X}{\sigma_Y}\{Y - E(Y)\}\right]\right]$$

$$= \sigma_X^2 + \frac{\sigma_X}{\sigma_Y}\text{Cov}(X, Y) + k\,\text{Cov}(X, Y) + k\frac{\sigma_X}{\sigma_Y}\cdot\sigma_Y^2$$

$$= (\sigma_X^2 + k\sigma_X\sigma_Y) + \left(\frac{\sigma_X}{\sigma_Y} + k\right)\text{Cov}(X, Y)$$

$$= \sigma_X(\sigma_X + k\sigma_Y) + \left(\frac{\sigma_X + k\sigma_Y}{\sigma_Y}\right)\text{Cov}(X, Y)$$

$$= (\sigma_X + k\sigma_Y)\left[\sigma_X + \frac{\text{Cov}(X, Y)}{\sigma_Y}\right] = (\sigma_X + k\sigma_Y)(1 + r)\sigma_X$$

U and V will be uncorrelated if $r(U, V) = 0 \Rightarrow \text{Cov}(U, V) = 0$

i.e., if $(\sigma_X + k\sigma_Y)(1 + r)\sigma_X = 0 \quad \Rightarrow \quad \sigma_X + k\sigma_Y = 0 \quad \therefore \quad k = -\frac{\sigma_X}{\sigma_Y}.$

Example 10·7. *The random variables X and Y are jointly normally distributed and U and V are defined by:*

$$U = X\cos\alpha + Y\sin\alpha, \quad V = Y\cos\alpha - X\sin\alpha.$$

Show that U and V will be uncorrelated if

$$\tan 2\alpha = \frac{2r\sigma_X\sigma_Y}{\sigma_X^2 - \sigma_Y^2},$$

where, $r = \text{corr.}(X, Y)$, $\sigma_X^2 = Var(X)$ and $\sigma_Y^2 = Var(Y)$. Are U and V then independent?

Solution. We have: $\text{Cov}(U, V) = E[\{U - E(U)\}\{V - E(V)\}]$

$$= E[\{X - E(X)\}\cos\alpha + \{Y - E(Y)\}\sin\alpha] \times [\{Y - E(Y)\}\cos\alpha - \{X - E(X)\}\sin\alpha]$$

$$= \cos^2\alpha\,\text{Cov}(X, Y) - \sin\alpha\cos\alpha\,\sigma_X^2 + \sin\alpha\cos\alpha\,\sigma_Y^2 - \sin^2\alpha\,\text{Cov}(X, Y)$$

$$= (\cos^2\alpha - \sin^2\alpha)\,\text{Cov}(X, Y) - \sin\alpha\cos\alpha(\sigma_X^2 - \sigma_Y^2)$$

$$= \cos 2\alpha\,\text{Cov}(X, Y) - \sin\alpha\cos\alpha(\sigma_X^2 - \sigma_Y^2)$$

U and V will be uncorrelated if and only if $r(U, V) = 0$, *i.e.*, if Cov $(U, V) = 0$

i.e., if $$\cos 2\alpha \, \text{Cov}(X, Y) - \sin \alpha \cos \alpha \left(\sigma_X^2 - \sigma_Y^2\right) = 0$$

or if $$\cos 2\alpha \, r\sigma_X\sigma_Y = \frac{\sin 2\alpha}{2}\left(\sigma_X^2 - \sigma_Y^2\right)$$

or if $$\tan 2\alpha = \frac{2r\sigma_X\sigma_Y}{\sigma_X^2 - \sigma_Y^2}$$

In general, $r(U, V) = 0$, does not imply that the variables U and V are independent. But since X and Y are given to be jointly normally distributed, in this case,

$$r(u, V) = 0 \quad \Rightarrow \quad X \text{ and } Y \text{ are inepentdent.}$$

[For detailed discussion, see Theorem 10·2, page 10·5.]

10·5. CALCULATION OF THE CORRELATION COEFFICIENT FOR A BIVARIATE FREQUENCY DISTRIBUTION

When the data are considerably large, they may be summarised by using a two-way table. Here, for each variable a suitable number of classes are taken, keeping in view the same considerations as in the univariate case. If there are m classes for X and n classes for Y, there will be in all $m \times n$ cells in the two-way table. By going through the pairs of values of X and Y, we can find the frequency for each cell. The whole set of cell frequencies will then define a *bivariate frequency distribution*. The column totals and row totals will give us the marginal distributions of X and Y, respectively. A particular column or row will be called the conditional distribution of Y for given X or of X for given Y respectively.

Suppose that the bivariate data on X and Y are presented in a two-way correlation table (shown on *page 10·13*) where there are n classes of Y placed along the horizontal lines and m classes of X along the vertical lines and f_{ij} is the frequency of individuals lying in the (i, j)th cell.

Here $\sum_x f(x, y) = g(y)$, is the sum of the frequencies along any row and

$\sum_y f(x, y) = f(x)$, is the sum of the frequencies along any column.

We observe that $$\sum_x \sum_y f(x, y) = \sum_y \sum_x f(x, y) = \sum_x f(x) = \sum_y g(y) = N$$

$$\therefore \quad \bar{x} = \frac{1}{N}\sum_x \sum_y x f(x, y) = \frac{1}{N}\left[\sum_x \left\{x \sum_y f(x, y)\right\}\right] = \frac{1}{N}\sum_x x f(x)$$

Similarly,

$$\bar{y} = \frac{1}{N}\sum_x \sum_y y f(x, y) = \frac{1}{N}\sum_y y\, g(y)$$

$$\sigma_X^2 = \frac{1}{N}\sum_x \sum_y x^2 f(x, y) - \bar{x}^2 = \frac{1}{N}\sum_x x^2 f(x) - \bar{x}^2$$

$$\sigma_Y^2 = \frac{1}{N}\sum_x \sum_y y^2 f(x, y) - \bar{y}^2 = \frac{1}{N}\sum_y y^2 g(y) - \bar{y}^2$$

BIVARIATE FREQUENCY TABLE (CORRELATION TABLE)

Y Series ↓		*X Series* → Classes / Mid Points: x_1 x_2 ... x_i ... x_m	Total of Frequencies of Y: $g(y)$
	y_1		
	y_2		
	⋮		
	y_j	$f(x, y)$	$g(y) = \sum_x f(x, y)$
	⋮		
	⋮		
	y_n		
Total of frequencies of X: f(x)		$f(x) = \sum_y f(x, y)$	$N \to \sum_x \sum_y f(x, y)$ ↓ $\sum_y \sum_x f(x, y)$

Example 10·8. *The following table gives, according to age, the frequency of marks obtained by 100 students in an intelligence test:*

Ages in years → *Marks ↓*	18	19	20	21	*Total*
10–20	4	2	2	–	8
20–30	5	4	6	4	19
30–40	6	8	10	11	35
40–50	4	4	6	8	22
50–60	–	2	4	4	1
60–70	–	2	3	1	6
Total	19	22	31	28	100

Calculate the correlation coefficient.

Solution.

CORRELATION TABLE

v	Mid-value	u Age (X)→ Marks (Y)↓	-1 18	0 19	1 20	2 21	Total $g(v)$	$vg(v)$	$v^2g(v)$	$\sum_u uvf(u, v)$
-2	15	10–20	(8) 4	(0) 2	(−4) 2		8	−16	32	4
-1	25	20–30	(5) 5	(0) 4	(−6) 6	(−8) 4	19	−19	19	−9
0	35	30–40	(0) 6	(0) 8	(0) 10	(0) 11	35	0	0	0
1	45	40–50	(−4) 4	(0) 4	(6) 6	(16) 8	22	22	22	18
2	55	50–60		(0) 2	(8) 4	(16) 4	10	20	40	24
3	65	60–70		(0) 2	(9) 3	(6) 1	6	18	54	15
		Total $f(u)$	19	22	31	28	100	25	167	52
		$u\,f(u)$	−19	0	31	56	68			
		$u^2 f(u)$	19	0	31	112	162			
		$\sum_v uvf(u, v)$	9	0	13	30	52			

Let $U = X - 19,$ $\quad V = [(Y - 35)/10]$

$$\bar{u} = \frac{1}{N}\sum_u u\,f(u) = \frac{68}{100} = 0{\cdot}68, \qquad \bar{v} = \frac{1}{N}\sum_v v\,g(v) = \frac{25}{100} = 0{\cdot}25$$

$$\text{Cov}\,(u, v) = \frac{1}{N}\sum_u\sum_v uv\,f(u, v) - \bar{u}\,\bar{v} = \frac{1}{100} \times 52 - 0{\cdot}68 \times 0{\cdot}25 = 0{\cdot}35$$

$$\sigma_U^2 = \frac{1}{N}\sum_u u^2 f(u) - \bar{u}^2 = \frac{162}{100} - (0{\cdot}68)^2 = 1{\cdot}1576$$

$$\sigma_V^2 = \frac{1}{N}\sum_v v^2 g(v) - \bar{v}^2 = \frac{167}{100} - (0\cdot25)^2 = 1\cdot6075$$

$$\therefore \quad r(U, V) = \frac{\text{Cov}(U, V)}{\sigma_U \sigma_V} = \frac{0\cdot35}{\sqrt{1\cdot1576 \times 1\cdot6075}} = 0\cdot25$$

Since correlation coefficient is independent of change of origin and scale,

$$r(X, Y) = r(U, V) = 0\cdot25$$

Remark. Figures in circles in the table are the product terms $uvf(u, v)$.

Example 10·9. *The joint probability distribution of X and Y is given below :*

Y \ X	−1	+1
0	$\frac{1}{8}$	$\frac{3}{8}$
1	$\frac{2}{8}$	$\frac{2}{8}$

Find the correlation coefficient between X and Y.

Solution.

COMPUTATION OF MARGINAL PROBABILITIES

Y \ X	−1	+1	g(y)
0	$\frac{1}{8}$	$\frac{3}{8}$	$\frac{4}{8}$
1	$\frac{2}{8}$	$\frac{2}{8}$	$\frac{4}{8}$
p(x)	$\frac{3}{8}$	$\frac{5}{8}$	1

We have

$$E(X) = \sum x\,p(x) = (-1)\times\frac{3}{8} + 1\times\frac{5}{8} = \frac{1}{4}, \quad E(X^2) = \sum x^2 p(x) = (-1)^2 \times \frac{3}{8} + 1^2 \times \frac{5}{8} = 1$$

$$\therefore \quad \text{Var}(X) = E(X^2) - [E(X)]^2 = 1 - \frac{1}{16} = \frac{15}{16}$$

$$E(Y) = \sum y\,g(y) = 0\times\frac{4}{8} + 1\times\frac{4}{8} = \frac{1}{2}, \quad E(Y^2) = \sum y^2 g(y) = 0^2 \times \frac{4}{8} + 1^2 \times \frac{4}{8} = \frac{1}{2}$$

$$\therefore \quad \text{Var}(Y) = E(Y^2) - [E(Y)]^2 = \frac{1}{2} - \frac{1}{4} = \frac{1}{4}$$

$$E(XY) = 0\times(-1)\times\frac{1}{8} + 0\times1\times\frac{3}{8} + 1\times(-1)\times\frac{2}{8} + 1\times1\times\frac{2}{8} = -\frac{2}{8} + \frac{2}{8} = 0$$

$$\therefore \text{Cov}(X, Y) = E(XY) - E(X)\,E(Y) = 0 - \frac{1}{4}\times\frac{1}{2} = -\frac{1}{8}$$

$$r(X, Y) = \frac{\text{Cov}(X, Y)}{\sigma_X \sigma_Y} = \frac{-\frac{1}{8}}{\sqrt{\frac{15}{16}\times\frac{1}{4}}} = \frac{-1}{\sqrt{15}} = \frac{-1}{3\cdot873} = -0\cdot2582.$$

10·6. PROBABLE ERROR OF CORRELATION COEFFICIENT

If r is the correlation coefficient in a sample of n pairs of observations, then its standard error (S.E.) is given by :

$$S.E(r) = \frac{1-r^2}{\sqrt{n}}.$$

Probable error (P.E.) of correlation coefficient is given by :

$$P.E.(r) = 0{\cdot}6745 \times S.E.(r) = 0{\cdot}6745 \times \frac{(1-r^2)}{\sqrt{n}} \qquad \text{...(10·6)}$$

Probable error is an old measure for testing the reliability of an observed correlation coefficient. The reason for taking the factor 0·6745 is that in a normal distribution, the range $\mu \pm 0{\cdot}6745\sigma$ covers 50% of the total area. According to Secrist, *"The probable error of correlation coefficient is an amount which if added to and subtracted from the mean correlation coefficient, produces amounts within which the chances are even that a coefficient of correlation from a series selected at random will fall."*

If $r < P.E.(r)$, correlation is not at all significant. If $r > 6\ P.E.\ (r)$, it is definitely significant. A rigorous method (t-test) of testing the significance of an observed correlation coefficient will be discussed later in *"test of significance"* in sampling.

Probable error also enables us to find the limits within which the population correlation coefficient can be expected to vary. The limits are : $r \pm P.E.\ (r)$.

10·7. RANK CORRELATION

Let us suppose that a group of n individuals is arranged in order of merit or proficiency in possession of two characteristics A and B. These ranks in the two characteristics will, in general, be different. For example, if we consider the relation between intelligence and beauty, it is not necessary that a beautiful individual is intelligent also. Let (x_i, y_i); $i = 1, 2, \ldots, n$ be the rank of the ith individual in two characteristics A and B, respectively. Pearsonian coefficient of correlation between the ranks x_i's and y_i's is called the *rank correlation coefficient* between A and B for that group of individuals.

10·7·1. Spearman's Rank Correlation Coefficient.

Assuming that no two individuals are bracketed equal in either classification, each of the variables X and Y takes the values $1, 2, \ldots, n$.

Hence,
$$\bar{x} = \bar{y} = \frac{1}{n}(1+2+3+\cdots+n) = \frac{n+1}{2}$$

$$\sigma_X^2 = \frac{1}{n}\sum_{i=1}^{n} x_i^2 - \bar{x}^2 = \frac{1}{n}\left(1^2+2^2+\cdots+n^2\right) - \left(\frac{n+1}{2}\right)^2 = \frac{n(n+1)(2n+1)}{6n} - \left(\frac{n+1}{2}\right)^2 = \frac{n^2-1}{12}$$

$$\therefore \quad \sigma_X^2 = \frac{n^2-1}{12} = \sigma^2{}_Y$$

In general: $x_i \neq y_i$. Let $d_i = x_i - y_i$ $\quad \therefore \quad d_i = (x_i - \bar{x}) - (y_i - \bar{y})$

Squaring and summing over i from 1 to n, we get

$$\sum_i d_i^2 = \sum_i [(x_i - \bar{x}) - (y_i - \bar{y})]^2 = \sum (x_i - \bar{x})^2 + \sum (y_i - \bar{y})^2 - 2\sum (x_i - \bar{x})(y_i - \bar{y})$$

Dividing both sides by n, we get

$$\frac{1}{n}\sum d_i^2 = \sigma_X^2 + \sigma_Y^2 - 2\,\text{Cov}(X, Y) = \sigma_X^2 + \sigma_Y^2 - 2\rho\,\sigma_X\sigma_Y,$$

where, ρ is the rank correlation coefficient between A and B.

$$\therefore \quad \frac{1}{n}\sum d_i^2 = 2\sigma_X^2 - 2\rho\sigma_X^2 \quad \Rightarrow \quad 1-\rho = \frac{\sum d_i^2}{2n\sigma_X^2} \qquad (\because \sigma_X = \sigma_Y)$$

$$\Rightarrow \quad \rho = 1 - \frac{\sum_{i=1}^{n} d_i^2}{2n\sigma_X^2} = 1 - \frac{6\sum_{i=1}^{n} d_i^2}{n(n^2-1)}, \qquad \text{[From (*)]} \qquad ...(10\cdot7)$$

which is the *Spearman's formula for the rank correlation coefficient.*

Remark. We always have

$$\sum d_i = \sum (x_i - y_i) = \sum x_i - \sum y_i = n(\bar{x} - \bar{y}) = 0 \qquad (\because \bar{x} = \bar{y})$$

This serves as a check on the calculations.

10·7·2. Tied or Repeated Ranks.

If some of the individuals receive the same rank in a ranking of merit, they are said to be tied. Let us suppose that m of the individuals, say, $(k+1)^{\text{th}}, (k+2)^{\text{th}}, \ldots, (k+m)^{\text{th}}$ are tied. Then each of these m individuals is assigned a common rank, which is arithmetic mean of the ranks $k+1, k+2, \ldots, k+m$.

Derivation of ρ (X, Y). We have :

$$\rho(X, Y) = \frac{\sum (X-\bar{X})(Y-\bar{Y})}{\left[\sum (X-\bar{X})^2 \cdot \sum (Y-\bar{Y})^2\right]^{1/2}} = \frac{\sum xy}{\sqrt{\sum x^2 \cdot \sum y^2}}, \qquad ...(i)$$

where, $\qquad x = X - \bar{X}, \quad y = Y - \bar{Y}.$

If X and Y each takes the values 1, 2, …, n, then $\bar{X} = [(n+1)/2] = \bar{Y}$

$$\text{and} \qquad n\sigma_X^2 = \sum x^2 = \frac{n(n^2-1)}{12} \quad \text{and} \quad n\sigma_Y^2 = \sum y^2 = \frac{n(n^2-1)}{12} \qquad ...(ii)$$

$$\text{Also} \qquad \sum d^2 = \sum (X-Y)^2 = \sum \left[(X-\bar{X}) - (Y-\bar{Y})\right]^2 = \sum (x-y)^2$$

$$\Rightarrow \qquad \sum d^2 = \sum x^2 + \sum y^2 - 2\sum xy \quad \Rightarrow \quad \sum xy = \frac{1}{2}\left(\sum x^2 + \sum y^2 - \sum d^2\right) \qquad ...(iii)$$

We shall now investigate the effect of common ranking (in case of ties), on the sum of squares of the ranks. Let S^2 and S_1^2 denote the sum of the squares of untied and tied ranks, respectively. Then, we have

$$S^2 = (k+1)^2 + (k+2)^2 + \ldots + (k+m)^2$$

$$= m\,k^2 + (1^2 + 2^2 + \dots + m^2) + 2k\,(1 + 2 + \dots + m)$$

$$= mk^2 + \frac{m(m+1)(2m+1)}{6} + mk(m+1)$$

$$S_1^2 = m\,(\text{Average rank})^2 = m\left[\frac{(k+1)+(k+2)+\cdots+(k+m)}{m}\right]^2$$

$$= m\left(k + \frac{m+1}{2}\right)^2 = mk^2 + \frac{m(m+1)^2}{4} + mk(m+1)$$

$$\therefore \qquad S^2 - S_1^2 = \frac{m(m+1)}{12}\left[2(2m+1) - 3(m+1)\right] = \frac{m(m^2-1)}{12}$$

Thus, the effect of tying m individuals (ranks) is to reduce the sum of the squares by $m\,(m^2 - 1)/12$, though the mean value of the ranks remains the same, *viz.*, $(n + 1)/2$.

Suppose that there are s such sets of ranks to be tied in the X-series so that the total sum of squares due to them is :

$$\frac{1}{12}\sum_{i=1}^{s} m_i(m_i^2 - 1) = \frac{1}{12}\sum_{i=1}^{s}(m_i^3 - m_i) = T_X, \text{ (say)} \qquad \dots(10\cdot7a)$$

Similarly, suppose that there are t such sets of ranks to be tied with respect to the other series Y so that sum of squares due to them is :

$$\frac{1}{12}\sum_{j=1}^{t} m'_j\,(m'^2_j - 1) = \frac{1}{12}\sum_{j=1}^{t}(m'^3_j - m'_j) = T_Y, \text{ (say)} \qquad \dots(10\cdot7b)$$

Thus, in the case of ties, the new sums of squares are given by :

$$n\,\text{Var}'(X) = \sum x^2 - T_X = \frac{n(n^2-1)}{12} - T_X, \quad n\,\text{Var}'(Y) = \sum y^2 - T_Y = \frac{n(n^2-1)}{12} - T_Y$$

and $n\,\text{Cov}'(X, Y) = \frac{1}{2}\left(\sum x^2 - T_X + \sum y^2 - T_Y - \sum d^2\right)$ [From (*iii*)]

$$= \frac{1}{2}\left[\frac{n(n^2-1)}{12} - T_X + \frac{n(n^2-1)}{12} - T_Y - \sum d^2\right]$$

$$= \frac{n(n^2-1)}{12} - \frac{1}{2}\left[(T_X + T_Y) + \sum d^2\right]$$

$$\rho(X, Y) = \frac{\dfrac{n(n^2-1)}{12} - \dfrac{1}{2}\left(T_X + T_Y + \sum d^2\right)}{\left[\dfrac{n(n^2-1)}{12} - T_X\right]^{1/2} \cdot \left[\dfrac{n(n^2-1)}{12} - T_Y\right]^{1/2}}$$

$$= \frac{\left[n(n^2-1)/6\right] - \left(\sum d^2 + T_X + T_Y\right)}{\left[\dfrac{n(n^2-1)}{6} - 2T_X\right]^{1/2}\left[\dfrac{n(n^2-1)}{6} - 2T_Y\right]^{1/2}} \quad ...(10\cdot7c)$$

where, T_X and T_Y are given by (10·7*a*) and (10·7*b*).

Remark. If we adjust only the covariance term, *i.e.*, Σxy and not the variances σ_X^2 (or Σx^2) and σ_Y^2 (or Σx^2) for ties, then the formula (10·7*c*) reduces to :

$$\rho(X, Y) = \frac{\{n(n^2-1)/6\} - \left(\sum d^2 + T_X + T_Y\right)}{n(n^2-1)/6} = 1 - \frac{6\left(\sum d^2 + T_X + T_Y\right)}{n(n^2-1)} \quad ...(10\cdot7d)$$

a formula which is commonly used in practice for numerical problems. For illustration, see Example 10·12.

Example 10·10. *The ranks of same 16 students in Mathematics and Statistics are as follows. Two numbers within brackets denote the ranks of the students in Mathematics and Statistics:*

(1, 1)	*(2, 10)*	*(3, 3)*	*(4, 4)*	*(5, 5)*	*(6, 7)*	*(7, 2)*	*(8, 6)*
(9, 8)	*(10, 11)*	*(11, 15)*	*(12, 9)*	*(13, 14)*	*(14, 12)*	*(15, 16)*	*(16, 13).*

Calculate the rank correlation coefficient for proficiencies of this group in Mathematics and Statistics.

Solution.

Ranks in Maths. (X)	1	2	3	4	5	6	7	8	9	10	11	12	13	14	15	16	Total
Ranks in Statistics (Y)	1	10	3	4	5	7	2	6	8	11	15	9	14	12	16	13	
$d = X - Y$	0	− 8	0	0	0	−1	5	2	1	−1	− 4	3	−1	2	−1	3	0
d^2	0	64	0	0	0	1	25	4	1	1	16	9	1	4	1	9	136

Rank correlation coefficient is given by :

$$\rho = 1 - \frac{6\sum d^2}{n(n^2-1)} = 1 - \frac{6 \times 136}{16 \times 255} = 1 - \frac{1}{5} = \frac{4}{5} = 0\cdot8$$

Example 10·11. *Ten competitors in a musical test were ranked by the three judges A, B and C in the following order:*

Ranks by A :	1	6	5	10	3	2	4	9	7	8
Ranks by B :	3	5	8	4	7	10	2	1	6	9
Ranks by C :	6	4	9	8	1	2	3	10	5	7

Using rank correlation method, discuss which pair of judges has the nearest approach to common likings in music.

Solution. Here $n = 10$

Ranks by A (X)	*Ranks by B* (Y)	*Ranks by C* (Z)	$d_1 = X - Y$	$d_2 = X - Z$	$d_3 = Y - Z$	d_1^2	d_2^2	d_3^2
1	3	6	−2	−5	−3	4	25	9
6	5	4	1	2	1	1	4	1
5	8	9	−3	−4	−1	9	16	1
10	4	8	6	2	−4	36	4	16
3	7	1	−4	2	6	16	4	36
2	10	2	−8	0	8	64	0	64
4	2	3	2	1	−1	4	1	1
9	1	10	8	−1	−9	64	1	81
7	6	5	1	2	1	1	4	1
8	9	7	−1	1	2	1	1	4
Total			$\Sigma d_1 = 0$	$\Sigma d_2 = 0$	$\Sigma d_3 = 0$	$\Sigma d_1^2 = 200$	$\Sigma d_2^2 = 60$	$\Sigma d_3^2 = 214$

$$\rho(X, Y) = 1 - \frac{6\sum d_1^2}{n(n^2-1)} = 1 - \frac{6 \times 200}{10 \times 99} = 1 - \frac{40}{33} = -\frac{7}{33}$$

$$\rho(X, Z) = 1 - \frac{6\sum d_2^2}{n(n^2-1)} = 1 - \frac{6 \times 60}{10 \times 99} = 1 - \frac{4}{11} = \frac{7}{11}$$

$$\rho(Y, Z) = 1 - \frac{6\sum d_3^2}{n(n^2-1)} = 1 - \frac{6 \times 214}{10 \times 99} = 1 - \frac{214}{165} = -\frac{49}{165}$$

Since, $\rho(X, Z)$ is maximum, we conclude that the pair of judges A and C has the nearest approach to common likings in music.

10·7·3. Repeated Ranks (Continued).

If any two or more individuals are bracketed equal in any classification with respect to characteristics A and B, or if there is more than one item with the same value in the series, then the Spearman's formula (10·7) for calculating the rank correlation coefficient breaks down, since in this case each of the variables X and Y does not assume the values $1, 2, \ldots, n$ and consequently, $\bar{x} \neq \bar{y}$.

In this case (10.7), common ranks are given to the repeated items. This common rank is the average of the ranks which these items would have assumed if they were slightly different from each other and the next item will get the rank next to the ranks already assumed. As a result of this, following adjustment or correction is made in the rank correlation formula [see (10·7*d*)].

In the formula, we add the factor $\dfrac{m(m^2-1)}{12}$ to $\sum d^2$, where m is the number of times an item is repeated. This correction factor is to be added for each repeated value in both the X-series and Y-series.

Example 10·12. *Obtain the rank correlation coefficient for the following data:*

X :	68	64	75	50	64	80	75	40	55	64
Y :	62	58	68	45	81	60	68	48	50	70

Solution.

CALCULATIONS FOR RANK CORRELATION

X	Y	*Rank X* (*x*)	*Rank Y* (*y*)	$d = x - y$	d^2
68	62	4	5	−1	1
64	58	6	7	−1	1
75	68	2·5	3·5	−1	1
50	45	9	10	−1	1
64	81	6	1	5	25
80	60	1	6	−5	25
75	68	2·5	3·5	−1	1
40	48	10	9	1	1
55	50	8	8	0	0
64	70	6	2	4	16
				$\Sigma d = 0$	$\Sigma d^2 = 72$

In the *X*-series we see that the value 75 occurs 2 times. The common rank given to these values is 2·5 which is the average of 2 and 3, the ranks which these values would have taken if they were different. The next value 68, then gets the next rank which is 4. Again we see that value 64 occurs thrice. The common rank given to it is 6 which is the average of 5, 6 and 7. Similarly in the *Y*-series, the value 68 occurs twice and its common rank is 3·5 which is the average of 3 and 4. As a result of these common rankings, the formula for 'ρ' has to be corrected. To Σd^2 we add $\dfrac{m(m^2-1)}{12}$ for each value repeated, where *m* is the number of times a value occurs. In the X-series the correction is to be applied twice, once for the value 75 which occurs twice ($m = 2$) and then for the value 64 which occurs thrice ($m = 3$), thus the correction for the X-series is $\dfrac{2(4-1)}{12} + \dfrac{3(9-1)}{12} = \dfrac{5}{2}$. Similarly, this correction for the Y-series is $\dfrac{2(4-1)}{12} = \dfrac{1}{2}$, as the value 68 occurs twice.

$$\therefore \qquad \rho = 1 - \frac{6\left(\sum d^2 + \frac{5}{2} + \frac{1}{2}\right)}{n(n^2-1)} = 1 - \frac{6(72+3)}{10 \times 99} = 0{\cdot}545$$

10·7·4. Limits for the Rank Correlation Coefficient.

Spearman's Rank correlation coefficient is given by :

$$\rho = 1 - \frac{\sum_{i=1}^{n} d_i^2}{n(n^2-1)}$$

'ρ' is maximum, if $\sum_{i=1}^{n} d_i^2$ is minimum, *i.e.*, if each of the deviations d_i is minimum. But the minimum value of d_i is zero in the particular case $x_i = y_i$, *i.e.*, if the ranks of the *i*th individual in the two characteristics are equal. Hence, the maximum value of ρ is +1, *i.e.*, $\rho \le 1$.

'ρ' is minimum if $\sum_{i=1}^{n} d_i^2$ is maximum, *i.e.*, if each of the deviations d_i is maximum which is so if the ranks of the *n* individuals in the two characteristics are in the opposite directions as given below:

x	1	2	3	...	...	$n-1$	n
y	n	$n-1$	$n-2$	...	...	2	1

...(*)

Case I. Suppose *n* is odd equal to $(2m + 1)$, then the values of *d* are :

d : $2m$, $2m-2$, $2m-4$, ..., 2, $0, -2, -4, \ldots$ $-(2m-2)$, $-2m$.

$$\therefore \quad \sum_{i=1}^{n} d_i^2 = 2\left[(2m)^2 + (2m-2)^2 + \ldots + 4^2 + 2^2\right]$$

$$= 8\left[m^2 + (m-1)^2 + \cdots + 2^2 + 1^2\right] = \frac{8m(m+1)(2m+1)}{6}$$

$$\text{Hence,} \qquad \rho = 1 - \frac{6\sum_{i=1}^{n} d_i^2}{n(n^2-1)} = 1 - \frac{8m(m+1)(2m+1)}{(2m+1)\,[(2m+1)^2-1]} = 1 - \frac{8m(m+1)}{4m(m+1)} = -1.$$

Case II. Let *n* be even and equal to $2m$, (say).

Then, the values of *d* are: $(2m-1), (2m-3), \ldots, 1, -1, -3, \ldots, -(2m-3), -(2m-1)$

$$\therefore \quad \sum d_i^2 = 2\left[(2m-1)^2 + (2m-3)^2 + \ldots + 1^2\right]$$

$$= 2\left[\left\{(2m)^2 + (2m-1)^2 + (2m-2)^2 + \cdots + 2^2 + 1^2\right\} - \left\{(2m)^2 + (2m-2)^2 + \cdots + 4^2 + 2^2\right\}\right]$$

$$= 2\left[1^2 + 2^2 + \cdots + (2m)^2 - \left\{2^2 m^2 + 2^2 (m-1)^2 + \cdots + 2^2\right\}\right]$$

$$= 2\left[\frac{2m(2m+1)(4m+1)}{6} - 4\frac{m(m+1)(2m+1)}{6}\right]$$

$$= \frac{2m}{3}\left[(2m+1)(4m+1) - 2(m+1)(2m+1)\right]$$

$$= \frac{2m}{3}\left[(2m+1)(4m+1-2m-2)\right] = \frac{2m}{3}\left\{(2m+1)(2m-1)\right\} = \frac{2m(4m^2-1)}{3}$$

$$\therefore \qquad \rho = 1 - \frac{6\sum d_i^2}{n(n^2-1)} = 1 - \frac{4m(4m^2-1)}{2m(4m^2-1)} = -1$$

Thus, the limits for rank correlation coefficient are given by: $\quad -1 \le \rho \le 1$.

Remarks on Spearman's Rank Correlation Coefficient.

1. $\Sigma d = \Sigma x - \Sigma y = n(\bar{x} - \bar{y}) = 0$, which provides a check for numerical calculations.

2. Since Spearman's rank correlation coefficient r is nothing but Pearsonian correlation coefficient between the ranks, it can be interpreted in the same way as the Karl Pearson's correlation coefficient.

3. Karl Pearson's correlation coefficient assumes that the parent population from which sample observations are drawn is normal. If this assumption is violated, then we need a measure which is distribution free (or non-parametric). A distribution-free measure is one which does not make any assumptions about the parameters of the population. Spearman's ρ is such a measure (*i.e.,* distribution-free), since no strict assumptions are made about the form of the population from which sample observations are drawn.

4. Spearman's formula is easy to understand and apply as compared with Karl Pearson's formula. The value obtained by the two formulae, *viz.,* Pearsonian r and Spearman's ρ, are generally different. The difference arises due to the fact that when ranking is used instead of full set of observations, there is always some loss of information. Unless many ties exist, the coefficient of rank correlation should be only slightly lower than the Pearsonian coefficient.

5. Spearman's formula is the only formula to be used for finding correlation coefficient if we are dealing with qualitative characteristics which cannot be measured quantitatively but can be arranged serially. It can also be used where actual data are given. In case of extreme observations, Spearman's formula is preferred to Pearson's formula.

6. Spearman's formula has its limitations also. It is not practicable in the case of bivariate frequency distribution (Correlation Table). For $n > 30$, this formula should not be used unless the ranks are given, since in the contrary case the calculations are quite time-consuming.

SUPPLEMENTARY EXAMPLES ON CORRELATION

Example 10·13. *If X, Y are standardized random variables and*

$$r(aX + bY, bX + aY) = \frac{1 + 2ab}{a^2 + b^2}, \quad \text{...(*)}$$

find r(X, Y), the coefficient of correlation between X and Y.

Solution. Since X and Y are standardised random variables, $E(X) = E(Y) = 0$

and $\quad \text{Var}(X) = \text{Var}(Y) = 1 \quad \Rightarrow \quad E(X^2) = E(Y^2) = 1$

and $\quad \text{Cov}(X, Y) = E(XY) \quad \Rightarrow \quad E(XY) = r(X, Y)\,\sigma_X\sigma_Y = r(X, Y)$ ...(**)

Also, we have

$$r[(aX + bY), (bX + aY)]$$

$$= \frac{E[(aX + bY)(bX + aY)] - E(aX + bY)\cdot E(bX + aY)}{[\text{Var}(aX + bY)\cdot \text{Var}(bX + aY)]^{1/2}}$$

$$= \frac{E\left[abX^2 + a^2XY + b^2YX + abY^2\right] - 0}{\left[\left\{a^2\,Var(X) + b^2\,\text{Var}(Y) + 2ab\,\text{Cov}(X, Y)\right\}\left\{b^2\,\text{Var}(X) + a^2\,\text{Var}(Y) + 2ba\,\text{Cov}(X, Y)\right\}\right]^{1/2}}$$

$$= \frac{a \cdot b \cdot 1 + a^2 \cdot r(X, Y) + b^2 \cdot r(X, Y) + ab \cdot 1}{\left[\left\{a^2 + b^2 + 2ab\, r(X, Y)\right\}\left\{b^2 + a^2 + 2ba\, r(X, Y)\right\}\right]^{1/2}} \qquad \text{[Using (**)]}$$

$$= \frac{2ab + (a^2 + b^2) \cdot r(X, Y)}{a^2 + b^2 + 2ab \cdot r(X, Y)} \qquad (***)$$

From (*) and (***), we get $\dfrac{1 + 2ab}{a^2 + b^2} = \dfrac{(a^2 + b^2) \cdot r(X, Y) + 2ab}{a^2 + b^2 + 2ab \cdot r(X, Y)}$

Cross multiplying, we get

$$(a^2 + b^2)(1 + 2ab) + 2ab \cdot r(X, Y)(1 + 2ab) = (a^2 + b^2)^2 \cdot r(X, Y) + 2ab\,(a^2 + b^2)$$

$$\Rightarrow \quad (a^4 + b^4 + 2a^2b^2 - 2ab - 4a^2b^2) \cdot r(X, Y) = (a^2 + b^2)$$

$$\Rightarrow \quad \left[(a^2 - b^2)^2 - 2ab\right] r(X, Y) = a^2 + b^2 \quad \Rightarrow \quad r(X, Y) = \frac{a^2 + b^2}{(a^2 - b^2)^2 - 2ab}$$

Example 10·14. *If X and Y are uncorrelated random variables with means zero and variances σ_1^2 and σ_2^2 respectively, show that*

$$U = X \cos \alpha + Y \sin \alpha, \qquad V = X \sin \alpha - Y \cos \alpha$$

have a correlation coefficient ρ given by : $\quad \rho = \dfrac{\sigma_1^2 - \sigma_2^2}{\left[\left(\sigma_1^2 - \sigma_2^2\right)^2 + 4\sigma_1^2\sigma_2^2 \operatorname{cosec}^2 2\alpha\right]^{1/2}}$

Solution. We are given that

$$r(X, Y) = 0 \quad \Rightarrow \quad \text{Cov}(X, Y) = 0, \quad \sigma_X^2 = \sigma_1^2 \quad \text{and} \quad \sigma_Y^2 = \sigma_2^2 \qquad \ldots(i)$$

We have

$$\sigma_U^2 = V(X \cos \alpha + Y \sin \alpha)$$
$$= \cos^2 \alpha\, V(X) + \sin^2 \alpha\, V(Y) + 2 \sin \alpha \cos \alpha \,\text{Cov}(X, Y)$$
$$= \cos^2 \alpha\, \sigma_1^2 + \sin^2 \alpha\, \sigma_2^2 \qquad \text{[Using (i)]}$$

Similarly,

$$\sigma_V^2 = V(X \sin \alpha - Y \cos \alpha) = \sin^2 \alpha\, \sigma_1^2 + \cos^2 \alpha\, \sigma_2^2$$

$$\text{Cov}(U, V) = E\left[\{U - E(U)\}\{V - E(V)\}\right]$$
$$= E\left[\{(X - E(X))\} \cos \alpha + \{(Y - E(Y)) \sin \alpha\} \times \{(X - E(X)) \sin \alpha - (Y - E(Y)) \cos \alpha\}\right]$$
$$= \sin \alpha \cos \alpha\, V(X) - \cos^2 \alpha\, \text{Cov}(X, Y) + \sin^2 \alpha\, \text{Cov}(X, Y) - \sin \alpha \cos \alpha\, V(Y)$$
$$= (\sigma_1^2 - \sigma_2^2) \sin \alpha \cos \alpha \qquad \text{[Using (i)]} \qquad \ldots(ii)$$

$$\rho = \frac{\text{Cov}(U, V)}{\sigma_U \sigma_V} \quad \Rightarrow \quad \rho^2 = \frac{[\text{Cov}(U, V)]^2}{\sigma_U^2 \cdot \sigma_V^2}, \qquad \ldots(iii)$$

where, $\sigma_U^2\sigma_V^2 = \left(\cos^2\alpha\,\sigma_1^2 + \sin^2\alpha\,\sigma_2^2\right)\cdot\left(\sin^2\alpha\,\sigma_1^2 + \cos^2\alpha\,\sigma_2^2\right)$

$$= \sin^2\alpha\cos^2\alpha\left(\sigma_1^4 + \sigma_2^4\right) + \sigma_1^2\sigma_2^2\left(\cos^4\alpha + \sin^4\alpha\right)$$

$$= \sin^2\alpha\cos^2\alpha\left(\sigma_1^4 + \sigma_2^4\right) + \sigma_1^2\,\sigma_2^2\left[\left(\sin^2\alpha + \cos^2\alpha\right)^2 - 2\sin^2\alpha\cos^2\alpha\right]$$

$$= \sin^2\alpha\cos^2\alpha\left(\sigma_1^4 + \sigma_2^4 - 2\sigma_1^2\sigma_2^2\right) + \sigma_1^2\sigma_2^2$$

$$= \sin^2\alpha\cos^2\alpha\left(\sigma_1^2 - \sigma_2^2\right)^2 + \sigma_1^2\sigma_2^2 \qquad \text{... } (iv)$$

$$\therefore \quad \rho^2 = \frac{\left(\sigma_1^2 - \sigma_2^2\right)^2\cdot\sin^2\alpha\cos^2\alpha}{\sigma_1^2\sigma_2^2 + \sin^2\alpha\cos^2\alpha\cdot\left(\sigma_1^2 - \sigma_2^2\right)^2} = \frac{\frac{1}{4}\left(\sigma_1^2 - \sigma_2^2\right)^2\sin^2 2\alpha}{\sigma_1^2\sigma_2^2 + \sin^2 2\alpha\cdot\frac{1}{4}\left(\sigma_1^2 - \sigma_2^2\right)^2}$$

[Using (*ii*) to (*iv*)]

$$= \frac{\left(\sigma_1^2 - \sigma_2^2\right)^2}{4\sigma_1^2\sigma_2^2\,\mathrm{cosec}^2 2\alpha + \left(\sigma_1^2 - \sigma_2^2\right)^2}$$

$$\Rightarrow \quad \rho = \frac{\sigma_1^2 - \sigma_2^2}{\left[\left(\sigma_1^2 - \sigma_2^2\right)^2 + 4\sigma_1^2\sigma_2^2\,\mathrm{cosec}^2 2\alpha\right]^{1/2}}.$$

Example 10·15. *If U = aX + bY and V = cX + dY, where X and Y are measured from their respective means and if r is the correlation coefficient between X and Y, and if U and V are uncorrelated, show that*

$$\sigma_U\sigma_V = (ad - bc)\,\sigma_X\sigma_Y(1 - r^2)^{1/2}.$$

Solution. We have

$$r = \frac{\mathrm{Cov}(X, Y)}{\sigma_X\sigma_Y} \quad \Rightarrow \quad 1 - r^2 = 1 - \frac{[\mathrm{Cov}(X, Y)]^2}{\sigma_X^2\cdot\sigma_Y^2}$$

$$\Rightarrow \quad \left(1 - r^2\right)\sigma_X^2\sigma_Y^2 = \sigma_X^2\,\sigma_Y^2 - [\mathrm{Cov}(X, Y)]^2 \qquad \text{...}(*)$$

[This step is suggested by the answer.]

Since X, Y are measured from their means,

$$\left.\begin{aligned} E(X) = 0 = E(Y) \quad &\Rightarrow \quad E(U) = 0 = E(V) \\ \text{and} \quad \sigma_U^2 = E(U)^2, \quad &\sigma_V^2 = E(V^2) \end{aligned}\right\} \qquad \text{...}(**)$$

Also $\qquad aX + bY - U = 0$ and $cX + dY - V = 0$

$$\Rightarrow \quad \frac{X}{-bV + dU} = \frac{Y}{-cU + aV} = \frac{1}{ad - bc} \quad \text{or} \quad \left.\begin{aligned} X &= \frac{1}{ad - bc}(dU - bV) \\ Y &= \frac{1}{ad - bc}(-cU + aV) \end{aligned}\right\} \qquad \text{...}(***)$$

$$\therefore \quad \text{Var}(X) = \frac{1}{(ad-bc)^2}\cdot\left[d^2\sigma_U^2 + b^2\sigma_V^2 - 2bd\,\text{Cov}(U,V)\right]$$

$$= \frac{1}{(ad-bc)^2}\left[d^2\sigma_U^2 + b^2\sigma_V^2\right]$$

[Since U, V are uncorrelated $\Rightarrow$ Cov $(U, V) = 0$]

Similarly, we have

$$\text{Var}(Y) = \frac{1}{(ad-bc)^2}\left[c^2\sigma_U^2 + a^2\sigma_V^2\right]$$

$$\text{Cov}(X, Y) = E(XY) - E(X)\,E(Y) = E(XY) \qquad [\because E(X) = 0 = E(Y)]$$

$$= \frac{1}{(ad-bc)^2}\,E\left[(dU - bV)(-cU + aV)\right] \qquad \text{[From (***)]}$$

$$= \frac{1}{(ad-bc)^2}\left[-cd\,\sigma_U^2 - ab\,\sigma_V^2\right] \qquad \text{[Using (**) and Cov } (U, V) = 0 \text{ given]}$$

$$= \frac{-1}{(ad-bc)^2}\left[cd\,\sigma_U^2 + ab\,\sigma_V^2\right]$$

Substituting in (*), we get

$$(1 - r^2)\sigma_X^2\sigma_Y^2 = \frac{1}{(ad-bc)^4}\left[(d^2\sigma_U^2 + b^2\sigma_V^2)(c^2\sigma_U^2 + a^2\sigma_V^2) - (cd\,\sigma_U^2 + ab\,\sigma_V^2)^2\right]$$

$$= \frac{1}{(ad-bc)^4}\left[c^2d^2\,\sigma_U^4 + a^2b^2\,\sigma_V^4 + (a^2d^2 + b^2c^2)\,\sigma_U^2\,\sigma_V^2 - c^2d^2\,\sigma_U^4 - a^2b^2\,\sigma_V^4 - 2abcd\,\sigma_U^2\,\sigma_V^2\right]$$

$$= \frac{1}{(ad-bc)^4}\left[a^2d^2 + b^2c^2 - 2abcd\right]\sigma_U^2\sigma_V^2 = \frac{1}{(ad-bc)^4}(ad-bc)^2\,\sigma_U^2\sigma_V^2$$

$$= \frac{\sigma_U^2\sigma_V^2}{(ad-bc)^2}.$$

Cross multiplying and taking square root, we get the required result.

Example 10·16. *Establish the formula:*

$$n r \sigma_X \sigma_Y = n_1 r_1 \sigma_{X_1}\sigma_{Y_1} + n_2 r_2\,\sigma_{X_2}\,\sigma_{Y_2} + n_1\,dx_1\,dy_1 + n_2\,dx_2\,dy_2$$

where n_1, n_2 and n are respectively the sizes of the first, second and combined sample $(\overline{x}_1, \overline{y}_1)$, $(\overline{x}_2, \overline{y}_2)$, $(\overline{x}, \overline{y})$, their means; r_1, r_2 and r their coefficients of correlation; $(\sigma_{X_1}, \sigma_{Y_1})$, $(\sigma_{X_2}, \sigma_{Y_2})$, (σ_X, σ_Y) their standard deviations, and

$$dx_1 = \overline{x}_1 - \overline{x} \qquad dy_1 = \overline{y}_1 - \overline{y}$$

$$dx_2 = \overline{x}_2 - \overline{x} \qquad dy_2 = \overline{y}_2 - \overline{y}$$

Solution. Let (x_{1i}, y_{1i}); $i = 1, 2, \ldots, n_1$ and (x_{2j}, y_{2j}); $j = 1, 2, \ldots, n_2$, be the two samples of sizes n_1 and n_2, respectively from the bivariate population. Then, with the given notations, we have

$$\bar{x} = \frac{n_1\bar{x}_1 + n_2\bar{x}_2}{n_1 + n_2}, \qquad \bar{y} = \frac{n_1\bar{y}_1 + n_2\bar{y}_2}{n_1 + n_2}$$

$$n\sigma_X^2 = n_1\left(\sigma_{X_1}^2 + dx_1^2\right) + n_2\left(\sigma_{X_2}^2 + dx_2^2\right) \qquad \text{...}(i)$$

$$n\sigma_Y^2 = n_1\left(\sigma_{Y_1}^2 + dy_1^2\right) + n_2\left(\sigma_{Y_2}^2 + dy_2^2\right)$$

$$r_1 = \frac{\sum_{i=1}^{n_1}(x_{1i} - \bar{x}_1)(y_{1i} - \bar{y}_1)}{n_1\sigma_{X_1}\sigma_{Y_1}}, \quad r_2 = \frac{\sum_{j=1}^{n_2}(x_{2j} - \bar{x}_2)(y_{2j} - \bar{y}_2)}{n_2\sigma_{X_2}\sigma_{Y_2}} \qquad \text{...}(ii)$$

For the pooled sample, we have

$$r = \frac{\sum_{i=1}^{n_1}(x_{1i} - \bar{x})(y_{1i} - \bar{y}) + \sum_{j=1}^{n_2}(x_{2j} - \bar{x})(y_{2j} - \bar{y})}{n\sigma_X\sigma_Y} \qquad \text{...}(iii)$$

$$\sum_{i=1}^{n_1}(x_{1i} - \bar{x})(y_{1i} - \bar{y}) = \sum_{i=1}^{n_1}\left[\{(x_{1i} - \bar{x}_1) + (\bar{x}_1 - \bar{x})\}\{(y_{1i} - \bar{y}_1) + (\bar{y}_1 - \bar{y})\}\right]$$

$$= \sum_{i=1}^{n_1}(x_{1i} - \bar{x}_1)(y_{1i} - \bar{y}_1) + (\bar{y}_1 - \bar{y})\sum_{i=1}^{n_1}(x_{1i} - \bar{x}_1)$$

$$+ (\bar{x}_1 - \bar{x})\sum_{i=1}^{n_1}(y_{1i} - \bar{y}_1) + n_1(\bar{x}_1 - \bar{x})(\bar{y}_1 - \bar{y})$$

But $\sum_{i=1}^{n_1}(x_{1i} - \bar{x}_1) = 0$ and $\sum_{i=1}^{n_1}(y_{1i} - \bar{y}_1) = 0,$

being the algebraic sum of the deviations from the mean.

$$\therefore \qquad \sum_{i=1}^{n_1}(x_{1i} - \bar{x})(y_{1i} - \bar{y}) = n_1 r_1 \sigma_{X_1}\sigma_{Y_1} + n_1\, dx_1\, dy_1 \qquad \text{[Using } (ii)]$$

Similarly, we will get

$$\sum_{j=1}^{n_2}(x_{2j} - \bar{x})(y_{2j} - \bar{y}) = n_2 r_2 \sigma_{X_2}\sigma_{Y_2} + n_2\, dx_2\, dy_2$$

Substituting in *(iii)*, we get the required formula.

Example 10·17. *The independent variables X and Y are defined by :*

$$f(x) = \begin{cases} 4ax, & 0 \le x \le r \\ 0, & \text{otherwise} \end{cases} \qquad \Bigg| \qquad f(y) = \begin{cases} 4by, & 0 \le y \le s \\ 0, & \text{otherwise} \end{cases}$$

Show that: $\quad r(U, V) = \dfrac{b - a}{b + a}, \quad$ *where* $\quad U = X + Y \quad$ *and* $\quad V = X - Y.$

Solution. Since the total area under probability curve is unity, we have

$$\int_0^r f(x)\,dx = 4a\int_0^r x\,dx = 1 \quad \Rightarrow \quad 2ar^2 = 1 \quad \Rightarrow \quad a = \frac{1}{2r^2} \qquad \text{...(1)}$$

$$\int_0^s f(y)\,dy = 4b\int_0^s y\,dy = 1 \quad \Rightarrow \quad 2bs^2 = 1 \quad \Rightarrow \quad b = \frac{1}{2s^2} \qquad \text{...(2)}$$

$$\therefore \qquad f(x) = 4ax = \frac{2x}{r^2}, \quad 0 \le x \le r; \quad \text{and} \quad f(y) = 4by = \frac{2y}{s^2}, 0 \le y \le s \qquad \text{...(3)}$$

Since X and Y are independent variates,

$$r(X, Y) = 0 \quad \Rightarrow \quad \text{Cov}\,(X, Y) = 0 \qquad \text{...(4)}$$

$$\begin{aligned}\text{Cov}\,(U, V) &= \text{Cov}\,(X + Y, X - Y)\\ &= \text{Cov}\,(X, X) - \text{Cov}\,(X, Y) + \text{Cov}\,(Y, X) - \text{Cov}\,(Y, Y)\\ &= \sigma_X^2 - \sigma_Y^2 \qquad \text{[Using (4)]}\end{aligned}$$

$$\begin{aligned}\text{Var}\,(U) &= \text{Var}\,(X + Y) = \text{Var}\,(X) + \text{Var}\,(Y) + 2\,\text{Cov}\,(X, Y)\\ &= \sigma_X^2 + \sigma_Y^2 \qquad \text{[Using (4)]}\end{aligned}$$

$$\begin{aligned}\text{Var}\,(V) &= \text{Var}\,(X - Y) = \text{Var}\,(X) + \text{Var}\,(Y) - 2\,\text{Cov}\,(X, Y)\\ &= \sigma_X^2 + \sigma_Y^2 \qquad \text{[Using (4)]}\end{aligned}$$

$$\therefore \qquad r(U, V) = \frac{\text{Cov}\,(U, V)}{\sigma_U \sigma_V} = \frac{\sigma_X^2 - \sigma_Y^2}{\sigma_X^2 + \sigma_Y^2} \qquad \text{...(5)}$$

$$\therefore \qquad E(X) = \int_0^r x\,f(x)\,dx = \frac{2}{r^2}\int_0^r x^2\,dx = \frac{2r}{3} \qquad \text{[From (3)]}$$

$$E(X^2) = \int_0^r x^2 f(x)\,dx = \frac{2}{r^2}\int_0^r x^3\,dx = \frac{r^2}{2}$$

$$\Rightarrow \qquad \text{Var}\,(X) = E(X^2) - [E(X)]^2 = \frac{r^2}{2} - \frac{4r^2}{9} = \frac{r^2}{18} = \frac{1}{36a} \qquad \text{[From (1)]}$$

Similarly, we shall get

$$E(Y) = \frac{2s}{3}, \quad E(Y^2) = \frac{s^2}{2} \quad \text{and} \quad \text{Var}\,(Y) = \frac{s^2}{18} = \frac{1}{36b}$$

Substituting in (5), we get

$$r(U, V) = \frac{(1/36a) - (1/36b)}{(1/36a) + (1/36b)} = \frac{b - a}{b + a}.$$

Example 10·18. *Let the random variable X have the marginal density :*

$$f_1(x) = 1, -\frac{1}{2} < x < \frac{1}{2}$$

and let the conditional density of Y be

$$f(y \mid x) = \begin{cases} 1, x < y < x+1, -\frac{1}{2} < x < 0 \\ 1, -x < y < 1-x, 0 < x < \frac{1}{2} \end{cases} \quad ...(*)$$

Show that the variables X and Y are uncorrelated.

Solution. We have

$$E(X) = \int_{-1/2}^{1/2} x\, f_1(x)\, dx = \int_{-1/2}^{1/2} x \cdot 1 dx = \left| \frac{x^2}{2} \right|_{-\frac{1}{2}}^{\frac{1}{2}} = 0$$

If $f(x, y)$ is the joint *p.d.f.* of X and Y, then

$$f(x, y) = f(y \mid x)\, f_1(x) = f(y \mid x) \qquad [\because f_1(x) = 1] \quad ...(**)$$

$$E(XY) = \int_{-1/2}^{0} \int_{x}^{x+1} xy\, dx\, dy + \int_{0}^{1/2} \int_{-x}^{1-x} xy\, dx\, dy$$

$$= \int_{-1/2}^{0} \left(x \int_{x}^{x+1} y\, dy \right) dx + \int_{0}^{1/2} \left(x \int_{-x}^{1-x} y\, dy \right) dx \quad \text{[From (*) and (**)]}$$

$$= \frac{1}{2} \int_{-1/2}^{0} x(2x+1)\, dx + \frac{1}{2} \int_{0}^{1/2} x(1-2x)\, dx$$

$$= \frac{1}{2} \left[\frac{2}{3}x^3 + \frac{x^2}{2} \right]_{-1/2}^{0} + \frac{1}{2} \left[\frac{1}{2}x^2 - \frac{2}{3}x^3 \right]_{0}^{1/2}$$

$$= \frac{1}{2} \left(\frac{1}{12} - \frac{1}{8} - \frac{1}{12} + \frac{1}{8} \right) = 0$$

$\therefore \quad \text{Cov}(XY) = E(XY) - E(X)\, E(Y) = 0 \quad \Rightarrow \quad r(X, Y) = 0$

Hence, the variables X and Y are uncorrelated.

CHAPTER CONCEPTS QUIZ

1. State whether the following statements are TRUE or FALSE:

(*i*) $r_{XY} = 0 \Rightarrow$ X and Y are independent.

(*ii*) If $r_{XY} > 0$, then $r_{X, -Y} > 0$, $r_{-X, Y} > 0$ and $r_{-X, -Y} > 0$.

(*iii*) $r_{XY} > 0 \quad \Rightarrow \quad E(XY) > E(X)\, E(Y)$.

(*iv*) Pearson's coefficient of correlation is independent of change of origin but not of scale.

(*v*) The numerical value of product moment correlation coefficient '*r*' between two variables *X* and *Y* cannot exceed unity.

(*vi*) If the correlation coefficient between the variables X and Y is zero then the correlation coefficient between X^2 and Y^2 is also zero.

(*vii*) If $r > 0$, then as X increases, Y also increases.

(*viii*) *r* measures every type of relationship between the two variables.

(*ix*) "High positive coefficient of correlation between increase in the sale of newspapers and increase in the number of crimes leads to the conclusion that newspaper reading may be responsible for the increase in the number of crimes."

(x) "A high positive value of r between the increase in cigarette smoking and increase in lung cancer establishes that cigarette smoking is responsible for lung cancer."

(xi) Let (X, Y) be jointly discrete random variables such that each X and Y has at most two mass points. Then X and Y are independent if and only if they are uncorrelated.

2. Fill in the blanks :

(i) If $r_{XY} = 1$, the line on the graph will extend from ... to ...

(ii) If $r_{XY} = 0$, it depicts ... association.

(iii) Correlation coefficient is a ... number

(iv) If r is the correlation coefficient, then $\sqrt{1-r^2}$ is termed as ... .

(v) The Karl Pearson coefficient of correlation between variables X and Y is ...

(vi) Two independent variables are ...

(vii) Limits for correlation coefficient are ...

(viii) If r be the correlation coefficient between the random variables X and Y, then the variance of $X + Y$ is ...

(ix) The absolute value of the product moment correlation coefficient is less than ...

(x) Correlation coefficient is invariant under changes of ... and ...

3. Indicate the correct answer :

(i) The coefficient of correlation will have positive sign when

(a) X is increasing and Y is decreasing, (b) both X and Y are increasing,

(c) X is decreasing and Y is increasing (d) there is no change in X and Y.

(ii) The coefficient of correlation (a) can take any value between – 1 and + 1, (b) is always less then – 1, (c) is always more than + 1, (d) cannot be zero.

(iii) The coefficient of correlation (a) cannot be positive, (b) cannot be negative, (c) is always positive, (d) can be both positive as well as negative.

(iv) Probable error of r is

(a) $0{\cdot}6745\dfrac{1-r^2}{\sqrt{n}}$, (b) $0{\cdot}6754\dfrac{1+r^2}{\sqrt{n}}$, (c) $0{\cdot}6547\dfrac{1-r^2}{n}$, (d) $0{\cdot}6754\dfrac{1-r^2}{n}$.

(v) The coefficient of correlation between X & Y is 0·6. Their covariance is 4·8. The variance of X is 9. Then the S.D. of Y is

(a) $\dfrac{4{\cdot}8}{3\times 0{\cdot}6}$, (b) $\dfrac{0{\cdot}6}{4{\cdot}8\times 3}$, (c) $\dfrac{3}{4{\cdot}8\times 0{\cdot}6}$, (d) $\dfrac{4{\cdot}8}{9\times 0{\cdot}6}$.

(vi) The coefficient of correlation is independent of (a) change of scale only, (b) change of origin only. (c) both change of scale and origin, (d) neither change of scale nor change of origin.

(vii) X, Y and Z are three uncorrelated variables having variances σ_X^2, σ_Y^2 and σ_Z^2, respectively, then the correlation coefficient between $X + Y$ and $Y + Z$ is :

(a) 0, (b) $\dfrac{1}{2}$, (c) 1, (d) None of the above.

(viii) Let X be normally distributed with mean 0 and variance σ_1^2 and Y be normally distributed with mean 0 and variance σ_2^2. Let ρ be the correlation coefficient between X and Y, then

$$(X/\sigma_1) + (Y/\sigma_2) \text{ and } (X/\sigma_1) - (Y/\sigma_2) :$$

(a) have correlation coefficient $= \rho$ (b) have correlation coefficient $\leq \rho$

(c) have correlation coefficient $\geq \rho$ (d) None of the above.

(ix) If $X_1, X_2, ..., X_n$ are n uncorrelated random variables following the same distribution, then the correlation coefficient between $\bar{X}$ and $X_i - \bar{X}$ is :

(a) 0, (b) 1, (c) $\frac{1}{2}$, (d) None of the above.

4. Prove or disprove :

(a) $r(X, Y) = 0 \Rightarrow r(X \mid Y) = 0$ (b) $r(X, Y) = 0, r(Y, Z) = 0 \Rightarrow r(X, Z) = 0$

DISCUSSION AND REVIEW QUESTIONS

1. Show that the coefficient of correlation r is independent of change of scale and origin of the variables. Also prove that for two independent variables $r = 0$. Show by an example that the converse is not true. State the limits between which r lies and give its proof.
2. Let ρ be the correlation coefficient between two jointly distributed random variables X and Y. Show that $|\rho| \le 1$ and that $|\rho| = 1$ if and only if X and Y are linearly related.
3. How can you use scatter diagram to obtain an idea of the extent and nature (direction) of the correlation coefficient?
4. Write a brief note on the correlation table.
5. Prove that Spearman's rank correlation coefficient is given by $1 - \frac{6\sum d_i^2}{n^3 - n}$, where d_i denotes the difference between the ranks of *ith* individuals.
6. Explain the difference between product moment correlation coefficient and rank correlation coefficient.
7. Prove that : $V(aX \pm bY) = a^2V(X) + b^2V(Y) \pm 2ab \operatorname{Cov}(X, Y)$.

   Hence deduce that if X and Y are independent, $V(X \pm Y) = V(X) + V(Y)$.
8. Prove that correlation coefficient between X and Y is positive or negative according as :
$$\sigma_{X+Y} > \text{ or } < \sigma_{X-Y}$$
9. Show that if X and Y are two random variables each assuming only two values and the correlation coefficient between them is zero, then they are independent. Indicate with justification whether the result is true in general.
10. Find the correlation coefficient between X and $a - X$, where X is any random variable and a is constant.

ASSORTED REVIEW PROBLEMS FOR SELF-ASSESSMENT

10·1. (a) Calculate the coefficient of correlation between X and Y for the following :

X :	1	3	4	5	7	8	10
Y :	2	6	8	10	14	16	20

Ans. $r(X, Y) = 1$

(b) By effecting suitable change of origin and scale, compute the product moment correlation coefficient for the following set of 5 observations on (X, Y):

X :	– 10	– 5	0	5	10
Y :	5	9	7	11	13

Ans. $r(X, Y) = 0{\cdot}34$

(c) The marks obtained by 10 students in Mathematics (X) and Statistics (Y) are given below. Find the coefficient of correlation between X and Y.

Roll No. :	1	2	3	4	5	6	7	8	9	10
X :	75	30	60	80	53	35	15	40	38	48
Y :	85	45	54	91	58	63	35	43	45	44

(d) The following table gives the distribution of items of production and also the relatively defective items among them, according to size-groups. Is there any correlation between size and defect in quality?

Size-group	:	15–16	16–17	17–18	18–19	19–20	20–21
No. of items	:	200	270	340	360	400	300
No. of defective items	:	150	162	170	180	180	120

Hint. Here we have to find the correlation coefficient between the size group (X) and the percentage defectives (Y) given below:

X	:	15·5	16·5	17·5	18·5	19·5	20·5
Y	:	75	60	50	50	45	40

Ans. $r(X, Y) = 0{\cdot}94$.

10·2. (*a*) From the following data, compute the coefficient of correlation between X and Y.

	X-series	Y-series
No. of items	15	15
Arithmetic mean	25	18
Sum of squares of deviations from mean	136	138

Summation of product of deviations of X and Y series from respective arithmetic means = 122.

(*b*) In two sets of variables X and Y with 50 observations each, the following data were observed :

$$\overline{X} = 10, \quad \sigma_X = 3, \quad \overline{Y} = 6, \quad \sigma_Y = 2 \quad \text{and} \quad r(X, Y) = 0{\cdot}3$$

But on subsequent verification it was found that one value of X (= 10) and the corresponding value of Y (= 6) were inaccurate and hence weeded out. With the remaining 49 pairs of values, how is the original value of r affected?

Hint. $\Sigma x = n\overline{X} = 500, \quad \Sigma Y = n\overline{Y} = 300$

$$\Sigma X^2 = n\left(\sigma_X^2 + \overline{X}^2\right) = 5450, \qquad \Sigma Y^2 = 50(4 + 36) = 2000$$

$$r\,\sigma_X\sigma_Y = \text{Cov}(X, Y) = \frac{\Sigma XY}{n} - \overline{X}\overline{Y}$$

$$\Rightarrow \quad 0{\cdot}3 \times 3 \times 2 = \frac{\Sigma XY}{50} - 10 \times 6 \quad \Rightarrow \quad \Sigma XY = 50(1{\cdot}8 + 60) = 3090.$$

After weeding out the incorrect pair of observation, *viz.*, ($X = 10$, $Y = 6$), the corrected values of ΣX, ΣY, ΣX^2, ΣY^2 and ΣXY for the remaining $50 - 1 = 49$, pairs of observations are given below:

Corrected Values:

$$\Sigma X = 50 - 10 = 490; \qquad \Sigma Y = 300 - 6 = 294$$

$$\Sigma XY = 3090 - 10 \times 6 = 3090 - 60 = 3030$$

$$\Sigma X^2 = 5450 - 10^2 = 5350, \qquad \Sigma Y^2 = 2000 - 6^2 = 1964$$

$$\therefore \quad r = \frac{\text{Corrected Cov}(X, Y)}{(\text{Corrected } \sigma_X) \times (\text{Corrected } \sigma_Y)} = \frac{90/49}{\sqrt{\dfrac{450}{49} \times \dfrac{200}{49}}} = 0{\cdot}3$$

Hence, the correlation coefficient is invariant in this case.

(*c*) Given the following information :

$r = 0{\cdot}8$, $\Sigma xy = 60$, $\sigma_Y = 2{\cdot}5$ and $\Sigma x^2 = 90$, where x and y are the deviations from the respective means, find the number of items (n)

Hint. Using the formula : $r(X, Y) = \dfrac{\text{Cov}(X, Y)}{\sigma_X\, \sigma_Y}$, we get

$$0{\cdot}8 = \frac{60/n}{\sqrt{(90/n)} \times 2{\cdot}5} \quad \Rightarrow \quad 2 = \frac{60}{n} \times \frac{\sqrt{n}}{\sqrt{90}} \quad \Rightarrow \quad n = 10$$

(*d*) The coefficient of correlation between two variables X and Y is 0·4 and their covariance is 10. If the variance of X series is 9, find the second moment about mean of Y series.

Hint. $r(X, Y) = \dfrac{\text{Cov}(X, Y)}{\sigma_X \sigma_Y} \Rightarrow 0{\cdot}4 = \dfrac{10}{3 \times \sigma_Y} \Rightarrow \sigma_Y = 8{\cdot}33$

The second moment about mean of Y series is

$$\mu_2 = \text{Var } Y = \sigma_Y^2 = 69{\cdot}39.$$

10·3. (*a*) X_1 and X_2 are independent variables with means 5 and 10 and standard deviations 2 and 3, respectively. Obtain $r(U, V)$, where $U = 3X_1 + 4X_2$ and $V = 3X_1 - X_2$.

Ans. $r(U, V) = 0$.

(*b*) If X and Y are normal and independent with zero means and standard deviations 9 and 12, respectively, and if $X + 2Y$ and $kX - Y$ are uncorrelated, find *k*.

Ans. $k = 32/9$

(*c*) If X_i ($i = 1, 2, 3$) are three uncorrelated variables having standard deviations σ_1, σ_2 and σ_3, respectively, obtain the coefficient of correlation between $(X_1 + X_2)$ and $(X_2 + X_3)$.

Ans. $\dfrac{\sigma_2^2}{\left[\left(\sigma_1^2 + \sigma_2^2\right)\left(\sigma_2^2 + \sigma_3^2\right)\right]^{1/2}}$

10·4. (*a*) Two random variables X and Y have zero means, the same variance σ^2 and zero correlation. Show that $U = X \cos \alpha + Y \sin \alpha$ and $V = X \sin \alpha - Y \cos \alpha$, have the same variance σ^2 and zero correlation.

(*b*) Let X and Y be uncorrelated random variables. If $U = X + Y$ and $V = X - Y$, prove that the coefficient of correlation between U and V is $\left(\sigma_X^2 - \sigma_Y^2\right)/\left(\sigma_X^2 + \sigma_Y^2\right)$, where σ_X^2 and σ_Y^2 are variances of X and Y respectively.

(*c*) If X_1 and X_2 are independent normal variates and U and V are defined by :

$$U = X_1 \cos \alpha + X_2 \sin \alpha, \quad V = X_2 \cos \alpha - X_1 \sin \alpha,$$

show that the correlation coefficient ρ between U and V is given by :

$$\rho^2 = 1 - \frac{4\sigma_1^2\sigma_2^2}{4\sigma_1^2\sigma_2^2 + \left(\sigma_1^2 - \sigma_2^2\right)\sin^2 2\alpha},$$

where σ_1^2 and σ_2^2 are variances of X_1 and X_2 respectively.

10·5. $X_1, X_2, \ldots, X_n$ are uncorrelated random variables, all with the same distribution and zero means. Let $\bar{X} = \sum_i X_i/n$. Find the correlation coefficient between (*i*) X_i and $\bar{X}$ and (*ii*) $X_i - \bar{X}$ and $\bar{X}$.

Hint. $$r(X_i, \bar{X}) = \frac{\sigma^2/n}{\sqrt{\sigma^2 \cdot \sigma^2/n}} = \frac{1}{\sqrt{n}}$$

$$\text{Cov}\left(X_i - \bar{X}, \bar{X}\right) = \text{Cov}(X_i, \bar{X}) - \text{Var}(\bar{X}) = (\sigma^2/n) - (\sigma^2/n) = 0$$

$$\therefore \quad r\left(X_i - \bar{X}, \bar{X}\right) = 0$$

10·6. $X_1, X_2, \ldots, X_n$ are random variables each with the same expected value μ and *s.d.* σ. The correlation coefficient between any two X's is ρ. Show that:

(*i*) $\text{Var}(\bar{X}) = \dfrac{\sigma^2}{n} + \left(1 - \dfrac{1}{n}\right)\rho\sigma^2$, (*ii*) $E\sum_{i=1}^{n}(X_i - \bar{X})^2 = (n-1)(1-\rho)\sigma^2$, and (*iii*) $\rho > -\dfrac{1}{n-1}$

10.7. If X and Y are independent random variables, show that

$$r(X + Y, X - Y) = r^2 (X, X + Y) - r^2(Y, X + Y).$$

where $r(X + Y, X - Y)$ denotes the coefficient of correlation between $(X + Y)$ and $(X - Y)$.

10·8. (*a*) X_1 and X_2 are two variables with zero means, variances σ_1^2 and σ_2^2, respectively and r is the correlation coefficient between them. Determine the values of the constants a and b which are independent of r such that $X_1 + aX_2$ and $X_1 + bX_2$ are uncorrelated.

10·9. (*a*) If $U = aX + bY$ and $V = bX - aY$, where X and Y are measured from their respective means and if U and V are uncorrelated, the coefficient of correlation (r) between X and Y is given by the equation:

$$\sigma_U \sigma_V = \left(a^2 + b^2\right)\sigma_X \sigma_Y \left(1 - r^2\right)^{1/2}.$$

(*b*) Let $U = aX + bY$ and $V = aX - bY$ where X, Y represent deviations from the means of two measurements on the same individual. The coefficient of correlation between X and Y is r. If U, V are uncorrelated, show that : $\sigma_U \sigma_V = 2\, a\, b\, \sigma_X \sigma_Y (1 - r^2)^{1/2}$

10·10. (*a*) Show that, if a and b are constants and r is the correlation coefficient between X and Y, then the correlation coefficient between aX and bY is equal to r if the signs of a and b are alike, and to $-r$ if they are different.

Also show that, if constants a, b and c are positive, the correlation coefficient between $(aX + bY)$ and cY is equal to

$$(a r \sigma_X + b \sigma_Y)/\sqrt{\left(a^2\sigma_X^2 + b^2\sigma_Y^2 + 2a\, b\, r\, \sigma_X\, \sigma_Y\right)}.$$

10·11. If $U = aX + bY$ and $V = bX - aY$, show that U and V are uncorrelated if

$$\frac{ab}{a^2 - b^2} = \frac{\rho \sigma_X \sigma_Y}{\sigma_X^2 - \sigma_Y^2},$$

where, ρ is the coefficient of correlation between X and Y. Show further that in this case

$$\sigma_U^2 + \sigma_V^2 = \left(a^2 + b^2\right)\left(\sigma_X^2 + \sigma_Y^2\right) \quad \text{and} \quad \sigma_U \sigma_V = \left(a^2 + b^2\right)\sigma_X \sigma_Y \sqrt{1 - \rho^2}$$

10·12. If $U = X + kY$ and $V = X + mY$ and r is the correlation coefficient between X and Y, find the correlation coefficient between U and V. Show that U and V are uncorrelated if

$$k = \frac{-\sigma_X(\sigma_X + rm\sigma_Y)}{\sigma_Y(r\sigma_X + m\sigma_Y)} \quad \text{and further if} \quad m = \frac{\sigma_X}{\sigma_Y}, \quad \text{then} \quad k = -\frac{\sigma_X}{\sigma_Y}.$$

10·13. (*a*) X_1, X_2, X_3 are three variables, each with variance σ^2 and the correlation coefficient between any two of them is r. If $\bar{X} = (X_1 + X_2 + X_3)/3$, show that $\text{Var}(\bar{X}) = \dfrac{\sigma^2}{3}(1 + 2r)$.

Deduce that $r \geq -\dfrac{1}{2}$.

(*b*) If X_1, X_2 and X_3 are three random variables measured from their respective means as origin and of equal variances, find the coefficient of correlation between $X_1 + X_2$ and $X_2 + X_3$ in terms of r_{12}, r_{13} and r_{23} and show that it is equal to.

(*i*) $\dfrac{r_{12} + 1}{2}$, if $r_{13} = r_{23} = 0$, and (*ii*) $\dfrac{r_{12} + 3}{4}$, if $r_{13} = r_{23} = 1$

10·14. (*a*) The following are the marks obtained by 24 students in a class test of Mathematics and Statistics:

Roll No. of Student	:	1	2	3	4	5	6	7	8	9	10	11	12
Marks in Mathematics	:	15	0	1	3	16	2	18	5	4	17	6	19
Marks in Statistics	:	13	1	2	7	8	9	12	9	17	16	6	18
Roll No. of Student	:	13	14	15	16	17	18	19	20	21	22	23	24
Marks in Mathematics	:	14	9	8	13	10	13	11	11	12	18	9	7
Marks in Statistics	:	11	3	5	4	10	11	14	7	18	15	15	3

Form a two-way bivariate frequency distribution and obtain the correlation table between marks in two subjects.

(*b*) Find the coefficient of correlation between the ages of 100 mothers and daughters:

Age of mothers in years (X)	*Age of daughters in years (Y)* 5–10	10–15	15–20	20–25	25–30	*Total*
15–25	6	3				9
25–35	3	16	10			29
35–45		10	15	7		32
45–55			7	10	4	21
55–65				4	5	9
Total	9	29	32	21	9	100

10·15. (*a*) Given the following frequency distribution of (X, Y), find the frequency distribution of (U, V) where

Y↓ X→	5	10	*Total*
10	30	20	50
20	20	30	50
Total	50	50	100

$$U = \frac{X - 7{\cdot}5}{2{\cdot}5}, \quad V = \frac{Y - 15}{5}$$

What shall be the relationship between the correlation coefficient between X, Y, and U, V?

(*b*) Find the coefficient of correlation between X and Y for the table below:

X↓ Y→	y_1	y_2	*Total*
x_1	p_{11}	p_{12}	P
x_2	p_{21}	p_{22}	Q
Total	P'	Q'	1

10·16. Consider the following probability distribution :

X↓ \ Y→	0	1	2
0	0·1	0·2	0·1
1	0·2	0·3	0·1

Calculate $E(X)$, Var (X), Cov (X, Y) and $r(X, Y)$.

10·17. Let the random variables X and Y have the joint *p.d.f.*

$$f(x, y) = \frac{1}{3}; (x, y) = (0, 0), (1, 1) (2, 0)$$

Compute $E(X)$, $V(X)$, $E(Y)$, $V(Y)$ and $r(X, Y)$. Are X and Y stochastically independent? Give reasons.

10·18. (*a*) Let (X, Y) have the probability distribution :

$$f(0, 0) = 0{\cdot}45, \quad f(0, 1) = 0{\cdot}05, \quad f(1, 0) = 0{\cdot}35, \quad f(1, 1) = 0{\cdot}15.$$

Evaluate $V(X)$, $V(Y)$ and $r(X, Y)$. Show that while X and Y are correlated, X and $X - 5Y$ are uncorrelated. Are X and $X - 5Y$ independent ?

(*b*) Give the bivariate probability distribution :

$f(-1, 0) = 1/15$, $f(-1, 1) = 3/15$, $f(-1, 2) = 2/15$

$f(0, 0) = 2/15$, $f(0, 1) = 2/15$, $f(0, 2) = 1/15$

$f(1, 0) = 1/15$, $f(1, 1) = 1/15$, $f(1, 2) = 2/15$

$f(x, y) = 0$ elsewhere.

Obtain :

(*i*) The marginal distributions of X and Y.

(*ii*) The conditional distributions of Y given $X = 0$.

(*iii*) $E(Y \mid X = 0)$.

(*iv*) The product moment correlation coefficient between X and Y.

Are X and Y independently distributed:?

10·19. (*a*) The joint *p.d.f.* of two variates X and Y is given by :

$$f(x, y) = k\left[(x + y) - \left(x^2 + y^2\right)\right]; \; 0 < (x, y) < 1$$
$$= 0; \text{ otherwise}$$

Show that X and Y are uncorrelated but not independent.

(*b*) If the random variables X and Y have the joint *p.d.f.*,

$$f(x, y) = \begin{cases} x + y; 0 < x < 1, 0 < y < 1 \\ 0, \text{ elsewhere} \end{cases}$$

then show that the correlation coefficient between X and Y is $-\frac{1}{11}$.

(*c*) The density function $f(x)$ of a random variable X is given by :

$$f(x) = \begin{cases} kx^2, & \text{if } -1 \le x \le 1 \\ 0, & \text{otherwise} \end{cases}$$

(*i*) What is the value of k ? What is the distribution function of X ?

(*ii*) Obtain the correlation coefficient between X and Y.

(*iii*) Are X and Y independently distributed ?

10·20. (*a*) Calculate the coefficient of correlation for ranks from the following data :

(X, Y) : (5, 8), (10, 3), (6, 2), (3, 9), (19, 12), (5, 3), (6, 17), (12, 18), (8, 22), (2, 12), (10, 17), (19, 20).

(*b*) Ten recruits were subjected to a selection test to ascertain their suitability for a certain course of training. At the end of training they were given a proficiency test.

The marks secured by recruits in the selection test (*X*) and in the proficiency test (*Y*) are given below:

Serial No.	1	2	3	4	5	6	7	8	9	10
X	10	15	12	17	13	16	24	14	22	20
Y	30	42	45	46	33	34	40	35	39	38

Calculate product moment correlation coefficient and rank correlation coefficient. Why are two coefficients different.

10·21. (*a*) Ten competitors in a beauty contest are ranked by three judges as follows :

	Competitors									
Judges	1	2	3	4	5	6	7	8	9	10
A	6	5	3	10	2	4	9	7	8	1
B	5	8	4	7	10	2	1	6	9	3
C	4	9	8	1	2	3	10	5	7	6

Discuss which pair of judges has the nearest approach to common tastes of beauty.

(*b*) A sample of 12 fathers and their eldest sons gave the following data about their height in inches:

Father	65	63	67	64	68	62	70	66	68	67	69	71
Son	68	66	68	65	69	66	68	65	71	67	68	70

Calculate coefficient of rank correlation.

(*c*) Obtain the rank correlation coefficient between the variables X and Y from the following pairs of observed values :

X	50	55	65	50	55	60	50	65	70	75
Y	110	110	115	125	140	115	130	120	115	160

Hint.

X	R_1	Y	R_2	$d = R_1 - R_2$	d^2
50	2	110	1·5	0·5	0·25
55	4·5	110	1·5	3.0	9·00
65	7·5	115	4	3·5	12·25
50	2	125	7	– 5·0	25·00
55	4·5	140	9	– 4·5	20·25
60	6	115	4	2·0	4·00
50	2	130	8	– 6·0	36·00
65	7·5	120	6	1·5	2.25
70	9	115	4	5·0	25·00
75	10	160	10	0	0
Total				$\Sigma d = 0$	$\Sigma d^2 = 134$

$$\rho = 1 - \frac{6\left\{134 + \frac{1}{12}(3^3 - 3) + \frac{1}{12}(2^3 - 2) + \frac{1}{12}(2^3 - 2) + \frac{1}{12}(2^3 - 2) + \frac{1}{12}(3^3 - 3)\right\}}{10\,(100 - 1)}$$

$$= 1 - \frac{6 \times 139{\cdot}5}{990} = 1 - 0{\cdot}845 = 0{\cdot}155.$$

10·22. (*a*) The coefficient of rank correlation between marks in Statistics and marks in Mathematics obtained by a certain group of students is 0·8. If the sum of the squares of the difference in ranks is given to be 33, find the number of students in the group.

(*b*) The coefficient of rank correlation between the marks obtained by 10 students in Mathematics and Statistics was found to be 0·5. It was later discovered that the difference in ranks in two subjects obtained by one of the students was wrongly taken as 3 instead of 7. Find the correct coefficient of rank correlation.

Hints. $0{\cdot}5 = 1 - \dfrac{6\Sigma d^2}{10 \times 99} \quad \Rightarrow \quad \Sigma d^2 = \dfrac{990}{6 \times 2} = 82{\cdot}5$

Since one difference was wrongly taken as 3 instead of 7, the correct value of Σd^2 is given by :

$$\text{Corrected } \Sigma d^2 = 82{\cdot}5 - (3)^2 + (7)^2 = 122{\cdot}5$$

$$\therefore \quad \text{Corrected } \rho = 1 - \frac{6 \times 122{\cdot}5}{10 \times 99} = 0{\cdot}2576$$

10·23. If d_i be the difference in the ranks of the *i*th individual in two different characteristics, then show that the maximum value of $\sum_{i=1}^{n} d_i^2$ is $\frac{1}{3}(n^3 - n)$. Hence, or otherwise, show that rank correlation coefficient lies between –1 and 1

10·24. Let $x_1, x_2, \ldots, x_n$ be the ranks of n individuals according to a character A and $y_1, y_2, \ldots, y_n$ be the corresponding ranks of the same individuals according to other character B. Obviously $(x_1, x_2, \ldots, x_n)$ and $(y_1, y_2, \ldots, y_n)$ are permutations of 1, 2, ..., n. It is given that $x_i + y_i = 1 + n$, for $i = 1, 2, \ldots, n$. Show that the value of the rank correlation coefficient ρ between the characters A and B is –1.

Hint. We are given

$$x_i + y_i = n + 1 \;\forall\; i = 1, 2, \ldots, n$$

Also $\quad x_i - y_i = d_i$

$$\therefore \quad 2x_i = n + 1 + d_i \quad \Rightarrow \quad d_i = 2x_i - (n + 1)$$

$$\therefore \quad \sum_{i=1}^{n} d_i^2 = \sum_{i=1}^{n} \left[4x_i^2 + (n+1)^2 - 2(n+1)\,2x_i\right]$$

$$= 4\frac{n(n+1)(2n+1)}{6} + n(n+1)^2 - \frac{4(n+1)\,n(n+1)}{2} = \frac{n(n^2-1)}{3}$$

$$\therefore \quad \rho = 1 - \frac{6\sum_{i=1}^{n} d_i^2}{n(n^2-1)} = -1$$

10·25. If X_1 and X_2 are two independent normal variates with a common mean zero and variances σ_1^2 and σ_2^2, respectively, show that the variates defined by :

$$U_1 = X_1 + X_2 \quad \text{and} \quad U_2 = -\frac{\sigma_2}{\sigma_1} X_1 + \frac{\sigma_1}{\sigma_2} X_2$$

are independent and that each is normally distributed with mean zero and common variance $(\sigma_1^2 + \sigma_2^2)$.

10·26. If X_1, X_2 and X_3 are uncorrelated variables with equal mean M and variances V_1^2, V_2^2 and V_3^2, respectively, prove that correlation coefficient ρ between $Z_1 = (X_1/X_3)$ and $Z_2 = (X_2/X_3)$ is given by:

$$\rho = \frac{V_3^2}{\sqrt{\{(V_1^2 + V_3^2)(V_2^2 + V_3^2)\}}}$$

10·27. If X is a standard normal variate and $Y = a + bX + cX^2$, where a, b, c are constants, find the correlation coefficient between X and Y. Hence or otherwise obtain the conditions when :
(i) X and Y are uncorrelated, and (ii) X and Y are perfectly correlated.

Ans: $r(X,Y) = \dfrac{b}{\sqrt{b^2 + 2c^2}}$; (i) $b = 0$, (ii) $c = 0$.

10·28. (a) If X_i, Y_j and Z_k are all independent random variables with mean zero and unit variance, find the correlation coefficient between $U = \sum_{i=1}^{m} X_i + \sum_{j=1}^{n} Y_j$ and $V = \sum_{i=i}^{m} X_i + \sum_{k=1}^{n} Z_k$.

Ans: $r(U,V) = \dfrac{m}{m+n}$

10·28. (b) *(Wedon's Dice Problem).* n white dice and m red dice are shaken together and thrown on a table.The sum of the dots on the upper faces are noted. The red dice are then picked up and thrown again among the white dice left on the table. The sum of the dice on the upper faces is again noted. What is the correlation coefficient between the first and the second sums?

Ans : $n / (m + n)$

10·29. If the variables $X_1, X_2, \ldots, X_{2n}$, all have the same variance σ^2 and the correlation coefficient between X_i and X_j $(i \neq j)$ has the same value, show that the correlation between $\sum_{i=1}^{n} X_i$ and $\sum_{j=n+1}^{2n} X_j$ is given by:

$$\left[\frac{n\rho}{1+(n-1)\rho}\right]$$

10·30. The means of independent *r.v.*'s $X_1, X_2, \ldots, X_n$ are zero and variances are equal, say, unity. The correlation coefficients between the sum of selected t $(< n)$ variables out of these variables and the sum of all n variables are found out. Prove that the sum of squares of all these correlation coefficients is ${}^{n-1}C_{t-1}$.

10·31. Two variables U and V are made up of the sum of a number of terms as follows :

$$U = X_1 + X_2 + \ldots + X_n + Y_1 + Y_2 + \ldots + Y_a,$$
$$V = X_1 + X_2 + \ldots + X_n + Z_1 + Z_2 + \ldots + Z_b,$$

where a and b are all suffixes and where X's, Y's and Z's are all uncorrelated standardised random variables. Show that the correlation coefficient between U and V is $\dfrac{n}{\sqrt{(n+a)(n+b)}}$.

Show further that $\xi = \sqrt{(n+b)} \cdot U + \sqrt{(n+a)} \cdot V$ and $\eta = \sqrt{(n+b)} \cdot U - \sqrt{(n+a)} \cdot V$ are uncorrelated.

10·32. Let the number X be chosen at random from among the integers 1, 2, 3, 4 and the number Y be chosen from among those at least as large as X. Prove that Cov (X, Y) = 5/8. Also find $r(X, Y)$.

Ans : $r(X,Y) = \dfrac{2}{\sqrt{11}}$.

10·33. Two ideal dice are thrown. Let X_1 be the score on the first die and X_2, the score on the second die. Let $Y = \max\,[X_1, X_2]$. Obtain the joint distribution of Y and X_1 and show that

$$\text{Corr}\,(Y, X_1) = \frac{3\sqrt{3}}{\sqrt{73}}$$

10·34. Three fair coins are tossed. Let X denote the number of heads on the first two coins and let Y denote the number of tails on the last two coins.

(i) Find the joint distribution of X and Y.

(*ii*) Find the conditional distribution of Y given that $X = 1$.
(*iii*) Find Cov. (X, Y) and $r(X, Y)$.

Ans. (*iii*) $\text{Cov}(X, Y) = -1/4$; $\sigma_x = \sigma_y = 1/\sqrt{2}$; $r(X,Y) = -0\cdot5$.

10·35. For the trinomial distribution of two random variables X and Y :

$$f(x, y) = \frac{n!}{x!\,y!\,(n-x-y)!}\, p^x q^y (1-p-q)^{n-x-y}$$

for $x, y = 0, 1, 2, \ldots, n$ and $x + y \le n$, $p \ge 0$, $q \ge 0$ and $p + q \le 1$.
(*i*) Obtain the marginal distribution of Y. (*ii*) Obtain $E(X \mid Y = y)$. (*iii*) Find $\rho(X, Y)$.

Ans. (*i*) $X \sim B(n,p),\ Y \sim B(n,q)$; $(X|Y = y) \sim B\left(n-y, \dfrac{p}{1-q}\right)$

(*ii*) $E(X \mid Y = y) = (n-y)\left(\dfrac{p}{1-q}\right); \begin{Bmatrix} x = 0,1,2,\cdots,n \\ y = 0,1,2,\cdots,n \end{Bmatrix}$

(*iii*) $\text{Cov}(X, Y) = -\,npq$; $\rho(x,y) = -\left[\dfrac{pq}{(1-p)(1-q)}\right]^{1/2}$

CHAPTER

11 Curve Fitting and Regression Analysis

LEARNING OBJECTIVES

Upon completion of this Chapter, you should be able to:

- Understand the meaning of regression, and its role in statistical analysis.
- Obtain the equations of the regression lines (lines of best fit), by using the concept of 'residuals' and, 'the principle of least squares'.
- Differentiate between linear and non-linear (curvi-linear) regression.
- Understand the concepts and meaning of regression coefficients, and their relationship with correlation coefficient.
- Demonstrate the meaning of curvilinear regression and fitting of non-linear functions, using the Principle of Least Squares.
- Explain the concept and computation of regression curves for means.
- Appreciate the use of regression analysis for estimation and prediction purposes.
- Understand the concepts of correlation ratio and intra-class correlation, their computation and interpretation.
- Explain:
  (i) The difference between correlation ratio and correlation analysis and
  (ii) Inter-relations between them.

CHAPTER OUTLINES

11·1. INTRODUCTION

The term *"regression"* linterally means *"stepping back towards the average"*. It was first used by British biometrician Sir Francis Galton (1822–1911), in connection with the inheritance of stature. Galton found that the offsprings of abnormally tall or short parents tend to *"regress"* or *"step back"* to the average population height. But the term *"regression"* as now used in Statistics is only a convenient term without having any reference to biometry.

Defintion. *Regression analysis is a mathematical measure of the average relationship between two or more variables in terms of the original units of the data.*

In regression analysis there are two types of variables. The variable whose value is influenced or is to be predicted is called *dependent variable* and the variable which influences the values or is used for prediction is called *independent variable.* In regression analysis independent variable is also known as *regressor or predictor or explanatory variable while* the dependent variable is also known as regressed or *explained variable.*

11·2. LINEAR REGRESSION

If the variables in a bivariate distribution are related, we will find that the points in the scatter diagram will cluster round some curve called the *"curve of regression"*. If the curve is a straight line, it is called the line of regression and there is said to be *linear regression* between the variables, otherwise regression is said to be *curvilinear.*

The line of regression is the line which gives the best estimate to the value of one variable for any specific value of the other variable. Thus the line of regression is the line of *"best fit"* and is obtained by the *principle of least squares.*

Let us suppose that in the bivariate distribution (x_i, y_i); $i = 1, 2, \ldots, n$; Y is dependent variable and X is independent variable. Let the line of regression of Y on X be

$$Y = a + bX \qquad \text{...(11·1)}$$

The above equation (11·1) represents a family of straight lines for different values of the arbitrary constants *'a'* and *'b'*. The problem is to determine *'a'* and *'b'* so that the line (11·1) is the line of *'best fit'*.

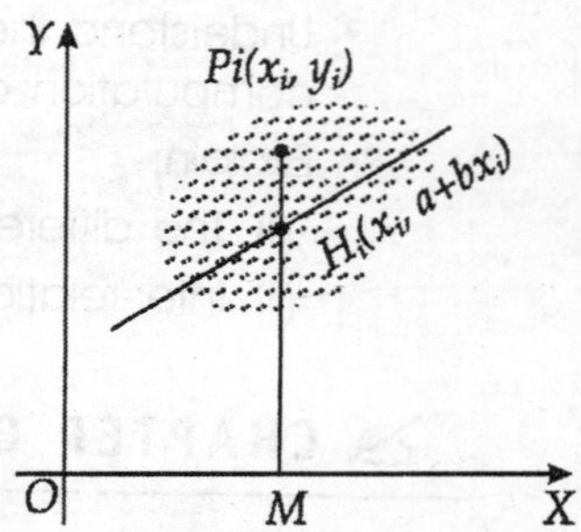

Fig. 11·1

The term *'best fit'* is interpreted in accordance with Legender's principle of least squares which consists in minimising the sum of the squares of the deviations of the actual values of y from their estimated values as given by the line of best fit.

Let $P_i(x_i, y_i)$ be any general point in the scatter diagram (§10·2). Draw $P_iM \perp$ to x-axis meeting the line, (11·1) in H_i. Abscissa of H_i is x_i and since H_i lies on (11·1), its ordinate is $a + bx_i$. Hence the co-ordinates of H_i are $(x_i, a + bx_i)$.

$$P_iH_i = P_iM - H_iM = y_i - (a + bx_i).$$

is called the *error of estimate* or the *residual* for y_i.

According to the principle of least squares, we have to determine a and b so that

$$\sum_{i=1}^{n} P_iH_i^2 = \sum_{i=1}^{n} (y_i - a - bx_i)^2$$

in minimum. From the principle of maxima and minima, the partial derivatives of E, *w.r.* to a and b should vanish separately, *i.e.,*

$$\frac{\partial E}{\partial a} = 0 = -2\sum_{i=1}^{n}(y_i - a - bx_i) \quad \text{and} \quad \frac{\partial E}{\partial b} = 0 = -2\sum_{i=1}^{n}x_i(y_i - a - bx_i)$$

$$\Rightarrow \quad \sum_{i=1}^{n} y_i = na + b\sum_{i=1}^{n} x_i \quad ...(11\cdot2) \quad \text{and} \quad \sum_{i=1}^{n} x_i y_i = a\sum_{i=1}^{n} x_i + b\sum_{i=1}^{n} x_i^2 \quad ...(11\cdot3)$$

Remark. Equations (11·2) and (11·3) are known as the *normal equations* for estimating *a* and *b*. All the quantities $\sum_{i=1}^{n} x_i, \sum_{i=1}^{n} x_i^2, \sum_{i=1}^{n} y_i$ and $\sum_{i=1}^{n} x_i y_i$, can be obtained from the given set of points (x_i, y_i); $i = 1, 2, ..., n$ and the equations (11·2) and (11·3) can be solved for *a* and *b*. With the values of *a* and *b* so obtained, equation (11·1) is the line of best fit to the given set of points (x_i, y_i); $i = 1, 2, ..., n$.

From (11·2) on dividing by *n*, we get

$$\overline{Y} = a + b\overline{X} \quad ...(11\cdot4)$$

Thus, the line of regression of *Y* on *X* passes through the point $(\overline{X}, \overline{Y})$.

$$\mu_{11} = \text{Cov}(X, Y) = \frac{1}{n}\sum_{i=1}^{n} x_i y_i - \overline{x}\,\overline{y} \quad \Rightarrow \quad \frac{1}{n}\sum_{i=1}^{n} x_i y_i = \mu_{11} + \overline{X}\,\overline{Y} \quad ...(11\cdot5)$$

Also

$$\sigma_X^2 = \frac{1}{n}\sum_{i=1}^{n} x_i^2 - \overline{X}^2 \quad \Rightarrow \quad \frac{1}{n}\sum_{i=1}^{n} x_i^2 = \sigma_X^2 + \overline{X}^2 \quad ...(11\cdot5a)$$

Dividing (11·3) by *n* and using (11·5) and (11·5*a*), we get

$$\mu_{11} + \overline{X}\,\overline{Y} = a\overline{X} + b(\sigma^2{}_X + \overline{X}^2) \quad ...(11\cdot6)$$

Multiplying (11·4) by $\overline{X}$ and then subtracting from (11·6), we get

$$\mu_{11} = b\,\sigma_X^2 \quad \Rightarrow \quad b = \frac{\mu_{11}}{\sigma_X^2} \quad ...(11\cdot7)$$

Since '*b*' is the slope of the line of regression of *Y* on *X* and since the line of regression passes through the point $(\overline{X}, \overline{Y})$, its equation is :

$$Y - \overline{Y} = b(X - \overline{X}) = \frac{\mu_{11}}{\sigma_X{}^2}(X - \overline{X})^* \quad ...(11\cdot8)^*$$

$$\Rightarrow \quad Y - \overline{Y} = r\frac{\sigma_Y}{\sigma_X}(X - \overline{X}) \quad ...(11\cdot8a)$$

Starting with the equation $X = A + BY$ and proceeding similarly or by simply interchanging the variables *X* and *Y* in (11·8) and (11·8*a*), the equation of the line of regression of *X* and *Y* becomes:

---

* The general problem in curve fitting is to find, if possible, an analytic expression of the form $y = f(x)$, for the functional relationship suggested by the given data. Fitting of curves to a set of numerical data is of considerable importance, theoretical as well as practical. Theoretically it is useful in the study of correlation and regression, *e.g.*, lines of regression can be regarded as fitting of linear curves to the given bivariate distribution. In practical statistics it enables us to represent the relationship between two variables by simple algebraic expressions, *e.g.*, polynomials, exponential or logarithmic functions. Moreover, it may be used to estimate the values of one variable which would correspond to the specified values of the other variable.

$$X - \bar{X} = \frac{\mu_{11}}{\sigma_{Y^2}}(Y - \bar{Y}) \qquad ...(11{\cdot}9)$$

$$\Rightarrow \qquad X - \bar{X} = r\frac{\sigma_X}{\sigma_Y} = r\frac{\sigma_X}{\sigma_Y}(Y - \bar{Y}) \qquad ...(11{\cdot}9a)$$

Aliter. The straight line defined by $Y = a + bX$ ...(*i*)
and satisfying the residual (least square) condition: $S = E\left[(Y - a - bX)^2\right] =$ Minimum ...(*ii*)
for variations in a and b, is called the *line of regression of Y on X*.

The necessary and sufficient conditions for a minima of S, subject to variations in a and b are :

$$(i)\ \frac{\partial S}{\partial a} = 0\ ;\ \frac{\partial S}{\partial b} = 0 \quad ...(*) \quad \text{and} \quad (ii)\ \Delta = \begin{vmatrix} \dfrac{\partial^2 S}{\partial a^2} & \dfrac{\partial^2 S}{\partial a\,\partial b} \\ \dfrac{\partial^2 S}{\partial b\,\partial a} & \dfrac{\partial^2 S}{\partial b^2} \end{vmatrix} > 0 \ \text{ and } \ \frac{\partial^2 S}{\partial a^2} > 0 \qquad ...(**)$$

Using (*), we get

$$\frac{\partial S}{\partial a} = -2\,E\,[Y - a - bX] = 0 \qquad ...(iii)$$

$$\frac{\partial S}{\partial b} = -2\,E[X\,(Y - a - bX)] = 0 \qquad ...(iv)$$

$$\Rightarrow \qquad E(Y) = a + bE(X) \qquad ...(v)$$

$$\text{and} \qquad E(XY) = aE(X) + bE(X^2) \qquad ...(vi)$$

Equation (*v*) implies that the line (*i*) of regression of Y on X passes through the mean value $[E(X), E(Y)]$.

Multiplying (*v*) by $E(X)$ and subtracting from (*vi*), we get

$$E(XY) - E(X)\,E(Y) = b\left[E(X^2) - \{E(X)\}^2\right]$$

$$\Rightarrow \qquad \text{Cov}(X, Y) = b\sigma_X^2 \quad \Rightarrow \quad b = \frac{\text{Cov}(X, Y)}{\sigma_X^2} = \frac{r\sigma_Y}{\sigma_X} \qquad ...(vii)$$

Subtracting (*v*) from (*i*) and using (*vii*), we obtain the equation of line of regression of Y on X as:

$$Y - E\,(Y) = \frac{\text{Cov}(X, Y)}{\sigma_X^2}[X - E(X)] \quad \Rightarrow \quad Y - E(Y) = \frac{r\sigma_Y}{\sigma_X}[X - E(X)]$$

Similarly, the straight line defined by $X = A + BY$ and satisfying the residual condition $E[X - A - BY]^2 =$ Minimum, is called the line of regression of X on Y.

Remarks:

1. We note that

$$\frac{\partial^2 S}{\partial a^2} = 2 > 0, \quad \frac{\partial^2 S}{\partial b^2} = 2E(X^2) \quad \text{and} \quad \frac{\partial^2 S}{\partial a\,\partial b} = 2E(X)$$

Substituting in (**), we have

$$\Delta = \frac{\partial^2 S}{\partial a^2}\cdot\frac{\partial^2 S}{\partial b^2} - \left(\frac{\partial^2 S}{\partial a\,\partial b}\right)^2 = 4\left[E(X^2) - \{E(X)\}^2\right] = 4\sigma_X^2 > 0$$

Hence, the solution of the least square equations (*iii*) and (*iv*), in fact provides a minima of S.

2. The regression equation (11·8*a*) implies that the line of regression of Y on X passes through the point $(\bar{X}, \bar{Y})$. Similarly (11·9*a*) implies that the line of regression of X on Y also passes through the point $(\bar{X}, \bar{Y})$. Hence, both the lines of regression pass through the point $(\bar{X}, \bar{Y})$. In other words, the mean values $(\bar{X}, \bar{Y})$. can be obtained as the point of intersection of the two regression lines.

3. *Why two lines of Regression?* There are always two lines of regression, one of Y on X and the other of X on Y. The line of regression of Y on X (11·8*a*) is used to estimate or predict the value of Y for any given value of X, *i.e.*, when Y is a dependent variable and X is an independent variable. The estimate so obtained will be best in the sense that it will have the minimum possible error as defined by the principle of least squares. We can also obtain an estimate of X for any given value of Y by using equation (11·8*a*) but the estimate so obtained will not be best since (11·8*a*) is obtained on minimising the sum of the squares of errors of estimates in Y and not in X. Hence to estimate or predict X for any given value of Y, we use the regression equation of X on Y (11·9*a*) which is derived on minimising the sum of the squares of errors of estimates in X. Here X is a dependent variable and Y is an independent variable. The two regression equations are not reversible or interchangeable because of the simple reason that the basis and assumptions for deriving these equations are quite different. The regression equation of Y on X is obtained on minimising the sum of the squares of the errors parallel to the Y-axis while the regression equation of X on Y is obtained on minimising the sum of squares of the errors parallel to the X-axis.

In a particular case of perfect correlation, positive or negative, *i.e.*, $r \pm 1$, the equation of line of regression of Y on X becomes :

$$Y - \bar{Y} = \pm \frac{\sigma_Y}{\sigma_X}(X - \bar{X}) \Rightarrow \frac{Y - \bar{Y}}{\sigma_Y} = \pm \left(\frac{X - \bar{X}}{\sigma_X}\right) \qquad \text{...(11·10)}$$

Similarly, the equation of the line of regression of X on Y becomes:

$$X - \bar{X} = \pm \frac{\sigma_X}{\sigma_Y}(Y - \bar{Y}) \Rightarrow \frac{Y - \bar{Y}}{\sigma_Y} = \pm \left(\frac{X - \bar{X}}{\sigma_X}\right)$$

which is same as (11·10).

Hence, in case of perfect correlation, $(r = \pm 1)$, both the lines of regression coincide. Therefore, in general we always have two lines of regression except in the particular case of perfect correlation when both the lines coincide and we get only one line.

4. Regression analysis has great practical utility in almost all disciplines and its applications are extended to natural, physical and social sciences.

11·2·1. Regression Coefficients.

'b', the slope of the line of regression of Y on X is also called the coefficient of regression of Y on X. It represents the increment in the value of dependent variable Y corresponding to a unit change in the value of independent variable X. More precisely, we write

$$b_{YX} = \text{Regression coefficient of } Y \text{ on } X = \frac{\mu_{11}}{\sigma_X^2} = r\frac{\sigma_Y}{\sigma_X} \qquad \text{...(11·11)}$$

Similarly, the coefficient of regression of X on Y indicates the change in the value of variable X corresponding to a unit change in the value of variable Y and is given by :

$$b_{XY} = \text{Regression coefficient of } X \text{ on } Y = \frac{\mu_{11}}{\sigma_Y^2} = r\frac{\sigma_X}{\sigma_Y} \qquad \text{...(11·11}a\text{)}$$

11·2·2. Properties of Regression Coefficients

(a) Correlation coefficient is the geometric mean between the regression coefficients.

Proof. Multiplying (11·11) and (11·11*a*), we get

$$b_{XY} \times b_{YX} = r\frac{\sigma_X}{\sigma_Y} \times r\frac{\sigma_Y}{\sigma_X} = r^2 \quad \Rightarrow \quad r = \pm\sqrt{b_{XY} \times b_{YX}} \quad \text{...(11·12)}$$

Remark. We have

$$r = \frac{\mu_{11}}{\sigma_X \sigma_Y}, \quad b_{YX} = \frac{\mu_{11}}{\sigma_X^2} \quad \text{and} \quad b_{XY} = \frac{\mu_{11}}{\sigma_Y^2}$$

It may be noted that the sign of correlation coefficient is the same as that of regression coefficients, since the sign of each depends upon the co-variance term μ_{11}. Thus, if the regression coefficients are positive, 'r' is positive and if the regression coefficients are negative 'r' is negative.

Hence, the sign to be taken before the square root in (11·12) is that of the regression coefficients.

(b) If one of the regression coefficients is greater than unity, the other must be less than unity.

Proof. Let one of the regression coefficients (say) b_{YX} be greater than unity, then we have to show that $b_{XY} < 1$.

$$b_{YX} > 1 \quad \Rightarrow \quad \frac{1}{b_{YX}} < 1 \quad \text{...(*)} \qquad \text{Also} \quad r^2 \le 1 \quad \Rightarrow \quad b_{YX} \cdot b_{XY} \le 1$$

Hence, $\quad b_{XY} \le \dfrac{1}{b_{XY}} < 1 \qquad$ [From (*)]

(c) The modulus value of the arithmetic mean of the regression coefficients is not less than the modulus value of the correlation coefficient r.

Proof. We have to prove that $\quad \left|\frac{1}{2}(b_{YX} + b_{XY})\right| \ge |r|$.

$$\Rightarrow \quad \left|\frac{1}{2}\left(r\frac{\sigma_Y}{\sigma_X} + r\frac{\sigma_X}{\sigma_Y}\right)\right| \ge |r| \qquad \Rightarrow \qquad \frac{\sigma_Y}{\sigma_X} + \frac{\sigma_X}{\sigma_Y} \ge 2 \qquad (\because |r| > 0)$$

$$\Rightarrow \quad \sigma_Y^2 + \sigma_X^2 - 2\sigma_X\sigma_Y \ge 0 \qquad \Rightarrow \qquad (\sigma_Y - \sigma_X)^2 \ge 0,$$

which is always true, since the square of a real quantity is ≥ 0.

(d) Regression coefficients are independent of the change of origin but not of scale.

Proof. Let $U = \dfrac{X-a}{h}$, $V = \dfrac{Y-b}{k} \quad \Rightarrow \quad X = a + hU, \; Y = b + kV$, which a, b, h (> 0) and k(> 0) are constants.

Then, $\quad \text{Cov}(X, Y) = hk\,\text{Cov}(U, V), \quad \sigma_X^2 = h^2\sigma_U^2 \quad \text{and} \quad \sigma_Y^2 = k^2\sigma_V^2$

$$\therefore \qquad b_{YX} = \frac{\text{Cov}(X, Y)}{\sigma_X^2} = \frac{hk\,\text{Cov}(U, V)}{h^2\sigma_U^2} = \frac{k}{h}\frac{\text{Cov}(U, V)}{\sigma_U^2} = \frac{k}{h}b_{VU}$$

Similarly, we can prove that: $b_{XY} = (h/k)\, b_{UV}$.

11·2·3. Angle Between Two Lines of Regression

Equations of the lines of regression of Y on X, and X on Y are :

$$Y - \bar{Y} = r \cdot \frac{\sigma_Y}{\sigma_X}(X - \bar{X}) \quad \text{and} \quad X - \bar{X} = r \cdot \frac{\sigma_X}{\sigma_Y}(Y - \bar{Y}) \quad \Rightarrow \quad Y - \bar{Y} = \frac{\sigma_Y}{r\sigma_X}(X - \bar{X})$$

Slopes of these lines are $r\dfrac{\sigma_Y}{\sigma_X}$ and $\dfrac{\sigma_Y}{r\sigma_X}$, respectively. If θ is the acute angle between the two lines of regression, then

$$\tan\theta = \left(\frac{r\dfrac{\sigma_Y}{\sigma_X} - \dfrac{\sigma_Y}{r\sigma_X}}{1 + r\dfrac{\sigma_Y}{\sigma_X}\cdot\dfrac{\sigma_Y}{r\sigma_X}}\right) = \left(\frac{r^2-1}{r}\right)\left(\frac{\sigma_X\sigma_Y}{\sigma_X^2+\sigma_Y^2}\right) = \frac{1-r^2}{|r|}\left(\frac{\sigma_X\sigma_Y}{\sigma_X^2+\sigma_Y^2}\right) \qquad (\because r^2 \le 1)$$

$$\therefore \qquad \theta = \tan^{-1}\left\{\frac{1-r^2}{|r|}\left(\frac{\sigma_X\sigma_Y}{\sigma_X^2+\sigma_Y^2}\right)\right\} \qquad \text{...(11·13)}$$

Case (i). ($r = 0$). If $r = 0$, $\tan\theta = \infty \Rightarrow \theta = \dfrac{\pi}{2}$. Thus, if the two variables are uncorrelated, the lines of regression become perpendicular to each other.

Case (ii). ($r = \pm 1$). If $r = \pm 1$, $\tan\theta = 0 \Rightarrow \theta = 0$ or π. In this case the two lines of regression either coincide or they are parallel to each other. But since both lines of regression pass through the point $(\bar{X}, \bar{Y})$, they cannot be parallel. Hence in the case of perfect correlation, positive or negative, the two lines of regression coincide.

Remarks:

1. *Whenever* two lines intersect, there are two angles between them, one acute angle and other obtuse angle. Further $\tan\theta > 0$ if $0 < \theta < \pi/2$, *i.e.,* θ is an acute angle and $\tan\theta < 0$ if $\pi/2 < \theta < \pi$, *i.e.,* θ is an obtuse a ngle and since $0 < r^2 < 1$, the acute angle (θ_1) and obtuse angle θ_2 between the two lines of regression are given by :

$$\theta_1 = \text{Acute angle} = \tan^{-1}\left\{\frac{\sigma_X\sigma_Y}{\sigma_X^2+\sigma_Y^2}\cdot\frac{1-r^2}{|r|}\right\}, \quad \text{and} \quad \theta_2 = \pi - \theta_1$$

2. When $r = 0$, *i.e.,* variables X and Y are uncorrelated, then the lines of regression of Y on X and X on Y are given respectively by:

[From (11·8*a*) and (11·9*a*)] $\quad Y = \bar{Y} \quad$ and $\quad X = \bar{X}$

as shown in the adjoining Fig. 11·2. Hence, in this case ($r = 0$), the lines of regression are perpendicular to each other and are parallel to X-axis and Y-axis respectively.

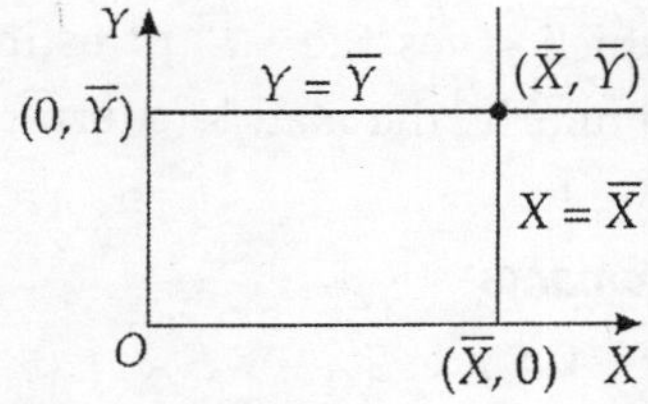

Fig. 11·2

3. The fact that if $r = 0$ (variables uncorrelated), the two lines of regression are perpendicular to each and if $r = \pm 1$, $\theta = 0$, *i.e.,* the two lines coincide, leads us to the conclusion that for higher degree of correlation between the variables, the angle between the lines is smaller, *i.e.,* the two lines of regression are nearer to each other. On the other hand, if the lines of regression make a larger angle, they indicate a poor degree of correlation between the variables and ultimately for $\theta = \pi/2$, $r = 0$, *i.e.,* the lines become perpendicular if no correlation exists between the variables. Thus, by plotting the lines of regression on a graph paper, we can have an approximate idea about the degree of correlation between the two variables under study. Consider the illustrations on page 11·8.

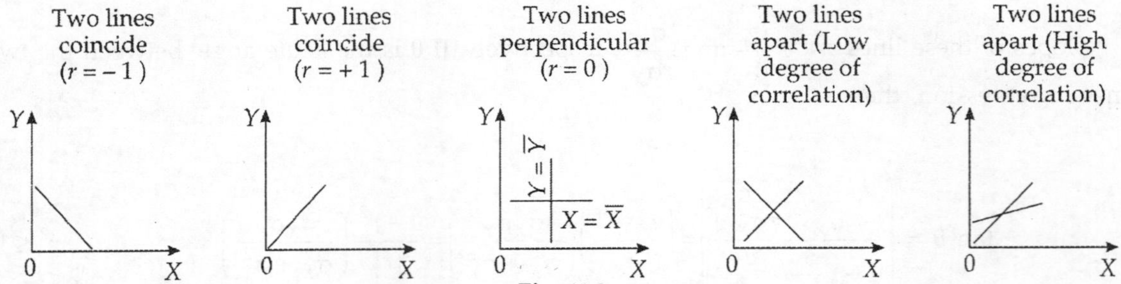

Fig. 11·3

11·2·4. Standard Error of Estimate or Residual Variance.

The regression equations enable us to estimate (predict) the value of the dependent variable for any given value of the independent variable. The estimates so obtained are, however, not perfect. A measure of the reliability of the estimates obtained from the fitted regression equation is provided by the *Standard Error (S.E.)* of the estimate which gives a measure of the scatter of the observations about the line of regression.

The equation of the line of regression of Y on X is :

$$Y - \bar{Y} = r\,\frac{\sigma_Y}{\sigma_X}(X - \bar{X}) \quad \Rightarrow \quad \frac{Y - \bar{Y}}{\sigma_Y} = r\left(\frac{X - \bar{X}}{\sigma_X}\right)$$

The residual variance S_{YX}^2 is the expected value of the squares of deviations of the observed values of Y from the expected values as given by the line of regression of Y on X. Thus,

$$S_{YX}^2 = E\left[Y - \hat{Y}\right]^2 = E\left[Y - \left\{\bar{Y} + (r\sigma_Y(X - \bar{X})/\sigma_X)\right\}\right]^2.$$

$$= \sigma_Y^2\,E\left[\frac{Y - \bar{Y}}{\sigma_Y} - r\left(\frac{X - \bar{X}}{\sigma_X}\right)\right]^2 = \sigma_Y^2\,E(Y^* - rX^*)^2,$$

where X^* and Y^* are standardised variates so that

$$E(X^{*2}) = 1 = E(Y^{*2}) \quad \text{and} \quad E(X^*\,Y^*) = r.$$

$$\therefore \quad S_{YX}^2 = \sigma_Y^2\left[E(Y^{*2}) + r^2\,E(X^{*2}) - 2r\,E(X^*\,Y^*)\right] = \sigma_Y^2(1 - r^2)$$

$$\Rightarrow \quad S_{YX} = \sigma_Y\,(1 - r^2)^{1/2} \qquad \text{...(11·13a)}$$

which gives the *S.E.* of estimate of Y for given X.

Similarly, the standard error of estimates of X, (for given Y), is given by :

$$S_{XY} = \sigma_X\,(1 - r^2)^{1/2} \qquad \text{...(11·13b)}$$

Remarks :

1. Since S_{XY}^2 or $S_{YX}^2 \geq 0$, it follows that $(1 - r^2) \geq 0 \;\Rightarrow\; |r| \leq 1$.

Hence, $\qquad -1 \leq r\,(X, Y) \leq 1$.

2. If $r = \pm 1$, $S_{XY} = S_{YX} = 0$ so that each deviation is zero, and the two lines of regression are coincident.

3. Since, as $r^2 \to 1$, S_{XY} and $S_{YX} \to 0$, it follows that departure of the value r^2 from unity indicates the departure of the relationship between the variables X and Y from linearity.

4. From the definition of linear regression, the minima condition implies that S_{YX}^2 or S_{XY}^2 is the minimum variance.

11·2·5. Explained and Unexplained Variation.

Consider the regression equation of Y on X for set of n points (X_i, Y_i); $i = 1, 2, \ldots, n$. Then, the total variation in Y is given by :

$$\sum_{i=1}^{n}(Y_i - \overline{Y})^2 = \sum_{i=1}^{n}\left[(Y_i - \hat{Y}_i) + (\hat{Y}_i - \overline{Y})\right]^2$$

$$\Rightarrow \quad \sum_{i=i}^{n}(Y_i - \overline{Y})^2 = \sum_{i=1}^{n}(Y_i - \hat{Y}_i)^2 + \sum_{i=1}^{n}(\hat{Y}_i - \overline{Y})^2 \quad \text{...(11·13}c\text{)}$$

the product term is zero, because

$$\sum_{i=1}^{n}(Y_i - \hat{Y}_i) = 0 \qquad \text{[By normal equation (11·2)]}$$

Thus, the total variation in Y can be split into two parts :

(*i*) **Un-explained Variation** $= \sum_{i=1}^{n}\left(Y_i - \hat{Y}_i\right)^2,$ ...(11·13*d*)

which is also known as the residual sum of squares, is the variation which is not explained by the linear regression of Y on X, and is due to some chance factors which are beyond our control and other characteristics not included in the study.

(*ii*) **Explained Variation** $= \sum_{i=1}^{n}(\hat{Y}_i - \overline{Y})^2,$ ...(11·13*e*)

i.e., the variation explained by linear regression of Y on X. This is also called *the sum of squares due to linear regression.*

The proportion of variability (in Y) explained by linear regression is given by :

$$\frac{\text{Explained Variation}}{\text{Total Variation}} = \frac{\sum_{i=1}^{n}(\hat{Y}_i - \overline{Y})^2}{\sum_{i=1}^{n}(Y_i - \overline{Y})^2} = r^2, \quad \text{...(11·13}f\text{)}$$

where 'r' is the correlation coefficient between X and Y.

r^2 is called the *coefficient of determination,* discussed below.

11·2·6. Coefficient of Determination.

A measure of linear relationship between two variables X and Y is provided by the correlation coefficient $r(X, Y)$. However, a more useful measure for this purpose is the *coefficient of determination,* which gives the percentage variation in the dependent variable that is accounted for by the independent variable.

Definition. *Coefficients of determination is given by the square of correlation coefficient.*

$$\text{Coefficient of determination} = r^2 = \frac{\text{Explained Variation}}{\text{Total Variation}}, \text{ as explained in (11·13}f\text{)}.$$

The coefficient of determination is a more useful measure for interpreting the value of r.

For example if $r = 0{\cdot}7$, we cannot say that 70% of the variation in dependent variable is explained by independent variable. But coefficient of determination $r^2 = 0{\cdot}7^2 = 0{\cdot}49$, which implies that only 49% of variation in dependent variable is explained by independent variable and the remaining 51% of the variation is due to other factors affecting the dependent variable but not included in the study.

Similarly, while comparing two correlation coefficients, say, $r_1 = 0{\cdot}3$ and $r_2 = 0{\cdot}6$, it will be misleading that correlation in the second case is double the correlation in the first case. The coefficients of determination in these cases are $r_1^2 = 0{\cdot}09$ and $r_2^2 = 0{\cdot}36$, from which we conclude that correlation in the second case is four times the correlation in the first case.

Thus, we observe that :

"The closeness of relationship between two variables as determined by correlation coefficient r is not proportional."

Example 11·1. *Obtain the equations of two lines of regression for the following data. Also obtain the estimate of X for Y = 70.*

X :	65	66	67	67	68	69	70	72
Y :	67	68	65	68	72	72	69	71

Solution. Let $U = X - 68$ and $V = Y - 69$, then

$$\bar{U} = 0,\ \bar{V} = 0,\ \sigma_U^2 = 4{\cdot}5,\ \sigma_V^2 = 5{\cdot}5,\ \text{Cov}(U, V) = 3 \text{ and } r(U, V) = 0{\cdot}6 \quad (c.f. \text{ page } 10{\cdot}8)$$

Since correlation coefficient is independent of change of origin, we get

$$r = r(X, Y) = r(U, V) = 0{\cdot}6$$

We know that if $U = \dfrac{X - a}{h},\ V = \dfrac{Y - b}{k}$, Then

$$\bar{X} = a + h\bar{U},\ \bar{Y} = b + k\bar{V},\ \sigma_X = h\sigma_U \text{ and } \sigma_Y = k\sigma_V$$

Here, we are given : $h = k = 1$, $a = 68$ and $b = 69$.

Thus, $\bar{X} = 68 + 0 = 68,\ \bar{Y} = 69 + 0 = 69$

$$\sigma_X = \sigma_U = \sqrt{4{\cdot}5} = 2{\cdot}12 \text{ and } \sigma_Y = \sigma_V = \sqrt{5{\cdot}5} = 2{\cdot}35$$

Equation of line of regression of Y on X is : $Y - \bar{Y} = r\dfrac{\sigma_Y}{\sigma_X}(X - \bar{X})$

i.e.,
$$Y = 69 + 0{\cdot}6 \times \frac{2{\cdot}35}{2{\cdot}12}(X - 68) \Rightarrow Y = 0{\cdot}665\,X + 23{\cdot}78$$

Equation of line of regression of X on Y is :

$$X - \bar{X} = r\frac{\sigma_X}{\sigma_Y}(Y - \bar{Y}) \Rightarrow X = 68 + 0{\cdot}6 \times \frac{2{\cdot}12}{2{\cdot}35}(Y - 69) \Rightarrow X = 0{\cdot}54Y + 30{\cdot}74$$

To estimate X for given Y, we use the line of regression of X on Y. If $Y = 70$, estimated value of X is given by : $\hat{X} = 0{\cdot}54 \times 70 + 30{\cdot}74 = 68{\cdot}54$, where $\hat{X}$ is estimate of X.

Example 11·2. *For a given set of bivariate data, the following results were obtained :*

$n = 100,\ \Sigma X = 5{,}000;\ \Sigma Y = 10{,}000;\ \Sigma X^2 = 2{,}60{,}000\ ;\ \Sigma Y^2 = 10{,}40{,}000$ *and* $\Sigma XY = 5{,}16{,}000$.

(i) Find the predicted value of Y when X = 60. (ii) What is the predicted value of X when Y = 80

(iii) Calculate the coefficient of correlation between X and Y.

Solution. Here, $\bar{X} = \dfrac{1}{n}\Sigma X = \dfrac{5000}{100} = 50,\quad \bar{Y} = \dfrac{1}{n}\Sigma Y = \dfrac{10000}{100} = 100$

$$b_{YX} = \text{Coefficient of regression of } Y \text{ on } X = r\,\frac{\sigma_Y}{\sigma_X} = \frac{\text{cov}(X, Y)}{\sigma_X^2}$$

$$= \frac{n\,\Sigma XY - (\Sigma X)(\Sigma Y)}{n\,\Sigma X^2 - (\Sigma X)^2} = \frac{100 \times 516000 - 5000 \times 10000}{100 \times 260000 - (5000)^2} = 1{\cdot}6,$$

$$\text{and } b_{XY} = \text{Coefficient of regression of } X \text{ on } Y = r\,\frac{\sigma_X}{\sigma_Y} = \frac{\text{cov}(X, Y)}{\sigma_X^2}$$

$$= \frac{n\,\Sigma XY - (\Sigma X)(\Sigma Y)}{n\,\Sigma Y^2 - (\Sigma Y)^2} = \frac{100 \times 516000 - 5000 \times 10000}{100 \times 1040000 - (10000)^2} = 0{\cdot}4$$

(*i*) Regression equation of Y on X is : $\quad Y - \bar{Y} = b_{YX}(X - \bar{X})$

$\Rightarrow \quad Y - 100 = 1{\cdot}6(X - 50) \quad \Rightarrow \quad Y = 100 + 1{\cdot}6X - 80 = 1{\cdot}6X + 20$

When $X = 60$, $\quad \hat{Y} = 1{\cdot}6 \times 60 + 20 = 116$

(*ii*) Regression equation of X on Y is : $\quad X - \bar{X} = b_{XY}(Y - \bar{Y})$

$\Rightarrow \quad X - 50 = 0{\cdot}4(Y - 100) \quad \Rightarrow \quad X = 50 + 0{\cdot}4Y - 40 = 0{\cdot}4Y + 10$

When $Y = 80$, $\quad \hat{X} = 0{\cdot}4 \times 80 + 10 = 42$

(*iii*) The correlation coefficient, r between X and Y is given by :

$$r^2 = b_{YX} \times b_{XY} = 1{\cdot}6 \times 0{\cdot}4 = 0{\cdot}64 \quad \Rightarrow \quad r = \pm\sqrt{0{\cdot}64} = \pm\,0{\cdot}8$$

Since the regression coefficients are positive, we take : $r = +\,0{\cdot}8$.

Example 11·3. *The following table shows the number of motor registrations in a certain territory for a term of 5 years and the sale of motor tyres by a firm in that territory for the same period.*

Find the regression equation to estimate the sale of tyres when motors registration is known.

Year	*Motor registrations*	*Number of Tyres sold*
1	600	1,250
2	630	1,100
3	720	1,300
4	750	1,350
5	800	1,500

Estimate sale of tyres when registration is 850.

Solution. Let the number of motor registrations and tyres sold be denoted by X and Y, respectively. Then it is required to find the regression equation of Y on X. In this problem, the origin as well as scale in both the X and Y series can be changed by taking :

$$U = \frac{X - a}{h} = \frac{X - 720}{10} \quad ; \quad V = \frac{Y - b}{k} = \frac{Y - 1350}{50}$$

CALCULATIONS FOR REGRESSION EQUATION

X	Y	U	V	U^2	UV
600	1250	– 12	– 2	144	24
630	1100	– 9	– 5	81	45
720	1300	0	– 1	0	0
750	1350	3	0	9	0
800	1500	8	3	64	24
Total		$\Sigma U = -10$	$\Sigma V = -5$	$\Sigma U^2 = 298$	$\Sigma UV = 93$

$$\overline{X} = a + h \cdot \frac{\Sigma U}{n} = 720 + \frac{10(-10)}{5} = 700$$

$$\overline{Y} = b + k \cdot \frac{\Sigma V}{n} = 1350 + \frac{50(-5)}{5} = 1300$$

Equation of regression of sale of tyres (Y) on the motor regressions (X) is :

$$Y - \overline{Y} = b_{YX} (X - \overline{X}) \quad \dots (*)$$

where $b_{YX} = \frac{k}{h} \cdot b_{VU} = \frac{k}{h}\left[\frac{n\Sigma UV - (\Sigma U)(\Sigma V)}{n\Sigma U^2 - (\Sigma U)^2}\right]$

$$= \frac{50}{10}\left[\frac{5 \times 93 - (-10) \times (-5)}{5 \times 298 - (-10)^2}\right] = \frac{5 \times 415}{1390} = 1{\cdot}4928$$

$\therefore \quad Y - 1300 = 1{\cdot}4928\ (X - 700) \quad \Rightarrow \quad Y = 1{\cdot}4928\ X + 255{\cdot}04 \quad$ (From (*))

The estimate of sales of tyres (Y) when registration X = 850, is given by :

$$Y = 1{\cdot}4928 \times 850 + 255{\cdot}04 = 1523{\cdot}92 \simeq 1524$$

Example 11·4. *Find the most likely price in Mumbai corresponding to the price of ₹70 at Kolkata from the following :*

	Kolkata	*Mumbai*
Average price	65	67
Standard deviation	2·5	3·5

Correlation coefficient between the prices of commodities in the two cities is 0·8.

Solution. Let the prices (in Rupees) in Kolkata and Mumbai be denoted by X and Y , respectively. Then we are given :

$\overline{X} = 65$, $\overline{Y} = 67$, $\sigma_X = 2{\cdot}5$, $\sigma_Y = 3{\cdot}5$ and $r = r(X, Y) = 0{\cdot}8$. We want Y for X = 70.

Line of regression of Y on X is :

$$Y - \overline{Y} = r\frac{\sigma_Y}{\sigma_X}(X - \overline{X}) \quad \Rightarrow \quad Y = 67 + 0{\cdot}8 \times \frac{3{\cdot}5}{2{\cdot}5}(X - 65)$$

When X = 70, $\quad \hat{Y} = 67 + 0{\cdot}8 \times \frac{3{\cdot}5}{2{\cdot}5}(70 - 65) = 72{\cdot}6$

Hence, the most likely price in Mumbai corresponding to price of ₹ 70 at Kolkata is ₹ 72·60.

Example 11·5. *Can $Y = 5 + 2{\cdot}8\ X$ and $X = 3 - 0{\cdot}5Y$ be the estimated regression equations of Y on X and X on Y, respectively ? Explain your answer with suitable theoretical arguments.*

Solution. Line of regression of Y on X is : $Y = 5 + 2{\cdot}8\ X \Rightarrow b_{YX} = 2{\cdot}8$ (*)

Line of regression of X on Y is : $X = 3 - 0{\cdot}5\ Y \Rightarrow b_{XY} = -\ 0{\cdot}5$ (**)

This is not possible, since each of the regression coefficients b_{YX} and b_{XY} must have the same sign, which is same as that of Cov(X, Y). If Cov (X, Y) is positive, then both the regression coefficients are positive and if Cov (X, Y) is negative, then both the regression coefficients are negative. Hence (*) and (**) cannot be the estimated regression equations of Y on X and X on Y respectively.

Example 11·6. *In a partially destroyed laboratory, record of an analysis of correlation, data, the following results only are legible :*

Variance of X = 9. Regression equations : $8X - 10Y + 66 = 0$, $40X - 18Y = 214$.

What are: (i) *the mean values X and Y,*

(ii) *the correlation coefficient between X and Y, and*

(iii) *the standard deviation of Y ?*

Solution. (i) Since both lines of regression pass through the point $(\overline{X}, \overline{Y})$, we have :

$$8\overline{X} - 10\overline{Y} + 66 = 0, \quad \text{and} \quad 40\overline{X} - 18\overline{Y} = 214. \text{ Solving, we get } \overline{X} = 13, \quad \overline{Y} = 17.$$

(ii) Let $8X - 10Y + 66 = 0$ and $40X - 18Y = 214$ be the lines of regression of Y on X and X on Y, respectively. These equations can be put in the form :

$$Y = \frac{8}{10}X + \frac{66}{10} \quad \text{and} \quad X = \frac{18}{40}Y + \frac{214}{40}$$

$$\therefore \qquad b_{YX} = \text{Regression coefficient of } Y \text{ on } X = \frac{8}{10} = \frac{4}{5}$$

$$\text{and} \qquad b_{XY} = \text{Regression coefficient of } X \text{ on } Y = \frac{18}{40} = \frac{9}{20}$$

$$\text{Hence,} \qquad r^2 = b_{YX} \cdot b_{XY} = \frac{4}{5} \cdot \frac{9}{20} = \frac{9}{25} \quad \Rightarrow \quad r = \pm \frac{3}{5} = \pm 0{\cdot}6$$

But since both the regression coefficients are positive, we take $r = +\ 0{\cdot}6$.

(iii) We have
$$b_{YX} = r \cdot \frac{\sigma_Y}{\sigma_X} \Rightarrow \frac{4}{5} = \frac{3}{5} \times \frac{\sigma_Y}{3} \Rightarrow \sigma_Y = 4. \quad \left[\because \sigma_X^2 = 9 \Rightarrow \sigma_X = 3\right]$$

Remarks:

1. It can be verified that the values of $\overline{X} = 13$ and $\overline{Y} = 17$ as obtained in part (i) satisfy both the regression equations. In numerical problems of this type, this check should invariably be applied to ascertain the correctness of the answer.

2. If we had assumed that $8X - 10Y + 66 = 0$, is the equation of the line of regression of X on Y and $40X - 18Y = 214$ is the equation of line of regression of Y on X, then we get, respectively :

$$8X = 10Y - 66 \quad \text{and} \quad 18Y = 40X - 214$$

$$\Rightarrow \quad X = \frac{10}{8}Y - \frac{66}{8} \quad \text{and} \quad Y = \frac{40}{18}X - \frac{214}{18}$$

$$\Rightarrow \quad b_{XY} = \frac{10}{8} \quad \text{and} \quad b_{YX} = \frac{40}{18}$$

$$\therefore \quad r^2 = b_{XY} \cdot b_{YX} = \frac{10}{8} \times \frac{40}{18} = 2{\cdot}78$$

But since r^2 always lies between 0 and 1, our supposition is wrong.

11·3. CURVILINEAR REGRESSION

One of the criteria set forth previously in our regression studies has been that the variables X and Y are related linearly. In many cases, this assumption may not be valid. For example, in a scatter diagram showing the relationship between X, the number of overtime hours, versus Y, the resulting number of additional units produced for a manufacturing firm, it appears by inspection that a curvilinear regression may explain more of the variability of Y than does a linear line. For this reason, we shall fit the regression line using the following formula :

$$Y = a + b_1X + b_2X^2 \qquad \text{...(11·14)}$$

Using the principle of least squares, we have to determine the constants, a, b_1 and b_2 so that :

$$E = \sum_{i=1}^{n}\left(y_i - a - b_1x_i - b_2x_i^2\right)^2 \text{ is minimum.}$$

Equating to zero the partial derivatives of E with respect to a, b and c separately, we get the normal equations for estimating a, b_1 and b_2 as

$$\left.\begin{aligned} \frac{\partial E}{\partial a} &= 0 = -2\sum_{t}\left(y_i - a - b_1x_i - b_2x_i^2\right) \\ \frac{\partial E}{\partial b_1} &= 0 = -2\sum_{i} x_i\left(y_i - a - b_1x_i - b_2x_i^2\right) \\ \frac{\partial E}{\partial b_2} &= 0 = -2\sum_{i} x_i^2\left(y_i - a - b_1x_i - b_2x_i^2\right) \end{aligned}\right\} \qquad \text{...(11·15)}$$

$$\Rightarrow \quad \left.\begin{aligned} \sum y_i &= na + b_1\sum x_i + b_2\sum x_i^2 \\ \sum x_i y_i &= a\sum x_i + b_1\sum x_i^2 + b_2\sum x_i^3 \\ \sum x_i^2 y_i &= a\sum x_i^2 + b_1\sum x_i^3 + b_2\sum x_i^4, \end{aligned}\right\} \qquad \text{...(11·15}a\text{)}$$

summation taken over i from 1 to n.

For given set of points (x_i, y_i); $i = 1, 2, \ldots, n$, equations (11·15a) can be solved for a, b_1 and b_2, and with these values of a, b_1 and b_2 , (11·14) is the parabola of best fit.

Remarks:

1. If $Y = a_0 + a_1 X + a_2X^2 + \cdots + a_kX^k$ is the kth degree polynomial of best fit to the set of points (x_i, y_i); $i = 1, 2, \ldots, n$ the constants $a_0, a_1, a_2, \ldots, a_k$ are to be obtained so that

$$E = \sum_{i=1}^{n}\left(y_i - a_0 - a_1x_i - a_2x_i^2 - \cdots - a_kx_i^k\right)^2$$

is minimum. Thus, the normal equations for estimating $a_0, a_1, \ldots, a_k$ are obtained on equating to zero the partial derivatives of E *w.r.to.* $a_0, a_1, \ldots, a_k$ separately, *i.e.,*

$$\left.\begin{aligned}\frac{\partial E}{\partial a_0} &= 0 = -2\sum\left(y_i - a_0 - a_1x_i - a_2x_i^2 - \cdots - a_kx_i^k\right)\\ \frac{\partial E}{\partial a_1} &= 0 = -2\sum x_i\left(y_i - a_0 - a_1x_i - a_2x_i^2 - \cdots - a_kx_i^k\right)\\ &\vdots\\ \frac{\partial E}{\partial a_k} &= 0 = -2\sum x_i^k\left(y_i - a_0 - a_1x_i - a_2x_i^2 - \cdots - a_kx_i^k\right)\end{aligned}\right\} \quad ...(11·17)$$

$$\Rightarrow \left.\begin{aligned}\sum y_i &= na_0 + a_1\sum x_i + a_2\sum x_i^2 + \cdots + a_k\sum x_i^k\\ \sum x_iy_i &= a_0\sum x_i + a_1\sum x_i^2 + a_2\sum x_i^3 + \cdots + a_k\sum x_i^{k+1}\\ &\vdots\\ \sum x_i^k y_i &= a_0\sum x_i^k + a_1\sum x_i^{k+1} + a_2\sum x_i^{k+2} + \cdots + a_k\sum x_i^{2k}\end{aligned}\right\} \quad ...(11·17a)$$

summation extended over i from 1 to n. These are $(k + 1)$ equations in $(k + 1)$ unknowns $a_0, a_1, a_2, \ldots, a_k$ and can be solved with the help of algebra.

2. Depending on the pattern suggested by the scatter diagram, there are many non-linear equations from which we can select a regression line. In such cases' the original data which is not in a linear form can be reduced to linear form by simple transformation of variables. This will be illustrated by considering the following curves.

(*a*) Fitting of a Power Curve $Y = aX^b$ to a set of n points.

$$Y = aX^b \quad ... (11.18)$$

Taking logarithm of each side, we get

$$\log Y = \log a + b \log X \quad \Rightarrow \quad U = A + bV,$$

where $U = \log Y$, $A = \log a$ and $V = \log X$.

This is a linear equation in V and U.

Normal equations for estimating A and b are :

$$\sum U = nA + b\sum V \quad \text{and} \quad \sum UV = A\sum V + b\sum V^2 \quad ...(11·18a)$$

These equations can be solved for A and b and consequently, we get a = antilog (A)

With the values of a and b so obtained, (11·18) is the curve of best fit to the set of n points.

(*b*) Fitting of Exponential Curves (*i*) $Y = ab^X$, (*ii*) $Y = a\,e^{bX}$, **to a set of *n* points.**

(*i*) $$Y = ab^X \quad ...(11·19)$$

Taking logarithm of each side, we get

$$\log Y = \log a + X \log b \quad \Rightarrow \quad U = A + BX,$$

where $U = \log Y$, $A = \log a$ and $B = \log b$

This is linear equation in X and U.

The normal equations for estimating A and B are :

$$\sum U = nA + B\sum X \quad \text{and} \quad \sum XU = A\sum X + B\sum X^2 \qquad ...(11{\cdot}19a)$$

Solving these equations for A and B, we finally get

$$a = \text{antilog }(A) \quad \text{and} \quad B = \text{antilog }(B)$$

With these values of a and b, (11·19) is the curve of best fit to the given set of n points.

(ii) $$Y = a\,e^{bX} \qquad ...(11{\cdot}20)$$

$$\log Y = \log a + bX \log e = \log a + (b \log e)\, X \quad \Rightarrow \quad U = A + BX,$$

where $U = \log Y, \quad A = \log a \quad \text{and} \quad B = b \log e.$

This is linear equation in X and U, and the normal equations for estimating A and B are:

$$\sum U = nA + B\sum X \quad \text{and} \quad \sum XU = A\sum X + B\sum X^2 \qquad ...(11{\cdot}20a)$$

From these we find A and B and consequently, a = Antilog A and $b = \dfrac{B}{\log e}$

Example 11·7. *For 10 randomly selected observations, the following data were recorded :*

Observation No.	:	1	2	3	4	5	6	7	8	9	10
Overtime hours (X)	:	1	1	2	2	3	3	4	5	6	7
Additional units (Y)	:	2	7	7	10	8	12	10	14	11	14

Determine the coefficients of regression and regression equation using the non-linear form :

$$Y = a + b_1X + b_2X^2$$

Solution.

S. No.	X	Y	X^2	X^3	X^4	XY	X^2Y
1	1	2	1	1	1	2	2
2	1	7	1	1	1	7	7
3	2	7	4	8	16	14	28
4	2	10	4	8	16	20	40
5	3	8	9	27	81	24	72
6	3	12	9	27	81	36	108
7	4	10	16	64	256	40	160
8	5	14	25	125	625	70	350
9	6	11	36	216	1296	66	396
10	7	14	49	343	2401	98	686
Total	34	95	154	820	4774	377	1849

Using normal equations (11·15a), we get

$10a + 34b_1 + 154\, b_2 = 95,\ 34a + 154\, b_1 + 820\, b_2 = 377$ and $154a + 820\, b_1 + 4774\, b_2 = 1849.$

The solutions to these three simultaneous equations are :

$$a = 1{\cdot}80, \quad b_1 = 3{\cdot}48 \quad \text{and} \quad b_2 = -\,0{\cdot}27$$

The regression equation, therefore, is : $\quad Y = 1{\cdot}80 + 3{\cdot}48X - 0{\cdot}27X^2$

Example 11·8. *Fit an exponential curve of the form $Y = ab^X$ to the following data :*

X :	1	2	3	4	5	6	7	8
Y :	1·0	1·2	1·8	2·5	3·6	4·7	6·6	9·1

Solution.

S. No.	X	Y	U = log Y	XU	X^2
1	1	1·0	0·0000	0·0000	1
2	2	1·2	0·0792	0·1584	4
3	3	1·8	0·2553	0·7659	9
4	4	2·5	0·3979	1·5916	16
5	5	3·6	0·5563	2·7815	25
6	6	4·7	0·6721	4·0326	36
7	7	6·6	0·8195	5·7365	49
8	8	9·1	0·9590	7·6720	64
Total	36	30·5	3·7393	22·7385	204

(11·19*a*) gives the normal equation as :

$$3{\cdot}7393 = 8A + 36B \quad \text{and} \quad 22{\cdot}7385 = 36A + 204B$$

Solving, we get

$$B = 0{\cdot}1408 \quad \text{and} \quad A = -\,0{\cdot}1662 = \bar{1}{\cdot}8338$$

$$\therefore \qquad b = \text{Antilog } B = 1{\cdot}383 \quad \text{and} \quad a = \text{Antilog } A = 0{\cdot}6821$$

Hence, the equation of the required curve is : $Y = ab^X = 0{\cdot}6821\ (1{\cdot}38)^X$.

11·4. REGRESSION CURVES

In modern terminology, the conditional mean $E(Y \mid X = x)$ for a continuous distribution is called the regression function of Y on X and the graph of this function of x is known as the *regression curve of Y on X* or sometimes the regression curve for the mean of Y. Geometrically, the regression function represents the y co-ordinate of the centre of mass of the bivariate probability mass in the infinitesimal vertical strip bounded by x and $x + dx$.

Similarly, the regression function of X on Y is $E(X \mid Y = y)$ and the graph of this function of y is called the *regression curve* (*of the mean*) of X on Y.

In case a regression curve is a straight line, the corresponding regression is said to be *linear.* If one of the regressions is linear, it does not however follow that the other is also linear.

Theorem. *Let (X, Y) be a two-dimensional random variable with* $E(X) = \overline{X}$, $E(Y) = \overline{Y}$, $V(X) = \sigma_X^2$ $V(Y) = \sigma_Y^2$ *and let* $r = r(X, Y)$ *be the correlation coefficient between X and Y. If the regression of Y on X is linear, then*

$$E(Y \mid X) = \overline{Y} + r\frac{\sigma_Y}{\sigma_X}(X - \overline{X}) \qquad \text{...(11·21}a\text{)}$$

Similarly, if the regression of X on Y is linear, then

$$E(X \mid Y) = \overline{X} + r\frac{\sigma_X}{\sigma_Y}(Y - \overline{Y}). \qquad \text{...(11·21}b\text{)}$$

Proof. Let the regression equation of Y on X be: $E(Y \mid X) = a + bX$ ...(*i*)

But by def., $$E(Y \mid X) = \int_{-\infty}^{\infty} y\, f(y \mid x)\, dy = \int_{-\infty}^{\infty} y \cdot \frac{f(x, y)}{f_X(x)}\, dy$$

$$\therefore \quad \frac{1}{f_X(x)} \int_{-\infty}^{\infty} y\, f(x, y)\, dy = a + bx \quad \text{...(ii)}$$

Multiplying each side of (*ii*) by $f_X(x)$ and integrating *w.r.to* x, we get

$$\int_{-\infty}^{\infty} \int_{-\infty}^{\infty} y\, f(x, y)\, dy\, dx = a \int_{-\infty}^{\infty} f_X(x) dx + b \int_{-\infty}^{\infty} x\, f_X(x)\, dx$$

$$\Rightarrow \quad \int_{-\infty}^{\infty} y \left[\int_{-\infty}^{\infty} f(x, y)\, dx \right] dy = a + bE(X)$$

$$\Rightarrow \quad \int_{-\infty}^{\infty} y\, f_Y(y) dy = a + bE(X) \Rightarrow E(Y) = a + bE(X) \Rightarrow \bar{Y} = a + b\bar{X} \quad \text{...(iii)}$$

Multiplying each side of (*ii*) by $x\, f_X(x)$ and integrating *w.r.to* x, we get

$$\int_{-\infty}^{\infty} \int_{-\infty}^{\infty} xy\, f(x, y) dy\, dx = a \int_{-\infty}^{\infty} x\, f_X(x)\, dx + b \int_{-\infty}^{\infty} x^2 f_X(x) dx$$

$$\Rightarrow \quad E(XY) = aE(X) + bE(X^2) \Rightarrow \mu_{11} + \bar{X}\,\bar{Y} = a\bar{X} + b(\sigma_X^2 + \bar{X}^2) \quad \text{...(iv)}$$

$$\left[\because \mu_{11} = E(XY) - E(X)E(Y) = E(XY) - \bar{X}\,\bar{Y}; \quad \sigma_X^2 = E(X^2) - \{E(X)\}^2 = E(X^2) - \bar{X}^2\right]$$

Solving (*iii*) and (*iv*) simultaneously, we get : $b = \dfrac{\mu_{11}}{\sigma_X^2}$ and $a = \bar{Y} - \dfrac{\mu_{11}}{\sigma_X^2} \cdot \bar{X}$

Substituting in (*i*) and on simplification we get the required equation of the line of regression of Y on X :

$$\Rightarrow \quad E(Y \mid X) = \bar{Y} + \frac{\mu_{11}}{\sigma_X^2}(X - \bar{X}) \quad \Rightarrow \quad E(Y \mid X) = \bar{Y} + r\frac{\sigma_Y}{\sigma_X}(X - \bar{X})$$

By starting with the line $E(X \mid Y) = A + BY$ and proceeding similarly, we shall obtain the equation of the line of regression of X on Y as :

$$E(X \mid Y) = \bar{X} + \frac{\mu_{11}}{\sigma_Y^2}(Y - \bar{Y}) = \bar{X} + r\frac{\sigma_X}{\sigma_Y}(Y - \bar{Y})$$

Example 11·9. *Given* $f(x, y) = x\, e^{-x(y+1)}$; $x \geq 0,\ y \geq 0$, *find the regression curve of Y on X.*

Solution. Marginal *p.d.f.* of X is given by :

$$f_1(x) = \int_0^{\infty} f(x, y)\, dy = \int_0^{\infty} x e^{-x(y+1)}\, dy$$

$$= xe^{-x} \int_0^{\infty} e^{-xy} dy = xe^{-x} \left| \frac{e^{-xy}}{-x} \right|_{y=0}^{\infty}$$

$$= e^{-x}, x \geq 0$$

Conditional *p.d.f.* of Y on X is given by

$$f(y \mid x) = \frac{f(x, y)}{f_1(x)} = \frac{xe^{-x(y+1)}}{e^{-x}}$$

$$= xe^{-xy}$$

The regression curve of Y for given X is given by :

$$y = E(Y|X=x) = \int_0^{\infty} y\, f(y|x)\, dy = \int_0^{\infty} y\, x\, e^{-xy}\, dy$$

$$= x\left[\left|\frac{ye^{-xy}}{-x}\right|_{y=0}^{\infty} + \int_0^{\infty}\frac{e^{-xy}}{x}\,dy\right] = 0 + \left|\frac{e^{-xy}}{-x}\right|_{y=0}^{\infty} = \frac{1}{x}$$

i.e., $\quad y = \dfrac{1}{x} \quad \Rightarrow \quad xy = 1,$ which is the equation of a rectangular hyperbola.

Hence, the regression of Y on X is *not* linear.

Example 11·10. *Obtain regression curve of Y on X for the following distribution :*

$$f(x, y) = \frac{y}{(1+x)^4}\, exp\left(-\frac{y}{1+x}\right); \quad x, y \geq 0$$

Solution. Marginal *p.d.f.* of X is given by :

$$f_1(x) = \int_0^{\infty} f(x, y)\, dy = \frac{1}{(1+x)^4}\int_0^{\infty} y\, e^{-y/(1+x)}\, dy$$

$$= \frac{1}{(1+x)^4}\cdot \overline{|(2)}\cdot (1+x)^2 \qquad \text{[Using Gamma Integral]}$$

$$= \frac{1}{(1+x)^2}; \quad x \geq 0$$

The conditional *p.d.f.* of Y (for given X) is given by :

$$f(y|x) = \frac{f(x, y)}{f_1(x)} = \frac{y}{(1+x)^2}\exp\left(-\frac{y}{1+x}\right); y \geq 0$$

Regression equation of Y on X is given by :

$$y = E(Y|X) = \int_0^{\infty} y\, f(y|x)\, dy = \frac{1}{(1+x)^2}\int_0^{\infty} y^2 e^{-y/(1+x)}\, dy$$

$$= \frac{1}{(1+x)^2}\cdot \overline{|(3)}\cdot (1+x)^3 \qquad \text{[Using Gamma Integral]}$$

$$\therefore \qquad y = 2(1+x) \qquad \left[\because \overline{|(3)} = 2! = 2\right]$$

Hence, the regression of Y on X is linear.

Example 11·11. *Let (X, Y) have the joint p.d.f. given by :*

$$f(x, y) = \begin{cases} 1, \text{ if } |y| < x, & 0 < x < 1 \\ 0, \text{ otherwise} \end{cases}$$

Show that the regression of Y on X is linear but regression of X on Y is not linear.

Solution. $|y| < x \;\Rightarrow\; -x < y < x$ and $x > |y|$.

The marginal *p.d.f.'s* $f_1(\cdot)$ of X and $f_2(\cdot)$ of Y are given by :

$$f_1(x) = \int_{-x}^{x} f(x, y)dy = \int_{-x}^{x} 1 \cdot dy = 2x \; ; 0 < x < 1$$

$$f_2(y) = \int_{|y|}^{1} f(x, y)dx = \int_{|y|}^{1} 1 \cdot dx = 1 - |y| \; ; -1 < y < 1$$

$$\therefore \quad f_3(x \mid y) = \frac{f(x, y)}{f_2(y)} = \frac{1}{1-|y|} \; ; \; -1 \le y < 1, \; 0 < x < 1$$

$$= \begin{cases} \dfrac{1}{1-y}, & 0 < y < 1 \; ; \; 0 < x < 1 \\ \dfrac{1}{1+y}, & -1 < y < 0 \; ; \; 0 < x < 1 \end{cases}$$

$$f_4(y \mid x) = \frac{f(x, y)}{f_1(x)} = \frac{1}{2x}, \quad 0 < x < 1; \; |y| < x$$

$$E(Y \mid X = x) = \int_{-x}^{x} y \cdot f_4(y \mid x)dy = \int_{-x}^{x} \frac{y}{2x} dy = \frac{1}{4x} \left| y^2 \right|_{-x}^{x} = 0$$

Hence, the curve of regression of Y on X is $y = 0$, which is a straight line.

$$E(X \mid Y = y) = \int_0^1 x f_3(x \mid y)dx$$

$$\therefore \quad E(X \mid Y = y) = \int_0^1 x \left(\frac{1}{1-y} \right) dx = \frac{1}{2(1-y)}, \quad 0 < y < 1$$

$$\text{and} \quad E(X \mid Y = y) = \int_0^1 x \left(\frac{1}{1+y} \right) dx = \frac{1}{2(1+y)}, \; -1 < y < 0$$

Hence, the curve of regression of X on Y is : $x = \begin{cases} \dfrac{1}{2(1-y)}, & 0 < y < 1 \\ \dfrac{1}{2(1+y)}, & -1 < y < 0, \end{cases}$

which is not a straight line.

Example 11·12. *The equations of two lines of regression, obtained in a correlation analysis between variables X and Y (not necessarily in that order), are as follows :*

$$2X + 3Y - 8 = 0 \quad \text{and} \quad 2Y + X - 5 = 0. \quad \text{The variance of } X = 4$$

Find : (i) variance of Y, (ii) coefficient of determination of X on Y,
(iii) standard error of estimate of X on Y and standard error of estimate of Y on X.

Solution. (*i*) Let $2X + 3Y - 8 = 0$, be regression of X on Y so that we may re-write :

$$X = -\frac{3}{2} Y + 4 \quad \Rightarrow \quad b_{XY} = -\frac{3}{2}.$$

Also, assume $2Y + X - 5 = 0$, is regression of Y on X so that

$$2Y = -X + 5 \quad \Rightarrow \quad Y = -\frac{1}{2}X + \frac{5}{2} \quad \Rightarrow \quad b_{YX} = -\frac{1}{2}$$

(*ii*) The coefficient of determination of X on Y is given by :

$$r^2 = b_{XY} \times b_{YX} = \left(-\frac{3}{2}\right) \times \left(-\frac{1}{2}\right) = 0{\cdot}75$$

$$\Rightarrow \quad r = -\sqrt{0{\cdot}75} = -0{\cdot}87 \qquad [\because \text{ Both } b_{XY} \text{ and } b_{YX} \text{ are negative}]$$

(*i*) $$b_{YX} = r\frac{\sigma_Y}{\sigma_X} \quad \Rightarrow \quad -\frac{1}{2} = -0{\cdot}87 \times \frac{\sigma_Y}{2} \quad \Rightarrow \quad \sigma_Y = 1{\cdot}15 \qquad [\because \sigma_X^2 = 4, \textit{(Given)}]$$

$$\therefore \sigma_Y^2 = (1{\cdot}15)^2 = 1{\cdot}3225$$

(*iii*) Standard error of estimate of X on Y is given by :

$$S_{XY} = \sigma_X \sqrt{1 - r^2} = 2\sqrt{0{\cdot}25} = 1$$

and Standard error of estimate of Y on X is given by :

$$S_{YX} = \sigma_Y \sqrt{1 - r^2} = 1{\cdot}15\sqrt{0{\cdot}25} = 0{\cdot}575.$$

SUPPLEMENTARY EXAMPLES

Example 11·13. *Show that the coefficient of correlation between the observed and the estimated values of Y obtained from the line of regression of Y on X, is the same as that between X and Y.*

Solution. The correlation coefficient between Y and the estimated value of Y, $\hat{Y} = \bar{Y} + r\frac{\sigma_Y}{\sigma_X}(X - \bar{X})$ *is given by :*

$$r(Y, \hat{Y}) = \frac{\{\text{Cov}(\hat{Y}), Y\}}{\sigma_Y \cdot \sigma_{\hat{Y}}}$$

We have $$E(\hat{Y}) = E\left[\bar{Y} + r\frac{\sigma_Y}{\sigma_X}(X - \bar{X})\right] = \bar{Y} + r\frac{\sigma_Y}{\sigma_X}E(X - \bar{X}) = \bar{Y}$$

$$\therefore \quad \text{Var}(\hat{Y}) = E[\hat{Y} - E(\hat{Y})]^2 = E\left[r\frac{\sigma_Y}{\sigma_X}(X - \bar{X})\right]^2 = r^2\sigma_Y^2$$

$$\Rightarrow \quad \sigma_{\hat{Y}} = r\,\sigma_Y$$

Also, $$\text{Cov}(Y, \hat{Y}) = E[\{Y - E(Y)\}\{\hat{Y} - E(\hat{Y})\}]$$

$$= E\left[b\{(X - E(X))\}\left\{r\frac{\sigma_Y}{\sigma_X}(X - \bar{X})\right\}\right]$$

$$= b \cdot r\frac{\sigma_Y}{\sigma_X}E[\{X - E(X)\}^2] = \left(r\frac{\sigma_Y}{\sigma_X}\right)^2 \sigma_X^2 = r^2\sigma_Y^2 \qquad \left(\because b = b_{YX} = r\frac{\sigma_Y}{\sigma_X}\right)$$

$$\therefore \quad r(Y, \hat{Y}) = \frac{r^2\sigma_Y^2}{\sigma_Y \cdot r\sigma_Y} = |r| = |r(X, Y)| \qquad [\because Cov(Y, \hat{Y}) = r^2\sigma_Y^2 > 0]$$

Hence, the correlation coefficient between observed and estimated value of Y is the same as the correlation coefficient between X and Y.

Example 11·14. *Variables X and Y have the joint p.d.f. given by:*

$$f(x, y) = \frac{1}{3}(x+y)\ ;\ 0 \le x \le 1, 0 \le y \le 2.$$

Find (i) The two regression curves for the means, (ii) r(X, Y), and (iii) The two lines of regression.

Solution. (*i*) The marginal *p.d.f.*'s of X and Y are given by :

$$f_1(x) = \int_0^2 f(x, y)\, dy = \frac{1}{3}\int_0^2 (x+y)\, dy = \frac{1}{3}\left| xy + \frac{y^2}{2} \right|_{y=0}^{y=2}$$

$$\Rightarrow \quad f_1(x) = \frac{2}{3}(1+x)\ ;\ 0 \le x \le 1 \qquad \text{...(*)}$$

$$f_2(y) = \int_0^1 f(x, y)\, dx = \frac{1}{3}\int_0^1 (x+y)\, dx = \frac{1}{3}\left| \frac{x^2}{2} + xy \right|_{x=0}^{x=1}$$

$$\Rightarrow \quad f_2(y) = \frac{1}{3}\left(\frac{1}{2}+y\right); 0 \le y \le 2 \qquad \text{...(**)}$$

The conditional distributions are given by :

$$f_3(y \mid x) = \frac{f(x, y)}{f_1(x)} = \frac{1}{2}\left(\frac{x+y}{1+x}\right);\quad f_4(x \mid y) = \frac{f(x, y)}{f_2(y)} = \frac{2(x+y)}{1+2y} \qquad \text{...(***)}$$

$$E(Y \mid X) = \int_0^2 y \cdot f_3(y \mid x)\, dy = \frac{1}{2(1+x)}\int_0^2 y(x+y)\, dy$$

$$= \frac{1}{2(1+x)}\left| \frac{xy^2}{2} + \frac{y^3}{3} \right|_{y=0}^{y=2} = \frac{3x+4}{3(x+1)}$$

and
$$E(X \mid Y) = \int_0^1 x\, f_4(x \mid y) dx = \frac{2}{1+2y}\int_0^1 (x^2 + xy) dx = \frac{2+3y}{3(1+2y)}$$

Hence, the regression curves for means are:

$$y = E(Y \mid X) = \frac{3x+4}{3(x+1)} \quad \text{and} \quad x = E(X \mid Y) = \frac{2+3y}{3(1+2y)}.$$

(*ii*) From the marginal distributions, we get

$$E(X) = \int_0^1 x\, f_1(x) dx = \frac{5}{9},\quad E(X^2) = \int_0^1 x^2 f_1(x) dx = \frac{7}{18}$$

$$\Rightarrow \quad \text{Var}(X) = \sigma_X^2 = \frac{7}{18} - \left(\frac{5}{9}\right)^2 = \frac{13}{162}$$

Similarly, we obtain

$$E(Y) = \frac{11}{9},\quad E(Y^2) = \frac{16}{9};\quad \sigma_Y^2 = \frac{16}{9} - \left(\frac{11}{9}\right)^2 = \frac{23}{81}$$

Also,
$$E(XY) = \int_0^1 \int_0^2 xy\, f(x, y)\, dx\, dy = \frac{1}{3}\int_0^1 \int_0^2 (x^2 y + xy^2)\, dx\, dy$$

$$= \frac{1}{3}\left\{\left(\int_0^1 x^2\,dx\right)\left(\int_0^2 y\,dy\right) + \left(\int_0^1 x\,dx\right)\left(\int_0^2 y^2 dy\right)\right\}$$

$$= \frac{1}{3}\left[\frac{1}{3} \times 2 + \frac{1}{2} \times \frac{8}{3}\right] = \frac{2}{3}$$

$$\text{Cov}(X, Y) = E(XY) - E(X)E(Y) = \frac{2}{3} - \frac{5}{9} \times \frac{11}{9} = -\frac{1}{81}$$

$$r(X, Y) = \frac{\text{Cov}(X, Y)}{\sigma_X \cdot \sigma_Y} = \frac{-\frac{1}{81}}{\sqrt{\frac{13}{162} \times \frac{23}{81}}} = -\left(\frac{2}{299}\right)^{1/2}$$

(*iii*) The two lines of regression are :

$$Y - E(Y) = \frac{\text{Cov}(X, Y)}{\sigma_X^2}[X - E(X)] \quad \Rightarrow \quad Y - \frac{11}{9} = -\frac{2}{13}\left(X - \frac{5}{9}\right)$$

and
$$X - E(X) = \frac{\text{Cov}(X, Y)}{\sigma_Y^2}[Y - E(Y)] \quad \Rightarrow \quad X - \frac{5}{9} = -\frac{1}{23}\left(Y - \frac{11}{9}\right).$$

Example 11·15. *By minimising* $\sum_{i=1}^{n} f_i(x_i \cos\alpha + y_i \sin\alpha - p)^2$ *for variation in* α *and* p, *show that there are two straight lines passing through the mean of the distribution for which the sum of squares of normal deviations has an extreme value. Prove also that their slopes are given by:*

$$\tan 2\alpha = \frac{2\mu_{11}}{\sigma_X^2 - \sigma_Y^2}.$$

Solution. We have to minimize: $S = \sum_{i=1}^{n} f_i(x_i \cos\alpha + y_i \sin\alpha - p)^2$ ... (*)

for variation in α and p.

Equating to zero, the partial derivatives of (*) *w.r.* to α and p, we have

$$\frac{\partial S}{\partial \alpha} = 0 = 2\sum_{i=1}^{n} f_i(x_i \cos\alpha + y_i \sin\alpha - p)(-x_i \sin\alpha + y_i \cos\alpha) \quad \text{... (ii)}$$

$$\frac{\partial S}{\partial p} = 0 = -2\sum_{i=1}^{n} f_i(x_i \cos\alpha + y_i \sin\alpha - p) \quad \text{... (iii)}$$

Equation (*iii*) can be written as:

$$\sum_{i=1}^{n} f_i(x_i \cos\alpha + y_i \sin\alpha - p) = 0 \quad \Rightarrow \quad \bar{x}\cos\alpha + \bar{y}\sin\alpha - p = 0 \quad \text{... (iv)}$$

From equation (*ii*), we get a quadratic equation which shows that there are two straight lines for extreme values of E.

From equation (*iv*), it becomes clear that both the straight lines pass through the point $(\bar{x}, \bar{y})$. Again equation (*ii*) can be written as:

$$\sum_{i=1}^{n} f_i(x_i \cos\alpha + y_i \sin\alpha - p)(y_i \cos\alpha - x_i \sin\alpha) = 0$$

$$\Rightarrow \quad \sum_{i=1}^{n} f_i\left[\{\cos\alpha(x_i - \bar{x}) + \sin\alpha(y_i - \bar{y})\}(y_i \cos\alpha - x_i \sin\alpha)\right] = 0 \qquad \text{[Using } (iv)\text{]}$$

$$\Rightarrow \quad \cos^2\alpha \sum_{i=1}^{n} f_i y_i (x_i - \bar{x}) - \sin\alpha\cos\alpha \sum_{i=1}^{n} f_i x_i (x_i - \bar{x})$$

$$+ \sin\alpha \cos\alpha \sum_{i=1}^{n} f_i y_i (y_i - \bar{y}) - \sin^2\alpha \sum_{i=1}^{n} f_i x_i (y_i - \bar{y}) = 0 \qquad \ldots (v)$$

We have
$$\mu_{11} = \frac{1}{N}\sum_i f_i (x_i - \bar{x})(y_i - \bar{y})$$

$$= \frac{1}{N}\sum_1 f_i x_i (y_i - \bar{y}) - \bar{x}\cdot\frac{1}{N}\sum_i f_i (y_i - \bar{y}) = \frac{1}{N}\sum_i f_i x_i (y_i - \bar{y})$$

Similarly,
$$\mu_{11} = \frac{1}{N}\sum_1 f_i y_i (x_i - \bar{x})$$

$$\sigma_X^2 = \frac{1}{N}\sum_1 f_i x_i (x_i - \bar{x}) \quad \text{and} \quad \sigma_Y^2 = \frac{1}{N}\sum_1 f_i y_i (y_i - \bar{y})$$

Substituting these values in (*v*), we get:

$$\mu_{11}(\cos^2\alpha - \sin^2\alpha) - \sin\alpha \cdot \cos\alpha(\sigma_X^2 - \sigma_Y^2) = 0$$

$$\Rightarrow \quad \mu_{11} \cdot \cos 2\alpha = \frac{1}{2}\sin 2\alpha \cdot (\sigma_X^2 - \sigma_Y^2) \quad \Rightarrow \quad \tan 2\alpha = \frac{2\mu_{11}}{\sigma_X^2 - \sigma_Y^2}$$

11·5. CORRELATION RATIO

As discussed earlier, when variables are linearly related, we have regression lines of one variable on another variable and correlation coefficient can be computed to tell us about the extent of association between them. However, if the variables are not linearly related but some sort of curvilinear relationship exists between them, the use of r which is a measure of the degree to which the relation approaches straight line "law" will be misleading. We might come across bivariate distributions where r may be very low or even zero but the regression may be strong, or even perfect. *Correlation ratio* 'η' is the appropriate measure of curvilinear relationship between the two variables. Just as r measures the concentration of points about the straight line of best fit, η measures the concentration of points about the curve of best fit. If regression is linear $\eta = r$, otherwise $|\eta| > |r|$ (c.f. Remark 2, 11·5·1).

11·5·1. Measure of Correlation Ratio

In the previous articles we have assumed that there is a single observed value Y corresponding to the given value x_i of X but sometimes there are more than one such value of Y.

Suppose corresponding to the value $x_i (i = 1, 2, \ldots, m)$ of the variable X, the variable Y takes the values y_{ij} with respective frequencies f_{ij}; $j = 1, 2, \ldots, n$.

Though all the x's in the ith vertical array have the same value, the y's are different. A typical pair of values in the ith array is (x_i, y_{ij}), with frequency f_{ij}. Thus the first suffix i indicates the vertical array while the second suffix j indicates the positions of y in that array. Let

$$\sum_{j=1}^{n} f_{ij} = n_i \quad \text{and} \quad \sum_{i=1}^{m}\sum_{j=1}^{n} f_{ij} = \sum_{i=1}^{m}\left(\sum_{j=1}^{n} f_{ij}\right) = \sum_{i=1}^{m} n_i = N, \qquad \text{(say).}$$

If $\bar{y}_i$ and $\bar{y}$ denote the means of the ith array and the overall mean, respectively,

then
$$\bar{y}_i = \frac{\sum_j f_{ij} y_{ij}}{\sum_j f_{ij}} = \frac{\sum_j f_{ij} y_{ij}}{n_i} = \frac{T_i}{n_i} \quad \text{and} \quad \bar{y} = \frac{\sum_i \sum_j f_{ij}\, y_{ij}}{\sum_i \sum_j f_{ij}} = \frac{\sum_i n_i\, \bar{y}_i}{\sum_i n_i} = \frac{T}{N}$$

In other words $\bar{y}$ is the weighted mean of all the array means, the weights being the array frequencies.

Definition. *The correlation ratio of Y on X, usually denoted by η_{YX} is given by :*

$$\eta_{YX}^2 = 1 - \frac{\sigma_{eY}^2}{\sigma_Y^2}, \qquad \text{...(11·22)}$$

where
$$\sigma_{eY}^2 = \frac{1}{N}\sum_i \sum_j f_{ij}(y_{ij} - \bar{y}_i)^2 \quad \text{and} \quad \sigma_Y^2 = \frac{1}{N}\sum_i \sum_j f_{ij}(y_{ij} - \bar{y})^2 \qquad \text{...(11·22}a\text{)}$$

A convenient expression for η_{YX} can be obtained in terms of standard deviation σ_{mY} of the means of vertical arrays, each mean being weighted by the array frequency.

We have

$$N\sigma_Y^2 = \sum\sum f_{ij}(y_{ij} - \bar{y})^2 = \sum_i \sum_j f_{ij}\left\{(y_{ij} - \bar{y}_i) + (\bar{y}_i - \bar{y})\right\}^2$$

$$= \sum_i \sum_j f_{ij}(y_{ij} - \bar{y}_i)^2 + \sum_i \sum_j f_{ij}\,(\bar{y}_i - \bar{y})^2 + 2\sum_i \sum_j f_{ij}(y_{ij} - \bar{y}_i)(\bar{y}_i - \bar{y})$$

The term $2\left[\sum_i (\bar{y}_i - \bar{y})\left\{\sum_j f_{ij}\,(y_{ij} - \bar{y}_i)\right\}\right]$ vanishes since $\sum_j f_{ij}\,(y_{ij} - \bar{y}_i) = 0$, being the algebraic sum of the deviations from mean.

$$\therefore \quad N\sigma_Y^2 = \sum_i \sum_j f_{ij}\,(y_{ij} - \bar{y}_i)^2 + \sum_i n_i(\bar{y}_i - \bar{y})^2$$

$$\Rightarrow \quad N\sigma_Y^2 = N\sigma_{eY}^2 + N\sigma_{mY}^2 \quad \Rightarrow \quad \sigma_Y^2 = \sigma_{eY}^2 + \sigma_{mY}^2$$

where $\sigma_{mY}^2 = \frac{1}{N}\sum n_i\,(\bar{y}_i - \bar{y})^2$, is variance of means of the vertical arrays.

$\Rightarrow \quad 1 - \dfrac{\sigma_{eY}^2}{\sigma_Y^2} = \dfrac{\sigma_{mY}^2}{\sigma_Y^2}$, which on comparison with (11·22), gives

$$\eta_{YX}^2 = \frac{\sigma_{mY}^2}{\sigma_Y^2} = \frac{\sum_i n_i(\bar{y}_i - \bar{y})^2}{\sum_i \sum_j f_{ij}\,(y_{ij} - \bar{y})^2} \qquad \text{...(11·23)}$$

We have : $$N\sigma_{mY}^2 = \sum_i n_i\,(\bar{y}_i - \bar{y})^2 = \sum_i n_i \bar{y}_i^2 - N\bar{y}^2 = \sum_i \frac{T_i^2}{n_i} - \frac{T^2}{N}$$

$$\therefore \qquad \eta_{YX}^2 = \left[\sum_i \left(\frac{T_i^2}{n_i}\right) - \frac{T^2}{N}\right] \Big/ N\sigma_Y^2 \qquad ...(11·24)$$

a formula, much more convenient for computational purposes.

Remarks:

1. (11·22) implies that : $\sigma_{eY}^2 = \sigma_Y^2 (1 - \eta_{YX}^2)$

Since σ_{eY}^2 and σ_Y^2 are non-negative, we have

$$1 - \eta_{YX}^2 \geq 0 \quad \Rightarrow \quad \eta_{YX}^2 \leq 1 \quad \Rightarrow \quad |\eta_{YX}| \leq 1 \qquad ...(11·24a)$$

2. Since the sum of squares of deviations in any array is minimum when measured from its mean, we have:

$$\sum_i \sum_j f_{ij} (y_{ij} - \bar{y}_i)^2 \leq \sum_i \sum_j f_{ij} (y_{ij} - \hat{y}_{ij})^2, \qquad ...(*)$$

where $\hat{y}_{ij}$ is the estimate of y_{ij} for given value of $X = x_{ij}$, say, as given by the line of regression of Y on X,

$$i.e., \ \hat{y}_{ij} = a + bx_i, (j = 1, 2, ..., n).$$

But $\sum_i \sum_j f_{ij} (y_{ij} - \bar{y}_i)^2 = N\sigma_{eY}^2 = N\sigma_Y^2 (1 - \eta_{YX}^2)$ and $\sum_i \sum_j f_{ij} (y_{ij} - a - bx_i)^2 = N\sigma_Y^2 (1 - r^2)$ ...(11·24b)

$$\therefore \qquad (*) \quad \Rightarrow \quad 1 - \eta_{YX}^2 \leq 1 - r^2 \quad \Rightarrow \quad \eta_{YX}^2 \geq r^2 \quad \Rightarrow \quad |\eta_{YX}| \geq |r|$$

Thus, the absolute value of the correlation ratio can never be less than the absolute of r, the correlation coefficient.

When the regression of Y on X is linear, straight line of means of arrays coincides with the line of regression and $\eta_{YX}^2 = r^2$. Thus $(\eta_{YX}^2 - r^2)$ *is the departure of regression from linearity.* It is also clear (from Remark 1) that the more nearly η_{YX}^2 approaches unity, the smaller is σ_{eY}^2 and, therefore, closer are the points to the curve of means of vertical arrays.

When $\eta_{YX}^2 = 1$, $\sigma_{eY}^2 = 0 \Rightarrow \sum \sum f_{ij} (y_{ij} - \bar{y}_i)^2 = 0$ or $y_{ij} = \bar{y}_i, \forall j = 1, 2, \cdots, n,$ *i.e.,*

all the points lie on the curve of means. This implies that there is a functional relationship between X and Y. η_{YX} *is, therefore, the measure of the degree to which the association between the variables approaches a functional relationship of the form* $Y = \phi(X)$, *where* $\phi(X)$ *is a single valued function of X,* $[\phi(X) = a + bX]$.

3. It is worth noting that the value of η_{YX} is not independent of the classification of the data. As the class intervals become narrower η_{YX} approaches unity, since in that case σ_{mY}^2 gets nearer to σ_Y^2. If the grouping is so fine that only one item appears in each row (related to each x - class), that item will constitute the mean of that column and thus in this case σ_{mY}^2 and σ_Y^2 become equal so that $\eta_{YX}^2 = 1$. On the other hand, a very coarse grouping tends to make the value of η_{YX} approach r. "Student" has given a formula for 'the correction' to be made in the correlation ratio 'for grouping' in Biometrika (Vol IX page 316–320)

4. It can be easily proved that η_{YX}^2 is independent of change of origin and scale of measurements.

5. η_{XY}^2, the second correlation ratio of X on Y depends upon the scatter of observations about the line of column means.

6. r_{XY} and r_{YX} are same but η_{YX} is, in general, different from η_{XY}.

7. In terms of expectation, correlation ratio is defined as follows:

$$\eta_{YX}^2 = \frac{E_X[E(Y|X) - E(Y)]^2}{E[Y - E(Y)]^2} = \frac{E_X[E(Y|X) - E(Y)]^2}{\sigma_Y^2}$$

and $$\eta_{XY}^2 = \frac{E_Y[E(X \mid Y) - E(X)]^2}{E[X - E(X)]^2} = \frac{E_Y[E(X \mid Y) - E(X)]^2}{\sigma_X^2}$$

8. We give below some diagrams, exhibiting the relationship between r and η_{YX}.

(*i*) For completely random scattering of the dots with no trend, both r and η are zero.

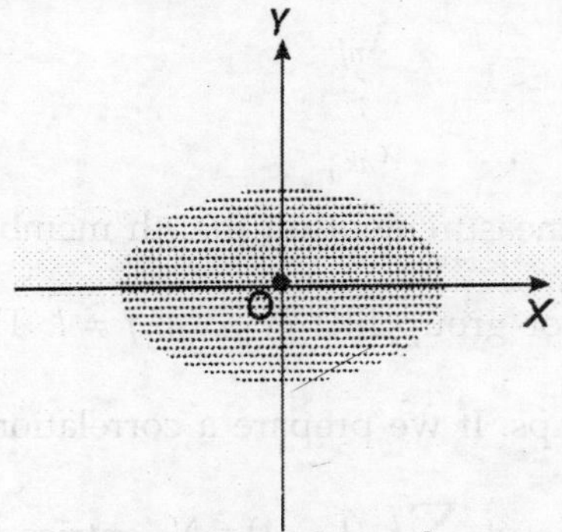

$r=0, \quad \eta_{YX} = \eta_{XY} = 0$

(*ii*) If dots lie precisely on a line, $r = 1$ and $\eta = 1$

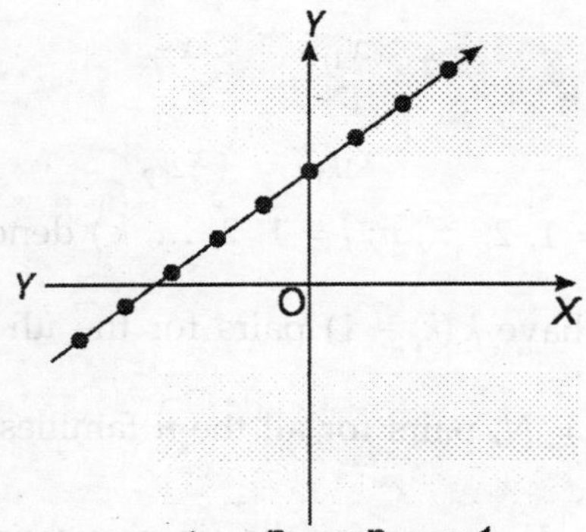

$r = 1, \quad \eta_{YX} = \eta_{XY} = 1$

(*iii*) If dots lie on a curve, such that no ordinate cuts it more than once, $\eta_{YX} = 1$ and if furthermore, the dots are symmetrically placed about Y-axis, then $\eta_{XY} = 0$, $r = 0$.

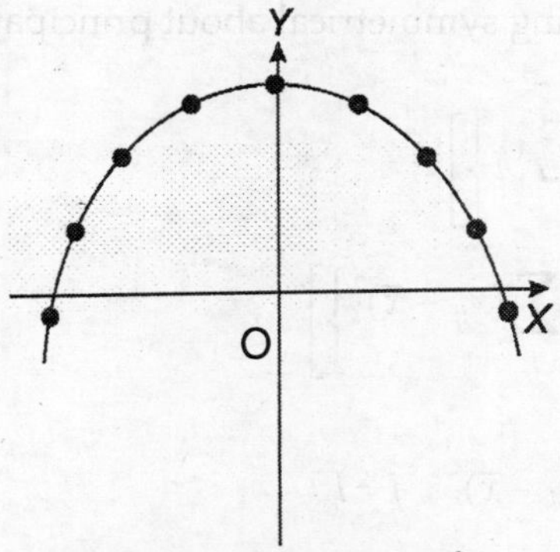

$r = 0, \quad \eta_{YX} = 1, \quad \eta_{XY} = 0$

(*iv*) If $\eta_{YX} > r$, the dots are scattered around a definitely curved trend line.

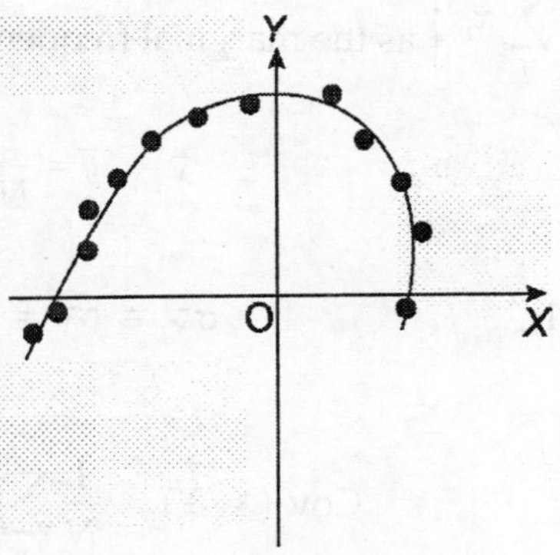

$\eta_{YX} > r$

11·6. INTRA-CLASS CORRELATION

Intra-class correlation means within class correlation. It is distinguishable from product moment correlation in as much as here both the variables measure the same characteristic. Sometimes specially in biological and agricultural study, it is of interest to know how the members of a family or group are correlated among themselves with respect to some one of their common characteristic. For example, we may require the correlation between the heights of brothers of a family or between yields of plots of an experimental block. In such cases both the variables measure the same characteristic, *e.g.*, height and height or weight and weight. There is nothing to distinguish one from the other so that one may be treated as X-variable and the other as the Y-variable.

Suppose we have n families $A_1, A_2, \ldots, A_n$ with $k_1, k_2, \ldots, k_n$ members, each of which may be represented as shown below :

$$\begin{array}{ccccccc}
A_1 & A_2 & \cdots & A_i & \cdots & A_n \\
x_{11} & x_{21} & \cdots & x_{i1} & \cdots & x_{n1} \\
x_{12} & x_{22} & \cdots & x_{i2} & \cdots & x_{n2} \\
\vdots & \vdots & & \vdots & & \vdots \\
x_{1j} & x_{2j} & \cdots & x_{ij} & \cdots & x_{nj} \\
\vdots & \vdots & & \vdots & & \vdots \\
x_{1k_1} & x_{2k_2} & \cdots & x_{ik_i} & \cdots & x_{nk_n}
\end{array}$$

and let x_{ij} $(i = 1, 2, \ldots, n;\ j = 1, 2, \ldots, k_i)$ denote the measurement on the jth member in the ith family.

We shall have $k_i(k_i - 1)$ pairs for the ith family or group like (x_{ij}, x_{il}), $j \neq l$. There will be $\sum_{i=1}^{n} k_i\ (k_i - 1) = N$, pairs for all the n families or groups. If we prepare a correlation table there will be $k_i\ (k_i - 1)$ entries for the ith group or family and $\sum_i k_i\,(k_i - 1) = N$ entries for all the n families or groups. The table is symmetrical about the principal diagonal. Such a table is called an *intra-class correlation table* and the correlation is called *intra-class correlation.*

In the bivariate table x_{i1} occurs $(k_i - 1)$ times, x_{i2} occurs $(k_i - 1)$ times, ..., x_{ik} occurs $(k_i - 1)$ times, *i.e.*, from the ith family we have $(k_i - 1) \sum_j x_{ij}$ and hence for all the n families we have $\sum_i \left[(k_i - 1) \sum_j x_{ij} \right]$ as the marginal frequency, the table being symmetrical about principal diagonal.

$$\therefore \qquad \bar{x} = \bar{y} = \frac{1}{N}\left[\sum_i \left\{ (k_i - 1) \sum_j x_{ij} \right\} \right]$$

Similarly,
$$\sigma_X^2 = \sigma_Y^2 = \frac{1}{N}\left[\sum_i \left\{ (k_i - 1) \sum_j (x_{ij} - \bar{x})^2 \right\} \right]$$

Further
$$\text{Cov}\,(X, Y) = \frac{1}{N} \sum_i \left[\sum_{j,l} (x_{ij} - \bar{x})\,(x_{il} - \bar{x}), \quad j \neq l \right.$$

$$= \frac{1}{N} \sum_i \left[\sum_{j=1}^{k_i} \sum_{l=1}^{k_i} (x_{il} - \bar{x})\,(x_{ij} - \bar{x}) - \sum_{j=1}^{k_i} (x_{ij} - \bar{x})^2 \right]$$

If we write
$$\bar{x}_i = \sum_j x_{ij}/k_i, \text{ then}$$

$$\sum_i \left[\sum_{j=1}^{k_i} \sum_{l=1}^{k_i} (x_{ij} - \bar{x})\,(x_{il} - \bar{x}) \right] = \sum_i \left[\sum_j (x_{ij} - \bar{x}) \sum_l (x_{il} - \bar{x}) \right] = \sum_i \left[k_i(\bar{x}_i - \bar{x})\, k_i(\bar{x}_i - \bar{x}) \right]$$

$$= \sum_i k_i^2\, (\bar{x}_i - \bar{x})^2$$

Therefore, intra-class correlation coefficient is given by :

$$r(X, Y) = \frac{\text{Cov}(X, Y)}{\sqrt{V(X)\,V(Y)}} = \frac{\sum_i k_i^2 (\bar{x}_i - \bar{x})^2 - \sum_i \sum_j (x_{ij} - \bar{x})^2}{\sum_i \sum_j (k_i - 1)(x_{ij} - \bar{x})^2} \qquad ...(11·25)$$

If we put $k_i = k$, *i.e.*, if all families have equal members then :

$$r = \frac{k^2 \sum_i (\bar{x}_i - \bar{x})^2 - \sum_i \sum_j (x_{ij} - \bar{x})^2}{(k-1) \sum_i \sum_j (x_{ij} - \bar{x})^2} = \frac{nk^2\sigma_m^2 - nk\sigma^2}{(k-1)\,nk\sigma^2}$$

$$= \frac{1}{(k-1)} \left\{ \frac{k\sigma_m^2}{\sigma^2} - 1 \right\} \qquad ...(11·25a)$$

where, σ^2 denotes the variance of X and σ^2_m the variance of means of families.

Limits. We have from (11·25*a*),

$$1 + (k-1)r = \frac{k\sigma_m^2}{\sigma^2} \geq 0 \quad \Rightarrow \quad r \geq -\frac{1}{(k-1)}$$

Also $\quad 1 + (k-1)r \leq k$ as the ratio $\dfrac{\sigma_m^2}{\sigma^2} \leq 1 \Rightarrow r \leq 1$

$$\therefore \quad -\frac{1}{(k-1)} \leq r \leq 1. \qquad ...(11·25b)$$

Interpretation. Intra-class correlation cannot be less than $\{-1/(k-1)\}$, though it may attain the value +1 on the positive side, so that it is a skew coefficient and a negative value has not the same significance as a departure from independence as an equivalent positive value.

CHAPTER CONCEPTS QUIZ

1. State, giving reasons, whether each of the following statements is true or false :
   (*i*) The regression lines of Y on X, and of X on Y do not intersect at all.
   (*ii*) In a bivariate regression, $b_{YX} = \frac{1}{5}$, $b_{XY} = 10$.
   (*iii*) The regression coefficient of Y on X is 3·2 and that of X on Y is 0·8.
   (*iv*) There is no relationship between correlation coefficient and regression coefficient.
   (*v*) Both the regression coefficients cannot exceed unity.
   (*vi*) The greater the value of '*r*', the better are the estimates obtained through regression analysis.
   (*vii*) If X and Y are negatively correlated variables, and (0, 0) is on the least squares line of Y on X, and if $X = 1$ is the observed value then predicted value of Y must be negative.
   (*viii*) Let the correlation between X and Y be perfect and positive. Suppose the points (3, 5) and (1, 4) are on the regression lines. With this knowledge it is possible to determine the least square line exactly.

(ix) If the lines of regression are $Y = \frac{1}{4}X$ and $X = \frac{1}{9}Y + 1$, then $\rho = \frac{1}{6}$ and $E(X \mid Y = 0) = 1$.

(x) In bivariate distribution, $b_{YX} = 2{\cdot}8$ and $b_{XY} = 0{\cdot}3$.

(xi) 'Estimating lines' are same as regression lines.

(xii) Both the regression lines for a set of data must have the same slope.

(xiii) The 'Standard Error of Estimate' indicates the error generally expected in computing the regression coefficients.

(xiv) It is possible to have the regression of Y on X constant (does not depend on X), but the regression of X on Y is not constant (does depend on Y).

(xv) $E(Y \mid X = x) = \text{constant} \Rightarrow r(X, Y) = 0$.

(xvi) Correlation ratio is used to measure the extent of relationship between non-linearly related variables.

(xvii) Correlation ratio lies between – 1 and 1.

(xviii) The absolute value of the correlation ratio can never be less than the absolute value of the correlation coefficient.

2. Fill in the blanks :

(i) The regression analysis measures ... between X and Y.

(ii) Lines of regression are ... if $r_{XY} = 0$ and they are ... if $r_{XY} = \pm 1$.

(iii) If the regression coefficients of X on Y and Y on X are $-0{\cdot}4$ and $-0{\cdot}9$, respectively then the correlation coefficient between X and Y is ...

(iv) If the two regression lines are $X + 3Y - 5 = 0$ and $4X + 3Y - 8 = 0$, then $r_{XY} = ...$

(v) If one of the regression coefficients is ... unity, the other must be ... unity.

(vi) The farther the two regression lines cut each other, the ... will be the degree of correlation.

(vii) When one regression coefficient is positive, the other would be ...

(viii) The sign of regression coefficient is ... as that of correlation coefficient.

(ix) Correlation coefficient is the ... between regression coefficients.

(x) When the correlation coefficient is zero, the two regression lines are ... and when it is ± 1, then the regression lines are ...

(xi) Correlation ratio is a suitable measure of correlation if the relation between X and Y is ...

(xii) In correlation ratio, for each value of X, there can be ... values of Y.

(xiii) Correlation ratio is defined as the ratio ... sum of squares to the ... sum of squares.

(xiv) If all observations are in one group, the value of correlation ratio is ...

(xv) Similarity among individuals within the groups can be measured through ... correlation.

3. Indicate the correct answer :

(i) The regression line of Y on X minimises (a) total of the squares of horizontal deviations, (b) total of the squares of the vertical deviations, (c) total of both vertical and horizontal deviations, (d) none of these.

(ii) The regression coefficients are b_2 and b_1. Then the correlation coefficients r is

(a) b_1/b_2, (b) b_2/b_1, (c) $b_1 b_2$, (d) $\pm\sqrt{b_1 b_2}$.

(iii) The farther the two regression lines cut each other (a) the greater will be the degree of correlation, (b) the lesser will be the degree of correlation, (c) does not matter.

(iv) If one regression coefficient is greater than unity, then the other must be
(a) greater than the first one, (b) equal to unity, (c) less than unity, (d) equal to zero.

(v) When the correlation coefficient $r = \pm 1$, then the two regression lines
(a) are perpendicular to each other, (b) coincide, (c) are parallel to each other, (d) do not exist.

(*vi*) The two lines of regression are given as $X + 2Y - 5 = 0$ and $2X + 3Y = 8$. Then the mean values of X and Y, respectively, are

(*a*) 2, 1 (*b*) 1, 2 (*c*) 2, 5 (*d*) 2, 3.

(*vii*) The tangent of the angle between two regression lines is given as 0·6 and the *s.d.* of Y is known to be twice that of X. Then the value of correlation coefficient between X and Y is :

(*a*) $-\frac{1}{2}$, (*b*) $\frac{1}{2}$, (*c*) 0·7, (*d*) 0·3.

(*viii*) If $e_i = y_i - \widehat{y_i}$ is the error in the *i*th estimate for a linear regression of Y on X for *n* pairs of (x_i, y_i)'s, then the standard error of estimate of Y is :

(*a*) $\sigma_y \sqrt{1-r^2}$, (*b*) σ_y, (*c*) $\sigma_y \sqrt{1-2r^2}$, (*d*) none of these

(*ix*) *Let* η_{YX} be the correlation ratio of Y on X then while the absolute value of the correlation ratio can never be less than the absolute value of the correlation coefficient, then η_{YX}^2

(*a*) like correlation coefficient, is independent of origin and scale.
(*b*) like regression coefficient, is independent of origin and not of scale.
(*c*) is independent of scale measurements and not of origin.
(*d*) is neither independent of origin nor scale.

(*x*) To find out the correlation with respect to a particular variate between members of particular group (having $k > 2$ members each) we go in for intra-class correlation, which :

(*a*) like Karl Pearson's correlation coefficient always lies between − 1 and 1.
(*b*) like correlation ratio always lies between − 1 and 1.
(*c*) lies always between $-1/k$ and 1.
(*d*) lies always between $-1/(k-1)$ and 1.

DISCUSSION AND REVIEW QUESTIONS

1. (*a*) Explain what are regression lines. Why are there two such lines? Also derive their equations.

(*b*) Define (*i*) Line of regression, and (*ii*) Regression coefficients. Find the equations to the lines of regression and show that the coefficient of correlation is the geometric mean of coefficients of regression.

(*c*) What equation is the equivalent mathematical statement for the following words :

"If the respective deviations in each series, X and Y, from their means were expressed in units of standard deviations, *i.e.*, if each were divided by the standard deviation of the series, to which it belongs and plotted to a scale of standard deviations, the slope of a straight line best describing the plotted points would be the correlation coefficient *r*."

2. (*a*) Obtain the equation of the line of regression of Y on X and show that the angle θ, between the two lines of regression is given by :

$$\tan\theta = \frac{1-\rho^2}{\rho} \times \frac{\sigma_1\sigma_2}{\sigma_1^2+\sigma_2^2},$$

where σ_1, σ_2 are the standard deviations of X and Y, respectively, and ρ is the correlation coefficient. between X and Y. Interpret the cases when $\rho = 0$ and $\rho = \pm 1$.

(*b*) If θ is the acute angle between the two regression lines with correlation coefficient *r*, show that :

$$\sin\theta \le 1 - r^2.$$

3. (*a*) Explain the term "Regression" by giving examples. Assuming that the regression of Y on X is linear, outline a method for the estimation of the coefficients in the regression line based on the random paired sample of X and Y, and show that the variance of the error of the estimate for Y for the regression line is $\sigma_Y^2(1-\rho^2)$, where σ_Y^2 is the variance of Y and ρ is the correlation coefficient between X and Y.

(b) Prove that X and Y are linearly related if and only if $r_{XY}^2 = 1$. Further show that the slope of the regression line is positive or negative according as $r > 0$ or $r < 0$.

(c) Let X and Y be two variates. Define $X^* = \dfrac{X-a}{b}, Y^* = \dfrac{Y-c}{d}$ for some constants a, b, c and d. Show that the regression line (least square) of Y on X can be obtained from that of Y* on X*.

4. Two variables X and Y are known to be related to each other by the relation : $Y = X/(aX + b)$. How is the theory of linear regression to be employed to estimate the constants a and b from a set of n pairs of observations (x_i, y_i); $i = 1, 2, ..., n$?

5. Derive the standard error of estimate of Y obtained from the linear regression equation of Y on X. What does this standard error measure ?

6. Prove that the error terms that are calculated using the method of least squares (difference between the actual value of the depandent variable and the value predicted by the regression equation) are uncorrelated with the values of the independent variable.

7. Define correlation coefficient and correlation ratio. When is the later a more suitable measure of correlation than the former? Show that the correlation ratio is never less than the correlation coefficient. What do you infer if the two are equal ? Further, show that none of these can exceed one.

8. Show that $1 \geq \eta_{YX}^2 \geq r_{XY}^2 \geq 0$.

Interpret each of the following statements.

(i) $r = 0$, (ii) $r^2 = 1$, (iii) $\eta^2 = 1$, (iv) $\eta^2 = r^2$ and (v) $\eta = 0$.

9. (a) When the correlation coefficient is equal to unity, show that the two correlation ratios are also equal to unity. Is the converse true?

(b) Define correlation ratio η_{XY} and prove that $1 \geq \eta_{XY}^2 \geq r^2$, where r is the coefficient of cor 'ation between X and Y. Show further that $(\eta_{XY}^2 - r^2)$ is a measure of non-linearity of regression.

10. What do you understand by intra-class correlation coefficient. If $x_1, x_2, ..., x_k$ be k variates with standard deviation σ and m be any number, prove that

$$k^2\sigma^2 = (k-1)\sum_{r=1}^{k}(x_r - m)^2 - \sum_{r=1}^{k}\sum_{s=1}^{k}(x_r - m)(x_s - m),\ r \neq s$$

Hence, deduce that the coefficient of intraclass correlation for n families with varying number of members in each family is : $1 - \dfrac{\sum_i k_i\sigma_i^2}{\sigma^2 \sum_i k_i(k_i - 1)}$, where k_i, σ_i^2 denote the number of members and the variance respectively in the ith family and σ^2 is the general variance.

ASSORTED REVIEW PROBLEMS FOR SELF-ASSESSMENT

11·1. Obtain the equations of the two lines of regression for the following data.

X	:	43	44	46	40	44	42	45	42	38	40	42	57
Y	:	29	31	19	18	19	27	27	29	41	30	26	10

Hence, obtain the value of correlation coefficient between X and Y.

Ans. Y on X : Y = – 1·22 X + 78·67 ; X on Y : X = – 0·44 Y + 54·8; r (X, Y) = – 0·73

11·2. (a) Why do we have, in general, two lines of regression ? Obtain the regression of Y on X, and X on Y from the following table and estimate the blood pressure when the age is 45 years :

Age in years (X)	:	56	42	72	36	63	47	55	49	38	42	68	60
Blood pressure (Y)	:	147	125	160	118	149	128	150	145	115	140	152	155

Ans. (a). Y = 1·138 X + 80·778, Y = 131·988 for X = 45.

(*b*) Suppose the observations on X and Y are given as :

X :	59	65	45	52	60	62	70	55	45	49
Y :	75	70	55	65	60	69	80	65	59	61

where $N = 10$ students, and Y = Marks in Maths, X = Marks in Economics. Compute the least square regression equations of Y on X and of X on Y.

If a student gets 61 marks in Economics, what would you estimate his marks in Maths to be ?

11·3. Will it be useful to fit a linear regression to the following data? Support your argument by computing the relevant measures.

X :	12	10	14	18	16
Y :	9	10	8	9	7

Ans. $r(X, Y) = 0{\cdot}5547$, which is a measure of linear relationship between X and Y, is not high (much less than 1). It may be concluded that it is not advisable to fit linear regression to the given data.

11·4. A panel of two Judges P and Q graded seven dramatic performances by independently awarding marks as follows :

Performance :	1	2	3	4	5	6	7
Marks by P :	46	42	44	40	43	41	45
Marks by Q :	40	38	36	35	39	37	41

The 8th performance, which Judge Q could not attend, was awarded 37 marks by Judge P. If Judge Q had also been present, how many marks would be expected to have been awarded by him to the 8th performance ?

Ans. $33{\cdot}5 \simeq 34$

11·5. The weight (in lbs.) of a new born calf is taken at weekly intervals. Below are the observations for 10 weeks :

Age X :	1	2	3	4	5	6	7	8	9	10
Weight Y :	52·5	58·7	65·0	70·2	75·4	81·1	87·2	85·5	102·2	108·4

Let $Y = a + bU$, where $U = 2X - 11$. Use normal equations to estimate a and b. (Given : The sum of the products of the form UY for these 10 observations = 1016·8). Hence obtain the line of best fit of Y on X. What is the average rate of growth of the calf per week ?

Ans. $a = 79{\cdot}62$, $b = 3{\cdot}08$; $Y = 6{\cdot}16X + 45{\cdot}74$; 6·16 lbs.

11·6. (*a*) Compare and contrast the roles of correlation and regression in studying the inter-dependence of two variables.

For 10 observations on price (X) and supply (Y) the following data were obtained (in appropriate units) :

$$\Sigma X = 130,\quad \Sigma Y = 220,\quad \Sigma X^2 = 2288,\quad \Sigma Y^2 = 5506 \quad \text{and} \quad \Sigma XY = 3467$$

Obtain the line of regression of Y on X and estimate the supply when the price is 16 units and find out the S.E.. of the estimate.

Ans. $Y = 8{\cdot}8 + 1{\cdot}015\,X$; 25·04

(*b*) Twenty-five pairs of value of variates X and Y led to the following results :

$$N = 25,\quad \Sigma X = 127,\quad \Sigma Y = 100,\quad \Sigma X^2 = 760\quad \Sigma Y^2 = 449 \quad \text{and} \quad \Sigma XY = 500$$

A subsequent scrutiny showed that two pairs of values were copied down as :

X	Y
8	14
8	6

instead of

X	Y
8	12
6	8

(*i*) Obtain the correct value of the correlation coefficient.

(*ii*) Hence or otherwise, find the correct equations of the two lines of regression.

(*iii*) Find the angle between the regression lines.

Ans. (*i*) $r(X, Y) = -(0{\cdot}64 \times 0{\cdot}15)^{1/2} = -0{\cdot}31$ (*ii*) $X = -0{\cdot}64Y + 7{\cdot}56$, $Y = -0{\cdot}15X + 4{\cdot}75$.

11·7. Using the following data, compute the two coefficients of regression and Karl Pearson's coefficient of correlation.

X \ Y	0–20	20–40	40–60
10–25	10	6	3
25–40	4	40	8
40–55	6	9	15

Ans. $b_{YX} = 0{\cdot}4375$, $b_{XY} = 0{\cdot}2511$; $r(X, Y) = 0{\cdot}3314$.

11·8. (*a*) The following data pertain to the marks in subjects A and B in a certain examination :

Mean marks in $A = 39{\cdot}5$; Mean marks in $B = 47{\cdot}5$

Standard deviation of marks in $A = 10{\cdot}8$; Standard deviation of marks in $B = 16{\cdot}8$

Coefficient of correlation between marks in A and marks in $B = 0{\cdot}42$.

Draw the two lines of regression and explain why there are two regression equations. Give the estimate of marks in B for candidate who secured 50 marks in A.

Ans. (*a*) $Y = 0{\cdot}65X + 21{\cdot}825$, $X = 0{\cdot}27Y + 26{\cdot}675$ and $Y = 54{\cdot}342$ for $X = 50$;

(*b*) You are given the following information about advertising expenditure and sales :

	Advertising Expenditure (X) (₹ *crores*)	*Sales (Y)* (₹ *crores*)
Mean	20	120
S.D.	15	25

Correlation coefficient = 0·8

(*i*) Calculate the two regression equations.

(*ii*) Find the likely sales when advertisement expenditure is ₹ 25 crores.

(*iii*) What should be the advertisement budget if the company wants to attain sales target of ₹ 150 crore ?

Ans. (*b*) (*i*) $X = 0{\cdot}16Y + 0{\cdot}8$, $Y = 4X + 40$; (*ii*) ₹ 140 crore; (*iii*) ₹ 24·8 crore

11·9. (*a*) The regression equations of X on Y, and Y on X (not necessarily in that order) are:

$10X + 3Y = 25$ and $6X + 5Y = 31$. The variance of X is 4.

Find: (*i*) The mean of X and Y

(*ii*) The predicted value of X when Y is 1 and the predicted value of Y when X = 2.

(*iii*) The correlation coefficient between X and Y.

(*iv*) The standard error of estimate in the regression of Y on X.

Ans. (*i*) $\bar{X} = 1$, $\bar{Y} = 5$; (*ii*) $(\hat{X})_{Y=1} = 2{\cdot}2$; $(\hat{Y})_{X=2} = 3.8$ (*iii*) $r(X, Y) = -0{\cdot}6$;

(*iv*) $S_{YX} = \sigma_Y (1 - r^2)^{1/2} = 3{\cdot}2$, $b_{YX} = \dfrac{r\sigma_Y}{\sigma_X} = \dfrac{-6}{5}$ $\Rightarrow$ $\sigma_Y = 4$ $[\because \sigma_X = 2]$

(*b*) Below is given daily family consumption expenditure and daily family income for 10 years.

Daily family consumption expenditure :	700	650	900	950	1,100	1,150,	1,200	1,400	1,550	1,500
Daily family income :	800	1,000	1,200	1,400	1,600	1,800	2,000	2,200	2,400	2,600

(*i*) Obtain regression of consumption expenditure on family income.

(*ii*) Compute r^2 and obtain explained and unexplained variation in consumption expenditure.

(*iii*) Calculate standard error of estimate of consumption expenditure on family income.

11·10. (*a*) Given that $X = 4Y + 5$ and $Y = kX + 4$, are the lines of regression of X on Y and Y on X, respectively, show that $0 < 4k < 1$. If $k = \dfrac{1}{16}$, find the means of the two variables and coefficient of correlation between them.

Hint. (a) $X = 4Y + 5 \Rightarrow b_{XY} = 4$ and $Y = kX + 4 \Rightarrow b_{YX} = k \Rightarrow r^2 = b_{YX} \cdot b_{XY} = 4k$ ...(*)

But $0 \le r^2 \le 1 \Rightarrow 0 \le 4k \le 1$. If $k = \frac{1}{16}$, then from (*), we get $r^2 = 4 \times \frac{1}{16} \Rightarrow r = +\frac{1}{2}$.

[Since both the regression coefficient are positive]

For $k = \frac{1}{16}$, the two lines of regression become $X = 4Y + 5$ and $Y = \frac{1}{16}X + 4$

Solving the two equations, we get. $\bar{Y} = 5{\cdot}75,\ \bar{X} = 28.$

(b) For 50 students of a class, the regression equation of marks in Statistics (X) on marks in Mathematics (Y) is $3Y - 5X + 180 = 0$. The mean marks in mathematics is 44 and variance of marks in Statistics is 9/16th of the variance of marks in Mathematics. Find the mean marks in Statistics and the coefficient of correlation between marks in two subjects.

Hint. (b)We are given : $n = 50$, $\bar{Y} = 44$ and $\sigma_X^2 = \frac{9}{16}\sigma_Y^2 \Rightarrow \frac{\sigma_X}{\sigma_Y} = \frac{3}{4}$

The equation of the line of regression of X on Y is given to be

$$3Y - 5X + 180 = 0 \quad \Rightarrow \quad X = \frac{3}{5}Y + \frac{180}{5}$$

$$\therefore \quad b_{XY} = r\frac{\sigma_X}{\sigma_Y} = \frac{3}{5} \quad \Rightarrow \quad r \cdot \frac{3}{4} = \frac{3}{5} \quad \Rightarrow \quad r = 0{\cdot}8$$

Since, the lines of regression pass through the point $(\bar{X}, \bar{Y})$, we get

$$\bar{X} = \frac{3}{5}\bar{Y} + 36 = \frac{3}{5} \times 44 + 36 = 62{\cdot}4$$

11·11. (a) The lines of regression in a bivariate distribution are : $X + 9Y = 7$ and $Y + 4X = \frac{49}{3}$.

Find (i) the coefficient of correlation

(ii) the ratios $\sigma_X^2 : \sigma_Y^2 : \text{Cov}(X, Y)$,

(iii) the means of the distribution and

(iv) $E(X \mid Y = 1)$.

(b) Estimate X when $Y = 10$, if the two lines of regression are:

$$X = -\frac{1}{18}Y + \lambda \quad \text{and} \quad Y = -2X + \mu,$$

(λ, μ) being unknown and the mean of the distribution is at $(-1, 2)$. Also compute r, λ and μ.

(c) Two lines of regression are : $3y - 2x - 10 = 0$ and $2y - x - 50 = 0$

It is known that $\sigma_Y^2 = 9$.

Obtain (i) the standard error of estimate of y on x.

(ii) the coefficient of determination.

Ans. (c) (i) $S_{YX} = 1{\cdot}5$, (ii) $r^2 = 0{\cdot}75$ ·

(d) In fitting a line of regression of Y on X, the following data are available :

Explained variation = 24; Un-explained variation = 36; $n = 9$

Find (i) S.E. of estimate of Y on X

(ii) The coefficient of determination.

Hint. Total variation $= \Sigma(Y - \bar{Y})^2 = 24 + 36 = 60$; $\sigma_Y^2 = \frac{60}{9} = \frac{20}{3}$

$$(i)\ S_{YX} = 2; \quad (ii)\ r^2 = \frac{\text{Explained variation}}{\text{Total variation}} = \frac{24}{60} = 0{\cdot}6$$

11·12. Suppose you have n observations :

$$(x_1, y_1), (x_2, y_2), \ldots, (x_n, y_n)$$

on two variables X and Y, and you have fitted a linear regression $Y = a + bX$ by the method of least squares. Denote the 'expected' value of Y by Y^*, and the residual $Y - Y^*$, by e. Find means and variances of Y^* and e, and the correlation coefficient between (*i*) X and e, (*ii*) Y and e and (*iii*) Y and Y^*. Use these results to bring out the significance and limitations of the correlation coefficient.

Ans. (*i*) $r(X, e) = 0$, (*ii*) $r(Y, e) = 0$, (*iii*) $r(Y, Y^*) = |r(X, Y)|$.

11·13. (*a*) The regression lines of Y on X and of X on Y are : $Y = aX + b$ and $X = cY + d$. Show that

(*i*) Means are $\overline{X} = (bc + d)/(1 - ac)$ and $\overline{Y} = (ad + b)/(1 - ac)$.

(*ii*) Correlation coefficient between X and Y is $\sqrt{ac}$.

(*iii*) The ratio of the standard deviations of X and Y is $\sqrt{c/a}$.

(*b*) For two random variables X and Y with same means, the two regression equations are $Y = aX + b$ and $X = \alpha Y + \beta$. Show that: $\dfrac{b}{\beta} = \dfrac{1-a}{1-\alpha}\cdot$ Find also the common mean.

(*c*) If the lines of regression of Y on X, and X on Y are respectively

$a_1X + b_1Y + c_1 = 0$ and $a_2X + b_2Y + c_2 = 0$, prove that: $a_1b_2 < a_2b_1$.

Hint. $r^2 = b_{YX} \cdot b_{XY} \le 1 \;\Rightarrow\; \left(-\dfrac{a_1}{b_1}\right)\left(-\dfrac{b_2}{a_2}\right) = \dfrac{a_1 b_2}{a_2 b_1} \le 1. \;\Rightarrow\; a_1 b_2 \le a_2 b_1$

11·14. (*a*) The job rating efficiency of an employee seems to be related to the number of weeks of employment. For a random sample of 10 employees, the following data were observed :

Job Efficiency (X)	:	55	50	20	55	75	80	90	30	75	70
Weeks of Employment (Y)	:	2	4	1	3	5	9	12	2	7	5

Determine the coefficients of regression and regression equation using the non-linear form

$$Y = a + b_1X + b_2X^2.$$

(*b*) The amount of money spent on research and development (R & D) by a large corporation is believed to have an effect on their gross sales. For the past 12 years, the following data have been recorded :

Year	*Amount Spent on R & D (in ₹ lakhs)*	*Gross Sales (in ₹ crores)*	*Year*	*Amount Spent on R & D (in ₹ lakhs)*	*Gross Sales (in ₹ crores)*
1	1·9	2·8	7	6·2	3·4
2	3·4	3·2	8	7·6	3·2
3	6·5	3·0	9	5·0	4·7
4	5·7	4·0	10	5·3	5·2
5	2·8	4·6	11	4·7	2·8
6	2·3	3·8	12	5·2	5·0

(*i*) Determine the coefficients of linear regression and coefficient of determination.

(*ii*) Determine the coefficients of regression and regression equation using the non-linear form

$$Y = a + b_1X + b_2X^2.$$

11·15. The expected remaining life of an electronic part is believed to be related to the age of the part. The ages of 10 of these parts that were in use on a certain date were reccorded in operating hours. When each part burned out, the elapsed time was recorded. The results were as follows :

Ages of Part (in hrs.)	:	40	65	90	5	30	10	80	85	70	25
Remaining Life (in hrs.)	:	30	20	10	80	40	65	15	15	20	50

Determine the coefficients of regression and regression curve using the non-linear form $Y = ab^X$.

11·16. Define curve of regression of Y on X.

The joint density function of X and Y is given by :

$$f(x, y) = \begin{cases} x+y, & 0<x<1, \quad 0<y<1 \\ 0, & \text{otherwise} \end{cases}$$

Find (*i*) the correlation coefficient between X and Y, (*ii*) the regression curves of Y on X, and X on Y.

11.17. Let
$$f(x, y) = \begin{cases} x+y, & 0<x<1, \quad 0<y<1 \\ 0, & \text{otherwise} \end{cases}$$

be the joint *p.d.f.* of X_1 and X_2. Find conditional means and variances. Also show that $\rho = \frac{1}{2}$.

11·18. Let the joint density function of X and Y be given by :

$$f(x, y) = \begin{cases} 8xy, & 0<x<y<1 \\ 0, & \text{otherwise} \end{cases}$$

Find: (*i*) $E(Y \mid X = x)$, (*ii*) $E(XY \mid X = x)$, (*iii*) Var $(Y \mid X = x)$.

Ans. (*i*) $E(Y|x) = \frac{2}{3}\left(\frac{1+x+x^2}{1+x}\right)$; (*ii*) $E(XY|x) = xE(Y|x)$, (*iii*) $E(Y^2|x) = \frac{1+x^2}{2}$;

$$\text{Var}(Y \mid X = x) = \frac{(x-1)^2\,(x^2+4x+1)}{18\,(x+1)^2}.$$

11·19. The joint p.d.f. of X, Y is : $f(x, y) = x \exp[-x\,(1+y)]$, $x \geq 0$, $y \geq 0$

Prove that the regression of (*a*) Y on X; and (*b*) X on Y, are both non-linear.

Ans. $E[X \mid Y = y] = \frac{2}{1+y}$ and $E[Y \mid X = x] = \frac{1}{x}$

11·20. Variables (X, Y) have joint *p.d.f.* :

$$f(x, y) = \begin{cases} 6(1-x-y), & x>0, \ y>0, \ x+y<1 \\ 0, & \text{otherwise} \end{cases}$$

Find $f_X(x)$, $f_Y(y)$ and Cov (X, Y). Are X and Y independent ? Obtain the regression curves for the means.

Ans. $f_1(x) = 3\,(1-x)^2,\ 0 < x < 1$; $f_2(y) = 3(1-y)^2,\ 0 < y < 1$; Cov $(X, Y) = -\frac{7}{48}$

X and Y are not independent. Regression curves for the means are :

$$y = E(Y \mid x) = \frac{1-3x}{3(1-x)^2} \quad \text{and} \quad x = E(X \mid y) = \frac{1-3y}{3(1-y)^2}.$$

11·21. For the joint *p.d.f.* : $f(x, y) = 3x^2 - 8xy + 6y^2,\ 0 < (x, y) \leq 1$,
find the least square regression lines and the regression curves for the means.

11·22. Let (X, Y) be jointly distributed with *p.d.f.* :

$$f(x, y) = \begin{cases} \exp(-y), & 0<x<y<\infty \\ 0, & \text{otherwise} \end{cases}$$

Prove that : $E(Y \mid X = x) = x + 1$ and $E(X \mid Y = y) = y/2$.

Hence, prove that: $r(X, Y) = \sqrt{1/2}$.

11·23. Let

$$f(x,y)=\begin{cases} e^{-y}(1-e^{-x}), & 0<x<y; \quad 0<y<\infty \\ e^{-x}(1-e^{-y}), & 0<y<x; \quad 0<x<\infty \end{cases}$$

(*i*) Show that $f(x, y)$ is a *p.d.f.*; (*ii*) Find marginal distributions of X and Y; (*iii*) Find $E(Y \mid X = x)$ for $x > 0$; (*iv*) Find $P(X \le 2, Y \le 2)$; (*v*) Find the correlation coefficient $r(X, Y)$; (*vi*) Find another joint p.d.f. having the same marginals.

Ans. (*ii*) $f_1(x) = x \cdot exp(-x),\ 0 < x < \infty;\quad f_2(y) = y\ exp(-y),\ 0 < y < \infty,$

(*iii*) $E(Y \mid x) = \dfrac{1-e^{x}}{x}[x+1] + \dfrac{1}{x}\left(\dfrac{x^2}{2} + xe^{-x} + e^{-x} - 1\right)$ (*iv*) $1 - \dfrac{1}{e^4} - \dfrac{4}{e^2}$,

(*v*) $r(x, y) = \dfrac{\text{Cov}(x, y)}{\sigma_x\, \sigma_y} = \dfrac{1}{\sqrt{2}\sqrt{2}} = \dfrac{1}{2}.$

Hint. (*vi*) $f(x, y, \alpha) = f_1(x)\, f_2(y)\left[1 + \alpha\{2F(x) - 1\}\{2F(y) - 1\}\right], |\alpha| < 1,$

has the same marginals $f_1(x)$ and $f_2(y)$.

11·24. Obtain regression equation of Y on X for the distributions :

(*a*) $f(x, y) = \dfrac{9}{2} \cdot \dfrac{1+x+y}{(1+x)^4\,(1+y)^4};\ x, y \ge 0$ (*b*) $f(x, y) = \dfrac{4}{5}(x+3y) \cdot \exp(-x-2y);\ x \ge 0,\ y \ge 0$

Ans. (*a*) $y = E(Y \mid X = x) = \dfrac{x+3}{2x+3};$ (*b*) $y = E(Y \mid X = x) = \dfrac{x+3}{2x+3}$

11·25. An item (say, a pen) from a production line can be acceptable, repairable or useless. Suppose a production is stable and let $p, q, r (p + q + r = 1)$, denote the probabilities for three possible conditions of an item. If the items are put into lots of 100 :

(*i*) Derive an expression for the probability function of (X, Y), where X and Y are the number of items in the lots that are respectively in the first two conditions.

(*ii*) Derive the moment generating function of X and Y.

(*iii*) Find the marginal distribution X.

(*iv*) Find the conditional distribution of Y given $X = 90$.

(*v*) Obtain the regression function of Y on X.

Ans. (*i*) $p(x, y) = \dfrac{n!}{x!\,y!\,(n-x-y)!} \cdot p^x q^y r^{n-x-y};\ [0 < (p, q) < 1;\ p + q + r = 1;\ x + y = n]$

(*ii*) $M_{X,Y}(t_1, t_2) = \left(pe^{t_1} + qe^{t_2} + r\right)^{100}$

(*iii*) $X \sim B(n = 100, p);$

(*iv*) $(Y \mid X = 90) \sim B\left(10, \dfrac{q}{1-p}\right);$ [Note that $1 - p = q + r \ne q$]

(*v*) $y = E(Y \mid X = x) = \dfrac{q\,(n-x)}{1-p}.$

11·26. A ball is drawn at random from an urn containing 3 white balls numbered 0,1,2; 2 red balls numbered 0, 1 and one black ball numbered 0. If the colours white, red and black are again numbered 0, 1, and 2 respectively, find the correlation coefficient between the variables: X, the colur number and Y, the number of the ball. Also write down least square regression lines of Y on X and X on Y.

Ans. $r(X,Y) = -1/2$; Lines of Regression are:

Y on X : $X + 2Y - 2 = 0$; X on Y : $2X + Y - 2 = 0$

CHAPTER

12 Multiple and Partial Correlation and Regression Analysis

LEARNING OBJECTIVES

Upon completion of this Chapter, you should be able to:

- Measure the relationship between two variables when influence of one or more other variables is held constant.
- Understand the concepts of multiple and partial correlation and regression.
- Derive expressions for multiple and partial correlation coefficients and regression coefficients in terms of total correlation coefficients and discuss the interrelationship between them.
- Understand the concept and interpretation of the coefficient of partial determination.
- Demonstrate how multiple correlation and regression analysis is used in statistical analysis.
- Measure the coefficient of multiple determination to understand the proportion of variation in the dependent variable which is explained by the independent variables.

CHAPTER OUTLINES

12·1. INTRODUCTION

In Chapters 10 (correlation) and 11 (Regression), our discussion was mainly confined to the simultaneous study of two variables. However, in most of the studies relating to social sciences, biological and physical sciences, the value of any one phenomenon (variable) is affected by multiplicity of factors (variables), many of which may be inter-related among themselves. Multiple correlation and multiple regression are the important tools for studying the relationship between three or more variables and also to measure the intensity of the relationship between them. For example,

(i) The weight (X_1) of an individual depends on the variables, height (X_2) and age (X_3) of the individual.

(ii) The yield of crop per acre, say, (X_1) depends upon quality of seed (X_2), fertility of soil (X_3), fertilizer used (X_4), irrigation facilities (X_5), weather conditions (X_6) and so on.

The dependent variable (X_1) is also called the *response variable* and the independent variables (X_2, X_3, ... etc.) are also called the *explanatory variables.*

Whenever we are interested in studying the joint effect of a group of variables upon a variable not included in that group, our study is that of *multiple correlation and multiple regression.*

Suppose in a trivariate or multi-variate distribution we are interested in the relationship between two variables only. There are two alternatives, *viz.*, (*i*) we consider only those two members of the observed data in which the other members have specified values or (*ii*) we may eliminate mathematically the effect of other variates on two variates. The first method has the disadvantage that it limits the size of the data and also it will be applicable to only the data in which the other variates have assigned values. In the second method it may not be possible to eliminate the entire influence of the variates but the linear effect can be easily eliminated. The correlation and regression between only two variates after eliminating the linear effect of other variates in them is called the *partial correlation and partial regression.*

12·2. MULTIPLE AND PARTIAL CORRELATION AND REGRESSION

Consider a multivariate distribution with n variables, say, $X_1, X_2, ..., X_n$. The study of the relationship between the dependent variable, say X_i and any one of the independent variable, say X_j, ($j \neq i = 1, 2, ..., n$), after eliminating the linear effect of the other independent variables on each of the variables X_i and X_j (or by keeping all other independent variables constant), is called the study of partial correlation. On the other hand, the study of relationship between the dependent variable, say X_1 and the joint effect of all the independent variables ($X_2, X_3, ..., X_n$) on X_1 is termed as the study of multiple correlation and multiple regression.

Multiple regression analysis is the extension of the two variable regression theory (in Chapter 11) to n variables $X_1, X_2, ..., X_n$. The basic objectives of multiple linear regression are:

(i) To fit the plane of regression of the dependent variable, say X_1 on the independent variables ($X_2, X_3, ..., X_n$) by the *principle of least squares* and use this plane to estimate the value of the dependent variable X_1 for a given set of values of the independent variables.

(ii) To estimate the error involved as a result of using this plane of regression for estimation, *i.e.*, to compute $(X_1 - \hat{X}_1)$. This is achieved by calculating the standard error of the estimate so obtained.

(iii) To determine how much variation in the dependent variable is accounted for or *explained* by the fitted plane of regression. This is achieved through the *multiple coefficient of determination.*

12·2·1. Yule's Notation

Let us consider a distribution involving three random variables X_1, X_2 and X_3. Then the equation of the plane of regression of X_1 on X_2 and X_3 is:

$$X_1 = a + b_{12\cdot3}\, X_2 + b_{13\cdot2}\, X_3 \quad \text{... (12·1)}$$

Without loss of generality we can assume that the variables X_1, X_2 and X_3 have been measured from their respective means, so that

$$E(X_1) = E(X_2) = E(X_3) = 0 \quad \text{... (12·1}a\text{)}$$

($\because$ Correlation coefficients and the regression coefficients are not affected by change of origin.)

Hence, on taking expectation of both sides in (12·1), we get $a = 0$.

Thus, the plane of regression of X_1 on X_2 and X_3 becomes

$$X_1 = b_{12\cdot3}\, X_2 + b_{13\cdot2}\, X_3 \quad \text{... (12·1}b\text{)}$$

The coefficients $b_{12\cdot3}$ and $b_{13\cdot2}$ are known as the *partial regression coefficients of* X_1 on X_2 and of X_1 on X_3, respectively.

$$e_{1\cdot23} = b_{12\cdot3}\, X_2 + b_{13\cdot2}\, X_3 \quad \text{... (12·1}c\text{)}$$

is called the estimate of X_1 as given by the plane of regression (12·1*b*) and the quantity

$$X_{1\cdot23} = X_1 - b_{12\cdot3}\, X_2 - b_{13\cdot2}\, X_3 = X_1 - e_{1\cdot23}, \quad \text{... (12·1}c\text{)}$$

is called the *error of estimate or residual.*

In the general case of n variable $X_1, X_2, \ldots, X_n$ the equation of the plane of regression of X_1 on $X_2, X_3 \ldots X_n$ becomes

$$X_1 = b_{12\cdot34\,\ldots\,n}\, X_2 + b_{13\cdot24\,\ldots\,n}\, X_3 + \ldots + b_{1n\cdot23\,\ldots\,(n-1)}\, X_n$$

The error of estimate or residual is given by:

$$X_{1\cdot23\,\ldots\,n} = X_1 - (b_{12\cdot34\,\ldots\,n}\, X_2 + b_{13\cdot24\,\ldots\,n}\, X_3 + \ldots + b_{1n\cdot23\,\ldots\,(n-1)}\, X_n) \quad \text{...(12·1}d\text{)}$$

The notations used here are due to Yule. The subscripts before the dot(·) are known as *primary subscripts* and those after the dot are called *secondary subscripts*. The order of a regression coefficient is determined by the number of secondary subscripts, *i.e.*, $b_{12\cdot3}$, $b_{12\cdot34}$, ... , $b_{12\cdot34\,\ldots\,n}$ are the regression coefficients of order 1, 2, ... , $(n-2)$ respectively. Thus, in general, a regression coefficient with *p*-secondary subscripts will be called a regression co-efficient of order '*p*'. It may be pointed out that the order in which the secondary subscripts are written is immaterial but the order of the primary subscripts is important, *e.g.*, in $b_{12\cdot34\,\ldots\,n}$, X_2 is independent while X_1 is dependent variable but in $b_{21\cdot34\,\ldots\,n}$, X_1 is independent while X_2 is dependent variable. Thus, of the two primary subscripts, former refers to dependent variable and the latter to independent variable.

The order of a residual is also determined by the number of secondary subscripts in it, *e.g.*, $X_{1\cdot23}$, $X_{1\cdot234}$, ..., $X_{1\cdot23\,\ldots\,n}$ are the residuals of order 2, 3, ..., $(n-1)$ respectively.

Remark. In the following sequences we shall assume that the variables under consideration have been measured from their respective means so that $E(X_i) = 0$; $i = 1, 2, \ldots n$.

12·3. PLANE OF REGRESSION

The equation of the plane of regression of X_1 and X_2 and X_3 is:

$$X_1 = b_{12\cdot3}\, X_2 + b_{13\cdot2}\, X_3 \quad \text{... (12·2)}$$

The constants *b's* in (12·2) are determined by the principle of least squares, *i.e.*, by minimising the sum of the squares of the residuals, *viz.*,

$$S = \Sigma X_{1\cdot23}^{\;2} = \Sigma (X_1 - b_{12\cdot3}\, X_2 - b_{13\cdot2}\, X_3)^2,$$

the summation being extended to the given values (N in number) of the variables.

Here, we make N observations on each of the variables X_1, X_2 and X_3.

The normal equations for estimating $b_{12.3}$ and $b_{13.2}$ are:

$$\left.\begin{aligned}\frac{\partial S}{\partial b_{12\cdot3}} &= 0 = -2\sum X_2\left(X_1 - b_{12\cdot3}X_2 - b_{13\cdot2}X_3\right)\\ \frac{\partial S}{\partial b_{13\cdot2}} &= 0 = -2\sum X_3\left(X_1 - b_{12\cdot3}X_2 - b_{13\cdot2}X_3\right)\end{aligned}\right\} \quad \text{... (12·3)}$$

i.e., $\Sigma X_2\, X_{1\cdot23} = 0$ and $\Sigma X_3\, X_{1\cdot23} = 0$ ... (12·3*a*)

$$\Rightarrow \left.\begin{aligned}\sum X_1X_2 - b_{12\cdot3}\sum X_2^{\,2} - b_{13\cdot2}\sum X_2X_3 &= 0\\ \sum X_1X_3 - b_{12\cdot3}\sum X_2X_3 - b_{13\cdot2}\sum X_3^{\,2} &= 0\end{aligned}\right\} \quad \text{... (12·3b)}$$

Since X_i's are measured from their respective means, we have

$$\sigma_i^2 = \frac{1}{N}\sum X_i^2, \quad \text{Cov}(X_i, X_j) = \frac{1}{N}\sum X_i X_j \quad \text{and} \quad r_{ij} = \frac{\text{Cov}(X_i, X_j)}{\sigma_i\,\sigma_j} = \frac{\sum X_i X_j}{N\sigma_i\,\sigma_j} \quad \text{... (12·3c)}$$

Hence, from (12·3*b*), we get

$$\left.\begin{aligned}r_{12}\,\sigma_1\sigma_2 - b_{12\cdot3}\,\sigma_2^{\,2} - b_{13\cdot2}\,r_{23}\,\sigma_2\sigma_3 &= 0\\ r_{13}\,\sigma_1\sigma_3 - b_{12\cdot3}\,r_{23}\,\sigma_2\sigma_3 - b_{13\cdot2}\,\sigma_3^{\,2} &= 0\end{aligned}\right\} \quad \text{... (12·3d)}$$

Solving the equations (12·3*d*) simultaneously for $b_{12\cdot3}$ and $b_{13\cdot2}$, we get

$$b_{12.3} = \frac{\begin{vmatrix} r_{12}\,\sigma_1 & r_{23}\,\sigma_3 \\ r_{13}\,\sigma_1 & \sigma_3\end{vmatrix}}{\begin{vmatrix}\sigma_2 & r_{23}\,\sigma_3\\ r_{23}\,\sigma_2 & \sigma_3\end{vmatrix}} = \frac{\sigma_1}{\sigma_2}\cdot\frac{\begin{vmatrix} r_{12} & r_{23}\\ r_{13} & 1\end{vmatrix}}{\begin{vmatrix}1 & r_{23}\\ r_{23} & 1\end{vmatrix}} = \frac{\sigma_1}{\sigma_2}\cdot\left(\frac{r_{12} - r_{13}\,r_{23}}{1 - r_{23}^2}\right) \quad \text{... (12·4)}$$

Similarly, we will get $$b_{13.2} = \frac{\sigma_1}{\sigma_3}\frac{\begin{vmatrix}1 & r_{12}\\ r_{23} & r_{13}\end{vmatrix}}{\begin{vmatrix}1 & r_{23}\\ r_{23} & 1\end{vmatrix}} = \frac{\sigma_1}{\sigma_3}\cdot\left(\frac{r_{13} - r_{12}\,r_{23}}{1 - r_{23}^2}\right) \quad \text{... (12·4a)}$$

Substituting these values in (12·2), we obtain the required equation of the plane of regression of X_1 on (X_2, X_3).

By symmetry of the results in (12·4) and (12·4*a*), we have

$$b_{ij.k} = \frac{\sigma_i}{\sigma_j}\left[\frac{r_{ij} - r_{ik}\,r_{jk}}{(1 - r_{jk}^2)}\right]; \ (i \neq j \neq k = 1, 2, 3) \quad \text{... (12·4b)}$$

If we write $$\omega = \begin{vmatrix}1 & r_{12} & r_{13}\\ r_{21} & 1 & r_{23}\\ r_{31} & r_{32} & 1\end{vmatrix} \quad \text{... (12·5)}$$

$$= 1\cdot\begin{vmatrix}1 & r_{23}\\ r_{32} & 1\end{vmatrix} - r_{12}\begin{vmatrix}r_{21} & r_{23}\\ r_{31} & 1\end{vmatrix} + r_{13}\begin{vmatrix}r_{21} & 1\\ r_{31} & r_{32}\end{vmatrix}$$

$$= \left(1 - r_{23}^2\right) - r_{12}\left(r_{12} - r_{13}r_{23}\right) + r_{13}\left(r_{12}r_{23} - r_{13}\right)$$

$$= 1 - r_{12}^2 - r_{13}^2 - r_{23}^2 + 2r_{12}\,r_{13}\,r_{23} \qquad [\because r_{ii} = 1 \text{ and } r_{ij},\ i \neq j = 1, 2, 3]$$

and ω_{ij} = (Cofactor of element in the ith row and jth column of ω) ... (12·6)

$= (-1)^{i+j} \times$ [Determinant obtained on deleting the ith row and jth column in ω]

Then ω_{11} = (Cofactor of element in 1st row and 1st column of ω)

$$= (-1)^{1+1}\begin{vmatrix} 1 & r_{23} \\ r_{32} & 1 \end{vmatrix} = 1 - r_{23}^2 \quad \text{... (12·6}a\text{)}$$

$$\omega_{12} = \omega_{21} = (-1)^{1+2}\begin{vmatrix} r_{21} & r_{23} \\ r_{31} & 1 \end{vmatrix} = -(r_{12} - r_{13} r_{23}) = (r_{13} r_{23} - r_{12}) \quad \text{... (12·6}b\text{)}$$

$$\omega_{13} = \omega_{31} = (-1)^{1+3}\begin{vmatrix} r_{21} & 1 \\ r_{31} & r_{32} \end{vmatrix} = r_{12} r_{23} - r_{13} \quad \text{... (12·6}c\text{)}$$

$$\omega_{22} = (-1)^{2+2}\begin{vmatrix} 1 & r_{13} \\ r_{31} & 1 \end{vmatrix} = 1 - r_{13}^2 \quad \text{... (12·6}d\text{)}$$

$$\omega_{33} = (-1)^{3+3}\begin{vmatrix} 1 & r_{12} \\ r_{21} & 1 \end{vmatrix} = 1 - r_{12}^2 \quad \text{... (12·6}e\text{)}$$

Substituting from (12·6a) to (12·6e) in (12·4) and (12·4a), the equation of the plane of regression of X_1 on (X_2 and X_3) becomes:

$$X_1 = -\frac{\sigma_1}{\sigma_2}\cdot\frac{\omega_{12}}{\omega_{11}}\cdot X_2 - \frac{\sigma_1}{\sigma_3}\cdot\frac{\omega_{13}}{\omega_{11}} X_3 \quad \Rightarrow \quad \frac{X_1}{\sigma_1}\cdot\omega_{11} + \frac{X_2}{\sigma_2}\cdot\omega_{12} + \frac{X_3}{\sigma_3}\cdot\omega_{13} = 0 \quad \text{... (12·7)}$$

Similarly, the equation of the plane of regression of X_3 on (X_1, X_2) will be given by:

$$\frac{X_1}{\sigma_1}\cdot\omega_{31} + \frac{X_2}{\sigma_2}\cdot\omega_{32} + \frac{X_3}{\sigma_3}\cdot\omega_{33} = 0 \quad \text{... (12·7}a\text{)}$$

In general, the equation of the plane of regression of X_i on (X_j, X_k) is given by:

$$X_i = b_{ij\,.\,k} X_j + b_{ik\,.\,j} X_k\ ;\ (i \neq j \neq k = 1, 2, 3), \quad \text{... (12·7}b\text{)}$$

$$\text{where } b_{ij\,.\,k} = \frac{\sigma_i}{\sigma_j}\left(\frac{r_{ij} - r_{ik} r_{jk}}{1 - r_{jk}^2}\right) = -\frac{\sigma_i}{\sigma_j}\cdot\frac{\omega_{ij}}{\omega_{ii}} \quad \text{and} \quad b_{ik\,.\,j} = \frac{\sigma_i}{\sigma_k}\left(\frac{r_{ik} - r_{ij} r_{kj}}{1 - r_{kj}^2}\right) = -\frac{\sigma_i}{\sigma_k}\cdot\frac{\omega_{ik}}{\omega_{ii}} \quad \text{... (12·7}c\text{)}$$

12·3·1. Generalisation of the Plane of Regression to *n* Variables

Consider n-variate distribution with variables $X_1, X_2, ..., X_n$. Assuming that X_i's; $i = 1, 2, ..., n$ are measured from their respective means, the equation of the plane of regression of X_1 on ($X_2, X_3, ..., X_n$) is given by:

$$X_1 = b_{12\cdot34\,...\,n} X_2 + b_{13\cdot24\,...\,n} X_3 + ... + b_{1n\cdot23\,...\,(n-1)} X_n \quad \text{... (12·8)}$$

where $b_{12\cdot34\,...\,n}$, $b_{13\cdot24\,...\,n}$, ..., $b_{1n\cdot23\,...\,(n-1)}$ are partial regression coefficients of X_1 on $X_2, X_3, ..., X_n$ respectively.

The sum of the squares of residuals is given by

$$S = \Sigma X^2_{1\cdot23\,...\,n} = \Sigma[X_1 - b_{12\cdot34\,...\,n} X_2 - b_{13\cdot24\,...\,n} X_3 - ... - b_{1n\cdot23\,...\,(n-1)} X_n]^2$$

Using the principle of least squares, the normal equations for estimating the $(n-1)$, b's are:

$$\left.\begin{aligned}
\frac{\partial S}{\partial b_{12\cdot34\ldots n}} &= 0 = -2\sum X_2\left(X_1 - b_{12\cdot34\ldots n} X_2 - b_{13\cdot24\ldots n} X_3 - \ldots - b_{1n\cdot23\ldots(n-1)} X_n\right)\\
\frac{\partial S}{\partial b_{13\cdot24\ldots n}} &= 0 = -2\sum X_3\left(X_1 - b_{12\cdot34\ldots n} X_2 - b_{13\cdot24\ldots n} X_3 - \ldots - b_{1n\cdot23\ldots(n-1)} X_n\right)\\
&\vdots\\
\frac{\partial S}{\partial b_{1n\cdot23\ldots(n-1)}} &= 0 = -2\sum X_n\left(X_1 - b_{12\cdot34\ldots n} X_2 - b_{13\cdot24\ldots n} X_3 - \ldots - b_{1n\cdot23\ldots(n-1)} X_n\right)
\end{aligned}\right\} \quad \ldots(12\cdot9)$$

i.e.,
$$\sum X_i\, X_{1\cdot23\ldots n} = 0,\ (i = 2, 3, \ldots, n) \quad \ldots(12\cdot9a)$$

which on simplification after using (12·3*c*), give

$$\left.\begin{aligned}
r_{12}\,\sigma_1\sigma_2 &= b_{12\cdot34\ldots n}\,\sigma_2^2 + b_{13\cdot24\ldots n}\, r_{23}\sigma_2\sigma_3 + \ldots + b_{1n\cdot23\ldots(n-1)}\, r_{2n}\sigma_2\sigma_n\\
r_{13}\,\sigma_1\sigma_3 &= b_{12\cdot34\ldots n}\, r_{23}\sigma_2\sigma_3 + b_{13\cdot24\ldots n}\,\sigma_3^2 + \ldots + b_{1n\cdot23\ldots(n-1)}\, r_{3n}\sigma_3\sigma_n\\
&\vdots\\
r_{1n}\,\sigma_1\sigma_n &= b_{12\cdot34\ldots n}\, r_{2n}\sigma_2\sigma_n + b_{13\cdot24\ldots n}\, r_{3n}\sigma_3\sigma_n + \ldots + b_{1n\cdot23\ldots(n-1)}\,\sigma_n^2
\end{aligned}\right\} \quad \ldots(12\cdot9b)$$

Hence, the eliminant of b's between (12·8) and (12·9*b*) is:

$$\begin{vmatrix}
X_1 & X_2 & X_3 & \cdots & X_n\\
r_{12}\sigma_1\sigma_2 & \sigma_2^2 & r_{23}\sigma_2\sigma_3 & \cdots & r_{2n}\sigma_2\sigma_n\\
r_{13}\sigma_1\sigma_3 & r_{23}\sigma_2\sigma_3 & \sigma_3^2 & \cdots & r_{3n}\sigma_3\sigma_n\\
\vdots & \vdots & \vdots & & \vdots\\
r_{1n}\sigma_1\sigma_n & r_{2n}\sigma_2\sigma_n & r_{3n}\sigma_3\sigma_n & \cdots & \sigma_n^2
\end{vmatrix} = 0$$

Dividing $C_1, C_2, \ldots, C_n$ by $\sigma_1, \sigma_2, \ldots, \sigma_n$ respectively and $R_2, R_3, \ldots, R_n$ by $\sigma_2, \sigma_3, \ldots, \sigma_n$ respectively, we get

$$\begin{vmatrix}
\dfrac{X_1}{\sigma_1} & \dfrac{X_2}{\sigma_2} & \dfrac{X_3}{\sigma_3} & \cdots & \dfrac{X_n}{\sigma_n}\\
r_{12} & 1 & r_{32} & \cdots & r_{2n}\\
r_{13} & r_{23} & 1 & \cdots & r_{3n}\\
\vdots & \vdots & \vdots & & \vdots\\
r_{1n} & r_{2n} & r_{3n} & \cdots & 1
\end{vmatrix} = 0 \quad \ldots(12\cdot10)$$

If we write

$$\omega = \begin{vmatrix}
1 & r_{12} & r_{13} & \cdots & r_{1n}\\
r_{21} & 1 & r_{23} & \cdots & r_{2n}\\
r_{31} & r_{32} & 1 & \cdots & r_{3n}\\
\vdots & \vdots & \vdots & & \vdots\\
r_{n1} & r_{n2} & r_{n3} & \cdots & 1
\end{vmatrix} \quad \ldots(12\cdot11)$$

and ω_{ij} is the cofactor of the element in the ith row and jth column of ω, we get from (12·10):

$$\frac{X_1}{\sigma_1}\omega_{11} + \frac{X_2}{\sigma_2}\omega_{12} + \frac{X_3}{\sigma_3}\omega_{13} + \ldots + \frac{X_n}{\sigma_n}\omega_{1n} = 0 \qquad \ldots (12\cdot12)$$

as the required equation of the plane of regression of X_1 and $X_2, X_3, \ldots, X_n$.

Equation (12·12) can be re-written as:

$$X_1 = -\frac{\sigma_1}{\sigma_2}\cdot\frac{\omega_{12}}{\omega_{11}}X_2 - \frac{\sigma_1}{\sigma_3}\cdot\frac{\omega_{13}}{\omega_{11}}X_3 - \ldots - \frac{\sigma_1}{\sigma_n}\cdot\frac{\omega_{1n}}{\omega_{11}}X_n \qquad \ldots (12\cdot12a)$$

Comparing (12·12*a*) with (12·8), we get

$$\left.\begin{aligned} b_{12\cdot34\ldots n} &= -\frac{\sigma_1}{\sigma_2}\cdot\frac{\omega_{12}}{\omega_{11}} \\ b_{13.24\ldots n} &= -\frac{\sigma_1}{\sigma_3}\cdot\frac{\omega_{13}}{\omega_{11}} \\ \vdots \quad & \quad \vdots \\ b_{1n\cdot23\ldots(n-1)} &= -\frac{\sigma_1}{\sigma_n}\cdot\frac{\omega_{1n}}{\omega_{11}} \end{aligned}\right\} \qquad \ldots (12\cdot13)$$

Remarks:

1. From the symmetry of the result obtained in (12·13) the equation of the plane of regression of X_i, (say), on the remaining variables X_j ($j \neq i = 1, 2, \ldots, n$), is given by:

$$\frac{X_1}{\sigma_1}\omega_{i1} + \frac{X_2}{\sigma_2}\omega_{i2} + \ldots + \frac{X_i}{\sigma_i}\omega_{ii} + \ldots + \frac{X_n}{\sigma_n}\omega_{in} = 0 \ ; \ i = 1, 2, \ldots, n \qquad \ldots (12\cdot14)$$

2. We have

$$b_{12\cdot34\ldots n} = -\frac{\sigma_1}{\sigma_2}\cdot\frac{\omega_{12}}{\omega_{11}} \quad \text{and} \quad b_{21.34\ldots n} = -\frac{\sigma_2}{\sigma_1}\cdot\frac{\omega_{21}}{\omega_{22}} \qquad \ldots (12\cdot14a)$$

Since each of $\sigma_1, \sigma_2, \omega_{11}$ and ω_{22} is non-negative and $\omega_{12} = \omega_{21}$, [*c.f.* Remarks 3 and 4 to § 12.5], the sign of each regression coefficient $b_{12.34\ldots n}$ and $b_{21.34\ldots n}$ depends on ω_{12}.

12·4. PROPERTIES OF RESIDUALS

Property 1. *The sum of the product of any residual of order zero with any other residual of higher order is zero, provided the subscript of the former occurs among the secondary subscripts of the latter.*

The normal equations for estimating *b*'s in trivariate and *n*-variate distributions, as obtained in equations (12.3*a*) and (12·9*a*), are:

$$\Sigma X_2 X_{1\cdot23} = 0, \quad \Sigma X_3 X_{1\cdot23} = 0 \quad \text{and} \quad \Sigma X_i X_{1\cdot23\ldots i\ldots n} = 0 \ ; \ i = 2, 3, \ldots, n$$

respectively. Here X_i, ($i = 1, 2, 3, \ldots, n$) can be regarded as a residual of order zero. Hence the result.

Property 2. *The sum of the product of any two residuals in which all the secondary subscripts of the first occur among the secondary subscripts of the second is unaltered if we omit any or all of the secondary subscripts of the first. Conversely, the product sum of any residual of order 'p' with a residual of order p + q, the 'p' subscripts being the same in each case is unaltered by adding to the secondary subscripts of the former any or all the 'q' additional subscripts of the latter.*

Let us consider:

$$\Sigma X_{1\cdot2}X_{1\cdot23} = \Sigma(X_1 - b_{12}X_2)X_{1\cdot23} = \Sigma X_1 X_{1\cdot23} - b_{12}\Sigma X_2 X_{1\cdot23}$$
$$= \Sigma X_1 X_{1\cdot23} \qquad \text{(Using Property 1)}$$

Also

$$\Sigma X^2_{1\cdot23} = \Sigma X_{1\cdot23}X_{1\cdot23} = \Sigma(X_1 - b_{12\cdot3}X_2 - b_{13\cdot2}X_3)X_{1\cdot23}$$
$$= \Sigma X_1 X_{1\cdot23} - b_{12\cdot3}\Sigma X_2 X_{1\cdot23} - b_{13\cdot2}\Sigma X_3 X_{1\cdot23}$$
$$= \Sigma X_1 X_{1\cdot23} \qquad \text{(Using Property 1)}$$

$$\therefore \qquad \Sigma X^2_{1\cdot 23} = \Sigma X_{1\cdot 2}\, X_{1\cdot 23} = \Sigma X_1\, X_{1\cdot 23}$$

Also, $\Sigma X_{1\cdot 34 \ldots n}\, X_{2\cdot 34 \ldots n} = \Sigma[(X_1 - b_{13\cdot 4 \ldots n}\, X_3 - b_{14\cdot 35 \ldots n}\, X_4 - \ldots - b_{1n\cdot 34 \ldots (n-1)}\, X_n)\, X_{2\cdot 34 \ldots n}]$

$= \Sigma X_1\, X_{2\cdot 34 \ldots n}$ (Using Property 1)

Hence, the property 2.

Property 3. *The sum of the product of two residuals is zero if all the subscripts (primary as well as secondary) of the one occur among the secondary subscripts of the other, e.g.,*

$$\Sigma X_{1\cdot 2}\, X_{3\cdot 12} = \Sigma(X_1 - b_{12}\, X_2)X_{3\cdot 12} = \Sigma X_1\, X_{3\cdot 12} - b_{12}\, \Sigma X_2\, X_{3\cdot 12} = 0 \quad \text{(Using Property 1)}$$

$$\Sigma X_{2\cdot 34 \ldots n}\, X_{1\cdot 23 \ldots n} = \Sigma[(X_2 - b_{23\cdot 4 \ldots n}\, X_3 - b_{24\cdot 35 \ldots n}\, X_4 - \ldots - b_{2n\cdot 34 \ldots (n-1)}\, X_n)\, X_{1\cdot 23 \ldots n}]$$

$$= \Sigma X_2\, X_{1\cdot 23 \ldots n} - b_{23\cdot 4 \ldots n}\, \Sigma X_3\, X_{1\cdot 234 \ldots n} - b_{24\cdot 35 \ldots n}\, \Sigma X_4\, X_{1\cdot 23 \ldots n} - \ldots$$

$$- b_{2n\cdot 34 \ldots (n-1)}\, \Sigma X_n\, X_{1\cdot 23 \ldots n} = 0 \quad \text{(Using Property 1)}$$

Hence, the property 3.

12·4·1. Variance of the Residual

Let us consider the plane of regression of X_1 on $X_2, X_3, \ldots, X_n$ *viz.,*

$$X_1 = b_{12\cdot 34 \ldots n}\, X_2 + b_{13\cdot 24 \ldots n}\, X_3 + \ldots + b_{1n\cdot 23 \ldots (n-1)}\, X_n.$$

Since all the X_i's are measured from their respective means, we have

$$E(X_i) = 0 \ ; \ i = 1, 2, \ldots, n \quad \Rightarrow \quad E(X_{1\cdot 23 \ldots n}) = 0$$

Hence, the variance of the residual is given by:

$$\sigma^2_{1\cdot 23 \ldots n} = \frac{1}{N}\Sigma\left[X_{1\cdot 23 \ldots n} - E(X_{1\cdot 23 \ldots n})\right]^2 = \frac{1}{N}\Sigma X^2_{1\cdot 23 \ldots n}$$

$$= \frac{1}{N}\Sigma X_{1\cdot 23 \ldots n}\, X_{1\cdot 23 \ldots n} = \frac{1}{N}\Sigma X_1\, X_{1\cdot 23 \ldots n} \quad \text{(Using Property 2 § 12·4)}$$

$$= \frac{1}{N}\Sigma X_1\left[X_1 - b_{12\cdot 34 \ldots n}\, X_2 - b_{13\cdot 24 \ldots n}\, X_3 - \ldots - b_{1n\cdot 23 \ldots (n-1)}\, X_n\right]$$

$$= \sigma_1^2 - b_{12\cdot 34 \ldots n}\, r_{12}\sigma_1\sigma_2 - b_{13\cdot 24 \ldots n}\, r_{13}\sigma_1\sigma_3 - \ldots - b_{1n\cdot 23 \ldots (n-1)}\, r_{1n}\sigma_1\sigma_n$$

$$\Rightarrow \quad \sigma_1^2 - \sigma^2_{1\cdot 23 \ldots n} = b_{12\cdot 34 \ldots n}\, r_{12}\sigma_1\sigma_2 + b_{13\cdot 24 \ldots n}\, r_{13}\sigma_1\sigma_3 + \ldots + b_{1n\cdot 23 \ldots (n-1)}\, r_{1n}\sigma_1\sigma_n \quad \ldots (12\cdot 15)$$

Eliminating the b's in equations (12·15) and (12·9*b*), we get

$$\begin{vmatrix} \sigma_1^2 - \sigma^2_{1\cdot 23 \ldots n} & r_{12}\sigma_1\sigma_2 & \ldots & r_{1n}\sigma_1\sigma_n \\ r_{12}\sigma_1\sigma_2 & \sigma_2^2 & \ldots & r_{2n}\sigma_2\sigma_n \\ \vdots & \vdots & & \vdots \\ r_{1n}\sigma_1\sigma_n & r_{2n}\sigma_2\sigma_n & \ldots & \sigma_n^2 \end{vmatrix} = 0$$

Dividing $R_1, R_2, \ldots, R_n$, by $\sigma_1, \sigma_2, \ldots, \sigma_n$ respectively and also $C_1, C_2, \ldots, C_n$, by $\sigma_1, \sigma_2, \ldots, \sigma_n$, respectively, we get

$$\begin{vmatrix} 1 - \dfrac{\sigma^2_{1\cdot 23 \ldots n}}{\sigma_1^2} & r_{12} & \ldots & r_{1n} \\ r_{12} & 1 & \ldots & r_{2n} \\ \vdots & \vdots & & \vdots \\ r_{1n} & r_{2n} & \ldots & 1 \end{vmatrix} = 0 \quad \Rightarrow \quad \begin{vmatrix} 1 - \dfrac{\sigma^2_{1\cdot 23 \ldots n}}{\sigma_1^2} & r_{12} & \ldots & r_{1n} \\ r_{12} + 0 & 1 & \ldots & r_{2n} \\ \vdots & \vdots & & \vdots \\ r_{1n} + 0 & r_{2n} & \ldots & 1 \end{vmatrix} = 0$$

$$\Rightarrow \begin{vmatrix} 1 & r_{12} & \cdots & r_{1n} \\ r_{12} & 1 & \cdots & r_{2n} \\ \vdots & \vdots & & \vdots \\ r_{1n} & r_{2n} & \cdots & 1 \end{vmatrix} - \begin{vmatrix} \dfrac{\sigma^2_{1\cdot 23 \ldots n}}{\sigma_1^2} & r_{12} & \cdots & r_{1n} \\ 0 & 1 & \cdots & r_{2n} \\ \vdots & \vdots & & \vdots \\ 0 & r_{2n} & \cdots & 1 \end{vmatrix} = 0$$

$$\Rightarrow \qquad \omega - \frac{\sigma^2_{1\cdot 23 \ldots n}}{\sigma_1^2}\,\omega_{11} = 0 \quad \Rightarrow \quad \sigma^2_{1\cdot 23 \ldots n} = \sigma_1^2 \frac{\omega}{\omega_{11}} \qquad \ldots (12\cdot 16)$$

Remark. In a tri-variate distribution, $\sigma^2_{1\cdot 23} = \sigma_1^2 \dfrac{\omega}{\omega_{11}}$ ... (12·16*a*)

where ω and ω_{11} are defined in (12·5) and (12.6).

12·5. COEFFICIENT OF MULTIPLE CORRELATION

In a tri-variate distribution in which each of the variables X_1, X_2 and X_3 has N observations, the multiple correlation coefficient of X_1 on X_2 and X_3, usually denoted by $R_{1\cdot 23}$, is the simple correlation coefficient between X_1 and the joint effect of X_2 and X_3 on X_1. In other words $R_{1\cdot 23}$ is the correlation coefficient between X_1 and its estimated value as given by the plane of regression of X_1 on X_2 and X_3 *viz.*,

$$e_{1\cdot 23} = b_{12\cdot 3}\, X_2 + b_{13\cdot 2}\, X_3$$

We have $\quad X_{1\cdot 23} = X_1 - b_{12\cdot 3}\, X_2 - b_{13\cdot 2}\, X_3 = X_1 - e_{1\cdot 23} \quad \Rightarrow \quad e_{1\cdot 23} = X_1 - X_{1\cdot 23}$

Since X_i's are measure from their respective means, we have

$$E(X_{1\cdot 23}) = 0 \text{ and } E(e_{1\cdot 23}) = 0 \qquad [\because E(X_i) = 0 \ ; \ i = 1, 2, 3] \qquad \ldots (*)$$

By def., $\quad R_{1\cdot 23} = \dfrac{\text{Cov}\,(X_1, e_{1\cdot 23})}{\sqrt{V(X_1)\, V(e_{1\cdot 23})}}$ ... (12·17)

$$\text{Cov}\,(X_1, e_{1\cdot 23}) = E\big[\{X_1 - E(X_1)\}\{e_{1\cdot 23} - E(e_{1\cdot 23})\}\big] = E(X_1\, e_{1\cdot 23}) \qquad \text{[From (*)]}$$

$$= \frac{1}{N}\Sigma X_1\, e_{1\cdot 23} = \frac{1}{N}\Sigma X_1\,(X_1 - X_{1\cdot 23})$$

$$= \frac{1}{N}\Sigma X_1^2 - \frac{1}{N}\Sigma X_1\, X_{1\cdot 23} = \frac{1}{N}\Sigma X_1^2 - \frac{1}{N}\Sigma X_{1\cdot 23}^2 \qquad \text{(Using property 2, § 12·4)}$$

$$= \sigma_1^2 - \sigma_{1\cdot 23}^2$$

Also $\quad V(e_{1\cdot 23}) = E\left(e_{1\cdot 23}^2\right) = \dfrac{1}{N}\Sigma e_{1\cdot 23}^2 = \dfrac{1}{N}\Sigma(X_1 - X_{1\cdot 23})^2$

$$= \frac{1}{N}\Sigma(X_1^2 + X_{1\cdot 23}^2 - 2X_1X_{1\cdot 23}) = \frac{1}{N}\Sigma X_1^2 + \frac{1}{N}\Sigma X_{1\cdot 23}^2 - \frac{2}{N}\Sigma X_1 X_{1\cdot 23}$$

$$= \frac{1}{N}\Sigma X_1^2 + \frac{1}{N}\Sigma X_{1\cdot 23}^2 - \frac{2}{N}\Sigma X_{1\cdot 23}^2 \qquad \text{(Using property 2, § 12·4)}$$

$$= \sigma_1^2 - \sigma^2_{1\cdot 23}$$

Substituting in (12·17), we get

$$R_{1\cdot 23} = \frac{\sigma_1^2 - \sigma_{1\cdot 23}^2}{\sqrt{\sigma_1^2(\sigma_1^2 - \sigma_{1\cdot 23}^2)}} \quad \Rightarrow \quad R_{1\cdot 23}^2 = \frac{\sigma_1^2 - \sigma_{1\cdot 23}^2}{\sigma_1^2} = 1 - \frac{\sigma_{1\cdot 23}^2}{\sigma_1^2}$$

$$\Rightarrow \qquad 1 - R^2_{1\cdot 23} = \frac{\sigma^2_{1\cdot 23}}{\sigma_1^2} = \frac{\omega}{\omega_{11}} \qquad \text{[From (12·16}a\text{)]} \qquad \text{... (12·18)}$$

where
$$\omega = \begin{vmatrix} 1 & r_{12} & r_{13} \\ r_{21} & 1 & r_{23} \\ r_{31} & r_{32} & 1 \end{vmatrix} = 1 - r_{12}^2 - r_{13}^2 - r_{23}^2 + 2r_{12}\, r_{13}\, r_{23} \qquad \text{(On simplification)}$$

and
$$\omega_{11} = \begin{vmatrix} 1 & r_{23} \\ r_{32} & 1 \end{vmatrix} = 1 - r_{23}^2$$

Hence, from (12·18), we get:

$$R^2_{1\cdot 23} = 1 - \frac{\omega}{\omega_{11}} = \frac{{r_{12}}^2 + {r_{13}}^2 - 2\, r_{12}\, r_{13}\, r_{23}}{1 - {r_{23}}^2} \qquad \text{... (12·18}a\text{)}$$

This formula expresses the multiple correlation coefficient in terms of the total correlation coefficients between the pairs of variables.

Remark. $R^2{}_{1\cdot 23} \le 1 \quad \Rightarrow \quad 1 - R^2{}_{1\cdot 23} = \dfrac{\omega}{\omega_{11}} \ge 0$ ... (*)

But $\omega_{11} = 1 - r^2{}_{23} > 0 \qquad (\because\ |r_{23}| \le 1)$

$\therefore$ From (*), $\omega \ge 0$ ... (**)

Also $\omega_{22} = (-1)^{2+2} \,.\, (1 - {r_{13}}^2) \ge 0$ and $\omega_{33} = (-1)^{3+3}\,(1 - {r_{12}}^2) \ge 0$

In general, $\omega_{i\,i} \ge 0\ ;\ i = 1, 2, 3$ ... (12·18*b*)

Generalisation. *In case of n-variate distribution, the multiple correlation coefficient of* X_1 *on* $X_2, X_3, ..., X_n$, *usually denoted by* $R_{1\cdot 23\,...\,n}$, *is the correlation coefficient between* X_1 *and* $e_{1\cdot 23\,...\,n} = X_1 - X_{1\cdot 23\,...\,n}$

$$\therefore \qquad R_{1\cdot 23\,...\,n} = \frac{\text{Cov}\,(X_1, e_{1\cdot 23\,...\,n})}{\sqrt{V(X_1)\,(e_{1\cdot 23\,...\,n})}}$$

$$\text{Cov}\left(X_1, e_{1\cdot 23\,...\,n}\right) = \frac{1}{N}\Sigma X_1\, e_{1\cdot 23\,...\,n} = \frac{1}{N}\Sigma X_1 (X_1 - X_{1\cdot 23\,...\,n})$$

$$= \frac{1}{N}\Sigma X_1^2 - \frac{1}{N}\Sigma X_1\, X_{1\cdot 23\,...\,n}$$

$$= \frac{1}{N}\Sigma X_1^2 - \frac{1}{N}\Sigma X^2_{1\cdot 23\,...\,n} = \sigma_1^2 - \sigma^2_{1\cdot 23\,...\,n} \qquad \text{... (*)}$$

$$V\left(e_{1\cdot 23\,...\,n}\right) = \frac{1}{N}\Sigma e^2_{1\cdot 23\,...\,n} = \frac{1}{N}\Sigma\left(X_1 - X_{1\cdot 23\,...\,n}\right)^2$$

$$= \frac{1}{N}\Sigma\left(X_1^2 + X^2_{1\cdot 23\,...\,n} - 2X_1 X_{1\cdot 23\,...\,n}\right)$$

$$= \frac{1}{N}\Sigma X_1^2 + \frac{1}{N}\Sigma X^2_{1\cdot 23\,...\,n} - 2\frac{1}{N}\Sigma X_1 X_{1\cdot 23\,...\,n}$$

$$= \frac{1}{N}\Sigma X_1^2 + \frac{1}{N}\Sigma X^2_{1\cdot 23\,...\,n} - \frac{2}{N}\Sigma X^2_{1\cdot 23\,...\,n}$$

$$= \sigma_1^2 - \sigma^2_{1\cdot 23\,...\,n}$$

$$\therefore \qquad R_{1\cdot 23\,...\,n} = \frac{{\sigma_1}^2 - \sigma^2{}_{1\cdot 23\,...\,n}}{\sqrt{{\sigma_1}^2({\sigma_1}^2 - \sigma^2{}_{1\cdot 23\,...\,n})}} = \left(\frac{{\sigma_1}^2 - \sigma^2{}_{1\cdot 23\,...\,n}}{{\sigma_1}^2}\right)^{1/2}$$

$$R^2_{1\cdot 23 \ldots n} = 1 - \frac{\sigma^2_{1\cdot 23 \ldots n}}{\sigma_1^2} = 1 - \frac{\omega}{\omega_{11}}, \quad \ldots (12\cdot 18c)$$

where ω *and* ω_{11} *are defined in* (12·11).

Remarks:

1. It may be pointed out here that multiple correlation coefficient can never be negative, because from (*) and (**), we get $\text{Cov}(X_1, e_{1\cdot 23 \ldots n}) = \sigma_1^2 - \sigma^2_{1\cdot 23 \ldots n} = \text{Var}(e_{1\cdot 23 \ldots n}) \geq 0$.
   Since the sign of $R_{1\cdot 23 \ldots n}$ depends upon the covariance term $\text{Cov}(X_1, e_{1\cdot 23 \ldots n})$ we conclude that $R_{1\cdot 23 \ldots n} \geq 0$.
2. Since $R^2_{1\cdot 23 \ldots n} \geq 0$, we have $\quad 1 - \dfrac{\omega}{\omega_{11}} \geq 0 \quad \Rightarrow \quad \omega \leq \omega_{11}$ ... (12·18*d*)
3. Also $R^2_{1\cdot 23 \ldots n} \leq 1 \quad \Rightarrow \quad 1 - R^2_{1\cdot 23 \ldots n} = \dfrac{\omega}{\omega_{11}} \geq 0 \quad \Rightarrow \quad \omega \geq 0$ ... (12·18*e*)
   From the above results, we get $\quad \omega_{11} \geq \omega \geq 0$ ... (12·18*f*)
   In general, we have $\quad \omega_{ii} \geq 0 \; ; \; i = 1, 2, \ldots, n$
4. Since ω is symmetric in r_{ij}'s, we have $\omega_{ij} = \omega_{ji} \; ; \; i \neq j = 1, 2, \ldots, n$ ... (12·18*g*)

12·5·1. Properties of Multiple Correlation Coefficient

1. Multiple correlation co-efficient measures the closeness of the association between the observed values and the expected values of a variable obtained from the multiple linear regression of that variable on other variables.
2. Multiple correlation coefficient between observed values and expected values, when the expected values are calculated from a linear relation of the variables determined by the method of least squares, is always greater than that where expected values are calculated from any other linear combination of the variables.
3. Since $R_{1\cdot 23}$ is the simple correlation between X_1 and $e_{1\cdot 23}$, it must lie between –1 and +1. But as seen in Remark 1 above, $R_{1\cdot 23}$ is a non-negative quantity, we conclude that $0 \leq R_{1\cdot 23} \leq 1$.
4. If $R_{1\cdot 23} = 1$, then association is perfect and all the regression residuals are zero, and as such $\sigma^2_{1\cdot 23} = 0$. In this case, since $X_1 = e_{1\cdot 23}$, the predicted value of X_1, the multiple linear regression equation of X_1 on X_2 and X_3 may be said to be a perfect prediction formula.
5. If $R_{1\cdot 23} = 0$, then all total and partial correlations involving X_1 are zero. [See *Example* 12·5]. So X_1 is completely uncorrelated with all the other variables in this case and the multiple regression equation fails to throw any light on the value of X_1 when X_2 and X_3 are known.
6. $R_{1\cdot 23}$ is not less tan any total correlation coefficient, *i.e.*, $R_{1\cdot 23} \geq r_{12}, r_{13}, r_{23}$.

12·5·2. Coefficient of Multiple Determination

The square of the multiple correlation coefficient is known as *coefficient of multiple determination.*

In particular, the coefficient of multiple determination corresponding to multiple correlation coefficient $R_{1\cdot 23}$ is:

$$R^2_{1\cdot 23} = \frac{\text{Explained Variation}}{\text{Total Variation}} \quad \ldots (12\cdot 19)$$

$$= \frac{\text{Sum of squares due to linear regression of } X_1 \text{ on } (X_2, X_3)}{\text{Total Variation}}$$

$$= 1 - \frac{\text{Sum of squares due to error}}{\text{Total Sum of squares}} = 1 - \frac{\sigma_{1\cdot23}^2}{\sigma_1^2} \quad \text{... (12·19a)}$$

$R_{12\cdot3}^2$ may be interpreted as the proportion of the total variation in the dependent variable that is explained by or is associated with the plane of regression of X_1 on (X_2 and X_3).

Remark. The value of $R_{1\cdot23}^2$ may also be interpreted as a measure of the closeness of fit of the plane of regression to the actual data values. The values of $R_{1\cdot23}^2$ nearer to 1 indicate that the given data values are closely scattered about the plane of regression and hence it is a better fit.

Example 12·1. *In a trivariate distribution, if $r_{12} = 0{\cdot}7$, $r_{13} = 0{\cdot}61$ and $r_{23} = 0{\cdot}4$, find all the multiple correlation coefficients. Also obtain the standard errors of estimates $\sigma_{1\cdot23}$, $\sigma_{2\cdot13}$ and $\sigma_{3\cdot12}$ given that:*

$$\sigma_1 = 1{\cdot}86,\ \sigma_2 = 2{\cdot}15 \text{ and } \sigma_3 = 3{\cdot}46.$$

Solution. (*i*)

$$R_{1\cdot23}^2 = \frac{r_{12}^2 + r_{13}^2 - 2\, r_{12}\, r_{13}\, r_{23}}{1 - r_{23}^2} = \frac{(0{\cdot}7)^2 + (0{\cdot}61)^2 - 2 \times 0{\cdot}7 \times 0{\cdot}61 \times 0{\cdot}4}{1 - (0{\cdot}4)^2}$$

$$= \frac{0{\cdot}49 + 0{\cdot}3721 - 0{\cdot}3416}{1 - 0{\cdot}16} = \frac{0{\cdot}5205}{0{\cdot}84} = 0{\cdot}6196$$

$$\Rightarrow \quad R_{1\cdot23} = \sqrt{0{\cdot}6196} = 0{\cdot}7871 \qquad [\because R_{1\cdot23} \geq 0]$$

(*ii*)

$$R_{2\cdot13}^2 = \frac{r_{21}^2 + r_{23}^2 - 2\, r_{12}\, r_{13}\, r_{23}}{1 - r_{13}^2} = \frac{0{\cdot}49 + 0{\cdot}16 - 0{\cdot}3416}{1 - (0{\cdot}61)^2} = \frac{0{\cdot}3084}{1 - 0{\cdot}3721} = 0{\cdot}4912$$

$$\Rightarrow \quad R_{2\cdot13} = \sqrt{0{\cdot}4912} = 0{\cdot}7009$$

(*iii*)

$$R_{3\cdot21}^2 = \frac{r_{32}^2 + r_{31}^2 - 2\, r_{12}\, r_{13}\, r_{23}}{1 - r_{21}^2} = \frac{0{\cdot}16 + 0{\cdot}3721 - 0{\cdot}3416}{1 - 0{\cdot}49} = 0{\cdot}3735$$

$$\Rightarrow \quad R_{3\cdot21} = \sqrt{0{\cdot}3735} = 0{\cdot}6111$$

Standard Error of Estimate can be obtained by using the formulae:

$$\sigma^2_{i\,\cdot\,jk} = \sigma^2_i\,(1 - R^2_{i\,\cdot\,jk})\ ;\ i \neq j \neq k = 1, 2, 3 \quad \text{... (*)}$$

$$\therefore \quad \sigma_{1\cdot23} = \sigma_1 \sqrt{(1 - R_{1\cdot23}^2)} = 1{\cdot}86\,\sqrt{1 - 0{\cdot}6196} = 1{\cdot}1472$$

$$\sigma_{2\cdot13} = \sigma_2 \sqrt{(1 - R_{2\cdot13}^2)} = 2{\cdot}15\,\sqrt{1 - 0{\cdot}4912} = 1{\cdot}5336$$

$$\sigma_{3\cdot21} = \sigma_3 \sqrt{(1 - R_{3\cdot21}^2)} = 3{\cdot}46\,\sqrt{1 - 0{\cdot}3735} = 2{\cdot}7386$$

Example 12·2. *If $r_{12} = r_{13} = r_{23} = r$, ($r \neq 1$), then prove that*

$$R_{1\cdot23} = R_{2\cdot31} = R_{3\cdot12} = r\sqrt{\frac{2}{1 + r}}.$$

Solution.

$$r_{12} = r_{13} = r_{23} = r \text{ (Given)}$$

$$\therefore \quad R_{1\cdot23}^2 = \frac{r_{12}^2 + r_{13}^2 - 2\, r_{12}\, r_{13}\, r_{23}}{1 - r_{23}^2} = \frac{r^2 + r^2 - 2r^3}{1 - r^2}$$

$$= \frac{2r^2(1 - r)}{(1 + r)(1 - r)} = \frac{2r^2}{1 + r}$$

$$\Rightarrow \qquad R_{1\cdot23} = \sqrt{\frac{2r^2}{1+r}} = r\sqrt{\frac{2}{1+r}}\,; \qquad (\because 0 \le R_{1\cdot23} \le 1)$$

Similarly, we get

$$R^2_{2\cdot31} = \frac{r_{23}^2 + r_{21}^2 - 2\,r_{12}\,r_{13}\,r_{23}}{1 - r_{13}^2} = \frac{2r^2}{1+r} \quad \Rightarrow \quad R_{2\cdot31} = r\sqrt{\frac{2}{1+r}}$$

and

$$R^2_{3\cdot12} = \frac{r_{31}^2 + r_{32}^2 - 2\,r_{12}\,r_{13}\,r_{23}}{1 - r_{12}^2} = \frac{2r^2}{1+r} \quad \Rightarrow \quad R_{3\cdot12} = r\sqrt{\frac{2}{1+r}}$$

$$\therefore \qquad R_{1\cdot23} = R_{2\cdot31} = R_{3\cdot12} = r\sqrt{\frac{2}{1+r}}$$

12·6. COEFFICIENT OF PARTIAL CORRELATION

Sometimes the correlation between two variables X_1 and X_2 may be partly due to the correlation of a third variable, X_3 with both X_1 and X_2. In such a situation, one may want to know what the correlation between X_1 and X_2 would be if the effect of X_3 on each of X_1 and X_2 were eliminated. This correlation is called the *partial correlation* and the correlation coefficient between X_1 and X_2 after the linear effect of X_3 on each of them has been eliminated is called the *partial correlation coefficient*.

The residual $X_{1\cdot3} = X_1 - b_{13}\,X_3$, may be regarded as that part of the variable X_1 which remains after the linear effect of X_3 has been eliminated. Similarly, the residual $X_{2\cdot3}$ may be interpreted as the part of the variable X_2 obtained after eliminating the linear effect of X_3. Thus the partial correlation coefficient between X_1 and X_2, usually denoted by $r_{12\cdot3}$, is given by:

$$r_{12\cdot3} = \frac{\text{Cov}\,(X_{1\cdot3}, X_{2\cdot3})}{\sqrt{\text{Var}(X_{1\cdot3})\,\text{Var}(X_{2\cdot3})}} \qquad \ldots (12\cdot20)$$

We have, assuming that X_i's, $(i = 1, 2, \ldots, n)$ are measured from their respective means,

$$\text{Cov}\,(X_{1\cdot3}\,,\,X_{2\cdot3}) = \frac{1}{N}\Sigma X_{1\cdot3}\,X_{2\cdot3} = \frac{1}{N}\Sigma X_1\,X_{2\cdot3} = \frac{1}{N}\Sigma X_1(X_2 - b_{23}X_3)$$

$$= \frac{1}{N}\Sigma X_1X_2 - b_{23}\frac{1}{N}\Sigma X_1X_3 = r_{12}\,\sigma_1\sigma_2 - r_{23}\,\frac{\sigma_2}{\sigma_3}\cdot(r_{13}\,\sigma_1\sigma_3)$$

$$= \sigma_1\sigma_2\,(r_{12} - r_{13}\,r_{23})$$

$$V(X_{1\cdot3}) = \frac{1}{N}\Sigma X^2_{1\cdot3} = \frac{1}{N}\Sigma X_{1\cdot3}\,X_{1\cdot3}$$

$$= \frac{1}{N}\Sigma X_1X_{1\cdot3} = \frac{1}{N}\Sigma X_1(X_1 - b_{13}\,X_3)$$

$$= \frac{1}{N}\Sigma X_1^2 - b_{13}\cdot\frac{1}{N}\Sigma X_1\,X_3 = \sigma_1^2 - r_{13}\,\frac{\sigma_1}{\sigma_3}\cdot(r_{13}\,\sigma_1\sigma_3)$$

$$= \sigma_1^2(1 - r_{13}^2)$$

Similarly, we shall get : $\qquad V(X_{2\cdot3}) = \sigma_2{}^2(1 - r_{23}{}^2)$

Substituting in (12·20), we get :

$$r_{12\cdot3} = \frac{\sigma_1\sigma_2\,(r_{12} - r_{13}\,r_{23})}{\sqrt{\sigma_1^2\,(1 - r_{13}^2)\,\sigma_2^2\,(1 - r_{23}^2)}} = \frac{r_{12} - r_{13}\,r_{23}}{\sqrt{(1 - r_{13}{}^2)\,(1 - r_{23}{}^2)}} \qquad \ldots (12\cdot20a)$$

Aliter. We have

$$0 = \Sigma X_{2\cdot3}\, X_{1\cdot23} = \Sigma X_{2\cdot3}\,(X_1 - b_{12\cdot3}\, X_2 - b_{13\cdot2}\, X_3)$$
$$= \Sigma X_1\, X_{2\cdot3} - b_{12\cdot3}\, \Sigma X_{2\cdot3}\, X_2 - b_{13\cdot2}\, \Sigma X_{2\cdot3}\, X_3 = \Sigma X_{1\cdot3}\, X_{2\cdot3} - b_{12\cdot3}\, \Sigma X_{2\cdot3}\, X_{2\cdot3}$$

$$\therefore \qquad b_{12\cdot3} = \frac{\Sigma X_{1\cdot3}\, X_{2\cdot3}}{\Sigma X_{2\cdot3}^2}$$

From this it follows that $b_{12\cdot3}$ is coefficient of regression of $X_{1\cdot3}$ on $X_{2\cdot3}$.

Similarly, $b_{21\cdot3}$ is the coefficient of regression of $X_{2\cdot3}$ on $X_{1\cdot3}$.

Since correlation coefficient is the geometric mean between regression coefficients, we have

$$r_{12\cdot3}^2 = b_{12\cdot3} \times b_{21\cdot3}$$

But by *def.*, $\qquad b_{12\cdot3} = -\dfrac{\sigma_1}{\sigma_2} \cdot \dfrac{\omega_{12}}{\omega_{11}}$ and $\qquad b_{21\cdot3} = -\dfrac{\sigma_2}{\sigma_1} \cdot \dfrac{\omega_{21}}{\omega_{22}}$

$$\therefore \qquad r_{12\cdot3}^2 = \left(-\frac{\sigma_1}{\sigma_2} \cdot \frac{\omega_{12}}{\omega_{11}}\right)\left(-\frac{\sigma_2}{\sigma_1} \cdot \frac{\omega_{21}}{\omega_{22}}\right) = \frac{\omega_{12}^{\ 2}}{\omega_{11}\, \omega_{22}} \qquad (\because \omega_{12} = \omega_{21})$$

$$\Rightarrow \qquad r_{12\cdot3} = -\frac{\omega_{12}}{\sqrt{\omega_{11}\, \omega_{22}}}, \qquad \text{... (12·20}b\text{)}$$

the negative sign being taken since the sign of regression coefficients is the same as that of $(-\omega_{12})$.

Substituting the values of ω_{12}, ω_{11} and ω_{22} from (12·6*a*), (12·6*b*) and (12·6*d*), we get

$$r_{12\cdot3} = \frac{r_{12} - r_{13}\, r_{23}}{\sqrt{(1 - r_{13}^2)\,(1 - r_{23}^2)}}$$

Remarks:

1. The expressions for $r_{13\cdot2}$ and $r_{23\cdot1}$ can be similarly obtained, to give

$$r_{13\cdot2} = \frac{r_{13} - r_{12}\, r_{32}}{\sqrt{(1 - r_{12}^2)\,(1 - r_{32}^2)}} \qquad \text{and} \qquad r_{23\cdot1} = \frac{r_{23} - r_{21}\, r_{31}}{\sqrt{(1 - r_{21}^2)\,(1 - r_{31}^2)}} \qquad \text{... (12·20}c\text{)}$$

2. If $r_{12\cdot3} = 0$, we have then $r_{12} = r_{13}\, r_{23}$. It means that r_{12} will not be zero if X_3 is correlated with both X_1 and X_2. Thus, although X_1 and X_2 may be uncorrelated when effect of X_3 is eliminated, yet X_1 and X_2 may appear to be correlated because they carry the effect of X_3 on them.
3. Partial correlation coefficient helps in deciding whether to include or not, an additional independent variable in regression analysis.
4. We know that $\sigma_1^2\,(1 - r_{12}^2)$ and $\sigma_1^2\,(1 - r_{13}^2)$ are the residual variances if X_1 is estimated from X_2 and X_3 individually, while $\sigma_1^2\,(1 - R_{1\cdot23}^2)$ is the residual variance if X_1 is estimated from X_2 and X_3 taken together. So from the above remark and $R_{1\cdot23}^2 \geq r_{12}^2$ and r_{13}^2, it follows that inclusion of an additional variable can only reduce the residual variance. Now inclusion of X_3 when X_2 has already been taken for predicting X_1, is worthwhile only when the resultant in the residual variance is substantial. This will be the case when $r_{13\cdot2}$ is sufficiently large. Thus in this respect partial correlation coefficient has its significance in regression analysis.

Generalisation. *In the case of n variables* $X_1, X_2, ..., X_n$ *the partial correlation coefficient* $r_{12\cdot34\,...\,n}$ *between* X_1 *and* X_2 *(after the linear effect of* $X_3, X_4, ..., X_n$ *on them has been eliminated), is given by:*

$$r^2_{12\cdot34\,...\,n} = b_{12\cdot34\,...\,n} \times b_{21\cdot34\,...\,n} \qquad \text{... (12·20}d\text{)}$$

But, we have

$$b_{12\cdot34\ldots n} = -\frac{\sigma_1}{\sigma_2}\cdot\frac{\omega_{12}}{\omega_{11}} \quad \text{and} \quad b_{21\cdot34\ldots n} = -\frac{\sigma_2}{\sigma_1}\cdot\frac{\omega_{21}}{\omega_{22}} \qquad \text{[c.f. Equation (12·14a)]}$$

$$\therefore \quad r^2_{12\cdot34\ldots n} = \left(-\frac{\sigma_1}{\sigma_2}\cdot\frac{\omega_{12}}{\omega_{11}}\right)\left(-\frac{\sigma_2}{\sigma_1}\cdot\frac{\omega_{21}}{\omega_{22}}\right) = \frac{\omega_{12}^2}{\omega_{11}\,\omega_{22}}$$

$$\Rightarrow \quad r_{12\cdot34\ldots n} = -\frac{\omega_{12}}{\sqrt{\omega_{11}\,\omega_{22}}}, \qquad \ldots (12.20e)$$

negative sign being taken since the sign of the regression coefficient is the same as that of $(-\omega_{12})$.

12.6.1. Coefficient of Partial Determination

A more useful measure for interpreting the value of partial correlation coefficient is the coefficient of partial determination, given by:

Coefficient of partial determination = Square of partial correlation coefficient.

Thus, if $r_{13\cdot2} = 0\cdot7$, then coefficient of partial determination = $r^2_{13\cdot2} = (0\cdot7)^2 = 0\cdot49$. This, implies that 49% of the variation in the variable X_1 is associated with the variable X_3 when the variable X_2 is held constant. In other words, 49% of the variation in X_1 is explained by the plane of regression of X_1 on X_2 and X_3 when the value of X_2 is held constant.

Example 12·3. *From the data relating to the yield of dry bark (X_1), height (X_2) and girth (X_3) for 18 cinchona plants, the following correlation coefficients were obtained:*

$$r_{12} = 0\cdot77, \quad r_{13} = 0\cdot72 \quad \text{and} \quad r_{23} = 0\cdot52.$$

Find the partial correlation coefficient $r_{12\cdot3}$ and multiple correlation coefficient $R_{1\cdot23}$.

Solution.

$$r_{12\cdot3} = \frac{r_{12} - r_{13}\,r_{23}}{\sqrt{(1-r_{13}^2)(1-r_{23}^2)}} = \frac{0\cdot77 - 0\cdot72 \times 0\cdot52}{\sqrt{(1-(0\cdot72)^2)(1-(0\cdot52)^2)}} = 0\cdot62$$

$$R^2_{1\cdot23} = \frac{r_{12}^2 + r_{13}^2 - 2\,r_{12}\,r_{13}\,r_{23}}{1-r_{23}^2} = \frac{(0\cdot77)^2 + (0\cdot72)^2 - 2\times0\cdot77\times0\cdot72\times0\cdot52}{1-(0\cdot52)^2} = 0\cdot7334$$

$$\therefore \quad R_{1\cdot23} = +\,0\cdot8564\ ; \text{ (since multiple correlation coefficient is non-negative).}$$

Example 12·4. *In a trivariate distribution:*

$\sigma_1 = 2,\quad \sigma_2 = \sigma_3 = 3,\quad r_{12} = 0\cdot7,\quad r_{23} = r_{31} = 0\cdot5.$

Find (i) $r_{23\cdot1}$, (ii) $R_{1\cdot23}$, (iii) $b_{12\cdot3}$, and (iv) $\sigma_{1\cdot23}$.

Solution. We have

$$(i) \quad r_{23\cdot1} = \frac{r_{23} - r_{21}\,r_{31}}{\sqrt{(1-r_{21}^2)(1-r_{31}^2)}} = \frac{0\cdot5 - (0\cdot7)(0\cdot5)}{\sqrt{(1-0\cdot49)(1-0\cdot25)}} = 0\cdot2425$$

$$(ii) \quad R^2_{1\cdot23} = \frac{r_{12}^2 + r_{13}^2 - 2\,r_{12}\,r_{13}\,r_{23}}{1-r_{23}^2} = \frac{0\cdot49 + 0\cdot25 - 2(0\cdot7)(0\cdot5)(0\cdot5)}{1-0\cdot25} = 0\cdot52$$

$$\Rightarrow \quad R_{1\cdot23} = \sqrt{0\cdot52} = 0\cdot7211$$

$$(iii) \quad b_{12\cdot3} = r_{12\cdot3}\,\frac{\sigma_{1\cdot3}}{\sigma_{2\cdot3}} \quad \text{and} \quad b_{13\cdot2} = r_{13\cdot2}\,\frac{\sigma_{1\cdot2}}{\sigma_{3\cdot2}} \qquad \ldots (*)$$

$$r_{12\cdot3} = \frac{r_{12} - r_{13}\, r_{23}}{\sqrt{(1-r_{13}^2)\,(1-r_{23}^2)}} = 0{\cdot}6, \qquad r_{13\cdot2} = \frac{r_{13} - r_{12}\, r_{32}}{\sqrt{(1-r_{12}^2)\,(1-r_{32}^2)}} = 0{\cdot}2425$$

$$\sigma_{1\cdot3} = \sigma_1\sqrt{(1-r_{13}^2)} = 2\sqrt{(1-0{\cdot}25)} = 1{\cdot}7320 \quad ; \qquad \sigma_{2\cdot3} = \sigma_2\sqrt{(1-r_{23}^2)} = 3\sqrt{(1-0{\cdot}25)} = 2{\cdot}5980$$

$$\sigma_{1\cdot2} = \sigma_1\sqrt{(1-r_{12}^2)} = 2\sqrt{(1-0{\cdot}49)} = 1{\cdot}4282 \quad ; \qquad \sigma_{3\cdot2} = \sigma_3\sqrt{(1-r_{32}^2)} = 3\sqrt{(1-0{\cdot}25)} = 2{\cdot}5980$$

Substituting in (*), we get:

$$b_{12\cdot3} = 0{\cdot}6 \times \frac{1{\cdot}7320}{2{\cdot}5980} = 0{\cdot}4 \quad \text{and} \quad b_{13\cdot2} = 0{\cdot}2425 \times \frac{1{\cdot}4282}{2{\cdot}5980} = 0{\cdot}1333$$

(*iv*)

$$\sigma_{1\cdot23} = \sigma_1\sqrt{\frac{\omega}{\omega_{11}}}$$

where

$$\omega = \begin{vmatrix} 1 & r_{12} & r_{13} \\ r_{21} & 1 & r_{23} \\ r_{31} & r_{32} & 1 \end{vmatrix} = 1 - r_{12}^2 - r_{13}^2 - r_{23}^2 + 2r_{12}\, r_{13}\, r_{23} = 0{\cdot}36$$

and

$$\omega_{11} = \begin{vmatrix} 1 & r_{23} \\ r_{32} & 1 \end{vmatrix} = 1 - r_{23}^2 = 1 - 0{\cdot}25 = 0{\cdot}75$$

$$\therefore \qquad \sigma_{1\cdot23} = 2\times\sqrt{\frac{0{\cdot}36}{0{\cdot}75}} = 2\times\sqrt{0{\cdot}48} = 2\times 0{\cdot}6928 = 1{\cdot}3856.$$

12·6·2. Relation Between Total, Multiple and Partial Correlation Coefficients

$$1 - R_{1\cdot23}^2 = \left(1 - r_{12}^2\right)\left(1 - r_{13\cdot2}^2\right) \qquad \text{... (12·21)}$$

Proof. We have

$$1 - R_{1\cdot23}^2 = 1 - \frac{r_{12}^2 + r_{13}^2 - 2\, r_{12}\, r_{13}\, r_{23}}{1 - r_{23}^2} = \frac{1 - r_{23}^2 - r_{12}^2 - r_{13}^2 + 2\, r_{12}\, r_{13}\, r_{23}}{1 - r_{23}^2}$$

Also

$$1 - r_{13\cdot2}^2 = 1 - \frac{(r_{13} - r_{12}\, r_{23})^2}{(1-r_{12}^2)\,(1-r_{23}^2)} = \frac{1 - r_{12}^2 - r_{23}^2 - r_{13}^2 + 2\, r_{12}\, r_{13}\, r_{23}}{(1-r_{12}^2)\,(1-r_{23}^2)}$$

Hence the result.

From the symmetry of the result in (12·21), we have in general for three variables X_1, X_2, X_3:

$$1 - R_{i\cdot jk}^2 = \left(1 - r_{ij}^2\right)\left(1 - r_{ik\cdot j}^2\right) \quad \text{or} \quad 1 - R_{i\cdot kj}^2 = \left(1 - r_{ik}^2\right)\left(1 - r_{ij.k}^2\right)$$

Since $R_{i\cdot jk} = R_{i\cdot kj}$, we have:

$$\left(1 - R_{i\cdot jk}^2\right) = \left(1 - r_{ij}^2\right)\left(1 - r_{ik\cdot j}^2\right) = \left(1 - r_{ik}^2\right)\left(1 - r_{ij\cdot k}^2\right) \;;\; i \neq j \neq k = 1, 2, 3 \qquad \text{...(12·21}a\text{)}$$

By symmetry, the result can be generalised to n variables to give:

$$1 - R_{1\cdot23\ldots n}^2 = \left(1 - r_{12}^2\right)\left(1 - r_{13\cdot2}^2\right)\left(1 - r_{14\cdot23}^2\right)\ldots\left(1 - r_{1n\cdot23\ldots\overline{n-1}}^2\right) \qquad \text{...(12·21}b\text{)}$$

Theorem. *Any standard deviation of order 'p' may be expressed in terms of a standard deviation of order (p – 1) and a partial correlation coefficient of order (p – 1).*

Proof. Let us consider the sum:

$$\Sigma X_{1\cdot23\cdots n}^2 = \Sigma\left[X_{1\cdot23\cdots n}\cdot X_{1\cdot23\cdots n}\right]$$

$$= \Sigma\left[X_{1\cdot23\cdots(n-1)}\cdot X_{1\cdot23\cdots n}\right] \qquad \text{(Using Property 2, § 12·4)}$$

$$= \Sigma[X_{1\cdot 23 \ldots (n-1)} (X_1 - b_{12\cdot 34 \ldots n} X_2 - \ldots - b_{1(n-1)\cdot 23 \ldots n} X_{n-1} - b_{1n\cdot 23 \ldots (n-1)} X_n)]$$

$$= \Sigma X_{1\cdot 23 \ldots (n-1)} X_1 - b_{1n\cdot 23 \ldots (n-1)} \Sigma X_{1\cdot 23 \ldots (n-1)} X_n \quad \text{(Using Property 1, § 12·4)}$$

$$= \Sigma X^2_{1\cdot 23 \ldots (n-1)} - b_{1n\cdot 23 \ldots (n-1)} \Sigma X_{1\cdot 23 \ldots (n-1)} X_{n\cdot 23 \ldots (n-1)}$$

Dividing both sides by N (total number of observations), we get

$$\sigma^2_{12\cdot 3 \ldots n} = \sigma^2_{1\cdot 23 \ldots (n-1)} - b_{1n\cdot 23 \ldots (n-1)} \text{Cov}\,[X_{1\cdot 23 \ldots (n-1)}, X_{n\cdot 23 \ldots (n-1)}] \quad \ldots (*)$$

The regression coefficient of $X_{n\cdot 23 \ldots (n-1)}$ on $X_{1\cdot 23 \ldots (n-1)}$ is given by:

$$b_{n1\cdot 23 \ldots (n-1)} = \frac{\text{Cov}\left[X_{1\cdot 23 \ldots (n-1)}, X_{n\cdot 23 \ldots (n-1)}\right]}{\sigma^2_{1\cdot 23 \ldots (n-1)}} \quad \ldots (**)$$

$$\therefore \quad \sigma^2_{12\cdot 3 \ldots n} = \sigma^2_{1\cdot 23 \ldots (n-1)} \left[1 - b_{1n\cdot 23 \ldots (n-1)} \cdot b_{n1\cdot 23 \ldots (n-1)}\right] \quad \text{[From (*) and (**)]}$$

$$= \sigma^2_{1\cdot 23 \ldots (n-1)} \left[1 - r^2_{n1\cdot 23 \ldots (n-1)}\right], \quad \ldots (12\cdot 22)$$

a formula which expresses the standard deviation of order $(n - 1)$ in terms of standard deviation of order $(n - 2)$ and partial correlation coefficient of order $(n - 2)$. If we take $p = (n - 1)$, the theorem is established.

***Cor.* 1.** From (12·22), we have $\sigma^2_{1\cdot 23 \ldots (n-1)} = \sigma^2_{1\cdot 23 \ldots (n-2)} \left[1 - r^2_{(n-1)1\cdot 23 \ldots (n-2)}\right]$ ... (12·22*a*)
and so on. Thus the repeated application of (12.22) gives

$$\sigma^2_{1\cdot 23 \ldots n} = \sigma^2_1 \left(1 - r^2_{12}\right)\left(1 - r^2_{13\cdot 2}\right)\left(1 - r^2_{14\cdot 32}\right) \ldots \left(1 - r^2_{1n.23 \ldots (n-1)}\right) \quad \ldots (12\cdot 22b)$$

Since partial correlation coefficients cannot exceed unity numerically, we get from (12·22), (12·22*a*) and so on:

$$\left.\begin{array}{ccc} \sigma^2_{1\cdot 23 \ldots n} & \leq & \sigma^2_{1\cdot 23 \ldots (n-1)} \\ \sigma^2_{1\cdot 23 \ldots (n-1)} & \leq & \sigma^2_{1\cdot 23 \ldots (n-2)} \\ \vdots & \vdots & \vdots \\ \sigma^2_{1\cdot 23} & \leq & \sigma^2_{1\cdot 2} \\ \sigma^2_{1\cdot 2} & \leq & \sigma^2_1 \end{array}\right\}$$

$$\Rightarrow \quad \sigma_1 \geq \sigma_{1\cdot 2} \geq \sigma_{1\cdot 23} \geq \ldots \geq \sigma_{1\cdot 23 \ldots n} \quad \ldots (12\cdot 22c)$$

***Cor.* 2.** Also, we have $\sigma^2_{1\cdot 23 \ldots n} = \sigma^2_1 \left(1 - R^2_{1\cdot 23 \ldots n}\right)$

On using (12·22*b*), we get

$$1 - R^2_{1\cdot 23 \ldots n} = \left(1 - r^2_{12}\right)\left(1 - r^2_{13\cdot 2}\right)\left(1 - r^2_{1n\cdot 23 \ldots (n-1)}\right) \quad \ldots (12\cdot 22d)$$

This is the generalisation of the result obtained in (12·21)

Since $| r_{ij\cdot(s)} | \leq 1;\ s = 0, 1, 2, 3, \ldots, (n - 1),$

where $r_{ij\cdot(s)}$ is a partial correlation coefficient of order s, we get from (12·22*d*)

$$1 - R^2_{1\cdot 23 \ldots n} \leq 1 - r^2_{12}, \quad 1 - R^2_{1\cdot 23 \ldots n} \leq 1 - r^2_{13\cdot 2} \quad \text{and so on.}$$

i.e.,

$$R^2_{1\cdot 23 \ldots n} \geq r^2_{12}, r^2_{13\cdot 2}, \ldots, r^2_{1n\cdot 23 \ldots (n-1)} \quad \ldots (12\cdot 22e)$$

Since $R_{1\cdot 23 \ldots n}$ is symmetric in its secondary subscripts, we have

$$\left.\begin{array}{l} R^2_{1\cdot 23 \ldots n} \geq r^2_{1i},\ (i = 2, 3, \ldots, n) \\ R^2_{1\cdot 23 \ldots n} \geq r^2_{1i.j},\ (i \neq j = 2, 3, \ldots, n) \end{array}\right\} \quad \ldots (12\cdot 22f)$$

and so on.

Example 12·5. *(i) If all the correlation coefficients of zero order are equal to ρ, prove that:*

$$r_{13\cdot 2} = \frac{\rho}{1+\rho}.$$

Hence, using the relationship between total, multiple and partial correlation coefficients, prove that:

$$1 - R_{1\cdot 23}^2 = \frac{(1-\rho)(1+2\rho)}{1+\rho}.$$

(ii) If $R_{1\cdot 23} = 0$, X_1 is uncorrelated with any of other variables, i.e., $r_{12} = r_{13} = 0$.

(iii) If $R_{1\cdot 23} = 1$, then linear regression of X_1 and X_2 and X_3 is perfect.

Solution. (*i*) Here, we are given that: $r_{12} = r_{13} = r_{23} = \rho$

$$\therefore \quad r_{13.2} = \frac{r_{13} - r_{12}\, r_{32}}{\sqrt{(1-r_{12}^2)(1-r_{32}^2)}} = \frac{\rho - \rho^2}{\sqrt{(1-\rho^2)(1-\rho^2)}} = \frac{\rho(1-\rho)}{(1-\rho^2)} = \frac{\rho}{1+\rho}$$

Using the relationship between total, multiple and partial correlation coefficients, viz ;

$$1 - R_{1\cdot 23}^2 = (1-r_{12}^2)(1-r_{13.2}^2), \text{ we get}$$

$$1 - R_{1\cdot 23}^2 = (1-\rho^2)\left[1 - \frac{\rho^2}{(1+\rho)^2}\right] = \frac{(1-\rho)(1+2\rho)}{(1+\rho)}$$

(*ii*) $$R_{1\cdot 23} = 0 \quad \Rightarrow \quad 1 = (1-r_{12}^2)(1-r_{13.2}^2) \qquad \text{... (*)}$$

Since $0 \le r_{12}^2 \le 1$ and $0 \le r_{13.2}^2 \le 1$, (*) will hold if and only if $r_{12} = 0$ and $r_{13.2} = 0$.

$$r_{13\cdot 2} = 0 \;\Rightarrow\; \frac{r_{13} - r_{12}\, r_{32}}{\sqrt{(1-r_{12}^2)(1-r_{32}^2)}} = 0 \;\Rightarrow\; \frac{r_{13}}{\sqrt{1-r_{32}^2}} = 0 \qquad (\because r_{12} = 0)$$

$$\therefore \quad r_{13} = 0$$

Thus if $R_{1\cdot 23} = 0$, then $r_{13} = r_{12} = 0$, *i.e.*, X_1 is uncorrelated with X_2 and X_3.

(*iii*) We have: $$1 - R_{1\cdot 23}^2 = \frac{\sigma_{1\cdot 23}^2}{\sigma_1^2}$$

$$\therefore \quad R_{1\cdot 23} = 1 \;\Rightarrow\; \sigma_{1\cdot 23}^2 = 0 \;\Rightarrow\; \Sigma(X_1 - e_{1\cdot 23})^2 = 0 \qquad \text{... (**)}$$

where $e_{1\cdot 23} = \hat{X}_1$, the estimated value of X_1 as given by plane of regression of X_1 on X_2 and X_3. (**) is true if and only if $X_1 = e_{1\cdot 23}$, *i.e.*, linear regression of X_1 on X_2 and X_3, is perfect.

12·7. EXPRESSION FOR REGRESSION COEFFICIENT IN TERMS OF REGRESSION COEFFICIENTS OF LOWER ORDER

Consider

$$\begin{aligned}
\Sigma X_{1\cdot 34\,\dots\,n} \cdot X_{2\cdot 34\,\dots\,n} &= \Sigma X_{1\cdot 34\,\dots\,(n-1)} \cdot X_{2\cdot 34\,\dots\,n} \\
&= \Sigma X_{1\cdot 34\,\dots\,(n-1)} (X_2 - b_{23\cdot 4\,\dots\,n} X_3 - \dots - b_{2n\cdot 34\,\dots\,(n-1)} X_n) \\
&= \Sigma X_{1\cdot 34\,\dots\,(n-1)} X_2 - b_{2n\cdot 34\,\dots\,(n-1)} \Sigma X_{1\cdot 34\,\dots\,(n-1)} X_n \\
&= \Sigma X_{1\cdot 34\,\dots\,(n-1)} X_{2\cdot 34\,\dots\,(n-1)} - b_{2n\cdot 34\,\dots\,(n-1)} \Sigma X_{1\cdot 34\,\dots\,(n-1)} X_{n\cdot 34\,\dots\,(n-1)}
\end{aligned}$$

Dividing both side by N, the total number of observations, we get

$$\text{Cov}(X_{1\cdot 34\,\dots\,n}, X_{2\cdot 34\,\dots\,n}) = \text{Cov}(X_{1\cdot 34\,\dots\,(n-1)}, X_{2\cdot 34\,\dots\,(n-1)}) - b_{2n\cdot 34\,\dots\,(n-1)} \text{Cov}(X_{1\cdot 34\,\dots\,(n-1)}, X_{n\cdot 34\,\dots\,(n-1)})$$

$$\Rightarrow b_{12\cdot 34\,\dots\,n}\, \sigma^2_{2\cdot 34\,\dots\,n} = b_{12\cdot 34\,\dots\,(n-1)}\, \sigma^2_{2\cdot 34\,\dots\,(n-1)} - b_{2n\cdot 34\,\dots\,(n-1)}\, b_{1n\cdot 34\,\dots\,(n-1)}\, \sigma^2_{n\cdot 34\,\dots\,(n-1)}$$

On using (12·22), we get

$$b_{12\cdot34\ldots n}\,\sigma^2_{2\cdot34\ldots(n-1)}\left[1-r^2_{2n\cdot34\ldots(n-1)}\right]$$

$$=\sigma^2_{2\cdot34\ldots(n-1)}\left[b_{12\cdot34\ldots(n-1)}-b_{2n\cdot34\ldots(n-1)}\,b_{1n\cdot34\ldots(n-1)}\times\frac{\sigma^2_{n\cdot34\ldots(n-1)}}{\sigma^2_{2\cdot34\ldots(n-1)}}\right] \quad \ldots (*)$$

In the case of two variables, we have

$$b_{ij}\,\sigma_j^2=b_{ji}\,\sigma^2_i=\text{Cov}(X_i,X_j)\ \Rightarrow\ b_{ij}=\frac{\sigma_i^2}{\sigma_j^2}\,b_{ji}$$

$$\therefore\quad b_{2n\cdot34\ldots(n-1)}\,\frac{\sigma^2_{n\cdot34\ldots(n-1)}}{\sigma^2_{2\cdot34\ldots(n-1)}}=b_{n2\cdot34\ldots(n-1)}$$

Hence, from (*), we get

$$b_{12\cdot34\ldots n}\left[1-r^2_{2n\cdot34\ldots(n-1)}\right]=\left[b_{12\cdot34\ldots(n-1)}-b_{1n\cdot34\ldots(n-1)}\,b_{n2\cdot34\ldots(n-1)}\right]$$

$$\Rightarrow\quad b_{12\cdot34\ldots n}=\left[\frac{b_{12\cdot34\ldots(n-1)}-b_{1n\cdot34\ldots(n-1)}\cdot b_{n2\cdot34\ldots(n-1)}}{1-r^2_{2n\cdot34\ldots(n-1)}}\right] \quad \ldots (12\cdot23)$$

$$\therefore\quad b_{12\cdot34\ldots n}=\left[\frac{b_{12\cdot34\ldots(n-1)}-b_{1n\cdot34\ldots(n-1)}\cdot b_{n2\cdot34\ldots(n-1)}}{1-b_{2n\cdot34\ldots(n-1)}\cdot b_{n2\cdot34\ldots(n-1)}}\right] \quad \ldots (12\cdot23a)$$

Hence, the result.

12·8. EXPRESSION FOR PARTIAL CORRELATION COEFFICIENT IN TERMS OF PARTIAL CORRELATION COEFFICIENTS OF LOWER ORDER

By def.,

$$b_{ij\cdot lm\ldots t}=r_{ij\cdot lm\ldots t}\times\frac{\sigma_{i\cdot lm\ldots t}}{\sigma_{j\cdot lm\ldots t}} \quad \ldots (12\cdot24)$$

$$\therefore\quad b_{1n\cdot34\ldots(n-1)}\cdot b_{n2\cdot34\ldots(n-1)}=r_{1n\cdot34\ldots(n-1)}\frac{\sigma_{1\cdot34\ldots(n-1)}}{\sigma_{n\cdot34\ldots(n-1)}}\times r_{n2\cdot34\ldots(n-1)}\frac{\sigma_{n\cdot34\ldots(n-1)}}{\sigma_{2\cdot34\ldots(n-1)}}$$

$$=r_{1n\cdot34\ldots(n-1)}\cdot r_{n2\cdot34\ldots(n-1)}\cdot\frac{\sigma_{1\cdot34\ldots(n-1)}}{\sigma_{2\cdot34\ldots(n-1)}} \quad \ldots (12\cdot24a)$$

Hence, from (12·23), on using (12·24) and (12·24a), we get

$$r_{12\cdot34\ldots n}\times\frac{\sigma_{1\cdot34\ldots n}}{\sigma_{2\cdot34\ldots n}}=\left[\frac{\left\{r_{12\cdot34\ldots(n-1)}-r_{1n\cdot34\ldots(n-1)}\,r_{n2\cdot34\ldots(n-1)}\right\}}{1-r^2_{2n\cdot34\ldots(n-1)}}\cdot\frac{\sigma_{1\cdot34\ldots(n-1)}}{\sigma_{2\cdot34\ldots(n-1)}}\right] \quad \ldots (12\cdot24b)$$

Also on using (12·22), we get

$$\frac{\sigma_{1\cdot34\ldots n}}{\sigma_{2\cdot34\ldots n}}=\frac{\sigma_{1\cdot34\ldots(n-1)}}{\sigma_{2\cdot34\ldots(n-1)}}\times\left[\frac{1-r^2_{1n\cdot34\ldots(n-1)}}{1-r^2_{2n\cdot34\ldots(n-1)}}\right]^{1/2}$$

Hence, from (12·24b), we get

$$r_{12\cdot34\ldots n}\left[\frac{1-r^2_{1n\cdot34\ldots(n-1)}}{1-r^2_{2n\cdot34\ldots(n-1)}}\right]^{\frac{1}{2}}=\left[\frac{r_{12\cdot34\ldots(n-1)}-r_{1n\cdot34\ldots(n-1)}\,r_{n2\cdot34\ldots(n-1)}}{1-r^2_{2n\cdot34\ldots(n-1)}}\right]$$

$$\Rightarrow\quad r_{12\cdot34\ldots n}=\frac{r_{12\cdot34\ldots(n-1)}-r_{1n\cdot34\ldots(n-1)}\,r_{n2\cdot34\ldots(n-1)}}{\left[1-r^2_{1n\cdot34\ldots(n-1)}\right]^{1/2}\left[1-r^2_{n2\cdot34\ldots(n-1)}\right]^{1/2}} \quad \ldots (12\cdot24c)$$

which is an expression for the correlation coefficient of order $p=(n-2)$ in terms of the correlation coefficient of order $(p-1)=(n-3)$.

SUPPLEMENTARY EXAMPLES

Example 12·6. *Five hundred students were examined in three subjects I, II and III, each subject carrying 100 marks. A student getting 120 or more but less than 150 marks was put in pass class. A student getting 150 or more but less than 180 marks was put in second class and a student getting 180 or more marks was put in the first class. The following marks were obtained:*

	I	II	III
Mean:	35·8	52·4	48·8
S.D.:	4·2	5·3	6·1
Correlation:	$r_{12} = 0·6$	$r_{13} = 0·7$	$r_{23} = 0·8$

Assuming that the marks are normally distributed,

(i) Find the number of students in each of the three classes.

(ii) Find the total number of students with total marks lying between 120 and 190.

(iii) Find the probability that a student gets more that 240 marks.

(iv) What should be the correlation between marks in subjects I and II among students who scored equal marks in subject III ?

(v) If r_{23} was not known, obtain the limits within which it may lie from the values of r_{12} and r_{13} (ignoring sampling errors.

Solution. If Z denotes the total marks of the students in the three subjects and X_1, X_2, X_3 the total marks of the students in subjects I, II and III respectively, then

$$Z = X_1 + X_2 + X_3 \quad \Rightarrow \quad E(Z) = E(X_1) + E(X_2) + E(X_3) = 35·8 + 52·4 + 48·8 = 137$$

$$V(Z) = V(X_1) + V(X_2) + V(X_3) + 2[\text{Cov}(X_1, X_2) + \text{Cov}(X_2, X_3) + \text{Cov}(X_3, X_1)]$$

$$= 17·64 + 28·09 + 37·21 + 26·712 + 35·868 + 51·728$$

$$= 197·248 \qquad [\text{Using Cov}(X_i, X_j) = r_{ij}\, \sigma_i \cdot \sigma_j]$$

$$\Rightarrow \qquad \sigma_Z = \sqrt{197·248} = 14·045$$

$$\xi = \frac{Z - E(Z)}{\sigma_Z} \sim N(0, 1)$$

Z	$\xi = \dfrac{Z-137}{14·045}$	$p = \int_{-\infty}^{\xi} p(\xi)\, d\xi$	Class	Area under the curve in this class (A)	Frequency 500 × (A)
120	–1.21050	0.11314	120 – 150	0.70937	354.685
150	0.92567	0.82251	150 – 180	0.17639	88.195
180	3.06180	0.99890	180 and over	0.00102	0.510
190	3.77400	0.99992	120 – 190	0.88678	443.390
240	7.33410	1.00000	240 and over	0.00000	0.000

(i) The number of students in the pass, second and first class respectively are 355, 88 and 0 (approx.)

(ii) Total number of students with total marks between 120 and 190 is 443.

(iii) Probability that a student gets more than 240 marks is zero.

(iv) The correlation coefficient between marks in subjects I and II of the students who secured equal marks in subject III is $r_{12·3}$ and is given by:

$$r_{12\cdot3} = \frac{r_{12} - r_{13}\, r_{23}}{\sqrt{(1-r_{13}^2)(1-r_{23}^2)}} = \frac{0\cdot04}{\sqrt{(1-0\cdot49)(1-0\cdot64)}} = 0\cdot0934.$$

(v) We have $\quad r_{12\cdot3}^2 = \dfrac{(r_{12} - r_{13}\, r_{23})^2}{(1-r_{13}^2)(1-r_{23}^2)} \le 1 \quad \Rightarrow \quad \dfrac{(0\cdot6 - 0\cdot7a)^2}{(1-0\cdot49)(1-a^2)} \le 1$, where $a = r_{23}$.

$\therefore \quad 0\cdot36 + 0\cdot49a^2 - 0\cdot84a \le 0\cdot51(1-a^2) \quad \Rightarrow \quad a^2 - 0\cdot84a - 0\cdot15 \le 0$

Thus 'a' lies between the roots of the equation: $a^2 - 0\cdot84a - 0\cdot15 = 0$, which are 0·99 and −0·15. Hence, r_{23} should lie between −0·15 and 0·99.

Example 12·7. *Show that the correlation coefficient between the residuals $X_{1\cdot23}$ and $X_{2\cdot13}$ is equal and opposite to that between $X_{1\cdot3}$ and $X_{2\cdot3}$.*

Solution. Assuming that $X_i'.s(i = 1,2,3)$, are measured from their respective means, the correlation coefficient between $X_{1\cdot23}$ and $X_{2\cdot13}$ is given by:

$$\frac{Cov\,(X_{1\cdot23}, X_{2\cdot13})}{\sigma_{1\cdot23}\,\sigma_{2\cdot13}} = \frac{\Sigma X_{1\cdot23}\, X_{2\cdot13}}{N\,\sigma_{1\cdot23}\,\sigma_{2\cdot13}} = \frac{\frac{1}{N}\Sigma X_{2\cdot13}\,(X_1 - b_{12\cdot3}\, X_2 - b_{13\cdot2}\, X_3)}{\sigma_{1\cdot23}\,\sigma_{2\cdot13}}$$

$$= -\, b_{12\cdot3}\frac{\Sigma X_{2\cdot13}\, X_2}{N\,\sigma_{1\cdot23}\,\sigma_{2\cdot13}} = -\, b_{12\cdot3}\frac{\Sigma X_{2\cdot13}^2}{N\,\sigma_{1\cdot23}\,\sigma_{2\cdot13}} \quad \text{(c.f. Property 1 and 2, § 12·4)}$$

$$= -\, b_{12\cdot3}\frac{\sigma_{2\cdot13}}{\sigma_{1\cdot23}} = -\, b_{12\cdot3}\frac{(\sigma_2\sqrt{\omega/\omega_{22}})}{(\sigma_1\sqrt{\omega/\omega_{11}})},$$

where $\omega = \begin{vmatrix} 1 & r_{12} & r_{13} \\ r_{21} & 1 & r_{23} \\ r_{31} & r_{32} & 1 \end{vmatrix}$, $\quad \omega_{11} = \begin{vmatrix} 1 & r_{23} \\ r_{32} & 1 \end{vmatrix} = 1 - r_{23}^2 \quad$ and $\quad \omega_{22} = \begin{vmatrix} 1 & r_{13} \\ r_{31} & 1 \end{vmatrix} = 1 - r_{13}^2$

$$\therefore \quad r\,(X_{1\cdot23}, X_{2\cdot13}) = -\, b_{12\cdot3}\,\frac{\sigma_2}{\sigma_1}\cdot\sqrt{\frac{1-r_{23}^2}{1-r_{13}^2}} = -\, b_{12\cdot3}\,\frac{\sigma_{2\cdot3}}{\sigma_{1\cdot3}}$$

[since $\sigma_{2\cdot3}^2 = \sigma_2^2(1-r_{23}^2)$ and $\sigma_{1\cdot3}^2 = \sigma_1^2(1-r_{13}^2)$]

$$\therefore \quad r\,(X_{1\cdot23}, X_{2\cdot13}) = -\frac{\text{Cov}\,(X_{1\cdot3}, X_{2\cdot3})}{\sigma_{2\cdot3}^2}\cdot\frac{\sigma_{2\cdot3}}{\sigma_{1\cdot3}} = -\frac{\text{Cov}\,(X_{1\cdot3}, X_{2\cdot3})}{\sigma_{2\cdot3}\,\sigma_{1\cdot3}} = -\, r\,(X_{1\cdot3}, X_{2\cdot3})$$

Hence, the result.

Example 12·8. *Show that if $X_3 = aX_1 + bX_2$, the three partial correlations are numerically equal to unity, $r_{13\cdot2}$ having the sign of a, $r_{23\cdot1}$, the sign of b and $r_{12\cdot3}$, the opposite sign of a/b.*

Solution. Here we may regard X_3 as dependent on X_1 and X_2 which may be taken as independent variables. Since X_1 and X_2 are independent, they are uncorrelated.

Thus, $\quad r(X_1, X_2) = 0 \quad \Rightarrow \quad \text{Cov}\,(X_1, X_2) = 0 \quad$ ... (*)

$$V(X_3) = V(aX_1 + bX_2) = a^2V(X_1) + b^2V(X_2) + 2ab\,\text{Cov}\,(X_1, X_2)$$

$$= a^2\sigma_1^2 + b^2\sigma_2^2, \qquad \text{[From (*)]}$$

where $\quad V(X_1) = \sigma_1^2, \quad V(X_2) = \sigma_2^2. \quad$ Also $X_1X_3 = X_1(aX_1 + bX_2) = a\,X_1^2 + bX_1X_2$

Assuming that X_i's are measured from their means, on taking expectations of both sides, we get

$$\text{Cov}\,(X_1, X_3) = a\sigma_1^2 + b\,\text{Cov}\,(X_1, X_2) = a\sigma_1^2 \qquad \text{[Using (*)]}$$

$$\therefore \quad r_{13} = \frac{\text{Cov}(X_1, X_3)}{\sqrt{V(X_1)V(X_3)}} = \frac{a\sigma_1^2}{\sqrt{\sigma_1^2(a^2\sigma_1^2 + b^2\sigma_2^2)}} = \frac{a\sigma_1}{k}, \text{ where } k^2 = a^2\sigma_1^2 + b^2\sigma_2^2. \qquad \text{... (**)}$$

Similarly, we will get: $r_{23} = \dfrac{\text{Cov}(X_2, X_3)}{\sqrt{V(X_2)V(X_3)}} = \dfrac{b\sigma_2}{k}.$

$$\therefore \quad r_{13\cdot 2} = \frac{r_{13} - r_{12}\, r_{32}}{\sqrt{(1 - r_{12}^2)\,(1 - r_{32}^2)}} = \frac{a\sigma_1}{k} \cdot \frac{k}{\sqrt{k^2 - b^2\sigma_2^2}} = \frac{a\sigma_1}{\sqrt{a^2\sigma_1^2}} = \frac{a\sigma_1}{|a|\sigma_1} = \pm 1, \qquad \text{[From (**)]}$$

according as 'a' is positive or negative. Hence, $r_{13.2}$ has the same sign as 'a'. Again

$$r_{23\cdot 1} = \frac{r_{23} - r_{21}\, r_{31}}{\sqrt{(1 - r_{21}^2)(1 - r_{31}^2)}} = \frac{b\sigma_2}{k} \cdot \frac{k}{\sqrt{k^2 - a^2\sigma_1^2}} = \frac{b\sigma_2}{\sqrt{b^2\sigma_2^2}} = \pm 1,$$

according as 'b' is positive or negative. Hence, $r_{23.1}$ has the same sign as 'b'.

$$r_{12\cdot 3} = \frac{r_{12} - r_{13}\, r_{23}}{\sqrt{(1 - r_{13}^2)(1 - r_{23}^2)}} = -\frac{a\sigma_1}{k} \cdot \frac{b\sigma_2}{k} \cdot \frac{k^2}{\sqrt{(k^2 - a^2\sigma_1^2)(k^2 - b^2\sigma_2^2)}}$$

$$= -\frac{ab\sigma_1\sigma_2}{\sqrt{b^2\sigma_2^2 \times a^2\sigma_1^2}} = -\frac{ab}{\sqrt{a^2b^2}} = -\frac{ab}{\sqrt{a^2/b^2}} = -\frac{(a/b)}{\pm|(a/b)|} = \mp 1,$$

according as a/b is positive or negative. Hence, $r_{12\cdot 3}$ has the sign opposite to that of (a/b).

Example 12·9. *If all the correlation coefficients of zero order in a set of p-variates are equal to ρ, show that*

(i) Every partial correlation of s'th order is $\rho/(1 + s\rho)$. ... (*)

(ii) The coefficient of multiple correlation R of a variate with the other $(p - 1)$ variates is given by:

$$1 - R^2 = (1 - \rho)\left[\frac{1 + (p-1)\rho}{1 + (p-2)\rho}\right].$$

Solution. We are given that: $r_{mn} = \rho, \ (m, n = 1, 2, ..., p \ ; m \neq n)$ ...(**)

We have: $$r_{ij\cdot k} = \frac{r_{ij} - r_{ik}\, r_{jk}}{\sqrt{(1 - r_{ik}^2)(1 - r_{jk}^2)}}; \ (i, j, k = 1, 2, ..., p \ ; i \neq j \neq k)$$

$$= \frac{\rho - \rho \cdot \rho}{\sqrt{(1 - \rho^2)(1 - \rho^2)}} = \frac{\rho}{1 + \rho} \qquad \text{[From (**)]}$$

Thus every partial correlation coefficient of first order is $\rho/(1 + \rho)$. ⇒ (*) is true for $s = 1$ ...(***)
The result will be established by the principle of mathematical induction. Let us suppose that every partial correlation coefficient of order s is given by $\rho/(1 + s\rho)$. Then the partial correlation coefficient of order $(s + 1)$ is given by:

$$r_{ij\cdot km\,...\,t} = \frac{r_{ij\cdot(s)} - r_{ik\cdot(s)}\, r_{jk\cdot(s)}}{\sqrt{(1 - r_{ik\cdot(s)}^2)(1 - r_{jk\cdot(s)}^2)}}$$

where $k, m, ..., t$ are $(s + 1)$ secondary subscripts and $r_{ij.(s)}$, $r_{ik.(s)}$, $r_{jk.(s)}$, are partial correlation coefficients of order s. Thus

$$r_{ij\cdot km\,...\,t} = \frac{\dfrac{\rho}{1+s\rho} - \left(\dfrac{\rho}{1+s\rho}\right)^2}{1 - \left(\dfrac{\rho}{1+s\rho}\right)^2} = \frac{\dfrac{\rho}{1+s\rho}\left(1 - \dfrac{\rho}{1+s\rho}\right)}{\left(1 - \dfrac{\rho}{1+s\rho}\right)\left(1 + \dfrac{\rho}{1+s\rho}\right)} = \frac{\rho}{1 + (s+1)\rho} \qquad \text{... (****)}$$

Using (***) and (****), the required result follows by induction.

(*ii*) We have $1 - R^2 = \dfrac{\omega}{\omega_{11}}$, where R is the multiple correlation coefficient of a variable with other $(p - 1)$ variables and

$$\omega = \begin{vmatrix} 1 & \rho & \rho & \cdots & \rho \\ \rho & 1 & \rho & \cdots & \rho \\ \rho & \rho & 1 & \cdots & \rho \\ \vdots & \vdots & \vdots & & \vdots \\ \rho & \rho & \rho & \cdots & 1 \end{vmatrix}_{p \times p} \quad \text{and} \quad \omega_{11} = \begin{vmatrix} 1 & \rho & \rho & \cdots & \rho \\ \rho & 1 & \rho & \cdots & \rho \\ \rho & \rho & 1 & \cdots & \rho \\ \vdots & \vdots & \vdots & & \vdots \\ \rho & \rho & \rho & \cdots & 1 \end{vmatrix}_{(p-1) \times (p-1)}$$

We have

$$\omega = [1 + (p-1)\rho] \cdot \begin{vmatrix} 1 & \rho & \rho & \rho & \cdots & \rho \\ 1 & 1 & \rho & \rho & \cdots & \rho \\ 1 & \rho & 1 & \rho & \cdots & \rho \\ \vdots & \vdots & \vdots & \vdots & & \vdots \\ 1 & \rho & \rho & \rho & \cdots & 1 \end{vmatrix}_{p \times p} \qquad \left(\text{On adding } C_2, C_3, \ldots, C_p \text{ to } C_1\right).$$

$$\Rightarrow \quad \omega = [1 + (p-1)\rho] \cdot \begin{vmatrix} 1 & \rho & \rho & \rho & \cdots & \rho \\ 0 & (1-\rho) & 0 & 0 & \cdots & 0 \\ 0 & 0 & (1-\rho) & 0 & \cdots & 0 \\ \vdots & \vdots & \vdots & & & \vdots \\ 0 & 0 & 0 & 0 & \cdots & (1-\rho) \end{vmatrix}; \qquad \left[\begin{array}{c}\text{On Operating } R_i \to R_i - R_1, \\ (i = 2, 3, \ldots p)\end{array}\right]$$

$$\therefore \qquad \omega = [1 + (p - 1)\rho]\,(1 - \rho)^{p-1}$$

Similarly, we will have: $\qquad \omega_{11} = [1 + (p - 2)\rho]\,(1 - \rho)^{p-2}$

$$\therefore \qquad 1 - R^2 = \frac{\omega}{\omega_{11}} = (1 - \rho)\left[\frac{1 + (p-1)\rho}{1 + (p-2)\rho}\right].$$

Example 12·10. *In a p-variate distribution, all the total (order zero) correlation coefficients are equal to* $\rho_0 \neq 0$. *Let* ρ_k *denote the partial correlation coefficient of order k and* R_k *be the multiple correlation coefficient of one variate on k other variates. Prove that*

(*i*) $\rho_0 \geq -\dfrac{1}{(p-1)}$, $\qquad$ (*ii*) $\rho_k - \rho_{k-1} = -\rho_k\,\rho_{k-1}$ $\qquad$ *and* $\qquad$ (*iii*) $R_k^{\,2} = \dfrac{k\,\rho_0^{\,2}}{1 + (k-1)\rho_0}$.

Solution. (*i*) We have proved in *Example* 12·9, that: $\qquad \rho_k = \dfrac{\rho_0}{1 + k\,\rho_0}$

In the case of *p*-variate distribution, the partial correlation coefficient of the highest order is ρ_{p-2} and is given by: $\qquad \rho_{p-2} = \dfrac{\rho_0}{1 + (p-2)\rho_0}$

Since $|\rho_{p-2}| \leq 1 \quad \Rightarrow \quad -1 \leq \rho_{p-2} \leq 1$, we have (on considering the lower limit)

$$-1 \leq \frac{\rho_0}{1 + (p-2)\rho_0} \quad \Rightarrow \quad -[1 + (p-2)\rho_0] \leq \rho_0 \quad \Rightarrow \quad \rho_0 \geq -\frac{1}{(p-1)}$$

(*ii*)
$$\text{L.H.S} = \rho_k - \rho_{k-1} = \frac{\rho_0}{1 + k\rho_0} - \frac{\rho_0}{1 + (k-1)\rho_0} = \rho_0\left[\frac{-\rho_0}{(1 + k\rho_0)\{1 + (k-1)\rho_0\}}\right]$$

$$= -\left(\frac{\rho_0}{1 + k\rho_0}\right)\left(\frac{\rho_0}{1 + (k-1)\rho_0}\right) = -\rho_k\,\rho_{k-1}$$

(*iii*) Taking $\rho = \rho_0$ and $k = p - 1$ in part (*ii*) Example 12.9, we get

$$1 - R_k^{\,2} = (1 - \rho_0)\left[\frac{1 + k\rho_0}{1 + (k-1)\rho_0}\right]$$

$$R_k^{\,2} = 1 - \frac{(1 - \rho_0)(1 + k\rho_0)}{1 + (k-1)\rho_0} = \frac{k\,\rho_0^{\,2}}{1 + (k-1)\rho_0}. \qquad \text{(On simplification)}$$

Example 12·11. *If r_{12} and r_{23} are given, show that r_{23} must lie in the range:*

$$r_{12}\, r_{13} \pm 1 - \left(r_{12}^2 - r_{13}^2 + r_{12}^2\, r_{13}^2\right)^{1/2}$$

If $r_{12} = k$ and $r_{13} = -k$, show that r_{23} will lie between -1 and $1 - 2k^2$.

Solution. We have $r_{12\cdot3} = \dfrac{r_{12} - r_{13}\, r_{23}}{\sqrt{\left(1 - r_{13}^2\right)\left(1 - r_{23}^2\right)}} \le 1 \quad \Rightarrow \quad (r_{12} - r_{13}\, r_{23})^2 \le (1 - r_{13}^2)(1 - r_{23}^2)$

$$\Rightarrow \quad r_{12}^2 + r_{13}^2\, r_{23}^2 - 2r_{12}\, r_{13}\, r_{23} \le 1 - r_{13}^2 - r_{23}^2 + r_{13}^2\, r_{23}^2 \quad \Rightarrow \quad r_{12}^2 + r_{13}^2 + r_{23}^2 - 2r_{12}\, r_{13}\, r_{23} \le 1 \qquad \dots (*)$$

This condition holds for consistent values of r_{12}, r_{13} and r_{23}. (*) may be rewitten as:

$$r_{23}^2 - (2r_{12}\, r_{13})\, r_{23} + \left(r_{12}^2 + r_{13}^2 - 1\right) \le 0.$$

Hence, for given values of r_{12} and r_{13}, r_{23} must lie between the roots of the quadratic (in r_{23}) equation:

$$r_{23}^2 - (2r_{12}\, r_{13})\, r_{23} + \left(r_{12}^2 + r_{13}^2 - 1\right) = 0,$$

which are given by: $\quad r_{23} = r_{12}\, r_{13} \pm \sqrt{r_{12}^2\, r_{13}^2 - \left(r_{12}^2 + r_{13}^2 - 1\right)}$

Hence, $\quad r_{12}\, r_{13} - \sqrt{1 - r_{12}^2 - r_{13}^2 + r_{12}^2\, r_{13}^2} \le r_{23} \le r_{12}\, r_{13} + \sqrt{1 - r_{12}^2 - r_{13}^2 + r_{12}^2\, r_{13}^2} \qquad \dots (**)$

In other words, r_{23} must lie in the range: $\quad r_{12}\, r_{13} \pm \sqrt{1 - r_{12}^2 - r_{13}^2 + r_{12}^2\, r_{13}^2}$

In particular, if $r_{12} = k$ and $r_{13} = -k$, we get from (**)

$$-k^2 - \sqrt{1 - k^2 - k^2 + k^4} \le r_{23} \le -k^2 + \sqrt{1 - k^2 - k^2 + k^4}$$

$$\Rightarrow \quad -k^2 - (1 - k^2) \le r_{23} \le -k^2 + (1 - k^2)$$

$$\therefore \quad -1 \le r_{23} \le 1 - 2k^2.$$

CHAPTER CONCEPTS QUIZ

1. Indicate the correct answer in each of the following:

(*i*) Let there be n observations on X_1, X_2 and X_3 in a tri-variate distribution. In the usual notations, if $r_{12\cdot3} = 0$, then we can say that:

(*a*) X_1 and X_2 may be uncorrelated when the effect of X_3 is eliminated.
(*b*) X_1 and X_2 may appear to be correlated because of the effect of X_3.
(*c*) r_{12} will not be zero if X_3 is correlated with both X_1 and X_2.
(*d*) all the above statements are false.

(*ii*) In the usual notations $r_{12\cdot3}$ is equal to:

(*a*) $\dfrac{r_{12} - r_{13}\, r_{23}}{\sqrt{1 - r_{13}^2}\,\sqrt{1 - r_{23}^2}}$ $\qquad$ (*b*) $\dfrac{r_{12}^2 - r_{13}\, r_{23}}{\sqrt{1 - r_{13}^2}\,\sqrt{1 - r_{23}^2}}$

(*c*) $\dfrac{r_{13}\, r_{23} - r_{12}^2}{\sqrt{1 - r_{13}^2}\,\sqrt{1 - r_{23}^2}}$ $\qquad$ (*d*) $\dfrac{r_{13}\, r_{23} - r_{12}}{\sqrt{1 - r_{13}^2}\,\sqrt{1 - r_{23}^2}}$

(*iii*) In a tri-variate distribution X_1, X_2 and X_3, if all the total correlation coefficients are equal, say ρ, then while the multiple correlation coefficient is never less then the total correlation coefficients, $1 - R_{1\cdot23}^2$ is equal to:

(*a*) $(1 - \rho)(1 + 2\rho)/(1 + \rho)$ $\qquad$ (*b*) $[(1 + \rho) - (1 - \rho)(1 + 2\rho)]/(1 + \rho)$
(*c*) $(1 - \rho)(1 - 2\rho)/(1 + \rho)$ $\qquad$ (*d*) $(1 + \rho)(1 - 2\rho)/(1 - \rho)$

(*iv*) In the usual notations, $R_{1\cdot23}^2$, is equal to:

(*a*) $\dfrac{(r_{12}^2 + r_{13}^2 - 2r_{12}\, r_{13}\, r_{23})}{(1 - r_{23}^2)}$ $\qquad$ (*b*) $\dfrac{(r_{12}^2 + r_{13}^2 - 2r_{12}\, r_{13})}{(1 - r_{23}^2)}$

(*c*) $\dfrac{(r_{12}^2 + r_{13}^2 - 2r_{23})}{(1 - r_{23}^2)}$ $\qquad$ (*d*) $\dfrac{(r_{12}^2 + r_{13}^2 - r_{12}\, r_{13}\, r_{23})}{(1 - 2r_{23})}$

(v) In the usual notations, $R^2_{1\cdot 23}$ can be expressed as:

(a) $R^2_{1\cdot 23} = 1 - (1 - r^2_{12})(1 - r^2_{13\cdot 2})$ (b) $R^2_{1\cdot 23} = 1 - (1 - r^2_{12})(1 - r_{13\cdot 2})$

(c) $R^2_{1\cdot 23} = (1 - r^2_{12})(1 - r^2_{13\cdot 2})$ (d) $R^2_{1\cdot 23} = (1 - r_{12})(1 - r_{13\cdot 2})$

(vi) If multiple correlation coefficient $R_{1\cdot 23} = 1$, then $R_{2\cdot 13}$ is

(a) 0 (b) –1 (c) 1 (d) None of these

(vii) The range of partial correlation coefficient $r_{12\cdot 3}$ is:

(a) –1 to 1 (b) 0 to ∞ (c) 0 to 1 (d) None of these

(viii) Which of the following relationship is true ?

(a) $R_{1\cdot 23} \le r_{12}$ (b) $R_{1\cdot 23} \ge r_{12}$

(c) $R_{1\cdot 23} = r_{12}$ (d) $R_{1\cdot 23} \ge -1$

(ix) The relationship between the multiple correlation coefficient of X_1 on X_2 and X_3 and the standard error of estimate is given by the expression:

(a) $R_{1\cdot 23} = \sqrt{1 - \frac{\sigma^2_{1\cdot 23}}{\sigma^2_1}}$ (b) $R_{1\cdot 23} = \sqrt{1 - \frac{\sigma^2_{1\cdot 23}}{\sigma^2_2}}$ (c) $R_{1\cdot 23} = \sqrt{1 - \frac{\sigma^2_{1\cdot 23}}{\sigma^2_3}}$ (d) None of these

2. Fill in the blanks:

(i) The order of a partial correlation coefficient is given by the number of ...... subscripts.

(ii) The second order partial correlation coefficient can be obtained in terms of partial correlation coefficients of ...... oder.

(iii) The orders of the partial correlation coefficients $r_{13\cdot 2}$, $r_{14\cdot 23}$ and $r_{23\cdot 145}$ are respectively ......, ..., ...

(iv) In a partial correlation coefficient, the ordering of the primary as well as the secondary subscripts is ......

(v) In partial regression coefficient $b_{ij.k}$, the first primary subscript refers to ...... variable and the second primary subscript refers to ...... variable.

(vi) The limits for multiple correlation coefficient $R_{i\,j.k}$ are ......

(vii) Partial correlation coefficient $r_{13.2}$ is the simple correlation between ......

(viii) $r_{12\cdot 3}$ is known as ...... order partial correlation coefficient.

(ix) The range of partial correlation coefficient is ......

(x) If the value of multiple correlation is near to unity, it indicates that multiple linear regression is a ...... fit.

DISCUSSION AND REVIEW QUESTIONS

1. Explain the concept of multiple regression. What is difference between simple linear and multiple linear regression ?

2. (a) Explain partial correlation and multiple correlation.

(b) Explain the concepts of multiple and partial correlation coefficients. Show that the multiple correlation coefficient $R_{1\cdot 23}$ is, in the usual notations given by:

$$R^2_{1\cdot 23} = 1 - (\omega/\omega_{11})$$

3. (a) In the usual notations, prove that $R^2_{1\cdot 23} = \dfrac{r^2_{12} + r^2_{13} - 2r_{12}\, r_{13}\, r_{23}}{1 - r^2_{23}} \le r^2_{12}$.

(b) If $R_{1\cdot 23} = 1$, prove that $r_{2\cdot 13}$ is also equal to 1. If $R_{1\cdot 23} = 0$, does it necessarily mean that $R_{2\cdot 13}$ is also zero?

4. (a) Obtain an expression for the variance of the residual $X_{1\cdot 23}$ in terms of the correlations r_{12}, r_{23} and r_{31} and deduce that $R_{1(23)} \ge r_{12}$ and r_{13}.

(b) Show that the standard deviation of order p may be expressed in terms of standard deviation of order $(p - 1)$ and a correlation coefficient of order $(p - 1)$. Hence deduce that:

(i) $\sigma_1 \ge \sigma_{1\cdot 2} \ge \sigma_{1\cdot 23} \ge \ldots \ge \sigma_{1\cdot 23\ldots n}$, (ii) $1 - R^2_{1\cdot 23\ldots n} = (1 - r^2_{12})(1 - r^2_{13.2}) \ldots (1 - r^2_{1n.23\ldots(n-1)})$

5. With the usual notations, prove that

$$\sigma^2_{1\cdot 234\,\ldots\, n} = \frac{\omega}{\omega_{11}}\sigma_1^2 = \sigma_1^2\left(1 - r_{12}^2\right)\left(1 - r_{13\cdot 2}^2\right)\ldots\left(1 - r_{1n\cdot 23\ldots(n-1)}^2\right)$$

ASSORTED REVIEW PROBLEMS FOR SELF-ASSESSMENT

12·1. For a trivariate distribution of three variables X_1, X_2 and X_3, the following results were obtained:

$$\overline{X}_1 = 40,\ \overline{X}_2 = 50,\ \overline{X}_3 = 20,\quad \sigma_1 = 3,\quad \sigma_2 = 4,\quad \sigma_3 = 2;\quad r_{12} = 0{\cdot}4,\quad r_{23} = 0{\cdot}5,\quad r_{31} = 0{\cdot}25,$$

where $\overline{X}_i$ = Mean of X_i, σ_i^2 = Var (X_i) and r_{ij} = Correlation coefficient between X_i and X_j, $(i, j) = 1, 2, 3$. Find:

(*i*) The multiple regression equation of X_3 on X_1 and X_2, and

(*ii*) Estimate the value of X_3 when $X_1 = 43$ and $X_2 = 54$.

Ans. (*i*) $X_3 - \overline{X}_3 = b_{31\cdot 2}(X_1 - \overline{X}_1) + b_{32\cdot 1}(X_2 - \overline{X}_2) \Rightarrow X_3 - 20 = 0{\cdot}04(X_1 - 40) + 0{\cdot}24(X_2 - 50)$

(*ii*) $\hat{X}_3 = 20 + 0{\cdot}04 \times 3 + 0{\cdot}24 \times 4 = 21{\cdot}08$

12·2. The following data relates to three variables X_1, X_2, X_3:

X_1:	4	6	7	9	13	15
X_2:	15	12	8	6	4	3
X_3:	30	24	20	14	10	4

Obtain: (*i*) Regression coefficients $b_{12\cdot 3}$ and $b_{13\cdot 2}$.

(*ii*) The equation of the plane of regression of X_1 on (X_2, X_3)

(*iii*) Estimate the value of X_1 when $X_2 = 15$ and $X_3 = 30$.

Ans. (*a*) $b_{12\cdot 3} = 0{\cdot}389$, $b_{13\cdot 2} = -\,0{\cdot}623$; (*b*) $X_1 = 16{\cdot}479 + 0{\cdot}389X_2 - 0{\cdot}623\,X_3$; (*c*) $\hat{X}_1 = 3{\cdot}624$.

12·3. An instructor of mathematics wishes to determine the relationship of grades on a final examination to grades on two quizzes given during the semester. Considering X_1, X_2 and X_3 the grades of a student on the first quiz, second quiz and final examination respectively, he made the following computations for a total of 120 students. Find:

(*i*) The least square regression equation of X_3 on X_1 and X_2.

(*ii*) Estimate the final grades of two students who scored respectively (9 and 7) and (4 and 5) on the two quizzes.

$$\overline{X}_1 = 6{\cdot}8 \qquad \overline{X}_2 = 7{\cdot}0 \qquad \overline{X}_3 = 74$$

$$\sigma_1 = 1{\cdot}0 \qquad \sigma_2 = 0{\cdot}80 \qquad \sigma_3 = 9{\cdot}0$$

$$r_{12} = 0{\cdot}60 \qquad r_{13} = 0{\cdot}70 \qquad r_{23} = 0{\cdot}65$$

Ans. (*i*) $X_3 = 4{\cdot}36X_1 + 4{\cdot}04X_2 + 16{\cdot}072$, (*ii*) $\hat{X}_3 = 84$ and $\hat{X}_3 = 54$.

12·4. (*a*) The simple correlation coefficients between temperature (X_1), corn yield (X_2) and rainfall (X_3) are:

$$r_{12} = 0{\cdot}59, \qquad r_{13} = 0{\cdot}46 \qquad \text{and} \qquad r_{23} = 0{\cdot}77.$$

Calculate the partial correlation coefficients $r_{12\cdot 3}$, $r_{23\cdot 1}$ and $r_{31\cdot 2}$. Also calculate $R_{1\cdot 23}$.

(*b*) If $r_{12} = 0{\cdot}80$, $r_{13} = -\,0{\cdot}40$ and $r_{23} = -\,0{\cdot}56$, find the values of $r_{12\cdot 3}$, $r_{13\cdot 2}$ and $r_{23\cdot 1}$. Calculate further $R_{1\cdot 23}$, $R_{2\cdot 13}$ and $R_{3\cdot 12}$.

12·5. From the data relating to the height (X_1), weight (X_2) and age (X_3) of 20 students of a college, the following correlation coefficients are obtained:

$$r_{12} = 0{\cdot}75,\ r_{13} = 0{\cdot}72 \quad \text{and} \quad r_{23} = 0{\cdot}50.$$

Find the partial correlation coefficient $r_{12\cdot 3}$ and multiple correlation coefficient $R_{3\cdot 12}$.

Ans. $r_{12\cdot 3} = 0.65$, $R_{3\cdot 12} = 0{\cdot}7225$

12·6. In a trivariate distribution, if $r_{12} = 0{\cdot}7$, $r_{13} = 0{\cdot}61$ and $r_{23} = 0{\cdot}4$, find the values of:

(*i*) All partial correlation coefficients and (*ii*) all multiple correlation coefficients.

Ans. $r_{12\cdot 3} = 0{\cdot}6279$, $r_{23\cdot 1} = -\,0{\cdot}0477$, $r_{31\cdot 2} = 0{\cdot}5042$; $R_{1\cdot 23} = 0{\cdot}79$, $R_{2\cdot 31} = 0{\cdot}70$, $R_{3\cdot 12} = 0{\cdot}61$

12·7. Prove that: $R^2_{1\cdot23} = b_{12\cdot3}\, r_{12}\dfrac{\sigma_2}{\sigma_1} + b_{13\cdot2}\, r_{13}\dfrac{\sigma_3}{\sigma_1}$.

12·8. (*a*) If $X_3 = aX_1 + bX_2$ for all sets of values of X_1, X_2 and X_3, find the value of $r_{23\cdot1}$.

(*b*) If the relation $aX_1 + bX_2 + cX_3 = 0$ holds for all sets of values X_1, X_2 and X_3, what must be the partial correlation coefficients ?

12·9. (*a*) If $r_{12} = r_{23} = r_{31} = \rho \neq 1$, then $r_{12\cdot3} = r_{23\cdot1} = r_{31\cdot2} = \dfrac{\rho}{1+\rho}$ and $R_{1\cdot23} = R_{2\cdot13} = R_{3\cdot12} = \dfrac{\rho\sqrt{2}}{\sqrt{1+\rho}}$.

(*b*) Y_1, Y_2, Y_3 are uncorrelated standard variates, $X_1 = Y_2 + Y_3$, $X_2 = Y_3 + Y_1$, and $X_3 = Y_1 + Y_2$. Find the multiple correlation coefficient between X_3 and (X_1 and X_2).

12·10. *X*, *Y*, *Z* are independent random variables with the same variance. If

$$X_1 = \frac{1}{\sqrt{2}}(X - Z), \qquad X_2 = \frac{1}{\sqrt{3}}(X + Y + Z), \qquad X_3 = \frac{1}{\sqrt{6}}(X + 2Y + Z),$$

show that X_1, X_2, X_3 have equal variances. Calculate $r_{12\cdot3}$ and $R_{1\cdot23}$.

12·11. (*a*) If X_1, X_2 and X_3 are three variables measured from their respective means as origin and if e_1 is the expected value of X_1 for given values of X_2 and X_3 from the linear regression of X_1 on X_2 and X_3, prove that $\text{Cov}(X_1, e_1) = \text{Var}(e_1) = \text{Var}(X_1) - \text{Var}(X_1 - e_1)$.

(*b*) If $r_{12} = k$ and $r_{23} = -k$, show that r_{13} will lie between -1 and $1 - 2k^2$.

12·12. If *R* denotes the multiple correlation co-efficient of X_1 on $X_2, X_3, \ldots, X_p$ in *p*-variate distribution, prove that

(*i*) $R^2 \geq R_0^2$, where R_0 is the correlation of X_1 with any arbitrary linear function of $X_2, X_3, \ldots, X_p$.

(*ii*) $R^2 \geq R_1^2$, where R_1 is the multiple correlation coefficient of X_1 with $X_2, X_3, \ldots, X_k$, $k < p$.

(*iii*) $1 - R^2 = \prod_{j=2}^{p}\left[1 - r^2_{1j\cdot234\ldots(j-1)}\right]$.

12·13. Let the maximum correlation between X_0 and any linear function of $X_1, X_2, \ldots, X_n$ be *R* and if $r_{01} = r_{02} = \ldots = r_{0n} = r$ and all other correlation coefficients are equal to *s* then show that:

$$R = r\left[\frac{n}{1 + (n-1)s}\right]^{1/2}$$

Hint: Note that *R* is the multiple correlation coefficient between X_0 and $(X_1, X_2, \ldots, X_{ni})$.

$$R^2 = 1 - \frac{\omega}{\omega_{11}}, \text{ where } \omega = \left|r_{ij}\right|_{(n+1)\times(n+1)} \;;\; (i, j) = 0, 1, 2, \ldots, n$$

12·14. If all the inter-correlations between $(p + 1)$ variates $X_0, X_1, X_2, \ldots, X_p$ are equal to *r*, show that each of the partial correlation co-efficients of order $p - 1$ is equal to $r/[1 + (p-1)r]$ and that the multiple correlation of X_0 on $X_1, X_2, \ldots, X_p$ is given by:

$$1 - R^2_{0\cdot(12\,\ldots\,p)} = \frac{(1-r)(1+pr)}{1+(p-1)r}.$$

Hint: Proceed as in Example 12·9.

12·15. X_1, X_2, X_3 have joint distribution as a multinomial distribution with parameters N, p_1, p_2, p_3. If r_{ij} is the correlation coefficient between X_i and X_j, find the expression for r_{12}, r_{23} and r_{31} and hence deduce the expression for the partial correlation coefficient $r_{12\cdot3}$.

$$\textbf{Hint: } r(X_i, X_j) = -\left[\frac{p_i\, p_j}{(1-p_i)(1-p_j)}\right]^{1/2} \;; i \neq j = 1, 2, 3.$$

12·16. Prove that the necessary and sufficient condition that the three planes of regression in case of trivariate distribution coincide is:

$$r_{12}^2 + r_{23}^2 + r_{31}^2 - 2r_{12}\, r_{13}\, r_{23} = 1.$$

CHAPTER

13 Theory of Attributes

LEARNING OBJECTIVES

Upon completion of this Chapter, you should be able to:

- Differentiate between variables and attributes.
- Understand the concept of class frequencies of different orders and obtain the relationship between them
- Understand the concept of consistency of data and examine the conditions for the consistency of given set of data.
- Understand criteria for the independence of attributes.
- Demonstrate the concept of association of attributes and obtain a quantitative measure of association between two attributes. (Yule's Coefficient of Association).
- Establish relationship between Yule's Coefficient of Association and Coefficient of Colligation.

CHAPTER OUTLINES

13·1. INTRODUCTION

Literally, an attribute means a *quality* or *characteristic.* Theory of attributes deals with qualitative characteristics which are not amenable to quantitative measurements and hence need slightly different statistical treatment from that of the variables. Examples of attributes are drinking, smoking, blindness, health, honesty, etc. An attribute may be marked by its presence (possession) or absence (dispossession) in a member of given population. It may be pointed out that the methods of statistical analysis applicable to the study of variables can also be used to a great extent in the theory of attributes and *vice-versa.* For example, the presence or absence of an attribute may be regarded as changes in the values of a variable which can possess only two values, *viz.,* 0 and 1. For the sake of clarity and simplicity, the theory of attributes has been developed independently.

In the study of attributes, if the universe (population) is divided into two sub-classes or complementary classes and no more, with respect to each of the attributes A, B, C, etc., the division or classification is said to be *'dichotomous classification'*. The classification is termed manifold if each class is further sub-divided.

13·2. NOTATIONS AND TERMINOLOGY

Suppose the population is divided into two classes according to the presence or absence of a single attribute. The *positive class* which denotes the presence of the attribute is generally written in capital Roman letters such as A, B, C, D, etc. and the *negative class,* denoting the absence of the attribute is written in corresponding small Greek letters such as α, β, γ, δ, etc. For example, if A represents the attribute sickness and B represents blindness, than α and β represent the attributes non-sickness (health) and non-blindness (sight), respectively. The two classes, *viz.,* A (possession of the attribute) and α (dispossession of the attribute) are said to be *complementary classes* and the attribute α used in the sense of not-A is often called the complementary attribute of A. Similarly β, γ, δ are the complementary attributes of B, C, D respectively.

The combinations of attributes are denoted by grouping together the letters concerned, *e.g.,* AB is the combination of the attributes A and B. Thus, for the attributes A (sickness) and B (smoking), AB would mean the simultaneous possession of sickness and smoking. Similarly $A\beta$ will represent sickness and non-smoking, αB non-sickness (health) and smoking, and $\alpha\beta$ non-sickness and non-smoking.

If a third attribute be included to represent, say male, then ABC will stand for sick males who are smokers. Similar interpretations can be given to $AB\gamma$, $A\beta C$, $A\beta\gamma$, etc.

13·3. CLASSES AND CLASS FREQUENCIES

Different attributes in themselves are called different classes and the number of observations assigned to them are called *class frequencies* which are denoted by bracketing the class symbols. Thus (A) stands for the frequency of A and (AB) for the number of objects possessing the attribute AB.

The Class frequencies of the type (A), (B), (AC), (BC), (ABC) etc. which involve only positive attributes, are called positive frequencies. The class frequencies of the type (α), (β), $(\alpha\gamma)$, $(\beta\gamma)$, $(\alpha\beta\gamma)$, etc., which involve only negative attributes, are called *negative frequencies.* The class frequencies of the type $(A\beta)$, (αB), $(AB\gamma)$, (αBC), etc., which involve the mixture of positive and negative attributes are called *contrary frequencies.*

13·3·1. Order of Classes and Class Frequencies.

A class represented by n attributes is called a class of nth order and the corresponding frequency as the frequency of the nth order. Thus (A) is a class frequency of order 1; (AB), (AC), $(\beta\gamma)$ etc. are class frequencies of second order; (ABC), $(A\beta\gamma)$, $(\alpha\beta C)$, etc. are frequencies of third order, and so on. N, the total number of members of the population, without any specification of attributes, is reckoned as frequency of zero-order.

Thus, in a dichotomous classification with respect to n attributes, the number of class frequencies of order 'r' is ${}^nC_r . 2^r$, since r attributes out of n can be selected in ${}^nC_r$ ways and each of the r attributes contributes two symbols, one representing the positive part (*e.g.*, A) and the other the negative part (*e.g.*, α). Thus the total number of class frequencies of all orders, for n attributes is:

$$\sum_{r=0}^{n} {}^nC_r 2^r = 1 + {}^nC_1\, 2 + {}^nC_2\, 2^2 + \ldots + {}^nC_n\, 2^n = (1+2)^n = 3^n \qquad \ldots (13\cdot1)$$

Remarks:

1. In particular, for n attributes, the total number of class frequencies of different orders are given as follows:

Order	0	1	2	...	r	...	n
No. of. frequencies	1	$2n$	${}^nC_2\, 2^2$	...	${}^nC_r\, 2^r$	...	2^n

2. Since in the case of n attributes, the positive class frequency of order r has ${}^nC_r$ elements, their total number is:

$$\sum_{r=0}^{n} {}^nC_r = {}^nC_0 + {}^nC_1 + \ldots + {}^nC_n = (1+1)^n = 2^n$$

3. In case of 3 attributes A, B and C, the total number of class frequencies is $3^3 = 27$, as given below:

Order	*Frequencies*				
0	N				
1.	(A), (α)	(B), (β)	(C), (γ)		... (13·2)
2.	(AB), (AC), (BC)	$(A\beta)$, $(A\gamma)$, $(B\gamma)$	(αB), (αC), (βC)	$(\alpha\beta)$, $(\alpha\gamma)$, $(\beta\gamma)$	... (13·2*a*)
3.	(ABC), (αBC)	$(AB\gamma)$, $(\alpha B\gamma)$	$(A\beta C)$, $(\alpha\beta C)$	$(A\beta\gamma)$, $(\alpha\beta\gamma)$	... (13·2*b*)

13·3·2. Relation Between Class Frequencies.

All the class frequencies of various orders are not independent of each other and any class frequency can always be expressed in terms of class frequencies of higher order. Thus

$$N = (A) + (\alpha) = (B) + (\beta) = (C) + (\gamma), \text{ etc.}$$

Also, since each of these A's or α' s can either be B's or β's, we have

$$(A) = (AB) + (A\beta) \quad \text{and} \quad (\alpha) = (\alpha B) + (\alpha\beta)$$

Similarly,

$$(B) = (AB) + (\alpha B) \quad \text{and} \quad (\beta) = (A\beta) + (\alpha\beta)$$

$$N = (A) + (\alpha) = (AB) + (A\beta) + (\alpha B) + (\alpha\beta) \qquad \ldots (13\cdot3)$$

Thus, in the case of two attributes, all the class frequencies can be expressed in terms of the ultimate class frequencies.

Also $(C) = (AC) + (\alpha C) = (ABC) = (A\beta C) + (\alpha BC) + (\alpha\beta C)$ ... (13·3a)

$$(C) = (BC) + (\beta C) = (ABC) + (\alpha BC) + (A\beta C) + (\alpha\beta C)$$

$$\left.\begin{aligned}(AB) &= (ABC) + (AB\gamma) \quad ; \quad (A\beta) = (A\beta C) + (A\beta\gamma)\\ (\alpha B) &= (\alpha BC) + (\alpha B\gamma) \quad ; \quad (\alpha\beta) = (\alpha\beta C) + (\alpha\beta\gamma)\end{aligned}\right\} \quad \text{... (13·3b)}$$

Substituting in (13·3), we get

$$N = (ABC) + (AB\gamma) + (A\beta C) + (A\beta\gamma) + (\alpha BC) + (\alpha B\gamma) + (\alpha\beta C) + (\alpha\beta\gamma) \quad \text{... (13·4)}$$

$$\left.\begin{aligned}(A) &= (ABC) + (AB\gamma) + (A\beta C) + (A\beta\gamma)\\ (B) &= (ABC) + (AB\gamma) + (\alpha BC) + (\alpha B\gamma)\end{aligned}\right\} \quad \text{... (13·4a)}$$

Similarly, we can express all other class frequencies in terms of the class frequencies of the third order.

The above discussion leads to the following important result.

"*Any class frequency can be expressed as the sum of some of the* 2^n ultimate class frequencies."

Remark. The set of ultimate class frequencies is not the only set which specify the data completely. In fact any set of class frequencies which are (*i*) 2^n in number and (*ii*) which are algebraically independent of each other, will specify the data completely. Such a set is called the *Fundamental Set*. For example, the positive class frequencies form such a set. Thus for $n = 2$, the set of positive class frequencies $2^2 = 4$, is N, (A), (B), (AB). If we are given these frequencies, then it is obvious from the table that the remaining frequencies, *viz.*, $(A\beta)$, $(\alpha\beta)$, (αB), (α) and (β) can be obtained by subtraction, as given below:

	A	α	Total
B	(AB)	–	(B)
β	–	–	(β)
Total	(A)	(α)	N

$(\alpha) = N - (A)$, $\quad (\beta) = N - (B)$

$(A\beta) = (A) - (AB)$, $\quad (\alpha B) = (B) - (AB)$

$(\alpha\beta) = (\alpha) - (\alpha B) = N - (A) - (B) + (AB)$

Example 13·1. *From the following set of positive frequencies, obtain all other class frequencies.*

$$N = 500, (AB = 140, (A) = 160, (B) = 200$$

Solution. We have:

$$(\alpha) = N - (A) = 500 - 160 = 340; \qquad (\beta) = N - (B) = 500 - 200 = 300$$

$$(A\beta) = (A) - (AB) = 160 - 140 = 20; \qquad (\alpha B) = (B) - (AB) = 200 - 140 = 60$$

$$(\alpha\beta) = (\alpha) - (\alpha B) = N - (A) - (B) + (AB) \qquad = 500 - 160 - 200 + 140 = 280$$

The remaining class frequencies can also be very conveniently found out on completing the nine square table, as given below.

Attribute	A	α	*Total*
B	$(AB) = $ **140**	$(\alpha B) = 240 - 140 = 60$	$(B) = $ **200**
β	$(A\beta) = 160 - 140 = 20$	$(\alpha\beta) = 300 - 20 = 280$	$(\beta) = 500 - 200 = 300$
Total	$(A) = $ **160**	$(\alpha) = 280 + 60 = 340$	$N = $ **500**

The figures in the bold are the given frequencies. The other frequencies are obtained on appropriate subtraction.

Example 13·2.

(a) *Given the following ultimate class frequencies, find the frequencies of positive class.*

$(ABC) = 149 \quad (AB\gamma) = 738, \quad (A\beta C) = 225, \quad (A\beta\gamma) = 1{,}196$

$(\alpha BC) = 204, \quad (\alpha B\gamma) = 1{,}762, \quad (\alpha\beta C) = 171$ *and* $(\alpha\beta\gamma) = 21{,}842$

(b) *Find the remaining class frequencies, given the follwoing data:*

$N = 23{,}713, \quad (A) = 1{,}618, \quad (B) = 2{,}015, \quad (C) = 770$

$(AB) = 587, \quad (AC) = 428 \quad (BC) = 335 \quad (ABC) = 156$

Solution.

(a) $(A) = (ABC) + (AB\gamma) + (A\beta C) + (A\beta\gamma) = 2{,}308$

$(B) = (ABC) + (AB\gamma) + (\alpha BC) + (\alpha B\gamma) = 2{,}853$

$(C) = (ABC) + (A\beta C) + (\alpha BC) + (\alpha\beta C) = 749$

$(AB) = (ABC) + (A\beta\gamma) = 887$

$(AC) = (ABC) + (A\beta C) = 374$

$(BC) = (ABC) + (\alpha BC) = 353$

and $N = \left[(ABC) + (AB\gamma) + (A\beta C) + (A\beta\gamma) + (\alpha BC) + (\alpha B\gamma) + (\alpha\beta C) + (\alpha\beta\gamma)\right] = 26{,}287$

(b) For three attributes, there are $3^3 = 27$, class frequencies in all. Thus we have to determine the remaining 19 class frequencies:

Order 1:

$(\alpha) = N - (A) = 22{,}095; \quad (\beta) = N - (B) = 21{,}698 \quad (\gamma) = N - (C) = 22{,}943$

Order 2:

$(A\beta) = (A) - (AB) = 1{,}031$

$(\alpha B) = (B) - (AB) = 1{,}428$

$(\alpha\beta) = (\alpha) - (\alpha B) = 20{,}667$

$(A\gamma) = (A) - (AC) = 1{,}190$

$(\alpha C) = (C) - (AC) = 342$

$(\alpha\gamma) = (\alpha) - (\alpha C) = 21{,}753$

$(B\gamma) = (B) - (BC) = 1{,}680$

$(\beta C) = (C) - (BC) = 435$

$(\beta\gamma) = (\beta) - (\beta C) = 21{,}263$

Order 3:

$(AB\gamma) = (AB) - (ABC) = 431$

$(A\beta C) = (AC) - (ABC) = 272$

$(A\beta\gamma) = (A\beta) - (A\beta C) = 759$

$(\alpha BC) = (BC) - (ABC) = 179$

$(\alpha B\gamma) = (\alpha B) - (\alpha BC) = 1{,}249$

$(\alpha\beta C) = (\beta C) - (A\beta C) = 163$

$(\alpha\beta\gamma) = (\alpha\beta) - (\alpha\beta C) = 20{,}504$

Example 13·3. *An investigation of 23,713 households was made in an urban and rual mixed locality. Of these 1,618 were farmers, 2,015 well-to-do and 770 families were having at least one graduate. Of these graduate families 335 were those of farmers and 428 were well-to-do, also 587 well-to-do families were those of farmers and out of them only 156 were having at least one of their family member as graduate. Obtain all the ultimate class frequencies.*

Solution. Let the attribute 'farming' be denoted by A, the attribute 'well-to-do' by B and 'having at least one graduate' by C. Then in the usual notations,

N	(A)	(B)	(C)
23,713	1,618	2,015	770
(AB)	(BC)	(AC)	(ABC)
587	428	335	156

For three attributes A, B, C the number of ultimate class frequencies is $2^3 = 8$, one of them being $(ABC) = 156$. The remaining frequencies are obtained below:

$$
\begin{aligned}
(AB\gamma) &= (AB) - (ABC) = 587 - 156 &&= 431 \\
(A\beta C) &= (AC) - (ABC) = 335 - 156 &&= 179 \\
(A\beta\gamma) &= (A) - (AB) - (AC) + (ABC) &&= 1{,}618 - 587 - 335 + 156 = 852 \\
(\alpha BC) &= (BC) - (ABC) = 428 - 156 &&= 272 \\
(\alpha B\gamma) &= (B) - (AB) - (BC) + (ABC) &&= 2{,}015 - 587 - 428 + 156 = 1{,}156 \\
(\alpha\beta C) &= (C) - (AC) - (BC) + (ABC) &&= 770 - 335 - 428 + 156 = 163 \\
(\alpha\beta\gamma) &= N - (A) - (B) - (C) + (AB) + (AC) + (BC) - (ABC) \\
&= 23{,}713 - 1{,}618 - 2{,}015 - 770 + 587 + 335 + 428 - 156 = 20{,}504.
\end{aligned}
$$

13·4. CONSISTENCY OF DATA

Any class frequencies which have been or might have been observed within one and the same population are said to be consistent if they conform with one another and do not in any way conflict. For example, the figures $(A) = 20$, $(AB) = 25$ are inconsistent as (AB) cannot be greater than (A), if they are observed from the same population.

'Consistency' of a set of class frequencies may be defined as the property that none of them is negative, otherwise, the data for class frequencies are said to be *'inconsistent'*.

Since any class frequency can be expressed as the sum of some of the ultimate class frequencies, it is necessarily non-negative if all the ultimate class frequencies are non-negative. This provides a criterion for testing the consistency of the data. In fact, we have the following theorem:

Theorem 13·1. *"The necessary and sufficient condition for the consistency of a set of independent class frequencies is that no ultimate class frequency is negative."*

Remark. We can test the consistency of a set of 2^n algebraically independent class frequencies by calculating the ultimate class frequencies. If any one of them is negative , the given data are inconsistent.

13·4·1. Conditions for Consistency of Data.

Criteria for consistency of class frequencies are obtained by using theorem 13·1. For a single attribute A we have conditions of consistency as follows:

$$
\left.
\begin{aligned}
&(i)\ (A) \geq 0 \\
&(ii)\ (\alpha) \geq 0 \Rightarrow (A) \leq N
\end{aligned}
\right\} \quad \text{... (13·5)}
$$

For two attributes A and B, the conditions of consistency are:

$$
\left.
\begin{aligned}
&(i)\ (AB) \geq 0 \\
&(ii)\ (A\beta) \geq 0 \Rightarrow (A) - (AB) \geq 0 \Rightarrow (AB) \leq (A) \\
&(iii)\ (\alpha B) \geq 0 \Rightarrow (B) - (AB) \geq 0 \Rightarrow (AB) \leq (B) \\
&(iv)\ (\alpha\beta) \geq 0 \Rightarrow N + (AB) - (A) - (B) \geq 0 \Rightarrow (AB) \geq (A) + (B) - N
\end{aligned}
\right\} \quad \text{... (13·6)}
$$

Conditions of consistency for three attributes A, B and C are:

$$
\left.
\begin{aligned}
&(i)\ (ABC) \geq 0 \\
&(ii)\ (AB\gamma) \geq 0 \Rightarrow (AB) - (ABC) \geq 0 \Rightarrow (ABC) \leq (AB) \\
&(iii)\ (A\beta C) \geq 0 \Rightarrow (AC) - (ABC) \geq 0 \Rightarrow (ABC) \leq (AC) \\
&(iv)\ (\alpha BC) \geq 0 \Rightarrow (BC) - (ABC) \geq 0 \Rightarrow (ABC) \leq (BC) \\
&(v)\ (A\beta\gamma) \geq 0 \Rightarrow (A) - (AB) - (AC) + (ABC) \geq 0 \\
&\qquad\qquad \Rightarrow (ABC) \geq (AB) + (AC) - (A) \\
&(vi)\ (\alpha B\gamma) \geq 0 \Rightarrow (ABC) \geq (AB) + (BC) - (B) \\
&(vii)\ (\alpha\beta C) \geq 0 \Rightarrow (ABC) \geq (AC) + (BC) - (C) \\
&(viii)\ (\alpha\beta\gamma) \geq 0 \Rightarrow (ABC) \geq (AB) + (BC) + (AC) - (A) - (B) - (C) + N
\end{aligned}
\right\} \quad \text{... (13·7)}
$$

(*i*) and (*viii*) in (13·7) give:

$$(AB)+(BC)+(AC) \leq (A)+(B)+(C)-(N)$$

Similarly,

$$\left.\begin{array}{lll} (ii) \text{ and } (vii) & \Rightarrow & (AC)+(BC)-(AB) \leq (C) \\ (iii) \text{ and } (vi) & \Rightarrow & (AB)+(BC)-(AC) \leq (B) \\ (iv) \text{ and } (v) & \Rightarrow & (AB)+(AC)-(BC) \leq (A) \end{array}\right\}$$

Remark. As already pointed out [*c.f.* Remarks (3) and (4), § 13·4·2)], 2^n algebraically independent class frequencies are necessary to specify the data completely, one such set being the set of ultimate class frequencies and the other being the set of positive class frequencies. If the data supplied are incomplete so that it is not possible to determine all the class frequencies, then the conditions (13·5), (13·6) and (13·8) for one, two and three attributes respectively, enable us to assign the limits within which an unknown class freqeuency can lie.

Example 13·4. *Examine the consistency of the following data:*
N = 4000, (A) = 3000, (B) = 200 and (AB) = 700, the symbols having their usual meaning.

Solution. To test the consistency of the data, we will find the ultimate class frequencies *viz.*, (αB), $(A\beta)$ and $(\alpha\beta)$.

$$(\alpha B)=(B)-(AB)=100-350=-250.$$

Since $(\alpha B) < 0$ *i.e.*, negative, the given data are inconsistent.

Example 13·5. *In a report on consumer preference, it was given that out of 1000 persons surveyed, 600 preferred variety A, 500 preferred variety B and 50 liked both A and B. Are the data consistent?*

Solution. In the usual notations, we are given: $N = 1000$, $(A) = 600$, $(B) = 500$, $(AB) = 50$

$$(\alpha\beta)=N-(A)-(B)+(AB)=1000-600-500+50=-50$$

Since each frequency has to be positive, the given data are inconsistent.

Example 13·6. *In a series of houses actually invaded by smallpox, 70% of the inhabitants are attacked and 85% have been vaccinated. What is the lowest percentage of the vaccinated that must have been attacked?*

Solution. Let *A* and *B* denote the attributes of the inhabitants being attacked and vaccinated respectively. Then we are given $N = 100$, $(A) = 70$ and $(B) = 85$.

Consistency condition gives: $(AB) \geq (A)+(B)-N \Rightarrow (AB) \geq 55$

Hence, the lowest percentage of inhabitants vaccinated, who have been attacked is given by:

$$\frac{(AB)}{(B)} \times 100 = \frac{55}{85} \times 100 = 64{\cdot}7\ \%.$$

Example 13·7. *Show that:*

(*i*) *If all A's are B's and all B's are C's, then all A's are C's,*

(*ii*) *If all A's are B's and no B's are C's then no A's are C's.*

Solution.

(*i*) All *A*'s are *B*'s $\Rightarrow$ $(AB) = (A)$ and all *B*'s are *C*'s $\Rightarrow$ $(BC) = (B)$ ... (*)

To prove $(AC) = (A)$

We have $(AB)+(BC)-(AC) \leq (B)$

$\Rightarrow$ $(A)+(B)-(AC) \leq (B)$ [Using (*)]

$\Rightarrow$ $(A) \leq (AC)$ $\Rightarrow$ $(AC) \geq (A)$

But since $(AC) \not> (A)$, we have $(AC) = (A)$, as desired.

(ii) We are given $(AB) = (A)$ and $(BC) = 0$ and we want to prove $(AC) = 0$.

We have $(AB) + (AC) - (BC) \le (A) \Rightarrow (A) + (AC) - 0 \le (A) \Rightarrow (AC) \le 0$

And since $(AC) \ge 0$, we must have $(AC) = 0$.

13·5. INDEPENDENCE OF ATTRIBUTES

Two attributes are said to be independent if there does not exist any kind of relationship between them.

13·5·1. Criterion of Independence of Two Attributes

First Criterion (Proportion Method). In this method, the presence or absence of given attribute in the other attribute is compared. If two attributes A and B are independent, then we would expect:

(a) the same proportion of A's amongst B's as amongst β's, and

(b) the proportion of B's amongst A's is same as that amongst the α's.

For example, if the attributes 'Insanity' and 'Deafness' are independent, then the proportion of deaf persons among insane and sane persons must be same.

Symbolically, if attributes A and B are independent, the criterion in (a) and (b) imply that:

$$\frac{(AB)}{(B)} = \frac{(A\beta)}{(\beta)} \quad \text{... (13·9)}$$

$$\Rightarrow \quad 1 - \frac{(AB)}{(B)} = 1 - \frac{(A\beta)}{(\beta)}$$

$$\Rightarrow \quad \frac{(B)-(AB)}{(B)} = \frac{(\beta)-(A\beta)}{\beta}$$

$$\Rightarrow \quad \frac{(\alpha B)}{(B)} = \frac{(\alpha\beta)}{(\beta)} \quad \text{... (13·9a)}$$

$$\frac{(AB)}{(A)} = \frac{(\alpha B)}{(\alpha)} \quad \text{... (13·10)}$$

$$\Rightarrow \quad 1 - \frac{(AB)}{(A)} = 1 - \frac{(\alpha B)}{(\alpha)}$$

$$\Rightarrow \quad \frac{(A)-(AB)}{(A)} = \frac{(\alpha)-(\alpha B)}{(\alpha)}$$

$$\Rightarrow \quad \frac{(A\beta)}{(A)} = \frac{(\alpha\beta)}{\alpha} \quad \text{... (13·10a)}$$

In fact (13·9) $\Rightarrow$ (13·10) *i.e.*, Criterion (*a*) $\Rightarrow$ Criterion (*b*), as explained below:

For example, (13·9) gives

$$\frac{(AB)}{(B)} = \frac{(AB)}{\beta} = \frac{(AB)+(A\beta)}{(B)+(\beta)} = \frac{(A)}{(N)} \quad \Rightarrow \quad \frac{(AB)}{(A)} = \frac{(B)}{N} = \frac{(B)-(AB)}{N-(A)} = \frac{(\alpha B)}{(\alpha)} \quad \text{...(13·10b)}$$

which is (13·10). Similarly, starting from (13·10), we would arrive at (13·9).

It becomes easier to grasp the nature of the above relations if the frequencies are supposed to be grouped into a table with two rows and two columns as follows:

Attributes	*A*	α	*Total*
B	(AB)	(αB)	(B)
β	(Aβ)	($\alpha\beta$)	(β)
Total	(A)	(α)	N

Second Criterion of independence may be obtained in terms of the class frequencies of first order. Rewriting (13·10*b*), we get

$$(AB) = \frac{(A)\times(B)}{N} \quad \Rightarrow \quad \frac{(AB)}{N} = \frac{(A)}{N}\cdot\frac{(B)}{N}, \quad \text{... (13·11)}$$

which leads to the following important fundamental rule:

> *"If the attributes A and B are independent, the proportion of AB's in the population is equal to the product of the proportions of A's and B's in the popualtion."*

Third Criterion of independence of two attributes is based on the class frequencies of second order.

If attributes A and B are independent, then by Criterion 1, we have

$$\frac{(AB)}{(B)} = \frac{(A\beta)}{(\beta)} \quad \text{and} \quad \frac{(\alpha B)}{(B)} = \frac{(\alpha\beta)}{(\beta)} \qquad \text{[From (13·9) and (13·9}a\text{)]}$$

Dividing these equations, we get

$$\frac{(AB)}{(\alpha B)} = \frac{(A\beta)}{(\alpha\beta)} \quad \Rightarrow \quad (AB)\cdot(\alpha\beta) = (A\beta)\cdot(\alpha B) \qquad \text{... (13·12)}$$

Aliter. *If A and B are independent, then (A and β), (α and B), and (α and β) are also independent. Hence, using the Criterion 2 in (13·11), we get*

$$(AB)\cdot(\alpha\beta) = \frac{(A)\times(B)}{N}\cdot\frac{(\alpha)\times(\beta)}{N} = \frac{(A)\times(\beta)}{N}\cdot\frac{(\alpha)\times(B)}{N} = (A\beta)\cdot(\alpha B)$$

a result obtained in (13·12).

Symbol $(AB)_0$ and (δ). Let $(AB)_0$ denote the expected value of the classe frequency (AB) under the hypothesis that the attributes A and B are independent so that

$$(AB)_0 = \frac{(A)\cdot(B)}{N} \qquad \text{... (13·13)}$$

Let
$$\delta = (AB) - (AB)_0 \qquad \text{... (13·14)}$$

denote the excess of (AB) over $(AB)_0$. Then

$$\delta = (AB) - \frac{(A)(B)}{N} = \frac{1}{N}[N(AB) - (A)(B)]$$

$$= \frac{1}{N}\Big[\{(AB) + (A\beta) + (\alpha B) + (\alpha\beta)\}\cdot(AB) - \{(AB) + (A\beta)\}\cdot\{(AB) + (\alpha B)\}\Big]$$

$$= \frac{1}{N}\Big[(AB)(\alpha\beta) - (A\beta)(\alpha B)\Big] \qquad \text{[On simplification]} \qquad \text{... (13·14}a\text{)}$$

Hence, on using (13·12), we get,

δ = 0, if and only if the attributes A and B are independent.

Example 13·8. *In an analysis of two attributes, if N = 1000, (A) = 100 and (B) = 300 find the frequencies of the remaining classes on the assumption that A and B are independent.*

Solution. Since the attributes A and B are independent (given),

$$(AB) = \frac{(A)\times(B)}{N} = \frac{100\times 300}{1000} = 30$$

The remaining frequencies can now be obtained on completing the nine-square table as given in the table.

	A	α	*Total*
B	30	300 − 30 = 270	(B) = **300**
β	100 − 30 = 70	700 − 70 = 630	1000 − 300 = 700
Total	(A) = **100**	1000 − 100 = 900	N = **1000**

∴ (α) = 900; (β) = 700, (αB) = 270, (Aβ) = 70, (αβ) = 630. [From Table].

13·6. ASSOCIATION OF ATTRIBUTES

Two attributes A and B are said to be associated if they are not independent but are related in some way or the other. They are said to be

$$\left.\begin{array}{llll} \textit{Positively associated if} & (AB) > \dfrac{(A)(B)}{N} & \textit{or} & \delta > 0 \\ \textit{Negatively associated if} & (AB) < \dfrac{(A)(B)}{N} & \textit{or} & \delta < 0 \\ \textit{and independent if} & (AB) = \dfrac{(A)(B)}{N} & \textit{or} & \delta = 0 \end{array}\right\} \quad \text{... (13·15)}$$

Similar expressions can be obtained for other ultimate class frequencies such as $(A\beta)$, (αB), and $(\alpha\beta)$.

Remarks:

1. Two attributes A and B are said to be *completely asosciated* if A cannot occur without B, though B may occur without A and *vice-versa*. In other words, for complete association either all A's are B's, *i.e.*, $(AB) = (A)$ or all B's are A's, *i.e.*, $(AB) = (B)$ according as either A's or B's are in a minority. Similarly, *complete dissociation* means that no A's are B's, *i.e.*, $(AB) = 0$ or no α's are β's, *i.e.*, $(\alpha\beta) = 0$ or more generally when either of these statements is true.
2. It should be carefully noted that the word 'association' used in Statistics is technically different from the general notion of association as used in day-to-day life. Ordinarily, two attributes are said to be associated if they occur together in a number of cases. But statistically two attributes are said to be associated if they occur together in a large number of cases than expected if they were independent, *i.e.*, if $\delta = (AB) - (A)(B) / N > 0$. In Statistics, the statement that "some A's are B's", however great the proportion, does not necessarily imply association between them. Thus to find out if two attributes are associated, we must know (A), (B), (AB) and N. Incomplete information will not enable us to conclude anything about association between them. For example, consider the following statment:

   "90 per cent of the people who drink alcohol die before reaching the age of 75 years. Hence drinking is bad for longevity of life."

   The inference drawn is not correct, since the given information is not complete for drawing any valid conclusions about association. It might happen that 95% of the people who do not drink alcohol, die before reaching 75 years of age. In that case drinking might be found good for longevity of life.
3. *Sampling fluctuations.* If $\delta \neq 0$ and its value is fairly small, then it is possible that this association is just by chance (or commonly termed as due to flutuations of sampling) and not really significant of any real association between the attributes. We should not, therefore, draw hasty conclusions about association or dissociation unless δ, the difference between (AB) and its expected value (under the hypothesis of independence) $(A)(B)/N$, is significant. The problem: 'how much difference is to be regarded as significant' will be discussed in detail in Chapters 14 (Large sample test for attributes) and 15 (Chi-square test of goodness of fit). This point has been raised here only as a precautionary measure to warn the reader against drawing hasty inferences.

13·6·1. Yule's Coefficient of Association.

As a measure of the intensity of association between two attributes A and B, G. Udny Yule gave the coefficient of association, denoted by Q (based on all the ultimate class frequencies), defined as follows:

$$Q = \frac{(AB)(\alpha\beta) - (A\beta)(\alpha B)}{(AB)(\alpha\beta) + (A\beta)(\alpha B)} = \frac{N\delta}{(AB)(\alpha\beta) + (A\beta)(\alpha B)} \quad \text{[From 13·14a)] ... (13·16)}$$

Limits for Q. Let : $(AB)(\alpha\beta) = a$ and $(A\beta)(\alpha B) = b$. Then, obviously $a > 0$ and $b > 0$, so that

$$|a-b| \le |a+b| \quad \Rightarrow \quad \left|\frac{a-b}{a+b}\right| \le 1 \qquad \text{... (*)}$$

$$\therefore \quad Q = \frac{a-b}{a+b} \quad \Rightarrow \quad |Q| = \left|\frac{a-b}{a+b}\right| \le 1 \qquad \text{[From (*)]... (13·17)}$$

$$\Rightarrow \qquad -1 \le Q \le 1$$

Hence, Yule's coefficient of association lies between –1 and 1, both inclusive.

(i) $Q = 1$, implies that there is perfect positive association between A and B.

(ii) $Q = -1 \Rightarrow$ There is perfect negative association or complete dis-association between A and B.

(iii) $Q = 0 \Rightarrow A$ and B are independent.

Any other value of Q between –1 and 1 gives us an idea of the intensity of association between A and B. A value of $Q \ge 0{\cdot}5$ indicates a fairly good degree of positive association and a value of Q lying between 0 and 0·5 indicates a low degree of positive association between A and B. Similar interpretations can be given to negative values of Q.

Perfect Positive and Perfect Negative Association. Q = +1 *i.e.*, there is perfect positive association between the two attributes A and B if and only if

$$b = (A\beta)(\alpha B) = 0 \Rightarrow (A\beta) = 0 \text{ or } (\alpha B) = 0$$

$$(A\beta) = 0 \Rightarrow (AB) = (A) \qquad [\because (A) = (AB) + (A\beta)]$$

or

$$(\alpha B) = 0 \Rightarrow (AB) = (B) \qquad [\because (B) = (AB) + (\alpha B)]$$

Hence, A and B are completely associated or there is perfect positive association between A and B if and only if $(AB) = (A)$ *i.e.*, all A's are B's or $(AB) = (B)$ *i.e.*, all B's are A's, whichever is less.

Q = –1, *i.e.*, there is perfect negative association between the two attributes A and B if and only if,

$$a = (AB)\cdot(\alpha\beta) = 0 \quad \Rightarrow \quad (AB) = 0 \quad \text{or} \quad (\alpha\beta) = 0$$

Hence, A and B are completely dis-associated or there is perfect negative association between A and B, if and only if $(AB) = 0$ *i.e.*, none of the A's is B or $(\alpha\beta) = 0$ *i.e.*, none of the α's is β.

Another important property of Q is that it is independent of the relative proportion of A's or α's in the data. Thus if all the terms containing any attribute, say A (or B or α or β) in Q are multiplied by a constant, k (say), its value remains unaltered. This property renders it specially useful to situations where the proportions are arbitrary *i.e., experiments.*

13·6·2. Coefficient of Colligation.

This is another measure of association suggested by Professor Yule and has the same properties as Q. It is called the coefficient of Colligation and is denoted by Y and is given by:

$$Y = \left\{\left(1 - \sqrt{\frac{(A\beta)(\alpha B)}{(AB)(\alpha\beta)}}\right)\right\} \Big/ \left\{\left(1 + \sqrt{\frac{(A\beta)(\alpha B)}{(AB)(\alpha\beta)}}\right)\right\}. \qquad \text{... (13·18)}$$

Relation Between Q and Y.

$$\text{Let} \quad \frac{(A\beta)(\alpha B)}{(AB)(\alpha\beta)} = k, \quad \text{so that} \quad Y = \frac{1-\sqrt{k}}{1+\sqrt{k}} \quad \Rightarrow \quad Y^2 = \frac{1+k-2\sqrt{k}}{1+k+2\sqrt{k}}$$

$$\therefore \qquad 1 + Y^2 = \frac{1 + k + 2\sqrt{k} + (1 + k - 2\sqrt{k})}{1 + k + 2\sqrt{k}} = \frac{2(1 + k)}{(1 + \sqrt{k})^2}$$

and
$$\frac{2Y}{1 + Y^2} = \frac{2(1 - \sqrt{k})(1 + \sqrt{k})}{2(1 + k)}$$

$$= \frac{1 - k}{1 + k} = \frac{1 - \dfrac{(A\beta)(\alpha B)}{(AB)(\alpha\beta)}}{1 + \dfrac{(A\beta)(\alpha B)}{(AB)(\alpha\,\beta)}} = \frac{(AB)(\alpha\beta) - (A\beta)(\alpha B)}{(AB)(\alpha\beta) + (A\beta)(\alpha B)} = Q$$

Hence, $Q = \dfrac{2Y}{1 + Y^2}$. ... (13·19)

The relation (13·19) can be used to establish the limits for Y.

We know that, $|Q| \le 1 \Rightarrow -1 \le Q \le 1$

From (13·19), we get: Attributes are independent if $Q = 0 \quad \Rightarrow \quad Y = 0$

There is perfect (positive) association between A and B if:

$$Q = 1 \Rightarrow 2Y = 1 + Y^2 \Rightarrow 1 + Y^2 - 2Y = 0 \Rightarrow (1 - Y)^2 = 0 \Rightarrow Y = 1$$
$$Q = -1 \Rightarrow 2Y = -(1 + Y^2) \Rightarrow 1 + Y^2 + 2Y = 0 \Rightarrow (1 + Y)^2 = 0 \Rightarrow Y = -1$$

Remark. From the above discussion, it is obvious that both Q and Y possess the same properties. Since Yule's coefficient of association (Q) is much easier to compute, it is more commonly used in practice than the coefficient of colligation (Y).

Example 13·9. *Find if A and B are independent, positively associated or negatively associated, in each of the following cases:*

(i) $N = 1000$, $(A) = 470$, $(B) = 620$, and $(AB) = 320$.
(ii) $(A) = 490$, $(AB) = 294$, $(\alpha) = 570$, and $(\alpha B) = 380$.
(iii) $(AB) = 256$, $(\alpha B) = 768$, $(A\beta) = 48$, and $(\alpha\beta) = 144$.

Solution.

(i) $\delta = (AB) - \dfrac{(A)\,(B)}{N} = 320 - \dfrac{470 \times 620}{1{,}000} = 320 - 291{\cdot}4 = 28{\cdot}6$

Since $\delta > 0$, A and B are positively associated.

(ii) We have $N = (A) + (\alpha) = 490 + 570 = 1{,}060$

$(B) = (AB) + (\alpha B) = 294 + 380 = 674$

$$\therefore \qquad \delta = (AB) - \frac{(A)\,(B)}{N} = 294 - \frac{490 \times 674}{1060} = 294 - 311{\cdot}6 < 0$$

Hence, A and B are negatively associated.

(iii) $(A) = (AB) + (A\beta) = 256 + 48 = 304$

$(B) = (AB) + (\alpha B) = 256 + 768 = 1024$

$N = (AB) + (A\beta) + (\alpha B) + (\alpha\beta) = 256 + 48 + 768 + 144 = 1{,}216$

$$\therefore \qquad \delta = (AB) - \frac{(A)\,(B)}{N} = 256 - \frac{304 \times 1024}{1216} = 0$$

Hence, A and B are independent.

Example 13·10. *Investigate the association between darkness of eye-colour in fathers and sons from the following data:*

Fathers with dark eyes and sons with dark eyes:	*50*
Fathers with dark eyes and sons with not dark eyes:	*79*

Fathers with not dark eyes and sons with dark eyes: 89
Fathers with not dark eyes and sons with not dark eyes: 782
Also tabulate for comparison the frequencies that would have been observed had there been no heredity.

Solution. Let A: Dark eye-colour of father and B: Dark eye-colour of son.

In the usual notations, we are given: $(AB) = 50,\ (A\beta) = 79,\ (\alpha B) = 89,\ (\alpha\beta) = 782$

$$\therefore \qquad Q = \frac{50 \times 782 - 79 \times 89}{50 \times 782 + 79 \times 89} = \frac{32069}{46131} = +0 \cdot 69$$

Hence, there is a fairly high degree of positive association between the eye colour of fathers and sons. Also:

$$(A) = (AB) + (A\beta) = 50 + 79 = 129\ ; \qquad (B) = (AB) + (\alpha B) = 50 + 89 = 139,$$
$$(\alpha) = (\alpha B) + (\alpha\beta) = 89 + 782 = 871\ ; \qquad (\beta) = (A\beta) + (\alpha\beta) = 79 + 782 = 861,$$

and $$N = (A) + (\alpha) = 129 + 871 = 1{,}000$$

Under the condition of *no heredity, i.e.,* independence of attributes A and B, we have

$$(AB)_0 = \frac{(A)(B)}{N} = \frac{129 \times 139}{1000} = 18\ ; \qquad (A\beta)_0 = \frac{(A)(\beta)}{N} = \frac{129 \times 861}{1000} = 111$$

$$(\alpha B)_0 = \frac{(\alpha)(B)}{N} = \frac{871 \times 139}{1000} = 121\ ; \qquad (\alpha\beta)_0 = \frac{(\alpha)(B)}{N} = \frac{871 \times 861}{1000} = 750.$$

Example 13·11. *Can vaccination be regarded as a preventive measure for small-pox from the data given below?*

'Of 1,482 persons in a locality exposed to small-pox, 368 in all were attacked.'

'Of 1,482 persons, 343 had been vaccinated and of these only 35 were attacked.'

Solution. Let A denote the attribute of vaccination and B that of attack by small-pox. Then in the usual notations, we are given:

$$N = 1482, \quad (A) = 368, \quad (B) = 343 \quad \text{and} \quad (AB) = 35$$
$$(\alpha\beta) = N - (A) - (B) + (AB) = 1482 - 368 - 343 + 35 = 806$$
$$(A\beta) = (A) - (AB) = 368 - 35 = 333; \qquad (\alpha B) = (B) - (AB) = 343 - 35 = 308$$

$$\therefore \qquad Q = \frac{(AB)(\alpha\beta) - (A\beta)(\alpha B)}{(AB)(\alpha\beta) + (A\beta)(\alpha B)} = \frac{35 \times 806 - 333 \times 308}{35 \times 806 + 333 \times 308} = -0 \cdot 57$$

Thus, there is negative association between A and B, *i.e.,* between 'attacked' and 'vaccinated'. In other words, there is positive association between not attacked and vaccinated. Hence vaccination can be regarded as a preventive measure for small pox.

Example 13·12. *800 candidates of both sexes appeared at an examination. The boys outnumbered the girls by 15% of the total. The number of candidates who passed exceed the number failed by 480. Equal number of boys and girls failed in the examination. Prepare a 2 × 2 table and find the coefficient of association. Comment.*

Solution. Let us define the attributes:

A: Boys, α: girls; B: Passed in the examination, β: Failed in the examination

Then, we are given: $N = 800$ and $(A\beta) = (\alpha B) = \frac{1}{2} \times 160 = 80$

$(A) - (\alpha) = 15\%$ of $N = 15/100 \times 800 = 120$

Also $(A) + (\alpha) = N = 800$

$\therefore$ Adding and subtracting, we get respectively:

$$2(A) = 920 \quad \Rightarrow \quad (A) = \frac{920}{2} = 460$$

$$2(\alpha) = 680 \quad \Rightarrow \quad (\alpha) = \frac{680}{2} = 340$$

$(B) - (\beta) = 480$

Also $(B) + (\beta) = N = 800$

$\therefore$ Adding and subtracting, we get respectively:

$$2(B) = 1280 \quad \Rightarrow \quad (B) = \frac{1280}{2} = 640$$

$$2(\beta) = 320 \quad \Rightarrow \quad (\beta) = \frac{320}{2} = 160$$

We can now present the above information in the form of a 2 × 2 table as follows:

Attributes	A	α	*Total*
B	380	260	640
β	80	80	160
Total	460	340	800

Yule's coefficient of association is:

$$Q = \frac{(AB)(\alpha\beta) - (A\beta)(\alpha B)}{(AB)(\alpha\beta) + (A\beta)\cdot(\alpha B)}$$

$$= \frac{380 \times 80 - 80 \times 260}{380 \times 80 + 80 \times 260} = +0{\cdot}1875.$$

$$= \frac{30400 - 20800}{30400 + 20800} = \frac{9600}{51200}$$

Thus, the coefficient shows positive association of a very low degree between sex and success in the examination.

Example 13·13. *The following table gives the distribution of students and also of regular players among them, according to age in completed years:*

Age in years	:	15	16	17	18	19	20
No. of students	:	250	200	150	120	100	80
Regular players	:	200	150	90	48	30	12

Calculate the coefficient of association between majority and playing habit, on the assumption that majority is attained in 18th year.

Solution. Let 'A' denote 'major' (≥ 18 years) and 'α' 'minor' (15 to 17 years) student and 'B' denote regular players and 'β' not regular players.

From the given data, we have

No. of minor students : $(\alpha) = 250 + 200 + 150 = 600$

No. of major students : $(A) = 120 + 100 + 80 = 300$

No. of minor regular players : $(\alpha B) = 200 + 150 + 90 = 440$

No. of major regular players : $(AB) = 48 + 30 + 12 = 90$

$$(A\beta) = (A) - (AB) = 300 - 90 = 210 \quad \text{and} \quad (\alpha\beta) = (\alpha) - (\alpha B) = 600 - 440 = 160$$

The coefficient of association (Q) between majority 'A' and playing habits 'B' is given by:

$$Q = \frac{(AB)(\alpha\beta) - (A\beta)(\alpha B)}{(AB)(\alpha\beta) + (A\beta)(\alpha B)} = \frac{(90 \times 160) - (210 \times 440)}{(90 \times 160) + (210 \times 440)} = -0.73$$

Example 13·14. *A survey was conducted in respect of marital status and success in examination. Out of 2,000 persons who appeared for an examination, 80% of them were boys, and the rest were girls. Among 300 married boys, 140 were successful. 1100 boys were successful among unmarried boys. In respect of 100 married girls, 40 were successful. 200 unmarried girls were successful. Construct two separate nine square tables, and find out the Yule's coefficient of association to discuss the association between marital status and success in examination.*

Solution. Total number of persons who appeared in the examination = 2000

Number of boys 80% of 2000 = $\frac{80}{100} \times 2000 = 1600$ and Number of girls = 2000 – 1600 = 400

Let us define the attributes:

A: Married person ; *B* : Success in the examination

α: Unmarried perosn ; β : Failure in the examination

Then, in the usual notations, we are given:

For Boys: $N = 1600$, $(A) = 300$, $(AB) = 140$, $(\alpha B) = 1100$

For Girls: $N = 400$, $(A) = 100$, $(AB) = 40$, $(\alpha B) = 200$

The above data can be expressed in two separate nine square tables for boys and girls as given below:

NINE SQUARE TABLES

TABLE (*a*): FOR BOYS

Attributes	*A*	α	*Total*
B	**140**	**1100**	140 + 1100 = 1240
β	300 – 140 = 160	360 – 160 = 200	1600 – 1240 = 360
Total	**300**	1600 – 300 = 1300	**1600**

TABLE (*b*): FOR GIRLS

Attributes	*A*	α	*Total*
B	**40**	**200**	40 + 200 = 240
β	100 – 40 = 60	300 – 200 = 100	400 – 240 = 160
Total	**100**	400 – 100 = 300	**400**

In the above Tables, the frequencies in the bold type are the given valus. The remaining frequencies can be obtained by appropriate additions or subtractions, as explained in the tables.

Yule's coefficient of association (Q) is given by: $Q = \frac{(AB)(\alpha\beta) - (\alpha B)(A\beta)}{(AB)(\alpha\beta) + (\alpha B)(A\beta)}$

FOR BOYS

$$Q = \frac{140 \times 200 - 1100 \times 160}{140 \times 200 + 1100 \times 160}$$

$$= \frac{28000 - 176000}{28000 + 176000} = -\frac{148000}{204000}$$

$$= -0.725$$

FOR GIRLS

$$Q = \frac{40 \times 100 - 200 \times 60}{40 \times 100 + 200 \times 60}$$

$$= \frac{4000 - 12000}{4000 + 12000} = -\frac{8000}{16000}$$

$$= -0.5$$

Hence, in each case (for boys and for girls) there is a fairly good degree of negative association between marital status and success in the examination, and the degree of association is more in boys.

SUPPLEMENTARY EXAMPLES

Example 13·15. *Among the adult population of a certain town 50 per cent are males, 60 per cent are wage earners and 50 per cent are 45 years of age or over, 10 per cent of the males are not wage-earners and 40 per cent of the males are under 45. Make the best possible inference about the limits within which the percentage of persons (male or female) of 45 years or over are wage-earners.*

Solution. Let $N = 100$. Then denoting males by *A*, wage-earners by *B* and 45 years of age or over by *C*, we are given: $N = 100$, $(A) = 50$, $(B) = 60$, $(C) = 50$

$$(A\beta) = \frac{10}{100} \times 50 = 5, \qquad (A\gamma) = \frac{40}{100} \times 50 = 20$$

$$(AB) = (A) - (A\beta) = 45, \qquad (AC) = (A) - (A\gamma) = 30$$

We are required to find the limits for (BC).

Conditions of consistency (13·8) give

$$(i) \quad (AB) + (BC) + (AC) \geq (A) + (B) + (C) - N$$

$$\Rightarrow \quad (BC) \geq 50 + 60 + 50 - 100 - 45 - 30 = -15$$

$$(ii) \quad (AB) + (AC) - (BC) \leq A \quad \Rightarrow \quad (BC) \geq (AB) + (AC) - (A) = 45 + 30 - 50 = 25$$

$$(iii) \quad (AB) + (BC) - (AC) \leq (B) \quad \Rightarrow \quad (BC) \leq (B) + (AC) - (AB) = 60 + 30 - 45 = 45$$

$$(iv) \quad (AC) + (BC) - (AB) \leq (C) \quad \Rightarrow \quad (BC) \leq (C) + (AB) - (AC) = 50 + 45 - 30 = 65$$

$$(i) \text{ to } (iv) \quad \Rightarrow \quad 25 \leq (BC) \leq 45$$

Hence, the percentage of wage-earning population of 45 years or over must lie between 25 and 45.

Example 13·16 *Show that if*

$$\frac{(A)}{N} = x, \quad \frac{(B)}{N} = 2x, \quad \frac{(C)}{N} = 3x \quad \text{and} \quad \frac{(AB)}{N} = \frac{(BC)}{N} = \frac{(CA)}{N} = y,$$

then the value of neither x nor y can exceed 1/4.

Solution. We are given:

$$\frac{(A)}{N} = x, \quad \frac{(B)}{N} = 2x, \quad \frac{(C)}{N} = 3x \quad \Rightarrow \quad (A) = Nx, \quad (B) = 2Nx, \quad (C) = 3Nx \qquad \dots (i)$$

$$\frac{(AB)}{N} = \frac{(BC)}{N} = \frac{(CA)}{N} = y \quad \Rightarrow \quad (AB) = (BC) = (CA) = Ny \qquad \dots (ii)$$

By the condition of consistency, we have

$$(AB) \leq (A) \quad \Rightarrow \quad Ny \leq Nx \quad \Rightarrow \quad y \leq x \qquad \text{[From } (i) \text{ and } (ii)] \dots (iii)$$

Also, $(BC) \geq (B) + (C) - N \quad \Rightarrow \quad Ny \geq N(2x + 3x) - N \quad \Rightarrow \quad y \geq 5x - 1$ ... (iv)

[From (i) and (ii)]

From (iv) and (iii), we get $\quad 5x - 1 \leq y \;\Rightarrow\; 5x - 1 \leq x \;\Rightarrow\; 4x \leq 1 \;\Rightarrow\; x \leq \frac{1}{4}$ ... (v)

From (iii) and (v), we get $\quad y \leq x \leq \frac{1}{4}$ as required.

Example 13·17. *Show that for n attributes* $A_1, A_2, A_3, \dots, A_n$

$$(A_1 A_2 A_3 \dots A_n) \geq (A_1) + (A_2) + (A_3) + \dots + (A_n) - (n-1) N, \qquad \dots (*)$$

where N is the total number of observations.

Solution. We have

$$(\alpha_1\alpha_2) = (\alpha_1) - (\alpha_1 A_2) = [N - (A_1)] - [(A_2) - (A_1A_2)] = N - (A_1) - (A_2) + (A_1A_2)$$

Since class frequency is always non-negative, we have

$$(\alpha_1 \alpha_2) \geq 0 \quad \Rightarrow \quad (A_1A_2) \geq (A_1) + (A_2) - N \qquad \dots (**)$$

It follows that (*) is true for 2 attributes.

Let us now suppose that (*) is true for r attributes $A_1, A_2, \dots, A_r$ so that

$$(A_1 A_2 A_3 \dots A_r) \geq (A_1) + (A_2) + (A_3) + \dots + (A_r) - (r-1) N$$

Replacing the attribute A_r by another compound attribute $A_r A_{r+1}$, we get

$$\begin{aligned}(A_1 A_2 A_3 \dots A_r A_{r+1}) &\geq (A_1) + (A_2) + (A_3) + \dots + (A_r A_{r+1}) - (r-1) N \\ &\geq (A_1) + (A_2) + (A_3) + \dots + \{(A_r) + A_{r+1}) - N\} - (r-1)N \qquad \text{[From (**)]} \\ &= (A_1) + (A_2) + \dots + (A_r) + (A_{r+1}) - rN\end{aligned}$$

This implies that if (*) is true for $n = r$, it is also true for $n = r + 1$ attributes. But we have seen in (**) that (*) is true for $n = 2$. Hence, by the principle of mathematical induction, the result is true for all positive integral values of n.

Example 13·18. *Show that if A occurs in a larger proportion of the cases where B is than where B is not, then B will occur in a larger proportion of cases where A is than where A is not.*

Solution. The problem can be restated as follows:

Given $\dfrac{(AB)}{(B)} > \dfrac{(A\beta)}{(\beta)}$, prove that $\dfrac{(AB)}{(A)} > \dfrac{(\alpha B)}{(\alpha)}$

$$\frac{(AB)}{(B)} > \frac{(A\beta)}{(\beta)} \quad \Rightarrow \quad \frac{(\beta)}{(B)} > \frac{(A\beta)}{(AB)}$$

$$\therefore \quad 1 + \frac{(\beta)}{(B)} > 1 + \frac{(A\beta)}{(AB)} \quad \Rightarrow \quad \frac{(N)}{(B)} > \frac{(A)}{(AB)} \quad \text{or} \quad \frac{N}{(A)} > \frac{(B)}{(AB)}$$

$$\Rightarrow \quad \frac{(A) + (\alpha)}{(A)} > \frac{(AB) + (\alpha B)}{(AB)} \quad \Rightarrow \quad 1 + \frac{(\alpha)}{(A)} > 1 + \frac{(\alpha B)}{(AB)}$$

$$\Rightarrow \quad \frac{(\alpha)}{(A)} > \frac{(\alpha B)}{(AB)} \quad \Rightarrow \quad \frac{(AB)}{(A)} > \frac{(\alpha B)}{(\alpha)}, \quad \text{as required.}$$

Example 13·19. *If* $\delta = (AB) - (AB)_0$, *then with usual notations, prove that*

(*i*) $[(A) - (\alpha)]\,[(B) - (\beta)] + 2N\delta = (AB)^2 + (\alpha\beta)^2 - (A\beta)^2 - (\alpha B)^2$

(*ii*) $\delta = \dfrac{(B)(\beta)}{N}\left[\dfrac{(AB)}{(B)} - \dfrac{(A\beta)}{(\beta)}\right] = \dfrac{(A)(\alpha)}{N}\left[\dfrac{(AB)}{(A)} - \dfrac{(\alpha B)}{(\alpha)}\right]$

Solution. (*i*) We have $\delta = (AB) - (AB)_0 = (AB) - \dfrac{(A)(B)}{N}$

$$\begin{aligned}\text{L.H.S.} &= [(A) - (\alpha)]\,[(B) - (\beta)] + 2N\delta \\ &= \Big[\{(AB) + (A\beta)\} - \{(\alpha B) + (\alpha\beta)\}\Big] \times \Big[\{(AB) + (\alpha B)\} - \{(A\beta) + (\alpha\beta)\}\Big] \\ &\qquad + 2N\left[(AB) - \frac{(A)\,(B)}{N}\right] \\ &= \Big[\{(AB) - (\alpha\beta)\} + \{(A\beta) - (\alpha B)\}\Big] \times \Big[\{(AB) - (\alpha\beta)\} - \{(A\beta) - (\alpha B)\}\Big] \\ &\qquad + 2\big[N \cdot (AB) - (A)\,(B)\big] \\ &= \big[(AB) - (\alpha\beta)\big]^2 - \big[(A\beta) - (\alpha B)\big]^2 + 2(AB)\big[(AB) + (A\beta) + (\alpha B) + (\alpha\beta)\big] \\ &\qquad - 2\big[\{(AB) + (A\beta)\} \times \{(AB) + (\alpha B)\}\big]\end{aligned}$$

$$= \left[(AB)^2 + (\alpha\beta)^2 - 2(AB)(\alpha\beta)\right] - \left[(A\beta)^2 + (\alpha B)^2 - 2(A\beta)(\alpha B)\right]$$
$$+ 2\left[(AB)(\alpha\beta) - (A\beta)(\alpha B)\right] \qquad \text{(On simplification)}$$
$$= (AB)^2 + (\alpha\beta)^2 - (A\beta)^2 - (\alpha B)^2 = \text{R.H.S}$$

(*ii*) $$\frac{(B)(\beta)}{(N)}\left[\frac{(AB)}{(B)} - \frac{(A\beta)}{(\beta)}\right] = \frac{1}{N}(\beta)(AB) - (B)(A\beta)$$
$$= \frac{1}{N}\left[(AB)\{N-(B)\} - (B)\{(A)-(AB)\}\right]$$
$$= \frac{1}{N}[N\cdot(AB) - (A)(B)] = (AB) - \frac{(A)\cdot(B)}{N} = \delta$$

Since δ is symmetric in A and B, by interchanging A and B, we will obtain the second result.

CHAPTER CONCEPTS QUIZ

1. State, giving reasons, whether each of the following statements is true or false:
   (*i*) There is no difference between correlation and association.
   (*ii*) All the class frequencies of various orders are independent of each other.
   (*iii*) If the attributes A and B are positively associated, then α and B are also positively associated.
   (*iv*) Square of Yule's coefficient of association cannot exceed 1.
   (*v*) Yule's coefficient of association cannot be negative.
   (*vi*) For two attributes A and B, the coefficient of association Q is 0·36. If each ultimate class frequency is doubled then Q is 0·72.
   (*vii*) If $(AB) = 10$, $(\alpha B) = 15$, $(A\beta) = 20$ and $(\alpha\beta) = 30$, then A and B are associated.
   (*viii*) If every item which possesses an attribute A possesses the attribute B as well, then the coefficient of association between A and B is 1.
2. Indicate the correct answer:
   (*i*) In case of two attributes A and B, the ultimate class frequencies are:
   (*a*): (A), (*b*): (AB), (*c*): (α), (*d*): (B).
   (*ii*) The condition for the consistency of a set of independent class frequencies is that no ultimate class frequency is: (*a*) zero, (*b*) positive, (*c*) negative.
   (*iii*) Attributes A and B are said to be independent if
   (*a*) $(AB) > \frac{(A)\times(B)}{N}$ (*b*) $(AB) = \frac{(A)\times(B)}{N}$ (*c*) $(AB) < \frac{(A)\times(B)}{N}$
   (*iv*) Attributes A and B are said to be positively associated if
   (*a*) $\frac{(AB)}{(B)} < \frac{(A\beta)}{(\beta)}$, (*b*) $\frac{(AB)}{(B)} = \frac{(A\beta)}{(\beta)}$, (*c*) $\frac{(AB)}{(B)} > \frac{(A\beta)}{(\beta)}$, (*d*) $\frac{(AB)}{(A)} < \frac{(A\beta)}{(\beta)}$
   (*v*) If $N = 50$, $(A) = 35$, $(B) = 25$, $(AB) = 15$, then the attributes A and B are said to be:
   (*a*) associated, (*b*) independent, (*c*) negatively associated, (*d*) positively associated.
   (*vi*) When there is a perfect positive association between two attributes, Q would be
   (*a*) zero, (*b*) $-0·9$, (*c*) -1, (*d*) $+1$.

DISCUSSION AND REVIEW QUESTIONS

1. Explain the following:
   (*i*) Order of a class, (*ii*) Ultimate classes, and (*iii*) Fundamental set of class frequencies.
2. What is meant by a class-frequency of (*i*) first order, (*ii*) third order? How would you express a class frequncy of first order in terms of class frequencies of third order?

3. What is dichotomy? Show that the continued dichotomy according to n attributes gives rise to 3^n classes.
4. In dichotomous classification of n attributes, prove that:
   (*a*) There are 2^n positive frequencies.
   (*b*) There are 2^n ultimate class frequencies.
5. Given the set of positive frequencies, explain how you will obtain all other class frequencies in dichotomous classification of (*a*) two attributes (*b*) three attributes.
6. If a collection contains N items, each of which is chracterised by one or more of the attributes A, B, C and D, show that with the usual notations
   (*a*) $(ABCD) \geq (A) + (B) + (C) + (D) - 3N$, and (*b*) $(ABCD) = (ABD) + (ACD) - (AD) + (AD\beta\gamma)$,
   where β and γ represent the characteristics of the absence of B and C, respectively.
7. Given $(A) = (\alpha) = (B) = (\beta) = \frac{1}{2}N$, show that $(AB) = (\alpha\beta)$, $(A\beta) = (\alpha B)$.
8. Given that $(A) = (\alpha) = (B) = (\beta) = (C) = (\gamma) = \frac{1}{2}N$ and also that $(ABC) = (\alpha\beta\gamma)$, show that $2(ABC) = (AB) + (AC) + (BC) - \frac{1}{2}N$.
9. What do you mean by independence of attributes? Give a criterion of independence for attributes A and B.
10. What are the various methods of finding whether two attributes are associated, dissociated or independent? Deduce any one such measure of association.
11. When are two attributes said to be (*i*) positively associated, and (*ii*) negatively associated? Also define complete association and dissociation of two attributes.
12. Distinguish between concepts of association and correlation and describe the situations in which each of them should be used. Illustrate your answer with examples.
13. (*a*) Define Yule's coefficient of association (Q) between two attributes A and B.
    (*b*) Prove that $-1 \leq Q \leq 1$.
    When are the limits +1 and –1 attained? Interpret them.
14. Define Yule's coefficient of association (Q) and coefficient of colligation (Y). Establish the following relation between Q and Y:

$$Q = \frac{2Y}{1+Y^2}$$

Hence, deduce the limits for Y and interpret them.

ASSORTED REVIEW PROBLEMS FOR SELF-ASSESSMENT

13·1. Given the following positive frequencies, obtain all other class frequencies:

(*i*) $N = 164$,	$(A) = 90$,	$(B) = 62$,	$(AB) = 6$
(*ii*) $N = 200$,	$(A) = 160$,	$(B) = 79$,	$(AB) = 69$
(*iii*) $N = 2500$,	$(A) = 800$,	$(B) = 1600$,	$(AB) = 400$

[**Ans.** (*i*) $(\alpha) = 74$, $(\beta) = 102$, $(A\beta) = 84$, $(\alpha B) = 56$, $(\alpha\beta) = 18$
(*ii*) $(\alpha) = 40$, $(\beta) = 121$, $(A\beta) = 91$, $(\alpha B) = 10$, $(\alpha\beta) = 30$
(*iii*) $(\alpha) = 1700$, $(\beta) = 900$, $(A\beta) = 400$, $(\alpha B) = 1200$, $(\alpha\beta) = 500$

13·2. Given the following ultimate class frequencies, obtain all other class frequencies.

(*i*) $(AB) = 47$,	$(A\beta) = 63$,	$(\alpha B) = 53$,	$(\alpha\beta) = 77$
(*ii*) $(AB) = 83$,	$(\alpha B) = 57$,	$(A\beta) = 45$,	$(\alpha\beta) = 68$

[**Ans.** (*i*) $(A) = 110$, $(\alpha) = 130$, $(B) = 100$, $(\beta) = 140$, $N = 240$
(*ii*) $(A) = 128$, $(\alpha) = 125$, $(B) = 140$, $(\beta) = 113$, $N = 253$]

13·3. Given the following ultimate frequencies , find the frequencies of the positive classes including the total number of observation N:

$(ABC) = 75,\quad (\alpha BC) = 98,\quad (AB\gamma) = 310,\quad (\alpha B\gamma) = 702$
$(A\beta C) = 106,\quad (\alpha\beta C) = 74,\quad (AB\gamma) = 489,\quad (\alpha\beta\gamma) = 8415.$

Ans. $N = 10269,\quad (A) = 980,\quad (B) = 1185,\quad (C) = 353$
$(AB) = 385,\quad (AC) = 181,\quad (BC) = 173,\quad (ABC) = 75.$

13·4. Given the following positive frequencies, obtain all the ultimate class frequencies.

$N = 12000,\quad (A) = 977,\quad (B) = 1185,\quad (C) = 596$
$(AB) = 453,\quad (AC) = 284,\quad (BC) = 250,\quad (ABC) = 127$

Ans. $(ABC) = 127,\quad (\alpha BC) = 123,\quad (AB\gamma) = 326,\quad (\alpha B\gamma) = 609,$
$(A\beta C) = 157,\quad (\alpha\beta C) = 189,\quad (A\beta\gamma) = 367,\quad (\alpha\beta\gamma) = 10102.$

13·5. At a competitive examination at which 600 graduates appeared, boys outnumbered girls by 96. Those qualifying for interview exceeded in number those failing to qualify by 310. The number of Science graduate boys interviewed was 300 while among the Arts graduate girls there were 25 who failed to qualify for interview. Alogether there were only 135 Arts graduates and 33 among them failed to qualify. Boys who failed to qualify numbered 18.

Find (*i*) the number of boys who qualified for interview,
(*ii*) the total number of Science graduate boys appearing, and
(*iii*) the number of Science graduate girls who qualified.

Ans. (*i*) = 330, (*ii*) 310 (*iii*) 53

13·6. (*a*) 100 children took three examinations *A*, *B* and *C*; 40 passed the first, 39 passed the second and 48 passed the third, 10 passed all the three, 21 failed all three, 9 passed the first two and failed the third, 19 failed the first two and passed the third. Find how many children passed at least two examinations. Show that for the question asked certain of the given frequences are not necessary. Whch are they?

Ans. 38. Only frequencies required are (C), $(\alpha\beta C)$ $(AB\gamma)$, 3%

(*b*) In a university examination, which was indeed very tough, 50% at least failed in "Statistics", 75% at least in "Topology", 82% at least in "Functional Analysis" and 96% at least in "Applied Mathematics". How many at least failed in all the four?

13·7. In a certain class it was found that 60% of the students passed in the terminal examination, 40% passed in the terminal and annual examinations, 25% passed in the annual but failed in the terminal examination. Find, the percentage of students who

(*i*) passed in the annual examination,
(*ii*) passed in the terminal examination but failed in the annual examination,
(*iii*) failed in both the examinations.

Ans.(*i*) 65%, (*ii*) 20%, (*iii*) 15%

13·8. A college submitted the following results to the university office:

Number of students appearing in all the three tests	= 350
Number of students passing the first test	= 125
Number of students passing the second test	= 135
Number of students passing the third test	= 145
Number of students passing all the three tests	= 20
Number of students failing in all the three tests	= 75
Number of students passing the first two but failing in the third	= 25
Number of students failing in the first two but passing in the third	= 60

From the information given above, find out the number of students passing at least two tests.

Ans.110.

13·9. (*a*) If a report gives the following frequencies as actually observed, show that three must be a misprint or mistake of some sort, and that possibly the misprint consists in the dropping of 1 before 85 given as the frequency (BC)

$N = 1000, (A) = 510, \quad (B) = 490, \quad (C) = 427, \quad (AB) = 189, \quad (AC) = 140, \quad (BC) = 85.$

(*b*) A student reported the results of survey in the following manner, in terms of the usual notations:

$N = 1000, \quad (A) = 525, \quad (B) = 312, \quad (C) = 470, \quad (AB) = 42, \quad (BC) = 86, \quad (AC) = 147,$ and $(ABC) = 25.$

Examine the consistency of the above data.

(*c*) Examine the consistency and adequacy of the following data to determine the frequencies of the remaining positive and ultimate classes.

$N = 10{,}000, \quad (A) = 1{,}087, \quad (B) = 286, \quad (C) = 877, \quad (CA\beta) = 281, \quad (C\alpha\beta) = 86, \quad (\gamma AB) = 78, \quad (ABC) = 57.$

13·10. (*a*) Given that $(A) = (B) = (C) = (N/2)$ and 80 per cent of A's are B's, 75 per cent of A's are C's, find the limits to the percentage of B' s that are C's.

(*b*) If $(A) = 50$, $(B) = 60$, $(C) = 50$, $(A\beta) = 5$, $(A\gamma) = 20$, $N = 100$, find the greatest and the least possible values of (BC) so that the data may be consistent.

13·11. (*a*) In a university examination 65% of the candidates passed in English, 90% passed in the second language and 60% passed in the optional subjects. Find how many at least should have passed the whole examination.

(*b*) A market investigator returns the following data. Of 1,000 people consulted 811 liked chocolates, 752 liked toffees and 418 liked boiled sweets, 570 liked both chocolates and toffees, 356 liked chocolates and boiled sweets and 348 liked toffees and boiled sweets, 297 liked all three. Show that this information as it stands must be incorrect.

(*c*) In a college 50 per cent of the students are boys 60 percent of the students are above 18 years and 80 percent receive scholarships. 35 per cent of the students are boys above 18 years of age, 45 per cent are boys receiving scholarships and 42 per cent are above 18 yeas and receive scholarships. Determine the limits to the proportion of boys above 18 years who are in receipt of sholarships.

Ans.(*a*) 15% (*b*) $(\alpha\beta\gamma) = -4$; Incorrect, (*c*) 30% to 32%

13·12. (*a*) The following summary appears in a report on a survey covering 1,000 fields. Scrutinise the numbers and point out if there is any mistake or misprint in them.

Manured fileds	510
Irrigated fields	490
Fields growing improved varieties	427
Fields both irrigated and manured	189
Fields both manured and growing improved varieties	140
Fields both irrigated and growing improved varieties	85

Hint. Let A: Manured fields, B: Irrigated fields, C: Growing Improved varieties, then $(\alpha\beta\gamma) < 0$.

(*b*) A social survey in a village revealed that there were more uneducated employed males than educated ones; there were more educated employed males than uneducated unemployed males. There were more educated unemployed under 35 years of age than employed uneducated males over 35 years of age. Show that there are more uneducated employed males under 35 years of age than educated unemployed males over 35 years of age.

13·13. Find if A & B are independent, positively associated or negatively associated, in each of the following cases:

(*i*)	$N = 1000$,	$(A) = 470$,	$(B) = 620$,	and	$(AB) = 320$
(*ii*)	$(A) = 490$,	$(AB) = 294$,	$(\alpha) = 570$,	and	$(\alpha B) = 380$
(*iii*)	$(AB) = 256$,	$(\alpha B) = 768$,	$(A\beta) = 48$,	and	$(\alpha\beta) = 144$

Ans. (*i*) Positively associated, (*ii*) Negatively associated, (*iii*) Independent.

13·14. The male population of a State is 250 lakhs. The number of literate males is 20 lakhs and total number of male criminals is 26 thousand. The number of literate male criminals is 2 thousand. Do you find any association between literacy and criminality?

Ans. Negatively associated.

13·15. Find the Association between Literacy and Unemplyment from the following figures:

Total Adults	10,000
Literates	1,290
Unemployed	1,390
Literate unemployed	820

Comment on the results.

Ans. Q = 0.923. Very high positive association.

13·16. Calculate the coefficient of association between extravagance in fathers and sons from the following data:

Extravagant fathers and extravagant sons	= 327
Extravagant fathers miserly sons	= 545
Miserly fathers with extravagant sons	= 741
Miserly fathers with miserly sons	= 235

Ans. Q = – 0·687

13·17. In a certain investigation carried on with regard to 500 graduates and 1,500 non-graduates, it was found that the number of employed graduates was 450 while the number of unemployed non-graduates was 300. In the second investigation 5,000 cases were examined. The number of non-graduates was 3,000 and the number of employed non-graduates was 2,500. The number of graduates who were found to be employed was 1,600.

Calculate the coefficient of association between graduation and employment in both the investigations. Can any definite conclusion be drawn from the coefficients ?

Ans. Q (First investigation) = + 0·38, Q (Second investigation) = – 0·11

13·18. Three apitude tests *A*, *B* and *C* were given to 200 apprentice trainees. From amongst them 80 passed test *A*, 78 passed test *B* and 96 passed the third test. While 20 passed all the three tests, 42 failed all the three, 18 passed *A* and *B* but failed *C* and 38 failed *A* and *B* but passed the third. Determine

(*i*) how many trainees passed at least two of the three tests, and

(*ii*) whether the performances in tests *A* and *B* are associated.

13·19. In a survey of a population of 12,000, information is gathered regarding three attributes *A*, *B* and *C*. In the usual notations,

$(A) = 980$, $(AB) = 450$, $(ABC) = 130$, $(B) = 1190$, $(AC) = 280$, $(C) = 600$ and $(BC) = 250$.

Find: (*i*) $(\alpha\beta\gamma)$ (*ii*) Q_{AB} = Coefficient of association between *A* and *B*.

Comment on your findings.

13·20. A group of 1,000 fathers was studied and it was found that 12·9% had dark eyes. Among them the ratio of those having sons with dark eyes to those having sons with not dark eyes was 1: 1·58. The number of cases where fathers and sons both did not have dark eyes was 782. Calculate coefficient of association between darkness of eye colour in father and son. Give the frequencies that would have been observed had there been completely no heredity.

CHAPTER

14 Large Sample Theory

LEARNING OBJECTIVES

Upon completion of this Chapter, you should be able to:

- Understand the meaning and utility of sampling in Statistics.
- Know about the various types of sampling.
- Demonstrate the concepts of parameter, statistic, sampling distribution of a statistic and its standard error, and their utility in Large Sample Tests.
- Develop a framework for the null hypothesis, alternative hypothesis, level of significance, one-tailed and two-tailed tests of significance and their importance in testing of hypothesis.
- Discuss the procedure for testing of hypothesis for large samples.
- Review and appreciate the various tests of significance (for large samples)
- (a) For Attributes *viz;*
  - (i) Test for Single Proportion
  - (ii) Test for Difference of Proportions.
- (b) For Variables *viz.;*
  - (i) Test for Single Mean
  - (ii) Test for Difference of Means
  - (iii) Test for Difference of Standard Deviations.

CHAPTER OUTLINES

14·1. INTRODUCTION

Before giving the notion of sampling we will first define *population*. In a statistical investigation the interest usually lies in the assessment of the general magnitude and the study of variation with respect to one or more characteristics relating to individuals belonging to a group. This group of individuals under study is called *population or universe*. Thus in statistics, population is an aggregate of objects, animate or inanimate, under study. The population may be *finite* or infinite.

It is obvious that for any statistical investigation complete enumeration of the population is rather impracticable. For example, if we want to have an idea of the average per capita (monthly) income of the people in India, we will have to enumerate all the earning individuals in the country, which is rather a very difficult task.

If the population is infinite, complete enumeration is not possible. Also if the units are destroyed in the course of inspection (*e.g.*, inspection of crackers, explosive materials, etc.), 100% inspection, though possible, is not at all desirable. But even if the population is finite or the inspection is not destructive, 100% inspection is not taken recourse to because of multiplicity of causes, *viz.*, administrative and financial implications, time factor, etc., and we take the help of *sampling*.

A finite subset of statistical individuals in a population is called a sample and the number of individuals in a sample is called the sample size.

For the purpose of determining population characteristics, instead of enumerating the entire population, the individuals in the sample only are observed. Then the sample characteristics are utilised to approximately determine or estimate the population. For example, on examining the sample of a particular stuff we arrive at a decision of purchasing or rejecting that stuff. The error involved in such approximation is known as *sampling error* and is inherent and unavoidable in any and every sampling scheme. But sampling results in considerable gains, especially in time and cost, not only in respect of making observations of characteristics but also in the subsequent handling of the data.

Sampling is quite often used in our day-to day practical life. For example, in a shop we assess the quality of sugar, wheat or any other commodity by taking a handful of it from the bag and then decide to purchase it or not. A housewife normally tests the cooked products to find if they are properly cooked and contain the proper quantity of salt.

14·2. TYPES OF SAMPLING

Some of the commonly known and frequently used types of sampling are :

(*i*) *Purposive sampling*, (*ii*) *Random sampling*,
(*iii*) *Stratified sampling*, (*iv*) *Systematic sampling*.

Below we will precisely explain these terms, without entering into detailed discussion.

14·2·1. Purposive Sampling

Purposive sampling is one in which the sample units are selected with definite purpose in view. For example, if we want to give the picture that the standard of living has increased in the city of New Delhi, we may take individuals in the sample from rich and posh localities like Defence Colony, South Extension, Golf Links, Jor Bagh, Chanakyapuri, Greater Kailash etc. and ignore the localities where low income group and the middle class families live. This sampling suffers from the drawback of favouritism and nepotism and does not give a representative sample of the population.

14·2·2. Random Sampling.

In this case the sample units are selected at random and the drawback of purposive sampling, *viz.*, favouritism or subjective element, is completely overcome. *A random sample is one in which each unit of population has an equal chance of being included in it.*

Suppose we take a sample of size n from a finite population of size N. Then there are ${}^N C_n$ possible samples. A sampling technique in which each of the ${}^N C_n$ samples has an equal chance of being selected is known as *random sampling* and the sample obtained by this technique is termed as a *random sample.*

Proper care has to be taken to ensure that the selected sample is random. Human bias, which varies from individual to individual, is inherent in any sampling scheme administered by human being. Fairly good random samples can be obtained by the use of *Tippet's random number tables* or by throwing of a dice, draw of a lottery, etc.

The simplest method, which is normally used is the *lottery system* which is illustrated below by means of an example.

Suppose we want to select 'r' candidates out of n. We assign numbers one to n, one number to each candidate and write these numbers (1 to n) on n slips which are made as homogeneous as possible in shape, size, etc. These slips are then put in a bag and thoroughly shuffled and then 'r' slips are drawn one by one. The 'r' candidates corresponding to the numbers on the slips drawn, will constitute the random sample.

Remark. *Tippet's Random Numbers.* L.H.C. Tippet's random number tables consist of 10400 four-digited numbers, giving in all 10400 × 4, *i.e.*, 41600 digits, taken from the British census reports. These tables have proved to be fairly random in character. Any page of the table is selected at random and the numbers in any row or column or diagonal selected at random may be taken to constitute the sample.

14·2·3. Simple Sampling

Simple sampling is random sampling in which each unit of the population has an equal chance, say p, of being included in the sample and that this probability is independent of the previous drawings. Thus a simple sample of size n from a population may be identified with series of n independent trials with constant probability 'p' of success for each trial.

Remark. It may be pointed out that random sampling does not necessarily imply simple sampling though, obviously, the converse is true. For example, if an urn contains 'a' white balls and 'b' black balls, the probability of drawing a white ball at the first draw is $[a/(a + b)] = p_1$, (say) and if this ball is not replaced the probability of getting a white ball in the second draw is $[(a - 1)/(a + b - 1)] = p_2 \neq p_1$, the sampling is not simple. But since in the first draw each white ball has the same chance, *viz.*, $a/(a + b)$, of being drawn and in the second draw again each white ball has the same chance, *viz.*, $(a - 1)/(a + b - 1)$, of being drawn, the sampling is random. Hence in this case, the sampling, though random, is not simple. To ensure that sampling is simple. it must be done with replacement, if population is finite. However, in case of infinite population no replacement is necessary.

14·2·4. Stratified Sampling

Here the entire heterogeneous population is divided into a number of homogeneous groups, usually termed as *strata,* which differ from one another but each of these groups is homogeneous within itself. Then units are sampled at random from each of these stratum, the sample size in each stratum varies according to the relative importance of the stratum in the population. The sample, which is the aggregate of the sampled units of each of the stratum, is termed as *stratified sample* and the technique of drawing this sample is known as *stratified sampling.* Such a sample is by far the best and can safely be considered as representative of the population from which it has been drawn.

14·3. PARAMETER AND STATISTIC

In order to avoid verbal confusion with the statistical constants of the population, *viz.*, mean (μ), variance σ^2, etc., which are usually referred to as *parameters*, statistical measures computed from the sample observations alone, *e.g.*, mean ($\bar{x}$), variance (s^2), etc., have been termed by Professor R.A. Fisher as *statistics*.

In practice parameter values are not known and the estimates based on the sample values are generally used. Thus, statistic which may be regarded as an estimate of parameter, obtained from the sample, is a function of the sample values only. It may be pointed out that a statistic, as it is based on sample values and as there are multiple choices of the samples that can be drawn from a population, varies from sample to sample. The determination or the characterisation of the variation (in the values of the statistic obtained from different samples) that may be attributed to chance or fluctuations of sampling, is one of the fundamental problems of the sampling theory.

Remarks:

1. Now onwards, μ and σ^2 will refer to the population mean and variance respectively while the sample mean and variance will be denoted by $\bar{x}$ and s^2 respectively.
2. *Unbiased Estimate.* A statistic $t = t(x_1, x_2, ..., x_n)$, a function of the sample values $x_1, x_2, ..., x_n$ is an unbiased estimate of the population parameter θ, if $E(t) = \theta$. In other words, if :

$$E\ (Statistic) = Parameter, \qquad ...\ (14·1)$$

then statistic is said to be an unbiased estimate of the parameter.

14·3·1. Sampling Distribution of a Statistic

If we draw a sample of size n from a given finite population of size N, then the total number of possible samples is:

$${}^N C_n = \frac{N!}{n!(N-n)!} = k, (say).$$

For each of these k samples we can compute some statistic $t = t(x_1, x_2, ..., x_n)$, in particular the mean $\bar{x}$, the variance s^2, etc., as given below.

Sample Number	Statistic t	$\bar{x}$	s^2
1	t_1	$\bar{x}_1$	s_1^2
2	t_2	$\bar{x}_2$	s_2^2
3	t_3	$\bar{x}_3$	s_3^2
⋮	⋮	⋮	⋮
k	t_k	$\bar{x}_k$	s_k^2

The set of the values of the statistic so obtained, one for each sample, constitutes what is called the *sampling distribution* of the statistic. For example, the values $t_1, t_2, t_3, ..., t_k$ determine the sampling distribution of the statistic t. In other words, statistic t may be regarded as a random variable which can take the values $t_1, t_2, t_3, ..., t_k$ and we can compute the various statistical constants like mean variance, skewness, kurtosis, etc., for its distribution. For example, the mean and variance of the sampling distribution of the statistic t are given by :

$$\bar{t} = \frac{1}{k}(t_1 + t_2 + ... + t_k) = \frac{1}{k}\sum_{i=1}^{k} t_i$$

and $$Var(t) = \frac{1}{k}\left[(t_1 - \bar{t}) + (t_2 - \bar{t})^2 + \text{L} + (t_k - \bar{t})^2\right] = \frac{1}{k}\sum_{i=1}^{k}(t_i - \bar{t})^2.$$

14·3·2. Standard Error

The standard deviation of the sampling distribution of a statistic is known as its *Standard Error*, abbreviated as S.E. The standard errors of some of the well-known statistics, *for large samples*, are given below, where n is the sample size, σ^2 the population variance, and P the population proportion, and $Q = 1 - P$; n_1 and n_2 represent the sizes of two independent random samples respectively drawn from the given population (*s*).

S. No.	*Statistic*	*Standard Error*
1.	Sample mean : $\bar{x}$	$\sigma/\sqrt{n}$
2.	Observed sample proportion 'p'	$\sqrt{PQ/n}$
3.	Sample standard deviation : (s)	$\sqrt{\sigma^2/2n}$
4.	Sample variance : s^2	$\sigma^2\sqrt{2/n}$
5.	Sample quartiles	$1{\cdot}36263\,\sigma/\sqrt{n}$
6.	Sample median	$1{\cdot}25331\,\sigma/\sqrt{n}$
7.	Sample correlation coefficient (r)	$(1-\rho^2)/\sqrt{n}$, ρ being the population correlation coefficient
8.	Sample moment : μ_3	$\sigma^3\sqrt{96/n}$
9.	Sample moment : μ_4	$\sigma^4\sqrt{96/n}$
10.	Sample coefficient of variation (υ)	$\frac{v}{\sqrt{2n}}\sqrt{1+\frac{2v^3}{10^4}}$; $\frac{v}{\sqrt{2n}}$
11.	Difference of two sample means : $(\bar{x}_1 - \bar{x}_2)$	$\sqrt{\frac{\sigma_1^2}{n_1}+\frac{\sigma_2^2}{n_2}}$
12.	Difference of two sample *s.d.'s* : $(s_1 - s_2)$	$\sqrt{\frac{\sigma_1^2}{2n_1}+\frac{\sigma_2^2}{2n_2}}$
13.	Difference of two sample proportions : $(p_1 - p_2)$	$\sqrt{\frac{P_1Q_1}{n_1}+\frac{P_2Q_2}{n_2}}$

14·3·3. Utility of Standard Error

S.E. plays a very important role in the large sample theory and forms the basis of the testing of hypothesis. If t is any statistic, then for large samples : $Z = \frac{t - E(t)}{\sqrt{V(t)}} \sim N(0,1)$ (*c.f.* § 14·4)

$$\Rightarrow \quad Z = \frac{t - E(t)}{S.E.(t)} \sim N(0,1), \text{ for large samples.}$$

Thus, if the discrepancy between the observed and the expected (hypothetical) value of a statistic is greater than z_α (*c f.* § 14·4·5) times its S.E., the null hypothesis is rejected at α level of significance. Similarly, if

$$|t - E(t)| \le z_\alpha \times S.E.(t),$$

the deviation is not regarded significant at level of significance. In other words, the deviation, $t - E(t)$, could have arisen due to fluctuations of sampling and the data do not provide us any evidence against the null hypothesis which may, therefore, be accepted at α level of significance. [For details see § *14·4·3*.]

(*i*) The magnitude of the standard error gives an index of the precision of the estimate of the parameter. The reciprocal of the standard error is taken as the measure of reliability or precision of the statistic.

$$S.E.(p) = \sqrt{PQ/n} \qquad \text{and} \qquad S.E.(\bar{x}) = \sigma/n$$

In other words, the standard errors of p and $\bar{x}$ vary inversely as the square root of the sample size. Thus in order to double the precision, which amounts to reducing the standard error to one half, the sample size has to be increased four times.

(*ii*) S.E. enables us to determine the probable limits within which the population parameter may be expected to lie. For example, the probable limits for population proportion P are given by: $p \pm 3pq/n$ (*c. f. Remark § 14·7·1*)

Remark. S.E. of a statistic may be reduced by increasing the sample size but this results in corresponding increase in cost, labour and time, etc.

14·4. TESTS OF SIGNIFICANCE

A very important aspect of the sampling theory is the study of the *tests of significance,* which enable us to decide on the basis of the sample results, if

(*i*) the deviation between the observed sample statistic and the hypothetical parameter value, or

(*ii*) the deviation between two independent sample statistics;

is significant or might be attributed to chance or the fluctuations of sampling.

Since, for large n, almost all the distributions, *e.g.*, Binomial, Poisson, Negative binomial, Hypergeometric (*c.f.* Chapter 8), t, F (Chapter 16), Chi-square (Chapter 15), can be approximated very closely by a normal probability curve, we use the *Normal Test of Significance* (*c. f.* § *14·7*) for large samples. Some of the well-known tests of significance for studying such differences for small samples are *t-test, F-test* and Fisher's *z-transformation.*

14·4·1. Null and Alternative Hypotheses

The technique of randomisation used for the selection of sample units makes the test of significance valid for us. For applying the test of significance we first set up a hypothesis—a definite statement about the population parameter. Such a hypothesis, which is usually a hypothesis of no difference, is called ***null hypothesis*** and is usually denoted by H_0. According to *Prof. R.A. Fisher, null hypothesis is the hypothesis which is tested for possible rejection under the assumption that it is true.*

For example, in case of a single statistic, H_0 will be that the sample statistic does not differ significantly from the hypothetical parameter value and in the case of two statistics, H_0 will be that the sample statistics do not differ significantly.

Having set up the null hypothesis we compute the probability P that the deviation between the observed sample statistic and the hypothetical parameter value might have occurred due to fluctuations of sampling. If the deviation comes out to be significant (as measured by a test of significance), null hypothesis is refuted or rejected at the particular level of significance adopted (*c.f.* § 14·4·3) and if the deviation is not significant, null hypothesis may be retained at that level.

Any hypothesis which is complementary to the null hypothesis is called an ***alternative hypothesis,*** usually denoted by H_1. For example, if we want to test the null hypothesis that the population has a specified mean μ_0, (say), *i.e.*, $H_0 : \mu = \mu_0$ then the alternative hypothesis could be :

(*i*) $H_1 : \mu \neq \mu_0$ (*i.e.*, $\mu > \mu_0$ or $\mu < \mu_0$) (*ii*) $H_1 : \mu > \mu_0$, (*iii*) $H_1 : \mu < \mu_0$

The alternative hypothesis in (*i*) is known as a *two-tailed alternative* and the alternatives in (*ii*) and (*iii*) are known as *right-tailed* and *left-tailed alternatives* respectively. The setting of alternative hypothesis is very important since it enables us to decide whether we have to use a single-tailed (right or left) or two-tailed test (*c.f.* § 14·4·4).

14·4·2. Errors in Sampling

The main objective in sampling theory is to draw valid inferences about the population parameters on the basis of the sample results. In practice we decide to accept or reject the lot after examining a sample from it. As such we are liable to commit the following two types of errors :

Type I Error *: Reject H_0 when it is true.*

Type II Error *: Accept H_0 when it is wrong, i.e., accept H_0 when H_1 is true.*

$$\left.\begin{aligned}&\text{If we write } P\{\text{Reject } H_0 \text{ when it is true}\} = P\left[\text{Reject } H_0 | H_0\right] = \alpha\\ &\text{and } P\{\text{Accept } H_0 \text{ when it is wrong}\} = P\left[\text{Accept } H_0 | H_1\right] = \beta\end{aligned}\right\} \quad \text{... (14·2)}$$

then α and β are called the *sizes of type I error and type II error,* respectively.

In practice, type I error amounts to rejecting a lot when it is good and type II error may be regarded as accepting the lot when it is bad.

$$\begin{aligned}\text{Thus} \quad & P\,(\text{Reject a lot when it is good}) = \alpha \quad \text{... (14·2a)}\\ \text{And} \quad & P\,(\text{Accept a lot when it is bad}) = \beta\end{aligned}$$

where α and β are referred to as *Producer's risk* and *Consumer's risk* respectively.

14·4·3. Critical Region and Level of Significance

A region (corresponding to a statistic (t)) in the sample space S which amounts to rejection of H_0 is termed as *critical region or region of rejection.* If ω is the critical region and if $t = t\,(x_1, x_2, ..., x_n)$ is the value of the statistic based on a random sample of size n, then

$$P\,(t \in \omega \mid H_0) = \alpha, \qquad P\,(t \in \bar{\omega} \mid H_1) = \beta \quad \text{... (14·2b)}$$

where $\bar{\omega}$, the complementary set of ω, is called the *acceptance region.*

We have $\omega \cup \bar{\omega} = S$ and $\omega \cap \bar{\omega} = \phi$

The probability 'α' that a random value of the statistic t belongs to the critical region is known as the *level of significance.* In other words, level of significance is the size of the type I error (or the maximum producer's risk). The levels of significance usually employed in testing of hypothesis are 5% and 1%. The level of significance is always fixed in advance before collecting the sample information.

14·4·4. One-tailed and Two-tailed Tests

In any test, the critical region is represented by a portion of the area under the probability curve of the sampling distribution of the test statistic.

A test of any statistical hypothesis where the alternatives hypothesis is one tailed (right-tailed or left-tailed) is called a *one-tailed test*. For example, a test for testing the mean of a population $H_0 : \mu = \mu_0$ against the alternative hypothesis :

$H_1 : \mu > \mu_0$ (*Right-tailed*) or $H_1 : \mu < \mu_0$ (*Left-tailed*), is a *single tailed test.*

In the right-tailed test ($H_1 : \mu > \mu_0$), the critical region lies entirely in the right tail of the sampling distribution of $\overline{x}$, while for the left-tailed test ($H_1 : \mu < \mu_0$), the critical region is entirely in the left tail of the distribution.

A test of statistical hypothesis where the alternative hypothesis is two-tailed such as : $H_0 : \mu = \mu_0$, against the alternative hypothesis $H_1 : \mu \neq \mu_0$ ($\mu > \mu_0$ or $\mu < \mu_0$), is known as *two tailed test* and in such a case the critical region is given by the portion of the area lying in both tails of the probability curve of the test statistic.

In a particular problem, whether one-tailed or two-tailed test is to be applied depends entirely on the nature of the alternative hypothesis. If the alternative hypothesis is two-tailed, we apply two-tailed test and if alternative hypothesis is one-tailed, we apply one tailed test.

For example, suppose that there are two population brands of bulbs, one manufactured by standard process (with mean life μ_1) and the other manufactured by some new technique (with mean life μ_2). If we want to test if the bulbs differ significantly, then our null hypothesis is $H_0 : \mu_1 = \mu_2$ and alternative will be $H_1 : \mu_1 \neq \mu_2$, thus giving us two-tailed test. However, if we want to test if the bulbs produced by new process have higher average life than those produced by standard process, then we have $H_0 : \mu_1 = \mu_2$ and $H_1 : \mu_1 < \mu_2$, thus giving us a left-tailed test. Similarly, for testing if the product of new process is inferior to that of standard process, then we have : $H_0 : \mu_1 = \mu_2$ and $H_1 : \mu_1 > \mu_2$, thus giving us a right-tailed test. Thus, the decision about applying a two-tailed test or a single-tailed (right or left) test will depend on the problem under study.

14·4·5. Critical Values or Significant Values

The value of test statistic which separates the critical (or rejection) *region* and the *acceptance region* is called the *critical value* or *significant value*. It depends upon :

(*i*) The level of significance used, and

(*ii*) The alternative hypothesis, whether it is two-tailed or single-tailed.

As has been pointed out earlier, for large samples, the standardised variable corresponding to the statistic t ,*viz.*,

$$Z = \frac{t - E(t)}{S.E(t)} \sim N(0, 1), \qquad \text{... (*)}$$

asymptotically as $n \to \infty$. The value of Z given by (*) under the null hypothesis is known as *test statistic*. The critical value of the test statistic at level of significance α for a two-tailed test is given by z_α , where z_α is determined by the equation :

$$P(|Z| > z_\alpha) = \alpha \qquad \text{... (14·2c)}$$

i.e., z_α is the value so that the total area of the critical region on both tails is α. Since normal probability curve is a symmetrical curve, from (14·2*c*), we get

$$P(Z > z_\alpha) + P(Z < -z_\alpha) = \alpha \quad \Rightarrow \quad P(Z > z_\alpha) + P(Z > z_\alpha) = \alpha \qquad \text{[By symmetry]}$$

$$\Rightarrow \quad 2P(Z > z_\alpha) = \alpha \quad \Rightarrow \quad P(Z > z_\alpha) = \alpha/2$$

In other words, the area of each tail is $\alpha/2$. Thus z_α is the value such that area to the right of z_α is $\alpha/2$ and to the left of $(-z_\alpha)$ is $\alpha/2$, as shown in the Fig 14.1 on Page 14.10.

TWO-TAILED TEST

(Level of Significance 'α')

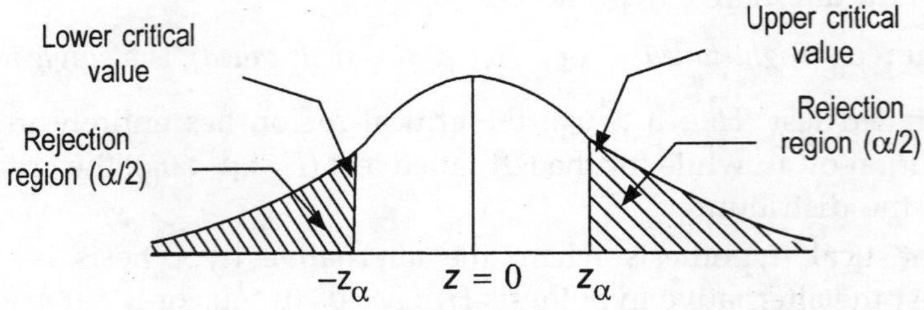

Fig. 14.1

In case of single-tail alternative, the critical value z_α is determined so that total area to the right of it (for right-tailed test) is α and for left-tailed test the total area to the left of $(-z_\alpha)$ is α (See Fig. 14.2 and Fig. 14.3 below), *i.e.*,

For Right-tailed Test : $P(Z > z_\alpha) = \alpha$ ... (14·2*d*)

For Left-tailed Test : $P(Z < -z_\alpha) = \alpha$ ... (14·2*e*)

RIGHT-TAILED TEST *(Level of Significance 'α')* — LEFT-TAILED TEST *(Level of Significance 'α')*

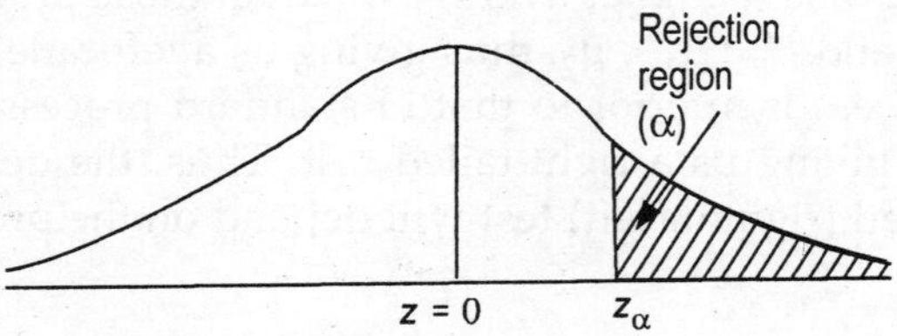

Fig. 14.2

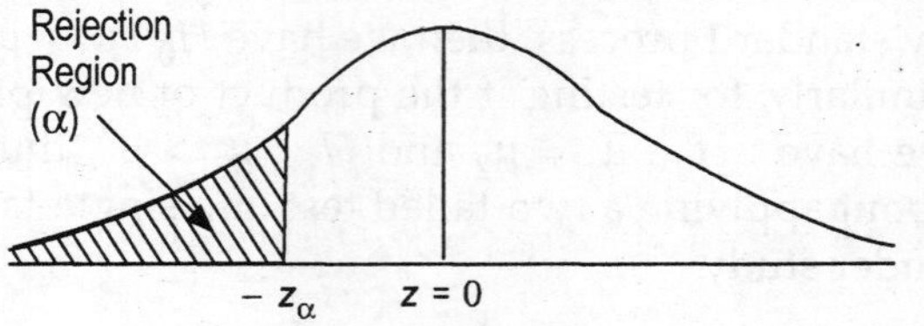

Fig. 14.3

Thus, *the significant* or *critical value of Z for a single-tailed test (left or right) at level of significance 'α' is same as the critical value of Z for a two-tailed test at level of significance '2α'.*

We give below, the critical values of Z at commonly used levels of significance for both two-tailed and single-tailed tests. These values have been obtained from equations (14·2*c*), (14·2*d*), (14·2*e*), on using the Normal Probability Tables as explained in § 14·6.

Critical value (z_α)	*Level of significance* (α)		
	1%	5%	10%
Two-tailed test	$\lvert z_\alpha \rvert = 2.58$	$\lvert z_\alpha \rvert = 1.96$	$\lvert z_\alpha \rvert = 1.645$
Right-tailed test	$z_\alpha = 2.33$	$z_\alpha = 1.645$	$z_\alpha = 1.28$
Left-tailed test	$z_\alpha = -2.33$	$z_\alpha = -1.645$	$z_\alpha = -1.28$

Remark. If *n* is small, then the sampling distribution of the test statistic Z will not be normal and in that case we can't use the above significant values which have been obtained from normal probability curves. In this case, *viz.*, *n* small (usually less than 30), we use the significant values based on the exact sampling distribution of the statistic Z. [defined in (*), § *14·4·5*, which turns out to be *t*, *F*, or χ^2 distribution [See Chapters 15 and 16]. These significant values have been tabulated for different values of *n* and α and are given in the Tables at the end of the book.

14·5. PROCEDURE FOR TESTING OF HYPOTHESIS

We now summarise the various steps in testing of a statistical hypothesis in a systematic manner.

1. *Null Hypothesis.* Set up the Null Hypothesis H_0.

2. *Alternative Hypothesis.* Set up the Alternative Hypothesis H_1. This will enable us to decide whether we have to use a single-tailed (right or left) test or two-tailed test.
3. *Level of Significance.* Choose the appropriate level of significance (α) depending on the reliability of the estimates and permissible risk. This is to be decided before sample is drawn, *i.e.*, *α is fixed in advance.*
4. *Test Statistic (or Test Criterion).* Compute the test statistic :
$$Z = \frac{t - E(t)}{S.E(t)}, \text{ under } H_0.$$
5. *Conclusion.* We compare the computed value of Z in *step 4* with the significant value (tabulated value) z_α at the given level of significance, 'α'.

If $|Z| < z_\alpha$, *i.e.*, if the calculated value of Z (in modulus value) is less than z_α we say it is not significant. By this we mean that the difference $t - E(t)$ is just due to fluctuations of sampling and the sample data do not provide us sufficient evidence against the null hypothesis which may, therefore, be accepted.

If $|Z| > z_\alpha$ *i.e.*, if the computed value of test statistic is greater than the critical or significant value, then we say that it is significant and the null hypothesis is rejected at level of significance α, *i.e.*, with confidence coefficient $(1 - \alpha)$.

14·6. TESTS OF SIGNIFICANCE FOR LARGE SAMPLES

In this section, we will discuss the tests of significance when samples are large. We have seen that for large values of *n*, the number of trials, almost all the distributions, *e.g.*, binomial, Poisson, negative binomial, etc., are very closely approximated by normal distribution. Thus in this case we apply the *normal test*, which is based upon the following fundamental property (*area property*) of the normal probability curve.

If $X \sim N(\mu, \sigma^2)$, then $$Z = \frac{X - \mu}{\sigma} = \frac{X - E(X)}{\sqrt{V(X)}} : N(0,1)$$

Thus, from the normal probability tables, we have

$$P(-3 \leq Z \leq 3) = 0.9973, \quad i.e., \quad P(|Z| \leq 3) = 0.9973$$

$$\Rightarrow \quad P(|Z| > 3) = 1 - P(|Z| \leq 3) = 0.0027 \qquad \text{... (14·3)}$$

i.e., in all probability we should expect a standard normal variate to lie between ± 3.

Also, from the normal probability tables, we get

$$P(-1.96 \leq Z \leq 1.96) = 0.95, \quad i.e., \quad P(|Z| \leq 1.96) = 0.95$$

$$\Rightarrow \quad P(|Z| > 1.96) = 1 - 0.95 = 0.05 \qquad \text{... (14·3a)}$$

and $$P(|Z| < 2.58) = 0.99 \quad \Rightarrow \quad P(|Z| > 2.58) = 0.01 \qquad \text{... (14·3b)}$$

Thus, the significant values of Z at 5% and 1% levels of significance for a two-tailed test are 1·96 and 2·58 respectively.

Thus the steps to be used in the normal test are as follows:

(*i*) Compute the test statistic Z under H_0.

(*ii*) If $|Z| > 3$, H_0 is always rejected.

(*iii*) If $|Z| \leq 3$, we test its significance at certain level of significance, usually at 5% and sometimes at 1% level of significance. Thus, for a two-tailed test if $|Z| > 1.96$, H_0 is rejected at 5% level of significance.

Similarly, if $|Z| > 2.58$, H_0 is contradicted at 1% level of significance and if $|Z| \leq 2.58$, H_0 may be accepted at 1% level of significance.

From the normal probability tables, we have :

$$P(Z > 1{\cdot}645) = 0{\cdot}5 - P(0 \le Z \le 1{\cdot}645) = 0{\cdot}5 - 0{\cdot}45 = 0{\cdot}5 - 0{\cdot}45 = 0{\cdot}05$$

$$P(Z > 2{\cdot}33) = 0{\cdot}5 - P(0 \le Z \le 2{\cdot}33) = 0{\cdot}5 - 0{\cdot}49 = 0{\cdot}01$$

Hence, for a single-tail test (Right-tail or Left-tail) we compare the computed value of $|Z|$ with 1·645 (at 5% level) and 2·33 (at 1% level) and accept or reject H_0 accordingly.

Important Remark. In the theoretical discussion that follows in the next sections, the samples under consideration are supposed to be large. For practical purposes, sample may be regarded as large if $n > 30$.

14·7. SAMPLING OF ATTRIBUTES

Here we shall consider sampling from a population which is divided into two mutually exclusive and collectively exhaustive classes — one class possessing a particular attribute, say A, and the other class not possessing that attribute, and then note down the number of persons in the sample of size n, possessing that attribute. The presence of an attribute in sampled unit may be termed as success and its absence as failure. In this case a sample of n observations is identified with that of a series of n independent Bernoulli trials with constant probability P of success for each trial. Then the probability of x successes in n trials, as given by the binomial probability distribution is : $p(x) = {}^nC_x\ P^x\ Q^{n-x}$; $x = 0, 1, 2, ..., n$.

14·7·1. Test of Significance for Single Proportion

If X is the number of successes in n independent trials with constant probability P of success for each trial, then

$$E(X) = nP \quad \text{and} \quad V(X) = nPQ, \quad \text{where} \quad Q = 1 - P, \text{ is the probability of failure.}$$

It has been proved that for large n, the binomial distribution tends to normal distribution. Hence, for large n, $X \sim N(nP, nPQ)$, *i.e.*,

$$Z = \frac{X - E(X)}{\sqrt{V(X)}} = \frac{X - nP}{\sqrt{nPQ}} \ :\ N(0,1) \qquad \text{... (14·4)}$$

and we can apply the normal test.

Remarks:

1. In a sample of size n, let X be the number of persons possessing the given attribute. Then

Observed proportion of successes $= X/n = p$, (say).

$$\therefore \quad E(p) = E(X/n) = \frac{1}{n}E(X) = \frac{1}{n}nP = P \qquad \text{... (14·4}a\text{)}$$

Thus, the sample proportion 'p' gives an unbiased estimate of the population proportion P.

Also, $$V(p) = V\left(\frac{X}{n}\right) = \frac{1}{n^2}V(X) = \frac{1}{n^2}\cdot nPQ = \frac{PQ}{n} \quad \Rightarrow \quad S.E.(p) = \sqrt{\frac{PQ}{n}} \qquad \text{... (14·4}b\text{)}$$

Since X and consequently X/n is asymptotically normal for large n, the normal test for the proportion of successes becomes:

$$Z = \frac{p - E(p)}{S.E.(p)} = \frac{p - P}{\sqrt{PQ/n}} \ :\ N(0,1) \qquad \text{... (14·4}c\text{)}$$

2. If we have sampling from a finite population of size N, then

$$S.E.(p) = \sqrt{\left(\frac{N-n}{N-1}\right)\cdot\frac{PQ}{n}} \qquad \text{... (14·4}d\text{)}$$

3. Since the probable limits for a normal variate X are $E(X) \pm 3\sqrt{V(X)}$, the probable limits for the observed proportion of successes are:

$$E(p) \pm 3\,S.E.(p), \quad i.e., \quad P \pm 3\sqrt{PQ/n}$$

If P is not known then taking p (the sample proportion) as an estimate of P, the probable limits for the proportion in the population are : $p \pm 3\sqrt{pq/n}$. ... (14·4e)

However, the limits for P at level of significance α are given by : $p \pm z_\alpha \sqrt{pq/n}$, ... (14·4f)

where z_α is the significant value of Z at level of significance α.

In particular : 95% confidence limits for P are given by : $p \pm 1{\cdot}96\sqrt{pq/n}$, ... (14·4g)

and 99% confidence limits for P are given by : $p \pm 2{\cdot}58\sqrt{pq/n}$. ... (14·4h)

Example 14·1. *A die is thrown 9,000 times and a throw of 3 or 4 is observed 3,240 times. Show that the die cannot be regarded as an unbiased one and find the limits between which the probability of a throw of 3 or 4 lies.*

Solution. If the coming of 3 or 4 is called a success, then in usual notations:

$$n = 9{,}000\ ;\quad X = \text{Number of successes} = 3{,}240$$

Under the null hypothesis (H_0) that the die is an unbiased one, we get

$$P = \text{Probability of success} = \text{Probability of getting a 3 or 4} = \frac{1}{6} + \frac{1}{6} = \frac{1}{3}$$

Alternative hypothesis, $H_1 : P \neq \frac{1}{3}$, (*i.e.,* die is biased).

We have $Z = \dfrac{X - nP}{\sqrt{nQP}} \sim N(0,1)$, since n is large.

$$Z = \frac{3240 - 9000 \times (1/3)}{\sqrt{9000 \times (1/3) \times (2/3)}} = \frac{240}{\sqrt{2000}} = \frac{240}{44{\cdot}73} = 5{\cdot}36$$

Since $|Z| > 3$, H_0 is rejected and we conclude that the die is almost certainly biased.

Since die is not unbiased, $P \neq \frac{1}{3}$. The probable limits for 'P' are given by:

$$\hat{P} \pm 3\sqrt{\hat{P}\hat{Q}/n} = p \pm 3\sqrt{pq/n},\ \text{where}\ \hat{P} = p = \frac{3{,}240}{9{,}000} = 0{\cdot}36\ \text{and}\ \hat{Q} = q = 1 - p = 0{\cdot}64$$

Probable limits for population proportion of successes may be taken as :

$$\hat{P} \pm 3\sqrt{\hat{P}\hat{Q}/n} = 0{\cdot}36 \pm 3\sqrt{\frac{0{\cdot}36 \times 0{\cdot}64}{9000}} = 0{\cdot}36 \pm 3 \times \frac{0{\cdot}6 \times 0{\cdot}8}{30\sqrt{10}} = 0{\cdot}345 \text{ and } 0{\cdot}375$$

Hence, the probability of getting 3 or 4 almost certainly lies between 0·345 and 0·375.

Example 14·2. *A random sample of 500 pineapples was taken from a large consignment and 65 were found to be bad. Show that the S.E. of the proportion of bad ones in a sample of this size is 0·015 and deduce that the percentage of bad pineapples in the consignment almost certainly lies between 8·5 and 17·5.*

Solution. Here we are given : $n = 500$

X = Number of bad pineapples in the sample = 65

p = Proportion of bad pineapples in the sample $= \dfrac{65}{500} = 0{\cdot}13 \Rightarrow q = 1 - p = 0{\cdot}87$

Since P, the proportion of bad pineapples in the consignment is not known, we may take (as in the last example) : $\hat{P} = p = 0{\cdot}13$, $\hat{Q} = q = 0{\cdot}87$.

$$\text{S.E. of proportion} = \sqrt{\hat{P}\hat{Q}/n} = \sqrt{0{\cdot}13 \times 0{\cdot}87/500} = 0{\cdot}015$$

Thus, the limits for the proportion of bad pineapples in the consignment are :

$$\hat{P} \pm 3\sqrt{\hat{P}\hat{Q}/n} = 0{\cdot}130 \pm 3 \times 0{\cdot}015 = 0{\cdot}130 \pm 0{\cdot}045 = (0{\cdot}085, 0{\cdot}175)$$

Hence, the percentage of bad pineapples in the consignment lies almost certainly between 8·5 and 17·5.

Example 14·3. *A random sample of 500 apples was taken from a large consignment and 60 were found to be bad. Obtain the 98% confidence limits for the percentage of bad apples in the consignment.*

Solution. We have :

$$p = \text{Proportion of bad apples in the sample} = \frac{60}{500} = 0{\cdot}12$$

Since significant value of Z at 98% confidence coefficient (level of significance 2%) is 2·33, [from Normal Tables], 98% confidence limits for population proportion are :

$$p \pm 2{\cdot}33\ \sqrt{pq/n} = 0{\cdot}12 \pm 2{\cdot}33\ \sqrt{0{\cdot}12 \times 0{\cdot}88/500} = 0{\cdot}12 \pm 2{\cdot}33 \times \sqrt{0{\cdot}0002112}$$

$$= 0{\cdot}12 \pm 2{\cdot}33 \times 0{\cdot}01453 = (0{\cdot}08615,\ 0{\cdot}15385)$$

Hence, 98% confidence limits for percentage of bad apples in the consignment are (8·61, 15·38).

Example 14·4. *In a sample of 1,000 people in Maharashtra, 540 are rice eaters and the rest are wheat eaters. Can we assume that both rice and wheat are equally popular in this State at 1% level of significance ?*

Solution. In the usual notations, we are given : $n = 1{,}000$

$$X = \text{Number of rice eaters} = 540$$

$$\therefore \qquad p = \text{Sample proportion of rice eaters} = \frac{X}{n} = \frac{540}{1000} = 0{\cdot}54$$

Null Hypothesis, H_0 : Both rice and wheat are equally popular in the State so that

$$P = \text{Population proportion of rice eaters in Maharashtra} = 0{\cdot}5 \quad \Rightarrow \quad Q = 1 - P = 0{\cdot}5.$$

Alternative Hypothesis, $H_1 : P \neq 0{\cdot}5$ (Two-tailed alternative)

Test Statistic. Under H_0, the test statistic is :

$$Z = \frac{p - P}{\sqrt{PQ/n}} \sim N(0, 1), \text{ (since } n \text{ is large).}$$

$$Z = \frac{0{\cdot}54 - 0{\cdot}50}{\sqrt{0{\cdot}5 \times 0{\cdot}5/1000}} = \frac{0{\cdot}04}{0{\cdot}0158} = 2{\cdot}532$$

Conclusion. The significant or critical value of Z at 1% level of significance for two-tailed test is 2·58. Since computed Z = 2·532 is less than 2·58, it is not significant at 1% level of significance. Hence, the null hypothesis is accepted and we may conclude that rice and wheat are equally popular in Maharashtra State.

Example 14·5. *Twenty people were attacked by a disease and only 18 survived. Will you reject the hypothesis that the survival rate, if attacked by this disease, is 85% in favour of the hypothesis that it is more, at 5% level. (Use Large Sample Test.)*

Solution. In the usual notations, we are given : $n = 20$.

$$X = \text{Number of persons who survived after attack by a disease} = 18$$

$$p = \text{Proportion of persons survived in the sample} = \frac{18}{20} = 0{\cdot}90$$

Null Hypothesis, $H_0 : P = 0{\cdot}85$, *i.e.*, the proportion of persons survived after attack by a disease in the lot is 85%.

Alternative Hypothesis, $H_1 : P > 0{\cdot}85$ (Right-tailed alternative).

Test Statistic. Under H_0, the test statistic is:

$$Z = \frac{p - P}{\sqrt{PQ/n}} \sim N(0,1), \text{ (since sample is large)}$$

$$Z = \frac{0{\cdot}90 - 0{\cdot}85}{\sqrt{0{\cdot}85 \times 0{\cdot}15/20}} = \frac{0{\cdot}05}{\sqrt{0{\cdot}006375}} = \frac{0{\cdot}05}{0{\cdot}0798} = 0{\cdot}63$$

Conclusion. Since the alternative hypothesis is one-sided (right-tailed), we shall apply right-tailed test for testing significance of Z. The significant value of Z at 5% level of significance for right-tailed test is $+ 1{\cdot}645$. Since computed value of $Z = 0{\cdot}63$ is less than $1{\cdot}645$, it is not significant and we may accept the null hypothesis at 5% level of significance.

Example 14·6. *Work sampling studies are conducted to find the utilization of a machine. Out of 200 observations made, only 40 observations indicated the machine to be idle. Find the number of observations to be made in order to satisfy 95% confidence to state the utilization of machine with expected accuracy of ± 5%.*

Solution. The sample size n for estimating the population proportion P with confidence coefficient $(1 - \alpha) = 0{\cdot}95$ is given by the equation:

$$Pr\left[\,|p-P| \le 1{\cdot}96\sqrt{PQ/n}\,\right] = 0{\cdot}95 \qquad \text{... (*)}$$

We want. $$Pr\,[\,|p - P| < 0{\cdot}05] = 0{\cdot}95. \qquad \text{... (**)}$$

From (*) and (**), we obtain $\quad 1{\cdot}96\sqrt{PQ/n} = 0{\cdot}05 \quad \Rightarrow \quad n = \dfrac{PQ(1{\cdot}96)^2}{(0{\cdot}05)^2}$

Since P is not known, we may use its sample estimate $\quad \hat{P} = p = \dfrac{40}{200} = 0{\cdot}2.$

$$\therefore \qquad n = \frac{0{\cdot}2 \times (1 - 0{\cdot}2) \times 3{\cdot}8416}{0{\cdot}0025} = \frac{0{\cdot}6147}{0{\cdot}0025} = 245{\cdot}88 \simeq 246.$$

14·7·2. Test of Significance for Difference of Proportions.

Suppose we want to compare two distinct populations with respect to the prevalence of a certain attribute, say A, among their members. Let X_1, X_2 be the number of persons possessing the given attribute A in random samples of sizes n_1 and n_2 from the two populations respectively. Then sample proportions are given by : $p_1 = X_1/n_1$ and $p_2 = X_2/n_2$. If P_1 and P_2 are population proportions, then

$$E\,(p_1) = P_1, \qquad E\,(p_2) = P_2 \qquad [\text{c.f. Equation (14·4}a)]$$

and $$V\,(p_1) = \frac{P_1Q_1}{n_1} \quad \text{and} \quad V\,(p_2) = \frac{P_2Q_2}{n_2}$$

Since for large samples, p_1 and p_2 are independently and asymptotically normally distributed, $(p_1 - p_2)$ is also normally distributed. Then the standard variable corresponding to the difference $(p_1 - p_2)$ is given by:

$$Z = \frac{(p_1 - p_2) - E(p_1 - p_2)}{\sqrt{V(p_1 - p_2)}} \sim N(0, 1) \qquad \text{...(*)}$$

Under the *null hypothesis*, $H_0 \;:\; P_1 = P_2$, *i.e.*, there is no significant difference between the sample proportions, we have

$$E\,(p_1 - p_2) = E\,(p_1) - E\,(p_2) = P_1 - P_2 = 0 \qquad (\text{Under } H_0)$$

Also $$V\,(p_1 - p_2) = V\,(p_1) + V\,(p_2),$$
the covariance term Cov (p_1, p_2) vanishes, since sample proportions are independent.

$$\Rightarrow \qquad V\,(p_1 - p_2) = \frac{P_1Q_1}{n_1} + \frac{P_2Q_2}{n_2} = PQ\left(\frac{1}{n_1} + \frac{1}{n_2}\right),$$

$[\because$ Under H_0, $P_1 = P_2 = P$ (say), and $Q_1 = Q_2 = Q]$.

Hence, under $H_0 : P_1 = P_2$, the test statistic for difference of proportions becomes :

$$Z = \frac{p_1 - p_2}{\sqrt{PQ\left(\frac{1}{n_1} + \frac{1}{n_2}\right)}} \sim N(0, 1) \qquad \text{... (14·5)}$$

In general, we do not have any information as to the proportion of A's in the populations from which the samples have been taken. Under $H_0 : P_1 = P_2 = P$ (say), an unbiased estimate of the population proportion P, based on both the samples is given by :

$$\hat{P} = \frac{n_1 p_1 + n_2 p_2}{n_1 + n_2} = \frac{X_1 + X_2}{n_1 + n_2} \qquad \text{... (14·5}a)$$

The estimate is unbiased, since

$$E(\hat{P}) = \frac{1}{n_1 + n_2} E[n_1 p_1 + n_2 p_2] = \frac{1}{n_1 + n_2}[n_1 E(p_1) + n_2 E(p_2)]$$

$$= \frac{1}{n_1 + n_2}(n_1 P_1 + n_2 P_2) = P$$

$$[\because \; P_1 = P_2 = P, \text{ under } H_0]$$

Thus (14·5) along with (14·5*a*) gives the required test statistic.

Remarks:

1. Suppose we want to test the significance of the difference between p_1 and p, where $p = \frac{(n_1 p_1 + n_2 p_2)}{(n_1 + n_2)}$ gives a pooled estimate of the population proportion on the basis of both the samples. We have

$$V(p_1 - p) = V(p_1) + V(p) - 2\,\text{Cov}(p_1, p).$$

Since p_1 and p are not independent, $\quad \text{Cov}(p_1, p) \neq 0.$ ... (*)

$$\text{Cov}(p_1, p) = E\big[\{p_1 - E(p_1)\}\{p - E(p)\}\big]$$

$$= E\left[\{p_1 - E(p_1)\}\left\{\frac{1}{n_1 + n_2}\{(n_1 p_1 + n_2 p_2) - E(n_1 p_1 + n_2 p_2)\}\right\}\right]$$

$$= \frac{1}{n_1 + n_2} E\big[\{p_1 - E(p_1)\}\{n_1(p_1 - E(p_1)) + n_2(p_2 - E(p_2))\}\big]$$

$$= \frac{1}{n_1 + n_2}\big[n_1 E\{p_1 - E(p_1)\}^2 + n_2 E\{(p_1 - E(p_1))(p_2 - E(p_2))\}\big]$$

$$= \frac{1}{n_1 + n_2}[n_1 V(p_1) + n_2\,\text{Cov}(p_1, p_2)] = \frac{1}{n_1 + n_2} \cdot n_1 V(p_1), \qquad [\because \; \text{Cov}(p_1, p_2) = 0]$$

$$= \frac{n_1}{n_1 + n_2} \cdot \frac{pq}{n_1} = \frac{pq}{n_1 + n_2}.$$

$$\text{Var}(p) = \text{Var}\left[\frac{n_1 p_1 + n_2 p_2}{n_1 + n_2}\right] = \frac{1}{(n_1 + n_2)^2} \cdot \text{Var}(n_1 p_1 + n_2 p_2)$$

$$= \frac{1}{(n_1 + n_2)^2}\left[n_1^2\,\text{Var}(p_1) + n_2^2\,\text{Var}(p_2)\right]$$

covariance term vanishes since p_1 and p_2 are independent.

$$\therefore \quad \text{Var}(p) = \frac{1}{(n_1 + n_2)^2}\left(n_1^2 \cdot \frac{pq}{n_1} + n_2^2 \cdot \frac{pq}{n_2}\right) = \frac{pq}{n_1 + n_2}$$

Substituting (*) and simplifying, we shall get

$$\text{Var}(p_1 - p) = \frac{pq}{n_1} + \frac{pq}{n_1 + n_2} - 2\frac{pq}{n_1 + n_2} = pq\left[\frac{n_2}{n_1(n_1 + n_2)}\right]$$

Also $\quad E(p_1 - p) = E(p_1) - E(p) = P - P = 0$

Thus, the test statistic in this case becomes :

$$Z = \frac{(p_1 - p) - E(p_1 - p)}{S.E.(p_1 - p)} = \frac{p_1 - p}{\sqrt{\left\{\frac{n_2}{(n_1 + n_2)} \cdot \frac{pq}{n_1}\right\}}} \sim N(0,1) \quad \text{... (14·5}b\text{)}$$

2. Suppose the population proportions P_1 and P_2 are given to be distinctly different, *i.e.*, $P_1 \neq P_2$ and we want to test if the difference $(P_1 - P_2)$ in population proportions is likely to be hidden in simple samples of sizes n_1 and n_2 from the two populations respectively.

We have seen that in the usual notations,

$$Z = \frac{(p_1 - p_2) - E(p_1 - p_2)}{S.E.(p_1 - p_2)} = \frac{(p_1 - p_2) - (P_1 - P_2)}{\sqrt{\left(\frac{P_1Q_1}{n_1} + \frac{P_2Q_2}{n_2}\right)}} \sim N(0,1)$$

Here sample proportions are not given. If we set up the *null hypothesis* $H_0 : p_1 = p_2$, *i.e.*, the samples will not reveal the difference in the population proportions or in other words the difference in population proportions is likely to be hidden in sampling, the test statistic becomes :

$$Z = \frac{|P_1 - P_2|}{\sqrt{\left(\frac{P_1Q_1}{n_1} + \frac{P_2Q_2}{n_2}\right)}} \sim N(0,1) \quad \text{... (14·5}c\text{)}$$

Example 14·7. *Random samples of 400 men and 600 women were asked whether they would like to have a flyover near their residence. 200 men and 325 women were in favour of the proposal. Test the hypothesis that proportions of men and women in favour of the proposal, are same against that they are not, at 5% level.*

Solution. *Null Hypothesis* $H_0 : P_1 = P_2 = P$ (say), *i.e.*, there is no significant difference between the opinions of men and women as far as proposal of flyover is concerned.

Alternative Hypothesis, $H_1 : P_1 \neq P_2$ (two-tailed).

We are given :

$n_1 = 400,\quad X_1$ = Number of men favouring the proposal = 200

$n_2 = 600,\quad X_2$ = Number of women favouring the proposal = 325

$\therefore\quad p_1$ = Proportion of men favouring the proposal in the sample $= \frac{X_1}{n_1} = \frac{200}{400} = 0{\cdot}5$

p_2 = Proportion of women favouring the proposal in the sample $= \frac{X_2}{n_2} = \frac{325}{600} = 0{\cdot}541$

Test Statistic. Since samples are large, the test statistic under the Null Hypothesis, H_0 is :

$$Z = \frac{p_1 - p_2}{\sqrt{\hat{P}\hat{Q}\left(\frac{1}{n_1} + \frac{1}{n_2}\right)}} \sim N(0,1), \quad \text{where}$$

$$\hat{P} = \frac{n_1p_1 + n_2p_2}{n_1 + n_2} = \frac{X_1 + X_2}{n_1 + n_2} = \frac{200 + 325}{400 + 600} = 0{\cdot}525 \Rightarrow \hat{Q} = 1 - \hat{P} = 1 - 0{\cdot}525 = 0{\cdot}475$$

$$\therefore\quad Z = \frac{0{\cdot}500 - 0{\cdot}541}{\sqrt{0{\cdot}525 \times 0{\cdot}475 \times \left(\frac{1}{400} + \frac{1}{600}\right)}} = \frac{-0{\cdot}041}{\sqrt{0{\cdot}001039}} = \frac{-0{\cdot}041}{0{\cdot}0323} = -1{\cdot}269$$

Conclusion. Since $|Z| = 1.269$ which is less than 1·96, it is not significant at 5% level of significance. Hence, H_0 may be accepted at 5% level of significance and we may conclude that men and women do not differ significantly as regards proposal of flyover is concerned.

Example 14·8. *In a large city A, 20 per cent of a random sample of 900 school children had defective eye-sight. In other large city B, 15 per cent of random sample of 1,600 children had the same defect. Is this difference between the two proportions significant ? Obtain 95% confidence limits for the difference in the population proportions.*

Solution. In usual notations : $n_1 = 900, \quad p_1 = 20\% = 0.20, \quad n_2 = 1600, \quad p_2 = 15\% = 0.15.$

Null hypothesis, $H_0 : P_1 = P_2.$

Alternative hypothesis, $H_1 : P_1 \neq P_2$ (Two-tailed).

Test Statistic. Under H_0, the test statistic is :

$$Z = \frac{p_1 - p_2}{S.E.\,(p_1 - p_2)} \sim N(0,1), \qquad \text{(Since the samples are large.)}$$

where
$$\hat{P} = \frac{900 \times 0.20 + 1600 \times 0.15}{900 + 1600} = 0.168 \Rightarrow \hat{Q} = 1 - \hat{P} = 0.832.$$

$$S.E.\,(p_1 - p_2) = \sqrt{\left\{0.168 \times 0.832\left(\frac{1}{900} + \frac{1}{1600}\right)\right\}} = \sqrt{0.0002427} = 0.0156$$

$\therefore$
$$Z = \frac{0.20 - 0.15}{0.0156} = 3.21$$

Conclusion. Since the calculated value of Z is greater than 1·96, it is significant at 5% level. We, therefore, reject the null hypothesis H_0 and conclude that the difference between the two proportions is significant.

The 95% confidence limits for the difference $P_1 - P_2$ are :

$$(p_1 - p_2) \pm 1.96\ S.E.\ \text{of}\ (p_1 - p_2),$$

where
$$S.E.\ of\ (p_1 - p_2) = \sqrt{\left(\frac{P_1Q_1}{n_1} + \frac{P_2Q_2}{n_2}\right)} \simeq \sqrt{\left(\frac{p_1q_1}{n_1} + \frac{p_2q_2}{n_2}\right)}$$

$$\simeq \sqrt{\left(\frac{0.20 \times 0.80}{900} + \frac{0.15 \times 0.85}{1600}\right)} = 0.016.$$

Hence, the 95% confidence limits for $P_1 - P_2$ are:

$$(0.20 - 0.15) \pm 1.96\,(0.016) = 0.05 \pm 0.031 = (0.019 \text{ and } 0.081).$$

Example 14·9. *A company has the head office at Kolkata and a branch at Mumbai. The personnel director wanted to know if the workers at the two places would like the introduction of a new plan of work and a survey was conducted for this purpose. Out of a sample of 500 workers at Kolkata, 62% favoured the new plan. At Mumbai out of a sample of 400 workers, 41% were against the new plan. Is there any significant difference between the two groups in their attitude towards the new plan at 5% level ?*

Solution. In the usual notations, we are given:

$$n_1 = 500, \quad p_1 = 0.62 \qquad \text{and} \qquad n_2 = 400, \quad p_2 = 1 - 0.41 = 0.59$$

Null hypothesis, $H_0 : P_1 = P_2$, *i.e.,* there is no significant difference between the two groups in their attitude towards the new plan.

Alternative Hypothesis, $H_1 : P_1 \neq P_2$ (Two-tailed).

Test Statistic. Under H_0, the test statistic for large samples is :

$$Z = \frac{p_1 - p_2}{S.E.(p_1 - p_2)} = \frac{p_1 - p_2}{\sqrt{\left\{\hat{P}\hat{Q}\left(\frac{1}{n_1} + \frac{1}{n_2}\right)\right\}}} \sim N(0,1), \quad \text{where}$$

$$\hat{P} = \frac{n_1 p_1 + n_2 p_2}{n_1 + n_2} = \frac{500 \times 0{\cdot}62 + 400 \times 0{\cdot}59}{500 + 400} = 0{\cdot}607 \quad \Rightarrow \quad \hat{Q} = 1 - \hat{P} = 0{\cdot}393$$

$$\therefore \quad Z = \frac{0{\cdot}62 - 0{\cdot}59}{\sqrt{\left\{0{\cdot}607 \times 0{\cdot}393 \times \left(\frac{1}{500} + \frac{1}{400}\right)\right\}}} = \frac{0{\cdot}03}{\sqrt{0{\cdot}00107}} = \frac{0{\cdot}03}{0{\cdot}0327} = 0{\cdot}917.$$

Critical region. At 5% level of significance, the critical value of Z for a two-tailed test is 1·96. Thus the critical region consists of all values of $Z \geq 1{\cdot}96$ or $Z \leq -1{\cdot}96$.

Conclusion. Since the calculated value of $|Z| = 0{\cdot}917$ is less than the critical value of $|Z|$ (1·96), it is not significant at 5% level of significance. Hence, the data do not provide us any evidence against the null hypothesis which may be accepted, and we may conclude that there is no significant difference between the two groups in their attitude towards the new plan.

Example 14·10. *Before an increase in excise duty on tea, 800 persons out of a sample of 1,000 persons were found to be tea drinkers. After an increase in duty, 800 people were tea drinkers in a sample of 1,200 people. Using standard error of proportion, state whether there is a significant decrease in the consumption of tea after the increase in excise duty ?*

Solution. In the usual notations, we have: $n_1 = 1{,}000$; $n_2 = 1{,}200$.

p_1 = Sample proportion of tea drinkers before increase in excise duty $= \frac{800}{1000} = 0{\cdot}80$

p_2 = Sample proportion of tea drinkers after increase in excise duty $= \frac{800}{1200} = 0{\cdot}67$

Null Hypothesis, $H_0 : P_1 = P_2$, *i.e.,* there is no significant difference in the consumption of tea before and after the increase in excise duty.

Alternative Hypothesis, $H_1 : P_1 > P_2$ (Right-tailed alternative).

Test Statistic. Under the null hypothesis, the test statistic is :

$$Z = \frac{p_1 - p_2}{\sqrt{\left\{\hat{P}\hat{Q}\left(\frac{1}{n_1} + \frac{1}{n_2}\right)\right\}}} \sim N(0,1) \qquad \text{(Since samples are large)}$$

where $\hat{P} = \frac{n_1 p_1 + n_2 p_2}{n_1 + n_2} = \frac{800 + 800}{1000 + 1200} = \frac{16}{22}$, and $\hat{Q} = 1 - \hat{P} = \frac{6}{22}$

$$\therefore \quad Z = \frac{0{\cdot}80 - 0{\cdot}67}{\sqrt{\left[\frac{16}{22} \times \frac{6}{22} \times \left(\frac{1}{1000} + \frac{1}{1200}\right)\right]}} = \frac{0{\cdot}13}{0{\cdot}019} = 6{\cdot}842$$

Conclusion. Since Z is much greater than 1·645 as well as 2·33 (since test is one-tailed), it is highly significant at both 5% and 1% levels of significance. Hence, we reject the null hypothesis H_0 and conclude that there is a significant decrease in the consumption of tea after increase in the excise duty.

Example 14·11. *A cigarette manufacturing firm claims that its brand A of the cigarettes outsells its brand B by 8%. If it is found that 42 out of a sample of 200 smokers prefer brand A and 18 out of* *another random sample of 100 smokers prefer brand B, test whether the 8% difference is a valid claim. (Use 5% level of significance.)*

Solution. In usual notations, we are given:

$$n_1 = 200, \quad X_1 = 42 \quad \Rightarrow \quad p_1 = \frac{X_1}{n_1} = \frac{42}{200} = 0{\cdot}21$$

$$n_2 = 100, \quad X_2 = 18 \quad \Rightarrow \quad p_2 = \frac{X_2}{n_2} = \frac{18}{100} = 0{\cdot}18$$

We set up the *Null Hypothesis* that 8% difference in the sale of two brands of cigarettes is a valid claim, *i.e.*, $H_0 : P_1 - P_2 = 0{\cdot}08$.

Alternative Hypothesis : $H_1 : P_1 - P_2 \neq 0{\cdot}08$ (Two-tailed).

Test Statistic. Under H_0, the test statistic is:

$$Z = \frac{(p_1 - p_2) - (P_1 - P_2)}{\sqrt{\left\{\hat{P}\hat{Q}\left(\frac{1}{n_1} + \frac{1}{n_2}\right)\right\}}} \sim N(0,1), \qquad \text{[Since samples are large]}$$

where $$\hat{P} = \frac{X_1 + X_2}{n_1 + n_2} = \frac{42 + 18}{200 + 100} = \frac{60}{300} = 0{\cdot}20 \Rightarrow \hat{Q} = 1 - \hat{P} = 0{\cdot}80$$

$$\therefore \quad Z = \frac{(0{\cdot}21 - 0{\cdot}18) - 0{\cdot}08}{\sqrt{\left\{0{\cdot}2 \times 0{\cdot}8\left(\frac{1}{200} + \frac{1}{100}\right)\right\}}} = \frac{-0{\cdot}05}{\sqrt{0{\cdot}16 \times 0{\cdot}015}} = \frac{-0{\cdot}05}{0{\cdot}04899} = -1{\cdot}02$$

Since $|Z| = 1{\cdot}02 < 1{\cdot}96$, it is not significant at 5% level of significance. Hence, null hypothesis may be retained at 5% level of significance and we may conclude that a difference of 8% in the sale of two brands of cigarettes is a valid claim by the firm.

Example 14·12. *On the basis of their total scores, 200 candidates of a civil service examination are divided into two groups, the upper 30 per cent and the remaining 70 per cent. Consider the first question of this examination. Among the first group, 40 had the correct answer, whereas among the second group, 80 had the correct answer. On the basis of these results, can one conclude that the first question is no good at discriminating ability of the type being examined here ?*

Solution. We are given:

n = Total number of candidates = 200

n_1 = The number of candidates in the upper 30% group = $\frac{30}{100} \times 200$ = 60

n_2 = The number of candidates in the remaining 70% group = $\frac{70}{100} \times 200$ = 140

X_1 = The number of candidates, with correct answer in the first group = 40

X_2 = The number of candidates, with correct answer in the second group = 80

$$\therefore \quad p_1 = \frac{X_1}{n_1} = \frac{40}{60} = 0{\cdot}6666 \quad \text{and} \quad p_2 = \frac{X_2}{n_2} = \frac{80}{140} = 0{\cdot}5714$$

Null Hypothesis, H_0 : There is no significance difference in the sample proportions, *i.e.*, $P_1 = P_2$. In other words, the first question is not good at discriminating the ability of the type being examined here.

Alternative Hypothesis, $H_1 : P_1 \neq P_2$.

Test Statistic. Under H_0 the test statistic is :

$$Z = \frac{p_1 - p_2}{\sqrt{\left\{\hat{P}\hat{Q}\left(\frac{1}{n_1} + \frac{1}{n_2}\right)\right\}}} \sim N(0,1) \qquad \text{(Since samples are large.)}$$

where $\hat{P} = \dfrac{X_1 + X_2}{n_1 + n_2} = \dfrac{40 + 80}{60 + 140} = 0{\cdot}6, \qquad \hat{Q} = 1 - \hat{P} = 0{\cdot}4$

$$\therefore \qquad Z = \frac{0{\cdot}6666 - 0{\cdot}5714}{\sqrt{\left\{0{\cdot}6 \times 0{\cdot}4\left(\frac{1}{60} + \frac{1}{140}\right)\right\}}} = \frac{0{\cdot}0953}{0{\cdot}0756} = 1{\cdot}258$$

Conclusion. Since $|Z| < 1{\cdot}96$, the data are consistent with the null hypothesis at 5% level of significance. Hence, we conclude that the first question is not good enough to distinguish between the ability of the two groups of candidates.

Example 14·13. *In a year there are 956 births in a town A, of which 52·5% were males, while in towns A and B combined, this proportion in a total of 1,406 births was 0·496. Is there any significant difference in the proportion of male births in the two towns ?*

Solution. In usual notations, we are given:

$n_1 = 956, \quad n_1 + n_2 = 1{,}406 \quad \Rightarrow \quad n_2 = 1{,}406 - 956 = 450$

p_1 = Proportion of males in the sample of town $A = 0{\cdot}525$.

Let p_2 be the proportion of males in the sample (of size n_2) of town B. Then

$\hat{P}$ = Proportion of males in both the samples combined $= \dfrac{n_1p_1 + n_2p_2}{n_1 + n_2} = 0{\cdot}496$ (Given)

$$\therefore \qquad \frac{956 \times 0{\cdot}525 + 450 \times p_2}{1{,}406} = 0{\cdot}496 \quad \Rightarrow \quad p_2 = 0{\cdot}434 \qquad \text{(On simplification).}$$

Null Hypothesis, $H_0 : P_1 = P_2$, *i.e.*, there is no significant difference in the proportion of male births in the two towns A and B.

Alternative Hypothesis, $H_1 : P_1 \neq P_2$ (two-tailed).

Test Statistic. Under H_0, the test statistic is:

$$Z = \frac{p_1 - p_2}{\sqrt{\left\{\hat{P}\hat{Q}\left(\frac{1}{n_1} + \frac{1}{n_2}\right)\right\}}} \sim N(0,1), \qquad \text{(Since samples are large.)}$$

where $\hat{P} = \dfrac{n_1p_1 + n_2p_2}{n_1 + n_2} = 0{\cdot}496 \quad \Rightarrow \quad \hat{Q} = 1 - \hat{P} = 0{\cdot}504$

$$\therefore \qquad Z = \frac{0{\cdot}525 - 0{\cdot}434}{\sqrt{\left\{0{\cdot}496 \times 0{\cdot}504\left(\frac{1}{956} + \frac{1}{450}\right)\right\}}} = \frac{0{\cdot}091}{0{\cdot}027} = 3{\cdot}368$$

Conclusion. Since $|Z| > 3$, the null hypothesis is rejected, *i.e.*, the data are inconsistent with the hypothesis $P_1 = P_2$ and we conclude that there is significant difference in the proportion of male births in the towns A and B.

Example 14·14. *In two large populations, there are 30 and 25 per cent respectively of blue-eyed people. Is this difference likely to be hidden in samples of 1,200 and 900 respectively from the two populations ?*

Solution. Here, we are given : $\quad n_1 = 1{,}200, \quad n_2 = 900.$

P_1 = Proportion of blue-eyed people in the first population $= 30\% = 0{\cdot}30$

P_2 = Proportion of blue-eyed people in the second population $= 25\% = 0{\cdot}25$

$\therefore \qquad Q_1 = 1 - P_1 = 0{\cdot}70 \quad \text{and} \quad Q_2 = 1 - P_2 = 0{\cdot}75$

We set up the *null hypothesis*, H_0 that $p_1 = p_2$, *i.e.*, the sample proportions are equal. In other words, the difference in population proportions is likely to be hidden in sampling.

Test Statistic. Under $H_0 : p_1 = p_2$, the test statistic is :

$$|Z| = \frac{|P_1 - P_2|}{\sqrt{\left(\frac{P_1Q_1}{n_1} + \frac{P_2Q_2}{n_2}\right)}} \sim N(0, 1) \qquad \text{(Since samples are large.)}$$

$$\therefore \qquad |Z| = \frac{0{\cdot}30 - 0{\cdot}25}{\sqrt{\left(\frac{0{\cdot}3 \times 0{\cdot}7}{1,200} + \frac{0{\cdot}25 \times 0{\cdot}75}{900}\right)}} = \frac{0{\cdot}05}{0{\cdot}0195} = 2{\cdot}56$$

Conclusion. Since $|Z| > 1{\cdot}96$, the null hypothesis ($p_1 = p_2$) is refuted at 5% level of significance and we conclude that the difference in population proportions is unlikely to be hidden in sampling. In other words, the samples will reveal the difference in the population proportions.

Example 14·15. *In a random sample of 400 students of the university teaching departments, it was found that 300 students failed in the examination. In another random sample of 500 students of the affiliated colleges, the number of failures in the same examination was found to be 300. Find out whether the proportion of failures in the university teaching departments is significantly greater than the proportion of failures in the university teaching departments and affiliated colleges taken together.*

Solution. Here we are given: $n_1 = 400$, $n_2 = 500$, $p_1 = \frac{300}{400} = 0{\cdot}75$, $p_2 = \frac{300}{500} = 0{\cdot}60$

$$\therefore \qquad q_1 = 1 - p_1 = 1 - 0{\cdot}75 = 0{\cdot}25 \quad \text{and} \quad q_2 = 1 - p_2 = 1 - 0{\cdot}60 = 0{\cdot}40$$

Here we set up the *null hypothesis*, H_0 that p_1 and $p = \hat{p}$, where $\hat{p}$ is the pooled estimate, *i.e.*, proportion of failures in the university teaching departments and affiliated colleges taken together, do not differ singificantly.

$$\text{S.E. of } (p - p_1) = \sqrt{\left(\frac{pq}{n_1 + n_2} \times \frac{n_2}{n_1}\right)} \qquad [\ c.f.\ (14{\cdot}5b)\ ] \qquad \ldots (*)$$

where $\quad p = \hat{P} = \dfrac{n_1p_1 + n_2p_2}{n_1 + n_2} = \dfrac{400 \times 0{\cdot}75 + 500 \times 0{\cdot}60}{400 + 500} = 0{\cdot}67 \Rightarrow q = 1 - 0{\cdot}67 = 0{\cdot}33$

$$\therefore \qquad \text{S.E. of } (p - p_1) = \sqrt{\left(\frac{0{\cdot}67 \times 0{\cdot}33}{400 + 500} \times \frac{500}{400}\right)} = 0{\cdot}018 \qquad \text{[Using (*)]}$$

Test Statistic. Under the null hypothesis, H_0, the test statistic is :

$$Z = \frac{p - p_1}{\text{S.E. of } (p - p_1)} \sim N(0,1), \qquad \text{(Since samples are large.)}$$

$$\therefore \qquad Z = \frac{0{\cdot}67 - 0{\cdot}75}{0{\cdot}018} = \frac{-0{\cdot}08}{0{\cdot}018} = -4{\cdot}08$$

Conclusion. Since the calculated value of $|Z|$ is greater than 3, it is significant. Hence the null hypothesis H_0 is rejected and we conclude that there is significant difference between p_1 and $p = \hat{p}$.

Example 14·16. *If for one-half of n events, the chance of success is p and the chance of failure is q, while for the other half the chance of success is q and the chance of failure is p, show that the standard deviation of the number of successes is the same as if the chance of successes were p in all the cases, i.e.,* $\sqrt{npq}$ *but that the mean of the number of successes is n/2 and not np.*

Solution. Let X_1 and X_2 denote the number of successes in the first half and the second half of n events respectively. Then according to the given conditions, we have:

$$\left.\begin{aligned} E(X_1) &= \frac{n}{2}p \\ V(X_1) &= \frac{n}{2}pq \end{aligned}\right\} \quad \text{and} \quad \left.\begin{aligned} E(X_2) &= \frac{n}{2}q \\ V(X_2) &= \frac{n}{2}qp \end{aligned}\right\}$$

The mean and variance of the number of successes in all the n events are given by:

$$E(X_1 + X_2) = E(X_1) + E(X_2) = \frac{n}{2}p + \frac{n}{2}q = \frac{n}{2} \qquad (\because p + q = 1)$$

and
$$V(X_1 + X_2) = V(X_1) + V(X_2) = \frac{n}{2}pq + \frac{n}{2}qp = npq$$

since the first and second half of events are independent.

Hence the variance and *s.d.* is the same as if the probability of success in all the n events is p.

14·8. SAMPLING OF VARIABLES

In the present section we will discuss in detail the sampling of variables such as height, weight, age, income, etc. In the case of sampling of variables each member of the population provides the value of the variable and the aggregate of these values forms the frequency distribution of the population. From the population, a random sample of size n can be drawn by any of the sampling methods discussed earlier, which is same as choosing n values of the given variable from the distribution.

14·8·1. Unbiased Estimate for Population Mean (μ) and Variance (σ²)

Let $x_1, x_2, \ldots, x_n$ be a random sample of size n from a large population $X_1, X_2, \ldots, X_N$ (of size N) with mean μ and variance σ^2. Then the sample mean ($\bar{x}$) and variance (s^2) are given by :–

$$\bar{x} = \frac{1}{n}\sum_{i=1}^{n} x_i, \quad \text{and} \quad s^2 = \frac{1}{n}\sum_{i=1}^{n}(x_i - \bar{x})^2$$

$$E(\bar{x}) = E\left(\frac{1}{n}\sum_{i=1}^{n} x_i\right) = \frac{1}{n}\sum_{i=1}^{n} E(x_i)$$

Since x_i is a sample observation from the population X_i, $(i = 1, 2, \ldots, N)$ it can take any one of the values $X_1, X_2, \ldots, X_N$ each with equal probability $1/N$.

$$\therefore \quad E(x_i) = \frac{1}{N}X_1 + \frac{1}{N}X_2 + \ldots + \frac{1}{N}X_N = \frac{1}{N}(X_1 + X_2 + \ldots + X_N) = \mu \qquad \ldots (1)$$

Hence,
$$E(\bar{x}) = \frac{1}{n}\sum_{i=1}^{n}(\mu) = \frac{1}{n}n\mu \quad \Rightarrow \quad E(\bar{x}) = \mu \qquad \ldots (14·6)$$

Thus, *the sample mean ($\bar{x}$) is an unbiased estimate of the population mean* (μ).

Now
$$E(s^2) = E\left\{\frac{1}{n}\sum_{i=1}^{n}(x_i - \bar{x})^2\right\} = E\left(\frac{1}{n}\sum_{i=1}^{n} x_i^2 - \bar{x}^2\right) = \frac{1}{n}\sum_{i=1}^{n} E\left(x_i^2\right) - E\left(\bar{x}^2\right) \qquad \ldots (2)$$

We have $V(x_i) = E[x_i - E(x_i)]^2 = E(x_i - \mu)^2$, [From (1)]

$$= \frac{1}{N}\left[(X_1 - \mu)^2 + (X_2 - \mu)^2 + \ldots + (X_N - \mu)^2\right] = \sigma^2 \qquad \ldots (3)$$

Also
$$V(x) = E(x^2) - [E(x)]^2 \quad \Rightarrow \quad E(x^2) = V(x) + [E(x)]^2 \qquad \ldots (4)$$

In particular
$$E(x_i^2) = V(x_i) + [E(x_i)]^2 = \sigma^2 + \mu^2 \qquad \ldots (5)$$

Also from (4), we obtain
$$E(\bar{x}^2) = V(\bar{x}) + [E(\bar{x})]^2$$

But $V(\bar{x}) = \dfrac{\sigma^2}{n}$, where σ^2 is the population variance. [*c.f.* § 14·8·2]

$$\therefore \qquad E\left(\overline{x}^2\right) = \frac{\sigma^2}{n} + \mu^2 \qquad \text{[Using (14·6)]} \qquad \ldots (5a)$$

Substituting from (5) and (5*a*) in (2), we get

$$E\left(s^2\right) = \frac{1}{n}\sum_{i=1}^{n}\left(\sigma^2 + \mu^2\right) - \left(\frac{\sigma^2}{n} + \mu^2\right)$$

$$= \frac{1}{n}\cdot n\left(\sigma^2 + \mu^2\right) - \left(\frac{\sigma^2}{n} + \mu^2\right) = \left(1 - \frac{1}{n}\right)\sigma^2 = \frac{n-1}{n}\sigma^2 \qquad \ldots (14·7)$$

Since $E(s^2) \neq \sigma^2$, *sample variance is not an unbiased estimate of population variance.*

From (14·7), we get $\dfrac{n}{n-1}E\left(s^2\right) = \sigma^2 \;\Rightarrow\; E\left(\dfrac{ns^2}{n-1}\right) = \sigma^2$

$$\Rightarrow \qquad E\left\{\frac{1}{n-1}\sum_{i=1}^{n}(x_i - \overline{x})^2\right\} = \sigma^2 \;\Rightarrow\; E\left(S^2\right) = \sigma^2 \qquad \ldots (14·8)$$

where $$S^2 = \frac{1}{n-1}\sum_{i=1}^{n}(x_i - \overline{x})^2 \qquad \ldots (14·8a)$$

$\therefore$ S^2 *is an unbiased estimate of the population variance* σ^2.

Aliter for $E(s^2)$

$$s^2 = \frac{1}{n}\left[\sum_{i=1}^{n}(x_i - \overline{x})^2\right] = \frac{1}{n}\left[\sum_{i=1}^{n}\{(x_i - \mu) - (\overline{x} - \mu)\}^2\right]$$

$$= \frac{1}{n}\left[\sum_{i=1}^{n}(x_i - \mu)^2 + n(\overline{x} - \mu)^2 - 2(\overline{x} - \mu)\sum_{i=1}^{n}(x_i - \mu)\right]$$

But $$\sum_i(x_i - \mu) = \sum_i x_i - n\mu = n\overline{x} - n\mu = n(\overline{x} - \mu)$$

$$\therefore \qquad s^2 = \frac{1}{n}\left\{\sum_{i=1}^{n}(x_i - \mu)^2 - n(\overline{x} - \mu)^2\right\} = \frac{1}{n}\sum_{i=1}^{n}(x_i - \mu)^2 - (\overline{x} - \mu)^2$$

$$\Rightarrow \qquad E(s^2) = \frac{1}{n}\sum_{i=1}^{n}E(x_i - \mu)^2 - E(\overline{x} - \mu)^2 = \frac{1}{n}\sum_{i=1}^{n}E\{x_i - E(x_i)\}^2 - E[\overline{x} - E(\overline{x})]^2$$

$$= \frac{1}{n}\sum_{i=1}^{n}V(x_i) - V(\overline{x}) = \left(1 - \frac{1}{n}\right)\sigma^2$$

Remarks:

1. Here we see that although sample mean is an unbiased estimate of population mean, sample variance is not an unbiased estimate of population variance. However, an unbiased estimate of σ^2 is given by S^2, defined in equation (14·8*a*).

   S^2 plays a very important role in sampling theory, particularly in small sampling theory. Whenever σ^2 is not known, its estimate S^2 given by (14·8*a*) is used for practical purposes.

2. We have $\quad s^2 = \dfrac{1}{n}\displaystyle\sum_{i=1}^{n}(x_i - \overline{x})^2 \quad$ and $\quad S^2 = \dfrac{1}{n-1}\displaystyle\sum_{i=1}^{n}(x_i - \overline{x})^2$

$$\therefore \qquad n\,s^2 = (n-1)S^2 \;\Rightarrow\; s^2 = \left(1 - \frac{1}{n}\right)S^2$$

   Hence, for large samples, *i.e.*, for $n \to \infty$, we have $s^2 \to S^2$.

   In other words, for large samples (*i.e.*, $n \to \infty$), we may take $\hat{\sigma}^2 = s^2$ $\qquad \ldots (14·8b)$

14·8·2. Standard Error of Sample Mean

The variance of the sample mean is σ^2/n, where σ is the population standard deviation and n is the size of the random sample. The S.E. of mean of a random sample of size n from a population with variance σ^2 is $\sigma/\sqrt{n}$.

Proof. Let x_i, $(i = 1, 2, ..., n)$ be a random sample of size n from a population with variance σ^2, then the sample mean $\bar{x}$ is : $\bar{x} = \frac{1}{n}(x_1 + x_2 + ... + x_n)$

$$V(\bar{x}) = V\left\{\frac{1}{n}(x_1 + x_2 + ... + x_n)\right\} = \frac{1}{n^2}V(x_1 + x_2 + ... + x_n)$$

$$\therefore \quad = \frac{1}{n^2}\{V(x_1) + V(x_2) + ... + V(x_n)\},$$

the covariance terms vanish since the sample observations are independent.

But $\quad V(x_i) = \sigma^2, (i = 1, 2, ..., n)$ [From (3) of § 14·8·1]

$$\therefore \quad V(\bar{x}) = \frac{1}{n^2}(n\sigma^2) = \frac{\sigma^2}{n} \quad \Rightarrow \quad S.E.(\bar{x}) = \sqrt{\frac{\sigma^2}{n}} = \frac{\sigma}{\sqrt{n}} \quad ... (14·9)$$

14·8·3. Test of Significance for Single Mean

We have proved that if x_i, $(i = 1, 2, ..., n)$ is a random sample of size n from a normal population with mean μ and variance σ^2, then the sample mean is distributed normally with mean μ and variance σ^2/n, *i.e.*, $\bar{x} \sim N(\mu, \sigma^2/n)$. However, this result holds, *i.e.*, $\bar{x} \sim N(\mu, \sigma^2/n)$, even in random sampling from non-normal population provided the sample size n is large [*c.f.* Central Limit Theorem]. Thus for large samples, the *standard normal variate* corresponding to $\bar{x}$ is :

$$Z = \frac{\bar{x} - \mu}{\sigma/\sqrt{n}}$$

Under the *null hypothesis* H_0, that the sample has been drawn from a population with mean μ and variance σ^2, *i.e.*, there is no significant difference between the sample mean ($\bar{x}$) and population mean (μ), the test statistic (for large samples), is :

$$Z = \frac{\bar{x} - \mu}{\sigma/\sqrt{n}} \sim N(0,1) \quad ... (14·9a)$$

Remarks:

1. If the population *s.d.* σ is unknown then we use its estimate provided by the sample variance given by [See (14·8*b*)]. $\hat{\sigma}^2 = s^2 \Rightarrow \hat{\sigma} = s$ (for large samples).
2. *Confidence limits for* μ. 95% confidence interval for μ are given by:

$$|Z| \le 1·96, \quad \Rightarrow \quad \left|\frac{\bar{x} - \mu}{\sigma/\sqrt{n}}\right| \le 1·96 \quad \Rightarrow \quad \bar{x} - 1·96(\sigma/\sqrt{n}) \le \mu \le \bar{x} + 1·96(\sigma/\sqrt{n}) \quad ... (14·10)$$

and $\bar{x} \pm 1·96\sigma/\sqrt{n}$ are known as 95% confidence limits for μ. Similarly, 99% confidence limits for μ are $\bar{x} \pm 2·58\sigma/\sqrt{n}$ and 98% confidence limits for μ are $\bar{x} \pm 2·33\sigma/\sqrt{n}$.

However, in sampling from a finite population of size N, the corresponding 95% and 99% confidence limits for μ are respectively

$$\bar{x} \pm 1·96\frac{\sigma}{\sqrt{n}}\sqrt{\frac{N-n}{N-1}} \quad \text{and} \quad \bar{x} \pm 2·58\frac{\sigma}{\sqrt{n}}\sqrt{\frac{N-n}{N-1}} \quad ... (14·10a)$$

3. The confidence limits for any parameter (P, μ, etc.) are also known as its *fiducial limits.*

Example 14·17. *A sample of 900 members has a mean 3·4 cms. and s.d. 2·61 cms. Is the sample from a large population of mean 3·25 cms. and s.d. 2·61 cms. ?*

If the population is normal and its mean is unknown, find the 95% and 98% fiducial limits of true mean.

Solution. *Null Hypothesis,* (H_0) : The sample has been drawn from the population with mean $\mu = 3·25$ cms. and S.D. $\sigma = 2·61$ cms.

Alternative Hypothesis, $H_1 : \mu \neq 3{\cdot}25$ (Two-tailed).

Test Statistic. Under H_0, the test statistic is : $Z = \dfrac{\overline{x} - \mu}{\sigma/\sqrt{n}} \sim N(0, 1)$, (Since n is large.)

Here, we are given: $\overline{x} = 3{\cdot}4$ cms., $n = 900$ cms., $\mu = 3{\cdot}25$ cms. and $\sigma = 2{\cdot}61$ cms.

$$\therefore \quad Z = \frac{3{\cdot}40 - 3{\cdot}25}{2{\cdot}61/\sqrt{900}} = \frac{0{\cdot}15 \times 30}{2{\cdot}61} = 1{\cdot}73$$

Since $|Z| < 1{\cdot}96$, we conclude that the data don't provide us any evidence against the null hypothesis (H_0) which may, therefore, be accepted at 5% level of significance.

95% *fiducial limits* for the population mean μ are :

$$\overline{x} \pm 1{\cdot}96\,(\sigma/\sqrt{n}) = 3{\cdot}40 \pm 1{\cdot}96\,(2{\cdot}61/\sqrt{900}) = 3{\cdot}40 \pm 0{\cdot}1705, \quad i.e., \quad 3{\cdot}5705 \text{ and } 3{\cdot}2295$$

98% *fiducial limits* for μ are given by :

$$\overline{x} \pm 2{\cdot}33\,\frac{\sigma}{\sqrt{n}} = 3{\cdot}40 \pm 2{\cdot}33 \times \frac{2{\cdot}61}{30} = 3{\cdot}40 \pm 0{\cdot}2027, \quad i.e., \quad (3{\cdot}6027 \text{ and } 3{\cdot}1973)$$

Remark. 2·33 is the value z_1 of Z from standard normal probability integrals, such that

$$P(|Z| > z_1) = 0{\cdot}98 \quad \Rightarrow \quad P(Z > z_1) = 0{\cdot}49$$

Example 14·18. *An insurance agent has claimed that the average age of policy-holders who insure through him is less than the average for all agents, which is 30·5 years.*

A random sample of 100 policy-holders who had insured through him gave the following age distribution :

Age last birthday :	*16—20*	*21—25*	*26—30*	*31—35*	*36—40*
Number of persons :	*12*	*22*	*20*	*30*	*16*

Calculate the arithmetic mean and standard deviation of this distribution and use these values to test his claim at the 5% level of significance. You are given that Z (1·645) = 0·95.

Solution. *Null Hypothesis*, $H_0 : \mu = 30{\cdot}5$ years, *i.e.*, the sample mean ($\overline{x}$) and population mean (μ) do not differ significantly.

Alternative Hypothesis, $H_1 : \mu < 30{\cdot}5$ years (Left-tailed alternative).

CALCULATIONS FOR SAMPLE MEAN AND S.D.

Age last birthday	*No. of persons* (f)	*Mid-point* x	$d = \dfrac{x-28}{5}$	fd	fd^2
16—20	12	18	−2	−24	48
21—25	22	23	−1	−22	22
26—30	20	28	0	0	0
31—35	30	33	1	30	30
36—40	16	38	2	32	64
Total	$N = 100$			$\Sigma fd = 16$	$\Sigma fd^2 = 164$

$$\overline{x} = 28 + \frac{5 \times 16}{100} = 28{\cdot}8 \text{ years}; \quad s = 5 \times \sqrt{\left[\frac{164}{100} - \left(\frac{16}{100}\right)^2\right]} = 6{\cdot}35 \text{ years}$$

Since the sample is large, $\hat{\sigma}\ ;\ s = 6{\cdot}35$ years.

Test Statistic. Under H_0, the test statistic is :

$$Z = \frac{\overline{x} - \mu}{\sqrt{s^2/n}} \sim N(0,1), \qquad \text{(Since sample is large.)}$$

$$Z = \frac{28\cdot8 - 30\cdot5}{\sqrt{(6\cdot35)^2/100}} = \frac{-1\cdot7}{0\cdot635} = -2\cdot681.$$

Conclusion. Since computed value of $Z = -2\cdot681 < -1\cdot645$ or $|Z| = 2\cdot681 > 1\cdot645$, it is significant at 5% level of significance. Hence, we reject the null hypothesis, H_0 (Accept H_1) at 5% level of significance and conclude that the insurance agent's claim, that the average age of policy-holders who insure through him is less than the average for all agents, is valid.

Example 14·19. *As an application of Central Limit Theorem, show that if E is such that* $P(|\bar{X} - \mu| < E) > 0\cdot95$, *then the minimum sample size n is given by* $n = \{(1\cdot96)^2 \sigma^2/E^2\}$ *where* μ *and* σ^2 *are the mean and variance respectively of the population and* $\bar{X}$ *is the mean of the random sample.*

Solution. By Central Limit Theorem, we know that $\bar{x} \sim N(\mu, \sigma^2/n)$ asymptotically, *i.e.*, for large n.

$$\therefore \qquad Z = \frac{\bar{x} - \mu}{\sigma/\sqrt{n}} \sim N(0, 1), \text{ asymptotically, i.e., for large } n.$$

From normal probability tables, we have $P(|Z| \le 1\cdot96) = 0\cdot95$

$$\Rightarrow \qquad P\left[\left|\frac{\bar{x} - \mu}{\sigma/\sqrt{n}}\right| \le 1\cdot96\right] = 0\cdot95 \quad \Rightarrow \quad P\left[|\bar{x} - \mu| \le 1\cdot96 \frac{\sigma}{\sqrt{n}}\right] = 0\cdot95 \qquad \ldots (*)$$

We are given that $\qquad P\left[|\bar{X} - \mu| < E\right] > 0\cdot95 \qquad \ldots (**)$

From (*) and (**), we have $\qquad E > \dfrac{1\cdot96\sigma}{\sqrt{n}} \quad \Rightarrow \quad n > \dfrac{(1\cdot96)^2\sigma^2}{E^2} = \dfrac{3\cdot84\sigma^2}{E^2}. \qquad \ldots$ (14·10 b)

Hence, minimum sample size n for estimating μ with 95% confidence coefficient is given by $n = 3\cdot84\, \sigma^2/E^2$, where E is the permissible error.

Remark. The minimum sample size for estimating μ with confidence coefficient $(1-\alpha)$ is given by $\sigma^2 z_\alpha^2/E^2$, where z_α is the significant value of Z at level of significance α and E is the permissible error in the estimate.

Arguing similarly, the minimum sample size for estimating population proportion P with confidence coefficient $(1 - \alpha)$ is given by $n = PQ\, z_\alpha^2/E^2$, where z_α is the significant value of Z at 'α' level of significance and E is the permissible error in the estimate. If P is unknown, we may use $\hat{P} = p$. ... (14·10 c)

Example 14·20. *The mean muscular endurance score of a random sample of 60 subjects was found to be 145 with a s.d. of 40. Construct a 95% confidence interval for the true mean. Assume the sample size to be large enough for normal approximation. What size of sample is required to estimate the mean within 5 of the true mean with a 95% confidence ?*

Solution. In usual notations, we are given : $n = 60$, $\bar{x} = 145$ and $s = 40$.

95% confidence limits for true mean (μ) are :

$$\bar{x} \pm 1\cdot96\, s/\sqrt{n}, \quad (\hat{\sigma} = s^2, \text{ since sample is large})$$

$$= 145 \pm \frac{1.96 \times 40}{\sqrt{60}} = 145 \pm \frac{78\cdot4}{7\cdot75} = 145 \pm 10\cdot12 = (134\cdot88 \text{ and } 155\cdot12)$$

Hence, 95% confidence interval for μ is (134·88 and 155·12).

In the notations of Example 14·19, we have

$$n = \left(\frac{z_\alpha \times \sigma}{E}\right)^2 = \left(\frac{1\cdot96 \times 40}{5}\right)^2 = (15\cdot68)^2 = 245\cdot86\,;\; 246.$$

$$[\because z_{0\cdot05} = 1\cdot96, \sigma = s = 40 \text{ and } |\bar{x} - \mu| < 5 = E]$$

Example 14·21. *The standard deviation of a population is 2·70 cms. Find the probability that in a random sample of size 66 (i) the sample mean will differ from the population mean by 0·75 cm. or more, and (ii) the sample mean will exceed the population mean by 0·75 cm. or more (given that the value of the standard normal probability integral from 0 to 2·25 is 0·4877).*

Solution. Here we are given : $n = 66, \quad \sigma = 2{\cdot}70$ cms.

Since n is large, the sample mean $\bar{x} \sim N(\mu, \sigma^2/n)$.

$$\therefore \qquad Z = \frac{\bar{x} - \mu}{\sigma/\sqrt{n}} \sim N(0, 1) \qquad \ldots (*)$$

We want

(i) $P[|\bar{x} - \mu| \geq 0{\cdot}75] = 1 - P[|\bar{x} - \mu| < 0{\cdot}75]$

$$= 1 - P\left(\left|\frac{\sigma Z}{\sqrt{n}}\right| < 0{\cdot}75\right) \qquad \text{[From (*)]}$$

$$= 1 - P\left(|Z| < 0{\cdot}75\frac{\sqrt{n}}{\sigma}\right) = 1 - 2P\left(0 < Z < 0{\cdot}75\frac{\sqrt{n}}{\sigma}\right)$$

$$= 1 - 2P\left(0 < Z < 0{\cdot}75\frac{\sqrt{66}}{2{\cdot}70}\right) = 1 - 2P\left(0 < Z < \frac{0{\cdot}75 \times 8{\cdot}124}{2{\cdot}70}\right)$$

$$= 1 - 2P(0 < Z < 2{\cdot}25) = 1 - 2 \times 0{\cdot}4877 = 0{\cdot}0246$$

(ii) $P(\bar{x} - \mu > 0{\cdot}75) = P(Z > 0{\cdot}75\ \sqrt{n}/\sigma) = P(Z > 2{\cdot}25)$

$$= 0{\cdot}5 - P(0 < Z < 2{\cdot}25) = 0{\cdot}5 - 0{\cdot}4877 = 0{\cdot}0123.$$

Example 14·22. *A normal population has a mean of 0·1 and standard deviation of 2·1. Find the probability that mean of a sample of size 900 will be negative.*

Solution. Here we are given that $X \sim N(\mu, \sigma^2)$, where $\mu = 0{\cdot}1$ and $\sigma = 2{\cdot}1$ and $n = 900$. Since $X \sim N(\mu, \sigma^2)$, the sample mean $\bar{x} \sim N(\mu, \sigma^2/n)$. The standard normal variate corresponding to $\bar{x}$ is given by :

$$Z = \frac{\bar{x} - \mu}{\sigma/\sqrt{n}} = \frac{\bar{x} - 0{\cdot}1}{2{\cdot}1/30} = \frac{\bar{x} - 0{\cdot}1}{0{\cdot}07} \quad \Rightarrow \quad \bar{x} = 0{\cdot}1 + 0{\cdot}07Z, \text{ where } Z \sim N(0,1)$$

The required probability p, that the sample mean is negative is given by :

$$p = P(\bar{x} < 0) = P(0{\cdot}1 + 0{\cdot}07Z < 0) = P\left(Z < \frac{-0{\cdot}10}{0{\cdot}07}\right)$$

$$= P(Z < -1{\cdot}43) = P(Z > 1{\cdot}43) = 0{\cdot}5 - P(0 < Z < 1{\cdot}43) = 0{\cdot}5 - 0{\cdot}4236 = 0{\cdot}0764.$$

(From Normal Probability Tables)

Example 14·23. *The guaranteed average life of a certain type of electric light bulbs is 1,000 hours with a standard deviation of 125 hours. It is decided to sample the output so as to ensure that 90 per cent of the bulbs do not fall short of the guaranteed average by more than 2·5 per cent. What must be the minimum size of the sample ?*

Solution. Here $\mu = 1{,}000$ hours, $\sigma = 125$ hours.

Since we do not want the sample mean to be less than the guaranteed average mean ($\mu = 1{,}000$) by more than 2·5%, we should have

$$\bar{x} > 1{,}000 - 2{\cdot}5\% \text{ of } 1{,}000 \quad \Rightarrow \quad \bar{x} > 1{,}000 - 25 = 975$$

Let n be the given sample size. Then

$$Z = \frac{\bar{x} - \mu}{\sigma/\sqrt{n}} \sim N(0, 1), \qquad \text{(since sample is large.)}$$

We want,

$$Z = \frac{\bar{x} - \mu}{\sigma/\sqrt{n}} > \frac{975 - 1{,}000}{125/\sqrt{n}} = -\frac{\sqrt{n}}{5} \qquad (\because \bar{x} > 975)$$

According to the given condition : $P(Z > -\sqrt{n}/5) = 0{\cdot}90 \quad \Rightarrow \quad P(0 < Z < \sqrt{n}/5) = 0{\cdot}40$

$$\Rightarrow \quad \frac{\sqrt{n}}{5} = 1{\cdot}28 \qquad \text{(From Normal Probability Tables)}$$

$$\therefore \quad n = 25 \times (1{\cdot}28)^2 = 41 \text{ (approx.)}$$

Example 14·24. *A survey is proposed to be conducted to know the annual earnings of the old Statistics graduates of Delhi University. How large should the sample be taken in order to estimate the mean monthly earnings within plus and minus ₹ 10,000 at 95% confidence level ? The standard deviation of the annual earnings of the entire population is known to be ₹ 30,000.*

Solution. We are given : σ = ₹ 30,000.

We want : $P[\,|\,\bar{x} - \mu\,| < 10{,}000] = 0{\cdot}95$ ... (*)

We know that, in sampling from normal population or for large samples from any population, $\bar{X} \sim N(\mu, \sigma^2/n)$. Hence from Normal Probability Tables, we have

$$P[\,|\,Z\,| \le 1{\cdot}96] = 0{\cdot}95 \quad \Rightarrow \quad P\left[\left|\frac{\bar{x} - \mu}{\sigma/\sqrt{n}}\right| \le 1{\cdot}96\right] = 0{\cdot}95$$

$$\Rightarrow \quad P\left[\,|\,\bar{x} - \mu\,| < 1{\cdot}96 \times (\sigma/\sqrt{n})\right] = 0{\cdot}95 \qquad \text{... (**)}$$

From (*) and (**), we get $\dfrac{1{\cdot}96 \times \sigma}{\sqrt{n}} = 10{,}000 \quad \Rightarrow \quad \dfrac{1{\cdot}96 \times 30{,}000}{\sqrt{n}} = 10{,}000$

$$\therefore \quad n = (1{\cdot}96 \times 3)^2 = (5{\cdot}88)^2 = 34{\cdot}56 \simeq 35$$

Aliter. Using Remark to Example 14·19,

$$n = \left(\frac{z_\alpha \times \sigma}{E}\right)^2 = \left(\frac{1{\cdot}96 \times 30{,}000}{10{,}000}\right)^2 \simeq 35\,.$$

14·8·4. Test of Significance for Difference of Means

Let $\bar{x}_1$ be the mean of a sample of size n_1 from a population with mean μ_1 and variance σ_1^2 and let $\bar{x}_2$ be the mean of an independent random sample of size n_2 from another population with mean μ_2 and variance σ_2^2. Then, since sample sizes are large,

$$\bar{x}_1 \sim N(\mu_1, \sigma_1^2/n_1) \quad \text{and} \quad \bar{x}_2 \sim N(\mu_2, \sigma_2^2/n_2)$$

Also $\bar{x}_1 - \bar{x}_2$, being the difference of two independent normal variates is also a normal variate. The value of Z (*S.N.V.*) corresponding to $\bar{x}_1 - \bar{x}_2$ is given by:

$$Z = \frac{(\bar{x}_1 - \bar{x}_2) - E(\bar{x}_1 - \bar{x}_2)}{S.E.(\bar{x}_1 - \bar{x}_2)} \sim N(0,1), \quad \text{for large samples.}$$

Under the null hypothesis, $H_0 : \mu_1 = \mu_2$, *i.e.*, there is no significant difference between the sample means, we get

$$E(\bar{x}_1 - \bar{x}_2) = E(\bar{x}_1) - E(\bar{x}_2) = \mu_1 - \mu_2 = 0\ ; \qquad V(\bar{x}_1 - \bar{x}_2) = V(\bar{x}_1) + V(\bar{x}_2) = \frac{\sigma_1^2}{n_1} + \frac{\sigma_2^2}{n_2},$$

the covariance term vanishes, since the sample means $\bar{x}_1$ and $\bar{x}_2$ are independent.

Thus, under $H_0 : \mu_1 = \mu_2$, the test statistic becomes (for large samples),

$$Z = \frac{\bar{x}_1 - \bar{x}_2}{\sqrt{(\sigma_1^2/n_1) + (\sigma_2^2/n_2)}} \sim N(0,1) \qquad \text{... (14·11)}$$

Remarks :

1. If $\sigma_1^2 = \sigma_2^2 = \sigma^2$, *i.e.*, if the samples have been drawn from the populations with common S.D. σ, then under $H_0 : \mu_1 = \mu_2$,

$$Z = \frac{\bar{x}_1 - \bar{x}_2}{\sigma\sqrt{\{(1/n_1) + (1/n_2)\}}} \sim N(0,1) \qquad \text{... [14·11}a\text{]}$$

2. If in (14·11*a*), σ is not known, then its estimate based on the sample variances is used. If the sample sizes are not sufficiently large, then an unbiased estimate of σ^2 is given by :

$$\hat{\sigma}^2 = \frac{(n_1-1)S_1^2+(n_2-1)S_2^2}{(n_1+n_2-2)}, \text{ since}$$

$$E(\hat{\sigma}^2) = \frac{1}{n_1+n_2-2}\left\{(n_1-1)E\left(S_1^2\right)+(n_2-1)E\left(S_2^2\right)\right\} = \frac{1}{n_1+n_2-2}\left[(n_1-1)\,\sigma^2+(n_2-1)\,\sigma^2\right] = \sigma^2$$

But since sample sizes are large, $S_1^2 \simeq s_1^2$, $S_2^2 \simeq s_2^2$, $n_1 - 1 \simeq n_1$, $n_2 - 1 \simeq n_2$. Therefore, in practice, for large samples, the following estimate of σ^2 without any serious error is used :

$$\hat{\sigma}^2 = \frac{n_1 s_1^2 + n_2 s_2^2}{n_1+n_2} \qquad \text{... [14·11}b\text{]}$$

However, if sample sizes are small, then an exact sample test, *t*-test for difference of means (*c.f.* Chapter 16) is to be used.

3. If $\sigma_1^2 \neq \sigma_2^2$ and σ_1 and σ_2 are not known, then they are estimated from sample values. This results in some error, which is practically immaterial, if samples are large. These estimates for large samples are given by : $\hat{\sigma}_1^2 = S_1^2 \simeq s_1^2$ and $\hat{\sigma}_2^2 = S_2^2 \simeq s_2^2$. (Since samples are large.)

In this case, (14·11) gives
$$Z = \frac{\bar{x}_1-\bar{x}_2}{\sqrt{\left(s_1^2/n_1\right)+\left(s_2^2/n_2\right)}} \sim N(0,1) \qquad \text{... [14·11(}c\text{)]}$$

Example 14·25. *The means of two single large samples of 1,000 and 2,000 members are 67·5 inches and 68·0 inches respectively. Can the samples be regarded as drawn from the same population of standard deviation 2·5 inches? (Test at 5% level of significance).*

Solution. In usual notations, we are given :

$n_1 = 1{,}000$, $n_2 = 2{,}000$, $\bar{x}_1 = 67{\cdot}5$ inches, $\bar{x}_2 = 68{\cdot}0$ inches.

Null hypothesis, $H_0 : \mu_1 = \mu_2$ and $\sigma = 2{\cdot}5$ inches, *i.e.*, the samples have been drawn from the same population of standard deviation 2·5 inches.

Alternative Hypothesis, $H_1 : \mu_1 \neq \mu_2$ (Two-tailed)

Test Statistic. Under H_0, the test statistic is:

$$Z = \frac{\bar{x}_1-\bar{x}_2}{\sqrt{\left\{\sigma^2\left(\frac{1}{n_1}+\frac{1}{n_2}\right)\right\}}} \sim N(0,1) \qquad \text{(Since samples are large)}$$

$$Z = \frac{67{\cdot}5-68{\cdot}0}{2{\cdot}5\times\sqrt{\left(\frac{1}{1000}+\frac{1}{2000}\right)}} = \frac{-0{\cdot}5}{2{\cdot}5\times 0{\cdot}0387} = -5{\cdot}1$$

Conclusion. Since $|Z| > 3$, the value is highly significant and we reject the null hypothesis and conclude that samples are certainly not from the same population with standard deviation 2·5.

Example 14·26. *In a survey of buying habits, 400 women shoppers are chosen at random in super market 'A' located in a certain section of the city. Their average weekly food expenditure is ₹ 250 with a standard deviation of ₹ 40. For 400 women shoppers chosen at random in super market 'B' in another section of, the city, the average weekly food expenditure is ₹ 220 with a standard deviation of ₹ 55. Test at 1% level of significance whether the average weekly food expenditure of the two populations of shoppers are equal.*

Solution. In the usual notations, we are given that

$n_1 = 400$, $\bar{x}_1 = ₹\ 250$, $s_1 = ₹\ 40$

$n_2 = 400$, $\bar{x}_2 = ₹\ 220$, $s_2 = ₹\ 55$

Null hypothesis, $H_0 : \mu_1 = \mu_2$, *i.e.*, the average weekly food expenditures of the two populations of shoppers are equal.

Alternative Hypothesis, $H_1 : \mu_1 \neq \mu_2$. (Two-tailed)

Test Statistic. Since samples are large, under H_0, the test statistic is :

$$Z = \frac{\bar{x}_1 - \bar{x}_2}{\sqrt{\left(\frac{\sigma_1^2}{n_1} + \frac{\sigma_2^2}{n_2}\right)}} \sim N(0,1)$$

Since σ_1 and σ_2, the population standard deviations are not known, we can take for large samples $\hat{\sigma}_1^2 = s_1^2$ and $\hat{\sigma}_2^2 = s_2^2$ and then Z is given by :

$$Z = \frac{\bar{x}_1 - \bar{x}_2}{\sqrt{\left(\frac{s_1^2}{n_1} + \frac{s_2^2}{n_2}\right)}} = \frac{250 - 220}{\sqrt{\left[\frac{(40)^2}{400} + \frac{(55)^2}{400}\right]}} = 8{\cdot}82 \text{ (approx.)}$$

Conclusion. Since $|Z|$ is much greater than 2·58, the null hypothesis ($\mu_1 = \mu_2$) is rejected at 1% level of significance and we conclude that the average weekly expenditures of two populations of shoppers in markets A and B differ significantly.

Example 14·27. *The average hourly wage of a sample of 150 workers in a plant 'A' was ₹ 2·56 with a standard deviation of ₹ 1·08. The average hourly wage of a sample of 200 workers in plant 'B' was ₹ 2·87 with a standard deviation of ₹ 1·28. Can an applicant safely assume that the hourly wages paid by plant 'B' are higher than those paid by plant 'A' ?*

Solution. Let X_1 and X_2 denote the hourly wages (in ₹) of workers in plant A and plant B, respectively. Then, in usual notations we are given :

$$\left.\begin{array}{lll} n_1 = 150, & \bar{x}_1 = 2{\cdot}56, & s_1 = 1{\cdot}08 = \hat{\sigma}_1 \\ n_2 = 200, & \bar{x}_2 = 2{\cdot}87, & s_2 = 1{\cdot}28 = \hat{\sigma}_2 \end{array}\right\} \quad \text{(Since samples are large.)}$$

Null Hypothesis, $H_0 : \mu_1 = \mu_2$, *i.e.,* there is no significant difference between the mean level of wages of workers in plant A and plant B.

Alternative Hypothesis, $H_1 : \mu_2 > \mu_1$ *i.e.,* $\mu_1 < \mu_2$ (Left-tailed test)

Test Statistic. Under H_0, the test statistic (for large samples) is :

$$Z = \frac{\bar{x}_1 - \bar{x}_2}{\sqrt{\left(\frac{\sigma_1^2}{n_1} + \frac{\sigma_2^2}{n_2}\right)}} = \frac{\bar{x}_1 - \bar{x}_2}{\sqrt{\left\{\frac{s_1^2}{n_1} + \frac{s_2^2}{n_1}\right\}}} \sim N(0,1)$$

$$\therefore \quad Z = \frac{2{\cdot}56 - 2{\cdot}87}{\sqrt{\left(\frac{(1{\cdot}08)^2}{150} + \frac{(1{\cdot}28)^2}{200}\right)}} = \frac{-0{\cdot}31}{\sqrt{0{\cdot}016}} = \frac{-0{\cdot}31}{0{\cdot}126} = -2{\cdot}46.$$

Critical region. For a one-tailed test, the critical value of Z at 5% level of significance is 1·645. The critical region for left-tailed test thus consists of all values of $Z \leq -1{\cdot}645$.

Conclusion. Since calculated value of Z (−2·46) is less than critical value (−1·645), it is significant at 5% level of significance. Hence, the null hypothesis is rejected at 5% level of significance and we conclude that the average hourly wages paid by plant 'B' are certainly higher than those paid by plant 'A'.

Example 14·28. *In a certain factory there are two independent processes manufacturing the same item. The average weight in a sample of 250 items produced from one process is found to be 120 ozs. with a standard deviation of 12 ozs. while the corresponding figures in a sample of 400 items from the other process are 124 and 14. Obtain the standard error of difference between the two sample means. Is this difference significant ? Also find the 99% confidence limits for the difference in the average weights of items produced by the two processes respectively.*

Solution. In usual notations, we are given:

$$n_1 = 250, \quad \bar{x}_1 = 120 \text{ oz}, \quad s_1 = 12 \text{ oz.} = \hat{\sigma}_1$$
$$n_2 = 400, \quad \bar{x}_2 = 124 \text{ oz}, \quad s_2 = 14 \text{ oz.} = \hat{\sigma}_1$$ (Since samples are large.)

$$S.E.(\bar{x}_1 - \bar{x}_2) = \sqrt{\left(\frac{\sigma_1^2}{n_1} + \frac{\sigma_2^2}{n_2}\right)} = \sqrt{\left(\frac{s_1^2}{n_1} + \frac{s_2^2}{n_2}\right)}$$ (Since samples are large.)

$$= \sqrt{\left(\frac{144}{250} + \frac{196}{400}\right)} = \sqrt{0·576 + 0·490} = 1·034$$

Null Hypothesis, $H_0 : \mu_1 = \mu_2$, *i.e.*, the sample means do not differ significantly.

Alternative Hypothesis, $H_1 : \mu_1 \neq \mu_2$ (Two-tailed).

Test Statistic. Under H_0, the test statistic is :

$$Z = \frac{\bar{x}_1 - \bar{x}_2}{S.E.(\bar{x}_1 - \bar{x}_2)} = \frac{120 - 124}{1·034} \sim N(0,1)$$

$$\therefore \quad |Z| = \frac{4}{1·034} = 3·87$$

Conclusion. Since $|Z| > 3$, the null hypothesis is rejected and we conclude that there is significant difference between the sample means.

99% confidence limits for $|\mu_1 - \mu_2|$, *i.e.*, for the difference in the average weights of items produced by two processes, are :

$$|\bar{x}_1 - \bar{x}_2| \pm 2·58 \; S.E.(\bar{x}_1 - \bar{x}_2) = 4 \pm 2·58 \times 1·034 = 4 \pm 2·67 \text{ (approx.)} = 6·67 \text{ and } 1·33$$

$$\therefore \quad 1·33 < |\mu_1 - \mu_2| < 6·67.$$

Example 14·29. *The mean height of 50 male students who showed above average participation in college athletics was 68·2 inches with a standard deviation of 2·5 inches; while 50 male students who showed no interest in such participation had a mean height of 67·5 inches with a standard deviation of 2·8 inches.*

(i) *Test the hypothesis that male students who participate in college athletics are taller than other male students.*

(ii) *By how much should the sample size of each of the two groups be increased in order that the observed difference of 0·7 inches in the mean heights be significant at the 5% level of significance.*

Solution. Let X_1 and X_2 denote the height (in inches) of athletic participants and non-athletic participants, respectively. In the usual notations, we are given :

$$n_1 = 50, \quad \bar{x}_1 = 68·2, \quad s_1 = 2·5 \quad ; \quad n_2 = 50, \quad \bar{x}_2 = 67·5, \quad s_2 = 2·8$$

Null Hypothesis, $H_0 : \mu_1 = \mu_2$.

Alternative Hypothesis, $H_1 : \mu_1 > \mu_2$ (Right-tailed).

Test Statistic. Under H_0, the test statistic for large samples is :

$$Z = \frac{\bar{x}_1 - \bar{x}_2}{\sqrt{\left(\frac{s_1^2}{n_1} + \frac{s_2^2}{n_2}\right)}} \sim N(0,1)$$

$$\therefore \quad Z = \frac{68·2 - 67·5}{\sqrt{\left\{\frac{(2·5)^2}{50} + \frac{(2·8)^2}{50}\right\}}} = \frac{0·7}{\sqrt{0·282}} = \frac{0·7}{0·53} = 1·32$$

For a right-tailed test, the critical (significant) value of Z at 5% level of significance is 1·645.

(i) Since the calculated value of Z(1·32) is less than the critical value (1·645), it is not significant at 5% level of significance. Hence the null hypothesis is accepted and we conclude that the college athletes are not taller than other male students.

(ii) The difference between the mean heights of two groups, each of size n will be significant at 5% level of significance if $Z \geq 1{\cdot}645$

$$\Rightarrow \quad \frac{68{\cdot}2 - 67{\cdot}5}{\sqrt{\left\{\dfrac{(2{\cdot}5)^2}{n} + \dfrac{(2{\cdot}8)^2}{n}\right\}}} \geq 1{\cdot}645 \quad \text{or} \quad \frac{0{\cdot}7}{\sqrt{14{\cdot}09/n}} \geq 1{\cdot}645, \quad i.e., \quad \frac{0{\cdot}7}{3{\cdot}754/\sqrt{n}} \geq 1{\cdot}645$$

$$\therefore \quad n \geq \left(\frac{1{\cdot}645 \times 3{\cdot}754}{0{\cdot}7}\right)^2 = (8{\cdot}8219)^2 = 77{\cdot}83 \simeq 78$$

Hence, the sample size of each of the two groups should be increased by at least 78– 50 = 28, in order that the difference between the mean heights of the two groups is significant.

14·8·5. Test of Significance for Difference of Standard Deviations

If s_1 and s_2 are the standard deviations of two independent samples, then under null hypothesis, $H_0 : \sigma_1 = \sigma_2$, *i.e.,* sample standard deviations don't differ significantly, the statistic :

$$Z = \frac{s_1 - s_2}{S.E.(s_1 - s_2)} \sim N(0,1), \quad \text{for large samples.}$$

But in case of large samples, the S.E. of the difference of the sample standard deviations is given by :

$$S.E.(s_1 - s_2) = \sqrt{\frac{\sigma_1^2}{2n_1} + \frac{\sigma_2^2}{2n_2}}$$

$$\therefore \quad Z = \frac{s_1 - s_2}{\sqrt{\left(\dfrac{\sigma_1^2}{2n_1} + \dfrac{\sigma_2^2}{2n_2}\right)}} \sim N(0,1), \qquad \ldots (14{\cdot}12)$$

σ_1^2 and σ_2^2 are usually unknown and for large samples, we use their estimates given by the corresponding sample variances. Hence, the test statistic reduces to

$$Z = \frac{s_1 - s_2}{\sqrt{\left(\dfrac{s_1^2}{2n_1} + \dfrac{s_2^2}{2n_2}\right)}} \sim N(0,1) \qquad \text{(for large samples)} \quad \ldots (14{\cdot}13)$$

Example 14·30. *Random samples drawn from two countries gave the following data relating to the heights of adult males :*

	Country A	*Country B*
Mean height (in inches)	67·42	67·25
Standard deviation (in inches)	2·58	2·50
Number in samples	1,000	1,200

(i) Is the difference between the means significant?

(ii) Is the difference between the standard deviations significant?

Solution. In usual notations, we are given:

$$n_1 = 1{,}000, \qquad \bar{x}_1 = 67{\cdot}42 \text{ inches}, \qquad s_1 = 2{\cdot}58 \text{ inches},$$
$$n_2 = 1{,}200, \qquad \bar{x}_2 = 67{\cdot}25 \text{ inches}, \qquad s_2 = 2{\cdot}50 \text{ inches}.$$

As in the previous examples (since sample sizes are large), we can take

$$\hat{\sigma}_1 = s_1 = 2{\cdot}58, \qquad \hat{\sigma}_2 = s_2 = 2{\cdot}50$$

(i) $H_0 : \mu_1 = \mu_2$, *i.e.*, the sample means do not differ significantly.

$H_1 : \mu_1 \neq \mu_2$ (Two tailed).

Under the *Null hypothesis*, H_0, the test statistic is :

$$Z = \frac{\bar{x}_1 - \bar{x}_2}{\sqrt{(s_1^2/n_1) + (s_2^2/n_2)}} \sim N(0,1), \text{ since samples are large.}$$

$$\therefore \quad Z = \frac{67{\cdot}42 - 67{\cdot}25}{\sqrt{\left\{\frac{(2{\cdot}58)^2}{1000} + \frac{(2{\cdot}50)^2}{1200}\right\}}} = \frac{0{\cdot}17}{\sqrt{\left(\frac{6{\cdot}66}{1000} + \frac{6{\cdot}25}{1200}\right)}} = 1{\cdot}56$$

Conclusion. Since $|Z| < 1{\cdot}96$, null hypothesis may be accepted at 5% level of significance and we may conclude that three is no significant difference between the sample means.

(ii) Under H_0 : that there is no significant difference between sample standard deviations,

$$Z = \frac{s_1 - s_2}{S.E.(s_1 - s_2)} \sim N(0,1), \quad \text{since samples are large.}$$

$$S.E.\,(s_1 - s_2) = \sqrt{\left(\frac{\sigma_1^2}{2n_1} + \frac{\sigma_2^2}{2n_2}\right)} \simeq \sqrt{\left(\frac{s_1^2}{2n_1} + \frac{s_2^2}{2n_2}\right)}$$

if σ_1 and σ_2 are not known and $\hat{\sigma}_1 = s_1$, $\hat{\sigma}_2 = s_2$.

$$\therefore \quad S.E.\,(s_1 - s_2) = \sqrt{\left\{\frac{(2{\cdot}58)^2}{2 \times 1000} + \frac{(2{\cdot}50)^2}{2 \times 1200}\right\}} = 0{\cdot}07746$$

Hence,
$$Z = \frac{2{\cdot}58 - 2{\cdot}50}{0{\cdot}07746} = \frac{0{\cdot}08}{0{\cdot}07746} = 1{\cdot}03$$

Conclusion. Since $|Z| < 1{\cdot}96$, the data don't provide us any evidence against the null hypothesis which may be accepted at 5% level of significance. Hence, the sample standard deviations do not differ significantly.

Example 14·31. *Two populations have their means equal, but S.D. of one is twice the other.*

(i) *Show that in the samples of size 2,000 each drawn under simple sampling conditions, the difference of means will, in all probability, not exceed $0{\cdot}15\sigma$, where σ is the smaller S.D.*

(ii) *What is the probability that the difference will exceed half this amount ?*

Solution. Let the standard deviations of the two populations be σ and 2σ respectively and let μ be the mean of each of the two populations. Also we are given $n_1 = n_2 = 2{,}000$. If $\bar{x}_1$ and $\bar{x}_2$ be the two sample means then, since samples are large,

$$Z = \frac{(\bar{x}_1 - \bar{x}_2) - E(\bar{x}_1 - \bar{x}_2)}{S.E.\,(\bar{x}_1 - \bar{x}_2)} \sim N(0,1) \qquad \ldots (*)$$

Now
$$E(\bar{x}_1 - \bar{x}_2) = E(\bar{x}_1) - E(\bar{x}_2) = \mu - \mu = 0$$

and
$$S.E.\,(\bar{x}_1 - \bar{x}_2) = \sqrt{\left\{\frac{\sigma^2}{n_1} + \frac{(2\sigma)^2}{n_2}\right\}} = \sigma \cdot \sqrt{\left(\frac{1}{2000} + \frac{4}{2000}\right)} = 0{\cdot}05\sigma$$

Substituting in (*), we get

$$Z = \frac{\bar{x}_1 - \bar{x}_2}{0{\cdot}05\sigma} \sim N(0,1) \qquad \ldots (**)$$

(i) Under simple sampling conditions, we should in all probability have

$$|Z| < 3 \;\Rightarrow\; |\bar{x}_1 - \bar{x}_2| < 3 \times 0{\cdot}05\sigma = 0{\cdot}15\sigma, \text{ as required.}$$

(ii) We want $p = P\left[\,|\,\bar{x}_1 - \bar{x}_2\,| > \frac{1}{2} \times 0{\cdot}15\sigma\right]$

$\therefore$ $p = P[0{\cdot}05\sigma\,|\,Z\,| > 0{\cdot}075\sigma]$ [From (**)]

$= P\,[\,|Z| > 1{\cdot}5] = 1 - P\,[\,|Z| \le 1{\cdot}5]$

$= 1 - 2{\cdot}P[0 \le Z \le 1{\cdot}5) = 1 - 2 \times 0{\cdot}4332 = 0{\cdot}1336.$

DISCUSSION AND REVIEW QUESTIONS

1. (a) There are two populations and P_1 and P_2 are the proportions of members in the two populations belonging to 'low-income' group. It is desired to test the hypothesis $H_0 : P_1 = P_2$. Explain clearly, the procedure that you would follow to carry out the above test at 5% level of significance.

State the theorem on which the above test is based.

In respect of the above two populations, if it is claimed that P_1, the proportion of 'low-income' group in the first population, is greater than P_2, how will you modify the procedure to test this claim (at 5% level) ?

(b) Take a concrete illustration and in relation to this illustration, explain the following terms:

(i) Null hypothesis and alternative hypothesis. (ii) Type *I* and Type *II* errors.

(iii) Critical region. (iv) Level of significance.

2. (a) Explain clearly the terms "Standard Error" and "Sampling Distribution". Show that in series of n independent trials with constant probability p of success, the standard error of the proportion of successes is $\sqrt{pq/n}$, where $q = 1 - p$.

(b) n individuals fall into one or the other two categories with probabilities p and q $(= 1 - p)$, the number in the two categories are x_1 and x_2, $(x_1 + x_2 = n)$. Show that covariance between x_1 and x_2 is $(-npq)$. Hence obtain the variance of the difference $[(x_1/n) - (x_2/n)]$, between the proportions.

(c) Explain clearly the procedure generally followed in testing of a hypothesis. Point out the difference between one-tail and two-tail tests.

(d) What do you mean by interval estimation and how would you set up the confidence limits for a parameter from a sample? Give the formula for 95% confidence limits for mean and proportion. What modifications do you have to make if the sampling is done from finite population, (i) without replacement and (ii) with replacement ?

3. P_1 and P_2 are the (unknown) proportions of students wearing glasses in two universities *A* and *B*. To compare P_1 and P_2, samples of sizes n_1 and n_2 are taken from the two populations and the number of students wearing glasses is found to be x_1 and x_2 respectively. Suggest an unbiased estimate of $(P_1 - P_2)$ and obtain its sampling distribution when n_1 and n_2 are large. Hence, explain how to test the hypothesis that $P_1 = P_2$.

4. (a) Define sampling distribution and standard error. Obtain standard error of mean when population is large.

(b) Find the standard error of a linear function of a number of variables. Deduce the standard error of the mean of n uncorrelated variables following the same distribution.

5. Derive the expressions for the standard error of

(i) the mean of a random sample of size n, and

(ii) the difference of the means of two independent random samples of sizes n_1 and n_2.

6. What is meant by a statistical hypothesis ? What are the two types of errors of decision that arise in testing a hypothesis ? Briefly explain how a statistical hypothesis is tested.

7. Obtain the expressions for the standard error of sampling distributions of: (*i*) sample mean ($\bar{x}$), and (*ii*) sample variance (s^2), in random sampling from a large population. Assume that n, the sample size, is large.

8. Let $X_1, X_2, \ldots, X_n$ be a random sample from a population which has a finite fourth moment $\mu_r = E(X_i - \mu)^r$, $r = 4$; $E(X_i) = \mu$ and $\text{Var}(X_i) = \sigma^2$; and let $\bar{X} = \frac{1}{n}\sum_{i=1}^{n} X_i$ and $S^2 = \frac{1}{n-1}\sum_{i=1}^{n}(X_i - \bar{X})^2$.

Show that :

(*i*) $S^2 = \dfrac{1}{2n(n-1)}\sum_{i=1}^{n}\sum_{j=1}^{n}(X_i - X_j)^2$, (*ii*) $\text{Var}\left(S^2\right) = \dfrac{1}{n}\left[\mu_4 - \left(\dfrac{n-3}{n-1}\right)\sigma^4\right]$, (*iii*) $\text{Cov}\left(\bar{X}, S^2\right) = \dfrac{\mu_3}{n}$.

ASSORTED REVIEW PROBLEMS FOR SELF-ASSESSMENT

14·1. (*a*) A coin is tossed 10,000 times and it turns up head 5,195 items. Discuss whether the coin may be regarded as unbiased one, explaining briefly the theoretical principles you would use for this purpose.

(*b*) A biased coin was thrown 400 times and head resulted 240 times. Find the standard error of the observed proportion of heads and deduce that the probability of getting a head in a single throw of the coin lies almost certainly between 0·53 and 0·67.

(*c*) Experience has shown that 20% of a manufactured product is of the top quality. In one day's production of 400 articles only 50 are of top quality. Show that either the production of the day taken was not a representative sample or the hypothesis of 20% was wrong.

Ans. (*a*) No, (*b*) 0·02445 (*c*) Z=3·75; Hypothesis of 20 % was worng.

14·2. (*a*) In a large consignment of oranges a random sample of 64 oranges revealed that 14 oranges were bad. Is it reasonable to assume that 20% of the oranges were bad ?

(*b*) In a random sample of 81 items taken from a large consignment some were found to be defective. If the standard error of the proportion of defective items in the sample is 1/18, find 95% confidence limits of the percentage of defective items in the consignment.

14·3. (*a*) In some dice throwing experiments Weldon threw die 75,145 times and of these 49,152 yielded a 4, 5 or 6. Is this consistent with the hypothesis that the die was unbiased ?

Hint. H_0 Dice is unbiased, *i.e.*, $P = \frac{3}{6} = \frac{1}{2}$, $H_1 : P \neq \frac{1}{2}$ (Two tailed).

Test Statistic. Under H_0 , $Z = \dfrac{p - P}{\sqrt{PQ/n}} = \dfrac{0{\cdot}654 - 0{\cdot}5}{\sqrt{(0{\cdot}5 \times 0{\cdot}5 / 75{,}145)}} = \dfrac{0{\cdot}154}{0{\cdot}0018} = 85{\cdot}56$

Ans. Z highly significant; No.

(*b*) 1,000 apples are taken from a large consignment and 100 are found to be bad. Estimate the percentage of bad apples in the consignment and assign the limits within which the percentage lies.

14·4. (*a*) A personnel manager claims that 80 per cent of all single women hired for secretarial job get married and quit work within two years after they are hired. Test this hypothesis at 5% level of significance if among 200 such secretaries, 112 got married within two years after they were hired and quit their jobs.

(*b*) A manufacturer claimed that at least 98% of the steel pipes which he supplied to a factory conformed to specifications. An examination of a sample of 500 pieces of pipes revealed that 30 were defective. Test this claim at a significance level of (*i*) 0·05, (*ii*) 0·01.

Hint. Let

X = No. of pipes conforming to specifications in the sample = 500 – 30 = 470

p = Sample proportion of pipes conforming to specifications $= \frac{470}{500} = 0{\cdot}94$

$H_0 : P = 0{\cdot}98;\ H_1 : P < 0{\cdot}98$ (Left-tail alternative)

Test Statistic: $$Z = \frac{p - P}{\sqrt{PQ/n}} = \frac{0{\cdot}94 - 0{\cdot}98}{\sqrt{0{\cdot}98 \times 0{\cdot}02/500}}$$

(*c*) An airline must allocate available seating space between first class passengers and economy class passengers. The Null hypothesis is that 20 per-cent of the passengers need first class, but management recognizes the probability that the percentage could be more or less. A random sample of 400 passengers includes 70 passengers holding first class tickets. Can the null hypothesis be rejected at the 10 per-cent level of significance ?

Ans. $H_0 : P = 0{\cdot}20;\ \ H_1: P \neq 0{\cdot}20;\ \ Z = -\ 1{\cdot}25;\ |Z| < 1{\cdot}645$, Not significant ; H_0 may be accepted.

(*d*) In a sample of 500 people in Kerala, 280 are tea drinkers and the rest are coffee drinkers. Can we assume that both coffee and tea are equally popular in this state at 1% level of significance ?

Ans. $H_0 : P = 1/2;\ \ H_1 : P \neq 1/2;\ \ Z = 2{\cdot}68;$ significant at 1% level.

14·5. In a random sample of 800 adults from the population of a certain large city, 600 are found to have dark hair. In a random sample of 1,000 adults from the habitants of another large city, 700 are dark haired. Show that the difference of the proportion of dark haired people is nearly 2·4 times the standard error of the difference for samples of above sizes.

14·6. (*a*) In a random sample of 100 men taken from village *A*, 60 were found to be consuming alcohol. In another sample of 200 men taken from village *B*, 100 were found to be consuming alcohol. Do the two villages differ significantly in respect of the proportion of men who consume alcohol ?

(*b*) In a random sample of 500 men from a particular district of U.P., 300 are found to be smokers. In one of 1,000 men from another district, 550 are smokers. Do the data indicate that the two districts are significantly different with respect to the prevalence of smoking among men ?

Ans. $H_0 : P_1 = P_2;\ \ H_1 : P_1 \neq P_2;\ \ Z = 1{\cdot}85;$ Not significant at 5% level.

14·7. (*a*) In a referendum submitted to the students body at a university, 850 men and 566 women voted. 530 of the men and 304 of the women voted yes. Does this indicate a significant difference of opinion on the matter at 1% level, between men and women students.

(*b*) In a certain district *A*, 450 persons were considered regular consumers of tea out of a sample of 1000 persons. In another district *B*, 400 were regular consumers of tea out of a sample of 800 persons. Do these facts reveal a significant difference between the two districts as far as tea drinking habit is oncerned ? Use 5% level.

Ans. $H_0 : P_1 = P_2\quad H_1 : P_1 \neq P_2;\ \ Z = 2{\cdot}08;$ Significant at 5% level.

14·8. The subject under investigation is the measure of dependence of Tamil on words of Sanskrit origin. One newspaper article reporting the proceedings of the constituent assembly contained 2,025 words of which 729 words were declared by literary critic to be of Sanskrit origin. A second article by the same author describing atomic research contained 1,600 words of which 640 words were declared by the same critic to be of Sanskrit origin. Assuming that simple sampling conditions hold, estimate the limits for the proportion of Sanskrit words in the writer's vocabulary and examine whether there is any significant difference in the dependence of this writer on words of Sanskrit origin in writing the two articles. Also find 95% confidence limits.

Hint. In usual notations; $n_1 = 2025,\ \ n_2 = 1600,\ \ p_1 = \frac{729}{2025} = 0{\cdot}36,\ \ p_2 = \frac{640}{1600} = 0{\cdot}40$

Null Hypothesis, $H_0 : P_1 = P_2$, *i.e.*, there is no significant difference in the proportion of Sanskrit words in the writer's vocabulary in the two articles. In other words, Tamil does not depend on the words of Sanskrit origin.

Alternative Hypothesis, $H_0 : P_1 \neq P_2$ (Two-tailed).

Test Statistic. Under H_0 (since samples are large).

$$Z = \frac{p_1 - p_2}{\sqrt{\hat{P}\hat{Q}\left(\frac{1}{n_1} + \frac{1}{n_2}\right)}} \sim N(0,1), \text{ where } \hat{P} = \frac{729 + 640}{2{,}025 + 1{,}600} = 0{\cdot}38 \quad \Rightarrow \quad \hat{Q} = 0{\cdot}62$$

$$\therefore \quad Z = \frac{0{\cdot}36 - 0{\cdot}40}{\sqrt{0{\cdot}38 \times 0{\cdot}62 \times \left(\frac{1}{2025} + \frac{1}{1600}\right)}} = \frac{-0{\cdot}04}{0{\cdot}0162} = -2{\cdot}469$$

since $|Z| = 2{\cdot}469$ is greater than 1·96, it is significant at 5% level of significance. Hence we reject the null hypothesis and conclude that the proportion of Sanskrit words in the writer's two articles differ significantly. In other words, we conclude (at 5% level of significance) that Tamil depends on the words of Sanskrit origin.

14·9. (*a*) A machine put out 16 imperfect items in a sample of 500. After the machine was overhauled, it put out 3 imperfect items in a batch of 100. Has the machine improved after overhauling ?

Hint. We are given : $n_1 = 500$, and $n_2 = 100$, $p_1 = \frac{16}{500} = 0{\cdot}032$; $p_2 = \frac{3}{100} = 0{\cdot}030$

Null Hypothesis, $H_0 : P_1 = P_2$, *i.e.*, there is no significicat difference in the machine before overhauling and after overhauling. In other words, the machine has not improved after overhauling.

Alternative Hypothesis, $H_1 : P_2 < P_1$ or $P_1 > P_2$. (Right-tail test)

$$\hat{p} = \frac{n_1p_1 + n_2p_2}{n_1 + n_2} = \frac{16 + 3}{500 + 100} = \frac{19}{600} = 0{\cdot}032$$

$$S.E.(p_1 - p_2) = \sqrt{0{\cdot}032 \times 0{\cdot}968\left(\frac{1}{500} + \frac{1}{100}\right)} = 0{\cdot}0193$$

$$Z = \frac{0{\cdot}032 - 0{\cdot}030}{0{\cdot}0193} = \frac{0{\cdot}002}{0{\cdot}0193} = 1{\cdot}04$$

Since $Z < 1{\cdot}645$ (Right-tailed test), it is not significant at 5% level of significance.

(*b*) A company is considering two different television advertisements for promotion of a new product. Management believed that the advertisement *A* is more effective than advertisement *B*. Two test market areas with virtually identical consumer characteristics are selected; *A* is used in one area and *B* in other area. In a random sample of 60 customers who saw *A*, 18 tried the product. In another random sample of 100 customers who saw *B*, 22 tried the product. Does this indicate that advertisement *A* is more effective than advertisement *B*, if a 5% level of significance is used ? Given critical value at 5% level is 1·96 and at 10% level of significance is 1·645.

14·10. (*a*) A candidate for election made a speech in city *A* but not in *B*. A sample of 500 voters from city *A* showed that 59·6% of the voters were in favour of him, whereas a sample of 300 voters from city *B* showed that 50% of the voters favoured him. Discuss whether his speech could produce any effect on voters in city *A*. Use 5% level.

Ans. $|Z| = 2{\cdot}67$, Yes.

(*b*) In a large city, 16 out of a random sample of 500 men were found to be drinkers. After the heavy increase in tax on intoxicants another random sample of 100 men in the same city included 3 drinkers. Was the observed decrease in the proportion of drinkers significant after tax increase ?

Ans. $H_0 : P_1 = P_2$, $H_1 : P_1 > P_2$; $Z = 1{\cdot}04$; Not Significant.

14·11. The business manager of a large company wants to check the inventory records against the physical inventories by a sample survey. He wants to almost assure that the maximum sampling error should not be more than 5% above or below the true proportion of the inaccurate records. The proportion of the inaccurate records is estimated at 35% from past experiences. Determine the sample size.

14·12. The sex ratio at birth is sometimes given by the ratio of male to female births instead of the proportion of male to total births. If z is the ratio, *i.e.*, $z = p/q$, show that the standard error of z is approximately $\frac{1}{1+z}\sqrt{z/n}$, n being large, so that deviations are small compared with mean.

14·13. (*a*) The manufacturer of television tubes knows from past experience that the average life of a tube is 2,000 hours with a standard deviation of 200 hours. A sample of 100 tubes has an average life of 1,950 hours. Test at the 0·05 level of significance if this sample came from a normal population of mean 2,000 hours.

State your null and alternative hypotheses and indicate clearly whether a one-tail or a two-tail test is used and why ? Is the result of the test significant ?

(*b*) A sample of 100 items, drawn from a universe with mean value 64 and S.D. 3, has a mean value 63·5. Is the difference in the means significant ? What will be your inference, if the sample had 200 items ?

(*c*) A sample of 400 individuals is found to have a mean height of 67·47 inches. Can it be reasonably regarded as a sample from a large population with mean height of 67·39 inches and standard deviation 1·30 inches?

(*d*) The mean breaking strength of cables supplied by a manufacturer is 1,800 with a standard deviation 100. By a new technique in the manufacturing process it is claimed that the breaking strength of the cables has increased. In order to test this claim a sample of 50 cables is tested. It is found that the mean breaking strength is 1,850. Can we support the claim at 0·01 level of significance ?

Ans. $H_0 : \mu = 1800$, $H_1 : \mu > 1800$; $Z = 3{\cdot}535$; Significant, No.

(*e*) An ambulance service claims that it takes on the average 8·9 minutes to reach its destination in emergency calls. To check on this claim, the agency which licenses ambulance services has them timed on 50 emergency calls, getting a mean of 9·3 minutes with a standard deviation of 1·6 minutes. What can they conclude at the level of significance $\alpha = 0{\cdot}05$?

14·14 A paper mill in U.P. has agreed to buy waste paper for recycling from a waste collection firm, under the agreement that the waste collection firm will supply the waste paper in packages of 300 kg each, for which the paper mill will pay by the package. To speed up their work the waste collection firm is making packages by some *approximation* procedure. The paper mill does not object to this procedure as long as it gets 300 kg. per package on the average. The waste collection firm has an interest not to exceed 300 kg. per package, because it is not being paid for more, and not to go under 300 kg. because the paper mill might terminate the agreement if it does. To estimate the mean weight of waste paper per package, the waste collection firm weighed 75 randomly selected packages and found that the mean weight was 290 kg and standard deviation was 15 kg. Can we infer that the mean weight per package in the entire supply was 300 kg.?

Ans. $H_0 : \mu = 300$ kg, $H_1 : \mu \neq 300$kg; $Z = \dfrac{290 - 300}{15/\sqrt{75}} = 5{\cdot}77$; Significant.

14·15. (*a*) A sample of 450 items is taken from a population whose standard deviation is 20. The mean of the sample is 30. Test whether the sample has come from a population with mean 29. Also calculate the 95% confidence limits for the population mean.

(*b*) A sample of 400 observations has mean 95 and standard deviation 12. Could it be a random sample from a population with mean 98 ? What can be the maximum value of the population mean?

14·16. (*a*) *ABC* autoparts manufacturing company plans on advertising the average lifetime of the autobatteries it sells. To find out what that average lifetime is, the company gives away 100 randomly chosen batteries to 100 randomly selected customers, with the provision that the customer use the battery continuously with no maintenance, except for adding water as needed and cleaning the terminals, until it fails. The lifetimes, in terms of the 100 batteries, turn out to have mean 2 years with a standard deviation of 0·156 year. Determine 90%, 95% and 99% confidence interval for the true mean lifetime of the batteries.

(*b*) A quality control manager in a Measure Ruler Company is examining her company's metric rulers. Each ruler is supposed to be 100 cms. long and each ruler must be within 0·2 cm. of this figure. The manager wants to examine the output to be sure that rulers are within the specifications. Her data for a sample of 100 units gave a sample mean 99·9 cms. and the sample variance of 0·0036 cm.^{2}.

(*i*) Construct 95% confidence interval for the length of metric ruler.

(*ii*) Should the manager stop the process and reset it?

(*iii*) Should manager's decision change if 99% confidence interval is used in place of 95% ?

14·17. The weights of 1,500 ball bearings are normally distributed with mean 22·40 and standard deviation 0·048. If 300 random samples of size 36 each are drawn from this population, determine the expected mean and standard deviation of the sampling distribution of means, if sampling is done with replacement. How many of the random samples in the above problem would have their means between 22·39 and 22·41?

Hint. $E(\bar{X}) = \mu = 22{\cdot}40;\quad S.E.(\bar{X}) = \sigma / \sqrt{n} = 0{\cdot}048 / \sqrt{36} = 0{\cdot}008.$

Required number of samples (out of 300) $= 300 \times P\ (22{\cdot}39 \leq \bar{X} < 22{\cdot}41)$

$$= 300 \times P\left(\frac{22{\cdot}39 - 22{\cdot}40}{0{\cdot}008} < Z < \frac{22{\cdot}41 - 22{\cdot}40}{0{\cdot}008}\right);\quad Z \sim N(0,1)$$

$$= 300 \times P(-1{\cdot}25 < Z < 1{\cdot}25) = 600 \times P(0 < Z < 1{\cdot}25)\ ;\ 237$$

14·18. (*a*) A normal distribution has mean 0·5 and standard deviation 2·5. Find :

(*i*) The probability that the mean of a random sample of size 16 from the population is positive.

(*ii*) The probability that the mean of a sample of size 90 from the population will be negative.

(*b*) The mean of a certain normal distribution is equal to the standard error of the mean of a random sample of 100 from that distribution. Find the probability that the mean of a sample of 25 from the distribution will be negative.

(*c*) The average value $\bar{x}$ of a random sample of observations from a certain population is normally distributed with mean 20 and standard deviation $5/\sqrt{n}$. How large a sample should be drawn in order to have a probability of at least 0·90 that $\bar{x}$ will lie between 18 and 22.

14·19. (*a*) From a population of 169 units it is desired to choose a simple random sample of size *n*. If the population standard deviation is 2, determine the smallest '*n*' for which the probability that the sample mean differs from the population mean by more than 0·75 is controlled at 0·05.

(*b*) An economist would like to estimate the mean income (μ) in a large city. He has decided to use the sample mean as an estimate of μ and would like to ensure that the error in estimation is not more than ₹ 100 with probability 0·90. How large a sample should he take if the standard deviation is known to be ₹ 1,000 ?

Hint. $n = \left(\frac{z_\alpha \sigma}{E}\right)^2 = \left(\frac{1{\cdot}645 \times 1{,}000}{100}\right)^2 = 270{\cdot}6 \cong 271$

14·20. The management of a manufacturing firm wishes to determine the average time required to complete a certain manual operation. There should be 0·95 confidence that the error in the estimate will not exceed 2 minutes.

What sample size is required if the standard deviation of the time needed to complete the manual operation is estimated by a time and motion study expert as (*i*) 10 minutes, (*ii*) 16 minutes ? Explain intuitively (without referring to the formula) why the sample size is large in (*ii*) than in (*i*). (*Given* $Z_{0.975} = 1.96$ and $Z_{0.95} = 1.645$.)

Hint. (*i*) $n_1 = \left(\frac{z_\alpha \sigma}{E}\right)^2 = \left(\frac{1.96 \times 10}{2}\right)^2 = 96,$ (*ii*) $n_2 = \left(\frac{1.96 \times 16}{2}\right)^2 = 246.$

14·21. (*a*) Two populations have the same mean, but the standard deviation of one is twice that of the other. Show that in samples of 500 each drawn under simple random conditions, the difference of the means will, in all probability, not exceed 0.3σ, where σ is the smaller standard deviation, and assuming the distribution of the difference of the means to be normal, find the probability that it exceeds half that amount.

(*b*) A simple sample of heights of 6,400 Englishmen has a mean of 67·85 inches and S.D. 2·56 inches, while a simple sample of heights of 1,600 Australians has a mean of 68·55 inches and S.D. of 2·52 inches. Do the data indicate that Australians are, on the average, taller than Englishmen ?

Ans. $H_0 : \mu_1 = \mu_2,$ $H_1 : \mu_1 < \mu_2$; $|Z| = 9.2$, Significant.

(*c*) In a random sample of 500, the mean is found to be 20. In another independent random sample of 400, the mean is 15. Could the samples have been drawn from the same population with standard deviation 4 ?

14·22. (*a*) The following table presents data on the values of a harvested crop stored in the open and inside a godown :

	Sample size	*Mean*	$\Sigma(x_i - \bar{x})^2$
Outside	40	117	8,685
Inside	100	132	27,315

Assuming that the two samples are random and they have been drawn from normal populations with equal variances, examine if the mean value of the harvested crop is affected by weather conditions.

(*b*) The mean consumption of foodgrains among 400 sampled middle class consumers is 380 grams per day per person with a standard deviation of 120 grams. A similar sample survey of 600 working class consumers gave a mean of 410 grams with a standard deviation of 80 grams. Are we justified in saying that the two classes consume the same quantity of foodgrains ? Use 5% level of significance for the test.

Ans. $H_0 : \mu_1 = \mu_2$; $H_1 : \mu_1 \neq \mu_2$; $|Z| = 4.3915$; Significant.

14·23. (*a*) The mean yield of two sets of plots and their variability are as given below. Examine :

(*i*) whether the difference in the mean yields of the two sets of plots is significant, and

(*ii*) whether the difference in the variability in yields is significant.

	Set of 40 plots	*Set of 60 plots*
Mean yield per plot	1,258 kg.	1243 kg.
S.D. per plot	34 kg.	28 kg.

(*b*) A large organisation produces electric light bulbs in each of its two factories. It is suspected that the efficiency in the factories is not the same, so a test is carried out by ascertaining the variability of the life of the bulbs produced by each factory. The results are as follows :

	Factory A	*Factory B*
No. of bulbs in sample	100	200
Average life	1,100 hours	900 hours
Standard deviation	240 hours	220 hours

From the above information, determine whether the difference between the variability of life of bulbs from each sample is significant (take 1% level of significance).

14·24. A farmer claims that the average yield of corn of variety *A* exceeds the average yield of variety *B* by at least 12 bushels per acre. To test this claim, 50 acres of each variety are planted and grown under similar conditions. Variety *A* yielded, on the average, 86·7 bushels per acre with a standard deviation of 6·28 bushels per acre; while variety *B* yielded, on the average, 77·8 bushels per acre with a standard deviation of 5·61 bushels per acre. Test the farmer's claim using a 0·02 level of significance.

Ans. $H_0 : \mu_1 - \mu_2 \geq 12, H_1 : \mu_1 - \mu_2 < 12$; It sufices to test $H_0 : \mu_1 - \mu_2 = 12$. $Z = 2{\cdot}603$; Significant.

14·25. The electric light tubes of manufacturer *A* have a lifetime of 1,400 hours, with a standard deviation of 200 hours, while of manufacturer *B* have a mean lifetime of 1,200 hours with a standard deviation of 100 hours. If random samples of 125 tubes of each batch are tested, what is the probability that the brand *A* tubes will have a mean time which is at least (*i*) 160 hours more than the brand *B* tubes, and (*ii*) 250 hours more than the brand *B* tubes?

Hint. Under the assumption of normal population, the sampling distribution of $(\bar{x}_1 - \bar{x}_2)$ would have mean; $\mu_1 - \mu_2 = 1{,}400 - 1{,}200 = 200$ hours and standard deviation:

$$S.E.\ (\bar{x}_1 - \bar{x}_2) = \sqrt{\left(\frac{\sigma_1^2}{n_1} + \frac{\sigma_2^2}{n_2}\right)} = \sqrt{\left\{\frac{(100)^2}{125} + \frac{(200)^2}{125}\right\}} = 20 \text{ hours}$$

(*i*) The required probability is given by:

$$P\{(\bar{x}_1 - \bar{x}_2) \geq 160\} = P\left[\frac{(\bar{x}_1 - \bar{x}_2) - (\mu_1 - \mu_2)}{S.E.(\bar{x}_1 - \bar{x}_2)} \geq \frac{160 - 200}{20}\right]$$

$$= P(Z \geq -2) = 0{\cdot}5 + P(-2 < Z < 0) = 0{\cdot}9772$$

(*ii*) The required probability is given by :

$$P\{(\bar{x}_1 - \bar{x}_2) \geq 250\} = P(Z \geq 2{\cdot}5) = 0{\cdot}5 - P(0 < Z < 2{\cdot}5) = 0{\cdot}5 - 0{\cdot}4938 = 0{\cdot}0062.$$

14·26. A random sample of 100 articles selected from a batch of 2,000 articles shows that the average diameter of the articles is 0·354 with a standard deviation 0·048. Find 95% confidence interval for the average of this batch of 2,000 articles.

Hint. We are given $n = 100$, $N = 2000$, $\bar{x} = 0{\cdot}354$, $s = 0{\cdot}048$

The Standard Error of sample mean $\bar{x}$ in random sampling from the batch of $N = 2{,}000$ is given by : [c.f. (14·10a)]

$$S.E.(\bar{x}) = \sqrt{\frac{N-n}{N-1}} \times \frac{\sigma}{\sqrt{n}} = \sqrt{\frac{N-n}{N-1}} \times \frac{s}{\sqrt{n}} \qquad (\because \hat{\sigma} = s, \text{ for large } n)$$

$$= \sqrt{\frac{2{,}000 - 100}{2{,}000 - 1}} \times \frac{0.048}{\sqrt{100}} = 0{\cdot}00468$$

Hence, 95% confidence limits for μ are given by:

$$\bar{x} \pm 1{\cdot}96\, S.E.\,(\bar{x}) = 0{\cdot}354 \pm 1{\cdot}96 \times 0{\cdot}00468 = (0{\cdot}3448, 0{\cdot}3632)$$

CHAPTER

15 Exact Sampling Distributions-I (Chi-square (χ^2) Distribution)

LEARNING OBJECTIVES

Upon completion of this Chapter, you should be able to:

- Derive Chi-square Distribution and learn about the shape of the Distribution for different degrees of freedom.
- Explain various concepts like m.g.f., characteristic function, etc., related to Chi-square Distribution.
- Discuss various theorems and properties of Chi-square Distribution.
- Learn about the various applications of Chi-square Distribution by developing the Chi-square tests for:
  (i) Hypothetical value of population variance.
  (ii) Goodness of fit.
  (iii) Independence of attributes.
  (iv) Equality of several population proportions.
  (v) Homogeneity of several correlation coefficients.
  (vi) Homogeneity of several independent estimates of the same population variance. (Bartlett's Test);

and the underlying assumptions in their use.

CHAPTER OUTLINES

15·1. INTRODUCTION

In the preceding chapter, some tests of significance based on the assumption that the samples were drawn from normally distributed population or, more appropriately, that the sample means were normally distributed were discussed in detail with a number of examples. These tests are known as *parametric tests* as the testing procedure involves the assumption about the type of population values or parameters. But, there are many situations in which it is not possible to make any dependable assumption about the form of the parent population from which samples have been drawn. This limitation has led to the development of a number of alternative techniques that are known as *non-parametric* or *distribution-free* methods.

Chi-square test (chi-pronounced as ki) is one of the most important non-parametric or distribution-free tests introduced in 1900 by Karl Pearson. While the test is distribution free (as it makes no assumptions abut the population from which the sample is drawn), there is a rigidly defined chi-square distribution with a mathematically expressed probability function. The chi-square distribution may be defined as the *'Sum of the squares of independent normally distributed variables with zero means and unit variances'*.

The square of a standard normal variate is known as a chi-square variate (*pronounced as Ki - Sky without S*) with 1 degree of freedom (*d.f.*).

Thus, if $X \sim N(\mu, \sigma^2)$, then $Z = \dfrac{X-\mu}{\sigma} \sim N(0, 1)$

and $$Z^2 = \left(\frac{X-\mu}{\sigma}\right)^2 \text{ is a chi-square variate with 1 d.f.} \qquad \ldots (15{\cdot}1)$$

In general if X_i $(i = 1, 2, \ldots, n)$ are n independent normal variates with means μ_i and variances σ_i^2, $(i = 1, 2, \ldots, n)$, then

$$\chi^2 = \sum_{i=1}^{n}\left(\frac{X_i-\mu_i}{\sigma_i}\right)^2, \text{ is a chi-square variate with } n \text{ } d.f. \qquad \ldots (15{\cdot}1a)$$

Note. For detailed discussion on degrees of freedom (*d.f.*), see Remark 2 to § 15·6·3.

15·2. DERIVATION OF THE CHI-SQUARE (χ^2) DISTRIBUTION

First Method. *Method of Moment Generating Function*

If X_i, $(i = 1, 2, \ldots, n)$ are independent $N(\mu_i, \sigma_i^2)$, we want the distribution of

$$\chi^2 = \sum_{i=1}^{n}\left(\frac{X_i-\mu_i}{\sigma_i}\right)^2 = \sum_{i=1}^{n} U_i^2, \text{ where } U_i = \frac{X_i-\mu_i}{\sigma_i} \sim N(0, 1)$$

Since X_i's are independent, U_i's are also independent. Therefore,

$$M_{\chi^2}(t) = M_{\Sigma U_i^2}(t) = \prod_{i=1}^{n} M_{U_i^2}(t) = \left[M_{U_i^2}(t)\right]^n, \qquad [\because U_i\text{'s are } i.i.d.\ N(0, 1)] \quad \ldots (*)$$

$$M_{U_i^2}(t) = E\left[\exp\left(tU_i^2\right)\right] = \int_{-\infty}^{\infty} \exp(tu_i^2)\, f(x_i)\,dx_i$$

$$= \int_{-\infty}^{\infty} \exp(tu_i^2) \frac{1}{\sigma_i\sqrt{2\pi}} \exp - \left\{(x_i - \mu_i)^2 / 2\sigma_i^2\right\} dx_i$$

$$= \frac{1}{\sqrt{2\pi}} \int_{-\infty}^{\infty} \exp(tu_i^2) \exp(-u_i^2/2)\, du_i, \qquad \left[u_i = \frac{x_i - \mu_i}{\sigma_i}\right]$$

$$= \frac{1}{\sqrt{2\pi}} \int_{-\infty}^{\infty} \exp\left\{-\left(\frac{1-2t}{2}\right)u_i^2\right\} du_i = \frac{1}{\sqrt{2\pi}} \cdot \frac{\sqrt{\pi}}{\left(\frac{1-2t}{2}\right)^{1/2}} = (1-2t)^{-1/2}$$

$$\left[\because \int_{-\infty}^{\infty} \exp(-a^2x^2)\, dx = \frac{\sqrt{\pi}}{a}\right]$$

$$\therefore \quad M_{\chi^2}(t) = (1-2t)^{-n/2}. \qquad \text{[From (*)]}$$

which is the *m.g.f.* of a Gamma variate with parameters $\left(\frac{1}{2} \text{ and } \frac{1}{2}n\right)$.

Hence, by uniqueness theorem of *m.g.f.*'s,

$$\chi^2 = \sum_i^n \left(\frac{X_i - \mu_i}{\sigma_i}\right)^2, \text{ is a Gamma variate with parameters } \left(\frac{1}{2} \text{ and } \frac{1}{2}n\right); \text{ and } p.d.f.,$$

$$dP(\chi^2) = \frac{\left(\frac{1}{2}\right)^{n/2}}{\Gamma(n/2)} \cdot \left[\exp\left(-\frac{1}{2}\chi^2\right)\right] (\chi^2)^{(n/2)-1}\, d\chi^2$$

$$\Rightarrow \quad p(\chi^2) = \frac{1}{2^{n/2}\Gamma(n/2)} \left[\exp(-\chi^2/2)\right] (\chi^2)^{(n/2)-1}, \; 0 \le \chi^2 < \infty \qquad \text{... (15·2)}$$

which is the required *p.d.f.* of chi-square distribution with *n* degrees of freedom.

Remarks:

1. If a *r.v.* X has a chi-square distribution with *n d.f.*, we write $X \sim \chi^2_{(n)}$ and its *p.d.f.* is:

$$f(x) = \frac{1}{2^{n/2}\Gamma(n/2)} e^{-x/2} x^{(n/2)-1}; \; 0 \le x < \infty \qquad \text{... (15·2a)}$$

2. If $X \sim \chi^2_{(n)}$, then $\frac{1}{2}X \sim \gamma\left(\frac{1}{2}n\right)$.

Proof. The *p.d.f.* of $Y = \frac{1}{2}X$, is given by:

$$g(y) = f(x) \cdot \left|\frac{dx}{dy}\right| = \frac{1}{2^{n/2}\Gamma(n/2)} e^{-y} \cdot (2y)^{(n/2)-1} \cdot 2 = \frac{1}{\Gamma(n/2)} e^{-y} y^{(n/2)-1}; \quad 0 \le y < \infty$$

$$\therefore \quad Y = \frac{1}{2}X \sim \gamma\left(\frac{1}{2}n\right).$$

Second Method. *Method of Induction*

If $X_i \sim N(0, 1)$, then $\frac{1}{2} X_i^2$ is a $\gamma\left(\frac{1}{2}\right)$ so that X_i^2 is a χ^2 variate with *d.f.* 1.

If X_1 and X_2 are independent standard normal variates then $X_1^2 + X_2^2$ is a chi-square variate with 2 *d.f.* which may be proved as follows:

The joint probability differential of X_1 and X_2 is given by:

$$dP(x_1, x_2) = f(x_1, x_2)dx_1\, dx_2 = f_1(x_1)\, f_2(x_2)\, dx_1\, dx_2$$

$$= \frac{1}{2\pi} \exp\left\{-(x_1^2 + x_2^2)/2\right\} dx_1 dx_2, -\infty < (x_1, x_2) < \infty$$

Let us transform to polar co-ordinates by substitution $x_1 = r \cos \theta$, $x_2 = r \sin \theta$. Jacobian of transformation J is given by:

$$J = \begin{vmatrix} \dfrac{\partial x_1}{\partial r} & \dfrac{\partial x_2}{\partial r} \\ \dfrac{\partial x_1}{\partial \theta} & \dfrac{\partial x_2}{\partial \theta} \end{vmatrix} = \begin{vmatrix} \cos \theta & \sin \theta \\ -r \sin \theta & r \cos \theta \end{vmatrix} = r$$

Also, we have $r^2 = x_1^2 + x_2^2$ and $\tan \theta = x_2/x_1$. As x_1 and x_2 range from $-\infty$ to $+\infty$, r varies from 0 to ∞ and θ from 0 to 2π. The joint probability differential of r and θ now becomes

$$dG(r, \theta) = \frac{1}{2\pi} \exp(-r^2/2)\, r\, dr\, d\theta\,; \quad 0 \le r \le \infty, \quad 0 \le \theta \le 2\pi$$

Integrating over 0, the marginal distribution of r is given by:

$$dG_1(r) = \int_0^{2\pi} dG(r, \theta) = r \exp(-r^2/2) dr \left| \frac{\theta}{2\pi} \right|_0^{2\pi} = \exp(-r^2 / 2) r \cdot dr$$

$$\Rightarrow \qquad dG_1(r^2) = \frac{1}{2} \exp(-r^2/2)\, dr^2 = \frac{1}{\Gamma(1)} \exp(-r^2/2)\, (r^2/2)^{1-1}\, d(r^2/2)$$

Thus, $\dfrac{r^2}{2} = \dfrac{X_1^2 + X_2^2}{2}$ is a γ (1) variate and hence $r^2 = X_1^2 + X_2^2$ is a χ^2-variate with 2 *d.f.*

For n variables X_i, (i = 1, 2, 3, ..., n), we transform $(X_1, X_2, \ldots, X_n)$ to $(\chi, \theta_1, \theta_2, \ldots, \theta_{n-1})$ by taking:

$$\left.\begin{aligned} x_1 &= \chi \cos \theta_1 \cos \theta_2 \ldots \cos \theta_{n-1} \\ x_2 &= \chi \cos \theta_1 \cos \theta_2 \ldots \cos \theta_{n-2} \sin \theta_{n-1} \\ x_3 &= \chi \cos \theta_1 \cos \theta_2 \ldots \cos \theta_{n-3} \sin \theta_{n-2} \\ &\vdots \\ &\vdots \\ x_j &= \chi \cos \theta_1 \cos \theta_2 \ldots \cos \theta_{n-j} \sin \theta_{n-j+1} \\ &\vdots \\ &\vdots \\ x_n &= \chi \sin \theta_1 \end{aligned}\right\} \qquad \ldots (15{\cdot}3)$$

where, $\chi > 0$, $-\pi < \theta_1 < \pi$ and $-\frac{1}{2}\pi < \theta_i < \frac{1}{2}\pi$; for $i = 2, 3, \ldots, (n-1)$.

Then $x_1^2 + x_2^2 + \cdots + x_n^2 = \chi^2$ and $|J| = \chi^{n-1} \cos^{n-2} \theta_1 \cos^{n-3} \theta_2 \ldots \cos \theta_{n-2}$

(*c.f.* Advanced Theory of Statistics Vol. I, by Kendall and Stuart.)

The joint distribution of $X_1, X_2, \ldots, X_n$, *viz.*,

$$dF(x_1, x_2, \ldots, x_n) = \left(\frac{1}{\sqrt{2\pi}}\right)^n \exp(-\Sigma x_i^2/2) \cdot \prod_{i=1}^{n} dx_i, \text{ transforms to}$$

$$dG(\chi, \theta_1, \theta_2, \ldots, \theta_{n-1}) = \exp\left(-\frac{1}{2}\chi^2\right) \chi^{n-1} \cos^{n-2} \theta_1 \cos^{n-3} \theta_2 \ldots \cos \theta_{n-2}\, d\chi\, d\theta_1\, d\theta_2, \ldots d\theta_{n-1}$$

Integrating over $\theta_1, \theta_2, \ldots, \theta_{n-1}$, we get the distribution of χ^2 as:

$$dP(\chi^2) = k \exp(-\chi^2/2)(\chi^2)^{(n/2)-1}\, d\chi^2, 0 \leq \chi^2 < \infty$$

The constant k is determined from the fact that total probability is unity, *i.e.*,

$$\int_0^\infty dP(\chi^2) = 1 \Rightarrow k \int_0^\infty \exp(-\chi^2/2)(\chi^2)^{\frac{n}{2}-1} d\chi^2 = 1 \Rightarrow k = \frac{1}{2^{n/2}\Gamma(n/2)}$$

$$\therefore \quad dP(\chi^2) = \frac{1}{2^{n/2}\Gamma(n/2)} \exp(-\chi^2/2).(\chi^2)^{\frac{n}{2}-1}, \quad 0 \leq \chi^2 < \infty$$

Hence, $\frac{1}{2}\chi^2 = \frac{1}{2}\sum_{i=1}^{n} X_i^2$ is a $\gamma(n/2)$ variate $\Rightarrow \chi^2 = \sum_{i=1}^{n} X_i^2$ is a chi-square variate with n degrees of freedom (*d.f.*) and (15.2) gives *p.d.f.* of chi-square distribution with n *d.f.*

Remarks:

1. If X_i, $i = 1, 2, \ldots, n$ are n independent normal variates with mean μ_i and S.D. σ_i, then

$$\sum_{i=1}^{n} \left(\frac{X_i - \mu_i}{\sigma_i}\right)^2 \text{ is a } \chi^2\text{-variate with } n \text{ } d.f.$$

2. In random sampling from a normal population with mean μ and S.D. σ, the sample mean $\bar{X}$ is distributed normally about the mean μ with S.D. $\sigma/\sqrt{n}$.

$$\therefore \quad \frac{\bar{X} - \mu}{\sigma/\sqrt{n}} \sim N(0, 1) \Rightarrow \left[\frac{\bar{X} - \mu}{\sigma/\sqrt{n}}\right]^2 \text{ is a } \chi^2\text{-variate with 1 } d.f.$$

3. Normal distribution is a particular case of χ^2-distribution when $n = 1$, since for $n = 1$,

$$dP(\chi^2) = \frac{1}{\sqrt{2}\,\Gamma(1/2)} \exp(-\chi^2/2)(\chi^2)^{\frac{1}{2}-1} d\chi^2, 0 \leq \chi^2 < \infty$$

$$= \frac{1}{\sqrt{2\pi}} \exp(-\chi^2/2)\, d\chi, \ -\infty \leq \chi < \infty$$

Thus, χ is a standard normal variate.

4. For $n = 2$, *i.e.*, if X is a χ^2–variate with 2 d.f., then

$$p(x) = \frac{1}{2^1 \cdot \Gamma(1)} \exp\left(-\frac{x}{2}\right) \cdot x^{1-1} \; ; \; x > 0 \quad \Rightarrow \quad p(x) = \frac{1}{2} \exp\left(\frac{-x}{2}\right), \quad x \geq 0$$

which is the *p.d.f.* of exponential distribution with mean 2.

15·3. M.G.F. OF CHI-SQUARE DISTRIBUTION

Let $X \sim \chi^2(n)$, then

$$M_X(t) = E(e^{tX}) = \int_0^\infty e^{tx} f(x)\, dx = \frac{1}{2^{n/2}\Gamma(n/2)} \int_0^\infty e^{tx} \cdot e^{-x/2}\, x^{(n/2)-1}\, dx$$

$$= \frac{1}{2^{n/2}\Gamma(n/2)} \int_0^\infty \exp\left[-\left(\frac{1-2t}{2}\right)x\right] \cdot x^{(n/2)-1}\, dx$$

$$= \frac{1}{2^{n/2}\Gamma(n/2)} \cdot \frac{\Gamma(n/2)}{[(1-2t)/2]^{n/2}} \qquad \text{[Using Gamma Integral]}$$

$$= (1-2t)^{-n/2}, |2t| < 1 \qquad \text{... (15·4)}$$

which is the required m.g.f. of a χ^2-variate with n *d.f.*

Remarks:

1. Using Binomial expansion for negative index, we get from (15.4)

$$M(t) = 1 + \frac{n}{2}(2t) + \frac{\frac{n}{2}\left(\frac{n}{2}+1\right)}{2!}(2t)^2 + \cdots + \frac{\frac{n}{2}\left(\frac{n}{2}+1\right)\left(\frac{n}{2}+2\right)\cdots\left(\frac{n}{2}+r-1\right)}{r!}(2t)^r + \cdots$$

$\therefore \quad \mu_r' =$ Coefficient of $\frac{t^r}{r!}$ in the expansion of $M(t)$.

$$= 2^r \frac{n}{2}\left(\frac{n}{2}+1\right)\left(\frac{n}{2}+2\right) \cdots \left(\frac{n}{2}+r-1\right)$$

$$= n(n+2)(n+4) \ldots (n+2r-2) \qquad \text{... (15·4}a\text{)}$$

2. If n is even so that $n/2$ is a positive integer, then

$$\mu_r' = 2^r \Gamma[(n/2)+r] / \Gamma(n/2) \qquad \text{... (15·4}b\text{)}$$

15.3.1. Cumulant Generating Function of χ^2-Distribution

If $X \sim \chi^2_{(n)}$, then

$$K_X(t) = \log M_X(t) = -\frac{n}{2}\log(1-2t) = \frac{n}{2}\left[2t + \frac{(2t)^2}{2} + \frac{(2t)^3}{3} + \frac{(2t)^4}{4} + \cdots\right]$$

$\therefore \quad \kappa_1 =$ Coefficient of t in $K(t) = n$, $\qquad \kappa_2 =$ Coefficient of $\frac{t^2}{2!}$ in $K(t) = 2n$,

$$\kappa_3 = \text{Coefficient of } \frac{t^3}{3!} \text{ in } K(t) = 8n, \quad \text{and} \quad \kappa_4 = \text{Coefficient of } \frac{t^4}{4!} \text{ in } K(t) = 48n$$

In general, $\kappa_r = \text{Coefficient of } \dfrac{t^r}{r!} \text{ in } K(t) = n.2^{r-1}(r-1)!$ ... (15·4c)

$$\left.\begin{aligned} \therefore \quad \text{Mean} &= \kappa_1 = n, & \text{Variance} &= \mu_2 = \kappa_2 = 2n \\ \mu_3 &= \kappa_3 = 8n, & \mu_4 &= \kappa_4 + 3\kappa_2^2 = 48n + 12n^2 \\ \beta_1 &= \frac{\mu_3^2}{\mu_2^3} = \frac{8}{n} \quad \text{and} & \beta_2 &= \frac{\mu_4}{\mu_2^2} = \frac{12}{n} + 3 \end{aligned}\right\} \quad \text{... (15·4d)}$$

15·3·2. Limiting Form of χ^2 Distribution for Large Degrees of Freedom

If $X \sim \chi^2_{(n)}$, then

$$M_X(t) = (1-2t)^{-n/2}, \quad |t| < \frac{1}{2}.$$

The *m.g.f.* of standard χ^2-variate Z is: $M_{(X-\mu)/\sigma}(t) = e^{-\mu t/\sigma} M_X(t/\sigma)$

$$M_Z(t) = e^{-\mu t/\sigma}(1 - 2t/\sigma)^{-n/2} = e^{-nt/\sqrt{2n}}\left(1 - \frac{2t}{\sqrt{2n}}\right)^{-n/2} \qquad (\because \mu = n, \sigma^2 = 2n)$$

$$\therefore \quad K_Z(t) = \log M_Z(t) = -t\sqrt{\frac{n}{2}} - \frac{n}{2}\log\left(1 - t\sqrt{\frac{2}{n}}\right)$$

$$= -t\sqrt{\frac{n}{2}} + \frac{n}{2}\left[t \cdot \sqrt{\frac{2}{n}} + \frac{t^2}{2}\cdot\frac{2}{n} + \frac{t^3}{3}\left(\frac{2}{n}\right)^{3/2} + \cdots\right]$$

$$= -t\sqrt{\frac{n}{2}} + t\cdot\sqrt{\frac{n}{2}} + \frac{t^2}{2} + O(n^{-1/2}) = \frac{t^2}{2} + O(n^{-1/2}),$$

where $O(n^{-1/2})$ are terms containing $n^{1/2}$ and higher powers of n in the denominator.

$$\therefore \quad \lim_{n\to\infty} K_Z(t) = \frac{t^2}{2} \quad \Rightarrow \quad M_Z(t) = e^{t^2/2} \text{ as } n \to \infty,$$

which is the *m.g.f. of* a standard normal variate. Hence, by uniqueness theorem of *m.g.f.* Z is asymptotically normal. In other words, standard χ^2 variate tends to standard normal variate as $n \to \infty$. Thus, χ^2 distribution tends to normal distribution for large *d.f.*

In practice for $n \geq 30$, the χ^2-approximation to normal distribution is fairly good. So whenever $n \geq 30$, we use the normal probability tables for testing the significance of the value of χ^2. That is why in the tables, the significant values of χ^2 have been tabulated till $n = 30$ only.

Remark. For the distribution of χ^2-variate for large values of n, see *Example* 15·22, page 15-50 and also *Remark* 2 to §15·6·1, page 15·23.

15·3·3. Characteristic Function of χ^2-Distribution

If $X \sim \chi^2_{(n)}$, then

$$\phi_X(t) = E\{\exp(itX)\} = \int_0^\infty \exp(itx)\, f(x)\, dx$$

$$= \frac{1}{2^{n/2}\Gamma(n/2)} \int_0^\infty \exp\left\{-\left(\frac{1-2it}{2}\right)x\right\}(x)^{\frac{n}{2}-1}\, dx = (1-2it)^{-n/2} \qquad \text{... (15·4e)}$$

15.3.4. Mode and Skewness of c²-Distribution

Let $X \sim \chi^2_{(n)}$, so that

$$f(x) = \frac{1}{2^{n/2}\Gamma(n/2)} e^{-x/2} x^{(n/2)-1}, 0 \le x < \infty \qquad \ldots (*)$$

Mode of the distribution is the solution of $f'(x) = 0$ and $f''(x) < 0$.

Logarithmic differentiation *w.r.to* x in (*) gives:

$$\frac{f'(x)}{f(x)} = 0 - \frac{1}{2} + \left(\frac{n}{2} - 1\right)\cdot\frac{1}{x} = \frac{n-2-x}{2x} \qquad \ldots (15·5)$$

Since $f(x) \ne 0$, $f'(x) = 0 \quad \Rightarrow \quad x = n - 2$.

It can be easily seen that at the point, $x = (n - 2)$, $f''(x) < 0$.

Hence, *mode of the chi-square distribution with n d.f. is (n – 2).*

Also Karl Pearson's coefficient of skewness is given by:

$$\text{Skewness} = \frac{\text{Mean} - \text{Mode}}{\text{S.D.}} = \frac{n-(n-2)}{\sqrt{2n}} = \sqrt{\frac{2}{n}} \qquad \ldots (15·6)$$

Since Pearson's coefficient of skewness is greater than zero for $n \ge 1$, the χ^2-distribution is positively skewed. Further since skewness is inversely proportional to the square root of *d.f.*, it rapidly tends to symmetry as the *d.f.* increases.

15·3·5. Additive Property of χ^2-variates

The sum of independent chi-square variates is also a χ^2-variate. More precisely, if X_i, $(i = 1,2, \ldots, k)$ are inde-pendent χ^2-variates with n_i *d.f.* respectively, then the sum $\sum_{i=1}^{k} X_i$ is also a chi-square variate with $\sum_{i=1}^{k} n_i$ d.f.

Proof. We have

$$M_{X_i}(t) = (1-2t)^{-n_i/2}; i = 1, 2, \cdots, k.$$

The *m.g.f.* of the sum $\sum_{i=1}^{k} X_i$ is given by:

$$M_{\Sigma X_i}(t) = M_{X_1}(t)\, M_{X_2}(t) \ldots M_{X_k}(t) \qquad [\because X_i\text{'s are independent}]$$

$$= (1-2t)^{-n_1/2}(1-2t)^{-n_2/2} \cdots (1-2t)^{-n_k/2} = (1-2t)^{-(n_1+n_2+\ldots+n_k)/2}$$

which is the *m.g.f.* of a χ^2-variate with $(n_1 + n_2 + \ldots + n_k)$ *d.f.* Hence, by uniqueness theorem of *m.g.f.*'s, $\sum_{i=1}^{k} X_i$ is a χ^2-variate with $\sum_{i=1}^{k} n_i$ *d.f.*

Remarks:

1. Converse is also true, *i.e.*, if X_i ; $i = 1, 2, \ldots, k$ are χ^2-variates with n_i; $i = 1, 2, \ldots, k.d.f.$ respectively and if $\sum_{i=1}^{k} X_i$ is a χ^2-variate with $\sum_{i=1}^{k} n_i$ *d.f.*, then X_i's are independent.

2. Another useful version of the converse is as follows:

If X and Y are independent non-negative variates such that $X + Y$ follows chi-square distribution with $n_1 + n_2$ *d.f.* and if one of them say X is a χ^2-variate with n_1 *d.f.* then the other, viz., Y, is a χ^2-variate with n_2 *d.f.*

Proof. Since X and Y are independent variates, $M_{X+Y}(t) = M_X(t)M_Y(t)$

$$\Rightarrow \quad (1-2t)^{-(n_1+n_2)/2} = (1-2t)^{-n_1/2} \cdot M_Y(t) \qquad \left[\because X+Y \sim \chi^2_{(n_1+n_2)} \text{ and } X \sim \chi^2_{(n_1)}\right]$$

$$\therefore \quad M_Y(t) = (1-2t)^{-n_2/2}$$

which is the *m.g.f.* of χ^2-variate with n_2 d.f. Hence by uniqueness theorem of *m.g.f.'s*, $Y \sim \chi^2_{(n_2)}$

3. Still another form of the above theorem is "*Cochran theorem*" which is as follows:

Let $X_1, X_2, \ldots, X_n$ be independently distributed as standard normal variates, *i.e.*, $N(0, 1)$.

Let $\sum_{i=1}^{n} X_i^2 = Q_1 + Q_2 + \cdots + Q_k$, where each Q_i is a sum of squares of linear combinations of $X_1, X_2, \ldots, X_n$ with n_i degrees of freedom. Then if $n_1 + n_2 + \ldots + n_k = n$, the quantities $Q_1, Q_2, \ldots, Q_k$ are independent χ^2-variates with $n_1, n_2, \ldots, n_k$ *d.f.* respectively.

15·3·6. Chi-square Probability Curve

We get from (15·5),

$$f''(x) = \left[\frac{n-2-x}{2x}\right] f(x) \qquad \ldots (15·7)$$

Since $x > 0$ and $f(x)$ being *p.d.f.* is always non-negative, we get from (15·7) :

$$f''(x) < 0 \quad \text{if } (n-2) \le 0, \text{ i.e., if } n \le 2,$$

for all values of x. Thus the χ^2-probability curve for 1 and 2 degrees of freedom is monotonically decreasing. When $n > 2$.

$$f''(x) = \begin{cases} > 0, \text{ if } x < (n-2) \\ = 0, \text{ if } x = n-2 \\ < 0, \text{ if } x > (n-2) \end{cases}$$

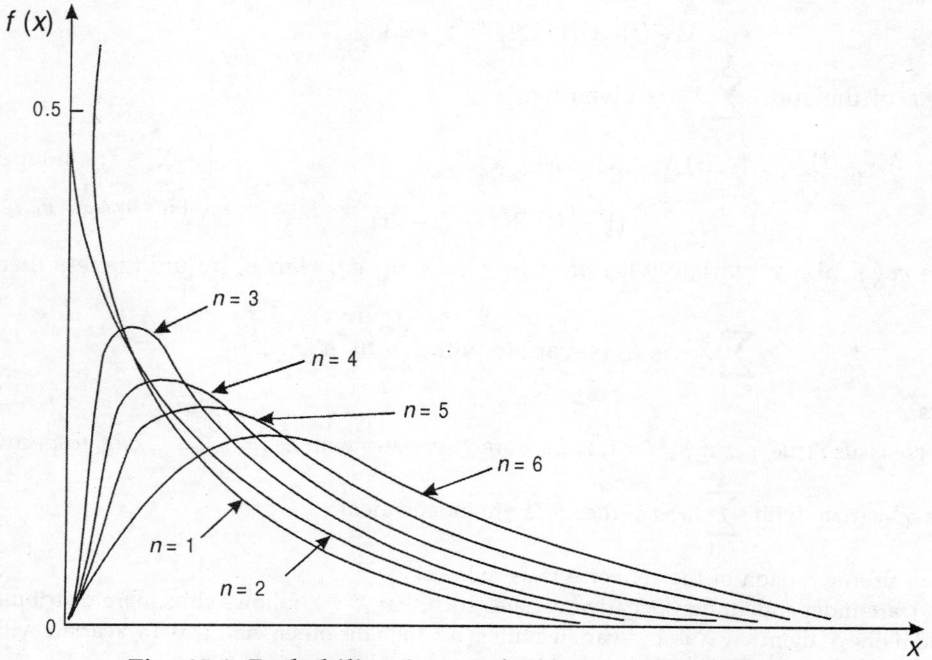

Fig. 15·1: Probability Curve of Chi-square Distribution

This implies that for $n > 2, f(x)$ is monotonically increasing for $0 < x < (n-2)$ and monotonically decreasing for $(n-2) < x < \infty$, while at $x = n - 2$, it attains the maximum value.

For $n \geq 1$, as x increases, $f(x)$ decreases rapidly and finally tends to zero as $x \to \infty$. Thus, for $n > 1$, the χ^2-probability curve is positively skewed [c.f. (15·6)] towards higher values of x. Moreover, x-axis is an asymptote to the curve. The shape of the curve for $n = 1, 2, 3, \ldots, 6$ is given in Fig. 15·1. For $n = 2$, the curve will meet $y = f(x)$ axis at $x = 0$, *i.e.*, at $f(x) = 0·5$. For $n = 1$, it will be an inverted J-shaped curve.

15·4. SOME THEOREMS ON CHI-SQUARE DISTRIBUTION

Theorem 1. *If X_1 and X_2 are two independent χ^2-variates with n_1 and n_2 d.f. respectively, then*

$$\frac{X_1}{X_2} \text{ is a } \beta_2\left(\frac{n_1}{2}, \frac{n_2}{2}\right) \text{ variate.}$$

Proof. Since X_1 and X_2 are independent χ^2 variates with n_1 and n_2 d.f. respectively, their joint probability differential is given by the compound probability theorem as:

$$dP(x_1, x_2) = dP_1(x_1)\, dP_2(x_2)$$

$$= \left[\frac{1}{2^{n_1/2}\Gamma(n_1/2)} \exp(-x_1/2)\cdot(x_1)^{(n_1/2)-1} dx_1\right] \times \left[\frac{1}{2^{n_2/2}\Gamma(n_2/2)} \exp(-x_2/2)\cdot(x_2)^{(n_2/2)-1} dx_2\right]$$

$$= \frac{1}{2^{(n_1+n_2)/2}\Gamma(n_1/2)\,\Gamma(n_2/2)} \exp\{-(x_1+x_2)/2\} \times (x_1)^{\frac{n_1}{2}-1} (x_2)^{\frac{n_2}{2}-1} dx_1 dx_2, \quad 0 \leq (x_1, x_2) < \infty$$

Let us make the transformation :

$$u = x_1/x_2 \quad \text{and} \quad v = x_2 \quad \text{so that} \quad x_1 = uv \quad \text{and} \quad x_2 = v.$$

Jacobian of transformation J is given by :

$$J = \frac{\partial(x_1, x_2)}{\partial(u, v)} = \begin{vmatrix} v & u \\ 0 & 1 \end{vmatrix} = v$$

Thus, the joint distribution of random variables U and V becomes:

$$dG(u, v) = \frac{1}{2^{(n_1+n_2)/2}\Gamma(n_1/2)\,\Gamma(n_2/2)} \exp\{-(1+u)v/2\} \times (uv)^{\frac{n_1}{2}-1} v^{\frac{n_2}{2}-1} |J|\, du\, dv,$$

$$= \frac{1}{2^{(n_1+n_2)/2}\Gamma(n_1/2)\,\Gamma(n_2/2)} \exp\{-(1+u)v/2\} \times u^{\frac{n_1}{2}-1} v^{\frac{n_1+n_2}{2}-1} du\, dv, \quad 0 \leq (u, v) < \infty$$

Integrating *w.r.to* v over the range 0 to ∞, we get marginal distribution of U as:

$$dG_1(u) = \int_0^\infty dG(u, v)$$

$$= \frac{1}{2^{(n_1+n_2)/2}\Gamma(n_1/2)\,\Gamma(n_2/2)} u^{(n_1/2)-1} du \times \int_0^\infty \exp\left\{-\left(\frac{1+u}{2}\right)v\right\} v^{\{(n_1+n_2)/2\}-1} dv$$

$$= \frac{u^{(n_1/2)-1}}{2^{(n_1+n_2)/2}\,\Gamma(n_1/2)\,\Gamma(n_2/2)} \cdot \frac{\Gamma\{(n_1+n_2)/2\}}{[(1+u)/2]^{(n_1+n_2)/2}} du$$

$$= \frac{1}{B\left(\frac{n_1}{2}, \frac{n_2}{2}\right)} \cdot \frac{u^{(n_1/2)-1}}{[1+u]^{(n_1+n_2)/2}} \, du, \; 0 \le u \le \infty$$

Hence, $$U = \frac{X_1}{X_2} \sim \beta_2\left(\frac{n_1}{2}, \frac{n_2}{2}\right) \text{ variate.}$$

Theorem 15·2. *If X_1 and X_2 are independent χ^2-variates with n_1 and n_2 d.f. respectively, then $U = \dfrac{X_1}{X_1 + X_2}$ and $V = X_1 + X_2$ are independently distributed, U as a $\beta_1\left(\frac{n_1}{2}, \frac{n_2}{2}\right)$ variate and V as a χ^2-variate with $(n_1 + n_2)$ d.f.*

Proof. From *Theorem 15·1*, we have

$$dP(x_1, x_2) = \frac{1}{2^{(n_1+n_2)/2}\,\Gamma(n_1/2)\,\Gamma(n_2/2)} \exp\{-(x_1 + x_2)/2\}$$

$$\times (x_1)^{(n_1/2)-1} (x_2)^{(n_2/2)-1} dx_1\, dx_2, \; 0 \le (x_1, x_2) < \infty$$

Let us transform to u and v defined as follows:

$$u = \frac{x_1}{x_1 + x_2} \quad \text{and } v = x_1 + x_2 \quad \text{so that} \quad x_1 = uv \quad \text{and} \quad x_2 = v - x_1 = (1-u)v$$

As x_1 and x_2 both range from 0 to ∞, u ranges from 0 to 1 and v from 0 to ∞.

Jacobian of transformation J is: $$J = \begin{vmatrix} v & u \\ -v & 1-u \end{vmatrix} = v$$

$$\therefore \quad dG(u, v) = \frac{1}{2^{(n_1+n_2)/2}\,\Gamma(n_1/2)\,\Gamma(n_2/2)} \exp(-v/2)$$

$$\times (uv)^{(n_1/2)-1} \times [(1-u)v]^{(n_2/2)-1} |J|\, du\, dv$$

$$= \frac{1}{2^{(n_1+n_2)/2}\,\Gamma(n_1/2)\,\Gamma(n_2/2)} u^{(n_1/2)-1} (1-u)^{(n_2/2)-1}$$

$$\times \exp(-v/2) \times v^{(n_1+n_2/2)-1} du\, dv$$

$$= \left[\frac{\Gamma\{(n_1+n_2)/2\}}{\Gamma(n_1/2)\,\Gamma(n_2/2)} u^{(n_1/2)-1} (1-u)^{(n_2/2)-1} du\right]$$

$$\times \left[\frac{1}{2^{(n_1+n_2)/2}\,\Gamma\{(n_1+n_2)/2\}} \exp(-v/2)\, v^{\{(n_1+n_2)/2\}-1} dv\right]$$

Since the joint probability differential of U and V is the product of their respective probability differentials, U and V are independently distributed, with

$$dG_1(u) = \frac{1}{B\left(\frac{n_1}{2}, \frac{n_2}{2}\right)} u^{(n_1/2)-1} (1-u)^{(n_2/2)-1} \, du, \; 0 \le u \le 1$$

and $$dG_2(v) = \frac{1}{2^{(n_1+n_2)/2}\,\Gamma\{(n_1+n_2)/2\}} \exp(-v/2)\, v^{\{(n_1+n_2)/2\}-1}\, dv,\ 0 \le v < \infty$$

i.e., U as a $\beta_1\left(\dfrac{n_1}{2}, \dfrac{n_2}{2}\right)$ variate and V as a χ^2 variate with $(n_1 + n_2)$ *d.f.*

Remark. The results in Theorems 15·1 and 15·2 can be summarised as follows:

If $X \sim \chi^2_{(n_1)}$ and $Y \sim \chi^2_{(n_2)}$ are independent chi-square variates, then

(*i*) $X + Y \sim \chi^2_{(n_1+n_2)}$ *i.e.,* the sum of two independent chi-square variates is also a chi-square variate.

(*ii*) $\dfrac{X}{Y} \sim \beta_2\left(\dfrac{n_1}{2}, \dfrac{n_2}{2}\right)$, *i.e.,* the ratio of two independent Chi-square variates is a β_2-variate.

(*iii*) $\dfrac{X}{X+Y} \sim \beta_1\left(\dfrac{n_1}{2}, \dfrac{n_2}{2}\right)$

Theorem 15·3. *In a random and large sample,* $$\chi^2 = \sum_{i=1}^{k}\left[\frac{(n_i - np_i)^2}{np_i}\right], \qquad \ldots (15·8)$$

follows Chi-square distribution approximately with (k – 1) degrees of freedom, where n_i is the observed frequency and np_i is the corresponding expected frequency of the ith class, (i = 1, 2, ..., k), $\sum_{i=1}^{k} n_i = n.$

Proof. Let us consider a random sample of size n, whose members are distributed at random in k classes or cells. Let p_i be the probability that sample observation will fall in the ith cell, (i = 1, 2, ..., k). Then the probability P of there being n_i members in the ith cell, (i = 1, 2, ..., k), respectively, is given by the multinomial probability law, by the expression:

$$P = \frac{n!}{n_1!\,n_2!\ldots n_k!}\, p_1^{n_1} p_2^{n_2} \ldots p_k^{n_k}, \text{ where } \sum_{i=1}^{k} n_i = n \text{ and } \sum_{i=1}^{k} p_i = 1.$$

If n is sufficiently large so that n_i, (i = 1, 2, ..., k) are not small then using Stirling's approximation to factorials for large n, *viz.,* $$\lim_{n\to\infty} (n!) \approx \sqrt{2\pi}\, e^{-n}\, n^{n+\frac{1}{2}},$$

$$P \approx \frac{\sqrt{2\pi}\, e^{-n}\, n^{n+\frac{1}{2}}}{(\sqrt{2\pi})^k\, e^{-(n_1+n_2+\cdots+n_k)}} \times \frac{p_1^{n_1} p_2^{n_2} \cdots p_k^{n_k}}{n_1^{n_1+\frac{1}{2}}\, n_2^{n_2+\frac{1}{2}} \cdots n_k^{n_k+\frac{1}{2}}}$$

$$\approx \frac{e^{-n} n^{n+\frac{1}{2}} \left(\dfrac{np_1}{n_1}\right)^{n_1+\frac{1}{2}} \left(\dfrac{np_2}{n_2}\right)^{n_2+\frac{1}{2}} \cdots \left(\dfrac{np_k}{n_k}\right)^{n_k+\frac{1}{2}}}{(\sqrt{2\pi})^{k-1} e^{-n} n^{n_1+n_2+\cdots+n_k+(k/2)} (p_1 p_2 \cdots p_k)^{1/2}} \approx C \prod_{i=1}^{k} \left(\frac{np_i}{n_i}\right)^{n_i+\frac{1}{2}}$$

where, $$C = \frac{1}{(2\pi)^{(k-1)/2}\, n^{(k-1)/2}\, (p_1 p_2 \cdots p_k)^{1/2}},$$ is a constant independent of n_i's.

$$\therefore \qquad \log P \approx \log C + \sum_{i=1}^{k}\left(n_i + \frac{1}{2}\right)\log\left(\frac{np_i}{n_i}\right) \quad \Rightarrow \quad \log(P/C) \approx \sum_{i=1}^{k}\left(n_i + \frac{1}{2}\right)\log\left(\frac{\lambda_i}{n_i}\right), \qquad \ldots (*)$$

where, $\lambda_i = np_i$ is the expected frequency for the ith cell, *i.e.*, $E(n_i) = np_i = \lambda_i$, $(i = 1, 2, \ldots, k)$.

Let us define $\quad \xi_i = \dfrac{n_i - \lambda_i}{\sqrt{\lambda_i}}$, so that $n_i - \lambda_i = \xi_i\sqrt{\lambda_i} \Rightarrow n_i = \lambda_i + \xi_i\sqrt{\lambda_i}$ $\qquad \ldots (**)$

Substituting in (*), we get

$$\log(P/C) \approx \sum_{i=1}^{k}\left(\lambda_i + \xi_i\sqrt{\lambda_i} + \frac{1}{2}\right)\cdot\log\left[\frac{\lambda_i}{\lambda_i + \xi_i\sqrt{\lambda_i}}\right]$$

$$\approx \sum_{i=1}^{k}\left(\lambda_i + \xi_i\sqrt{\lambda_i} + \frac{1}{2}\right)\cdot\log\left[1/\{1 + \xi_i/\sqrt{\lambda_i}\}\right]$$

$$\approx -\sum_{i=1}^{k}\left(\lambda_i + \xi_i\sqrt{\lambda_i} + \frac{1}{2}\right)\log\{1 + (\xi_i/\sqrt{\lambda_i})\}$$

If we assume that ξ_i is small compared with λ_i, the expansion of $\log\{1 + (\xi_i/\sqrt{\lambda_i})\}$ in ascending powers of $\xi_i/\sqrt{\lambda_i}$ is valid.

$$\therefore \qquad \log(P/C) \approx -\sum_{i=1}^{k}\left(\lambda_i + \xi_i\sqrt{\lambda_i} + \frac{1}{2}\right)\left[\frac{\xi_i}{\sqrt{\lambda_i}} - \frac{1}{2}\frac{\xi_i^2}{\lambda_i} + O\left(1/\lambda_i^{3/2}\right)\right]$$

$$\approx -\sum_{i=1}^{k}\left[\xi_i\sqrt{\lambda_i} - \frac{1}{2}\xi_i^2 + \xi_i^2 + O\left(\lambda_i^{-1/2}\right)\right],$$

(neglecting higher powers of $\xi_i/\sqrt{\lambda_i}$ if ξ_i is small compared with λ_i)

Since n is large, so is $\lambda_i = np_i$. Hence $O\left(\lambda_i^{-1/2}\right) \to 0$ for large n.

$$\text{Also,} \qquad \sum_{i=1}^{k}\xi_i\sqrt{\lambda_i} = \sum_{i=1}^{k}(n_i - \lambda_i) = \sum_{i=1}^{k}n_i - \sum_{i=1}^{k}\lambda_i = \sum_{i=1}^{k}n_i - n\sum_{i=1}^{k}p_i = n - n = 0$$

$$(\because \Sigma n_i = n,\ \Sigma p_i = 1)$$

$$\therefore \qquad \log(P/C) \approx -\left[\sum_{i=1}^{k}\xi_i\sqrt{\lambda_i} + \frac{1}{2}\sum_{i=1}^{k}\xi_i^2 + O(\lambda_i^{-1/2})\right] \approx -\frac{1}{2}\sum_{i=1}^{k}\xi_i^2$$

$$\Rightarrow \qquad P \approx C\cdot\exp\left(-\frac{1}{2}\sum_{i=1}^{k}\xi_i^2\right),$$

which shows that ξ_i, ($i = 1, 2, ..., k$) are distributed as independent standard normal variates.

Hence, $\sum_{i=1}^{k} \xi_i^2 = \sum_{i=1}^{k} \left[\frac{(n_i - \lambda_i)^2}{\lambda_i} \right]$, being the sum of the squares of k independent standard normal variates is a χ^2-variate with $(k - 1)$ *d.f.*, one *d.f.* being lost because of the linear constraint

$$\sum_{i=1}^{k} \xi_i \sqrt{\lambda_i} = \sum (n_i - \lambda_i) = 0 \quad \Rightarrow \quad \sum_{i=1}^{k} n_i = \sum_{i=1}^{k} \lambda_i \qquad ... (***)$$

Remarks:

1. If O_i and E_i ($i = 1, 2, ..., k$), be a set of observed and expected frequencies, then

$$\chi^2 = \sum_{i=1}^{k} \left[\frac{(O_i - E_i)^2}{E_i} \right], \quad \left(\sum_{i=1}^{k} O_i = \sum_{i=1}^{k} E_i \right) \qquad ... (15·8a)$$

follows chi-square distribution with $(k - 1)$ *d.f.*

Another convenient form of this formula is as follows:

$$\chi^2 = \sum_{i=1}^{k} \left(\frac{O_i^2 + E_i^2 - 2O_i E_i}{E_i} \right) = \sum_{i=1}^{k} \left(\frac{O_i^2}{E_i} + E_i - 2O_i \right)$$

$$= \sum_{i=1}^{k} (O_i^2/E_i) + \sum_{i=1}^{k} E_i - 2\sum_{i=1}^{k} O_i = \sum_{i=1}^{k} (O_i^2/E_i) - N \qquad ... (15·8b)$$

where, $\sum_{i=1}^{k} O_i = \sum_{i=1}^{k} E_i = N$ (say), is the total frequency.

2. Conditions for the Validity of χ^2-test. χ^2-test is an approximate test for large values of n. For the validity of chi-square test of '*goodness of fit*' between theory and experiment, the following conditions must be satisfied:

(*i*) The sample observations should be independent.

(*ii*) Constraints on the cell frequencies, if any, should be linear, *e.g.*, $\Sigma n_i = \Sigma \lambda_i$ or $\Sigma O_i = \Sigma E_i$.

(*iii*) N, the total frequency should be reasonably large, say, greater than 50.

(*iv*) No theoretical cell frequency should be less than 5. (The chi-square distribution is essentially a continuous distribution but it cannot maintain its character of continuity if cell frequency is less than 5.) If any theoretical cell frequency is less than 5, then for the application of χ^2-test, it is pooled with the preceding or succeeding frequency so that the pooled frequency is more than 5 and finally adjust for the *d.f.* lost in pooling.

3. It may be noted that the χ^2-test depends only on the set of observed and expected frequencies and on degrees of freedom (*d.f.*). It does not make ny assumptions regarding the parent population from which the observations are taken. Since χ^2 defined in (15.8) does not involve any population parameters, it is termed as a statistic and the test is known as *Non-Parametric Test* or *Distribution-Free Test.*

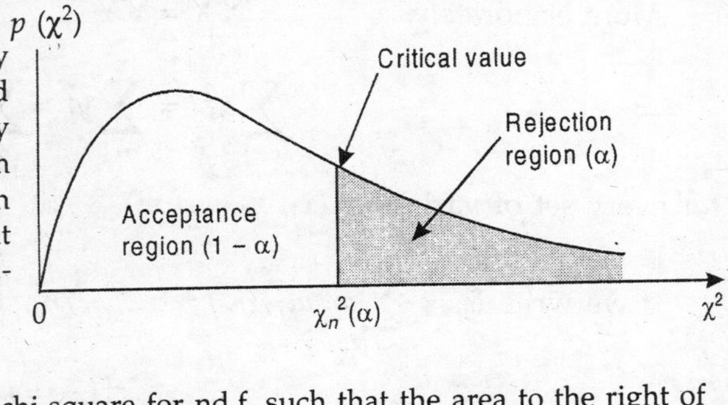

4. *Critical Value*. Let $\chi_n^2(\alpha)$ denote the value of chi-square for nd.f. such that the area to the right of this point is α., *i.e.*,

$$P\left[\chi^2 > \chi_n^2(\alpha)\right] = \alpha, \qquad ... (15·8c)$$

The value $\chi_n^2(\alpha)$ defined in (15·8c) is known as the *'Upper (right-tailed) α-point or Critical Value or Significant Value of Chi-square for n d.f.'* and has been tabulated for different values of n and α in the Table VII given at the end of the book. From these tables we observe that the critical values of χ^2 increase as n (*d.f.*) increases and level of significance (α) decreases.

15·5. LINEAR TRANSFORMATION

Let us suppose that the given set of variables $X' = (x_1, x_2, \ldots, x_n)$ is transformed to a new set of variables $\mathbf{Y}' = (y_1, y_2, \ldots, y_n)$ by means of the linear transformation:

$$\left.\begin{aligned} y_1 &= a_{11}x_1 + a_{12}x_2 + \cdots + a_{1n}x_n \\ y_2 &= a_{21}x_1 + a_{22}x_2 + \cdots + a_{2n}x_n \\ &\vdots \\ y_n &= a_{n1}x_1 + a_{n2}x_2 + \cdots + a_{nn}x_n \end{aligned}\right\} \quad \ldots (15·9)$$

i.e., $y_i = a_{i1}x_1 + a_{i2}x_2 + \ldots + a_{in}\, x_n \,;\, i = 1, 2, \ldots, n$

In matrix notation, this system of linear equations can be expressed symbolically as:

$$\mathbf{Y} = \mathbf{AX} \quad \ldots (15·10)$$

where $\mathbf{Y} = \begin{pmatrix} y_1 \\ y_2 \\ \vdots \\ y_n \end{pmatrix}$, $\mathbf{X} = \begin{pmatrix} x_1 \\ x_2 \\ \vdots \\ x_n \end{pmatrix}$, $\mathbf{A} = \begin{pmatrix} a_{11} & a_{12} & \cdots & a_{1n} \\ a_{21} & a_{22} & \cdots & a_{2n} \\ \vdots & \vdots & & \vdots \\ a_{n1} & a_{n2} & \cdots & a_{nn} \end{pmatrix}$

From matrix theory, we know that the system (15·10) has a unique solution iff $|\mathbf{A}| \neq 0$. In other words, we can express **X** uniquely in terms of **Y** if **A** is non-singular and the solution is given by:

$$\mathbf{X} = \mathbf{A}^{-1}\mathbf{Y} \quad \ldots (15·10a)$$

where $\mathbf{A}^{-1}$ is the inverse of the square matrix **A**.

The linear transformation defined in (15·9) or (15·10) is said to be *orthogonal* if

$$\mathbf{X'X} = \mathbf{Y'Y} \quad \ldots (15·11)$$

$$\Rightarrow \quad \mathbf{X'X} = \mathbf{(AX)'\, AX} = \mathbf{X'(A'A)X}$$

$$\Rightarrow \quad \mathbf{A'A} = \mathbf{I}_n \quad \ldots (15·11a)$$

$\therefore$ **A** is an orthogonal matrix.

More elaborately, $\mathbf{X'X} = \mathbf{Y'Y}$

$$\Rightarrow \quad \sum_{i=1}^{n} x_i^2 = \sum_{i=1}^{n} y_i^2 = \sum_{i=1}^{n} (a_{i1}x_1 + a_{i2}x_2 + \cdots + a_{in}x_n)^2, \quad \ldots (*)$$

for every set of variables, $(x_1, x_2, \ldots, x_n)$.

If we write $\delta_{ij} = \sum_{k=1}^{n} a_{ik}\, a_{kj}, (i, j = 1, 2, \ldots, n)$, then (*) implies that δ_{ij} is a *Kronecker delta*

so that
$$\delta_{ij} = \begin{cases} 1, & i = j \\ 0, & i \neq j \end{cases} \quad \ldots (15·12)$$

whence it follows that **A** is an orthogonal matrix.

Linear Orthogonal Transformation. Def.: *A linear transformation* $\mathbf{Y} = \mathbf{AX}$, *is said to be orthogonal if A is an orthogonal matrix.*

Remarks:

1. It is very easy to verify the equivalence of the following two definitions of an orthogonal matrix.

Def. 1. A square matrix $\mathbf{A}(n \times n)$ is said to be orthogonal if $\mathbf{A'A} = \mathbf{AA'} = \mathbf{I}_n$.

Def. 2. A square matrix $\mathbf{A}$ is said to be orthogonal if the transformation $\mathbf{Y} = \mathbf{AX}$ transforms $\mathbf{X'X}$ to $\mathbf{Y'Y}$.

2. If $\mathbf{Y} = \mathbf{AX}$ is orthogonal transformation, then $\mathbf{Y'Y} = \mathbf{X'X}$ and $\mathbf{A'A} = \mathbf{AA'} = \mathbf{I}_n$.

Theorem 15·4. *(Fisher's Lemma). If* X_i, $(i = 1, 2, \ldots, n)$ *are independent* $N(0, \sigma^2)$ *and they are transformed to a new set of variables* Y_i, $(i = 1, 2, \ldots, n)$, *by means of a linear orthogonal transformation, then* Y_i, $(i = 1, 2, \ldots, n)$ *are also independent* $N(0, \sigma^2)$.

Proof. Let the linear orthogonal transformation be $\mathbf{Y} = \mathbf{AX}$ so that

$$\mathbf{Y'Y} = \mathbf{X'X} \quad \text{and} \quad \mathbf{A'A} = \mathbf{I}_n$$

Since X_i, $(i = 1, 2, \ldots, n)$ are independent $N(0, \sigma^2)$, their joint probability density function is given by:

$$f(x_1, x_2, \ldots, x_n) = \left(\frac{1}{\sigma\sqrt{2\pi}}\right)^n \cdot \exp\left(-\sum_{i=1}^{n} x_i^2/2\sigma^2\right), -\infty < (x_1, x_2, \ldots, x_n) < \infty$$

$$= \left(\frac{1}{\sigma\sqrt{2\pi}}\right)^n \exp\left(-\mathbf{X'X}/2\sigma^2\right)$$

The joint density of $(Y_1, Y_2, \ldots, Y_n)$ becomes

$$g(y_1, y_2, \ldots, y_n) = \left(\frac{1}{\sigma\sqrt{2\pi}}\right)^n \exp\left(-\mathbf{Y'Y}/2\sigma^2\right)|J|, \quad \text{where } \frac{1}{J} = \frac{\partial(y_1, y_2, \ldots, y_n)}{\partial(x_1, x_2, \ldots, x_n)} = |\mathbf{A}|$$

$$\mathbf{A'A} = \mathbf{I}_n \quad \Rightarrow \quad |\mathbf{A'A}| = |\mathbf{I}_n| = 1 \quad \Rightarrow \quad |\mathbf{A'}|\ |\mathbf{A}| = 1$$

$$\Rightarrow \quad |\mathbf{A}|^2 = 1 \qquad (\because |\mathbf{A'}| = |\mathbf{A}|)$$

$$\Rightarrow \quad |\mathbf{A}| = \pm 1, \quad \text{i.e.,} \quad |J| = |\pm 1| = 1$$

$$\therefore \quad g(y_1, y_2, \ldots, y_n) = \left(\frac{1}{\sigma\sqrt{2\pi}}\right)^n \exp(-\mathbf{Y'Y}/2\sigma^2) = \prod_{i=1}^{n}\left[\frac{1}{\sigma\sqrt{2\pi}}\exp\left(-y_i^2/2\sigma^2\right)\right]$$

Hence, Y_i, $(i = 1, 2, \ldots, n)$ are independent $N(0, \sigma^2)$.

Theorem 15.5. *Let* $X_1, X_2, \ldots, X_n$ *be a random sample from a normal population with mean* μ *and variance* σ^2. *Then*

(i) $\overline{X} \sim N(\mu, \sigma^2/n)$, *(ii)* $\sum_{i=1}^{n}\left(\frac{X_i - \overline{X}}{\sigma}\right)^2$ *is a* χ^2*-variate with* $(n - 1)$ *d.f., and*

(iii) $\overline{X} = \frac{1}{n}\sum_{i=1}^{n} X_i$ and $\frac{ns^2}{\sigma^2} = \sum_{i=1}^{n}\left(\frac{X_i - \overline{X}}{\sigma}\right)^2$ *are independently distributed;*

$\overline{X}$ *as* $N\left(\mu, \frac{\sigma^2}{n}\right)$ *variate and* $\frac{ns^2}{\sigma^2}$ *as* $\chi^2_{(n-1)}$ *variate.*

Proof. The joint probability differential of $X_1, X_2, \ldots, X_n$ is given by:

$$dP(x_1, x_2, \ldots, x_n) = \left(\frac{1}{\sigma\sqrt{2\pi}}\right)^n \cdot \exp\left[-\frac{1}{2\sigma^2}\sum_{i=1}^{n}(x_i - \mu)^2\right] dx_1\, dx_2 \ldots dx_n\ ;$$

$$-\infty < (x_1, x_2, \ldots, x_n) < \infty$$

Let us transform to the variables Y_i, $(i = 1, 2, \ldots, n)$ by means of a linear orthogonal transformation $\mathbf{Y} = \mathbf{AX}$. Let us choose in particular:

$$a_{11} = a_{12} = \ldots = a_{1n} = \frac{1}{\sqrt{n}} \quad \Rightarrow \quad y_1 = \frac{1}{\sqrt{n}}(x_1 + x_2 + \ldots + x_n) = \sqrt{n}\,\bar{x} \qquad \ldots (*)$$

It can be easily seen that the above choice of $a_{11}, a_{12}, \ldots, a_{1n}$ satisfieds the condition of orthogonality, *viz.*, $\sum_{j=1}^{n} a_{ij}^2 = 1$. Since the transformation is orthogonal, we have

$$\sum_{i=1}^{n} y_i^2 = \sum_{i=1}^{n} x_i^2 = \sum_{i=1}^{n}(x_i - \bar{x})^2 + n\bar{x}^2 = \sum_{i=1}^{n}(x_i - \bar{x})^2 + y_1^2 \qquad \text{[From (*)]}$$

$$\Rightarrow \qquad \sum_{i=2}^{n} y_i^2 = \sum_{i=1}^{n}(x_i - \bar{x})^2 \qquad \ldots (**)$$

Also, $$\sum_{i=1}^{n}(x_i - \mu)^2 = \sum_{i=1}^{n}(x_i - \bar{x} + \bar{x} - \mu)^2 = \sum_{i=1}^{n}(x_i - \bar{x})^2 + n(\bar{x} - \mu)^2 = \sum_{i=2}^{n} y_i^2 + n(\bar{x} - \mu)^2$$

[From (**)]

As in Theorem 15·4, the Jacobian of transformation $J = \pm 1$. Thus, the joint density function of $X_1, X_2, \ldots, X_n$ transforms to:

$$dG(y_1, y_2, \ldots, y_n) = \left(\frac{1}{\sigma\sqrt{2\pi}}\right)^n \exp\left[-\frac{1}{2\sigma^2}\left\{\sum_{i=2}^{n} y_i^2 + n(\bar{x} - \mu)^2\right\}\right] \times |J|\, dy_1\, dy_2 \ldots dy_n$$

$$= \left[\frac{1}{\sqrt{2\pi}\,(\sigma/\sqrt{n})} \exp\left\{-\frac{n}{2\sigma^2}(\bar{x} - \mu)^2\right\} d\bar{x}\right]$$

$$\times \left[\left(\frac{1}{\sigma\sqrt{2\pi}}\right)^{n-1} \exp\left\{-\sum_{i=2}^{n} \frac{y_i^2}{2\sigma^2}\right\} dy_2\, dy_3 \ldots dy_n\right] \qquad (\because\ dy_1 = \sqrt{n}\, d\bar{x})$$

Thus, $\bar{X}$ and $\sum_{i=2}^{n} Y_i^2 = \sum_{i=1}^{n}(X_i - \bar{X})^2 = ns^2$, (where s^2 is the sample variance), are independently distributed, which establishes part (*iii*) of the theorem. Moreover, $\bar{X} \sim N(\mu, \sigma^2/n)$ and Y_i $(i = 2, 3, \ldots, n)$ are independent $N(0, \sigma^2)$.

Hence,
$$\sum_{i=2}^{n} \frac{Y_i^2}{\sigma^2} = \sum_{i=1}^{n}\left(\frac{X_i - \bar{X}}{\sigma}\right)^2 = \frac{ns^2}{\sigma^2}$$

being the sum of squares of $(n - 1)$ independent standard normal variates is distributed as χ^2-variate with $(n - 1)$ *d.f.*

Aliter. The alternative proof of the above Theorem is based on the use of *m.g.f.'s* and is given below. We shall first prove that:

$$\bar{X} = \frac{1}{n}\sum_{i=1}^{n} X_i \text{ and } X_i - \bar{X}\ ;\ i = 1, 2, \ldots, n \text{ are independently distributed.} \quad \ldots (i)$$

The joint *m.g.f.* of $\bar{X}$ and $(X_i - \bar{X})$ is given by:

$$M(t_1, t_2) = E\left[e^{t_1\bar{X} + t_2(X_i - \bar{X})}\right] = E\left[e^{(t_1 - t_2)\bar{X} + t_2 X_i}\right] = E\left[\exp\left\{\frac{t_1 - t_2}{n}\cdot\sum_{i=1}^{n} X_i + t_2 X_i\right\}\right]$$

$$= E\left[\exp\left\{\left(\frac{t_1 - t_2}{n} + t_2\right)X_i + \frac{t_1 - t_2}{n}\sum_{\substack{j=1\\(j\neq i)}}^{n} X_j\right\}\right]$$

$$= E\left[\exp\left\{\left(\frac{t_1 - t_2}{n} + t_2\right)X_i\right\}\right]\cdot E\left[\exp\left(\frac{t_1 - t_2}{n}\sum_{\substack{j=1\\(j\neq i)}}^{n} X_j\right)\right] \quad \ldots (ii)$$

($\because X_1, X_2, \ldots, X_n$ are independent)

Now $U = \sum_{\substack{j=1\\(j\neq i)}}^{n} X_j$, being the sum of $(n - 1)$ *i.i.d.* $N(\mu, \sigma^2)$ variates is a $N\{(n-1)\mu, (n-1)\sigma^2\}$ variate.

$$\therefore \quad M_U(t) = \exp\left[t\cdot(n-1)\,\mu + t^2\cdot(n-1)\,\sigma^2/2\right] \quad \ldots (iii)$$

Hence,
$$E\left[\exp\left\{\frac{t_1 - t_2}{n}\sum_{\substack{j=1\\(j\neq i)}}^{n} X_j\right\}\right] = E\left[\exp\left(\frac{t_1 - t_2}{n}\cdot U\right)\right] = M_U\left[(t_1 - t_2)/n\right]$$

$$= \exp\left[\left(\frac{t_1 - t_2}{n}\right)(n-1)\,\mu + \left(\frac{t_1 - t_2}{n}\right)^2 (n-1)\frac{\sigma^2}{2}\right] \quad \text{[On using (iii)]} \quad \ldots (iv)$$

and
$$E\left[\exp\left\{\left(\frac{t_1 - t_2}{n} + t_2\right)X_i\right\}\right] = M_{X_i}\left(\frac{t_1 - t_2}{n} + t_2\right) = \exp\left[\left(\frac{t_1 - t_2}{n} + t_2\right)\mu + \left(\frac{t_1 - t_2}{n} + t_2\right)^2\frac{\sigma^2}{2}\right]$$

$$\left[\because X_i \sim N(\mu, \sigma^2)\right] \quad \ldots (v)$$

Substituting from (*iv*) and (*v*) in (*ii*), we get

$$M(t_1, t_2) = \exp\left[\left\{\left(\frac{t_1 - t_2}{n}\right)(n-1) + \left(\frac{t_1 - t_2}{n} + t_2\right)\right\}\mu\right]$$

$$\times \exp\left[\left\{\left(\frac{t_1 - t_2}{n}\right)^2 \cdot (n-1) + \left(\frac{t_1 - t_2}{n} + t_2\right)^2\right\}\frac{\sigma^2}{2}\right]$$

$$= \exp\left[t_1 \mu + \frac{1}{2} t_1^2 \left(\frac{\sigma^2}{n}\right)\right] \times \exp\left[\frac{1}{2} t_2^2 \left(\frac{n-1}{n}\right)\sigma^2\right] \text{ (On simplification)}$$

$$= M(t_1) \cdot M(t_2)$$

$\Rightarrow$ (*a*) $\bar{X}$ and $(X_i - \bar{X})$; $i = 1, 2, \ldots, n$ are independently distributed ... (*vi*)

and (*b*) $\bar{X} \sim N(\mu, \sigma^2/n)$ ... (*vii*)

and $X_i - \bar{X} \sim N\left(0, \frac{n-1}{n}\sigma^2\right)$ ... (*viii*)

Since $\bar{X}$ and $X_i - \bar{X}$ are independently distributed, $\bar{X}$ and $s^2 = \frac{1}{n}\sum_{i=1}^{n}(X_i - \bar{X})^2$, are independently distributed .... (*viii a*)

To derive the distribution of s^2, we note that:

$$\sum_{i=1}^{n}(X_i - \mu)^2 = \sum_{i=1}^{n}(X_i - \bar{X} + \bar{X} - \mu)^2 = \sum_{i=1}^{n}(X_i - \bar{X})^2 + n(\bar{X} - \mu)^2,$$

the product term vanishes since $\sum_{i=1}^{n}(X_i - \bar{X}) = 0$.

$$\therefore \quad \sum_{i=1}^{n}\left(\frac{X_i - \mu}{\sigma}\right)^2 = \sum_{i=1}^{n}\frac{(X_i - \bar{X})^2}{\sigma^2} + \left[\frac{\bar{X} - \mu}{\sigma/\sqrt{n}}\right]^2 \quad \Rightarrow \quad V = W + Z, \quad \text{... (ix)}$$

where $V = \sum_{i=1}^{n}\left(\frac{X_i - \mu}{\sigma}\right)^2$, being the sum of squares of n independent standard normal variates is a $\chi^2_{(n)}$ variate. Hence,

$$M_V(t) = (1 - 2t)^{-n/2} \; ; \; |t| < \frac{1}{2}. \quad \text{... (x)}$$

Also $\bar{X} \sim N(\mu, \sigma^2/n) \quad \Rightarrow \quad \frac{\bar{X} - \mu}{\sigma/\sqrt{n}} \sim N(0, 1) \quad \Rightarrow \quad Z = \left[\frac{\bar{X} - \mu}{\sigma/\sqrt{n}}\right]^2 \sim \chi^2_{(1)}$

$$\therefore \quad M_Z(t) = (1 - 2t)^{-1/2} \quad \text{... (xi)}$$

Further, since $\bar{X}$ and s^2 are independent, [see *viii(a)*], *W* and *Z* are independently distributed.

$\therefore \quad M_V(t) = M_{W+Z}(t) = M_W(t) \cdot M_Z(t) \quad$ ($\because$ *W* and *Z* are independent)

$\Rightarrow \quad (1-2t)^{-n/2} = M_W(t) \cdot (1-2t)^{-1/2} \quad$ [From (*x*) and (*xi*)]

$\Rightarrow \quad M_W(t) = (1-2t)^{-(n-1)/2}, \ |t| < \dfrac{1}{2}$

which is *m.g.f.* of χ^2-variate with $(n-1)$ *d.f.* Hence, by uniqueness theorem of *m.g.f.*

$$W = \sum_{i=1}^{n} \frac{(X_i - \bar{X})^2}{\sigma^2} = \frac{ns^2}{\sigma^2} \sim \chi^2_{(n-1)} \qquad \text{... } (xii)$$

Remarks:

1. p.d.f. of the sample variance $s^2 = n^{-1} \sum_{i=1}^{n} (X_i - \bar{X})^2$.

Since $\sum_{i=1}^{n} \left(\dfrac{X_i - \bar{X}}{\sigma} \right)^2 = \dfrac{ns^2}{\sigma^2}$, is a χ^2-variate with $(n-1)$ *d.f.*, we have

$$dP\,(ns^2/\sigma^2) = \frac{1}{2^{(n-1)/2}\,\Gamma[(n-1)/2]} \cdot \exp\left(-\frac{ns^2}{2\sigma^2} \right) (ns^2/\sigma^2)^{\frac{n-1}{2}-1}\ d(ns^2/\sigma^2)$$

$$\Rightarrow \quad dP(s^2) = \frac{(n/2\sigma^2)^{(n-1)/2}}{\Gamma[(n-1)/2]} \cdot \exp\left(-\frac{ns^2}{2\sigma^2} \right) (s^2)^{(n-3)/2}\, ds^2, \ 0 < s^2 < \infty.$$

2. We have $\dfrac{ns^2}{\sigma^2} \sim \chi^2_{n-1}$

$$\therefore \quad E\left(\frac{ns^2}{\sigma^2} \right) = n - 1 \quad \Rightarrow \quad E(s^2) = \left(\frac{n-1}{n} \right) \sigma^2 = \left(1 - \frac{1}{n} \right) \sigma^2 \simeq \sigma^2, \ \text{for large } n. \qquad \text{... (*)}$$

$$\text{Also} \quad \text{Var}\left(\frac{ns^2}{\sigma^2} \right) = 2(n-1) \quad \Rightarrow \quad \text{Var}\,(s^2) = \frac{2}{n}\left(1 - \frac{1}{n} \right) \sigma^4 \simeq \frac{2\sigma^4}{n}, \ \text{for large } n. \qquad \text{... (**)}$$

$$\therefore \quad \text{S.E.}(s^2) = \sigma^2 \times \sqrt{2/n}, \ (\text{For large } n) \qquad \text{... (***)}$$

Theorem 15·6. *Let X_i, $(i = 1, 2, \ldots, n)$ be independent N(0, 1) variates. Then, the conditional distribution of $\chi^2 = \sum_{i=1}^{n} X_i^2$, subject to m (< n), (say), independent homogeneous linear constraints, viz.,*

$$\left.\begin{aligned} c_{11}X_1 + c_{12}X_2 + \cdots + c_{1n}X_n &= 0 \\ c_{21}X_1 + c_{22}X_2 + \ldots + c_{2n}X_n &= 0 \\ \vdots \quad\quad \vdots \quad\quad\quad \vdots \quad & \\ c_{m1}X_1 + c_{m2}X_2 + \ldots + c_{mn}X_n &= 0 \end{aligned}\right\} \qquad \text{... (15·13)}$$

is also a χ^2-distribution with $(n - m)$ degrees of freedom.

Proof. Equivalently, the constraints (15·13) can be expressed as:

$$\left.\begin{aligned} a_{11}X_1 + a_{12}X_2 + \cdots + a_{1n}X_n &= 0 \\ a_{21}X_1 + a_{22}X_2 + \ldots + a_{2n}X_n &= 0 \\ \vdots \quad \vdots \qquad\quad \vdots \\ a_{m1}X_1 + a_{m2}X_2 + \ldots + a_{mn}X_n &= 0 \end{aligned}\right\} \qquad \ldots (15·13a)$$

where $a_i = (a_{i1}, a_{i2}, \ldots, a_{in})$; $i = 1, 2, \ldots, m$ are m unitary, mutually orthogonal vectors.

Let us now transform the variables:

$$(X_1, X_2, \ldots, X_m, X_{m+1}, \ldots, X_n) \quad \text{to} \quad (Y_1, Y_2, \ldots, Y_m, Y_{m+1}, \ldots, Y_n)$$

by means of a linear orthogonal transformation: $\mathbf{Y} = \mathbf{AX}$, ... (15·13*b*)

where

$$\mathbf{Y} = \begin{pmatrix} Y_1 \\ Y_2 \\ \vdots \\ Y_m \\ Y_{m+1} \\ \vdots \\ Y_n \end{pmatrix}, \quad \mathbf{A} = \begin{pmatrix} a_{11} & a_{12} & \cdots & a_{1n} \\ a_{21} & a_{22} & \cdots & a_{2n} \\ \vdots & & & \\ a_{m1} & a_{m2} & \cdots & a_{mn} \\ a_{m+1,1} & a_{m+1,2} & \cdots & a_{m+1,n} \\ \vdots & & & \\ a_{n1} & a_{n2} & \cdots & a_{nn} \end{pmatrix} \quad \text{and} \quad \mathbf{X} = \begin{pmatrix} X_1 \\ X_2 \\ \vdots \\ X_m \\ X_{m+1} \\ \vdots \\ X_n \end{pmatrix}$$

(15.13*b*) implies that the constraints (15·13*a*) are equivalent to

$$Y_i = 0, \ (i = 1, 2, \ldots, m) \qquad \ldots (15·13c)$$

By Fisher's Lemma (Theorem 15·4) Y_i, $(i = 1, 2, \ldots, n)$ are also independent $N(0, 1)$ variates and

$$\sum_{i=1}^{n} X_i^2 = \sum_{i=1}^{n} Y_i^2 \qquad [\because \text{ Transformation (15·13}b\text{) is orthogonal}]$$

$$= \sum_{i=m+1}^{n} Y_i^2 \qquad [\text{Using (15·13}c\text{)}]$$

Thus, the conditional distribution of $\sum_{i=1}^{n} X_i^2$ subject to the conditions (15·13) is same as the unconditional distribution of $\sum_{i=m+1}^{n} Y_i^2$, where Y_i $(i = m + 1, m + 2, \ldots, n)$ are independent standard normal variates without any constraints on them. Hence,

$$\chi^2 = \sum_{i=1}^{n} X_i^2 = \sum_{i=m+1}^{n} Y_i^2,$$

being the sum of squares of $(n - m)$ independent standard normal variates follows χ^2-distribution with $(n - m)$ degrees of freedom.

15·6. APPLICATIONS OF CHI-SQUARE DISTRIBUTION

χ^2-distribution has a large number of applications in Statistics, some of which are enumerated below:

(*i*) To test if the hypothetical value of the population variance is $\sigma^2 = \sigma^2_0$ (say).

(*ii*) To test the 'goodness of fit'.

(*iii*) To test the independence of attributes.

(*iv*) To test the homogeneity of independent estimates of the population variance.

(*v*) To test the equality of several population proportions.

(*vi*) To combine various probabilities obtained from independent experiments to give a single test of significance.

(*vii*) To test the homogeneity of independent estimates of the population correlation coefficient.

In this section we will introduce various hypothesis-testing procedures based on the use of the chi-square distribution. As with other hypothesis-testing procedures, these tests compare the sample results with those that are expected when the null hypothesis is true. The acceptance or rejection of the null hypothesis is based upon how 'close' the sample or observed results are to the expected results. For detailed discussion on Testing of Hypothesis, see Chapter 17.

15·6·1. Inferences About a Population Variance

Suppose we want to test if a random sample x_i, $(i = 1, 2, \ldots, n)$ has been drawn from a normal population with a specified variance $\sigma^2 = \sigma_0^2$ (say).

Under the null hypothesis that the population variance is $\sigma^2 = \sigma_0^2$, the statistic

$$\chi^2 = \sum_{i=1}^{n}\left[\frac{(x_i - \bar{x})^2}{\sigma_0^2}\right] = \frac{1}{\sigma_0^2}\left[\sum_{i=1}^{n} x_i^2 - \frac{(\Sigma x_i)^2}{n}\right] = \frac{ns^2}{\sigma_0^2} \qquad \ldots (15\cdot14)$$

follows chi-square distribution with $(n - 1)$ *d.f.*

By comparing the calculated value with the tabulated value of χ^2 for $(n - 1)$ *d.f.* at certain level of significance (usually 5%), we may retain or reject the null hypothesis.

Remarks:

1. The above test (15·14) can be applied only if the population from which the sample is drawn is normal.

2. If the sample size n is large (> 30), then we can use Fisher's approximation

$$\sqrt{2\chi^2} \sim N\left(\sqrt{2n-1}, 1\right), \quad i.e., \quad Z = \sqrt{2\chi^2} - \sqrt{2n-1} \sim N(0, 1) \qquad \ldots (15\cdot14a)$$

and apply Normal Test.

Example 15·1a. *It is believed that the precision (as measured by the variance) of an instrument is no more than 0·16. Write down the null and alternative hypothesis for testing this belief. Carry out the test at 1% level given 11 measurements of the same subject on the instrument:*

2·5, 2·3, 2·4, 2·3, 2·5, 2·7, 2·5, 2·6, 2·6, 2·7, 2·5

Solution.

Null Hypothesis, H_0: $\sigma^2 = 0\cdot16$
Alternative Hypothesis, H_1: $\sigma^2 > 0\cdot16$

Table 15·1: COMPUTATION OF SAMPLE VARIANCE

X	$X - \bar{X}$	$(X - \bar{X})^2$
2·5	– 0·01	0·0001
2·3	– 0·21	0·0441
2·4	– 0·11	0·0121
2·3	– 0·21	0·0441
2·5	– 0·01	0·0001
2·7	+ 0·19	0·0361
2·5	– 0·01	0·0001
2·6	+ 0·09	0·0081
2·6	+ 0·09	0·0081
2·7	+ 0·19	0·0361
2·5	– 0·01	0·0001
$\bar{X} = \frac{27·6}{11} = 2·51$		$\Sigma(X - \bar{X})^2 = 0·1891$

Under the null hypothesis, H_0: $\sigma^2 = 0·16$, the test statistic is:

$$\chi^2 = \frac{ns^2}{\sigma^2} = \frac{\Sigma(X-\bar{X})^2}{\sigma^2} = \frac{0·1891}{0·16} = 1·182,$$

which follows χ^2-distribution with *d.f.* $n - 1 = (11 - 1) = 10$.

Since the calculated value of χ^2 is less than the tabulated value 23·2 of χ^2 for 10 *d.f.* at 1% level of significance, it is not significant. Hence H_0 may be accepted and we conclude that the data are consistent with the hypothesis that the precision of the instrument is 0·16.

Example 15·1b. *Test the hypothesis that $\sigma = 10$, given that the sample standard deviation $s = 15$ for a random sample of size 50 from a normal population.*

Solution. Null Hypothesis, H_0: $\sigma = 10$.

We are given $n = 50, \ s = 15.$

$$\therefore \quad \chi^2 = \frac{ns^2}{\sigma^2} = \frac{50 \times 225}{100} = 112·5$$

Since n is large, using (15·14*a*), the test statistic is: $Z = \sqrt{2\chi^2} - \sqrt{2n-1} \sim N(0,1)$

$$\therefore \quad Z = \sqrt{225} - \sqrt{99} = 15 - 9·95 = 5·05$$

Since $|Z| > 3$, it is significant at all levels of significance and hence H_0 is rejected and we conclude that $\sigma \neq 10$.

15·6·2. Goodness of Fit Test. A very powerful test for testing the significance of the discrepancy between theory and experiment was given by Prof. Karl Pearson in 1900 and is known as "*Chi-square test of goodness of fit*". It enables us to find if the deviation of the experiment from theory is just by chance or is it really due to the inadequacy of the theory to fit the observed data.

If f_i $(i = 1, 2, \ldots, n)$ is a set of observed (experimental) frequencies and e_i $(i = 1, 2, \ldots, n)$ is the corresponding set of expected (theoretical or hypothetical) frequencies, then Karl Pearson's chi-square, given by:

$$\chi^2 = \sum_{i=1}^{n}\left[\frac{(f_i - e_i)^2}{e_i}\right], \quad \left(\sum_{i=1}^{n} f_i = \sum_{i=1}^{n} e_i\right) \qquad \text{... (15·15)}$$

follows chi-square distribution with $(n - 1)$ *d.f.*

Remark. This is an approximate test for large values of n. Conditions for the validity of the χ^2-test of goodness of fit have already been given in § 15·4, Remark 2 on page 15·15.

The goodness of fit test uses the chi-square distribution to determine if a hypothesized probability distribution for a population provides a good fit. Acceptance or rejection of the hypothesized population distribution is based upon differences between observed frequencies (f_i's) in a sample and the expected frequencies (e_i's) obtained under null hypothesis H_0.

Decision rule: Accept H_0 if $\chi^2 \le \chi^2_\alpha\,(n - 1)$ and reject H_0 if $\chi^2 > \chi^2_\alpha\,(n - 1)$, where χ^2 is the calculated value of chi-square obtained on using (15·15) and $\chi^2_\alpha\,(n - 1)$ is the tabulated value of chi-square for $(n - 1)$ *d.f.* and level of significance α.

Example 15·2. *The demand for a particular spare part in a factory was found to vary from day-to-day. In a sample study the following information was obtained:*

Days :	*Mon.*	*Tues.*	*Wed.*	*Thurs.*	*Fri.*	*Sat.*
No. of parts demanded :	*1124*	*1125*	*1110*	*1120*	*1126*	*1115*

Test the hypothesis that the number of parts demanded does not depend on the day of the week. (Given: the values of chi-square significance at 5, 6, 7, d.f. are respectively 11·07, 12·59, 14·07 at the 5% level of significance).

Solution. Here we set up the *null hypothesis,* H_0 that the number of parts demanded does not depend on the day of week.

Under the null hypothesis, the expected frequencies of the spare parts demanded on each of the six days would be:

$$\frac{1}{6}(1124 + 1125 + 1110 + 1120 + 1126 + 1115) = \frac{6720}{6} = 1120$$

TABLE 15·2: CALCULATIONS FOR χ^2

	Frequency			
Days	*Observed* (f_i)	*Expected* (e_i)	$(f_i - e_i)^2$	$\frac{(f_i - e_i)^2}{e_i}$
Mon.	1124	1120	16	0·014
Tues.	1125	1120	25	0·022
Wed.	1110	1120	100	0·089
Thurs.	1120	1120	0	0
Fri.	1126	1120	36	0·032
Sat.	1115	1120	25	0·022
Total	6720	6720		0·179

$$\chi^2 = \sum \frac{(f_i - e_i)^2}{e_i} = 0{\cdot}179$$

The number of degrees of freedom = 6 – 1 = 5 (since we are given 6 frequencies subjected to only one linear constraint: $\Sigma f_i = \Sigma e_i = 6720$)

The tabulated $\chi^2_{0.05}$ for 5 *d.f.* = 11·07.

Since calculated value of χ^2 is less than the tabulated value, it is not significant and the null hypothesis may be accepted at 5% level of significance. Hence, we conclude that the number of spare parts demanded are same over the 6-day period.

Example 15·3. *The following figures show the distribution of digits in numbers chosen at random from a telephone directory:*

Digits	*0*	*1*	*2*	*3*	*4*	*5*	*6*	*7*	*8*	*9*	*Total*
Frequency	*1026*	*1107*	*997*	*966*	*1075*	*933*	*1107*	*972*	*964*	*853*	*10,000*

Test whether the digits may be taken to occur equally frequently in the directory.

Solution. We set up the *null hypothesis* that the digits occur equally frequently in the directory.

Under the null hypothesis, the expected frequency for each of the digits 0, 1, 2, ..., 9 is (10,000/10) = 1000. The value of χ^2 is computed as follows:

TABLE 15·3: CALCULATIONS FOR χ^2

Digits	*Frequency* Observed (f_i)	Expected (e_i)	$(f_i - e_i)^2$	$\frac{(f_i - e_i)^2}{e_i}$
0	1026	1000	676	0·676
1	1107	1000	11449	11·449
2	997	1000	9	0·009
3	966	1000	1156	1·156
4	1075	1000	5625	5·625
5	933	1000	4489	4·489
6	1107	1000	11149	11·449
7	972	1000	784	0·784
8	964	1000	1296	1·296
9	853	1000	21609	21·609
Total	10,000	10,000		58·542

$$\chi^2 = \sum \frac{(f_i - e_i)^2}{e_i} = 58{\cdot}542$$

The number of degrees of freedom

= Number of observations – Number of independent constraints

= 10 – 1 = 9

Tabulated $\chi^2_{0.05}$ for 9 *d.f.* = 16·919

Since the calculated value of χ^2 is much greater than the tabulated value, it is highly significant and we reject the null hypothesis. Thus, we conclude that the digits are not uniformly distributed in the directory.

Example 15·4. *A sample analysis of examination results of 200 MBA's was made. It was found that 46 students had failed, 68 secured a third division, 62 secured a second division and the rest were placed in first division. Are these figures commensurate with the general examination result which is in the ratio of 4 : 3 : 2 : 1 for various categories respectively.*

Solution. Set up the *null hypothesis* that the observed figures do not differ significantly from the hypothetical frequencies which are in the ratio 4 : 3 : 2 : 1. In other words the *given data are commensurate with the general examination result* which is in the ratio of 4 : 3 : 2 : 1 for the various categories.

Under the null hypothesis, the expected frequencies can be computed as shown in the following Table:

Category	*Frequency*	
	Observed (f_i)	*Expected* (e_i)
Failed	46	$\frac{4}{10} \times 200 = 80$
III Division	68	$\frac{3}{10} \times 200 = 60$
II Division	62	$\frac{2}{10} \times 200 = 40$
I Division	24	$\frac{1}{10} \times 200 = 20$
Total	200	200

TABLE 15·4: CALCULATIONS FOR χ^2

Category	*Frequency*		$(f_i - e_i)^2$	$\frac{(f_i - e_i)^2}{e_i}$
	Observed (f_i)	*Expected* (e_i)		
Failed	46	80	1156	14·450
III Division	68	60	64	1·067
II Division	62	40	484	12·100
I Division	24	20	16	0·800
Total	200	200		28·417

$$\chi^2 = \sum_i \frac{(f_i - e_i)^2}{e_i} = 28{\cdot}417$$

d.f. = 4 – 1 = 3, Tabulated $\chi^2_{0\cdot05}$ for 3 *d.f.* = 7·815.

Since the calculated value of χ^2 is greater than the tabulated value, it is significant and the null hypothesis is rejected at 5% level of significance. Hence, we may conclude that data are not commensurate with the general examination result.

Example 15·5. *A survey of 800 families with four children each revealed the following distribution:*

No. of boys	*0*	*1*	*2*	*3*	*4*
No. of girls	*4*	*3*	*2*	*1*	*0*
No. of families	*32*	*178*	*290*	*236*	*64*

Is this result consistent with the hypothesis that male and female births are equally probable ?

Solution. Let us set up the null hypothesis that the data are consistent with the hypothesis of equal probability for male and female births. Then under the null hypothesis:

$$p = \text{Probability of male birth} = \frac{1}{2} = q$$

$$p(r) = \text{Probability of 'r' male births in a family of } 4 = {}^4C_r \left(\frac{1}{2}\right)^r \left(\frac{1}{2}\right)^{4-r} = {}^4C_r \left(\frac{1}{2}\right)^4$$

The frequency of *families with r* male births is given by:

$$f(r) = N \cdot p(r) = 800 \times {}^4C_r \left(\frac{1}{2}\right)^4 = 50 \times {}^4C_r \; ; \; r = 0, 1, 2, 3, 4. \qquad \text{... (*)}$$

Substituting r = 0, 1, 2, 3, 4 successively in (*), we get the expected frequencies as follows:

$f(0) = 50 \times 1 = 50.$ $\quad f(1) = 50 \times {}^4C_1 = 200$ $\quad f(2) = 50 \times {}^4C_2 = 300,$

$f(3) = 50 \times {}^4C_3 = 200,$ $\quad f(4) = 50 \times {}^4C_4 = 50.$

TABLE 15·5: CALCULATIONS FOR χ^2

No. of male births	*Frequency (Families)*		$(f_i - e_i)^2$	$\dfrac{(f_i - e_i)^2}{e_i}$
	Observed (f_i)	*Expected* (e_i)		
0	32	50	324	6·48
1	178	200	484	2·42
2	290	300	100	0·33
3	236	200	1296	6·48
4	64	50	196	3·92
Total	800	800		19·63

$$\chi^2 = \sum\left\{\frac{(f_i - e_i)^2}{e_i}\right\} = 19{\cdot}63$$

Tabulated $\chi^2_{0\cdot05}$ for 5 – 1 = 4 *d.f.* is 9·488

Since calculated value of χ^2 is greater than tabulated value, it is significant at 5% level of significance. Hence, we reject the null hypothesis and conclude that male and female births are not equally probable.

Example 15·6. *When the first proof of 392 pages of a book of 1200 pages were read, the distribution of printing mistakes were found to be as follow:*

No. of mistakes in a page (x)	0	1	2	3	4	5	6
No. of pages (f)	275	72	30	7	5	2	1

Fit a Poisson distribution to the above data and test the goodness of fit.

Solution. Mean of the given distribution is: $\bar{X} = \frac{1}{N}\sum f x = \frac{189}{392} = 0{\cdot}482$

In order to fit a Poisson distribution to the given data, we take the mean (parameter) *m* of the Poisson distribution equal to the mean of the given distribution, *i.e.*, we take:

$$m = \bar{X} = 0{\cdot}482$$

The frequency of *r* mistakes per page is given by the Poisson Probability law as follows:

$$f(r) = N p(r) = 392 \times \frac{e^{-0\cdot482}(0{\cdot}482)^r}{r!}\ ;\ r = 0, 1, 2, \ldots, 6$$

Now $f(0) = 392 \times e^{-0\cdot482} = 392 \times \text{Antilog}\,(-\,0{\cdot}482 \log_{10} e)$

$= 392 \times \text{Antilog}\,(-\,0{\cdot}482 \times \log_{10} 2{\cdot}7183)$ $\quad (\because\ e = 2{\cdot}7183)$

$= 392 \times \text{Antilog}\,(-\,0{\cdot}482 \times 0{\cdot}4343) = 392 \times \text{Antilog}\,(-\,0{\cdot}2093)$

$= 392 \times \text{Antilog}\,(\bar{1}{\cdot}7907) = 392 \times 0{\cdot}6176 = 242{\cdot}1$

$f(1) = m \times f(0) = 0{\cdot}482 \times 242{\cdot}1 = 116{\cdot}69,$ $\quad f(2) = \frac{m}{2} \times f(1) = 0{\cdot}241 \times 116{\cdot}69 = 28{\cdot}12$

$f(3) = \frac{m}{3} \times f(2) = \frac{0{\cdot}482}{3} \times 28{\cdot}12 = 4{\cdot}518,$ $\quad f(4) = \frac{m}{4} \times f(3) = \frac{0{\cdot}482}{4} \times 4{\cdot}51 = 0{\cdot}544$

$f(5) = \frac{m}{5} \times f(4) = \frac{0{\cdot}482}{5} \times 0{\cdot}544 = 0{\cdot}052,$ $\quad f(6) = \frac{m}{6} \times f(5) = \frac{0{\cdot}482}{6} \times 0{\cdot}052 = 0{\cdot}004$

Hence, the theoretical Poisson frequencies correct to one decimal place are as given below:

X	:	0	1	2	3	4	5	6	*Total*
Expected Frequency	:	242·1	116·7	28·1	4·5	0·5	0·1	0	392

TABLE 15·6: CALCULATIONS FOR χ^2

Mistakes per page (X)	*Frequency* Observed (f_i)	Expected (e_i)	$(f_i - e_i)^2$	$\frac{(f_i - e_i)^2}{e_i}$
0	275	242·1	1082·41	4·471
1	72	116·7	1998·09	17·121
2	30	28·1	3·61	0·128
3	7	4·5		
4	5	0·5		
	} 15	} 5·1	98·01	19·217
5	2	0·1		
6	1	0		
Total	392	392		40·937

Ho: Posson distribution is a good fit to the given data. Under *Ho*, the test statistic is:

$$\chi^2 = \sum_i \left\{ \frac{(f_i - e_i)^2}{e_i} \right\} = 40{\cdot}937$$

d.f. = 7 – 1 – 1 – 3 = 2

(One *d.f.* being lost because of the linear constraint $\Sigma f_i = \Sigma e_i$; 1 *d.f.* is lost because the parameter *m* has been estimated from the given data and is then used for computing the expected frequencies; 3 *d.f.* are lost because of pooling the last four expected cell frequencies which are less than five.)

Tabulated value of χ^2 for 2 *d.f.* at 5% level of significance is 5·99.

Conclusion. Since calculated value of χ^2(40·937) is much greater than 5.99, it is highly significant and we reject the null hypothesis H_0. Hence, we conclude that Poisson distribution is not a good fit to the given data.

Example 15·7. *The figures given below in the first row (A) are the theoretical frequencies of a distribution and in the second row (B) are the frequencies of the normal distribution having the same mean, variance and the total frequencies. Do you think that the normal distribution so fitted provides a good fit to the data?*

A :	1	12	66	220	495	792	924	792	495	220	66	12	1
B :	2	5	66	210	484	797	943	799	484	210	66	15	2

Solution. Let us set up the null hypothesis, H_0: *"Normal distribution is a good fit to the given frequency distribution"*. In other words, the observed frequencies (f_i) and the expected (theoretical) normal frequencies (e_i) do not differ significantly.

TABLE 15·7: CALCULATIONS FOR χ^2.

f_i	e_i	$f_i - e_i$	$(f_i - e_i)^2$	$\frac{(f_i - e_i)^2}{e_i}$
1, 12 } = 13	2, 5 } = 7	6	36	5·14
66	66	0	0	0
220	210	10	100	0·48
495	484	11	121	0·25

	792	797	– 5	25	0·03
	924	943	– 19	361	0·38
	792	799	– 7	49	0·06
	495	484	11	121	0·25
	220	210	10	100	0·48
	66	66	0	0	0
	$\left.\begin{matrix}12\\1\end{matrix}\right\} = 13$	$\left.\begin{matrix}15\\2\end{matrix}\right\} = 17$	– 4	16	0·94
Total	4096	4096	0		8·01

$$\chi^2 = \sum_i \left\{ \frac{(f_i - e_i)^2}{e_i} \right\} = 8{\cdot}01$$

$$d.f. = 13 - 1 - 2 - 2 = 8$$

One *d.f.* being lost because of the linear constant $\Sigma f_i = \Sigma e_i$; 2 *d.f.* are lost because the parameters μ and σ^2 have been estimated from the given data and are then used for computing the expected frequencies; 2 *d.f.* are lost in pooling because the first and the last frequencies are less than 5 and they are pooled with succeeding and preceding frequency respectively.

Tabulated value of χ^2 for 8 *d.f.* at 5% level of significance is 15·507.

Conclusion. Since calculated value of χ^2 (8·01) is much less than the tabulated value (15·507), it is not significant. Hence, we may accept the null hypothesis at 5% level of significance and conclude that the normal distribution is a good fit to the given distribution.

15·6·3. Test of Independence of Attributes — Contingency Tables

Let us consider two attributes A and B, A divided into r classes $A_1, A_2, \ldots, A_r$ and B divided into s classes $B_1, B_2, \ldots, B_s$. Such a classification in which attributes are divided into more than two classes is known as *manifold classification*. The various cell frequencies can be expressed in the following table known as *r × s manifold contingency table* where (A_i) is the number of persons possessing the attribute A_i, $(i = 1, 2, \ldots, r)$, (B_{Ji}) is the number of persons possessing the attribute B_j $(j = 1, 2, \ldots, s)$ and (A_iB_j) is the number of persons possessing both the attributes A_i and B_j, $(i = 1, 2, \ldots, r; j = 1, 2, \ldots, s)$.

Also, $$\sum_{i=1}^{r} (A_i) = \sum_{j=1}^{s} (B_j) = N, \text{ where } N \text{ is the total frequency.}$$

TABLE 15.8: $r \times s$ CONTINGENCY TABLE

B \ A	A_1	A_2	…	A_i	…	A_r	Total
B_1	(A_1B_1)	(A_2B_1)	…	(A_iB_1)	…	(A_rB_1)	(B_1)
B_2	(A_1B_2)	(A_2B_2)	…	(A_iB_2)	…	(A_rB_2)	(B_2)
⋮	⋮	⋮	⋮	⋮	⋮	⋮	⋮
B_j	(A_1B_j)	(A_2B_j)	…	(A_iB_j)	…	(A_rB_j)	(B_j)
⋮	⋮	⋮	⋮	⋮	⋮	⋮	⋮
B_s	(A_1B_s)	(A_2B_s)	…	(A_iB_s)	…	(A_rB_s)	(B_s)
Total	(A_1)	(A_2)	…	(A_i)	…	(A_r)	N

The problem is to test if the two attributes A and B under consideration are independent or not.

Under the *null hypothesis* that the attributes are independent, the theoretical cell frequencies are calculated as follows:

$P[A_i]$ = Probability that a person possesses the attribute $A_i = \dfrac{(A_i)}{N}$; $i = 1, 2, \ldots, r$

$P[B_j]$ = Probability that a person possesses the attribute $B_j = \dfrac{(B_j)}{N}$; $j = 1, 2, \ldots, s$

$P[A_iB_j]$ = Probability that a person possesses the attributes A_i and $B_j = P(A_i)\ P(B_j)$

[By compound probability theorem, since the attributes A_i and B_j are independent, under the null hypothesis].

$$\therefore \quad P[A_iB_j] = \frac{(A_i)}{N}\cdot\frac{(B_j)}{N}; \quad i = 1, 2, \ldots, r; \quad j = 1, 2, \ldots, s$$

and $(A_iB_j)_0$ = Expected number of persons possessing both the attributes A_i and B_j

$$= N \cdot P[A_iB_j] = \frac{(A_i)(B_j)}{N}$$

$$\Rightarrow \quad (A_iB_j)_0 = \frac{(A_i)(B_j)}{N}, (i = 1, 2, \ldots, r;\ j = 1, 2, \ldots, s) \qquad \ldots (15{\cdot}16)$$

By using this formula, we can find out expected frequencies for each of the cell-frequencies $(A_i\, B_j)$ $(i = 1, 2, \ldots, r; j = 1, 2, \ldots, s)$, under the null hypothesis of independence of attributes.

The exact test for the independence of attributes is very complicated but a fair degree of approximation is given, for large samples, (large N), by the χ^2-test of goodness of fit, *viz.*,

$$\chi^2 = \sum_{i=1}^{r}\sum_{j=1}^{s}\left[\frac{\{(A_iB_j) - (A_iB_j)_0\}^2}{(A_iB_j)_0}\right] = \sum_i\sum_j \frac{(f_{ij} - e_{ij})^2}{e_{ij}} \qquad \ldots (15{\cdot}16a)$$

where f_{ij} = observed frequency for contingency table category in column i and row j,

e_{ij} = expected frequency for contingency table category in column i and row j,

which is distributed as χ^2-variate with $(r - 1)\ (s - 1)$ *d.f.* [*See Remark* 2 on degrees of freedom].

Remarks:

1. $\phi^2 = \chi^2/N$ is known as *mean-square contingency.*

Since the limits of χ^2 and ϕ^2 vary in different cases, they cannot be used for establishing the closeness of the relationship between qualitative characters under study. Prof. Karl Pearson suggested another measure, known as *"coefficient of mean square contingency"* which is denoted by C and is given by:

$$C = \sqrt{\frac{\chi^2}{\chi^2 + N}} = \sqrt{\frac{\phi^2}{1 + \phi^2}} \qquad \ldots (15{\cdot}17)$$

Obviously C is always less than unity. The maximum value of C depends on r and s, the number of classes into which A and B are divided. In a $r \times r$ contingency table, the maximum value of $C = \sqrt{(r-1)/r}$. Since the maximum value of C differs for different classification, *viz.*, $r \times r$ $(r = 2, 3, 4, \ldots)$, strictly speaking, the values of C obtained from different types of classification are not comparable.

2. ***Note on Degrees of Freedom (d.f.).*** The number of independent variates which make up the statistic (*e.g.*, χ^2) is known as the *degrees of freedom (d.f.)* and is usually denoted by ν (the letter 'Nu' of the Greek alphabet).

The number of degrees of freedom, in general, is the total number of observations less the number of independent constraints imposed on the observations. For example, if k is the number of independent constraints in a set of data of n observations than $\nu = (n - k)$.

Thus in a set of n observations usually, the degrees of freedom for χ^2 are $(n - 1)$, one *d.f.* being lost because of the linear constraint $\sum_i f_i = \sum_i e_i = N$, on the frequencies (*c.f. Theorem 15·3*). If 'r' independent linear constraints are imposed on the cell frequencies, then the *d.f.* are reduced by 'r'. [*c.f.* Theorem 15·6].

In addition, if any of the population parameters(s) is (are) calculated from the given data and used for computing the expected frequencies then in applying χ^2-test of goodness of fit, we have to subtract one *d.f.* for each parameter calculated. Thus, if 's' is the number of population parameters estimated from the sample observations (n in number), then the required number of degrees of freedom for χ^2-test is $(n - s - 1)$.

If any one or more of the theoretical frequencies is less than 5, then in applying χ^2-test we have also to subtract the degrees of freedom lost in pooling these frequencies with the preceding or succeeding frequency (or frequencies).

In a $r \times s$ contingency table, in calculating the expected frequencies, the row totals, the column totals and the grand totals remain fixed. The fixation of 'r' column totals and 's' row totals imposes $(r + s)$ constraints on the cell frequencies. But since $\sum_{i=1}^{r} (A_i) = \sum_{j=1}^{s} (B_j) = N$, the total number of independent constraints is only $(r + s - 1)$. Further, since the total number of cell-frequencies is $r \times s$, the required number of *d.f.* is:

$$\nu = rs - (r + s - 1) = (r - 1)(s - 1) \qquad \text{... (15.17a)}$$

Example 15·8. *Two sample polls of votes for two candidates A and B for a public office are taken, one from among the residents of rural areas. The results are given in the table below. Examine whether the nature of the area is related to voting preference in this election.*

Area	*Votes for*		*Total*
	A	*B*	
Rural	*620*	*380*	*1000*
Urban	*550*	*450*	*1000*
Total	*1170*	*830*	*2000*

Solution. Under the *null hypothesis* that the nature of the area is independent of the voting preference in the election, we get the expected frequencies as follows:

$$E(620) = \frac{1170 \times 1000}{2000} = 585, \qquad E(380) = \frac{830 \times 1000}{2000} = 415,$$

$$E(550) = \frac{1170 \times 1000}{2000} = 585, \qquad \text{and} \qquad E(450) = \frac{830 \times 1000}{2000} = 415$$

Aliter. In a 2×2 contingency table, since *d.f.* = $(2 - 1)(2 - 1) = 1$, only one of the cell frequencies can be filled up independently and the remaining will follow immediately, since the observed and theoretical marginal totals are fixed. Thus having obtained any one of the theoretical frequencies (say) $E(620) = 585$, the remaining theoretical frequencies can be easily obtained as follows:

$$E(380) = 10000 - 585 = 415, \; E(550) = 1170 - 585 = 585, \quad \text{and} \quad E(450) = 1000 - 585 = 415$$

$$\therefore \quad \chi^2 = \sum_i \left[\frac{(f_i - e_i)^2}{e_i}\right] = \frac{(620-585)^2}{585} + \frac{(380-415)^2}{415} + \frac{(550-585)^2}{585} + \frac{(450-415)^2}{415}$$

$$= (35)^2 \left(\frac{1}{585} + \frac{1}{415} + \frac{1}{585} + \frac{1}{415}\right) = (1225)\,[2 \times 0{\cdot}002409 + 2 \times 0{\cdot}001709] = 10{\cdot}0891$$

Tabulated $\chi^2_{0\cdot05}$ for (2 –1) (2 – 1) = 1 *d.f.* is 3·841. Since calculated χ^2 is much greater than the tabulated value, it is highly significant and null hypothesis is rejected at 5% level of significance. Thus we conclude that nature of area is related to voting preference in the election.

Example 15·9. *(2 × 2 Contingency Table). For the 2 × 2 table,*

a	*b*
c	*d*

prove that chi-square test of independence gives :

$$\chi^2 = \frac{N(ad-bc)^2}{(a+c)\,(b+d)\,(a+b)\,(c+d)}, \; N = a+b+c+d \qquad \text{... (15·18)}$$

Solution.

a	*b*	*a + b*
c	*d*	*c + d*
a + c	*b + d*	*N*

Under the hypothesis of independence of attributes,

$$E(a) = \frac{(a+b)\,(a+c)}{N}; \qquad E(b) = \frac{(a+b)\,(b+d)}{N}$$

$$E(c) = \frac{(a+c)\,(c+d)}{N}; \qquad E(d) = \frac{(b+d)\,(c+d)}{N}$$

$$\therefore \quad \chi^2 = \frac{[a-E(a)]^2}{E(a)} + \frac{[b-(E(b)]^2}{E(b)} + \frac{[c-E(c)]^2}{E(c)} + \frac{[d-E(d)]^2}{E(d)} \qquad \text{... (*)}$$

$$a - E(a) = a - \frac{(a+b)\,(a+c)}{N} = \frac{a(a+b+c+d) - (a^2+ac+ab+bc)}{N} = \frac{ad-bc}{N}$$

Similarly, we will get:

$$b - E(b) = -\frac{ad-bc}{N} = c - E(c); \qquad d - E(d) = \frac{ad-bc}{N}$$

Substituting in (*), we get

$$\chi^2 = \frac{(ad-bc)^2}{N^2}\left[\frac{1}{E(a)} + \frac{1}{E(b)} + \frac{1}{E(c)} + \frac{1}{E(d)}\right]$$

$$= \frac{(ad-bc)^2}{N}\left[\left\{\frac{1}{(a+b)\,(a+c)} + \frac{1}{(a+b)\,(b+d)}\right\} + \left\{\frac{1}{(a+c)\,(c+d)} + \frac{1}{(b+d)\,(c+d)}\right\}\right]$$

$$= \frac{(ad-bc)^2}{N}\left[\frac{b+d+a+c}{(a+b)\,(a+c)\,(b+d)} + \frac{b+d+a+c}{(a+c)\,(c+d)\,(b+d)}\right]$$

$$= (ad - bc)^2 \left[\frac{c + d + a + b}{(a + b)(a + c)(b + d)(c + d)} \right] = \frac{N(ad - bc)^2}{(a + b)(a + c)(b + d)(c + d)}$$

Remark. We can calculate the value of χ^2 for 2 × 2 contingency table by using (15·18) directly. The reader is advised to obtain the value of χ^2 in *Example 15·8* by using (15·18).

Example 15·10. *Out of 8,000 graduates in a town 800 are females, out of 1,600 graduate employees 120 are females. Use χ^2 to determine if any distinction is made in appointment on the basis of sex. Value of χ^2 at 5% level for one degree of freedom is 3·84.*

Solution. We set up the *Null hypothesis* that no distinction is made in appointment on the basis of sex,

The observed and expected frequencies are shown in the following table:

	Observed Frequencies			*Expected Frequencies*		
	Employed	*Not employed*	*Total*	*Employed*	*Not employed*	*Total*
Male	1480	5720	7200	$\frac{7200 \times 1600}{8000} = 1440$	7200 – 1440 = 5760	7200
Female	**120**	680	**800**	1600 – 1440 = 160	6400 – 5760 = 640	800
Total	**1600**	6400	**8000**	1600	6400	8000

The frequencies in bold are the given frequencies. Rest of the frequencies are obtained on subtraction, from row and column totals.

TABLE 15.9: CALCULATIONS FOR χ^2

Class	*Frequency* Observed (f_i)	Expected (e_i)	($f_i - e_i$)	$\frac{(f_i - e_i)^2}{e_i}$
Male employed	1480	1440	40	$\frac{1600}{1440} = 1{\cdot}11$
Male unemployed	5720	5760	– 40	$\frac{1600}{5760} = 0{\cdot}28$
Female employed	120	160	– 40	$\frac{1600}{160} = 10{\cdot}00$
Female unemployed	680	640	40	$\frac{1600}{640} = 2{\cdot}50$

$$\chi^2 = \sum \frac{(f_i - e_i)^2}{e_i} = 13{\cdot}89$$

$$d.f. = (2 - 1)(2 - 1) = 1$$

Tabulated $\chi^2_{0\cdot05}$ for 1 *d.f.* = 3·841.

Conclusion. Since the calculated value of χ^2 (13·89) is much greater than the tabulated value of χ^2 (3·841), the value of χ^2 is highly significant and null hypothesis is rejected. Hence, we conclude that distinction is made in appointment on the basis of sex.

Example 15·11. *A random sample of students of XYZ University was selected and asked their opinions about 'autonomous colleges'. The results are given below. The same number of each sex was included within each class-group. Test the hypothesis at 5% level that opinions are independent of the class groupings:*

Class	*Numbers*		*Total*
	Favouring 'autonomous colleges'	*Opposed to 'autonomous' colleges'*	
B.A./B.Sc./B.Com. Part I	*120*	*80*	*200*
B.A./B.Sc./B.Com. Part II	*130*	*70*	*200*
B.A./B.Sc./B.Com. Part III	*70*	*30*	*100*
M.A./M.Sc./M.Com.	*80*	*20*	*100*
Total	*400*	*200*	*600*

Solution. We set up the null hypothesis that the opinions about autonomous colleges are independent of the class-groupings.

Here the frequencies are arranged in the form of a 4 × 2 contingency table. Hence the *d.f.* are $(4 - 1) \times (2 - 1) = 3 \times 1 = 3$. Hence we need to compute independently only three expected frequencies and the remaining expected frequencies can be obtained by subtraction from the row and column totals.

Under the null hypothesis of independence:

$$E(120) = \frac{400 \times 200}{600} = 133,33\,; \quad E(130) = \frac{400 \times 200}{600} = 133,33\,; \quad E(70) = \frac{400 \times 100}{600} = 66,67$$

Now the table of expected frequencies can be completed as shown below:

Class	*Numbers*		*Total*
	Favouring 'autonomous colleges'	*Opposed to 'autonomous' colleges'*	
B.A./B.Sc./B.Com. Part I	*133·33*	*200 – 133·33 = 66·67*	*200*
B.A./B.Sc./B.Com. Part II	*133·33*	*200 – 133·33 = 66·67*	*200*
B.A./B.Sc./B.Com. Part III	*66·67*	*100 – 66·67 = 33·33*	*100*
M.A./M.Sc./M.Com.	*66·67*	*100 – 66·67 = 33·33*	*100*
Total	*400*	*200*	*600*

TABLE 15.10: CALCULATIONS FOR CHI-SQUARE

f_i	e_i	$(f_i - e_i)$	$(f_i - e_i)^2$	$\frac{(f_i - e_i)^2}{e_i}$
120	113·33	– 13·33	177·6889	1·3327
130	133·33	– 3·33	11·0889	0·0832
70	66·67	3·33	11·0889	0·1663

80	66·67	13·33	177·6889	2·6652
80	66·67	13·33	177·6889	2·6652
70	66·67	3·33	11·0889	0·1663
30	33·33	– 3·33	11·0889	0·3327
20	33·33	– 13·33	177·6889	5·3312
Total 400	400			12·7428

$$\chi^2 = \sum_i \frac{(f_i - e_i)^2}{e_i} = 12{\cdot}7428$$

Tabulated value of χ^2 for $(4 - 1) \times (2 - 1) = 3$ *d.f.* at 5% level of significance is 7·815.

Conclusion. Since calculated value of χ^2 is greater than the tabulated value, it is significant at 5% level of significance and we reject the null hypothesis. Hence, we conclude that the opinions about autonomous colleges are dependent on the class-groupings.

Example 15·12. *Two researchers adopted different sampling techniques while investigating the same group of students to find the number of students falling in different intelligence levels. The results are as follows:*

Researcher	*No. of students in each level*				*Total*
	Below Average	*Average*	*Above Average*	*Genius*	
X	86	60	44	10	200
Y	40	33	25	2	100
Total	126	93	69	12	300

Would you say that the sampling techniques adopted by the two researchers are significantly different? (Given 5% value of χ^2 for 2 d.f. and 3 d.f. are 5·991 and 7·82 respectively.)

Solution. We set up the *null hypothesis* that the data obtained are independent of the sampling techniques adopted by the two researchers. In other words, the null *hypothesis* is that there is no significant difference between the sampling techniques used by the two researchers for collecting the required data.

Here we have a 4 × 2 contingency table and *d.f.* $= (4 - 1) \times (2 - 1) = 3 \times 1 = 3$. Hence we need to compute only 3 independent expected frequencies and the remaining expected frequencies can be obtained by subtraction from the marginal row and column totals.

Under the null hypothesis of independence, we have

$$E(86) = \frac{126 \times 200}{300} = 84\ ; \quad E(60) = \frac{93 \times 200}{300} = 62\ ; \quad E(44) = \frac{69 \times 200}{300} = 46.$$

The table of expected frequencies can now be completed as shown below:

Researcher	*No. of students in each level*				*Total*
	Below Average	*Average*	*Above Average*	*Genius*	
X	84	62	46	200 – 192 = 8	200
Y	126 – 84 = 42	93 – 62 = 31	69 – 46 = 23	12 – 8 = 4	100
Total	126	93	69	12	300

Since we cannot apply the χ^2-test straightway here as the last expected frequency is less than 5, we should use the technique of pooling in this case as given below:

TABLE 15·11: COMPUTATION OF THE VALUE OF χ^2

Researchers	*Type of Students*	f_i	e_i	$f_i - e_i$	$(f_i - e_i)^2$	$\frac{(f_i - e_i)^2}{e_i}$
X	Below average	86	84	2	4	0·048
	Average	60	62	– 2	4	0·064
	Above average	44	46	– 2	4	0·087
	Genius	10	8	2	4	0·500
Y	Below average	40	42	– 2	4	0·095
	Average	33	31	2	4	0·129
	Above average	25 } 27	23 } 27	0	0	0
	Genius	2	4			
Total		300	300			0·923

$$\chi^2 = \sum_i \frac{(f_i - e_i)^2}{e_i} = 0{\cdot}923$$

and the *d.f.* = (4 – 1) × (2 – 1) – 1 = 3 – 1 = 2, since 1 *d.f.* is lost in the method of pooling. Tabulated value of χ^2 for 2 *d.f.* at 5% level of significance is 5·991.

Conclusion. Since calculated value is less than the tabulated value, null hypothesis may be accepted at 5% level of significance and we may conclude that there is no significant difference in the sampling techniques used by the two researchers.

15·6·4. Yate's Correction (for 2 × 2 Contingency Table)

In a 2 × 2 contingency table, the number of *d.f.* is (2 – 1) (2 – 1) = 1. If any one of the theoretical cell frequencies is less than 5, then use of pooling method for χ^2-test results in χ^2 with 0 *d.f.* (since 1 *d.f.* is lost in pooling) which is meaningless. In this case we apply a correction due to F. Yates (1934), which is usually known as *"Yate's Correction for Continuity"* [As already pointed out, χ^2 is a continuous distribution and it fails to maintain its character of continuity if any of the expected frequency is less than 5; hence the name 'Correction for Continuity'.] This consists in adding 0·5 to the cell frequency which is less than 5 and then adjusting for the remaining cell frequencies accordingly. The χ^2-test of goodness of fit is then applied without pooling method.

For a 2 × 2 contingency table,

a	*b*
c	*d*

we have $$\chi^2 = \frac{N(ad - bc)^2}{(a + c)(b + d)(a + b)(c + d)}$$

According to Yate's correction, as explained above, we subtract (or add) $\frac{1}{2}$ from a and d and add (subtract) $\frac{1}{2}$ to b and c so that the marginal totals are not disturbed at all. Thus, corrected value of χ^2 is given as:

$$\chi^2 = \frac{N\left[\left(a \mp \frac{1}{2}\right)\left(d \mp \frac{1}{2}\right) - \left(b \pm \frac{1}{2}\right)\left(c \pm \frac{1}{2}\right)\right]^2}{(a+c)(b+d)(a+b)(c+d)}$$

$$\text{Numerator} = N\left[(ad - bc) \mp \frac{1}{2}(a+b+c+d)\right]^2 = N\left[|ad - bc| - \frac{N}{2}\right]^2$$

$$\therefore \quad \chi^2 = \frac{N[| ad - bc | - (N/2)]^2}{(a+c)(b+d)(a+b)(c+d)} \qquad \text{... (15·18}a)$$

Remarks:

1. If N is large, the use of Yate's correction will make very little difference in the value of χ^2. If, however, N is small, the application of Yate's correction may overstate the probability.

2. It is recommended by many authors and it seems quite logical in the light of the above discussion that Yate's correction be applied to every 2 × 2 table, even if no theoretical cell frequency is less than 5.

15·6·5. Brandt and Snedecor Formula for 2 × *k* Contingency Table

Let the observations a_{ij}, $(i = 1, 2; j = 1, 2, \ldots, k)$ be arranged in a 2 × k contingency table as follows:

B \ A	A_1	A_2	...	A_i	...	A_k	Total
B_1	a_{11}	a_{12}	...	a_{1i}	...	a_{1k}	m_1
B_2	a_{21}	a_{22}	...	a_{2i}	...	a_{2k}	m_2
Total	n_1	n_2	...	n_i	...	n_k	N

Under the hypothesis of independence of attributes, we have

$$E(a_{1i}) = \frac{n_i \times m_1}{N}; \qquad E(a_{2i}) = \frac{n_i \times m_2}{N}; \qquad i = 1, 2, \ldots, k$$

$$\therefore \quad \chi^2 = \sum_{i=1}^{k}\left[\frac{\{a_{1i} - E(a_{1i})\}^2}{E(a_{1i})} + \frac{\{a_{2i} - E(a_{2i})\}^2}{E(a_{2i})}\right] = \sum_{i=1}^{k}\left\{\frac{\left(a_{1i} - \frac{m_1 n_i}{N}\right)^2}{\left(\frac{m_1 n_i}{N}\right)} + \frac{\left(a_{2i} - \frac{m_2 n_i}{N}\right)^2}{\left(\frac{m_2 n_i}{N}\right)}\right\}$$

$$= \sum_{i=1}^{k}\left[\frac{Nn_i}{m_1}\left(\frac{a_{1i}}{n_i} - \frac{m_1}{N}\right)^2 + \frac{Nn_i}{m_2}\left(\frac{a_{2i}}{n_i} - \frac{m_2}{N}\right)^2\right] = \sum_{i=1}^{k}\left[\frac{n_i}{p}(p_i - p)^2 + \frac{n_i}{q}(q_i - q)^2\right]$$

where $p_i = \dfrac{a_{1i}}{n_i}$, $\quad q_i = 1 - p_i = \dfrac{a_{2i}}{n_i}$ $\quad$ *and* $\quad p = \dfrac{m_1}{N}$, $\quad q = 1 - p = \dfrac{m_2}{N}$

$$\therefore \quad \chi^2 = \sum_{i=1}^{k}\left[\frac{n_i}{p}(p_i - p)^2 + \frac{n_i}{q}\{(1 - p_i) - (1 - p)\}^2\right] = \sum_{i=1}^{k} n_i (p_i - p)^2 \left(\frac{1}{p} + \frac{1}{q}\right)$$

$$= \sum_{i=1}^{k} n_i (p_i - p)^2 \frac{1}{pq} = \frac{1}{pq} \sum_{i=1}^{k} n_i (p_i^2 + p^2 - 2p_i p) \qquad (\because \ p + q = 1)$$

$$= \frac{1}{pq} \left(\sum_{i=1}^{k} n_i p_i^2 + p^2 \sum_{i=1}^{k} n_i - 2p \sum_{i=1}^{k} p_i n_i \right)$$

$$= \frac{1}{pq} \left(\sum_{i=1}^{k} n_i p_i^2 + p^2 N - 2p \cdot Np \right) = \frac{1}{pq} \left(\sum_{i=1}^{k} n_i p_i^2 - Np^2 \right)$$

But $\displaystyle\sum_{i=1}^{k} n_i p_i^2 = \sum_{i=1}^{k} n_i p_i \cdot p_i = \sum_{i=1}^{k} a_{1i} p_i$

$$\therefore \qquad \chi^2 = \frac{1}{pq} \left(\sum_{i=1}^{k} a_{1i} p_i - Np^2 \right) = \frac{1}{pq} \left(\sum_{i=1}^{k} \frac{a_{1i}^2}{n_i} - Np^2 \right) \qquad \text{... (15·19)}$$

$$= \frac{1}{pq} \left(\sum_{i=1}^{k} \frac{a_{1i}^2}{n_i} - m_1 p \right) = \frac{1}{pq} \left(\sum_{i=1}^{k} \frac{a_{1i}^2}{n_i} - \frac{m_1^2}{N} \right) \qquad \text{... [15·19(a)]}$$

Example 15·13. *The following table shows three age-groups of boys and girls, (a) the number of children affected by a non-infectious disease, and (b) the total number of children exposed to risk.*

	Boys			*Girls*		
	I	*II*	*III*	*I*	*II*	*III*
(a)	60	25	48	96	18	42
(b)	240	470	350	530	200	210

(i) Test whether there are differences between the incidence rates in the three age-groups of boys.

(ii) Test whether the boys and girls are equally susceptible or not.

Solution. (*i*) We set up the *null hypothesis* (H_0) *that there is no significant difference between the incidence rates in the three age-groups of boys.*

In the notations of the above article § 15·6·5, we have

$$\left.\begin{matrix} a_{11} = 60, & a_{12} = 25, & a_{13} = 48, & m_1 = 133 \\ n_1 = 240, & n_2 = 470, & n_3 = 350, & N = 1060 \end{matrix}\right\} \qquad p = \frac{m_1}{N} = \frac{133}{1060} = 0·1255, \ q = 1 - p = 0·8745$$

Substituting these values in (15·19*a*), we get

$$\chi^2 = \frac{1}{(0·1255)(0·8745)} (15·00 + 1·33 + 6·58 - 133 \times 0·1255) = \frac{6·2187}{0·1097} = 56·688$$

Here *d.f.* = (3 – 1) (2 – 1) = 2.

The tabulated value of χ^2 for 2 degrees of freedom at 5% level of significance is 5·991. Since calculated value of χ^2 is much greater than tabulated value, we reject the null hypothesis and conclude that the incidence rates in the three age-groups of boys differ significantly.

(*ii*) Here we set up the *null hypothesis* that the boys and girls are equally susceptible to the disease. In the usual notations, we have

$a_{11} = 60 + 25 + 48 = 133$ and $a_{12} = 96 + 18 + 42 = 156$, $m_1 = 289$

$n_1 = 240 + 470 + 350 = 1,060$ and $n_2 = 530 + 200 + 210 = 940$, $N = 2,000$

$$\therefore \quad p = \frac{m_1}{N} = \frac{289}{2000} = 0{\cdot}1445, \quad q = 1 - p = 0{\cdot}8555 \quad \text{and} \quad Np^2 = m_1 \cdot p = 289 \times 0{\cdot}1445 = 41{\cdot}76$$

$$\therefore \quad \chi^2 = \frac{1}{(0{\cdot}1445)\,(0{\cdot}8555)}(16{\cdot}69 + 25{\cdot}89 - 41{\cdot}76) = 6{\cdot}605 \quad [From (15{\cdot}19a)]$$

d.f. = (2 – 1) (2 – 1) = 1

From the tables, the value of χ^2 for 1 degree of freedom at 5% level of significance is 3·841 which is much less than the calculated value. We, therefore, reject the null hypothesis and conclude that boys and girls are not equally susceptible to the disease.

15·6·6. Test for Equality of Several Population Proportions. In order to test the equality of *several population proportions, i.e.*, $P_1 = P_2 = \ldots = P_k$; independent random samples of predetermined sample sizes, say $n_1, n_2, \ldots, n_k$ are selected from these populations respectively and their responses are recorded into one of the two categories (dichotomous classification) *e.g.*, (literate; illiterate), (trained; untrained) and so on. This will give data in the form of a two-way $2 \times k$ contingency table, in which one classification refers to the populations and the other classification refers to response under study.

The test statistic for testing the equality of several population proportions, *i.e.*, to test $H_0 : P_1 = P_2 = \ldots = P_k$, one can use the chi-square statistic for testing independence of two attributes for $2 \times k$ contingency table, *viz.*,

$$\chi^2 = \sum_i \left\{ \frac{(f_i - e_i)^2}{e_i} \right\} \sim \chi^2_{(2-1)(k-1)} = \chi^2_{(k-1)} \qquad \ldots (15{\cdot}19b)$$

where

$$\text{Expected frequency for any cell frequency} = \frac{(\text{Row total}) \times (\text{Column total})}{\text{Grand total}}$$

Note. The proof for the equivalence of the two tests *i.e.*,

(*i*) χ^2-test for independence of attributes.

(*ii*) χ^2-test for equality of several proportions,

is beyond the scope of the book.

Remarks:

1. Since the response data are in the form of a 2 × k contingency table, we can conveniently compute the value of chi-square statistic by using 'Brandt and Snedecor' formula (15·19*a*).

2. In the usual contingency table for 'Independence of two attributes', one random sample of size n is classified simultaneously according to the attributes into, say r class *w.r.t.* one and s classes w.r.t. other. Here, both the row and column totals are the result of random sampling and none of these were fixed in advance.

However, in testing $H_0 : P_1 = P_2 = \ldots = P_k$, independent random samples of sizes $n_1, n_2, \ldots, n_k$ are drawn from the populations and two responses are recorded in each sample. It may be noted that, here row (or column) totals $n_1, n_2, \ldots, n_k$ are not the outcome of random sampling but they are fixed (pre-determined) sample sizes. However, the other marginal totals in the table are the results of random sampling. This type of classification is also referred to as $2 \times k$ contingency table with fixed marginals totals *w.r.t.* one characteristic.

Example 15·14. *The following data gives the number of users of a new shampoo in random samples of size 100, from each of the four regions NW, NE, SW and SE in a city.*

Region	*NW*	*NE*	*SW*	*SE*	*Total*
Do not buy	60	45	55	50	210
Buy	40	55	45	50	190
Total	100	100	100	100	400

What conclusion can you draw about the pattern of demand in different regions ?

Solution. We set up the null hypothesis, H_o: "Proportions of users of a new shampoo is same in all the four regions," *i.e.*, $H_o : P_1 = P_2 = P_3 = P_4$, where P_1, P_2, P_3 and P_4 represent the true population proportions of users. Under H_o, the expected frequencies are computed as shown in the following table.

TABLE 15·13: COMPUTATION OF EXPECTED FREQUENCIES

Region	*NW*	*NE*	*SW*	*SE*	*Total*
Do not buy	$\frac{210 \times 100}{400} = 52{\cdot}5$	$\frac{210 \times 100}{400} = 52{\cdot}5$	$\frac{210 \times 100}{400} = 52{\cdot}5$	52·5	210
Buy	100 − 52.5 = 47·5	100 − 52·5 = 47·5	100 − 52.5 = 47.5	47·5	190
Total	100	100	100	100	400

TABLE 15·13(*a*): CALCULATIONS FOR CHI-SQUARE

f_i	e_i	$f_i - e_i$	$(f_i - e_i)^2$	$\frac{(f_i - e_i)^2}{e_i}$
60	52·5	7·5	56·25	1·071
45	52·5	− 7·5	56·25	1·071
55	52·5	2·5	6·25	0·119
50	52·5	− 2·5	6·25	0·119
40	47·5	− 7·5	56·25	1·184
55	47·5	7·5	56·25	1·184
45	47·5	− 2·5	6·25	0·132
50	47·5	2·5	6·25	0·132
Total				5·012

$$\chi^2 = \sum_i \left\{ \frac{(f_i - e_i)^2}{e_i} \right\} = 5{\cdot}012$$

Tabulated $\chi^2_{0{\cdot}05}$ for (2 − 1) (4 − 1) = 3 *d.f.* is 7·815

Since the calculated value of χ^2 is less than the tabulated value, it is not-significant at 5% level of significance. Hence, the null hypothesis may be accepted and it may be concluded that the regions do not differ significantly as regards the users of a new shampoo.

Aliter. **For computing χ^2.** Using the 'Brandt and Snedecor' formula (15·19*a*), we have

$$\chi^2 = \frac{(400)^2}{210 \times 190} \left[\frac{60^2 + 45^2 + 55^2 + 50^2}{100} - \frac{(210)^2}{400} \right]$$

$$= \frac{1600}{21 \times 19}\left[\frac{11150}{100} - \frac{441}{4}\right] = 4{\cdot}01 \times 1{\cdot}25 = 5{\cdot}0125,$$

the value, which is approximately the same as obtained in (*).

15·6·7. χ^2-test of Homogeneity of Correlation Coefficients. Let $r_1, r_2, \ldots, r_k$ be k correlation coefficients from independent samples of sizes $n_1, n_2, \ldots, n_k$, respectively.

We want to test the hypothesis that these sample correlation coefficients are the estimates of the same correlation coefficient ρ from a bivariate normal population. Obtain the values of z_1, z_2, ..., z_k from the Table of Fisher's z-transformation [TABLE X, given at end of the book] or from

$$z_i = \frac{1}{2}\log_e\left(\frac{1+r_i}{1-r_i}\right) = \tan h^{-1} r_i;\ i = 1, 2, \ldots, k \qquad \text{... (15·20)}$$

These z_i's are normally distributed about a common mean

$$\xi = \frac{1}{2}\log_e\left(\frac{1+\rho}{1-\rho}\right) \text{ and variance } = \frac{1}{n_i - 3} \qquad \text{... (15·21)}$$

The minimum variance estimate $\bar{z}$ of the common mean ξ of Z's is obtained by weighting the values z_i's inversely with their respectively variances. The estimate of ξ is, therefore,

$$\bar{z} = \frac{\sum_i z_i(n_i - 3)}{\sum_i (n_i - 3)},$$ so that $(z_i - \bar{z})\ \sqrt{n_i - 3}\ ;\ i = 1, 2, \ldots, k$ are independent standard normal variates.

$$\text{Hence,} \qquad \sum_{i=1}^{k} (n_i - 3)(z_i - \bar{z})^2 \sim \chi^2_{(k-1)} \qquad \text{...(15·21}a\text{)}$$

[By additive property of χ^2-distribution, one *d.f.* being lost since $\bar{z}$ has been determined from the data.]

If χ^2 value thus obtained is greater than the tabulated 5 per cent value of χ^2 for $(k-1)$ *d.f.*, the *null hypothesis of homogeneity of correlation coefficients* is rejected. If not, the correlation coefficients are supposed to be homogeneous in which case we combine the sample correlation coefficients to find the estimate $\hat{\rho}$ of the population correlation coefficients ρ.

$$\text{We have,} \qquad \bar{z} = \frac{1}{2}\log\left(\frac{1+\hat{\rho}}{1-\hat{\rho}}\right) \quad \Rightarrow \quad (1+\hat{\rho}) = (1-\hat{\rho})\,e^{2\bar{z}}$$

$$\Rightarrow \qquad \left(1 + e^{2\bar{z}}\right)\hat{\rho} = e^{2\bar{z}} - 1 \quad \Rightarrow \quad \hat{\rho} = \frac{e^{2\bar{z}} - 1}{e^{2\bar{z}} + 1} = \tanh \bar{z} \qquad \text{... (15·22)}$$

Remark. For testing the homogeneity of independent estimates of the parent partial correlation coefficient, the above formulae hold with the only difference that for a partial correlation coefficient of order s, n_i will be replaced by $n_i - s$.

Example 15·15. *The correlation coefficient between daily ration of green grass and rate of growing calves on the basis of observations taken on 10, 14, 16, 20, 25 and 28 cows at six farms were found to be 0·318, 0·106, 0·253, 0·340, 0·116 and 0·112. Can these be considered homogeneous ? If so, estimate the common correlation coefficient.*

Solution. H_0: The given values of sample correlation coefficients are homogeneous or the samples are from equally correlated populations.

Using (15·20), (or From Table X in the Appendix) we get

$z_1 = 0{\cdot}3294, \quad z_2 = 0{\cdot}1063, \quad z_3 = 0{\cdot}2586, \quad z_4 = 0{\cdot}3541, \quad z_5 = 0{\cdot}1165$, and $z_6 = 0{\cdot}1125$

$$\therefore \qquad \bar{z} = \sum_i z_i(n_i - 3) \Big/ \sum_i (n_i - 3) = 0{\cdot}1919$$

From (15·21*a*) $\qquad \chi^2 = \sum (n_i - 3)(z_i - \bar{z})^2 = 0{\cdot}1008$

Tabulated value of χ^2 for $(6 - 1) = 5$ *d.f.* at 5% level of significance is 11·070.

Since the calculated value of χ^2 is less than the tabulated value, we may accept the null hypothesis that the sample correlation coefficients are homogeneous.

If $\hat{\rho}$ is the pooled estimate of the population correlation coefficient, then using (15·22),

$$\hat{\rho} = \frac{e^{2\bar{z}} - 1}{e^{2\bar{z}} + 1} = \frac{1{\cdot}468 - 1}{1{\cdot}468 + 1} = 0{\cdot}1894$$

15·6·8. Bartlett's Test for Homogeneity of Several Independent Estimates of the Same Population Variance. Let

$$s_i^2 = \frac{1}{n_i - 1} \sum_{j=1}^{n_i} (X_{ij} - \bar{X}_i)^2, \; (i = 1, 2, \ldots, k) \qquad \ldots (15{\cdot}23)$$

be the unbiased estimate of the population variance, obtained from the ith sample X_{ij}, $(j = 1, 2, \ldots, n_i)$ and based on $\nu_i = (n_i - 1)$ *d.f.*, all the k samples being independent.

Under the *null hypothesis* that the samples come from the same population with variance σ^2, *i.e.*, the independent estimates s_i^2, $(i = 1, 2, \ldots, k)$ of σ^2 are homogeneous, Bartlett proved that the statistic

$$\chi^2 = \sum_{i=1}^{k} \left(\nu_i \log_e \frac{s^2}{s_i^2} \right) \Bigg/ \left[\left(1 + \frac{1}{3(k-1)} \right) \left\{ \sum_i \left(\frac{1}{\nu_i} \right) - \frac{1}{\nu} \right\} \right] \qquad \ldots (15{\cdot}23a)$$

where
$$s^2 = \frac{\sum \nu_i s_i^2}{\sum \nu_i} = \frac{\sum \nu_i s_i^2}{\nu}, \qquad \sum_{i=1}^{k} \nu_i = \nu, \qquad \ldots (*)$$

follows chi-square distribution with $(k - 1)$ degrees of freedom.

Remarks:

1. S^2 defined in (*) is also an unbiased estimate of σ^2, since

$$E(S^2) = \frac{\sum \nu_i E(S_i^2)}{\sum \nu_i} = \frac{\left(\sum \nu_i\right) \sigma^2}{\sum \nu_i} = \sigma^2 \qquad \left[\because \; E(S_i^2) = \sigma^2, i = 1, 2, \ldots, k \right]$$

2. Let s_i^2 and s_j^2; $i \neq j$, $1 < (i, j) \leq k$ be the smallest and the largest values of the unbiased estimates of s^2, respectively. If on the basis of F-test (*c.f.* Chapter 16), these do not differ significantly, then all the estimates s_i^2 which lie between s_i^2 and s_j^2 won't differ significantly either and consequently all the estimates can be reasonably regarded as homogeneous, coming from the same population. In this case, therefore, there is no need to apply Bartlett's test.

SUPPLEMENTARY EXAMPLES ON CHI-SQUARE DISTRIBUTION

Example 15·16. *(a) Show that for 2 d.f. the probability P of a value of χ^2 greater than χ_0^2 is*

$$\exp\left(-\frac{1}{2}\chi_0^2\right), \text{ and hence that: } \chi_0^2 = 2\log_e(1/P)$$

Deduce the value of χ_0^2 when $P = 0{\cdot}05$.

(b) Given different probabilities $P_1, P_2, \ldots, P_n$ obtained from n independent tests of significance, explain how you will pool them to get a single probability in order to decide about the significance of the aggregate of these tests.

Solution. (*a*) The *p.d.f.* of χ^2-distribution with 2 *d.f.* is:

$$f(\chi^2) = \left[\frac{1}{2^{n/2}\Gamma(n/2)}\exp(-\chi^2/2)\cdot(\chi^2)^{(n/2)-1}\right]_{n=2} = \frac{1}{2}\exp(-\chi^2/2), 0 \le \chi^2 < \infty$$

$$\therefore \quad P = P(\chi^2 > \chi_0^2) = \int_{\chi_0^2}^{\infty}\frac{1}{2}\exp(-\chi^2/2)\,d\chi^2 \qquad \ldots (*)$$

$$= \frac{1}{2}\left|\frac{\exp(-\chi^2/2)}{-\frac{1}{2}}\right|_{\chi_0^2}^{\infty} = \exp(-\chi_0^2/2)$$

$$\therefore \quad \log_e P = -\chi_0^2/2 \quad \Rightarrow \quad \chi_0^2 = -2\log_e P = 2\log_e(1/P)$$

When $P = 0{\cdot}05$, we get: $\chi_0^2 = 2\log_e 20 = 3{\cdot}012$

Remark. The value χ_0^2 of χ^2 defined in (*) is known as the significant or critical value [*c.f. Remark* 4, Page 15·15, to *Theorem 15·3*,] of χ^2 corresponding to the probability level *P*. Thus, if *P* is the significant probability, then

$$\chi^2 = -2\log_e P = 2\log_e(1/P) \qquad \ldots (15{\cdot}24)$$

is a χ^2-variate with 2 *d.f.*

(*b*) $-2\log_e P_i$ ($i = 1, 2, \ldots, n$) are *independent* χ^2-variates each with 2 *d.f.* (*c.f.* Remark and the fact that P_i's obtained from independent tests of significance are independent). Hence, by additive property of chi-square distribution:

$$\chi^2 = \sum_{i=1}^{n}(-2\log_e P_i) = 2\log_e\left(\frac{1}{P_1 P_2 \ldots P_n}\right) \qquad \ldots (15{\cdot}25)$$

is a chi-square variate with 2*n d.f.*

If $\chi^2 > \chi_{0\cdot05}^2$ for 2*n d.f.*, then we conclude that the pooled result (aggregate of the tests) is significant at 5% level of significance.

Example 15.17. *(Pearson's P_l – Statistic). The variables $X_1, X_2, \ldots, X_n$ are independently distributed in the rectangular form: $dF = dx$, $0 \le x \le 1$. Then if $P = X_1 X_2 \ldots X_n$, show that $-2\log_e P$ has χ^2-distribution with 2n degrees of freedom.*

Solution. $-2\log_e P = -2\log_e(X_1 X_2 \ldots X_n) = \xi_1 + \xi_2 + \ldots + \xi_n = \sum_{i=1}^{n}\xi_i,$

where $\xi_i = -2 \log X_i \quad \Rightarrow \quad X_i = \exp(-\xi_i/2)$

The probability function of ξ_i is given by: $g(\xi_i) = f(x_i)\left|\dfrac{dx_i}{d\xi_i}\right|$

Since, $dF(x) = dx, \quad f(x) = 1 \;\forall\; x$ in $[0, 1]$

$$\therefore \quad g(\xi_i) = 1\cdot\left|\exp(-\xi_i/2)\times\left(-\frac{1}{2}\right)\right| = \frac{1}{2}\exp(-\xi_i/2),$$

which is the probability function of χ^2-distribution with 2 *d.f.*

$\therefore$ ξ_i, $(i = 1, 2, \ldots, n)$ are independent χ^2-variates each with 2 d.f. Hence, by additive property of χ^2-distribution,

$$-2 \log_e P = \sum_{i=1}^{n} \xi_i, \text{ is a } \chi^2\text{-variate with } 2n \; d.f.$$

Remark. The significance of this result lies in testing of hypothesis as explained in Example 15·16.

Example 15·18. *Show that if v is even,*

$$P = \frac{1}{2^{(v-2)/2}\,\Gamma(v/2)}\int_{\chi}^{\infty}\exp(-\chi^2/2)\cdot\chi^{v-1}d\chi$$

$$= \exp(-\chi^2/2)\left[1 + \frac{\chi^2}{2} + \frac{\chi^4}{2.4} + \cdots + \frac{\chi^{v-2}}{2.4\cdots(v-2)}\right]$$

and hence the values of P for a given χ^2 can be derived from tables of Poisson's exponential limit.

Solution. Let us consider the incomplete Gamma integral : $I_r = \dfrac{1}{r!}\displaystyle\int_{\beta}^{\infty} e^{-t}\, t^r dt,$

where r is a positive integer. Integrating by parts, we get

$$I_r = \left|-\frac{e^{-t}\cdot t^r}{r!}\right|_{\beta}^{\infty} + \frac{1}{(r-1)!}\int_{\beta}^{\infty} e^{-t}t^{r-1}dt = \frac{e^{-\beta}\cdot\beta^r}{r!} + I_{r-1},$$

which is a reduction formula. Repeated application of this gives

$$I_r = \frac{e^{-\beta}\beta^r}{r!} + \frac{e^{-\beta}\beta^{r-1}}{(r-1)!} + \cdots + \frac{e^{-\beta}\cdot\beta^2}{2!} + \frac{e^{-\beta}\cdot\beta}{1!} + I_0$$

But $\quad I_0 = \displaystyle\int_{\beta}^{\infty} e^{-t}\,dt = \left|-e^t\right|_{\beta}^{\infty} = e^{-\beta}$

$$\therefore \quad \frac{1}{r!}\int_{\beta}^{\infty} e^{-t}t^r\,dt = e^{-\beta}\left[1 + \beta + \frac{\beta^2}{2!} + \cdots + \frac{\beta^{r-1}}{(r-1)!} + \frac{\beta^r}{r!}\right]$$

Putting $\beta = \dfrac{1}{2}\chi^2$ and $r = \dfrac{1}{2}(v-2) = \dfrac{1}{2}v - 1$, (since r is an integer, $v = 2r + 2$ must be even), in (*), we get

$$\frac{1}{\{(v/2)-1\}!}\int_{\chi^2/2}^{\infty} e^{-t}t^{(v/2)-1}dt = \exp(-\chi^2/2)\left[1 + \frac{\chi^2}{2} + \frac{\chi^4}{2.4} + \frac{\chi^6}{2.4.6} + \cdots + \frac{\chi^{v-2}}{2.4.6\ldots(v-2)}\right] \quad \ldots (**)$$

Taking $t = \frac{1}{2}\chi^2$ in the integral on the L.H.S., we get

$$\text{L.H.S.} = \frac{1}{\{(v/2)-1\}!}\int_{\chi}^{\infty} \exp(-\chi^2/2)\,(\chi^2/2)^{(v/2)-1}\cdot d(\chi^2/2)$$

$$= \frac{1}{2^{(v-2)/2}\,\Gamma(v/2)}\int_{\chi}^{\infty} \exp(-\chi^2/2)\cdot\chi^{v-1}d\chi \qquad \ldots (***)$$

From (**) and (***), we get the required result.

Let the given value of χ^2 be χ_0^2, then

$$P = P(\chi^2 > \chi_0^2) = \frac{1}{2^{(v-2)/2}\,\Gamma(v/2)}\int_{\chi_0}^{\infty} \exp(-\chi^2/2)\cdot\chi^{v-1}d\chi$$

$$= \exp(-\chi_0^2/2)\left[1+\frac{\chi_0^2}{2}+\frac{\chi_0^4}{2.4}+\cdots+\frac{\chi_0^{v-2}}{2.4\ldots(v-2)}\right]$$

$$= e^{-\lambda}\left[1+\lambda+\frac{\lambda^2}{2!}+\frac{\lambda^3}{3!}+\cdots+\frac{\lambda^{\frac{v}{2}-1}}{\{(v/2)-1\}!}\right], \text{ where } \lambda = \frac{1}{2}\chi_0^2$$

The terms on the right-hand side, *viz.*, $e^{-\lambda}$, $\lambda e^{-\lambda}$, $\frac{\lambda^2}{2!}e^{-\lambda}$..., etc. are the successive terms of the Poisson distribution with parameter $\lambda = \frac{1}{2}\chi_0^2$.

Hence, the result.

Example 15·19. *If X and Y are independent normal variates with means μ_1, μ_2 and variances σ_1^2, σ_2^2 respectively, derive the distribution of: $Z = (X-\mu_1)/(Y-\mu_2)$.*

What is the name of the distribution so obtained? Mention one important property of this distribution.

Solution. Here $$Z^2 = \frac{(X-\mu_1)^2}{(Y-\mu_2)^2} \Rightarrow \frac{\sigma_2^2}{\sigma_1^2}\cdot Z^2 = \frac{[(X-\mu_1)/\sigma_1]^2}{[(Y-\mu_2)/\sigma_2]^2}$$

But $\{(X-\mu_1)/\sigma_1\}^2$ and $\{(Y-\mu_2)/\sigma_2\}^2$, being the squares of independent standard normal variates, are independent χ^2-variates with 1 *d.f.* each.

Thus $\frac{\sigma_2^2 Z^2}{\sigma_1^2}$ being the quotient of two independent χ^2-variates each with 1 *d.f.* is a $\beta_2\left(\frac{1}{2},\frac{1}{2}\right)$ variate (*c.f. Theorem 15·1*).

Hence, its probability differential is given by:

$$dF\left(\frac{\sigma_2^2}{\sigma_1^2}z^2\right) = \frac{1}{B\left(\frac{1}{2},\frac{1}{2}\right)}\times\frac{\left(\sigma_2^2 z^2/\sigma_1^2\right)^{(1/2)-1}}{\left(1+\frac{\sigma_2^2 z^2}{\sigma_1^2}\right)^{\frac{1}{2}+\frac{1}{2}}}\,d(\sigma_2^2 z^2/\sigma_1^2)$$

$$= \frac{\Gamma\left(\frac{1}{2}+\frac{1}{2}\right)}{\Gamma\left(\frac{1}{2}\right)\Gamma\left(\frac{1}{2}\right)} \cdot \frac{\left(\frac{\sigma_2^2 z^2}{\sigma_1^2}\right)^{\frac{1}{2}-1}}{\left(1+\frac{\sigma_2^2 z^2}{\sigma_1^2}\right)} \cdot \frac{\sigma_2^2}{\sigma_1^2} d(z^2)$$

$$= \frac{\sigma_1 \sigma_2}{\pi\left(\sigma_1^2+\sigma_2^2 z^2\right) z} dz^2, 0 < z^2 < \infty \qquad \left[\because \Gamma(1/2) = \sqrt{\pi}\right]$$

Thus, the probability differential of Z is given by:

$$dF(z) = f(z)\,dz = \frac{\sigma_1 \sigma_2}{\pi\left(\sigma_1^2+\sigma_2^2 z^2\right)} dz, -\infty < z < \infty$$

If $\sigma_1 = \sigma_2 = 1$, then it conforms to standard Cauchy distribution,

$$dF(z) = \frac{1}{\pi} \cdot \frac{dz}{\left(1+z^2\right)}, -\infty < z < \infty$$

[For its properties see Chapter 9.]

Aliter. $\quad Z = \dfrac{X-\mu_1}{Y-\mu_2} \quad \Rightarrow \quad \dfrac{\sigma_2}{\sigma_1} Z = \dfrac{(X-\mu_1)/\sigma_1}{(Y-\mu_2)/\sigma_2}$

$\dfrac{\sigma_2}{\sigma_1} Z$, being the ratio of two independent standard normal variates is a standard Cauchy variate.

$$\therefore \quad dF\left(\frac{\sigma_2}{\sigma_1} z\right) = \frac{d\left(\frac{\sigma_2}{\sigma_1} z\right)}{\pi\left[1+\left(\frac{\sigma_2}{\sigma_1} z\right)^2\right]} = \frac{\sigma_1 \sigma_2}{\pi\left(\sigma_1^2+\sigma_2^2 z^2\right)} dz, -\infty < z < \infty$$

Example 15·20. *X_i, $(i = 1, 2, \ldots, n)$ are independently and normally distributed with zero mean and common variance σ^2.*

Let $\xi_i = \sum_{j=1}^{n} c_{ij} X_j, i = 1, 2, \ldots, n,$ *where* $\sum_{j=1}^{n} c_{ij} c_{i'j} = \delta_{i i'}$ *is Kroneker delta.*

Show that, $\dfrac{1}{\sigma^2}\left[\sum_{i=1}^{n} X_i^2 - \sum_{i=1}^{p} \xi_i^2\right]$ *is distributed as χ^2 variate with $(n-p)$ d.f.*

Solution. Since $\delta_{ii'} = \sum_{j=1}^{n} c_{ij} c_{i'j}$ is a Kroneker delta, we have $\sum_{j=1}^{n} c_{ij} c_{i'j} = \begin{cases} 0, & i \neq i' \\ 1, & i = i' \end{cases}$

i.e. X_i's are transformed to ξ_i's by means of a linear orthogonal transformation. Hence, by Fisher's Lemma, ξ_i $(i = 1, 2, \ldots, n)$ are independent normal variates with zero mean and common variance σ^2.

Since the transformation is orthogonal, we have: $\quad \sum_{i=1}^{n} X_i^2 = \sum_{i=1}^{n} \xi_i^2 \qquad \ldots (*)$

Thus, $$\frac{\left[\sum_{i=1}^{n} X_i^2 - \sum_{i=1}^{p} \xi_i^2\right]}{\sigma^2} = \frac{\sum_{i=1}^{n} \xi_i^2 - \sum_{i=1}^{p} \xi_i^2}{\sigma^2} = \frac{\sum_{i=p+1}^{n} \xi_i^2}{\sigma^2} \qquad \text{[From (*)]}$$

$$\frac{\sum_{i=p+1}^{n} \xi_i^2}{\sigma^2} = \sum_{i=p+1}^{n} (\xi_i/\sigma)^2,$$ being the sum of the squares of $(n-p)$ independent standard normal variates is a χ^2-variate with $(n-p)$ degrees of freedom. Hence the result.

Example 15·21. *Show that the m.g.f. of $Y = \log \chi^2$, where χ^2 follows chi-square distribution with n d.f., is given by*

$$M_Y(t) = 2^t\, \Gamma\left(\frac{n}{2}+t\right) \Big/ \Gamma(n/2)$$

If χ_1^2 and χ_2^2 are independent χ^2-variates each with n d.f. and $U = \chi_1^2/\chi_2^2$, deduce that for positive integer k,

$$E(U^k) = \Gamma\left(\frac{n}{2}+k\right)\Gamma\left(\frac{n}{2}-k\right) \Big/ \left[\Gamma\left(\frac{n}{2}\right)\right]^2$$

Solution. $y = \log \chi^2 \quad \Rightarrow \quad \chi^2 = e^y \quad$ or $\quad d\chi^2 = e^y\, dy$

The probability differential of χ^2, *viz.*,

$$dP(\chi^2) = \frac{1}{2^{n/2}\,\Gamma(n/2)}\, e^{-\chi^2/2}\, (\chi^2)^{\frac{n}{2}-1}\, d\chi^2, \; 0 < \chi^2 < \infty$$

transforms to

$$dG(y) = \frac{1}{2^{n/2}\,\Gamma(n/2)} \exp\left(-\frac{1}{2}e^y + \frac{ny}{2}\right) dy, \; -\infty < y < \infty$$

$$\therefore \quad M_Y(t) = \frac{1}{2^{n/2}\,\Gamma(n/2)} \int_{-\infty}^{\infty} \exp\left(-\frac{1}{2}e^y + \frac{ny}{2} + ty\right) dy$$

$$= \frac{1}{2^{n/2}\,\Gamma(n/2)} \int_0^{\infty} e^{-z} (2z)^{\frac{n}{2}+t}\, \frac{dz}{z}, \qquad (\, 2z = e^{\,y}\,)$$

$$= \frac{2^t}{\Gamma(n/2)} \int_0^{\infty} e^{-z}\, z^{\frac{n}{2}+t-1}\, dz = 2^t\, \Gamma\left(\frac{n}{2}+t\right) \Big/ \Gamma\left(\frac{n}{2}\right) \qquad \dots (*)$$

$$E(U^k) = E\left[\left(\frac{\chi_1^2}{\chi_2^2}\right)^k\right] = E\left[\exp\left\{\log\left(\frac{\chi_1^2}{\chi_2^2}\right)^k\right\}\right] = E\left(e^{k \log \chi_1^2 - k \log \chi_2^2}\right)$$

$$= E\left(e^{k \log \chi_1^2}\right) \cdot E\left(e^{-k \log \chi_2^2}\right) \qquad \left[\because \chi_1^2 \text{ and } \chi_2^2 \text{ are independent.}\right]$$

$$= M_{\log \chi_1^2}(k) \cdot M_{\log \chi_2^2}(-k) = \frac{2^k \Gamma\left(\frac{n}{2}+k\right)}{\Gamma(n/2)} \cdot \frac{2^{-k} \Gamma\left(\frac{n}{2}-k\right)}{\Gamma(n/2)} \qquad \text{[From (*)]}$$

$$= \Gamma\left(\frac{n}{2}+k\right)\Gamma\left(\frac{n}{2}-k\right)\Big/\left[\Gamma\left(\frac{n}{2}\right)\right]^2$$

Example 15·22. *If X is chi-square variate with n d.f., then prove that for large n,*

$$\sqrt{2X} \sim N\left(\sqrt{2n}, 1\right). \qquad \text{... (15·26)}$$

Solution. Since $X \sim \chi^2_{(n)}$, we have $E(X) = n$, Var $(X) = 2n$

$$\therefore \qquad Z = \frac{X - E(X)}{\sigma_X} = \frac{X - n}{\sqrt{2n}} \sim N(0, 1), \text{ for large } n.$$

Consider, $P\left(\frac{X-n}{\sqrt{2n}} \le z\right) = P\left(X \le n + z\sqrt{2n}\right) = P\left[\sqrt{2X} \le \left(2n + 2z\sqrt{2n}\right)^{1/2}\right]$

$$= P\left[\sqrt{2X} \le \sqrt{2n}\left(1 + z\sqrt{\frac{2}{n}}\right)^{1/2}\right] = P\left[\sqrt{2X} \le \sqrt{2n}\left(1 + \frac{z}{\sqrt{2n}} - \frac{z^2}{4n} + \cdots\right)\right]$$

$$\approx P\left(\sqrt{2X} \le \sqrt{2n} + z\right), \text{ for large } n.$$

$$= P\left[\sqrt{2X} - \sqrt{2n} \le z\right], \text{ for large } n. \qquad \text{... (*)}$$

Since for large n, $(X - n)/\sqrt{2n} \sim N(0, 1)$, from (*) we conclude that

$$\sqrt{2X} - \sqrt{2n} \sim N(0, 1) \text{ for large } n \quad \Rightarrow \quad \sqrt{2X} \text{ is asymptotically } N\left(\sqrt{2n}, 1\right).$$

Remark. This approximation is often used for the value of n larger than 30. This result does not reflect anything as to how good the approximation is, for moderate values of n. R.A. Fisher has proved that the approximation is improved by taking $\sqrt{(2n-1)}$ instead of $\sqrt{2n}$.

A still better approximation is: $\left(\frac{\chi^2}{n}\right)^{1/3} \sim N\left(1 - \frac{2}{9n}, \frac{2}{9n}\right)$ ... (15.26*a*)

Example 15·23. *For a chi-square distribution with n d.f. establish the following recurrence relation between the moments:*

$$\mu_{r+1} = 2\,r\,(\mu_r + n\,\mu_{r-1}),\ r \ge 1. \textit{ Hence find } \beta_1 \textit{ and } \beta_2.$$

Solution. If $X \sim \chi^2_{(n)}$, then its *m.g.f.* about origin is:

$$M_X(t) = E\left(e^{tX}\right) = (1 - 2t)^{-n/2};\ t < \frac{1}{2} \qquad \text{... (*)}$$

Also $\qquad E(X) = n = \mu$ (say).

Hence, *m.g.f.* about mean, say $M(t)$ is:

$$M(t) = M_{X-\mu}(t) = E\left[e^{t(X-\mu)}\right] = e^{-\mu t} \cdot E(e^{tX}) = e^{-nt}(1-2t)^{-n/2}$$

Taking logarithms of both sides, $\log M(t) = -nt - \frac{n}{2}\log(1-2t)$ [Using (*)]

Differentiating *w.r.* to t, we have

$$\frac{M'(t)}{M(t)} = -n + \frac{n}{2}\cdot\frac{2}{(1-2t)} = \frac{2nt}{1-2t} \quad \Rightarrow \quad (1-2t)\,M'(t) = 2nt\,M(t)$$

Differentiating r times *w.r.to* t by Leibnitz theorem, we get

$$(1-2t)\,M^{r+1}(t) + r(-2)\,M^r(t) = 2nt\,M^r(t) + 2nr\,M^{r-1}(t)$$

Putting $t = 0$ and using the relation, $\mu_r = \left[\frac{d^r}{dt^r}M(t)\right]_{t=0} = M^r(0)$, we get

$$\mu_{r+1} - 2r\mu_r = 2nr\mu_{r-1} \quad \Rightarrow \quad \mu_{r+1} = 2r\left(\mu_r + n\mu_{r-1}\right),\ r \geq 1 \qquad \text{...(15·27)}$$

Taking $r = 1, 2, 3$, in (15·27), we get

$$\mu_2 = 2n\mu_0 = 2n, \qquad \mu_3 = 4(\mu_2 + n\mu_1) = 8n \qquad [\because \mu_1 = 0 \text{ and } \mu_0 = 1]$$

$$\mu_4 = 6(\mu_3 + n\mu_2) = 48n + 12n^2$$

$$\therefore \quad \beta_1 = \frac{\mu_3^2}{\mu_2^3} = \frac{8}{n} \quad \text{and} \quad \beta_2 = \frac{\mu_4}{\mu_2^2} = 3 + \frac{12}{n}$$

Example 15·24. *Two samples of sizes N_1, N_2 have respectively frequencies $f_1, f_2, \ldots, f_n$ and $f_1', f_2', \ldots, f_n'$ under the same headings. Show that χ^2 for such a distribution is equal to*

$$\sum_{r=1}^{n} N_1 N_2 \left[\frac{\{(f_r/N_1) - (f_r'/N_2)\}^2}{f_r + f_r'}\right]$$

Solution. The $2 \times n$ contingency table for which χ^2 is to be calculated is given below:

B \ A	A_1	A_2	…	A_r	…	A_n	Total
B_1	f_1	f_2	…	f_r	…	f_n	N_1
B_2	f_1'	f_2'	…	f_r'	…	f_n'	N_2

Under the hypothesis of independence of attributes, we have

$$E(f_r) = \frac{N_1(f_r + f_r')}{N_1 + N_2}, \qquad E(f_r') = \frac{N_2(f_r + f_r')}{N_1 + N_2}$$

$$\chi^2 = \sum_{i=1}^{n}\left[\frac{\{f_r - E(f_r)\}^2}{E(f_r)} + \frac{\{f_r' - E(f_r')\}^2}{E(f_r')}\right]$$

$$= \sum_{r=1}^{n}\left[\frac{(N_1+N_2)}{N_1(f_r+f_r')}\cdot\left\{f_r - \frac{N_1(f_r+f_r')}{N_1+N_2}\right\}^2 + \frac{(N_1+N_2)}{N_2(f_r+f_r')}\left\{f_r' - \frac{N_2(f_r+f_r')}{N_1+N_2}\right\}^2\right]$$

$$= \sum_{r=1}^{n}\left[\frac{(N_2 f_r - N_1 f_r')^2}{N_1(N_1+N_2)(f_r+f_r')} + \frac{(N_1 f_r' - N_2 f_r)^2}{N_2(N_1+N_2)(f_r+f_r')}\right]$$

$$= \sum_{r=1}^{n}\frac{(N_2 f_r - N_1 f_r')^2}{(N_1+N_2)(f_r+f_r')}\left(\frac{1}{N_1}+\frac{1}{N_2}\right) = \sum_{r=1}^{n}\left[\frac{N_1N_2}{f_r+f_r'}\left(\frac{f_r}{N_1}-\frac{f_r'}{N_2}\right)^2\right]$$

15·7. NON-CENTRAL χ^2-DISTRIBUTION

The χ^2-distribution defined as the sum of the squares of independent standard normal variates is often referred to as the *central* χ^2 - distribution. The distribution of the sum of the squares of independent normal variates each having unit variance but with possibly non-zero means is known as *non-central chi-square distribution.* Thus, if X_i, $(i = 1, 2, \ldots, n)$ are independent $N(\mu_i, 1)$, *r.v.*'s then:

$$\chi'^2 = \sum_{i=1}^{n} X_i^2, \qquad \ldots (15\cdot28)$$

has the non-central χ^2 distribution with n *d.f.* Intuitively, this distribution would seem to depend upon the n parameters $\mu_1, \mu_2, \ldots, \mu_n$ but it will be seen that it depends on these parameters only through the *non-centrality parameter*

$$\lambda = \frac{1}{2}\left(\mu_1^2 + \mu_2^2 + \cdots + \mu_n^2\right) \qquad \ldots (15\cdot28a)$$

and we write $\chi'^2 \sim \chi'^2(n, \lambda)$.

Non-central χ^2-distribution with Non-centrality Parameter λ. The *p.d.f.* is given by

$$f\left(\chi_n'^2, \lambda\right) = \sum_{j=0}^{\infty}\left[\frac{e^{-\lambda}\lambda^j}{j!}\cdot p\left(\chi_{n+2j}^2\right)\right] \qquad \ldots (15\cdot29)$$

where $p\left(\chi_{n+2j}^2\right)$ is the *p.d.f.* of (central) χ^2-variate with $(n + 2j)$ *d.f.*

Thus, $f\left(\chi_n'^2, \lambda\right)$ is the mixture of central χ^2-distributions with *d.f.* $n, n + 2, n + 4, \ldots$; the corresponding weights being the successive terms of the Poisson distribution with parameter λ.

Derivation of p.d.f. of χ'^2. We shall obtain the *p.d.f* of non-central χ^2-distribution through moment generating function (*m.g.f.*), by using the uniqueness theorem of *m.g.f.*

15·7·1. Moment Generating Function of Non-central χ^2-Distribution. If $X \sim N(\mu, 1)$ then

$$M_{X^2}(t) = \frac{1}{\sqrt{2\pi}} \int_{-\infty}^{\infty} e^{tx^2} \cdot e^{-(x-\mu)^2/2} dx \qquad \ldots (*)$$

$$\exp\left[tx^2 - \frac{1}{2}(x-\mu)^2\right] = \exp\left[-\left\{\left(\frac{1}{2} - t\right)x^2 - \mu x + \frac{\mu^2}{2}\right\}\right]$$

$$= \exp\left[-\left(\frac{1-2t}{2}\right)\left\{x^2 - \frac{2\mu x}{1-2t} + \frac{\mu^2}{1-2t}\right\}\right]$$

$$= \exp\left[-\left(\frac{1-2t}{2}\right)\left\{\left(x - \frac{\mu}{1-2t}\right)^2 + \frac{\mu^2}{1-2t} - \frac{\mu^2}{(1-2t)^2}\right\}\right]$$

$$= \exp\left(\frac{t\mu^2}{1-2t}\right) \cdot \exp\left[-\left(\frac{1-2t}{2}\right)\left(x - \frac{\mu}{1-2t}\right)^2\right]$$

$$\therefore \quad M_{X^2}(t) = \exp\left(\frac{t\mu^2}{1-2t}\right) \frac{1}{\sqrt{2\pi}} \int_{-\infty}^{\infty} \exp\left[-\left(\frac{1-2t}{2}\right)\left(x - \frac{\mu}{1-2t}\right)^2\right] dx \qquad \text{[From (*)]}$$

$$= \exp\left(\frac{t\mu^2}{1-2t}\right) \frac{1}{\sqrt{2\pi}} \int_{-\infty}^{\infty} \exp\left(-\frac{1}{2}u^2\right) \frac{du}{(1-2t)^{1/2}}$$

$$= (1-2t)^{-1/2} \exp\left(\frac{t\mu^2}{1-2t}\right); \; 1 - 2t > 0 \;\Rightarrow\; t < \frac{1}{2} \qquad \ldots (**)$$

If X_i $(i = 1, 2, \ldots, n)$, are independent $N(\mu_i, 1)$ then the *m.g.f.* of the non-central χ^2-variate

$\chi'^2 = \sum_{i=1}^{n} X_i^2$, is given by

$$M_{\chi'^2}(t) = M_{\sum_{i=1}^{n} X_i^2}(t) = \prod_{i=1}^{n} M_{X_i^2}(t) \qquad \left[\text{Since } X_i^2\text{'s are independent}\right]$$

$$= \prod_{i=1}^{n}\left[(1-2t)^{-1/2} \exp\left(\frac{t\mu_i^2}{1-2t}\right)\right] \qquad \text{[From (**)]}$$

$$= (1-2t)^{-n/2} \exp\left[\frac{t}{(1-2t)} \sum_{i=1}^{n} \mu_i^2\right]$$

$$= (1-2t)^{-n/2} \exp\left[2\lambda t / (1-2t)\right], \; t < \frac{1}{2} \qquad \ldots (15\cdot30)$$

where, $\lambda = \frac{1}{2} \sum_{i=1}^{n} \mu_i^2$, is the non-centrality parameter.

(15·30) can be re-written as:

$$M_{\chi'^2}(t) = (1-2t)^{-n/2}\exp\left[\lambda\left(-1+\frac{1}{1-2t}\right)\right] = (1-2t)^{-n/2}e^{-\lambda}\exp\left(\frac{\lambda}{1-2t}\right)$$

$$= (1-2t)^{-n/2}e^{-\lambda}\sum_{r=0}^{\infty}\left(\frac{\lambda}{1-2t}\right)^r \times \frac{1}{r!}$$

$$= \sum_{r=0}^{\infty}\frac{e^{-\lambda}\lambda^r}{r!}(1-2t)^{-\left(r+\frac{n}{2}\right)}\ ;\ t<\frac{1}{2} \qquad \text{... (15·30}a\text{)}$$

Thus, the m.g.f. of a non-central χ^2 distribution is seen to be a *convex combination* of χ^2 *m.g.f.'s* with *d.f.* $n, n+2, n+4, \ldots$ The coefficients appearing in the convex combination are merely the Poisson probabilities.

Hence, by the uniqueness theorem of *m.g.f.*, the *p.d.f.* of non-central χ^2-distribution with n *d.f.* and with non-centrality parameter λ is given by

$$f(\chi_n'^2, \lambda) = \sum_{r=0}^{\infty}\frac{e^{-\lambda}\lambda^r}{r!} \times p\left(\chi_{n+2r}^2\right),$$

where, $$p\left(\chi_{n+2r}^2\right) = \frac{1}{2^{(n+2r)/2}\Gamma\left(\frac{n+2r}{2}\right)}\cdot e^{-\frac{1}{2}\chi^2}(\chi^2)^{\frac{n}{2}+r-1}\ ;\ 0\le\chi^2<\infty$$

is the *p.d.f.* of central χ^2-distribution with $(n+2r)$ *d.f.*

Remarks:

1. We can also write the *m.g.f.* of non-central χ^2 distribution with non-centrality parameter λ as $E\left[(1-2t)^{-\frac{n}{2}-Y}\right]$, where Y is the Poisson variate with parameter λ.

2. If we take $\lambda = 0 \Rightarrow \mu_i = 0\ \forall\ i = 1, 2, \ldots, n$; the *m.g.f.* of the non-central χ^2-distribution reduces to the *m.g.f.* of central χ^2 distribution; *viz.*, $(1-2t)^{-n/2}$. [From (15·30)]

3. Taking $\lambda = 0$ in the *p.d.f.* of non-central χ^2-variate, *i.e.*, in (15·29), we get

$$f(\chi_n'^2) = p(\chi_n^2) = \frac{1}{2^{n/2}\Gamma(n/2)}e^{-\frac{\chi^2}{2}}\cdot(\chi^2)^{\frac{n}{2}-1},\quad 0\le\chi^2<\infty$$

[$\because$ we get contribution only when $j = 0$, the other terms vanish when $\lambda = 0$]; which is *p.d.f.* of central χ^2-distribution with n *d.f.*

15·7·2. Additive or Re-productive Property of Non-central Chi-Square Distribution

If Y_i, $(i = 1, 2, \ldots, k)$, are independent non-central χ^2-variates with n_i *d.f.* and non-centrality parameter λ_i, then $\sum_{i=1}^{k} Y_i$ is also a non-central χ^2-variate with $\sum_{i=1}^{k} n_i$ *d.f.* and non-centrality parameter $\lambda = \sum_{i=1}^{k}\lambda_i$.

Proof. We have from (15·30)

$$M_{Y_i}(t) = (1-2t)^{-n_i/2} \exp[2t\lambda_i/(1-2t)], (i = 1, 2, \ldots, k)$$

$$\therefore \quad M_{\Sigma Y_i}(t) = \prod_{i=1}^{k} M_{Y_i}(t) = (1-2t)^{-\Sigma n_i/2} \exp\left[2t\sum_i \lambda_i/(1-2t)\right]$$

which is the *m.g.f.* of a non central χ^2-variate with $\sum_i n_i$ *d.f.* and non-centrality parameter $\lambda = \sum_i \lambda_i$ Hence, by uniqueness theorem of *m.g.f.'s*

$$\sum_{i=1}^{k} Y_i \sim \chi'^2_{\Sigma n_i}\left(\sum_i \lambda_i\right)$$

15·7·3. Cumulants of Non-central Chi-square Distribution.

Cumulant generating function is given by

$$K_{\chi'^2}(t) = \log M_{\chi'^2}(t) = -\frac{n}{2}\log(1-2t) + 2t\lambda(1-2t)^{-1}$$

$$= \frac{n}{2}\left[2t + \frac{(2t)^2}{2} + \cdots + \frac{(2t)^r}{r} + \cdots\right] + 2\lambda t\left[1 + 2t + (2t)^2 + \cdots + (2t)^r + \cdots\right]$$

the expansion being valid for $t < 1/2$.

$$\therefore \quad K_{\chi'^2}(t) = (n+2\lambda)t + (n+4\lambda)t^2 + \cdots + \left(\frac{2^{r-1}}{r}\cdot n + 2\lambda\, 2^{r-1}\right)t^r + \cdots$$

$$\Rightarrow \quad \kappa_r = \text{Coefficient of } \frac{t^r}{r!} \text{ in } K_{\chi'^2}(t) = r!\left(\frac{n}{r} + 2\lambda\right)2^{r-1}$$

$$= 2^{r-1}(r-1)!\,(n+2\lambda r) \qquad \ldots (15·31)$$

$$\therefore \quad \kappa_{r-1} = 2^{r-2}(r-2)!\,[n + 2\lambda(r-1)]$$

$$\therefore \quad \frac{d}{d\lambda}\{\kappa_{r-1}\} = 2^{r-2}(r-2)!\,2(r-1) = 2^{r-1}(r-1)! = \frac{\kappa_r}{(n+2\lambda r)} \qquad \text{[From (15·31)]}$$

$$\Rightarrow \quad \kappa_r = (n+2\lambda r)\frac{d}{d\lambda}(\kappa_{r-1}) \qquad \ldots (15·32)$$

CHAPTER CONCEPTS QUIZ

1. Choose the correct answer from *B* and match it with each item in *A*:

A	*B*
(*a*) β_2 for a chi-square distribution	1. $(1-2it)^{-n/2}$
(*b*) β_1 for a chi-square distribution	2. $8/n$
(*c*) Mean for a chi-square distribution	3. $2n$
(*d*) Variance for a chi-square distribution	4. $(1-2t)^{-n/2}$
(*e*) Characteristic function for χ^2-distribution	5. $(12/n) + 3$

(f) Mode of χ^2-distribution — 6. $\sqrt{2/n}$

(g) M.G.F. of χ^2-distribution — 7. n

(h) Skewness of χ^2-distribution — 8. $(n-2)$

Note. Here n represents the degrees of freedom of the chi-square distribution.

2. State which of the following statements are True and which are False. In case of false statements, give the correct statement:

(i) Normal distribution is particular case of χ^2-distribution for one *d.f.*

(ii) For large degrees of freedom, chi-square distribution tends to normal distribution.

(iii) The sum of independent chi-square variates is also a chi-square variate.

(iv) For the validity of χ^2-test, it is always necessary that the sample observations should be independent.

(v) The chi-square distribution maintains its character of continuity if cell frequency is less than 5.

(vi) Each linear constraint reduces the number of degrees of freedom of chi-square by unity.

(vii) In chi-square test of goodness of fit, if the calculated value of χ^2 is zero then the fit is a bad fit.

3. Mention the correct answer:

(i) The mean of a chi-square distribution with n *d.f.* is:

(a) $2n$, (b) n^2, (c) $\sqrt{n}$, (d) n

(ii) The characteristic function of chi-square distribution with n *d.f.* is:

(a) $(1-2it)^{n/2}$, (b) $(1+2it)^{n/2}$, (c) $(1-2it)^{-n/2}$ (d) None of these

(iii) The range of χ^2-variate is:

(a) $-\infty$ to $+\infty$, (b) 0 to ∞, (c) 0 to 1, (d) $-\infty$ to 0

(iv) The skewness in a chi-square distribution will be zero if

(a) $n \to \infty$, (b) $n = 0$ (c) $n = 1$ (d) $n < 0$

(v) The moment generating function of a χ^2-distribution with n degrees of freedom is:

(a) $(1-t)^{-n/2}$ (b) $(1-2t)^{-n/2}$ (c) $(1-3t)^{-n/2}$, (d) $(1-2t)^{n/2}$

(vi) Chi-square distribution is

(a) Continuous, (b) multimodal, (c) symmetrical.

4. Mention some prominent features of the chi-square distribution with n degrees of freedom.

5. If X and Y are independent random variables having chi-square distribution with m and n degrees of freedom respectively, write down the distributions of

(i) $X+Y$, (ii) X/Y, (iii) $X/(X+Y)$.

6. (a) For how many degrees of freedom does the χ^2-distribution reduce to negative exponential distribution ?

(b) Give an example of two independent variates none of which is a chi-square variate, although their sum is a chi-square variate.

DISCUSSION AND REVIEW QUESTIONS

1. Derive the *p.d.f.* of chi-square distribution with n degrees of freedom.

2. If X has a chi-square distribution with n *d.f.*, find *m.g.f.* $M_X(t)$.

Deduce that:

(i) $\mu_r' = E(X^r) = 2^r\Gamma[(n/2)+r]/\Gamma(n/2)$

(ii) κ_r = rth cumulant $= n\,2^{r-1}\,(r-1)!$

(iii) $\kappa_1\,\kappa_3 = 2\kappa_2^2$; $2\beta_2 - 3\beta_1 - 6 = 0$.

3. If $X \sim \chi^2_{(n)}$, show that:

(i) Mode is at $x = n - 2$.

(ii) The points of inflexion are equidistant from the mode.

4. (a) If $X \sim \chi^2_{(n)}$, obtain the *m.g.f.* of X. Hence, find the *m.g.f.* of standard chi-square variate and obtain its limiting form as $n \to \infty$. Also interpret the result.

(b) Let X and Y be two independent random variables having chi-square distribution with degrees of freedom m and n, respectively.

(i) Obtain the distribution of $U = X/(X + Y)$.

(ii) when $m = n$, show that the distribution of U is symmetrical about $\frac{1}{2}$.

Hence, or otherwise derive the rth moment about the mean of U when $m = n$.

(iii) Deduce the distribution of U when $m = n = 1$.

5. If X and Y are independently distributed chi-square variates with m and n degrees of freedom, respectively, show that $U = X + Y$ nd $V = X/Y$ are independently distributed. Also identify their distributions.

6. If χ_1^2 and χ_2^2 are independent χ^2 variates with n_1 & n_2 degrees of freedom, respectively, then show that:

(i) $\chi^2 = \chi_1^2 + \chi_2^2$ is a χ^2-variate with $(n_1 + n_2)$ degrees of freedom.

(ii) $U = \dfrac{\chi_1^2}{\chi_2^2}$ is a $\beta_2\left(\dfrac{n_1}{2}, \dfrac{n_2}{2}\right)$ variate.

7. If X_i $(i = 1, 2, \ldots, n)$ are n independent normal variates with zero means and unit variances, show that $\sum_{i=1}^{n} X_i$ and $\sum_{i=1}^{n} (X_i - \bar{X})^2$ are independently distributed. Hence, or otherwise obtain the distribution of

$$U = \sum_{i=1}^{n} X_i \Big/ \sqrt{\sum_{i=1}^{n} (X_i - \bar{X})^2}.$$

8. Prove that in sampling from a normal population ns^2/σ^2 is distributed like χ^2 with $(n - 1)$ *d.f.*, where s^2 and σ^2 are the variances of sample (of size n) and the population, respectively.

9. Let X/σ_1^2 and Y/σ_2^2 be two independent chi-square variates with n and m degrees of freedom, respectively. Find an unbiased estimate of $(\sigma_1/\sigma_2)^2$ and find its variance.
Show that X/Y and $(X/\sigma_1^2) + (Y/\sigma_2^2)$ are independently distributed and identify their distributions.

10. (a) Describe the chi-square test for testing a hypothesis that a normal population has a specified variance σ^2.

(b) Give the approximation to the test statistic in (a) if n, the sample size, is sufficiently large.

11. (a) Define chi-square and obtain its sampling distribution. Mention some prominent features of its frequency curve. Obtain the mean and the variance of the chi-square distribution with n degrees of freedom.

(b) Show that the sum of two independent variates having chi-square distributions, has a chi-square distribution.

12. (a) Write a short note on the chi-square test of goodness of fit of a random sample to a hypothetical distribution.

(b) Describe the chi-square test of significance and state various uses to which it can be put.

(c) Discuss the χ^2-test of goodness of fit of a theoretical distribution to an observed frequency distribution. How are the degrees of freedom ascertained when some parameters of the theoretical distribution have to be estimated from the data?

13. (*a*) What is contingency table? Describe how the χ^2 distribution may be used to test whether the two criteria of classification in an $m \times n$ contingency table are independent.

(*b*) State the hypothesis you test using the chi-square statistic in a contingency table.

(*c*) Describe the χ^2-test for independence of attributes, stating clearly the conditions for its validity. Give a rule for calculating the number of degrees of freedom to be assigned to χ^2. Illustrate your answer with an $m \times n$ contingency table, explaining the null hypothesis that is being tested.

14. Of '*A*' candidates taking a certain paper, '*a*' are successful; of '*B*' taking another paper, '*b*' are successful. Show how the significance of the difference between the ratios a/A, b/B may be tested (*i*) by a χ^2 test on contingency table and (*ii*) by comparing the difference with its standard error assessed by means of a binomial distribution. (You may assume all frequencies are sufficiently large.) Prove algebraically that the value of χ^2 is the square of the ratio of $(a/A) - (b/B)$ to its standard error.

15. (*a*) Show that for the entries in the following $2 \times r$ contingency table,

	A_1	A_2	...	A_i	...	A_r	Total
B_1	a_1	a_2	...	a_i	...	a_r	a
B_2	b_1	b_2	...	b_i	...	b_r	b
Total	n_1	n_2	...	n_i	...	n_r	n

the value of χ^2 is: $\chi^2 = \sum\limits_{i=1}^{r} \omega_i\,(p_i - p)^2$, where $p_i = \dfrac{a_i}{n_i},\ p = \dfrac{a}{n},\ \omega_i = \dfrac{n_i}{pq}$ and $q_i = 1 - p_i,\ q = \dfrac{b}{n}$.

(*b*) Given χ^2 contingency table representing two independent samples:

					Total
Sample I	μ_1	μ_2	...	μ_r	m
Sample II	ν_1	ν_2	...	ν_r	n
Total	$\mu_1 + \nu_1$	$\mu_2 + \nu_2$	...	$\mu_r + \nu_r$	$m + n$

Show that

$$\chi^2 = \frac{1}{\omega(1-\omega)}\left[\sum_{i=1}^{r} \mu_i\,\omega_i - m\,\omega\right], \text{ where } \omega_i = \frac{\mu_i}{\mu_i + \nu_i} \text{ and } \omega = \frac{m}{m+n},$$

can be used to test whether the samples are drawn from the same population. Clearly state the underlying assumptions, and give the number of degrees of freedom.

(*c*) In a 2×3 contingency table if $N = x + y + z$, $N' = x' + y' + z'$ and $N = N'$, show that

$$\chi^2 = \frac{(x-x')^2}{x+x'} + \frac{(y-y')^2}{y+y'} + \frac{(z-z')^2}{z+z'}$$

16. (*a*) Show that for a 2×2 table,

a	b
c	d

the value of χ^2, after applying Yate's correction for continuity is: $\dfrac{N}{D}\left(ad - bc - \dfrac{N}{2}\right)^2$ or $\dfrac{N}{D}\left(ad - bc + \dfrac{N}{2}\right)^2$ according as ad – bc > 0 or < 0 respectively, where $D = (a + b)\,(a + c)\,(b + d)\,(c + d)$.

(*b*) What is Yates' correction? Show that for a 2×2 contingency table in Part (*a*), the value of χ^2 after applying this correction is: $\chi^2 = \dfrac{N\left[|ad - bc| - N/2\right]^2}{(a+b)\,(a+c)\,(b+d)\,(c+d)}$.

17. Consider the following 2 × 2 table of observed frequencies based on random samples (with replacement) of sizes n_1 and n_2 from two populations:

	Population I	*Population II*	*Total*
Class A	n_{11}	n_{12}	$n_{1.}$
Class B	n_{21}	n_{22}	$n_{2.}$
Total	$n_{.1}$	$n_{.2}$	n

(*i*) Define the χ^2-statistic to be used for test of homogeneity of the two populations.

(*ii*) Show that $$\chi^2 = \frac{n(n_{11}n_{22} - n_{12}n_{21})^2}{(n_{1\cdot}\,n_{2\cdot}\,n_{\cdot 1}\,n_{\cdot 2})}$$

18. Show that for $r \times s$ contingency table:

(*a*) Number of degrees of freedom is $(r - 1) \times (s - 1)$

(*b*) $\chi^2 = N(s - 1)$ or $\chi^2 = N(r - 1)$, whichever is less.

(*c*) $E(\chi^2) = N(r - 1)(s - 1)/(N - 1)$

(*d*) $\max(C) = [(s - 1)/s]^{1/2}$, $r = s$,

where C is the coefficient of contingency and N is the total frequency.

19. Let X and Y denote the number of successes and failures respectively in n independent Bernoulli trials with p as the probability of success in each trial. Show that:

$$\frac{(X - np)^2}{np} + \frac{[Y - n(1-p)]^2}{n(1-p)}$$

can be approximated by a chi-square distribution with one degree of freedom when n is large.

20. (*a*) Define χ^2-statistic. What are its uses? What is Fisher's *z*-transformation for correlation coefficient and what are its properties? How is it used to combine the correlation coefficients between two random variables computed independently from different sources.

(*b*) Explain the use of the chi-square statistic for testing the homogeneity of several independent estimates of population correlation coefficient, clearly stating the underlying assumptions.

ASSORTED REVIEW PROBLEMS FOR SELF-ASSESSMENT

15·1. (*a*) If $X \sim \chi^2_{(n)}$ obtain the *m.g.f.* of X. Hence find the *m.g.f.* of standard chi-square variate and obtain its limiting form as $n \to \infty$. Also interpret the result.

(*b*) The *p.d.f.* of a *r. v.* X is given by:

$$f(x) = \frac{1}{4} x e^{-x/2}, \quad 0 < x < \infty$$
$$= 0 \quad , \text{ elsewhere}$$

Find mean, variance and *m.g.f.*

Hint. $X \sim \chi^2_{n=4}$. Mean = n = 4; Variance = $2n$ = 8; and $M_X(t) = (1 - 2t)^{-n/2} = (1 - 2t)^{-2}$; $t < 1/2$.

15·2. Let $X \sim N(0, 1)$ and $Y = X^2$. Calculate $E(Y)$ in two different ways.

Hint. $E(Y) = 1$. Use normal distribution and chi-square distribution.

15·3. Let X_1 and X_2 be independent standard normal variates and let $Y = (X_2 - X_1)^2/2$. Find the distribution of Y.

Ans. $Y \sim \chi^2(1)$.

15.4. If $X_1, X_2, ..., X_n$ are *i.i.d.* exponential variates with parameter λ, prove that

$$2\lambda \sum_{i=1}^{n} X_i \sim X^2_{(2n)}$$

15·5. (*a*) If $X \sim U[0, 1]$, show that: $\quad -2 \log X \sim \chi^2_{(2)}$

Hence, show that if $X_1, X_2, ..., X_n$, are *i.i.d.* $U[0, 1]$ variates, and if $P = X_1 X_2 ... X_n$ then

$$-2 \log_e P \sim \chi^2_{(2n)}$$

Hint. Find *m.g.f.* of $(-2 \log X)$.

(*b*) If $X_1, X_2, ..., X_n$ are independent random variables with continuous distribution functions $F_1, F_2, ..., F_n$ respectively, show that

$$-2\log\left[F_1(X_1)\cdot F_2(X_2)\ldots F_n(X_n)\right] \sim \chi^2_{(2n)}$$

Hint. Use $F(X) \sim U[0, 1]$ and Part (*a*).

15·6. X denotes the random variable with chi-square distribution having n degrees of freedom. Show that for suitably chosen constants a_n and b_n, the moment generating function of $(X - a_n)/b_n$ tends to that of the standard normal distribution as $n \to \infty$. From this what would you conclude about the behaviour, for large n, of: $\quad P\left(\dfrac{X - a_n}{b_n} \le x\right)$?

15·7. Show that $P\{\chi^2_{2v+2} \ge 2\lambda\} = \dfrac{1}{v!}\displaystyle\int_{\lambda}^{\infty} e^{-y} y^v dy = \sum_{r=0}^{v} \frac{e^{-\lambda}\lambda^r}{r!}$, where $\lambda = \dfrac{1}{2}\chi_0^2$.

Explain the uses of the result.

15·8. X is a Poisson variate with parameter λ and χ^2 is a chi-square variate with $2k$ *d.f.* Prove that for all positive integers k, $\quad P(X \le k - 1) = P(\chi^2 > 2\lambda)$.

Hint. $$P(\chi^2 > 2\lambda) = \frac{1}{2^k \Gamma(k)}\int_{2\lambda}^{\infty} \exp\left(-\frac{1}{2}\chi^2\right)(\chi^2)^{k-1}\,(d\chi^2)$$

$$= \frac{1}{(k-1)!}\int_{\lambda}^{\infty} e^{-y} y^{k-1} dy = \frac{1}{(k-1)!}\left\{\lambda^{k-1}e^{-\lambda} + (k-1)\int_{\lambda}^{\infty} e^{-y} y^{k-2}\, dy\right\}$$

By repeated integration, we get the required result.

15·9. Let $X_1, X_2, ..., X_m$ and $Y_1, Y_2, ..., Y_n$ be independent random samples from a normal population with mean zero and variance σ^2. Let their means be $\overline{X}$ and $\overline{Y}$ and their variances be S_X^2 and S_Y^2 respectively. Let the pooled variance S_p^2 be defined as: $S_p^2 = \dfrac{(m-1)\, S_X^2 + (n-1)\, S_Y^2}{(m+n-2)}$.

Prove that $(\overline{X} - \overline{Y})$ and $(m + n - 2)\; S_p^2/\sigma^2$ are independently distributed, the former as a normal variate with zero mean and variance $\sigma^2\left(\dfrac{1}{m} + \dfrac{1}{n}\right)$ and the latter as a chi-square variate with $(m + n - 2)$ *d.f.*

15·10. If X is a random variable following Poisson distribution with parameter λ, and λ is also a random variable so that $2\alpha\lambda$ is a chi-square variate with $2p$ degrees of freedom, obtain the unconditional distribution of X. Give the name of this distribution and find its mean.

Ans. $X \sim$ Negative Binomial $\left(p, \dfrac{\alpha}{1+\alpha}\right)$. Mean $= (p / \alpha)$

15·11. X_1, X_2, and X_3 denote independent chi-square variates with v_1, v_2 and v_3 *d.f.* respectively.

(*i*) Show that $X_1/(X_1 + X_2)$ is independently distributed of $(X_1 + X_2)/(X_1 + X_2 + X_3)$.

(*ii*) Obtain the joint density function of the distribution of:

$$X = X_1/(X_1 + X_2 + X_3) \quad \text{and} \quad Y = X_2/(X_1 + X_2 + X_3)$$

(*iii*) Hence, or otherwise obtain the mean and variance of X and Y and Cov (X, Y)

15·12. Prove that each linear constraint on $\{f_i\}$, $i = 1, 2, \ldots, n$ reduces by unity the number of degrees of freedom of the chi-square, $\sum_{i=1}^{n} \{(f_i - e_i)^2/e_i\}$, where $e_i = E(f_i)$.

15·13. If X follows a chi-square distribution with n *d.f.* show that $E(X) = n$ and $V(X) = 2n$. Prove that $(X - n)/\sqrt{2n}$ is a $N(0, 1)$, for large n.

15·14. If $y \cdot dx$ is the probability that X lies between x and $x + dx$ and y is given by the solution of the differential equation: $\dfrac{dy}{dx} = \dfrac{y(a-x)}{bx + c}$, show that (for suitable values of the constants a, b and c), a certain linear function of X has the χ^2-distribution with n degrees of freedom, where $n = 2\left(1 + \dfrac{a}{b} + \dfrac{c}{b^2}\right)$.

15·15. If X_1 and X_2 are independently distributed, each as χ^2-variate with 2 *d.f.*, show that the density function of $Y = \frac{1}{2}(X_1 - X_2)$ is: $g(y) = \frac{1}{2}\exp(-|y|), -\infty < y < \infty$.

15·16. If X and Y are independent *r.v.*'s having rectangular distribution in the interval (0, 1), show that: $U = \sqrt{-2 \log X} \cos 2\pi Y$ and $V = \sqrt{-2 \log X} \sin 2\pi Y$, are independently distributed as $N(0, 1)$. Hence, show that U^2 and V^2 are independently distributed as χ^2-variates, each with 1 *d.f.*

15·17. Find the p.d.f. of $\chi_n = +\sqrt{\chi_n^2}$, where χ_n^2 is a χ^2-variate with n *d.f.* and show that

$$\mu_r' = E(\chi_n^r) = 2^{r/2} \frac{\Gamma[(n+r)/2]}{\Gamma(n/2)}.$$ Hence, establish that for large n, $E(\chi_n^2) \simeq [E(\chi_n)]^2$.

15·18. Let $P_x = \dfrac{1}{2^{n/2}\Gamma(n/2)} \displaystyle\int_0^x w^{(n-2)/2} \cdot e^{-w/2}\, dw, x > 0$. Show that $x < \dfrac{n}{1 - P_x}$.

15·19. Let $X_1, X_2, \ldots, X_n$ be a random sample from $N(\mu, \sigma^2)$, and k be a positive integer. Find $E(S^{2k})$. In particular, find $E(S^2)$ and Var (S^2).
$$\left[S^2 = (n-1)^{-1} \sum_{i=1}^{n} (X_i - \bar{X})^2 \; ; \; \bar{X} = n^{-1} \sum_{i=1}^{n} X_i\right]$$

Ans. $E(S^{2k}) = \left(\dfrac{2\sigma^2}{n-1}\right)^k \cdot \dfrac{\Gamma[k + (n-1)/2]}{\Gamma\left(\frac{n-1}{2}\right)}$; $k > 0, n > 1$ and $E(S^2) = \sigma^2$, Var $(S^2) = 2\sigma^4/(n-1)$.

15·20. Let X_1 and X_2 be independent random variables, each $N(0, 1)$. Find the joint distribution of $Y_1 = X_1^2 + X_2^2$ and $Y_2 = X_1/X_2$. Find the marginal distributions of Y_1 and Y_2. Are Y_1 and Y_2 independent ?

15·21. Let X_1 and X_2 be independent standard normal variates.

Let $Y_1 = X_1 + X_2$ and $Y_2 = X_1^2 + X_2^2$.

(*i*) Show that the joint *m.g.f.* of Y_1 and Y_2 is:

$$M(t_1, t_2) = \frac{1}{1 - 2t_2} \exp\left[\frac{t_1^2}{1 - 2t_2}\right] - \infty < t_1 < \infty, -\infty < t_2 < \frac{1}{2}.$$

(*ii*) Hence or otherwise, show that Y_1 is a normal variate and Y_2 is a chi-square variate.

(*iii*) Are Y_1 and Y_2 independent? If not, find the correlation coefficient between Y_1 and Y_2.

Ans. (Y_1, Y_2) are not independent. $\rho(Y_1, Y_2) = 0$.

15·22. If $X_1, X_2, \ldots, X_n$ are independently and normally distributed with the same mean μ but different variances $\sigma_1^2, \sigma_2^2, \ldots, \sigma_n^2$, respectively, and assuming that

$$U = \frac{\sum_i (X_i/\sigma_i^2)}{\sum_i (1/\sigma_i^2)} \quad \text{and} \quad V = \sum_{i=1}^{n} \left[\frac{(X_i - U)^2}{\sigma_i^2}\right]$$

are independently distributed, show that $U \sim N\left\{\mu, 1/\sum_i \left(1/\sigma_i^2\right)\right\}$ and V has χ^2-distribution with $(n-1)$ *d.f.*

15·23. (*a*) If $X_1, X_2, \ldots, X_n$ is a random sample from $N(\mu, \sigma^2)$, find the mean and variance of

$$S = \left[\sum_{i=1}^{n} (X_i - \overline{X})^2/(n-1)\right]^{1/2}$$

(*b*) If $X \sim NB(r, p)$, then prove that

$$2pX \xrightarrow{L} Y, \text{ as } p \to 0, \text{ where } Y \sim \psi^2_{(2r)}$$

(*c*) Let the sequence of random variables $\{Z_n\}$ be such that Z_n follows chi-square distribution with n degrees of freedom. Define $W_n = Z_n/n^2$. Find the limiting distribution of W_n.

15·24. Let $X_1, X_2, \ldots, X_n$ be a random sample from $N(0, 1)$.

Define: $$\overline{X}_k = \frac{1}{k}\sum_{i=1}^{k} X_i \quad \text{and} \quad \overline{X}_{n-k} = \frac{1}{n-k}\sum_{i=k+1}^{n} X_i.$$

(*a*) What is the distribution of $\frac{1}{2}\left(\overline{X}_k + \overline{X}_{n-k}\right)$?

(*b*) What is the distribution of $k\overline{X}_k^2 + (n-k)\,\overline{X}_{n-k}^2$?

(*c*) What is the distribution of $X_i^2 / X_j^2, i \neq j$?

(*d*) What is the distribution of X_i / X_j, $i \neq j$?

Ans. (*a*) $N\left(0, \dfrac{n}{4k(n-k)}\right)$, (*b*) $\psi^2_{(2)}$ (*c*) $\beta_2\left(\dfrac{1}{2}, \dfrac{1}{2}\right)$ (*d*) Standard Cauchy Distribution.

15·25. (*a*) A sample of 15 values shows that the s.d. is 6·4. Is this compatible with the hypothesis that the sample is from a normal population with *s.d.* 5 ?

Ans. $H_0: \sigma = 5$, $\chi^2 = 24·58$; Significant.

(*b*) Test the hypothesis that $\sigma = 8$, given that $s = 10$, for a random sample of size 51 from a normal population.

Ans. $Z = \sqrt{2\chi^2} - \sqrt{2n-1} = \sqrt{2 \times 79·69} - \sqrt{101} = 2·57$; Significant.

(*c*) A manufacturer claims that the life time of a certain brand of batteries produced by his factory has a variance of 5,000 (hours)^2. A sample of size 26 has a variance of 7,200 (hours)^2. Assuming that it is reasonable to treat these data as a random sample from a normal population, test the manufacturer's claim at the $\alpha = 0·02$ level.

(*d*) A manufacturer recorded the cut-off bias (Volt) of a sample of 10 tubes as follows:

12·1, 12·3, 11·8, 12·0, 12·4, 12·0, 12·1, 11·9, 12·2, 12·2

The variability of cut-off bias for tubes of a standard type as measured by the standard deviation is 0·208 volts. Is the variability of the new tube with respect to cut-off bias less than that of the standard type?

15·26. (*a*) The following table gives the number of aircraft accidents that occurred during the seven days of the week. Find whether the accidents are uniformly distributed over the week.

Days	Mon.	Tues.	Wed.	Thurs.	Fri.	Sat.	Total
No. of accidents	14	18	12	11	15	14	84

Ans. $\chi^2 = 2{\cdot}143$

(*b*) A die is thrown 60 times with the following results.

Face	1	2	3	4	5	6
Frequency	8	7	12	8	14	11

Test at 5% level of significance if the die is unbiased, assuming that $P(\chi^2 > 11{\cdot}1) = 0{\cdot}05$ with 5 *d.f.*

(*c*) 200 digits are chosen at random from a set of tables. The frequencies of the digits were:

Digits	0	1	2	3	4	5	6	7	8	9
Frequency	18	19	23	21	16	25	22	20	21	15

Use χ^2 test to assess the correctness of hypothesis that the digits were distributed in equal numbers in the table, given that the values of χ^2 are respectively 16·9, 18·3 and 19·7 for 9, 10 and 11 degrees of freedom at 5% level of significance.

Ans. $\chi^2 = 4{\cdot}3$

15·27. (*a*) Among 64 offsprings of a certain cross between pigs, 34 were red, 10 were black and 20 were white. According to genetic model these numbers should be in the ratio 9 : 3 : 4. Are the data consistent with the model at 5 per cent level ?

[You are given that the value of χ^2 with the probability 0·05 being exceeded is 5·99 for 2 *d.f.* and 3·84 for 1 *d.f.*]

(*b*) In an experiment on pea-breeding, Mendal obtained the following frequencies of seeds: 315 round and yellow, 101 wrinkled and yellow; 108 round and green, 32 wrinkled and green. Total 556. Theory predicts that the frequencies should be in the proportion 9 : 3 : 3 : 1 respectively. Set up proper hypothesis and test it at 10% level of significance.

Ans. $\chi^2 = 0{\cdot}51$

(*c*) Selfed progenies of a cross between pure strains of plant segregated as follows:

		Early flowering	*Late flowering*
Tall	:	120	48
Short	:	36	13

Do the results agree with the theoretical frequencies which specify a 9 : 3 : 3 : 1 ratio?

Ans. $\chi^2 = 0{\cdot}0782$

(*d*) A genetical law says that children having one parent of blood group *M* and the other parent of blood group *N* will always be one of the three blood groups *M*, *MN*, *N*; and that the average number of children in these groups will be in the ratio 1 : 2 : 1. The report on an experiment states as follows: "Of 162 children having one *M* parent, and one *N* parent, 28·4% were found to be of group *M*, 42% of group *MN* and the rest of the group *N*." Do the data in the report conform to the expected genetic ratio 1 : 2 : 1?

(*e*) A chemical extraction plant processes sea water to collect sodium chloride and magnesium. It is known that sea water contains sodium chloride, magnesium and other elements in the ratio of 62 : 4 : 34. A sample of 200 tonnes of sea water has resulted in 130 tonnes of sodium chloride and 6 tonnes of magnesium. Are these data consistent with the known composition of sea water at 5% level?

Ans. $\chi^2 = 1{\cdot}025$

15.28. A bird watcher sitting in a park has spotted a number of birds belonging to 6 categories. The exact classification is given below:

Category	1	2	3	4	5	6
Frequency	6	7	13	17	6	5

Test at 5% level of significance whether or not the data is compatible with the assumption that this particular park is visited by birds belonging to these six categories in the proportion 1 : 1 : 2 : 3 : 1 : 1.
[Given $P(\chi^2 \geq 11{\cdot}07) = 0{\cdot}05$ for 5 degrees of freedom.]

15·29. (*a*) Five unbiased dice were thrown 96 times and the number of times 4, 5 or 6 was obtained is given below.

No. of dice showing 4, 5 or 6	5	4	3	2	1	0
Frequency	8	18	35	24	10	1

Fit a suitable distribution and test for the goodness of fit as far as you can proceed without the use of any tables and state how you would proceed further.

(*b*) In 120 throws of a single die, the following distribution of faces was obtained:

Face	1	2	3	4	5	6	Total
Frequency	30	25	18	10	22	15	120

Compute the statistic you would use to test whether the results constituted refutation of the "equal probability" (null hypothesis). Also state how you would proceed further.

(*c*) Given below is the number of male and female births in 1,000 families having five children:

Number of male births	*Number of female births*	*Number of families*
0	5	40
1	4	300
2	3	250
3	2	200
4	1	30

Test whether the given data is consistent with the hypothesis that the binomial law holds if the chance of a male birth is equal to that of a female birth.

(*d*)

Five-pig litters		*Six-pig litters*	
Number of males in litter	*Number of litters*	*Number of males in litter*	*Number of litters*
0	2	0	3
1	20	1	16
2	41	2	53
3	35	3	78
4	14	4	53
5	4	5	18
		6	0

Test whether each of the above two samples is a binomial sample: (*i*) with $p = 0{\cdot}5$, given a priori, and (*ii*) with p determined from the data. Test the significance of the difference between the two sample p's.

15·30. (*a*) The following table gives the count of yeast cells in square of a cyclometer. A square millimeter is divided into 400 equal squares and the number of these squares containing 0, 1, 2, ... cells are recorded:

Number of cells	0	1	2	3	4	5	6	7	8
Frequency	0	20	43	53	86	70	54	37	18
Number of cells	9	10	11	12	13	14	15	16	
Frequency	10	5	2	2	0	0	0	0	

Fit a Poisson distribution to the data and test the goodness of fit.

(*b*) The following is the distribution of the hourly number of trucks arriving at a company's warehouse:

Trucks arriving per hour	0	1	2	3	4	5	6	7	8
Frequency	52	151	130	102	45	12	5	1	2

Find the mean of the distribution and using this mean, (rounded to one decimal) as the parameter λ, fit a Poisson distribution. Test for goodness of fit at 5% level of significance.

15.31. Obtain the equation of the normal curve that may be fitted to the following data:

Class :	60–65	65–70	70–75	75–80	80–85	85–90	90–95	95–100
Frequency :	3	21	150	335	326	135	26	4

Obtain the expected normal frequencies and test the goodness of fit.

15.32. The following independent observations were made on the price of grain in 10 consecutive months:

Month	1	2	3	4	5	6	7	8	9	10
Price (in ₹)	115	118	120	140	135	137	139	142	144	150

Test the hypothesis that the expected price in the *i*th month is ₹ $(100 + 3i)$, $i = 1, 2, ..., 10$ with a standard deviation of ₹ 5 under the assumption that the prices are normally distributed.

15.33. To test a hypothesis H_0, an experiment is performed 3 times. The resulting values of chi-square are 2·37, 1·86 and 3·54, each of which corresponds to one degree of freedom. Show that while H_0 cannot be rejected at 5% level on the basis of any individual experiment, it can be rejected when the three experiments are collectively counted.

Hint. Use additive property of chi-square variates.

15·34. (*a*) In an experiment on immunization of cattle from tuberculosis, the following results were obtained:

	Affected	*Unaffected*
Inoculated	12	28
Not inoculated	13	7

Examine the effect of vaccine in controlling the incidence of the disease.

(*b*) The following data is collected on two characters:

	Cinegoers	*Non-cinegoers*
Literate	83	57
Illiterate	45	68

Based on this, can you conclude that there is no relation between the habit of cinema going and literacy?

(*c*) To find whether a certain vaccination prevents a certain disease or not, an experiment was conducted and the figures in various classes were obtained, as given in the table on page 15·66 top; *A* showing vaccination and *B* attacked by the disease.

	A	α	Total
B	69	10	79
β	91	30	121
Total	160	40	200

Using χ^2-test, analyse the results of the experiment for independence between *A* and *B*; examine whether Yate's correction modifies the conclusion or not. Test also the significance of the difference between the proportions of persons attacked by the disease among vaccinated and non-vaccinated which are 69/160 and 10/40.

15·35. (*a*) In an epidemic of certain disease 92 children contracted the disease. Of these, 41 received no treatment and of these 10 showed after-effects. Of the remainder who did receive treatment, 17 showed after-effects. Test the hypothesis that treatment was not effective.

(*b*) Can vaccination be regarded as a preventive measure of small-pox as evidenced by the following data?

"Of 1,482 persons exposed to small-pox in a locality, 368 in all were attacked. Of these 1,482 persons, 343 were vaccinated and of these, only 35 were attacked."

(*c*) In a certain sample of 2,000 families; 1,400 families are consumers of tea. Out of 1,800 Hindu families, 1,236 families consume tea. Use χ^2-test and state whether there is any significant difference between consumption of tea among Hindu and non-Hindu families.

Ans. $\chi^2 = 13{\cdot}89$

(*d*) Two groups *A* and *B*, of certain type of patients each consisting of 200 people are used to test the effectiveness of a new serum. Both groups are treated identically except that Group *A* is given the serum while Group *B* is not. It is found that 140 and 120 of Groups *A* and *B*, respectively, recover from the disease. Does this observed result give sufficient evidence for the conclusion that the new serum helps to cure the disease if we were willing to assume an α-risk of 0·01?

Ans. $\chi^2 = 4{\cdot}4$

(*e*) The following results were obtained when two sets of items were subjected to two different treatments *X* and *Y* to enhance their tensile strength:

Treatment *X* was applied on 400 items and 80 were found to have gained in strength. Treatment *Y* was applied on 400 items and 20 were found to have gained in strength. Is treatment *X* superior to treatment *Y*? Use the χ^2-test.

Ans. $\chi^2 = 41{\cdot}14$

15·36. (*a*) 1,072 college students were classified according to their intelligence and economic conditions. Test whether there is any association between intelligence and economic conditions.

		Intelligence			
		Excellent	*Good*	*Mediocre*	*Dull*
Economic Conditions	Good	48	199	181	82
	Not good	81	185	190	106

(*b*) Below is given the distribution of hair colours for either sex in a university:

Hair colour	(1) *Fair*	(2) *Red*	(3) *Medium*	(4) *Dark*	(5) *Jet black*	*Total*
Boys	592	119	849	504	36	2,100
Girls	544	97	677	451	14	1,783
Total	1,136	216	1,526	955	50	3,883

Test the homogeneity of hair colour for either sex. If the result is significant at 5 per cent level, explain the reason why it should be so.

15·37. (*a*) Bricks made in two kilns have been graded as facings (high quality), seconds (medium quality) and commons (rather poor quality).

The production of bricks in a particular period was as follows:

	Facings '000s	*Seconds* '000s	*Commons* '000s	*Total* '000s
Kiln A	24	43	13	80
Kiln B	31	57	32	120

It is the policy of the business to produce bricks of the highest quality and the number of seconds and commons produced is due to the inefficiency of the process. Use the χ^2-test to determine whether the manager of kiln *A* is justified in claiming that he produces bricks of a higher quality than his colleague at kiln *B*.

(*b*) The following data are for a sample of 300 car owners who were classified with respect to age and the number of accidents they had during the past two years. Test whether there is any relationship between these two variables.

		Accidents		
		0	1 or 2	3 or more
	≤ 21	8	23	14
Age	22 – 26	21	42	12
	≥ 27	71	90	19

(*c*) For the data in the following table, test for independence between a person's ability in Mathematics and interest in Economics:

		Ability in Mathematics		
		Low	*Average*	*High*
Interest	Low	63	42	15
in	Average	58	61	31
Economics	High	14	47	29

State clearly the assumptions underlying your test procedure.

15·38. (*a*) The following table gives for a sample of married women, the level of education and marriage adjustment score:

		Marriage adjustment score			
		Very low	*Low*	*High*	*Very high*
Level of	College	24	97	62	58
Education	High School	22	28	30	41
	Middle School	32	10	11	20

Can you conclude from the above, 'the higher the level of education, the greater is the degree of adjustment in marriage'?

(*b*) Test the hypothesis that there is no difference in the quality of the four kinds of tyres *A*, *B*, *C* and *D* based on the data given below. Use 5% level of significance.

	Tyre Brand			
	A	*B*	*C*	*D*
Failed to last 40,000 kms.	26	23	15	32
Lasted from 40,000 kms. to 60,000 km.s	118	93	116	121
Lasted more than 60,000 kms.	56	84	69	47

(c) The result of a survey regarding radio listeners' preference for different types of music are given in the following table, with listeners classified by age group. Is preference of type of music influenced by age?

	Age group		
Type of music preferred	19–25	26–35	Above 36/
National music	80	60	9
Foreign music	210	325	44
Indifferent	16	45	132

15·39. (a) Explain the chi-square test for testing the equality of several population proportions.

(b) The following data gives the HDL-levels in random samples of sizes 120, 200, 150 and 130 from the adult population of four cities *A, B, C* and *D* respectively.

City →	A	B	C	D
High HDL	53	80	68	57
Not High HDL	67	120	82	73

Test the equality of proportion of adults with high HDL cholesterol in these four cities at $\alpha = 0{\cdot}025$.

Ans. $H_0 : P_A = P_B = P_C = P_D$; $H_1 = P_A, P_B, P_C$ and P_D are not all equal.

(c) From the adult male population of four large cities *P, Q, R* and *S*, random samples of sizes 169, 221, 208 and 182 were taken and the number of married and single men recorded. Can we say at 5% level of significance, that the proportion of married men is same in all the four cities.

City →	P	Q	R	S	*Total*
Married	137	164	152	147	600
Single	32	57	56	35	180
Total	169	221	208	182	780

Ans. H_0: $P_1 = P_2 = P_3 = P_4$; H_1: All P_i's are not equal; $\chi^2 = 5{\cdot}801 < \chi^2_{(3)}\ (0{\cdot}05) = 7{\cdot}815$; Not significant.

15·40. (a) The correlation coefficients between wing length and tongue length were estimated from 2 independent samples each of size 44 to be 0·731 and 0·690. Test whether the correlation coefficients are significantly different or not. If not, obtain the best estimate of the common correlation coefficient.

(b) Test for equality of the correlation coefficients between the scores in two halves of a psychological test applied to different groups of sizes 30, 20 and 25 if the corresponding sample values are 0·63, 0·48, 0·71, respectively.

15·41. (a) Independent samples of 21, 30, 39, 26 and 35 pairs of values yielded correlation coefficients 0·39, 0·61, 0·43, 0·54 amd 0·48, respectively. Can these estimates be regarded as homogeneous? If so, find an estimate of the correlation coefficient in the population.

(b) Test whether the following set of correlation coefficients between stature and sitting heights obtained for persons from 8 districts can be regarded as homogeneous:

Sample size :	130	60	338	78	125	299	170	139
Correlation coefficient :	0·718	0·961	0·825	0·685	0·700	0·548	0·793	0·687

(c) The correlation coefficients between fibre weight and staple length in six cotton crosses were estimated as: – 0·129, 0·1138, – 0·2780, 0·0033, 0·2331 and 0·0550

based on samples of sizes 73, 81, 67, 83, 71, 57 respectively. Test the homogeneity of r_i's and obtain their best estimate.

CHAPTER

16 Exact Sampling Distributions-II

(t, F and z Distributions)

LEARNING OBJECTIVES

Upon completion of this Chapter, you should be able to:

- Understand and derive the sampling distribution of mean for small samples ($n \leq 30$) and know about Student's t-statistic and its sampling distribution.
- Define Fisher's t, F and Z statistics and derive their probability distributions.
- Obtain the various constants of t, F and Z distributions and discuss their important properties.
- Understand and appreciate various applications of t and F distributions in Statistics.
- Understand and derive the relationship between t, F and chi-square distributions.
- Define the non-central 't' and non-central 'F' and derive their probability distributions.
- Derive the sampling distribution of correlation coefficient in a random sample from an uncorrelated bivariate normal population.
- Define Fisher's Z-transformation for correlation coefficient and discuss its various applications in Statistics.

CHAPTER OUTLINES

16·1. INTRODUCTION

The entire large sample theory was based on the application of *"Normal Test"* (*c.f.* § 14·4). However, if the sample size n is small,the distribution of the various statistics, *e.g.*, $Z = \dfrac{\bar{x} - \mu}{\sigma / \sqrt{n}}$ or $Z = (X - nP)/\sqrt{nPQ}$ etc., are far from normality and as such *'normal test'* cannot be applied if n is small. In such cases exact sample tests, pioneered by W.S. Gosset (1908) who wrote under the pen name of Student, and later on developed and extended by Prof. R.A. Fisher (1926), are used. In the following sections we shall discuss : *(i) t-test, (ii) F-test,* and *(iii) Fisher's Z-transformation.*

The exact sample tests can, however, be applied to large samples also though the converse is not true. In all the exact sample tests, the basic assumption is that *"the population(s) from which sample(s) is (are) drawn is (are) normal, i.e., the parent population(s) is (are) normally distributed."*

16·2. STUDENT'S '*t*' DISTRIBUTION

Let x_i $(i = 1, 2, \ldots, n)$ be a random sample of size n from a normal population with mean μ and variance σ^2. Then Student's t statistic is defined as :

$$t = \frac{\bar{x} - \mu}{S/\sqrt{n}} \qquad \ldots (16·1)$$

where $\bar{x} = \dfrac{1}{n}\sum_{i=1}^{n} x_i$, is the sample mean and $S^2 = \dfrac{1}{n-1}\sum_{i=1}^{n}(x_i - \bar{x})^2$, ... (16·1*a*)

is an unbiased estimate of the population variance σ^2, and it follows Student's t-distribution with $\nu = (n - 1)$ *d.f.* with probability density function :

$$f(t) = \frac{1}{\sqrt{\nu}\cdot B\left(\frac{1}{2}, \frac{\nu}{2}\right)} \cdot \frac{1}{\left(1 + \frac{t^2}{\nu}\right)^{(\nu+1)/2}}\ ; -\infty < t < \infty \qquad \ldots (16·2)$$

Remarks:

1. A statistic t following Student's t-distribution with n *d.f.* will be abbreviated as $t \sim t_n$.
2. If we take $\nu = 1$ in (16·2), we get

$$f(t) = \frac{1}{B\left(\frac{1}{2}, \frac{1}{2}\right)} \cdot \frac{1}{(1 + t^2)} = \frac{1}{\pi} \cdot \frac{1}{(1 + t^2)}; \ -\infty < t < \infty$$

which is the *p.d.f.* of standard Cauchy distribution. Hence, when $\nu = 1$ *i.e.*, for 1 *d.f.* Student's t distribution reduces to Cauchy distribution.

16·2·1. Derivation of Student's *t*-distribution

The expression (16·1) can be re-written as :

$$t^2 = \frac{n(\bar{x} - \mu)^2}{S^2} = \frac{n(\bar{x} - \mu)^2}{ns^2/(n-1)} \Rightarrow \frac{t^2}{(n-1)} = \frac{(\bar{x} - \mu)^2}{\sigma^2/n} \cdot \frac{1}{ns^2/\sigma^2} = \frac{(\bar{x} - \mu)^2/(\sigma^2/n)}{ns^2/\sigma^2}$$

Since x_i $(i = 1, 2, \ldots, n)$ is a random sample from the normal population with mean μ and variance σ^2,

$$\bar{x} \sim N(\mu, \sigma^2/n) \Rightarrow \frac{(\bar{x} - \mu)}{\sigma/\sqrt{n}} \sim N(0,1)$$

Hence, $\dfrac{(\bar{x} - \mu)^2}{\sigma^2/n}$, being the square of a standard normal variate is a chi-square variate with 1 *d.f.*

Also, $\dfrac{ns^2}{\sigma^2}$ is a χ^2-variate with $(n - 1)$ *d.f.* (*c.f. Theorem 15·5*).

Further since $\bar{x}$ and s^2 are independently distributed (*c.f. Theorem* 15·5), $t^2/(n-1)$ being the ratio of two independent χ^2-variates with 1 and $(n-1)$ *d.f.* respectively, is a $\beta_2\left(\frac{1}{2}, \frac{n-1}{2}\right)$ variate and its distribution is given by :

$$dF(t) = \frac{1}{B\left(\frac{1}{2}, \frac{\nu}{2}\right)} \cdot \frac{(t^2/\nu)^{\frac{1}{2}-1}}{\left(1+\frac{t^2}{\nu}\right)^{(\nu+1)/2}} \cdot d(t^2/\nu),\ 0 \le t^2 < \infty \quad [\text{where } \nu = (n-1)]$$

$$= \frac{1}{\sqrt{\nu}\,B\left(\frac{1}{2}, \frac{\nu}{2}\right)} \cdot \frac{1}{\left(1+\frac{t^2}{\nu}\right)^{(\nu+1)/2}} dt\ ;\ -\infty < t < \infty$$

the factor 2 disappearing since the integral from $-\infty$ to ∞ must be unity. This is the required probability density function as given in (16·2) of Student's *t*-distribution with $\nu = (n-1)$ *d.f.*

Remarks on Student's '*t*'.

1. ***Importance of Student's t-distribution in Statistics.*** W.S. Gosset, who wrote under pseudonym (pen-name) of Student defined his t in a slightly different way, *viz.*, $t = (\bar{x} - \mu)/s$ and investigated its sampling distribution, somewhat empirically, in a paper entitled '*The Probable Error of the Mean*', published in 1908. Prof. R.A. Fisher, later on defined his own '*t*' and gave a rigorous proof for its sampling distribution in 1926. The salient feature of '*t*' is that both the statistic and its sampling distribution are functionally independent of σ, the population standard deviation.

   The discovery of '*t*' is regarded as a landmark in the history of statistical inference. Before Student gave his '*t*', it was customary to replace σ^2 in $Z = \dfrac{\bar{x}-\mu}{\sigma/\sqrt{n}}$, by its unbiased estimate S^2 to give $t = \dfrac{\bar{x}-\mu}{S/\sqrt{n}}$ and then normal test was applied even for small samples. It has been found that although the distribution of t is asymptotically normal for large n (*c.f.* § 16·2·5), it is far from normality for small samples. The Student's t ushered in an era of exact sample distributions (and tests) and since its discovery many important contributions have been made towards the development and extension of small (exact) sample theory.

2. ***Confidence or Fiducial Limits for μ.*** If $t_{0\cdot05}$ is the tabulated value of t for $\upsilon = (n-1)$ *d.f.* at 5% level of significance, *i.e.*, $P(|t| > t_{0\cdot05}) = 0\cdot05 \Rightarrow P(|t| \le t_{0\cdot05}) = 0\cdot95$, $\Rightarrow$ the 95% confidence limits for μ are given by :

$$|t| \le t_{0\cdot05},\ \textit{i.e.},\ \left|\frac{\bar{x}-\mu}{S/\sqrt{n}}\right| \le t_{0\cdot05} \Rightarrow \bar{x} - t_{0\cdot05}\frac{S}{\sqrt{n}} \le \mu \le \bar{x} + t_{0\cdot05}\frac{S}{\sqrt{n}}$$

Thus, 95% confidence limits for μ are : $\bar{x} \pm t_{0\cdot05} \cdot \left(S/\sqrt{n}\right)$ ...[16·2*a*]

Similarly, 99% confidence limits for μ are : $\bar{x} \pm t_{0\cdot01} \cdot \left(S/\sqrt{n}\right)$ ...[16·2*b*]

where $t_{0\cdot01}$ is the tabulated value of t for $\nu = (n-1)$ *d.f.* at 1% level of significance.

16·2·2. Fisher's '*t*' (*Definition*)

It is the ratio of a standard normal variate to the square root of an independent chi-square variate divided by its degrees of freedom. If ξ is a $N(0, 1)$ and χ^2 is an independent chi-square variate with n *d.f.*, then Fisher's t is given by :

$$t = \frac{\xi}{\sqrt{\chi^2/n}} \qquad \text{... (16·3)}$$

and it follows Student's '*t*' distribution with n degrees of freedom.

16·2·3. Distribution of Fisher's '*t*'

Since ξ and χ^2 are independent, their joint probability differential is given by :

$$dF\left(\xi,\chi^2\right)=\frac{1}{\sqrt{2\pi}}\exp\left(-\xi^2/2\right)\cdot\frac{\exp\left(-\chi^2/2\right)\left(\chi^2\right)^{\frac{n}{2}-1}}{2^{n/2}\Gamma(n/2)}d\xi\, d\chi^2$$

Let us transform to new variates *t* and *u* by the substitution :

$$t=\frac{\xi}{\sqrt{\chi^2/n}}\quad\text{and}\quad u=\chi^2\quad\Rightarrow\quad \xi=t\sqrt{u/n}\ \text{and}\ \chi^2=u$$

Jacobian of transformation *J* is given by :

$$J=\frac{\partial\left(\xi,\chi^2\right)}{\partial(t,u)}=\begin{vmatrix}\sqrt{u/n} & t/\left(2\sqrt{un}\right)\\ 0 & 1\end{vmatrix}=\sqrt{\frac{u}{n}}$$

The joint *p.d.f* $g(t, u)$ of *t* and *u* becomes :

$$g(t,u)=\frac{1}{\sqrt{2\pi}\,2^{n/2}\Gamma(n/2)\sqrt{n}}\exp\left\{-\frac{u}{2}\left(1+\frac{t^2}{n}\right)\right\}u^{\frac{n}{2}-\frac{1}{2}}$$

Since, $\chi^2\geq 0$ and $-\infty<\xi<\infty, u\geq 0$ and $-\infty<t<\infty$.

Integrating *w.r.t.* '*u*' over the range 0 to ∞, the marginal *p.d.f.* $g_1(.)$ of *t* becomes :

$$g_1(t)=\frac{1}{\sqrt{2\pi}\,2^{n/2}\Gamma(n/2)\sqrt{n}}\left[\int_0^\infty \exp\left\{-\frac{u}{2}\left(1+\frac{t^2}{n}\right)\right\}u^{(n-1)/2}du\right]$$

$$=\frac{1}{\sqrt{2\pi}\,2^{n/2}\Gamma(n/2)\sqrt{n}}\cdot\frac{\Gamma[(n+1)/2]}{\left[\frac{1}{2}\left(1+\frac{t^2}{n}\right)\right]^{(n+1)/2}}$$

$$=\frac{\Gamma\,[(n+1)/2]}{\sqrt{n}\,\Gamma(n/2)\,\Gamma(1/2)}\cdot\frac{1}{\left(1+\frac{t^2}{n}\right)^{(n+1)/2}},\ -\infty<t<\infty$$

$$=\frac{1}{\sqrt{n}\,B\left(\frac{1}{2},\frac{n}{2}\right)\left(1+\frac{t^2}{n}\right)^{(n+1)/2}},\ -\infty<t<\infty$$

which is the probability density function of Student's *t*-distribution with *n d.f.*

Remarks:

1. In Fisher's '*t*' the *d.f.* is the same as the *d.f.* of chi-square variate.
2. Student's '*t*' may be regarded as a particular case of Fisher's '*t*' as explained below.

Since $\bar{x}\sim N(\mu,\sigma^2/n)$, $\xi=\dfrac{\bar{x}-\mu}{\sigma/\sqrt{n}}\sim N(0,1)$ and $\chi^2=\dfrac{ns^2}{\sigma^2}=\sum\limits_{i=1}^{n}\left[(x_i-\bar{x})^2/\sigma^2\right]$ (**)

is independently distributed as chi-square variate with $(n-1)$ *d.f.* Hence Fisher's *t* is given by

$$t=\frac{\xi}{\sqrt{\chi^2/(n-1)}}=\frac{\sqrt{n}(\bar{x}-\mu)}{\sigma}\cdot\frac{\sigma}{\sqrt{\Sigma(x_i-\bar{x})^2/(n-1)}}$$

$$=\frac{\sqrt{n}(\bar{x}-\mu)}{S}=\frac{\bar{x}-\mu}{S/\sqrt{n}}\qquad\ldots(***)$$

and it follows Student's *t*-distribution with $(n-1)$ *d.f.* (*c.f.* Remark 1 above.)

Now, (***) is same as Student's '*t*' defined in (16·1). Hence, Student's '*t*' is a particular case of Fisher's '*t*'.

16·2·4. Constants of *t*-distribution

Since $f(t)$ is symmetrical about the line $t = 0$, all the moments of odd order about origin vanish, *i.e.*,

$$\mu'_{2r+1} \text{ (about origin)} = 0\ ;\ r = 0, 1, 2, \ldots$$

In particular, μ_1' (about origin) = 0 = Mean

Hence, central moments coincide with moments about origin.

$$\therefore \qquad \mu_{2r+1} = 0, \ (r = 0, 1, 2, \ldots) \qquad \ldots (16·4)$$

The moments of even order are given by :

$$\mu_{2r} = \mu'_{2r} \text{(about origin)} = \int_{-\infty}^{\infty} t^{2r} f(t)\,dt = 2\int_0^{\infty} t^{2r} f(t)\,dt$$

$$= 2 \cdot \frac{1}{B\left(\frac{1}{2}, \frac{n}{2}\right)\sqrt{n}} \int_0^{\infty} \frac{t^{2r}}{\left(1 + \frac{t^2}{n}\right)^{(n+1)/2}}\,dt$$

This integral is absolutely convergent if $2r < n$.

Put $\qquad 1 + \dfrac{t^2}{n} = \dfrac{1}{y} \quad \Rightarrow \quad t^2 = \dfrac{n(1-y)}{y} \quad \Rightarrow \quad 2t\,dt = -\dfrac{n}{y^2}\,dy$

When $t = 0$, $y = 1$ and when $t = \infty$, $y = 0$. Therefore,

$$\mu_{2r} = \frac{2}{\sqrt{n}B\left(\frac{1}{2}, \frac{n}{2}\right)} \int_1^0 \frac{t^{2r}}{(1/y)^{(n+1)/2}} \cdot \frac{-n}{2ty^2}\,dy$$

$$= \frac{n}{\sqrt{n}B\left(\frac{1}{2}, \frac{n}{2}\right)} \int_0^1 \left(t^2\right)^{(2r-1)/2} y^{\{(n+1)/2\}-2}\,dy$$

$$= \frac{\sqrt{n}}{B\left(\frac{1}{2}, \frac{n}{2}\right)} \int_0^1 \left[n\left(\frac{1-y}{y}\right)\right]^{r-\frac{1}{2}} y^{[(n+1)/2]-2}\,dy$$

$$= \frac{n^r}{B\left(\frac{1}{2}, \frac{n}{2}\right)} \int_0^1 y^{\frac{n}{2}-r-1} (1-y)^{r-\frac{1}{2}}\,dy = \cdot \frac{n^r}{B\left(\frac{1}{2}, \frac{n}{2}\right)} \cdot B\left(\frac{n}{2} - r, r + \frac{1}{2}\right), \ n > 2r.$$

$$= n^r \cdot \frac{\Gamma[(n/2) - r]\,\Gamma\left(r + \frac{1}{2}\right)}{\Gamma(1/2)\,\Gamma(n/2)} \qquad \ldots (16·4a)$$

$$= n^r \cdot \frac{\left(r - \frac{1}{2}\right)\left(r - \frac{3}{2}\right) \cdots \frac{3}{2} \cdot \frac{1}{2}\,\Gamma\left(\frac{1}{2}\right)\Gamma\left(\frac{n}{2} - r\right)}{\Gamma(1/2)\ [(n/2) - 1]\ [(n/2) - 2]\ \ldots\ [(n/2) - r]\Gamma[(n/2) - r]}$$

$$= n^r \frac{(2r-1)(2r-3)\ldots3.1}{(n-2)(n-4)\ldots(n-2r)}, \ \frac{n}{2} > r \qquad \ldots (16·4b)$$

In particular,

$$\mu_2 = n \cdot \frac{1}{(n-2)} = \frac{n}{n-2}, \ (n > 2) \qquad \ldots (16·4c)$$

and

$$\mu_4 = n^2 \frac{3.1}{(n-2)(n-4)} = \frac{3n^2}{(n-2)(n-4)}, \ (n > 4) \qquad \ldots (16·4d)$$

Hence, $\beta_1 = \dfrac{\mu_3^2}{\mu_2^3} = 0$ and $\beta_2 = \dfrac{\mu_4}{\mu_2^2} = 3\left(\dfrac{n-2}{n-4}\right)$; $(n > 4)$.

Remarks:

1. As $n \to \infty$, $\beta_1 = 0$ and $\beta_2 = \lim\limits_{n\to\infty} 3\left(\dfrac{n-2}{n-4}\right) = 3 \lim\limits_{n\to\infty}\left[\dfrac{1-(2/n)}{1-(4/n)}\right] = 3$ ... (16·4e)

2. Changing r to $(r-1)$ in [16·4(b)], dividing and simplifying, we shall get the recurrence relation for the moments as $\dfrac{\mu_{2r}}{\mu_{2r-2}} = \dfrac{n(2r-1)}{(n-2r)}$, $\dfrac{n}{2} > r$ ... (16·4f)

3. ***Moment Generating Function of t-distribution.*** From [16·4(b)] we observe that if $t \sim t_n$, then all the moments of order $2r < n$ exist but the moments of order $2r \geq n$ do not exist. Hence, the *m.g.f.* of t-distribution does not exist.

Example 16·1. *Express the constants y_0, a and m of the distribution :*

$$dF(x) = y_0\left(1-\frac{x^2}{a^2}\right)^m dx, -a \le x \le a \qquad \dots (*)$$

in terms of its μ_2 and β_2.

Show that if x is related to a variable t by the equation: $\quad x = \dfrac{at}{\{2(m+1)+t^2\}^{1/2}}$, ... (**)

then t has Student's distribution with $2(m+1)$ degrees of freedom. Use the transformation to calculate the probability that $t \geq 2$ when the degrees of freedom are 2 and also when 4.

Solution. First of all, we shall determine the constant y_0 from the consideration that total probability is unity.

$$\therefore \quad y_0\int_{-a}^{a}\left(1-\frac{x^2}{a^2}\right)^m dx = 1 \;\Rightarrow\; 2y_0\int_0^a\left(1-\frac{x^2}{a^2}\right)^m dx = 1$$

($\because$ Integrand is an even function of x.)

$$\Rightarrow \quad 2y_0\int_0^{\pi/2}\cos^{2m}\theta \cdot a\cos\theta\, d\theta = 1, \quad (x = a\sin\theta)$$

$$\Rightarrow \quad 2ay_0\int_0^{\pi/2}\cos^{2m+1}\theta\, d\theta = 1$$

But we have the Beta integral, $2\displaystyle\int_0^{\pi/2}\sin^p\theta\cos^q\theta\, d\theta = B\left(\frac{p+1}{2}, \frac{q+1}{2}\right)$ ... (1)

$$\therefore \quad ay_0\cdot 2\int_0^{\pi/2}\cos^{2m+1}\theta\sin^0\theta\, d\theta = 1 \;\Rightarrow\; ay_0B\left(m+1, \frac{1}{2}\right) = 1 \qquad \text{[Using (1)]}$$

$$\Rightarrow \quad y_0 = \frac{1}{aB\left(m+1, \frac{1}{2}\right)} \qquad \dots (2)$$

Since the given probability function is symmetrical about the line $x = 0$, we have as in § 16·2·4, $\mu_{2r+1} = \mu'_{2r+1} = 0$; $r = 0, 1, 2, \dots$ [$\because$ Mean = Origin]

The moments of even order are given by :

$$\mu_{2r} = \mu'_{2r}(\text{about origin}) = \int_{-a}^{a} x^{2r} f(x)dx = y\int_{-a}^{a} x^{2r}\left(1-\frac{x^2}{a^2}\right)^m dx$$

$$\therefore \quad \mu_{2r} = 2y_0 \int_0^a x^{2r}\left(1-\frac{x^2}{a^2}\right)^m dx = 2y_0 \int_0^{\pi/2} (a\sin\theta)^{2r}\cos^{2m}\theta \cdot a\cos\theta\, d\theta, \quad (x = a\sin\theta)$$

$$= y_0 a^{2r+1}\cdot 2\int_0^{\pi/2} \sin^{2r}\theta\cdot\cos^{2m+1}\theta d\theta$$

$$= y_0 a^{2r+1} B\left(r+\frac{1}{2}, m+1\right) \qquad \text{[Using (1)]}$$

$$= a^{2r}\frac{B\left(r+\frac{1}{2}, m+1\right)}{B\left(m+1, \frac{1}{2}\right)} = a^{2r}\cdot\frac{\Gamma\left(r+\frac{1}{2}\right)\Gamma\left(m+\frac{3}{2}\right)}{\Gamma\left(m+r+\frac{3}{2}\right)\Gamma\left(\frac{1}{2}\right)} \qquad \ldots(***)$$

In particular, $$\mu_2 = a^2\cdot\frac{\Gamma\{m+(3/2)\}\cdot\frac{1}{2}\Gamma(1/2)}{\{m+(3/2)\}\Gamma\{m+(3/2)\}\Gamma(1/2)} = \frac{a^2}{2m+3}$$

$$\Rightarrow \quad a^2 = (2m+3)\mu_2 \qquad \ldots (3)$$

Also, $$\mu_4 = a^4\,\frac{\Gamma(5/2)}{\Gamma\{m+(7/2)\}}\times\frac{\Gamma\{m+(3/2)\}}{\Gamma(1/2)} = \frac{3a^4}{(2m+5)(2m+3)}$$

(On simplification)

$$\therefore \quad \beta_2 = \frac{\mu_4}{\mu_2^2} = \frac{3(2m+3)}{(2m+5)} \Rightarrow m = \frac{9-5\beta_2}{2(\beta_2-3)} \quad \text{(On simplification)} \ldots (4)$$

Equations (2), (3) and (4) express the constants y_0, a and m in terms of μ_2 and β_2.

$$x = \frac{at}{\left[2(m+1)+t^2\right]^{1/2}} \Rightarrow \frac{x^2}{a^2} = \frac{t^2}{2(m+1)+t^2}$$

i.e., $$1-\frac{x^2}{a^2} = \frac{2(m+1)}{2(m+1)+t^2} = \left(1+\frac{t^2}{n}\right)^{-1}, \quad (n = 2m+2)$$

Also, $$dx = a\left[\frac{dt}{\left(n+t^2\right)^{1/2}} - t\cdot\frac{1}{2}\frac{2t\,dt}{\left(n+t^2\right)^{3/2}}\right] = a\frac{1}{\left(n+t^2\right)^{1/2}}\left(1-\frac{t^2}{n+t^2}\right)dt$$

$$= \frac{an}{\left(n+t^2\right)^{3/2}}dt = \frac{a}{\sqrt{n}}\cdot\frac{1}{\left[1+\left(t^2/n\right)\right]^{3/2}}dt$$

Hence, the *p.d.f.* of X transforms to

$$dF(t) = y_0\frac{1}{\left(1+\frac{t^2}{n}\right)^m}\cdot\frac{a}{\sqrt{n}}\cdot\frac{dt}{\left(1+\frac{t^2}{n}\right)^{3/2}}$$

$$= \frac{1}{aB\left(m+1, \frac{1}{2}\right)}\cdot\frac{a}{\sqrt{n}}\cdot\frac{dt}{\left(1+\frac{t^2}{n}\right)^{m+(3/2)}}$$

$$= \frac{1}{\sqrt{n}B\left(\frac{n}{2}, \frac{1}{2}\right)}\cdot\frac{dt}{\left(1+\frac{t^2}{n}\right)^{(n+1)/2}}, -\infty < t < \infty \qquad \ldots (5)$$

which is the probability differential of Student's t-distribution with $n = 2(m+1)$ d.f. Hence the result.

For 2 d.f., i.e., $n = 2$, we get $2(m+1) = 2 \Rightarrow m = 0$. Hence from (**), we get (for $m = 0$),

$$x = \frac{at}{\left(2+t^2\right)^{1/2}} \quad \Rightarrow \quad x = \frac{\sqrt{2}}{\sqrt{3}}a, \text{ when } t = 2.$$

$$\therefore \qquad P(t \ge 2) = P\left[X \ge \sqrt{(2/3)}\, a\right] = \int_{a\sqrt{(2/3)}}^{a} dF(x) = \int_{a\sqrt{(2/3)}}^{a} \frac{1}{aB\left(1, \frac{1}{2}\right)} dx$$

[From (*), since $m = 0$]

$$= \frac{1}{2a}\left(a - \frac{\sqrt{2}}{\sqrt{3}}a\right) = \frac{\sqrt{3}-\sqrt{2}}{2\sqrt{3}} \qquad \left[\because B\left(1, \frac{1}{2}\right) = \frac{\Gamma 1 \Gamma(1/2)}{\Gamma(3/2)} = \frac{\Gamma(1/2)}{(1/2)\Gamma(1/2)} = 2\right]$$

For 4 d.f., i.e., $n = 2$, we get $m = 1$. Proceeding exactly similarly we shall obtain

$$P(t \ge 2) = \frac{1}{2} - \frac{5\sqrt{2}}{16}.$$

Example 16·2. *Show that for t-distribution with n d.f., mean deviation about mean is given by :*

$$\sqrt{n}\,\Gamma[(n-1)/2]/\sqrt{\pi}\,\Gamma(n/2)$$

Solution. $E(t) = 0$.

$$\text{M.D. (about mean)} = \int_{-\infty}^{\infty} |t|\, f(t)dt = \frac{1}{\sqrt{n}\, B\left(\frac{1}{2}, \frac{n}{2}\right)} \int_{-\infty}^{\infty} \frac{|t|\, dt}{\left(1 + \frac{t^2}{n}\right)^{(n+1)/2}}$$

$$= \frac{2}{\sqrt{n}\, B\left(\frac{1}{2}, \frac{n}{2}\right)} \int_0^{\infty} \frac{tdt}{\left(1 + \frac{t^2}{n}\right)^{(n+1)/2}} = \frac{\sqrt{n}}{B\left(\frac{1}{2}, \frac{n}{2}\right)} \int_0^{\infty} \frac{dy}{(1+y)^{(n+1)/2}}, \qquad \left(\frac{t^2}{n} = y\right)$$

$$= \frac{\sqrt{n}}{B\left(\frac{1}{2}, \frac{n}{2}\right)} \int_0^{\infty} \frac{y^{1-1}}{(1+y)^{\frac{n-1}{2}+1}} dy = \frac{\sqrt{n}}{B\left(\frac{1}{2}, \frac{n}{2}\right)} \cdot B\left(\frac{n-1}{2}, 1\right) = \frac{\sqrt{n}\,\Gamma[(n-1)/2]}{\sqrt{\pi}\,\Gamma(n/2)}$$

Example 16·3· *If $I_x(p, q)$ represents the incomplete Beta function defined by:*

$$I_x(p,q) = \frac{1}{B(p,q)} \int_0^x t^{p-1}(1-t)^{q-1} dt \; ; \; p > 0, q > 0 \qquad \ldots (*)$$

show that the distribution function F(.) of Student's t-distribution with n d.f. is given by :

$$F(t) = 1 - \frac{1}{2} I_x\left(\frac{n}{2}, \frac{1}{2}\right), \quad \textit{where} \quad x = \left(1 + \frac{t^2}{n}\right)^{-1}$$

Solution. *If f(.) is p.d.f. of Student's t-distribution with n d.f., then*

$$F(t) = \int_{-\infty}^{t} f(u)du = 1 - \int_t^{\infty} f(u)du$$

$$= 1 - \frac{1}{\sqrt{n}B\left(\frac{1}{2}, \frac{n}{2}\right)} \int_t^{\infty} \left(1 + \frac{u^2}{n}\right)^{-(n+1)/2} du \qquad \ldots (**)$$

Put $$1+\frac{u^2}{n}=\frac{1}{z} \quad \Rightarrow \quad u=\sqrt{n}\left(\frac{1-z}{z}\right)^{1/2}$$

Also $$\frac{2u\,du}{n}=\frac{-dz}{z^2} \quad \Rightarrow \quad du=-\frac{ndz}{2uz^2}=-\frac{\sqrt{n}}{2}\left(\frac{z}{1-z}\right)^{1/2}\cdot\frac{dz}{z^2}$$

Substituting in (**), we get

$$F(t)=1-\frac{1}{\sqrt{n}\,B\left(\frac{1}{2},\frac{n}{2}\right)}\int_{\left(1+\frac{t^2}{n}\right)^{-1}}^{0} z^{(n+1)/2}\left\{-\frac{\sqrt{n}}{2}z^{-3/2}(1-z)^{-1/2}\right\}dz$$

$$=1+\frac{1}{2B\left(\frac{1}{2},\frac{n}{2}\right)}\int_{\left(1+\frac{t^2}{n}\right)^{-1}}^{0} z^{(n/2)-1}(1-z)^{-1/2}\,dz$$

$$=1-\frac{1}{2B\left(\frac{1}{2},\frac{n}{2}\right)}\int_0^x z^{(n/2)-1}(1-z)^{(1/2)-1}\,dz \quad ; \quad \left[\text{where } x=\left(1+\frac{t^2}{n}\right)^{-1}\right]$$

$$=1-\frac{1}{2}I_x\left(\frac{n}{2},\frac{1}{2}\right), \qquad \left[x=\left(1+\frac{t^2}{n}\right)^{-1}\right]$$

where, $I_x\ (p,\ q)$ is defined in (*).

Example 16·4. *If the random variables X_1 and X_2 are independent and follow chi-square distribution with n d.f., show that $\sqrt{n}(X_1-X_2)/2\sqrt{X_1X_2}$ is distributed as Student's t with n d.f., independently of (X_1+X_2).*

Solution. Since X_1 and X_2 are independent chi-square variates each with n d.f., their joint p.d.f. is given by:

$$p(x_1,x_2)=p_1(x_1)\times p_2(x_2)$$
$$=\frac{1}{2^n[\Gamma(n/2)]^2}\cdot e^{-(x_1+x_2)/2}\,x_1^{(n/2)-1}x_2^{(n/2)-1};\ 0\le x_1<\infty, 0\le x_2<\infty$$

Put $$u=\frac{\sqrt{n}(x_1-x_2)}{2\sqrt{x_1x_2}} \quad \text{and} \quad v=x_1+x_2$$

$$\Rightarrow \quad x_1=\frac{v}{2}\left[1+\frac{1}{\sqrt{\{1+(n/u^2)\}}}\right], \quad x_2=\frac{v}{2}\left[1-\frac{1}{\sqrt{\{1+(n/u^2)\}}}\right]$$

Jacobian of transformation is : $$J=\frac{\partial(x_1,x_2)}{\partial(u,v)}=\frac{v}{2\sqrt{n}\left(1+\frac{u^2}{n}\right)^{3/2}}$$

The joint *p.d.f.* of U and V becomes

$$g(u,\ v)=p(x_1,x_2)\,|\,J\,|=\frac{1}{2^{2n-1}\Gamma(n/2)\ \Gamma(n/2)\sqrt{n}}\times\frac{e^{-v/2}v^{n-1}}{\left(1+\frac{u^2}{n}\right)^{(n+1)/2}};\ -\infty<u<\infty,\ 0\le v<\infty$$

Using Legender's duplication formula, *viz.,*

$$\Gamma n=2^{n-1}\Gamma(n/2)\Gamma\left(\frac{n+1}{2}\right)\Big/\sqrt{\pi} \quad \Rightarrow \quad \Gamma(n/2)=\frac{\Gamma n\sqrt{\pi}}{2^{n-1}\Gamma\left(\frac{n+1}{2}\right)}, \quad \text{we get}$$

$$2^{2n-1}\Gamma(n/2)\Gamma(n/2)\sqrt{n}=\frac{2^{2n-1}\cdot\Gamma n\sqrt{\pi}}{2^{n-1}\Gamma\left(\frac{n+1}{2}\right)}\Gamma\left(\frac{n}{2}\right)\sqrt{n}=2n\,\Gamma n\sqrt{n}\,B\left(\frac{1}{2},\frac{n}{2}\right)\qquad\left[\because\sqrt{\pi}=\Gamma\left(\frac{1}{2}\right)\right].$$

$$\therefore\qquad g(u,v)=\left(\frac{1}{2^n\Gamma n}e^{-v/2}v^{n-1}\right)\left[\frac{1}{\sqrt{n}\,B\left(\frac{1}{2},\frac{n}{2}\right)}\cdot\frac{1}{\left(1+\frac{u^2}{n}\right)^{(n+1)/2}}\right];\quad 0<v<\infty,-\infty<u<\infty.$$

$$\Rightarrow\qquad g(u,v)=g_1(u)g_2(v),\qquad\ldots(i)$$

where
$$g_1(u)=\frac{1}{\sqrt{n}\,B\left(\frac{1}{2},\frac{n}{2}\right)}\cdot\frac{1}{\left(1+\frac{u^2}{n}\right)^{(n+1)/2}};-\infty<u<\infty\qquad\ldots(ii)$$

and
$$g_2(v)=\frac{1}{2^n\Gamma n}e^{-v/2}v^{n-1},\quad 0<v<\infty\qquad\ldots(iii)$$

(*i*) $\Rightarrow U=\sqrt{n}(X_1-X_2)/2\sqrt{X_1X_2}$ and $V=X_1+X_2$ are independently distributed.

(*ii*) $\Rightarrow U=\sqrt{n}(X_1-X_2)/2\sqrt{X_1X_2}\sim t_n$, and (*iii*) $\Rightarrow V=X_1-X_2\sim\gamma\left(a=\frac{1}{2},n\right)$

16·2·5. Limiting Form of *t*-distribution

As $n\to\infty$, the p.d.f. of t-distribution with n d.f. viz.,

$$f(t)=\frac{1}{\sqrt{n}B\left(\frac{1}{2},\frac{n}{2}\right)}\left(1+\frac{t^2}{n}\right)^{-(n+1)/2}\longrightarrow\frac{1}{\sqrt{2\pi}}\cdot\exp\left(-\frac{1}{2}t^2\right),-\infty<t<\infty$$

Proof.
$$\lim_{n\to\infty}\frac{1}{\sqrt{n}\,B\left(\frac{1}{2},\frac{n}{2}\right)}=\lim_{n\to\infty}\frac{1}{\sqrt{n}}\frac{\Gamma[(n+1)/2]}{\Gamma(1/2)\,\Gamma(n/2)}=\frac{1}{\sqrt{n}}\cdot\frac{1}{\sqrt{\pi}}\left(\frac{n}{2}\right)^{1/2}=\frac{1}{\sqrt{2\pi}}$$

$$\left[\because\Gamma(1/2)=\sqrt{\pi}\text{ and }\lim_{n\to\infty}\frac{\Gamma(n+k)}{\Gamma(n)}=n^k,\ (c.f.\text{ Remark to § 16·9, page 16·42})\right]$$

$$\therefore\qquad\lim_{n\to\infty}f(t)=\lim_{n\to\infty}\frac{1}{\sqrt{n}B\left(\frac{1}{2},\frac{n}{2}\right)}\cdot\lim_{n\to\infty}\left[\left(1+\frac{t^2}{n}\right)^n\right]^{-\frac{1}{2}}\times\lim_{n\to\infty}\left(1+\frac{t^2}{n}\right)^{-\frac{1}{2}}$$
$$=\frac{1}{\sqrt{2\pi}}\exp(-t^2/2),-\infty<t<\infty$$

Hence, for large d.f. *t*-distribution tends to standard normal distribution.

16·2·6. Graph of *t*-distribution

The *p.d.f.* of *t*-distribution with *n d.f.* is :

$$f(t)=C\cdot\left(1+\frac{t^2}{n}\right)^{-(n+1)/2},-\infty<t<\infty$$

Since $f(-t)=f(t)$, the probability curve is symmetrical about the line $t=0$. As t increases, $f(t)$ decreases rapidly and tends to zero as $t\to\infty$, so that t-axis is an asymptote to the curve.

We have shown that

$$\mu_2 = \frac{n}{n-2}, \ n > 2 \quad ; \quad \beta_2 = \frac{3(n-2)}{(n-4)}, \ n > 4$$

Hence, for $n > 2$, $\mu_2 > 1$ *i.e.*, the variance of *t*-distribution is greater than that of standard normal distribution and for $n > 4$, $\beta_2 > 3$ and thus *t*-distribution is more flat on the top than the normal curve. In fact, for small *n*, we have

$$P(|t| \geq t_0) \geq P(|Z| \geq t_0), \ Z \sim N(0,1)$$

i.e., the tails of the *t*-distribution have a greater probability (area) than the tails of standard normal distribution. Moreover, we have also seen [§ 16·2·5], that for large *n* (*d.f.*), *t*-distribution tends to standard normal distribution.

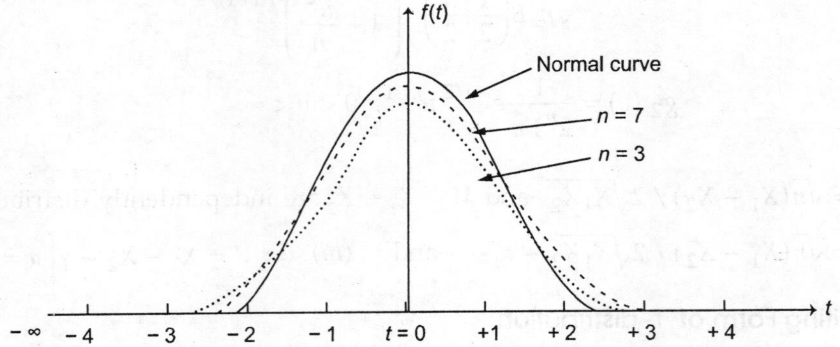

Fig. 16·1: *Graph of t-distribution*

16·2·7. Critical Values of *t*

The critical (or significant) values of *t* at level of significance α and *d.f.* *v* for *two-tailed test* are given by the equation :

$$P[|t| > t_v(\alpha)] = \alpha \qquad \ldots(16·5)$$

$$\Rightarrow \quad P[|t| \leq t_v(\alpha)] = 1 - \alpha \qquad \ldots(16·5a)$$

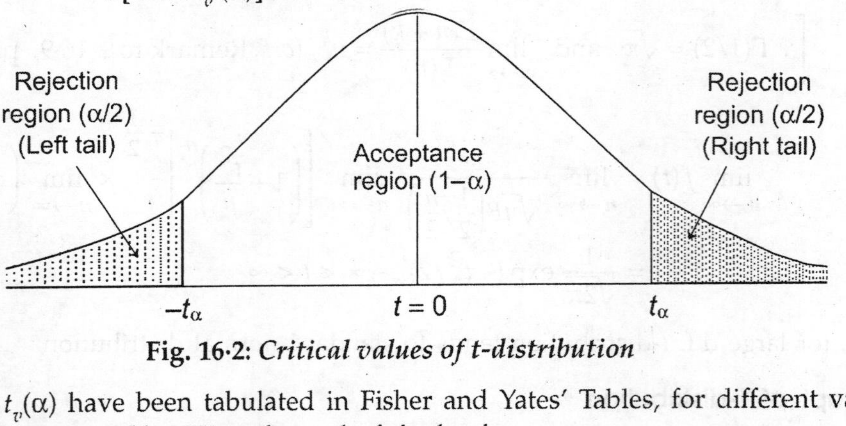

Fig. 16·2: *Critical values of t-distribution*

The values $t_v(\alpha)$ have been tabulated in Fisher and Yates' Tables, for different values of α and *v* and are given in *Table VIII* at the end of the book.

Since *t*-distribution is symmetric about $t = 0$, we get from (16·5)

$$P[t > t_v(\alpha)] + P[t < -t_v(\alpha)] = \alpha \quad \Rightarrow \quad 2P[t > t_v(\alpha)] = \alpha$$

$$\Rightarrow \quad P[t > t_v(\alpha)] = \alpha/2 \qquad \therefore \quad P[t > t_v(2\alpha)] = \alpha \qquad \ldots(16·5b)$$

$t_v(2\alpha)$ (from the Tables at the end of the book) gives the significant value of *t* for a single-tail test [Right-tail or Left-tail-since the distribution is symmetrical], at level of significance α and ν *d.f.*

Hence, the significant values of t at level of significance 'α' for a single-tailed test can be obtained from those of two-tailed test by looking the values at level of significance 2α, i.e., $t_\nu(\alpha)$ for single-tail test = (2) for two-tailed test.

For example,

t_8 (0·05) for single-tail test = t_8 (0·10) for two-tail test = 1·86

t_{15}(0·01) for single-tail test = t_{15} (0·02) for two-tail test = 2·60.

16·3· APPLICATIONS OF *t*-DISTRIBUTION

The *t*-distribution has a wide number of applications in Statistics, some of which are enumerated below.

(*i*) To test if the sample mean ($\overline{x}$) differs significantly from the hypothetical value μ of the population mean.

(*ii*) To test the significance of the difference between two sample means.

(*iii*) To test the significance of an observed sample correlation coefficient and sample regression coefficient.

(*iv*) To test the significance of observed partial correlation coefficient.

In the following sections we will discuss these applications in detail, one by one.

16·3·1. *t*-Test for Single Mean

Suppose we want to test :

(*i*) if a random sample x_i (i = 1, 2, ..., n) of size n has been drawn from a normal population with a specified mean, say μ_0, or

(*ii*) if the sample mean differs significantly from the hypothetical value μ_0 of the population mean.

Under the null hypothesis, H_0 :

(*i*) *The sample has been drawn from the population with mean μ_0 or*

(*ii*) *There is no significant difference between the sample mean $\overline{x}$ and the population mean μ_0,*

the statistic
$$t = \frac{\overline{x} - \mu_0}{S/\sqrt{n}}, \qquad \ldots (16·6)$$

where
$$\overline{x} = \frac{1}{n}\sum_{i=1}^{n} x_i \quad \text{and} \quad S^2 = \frac{1}{n-1}\sum_{i=1}^{n}(x_i - \overline{x})^2, \qquad \ldots (16·6a)$$

follows Student's *t*-distribution with $(n - 1)$ *d.f.*

We now compare the calculated value of t with the tabulated value at certain level of significance. If calculated $|t|$ > tabulated t, null hypothesis is rejected and if calculated $|t|$ < tabulated t, H_0 may be accepted at the level of significance adopted.

Remarks:

1. ***On computation of S^2 for numerical problems.*** If $\overline{x}$ comes out in integers, the formula (16·6*a*) can be conveniently used for computing S^2. However, if comes in fractions then the formula (16·6*a*) for computing S^2 is very cumbersome and is not recommended. In that case, step deviation method, given below, is quite useful.

If we take $d_i = x_i - A$, where A is any arbitrary number, then

$$S^2 = \frac{1}{n-1}\left[\Sigma(x_i - \overline{x})^2\right] = \frac{1}{n-1}\left[\Sigma x_i^2 - \frac{(\Sigma x_i)^2}{n}\right] \qquad \ldots (16·6b)$$

$$= \frac{1}{n-1}\left[\Sigma d_i^2 - \frac{(\Sigma d_i)^2}{n}\right], \text{ since variance is independent of change of origin} \qquad \ldots (16·6c)$$

Also, in this case $\overline{x} = A + \dfrac{\Sigma d_i}{n}$. ... (16·6*d*)

2. The sample variance : $s^2 = \dfrac{1}{n}\underset{i}{\Sigma}(x_i - \overline{x})^2 \Rightarrow ns^2 = (n-1)S^2 \Rightarrow \dfrac{S^2}{n} = \dfrac{s^2}{n-1}$ ... (16·6*e*)

Hence, for numerical problems, the test statistic (16·6) on using [16·6*e*] becomes

$$t = \frac{\overline{x} - \mu_0}{\sqrt{S^2/n}} = \frac{\overline{x} - \mu_0}{\sqrt{s^2/(n-1)}} \sim t_{(n-1)} \quad \text{... (16·6f)}$$

3. ***Assumptions for Student's t-test.*** The following assumptions are made in the Student's *t*-test :

(*i*) The parent population from which the sample is drawn is normal.

(*ii*) The sample observations are independent, *i.e.*, the sample is random.

(*iii*) The population standard deviation σ is unknown.

Example 16·5. *A machinist is making engine parts with axle diameters of 0·700 inch. A random sample of 10 parts shows a mean diameter of 0·742 inch with a standard deviation of 0·040 inch. Compute the statistic you would use to test whether the work is meeting the specifications. Also state how you would proceed further.*

Solution. Here we are given :

$\mu = 0{\cdot}700$ inch ; $\overline{x} = 0{\cdot}742$ inch, $s = 0{\cdot}040$ inch and $n = 10$

Null Hypothesis, H_0: $\mu = 0{\cdot}700$, *i.e., the product is conforming to specifications.*

Alternative Hypothesis, H_1: $\mu \neq 0{\cdot}700$ (Two-tailed)

Test Statistic. Under H_0, the test statistic is : $t = \dfrac{\overline{x} - \mu}{\sqrt{S^2/n}} = \dfrac{\overline{x} - \mu}{\sqrt{s^2/(n-1)}} \sim t_{(n-1)}$

$$\therefore \quad t = \frac{\sqrt{9}\,(0{\cdot}742 - 0{\cdot}700)}{0{\cdot}040} = 3{\cdot}15$$

How to proceed further. Here the test statistic '*t*' follows Student's *t*-distribution with $10 - 1 = 9$ *d.f.* We will now compare this calculated value with the tabulated value of *t* for 9 *d.f.* and at certain level of significance, say 5% for a two tailed test. Let this tabulated value be denoted by t_0.

(*i*) If calculated '*t*', *viz.*, $3{\cdot}15 > t_0$, we say that the value of *t* is significant. This implies that $\overline{x}$ differs significantly from μ and H_0 is rejected at this level of significance and we conclude that the product is not meeting the specifications.

(*ii*) If calculated $t < t_0$, we say that the value of *t* is not significant, *i.e.*, there is no significant difference between $\overline{x}$ and μ. In other words, the deviation $(\overline{x} - \mu)$ is just due to fluctuations of sampling and null hypothesis H_0 may be retained at 5% level of significance, *i.e.*, we may take the product conforming to specifications.

Example 16·6. *The mean weekly sales of soap bars in departmental stores was 146·3 bars per store. After an advertising campaign the mean weekly sales in 22 stores for a typical week increased to 153·7 and showed a standard deviation of 17·2. Was the advertising campaign successful ?*

Solution. We are given : $n = 22$, $\overline{x} = 153{\cdot}7$, $s = 17{\cdot}2$.

Null Hypothesis. The advertising campaign is not successful, *i.e.*, H_0: $\mu = 146{\cdot}3$

Alternative Hypothesis, H_1: $\mu > 146{\cdot}3$ (Right-tail).

Test Statistic. Under H_0, the test statistic is: $t = \dfrac{\overline{x} - \mu}{\sqrt{s^2/(n-1)}} \sim t_{22-1} = t_{21}$

$$\therefore \quad t = \frac{153{\cdot}7 - 146{\cdot}3}{\sqrt{(17{\cdot}2)^2/21}} = \frac{7{\cdot}4 \times \sqrt{21}}{17{\cdot}2} = 9{\cdot}03$$

Conclusion. Tabulated value of t for 21 *d.f.* at 5% level of significance for *single-tailed test* is 1·72. Since calculated value is much greater than the tabulated value, it is highly significant. Hence, we reject the null hypothesis and conclude that the advertising campaign was definitely successful in promoting sales.

Example 16·7. *A random sample of 10 boys had the following I.Q.'s:*

70, 120, 110, 101, 88, 83, 95, 98, 107, 100.

Do these data support the assumption of a population mean I.Q. of 100? Find a reasonable range in which most of the mean I.Q. values of samples of 10 boys lie.

Solution. *Null hypothesis,* H_0: The data are consistent with the assumption of a mean I.Q. of 100 in the population, *i.e.,* $\mu = 100$.

Alternative hypothesis, H_1: $\mu \neq 100$. (Two Tailed)

Test Statistic. Under H_0, the test statistic is : $t = \dfrac{(\bar{x} - \mu)}{\sqrt{S^2/n}} \sim t_{(n-1)}$,

where, $\bar{x}$ and S^2 are to be computed from the sample values of I.Q.'s.

TABLE 16·1: CALCULATIONS FOR SAMPLE MEAN AND S.D.

x	$(x - \bar{x})$	$(x - \bar{x})^2$
70	– 27·2	739·84
120	22·8	519·84
110	12·8	163·84
101	3·8	14·44
88	– 9·2	84·64
83	– 14·2	201·64
95	– 2·2	4·84
98	0·8	0·64
107	9·8	96·04
100	2·8	7·84
Total 972	0	1833·60

Here, $n = 10$

$$\bar{x} = \frac{972}{10} = 97{\cdot}2$$

$$S^2 = \frac{1833{\cdot}6}{9} = 203{\cdot}73$$

$$\therefore \quad |\,t\,| = \frac{|\,97{\cdot}2 - 100\,|}{\sqrt{203{\cdot}73/10}} = \frac{2{\cdot}8}{\sqrt{20{\cdot}37}} = \frac{2{\cdot}8}{4{\cdot}514} = 0{\cdot}62$$

Tabulated $t_{0{\cdot}05}$ for (10 – 1), *i.e.,* 9 *d.f.* for two-tailed test is 2·262.

Conclusion. Since calculated t is less than tabulated $t_{0{\cdot}05}$ for 9 *d.f.,* H_0 may be accepted at 5% level of significance and we may conclude that the data are consistent with the assumption of mean I.Q. of 100 in the population.

The 95% confidence limits within which the mean I.Q. values of samples of 10 boys will lie are given by:

$$\bar{x} \pm t_{0{\cdot}05} S/\sqrt{n} = 97{\cdot}2 \pm 2{\cdot}262 \times 4{\cdot}514 = 97{\cdot}2 \pm 10{\cdot}21 = 107{\cdot}41 \quad \text{and} \quad 86{\cdot}99$$

Hence, the required 95% confidence interval for mean I.Q. is [86·99, 107·41].

Remark. Aliter for computing $\bar{x}$ and S^2.

Here we see that $\bar{x}$ comes in fractions and as such the computation of $(x - \bar{x})^2$ is quite laborious and time consuming. In this case we use the method of step deviations to compute $\bar{x}$ and S^2.

x	70	120	110	101	88	83	95	98	107	100	Total
$d = x - 90$	–20	30	20	11	–2	–7	5	8	17	10	$\Sigma d = 72$
d^2	400	900	400	121	4	49	25	64	289	100	$\Sigma d^2 = 2,352$

Here, $d = x - A$, where $A = 90$.

$\therefore$ $$\bar{x} = A + \frac{1}{n}\Sigma d = 90 + \frac{72}{10} = 97{\cdot}2 \text{ and}$$

$$S^2 = \frac{1}{n-1}\left[\Sigma d^2 - \frac{(\Sigma d)^2}{n}\right] = \frac{1}{9}\left[2352 - \frac{(72)^2}{10}\right] = 203{\cdot}73$$

Example 16·8. *The heights of 10 plants of a given garden are found to be 70, 67, 62, 68, 61, 68, 70, 64, 64, 66 cms. Is it reasonable to believe that the average height is greater than 64 cms.? Test at 5% significance level assuming that for 9 degrees of freedom $P(t > 1{\cdot}83) = 0{\cdot}05$.*

Solution. *Null Hypothesis,* H_0: $\mu = 64$ cms; *Alternative Hypothesis,* H_1: $\mu > 64$ *cms. (Right Tailed)*

TABLE 16·2: CALCULATIONS FOR SAMPLE MEAN AND S.D.

x	70	67	62	68	61	68	70	64	64	66	Total 660
$x - \bar{x}$	4	1	– 4	2	– 5	2	4	– 2	– 2	0	0
$(x - \bar{x})^2$	16	1	16	4	25	4	16	4	4	0	90

$$\bar{x} = \frac{\Sigma x}{n} = \frac{660}{10} = 66; \quad S^2 = \frac{1}{n-1}\Sigma(x - \bar{x})^2 = \frac{90}{9} = 10$$

Test Statistic. Under H_0, the test statistic is : $$t = \frac{\bar{x} - \mu}{\sqrt{S^2/n}} = \frac{66 - 64}{\sqrt{10/10}} = 2,$$

which follows Student's *t*-distribution with $(10 - 1) = 9$ *d.f.*

Tabulated value of t for 9 *d.f.* at 5% level of significance for single (right) tail-test is 1·833. (This is the value $t_{0{\cdot}10}$ for 9 *d.f.* in the two-tailed tables given at the end of the book.)

Conclusion. Since calculated value of t is greater than the tabulated value, it is significant. Hence, H_0 is rejected at 5% level of significance and we conclude that the average height is greater than 64 cms.

Example 16·9. *A random sample of 16 values from a normal population showed a mean of 41·5 cms. and the sum of squares of deviations from this mean equal to 135 square cms. Show that the assumption of a mean of 43·5 cms. for the population is not reasonable. Obtain 95 per cent and 99 per cent fiducial limits for the same.*

You may use the following information from statistical tables :

$$v = 15, \begin{cases} P = 0{\cdot}05, t = 2{\cdot}131 \\ P = 0{\cdot}01, t = 2{\cdot}947 \end{cases}$$

Solution. We are given $n = 16$, $\bar{x} = 41{\cdot}5$ cms. and $\Sigma(x - \bar{x})^2 = 135$ sq. cms.

$\therefore$ $$S^2 = \frac{1}{n-1}\Sigma(x - \bar{x})^2 = \frac{135}{15} = 9 \quad \Rightarrow \quad S = 3$$

Null Hypothesis, H_0: $\mu = 43{\cdot}5$ cms., *i.e.*, the data are consistent with the assumption that the mean height in the population is 43·5 cms.

Alternative Hypothesis, H_1: $\mu \neq 43{\cdot}5$ cms. (Two Tailed)

Test Statistic. Under H_0, the test statistic is : $$t = \frac{\bar{x} - \mu}{S/\sqrt{n}} \sim t_{(n-1)}$$

$$\therefore \qquad |t| = \frac{|41{\cdot}5 - 43{\cdot}5|}{3/4} = \frac{8}{3} = 2{\cdot}667$$

Here, number of degrees of freedom is (16 – 1) = 15.

We are given: $t_{0{\cdot}05}$ for 15 *d.f.* = 2·131 and $t_{0{\cdot}01}$ for 15 *d.f.* = 2·947.

Conclusion. Since calculated $|t|$ is greater than 2·131, null hypothesis is rejected at 5% level of significance and we conclude that the assumption of mean of 43·5 cms. for the population is not reasonable.

Remark. Since calculated $|t|$ is less than 2·947, null hypothesis (μ = 43·5) may be accepted at 1% level of significance.

95% fiducial limits for μ : (*d.f.* = 15)

$$\bar{x} \pm t_{0{\cdot}05} \times \frac{S}{\sqrt{n}} = 41{\cdot}5 \pm 2{\cdot}131 \times \frac{3}{4} = 41{\cdot}5 \pm 1{\cdot}598 \quad \Rightarrow \quad 39{\cdot}902 < \mu < 43{\cdot}098$$

99% fiducial limits for μ: (*d.f.* = 15)

$$\bar{x} \pm t_{0{\cdot}01} \times \frac{S}{\sqrt{n}} = 41{\cdot}5 \pm 2{\cdot}947 \times \frac{3}{4} = 43{\cdot}71 \text{ and } 39{\cdot}29 \quad \Rightarrow \quad 39{\cdot}29 < \mu < 43{\cdot}71$$

16·3·2. *t*-Test for Difference of Means

Suppose we want to test if two independent samples x_i, (i = 1, 2, ..., n_1) and y_j, (j = 1, 2, ..., n_2) of sizes n_1 and n_2 have been drawn from two normal populations with means μ_X and μ_Y respectively.

Under the null hypothesis (H_0) *that the samples have been drawn from the normal populations with means* μ_X *and* μ_Y *and under the assumption that the population variance are equal, i.e.,* $\sigma_X^2 = \sigma_Y^2 = \sigma^2$ (say), the statistic

$$t = \frac{(\bar{x} - \bar{y}) - (\mu_X - \mu_Y)}{S\sqrt{\left(\dfrac{1}{n_1} + \dfrac{1}{n_2}\right)}} \qquad \ldots (16{\cdot}7)$$

where
$$\bar{x} = \frac{1}{n_1}\sum_{i=1}^{n_1} x_i, \qquad \bar{y} = \frac{1}{n_2}\sum_{j=1}^{n_2} y_j$$

and
$$S^2 = \frac{1}{n_1 + n_2 - 2}\left[\sum_i (x_i - \bar{x})^2 + \sum_j (y_j - \bar{y})^2\right] \qquad \ldots (16{\cdot}7a)$$

is an unbiased estimate of the common population variance σ^2, follows Student's *t*-distribution with $(n_1 + n_2 - 2)$ *d.f.*

Proof. *Distribution of t defined in (16·7).*

$$\xi = \frac{(\bar{x} - \bar{y}) - E(\bar{x} - \bar{y})}{\sqrt{V(\bar{x} - \bar{y})}} \sim N(0, 1)$$

But
$$E(\bar{x} - \bar{y}) = E(\bar{x}) - E(\bar{y}) = \mu_X - \mu_Y$$

$$V(\bar{x} - \bar{y}) = V(\bar{x}) + V(\bar{y}) = \frac{\sigma_X^2}{n_1} + \frac{\sigma_Y^2}{n_2} = \sigma^2\left(\frac{1}{n_1} + \frac{1}{n_2}\right) \qquad \text{(By assumption)}$$

[The covariance term vanishes since samples are independent.]

$$\therefore \qquad \xi = \frac{(\bar{x} - \bar{y}) - (\mu_X - \mu_Y)}{\sqrt{\sigma^2\left(\dfrac{1}{n_1} + \dfrac{1}{n_2}\right)}} \sim N(0, 1) \qquad \ldots (*)$$

Let,
$$\chi^2 = \frac{1}{\sigma^2}\left[\sum_{i=1}^{n_1}(x_i-\bar{x})^2 + \sum_{j=1}^{n_2}(y_j-\bar{y})^2\right]$$
$$= \left[\sum_i (x_i-\bar{x})^2/\sigma^2\right] + \left[\sum_j (y_j-\bar{y})^2/\sigma^2\right] = \frac{n_1 s_X^2}{\sigma^2} + \frac{n_2 s_Y^2}{\sigma^2} \quad \ldots (**)$$

Since $n_1 s_X^2/\sigma^2$ and $n_2 s_Y^2/\sigma^2$ are independent χ^2-variates with $(n_1 - 1)$ and $(n_2 - 1)$ *d.f.* respectively, by the additive property of chi-square distribution, χ^2 defined in (**) is a χ^2-variate with $(n_1 - 1) + (n_2 - 1)$, *i.e.*, $n_1 + n_2 - 2$ *d.f.* Further, since sample mean and sample variance are independently distributed, ξ and χ^2 are independent random variables. Hence, Fisher's t statistic is given by

$$t = \frac{\xi}{\sqrt{\dfrac{\chi^2}{n_1+n_2-2}}}$$
$$= \frac{(\bar{x}-\bar{y})-(\mu_X-\mu_Y)}{\sqrt{\sigma^2\left(\dfrac{1}{n_1}+\dfrac{1}{n_2}\right)}} \times \frac{1}{\left[\dfrac{1}{n_1+n_2-2}\left\{\sum_i(x_i-\bar{x})^2+\sum_j(y_j-\bar{y})^2\right\}/\sigma^2\right]^{1/2}}$$
$$= \frac{(\bar{x}-\bar{y})-(\mu_X-\mu_Y)}{S\sqrt{\left(\dfrac{1}{n_1}+\dfrac{1}{n_2}\right)}}, \quad \text{where } S^2 = \frac{1}{n_1+n_2-2}\left[\sum_i(x_i-\bar{x})^2+\sum_j(y_j-\bar{y})^2\right]$$

and it follows Student's t-distribution with (n_1+n_2-2) *d.f.* (*c.f. Remark 1, § 16·2·3*).

Remarks:

1. S^2, defined in (16·7*a*) is an unbiased estimate of the common population variance σ^2, since

$$E(S^2) = \frac{1}{n_1+n_2-2}E\left[\sum_i(x_i-\bar{x})^2+\sum_i(y_i-\bar{y})^2\right] = \frac{1}{n_1+n_2-2}E\left[(n_1-1)S_X^2+(n_2-1)S_Y^2\right]$$
$$= \frac{1}{n_1+n_2-2}\left[(n_1-1)E\left(S_X^2\right)+(n_2-1)E\left(S_Y^2\right)\right] = \frac{1}{n_1+n_2-2}\left[(n_1-1)\sigma^2+(n_2-1)\sigma^2\right] = \sigma^2$$

2. An important deduction which is of much practical utility is discussed below :
Suppose we want to test if :

(*a*) two independent samples x_i $(i = 1, 2, \ldots, n_1)$, and y_j $(j = 1, 2, \ldots, n_2)$, have been drawn from the populations with same means, or

(*b*) the two sample means $\bar{x}$ *and* $\bar{y}$ differ significantly or not.

Under the null hypothesis, H_0 that (*a*) *samples have been drawn from the populations with same means, i.e.,* $\mu_X = \mu_Y$, *or (b) the sample means* $\bar{x}$ *and* $\bar{y}$ *do not differ significantly,* the statistic :

$$t = \frac{\bar{x}-\bar{y}}{S\sqrt{\left(\dfrac{1}{n_1}+\dfrac{1}{n_2}\right)}} \qquad [\because\ \mu_X = \mu_Y, \text{ under } H_0] \qquad \ldots (16·8)$$

where symbols are defined in (16·7*a*), follows Student's t-distribution with $(n_1 + n_2 - 2)$ *d.f.*

3. *On the assumption of t-test for difference of means.* Here we make the following three fundamental assumptions:

(*i*) Parent populations, from which the samples have been drawn are normally distributed.

(*ii*) The population variances are equal and unknown, *i.e.*, $\sigma_X^2 = \sigma_Y^2 = \sigma^2$ (say), where σ^2 is unknown.

(*iii*) The two samples are random and independent of each other.

Thus, before applying t-test for testing the equality of means it is theoretically desirable to test the equality of population variances by applying F-test. (*c.f* § 16·7·1) If the variances do not come out to be equal then t-test becomes invalid and in that case Behren's 'd'-test based on fiducial intervals is used. For practical problems, however, the assumptions (i) and (ii) are taken for granted.

16·3·3. Paired t-test for Difference of Means.

Let us now consider the case when (i) the sample sizes are equal, *i.e.*, $n_1 = n_2 = n$ (say), and (ii) the two samples are not independent but the sample observations are paired together, *i.e.*, the pair of observations (x_i, y_i), $(i = 1, 2, \ldots, n)$ corresponds to the same (ith) sample unit. The problem is to test if the sample means differ significantly or not.

For example, suppose we want to test the efficacy of a particular drug, say, for inducing sleep. Let x_i and y_i $(i = 1, 2, \ldots, n)$ be the readings, in hours of sleep, on the ith individual, before and after the drug is given respectively. Here instead of applying the difference of the means test discussed in § 16·3·2, we apply the paired t-test given below.

Here we consider the increments, $d_i = x_i - y_i$, $(i = 1, 2, \ldots, n)$.

Under the null hypothesis, H_0 that *increments are due to fluctuations of sampling, i.e., the drug is not responsible for these increments*, the statistic:

$$t = \frac{\bar{d}}{S/\sqrt{n}} \quad \ldots (16·9)$$

where

$$\bar{d} = \frac{1}{n}\sum_{i=1}^{n} d_i \text{ and } S^2 = \frac{1}{n-1}\sum_{i=1}^{n} (d_i - \bar{d})^2, \quad \ldots (16·9a)$$

follows Student's t-distribution with $(n - 1)$ *d.f.*

Example 16·10. *Below are given the gain in weights (in kgs.) of pigs fed on two diets A and B.*

Gain in weight

Diet A: 25, 32, 30, 34, 24, 14, 32, 24, 30, 31, 35, 25

Diet B: 44, 34, 22, 10, 47, 31, 40, 30, 32, 35, 18, 21, 35, 29, 22

Test, if the two diets differ significantly as regards their effect on increase in weight.

Solution. *Null hypothesis,* H_0: $\mu_X = \mu_Y$, *i.e., there is no significant difference between the mean increase in weight due to diets A and B.*

Alternative hypothesis, H_1: $\mu_X \neq \mu_Y$ (two-tailed).

Diet A			Diet B		
x	$x - \bar{x}$	$(x - \bar{x})^2$	y	$y - \bar{y}$	$(y - \bar{y})^2$
25	−3	9	44	14	196
32	4	16	34	4	16
30	2	4	22	− 8	64
34	6	36	10	− 20	400
24	− 4	16	47	17	289
14	−14	196	31	1	1
32	4	16	40	10	100
24	− 4	16	30	0	0
30	2	4	32	2	4
31	3	9	35	5	25
35	7	49	18	−12	144
25	− 3	9	21	− 9	81
			35	5	25
			29	−1	1
			22	− 8	64
$\Sigma x = 336$	$\Sigma(x - \bar{x}) = 0$	$\Sigma(x - \bar{x})^2 = 380$	$\Sigma y = 450$	$\Sigma(y - \bar{y}) = 0$	$\Sigma(y - \bar{y})^2 = 1{,}410$

$\bar{x} = \dfrac{336}{12} = 28, \bar{y} = \dfrac{450}{15} = 30, S^2 \dfrac{1}{n_1 + n_2 - 2}\left[\Sigma(x - \bar{x})^2 + \Sigma(y - \bar{y})^2\right] = 71{\cdot}6$; and $n_1 = 12, n_2 = 15$

Under null hypothesis (H_0):

$$t = \frac{\bar{x} - \bar{y}}{\sqrt{S^2\left(\dfrac{1}{n_1} + \dfrac{1}{n_2}\right)}} : t_{n_1+n_2-2}$$

$$\therefore \quad t = \frac{28 - 30}{\sqrt{71{\cdot}6\left(\dfrac{1}{12} + \dfrac{1}{15}\right)}} = \frac{-2}{\sqrt{10{\cdot}74}} = -0{\cdot}609$$

Tabulated $t_{0{\cdot}05}$ for (12 + 15 – 2) = 25 *d.f.* for two tailed test is 2·06.

Conclusion. Since calculated $|t|$ is less than tabulated t, H_0 may be accepted at 5% level of significance and we may conclude that the two diets do not differ significantly as regards their effect on increase in weight.

Remark. Here $\bar{x}$ and $\bar{y}$ come out to be integral values and hence the direct method of computing $\Sigma(x - \bar{x})^2$ and $\Sigma(y - \bar{y})^2$ is used. In case $\bar{x}$ and (or) $\bar{y}$ comes out to be fractional, then the step deviation method is recommended for computation of $\Sigma(x - \bar{x})^2$ and $\Sigma(y - \bar{y})^2$.

Example 16·11. *Samples of two types of electric light bulbs were tested for length of life and following data were obtained :*

	Type I	*Type II*
Sample No.	$n_1 = 8$	$n_2 = 7$
Sample Means	$\bar{x}_1$ = *1,234 hrs.*	$\bar{x}_2$ = *1,036 hrs.*
Sample S.D.'s	s_1 = *36 hrs.*	s_2 = *40 hrs.*

Is the difference in the means sufficient to warrant that type I is superior to type II regarding length of life?

Solution. *Null Hypothesis,* H_0: $\mu_X = \mu_Y$, *i.e.,* the two types I and II of electric bulbs are indentical.

Alternative Hypothesis, H_1: $\mu_X > \mu_Y$, *i.e.,* type I is superior to type II.

Test Statistic. Under H_0, the test statistic is :

$$t = \frac{\bar{x}_1 - \bar{x}_2}{\sqrt{S^2\left(\dfrac{1}{n_1} + \dfrac{1}{n_2}\right)}} : t_{n_1+n_2-2} = t_{13}$$

where
$$S^2 = \frac{1}{n_1 + n_2 - 2}\left[\Sigma(x_1 - \bar{x}_1)^2 + \Sigma(x_2 - \bar{x}_2)^2\right]$$

$$= \frac{1}{n_1 + n_2 - 2}\left(n_1 s_1^2 + n_2 s_2^2\right) = \frac{1}{13}\left[8 \times (36)^2 + 7 \times (40)^2\right] = 1{,}659{\cdot}08$$

$$\therefore \quad t = \frac{1234 - 1036}{\sqrt{1659{\cdot}08\left(\dfrac{1}{8} + \dfrac{1}{7}\right)}} = \frac{198}{\sqrt{1659{\cdot}08 \times 0{\cdot}2679}} = 9{\cdot}39$$

Tabulated value of t for 13 *d.f.* at 5% level of significance for right (single)-tailed test is 1·77. [This is the value of $t_{0{\cdot}10}$ for 13 *d.f.* from two-tail tables given at the end of the chapter.]

Conclusion. Since calculated 't' is much greater than tabulated 't', it is highly significant and H_0 is rejected. Hence, the two types of electric bulbs differ significantly. Further, since $\bar{x}_1$ is much greater than $\bar{x}_2$, we conclude that type I is definitely superior to type II.

Example 16·12. *The heights of six randomly chosen sailors are (in inches): 63, 65, 68, 69, 71, and 72. Those of 10 randomly chosen soldiers are 61, 62, 65, 66, 69, 69, 70, 71, 72 and 73. Discuss, the light that these data throw on the suggestion that sailors are on the average taller than soldiers.*

Solution. If the heights of sailors and soldiers be represented by the variables X and Y respectively then the *Null Hypothesis* is, H_0: $\mu_X = \mu_Y$, *i.e., the sailors are not on the average taller than the soldiers.*

Alternative Hypothesis, H_1: $\mu_X > \mu_Y$ (Right-tailed).

Under H_0, the test statistic is : $t = \dfrac{\bar{x}-\bar{y}}{\sqrt{S^2\left(\dfrac{1}{n_1}+\dfrac{1}{n_2}\right)}} \sim t_{n_1+n_2-2} = t_{14}$

Sailors			*Soldiers*		
X	$d = X - A$ $= X - 68$	d^2	Y	$D = Y - B$ $= Y - 66$	D^2
63	− 5	25	61	− 5	25
65	− 3	9	62	− 4	16
68	0	0	65	− 1	1
69	1	1	66	0	0
71	3	9	69	3	9
72	4	16	69	3	9
			70	4	16
			71	5	25
Total	0	60	72	6	36
			73	7	49
			Total	18	186

$\therefore \quad \bar{x} = A + \dfrac{\Sigma d}{n_1} = 68 + 0 = 68 \qquad\qquad \bar{y} = B + \dfrac{\Sigma D}{n_2} = 66 + \dfrac{18}{10} = 67{\cdot}8$

and $\quad \Sigma(x-\bar{x})^2 = \Sigma d^2 - \dfrac{(\Sigma d)^2}{n_1} = 60 - 0 = 60 \qquad$ and $\quad \Sigma(y-\bar{y})^2 = \Sigma D^2 - \dfrac{(\Sigma D)^2}{n_2} = 186 - \dfrac{324}{10} = 153{\cdot}6$

$$S^2 = \frac{1}{n_1+n_2-2}\left[\Sigma(x-\bar{x})^2 + \Sigma(y-\bar{y})^2\right] = \frac{1}{14}(60 + 153{\cdot}6) = 15{\cdot}2571$$

$$\therefore \quad t = \frac{68 - 67{\cdot}8}{\sqrt{15{\cdot}2571\left(\dfrac{1}{6}+\dfrac{1}{10}\right)}} = \frac{0{\cdot}2}{\sqrt{15{\cdot}2571 \times 0{\cdot}2667}} = 0{\cdot}099$$

Tabulated $t_{0{\cdot}05}$ for 14 *d.f.* for single-tail test is 1·76.

Conclusion. Since calculated t is much less than 1·76, it is not at all significant at 5% level of significance. Hence, null hypothesis may be retained at 5% level of significance and we conclude that the data are inconsistent with the suggestion that the sailors are on the average taller than soldiers.

Example 16·13. *To test the claim that the resistance of electric wire can be reduced by at least 0·05 ohm by alloying, 25 values obtained for each alloyed wire and standard wire produced the following results:*

	Mean	*Standard deviation*
Alloyed wire	*0·083 ohm*	*0·003 ohm*
Standard wire	*0·136 ohm*	*0·002 ohm*

Test at 5% level whether or not the claim is substantiated.

Solution. *Null Hypothesis* H_0: $\mu_1 - \mu_2 \geq 0{\cdot}05$, [*i.e.*, the claim is substantiated]

Alternative Hypothesis H_1: $\mu_1 - \mu_2 < 0{\cdot}05$ (Left-tailed, test)

Test Statistic. Under H_0, the test statistic is :

$$t = \frac{(\overline{X}_1 - \overline{X}_2) - (\mu_1 - \mu_2)}{\sqrt{S^2\left(\dfrac{1}{n_1} + \dfrac{1}{n_2}\right)}} \sim N(0,1)$$

where $$S^2 = \frac{n_1 s_1^2 + n_2 s_2^2}{n_1 + n_2 - 2} = \frac{25 \times (0{\cdot}003)^2 + 25 \times (0{\cdot}002)^2}{25 + 25 - 2} = \frac{0{\cdot}000225 + 0{\cdot}0001}{48} = 0{\cdot}0000067$$

$$\therefore \quad t = \frac{(0{\cdot}083 - 0{\cdot}136) - 0{\cdot}05}{\sqrt{\left\{0{\cdot}0000067\left(\dfrac{1}{25} + \dfrac{1}{25}\right)\right\}}} = -\frac{0{\cdot}103}{0{\cdot}00071} = -145{\cdot}07$$

The (critical) tabulated value of t for 48 *d.f.*, at 5% level of significance for left-tailed test is $-1{\cdot}645$.

Conclusion. Since calculated value of t is much less than tabulated value of t, it falls in the rejection region. We, therefore, reject the null hypothesis and conclude that the claim is not substantiated.

Example 16·14. *A certain stimulus administered to each of the 12 patients resulted in the following increase of blood pressure:*

5, 2, 8, −1, 3, 0, −2, 1, 5, 0, 4 and 6

Can it be concluded that the stimulus will, in general, be accompanied by an increase in blood pressure ?

Solution. Here we are given the increments in blood pressure, *i.e.*, $d_i \; (= x_i - y_i)$.

Null Hypothesis, H_0: $\mu_X = \mu_Y$, *i.e.*, there is no significant difference in the blood pressure readings of the patients before and after the drug. In other words, the given increments are just by chance (fluctuations of sampling) and not due to the stimulus.

Alternative Hypothesis, H_1: $\mu_X < \mu_Y$, *i.e.*, the stimulus results in an increase in blood pressure.

Test Statistic. Under H_0, the test statistic is: $t = \dfrac{\overline{d}}{S/\sqrt{n}} \sim t_{(n-1)}$

d	5	2	8	−1	3	0	−2	1	5	0	4	6	31
d^2	25	4	64	1	9	0	4	1	25	0	16	36	185

$$\overline{d} = \frac{1}{n}\sum d = 2{\cdot}58 \quad \text{and} \quad S^2 = \frac{1}{n-1}\left[\sum d^2 - \frac{(\sum d)^2}{n}\right] = \frac{1}{11}\left[185 - \frac{(31)^2}{12}\right] = 9{\cdot}5382$$

$$\therefore \quad t = \frac{\overline{d}}{S/\sqrt{n}} = \frac{2{\cdot}58 \times \sqrt{12}}{\sqrt{9{\cdot}5382}} = \frac{2{\cdot}58 \times 3{\cdot}464}{3{\cdot}09} = 2{\cdot}89$$

Tabulated $t_{0{\cdot}05}$ for 11 *d.f.* for single-tail test is 1·80. [This is the value of $t_{0{\cdot}10}$ for 11 *d.f.* in the table for two-tail test given at the end of the book].

Conclusion. Since calculated $t > t_{0\cdot05}$, H_0 is rejected at 5% level of significance. Hence, we conclude that the stimulus will, in general, be accompanied by an increase in blood pressure.

Example 16·15. *In a certain experiment to compare two types of animal foods A and B, the following results of increase in weights were observed in animals.*

Animal number →		1	2	3	4	5	6	7	8	Total
Increase	*Food A*	49	53	51	52	47	50	52	53	407
weight in lb	*Food B*	52	55	52	53	50	54	54	53	423

(*i*) *Assuming that the two samples of animals are independent, can we conclude that food B is better than food A?*

(*ii*) *Also examine the case when the same set of eight animals were used in both the foods.*

Solution. *Null Hypothesis,* H_0: If the increase in weights due to foods A and B are denoted by X and Y respectively, then $H_0: \mu_X = \mu_Y$, *i.e.*, there is no significant difference in increase in weights due to diets A and B.

Alternative Hypothesis, $H_1: \mu_X < \mu_Y$ (Left-tailed).

(*i*) If the two samples of animals be assumed to be independent, then we will apply t-test for difference of means to test H_0.

Test Statistic. Under $H_0: \mu_X = \mu_Y$, the test criterion is :

$$t = \frac{\bar{x} - \bar{y}}{\sqrt{S^2\left(\frac{1}{n_1} + \frac{1}{n_2}\right)}} \sim t_{n_1+n_2-2}$$

	Food A			Food B	
X	$d = X - 50$	d^2	Y	$D = Y - 52$	D^2
49	–1	1	52	0	0
53	3	9	55	3	9
51	1	1	52	0	0
52	2	4	53	1	1
47	–3	9	50	–2	4
50	0	0	54	2	4
52	2	4	54	2	4
53	3	9	53	1	1
Total	7	37		7	23

$$\therefore \quad \bar{x} = 50 + \frac{7}{8} = 50{\cdot}875 \qquad \bar{y} = 52 + \frac{7}{8} = 52{\cdot}875$$

and

$$\Sigma(x - \bar{x})^2 = \Sigma d^2 - \frac{(\Sigma d)^2}{n_1} = 37 - \frac{49}{8} = 30{\cdot}875 \qquad \Sigma(y - \bar{y})^2 = \Sigma D^2 - \frac{(\Sigma D)^2}{n_2} = 23 - \frac{49}{8} = 16{\cdot}875$$

$$S^2 = \frac{1}{n_1 + n_2 - 2}\left[\Sigma(x-\bar{x})^2 + \Sigma(y-\bar{y})^2\right] = \frac{1}{14}(30{\cdot}875 + 16{\cdot}875) = 3{\cdot}41$$

$$\therefore \quad t = \frac{\bar{x} - \bar{y}}{\sqrt{S^2\left(\frac{1}{n_1} + \frac{1}{n_2}\right)}} = \frac{50{\cdot}875 - 52{\cdot}875}{\sqrt{3{\cdot}41\left(\frac{1}{8} + \frac{1}{8}\right)}} = -2{\cdot}17$$

Tabulated $t_{0\cdot05}$ for $(8 + 8 - 2)$ for 14 *d.f.* for one-tail test is 1·76.

Conclusion. The critical region for the left-tail test is $t < -1{\cdot}76$. Since calculated t is less than $-1{\cdot}76$, H_0 is rejected at 5% level of significance. Hence, we conclude that the foods A and B differ significantly as regards their effect on increase in weight. Further, since $\bar{y} > \bar{x}$, food B is superior to food A.

(*ii*) If the same set of animals is used in both the cases, then the readings X and Y are not independent but they are paired together and we apply the paired t-test for testing H_0.

Under H_0: $\mu_X = \mu_Y$, the test statistic is: $t = \dfrac{\bar{d}}{S/\sqrt{n}} \sim t_{(n-1)}$

X	49	53	51	52	47	50	52	53	*Total*
Y	52	55	52	53	50	54	54	53	
$d = X - Y$	−3	−2	−1	−1	−3	− 4	−2	0	−16
d^2	9	4	1	1	9	16	4	0	44

$$\bar{d} = \frac{\Sigma d}{n} = \frac{-16}{8} = -2 \quad \text{and} \quad S^2 = \frac{1}{n-1}\left[\Sigma d^2 - \frac{(\Sigma d)^2}{n}\right] = \frac{1}{7}\left(44 - \frac{256}{8}\right) = 1{\cdot}7143$$

$$\therefore \qquad |t| = \frac{|\bar{d}|}{\sqrt{S^2/n}} = \frac{2}{\sqrt{1{\cdot}7143/8}} = \frac{2}{0{\cdot}4629} = 4{\cdot}32$$

Tabulated $t_{0{\cdot}05}$ for $(8-1) = 7$ *d.f.* for one-tail test is 1·90.

Conclusion. Here also the observed value of 't' is significant at 5% level of significance and we conclude that food B is superior to food A.

Example 16·16. *Two laboratories carry out independent estimates of a particular chemicals in a medicine produced by a certain firm. A sample is taken from each batch, halved and the separate halves sent to the two laboratories. The following data is obtained:*

No. of samples	10
Mean value of the difference of estimates	0·6
Sum of the squares of the differences (from their means)	20

Is the difference significant ? (Value of t at 5% level for 9 d.f. is 2·262.)

Solution. Let d stand for the difference between the estimates of the chemical between the two halves of each batch, and $\bar{d}$ the mean value of the difference of estimates. In usual notations, we are given :

$$n = 10, \quad \bar{d} = 0{\cdot}6, \quad \Sigma(d - \bar{d})^2 = 20$$

Null hypothesis, $H_0 : \mu_1 = \mu_2$, *i.e.*, the difference is insignificant.

Alternative hypothesis, H_1: $\mu_1 \neq \mu_2$.

Test Statistic. Under H_0, the test statistic is: $t = \dfrac{\bar{d}}{\sqrt{S^2/n}} \sim t_{10-1}$

where $$S^2 = \frac{1}{n-1}\Sigma(d - \bar{d})^2 = \frac{20}{9} = 2{\cdot}22 \qquad \therefore \; t = \frac{0{\cdot}6}{\sqrt{2{\cdot}22/10}} = \frac{0{\cdot}6}{0{\cdot}471} = 1{\cdot}274.$$

The tabulated value of t at 5% level for 9 *d.f.*, is 2·262 (given).

Conclusion. Since calculated value of t is less than tabulated value of t, it is not significant. Hence, we may accept the null hypothesis and conclude that the difference is not significant.

16·3·4. t-test for Testing the Significance of an Observed Sample Correlation Coefficient

If r is the observed correlation coefficient in a sample of n pairs of observations from a bivariate normal population, then Prof. Fisher proved that under the null hypothesis, H_0: $\rho = 0$, *i.e., population correlation coefficient is zero,* the statistic :

$$t = \frac{r}{\sqrt{(1-r^2)}}\sqrt{(n-2)} \qquad \ldots (16\cdot10)$$

follows Student's t-distribution with $(n - 2)$ *d.f.* (*c.f.* Deduction in § *16·4*).

If the value of t comes out to be significant, we reject H_0 at the level of significance adopted and conclude that $\rho \neq 0$, *i.e.,* 'r' is significant of correlation in the population.

If t comes out to be non-significant, then H_0 may be accepted and we conclude that variables may be regarded as uncorrelated in the population.

Example 16·17.

(*a*) *A random sample of 27 pairs of observations from a normal population gave a correlation coefficient of 0·6. Is this significant of correlation in the population ?*

(*b*) *Find the least value of r in a sample of 18 pairs of observations from a bi-variate normal population, significant at 5% level of significance.*

Solution.

(*a*) We set up the null hypothesis, H_0: $\rho = 0$, *i.e.,* the *observed sample correlation coefficient is not significant of any correlation in the population.*

Under H_0 : $t = \dfrac{r\sqrt{(n-2)}}{\sqrt{(1-r^2)}} \sim t_{(n-2)}$ $\quad\therefore\quad$ $t = \dfrac{0\cdot6\sqrt{27-2}}{\sqrt{(1-0\cdot36)}} = \dfrac{3}{\sqrt{0\cdot64}} = 3\cdot75.$

Tabulated $t_{0\cdot05}$ for $(27 - 2) = 25$ *d.f.* is 2·06.

Conclusion. Since calculated t is much greater than the tabulated t, it is significant and hence H_0 is discredited at 5% level of significance. Thus, we conclude that the variables are correlated in the population.

(*b*) Here $n = 18$. From the tables $t_{0\cdot05}$ for $(18 - 2) = 16$ *d.f.* is 2·12.

Under H_0: $\rho = 0$, $\quad t = \dfrac{r\sqrt{(n-2)}}{\sqrt{(1-r^2)}} \sim t_{(n-2)} = t_{16}$

In order that the calculated value of t is significant at 5% level of significance, we should have

$$\left|\frac{r\sqrt{(n-2)}}{\sqrt{(1-r^2)}}\right| > t_{0\cdot05} \quad\Rightarrow\quad \left|\frac{r\sqrt{16}}{\sqrt{(1-r^2)}}\right| > 2\cdot12$$

$$\Rightarrow \quad 16r^2 > (2\cdot12)^2(1-r^2) \quad or \quad 20\cdot493r^2 > 4\cdot493 \quad or \quad r^2 > \frac{4\cdot493}{20\cdot493} = 0\cdot2192$$

Hence, $\quad | \, r \, | > 0\cdot4682.$

Example 16·18. *A coefficient of correlation of 0·2 is derived from a random sample of 625 pairs of observations.*

(*i*) *Is this value of r significant?*

(*ii*) *What are the 95% and 99% confidence limits to the correlation coefficient in the population ?*

Solution. *Under the null hypothesis* H_0: $\rho = 0$, *i.e., the value of* $r = 0\cdot2$ *is not significant, the test statistic is :*

$$t = \frac{r\sqrt{n-2}}{\sqrt{1-r^2}} \sim t_{n-2}$$

$$\therefore \quad t = \frac{0\cdot2 \times \sqrt{(625-2)}}{\sqrt{(1-0\cdot04)}} = 5\cdot09$$

Since *d.f.* = 625 − 2 = 623, the significant values of t are same as in the case of normal distribution, *viz.*, $t_{0.05} = 1.96$ and $t_{0.01} = 2.58$. Since calculated t is much greater than these values; it is highly significant. Hence H_0: $\rho = 0$ is rejected and we conclude that the sample correlation is significant of correlation in the population.

95% Confidence Limits for ρ (population correlation coefficient) are :

$$r \pm 1.96 \text{ S.E. } (r) = r \pm 1.96\ (1 - r^2)/\sqrt{n} \qquad \text{[Since } n \text{ large]}$$

$$= 0.2 \pm (1.96 \times 0.96/\sqrt{625}\,) = 0.2 \pm 0.075 = (0.125,\ 0.275)$$

99% Confidence Limits for ρ are:

$$0.2 \pm 2.58 \times 0.0384 = 0.2 \pm 0.099 = (0.101,\ 0.299)$$

16·3·5. *t*-test for Testing the Significance of an Observed Regression Coefficient

Here the problem is to test if a random sample (x_i, y_i), $(i = 1, 2, \ldots, n)$ has been drawn from a bivariate normal population in which regression coefficient of Y on X is β.

The line of regression of Y on X (for the given sample) is :

$$Y - \bar{y} = b(X - \bar{x}), \quad b = \frac{\mu_{11}}{\sigma_X^2} \qquad (*)$$

The estimate of Y for a given value (say) x_i of X as given by line (*) is :

$$\hat{y}_i = \bar{y} + b\,(x_i - \bar{x})$$

Under the null hypothesis H_0, that the *population regression coefficient* is β, Prof. R.A. Fisher proved that the statistic,

$$t = (b - \beta)\left[\frac{(n-2)\sum_i (x_i - \bar{x})^2}{\sum_i (y_i - \hat{y}_i)^2}\right]^{1/2} \qquad \ldots (16{\cdot}11)$$

follows Student's t-distribution with $(n - 2)$ *d.f.*

16·3·6. *t*-test for Testing the Significance of an Observed Partial Correlation Coefficient

Let $r_{12\cdot34\ldots(k+2)}$ be the partial correlation coefficient of order k, observed in a sample of size n from a multivariate normal population. Prof. R.A. Fisher proved that under the null hypothesis $H_0 : \rho_{12\cdot34\ldots(k+2)} = 0$, *i.e.*, the population correlation coefficient is zero, the sampling distribution of $r_{12\cdot34\ldots(k+2)}$ is same as that of r_{12} under the null hypothesis $\rho_{12} = 0$, with sample size n reduced by k. Hence, the test statistic for H_0 becomes :

$$t = \frac{r_{12\cdot34\ldots(k+2)}}{\left[1 - r^2_{12\cdot34\ldots(k+2)}\right]^{1/2}}\sqrt{(n-k-2)} \qquad \ldots (16{\cdot}11a)$$

which follows Student's t-distribution with $(n - k - 2)$ *d.f.*

16·4. DISTRIBUTION OF SAMPLE CORRELATION COEFFICIENT (*r*) WHEN POPULATION CORRELATION COEFFICIENT, ρ = 0 (SAWKIN'S METHOD)

The sample correlation coefficient, r in a random sample of size n from an uncorrelated bivariate normal population (ρ = 0) has the following probability density function:

$$f(r) = \frac{1}{B\left(\frac{1}{2}\cdot\frac{n-2}{2}\right)}\cdot(1 - r^2)^{(n-4)/2}, \; -1 \le r \le 1 \qquad \ldots (16{\cdot}12)$$

Proof. Let (x_i, y_i), $(i = 1, 2, \ldots, n)$ be a random sample of size n drawn from an uncorrelated bivariate normal population $(\rho = 0)$ in which $E(X) = E(Y) = 0$ and $V(X) = \sigma_X^2$, $V(Y) = \sigma_Y^2$. Let the variable Y be transformed to the variable Z by means of a linear orthogonal transformation, *viz.*, $\mathbf{Z} = \mathbf{CY}$ where $\mathbf{Z}_{n \times 1} = (z_1, z_2, \ldots, z_n)'$,

$$\mathbf{Y}_{n \times 1} = (y_1, y_2, \ldots, y_n)' \quad \text{and} \quad C_{n \times n} = (c_{ij}), \text{ C is an orthogonal matrix.}$$

Let us in particular take $c_{11} = c_{12} = \ldots = c_{1n} = 1/\sqrt{n}$, so that

$$z_1 = \frac{1}{\sqrt{n}}(y_1 + y_2 + \ldots + y_n) = \sqrt{n}\,\bar{y}$$

Now proceeding as in (*Theorem 15·5*), we get $\quad \sum_{i=2}^{n} z_i^2 = \sum_{i=1}^{n} (y_i - \bar{y})^2 = ns_Y^2 \quad \ldots (*)$

Since in a bivariate normal distribution, the marginal distributions of X and Y are also normal, we have $Y \sim N(0, \sigma_Y^2)$. Hence, by Fisher's Lemma (*Theorem 15·4*) z_i, $(i = 1, 2, \ldots, n)$ are independent $N(0, \sigma_Y^2)$.

$$\therefore \quad r = \frac{Cov(X,Y)}{s_X s_Y} = \frac{\sum_{i=1}^{n}(x_i - \bar{x})(y_i - \bar{y})}{ns_X s_Y} = \frac{\sum_{i=1}^{n}(x_i - \bar{x})y_i - \bar{y}\sum_{i=1}^{n}(x_i - \bar{x})}{ns_X s_Y} = \frac{\sum_{i=1}^{n}(x_i - \bar{x})y_i}{ns_X s_Y}$$

$$\Rightarrow \quad \sqrt{n}\, s_Y r = \frac{\Sigma(x_i - \bar{x})y_i}{\sqrt{n}s_X} = z_2, \ (say) \quad \ldots (**)$$

[Since the sum of the squares of coefficients of $y_1, y_2, \ldots, y_n$ in (**) is unity.]

From (*) and (**), we get

$$ns_Y^2 = \sum_{i=2}^{n} z_i^2 = \sum_{i=3}^{n} z_i^2 + z_2^2 = \sum_{i=3}^{n} z_i^2 + nr^2 s_Y^2 \quad \Rightarrow \quad (1 - r^2)ns_Y^2 = \sum_{i=3}^{n} z_i^2 \quad \ldots (***),$$

Since z_i ; $i = 1, 2, \ldots, n$, are independent $N(0, \sigma_Y^2)$, (Z_i/σ_Y) are independent $N(0, 1)$.

Hence, from (**), $\quad U = \dfrac{z_2^2}{\sigma_Y^2} = \dfrac{nr^2 s_Y^2}{\sigma_Y^2}$

being the square of a standard normal variate is a χ^2-variate with 1 *d.f.* and from (***),

$$V = \sum_{i=3}^{n} z_i^2/\sigma_Y^2 = \sum_{i=3}^{n}(z_i/\sigma_Y)^2 = \frac{(1 - r^2)ns_Y^2}{\sigma_Y^2}$$

being the sum of squares of $(n - 2)$ independent standard normal variates is an independent χ^2-variate with $(n - 2)$ *d.f.*

Further, since z_2 and $(z_3, z_4, \ldots, z_n)$ are independent *r.v.'s*, U and V are independent chi-square variates with 1 and $(n - 2)$ *d.f.* respectively.

$$\therefore \quad \frac{U}{U + V} = \frac{nr^2 s_Y^2/\sigma_Y^2}{\left[nr^2 s_Y^2 + (1 - r^2)ns_Y^2\right]/\sigma_Y^2} \sim \beta_1\left(\frac{1}{2}, \frac{n-2}{2}\right) \quad \text{[c.f. Theorem 15·2]}$$

$$\Rightarrow \quad r^2 \sim \beta_1\left(\frac{1}{2}, \frac{n-2}{2}\right)$$

Hence, the probability density function of r^2 is given by :

$$dF(r^2) = \frac{1}{B\left(\frac{1}{2}, \frac{n-2}{2}\right)} \cdot (r^2)^{(1/2)-1}(1 - r^2)^{\frac{n-2}{2}-1} d(r^2), \ 0 \le r^2 \le 1$$

$$\Rightarrow \quad f(r) = \frac{1}{B\left(\frac{1}{2}, \frac{n-2}{2}\right)} \cdot (1 - r^2)^{(n-4)/2}, -1 \le r \le 1$$

the factor 2 disappearing from the fact that total probability in the range $-1 \le r \le 1$ must be unity.

Deduction. *If $\rho = 0$, then $t = \dfrac{r}{\sqrt{1-r^2}}\sqrt{(n-2)}$ is distributed as Student's t with $(n-2)$ d.f.*

Proof.
$$t = \frac{r\sqrt{n-2}}{\sqrt{1-r^2}} \qquad \ldots (i)$$

$$\Rightarrow \quad \frac{t^2}{n-2} = \frac{r^2}{(1-r^2)} = \frac{1}{1-r^2} - 1 \quad \Rightarrow \quad (1-r^2) = \left[1 + \frac{t^2}{(n-2)}\right]^{-1} \qquad \ldots (ii)$$

From (*i*), we get

$$\Rightarrow \quad dt = \sqrt{(n-2)}\, d\left[r/\sqrt{(1-r^2)}\right] \quad \Rightarrow \quad dt = \sqrt{(n-2)}\left[\frac{dr}{\sqrt{(1-r^2)}} + \left(-\frac{r}{2}\right)\frac{(-2r)dr}{(1-r^2)^{3/2}}\right]$$

$$\Rightarrow \quad dt = \sqrt{(n-2)} \times \frac{dr}{\sqrt{(1-r^2)}}\left(1 + \frac{r^2}{1-r^2}\right)$$

$$\Rightarrow \quad dt = \sqrt{(n-2)} \times \frac{dr}{(1-r^2)^{3/2}} \quad \Rightarrow \quad dr = \frac{1}{\sqrt{n-2}}(1-r^2)^{3/2}\, dt$$

As r ranges from -1 to 1, from (*i*), t ranges from $-\infty$ to ∞.
When $\rho = 0$, the *p.d.f.* of 'r' is given by (16·12) and it transforms to

$$dG(t) = \frac{1}{B\left(\frac{1}{2}, \frac{n-2}{2}\right)}(1-r^2)^{(n-4)/2}\frac{1}{\sqrt{(n-2)}}(1-r^2)^{3/2}\, dt$$

$$\Rightarrow \quad g(t) = \frac{1}{\sqrt{(n-2)}\, B\left(\frac{1}{2}, \frac{n-2}{2}\right)} \cdot \frac{1}{\left(1 + \frac{t^2}{n-2}\right)^{(n-1)/2}} \qquad \text{[From } (ii)]$$

$$= \frac{1}{\sqrt{(n-2)}\, B\left(\frac{1}{2}, \frac{n-2}{2}\right)} \cdot \frac{1}{\left(1 + \frac{t^2}{n-2}\right)^{(n-2+1)/2}}; \; -\infty < t < \infty$$

which is the *p.d.f.* of *t*-distribution with $(n-2)$ *d.f.*

Hence,
$$t = \frac{r}{\sqrt{(1-r^2)}} \cdot \sqrt{(n-2)} \;:\; t_{(n-2)}$$

Example 16·19.

(*a*) *If (x_i, y_i) is a random sample drawn from an uncorrelated bivariate normal population, derive the distribution of :*

$$r = \frac{\Sigma(x_i - \bar{x})(y_i - \bar{y})}{\sqrt{\Sigma(x_i - \bar{x})^2\, \Sigma(y_i - \bar{y})^2}}$$

(*b*) *Further, when $n = 5$ and if $P(|r| \ge C) = \alpha$, show that C is a root of the equation,*

$$C\sqrt{(1-C^2)} + \sin^{-1} C + \frac{\pi(\alpha - 1)}{2} = 0.$$

Solution.

(*a*) *c.f.* § 16·4, page 16·26.

(*b*) $P\ (|r| \geq C) = 1 - P\ (|r| \leq C) = 1 - P\ (-C \leq r \leq C)$

$$= 1 - 2\,P(0 \leq r \leq C) = 1 - 2\int_0^C f(r)dr \qquad [\because\ f(r) \text{ is symmetrical about } r = 0]$$

When $n = 5$, $$f(r) = \frac{1}{B\left(\frac{1}{2}, \frac{3}{2}\right)} \cdot (1 - r^2)^{1/2} \qquad [c.f. \text{ Equation (16·12), page 16·26}]$$

$$\therefore \quad P(|r| \geq C) = 1 - 2\frac{\Gamma(2)}{\Gamma(1/2)\Gamma(3/2)}\int_0^C (1 - r^2)^{1/2} dr$$

$$= 1 - 2 \times \frac{1}{\frac{1}{2}\cdot\pi}\left[\frac{1}{2}r(1 - r^2)^{1/2} + \frac{1}{2}\sin^{-1} r\right]_0^C \qquad \left[\because\ \Gamma\left(\frac{1}{2}\right) = \sqrt{\pi}\right]$$

$$= 1 - \frac{4}{\pi}\left[\frac{1}{2}C(1 - C^2)^{1/2} + \frac{1}{2}\sin^{-1} C\right] = \alpha \qquad \text{(Given)}$$

$$\therefore \quad 1 - \frac{2}{\pi}\left[\ \cdot (1 - \ \ ^2)^{\frac{1}{2}} + \sin^{-1}\ \ \right] = \alpha \quad \Rightarrow \quad \cdot (1 - \ \ ^2)\ \ + \sin^{-1}\ \ + (\alpha - 1)\frac{\pi}{2} = 0.$$

16·5. NON-CENTRAL *t*-DISTRIBUTION

The non-central *t*-distribution is the distribution of the ratio of a normal variate with possibly non-zero mean and variance unity, to the square root of an independent χ^2-variate divided by its degrees of freedom. If $X \sim N(\mu, 1)$ and Y is an independent χ^2-variate with *n d.f.* then

$$t' = \frac{X}{\sqrt{Y/n}} \qquad ...(16·12a)$$

is said to have a *non-central t-distribution* with *n d.f.* and non-centrality parameter μ. Non-central *t*-distribution is requireed for the power functions of certain tests concerning normal population.

p.d.f. of '*t*' Since $X \sim N\ (\mu, 1)$, its *p.d.f.* $f(.)$ is :

$$f(x) = \frac{1}{\sqrt{2\pi}}\exp\left[-\frac{1}{2}(x - \mu)^2\right]$$

$$= \frac{1}{\sqrt{2\pi}}\exp\left[-\frac{1}{2}\left(\mu^2 + x^2\right)\right]\sum_{i=0}^{\infty}\frac{(\mu x)^i}{i!}, \quad -\infty < x < \infty$$

Since, $Y \sim \chi^2_{(n)}$, its *p.d.f.* $g(.)$ is :

$$g(y) = \frac{1}{2^{n/2}\Gamma(n/2)} e^{-y/2} y^{(n/2)-1}, \quad 0 < y < \infty$$

Since, X and Y are independent, their joint p.d.f. becomes

$$f(x, y) = \frac{1}{\sqrt{2\pi}\ 2^{n/2}\Gamma(n/2)}\exp\left[-\frac{1}{2}(\mu^2 + x^2 + y)\right]\cdot y^{(n/2)-1} . \sum_{i=0}^{\infty}\frac{(\mu x)^i}{i!}$$

Let us transform to new variables t' and z by the substitution:

$$t' = \frac{x}{\sqrt{y/n}} = \frac{\sqrt{n}.x}{\sqrt{y}}, \quad z = +\sqrt{y} \quad \Rightarrow \quad x = \frac{zt'}{\sqrt{n}}, \quad y = z^2$$

Jacobian of transformation J is

$$J = \begin{vmatrix} \dfrac{\partial x}{\partial t'} & \dfrac{\partial x}{\partial z} \\[2ex] \dfrac{\partial y}{\partial t'} & \dfrac{\partial y}{\partial z} \end{vmatrix} = \begin{vmatrix} \dfrac{z}{\sqrt{n}} & \dfrac{t'}{\sqrt{n}} \\[2ex] 0 & 2z \end{vmatrix} = \frac{2z^2}{\sqrt{n}}$$

The joint p.d.f. of t' and z becomes [From (*)],

$$h(t',z)=\frac{\exp\left(-\mu^2/2\right)}{\sqrt{2\pi}\,2^{n/2}\Gamma(n/2)}(z^2)^{\frac{n}{2}-1}\sum_{i=0}^{\infty}\frac{\left(\mu z t'/\sqrt{n}\right)^i}{i!}\cdot\frac{2z^2}{\sqrt{n}}$$

$$\times\exp\left[-\frac{1}{2}\left(1+\frac{t'^2}{n}\right)z^2\right];\qquad -\infty<t'<\infty, 0<z<\infty$$

$$=\frac{\exp(-\mu^2/2)}{\sqrt{\pi}\,2^{(n-1)/2}\Gamma(n/2)}\sum_{i=0}^{\infty}\left[\frac{(\mu t)^i}{i\,!\,n^{(i+1)/2}}\cdot\exp\left\{-\frac{1}{2}\left(1+\frac{t'^2}{n}\right)z^2\right\}z^{n+i}\right]$$

Integrating w.r.t z in the range 0 to ∞ , we get the *p.d.f.* of t'.

$$h_1(t')=\frac{\exp(-\mu^2/2)}{\sqrt{\pi}\;2^{(n-1)/2}\Gamma(n/2)}\times\sum_{i=0}^{\infty}\left[\frac{(\mu t')^i}{i!n^{(i+1)/2}}\int_0^{\infty}\exp\left\{-\frac{1}{2}\left(1+\frac{t'^2}{n}\right)z^2\right\}z^{n+i}dz\right]$$

$$=\frac{\exp\,(-\mu^2/2)}{\sqrt{\pi}2^{(n-1)/2}\Gamma(n/2)}\times\sum_{i=0}^{\infty}\left[\frac{(\mu t')^i}{i!n^{(i+1)/2}}\int_0^{\infty}\exp\left\{-\left(1+\frac{t'^2}{n}\right)v\right\}\cdot(2v)^{(n+i-1)/2}dv\right]$$

$$=\frac{\exp\,(-\mu^2/2)}{\sqrt{\pi}\,\Gamma(n/2)}\sum_{i=0}^{\infty}\left[\frac{\mu^i 2^{i/2}\Gamma\left(\dfrac{n+i+1}{2}\right)}{i!n^{(i+1)/2}}\frac{(t')^i}{\left(1+\dfrac{t'^2}{n}\right)^{(n+i+1)/2}}\right]\qquad \text{...(16·12}b\text{)}$$

which is the *p.d.f.* of non-central *t*-distribution with *n d.f.* and non-centrality element μ.

Remark. If $\mu = 0$, we get from (16·12b),

$$h_1(t')=\frac{1}{\sqrt{\pi}\,\Gamma\,(n/2)}\cdot\frac{\Gamma\left[(n+1)/2\right]}{\sqrt{n}}\cdot\frac{1}{\left[1+\dfrac{t'^2}{n}\right]^{(n+1)/2}}$$

since, for $\mu=0$, we get contribution from the sum only when $i=0$ and other terms vanish.

$$\therefore\qquad h_1(t')=\frac{1}{\sqrt{n}B\left(\dfrac{1}{2},\dfrac{n}{2}\right)}\left[1+\frac{t'^2}{n}\right]^{-(n+1)/2},\quad -\infty<t'<\infty$$

which is the *p.d.f.* of central *t*-distribution with *n d.f.*

16·6. *F*-DISTRIBUTION

Definition. *If X and Y are two independent chi-square variates with ν_1 and ν_2 d.f. respectively, then F-statistic is defined by :*

$$F=\frac{X/\nu_1}{Y/\nu_2}\qquad \text{... (16·13)}$$

In other words, F is defined as the ratio of two independent chi-square variates divided by their respective degrees of freedom and it follows Snedecor's F-distribution with (υ_1, υ_2) *d.f.* with probability function given by :

$$f(F)=\frac{\left(\dfrac{\nu_1}{\nu_2}\right)^{\frac{\nu_1}{2}}}{B\left(\dfrac{\nu_1}{2},\dfrac{\nu_2}{2}\right)}\cdot\frac{F^{(\nu_1/2)-1}}{\left(1+\dfrac{\nu_1}{\nu_2}F\right)^{(\nu_1+\nu_2)/2}},\quad 0\le F<\infty\qquad \text{...[16·13}a\text{]}$$

***Remarks*:**

1. The sampling distribution of *F*-statistic does not involve any population parameters and depends only on the degrees of freedom υ_1 and υ_2.
2. A statistic *F* following Snedecor's *F*-distribution with (υ_1, υ_2) *d.f.* will be denoted by $F \sim F(\upsilon_1, \upsilon_2)$.

16·6·1 Derivation of Snedecor's *F*-distribution

Since *X* and *Y* are independent chi-square variates with ν_1 and ν_2 *d.f.* respectively, their joint probability density function is given by :

$$f(x,y) = \left\{\frac{1}{2^{\nu_1/2}\Gamma(\nu_1/2)}\exp(-x/2)\cdot x^{(\nu_1/2)-1}\right\} \times \left\{\frac{1}{2^{\nu_2/2}\Gamma(\nu_2/2)}\exp(-y/2)\cdot y^{(\nu_2/2)-1}\right\}$$

$$= \frac{1}{2^{(\nu_1+\nu_2)/2}\Gamma(\nu_1/2)\Gamma(\nu_2/2)}\exp\{-(x+y)/2\} \times x^{(\nu_1/2)-1} y^{(\nu_2/2)-1}, 0 \le (x,y) < \infty$$

Let us make the following transformation of variables :

$$F = \frac{x/\nu_1}{y/\nu_2} \text{ and } u = y, \text{ so that } 0 \le F < \infty,\ 0 < u < \infty \qquad \therefore \quad x = \frac{\nu_1}{\nu_2}Fu \quad \text{and} \quad y = u$$

Jacobian of transformation *J* is given by : $J = \dfrac{\partial(x,y)}{\partial(F,u)} = \begin{vmatrix} \frac{\nu_1}{\nu_2}u & 0 \\ \frac{\nu_1}{\nu_2}F & 1 \end{vmatrix} = \dfrac{\nu_1 u}{\nu_2}$

Thus, the joint *p.d.f.* of the transformed variables is :

$$g(F,u) = \frac{1}{2^{(\nu_1+\nu_2)/2}\Gamma(\nu_1/2)\Gamma(\nu_2/2)}\exp\left\{-\frac{u}{2}\left(1+\frac{\nu_1}{\nu_2}F\right)\right\} \times \left(\frac{\nu_1}{\nu_2}Fu\right)^{(\nu_1/2)-1} u^{(\nu_2/2)-1}|J|$$

$$= \frac{(\nu_1/\nu_2)^{\nu_1/2}}{2^{(\nu_1+\nu_2)/2}\Gamma(\nu_1/2)\Gamma(\nu_2/2)}\exp\left\{-\frac{u}{2}\left(1+\frac{\nu_1}{\nu_2}F\right)\right\} \times u^{\{(\nu_1+\nu_2)/2\}-1} F^{(\nu_1/2)-1}; 0 < u < \infty, 0 \le F < \infty$$

Integrating *w.r. to u* over the range 0 to ∞, the *p.d.f.* of *F* becomes :

$$g_1(F) = \frac{(\nu_1/\nu_2)^{(\nu_1/2)}F^{(\nu_1/2)-1}}{2^{(\nu_1+\nu_2)/2}\Gamma(\nu_1/2)\Gamma(\nu_2/2)} \times \left[\int_0^\infty \exp\left\{-\frac{u}{2}\left(1+\frac{\nu_1}{\nu_2}F\right)\right\}u^{\{\nu_2+\nu_2/2\}-1}\,du\right]$$

$$= \frac{(\nu_1/\nu_2)^{(\nu_1/2)}F^{(\nu_1/2)-1}}{2^{(\nu_1+\nu_2)/2}\Gamma(\nu_1/2)\Gamma(\nu_2/2)} \times \frac{\Gamma[(\nu_1+\nu_2)/2]}{\left[\frac{1}{2}\left(1+\frac{\nu_1}{\nu_2}F\right)\right]^{(\nu_1+\nu_2)/2}}$$

$$\therefore \qquad g_1(F) = \frac{(\nu_1/\nu_2)^{\nu_1/2}}{B\left(\frac{\nu_1}{2},\frac{\nu_2}{2}\right)} \cdot \frac{F^{(\nu_1/2)-1}}{\left(1+\frac{\nu_1}{\nu_2}F\right)^{(\nu_1+\nu_2)/2}}, 0 \le F < \infty$$

which is the required probability function of *F*-distribution with (ν_1, ν_2) *d.f.*

Aliter. $F = \dfrac{x/\nu_1}{y/\nu_2}$

$\therefore \ \dfrac{\nu_1}{\nu_2}F = \dfrac{x}{y}$, being the ratio of two independent chi-square variates with υ_1 and υ_2 *d.f.*, respectively is a $\beta_2\left(\dfrac{\nu_1}{2},\dfrac{\nu_2}{2}\right)$ variate. Hence, the probability function of *F* is given by:

$$dP(F) = \frac{1}{B\left(\frac{\nu_1}{2}, \frac{\nu_2}{2}\right)} \cdot \frac{\left(\frac{\nu_1}{\nu_2} F\right)^{(\nu_1/2)-1}}{\left(1 + \frac{\nu_1}{\nu_2} F\right)^{(\nu_1+\nu_2)/2}} d\left(\frac{\nu_1}{\nu_2} F\right)$$

$$\Rightarrow \qquad f(F) = \frac{\left(\frac{\nu_1}{\nu_2}\right)^{\nu_1/2}}{B\left(\frac{\nu_1}{2}, \frac{\nu_2}{2}\right)} \cdot \frac{F^{(\nu_1/2)-1}}{\left(1 + \frac{\nu_1}{\nu_2} F\right)^{(\nu_1+\nu_2)/2}}, \; 0 \le F < \infty$$

16·6·2. Constants of F-distribution

$$\mu'_r \text{ (about origin)} = E(F^r) = \int_0^\infty F^r f(F) dF$$

$$= \frac{(\nu_1/\nu_2)^{\nu_1/2}}{B\left(\frac{\nu_1}{2}, \frac{\nu_2}{2}\right)} \int_0^\infty F^r \cdot \frac{F^{(\nu_1/2)-1}}{\left(1 + \frac{\nu_1}{\nu_2} F\right)^{(\nu_1+\nu_2)/2}} dF \qquad \ldots (*)$$

To evaluate the integral, put : $\frac{\nu_1}{\nu_2} F = y$, so that $dF = \frac{\nu_2}{\nu_1} dy$

$$\therefore \qquad \mu_r' = \frac{[\nu_1/\nu_2]^{\nu_1/2}}{B\left(\frac{\nu_1}{2}, \frac{\nu_2}{2}\right)} \int_0^\infty \frac{\left(\frac{\nu_2}{\nu_1} y\right)^{r+(\nu_1/2)-1}}{(1+y)^{(\nu_1+\nu_2)/2}} \left(\frac{\nu_2}{\nu_1}\right) dy$$

$$\mu_r' = \frac{(\nu_2/\nu_1)^r}{B\left(\frac{\nu_1}{2}, \frac{\nu_2}{2}\right)} \cdot \int_0^\infty \frac{y^{r+(\nu_1/2)-1}}{(1+y)^{(\nu_1/2)+r+\{(\nu_2/2)-r\}}} dy$$

$$= \left(\frac{\nu_2}{\nu_1}\right)^r \cdot \frac{1}{B\left(\frac{\nu_1}{2}, \frac{\nu_2}{2}\right)} \cdot B\left(r + \frac{\nu_1}{2}, \frac{\nu_2}{2} - r\right), \nu_2 > 2r \qquad \ldots (16·14)$$

Aliter for (16·14). (16·14) could also be obtained by substituting $\frac{\nu_2}{\nu_1} F = \tan^2 \theta$ (*) and using the Beta integral : $2\int_0^{\pi/2} \sin^p \theta \cos^q \theta \, d\theta = B\left(\frac{p+1}{2}, \frac{q+1}{2}\right)$

$$\therefore \qquad \mu_r' = \left(\frac{\nu_2}{\nu_1}\right)^r \cdot \frac{\Gamma[r + (\nu_1/2)] \cdot \Gamma[(\nu_2/2) - r]}{\Gamma(\nu_1/2) \cdot \Gamma(\nu_2/2)}; r < \frac{\nu_2}{2} \Rightarrow \nu_2 > 2r \qquad \ldots (16·15)$$

In particular

$$\mu_1' = \frac{\nu_2}{\nu_1} \cdot \frac{\Gamma[1 + (\nu_1/2)] \cdot \Gamma[(\nu_2/2) - 1]}{\Gamma(\nu_1/2) \cdot \Gamma(\nu_2/2)} = \frac{\nu_2}{\nu_2 - 2}, \nu_2 > 2 \qquad \ldots (16·15a)$$

$$[\because \Gamma(r) = (r-1)\,\Gamma(r-1)]$$

Thus, *the mean of F-distribution is independent of* ν_1.

$$\mu_2' = \left(\frac{\nu_2}{\nu_1}\right)^2 \cdot \frac{\Gamma\,[(\nu_1/2) + 2] \cdot \Gamma\,[(\nu_2/2) - 2]}{\Gamma(\nu_1/2) \cdot \Gamma\,(\nu_2/2)}$$

$$= \left(\frac{\nu_2}{\nu_1}\right)^2 \cdot \frac{[(\nu_1/2) + 1]\,(\nu_1/2)}{[(\nu_2/2) - 1] \cdot [(\nu_2/2) - 2]} = \frac{\nu_2^2\,(\nu_1 + 2)}{\nu_1(\nu_2 - 2)\,(\nu_2 - 4)}, \quad \nu_2 > 4.$$

$$\therefore \quad \mu_2 = \mu_2' - \mu_1'^2 = \frac{\nu_2^2(\nu_1+2)}{\nu_1(\nu_2-2)(\nu_2-4)} - \frac{\nu_2^2}{(\nu_2-2)^2} = \frac{2\nu_2^2(\nu_2+\nu_1-2)}{\nu_1(\nu_2-2)^2(\nu_2-4)}, \nu_2 > 4 \quad \ldots (16{\cdot}15b)$$

Similarly, on putting $r = 3$ and 4 in μ_r', we get μ_3' and μ_4'respectively, from which the central moments μ_3 and μ_4 can be obtained.

16·6·3. Mode and Points of Inflexion of *F*-distribution

We have $\quad \log f(F) = C + \{(\nu_1/2) - 1\}\log F - \left(\frac{\nu_1+\nu_2}{2}\right)\log\{1 + (\nu_1/\nu_2)F\},$

where C is a constant independent of *F*.

$$\frac{\partial}{\partial F}[\log f(F)] = \left(\frac{\nu_1}{2} - 1\right)\cdot\frac{1}{F} - \frac{(\nu_1+\nu_2)}{2}\cdot\frac{1}{\left(1+\frac{\nu_1}{\nu_2}F\right)}\cdot\frac{\nu_1}{\nu_2}$$

$$f'(F) = \frac{\partial}{\partial F}f(F) = 0 \quad \Rightarrow \quad \frac{\nu_1-2}{2F} - \frac{\nu_1(\nu_1+\nu_2)}{2(\nu_2+\nu_1 F)} = 0$$

Hence, $\quad F = \dfrac{\nu_2(\nu_1-2)}{\nu_1(\nu_2+2)}$

It can be easily verified that at this point $f''(F) < 0$. Hence, Mode $= \dfrac{\nu_2(\nu_1-2)}{\nu_1(\nu_2+2)}$ ... (16·16)

Remarks:

1. Since $F > 0$, mode exists if and only if $\nu_1 > 2$.
2. $\quad$ Mode $= \left(\dfrac{\nu_2}{\nu_2+2}\right)\cdot\left(\dfrac{\nu_1-2}{\nu_1}\right)$

   Hence, mode of *F*-distribution is always less than unity.
3. The points of inflexion of *F*-distribution exist for $\upsilon_1 > 4$ and are equidistant from mode.

Proof. We have $\quad \dfrac{\nu_1}{\nu_2}F = \dfrac{X}{Y} \sim \beta_2(l, m),$ ... (*)

where $l = \upsilon_1/2$ and $m = \upsilon_2/2$. We now find the points of inflexion of Beta distribution of second kind with parameters l and m. If $X \sim \beta_2\,(l, m)$, its *p.d.f.* is :

$$f(x) = \frac{1}{\beta(l,m)}\cdot\frac{x^{l-1}}{(1+x)^{l+m}};\quad 0 \le x < \infty \quad \ldots (**)$$

Points of inflexion are the solution of $f''(x) = 0$ and $f'''(x) \neq 0$.
From (**), $\quad \log f(x) = -\log\beta(l, m) + (l-1)\log x - (l+m)\log(1+x)$
Differentiating twice *w.r. to x*, we get

$$\frac{f'(x)}{f(x)} = \frac{l-1}{x} - \frac{l+m}{1+x} \quad \ldots (***)$$

and

$$\frac{f(x)f''(x) - [f'(x)]^2}{[f(x)]^2} = -\left(\frac{l-1}{x^2}\right) + \frac{l+m}{(1+x)^2}$$

If $f''(x) = 0$, then $-\left[\dfrac{f'(x)}{f(x)}\right]^2 = -\left(\dfrac{l-1}{x^2}\right) + \dfrac{l+m}{(1+x)^2} \Rightarrow -\left[\dfrac{l-1}{x} - \dfrac{l+m}{1+x}\right]^2 = -\left(\dfrac{l-1}{x^2}\right) + \dfrac{l+m}{(1+x)^2}$

[On using (***)]

$$\Rightarrow \quad \frac{l-1}{x^2}(l-1-1) - 2\frac{(l-1)(l+m)}{x(1+x)} + \frac{l+m}{(1+x)^2}\times(l+m+1) = 0$$

$$\Rightarrow \quad (l-1)(l-2)(1+x)^2 - 2x(l+x)(l-1)(l+m) + x^2(l+m)(l+m+1) = 0 \quad \ldots (****)$$

which is a quadratic in x. It can be easily verified that at these values of x, $f'''(x) \neq 0$, if $l > 2$. The roots of (****) give the points of inflexion of $\beta_2(l, m)$ distribution. The sum of the points of inflexion is equal to the sum of roots of (****) and is given by :

$$-\left[\frac{\text{Coeff. of } x \text{ in (****)}}{\text{Coeff. of } x^2 \text{ in (****)}}\right] = -\left[\frac{2(l-1)(l-2)-2(l-1)(l+m)}{(l-1)(l-2)-2(l-1)(l+m)+(l+m)(l+m+1)}\right]$$

$$= \frac{2(l-1)\,[(l+m)-(l-2)]}{(l-1)(l-2)-(l-1)(l+m)-(l-1)(l+m)+(l+m)(l+m+1)}$$

$$= \frac{2(l-1)(m+2)}{(l-1)[(l-2-l-m)]+(l+m)(l+m+1-l+1)}$$

$$= \frac{2(l-1)(m+2)}{-(l-1)(m+2)+(l+m)(m+2)} = \frac{2(l-1)}{l+m-l+1} = \frac{2(l-1)}{(m+1)}$$

$\therefore$ Sum of points of inflexion of $\left(\frac{\nu_1}{\nu_2}F\right)$ distribution $= \dfrac{2(l-1)}{(m+1)} = \dfrac{2\left(\frac{\nu_1}{2}-1\right)}{\left(\frac{\nu_2}{2}+1\right)} = \dfrac{2(\nu_1-2)}{(\nu_2+2)}$

$\Rightarrow$ Sum of points of inflexion of $F(\upsilon_1, \upsilon_2)$ distribution

$$= \left[\frac{\nu_2}{\nu_1}\cdot\times\frac{2(\nu_1-2)}{(\nu_2+2)},\ \text{provided } l = \frac{\nu_1}{2} > 2\right] = \left[2\cdot\frac{\nu_2\,(\nu_1-2)}{\nu_1\,(\nu_2+2)} = 2\,.\,\text{Mode, provided } \nu_1 > 4\right]$$

Hence, *the points of inflexion of $F(\nu_1, \nu_2)$ distribution, when they exist, (i.e., when $\nu_1 > 4$), are equidistant from the mode.*

4. Karl Pearson's coefficient of skewness is given by : $S_k = \dfrac{\text{Mean} - \text{Mode}}{\sigma} > 0,$

since mean > 1 and mode < 1. Hence F-distribution is highly positively skewed.

5. The probability $f(F)$ increases steadily at first until it reaches its peak (corresponding to the modal value which is less than 1) and then decreases slowly so as to become tangential at $F = \infty$, *i.e.*, F-axis is an asymptote to the right tail.

Example 16·20. *When $\nu_1 = 2$, show that the significance level of F corresponding to a significant probability p is :* $F = \dfrac{\nu_2}{2}\left[p^{-(2/\nu_2)} - 1\right]$ *where υ_1 an υ_2 have their usual meanings.*

Solution. When $\nu_1 = 2$, the p.d.f of $F(\upsilon_1\ \upsilon_2)$ distribution is:

$$f(F) = \frac{1}{\left(1, \frac{\nu_2}{2}\right)}\cdot\frac{2}{\nu_2}\cdot\frac{1}{\left(1+\frac{2}{\nu_2}F\right)^{(\nu_2/2)+1}}\ ;\ 0 < F < \infty$$

$$= \frac{\Gamma\left(\frac{\nu_2}{2}+1\right)}{\Gamma(1)\Gamma(\nu_2/2)} \times \frac{2/\nu_2}{\left(\frac{2}{\nu_2}\right)^{(\nu_2/2)+1}\left(F+\frac{\nu_2}{2}\right)^{(\nu_2/2)+1}} = \frac{\left(\frac{\nu_2}{2}\right)^{(\nu_2/2)+1}}{\left(F+\frac{\nu_2}{2}\right)^{(\nu_2/2)+1}}$$

Hence, $$p = \int_F^\infty f(F)dF = \left[\left(\frac{\nu_2}{2}\right)\right]^{(\nu_2/2)+1} \times \int_F^\infty \frac{dF}{\left(F+\frac{\nu_2}{2}\right)^{(\nu_2/2)+1}}$$

$$= \left(\frac{\nu_2}{2}\right)^{(\nu_2/2)+1} \times \left|\frac{\left(F+\frac{\nu_2}{2}\right)^{-(\nu_2/2)}}{-\frac{\nu_2}{2}}\right|_F^\infty = \left[\frac{\left(\frac{\nu_2}{2}\right)}{F+\frac{\nu_2}{2}}\right]^{(\nu_2/2)} = \frac{1}{\left(1+\frac{2}{\nu_2}F\right)^{(\nu_2/2)}}$$

$$\Rightarrow \qquad p^{-(2/\nu_2)} = 1 + \frac{2F}{\nu_2} \qquad \Rightarrow \qquad F = \frac{\nu_2}{2}\left[p^{-(2/\nu_2)} - 1\right].$$

Example 16·21 (a). *If $F(n_1, n_2)$ represents an F-variate with n_1 and n_2 d.f., prove that $F(n_2, n_1)$ is distributed as $1/F(n_1, n_2)$ variate. Deduce that:*

$$P\left[F(n_1, n_2) \geq c\right] = P\left[F(n_2, n_1) \leq \frac{1}{c}\right]$$

(b). *Show how probability points of $F(n_2, n_1)$ can be obtained from those of $F(n_1, n_2)$.*

Solution (a). Let X and Y be independent chi-square variates with n_1, and n_2 d.f. respectively. Then, by definition, we have

$$F = \frac{(X/n_1)}{(Y/n_2)} \sim F(n_1, n_2) \quad \text{and} \quad \frac{1}{F} = \frac{(Y/n_2)}{(X/n_1)} \sim F(n_2, n_1) \qquad \text{...(*)}$$

Hence, the result.

We have :
$$P\left[F(n_1, n_2) \geq c\right] = P\left[\frac{1}{F(n_1, n_2)} \leq \frac{1}{c}\right] = P\left[F(n_2, n_1) \leq \frac{1}{c}\right] \qquad \text{[From(*)]}$$

(b). *Probability Points of $F(n_2, n_1)$ from Those of $F(n_1, n_2)$ Distribution.*

Let $P\left[F(n_1, n_2) \geq c\right] = \alpha$, *i.e.*, let c be the upper α-significant point of $F(n_1, n_2)$ distribution.

$$\therefore \qquad 1 - \alpha = 1 - P\left[F(n_1, n_2) \geq c\right] = 1 - P\left[\frac{1}{F(n_1, n_2)} \leq \frac{1}{c}\right]$$

$$\Rightarrow \qquad \alpha = P\left[F(n_2, n_1) \leq \frac{1}{c}\right] = 1 - P\left[F(n_2, n_1) \geq \frac{1}{c}\right] \qquad \text{[From(*)]}$$

$$\therefore \qquad P\left[F(n_2, n_1) \geq \frac{1}{c}\right] = 1 - \alpha$$

Thus, $(1 - \alpha)$ significant points of $F(n_2, n_1)$ distribution are the reciprocal of α-significant points of $F(n_1, n_2)$ distribution, e.g., $\quad F_{8,4}(0{\cdot}05) = 6{\cdot}04 \quad \Rightarrow \quad F_{4,8}(0{\cdot}95) = \dfrac{1}{6{\cdot}04}$.

Example 16·22. *Prove that if $n_1 = n_2$, the median of F-distribution is at F = 1 and that the quartiles Q_1 and Q_3 satisfy the condition:* $Q_1 Q_3 = 1$.

Solution. Since $n_1 = n_2 = n$, (say), the median (M) of $F(n_1, n_2) = F(n, n)$ distribution is given by :

$$P\left[F(n, n) \leq M\right] = 0{\cdot}5 \qquad \text{...(*)}$$

$$\Rightarrow \qquad P\left[\frac{1}{F(n, n)} \geq \frac{1}{M}\right] = 0{\cdot}5 \Rightarrow P\left[F(n, n) \geq \frac{1}{M}\right] = 0{\cdot}5 \qquad \left[\because \frac{1}{F(m, n)} = F(n, m)\right]$$

$$\Rightarrow \qquad P\left[F(n, n) \leq \frac{1}{M}\right] = 1 - P\left[F(n, n) \geq \frac{1}{M}\right] = 1 - 0{\cdot}5 = 0{\cdot}5 \qquad \text{...(**)}$$

From (*) and (**), we get $M = \dfrac{1}{M} \Rightarrow M^2 = 1$ or $M = 1$, the negative value $M = -1$, is discarded since $F > 0$.

Hence, the median of $F(n, n)$ distribution is at $F = 1$.

Similarly by definition of Q_1 and Q_3, we have

$$P\left[F(n, n) \leq Q_1\right] = 0{\cdot}25 \qquad \text{...(**)}$$

and
$$P[F(n,n) \geq Q_3] = 0{\cdot}25 \Rightarrow P\left[\frac{1}{F(n,n)} \leq \frac{1}{Q_3}\right] = 0{\cdot}25$$

$$\Rightarrow \quad P\left[F(n,n) \leq \frac{1}{Q_3}\right] = 0{\cdot}25 \qquad \left[\because \frac{1}{F(m,n)} = F(n,m)\right] \qquad ...(****)$$

From (***) and (****), we get $Q_1 = \dfrac{1}{Q_3} \Rightarrow Q_1 Q_3 = 1.$

Example 16·23. *Let $X_1, X_2, ..., X_n$ be a random sample from N(0, 1).*

Define $\bar{X}_k = \frac{1}{k}\sum_{1}^{k} X_i$ and $\bar{X}_{n-k} = \frac{1}{n-k}\sum_{k+1}^{n} X_i$. *Find the distribution of :*

(a) $\frac{1}{2}\left(\bar{X}_k + \bar{X}_{n-k}\right)$, (b) $k\bar{X}_k^2 + (n-k)\bar{X}_{n-k}^2$, (c) X_1^2/X_2^2, and (d) X_1/X_2.

Solution.

(a) Since $X_1, X_2, ..., X_n$ *is a random sample from N (0, 1),*

$$\bar{X}_k \sim N\left(0, \frac{1}{k}\right) \text{ and } \bar{X}_{n-k} \sim N\left(0, \frac{1}{n-k}\right) \qquad ...(*)$$

Further, since $(X_1, X_2, ..., X_k)$ and $(X_{k+1}, X_{k+2}, ..., X_n)$ are independent, $\bar{X}_k$ and $\bar{X}_{n-k}$ are independent. Hence,

$$\frac{1}{2}\left(\bar{X}_k + \bar{X}_{n-k}\right) = \frac{1}{2}\bar{X}_k + \frac{1}{2}\bar{X}_{n-k} \sim N\left\{0, \frac{1}{4k} + \frac{1}{4(n-k)}\right\}$$

$$\Rightarrow \quad \frac{1}{2}\left(\bar{X}_k + \bar{X}_{n-k}\right) \sim N\left\{0, \frac{n}{4k(n-k)}\right\}$$

(b) From (*), we get

$$\frac{\bar{X}_k}{\sqrt{1/k}} \sim N(0,1) \quad \text{and} \quad \frac{\bar{X}_{n-k}}{\sqrt{1/(n-k)}} \sim N(0,1)$$

$$k\bar{X}_k^2 \sim \chi^2_{(1)} \quad \text{and} \quad (n-k)\bar{X}_{n-k}^2 \sim \chi^2_{(1)}$$

Since $\bar{X}_k$ and $\bar{X}_{n-k}$ are independent, by additive property of chi-square distribution,

$$k\bar{X}_k^{\,2} + (n-k)\bar{X}_{n-k}^2 \sim \chi^2_{(1+1)} = \chi^2_{(2)}$$

(c) Since $X_1 \sim N(0, 1)$ and $X_2 \sim N(0, 1)$ are independent, $X_1^2 \sim \chi^2_{(1)}$ and $X_2^2 \sim \chi^2_{(1)}$, are also independent. Hence by definition of F-statistic,

$$\frac{X_1^2/1}{X_2^2/1} \sim F_{1,1} \quad \Rightarrow \quad \frac{X_1^2}{X_2^2} \sim F_{(1,1)}$$

(d) X_1/X_2, being the ratio of two independent standard normal variates is a standard Cauchy variate.

Example 16·24. *X is a binomial variate with parameters n and p and F_{ν_1,ν_2} is an F-statistic with v_1 and v_2 d.f. Prove that :*

$$P(X \leq k-1) = P\left[F_{2k,2(n-k+1)} > \frac{(n-k+1)}{k} \cdot \frac{p}{1-p}\right]$$

Solution. If $X \sim B(n, p)$, then we have

$$P(X \le k-1) = (n-k+1)\cdot\binom{n}{k-1}\int_0^q t^{n-k}(1-t)^{k-1}\,dt$$
$$= \frac{1}{B(k, n-k+1)}\int_0^q t^{n-k}(1-t)^{k-1}\,dt \quad \ldots (*)$$

[See Chapter 8 - Binomial distribution.]

$$P = P_r\left[F_{2k, 2(n-k+1)} > \frac{n-k+1}{k}\left(\frac{p}{1-p}\right)\right] = \int_{\frac{n-k+1}{k}\cdot\frac{p}{q}}^{\infty} p\left[F_{2k,2(n-k+1)}\right]dF$$

$$= \frac{1}{B(k, n-k+1)}\int_{\frac{n-k+1}{k}\cdot\frac{p}{q}}^{\infty}\frac{[k/(n-k+1)]^k\cdot F^{k-1}}{\left(1+\dfrac{kF}{n-k+1}\right)^{n+1}}\,dF \quad \ldots (**)$$

Put $\quad 1+\dfrac{kF}{n-k+1} = \dfrac{1}{y} \;\Rightarrow\; F = \dfrac{n-k+1}{k}\left(\dfrac{1-y}{y}\right)$ and $dF = \dfrac{n-k+1}{k}\cdot\dfrac{-dy}{y^2}$

$$F = \infty \;\Rightarrow\; y = 0 \text{ and } F = \frac{n-k+1}{k}\cdot\frac{p}{q} \;\Rightarrow\; \frac{1}{y} = \frac{q+p}{q} = \frac{1}{q} \;\Rightarrow\; y = q$$

Substituting in (**), we get

$$P = \frac{1}{B(k, n-k+1)}\int_q^0\left(\frac{1-y}{y}\right)^{k-1}\cdot y^{n+1}\left(\frac{-dy}{y^2}\right)$$
$$= \frac{1}{B(k, n-k+1)}\int_0^q y^{n-k}(1-y)^{k-1}\,dy \quad \ldots (***)$$

From (*), (**) and (***), we get the result.

16·7. APPLICATIONS OF *F*-DISTRIBUTION

F-distribution has the following applications in statistical theory.

16·7·1. *F*-test for Equality of Two Population Variances

Suppose we want to test (*i*) whether two independent samples x_i, $(i = 1, 2, \ldots, n_1)$ and y_j, $(j = 1, 2, \ldots, n_2)$ have been drawn from the normal populations with the same variance σ^2 (say), or (*ii*) whether the two independent estimates of the population variance are homogeneous or not.

Under the null hypothesis (H_0) that (*i*) $\sigma_X^2 = \sigma_Y^2 = \sigma^2$, *i.e., the population variances are equal, or (ii) Two independent estimates of the population variance are homogeneous,* the statistic *F* is given by :

$$F = \frac{S_X^2}{S_Y^2} \quad \ldots (16\cdot17)$$

where $\quad S_X^2 = \dfrac{1}{n_1-1}\sum_{i=1}^{n_1}(x_i-\bar{x})^2 \quad$ and $\quad S_Y^2 = \dfrac{1}{n_2-1}\sum_{j=1}^{n_2}(y_j-\bar{y})^2 \quad \ldots (16\cdot17a)$

are unbiased estimates of the common population variance σ^2 obtained from two independent samples and it follows Snedecor's *F*-distribution with (ν_1, ν_2) *d.f.* where $\nu_1 = n_1 - 1$ and $\nu_2 = n_2 - 1$.

Proof.

$$F = \frac{S_X^2}{S_Y^2} = \frac{\left[\dfrac{n_1}{n_1-1}s_X^2\right]}{\left[\dfrac{n_2}{n_2-1}s_Y^2\right]} = \frac{\left[\dfrac{n_1 s_X^2}{\sigma_X^2}\cdot\dfrac{1}{(n_1-1)}\right]}{\left[\dfrac{n_2 s_Y^2}{\sigma_Y^2}\cdot\dfrac{1}{(n_2-1)}\right]} \qquad (\because \sigma_X^2 = \sigma_Y^2 = \sigma^2, \text{ under } H_0)$$

Since $\frac{n_1 s_X^2}{\sigma_X^2}$ and $\frac{n_2 s_Y^2}{\sigma_Y^2}$ are independent chi-square variates with $(n_1 - 1)$ and $(n_2 - 1)$ *d.f.* respectively, F follows Snedecor's F-distribution with $(n_1 - 1, n_2 - 1)$ *d.f.* (*c.f.* § 16·5).

Remarks:

1. In (16·17), greater of the two variances S_X^2 and S_Y^2 is to be taken in the numerator and n_1 corresponds to the greater variance.
   By comparing the calculated value of F obtained by using (16·17) for the two given samples, with the tabulated value of F for (n_1, n_2) *d.f.* at certain level of significance (5% or 1%), H_0 is either rejected or accepted.
2. *Critical values of F-distribution.* The available F-tables (given in Tables IX and IXA at the end of the book) give the critical values of F for the right-tailed test, *i.e.,* the critical region is determined by the right-tail areas. Thus the significant value $F_\alpha(n_1, n_2)$ at level of significance α and (n_1, n_2) *d.f.* is determined by :

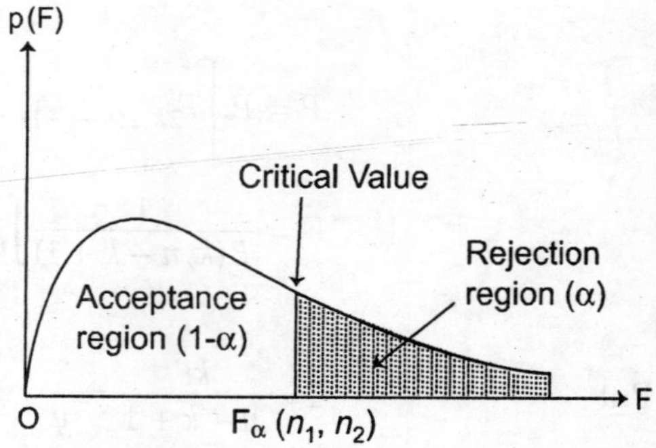

Fig. 16·3 : Critical Values of F-Distribution

$$P[F > F_\alpha(n_1, n_2)] = \alpha, \quad \ldots (*)$$

as shown in Fig. 16·3·

From the *Remark to Example 16·21*, we have the following reciprocal relation between the upper and lower α-significant points of F-distribution :

$$F_\alpha(n_1, n_2) = \frac{1}{F_{1-\alpha}(n_2, n_1)} \quad \Rightarrow \quad F_\alpha(n_1, n_2) \times F_{1-\alpha}(n_2, n_1) = 1 \quad \ldots (**)$$

The critical values of F for left tail–test $H_0 : \sigma_1^2 = \sigma_2^2$ against $H_1 : \sigma_1^2 < \sigma_2^2$ are given by $F < F_{n_1-1, n_2-1}(1-\alpha)$, and for the two – tailed test, $H_0 : \sigma_1^2 = \sigma_2^2$ against $H_1 : \sigma_1^2 \neq \sigma_2^2$ are given by $F > F_{n_1-1, n_2-1}(\alpha/2)$ and $F < F_{n_1-1, n_2-1}(1-\alpha/2)$ [For details, see Chapter 18]

Example 16·25. *Pumpkins were grown under two experimental conditions. Two random samples of 11 and 9 pumpkins show the sample standard deviations of their weights as 0·8 and 0·5 respectively. Assuming that the weight distributions are normal, test the hypothesis that the true variances are equal, against the alternative that they are not, at the 10% level.*

[Assume that $P(F_{10, 8} \geq 3{\cdot}35) = 0{\cdot}05$ and $P(F_{8, 10} \geq 3{\cdot}07) = 0{\cdot}05$.]

Solution. We want to test *Null Hypothesis,* $H_0 : \sigma_X^2 = \sigma_Y^2$ against the *Alternative Hypothesis,*

$H_1 : \sigma_X^2 \neq \sigma_Y^2$ (Two-tailed).

We are given : $n_1 = 11, \quad n_2 = 9, \quad s_X = 0{\cdot}8$ and $s_Y = 0{\cdot}5$.

Under the null hypothesis, $H_0 : \sigma^2_X = \sigma^2_Y$, the statistic :

$F = \frac{S_X^2}{S_Y^2}$, follows F distribution with $(n_1 - 1, n_2 - 1)$ *d.f.*

$$n_1 s_X^2 = (n_1 - 1) S_X^2 \Rightarrow S_X^2 = \left(\frac{n_1}{n_1 - 1}\right) s_X^2 = \left(\frac{11}{10}\right) \times (0{\cdot}8)^2 = 0{\cdot}704$$

Similarly,
$$S_Y^2 = \left(\frac{n_2}{n_2 - 1}\right) s_Y^2 = \left(\frac{9}{8}\right) \times (0{\cdot}5)^2 = 0{\cdot}28125$$

$\therefore$
$$F = \frac{0{\cdot}704}{0{\cdot}28125} = 2{\cdot}5$$

The significant values of F for two-tailed test at level of significance $\alpha = 0{\cdot}10$ are :

$$F > F_{10,8}\,(\alpha/2) = F_{10,8}\,(0{\cdot}05)$$

and $$F < F_{10,8}(1-\alpha/2) = F_{10,8}\,(0{\cdot}95) \qquad \ldots (*)$$

We are given the tabulated (significant) values :

$$P\,(F_{10,8} \geq 3{\cdot}35) = 0{\cdot}05 \quad \Rightarrow \quad F_{10,8}\,(0{\cdot}05) = 3{\cdot}35 \qquad \ldots (**)$$

Also $$P(F_{8,\,10} \geq 3{\cdot}07) = 0{\cdot}05 \quad \Rightarrow \quad P\left(\frac{1}{F_{8,10}} \leq \frac{1}{3{\cdot}07}\right) = 0{\cdot}05$$

$$\Rightarrow \quad P\,(F_{10,\,8} \leq 0{\cdot}326) = 0{\cdot}05 \quad \Rightarrow \quad P(F_{10,8} \geq 0{\cdot}326) = 0{\cdot}95 \qquad \ldots (***)$$

Hence, from (*), (**) and (***), the critical values for testing $H_0 : \sigma_X{}^2 = \sigma_Y{}^2$, against $H_1 : \sigma_X^2 \neq \sigma_Y^2$ at level of significance $\alpha = 0{\cdot}10$ are given by :

$$F > 3{\cdot}35 \text{ and } F < 0{\cdot}326 = 0{\cdot}33$$

Since, the calculated value of F (=2·5) lies between 0·33 and 3·35, it is not significant and hence null hypothesis of equality of population variances may be accepted at level of significance $\alpha = 0{\cdot}10$.

Example 16·26. *In one sample of 8 observations, the sum of the squares of deviations of the sample values from the sample mean was 84·4 and in the other sample of 10 observations it was 102·6. Test whether this difference is significant at 5 per cent level, given that the 5 per cent point of F for $n_1 = 7$ and $n_2 = 9$ degrees of freedom is 3·29.*

Solution. Here $n_1 = 8, n_2 = 10$ and $\Sigma(x - \bar{x})^2 = 84{\cdot}4$, $\Sigma(y - \bar{y})^2 = 102{\cdot}6$

$$\therefore \quad S_X^2 = \frac{1}{n_1 - 1}\Sigma(x-\bar{x})^2 = \frac{84{\cdot}4}{7} = 12{\cdot}057$$

$$S_Y^2 = \frac{1}{n_2 - 1}\Sigma(y-\bar{y})^2 = \frac{102{\cdot}6}{9} = 11{\cdot}4$$

Under $H_0 : \sigma_X{}^2 = \sigma_Y{}^2 = \sigma^2$, *i.e.*, the estimates of σ^2 given by the samples are homogeneous, the test statistic is :

$$F = \frac{S_X^2}{S_Y^2} = \frac{12{\cdot}057}{11{\cdot}4} = 1{\cdot}057$$

Tabulated $F_{0{\cdot}05}$ for (7, 9) *d.f.* is 3·29) (given).

Since calculated $F <$ tabulated $F_{0{\cdot}05}$, H_0 may be accepted at 5% level of significance and it may be concluded, that difference in sample variability is not significant and is due to fluctuations of sampling.

Example 16·27. *Two random samples gave the following results :*

Sample	*Size*	*Sample mean*	*Sum of squares of deviations from the mean*
1	10	15	90
2	12	14	108

Test whether the samples come from the same normal population at 5% level of significance.

[*Given* : $F_{0{\cdot}05}\,(9, 11) = 2{\cdot}90$, $F_{0{\cdot}05}(11, 9) = 3{\cdot}10$ (approx.) and $t_{0{\cdot}05}(20) = 2{\cdot}086$, $t_{0{\cdot}05}(22) = 2{\cdot}07$]

Solution. A normal population has two parameters, *viz.*, mean μ and variance σ^2. To test if two independent samples have been drawn from the same normal population, we have to test (*i*) the equality of population means, and (*ii*) the equality of population variances.

Null Hypothesis : The two samples have been drawn from the same normal population, *i.e.*,

$$H_0 : \mu_1 = \mu_2 \quad \text{and} \quad \sigma_1^2 = \sigma_2^2.$$

Equality of means will be tested by applying t-test and equality of variances will be tested by applying F-test. Since t-test assumes $\sigma_1^2 = \sigma_2^2$, we shall first apply F-test and then t-test. In usual notations, we are given :

$$n_1 = 10, \quad n_2 = 12; \quad \bar{x}_1 = 15, \quad \bar{x}_2 = 14, \quad \Sigma(x_1 - \bar{x}_1)^2 = 90, \quad \Sigma(x_2 - \bar{x}_2)^2 = 108.$$

***F*-test** : Here,

$$S_1^2 = \frac{1}{n_1 - 1}\Sigma(x_1 - \bar{x}_1)^2 = \frac{90}{9} = 10, \qquad S_2^2 = \frac{1}{n_2 - 1}\Sigma(x_2 - \bar{x}_2)^2 = \frac{108}{11} = 9{\cdot}82$$

Since $S_1^2 > S_2^2$, under $H_0 : \sigma_1^2 = \sigma_2^2$, the test statistic is

$$F = \frac{S_1^2}{S_2^2} \sim F(n_1 - 1, n_2 - 1) = F(9,11)$$

$$\therefore \qquad F = \frac{S_1^2}{S_2^2} = \frac{10}{9{\cdot}82} = 1{\cdot}018$$

Tabulated $F_{0{\cdot}05}$ (9, 11) = 2·90. Since calculated F is less than tabulated F, it is not significant. Hence, null hypothesis of equality of population variances may be accepted.

Since $\sigma_1^2 = \sigma_2^2$, we can now apply t test for testing $H_0 : \mu_1 = \mu_2$.

***t*-test** : Under $H_0' : \mu_1 = \mu_2$, against alternative hypothesis, $H_1' : \mu_1 \neq \mu_2$, the test statistic is :

$$t = \frac{\bar{x}_1 - \bar{x}_2}{\sqrt{S^2\left(\frac{1}{n_1} + \frac{1}{n_2}\right)}} \sim t_{n_1+n_2-2} = t_{20}$$

where
$$S^2 = \frac{1}{n_1 + n_2 - 2}\left[\Sigma(x_1 - \bar{x}_1)^2 + \Sigma(x_2 - \bar{x}_2)^2\right] = \frac{1}{20}(90 + 108) = 9{\cdot}9$$

$$\therefore \qquad t = \frac{15 - 14}{\sqrt{9{\cdot}9\left(\frac{1}{10} + \frac{1}{12}\right)}} = \frac{1}{\sqrt{9{\cdot}9 \times \frac{11}{60}}} = \frac{1}{\sqrt{1{\cdot}815}} = 0{\cdot}742$$

Tabulated $t_{0{\cdot}05}$ for 20 *d.f.* = 2·086. Since $|t| < t_{0{\cdot}05}$, it is not significant. Hence, the hypothesis $H_0' : \mu_1 = \mu_2$ may be accepted. Since both the hypotheses, *i.e.*, $H_0' : \mu_1 = \mu_2$ and $H_0 : \sigma_1^2 = \sigma_2^2$ are accepted, we may regard that the given samples have been drawn from the same normal population.

16·7·2. F-test for Testing the Significance of an Observed Multiple Correlation Coefficient

If R is the observed multiple correlation coefficient of a variate with k other variates in a random sample of size n from a $(k + 1)$ variate population, then Prof. R.A. Fisher proved that under the null hypothesis (H_0) *that the multiple correlation coefficient in the population is zero, the statistic* :

$$F = \frac{R^2}{1 - R^2} \cdot \frac{n - k - 1}{k} \qquad \ldots (16{\cdot}18)$$

conforms to F-distribution with $(k, n - k - 1)$ *d.f.*

16·7·3. F-test for Testing the Significance of an Observed Sample Correlation Ratio η_{YX}

Under the null hypothesis that *population correlation ratio is zero,* the test statistic is :

$$F = \frac{\eta^2}{1 - \eta^2} \cdot \frac{N - h}{h - 1} \sim F(h - 1, N - h) \qquad \ldots (16{\cdot}19)$$

where N is the size of the sample (from a bi-variate normal population) arranged in h arrays.

16·7·4. F-test for Testing the Linearity of Regression

For a sample of size N arranged in h arrays, from a bi-variate normal population, the test statistic for testing the hypothesis of linearity of regression is :

$$F = \frac{\eta^2 - r^2}{1-\eta^2} \cdot \frac{N-h}{h-2} \sim F(h-2, N-h) \qquad \text{... (16·20)}$$

16·7·5. F-test for Equality of Several Means

This test is carried out by the technique of Analysis of Variance, which plays a very important and fundamental role in Design of Experiments in Agricultural Statistics.

[For a detailed discussion of the Analysis of Variance Technique, see Fundamentals of Applied Statistics by the same authors.]

16·8. RELATION BETWEEN *t* AND *F* DISTRIBUTIONS

In F-distribution with (υ_1, υ_2) *d.f.* [*c.f.* 16·13 (*a*)], take $\nu_1 = 1$, $\nu_2 = v$ and $t^2 = F$, *i.e.*, $dF = 2t\, dt$. Thus, the probability differential of F transforms to :

$$dG(t) = \frac{(1/\nu)^{1/2}}{B\left(\frac{1}{2}, \frac{\nu}{2}\right)} \cdot \frac{\left(t^2\right)^{(1/2)-1}}{\left(1+\frac{t^2}{\nu}\right)^{(\nu+1)/2}} \, 2t\, dt, \quad 0 \le t^2 < \infty$$

$$= \frac{1}{\sqrt{\nu}\, B\left(\frac{1}{2}, \frac{\nu}{2}\right)} \cdot \frac{1}{\left(1+\frac{t^2}{\nu}\right)^{(\nu+1)/2}} dt, \quad -\infty < t < \infty$$

the factor 2 disappearing since the total probability in the range $(-\infty, \infty)$ is unity. This is the probability function of Student's t-distribution with ν *d.f.* Hence, we have the following relation between t and F distributions.

If a statistic t follows Student's t distribution with n d.f., then t^2 follows Snedecor's F-distribution with (1, n) d.f. Symbolically,

$$\left.\begin{array}{ll} \text{if} & t \sim t_{(n)} \\ \text{then} & t^2 \sim F_{(1,n)} \end{array}\right\} \qquad \text{... (16·21)}$$

Aliter Proof of (16·21). *If $\xi \sim N(0, 1)$ and $X \sim \chi^2_{(n)}$ are independent r.v.'s, then :*

$$U = \xi^2 \sim \chi^2_{(1)} \qquad \text{[Square of a } S.N.V.\text{]}$$

and

$$t = \frac{\xi}{\sqrt{X/n}} \sim t_{(n)} \Rightarrow t^2 = \frac{\xi^2}{(X/n)} = \frac{(\xi^2/1)}{(X/n)},$$

being the ratio of two independent chi-square variates divided by their respective degrees of freedom is $F(1, n)$ variate. Hence, $t^2 \sim F(1, n)$.

With the help of relation (16·21), all the uses of t-distribution can be regarded as the applications of F-distribution also, *e.g.*, for test for a single mean, instead of computing $t = \frac{\bar{x}-\mu}{S/\sqrt{n}}$, we may compute $F = t^2 = \frac{n(\bar{x}-\mu)^2}{S^2}$, and then apply F-test with $(1, n)$ *d.f.*, and so on.

Similarly, we can write the test statistic F from § 16·3·4, § 16·3·5 and § 16·3·6 for testing the significance of an observed sample correlation coefficient, regression coefficient and partial correlation coefficient respectively.

Example 16·28. *Given :* $P[F(10, 12) > 2{\cdot}753] = 0{\cdot}05 = P[F(1, 12) > 4{\cdot}747]$,

find $P[F(12, 10) > (2{\cdot}753)^{-1}]$, *and* $P(-\sqrt{4{\cdot}747} < t_{12} < \sqrt{4{\cdot}747})$

Solution.

$$P[F(12, 10) > (2{\cdot}753)^{-1}] = P\left[\frac{1}{F(12,10)} < 2{\cdot}753\right] = P\,[F(10, 12) < 2{\cdot}753]$$

$$= 1 - P[F(10, 12) > 2{\cdot}753] = 1 - 0{\cdot}05 = 0{\cdot}95$$

$$P(-\sqrt{4{\cdot}747} < t_{12} < \sqrt{4{\cdot}747}) = P(t^2_{12} < 4{\cdot}747) = P[F(1, 12) < 4{\cdot}747]$$

$$= 1 - P[F(1, 12) > 4{\cdot}747] = 1 - 0{\cdot}05 = 0{\cdot}95$$

16·9. RELATION BETWEEN F AND χ^2 DISTRIBUTION

In $F\ (n_1, n_2)$ distribution if we let $n_2 \to \infty$, then $\chi^2 = n_1 F$ follows χ^2-distribution with n_1 *d.f.*

Proof.

$$f(F) = \frac{(n_1/n_2)^{n_1/2}\, F^{(n_1/2)-1}}{\Gamma(n_1/2)\Gamma(n_2/2)} \cdot \frac{\Gamma[(n_1+n_2)/2]}{\left(1+\dfrac{n_1}{n_2}F\right)^{(n_1+n_2)/2}},\quad 0 < F < \infty \qquad \dots(16{\cdot}22)$$

In the limit as $n_2 \to \infty$, we have

$$\frac{\Gamma[(n_1+n_2)/2]}{{n_2}^{n_1/2}\Gamma(n_2/2)} \to \frac{(n_2/2)^{n_1/2}}{{n_2}^{n_1/2}} = \frac{1}{2^{n_1/2}} \qquad \dots(*)$$

$$\left[\because \frac{\Gamma(n+k)}{\Gamma(n)} \to n^k \text{ as } n \to \infty. (c.f.\ \text{Remark below})\right]$$

Also

$$\lim_{n_2\to\infty}\left(1+\frac{n_1}{n_2}F\right)^{(n_1+n_2)/2} = \lim_{n_2\to\infty}\left[\left(1+\frac{n_1}{n_2}F\right)^{n_2}\right]^{1/2} \times \lim_{n_2\to\infty}\left(1+\frac{n_1}{n_2}F\right)^{n_1/2} \qquad \dots(**)$$

$$= \exp\,(n_1F/2) = \exp(\chi^2/2) \qquad (\because\ n_1F = \chi^2)$$

Hence, in the limit, on using (*) and (**), the *p.d.f.* of $\chi^2 = n_1F$ becomes :

$$dP(\chi^2) = \frac{(n_1/2)^{n_1/2} e^{-\chi^2/2}}{\Gamma(n_1/2)} \cdot \left(\frac{\chi^2}{n_1}\right)^{(n_1/2)-1} d\left(\frac{\chi^2}{n_1}\right)$$

$$= \frac{1}{2^{n_1/2}\Gamma(n_1/2)} \cdot e^{-\chi^2/2}(\chi^2)^{(n_1/2)-1} d\chi^2,\ 0 < \chi^2 < \infty$$

which is the p.d.f. of chi-square distribution with n_1 *d.f.*

Remark.

$$\lim_{n\to\infty}\frac{\Gamma(n+k)}{\Gamma(n)} = \lim_{n\to\infty}\frac{(n+k-1)!}{(n-1)!},\ \text{(for large } n) \simeq \lim_{n\to\infty}\frac{\sqrt{2\pi}\, e^{-(n+k-1)}(n+k-1)^{n+k-(1/2)}}{\sqrt{2\pi}\, e^{-(n-1)}(n-1)^{n-(1/2)}}$$

(On using Stirling's approximation for $n\,!$ as $n \to \infty$.)

$$= e^{-k}\lim_{n\to\infty}\frac{n^{n+k-\frac{1}{2}}\left(1+\dfrac{k-1}{n}\right)^{n+k-\frac{1}{2}}}{n^{n-\frac{1}{2}}\left(1-\dfrac{1}{n}\right)^{n-\frac{1}{2}}} = e^{-k}n^k\,\frac{\lim\limits_{n\to\infty}\left(1+\dfrac{k-1}{n}\right)^n \lim\limits_{n\to\infty}\left(1+\dfrac{k-1}{n}\right)^{k-\frac{1}{2}}}{\lim\limits_{n\to\infty}\left(1-\dfrac{1}{n}\right)^n \lim\limits_{n\to\infty}\left(1-\dfrac{1}{n}\right)^{-\frac{1}{2}}}$$

$$= e^{-k} n^k \left[\frac{e^{(k-1)} \cdot 1}{e^{-1} 1} \right] = n^k$$

$$\therefore \quad \lim_{n \to \infty} \frac{\Gamma(n+k)}{\Gamma n} = n^k \qquad \ldots (16{\cdot}22a)$$

16·10 NON-CENTRAL F-DISTRIBUTION

The ratio of two independent χ^2 variates each divided by the corresponding d.f. has a non-central F-distribution if the numerator has non-central χ^2 -distribution and the denominator has a central χ^2 -distribution. Thus, if X has a non-central χ^2 -distribution with n_1 d.f. and non-centrality parameter λ, *i.e.*, if $X \sim \chi'^2_{n_1}$ and Y is an independent (central) χ^2 -variate with n_2 d.f. *i.e.*, if $Y \sim \chi^2_{n_2}$, then the non-central F-statistic is defined as : $F' = \dfrac{X/n_1}{Y/n_2} = \dfrac{n_2 X}{n_1 Y}$ ... (16.23)

p.d.f. of F′. Since X and Y are independent, their joint p.d.f. is given by:

$$p(x,y) = p_1(x) \cdot p_2(y) = \left[\sum_{i=0}^{\infty} \frac{e^{-\lambda} \lambda^i}{i!} \cdot \frac{e^{-x/2} x^{(n_1/2)+i-1}}{2^{(n_1+2i)/2} \Gamma[(n_1+2i)/2]} \right] \times \frac{e^{-y/2} y^{(n_2/2)-1}}{2^{n_2/2} \Gamma(n_2/2)}; \; 0 \le (x,y) < \infty$$

Let us transform to the new set of *r.v.'s* F' and U defined by the transformation :

$$F' = \frac{n_2 x}{n_1 y} \quad \text{and} \quad u = y \quad \Rightarrow \quad y = u, \; x = \frac{n_1 u F'}{n_2}$$

$$J = \frac{\partial(x,y)}{\partial(F',u)} = \begin{vmatrix} \frac{n_1}{n_2} u & \frac{n_1}{n_2} F' \\ 0 & 1 \end{vmatrix} = \frac{n_1}{n_2} u$$

The joint p.d.f. of F' and U is given by

$$g(F',u) = \left\{ \sum_{i=0}^{\infty} \frac{e^{-\lambda} \lambda^i}{i!} \frac{\exp\left[-\frac{n_1 u F'}{2n_2} \right] \cdot \left(\frac{n_1 u F'}{n_2} \right)^{(n_1/2)+i-1}}{2^{i+(n_1+n_2)/2} \Gamma[(n_2/2)] \, \Gamma[(n_1+2i)/2]} \right\} \times e^{-u/2} u^{(n_2/2)-1} \cdot \left(\frac{n_1}{n_2} \right) u$$

$$= \frac{n_1}{n_2} \sum_{i=0}^{\infty} \left\{ \frac{e^{-\lambda} \lambda^i}{i!} \cdot \frac{\left(\frac{n_1}{n_2} F' \right)^{(n_1/2)+i-1}}{2^{i+(n_1+n_2)/2} \, \Gamma[(n_2/2)] \Gamma[(n_1+2i)/2]} \right.$$

$$\left. \times \exp\left[-\frac{u}{2}\left(1 + \frac{n_1}{n_2} F' \right) \right] \cdot u^{\frac{n_1+n_2}{2}+i-1} \right\}; \; = 0 \le F' < \infty, \; 0 < u < \infty$$

Integrating it *w.r.t.* u between the limits 0 to ∞ and using Gamma Integral, we obtain the marginal p.d.f. of F' *as :*

$$g_1(F') = \frac{n_1}{n_2} \sum_{i=0}^{\infty} \left\{ \frac{e^{-\lambda} \lambda^i}{i!} \times \frac{\left(\frac{n_1}{n_2} F' \right)^{(n_1/2)+i-1}}{2^{i+(n_1+n_2)/2} \Gamma[n_2/2] \cdot \Gamma[(n_1+2i)/2]} \times \frac{\Gamma\left(\frac{n_1+n_2}{2} + i \right)}{\left[\frac{1}{2}\left(1 + \frac{n_1}{n_2} F' \right) \right]^{i+(n_1+n_2)/2}} \right\}$$

$$= \frac{n_1}{n_2} \sum_{i=0}^{\infty} \left\{ \frac{e^{-\lambda}\lambda^i}{i!} \times \frac{\left(\frac{n_1}{n_2}F'\right)^{(n_1/2)+i-1}}{B\left(\frac{n_1}{2}+i, \frac{n_2}{2}\right)} \times \frac{1}{\left(1+\frac{n_1}{n_2}F'\right)^{i+(n_1+n_2)/2}} \right\}; \; 0 \le F' < \infty \qquad ...(16·24)$$

which is p.d.f. of non-central F distribution with non-centrality parameter λ.

Remarks:

1. Putting $\lambda = 0$ in (16·24), we get

$$g_1(F') = \frac{n_1}{n_2} \frac{1}{B\left(\frac{n_1}{2}, \frac{n_2}{2}\right)} \frac{\left(\frac{n_1}{n_2}F'\right)^{(n_1/2)-1}}{\left(1+\frac{n_1}{n_2}F'\right)^{(n_1+n_2)/2}} \; ; \; 0 \le F' < \infty, \qquad ...(16·24a)$$

since, for $\lambda = 0$, we get the contribution from the sum only when $i = 0$ and other terms vanish. Thus, for $\lambda = 0$, $g_1(F')$ reduces to the p.d.f. of central F-distribution with (n_1, n_2) d.f.

2. The hyper-geometric function of first kind is defined by

$${}_1F_1(a,b,y) = \sum_{i=0}^{\infty} \frac{\Gamma(a+i)\cdot\Gamma b}{\Gamma a \cdot \Gamma(b+i)} \cdot \frac{y^i}{i!} \qquad ...(16·25)$$

$$\therefore \quad {}_1F_1\left(\frac{n_1+n_2}{2}, \frac{n_1}{2}, \frac{\lambda n_1 F'}{n_2\left(1+\frac{n_1}{n_2}F'\right)}\right) = \sum_{i=0}^{\infty} \frac{\Gamma\left(\frac{n_1+n_2}{2}+i\right)\Gamma\left(\frac{n_1}{2}\right)}{\Gamma\left(\frac{n_1+n_2}{2}\right)\cdot\Gamma\left(\frac{n_1}{2}+i\right)} \times \frac{(\lambda n_1 F')^i}{\left[n_2\left(1+\frac{n_1}{n_2}F'\right)\right]^i} \times \frac{1}{i!}$$

$$= \sum_{i=0}^{\infty} \frac{B\left(\frac{n_1}{2}, \frac{n_2}{2}\right)}{B\left(\frac{n_1}{2}+i, \frac{n_2}{2}\right)} \cdot \frac{\lambda^i}{i!} \cdot \frac{\left(\frac{n_1}{n_2}\right)^i (F')^i}{\left(1+\frac{n_1}{n_2}F'\right)^i}$$

$$\therefore \quad g_1(F') = \frac{e^{-\lambda}\left(\frac{n_1}{n_2}\right)^{n_1/2} (F')^{\frac{n_1}{2}-1}}{B\left(\frac{n_1}{2}, \frac{n_2}{2}\right)\cdot\left(1+\frac{n_1}{n_2}F'\right)^{\frac{n_1+n_2}{2}}} \times {}_1F_1\left(\frac{n_1+n_2}{2}, \frac{n_1}{2}, \frac{\lambda n_1 F'}{n_2\left(1+\frac{n_1}{n_2}F'\right)}\right) \qquad ...(16·25a)$$

3. It can be proved that the mean of $F'(n_1, n_2)$ distribution is given by:

$$E(F') = \int_0^{\infty} F' \cdot g_1(F')\, dF'$$

$$= \sum_{i=0}^{\infty} \left[\frac{e^{-\lambda}\lambda^i}{i!} \cdot \frac{n_2(n_1+2i)}{n_1(n_2-2)}\right] \; ; \; n_2 > 2. \qquad ...(16·26)$$

If $\lambda = 0$ (in which case we get contribution from the sum only when $i = 0$), we get:

$$E(F') = \frac{n_2}{n_2-2}, \qquad ...(16·27)$$

which is the mean of central F-distribution with (n_1, n_2) *d.f.*

16·11. FISHER'S *Z*- DISTRIBUTION

In G.W. Snedecor's *F*-distribution with (ν_1, ν_2) *d.f.*, if we put

$$F = \exp(2Z) \quad \Rightarrow \quad Z = \frac{1}{2}\log_e F, \qquad \ldots (16\cdot28)$$

the distribution of *Z* becomes:

$$g(z) = p(F)\cdot\left|\frac{dF}{dz}\right| = \frac{(\nu_1/\nu_2)^{\nu_1/2}}{B\left(\frac{\nu_1}{2}, \frac{\nu_2}{2}\right)}\cdot\frac{(e^{2z})^{(\nu_1/2)-1}\cdot 2e^{2z}}{\left[1+\frac{\nu_1}{\nu_2}e^{2z}\right]^{(\nu_1+\nu_2)/2}}$$

$$= 2\frac{(\nu_1/\nu_2)^{\nu_1/2}}{B\left(\frac{\nu_1}{2}, \frac{\nu_2}{2}\right)}\cdot\frac{e^{\nu_1 z}}{\left[1+\frac{\nu_1}{\nu_2}e^{2z}\right]^{(\nu_1+\nu_2)/2}} \; ; -\infty < z < \infty \qquad \ldots (16\cdot29)$$

which is the probability function of Fisher's *z*-distribution with (ν_1, ν_2) *d.f.* The tables of significant values z_0 of *z* which will be exceeded in random sampling with probabilities 0·05 and 0·01, *i.e.*, $P(z > z_0) = 0\cdot05$ and $P(z > z_0') = 0\cdot01$ corresponding to various *d.f.* (ν_1, ν_2) were published by Fisher (*c.f.* Statistical Methods for Research Workers) in 1925. From these tables, Snedecor (1934–38) by using (16·28) deduced the tables of significant values of the variance ratio which he denoted by *F* in honour of Prof. R.A. Fisher.

Remark. With the help of relation (16·28), all the applications of *F*-distribution may be regarded as the applications of *Z*-distribution also.

16·11·1. Moment Generating Function of *Z*-distribution

$$M_Z(t) = E(e^{tZ}) = \int_{-\infty}^{\infty} e^{tz} g(z)\, dz = \int_0^{\infty} F^{t/2} f(F) dF \qquad [\because e^{2z} = F]$$

Since μ_r' (about origin) for *F*-distribution is $\int_0^{\infty} F^r f(F) dF$, we can find *m.g.f.* of the *Z*-distribution by putting $r = t/2$ in the expression for μ_r' for *F*-distribution.

$$\text{Hence, } M_Z(t) = \left(\frac{v_2}{v_1}\right)^{t/2}\frac{\Gamma\{(v_1+t)/2\}\cdot\Gamma\{(v_2-t)/2\}}{\Gamma(v_1/2)\cdot\Gamma(v_2/2)} \qquad [c.f. \text{ Equation } (16\cdot15)] \qquad \ldots (16\cdot29a)$$

$$\Rightarrow \quad K_Z(t) = \log M_Z(t)$$

$$= \frac{t}{2}(\log v_2 - \log v_1) + \log\Gamma\{(v_1+t)/2\} + \log\Gamma\{(v_2-t)/2\} - \log\Gamma(v_1/2) - \log\Gamma(v_2/2)$$

Using Stirling's approximation for *n* !, when *n* is large, *viz.*,

$$\lim_{n\to\infty}\Gamma(n+1) = \lim_{n\to\infty} n! \approx \sqrt{2\pi}\, e^{-n} n^{n+\frac{1}{2}}$$

$$\Rightarrow \quad \log\Gamma(n+1) \simeq \left(n+\frac{1}{2}\right)\log n - n + \log\sqrt{2\pi}, \text{ we get}$$

$$\left.\begin{aligned}
\kappa_1 &= \mu_1' = \frac{1}{2}\left(\frac{1}{\nu_2} - \frac{1}{\nu_1}\right), \quad \kappa_2 = \mu_2 = \frac{1}{2}\left(\frac{1}{\nu_2} + \frac{1}{\nu_1} + \frac{1}{\nu_1^2} + \frac{1}{\nu_2^2}\right),\\
\kappa_3 &= \mu_3 = \frac{1}{2}\left[\left(\frac{1}{\nu_2^2} - \frac{1}{\nu_1^2}\right) + \left(\frac{1}{\nu_2^3} + \frac{1}{\nu_1^3}\right)\right]\\
\kappa_4 &= \mu_4 - 3\mu_2^2 = \frac{1}{\nu_1^3} + \frac{1}{\nu_2^3} + 3\left(\frac{1}{\nu_1^4} + \frac{1}{\nu_2^4}\right),
\end{aligned}\right\} \qquad (16\cdot29b)$$

whence β_1 and β_2 can be found.

Remark. Z-distribution tends to normal distribution with mean $\frac{1}{2}\left(\frac{1}{\nu_2}-\frac{1}{\nu_1}\right)$ and variance $\frac{1}{2}\left(\frac{1}{\nu_1}+\frac{1}{\nu_2}\right)$, as ν_1 and ν_2 become large.

16·12. FISHER'S Z-TRANSFORMATION

To test the significance of an observed sample correlation coefficient from an uncorrelated bivariate normal population, *t*-test is used. But in random sample of size *n* from a bivariate normal population in which $\rho \neq 0$, Prof. R.A. Fisher proved that the distribution of '*r*' is by no means normal and in the neighbourhood of $\rho = \pm 1$, its probability curve is extremely skewed even for large *n*. If $\rho \neq 0$, Fisher suggested the following transformation :

$$Z = \frac{1}{2}\log_e\left(\frac{1+r}{1-r}\right) = \tanh^{-1} r \qquad \text{... (16·30)}$$

and proved that even for small samples, the distribution of Z is approximately normal with mean:

$$\xi = \frac{1}{2}\log_e\left(\frac{1+\rho}{1-\rho}\right) = \tanh^{-1}\rho \qquad \text{... (16·30a)}$$

and variance $1/(n-3)$ and for large values of *n*, say > 50, the approximation is fairly good.

Remark. The values of Z have been tabulated for different values of *r* and are given in Table X at the end of the book.

16·12·1. Applications of Z-Transformation.

Z-transformation has the following applications in Statistics.

(1) To test if an observed value of 'r' differs significantly from a hypothetical value ρ of the population correlation coefficient.

H_0 : *There is no significant difference between r and ρ. In other words, the given sample has been drawn from a bivariate normal population with correlation coefficient ρ.* If we take

$$Z = \frac{1}{2}\log_e\{(1+r)/(1-r)\} \quad \text{and} \quad \xi = \frac{1}{2}\log_e\{(1+\rho)/(1-\rho)\}, \text{ then under } H_0,$$

$$Z \sim N\left(\xi, \frac{1}{n-3}\right) \Rightarrow \frac{Z-\xi}{\sqrt{1/(n-3)}} \sim N(0, 1) \qquad \text{...(16·30b)}$$

Thus, if $(Z-\xi)\sqrt{(n-3)} > 1{\cdot}96$, H_0 is rejected at 5% level of significance and if it is greater than 2·58, H_0 is rejected at 1% level of significance, where Z and ξ are defined in (16·30) and (16·30*a*).

Remark. Z defined in equation (16·30) should not be confused with the Z used in Fisher's Z-distribution (*c.f.* § 16·11).

Example 16·29. *A correlation coefficient of 0·72 is obtained from a sample of 29 pairs of observations.*

(i) Can the sample be regarded as drawn from a bivariate normal population in which true correlation coefficient is 0·8?

(ii) Obtain 95% confidence limits for ρ in the light of the information provided by the sample.

Solution.

(*i*) H_0 : There is no significant difference between $r = 0{\cdot}72$; and $\rho = 0{\cdot}80$, *i.e.*, the sample can be regarded as drawn from the bivariate normal population with $\rho = 0{\cdot}8$. Here

$$Z = \frac{1}{2}\log_e\left(\frac{1+r}{1-r}\right) = 1{\cdot}1513\log_{10}\left(\frac{1+r}{1-r}\right) = 1{\cdot}1513\log_{10} 6{\cdot}14 = 0{\cdot}907$$

$$\xi = \frac{1}{2}\log_e\left(\frac{1+\rho}{1-\rho}\right) = 1{\cdot}1513\log_{10}\frac{(1+0{\cdot}8)}{(1-0{\cdot}8)} = 1{\cdot}1513 \times 0{\cdot}9541 = 1{\cdot}1$$

$$\text{S.E. (Z)} = \frac{1}{\sqrt{n-3}} = \frac{1}{\sqrt{29-3}} = \frac{1}{\sqrt{26}} = 0{\cdot}196$$

Under H_0, the test statistic is : $$U = \frac{Z - \xi}{1/\sqrt{n-3}} \sim N(0,1)$$

$$\therefore \qquad U = \frac{(0{\cdot}907 - 1{\cdot}100)}{0{\cdot}196} = -0{\cdot}985$$

Since $|U| < 1{\cdot}96$, it is not significant at 5% level of significance and H_0 may be accepted. Hence, the sample may be regarded as coming from a bivariate normal population with $\rho = 0{\cdot}8$.

(*ii*) 95% confidence limits for ρ on the basis of the information supplied by the sample, are given by :

$$|U| \le 1{\cdot}96 \quad \Rightarrow \quad |Z - \xi| \le 1{\cdot}96 \times \frac{1}{\sqrt{n-3}} = 1{\cdot}96 \times 0{\cdot}196$$

$$\Rightarrow |0{\cdot}907 - \xi| \le 0{\cdot}384 \quad \text{or} \quad 0{\cdot}907 - 0{\cdot}384 \le \xi \le 0{\cdot}907 + 0{\cdot}384$$

$$\Rightarrow 0{\cdot}523 \le \xi \le 1{\cdot}291 \qquad \text{or} \quad 0{\cdot}523 \le \frac{1}{2}\log_e\left(\frac{1+\rho}{1-\rho}\right) \le 1{\cdot}291$$

$$\Rightarrow 0{\cdot}523 \le 1{\cdot}1513 \log_{10}\left(\frac{1+\rho}{1-\rho}\right) \le 1{\cdot}291 \qquad \text{or} \quad \frac{0{\cdot}523}{1{\cdot}1513} \le \log_{10}\left(\frac{1+\rho}{1-\rho}\right) \le \frac{1{\cdot}291}{1{\cdot}1513}$$

$$\therefore \qquad 0{\cdot}4543 \le \log_{10}\left(\frac{1+\rho}{1-\rho}\right) \le 1{\cdot}1213 \qquad \ldots (*)$$

$$\log_{10}\left(\frac{1+\rho}{1-\rho}\right) = 0{\cdot}4543$$

$$\Rightarrow \frac{1+\rho}{1-\rho} = \text{Antilog}\,(0{\cdot}4543) = 2{\cdot}846$$

$$\therefore \quad \rho = \frac{2{\cdot}846 - 1}{2{\cdot}846 + 1} = \frac{1{\cdot}846}{3{\cdot}846} = 0{\cdot}4799$$

and

$$\log_{10}\left(\frac{1+\rho}{1-\rho}\right) = 1{\cdot}1213$$

$$\Rightarrow \frac{1+\rho}{1-\rho} = \text{Antilog}\,(1{\cdot}1213) = 13{\cdot}22$$

$$\therefore \quad \rho = \frac{13{\cdot}22 - 1}{13{\cdot}22 + 1} = \frac{12{\cdot}22}{14{\cdot}22} = 0{\cdot}86$$

Hence, substituting in (*), we get: $$0{\cdot}48 \le \rho \le 0{\cdot}86$$

(2) To test the significance of the difference between two independent sample correlation coefficients. Let r_1 and r_2 be the sample correlation coefficients observed in two independent samples of sizes n_1 and n_2 respectively, then

$$Z_1 = \log_e\left(\frac{1+r_1}{1-r_1}\right) \quad \text{and} \quad Z_2 = \frac{1}{2}\log_e\left(\frac{1+r_2}{1-r_2}\right)$$

Under the null hypothesis, H_0 : *that sample correlation coefficients do not differ significantly, i.e., the samples are drawn from the same bivariate normal population or from different populations with same correlation coefficient* ρ, *(say), the statistic :*

$$Z = \frac{(Z_1 - Z_2) - E(Z_1 - Z_2)}{S.E.(Z_1 - Z_2)} \sim N(0,1) \qquad \ldots(16{\cdot}30c)$$

$$E(Z_1 - Z_2) = E(Z_1) - E(Z_2) = \xi_1 - \xi_2 = 0 \qquad \left[\because \xi_1 = \xi_2 = \frac{1}{2}\log_e \frac{1+\rho}{1-\rho} (\text{under } H_0)\right]$$

and $$\text{S.E. }(Z_1 - Z_2) = \sqrt{V(Z_1) + V(Z_2)} = \sqrt{\left\{\frac{1}{n_1 - 3} + \frac{1}{n_2 - 3}\right\}}$$

[Covariance term vanishes since samples are independent.]

Under H_0, the test statistic is :

$$Z = \frac{Z_1 - Z_2}{\sqrt{\left\{\frac{1}{n_1 - 3} + \frac{1}{n_2 - 3}\right\}}} \sim N(0, 1)$$

By comparing this value with 1·96 or 2·58, H_0 may be accepted or rejected at 5% and 1% levels of significance respectively.

(3) To obtain pooled estimate of ρ. Let $r_1, r_2, \ldots, r_k$ be observed correlation coefficients in k-independent samples of sizes $n_1, n_2, \ldots, n_k$ respectively from a bivariate normal population. The problem is to combine these estimates of ρ to get a pooled estimate for the parameter. If we take $Z_i = \frac{1}{2}\log_e\left(\frac{1+r_i}{1-r_i}\right); i = 1, 2, \ldots, k$; then Z_i ; $i = 1, 2, \ldots, k$ are independent normal variates with variances $\frac{1}{(n_i - 3)}$; $i = 1, 2, \ldots, k$ and common mean $\xi = \frac{1}{2}\log_e\left(\frac{1+\rho}{1-\rho}\right)$

The weighted mean, (say $\overline{Z}$), of these Z_i's is given by :

$$\overline{Z} = \sum_{i=1}^{k} w_i Z_i \Big/ \sum_{i=1}^{k} w_i; \qquad \text{where } w_i \text{ is the weight of } Z_i.$$

$\overline{Z}$ is also an unbiased estimate of ξ, since

$$E(\overline{Z}) = \frac{1}{(\Sigma w_i)}\left(E \sum_{i=1}^{k} w_i Z_i\right) = \frac{1}{\Sigma w_i}\left[\sum_i w_i E(Z_i)\right] = \frac{1}{\Sigma w_i}\left(\sum_i w_i \xi\right) = \xi$$

and

$$V(\overline{Z}) = \frac{1}{(\Sigma w_i)^2} V(\Sigma w_i Z_i) = \frac{1}{(\Sigma w_i)^2}\left[\Sigma w_i^2 V(Z_i)\right] \qquad \ldots (*)$$

The weights w_i's, $(i = 1, 2, \ldots, n)$ are so chosen that $\overline{Z}$ has minimum variance.

In order that $V(\overline{Z})$ is minimum for variations in w_i, we should have

$$\frac{\partial}{\partial w_i} V(\overline{Z}) = 0 ; \quad i = 1, 2, \ldots, k.$$

$$\Rightarrow \quad \frac{(\Sigma w_i)^2 2 w_i V(Z_i) - \left[\sum_i w_i^2 V(Z_i)\right] 2\left(\sum_i w_i\right)}{(\Sigma w_i)^4} = 0 \quad \text{or} \quad w_i V(Z_i) = \frac{\Sigma w_i^2 V(Z_i)}{\Sigma w_i}, \text{ a constant.}$$

$$\therefore \quad w_i \propto \frac{1}{V(Z_i)} = (n_i - 3) ; i = 1, 2, \ldots, k. \qquad \ldots (**)$$

Hence, the minimum variance estimate of ξ is given by :

$$\overline{Z} = \frac{\sum_{i=1}^{k} w_i Z_i}{\sum_{i=1}^{k} w_i} = \frac{\sum_{i=1}^{k} (n_i - 3) Z_i}{\sum_{i=1}^{k} (n_i - 3)} \qquad \text{[On using (**)]}$$

and the best estimate of ρ is then given by

$$\overline{Z} = \frac{1}{2}\log_e \frac{1+\rho}{1-\rho} \quad \Rightarrow \quad \hat{\rho} = \frac{e^{2\overline{Z}} - 1}{e^{2\overline{Z}} + 1} = \tanh \overline{Z} = \tanh\left[\frac{\Sigma(n_i - 3)Z_i}{\Sigma(n_i - 3)}\right]$$

Remark. Minimum variance of $\bar{Z}$ is given by :

$$\left[V(\bar{Z})\right]_{\min} = \frac{\sum_i \left\{(n_i-3)^2\left(\frac{1}{n_i-3}\right)\right\}}{\{\Sigma(n_i-3)\}^2} = \frac{\Sigma(n_i-3)}{\{\Sigma(n_i-3)\}^2} = \frac{1}{\sum_{i=1}^{k}(n_i-3)} \qquad \text{[From (*)]}$$

Example 16·30. *Two independent samples have 28 and 19 pairs of observations with correlation coefficients 0·55 and 0·75 respectively. Are these values of correlation coefficients, consistent with the hypothesis that both the samples are drawn from the same population ?*

Solution. *Null Hypothesis,* $H_0 : \rho_1 = \rho_2$, *i.e.*, samples have been drawn from the same population.

Allernative Hypothesis, $H_1 : \rho_1 \neq \rho_2$ (Two-tailed).

$$Z_1 = \frac{1}{2}\log_e\left(\frac{1+r_1}{1-r_1}\right) = 1{\cdot}1513\log_{10}\left(\frac{1+r_1}{1-r_1}\right) = 1{\cdot}1513\log_{10}\left(\frac{1+0{\cdot}55}{1-0{\cdot}55}\right)$$

$$= 1{\cdot}1513\log_{10}(3{\cdot}444) = 1{\cdot}1513\times 0{\cdot}5371 = 0{\cdot}6184$$

$$Z_2 = 1{\cdot}1513\log_{10}\left(\frac{1+0{\cdot}75}{1-0{\cdot}75}\right) = 1{\cdot}1513\log_{10}7 = 1{\cdot}1513\times 0{\cdot}8451 = 0{\cdot}9730$$

Under H_0, the test statistic is :

$$Z = \frac{Z_1 - Z_2}{\sqrt{\left\{\frac{1}{(n_1-3)}+\frac{1}{(n_2-3)}\right\}}} \sim N(0,1)$$

$$\therefore \quad Z = \frac{0{\cdot}6184 - 0{\cdot}9730}{\sqrt{\{(1/25)+(1/16)\}}} = \frac{-7{\cdot}092}{6{\cdot}403} = -1{\cdot}11$$

Since $|Z| < 1{\cdot}96$, it is not significant. Hence, the data donot provide any evidence against the null hypothesis which may, therefore, be accepted; and it may be concluded that the samples have been drawn from the same population.

DISCUSSION AND REVIEW QUESTIONS

1. (*a*) Given that

   (*i*) u is normally distributed with zero mean and unit variance,

   (*ii*) v^2 has a chi-square distribution with n degrees of freedom, and

   (*iii*) u and v are independently distributed,

   find the distribution of the variable $t = \dfrac{u\sqrt{n}}{v}$.

   (*b*) Find the variance of the t-distribution with n degrees of freedom, $(n > 2)$.

   (*c*) If the variable t has Student's t-distribution with 2 degrees of freedom, prove that

   $$P(t \geq 2) = \frac{3-\sqrt{6}}{6}$$

2. (*a*) State (without proof) the sampling distribution of Student's t. Who discovered it ?

(b) "Discovery of Student's t is regarded as a landmark in the history of statistical inference." Elucidate.

(c) Let t have Student's t-distribution with 2 $d.f.$ Find the probability $P(-\sqrt{2} \le t \le \sqrt{2})$.

3. (a) Show that

$$E(T) = \begin{cases} \dfrac{k^{r/2}\Gamma[(1+r)/2]\cdot\Gamma[(k-r)/2]}{\Gamma[(1/2)\cdot\Gamma(k/2)]}, & \text{if } r \text{ is even for } -1 < r < k \\ 0, \text{ if } r \text{ is odd} \end{cases}$$

where T has Student's t-distribution with k degrees of freedom.

(b) For the t-distribution with n $d.f.$, establish the recurrence relation :

$$\mu_{2r} = \frac{n(2r-1)}{(n-2r)}\cdot\mu_{2r-2}, \; n > 2r$$

(c) For how many $d.f.$ does

(i) χ^2-distribution reduce to negative exponential distribution, and

(ii) t-distribution reduce to Cauchy distribution ?

4. (a) Write a short note on Student's t-distribution and point out its uses.

(b) Show how the t-distribution has been found useful in testing whether the mean of small sample is significantly different from a hypothetical value.

(c) It is desired to test the hypothesis that the mean of a normal population is $\mu = \mu_0$ against the alternative that $\mu \ne \mu_0$. Explaining the assumptions involved, develop the statistic suitable for testing this hypothesis if the size of the sample is small. What modification do you suggest when the sample size is large ?

5. What is a test of significance ?

To test the hypothesis that the mean of a normal distribution is zero, two independent observations x_1 and x_2 are taken from the distribution. Show that the hypothesis is rejected at 10% level of significance, using t-test with equal tail ends, if $|x_1 + x_2| > |x_1 - x_2| \tan 81°$.

6. It is required to test that the mean of a normal population is zero. A random sample drawn from the population gives the values $x_1, x_2, \ldots, x_n$. Show that the t-test for acceptance of the hypothesis reduces to

$$\left(\sum_{i=1}^{n} x_i\right)^2 \le \frac{n\cdot t_\alpha^2}{t_\alpha^2 + (n-1)}\left(\sum_{i=1}^{n} x_i^2\right),$$

where t_α is the value of Student's t at the desired level of significance α for $(n-1)$ $d.f.$

7. Explain, stating clearly the assumptions involved, the t-test for testing the significance of the difference between the two sample means.

8. (a) Derive the distribution of $F = S_1^2/S_2^2$, where S_1^2 and S_2^2 are two independent unbiased estimates of the common population variance σ^2, defined by

$$S_1^2 = \frac{1}{n_1 - 1}\sum_{i=1}^{n_1}(x_{1i} - \bar{x}_1)^2; \quad S_2^2 = \frac{1}{n_2 - 1}\sum_{j=1}^{n_2}(x_{2j} - \bar{x}_2)^2$$

(b) Find the limiting form when the degrees of freedom of the χ^2 in the denominator tend to infinity and give an intuitive justification of the result.

9. (a) If $X_1, X_2, \ldots, X_m, X_{m+1}, \ldots, X_{m+n}$ are independent normal variates with zero mean and standard deviation σ, obtain the distribution of $\sum_{i=1}^{m} X_i^2 \Big/ \sum_{i=m+1}^{m+n} X_i^2$

(b) If X has an F-distribution with n_1 and n_2 $d.f.$, find the distribution of $1/X$ and give one use of this result.

(c) If X is t-distributed, show that X^2 is F-distributed.

10. (a) Derive the distribution of the F-statistic on (n_1, n_2) degrees of freedom and show that the statistic

$\left(1+\frac{n_1}{n_2}F\right)^{-1}$ has a Beta-distribution.

(b) Show that the probability curve of the distribution of F is positively skewed.

11. Prove the following :

(i) $F_{n_1,n_2} = \dfrac{1}{F_{n_2,n_1}}$ and (ii) $F_{n_1,n_2} = \dfrac{n_2}{n_1}\cdot\dfrac{x}{1-x}$, where x has Beta-distribution.

12. If X and Y are independent chi-square variates with v_1 and v_2 $d.f.$ respectively, show that $U = X + Y$ and $V = \dfrac{v_2 X}{v_1 Y}$ are independently distributed. Find the distribution of V.

13. Prove that if X has the F-distribution with (m, n) $d.f.$ and Y has the F-distribution with (n, m) $d.f.$, then for every $a > 0$, $P(X \le a) + P\left(Y \le \frac{1}{a}\right) = 1$.

14. Show that the mode of the F-distribution with v_1 (≥ 2), v_2 $d.f.$ is given by $\dfrac{v_2(v_1-2)}{v_1(v_2+2)}$ and is always less than unity.

15. X is F-variate with 2 and n $(n \ge 2)$ degrees of freedom. Show that $P(F \ge k) = \left(1+\frac{2k}{n}\right)^{-n/2}$. Deduce the significance level of F corresponding to the significance level of probability P.

16. Let X_1, X_2 be independent random variables following density law : $f(x) = \exp(-x)$, $0 < x < \infty$. Show that $Z = X_1/X_2$, has an F-distribution.

17. (a) If $X \sim F(n_1, n_2)$, show that its mean is independent of n_1.

(b) Obtain the mode of F-distribution with (n_1, n_2) $d.f.$ and show that it lies between 0 and 1.

(c) Show that for F-distribution with n_1 and n_2 $d.f.$, the points of inflexion exist if $n_1 > 4$ and are equidistant from the mode.

18. If $X \sim F(n_1, n_2)$ distribution, show that : $U = \dfrac{n_1 X}{n_2 + n_1 X} \sim \beta_1\left(\dfrac{n_1}{2}, \dfrac{n_2}{2}\right)$

Hence obtain the distribution function of X.

19. If χ_1^2 and χ_2^2 are independent chi-square variates with n_1 and n_2 $d.f.$, obtain the probability density function of F-statistic defined by : $F = \dfrac{(\chi_1^2/n_1)}{(\chi_2^2/n_2)}$.

Mention the types of hypotheses which are tested with the help of this statistic.

20. Explain why the larger variance is placed in the numerator of the statistic F. Discuss the application of F-test in testing if two variances are homogeneous.

ASSORTED REVIEW PROBLEMS FOR SELF-ASSESSMENT

16·1. Suppose $X_1, X_2, \ldots, X_n$ $(n > 1)$ are independent variates each distributed as $N(0, \sigma^2)$. Find the $p.d.f.$ of

$$W = X_1 \Big/ \left(\frac{1}{n}\sum_{i=1}^{n} X_i^2\right)^{1/2}$$

Why does not W follow the t-distribution?

16·2. Let $x_1, x_2, \ldots, x_n$ be independent observations from a normal universe with mean μ and variance σ^2 and let $\bar{x}$ and s^2 be the sample mean and sum of the squares of the deviations from the mean respectively. Let x' be one more observation independent of previous ones. Show that :

$\dfrac{x'-\bar{x}}{s}\left[\dfrac{n(n-1)}{n+1}\right]^{1/2}$ has a Student t-distribution with $(n-1)$ degrees of freedom.

16·3. (*a*) Let X_1 and X_2 be two independent normal variates with the same normal distribution $N(\mu, \sigma^2)$.

Obtain the distribution of $$Y = \frac{X_1+X_2-2\mu}{\sqrt{|X_1-X_2|^2}}.$$

(*b*) If X is *t*-distributed with *k d.f.*, show that $\dfrac{1}{1+(X^2/k)}$ has a beta distribution.

16·4. Define Student's *t*-statistic and state its probability density function.

If x_i $(i = 1, 2, ..., n)$ is a random sample of n independent observations from a normal population with mean μ and variance σ^2, show that

$$U = \frac{(\bar{x}-\mu)\sqrt{n(n-1)}}{\sqrt{\sum_{i=1}^{n}(x_i-\bar{x})^2}}, \quad \text{where} \quad \bar{x} = \frac{1}{n}\sum_{i=1}^{n} x_i$$

conforms to Student's *t*-variate. If x is an additional observation drawn independently from the same normal population, show that :

$$W = \frac{(x-\bar{x})}{\sqrt{\sum_{i=1}^{n}(x_i-\bar{x})^2}} \times \sqrt{\frac{n(n-1)}{n+1}}$$

also conforms to Student's *t*-variate.

16·5. Let $X_1, X_2, ..., X_n$ be a random sample from $N(\mu, \sigma^2)$, and $\bar{X}$ and S^2, respectively, be the sample mean and sample variance. Let $X_{n+1} \sim N(\mu, \sigma^2)$, and assume that $X_1, X_2, ..., X_n, X_{n+1}$ are independent. Obtain the sampling distribution of :

$$U = \frac{(X_{n+1}-\bar{X})}{S}\cdot\sqrt{\frac{n}{n+1}} \; ; \quad \left\{S^2 = \frac{1}{n-1}\sum_{i=1}^{n}(X_i-\bar{X})^2\right\}$$

16·6. Let $X_1, X_2, ..., X_m$ and $Y_1, Y_2, ..., Y_n$ be independent random samples from $N(\mu_1, \sigma^2)$ and $N(\mu_2, \sigma^2)$, respectively. If $\bar{X}$ and $\bar{Y}$ denote the corresponding sample means and if

$(m-1)S_1^2 = \sum_{i=1}^{m}(X_i-\bar{X})^2$, $(n-1)S_2^2 = \sum_{j=1}^{n}(Y_j-\bar{Y})^2$, obtain the sampling distribution of

$$\frac{a(\bar{X}-\mu_1)+b(\bar{Y}-\mu_2)}{\left[\left\{\frac{(m-1)S_1^2+(n-1)S_2^2}{(m+n-2)}\right\}\left\{\frac{a^2}{m}+\frac{b^2}{n}\right\}\right]^{1/2}},$$ where a and b are two fixed real numbers.

16·7. If $X \sim t_{(n)}$, show that $\left(n-\frac{1}{2}\right)\cdot\log\left(1+\frac{x^2}{n}\right) \sim \chi^2_{(1)}$, for large n.

You may assume that for large n, $$\frac{\Gamma\left(\frac{n}{2}+\frac{1}{2}\right)}{\Gamma\left(\frac{1}{2}+n\right)\sqrt{\frac{1}{2}n}} \simeq \left(1-\frac{1}{4n}\right).$$

16·8. If $\bar{X}$ and $\hat{\sigma}^2 = S^2$ be the usual sample mean and sample variance based on a random sample of n observations from $N(\mu, \sigma^2)$, and if $T = (\bar{X}-\mu)\sqrt{n}/S$, prove that

(*i*) $\text{Var}(T) = (n-1)/(n-3)$; (*ii*) $\text{Cov}(\bar{X}, T) = \sigma\dfrac{\sqrt{n-1}\,\Gamma[(n-2)/2]}{\sqrt{2n}\cdot\Gamma[(n-1)/2]}$;

(*iii*) $r(\bar{X}, T) = \left[\frac{1}{2}(n-3)\right]^{1/2} \Gamma\left[\frac{1}{2}(n-2)\right] / \Gamma\left[\frac{1}{2}(n-1)\right].$

16·9. (*a*) Find the Student's t for the variate values in a sample of eight : – 4, –2, –2, 0, 2, 2, 3, 3, taking the mean of the universe to be zero. How would you proceed further ?

(*b*) Ten individuals are chosen at random from a normal population and their heights are found to be 63, 63, 66, 67, 68, 69, 70, 70, 71, 71 inches. Test if the sample belongs to the population whose mean heights is 66 inches. [Given $t_{0\cdot05} = 2\cdot62$ for 9 *d.f.*]

(*c*) A random sample of 9 experimental animals under a certain diet gave the following increase in weight : $\Sigma x_i = 45$ lbs., $\Sigma x_i^2 = 279$ lbs., where x_i denotes the increase in weight of the *i*th animal. Assuming that the increase in weight is normally distributed as $N(\mu, \sigma^2)$ variate, test $H_0 : \mu = 1$ against $H_1 : \mu \neq 1$ at 5% level. Given $P(|t| > 2\cdot306) = 0\cdot05$ for 8 degrees of freedom.

(*d*) A random sample of eight cigarettes of a certain brand has an average nicotine content of 18·6 milligrams and a standard deviation of 2·4 milligrams. Is this in line with the manufacturer's claim that average nicotine content does not exceed 17·5 milligrams?Use a 0·01 level of significance and assume the distribution of nicotine contents to be normal.

16·10. (*a*) Ten specimens of copper wires drawn from a large lot have the following breaking strength (in kg. weight):

578 572 570 568 572 571 570 572 596 548

Test whether the mean breaking strength of the lot may be taken to be 578 kg. *wt*.

(*b*) A manufacturer of gunpowder has developed a new powder which is designed to produce a muzzle velocity equal to 3,000 ft/sec. Seven shells are loaded with the charge and the muzzle velocities measured.

The resulting velocities are: 3,005; 2,935; 2,965; 2,995; 3,905; 2,935; and 2,905. Do these data present sufficient evidence to indicate that the average velocity differs from 3,000 ft./sec?

(*c*) The average length of time for students to register for summer classes at a certain college has been 50 minutes with *s.d.* of 10 minutes. A new registration procedure using modern computing machines is being tried. If a random sample of 12 students had an average registration time of 42 minutes with *s.d.* of 11·9 minutes under the new system, test the hypothesis that the population mean has not changed, using 0·05 as level of significance.

(*d*) A time study engineer developed a new sequence of operation elements that he hopes will reduce the mean cycle time of a certain production process. The results of a time study of 20 cycles are given below :

Cycle time in minutes

12·25	11·97	12·15	12·08	12·31	12·28	11·94	11·89	12·16	12·04
12·09	12·15	12·14	12·47	11·98	12·04	12·11	12·25	12·15	12·34

If the present mean cycle time is 12·5 minutes, should he adopt the new sequence?

16·11. (*a*) The life (in hours) of a 100 watt bulb is known to be normally distributed with standard deviation of 36 hours. A random sample of 15 bulbs yielded the following results :

Life in hours

2,216	2,237	2,249	2,204	2,225	2,301	2,281	
2,263	2,318	2,255	2,275	2,295	2,250	2,295	2,300

Construct a 95% two-sided confidence interval so that the actual mean of the life of bulbs falls within this interval.

(b) The average breaking strength of steel rods is specified to be 18·5 thousand pounds. To test this a sample of 14 rods was tested. The mean and standard deviations obtained were 17·85 and 1·955 thousand pounds respectively. Is the result of the experiment significant? Also obtain the 95 per cent fiducial limits from the sample for the average breaking strength of steel rods.

(c) A sample of 9 shafts is inspected from a production line. The following measurements are the diameters (in mm.) of shafts : 45·010, 45·020, 45·021, 45·015, 45·019, 45·018, 45·020, 45·023 and 45·005. If the production line meets the specifications laid by the I.S.I., with S.D. 0·006 mm, estimate the 95% confidence interval within which the true diameter of the shaft lies.

(d) A random sample of 8 envelopes is taken from letter box of a post office and their weights in grams are found to be 12·1, 11·9, 12·4, 12·3, 11·9, 12·1, 12·4, 12·1.

(i) Find 99% confidence limits for the mean weight of the envelopes received at that post office.

(ii) Using the result of part (i), does this sample indicate at 1% level that the average weight of envelopes received at that post office is 12·35 gms ?

16·12. (a) Two independent samples of 8 and 7 items respectively had the following values :

Sample I :	9	11	13	11	15	9	12	14
Sample II :	10	12	10	14	9	8	10	

Is the difference between the means of samples significant ?

(b) Two horses A and B were tested according to the time (in seconds) to run a particular track with the following results :

Horse A :	28	30	32	33	33	29	34
Horse B :	29	30	30	24	27	29	

Test whether the two horses have the same running capacity.

[5 per cent values of t for 11 and 12 degrees of freedom respectively are 2·20 and 2·18.]

(c) The gain in weight of two random samples of rats fed on two different diets A and B are given below. Examine whether the difference in mean increases in weight is significant.

Diet A :	13	14	10	11	12	16	10	8	
Diet B :	7	10	12	8	10	11	9	10	11

16·13. (a) Show how you would use Student's t-test to decide whether the two sets of observations [17, 27, 18, 25, 27, 29, 27, 23, 17] and [16, 16, 20, 16, 20, 17, 15, 21] indicate samples drawn from the same universe.

(b) A reading test is given to an elementary school class that consists of 12 Anglo-American children and 10 Mexican-American children. The results of the test are :

Anglo-American	*Mexican-American*
$\bar{x}_1 = 74;\ s_1 = 8$	$\bar{x}_2 = 70;\ s_2 = 10$

Is the difference between the means of the two groups significant at the 0·05 level?

(c) For a random sample of 10 pigs, fed on a diet A, the increases in weight in a certain period were : 10, 6, 16, 17, 13, 12, 8, 14, 15, 9 lbs.

For another random sample of 12 pigs fed on diet B, the increases in the same period were :

7, 13, 22, 15, 12, 14, 18, 8, 21, 23, 10, 17 lbs.

Find if the two samples are significantly different regarding the effect of diet, given that for $d.f.$ $v = 20, 21, 22$, the five per cent values of t are respectively 2·09, 2·07, 2·06.

(d) Two independent samples of rats chosen among both the sexes had the following increase in weights when fed on a diet. Can you say that the mean increase in weight differs significantly with sex?

Male :	96,	88,	97.	89,	92,	95	and	90		
Female :	112,	80,	98,	100,	84,	82,	89,	95,	100	and 96.

16·14. (*a*) Ten soldiers visit a riffle range for two consecutive weeks. For the first week their scores are: 67, 24, 57, 55, 63, 54, 56, 68, 33, 43

and during the second week they score in the same order—

70, 38, 58, 58, 56, 67, 68, 72, 42, 38

Examine if there is any significant difference in their performance.

(*b*) Two independent groups of 10 children were tested to find how many digits they could repeat from memory after hearing them. The results are as follows :

Group *A* :	8	6	5	7	6	8	7	4	5	6
Group *B* :	10	6	7	8	6	9	7	6	7	7

Is the difference between the mean scores of the two groups significant ?

(*c*) Measurements of the fat content of two kinds of ice cream, Brand *A* and Brand *B*, yielded the following sample data :

Brand *A* :	13·5	14·0	13·6	12·9	13·0
Brand *B* :	12·9	13·0	12·4	13·5	12·7

Test the null hypothesis $\mu_1 = \mu_2$, (where μ_1 and μ_2 are the respective true average fat contents of the two kinds of ice cream), against the alternative hypothesis $\mu_1 \neq \mu_2$ at the level of significance $\alpha = 0{\cdot}05$.

16·15. (*a*) A random sample of 16 values from a normal population has a mean of 41·5 inches and sum of squares of deviations from the mean is equal to 135 inches. Another sample of 20 values from an unknown population has a mean of 43·0 inches and sum of squares of deviations from their mean is equal to 171 inches. Show that the two samples may be regarded as coming from the same normal population.

(*b*) A company is interested in knowing if there is a difference in the average salary received by foremen in two divisions. Accordingly, samples of 12 foremen in the first division and 10 foremen in the second division are selected at random. Based upon experience, foremen's salaries are known to be approximately normally distributed, and the standard deviations are about the same.

	First Division	*Second Division*
Sample size	12	10
Average daily salary of foremen (₹)	1,050	980
Standard deviation of salaries (₹)	68	74

The table value of *t* for 20 *d.f.* at 5% level of significance is 2·086.

Ans. $t = 2{\cdot}2$; Reject $H_0 : \mu_1 = \mu_2$

16·16. (*a*) Two different machines are used for manufacturing automobile horns in a factory. The number produced in an hour on each machine is noted for hours chosen at random over a period of several months. Notes on the first machine, taken over 10 hours, show a mean production per hour of 100 horns with a standard deviation of 6·0. The second machine, noted also for 10 hours, has a mean production of 96 horns with a standard deviation of 5·0. Does the first machine produce at least 5 more horns per hour at a 0·05 level of significance?

(*b*) The average number of articles produced by two machines per day are 200 and 250 with standard deviations 20 and 25 respectively on the basis of records of 25 days production. Can you regard both the machines equally efficient at 10% level of significance ?

Ans. $t = -7{\cdot}65$

16·17. Eleven college boys were given a test in Statistics. They were given a month's tuition and a second test was held at the end of it. Do the marks give evidence that the students have benefited by the extra coaching?

Boys	1	2	3	4	5	6	7	8	9	10	11
Marks in 1st test	23	20	19	21	18	20	18	17	23	16	19
Marks in 2nd test	24	19	22	18	20	22	20	20	23	20	18

Hint: $H_0 = \mu_1 = \mu_2$ H_1: $\mu_1 < \mu_2$. Paired $|t| = 1{\cdot}483$; Not significant. Hence students have not benefited from extra coaching.

16·18. (*a*) The following table gives the additional hours of sleep gained by 10 patients in an experiment to test the effect of a drug. Do these data give evidence that the drug produces additional hours of sleep?

Patients :	1	2	3	4	5	6	7	8	9	10
Hours gained :	0·7	0·1	0·2	1·2	0·31	0·4	3·7	0·8	3·8	2·0

(*b*) A drug was administered to 10 patients, and the increments in their blood pressure were recorded to be 6, 3, – 2, 4, – 3, 4, 4, 6, 0, 3, 2. Is it reasonable to believe that the drug has no effect on change of blood pressure? Use 5% significance level, and assume that for 9 degrees of freedom, $P(t > 2{\cdot}26) = 0{\cdot}025$.

(*c*) The scores of 10 candidates prior and after training are given below :

Prior :	84	48	36	37	54	69	83	96	90	65
After :	90	58	56	49	62	81	84	86	84	75

Is the training effective?

16·19. The following table gives measurements of blood pressure on subjects by two investigators :

Subject No. :	1	2	3	4	5	6	7	8	9	10
Investigator I :	70	68	56	75	80	90	68	75	56	58
Investigator II :	68	70	52	73	75	78	67	70	54	55

No other details of the experiment were given.

(*i*) If a valid inference has to be drawn about the difference between the investigators, mention the precautions that should have been taken in conducting the experiment with respect to the time of measurement, interval between the first and second measurements, the order in which the investigators measure, etc.

(*ii*) After the experiment was conducted it was discovered that all the subjects were unrelated except that No. 10 was the father of No. 9. Assuming that all the precautions you mention in (*a*) are satisfied, analyse the data to draw an inference on the difference between the investigators. 5 per cent values of the *t*-statistic corresponding to various degrees of freedom are as follows :

5 per cent values of *t* ...	2·40	2·31	2·26	2·23	2·10	2·09
Degrees of freedom ...	7	8	9	10	18	19

16·20. The following are the values of the cephalic index found in two samples of skulls, one consisting of 15 and the other of 13 individuals :

Sample I : 74·1 77·7 74·0 74·4 73·8 79·3 75·8 82·8 72·2 75·2 78·2 77·1 78·4 76·3 76·8

Sample II : 70·8 74·9 74·2 70·4 69·2 72·2 76·8 72·4 77·4 78·1 72·8 74·3 74·7

(*i*) Test the hypothesis that the means of population I and population II could be equal.

(*ii*) Is it possible that the sample II has come from a population of mean 72·0 ?

(*iii*) Obtain confidence limits for the mean of population I and for the mean of population II.

(Assume that the distribution of cephalic indices for a homogeneous population is normal.)

16·21. (*a*) The following table gives the gain in weight in decagrams in a feeding experiment with pigs on the relative value of limestone and bone meal for bone development :

Limestone	:	49·2	53·3	50·6	52·0	46·8	50·5	52·1	53·0
Bone meal	:	51·5	54·9	52·2	53·3	51·6	54·1	54·2	53·3

Test for the significance of difference between the means in two ways :

(*i*) by assuming that the values are paired, (*ii*) by assuming that the values are not paired.

(*b*) The following table shows the mean number of bacterial colonies per plate obtainable by four slightly different methods from soil samples taken at 4 P.M. and 8 P.M. respectively :

Method	*A*	*B*	*C*	*D*
4 P.M.	29·75	27·50	30·25	27·80
8 P.M.	39·20	40·60	36·20	42·40

Are there significantly more bacteria at 8 P.M. than at 4 P.M.

[Given $t_{0·05}$ (3)= 3·18 and $t_{0·01}$ (3) = 5·84]

16·22. (*a*) It is believed that glucose treatment will extend the sleep time of mice. In an experiment to test this hypothesis ten mice selected at random are given glucose treatment and are found to have a mean hexabarbital sleep time of 47·2 min with a standard deviation of 9·3 min. A further sample of ten untreated mice are found to have a mean hexabarbital sleep time of 28·5 min with a standard deviation of 7·2 min. Are these results significant evidence in favour of the hypothesis?

Find 95% confidence limits for the population mean difference in sleep time. State any assumptions made concerning the data in carrying out the test and finding the limits.

(*b*) An experiment was performed to compare the abrasive wear of two different laminated materials. Twelve pieces of material I were tested, by exposing each piece to a machine measuring wear. Ten pieces of material II were similarly tested. In each case the depth of wear was observed. The sample of material I gave an average (coded) wear 8·5 units with a standard deviation of 0·4 while the sample of material II gave an average of 8·1 and a standard deviation of 0·5. Test the hypothesis that the two types of material exhibit the same mean abrasive wear at the 0·10 level of significance. Assume the populations to be approximately normal with equal variances.

If the level of significance is 0·01, what will be your conclusion ?

16·23. A restaurant owner ranked his 17 waiters in terms of their speed and efficiency on the job. He correlated these ranks with the total amount of tips each of these waiters received for a one-week period. The obtained value of correlation coefficient is 0·438. What do you conclude ?

16·24. Test the significance of the values of correlation coefficient 'r' obtained from samples of size n pairs from a bivariate normal population.

(*i*) $r = 0·6$, $n = 38$ (*ii*) $r = 0·5$, $n = 11$

16·25. (*a*) A random sample of 27 pairs of observations from a normal population gives a correlation coefficient of 0·42. Is it likely that the variables in the population are uncorrelated ?

Ans: $t = 2·314$; Significant at 5% level; $H_0 : \rho = 0$, rejected.

(*b*) Find the least value of r in a sample of 27 pairs from a bivariate normal population significant at 5% level.

Ans: $|r| > 0·3809$.

16·26. Let X_1 and X_2 be a random sample of size 2 from $N(0, 1)$ and Y_1 and Y_2 be a random sample of size 2 from $N(1, 1)$, and let the Y_i's be independent of the X_i's. Find the distribution of the following :

(i) $(X_1 - X_2)/\sqrt{2}$ (ii) $(X_1 + X_2)^2/(X_2 - X_1)^2$

(iii) $\overline{X} + \overline{Y}$ (iv) $(Y_1 + Y_2 - 2)^2/(X_2 - X_1)^2$

(v) $(X_1 + X_2)/\sqrt{[(X_2 - X_1)^2 + (Y_2 - Y_1)^2]/2}$ (vi) $[(Y_1 - Y_2)^2 + (X_1 - X_2)^2 + (X_1 + X_2)^2]/2$

16·27. Let $X_i \sim N(i, i^2)$; $i = 1, 2, 3$ be independent random variables. Using only the three random variables X_1, X_2 and X_3, give an example of a statistic that has :

(i) A chi-square distribution with 3 d.f (ii) An F-distribution with (1, 2) d.f.

(iii) A t-distribution with 2 d.f.

Hint: $Z_i = (x_i - i)/i$; $i = 1,2,3$ are $i,i,d, N(0,1)$.

(i) $\sum_{i=1}^{3} Z_i^2 \sim \chi^2_{(3)}$, (ii) $\dfrac{2Z_1^2}{Z_2^2 + Z_3^2} \sim F(1,2)$, (iii) $\dfrac{Z_1}{\left[\left(Z_2^2 + Z_3^2\right)/2\right]^{1/2}} \sim t_{(2)}$

16·28 (a) Let $X_1, X_2, \ldots, X_n$ be a random sample from $N(\mu, \sigma^2)$. Define :

$$\overline{X}_k = \frac{1}{k}\sum_1^k X_i, \quad \overline{X}_{n-k} = \frac{1}{n-k}\sum_{k+1}^{n} X_i, \quad \overline{X} = \frac{1}{n}\sum_1^n X_i$$

$$S_k^2 = \frac{1}{k-1}\sum_1^k\left(X_i - \overline{X}_k\right)^2, \quad S_{n-k}^2 = \frac{1}{n-k-1}\sum_{k+1}^{n}\left(X_i - \overline{X}_{n-k}\right)^2, \quad \text{and} \quad S^2 = \frac{1}{n-1}\sum_1^n\left(X_i - \overline{X}\right)^2$$

Answer the following questions, giving reasons.

(i) What is the distribution of $\sigma^{-2}\left[(k-1)\, S_k^2 + (n-k-1)\, S_{n-k}^2\right]$?

(ii) What is the distribution of S_k^2/S_{n-k}^2 ?

(iii) What is the distribution of $\left(\overline{X} - \mu\right)\sqrt{n}/S$?

(iv) What is the distribution of $\frac{1}{2}\left(\overline{X}_k + \overline{X}_{n-k}\right)$?

(v) What is the distribution of $(X_i - \mu)^2/\sigma^2$?

(b) If (X_1, X_2) is a random sample from $N(0, 1)$, what is the distribution of :

(i) $(X_2 - X_1)^2/2$ (ii) $(X_1 + X_2)^2/(X_2 - X_1)^2$

(iii) $(X_2 + X_1)/\sqrt{(X_1 - X_2)^2}$ (iv) $(1/Z)$ if $Z = X_1^2/X_2^2$

Give reasons for your answer.

16·29. If $X \sim F(1, n)$, show that $\left(n - \frac{1}{2}\right)\log\left[1 + \frac{X}{n}\right] \sim \chi^2_{(1)}$, for large n.

16·30. If X_1, X_2, X_3 and X_4 are independent observations from $N(0, 1)$ population, state giving reasons, the sampling distributions of (i) $U = \dfrac{\sqrt{2}X_3}{\sqrt{X_1^2 + X_2^2}}$ and (ii) $V = \dfrac{3X_4^2}{X_1^2 + X_2^2 + X_3^2}$

Ans. (i) $U \sim t_{(2)}$ (ii) $V \sim F(1,3)$

16·31. An investigator, newly appointed, was made to take ten independent measurements on the maximum internal diameter of a pot at specified equal intervals of time and the standard deviation of these ten observations was found to be 0·0345 mm. After he had been some time on similar jobs, he was asked to repeat this experiment an equal number of times and the standard deviation of the new set of ten observations was found to be 0·0285 mm. Can it be concluded that the investigator has become more consistent (i.e. less variable) with practice?

16·32. (a) Two independent samples of 8 and 7 items respectively had the following values of the variables :

Sample I :	9	11	13	11	15	9	12	14
Sample II :	10	12	10	14	9	8	10	

Do the estimates of population variance differ significantly?

(b) Five measurements of the output of two units have given the following results (in kilograms of material per one hour of operation).

Unit A :	14·1	10·1	14·7	13·7	14·0
Unit B :	14·0	14·5	13·7	12·7	14·1

Assuming that both samples have been obtained from normal populations, test at 10% significance level if two populations have the same variance, it being given that $F_{0.95}$ (4, 4) = 6·39.

(c) In one sample of 10 observations from a normal population, the sum of the squares of the deviations of the sample values from the sample mean is 102·4 and in another sample of 12 observations from another normal population, the sum of the squares of the deviations of the sample values from the sample mean is 120·5. Examine whether the two normal populations have the same variance.

16·33. (a) Two random samples of sizes 8 and 11, drawn from two normal populations, are characterised as follows :

Population from which the sample is drawn	*Size of sample*	*Sum of observations*	*Sum of squares of observations*
I	8	9·6	61·52
II	11	16·5	73·26

You are to decide if the two populations can be taken to have same variance. What test function would you use? How is it distributed and what value it has in this sampling experiment ?

(b) The following are the values in thousands of an inch obtained by two engineers in 10 successive measurements with the same micrometer. Is one engineer significantly more consistent than the other?

Engineer A :	503,	505,	497,	505,	495,	502,	499,	493,	510,	501
Engineer B :	502,	497,	492,	498,	499,	495,	497,	496,	498	

Ans. $F = 2{\cdot}4$

16·34. The nicotine content (in miligrams) of two samples of tobacco were found to be as follows :

Sample A :	24	27	26	21	25	
Sample B :	27	30	28	31	22	36

Can it be said that the two samples come from the same normal population?

Ans. $H_0 : \mu_1 = \mu_2;\ t = 1{\cdot}9,$ Not significant;

$H_0' : \sigma_1^2 = \sigma_2^2,\ F = 4{\cdot}08 < 6{\cdot}26\ [F_{0.05}(5,4)];$ Not significant.

16·35. (a) Two random samples drawn from two normal populations are :

Sample I :	20,	16,	26,	27,	23,	22,	18,	24,	25,	19		
Sample II :	27,	33,	42,	35,	32,	34,	38,	28,	41,	43,	30,	37

Obtain estimates of the variances of the populations and test whether the populations have same variances.

(b) Test $H_0 : \sigma^2_1 = \sigma^2_2$ against $H_1 : \sigma^2_1 \neq \sigma^2_2$,

given $n_1 = 25$, $\Sigma(x_i - \overline{x})^2 = 164 \times 24$, $n_2 = 22$, $\Sigma(y_j - \overline{y})^2 = 190 \times 21$.

Make necessary assumptions, stating them.

(c) The diameters of two random samples, each of size 10, of bullets produced by two machines have standard deviations $s_1 = 0{\cdot}01$ and $s_2 = 0{\cdot}015$. Assuming that the diameters have independent distributions $N(\mu_1, \sigma_1^2)$ and $N(\mu_2, \sigma_2^2)$, test the hypothesis that, the two machines are equally good by testing : $H_0 : \sigma_1 = \sigma_2$ against $H_1 : \sigma_1 \neq \sigma_2$.

16·36. The following table shows the yield of corn in bushels per plot in 20 plots, half of which are treated with phosphate as fertiliser:

Treated :	5	0	8	3	6	1	0	3	3	1
Untreated :	1	4	1	2	3	2	5	0	2	0

Test whether the treatment by phosphate has

(*i*) changed the variability of the plot yields, (*ii*) improved the average yield of corn.

16·37. The following figures give the prices in rupees of a certain commodity in a sample of 15 shops selected at random from a city *A* and those in a sample of 13 shops from another city *B* :

City *A* :	7·41	7·77	7·44	7·40	7·38	7·93	7·58	8·28
	7·23	7·52	7·82	7·71	7·84	7·63	7·68	
City *B* :	7·08	7·49	7·42	7·04	6·92	7·22	7·68	
	7·24	7·74	7·81	7·28	7·43	7·47		

Assuming that the distribution of prices in the two cities is normal, answer the following :

(*i*) Is it possible that the average price of city *B* is Rs. 7·20 ?

(*ii*) Is the observed variance in the first sample consistent with the hypothesis that the standard deviation of prices in city *A* is Rs. 0·30 ?

(*iii*) Is it reasonable to say that the variability of prices in the two cities is the same ?

(*iv*) Is it reasonable to say that the average prices are the same in the two cities ?

CHAPTER 17 Statistical Inference – I (Theory of Estimation)

LEARNING OBJECTIVES

Upon completion of this Chapter, you should be able to:

- Differentiate between Point Estimate and Interval Estimate.
- Understand and discuss various characteristics of estimators like:
  (i) Unbiasedness (ii) Consistency
  (iii) Efficiency (iv) Sufficiency,
  along with their importance in estimation theory.
- Know and understand about Cramer-Rao Inequality regarding the lower bound to the variance of an unbiased estimator and obtain the conditions under which this lower bound is attained.
- Learn about the concept of complete statistic and complete family of distributions.
- Distinguish between Minimum Variance Unbiased (MVU) Estimator and Minimum Variance Bound (MVB) Unbiased Estimator.
- Understand the various methods of estimation like methods of :
  (i) 'Maximum Likelihood Estimation'
  (ii) 'Minimum Variance'
  (iii) 'Moments'
  (iv) 'Least Squares'
  (v) 'Minimum Chi-square'
  along with their important properties.
- Explain what is meant by Confidence Interval and Confidence Limits.

CHAPTER OUTLINES

17·1. INTRODUCTION

One of the main objectives of Statistics is to draw inferences about a population from the analysis of a sample drawn from that population. Two important problems in statistical inference are:

(*i*) Estimation and (*ii*) Testing of Hypothesis.

The theory of estimation was founded by Prof. R.A. Fisher in a series of fundamental papers round about 1930.

Parameter Space. Let us consider a random variable X with *p.d.f.* $f(x, \theta)$. In most common applications, though not always, the functional form of the population distribution is assumed to be known except for the value of some unknown parameter(s) θ which may take any value on a set Θ. This is expressed by writing the *p.d.f.* in the form $f(x, \theta), \theta \in \Theta$. The set Θ, which is the set of all possible values of θ is called the *parameter space*. Such a situation gives rise not to one probability distribution but a family of probability distributions which we write as $\{f(x, \theta), \theta \in \Theta\}$, *e.g.*, if $X \sim N(\mu, \sigma^2)$, then the parameter space is:

$$\Theta = \{(\mu, \sigma^2) : \varepsilon \infty < \mu < \infty; 0 < \sigma < \infty\}.$$

In particular, for $\sigma^2 = 1$, the family of probability distributions is given by :

$$\{N(\mu, 1) ; \mu \in \Theta\}, \text{ where } \Theta = \{\mu : -\infty < \mu < \infty\}$$

In the following discussion we shall consider a general family of distributions :

$$\{f(x ; \theta_1, \theta_2, \ldots, \theta_k) : \theta_i \varepsilon \Theta, i = 1, 2, \ldots, k\}.$$

Let us consider a random sample $x_1, x_2, \ldots, x_n$ of size n from a population, with probability function $f(x ; \theta_1, \theta_2, \ldots, \theta_k)$, where $\theta_1, \theta_2, \ldots, \theta_k$ are the unknown population parameters. There will then always be an infinite number of functions of sample values, called statistics, which may be proposed as estimates of one or more of the parameters.

Evidently, the best estimate would be one that falls nearest to the true value of the parameter to be estimated. In other words, the statistic whose distribution concentrates as closely as possible near the true value of the parameter may be regarded the best estimate. Hence the basic problem of the estimation in the above case, can be formulated as follows :

We wish to determine the functions of the sample observations :

$$T_1 = \hat{\theta}_1 (x_1, x_2, \ldots, x_n), \; T_2 = \hat{\theta}_2 (x_1, x_2, \ldots, x_n), \ldots, \; T_k = \hat{\theta}_k (x_1, x_2, \ldots, x_n),$$

such that their distribution is concentrated as closely as possible near the true value of the parameter. The estimating functions are then referred to as *estimators.*

Definition. *Any function of the random sample $x_1, x_2, \ldots, x_n$ that are being observed, say $T_n (x_1, x_2, \ldots, x_n)$ is called a statistic. Clearly, a statistic is a random variable. If it is used to estimate an unknown parameter θ of the distribution, it is called an estiamtor. A particular value of the estimator, say, $T_n (x_1, x_2, \ldots, x_n)$ is called an estimate of θ.*

We shall, however, use the terms *estimator* and *estimate*, somewhat loosely, their actual implications being clear from the context.

17·2. CHARACTERISTICS OF ESTIMATORS

The following are some of the criteria that should be satisfied by a good estimator.

(*i*) Unbiasedness, (*ii*) Consistency,

(*iii*) Efficiency, and (*iv*) Sufficiency.

We shall now, briefly, explain these terms one by one.

17·2·1. Unbiasedness

Definition. *An estimator $T_n = T(x_1, x_2, ..., x_n)$ is said to be an unbaised estimator of $\gamma(\theta)$ if* $E(T_n) = \gamma(\theta)$, *for all* $\theta \in \Theta$ ... (17·1)

We have seen in chapter 14 that in sampling from a population with mean μ and variance σ^2, $E(\bar{x}) = \mu$ and $E(s^2) \neq \sigma^2$ but $E(S^2) = \sigma^2$. Hence, there is a reason to prefer

$$S^2 = \frac{1}{n-1}\sum_{i=1}^{n}(x_i - \bar{x})^2, \quad \text{to the sample variance} \quad s^2 = \frac{1}{n}\sum_{i=1}^{n}(x_i - \bar{x})^2,$$

Remark. If $E(T_n) > \gamma(\theta)$, T_n is said to be positively biased and if $E(T_n) < \gamma(\theta)$, it is said to be negatively biased, the amount of bias $b(\theta)$ being given by: $b(\theta) = E(T_n) - \gamma(\theta),\ \theta \in \Theta$ ... (17·1*a*)

Example 17·1. *$x_1, x_2, \ldots, x_n$ is a random sample from a normal population $N(\mu, 1)$.*

Show that: $t = \frac{1}{n}\sum_{i=1}^{n} x_i^2$, *is an unbiased estimator of $(\mu^2 + 1)$.*

Solution. (*a*) We are given : $E(x_i) = \mu,\ V(x_i) = 1\ \forall\ i = 1, 2, \ldots, n$ ... (*)

$$E(x_i^2) = V(x_i) + \{E(x_i)\}^2 = 1 + \mu^2 \qquad \text{[From (*)]}$$

$$\therefore \quad E(t) = E\left(\frac{1}{n}\sum_{i=1}^{n} x_i^2\right) = \frac{1}{n}\sum_{i=1}^{n} E(x_i^2) = \frac{1}{n}\sum_{i=1}^{n}(1+\mu^2) = 1 + \mu^2$$

Hence, t is an unbiased estimator of $(1 + \mu^2)$.

Example 17·2. *If T is an unbiased estimator for θ, show that T^2 is a biased estimator for θ^2.*

Solution. Since T is an unbiased estimator for θ, we have: $E(T) = \theta$

Also $\text{Var}(T) = E(T^2) - \{E(T)\}^2 = E(T^2) - \theta^2 \Rightarrow E(T^2) = \theta^2 + \text{Var}(T)$, $(\text{Var}\ T > 0)$.

Since $E(T^2) \neq \theta^2$, T^2 is a biased estimator for θ^2.

Example 17·3. *Show that* $\frac{[\Sigma x_i(\Sigma x_i - 1)]}{n(n-1)}$ *is an unbiased estimate of θ^2, for the sample $x_1, x_2, \ldots, x_n$ drawn on X which takes the values 1 or 0 with respective probabilities θ and $(1 - \theta)$.*

Solution. Since $x_1, x_2, \ldots, x_n$ is a random sample from Bernoulli population with parameter θ,

$$T = \sum_{i=1}^{n} x_i \sim B(n, \theta) \Rightarrow E(T) = n\theta \ \text{and}\ Var(T) = n\theta(1-\theta)$$

$$\therefore \quad E\left\{\frac{\Sigma x_i(\Sigma x_i - 1)}{n(n-1)}\right\} = E\left\{\frac{T(T-1)}{n(n-1)}\right\} = \frac{1}{n(n-1)}\left\{E(T^2) - E(T)\right\}$$

$$= \frac{1}{n(n-1)}\left[\text{Var}(T) + \{E(T)\}^2 - E(T)\right]$$

$$= \frac{1}{n(n-1)}\left\{n\theta(1-\theta) + n^2\theta^2 - n\theta\right\} = \frac{n\theta^2(n-1)}{n(n-1)} = \theta^2$$

$\Rightarrow \quad \{\Sigma x_i(\Sigma x_i - 1)\} / \{n(n-1)\}$ is an unbiased estimator of θ^2.

Example 17·4. *Let X be distributed in the Poisson form with parameter θ. Show that only unbiased estimator of* $\exp\{-(k+1)\theta\}$, $k > 0$, *is*

$$T(X) = (-k)^X \text{ so that } T(x) > 0 \text{ if } x \text{ is even and } T(x) < 0 \text{ if } x \text{ is odd.}$$

Solution.

$$E\{T(X)\} = E\left\{(-k)^X\right\}, k > 0 = \sum_{x=0}^{\infty} (-k)^x \left(\frac{e^{-\theta}\theta^x}{x!}\right)$$

$$= e^{-\theta} \sum_{x=0}^{\infty} \left\{\frac{(-k\theta)^x}{x!}\right\} = e^{-\theta} e^{-k\theta} = e^{-(1+k)\theta}$$

$\Rightarrow$ $T(X) = (-k)^X$ is an unbiased estimator for $\exp\{-(1+k)\theta\}$, $k > 0$.

17·2·2. Consistency

Definition. *An estimator $T_n = T(x_1, x_2, \ldots, x_n)$, based on a random sample of size n, is said to be consistent estimator of $\gamma(\theta)$, $\theta \in \Theta$, the parameter space, if T_n converges to $\gamma(\theta)$ in probability, i.e., if* $T_n \xrightarrow{p} \gamma(\theta)$ *as* $n \to \infty$.

In other words, T_n is a consistent estimator of $\gamma(\theta)$ if for every $\varepsilon > 0$, $\eta > 0$, there exists a positive integer $n \geq m(\varepsilon, \eta)$ such that

$$P\{|T_n - \gamma(\theta)| < \varepsilon\} \to 1 \text{ as } n \;\Rightarrow\; P\{|T_n - \gamma(\theta)| < \varepsilon\} > 1 - \eta\,;\; \forall\, n \geq m \quad \ldots (17\cdot 2)$$

where m is some very large value of n.

Remarks:

1. If $X_1, X_2, \ldots, X_n$ is a random sample from population with finite mean $E(X_i) = \mu < \infty$, then by Khinchine's weak law of large numbers (W.L.L.N), we have

$$\bar{X}_n = \frac{1}{n}\sum_{i=1}^{n} X_i \xrightarrow{p} E(X_i) = \mu, \text{ as } n \to \infty.$$

Hence, sample mean $(\bar{X}_n)$ is always a consistent estimator of the population mean (μ).

2. Obviously consistency is a property concerning the behaviour of an estimator for indefinitely large values of the sample size n, *i.e.*, as $n \to \infty$. Nothing is regarded of its behaviour for finite n.

Moreover, if there exists a consistent estimator, say, T_n of $\gamma(\theta)$, then infinitely many such estimators can be constructed, *e.g.*,

$$T_n' = \left(\frac{n-a}{n-b}\right) T_n = \left[\frac{1-(a/n)}{1-(b/n)}\right] T_n \;\to\; T_n \xrightarrow{p} \gamma(\theta), \text{ as } n \to \infty$$

and hence, for different values of a and b, T_n' is also consistent for $\gamma(\theta)$.

Invariance Property of Consistent Estimators

Theorem 17·1. *If T_n is a consistent estimator of $\gamma(\theta)$ and $\psi\{\gamma(\theta)\}$ is a continuous function of $\gamma(\theta)$, then $\psi(T_n)$ is a consistent estimator of $\psi\{\gamma(\theta)\}$.*

Proof. Since T_n is a consistent estimator of $\gamma(\theta)$, $T_n \xrightarrow{p} \gamma(\theta)$ as $n \to \infty$, *i.e.*, for every $\varepsilon > 0$, $\eta > 0$, $\exists$ a positive integer $n \geq m(\varepsilon, \eta)$ such that

$$P\{|T_n - \gamma(\theta)| < \varepsilon\} > 1 - \eta, \;\forall\, n \geq m \quad \ldots (*)$$

Since $\psi(\cdot)$ is a continuous function, for every $\varepsilon > 0$, however small, $\exists$ a positive number ε_1 such that $|\psi(T_n) - \psi\{\gamma(\theta)\}| < \varepsilon_1$, whenever $|T_n - \gamma(\theta)| < \varepsilon$, *i.e.*,

$$|T_n - \gamma(\theta)| < \varepsilon \;\Rightarrow\; |\psi(T_n) - \psi\{\gamma(\theta)\}| < \varepsilon_1 \quad \ldots (**)$$

For two events A and B, if $A \Rightarrow B$, then

$$A \subseteq B \quad \Rightarrow \quad P(A) \le P(B) \quad \text{or} \quad P(B) \ge P(A) \qquad \ldots (***)$$

From (**) and (***), we get

$$P[\,|\psi(T_n) - \psi\{\gamma(\theta)\}\,| < \varepsilon_1] \ge P[\,|T_n - \gamma(\theta)| < \varepsilon]$$

$$\Rightarrow \quad P[\,|\psi(T_n) - \psi\{\gamma(\theta)\}\,| < \varepsilon_1] \ge 1 - \eta \,;\ \forall\, n \ge m \qquad \text{[Using (*)]}$$

$\Rightarrow \quad \psi(T_n) \xrightarrow{p} \psi\{\gamma(\theta)\}$, as $n \to \infty$ or $\psi(T_n)$ is a consistent estimator of $\psi\{\gamma(\theta)\}$.

Sufficient Conditions for Consistency

Theorem 17·2. *Let $\{T_n\}$ be a sequence of estimators such that for all $\theta \in \Theta$,*

$$\text{(i) } E_\theta(T_n) \to \gamma(\theta), \text{ as } n \to \infty \quad \text{and} \quad \text{(ii) } Var_\theta(T_n) \to 0, \text{ as } n \to \infty.$$

Then, T_n is a consistent estimator of $\gamma(\theta)$.

Proof. We have to prove that T_n is a consistent estimator of $\gamma(\theta)$

i.e., $\quad T_n \xrightarrow{p} \gamma(\theta)$, as $n \to \infty$

i.e., $$P[\,|T_n - \gamma(\theta)| < \varepsilon] > 1 - \eta \,;\ \forall\, n \ge m(\varepsilon, \eta) \qquad \ldots (17·3)$$

where ε and η are arbitrarily small positive numbers and m is some large value of n.

Applying Chebychev's inequality to the statistic T_n, we get

$$P[\,|T_n - E_\theta(T_n)| \le \delta] \ge 1 - \frac{\text{Var}_\theta(T_n)}{\delta^2} \qquad \ldots (17·4)$$

We have

$$|T_n - \gamma(\theta)| = |T_n - E(T_n) + E(T_n) - \gamma(\theta)| \le |T_n - E_\theta(T_n)| + |E_\theta(T_n) - \gamma(\theta)| \qquad \ldots (17·5)$$

Now $$|T_n - E_\theta(T_n)| \le \delta \Rightarrow |T_n - \gamma(\theta)| \le \delta + |E_\theta(T_n) - \gamma(\theta)| \qquad \ldots (17·6)$$

Hence, on using (***) of Theorem 17·1, we get:

$$P\{\,|T_n - \gamma(\theta)| \le \delta + |E_\theta(T_n) - \gamma(\theta)|\,\} \ge P\{\,|T_n - E_\theta(T_n)| \le \delta\}$$

$$\ge 1 - \frac{\text{Var}_\theta(T_n)}{\delta^2} \qquad \text{[From (17·4)]} \quad \ldots (17·7)$$

We are given: $\quad E_\theta(T_n) \to \gamma(\theta)\ \forall\ \theta \in \Theta$ as $n \to \infty$

Hence, for every $\delta_1 > 0$, $\exists$ a positive integer $n \ge n_0(\delta_1)$ such that

$$|E_\theta(T_n) - \gamma(\theta)| \le \delta_1,\ \forall\, n \ge n_0(\delta_1) \qquad \ldots (17·8)$$

Also $\text{Var}_\theta(T_n) \to 0$ as $n \to \infty$, (Given) $\quad \therefore \dfrac{\text{Var}_\theta(T_n)}{\delta^2} \le \eta,\ \forall\, n \ge n_0'(\eta)$, $\qquad \ldots (17·9)$

where η is arbitrarily small positive number.

Substituting from (17·8) and (17·9) in (17·7), we get

$$P[\,|T_n - \gamma(\theta)| \le \delta + \delta_1] \ge 1 - \eta \,;\ n \ge m(\delta_1, \eta)$$

$$\Rightarrow \quad P[\,|T_n - \gamma(\theta)| \le \varepsilon] \ge 1 - \eta;\ n \ge m,$$

where $m = \max(n_0, n_0')$ and $\varepsilon = \delta + \delta_1 > 0$.

$$\Rightarrow \quad T_n \xrightarrow{p} \gamma(\theta) \text{ as } n \to \infty \qquad \text{[Using (17·3)]}$$

$\therefore\ T_n$ is a consistent estimator of $\gamma(\theta)$.

Example 17·5.

(a) *Prove that in sampling from a $N(\mu, \sigma^2)$ population, the sample mean is a consistent estimator of μ.*

(b) *Prove that for Cauchy's distribution, not sample mean but sample median is a consistent estimator of the population mean.*

Solution.

(a) In sampling from a $N(\mu, \sigma^2)$ population, the sample mean $\bar{x}$ is also normally distributed as $N(\mu, \sigma^2/n)$, *i.e.*, $E(\bar{x}) = \mu$ and $V(\bar{x}) = \sigma^2/n$.

Thus, as $n \to \infty$, $E(\bar{x}) = \mu$ and $V(\bar{x}) \to 0$.

Hence, by *Theorem 17·2*, $\bar{x}$ is a consistent estimator for μ.

(b) Cauchy's population is given by the probability function :

$$dF(x) = \frac{1}{\pi} \cdot \frac{dx}{1 + (x - \mu)^2}, -\infty \le x \le \infty \qquad \ldots (*)$$

The mean of the distribution, if we conventionally agree to assume that it exists, is at $x = \mu$. If $\bar{x}$, the sample mean is taken as an estimator of μ, then the sampling distribution of $\bar{x}$ is given by :

$$dF(\bar{x}) = \frac{1}{\pi} \cdot \frac{d\bar{x}}{1 + (\bar{x} - \mu)^2}; -\infty < \bar{x} < \infty$$

because in Cauchy's distribution, the distribution of $\bar{x}$ is same as the distribution of x.

Since in this case, the distribution of $\bar{x}$ is same as distribution of any single sample observation, it does not increase in accuracy with increasing n. In other words

$$E(\bar{x}) = \mu \quad \text{but} \quad V(\bar{x}) = V(x) \ne 0, \text{ as } n \to \infty$$

Hence, by *Theorem 17·2*, $\bar{x}$ is not a consistent estimator of μ in this case.

Consideration of symmetry of (*) is enough to show that the sample median Md is an unbiased estimate of the population mean, which of course is same as the population median. Therefore,

$$E(Md) = \mu. \qquad (**)$$

For large n, the sampling distribution of median is asymptotically normal and is given by

$$dF \propto \exp\{-2n f_1^2 (x - \mu)^2\} dx,$$

where f_1 is the median ordinate of the parent population, *i.e.*,

$$dF \propto \exp\left\{-\frac{(x-\mu)^2}{1/(2nf_1^2)}\right\} \qquad \ldots (***)$$

But f_1 = Median ordinate of (*) = Modal ordinate of (*) [Because of symmetry]

$$= [f(x)]_{x=\mu} = \frac{1}{\pi}$$

Hence, from (***), the variance of the sampling distribution of median is :

$$V(Md) = \frac{1}{4nf_1^2} = \frac{1}{4n(1/\pi)^2} = \frac{\pi^2}{4n} \to 0 \text{ as } n \to \infty \qquad \ldots (****)$$

Hence, from (**) and (****), using *Theorem 17·2*, we conclude that for Cauchy's distribution, median is a consistent estimator for μ.

Example 17·6. *If $X_1, X_2, \ldots, X_n$ are random observations on a Bernoulli variate X taking the value 1 with probability p and the value 0 with probability $(1 - p)$, show that :*

$$\frac{\Sigma x_i}{n}\left(1 - \frac{\Sigma x_i}{n}\right) \text{ is a consistent estimator of } p(1-p).$$

Solution. Since $X_1, X_2, \ldots, X_n$ are *i.i.d* Bernoulli variates with parameter 'p',

$$T = \sum_{i-1}^{n} x_i \sim B(n, p) \;\Rightarrow\; E(T) = np \text{ and } Var(T) = npq \qquad \ldots (i)$$

$$\overline{X} = \frac{1}{n}\sum_{i-1}^{n} x_i = \frac{T}{n} \;\Rightarrow\; E(\overline{X}) = \frac{1}{n}E(T) = \frac{1}{n}\cdot np = p \qquad \text{[From } (i)]$$

and $$\text{Var}(\overline{X}) = \text{Var}\left(\frac{T}{n}\right) = \frac{1}{n^2}\cdot \text{Var}(T) = \frac{pq}{n} \to 0 \text{ as } n \to \infty \qquad \text{[From } (i)]$$

Since $E(\overline{X}) \to p$ and $\text{Var}(\overline{X}) \to 0$, as $n \to \infty$; $\overline{X}$ is a consistent estimator of p. Also $\frac{\Sigma x_i}{n}\left(1 - \frac{\Sigma x_i}{n}\right) = \overline{X}(1 - \overline{X})$, being a polynomial in $\overline{X}$, is a continuous function of $\overline{X}$.

Since $\overline{X}$ is consistent estimator of p, by the invariance property of consistent estimators (*Theorem 17·1*), $\overline{X}(1 - \overline{X})$ is a consistent estimator of $p(1-p)$.

17·2·3. Efficient Estimators

Efficiency. Even if we confine ourselves to unbiased estimates, there will, in general, exist more than one consistent estimator of a parameter. For example, in sampling from a normal population $N(\mu, \sigma^2)$, when σ^2 is known, sample mean $\overline{x}$ is an unbiased and consistent estimator of μ [*c.f.* Example 17·5*a*].

From symmetry it follows immediately that sample median (*Md*) is an unbiased estimate of μ, which is same as the population median. Also for large n,

$$V(Md) = \frac{1}{4nf_1^2} \qquad [c.f. \text{ Example 17·5}(b)]$$

Here, $$f_1 = \text{Median ordinate of the parent distribution.}$$
$$= \text{Modal ordinate of the parent distribution.}$$
$$= \left[\frac{1}{\sigma\sqrt{2\pi}}\exp\left\{-(x-\mu)^2/2\sigma^2\right\}\right]_{x=\mu} = \frac{1}{\sigma\sqrt{2\pi}}$$

$$\therefore \qquad V(Md) = \frac{1}{4n}\cdot 2\pi\sigma^2 = \frac{\pi\sigma^2}{2n}$$

Since $\quad E(Md) = \mu \quad$ and $\quad V(Md) \to 0$, as $n \to \infty$,

median is also an unbiased and consistent estimator of μ.

Thus, there is a necessity of some further criterion which will enable us to choose between the estimators with the common property of consistency. Such a criterion which is based on the variances of the sampling distribution of estimators is usually known as *efficiency*.

If, of the two consistent estimators T_1, T_2 of a certain parameter θ, we have

$$V(T_1) < V(T_2), \quad \textit{for all } n \qquad \ldots (17·10)$$

then T_1 is more efficient than T_2 for all samples sizes.

We have seen above :

For all n, $V(\overline{x}) = \frac{\sigma^2}{n}$ and for large n, $V(Md) = \frac{\pi\sigma^2}{2n} = 1{\cdot}57\frac{\sigma^2}{n}$

Since $V(\overline{x}) < V(Md)$, we conclude that for normal distribution, sample mean is more efficient estimator for μ than the sample median, for large samples at least.

Most Efficient Estimator. *If in a class of consistent estimators for a parameter, there exists one whose sampling variance is less than that of any such estimator, it is called the most efficient estimator. Whenever such an estimator exists, it provides a criterion for measurement of efficiency of the other estimators.*

Efficiency *(Definition) If T_1 is the most efficient estimator with variance V_1 and T_2 is any other estimator with variance V_2, then the efficiency E of T_2 is defined as :*

$$E = \frac{V_1}{V_2} \quad \ldots (17 \cdot 11)$$

Obviously, E cannot exceed unity, i.e. $0 \leq E \leq 1$.

If $T, T_1, T_2, \ldots, T_n$ are all estimators of $\gamma(\theta)$ and Var(T) is minimum, then the efficiency E_i of T_i, $(i = 1, 2, \ldots, n)$ is defined as :

$$E_i = \frac{\text{Var } T}{\text{Var } T_i} \; ; i = 1, 2, \ldots, n \quad \ldots (17 \cdot 11a)$$

Obviously $E_i \leq 1; i = 1, 2, \ldots n$. For example, in the normal samples, since sample mean $\overline{x}$ is the most efficient estimator of μ [*c.f. Remark to Example 17·31*], the efficiency E of *Md* for such samples, (for large *n*), is :

$$E = \frac{V(\overline{x})}{V(Md)} = \frac{\sigma^2/n}{\pi\sigma^2/(2n)} = \frac{2}{\pi} = 0 \cdot 637.$$

Example 17·7. *A random sample $(X_1, X_2, X_3, X_4, X_5)$ of size 5 is drawn from a normal population with unknown mean μ. Consider the following estimators to estimate μ :*

$$(i)\ t_1 = \frac{X_1 + X_2 + X_3 + X_4 + X_5}{5}, \qquad (ii)\ t_2 = \frac{X_1 + X_2}{2} + X_3 \qquad (iii)\ t_3 = \frac{2X_1 + X_2 + \lambda X_3}{3},$$

where λ is such that t_3 is an unbiased estimator of μ.

Find λ. Are t_1 and t_2 unbiased? State giving reasons, the estimator which is best among t_1, t_2 and t_3.

Solution. We are given :

$$E(X_i) = \mu, \qquad \text{Var}(X_i) = \sigma^2, \text{(say)} ; \qquad \text{Cov}(X_i, X_j) = 0, (i \neq j = 1, 2, \ldots, n) \quad \ldots (*)$$

(i) $E(t_1) = \frac{1}{5}\sum_{i=1}^{5} E(X_i) = \frac{1}{5}\sum_{i=1}^{5} \mu = \frac{1}{5} \cdot 5\mu = \mu \quad \Rightarrow \quad t_1$ is an unbiased estimator of μ.

(ii) $E(t_2) = \frac{1}{2}E(X_1 + X_2) + E(X_3) = \frac{1}{2}(\mu + \mu) + \mu = 2\mu$ [Using (*)]

$\Rightarrow$ t_2 is *not* an unbiased estimator of μ.

(iii) $E(t_3) = \mu \Rightarrow \frac{1}{3}E(2X_1 + X_2 + \lambda X_3) = \mu$ ($\because$ t_3 is unbiased estimator of μ)

$\therefore \quad 2E(X_1) + E(X_2) + \lambda E(X_3) = 3\mu \quad \Rightarrow \quad 2\mu + \mu + \lambda\mu = 3\mu \Rightarrow \lambda = 0.$

Using (*), we get

$$V(t_1) = \frac{1}{25}\{V(X_1) + V(X_2) + V(X_3) + V(X_4) + V(X_5)\} = \frac{1}{5}\sigma^2$$

$$V(t_2) = \frac{1}{4}[V(X_1) + V(X_2)] + V(X_3) = \frac{1}{2}\sigma^2 + \sigma^2 = \frac{3}{2}\sigma^2$$

$$V(t_3) = \frac{1}{9}[4V(X_1) + V(X_2)] = \frac{1}{9}\left(4\sigma^2 + \sigma^2\right) = \frac{5}{9}\sigma^2 \qquad (\because \lambda = 0)$$

Since $V(t_1)$ is least, t_1 is the best estimator (in the sense of least variance) of μ.

Example 17·8. *X_1, X_2, and X_3 is a random sample of size 3 from a population with mean value μ and variance σ^2. T_1, T_2, T_3 are the estimators used to estimate mean value μ, where*

$$T_1 = X_1 + X_2 - X_3, \quad T_2 = 2X_1 + 3X_3 - 4X_2, \quad \text{and} \quad T_3 = (\lambda X_1 + X_2 + X_3)/3.$$

(i) Are T_1 and T_2 unbiased estimators?

(ii) Find the value of λ such that T_3 is unbiased estimator for μ.

(iii) With this value of λ, is T_3 a consistent estimator?

(iv) Which is the best estimator?

Solution. Since X_1, X_2, X_3 is a random sample from a population with mean μ and variance σ^2,

$$E(X_i) = \mu, \text{ Var } (X_i) = \sigma^2 \text{ and } \text{Cov } (X_i, X_j) = 0, (i \neq j = 1, 2, \ldots, n) \quad \ldots (*)$$

(i) We have [On using (*)],

$$E(T_1) = E(X_1) + E(X_2) - E(X_3) = \mu \quad \Rightarrow \quad T_1 \text{ is an unbiased estimator of } \mu$$

$$E(T_2) = 2E(X_1) + 3E(X_3) - 4E(X_2) = \mu \quad \Rightarrow \quad T_2 \text{ is an unbiased estimator of } \mu.$$

(ii) We are given : $E(T_3) = \mu \quad \Rightarrow \frac{1}{3}\{\lambda E(X_1) + E(X_2) + E(X_3)\} = \mu$

$$\Rightarrow \quad \frac{1}{3}(\lambda\mu + \mu + \mu) = \mu \quad \Rightarrow \quad \lambda + 2 = 3 \Rightarrow \lambda = 1. \quad \text{[From (*)]}$$

(iii) With $\lambda = 1$, $T_3 = \frac{1}{3}(X_1 + X_2 + X_3) = \bar{X}$. Since sample mean is a consistent estimator of population mean μ, by Weak Law of Large Numbers, T_3 is a consistent estimator of μ.

(iv) We have [on using (*)] :

$$\text{Var}(T_1) = \text{Var}(X_1) + \text{Var}(X_2) + \text{Var}(X_3) = 3\sigma^2$$

$$\text{Var}(T_2) = 4\text{Var}(X_1) + 9\text{Var}(X_3) + 16\text{Var}(X_2) = 29\sigma^2$$

$$\text{Var}(T_3) = \frac{1}{9}[\text{Var}(X_1) + \text{Var}(X_2) + \text{Var}(X_3)] = \frac{1}{3}\sigma^2 \quad (\because \lambda = 1)$$

Since Var(T_3) is minimum, T_3 is the best estimator of μ in the sense of minimum variance.

Definition. Minimum Variance Unbiased (M.V.U.) Estimators

If a statistic $T = T(x_1, x_2, \ldots, x_n)$, based on sample of size n is such that :

(i) T is unbiased for $\gamma(\theta)$, for all $\theta \in \Theta$ and

(ii) It has the smallest variance among the class of all unbiased estimators of $\gamma(\theta)$,

then T is called the minimum variance unbiased estimator (MVUE) of $\gamma(\theta)$. ... (17·12)

More precisely, T is MVUE of $\gamma(\theta)$ if

$$E_\theta(T) = \gamma(\theta) \text{ for all } \theta \in \Theta \quad \ldots (17·13)$$

and

$$\text{Var}_\theta(T) \leq \text{Var}_\theta(T') \text{ for all } \theta \in \Theta \quad \ldots (17·14)$$

where T' is any other unbiased estimator of $\gamma(\theta)$.

We give below some important Theorems concerning MVU estimators.

Theorem 17·3. *An M.V.U. is unique in the sense that if T_1 and T_2 are M.V.U. estimators for $\gamma(\theta)$, then $T_1 = T_2$, almost surely.*

Proof. We are given that

$$\left.\begin{aligned} E_\theta(T_1) = E_\theta(T_2) &= \gamma(\theta), \quad \text{for all } \theta \in \Theta \\ \text{and} \quad \text{Var}_\theta(T_1) = \text{Var}_\theta(T_2)&, \quad \text{for all } \theta \in \Theta \end{aligned}\right\} \quad \ldots (17·15)$$

Consider a new estimator, $T = \frac{1}{2}(T_1 + T_2)$ which is also unbiased since,

$$E(T) = \frac{1}{2}\{E(T_1) + E(T_2)\} = \gamma(\theta) \qquad \text{[From(17·15)]}$$

$$\text{Var}(T) = \text{Var}\left\{\frac{1}{2}(T_1 + T_2)\right\} = \frac{1}{4}\text{Var}(T_1 + T_2)$$

$$= \frac{1}{4}\{\text{Var}(T_1) + \text{Var}(T_2) + 2\,\text{Cov}(T_1, T_2)\} \qquad [\because \text{Var}(CX) = C^2\,\text{Var}(X)]$$

$$= \frac{1}{4}\left\{\text{Var}(T_1) + \text{Var}(T_2) + 2\rho\sqrt{\text{Var}(T_1)\text{Var}(T_2)}\right\}$$

$$= \frac{1}{2}\text{Var}(T_1)(1 + \rho), \qquad \text{[From (17·15)]}$$

where ρ is Karl Pearson's co-efficient of correlation between T_1 and T_2.

Since T_1 is the *MVU* estimator,

$$\text{Var}(T) \geq \text{Var}(T_1) \quad \Rightarrow \frac{1}{2}\text{Var}(T_1)(1 + \rho) \geq \text{Var}(T_1) \quad \Rightarrow \quad \frac{1}{2}(1 + \rho) \geq 1 \quad \Rightarrow \quad \rho \geq 1$$

Since $|\rho| \leq 1$, we must have $\rho = 1$, *i.e.*, T_1 and T_2 must have a linear relation of the form:

$$T_1 = \alpha + \beta T_2, \qquad \text{... (17·16)}$$

where α and β are constants, independent of $x_1, x_2, \ldots, x_n$ but may depend on θ, *i.e.*, we may have

$$\alpha = \alpha(\theta) \quad \text{and} \quad \beta = \beta(\theta).$$

Taking expectation of both sides in (17·16) and using (17·15), we get

$$\theta = \alpha + \beta\theta \qquad \text{... (17·17)}$$

Also from (17·16), we get: $\text{Var}(T_1) = \text{Var}(\alpha + \beta T_2) = \beta^2\,\text{Var}(T_2)$

$$\Rightarrow \quad 1 = \beta^2 \quad \Rightarrow \quad \beta = \pm 1 \qquad \text{... [From (17·15)]}$$

But since $\rho(T_1, T_2) = +1$, the coefficient of regression of T_1 on T_2 must be positive.

$$\therefore \quad \beta = 1 \quad \Rightarrow \quad \alpha = 0 \qquad \text{...[From 17·17)]}$$

Substituting in (17·16), we get $T_1 = T_2$, as desired.

Theorem 17·4. *Let T_1 and T_2 be unbiased estimators of $\gamma(\theta)$ with efficiencies e_1 and e_2 respectively and $\rho = \rho_\theta$ be the correlation coefficient between them. Then*

$$\sqrt{e_1e_2} - \sqrt{(1-e_1)(1-e_2)} \leq \rho \leq \sqrt{e_1e_2} + \sqrt{(1-e_1)(1-e_2)}$$

Proof. Let T be minimum variance unbiased estimator of $\gamma(\theta)$. Then we are given :

$$E_\theta(T_1) = \gamma(\theta) = E_\theta(T_2) \;\; \forall\, \theta \in \Theta \qquad \text{... (17·18)}$$

and

$$e_1 = \frac{V_\theta(T)}{V_\theta(T_1)} = \frac{V}{V_1}, \text{(say)} \quad \Rightarrow \quad V_1 = \frac{V}{e_1} \qquad \text{... (17·19)}$$

$$e_2 = \frac{V_\theta(T)}{V_\theta(T_2)} = \frac{V}{V_2}, \text{(say)} \quad \Rightarrow \quad V_2 = \frac{V}{e_2} \qquad \text{... (17·20)}$$

Let us consider another estimator: $\quad T_3 = \lambda T_1 + \mu T_2, \qquad$ ... (17·21)

which is also unbiased estimator of $\gamma(\theta)$ *i.e.*,

$$E(T_3) = (\lambda + \mu)\,\gamma(\theta) \;=\; \gamma(\theta) \quad \text{[Using (17·18)]} \;\Rightarrow\; \lambda + \mu = 1 \qquad \text{... (17·22)}$$

$$V_\theta(T_3) = V(\lambda T_1 + \mu T_2) \;=\; \lambda^2 V(T_1) + \mu^2 V(T_2) + 2\lambda\,\mu\,\text{Cov}(T_1, T_2)$$

$$= V\left(\frac{\lambda^2}{e_1} + \frac{\mu^2}{e_2} + 2 \cdot \frac{\lambda\,\mu\,\rho}{\sqrt{e_1e_2}}\right) \qquad \text{[Using (17·19) and (17·20)]}$$

But $V_\theta(T_3) \geq V$, since V is the minimum variance.

$$\therefore \quad \frac{\lambda^2}{e_1} + \frac{\mu^2}{e_2} + \frac{2\rho\lambda\mu}{\sqrt{e_1 e_2}} \geq 1 = (\lambda + \mu)^2 \qquad \text{[Using (17·22)]}$$

$$\Rightarrow \quad \left(\frac{1}{e_1} - 1\right)\lambda^2 + \left(\frac{1}{e_2} - 1\right)\mu^2 + 2\lambda\mu\left(\frac{\rho}{\sqrt{e_1 e_2}} - 1\right) \geq 0$$

$$\Rightarrow \quad \left(\frac{1}{e_1} - 1\right)\left(\frac{\lambda}{\mu}\right)^2 + 2\left(\frac{\rho}{\sqrt{e_1 e_2}} - 1\right)\left(\frac{\lambda}{\mu}\right) + \left(\frac{1}{e_2} - 1\right) \geq 0 \qquad \text{... (17·23)}$$

which is quadratic expression in (λ/μ).

Note that : $\quad e_i < 1 \Rightarrow \frac{1}{e_i} > 1 \Rightarrow \left(\frac{1}{e_i} - 1\right) > 0, i = 1, 2$

We know that $AX^2 + BX + C \geq 0 \ \forall X$, $A > 0$, $C > 0$; if and only if

$$\text{Discriminant} = B^2 - 4AC \leq 0 \qquad \text{... (17·24)}$$

Using (17·24), we get from (17·23) :

$$\left(\frac{\rho}{\sqrt{e_1 e_2}} - 1\right)^2 - \left(\frac{1}{e_1} - 1\right)\left(\frac{1}{e_2} - 1\right) \leq 0 \quad \Rightarrow \quad \left(\rho - \sqrt{e_1 e_2}\right)^2 - (1 - e_1)(1 - e_2) \leq 0$$

$$\therefore \quad \rho^2 - 2\sqrt{e_1 e_2}\,\rho + (e_1 + e_2 - 1) \leq 0$$

This implies that ρ lies between the roots of the equation : $\rho^2 - 2\sqrt{e_1 e_2}\,\rho + (e_1 + e_2 - 1) = 0$,

which are given by : $\quad \frac{1}{2}\left\{2\sqrt{e_1 e_2} \pm 2\sqrt{e_1 e_2 - (e_1 + e_2 - 1)}\right\} = \sqrt{e_1 e_2} \pm \sqrt{(e_1 - 1)(e_2 - 1)}$

Hence, we have :

$$\sqrt{e_1 e_2} - \sqrt{(e_1 - 1)(e_2 - 1)} \leq \rho \leq \sqrt{e_1 e_2} + \sqrt{(e_1 - 1)(e_2 - 1)}$$

$$\Rightarrow \quad \sqrt{e_1 e_2} - \sqrt{(1 - e_1)(1 - e_2)} \leq \rho \leq \sqrt{e_1 e_2} + \sqrt{(1 - e_1)(1 - e_2)} \qquad \text{... (17·25)}$$

Corollary. If we take $e_1 = 1$ and $e_2 = e$ in (17·25), we get $\sqrt{e} \leq \rho \leq \sqrt{e} \Rightarrow \rho = \sqrt{e}$.
This leads to the following important result, which we state in the form of a theorem.

Theorem 17·5. *If T_1 is an MVU estimator of $\gamma(\theta)$, $\theta \in \Theta$ and T_2 is any other unbiased estimator of $\gamma(\theta)$ with efficiency $e = e_\theta$, then the correlation coefficient between T_1 and T_2 is given by $\rho = \sqrt{e}$, i.e., $\rho = \sqrt{e_\theta}, \forall \ \theta \in \Theta$.*

For an alternate proof, see Examples 17·9 and 17·10.

Theorem 17·6. *If T_1 is an MVUE of $\gamma(\theta)$ and T_2 is any other unbiased estimator of $\gamma(\theta)$ with efficiency $e < 1$, then no unbiased linear combination of T_1 and T_2 can be an MVUE of $\gamma(\theta)$.*

Proof. A linear combination: $\quad T = l_1 T_1 + l_2 T_2 \qquad$ ... (17·27)

will be unbiased estimator of $\gamma(\theta)$ if

$$E(T) = l_1 E(T_1) + l_2 E(T_2) = \gamma(\theta), \text{ for all } \theta \in \Theta \quad \Rightarrow \quad l_1 + l_2 = 1, \qquad \text{... (17·27a)}$$

since we are given $E(T_1) = E(T_2) = \gamma(\theta)$.

We have $\quad e = \frac{\text{Var } T_1}{\text{Var } T_2} \quad \Rightarrow \quad \text{Var } T_2 = \frac{\text{Var } T_1}{e} \qquad$ ... (17·28)

and $\quad \rho = \rho(T_1, T_2) = \sqrt{e}$ [*c.f.* Theorem 17·5]

From (17·27), on using (17·28), we get

$$\text{Var } T = l_1^2 \text{Var}(T_1) + l_2^2 \text{Var}(T_2) + 2l_1 l_2 \text{Cov}(T_1, T_2)$$

$$= l_1^2 \text{Var}(T_1) + l_2^2 \text{Var}(T_2) + 2l_1 l_2\, \rho \sqrt{\text{Var}(T_1)\text{Var}(T_2)}$$

$$= \text{Var}(T_1)\left(l_1^2 + \frac{l_2^2}{e} + 2l_1 l_2 \frac{\rho}{\sqrt{e}}\right) = \text{Var}(T_1)\left(l_1^2 + 2l_1 l_2 + \frac{l_2^2}{e}\right) \qquad (\because \rho = \sqrt{e})$$

$$> \text{Var}(T_1).\left(l_1^2 + 2l_1 l_2 + l_2^2\right) \qquad \left(\because 0 < e < 1 \Rightarrow \frac{1}{e} > 1\right)$$

$$\Rightarrow \qquad \text{Var } T > \text{Var } T_1 \cdot (l_1 + l_2)^2 = \text{Var}(T_1) \qquad \text{[From (17·27a)]}$$

$\therefore$ T cannot be an MVU estimator.

Example 17·9. *If T_1 and T_2 be two unbiased estimators of $\gamma(\theta)$ with variances σ_1^2, σ_2^2 and correlation ρ, what is the best unbiased linear combination of T_1 and T_2 and what is the variance of such a combination?*

Solution. Let T_1 and T_2 be two unbiased estimators of $\gamma(\theta) \Rightarrow E(T_1) = E(T_2) = \gamma(\theta)$ ... (1)

Let T be a linear combination of T_1 and T_2 given by $T = l_1 T_1 + l_2 T_2$, ... (*)

where l_1, l_2 are arbitrary constants.

$$E(T) = l_1 E(T_1) + l_2 E(T_2) = (l_1 + l_2)\, \gamma(\theta) \qquad \text{[From (1)]}$$

$\therefore$ T is also an unbiased estimator of $\gamma(\theta)$ if and only if : $l_1 + l_2 = 1$ ... (2)

$$V(T) = V(l_1 T_1 + l_2 T_2) = l_1^2 V(T_1) + l_2^2 V(T_2) + 2l_1 l_2 \text{ Cov}(T_1, T_2)$$

$$= l_1^2 \sigma_1^2 + l_2^2 \sigma_2^2 + 2l_1 l_2 \rho\, \sigma_1 \sigma_2 \qquad \text{... (3)}$$

We want the minimum value of (3) for variations in l_1 and l_2, subject to the condition (2).

$$\therefore \qquad \frac{\partial}{\partial l_1}[V(T)] = 0 = l_1 \sigma_1^2 + l_2 \rho \sigma_1 \sigma_2 \ ; \quad \frac{\partial}{\partial l_2}[V(T)] = 0 = l_2 \sigma_2^2 + l_1 \rho \sigma_1 \sigma_2$$

Substracting, we get $\quad l_1(\sigma_1^2 - \rho\sigma_1\sigma_2) = l_2(\sigma_2^2 - \rho\sigma_1\sigma_2)$

$$\Rightarrow \qquad \frac{l_1}{\sigma_2^2 - \rho\sigma_1\sigma_2} = \frac{l_2}{\sigma_1^2 - \rho\sigma_1\sigma_2} = \frac{l_1 + l_2}{\sigma_1^2 + \sigma_2^2 - 2\rho\sigma_1\sigma_2} = \frac{1}{\sigma_1^2 + \sigma_2^2 - 2\rho\sigma_1\sigma_2} \qquad \text{[From (2)]}$$

$$\therefore \qquad l_1 = \frac{\sigma_2^2 - \rho\sigma_1\sigma_2}{\sigma_1^2 + \sigma_2^2 - 2\rho\sigma_1\sigma_2} \quad \text{and} \quad l_2 = \frac{\sigma_1^2 - \rho\sigma_1\sigma_2}{\sigma_1^2 + \sigma_2^2 - 2\rho\sigma_1\sigma_2} \qquad \text{... (4)}$$

With these values of l_1 and l_2, T given by (*) is the best unbiased linear combination of T_1 and T_2 and its variance is given by (3).

Example 17·10. *Suppose T_1 in the above Example 17·9. is an unbiased minimum variance estimate and T_2 is any other unbiased estimate with variance σ^2/e. Then prove that the correlation coefficient between T_1 and T_2 is $\sqrt{e}$.*

Solution. The coefficients of the best linear unbiased combination of T_1 and T_2, given by (*) in Example 17·9 are given by (4). We are given that $\sigma_1^2 = V(T_1) = \sigma^2$

and
$$e = \frac{V(T_1)}{V(T_2)} = \frac{\sigma^2}{V(T_2)} \quad \Rightarrow \quad V(T_2) = \sigma_2^2 = \frac{\sigma^2}{e},$$

Substituting in (4) of Example 17·9, we get :

$$\left.\begin{aligned} l_1 &= \frac{1-\rho\sqrt{e}}{D} \\ l_2 &= \frac{e-\rho\sqrt{e}}{D} \end{aligned}\right\}, \text{where } D = 1+e-2\rho\sqrt{e} \qquad \ldots (5)$$

Hence, from (*), the unbiased statistic is : $T = \dfrac{\left[(1-\rho\sqrt{e})T_1 + (e-\rho\sqrt{e})T_2\right]}{D}$
and from (3) the minimum variance is :

$$\begin{aligned} V(T) &= \frac{1}{D^2}\left\{\left(1-\rho\sqrt{e}\right)^2\sigma^2 + \left(e-\rho\sqrt{e}\right)^2\frac{\sigma^2}{e} + 2\left(1-\rho\sqrt{e}\right)\left(e-\rho\sqrt{e}\right)\cdot\rho\cdot\sigma\cdot\sigma/\sqrt{e}\right\} \\ &= \frac{\sigma^2}{D^2}\left\{\left(1+\rho^2 e - 2\rho\sqrt{e}\right) + \frac{1}{e}\left(e^2+\rho^2 e - 2\rho e\sqrt{e}\right) + 2\left(1-\rho\sqrt{e}\right)\left(\sqrt{e}-\rho\right)\rho\right\} \\ &= \frac{\sigma^2}{D^2}\left\{1+\rho^2 e - 2\rho\sqrt{e} + e + \rho^2 - 2\rho\sqrt{e} + 2\left(\rho\sqrt{e} - \rho^2 e - \rho^2 + \rho^3\sqrt{e}\right)\right\} \\ &= \frac{\sigma^2}{D^2}\left(1-\rho^2 e + e - \rho^2 - 2\rho\sqrt{e} + 2\rho^3\sqrt{e}\right) \\ &= \frac{\sigma^2}{D^2}\left\{\left(1+e-2\rho\sqrt{e}\right) - \rho^2\left(e+1-2\rho\sqrt{e}\right)\right\} \\ &= \frac{\sigma^2\left(1-\rho^2\right)\left(1+e-2\rho\sqrt{e}\right)}{\left(1+e-2\rho\sqrt{e}\right)^2} = \frac{\sigma^2\left(1-\rho^2\right)}{1+e-2\rho\sqrt{e}} = \frac{\sigma^2\left(1-\rho^2\right)}{\left(1-\rho^2\right)+\left(\sqrt{e}-\rho\right)^2} \end{aligned}$$

$$\therefore \qquad \frac{V(T)}{\sigma^2} = \frac{1-\rho^2}{\left(1-\rho^2\right)+\left(\sqrt{e}-\rho\right)^2} \le 1 \qquad \ldots (6)$$

Since T_1 is the most efficient estimator, $V(T) \not< \sigma^2 \;\Rightarrow\; \dfrac{V(T)}{\sigma^2} \ge 1$ ... (7)

From (6) and (7), we get

$$\frac{V(T)}{\sigma^2} = 1 \;\Rightarrow\; \frac{1-\rho^2}{(1-\rho^2)+\left(\sqrt{e}-\rho\right)^2} = 1 \;\Rightarrow\; \left(\sqrt{e}-\rho\right)^2 = 0 \;\Rightarrow\; \rho = \sqrt{e}.$$

Aliter. *From (5) onwards.* Since T_1 is given to be the most efficient estimator, it cannot be improved upon (*c.f.* Theorem 17·6). Hence, in order that T defined in (*) is minimum variance unbiased estimator we must have

$$\left.\begin{aligned} l_1 &= 1 \\ l_2 &= 0 \end{aligned}\right\} \Rightarrow \rho = \sqrt{e} \qquad \ldots \text{[From (5)]}$$

Remark. This problem leads to the following very important result :
"The correlation coefficient between a most efficient estimator and any other estimator with efficiency e is $\sqrt{e}$*."*

Example 17·11.

(a) *Show that if a most efficient estimator A and a less efficient estimator B with efficiency e tend to joint normality for large samples, B – A tends to zero correlation with A.*

(b) *Show that the error in B may be regarded as composed (for large samples) of two parts which are independent, the error in A and the error in (B –A).*

(c) *Show further that* $V(A-B) = \left(\dfrac{1}{e}-1\right) V(A)$.

Solution.

(*a*) We have to prove that $r\{A, (B-A)\} = 0 \Rightarrow \text{Cov}\,(A, B-A) = 0$

$$\text{Cov}\,\{A, (B-A)\} = \text{Cov}\,(A, B) - V(A) = \rho\sigma_A\sigma_B - \sigma_A^2,$$

where ρ is the correlation coefficient between A and B.

If we take $\sigma_A = \sigma$, then $\sigma_B = \dfrac{\sigma}{\sqrt{e}}$ and $\rho = \sqrt{e}$ (*c.f.* Theorem 17·5)

$\therefore$ $\text{Cov}\,(A, B-A) = \sqrt{e}\cdot\sigma\cdot\dfrac{\sigma}{\sqrt{e}} - \sigma^2 = 0$. Hence $(B-A)$ has zero correlation with A.

(*b*) We have $B = A + (B - A)$

$$\therefore\quad V(B) = V[A + (B-A)] = V(A) + V(B-A) + 2\,\text{Cov}\,(A, B-A)$$

$$= V(A) + V(B-A) \qquad \text{[Using part } (a)]$$

$\Rightarrow$ Error in B = Error in A + Error in $(B-A)$

and since A and $(B-A)$ are independent, [*c.f.* part (*a*) *viz.*, $r(A, B-A) = 0$ and A and B tend to joint normality], the result follows.

(*c*) $V(A - B) = V(A) + V(B) - 2\,\text{Cov}\,(A, B) = \sigma_A^2 + \sigma_B^2 - 2\,\rho\,\sigma_A\,\sigma_B$

$$= \sigma^2 + \frac{\sigma^2}{e} - 2\sqrt{e}\,\sigma\frac{\sigma}{\sqrt{e}} = \frac{\sigma^2}{e} - \sigma^2 = \left(\frac{1}{e} - 1\right)\sigma^2.$$

Example 17·12. *If T_1 and T_2 are two unbiased estimators of $\gamma(\theta)$, having the same variance and ρ is the correlation between them, then show that $\rho \geq 2e - 1$, where e is the efficiency of each estimator.*

Solution. Let T be MVUE of $\gamma(\theta)$. Then, since $V(T_1) = V(T_2)$, the efficiency e of each estimator is given by :

$$e = \frac{V(T)}{V(T_1)} = \frac{V(T)}{V(T_2)} \qquad \ldots (*)$$

Consider another unbiased estimator of $\gamma(\theta)$ *viz.*, $T_3 = \frac{1}{2}(T_1 + T_2)$

$$\Rightarrow\quad V(T_3) = \frac{1}{4}\left[V(T_1) + V(T_2) + 2\,\text{Cov}\,(T_1, T_2)\right] = \frac{1}{4}\left\{\frac{V(T)}{e} + \frac{V(T)}{e} + 2\rho\sqrt{\frac{V(T)}{e}\cdot\frac{V(T)}{e}}\right\} \qquad \text{[From (*)]}$$

$$= \frac{V(T)}{4e}(1 + 1 + 2\rho) = \frac{(1+\rho)V(T)}{2e}$$

Since $V(T)$ is the minimum variance, $V(T_3) = \dfrac{(1+\rho)\cdot V(T)}{2e} \geq V(T) \Rightarrow \rho \geq (2e - 1)$.

Aliter. **Deduction from (17·25), page 17·12.** If T_1 and T_2 have same variances/efficiencies *i.e.*, $e_1 = e_2 = e$, (say), then (17·25) gives

$$e - (1-e) \leq \rho \leq e + (1-e) \quad\Rightarrow\quad \rho \geq 2e - 1.$$

17·2·4. Sufficiency

An estimator is said to be sufficient for a parameter, if it contains all the information in the sample regarding the parameter.

If $T = t(x_1, x_2, \ldots, x_n)$ is an estimator of a parameter θ, based on a sample $x_1, x_2, \ldots, x_n$ of size n from the population with density $f(x, \theta)$ such that the conditional distribution of $x_1, x_2, \ldots, x_n$ given T, is independent of θ, then T is sufficient estimator for θ.

Illustration. Let $x_1, x_2, \ldots, x_n$ be a random sample from a Bernoulli population with parameter 'p', $0 < p < 1$, *i.e.,*

$$x_i = \begin{cases} 1, \text{ with probability } p \\ 0, \text{ with probability } q = (1-p) \end{cases}$$

Then $$T = t(x_1, x_2, \ldots, x_n) = x_1 + x_2 + \ldots + x_n \sim B(n, p)$$

$$\therefore \quad P(T = k) = \binom{n}{k} p^k (1-p)^{n-k};\ k = 0, 1, 2, \ldots, n$$

The conditional distribution of $(x_1, x_2, \ldots, x_n)$ given T is :

$$P(x_1 \cap x_2 \cap \ldots \cap x_n \mid T = k) = \frac{P(x_1 \cap x_2 \cap \ldots x_n \cap T = k)}{P(T = k)} = \begin{cases} \dfrac{p^k (1-p)^{n-k}}{\binom{n}{k} p^k (1-p)^{n-k}} = \dfrac{1}{\binom{n}{k}} \\ 0, \text{ if } \sum\limits_{i=1}^{n} x_i \neq k \end{cases}$$

Since this does not depend on 'p', $T = \sum\limits_{i=1}^{n} x_i$, is sufficient for '$p$'.

Theorem 17·7. FACTORIZATION THEOREM (Neyman). *The necessary and sufficient condition for a distribution to admit sufficient statistic is provided by the 'factorization theorem' due to Neyman.*

Statement *$T = t(x)$ is sufficient for θ if and only if the joint density function L (say), of the sample values can be expressed in the form :*

$$L = g_\theta\,[t(x)] \cdot h(x) \qquad \ldots (17·29)$$

where (as indicated) $g_\theta\,[t(x)]$ depends on θ and x only through the value of $t(x)$ and $h(x)$ is independent of θ.

Remarks:

1. It should be clearly understood that by 'a function independent of θ' we not only mean that it does not involve θ but also that its domain does not contain θ. For example, the function :

$$f(x) = \frac{1}{2a},\ a - \theta < x < a + \theta\ ;\ -\infty < \theta < \infty,$$

depends on θ.

2. It should be noted that the original sample $\mathbf{X} = (X_1, X_2, \ldots, X_n)$, is always a sufficient statistic.
3. The most general form of the distributions admitting sufficient statistic is *Koopman's form* and is given by :

$$L = L(\mathbf{x}, \theta) = g(\mathbf{x}).h(\theta).\exp\{a(\theta)\psi(\mathbf{x})\} \qquad \ldots (17·30)$$

where $h(\theta)$ and $a(\theta)$ are functions of the parameter θ only and $g(\mathbf{x})$ and $\psi(\mathbf{x})$ are the functions of the sample observations only.

Equation (17·30) represents the famous *exponential family of distributions,* of which most of the common distributions like the binomial, the Poisson and the normal with unknown mean and variance, are the members.

4. **Invariance Property of Sufficient Estimator.** *If T is a sufficient estimator for the parameter θ and if $\psi(T)$ is a one to one function of T, then $\psi(T)$ is sufficient for $\psi(\theta)$.*

5. **Fisher-Neyman Criterion.** *A statistic $t_1 = t(x_1, x_2, \ldots, x_n)$ is sufficient estimator of parameter θ if and only if the likelihood function (joint p.d.f. of the sample) can be expressed as :*

$$L = \prod_{i=1}^{n} f(x_i, \theta) = g_1(t_1, \theta) . k(x_1, x_2, \ldots, x_n) \qquad \ldots (17\cdot31)$$

where $g_1(t_1, \theta)$ is the p.d.f. of the statistic t_1 and $k(x_1, x_2, \ldots, x_n)$ is a function of sample observations only, independent of θ.

Note that this method requires the working out of the *p.d.f.* (*p.m.f.*) of the statistic $t_1 = t(x_1, x_2, \ldots, x_n)$, which is not always easy.

Example 17·13. *Let $x_1, x_2, \ldots, x_n$ be a random sample from a uniform population on $[0, \theta]$. Find a sufficient estimator for θ.*

Solution. We are given : $f_\theta(x_i) = \begin{cases} \dfrac{1}{\theta}, 0 \le x_i \le \theta \\ 0, \text{otherwise} \end{cases}$

Let $k(a,b) = \left.\begin{matrix} 1, \text{if } a \le b \\ 0, \text{if } a > b \end{matrix}\right\}$, then $f_\theta(x_i) = \dfrac{k(0, x_i)\, k(x_i, \theta)}{\theta}$

$$L = \prod_{i=1}^{n} f_\theta(x_i) = \prod_{i=1}^{n}\left[\frac{k(0,x_i).\,k(x_i,\theta)}{\theta}\right] = \frac{k\left(0, \min\limits_{1\le i\le n} x_i\right)\cdot k\left(\max\limits_{1\le i\le n} x_i, \theta\right)}{\theta^n} = g_\theta\{t(x)\}h(x)$$

where $\quad g_\theta[t(\mathbf{x})] = \dfrac{k\{t(\mathbf{x}),\theta\}}{\theta^n}, \quad t(\mathbf{x}) = \max\limits_{1\le i\le n} x_i \;$ and $\; h(\mathbf{x}) = k\left(0, \min\limits_{1\le i\le n} x_i\right)$

Hence, by Factorization theorem, $T = \max\limits_{1\le i\le n} x_i,$ is sufficient statistic for θ.

Aliter. We have
$$L = \prod_{i=1}^{n} f(x_i,\theta) = \frac{1}{\theta^n};\, 0 < x_i < \theta \qquad \ldots (*)$$

If $t = \max(x_1, x_2, \ldots, x_n) = x_{(n)}$, then *p.d.f.* of t is given by :
$$g(t, \theta) = n\{F(x_{(n)})\}^{n-1} . f(x_{(n)}) \qquad \ldots (**)$$

We have
$$F(x) = P(X \le x) = \int_0^x f(x,\theta)dx = \int_0^x \frac{1}{\theta}\cdot dx = \frac{x}{\theta}$$

$$g(t, \theta) = n\left[\frac{x_{(n)}}{\theta}\right]^{n-1} \cdot \frac{1}{\theta} = \frac{1}{\theta^n}\left[x_{(n)}\right]^{n-1} \qquad \ldots [\text{From } (**)]$$

Rewriting (*),
$$L = \frac{n\left[x_{(n)}\right]^{n-1}}{\theta^n} \cdot \frac{1}{n\left[x_{(n)}\right]^{n-1}} = g(t,\theta)\cdot h(x_1, x_2, \ldots, x_n)$$

Hence, by Fisher – Neyman criterion, the statistic $t = x_{(n)}$, is sufficient estimator for θ.

Example 17·14. *Let $x_1, x_2, \ldots, x_n$ be a random sample from $N(\mu, \sigma^2)$ population. Find sufficient estimators for μ and σ^2.*

Solution. Let us write $\theta = (\mu, \sigma^2)$; $-\infty < \mu < \infty$, $0 < \sigma^2 < \infty$.

Then
$$L = \prod_{i=1}^{n} f(x_i,\theta) = \left\{\frac{1}{\sigma\sqrt{2\pi}}\right\}^n \cdot \exp\left\{-\frac{1}{2\sigma^2}\sum_{i=1}^{n}(x_i - \mu)^2\right\}$$

$$= \left(\frac{1}{\sigma\sqrt{2\pi}}\right)^n \exp\left\{-\frac{1}{2\sigma^2}\left(\sum_{i=1}^{n} x_i^2 - 2\mu\sum x_i + n\mu^2\right)\right\}$$

$$= g_\theta[t(\mathbf{x})] \cdot h(\mathbf{x})$$

where, $$g_\theta[t(\mathbf{x})] = \left(\frac{1}{\sigma\sqrt{2\pi}}\right)^n \exp\left[-\frac{1}{2\sigma^2}\left\{t_2(\mathbf{x}) - 2\mu t_1(\mathbf{x}) + n\mu^2\right\}\right],$$

$$t(\mathbf{x}) = \{t_1(\mathbf{x}), t_2(\mathbf{x})\} = \left(\sum x_i, \sum x_i^2\right) \text{ and } h(\mathbf{x}) = 1$$

Thus, $t_1(\mathbf{x}) = \sum x_i$ is sufficient for μ and $t_2(\mathbf{x}) = \sum x_i^2$, is sufficient for σ^2.

Example 17·15. *Let $X_1, X_2, \ldots, X_n$ be a random sample from a distribution with p.d.f.:*

$$f(x, \theta) = e^{-(x-\theta)}, \theta < x < \infty ; -\infty < \theta < \infty$$

Obtain sufficient statistic for θ.

Solution. Here,

$$L = \prod_{i=1}^{n} f(x_i, \theta) = \prod_{i=1}^{n}\left\{e^{-(x_i-\theta)}\right\} = \exp\left(-\sum_{i=1}^{n} x_i\right) \times \exp(n\theta) \qquad \ldots (*)$$

Let $Y_1, Y_2, \ldots, Y_n$ denote the order statistics of the random sample such that $Y_1 < Y_2 < \ldots < Y_n$. The *p.d.f.* of the smallest observation Y_1 is given by :

$$g_1(y_1, \theta) = n[1 - F(y_1)]^{n-1} f(y_1, \theta),$$

where $F(\cdot)$ is the distribution function corresponding to *p.d.f.* $f(\cdot)$.

Now, $$F(x) = \int_\theta^x e^{-(u-\theta)} du = \left|\frac{e^{-(u-\theta)}}{-1}\right|_{u=\theta}^{x} = 1 - e^{-(x-\theta)}$$

$\therefore$ $$g_1(y_1, \theta) = n\left[e^{-(y_1-\theta)}\right]^{n-1} \cdot e^{-(y_1-\theta)} = \begin{cases} ne^{-n(y_1-\theta)} & , 0<y_1<\infty \\ 0, \text{otherwise} \end{cases}$$

Thus, the likelihood function (*) of $X_1, X_2, \ldots, X_n$ may be expressed as

$$L = e^{n\theta} \exp\left(-\sum_{i=1}^{n} x_i\right) = n\exp\{-n(y_1 - \theta)\}\left\{\frac{\exp\left(-\sum_{i=1}^{n} x_i\right)}{n\exp(-ny_1)}\right\}$$

$$= g_1(\min x_i, \theta)\left\{\frac{\exp\left(-\sum_{i=1}^{n} x_i\right)}{n\exp(-n\min x_i)}\right\}$$

Hence, by Fisher-Neyman criterion, the first order statistic $Y_1 = \min(X_1, X_2, \ldots, X_n)$ is a sufficient statistic for θ.

Example 17·16. *Let $X_1, X_2, \ldots, X_n$ be a random sample from a population with p.d.f :*

$$f(x, \theta) = \theta x^{\theta-1}; 0 < x < 1, \theta > 0.$$

Show that $t_1 = \prod_{i=1}^{n} X_i$, *is sufficient for θ.*

Solution. $L(\mathbf{x}, \theta) = \prod_{i=1}^{n} f(x_i, \theta) = \theta^n \prod_{i=1}^{n} (x_i^{\theta - 1})$

$$= \theta^n \left(\prod_{i=1}^{n} x_i\right)^{\theta} \cdot \frac{1}{\left(\prod_{i=1}^{n} x_i\right)} = g(t_1, \theta).\, h(x_1, x_2, \ldots, x_n), \text{ (say)}.$$

Hence, by Factorisation Theorem, $t_1 = \prod_{i=1}^{n} (X_i)$, is sufficient estimator for θ.

Example 17·17. *Let $X_1, X_2, \ldots, X_n$ be a random sample from Cauchy population :*

$$f(x,\theta) = \frac{1}{\pi} \cdot \frac{1}{1 + (x - \theta)^2}\,;\ -\infty < x < \infty\,;\ -\infty < \theta < \infty$$

Examine if there exists a sufficient statistic for θ.

Solution. $L(\mathbf{x},\theta) = \prod_{i=1}^{n} f(x_i, \theta) = \frac{1}{\pi^n} \cdot \prod_{i=1}^{n} \left\{\frac{1}{1 + (x_i - \theta)^2}\right\} \neq g\,(t_1, \theta)\,.\, h(x_1, x_2, \ldots, x_n)$

Hence, by Factorisation Theorem, there is no single statistic, which alone, is sufficient estimator of θ.

However, $L(\mathbf{x}, \theta) = k_1 (X_1, X_2, \ldots, X_n, \theta).\ k_2 (X_1, X_2, \ldots, X_n)$

$\Rightarrow$ The whole set $(X_1, X_2, \ldots, X_n)$ is jointly sufficient for θ.

17·3. CRAMER-RAO INEQUALITY

Definition. *If t is an unbiased estimator for $\gamma(\theta)$, a function of parameter θ, then*

$$Var(t) \geq \frac{\left\{\dfrac{d}{d\theta} \cdot \gamma(\theta)\right\}^2}{E\left(\dfrac{\partial}{\partial\theta} \log L\right)^2} = \frac{\{\gamma'(\theta)\}^2}{I(\theta)} \qquad \ldots (17·32)$$

where $I(\theta)$ is the information on θ, supplied by the sample.

In other words, Cramer-Rao inequality provides a lower bound $\{\gamma'(\theta)\}^2/I(\theta)\}$, to the variance of an unbiased estimator of $\gamma(\theta)$.

Proof. In proving this result, we assume that there is only a single parameter θ which is unknown. We also take the case of continuous *r.v.* The case of descrete random variables can be dealt with similarly on replacing the multiple integrals by appropriate multiple sums.

We further make the following assumptions, which are known as the *Regularity conditions* for *Cramer-Rao Inequality.*

(1) The parameter space Θ is a non-degenerate open interval on the real line R^1 $(-\infty, \infty)$.

(2) For almost all $\mathbf{x} = (x_1, x_2, \ldots, x_n)$, and for all $\theta \in \Theta$, $\frac{\partial}{\partial\theta} L(\mathbf{x}, \theta)$ exists, the exceptional set, if any, is independent of θ.

(3) The range of integration is independent of the parameter θ, so that $f(x, \theta)$ is differentiable under integral sign.

If range is not independent of θ and f is zero at the extremes of the range, *i.e.*,

$$f(a, \theta) = 0 = f(b, \theta), \text{ then}$$

$$\frac{\partial}{\partial\theta}\int_a^b f dx = \int_a^b \frac{\partial f}{\partial\theta} dx - f(a,\theta)\frac{\partial a}{\partial\theta} + f(b,\theta)\frac{\partial b}{\partial\theta}$$

$$\Rightarrow \quad \frac{\partial}{\partial\theta}\int_a^b f dx = \int_a^b \frac{\partial f}{\partial\theta} dx, \text{ since } f(a,\theta) = 0 = f(b,\theta).$$

(4) The conditions of uniform convergence of integrals are satisfied so that differentiation under the integral sign is valid.

(5) $I(\theta) = E\left[\left\{\frac{\partial}{\partial\theta}\log L(x,\theta)\right\}^2\right]$, exists and is positive for all $\theta \in \Theta$.

Let X be a *r.v.* following the *p.d.f.* $f(x, \theta)$ and let L be the likelihood function of the random sample $(x_1, x_2, \ldots, x_n)$ from this population. Then

$$L = L(\mathbf{x}, \theta) = \prod_{i=1}^{n} f(x_i, \theta)$$

Since L is the joint *p.d.f.* of $(x_1, x_2, \ldots, x_n)$, $\int L(\mathbf{x}, \theta)\, d\mathbf{x} = 1$,

where $\int d\mathbf{x} = \int\int \ldots \int dx_1\, dx_2 \ldots dx_n$.

Differentiating w.r. to θ and using regularity conditions given above, we get

$$\int \frac{\partial}{\partial\theta} L\, d\mathbf{x} = 0 \quad \Rightarrow \quad \int\left(\frac{\partial}{\partial\theta}\log L\right) L\, d\mathbf{x} = 0 \quad \Rightarrow \quad E\left(\frac{\partial}{\partial\theta}\log L\right) = 0 \qquad \ldots (17\cdot33)$$

Let $t = t(x_1, x_2, \ldots, x_n)$ be an unbiased estimator of $\gamma(\theta)$ such that

$$E(t) = \gamma(\theta) \quad \Rightarrow \quad \int t \,.\, L\, d\mathbf{x} = \gamma(\theta) \qquad \ldots (17\cdot34)$$

Differentiating w.r. to θ, we get $\int t \cdot \frac{\partial L}{\partial\theta} d\mathbf{x} = \gamma'(\theta) \quad \Rightarrow \quad \int t\left(\frac{\partial}{\partial\theta}\log L\right) L\, d\mathbf{x} = \gamma'(\theta)$

$$\Rightarrow \quad E\left(t \cdot \frac{\partial}{\partial\theta}\log L\right) = \gamma'(\theta) \qquad \ldots (17.35)$$

$$\text{Cov}\left(t, \frac{\partial}{\partial\theta}\log L\right) = E\left(t \cdot \frac{\partial}{\partial\theta}\log L\right) - E(t) \cdot E\left(\frac{\partial}{\partial\theta}\log L\right)$$

$$= \gamma'(\theta) \qquad \text{[From (17·33) and (17·35)]} \qquad \ldots (17\cdot35a)$$

We have : $\{r(X, Y)\}^2 \leq 1 \quad \Rightarrow \quad \{\text{Cov}(X, Y)\}^2 \leq \text{Var}(X). \text{Var}(Y)$

$$\therefore \quad \left\{\text{Cov}\left(t, \frac{\partial}{\partial\theta}\log L\right)\right\}^2 \leq \text{Var}\, t.\ \text{Var}\left(\frac{\partial}{\partial\theta}\log L\right)$$

$$\Rightarrow \quad \{\gamma'(\theta)\}^2 \leq \text{Var}\, t \left[E\left(\frac{\partial}{\partial\theta}\log L\right)^2 - \left\{E\left(\frac{\partial}{\partial\theta}\log L\right)\right\}^2\right]$$

$$\{\gamma'(\theta)\}^2 \leq \text{Var}\, t \cdot E\left\{\left(\frac{\partial}{\partial\theta}\log L\right)^2\right\} \qquad \text{[Using (17·33)]} \ldots (17\cdot36)$$

$$\Rightarrow \quad \text{Var}(t) \geq \frac{\{\gamma'(\theta)\}^2}{E\left\{\left(\frac{\partial}{\partial\theta}\log L\right)^2\right\}}, \qquad \ldots (17\cdot36a)$$

which is Cramer-Rao Inequality.

Corollary. If t is an unbiased estimator of parameter θ, *i.e.*,

$$E(t) = \theta \;\Rightarrow\; \gamma(\theta) = \theta \;\Rightarrow\; \gamma'(\theta) = 1,$$

then from (17·36*a*), we get

$$\text{Var}(t) \geq \frac{1}{E\left\{\left(\frac{\partial}{\partial\theta}\log L\right)^2\right\}} = \frac{1}{I(\theta)}, \quad \text{where} \quad I(\theta) = E\left\{\left(\frac{\partial}{\partial\theta}\log L\right)^2\right\} \qquad \ldots (17·37)$$

is called by R.A. Fisher as the *amount of information* on θ supplied by the sample $(x_1, x_2, \ldots, x_n)$ and its reciprocal $1/I(\theta)$, as the *information limit* to the variance of estimator $t = t(x_1, x_2, \ldots, x_n)$.

Remarks:

1. An unbiased estimator t of $\gamma(\theta)$ for which Cramer-Rao lower bound in (17·32) is attained is called a *minimum variance bound (MVB) estimator.*
2. We have :

$$I(\theta) = E\left\{\left(\frac{\partial}{\partial\theta}\log L\right)^2\right\} = -E\left(\frac{\partial^2}{\partial\theta^2}\log L\right) \qquad \ldots (17·38)$$

and

$$I(\theta) = n\left\{\left(\frac{\partial}{\partial\theta}\log f(x,\theta)\right)\right\}^2 = -n\left(\frac{\partial^2}{\partial\theta^2}\log f\right) \qquad \ldots (17·38a)$$

Proof. We have proved in (17·33), $\quad E\left(\frac{\partial}{\partial\theta}\log L\right) = 0 \qquad \ldots (*)$

$$\left(\frac{\partial^2}{\partial\theta^2}\log L\right)L = \frac{\partial}{\partial\theta}\left\{\left(\frac{\partial}{\partial\theta}\log L\right)\cdot L\right\} - \left(\frac{\partial}{\partial\theta}\log L\right)\cdot\frac{\partial L}{\partial\theta}$$

$$= \frac{\partial}{\partial\theta}\left\{\left(\frac{\partial}{\partial\theta}\log L\right)\cdot L\right\} - \left(\frac{\partial}{\partial\theta}\log L\right)^2\cdot L$$

Integrating both sides w.r. to $\mathbf{x} = (x_1, x_2, \ldots, x_n)$, we get

$$E\left(\frac{\partial}{\partial\theta}\log L\right) = \frac{\partial}{\partial\theta}\cdot E\left(\frac{\partial}{\partial\theta}\log L\right) - E\left(\frac{\partial}{\partial\theta}\log L\right)^2 = -E\left(\frac{\partial}{\partial\theta}\log L\right)^2 \quad [\text{Using } (*)]$$

$$\Rightarrow \qquad I(\theta) = E\left(\frac{\partial}{\partial\theta}\log L\right) = -E\left(\frac{\partial}{\partial\theta}\log L\right),$$

a form which is more convenient to use in practice.

Also $\quad I(\theta) = E\left\{\left(\frac{\partial}{\partial\theta}\log L\right)^2\right\} = E\left\{\sum_{i=1}^{n}\frac{\partial}{\partial\theta}\log f(x_i,\theta)\right\}^2$

$$= E\left[\sum_{i=1}^{n}\left\{\frac{\partial}{\partial\theta}\log f(x_i,\theta)\right\}^2 + \sum_{i\neq j=1}^{n}\left\{\left(\frac{\partial}{\partial\theta}\log f(x_i,\theta)\right)\left(\frac{\partial}{\partial\theta}\log f(x_j,\theta)\right)\right\}\right]$$

$$= n\cdot E\left\{\frac{\partial}{\partial\theta}\log f(x_i,\theta)\right\}^2 \qquad [\text{On using } (*)], \text{ since } x_i\text{'s}; i = 1, 2, \ldots, n \text{ are } i.i.d.\ r.v.\text{'s.}]$$

17·3·1. Conditions for the Equality Sign in Cramer-Rao Inequality

In proving (17·32) we used [*c.f.* (17·36)] that

$$[\gamma'(\theta)]^2 \leq E[t - \gamma(\theta)]^2\cdot E\left[\left(\frac{\partial}{\partial\theta}\log L\right)^2\right] \qquad \ldots (17·39)$$

The sign of equality will hold in C.R. inequality if only if the sign of equality holds in (17·39). The sign of equality will hold in (17·39) by Cauchy Schwartz Inequality, if and only if the variables $[t - \gamma(\theta)]$ and $\left(\frac{\partial}{\partial\theta}\log L\right)$ are proportional to each other, *i.e.*,

$$\frac{t - \gamma(\theta)}{\frac{\partial}{\partial\theta}\log L} = \lambda = \lambda(\theta),$$

where λ is a constant independent of $(x_1, x_2, \ldots, x_n)$ but may depend on θ.

$$\therefore \quad \frac{\partial}{\partial\theta}\log L = \frac{t - \gamma(\theta)}{\lambda(\theta)} = [t - \gamma(\theta)]A(\theta), \quad \ldots (17{\cdot}40)$$

where $\quad A = A(\theta) \; = \; 1/[\lambda(\theta)]$, say.

Hence, *a necessary and sufficient condition for an unbiased estimator t to attain the lower bound of its variance is given by (17·40).*

Further, the C-R minimum variance bound is given by :

$$\text{Var}(t) = [\gamma'(\theta)]^2 / E\left(\frac{\partial}{\partial\theta}\log L\right)^2 \quad \ldots (17{\cdot}41)$$

But

$$E\left(\frac{\partial}{\partial\theta}\log L\right)^2 = E[A(\theta)\cdot\{t - \gamma(\theta)\}]^2 \quad \text{[From (17·40)]}$$

$$= \{A(\theta)\}^2 \cdot E\{t - \gamma(\theta)\}^2 = \{A(\theta\}^2 \cdot \text{Var}(t)$$

Substituting in (17·41), we get: $\quad \text{Var}(t) = \dfrac{\{\gamma'(\theta)\}^2}{\{A(\theta)\}^2 \cdot \text{Var}(t)}$

$$\Rightarrow \quad \text{Var}(t) = \left|\frac{\gamma'(\theta)}{A(\theta)}\right| = |\gamma'(\theta)\lambda(\theta)| \quad \ldots (17{\cdot}42)$$

Hence, if the likelihood function L is expressible in the form (17·40) then

(*i*) t is an unbiased estimator of $\gamma(\theta)$,

(*ii*) Minimum Variance Bound (MVB) estimator (t) for $\gamma(\theta)$ exists, and

(*iii*) $\text{Var}(t) = \left|\dfrac{\gamma'(\theta)}{A(\theta)}\right| = |\gamma'(\theta)\lambda(\theta)|$

The importance of this result lies in the fact that C.R. inequality, in addition to find if MVBU estimator for $\gamma(\theta)$ exists, also gives us the variance of such an estimator, which is given by (17·42).

Remarks:

1. If $\gamma(\theta) = \theta$, *i.e.*, if t is an unbiased estimator of θ, then (17·40) can be written as :

$$\frac{\partial}{\partial\theta}\log L = \frac{t - \theta}{\lambda} \quad \ldots (17{\cdot}43)$$

Hence, if (17·43) holds, then t is an MVB estimator for θ with

$$\text{Var}(t) = |\lambda(\theta)| = 1/|A(\theta)| \quad \ldots (17{\cdot}43a)$$

2. We have seen in (17·40) that MVB estimator exists for $\gamma(\theta)$ if

$$\frac{\partial}{\partial\theta}\log L = \frac{t - \gamma(\theta)}{\lambda} = \{t - \gamma(\theta)\}\cdot\frac{1}{\lambda}, \text{ where } \lambda = \lambda(\theta), \text{ say.} \quad \ldots (*)$$

If we write $\int \frac{1}{\lambda}\, d\theta = \alpha(\theta)$, then integrating (*) w.r.t. θ (by parts), we get

$$\log L = \{t - \gamma(\theta)\}\,\alpha(\theta) + \int \alpha(\theta).\,\gamma'(\theta)\,d\theta + k(\mathbf{x}) \;\Rightarrow\; \log L = [t - \gamma(\theta)]\,\alpha(\theta) + \beta(\theta) + k(\mathbf{x}) \quad \ldots (17{\cdot}44))$$

where $\alpha(\theta)$ and $\beta(\theta)$ are arbitrary functions of θ and $k(\mathbf{x}) = k(x_1, x_2, \ldots, x_n)$, is an arbitrary function of x_i's independent of θ. Hence,

$$\log f(x, \theta) = [t - \gamma(\theta)]\,A_1(\theta) + B_1(\theta) + k_1(\mathbf{x}) \;\Rightarrow\; f(x, \theta) = g(\mathbf{x})\,.\,h(\theta)\,.\,\exp\{a(\theta)\,.\,\psi(\mathbf{x})\} \quad \ldots (15{\cdot}44a)$$

which is the necessary and sufficient condition for the existence of a sufficient statistic [*c.f.* Koopman's form, Equation (17·30) in Remark 3 to § 17·2·4)]. Hence, *an MVB estimator for $\gamma(\theta)$ exists if and only if there exists a sufficient estimator for $\gamma(\theta)$.*

This suggests that in our search for an MVB estimator for $\gamma(\theta)$, we need to confine ourselves to sufficient estimators of $\gamma(\theta)$ alone. This explains why the method failed in the case of Cauchy population [*c.f.* Example 17·19], where no sufficient estimator exists and its success in the case of normal population [*c.f.* Examples 17·18], where $\bar{x}$ is sufficient for μ and in Example 17·20, $\sum_{i=1}^{n} x_i^2/n$ is sufficient for σ^2].

Example 17·18. *Obtain the MVB estimator for μ in normal population $N(\mu, \sigma^2)$, where σ^2 is known.*

Solution. If $x_1, x_2, ..., x_n$ is a random sample of size n from the normal population, then

$$L = \prod_{i=1}^{n} f(x_i, \mu) = \left(\frac{1}{\sigma\sqrt{2\pi}}\right)^n \cdot \exp\left\{-\sum_{i=1}^{n}(x_i-\mu)^2/2\sigma^2\right\}$$

$$\Rightarrow \quad \log L = -n\log\left(\sqrt{2\pi}\,\sigma\right) - \frac{1}{2\sigma^2}\sum_{i=1}^{n}(x_i-\mu)^2 = k - \frac{1}{2\sigma^2}\sum_{i=1}^{n}(x_i-\mu)^2,$$

where k is a constant independent of μ, (σ being known).

$$\frac{\partial}{\partial\mu}\log L = -\frac{1}{2\sigma^2}\sum_{i=1}^{n}\{2(x_i-\mu)(-1)\} = \frac{\sum_{i=1}^{n}(x_i-\mu)}{\sigma^2} = \frac{\sum x_i - n\mu}{\sigma^2} = \frac{(\bar{x}-\mu)}{\sigma^2/n}$$

which is of the form (17·40).

Hence, $\bar{x}$ is a MVB unbiased estimator for μ and $V(\hat{\mu}) = V(\bar{x}) = \dfrac{\sigma^2}{n}$.

Example 17·19. *Find if MVB estimator exists for θ in the Cauchy's population :*

$$dF(x,\theta) = \frac{1}{\pi}\cdot\frac{1}{1+(x-\theta)^2}\;;\; -\infty < x < \infty$$

Solution.

$$L = \prod_{i=1}^{n} f(x_i, \theta) = \left(\frac{1}{\pi}\right)^n \prod_{i=1}^{n}\left\{\frac{1}{1+(x_i-\theta)^2}\right\}$$

$$\Rightarrow \quad \log L = -n\log\pi - \sum_{i=1}^{n}\log\left\{1+(x_i-\theta)^2\right\} \quad \Rightarrow \quad \frac{\partial}{\partial\theta}\log L = 2\sum_{i=1}^{n}\left\{\frac{(x_i-\theta)}{1+(x_i-\theta)^2}\right\}$$

Since this cannot be expressed in form (17·40), MVB estimator does not exist for θ, in the Cauchy's population and so Cramer Rao lower bound is not attainable by the variance of any unbiased estimator of θ.

Example 17·20. *A random sample $x_1, x_2, ..., x_n$ is taken from a normal population with mean zero and variance σ^2. Examine if $\sum_{i=1}^{n} x_i^2/n$ is a MVB estimator for σ^2.*

Solution. Since $X \sim N(0, \sigma^2)$, $f(x, \sigma^2) = \dfrac{1}{\sigma\sqrt{2\pi}} \cdot \exp(-x^2/2\sigma^2), -\infty < x < \infty$

$$L = \prod_{i=1}^{n} f(x_i, \sigma^2) = \left(\frac{1}{\sigma\sqrt{2\pi}}\right)^n \exp\left\{-\sum_{i=1}^{n}\left(x_i^2/2\sigma^2\right)\right\}$$

$$\Rightarrow \quad \log L = -\frac{n}{2}\log(2\pi) - \frac{n}{2}\log\sigma^2 - \frac{1}{2\sigma^2}\sum_{i=1}^{n} x_i^2$$

$$\therefore \quad \frac{\partial}{\partial\sigma^2}\log L = -\frac{n}{2\sigma^2} + \frac{1}{2\sigma^4}\sum_{i=1}^{n} x_i^2 = \frac{\sum_{i=1}^{n} x_i^2 - n\sigma^2}{2\sigma^4} = \frac{\left(\sum_{i=1}^{n} x_i^2/n\right) - \sigma^2}{(2\sigma^4/n)},$$

which is of the form (17·40).

Hence, $\quad \hat{\sigma}^2 = \sum_{i=1}^{n} \frac{x_i^2}{n}$, is an MVB estimator and $V(\hat{\sigma}^2) = \frac{2\sigma^4}{n}$.

Example 17·21. *Show that* $\bar{X} = \sum_{i=1}^{n} x_i/n$, *in random sampling from*

$$f(x,\theta) = \begin{cases} \frac{1}{\theta} exp(-x/\theta), \; 0 < x < \infty \\ 0, \; otherwise \end{cases} \qquad \ldots (*)$$

where $0 < \theta < \infty$, *is an MVB estimator of* θ *and has variance* θ^2/n.

Solution. Let $x_1, x_2, \ldots, x_n$ be a random sample of size n from population with *p.d.f.* in (*). Then

$$L = \prod_{i=1}^{n} f(x_i,\theta) = \frac{1}{\theta^n}\cdot exp\left(-\sum_{i=1}^{n} x_i/\theta\right) \quad \Rightarrow \quad \log L = -n\log\theta - \frac{1}{\theta}\cdot\sum_{i=1}^{n} x_i$$

$$\therefore \quad \frac{\partial}{\partial\theta}\log L = -\frac{n}{\theta} + \frac{1}{\theta^2}\cdot\sum_{i=1}^{n} x_i \quad \Rightarrow \quad \frac{\partial^2}{\partial\theta^2}\log L = \frac{n}{\theta^2} - \frac{2}{\theta^3}\sum_{i=1}^{n} x_i$$

Thus, $$I(\theta) = -E\left(\frac{\partial^2}{\partial\theta^2}\log L\right) = -\frac{n}{\theta^2} + \frac{2}{\theta^3}\sum_{i=1}^{n} E(x_i)$$

In sampling from exponential population (*), we have $E(X) = \theta$ and $\text{Var}(X) = \theta^2$ ... (**)

$$\therefore \quad I(\theta) = -\frac{n}{\theta^2} + \frac{2}{\theta^3}\sum_{i=1}^{n}(\theta) = -\frac{n}{\theta^2} + \frac{2}{\theta^3}\cdot n\theta = \frac{n}{\theta^2} \qquad (\because x_i\text{'s are } i.i.d)$$

Also $\quad \gamma(\theta) = \theta \quad \Rightarrow \quad \gamma'(\theta) = 1.$

Hence, Cramer Rao lower bound to the variance of an unbiased estimator of θ is :

$$\frac{[\gamma'(\theta)]^2}{I(\theta)} = \frac{1}{(n/\theta^2)} = \frac{\theta^2}{n} \qquad \ldots (***)$$

Consider the estimator : $\bar{X} = \frac{1}{n}\sum_{i=1}^{n} x_i$. We have

$$E(\bar{X}) = \frac{1}{n}\sum_{i=1}^{n} E(x_i) = \frac{1}{n}\sum_{i=1}^{n}(\theta) = \theta \quad \Rightarrow \quad \bar{X} \text{ is an unbiased estimator of } \theta.$$

Also $$\text{Var}(\bar{X}) = \frac{\sigma^2}{n} = \frac{\text{Var } X}{n} = \frac{\theta^2}{n} \qquad \text{[From (**)]}$$

Thus, we see that Var ($\bar{X}$) coincides with the Cramer-Rao lower bound obtained in (***). Hence $\bar{X}$, the sample mean is an MVB unbiased estimator, for θ.

Aliter. A more convenient way of doing this problem is as follows :

We have $$\frac{\partial}{\partial\theta}\log L = -\frac{n}{\theta} + \frac{1}{\theta^2}\sum_{i=1}^{n} x_i = \frac{\sum_{i=1}^{n} x_i - n\theta}{\theta^2} = \frac{\bar{X}-\theta}{(\theta^2/n)} = \frac{\bar{X}-\theta}{\lambda(\theta)}, \text{(say)}$$

which is of the form (17·40).

Hence, $\bar{X}$ is an MVB unbiased estimator of θ and Var ($\bar{X}$) = λ (θ) = (θ^2/n) .

Example 17·22. *Given the probability density function :*

$$f(x\,;\theta) = [\pi\{1+(x-\theta)^2\}]^{-1}\,; -\infty < x < \infty, -\infty < \theta < \infty \qquad \ldots (*)$$

show that the Cramer-Rao lower bound of variance of an unbiased estimator of θ is 2/n, where n is the size of the random sample from this distribution.

Solution. $$\log f = -\log\pi - \log\{1+(x-\theta)^2\} \Rightarrow \frac{\partial \log f}{\partial\theta} = \frac{2(x-\theta)}{\{1+(x-\theta)^2\}}$$

$$E\left(\frac{\partial \log f}{\partial\theta}\right)^2 = \int_{-\infty}^{\infty} \frac{4(x-\theta)^2}{\{1+(x-\theta)^2\}^2} f(x,\theta)dx = \frac{1}{\pi}\int_{-\infty}^{\infty}\frac{4(x-\theta)^2}{\{1+(x-\theta)^2\}^3}dx$$

Put $x - \theta = \tan\phi$ so that $dx = \sec^2\phi\, d\phi.$

$$\therefore \quad E\left(\frac{\partial}{\partial\theta}\log f\right)^2 = \frac{2}{\pi}\int_0^{\pi/2}\frac{4\tan^2\phi}{\sec^6\phi}\sec^2\phi\, d\phi = \frac{2}{\pi}\int_0^{\pi/2}\frac{4\sin^2\phi}{\cos^2\phi}\cos^4\phi\, d\phi$$

$$= \frac{2}{\pi}\int_0^{\pi/2} 4\sin^2\phi\cos^2\phi\, d\phi = \frac{8}{\pi}\int_0^{\pi/2}(\cos^2\phi - \cos^4\phi)\, d\phi$$

$$= \frac{8}{\pi}\left(\frac{1}{2}\cdot\frac{\pi}{2} - \frac{3.1}{4.2}\cdot\frac{\pi}{2}\right) = \frac{8}{\pi}\left(\frac{\pi}{4} - \frac{3\pi}{16}\right) = \frac{1}{2}.$$

Hence, $$\text{Cramer-Rao lower bound to Var}(\hat{\theta}) = \frac{1}{I(\theta)} \Re \frac{1}{nE\left(\frac{\partial}{\partial\theta} f\right)^2} \quad \frac{1}{n\left(\frac{1}{2}\right)} \quad \frac{2}{n}.$$

Example 17·23. *Prove that under certain general conditions of regularity to be stated clearly the mean square deviation $E(\hat{\theta}-\theta)^2$ of an estimator $\hat{\theta}$ of the parameter θ, can never fall below a positive limit depending only on the density function f (x, θ), the size of the sample and the bias of the estimate.*

Solution. We have proved Cramer-Rao's inequality : $V(\hat{\theta}) \geq \frac{\{\psi'(\theta)\}^2}{I(\theta)}$, where $E(\hat{\theta}) = \Psi(\theta)$ ... (*)

$$E(\hat{\theta}-\theta)^2 = E\{\hat{\theta}-\psi(\theta)+\psi(\theta)-\theta\}^2$$

$$= E\{\hat{\theta}-\psi(\theta)\}^2 + \{\theta-\psi(\theta)\}^2 + 2[\psi(\theta)-\theta]\,.\,E[\hat{\theta}-\psi(\theta)]$$

$$= V(\hat{\theta}) + \{\theta-\psi(\theta)\}^2 \qquad \text{[From (*)]}$$

$$\therefore \quad E(\hat{\theta}-\theta)^2 \geq \frac{\{\psi'(\theta)\}^2}{I(\theta)} + \{\theta-\psi(\theta)\}^2 \qquad \text{[Using (*)]} \qquad \ldots (**)$$

Let $\hat{\theta}$ be a *'biased'* estimate of θ with bias given by $b(\theta)$, *i.e.,*

$$E(\hat{\theta}) = \theta + b(\theta) = \psi(\theta), \text{(say)} \quad \Rightarrow \quad \psi(\theta) - \theta = b(\theta) \qquad \ldots (***)$$

From (**) and (***), we get

$$E(\hat{\theta}-\theta)^2 \geq \frac{\left\{1+\frac{\partial}{\partial\theta}b(\theta)\right\}^2}{I(\theta)} + [b(\theta)]^2 > 0$$

where, $$I(\theta) = n\int_{-\infty}^{\infty}\left(\frac{\partial}{\partial\theta}\log f\right)^2 f(x,\theta)dx > 0.$$

This proves the result.

17·4. COMPLETE FAMILY OF DISTRIBUTIONS

Consider a statistic $T = t(x_1, x_2, \ldots, x_n)$, based on a random sample of size n from the population $f(x, \theta)$, $\theta \in \Theta$. The distribution of the statistic T will, in general, depend on θ. Hence, corresponding to T, we again have a family of distributions, say, $\{g(t, \theta), \theta \in \Theta\}$.

Definition. *The statistic $T = t(\mathbf{x})$, or more precisely the family of distributions $\{g(t, \theta), \theta \in \Theta\}$ is said to be complete for θ if*

$$E_\theta [h(T)] = 0 \text{ for all } \theta \quad \Rightarrow \quad P_\theta [h(T) = 0] = 1 \qquad \ldots (17·45)$$

$$\left.\begin{aligned} &\textit{i.e.,} \quad \int h(t)\, g(t,\theta)\, dt = 0 \text{ for all } \theta \in \Theta \\ &\textit{or} \quad \sum_t h(t)\, g(t,\theta) = 0 \text{ for all } \theta \in \Theta \end{aligned}\right\} \qquad \ldots (17·45a)$$

$$\Rightarrow \quad h(T) = 0, \text{ for all } \theta \in \Theta, \text{ almost surely (a.s.)}. \qquad \ldots (17·45b)$$

The concept of complete sufficient statistic is specially useful in Rao-Blackwell Theorem [*c.f.* Theorem 17·9].

Example 17·24. *Let $X_1, X_2, \ldots, X_n$ be a random sample from Bernoulli distribution :*

$$f(x,\theta) = \begin{cases} \theta^x (1-\theta)^{1-x} ; x = 0,1 \\ 0 \quad , \quad \text{otherwise} \end{cases}$$

Show that $\sum_{i=1}^{n} X_i$ is a complete sufficient statistic for θ.

Solution. The likelihood function of the sample $(X_1, X_2, \ldots, X_n)$ is given by :

$$L = \prod_{i=1}^{n} f(x_i, \theta) = \left\{\theta^{\sum_i x_i} (1-\theta)^{n - \sum_i x_i}\right\} \times 1 = g\{t(\mathbf{x}), \theta\} \,.\, h(x_1, x_2 \ldots, x_n)$$

where, $\quad t(\mathbf{x}) = \sum_{i=1}^{n} x_i$ and $h(x_1, x_2, \ldots, x_n) = 1$

Hence, by Factorisation theorem, $T = \sum_{i=1}^{n} X_i$, is sufficient estimator of θ.

Since X_i's are *i.i.d.* Bernoulli variates with parameter θ, $T = \sum_{i=1}^{n} X_i \sim B(n, \theta)$,

with *p.m.f.*

$$P(T = k) = \begin{cases} {}^nC_k \theta^k (1-\theta)^{n-k} ; \ k = 0,1,2\ldots, n \\ 0, \quad \text{otherwise} \end{cases}$$

$$E_\theta\{h(T)\} = \sum_{k=0}^{n} h(k) \,.\, P(T = k) = \sum_{k=0}^{n} h(k) .\, {}^nC_k\, \theta^k (1-\theta)^{n-k}$$

$$= \sum_{k=0}^{n} A(k) \,.\, \theta^k (1-\theta)^{n-k} ; \qquad A(k) = h(k) .\, {}^nC_k \qquad \ldots (*)$$

$$= A(0)(1-\theta)^n + A(1)\,\theta(1-\theta)^{n-1} + \ldots + A(n) .\, \theta^n$$

Now $\quad E_\theta\{h(T)\} = 0$ for all $\theta \in \Theta = \{\theta : 0 < \theta < 1\}$

$$\Rightarrow \quad A(0)(1-\theta)^n + A(1)\,\theta\,(1-\theta)^{n-1} + \ldots + A(n)\,\theta^n = 0, \ \forall\, \theta$$

$$\Rightarrow \quad A(0) + A_1 [\theta/(1-\theta)] + \ldots + A(n)\{\theta/(1-\theta)\}^n = 0 \ \forall\, \theta \in [0, 1] \qquad \ldots (**)$$

$$\Rightarrow \quad A(0) = A(1) = A(2) = \dots = A(n) = 0,$$

since a polynomial of degree n in x is identically zero (for all x), if all the coefficients are zero.

From (*) and (**), we get: $h(k) = 0\ ;\ k = 0, 1, 2, \dots, n \quad \Rightarrow \quad h(t) = 0\ ;\ t = 0, 1, 2, \dots, n$

Hence, T is a complete (sufficient) statistic for θ.

Example 17·25. *Let $X_1, X_2, \dots, X_n$ be a random sample of size n from $N(\theta, 1)$ population. Examine if $T = t(x) = X_1$ is complete for θ.*

Solution. We have $\quad T = X_1\ ;\ \Theta = \{\theta : -\infty < \theta < \infty\}$ so that $E_\theta\,[h(T)] = 0,\ \forall\ \theta.$

$$\Rightarrow \quad \int_{-\infty}^{\infty} h(u) e^{-(u-\theta)^2/2} du = 0, \forall \theta \in \Theta \qquad [\because T = X_1 \sim N(\theta, 1)]$$

$$\Rightarrow \quad \int_{-\infty}^{\infty} \left\{ h(u) e^{-u^2/2} \right\} e^{\theta u} \cdot du = 0, \forall \theta \in \Theta$$

This is a bilateral Laplace transform in θ. Since these are unique :

$$h(u) \cdot e^{-u^2/2} = 0, a.s. \quad \Rightarrow \quad h(u) = 0,\ a.s.$$

$$\Rightarrow \quad P\,\{h(T) = 0\} = 1,\ \forall\ \theta\ \Theta \qquad \therefore \qquad T = X_1 \text{ is complete statistic for } \theta.$$

Remark. It can be easily seen that $T_1 = \sum_{i=1}^{n} X_i$, is a sufficient estimator of θ and since $T_1 \sim N(n\theta, n)$, by proceeding as in the above problem, we can prove that $T_1 = \sum_{i=1}^{n} X_i$, is a complete sufficient statistic for θ and the family of distributions $\{g_1(t_1, \theta), \theta \in \Theta\}$, is complete.

Example 17·26. *Let $X_1, X_2, \dots, X_n$ be a random sample from $N(0, \theta)$. Prove that $T = X_1$ is not a complete statistic for θ but $T_1 = X_1^{\,2}$ is complete for θ.*

Solution. Here $\quad T = t(\mathrm{x}) = X_1\ ;\ \Theta = \{\theta\ ;\ 0 < \theta < \infty\}$

$$E_\theta\,[h(T)] = 0,\ \forall\ \theta \in \Theta$$

$$\Rightarrow \quad \int_{-\infty}^{\infty} h(u) \exp\left\{-u^2/(2\theta)\right\} du = 0,\ \forall\ \theta \in \Theta$$

This holds only for all odd functions $h(u)$ of u, for which the integral exists, *i.e.*, for all functions *s.t.* $h(u) = -h(-u)$; for all u.

$$\Rightarrow \quad h(u) \neq 0, \text{ a.s.} \quad \text{or} \quad T = X_1 \text{ is not complete statistic for } \theta.$$

Let us now consider the statistic $T_1 = X_1^{\,2}$.

$$E_\theta\left[h(T_1)\right] = 0,\ \forall\ \theta \in \Theta \quad \Rightarrow \quad \int_{-\infty}^{\infty} h(x^2) \exp\,(-x^2/2\theta)\, dx = 0,\ \forall\ \theta \in \Theta$$

$$\Rightarrow \quad \int_{-\infty}^{\infty} \frac{h(u)}{\sqrt{u}} \exp\,(-u/2\theta)\, du = 0,\ \forall\ \theta \in \Theta$$

This being a Laplace transform in $(1/\theta)$, we have

$$\frac{h(u)}{\sqrt{u}} = 0,\ a.s. \quad \Rightarrow \quad h(u) = 0,\ a.s.$$

$$\therefore \quad T_1 = X_1^{\,2}, \text{ is complete statistic for } \theta.$$

Remark. We can easily see that $T_1 = X_1^{\,2}$, is sufficient statistic for θ. Hence $T_1 = X_1^{\,2}$ is a complete sufficient statistic for θ.

Example 17·27. *Let $X_1, X_2, \ldots, X_n$ be a random sample from uniform $U[0, \theta]$, $\theta > 0$ population. Show that $T = \max_{1 \le i \le n} (X_i) = X_{(n)}$, is a complete sufficient statistic for θ.*

Solution. $T = X_{(n)}$ has *p.d.f.* : $g(t, \theta) = \begin{cases} \dfrac{nt^{n-1}}{\theta^n}, 0 \le t \le \theta \\ 0, \quad otherwise \end{cases}$

$$E_\theta [h(T)] = 0, \ \forall \ \theta \in \Theta = \{\theta : 0 < \theta < \infty\} \quad \Rightarrow \quad \frac{n}{\theta^n} \int_0^\theta h(u).\, u^{n-1}\, du = 0, \forall\ \theta \in \Theta$$

Differentiating *w.r. to* θ, we get from the fundamental theorem of integral calculus :

$$h(\theta) \,.\, \theta^{n-1} = 0, \ \forall\ \theta \in \Theta \quad \Rightarrow \quad h(T) = 0, \ a.s.$$

$$\therefore \qquad T = \max (X_1, X_2, \ldots, X_n) = X_{(n)}, \text{ is complete for } \theta.$$

We have also proved in *Example 17·13*, that $T = X_{(n)}$, is sufficient for θ.

Hence, $T = X_{(n)}$, is complete sufficient statistic for θ.

17·5. MVUE AND BLACKWELLISATION

Cramer-Rao inequality (c.f. § 17·3) provides us a technique of finding if the unbiased estimator is also an *MVU* estimator or not. Here, since the regularity conditions are very strict, its applications become quite restrictive. Moreover *MVB* estimator is not the same as an *MVU* estimator since the Cramer-Rao lower bound may not always be attained. Moreover, if the regularity conditions are violated, then the least attainable variance may be less than the Cramer-Rao bound. [For illustration see Example 17·30]. In this section we shall discuss *how to obtain MVU estimator from any unbiased estimator through the use of sufficient statistic.* This technique is called Blackwellisation after *D.* Blackwell. The result is contained in the following Theorem due to C.R. Rao and *D.* Blackwell.

Theorem 17·9. *(RAO-BLACKWELL THEOREM). Let X and Y be random variables such that*

$$E(Y) = \mu \text{ and } Var(Y) = \sigma_Y^{\,2} > 0$$

Let $E(Y \mid X = x) = \phi(x)$. Then (i) $E[\phi(X)] = \mu$, and (ii) $Var[\phi(X)] \le Var(Y)$.

Proof. Let $f_{XY}(x, y)$ be the joint p.d.f. of random variables X and Y, $f_1(.)$ and $f_2(.)$ the marginal p.d.f.'s of X and Y respectively and $h(y \mid x)$ be the conditional p.d.f. of Y for given $X = x$ such that $h(y \mid x) = \{f(x, y)/f_1(x)\}$.

$$E(Y \mid X = x) = \int_{-\infty}^{\infty} y.\, h(y \mid x)\, dy = \int_{-\infty}^{\infty} y \,.\, \frac{f(x,y)}{f_1(x)}\, dy$$

$$= \frac{1}{f_1(x)} \int_{-\infty}^{\infty} y\, f(x, y)\, dy = \phi(x), \text{ (say)} \qquad \ldots (17·46)$$

$$\Rightarrow \qquad \int_{-\infty}^{\infty} y\, f(x, y)\, dy = \phi(x).\, f_1(x) \qquad \ldots (17·46a)$$

From (17·46) we observe that the conditional distribution of Y given $X = x$ does not depend on the parameter μ. Hence, X is sufficient statistic for μ. Also

$$E\{\phi(X)\} = E\{E(Y \mid X)\} = E(Y) = \mu, \qquad \ldots (17·47)$$

which establishes part (*i*) of the theorem.

Now $\text{Var}(Y) = E\big[Y - E(Y)\big]^2 = E\big[Y - \mu\big]^2 = E\big[Y - \phi(X) + \phi(X) - \mu\big]^2$

$$= E\big[Y - \phi(X)\big]^2 + E\big[\phi(X) - \mu\big]^2 + 2E\big[\{Y - \phi(X)\}\{\phi(X) - \mu\}\big] \quad \ldots (17{\cdot}48)$$

The product term gives

$$E\big[\{Y - \phi(X)\}\{\phi(X) - \mu\}\big] = \int_{-\infty}^{\infty}\int_{-\infty}^{\infty} \{(y - \phi(x)\}\{\phi(x) - \mu\} f(x, y)\, dx\, dy$$

$$= \int_{-\infty}^{\infty}\int_{-\infty}^{\infty} \{y - \phi(x)\}\{\phi(x) - \mu\} f_1(x)\, h(y \mid x)\, dx\, dy$$

$$= \int_{-\infty}^{\infty} \{\phi(x) - \mu\}\left[\int_{-\infty}^{\infty} \{y - \phi(x)\} h(y \mid x)\, dy\right] f_1(x) dx$$

But $\displaystyle\int_{-\infty}^{\infty} \{y - \phi(x)\}\, h(y \mid x)\, dy = 0$ $\quad [\because E(Y \mid X = x) = \phi(x)]$

$\therefore$ $E\big[\{Y - \phi(X)\}\{\phi(X) - \mu\}\big] = 0.$ Substituting in (17·48), we get

$$\text{Var}(Y) = E\{Y - \phi(X)\}^2 + \text{Var}\{\phi(X)\} \quad \ldots (17{\cdot}49)$$

$\Rightarrow$ $\text{Var}\, Y \geq \text{Var}\,[\phi(X)]$ $\quad [\because E\{Y - \phi(X)\}^2 \geq 0]$

$\therefore$ $\text{Var}\{\phi(X)\} \leq \text{Var}\, Y,$ $\quad \ldots (17{\cdot}49a)$

which completes the proof of the theorem.

Remarks:

1. From (17·49), it is obvious that the sign of equality holds in (17·49*a*) iff

$$E\{Y - \phi(X)\}^2 = 0 \quad \Rightarrow \quad Y - \phi(X) = 0, \text{ almost surely}$$

i.e., iff $\quad P\big[(x, y) : y - \phi(x) = 0\big] = 1 \quad \ldots (17{\cdot}50)$

2. Here we have proved the theorem for continuous *r.v.*'s. The result can be similarly proved for discrete case, replacing integration by summation.
3. Rao-Blackwell theorem enables us to obtain *MVU* estimators through sufficient statistic. If a sufficient estimator exists for a parameter, then in our search for *MVU* estimator we may restrict ourselves to functions of the sufficient statistic. The theorem can be stated slightly differently as follows :

Let $U = U(x_1, x_2, \ldots, x_n)$ be an unbiased estimator of parameter $\gamma(\theta)$ and let $T = T(x_1, x_2, \ldots, x_n)$ be sufficient statistic for $\gamma(\theta)$. Consider the function $\phi(T)$ of the sufficient statistic defined as

$$\phi(t) = E\,(U \mid T = t) \quad \ldots (17{\cdot}51)$$

which is independent of θ (since T is sufficient for θ). Then

$$E[\phi(T)] = \gamma(\theta) \quad \textit{and} \quad Var\,[\phi(T)] \leq Var\,(U) \quad \ldots (17{\cdot}52)$$

This result implies that starting with an unbiased estimator *U*, we can improve upon it by defining a function $\phi(T)$ of the sufficient statistics as given in (17·51). This technique of obtaining improved estimators is called Blackwellisation.

If in addition, the sufficient statistic *T* is also complete, then the estimator $\phi(T)$ discussed above will not only be an improved estimator over *U* but also the *best (unique)* estimator. We state below the relevant theorem.

Theorem 17·10. *Let T be a complete sufficient statistic for $\gamma(\theta)$, $\theta \in \Theta$. Then $\phi(T)$, the function of T defined in (17·51) is the unique unbiased estimator of $\gamma(\theta)$.*

Combining the results of the two *Theorems 17·9 and 17·10*, we have the following result.

Corollary. *If T is a complete sufficient statistic for $\gamma(\theta)$ and if we can find some function of T, say g(T), which is unbiased estimator of $\gamma(\theta)$, then g(T) is the MVU estimator of $\gamma(\theta)$.*

Example 17·28. *Let $X_1, X_2, ..., X_n$ be a random sample from N(θ, 1). Obtain MVUE of θ.*

Solution. It can be easily proved [*c.f.* Example 17·25] that the statistic :

$$T = X_1 + X_2 + ... + X_n = \sum_{i=1}^{n} X_i, \text{ is complete sufficient statistic for } \theta.$$

Consider $\overline{X}_n = \frac{1}{n}\sum_{i=1}^{n} X_i = \frac{T}{n} = g(T)$, (say), Since $\overline{X}_n = g(T)$, is unbiased estimator of θ, by corollary to Theorem 17·10, $\overline{X}_n$ is *MVUE* of θ.

Example 17·29. *Let $X_1, X_2, ..., X_n$ be a random sample from U[0, θ] population. Obtain MVUE for θ.*

Solution. We have seen that in sampling from *U* [0, θ] population, the statistic :
$T = X_{(n)} = \max_{1\le i\le n} (X_i)$ is sufficient (Example 17·13) and complete (Example 17·27) for θ.

Also, $$E(T) = E[X_{(n)}] = \left(\frac{n}{n+1}\right)\theta \Rightarrow E\left\{\frac{(n+1)T}{n}\right\} = \theta \quad \text{[See Example 17·30]}$$

Hence, by corollary to Theorem 17·10,

$[(n+1)T/n] = [(n+1)X_{(n)}/n]$ is an *MVU* estimator of θ.

Example 17·30. *Given :* $$f(x,\theta) = \begin{cases} \frac{1}{\theta}, 0 < x < \theta, \theta > 0 \\ 0, \quad \textit{elsewhere} \end{cases} \quad ... (*)$$

compute the reciprocal of $n\cdot E\left[\left\{\frac{\partial \log f(x,\theta)}{\partial\theta}\right\}^2\right]$ *and compare this with the variance of (n + 1) Y_n/n, where Y_n is the largest item of a random sample of size n from this distribution. Comment on the result.*

Solution. $\log f(x,\theta) = -\log\theta \Rightarrow \frac{\partial}{\partial\theta}\log f = -\frac{1}{\theta} \Rightarrow n.E\left(\frac{\partial}{\partial\theta}\log f\right) = nE\left(\frac{1}{\theta^2}\right) = \frac{n}{\theta^2}$

Hence, $$\text{Reciprocal of } n.E\left[\left\{\frac{\partial}{\partial\theta}\log f(x,\theta)\right\}^2\right] = \frac{\theta^2}{n} \quad ... (**)$$

For the rectangular population (*), the *p.d.f.* of *n*th order statistic (the largest sample observation), Y_n is : $g(y) = n\,.\,\{F(y,\theta)\}^{n-1}\,.\,f(y,\theta)$,

where $$F(x,\theta) = P(X \le x) = \int_0^x f(u)du = \int_0^x \frac{1}{\theta} = \frac{x}{\theta}$$

$$\therefore \quad g(y) = n\left(\frac{y}{\theta}\right)^{n-1}\frac{1}{\theta} = \frac{n}{\theta^n}\cdot y^{n-1}; 0 \le y < \theta$$

$$E(Y_n^{\,r}) = \int_0^\theta y^r \cdot g(y)dy = \frac{n}{\theta^n}\int_0^\theta y^{r+n-1}dy = \frac{n\theta^r}{n+r}$$

Taking *r* = 1 and 2 ; $$E(Y_n) = \frac{n\theta}{n+1}; \quad E(Y_n^2) = \frac{n\theta^2}{n+2} \quad ... (***)$$

$$\therefore \quad E\left(\frac{n+1}{n}\cdot Y_n\right) = \frac{n+1}{n}E(Y_n) = \theta \quad \text{[Using ***]}$$

$\Rightarrow$ $(n+1)Y_n/n$ is an unbiased estimator of θ.

$$\text{Var}\left(\frac{n+1}{n}Y_n\right) = \left(\frac{n+1}{n}\right)^2 \cdot \text{Var}(Y_n) = \left(\frac{n+1}{n}\right)^2 \left\{EY_n^2 - \left(EY_n\right)^2\right\}$$

$$= \left(\frac{n+1}{n}\right)^2 \left\{\frac{n\theta^2}{n+2} - \frac{n^2\theta^2}{(n+1)^2}\right\} = \theta^2 \left\{\frac{(n+1)^2}{n(n+2)} - 1\right\} = \frac{\theta^2}{n(n+2)} < \frac{\theta^2}{n}$$

$\Rightarrow$ $$\text{Var}\left(\frac{n+1}{n} \cdot Y_n\right) \le 1 \Big/ \left\{nE\left(\frac{\partial}{\partial\theta}\log f\right)^2\right\}.$$ Hence $(n + 1)Y_n/n$ is an *MVUE*.

Remark. This example illustrates that if the regularity conditions underlying Cramer-Rao inequality are violated, then the least attainable variance may be less than the Cramer-Rao lower bound.

17·6. METHODS OF ESTIMATION

So far we have been discussing the requisites of a good estimator. Now we shall briefly outline some of the important methods for obtaining such estimators. Commonly used methods are :

(*i*) *Method of Maximum Likelihood Estimation.*
(*ii*) *Method of Minimum Variance.*
(*iii*) *Method of Moments.*
(*iv*) *Method of Least Squares.*
(*v*) *Method of Minimum Chi-square.*
(*vi*) *Method of Inverse Probability.*

In the following sections, we shall discuss briefly the first five methods only.

17·6·1. Method of Maximum Likelihood Estimation

From theoretical point of view, the most general method of estimation known is the method of *Maximum Likelihood Estimators* (M.L.E.) which was initially formulated by C.F. Gauss but as a general method of estimation was first introduced by Prof. R.A. Fisher and later on develoed by him in a series of papers. Before introducing the method, we will first define *Likelihood Function.*

Likelihood Function. *Definition. Let $x_1, x_2, \ldots, x_n$ be a random sample of size n from a population with density function $f(x, \theta)$. Then the likelihood function of the sample values $x_1, x_2, \ldots, x_n$, usually denoted by $L = L(\theta)$ is their joint density function, given by :*

$$L = f(x_1, \theta) f(x_2, \theta) \ldots f(x_n, \theta) = \prod_{i=1}^{n} f(x_i, \theta). \qquad \ldots (17·53)$$

L gives the relative likelihood that the random variables assume a particular set of values $x_1, x_2, \ldots, x_n$. For a given sample $x_1, x_2, \ldots, x_n$, L becomes a function of the variable θ, the parameter.

The principle of maximum likelihood consists in finding an estimator for the unknown parameter $\theta = (\theta_1, \theta_2, \ldots, \theta_k)$, say, which maximises the likelihood function $L(\theta)$ for variations in parameter, *i.e.*, we wish to find $\hat{\theta} = (\hat{\theta}_1, \hat{\theta}_2, \ldots, \hat{\theta}_k)$ so that

$$L(\hat{\theta}) > L(\theta) \quad \forall \quad \theta \in \Theta, \quad i.e., \quad L(\hat{\theta}) = \text{Sup } L(\theta) \ \forall \ \theta \in \Theta.$$

Thus, if there exists a function $\hat{\theta} = \hat{\theta}(x_1, x_2, \ldots, x_n)$ of the sample values which maximises L for variations in θ, then $\hat{\theta}$ is to be taken as an estimator of θ. $\hat{\theta}$ is usually called *Maximum Likelihood Estimator (M.L.E.)*. Thus $\hat{\theta}$ is the solution, if any, of

$$\frac{\partial L}{\partial\theta} = 0 \quad \text{and} \quad \frac{\partial^2 L}{\partial\theta^2} < 0 \qquad \ldots (17·54)$$

Since $L > 0$, and $\log L$ is a non-decreasing function of L ; L and $\log L$ attain their extreme values (maxima or minima) at the same value of $\hat{\theta}$. The first of the two equations in (17·54) can be rewritten as:

$$\frac{1}{L}\cdot\frac{\partial L}{\partial \theta}=0 \quad \Rightarrow \quad \frac{\partial \log L}{\partial \theta}=0 \qquad \ldots (17\cdot54a)$$

a form which is much more convenient from practical point of view.

If θ is vector valued parameter, then $\hat{\theta}=\hat{\theta}_1,\hat{\theta}_2,\ldots,\hat{\theta}_k$, is given by the solution of simultaneous equations :

$$\frac{\partial}{\partial \theta_i}\log L=\frac{\partial}{\partial \theta_i}\log L\,(\theta_1,\theta_2,\ldots,\theta_k)=0\,;\; i=1,2,\ldots,k \qquad \ldots (17\cdot54b)$$

The above equations (17·54 *a*) and (17·54 *b*) are usually referred to as the *Likelihood Equations* for estimating the parameters.

Note. For the solution $\hat{\theta}$ of the likelihood equations, we have to see that the second derivative of *L w.r. to* θ is negative. If θ is vector valued, then for *L* to be maximum, the matrix of derivatives $\left(\frac{\partial^2}{\partial\theta_i\partial\theta_j}\log L\right)_{\theta=\hat{\theta}}$ should be negative definite.

Properties of Maximum Likelihood Estimators. We make the following assumptions, known as the *Regularity Conditions* :

(*i*) The first and second order derivatives, *viz.*, $\frac{\partial \log L}{\partial \theta}$ and $\frac{\partial^2 \log L}{\partial \theta^2}$ exist and are continuous functions of θ in a range *R* (including the true value θ_0 of the parameter) for almost all *x*. For every θ in *R*, $\left|\frac{\partial}{\partial\theta}\log L\right|<F_1(x)$ and $\left|\frac{\partial^2}{\partial\theta^2}\log L\right|<F_2(x)$, where $F_1(x)$ and $F_2(x)$ are integrable functions over $(-\infty,\infty)$.

(*ii*) The third order derivative $\frac{\partial^3}{\partial\theta^3}$ exists such that $\left|\frac{\partial^3}{\partial\theta^3}\log L\right|<M(x)$, where $E[M(x)]<K$, a positive quantity.

(*iii*) For every θ in *R*, $E\left(-\frac{\partial^2}{\partial\theta^2}\log L\right)=\int_{-\infty}^{\infty}\left(-\frac{\partial^2}{\partial\theta^2}\log L\right)L\,dx=I(\theta)$, is finite and non-zero.

(*iv*) The range of integration is independent of θ. But if the range of integration depends on θ, then $f(x,\theta)$ vanishes at the extremes depending on θ.

This assumption is to make the differentiation under the integral sign valid.

Under the above assumptions M.L.E. possesses a number of important properties, which will be stated in the form of theorems.

Theorem 17·11. (*Cramer-Rao Theorem*). *"With probability approaching unity as $n\to\infty$, the likelihood equation $\frac{\partial}{\partial\theta}\log L=0$, has a solution which converges in probability to the true value θ_0". In other words, M.L.E.'s are consistent.*

Remark. *MLE's* are always consistent estimators but need not be unbiased. For example in sampling from $N(\mu,\sigma^2)$ population, [c.f. Example 17·31],

MLE(μ) = $\bar{x}$ (sample mean), which is both unbiased and consistent estimator of μ.

MLE(σ^2) = s^2 (sample variance), which is consistent but not unbiased estimator of σ^2.

Theorem 17·12. (*Hazoor Bazar's Theorem*). *Any consistent solution of the likelihood equation provides a maximum of the likelihood with probability tending to unity as the sample size (n) tends to infinity.*

Theorem 17·13. (*ASYMPTOTIC NORMALITY OF MLE'S*). *A consistent solution of the likelihood equation is asymptotically normally distributed about the true value θ_0. Thus, $\hat{\theta}$ is asymptotically $N\left(\theta_0,\frac{1}{I(\theta_0)}\right)$, as $n\to\infty$.*

Remark. Variance of M.L.E. is given by :

$$V(\hat{\theta}) = \frac{I}{I(\theta)} = \frac{1}{\left\{E\left(-\frac{\partial^2}{\partial\theta^2}\log L\right)\right\}} \quad \ldots (17\cdot55)$$

Theorem 17·14. *If M.L.E. exists, it is the most efficient in the class of such estimators.*

Theorem 17·15. *If a sufficient estimator exists, it is a function of the Maximum Likelihood Estimator.*

Proof. *If $t = t(x_1, x_2, \ldots, x_n)$ is a sufficient estimator of θ, then Likelihood Function can be written as (c.f. Theorem 17·7) : $L = g(t, \theta) . h(x_1, x_2, x_3, \ldots, x_n \mid t)$, where $g(t, \theta)$ is the density function of t and θ and $h(x_1, x_2, \ldots, x_n \mid t)$ is the density function of the sample, given t, and is independent of θ.*

$$\therefore \quad \log L = \log g(t, \theta) + \log h(x_1, x_2, \ldots, x_n \mid t)$$

Differentiating *w.r. to* θ, we get : $\dfrac{\partial \log L}{\partial\theta} = \dfrac{\partial}{\partial\theta} \log g(t, \theta) = \psi(t, \theta)$, (say), ...(17·56)

which is a function of t and θ only.

M.L.E. of θ is given by $\dfrac{\partial \log L}{\partial\theta} = 0 \quad \Rightarrow \quad \psi(t, \theta) = 0$

$\therefore \quad \hat{\theta} = \eta(t) =$ Some function of sufficient statistic.

$\Rightarrow \quad t = \xi(\hat{\theta}) =$ Some function of M.L.E.

Hence, the theorem.

Remark. This theorem is quite helpful in finding if a sufficient estimator exists or not. If $\frac{\partial}{\partial\theta}\log L$ can be expressed in the form (17·56), *i.e.*, as a function of a statistic and parameter alone, then the statistic is regarded as a sufficient estimator of the parameter. If $\frac{\partial}{\partial\theta}\log L$ cannot be expressed in the form (17·56), no sufficient estimator exists in that case.

Theorem 17·16. *If for a given population with p.d.f. $f(x, \theta)$, an MVB estimator T exits for θ, then likelihood equation will have a solution equal to the estimator T.*

Proof. Since T is an MVB estimator of θ, we have *[c.f. (17·40)]*,

$$\frac{\partial}{\partial\theta}\log L = \frac{T-\theta}{\lambda(\theta)} = (T-\theta)\,A(\theta)$$

MLE for θ is the solution of the likelihood equation :

$$\frac{\partial}{\partial\theta}\log L = 0 \quad \Rightarrow \quad \hat{\theta} = T, \text{ as required.}$$

Theorem 17·17. (INVARIANCE PROPERTY OF MLE). *If T is the MLE of θ and $\psi(\theta)$ is one to one function of θ, then $\psi(T)$ is the MLE of $\psi(\theta)$.*

Example 17·31. *In random sampling from normal population $N(\mu, \sigma^2)$, find the maximum likelihood estimators for*

(i) μ when σ^2 is known,

(ii) σ^2 when μ is known, and

(iii) the simultaneous estimation of μ and σ^2.

Solution. $X \sim N(\mu, \sigma^2)$, then

$$L = \prod_{i=1}^{n}\left[\frac{1}{\sigma\sqrt{2\pi}}\exp\left\{-\frac{1}{2\sigma^2}(x_i-\mu)^2\right\}\right] = \left(\frac{1}{\sigma\sqrt{2\pi}}\right)^n \exp\left\{-\sum_{i=1}^{n}(x_i-\mu)^2/2\sigma^2\right\}$$

$$\log L = -\frac{n}{2}\log(2\pi) - \frac{n}{2}\log\sigma^2 - \frac{1}{2\sigma^2}\sum_{i=1}^{n}(x_i-\mu)^2$$

Case (i). When σ^2 is known, the likelihood equation for estimating μ is :

$$\frac{\partial}{\partial \mu}\log L = 0 \quad \Rightarrow \quad -\frac{1}{2\sigma^2}\sum_{i=1}^{n} 2(x_i - \mu)(-1) = 0$$

$$\Rightarrow \quad \sum_{i=1}^{n}(x_i - \mu) = 0 \quad \Rightarrow \quad \hat{\mu} = \frac{1}{n}\sum_{i=1}^{n} x_i = \overline{x} \qquad \ldots (*)$$

Hence, M.L.E. for μ is the sample mean $\overline{x}$.

Case (ii). When μ is known, the likelihood equation for estimating σ^2 is :

$$\frac{\partial}{\partial \sigma^2}\log L = 0 \quad \Rightarrow \quad -\frac{n}{2}\times\frac{1}{\sigma^2} + \frac{1}{2\sigma^4}\sum_{i=1}^{n}(x_i - \mu)^2 = 0$$

$$\Rightarrow \quad n - \frac{1}{\sigma^2}\sum_{i=1}^{n}(x_i - \mu)^2 = 0 \quad \Rightarrow \hat{\sigma}^2 = \frac{1}{n}\sum_{i=1}^{n}(x_i - \mu)^2 \qquad \ldots (**)$$

Case (iii). The likelihood equations for simultaneous estimation of μ and σ^2 are :

$$\frac{\partial}{\partial \mu}\log L = 0 \text{ and } \frac{\partial}{\partial \sigma^2}\log L = 0, \text{ thus giving } \quad \hat{\mu} = \overline{x} \qquad \text{[From (*)]}$$

and $\quad \hat{\sigma}^2 = \frac{1}{n}\sum_{i=1}^{n}(x_i - \hat{\mu})^2 = \frac{1}{n}\sum_{i=1}^{n}(x_i - \overline{x})^2 = s^2,$ the sample variance.

Important Note. It may be pointed out here that though

$$E(\hat{\mu}) = E(\overline{x}) = \mu, \qquad E(\hat{\sigma}^2) = E(s^2) \neq \sigma^2$$

Hence, the maximum likelihood estimators (M.L.Es.) need not necessarily be unbiased. Another illustration is given in *Example 17·32.*

Remark. Since M.L.E. is the most efficient, we conclude that in sampling from a normal population, the sample mean $\overline{x}$ is the most efficient estimator of the population mean μ.

Example 17·32. *Prove that the maximum likelihood estimate of the parameter α of a population having density function : $\frac{2}{\alpha^2}(\alpha - x)$, $0 < x < \alpha$, for a sample of unit size is $2x$, x being the sample value. Show also that the estimate is biased.*

Solution. For a random sample of unit size ($n = 1$), the likelihood function is :

$$L(\alpha) = f(x, \alpha) = \frac{2}{\alpha^2}(\alpha - x)\,; 0 < x < \alpha$$

Likelihood equation gives : $\frac{d}{d\alpha}\log L = \frac{d}{d\alpha}\{\log 2 - 2\log\alpha + \log(\alpha - x)\} = 0$

$$\Rightarrow \quad -\frac{2}{\alpha} + \frac{1}{\alpha - x} = 0 \quad \Rightarrow \quad 2(\alpha - x) - \alpha = 0 \quad \Rightarrow \quad \alpha = 2x$$

Hence, MLE of α is given by : $\quad \hat{\alpha} = 2x.$

$$E(\hat{\alpha}) = E(2X) = 2\int_0^{\alpha} x f(x, \alpha)\,dx = \frac{4}{\alpha^2}\int_0^{\alpha} x(\alpha - x)\,dx = \frac{4}{\alpha^2}\left|\frac{\alpha x^2}{2} - \frac{x^3}{3}\right|_0^{\alpha} = \frac{2}{3}\alpha$$

Since $E(\hat{\alpha}) \neq \alpha$, $\hat{\alpha} = 2x$ is not an unbiased estimate of α.

Example 17·33.

(a) *Find the maximum likelihood estimate for the parameter λ of a Poisson distribution on the basis of a sample of size n. Also find its variance.*

(b) *Show that the sample mean $\overline{x}$, is sufficient for estimating the parameter λ of the Poisson distribution.*

Solution. The probability function of the Poisson distribution with parameter λ is given by :

$$P(X = x) = f(x, \lambda) = \frac{e^{-\lambda}\lambda^x}{x!} \; ; x = 0, 1, 2, \ldots$$

Likelihood function of random sample $x_1, x_2, \ldots, x_n$ of n observations from this population is :

$$L = \prod_{i=1}^{n} f(x_i, \lambda) = \frac{e^{-n\lambda}\lambda^{\sum_{i=1}^{n} x_i}}{x_1!x_2!\ldots x_n!}$$

$$\therefore \quad \log L = -n\lambda + n\bar{x}\log\lambda - \sum_{i=1}^{n}\log(x_i!)$$

The likelihood equation for estimating λ is :

$$\frac{\partial}{\partial\lambda}\log L = 0 \quad \Rightarrow \quad -n + \frac{n\bar{x}}{\lambda} = 0 \quad \Rightarrow \quad \lambda = \bar{x}$$

Thus, the M.L.E. for λ is the sample mean $\bar{x}$. The variance of estimate is given by :

$$\frac{1}{V(\hat{\lambda})} = E\left\{-\frac{\partial^2}{\partial\lambda^2}(\log L)\right\} \qquad [c.f.\ (17\cdot55)]$$

$$= E\left[-\frac{\partial}{\partial\lambda}\left(-n + \frac{n\bar{x}}{\lambda}\right)\right] = E\left[-\left(-\frac{n\bar{x}}{\lambda^2}\right)\right] = \frac{n}{\lambda^2}E(\bar{x}) = \frac{n}{\lambda} \qquad [\because E(\bar{x}) = \lambda]$$

$$\therefore \quad V(\hat{\lambda}) = \lambda/n$$

(*b*) For the Poisson distribution with parameter λ, we have

$$\frac{\partial}{\partial\lambda}\log L = -n + \frac{n\bar{x}}{\lambda} = n\left(\frac{\bar{x}}{\lambda} - 1\right) = \psi(\bar{x}, \lambda),$$ a function of $\bar{x}$ and λ only.

Hence, (*c.f.* Remark to *Theorem 17·15*), $\bar{x}$ is sufficient for estimating λ.

Example 17·34. *Let $x_1, x_2, \ldots, x_n$ denote random sample of size n from a uniform population with*

p.d.f. : $\quad f(x, \theta) = 1 \; ; \theta - \frac{1}{2} \leq x \leq \theta + \frac{1}{2}, -\infty < \theta < \infty.$

Obtain M.L.E. for θ.

Solution. $L = L(\theta; x_1, x_2, \ldots, x_n) = \begin{cases} 1, & \theta - \frac{1}{2} \leq x_i \leq \theta + \frac{1}{2} \\ 0, & \textit{otherwise} \end{cases}$

If $x_{(1)}, x_{(2)}, \ldots, x_{(n)}$ is the ordered sample, then

$$\theta - \frac{1}{2} \leq x_{(1)} \leq x_{(2)} \leq \ldots \leq x_{(n)} \leq \theta + \frac{1}{2}$$

Thus, L attains the maximum if

$$\theta - \frac{1}{2} \leq x_{(1)} \quad \text{and} \quad x_{(n)} \leq \theta + \frac{1}{2} \quad \Rightarrow \quad \theta \leq x_{(1)} + \frac{1}{2} \quad \text{and} \quad x_{(n)} - \frac{1}{2} \leq \theta$$

Hence, every *statistic* $t = t(x_1, x_2, \ldots, x_n)$ such that

$$x_{(n)} - \frac{1}{2} \leq t(x_1, x_2, \ldots, x_n) \leq x_{(1)} + \frac{1}{2},$$ provides an M.L.E. for θ.

Remark. This example illustrates that M.L.E. for a parameter need not be unique.

Example 17·35. *Find the M.L.E. of the parameters α and λ, (λ being large), of the*

distribution : $\quad f(x; \alpha, \lambda) = \frac{1}{\Gamma(\lambda)}\left(\frac{\lambda}{\alpha}\right)^{\lambda} \exp\left(-\frac{\lambda x}{\alpha}\right) x^{\lambda-1} \; ; \; 0 \leq x < \infty, \; \lambda > 0.$

You may use that for large values of λ,

$$\psi(\lambda) = \frac{\lambda}{\partial\lambda}\log\Gamma(\lambda) = \log\lambda - \frac{1}{2\lambda} \quad\text{and}\quad \Psi'(\lambda) = \frac{1}{\lambda} + \frac{1}{2\lambda^2}. \qquad \ldots (*)$$

Solution. Let $x_1, x_2, \ldots, x_n$ be a random sample of size n from the given population.

Then $$L = \prod_{i=1}^{n} f(x_i;\alpha,\lambda) = \left(\frac{1}{\Gamma(\lambda)}\right)^n \cdot \left(\frac{\lambda}{\alpha}\right)^{n\lambda} \cdot \exp\left(-\frac{\lambda}{\alpha}\sum_{i=1}^{n} x_i\right) \cdot \prod_{i=1}^{n}\left(x_i^{\lambda-1}\right)$$

$$\therefore \quad \log L = -\,n \log \Gamma(\lambda) + n\lambda(\log\lambda - \log\alpha) - \frac{\lambda}{\alpha}\sum_{i=1}^{n} x_i + (\lambda - 1)\sum_{i=1}^{n}\log x_i$$

If G is the geometric mean of $x_1, x_2, \ldots, x_n$, then

$$\log G = \frac{1}{n}\sum_{i=1}^{n}\log x_i \quad \Rightarrow \quad n\log G = \sum_{i=1}^{n}\log x_i$$

$$\therefore \quad \log L = -\,n \log \Gamma(\lambda) + n\lambda(\log\lambda - \log\alpha) - \frac{\lambda}{\alpha} n\bar{x} + (\lambda - 1).n\log G,$$

where G is independent of λ and α.

The likelihood equations for the simultaneous estimation of α and λ are :

$$\frac{\partial}{\partial\alpha}\log L = 0 \quad \ldots(1) \qquad \text{and} \qquad \frac{\partial}{\partial\lambda}\log L = 0 \qquad \ldots (2)$$

(1) gives : $$-\frac{n\lambda}{\alpha} + \frac{\lambda}{\alpha^2}\cdot n\bar{x} = 0 \quad \Rightarrow \quad -1 + \frac{\bar{x}}{\alpha} = 0 \Rightarrow \hat{\alpha} = \bar{x} \qquad \ldots (3)$$

(2) gives (for large values of λ), on using (*) :

$$-n\left(\log\lambda - \frac{1}{2\lambda}\right) + n\left\{1\cdot(\log\lambda - \log\alpha) + \lambda\cdot\frac{1}{\lambda}\right\} - \frac{n\bar{x}}{\alpha} + n\log G = 0$$

$$\Rightarrow \quad \frac{1}{2\lambda} + \left(1 - \log\alpha + \log G - \frac{\bar{x}}{\alpha}\right) = 0 \qquad \text{[From (3)]}$$

$$\Rightarrow \quad 1 + 2\lambda(\log G - \log\bar{x}) = 0$$

$$\Rightarrow \quad 1 - 2\lambda\log\left(\frac{\bar{x}}{G}\right) = 0 \quad \Rightarrow \quad \hat{\lambda} = \frac{1}{2\log(\bar{x}/G)}$$

Hence, the M.L.E. for α and λ are given by : $\hat{\alpha} = \bar{x}$ and $\hat{\lambda} = \dfrac{1}{2\log(\bar{x}/G)}$.

Example 17·36. *In sampling from a power series distribution with p.d.f. :*

$$f(x, \theta) = a_x\theta^x/\psi(\theta)\ ;\ x = 0, 1, 2, \ldots$$

where a_x *may be zero for some* x, *show that MLE of* θ *is a root of the equation :*

$$\bar{X} = \frac{\theta\,\psi'(\theta)}{\psi(\theta)} = \mu(\theta), \quad \textit{where} \quad \mu(\theta) = E(X) \qquad \ldots (*)$$

Solution. Likelihood function is given by :

$$L = \prod_{i=1}^{n} f(x_i,\theta) = \prod_{i=1}^{n}\left(\frac{a_{x_i}\theta^{x_i}}{\psi(\theta)}\right) = \left(\prod_{i=1}^{n} a_{x_i}\right)\frac{\theta^{\Sigma x_i}}{[\psi(\theta)]^n}$$

$$\Rightarrow \quad \log L = \sum_{i=1}^{n}\log a_{x_i} + \log\theta\cdot\sum_{i=1}^{n} x_i - n\log\psi(\theta)$$

Likelihood equation for estimating θ gives :

$$\frac{\partial}{\partial\theta}\log L = 0 = \frac{\Sigma x_i}{\theta} - \frac{n\psi'(\theta)}{\psi(\theta)} \quad \Rightarrow \quad \bar{X} = \frac{\Sigma x_i}{n} = \frac{\theta\psi'(\theta)}{\psi(\theta)} = \mu(\theta), \text{ (say)}. \qquad \ldots (i)$$

Hence, MLE of θ is a root of equation (*). We have

$$E(X) = \sum_{x=0}^{\infty} x f(x,\theta) = \sum_{x=0}^{\infty} \left[x \left\{ \frac{a_x \theta^x}{\psi(\theta)} \right\} \right] \qquad \text{... (ii)}$$

$$\sum_{x=0}^{\infty} f(x,\theta) = 1 \Rightarrow \sum_{x=0}^{\infty} \frac{a_x \theta^x}{\psi(\theta)} = 1 \Rightarrow \sum_{x=0}^{\infty} a_x \theta^x = \psi(\theta)$$

Differentiating *w.r. to* θ, we get

$$\sum_x \left[a_x \cdot x\theta^{x-1} \right] = \psi'(\theta) \quad \Rightarrow \quad \sum_x \left\{ a_x \cdot \frac{x\theta^x}{\psi(\theta)} \right\} = \frac{\theta \cdot \psi'(\theta)}{\psi(\theta)}$$

$$\therefore \quad E(X) = \mu(\theta) = \overline{X}, \qquad \text{[From (ii) and (i)]}$$

as required.

Example 17·37.

(a) Let $x_1, x_2, \ldots, x_n$ be a random sample from the uniform distribution with p.d.f.:

$$f(x,\theta) = \begin{cases} \dfrac{1}{\theta}, & 0 < x < \infty, \quad \theta > 0 \\ 0, & \textit{elsewhere} \end{cases}$$

Obtain the maximum likelihood estimator for θ.

(b) Obtain the M.L.Es. for α and β for the rectangular population :

$$f(x;\alpha,\beta) = \begin{cases} \dfrac{1}{\beta-\alpha}, & \alpha < x < \beta \\ 0, & \textit{elsewhere} \end{cases}$$

Solution.

(a)
$$L = \prod_{i=1}^{n} f(x_i,\theta) = \frac{1}{\theta}\cdot\frac{1}{\theta} \;\cdots\; \frac{1}{\theta} = \left(\frac{1}{\theta}\right)^n \qquad \text{... (*)}$$

Likelihood equation, *viz.*, $\dfrac{\partial}{\partial\theta}\log L = 0$, gives :

$$\frac{\partial}{\partial\theta}(-n\log\theta) = 0 \quad \Rightarrow \quad \frac{-n}{\theta} = 0 \text{ or } \hat{\theta} = \infty, \text{ obviously an absurd result.}$$

In this case we locate M.L.E. as follows : We have to choose θ so that L in (*) is maximum. L is maximum if θ is minimum.

Let $x_{(1)}, x_{(2)}, \ldots, x_{(n)}$ be the *ordered* random sample of n independent observations from the given population so that $0 \le x_{(1)} \le x_{(2)} \le \ldots \le x_{(n)} \le \theta \quad \Rightarrow \quad \theta \ge x_{(n)}$

Since the minimum value of θ consistent with the sample is $x_{(n)}$, the largest sample observation, $\hat{\theta} = x_{(n)}$.

∴ M.L.E. for $\theta = x_{(n)}$ = The largest sample observation.

(b) Here,
$$L = \left(\frac{1}{\beta-\alpha}\right)^n \quad \Rightarrow \quad \log L = -n\log(\beta-\alpha) \qquad \text{... (**)}$$

The likelihood equations for α and β give :

$$\left.\begin{aligned} \frac{\partial}{\partial\alpha}\log L = 0 &= \frac{n}{\beta-\alpha} \\ \frac{\partial}{\partial\beta}\log L = 0 &= \frac{-n}{\beta-\alpha} \end{aligned}\right\}$$

and

Each of these equations gives $\beta - \alpha = \infty$, an obviously negative result. So, we find M.L.Es for α and β by some other means.

L in (**) is maximum if $(\beta - \alpha)$ is minimum, *i.e.*, if β takes the minimum possible value and α takes the maximum possible value.

As in part (*a*), if $x_{(1)}, x_{(2)}, \ldots, x_{(n)}$ is an *ordered* random sample from this population, then $\alpha \le x_{(1)} \le x_{(2)} \le \ldots \le x_{(n)} \le \beta$. Thus $\beta \ge x_{(n)}$ and $\alpha \le x_{(1)}$. Hence the minimum possible value of β consistent with the sample is $x_{(n)}$ and the maximum possible value of α consistent with the sample is $x_{(1)}$. Hence L is maximum if $\beta = x_{(n)}$ and $\alpha = x_{(1)}$.

$\therefore$ M.L.E. for α and β are given by :

$$\hat{\alpha} = x_{(1)} = \text{The smallest sample observation,}$$

and $$\hat{\beta} = x_{(n)} = \text{The largest sample observation.}$$

Example 17·38. *State as precisely as possible the properties of the M.L.E. Obtain the M.L.Es. of α and β for a random sample from the exponential population :*

$$f(x\,;\,\alpha, \beta) = y_0\, e^{-\beta(x-\alpha)},\ \alpha \le x \le \infty,\ \beta > 0 \text{ and } y_0 \text{ being a constant.}$$

Solution. Here first of all we shall determine the constant y_0 from the consideration that the total area under a probability curve is unity.

$$y_0 \int_\alpha^\infty \exp[-\beta(x-\alpha)]\,dx = 1 \;\Rightarrow\; y_0 \left| \frac{e^{-\beta(x-\alpha)}}{-\beta} \right|_\alpha^\infty = 1 \;\Rightarrow\; -\frac{y_0}{\beta}(0-1) = 1 \;\Rightarrow\; y_0 = \beta$$

$$\therefore \qquad f(x\,;\,\alpha, \beta) = \beta\, e^{-\beta(x-\alpha)},\ \alpha \le x < \infty$$

If $x_1, x_2, \ldots, x_n$ is a random sample of n observations from this population, then

$$L = \prod_{i=1}^n f(x_i;\alpha,\beta) = \beta^n \exp\left\{-\beta \sum_{i=1}^n (x_i - \alpha)\right\} = \beta^n \exp\left[-n\beta(\bar{x} - \alpha)\right]$$

$$\therefore \qquad \log L = n \log \beta - n\beta(\bar{x} - \alpha) \qquad \ldots (*)$$

The likelihood equations for estimating α and β give

$$\frac{\partial}{\partial \alpha} \log L = 0 = n\beta \qquad \ldots (**)$$

and $$\frac{\partial}{\partial \beta} \log L = 0 = \frac{n}{\beta} - n(\bar{x} - \alpha) \qquad \ldots (***)$$

Equation (**) gives $\beta = 0$, which is obviously inadmissible and this on substitution in (***) gives $\alpha = \infty$, a nugatory result. Thus, the likelihood equations fail to give us valid estimates of α and β and we try to locate M.L.Es. for α and β by maximising L directly.

$$L \text{ is maximum} \;\Rightarrow\; \log L \text{ is maximum.}$$

From (*), $\log L$ is maximum (for any value of β), if $(\bar{x} - \alpha)$ is minimum, which is so if α is maximum.

If $x_{(1)}, x_{(2)}, \ldots, x_{(n)}$ is ordered sample from this population then $\alpha \le x_{(1)} \le x_{(2)} \le \ldots \le x_{(n)} < \infty$, so that the maximum value of α consistent with the sample is $x_{(1)}$, the smallest sample observation, *i.e.*, $\hat{\alpha} = x_{(1)}$.

Consequently, (***) gives $\dfrac{1}{\beta} = \bar{x} - \hat{\alpha} = \bar{x} - x_{(1)} \;\Rightarrow\; \hat{\beta} = \dfrac{1}{\bar{x} - x_{(1)}}$

Hence, M.L.Es. for α and β are given by : $\hat{\alpha} = x_{(1)}$ and $\hat{\beta} = \dfrac{1}{\bar{x} - x_{(1)}}$

Remarks:

1. Whenever the given probability function involves a constant and the range of the variable is dependent on the parameter(s) to be estimated, first of all we should determine the constant by taking the total probability as unity and then proceed with the estimation part.
2. From the last two examples, it is obvious that whenever the range of the variable involves the parameter(s) to be estimated, the likelihood equations fail to give us valid estimates and in this case M.L.Es are obtained by adopting some other approach of maximising L or $\log L$ directly.

Example 17·39. *Obtain maximum likelihood estimate of θ in $f(x, \theta) = (1 + \theta)\, x^{\theta}$, $0 < x < 1$, based on an independent sample of size n. Examine whether this estimate is sufficient for θ.*

Solution. $L(x,\theta) = \prod_{i=1}^{n} f(x_i,\theta) = (1+\theta)^n \left(\prod_{i=1}^{n} x_i\right)^{\theta} \Rightarrow \log L = n\log(1+\theta) + \theta.\sum_{i=1}^{n} \log x_i$

$$\frac{\partial}{\partial\theta}\log L = \frac{n}{1+\theta} + \sum_{i=1}^{n} \log x_i = 0 \quad \Rightarrow \quad n + \theta\sum_i \log x_i + \sum_i \log x_i = 0$$

$$\therefore \quad \hat{\theta} = \frac{-n}{\sum_{i=1}^{n} \log x_i} - 1 = \frac{-n}{\log\left(\prod_{i=1}^{n} x_i\right)} - 1 \qquad \dots (*)$$

Also $\quad L(x,\theta) = \left\{(1+\theta)^n \cdot \left(\prod_{i=1}^{n} x_i\right)^{\theta-1}\right\} \cdot \left(\prod_{i=1}^{n} x_i\right)$

Hence, by Factorisation theorem, $T = \left(\prod_{i=1}^{n} x_i\right)$ is a sufficient statistic for θ, and $\hat{\theta}$ being a one to one function of sufficient statistic $\left(\prod_{i=1}^{n} x_i\right)$, is also sufficient for θ.

Example 17·40.

(a) Obtain the most general form of distribution differentiable in θ, for which the sample mean is the M.L.E.

(b) Show that the most general continuous distribution for which the M.L.E. of a parameter θ is the geometric mean of the sample is

$$f(x,\theta) = \left(\frac{x}{\theta}\right)^{\theta\cdot\frac{\partial\psi}{\partial\theta}} \exp\{\psi(\theta) + \xi(x)\},$$

where $\psi(\theta)$ and $\xi(x)$ are arbitrary functions of θ and x respectively.

Solution.

(a) We have $\quad L = \prod_{i=1}^{n} f(x_i, \theta) \Rightarrow \log L = \sum_{i=1}^{n} \log f(x_i, \theta) = \sum_x \log f$, $[f = f(x, \theta)]$, the summation extending to all the values of $x = (x_1, x_2, \dots, x_n)$in the sample. The likelihood equation is :

$$\frac{\partial}{\partial\theta}\log L = 0, \quad i.e., \quad \frac{\partial}{\partial\theta}\left(\sum_x \log f\right) = 0$$

$$\Rightarrow \quad \sum_x \frac{\partial}{\partial\theta}\log f = 0 \quad \Rightarrow \quad \sum_x \frac{1}{f}\cdot\frac{\partial f}{\partial\theta} = 0 \qquad \dots (*)$$

We are given that the solution of (*) is : $\theta = \frac{1}{n}\sum x \Rightarrow n\theta = \sum x \Rightarrow \sum_x (x - \theta) = 0$ ...(**)

Since this is true for all values of x and θ, we get from (*) and (**),

$\frac{1}{f}\cdot\frac{\partial f}{\partial\theta} = A(x - \theta)$, where A is independent of x but may be function of θ.

Let us take $A = \frac{\partial^2\psi}{\partial\theta^2}$, where $\psi = \psi(\theta)$ is any arbitrary function of θ. Thus $\frac{\partial}{\partial\theta}\log f = \frac{\partial^2\psi}{\partial\theta^2}(x - \theta)$.

Integrating *w.r.t.* θ (partially), we get

$$\log f = (x - \theta) . \frac{\partial \psi}{\partial \theta} - \int \frac{\partial \psi}{\partial \theta} (-1)\, d\theta + \xi(x) + k,$$

where $\xi(x)$ is an arbitrary function of x and k is arbitrary constant.

$$\therefore \qquad \log f = (x - \theta) . \frac{\partial \psi}{\partial \theta} + \psi(\theta) + \xi(x) + k$$

Hence, $$f = \text{Const.} \exp \left\{ (x - \theta) \frac{\partial \psi}{\partial \theta} + \psi(\theta) + \xi(x) \right\},$$

which is the probability function of the required distribution.

Note In particular, if we take. $\psi(\theta) = \dfrac{\theta^2}{2}$ and $\xi(x) = -\dfrac{x^2}{2}$, then

$$f = \text{Const.} \exp \left\{ (x - \theta) \cdot \theta + \frac{\theta^2}{2} - \frac{x^2}{2} \right\}$$

$$= \text{Const.} \exp \left\{ -\frac{1}{2}(x^2 + \theta^2 - 2\theta x) \right\} = \text{Const.} \exp \left\{ -\frac{1}{2}(x - \theta)^2 \right\}$$

which is the probability function of the normal distribution with mean θ and unit variance.

(*b*) Here, the solution of the likelihood equation

$$\frac{\partial}{\partial \theta} \log L = \sum_x \frac{\partial}{\partial \theta} \log f = 0 \qquad \ldots (*)$$

is $\theta = (x_1 \, x_2 \ldots x_n)^{1/n} \quad \Rightarrow \quad \log \theta = \dfrac{1}{n} \sum_x \log x \quad \Rightarrow \quad \sum_x (\log x - \log \theta) = 0 \qquad \ldots (**)$

Since this is true for all x and all θ, we get from (*) and (**),

$$\frac{\partial}{\partial \theta} \log f = (\log x - \log \theta)\, A(\theta),$$

where $A(\theta)$ is an arbitrary function of θ and is independent of x.

Integrating *w.r. to* θ (partially),

$$\log f = \log x \int A(\theta)\, d\theta - \int A(\theta) \log \theta \, d\theta + \xi(x),$$

where $\xi(x)$ is an arbitrary function of x alone.

If we take $\int A(\theta)\, d\theta = A_1(\theta)$, then

$$\log f = \log x \cdot A_1(\theta) - \left\{ A_1(\theta) \log \theta - \int A_1(\theta) \cdot \frac{1}{\theta} d\theta \right\} + \xi(x)$$

$$= A_1(\theta) \log(x/\theta) + \int \frac{A_1(\theta)}{\theta} d\theta + \xi(x)$$

Let us take $A_1(\theta) = \theta \dfrac{\partial \psi}{\partial \theta}$, (suggested by the answer), where $\psi = \psi(\theta)$ is an arbitrary function of θ alone.

$$\therefore \qquad \log f = \theta \frac{\partial \psi}{\partial \theta} \log(x/\theta) + \int \frac{\partial \psi}{\partial \theta} d\theta + \xi(x)$$

$$= \theta \frac{\partial \psi}{\partial \theta} \cdot \log(x/\theta) + \psi(\theta) + \xi(x) = \log \left\{ \left(\frac{x}{\theta} \right)^{\theta \frac{\partial \psi}{\partial \theta}} \right\} + \psi(\theta) + \xi(x)$$

Hence, $$f = f(x, \theta) = \left(\frac{x}{\theta} \right)^{\theta \frac{\partial \psi}{\partial \theta}} \cdot \exp\{\psi(\theta) + \xi(x)\}.$$

***Example 17·41.** A sample of size n is drawn from each of the four normal populations which have the same variance σ^2. The means of the four populations are $a + b + c$, $a + b - c$, $a - b + c$ and $a - b - c$. What are the M.L.Es. for a, b, c, and σ^2 ?*

Solution. Let the sample observations be denoted by x_{ij} , $i = 1, 2, 3, 4; j = 1, 2, ..., n$. Since the four samples, from the four normal populations are independent, the likelihood function L of all the sample observations x_{ij}, $(i = 1, 2, 3, 4 ; j = 1, 2, ..., n)$, is given by :

$$L = \left(\frac{1}{\sigma\sqrt{\sqrt{2\pi}}}\right)^{4n} \cdot \exp\left\{-\frac{1}{2\sigma^2}\sum_{i=1}^{4}\sum_{j=1}^{n}(x_{ij} - \mu_i)^2\right\},$$

where μ_i , $(i = 1, 2, 3, 4)$ is mean of the ith population. Therefore,

$$L = \left(\frac{1}{\sigma\sqrt{2\pi}}\right)^{4n} \cdot \exp\left[-\frac{1}{2\sigma^2}\left\{\sum_j(x_{1j} - \mu_1)^2 + \sum_j(x_{2j} - \mu_2)^2 + \sum_j(x_{3j} - \mu_3)^2 + \sum_j(x_{4j} - \mu_4)^2\right\}\right]$$

$$\Rightarrow \log L = k - 2n\log\sigma^2 - \frac{1}{2\sigma^2}\left\{\sum_j(x_{1j} - a - b - c)^2 + \sum_j(x_{2j} - a - b + c)^2\right.$$

$$\left. + \sum_j(x_{3j} - a + b - c)^2 + \sum_j(x_{4j} - a + b + c)^2\right\},$$

where k is a constant w.r. to a, b, c and σ^2. The M.L.Es. for a, b, c and σ^2 are the solutions of the simultaneous equations (maximum likelihood equations for estimating a, b, c and σ^2) :

$$\frac{\partial}{\partial a}\log L = 0 \quad \ldots(1) \qquad \frac{\partial}{\partial b}\log L = 0 \quad \ldots(2)$$

$$\frac{\partial}{\partial c}\log L = 0 \quad \ldots(3) \qquad \frac{\partial}{\partial \sigma^2}\log L = 0 \quad \ldots(4)$$

(1) gives :

$$-\frac{1}{2\sigma^2}\left\{\sum_j(x_{1j} - a - b - c)(-2) + \sum_j(x_{2j} - a - b + c)(-2)\right.$$

$$\left. + \sum_j(x_{3j} - a + b - c)(-2) + \sum(x_{4j} - a + b + c)(-2)\right\} = 0$$

$$\Rightarrow \quad \sum_j(x_{1j} + x_{2j} + x_{3j} + x_{4j}) + n[(-a - b - c) + (-a - b + c) + (-a + b - c) + (-a + b + c)] = 0$$

$$\Rightarrow \quad \sum_{j=1}^{n}\left(\sum_{i=1}^{4} x_{ij}\right) + n(-4a) = 0 \qquad \Rightarrow \qquad \hat{a} = \frac{1}{4n}\sum_{i=1}^{4}\sum_{j=1}^{n} x_{ij} = \overline{x}$$

(2) gives :

$$-\frac{1}{2\sigma^2}\left\{\sum_j(x_{1j} - a - b - c)(-2) + \sum_j(x_{2j} - a - b + c)(-2)\right.$$

$$\left. + \sum_j(x_{3j} - a + b - c)(2) + \sum_j(x_{4j} - a + b + c)(2)\right\} = 0$$

$$\Rightarrow \quad \sum_j x_{1j} + \sum_j x_{2j} - \sum_j x_{3j} - \sum_j x_{4j} + n[(-a - b - c) + (-a - b + c) - (-a + b - c) - (-a + b + c)] = 0$$

$$\Rightarrow \quad \sum x_{1j} + \sum x_{2j} - \sum x_{3j} - \sum x_{4j} - 4nb = 0$$

$$\therefore \quad \hat{b} = \frac{1}{4}\left(\frac{1}{n}\sum x_{1j} + \frac{1}{n}\sum x_{2j} - \frac{1}{n}\sum x_{3j} - \frac{1}{n}\sum x_{4j}\right) \Rightarrow \hat{b} = (\overline{x}_1 + \overline{x}_2 - \overline{x}_3 - \overline{x}_4)/4$$

where $\overline{x}_i$ is the mean of the ith sample.

Similarly (3) will give : $\hat{c} = (\bar{x}_1 - \bar{x}_2 + \bar{x}_3 - \bar{x}_4)/4$

(4) gives : $$-\frac{2n}{\sigma^2} + \frac{1}{2\sigma^4}\Bigg\{\sum_j (x_{1j} - a - b - c)^2 + \sum_j (x_{2j} - a - b + c)^2$$
$$+ \sum_j (x_{3j} - a + b - c)^2 + \sum_j (x_{4j} - a + b + c)^2\Bigg\} = 0$$

$$\therefore \quad \hat{\sigma}^2 = \frac{1}{4n}\Bigg\{\sum_j (x_{1j} - \hat{a} - \hat{b} - \hat{c})^2 + \sum_j (x_{2j} - \hat{a} - \hat{b} + \hat{c})^2 + \sum_j (x_{3j} - \hat{a} + \hat{b} - \hat{c})^2 + \sum_j (x_{4j} - \hat{a} + \hat{b} + \hat{c})^2\Bigg\}$$

Example 17·42. *The following table gives probabilities and observed frequencies in four classes AB, Ab, aB and ab in a genetical experiment. Estimate the parameter θ by the method of maximum likelihood and find its standard error.*

Class	:	*AB*	*Ab*	*aB*	*ab*
Probability	:	$\frac{1}{4}(2+\theta)$	$\frac{1}{4}(1-\theta)$	$\frac{1}{4}(1-\theta)$	$\frac{1}{4}\theta$
Observed frequency	:	108	27	30	8

Solution. Using multinomial probability distribution, we have

$$L = L(x, \theta) = \frac{n\,!}{n_1!\, n_2!\, n_3!\, n_4!} \cdot p_1^{n_1} p_2^{n_2} p_3^{n_3} p_4^{n_4}, \quad \sum p_i = 1, \quad \sum n_i = n$$

$$\Rightarrow \qquad \log L = C + n_1 \log p_1 + n_2 \log p_2 + n_3 \log p_3 + n_4 \log p_4,$$

where $$C = \log\left[\frac{n\,!}{n_1!\, n_2!\, n_3!\, n_4!}\right], \text{ is a constant.}$$

$$\therefore \quad \log L = C + n_1 \log\left[(2+\theta)/4\right] + n_2 \log\left[(1-\theta)/4\right] + n_3 \log\left[(1-\theta)/4\right] + n_4 \log(\theta/4)$$

Likelihood equation gives :

$$\frac{\partial \log L}{\partial \theta} = \frac{n_1}{2+\theta} - \frac{n_2}{1-\theta} - \frac{n_3}{1-\theta} + \frac{n_4}{\theta} = 0 \Rightarrow \frac{n_1}{2+\theta} - \frac{(n_2+n_3)}{1-\theta} + \frac{n_4}{\theta} = 0 \qquad \ldots (*)$$

Taking $n_1 = 108$, $n_2 = 27$, $n_3 = 30$ and $n_4 = 8$, we get $\dfrac{108}{2+\theta} - \dfrac{(27+30)}{1-\theta} + \dfrac{8}{\theta} = 0$

$$\Rightarrow \qquad 108\theta(1-\theta) - 57\theta(2+\theta) + 8(1-\theta)(2+\theta) = 0 \quad \Rightarrow \quad 173\theta^2 + 14\theta - 16 = 0$$

$$\therefore \qquad \theta = \frac{-14 \pm \sqrt{196 + 11072}}{346} = -0{\cdot}34 \text{ and } 0{\cdot}26$$

But θ, being the probability cannot be negative. Hence,

M.L.E. of θ is given by : $\hat{\theta} = 0{\cdot}26$ ... (**)

Differentiating (*) again partially w.r.t. θ, we get

$$\frac{\partial^2 \log L}{\partial \theta^2} = \frac{-n_1}{(2+\theta)^2} - \frac{(n_2+n_3)}{(1-\theta)^2} - \frac{n_4}{\theta^2}$$

$$\therefore \quad -E\left(\frac{\partial^2 \log L}{\partial \theta^2}\right) = \frac{E(n_1)}{(2+\theta)^2} + \frac{E(n_2)+E(n_3)}{(1-\theta)^2} + \frac{E(n_4)}{\theta^2}$$

$$= \frac{np_1}{(2+\theta)^2} + \frac{n(p_2+p_3)}{(1-\theta)^2} + \frac{np_4}{\theta^2} = \frac{n(2+\theta)}{4(2+\theta)^2} + \frac{n(1-\theta)}{2(1-\theta)^2} + \frac{n\theta}{4\theta^2}$$

$$\Rightarrow \quad I(\theta) = \frac{n}{4(2+\hat{\theta})} + \frac{n}{2(1-\hat{\theta})} + \frac{n}{4\hat{\theta}}\ ; \ n = \sum n_i = 173$$

$$= 173\left(\frac{1}{4\times 2{\cdot}26} + \frac{1}{2\times 0{\cdot}74} + \frac{1}{4\times 0{\cdot}26}\right) = 301{\cdot}02$$

$$S.E.(\hat{\theta}) = \sqrt{I/I(\theta)} = \frac{1}{\sqrt{301{\cdot}02}} = 0{\cdot}0576$$ [*c.f.* Theorem (17·13)]

17·6·2. Method of Minimum Variance

{Minimum Variance Unbiased Estimates (M.V.U.E.)}. In this section we shall look for estimates which (*i*) are unbiased and (*ii*) have minimum variance.

If $L = \prod_{i=1}^{n} f(x_i, \theta)$, is the likelihood function of a random sample of n observations $x_1, x_2, \ldots, x_n$ from a population with probability function $f(x, \theta)$, then the problem is to find a statistic $t = t\ (x_1, x_2, \ldots, x_n)$, such that

$$E(t) = \int_{-\infty}^{\infty} t\,L\,dx = \gamma(\theta) \quad \Rightarrow \quad \int_{-\infty}^{\infty} \{t-\gamma(\theta)\}L\,dx = 0 \qquad \ldots (17{\cdot}57)$$

and
$$V(t) = \int_{-\infty}^{\infty} [t-E(t)]^2 L\,dx = \int_{-\infty}^{\infty} [t-\gamma(\theta)]^2 L\,dx \qquad \ldots (17{\cdot}58)$$

is minimum where

$\int_{-\infty}^{\infty} dx$ represents the n-fold integration $\int_{-\infty}^{\infty}\int_{-\infty}^{\infty}\cdots\int_{-\infty}^{\infty} dx_1\,dx_2\ldots dx_n$

In other words, we have to minimise (17·58) subject to the condition (17·57).

For detailed discussion of this method see *MVU* Estimators (§ 17·5) and Cramer-Rao Inequality (§ 17·3).

17·6·3. Method of Moments

This method was discovered and studied in detail by Karl Pearson.

Let $f(x\ ;\ \theta_1, \theta_2, \ldots, \theta_k)$ be the density function of the parent population with k parameters $\theta_1, \theta_2, \ldots, \theta_k$. If μ_r' denotes the rth moment about origin, then

$$\mu_r' = \int_{-\infty}^{\infty} x^r f(x\ ;\theta_1, \theta_2, \ldots, \theta_k)\,dx, \ (r = 1, 2, \ldots, k) \qquad \ldots (17{\cdot}59)$$

In general $\mu_1', \mu_2', \ldots, \mu_k'$ will be function of the parameters $\theta_1, \theta_2, \ldots, \theta_k$.

Let x_i, $i = 1, 2, \ldots, n$ be a random sample of size n from the given population. The method of moments consists in solving the k-equations (17·59) for $\theta_1, \theta_2, \ldots, \theta_k$ in terms of $\mu_1', \mu_2', \ldots, \mu_k'$ and then replacing these moments μ_r' ; $r = 1, 2, \ldots, k$ by the sample moments, *e.g.*,

$$\hat{\theta}_i = \theta_i\ (\hat{\mu}_1', \hat{\mu}_2', \ldots, \hat{\mu}_k') = \theta_i\ (m_1', m_2', \ldots, m_k')\ ;\ i = 1, 2, \ldots, k,$$

where m_i' is the ith moment about origin in the sample.

Then, by the method of moments $\hat{\theta}_1, \hat{\theta}_2, \ldots, \hat{\theta}_k$ are the required estimators of $\theta_1, \theta_2, \ldots, \theta_k$. respectively.

Remarks:

1. Let $(x_1, x_2, \ldots, x_n)$ be a random sample of size n from a population with *p.d.f.* $f(x, \theta)$. Then X_i, $(i = 1, 2, \ldots, n)$ are *i.i.d.* $\Rightarrow X_i^r$, $(i = 1, 2, \ldots, n)$ are *i.i.d.* Hence if $E\ (X_i^r)$ exists, then by W.L.L.N., we get

$$\frac{1}{n}\sum_{i=1}^{n} x_i^r \xrightarrow{p} E(X_1^r) \quad \Rightarrow \quad m_r' \xrightarrow{p} \mu_r' \qquad \text{... (17·60)}$$

Hence, the sample moments are consistent estimators of the corresponding population moments.

2. It has been shown that under quite general conditions, the estimates obtained by the method of moments are asymptotically normal but not, in general, efficient.
3. Generally the method of moments yields less efficient estimators than those obtained from the principle of maximum likelihood. The estimators obtained by the method of moments are identical with those given by the method of maximum likelihood if the probability mass function or probability density function is of the form :

$$f(x, \theta) = \exp\left(b_0 + b_1 x + b_2 x^2 + \dots\right), \qquad \text{... (17·61)}$$

where b's are independent of x but may depend on $\theta = (\theta_1, \theta_2, \dots)$.

(17·61) implies that : $L\ (x_1, x_2, \dots, x_n\ ;\ \theta) = \exp\left(nb_0 + b_1\Sigma x_i + b_2\Sigma x_i^2 + \dots\right)$

$$\Rightarrow \qquad \frac{\partial}{\partial\theta}\log L = a_0 + a_1\,\Sigma\, x_i + a_2\,\Sigma\, x_i^2 + a_3\,\Sigma x_i^3 + \dots \qquad \text{... (17·61a)}$$

where $a_i = \dfrac{\partial}{\partial\theta}\,(b_i),\ (i = 1, 2, \dots\dots)$ and $a_0 = n\,\dfrac{\partial b_0}{\partial\theta}$

Thus, both the methods yield identical estimators if *MLE's* are obtained as linear functions of the moments.

Example 17·43. *Estimate α and β in the case of Pearson's Type III distribution by the method of moments :*

$$f(x;\, \alpha,\, \beta) = \frac{\beta^\alpha}{\Gamma(\alpha)}\, x^{\alpha-1}\, e^{-\beta x},\ 0 \le x < \infty$$

Solution. We have

$$\mu_r' = \frac{\beta^\alpha}{\Gamma(\alpha)}\int_0^\infty x^r x^{\alpha-1} e^{-\beta x} dx = \frac{\beta^\alpha}{\Gamma(\alpha)}\cdot\frac{\Gamma(\alpha+r)}{\beta^{\alpha+r}} = \frac{\Gamma(\alpha+r)}{\Gamma(\alpha)\beta^r}$$

$$\mu_1' = \frac{\Gamma(\alpha+1)}{\Gamma(\alpha)\cdot\beta} = \frac{\alpha}{\beta}, \qquad \mu_2' = \frac{\Gamma(\alpha+2)}{\Gamma(\alpha)\beta^2} = \frac{(\alpha+1)\alpha}{\beta^2}$$

$$\frac{\mu_2'}{\mu_1'^2} = \frac{\alpha+1}{\alpha} = \frac{1}{\alpha} + 1 \quad \Rightarrow \quad \alpha = \frac{\mu_1'^2}{\mu_2' - \mu_1'^2}, \quad \beta = \frac{\alpha}{\mu_1'} = \frac{\mu_1'}{\mu_2' - \mu_1'^2}$$

Hence, $\hat{\alpha} = \dfrac{m_1'^2}{m_2' - m_1'^2}$ and $\hat{\beta} = \dfrac{m_1'}{m_2' - m_1'^2}$, where m_1' and m_2' are sample moments.

Example 15·44. *For the double Poisson distribution :*

$$p(x) = P(X = x) = \frac{1}{2}\cdot\frac{e^{-m_1} m_1^x}{x!} + \frac{1}{2}\cdot\frac{e^{-m_2}\cdot m_2^x}{x!}\ ;\ x = 0, 1, 2, \dots$$

show that the estimates for m_1 and m_2 by the method of moments are : $\mu_1' \pm \sqrt{\mu_2' - \mu_1' - \mu_1'^2}$.

Solution. We have

$$\mu_1' = \sum_{x=0}^{\infty} x\cdot p(x) = \frac{1}{2}\sum_{x=0}^{\infty} x\cdot\frac{e^{-m_1} m_1^x}{x!} + \frac{1}{2}\sum_{x=0}^{\infty} x\cdot\frac{e^{-m_2}\cdot m_2^x}{x!} = \frac{1}{2}m_1 + \frac{1}{2}m_2 \qquad \text{... (*)}$$

(since the first and second summations are the means of Poisson distributions with parameters m_1 and m_2, respectively).

$$\mu_2' = \sum_{x=0}^{\infty} x^2 \cdot p(x) = \frac{1}{2}\left\{\sum_{x=0}^{\infty} x^2 \cdot \left(\frac{e^{-m_1} m_1^x}{x!}\right) + \sum_{x=0}^{\infty} x^2 \cdot \left(\frac{e^{-m_2} \cdot m_2^x}{x!}\right)\right\}$$

$$= \frac{1}{2}\{(m_1^2 + m_1) + (m_2^2 + m_2)\}$$

$$\Rightarrow \qquad \mu_2' = \frac{1}{2}\{(m_1 + m_2) + (m_1^2 + m_2^2)\} \qquad \ldots (**)$$

$$= \frac{1}{2}\{2\mu_1' + m_1^2 + (2\mu_1' - m_1)^2\} \qquad \text{[Using (*)]}$$

$$= \frac{1}{2}\left(2\mu_1' + m_1^2 + 4\mu_1'^2 + m_1^2 - 4m_1\mu_1'\right)$$

$$\Rightarrow \qquad \mu_2' = \mu_1' + m_1^2 + 2\mu_1'^2 - 2\mu_1' m_1 \quad \Rightarrow \quad m_1^2 - 2m_1\mu_1' + (2\mu_1'^2 + \mu_1' - \mu_2') = 0$$

$$\therefore \qquad \hat{m}_1 = \frac{2\mu_1' \pm \sqrt{4\mu_1'^2 - 4(2\mu_1'^2 + \mu_1' - \mu_2')}}{2} = \mu_1' \pm \sqrt{\mu_2' - \mu_1' - \mu_1'^2}$$

Similarly, on substituting for m_1 in terms of m_2 from (*) in (**), we get

$$m_2^2 - 2m_2\mu_1' + (2\mu_1'^2 + \mu_1' - \mu_2') = 0$$

Solving for m_2, we get $\quad \hat{m}_2 = \mu_1' \pm \sqrt{\mu_2' - \mu_1' - \mu_1'^2}$

Example 17·45. *A random variable X takes the values, 0, 1, 2, with respective probabilities* $\dfrac{\theta}{4N} + \dfrac{1}{2}\left(1 - \dfrac{\theta}{N}\right)$, $\dfrac{\theta}{2N} + \dfrac{\alpha}{2}\left(1 - \dfrac{\theta}{N}\right)$ *and* $\dfrac{\theta}{4N} + \dfrac{1-\alpha}{2}\left(1 - \dfrac{\theta}{N}\right)$, *where N is a known number and* α, θ *are unknown parameters. If 75 independent observations on X yielded the values 0, 1, 2 with frequencies 27, 38, 10 respectively, estimate* θ *and* α *by the method of moments.*

Solution.

$$E(X) = 0 \cdot \left\{\frac{\theta}{4N} + \frac{1}{2}\left(1 - \frac{\theta}{N}\right)\right\} + 1 \cdot \left\{\frac{\theta}{2N} + \frac{\alpha}{2}\left(1 - \frac{\theta}{N}\right)\right\} + 2\left\{\frac{\theta}{4N} + \frac{1-\alpha}{2}\left(1 - \frac{\theta}{N}\right)\right\}$$

$$= \frac{\theta}{N} + \left(1 - \frac{\theta}{N}\right)\left[\frac{\alpha}{2} + (1 - \alpha)\right] \qquad \text{[On simplification]}$$

$$\Rightarrow \qquad \mu_1' = \frac{\theta}{N} + \left(1 - \frac{\theta}{N}\right)\left(1 - \frac{\alpha}{2}\right) = 1 - \frac{\alpha}{2}\left(1 - \frac{\theta}{N}\right) \qquad \ldots (*)$$

$$E(X^2) = 1^2 \cdot \left\{\frac{\theta}{2N} + \frac{\alpha}{2}\left(1 - \frac{\theta}{N}\right)\right\} + 2^2 \cdot \left\{\frac{\theta}{4N} + \frac{1-\alpha}{2}\left(1 - \frac{\theta}{N}\right)\right\}$$

$$= \frac{3\theta}{2N} + \left(1 - \frac{\theta}{N}\right)\left[\frac{\alpha}{2} + 2(1 - \alpha)\right] = \frac{3\theta}{2N} + \left(1 - \frac{\theta}{N}\right)\left(2 - \frac{3\alpha}{2}\right)$$

$$\mu_2' = 2 - \frac{\theta}{2N} - \frac{3}{2}\alpha\left(1 - \frac{\theta}{N}\right) \qquad \ldots (**)$$

The sample frequency distribution is :

x	0	1	2
f	27	38	10

$$\mu_2' = \frac{1}{N}\Sigma fx = \frac{1}{75}(38 + 20) = \frac{58}{75}, \qquad \mu_2' = \frac{1}{N}\Sigma fx^2 = \frac{1}{75}(38 + 40) = \frac{78}{75}$$

Equating the sample moments to theoretical moments, we get

$$1-\frac{\alpha}{2}\left(1-\frac{\theta}{N}\right)=\frac{58}{75} \Rightarrow \frac{\alpha}{2}\left(1-\frac{\theta}{N}\right)=1-\frac{58}{75}=\frac{17}{75}. \quad \ldots (***)$$

Substituting in (**), we get $\quad 2-\frac{\theta}{2N}-3\times\frac{17}{75}=\frac{78}{75} \quad \Rightarrow \quad \hat{\theta}=\frac{42}{75}N$

Substituting in (***), we get $\quad \frac{\alpha}{2}\left(1-\frac{42}{75}\right)=\frac{17}{75} \quad \Rightarrow \quad \hat{\alpha}=\frac{34}{33}$

17·6·4. Method of Least Squares.

The principle of least squares is used to fit a curve of the form :

$$y=f(x, a_0, a_1, \ldots, a_k) \quad \ldots (17\cdot62)$$

where a_i's are unknown parameters, to a set of n sample observations (x_i, y_i); $i = 1, 2, \ldots, n$ from a bivariate population. It consists in minimising the sum of squares of residuals, *viz.*,

$$E = \Sigma \; \{y_i - f(x_i, a_0, a_1, \ldots, a_k)\}^2 \quad \ldots (17\cdot63)$$

subject to variations in $a_0, a_1, \ldots, a_k$.

The normal equations for estimating $a_0, a_1, \ldots, a_k$ are given by :

$$\frac{\partial E}{\partial a_i}=0 \; ; \; i = 0, 1, 2, \ldots, k \quad \ldots (17\cdot64)$$

Remarks:

1. In chapter 11, we have discussed in detail the method of least squares for fitting linear regression, polynomial regression and the exponential family of curves reducible to linear regression. In chapter 12, we have discussed the method of fitting multiple linear regression.
2. If we are estimating $f(x, a_0, a_1, \ldots, a_k)$ as a linear function of the parameters $a_0, a_1, \ldots, a_k$, the x's being known given values, the least square estimators obtained as linear functions of the y's will be MVU estimators.

17·6·5. Method of Minimum Chi-Square

In order that χ^2-test for compatibility of observed and theoretical frequencies or of 'goodness of fit', may be applicable, it is clear from the proof of Theorem 15·3 and further discussion of the same by Prof. R. A. Fisher that the following conditions must be satisfied.

Conditions (i) to (iv): [See Remark 2, to Theorem 15·3]

(v) In case parameter values have to be estimated from the data, the method of estimation should be such as to minimize χ^2. This is, in large samples, equivalent to choosing *maximum likelihood estimates.*

If s is the number of parameters (estimated from the data) contained in p_i *i.e.*, $p_i = p_i(\theta_1, \theta_2, \ldots, \theta_s)$ and $\hat{m}_i = E(n_i) = n\hat{p}_i \; ; i = 1, 2, \ldots, k$, then

$$\chi^2 = \sum_{i=1}^{k} \frac{(n_i-\hat{m}_i)^2}{\hat{m}_i} \sim \chi^2_{k-1-s} \quad (17\cdot65)$$

The likelihood function L is given by :

$$L=\frac{n!}{n_1!n_2!\ldots n_k!}\,p_1^{n_1}p_2^{n_2}\ldots p_k^{n_k}\;;\;\sum_i n_i = n,\;\; \sum_i p_i = 1 \;\text{ and }\; p_i = p_i(\theta_1, \theta_2, \ldots, \theta_s)$$

$$\Rightarrow \quad \log L = \log C + \sum_{i=1}^{k} n_i \log p_i,$$

where C is a constant independent of p_i' s. The likelihood equations are given by:

$$\frac{\partial \log L}{\partial \theta_j} = \sum_{i=1}^{k}\left[\frac{n_i}{p_i}\left(\frac{\partial p_i}{\partial \theta_j}\right)\right] = 0\ ;\ j = 1, 2, \ldots, s \qquad \ldots (17\cdot65a)$$

Solving these 's' equations simultaneously, we obtain the MLE's $\mathfrak{R}_1, \hat{}_2, \ldots, \hat{}_s$ of $\theta_1, \theta_2, \ldots, \theta_s$ respectively.

Now, we obtain the Minimum Chi-square estimates of $\theta_j ; j = 1, 2, \ldots, s$.

$$\chi^2 = \sum_{i=1}^{k}\frac{(n_i - \hat{m}_i)^2}{\hat{m}_i} = \sum_{i=1}^{k}\frac{(n_i - np_i)^2}{np_i};\ p_i = p_i(\theta_1, \theta_2, \ldots, \theta_s)$$

$$\frac{\partial \chi^2}{\partial \theta j} = \sum_{i=1}^{k}\left[\frac{-2n(n_i - np_i)}{np_i}\cdot\frac{\partial p_i}{\partial \theta_j} - \frac{(n_i - np_i)^2}{np_i^2}\cdot\frac{\partial p_i}{\partial \theta_j}\right]$$

$$\Rightarrow \qquad -\frac{1}{2}\cdot\frac{\partial \chi^2}{\partial \theta_j} = \sum_{i=1}^{k}\left[\left\{\frac{n_i - np_i}{p_i} + \frac{(n_i - np_i)^2}{2np_i^2}\right\}\cdot\frac{\partial p_i}{\partial \theta j}\right]$$

The second term in the bracket { }, is of order $(1/n)$, *i.e.*, $O(n^{-1})$ tends to zero for large values of n.

$$\therefore \qquad -\frac{1}{2}\cdot\frac{\partial \chi^2}{\partial \theta_j} = \sum_{i=1}^{k}\left[\frac{n_i - np_i}{p_i}\right]\cdot\frac{\partial p_i}{\partial \theta_j}, \text{for large } n.$$

$$= \sum_{i=1}^{k}\left[\left(\frac{n_i}{p_i}\right) - n\right]\cdot\frac{\partial p_i}{\partial \theta_j}$$

$$= \sum_{i=1}^{k}\frac{n_i}{p_i}\cdot\frac{\partial p_i}{\partial \theta_j} - n\sum_{i=1}^{k}\frac{\partial p_i}{\partial \theta_j}$$

Since $\qquad \sum_{i=1}^{k} p_i = 1, \quad \sum_{i=1}^{k}\frac{\partial p_i}{\partial \theta_j} = \frac{\partial}{\partial \theta_j}\left(\sum_{i=1}^{k} p_i\right) = \frac{\partial (1)}{\partial \theta_j} = 0.$

$$\therefore \qquad -\frac{1}{2}\cdot\frac{\partial \chi^2}{\partial \theta_j} = \sum_{i=1}^{k}\left[\frac{n_i}{p_i}\cdot\frac{\partial p_i}{\partial \theta_j}\right], \qquad \text{for large } n.$$

For minimum value of χ^2, we have

$$\frac{\partial \chi^2}{\partial \theta_j} = 0 \quad \Rightarrow \quad \sum_{i=1}^{k}\left[\frac{n_i}{p_i}\cdot\frac{\partial p_i}{\partial \theta_j}\right] = 0, (\text{for large } n);\ j = 1, 2, \ldots, s \qquad \ldots (17\cdot65b)$$

Solving these s-equations, we obtain the minimum χ^2 estimates $(\hat{\theta}_1, \hat{\theta}_2, \ldots, \hat{\theta}_s)$ of $(\theta_1, \theta_2, \ldots, \theta_s)$. From (17·65a) and (17·65b), we conclude that for large samples, *Maximum likelihood Estimates and Minimum Chi-square Estimates are equal.*

17·7. CONFIDENCE INTERVAL AND CONFIDENCE LIMITS

Let x_i, $(i = 1, 2, \ldots, n)$ be a random sample of n observations from a population involving a single unknown parameter θ, (say). Let $f(x, \theta)$ be the probability function of the parent distribution from which the sample is drawn and let us suppose that this distribution is continuous. Let $t = t(x_1, x_2, \ldots, x_n)$, *a* function of the sample values be an estimate of the population parameter θ, with the sampling distribution given by $g(t, \theta)$.

Having obtained the value of the statistic t from a given sample, the problem is, "Can we make some reasonable probability statements about the unknown parameter θ in the population, from which the sample has been drawn?" This question is very well answered by the technique of *Confidence interval* due to Neyman and is obtained below :

We choose once for all some small value of α (5% or 1%) and then determine two constants say, c_1 and c_2 such that :

$$P(c_1 < \theta < c_2 \mid t) = 1 - \alpha \qquad \ldots (17{\cdot}66)$$

The quantities c_1 and c_2, so determined, are known as the *confidence limits* or *fiducial limits* and the interval $[c_1, c_2]$ within which the unknown value of the population parameter is expected to lie, is called the *confidence interval* and $(1 - \alpha)$ is called the *confidence coefficient.*

Thus if we take $\alpha = 0{\cdot}05$ (or $0{\cdot}01$), we shall get 95% (or 99%) confidence limits.

How to find c_1 and c_2? Let T_1 and T_2 be two statistics such that

$$P(T_1 > \theta) = \alpha_1 \qquad \ldots (17{\cdot}67)$$

and

$$P(T_2 < \theta) = \alpha_2 \qquad \ldots (17{\cdot}67a)$$

where α_1 and α_2 are constants independent of θ. (17·67) and (17·67*a*) can be combined to give:

$$P(T_1 < \theta < T_2) = 1 - \alpha, \qquad \ldots (17{\cdot}67b)$$

where $\alpha = \alpha_1 + \alpha_2$. Statistics T_1 and T_2 defined in (17·67) and (17·67*a*) may be taken as c_1 and c_2 defined in (17·66).

For example, if we take a large sample from a normal population with mean μ and standard deviation σ, then

$$Z = \frac{\bar{x}-\mu}{\sigma/\sqrt{n}} \sim N(0, 1)$$

and

$$P(-1{\cdot}96 \le Z \le 1{\cdot}96) = 0{\cdot}95 \qquad \text{(From Normal Probability Tables)}$$

$$\Rightarrow P\left(-1{\cdot}96 \le \frac{\bar{x}-\mu}{\sigma/\sqrt{n}} \le 1{\cdot}96\right) = 0{\cdot}95 \quad \Rightarrow \quad P\left(\bar{x} - 1{\cdot}96\frac{\sigma}{\sqrt{n}} \le \mu \le \bar{x} + 1{\cdot}96\frac{\sigma}{\sqrt{n}}\right) = 0{\cdot}95$$

Thus, $\left(\bar{x} \pm 1{\cdot}96\frac{\sigma}{\sqrt{n}}\right)$ are 95% confidence limits for the unknown parameter μ, the population mean and the interval $\left(\bar{x} - 1\cdot 96\frac{\sigma}{\sqrt{n}}, \bar{x} + 1\cdot 96\frac{\sigma}{\sqrt{n}}\right)$ is called the 95% confidence interval for μ.

Also, $P(-2{\cdot}58 \le Z \le 2{\cdot}58) = 0{\cdot}99$ or $P\left(-2{\cdot}58 \le \frac{\bar{x}-\mu}{\sigma/\sqrt{n}} \le 2{\cdot}58\right) = 0{\cdot}99$

$$\Rightarrow \quad P\left(\bar{x} - 2{\cdot}58\frac{\sigma}{\sqrt{n}} \le \mu \le \bar{x} + 2{\cdot}58\frac{\sigma}{\sqrt{n}}\right) = 0{\cdot}99$$

Hence, 99% confidence limits for μ are : $\bar{x} \pm 2{\cdot}58\frac{\sigma}{\sqrt{n}}$

and 99% confidence interval for μ is: $\left(\bar{x} - 2{\cdot}58\frac{\sigma}{\sqrt{n}}, \quad \bar{x} + 2{\cdot}58\frac{\sigma}{\sqrt{n}}\right).$

Remarks:

1. Usually σ^2 is not known and its unbiased estimate S^2 obtained from the samples, is used. However if n is *small*, $Z = \dfrac{\bar{x}-\mu}{S/\sqrt{n}}$ is not N (0, 1) and in this case the confidence limits and confidence intervals for μ are obtained by using Student's 't' distribution.
2. It can be seen that in many cases there exist more than one set of confidence intervals with the same confidence coefficient. Then the problem arises as to which particular set is to be regarded as better than the others in some useful sense and in such cases we look for the shortest of all the intervals.

Example 17·46. *Obtain 100 (1 – α)% confidence intervals for the parameters (a) θ and (b) σ², of the normal distribution :*

$$f(x, \theta; \sigma) = \frac{1}{\sigma\sqrt{2\pi}} \, exp \left\{-\frac{1}{2}\left(\frac{x-\theta}{\sigma}\right)^2\right\}, -\infty < x < \infty$$

Solution. Let X_i, $(i = 1, 2, \ldots, n)$ be a random sample of size n from the density $f(x ; \theta, \sigma)$ and let :

$$\overline{X} = \frac{1}{n}\sum_{i=1}^{n} X_i, \quad s^2 = \frac{1}{n}\sum_{i=1}^{n}(X_i - \overline{X})^2, \quad S^2 = \frac{1}{n-1}\sum_{i=1}^{n}(X_i - \overline{X})^2,$$

(*a*) The statistic $t = \dfrac{\overline{X} - \theta}{S/\sqrt{n}}$ follows student's t-distribution with $(n - 1)$ degrees of freedom. Hence 100(1 – α)% confidence limits for θ are given by :

$$P(|t| \le t_\alpha) = 1 - \alpha \quad \Rightarrow \quad P\left(\left|\overline{X} - \theta\right| \le \frac{S}{\sqrt{n}} t_\alpha\right) = 1 - \alpha$$

$$\therefore \quad P\left(\overline{X} - t_\alpha \cdot \frac{S}{\sqrt{n}} \le \theta \le \overline{X} + t_\alpha \cdot \frac{S}{\sqrt{n}}\right) = 1 - \alpha \qquad \ldots (17{\cdot}68)$$

where t_α is the tabulated value of t for $(n - 1)$ *d.f.* at significance level 'α'. Hence, the required confidence interval for θ is : $\left(\overline{X} - t_\alpha \dfrac{S}{\sqrt{n}}, \overline{X} + t_\alpha \dfrac{S}{\sqrt{n}}\right)$

(*b*) *Case (i) θ is known and equal to μ (say).*

Then $\dfrac{\Sigma(X_i - \mu)^2}{\sigma^2} = \dfrac{ns^2}{\sigma^2} \sim \chi^2_n$

If we define χ^2_α as the value of χ^2 such that : $P(\chi^2 > \chi_\alpha^2) = \displaystyle\int_{\chi^2_\alpha}^{\infty} p(\chi^2)d\chi^2 = \alpha$ ... (*)

where $p(\chi^2)$ is the *p.d.f.* of χ^2-distribution with n *d.f.,* then the required confidence interval is given by:

$$P\left\{\chi^2_{1-(\alpha/2)} \le \chi^2 \le \chi^2_{\alpha/2}\right\} = 1 - \alpha \quad \Rightarrow \quad P\left\{\chi^2_{1-(\alpha/2)} \le \frac{ns^2}{\sigma^2} \le \chi^2_{\alpha/2}\right\} = 1 - \alpha \qquad \ldots (^{**})$$

$$\frac{ns^2}{\sigma^2} \le \chi^2_{\alpha/2} \quad \Rightarrow \quad \frac{ns^2}{\chi^2_{\alpha/2}} \le \sigma^2 \quad \text{and} \quad \chi^2_{1-(\alpha/2)} \le \frac{ns^2}{\sigma^2} \quad \Rightarrow \quad \sigma^2 \le \frac{ns^2}{\chi^2_{1-(\alpha/2)}}$$

Hence, (**) gives :

$$P\left\{\frac{ns^2}{\chi^2_{\alpha/2}} \le \sigma^2 \le \frac{ns^2}{\chi^2_{1-(\alpha/2)}}\right\} = 1 - \alpha, \qquad \ldots (^{***})$$

where, $\chi^2_{\alpha/2}$ and $\chi^2_{1-(\alpha/2)}$ are obtained from (*) by using n *d.f.*

Thus, *e.g.*, 95% confidence interval for σ^2 is : $P\left(\dfrac{ns^2}{\chi^2_{0\cdot025}} \le \sigma^2 \le \dfrac{ns^2}{\chi^2_{0\cdot975}}\right) = 0{\cdot}95 \qquad (\because \alpha = 0{\cdot}05)$

Case (*ii*). θ is unknown. In this case the statistic : $\dfrac{\Sigma(X_i - \overline{X})^2}{\sigma^2} = \dfrac{ns^2}{\sigma^2} \sim \chi^2_{(n-1)}$.

Here also confidence interval for σ^2 is given by (***), where now χ^2_α is the significant value of χ^2 [as defined in (*)] for $(n - 1)$ *d.f.* at the significance level 'α'.

Example 17·47. *Show that the largest observations L of a sample of n observations from a rectangular distribution with density function :*

$$f(x,\theta)=\begin{cases}\dfrac{1}{\theta}, & 0\le x\le\theta\\ 0, & otherwise\end{cases} \qquad \ldots (*)$$

has the distribution : $dG(L)=n\left(\dfrac{L}{\theta}\right)^{n-1}\cdot\dfrac{dL}{\theta}, 0\le L\le\theta$

Show that the distribution of $V=L/\theta$ *is given by p.d.f. :* $h(v)=nv^{n-1}, 0\le v\le 1$

Deduce that the confidence limits for θ *corresponding to confidence coefficient* α *are:*

$$L \text{ and } \frac{L}{(1-\alpha)^{1/n}}.$$

Solution. Let $X_1, X_2, \ldots, X_n$ be a random sample of size n from the population (*) and let $L=\max(X_1, X_2, \ldots, X_n)$. The distribution of L is given by :

$dG(L)=n[F(L)]^{n-1}\cdot f(L)\,dL$, where $F(.)$ is the distribution function of X given by :

$$F(L)=\int_0^L f(x,\theta)dx=\frac{L}{\theta} \qquad \therefore \qquad dG(L)=n\left(\frac{L}{\theta}\right)^{n-1}\cdot\frac{dL}{\theta}, 0\le L\le\theta$$

If we take $V=L/\theta$, the Jacobian of transformation is θ. Hence *p.d.f.* $h(.)$ of V is :

$$h(v)=nv^{n-1}\cdot\frac{1}{\theta}\,|\,J\,|=n\,v^{n-1}, 0\le v\le 1,$$

which is independent of θ.

To obtain the confidence limits for θ, with confidence coefficient α, let us define v_α such that

$$P(v_\alpha<V<1)=\alpha \quad\Rightarrow\quad \int_{v_\alpha}^1 h(v)\,dv=\alpha \qquad \ldots (**)$$

$$\Rightarrow \quad n\int_{v_\alpha}^1 v^{n-1}dv=\alpha \quad\Rightarrow\quad 1-v_\alpha^n=\alpha \quad\Rightarrow\quad v_\alpha=(1-\alpha)^{1/n} \qquad \ldots (***)$$

From (**) and (***), $P\{(1-\alpha)^{1/n}<V<1\}=\alpha \quad\Rightarrow\quad P\left\{(1-\alpha)^{1/n}<\dfrac{L}{\theta}<1\right\}=\alpha$

$$\therefore \qquad P\left\{L<\theta<\frac{L}{(1-\alpha)^{1/n}}\right\}=\alpha$$

Hence, the required confidence limits for θ are L and $L/(1-\alpha)^{1/n}$.

Example 17·48. *Given a random sample from a population with p.d.f. :*

$$f(x,\theta)=\frac{1}{\theta}, 0\le x\le\theta$$

show that 100 $(1-\alpha)$% *confidence interval for* θ *is given by R and* R/ψ*, where* ψ *is given by*

$$\psi^{n-1}[n-(n-1)\psi]=\alpha, \text{ and } R \text{ is the sample range.}$$

Solution. The joint *p.d.f.* of $x_1, x_2, \ldots, x_n$ is given by : $L=\dfrac{1}{\theta^n}, 0\le x_i\le\theta$.

If $x_{(1)}, x_{(2)}, \ldots, x_{(n)}$ is the ordered sample then the joint *p.d.f.* of $x_{(n)}$ and $x_{(1)}$ is :

$$g[x_{(1)}, x_{(n)}]=\frac{n(n-1)}{\theta^n}[x_{(n)}-x_{(1)}]^{n-2}, 0\le x_{(1)}\le x_{(n)}\le\theta.$$

To obtain the distribution of the sample range R, let us make the transformation of variables :

$$R = x_{(n)} - x_{(1)} \text{ and } v = x_{(1)} \quad \Rightarrow \quad v = x_{(n)} - R \le \theta - R.$$

The Jacobian of transformation is $|J| = 1$ and the joint *p.d.f.* of R and V becomes :

$$h(R, v) = \frac{n(n-1)}{\theta^n} R^{n-2}, \quad 0 < v < \theta - R$$

The marginal density of R is given by :

$$h_1(R) = \int_0^{\theta - R} \frac{n(n-1)}{\theta^n} \cdot R^{n-2} dv = \frac{n(n-1)R^{n-2}(\theta - R)}{\theta^n}, \quad 0 \le R \le \theta$$

The *p.d.f.* $h_2(.)$ of $U = R/\theta$ is :

$$h_2(u) = h_1(R)\left|\frac{dR}{du}\right| = \frac{n(n-1)R^{n-2}(\theta - R)}{\theta^n} \cdot \theta = n(n-1)\,u^{n-2}\,(1-u), 0 \le u \le 1$$

100 $(1-\alpha)$% confidence interval for θ is given by : $P(\psi \le U \le 1) = 1 - \alpha$ ... (*)

where ψ is obtained from the equation $\displaystyle\int_0^{\psi} h_2(u)du = \alpha$

$$\Rightarrow \quad n(n-1)\int_0^{\psi} u^{n-2}(1-u)du = \alpha \quad \Rightarrow \quad \left| nu^{n-1} - (n-1)u^n \right|_0^{\psi} = \alpha$$

$$\therefore \quad \psi^{n-1}\{n - (n-1)\psi\} = \alpha \qquad \text{... (**)}$$

From (*), we get $P\left(\psi \le \dfrac{R}{\theta} \le 1\right) = 1 - \alpha \quad \Rightarrow \quad P\left(R \le \theta \le \dfrac{R}{\psi}\right) = 1 - \alpha$

Hence, the required limits for θ are given by R and R/ψ where ψ is given by (**).

Example 17·49. *Given one observation from a population with p.d.f. :*

$$f(x, \theta) = \frac{2}{\theta^2}(\theta - x), \quad 0 \le x \le \theta,$$

obtain 100 $(1-\alpha)$% confidence interval for θ.

Solution. The *p.d.f* of $u = x/\theta$ is given by :

$$g(u) = f(x, \theta) \cdot \left|\frac{dx}{du}\right| = \frac{2}{\theta^2}(\theta - x) \cdot \theta = 2(1-u), \quad 0 \le u \le 1$$

To obtain 100 $(1-\alpha)$% confidence interval for θ, we choose two quantities u_1 and u_2 such that

$$\left.\begin{aligned} &P(u_1 \le u \le u_2) = 1 - \alpha \\ \text{and} \quad &P(u < u_1) = P(u > u_2) = \tfrac{1}{2}\alpha \end{aligned}\right\} \qquad \text{... (*)}$$

$$P(u < u_1) = \frac{\alpha}{2} \quad \Rightarrow \quad \int_0^{u_1} 2(1-u)\,du = \frac{\alpha}{2} \quad \Rightarrow \quad u_1^2 - 2u_1 + \frac{\alpha}{2} = 0 \qquad \text{... (**)}$$

and $$P(u > u_2) = \frac{1}{2}\alpha \quad \Rightarrow \quad \int_{u_2}^{1} 2(1-u)\,du = \frac{\alpha}{2} \quad \Rightarrow \quad u_2^2 - 2u_2 + \left(1 - \frac{\alpha}{2}\right) = 0 \qquad \text{... (***)}$$

From (*), we get $P\left(u_1 \le \dfrac{x}{\theta} \le u_2\right) = 1 - \alpha \quad \Rightarrow \quad \left(\dfrac{x}{u_2} \le \theta \le \dfrac{x}{u_1}\right) = 1 - \alpha$

Hence, the required interval for θ is $\left(\dfrac{x}{u_2}, \dfrac{x}{u_1}\right)$, where u_1 and u_2 are given by (**) and (***) respectively.

17·7·1. Confidence Intervals for Large Samples

It has been proved that under certain regularity conditions, the first derivative of the logarithm of the likelihood function w.r.to parameter θ *viz.*, $\frac{\partial}{\partial\theta}\log L$, is asymptotically normal with mean zero and variance given by :

$$\text{Var}\left(\frac{\partial}{\partial\theta}\log L\right) = E\left(\frac{\partial}{\partial\theta}\log L\right)^2 = E\left(-\frac{\partial^2}{\partial\theta^2}\log L\right)$$

Hence, for large n,
$$Z = \frac{\frac{\partial}{\partial\theta}\log L}{\sqrt{\text{Var}\left(\frac{\partial}{\partial\theta}\log L\right)}} \sim N(0,1). \qquad \ldots (17\cdot69)$$

The result enables us to obtain confidence interval for the parameter θ in large samples. Thus, for large samples, the confidence interval for θ with confidence coefficient $(1-\alpha)$ is obtained by converting the inequalities in

$$P(|Z| \le \lambda_\alpha) = 1-\alpha \qquad \ldots (17\cdot70)$$

where λ_α is given by :
$$\frac{1}{\sqrt{2\pi}}\int_{-\lambda_\alpha}^{\lambda_\alpha} \exp(-u^2/2)\, du = 1-\alpha \qquad \ldots[17\cdot70a]$$

Example 17·50. *Obtain 100 $(1-\alpha)$% confidence limits (for large samples) for the parameter λ of the Poisson distribution :*

$$f(x, \lambda) = \frac{e^{-\lambda}\cdot\lambda^x}{x!}\ ; x = 0, 1, 2, \ldots$$

Solution. We have

$$\frac{\partial}{\partial\lambda}\log L = \frac{\partial}{\partial\lambda}\left\{-n\lambda + \left(\sum_{i=1}^n x_i\right)\log\lambda - \sum_{i=1}^n \log(x_i)\right\} = -n + \frac{\Sigma x_i}{\lambda} = n\left(\frac{\bar{x}}{\lambda} - 1\right)$$

$$\text{Var}\left(\frac{\partial}{\partial\lambda}\log L\right) = E\left(-\frac{\partial^2}{\partial\lambda^2}\log L\right) = E\left(\frac{n\bar{x}}{\lambda^2}\right) = \frac{n}{\lambda^2}E(\bar{x}) = \frac{n}{\lambda} \qquad [\because E(\bar{x}) = \lambda]$$

$$Z = \frac{n\left(\frac{\bar{x}}{\lambda} - 1\right)}{\sqrt{n/\lambda}} = \sqrt{(n/\lambda)}\,(\bar{x}-\lambda) \sim N(0,1) \qquad \text{[Using (17·69)]}$$

Thus, 100 $(1-\alpha)$% confidence interval for λ is given by (for large samples)

$$P\left\{\left|\sqrt{(n/\lambda)}\,(\bar{x}-\lambda)\right| \le \lambda_\alpha\right\} = 1-\alpha$$

Thus, the required limits for λ are the roots of the equation :

$$\left|\sqrt{n/\lambda}\,(\bar{x}-\lambda)\right| = \lambda_\alpha \quad \Rightarrow \quad n(\bar{x}-\lambda)^2 - \lambda\cdot\lambda_\alpha^2 = 0$$

$$\Rightarrow \quad \lambda^2 - \lambda\left(2\bar{x} + \frac{\lambda_\alpha^2}{n}\right) + \bar{x}^2 = 0 \quad \Rightarrow \quad \lambda = \frac{\left(2\bar{x} + \frac{\lambda_\alpha^2}{n}\right) \pm \left\{\left(2\bar{x} + \frac{\lambda_\alpha^2}{n}\right)^2 - 4\bar{x}^2\right\}^{1/2}}{2} \qquad \ldots (*)$$

For example, 95% confidence interval for λ is given by taking $\lambda_\alpha = 1\cdot96$ in (*), thus giving :

$$\lambda = \frac{1}{2}\left(2\bar{x} + \frac{3\cdot84}{n}\right) \pm \left(\frac{3\cdot84\bar{x}}{n} + \frac{3\cdot69}{n^2}\right)^{1/2} = \bar{x} \pm 1\cdot96\sqrt{\bar{x}/n}\text{, to the order } n^{-1/2}.$$

Example 17·51. *Show that for the distribution : $dF(x) = \theta \exp(-x\theta)$; $0 < x < \infty$, central confidence limits for large samples with 95% confidence coefficient are given by*

$$\theta = \left(1 \pm \frac{1\cdot 96}{\sqrt{n}}\right) \Big/ \bar{x}.$$

Solution. Here, $L = \theta^n \exp\left(-\theta \sum_{i=1}^{n} x_i\right)$

$$\frac{\partial}{\partial\theta}\log L = \frac{\partial}{\partial\theta}(n\log\theta - \theta\Sigma x_i) = \frac{n}{\theta} - \sum_{i=1}^{n} x_i = n\left(\frac{1}{\theta} - \bar{x}\right)$$

and $$\frac{\partial^2}{\partial\theta^2}\log L = -\frac{n}{\theta^2} \quad \Rightarrow \quad \text{Var}\left(\frac{\partial}{\partial\theta}\log L\right) = E\left(-\frac{\partial^2}{\partial\theta^2}\log L\right) = \frac{n}{\theta^2}$$

Hence, for large samples, using (17·69), we have

$$Z = \frac{n\left(\frac{1}{\theta} - \bar{x}\right)}{\sqrt{n/\theta^2}} \sim N(0,1) \quad \Rightarrow \quad \sqrt{n}(1 - \theta\bar{x}) \sim N(0,1)$$

Hence, 95% confidence limits for θ are given by :

$$P\left[-1\cdot 96 \le \sqrt{n}\ (1 - \theta\ \bar{x}) \le 1\cdot 96\right] = 0\cdot 95 \qquad \ldots (*)$$

$$\sqrt{n}\ (1 - \theta\ \bar{x}) \le 1\cdot 96 \quad \Rightarrow \quad \left(1 - \frac{1\cdot 96}{\sqrt{n}}\right)\frac{1}{\bar{x}} \le \theta \qquad \ldots (**)$$

and $$-1\cdot 96 \le \sqrt{n}\ (1 - \theta\bar{x}) \quad \Rightarrow \quad \theta \le \left(1 + \frac{1\cdot 96}{\sqrt{n}}\right)\frac{1}{\bar{x}} \qquad \ldots (***)$$

Hence, from (*), (**) and (***), the central 95% confidence limits for θ are given by:

$$\theta = \left(1 \pm \frac{1\cdot 96}{\sqrt{n}}\right) \cdot \frac{1}{\bar{x}}.$$

CHAPTER CONCEPTS QUIZ

1. Comment on the following statements :

(*i*) In the case of Poisson distribution with parameter λ, $\bar{x}$ is sufficient for λ.

(*ii*) If $(X_1, X_2, \ldots X_n)$ be a sample of independent observation from the uniform distribution on $(\theta, \theta + 1)$, then the maximum likelihood estimator of θ is unique.

(*iii*) A maximum likelihood estimator is always unbiased.

(*iv*) Unbiased estimator is necessarily consistent

(*v*) A consistent estimator is also unbiased.

(*vi*) An unbiased estimator whose variance tends to zero as the sample size increases is consistent.

(*vii*) If t is a sufficient statistic for θ then $f(t)$ is a sufficient statistic for $f(\theta)$.

(*viii*) If t_1 and t_2 are two independent estimators of θ, then $t_1 + t_2$ is less efficient than both t_1 and t_2.

(*ix*) If T is consistent estimator of a parameter θ, then $aT + b$ is a consistent estimator of $a\theta + b$, where a and b are constants.

(*x*) If x is the number of successes in n independent trials with a constant probability p of success in each trial, then x/n is a consistent estimator of p.

2. Fill in the blanks :

(*i*) In a random sample of size n from a population with mean μ, the sample mean ($\overline{x}$) is ... estimate of ...

(*ii*) The sample median is ... estimate for the mean of normal population.

(*iii*) An estimator $\hat{\theta}$ of a parameter θ is said to be unbiased if...

(*iv*) The variance s^2 of a sample of size n is a ... estimator of population variance σ^2.

(*v*) If a sufficient estimator exists, it is a function of the ... estimator.

(*vi*) ...estimate may not be unique.

3. (*a*) Give example of a statistic t which is unbiased for a parameter θ but t^2 is not unbiased for θ^2.

(*b*) Give example of an M.L. estimator which is not unbiased.

4. (*i*) If $\overline{x}$ is an unbiased estimator for the population mean μ, state which of the following are unbiased estimators for μ^2 :

$$(a)\ \overline{x}^2,\ (b)\ \overline{x}^2 - \frac{\sigma^2}{n}\ (\sigma^2 \text{ is known/unknown}).$$

(*ii*) If t is the maximum likelihood estimator for θ, state the condition under which $f(t)$ will be the maximum likelihood estimator for $f(\theta)$.

(*iii*) Write down the condition for the Cramer-Rao lower bound for the variance of the estimator to be attained.

(*iv*) Write down the general form of the distribution admitting sufficient statistic.

5. A random variable X takes the values 1, 2, 3 and 4, each with probability $\frac{1}{4}$. A random sample of three values of x is taken, $\overline{x}$ is the mean and m is the median of this sample. Show that both $\overline{x}$ and m are unbiased estimators of the mean of the population, but $\overline{x}$ is more efficient than m. Compare their efficiencies.

6. Give an example of estimates which are (*i*) Unbiased and efficient, (*ii*) Unbiased and inefficient.

7. Mark the correct alternative :

(*i*) Let T_n be an estimator, based on a sample $x_1, x_2, ..., x_n$, of the parameter θ. Then T is a consistent estimator of θ if

(*a*) $P(T_n - \theta > \varepsilon) = 0\ \forall\ \varepsilon > 0$, (*b*) $P(|T_n - \theta| < \varepsilon) = 0$,

(*c*) $\lim_{n\to\infty} P(|T_n - \theta| > \varepsilon) = 0\ \forall\ \varepsilon > 0$, (*d*) $\lim_{n\to\infty} P(T_n - \theta > \varepsilon) = 0\ \forall\ \varepsilon > 0$

(*ii*) Let $E(T_1) = \theta = E(T_2)$, where T_1 and T_2 are the linear functions of the sample observations. If $V(T_1) \le V(T_2)$ then :

(*a*) T_1 is an unbiased linear estimator. (*b*) T_1 is the best linear unbiased estimator.

(*c*) T_1 is a consistent linear unbiased estimator. (*d*) T_1 is a consistent best linear unbiased estimator.

(*iii*) Let X be a random variable with $E(X) = \mu$ and $V(X) = \sigma^2$. Let $\overline{X}$ be the sample mean based on a random sample of size n, then $\overline{X}$ is :

(*a*) the best linear unbiased estimator of μ; (*b*) an unbiased and consistent estimator of μ.

(*c*) an unbiased and linear estimator of μ; (*d*) the best linear consistent estimator of μ.

(*iv*) Let θ be an unknown parameter and T_1 be an unbiased estimator of θ. If $Var(T_1) \le Var(T_2)$, for T_2 to be any other unbiased estimator, then T_1 is known as :

(*a*) minimum variance unbiased estimator; (*b*) unbiased and efficient estimator.

(*c*) consistent and efficient estimator; (*d*) unbiased, consistent and minimum variance estimator.

(*v*) Which one of the following statements is false ?

(*a*) In sufficiency we reduce original random variables to a few statistics for the purpose of drawing inference about parameters.

(b) $T_1, T_2, ..., T_k$ will be called sufficient for θ if the conditional distribution of $T_1, T_2, ..., T_k$ given $X_1, X_2, ..., X_n$ is independent of θ.

(c) A statistic is said to be sufficient if and only if the joint probability density function of random variables can be expressed as product of two factors, where one factor depends on parameters(s) and observations through the statistic, while the other is independent of parameter(s).

(vi) The most important of all the methods of estimation is the method of maximum likelihood. It generally yields good estimator as judged from various criteria. Which one of the following statements about maximum likelihood estimates is not true ?

(a) Maximum likelihood estimates are consistent.
(b) If maximum likelihood estimate exists, it is most efficient in the class of estimates.
(c) Maximum likelihood estimates are sufficient.
(d) Maximum likelihood estimates are unbiased.

(vii) The maximum likelihood estimates, which are obtained by maximizing the function of joint density of random variables, are generally :

(a) unbiased and inconsistent,
(b) unbiased and consistent,
(c) consistent and invariant, and
(d) invariant and unbiased.

DISCUSSION AND REVIEW QUESTIONS

1. What do you understand by Point Estimation? Define the following terms and give one example for each :
(i) Consistent Statistic, (ii) Unbiased Statistic, (iii) Sufficient Statistic, (iv) Efficiency.
2. What do you understand by Point Estimation? When would you say that estimate of a parameter is good? In particular, discuss the requirements of consistency and unbiasedness of an estimate. Give an example to show that a consistent estimate need not be unbiased.
3. Discuss the terms (i) estimate, (ii) consistent estimate, (iii) unbiased estimate, of a parameter and show that sample mean is both consistent and unbiased estimate of the population mean.
4. Give an example of an estimator :
(i) which is consistent but not unbiased, (ii) which is unbiased but not consistent.
5. (a) State and prove a sufficient condition for the consistency of an estimator. Define the invariance property of a consistent estimator and establish it.
(b) Given a random sample $X_1, X_2, ..., X_n$ from a normal $N(\mu, \sigma^2)$ distribution, examine unbiasedness and consistency of (i) $\bar{X}$ for μ, (ii) $\frac{1}{n}\Sigma\ (X_i - \bar{X})^2$ for σ^2.
6. (a) When would you say that estimate of a parameter is good? In particular, discuss the requirements of consistency and unbiasedness of an estimate. Give an example to show that a consistent estimate need not be unbiased.
Show that an unbiased estimator whose variance tends to zero as the sample size increases to infinity is consistent.
(b) Define unbiasedness and consistency of estimators. Let $X_1, X_2, ..., X_n$ be a random sample from the $N(\mu, \sigma^2)$ distribution. Propose three estimators of μ based on this random sample such that the first is unbiased but not consistent, the second is consistent but not unbiased, and the third is both unbiased and consistent.
7. Define an unbiased and consistent estimate of a parameter in a distribution.
Prove that for a sample of size n from a normal $(m, 1)$ population, the arithmetic mean is an unbiased estimate of m and by Chebyshev's inequality or otherwise, show that the estimate is consistent too.
8. (a) Define a consistent estimator. Let $T_{1,n}$ and $T_{2,n}$ be consistent estimators of $g_1(\theta)$ and $g_2(\theta)$ respectively. Prove that $aT_{1,n} + b\,T_{2,n}$ is a consistent estimator of $ag_1(\theta) + bg_2(\theta)$, where a and b are constants independent of θ.

(b) Define consistent estimator. If the estimator t_n based on a random sample of size n is such that $E(t_n) \to \theta$ and $V(t_n) \to 0$, as $n \to \infty$, then prove that t_n is a consistent estimator for θ. Hence, prove that sample mean is always a consistent estimate for population mean.

(c) If t_n is a biased estimate of paramater θ based on a random sample of size n, and $E(t_n) = \theta + b_n$ and if $b_n \to 0$ and $V(t_n) \to 0$ as $n \to \infty$, show that t_n is consistent estimator of θ.

9. Define a consistent estimator of parameter θ. If T is a consistent estimator of θ and if ϕ is any continuous function of its argument, show that $\phi(T)$ is a consistent estimator of $\phi(\theta)$.

10. Define *MVU* estimator. If T_1 and T_2 are two unbiased estimators of a parameter θ, with variances σ_1^2 and σ_2^2 and correlation coefficient ρ, then obtain the best unbiased linear combination of T_1 and T_2. Also obtain its variance.

11. (a) Define *MVU* estimator. Show that an *MVU* estimator is unique.

(b) If T_1 and T_2 are two unbiased statistics having the same variance and ρ is the correlation between them then show that $\rho \geq 2e - 1$, where e is the ratio of the variance of the best estimator to the common variance of T_1 and T_2.

12. (a) Let T be an *MVU* estimate for $\gamma(\theta)$ and T_1, T_2 be two other unbiased estimators of $\gamma(\theta)$ with efficiencies e_1 and e_2 respectively. If ρ_θ is the correlation coefficient between T_1 and T_2, then

$$(e_1e_2)^{1/2} - \{(1-e_1)(1-e_2)\}^{1/2} \leq \rho_\theta \leq (e_1e_2)^{1/2} + \{(1-e_1)(1-e_2)\}^{1/2}.$$

(b) Let t_1 and t_2 be two unbiased estimates of θ with variances σ_1^2 and σ_2^2, (both known) and correlation ρ (known). Consider the estimate $\hat{\theta} = \alpha\, t_1 + (1 - \alpha)\, t_2$.

Show that $\hat{\theta}$ is unbiased. Find α such that $\hat{\theta}$ has minimum variance.

13. (a) Prove that if an unbiased estimator and a sufficient statistic exist for $\psi(\theta)$ and the density function $f(x, \theta)$ satisfies certain regularity conditions (to be stated by you), then the best unbiased estimate of $\psi(\theta)$ is an explicit function of the sufficient statistic.

14. Prove that under certain regularity conditions to be stated by you, the variance of an unbiased estimator T for $\gamma(\theta)$, satisfies the inequality :

$$\text{Var}_\theta(T) \geq \frac{[\gamma'(\theta)]^2}{E_\theta\left[\frac{\partial}{\partial\theta}\{\log f_\theta(X_1, X_2, ... X_n)\}\right]^2}$$

15. (a) If T is an unbiased estimator of a parameter θ, based on a random sample of size n, prove that $\text{Var}(T) \geq 1/[nI(\theta)]$, where $I(\theta)$ is the information function.

(b) Show that under certain regularity conditions, an unbiased estimate T of a parametric function $\psi(\theta)$ attains a Cramer-Rao bound for the variance of unbiased estimator of $\psi(\theta)$, if and only if T satisfies the relation : $\dfrac{\partial \log L}{\partial\theta} = \dfrac{nI(\theta)}{\psi'(\theta)}\{T - \psi(\theta)\}$, where L is the likelihood function of a sample of n observations and $nI(\theta) = E\left(\dfrac{\partial \log L}{\partial\theta}\right)^2$.

What is the variance of T in such a case? Show that an estimator T satisfying the above relation is unique when it exists. Further a parametric function admitting such an estimator T is unique except for an additive and multiplicative constant.

16. (a) Define Minimum Variance unbiased estimator and Minimum Variance Bound unbiased estimator and explain clearly the difference between them. Prove that minimum variance unbiased estimator is essentially unique.

(b) How is Cramer-Rao inequality useful in obtaining *MVUE*? Derive this inequality.

17. (a) If $b(\theta)$ is the bias in the estimator T of θ, then show that (under conditions to be stated by you),

$$E(T - \theta)^2 > \frac{\{1 + b'(\theta)\}^2}{I(\theta)} + \{b(\theta)\}^2,$$

where $I(\theta)$ is the information on θ supplied by a sample of n observations.

(*b*) Prove the following result :

$$\int_{-\infty}^{\infty}(x-\theta)^2\cdot g(x,\theta)\,dx\int_{-\infty}^{\infty}\left(\frac{\partial \log g}{\partial\theta}\right)^2 g(x,\theta)dx \ge \left(\frac{d\psi}{d\theta}\right)^2$$

where $g(x, \theta)$ is the frequency function of x having the first moment $\psi(\theta)$ and finite second moment. Discuss when the equality sign holds.

18. (*a*) Define completeness of a statistic *T*. Let $X_1, X_2, \ldots, X_n$ be a random sample from uniform population $U[0, \theta]$. Obtain sufficient statistic for θ. Show that it is complete. Hence obtain *MVU* estimator for θ.

(*b*) State and prove Rao-Blackwell theorem and explain its significance in point estimation.

19. (*a*) State and explain the principle of maximum likelihood (M.L.) for estimation of population parameter.

(*b*) (*i*) Describe the M.L. method of estimation and discuss five of its optimal properties.

(*ii*) Examine a situation when M.L. method fails and explain how you tackle such situations.

(*c*) Define the likelihood function for a random sample drawn from

(*i*) a discrete population, (*ii*) a continuous population.

(*d*) Find the likelihood function for a random sample of size n from each of the following populations :

(*i*) Normal (m, σ^2), (*ii*) Binomial (n, p), (*iii*) Poisson (μ), (*iv*) Uniform on (a, b).

20. Explain the technique of maximum likelihood estimation and give a formula for the large sample standard error of the maximum-likelihood estimator.

21. Explain the terms :

(*i*) sufficient estimator, (*ii*) efficient estimator,

(*iii*) Cramer-Rao lower bound to the variance of an estimator,

(*iv*) maximum likelihood estimator;

and describe the relations amongst these four concepts.

22. Describe the method of moments for estimating the parameters. What are the properties of the estimates obtained by this method?

23. Explain the methods of estimation : (*i*) Method of moments and (*ii*) Maximum likelihood.

Do these lead to the same estimates in respect of the standard deviation of a normal population? Examine the properties of the estimates from the point of view of consistency and unbiasedness.

24. Discuss the concept of interval estimation and provide suitable illustration.

25. (*a*) Critically examine how interval estimation differs from point estimation. Give the 95% confidence interval for the mean of the normal distribution, when its variance is known.

(*b*) What are confidence intervals ? How are they constructed using *t*-distribution ?

26. Describe the general method of constructing the confidence interval for large samples.

If $X_1, X_2, \ldots, X_n$ is a random sample from an exponential distribution with mean θ, obtain 95% confidence interval for θ when n is large.

27. Discuss the problem of interval estimation. Obtain the minimum confidence interval for the variance for a random sample of size n from a normal population with unknown mean.

28. Give a method of determining the confidence limits for a single unknown parameter, stating the conditions of validity. From amongst intervals of Confidence Coefficient α, how will you decide one as being superior to another ?

ASSORTED REVIEW PROBLEMS FOR SELF-ASSESSMENT

17·1. (*a*) If $s_1^2, s_2^2, \ldots, s_r^2$ are r sample variances based on random samples of sizes $n_1, n_2, \ldots, n_r$ respectively, and if T is some statistic given by : $T = \frac{1}{a}(n_1 s_1^2 + n_2 s_2^2 + \ldots + n_r s_r^2)$, for estimating σ^2 as an unbiased estimator, find the value, of a, supposing population is very large and for every sample of size n, $s^2 = \frac{1}{n}\Sigma(x_i - \overline{x})^2$.

Ans. $a = (n_1 + n_2 + \ldots + n_r) - r$.

(*b*) If $\overline{X}_1, \overline{X}_2, \overline{X}_3, \ldots, \overline{X}_r$ are the sample means based on samples of sizes $n_1, n_2, n_3, \ldots, n_r$ respectively, an unbiased estimator : $t = \frac{1}{k}\left(n_1 \overline{X}_1 + n_2 \overline{X}_2 + \ldots + n_r \overline{X}_r\right)$ has been defined to estimate μ. Find the value of k.

Ans. $k = n_1 + n_2 + \ldots + n_r$.

17·2. (*a*) For the geometric distribution, $f(x, \theta) = \theta(1-\theta)^{x-1}$, $(x = 1, 2, \ldots)$, $0 < \theta < 1$, obtain an unbiased estimator of $1/\theta$.

(*b*) The random variable X takes the values 1 and 0 with respective probabilities θ and $1 - \theta$. Independent observations $X_1, X_2, \ldots, X_n$ on X are available. Write $\xi = X_1 + X_2 + \ldots + X_n$. Show that $\xi(n - \xi)/n(n-1)$ is an unbiased estimate of $\theta(1-\theta)$.

17·3. Show that if T is an unbiased estimator of a parameter θ, then $\lambda_1 T + \lambda_2$ is an unbiased estimator of $\lambda_1\theta + \lambda_2$, where λ_1 and λ_2 are known constants, but T^2 is a biased estimator of θ^2.

17·4. For the following cases determine if the given estimator is unbiased for the parametric function. When it is biased, derive an unbiased estimator from it. $\overline{x}$ is the sample mean.

(*a*) $x_1, \ldots, x_n$ is a random sample from a distribution with variance σ^2. The estimator $\frac{1}{n}\{(x_1 - \overline{x})^2 + \ldots + (x_n - \overline{x})^2\}$ is used to estimate σ^2.

(*b*) $x_1, x_2, \ldots, x_n$ is a random sample from an exponential distribution with mean θ. The estimator $\left[1 - \{1/(n\overline{X})\}\right]^{n-1}$ is used to estimate $\exp(-1/\theta)$ when $n\overline{X} > 1$ and zero is used when $n\overline{X} < 1$.

17·5. $f(x\,;\mu, \sigma) = \frac{1}{\sigma}\exp\{-(x-\mu)/\sigma\}$; $\mu \le x < \infty$, $-\infty < \mu < \infty$ and $0 < \sigma < \infty$.

Obtain (*i*) an unbiased estimate of μ when σ is known, (*ii*) an unbiased estimate of σ when μ is known, and (*iii*) two unbiased estimators of σ^2 when μ is known.

Hence, obtain an infinity of unbiased estimators of σ^2 in this case.

17·6. Suppose X and Y are independent random variables with the same unknown means μ. Both X and Y have variance as 36. Let $T = aX + bY$ be an estimator of μ.

(*i*) Show that T is an unbiased estimator of μ if $a + b = 1$. (*ii*) If $a = \frac{1}{3}$ and $b = \frac{2}{3}$, what is the variance of T? (*iii*) If $a = \frac{1}{2}$ and $b = \frac{1}{2}$, what is the variance of T? (*iv*) What choice of a and b minimizes the variance of T subject to the requirement that T is an unbiased estimate of μ?

17·7. (*a*) Examine the unbiasedness of the following estimates :

(*i*) $s_1^2 = \frac{1}{n}\sum_{i=1}^{n}(x_i - \overline{x})^2$ (*ii*) $s_2^2 = \frac{1}{n}\sum_{i=1}^{n}(x_i - \mu)^2$ (where μ is known), for σ^2, the population variance.

Ans. (*i*) $E(s_1^2) = \frac{1}{n}(n-1)\sigma^2 \ne \sigma^2$, (*ii*) $E(s_2^2) = \sigma^2$.

(*b*) Let $x_1, x_2, \ldots, x_n$ be a random sample of size n drawn from a population with mean μ and variance σ^2. Obtain an unbiased estimator for μ^2.

17·8. If $X_1, X_2, \ldots, X_n$ is a random sample of size n from $N(\mu, \sigma^2)$, where μ is known and if

$$T = \frac{1}{n}\sum_{i=1}^{n} | X_i - \mu |.$$

examine if T is unbiased for σ. If not, obtain an unbiased estimator of σ.

Hint. $E(T) = \frac{1}{n}\sum_{i=1}^{n} E\,|\,X_i - \mu\,| = \sqrt{(2/\pi)}.\sigma$, since, for $N(\mu, \sigma^2)$, Mean Deviation about mean $= \sqrt{2/\pi}\ \sigma$.

Ans. No; $\sqrt{(\pi/2)}\ T$.

17·9. If $x_1, x_2, \ldots, x_n$ is a random sample from the population

$$f(x, \theta) = (\theta + 1)\, x^{\theta}\ ;\ 0 < x < 1\ ;\ \theta > -1 \qquad \ldots (*)$$

show that $\left[\dfrac{-(n-1)}{\Sigma \log x_i} - 1\right]$ is unbiased estimator of θ.

Hint. In sampling from *p.d.f.* in (*), $U = -\log X$ has an exponential distribution with parameter $(\theta+1)$

$$\Rightarrow \qquad U_i = -\log X_i \overset{i.i.d}{\sim} \gamma(\theta + 1, 1);\ i = 1, 2, \ldots, n$$

$$\Rightarrow \qquad Y = \sum_{i=1}^{n} \log X_i \sim \gamma(\theta + 1, n); \quad E[1/Y] = \frac{\theta + 1}{n-1}.$$

17·10. Suppose X has a truncated Poisson distribution with p.m.f.

$$f(x, \theta) = \begin{cases} \dfrac{\exp(-\theta)\cdot\theta^x}{[1-\exp(-\theta)]\ x!}\ ; & x = 1, 2, 3\ldots \\ 0 & \text{, otherwise} \end{cases}$$

Show that the only unbiased estimator of $[1 - \exp(-\theta)]$ based on X is the statistic T, defined as :

$$T(x) = \begin{cases} 0, & \text{when } x \text{ is odd} \\ 2, & \text{when } x \text{ is even} \end{cases}$$

Note. This is an Example of *absurd unbiased estimator.*

17·11. Consider a random sample X_1, X_2, X_3 of size 3 from uniform p.d.f. : $f(x, \theta) = \begin{cases} \frac{1}{\theta}, & 0 < x < \theta \\ 0, & \text{otherwise} \end{cases}$

Show that each of the statistics $4X_{(1)}$, $2X_{(2)}$ and $\frac{1}{3}\ X_{(3)}$, where $X_{(i)}$ is the ith order statistic is an unbiased estimator for θ. Find the variance and hence the efficiency of each.

17·12. Obtain an unbiased estimator for (*i*) θ, and (*ii*) θ^2, in case of binomial probability distribution :

$$f(x, \theta) = {}^nC_x\, \theta^x (1 - \theta)^{n-x};\ x = 0, 1, 2, \ldots, n\ ; 0 < \theta < 1.$$

Hint. $E\left(\dfrac{x}{n}\right) = \theta; \qquad E\left\{\dfrac{x(x-1)}{n(n-1)}\right\} = \theta^2$

If we write $T = x/n$, the observed proportion of successes, then

$$E(T) = \theta; \qquad E(T^2) = \frac{\theta^2}{n} + \left(\frac{n-1}{n}\right).\theta^2 \neq \theta^2.$$

This illustrates that we may have t_n unbiased for θ but t_n^2 not unbiased for θ^2.

17·13. X is a uniform *r.v.* with range $[0, \theta]$. $x_1, x_2, \ldots, x_n$ are independent observations on X.

Define : $\hat{\theta}_1 = \dfrac{2}{n}\ (x_1 + x_2 \ldots + x_n) \quad ; \quad \hat{\theta}_2 = \left\{\dfrac{(n+1)}{n}\right\}.\ \max\ (x_1, x_2, \ldots, x_n).$

Show that $\hat{\theta}_1$ and $\hat{\theta}_2$ are unbiased for θ. Evaluate their relative efficiency.

17·14. The observations $x_1, x_2, \ldots, x_n$ represent a random sample from a uniform distribution over the interval $(0, \theta)$, where θ is an unknown parameter. The statistics $\bar{X}$, m and M are the mean, the smallest value and the largest value respectively for the sample. Find values for k so that, kt is an unbiased estimator for θ where (*a*) $t = \bar{X}$ (*b*) $t = M$, (*c*) $t = M - m$,

Of the three unbiased estimators which is the best? Give your reasons.

17·15. Let $X_1, X_2, \ldots, X_n$, $(n > 2)$ be a random sample of size n from the distribution having density function :

$$f(x\,;\theta) = \theta x^{\theta-1}, 0 < x < 1,\ \theta > 0$$

If $Z = -\sum_{t=1}^{n} \log X_i$, show that $(n-1)/Z$ is an unbiased estimator for θ and its efficiency is $(n-2)/n$.

17·16. (*a*) Suppose $X_1, X_2, \ldots, X_n$ are sample values independently drawn from population with mean m and variance σ^2. Consider the estimates :

$$Y_n = \frac{X_1 + X_2 + \ldots + X_n}{n+1}, \quad Z_n = \frac{X_1 + 2X_2 + 3X_3 + \ldots + nX_n}{n^2}$$

Discuss whether they are unbiased, consistent for m. What is the efficiency of Y_n over Z_n?

(*b*) Let X_1, X_2, X_3 and X_4 be independent random variables such that

$$E(X_i) = \mu \quad \text{and} \quad \text{Var}(X_i) = \sigma^2 \text{ for } i = 1, 2, 3, 4.$$

If $$Y = \frac{X_1 + X_2 + X_3 + X_4}{4}, \quad Z = \frac{X_1 + X_2 + X_3 + X_4}{5} \quad \text{and} \quad T = \frac{X_1 + 2X_2 + X_3 - X_4}{4},$$

examine whether Y, Z and T are unbiased estimators of μ? What is the efficiency of Y relative to Z?

17·17. Let x, the earnings of a commercial bank, be a random variable with mean μ and variance σ^2. A random sample of earnings of n banks is denoted by $x_1, x_2, \ldots, x_n$. However, because of the disclosure laws, individual bank earnings are not disclosed and only the following average values are made available to the researcher :

$a_1 = \frac{1}{2}(x_1 + x_2)$, $a_2 = \frac{1}{2}(x_3 + x_4)$, ... , $a_m = \frac{1}{2}(x_{n-1} + x_n)$ where n is an even number and $m = n/2$.

(*i*) Obtain the best linear unbiased estimator of μ, given the available information. What is the variance of the proposed estimator?

(*ii*) Obtain an unbiased estimator of σ^2.

17·18. If $X_1, X_2, \ldots, X_n$ is a random sample obtained from the density function :

$$f(x,\theta) = \begin{cases} 1, & \theta < x < \theta + 1 \\ 0, & \text{elsewhere} \end{cases}$$

show that the sample mean $\overline{X}$ is an unbiased and consistent estimator of $\left(\theta + \frac{1}{2}\right)$.

17·19. (*a*) Show that $\overline{x} = \frac{1}{n}\sum_{i=1}^{n} x_i$ and $s^2 = \frac{1}{n}\sum_{i=1}^{n}(x_i - \overline{x})^2$, are joint consistent estimators for μ and σ^2 respectively, if $x_1, x_2, \ldots, x_n$ is a random sample from a normal population $N(\mu, \sigma^2)$. Also find the efficiency of $ns^2/(n-1)$.

(*b*) Show that if t is a consistent estimator of a parameter θ, then e^t is a consistent estimator of e^{θ}.

(*c*) Prove that in case of Binomial distribution with parameter θ, t_n defined as r/n is a consistent unbiased estimator for θ, but T_n defined as $(r/n)^2$ is consistent but not unbiased estimator for θ^2.

17·20. Show that in sampling from Cauchy distribution:

$$f(x,\theta) = \frac{1}{\pi[1 + (x-\theta)^2]}, \ -\infty < x < \infty,\ \theta > 0;$$

(*i*) Sample mean $\overline{X}$ is not a consistent estimator of θ.

(*ii*) Sample median is a consistent estimator of θ and its asymptotic efficiency is $8/\pi^2$.

17·21. (*a*) If T_1 and T_2 are consistent estimators of $\gamma(\theta)$, show that $a_1T_1 + a_2T_2$, such that $a_1 + a_2 = 1$, is also consistent for $\gamma(\theta)$.

(*b*) For a Poisson distribution with parameter θ, show that $1/\overline{X}$ is consistent estimator of $1/\theta$, where $\overline{X}$ is the mean of a random sample from the given population.

17·22. (*a*) Let T_1 and T_2 be two unbiased estimators of $\gamma(\theta)$ having the same variance. Show that their correlation coefficient ρ_0 cannot be smaller than $(2e_0 - 1)$, where e_0 is the efficiency of each estimator.

Further show that if T_1 is *MVU* estimator and T_2 is any unbiased estimator with efficiency *e*, then :

$$V(T_1 - T_2) = \left(\frac{1}{e} - 1\right) V(T_1)$$

(*b*) If T_1 is a *MVU* for θ and T_2 is any other unbiased estimator of θ with efficiency e_θ, then prove that the correlation between T_1 and T_2 is $\sqrt{e_\theta}$.

17·23. Suppose *X* and *Y* are independent unbiased estimates of μ. It is known that the variance of *X* is 12 and the variance of *Y* is 4. It is desired to combine two estimators in order to obtain a more efficient estimator. Let $T = aX + bY$, be the new estimator.

(*i*) In order that *T* be an unbiased estimator of μ, what conditions must be imposed on *a* and *b* ?

(*ii*) Find the values of *a* and *b* that minimize the variance of *T* subject to the condition that *T* be an unbiased estimator.

17·24. (*a*) What is an efficient estimator?

If T_1, T_2 are both efficient estimators with variance *v* and if $T = \frac{1}{2}(T_1 + T_2)$, show that variance of *T* is $(v/2)(1 + \rho)$, where ρ is the coefficient of correlation between T_1 and T_2. Deduce that $\rho = 1$ and that *T* is also efficient.

(*b*) If *T* and *T* ′ be two consistent estimators of which *T* is the most efficient, prove that the correlation coefficient between them is $\sqrt{\dfrac{V(T)}{V(T')}}$ where $V(T)$ and $V(T')$ are the variance of *T* and *T*′ respectively.

Show also that the correlation coefficient between two most efficient estimators is unity.

17·25. Define a sufficient statistic. Explain the method of finding sufficient estimator. If $(x_1, x_2, ..., x_n)$ is a random sample from a distribution :

$$f(x, p) = p^x (1 - p)^{1-x} \;;\; x = 0, 1 \text{ and } 0 \le p \le 1, \text{ find the sufficient estimator of } p.$$

17·26. (*a*) State the factorisation theorem on sufficiency. Obtain a sufficient statistic for the parameter θ in the following distribution : $f(x, \theta) = \frac{1}{\theta}$, $0 < x < \theta$.

(*b*) Define a sufficient statistic.

$x_1, x_2, ..., x_n$ is a random sample from a distribution : $f(x, \theta) = \theta^x (1 - \theta)^x \;;\; x = 0, 1, 0 < \theta < 1$

$$= 0, \quad \text{elsewhere.}$$

Show that $Y_1 = x_1 + x_2 + ... + x_n$ is a sufficient statistic for θ.

(*c*) Let $x_1, x_2, ..., x_n$ denote a random sample from a population with p.d.f

$$f(x, \theta) = \theta\, x^{\theta - 1}, 0 < x < 1.$$

Show that $Y = x_1 x_2 ... x_n$ is a sufficient statistic for θ.

17·27. (*a*) Let *X* be a random sample of size one from a normal distribution $N(0, \sigma^2)$.

(*i*) Is X a sufficient statistic for σ^2?; (*ii*) Is $|X|$ a sufficient statistic for σ^2?

(*iii*) Is X^2 a sufficient statistic for σ^2?

(*b*) Examine which of the following distributions admit sufficient estimators for their parameters

(*i*) $f(x, \theta) = \theta x^{\theta - 1}$, $0 \le x \le 1$, $\theta > 0$.

(*ii*) $f(x\ y, \rho) = \dfrac{1}{2\pi\sqrt{(1-\rho^2)}} \exp\left\{-\dfrac{1}{2(1-\rho^2)}\left(x^2 - 2\rho xy + y^2\right)\right\}$

17·28. (*a*) Do the following distributions admit of sufficient estimators?

(*i*) $f(x, \theta) = \frac{1}{\theta}$; $k\theta \le x \le (k + 1)\theta$, where *k* is an integer ; (*ii*) $f(x, \theta) = \dfrac{1 + \theta}{(x + \theta)^2}, 1 \le x < \infty$

(*b*) Examine if the following distribution admits a sufficient statistic for the parameter θ.

$$f(x, \theta) = (1 + \theta)\, x^\theta \quad ; \quad 0 \le x \le 1, \quad \theta > 0$$

(c) Discuss if a sufficient statistic exists for the parameter θ, in sampling from double exponential distribution with *p.d.f.* : $f(x, \theta) = \frac{1}{2} \exp\left(-|x-\theta|\right), -\infty < x < \infty$.

(d) Obtain jointly sufficient estimators for α and β in a random sample $X_1, X_2, \ldots, X_n$ from the uniform population with p.d.f.

$$f(x,\alpha,\beta) = \begin{cases} \frac{1}{\beta-\alpha}, & \alpha \le x \le \beta \\ 0, & \text{otherwise} \end{cases}$$

17·29. (a) Show that a necessary and sufficient condition for a statistic T to be sufficient for θ is that the probability function should belong to an exponentially family.

(b) Let $x_1, x_2, \ldots x_n$ be a random sample from a distribution with p.d.f. $f(x, \theta) = e^{-(x-\theta)}$, $x \ge \theta$, $-\infty < \theta < \infty$. Obtain a sufficient statistic for θ.

17·30. (a) Let (X_1, X_2, X_3) be a random sample from the probability mass function :

$$P(X = x) = \theta^x (1-\theta)^{1-x}, \ (x = 0, 1; 0 < \theta < 1).$$

If $t = X_1 + X_2 + X_3$, show that the conditional distribution of the random sample given $t = r$ does not depend on θ. Interpret this result in the light of sufficiency-concept.

(b) Let (X_1, X_2) be a random sample from a Poisson distribution with parameter θ. Prove that

$$t = X_1 + 2X_2 \text{ is not sufficient for } \theta.$$

(c) Let (X_1, X_2) be a random sample from $N(\theta, 1)$. If $T = X_1 + X_2$ and $U = X_2 - X_1$, show that the conditional distribution of U given $T = t$ does not depend on θ. Interpret this result in the light of sufficiency-concept.

(d) For a random sample X_i $(i = 1, 2, \ldots, n)$ from an exponential distribution with *p.d.f* :

$$f(x, \theta) = \frac{1}{\theta} \exp\left(-\frac{x}{\theta}\right), \ x > 0, \ \theta > 0,$$

obtain an unbiased and sufficient estimator for θ.

17·31. Let $X_1, X_2, \ldots, X_n$ be a random sample from a population with *p.d.f.* :

$$f(x, \theta) = \theta \cdot \exp(-\theta x)\ ; \ x > 0, \theta > 0.$$

Find Cramer-Rao lower bound for the variance of the unbiased estimator of θ.

17·32. $f(x, \theta)$ is a probability density function and $(x_1, x_2, \ldots, x_n)$ is a random sample from it. Prove that if an unbiased minimum variance bound (*MVB*) estimator T exists, it must be of the form :

$$T = \theta + \lambda \sum \frac{\partial}{\partial\theta} \log f(x_i, \theta), \text{ in which } \lambda \text{ does not depend on sample values.}$$

Show that the variance of T is λ and is given by :

$$\frac{1}{\text{Var}T} = nE\left\{-\frac{\partial}{\partial\theta^2} \log f(x_i, \theta)\right\}$$

17·33. (a) Verify that there exists an M.V.B. estimator for the parameter θ of the distribution :

$$f(x, \theta) = \frac{e^{-\theta}\theta^x}{x!}\ ; x = 0, 1, 2, \ldots$$

and hence obtain the value of M.V.B.

(b) Show that there exists a parameter function $\psi(\theta)$ in the case of the geometric distribution :

$$f(x, \theta) = (1-\theta)\,\theta^x\ ; x = 0, 1, 2, \ldots\ ; \ 0 < \theta < 1$$

such that there exists an M.V.B. unbiased estimator T of $\psi(\theta)$. Obtain $\psi(\theta)$, T and $V(T)$.

17·34. Obtain minimum variance unbiased estimator of θ from a sample of n independent observations $x_1, x_2, \ldots, x_n$ drawn from the binomial $B(N, \theta)$ population having probability function :

$$f(x\ ; \theta) = {}^N C_x\, \theta^x (1-\theta)^{N-x};\ x = 0, 1, 2, \ldots, N$$

Also obtain variance of this estimator of θ.

17·35. For the gamma distribution :

$$f(x, \theta) = \frac{1}{\theta^p \Gamma p} x^{p-1} \exp(-x/\theta) \,;\; 0 \le x < \infty, \theta > 0,\; p \text{ (known)},$$

find the expectation of X^2. Use it to obtain an unbiased estimator T of θ^2. Find $V(T)$:
Evaluate Fisher's information function $I(\theta)$ about θ^2 and verify the truth of the inequality: $V(T) > [nI(\theta^2)]^{-1}$.

17·36. (*a*) Let $x_1, x_2, \ldots, x_n$ be a random sample from Poisson distribution with parameter λ. Obtain Cramer-Rao lower bound to the variance of an unbiased estimator for λ. Hence find the M.V.U.E. for λ.

(*b*) Let $X_1, X_2, \ldots, X_n$ be a random sample from a rectangular distribution with *p.d.f.* $f(x, \theta) = \frac{1}{\theta}, 0 \le x \le \theta$.

Find MVU estimators of θ and $3\theta + 5$.

17·37. If T is a complete sufficient statistic for $\gamma(\theta)$, and $E[\phi(T)] = \gamma(\theta)$, then show that $\phi(T)$ is the unique *MVUE* of $\gamma(\theta)$.

Use this property to obtain *MVU* estimator of θ based on a random sample $X_1, X_2, \ldots, X_n$ from the distribution with *p.m.f.*

$$f(x,\theta) = \begin{cases} \theta^x (1-\theta)^{1-x} \,;\; x=0, 1 \\ 0, \quad \text{elsewhere} \end{cases}$$

17·38. Show that the family $\{f(x, \theta), \theta \in (0, 1)\}$ with $f(x, \theta) = {}^2C_x\, \theta^x (1-\theta)^{2-x},\; x = 0, 1, 2$, is complete.

17·39. Let $X_1, X_2, \ldots, X_n$ be a random sample from

$$f_\theta(x) = \frac{1}{\theta}, 0 < x < \theta \text{ for all } \theta \in \Theta$$

Show that $X_{(n)} = \text{Max}(X_1, X_2, \ldots, X_n)$ is sufficient for θ and $\frac{(n+1)}{n} X_{(n)}$ is an unbiased estimator for θ. Comment on the result.

17·40. Let the random variables X and Y have the joint *p.d.f.* :

$$f(x, y) = \frac{2}{\theta^2} \exp\left\{-\frac{(x+y)}{\theta}\right\}, 0 < x < y < \infty$$

and zero elsewhere.

(*i*) Show that : $\quad E(Y \mid x) = x + \theta$

Obtain the expected value of $X + \theta$ and compare the variance of $X + \theta$ with that of Y.

(*ii*) Show that : $\quad E(Y) = \frac{3}{2}\theta, \; \text{Var}(Y) = \frac{5}{4}\theta^2.$

17·41. (*a*) A random sample of size n is drawn from a Poisson population with parameters λ. Obtain the minimum variance unbiased estimator of λ.

(*b*) Establish a necessary and sufficient condition for an unbiased estimator to be an *MVU* estimator.

17·42. Define sufficiency of an estimator.

Let $Y_1 < Y_2 < Y_3 < Y_4 < Y_5$ be the order statistics of a random sample of size 5 from the uniform distribution with *p.d.f.* :

$$f(x,\theta) = \begin{cases} \frac{1}{\theta}; & 0<x<\theta,\; 0<\theta<\infty \\ 0, & \text{elsewhere} \end{cases}$$

Show that $2Y_3$ is an unbiased estimator of θ. Find the conditional expectation

$E[2Y_3 \mid Y_5] = \phi(Y_5)$, say. Compare the variances of $2Y_3$ and $\phi(Y_5)$.

17·43. Let $X_1, X_2, \ldots, X_n$ be a random sample from the Bernoulli population with parameter θ, $0 < \theta < 1$. Obtain a sufficient statistic for θ and show that it is complete. Hence obtain *MVU* estimator of θ.

17·44. Show that $T = \sum_{i=1}^{n} X_i$, is a complete sufficient statistic for the parameter θ in a random sample $X_1, X_2, \ldots, X_n$ drawn from the population with *p.d.f.*

(*a*) $$f(x,\theta) = \begin{cases} \theta^x (1-\theta)^{1-x}; x=0,1 \\ 0, \quad \text{elsewhere} \end{cases}$$

(*b*) $$f(x,\theta) = \begin{cases} e^{-\theta}\, \theta^x / x!\,; x=0, 1, 2, \ldots \\ 0, \quad \text{elsewhere} \end{cases}$$

17·45. If $X_1, X_2, \ldots, X_n$ is a random sample from $N(\mu, \sigma^2)$, show that :

(*a*) $T = \overline{X}$, is complete sufficient statistic for μ, $(-\infty < \mu < \infty)$, when σ^2 is known.

(*b*) $T = \sum_{i=1}^{n} (X_i - \mu)^2$, is complete sufficient statistic for σ^2, $(0 < \sigma^2 < \infty)$, when μ is known.

17·46. (*a*) A random variable X takes the values 0 and 1 with respectively probabilities p and $1 - p$. Obtain on the basis of random sample of size n, the maximum likelihood estimator of p.

(*b*) Obtain the maximum likelihood estimator of θ in the following cases :

(*i*) $f(x, \theta) = \frac{1}{\theta} . \exp(-x/\theta)$; $x \geq 0, \theta > 0$ (*ii*) $f(x, \theta) = {}^nC_x\, \theta^x (1-\theta)^{n-x}$; $x = 0, 1, 2, \ldots, n$

(*c*) Suppose that X has a normal distribution $N(\mu, \sigma^2)$.

Using M.L. estimation, determine estimates of μ and σ^2. What conclusions do you draw on the nature of the result so obtained ?

17·47. (*a*) For the distribution with p.d.f. $f(x, \theta) = \theta\, exp(-\theta x)$, $(x \geq 0; \theta > 0)$, find the maximum likelihood estimators of θ and $E(X)$, and obtain their large sample standard errors.

(*b*) X is a random variable such that $P(X \leq x) = \begin{cases} 0, & \text{for } x < 0 \\ 1 - exp(-\theta x), & \text{for } x \geq 0 \end{cases}$

Based on n independent observations on X, obtain the maximum likelihood estimator of $E(X)$.

17·48. (*a*) Let $X_1, X_2, \ldots, X_n$ be a random sample from the distribution with probability density function :

$$f(x, \theta) = \frac{1}{\theta} \exp(-x/\theta)\,; 0 < x < \infty,\ 0 < \theta < \infty$$

Find the maximum likelihood estimator of θ.

(*b*) For the distribution,

$$dF(x) = \frac{1}{\theta^p \Gamma(p)} \exp(-x/\theta)\, x^{p-1}; 0 \leq x < \infty, p > 1, \theta > 0$$

where p is known, find out the maximum likelihood estimate of θ on the basis of a random sample of size n from the distribution. Find the variance of the estimate.

17·49. (*a*) If x_i $(i = 1, 2, \ldots n)$ is an observed random sample from the distribution having *p.d.f.* :

$$f(x,\lambda) = \frac{\lambda^{k+1} x^k \exp(-\lambda x)}{\Gamma(k+1)}, x > 0,$$

where $\lambda > 0$ and k is a known constant, show that the ML estimator $\hat{\lambda}$ for λ is $(k + 1)/\overline{x}$. Show that the corresponding estimator is biased but consistent and that its asymptotic distribution for large n is : $N\{\lambda, \lambda^2/[n(k + 1)]\}$.

(*b*) Derive the *MLE* of the mean $[\alpha/(\alpha+2)]$ of the beta distribution :

$$f(x) = [B(\alpha, 2)]^{-1} x^{\alpha-1} (1-x), 0 < x \leq 1, \alpha > 0.$$

17·50. (*a*) From a sample of size n from the population of X, determine the maximum likelihood estimates of the parameters a and b of the probability density

$$f(x) = \text{Constant.} \exp\{-(x-a)/b\} \;; x \geq a, b > 0, -\infty < a < \infty$$

(*b*) Let $X_1, X_2, \ldots, X_n$ be a random sample from the distribution with p.d.f.

$$f(x;\theta_1,\theta_2)=\begin{cases}\dfrac{1}{\theta_2}\exp\left[-\dfrac{(x-\theta_1)}{\theta_2}\right]; & x\geq\theta_1, -\infty<\theta_1<\infty, \theta_2>0\\ 0, & \text{elsewhere}\end{cases}$$

Obtain the maximum likelihood estimators of:

(*i*) θ_1 when θ_2 is known; (*ii*) θ_2 when θ_1 is known; (*iii*) θ_1 and θ_2 (Joint).

17·51. (*a*) A random variable X has a distribution with density function

$$f(x)=\begin{cases}(\alpha+1)\,x^{\alpha}, & (0\leq x\leq 1, \alpha>-1)\\ 0, & \text{otherwise}\end{cases}$$

and a random sample of size 8 produces the data :

0·2, 0·4, 0·8, 0·5, 0·7, 0·9, 0·8, 0·9.

Find the maximum likelihood estimate of the unknown parameter α, it being given that ln (0·0145152) $\simeq$ – 4·2326 (ln denotes natural logarithm).

(*b*) Find the MLE of θ for a random sample of size n from the distribution:

$$f(x,\theta)=\begin{cases}(\theta+1)x^{\theta}, & 0\leq x\leq 1, \theta>-1\\ 0, & \text{otherwise}\end{cases}$$

Show that it is also sufficient statistic for θ.

Also obtain an unbiased estimator of $(\theta+1)/(\theta+2)$ when $\theta\neq-2$.

17·52. (*a*) Obtain the MLE for the parameter θ in a random sample of size n from the uniform population $U[0, \theta]$.

(*b*) Show by means of an example, that MLE are not, in general unique.

(*c*) Show that in a random sample from a distribution with *p.d.f.* $f(x, \theta) = \theta \exp(-\theta x)$, $x \geq 0$; $1/\overline{X}$ is the MLE for θ and has greater variance than the unbiased estimator $(n-1)/(n\overline{X})$.

17·53. (*a*) Let $X_1, X_2, \ldots, X_n$ be a random sample from a population with density.

$$f(x, \theta)=\frac{1}{2}\exp(-|x-\theta|), -\infty<x<\infty.$$

Find the estimator for θ based on the method of maximum likelihood.

Ans. $\hat{\theta}$ = Median $(x_1, x_2, \ldots, x_n)$.

(*b*) Obtain the maximum likelihood estimator of θ based on a random sample of size n from the population with p.d.f.

(*i*) $f(x, \theta) = exp[-(x-\theta)]$, $\theta \leq x < \infty$, $-\infty < \theta < \infty$

(*ii*) $f(x, \theta) = \theta x^{\theta-1}$, $0 < x < 1$, $0 < \theta < \infty$.

Examine in each case, whether θ is unbiased.

Ans. (*i*) $\hat{\theta} = x_{(1)}$.

17·54. (*a*) Explain what is meant by an estimate of a population parameter. Find the maximum likelihood estimate of the parameter θ of a population having density function

$$2(\theta-x)/\theta^2, (0<x<\theta)$$

for a sample of unit size and examine whether the estimate so obtained is biased or not.

Ans. $\hat{\theta} = 2x$, biased.

(*b*) Obtain Maximum Likelihood Estimator of θ for the distribution :

$$f(x,\theta)=\frac{C_o\theta^x}{x!}\ ; x=0, 1, 2, \ldots ; \theta>0.$$

where C_o is a constant. Also write the Maximum Likelihood Estimator of $3\theta^2 + 4\theta + 5$.

Hint. For MLE of $3\theta^2 + 4\theta + 5$, use Invarince Property of MLE.

17·55. (*a*) Consider a population made up of 3 different types of individuals occurring in the population with probabilities θ^2, $2\theta(1-\theta)$ and $(1-\theta)^2$, respectively where $0 < \theta < 1$. Let n_1, n_2 and n_3 denote the respective random sample sizes of the above three types of individuals. Determine the maximum likelihood estimator for θ.

(*b*) Obtain the maximum likelihood estimate of θ, if the variable takes the values 1, 2, 3 and 4 with probabilities $(1-\theta)/2$, $(1-\theta)/2$, $\theta(1-\theta)$ and θ^2 respectively and the observed frequencies are n_1, n_2, n_3 and n_4 respectively.

17·56. In life-testing it is sometimes assumed that the life-time of an item is a random variable which is greater than or equal to x with probability : $\exp\{-(x/\theta)^m\}$, $x \geq 0$, $m > 0$ is known and $\theta > 0$ is unknown. Suppose n such items are tested and yield $X_1, X_2, \ldots, X_n$ as their times of "death".

Find the maximum likelihood estimate of θ.

17·57. X_1, X_2, X_3, X_4 are independent normal random variables with means $\alpha+\beta$, $\alpha-\beta$, $\alpha+2\beta$, $\alpha-2\beta$ respectively and a common variance σ^2, on the basis of one observation on each X_i. Obtain the maximum likelihood estimators of α, β and σ^2.

17·58. (*a*) For the bivariate normal distribution $N(\mu_1, \mu_2, \sigma_1^2, \sigma_2^2, \rho)$, find the maximum likelihood estimators :

(*i*) of σ_1^2, σ_2^2 and ρ when μ_1 and μ_2 are known,

(*ii*) of all five parameters of the distribution.

(*b*) If (x_i, y_i), $(i = 1, 2, \ldots, n)$ come from a bivariate normal population with zero means, unit variances and co-efficient of correlation ρ, obtain the maximum likelihood estimator of ρ.

17·59. (*a*) Show that the most general continuous distribution for which the M.L.E. of a parameter θ is the sample harmonic mean is :

$$f(x,\theta) = \exp\left[\frac{1}{x}\left\{\theta\frac{\partial\psi}{\partial\theta} - \psi(\theta)\right\} - \frac{\partial\psi}{\partial\theta} + \xi(x)\right]$$

where $\psi(\theta)$ and $\xi(x)$ are arbitrary function of θ and x respectively.

(*b*) Show that the most general form of the distribution for which the sample arithmetic mean $\overline{X}$ is the MLE of θ has the *p.d.f.* : $f(x,\theta) = \exp\left[(x-\theta)\,A'(\theta) + A(\theta) + B(x)\right]$.

17·60. (*a*) Suppose that distribution of X is represented by the function :

$$P(X = x) = e^{-\lambda}\,\frac{\lambda^x}{x!}\,;\ x = 0, 1, 2, \ldots$$

where $\lambda > 0$. Given *a* random sample of size n, show that the sample mean is the maximum likelihood estimate of λ. Show further that this estimate is (*i*) best unbiased, and (*ii*) consistent.

(*b*) Consider the estimation of the Poisson parameter from a random sample.

(*i*) Work out the maximum likelihood estimator and its variance.

(*ii*) Work out the Cramer-Rao lower bound and show that it is equal to the variance worked out in (*i*). Comment on the significance of this result.

17·61. X is a discrete random variable and

$$P(X = r) = (1-p)\,p^{r-1};\ r = 1, 2, 3, \ldots$$

Find the MLE of p based on a random sample of n observations and its variance in large samples. Show that the variance attains the lower bound of C.R. inequality.

17·62 (*a*) Let $(X_1, X_2, \ldots, X_n)$ be a random sample from the *p.d.f.* :

$$f(x,\theta) = \begin{cases} \theta\, exp(-\theta x), & 0 < x < \infty,\ \theta > 0. \\ 0, & \text{elsewhere} \end{cases}$$

Estimate θ using the method of moments.

(*b*) $X_1, X_2, \ldots, X_n$ is a random sample from $f(x, a,b) = \begin{cases} \frac{1}{b-a} & a<x<b \\ 0, & \text{elsewhere} \end{cases}$

Find estimates of a and b by the method of moments.

(*c*) Estimate θ in the density function : $f(x, \theta) = (1 + \theta)x^{\theta}$; $\theta < x < 1$, by the method of moments and obtain the standard error of the estimator.

17·63. The sample values from population with *p.d.f.*: $f(x) = (1 + \theta)\, x^{\theta}$, $0 < x < 1$, $\theta > 0$, are given below :

0·46, 0·38, 0·61, 0·82, 0·59, 0·53, 0·72, 0·44, 0·59, 0·60

Find the estimate of θ by method of (*i*) moments and (*ii*) maximum likelihood estimation.

17·64. (*a*) For the distribution with density function: $f(x, \theta) = \dfrac{e^{-\theta}\theta^x}{x!(1-e^{-\theta})}$; $x = 1, 2, 3, \ldots$

obtain the estimate of the θ by method of moments.

(*b*) For the following probability function : $f(x,p) = \dbinom{3}{x}\dfrac{p^x(1-p)^{3-x}}{1-(1-p)^3}$; $x = 1,2,3;$

obtain the estimator of p by the method of moments, if the frequencies at x = 1, 2 and 3 are respectively 22, 20 and 18.

17·65. (*a*) Let $x_1, x_2, \ldots, x_n$ be a random sample from a distribution with density function :

$$f_{\theta}(x) = \theta(\theta+1)\, x^{\theta-1}(1-x)\ ;\ 0<x<1, \theta>0$$

Determine the estimate of θ by the method of moments.

(*b*) Describe the method of moments and discuss when the estimates obtained by the method of moments are identical with those of maximum likelihood estimates.

Estimate α and β by the method of moments for the distribution

$$f(x, \alpha, \beta) = \frac{\beta^{\alpha}}{\Gamma(\alpha)}\, x^{\alpha-1}\, exp(-\beta x)\ , 0 \le x < \infty.$$

17·66. Examine if the MLE's of the parameter(s) are identical with those obtained by the method of moments in random sampling from the following distributions:

(*i*) $f(x, \theta) = \dfrac{1}{\theta} \cdot \exp\left(-\dfrac{x}{\theta}\right)$; $0 < x < \infty$ (*ii*) $f(x, \mu, \sigma^2) = \dfrac{1}{\sigma\sqrt{2\pi}} \exp\{-(x-\mu)^2/2\sigma^2\}$; $-\infty < x < \infty$.

17·67. Independent samples of sizes n_1 and n_2 are taken from two normal populations with equal means μ and variances respectively equal to $\lambda\sigma^2$, σ^2. Find the maximum likelihood estimator of μ based on $n_1 + n_2$ sample observations and show that its large sample variance is :

$$\text{Var}(\hat{\mu}) = \sigma^2 \Big/ \left(\frac{n_1}{\lambda} + n_2\right)$$

Hence, show that the unbiased estimator, $t = (n_1\bar{x}_1 + n_2\bar{x}_2)/(n_1 + n_2)$ has efficiency,

$$\frac{\lambda(n_1+n_2)^2}{(n_1\lambda + n_2)(n_1 + n_2\lambda)}$$, which attains the value 1 if and only if $\lambda = 1$.

17·68. The random variable X is uniformly distributed in $(a, a + 2)$. Obtain limits x_1 and x_2 such that :

$$P(X \le x_1) = P(X \ge x_2) = 0{\cdot}025$$

The random variable is observed once, the value being x_0. Give a method of obtaining an interval estimate for '*a*' which you expect to be correct in 95% of trials.

17·69. Obtain 100 $(1 - \alpha)$% confidence interval:

(*i*) for the unknown parameter p of a binomial distribution when the parameter n is known, and

(*ii*) for the population correlation coefficient when the population is Normal.

17·70. Let $$f_\theta(x) = \frac{1}{\theta}, 0 \le x \le \theta$$

and let L be the largest observation of a sample of size n from the above distribution.

Obtain the distribution of L/θ and hence deduce that the confidence limits corresponding to confidence coefficient α are L and $\frac{L}{(1-\alpha)^{1/n}}$ respectively.

17·71. What are confidence intervals? y is the largest observation in a sample of size n drawn from a rectangular population in $(0, \theta)$. Find the confidence coefficient corresponding to the confidence interval $\{y, y/(1-\alpha)^{1/n}\}$, where '$\alpha$' is the level significance.

17·72. Develop a general method for constructing confidence intervals. Consider a random sample of size n from the exponential distribution with *p.d.f.*:

$$f(x, \theta) = \exp\left[-(x-\theta)\right], \theta \le x < \infty, \ -\infty < \theta < \infty.$$

Show that: $$P\left\{X_{(1)} - \frac{1}{n}\log\alpha \le \theta \le X_{(1)}\right\} = 1-\alpha$$

where symbols have their usual meanings. Also interpret the result.

17·73. (*a*) Consider a random sample $X_1, X_2, \ldots, X_n$ from an $U[0, \theta]$ population. Show that R and R/ξ are the confidence limits for θ with confidence coefficient $(1-\alpha)$, where R is the sample range and ξ satisfies the equation:

$$\xi^{n-1}\{n-(n-1)\xi\} = \alpha$$

(*b*) Obtain 100 $(1-\alpha)$% confidence interval for the population correlation coefficient 'ρ' when the random sample of size n has been drawn from bivariate normal population.

17·74. (*a*) Obtain a large sample 100 $(1-\alpha)$% confidence interval for the parameter θ in random sampling from the population :

$$dF(x) = \theta\exp(-\theta x)\,dx\,; x > 0, \ \theta > 0$$

(*b*) Develop a general method for obtaining confidence intervals. Obtain a 100$(1-\alpha)$% confidence interval for large sample size for the parameter θ of the Poisson distribution.

(*c*) Show that for the exponential distribution : $dF(x) = \theta\, e^{-\theta x}\, dx, \ x \ge 0,$

central confidence limits for θ for large samples of size n and 95% confidence coefficient are :

$$\{1 \pm 1{\cdot}96/\sqrt{n}\}/\bar{x}$$

where $\bar{x}$ is the mean of the sample observations $x_1, x_2, \ldots, x_n$ drawn randomly from the exponential population.

17·75. (*a*) Let $X_1, X_2, \ldots, X_n$ be a random sample from a distribution with *p.d.f.*:

$$f(x, \theta) = \theta\, exp(-\theta x), 0 \le x < \infty, \theta > 0$$

Find a 100 $(1-\alpha)$, (when $0 < \alpha < 1$), per cent confidence interval for the mean of this population, for large samples.

(*b*) Consider a random sample $X_1, X_2, \ldots, X_n$ from the exponential distribution with p.d.f.

$$f(x,\theta,p) = \begin{cases} \dfrac{\exp(-x/\theta)\cdot x^{p-1}}{\Gamma p\, \theta^p}, & x > 0 \\ 0, & \text{otherwise} \end{cases}$$

If p is known, obtain a confidence interval for θ, starting from the sufficient statistic $\bar{x}/p$.

CHAPTER

18 Statistical Inference-II (Testing of Hypothesis and Non-parametric Methods)

LEARNING OBJECTIVES

Upon completion of this Chapter, you should be able to:

- Understand the various concepts related to the testing of Hypothesis.
- Explain the various steps involved in Testing of Hypothesis problem.
- Understand the concept of optimum test under different situations.
- State and prove Neyman and Pearson Lemma developing the most powerful (M.P.) test for testing simple hypothesis against simple alternative hypothesis.
- Understand the concept of Likelihood Ratio Test and learn about:

(a) Its important properties.

(b) Its use in developing various tests of significance for mean(s) and variance(s) of normal population(s).

(c) Differentiate between Parametric Tests and Non-parametric Tests.

(d) Understand the advantages and limitations of Non-parametric Methods over Parametric Methods.

(e) Develop various Non-parametric Tests like:

  (i) Wald-Wolfowitz Run Test.

  (ii) Median Test.

  (iii) Sign Test.

  (iv) Mann-Whitney-Wilcoxon U-Test.

(f) Learn and develop A. Wald's Sequential Probability Ratio Test (S.P.R.T.) and understand its difference from the Neyman-Pearson Theory of Testing of Hypothesis.

(g) Obtain its S.P.R.T's:

  (i) Operating Characteristic Function (O.C. Function) and

  (ii) Average Sample Number (A.S.N.).

CHAPTER OUTLINES

18·1. INTRODUCTION

The main problems in statistical inference can be broadly classified into two areas :

(*i*) The area of estimation of population parameters(s) and setting up of confidence intervals for them, *i.e.*, the area of *point and interval estimation* and,

(*ii*) *Tests of statistical hypothesis.*

The first topic has already been discussed in Chapter 17. In this chapter we shall discuss:

(*a*) The theory of testing of hypothesis initiated by J. Neyman and E.S. Pearson (§ 18.2)

(*b*) Sequential analysis propounded by Abraham Wald (§ 18·8) and

(*c*) Non-parametric tests (§ 18·7).

In Neyman-Pearson theory, we use statistical methods to arrive at decisions in certain situations where there is lack of certainty on the basis of a sample whose size is fixed in advance while in Wald's sequential theory the sample size is not fixed but is regarded as a random variable. Before taking up a detailed discussion of the topics in (*a*), (*b*) and (*c*), we shall explain below certain concepts which are of fundamental importance.

18·2. STATISTICAL HYPOTHESIS — SIMPLE AND COMPOSITE

A statistical hypothesis is some statement or assertion about a population or equivalently about the probability distribution characterising a population, which we want to verify on the basis of information available from a sample. If the statistical hypothesis specifies the population completely then it is termed as *a simple statistical hypothesis* otherwise it is called a *composite statistical hypothesis.*

For example, if $X_1, X_2, ..., X_n$ is a random sample of size n from a normal population with mean μ and variance σ^2, then the hypothesis $H_0 : \mu = \mu_0, \sigma^2 = \sigma_0^2$ is a simple hypothesis, whereas each of the following hypotheses is a composite hypothesis:

(*i*) $\mu = \mu_0$, (*ii*) $\sigma^2 = \sigma_0^2$, (*iii*) $\mu < \mu_0,\ \sigma^2 = \sigma_0^2$,

(*iv*) $\mu > \mu_0, \sigma^2 = \sigma_0^2$ (*v*) $\mu = \mu_0,\ \sigma^2 < \sigma_0^2$, (*vi*) $\mu = \mu_0,\ \sigma^2 > \sigma_0^2$,

(*vii*) $\mu < \mu_0, \sigma^2 > \sigma_0^2$.

A hypothesis which does not specify completely '*r*' parameters of a population is termed as a composite hypothesis with *r* degrees of freedom.

18·2·1. Test of a Statistical Hypothesis

A test of a statistical hypothesis is a two action decision problem after the experimental sample values have been obtained, the two actions being the acceptance or rejection of the hypothesis under consideration.

18·2·2. Null Hypothesis

In hypothesis testing, a statistician or decision maker should not be motivated by prospects of profit or loss resulting from the acceptance or rejection of the hypothesis. He should be completely impartial and should have no brief for any party or company nor should he allow his personal views to influence the decision. *Much, therefore, depends upon how the hypothesis is framed.* For example, let us consider the 'light-bulbs' problem. Let us suppose that the bulbs manufactured under some standard manufacturing process have an average life of μ hours and it is proposed to test a new procedure for manufacturing light bulbs. Thus, we have two populations of bulbs, those manufactured by standard process and those manufactured by the new process. In this problem the following three hypotheses may be set up:

(*i*) New process is better than standard process.

(*ii*) New process is inferior to standard process.

(*iii*) There is no difference between the two processes.

The first two statements appear to be based since they reflect a preferential attitude to one or the other of the two processes. Hence, the best course is to adopt the hypothesis of *no difference,* as stated in (*iii*). This suggests that the *statistician should take up the neutral or null attitude regarding the outcome of the test.* His attitude should be on the null or zero line in which the experimental data has the due importance and complete say in the matter. *This neutral or non-committal attitude of the statistician or decision-maker before the sample observations are taken is the keynote of the null hypothesis.*

Thus in the above example of light bulbs if μ_0 is the mean life (in hours) of the bulbs manufactured by the new process then the null hypothesis which is usually denoted by H_0, can be stated as follows: H_0: $\mu = \mu_0$.

As another example let us suppose that two different concerns manufacture drugs for inducing sleep, drug *A* manufactured by first concern and drug *B* manufactured by second concern. Each company claims that its drug is superior to that of the other and it is desired to test which is a superior drug *A* or *B*? To formulate the statistical hypothesis let *X* be a random variable which denotes the additional hours of sleep gained by an individual when drug *A* is given and let the random variable *Y* denote the additional hours of sleep gained when drug *B* is used. Let us suppose that *X* and *Y* follow the probability distributions with means μ_X and μ_Y respectively. Here our *null hypothesis would be that there is no difference between the effects of two drugs.* Symbolically, $H_0 : \mu_X = \mu_Y$.

18·2·3. Alternative Hypothesis

It is desirable to state what is called an *alternative hypothesis* in respect of every statistical hypothesis being tested because the acceptance or rejection of null hypothesis is meaningful only when it is being tested against a rival hypothesis which should rather be explicitly mentioned. *Alternative hypothesis* is usually denoted by H_1. For example, in the example of light bulbs, alternative hypothesis could be $H_1 : \mu > \mu_0$ or $\mu < \mu_0$ or $\mu \neq \mu_0$. In the example of drugs, the alternative hypothesis could be $H_1 : \mu_X > \mu_Y$ or $\mu_X < \mu_Y$ or $\mu_X \neq \mu_Y$.

In both the cases, the first two of the alternative hypotheses give rise to what are called '*one tailed*' tests and the third alternative hypothesis results in '*two tailed*' tests.

Remarks:

1. In the formulation of a testing problem and devising a 'test of hypothesis' the roles of H_0 and H_1 are not at all symmetric. In order to decide which one of the two hypotheses should be taken as null hypothesis H_0 and which one as alternative hypothesis H_1, the intrinsic difference between the roles and the simplifications of these two terms should be clearly understood.
2. If a particular problem cannot be stated as a test between two simple hypotheses, *i.e.,* simple null hypothesis against a simple alternative hypothesis, then the next best alternative is to formulate the problem as the test of a simple null hypothesis against a composite alternative hypothesis. In other words, one should try to structure the problem so that null hypothesis is simple rather than composite.
3. Keeping in mind the potential losses due to wrong decisions (which may or may not be measured in terms of money), the decision maker is somewhat conservative in holding the null hypothesis as true unless there is a strong evidence from the experimental sample observations that it is false. To him, the *consequences of wrongly rejecting a null hypothesis seem to be more severe than those of wrongly accepting it.* In most of the cases, the statistical hypothesis is in the form of a claim that a particular

product or product process is superior to some existing standard. The null hypothesis H_0 in this case is that there is no difference between the new product or production process and the existing standard. In other words, null hypothesis nullifies this claim. The rejection of the null hypothesis wrongly which amounts to the acceptance of claim wrongly involves huge amount of pocket expenses towards a substantive overhaul of the existing set-up. The resulting loss is comparatively regarded as more serious than the opportunity loss in wrongly accepting H_0 which amounts to wrongly rejecting the claim, *i.e.*, in sticking to the less efficient existing standard. In the light-bulbs problem discussed earlier, suppose the research division of the concern, on the basis of the limited experimentation, claims that its brand is more effective than that manufactured by standard process. If in fact, the brand fails to be more effective the loss incurred by the concern due to an immediate *obsolescence* of the product, decline of the concern's image, etc., will be quite serious. On the other hand, the failure to bring out a superior brand in the market is an opportunity loss and is not a consideration to be as serious as the other loss.

18·2·4. Critical Region

Let $x_1, x_2, ..., x_n$ be the sample observations denoted by O. All the values of O will be aggregate of a sample and they constitute a space, called the *sample space,* which is denoted by S.

Since the sample values $x_1, x_2, ..., x_n$ can be taken as a point in n-dimensional space, we specify some region of the n-dimensional space and see whether this point lies within this region or outside this region. We divide the whole sample space S into two disjoint parts W and $S - W$ or $\bar{W}$ or W'. The null hypothesis H_0 is rejected if the observed sample point falls in W and if it falls in W' we reject H_1 and accept H_0. *The region of rejection of H_0 when H_0 is true is that region of the outcome set where H_0 is rejected if the sample point falls in that region and is called critical region.* Evidently, the size of the critical region is α, the probability of committing type 1 error (discussed below).

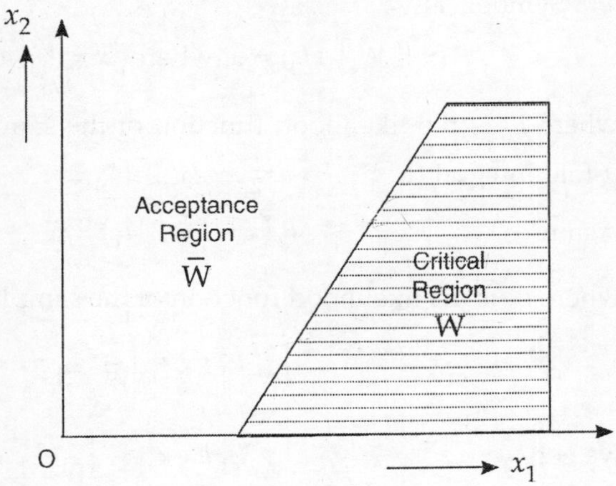

Suppose if the test is based on a sample of size 2, then the outcome set or the sample space is the first quadrant in a two-dimensional space and a test criterion will enable us to separate our outcome set into two complementary subsets, W and $\bar{W}$. If the sample point falls in the subset W, H_0 is rejected, otherwise H_0 is accepted. This is shown in the above diagram.

18·2·5. Two Types of Errors in Testing of Hypothesis

The decision to accept or reject the null hypothesis H_0 is made on the basis of the information supplied by the observed sample observations. The conclusion drawn on the basis of a particular sample may not always be true in respect of the population. The four possible situations that arise in any test procedure are given in the table on page 18·6.

DOUBLE DICHOTOMY RELATING TO DECISION AND HYPOTHESIS

		Decision From Sample	
		Reject H_0	*Accept H_0*
True State	*H_0 True*	Wrong (Type I Error)	Correct
	H_0 False (*H_1 True*)	Correct	Wrong (Type II Error)

From the above table it is obvious that in any testing problem we are liable to commit two types of errors.

Errors of Type I and Type II. The error of rejecting H_0 (accepting H_1) when H_0 is true is called *Type I error* and the error of accepting H_0 when H_0 is false (H_1 is true) is called *Type II* error. The probabilities of type I and type II errors are denoted by α and β respectively. Thus

α = Probability of type I error
= Probability of rejecting H_0 when H_0 is true.

β = Probability of type II error
= Probability of accepting H_0 when H_0 is false.

Symbolically :

$$P(\mathbf{x} \in W \mid H_0) = \alpha, \text{ where } \mathbf{x} = (x_1, x_2, \ldots, x_n) \quad \Rightarrow \quad \int_W L_0 \, d\mathbf{x} = \alpha \qquad \ldots (18{\cdot}1)$$

where L_0 is the likelihood function of the sample observations under H_0 and $\int d\mathbf{x}$ represents the n-fold integral $\iint \ldots \int dx_1 \, dx_2 \ldots dx_n$.

Again
$$P(\mathbf{x} \in \overline{W} \mid H_1) = \beta \quad \Rightarrow \quad \int_{\overline{W}} L_1 \, d\mathbf{x} = \beta \qquad \ldots (18{\cdot}2)$$

where L_1 is the likelihood function of the sample observations under H_1. Since

$$\int_W L_1 \, d\mathbf{x} + \int_{\overline{W}} L_1 \, d\mathbf{x} = 1,$$

we get
$$\int_W L_1 \, d\mathbf{x} = 1 - \int_{\overline{W}} L_1 \, d\mathbf{x} = 1 - \beta \qquad \ldots (18{\cdot}2a)$$

$$\Rightarrow \quad P(\mathbf{x} \in W \mid H_1) = 1 - \beta \qquad \ldots (18{\cdot}2b)$$

18·2·6. Level of Significance

α, the probability of type 1 error, is known as the level of significance of the test. It is also called the *size of the critical region.*

Remarks:

1. In practice, most commonly used levels of significance are 0·05 (5%) and 0·01 (1%).
2. Normally, level of significance is always fixed in advance before collecting the sample information.

18·2·7. Power of the Test

1 – β, defined in (18·2*a*) and (18·2*b*) is called the *power function of the test hypothesis H_0 against the alternative hypothesis H_1*. The value of the power function at a *parameter* point is called the *power of the test* at that point.

Remarks:

1. In quality control terminology, α and β are termed as *producer's risk and consumer's risk,* respectively.

2. An ideal test would be the one which properly keeps under control both the types of errors. But since the commission of an error of either type is a random variable, equivalently an ideal test should minimise the probability of both the types of errors, *viz.*, α and β. But unfortunately, for a fixed sample size n, α and β are so related (like producer's and consumer's risk in sampling inspection plans), that the reduction in one results in an increase in the other. Consequently, the simultaneous minimising of both the errors is not possible. Since type I error is deemed to be more serious than the type II error (*c.f.* Remark 3 § 18·2·3) the usual practice is to control α at a predetermined low level and subject to this constraint on the probabilities of type I error, *choose a test which minimises β or maximises the power function* $(1 - \beta)$. Generally, we choose $\alpha = 0{\cdot}05$ or $0{\cdot}01$.

18·3. STEPS IN SOLVING TESTING OF HYPOTHESIS PROBLEM

The major steps involved in the solution of a 'testing of hypothesis' problem may be outlined as follows:

1. Explicit knowledge of the nature of the population distribution and the parameter(s) of interest, *i.e.*, the parameter(s) about which the hypotheses are set up.
2. Setting up of the null hypothesis H_0 and the alternative hypothesis H_1 in terms of the range of the parameter values each one embodies.
3. The choice of a suitable statistic $t = t\,(x_1, x_2, \ldots, x_n)$ called the *test statistic*, which will best reflect upon the probability of H_0 and H_1.
4. Partitioning the set of possible values of the test statistic t into two disjoint sets W (called the *rejection region* or *critical region*) and $\overline{W}$ (called the *acceptance region*) and framing the following test:
   (*i*) Reject H_0 (*i.e.*, accept H_1) if the value of t falls in W.
   (*ii*) Accept H_0 if the value of t falls in $\overline{W}$.
5. After framing the above test, obtain experimental sample observations, compute the appropriate test statistic and take action accordingly.

18·4. OPTIMUM TEST UNDER DIFFERENT SITUATIONS

The discussion in § 18·3 and Remark 2, § 18·2·7 enables us to obtain the so called best test under different situations. In any testing problem the first two steps, *viz.*, the form of the population distribution, the parameter(s) of interest and the framing of H_0 and H_1 should be obvious from the description of the problem. The most crucial step is the choice of the '*best test, i.e.*, the best statistic 't' and the critical region W where *by best test we mean one which in addition to controlling α at any desired low level has the minimum type II error β or maximum power $(1 - \beta)$, compared to β of all other tests having this α.* This leads to the following definition.

18·4·1. Most Powerful Test (MP Test)

Let us consider the problem of testing a simple hypothesis :

$$H_0 : \theta = \theta_0 \text{ against a simple alternative hypothesis : } H_1 : \theta = \theta_1.$$

Definition. *The critical region W is the most powerful (MP) critical region of size α (and the corresponding test a most powerful test of level α) for testing $H_0 : \theta = \theta_0$ against $H_1 : \theta = \theta_1$ if*

$$P\,(\mathbf{x} \in W \mid H_0) = \int_W L_0\, d\mathbf{x} = \alpha \qquad \ldots (18{\cdot}3)$$

and

$$P\,(\mathbf{x} \in W \mid H_1) \geq P\,(\mathbf{x} \in W_1 \mid H_1) \qquad \ldots (18{\cdot}3a)$$

for every other critical region W_1 satisfying (18·3).

18·4·2. Uniformly Most Powerful Test (UMP Test)

Let us now take up the case of testing a simple null hypothesis against a composite alternative hypothesis, *e.g.*, of testing $H_0 : \theta = \theta_0$ against the alternative $H_1 : \theta \neq \theta_0$

In such a case, for a predetermined α, the best test for H_0 is called the *uniformly most powerful test of level* α.

Definition. *The region W is called uniformly most powerful (UMP) critical region of size α [and the corresponding test as uniformly most powerful (UMP) test of level α] for testing $H_0 : \theta = \theta_0$ against $H_1 : \theta \neq \theta_0$ i.e., $H_1 : \theta = \theta_1 \neq \theta_0$ if*

$$P(\mathbf{x} \in W \mid H_0) = \int_W L_0 \, dx = \alpha \quad \text{......(18·4)}$$

and

$$P(\mathbf{x} \in W \mid H_1) \geq P(\mathbf{x} \in W_1 \mid H_1) \text{ for all } \theta \neq \theta_0, \quad \text{.....(18·4a)}$$

whatever the region W_1 satisfying (18·4) may be.

18·5. NEYMAN J. AND PEARSON, E.S. LEMMA

This Lemma provides the most powerful test of simple hypothesis against a simple alternative hypothesis. The theorem, known as Neyman-Pearson Lemma, will be proved for density function $f(x, \theta)$ of a single continuous variate and a single parameter. However, by regarding x and θ as vectors, the proof can be easily generalised for any number of random variables $x_1, x_2, ..., x_n$ and any number of parameters $\theta_1, \theta_2, ..., \theta_k$. The variables $x_1, x_2 .... x_n$ occurring in this theorem are understood to represent a random sample of size n from the population whose density function is $f(x, \theta)$. The lemma is concerned with a simple hypothesis $H_0 : \theta = \theta_0$ and a simple alternative $H_1 : \theta = \theta_1$.

Neyman Pearson Lemma. *Let $k > 0$, be a constant and W be a critical region of size α such that*

$$W = \left\{ \mathbf{x} \in S : \frac{f(x, \theta_1)}{f(x, \theta_0)} \geq k \right\}$$

$$\Rightarrow \quad W = \left\{ \mathbf{x} \in S : \frac{L_1}{L_0} \geq k \right\} \quad \text{... (18·5)}$$

and

$$\bar{W} = \left\{ \mathbf{x} \in S : \frac{L_1}{L_0} < k \right\} \quad \text{... (18·5a)}$$

where L_0 and L_1 are the likelihood functions of the sample observations $\mathbf{x} = (x_1, x_2, ..., x_n)$ under H_0 and H_1 respectively. Then, W is the most powerful critical region of the test hypothesis $H_0 : \theta = \theta_0$ against the alternative $H_1 : \theta = \theta_1$.

Proof. We are given $P(\mathbf{x} \in W \mid H_0) = \int_W L_0 \, dx = \alpha$ ... (18·6)

The power of the region is: $P(\mathbf{x} \in W \mid H_1) = \int_W L_1 \, dx = 1 - \beta$, (say). ... (18·6a)

In order to establish the lemma, we have to prove that there exists no other critical region, of size less than or equal to α, which is more powerful than W. Let W_1 be another critical region of size $\alpha_1 \leq \alpha$ and power $1 - \beta_1$ so that we have

$$P(\mathbf{x} \in W_1 \mid H_0) = \int L_0 \, dx = \alpha_1 \quad \text{... (18·7)}$$

and

$$P(\mathbf{x} \in W_1 \mid H_1) = \int_{W_1} L_1 \, dx = 1 - \beta_1 \quad \text{... (18·7a)}$$

We have to prove that : $1-\beta \geq 1-\beta_1$

Let $W = A \cup C$ and $W_1 = B \cup C$

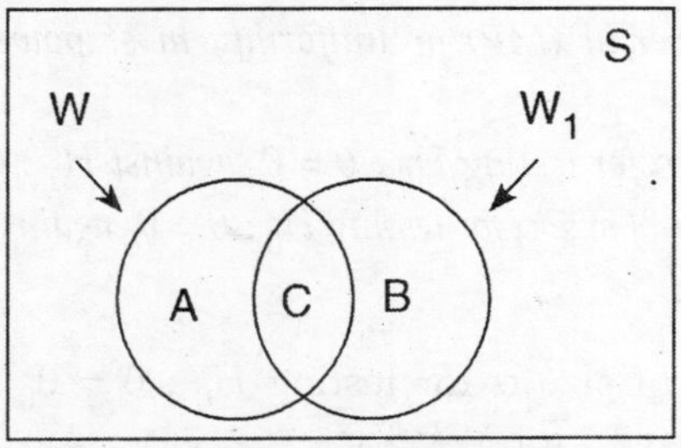

(C may be empty, *i.e.*, W and W_1 may be disjoint).

If $\alpha_1 \leq \alpha$, we have

$$\int_{W_1} L_0\, d\mathbf{x} \leq \int_W L_0\, d\mathbf{x}$$

$$\Rightarrow \quad \int_{B\cup C} L_0\, d\mathbf{x} \leq \int_{A\cup C} L_0\, d\mathbf{x}$$

$$\Rightarrow \quad \int_B L_0\, d\mathbf{x} \leq \int_A L_0\, d\mathbf{x}$$

$$\Rightarrow \quad \int_A L_0\, d\mathbf{x} \geq \int_B L_0\, d\mathbf{x} \qquad \text{... (18·8)}$$

Since $A \subset W$,

$$\text{(18·5)} \quad \Rightarrow \quad \int_A L_1\, d\mathbf{x} > k\int_A L_0\, d\mathbf{x} \geq k\int_B L_0\, d\mathbf{x} \qquad [\text{Using (18·8)}] \qquad \text{... (18·8a)}$$

Also [18·5 (*a*)] implies

$$\frac{L_1}{L_0} \leq k \;\forall\; x \in \overline{W} \quad \Rightarrow \quad \int_{\overline{W}} L_1\, d\mathbf{x} \leq k\int_{\overline{W}} L_0\, d\mathbf{x}$$

This result also holds for any subset of $\overline{W}$, say $\overline{W} \cap W_1 = B$. Hence,

$$\int_B L_1\, d\mathbf{x} \leq k\int_B L_0\, d\mathbf{x} \leq \int_A L_1\, d\mathbf{x} \qquad [\text{From (18·8a)}]$$

Adding $\int_C L_1\, d\mathbf{x}$ to both sides, we get

$$\int_{W_1} L_1\, d\mathbf{x} \leq \int_W L_1\, d\mathbf{x} \quad \Rightarrow \quad 1-\beta \geq 1-\beta_1$$

Hence, the Lemma.

Remark. Let W defined in (18·5) of the above theorem be the most powerful critical region of size α for testing $H_0 : \theta = \theta_0$ against $H_1 : \theta = \theta_1$, and let it be independent of $\theta_1 \in \Theta_1 = \Theta - \Theta_0$, where Θ_0 is the parameter space under H_0. Then we say that C.R. W is the UMP CR of size α for testing : $H_0 : \theta = \theta_0$, against $H_1 : \theta \in \Theta_1$.

18·5·1. Unbiased Test and Unbiased Critical Region

Let us consider the testing of $H_0 : \theta = \theta_0$ against $H_1 : \theta = \theta_1$. *The critical region W and consequently the test based on it is said to be unbiased if the power of the test exceeds the size of the critical region, i.e., if*

$$\text{Power of the test} \geq \text{Size of the C.R.} \qquad \text{... (18.9)}$$

$$\Rightarrow \quad 1-\beta \geq \alpha$$

$$\Rightarrow \quad P_{\theta_1}(W) \geq P_{\theta_0}(W)$$

$$\Rightarrow \quad P[\mathbf{x} : \mathbf{x} \in W \mid H_1] \geq P[\mathbf{x} : \mathbf{x} \in W \mid H_0] \qquad \text{... (18·9a)}$$

In other words, the critical region W is said to be unbiased if

$$P_\theta(W) \geq P_{\theta_0}(W),\ \forall\ \theta\ (\neq \theta_0) \in \Theta \qquad \ldots (18\cdot 9b)$$

Theorem 18·2. *Every most powerful (MP) or uniformly most powerful (UMP) critical region (CR) is necessarily unbiased.*

(*a*) *If W be an MPCR of size α for testing $H_0 : \theta = \theta_0$ against $H_1 : \theta = \theta_1$, then it is necessarily unbiased.*

(*b*) *Similarly if W be UMPCR of size α for testing $H_0 : \theta = \theta_0$ against $H_1 : \theta \in \Theta_1$, then it is also unbiased.*

Proof.

(*a*) Since W is the MPCR of size α for testing $H_0 : \theta = \theta_0$ against $H_1 : \theta = \theta_1$, by Neyman Pearson Lemma, we have ; for $\forall\ k > 0$,

$$W = \{\mathbf{x} : L(\mathbf{x}, \theta_1) \geq k\, L(\mathbf{x}, \theta_0\} = \{\mathbf{x} : L_1 \geq k\, L_0\}$$

and $$W' = \{\mathbf{x} : L(\mathbf{x}, \theta_1) < k\, L(\mathbf{x}, \theta_0)\} = \{\mathbf{x} : L_1 < k\, L_0\},$$

where k is determined so that the size of the test is α *i.e.*,

$$P_{\theta_0}(W) = P[\mathbf{x} \in W \mid H_0] = \int_W L_0\, d\mathbf{x} = \alpha \qquad \ldots (i)$$

To prove that W is unbiased, we have to show that:

$$\text{Power of } W \geq \alpha \quad i.e., \quad P_{\theta_1}(W) \geq \alpha \qquad \ldots (ii)$$

We have : $$P_{\theta_1}(W) = \int_W L_1\, d\mathbf{x} \geq k \int_W L_0\, d\mathbf{x} = k\alpha \quad [\because \text{On } W,\ L_1 \geq k\, L_0 \text{ and Using } (i)]$$

i.e., $$P_{\theta_1}(W) \geq k\alpha,\ \forall\ k > 0 \qquad \ldots (iii)$$

Also

$$1 - P_{\theta_1}(W) = 1 - P(\mathbf{x} \in W \mid H_1) = P(\mathbf{x} \in W' \mid H_1) = \int_{W'} L_1\, d\mathbf{x}$$

$$< k \int_{W'} L_0\, d\mathbf{x} = k\, P(\mathbf{x} : \mathbf{x} \in W' \mid H_0) \qquad [\because \text{On } W',\ L_1 < k\, L_0]$$

$$= k[1 - P(\mathbf{x} : \mathbf{x} \in W \mid H_0)] = k(1 - \alpha) \qquad [\text{Using } (i)]$$

i.e., $$1 - P_{\theta_1}(W) \leq k(1 - \alpha),\ \forall\ k > 0 \qquad \ldots (iv)$$

Case (*i*) $k \geq 1$. If $k \geq 1$, then from (*iii*), we get

$$P_{\theta_1}(W) \geq k\alpha \geq \alpha \quad \Rightarrow \quad W \text{ is unbiased CR.}$$

Case (*ii*) $0 < k < 1$. If $0 < k < 1$, then from (*iv*), we get :

$$1 - P_{\theta_1}(W) < 1 - \alpha \quad \Rightarrow \quad P_{\theta_1}(W) > \alpha \quad \Rightarrow \quad W \text{ is unbiased C.R.}$$

Hence, MP critical region is unbiased.

(*b*) If W is UMPCR of size α then also the above proof holds if for θ_1 we write θ such that $\theta \in \Theta_1$. So we have

$$P_\theta(W) > \alpha,\ \forall\ \theta \in \Theta_1 \quad \Rightarrow \quad W \text{ is unbiased CR.}$$

18·5·2. Optimum Regions and Sufficient Statistics

Let $x_1, x_2, \ldots, x_n$ be a random sample of size n from a population with p.m.f. or p.d.f. $f(x, \theta)$, where the parameter θ may be a vector. Let T be a sufficient statistic for θ. Then by Factorization Theorem,

$$L(\mathbf{x}, \theta) = \prod_{i=1}^{n} f(x_i, \theta) = g_\theta\{t(\mathbf{x})\}.\, h(\mathbf{x}) \quad \ldots (*)$$

where $g_\theta\{t(\mathbf{x})\}$ is the marginal distribution of the statistic $T = t(\mathbf{x})$.

By Neyman Pearson Lemma, the MPCR for testing $H_0 : \theta = \theta_0$ against $H_1 : \theta = \theta_1$ is given by:

$$W = \{\mathbf{x} : L(\mathbf{x}, \theta_1) \geq k\, L(\mathbf{x}, \theta_0)\}, \forall\, k > 0 \quad \ldots (**)$$

From (*) and (**), we get

$$W = \{\mathbf{x} : g_{\theta_1}(t(\mathbf{x})) \,.\, h(\mathbf{x}) \geq k.\, g_{\theta_0}(t(\mathbf{x}))\,.\, h(\mathbf{x})\}, \forall\, k > 0$$

$$= \{\mathbf{x} : g_{\theta_1}(t(\mathbf{x})) \geq k \,.\, g_{\theta_0}(t(\mathbf{x}))\}, \forall\, k > 0$$

Hence, if $T = t(\mathbf{x})$ is sufficient statistic for θ then the MPCR for the test may be defined in terms of the marginal distribution of $T = t(\mathbf{x})$, rather than the joint distribution of $x_1, x_2, \ldots, x_n$.

Example 18·1. *Given the frequency function:* $f(x,\theta) = \begin{cases} \dfrac{1}{\theta}, & 0 \leq x \leq \theta \\ 0, & elsewhare \end{cases}$

and that you are testing the null hypothesis $H_0 : \theta = 1$ against $H_1 : \theta = 2$, by means of a single observed value of x. What would be the sizes of the type 1 and type II errors, if you choose the interval (i) $0{\cdot}5 \leq x$, (ii) $1 \leq x \leq 1{\cdot}5$ as the critical regions? Also obtain the power function of the test.

Solution. We want to test $H_0 : \theta = 1$, against $H_1 : \theta = 2$.

(*i*) We are given: $W = \{x : 0{\cdot}5 \leq x\} = \{x : x \geq 0{\cdot}5\}$ and $\bar{W} = \{x : x < 0{\cdot}5\}$

$$\alpha = P\{x \in W \mid H_0\} = P\{x \geq 0{\cdot}5 \mid \theta = 1\} = P\{0{\cdot}5 \leq x \leq \theta \mid \theta = 1\}$$

$$= P\{0{\cdot}5 \leq x \leq 1 \mid \theta = 1\} = \int_{0\cdot5}^{1} [f(x,\theta)]_{\theta=1}\, dx = \int_{0\cdot5}^{1} 1{\cdot}dx = 0{\cdot}5$$

Similarly, $\beta = P\{x \in \bar{W} \mid H_1\} = P\{x \leq 0{\cdot}5 \mid \theta = 2\}$

$$= \int_0^{0\cdot5} [f(x,\theta)]_{\theta=2}\, dx = \int_0^{0\cdot5} \frac{1}{2} dx = 0{\cdot}25$$

Thus, the sizes of type I and type II errors are respectively: $\alpha = 0{\cdot}5$ and $\beta = 0{\cdot}25$ and power function of the test $= 1 - \beta = 0{\cdot}75$

(*ii*) $W = \{x : 1 \leq x \leq 1{\cdot}5\} \Rightarrow \alpha = P\{x \in W \mid \theta = 1\} = \int_1^{1\cdot5} [f(x,\theta)]_{\theta=1}\, dx = 0,$

since under $H_0 : \theta = 1, f(x, \theta) = 0$, for $1 \leq x \leq 1{\cdot}5$.

$$\beta = P\{x \in \bar{W} \mid \theta = 2\} = 1 - P\{x \in W \mid \theta = 2\} = 1 - \int_1^{1\cdot5} [f(x,\theta)]_{\theta=2}\, dx = 1 - \left|\frac{x}{2}\right|_1^{1\cdot5} = 0{\cdot}75$$

$\therefore$ Power Function $= 1 - \beta = 1 - 0{\cdot}75 = 0{\cdot}25$

Example 18·2. *If $x \geq 1$ is the critical region for testing $H_0 : \theta = 2$ against the alternative $\theta = 1$, on the basis of the single observation from the population,*

$$f(x, \theta) = \theta \exp(-\theta x),\ 0 \leq x < \infty,$$

obtain the values of type I and type II errors.

Solution. Here $W = \{x : x \geq 1\}$ and $\bar{W} = \{x : x < 1\}$ and $H_0 : \theta = 2$, $H_1 : \theta = 1$.

$$\alpha = \text{Size of Type I error} = P[x \in W \mid H_0] = P[x \geq 1 \mid \theta = 2]$$

$$= \int_1^\infty [f(x, \theta)]_{\theta = 2} dx = 2 \int_1^\infty e^{-2x} dx = 2 \left| \frac{e^{-2x}}{-2} \right|_1^\infty = \frac{1}{e^2}$$

β = Size of type II error $= P[x \in \overline{W} \mid H_1] = P\{x < 1 \mid \theta = 1\}$

$$= \int_0^1 e^{-x} dx = \left| \frac{e^{-x}}{-1} \right|_0^1 = (1 - e^{-1}) = \frac{e-1}{e}.$$

Example 18·3. *Let p be the probability that a coin will fall head in a single toss in order to test $H_0 : p = \frac{1}{2}$ against $H_1 : p = \frac{3}{4}$. The coin is tossed 5 times and H_0 is rejected if more than 3 heads are obtained. Find the probability of type I error and power of the test.*

Solution. Here $H_0 : p = \frac{1}{2}$ and $H_1 : p = \frac{3}{4}$.

If the *r.v.* X denotes the number of heads in *n* tosses of a coin then $X \sim B(n, p)$ so that

$$P(X = x) = {}^nC_x\, p^x (1-p)^{n-x} = {}^5C_x\, p^x (1-p)^{5-x} ; x = 0, 1, 2, \ldots, 5 \quad \ldots (*)$$

The critical region is given by: $W = \{x : x \ge 4\} \Rightarrow \overline{W} = \{x : x \le 3\}$

α = Probability of type I error $= P[X \ge 4 \mid H_0]$

$$= P\left(X = 4 \mid p = \frac{1}{2}\right) + P\left(X = 5 \mid p = \frac{1}{2}\right) = {}^5C_4 \left(\frac{1}{2}\right)^4 \left(\frac{1}{2}\right)^{5-4} + {}^5C_5 \left(\frac{1}{2}\right)^5 \quad \text{[From (*)]}$$

$$= 5\left(\frac{1}{2}\right)^5 + \left(\frac{1}{2}\right)^5 = 6\left(\frac{1}{2}\right)^5 = \frac{3}{16}$$

β = Probability of Type II error $= P(x \in \overline{W} \mid H_1) = 1 - P(x \in W \mid H_1)$

$$= 1 - \left[\left\{P\left(X = 4 \mid p = \frac{3}{4}\right)\right\} + \left\{P\left(X = 5 \mid p = \frac{3}{4}\right)\right\}\right] = 1 - \left\{{}^5C_4 \left(\frac{3}{4}\right)^4 \left(\frac{1}{4}\right) + {}^5C_5 \left(\frac{3}{4}\right)^5\right\}$$

$$= 1 - \left(\frac{3}{4}\right)^4 \left\{\frac{5}{4} + \frac{3}{4}\right\} = 1 - \frac{81}{128} = \frac{47}{128}$$

$\therefore$ Power of the test $= 1 - \beta = \frac{81}{128}$

Example 18·4. *Let $X \sim N(\mu, 4)$, μ unknown. To test $H_0 : \mu = -1$ against $H_1 : \mu = 1$, based on a sample of size 10 from this population, we use the critical region : $x_1 + 2x_2 + \ldots + 10x_{10} \ge 0$. What is its size? What is the power of the test?*

Solution. Critical Region $W = \{\mathbf{x} : x_1 + 2x_2 + \ldots + 10x_{10} \ge 0\}$.

Let $U = x_1 + 2x_2 + \ldots + 10x_{10}$

Since x_i's are *i.i.d.* $N(\mu, 4)$,

$$U \sim N[(1 + 2 + \ldots + 10)\mu, (1^2 + 2^2 + \ldots + 10^2)\sigma^2] = N(55\mu, 385\sigma^2)$$

$$\Rightarrow \quad U \sim N(55\mu, 385 \times 4) = N(55\mu, 1540) \quad \ldots (*)$$

The size 'α' of the critical region is : $\alpha = P(\mathbf{x} \in W \mid H_0) = P(U \ge 0 \mid H_0) \quad \ldots (**)$

Under $H_0 : \mu = -1$, $U \sim N(-55, 1540)$ [From (*)] $\Rightarrow Z = \dfrac{U - E(U)}{\sigma_U} = \dfrac{U + 55}{\sqrt{1540}}$

$\therefore$ Under H_0, when $U = 0$, $Z = \dfrac{55}{\sqrt{1540}} = \dfrac{55}{39{\cdot}2428} = 1{\cdot}4015$

$\therefore$ $\alpha = P(Z \geq 1{\cdot}4015) = 0{\cdot}5 - P(0 \leq Z \leq 1{\cdot}4015)$ [From (**)]

$= 0{\cdot}5 - 0{\cdot}4192 = 0{\cdot}1808$ (From Normal Probability Tables)

Alternatively, $\alpha = 1 - P(Z \leq 1{\cdot}4015) = 1 - \Phi(1{\cdot}4015)$,

where $\Phi(\cdot)$ is the distribution function of standard normal variate.

Power of the test is : $1 - \beta = P(\mathbf{x} \in W \mid H_1) = P(U \geq 0 \mid H_1)$

Under $H_1 : \mu = 1$, $U \sim N(55, 1540)$

$\Rightarrow$ $Z = \dfrac{U - E(U)}{\sigma_U} = \dfrac{-55}{\sqrt{1540}} = -1{\cdot}40$, when $U = 0$

$\therefore$ $1 - \beta = P(Z \geq -1{\cdot}40) = P(-1{\cdot}4 \leq Z \leq 0) + 0{\cdot}5$

$= P(0 \leq Z \leq 1{\cdot}4) + 0{\cdot}5$ (By symmetry)

$= 0{\cdot}4192 + 0{\cdot}5 = 0{\cdot}9192$

Alternatively, $1 - \beta = 1 - P(Z \leq -1{\cdot}40) = 1 - \Phi(-1{\cdot}40)$,

where $\Phi(.)$ is the distribution function of standard normal variate.

Example 18·5. *Let X have a p.d.f. of the form:* $f(x, \theta) = \begin{cases} \dfrac{1}{\theta} e^{-x/\theta} ; 0 < x < \infty, \theta > 0 \\ 0; \quad \textit{elsewhere.} \end{cases}$

To test $H_0 : \theta = 2$, *against* $H_1 : \theta = 1$, *use the random sample* x_1, x_2 *of size 2 and define a critical region :*

$$W = \{(x_1, x_2) : 9{\cdot}5 \leq x_1 + x_2\}.$$

Find : (i) *Power of the test.*

(ii) *Significance level of the test.*

Solution. We are given the critical region: $W = \{(x_1, x_2) : 9{\cdot}5 \leq x_1 + x_2\} = \{(x_1\ x_2) : x_1 + x_2 \geq 9{\cdot}5\}$

Size of the critical region *i.e.*, the significance level of the test is given by:

$$\alpha = P(\mathbf{x} \in W \mid H_0) = P[x_1 + x_2 \geq 9{\cdot}5 \mid H_0] \quad \dots (*)$$

In sampling from the given exponential distribution,

$\dfrac{2}{\theta} \sum_{i=1}^{n} x_i : \chi^2_{(2n)}$ $\Rightarrow$ $U = \dfrac{2}{\theta}(x_1 + x_2) \sim \chi^2_{(4)}, (n = 2)$ [*c.f.* Example 18·8]

$\therefore$ $\alpha = P\left[\dfrac{2}{\theta}(x_1 + x_2) \geq \dfrac{2}{\theta} \times 9{\cdot}5 \mid H_0\right]$ [From (*)]

$= P\left[\chi^2_{(4)} \geq 9{\cdot}5\right]$ ($\because$ Under H_0, $\theta = 2$)

$\Rightarrow$ $\alpha = 0{\cdot}05$ [From Probability Tables of χ^2 - distribution]

Power of the test is given by

$$1 - \beta = P(\mathbf{x} \in W \mid H_1) = P(x_1 + x_2 \geq 9{\cdot}5 \mid H_1)$$

$= P\left[\dfrac{2}{\theta}(x_1 + x_2) \geq \dfrac{2}{\theta} \times 9{\cdot}5 \mid H_1\right] = P\left[\chi^2_{(4)} \geq 19\right]$ ($\because$ Under H_1, $\theta = 1$)

Example 18·6. *Use the Neyman-Pearson Lemma to obtain the region for testing* $\theta = \theta_0$ *against* $\theta = \theta_1 > \theta_0$ *and* $\theta = \theta_1 < \theta_0$, *in the case of a normal population* $N(\theta, \sigma^2)$, *where* σ^2 *is known. Hence find the power of the test.*

Solution.

$$L = \prod_{i=1}^{n} f(x_i, \theta) = \left(\frac{1}{\sigma\sqrt{2\pi}}\right)^n \exp\left\{-\frac{1}{2\sigma^2}\sum_{i=1}^{n}(x_i - \theta)^2\right\}$$

Using Neyman-Pearson Lemma, best critical region (B.C.R.) is given by (for $k > 0$)

$$\frac{L_1}{L_0} = \frac{\exp\left\{-\frac{1}{2\sigma^2}\sum_{i=1}^{n}(x_i - \theta_1)^2\right\}}{\exp\left\{-\frac{1}{2\sigma^2}\sum_{i=1}^{n}(x_i - \theta_0)^2\right\}} \geq k$$

$$\Rightarrow \qquad \exp\left[-\frac{1}{2\sigma^2}\left\{\sum_{i=1}^{n}(x_i - \theta_1)^2 - \sum_{i=1}^{n}(x_i - \theta_0)^2\right\}\right] \geq k$$

$$\Rightarrow \qquad \exp\left[-\frac{n}{2\sigma^2}\left(\theta_1^2 - \theta_0^2\right) + \frac{1}{\sigma^2}(\theta_1 - \theta_0)\sum_{i=1}^{n} x_i\right] \geq k$$

$$\Rightarrow \qquad -\frac{n}{2\sigma^2}\left(\theta_1^2 - \theta_0^2\right) + \frac{1}{\sigma^2}(\theta_1 - \theta_0)\sum_{i=1}^{n} x_i \geq \log k$$

(Since $\log x$ is an increasing function of x)

$$\Rightarrow \qquad \overline{x}\left(\theta_1 - \theta_0\right) \geq \frac{\sigma^2}{n}\log k + \frac{\theta_1^2 - \theta_0^2}{2}$$

Case (*i*) If $\theta_1 > \theta_0$, the B.C.R. is determined by the relation (right-tailed test):

$$\overline{x} \geq \frac{\sigma^2}{n} \cdot \frac{\log k}{\theta_1 - \theta_0} + \frac{\theta_1 + \theta_0}{2} \quad \Rightarrow \quad \overline{x} \geq \lambda_1 \text{ (say)}.$$

$\therefore$ B.C.R. is : $\quad W = \{\mathbf{x} : \overline{x} \geq \lambda_1\}$ ... (18·10)

Case (*ii*) If $\theta_1 < \theta_0$, the B.C.R. is given by the relation (left handed test)

$$\overline{x} \leq \frac{\sigma^2}{n} \cdot \frac{\log k}{\theta_1 - \theta_0} + \frac{\theta_1 + \theta_0}{2} = \lambda_2, \text{ (say)}.$$

Hence, B.C.R. is : $\quad W_1 = \{\mathbf{x} : \overline{x} \leq \lambda_2\}$ ... (18·11)

The constants λ_1 and λ_2 are so chosen as to make the probability of each of the relations (18·10) and (18·11) equal to α when the hypothesis H_0 is true. The sampling distribution of $\overline{x}$, when H_i is true is $N\left(\theta_i, \frac{\sigma^2}{n}\right)$, ($i$ = 0, 1). Therefore, the constants λ_1 and λ_2 are determined from the relations :

$$P[\overline{x} \geq \lambda_1 \mid H_0] = \alpha \quad \text{and} \quad P[\overline{x} \leq \lambda_2 \mid H_0] = \alpha$$

$$\therefore \qquad P(\overline{x} \geq \lambda_1 \mid H_0) = P\left[Z \geq \frac{\lambda_1 - \theta_0}{\sigma/\sqrt{n}}\right] = \alpha \ ; \ Z \sim N(0, 1)$$

$$\Rightarrow \qquad \frac{\lambda_1 - \theta_0}{\sigma/\sqrt{n}} = z_\alpha \quad \Rightarrow \quad \lambda_1 = \theta_0 + \frac{\sigma}{\sqrt{n}} z_\alpha \qquad \text{... (18·12)}$$

where z_α is the upper α-point of the standard normal variate given by: $P(Z > z_\alpha) = \alpha$ ... (*)

Also $\quad P(\overline{x} \leq \lambda_2 \mid H_0) = \alpha \quad \Rightarrow \quad P(\overline{x} > \lambda_2 \mid H_0) = 1 - \alpha$

$$\Rightarrow \quad P\left(Z > \frac{\lambda_2 - \theta_0}{\sigma/\sqrt{n}}\right) = 1 - \alpha \quad \Rightarrow \quad \frac{\lambda_2 - \theta_0}{\sigma/\sqrt{n}} = z_{1-\alpha} \quad \Rightarrow \quad \lambda_2 = \theta_0 + \frac{\sigma}{\sqrt{n}} \cdot z_{1-\alpha} \qquad \text{... (18·12a)}$$

Remark. By symmetry of normal distribution, we have: $\quad z_{1-\alpha} = -z_\alpha$.

Power of the test. By definition, the power of the test in case (*i*) is:

$$1-\beta = P[\mathbf{x} \in W \mid H_1] = P[\,\bar{x} \geq \lambda_1 \mid H_1]$$

$$= P\left(Z \geq \frac{\lambda_1-\theta_1}{\sigma/\sqrt{n}}\right) \qquad \left[\because \text{Under } H_1, Z = \frac{\bar{x}-\theta_1}{\sigma/\sqrt{n}} \sim N(0,1)\right]$$

$$= P\left(Z \geq \frac{\theta_0 + \frac{\sigma}{\sqrt{n}} z_\alpha - \theta_1}{\sigma/\sqrt{n}}\right) = P\left(Z \geq z_\alpha - \frac{\theta_1-\theta_0}{\sigma/\sqrt{n}}\right) \qquad [\text{Using (18·12)}] \text{ and } \theta_1 > \theta_0$$

$$= 1 - P(Z \leq \lambda_3) = 1 - \Phi(\lambda_3) \qquad \left\{\lambda_3 = z_\alpha - \frac{\theta_1-\theta_0}{\sigma/\sqrt{n}}, \text{say}\right\} \qquad ...(18·13)$$

where Φ (.) is the distribution function of standard normal variate.

Similarly in case (*ii*), $(\theta_1 < \theta_0)$, the power of the test is

$$1-\beta = P(\bar{x} < \lambda_2 \mid H_1) = P\left(Z < \frac{\lambda_2-\theta_1}{\sigma/\sqrt{n}}\right) = P\left(Z < \frac{\theta_0 + \frac{\sigma}{\sqrt{n}} z_{1-\alpha} - \theta_1}{\sigma/\sqrt{n}}\right) \qquad [\text{Using (18·12}a)]$$

$$= P\left(Z < z_{1-\alpha} + \frac{\theta_0-\theta_1}{\sigma/\sqrt{n}}\right) = \Phi(\lambda_4), \qquad (\because \theta_0 > \theta_1) \qquad ...(18·13a)$$

where

$$\lambda_4 = z_{1-\alpha} + \frac{\sqrt{n}\,(\theta_0-\theta_1)}{\sigma} = \frac{\sqrt{n}(\theta_0-\theta_1)}{\sigma} - z_\alpha \qquad ...(18·13b)$$

UMP Critical Region. (18·10) provides best critical region for testing $H_0 : \theta = \theta_0$ against the hypothesis $\theta = \theta_1$, provided $\theta_1 > \theta_0$ while (18·11) defines the best critical region for testing $H_0 : \theta = \theta_0$ against $H_1 : \theta = \theta_1$, provided $\theta_1 < \theta_0$. Thus, the best critical region for testing simple hypothesis $H_0 : \theta = \theta_0$ against the simple hypothesis $\theta = \theta_1 + c$, $c > 0$ will not serve as best critical region for testing simple hypothesis $H_0 : \theta = \theta_0$ against simple alternative hypothesis $H_1 : \theta = \theta_0 - c$, $c > 0$.

Hence in this problem, no uniformly most powerful test exists for testing the simple hypothesis, $H_0 : \theta = \theta_0$ against the composite alternative hypothesis, $H_1 : \theta \neq \theta_0$.

However, for each alternative hypothesis, $H_1 : \theta = \theta_1 > \theta_0$ or $H_1 : \theta = \theta_1 < \theta_0$, *a UMP* test exists and is given by (18·10) and (18·11) respectively.

Remark. In particular, if we take $n = 2$, then the B.C.R. for testing $H_0 : \theta = \theta_0$, against $H_1 : \theta = \theta_1\ (> \theta_0)$ is given by : [From (18·10) and (18·12)]

$$W = \{\mathbf{x} : (x_1 + x_2)/2 \geq \theta_0 + \sigma z_\alpha/2\} \qquad [\because\ \bar{x} = (x_1+x_2)/2]$$

$$= [\mathbf{x} : x_1 + x_2 \geq 2\theta_0 + \sqrt{2}\ \sigma z_\alpha] = [\mathbf{x} : x_1 + x_2 \geq C], \text{ (say)} \qquad ... (**)$$

where

$$C = 2\theta_0 + \sqrt{2}\ \sigma z_\alpha = 2\theta_0 + \sqrt{2}\ \sigma \times 1{\cdot}645, \text{ if } \alpha = 0{\cdot}05.$$

Similarly, the B.C.R. for testing $H_0 : \theta = \theta_0$ against $H_1 : \theta = \theta_1\ (< \theta_0)$ with $n = 2$ and $\alpha = 0{\cdot}05$ is given by [From (18·11) and (18·12*a*)] :

$$W_1 = [\mathbf{x} : (x_1 + x_2)/2 \leq \theta_0 - \sigma z_\alpha/\sqrt{2}\,]$$

$$= [\mathbf{x} : (x_1 + x_2) \leq 2\theta_0 - \sqrt{2}\ \sigma \times 1{\cdot}645]$$

$$= [\mathbf{x} : x_1 + x_2 \leq C_1], \text{ (say)}, \qquad ... (***)$$

where

$$C_1 = 2\theta_0 - \sqrt{2}\ \sigma z_\alpha = 2\theta_0 - \sqrt{2}\ \sigma \times 1{\cdot}645, \text{ if } \alpha = 0{\cdot}05$$

The B.C.R. for testing $H_0 : \theta = \theta_0$ against the two tailed alternative $H_1 : \theta = \theta_1 \ (\neq \theta_0)$, is given by :

$$W_2 = \{\mathbf{x} : (x_1 + x_2 \geq C) \cup (x_1 + x_2 \leq C_1)\} \qquad \text{... (****)}$$

The regions in (**), (***), and (****) are given by the shaded portions in the following figures (*i*), (*ii*) and (*iii*) respectively.

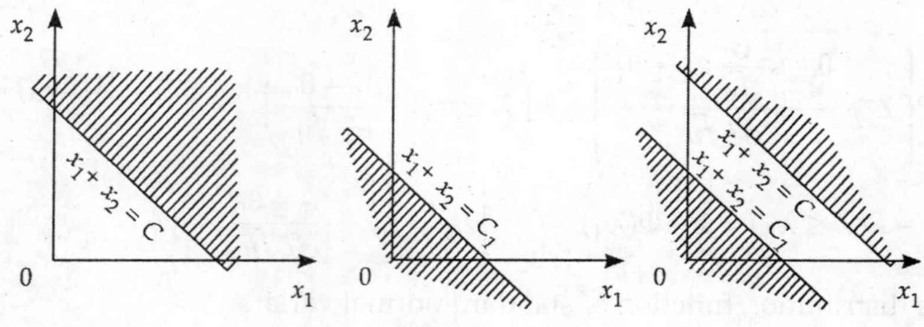

Fig. (*i*) Fig. (*ii*) Fig. (*iii*)

BCR for $\left.\right\} : \begin{matrix} H_0 : \theta = \theta_0 \\ H_1 : \theta = \theta_1 (> \theta_0) \end{matrix}$ BCR for $\left.\right\} : \begin{matrix} H_0 : \theta = \theta_0 \\ H_1 : \theta = \theta_1 (< \theta_0) \end{matrix}$ BCR for $\left.\right\} : \begin{matrix} H_0 : \theta = \theta_0 \\ H_1 : \theta = \theta_1 (\neq \theta_0) \end{matrix}$

Example 18·7. *Show that for the normal distribution with zero mean and variance σ^2, the best critical region for $H_0 : \sigma = \sigma_0$ against the alternative $H_1 : \sigma = \sigma_1$ is of the form :*

$$\sum_{i=1}^{n} x_i^2 \leq a_\alpha \text{ , for } \sigma_0 > \sigma_1 \qquad \text{and} \qquad \sum_{i=1}^{n} x_i^2 \geq b_\alpha \text{ , for } \sigma_0 < \sigma_1$$

Show that the power of the best critical region when $\sigma_0 > \sigma_1$ is $F\left(\dfrac{\sigma_0^2}{\sigma_1^2} \chi^2_{\alpha,n}\right)$, where $\chi^2_{\alpha,n}$ is lower 100 α-percent point and F is the distribution function of the χ^2-distribution with n degrees of freedom.

Solution. Here we are given:

$$f(x, \sigma) = \frac{1}{\sigma\sqrt{2\pi}} \exp\left(-\frac{x^2}{2\sigma^2}\right); -\infty < x < \infty$$

The best critical region (B.C.R.), according to Neyman-Pearson Lemma, is given by (for $k_\alpha > 0$),

$$\frac{L_0}{L_1} \leq \frac{1}{k_\alpha} = A_\alpha, \text{ (say)}.$$

$$\Rightarrow \qquad \left(\frac{\sigma_1}{\sigma_0}\right)^n \exp\left\{-\frac{1}{2}\sum_{i=1}^{n} x_i^2\left(\frac{1}{\sigma_0^2} - \frac{1}{\sigma_1^2}\right)\right\} \leq A_\alpha$$

$$\Rightarrow \qquad n \log\left(\frac{\sigma_1}{\sigma_0}\right) - \frac{1}{2}\sum_{i=1}^{n} x_i^2 \left(\frac{\sigma_1^2 - \sigma_0^2}{\sigma_0^2 \sigma_1^2}\right) \leq \log A_\alpha$$

(Since $\log x$ is an increasing function of x).

$$\Rightarrow \qquad \frac{\sigma_0^2 - \sigma_1^2}{2\sigma_0^2\, \sigma_1^2} \sum_{i=1}^{n} x_i^2 \leq \left\{\log A_\alpha - n \log\left(\frac{\sigma_1}{\sigma_0}\right)\right\} \qquad \text{... (*)}$$

Case (i). If $\sigma_0 > \sigma_1 \quad \Rightarrow \quad \sigma_1 < \sigma_0,$ then B.C.R. is given by

$$\sum_{i=1}^{n} x_i^2 \leq \left[\log A_\alpha - n \log\left(\frac{\sigma_1}{\sigma_0}\right)\right] \frac{2\sigma_0^2\sigma_1^2}{\sigma_0^2 - \sigma_1^2} = a_\alpha, \text{ (say), i.e., } W = \left(x : \sum_{i=1}^{n} x_i^2 \leq a_\alpha\right) \qquad \text{... (18·14)}$$

Case (ii). If $\sigma_0 < \sigma_1$, then B.C.R. is given by:

$$\sum_{i=1}^{n} x_i^2 \geq \left[\log A_\alpha - n\log\left(\frac{\sigma_1}{\sigma_0}\right)\right] \cdot \frac{2\sigma_0^2\sigma_1^2}{\sigma_0^2 - \sigma_1^2} = b_\alpha, \text{(say)}, \quad i.e., W_1 = \left(\mathbf{x} : \sum_{i=1}^{n} x_i^2 \geq b_\alpha\right) \qquad \ldots (18\cdot14a)$$

The constants a_α and b_α are so chosen that the size of the critical region is α.

Thus, a_α is determined so that

$$P[\mathbf{x} \in W \mid H_0] = \alpha \quad \Rightarrow \quad P\left[\sum_{i=1}^{n} x_i^2 \leq a_\alpha | H_0\right] = \alpha \Rightarrow P\left(\sum_{i=1}^{n} \frac{x_i^2}{\sigma_0^2} \leq \frac{a_\alpha}{\sigma_0^2} | H_0\right) = \alpha \qquad \ldots (*)$$

Since under H_0, $\quad \chi_{(n)}^2 = \sum_{i=1}^{n} \frac{x_i^2}{\sigma_0^2}$, is a χ^2-variate with n $d.f.$,

$$\therefore \qquad P\left[\chi_{(n)}^2 \leq \frac{a_\alpha}{\sigma_0^2}\right] = \alpha \Rightarrow \frac{a_\alpha}{\sigma_0^2} = \chi_{\alpha,n}^2 \Rightarrow \sigma_0^2\chi_{\alpha,n}^2 = a_\alpha \qquad \ldots (18\cdot15)$$

where $\chi^2_{\alpha,\, n}$ is the lower 100 α-per cent point of chi-square distribution with n $d.f.$ given by:

$$P(\chi^2 \leq \chi^2_{\alpha,\, n}) = \alpha \qquad \ldots (18\cdot15a)$$

Hence, the B.C.R. for testing $H_0 : \sigma = \sigma_0$ against $H_1 : \sigma = \sigma_1\ (< \sigma_0)$, is given by [From (18·14) and (18·15)]:

$$W = \left[\mathbf{x} : \sum_{i=1}^{n} x_i^2 \leq \sigma_0^2\, \chi_{\alpha,n}^2\right] \qquad \ldots (18\cdot15b)$$

where $\chi^2_{\alpha,\, n}$ is defined in (18·15a).

Also by definition, the power of the test is:

$$1 - \beta = P[x \in W | H_1] = P\left(\sum_{i=1}^{n} x_i^2 \leq a_\alpha | H_1\right)$$

$$= P\left(\frac{\sum_{i=1}^{n} x_i^2}{\sigma_0^2} \leq \frac{a_\alpha}{\sigma_0^2} | H_1\right) = P\left(\frac{\sum_{i=1}^{n} x_i^2}{\sigma_0^2} \leq \chi_{\alpha,n}^2 | H_1\right) \qquad \text{[From (18·15)]}$$

$$= P\left(\frac{\sum_{i=1}^{n} x_i^2}{\sigma_1^2} \leq \frac{\sigma_0^2}{\sigma_1^2}\chi_{\alpha,n}^2 | H_1\right) = P\left(\chi_{(n)}^2 \leq \frac{\sigma_0^2}{\sigma_1^2}\chi_{\alpha,n}^2\right),$$

since under H_1, $\Sigma x_i^2/\sigma_1^2$ is a χ^2-variate with n $d.f.$

Hence, $\qquad$ Power of the test $= F\left(\frac{\sigma_0^2}{\sigma_1^2} \cdot \chi_{\alpha,n}^2\right)$, $\qquad \ldots (18\cdot15c)$

where $F(\cdot)$ is the distribution function of chi-square distribution with n $d.f.$

Remarks:

1. Similarly, for testing $H_0 : \sigma = \sigma_0$ against $H_1 : \sigma = \sigma_1\ (> \sigma_0)$, b_α in (16·14a) is determined so that:

$$P\,[\mathbf{x} \in W_1 \mid H_0] = \alpha \quad \Rightarrow \quad P\left(\mathbf{x} : \sum_{i=1}^{n} x_i^2 \geq b_\alpha | H_0\right) = \alpha$$

$$\Rightarrow \quad P\left(\mathbf{x} : \frac{\Sigma x_i^2}{\sigma_0^2} \geq \frac{b_\alpha}{\sigma_0^2} | H_0\right) = \alpha \quad \Rightarrow \quad P\left[\mathbf{x} : \chi_{(n)}^2 \geq \frac{b_\alpha}{\sigma_0^2}\right] = \alpha \quad \Rightarrow \quad P\left(\mathbf{x} : \chi_{(n)}^2 \leq \frac{b_\alpha}{\sigma_0^2}\right) = 1 - \alpha$$

$$\therefore \qquad \frac{b_\alpha}{\sigma_0^2} = \chi^2_{1-\alpha,n} \quad \Rightarrow \quad b_\alpha = \sigma_0^2 \cdot \chi^2_{1-\alpha,n} \qquad \ldots (18·16)$$

where $\chi^2_{\alpha,\,n}$ is defined in (18·15*a*).

Hence, the B.C.R. for testing $H_0 : \sigma = \sigma_0$ against $H_1 : \sigma = \sigma_1\ (> \sigma_0)$, is given by :

$$W_1 = \left(x : \sum_{i=1}^{n} x_i^2 \geq \sigma_0^2 \cdot \chi^2_{1-\alpha,n} \right) \qquad \ldots (18·16a)$$

The power of the test in this case is given by:

$$1 - \beta = P(x \in W_1 \mid H_1) = P\left[\sum_{i=1}^{n} x_i^2 \geq \sigma_0^2 \chi^2_{1-\alpha,n} \mid H_1 \right]$$

$$= P\left(\frac{\sum_{i=1}^{n} x_i^2}{\sigma_1^2} \geq \frac{\sigma_0^2}{\sigma_1^2} \chi^2_{1-\alpha,n} \middle| H_1 \right) = P\left(\chi^2_{(n)} \geq \frac{\sigma_0^2}{\sigma_1^2} \cdot \chi^2_{1-\alpha,n} \right), \qquad \ldots (18·16b)$$

since under H_1, $\sum_{i=1}^{n} \left(x_i^2 / \sigma_1^2 \right)$ is a χ^2-variate with *n.d.f.*

$$\therefore \qquad 1 - \beta = 1 - P\left\{ \chi^2_{(n)} \leq \frac{\sigma_0^2}{\sigma_1^2} \cdot \chi^2_{1-\alpha,n} \right\} = 1 - F\left(\frac{\sigma_0^2}{\sigma_1^2} \cdot \chi^2_{1-\alpha,n} \right), \qquad \ldots (18·16c)$$

where $F(.)$ is the distribution function of chi-square distribution with *n.d.f.*

2. *Graphical representation of the B.C.R. for the particular case n = 2.*

For $n = 2$, the B.C.R. for testing $H_0 : \sigma = \sigma_0$, against $H_1 : \sigma = \sigma_1\ (< \sigma_0)$ is given by [From (18·15*b*)]

$$W = \left\{ \mathbf{x} : \sum_{i=1}^{2} x_i^2 \leq \sigma_0^2 \cdot \chi^2_{\alpha,2} \right\} = \left\{ \mathbf{x} : x_1^2 + x_2^2 \leq a^2 \right\},$$

where $a^2 = \sigma_0^2\, \chi^2_{\alpha,\,2}$. Thus, the B.C.R. is the interior of the circle with centre (0, 0) and radius '*a*' and is shown as the shaded region in Figure (*i*) below:

Similarly, from (18·16*a*), the B.C.R. for testing $H_0 : \sigma = \sigma_0$, against $H_1 : \sigma = \sigma_1\ (> \sigma_0)$ for $n = 2$ is given by :

$$W_1 = \left\{ \mathbf{x} : x_1^2 + x_2^2 \geq \sigma_0^2\, \chi^2_{1-\alpha,\,2} \right\} = \left\{ \mathbf{x} : x_1^2 + x_2^2 \geq b^2 \right\}$$

where $b^2 = \sigma_0^2 \cdot \chi^2_{1-\alpha,\,2}$. Thus, B.C.R. is the exterior of the circle with centre (0, 0) and radius b and is shown as the shaded region in Figure (*ii*) below:

Similarly, the B.C.R. for testing $H_0 : \sigma = \sigma_0$ against two tailed alternative

$$H_1 : \sigma = \sigma_1\ (\neq \sigma_0), \text{ for } n = 2 \text{ is given by:}$$

$$W_2 = W \cup W_1 = \{\mathbf{x} : x_1^2 + x_2^2 \leq a^2\} \cup \{\mathbf{x} : x_1^2 + x_2^2 \geq b^2\}$$

and is shown as the shaded region in the Figure (*iii*) below.

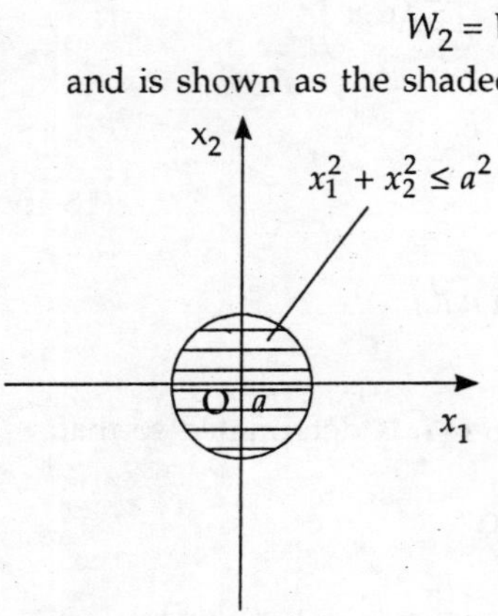

Fig. (*i*)

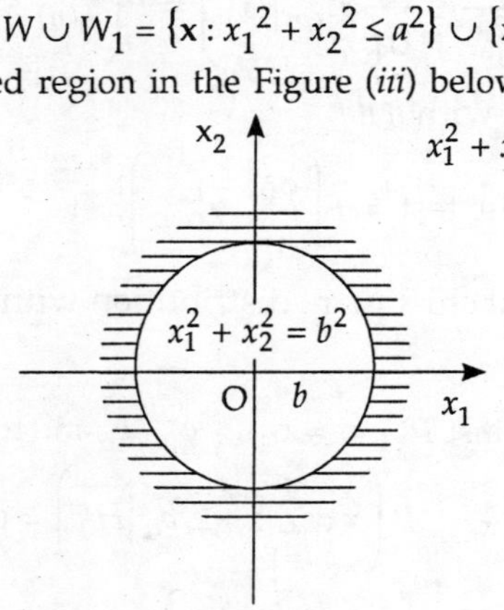

Fig. (*ii*)

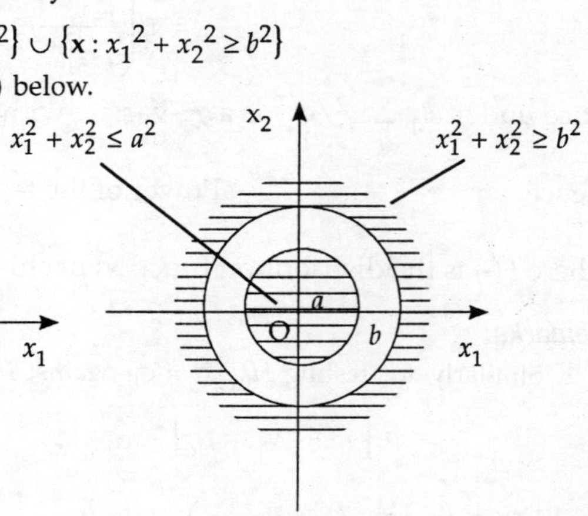

Fig (*iii*)

3. (18·14) defines an *UMP* test for testing simple hypothesis $H_0 : \sigma = \sigma_0$ against simple alternative hypothesis $H_1 : \sigma = \sigma_1$ ($< \sigma_0$) whereas (18·14*a*) defines an *UMP* test for testing simple hypothesis H_0 against the simple alternative hypothesis $H_1 : \sigma = \sigma_1$ ($> \sigma_0$). However no *UMP* test exists for testing simple hypothesis $H_0 : \sigma = \sigma_0$ against the composite alternative hypothesis $H_1 : \sigma \neq \sigma_0$.

Example 18·8. *Given a random sample $x_1, x_2, ..., x_n$ from the distribution with p.d.f.*

$$f(x, \theta) = \theta\, e^{-\theta x}, x > 0$$

show that there exists no UMP test for testing $H_0 : \theta = \theta_0$ against $H_1 : \theta \neq \theta_0$.

Solution.

$$L = \prod_{i=1}^{n} f(x_i, \theta) = \theta^n \cdot \exp\left[-\theta \sum_{i=1}^{n} x_i\right]$$

Consider $\qquad H_1 : \theta = \theta_1, (\theta_1 \neq \theta_0)$

The best critical region, using Neyman-Pearson Lemma is given by:

$$\theta_1^{\,n} \exp[-\theta_1 \Sigma x_i] \geq k \cdot \theta_0^{\,n} \exp[-\theta_0 \Sigma x_i]\ ; k > 0$$

$$\Rightarrow \qquad \exp[(\theta_0 - \theta_1)\Sigma x_i] \geq k \cdot \left(\frac{\theta_0}{\theta_1}\right)^n$$

$$\Rightarrow \qquad (\theta_0 - \theta_1)\Sigma x_i \geq \log\left\{k \cdot \left(\frac{\theta_0}{\theta_1}\right)^n\right\} = k_1, \text{ (say)}. \qquad ...(*)$$

Case (*i*) If $\theta_1 > \theta_0$, then *B.C.R.* is given by: $\Sigma x_i \leq \dfrac{k_1}{\theta_0 - \theta_1} = \lambda_1$, (say).

Case (*ii*) If $\theta_1 < \theta_0$, then *B.C.R.* is given by: $\Sigma x_i \geq \dfrac{k_1}{\theta_0 - \theta_1} = \lambda_2$, (say).

The constants λ_1 and λ_2 are so determined that

$$\left.\begin{aligned} &P[\Sigma x_i \leq \lambda_1 \mid H_0] = \alpha \\ \Rightarrow\ &P[2\theta \Sigma x_i \leq 2\theta\lambda_1 \mid H_0] = \alpha \end{aligned}\right\} ...(**) \quad \text{and} \quad \left.\begin{aligned} &P[\Sigma x_i \geq \lambda_2 \mid H_0] = \alpha \\ \Rightarrow\ &P[2\theta \Sigma x_i \geq 2\theta\lambda_2 \mid H_0] = \alpha \end{aligned}\right\} ...(***)$$

But in random sampling from the given exponential distribution,

$$M_{\Sigma X_i}(t) = \prod_{i=1}^{n} M_{X_i}(t) = \{M_{X_i}(t)\}^n = \left(1 - \frac{t}{\theta}\right)^{-n} \Rightarrow M_{2\theta\Sigma X_i}(t) = M_{\Sigma X_i}(2t\theta) = (1 - 2t)^{-n}$$

which is the *m.g.f.* of a χ^2-variate with $2n$ *d.f.* Hence by uniqueness theorem of *m.g.f.'s*,

$$2\theta \sum_{i=1}^{n} X_i \sim \chi^2_{(2n)}$$

Using this result in (**): $P[2\theta_0 \Sigma x_i \leq \mu_1] = P[\chi^2_{(2n)} \leq \mu_1] = \alpha \quad \Rightarrow \quad \mu_1 = \chi^2_{1-\alpha\,,\,2n}$

where $\chi^2_{\alpha, n}$ is the upper 'α' point of χ^2-distribution with *n.d.f.* given by $P(\chi^2 > \chi^2_{\alpha, n}) = \alpha$ ... (****)

Hence, B.C.R. for testing $H_0 : \theta = \theta_0$ against $H_1 : \theta = \theta_1$ ($> \theta_0$) is given by

$$W_0 = \left\{\mathbf{x} : 2\theta_0 \Sigma x_i \leq \chi^2_{1-\alpha, 2n}\right\} = \left\{\mathbf{x} : \Sigma x_i \leq \frac{1}{2\theta_0}\chi^2_{1-\alpha, 2n}\right\}$$

and since it is independent of θ_1, W_0 is *UMP C.R.* for $H_0 : \theta = \theta_0$ against $H_1 : \theta = \theta_1$ ($> \theta_0$).

Similarly, from (***) and (****), we get

$$P[2\theta_0 \Sigma x_i \geq \mu_2] = P(\chi^2_{(2n)} \geq \mu_2) = \alpha \quad \Rightarrow \quad \mu_2 = \chi^2_{\alpha,\, 2n}$$

Hence, B.C.R. for testing H_0 against $H_1 : \theta = \theta_1 \ (< \theta_0)$ is given by:

$$W_1 = \left\{ \mathbf{x} : 2\theta_0 \, \Sigma x_i \geq \chi^2_{\alpha, 2n} \right\} = \left\{ \mathbf{x} : \Sigma x_i \geq \frac{1}{2\theta_0} \chi^2_{\alpha, 2n} \right\}$$

and since it is independent of θ_1, W_1 is also *UMP C.R.* for $H_0 : \theta = \theta_0$ against $H_1 : \theta = \theta_1 \ (< \theta_0)$.

However, since the two critical regions W_0 and W_1 are different, there exists no critical region of size α which is *UMP.* for $H_0 : \theta = \theta_0$ against the two tailed alternative, $H_1 : \theta \neq \theta_0$.

Power of the test. The power of the test for testing $H_0 : \theta = \theta_0$, against $H_1 : \theta = \theta_1 \ (> \theta_0)$ is given by

$$1 - \beta = P[\mathbf{x} \in W_0 \mid H_1] = P\left(\sum_{i=1}^{n} x_i \leq \frac{1}{2\theta_0} \chi^2_{1-\alpha,\, 2n} \middle| H_1 \right)$$

$$= P\left(2\theta_1 \sum_{i=1}^{n} x_i \leq \frac{\theta_1}{\theta_0} \chi^2_{1-\alpha, 2n} \middle| H_1 \right) = P\left\{ \chi^2_{(2n)} \leq \frac{\theta_1}{\theta_0} \chi^2_{1-\alpha, 2n} \right\}, \qquad \dots (*)$$

since under H_1, $2\theta_1 \sum_{i=1}^{n} x_i \sim \chi^2_{(2n)}$.

Similarly, the power of the test for testing $H_0 : \theta = \theta_0$, against $H_1 : \theta = \theta_1 \ (< \theta_0)$ is given by :

$$1 - \beta = P(\mathbf{x} \in W_1 \mid H_1) = P\left(\sum_{i=1}^{n} x_i \geq \frac{1}{2\theta_0} \chi^2_{\alpha, 2n} \middle| H_1 \right)$$

$$= P\left(2\theta_1 \sum_{i=1}^{n} x_i \geq \frac{\theta_1}{\theta_0} \chi^2_{\alpha, 2n} \middle| H_1 \right) = P\left\{ \chi^2_{(2n)} \geq \frac{\theta_1}{\theta_0} \chi^2_{\alpha, 2n} \right\}, \qquad \dots (**)$$

Remark. The graphic representation of the B.C.R. for $H_0 : \theta = \theta_0$ against different alternatives $H_1 : \theta = \theta_1$ $(> \theta_0)$, $H_1 : \theta = \theta_1 \ (< \theta_0)$ and $H_1 : \theta = \theta_1 \ (\neq \theta_0)$ for $n = 2$, can be done similarly as in Example 18·6, for the mean of normal distribution.

Example 18·9. *For the distribution*

$$dF = \begin{cases} \beta exp\{-\beta(x - \gamma)\} dx, \ x \geq \gamma \\ 0, x < \gamma \end{cases}$$

show that for a hypothesis H_0 *that* $\beta = \beta_0$, $\gamma = \gamma_0$ *and an alternative* H_1 *that* $\beta = \beta_1$, $\gamma = \gamma_1$, *the best critical region is given by*

$$\bar{x} \leq \frac{1}{\beta_1 - \beta_0} \left\{ \gamma_1 \beta_1 - \gamma_0 \beta_0 - \frac{1}{n} \log k + \log \frac{\beta_1}{\beta_0} \right\}$$

provided that the admissible hypothesis is restricted by the condition $\gamma_1 \leq \gamma_0$, $\beta_1 \geq \beta_0$.

Solution. $f(x \,;\, \beta, \gamma) = \begin{cases} \beta \exp\{-\beta(x - \gamma)\}, x \geq \gamma \\ 0, \text{otherwise} \end{cases}$

$$\therefore \qquad \prod_{i=1}^{n} f(x_i ; \beta, \gamma) = \begin{cases} \beta^n \exp\left\{ -\beta \sum_{i=1}^{n} (x_i - \gamma) \right\} \ ; x_1, x_2, \dots, x_n \geq \gamma \\ 0, \text{otherwise} \end{cases}$$

Using Neyman-Pearson Lemma, B.C.R. for $k > 0$, is given by

$$\frac{\beta_1^n \exp\left\{-\beta_1 \sum_{i=1}^{n} (x_i - \gamma_1)\right\}}{\beta_0^n \exp\left\{-\beta_0 \sum_{i=1}^{n} (x_i - \gamma_0)\right\}} \geq k$$

$$\Rightarrow \quad \left(\frac{\beta_1}{\beta_0}\right)^n \exp\left\{-\beta_1 \sum_{i=1}^{n} (x_i - \gamma_1) + \beta_0 \sum_{i=1}^{n} (x_i - \gamma_0)\right\} \geq k$$

$$\Rightarrow \quad \left(\frac{\beta_1}{\beta_0}\right)^n \exp[-\beta_1 n(\bar{x} - \gamma_1) + \beta_0 n(\bar{x} - \gamma_0)] \geq k$$

$$\Rightarrow \quad n \log(\beta_1 / \beta_0) - n\bar{x}(\beta_1 - \beta_0) + n\beta_1\gamma_1 - n\beta_0\gamma_0 \geq \log k$$

(Since $\log x$ is an increasing function of x).

$$\Rightarrow \quad \bar{x}(\beta_1 - \beta_0) \leq \left\{\gamma_1\beta_1 - \gamma_0\beta_0 - \frac{1}{n}\log k + \log\left(\frac{\beta_1}{\beta_0}\right)\right\}$$

$$\therefore \quad \bar{x} \leq \frac{1}{\beta_1 - \beta_0}\left\{\gamma_1\beta_1 - \gamma_0\beta_0 - \frac{1}{n}\log k + \log\left(\frac{\beta_1}{\beta_0}\right)\right\} \text{provided } \beta_1 > \beta_0.$$

Example 18·10. *Examine whether a best critical region exists for testing the null hypothesis $H_0 : \theta = \theta_0$ against the alternative hypothesis $H_1 : \theta > \theta_0$ for the parameter θ of the distribution :*

$$f(x, \theta) = \frac{1 + \theta}{(x + \theta)^2}, 1 \leq x < \infty$$

Solution. $\prod_{i=1}^{n} f(x_i, \theta) = (1 + \theta)^n \prod_{i=1}^{n} \frac{1}{(x_i + \theta)^2}\ ; 1 \leq x < \infty$

By Neyman-Pearson Lemma, the B.C.R. for $k > 0$, is given by

$$(1 + \theta_1)^n \prod_{i=1}^{n} \frac{1}{(x_i + \theta_1)^2} \geq k(1 + \theta_0)^n \prod_{i=1}^{n} \frac{1}{(x_i + \theta_0)^2}$$

$$\Rightarrow \quad n \log (1 + \theta_1) - 2 \sum \log (x_i + \theta_1) \geq \log k + n \log (1 + \theta_0) - 2 \sum_{i=1}^{n} \log (x_i + \theta_0)$$

$$\Rightarrow \quad 2 \sum_{i=1}^{n} \log\left(\frac{x_i + \theta_0}{x_i + \theta_1}\right) \geq \log k + n \log\left(\frac{1 + \theta_0}{1 + \theta_1}\right).$$

Thus, the test criterion is: $\sum_{i=1}^{n} \log\left(\frac{x_i + \theta_0}{x_i + \theta_1}\right)$, which cannot be put in the form of a function of the sample observations, not depending on the hypothesis. Hence, no B.C.R. exists in this case.

18·6. LIKELIHOOD RATIO TEST

Neyman-Pearson Lemma based on the magnitude of the ratio of two probability density functions provides best test for testing simple hypothesis against simple alternative hypothesis. The best test in any given situation depends on the nature of the population distribution and the form of the alternative hypothesis being considered. In this section we shall discuss a general method of test construction called the *Likelihood Ratio* (L.R.) *Test* introduced by Neyman and Pearson for testing a hypothesis, simple or composite, against a simple or composite alternative hypothesis. This test is related to the maximum likelihood estimates.

Before defining the test, we give below some notations and terminology.

Parameter Space. Let us consider a random variable X with *p.d.f.* $f(x, \theta)$. In most common applications, though not always, the functional form of the population distribution is assumed to be known except for the value of some unknown parameter(s) θ which may take any value on a set Θ. This is expressed by writing the *p.d.f.* in the form $f(x, \theta)$, $\theta \in \Theta$. The set Θ, which is the set of all possible values of θ is called the *parameter space.* Such a situation gives rise not to one probability distribution but a family of probability distributions which we write as $\{f(x, \theta) = \theta \in \Theta\}$. For example if $X \sim N(\mu, \sigma^2)$, then the parameter space is :

$$\Theta = \{(\mu, \sigma^2) : -\infty < \mu < \infty, 0 < \sigma < \infty\}$$

In particular, for $\sigma^2 = 1$, the family of probability distributions is given by

$$\{N(\mu, 1); \mu \in \Theta\}, \text{ where } \Theta = \{\mu : -\infty < \mu < \infty\}$$

In the following discussion we shall consider a general family of distributions:

$$\{f(x : \theta_1, \theta_2, \ldots, \theta_k) : \theta_i \in \Theta, i = 1, 2, \ldots, k\}$$

The null hypothesis H_0 will state that the parameters belong to some subspace Θ_0 of the parameter space Θ.

Let $x_1, x_2, \ldots, x_n$ be a random sample of size $n > 1$ from a population with p.d.f. $f(x, \theta_1, \theta_2, \ldots, \theta_k)$, where Θ, the parameter space is the totality of all points that $(\theta_1, \theta_2, \ldots, \theta_k)$ can assume. We want to test the null hypothesis :

$$H_0 : (\theta_1, \theta_2, \ldots, \theta_k) \in \Theta_0$$

against all alternative hypotheses of the type:

$$H_1 : (\theta_1, \theta_2, \ldots, \theta_k) \in \Theta - \Theta_0$$

The likelihood function of the sample observations is given by

$$L = \prod_{i=1}^{n} f(x_i ; \theta_1, \theta_2, \ldots, \theta_k) \qquad \ldots (18\cdot16)$$

According to the principle of maximum likelihood, the likelihood equation for estimating any parameter θ_i is given by

$$\frac{\partial L}{\partial \theta_i} = 0, (i = 1, 2, \ldots, k) \qquad \ldots (18\cdot17)$$

Using (18·17), we can obtain the maximum likelihood estimates for the parameters $(\theta_1, \theta_2, \ldots, \theta_k)$ as they are allowed to vary over the parameter space Θ and the subspace Θ_0. Substituting these estimates in (18·16), we obtain the maximum values of the likelihood function for variation of the parameters in Θ and Θ_0 respectively. Then the criterion for the likelihood ratio test is defined as the quotient of these two maxima and is given by

$$\lambda = \lambda(x_1, x_2, \ldots, x_n) = \frac{L(\hat{\Theta}_0)}{L(\hat{\Theta})} = \frac{\sup_{\theta \in \Theta_0} L(x, \theta)}{\sup_{\theta \in \Theta} L(x, \theta)}, \qquad \ldots (18\cdot18)$$

where $L(\hat{\Theta}_0)$ and $L(\hat{\Theta})$ are the maxima of the likelihood function (18·16) with respect to the parameters in the regions Θ_0 and Θ respectively.

The quantity λ is a function of the sample observations only and does not involve parameters. Thus, λ being a function of the random variables, is also a random variable. Obvious $\lambda > 0$. Further

$$\Theta_0 \subset \Theta \quad \Rightarrow \quad L(\Theta_0) \leq L(\Theta) \quad \Rightarrow \quad \lambda \leq 1$$

Hence, we get

$$0 \leq \lambda \leq 1 \qquad \ldots (18\cdot19)$$

The critical region for testing H_0 (against H_1) is an interval

$$0 < \lambda < \lambda_0 \qquad \text{... (18·20)}$$

where λ_0 is some number (< 1) determined by the distribution of λ and the desired probability of type 1 error, *i.e.*, λ_0 is given by the equation :

$$P(\lambda < \lambda_0 \mid H_0) = \alpha \qquad \text{... (18·21)}$$

For example, if $g(.)$ is the p.d.f. of λ then λ_0 is determined from the equation :

$$\int_0^{\lambda_0} g(\lambda \mid H_0) d\lambda = \alpha \qquad \text{... (18·21}a\text{)}$$

A test that has critical region defined in (18·20) and (18·21) is a *likelihood ratio test* for testing H_0.

Remark. Equations (18·20) and (18·21) define the critical region for testing the hypothesis H_0 by the likelihood ratio test. Suppose that the distribution of λ is not known but the distribution of some function of λ is known, then this knowledge can be utilized as given in the following theorem.

Theorem 18·3. *If λ is the likelihood ratio for testing a simple hypothesis H_0 and if $U = \phi(\lambda)$ is a monotonic increasing (decreasing) function of λ then the test based on U is equivalent to the likelihood ratio test. The critical region for the test based on U is :*

$$\phi(0) < U < \phi(\lambda_0),\ [\phi(\lambda_0) < U < \phi(0)] \qquad \text{... (18·22)}$$

Proof. The critical region for the likelihood ratio test is given by $0 < \lambda < \lambda_0$, where λ_0 is determined by

$$\int_0^{\lambda_0} g(\lambda \mid H_0) d\lambda = \alpha \qquad \text{... (*)}$$

Let $U = \phi(\lambda)$ be a monotonically increasing function of λ. Then (*) gives

$$\alpha = \int_0^{\lambda_0} g(\lambda \mid H_0) d\lambda = \int_{\phi(0)}^{\phi(\lambda_0)} h(u \mid H_0) du$$

where $h(u \mid H_0)$ is the p.d.f. of U when H_0 is true. Here the critical region $0 < \lambda < \lambda_0$ transforms to $\phi(0) < U < \phi(\lambda_0)$. However if $U = \phi(\lambda)$ is a monotonic decreasing function of λ, then the inequalities are reversed and we get the critical region as $\phi(\lambda_0) < U < \phi(0)$.

If we are testing a simple null hypothesis H_0 then there is a unique distribution determined for λ. But if H_0 is composite, then the distribution of λ may or may not be unique. In such a case the distribution of λ may possibly be different for different parameter points in Θ_0 and then λ_0 is to be chosen such that

$$\int_0^{\lambda_0} \cdot\cdot \lambda \quad {}_0 \quad \lambda \le \alpha \qquad \text{... (18·23)}$$

for all values of the parameters in Θ_0.

However, if we are dealing with large samples, a fairly satisfactory situation to this testing of hypothesis problem exists as stated (without proof) in the following theorem.

Theorem 18·4. *Let $x_1, x_2, \ldots, x_n$ be a random sample from a population with p.d.f. $f(x ; \theta_1, \theta_2, \ldots, \theta_k)$ where the paramcter space Θ is k-dimensional. Suppose we want to test the composite hypothesis*

$$H_0 : \theta_1 = \theta_1', \quad \theta_2 = \theta_2', \ldots, \quad \theta_r = \theta_r' \quad ; \quad r < k$$

where $\theta_1', \theta_2', \ldots, \theta_r'$ are specified numbers. When H_0 is true, $-2 \log_e \lambda$ is asymptotically distributed as chi-square variate with r degrees of freedom, i.e., under H_0,

$$-2 \log_e \lambda \sim \chi^2_{(r)}, \qquad \text{if } n \text{ is large.} \qquad \text{... (18·24)}$$

Since $0 \le \lambda \le 1$, $(-2 \log_e \lambda)$ is an increasing function of λ and approaches infinity when $\lambda \to 0$, the critical region for $-2 \log \lambda$ being the right hand tail of the chi-square distribution. Thus at the level of significance 'α', the test may be given as follows:

$$\text{Reject } H_0 \text{ if } -2 \log_e \lambda > \chi_r^2(\alpha)$$

where $\chi_r^2(\alpha)$ is the upper α-point of the chi-square distribution with r *d.f.* given by:

$$P[\chi^2 > \chi_{(r)}^2(\alpha)] = \alpha,$$

otherwise H_0 may be accepted.

18·6·1. Properties of Likelihood Ratio Test

Likelihood ratio (*L.R.*) test principle is an intuitive one. If we are testing a simple hypothesis H_0 against a simple alternative hypothesis H_1 then the *LR* principle leads to the same test as given by the Neyman-Pearson Lemma. This suggests that *LR* test has some desirable properties, specially large sample properties.

In *L.R.* test, the probability of type I error is controlled by suitably choosing the cut off point λ_0. *LR* test is generally *UMP* if an *UMP* test at all exists. We state below, the two asymptotic properties of *LR* tests.

1. Under certain conditions, $-2 \log_e \lambda$ has an asymptotic chi-square distribution.
2. Under certain assumptions, *L.R.* test is consistent.

Now we shall illustrate how the likelihood ratio criterion can be used to obtain various standard tests of significance in Statistics.

18.6.2. Test For the Mean of a Normal Population

Let us take the problem of testing if the mean of a normal population has a specified value. Let $(x_1, x_2, \ldots, x_n)$ be a random sample of size n from the normal population with mean μ and variance σ^2, where μ and σ^2 are unknown. Suppose we want to test the (composite) null hypothesis:

$$H_0 : \mu = \mu_0 \text{ (specified)},\ 0 < \sigma^2 < \infty$$

against the composite alternative hypothesis : $\quad H_1 : \mu \ne \mu_0\ ; 0 < \sigma^2 < \infty$

In this case the parameter space Θ is given by

$$\Theta = \{(\mu, \sigma^2) : -\infty < \mu < \infty,\ 0 < \sigma^2 < \infty\}$$

and the subspace Θ_0 determined by the null hypothesis H_0 is given by

$$\Theta_0 = \{(\mu, \sigma^2) : \mu = \mu_0,\ 0 < \sigma^2 < \infty\}$$

The likelihood function of the sample observations $x_1, x_2, \ldots, x_n$ is given by

$$L = \left(\frac{1}{2\pi\sigma^2}\right)^{n/2} \cdot \exp\left\{-\frac{1}{2\sigma^2} \sum_{i=1}^{n} (x_i - \mu)^2\right\} \quad \ldots (18·25)$$

The maximum likelihood estimates of μ and σ^2 are given by:

$$\hat{\mu} = \frac{1}{n} \sum_{i=1}^{n} x_i = \bar{x},\quad \hat{\sigma}^2 = \frac{1}{n} \sum_{i=1}^{n} (x_i - \bar{x})^2 = s^2 \quad \ldots (18·26)$$

Hence, substituting in (18·25), the maximum of L in the parameter space Θ is given by

$$L(\hat{\Theta}) = \left(\frac{1}{2\pi s^2}\right)^{n/2} . \exp(-n/2) \quad \ldots (18·27)$$

In Θ_0, the only variate parameter is σ^2 and *MLE* of σ^2 for given $\mu = \mu_0$ is given by

$$\hat{\sigma}^2 = \frac{1}{n} \sum_i (x_i - \mu_0)^2 = s_0^2, \text{ (say)}. \quad \ldots (18·28)$$

$$= \frac{1}{n} \sum_i (x_i - \bar{x} + \bar{x} - \mu_0)^2 = \frac{1}{n} \sum_i (x_i - \bar{x})^2 + (\bar{x} - \mu_0)^2,$$

the product term vanishes, since $\sum_i (x_i - \bar{x})(\bar{x} - \mu_0) = (\bar{x} - \mu_0) \sum_i (x_i - \bar{x}) = 0$

$$\therefore \quad \hat{\sigma}^2 = s^2 + (\bar{x} - \mu_0)^2 = s_0^2, \text{ (say)}. \qquad \ldots (18{\cdot}28a)$$

Hence, substituting in (18·25),

$$L(\theta_0) = \left(\frac{1}{2\pi s_0^2} \right)^{n/2} \cdot \exp(-n/2) \qquad \ldots (18{\cdot}28b)$$

The ratio of (18·28*b*) and (18·27) gives the likelihood ratio criterion

$$\lambda = \frac{L(\hat{\Theta}_0)}{L(\hat{\Theta})} = \left(\frac{s^2}{s_0^2} \right)^{n/2} \qquad \ldots (18{\cdot}29)$$

$$= \left\{ \frac{s^2}{s^2 + (\bar{x} - \mu_0)^2} \right\}^{n/2} = \left\{ \frac{1}{1 + \left[(\bar{x} - \mu_0)^2 / s^2 \right]} \right\}^{n/2} \quad [\text{ From } 18{\cdot}28(a)\ ] \qquad \ldots (18{\cdot}29a)$$

We have proved earlier (§ *16·2*) that under H_0, the statistic

$$t = \frac{\bar{x} - \mu_0}{S/\sqrt{n}}, \quad \text{where} \quad S^2 = \frac{1}{n-1} \sum_i (x_i - \bar{x})^2 = \frac{ns^2}{n-1},$$ follows Student's *t*-distribution with $(n - 1)$ *d.f.*

Thus,
$$t = \frac{\bar{x} - \mu_0}{S/\sqrt{n}} = \frac{\bar{x} - \mu_0}{s / \sqrt{n-1}} \sim t_{n-1} \qquad \ldots (18{\cdot}30)$$

Substituting in (18·29*a*), we get:
$$\lambda = \frac{1}{\left(1 + \dfrac{t^2}{n-1} \right)^{n/2}} = \phi(t^2), \textit{ (say)}. \qquad \ldots (18{\cdot}31)$$

The likelihood ratio test for testing H_0 against H_1 consists in finding a critical region of the type $0 < \lambda < \lambda_0$, where λ_0 is given by (18·21*a*), which requires the distribution of λ under H_0. In this case, it is not necessary to obtain the distribution of λ since $\lambda = \phi(t)$ is a monotonic function of t^2 and the test can well be carried on with t^2 as a criterion as with λ [*c.f.* Theorem 18·1]. Now $t^2 = 0$ when $\lambda = 1$ and t^2 becomes infinite when $\lambda = 0$. The critical region of the *LR* test *viz.*, $0 < \lambda < \lambda_0$, on using (18·31) is equivalent to

$$\left(1 + \frac{t^2}{n-1} \right)^{-n/2} \le \lambda_0 \quad \Rightarrow \quad \left(1 + \frac{t^2}{n-1} \right)^{n/2} \ge \lambda_0^{-1}$$

$$\Rightarrow \quad \frac{t^2}{n-1} \ge (\lambda_0)^{-2/n} - 1 \quad \Rightarrow \quad t^2 \ge (n-1)\left[\lambda_0^{-2/n} - 1 \right] = A^2, \textit{ (say)}.$$

Thus, the critical region may well be defined by:
$$|\, t \,| = \left| \frac{\sqrt{n}(\bar{x} - \mu_0)}{S} \right| \ge A \qquad \ldots (18{\cdot}32)$$

where the constant A is determined such that: $P[\,|\, t \,| \ge A \mid H_0] = \alpha$ $\quad \ldots (18{\cdot}33)$

Since under H_0, the statistic t follows Student's t distribution with $(n - 1)$ *d.f.*,

$A = t_{n-1}(\alpha/2)$ where the symbol $t_n(\alpha)$ stands for the *right tail* 100 α% point of the *t*-distribution with n d.f. given by :

$$P\{t > t_n(\alpha)\} = \int_{t_n(\alpha)}^{\infty} f(t)dt = \alpha, \qquad \text{... (18·33a)}$$

where $f(\cdot)$ is the *p.d.f.* of Student's t with *n.d.f.* The critical region is shown in the adjoining diagram.

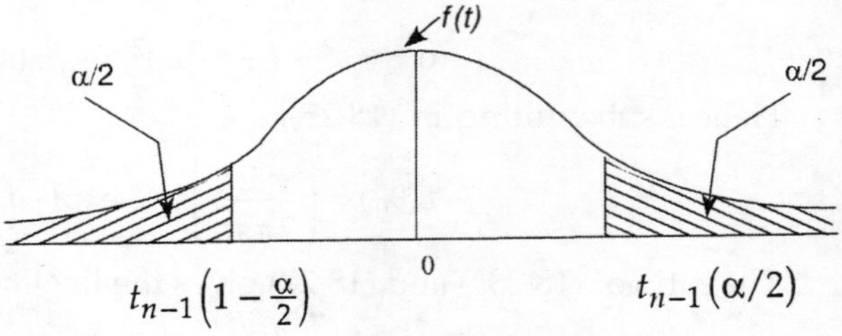

Thus, for testing $H_0 : \mu = \mu_0$ against $\mu \neq \mu_0$ (σ^2-unknown), we have the two-tailed t-test defined as follows :

If $|t| = \left|\dfrac{\sqrt{n}(\bar{x} - \mu_0)}{S}\right| > t_{n-1}(\alpha/2)$, *reject* H_0

and if $|t| < t_{n-1}(\alpha/2)$, H_0 *may be accepted.*

Remarks:

1. Let us now consider the problem of testing the hypothesis: $H_0 : \mu = \mu_0,\ 0 < \sigma^2 < \infty$ against the alternative hypothesis

$$H_1 : \mu > \mu_0,\ 0 < \sigma^2 < \infty$$

Here $\Theta = \{(\mu, \sigma^2) : -\infty < \mu < \infty,\ 0 < \sigma^2 < \infty\}$ and $\Theta_0 = \{(\mu, \sigma^2) : \mu = \mu_0,\ 0 < \sigma^2 < \infty\}$

The maximum likelihood estimates of μ and σ^2 belonging to Θ are given by:

$$\hat{\mu} = \begin{cases} \bar{x}, & \text{if } \bar{x} \geq \mu_0 \\ \mu_0, & \text{if } \bar{x} < \mu_0 \end{cases} \quad \text{... (18·34)} \quad \text{and} \quad \hat{\sigma}^2 = \begin{cases} s^2, & \text{if } \bar{x} \geq \mu_0 \\ s_0^2, & \text{if } \bar{x} < \mu_0 \end{cases} \qquad \text{... (18·34a)}$$

where $$s_0^2 = \frac{1}{n}\sum_{i=1}^{n}(x_i - \mu_0)^2 \qquad \text{... (18·34b)}$$

Thus, $$L(\hat{\Theta}) = \begin{cases} \left(\dfrac{1}{2\pi s^2}\right)^{n/2} \cdot \exp(-n/2), & \text{if } \bar{x} \geq \mu_0 \\ \left(\dfrac{1}{2\pi s_0^2}\right)^{n/2} \cdot \exp(-n/2), & \text{if } \bar{x} < \mu_0 \end{cases} \qquad \text{... (18·35)}$$

In Θ_0, the only unknown parameter is σ^2 whose *MLE* is given by $\hat{\sigma}^2 = s_0^2$. Thus

$$L(\hat{\Theta}_0) = \left(\frac{1}{2\pi s_0^2}\right)^{n/2} \cdot \exp(-n/2) \qquad \text{... (18·36)}$$

$$\therefore \quad \lambda = \frac{L(\hat{\Theta}_0)}{L(\hat{\Theta})} = \begin{cases} \left(s^2/s_0^2\right)^{n/2}, & \text{if } \bar{x} \geq \mu_0 \\ 1, & \text{if } \bar{x} < \mu_0 \end{cases} \qquad \text{... (18·37)}$$

Thus, the sample observations $(x_1, x_2, \ldots, x_n)$ for which $\bar{x} < \mu_0$ are to be included in the acceptance region. Hence, for the sample observations for which $\bar{x} \geq \mu_0$, the likelihood ratio criterion becomes

$$\lambda = (s^2/s_0^2)^{n/2},\ \bar{x} \geq \mu_0 \qquad \text{... (18·37a)}$$

which is the same as the expression obtained in (18·29). Proceeding similarly as in the above problem, the critical region of the form $0 < \lambda < \lambda_0$ will be equivalently given by [*c.f.* (18·32)]

$$t^2 = \frac{n(\bar{x} - \mu_0)^2}{S^2} \geq A^2 \quad \text{or} \quad \text{by} \quad t = \frac{\sqrt{n}(\bar{x} - \mu_0)}{S} \geq A \ (\because \bar{x} \geq \mu_0) \qquad \text{... (18·38)}$$

where t follows Student's t-distribution with $(n - 1)$ d.f. The constant A is to be determined so that

$$P(t > A) = \alpha \quad \Rightarrow \quad A = t_{n-1}(\alpha) \qquad \text{... (18·39)}$$

Hence, for testing $H_0 : \mu = \mu_0$ against $H_1 : \mu > \mu_0$, we have the right tailed t-test defined as follows:

Reject H_0 if $t = \dfrac{\sqrt{n}(\bar{x} - \mu_0)}{S} > t_{n-1}(\alpha)$ *and if* $t < t_{n-1}(\alpha)$*, H_0 may be accepted.*

2. If we want to test $H_0 : \mu = \mu_0,\ 0 < \sigma^2 < \infty$ against the alternative hypothesis:

$$H_1 : \mu < \mu_0,\ 0 < \sigma^2 < \infty,$$

then proceeding exactly similarly as in Remark 1 above, we shall get the critical region given by:

$$t < -t_{n-1}(\alpha) \qquad \text{... (18·40)}$$

In this case we have the left tailed t-test defined as follows :

If $t = \dfrac{\sqrt{n}(\bar{x} - \mu_0)}{S} < -t_{n-1}(\alpha)$*, reject H_0, otherwise H_0 may be accepted.*

3. We summarise below in a tabular form the test criterion, along with the confidence interval for the parameter for testing the hypothesis $H_0 : \mu = \mu_0$ against various alternatives for *the normal population when σ^2 is not known.*

[Here $t_n(\alpha)$ is upper α-point of the t-distribution with n *d.f.* as defined in (18·33*a*)].

NORMAL POPULATION $N(\mu, \sigma^2)$; σ^2 UNKNOWN

Serial No.	*Hypothesis*	*Test*	*Test Statistic*	*Reject H_0 at Level of Significance, α, iff*	*(1 – α) confidence interval for μ*
1.	$H_0 : \mu = \mu_0$ $H_1 : \mu \neq \mu_0$	Two tailed test	$t = \dfrac{\bar{x} - \mu_0}{S/\sqrt{n}}$	$\lvert t \rvert > t_{n-1}(\alpha/2)$	$\bar{x} - \dfrac{S}{\sqrt{n}} t_{n-1}(\alpha/2) \leq \mu$ $\leq \bar{x} + \dfrac{S}{\sqrt{n}} t_{n-1}(\alpha/2)$
2.	$H_0 : \mu = \mu_0$ $H_1 : \mu > \mu_0$	Right tailed test	— do —	$t > t_{n-1}(\alpha)$	$\mu \geq \bar{x} - \dfrac{S}{\sqrt{n}} t_{n-1}(\alpha)$
3.	$H_0 : \mu = \mu_0$ $H_1 : \mu < \mu_0$	Left tailed test	— do —	$t < -t_{n-1}(\alpha)$	$\mu \leq \bar{x} + \dfrac{S}{\sqrt{n}} t_{n-1}(\alpha)$

18·6·3. Test For the Equality of Means of Two Normal Populations

Let us consider two independent random variables X_1 and X_2 following normal distributions $N(\mu_1, \sigma_1^2)$ and $N(\mu_2, \sigma_2^2)$ respectively where the means μ_1, μ_2 and the variances σ_1^2, σ_2^2 are unspecified. Suppose we want to test the hypothesis :

$$H_0 : \mu_1 = \mu_2 = \mu, \text{ (say), (unspecified)};\ 0 < \sigma_1^2 < \infty,\ 0 < \sigma_2^2 < \infty,$$

against the alternative hypothesis:

$$H_1 : \mu_1 \neq \mu_2,\ \sigma_1^2 > 0,\ \sigma_2^2 > 0.$$

Case (i) Population variances are unequal.

$$\Theta = \{(\mu_1, \mu_2, \sigma_1^2, \sigma_2^2) : -\infty < \mu_i < \infty,\ \sigma_i^2 > 0,\ i = 1, 2\}$$

and $$\Theta_0 = \{(\mu, \sigma_1^2, \sigma_2^2) : -\infty < \mu < \infty,\ \sigma_i^2 > 0,\ i = 1, 2\}$$

Let x_{1i} $(i = 1, 2, ..., m)$ and x_{2j} $(j = 1, 2, ..., n)$ be two independent random samples of sizes m and n from the populations $N(\mu_1, \sigma_1^2)$ and $N(\mu_2, \sigma_2^2)$ respectively. Then the likelihood function is given by

$$L(\Theta) = \left(\frac{1}{2\pi\sigma_1^2}\right)^{m/2} \cdot \exp\left\{-\frac{1}{2\sigma_1^2}\sum_{i=1}^{m}(x_{1i}-\mu_1)^2\right\} \times \left(\frac{1}{2\pi\sigma_2^2}\right)^{n/2} \cdot \exp\left\{-\frac{1}{2\sigma_2^2}\sum_{j=1}^{n}(x_{2j}-\mu_2)^2\right\} \quad \ldots (18\cdot41)$$

The maximum likelihood estimates for μ_1, μ_2, σ_1^2 and σ_2^2 are given by the equations:

$$\left.\begin{aligned}
&\frac{\partial}{\partial \mu_1}\log L = 0 \;\Rightarrow\; \hat{\mu}_1 = \frac{1}{m}\sum_{i=1}^{n} x_{1i} = \bar{x}_1\\
&\frac{\partial}{\partial \mu_2}\log L = 0 \;\Rightarrow\; \hat{\mu}_2 = \frac{1}{n}\sum_{j=1}^{n} x_{2j} = \bar{x}_2\\
&\frac{\partial}{\partial \sigma_1^2}\log L = 0 \;\Rightarrow\; \hat{\sigma}_1^2 = \frac{1}{m}\sum_{i=1}^{m} (x_{1i}-\bar{x}_1)^2 = s_1^2, \text{(say)}.\\
\text{and } &\frac{\partial}{\partial \sigma_2^2}\log L = 0 \;\Rightarrow\; \hat{\sigma}_2^2 = \frac{1}{n}\sum_{j=1}^{n} (x_{2j}-\bar{x}_2)^2 = s_2^2, \text{(say)}.
\end{aligned}\right\} \quad \ldots (18\cdot41a)$$

Substituting in (18·41), we get: $L(\hat{\Theta}) = \left(\dfrac{1}{2\pi s_1^2}\right)^{m/2} \cdot \left(\dfrac{1}{2\pi s_2^2}\right)^{n/2} \cdot e^{-(m+n)/2}$ ... (18·42)

In Θ_0, we have $\mu_1 = \mu_2 = \mu$ and the likelihood function is given by:

$$L(\Theta_0) = \left(\frac{1}{2\pi\sigma_1^2}\right)^{m/2} \cdot \exp\left\{-\frac{1}{2\sigma_1^2}\sum_{i=1}^{m}(x_{1i}-\mu)^2\right\} \times \left(\frac{1}{2\pi\sigma_2^2}\right)^{n/2} \cdot \exp\left\{-\frac{1}{2\sigma_2^2}\sum_{i=1}^{n}(x_{2j}-\mu)^2\right\}$$

To obtain the maximum value of $L(\Theta_0)$ for variations in μ, σ_1^2 and σ_2^2, it will be seen that estimate of μ is obtained as the root of a cubic equation

$$\frac{m^2(\bar{x}_1-\mu)}{\sum_{i=1}^{m}(x_{1i}-\hat{\mu})^2} + \frac{n^2(\bar{x}_2-\mu)}{\sum_{j=1}^{m}(x_{2j}-\hat{\mu})^2} \quad \ldots (18\cdot43)$$

and is thus a complicated function of the sample observations. Consequently the likelihood ratio criterion λ will be a complex function of the observations and its distribution is quite tedious since it involves the ratio of two variances. Consequently, it is impossible to obtain the critical region $0 < \lambda < \lambda_0$ for given α since the distribution of the population variances is ordinarily unknown. However, in any given instance the cubic equation (16·43) can be solved for μ by numerical analysis technique and thus λ can be computed. Finally, as an approximate test, $-2 \log_e \lambda$ can be regarded as a χ^2-variate with 1 *d.f.* (*c.f. Theorem 18·2*).

***Case (ii)* Population Variances are equal**, *i.e.*, $\sigma_1^2 = \sigma_2^2 = \sigma^2$, (say). In this case

$$\Theta = \{(\mu_1, \mu_2, \sigma^2) : -\infty < \mu_i < \infty, \sigma^2 > 0, (i = 1, 2)\}$$

$$\Theta_0 = \{(\mu, \sigma^2) : -\infty < \mu < \infty, \sigma^2 > 0\}$$

The likelihood function is then given by :

$$L = \left(\frac{1}{2\pi\sigma^2}\right)^{(m+n)/2} \cdot \exp\left[-\frac{1}{2\sigma^2}\left\{\sum_{i=1}^{m}(x_{1i}-\mu_1)^2 + \sum_{j=1}^{n}(x_{2j}-\mu_2)^2\right\}\right] \quad \ldots (18\cdot44)$$

For $\mu_1, \mu_2, \sigma^2 \in \Theta$, the maximum likelihood equations are given by:

$$\frac{\partial}{\partial \mu_1} \log L = 0 \quad \Rightarrow \quad \hat{\mu}_1 = \bar{x}_1 \quad \text{and} \quad \frac{\partial}{\partial \mu_2} \log L = 0 \quad \Rightarrow \quad \hat{\mu}_2 = \bar{x}_2 \qquad \ldots (18\cdot45)$$

and

$$\frac{\partial}{\partial \sigma^2} \log L = 0 \quad \Rightarrow \quad \hat{\sigma}^2 = \frac{1}{m+n}\left\{\Sigma(x_{1i} - \hat{\mu}_1)^2 + \Sigma(x_{2j} - \hat{\mu}_2)^2\right\}$$

$$\Rightarrow \quad \hat{\sigma}^2 = \frac{1}{m+n}\left\{\sum (x_{1i} - \bar{x}_1)^2 + \sum (x_{2j} - \bar{x}_2)^2\right\} = \frac{1}{m+n}\left(ms_1^2 + ns_2^2\right) \qquad \ldots (18\cdot45a)$$

Substituting the values from (18·45) and (18·45*a*) in (18·44), we get:

$$L(\hat{\Theta}) = \left\{\frac{(m+n)}{2\pi\left(ms_1^2 + ns_2^2\right)}\right\}^{(m+n)/2} \cdot \exp\left\{-\frac{1}{2}(m+n)\right\} \qquad \ldots (18\cdot46)$$

In Θ_0, $\mu_1 = \mu_2 = \mu$ (say) and we get

$$L(\Theta_0) = \left(\frac{1}{2\pi\sigma^2}\right)^{(m+n)/2} \cdot \exp\left[-\frac{1}{2\sigma^2}\left\{\sum_{i=1}^{m}(x_{1i} - \mu)^2 + \sum_{j=1}^{n}(x_{2j} - \mu)^2\right\}\right] \qquad \ldots (18\cdot47)$$

$$\Rightarrow \quad \log L(\Theta_0) = C - \frac{m+n}{2}\log \sigma^2 - \frac{1}{2\sigma^2}\left\{\sum_i (x_{1i} - \mu)^2 + \sum_j (x_{2j} - \mu)^2\right\},$$

where C is a constant independent of μ and σ^2.

The likelihood equation for estimating μ gives:

$$\frac{\partial}{\partial \mu}\log L = \frac{1}{\sigma^2}\left\{\sum_{i=1}^{m}(x_{1i} - \mu) + \sum_{j=1}^{n}(x_{2j} - \mu)\right\} = 0 \Rightarrow (m\bar{x}_1 + n\bar{x}_2) - (m+n)\mu = 0$$

$$\Rightarrow \quad \hat{\mu} = \frac{1}{m+n}[m\bar{x}_1 + n\bar{x}_2] \qquad \ldots (18\cdot48)$$

Also

$$\frac{\partial}{\partial \sigma^2}\log L = 0 \quad \Rightarrow \quad -\frac{(m+n)}{2\sigma^2} + \frac{1}{2\sigma^4}\left[\Sigma(x_{1i} - \mu)^2 + \Sigma(x_{2j} - \mu)^2\right] = 0$$

$$\Rightarrow \quad \hat{\sigma}^2 = \frac{1}{m+n}\left\{\Sigma(x_{1i} - \hat{\mu})^2 + \Sigma(x_{2j} - \hat{\mu})^2\right\} \qquad \ldots (18\cdot49)$$

But

$$\sum_{i=1}^{m}(x_{1i} - \hat{\mu})^2 = \sum_{i=1}^{m}(x_{1i} - \bar{x}_1 + \bar{x}_1 - \hat{\mu})^2 = \Sigma(x_{1i} - \bar{x}_1)^2 + m(\bar{x}_1 - \hat{\mu})^2$$

the product term vanishes since $\sum_i (x_{1i} - \bar{x}_1) = 0.$

$$\therefore \quad \sum_{i=1}^{m}(x_{1i} - \hat{\mu})^2 = ms_1^2 + m\left(\bar{x}_1 - \frac{m\bar{x}_1 + n\bar{x}_2}{m+n}\right)^2 = ms_1^2 + \frac{m.n^2(\bar{x}_1 - \bar{x}_2)^2}{(m+n)^2}$$

Similarly, we shall get: $\sum_{j=1}^{n}\left(x_{2j} - \hat{\mu}\right)^2 = ns_2^2 + \dfrac{nm^2(\bar{x}_2 - \bar{x}_1)^2}{(m+n)^2}$

Substituting in (18·49), we get:

$$\hat{\sigma}^2 = \frac{1}{m+n}\left\{ms_1^2 + ns_2^2 + \frac{mn}{m+n}(\bar{x}_1 - \bar{x}_2)^2\right\} \qquad \ldots (18\cdot49a)$$

Substituting from (18·48) and (18·49*a*) in (18·47), we get:

$$L(\hat{\Theta}_0) = \left\{\frac{(m+n)}{2\pi\left(ms_1^2 + ns_2^2 + \frac{mn}{m+n}(\bar{x}_1 - \bar{x}_2)^2\right)}\right\}^{(m+n)/2} \times \exp\left(-\frac{m+n}{2}\right) \qquad \ldots (18\cdot50)$$

$$\therefore \qquad \lambda = \frac{L(\hat{\Theta}_0)}{L(\hat{\Theta})} = \left\{ \frac{ms_1^2 + ns_2^2}{ms_1^2 + ns_2^2 + \frac{mn}{m+n}(\bar{x}_1 - \bar{x}_2)^2} \right\}^{(m+n)/2}$$

$$= \left[\frac{1}{\left\{ 1 + \frac{mn(\bar{x}_1 - \bar{x}_2)^2}{(m+n)\left(ms_1^2 + ns_2^2\right)} \right\}} \right]^{(m+n)/2} \qquad \dots (18{\cdot}51)$$

We know that (c.f. § 16·3·3), under the null hypothesis $H_0 : \mu_1 = \mu_2$, the statistic:

$$t = \frac{\bar{x}_1 - \bar{x}_2}{S\sqrt{\frac{1}{m} + \frac{1}{n}}}, \qquad \dots (18{\cdot}52)$$

where $$S^2 = \frac{1}{m+n-2}\left(ms_1^2 + ns_2^2\right) \qquad \dots (18{\cdot}52a)$$

follows Student's t-distribution with $(m + n - 2)$ *d.f.* Thus, in terms of t, we get:

$$\lambda = \left(1 + \frac{t^2}{m+n-2}\right)^{-(m+n)/2} \qquad \dots (18{\cdot}53)$$

As in § 18·6·2, the test can as well be carried with t rather than with λ. The critical region $0 < \lambda < \lambda_0$ transforms to the critical region of the type

$$t^2 > (m+n-2)\left[\frac{1}{\lambda_0^2/(m+n)} - 1\right] = A^2, \text{ (say)},$$

i.e., by $\qquad | t | > A,$

where A is determined so that : $\qquad P[\,| t | > A \mid H_0] = \alpha \qquad \dots (18{\cdot}55)$

Since under H_0, the statistic t follows Student's t-distribution with $(m + n - 2)$ *d.f.*, we get from (18·55):

$$A = t_{m+n-2}\,(\alpha/2) \qquad \dots (18{\cdot}56)$$

where, $t_n(\alpha)$ is the right $100\,\alpha\%$ point of the t-distribution with $n.d.f.$

Thus, for testing the null hypothesis $\qquad H_0 : \mu_1 = \mu_2 \;;\quad \sigma_1{}^2 = \sigma_2{}^2 = \sigma^2 > 0$

against the alternative : $\qquad H_1 : \mu_1 \neq \mu_2 \;,\quad \sigma_1{}^2 = \sigma_2{}^2 = \sigma^2 > 0,$

we have the two-tailed t-test defined as follows ;

If $| t | = \left| \dfrac{\bar{x}_1 - \bar{x}_2}{S\sqrt{\frac{1}{m} + \frac{1}{n}}} \right| > t_{m+n-2}(\alpha/2)$, *reject* H_0, *otherwise* H_0 *may be accepted.*

Remarks:

1. Proceeding similarly as in Remarks to § 18·7·1, we can obtain the critical regions for testing

$$H_0 : \mu_1 = \mu_2 \,;\, \sigma_1{}^2 = \sigma_2{}^2 = \sigma^2 > 0$$

against the alternative hypothesis

$$H_1 : \mu_1 > \mu_2 \,;\, \sigma_1{}^2 = \sigma_2{}^2 = \sigma^2 > 0$$

or $$H_1' : \mu_1 < \mu_2 \,;\, \sigma_1{}^2 = \sigma_2{}^2 = \sigma^2 > 0$$

We give on page 18.31, in a tabular form the critical region, the test statistic and the confidence interval for testing the hypothesis

$$H_0 : \delta = \mu_1 - \mu_2 = \delta_0, \text{ (say)},$$

against various alternatives, *viz.*, $\delta > \delta_0$, $\delta < \delta_0$ or $\delta \neq \delta_0$.

2. For testing $H_0 : \delta = \delta_0$ against the alternative $H_1 : \delta < \delta_0$, the roles of x_1 and x_2 are interchanged and the case 1 of the table is applied.

3. If $\delta_0 = 0$, the above test reduces to testing $H_0 : \mu_1 = \mu_2$, *i.e.*, the equality of two population means.

4. If the two population variances are not equal, then for testing $H_0 : \delta = \delta_0$, we use Fisher-Behrens' d-test.

S. No.	*Alternative Hypothesis*	*Test*	*Test statistic*	*Reject H_0 at level of significance α if*	*(1 – α) Confidence interval of δ*
1.	$\delta > \delta_0$	Right, tailed	$t = \dfrac{(\bar{x} \cdot \Re\bar{x})}{S\sqrt{\dfrac{1}{m}+\dfrac{1}{n}}}$	$t > t_{m+n-2}(\alpha)$ $= t_1$, (say)	$\delta \ge (\bar{x}_1 - \bar{x}_2) - t_1 S\sqrt{\dfrac{1}{m}+\dfrac{1}{n}}$
2.	$\delta \ne \delta_0$	Two tailed	— do —	$\lvert t \rvert > t_{m+n-2}(\alpha/2)$ $= t_2$, (say)	$(\bar{x}_1 - \bar{x}_2) - t_2 S\sqrt{\dfrac{1}{m}+\dfrac{1}{n}}$ $\le \delta \le (\bar{x}_1 - \bar{x}_2) + t_2 S\sqrt{\dfrac{1}{m}+\dfrac{1}{n}}$

18·6·4. Test For the Equality of Means of Several Normal Populations

Let X_{ij}, $(j = 1, 2, \ldots, n_i;\ i = 1, 2, \ldots k)$ be k independent random samples from k normal populations with means $\mu_1, \mu_2, \ldots, \mu_k$ respectively and unknown but common variance σ^2. In other words, the k normal populations are supposed to be *homoscedastic.* We want to test the null hypothesis

$$H_0 : \mu_1 = \mu_2 = \ldots = \mu_k = \mu \text{ (say), (unspecified)}$$

$$\sigma_1^2 = \sigma_2^2 = \ldots = \sigma_k^2 = \sigma^2 \text{ (say), (unspecified)}$$

against the alternative hypothesis

$$H_1 : \mu_i\text{'s are not all equal, and}$$

$$\sigma_1^2 = \sigma_2^2 = \ldots = \sigma_k^2 = \sigma^2, \text{ (unspecified)}$$

Thus we have

$$\Theta = \left\{(\mu_1, \mu_2, \ldots, \mu_k, \sigma^2)\ ; -\infty < \mu_i < \infty, (i = 1, 2, \ldots, k), \sigma^2 > 0\right\}$$

and

$$\Theta_0 = \left\{(\mu_1, \mu_2, \ldots, \mu_k, \sigma^2)\ ; -\infty < \mu_i = \mu < \infty, (i = 1, 2, \ldots, k), \sigma^2 > 0\right\}$$

The likelihood function of the sample observations is given by

$$L(\Theta) = \left(\frac{1}{2\pi\sigma^2}\right)^{n/2} \cdot \exp\left\{-\frac{1}{2\sigma^2}\sum_{i=1}^{k}\sum_{j=1}^{n_i}(x_{ij} - \mu_i)^2\right\} \quad \text{where } n = \sum_{i=1}^{k} n_i. \qquad \ldots (18\cdot57)$$

For variations of μ_i, $(i = 1,2, \ldots, k)$ and σ^2 in Θ, the maximum likelihood estimates are given by

$$\frac{\partial}{\partial \mu_i}\log L(\Theta) = 0 \quad \Rightarrow \quad \sum_j (x_{ij} - \mu_i) = 0 \quad \Rightarrow \quad \hat{\mu}_i = \frac{1}{n_i}\sum_{j=1}^{n_i} x_{ij} = \bar{x}_i \qquad \ldots (18\cdot58)$$

$$\frac{\partial}{\partial \sigma^2}\log L(\Theta) = 0 \quad \Rightarrow \quad \hat{\sigma}^2 = \frac{1}{n}\sum_i\sum_j (x_{ij} - \hat{\mu}_i)^2$$

$$\Rightarrow \qquad \hat{\sigma}^2 = \frac{1}{n}\sum_i\sum_j (x_{ij} - \bar{x}_i)^2 = \frac{S_W}{n}, \text{(say)}, \qquad \ldots (18\cdot58a)$$

where in *ANOVA* (Analysis of Variance) terminology, S_W is called *within sample sum of squares (W.S.S.).*

In Θ_0, the only variable parameters are μ and σ^2 and we have:

$$L(\Theta\ ) = \left(\frac{1}{2\pi\sigma^2}\right)^{/2} \cdot \exp\left\{-\frac{1}{2\sigma^2}\sum_i\sum_j (x_{ij} - \mu)\right\} \qquad \ldots (18{\cdot}59)$$

The *MLE's* of μ and σ^2 are given by:

$$\frac{\partial}{\partial\mu}\log L(\Theta\ ) = 0 \quad \Rightarrow \quad \sum_i\sum_j (x_{ij} - \mu) = 0 \quad \Rightarrow \quad \mu = -\sum_i\sum_j x_{ij} = \bar{x} \qquad \ldots (18{\cdot}60)$$

and $$\frac{\partial}{\partial\sigma^2}\log L(\Theta_0) = 0 \quad \Rightarrow \hat{\sigma}^2 = \frac{1}{n}\sum\sum (x_{ij} - \hat{\mu})^2$$

$$\hat{\sigma}^2 = \frac{1}{n}\sum\sum (x_{ij} - \bar{x})^2 = \frac{S_T}{n}, \text{(say)}, \qquad \ldots (18.60\text{a})$$

where in *ANOVA* terminology, S_T, is called *total sum of squares (T.S.S.)*

Substituting from (18·58) and (18·58*a*) in (18·57) and from (18·60) and (18·60*a*) in (18·59), we get respectively,

$$L(\hat{\Theta}) = \left(\frac{n}{2\pi S_W}\right)^{n/2} \cdot \exp\left(-\frac{n}{2}\right) \ldots (18{\cdot}61) \quad \text{and} \quad L(\hat{\Theta}_0) = \left(\frac{n}{2\pi S_T}\right)^{n/2} \cdot \exp\left(-\frac{n}{2}\right) \ldots (18{\cdot}62)$$

$$\therefore \qquad \lambda = \frac{L(\hat{\Theta}_0)}{L(\hat{\Theta})} = \left(\frac{S_W}{S_T}\right)^{n/2} \qquad \ldots (18{\cdot}63)$$

We have $$S_T = \sum_i\sum_j (x_{ij} - \bar{x})^2 = \sum_i\sum_j (x_{ij} - \bar{x}_i + \bar{x}_i - \bar{x})^2$$

$$= \sum_i\sum_j (x_{ij} - \bar{x}_i)^2 + \sum_i\sum_j (\bar{x}_i - \bar{x})^2 + 2\sum_i\left[(\bar{x}_i - \bar{x})\sum_j (\bar{x}_{ij} - \bar{x}_i)\right]$$

But $\sum_{j=1}^{n_i} (x_{ij} - \bar{x}_i) = 0$, being the algebraic sum of the deviations of the observations of the *i*th sample from its mean.

$$\therefore \qquad S_T = \sum_i\sum_j (x_{ij} - \bar{x}_i)^2 + \sum_i n_i(\bar{x}_i - \bar{x})^2 = S_W + S_B, \text{(say)}, \qquad \ldots (18{\cdot}63a)$$

where $S_B = \sum_i n_i(\bar{x}_i - \bar{x})^2$, in *ANOVA* terminology is called *between samples sum of squares (B.S.S.)*.

Substituting in (18·63), we get: $$\lambda = \left(\frac{S_W}{S_W + S_B}\right)^{n/2} = \frac{1}{\left(1 + \dfrac{S_B}{S_W}\right)^{n/2}} \qquad \ldots (18{\cdot}64)$$

We know that under H_0, the statistic : $$F = \frac{S_B/(k-1)}{S_W/(n-k)} \qquad \ldots (18{\cdot}65)$$

follows *F*-distribution with $(k - 1, n - k)$ d.f.

Substituting in (18·64), the likelihood ratio criterion λ in terms of *F* is given by

$$\lambda = \left(1 + \frac{k-1}{n-k}F\right)^{-n/2} \qquad \ldots (18{\cdot}66)$$

Since λ is a monotonic function of *F*, the test can well be carried on with *F* as test statistic rather than with λ. The critical region for testing H_0 against H_1, *viz.*, $0 < \lambda < \lambda_0$ is equivalently given by:

$$\left(1+\frac{k-1}{n-k}F\right)^{n/2} > \lambda_0^{-1}$$

$$\Rightarrow \qquad F > \frac{n-k}{k-1}\left\{(\lambda_0)^{-2/n}-1\right\} = A, \text{(say)}, \qquad \ldots (18·67)$$

where A is determined from the equation: $\quad P[F > A \mid H_0] = \alpha \qquad \ldots (18·67a)$

Since F follows F-distribution with $(k-1, n-k)$ *d.f.*, $\quad A = F_{k-1,\, n-k}(\alpha)$

where $F_{k-1,\, n-k}(\alpha)$ denotes the upper α-point of the F-distribution with $(k-1, n-k)$ *d.f.*

Hence, the test for testing

$$H_0 : \mu_1 = \mu_2 = \ldots = \mu_k = \mu,\ \sigma_1^2 = \sigma_2^2 = \ldots = \sigma_k^2 = \sigma^2 > 0$$

against the alternative hypothesis

$$H_1 : \mu_i\text{'s are not all equal},\ \sigma_1^2 = \sigma_2^2 = \ldots = \sigma_k^2 = \sigma^2 > 0$$

is defined as follows :

Reject H_0 if $F > F_{k-1,\, n-k}(\alpha)$, otherwise H_0 may be accepted, where F is defined in (18·65).

Remark. In *ANOVA* terminology, $S_B/(k-1)$ is called Between Samples Mean Sum of Squares (M.S.S.) while $S_W/(n-k)$ is called Within Samples (or Error) Mean Sum of Squares and thus F is defined as

$$F = \frac{\text{Between Samples M.S.S.}}{\text{Within Samples M.S.S.}} \qquad \ldots (18·67b)$$

18·6·5. Test For the Variance of a Normal Population

Let us now consider the problem of testing if the variance of a normal population has a specified value σ_0^2 on the basis of a random sample $x_1, x_2, \ldots, x_n$ of size n from normal population $N(\mu, \sigma^2)$.

We want to test the hypothesis : $\quad H_0 : \sigma^2 = \sigma_0^2$, (specified),

against the alternative hypothesis : $\quad H_1 : \sigma^2 \neq \sigma_0^2$

Here, we have: $\quad \Theta = \{(\mu, \sigma^2) : -\infty < \mu < \infty,\ \sigma^2 > 0\}$

and $\quad \Theta_0 = \{(\mu, \sigma^2) : -\infty < \mu < \infty,\ \sigma^2 = \sigma_0^2\}$

The likelihood function of the sample observations is given by:

$$L = \left(\frac{1}{2\pi\sigma^2}\right)^{n/2} \cdot \exp\left\{-\frac{1}{2\sigma^2}\sum_{i=1}^{n}(x_i-\mu)^2\right\} \qquad \ldots (18·68)$$

As in § (18·6·2), [c.f. (18·27)], we shall get:

$$L(\hat{\Theta}) = \left(\frac{1}{2\pi s^2}\right)^{n/2} \exp\left(-\frac{n}{2}\right) \qquad \ldots (18·69)$$

In Θ_0, we have only one variable parameter, *viz.*, μ and

$$L(\Theta_0) = \left(\frac{1}{2\pi\sigma_0^2}\right)^{n/2} . \exp\left\{-\frac{1}{2\sigma_0^2}\sum_{i=1}^{n}(x_i-\mu)^2\right\} \qquad \ldots (18·70)$$

The *MLE* for μ is given by: $\quad \dfrac{\partial}{\partial\mu}\log L = 0 \quad \Rightarrow \quad \hat{\mu} = \bar{x}$

$$\therefore \qquad L(\hat{\Theta}_0) = \left(\frac{1}{2\pi\sigma_0^2}\right)^{n/2} . \exp\left\{-\frac{1}{2\sigma_0^2}\sum_{i=1}^{n}(x_i-\bar{x})^2\right\}$$

$$= \left(\frac{1}{2\pi\sigma_0^2}\right)^{n/2} . \exp\left(-ns^2/2\sigma_0^2\right) \qquad \ldots (18·71)$$

The likelihood ratio criterion is given by:

$$\lambda = \frac{L(\hat{\Theta}_0)}{L(\hat{\Theta})} = \left(\frac{s^2}{\sigma_0^2}\right)^{n/2} \exp\left\{-\frac{1}{2}\left(\frac{ns^2}{\sigma_0^2} - n\right)\right\} \qquad \text{... (18·72)}$$

We know that under H_0, the statistic : $\chi^2 = \dfrac{ns^2}{\sigma_0^2}$ follows chi-square distribution with $(n-1)$ *d.f.*

In terms of χ^2, we have

$$\lambda = \left[\frac{\chi^2}{n}\right]^{n/2} \cdot \exp\left[-\frac{1}{2}(\chi^2 - n)\right] \qquad \text{... (18·73)}$$

Since λ is a monotonic function of χ^2, the test may be done using χ^2 as a criterion. The critical region $0 < \lambda < \lambda_0$ is now equivalent to

$$\left(\chi^2/n\right)^{n/2} \exp\left[-\frac{1}{2}(\chi^2 - n)\right] < \lambda_0$$

or $$\exp\left(-\frac{1}{2}\chi^2\right)(\chi^2)^{n/2} < \lambda_0.(ne^{-1})^{n/2} = B, \text{ (say)}. \qquad \text{... (18·74)}$$

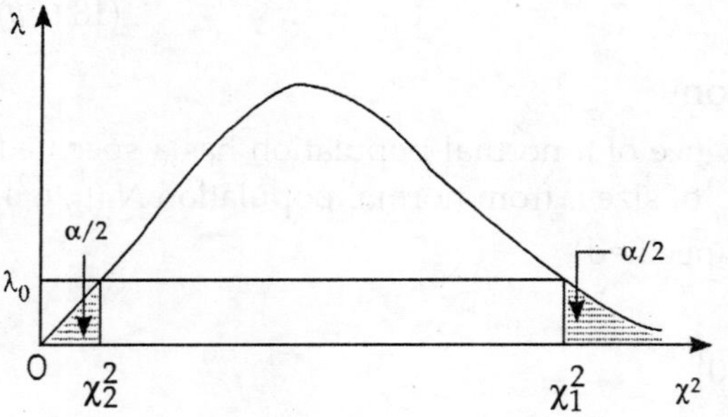

Since χ^2 has chi-square distribution with $(n-1)$ d.f., the critical region (18·74) is determined by a pair of intervals $0 < \chi^2 < \chi_2^2$ and $\chi_1^2 < \chi^2 < \infty$, where χ_1^2 and χ_2^2 are to be determined such that the ordinates of (18·73) are equal, *i.e.*,

$$(\chi_1^2)^{n/2} \exp\left(-\frac{1}{2}\chi_1^2\right) = (\chi_2^2)^{n/2} \exp\left(-\frac{1}{2}\chi_2^2\right).$$

Critical region is shown as shaded region in the adjacent diagram.

In other words, χ_1^2 and χ_2^2 are defined by the equations

$$P(\chi^2 > \chi_1^2) = \frac{\alpha}{2} \quad \text{and} \quad P(\chi^2 > \chi_2^2) = 1 - \frac{\alpha}{2} \qquad \text{... (18·75)}$$

In other words, $\chi_1^2 = \chi^2_{n-1}(\alpha/2)$ and $\chi_2^2 = \chi^2_{n-1}(1-\alpha/2)$,

where $\chi^2_{(n-1)}(\alpha)$ is the upper α-point of the chi-square distribution with $(n-1)$ d.f. Thus, the *critical region for testing* $H_0 : \sigma^2 = \sigma_0^2$ *against* $H_1 : \sigma^2 \neq \sigma_0^2$, *is a two-tailed region given by* :

$$\chi^2 > \chi^2_{n-1}(\alpha/2) \quad \text{and} \quad \chi^2 < \chi^2_{n-1}(1-\alpha/2) \qquad \text{... (18·76)}$$

Thus, in this case we have a two-tailed test.

Remarks:

1. If we want to test $H_0 : \sigma^2 = \sigma_0^2$ against the alternative hypothesis $H_1 : \sigma^2 < \sigma_0^2$, we get a one-tailed (left-tail) test with critical region $\chi^2 < \chi^2_{(n-1)}(\alpha)$ while for testing H_0 against $H_1 : \sigma^2 > \sigma_0^2$, we have a right tail test with critical region $\chi^2 > \chi^2_{(n-1)}(\alpha)$.

   We give on page 18·35 in a tabular form, the test statistic, the test criterion and the confidence interval for the parameter σ^2 for testing $H_0 : \sigma^2 = \sigma_0^2$, μ (unknown) against various alternative hypotheses.

NORMAL POPULATION $N(\mu, \sigma^2)$; μ UNKNOWN; $H_0 : \sigma^2 = \sigma_0^2$

S. No.	*Alternative Hypothesis*	*Test*	*Test Statistic*	*Reject H_0 at 'α' level of significance if*	*Confidence (1 – α) interval for σ^2*
1.	$\sigma^2 > \sigma_0^2$	Right-tailed test	$\chi^2 = \dfrac{ns^2}{\sigma_0^2}$	$\chi^2 > \chi^2_{n-1}(\alpha)$	$\sigma^2 \geq \dfrac{ns^2}{\chi^2_{n-1}(\alpha)}$
2.	$\sigma^2 < \sigma_0^2$	Left-tailed test	— do —	$\chi^2 < \chi^2_{n-1}(1-\alpha)$	$\sigma^2 \leq \dfrac{ns^2}{\chi^2_{n-1}(1-\alpha)}$
3.	$\sigma^2 \neq \sigma_0^2$	Two-tailed test	— do —	$\chi^2 > \chi^2_{n-1}(\alpha/2)$ and $\chi^2 < \chi^2_{n-1}(1-\alpha/2)$	$\dfrac{ns^2}{\chi^2_{n-1}(\alpha/2)} \leq \sigma^2 \leq \dfrac{ns^2}{\chi^2_{n-1}(1-\alpha/2)}$

2. If we want to test the null hypothesis $H_0 : \sigma^2 = \sigma_0^2$ against the various alternative hypotheses, *viz.*, $\sigma^2 > \sigma_0^2$ or $\sigma^2 < \sigma_0^2$ or $\sigma^2 \neq \sigma_0^2$ for the normal population $N(\mu, \sigma^2)$ where μ is known then the test statistic, the critical region and the confidence interval for σ^2 can be obtained from the table given above on replacing $(n-1)$ by n and ns^2 by the expression $\sum_{i=1}^{n} (x_i - \mu)^2$.

18·6·6. Test for the Equality of Variances of Two Normal Populations

Consider two normal populations $N(\mu_1, \sigma_1^2)$ and $N(\mu_2, \sigma_2^2)$ where the means μ_1 and μ_2 and variances σ_1^2, σ_2^2 are unspecified. We want to test the hypothesis :

$$H_0 : \sigma_1^2 = \sigma_2^2 = \sigma^2 \text{ (unspecified) with } \mu_1 \text{ and } \mu_2 \text{ (unspecified)}$$

against the alternative hypothesis : $H_1 : \sigma_1^2 \neq \sigma_2^2$; μ_1 and μ_2 (unspecified).

If x_{1i}, $(i = 1, 2, \ldots, m)$ and x_{2j}, $(j = 1, 2, \ldots, n)$ be independent random samples of sizes m and n from $N(\mu_1, \sigma_1^2)$ and $N(\mu_2, \sigma_2^2)$ respectively then

$$L = \left(\frac{1}{2\pi\sigma_1^2}\right)^{m/2} \exp\left\{-\frac{1}{2\sigma_1^2}\sum_{i=1}^{m}(x_{1i}-\mu_1)^2\right\} \times \left(\frac{1}{2\pi\sigma_2^2}\right)^{n/2} \exp\left\{-\frac{1}{2\sigma_2^2}\sum_{j=1}^{n}(x_{2j}-\mu_2)^2\right\} \quad \ldots (18·77)$$

In this case $\quad \Theta = \{(\mu_1, \mu_2, \sigma_1^2, \sigma_2^2) : -\infty < \mu_i < \infty;\ \sigma_i^2 > 0,\ (i = 1, 2)\}$

and $\quad \Theta_0 = \{(\mu_1, \mu_2, \sigma^2) : -\infty < \mu_i < \infty;\ (i = 1, 2),\ \sigma^2 > 0\}$

As in § 18·6·3 [c.f. (18·42)],

$$L(\hat{\Theta}) = \left(\frac{1}{2\pi s_1^2}\right)^{m/2} \cdot \left(\frac{1}{2\pi s_2^2}\right)^{n/2} \cdot \exp\left\{-\frac{1}{2}(m+n)\right\} \quad \ldots (18·78)$$

where s_1^2 and s_2^2 are as defined in (18·41*a*).

In Θ_0, the likelihood function (18·77) is given by

$$L(\Theta_0) = \left(\frac{1}{2\pi\sigma^2}\right)^{(m+n)/2} \cdot \exp\left[-\frac{1}{2\sigma^2}\left\{\sum_i (x_{1i}-\mu_1)^2 + \sum_j (x_{2j}-\mu_2)^2\right\}\right] \quad \ldots (18·79)$$

and the MLE's for μ_1, μ_2 and σ^2 are now given by $\hat{\mu}_1 = \bar{x}_1,\ \hat{\mu}_2 = \bar{x}_2$ $\quad \ldots (18·80)$

$$\hat{\sigma}^2 = \frac{1}{(m+n)}\left\{\sum_i (x_{1i}-\hat{\mu}_1)^2 + \sum_j (x_{2j}-\hat{\mu}_2)^2\right\}$$

and

$$= \frac{1}{m+n}\left\{\sum_i (x_{1i}-\bar{x}_1)^2 + \sum_j (x_{2j}-\bar{x}_2)^2\right\} = \frac{ms_1^2 + ns_2^2}{m+n} \quad \ldots (18·80a)$$

Substituting from (18·80) and (18·80*a*) in (18·79), we get:

$$L(\hat{\Theta}_0) = \left\{\frac{m+n}{2\pi\left(ms_1^2 + ns_2^2\right)}\right\}^{(m+n)/2} \cdot \exp\left\{-\frac{1}{2}(m+n)\right\} \quad \ldots (18{\cdot}81)$$

$$\therefore \quad \lambda = \frac{L(\hat{\Theta}_0)}{L(\hat{\Theta})} = (m+n)^{(m+n)/2}\left\{\frac{\left(s_1^2\right)^{m/2}\left(s_2^2\right)^{n/2}}{\left[ms_1^2 + ns_2^2\right]^{(m+n)/2}}\right\}$$

$$= \frac{(m+n)^{(m+n)/2}}{m^{m/2}\cdot n^{n/2}}\left\{\frac{\left(ms_1^2\right)^{m/2}\left(ns_2^2\right)^{n/2}}{\left[ms_1^2 + ns_2^2\right]^{(m+n)/2}}\right\} \quad \ldots (18{\cdot}82)$$

We know that under H_0, the statistic $F = \dfrac{\Sigma(x_{1i} - \bar{x}_1)^2 / (m-1)}{\Sigma(x_{2j} - \bar{x}_2)^2 / (n-1)} = \dfrac{S_1^2}{S_2^2}$ ... (18·83)

follows *F*-distribution with $(m-1, n-1)$ d.f. (18·83) also implies

$$F = \frac{m(n-1)s_1^2}{n(m-1)s_2^2} \Rightarrow \left(\frac{m-1}{n-1}\right)F = \frac{ms_1^2}{ns_2^2} \quad \ldots (18{\cdot}83a)$$

Substituting in (18·82) and simplifying, we get:

$$\lambda = \frac{(m+n)^{(m+n)/2}}{m^{m/2}n^{n/2}}\left\{\frac{\left(\frac{m-1}{n-1}F\right)^{m/2}}{\left(1+\frac{m-1}{n-1}F\right)^{(m+n)/2}}\right\} \quad \ldots (18{\cdot}84)$$

Thus λ is a monotonic function of *F* and hence the test can be carried on with *F*, defined in (18·83) as test statistic. The critical region $0 < \lambda < \lambda_0$ can be equivalently seen to be given by pair of intervals $F \le F_1$ and $F \ge F_2$, where F_1 and F_2 are determined so that under H_0

$$P(F \ge F_2) = \alpha/2 \quad \text{and} \quad P(F \ge F_1) = 1 - \alpha/2$$

Since, under H_0, *F* follows Snedecor's *F*-distribution with $(m-1, n-1)$ d.f., we have

$$F_2 = F_{m-1,n-1}(\alpha/2) \quad \text{and} \quad F_1 = F_{m-1,n-1}(1-\alpha/2)$$

where $F_{m,n}(\alpha)$ is the upper α-point of *F*-distribution with (m, n) d.f. Consequently, for testing $H_0 : \sigma_1^2 = \sigma_2^2$ against the alternative hypothesis $H_1 : \sigma_1^2 \ne \sigma_2^2$, we have a two-tailed *F*-test, the *critical region* being given by:

$$F > F_{m-1,n-1}(\alpha/2) \quad \text{and} \quad F < F_{m-1,n-1}(1-\alpha/2) \quad \ldots (18{\cdot}85)$$

where *F* is defined in (18·83) or (18·83*a*).

Remark. Let us suppose that we want to test the hypothesis $H_0 : \dfrac{\sigma_1^2}{\sigma_2^2} = \delta_0^2$.

Without loss of generality, we can assume that $S_1^2 > S_2^2$, where S_1^2 and S_2^2 are unbiased estimates of σ_1^2 and σ_2^2 respectively. We know that the statistic

$$F = \frac{S_1^2/\sigma_1^2}{S_2^2/\sigma_2^2} = \frac{S_1^2}{S_2^2}\cdot\frac{1}{\delta_0^2} \quad \text{(under } H_0\text{)},$$

follows *F*-distribution with $(m-1, n-1)$ d.f. The test-statistic, the test criterion and $(1-\alpha)$ confidence interval for the parameter for various alternative hypotheses are given in the table on page 18.37.

If $\delta_0 = 1$, the above test reduces to testing the equality of population variances.

NORMAL POPULATION; $H_0 : \frac{\sigma_1^2}{\sigma_2^2} = \delta_0^2$

S. No.	*Alternative Hypothesis*	*Test*	*Test Statistic*	*Critical region at level of significance 'α'*	*(1 – α) confidence coefficient for* $\frac{\sigma_1^2}{\sigma_2^2}$
1.	$\frac{\sigma_1^2}{\sigma_2^2} > \delta_0^2$	Right-tailed	$F = \frac{S_1^2}{S_2^2} \cdot \frac{1}{\delta_0^2}$	$F > F_{m-1,\, n-1}(\alpha)$	$\frac{\sigma_1^2}{\sigma_2^2} \geq \frac{S_1^2}{S_2^2} \times \frac{1}{F_{m-1,\, n-1}(\alpha)}$
2.	$\frac{\sigma_1^2}{\sigma_2^2} < \delta_0^2$	Left-tailed	— do —	$F < F_{m-1,\, n-1}(1-\alpha)$	$\frac{\sigma_1^2}{\sigma_2^2} \leq \frac{S_1^2}{S_2^2} \times \frac{1}{F_{m-1,n-1}(1-\alpha)}$
3.	$\frac{\sigma_1^2}{\sigma_2^2} \neq \delta_0^2$	Two-tailed	— do —	$F > F_{m-1,\, n-1}(\alpha/2)$ and $F < F_{m-1,\, n-1}(1-\alpha/2)$	$\frac{S_1^2}{S_2^2} \times \frac{1}{F_{m-1,n-1}(\alpha/2)} \leq \frac{\sigma_1^2}{\sigma_2^2} \leq \frac{S_1^2}{S_2^2} \cdot \frac{1}{F_{m-1,\, n-1}(1-\alpha/2)}$

18·6·7. Test for the Equality of Variances of Several Normal Populations

Let X_{ij}, $(j = 1, 2, \ldots, n_i)$ be a random sample of size n_i from the normal population $N(\mu_i, \sigma_i^2)$, $i = 1, 2, \ldots, k$. We want to test the null hypothesis :

$H_0 : \sigma_1^2 = \sigma_2^2 = \ldots = \sigma_k^2 = \sigma^2$ (unspecified), with $\mu_1, \mu_2, \ldots, \mu_k$ (unspecified), against the alternative hypothesis :

$H_1 : \sigma_i^2$ *(1, 2, ..., k)*, are not all equal; $\mu_1, \mu_2, \ldots, \mu_k$ (unspecified).

Here we have

$$\Theta = \{(\mu_1, \mu_2, \ldots, \mu_k ; \sigma_1^2, \sigma_2^2, \ldots, \sigma_k^2) : -\infty < \mu_i < \infty, \sigma_i^2 > 0 \ (i = 1, 2, \ldots, k)\}$$

and

$$\Theta_0 = \{(\mu_1, \mu_2, \ldots, \mu_k ; \sigma_1^2, \sigma_2^2, \ldots, \sigma_k^2) : -\infty < \mu_i < \infty, \sigma_i^2 = \sigma^2 > 0, \ (i = 1, 2, \ldots, k)\}$$

The likelihood function of the sample observations x_{ij}, $(j = 1, 2, \ldots, n_i : i = 1, 2, \ldots, k)$ is given by:

$$L = \prod_{i=1}^{k} \left[\left(\frac{1}{2\pi\sigma_i^2} \right)^{n_i/2} \cdot \exp\left\{ -\frac{1}{2\sigma_i^2} \sum_{j=1}^{n_i} (x_{ij} - \mu_i)^2 \right\} \right] \quad \ldots (18·86)$$

It can be easily seen that in Θ the MLE's of μ_i's and σ_i's are given by:

$$\hat{\mu}_i = \bar{x}_i \text{ and } \hat{\sigma}_i^2 = \frac{1}{n_i} \sum_{j=1}^{n_i} (x_{ij} - \bar{x}_i)^2 = s_i^2 ; \ i = 1, 2, \ldots, k \quad \ldots (18·87)$$

$$\therefore \quad L(\hat{\Theta}) = \prod_{i=1}^{k} \left\{ \left(\frac{1}{2\pi s_i^2} \right)^{n_i/2} \cdot \exp\left(-\frac{n_i}{2} \right) \right\}$$

$$= \exp\left(-\frac{n}{2} \right) \cdot \prod_{i=1}^{k} \left\{ \left(\frac{1}{2\pi s_i^2} \right)^{n_i/2} \right\}, \quad \text{where } n = \sum n_i \quad \ldots (18·88)$$

In Θ_0, $\sigma_1^2 = \sigma_2^2 = \ldots = \sigma_k^2 = \sigma^2$ and therefore,

$$L(\Theta_0) = \left(\frac{1}{2\pi\sigma^2}\right)^{n/2} \cdot \exp\left\{-\frac{1}{2\sigma^2}\sum_j\sum_i (x_{ij} - \mu_i)^2\right\} \quad \ldots (18{\cdot}89)$$

The *MLE*'s of μ_i's and σ^2 are given by

$$\hat{\mu}_i = \bar{x}_i \quad \text{and} \quad \hat{\sigma}^2 = \frac{1}{n}\sum_i\sum_j (x_{ij} - \bar{x})^2 = \frac{1}{n}\sum_i n_i s_i^2 \quad \ldots (18{\cdot}90)$$

Substituting from (18·90) in (18·89), we get : $L(\hat{\Theta}_0) = \left(\dfrac{n}{2\pi \sum_i n_i s_i^2}\right)^{n/2} \cdot \exp\left(-\dfrac{n}{2}\right)$ ... (18·91)

$$\therefore \quad \lambda = \frac{L(\hat{\Theta}_0)}{L(\hat{\Theta})} = \frac{n^{n/2} \prod_{i=1}^{k}\left[\left(s_i^2\right)^{n_i/2}\right]}{\left\{\sum_{i=1}^{k} n_i s_i^2\right\}^{n/2}}$$

$$= \frac{\prod_{i=1}^{k}\left\{\left(s_i^2\right)^{n_i/2}\right\}}{\left(s^2\right)^{\Sigma n_i/2}}, \quad \text{where } s^2 = \frac{1}{n}\sum_i n_i s_i^2$$

$$\therefore \quad = \prod_{i=1}^{k}\left\{\left(\frac{s_i^2}{s^2}\right)^{n_i/2}\right\} \quad \ldots (18{\cdot}92)$$

λ is thus a complicated function of sample observations and it is not easy to obtain its distribution. However, if n_i's are large ($i = 1, 2, \ldots, k$), Theorem 18·2 provides an approximate test defined as follows :

For large n_i's, the quantity $(-2 \log_e \lambda)$ is approximately distributed as a chi-square variate with $2k - (k + 1) = k - 1$ d.f.

The test can, however, be made even if n_i's are not large. It has been investigated and found that the distribution of $(-2 \log_e \lambda)$ is approximately a χ^2-distribution with $(k - 1)$ d.f. even for small n_i's. However, a better approximation is provided by the Bartlett's test statistic :

$$\chi^2 = \frac{-2 \log \lambda'}{1 + \dfrac{1}{3(k-1)}\left\{\sum_i \left(\dfrac{1}{n_i}\right) - \dfrac{1}{\sum_i n_i}\right\}}$$

where λ' is obtained from λ on replacing n_i by $(n_i - 1)$ in (18·92), which follows χ^2-distribution with $(k - 1)$ d.f. Thus, the test statistic, under H_0 is given by

$$\chi^2 = \frac{\sum_{i=1}^{k}\left\{(n_i - 1)\log_e\left(\dfrac{s^2}{s_i^2}\right)\right\}}{1 + \dfrac{1}{3(k-1)}\left\{\sum_i\left(\dfrac{1}{n_i}\right) - \dfrac{1}{\Sigma n_i}\right\}} \sim \chi^2_{k-1} \quad \ldots (18{\cdot}93)$$

The critical region for test is, of course, the right-tail of the χ^2-distribution given by :

$$\chi^2 > \chi^2_{(k-1)}(\alpha), \quad \ldots (18{\cdot}94)$$

where χ^2 is defined in (18·93).

18·7. NON-PARAMETRIC METHODS

Most of the statistical tests that we have discussed so far had the following two features in common.

(*i*) The form of the frequency function of the parent population from which the samples have been drawn is assumed to be known, and

(*ii*) They were concerned with testing statistical hypothesis about the parameters of this frequency function or estimating its parameters.

For example, almost all the exact (small) sample tests of significance are based on the fundamental assumption that the parent population is normal and are concerned with testing or estimating the means and variances of these populations. Such tests, which deal with the parameters of the population are known as *Parametric Tests*. Thus, a parametric statistical test is a test whose model specifies certain conditions about the parameters of the population from which the samples are drawn.

On the other hand, a *Non-parametric (N.P.) Test* is a test that does not depend on the particular form of the basic frequency function from which the samples are drawn. In other words, non-parametric test does not make any assumption regarding the form of the population.

However, certain assumptions associated with N.P. tests are :

(*i*) Sample observations are independent.

(*ii*) The variable under study is continuous.

(*iii*) p.d.f. is continuous.

(*iv*) Lower order moments exist.

Obviously these assumptions are fewer and much weaker than those associated with parametric tests.

18·7·1. Advantages and Drawbacks of Non-parametric Methods over Parametric Methods

Below we shall give briefly the comparative study of parametric and non-parametric methods and their relative merits and demerits.

Advantages

1. N.P. methods are readily comprehensible, very simple and easy to apply and do not require complicated sample theory.
2. No assumption is made about the form of the frequency function of the parent population from which sampling is done.
3. No parametric technique will apply to the data which are mere classification (*i.e.*, which are measured in nominal scale), while N.P. methods exist to deal with such data.
4. Since the socio-economic data are not, in general, normally distributed, N.P. tests have found applications in Psychometry, Sociology and Educational Statistics.

Drawbacks

1. N.P. tests can be used only if the measurements are nominal or ordinal. Even in that case, if a parametric test exists it is more powerful than the N.P. test. In other words, if all the assumptions of a statistical model are satisfied by the data and if the measurements are of required strength, then the N.P. tests are wasteful of time and data.
2. So far, no N.P. methods exist for testing interactions in 'Analysis of Variance' model unless special assumptions about the additivity of the model are made.

5. N.P. tests are available to deal with the data which are given in ranks or whose seemingly numerical scores have the strength of ranks. For instance, no parametric test can be applied if the scores are given in grades such as A^+, A^-, B, A, B^+, etc.

3. N.P. tests are designed to test statistical hypothesis only and not for estimating the parameters.

Remarks:

1. Since no assumption is made about the parent distribution, the N.P. methods are sometimes referred to as *Distribution Free* methods. These tests are based on the 'Order Statistic' theory. In these tests we shall be using median, range, quartil, inter-quartile range, etc., for which an ordered sample is desirable. By saying that $x_1, x_2, ..., x_n$ is an ordered sample we mean $x_1 \le x_2 \le ... \le x_n$.
2. The whole structure of the N.P. methods rests on a simple but fundamental property of order statistic, *viz. "The distribution of the area under the density function between any two ordered observations is independent of the form of the density function"*, which we shall now prove.

18·7·2. Basic Distribution

Let Z be a continuous random variable with a p.d.f. $f(\cdot)$. Let $Z_1, Z_2, ..., Z_n$ be a random sample of size n from $f(\cdot)$ and let $x_1, x_2, ..., x_n$ be the corresponding ordered sample. Then the joint density of $x_1, x_2, ..., x_n$ is given by

$$g(x_1, x_2, ..., x_n) = n\,!\,f(x_1)\,f(x_2)\,...\,f(x_n), -\infty < x_1 < x_2 < ... < x_n < \infty \qquad ...(18·95)$$

the factor $n\,!$ appearing since there are $n\,!$ permutations of the sample observations and each gives rise to the same ordered sample.

Let us define

$$U_i = \int_{-\infty}^{x_i} f(z)\,dz = F(x_i), (i = 1, 2, ..., n) \qquad ...(18·96)$$

where $F(\cdot)$ is the distribution function of Z. But since $F(x_i)$ is a uniform random variable on $[0, 1]$, U_i, $(i = 1, 2, ..., n)$, defined in (18·96) are random variable following uniform distribution on $[0, 1]$. Thus, the joint density $k(\cdot)$ of the random variables U_i, $(i = 1, 2, ..., n)$ is given by

$$k(u_1, u_2, ..., u_n) = n\,!, 0 \le u_1 < u_2 < u_3 ... < u_n \le 1 \qquad ...(18·97)$$

and does not depend on $f(.)$.

$$E(U_i) = \int_0^1 ... \int_0^{u_3} \int_0^{u_2} u_i n!\; du_1\, du_2 ... du_n$$

$$= \frac{i}{n+1} \quad \text{(On simplification)} \qquad ...(18·98)$$

Thus, the expected area under $f(.)$ between two successive ordered observations is given by

$$E(U_i) - E(U_{i-1}) = \frac{i}{n+1} - \frac{i-1}{n+1} = \frac{1}{n+1}, \qquad (18·98a)$$

which is independent of $f(.)$

18·7·3. Wald-Wolfowitz Run Test

Suppose $x_1, x_2, ..., x_{n1}$ is an ordered sample from a population with density $f_1(.)$ and let $y_1, y_2, ..., y_{n2}$ be an independent ordered sample from another population with density $f_2(.)$. We want to test if the samples have been drawn from the same population or from populations with the same density functions, *i.e.*, if $f_1(\cdot) = f_2(\cdot)$.

Let us combine the two samples and arrange the observations in order of magnitude to give the combined ordered sample as, (say),

$$x_1\, x_2\, y_1\, y_2\, y_3\, x_3\, y_4\, x_4\, x_5\, y_5 \ldots \qquad \ldots (18{\cdot}99)$$

Run *(Definition). A run is defined as a sequence of letters of one kind surrounded by a sequence of letters of the other kind, and the number of elements in a run is usually referred to a the length (l) of the run.*

Thus in (18·99), we have in order, a run of x $(l = 2)$, a run of y $(l = 3)$, a run of x $(l = 1)$, a run of y $(l = 1)$ etc.

If both the samples come from the same population then there would be thorough mingling of x's and y's and consequently the number of runs in the combined sample would be large. On the other hand if the samples come from two different populations so that their ranges do not overlap, then there would be only two runs of the type $x_1, x_2, \ldots, x_{n1}$ and $y_1, y_2, \ldots, y_{n2}$. Generally, any difference in mean and variance would tend to reduce the number of runs. Thus the *alternative hypothesis will entail too few runs.*

Procedure. In order to test the Null Hypothesis $H_0 : f_1(.) = f_2(.)$ *i.e.*, the samples have come from the same population, we count the number of runs 'U' in the combined ordered sample.

Null hypothesis is rejected if $U < u_0$, where the value of u_0 for given level of significance is determined from considering the distribution of U under H_0.

First of all let us find the probability of obtaining a specific arrangement (18·99) under $H_0 : f_1(.) = f_2(.) = f(.)$, (say).

If X's and Y's are transformed to U's and V's by the relation :

$$U_i = \int_{-\infty}^{x_i} f(z)dz, \quad V_i = \int_{-\infty}^{y_i} f(z)dz,$$

then the joint *p.d.f.* of U's and V's becomes

$$g(u_1, u_2, \ldots, u_n, v_1, v_2, \ldots, v_n) = n_1\,!\, n_2\,! \qquad \ldots (18{\cdot}100)$$

The probability of an arrangement (18·99) is obtained on integrating (18·100) over the region defined by:

$$0 < u_1 < u_2 < v_1 < v_2 < v_3 < u_3 < v_4 < \ldots < 1$$

i.e., integrating u_1 over 0 to u_2 ; then u_2 over 0 to v_1 and so on. The value of the integral will, on simplification, come out to be

$$\frac{n_1!n_2!}{(n_1+n_2)!} = \frac{1}{\binom{n_1+n_2}{n_1}}$$

Since, there are exactly $\binom{n_1+n_2}{n_1}$ arrangements of n_1, x's and n_2, y's, it follows that all the arrangements of x's and y's are equally likely.

Since, under H_0 , all the $\binom{n_1+n_2}{n_1}$ arrangements of n_1 x's and n_2 y's are equally likely, to obtain the distribution of U under H_0, it is necessary to count all the arrangements with exactly 'u' runs. Let us first take the case of even number of runs, *i.e.*, $u = 2k$. In this case we should have k runs of x's and k runs of y's.

n_1 x's will give k runs if they are separated by $(k - 1)$ vertical bars in distinct spaces between the x's. In other words, $(k - 1)$ spaces are to come out of the total number of $(n_1 - 1)$ spaces between the n_1 x's and this can happen in $\binom{n_1-1}{k-1}$ ways. Hence k runs of x's can be obtained in $\binom{n_1-1}{k-1}$ ways. Similarly, k runs of y's can be obtained in $\binom{n_2-1}{k-1}$ ways.

The same result holds if the sequence of runs in (18·99) starts with x or with y. Since a sequence of type (18·99) may start with x or y, we get

$$P(U = 2k) = \frac{2\binom{n_1 - 1}{k-1}\binom{n_2 - 1}{k-1}}{\binom{n_1 + n_2}{n_1}}$$

If the number of runs in (18·99) is odd, *i.e.*, $u = 2k + 1$ then we should have either (*i*) $(k + 1)$ runs of x and k runs of y or (*ii*) k runs of x and $(k + 1)$ runs of y. Hence,

$$P\,(U = 2k + 1) = P\,(i) + P\,(ii)$$

$$= \frac{\binom{n_1 - 1}{k}\binom{n_2 - 1}{k-1} + \binom{n_1 - 1}{k-1}\binom{n_2 - 1}{k}}{\binom{n_1 + n_2}{n_1}}$$

Hence, the distribution of U under H_0 is given by

$$\left.\begin{aligned} P(U = 2k) &= \frac{2\binom{n_1 - 1}{k-1}\binom{n_2 - 1}{k-1}}{\binom{n_1 - n_2}{n_1}}, \\ \text{and} \quad P(U = 2k + 1) &= \frac{\binom{n_1 - 1}{k}\binom{n_2 - 1}{k-1} + \binom{n_1 - 1}{k-1}\binom{n_2 - 1}{k}}{\binom{n_1 + n_2}{n_1}} \end{aligned}\right\} \quad \text{... (18·101)}$$

If the probability of type I error is fixed as α, then u_0 is determined from the equation :

$$\sum_{u=2}^{u_0} h(u) = \alpha \quad \text{... (18·102)}$$

where $h(u)$ is the probability function of U given by (18·101).

Calculation of u_0 from (18·102) is quite tedious and cumbersome unless n_1 and n_2 are large in which case under H_0, U is asymptotically normal with

$$E(U) = \frac{2n_1 n_2}{n_1 + n_2} + 1 \quad \text{... (18·103)}$$

$$\text{Var}\,(U) = \frac{2n_1 n_2(2n_1 n_2 - n_1 - n_2)}{(n_1 + n_2)^2(n_1 + n_2 - 1)} \quad \text{... (18·104)}$$

and we can use the normal test

$$Z = \frac{U \quad E(U)}{\sqrt{\text{Var}(\quad)}} \sim (0, 1), \quad \text{asymptotically.} \quad \text{... (18·105)}$$

This approximation is fairly good if each of n_1 and n_2 is greater than 10. Since the alternative hypothesis is 'too few runs', the test is ordinarily one-tailed with only negative values leading to the rejection of H_0.

18·7·4. Test for Randomness

Another application of the 'run' theory is in testing the randomness of a given set of observations. Let $x_1, x_2, \ldots, x_n$ be the set of observations arranged in the order in which they occur, *i.e.*, x_i is the *ith* observation in the outcome of an experiment. Then, for each of the observations, we see if it is above or below the value of the median of the observations and write A if the observation is above and B if it is below, the median value. Thus, we get a sequence of A's and B's of the type, (say),

$$A\,B\,B\,A\,A\,A\,B\,A\,B\,B \qquad \ldots (*)$$

Under the *null hypothesis* H_0 *that the set of observations is random,* the number of runs U in (*) is a *r.v.* with

$$E(U) = \frac{n+2}{2} \quad \text{and} \quad \text{Var}(U) = \frac{n}{4}\left(\frac{n-2}{n-1}\right) \qquad \ldots (18{\cdot}106)$$

For large n (say, > 25), U may be regarded as asymptotically normal and we may use the normal test.

18·7·5. Median Test

Median test is a statistical procedure for testing if two independent ordered samples differ in their central tendencies. In other words, it gives information if two independent samples are likely to have been drawn from the populations with the same median.

As in 'run' test, let $x_1, x_2, \ldots, x_{n1}$ and $y_1, y_2, \ldots, y_{n2}$ be two independent ordered samples from the populations with p.d.f.'s $f_1(.)$ and $f_2(.)$ respectively. The measurements must be at least ordinal. Let $z_1, z_2, \ldots, z_{n1+n2}$ be the combined *ordered* sample. Let m_1 be the number of x's and m_2 the number of y's exceeding the median value M, (say), of the combined sample.

Then, under the null hypothesis that the samples come from the same population or from different populations with the same median, *i.e.*, under $H_0 : f_1(.) = f_2(.)$, the joint distribution of m_1 and m_2 is the hypergeometric distribution with probability function :

$$p(m_1, m_2) = \frac{\binom{n_1}{m_1}\binom{n_2}{m_2}}{\binom{n_1+n_2}{m_1+m_2}} \qquad \ldots (18{\cdot}107)$$

If $m_1 < n_1/2$, then the critical region corresponding to the size of type 1 error α, is given by $m_1 < m_1'$ where m_1' is computed from the equation :

$$\sum_{m_1=1}^{m_1'} p(m_1, m_2) = \alpha \qquad \ldots (18{\cdot}108)$$

The distribution of m_1 under H_0 is also hyper-geometric with

$$E(m_1) = \begin{cases} \dfrac{n_1}{2}, \text{if } N = (n_1+n_2) \text{ is even} \\[2ex] \dfrac{n_1}{2}\cdot\left(\dfrac{N-1}{N}\right), \text{ if } N \text{ is odd} \end{cases}$$

$$\text{and} \quad \text{Var}(m_1) = \begin{cases} \dfrac{n_1 n_2}{4(N-1)}, & \text{if } N \text{ is even} \\[2ex] \dfrac{n_1 n_2 (N+1)}{4N^2}, & \text{if } N \text{ is odd} \end{cases} \qquad \ldots (18{\cdot}109)$$

This distribution is most of the times quite inconvenient to use. However, for large samples, we may regard m_1 to be asymptotically normal and use normal test, *viz.*,

$$Z = \frac{m_1 - E(m_1)}{\sqrt{Var(m_1)}} \sim N(0, 1), \text{ asymptotically.} \qquad \ldots (18{\cdot}110)$$

Remarks:

1. The observations m_1 and m_2 can be classified into the following 2×2 contingency table.

	Sample 1	*Sample 2*	*Total*
No. of observations > M	m_1	m_2	$m_1 + m_2$
No. of observations < M	$n_1 - m_1$	$n_2 - m_2$	$n_1 + n_2 - m_1 - m_2$
Total	n_1	n_2	$n_1 + n_2 = N$

If frequencies are small we can compute the exact probabilities from (18·107), rather than approximate them. However, if frequencies are large we may use χ^2-test with 1 *d.f.* (for a 2×2 contingency table) for testing H_0.

The approximation is fairly good if both n_1 and n_2 exceed 10.

2. Median test is sensitive to the differences in location between $f_1(x)$ and $f_2(y)$ but not to differences in their shapes. Thus, if $f_1(x)$ and $f_2(y)$ have the same median, we would expect $H_0 : f_1(.) = f_2(.)$ to be accepted ordinarily even though their shapes are quite different.

3. Generally, the median test makes the correct decision with a little more assurance than does the sign test (*c.f.* §. 18·7·6) but not as decisively as the *t*-test.

18·7·6. Sign Test

Consider a situation where it is desired to compare two things or materials under various sets of conditions. An experiment is thus conducted under the following circumstances :

(*i*) When there are pairs of observations on two things being compared.

(*ii*) For any given pair, each of the two observations is made under similar extraneous conditions.

(*iii*) Different pairs are observed under different conditions.

Condition (*iii*) implies that the differences $d_i = x_i - y_i \,; i = 1, 2, \ldots, n$ have different variances and thus renders the paired *t*-test (Chapter 16) invalid, which would have otherwise been used unless there was obvious non-normality. So, in such a case we use the '*Sign Test*', named so since it is based on the signs (plus or minus) of the deviations $d_i = x_i - y_i$. No assumptions are made regarding the parent population. The only assumptions are :

(*i*) Measurements are such that the deviations $d_i = x_i - y_i$, can be expressed in terms of positive or negative signs.

(*ii*) Variables have continuous distribution.

(*iii*) d_i's are independent.

Different pairs (x_i, y_i) may be from different populations (say *w.r.t.* age, weight, stature, education, etc.). The only requirement is that within each pair, there is matching *w.r.t* relevant extraneous factors.

Procedure. Let (x_i, y_i), $i = 1, 2, \ldots, n$ be n paired sample observations drawn from the two populations with *p.d.f.*'s $f_1(.)$ and $f_2(.)$. We want to test the null hypothesis $H_0 : f_1(.) = f_2(.)$. To test H_0, consider $d_i = x_i - y_i$, $(i = 1, 2, \ldots, n)$. When H_0 is true, x_i and y_i constitute a random sample of size 2 from the same population. Since the probability that the first of the two sample observations exceeds the second is same as the probability that the second exceeds the first and since hypothetically the probability of a tie is zero, H_0 may be restated as:

$$H_0 : \quad P[X - Y > 0] = \frac{1}{2} \quad \text{and} \quad P[X - Y < 0] = \frac{1}{2}$$

Let us define:

$$U_i = \begin{cases} 1, & \text{if } x_i - y_i > 0 \\ 0, & \text{if } x_i - y_i < 0 \end{cases}$$

U_i is a Bernoulli variate with $p = P(x_i - y_i > 0) = \frac{1}{2}$. Since U_i's, $i = 1, 2, ..., n$ are independent, $U = \sum_{i=1}^{n} U_i$, the total number of positive deviations, is a Binomial variate with parameters n and $p\left(=\frac{1}{2}\right)$. Let the number of positive deviations be k. Then

$$P(U \le k) = \sum_{r=0}^{k} \binom{n}{r} p^r q^{n-r}, \left(p = q = \frac{1}{2} \text{ under } H_0\right)$$

$$= \left(\frac{1}{2}\right)^n \sum_{r=0}^{k} \binom{n}{r} = p', \text{(say)} \qquad \text{... (18·111)}$$

If $p' \le 0{\cdot}05$, we reject H_0 at 5% level of significance and if $p' > 0{\cdot}05$, we conclude that the data do not provide any evidence against the null hypothesis, which may therefore, be accepted.

For large samples, $(n \ge 30)$, (under H_0), we may regard U to be asymptotically normal with,

$$E(U) = np = n/2 \quad \text{and} \quad \text{Var}(U) = npq = n/4$$

$$\therefore \quad Z = \frac{U - E(U)}{\sqrt{\text{Var}(U)}} = \frac{U - n/2}{\sqrt{(n/4)}}, \text{ is asymptotically } N(0, 1), \qquad \text{... (18·112)}$$

and we may use normal test.

18·7·7. Mann-Whitney-Wilcoxon U-Test

This non-parametric test for two samples was described by Wilcoxon and studied by Mann and Whitney. It is the most widely used test as an alternative to the t-test when we do not make the t-test assumptions about the parent population.

Let x_i $(i = 1, 2, ..., n_1)$ and y_j $(j = 1, 2, ..., n_2)$ be independent ordered samples of size n_1 and n_2 from the populations with *p.d.f.* $f_1(.)$ and $f_2(.)$ respectively. We want to test the null hypothesis $H_1 : f_1(.) = f_2(.)$. Like the run test, Mann-Whitney test is based on the pattern of the x's and y's in the combined ordered sample. Let T *denote the sum of ranks of the y's in the combined ordered sample.* For example, for the pattern (18·99) on page 18·41 of combined ordered sample the ranks of y observations are respectively 3, 4, 5, 7, 10 etc. and $T = 3 + 4 + 5 + 7 + 10 + ...$ The test statistic U is then defined in terms of T as follows :

$$U = n_1 n_2 + \frac{n_2(n_2+1)}{2} - T \qquad \text{... (18·113)}$$

If T is significantly large or small then $H_0 : f_1(.) = f_2(.)$ is rejected. The problem is to find the distribution of T under H_0. Unfortunately, it is very troublesome to obtain the distribution of T under H_0. However, Mann and Whitney have obtained the distribution of T for small n_1 and n_2; have found the moments of T in general and shown that T is asymptotically normal. It has been established that under H_0, U is asymptotically normally distributed as $N(\mu, \sigma^2)$, where

$$\mu = E(U) = \frac{n_1 n_2}{2} \quad \text{and} \quad \sigma^2 = \text{Var}(U) = \frac{n_1 n_2 (n_1 + n_2 + 1)}{12} \qquad \text{... (18·114)}$$

Hence,
$$Z = \frac{U - \mu}{\sigma} \sim N(0,1), \text{ asymptotically,} \qquad \text{... (18·114}a\text{)}$$

and normal test can be used. The approximation is fairly good if both n_1 and n_2 are greater than 8.

Remark: The asymptotic relative efficiency (*ARE*) of Mann-Whitney's U-test relative to two samples t-test is greater than or equal to 0·864. For a normal population, this $ARE = 3/\pi = 0{\cdot}955$. Accordingly, Mann-

Whitney's *U*-test is regarded as the best non-parametric test for location.

18·8. SEQUENTIAL ANALYSIS – INTRODUCTION

We have seen that in Neyman-Pearson theory of testing of hypothesis, n, the sample size is regarded as a fixed constant and keeping α fixed, we minimise β. But in the sequential analysis theory propounded by *A. Wald n*, the sample size is not fixed but is regarded as a random variable whereas both α and β are fixed constants.

18·8·1. Sequential Probability Ratio Test (S.P.R.T.)

The best known procedure in sequential testing is the *Sequential Probability Ratio Test* (S.P.R.T.) developed by A. Wald discussed below:

Suppose we want to test the hypothesis $H_0 : \theta = \theta_0$ against the alternative $H_1 : \theta = \theta_1$ for a distribution with p.d.f. $f(x, \theta)$. For any positive integer m, the likelihood function of a sample $x_1, x_2, \ldots, x_m$ from the population with *p.d.f. (p.m.f.)* $f(x, \theta)$ is given by:

$$L_{1m} = \prod_{i=1}^{m} f(x_i, \theta_1) \text{ when } H_1 \text{ is true, and by } L_{0m} = \prod_{i=1}^{m} f(x_i, \theta_0) \text{ when } H_0 \text{ is true,}$$

and the likelihood ratio λ_m is given by:

$$\lambda_m = \frac{L_{1m}}{L_{0m}} = \frac{\prod_{i=1}^{m} f(x_i,\theta_1)}{\prod_{i=1}^{m} f(x_i,\theta_0)} = \prod_{i=1}^{m} \frac{f(x_i,\theta_1)}{f(x_i,\theta_0)}, (m = 1, 2, \ldots) \quad \ldots (18·115)$$

The S.P.R.T. for testing H_0 against H_1 is defined as follows:

At each stage of the experiment (at the mth trial for any integral value m), the likelihood ratio λ_m, $(m = 1, 2, \ldots)$ is computed.

(*i*) If $\lambda_m \geq A$, we terminate the process with the rejection of H_0,
(*ii*) If $\lambda_m \leq B$, we terminate the process with the acceptance of H_0, and
(*iii*) If $B < \lambda_m < A$, we continue sampling by taking an additional observation. ... (18·116)

Here, A and B, $(B < A)$, are the constants which are determined by the relation

$$A = \frac{1-\beta}{\alpha}, \quad B = \frac{\beta}{1-\alpha} \quad \ldots (18·117)$$

where α and β are the probabilities of type I error and type II error respectively.

From computational point of view, it is much convenient to deal with $\log \lambda_m$ rather than λ_m, since

$$\log \lambda_m = \sum_{i=1}^{m} \log\left[\frac{f(x_i,\theta_1)}{f(x_i,\theta_0)}\right] = \sum_{i=1}^{m} z_i \quad \ldots (18·118)$$

where

$$z_i = \log\left[\frac{f(x_i,\theta_1)}{f(x_i,\theta_0)}\right] \quad \ldots (18·118a)$$

In terms of z_i s, *S.P.R.T.* is defined as follows:

(*i*) If $\Sigma z_i \geq \log A$, reject H_0
(*ii*) If $\Sigma z_i \leq \log B$, reject H_1 (Accept H_0)
(*iii*) If $\log B < \Sigma z_i < \log A$, continue sampling by taking an additional observation. ... (18·119)

Remarks:

1. In *S.P.R.T.*, we continue taking additional observations unless the inequality
$$B < \lambda_m < A \quad \Rightarrow \quad \log B < \Sigma z_i < \log A,$$
is violated at either end. It has been proved that S.P.R.T. eventually terminates with probability one.
2. Sequential schemes provide for a minimum amount of sampling and thus result is considerable saving in terms of inspection, time and money. As compared with single sampling, sequential scheme requires on the average 33% to 50% less inspection for the same degree of protection *i.e.,* for the same values of α and β.

18·8·2. Operating Characteristic (O.C.) Function of S.P.R.T.

The O.C. function $L(\theta)$ is defined as

$L(\theta)$ = Probability of accepting $H_0 : \theta = \theta_0$, when θ is the true value of the parameter,

and since the power function

$P(\theta)$ = Probability of rejecting H_0 where θ is the true value, we get
$$L(\theta) = 1 - P(\theta) \qquad \text{... (18·120)}$$

The O.C. function of a S.P.R.T. for testing $H_0 : \theta = \theta_0$ against the alternative $H_1 : \theta = \theta_1$, in sampling from a population with density function $f(x, \theta)$ is given by
$$L(\theta) = \frac{A^{h(\theta)} - 1}{A^{h(\theta)} - B^{h(\theta)}} \qquad \text{... (18·121)}$$
where for each value of θ, the value of $h(\theta) \neq 0$ is to be determined so that
$$E\left[\frac{f(x,\theta_1)}{f(x,\theta_0)}\right]^{h(\theta)} = 1 \qquad \text{... (18·122)}$$
where the constant A and B have already been defined in (18·117). It has been proved that under very simple conditions on the nature of the function $f(x, \theta)$, there exists a unique value of $h(\theta) \neq 0$ such that (18·122) is satisfied.

18·8·3. Average Sample Number (A.S.N.) of S.P.R.T.

The sample size n in sequential testing is a random variable which can be determined in terms of the true density function $f(x, \theta)$. The A.S.N. function for the S.P.R.T. for testing $H_0 : \theta = \theta_0$ against $H_1 : \theta = \theta_1$ is given by
$$E(n) = \frac{L(\theta)\log B + [1 - L(\theta)]\log A}{E(Z)} \qquad \text{... (18·123)}$$
where
$$Z = \log\left(\frac{f(x,\ \ )}{f(x,\theta\ )}\right), \quad A = \frac{-\beta}{\alpha}, \quad B = \frac{\beta}{1-\alpha} \qquad \text{... (18·123a)}$$

Example 18·11. *Give the S.P.R.T. for testing $H_0 : \theta = \theta_0$ against $H_1 : \theta = \theta_1$ ($> \theta_0$) in sampling from a normal density :*
$$f(x, \theta) = \frac{1}{\sigma\sqrt{2\pi}} \exp\left[-\frac{1}{2}\left(\frac{x-\theta}{\sigma}\right)^2\right], -\infty < x < \infty$$
where σ is known. Also obtain its O.C. function and A.S.N. function.

Solution.
$$\frac{f(x_i, \theta_1)}{f(x_i, \theta_0)} = \exp\left[-\frac{1}{2\sigma^2}\left\{(x_i - \theta_1)^2 - (x_i - \theta_0)^2\right\}\right]$$
$$= \exp\left[-\frac{1}{2\sigma^2}\left\{(\theta_0 - \theta_1)(2x_i - \theta_0 - \theta_1)\right\}\right] \qquad \text{...(*)}$$

$$\therefore \qquad z_i = \log\frac{f(x_i, \theta_1)}{f(x_i, \theta_0)} = \frac{\theta_1 - \theta_0}{\sigma^2}\left(x_i - \frac{\theta_0 + \theta_1}{2}\right) \qquad \ldots (**)$$

$$\Rightarrow \qquad \log\lambda_m = \sum_{i=1}^{m} z_i = \frac{\theta_1 - \theta_0}{\sigma^2}\left[\sum_i x_i - \frac{m(\theta_0 + \theta_1)}{2}\right]$$

Hence the S.P.R.T. for $H_0 : \theta = \theta_0$ against $H_1 : \theta = \theta_1$ is given by [*c.f.* (18·119)] :

(*i*) Reject H_0 if

$$\frac{\theta_1 - \theta_0}{\sigma^2}\left[\sum_i x_i - \frac{m(\theta_0 + \theta_1)}{2}\right] \geq \log\left(\frac{1-\beta}{\alpha}\right)$$

$$\Rightarrow \qquad \sum_{j=1}^{m} x_i \geq \frac{\sigma^2}{\theta_1 - \theta_0}\log\left(\frac{1-\beta}{\alpha}\right) + \frac{m(\theta_0 + \theta_1)}{2}; (\theta_1 > \theta_0)$$

(*ii*) Accept H_0 if

$$\frac{\theta_1 - \theta_0}{\sigma^2}\left[\sum_i x_i - \frac{m(\theta_0 + \theta_1)}{2}\right] \leq \log\left(\frac{\beta}{1-\alpha}\right)$$

$$\Rightarrow \qquad \sum_{i=1}^{m} x_i \leq \frac{\sigma^2}{\theta_1 - \theta_0}\log\left(\frac{\beta}{1-\alpha}\right) + \frac{m(\theta_0 + \theta_1)}{2}; (\theta_1 > \theta_0)$$

and (*iii*) Continue taking additional observations as long as:

$$\log\left(\frac{\beta}{1-\alpha}\right) < \frac{\theta_1 - \theta_0}{\sigma^2}\left[\sum_i x_i - \frac{m(\theta_0 + \theta_1)}{2}\right] < \log\left(\frac{1-\beta}{\alpha}\right)$$

$$\Rightarrow \qquad \frac{\sigma^2}{\theta_1 - \theta_0}\log\left(\frac{\beta}{1-\alpha}\right) + \frac{m(\theta_0 + \theta_1)}{2} < \sum_i x_i < \frac{\sigma^2}{\theta_1 - \theta_0}\log\left(\frac{1-\beta}{\alpha}\right) + \frac{m(\theta_0 + \theta_1)}{2}; \theta_1 > \theta_0$$

O.C. Function. First of all we shall determine $h = h(\theta) \neq 0$ from (18·122) *i.e.*, from

$$\int_{-\infty}^{\infty}\left[\frac{f(x, \theta_1)}{f(x, \theta_0)}\right]^h f(x, \theta)dx = 1$$

$$\Rightarrow \qquad \frac{1}{\sigma\sqrt{2\pi}}\int_{-\infty}^{\infty}\exp\left[-\frac{1}{2}\left(\frac{x-\theta}{\sigma}\right)^2\right]\cdot\left[\exp\left\{-\frac{1}{2\sigma^2}(\theta_1 - \theta_0) \times (-2x + \theta_0 + \theta_1)\right\}\right]^h dx = 1$$

[on using (*)]

$$\Rightarrow \qquad \frac{1}{\sigma\sqrt{2\pi}}\int_{-\infty}^{\infty}\exp\left[-\frac{1}{2\sigma^2}\left\{x^2 - 2x\left[\theta_1 - \theta_0)h + \theta\right] + \theta^2 + \left(\theta_1^2 - \theta_0^2\right)h\right\}\right]dx = 1$$

If we take

$$\left.\begin{aligned}\lambda &= (\theta_1 - \theta_0)h + \theta \\ \lambda^2 &= \left(\theta_1^2 - \theta_0^2\right)h + \theta^2\end{aligned}\right\} \qquad \ldots (***)$$

then L.H.S. becomes: $\dfrac{1}{\sigma\sqrt{2\pi}}\displaystyle\int_{-\infty}^{\infty}\exp\left[-\frac{1}{2\sigma^2}(x-\lambda)^2\right]dx,$

which being the total area under normal probability curve with mean λ and variance σ^2 is always unity, as desired. Thus $h = h(\theta)$ is the solution of (***) and is given by:

$$(\theta_1{}^2-\theta_0{}^2)\,h+\theta^2=\{(\theta_1-\theta_0)h+\theta\}^2 \quad\Rightarrow\quad (\theta_1{}^2-\theta_0{}^2)h=(\theta_1-\theta_0)^2h^2+2\theta\,(\theta_1-\theta_0)\,h$$

Since $h = h(\theta) \neq 0$ and $\theta_1 \neq \theta_0$, on dividing throughout by $(\theta_1-\theta_0)\,h$, we get

$$(\theta_1+\theta_0)=(\theta_1-\theta_0)\,h+2\theta \quad\Rightarrow\quad h(\theta)=\frac{\theta_1+\theta_0-2\theta}{\theta_1-\theta_0}$$

Substituting for $h(\theta)$ in (18·121), we get the required expression for the O.C. function.

A.S.N. function. We have:

$$Z=\log\frac{f(x,\theta_1)}{f(x,\theta_0)}=\frac{\theta_1-\theta_0}{\sigma^2}\left(x-\frac{\theta_0+\theta_1}{2}\right), \qquad \text{[From (**)]}$$

$$\therefore \qquad E(Z)=\frac{\theta_1-\theta_0}{2\sigma^2}\left[2E(x)-\theta_0-\theta_1\right]=\frac{\theta_1-\theta_0}{2\sigma^2}(2\theta-\theta_0-\theta_1)$$

Substituting in (18·123), we get the required A.S.N. function.

Example 18·12. *Let X have the distribution*

$$f(x,\theta)=\theta^x(1-\theta)^{1-x}\,;\; x=0,1;\; 0<\theta<1$$

For testing $H_0:\theta=\theta_0$ against $H_1:\theta=\theta_1$, construct S.P.R.T. and obtain its A.S.N. and O.C. Function.

Solution. We have

$$\lambda_m=\frac{L(x_1,x_2,\ldots,x_m\mid H_1)}{L(x_1,x_2,\ldots,x_m\mid H_0)}$$

$$=\left[\theta_1^{\sum_{i=1}^m x_i}(1-\theta_1)^{m-\sum_{i=1}^m x_i}\right]\div\left(\theta_0^{\sum x_i}(1-\theta_0)^{m-\sum x_i}\right)$$

$$=\left(\frac{\theta_1}{\theta_0}\right)^{\sum_{i=1}^m x_i}\cdot\left(\frac{1-\theta_1}{1-\theta_0}\right)^{m-\sum_{i=1}^m x_i}$$

$$\log\lambda_m=(\Sigma x_i)\log(\theta_1/\theta_0)+(m-\Sigma x_i)\log\left(\frac{1-\theta_1}{1-\theta_0}\right)$$

$$=\sum_{i=1}^m x_i\log\left[\frac{\theta_1(1-\theta_0)}{\theta_0(1-\theta_1)}\right]+m\log\left(\frac{1-\theta_1}{1-\theta_0}\right)$$

Hence, S.P.R.T. for testing $H_0:\theta=\theta_0$ against $H_1:\theta=\theta_1$ is given by [*c.f.* (18·119)] :

(*i*) Accept H_0 if $\log\lambda_m\le\log\left(\dfrac{\beta}{1-\alpha}\right)=b$, (say)

i.e., if $\displaystyle\sum_{i=1}^m x_i\le\frac{b-m\log[(1-\theta_1)/(1-\theta_0)]}{\log[\theta_1(1-\theta_0)/\theta_0(1-\theta_1)]}=a_m$, (say).

(*ii*) Reject H_0 (Accept H_1) if $\log\lambda_m\ge\log\left(\dfrac{1-\beta}{\alpha}\right)=a$, (say)

i.e., if $\displaystyle\sum_{i=1}^m x_i\ge\frac{a-m\log[(1-\theta_1)/(1-\theta_0)]}{\log[\theta_1(1-\theta_0)/\theta_0(1-\theta_1)]}=r_m$, (say).

(*iii*) Continue sampling if

$$b<\log\lambda_m<a \quad\Rightarrow\quad a_m<\Sigma x_i<r_m$$

O.C. Function. *O.C.* function is given by:

$$L(\theta) = [A^{h(\theta)} - 1]/[A^{h(\theta)} - B^{h(\theta)}] \qquad [c.f. \ (18\cdot121)] \ldots (i)$$

where for each value of θ, $h(\theta) \neq 0$ is to be determined such that

$$\Rightarrow \qquad E\left[\frac{f(x,\theta_1)}{f(x,\theta_0)}\right]^{h(\theta)} = 1 \qquad [c.f. \ (18\cdot122)]$$

$$\Rightarrow \qquad \sum_{x=0}^{1}\left[\frac{f(x,\theta_1)}{f(x,\theta_0)}\right]^{h(\theta)} f(x,\theta) = 1$$

$$\Rightarrow \qquad \sum_{x=0}^{1}\left[\left(\frac{\theta_1}{\theta_0}\right)^x\left(\frac{1-\theta_1}{1-\theta_0}\right)^{1-x}\right]^{h(\theta)} \theta^x(1-\theta)^{1-x} = 1$$

$$\Rightarrow \qquad \left(\frac{1-\theta_1}{1-\theta_0}\right)^{h(\theta)}(1-\theta) + \left(\frac{\theta_1}{\theta_0}\right)^{h(\theta)}\theta = 1 \qquad \ldots (ii)$$

The solution of this equation for $h = h(\theta)$ is very tedious. From practical point of view, instead of solving *(ii)* for h we regard h as a parameter and solve it for θ; thus giving

$$\theta\left[\left(\frac{\theta_1}{\theta_0}\right)^{h(\theta)} - \left(\frac{1-\theta_1}{1-\theta_0}\right)^{h(\theta)}\right] = 1 - \left(\frac{1-\theta_1}{1-\theta_0}\right)^{h(\theta)}$$

$$\Rightarrow \qquad \theta = \frac{1-[(1-\theta_1)/(1-\theta_0)]^{h(\theta)}}{(\theta_1/\theta_0)^{h(\theta)} - [(1-\theta_1)/(1-\theta_0)]^{h(\theta)}} = \theta(h), \text{(say)}. \qquad \ldots (iii)$$

Using *(i)*, we have:

$$L(\theta) = \frac{[(1-\beta)/\alpha]^h - 1}{[(1-\beta)/\alpha]^h - [\beta/(1-\alpha)]^h} = L(\theta, h), \text{(say)}. \qquad \ldots (iv)$$

Various points on the *O.C.* curve are obtained by assigning arbitrary values to 'h' and computing the corresponding values of θ and $L(\theta)$ from *(iii)* and *(iv)* respectively.

A.S.N. Function

$$Z = \log\left[\frac{f(x,\theta_1)}{f(x,\theta_0)}\right], \ A = \frac{1-\beta}{\alpha}, \ B = \frac{\beta}{1-\alpha}$$

$$\therefore \qquad E(Z) = \sum_{x=0}^{1}\log\left[\frac{f(x,\theta_1)}{f(x,\theta_0)}\right]\cdot f(x,\theta)$$

$$= \sum_{x=0}^{1}\log\left[\left(\frac{\theta_1}{\theta_0}\right)^x\left(\frac{1-\theta_1}{1-\theta_0}\right)^{1-x}\right]\cdot\theta^x(1-\theta)^{1-x}$$

$$= (1-\theta)\log\left(\frac{1-\theta_1}{1-\theta_0}\right) + \theta\cdot\log\left(\frac{\theta_1}{\theta_0}\right)$$

$$= \theta\log\left[\frac{\theta_1(1-\theta_0)}{\theta_0(1-\theta_1)}\right] + \log\left(\frac{1-\theta_1}{1-\theta_0}\right) \qquad \ldots (v)$$

A.S.N. is given by

$$E(n) = \frac{L(\theta)\log B + [1 - L(\theta)].\log A}{E(Z)} \qquad \dots (vi)$$

Substituting the values of $E(Z)$ and $L(\theta)$ from (*v*) and (*iv*) in (*vi*), we get the *A.S.N.* function.

Remark: If h assumes negative values *i.e.*, if instead of h we take $(-h)$ where $h > 0$, then

$$L(\theta, -h) = \frac{A^{-h} - 1}{A^{-h} - B^{-h}} = \left(\frac{1 - A^h}{B^h - A^h}\right) B^h = \left(\frac{A^h - 1}{A^h - B^h}\right) . B^h \qquad \dots (vii)$$

$$\Rightarrow \qquad L(\theta, -h) = B^h \cdot L(\theta, h)$$

and

$$\theta(-h) = \frac{[(1-\theta_1)/(1-\theta_0)]^h - 1}{[(1-\theta_1)/(1-\theta_0)]^h - (\theta_1/\theta_0)^h} \cdot \left(\frac{\theta_1}{\theta_0}\right)^h \qquad [\text{From } (iii)]$$

$$= \theta(h) \cdot \left(\frac{\theta_1}{\theta_0}\right)^h \qquad \dots (viii)$$

Formulae (*vii*) and (*viii*) are very convenient to use for obtaining the points on *O.C.* curve for arbitrary negative values of h.

CHAPTER CONCEPTS QUIZ

1. Fill in the blanks:

(*i*) The theory of testing of hypothesis was propounded by .... in 1933

(*ii*) Statistical hypothesis under test in called ....

(*iii*) A wrong decision about the null hypothesis leads to .... types of errors.

(*iv*) The probability of type 1 error is given by ....

(*v*) The probability of type 1 error is also called ....

(*vi*) The probability of type II error is given by ....

(*vii*) If β is the probability of type II error, then $(1 - \beta)$ is called .... of the test.

(*viii*) In any testing problem, the type .... error is considered more serious than type .... error.

(*ix*) The power function of a test is related to type .... error and is given by ....

(*x*) The level of significance of a test is related to type .... error and is given by ....

(*xi*) The most powerful (MP) test consists in minimising .... or maximising .... for fixed ....

(*xii*) The Neyman-Pearson lemma provides the best critical region (B.C.R.) for testing .... hypothesis against .... alternative hypothesis

(*xiii*) The L-R. Criterion gives the B.C.R. for testing .... against ....

(*xiv*) Uniformly most powerful test is one ....

(*xv*) SPRT was given by ....

(*xvi*) Critical region provides us the criterion for .... of null hypothesis.

(*xvii*) A test, which for fixed α, maximises $(1 - \beta)$ is called ....

(*xviii*) The choice of one tailed or two tailed test depends on the ....

2. What do you understand by non-parametric methods of testing of hypothesis? What are their advantages and drawbacks over the parametric methods?
3. Outline the testing of hypothesis approach by J. Neyman and the S.P.R.T. approach by A. Wald. What are the basic differences between the two approaches?

DISCUSSION AND REVIEW QUESTIONS

1. (*a*) What are simple and composite statistical hypotheses? Give examples. Define null and alternative hypotheses. How is a statistical hypothesis tested?

   (*b*) Explain the following terms :

   (*i*) Errors of first and second kinds. (*ii*) The best critical region.
   (*iii*) Power function of a test. (*iv*) Level of significance.
   (*v*) Simple and composite hypotheses. (*vi*) Most powerful test.
   (*vii*) Uniformly most powerful test.

   (*c*) Identify the composite hypotheses in the following, where μ is the mean and σ^2 is the variance of a distribution.

   (*i*) $H_0 : \mu \le 0, \sigma^2 = 1$ (*ii*) $H_0 : \mu = 0, \sigma^2 = 0$
   (*iii*) $H_0 : \mu \le 0, \sigma^2 =$ arbitrary (*iv*) $H_0 : \sigma^2 = \sigma^2_0$ (a given value), μ arbitrary.

   (*d*) (*i*) Explain the concepts of Type I and Type II errors, with examples and bring out their importance in Neyman and Pearson testing theory.
2. "In every hypothesis testing, the two types of errors are always present." If this is true then explain what is the use of hypothesis testing.
3. What is a statistical hypothesis? Define (*i*) two types of errors, (*ii*) power of a test, with reference to testing of a hypothesis. Explain how the best critical region is determined. State clearly the theorem which is used to determine the best critical region for simple hypothesis at a given significance level.
4. Explain the concept of most powerful test and discuss how the Neyman-Pearson lemma enables us to obtain the most powerful critical region for testing a simple hypothesis against a simple alternative.
5. What is meant by a statistical hypothesis? Explain the concepts of type I and type II errors. Show that a most powerful test is necessarily unbiased.
6. What are simple and composite statistical hypotheses? State and prove Neyman-Pearson Fundamental Lemma for testing a simple hypothesis against a simple alternative.
7. (*a*) Explain the basic concepts of a statistical hypothesis. Discuss the problems associated with the testing of simple and composite hypotheses. Show that a most powerful test is necessarily unbiased.

   (*b*) State Neyman-Pearson Lemma.

   Prove that if W is an *MP* region for testing $H_0 : \theta = \theta_0$ against $H_1 : \theta = \theta_1$, then it is necessarily unbiased. Also prove that the same holds good if W is an *UMP* region.

   (*c*) Define M.P. region and U.M.P. region. Show that an M.P. region is necessarily unbiased.
8. (*a*) Define 'Likelihood Ratio Test'. Under what circumstance would you recommend this test?

   (*b*) Let $p_\theta(x)$ be the density of a random variable with the mixed second derivative $\dfrac{\partial^2 \log p_\theta(x)}{\partial\theta\,\partial x} \ge 0$ for all x and θ. Then show that the family has monotone likelihood ratio in x.

   (*c*) Discuss the general method of construction of likelihood ratio test.

   (*d*) Show that the likelihood ratio test for testing the equality of variances of two normal distributions is the usual *F*-test.

9. Show that the likelihood ratio principle leads to the same test, when testing a simple hypothesis against an alternative simple hypothesis, as that given by Neyman-Pearson theorem.
10. Explain what is meant by non-parametric methods. How do they differ from parametric methods? Illustrate your answer by considering a suitable non-parametric test for the hypothesis that two independent samples have come from the same population.
11. (*a*) Derive the sign test, stating clearly the assumptions made.
(*b*) Describe the median test for the two-sample location problem. Find the distribution of the test statistic and compute its mean and variance under the null hypothesis. How is the test carried out in case of large samples?
12. (*a*) Explain the main difference between parametric and non-parametric approaches to the theory of statistical inference. Derive the sign test for two sample problem.
(*b*) Describe the procedure in median test when there are two independent samples. What non-parametric test would you use when the two samples are related.
(*c*) What are the advantages and drawbacks of non-parametric methods over parametric methods? Develop the following non-parametric tests, stating the underlying assumptions and the null hypotheses :
(*i*) Median test (*ii*) Mann-Whitney-Wilcoxon test.
13. (*a*) Distinguish between 'sign test' and 'Wilcoxon signed rank test'. Describe the sign test for testing that the population median is M_0 against the alternative that the median is M_1 ($> M_0$).
(*b*) Develop the Mann-Whitney-Wilcoxon test and obtain the mean and variance of the test statistic T. How is the test carried out for large samples?
14. (*a*) Describe Wald's Sequential Probability Ratio Test.
(*b*) Explain how the sequential test procedure differs from the Neyman-Pearson test procedure.
(*c*) Define the *O.C.* function and *A.S.N.* function in sequential analysis. Derive their approximate expressions for the sequential probability ratio test of a simple hypothesis against a simple alternative.
15. (*a*) What is a sequential test? How will you develop an optimum test of a specified strength for a simple null hypothesis versus a simple alternative?
(*b*) Find expressions for the sample size expected for termination of *S.P.R.T.* both under H_0 and H_1. Clearly state all the assumptions made.

ASSORTED REVIEW PROBLEMS FOR SELF-ASSESSMENT

18·1. (*a*) Let p denote the probability of getting a head when a given coin is tossed once. Suppose that the hypothesis $H_0 : p = 0{\cdot}5$ is rejected in favour of $H_1 : p = 0{\cdot}6$ if 10 trials result in 7 or more heads. Calculate the probabilities of type *I* and type *II* error.
(*b*) An urn contains 6 marbles of which θ are white and the others black. In order to test the null hypothesis $H_0 : \theta = 3$, against the alternative $H_1 : \theta = 4$, two marbles are drawn at random (without replacement) and H_0 is rejected if both the marbles are white; otherwise H_0 is accepted. Find the probabilities of committing type *I* and type *II* errors.

If it is decided to reject H_0 when both marbles are black and to accept it otherwise, find the probabilities of rejecting H_0 (*i*) when H_0 is true and (*ii*) when H_1 is true. Comment on your results.

18·2. (*a*) p is the probability that a given die shows even number. To test $H_0 : p = \frac{1}{2}$ against $H_1 : p = \frac{1}{3}$, following procedure is adopted. "Toss the die twice and accept H_0 if both times it shows even number" Find probabilities of type *I* and type *II* errors.

(*b*) Let p be the probability that a coin will fall head in a single toss. In order to test the hypothesis $H_0 : p = \frac{1}{2}$, the coin is tossed 6 times and the hypothesis H_0 is rejected if more than 4 heads are obtained. Find the probability of the error of first kind. If the alternative hypothesis is $H_1 : p = \frac{3}{4}$, find the probability of the error of second kind.

(*c*) In a Bernoulli distribution with parameter p, $H_0 : p = \frac{1}{2}$ against $H_1 : p = \frac{2}{3}$ is rejected if more than 3 heads are obtained out of 5 throws of a coin. Find the probabilities of Type I and Type II errors.

18·3. (*a*) Let $X_1, X_2, \ldots, X_9$ be a random sample from $N(\theta, 25)$. If, for testing $H_0 : \theta = 20$, against $H_1 : \theta = 26$, the critical region W is defined by : $W = \{\mathbf{x} \mid \bar{x} > 23{\cdot}266\}$, then find the size of critical region and the power of the test.

(*b*) Let $X \sim N(\mu, 4)$, μ unknown. To test $H_0 : \mu = -1$ against $H_1 : \mu = 1$, based on a sample of size 10 from this population we use the critical region $x_1 + 2x_2 + \ldots + 10x_{10} \geq 0$. What is its size? What is the power of the test?

(*c*) A sample of size 16 is drawn from a normal population with mean μ and standard deviation σ for testing the hypothesis $H_0 : \mu = \sigma = 1$ against the alternative hypothesis $H_1 : \mu = \sigma = 2$. It is decided to reject the hypothesis H_0 if the sample mean exceeds 1·5 and otherwise accept it. Calculate the probabilities of errors of the first and second kind in this procedure.

$$\left(\text{Given } \int_{-\infty}^{1} \frac{1}{\sqrt{2\pi}} \exp\left(-\frac{1}{2}t^2\right) dt = 0{\cdot}8413 \text{ and } \int_{-\infty}^{2} \frac{1}{\sqrt{2\pi}} \exp\left(-\frac{1}{2}t^2\right) dt = 0{\cdot}9773\right)$$

(*d*) A single value x is drawn from a normal population with mean m and variance 25. The null hypothesis $H_0 : m = 50$ is accepted if $x \leq 75$, otherwise $H_1 : m = 60$ is considered true. Evaluate the probabilities of type *I* and type *II* errors.

(*e*) Let p be the proportion of smokers in a certain city. You desire to test the hypothesis $H_0 : p = \frac{1}{2}$ against $H_1 : p = \frac{3}{4}$. If you reject H_0 when 60 persons or more are found smokers in a sample of 100 persons, compute the significance level and power of the test.

(*f*) Let p be the probability that a given die shows an even number. To test $H_0 : p = \frac{1}{2}$ against $H_1 : p = \frac{1}{3}$; the following procedure is adopted. Toss the die twice and accept H_0 if both times it shows even number. Find the probabilities of type *I* and type *II* errors.

18·4. (*a*) The hypothesis $\mu = 50$, is rejected if the mean of a sample of size 25 is either greater than 70·54 or less than 31·19. Assuming the distribution to be normal with s.d. 50, find the level of significance. Obtain the power function for the test and sketch the power curve with two values above 50 and two values below 50.

(*b*) Calculate the size of the type II error if the type I error is chosen to be $\alpha = 0{\cdot}16$ if you are testing $H_0 : \mu = 7$ against $H_1 : \mu = 6$, for a normal distribution with $\sigma = 2$, by means of a sample of size 25 and if the proper tail of the χ^2 distribution is used as the critical region.

18·5. (*a*) Given the frequency function:

$$f(x,\theta) = \begin{cases} \frac{1}{\theta}, 0 \leq x \leq \theta \\ 0, \text{elsewhere} \end{cases}$$

and that you are testing the hypothesis $H_0 : \theta = 1{\cdot}5$ against $H_1 : \theta = 2{\cdot}5$, by means of a single observed value of x, what would be the sizes of the type I and type II errors, if you choose the interval $0{\cdot}8 \leq x$, as the critical region? Also obtain the power function of the test.

(*b*) It is desired to test the hypothesis $H_0 : \theta = 0$ against $H_1 : \theta > 0$, by observing a random variable X which is uniformly distributed on $[\theta, \theta + 1]$. Given only one observation, sketch the power function of the test whose critical region is defined by $\{x > c\}$. What value of c would you choose?

Given n observations, derive the general formula of the power function of the test whose critical region is defined by : {at least one x is greater than c}, and indicate how you would construct a confidence interval for θ.

18·6. (*a*) Let X have a p.d.f. of the form:

$$f(x,\theta) = \begin{cases} \dfrac{1}{\theta}\exp(-x/\theta), & 0 < x < \infty, \theta > 0 \\ 0, & \text{elsewhere} \end{cases}$$

To test $H_0 : \theta = 2$ against $H_1 : \theta = 1$, use a random sample X_1, X_2 of size 2 and define a critical region $C = \{(x_1, x_2) : 9{\cdot}5 \le x_1 + x_2\}$.

Find (*i*) Power function of the test (*ii*) Significance level of the test.

(*b*) Let X have a p.d.f. of the form:

$$f(x,\theta) = \begin{cases} \theta\, x^{\theta-1}, & 0 < x < 1 \\ 0, & \text{elsewhere} \end{cases}$$

To test the simple hypothesis $H_0 : \theta = 1$ against the alternative simple hypothesis $H_1 : \theta = 2$, use a random sample X_1, X_2 of size $n = 2$ and define the critical region to be: $C = \{(x_1, x_2) : \frac{3}{4} \le x_1 x_2\}$ where x_1, x_2 are the values assumed by a sample. Obtain the power function of the test.

18·7. (*a*) Give a working rule of finding the best critical region for testing a simple hypothesis against a simple alternative.

For a normal (m, σ^2) population with known σ, construct a test for the null hypothesis $H_0 : m = m_0$ against the alternative $m > m_0$.

(*b*) Let $(x_1, x_2, ..., x_n)$ be a random sample from $N(\theta, \sigma^2)$, where σ^2 is known. Obtain an *UMP* test for testing $H_0 : \theta = \theta_0$ against $H_1 : \theta > \theta_0$. Also find the power function of the test and examine if the test is unbiased.

(*c*) Obtain the most powerful test for testing the mean $\mu = \mu_0$ against $\mu = \mu_1$, $(\mu_1 > \mu_0)$ when $\sigma^2 = 1$ in normal population.

(*d*) Obtain the most powerful test of size α for $H_0 : \mu = \mu_0$ against $H_1 : \mu = \mu_1$ when $\mu_1 > \mu_0$, if the probability density function of the random variable X is

$$f(x,\mu) = \frac{1}{\sqrt{8\pi}} \cdot \exp\left\{-\frac{1}{8}(x-\mu)^2\right\}, -\infty < x < \infty$$

18·8. (*a*) Let $x_1, x_2, ..., x_n$ denote a random sample from the distribution that has p.d.f.

$$f(x,\mu) = \frac{1}{\sqrt{2\pi}} \cdot \exp\left[-\frac{1}{2}(x-\mu)^2\right], -\infty < x < \infty$$

It is desired to test $H_0 : \mu = 0$ against $H_1 : \mu = 1$.

(*b*) Let $X_1, X_2, ..., X_n$ denote a random sample from the normal distribution $N(\theta, 1)$, θ is unknown. Show that there is no uniformly most powerful test of the simple hypothesis $H_0 : \theta = \theta_0$, where θ_0 is a fixed number against the alternative composite hypothesis $H_1 : \theta \ne \theta_0$.

(*c*) Let $x_1, x_2, ..., x_n$ be a random sample from $N(\mu, \theta)$, where μ is known. Obtain an UMP test for testing $H_0 : \theta = \theta_0$ against $H_1 : \theta < \theta_0$. Also find the power function of the test when,

(*i*) $\theta_1 > \theta_0$ and (*ii*) $\theta_1 < \theta_0$.

(*d*) Obtain M.P. regions of size α for testing:

$$H_0 : \theta = \theta_0 \text{ against } H_1 : \theta = \theta_1,$$

for $N(\mu, \theta)$, where μ is known.

Show that the tests in (*i*) and (*ii*) are U.M.P. against one-sided alternative.

18·9. (*a*) Let $x_1, x_2, \ldots, x_n$ be a random sample from a normal distribution $N(0, \sigma^2)$. Show that there exists a uniformly most powerful test with significance level α for testing

$$H_0 : \sigma^2 = \sigma_1^2 \text{ against } H_1 : \sigma^2 < \sigma_1^2.$$

If $n = 15$, $\alpha = 0{\cdot}05$ and $\sigma_1^2 = 3$, determine the best critical region and the power function of the above test.

(*b*) State Neyman and Pearson's fundamental lemma and apply it to obtain the test for testing $\sigma^2 = 1$ against $\sigma^2 > 1$, when the sample is from $N(0, \sigma^2)$. Is this test UMP? Is it unbiased? Give reasons.

18·10. (*a*) A sample of size 25 is drawn from a normal population with unknown mean μ and variance 16. It is required to test the hypothesis $H_0 : \mu = 1{\cdot}0$ against the alternative $H_1 : \mu = 3{\cdot}0$ at 5% level of significance. Obtain the most powerful test for testing H_0 against H_1 and state how you will find its power. Is the test uniformly most powerful?

(*b*) Let $X_1, X_2, \ldots, X_{20}$ be a random sample of size 20 from a Poisson distribution with mean θ. Show that the critical region defined by $\sum_{i=1}^{20} x_i \geq 5$, is uniformly most powerful critical region for testing $H_0 : \theta = \frac{1}{10}$ against $H_1 : \theta > \frac{1}{10}$.

(*c*) Let $X_1, X_2, \ldots, X_n$ denote a random sample from a normal distribution $N(\theta, 16)$. Find the sample size n and a uniformly most powerful test for $H_0 : \theta = 25$ against $H_1 : \theta < 25$ with power function $K(\theta)$ so that approximately $K(25) = 0{\cdot}10$ and $K(23) = 0{\cdot}90$.

18·11. Explain the statistical procedure of testing the following hypothesis regarding the standard deviation (σ) of normal population :

$$H_0 : \sigma = \sigma_0, \quad H_1 : \sigma = \sigma_1 > \sigma_0.$$

Will the test criterion remain the same when σ_1 is changed to $\sigma \neq \sigma_0$?

18·12. (*a*) State and prove Neyman-Pearson Lemma. If $x \geq 1$ is the critical region for testing $H_0 : \theta = 2$ against the alternative $H_1 : \theta = 1$, on the basis of a single observation from the population :

$$f(x, \theta) = \theta\, e^{-\theta x}, 0 \leq x < \infty,$$

obtain the values of type I and type II errors and the power function of the test.

(*b*) Given a random sample $X_1, X_2, \ldots, X_n$ of size n from the distribution with *p.d.f.*

$$f(x, \theta) = \theta\, e^{-\theta x}\ ; x > 0, 0 < \theta < \infty,$$

show that UMP test for testing $H_0 : \theta = \theta_0$ against $H_1 : \theta < \theta_0$ is given by

$$\left\{ x \mid \sum x_i \geq \frac{1}{2\theta_0} \chi^2_{\alpha, 2n} \right\}\ ; \ (\alpha = \text{Level of significance})$$

18·13. (*a*) Two independent observations x_1, x_2 are made on a random variable X with density function :

$$f(x, \theta) = \frac{1}{\theta} \exp(-x/\theta);\ 0 < x < \infty, \theta > 0$$

Test the null hypothesis $H_0 : \theta = 2$ against the alternative $H_1 : \theta = 4$. If H_0 is accepted when $x_1 + x_2 < 9{\cdot}5$, and rejected otherwise, obtain the level of significance and power of the test.

(*b*) On the basis of a single observation x from the following *p.p.f.*

$$f(x, \theta) = \frac{1}{\theta} \exp\left(-\frac{x}{\theta}\right), (x > 0; \theta > 0),$$

the null hypothesis $H_0 : \theta = 1$ against the alternative hypothesis $H_1 : \theta = 4$ is tested by using a set $C = \{x : x > 3\}$ as the critical region. Prove that the critical region C provides a most powerful test of its size. What is the power of the test?

18·14. (*a*) Obtain the statistic for testing the hypothesis that the mean of a Poisson population is 2 against the alternative that it is 3, on the basis of n independent observations.

(*b*) Suppose you are testing $H_0 : \lambda = 2$ against $H_1 : \lambda = 1$, where λ is the parameter of the Poisson distribution. Obtain the best critical region of the test.

18·15. (*a*) Suppose a random sample of size n is taken from the Poisson population $\left(\frac{\exp(-\lambda)\cdot\lambda^x}{x!}\right)$, $x = 0, 1, 2, \ldots$. Give the most powerful critical region of size α for testing the hypothesis $\lambda = \lambda_0$ against $\lambda = \lambda_1, (\lambda_1 > \lambda_0)$.

How can you use the above result to find a confidence interval for λ?

Write an expression for the power function of the test for the hypothesis $\lambda = \lambda_0$ against $\lambda > \lambda_0$.

(*b*) $x_1, x_2, \ldots, x_{10}$ is a random sample of size 10 from a Poisson distribution with mean θ. Show that the critical region C defined by $\sum_{i=1}^{10} x_i \geq 3$, is the best critical region for testing :

$$H_0 : \theta = 0{\cdot}1 \text{ against } H_1 : \theta = 0{\cdot}5.$$

18·16. (*a*) Let $X_1, X_2, \ldots, X_n$ denote a random sample from a distribution having *p.d.f.* :

$$f(x,p) = \begin{cases} p^x(1-p)^{1-x} & ; x = 0,1 \\ 0, & \text{elsewhere} \end{cases}$$

It is desired to test $H_0 : p = \frac{1}{2}$ against $H_1 : p = \frac{1}{3}$. Obtain the BCR.

(*b*) Suppose X has Bernoulli distribution with probability of success θ. On the basis of a random sample of size n it is proposed to reject the null hypothesis, $H_0 : \theta = \frac{1}{2}$

$$\text{if } (X_1 + X_2 + \cdots + X_n) \geq \frac{3}{8} \quad \text{or} \quad \leq \frac{5}{8}$$

For $n = 5$, find the level of significance of the test.

18·17. Let $x_1, x_2, \ldots, x_n$ be a random sample from a Bernoulli distribution with density :

$$f(x ; \theta) = \theta^x (1-\theta)^{1-x} ; \quad x = 0, 1$$

Obtain a uniformly most powerful size – α test for $H_0 : \theta = \theta_0$ against $H_1 : \theta > \theta_0$. Would you modify the test if $H_1 : \theta < \theta_0$?

18·18. Derive a most powerful test of the hypothesis $\theta = \frac{1}{4}$ against the alternative $\theta = \frac{1}{2}$ for the parameter θ in a geometric distribution $\theta\,(1-\theta)^x$, $x = 0, 1, 2, \ldots$ based on a random sample of size 2.

18·19. Describe the method for finding the best critical region of size α while testing a simple hypothesis against simple alternative one. Illustrate it by finding BCR for testing $H_0 : \theta = 0$ against $H_1 : \theta = 1$, for the Cauchy distribution :

$$dF(x) = \frac{dx}{\pi\left[1 + (x-\theta)^2\right]}, -\infty < x < \infty$$

based on a random sample of size 1.

18·20. The distribution of x is :

$$f(x,\theta) = \begin{cases} \frac{1}{2}, \theta - 1 \leq x \leq \theta + 1 \\ 0, \text{ elsewhere} \end{cases}$$

If $H_0 : \theta = 4$ and $H_1 : \theta = 5$, determine the critical region on the right hand tail of the distribution corresponding to $\alpha = 0{\cdot}25$. Also calculate the type II error.

18·21. (*a*) Let $X_1, X_2, ..., X_n$ be a random sample of size n from p.d.f.

$$f(x, \theta) = \theta x^{\theta-1}, 0 < x < 1, \theta > 0.$$

Obtain the U.M.P. region of size α for testing $H_0 : \theta = \theta_0$ against $H_1 : \theta > \theta_0$. Also find the power function of the test.

(*b*) Let $x_1, x_2, ..., x_n$ be n independent observations on a random variable X with density function

$$f(x, \theta) = \theta x^{\theta-1}; 0 < x < 1, \theta > 0$$

Show that the best critical region for testing $H_0 : \theta = 1$ against $H_1 : \theta = 2$, can be defined in terms of the geometric mean of $x_1, x_2, ..., x_n$.

(*c*) Let $X_1, X_2, ..., X_n$ be a random sample from a distribution with p.d.f.

$$f(x,\theta) = \begin{cases} \theta x^{\theta-1}, & if\ 0 < x < 1 \\ 0, & \text{otherwise} \end{cases}$$

where $0 < \theta < \infty$. Show that the M.P. test of level α for testing $H_0 : \theta = 1$ against the alternative $H_1 : \theta = 2$, is given by the critical region:

$$\left\{ x \,\middle|\, \prod_{i=1}^{n} x_i > \exp\left(-\frac{1}{2}\chi^2_{1-\alpha, 2n}\right) \right\}$$

where $\chi^2_{1-\alpha,\,2n}$ is the lower α = point of the χ^2-distribution with $2n$ *d.f.*

18·22. In testing $H_0 : \sigma = \sigma_0$ against $H_1 : \sigma = \sigma_1\ (\neq \sigma_0)$ for the distribution

$$f(x) = \frac{1}{\sigma} \exp\left[-\left(\frac{x-\theta}{\sigma}\right)\right], (\theta \le x < \infty, \sigma > 0)$$

show that the UMP test is of the form :

$$\Sigma x_i \ge \text{constant and } \Sigma x_i \le \text{constant.}$$

18·23. Let X be a single observation from the density : $f(x\,;\,\theta) = 2\theta x + 1 - \theta$, $0 < x < 1$, $|\theta| \le 1$, zero otherwise. Find the best critical region of size α, for testing $H_0 : \theta = 0$ against $H_1 : \theta < 0$. Express the power function of this test in terms of α. Is the test uniformly most powerful? Explain.

18·24. $X_1, X_2, ..., X_n$ is a random sample of size n from $N(\theta, 100)$. For testing $H_0 : \theta = 75$ against $H_1 : \theta > 75$ the following test procedure is proposed:

Reject H_0 if $\overline{x} \ge c$, Accept H_0 if $\overline{x} < c$.

Determine n and c so that the power function $P(\theta)$ of the test satisfies

$$P(75) = 0{\cdot}159 \text{ and } P(77) = 0{\cdot}841.$$

18·25. Let $x_1, x_2, ..., x_n$ be a random sample from a distribution having *p.d.f.*

$$f(x,\theta) = \begin{cases} \dfrac{[x(1-x)]^{\theta-1}}{\beta(\theta,\theta)}, & 0 < x < 1,\ \theta > 0 \\ 0, & \text{elsewhere} \end{cases}$$

Show that the best critical region for testing $H_0 : \theta = 1$ against $H_1 : \theta = 2$ is:

$$C = \left\{ (x_1, x_2, ..., x_n) : c \le \sum_{i=1}^{n} x_i(1 - x_i) \right\}.$$

18·26. Let $(x_1, x_2, ..., x_9)$ be a random sample from $N(\mu, 9)$. To test the hypothesis $H_0 : \mu = 40$ against $H_1 : \mu \neq 40$, consider the following two critical regions:

$$C_1 = \{\bar{x} : \bar{x} \geq a_1\}, \qquad C_2 = \{\bar{x} : |\bar{x} - 40| \geq a_2\}$$

(*i*) Obtain the values of a_1 and a_2 so that the size of each critical region is 0·05

(*ii*) Calculate the power of the two critical regions when $\mu = 39$ and $\mu = 41$ and comment on the results.

18·27. (*a*) For a sample of size 25 from a normal population $N(\mu, 25)$, $\bar{X} = 11{\cdot}5$. Test the hypothesis H_0: $\mu = 10$ against the alternative $H_1 : \mu > 10$. Calculate the power of the test for $\mu = 11$.

(*b*) Let $X \sim N(\mu, 25)$. The null and alternative hypotheses are $H_0 : \mu = 10$ and $H_1 : \mu < 10$.

(*i*) Give the best test of size $\alpha = 0{\cdot}05$ for a sample of size 25. (No derivation is expected).

(*ii*) Calculate the power of the test for $\mu = 8$.

(*c*) X is normally distributed with $\sigma = 5$ and it is desired to test $H_0 : \mu = 105$ against $H_1 : \mu = 110$. How large a sample should be taken if the probability of accepting H_0 when H_1 is true is 0·02 and if a critical region of size 0·05 is used?

18·28. The *p.d.f.* of x is given by $f(x) = 1/\theta,\ 0 < x < \theta$. Let the null hypothesis be $H_0 : \theta = 4/3$ against the alternative hypothesis $H_A : \theta > 4/3$. We have a random sample of one observation. The critical region is defined by $C = \{x : x > 1\}$.

(*i*) Find the significance level of the test.

(*ii*) Find the power of the test for $\theta = 7/3$ and $\theta = 10/3$.

18·29. Let $X_1, X_2, ..., X_n$ be an random sample from discrete distribution with probability function $f(x)$ for which x takes non-negative integral values 0, 1, 2, ...

$$\text{According to } H_0 : \quad f(x) = \begin{cases} \dfrac{e^{-1}}{x!}; & x = 0,1,2,... \\ 0, & \text{otherwise} \end{cases}$$

$$\text{According to } H_1 : \quad f(x) = \begin{cases} \dfrac{1}{2^{x+1}}; & x = 0,1,2,... \\ 0, & \text{otherwise} \end{cases}$$

Obtain the critical region of the most powerful test of level α for testing H_0 against H_1.

18·30. H_0 denotes the null hypothesis that a given distribution has the *p.d.f* :

$$\frac{1}{\sqrt{2\pi}} e^{-\frac{1}{2}x^2}, -\infty < x < \infty$$

and H_1 denotes the alternative hypothesis that the distribution has the *p.d.f.*

$$\frac{1}{2}\exp(-|x|), -\infty < x < \infty$$

Obtain the most powerful test for testing H_0 against H_1 with level of significance α.

18·31. It is required to test H_0 against H_1 from a single observation x, where H_0 is the hypothesis that the *p.d.f.* is:

$$f(x) = \frac{1}{\sqrt{2\pi}} \exp\left(-\frac{1}{2}x^2\right), (-\infty < x < \infty)$$

and H_1 is the hypothesis that the *p.d.f.* is : $\quad f(x) = \dfrac{2}{\Gamma(1/4)} \exp(-x^4),\ (-\infty < x < \infty)$.

Obtain the most powerful test with level of significance α in this case.

18·32. State Neyman-Pearson fundamental lemma. With the usual notation, if β is the power of the most powerful test of size α for testing $H_0 : p = p_0$ against $H_1 : p = p_1$, show that $\alpha < \beta$ unless $p_0 = p_1$. If

$$p_0(x) = \frac{1}{\sqrt{2\pi}} \exp\left[-\frac{1}{2}(x-\mu)^2\right], -\infty < x < \infty, \text{ and } \quad p_1(x) = \frac{1}{\pi} \cdot \frac{1}{1+(x-\mu)^2}, -\infty < x < \infty$$

and μ is known, determine the most powerful test of size α. Calculate its power, if α and μ have specified values.

18·33. Discuss the general method of construction of likelihood ratio test. Consider n Bernoullian trials with probability of success p for each trial. Derive the likelihood-ratio test for testing

$$H_0 : p = p_0 \text{ against } H_1 : p > p_0.$$

18·34. (*a*) Let $X_1, X_2, ..., X_n$ be a random sample from a Poisson distribution with parameter θ. Derive the likelihood ratio test for $H_0 : \theta = \theta_0$ against $H_1 : \theta > \theta_0$. Show that this is identical with the corresponding *UMP* test.

(*b*) Let $X_1, X_2, ..., X_n$ be a random sample from a normal population with unknown mean μ and, known variance σ^2. Develop the likelihood ratio test for testing $H_0 : \mu = \mu_0$ (specified) against

(*i*) $H_1 : \mu > \mu_0$ and (*ii*) $H_1 : \mu < \mu_0$.

(*c*) Let $X_1, X_2, ..., X_n$ be a random sample from a $N(\mu, \theta)$, θ is the unknown variance and μ is known. Obtain a likelihood ratio test for testing a simple $H_0 : \theta = \theta_0$ against $H_1 : \theta > \theta_0$.

(*d*) Let $x_1, x_2, ..., x_n$ be a random sample from a normal population with mean μ and variance σ^2, μ and σ^2 being unknown. We wish to test $H_0 : \mu = \mu_0$ (specified) against $H_1 : \mu \neq \mu_0$, $0 < \sigma^2 < \infty$. Show that the Likelihood Ratio Test is the same as the two tailed t-test.

18·35. Describe the likelihood ratio test.

The random variable X follows normal distribution with mean θ_1 and variance θ_2. The parameter space is $\quad \Theta = \{\theta_1, \theta_2\} : -\infty < \theta_1 < \infty, 0 < \theta_2 < \infty$

Let $\quad \Theta_0 = \{(\theta_1, \theta_2) : \theta_1 = 0, 0 < \theta_2 < \infty\}$.

Test the hypothesis $H_0 : \theta_1 = 0, \theta_2 > 0$ against the alternative composite hypothesis $H_1 : \theta_1 \neq 0, \theta_2 > 0$.

18·36. Discuss the general method of construction of likelihood ratio test. Given $N(\mu_1, \sigma_1^2)$ and $N(\mu_2, \sigma_2^2)$, where all the parameters μ_1, μ_2, σ_1^2 and σ_2^2 are unspecified, develop the *LR* test for testing $H_0 : \sigma_1^2 = \sigma_2^2$ against $H_1 : \sigma_1^2 \neq \sigma_2^2$.

18·37. Describe likelihood ratio test and state its important properties.

Let X_1 and X_2 be $N(\mu_1, \sigma^2)$ and $N(\mu_2, \sigma^2)$ respectively where the means and variance are unspecified. Develop *LR* test for testing $H_0 : \mu_1 = \mu_2$ against $H_1 : \mu_1 \neq \mu_2$.

OR

Construct *LR* test for testing $H_0 : \theta = \theta_0$ against all its alternatives in $N(\theta, \sigma^2)$, where σ^2 is known.

18·38. Show that the likelihood ratio test for testing $H_0 : \alpha = 0$ against $H_1 : \alpha \neq 0$, based on a random sample of size n from

$$f(x ; \alpha; \beta) = \frac{1}{2\beta} ; \alpha - \beta \leq x \leq \alpha + \beta$$

is $(R/2Z)^n$ where $R = X_{(n)} - X_{(1)}$ and $Z = \max\left[- X_{(1)}, X_{(n)}\right]$.

18·39. $X_1, X_2, ..., X_{10}$ is a random sample of size 10 from a population having distribution function $F(x)$. Test the hypothesis $H_0 : F(72) = \frac{1}{2}$ against the alternative hypothesis, $H_1 : F(72) > \frac{1}{2}$.

18·40. The observations of a random sample of size 10 from a distribution which is symmetric about $K_{.5}$ are 20·2, 24·1, 21·3, 17·2, 19·8, 16·5, 21·8, 18·7, 17·1, 19·9. Use Wilcoxon's Test to test the hypothesis $H_0 : K_{.5} = 18$ against $H_1 : K_{.5} > 18$ if $\alpha = 0{\cdot}05$. You may use the normal approximation.

18·41. Thirty observations as given below are obtained :

24,	35,	12,	50,	60,	70,	68,	49,	80,	25,	69,	28,	28,	11,	83,
31,	37,	34,	54,	75,	45,	95,	75,	26,	43,	57,	94,	48,	63,	45

Test their randomness by considering the sequence of positive and negative signs.

18·42. If R_1 and R_2 denote the number of runs of n_1 objects of one type and n_2 objects of another type in a sample of size $n_1 + n_2$, then find the probability that $R_1 + R_2 = r$, for r even and r odd, and also the mean and variance of $R_1 + R_2$ when all these $n_1 + n_2$ observations arise from the same distribution.

18·43. In the median test with samples of size 9 and 7 respectively, from two populations find the probability density function of the random variable representing the number of values of the samples from the first population in the lower half of the combined sample.

18·44. The win-lose record of a certain basketball team for their last 50 consecutive games was as follows:—

W W W W W W L W W W W W W L W L W W W L L W W W W

L W W W L L W W W W W W L L W W L L L W W L W W W

Apply run test to test that sequence of wins and losses is random.

18·45. At the beginning of the year a first grade class was randomly divided into two groups. One group was taught to read using a uniform method, where all students progressed from one stage to the next at the same time, following the teacher's direction. The second group was taught to read using an individual method, where each student progressed at his own rate according to a programmed work book, under supervision of the teacher. At the end of the year each student was given a reading ability test with the following results :

First Group			*Second Group*		
227	55	184	202	271	63
176	234	147	14	151	284
252	194	88	165	235	53
149	247	161	171	147	228
16	92	171	292	99	271

Use the Wald-Wolfowitz run-test to test for the equality of the distribution functions of the two groups.

18·46. Using the number of runs above and below the median, test for randomness the following set of a table of 2-digit numbers :

15,	77,	01,	65,	69,	69,	58,	40,	81,	16,	16,	20,	00,	84,	22,
28,	26,	46,	66,	36,	86,	66,	17,	43,	49,	85,	40,	51,	40,	10.

18·47. (*a*) Describe Wald's S.P.R.T. Let X be a Bernoulli variate with probability function:

$$f(x\,;\theta) = \theta^x (1-\theta)^{1-x}\,; \quad x = 0, 1; \quad 0 \le \theta \le 1.$$

Employ S.P.R.T. for testing $H_0 : \theta = \theta_0$ against $\theta = \theta_1$ and obtain its A.S.N. and O.C. functions.

(*b*) Develop the S.P.R.T. for testing $H_0 : \pi = \pi_0$ against $H_1 : \pi = \pi_1$ based on a random sample from a Binomial population with parameters (n, π), n being known. Obtain its O.C. and A.S.N. functions.

(*c*) Obtain the sequential probability ratio test of the hypothesis

$$H_0 : \theta = \frac{1}{3} \text{ against } H_1 : \theta = \frac{2}{3}, \text{ for the distribution :}$$

$$f(x;\theta) = \begin{cases} \theta^x (1-\theta)^{1-x}, & \text{for } x = 0,1 \\ 0, & \text{otherwise} \end{cases}$$

18·48. (*a*) Develop S.P.R.T. for testing $H_0 : \theta = \theta_0$ against $H_1 : \theta = \theta_1$, $(\theta_1 > \theta_0)$, where θ is the parameter of a Poisson distribution. Find approximate expressions for *O.C.* function and the *A.S.N.* function of the test.

(*b*) Describe S.P.R.T., its *O.C.* and *A.S.N.* functions.

Construct S.P.R.T. for testing $H_0 : \theta = \theta_0$ against $H_1 : \theta = \theta_1$, $(0 < \theta_0 < \theta_1)$ on the basis of a random sample drawn from the Pareto distribution with density function

$$f(x, \theta) = \frac{\theta a^{\theta}}{x^{\theta+1}}, \quad x \geq a$$

Also obtain its O.C. function and A.S.N. Function.

(*c*) Develop the S.P.R.T. for testing $H_0 : \theta = \theta_0$ against $H_1 : \theta = \theta_1$ $(> \theta_0)$ based on a random sample of size n from a population with p.d.f.

$$f(x, \theta) = \frac{1}{\theta} e^{-x/\theta}, x > 0, \theta > 0.$$

Also obtain its *A.S.N.* and *O.C.* functions.

(*d*) Let X have the p.d.f.

$$f(x; \theta) = \begin{cases} \theta e^{-\theta x} ; x \geq 0, \theta > 0 \\ 0, \text{ elsewhere} \end{cases}$$

For testing $H_0 : \theta = \theta_0$ against $H_1 : \theta = \theta_1$, construct the *S.P.R.T.* and obtain its *A.S.N.* and *O.C.* functions.

(*e*) A random variable follows the normal distribution $N[\theta, \sigma^2]$, where σ^2 is known. Derive the *S.P.R.T.* for testing $H_0 : \theta = \theta_0$ against $H_1 : \theta = \theta_1$. Obtain the approximate expression for the *O.C.* function.

18·49. To test sequentially the hypothesis H_0 that the distribution is given by $P(X = -1) = P(X = 1) = P(X = 2) = \frac{1}{3}$ against the alternative H_1 that it is given by $P(X = -1) = P(X = 1) = \frac{1}{4}$; $P(X = 2) = \frac{1}{2}$, it is decided to continue sampling as long as $-\frac{(n+1)}{2} < S_n < \frac{n+2}{2}$ where $S_n = X_1 + X_2 + \ldots + X_n$, the X_k's being the successive observations. Compute the probability under H_0 and under H_1 that the procedure will terminate with the fourth observation or earlier.

18·50. $X_1, X_2, \ldots, X_n$ is a sequence of *i.i.d.* observations from $N(\mu, \sigma^2)$ where μ is known, σ^2 being unknown. Obtain the *S.P.R.T.* for testing $H_0 : \sigma^2 = \sigma_0^2$, against $H_1 : \sigma^2 = \sigma_1^2$ $(> \sigma_0^2)$. Also obtain its *O.C.* function and *A.S.N.* function.

CHAPTER

19 Finite Differences and Numerical Analysis

LEARNING OBJECTIVES

Upon completion of this Chapter, you should be able to:

- Understand the meaning and uses of interpolation and extrapolation.
- Define and understand the use of different operators like:
  - (i) Δ (Forward Difference Operator),
  - (ii) E (Shift Operator),
  - (iii) ∇ (Backward Difference Operator),
  - (iv) $\mathbf{\Delta}$ (Divided Difference Operator),
  - (v) D (Differential Operator),

  and the inter-relationships between them.
- Derive various formulae of interpolation for equal intervals like Newton's forward difference and backward difference formulae and understand their use.
- Derive various formulae of interpolation for unequal intervals like Newton's divided difference formula and Lagrange's formula and understand their use.
- Understand the meaning and use of Inverse Interpolation.
- Understand the meaning of numerical integration.
- Derive various formulae for numerical integration like:
  - (i) General Quadrature Formula,
  - (ii) Trapezoidal Rule,
  - (iii) Simpson's One-third Rule
  - (iv) Simpson's Three-eighth Rule,
  - (v) Weddles Rule,

  along with their applications.

CHAPTER OUTLINES

19·1. Introduction

19·2. Finite Difference Table

- 19·2·1. Forward Difference Operator Δ
- 19·2·2. The Shift Operator E
- 19·2·3. Properties of Operators Δ and E
- 19·2·4. Fundamental Theorem of Finite Differences
- 19·2·5. Relation between E and D, where D = $\frac{d}{dx}$
- 19·2·6. Missing Terms (Equal Intervals)
- 19·2·7. Factorial Notation

19·3. Interpolation

- 19·3·1. Assumptions of Interpolation
- 19·3·2. Uses of Interpolation
- 19·3·3. Method of Parabolic Curve Fitting
- 19·3·4. Gregory Newton's Forward Interpolation Formula for Equal Intervals
- 19·3·5. Gregory Newton's Backward Interpolation Formula for Equal Intervals

19·4. Interpolation with Arguments at Unequal Intervals

- 19·4·1. Divided Difference Operator Δ.
- 19·4·2. Divided Difference Table
- 19·4·3. Theorems on Divided Differences
- 19·4·4. Newton's Divided Difference Formula
- 19·4·5. Lagrange's Interpolation Formula

19·5. Numerical integration

- 19·5·1. General Quadrature Formula for Equidistant Ordinates
- 19·5·2. Trapezoidal Rule
- 19·5·3. Simpson's One-third Rule
- 19·5·4. Simpson's Three-eighth Rule
- 19.5.5. Weddle's Rule

19·1. INTRODUCTION

The calculus of finite differences, in its broad meaning, deals with the changes that take place in the value of the function, the dependent variable, due to changes in the independent variable. It is a study of the relations that exist between the values assumed by the function whenever the independent variable changes by finite jumps whether equal or unequal. In infinitesimal calculus we study, on the other hand, those changes which occur when the independent variable changes continuously in a given interval. In the present chapter, we study the variations in the function when the independent variable changes by (*i*) equal intervals and (*ii*) unequal intervals.

19·2. FINITE DIFFERENCE TABLE

Let us consider a function $y = U_x$ where x is an independent variable and $y = U_x$ is a dependent variable. Suppose we are given equidistant values (finite in number) a, $a + h$, $a + 2h$, $a + 3h$,... of the variable x at an interval of 'h'. Then the corresponding values of the variable $y = U_x$ are U_a, U_{a+h}, U_{a+2h}, U_{a+3h} ...

The values of the independent variable x are known as *arguments* and the corresponding values of the dependent variable y are called *entries, e.g.*, U_{a+h} is the entry corresponding to the argument $a + h$, and so on.

19·2·1. Forward Difference Operator Δ

Now we shall define the *forward difference operator* Δ, which is of fundamental importance in the calculus of finite differences. By *def.*,

$$\Delta U_x = U_{x+h} - U_x \; ; x = a, a + h, a + 2h, \ldots \qquad \ldots (19{\cdot}1)$$

The equal interval of the arguments, *viz.*, 'h' is known as the *interval of differencing*. Thus

$$\Delta U_a = U_{a+h} - U_a , \quad \Delta U_{a+h} = U_{a+h+h} - U_{a+h} = U_{a+2h} - U_{a+h}, \text{ and so on.} \qquad \ldots[19{\cdot}1(a)]$$

The differences defined in (19·1) or [19·1 (*a*)] are known as *first order differences*. If the operation of Δ is again performed on differences obtained in (19·1) or (19·1 (*a*)], we get the second order differences, which are denoted by Δ^2. Thus

$$\Delta^2 U_x = \Delta (\Delta U_x) = \Delta (U_{x+h} - U_x) = (U_{x+2h} - U_{x+h}) - (U_{x+h} - U_x)$$
$$= \Delta U_{x+h} - \Delta U_x \; ; \; x = a, a + h, a + 2h, \ldots \qquad \ldots [19{\cdot}1(b)]$$

In particular, $\Delta^2 U_a = \Delta U_{a+h} - \Delta U_a$, $\quad \Delta^2 U_{a+h} = \Delta U_{a+2h} - \Delta U_{a+h}$ and so on $\qquad \ldots [19{\cdot}1(c)]$

The third order difference denoted by Δ^3 are obtained by taking the differences of the second order differences obtained in [19·1 (*b*)] or [19·1 (*c*)]. Proceeding similarly the differences of higher order, *viz.*, Δ^4, Δ^5, etc. can be obtained. These differences of various orders can be conveniently expressed in the form of a table, known, as *Finite Differences* (*Forward*) Table.

DIFFERENCE TABLE

Argument x	Entry U_x	First Differences ΔU_x	Second Differences $\Delta^2 U_x$	Third Differences $\Delta^3 U_x$
a	U_a			
→	→	$U_{a+h} - U_a = \Delta U_a$		
$a + h$	U_{a+h}	→	$\Delta U_{a+h} - \Delta U_a = \Delta^2 U_a$	
→	→	$U_{a+2h} - U_{a+h} = \Delta U_{a+h}$	→	
$a + 2h$	U_{a+2h}	→	$\Delta U_{a+2h} - \Delta U_{a+h} = \Delta^2 U_{a+h}$	

Contd...

Argument x	Entry U_x	First Differences ΔU_x	Second Differences $\Delta^2 U_x$	Third Differences $\Delta^3 U_x$
		$\longrightarrow U_{a+3h} - U_{a+2h} = \Delta U_{a+2h}$		$\Delta^2 U_{a+h} - \Delta^2 U_a = \Delta^3 U_a$
$a+3h$	U_{a+3h}	$\longrightarrow$	$\Delta U_{a+3h} - \Delta U_{a+2h} = \Delta^2 U_{a+2h}$	
		$\longrightarrow U_{a+4h} - U_{a+3h} = \Delta_{a+3h}$	$\longrightarrow$	
$a+4h$	U_{a+4h}			

U_a is known as the *first entry or the leading term* in the difference table and h is called the *interval of differencing*. ΔU_a, $\Delta^2 U_a$, $\Delta^3 U_a$ etc. in the difference table are known as *leading differences.*

The *Backward differences* usually denoted by ∇, are defined as follows :

$$\nabla U_{x+h} = U_{x+h} - U_x = \Delta U_x \; ; x = a, a+h, a+2h, \ldots$$

Thus the backward difference of U_{x+h} is the same as the forward difference of U_x and the *backward difference table* may be expressed as follows:

x	U_x	ΔU_x	$\Delta^2 U_x$	$\Delta^3 U_x$	$\Delta^4 U_x$
a	U_a				
		∇U_{a+h}			
$a+h$	U_{a+h}		$\nabla^2 U_{a+2h}$		
		∇U_{a+2h}		$\nabla^3 U_{a+3h}$	
$a+2h$	U_{a+2h}		$\nabla^2 U_{a+3h}$		$\nabla^4 U_{a+4h}$
		∇U_{a+3h}		$\nabla^3 U_{a+4h}$	
$a+3h$	U_{a+3h}		$\nabla^2 U_{a+4h}$		
		∇U_{a+4h}			
$a+4h$	U_{a+4h}				

Remarks:

1. Like the operator D^2 in differential calculus, the operators Δ^2 or ∇^2 does not mean the square of a quantity Δ or ∇ but it means that the operator Δ or ∇ is to be performed twice.
2. Unless otherwise stated, the interval of differencing is always taken as unity.

Example 19·1. *Given $U_0 = 3$, $U_1 = 12$, $U_2 = 81$, $U_3 = 200$, $U_4 = 100$ and $U_5 = 8$, find $\Delta^5 U_0$.*

Solution. The difference table is given below:

x	U_x	ΔU_x	$\Delta^2 U_x$	$\Delta^3 U_x$	$\Delta^4 U_x$	$\Delta^5 U_x$
0	3					
		9				
1	12		60			
		69		– 10		
2	81		50		– 259	
		119		– 269		755
3	200		– 219		496	
		– 100		227		
4	100		8			
		– 92				
5	8					

Hence, $\Delta^5 U_0 = 755.$

Example 19·2. *If $y = 2x^3 - x^2 + 3x + 1$, calculate the values of y corresponding to $x = 0, 1, 2, 3, 4, 5$ and form the table of differences. Prove theoretically that the second difference is $12x + 10$. Verify this numerically.*

Solution. Here $y = 2x^3 - x^2 + 3x + 1 = U_x$ (say).

The values of y corresponding to $x = 1, 2, 3, 4$ and 5 are respectively 1, 5, 19, 55, 125 and 241.

The difference table is given below :

x	U_x	ΔU_x	$\Delta^2 U_x$	$\Delta^3 U_x$	$\Delta^4 U_x$
0	1				
		4			
1	5		10		
		14		12	
2	19		22		0
		36		12	
3	55		34		0
		70		12	
4	125		46		
		116			
5	241				

Theoretically

$$\begin{aligned}\Delta U_x &= U_{x+1} - U_x \\ &= [2(x+1)^3 - (x+1)^2 + 3(x+1) + 1] \\ &\quad - (2x^3 - x^2 + 3x + 1) \\ &= 6x^2 + 4x + 4\end{aligned}$$

and $\Delta^2 U_x = [6(x+1)^2 + 4(x+1) + 4] - (6x^2 + 4x + 4)$

$= 12x + 10$

Verification : $\Delta^2 U_x = 12x + 10$

$\therefore\ \Delta^2 U_0 = 10, \qquad \Delta^2 U_1 = 12 \times 1 + 10 = 22$

$\Delta^2 U_2 = 12 \times 2 + 10 = 34$ and $\Delta^2 U_3 = 12 \times 3 + 10 = 46$

These values are same as obtained in the above difference table.

19·2·2. The Shift Operator E

The operator E is defined as:

$$EU_x = U_{x+h} \qquad \ldots (19·2)$$

i.e., it results in increasing the argumment by the interval of differencing.

The operator E^2 implies that the operator E is to be performed twice.

Thus, $E^2 U_x = E[EU_x] = E[U_{x+h}] = U_{x+2h}$ ... [19·2 (*a*)]

In general, $E^r U_x = U_{x+rh}$ ... [19·2 (*b*)]

a result, which is of much practical utility.

Remarks:

1. If the interval of differencing is taken as unity, *i.e.*, $h = 1$ then

$$E^r U_x = U_{x+r} \qquad \ldots [19·2 (c)]$$

In particular $E^5 U_0 = U_{0+5} = U_5$, $\quad E^4 U_2 = U_{2+4} = U_6$, $\quad E^4 U_{-1} = U_{-1+4} = U_3$

Conversely, $U_1 = EU_0 = E^2 U_{-1} = E^3 U_{-2}$ and $U_2 = E^2 U_0 = E\,U_1 = E^3 U_{-1}$ and so on.

Thus by using the operator E, the suffix in U can be conveniently changed. This result is of more practical utility. From (19·1), we have

$$\Delta U_x = U_{x+h} - U_x = EU_x - U_x = (E - 1)\,U_x$$

Since U_x is arbitrary, $\Delta \equiv E - 1 \Rightarrow E \equiv 1 + \Delta$, ... (19·3)

which is a very important relation between operators E and Δ.

2. *Ralation between* E *and* ∇. By *def.*

$\nabla U_{x+h} = U_{x+h} - U_x \quad \Rightarrow \quad U_x = U_{x+h} - \nabla U_{x+h} = (1 - \nabla)\,U_{x+h}$ ... (*)

Also $\quad U_x = E^{-1} U_{x+h}$ ... (**)

From (*) and (**), we get: $\quad 1 - \nabla \equiv E^{-1} \quad \Rightarrow \quad E = (1 - \nabla)^{-1}$ ... (19·3*a*)

19·2·3. Properties of Operators D and E

The operators E and Δ satisfy the basic laws of algebra, *viz.*, distributive law, commutative law and law of indices.

(a) *The operators* E and Δ *are commutative in operation w.r.t. constants.* For example, if C is a constant, then

$$\Delta (CU_x) = CU_{x+h} - CU_x = C(U_{x+h} - U_x) = C\,\Delta U_x \quad \text{and} \quad E(CU_x) = CU_{x+h} = C.EU_x$$

(b) *The operators* Δ *and* E *are distributive, w.r.t. addition, e.g.,*

$$\Delta (U_x + V_x) = (U_{x+h} + V_{x+h}) - (U_x + V_x) = (U_{x+h} - U_x) + (V_{x+h} - V_x) = \Delta U_x + \Delta V_x$$

Similarly, $E(U_x + V_x) = U_{x+h} + V_{x+h} = EU_x + EV_x$

(c) *The operators* Δ *and* E *are linear, i.e.,* if a and b are constants, then

$$\Delta(aU_x + bV_x) = a\,\Delta U_x + b\,\Delta\, V_x \quad \text{and} \quad E(aU_x + bV_x) = aEU_x + bEV_x$$

These results are obvious from the properties (*a*) and (*b*).

(d) E *and* Δ *obey the law of indices, i.e.,*

$$E^m E^n U_x = E^{m+n} U_x \quad \text{and} \quad \Delta^m \Delta^n U_x = \Delta^{m+n} U_x$$

Proof. $E^m (E^n U_x) = E^m U_{x+nh} = U_{x+nh+mh} = U_{x+(m+n)h} = E^{m+n} U_x$

and

$$\Delta^m \Delta^n U_x = (E-1)^m (E-1)^n U_x$$
$$= [\{E^m - {}^mC_1 E^{m-1} + {}^mC_2 E^{m-2} - \dots + (-1)^m\}$$
$$\times \{E^n - {}^nC_1 E^{n-1} + {}^nC_2 E^{n-2} + \dots + (-1)^n\}]U_x$$
$$= [E^{m+n} - ({}^mC_1 + {}^nC_1)\,E^{m+n-1} + ({}^mC_2 + {}^nC_2 + {}^mC_1{}^nC_1)\,E^{m+n-2}$$
$$- ({}^mC_3 + {}^nC_3 + {}^mC_1\,{}^nC_2 + {}^mC_2\,{}^nC_1)\,E^{m+n-3} + \dots + (-1)^{m+n}]U_x$$

$${}^mC_2 + {}^nC_2 + {}^mC_1{}^nC_1 = \frac{m(m-1)}{2!} + \frac{n(n-1)}{2!} + mn$$
$$= \frac{m^2 - m + n^2 - n + 2mn}{2} = \frac{(m+n)^2 - (m+n)}{2}$$
$$= \frac{(m+n)(m+n-1)}{2!} = {}^{m+n}C_2$$

Similarly,

$${}^mC_3 + {}^nC_3 + {}^mC_1\,{}^nC_2 + {}^mC_2\,{}^nC_1$$
$$= \frac{m(m-1)(m-2)}{3!} + \frac{n(n-1)(n-2)}{3!} + \frac{mn(n-1)}{2!} + \frac{nm(m-1)}{2!} = {}^{m+n}C_3 \text{ (On simplification)}$$

$$\therefore \quad \Delta^m \Delta^n U_x = [E^{m+n} - {}^{m+n}C_1\,E^{m+n-1} + {}^{m+n}C_2 E^{m+n-2} - {}^{m+n}C_3 E^{m+n-3} + \dots + (-1)^{m+n}]\,U_x$$
$$= (E-1)^{m+n}\,U_x = \Delta^{m+n} U_x$$

Remarks:

1. The operators E and Δ are not cummutative with respect to a variable, *i.e.*,

$$E[U_x V_x] \neq U_x E V_x \quad \text{and} \quad \Delta\,[U_x V_x] \neq U_x\,\Delta V_x$$

Proof. $E[U_x V_x] = U_{x+h}\,V_{x+h} = EU_x\,EV_x$

and $\Delta U_x V_x = U_{x+h}\,V_{x+h} - U_x V_x = (U_{x+h} - U_x)\,V_{x+h} + U_x\,(V_{x+h} - V_x)$

$$= V_{x+h}\,\Delta U_x + U_x\,\Delta V_x$$

2. The operator Δ, operated on the quotient gives

$$\Delta\left(\frac{U_x}{V_x}\right) = \frac{U_{x+h}}{V_{x+h}} - \frac{U_x}{V_x} = \frac{U_{x+h}V_x - V_{x+h}U_x}{V_{x+h}V_x}$$
$$= \frac{(U_{x+h} - U_x)V_x - U_x(V_{x+h} - V_x)}{V_{x+h}V_x} = \frac{V_x\Delta U_x - U_x\Delta V_x}{V_x V_{x+h}}$$

3. Δ (Constant) = 0

Let U_x = (Constant) = k (say). Then, $\Delta U_x = U_{x+h} - U_x = k - k = 0$

Hence, $\Delta\,(\text{Constant}) = 0.$

4. We have

$$\Delta^n U_x = (\mathrm{E}-1)^n U_x = \left[\sum_{r=0}^{n} {}^n C_r E^r (-1)^{n-r}\right] U_x = \sum_{r=0}^{n}\left[(-1)^{n-r}\cdot {}^n C_r\, U_{x+rh}\right]$$

This result, known as the *"Lagrange form of the differences"*, enables us to compute the nth difference without completing the entire difference table.

19·2·4. Fundamental Theorem of Finite Differences

If U_x is a polynomial of nth degree in x , then

$$\Delta^r U_x = \begin{cases} \text{Constant}, & r = n \\ 0, & r > n \end{cases} \qquad \ldots (19\cdot4)$$

In other words, *the nth order difference of a polynomial of nth degree is constant and higher differences are zero.*

Proof. Let U_x be a polynomial of nth degree given by

$$U_x = a_0 x^n + a_1 x^{n-1} + a_2 x^{n-2} + \ldots + a_{n-1} x + a_n, \ (a_0 \neq 0)$$

where $a_0, a_1, a_2, \ldots, a_n$ are constants.

By def.,

$$\begin{aligned}
\Delta U_x &= U_{x+h} - U_x \\
&= \left[a_0 (x+h)^n + a_1 (x+h)^{n-1} + a_2 (x+h)^{n-2} + \ldots + a_{n-1} (x+h) + a_n\right] \\
&\qquad - (a_0 x^n + a_1 x^{n-1} + a_2 x^{n-2} + \ldots + a_{n-1} x + a_n) \\
&= a_0 (x^n + {}^n C_1 x^{n-1} h + {}^n C_2 x^{n-2} h^2 + \ldots + h^n) \\
&\quad + a_1 (x^{n-1} + {}^{n-1} C_1 x^{n-2} h + {}^{n-1} C_2 x^{n-3} h^2 + \ldots + h^{n-1}) \\
&\quad \vdots \\
&\quad + a_{n-2} (x^2 + 2xh + h^2) + a_{n-1} (x+h) + a_n - \{a_0 x^n + a_1 x^{n-1} + \ldots + a_{n-1} x + a_n\} \\
&= a_0 n h x^{n-1} + b_2 x^{n-2} + b_3 x^{n-3} + \ldots + b_{n-1} x + b_n,
\end{aligned}$$

where $b_2, b_3, \ldots, b_n$ are constants independent of x.

$\Rightarrow$ ΔU_x = A polynomial in x of degree $(n-1)$.

Thus, the first order difference of a polynomial of degree n is a polynomial of degree $(n-1)$.

$$\begin{aligned}
\Delta^2 U_x &= \left[a_0 nh (x+h)^{n-1} + b_2 (x+h)^{n-2} + \ldots + b_{n-1} (x+h) + b_n\right] \\
&\qquad - \left[a_0 nh x^{n-1} + b_2 x^{n-2} + \ldots + b_{n-1} x + b_n\right] \\
&= \{a_0 nh (x^{n-1} + {}^{n-1} C_1 x^{x-2} h + \ldots + h^{n-1}) \\
&\quad + b_2 (x^{n-2} + {}^{n-2} C_1 x^{n-3} h + \ldots + h^{n-2}) + \ldots + b_{n-1} (x+h) + b_n\} \\
&\qquad - \{a_0 nh x^{n-1} + b_2 x^{n-2} + \ldots + b_{n-1} x + b_n\} \\
&= a_0 n (n-1) h^2 x^{n-2} + c_3 x^{n-3} + c_4 x^{n-4} + \ldots + c_{n-1} x + c_n,
\end{aligned}$$

where $c_3, c_4, \ldots, c_n$ are constants independent of x.

Thus, $\Delta^2 U_x$ = A polynomial in x of $(n-2)^{\text{th}}$ degree.

Similarly proceeding, we shall get:

$$\Delta^n U_x = a_0 n (n-1)(n-2) \ldots 2.1\, h^n x^{n-n} = a_0 (n!) h^n = \text{Constant} = k \text{ (say)}, \qquad \ldots [19\cdot4(a)]$$

i.e., the nth order difference of a polynomial of nth degree is constant .

Then, $\Delta^{n+1}U_x = \Delta(k) = 0$ and $\Delta^{n+2}U_x = \dots = 0$

Hence, the theorem.

Remark.
$$\begin{aligned}&\Delta^n\left(a_0x^n + a_1x^{n-1} + \dots + a_{n-1}x + a_n\right)\\ &= \Delta^n(a_0x^n) + a_1\Delta^n x^{n-1} + \dots + a_{n-1}\Delta^n x + \Delta^n a_n\\ &= \Delta^n(a_0x^n) && \text{[Using (19·4)]}\\ &= a_0(n!)h^n && \text{[Using (19·4a)]}\end{aligned}$$

In particular, if the interval of differencing is unity, then

$$\Delta^n(a_0x^n) = a_0(n!) \quad\Rightarrow\quad a_0\Delta^n(x^n) = a_0(n!) \quad\Rightarrow\quad \Delta^n(x^n) = n! \qquad \text{[19·4 (b)]}$$

19·2·5. Relation Between E and D, where D = $\dfrac{d}{dx}$

By *def.*, we have

$$\begin{aligned}\mathrm{E}U_x &= U_{x+h} = U_x + h\,U_x' + \frac{h^2}{2!}U_x'' + \frac{h^3}{3!}U_x''' + \dots && \text{[By Taylor's theorem]}\\ &= U_x + h\,DU_x + \frac{h^2}{2!}D^2U_x + \frac{h^3}{3!}D^3U_x + \dots\\ &= \left\{1 + hD + \frac{(hD)^2}{2!} + \frac{(hD)^3}{3!} + \dots\right\}U_x = e^{hD}U_x\end{aligned}$$

Hence, $\mathrm{E} \equiv 1 + \Delta \equiv e^{hD}$ ... (19·5)

If the interval of differencing is taken as unity, then

$$\mathrm{E} \equiv 1 + \Delta \equiv e^D \quad\Rightarrow\quad D \equiv \log(1+\Delta) = \Delta - \frac{1}{2}\Delta^2 + \frac{1}{3}\Delta^3 + \dots \qquad \text{... [(19·(5a)]}$$

Example 19·3. *Evaluate*

(a) $\Delta^n(ax^n + bx^{n-1})$ (b) $\Delta^{10}\left[(1-ax)(1-bx^2)(1-cx^3)(1-dx^4)\right]$

(c) $\Delta\tan^{-1}x$, (d) $\Delta\log f(x) = \log\left\{1 + \dfrac{\Delta f(x)}{f(x)}\right\}$.

Solution:

(a) $\Delta^n(ax^n + bx^{n-1}) = \Delta^n(ax^n) + \Delta^n(bx^{n-1}) = a\Delta^n(x^n) + b\Delta^n(x^{n-1})$
$= a.n! + b.0 = a.n!$ [Using (19·4) and (19·4a)]

(b) Since for a polynomial of nth degree, $\Delta^r U_x = 0$, $(r > n)$, we get

$$\Delta^{10}\left[(1-ax)(1-bx^2)(1-cx^3)(1-dx^4)\right] = \Delta^{10}(abcd.x^{10}) = abcd\,\Delta^{10}(x^{10}) = abcd\,(10!)$$

(c) $$\Delta\tan^{-1}x = \tan^{-1}(x+h) - \tan^{-1}x = \tan^{-1}\left\{\frac{(x+h)-x}{1+(x+h)x}\right\} = \tan^{-1}\left(\frac{h}{1+hx+x^2}\right)$$

(d) $$\begin{aligned}\Delta\log f(x) &= \log f(x+h) - \log f(x)\\ &= \log\left\{\frac{f(x+h)}{f(x)}\right\} = \log\left\{\frac{\mathrm{E}f(x)}{f(x)}\right\} = \log\left\{\frac{(1+\Delta)f(x)}{f(x)}\right\}\\ &= \log\left\{\frac{f(x)+\Delta f(x)}{f(x)}\right\} = \log\left\{1 + \frac{\Delta f(x)}{f(x)}\right\}.\end{aligned}$$

Example 19·4. *Evaluate* $\frac{\Delta^2}{\mathrm{E}} \sin(x+h) + \frac{\Delta^2 \sin(x+h)}{\mathrm{E}\sin(x+h)}$ *, where interval of differencing is h.*

Solution. $\frac{\Delta^2}{\mathrm{E}} \sin[(x+h)] + \frac{\Delta^2 \sin(x+h)}{\mathrm{E}\sin(x+h)}$

$$= \left[\frac{(\mathrm{E}-1)^2}{\mathrm{E}}\right] \cdot \sin(x+h) + \frac{(\mathrm{E}-1)^2 \sin(x+h)}{\mathrm{E}\sin(x+h)}$$

$$= (\mathrm{E} - 2 + \mathrm{E}^{-1}) \sin(x+h) + \frac{(\mathrm{E}^2 - 2\mathrm{E} + 1)\sin(x+h)}{\mathrm{E}\sin(x+h)}$$

$$= [\sin(x+2h) - 2\sin(x+h) + \sin x] + \left[\frac{\sin(x+3h) - 2\sin(x+2h) + \sin(x+h)}{\sin(x+2h)}\right]$$

$$= 2\sin(x+h)(\cos h - 1) + \frac{2\sin(x+2h)(\cos h - 1)}{\sin(x+2h)}$$

$$= 2(\cos h - 1)\{\sin(x+h) + 1\}.$$

Example 19·5. *Use the method of separation of symbols to prove the following identities :*

(*a*) $U_0 + U_1 + \ldots + U_n = \binom{n+1}{1} U_0 + \binom{n+1}{2} \Delta U_0 + \binom{n+1}{3} \Delta^2 U_0 + \ldots + \binom{n+1}{n+1} \Delta^n U_0$ ·

(*b*) $U_1 x + U_2 x^2 + U_3 x^3 + \ldots = \frac{x}{(1-x)} U_1 + \frac{x^2}{(1-x)^2} \Delta U_1 + \frac{x^3}{(1-x)^3} \Delta^2 U_1 + \ldots$

(*c*) $U_0 + \frac{U_1 x}{1!} + \frac{U_2 x^2}{2!} + \frac{U_3 x^3}{3!} + \ldots = e^x \left(U_0 + x\Delta U_0 + \frac{x^2}{2!}\Delta^2 U_0 + \frac{x^3}{3!}\Delta^3 U_0 + \ldots\right)$.

Solution.

(*a*) L.H.S. $= U_0 + U_1 + U_2 + \ldots + U_n$

$$= U_0 + \mathrm{E}U_0 + \mathrm{E}^2 U_0 + \ldots + \mathrm{E}^n U_0 = (1 + \mathrm{E} + \mathrm{E}^2 + \ldots + \mathrm{E}^n)\, U_0$$

$$= \left(\frac{\mathrm{E}^{n+1} - 1}{\mathrm{E} - 1}\right) U_0 = \frac{1}{\Delta}\{(1+\Delta)^{n+1} - 1\} U_0$$

$$= \frac{1}{\Delta}\left[\left\{1 + \binom{n+1}{1}\Delta + \binom{n+1}{2}\Delta^2 + \binom{n+1}{3}\Delta^3 + \ldots + \Delta^{n+1}\right\} - 1\right] U_0$$

$$= \binom{n+1}{1} U_0 + \binom{n+1}{2} \Delta U_0 + \binom{n+1}{3} \Delta^2 U_0 + \ldots + \binom{n+1}{n+1} \Delta^n U_0$$

= R.H.S.

(*b*) L.H.S $= U_1 x + U_2 x^2 + U_3 x^3 + \ldots = U_1 x + x^2\, \mathrm{E} U_1 + x^3 \mathrm{E}^2 U_1 + \ldots$

$$= (x + x^2\mathrm{E} + x^3\mathrm{E}^2 + \ldots)\, U_1 = x\,\{1 + x\,\mathrm{E} + (x\,\mathrm{E})^2 + \ldots\}\, U_1$$

$$= x\left(\frac{1}{1 - x\mathrm{E}}\right) U_1 = x\,(1 - x\mathrm{E})^{-1}\, U_1$$

$$= x[1 - x(1+\Delta)]^{-1} U_1 = \frac{x}{1-x}\left(\frac{1 - x - x\Delta}{1-x}\right)^{-1} U_1$$

$$= \frac{x}{1-x}\left(1 - \frac{x\Delta}{1-x}\right)^{-1} U_1$$

$$= \frac{x}{1-x}\left\{1+\frac{x}{1-x}\Delta+\frac{x^2}{(1-x)^2}\Delta^2+\cdots\right\}U_1$$

$$= \frac{x}{1-x}U_1+\frac{x^2}{(1-x)^2}\Delta U_1+\frac{x^3}{(1-x)^3}\Delta^2U_1+\cdots = \text{R.H.S.}$$

(c) L.H.S. $= U_0+\frac{U_1x}{1!}+\frac{U_2x^2}{2!}+\frac{U_3x^3}{3!}+\cdots$

$$= U_0 + x\,\text{E}U_0 + \frac{x^2}{2!}\text{E}^2U_0 + \frac{x^3}{3!}\text{E}^3U_0+\ldots$$

$$= \left\{1+x\text{E}+\frac{(x\text{E})^2}{2!}+\frac{(x\text{E})^3}{3!}+\cdots\right\}U_0$$

$$= \{\exp(x\text{E})\}\,U_0 = \exp\{x(1+\Delta)\}U_0 = e^x\{\exp(x\Delta)\}U_0$$

$$= e^x\left\{1+x\Delta+\frac{(x\Delta)^2}{2!}+\frac{(x\Delta)^3}{3!}+\cdots\right\}U_0$$

$$= e^x\left\{U_0+x\,\Delta U_0+\frac{x^2}{2!}\Delta^2U_0+\frac{x^3}{3!}\Delta^3U_0+\ldots\right\} = \text{R.H.S.}$$

Example 19·6. *Use the method of separation of symbols to prove the following identities :*

(a) $U_x - U_{x+1} + U_{x+2} - U_{x+3} + \ldots$

$$= \frac{1}{2}\left[U_{x-(1/2)} - \frac{1}{8}\Delta\,U_{x-(3/2)} + \frac{1\cdot3}{2!}\left(\frac{1}{8}\right)^2\Delta\,U_{x-(5/2)} - \frac{1\cdot3\cdot5}{3!}\left(\frac{1}{8}\right)^3\Delta\,U_{x-(7/2)}+\cdots\right]$$

(b) $\Delta x^m - \frac{1}{2}\Delta^2x^m + \frac{1\cdot3}{2\cdot4}\Delta^3x^m - \frac{1\cdot3\cdot5}{2\cdot4\cdot6}\Delta^4x^m + \ldots$ *m terms* $= \left(x+\frac{1}{2}\right)^m - \left(x-\frac{1}{2}\right)^m$.

(c) $\Delta^nU_x = U_{x+n} - {}^nC_1U_{x+n-1} + {}^nC_2U_{x+n-2} - \ldots + (-1)^nU_x$.

Solution.

(a) R.H.S. $= \frac{1}{2}\left[\text{E}^{-1/2} - \frac{1}{8}\cdot\Delta^2\text{E}^{-3/2} + \frac{1\cdot3}{2!}\left(\frac{1}{8}\right)^2\cdot\Delta^4\text{E}^{-5/2} - \frac{1\cdot3\cdot5}{3!}\left(\frac{1}{8}\right)^3\cdot\Delta^6\text{E}^{-7/2}+\cdots\right]U_x$

$$= \frac{1}{2}\left[1-\frac{1}{8}\Delta^2\text{E}+\frac{1\cdot3}{2!}\cdot\left(\frac{1}{8}\right)^2(\Delta^2\text{E}^{-1})^2 - \frac{1\cdot3\cdot5}{3!}\left(\frac{1}{8}\right)^3(\Delta^2\text{E}^{-1})^3+\cdots\right]\text{E}^{-1/2}U_x$$

$$= \frac{1}{2}\left[1-\frac{1}{2}a+\frac{1\cdot3}{2!}\frac{a^2}{4}-\frac{1\cdot3\cdot5}{3!}\cdot\frac{a^3}{8}+\cdots\right]\text{E}^{-1/2}U_{x'}, \qquad \text{where } a = \frac{1}{4}\Delta^2\text{E}^{-1}$$

$$= \frac{1}{2}[(1+a)^{-1/2}]\text{E}^{-1/2}U_x = \frac{1}{2}\left[\left\{1+\frac{\Delta^2\text{E}^{-1}}{4}\right\}^{-1/2}\right]\text{E}^{-1/2}U_x$$

$$= \frac{1}{2}\left[\frac{4\text{E}+\Delta^2}{4\text{E}}\right]^{-1/2}\text{E}^{-1/2}U_x = \left[4(1+\Delta)+\Delta^2\right]^{-1/2}U_x$$

$$= \left[(2+\Delta)^2\right]^{-1/2}U_x = (1+\text{E})^{-1}\,U_x = (1-\text{E}+\text{E}^2-\text{E}^3+\ldots)\,U_x$$

$$= U_x - U_{x+1} + U_{x+2} - U_{x+3} + \ldots = \text{L.H.S.}$$

(b) Since $\Delta^r x^m = 0,\ r > m,$

the sum of the series up to m terms is the same as the sum to infinity. ... (*)

$\therefore$ L.H.S. $= \Delta x^m - \frac{1}{2}\Delta^2x^m + \frac{1\cdot3}{2\cdot4}\Delta^3x^m - \frac{1\cdot3\cdot5}{2\cdot4\cdot6}\Delta^4x^m + \ldots$ m terms

$$= \Delta\left(1 - \frac{\Delta}{2} + \frac{1\cdot3}{2\cdot4}\Delta^2 - \frac{1\cdot3\cdot5}{2\cdot4\cdot6}\Delta^3 + \cdots \text{ upto } m \text{ terms}\right)x^m$$

$$= \Delta\left(1 - \frac{\Delta}{2} + \frac{1\cdot3}{2\cdot4}\Delta^2 - \frac{1\cdot3\cdot5}{2\cdot4\cdot6}\Delta^3 + \ldots \textit{ ad infinity}\right)x^m \qquad \text{[Using (*)]}$$

$$= \Delta(1+\Delta)^{-1/2}\, x^m = \Delta E^{-1/2}\, x^m$$

$$= \Delta\left[\left(x - \frac{1}{2}\right)^m\right], \text{ the interval of differencing being unity.}$$

$$= \left(x + \frac{1}{2}\right)^m - \left(x - \frac{1}{2}\right)^m$$

(c) R.H.S. $= U_{x+n} - {}^nC_1 \cdot U_{x+n-1} + {}^nC_2 \cdot U_{x+n-2} - \ldots + (-1)^n U_x$

$= [E^n - {}^nC_1 E^{n-1} + {}^nC_2\, E^{n-2} + \ldots + (-1)^n]\, U_x$

$= (E-1)^n\, U_x = \Delta^n U_x =$ L.H.S.

Example 19·7. *Find the function U_x whose first difference is ${}^xC_n$.*

Solution. $\Delta\,({}^xC_n) = {}^{x+1}C_n - {}^xC_n$, the interval of differencing being unity.

But $\quad {}^xC_n + {}^xC_{n-1} = {}^{x+1}C_n \quad \Rightarrow \quad {}^{x+1}C_n - {}^xC_n = {}^xC_{n-1} \quad \Rightarrow \quad \Delta\,({}^xC_n) = {}^xC_{n-1}$

Thus, the operation of Δ on ${}^xC_n$ diminishes the lower suffix by unity.

Hence, $\quad \Delta\,({}^xC_{n+1}) = {}^xC_n$

and the required function is given by $\quad U_x = {}^xC_{n+1}.$

Example 19·8. *If p, q, r and s be the successive entries corresponding to equidistant arguments in a difference table, show that when third difference are taken into account, the entry corresponding to the argument halfway between the arguments of q and r is $A + \frac{1}{24}B$, where A is the arithmetic mean of q and r and B is the arithmetic mean of $3q - 2p - s$ and $3r - 2s - p$.*

Solution. Let the equidistant arguments be a, $a + h$, $a + 2h$ and $a + 3h$, h being the interval of differencing. Then the difference table is :

x	U_x	ΔU_x	$\Delta^2 U_x$	$\Delta^3 U_x$
a	p			
		$q-p$		
$a+h$	q		$r-2q+p$	
		$r-q$		$s-3r+3q-p$
$a+2h$	r		$s-2r+q$	
		$s-r$		
$a+3h$	s			

The argument half way between the arguments of q and r is $\frac{1}{2}(a + h + a + 2h)$, *i.e.*, $a + \frac{3}{2}h$.
Hence, the required entry is given by

$$U_{a+(3/2)h} = E^{3/2}\, U_a = (1+\Delta)^{3/2}\, U_a$$

$$= \left[1 + \frac{3}{2}\cdot\Delta + \frac{3}{2}\cdot\frac{1}{2}\cdot\frac{1}{2!}\Delta^2 + \left(\frac{3}{2}\right)\left(\frac{1}{2}\right)\left(-\frac{1}{2}\right)\cdot\frac{1}{3!}\Delta^3\right]U_a$$

higher differences being neglected.

$$\therefore \quad U_{a+(3/2)h} = U_a + \frac{3}{2}\,\Delta U_a + \frac{3}{8}\,\Delta^2 U_a - \frac{1}{16}\,\Delta^3 U_a$$

$$= p + \frac{3}{2}(q-p) + \frac{3}{8}(r-2q+p) - \frac{1}{16}(s-3r+3q-p)$$

$$= p\left(1 - \frac{3}{2} + \frac{3}{8} + \frac{1}{16}\right) + q\left(\frac{3}{2} - \frac{3}{4} - \frac{3}{16}\right) + r\left(\frac{3}{8} + \frac{3}{16}\right) - \frac{1}{16}s$$

$$= -\frac{1}{16}p + \frac{9}{16}q + \frac{9}{16}r - \frac{1}{16}s = -\frac{1}{16}p + (q+r)\left(\frac{1}{16} + \frac{1}{2}\right) - \frac{1}{16}s$$

$$= \frac{1}{2}(q+r) + \frac{1}{16}(q+r-p-s) \qquad \ldots (*)$$

Now, $\quad A$ = Arithmetic mean of q and $r = \dfrac{q+r}{2}$

and $\quad B$ = Arithmetic mean of $3q - 2p - s$ and $3r - 2s - p$

$$= \frac{1}{2}[3q - 2p - s + 3r - 2s - p] = \frac{3}{2}(q+r-p-s)$$

$$\therefore \quad A + \frac{1}{24}B = \frac{q+r}{2} + \frac{1}{16}(q+r-p-s)$$

Subsituting in (*), we get the required result.

Example 19·9. *Taking fifth order differences of U_x to be constant and given $U_0, U_1, U_2, \ldots, U_5$, prove that*

$$U_{2\frac{1}{2}} = \frac{1}{2}c + \frac{25(c-b) + 3(a-c)}{256}, \quad \text{where } a = U_0 + U_5, \quad b = U_1 + U_4, \quad c = U_2 + U_3.$$

Solution.

$$U_{2\frac{1}{2}} = U_{5/2} = \mathrm{E}^{5/2}U_0 = (1+\Delta)^{5/2}U_0$$

$$= \left\{1 + \frac{5}{2}\Delta + \frac{5}{2}\cdot\frac{3}{2}\cdot\frac{1}{2!}\cdot\Delta^2 + \frac{5}{2}\cdot\frac{3}{2}\cdot\frac{1}{2}\cdot\frac{1}{3!}\Delta^3 + \frac{5}{2}\cdot\frac{3}{2}\cdot\frac{1}{2}\left(-\frac{1}{2}\right)\frac{1}{4!}\Delta^4 + \frac{5}{2}\cdot\frac{3}{2}\cdot\frac{1}{2}\left(-\frac{1}{2}\right)\left(-\frac{3}{2}\right)\frac{1}{5!}\Delta^5\right\}U_0 \quad \text{(Neglecting higher differences)}$$

$$\therefore \quad U_{2\frac{1}{2}} = U_0 + \frac{5}{2}\Delta U_0 + \frac{15}{8}\Delta^2 U_0 + \frac{5}{16}\Delta^3 U_0 - \frac{5}{128}\Delta^4 U_0 + \frac{3}{256}\Delta^5 U_0$$

$$= U_0 + \frac{5}{2}(\mathrm{E}-1)U_0 + \frac{15}{8}(\mathrm{E}-1)^2 U_0 + \frac{5}{16}(\mathrm{E}-1)^3 U_0 - \frac{5}{128}(\mathrm{E}-1)^4 U_0 + \frac{3}{256}(\mathrm{E}-1)^5 U_0$$

$$= U_0 + \frac{5}{2}(U_1 - U_0) + \frac{15}{8}(U_2 - 2U_1 + U_0) + \frac{5}{16}(U_3 - 3U_2 + 3U_1 - U_0) - \frac{5}{128}(U_4 - 4U_3 + 6U_2 - 4U_1 + U_0) + \frac{3}{256}(U_5 - 5U_4 + 10U_3 - 10U_2 + 5U_1 - U_0)$$

$$= \frac{3}{256}(U_0 + U_5) - \frac{25}{256}(U_1 + U_4) + \frac{75}{128}(U_2 + U_3) \qquad \text{(On simplification)}$$

$$= \frac{3a}{256} - \frac{25b}{256} + \frac{75c}{128} = \frac{3a}{256} - \frac{25b}{256} + \left(\frac{1}{2} + \frac{11}{128}\right)c$$

$$= \frac{1}{2}c + \frac{3a - 25b + 22c}{256} = \frac{1}{2}c + \frac{3a - 25b + 25c - 3c}{256}$$

$$= \frac{1}{2}c + \frac{3(a-c) + 25(c-b)}{256} \quad \text{where, } a = U_0 + U_5, \quad b = U_1 + U_4, \quad c = U_2 + U_3.$$

19·2·6. Missing terms (Equal intervals)

Sometimes we may be given a set of equidistant terms with some terms (one or two or more) missing. The problem of estimating such terms can be easily tackled by the use of the operators E and Δ.

Let us suppose that we are given $(n + 1)$ equidistant arguments, $(x = 0, 1, 2, \ldots, n$, say) but the entry (U_r) corresponding to any one of them say $(r + 1)^{th}$ argument is not given and we want to estimate that. Since we are given n entries, the data can be represented by a polynomial of $(n - 1)^{th}$ degree. Hence we may take U_x as a polynomial of degree $(n - 1)$.

$$\therefore \quad \Delta^{n-1} U_x = \text{constant} \quad \text{and} \quad \Delta^n U_x = 0 \,;\, x = 0, 1, 2, \ldots, n \qquad (*)$$

In particular $\Delta^n U_0 = 0$, *i.e.*, $(E - 1)^n U_0 = 0$

$$\Rightarrow \quad [E^n - {}^nC_1 E^{n-1} + {}^nC_2 E^{n-2} + \ldots (-1)^n] U_0 = 0$$

$$\Rightarrow \quad U_n - {}^nC_1 U_{n-1} + {}^nC_2 U_{n-2} + \ldots + (-1)^n U_0 = 0$$

From this equation, the missing entry can be easily calculated.

If, in a set of $(n + 2)$ equidistant arguments, two entries U_r and U_s are missing, then we have from (*),

$$\Delta^n U_0 = 0 \quad \text{and} \quad \Delta^n U_1 = 0$$

$$\Rightarrow \quad (E - 1)^n U_0 = 0 \quad \text{and} \quad (E - 1)^n U_1 = 0 \qquad (**)$$

Expanding and simplifying as above, the two missing terms can be estimated by solving equations (**).

Similarly, the three missing terms can be estimated by solving the equations :

$$\Delta^n U_0 = 0, \quad \Delta^n U_1 = 0 \quad \text{and} \quad \Delta^n U_2 = 0 \text{ and so on.}$$

Example 19·10. *Estimate U_2 from the following table :*

x :	1	2	3	4	5
U_x :	7	?	13	21	37

and explain why the value obtained is different from that obtained by putting $x = 2$ in the expression $2^x + 5$.

Solution. Since we are given four entries, *viz.*, U_1, U_3, U_4 and U_5, the function U_x can be represented by a third degree polynomial.

Hence, $\Delta^3 U_x = \text{constant}$ and $\Delta^4 U_x = 0$ (*)

In particular : $\Delta^4 U_1 = 0$, *i.e.*, $(E - 1)^4 U_1 = 0$

$$\Rightarrow \quad (E^4 - 4E^3 + 6E^2 - 4E + 1) U_1 = 0$$

$$\Rightarrow \quad U_5 - 4U_4 + 6U_3 - 4U_2 + U_1 = 0$$

$$\Rightarrow \quad 37 - 4 \times 21 + 6 \times 13 - 4U_2 + 7 = 0 \quad \Rightarrow U_2 = \frac{38}{4} = 9{\cdot}5.$$

Aliter. *From (*) onwards*

Let $U_2 = a$. The difference table is as follows :

x	U_x	ΔU_x	$\Delta^2 U_x$	$\Delta^3 U_x$	$\Delta^4 U_x$
1	7				
		$a - 7$			
2	a		$20 - 2a$		
		$13 - a$		$3a - 25$	
3	13		$a - 5$		$38 - 4a$
		8		$13 - a$	
4	21		8		
		16			
5	37				

Since $\Delta^4 U_1 = 0$, $\quad 38 - 4a = 0 \quad \Rightarrow \quad a = \frac{38}{4} = 9{\cdot}5.$

Verification. $\quad U_x = 2^x + 5 \Rightarrow U_2 = 2^2 + 5 = 9$

The estimated value of U_2, *viz.*, 9·5 is different from the actual value, *i.e.*, 9. The reason for this is given below.

Here we are given that the function U_x is of the form $U_x = a^x + b$

$$\therefore \quad \Delta U_x = U_{x+1} - U_x = (a^{x+1} + b) - (a^x + b) = a^x (a - 1)$$

$$\Delta^2 U_x = \Delta \left[a^x (a-1)\right] = (a-1)\,\Delta (a^x) = (a-1)\left[a^x (a-1)\right] = (a-1)^2 a^x$$

In general, we get : $\quad \Delta^r U_x = (a-1)^r a^x \quad$ (**)

While estimating U_2, we assumed that U_x can be expressed as a polynomial of third degree and hence fourth difference is zero. But in the given case U_x is not a polynomial but an exponential function and $\Delta^r U_x \neq 0$ for any r [c.f. (**)]. Since the basic assumption of interpolation (c.f. § 19·7) is violated, we get the difference in the actual and estimated value of U_2.

Example 19·11.

(*a*) *The number of members of a certain society are as given in the following table :*

Year x :	*1990*	*1991*	*1992*	*1993*	*1994*	*1995*
Number U_x :	*845*	*867*	—	*846*	*821*	*772*
Year x :	*1996*	*1997*	*1998*	*1999*		
Number U_x :	—	*757*	*761*	*796*		

Make the best estimate you can, of the member in 1992 and 1996

(*b*) *Given :*

$U_o + U_8 = 1{\cdot}9243, \quad U_1 + U_7 = 1{\cdot}9590, \quad U_2 + U_6 = 1{\cdot}9843, \quad U_3 + U_5 = 1{\cdot}9956,$ *find* U_4.

Solution.

(*a*) Since we are given 8 entries, U_x can be represented by a seventh degree polynomial and hence $\quad \Delta^7 U_x$ = constant and $\Delta^8 U_x = 0. \quad \forall\, x$ ... (1)

If we take 1990 as origin, then we have

$U_0 = 845, \quad U_1 = 867, \quad U_2 = ?, \quad U_3 = 846, \quad U_4 = 821,$

$U_5 = 772, \quad U_6 = ?, \quad U_7 = 757, \quad U_8 = 761 \quad$ and $\quad U_9 = 796$

From (1), we have in particular : $\Delta^8 U_0 = 0$ and $\Delta^8 U_1 = 0$

$\Delta^8 U_0 = 0, \quad$ gives $\quad (\mathrm{E} - 1)^8 U_0 = 0$ ... (2)

$$\Rightarrow \left[\mathrm{E}^8 - {}^8C_1\mathrm{E}^7 + {}^8C_2\mathrm{E}^6 - {}^8C_3\mathrm{E}^5 + {}^8C_4\mathrm{E}^4 - {}^8C_5\mathrm{E}^3 + {}^8C_6\mathrm{E}^2 - {}^8C_7\mathrm{E} + 1\right] U_0 = 0$$

$$\Rightarrow \left[\mathrm{E}^8 - 8\mathrm{E}^7 + 28\mathrm{E}^6 - 56\mathrm{E}^5 + 70\mathrm{E}^4 - 56\mathrm{E}^3 + 28\mathrm{E}^2 - 8\mathrm{E} + 1\right] U_0 = 0 \quad \text{... (3)}$$

$$\Rightarrow (U_8 + U_0) - 8\,(U_7 + U_1) + 28\,(U_6 + U_2) - 56\,(U_5 + U_3) + 70U_4 = 0 \quad \text{... (4)}$$

$$\Rightarrow 28\,(U_2 + U_6) - 44524 = 0 \quad \textit{(On simplification)} \quad \text{... (5)}$$

Similarly, $\quad \Delta^8 U_1 = 0$, gives

$$(U_9 + U_1) - 8\,(U_8 + U_2) + 28\,(U_7 + U_3) - 56\,(U_6 + U_4) + 70U_5 = 0$$

$$\Rightarrow 8U_2 + 56U_6 - 48523 = 0 \quad \textit{(On simplification)} \quad \text{... (6)}$$

Solving (5) and (6) for U_2 and U_6, we get : $\quad U_2 = 844{\cdot}2708$ and $U_6 = 745{\cdot}8720$

Hence, the number of members of the society for the years 1992 and 1996 are 844 and 746 , respectively. [Note that the number of members can not be in fractions].

(*b*) Using the given data and the equation (4) in Part (*a*), we get

$$1{\cdot}9243 - 8 \times 1{\cdot}9590 + 28 \times 1{\cdot}9823 - 56 \times 1{\cdot}9956 + 70U_4 = 0$$

$$\Rightarrow \quad -69{\cdot}9969 + 70U_4 = 0 \quad \Rightarrow \quad U_4 = 0{\cdot}9999557.$$

19·2·7. Factorial Notation. The product of n consecutive factors each at a constant difference, say, h, the first factor being x, is called a *factorial function* and is denoted by $x^{(n)}$. Thus

$$x^{(n)} = x(x-h)(x-2h)\ldots\left(x-\overline{n-1}\,h\right) \qquad \ldots (19·6)$$

In particular, if $h = 1$

$$x^{(n)} = x(x-1)(x-2)\ldots(x-n+1) \qquad \ldots [19·6\,(a)]$$

$\Delta x^{(n)} = (x+h)^{(n)} - x^{(n)}$, h being the interval of differencing.

$$\therefore \quad \Delta x^{(n)} = (x+h)x(x-h)\ldots\left(x+h-\overline{n-1}\,h\right) - x(x-h)\ldots\left(x-\overline{n-1}\,h\right)$$

$$= x(x-h)(x-2h)\ldots\left(x+h-\overline{n-1}\,h\right)\left[(x+h)-\left(x-\overline{n-1}\,h\right)\right]$$

$$= nh.x(x-h)(x-2h)\ldots\left(x-\overline{n-2}\,h\right) = nhx^{(n-1)} \qquad \ldots (19·7)$$

Similarly,

$$\Delta^2 x^{(n)} = \Delta\left[\Delta x^{(n)}\right] = \Delta\left[nh\,x^{(n-1)}\right] = nh\left[\Delta x^{(n-1)}\right] = nh(n-1)\,h\cdot x^{(n-2)} \qquad \ldots [19·7\,(a)]$$

In general, we get

$$\Delta^n x^{(n)} = n(n-1)(n-2)\ldots 3.2.1\,h^n = (n\,!)\,h^n \qquad \ldots [19·7\,(b)]$$

In particular if the interval of differing is taken as unity, then from (19·7), (19·7*a*) and (19·7*b*), we get respectively

$$\Delta x^{(n)} = nx^{(n-1)}, \quad \Delta^2 x^{(n)} = n(n-1)\,x^{(n-2)}, \ldots, \quad \text{and} \quad \Delta^n x^{(n)} = n\,!,$$

These results lead to the following very important conclusion :

"*In case of factorial notation, the operator Δ is equivalent to differentiation if the interval of differencing is unity." i.e.,*

$$\text{if } h = 1 \text{ then } \Delta \equiv D, \text{ where } D = \frac{d}{dx}. \qquad \ldots (19·7c)$$

Also, we have

$$x^{(n)} = x(x-h)(x-2h)\ldots\left(x-\overline{n-2}\,h\right)\left(x-\overline{n-1}\,h\right) \Rightarrow x^{(n)} = \left(x-\overline{n-1}\,h\right)x^{(n-1)} \qquad \ldots (19·8)$$

In particular, for $n = 0$, $\quad x^{(0)} = (x+h)\,x^{(-1)}$

But by convention

$$x^{(0)} = 1 \quad \Rightarrow \quad x^{(-1)} = \frac{1}{x+h} \qquad \ldots [19·8\,(a)]$$

Again, from (19·8) for $n = -1$, $\quad x^{(-1)} = (x+2h)\,x^{(-2)}$

$$\therefore \quad x^{(-2)} = \frac{x^{(-1)}}{x+2h} = \frac{1}{(x+h)(x+2h)} \text{[From 19·8}(a)] \qquad \ldots [19·8\,(b)]$$

In general, we will get

$$x^{(-n)} = \frac{1}{(x+h)(x+2h)\cdots(x+nh)} \qquad \ldots [19·8\,(c)]$$

If the interval of differencing is unity, then we have from [(19·8(c)]

$$x^{(-n)} = \frac{1}{(x+1)(x+2)\cdots(x+n)} \qquad \ldots [19·8\,(d)]$$

Also

$$\Delta x^{(-n)} = (x+h)^{(-n)} - x^{(-n)}$$

$$= \frac{1}{(x+2h)(x+3h)\cdots(x+nh)(x+(n+1)h)} - \frac{1}{(x+h)(x+2h)(x+3h)\cdots(x+nh)}$$

$$= \frac{1}{(x+2h)(x+3h)\cdots(x+nh)}\left[\frac{1}{(x+\overline{n+1}h)} - \frac{1}{x+h}\right]$$

$$= \frac{-nh}{(x+h)(x+2h)\cdots(x+nh)(x+\overline{n+1}h)}$$

$$\Rightarrow \quad \Delta x^{(-n)} = -\,-nh\,x^{(-\overline{n+1})} = -\,nh\,x^{(-n-1)} \qquad \ldots (19{\cdot}9)$$

$$\Delta^2 x^{(-n)} = \Delta\left[-nhx^{(-n-1)}\right] = (-nh)\,\Delta\left[x^{(-n-1)}\right] = (-nh)\,(-n-1)\,h\,x^{(-n-2)}$$

$$= (-1)^2\,n\,(n+1)\,h^2 x^{(-n-2)} \qquad \ldots [19{\cdot}9a]$$

and so on

If $h = 1$, we get from (19·9) and (19·9*a*)

$$\Delta x^{(-n)} = -n\,x^{(-n-1)}\ , \qquad \Delta^2 x^{(-n)} = (-n)\,(-n-1)\,x^{(-n-2)}, \text{ and so on.} \qquad \ldots [19{\cdot}9\,(b)]$$

i.e., In the case of negative index for factorial function also, the operator $\Delta \equiv D$, the result obtained in (19·7*c*).

Example 19·12.

(*a*) *Represent the function :* $x^4 - 12x^3 + 42x^2 - 30x + 9$ *and its successive differences in factorial notation.*

(*b*) *Obtain the function whose first difference is :* $x^3 + 4x^2 + 9x + 12$

Solution. (*a*) Using the method of synthetic division (method of detached co-efficients), we get

1	1	−12	42	−30	9
	0	1	−11	31	
2	1	−11	31	1	
	0	2	−18		
3	1	−9	13		
	0	3			
4	1	−6			
	0				
	1				

Thus,

$$U_x = x^4 - 12x^3 + 42x^2 - 30x + 9$$

$$\equiv x^{(4)} - 6x^{(3)} + 13x^{(2)} + x^{(1)} + 9$$

Since in the factorial notation, the operator Δ is equivalent to differentiation, we get:

$$\Delta U_x = 4x^{(3)} - 18x^{(2)} + 26x + 1$$

$$= 4x\,(x-1)\,(x-2) - 18x\,(x-1) + 26x + 1$$

$$= 4x^3 - 30x^2 + 52x + 1$$

$$\Delta^2 U_x = 12x^{(2)} - 36x + 26 = 12x^2 - 48x + 26$$

$$\Delta^3 U_x = 24x - 36 \quad \text{and} \quad \Delta^4 U_x = 24$$

Remarks:

1. While using the method of synthetic division (method of detached coefficients), we should first of all see whether the given expression is complete or not. If any power of x is missing, we should first make the expression complete by taking its coefficient zero and then employ the method of synthetic division.

2. The reader is advised to use the direct method, *viz.*, by taking

$$U_x = x^4 - 12x^3 + 42x^2 - 30x + 9 \equiv x^{(4)} + Ax^{(3)} + Bx^{(2)} + Cx + 9$$

and calculating the coefficients A, B and C by comparing coefficients of like powers of x on both sides. It may be pointed out that the coefficients of the highest power and the constant remain unchanged while transforming to factorial functions. This point reduces algebra to some extent.

(*b*) We are given : $\Delta U_x = x^3 + 4x^2 + 9x + 12$

$\therefore \quad \Delta U_x = x^{(3)} + 7x^{(2)} + 14x^{(1)} + 12$

1	1	4	9	12
	0	1	5	
2	1	5	14	
	0	2		
3	1	7		
	0			
	1			

Since in factorial notation,

$\Delta \equiv$ differentiation, $1/\Delta \equiv$ integration.

Hence, $U_x = \dfrac{1}{\Delta}\{x^{(3)} + 7x^{(2)} + 14x^{(1)} + 12\}$

$$= \frac{x^{(4)}}{4} + 7\frac{x^{(3)}}{3} + 14\frac{x^{(2)}}{2} + 12x + k,$$

where k is an arbitrary constant.

Example 19·13. *Find the relation between α, β, γ in order that $\alpha + \beta x + \gamma x^2$ may be expressible in one term in the factorial notation.*

Solution. Let $\alpha + \beta x + \gamma x^2 \equiv \gamma\,(x + A)^{(2)} \equiv \gamma\,(x + A)\,(x + A - 1)$

$$\equiv \gamma\,[x^2 + (2A - 1)x + A\,(A - 1)]$$

Comparing co-efficients, of like powers of x on both sides we get

$$\beta = (2A - 1)\gamma \quad \Rightarrow \quad (2A-1) = \frac{\beta}{\gamma} \quad \Rightarrow \quad A = \frac{\beta + \gamma}{2\gamma}$$

and $\quad \alpha = A\,(A - 1)\,\gamma \quad \Rightarrow \quad \alpha = \dfrac{\beta + \gamma}{2\gamma} \times \dfrac{\beta - \gamma}{2\gamma} = \dfrac{\beta^2 - \gamma^2}{4\gamma}$

Hence, $\quad \beta^2 - \gamma^2 = 4\alpha\gamma$, is the required relation.

19·3. INTERPOLATION

Suppose we are given the values of a function U_x, *i.e.*, the entries, for a set of values of the independent variable x, *i.e.*, the arguments. The *interpolation* is defined as *the technique of estimating the value of U_x for any intermediate value of the argument.* Theile defines interpolation as *"the art of reading between the lines of a table."*

Let us suppose we are given the census figures for the population of India for four years 1961 , 1971, 1981 and 1991 and we want to estimate the figure for any intermediate year, *e.g.*, for 1975 or 1988, etc. This can be done by applying the technique of interpolation. The interpolation may be defined as the *"technique of obtaining the most likely estimate of a certain quantity under certain assumptions."*

If the form of the function U_x is known, we can evaluate U_x for any value of x by simple substitution. But, in case the form of the function is unknown or if it is quite complicated, then the problem of determining the nature of the function or replacing the given function by a comparatively simpler one is also the problem of what is usually called *parabolic or polynomial interpolation*.

Remarks:

1. If we have to estimate the value of U_x for any value of the argument outside the given range of the arguments, the technique is known as *extrapolation*. For example, the estimation of the population figure for the year 1995 in the above example falls in the domain of extrapolation.

2. It should be clearly understood that there is no difference between interpolation and extrapolation as far as estimation methods and underlying assumptions are concerned. The only difference between the two is that interpolation relates to estimation of a value within the given range of the series while extrapolation deals with obtaining the forecast or projections (in the past or future) beyond the given range of the series.

"Interpolation is the estimation of a most likely estimate in given conditions. The technique of estimating a past figure is termed as interpolation, while that of estimating a probable figure for the future is called extrapolation."
— *Hirsch*

"There is no difference between interpolation and extrapolation so far as the methods are concerned but for distinguishing past from the future we give them two different names. Interpolation relates to the past where as extrapolation gives us the forecast for the future."
— *Frederick Mills*

19·3·1. Assumptions of Interpolation

The following are the fundamental assumptions in any method of interpolation :

(i) *There are no sudden jumps or falls in the values of the entries for the period under consideration.* In other words, the given data does not refer to abnormal periods such as period of famines, wars, epidemics, etc., which may result in sudden changes in the values of U_x. Mathematically, it means that the data can be represented by a smooth or continuous curve which implies that given data can be represented by a polynomial of certain degree, which is determined by the following theorem.

One and only one polynomial curve of degree less than or equal to n passes through a given set of (n + 1) distinct points.

All the formulae of interpolation are based upon the fundamental assumption that the data is expressible or can be expressed as a polynomial function, with fair degree of accuracy.

(ii) *In the absence of the evidence to the contrary, the rise or fall in the data has been uniform.*

19·3·2. Uses of Interpolation

1. The need for interpolating missing observations or making forecasts or projections arises in a number of disciplines like economics, business, social sciences, actuarial work, population studies, etc. Some of the examples are ascertaining the most likely prices, business changes, mid-term intercensal population figures, mid-term figures of industrial production from their totals and so on.
2. The interpolation technique has been used to derive the formulae for the computation of median and mode in case of continuous frequency distribution. (See formulae in Chapter 2.)
3. Interpolation techniques are used to fill in the gaps in the statistical data for the sake of continuity of information. These gaps in the data may be due to the following reasons :
   (i) Due to certain financial and organisational difficulties, data may not be collected on census basis and sampling techniques may be used to obtain the relevant information. The intermediate gaps are then filled by interpolation methods.
   (ii) Data for some periods may not be collected due to certain unavoidable circumstances.
   (iii) Figures of some of the periods may be erased, destroyed or lost due to certain reasons like improper handling or random and natural causes like fire, floods, etc.

Interpolation techniques help us to obtain the best (most likely) substitutes for the original missing values under certain assumptions discussed in § 19·3·1 and these methods are entirely different from those by which actual data are obtained.

Remark. *Accuracy of Estimates.* Since the interpolation techniques are based on certain assumptions which may not hold good in practice, the estimates so obtained may not always be accurate or reliable. It is not possible to ascertain the error of estimate. The accuracy of the interpolated value depends on :

(*i*) A knowledge of possible fluctuations in the values of the phenomenon under study, which is provided by the available data at our disposal.

(*ii*) A knowledge about the course of events which affect the value of the phenomenon under investigation. If we know that the estimated value of the given phenomenon at a particular period is affected by random factors like political riots, floods, etc., then the interpolated value should be modified in the light of this information and a better estimate may be obtained.

19·3·3. Method of Parabolic Curve Fitting

The form of function $y = f(x)$ or its estimate for any given value of x can be obtained by fitting a polynomial curve to the given set of observations provided the values of x (arguments) are at equal intervals. The method is based on the fundamental theorem of algebra *viz.*, " *One and only one polynomial curve of degree less than or equal n passes through a given set of (n + 1) distinct points.*" Thus if we are given $(n + 1)$ equidistant arguments and entries then we can represent the function $y = f(x)$ by a polynomial of nth degree,

viz.,
$$y = f(x) = a_0 x^n + a_1 x^{n-1} + a_2 x^{n-2} + \ldots + a_{n-1} x + a_n \quad \ldots (*)$$

where $a_0, a_1, \ldots, a_n$ are $(n + 1)$ constants whose values are to be determined from the $(n+1)$ equations obtained on substituting the given values of x and $y = f(x)$ in (*). Solving the $(n+1)$ equations so obtained and substituting the values of $a_0, a_1, \ldots, a_n$ in (*) we get the required form of function $y = f(x)$, which can then be used to estimate y for any given value of x. We shall explain the technique by means of an example.

Remark. The method of parabolic curve fitting is quite time consuming and tedious, particularly when the number of entries given is large. For example, if we are given 5 entries, then $y = f(x)$ can be represented by a polynomial of 4th degree which involves 5 unknown constants. To determine these 5 constants we have to solve simultaneous equations in the 5 unknowns which is quite a difficult job. Moreover, this method is applicable only if the arguments are at equal intervals and cannot be applied for unequal intervals.

Example 19·14. *Find $f(x)$, given that $f(0) = -3$, $f(1) = 6$, $f(2) = 8$, $f(3) = 12$. State your assumption, if any. Hence find $f(6)$.*

Solution. Since we are given 4 entries, we can approximate $f(x)$ by a polynomial of 3rd degree, say,
$$f(x) = ax^3 + bx^2 + cx + d. \quad \ldots (*)$$

This involves 4 known constants, a, b, c and d.

Putting $x = 0, 1, 2,$ and 3 in (*) we get respectively :

$$f(0) = d = -3, \qquad f(1) = a + b + c + d = 6$$

$$f(2) = 8a + 4b + 2c + d = 8, \qquad f(3) = 27a + 9b + 3c + d = 12$$

Solving the above equations, we get $a = \frac{3}{2}$, $b = -8$, $c = \frac{31}{2}$ and $d = -3$.

Finally, substituting the values of a, b, c and (d) in (*), we get the form of function $f(x)$ as :

$$f(x) = \frac{3}{2} x^3 - 8x^2 + \frac{31}{2} x - 3 \quad \ldots (**)$$

Putting $x = 6$ in (**), we get $f(6) = \frac{3}{2} \times 6^3 - 8 \times (6)^2 + \frac{31}{2} \times 6 - 3 = 126.$

19·3·4. Gregory Newton's Forward Interpolation Formula for Equal Intervals.

$$U_{a+xh} = U_a + {}^xC_1\,\Delta U_a + {}^xC_2\,\Delta^2 U_a + \ldots + {}^xC_r\,\Delta^r U_a + \ldots \qquad \ldots (19{\cdot}10)$$

where 'a' is the first argument and h is the common interval of differencing, last term depending on the degree of the polynomial U_x.

Proof. By def.,

$$\begin{aligned} U_{a+xh} &= E^x U_a = (1+\Delta)^x\, U_a \\ &= (1 + {}^xC_1\Delta + {}^xC_2\Delta^2 + {}^xC_3\Delta^3 + \ldots + {}^xC_r\Delta^r + \ldots + \Delta^x)\,U_a \\ &= U_a + {}^xC_1\Delta U_a + {}^xC_2\Delta^2 U_a + \ldots + {}^xC_r\Delta^r U_a + \ldots + \Delta^x U_a \end{aligned}$$

which is the required formula.

Aliter. Assuming that the entry U_{a+xh} can be approximated by a polynomial of nth degree and remembering that every polynomial can be expressed as a factorial function, we may take

$$U_{a+xh} = A_0 + A_1 x^{(1)} + A_2 x^{(2)} + A_3^{(3)} + \ldots + A_r x^{(r)} + \ldots + A_n x^{(n)}, \qquad (*)$$

where $\quad x^{(r)} = x(x-1)(x-2)\ldots(x-r+1)$

$$\therefore \quad \left[\Delta^r (U_{a+xh})\right]_{x=0} = \Delta^r [A_r x^{(r)}] = A_r\,(r\,!) \qquad (\because\ \Delta \equiv D \text{ in case of factorial notation})$$

$$\therefore \quad A_r = \frac{\Delta^r U_a}{r!}\,;\ r = 1, 2, \ldots, n$$

Also from (*) on putting $x = 0$, we get: $U_a = A_0$

$$\text{Hence,} \quad U_{a+xh} = U_a + x^{(1)}\,\Delta\, U_a + \frac{x^{(2)}}{2!}\,\Delta^2 U_a + \ldots + \frac{(\ )}{\ }\,\Delta^r U_a + \ldots$$

$$= U_a + {}^xC_1\Delta U_a + {}^xC_2\,\Delta^2 U_a + {}^xC_3\,\Delta^3 U_a + \ldots + {}^xC_r\Delta^r U_a + \ldots \qquad \ldots [19{\cdot}10a]$$

Remarks:

1. The formula (19·10) is mainly used while interpolating the value of U_x for the value of x slightly greater than a or for extrapolating U_x for value of x slightly less than a.
2. In particular, if in (19·10) we take $a = 0$ and $h = 1$, we get

$$U_x = U_0 + {}^xC_1\,\Delta U_0 + {}^xC_2\,\Delta^2 U_0 + \ldots + {}^xC_n\,\Delta^n U_0 + \ldots \qquad \ldots [19{\cdot}10\,b]$$

a form known as the *Gregory-Newton's Advancing (Forward) Differencing Formula.*

Example 19·15. *From the following table of yearly premiums for policies maturing at quinquenial ages, estimate the premiums for policies maturing at the ages 46 years.*

Age (x) :	45	50	55	60	65
Premium (U_x) :	2·871	2·404	2·083	1·862	1·712

Solution.

TABLE OF DIFFERENCES

S. No.	*Age x*	*Premium* U_x	ΔU_x	$\Delta^2 U_x$	$\Delta^3 U_x$	$\Delta^4 U_x$
1	45	2·871				
			– 0·467			
2	50	2·404		0·146		
			– 0·321		– 0·046	
3	55	2·083		0·100		0·017
			– 0·221		– 0·029	
4	60	1·862		0·071		
			– 0·150			
5	65	1·712				

We want $U_{46} = U_{a+xh}$, (say).

$$\therefore \quad a + xh = 46 \quad \Rightarrow \quad 45 + x \times 5 = 46, \quad \Rightarrow \quad x = \frac{1}{5} = 0{\cdot}2$$

Substituting $x = \frac{1}{5}$ in (19·10), we get:

$$U_{46} = U_{45} + \frac{1}{5}\,\Delta U_{45} + \frac{1}{5}\left(-\frac{4}{5}\right)\frac{1}{2!}\,\Delta^2 U_{45} + \frac{1}{5}\left(-\frac{4}{5}\right)\left(-\frac{9}{5}\right) \times \frac{1}{3!}\,\Delta^3 U_{45}$$

$$+ \frac{1}{5}\left(-\frac{4}{5}\right)\left(-\frac{9}{5}\right)\left(-\frac{14}{5}\right)\frac{1}{4!}\,\Delta^4 U_{45}$$

$$= 2{\cdot}871 + \frac{1}{5}\,(-0{\cdot}467) - \frac{2}{25}\,(0{\cdot}146) + \frac{6}{125}\,(-0{\cdot}046) - \frac{21}{625}\,(0{\cdot}017)$$

$$= 2{\cdot}871 - 0{\cdot}934 - 0{\cdot}02336 - 0{\cdot}0013248 - 0{\cdot}000506 = 2{\cdot}753 \text{ (approx.)}$$

Example 19·16. *From the following table, estimate the number of students who obtained marks between 40 and 45.*

Marks :	*30–40*	*40–50*	*50–60*	*60–70*	*70–80*
No. of students :	*31*	*42*	*51*	*35*	*31*

Solution. We first prepare the cumulative frequency table, as given below:

Marks less than (x) :	40	50	60	70	80
No. of students (U_x) :	31	73	124	159	190

TABLE OF DIFFERENCES

x	U_x	ΔU_x	$\Delta^2 U_x$	$\Delta^3 U_x$	$\Delta^4 U_x$
40	31				
		42			
50	73		9		
		51		–25	
60	124		– 16		37
		35		12	
70	159		– 4		
		31			
80	190				

We will find : U_{45} = Number of students with marks less than 45.

Taking $a + xh = 45$, we get : $40 + x \times 10 = 45 \quad \Rightarrow \quad x = \frac{1}{2}$

$\therefore$ Using Newton – Gregory formula, we get

$$U_{45} = 31 + \frac{1}{2} \times 42 + \frac{1}{2}\left(-\frac{1}{2}\right)\frac{1}{2!}\cdot 9 + \frac{1}{2}\left(-\frac{1}{2}\right)\left(-\frac{3}{2}\right)\cdot\frac{1}{3!}\,(-25) + \frac{1}{2}\left(-\frac{1}{2}\right)\left(-\frac{3}{2}\right)\left(-\frac{5}{2}\right)\frac{1}{4!} \times 37$$

$= 47{\cdot}868$ (On simplification)

Hence, the number of students with marks less than 45 is 47·868, *i.e.*, 48.

But the number of students with marks less than 40 is 31.

$\therefore$ Required number of students getting marks between 40 and 45 is 48 – 31 = 17.

Example 19·17. *Given :*

$$\sum_{1}^{10} U_x = 500426, \quad \sum_{4}^{10} U_x = 329240, \quad \sum_{7}^{10} U_x = 175212 \quad \text{and} \quad U_{10} = 40365.$$

Find U_1.

Solution. Here we are given the cumulative function, say $F(x)$. Then

$$F(1) = \sum_{1}^{10} U_x = 500426, \qquad F(4) = \sum_{4}^{10} U_x = 329240$$

$$F(7) = \sum_{7}^{10} U_x = 175212 \quad \text{and} \quad F(10) = U_{10} = 40365$$

The difference table for $F(x)$ becomes :

x	$F(x)$	Δ	Δ^2	Δ^3
1	500426			
		–171186		
4	329240		17158	
		– 154028		2023
7	175212		19181	
		– 134847		
11	40365			

Now we will find $F(2) = \sum_{2}^{10} U_x$

Taking $a + xh = 2$, we get $\quad 1 + x \times 3 = 2 \quad \Rightarrow \quad x = \frac{1}{3}$

$$\therefore \quad F(2) = F(1) + x\,\Delta F(1) + {}^xC_2\,\Delta^2 F(1) + {}^xC_3\,\Delta^3 F(1) + \ldots$$

$$= 500426 + \frac{1}{3}(-171186) + \frac{1}{2!}\left(-\frac{2}{3}\right)\cdot\frac{1}{3!}\cdot 17158 + \frac{1}{3}\left(-\frac{2}{3}\right)\left(-\frac{5}{3}\right)\cdot\frac{1}{3!}\cdot 2023$$

$$= 500426 - 57062 - 1906{\cdot}4444 + 124{\cdot}8765 = 441582{\cdot}432$$

Hence, $\quad U_1 = \sum_{1}^{10} U_x - \sum_{2}^{10} U_x = F(1) - F(2) = 500426 - 441582{\cdot}432 = 58843{\cdot}568.$

Example 19·18. *State Newton's forward interpolation formula and use it to obtain $\sqrt{5{\cdot}5}$, given that $\sqrt{5} = 2{\cdot}236$, $\sqrt{6} = 2{\cdot}449$, $\sqrt{7} = 2{\cdot}646$ and $\sqrt{8} = 2{\cdot}828$.*

Solution. Here $U = U_x = \sqrt{x}$ and we are given:

x :	5	6	7	8
U_x :	2·236	2·449	2·646	2·828

Newton's forward interpolation formula is :

$$U_{a+xh} = U_a + x\,\Delta U_a + \frac{x(x-1)}{2!}\,\Delta^2 U_a + \frac{x(x-1)(x-2)}{3!}\,\Delta^3 U_a + \ldots \qquad \ldots (*)$$

TABLE OF DIFFERENCES

x	U_x	ΔU_x	$\Delta^2 U_x$	$\Delta^3 U_x$
5	2·236			
		0·213		
6	2·449		– 0·016	
		0·197		0·001
7	2·646		– 0·015	
		0·182		
8	2·828			

Now $\quad a + xh = 5{\cdot}5 \quad \Rightarrow \quad 5 + x \times 1 = 5{\cdot}5 \quad \Rightarrow \quad x = 0{\cdot}5 \qquad \ldots (**)$

Hence, on using the difference table, we get from (*) and (**) :

$$U_{5\cdot5} = 2{\cdot}236 + (0{\cdot}5)(0{\cdot}213) + \frac{(0{\cdot}5)(0{\cdot}5-1)}{2}(-0{\cdot}016) + \frac{\{0{\cdot}5\,(0{\cdot}5-1)(0{\cdot}5-2)\}}{6}(0{\cdot}001)$$

$$= 2{\cdot}236 + 0{\cdot}1065 + 0{\cdot}002 + 0{\cdot}0000625 = 2{\cdot}3445626.$$

19·3·4. Gregory-Newton's Backward Interpolation Formula

This formula is based on the backward differences ∇.

If $b = a + nh$, is the last argument in the difference table, then by *def.*, we have

$$U_{b+xh} = \mathrm{E}^x U_b = (1-\nabla)^{-x} U_b \qquad [\because \mathrm{E} \equiv (1-\nabla)^{-1}]$$

$$= U_b + {}^xC_1 \nabla U_b + \frac{x(x+1)}{2!}\nabla^2 U_b + \frac{x(x+1)(x+2)}{3!}\nabla^3 U_b + \ldots \qquad \ldots(19{\cdot}11)$$

the last term depending on the degree of the polynomial U_{a+xh}. If U_{a+xh} is a polynomial of nth degree, $\Delta^r U_a = 0$, $r > n$ and hence the series (19·11) terminates after $(n+1)$ terms, thus giving

$$U_{b+xh} = U_b + x\,\nabla U_b + \frac{x\ x +}{2!}\nabla^2 U_b + \frac{x(x+1)(x+2)}{3!}\nabla^3 U_b$$

$$+ \ldots + \frac{x(x+1)(x+2)\cdots(x+n-1)}{n!}\nabla^n U_b, \qquad \ldots(19{\cdot}11a)$$

which is the required Gregory-Newton's Backward Difference formula.

Remarks:

1. Here

$$x = \frac{\text{Period of interpolation} - \text{Last argument}}{\text{Interval of Differencing}}$$

2. The formula (19·11) or (19·11*a*) is mainly helpful if we have to interpolate U_x for the value of x near the extreme end (*i.e.*, $x < a + nh$) of a st of tabulated values or when we have to extrapolate U_x for x slightly greater than $a + nh$ (*i.e.*, $x > a + nh$).

Example 19·19. *The following table gives the census population of a town for the years 1976 to 2016. Estimate the population for the year 2010 by using an appropriate interpolation formula.*

Year	:	*1976*	*1986*	*1996*	*2006*	*2016*
Population (in lakhs)	:	*46*	*66*	*81*	*93*	*101*

Solution. Since the value to be interpolated lies towards the end of the given data, we shall use Newton's backward difference formula.

TABLE OF DIFFERENCES

x	$f(x)$	$\nabla f(x)$	$\nabla^2 f(x)$	$\nabla^3 f(x)$	$\nabla^4 f(x)$
1976	46				
		20			
1986	66		– 5		
		15		2	
1996	81		– 3		– 3
		12		– 1	
2006	93		– 4		
		8			
2016	101				

In the notations of formula (19·11), we have $x = \frac{2010 - 2016}{10} = -0{\cdot}6$

The leading backward differences of last entry are given respectively by 8, – 4, –1 and –3. Substituting in (19·11), we get

$$U_{2010} = 101 - (0{\cdot}6)\times 8 + \frac{(-0{\cdot}6)(-0{\cdot}6+1)}{2!}\times(-4) + \frac{(-0{\cdot}6)(-0{\cdot}6+1)(-0{\cdot}6+2)}{3!}\times(-1)$$

$$\frac{(-0{\cdot}6)(-0{\cdot}6+1)(-0{\cdot}6+2)(-0{\cdot}6+3)}{4!}\times(-3)$$

$$= 101 - 4{\cdot}8 + 0{\cdot}48 + 0{\cdot}056 + 0{\cdot}1008 = 96{\cdot}837 \text{ lakhs}$$

19·4. INTERPOLATION WITH ARGUMENTS AT UNEQUAL INTERVALS

The operates E, ∇, Δ and Newton's-Gregory formulae of interpolation (Forward and Backward) can be used only if we are given the entries corresponding to equidistant values of the argument. If we are given the entries corresponding to the arguments at unequal intervals, the operators E and Δ don't serve our purpose and we define a more general class of differences, known as *divided differences*.

19·4.1. Divided Differences

Let $U_{a_0}, U_{a_1}, U_{a_2}, \ldots, U_{a_n}$ be the entries corresponding to the arguments $a_0, a_1, a_2 \ldots, a_n$ which need not necessarily be equidistant. Then the first order divided difference of U_x for the arguments a_0, a_1 denoted by $U(a_0, a_1)$ or $\underset{a_1}{\Delta} U_{a_0}$ is given by :

$$\left.\begin{aligned} \underset{a_1}{\Delta} U_{a_0} &= U(a_0, a_1) = \frac{U_{a_1} - U_{a_0}}{a_1 - a_0} \\ \text{and} \quad \underset{a_0}{\Delta} U_{a_1} &= U(a_1, a_0) = \frac{U_{a_0} - U_{a_1}}{a_0 - a_1} \end{aligned}\right\} \quad \ldots (19{\cdot}12)$$

Obviously,

$$\underset{a_1}{\Delta} U_{a_0} = \underset{a_0}{\Delta} U_{a_1} \quad \Rightarrow \quad U(a_0, a_1) = U(a_1, a_0) \quad \ldots [19{\cdot}12(a)]$$

Similarly,

$$\left.\begin{aligned} \underset{a_2}{\Delta} U_{a_1} &= \frac{U_{a_2} - U_{a_1}}{a_2 - a_1} = \underset{a_1}{\Delta} U_{a_2} \\ &\vdots \\ \underset{a_n}{\Delta} U_{a_{n-1}} &= \frac{U_{a_n} - U_{a_{n-1}}}{a_n - a_{n-1}} = \underset{a_{n-1}}{\Delta} U_{a_n} \end{aligned}\right\} \quad \ldots [19{\cdot}12\,(b)]$$

The second order divided difference of U_x for the arguments a_0, a_1 and a_2 is defined as

$$\underset{a_1 a_2}{\Delta^2} U_{a_0} = U(a_0, a_1, a_2) = \frac{\underset{a_2}{\Delta} U_{a_1} - \underset{a_1}{\Delta} U_{a_0}}{a_2 - a_0} = \frac{1}{a_2 - a_0}\left(\frac{U_{a_2} - U_{a_1}}{a_2 - a_1} - \frac{U_{a_1} - U_{a_0}}{a_1 - a_0}\right) \quad \ldots (19{\cdot}13)$$

$$= \frac{1}{a_2 - a_0}\left[\frac{U_{a_2}}{a_2 - a_1} + \frac{U_{a_0}}{a_1 - a_0} - \frac{(a_2 - a_0)U_{a_1}}{(a_2 - a_1)(a_1 - a_0)}\right]$$

$$= \frac{U_{a_0}}{(a_0 - a_1)(a_0 - a_2)} + \frac{U_{a_1}}{(a_1 - a_0)(a_1 - a_2)} + \frac{U_{a_2}}{(a_2 - a_0)(a_2 - a_1)} \quad \ldots [19{\cdot}13\,(a)]$$

Similarly, we have

$$\underset{a_0 a_2}{\Delta} U = U(a_1, a_0, a_2) = \frac{\underset{a_2}{\Delta} U_{a_0} - \underset{a_0}{\Delta} U_{a_1}}{a_2 - a_1} = \frac{1}{a_2 - a_1}\left(\frac{U_{a_2} - U_{a_0}}{a_2 - a_0} - \frac{U_{a_0} - U_{a_1}}{a_0 - a_1}\right)$$

$$= \frac{U_{a_2}}{(a_2 - a_0)(a_2 - a_1)} + \frac{U_{a_1}}{(a_1 - a_0)(a_1 - a_2)} + \frac{}{(a_0 - a_1)(a_0 - a_2)} \quad \ldots [19{\cdot}13\,(b)]$$

and so on.

From the symmetry of the results in [19·13 (*a*)] and [19·13 (*b*)], we see that

$$\underset{a_1 a_2}{\Delta^2} U_{a_0} = \underset{a_2 a_1}{\Delta^2} U_{a_0} = \underset{a_0 a_2}{\Delta^2} U_{a_1} = \underset{a_2 a_0}{\Delta^2} U_{a_1} = \underset{a_1 a_0}{\Delta^2} U_{a_2} = \underset{a_0 a_1}{\Delta^2} U_{a_2} \quad \text{... [19·13 (c)]}$$

$$\Rightarrow \quad U(a_0, a_1, a_2) = U(a_0, a_2, a_1) = U(a_1, a_0, a_2)$$

$$= U(a_1, a_2, a_0) = U(a_2, a_1, a_0) = U(a_2, a_0, a_1) \quad \text{... [19·13 (d)]}$$

The nth order divided difference of the arguments $a_0, a_1, ..., a_n$ is defined as

$$\left.\begin{aligned} \underset{a_1 a_2 \cdots a_n}{\Delta^n} U_{a_0} &= \frac{\underset{a_2 a_3 \cdots a_n}{\Delta^{n-1}} U_{a_1} - \underset{a_1 a_2 \cdots a_{n-1}}{\Delta^{n-1}} U_{a_0}}{a_n - a_0} \\ \Rightarrow \quad U(a_0, a_1, a_2 \cdots, a_n) &= \frac{U(a_1, a_2, \cdots, a_n) - U(a_0, a_1 \cdots, a_{n-1})}{a_n - a_0} \end{aligned}\right\} \quad \text{... (19·14)}$$

The result obtained in (19·13*a*) or (19·13*b*) and (19·13*c*) can be easily generalised by mathematical induction, to give

$$\underset{a_1 a_2 \cdots a_n}{\Delta^n} U_{a_0} = \frac{U_{a_0}}{(a_0 - a_1)(a_0 - a_2)\cdots(a_0 - a_n)} + \frac{U_{a_1}}{(a_1 - a_0)(a_1 - a_2)\cdots(a_1 - a_n)}$$

$$+ \cdots + \frac{U_{a_n}}{(a_n - a_0)(a_n - a_1)\cdots(a_n - a_{n-1})} \quad \text{... [19·14(a)]}$$

and

$$\underset{a_1 a_2 \cdots a_n}{\Delta^n} U_{a_0} = \underset{a_0 a_2 \cdots a_n}{\Delta^n} U_{a_1} = \cdots etc. \quad \text{... [(19·14 (b)]}$$

Remarks:

1. We have $\Delta U_{a_0} = \frac{U_{a_1} - U_{a_0}}{a_1 - a_0}$ *i.e.,* the ordinary difference $\left(U_{a_1} - U_{a_0}\right)$, is divided by the difference of the argument $a_1 - a_0$. Hence the name divided difference.
2. The order of a divided difference is one less than the number of arguments required for its definition.
3. The results obtained in [19·12*a*] or [19·13*a*] and [19·14*b*] lead to the following very important conclusion. *"Divided difference are symmetric functions of their arguments, i.e., the divided differences are independent of the order of the arguments."*
   Relation [19·14 (*a*)] expresses the divided difference of nth order in terms of n partial fractions.
4. It can be easily verified that

   (*i*) $\Delta\left[U_x \pm V_x\right] = \Delta U_x \pm \Delta V_x$, (*ii*) $\Delta\left[CU_x\right] = C\Delta U_x$, C being arbitrary constant

19·4·2. Divided Difference Table

Divided differences of various orders can be conveniently arranged as shown in *divided difference table on page 19·26.*

DIVIDED DIFFERENCES TABLE

x	U_x	ΔU_x	$\Delta^2 U_x$	$\Delta^3 U_x$	$\Delta^4 U_x$
a_0	U_{a_0}				
	$\longrightarrow$	$\dfrac{U_{a_1} - U_{a_0}}{a_1 - a_0} = \underset{a_1}{\Delta} U_{a_0}$			
a_1	U_{a_1}	$\longrightarrow$	$\dfrac{\underset{a_2}{\Delta} U_{a_1} - \underset{a_1}{\Delta} U_{a_0}}{a_2 - a_0} = \Delta^2 U_{a_0}$		
	$\longrightarrow$	$\dfrac{U_{a_2} - U_{a_1}}{a_2 - a_1} = \underset{a_2}{\Delta} U_{a_1}$	$\longrightarrow$	$\dfrac{\Delta^2 U_{a_1} - \Delta^2 U_{a_0}}{a_3 - a_0} = \Delta^3 U_{a_0}$	
a_2	U_{a_2}	$\longrightarrow$	$\dfrac{\underset{a_3}{\Delta} U_{a_2} - \underset{a_2}{\Delta} U_{a_1}}{a_3 - a_1} = \Delta^2 U_{a_1}$		$\dfrac{\Delta^3 U_{a_1} - \Delta^3 U_{a_0}}{a_4 - a_0} = \Delta^4 U_{a_0}$
	$\longrightarrow$	$\dfrac{U_{a_3} - U_{a_2}}{a_3 - a_2} = \underset{a_3}{\Delta} U_{a_2}$		$\dfrac{\Delta^2 U_{a_2} - \Delta^2 U_{a_1}}{a_4 - a_1} = \Delta^3 U_{a_1}$	
a_3	U_{a_3}		$\dfrac{\underset{a_4}{\Delta} U_{a_3} - \underset{a_3}{\Delta} U_{a_2}}{a_4 - a_2} = \Delta^2 U_{a_2}$		
	$\longrightarrow$	$\dfrac{U_{a_4} - U_{a_3}}{a_4 - a_3} = \underset{a_4}{\Delta} U_{a_3}$			
a_4	U_{a_4}				

Example 19·20. *If* $f(x) = \frac{1}{x^2}$, *find the divided differences* $f(a, b), f(a, b, c)$ *and* $f(a, b, c, d)$.

Solution. $$f(a, b) = \frac{f(b) - f(a)}{b - a} = \frac{\frac{1}{b^2} - \frac{1}{a^2}}{b - a} = -\frac{a + b}{a^2b^2}$$

Similarly, we will get $$f(b, c) = -\frac{b + c}{b^2c^2}$$

$$f(a, b, c) = \frac{f(b,c) - f(a,b)}{c - a} = -\frac{1}{(c - a)}\left(\frac{b + c}{b^2c^2} - \frac{a + b}{a^2b^2}\right)$$

$$= -\frac{1}{(c - a)}\left\{\frac{a^2(b + c) - c^2(a + b)}{a^2b^2c^2}\right\} = -\frac{1}{(c - a)}\left\{\frac{a^2b + a^2c - c^2a - c^2b}{a^2b^2c^2}\right\}$$

$$= -\left\{\frac{b(a^2 - c^2) + ac(a - c)}{(c - a)a^2b^2c^2}\right\} = \frac{ab + bc + ca}{a^2b^2c^2} \qquad \text{(On simplification)}$$

By symmetry, we shall get : $$f(b, c, d) = \frac{bc + cd + db}{b^2c^2d^2}$$

$$f(a, b, c, d) = \frac{f(b,c,d) - f(a,b,c)}{d - a} = \frac{1}{d - a}\left(\frac{bc + cd + db}{b^2c^2d^2} - \frac{ab + bc + ca}{a^2b^2c^2}\right)$$

$$= \frac{1}{d - a} \cdot \left\{\frac{a^2(bc + cd + db) - d^2(ab + bc + ca)}{a^2b^2c^2d^2}\right\} \qquad \text{... (*)}$$

Numerator $= a^2bc + a^2cd + a^2db - d^2ab - d^2bc - d^2ca$

$= bc\,(a^2 - d^2) + acd\,(a - d) + abd\,(a - d)$

$= (a - d)\,\{bc\,(a + d) + acd + abd\}$

$= (a - d)\,(abc + abd + acd + bcd)$

Hence, from (*) $$f(a, b, c, d) = -\left[\frac{abc + abd + acd + bcd}{a^2b^2c^2d^2}\right].$$

19·4·3. Theorems on Divided Differences

Theorem 19·1. *The nth order divided difference of a polynomical of nth degree is constant.*

Proof. It is sufficient to consider the function $U_x = x^n$.

Let this function be defined for the arguments $a_0, a_1, a_2, \ldots, a_n$. Then,

$$\underset{a_1}{\Delta} U_{a_0} = \frac{U_{a_1} - U_{a_0}}{a_1 - a_0} = \frac{a_1^n - a_0^n}{a_1 - a_0}$$

$$= (a_1^{n-1} + a_1^{n-2}\,a_0 + a_1^{n-3}\,a_0^2 + \ldots + a_0^{n-1}) \qquad \text{...(*)}$$

$$= \text{Co-eff. of } x^{n-1} \text{ in } \left[(1 + a_0x + a_0^2x^2 + a_0^3x^3 + \ldots) \times (1 + a_1x + a_1^2x^2 + a_1^3x^3 + \ldots)\right]$$

$$= \text{Co-eff. of } x^{n-1} \text{ in } \left\{(1 - a_0x)^{-1}\,(1 - a_1x)^{-1}\right\}$$

$$= \text{Co-eff. of } x^{n-1} \text{ in } \left[(1 - a_0x)\,(1 - a_1x)\right]^{-1}$$

From (*), we see that the first order divided difference is homogeneous function of order $(n - 1)$ of arguments a_0 and a_1.

Similarly : $\underset{a_2}{\Delta} U_{a_1} = \text{Co-eff. of } x^{n-1} \text{ in } \left[(1-a_1x)(1-a_2x)\right]^{-1}$

The divided difference of the second order is given by:

$$\underset{a_1 a_2}{\Delta}{}^2 U_{a_0} = \frac{\underset{a_2}{\Delta} U_{a_1} - \underset{a_1}{\Delta} U_{a_0}}{a_2 - a_0}$$

$$= \frac{1}{a_2 - a_0} \times \text{Co-eff. of } x^{n-1} \text{ in } \left\{\frac{1}{(1-a_1x)(1-a_2x)} - \frac{1}{(1-a_0x)(1-a_1x)}\right\}$$

$$= \frac{1}{a_2 - a_0} \times \text{Co-eff. of } x^{n-1} \text{ in } \left\{\frac{(1\ \Re a_0 x) \quad (1 \quad a_2 x)}{(1\ \Re a\cdot x)(1 \quad a\ x)(1 \quad a\ x)}\right\}$$

$$= \text{Co-eff. of } x^{n-1} \text{ in } \left[\frac{x}{(1-a_0x)(1-a_1x)(1-a_2x)}\right]$$

$$= \text{Co-eff. of } x^{n-2} \text{ in } \left[(1-a_0x)\,(1-a_1x)\,(1-a_2x)\right]^{-1}$$

By mathematical induction the result can be easily generalised to the case of nth order divided difference to give

$$\underset{a_1 a_2 \cdots a_n}{\Delta^n} U_{a_0} = \text{Co-eff. of } x^{n-n} \text{ in } \left[(1-a_0x)\,(1-a_1x)\ldots(1-a_nx)\right]^{-1}$$

$$= \text{Constant term in } \left[(1-a_0x)\,(1-a_1x)\ldots(1-a_nx)\right]^{-1} = 1$$

Thus $\qquad \Delta^n x^n = 1 \qquad$ ...(19·16)

In general, we will get $\qquad \Delta^n (Cx^n) = C\ \Delta^n x^n = C \qquad$ ...[19·16 (*a*)]

Remarks:

1. The divided differences of order higher than n for the function $U_x = x^n$, all vanish, *i.e.*,

$$\Delta^r (x^n) = 0 \ ; \ r = n+1, \ n+2, \ldots$$

$\Rightarrow \qquad \Delta^r (x^n) = 0, \ r > n \qquad$ ...[19·16 (*b*)]

2. The nth order divided difference of a polynomial of nth degree, *viz.*,

$$U_x = A_0 x^n + A_1 x^{n-1} + A_2 x^{n-2} + \ldots + A_{n-1} x + A_n, \quad (A_0 \neq 0)$$

where $A_0, A_1, \ldots, A_n$ are constants, is given by :

$$\Delta^n U_x = \Delta^n \left[A_0 x^n + A_1 x^{n-1} + A_2 x^{n-2} + \ldots + A_{n-1} x + A_n\right]$$

$$= \Delta^n (A_0 x^n) = A_0\ \Delta^n x^n = A_0 \qquad \left[\text{Using 19·16}(b) \text{ and } (19·16)\right]$$

Thus, the *nth* divided difference of a polynomial of degree n is constant and is equal to the co-efficient of x^n.

Theorem 19·2. *The nth order divided difference of a function can be expressed as the quotient of two determinants, each of order (n + 1).*

(For proof see Q 56 at the end of the chapter).

19·4·4. Newton's Divided Difference Formula

By definition., we have

$$\underset{x}{\Delta} U_a = U(x, a) = \frac{U_x - U_a}{x - b} \quad \Rightarrow \quad U_x = U_a + (x-a)\underset{x}{\Delta} U_a \qquad \ldots(*)$$

The second order divided difference for arguments x, a, b is given by :

$$\underset{xb}{\Delta}^2 U_a = U(x,a,b) = \frac{U(x,a) - U(a,b)}{x-b} = \frac{\underset{x}{\Delta} U_a - \underset{b}{\Delta} U_a}{x-b}$$

$\therefore$
$$\underset{x}{\Delta} U_a = \underset{b}{\Delta} U_a + (x-b)\underset{xb}{\Delta}^2 U_a$$

Substituting in (*), we get

$$U_x = U_a + (x-a)\left[\underset{b}{\Delta} U_a + (x-b)\underset{xb}{\Delta} U_a\right]$$
$$= U_a + (x-a)\underset{b}{\Delta} U_a + (x-a)(x-b)\underset{xb}{\Delta}^2 U_a \quad \ldots(**)$$

The third order divided difference for the arguments x, a, b, c is given by

$$\underset{xbc}{\Delta}^3 U_a = U(x,a,b,c) = \frac{U(x,a,b) - U(a,b,c)}{x-c} = \frac{\underset{xb}{\Delta}^2 U_a - \underset{bc}{\Delta}^2 U_a}{x-c}$$

$\therefore$
$$\underset{xb}{\Delta}^2 U_a = \underset{bc}{\Delta}^2 U_a + (x-c)\underset{xbc}{\Delta}^3 U_a$$

Substituting in (**), we get

$$U_x = U_a + (x-a)\underset{b}{\Delta} U_a + (x-a)(x-b)\underset{bc}{\Delta}^2 U_a + (x-a)(x-b)(x-c)\underset{xbc}{\Delta}^3 U_a$$
$$= U_a + (x-a)\underset{b}{\Delta} U_a + (x-a)(x-b)\underset{bc}{\Delta}^2 U_a + (x-a)(x-b)(x-c)\underset{bcd}{\Delta}^3 U_a$$
$$+ (x-a)(x-b)(x-c)(x-d)\underset{xbcd}{\Delta}^4 U_a$$

Similarly proceeding, we shall get

$$U_x = U_a + (x-a)\Delta U_a + (x-a)(x-b)\overset{2}{\Delta} U_a + (x-a)(x-b)(x-c)\overset{3}{\Delta} U_a$$
$$+ (x-a)(x-b)(x-c)(x-d)\overset{4}{\Delta} U_a + \ldots, \quad \ldots(19{\cdot}17)$$

the last term depending on the degree of the polynomial U_x.

If U_x is a polynomial of nth degree then, $\overset{r}{\Delta} U_x = 0, r > n$ and the series (19·17) terminates after $(n+1)$ terms.

Formula (19·17) is the required Newton's divided difference interpotation formula.

Remarks :

1. Relation between Δ and Δ. Let us take the arguments a, b, c, d, etc. to be at equal interval of h, (say), so that

$$b = a + h, \quad c = a + 2h, \quad d = a + 3h, \text{ and so on. Then by def.,}$$

$$\underset{b}{\Delta} U_a = \frac{U_b - U_a}{b-a} = \frac{U_{a+h} - U_a}{h} = \frac{1}{h}\Delta U_a \quad \ldots(*)$$

Similarly,

$$\underset{c}{\Delta} U_b = \frac{U_c - U_b}{c-b} = \frac{U_{a+2h} - U_{a+h}}{h} = \frac{1}{h}\Delta U_{a+h} \quad \ldots(**)$$

$$\underset{bc}{\Delta}^2 U_a = \frac{\underset{c}{\Delta} U_b - \underset{b}{\Delta} U_a}{c-a} = \frac{1}{2h}\left[\frac{1}{h}\{\Delta U_{a+h} - \Delta U_a\}\right] \quad \text{(Using (*) and (**))}$$

$$= \frac{1}{2h^2}\Delta[U_{a+h} - U_a] = \frac{1}{2h^2}\Delta(\Delta U_a) = \frac{1}{2h^2}\Delta^2 U_a$$

Similarly, $$\underset{cd}{\Delta}^2 U_b = \frac{1}{2h^2}\Delta^2 U_b = \frac{1}{2h^2}\Delta^2 U_{a+h}$$

$$\underset{bcd}{\Delta^3} U_a = \frac{\underset{cd}{\Delta^2} U_b - \underset{bc}{\Delta^2} U_a}{d-a} = \frac{1}{3h}\left[\frac{1}{2h^2}(\Delta^2 U_{a+h} - \Delta^2 U_a)\right] = \frac{1}{3!\cdot h^3}\Delta^2 (U_{a+h} - U_a) = \frac{1}{3!\cdot h^3}\Delta^3 U_a$$

In general, we shall get : $\Delta^r \equiv \frac{1}{r!\cdot h^r}\Delta^r, (r = 1, 2, 3, \cdots)$ ...(19·18)

which is the required relation between Δ and Δ .

2. Gregory-Newton's (Forward Interpolation) Formula as a Particular Case of Newton's Divided Difference Formula.

In (19·17), let us take $b = a + h$, $c = a + 2h$, $d = a + 3h$ and so on and $x = a + Xh$, so that

$(x - a) = (a + Xh) - a = Xh$; $(x - b) = (a + Xh) - (a + h) = h(X - 1)$

$(x - c) = (a + Xh) - (a + 2h) = h(X - 2)$; $(x - d) = (a + Xh) - (a + 3h) = h(X - 3)$, and so on.

Hence, on using (19·18), we get from (19·17)

$$U_{a+Xh} = U_a + Xh \cdot \frac{1}{h}\Delta U_a + hX\cdot h(X-1)\,.\,\frac{1}{2!h^2}\Delta^2 U_a + hX\cdot h(X-1).h(X-2)\cdot\frac{1}{3!h^3}\Delta^3 U_a + \ldots$$

$$= U_a + X\Delta U_a + \frac{X(X-1)}{2!}\Delta^2 U_a + \frac{X(X-1)(X-2)}{3!}\Delta^3 U_a + \ldots$$

$$= U_a + {}^X C_1\,\Delta U_a + {}^X C_2\,\Delta^2 U_a + {}^X C_3 \Delta^3 U_a + \ldots$$

which is the Newton-Gregory forward Interpolation formula.

Example 19·19. *The observed values of a function are respectively 168, 120, 72 and 63 at the four positions 3, 7, 9 and 10 of the independent variable. What is the best estimate you can give for the value of the function at the position 6 of the independent variable ?*

Solution. The divided difference table is given by:

x	U_x	ΔU_x	$\Delta^2 U_x$	$\Delta^3 U_x$
3	168			
		$\frac{120-168}{7-3} = -12$		
7	120		$\frac{-24+12}{9-3} = -2$	
		$\frac{72-120}{9-7} = -24$		$\frac{5+2}{10-3} = 1$
9	72		$\frac{-9+24}{10-7} = +5$	
		$\frac{63-72}{10-9} = -9$		
10	63			

Hence, using Newton's divided difference interpolation formula, we get :

$$U_x = U_a + (x-a)\,\Delta U_a + (x-a)(x-b)\,\Delta^2 U_a + (x-a)(x-b)(x-c)\,\Delta^3 U_a$$

$$= 168 - 12(x-3) - 2(x-3)(x-7) + 1(x-3)(x-7)(x-9)$$

$$= x^3 - 21x^2 + 119x - 27$$ (On simplification)

When $x = 6$, the estimate of the function is given by :

$$U_6 = 6^3 - 21 \times 6^2 + 119 \times 6 - 27 = 147.$$

Example 19·20. *Given the values :*

x:	4	5	7	10	11	13
U_x:	48	100	294	900	1210	2028

form the table of divided differences and extend it to include the values $x = 2$ and $x = 15$.

Solution.

DIVIDED DIFFERENCE TABLE

x	U_x	ΔU_x	$\Delta^2 U_x$	$\Delta^3 U_x$	$\Delta^4 U_x$	$\Delta^5 U_x$
2	**4**					
		22				
4	48		**10**			
		52		**1**		
5	100		15			
		97		1	**0**	
7	294		21			**0**
		202		1	0	
10	900		27			**0**
		310		1	**0**	
11	1210		33			
		409		**1**		
13	2028		**38**			
		561				
15	**3150**					

$\therefore\quad U_x = 48 + 52\,(x-4) + 15\,(x-4)\,(x-5) + 1.\,(x-4)(x-5)(x-7)$

$= 48 + 52\,(x-4) + 15\,(x^2 - 9x + 20) + (x^3 - 16x^2 + 83x - 140)$

$= x^3 - x^2 = x^2\,(x-1)$

$\therefore\quad U_2 = 2^3 - 2^2 = 4 \qquad$ and $\qquad U_{15} = 15^3 - 15^2 = 3150$

The original divided difference table is for the argument 4 to 13, leaving 2 and 15 and the figures in antique (bold).

The complete table, with arguments 2 and 15 and the figures in antique, gives the original divided difference table extended to include the values $x = 2$ and $x = 15$.

Example 19·23. *The following are the mean temperatures (Fahrenheit) on three days, 30 days apart round the periods of summer and winter. Estimate the approximate dates and values of the maximum and minimum temperature.*

	SUMMER		WINTER	
Day	*Date*	*Temperature*	*Date*	*Temperature*
0	*15th June*	*58·8*	*16th Dec.*	*40·7*
30	*15th July*	*63·4*	*15th Jan.*	*38·1*
60	*14th Aug.*	*62·5*	*14th Feb.*	*39·3*

Solution. Obviously the dates and values of maximum and minimum temperatures are to be obtained from the summer and winter records respectively.

If we choose unit 1 ≡ 30 days, then we are given the entries (temperatures) corresponding to equidistant arguments (dates), the interval of differencing being unity.

SUMMER

x	*Temperature* (U_x)	ΔU_x	$\Delta^2 U_x$
0	58·8		
		4·6	
1	63·4		$-\dfrac{5·5}{2} = -2·75$
		$-0·9$	
2	62·5		

$$U_x = U_0 + (x-0)\,\Delta U_0 + (x-0)(x-1)\,\Delta^2 U_0$$
$$= 58{\cdot}8 + x \times 4{\cdot}6 + x(x-1)(-2{\cdot}75) = -2{\cdot}75x^2 + 7{\cdot}35x + 58{\cdot}8 \qquad \ldots(*)$$

For maxima or minima, we should have

$$U'(x) = -5{\cdot}5x + 7{\cdot}35 = 0 \quad \Rightarrow \quad x = 1{\cdot}342$$
$$U''(x) = -5{\cdot}5 < 0$$

Hence, U_x is maximum at $x = 1{\cdot}342$.

Since, Unit 1 ≡ 30 days, 1·342 units ≡ 30 × 1·342 = 40·26 days.

Thus the maximum temperature was on 15th June + 40 days, *i.e.*, on 24th July and the value of maximum temperature is :

$$(U_x)_{max} = [U_x]_{x=1{\cdot}342}$$
$$= -2{\cdot}75 \times (1{\cdot}342)^2 + 7{\cdot}35 \times 1{\cdot}342 + 58{\cdot}8 \qquad \text{[From (*)]}$$
$$= 63{\cdot}72° \text{ F}$$

WINTER

x	Temperature (U_x)	ΔU_x	$\Delta^2 U_x$
0	40·7		
		– 2·6	
1	38·1		1·9
		1·2	
2	39·3		

$$\therefore \quad U_x = U_0 + (x-0)\,\Delta U_0 + x(x-1)\,\Delta^2 U_0$$
$$= 40{\cdot}7 - 2{\cdot}6x + 1{\cdot}9(x^2 - x) = 1{\cdot}9x^2 - 4{\cdot}5x + 40{\cdot}7 \qquad \ldots(**)$$
$$U'(x) = 3{\cdot}8x - 4{\cdot}5 = 0 \quad \Rightarrow \quad x = 1{\cdot}184$$
$$U''(x) = 3{\cdot}8 > 0$$

Hence, U_x is minimum at $x = 1{\cdot}184$.

Since, 1 unit ≡ 30 days, 1·184 units = 30 × 1·184 = 35·520 days.

Thus, the date of minimum temperature is 16th December + 35·5 days, *i.e.*, the mid-night of 19th January.

Minimum temperature $= (U_x)_{x=1{\cdot}184}$

$$= 1{\cdot}9(1{\cdot}184)^2 - 4{\cdot}5 \times 1{\cdot}184 + 40{\cdot}7 \qquad \text{[From (**)]}$$
$$= 38{\cdot}036° \text{ F}$$

19·4·5. Lagrange's Interpolation Formula

Let U_{a_0}, U_{a_1}, ... , U_{a_n} be the $(n+1)$ entries corresponding to the arguments $a_0, a_1, \ldots, a_n$ which are not necessarily equally spaced. Then the function U_x can be approximated by a polynomial of the nth degree.

Let us consider the divided difference $\underset{x a_1 a_2 \cdots a_n}{\Delta^{n+1}} U_{a_0}$ corresponding to $(n+2)$ arguments $x, a_0, a_1, \ldots, a_n$. [Since the order of a divided difference is one less than the number of arguments required for its definition, $\underset{x a_1 \cdots a_n}{\Delta} U_{a_0}$ is a divided difference of order $(n+1)$]. Further since U_x is a polynomial of degree n, $\Delta^r U_x = 0$, $r > n$.

Hence,
$$\underset{x a_1 a_2 \cdots a_n}{\Delta^{n+1}} U_{a_0} = 0$$

$$\Rightarrow \frac{U_{a_0}}{(a_0 - x)(a_0 - a_1)(a_0 - a_2) \cdots (a_0 - a_n)} + \frac{U_{a_1}}{(a_1 - x)(a_1 - a_0)(a_1 - a_2) \cdots (a_1 - a_n)}$$

$$+ \frac{U_{a_2}}{(a_2 - x)(a_2 - a_0)(a_2 - a_1) \cdots (a_2 - a_n)} + \cdots + \frac{U_{a_n}}{(a_n - x)(a_n - a_0)(a_n - a_1) \cdots (a_n - a_{n-1})}$$

$$+ \frac{U_x}{(x - a_0)(x - a_1)(x - a_2) \cdots (x - a_n)} = 0$$

$$\Rightarrow \frac{U_x}{(x - a_0)(x - a_1) \cdots (x - a_n)} = \left[\frac{U_{a_0}}{(x - a_0)(a_0 - a_1)(a_0 - a_2) \cdots (a_0 - a_n)} \right.$$

$$+ \frac{U_{a_1}}{(x - a_1)(a_1 - a_0)(a_1 - a_2) \cdots (a_1 - a_n)} + \cdots$$

$$\left. + \frac{U_{a_n}}{(x - a_n)(a_n - a_0)(a_n - a_1) \cdots (a_n - a_{n-1})} \right] \quad \ldots (19{\cdot}19)$$

$$\Rightarrow \quad U_x = \frac{(x - a_1)(x - a_2) \cdots (x - a_n)}{(a_0 - a_1)(a_0 - a_2) \cdots (a_0 - a_n)} U_{a_0} + \frac{(x - a_0)(x - a_2) \cdots (x - a_n)}{(a_1 - a_0)(a_1 - a_2) \cdots (a_1 - a_n)} U_{a_1}$$

$$+ \cdots + \frac{(x - a_0)(x - a_1) \cdots (x - a_{n-1})}{(a_n - a_0)(a_n - a_1) \cdots (a_n - a_{n-1})} U_{a_n} \quad \ldots [19{\cdot}19\,(a)]$$

Formula (19·19) or [19·19*a*] is the Lagrange's formula of interpolation.

Remarks:

1. Formula (19·19) is same as splitting the function

$$\frac{U_x}{(x - a_0)(x - a_1) \cdots (x - a_n)}, \text{ into linear fractions.}$$

2. While dealing with arguments not at equal intervals, any one of the two formulae of interpolation, *viz.*, Newton's Divided Difference formula or Lagrange's formula, can be used.

These formulae also enable us to determine the form of the function U_x.

3. Lagrange's formula (19·19*a*) can also be expressed as follows :

$$U_x = \sum_{r=0}^{n} \left[\frac{\phi(x)\, U_{a_r}}{(x - a_r)\, \phi'(a_r)} \right], \quad \text{where } \phi(x) = \prod_{r=0}^{n} (x - a_r) \quad \ldots (19{\cdot}19b)$$

For proof of (19·19*b*), see *Example 19·30*.

Example 19·24. *Given* $\log_{10} 654 = 2{\cdot}8156$, $\log_{10} 658 = 2{\cdot}8182$, $\log_{10} 659 = 2{\cdot}8189$, $\log_{10} 661 = 2{\cdot}8202$, *find by Lagrange's interpolation formula* $\log_{10} 656$. *(Retain for decimal places in your answer.)*

Solution. Here $U(x) = \log_{10} x$. The values of arguments and the corresponding entries are as given bellow :

x	:	$a_0 = 654$	$a_1 = 658$	$a_2 = 659$	$a_3 = 661$
$U_x = \log_{10} x$	:	2·8156	2·8182	2·8189	2·8202

We want U_x when $x = 656$, *viz.*, $U_{656} = \log_{10} 656$.

Taking $x = 656$ in Lagrange's formula (19·19), and using the above table, we get

$$\log_{10} 656 = U_{656}$$

$$= \frac{(656-658)(656-659)(656-661)}{(654-658)(654-659)(654-661)} \times 2{\cdot}8156$$

$$+ \frac{(656-654)(656-659)(656-661)}{(658-654)(658-659)(658-661)} \times 2{\cdot}8182$$

$$+ \frac{(656-654)(656-658)(656-661)}{(659-654)(659-658)(659-661)} \times 2{\cdot}8189$$

$$+ \frac{(656-654)(656-658)(656-659)}{(661-654)(661-658)(661-659)} \times 2{\cdot}8202$$

$$= \frac{(-2)(-3)(-5)}{(-4)(-5)(-7)} \times 2{\cdot}8156 + \frac{2(-3)(-5)}{4(-1)(-3)} \times 2{\cdot}8182$$

$$+ \frac{2\times(-2)\times(-5)}{5\times 1\times(-2)} \times 2{\cdot}8189 + \frac{2\times(-2)\times(-3)}{7\times 3\times 2} \times 2{\cdot}8202$$

$$= \frac{3}{14} \times 2{\cdot}8156 + \frac{5}{2} \times 2{\cdot}8182 - 2\times 2{\cdot}8189 + \frac{2}{7} \times 2{\cdot}8202$$

$$= 0{\cdot}6033 + 7{\cdot}0455 - 5{\cdot}6378 + 0{\cdot}8058 = 2{\cdot}8168.$$

Example 19·25. *The mode of a certain frequency curve $y = f(x)$ is attained at $x = 9{\cdot}1$ and the values of the frequency function $f(x)$ for $x = 8{\cdot}9$, $9{\cdot}0$ and $9{\cdot}3$ are respectively equal to $0{\cdot}30$, $0{\cdot}35$, $0{\cdot}25$. Calculate the approximate value of $f(x)$ at the mode.*

Solution. We are given :

x	:	$a_0 = 8{\cdot}9$	$a_1 = 9{\cdot}0$	$a_2 = 9{\cdot}3$
$f(x)$	:	0·30	0·35	0·25

Since mode of the frequency distribution is attained at $x = 9{\cdot}1$ and we want the value of $f(x)$ at mode, we have to obtain approximate value of $f(9{\cdot}1)$. Using Lagrange's formula of interpolation we get :

$$f(x) = \frac{(x-9)(x-9{\cdot}3)}{(8{\cdot}9-9)(8{\cdot}9-9{\cdot}3)}(0{\cdot}30) + \frac{(x-8{\cdot}9)(x-9{\cdot}3)}{(9{\cdot}0-8{\cdot}9)(9{\cdot}0-9{\cdot}3)}(0{\cdot}35) + \frac{(x-8{\cdot}9)(x-9{\cdot}0)}{(9{\cdot}3-8{\cdot}9)(9{\cdot}3-9{\cdot}0)}(0{\cdot}25) \quad \ldots (*)$$

Putting $x = 9{\cdot}1$ in (*), we get :

$$f(9{\cdot}1) = \frac{(9{\cdot}1-9{\cdot}0)(9{\cdot}1-9{\cdot}3)}{(8{\cdot}9-9)(8{\cdot}9-9{\cdot}3)} \times 0{\cdot}30 + \frac{(9{\cdot}1-8{\cdot}9)(9{\cdot}1-9{\cdot}3)}{(9{\cdot}0-8{\cdot}9)(9{\cdot}0-9{\cdot}3)} \times 0{\cdot}35 + \frac{(9{\cdot}1-8{\cdot}9)(9{\cdot}1-9{\cdot}0)}{(9{\cdot}3-8{\cdot}9)(9{\cdot}3-9{\cdot}0)} \times (0{\cdot}25)$$

$$= \frac{0{\cdot}1\times(-0{\cdot}2)}{(-0{\cdot}1)\times(-0{\cdot}4)} \times 0{\cdot}30 + \frac{0{\cdot}2\times(-0{\cdot}2)}{(0{\cdot}1)\times(-0{\cdot}3)} \times 0{\cdot}35 + \frac{0{\cdot}2\times 0{\cdot}1}{0{\cdot}4\times 0{\cdot}3} \times 0{\cdot}25$$

$$= -0{\cdot}15 + 0{\cdot}47 + 0{\cdot}04 = 0{\cdot}36.$$

Example 19·27. *If four equidistant values u_{-1}, u_0, u_1, and u_2 are given and a value u_x is interpolated by Lagrange's formula, show that*

$$u_x = yu_0 + xu_1 + \frac{y(y^2-1)}{3!}\,\Delta^2 u_{-1} + \frac{x(x^2-1)}{3!}\,\Delta^2 u_0 \quad \ldots (*)$$

where $x + y = 1$.

Solution. For four equidistant arguments − 1, 0, 1 and 2, Lagrange's formula gives :

$$u_x = \left\{\frac{(x-0)(x-1)(x-2)}{(-1-0)(-1-1)(-1-2)}u_{-1} + \frac{(x+1)(x-1)(x-2)}{(0+1)(0-1)(0-2)}u_0\right.$$

$$\left.+\frac{(x+1)(x-0)(x-2)}{(1+1)(1-0)(1-2)}u_1 + \frac{(x+1)(x-0)(x-1)}{(2+1)(2-0)(2-1)}u_2\right\}$$

$$= \left\{-\frac{x(x-1)(x-2)}{6}u_{-1} + \frac{x^3-2x^2-x+2}{2}u_0 - \frac{x^3-x^2-2x}{2}u_1 + \frac{x(x^2-1)}{6}u^2\right\} \quad ...(**)$$

$$x+y=1 \quad \Rightarrow \quad y = 1-x$$

∴ R.H.S. of (*) (in the question)

$$= (1-x)u_0 + xu_1 + \frac{(1-x)[(1-x)^2-1]}{3!}\cdot \Delta^2 u_{-1} + \frac{x(x^2-1)}{6}\Delta^2 u_0$$

$$= (1-x)\,u_0 + xu_1 - \frac{x(x-1)(x-2)}{3!}\cdot(\text{E}-1)^2 u_{-1} + \frac{x(x^2-1)}{3!}(\text{E}-1)^2 u_0$$

$$= (1-x)\,u_0 + xu_1 - \frac{x(x-1)(x-2)}{3!}(u_1 - 2u_0 + u_{-1}) + \frac{x(x^2-1)}{3!}(u_2 - 2u_1 + u_0)$$

$$= \left\{-\frac{x(x-1)(x-2)}{6}u_{-1} + \frac{x^3-2x^2-x+2}{2}u_0 - \frac{x^3-x^2-2x}{2}u_1 + \frac{x(x^2-1)}{6}u^2\right\}$$

[On simplification]

$= u_x$ [From (**)].

Example 19·28. *The values of f(x) are given at a, b and c. Show that (under certain conditions to be mentioned) the maximum or minimum of f(x) is attained at*

$$x = \frac{f(a)(b^2-c^2)+f(b)(c^2-a^2)+f(c)(a^2-b^2)}{2\{f(a)(b-c)+f(b)(c-a)+f(c)(a-b)\}}\ .$$

Solution.

(*a*) By Lagrange's formula, the function $f(x)$ for the arguments a, b and c is given by :

$$f(x) = \frac{(x-b)(x-c)}{(a-b)(a-c)}f(a) + \frac{(x-a)(x-c)}{(b-a)(b-c)}f(b) + \frac{(x-a)(x-b)}{(c-a)(c-b)}f(c)$$

$$= \frac{x^2-(b+c)x+bc}{(a-b)(a-c)}f(a) + \frac{x^2-(a+c)x+ac}{(b-a)(b-c)}f(b) + \frac{x^2-(a+b)x+ab}{(c-a)(c-b)}f(c)$$

For an extremum, (maxima or minima) of $f(x)$, we have

$f'(x) = \dfrac{\partial}{\partial x}f(x) = 0$, which gives

$$\frac{2x-(b+c)}{(a-b)(a-c)}f(a) + \frac{2x-(a+c)}{(b-a)(b-c)}f(b) + \frac{2x-(a+b)}{(c-a)(c-b)}f(c) = 0$$

$$\Rightarrow \quad \frac{2x-(b+c)}{(a-b)(c-a)}f(a) + \frac{2x-(a+c)}{(a-b)(b-c)}f(b) + \frac{2x-(a+b)}{(b-c)(c-a)}f(c) = 0$$

$$\Rightarrow \quad 2x\,\{(b-c)\,f(a) + (c-a)\,f(b) + (a-b)\,f(c)\}$$

$$-\{(b^2-c^2)\,f(a) + (c^2-a^2)\,f(b) + (a^2-b^2)\,f(c)\} = 0$$

Hence, extremum is attained at

$$x = \frac{f(a)(b^2-c^2)+f(b)(c^2-a^2)+f(c)(a^2-b^2)}{2\{f(a)(b-c)+f(b)(c-a)+f(c)(a-b)\}}$$

Example 19·29. *If U_x is a rational function of the second degree, prove that :*

$$\left(\frac{\partial}{\partial x}\right) = -\left\{\frac{b+c}{(a-b)(a-c)}\cdot U_a + \frac{(c \quad a)}{(b-c)(b-a)}\cdot U_b + \frac{a+b}{(c-a)(c-b)}\cdot U_c\right\}$$

Hence, evaluate :

$$a^2(b+c)(b-c) - b^2(a+c)(a-c) + c^2(a+b)(a-b).$$

Solution. By Lagrange's formula, a rational function of second degree for the arguments a, b and c is given by : [See *Example 19·28*]

$$U_x = \frac{x^2-(b+c)x+bc}{(a-b)(a-c)}\cdot U_a + \frac{x^2-(a+c)x+ac}{(b-c)(b-c)}\cdot U_b + \frac{x^2-(a+b)+ab}{(c-a)(c-b)}\cdot U_c$$

$$\therefore \quad \left(\frac{\partial U_x}{\partial x}\right)_{x=0} = \left\{\frac{2x-(b+c)}{(a-b)(a-c)}U_a + \frac{2x-(a+c)}{(b-a)(b-c)}U_b + \frac{2x-(a+b)}{(c-a)(c-b)}U_c\right\}_{x=0} \qquad \ldots (*)$$

$$= -\left[\frac{b+c}{(a-b)(a-c)}U_a + \frac{c+a}{(b-a)(b-c)}U_b + \frac{a+b}{(c-a)(c-b)}\cdot U_c\right]$$

If we take $U_x = x^2$, then, $\frac{\partial U_x}{\partial x} = 2x \Rightarrow \left(\frac{\partial U_x}{\partial x}\right)_{x=0} = 0$

Also $U_a = a^2,\ U_b = b^2,\ U_c = c^2.$

Hence, from (*), we get:

$$\frac{a^2(b+c)}{(a-b)(a-c)} + \frac{b^2(c+a)}{(b-a)(b-c)} + \frac{c^2(a+b)}{(c-a)(c-b)} = 0$$

$$\Rightarrow \quad \frac{a^2(b+c)}{(a-b)(c-a)} + \frac{b^2(c+a)}{(a-b)(b-c)} + \frac{c^2(a+b)}{(b-c)(c-a)} = 0$$

$$\Rightarrow \quad a^2(b+c)(b-c) - b^2(a+c)(a-c) + c^2(a+b)(a-b) = 0$$

$$\Rightarrow \quad a^2(b+c)(b-c) + b^2(c+a)(c-a) + c^2(a+b)(a-b) = 0$$

$$\therefore \quad \Sigma a^2(b+c)(b-c) = 0.$$

Verification:

$$\Sigma a^2(b+c)(b-c) = \Sigma a^2(b^2-c^2) = a^2(b^2-c^2) + b^2(c^2-a^2) + c^2(a^2-b^2) = 0.$$

Example 19·30. *Prove that Lagrange's formula for the arguments $x_1, x_2, \ldots, x_n$ can be put in the form:*

$$U_x = \sum_{r-1}^{n}\left[\frac{\phi(x)\cdot U_{x_r}}{(x-x_r)\phi'(x_r)}\right], \text{ where } \phi(x) = \prod_{r=1}^{n}(x-x_r) \qquad \ldots(19·20)$$

Solution. Lagrange's formula for the arguments $x_1, x_2, \ldots\ldots, x_n$ is given by

$$U_x = \sum_r\left\{\frac{(x-x_1)(x-x_2)\cdots(x-x_{r-1})(x-x_{r+1})\cdots(x-x_n)}{(x_r-x_1)(x_r-x_2)\cdots(x_r-x_{r-1})(x_r-x_{r+1})\cdots(x_r-x_n)}\cdot U_{x_r}\right\}$$

$$= \sum_{r=1}^{n}\left\{\frac{\phi(x)}{(x-x_r)(x_r-x_1)(x_r-x_2)\cdots(x_r-x_{r-1})(x_r-x_{r-1})\cdots(x_r-x_n)}\cdot U_{x_r}\right\}, \qquad \ldots(*)$$

where $\phi(x) = \prod_{r=1}^{n}(x-x_r)$

$$\therefore \qquad \log \phi(x) = \sum_{r=1}^{n} \log(x - x_r) \quad \Rightarrow \quad \frac{\phi'(x)}{\phi(x)} = \sum_{r=1}^{n}\left(\frac{1}{x - x_r}\right)$$

i.e.

$$\phi'(x) = \phi(x)\left(\frac{1}{x-x_1} + \frac{1}{x-x_2} + \cdots + \frac{1}{x-x_{r-1}} + \frac{1}{x-x_r} + \frac{1}{x-x_{r+1}} + \cdots + \frac{1}{x-x_n}\right)$$

$$\begin{aligned} = \Big\{ & (x-x_2)(x-x_3)\ldots(x-x_r)\ldots(x-x_n) \\ & + (x-x_1)(x-x_3)\ldots(x-x_r)\ldots(x-x_n) \\ & \quad \vdots \qquad \vdots \qquad \vdots \\ & + (x-x_1)(x-x_2)\ldots(x-x_{r-2})(x-x_r)\ldots(x-x_n) \\ & + (x-x_1)(x-x_2)\ldots(x-x_{r-1})(x-x_{r+1})\ldots(x-x_n) \\ & + (x-x_1)(x-x_2)\ldots(x-x_{r-1})(x-x_r)(x-x_{r+2})\ldots(x-x_n) \\ & \quad \vdots \qquad \vdots \qquad \vdots \\ & \qquad \ldots + (x-x_1)(x-x_2)\ldots(x-x_r)\ldots(x-x_{n-1}) \Big\} \end{aligned}$$

$$\begin{aligned} \therefore \qquad \phi'(x_r) &= \left[\phi'(x)\right]_{x=x_r} \\ &= (x_r - x_1)(x_r - x_2)\ldots(x_r - x_{r-1})(x_r - x_{r+1})\ldots(x_r - x_n) \end{aligned}$$

Substituting in (*), we get

$$U_x = \sum_{r=1}^{n}\left[\frac{\phi(x)\cdot U_{x_r}}{(x - x_r)\cdot \phi'(x_r)}\right],$$

which is the required result.

19·4·5. Inverse Interpolation

So far we were given a set of values of x (arguments) and the corresponding values of $y = U_x$, (entries) and we were required to estimate $y = U_x$ for some specified value of x. Let us now consider the reverse problem stated below :

"Given a set of values of x and $y = U_x$, we are interested to find the value of x for a certain value of y". This is termed as *inverse interpolation.*

The formula for inverse interpolation is obtained from Lagrange's interpolation formula by interchanging the variables x and $y = U_x$.

Thus, for $n + 1$ arguments $a_0, a_1, a_2, \ldots a_n$, the value of x for given value of U_x is given by the formula :

$$x = \frac{(U_x - U_{a_1})(U_x - U_{a_2})\cdots(U_x - U_{a_n})}{(U_{a_0} - U_{a_1})(U_{a_0} - U_{a_2})\cdots(U_{a_0} - U_{a_n})} \times a_0 + \frac{(U_x - U_{a_0})(U_x - U_{a_2})\cdots(U_x - U_{a_n})}{(U_{a_1} - U_{a_0})(U_{a_1} - U_{a_2})\cdots(U_{a_1} - U_{a_n})} \times a_1$$

$$+ \cdots + \frac{(U_x - U_{a_0})(U_x - U_{a_1})\cdots(U_x - U_{a_{n-1}})}{(U_{a_n} - U_{a_0})(U_{a_n} - U_{a_1})\cdots(U_{a_n} - U_{a_{n-1}})} \times a_n \qquad (19·21)$$

Let us write : $y = U_x$ and $y_r = (U_x)_{x = a_r}$; $r = 0, 1, 2, \ldots, n$, so that :

Argument (x) :	a_0	a_1	a_2	a_3	...	a_n
Entry : $y = U_x$:	y_0	y_1	y_2	y_n	...	y_n

Then, the formula (19·21) can be expressed as :

$$: = \frac{(y - y_1)(y - y_2)\cdots(y - y_n)}{(y_0 - y_1)(y_0 - y_2)\cdots(y_0 - y_n)} \times a_0 + \frac{(y - y_0)(y - y_2)\cdots(y - y_n)}{(y_1 - y_0)(y_1 - y_2)\cdots(y_1 - y_n)} \times a_{\cdot}$$

$$+ \cdots + \frac{(y - y_0)(y - y_1)\cdots(y - y_{n-1})}{(y_n - y_0)(y_n - y_1)\cdots(y_n - y_{n-1})} \times a_n$$

which is more convenient form to use.

Example 19·22. *The values of x and* $y = U_x$ *are given below :*

x :	5	6	9	11
U_x :	12	13	14	16

Find the value of x when $U_x = 15$

Solution. In the usual notations of Lagranges formula, we have :

x :	$a_0 = 5$	$a_1 = 6$	$a_2 = 9$	$a_3 = 11$
$y = U_x$:	12	13	14	16

We want x when $U_x = 15$.

Using inverse interpolation formula (19·21), we get

$$x = \frac{(15-13)(15-14)(15-16)}{(12-13)(12-14)(12-16)} \times 5 + \frac{(15-12)(15-14)(15-16)}{(13-12)(13-14)(13-16)} \times 6$$

$$+ \frac{(15-12)(15-13)(15-16)}{(14-12)(14-13)(14-16)} \times 9 + \frac{(15-12)(15-13)(15-14)}{(16-12)(16-13)(16-14)} \times 11$$

$$= \frac{2 \times 1 \times (-1)}{(-1)(-2)(-4)} \times 5 + \frac{(3)(1)(-1)}{1(-1)(-3)} \times 6 + \frac{3(2)(-1)}{2(1)(-2)} \times 9 + \frac{3(2)(1)}{4(3)(2)} \times 11$$

$$= \frac{5}{4} - 6 + \frac{27}{2} + \frac{11}{4} = 11{\cdot}5.$$

19·5. NUMERICAL INTEGRATION

If the form of the function U_x is completely known, then the definite integral $\int_a^b U_x dx$, can be evaluated, provided it exists.

Now suppose that the form of the function is not completely known but we are given the values of the function U_x for a set of equidistant values of the arguments x.

Then the technique of approximately calculating the value of the integral $\int_a^b U_x dx$, is known as *numerical integration*. If U_x is a single valued function then this technique is often termed as *quadrature.*

19·5·1. General Quadrature Formula for Equidistant Ordinates

Let U_a, U_{a+h}, U_{a+2h}, ..., U_{a+nh} be $(n + 1)$ values of the entries corresponding to the equidistant arguments a, $a + h$, $a + 2h$, ..., $a + nh = b$ respectively, h being the interval of differencing, so that $b - a = nh$.

The general problem in quadrature is to represent the function U_x by a polynomial of certain degree (depending on the number of entries given) and then to evaluate the integral between desired limits.

Here, since we are given $(n+1)$ entries, U_x can be approximated by a polynomial of nth degree, so that

$$\Delta^r U_x = \begin{cases} \text{constant}, & r = n \\ 0, & r > n \end{cases} \qquad \ldots (*)$$

$$\therefore \quad I = \int_a^b U_x \, dx = \int_a^{a+nh} U_x \, dx$$

Let us take: $\quad t = \dfrac{x-a}{h}$, so that $\quad dx = h\, dt$

$$\therefore \quad I = h\int_0^n U_{a+th}\,dt = h\int_0^n E^t U_a\,dt = \int_0^n (1+\Delta)^t U_a\,dt$$

$$= h\int_0^n \left[1 + {}^tC_1\,\Delta + {}^tC_2\,\Delta^2 + \ldots + {}^tC_n\,\Delta^n\right] U_a\,dt,$$

higher differences vanish because of (*).

$$I = h\int_0^n \left[U_a + (\Delta U_a)t + \Delta^2 U_a \cdot \frac{t(t-1)}{2} + \frac{t(t-1)(t-2)}{3!}\Delta^3 U_a + \cdots \text{up to } (n+1) \text{ terms}\right] dt$$

$$= h\left[tU_a + \frac{t^2}{2}\Delta U_a + \frac{\frac{t^3}{3} - \frac{t^2}{2}}{2!}\Delta^2 U_a + \frac{\left(\frac{t^4}{4} - 3\frac{t^3}{3} + 2\cdot\frac{t^2}{2}\right)}{3!}\Delta^3 U_a + \cdots + \cdots \text{up to } (n+1) \text{ terms}\right]_0^n$$

$$\Rightarrow \quad \int_a^{a+nh} U_x dx = h\left\{nU_a + \frac{n^2}{2}\Delta U_a + \frac{\frac{n^3}{3} - \frac{n^2}{2}}{2!}\Delta^2 U_a + \frac{\frac{n^4}{4} - n^3 + n^2}{3!}\Delta^3 U_a + \cdots \text{up to } (n+1) \text{ terms}\right\} \quad \ldots (19{\cdot}22)$$

which is the *general quadrature formula.* A number of important deductions, *viz., Trapezoidal rule, Simpson's 1/3rd and 3/8th rule, Weddle's rule, etc.,* can be immediately obtained by putting $n = 1, 2, 3, 6$ respectively in the above formula (19·22).

19·5·2. Trapezoidal Rule

Taking $n = 1$ in (19·22), so that second and higher differences are neglected, we get

$$\int_a^{a+h} U_x dx = h\left[U_a + \frac{1}{2}\Delta U_a\right] = h\left[U_a + \frac{1}{2}(E-1)U_a\right]$$

$$\therefore \quad \int_a^{a+h} U_x dx = \frac{h}{2}(U_a + U_{a+h}) \quad \ldots (19{\cdot}23)$$

which is the Trapezoidal rule for the two ordinates.

Generalisation of (19·23) to (n + 1) Ordinates

Changing the origin in (19·23), we get

$$\int_{a+h}^{a+2h} U_x dx = \frac{h}{2}(U_{a+h} + U_{a+2h})$$

$$\int_{a+2h}^{a+3h} U_x dx = \frac{h}{2}(U_{a+2h} + U_{a+3h})$$

$$\vdots \qquad \vdots \qquad \vdots$$

$$\int_{a+(n-1)h}^{a+nh} U_x dx = \frac{h}{2}(U_{a+(n-1)h} + U_{a+nh}) \quad \ldots (19.23a)$$

Adding the n integrals in (19·23) and (19·23a), we get

$$\int_a^{a+nh} U_x dx = \frac{h}{2}\left[U_a + 2(U_{a+h} + U_{a+2h} + \cdots + U_{a+\overline{n-1}h}) + U_{a+nh}\right]$$

$= h$ [Mean of two extreme ordinates

+ Sum of all the intermediate ordinates], ... (19·23b)

correct upto first difference only.

(19·23b) is known as Trapezoidal rule extended to $(n + 1)$ ordinates.

***Remark.* Geometric Significance of Trapezoidal Rule**

The right hand side in (19·23) represents the area of a trapezium with parallel sides of magnitude U_a and U_{a+h}, h being the distance (altitude) between the parallel sides. Hence the name *'Trapezoidal Rule'* to the formula (19·23).

19·5·3. Simpson's One-third Rule

Taking $n = 2$ in (19·22), so that third and higher differences vanish, we get

$$\int_a^{a+2h} U_x dx = h\left[2U_a + 2\Delta U_a + \left(\frac{8}{3} - 2\right)\frac{1}{2!}\Delta^2 U_a\right]$$

$$= h\left[2U_a + 2(E-1)U_a + \frac{1}{3}(E^2 - 2E + 1)U_a\right]$$

$$= h\left[2U_a + 2U_{a+h} - 2U_a + \frac{1}{3}(U_{a+2h} - 2U_{a+h} + U_a)\right]$$

$$= \frac{h}{3}(U_a + 4U_{a+h} + U_{a+2h}), \quad \ldots (19\cdot24)$$

which is *Simpson's one-third rule.*

Changing the origin, we get

$$\left.\begin{aligned}
\int_{a+2h}^{a+4h} U_x dx &= \frac{h}{3}(U_{a+2h} + 4U_{a+3h} + U_{a+4h}) \\
\int_{a+4h}^{a+6h} U_x dx &= \frac{h}{3}(U_{a+4h} + 4U_{a+5h} + U_{a+6h}) \\
\vdots \qquad & \qquad \vdots \qquad\qquad \vdots \\
\int_{a+(2m-2)h}^{a+2mh} U_x dx &= \frac{h}{3}(U_{a+(2m-2)h} + 4U_{a+(2m-1)h} + U_{a+2mh})
\end{aligned}\right\} \quad \ldots [19\cdot24(a)]$$

Adding the n-integrals in (19·24) and (19·24a), we get

$$\int_a^{a+2mh} U_x dx = \frac{h}{3}\Big[(U_a + U_{a+2nh}) + 4(U_{a+h} + U_{a+3h} + \cdots + U_{a+(2m-1)h})$$

$$+ \left(U_{a+2h} + U_{a+4h} + \cdots + U_{a+(2m-1)h}\right) \quad \ldots [19\cdot24\,(b)]$$

$$= \frac{h}{3}\Big[\text{(sum of extreme ordinates)} + 4\,\text{(sum of odd ordinates)}$$

$$+ 2\text{(sum of even ordinates)}\Big] \quad \ldots [19\cdot24\,(c)]$$

which is the generalisation of Simpson's one-third rule to$(2m + 1)$ ordinates, correct to second differences. In particular if $a = 0$ and $h = 1$, we get

$$\int_0^{2m} U_x\, dx = \frac{1}{3}\left[(U_0 + U_{2m}) + 4\,(U_1 + U_3 + \ldots + U_{2m-1}) + 2\,(U_2 + U_4 + \ldots + U_{2m-2})\right] \quad \ldots [19\cdot24\,(d)]$$

***Remark* Geometric Significance of Simpson's 1/3rd Rule**. The geometric significance of Simpson's 1/3rd rule extended to $n + 1$ ($= 2m + 1$) ordinates is that we replace graph of the given function U_x by $(n/2) = m$ arcs of the second degree polynomials (parabolas with vertical axes).

19·5·4. Simpson's Three-eighth Rule

If we take $n = 3$ in (19·22), so that $\Delta^r U_x = 0,\ r \geq 4$, we get

$$\int_a^{a+3h} U_x dx = h\left[3U_a + \frac{9}{2}\Delta U_a + \frac{1}{2}\left(9 - \frac{9}{2}\right)\Delta^2 U_a + \frac{1}{6}\left(\frac{81}{4} - 27 + 9\right)\Delta^3 U_a\right]$$

$$= h\left[3U_a + \frac{9}{2}(E-1)\,U_a + \frac{9}{4}(E^2 - 2E + 1)\,U_a + \frac{3}{8}\,(E^3 - 3E^2 + 3E - 1)U_a\right]$$

$$= h\left[3U_a + \frac{9}{2}(U_{a+h} - U_a) + \frac{9}{4}(U_{a+2h} - 2\,U_{a+h} + U_a)\right.$$
$$\left. + \frac{3}{8}\,(U_{a+3h} - 3U_{a+2h} + 3U_{a+h} - U_a)\right]$$

$$= \frac{3h}{8}\,(U_a + 3U_{a+h} + 3U_{a+2h} + U_{a+3h}) \qquad \text{... (19·25)}$$

which is Simpson's 3/8 th rule for four ordinates.

Changing the origin in (19·25), we get

$$\left.\begin{aligned} \int_{a+3h}^{a+6h} U_x dx &= \frac{3h}{8}[U_{a+3h} + 3U_{a+4h} + 3U_{a+5h} + U_{a+6h}] \\ &\vdots \qquad\qquad \vdots \qquad\qquad \vdots \\ \int_{a+r-3h}^{a+rh} U_x dx &= \frac{3h}{8}\left[U_{a+(r-3)h} + 3U_{a+(r-2)h} + 3U_{a+(r-1)h} + U_{a+rh}\right] \end{aligned}\right\} \quad \text{... [(19·25(a)]}$$

where r is a multiple of 3.

Adding all these integrals, we get

$$\int_a^{a+rh} U_x dx = \frac{3h}{8}\left[(U_a + U_{a+rh}) + 3\{U_{a+h} + U_{a+4h} + U_{a+5h} + \ldots + U_{a+(r-2)h} + U_{a+(r-1)h}\}\right.$$
$$\left. + 2\{U_{a+3h} + U_{a+6h} + \cdots + U_{a+(r-3)h}\}\right] \quad \text{... [19·25 (b)]}$$

which is the generalisation of Simpson's 3/8th rule, correct to third differences.

In particular if we take $a = 0$ and $h = 1$ and $r = 3$, then

$$\int_0^3 U_x dx = \frac{3}{8}\,[U_0 + 3U_1 + 3U_2 + U_3] \qquad \text{... [19·25(c)]}$$

and in general

$$\int_0^{3m} U_x dx = \frac{3}{8}[(U_0 + U_{3m}) + 3\{U_1 + U_2 + U_4 + U_5 + \ldots U_{3m-2} + U_{3m-1}\}$$
$$+ 2\{U_3 + U_6 + \ldots + U_{3m-3}\}] \quad \text{... [19·25(d)]}$$

Remarks:

1. Simpson's 1/3rd rule can be used for an odd number, (say, 3, 5, 7, ...etc.) of equidistant ordinates and Simpson's 3/8th rule can be used if the number of equidistant ordinates is 4, 7, 10, 13, etc. Thus for 7, 13, 19, etc. equidistant ordinates, Simpson's 1/3rd as well as 3/8th rules can be used. But the later will give better approximation since in the case of 1/3rd rule differences up to 2nd order are used while 3/8th rule takes into account third order differences also.
2. **Geometric Significance of Simpson's 3/8th Rule**

   The geometric significance of Simpson's 3/8th Rule extended to $n + 1$ (= $3m + 1$) ordinates is that we replace the graph of the given function U_x by $(n/3) = m$ arcs of third degree polynomials.
3. In the above sections, we have deduced Simpson's 1/3 and 3/8th rules from the general quadrature formula for equidistant arguments. We shall now obtain Simpson's 1/3rd and 3/8th rules independently. [See Remarks 4 and 5]

4. Aliter. Simpson's One-third Rule. Let us suppose we are given three entries U_0, U_1 and U_2 corresponding to equidistant arguments 0, 1 and 2 at differencing interval of unity. Then U_x can be approximated by a second degree polynomial. Let $U_x = a + bx + cx^2$ ... (*)

Then
$$\int_0^2 U_x\, dx = \int_0^2 (a+ bx + cx^2)\, dx = 2a + 2b + \frac{8}{3}c \qquad \text{... (**)}$$

Now, the problem is to determine a, b and c in terms of U_0, U_1 and U_2. We have from (*),
$$U_0 = a, \quad U_1 = a + b + c, \quad U_2 = a + 2b + 4c$$

Solving for a, b and c, we get
$$a = U_0, \quad b = \frac{1}{2}(4U_1 - 3U_0 - U_2), \quad c = \frac{1}{2}(U_2 - 2U_1 + U_0)$$

Substituting in (**), we get
$$\int_0^2 U_x\, dx = \left[2U_0 + (4U_1 - 3U_0 - U_2) + \frac{8}{3}\cdot\frac{1}{2}(U_2 - 2U_1 + U_0)\right]$$
$$= \frac{1}{3}(U_0 + 4U_1 + U_2)$$

which is Simpson's one-third rule.

5. Aliter. Simpson's Three-eighth Rule. Here we shall consider four entries U_0, U_1, U_2 and U_3 corresponding to equidistant arguments 0, 1, 2 and 3, and we want to evaluate $\int_0^3 U_x\, dx$ in terms of U_0, U_1, U_2, and U_3.

Doubling the scale, we shall get $\int_0^6 U_x\, dx$ in terms of U_0, U_2, U_4 and U_6 and on shifting the origin to −3, we get
$$\int_{-3}^3 U_x\, dx \equiv A(U_{-1} + U_1) + B(U_{-3} + U_3), \qquad \text{... (*)}$$
where A and B are constants to be determined.

Since we are given four entries, we may take
$$U_x = a + bx + cx^2 + dx^3$$
$$\therefore \quad U_{-1} + U_1 = 2(a + c) \quad \text{and} \quad U_{-3} + U_3 = 2(a + 9c)$$

Hence, from (*), we get
$$\int_{-3}^3 (a + bx + cx^2 + dx^3)\, dx \equiv A.\, 2(a + c) + B.\, 2(a + 9c)$$
$$\Rightarrow \quad \left| ax + \frac{bx^2}{2} + \frac{cx^3}{3} + \frac{dx^4}{4} \right|_{-3}^{3} \equiv A.\, 2(a + c) + B.\, 2(a + 9c)$$
$$\Rightarrow \quad 2(3a + 9c) \equiv A.\, 2(a + c) + B.\, 2(a + 9c)$$

Comparing co-efficient of a and c on both sides, we get: $3 = A + B$ and $9 = A + 9B$

Solving, we get $A = \dfrac{9}{4}$ and $B = \dfrac{3}{4}$.

Substituting in (*), we get
$$\int_{-3}^3 U_x\, dx = \frac{3}{4}[3(U_{-1} + U_1) + (U_{-3} + U_3)]$$
$$\Rightarrow \quad \int_0^6 U_x\, dx = \frac{3}{4}[3(U_2 + U_4) + (U_0 + U_6)] \qquad \text{(Changing the origin)}$$
$$\Rightarrow \quad \int_0^3 U_x\, dx = \frac{3}{8}[3(U_1 + U_2) + (U_0 + U_3)] \qquad \left[\text{Making scale } \frac{1}{2}\right]$$
$$= \frac{3}{8}[U_0 + 3U_1 + 3U_2 + U_3],$$

which is the required formula for Simpson's 3/8th rule.

19·5·5. Weddle's Rule

Taking $n = 6$ in (19·22), we get:

$$\int_a^{a+6h} U_x\,dx = h\left[6U_a + 18\Delta U_a + 27\Delta^2 U_a + 24\Delta^3 U_a + \frac{123}{10}\Delta^4 U_a + \frac{33}{10}\Delta^5 U_a + \frac{41}{140}\Delta^6 U_a + \ldots\right]$$

[Since only seven entries are given, $\Delta^6 U_a$ will be constant and higher differences will be zero.]

If we take $(41/140) = (3/10)$, in the last term, the error involved is $(h/140)$. $\Delta^6 U_a$ which will be negligible if the interval of differencing h is such that sixth differences are small. Hence, the above result can be approximated by:

$$\int_a^{a+6h} U_x dx = \left[6U_a + 18\Delta U_a + 27\Delta^2 U_a + 24\Delta^3 U_a + \frac{123}{10}\Delta^4 U_a + \frac{33}{10}\Delta^5 U_a + \frac{3}{10}\Delta^6 U_a + \ldots\right]$$

$$= \frac{3h}{10}\left[U_a + 5U_{a+h} + U_{a+2h} + 6U_{a+3h} + U_{a+4h} + 5U_{a+5h} + U_{a+6h}\right] \quad \text{(On simplification)} \quad \ldots(19{\cdot}26)$$

Generalisation. Changing the origin from a to $a+6h$, $a+12h$, ..., $a + (r-6)h$ successively in (19·26) and adding the resulting equations we get:

$$\int_a^{a+rh} U_x dx = \frac{3h}{10}[U_a + 5U_{a+h} + U_{a+2h} + 6U_{a+3h} + U_{a+4h} + 5U_{a+5h} + 2U_{a+6h} + U_{a+7h} + \ldots + U_{a+rh}] \quad (19{\cdot}27)$$

where r is a multiple of 6. Formulae (19·26) and (19·27) and known as Weddle's rule and give more accurate results than Simpson's formulae but these can be used only for 7, 13, 19, 25, ... equidistant ordinates.

Example 19·33. *Use Simpson's one-third rule to estimate approximately the area of the cross-section of a river 80 feet wide, the depth d (in feet) at a distance x from one bank being given by the following table :*

x :	0	10	20	30	40	50	60	70	80
d :	0	4	7	9	12	15	14	8	3

Solution. Area of the cross-section of a river 80 feet wide is given by $\int_0^{80} U_x dx$, *where* $U_x = d =$ depth at a distance x from one bank. Here we are given nine equidistant arguments at an interval of $h = 10$. Hence, using extended form of Simpson's $\frac{1}{3}$rd rule, we get

$$\int_0^{80} U_x\,dx = \frac{h}{3}\left[(\text{Sum of extreme ordinates}) + 4\,(\text{sum of odd ordinates}) + 2\,(\text{sum of even ordinates})\right]$$

$$= \frac{10}{3}\{(0+3) + 4\,(4+9+15+8) + 2(7+12+14)\}$$

$$= 10\,[1 + 48 + 22] = 710 \text{ sq. ft. approx.}$$

Example 19·34. *Find an approximate value of* $\int_{1.00}^{1.04} y dx$, *given the following values :*

x:	1·00	1·01	1·02	1·03	1·04
y:	3·953	4·066	4·182	4·300	4·421

Solution. If we take $y = U_x$, then using Simpson's $\frac{1}{3}$rd rule with $h = 0{\cdot}01$, we get

$$\int_{1.00}^{1.04} U_x dx = \frac{0\cdot 01}{3}\left[(3{\cdot}953 + 4{\cdot}421) + 4 \times (4{\cdot}066 + 4{\cdot}300) + 2 \times 4{\cdot}182)\right]$$

$$= \frac{0{\cdot}01 \times 50{\cdot}202}{3} = 0{\cdot}16734$$

Example 19·35. *Evaluate* $\log_e 7$ *by Simpson's (i)* $\frac{1}{3}$*rd, and (ii)* $\frac{3}{8}$*th rules.*

Solution. Let us consider the integral,

$$\int_0^6 \frac{1}{1+x}\,dx = \left|\log_e(1+x)\right|_0^6 = \log_e 7$$

$$\therefore \qquad \log_e 7 = \int_0^6 U_x dx, \text{ where } U_x = \frac{1}{1+x}$$

$$\Rightarrow \quad U_0 = 1, \quad U_1 = \frac{1}{2}, \quad U_2 = \frac{1}{3}, \quad U_3 = -, \quad U_4 = \frac{1}{5}, \quad U_5 = \frac{1}{6}, \quad \text{and} \quad U_6 = \frac{1}{7}$$

Hence, using *Simpson's* $\frac{1}{3}$rd *rule* for 7 ordinates at equal interval of unity, we get

$$\log_e 7 = \int_0^6 U_x dx = \frac{1}{3}\{(U_0 + U_6) + 4(U_1 + U_3 + U_5) + 2(U_2 + U_4)\}$$

$$= \frac{1}{3}\left\{\left(1+\frac{1}{7}\right) + 4\left(\frac{1}{2}+\frac{1}{4}+\frac{1}{6}\right) + 2\left(\frac{1}{3}+\frac{1}{5}\right)\right\} = 1{\cdot}9588$$

By Simpson's $\frac{3}{8}$ th rule for seven ordinates, we get

$$\int_0^6 U_x dx = \int_0^3 U_x dx + \int_3^6 U_x dx$$

$$= \frac{3}{8}(U_0 + 3U_1 + 3U_2 + U_3) + \frac{3}{8}(U_3 + 3U_4 + 3U_5 + U_6)$$

$$= \frac{3}{8}\left[(U_0 + U_6) + 2U_3 + 3(U_1 + U_2 + U_4 + U_5)\right]$$

$$= \frac{3}{8}\left[\left(1+\frac{1}{7}\right) + 2\times\frac{1}{4} + 3\left(\frac{1}{2}+\frac{1}{3}+\frac{1}{5}+\frac{1}{6}\right)\right] = 1{\cdot}966.$$

Example 19·36. *If* $U_x = a + bx + cx^2$, *prove that*

$$\int_1^3 U_x dx = 2U_2 + \frac{1}{12}(U_0 - 2U_2 + U_4)$$

and hence find an approximate value for

$$\int_{-\frac{1}{2}}^{\frac{1}{2}} \exp(-x^2/10)\,dx$$

Solution. Shifting the origin, we have to prove that:

$$\int_{-1}^{1} U_x dx = 2U_0 + \frac{1}{12}(U_{-2} - 2U_0 + U_2) \qquad \ldots (*)$$

$$\text{L.H.S. of } (*) = \int_{-1}^{1}(a + bx + cx^2)dx = \left|ax + \frac{bx^2}{2} + \frac{cx^3}{3}\right|_{-1}^{1} = 2\left(a + \frac{c}{3}\right)$$

$$U_x = a + bx + cx^2$$

$$U_0 = a, \qquad U_{-2} = a - 2b + 4c, \qquad U_2 = a + 2b + 4c$$

$$\therefore \quad \text{R.H.S. of } (*) = 2U_0 + \frac{1}{12}(U_{-2} - 2U_0 + U_2) = 2a + \frac{1}{12}\left[2(a + 4c) - 2a\right]$$

$$= 2a + \frac{8}{12}c = 2\left(a + \frac{c}{3}\right) = \text{L.H.S.}$$

Changing the scale to 1/2 in (*), we get

$$\int_{-\frac{1}{2}}^{\frac{1}{2}} U_x\,dx = \frac{1}{2}\left[2U_0 + \frac{1}{12}(U_{-1} - 2U_0 + U_1)\right]$$

Taking $U_x = \exp(-x^2/10)$, we get

$$\int_{-\frac{1}{2}}^{\frac{1}{2}} \exp(-x^2/10)\, dx = 1 + \frac{1}{24}\left[2e^{-0.1} - 2\right].$$

Example 19·37. *If third differences are constant, prove that*

$$\int_{-1}^{1} f(x)\, dx = \frac{2}{3}\left[f(0) + f\left(1/\sqrt{2}\right) + f\left(-1/\sqrt{2}\right)\right]$$

Solution. Since third differences are constant, $f(x)$ can be taken as a 3rd degree polynomial in x. Let.

$$f(x) = a + bx + cx^2 + dx^3 \qquad \text{... (*)}$$

$$\text{L.H.S.} = \int_{-1}^{1} f(x)\, dx = \int_{-1}^{1} (a + bx + cx^2 + dx^3)\, dx = 2\left(a + \frac{c}{3}\right)$$

$$\text{R.H.S.} = \frac{2}{3}\left[a + \left(a + \frac{b}{\sqrt{2}} + \frac{c}{2} + \frac{d}{2\sqrt{2}}\right) + \left(a - \frac{b}{\sqrt{2}} + \frac{c}{2} - \frac{d}{2\sqrt{2}}\right)\right] \qquad \text{[Using (*)]}$$

$$= \frac{2}{3}\left[a + 2\left(a + \frac{c}{2}\right)\right] = \frac{2}{3}(3a + c) = 2\left(a + \frac{c}{3}\right) = \text{L.H.S.}$$

Example 19·38. *If the third order difference are constant, prove that*

$$\int_0^2 U_x dx = \frac{1}{24}\left(U_{-1/2} + 23U_{1/2} + 23U_{3/2} + U_{5/2}\right) \qquad \text{... (*)}$$

Solution. Since the 3rd order differences are given to be constant, we may take U_x as a polynomial of 3rd degree in x

Let $\qquad U_x = a + bx + cx^2 + dx^3 \qquad$ ... (**)

Doubling the scale in (*), we have to prove

$$\int_0^4 U_x\, dx = \frac{1}{12}\left(U_{-1} + 23\, U_1 + 23\, U_3 + U_5\right)$$

Now, changing the origin to 2, we have to prove that

$$\int_{-2}^{2} U_x dx = \frac{1}{12}\left(U_{-3} + 23\, U_{-1} + 23\, U_1 + U_3\right) \qquad \text{... (***)}$$

$$\text{R.H.S. of (***)} = \frac{1}{12}\{(U_{-3} + U_3) + 23\,(U_{-1} + U_1)\}$$

$$= \frac{1}{12}\{2(a + 9c) + 23 \times 2\,(a + c)\} \qquad \text{[Using (**)]}$$

$$= \frac{1}{6}\left(24a + 32c\right) = 4a + \frac{16}{3}c$$

$$\text{L.H.S. of (***)} = \int_{-2}^{2} (a + bx + cx^2 + dx^3)\, dx = \left| ax + b\frac{x^2}{2} + c\cdot\frac{x^3}{3} + d\cdot\frac{x^4}{4}\right|_{-2}^{2}$$

$$= 2\left(2a + \frac{8c}{3}\right) = 4a + \frac{16}{3}c = \text{R.H.S.}$$

Example 19·39. *Obtain the approximate quadrature formula:*

$$\int_{-1}^{1} U_x\, dx = \frac{1}{12}\left[13\,(U_1 + U_{-1}) - (U_3 + U_{-3})\right]$$

Solution. Since, on the right hand side we are given four entries *viz.*, U_{-3}, U_{-1}, U_1 and U_3, U_x can be taken as a 3rd degree polynomial.

Let, $\qquad U_x = a + bx + cx^2 + dx^3$

$$\therefore \quad \text{L.H.S.} = \int_{-1}^{1} U_x\,dx = \int_{-1}^{1} (a + bx + cx^2 + dx^3)\,dx$$

$$= \left| ax + b \cdot \frac{x^2}{2} + c \cdot \frac{x^3}{3} + d \cdot \frac{x^4}{4} \right|_{-1}^{1} = 2\left(a + \frac{c}{3}\right)$$

$$\text{R.H.S.} = \frac{1}{12}\left[13\,(U_1 + U_{-1}) - (U_3 + U_{-3})\right] = \frac{1}{12}\left[13\,(a + c)\,2 - (a + 9c)\,2\right]$$

$$= \frac{1}{12}(24a + 8c) = 2\left(a + \frac{c}{3}\right) = \text{L.H.S.}$$

Example 19·40. *If U_x is a function whose fifth differences are constant and if $\int_{-1}^{1} U_x\,dx$ can be expressed in the form $pu_{-\alpha} + qu_0 + pu_{\alpha}$, find p, q and α. Use the formula to find $\log_e 2$ to four decimal places from the integral $\int_0^1 \frac{dx}{1+x}$.*

Solution. Let us take

$$U_x = a + bx + cx^2 + dx^3 + ex^4 + fx^5, \quad \text{... (1)}$$

so that fifth differences are constant.

Let
$$\int_{-1}^{1} U_x\,dx = p\,u_{-\alpha} + q\,u_0 + pu_{\alpha} = p\,(u_{-\alpha} + u_{\alpha}) + qu_0 \quad \text{... (2)}$$

$$\therefore \quad \left| ax + b\frac{x^2}{2} + c.\frac{x^3}{3} + d.\frac{x^4}{4} + e.\frac{x^5}{5} + f.\frac{x^6}{6} \right|_{-1}^{1} = p\left[2(a + c\alpha^2 + e\alpha^4)\right] + qa$$

$$\Rightarrow \quad 2\,(a + c/3 + e/5) = (2p + q)\,a + 2p\alpha^2 c + 2p\alpha^4 e$$

Comparing co-efficients of a, c and e on both sides, we get

$$2p + q = 2 \quad \text{... (3)} \qquad 2p\alpha^2 = \frac{2}{3} \quad \text{... (4)} \qquad 2p\alpha^4 = \frac{2}{5} \quad \text{... (5)}$$

Dividing (5) by (4), we get

$$\alpha^2 = \frac{3}{5} = 0{\cdot}6 \quad \Rightarrow \quad \alpha = +\sqrt{0{\cdot}6} \quad \text{or} \quad \alpha = -\sqrt{0{\cdot}6}$$

Substituting in (4), we get $\quad 2p \cdot \frac{3}{5} = \frac{2}{3} \quad \Rightarrow \quad p = \frac{5}{9}$

From (3), $\quad q = 2 - 2p = \frac{8}{9}$

$$\therefore \quad \int_{-1}^{1} U_x dx = \frac{5}{9}\left(U_{\sqrt{0\cdot6}} + U_{-\sqrt{0\cdot6}}\right) + \frac{8}{9}U_0 \quad \text{... (6)}$$

Changing the origin, we get

$$\int_0^2 U_x dx = \frac{5}{9}\left(U_{1+\sqrt{0\cdot6}} + U_{1-\sqrt{0\cdot6}}\right) + \frac{8}{9}U_1$$

Making the scale $\frac{1}{2}$, we have

$$\int_0^1 U_x dx = \frac{1}{2}\left[\frac{5}{9}\left\{\left(U_{\frac{1+\sqrt{0\cdot6}}{2}} + U_{\frac{1-\sqrt{0\cdot6}}{2}}\right) + \frac{8}{9}U_{0.5}\right\}\right] \quad \text{... (7)}$$

We have $\quad \int_0^1 \frac{1}{1+x}dx = \left|\log_e(1+x)\right|_0^1 = \log_e 2$

Taking $\quad U_x = \frac{1}{1+x}$ in (7) and simplifying, we get

$$\log_e 2 = \int_0^1 \frac{dx}{1+x} = \frac{4}{9}U_{0.5} + \frac{5}{18}\left(U_{\frac{1+\sqrt{0\cdot6}}{2}} + U_{\frac{1-\sqrt{0\cdot6}}{2}}\right) = 0{\cdot}694.$$

DISCUSSION AND REVIEW QUESTIONS

1. Define the operators E and Δ and show that

(i) $E \equiv 1 + \Delta$ (ii) $E^n f(a) = f(a + nh)$

(iii) $E^n \equiv (1 + \Delta)^n$ (iv) $E\Delta \equiv \Delta E$

2. Define the operators D, E and Δ, used in the calculus of finite differences and prove that

$$E = 1 + \Delta = e^D$$

Evaluate : $\left(\frac{\Delta^2}{E}\right)e^x \times \frac{Ee^x}{\Delta^2 e^x}$, interval of differencing being h.

3. What is the difference between $\left(\frac{\Delta U_x}{EU_x}\right)^2$ and $\left(\frac{\Delta^2 U_x}{E^2 U_x}\right)$

If $U_x = x^2$ and interval of differencing 1, find out the expressions for both.

4. Explain the difference between $\left(\frac{\Delta}{E}\right)^2 U_x$ and $\left(\frac{\Delta^2 U_x}{E^2 U_x}\right)$ and find the values of these functions when $U_x = x^3$.

5. What do you understand by interpolation ? What are the underlying assumptions for the validity of the various methods used for interpolation.

6. Prove Newton's interpolation formula for equidistant interval of units, *viz.*,

$$U_{x+n} = U_x + {}^nC_1 \Delta U_x + {}^nC_2 \Delta^2 U_x + \ldots + {}^nC_n \Delta^n U_x,$$

where n is a positive integer.

7. Establish Newton's Forward difference interpolation formula

$$U_x = U_0 + x\Delta U_0 + \frac{x(x-1)}{2!}\Delta^2 U_0 + \ldots + \frac{x(x-1)\cdots(x-n+1)}{n!}\Delta^n U_0$$

(i) Discuss the suitability of using the above formula.

(ii) If $U_x = \sin \pi x$, then do you expect the above interpolation formula to be good one ? If not, why not ?

(iii) If U_x were known to be a polynomial of degree not exceeding 5, then how many terms in the right-hand expression will give you an exact fit?

8. Show that Newtons-Gregory interpolation formula can be put in the form,

$$U_x = U_0 + x\Delta U_0 - xa\, \Delta^2 U_0 + xab\, \Delta^3 U_0 - xabc\, \Delta^4 U_0 + \ldots,$$

where, $a = 1 - \frac{1}{2}(x+1),\quad b = 1 - \frac{1}{3}(x+1),\quad c = 1 - \frac{1}{4}(x+1)$ etc.

9. Establish Newton's backward formula for interpolation. In which situation is it most appropriate ?

10. What are divided differences ? Prove that divided differences of order n of polynomial of nth degree are constant.

11. (a) Show that:

(i) a divided difference is a symmetric function of its arguments;

(ii) if all the intervals are equal to h, then

$$\Delta^n \equiv \frac{1}{h^n n!}\Delta^n$$

(b) The nth order divided difference $U[x_1\, x_2 \ldots x_{n+1}]$ for $U_x = \frac{1}{x}$ is equal to $\frac{(-1)^n}{x_1 x_2 \ldots x_{n+1}}$

12. (a) State Newtons divided difference interpolation formula and deduce thereform the Newton-Gregory interpolation formula.

(*b*) Derive Newton's divided difference formula *viz.*,

$$U_x = U_a + (x-a)\underset{b}{\Delta} U_a + (x-a)(x-b)\underset{bc}{\Delta^2} U_a + \ldots + (x-a)(x-b)\ldots\ldots(x-k)\underset{bc\ldots kl}{\Delta^n} U_a.$$

where $a, b, c, \ldots, k, l$ are $(n+1)$ arguments.

13. Establish Lagrange's formula for interpolation indicating the circumstances when it is most appropriate. Is there any difference between this and Newton's divided difference formula based on the same number of points.

14. Obtain the quadrature formula

$$\int_{x_0}^{x_0+nh} y_x\,dx = h\left[ny_0 + \frac{n^2}{2}\Delta y_0 + \left(\frac{n^3}{3} - \frac{n^2}{2}\right)\frac{\Delta^2 y_0}{2!} + \left(\frac{n^4}{4} - n^3 + n^2\right)\frac{\Delta^3 y_0}{3!} + \cdots \text{ to } (n+1) \text{ terms}\right],$$

where, $\quad y_0 = (y_x)_{x\,=\,x_0}$

15. Discuss briefly the (*i*) Trapezoidal rule and (*ii*) Simpsons's one-third and three-eighth rules, for numerical integration and give their geometrical significance.

ASSORTED REVIEW PROBLEMS FOR SELF-ASSESSMENT

19·1. (*a*) By constructing the difference table as well as using the relation E = 1 + Δ find

(*i*) 5th term of the series 6, 10, 17, 27, ...

(*ii*) 6th term and the 7th term of the series 1, 1, 3, 7, 13 ...

(*iii*) First term of the series whose second term and the subsequent terms are 8, 3, 0, – 1, 0, ...

(*b*) Discuss that E and Δ operators of interpolation theory obey the distributive, the commutative, the associative and indices laws of algebra.

19·2. (*a*) Simplify (*i*) $\Delta^2 x^3$ (*ii*) $\Delta^2 (3e^x)$ (*iii*) $\Delta^2 \cos 2x$.

(*b*) Prove that (*i*) $\Delta ab^{cx} = (b^c - 1)\, ab^{cx}$, (*ii*) $\Delta x^{(r)} = r x^{(r-1)}$

(*c*) Evaluate (*i*) $\Delta^2 (ab^{cx})$; (*ii*) $\Delta^n e^x$; (*iii*) $\Delta^n e^{ax+b}$; (*iv*) $\Delta^2 (ab)^{cx}$

19·3. Find

(*i*) $\Delta^3 \left(ax^3 + bx^2 + cx + d\right)$, interval of differencing is h.

(*ii*) $\Delta^2\left\{\dfrac{7x+10}{2x^2+5x+9}\right\}, h = 1$ (*iii*) $\Delta^2\left\{\dfrac{a^{2x}+a^{4x}}{\left(a^2-1\right)^2}\right\}, h = 1$

19·4. (*a*) If $U_x = \sin x$, show that $\Delta^2 U_x = k\mathrm{E}U_x$, where k is a constant and $\Delta x = h$.

(*b*) Prove that if the interval of differencing is h,

$$\Delta \sin x = 2\sin\frac{h}{2}\cdot\cos\left(x+\frac{h}{2}\right), \qquad \Delta \cos x = -2\sin\frac{h}{2}\cdot\sin\left(x+\frac{h}{2}\right)$$

(*c*) Prove that $\quad \Delta \cot (a + bx) = \dfrac{-\sin b}{\sin(a+bx)\sin(a+b+bx)}$

(*d*) If $U_x = \sin^{-1} x$, and the interval of differencing is $\dfrac{14}{65}$, prove that $\Delta U_{5/13} = \sin^{-1}\left(\dfrac{16}{65}\right)$

19·5. (*a*) Prove that the rth difference of a polynomial of degree n is a polynomial of degree $(n-r)$, if $r < n$. Find out the value of the rth difference for $r = n$ and for $r > n$. Hence obtain :

(*i*) $\Delta^6 (k - ax)(k - bx^2)(k - cx^3)$ (*ii*) $\Delta^{12}\left[(a-kx)(b-kx^2)(c-kx^3)(d-kx^4)\right]$

(*b*) Define $x^{(m)}$ and deduce that

(*i*) $\Delta^m x^{(m)} = m\,!$ (*ii*) $\Delta x^{(-m)} = -m x^{(-m-1)}$

19·6. (*a*) Express $(x^3 - 3x + 1)$ in factorials. Hence or otherwise find its 3rd difference.

(*b*) Express $U_x = x^3 + 3x^2 + 5x + 12$, in the factorial notation. Hence or otherwise find the value of ΔU_2.

(c) Represent the function $x^4 - 12x^3 + 42x^2 - 30x + 9$ and its successive differences in factorial notation.

(d) Obtain a function whose first difference is $x^3 + 3x^2 + 5x + 12$.

19·7. Use the method of separation of symbols to prove the following identities :

(a) $U_x = U_{x-1} + \Delta U_{x-2} + \Delta^2 U_{x-3} + \ldots + \Delta^{n-1} U_{x-n} + \Delta^n U_{x-n}$.

(b) $U_0 + {}^xC_1\, \Delta U_1 + \binom{x}{2}\Delta^2 U_2 + \ldots = U_x + {}^xC_1\, \Delta^2 U_{x-1} + {}^xC_2\, \Delta^4 U_{x-2} + \ldots$

(c) $U_{x+n} = U_n + {}^xC_1\, \Delta U_{n-1} + {}^{x+1}C_2\, \Delta^2 U_{n-2} + {}^{x+1}C_3\, \Delta^3\, U_{n-3} + \ldots$

(d) $\Delta U_x + \Delta U_{x+1} + \Delta U_{x+2} + \cdots + \Delta U_{x+n-1} = U_{x+n} - U_x$

(e) Prove that:

$$\Delta^n f(x) = \sum_{r=0}^{n} (-1)^{n-r} \binom{n}{r} f(x + rh),$$

where h is the interval of differencing.

Hint: Simplify R.H.S.

19·8 (a) If U_x is a function whose differences when the increment of x is unity are denoted by δU_x, $\delta^2 U_x$, ..., and by ΔU_x, $\Delta^2 U_x$,..., when the increment of x is n, then if $\delta^2 U_x$, $\delta^2 U_{x+1}$, $\delta^2 U_{x+2}$, ... are in G. P. with common ratio q, show that:

$$\frac{\Delta U_x - n\delta U_x}{(q^n - 1) - n(q-1)} = \frac{\delta^2 U_x}{(q-1)^2}$$

(b) Find the value of $x^{(m)} - 2\,\Delta x^{(m)} + 3\Delta^2 x^{(m)} - 4\Delta^3 x^{(m)} + \ldots$ $(m + 1)$ terms.

19·9. (i) Estimate the missing figure in the following tables :

x:	1	2	3	4	5
$f(x)$:	2	5	7	—	32

(ii)

Year :	1980	1985	1990	1995	2000
Sale of cloth in lakh yards :	250	285	328	—	444

19·10. (a) Complete the following table :

x:	10	15	20	25	30	35
$f(x)$:	43	—	29	32	—	77

(b) Find out the missing values in the following table:

x:	2.0	2.1	2.2	2.3	2.4	2.5	2.6
$f(x)$:	0·135	—	0·111	0·100	—	0·082	0·074

19·11. (a) If $U_0, U_1, \ldots, U_6$ be the consecutive terms of a series, prove that

$$U_3 = 0{\cdot}05\,(U_6 + U_0) - 0{\cdot}3\,(U_1 + U_5) + 0{\cdot}75\,(U_2 + U_4)$$

(b) Given $U_0 + U_8 = x$, $U_1 + U_7 = y$, $U_2 + U_6 = z$ and $U_3 + U_5 = t$, find U_4.

(c) The function $U_x = 3^x$, takes (as it should), the values 1, 3, 9 and 81 for $x = 0, 1, 2,$ and 4 respectively. Use any method, of finite differences to estimate U_x for $x = 3$. Explain why this result differs from the actual value, *viz.,* 3^3, *i.e,* 27.

19·12. (a) Prove that

(i) $(x + 5)^4 = x^4 + 5\,\Delta x^4 + 10\,\Delta^2 x^4 + 10\Delta^3 x^4 + 5\Delta^4 x^4 + \Delta^5 x^4$

(ii) $\Delta x(x + 1) = 2(x + 1)$,

(iii) $\Delta\,[x\,(x + 1)\,(x + 2)] = 3(x + 2)\,(x + 1)$

(b) Prove that

(i) $U_4 = U_3 + \Delta U_2 + \Delta^2 U_1 + \Delta^3 U_1$

(ii) $U_4 = U_3 + 4\Delta\, U_0 + 6\Delta^2 U_{-1} + 10\Delta^2 U_{-1}$ as far as third differences.

19·13. (a) A second degree polynomial passes through (0,1), (1, 3), (2, 7) and (3,13). Find the polynomial.

(b) Find the polynomial of 4th degree from the following data :

$$U_0 = 1, \quad U_1 = 1, \quad U_2 = 11, \quad U_3 = 61, \quad U_4 = 205, \quad U_5 = 521$$

(c) How do you interpolate a value of the given function by using difference table when the values of the given function are given at non-equidistant values of the argument ?

(d) Find the polynomial $f(x)$ of the lowest possible degree which assumes the values 3, 12, 15 and – 20 when x has the values 3, 9, 1, and – 1, respectively.

19·14. (a) If n is a positive integer, prove that: (i) $\Delta^n x^{(n)} = n\,!$, (ii) $\Delta^{n+1} x^{(n)} = 0$

(b) Evaluate (i) $\Delta^n \left(\frac{1}{x}\right)$ (ii) $\Delta^2 \left[\frac{a^{2x} + a^{4x}}{(a^2-1)^2}\right]$, (iii) $\Delta^2 \left[\frac{4^{2x} + 4^{4x}}{225}\right]$

19.15. (a) Express the function $f(x) = 10x^5 + 15x^3 - 12x^2 + 10$ and its successive differences in factorial notation.

(b) Obtain a function whose first difference is $x^2 + 3x^3 + 5x + 12$.

(c) If $U_x = (1 + a_1 x)(1 + a_2 x) \ldots (1 + a_n x)$, then prove that $\Delta^n U_x = a_1 a_2 \ldots a_n\, n!$

(d) If $U_x = (a_1 + x)(a_2 + x) \ldots (a_n + x)$, then prove that $\Delta^n U_x = n!$

19·16. (a) Use the method of separation of symbols to prove the following identities:

(i) $U_0 + U_1 + U_2 + \ldots + U_n = {}^{n+1}C_1 U_0 + {}^{n+1}C_2 \Delta U_0 + {}^{n+1}C_3 \Delta^2 U_0 + \ldots + \Delta^n U_0$

(ii) $U_{x+n} = U_n + {}^{x}C_1 \Delta U_{n-1} + {}^{x+1}C_2 \Delta^2 U_{n-2} + {}^{x+2}C_3 \Delta^3 U_{n-3} + \ldots$

(b) $U_0 + {}^{x}C_1 \Delta U_1 + {}^{x}C_2 \Delta^2 U_2 + {}^{x}C_3 \Delta^3 U_3 + \ldots = U_x + {}^{x}C_1 \Delta^2 U_{x-1} + {}^{x}C_2 \Delta^4 U_{x-2} + \ldots$

19·17. Prove the following identity.

$$\sum_{x=0}^{\infty} U_{2x} = \frac{1}{2}\sum_{x=0}^{\infty} U_x + \frac{1}{4}\left(1 - \frac{\Delta}{2} + \frac{\Delta^2}{2} - \cdots\right) U_0$$

19·18. Prove that:

$$U_{2n} - {}^{n}C_1 2 . U_{2n-1} + {}^{n}C_2 2^2 . U_{2n-2} + \cdots + (-2)^n U_n = (-1)^n (c - 2an),$$

where $U_n = an^2 + bn + c$ and $\Delta n = 1$.

19·19. Prove that

$$\sum_{r=0}^{n-1} U_r x^r = \frac{U_0 - x^n U_n}{1-x} + \frac{x}{(1-x)^2}\left(\Delta U_0 - x^n \Delta U_n\right) + \frac{x^2}{(1-x)^3}\left(\Delta^2 U_0 - x^n \Delta^2 U_n\right) + \cdots$$

19·20. Prove that

$$(U_1 - U_0) - x(U_2 - U_1) + x^2(U_3 - U_2) - x^3(U_4 - U_3) + \ldots = \frac{\Delta U_0}{1+x} - \frac{x\Delta^2 U_0}{(1+x)^2} + \frac{x^2 \Delta^2 U_0}{(1+x)^3} + \cdots$$

19·21. If $y_n = (-1)^{n-1}\left\{\left(\frac{\frac{1}{2}}{n+1}\right) + \left(\frac{-\frac{1}{2}}{n+1}\right)\right\}$, prove that $\Delta^n y_n = 0$

19·22. Assuming U_x to be a polynomial of seventh degree in x, prove that

$$U_5 = \frac{4}{5}(U_4 + U_6) - \frac{2}{5}(U_3 + U_7) + \frac{4}{35}(U_2 + U_8) - \frac{1}{70}(U_1 + U_9)$$

19·23. (a) (i) $U_0 = U_1 = b$, $U_2 = c$, then prove that

$$U_x = a + x(b-a) + \frac{x(x-1)}{2}(c - 2b + a)$$

(ii) If $U_0 = a$, $U_2 = b$, $U_4 = c$, then prove that

$$U_x = a + \frac{x}{2}(b-a) + \frac{(x-2)}{8}(c - 2b + a)$$

(*iii*) $U_0 = a,\quad U_3 = b,\quad U_6 = c$, then prove that

$$U_x = a + \frac{x}{3}(b-a) + \frac{x(x-3)}{18}(c-2b+a)$$

(*b*) If $U_0 + U_6 = p,\quad U_1 + U_5 = q$ and $U_2 + U_4 = r$, then prove that $U_3 = \frac{1}{2}(p - 3q + 3r)$

(*c*) For a certain polynomial function y_x it is known that

$$y_1 = -1,\quad y_2 + y_3 = -1,\quad y_4 + y_5 + y_6 = 61.$$

Find the value of y_3.

19·24. In usual notations prove that:

(*i*) $D = \frac{1}{h}\left[\Delta - \frac{1}{2}\Delta^2 + \frac{1}{3}\Delta^3 - \frac{1}{4}\Delta^4 + \frac{1}{5}\Delta^5 - \frac{1}{6}\Delta^6 + \cdots\right]$

(*ii*) $D^2 = \frac{1}{h^2}\left[\Delta^2 - \Delta^3 + \frac{11}{12}\Delta^4 - \frac{5}{6}\Delta^5 + \frac{137}{180}\Delta^6 - \cdots\right]$

19·25. Evaluate.

(*i*) $\Delta^{-1}\left[1/x(x+3)\right]$, (*ii*) $\Delta^{-1}\left[(x+3)/x(x+1)(x+2)\right]$, (*iii*) $\Delta^{-1}(xa^x)$ and (*iv*) $\Delta^{-1}\cos ax$.

19·26. (*a*) Given $U_1 = 386$, $U_3 = 530$ and $U_5 = 810$. Determine U_2 and U_4.

(*b*) Given, $U_0 = 78{\cdot}49$, $U_1 = 68{\cdot}74$, $U_3 = 56{\cdot}46$ and $U_4 = 52{\cdot}32$. Interpolate U_2.

(*c*) Find approximately the value of sin (0·101) from the following table :

x :	0·100	0·102	0·104	0·106
sin x :	0·0998	0·1018	0·1038	0·1058

19·27. (*a*) Using Newton's formula, compute l_x when $x = 24$, given

x :	15	20	25	30	35	40	45	50
l_x:	962	952	937	920	901	878	847	804

(*b*) The following table gives the value of $\log_{10} x$ for certain values of x. Find $\log_{10} 1{\cdot}33$

x:	1·1	1·2	1·3	1·4	1·5
$\log_{10} x$:	0·04139	0·07918	0·1139	0·14613	0·17809

(*c*) The following table gives the areas of the normal distribution:

t	*Area*
0·00	0·0000
·10	0·0398
·20	0·0793
·30	0·1179
·40	0·1554
·50	0·1915

Calculate the area corresponding to $t = 0{\cdot}15$.

(*d*) The following marks were obtained by 492 candidates in a certain examination.

Not more than 40 marks	212 candidates
" " " 45 "	296 "
" " " 50 "	368 "
" " " 55 "	429 "
" " " 60 "	460 "
" " " 65 "	481 "
" " " 70 "	490 "
" " " 75 "	492 "

Find out the number of candidates who secured more than 42 but not more than 45 marks.

(*e*) Based on the following table, find a polynomial of 3rd degree using Newton's divided difference formula ;

x :	3	4	5	6
y :	6	24	60	120

19·28. (*a*) Find log (57·4) given the following values of log x.

x :	55	56	57	58
log x :	1·7404	1·7482	1·7559	1·7634

(*b*) Given the following table, construct a difference table and from it estimate y when $x = 0·35$ by using Newton's backward difference interpolation formula.

x :	0	0·1	0·2	0·3	0·4
y :	1	1·095	1·179	1·251	1·310

(Answer to be given correct to 3 decimal places)

19·29. Given $U_0, U_1, U_2, \ldots, U_{n-1}$, show that U_x can be written as

$$U_x = \frac{x!}{(x-n)(n-1)!}\left\{\frac{U_{n-1}}{x-n+1} - \binom{n-1}{1}\frac{U_{n-2}}{x-n+2} + \binom{n-1}{2}\frac{U_{n-3}}{x-n+3} - \cdots \pm \binom{n-1}{n-1}\frac{U_0}{x}\right\}$$

19·30. (*a*) Given that :

$$\underset{b}{\Delta} U_a = \frac{U_a}{a-b} + \frac{U_b}{b-a}, \text{ derive a symmetrical expression for } \underset{bc}{\Delta^2} U_a.$$

(*b*) If the data are U_0, U_3, U_4, U_7, U_{11} and interpolation formula is

$$U_x = U_4 + C_1 \underset{3}{\Delta} U_4 + C_2 \underset{4,7}{\Delta^2} U_3 + C_3 \underset{3,4,7}{\Delta^3} U_0 + C_4 \underset{3,4,7,11}{\Delta^4} U_0,$$

find the values of C_1, C_2, C_3 and C_4 where $\underset{3,4,7,11}{\Delta^4} U_0$ is the fourth order divided differences for the arguments 3, 4, 7, 11, 0.

19·31(a) Prove Lagrange's interpolation formula for the function U_x for given arguments $x_0, x_1, \ldots, x_n$; and show that the sum of Lagrange's coefficients is unity *i.e.*, one.

(b) Show that for the arguments 0, α, 1, ; Lagransge's interpolation formula for $f(x)$ gives:

$$f(x) = \left[1 - \frac{x(x-\alpha)}{1-\alpha}\right]f(0) + \frac{x(1-x)}{1-\alpha}\cdot\left[\frac{f(\alpha)-f(0)}{\alpha}\right] + \frac{x(x-\alpha)}{1-\alpha}.f(1)$$

Also show that if $\alpha \to 0$, it reduces to:

$$f(x) = \left(1-x^2\right)f(0) + x(1-x)f'(0) + x^2 f(1)$$

(c) Derive Lagrange's interpolation formula for expressing U_x in terms of U_a, U_b, U_c, and U_d. Use it to prove that

$$U_{x+\frac{1}{2}} = \frac{1}{16}\{9(U_x + U_{x+1}) - (U_{x-1} + U_{x+2})\}$$

19·32. (*a*) If $U_{4·50} = 134·5$, $U_{4·55} = 1470$, $U_{4·70} = 2010$, $U_{4·90} = 3815$, $U_{5·15} = 10695$, find by the method of divided differences the value of U_5.

(*b*) Determine the polynomial $f(x)$ and find the value of $f(3)$ from the following data :

x :	1	2	4	5	7	9
$f(x)$:	−96	0	120	0	−780	0

19·33. (*a*) Find the form of U_x given that $U_0 = 8$, $U_1 = 11$, $U_4 = 68$ and $U_5 = 123$.

(*b*) The following table gives the percentage points of χ^2 distribution for 10 Degrees of freedom. Using Lagrange's formula, obtain the value of χ^2 for $p = 0·7$.

p :	0·05	0·10	0·50	0·90
χ^2 :	18·307	15·987	9·342	4·845

(c) Find the form of the function U_x given that : $U_{-1} = -25$, $U_1 = 27$, $U_4 = 0$ and $U_6 = 1280$.

Indicate briefly how your answer will change if you are given an additional value, $U_9 = 2475$.

19·34. (a) Given $U_0 = -4$, $U_1 = -2$, $U_4 = 220$, $U_5 = 546$, $U_6 = 1148$, find U_2 and U_3.

(b) Given that $U_{1\cdot5} = 21\cdot3$, $U_{4\cdot2} = 37\cdot8$, $U_6 = 42\cdot7$, $U_{7\cdot5} = 50\cdot1$ and $U_8 = 51\cdot6$, determine U_5.

19·35. Find the value of $f(8)$ and $f(15)$ from the following table using Newton's divided difference formula:

x :	4	5	7	10	11	13
$f(x)$:	48	100	294	900	1210	2028

19·36. (a) Use Newton's formula to find the value of $f(8)$ from the following set of values :

x :	4	5	7	10	11	13
$f(x)$:	2	4	8	104	114	452

(b) Using Lagrange's interpolation formula, find the form of the function $y = f(x)$, given

x:	0	2	3	6
$f(x)$:	659	705	729	804

19·37. If $y_0, y_1, \ldots, y_6$ be consecutive terms of a series, prove that

$$y_3 = 0\cdot05\,(y_0 + y_6) - 0\cdot3\,(y_1 + y_5) + 0\cdot75\,(y_2 + y_4)$$

Find y_3, given the following :

$y_0 = 72795$ $\qquad y_4 = 67919$

$y_1 = 71651$ $\qquad y_5 = 66566$

$y_2 = 70458$ $\qquad y_6 = 65152$

19·38. (a) A function $f(x)$ takes the values as given in the following table :

x :	1	3	4
$f(x)$:	4	12	19

Find the value of x so that $f(x) = 7$.

(b) Find the value of x (up to first decimal place) corresponding to $f(x) = 13\cdot6$, given :

$f(x)$:	15·9,	14·9,	14·1,	13·3,	12·5
x :	30,	35,	40,	45,	50

(c) Given the following table, find the value of $\log_e x$ at $x = 505$;

x :	500	510	520	530	540	550
$\log_e x$:	6·215	6·234	6·254	6·273	6·292	6·310

19·39. If the data are U_0, U_1, U_3, U_7, and U_{10} and the interpolation formula is

$$U_x = U_3 + (x-3)\ \Delta\ U + (x-1)(x-3)\ \Delta^2 U + (x-1)(x-3)(x-7)\Delta^3 U$$
$$+\ x(x-1)(x-3)(x-7)\ \Delta^4 U,$$

find the missing suffixes of operators and subscripts of U.

19·40. If four values U_{-5}, U_{-3}, U_3 and U_5 are given and a value U_x is interpolated by Lagrange's formula, show that

$$U_x = \frac{x^3 - 9}{160}\left[(x+5)U_5 - (x-5)U_{-5}\right] + \frac{x^2 - 25}{96}\left[(x-3)U_{-3} - (x+3)U_3\right]$$

Hence, or otherwise prove that

$$U_1 = U_3 - \frac{3}{10}\left(U_5 - U_{-3}\right) + \frac{2}{10}\left(U_{-3} - U_{-5}\right)$$

19·41. By means of Lagrange's formula, prove that

(a) $U_0 = \frac{1}{2}(U_1 + U_{-1}) - \frac{1}{8}\left[\frac{1}{2}(U_3 - U_1) - \frac{1}{2}(U_{-1} - U_{-3})\right]$

(b) $U_1 = U_3 - \cdot 3\,[U_5 - U_{-3}] + \cdot 2\,[U_{-3} - U_{-5}]$

19·42. (a) Use Simpson's one-third rule to estimate the value of $\int_1^5 f(x)dx$

given

x:	1	2	3	4	5
$f(x)$:	10	50	70	80	100

(b) Tabulate the values of the function $(2 + x^2)^{-1}$ for nine equally spaced values of x over the interval (3, 5) correct to three decimal places and hence compute the value of

$$\int_3^5 \frac{4}{(2 + x^2)} dx$$

using an appropriate quadrature formula.

(c) The following values of U_x are given :

x:	0	1	2	3	4	5	6
U_x:	0·146	0·161	0·176	0·190	0·204	0·217	0·230

Use an approximate integration formula to find the value of $\int_0^6 U_x dx$. .

19·43. (a) Use Simpson's one-third rule to find the approximate value of $\log_e 2$ from the integral

$$\int_0^1 \frac{dx}{1 + x}, \text{ taking 5 ordinates.}$$

(b) Taking four equal parts and using Simpson's one-third rule, find the approximate value of $\log_e 2$ from

$$\int_0^1 \frac{x^2}{1 + x^3} dx \text{ and find the error.}$$

(c) Evaluate $\int_0^6 \frac{1}{(1 + x)^2} dx$, using Simpson's one-third rule.

(d) Dividing the interval (2, 3) into six equal parts, find approximately the value of

$$\int_1^3 \frac{dx}{\sqrt{3 + 2x - x^2}}$$

(e) Taking 10 equal parts, show by Simpsons's rule that $\int_0^1 \frac{\log(1 + x^2)}{\sqrt{(1 + x^2)}} dx = 0{\cdot}1730$

19·44. (a) Using Simpson's 3/8th rule, obtain approximately the value of $\int_0^6 \frac{dx}{(1 + x^2)}$

(b) Compute the value of the definite integral $\int_4^{5\cdot 2} \log x\, dx$, by any two methods and hence show their relative accuracy.

19·45. (*a*) Calculate by Simpson's rule an approximate value of $\int_{-3}^{3} x^4 dx$,, by taking seven equidistant ordinates. Compare it with the exact value and the value obtained by using Trapezoidal rule.

(*b*) Find by Simpson's rule, an approximate value of the integral $\int_0^1 \sqrt{x}(1-x)dx$

19·46. Establish Simpson's rule for numerical integration and hence deduce that

$$\int_0^{2n} U_x dx = \frac{1}{3}\left[U_0 + 4(U_1 + U_3 + \cdots + U_{2n-1}) + 2(U_2 + U_4 + \cdots + U_{2n-2}) + U_{2n}\right]$$

19·47. (*a*) If U_x is a polynomial of second degree in x, prove that

$$\int_0^1 U_x dx = \frac{1}{12}(5U_0 + 8U_1 - U_2)$$

(*b*) Obtain the approximate quadrature formula

$$\int_{-1}^{1} U_x \, dx = \frac{1}{12}\left[13\,(U_1 + U_{-1}) - (U_3 + U_{-3})\right]$$

19·48. (*a*) Obtain the approximate quadrature formulae :

(*i*) $\int_{1\cdot5}^{2\cdot5} U_x dx = \frac{1}{24}\left(U_1 + 22U_2 + U_3\right)$

(*ii*) $\int_0^1 U_x dx = \frac{1}{12}\left(5U_1 + 8U_0 - U_{-1}\right)$

(*iii*) $\int_0^n U_x dx = n\left[\frac{3}{8}U_0 + \frac{1}{24}\{19U_n - 5U_{2n} + U_{3n}\}\right]$

(*b*) If third differences are constant, prove that

$$\int_{-1/2}^{1/2} f(x)\, dx = \frac{1}{2}\left[\{f(-1/2) + f(1/2)\} + \frac{1}{24}\{\Delta f(-3/2) - \Delta f(1/2)\}\right]$$

(*c*) Prove that

$$\int_{-\frac{1}{2}}^{\frac{1}{2}} f(x)dx = \frac{1}{24}\left[13f\left(-\frac{1}{2}\right) + 13f\left(\frac{1}{2}\right) - f\left(\frac{3}{2}\right) - f\left(-\frac{3}{2}\right)\right], \text{ (approx.)}$$

19·49. If U_0, U_2 and U_4 are the values of $U_x = a + bx + cx^2$ at $x = 0$, 2 and 4 respectively,

prove that: $$\int_1^3 Ux\, dx = 2U_2 + \frac{1}{12}(U_0 - 2U_2 + U_4).$$

and hence find the approximate value for $\int_{-1/2}^{1/2} \exp\left(-\frac{1}{10}x^2\right)dx$.

19·50. Prove that $\int_{-3}^{3} U_x\, dx = \frac{3}{10}\left[11\,(U_{-2} + U_2) - 14\,(U_{-1} + U_1) + 26U_0\right]$

19·51. U_x is a polynomial in x of the third degree, find an expression for $\int_0^1 U_x\, dx$ in terms of U_0, U_1, U_2 and U_3. Hence deduce that :

$$\int_1^2 U_x\, dx = \frac{1}{24}\left(-U_0 + 13U_1 + 13U_2 - U_3\right).$$

19·52. Obtain on approximate formula for $\int_{-3}^{3} U_x\, dx$ in the form

$a\,(U_{-2} + U_2) + b\,(U_{-3} + U_3)$ and find the values of a and b.

19·53. If U_x is of the form $a + bx + cx^2$, show that :

(*i*) $\int_0^1 U_x\, dx = \frac{1}{36}\left(5U_{-1} + 39U_1 - 8U_2\right)$ (*ii*) $\int_0^1 U_x\, dx = \frac{1}{12}\left(5U_0 + 8U_1 - U_2\right)$

19·54. If 4th and higher differences are zero and if

$$\int_{-1}^{1} f(x)dx = \frac{2}{3}\left[f(x_1) + f(x_2) + f(x_3)\right],$$

find the values of x_1, x_2 and x_3.

Ans: $0, \pm\left(1/\sqrt{2}\right)$

19·55. If $$\int_0^m U_x dx = a_1 U_{x_1} + a_2 U_{x_2} + \cdots + a_n U_{x_n}$$

where U_x is polynomial in x of the $(n - 1)$th degree $(n > 2)$, prove that:

$$3\sum_{r=1}^{m} a_r x_r^2 = m^2$$

19·56. Prove that the nth order divided difference of a function can be expressed as the quotient of two determinants, each of order $(n + 1)$.

Hint: For proving this result, we use the concepts of :

(*i*) Difference Product (D.P.) and (*ii*) Vander Monde's determinant, which are explained below.

Difference Product. The difference production (D.P.) of three numbers a_0, a_1 and a_2 is defined as:

$$\text{D.P.}\,(a_0, a_1, a_2) = (a_0 - a_1)(a_0 - a_2)(a_1 - a_2) = \prod_{i<j=0}^{2} (a_i - a_j)$$

Similarly, $$\text{D. P.}\,(a_0, a_1, a_2, a_3) = (a_0 - a_1)(a_0 - a_2)(a_0 - a_3)(a_1 - a_2)(a_1 - a_3)(a_2 - a_3)$$

$$= \prod_{i<j=0}^{3} (a_i - a_j)$$

In general, $$\text{D. P.}\,(a_0, a_1, a_2, \ldots, a_n) = \prod_{i<j=0}^{n} (a_i - a_j).$$

Vander-Monde's Determinant. We know that:

$$\begin{vmatrix} 1 & 1 & 1 \\ a & b & c \\ a^2 & b^2 & c^2 \end{vmatrix} = (a-b)(b-c)(c-a) \qquad \ldots (*)$$

$$\therefore \quad \begin{vmatrix} a_0^2 & a_1^2 & a_2^2 \\ a_0 & a_1 & a_2 \\ 1 & 1 & 1 \end{vmatrix} = (-1)^2 \begin{vmatrix} 1 & 1 & 1 \\ a_0^2 & a_1^2 & a_2^2 \\ a_0 & a_1 & a_2 \end{vmatrix} = -\begin{vmatrix} 1 & 1 & 1 \\ a_0 & a_1 & a_2 \\ a_0^2 & a_1^2 & a_2^2 \end{vmatrix}$$

$$= -(a_0-a_1)(a_1-a_2)(a_2-a_0) \qquad \text{[Using (*)]}$$

$$= (a_0-a_1)(a_0-a_2)(a_1-a_2) = \text{D.P.}\,(a_0, a_1, a_2)$$

In general, we have:

$$\begin{vmatrix} a_0^n & a_1^n & a_2^n & \ldots & a_n^n \\ a_0^{n-1} & a_1^{n-1} & a_2^{n-1} & \ldots & a_n^{n-1} \\ \vdots & \vdots & \vdots & & \vdots \\ a_0^2 & a_1^2 & a_2^2 & \ldots & a_n^2 \\ a_0 & a_1 & a_2 & \ldots & 1 \\ 1 & 1 & 1 & 1 & 1 \end{vmatrix} = \prod_{i<j=0}^{n} (a_i - a_j) = \text{D.P.}\,(a_0, a_1, \ldots, a_n) \qquad \ldots (**)$$

(**) is known as Vader-Monde's determinant

Proof of Theorem. For four arguments a_0, a_1, a_2 and a_3, the divided different of third order is given by:

$$\underset{a_1, a_2, a_3}{\Delta^3} U_{a_0} = \frac{U_{a_0}}{(a_0-a_1)(a_0-a_2)(a_0-a_3)} + \frac{U_{a_1}}{(a_1-a_0)(a_1-a_2)(a_1-a_3)}$$

$$+ \frac{U_{a_2}}{(a_2-a_0)(a_2-a_1)(a_2-a_3)} + \frac{U_{a_3}}{(a_3-a_0)(a_3-a_1)(a_3-a_2)}$$

$$= \frac{U_{a_0}}{(a_0-a_1)(a_0-a_2)(a_0-a_3)} - \frac{U_{a_1}}{(a_0-a_1)(a_1-a_2)(a_1-a_3)}$$

$$+ \frac{U_{a_2}}{(a_0-a_2)(a_1-a_2)(a_2-a_3)} - \frac{U_{a_3}}{(a_0-a_3)(a_1-a_3)(a_2-a_3)}$$

[Write each term as $a_i - a_j$, $i < j$]

$$= \frac{1}{\left\{\begin{matrix}(a_0-a_1)(a_0-a_2)(a_0-a_3) \\ \times (a_1-a_2)(a_1-a_3)\times(a_2-a_3)\end{matrix}\right\}} \times \begin{bmatrix} U_{a_0}\,\text{D.P.}\,(a_1, a_2, a_3) - U_{a_1}\text{D.P.}\,(a_0, a_2, a_3) \\ + U_{a_2}\text{D.P.}\,(a_0, a_1, a_3) - U_{a_3}\text{D.P.}\,(a_0, a_1, a_2) \end{bmatrix}$$

$$= \frac{1}{\text{D.P.}\,(a_0, a_1, a_2, a_3)} \times \left[\begin{vmatrix} a_1^2 & a_2^2 & a_3^2 \\ a_1 & a_2 & a_3 \\ 1 & 1 & 1 \end{vmatrix} U_{a_0} - \begin{vmatrix} a_0^2 & a_2^2 & a_3^2 \\ a_0 & a_2 & a_3 \\ 1 & 1 & 1 \end{vmatrix} U_{a_1} + \begin{vmatrix} a_0^2 & a_1^2 & a_3^2 \\ a_0 & a_1 & a_3 \\ 1 & 1 & 1 \end{vmatrix} U_{a_2} - \begin{vmatrix} a_0^2 & a_1^2 & a_2^2 \\ a_0 & a_1 & a_2 \\ 1 & 1 & 1 \end{vmatrix} U_{a_3} \right]$$

[Using Vander-Mode's determinant (**)]

$$= \begin{vmatrix} U_{a_0} & U_{a_1} & U_{a_2} & U_{a_3} \\ a_0^2 & a_1^2 & a_2^2 & a_3^2 \\ a_0 & a_1 & a_2 & a_3 \\ 1 & 1 & 1 & 1 \end{vmatrix} \div \begin{vmatrix} a_0^3 & a_1^3 & a_2^3 & a_3^3 \\ a_0^2 & a_1^2 & a_2^2 & a_3^2 \\ a_0 & a_1 & a_2 & a_3 \\ 1 & 1 & 1 & 1 \end{vmatrix}$$

Expanding the first determinant by first row and using Vander Monde's determinant (**).

Thus, the third order divided difference is expressed as the ratio of two determinants, each of order 4.

Proceeding similarly, the divided difference of order n of the arguments $a_0, a_1, a_2, \ldots, a_n$ is given by:

$$\underset{a_1, a_2, \ldots a_n}{\Delta^n} U_{a_0} = \sum_{r=0}^{n} \frac{U_{a_r}}{(a_r - a_0)(a_r - a_1) \ldots (a_r - a_n)}$$

$$= \frac{1}{\text{D.P.}\,(a_0, a_1, \ldots, a_n)} \times \left[\begin{array}{c} \text{D.P.}\,(a_1, a_2, \ldots, a_n)\, U_{a_0} - \text{D.P.}\,(a_0, a_2, \ldots, a_n) U_{a_1} \\ + \text{D.P.}\,(a_0, a_1, a_3, \ldots, a_n)\, U_{a_2} - \text{D.P.}\,(a_0, a_1, a_2, a_4, \ldots, a_n)\, U_{a_3} \\ + \ldots + (-1)^n\, \text{D.P.}\,(a_0, a_1, a_2, \ldots, a_{n-1})\, U_{a_n} \end{array} \right]$$

$$= \begin{vmatrix} U_{a_0} & U_{a_1} & U_{a_2} & \cdots & U_{a_n} \\ a_0^{n-1} & a_1^{n-1} & a_2^{n-1} & \cdots & a_n^{n-1} \\ \vdots & \vdots & \vdots & & \vdots \\ a_0 & a_1 & a_2 & \cdots & a_n \\ 1 & 1 & 1 & \cdots & 1 \end{vmatrix} \div \begin{vmatrix} a_0^n & a_1^n & a_2^n & \cdots & a_n^n \\ a_0^{n-1} & a_1^{n-1} & a_2^{n-1} & \cdots & a_n^{n-1} \\ \vdots & \vdots & \vdots & \cdots & \vdots \\ a_0 & a_1 & a_2 & \cdots & a_n \\ 1 & 1 & 1 & \cdots & 1 \end{vmatrix},$$

which is the quotient of two determinants, each of order $(n + 1)$.

Hence the result.

CHAPTER

20 Bivariate and Multivariate Normal Distributions

LEARNING OBJECTIVES

Upon completion of this Chapter, you should be able to:

- Derive the p.d.f. of Bivariate Normal (BVN) Distribution.
- Obtain the Moment Generating Function (m.g.f.) of BVN $\left(\mu_X, \mu_Y, \sigma_X^2, \sigma_Y^2, \rho\right)$ Distribution and discuss its important properties.
- Obtain the Marginal and Conditional Distributions of the BVN Distribution.
- Derive the p.d.f. of Multivariate Normal (MVN) Distribution.
- Obtain the m.g.f. and Characteristic Function of MVN Distribution.
- Obtain the Marginal and Conditional Distributions for MVN $\left[N_p(\mu, \Sigma)\right]$ Distribution.

CHAPTER OUTLINES

20·1. BIVARIATE NORMAL DISTRIBUTION

The bivariate normal distribution is a generalization of a normal distribution for a single variate. Let X and Y be two normal correlated variables with correlation coefficient ρ and $E(X) = \mu_1$, $\text{Var}(X) = \sigma_1^2$; $E(Y) = \mu_2$, $\text{Var}(Y) = \sigma_2^2$. In deriving the bivariate normal distribution we make the following three assumptions :

(i) *The regression of Y on X is linear.* Since the mean of each array is on the line of regression $Y = \rho\,(\sigma_2 / \sigma_1)\,X$, the mean or expected value of Y is $\rho\,(\sigma_2 / \sigma_1)\,X$, for different values of X.

(ii) *The arrays are homoscedastic, i.e.,.* variance in each array is same. The common variance of estimate of Y in each array is then given by $\sigma_2^2\,(1-\rho^2)$, ρ being the correlation coefficient between variables X and Y, and is independent of X.

(iii) *The distribution of Y in different arrays is normal.* Suppose that one of the variates, say X, is distributed normally with mean 0 and standard deviation σ_1 so that the probability that a random value of X will fall in the small interval dx is

$$g(x)\,dx = \frac{1}{\sigma_1\sqrt{(2\pi)}}\exp\left(-x^2 / 2\sigma_1^2\right)dx$$

The probability that a value of Y, taken at random in an assigned vertical array will fall in the interval dy is :

$$h(y \mid x)\,dy = \frac{1}{\sigma_2\sqrt{2\pi\,(1-\rho^2)}}\,.\exp\left\{-\frac{1}{2\sigma_2^2(1-\rho^2)}\left(y-\rho\frac{\sigma_2}{\sigma_1}x\right)^2\right\}$$

The joint probability differential of X and Y is given by:

$$dP(x,y) = g(x)\,h(y \mid x)\,dxdy$$

$$= \frac{1}{2\pi\sigma_1\sigma_2\sqrt{(1-\rho^2)}}\exp\left(-\frac{1}{2\sigma_1^2}\,.\,x^2\right).\exp\left[-\frac{1}{2\sigma_2^2(1-\rho^2)}\left(y-\rho\frac{\sigma_2}{\sigma_1}x\right)^2\right]dx\,dy$$

$$= \frac{1}{2\pi\sigma_1\sigma_2\sqrt{(1-\rho^2)}}\,.\exp\left\{-\frac{1}{2(1-\rho^2)}\left(\frac{x^2}{\sigma_1^2}-\frac{2\rho x y}{\sigma_1\sigma_2}+\frac{y^2}{\sigma_2^2}\right)\right\}dx\,dy$$

Shifting the origin to (μ_1, μ_2), we get

$$f_{XY}(x, y) = \frac{1}{2\pi\,\sigma_1\sigma_2\sqrt{(1-\rho^2)}}\,.\exp\left[-\frac{1}{2(1-\rho^2)}\left\{\frac{(x-\mu_1)^2}{\sigma_1^2}-2\rho\frac{(x-\mu_1)(y-\mu_2)}{\sigma_1\sigma_2}+\frac{(y-\mu_2)^2}{\sigma_2^2}\right\}\right];$$

$$(-\infty < x < \infty, -\infty < y < \infty) \quad \ldots (20{\cdot}1)$$

where μ_1, μ_2, $\sigma_1\,(>0)$, $\sigma_2\,(>0)$, and $\rho\,(-1<\rho<1)$ are the five parameters of the distribution.

Normal Correlation Surface : $z = f(x,y)$

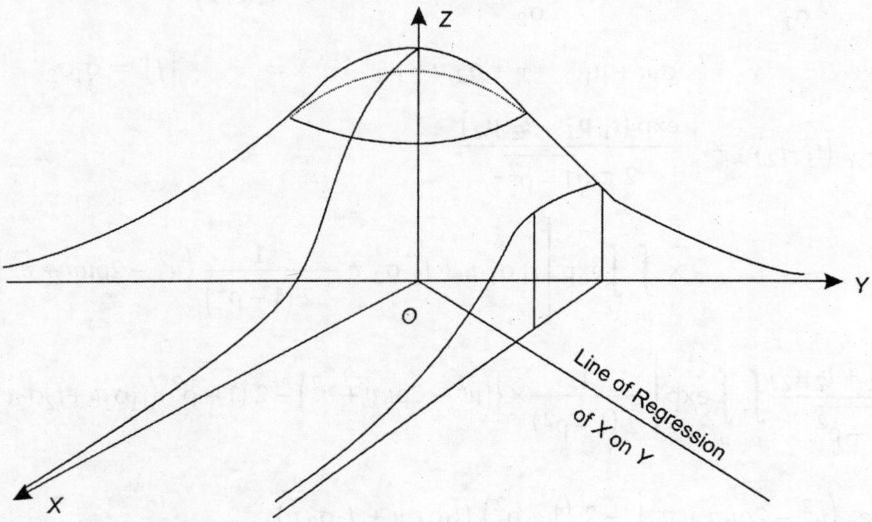

Fig. 20·1

(20·1) gives the density function of a bivariate normal distribution. The variables X and Y are said to be normally correlated and the surface $z = f(x,y)$ is known as the *normal correlation surface.* The nature of the normal correlation surface is indicated in the Fig. 20.1.

Remarks:

1. The vector $(X,Y)'$ following the joint *p.d.f.* $f(x,y)$ as given in (20.1), will be abbreviated as $(X,Y) \sim N\left(\mu_1, \mu_2, \sigma_1^2, \sigma_2^2, \rho\right)$ or BVN $\left(\mu_1, \mu_2, \sigma_1^2, \sigma_2^2, \rho\right)$. If in particular $\mu_1 = \mu_2 = 0$ and $\sigma_1 = \sigma_2 = 1$ then $(X,Y) \sim N\ (0, 0, 1, 1, \rho)$ or BVN $(0, 0, 1, 1, \rho)$.
2. The curve $Z = f(x, y)$ which is the equation of a surface in three dimensions, is called the 'Normal Correlation Surface'.
3. Consider the family of ellipses given by:

$$\left(\frac{x-\mu_x}{\sigma_x}\right)^2 - 2\rho\left(\frac{x-\mu_x}{\sigma_x}\right)\left(\frac{y-\mu_Y}{\sigma_y}\right) + \left(\frac{y-\mu_Y}{\sigma_y}\right)^2 = \lambda^2 \qquad \text{... (20·1 } a\text{)}$$

λ being the parameter of the family.

On any one of these ellipses, the density function is constant, and hence these are called *equi-probability ellipses.*

On the ellipse λ , the p.d.f. becomes :

$$f(x,y) = \frac{1}{2\pi\,\sigma_x\,\sigma_y\,\sqrt{1-\rho^2}} \times \exp\left[-\frac{\lambda^2}{2\left(1-\rho^2\right)}\right]; -\infty < (x,y) < \infty \qquad \text{... (21·1 } b\text{)}$$

20·1·1. Moment Generating Function of Bivariate Normal Distribution

Let $(X, Y) \sim$ BVN $\left(\mu_1, \mu_2, \sigma_1^2, \sigma_2^2, \rho\right)$. By *def.*

$$M_{XY}\,(t_1, t_2) = E\left[e^{t_1X+t_2Y}\right] = \int_{-\infty}^{\infty}\int_{-\infty}^{\infty} \exp\left(t_1x + t_2y\right).\,f\,(x, y)\,dxdy$$

Put $\dfrac{x-\mu_1}{\sigma_1} = u, \quad \dfrac{y-\mu_2}{\sigma_2} = v, \qquad -\infty < (u, v) < \infty$

i.e., $x = \sigma_1 u + \mu_1, \quad y = \sigma_2 v + \mu_2 \quad \Rightarrow \quad |J| = \sigma_1 \sigma_2.$

$$\therefore \quad M_{X,Y}(t_1, t_2) = \frac{\exp(t_1 \mu_1 + t_2 \mu_2)}{2\pi\sqrt{1-\rho^2}}$$

$$\times \int_u \int_v \exp\left[t_1 \sigma_1 u + t_2 \sigma_2 v - \frac{1}{2(1-\rho^2)}\left\{u^2 - 2\rho uv + v^2\right\}\right] du\, dv$$

$$= \frac{\exp(t_1 \mu_1 + t_2 \mu_2)}{2\pi\sqrt{1-\rho^2}} \int_u \int_v \exp\left[\frac{-1}{2(1-\rho^2)} \times \left\{(u^2 - 2\rho uv + v^2) - 2(1-\rho^2)(t_1\sigma_1 u + t_2 \sigma_2 v)\right\}\right] du\, dv$$

We have, $(u^2 - 2\rho uv + v^2) - 2(1-\rho^2)(t_1 \sigma_1 u + t_2 \sigma_2 v)$

$$= \left\{(u - \rho v) - (1-\rho^2) t_1 \sigma_1\right\}^2 + (1-\rho^2)\left\{(v - \rho t_1 \sigma_1 - t_2 \sigma_2)^2 - t_1^2 \sigma_1^2 - t_2^2 \sigma_2^2 - 2\rho t_1 t_2 \sigma_1 \sigma_2\right\} \quad \ldots (*)$$

By taking

$$\left.\begin{aligned} u - \rho v - (1-\rho^2) t_1 \sigma_1 &= \omega (1-\rho^2)^{1/2} \\ v - \rho t_1 \sigma_1 - t_2 \sigma_2 &= z \end{aligned}\right\} \Rightarrow du\, dv = \sqrt{1-\rho^2}\; dw dz$$

and using (*), we get

$$M_{X,Y}(t_1, t_2) = \exp\left[t_1 \mu_1 + t_2 \mu_2 + \frac{1}{2}\left(t_1^2 \sigma_1^2 + t_2^2 \sigma_2^2 + 2\rho t_1 t_2 \sigma_1 \sigma_2\right)\right]$$

$$\times \left[\frac{1}{\sqrt{2\pi}} \int_{-\infty}^{\infty} e^{-\omega^2/2}\, dw\right] \times \left[\frac{1}{\sqrt{2\pi}} \int_{-\infty}^{\infty} e^{-z^2/2}\, dz\right]$$

$$= \exp\left[t_1 \mu_1 + t_2 \mu_2 + \frac{1}{2}\left(t_1^2 \sigma_1^2 + t_2^2 \sigma_2^2 + 2\rho t_1 t_2 \sigma_1 \sigma_2\right)\right] \quad \ldots (20·2)$$

In particular if $(X, Y) \sim BVN(0, 0, 1, 1, \rho)$, then

$$M_{X,Y}(t_1, t_2) = \exp\left[\frac{1}{2}\left(t_1^2 + t_2^2 + 2\rho t_1 t_2\right)\right] \quad \ldots (20·2a)$$

Theorem 20.1: Let $(X,Y) \sim BVN(\mu_1, \mu_2, \sigma_1^2, \sigma_2^2, \rho)$. Then X and Y are independent if and only if $\rho = 0$.

Proof.

(*a*) If $(X, Y) \sim BVN(\mu_1, \mu_2, \sigma_1^2, \sigma_2^2, \rho)$ and $\rho = 0$, then X and Y are independent.

If $\rho = 0$, then from (20·1), we get:

$$f(x, y) = \frac{1}{\sigma_X \sqrt{2\pi}} \exp\left[-\frac{1}{2}\left(\frac{x-\mu_x}{\sigma_X}\right)^2\right] \times \frac{1}{\sigma_Y \sqrt{2\pi}} \exp\left[-\frac{1}{2}\left(\frac{y-\mu_Y}{\sigma_Y}\right)^2\right]$$

$$\Rightarrow \quad f(x,y) = f_1(x) . f_2(y)$$

$\therefore$ X and Y are independent, $X \sim N\left(\mu_X, \sigma_X^2\right)$ and $Y \sim N\left(\mu_Y, \sigma_Y^2\right)$.

Aliter. $(X, Y) \sim BVN\left(\mu_1, \mu_2, \sigma_1^2, \sigma_2^2, \rho\right)$

$$\therefore \quad M_{X,Y}(t_1, t_2) = \exp\left\{t_1\mu_1 + t_2\mu_2 + \frac{1}{2}\left(t_1^2\sigma_1^2 + 2\rho t_1 t_2 \sigma_1\sigma_2 + t_2^2\sigma_2^2\right)\right\}$$

If $\rho = 0$, then

$$M_{X,Y}(t_1, t_2) = \exp\left\{t_1\mu_1 + \frac{1}{2}t_1^2\sigma_1^2\right\} . \exp\left\{t_2\mu_2 + \frac{1}{2}t_2^2\sigma_2^2\right\} = M_X(t_1) . M_Y(t_2). \quad \ldots (*)$$

$\left[\because \text{ If } (X, Y) \sim BVN\left(\mu_1, \mu_2, \sigma_1^2, \sigma_2^2, \rho\right), \text{ then the marginal p.d.f.'s of } X \text{ and } Y \text{ are normal } i.e., X \sim N(\mu_1, \sigma_1^2) \text{ and } Y \sim N(\mu_2, \sigma_2^2)\right.$ {See § 20·1·2.}$]$

$\therefore$ (*) $\Rightarrow$ X and Y are independent

(*b*) Conversely, if X and Y are independent, then we have to prove that $\rho = 0$.

Proof. If X and Y are independent, then

$$M_{X,Y}(t_1, t_2) = M_X(t_1) . M_Y(t_2)$$

$$\Rightarrow \quad \exp\left[t_1\mu_1 + t_2\mu_2 + \frac{1}{2}\left(t_1^2\sigma_1^2 + 2\rho\, t_1 t_2 \sigma_1\sigma_2 + t_2^2\sigma_2^2\right)\right]$$

$$= \exp\left[t_1\mu_1 + \frac{1}{2}t_1^2\sigma_1^2\right] . \exp\left[t_2\mu_2 + \frac{1}{2}t_2^2\sigma_2^2\right]$$

$\left[\because X \sim N\left(\mu_1, \sigma_1^2\right) \text{ and } Y \sim N\left(\mu_2, \sigma_2^2\right); \ c.f.\ \S\ 20{\cdot}1{\cdot}2\right]$

$$\Rightarrow \quad \exp\left[\rho\, t_1 t_2\, \sigma_1 \sigma_2\right] = 1$$

Since, $\sigma_1 > 0$, $\sigma_2 > 0$, we get $\rho = 0$, as desired.

Aliter : Refer to Theorem 10·2.

Theorem 20·2: *A random vector (X,Y) possesses a bivariate normal distribution if and only if every linear combination of X and Y viz., (aX + bY), a ≠ 0, b ≠ 0, is a normal variate.*

Proof. If Part. Let, $Z = aX + bY$, $a \neq 0$, $b \neq 0$ be a normal variate.

Then we have to prove that (X,Y) has a bivariate normal distribution.

Let $$Z = aX + bY \sim N\left(\mu, \sigma^2\right),$$

where $$\mu = E(Z) = E(aX + bY) = a\mu_X + b\mu_Y$$

and $$\sigma^2 = \text{Var } Z = \text{Var}\,(aX + bY) = a^2\sigma_X{}^2 + 2ab\rho\sigma_X\sigma_Y + b^2\sigma_Y{}^2$$

$$\therefore \quad M_Z(t) = \exp\left[t\mu + t^2\sigma^2/2\right]$$

$$= \exp\left[t\,(a\mu_X + b\mu_Y) + \frac{t^2}{2}\left(a^2\sigma_X{}^2 + 2ab\rho\sigma_X\sigma_Y + b^2\sigma_Y{}^2\right)\right]$$

$$= \exp\left[t_1\mu_X + t_2\mu_Y + \frac{1}{2}\left(t_1{}^2\sigma_X{}^2 + 2\rho t_1 t_2\sigma_X\sigma_Y + t_2{}^2\sigma_Y{}^2\right)\right] \quad \ldots(*)$$

where $t_1 = at$ and $t_2 = bt$.

But (*) is the m.g.f. of *BVN* distribution with parameters $\left(\mu_X, \mu_Y, \sigma_X{}^2, \sigma_Y{}^2, \rho\right)$. Hence, by uniqueness theorem of m.g.f.,

$$(X, Y) \sim BVN\left(\mu_X, \mu_Y, \sigma_X{}^2, \sigma_Y{}^2, \rho\right)$$

Only if Part: Let $(X, Y) \sim BVN\left(\mu_X, \mu_Y, \sigma_X{}^2, \sigma_Y{}^2, \rho\right)$ then, we shall prove that $Z = aX + bY$, is a normal variate.

$$M_{X,Y}(t_1, t_2) = E\left[\exp(t_1X + t_2Y)\right] \quad \ldots(**)$$

$$= \exp\left[t_1\mu_X + t_2\mu_Y + \frac{1}{2}\left(t_1^2\sigma_X^2 + t_2^2\sigma_Y^2 + 2t_1t_2\rho\sigma_X\sigma_Y\right)\right]$$

$$M_Z(t) = E\left[\exp(tZ)\right] = E\left[\exp\left\{t\,(aX + bY)\right\}\right]$$

$$= E\left[\exp\left\{(at)X + (bt)Y\right\}\right] = M_{X,Y}(at, bt)$$

$$= \exp\left[(at\,\mu_X + bt\,\mu_Y) + \frac{1}{2}\left(a^2t^2\,\sigma_X{}^2 + b^2t^2\,\sigma_Y{}^2 + 2abt^2\rho\sigma_X\sigma_Y\right)\right]$$

$$= \qquad \left[\text{Taking } t_1 = at \text{ and } t_2 = bt \text{ in } (**)\right]$$

$$= \exp\left[(a\mu_X + b\mu_Y)\,t + \frac{1}{2}\left(a^2\sigma_X{}^2 + b^2\sigma_Y{}^2 + 2ab\,\rho\sigma_X\sigma_Y\right)t^2\right]$$

which is the m.g.f. of normal distribution with :

$$\text{Mean }(\mu) = a\mu_X + b\mu_Y \text{ and Variance }\left(\sigma^2\right) = a^2\sigma_X{}^2 + b^2\sigma_Y{}^2 + 2ab\rho\sigma_X\sigma_Y \quad \ldots(***)$$

Hence, by uniqueness theorem of m.g.f.

$$Z = aX + bY \sim N\left(\mu, \sigma^2\right), \text{ where } \mu \text{ and } \sigma^2 \text{ are given in } (***)$$

20·1·2. Marginal Distribution of Bivariate Normal Distribution

The marginal distribution of random variable X is given by:

$$f_X(x) = \int_{-\infty}^{\infty} f_{X,Y}(x,y)\,dy \text{ , where } f(x, y) \text{ is defined in (20·1)}$$

Put $\dfrac{y-\mu_2}{\sigma_2} = u$, then $dy = \sigma_2\, du$. Therefore,

$$f_X(x) = \frac{1}{2\pi\sigma_1\sigma_2\sqrt{(1-\rho^2)}} \int_{-\infty}^{\infty} \exp\left[-\frac{1}{2(1-\rho^2)}\left\{\left(\frac{x-\mu_1}{\sigma_1}\right)^2 - 2\rho u\left(\frac{x-\mu_1}{\sigma_1}\right) + u^2\right\}\right]\sigma_2\, du$$

$$= \frac{1}{2\pi\sigma_1\sqrt{(1-\rho^2)}} \exp\left[-\frac{1}{2}\left(\frac{x-\mu_1}{\sigma_1}\right)^2\right] \cdot \int_{-\infty}^{\infty} \exp\left[-\frac{1}{2(1-\rho^2)}\left\{u-\rho\left(\frac{x-\mu_1}{\sigma_1}\right)\right\}^2\right] du$$

Put $\dfrac{1}{\sqrt{1-\rho^2}}\left[u-\rho\left(\dfrac{x-\mu_1}{\sigma_1}\right)\right] = t$, then $du = \sqrt{(1-\rho^2)}\, dt$

$$\therefore\ f(x) = \frac{1}{2\pi\sigma_1}.\exp\left[-\frac{1}{2}\left(\frac{x-\mu_1}{\sigma_1}\right)^2\right] \int_{-\infty}^{\infty} \exp\left(-\frac{t^2}{2}\right) dt$$

$$= \frac{1}{2\pi\sigma_1}.\exp\left[-\frac{1}{2}\left(\frac{x-\mu_1}{\sigma_1}\right)^2\right] \sqrt{2\pi}$$

$$= \frac{1}{\sigma_1\sqrt{2\pi}} \exp\left[-\frac{1}{2}\left(\frac{x-\mu_1}{\sigma_1}\right)^2\right] ;\ -\infty < x < \infty \qquad \text{... (20·3)}$$

Similarly, we shall get

$$f_Y(y) = \int_{-\infty}^{\infty} f_{X,Y}(x,y)\,dx = \frac{1}{\sigma_2\sqrt{2\pi}} \exp\left[-\frac{1}{2}\left(\frac{y-\mu_2}{\sigma_2}\right)^2\right] ;-\infty < y < \infty \qquad \text{... (20·3}a)$$

Hence, $X \sim N(\mu_1, \sigma_1^2)$ and $Y \sim N(\mu_2, \sigma_2^2)$. ... (20·3$b$)

Aliter. $(X, Y) \sim BVN(\mu_1, \mu_2, \sigma_1^2, \sigma_2^2, \rho)$

$$\Rightarrow \quad M_{X,Y}(t_1, t_2) = \exp\,(t_1\mu_1 + t_2\mu_2) + \frac{1}{2}\left(t_1^2\sigma_1^2 + t_2^2\sigma_2^2 + 2\rho t_1 t_2\,\sigma_1\sigma_2\right)$$

$$\therefore \quad M_X(t_1) = M_{X,Y}(t_1, 0) = \exp\left[t_1\mu_1 + \frac{1}{2}t_1^2\sigma_1^2\right] \quad \Rightarrow \quad X \sim N(\mu_1, \sigma_1^2)$$

and
$$M_Y(t_2) = M_{X,Y}(0, t_2) = \exp\left[t_2\mu_2 + \frac{1}{2}t_2^2\sigma_2^2\right] \Rightarrow Y \sim N(\mu_2, \sigma_2^2)$$

Remark. We have proved that if $(X, Y) \sim BVN(\mu_1, \mu_2, \sigma_1^2, \sigma_2^2, \rho)$, then the marginal p.d.f.'s of X and Y are also normal. However, the converse is not true, *i.e.*, we may have joint p.d.f. $f(X, Y)$ of (X, Y) which is not normal but the marginal p.d.f's may still be normal as discussed in the illustration on page 20.8.

Consider the joint distribution of X and Y given by:

$$f(x, y) = \frac{1}{2}\left[\frac{1}{2\pi\left(1-\rho^2\right)^{1/2}} \exp\left\{\frac{-1}{2\left(1-\rho^2\right)}\left(x^2 - 2\rho xy + y^2\right)\right\} + \frac{1}{2\pi\left(1-\rho^2\right)^{1/2}} \exp\left\{\frac{-1}{2\left(1-\rho^2\right)}\left(x^2 + 2\rho xy + y^2\right)\right\}\right]$$

$$= \frac{1}{2}\left[f_1(x, y) + f_2(x, y)\right]; -\infty < (x, y) < \infty \qquad \text{... (20·3c)}$$

where $f_1(x, y)$ is the p.d.f. of $BVN(0, 0, 1, 1, \rho)$ distribution and $f_2(x, y)$ is the p.d.f. of $BVN(0, 0, 1, 1, -\rho)$ distribution. It can be easily verfied that $f(x, y)$ is the joint p.d.f. of (X, Y) and obviousaly $f(x, y)$ is not the p.d.f. of bivariate normal distribution.

Marginal distribution of X in (20·3c).

$$f_X(x) = \frac{1}{2}\left[\int_{-\infty}^{\infty} f_1(x, y)\, dy + \int_{-\infty}^{\infty} f_2(x, y)\, dy\right]$$

But, $\int_{-\infty}^{\infty} f_1(x, y)\, dy$ is the marginal p.d.f. of X,

where $(X, Y) \sim BVN(0, 0, 1, 1, \rho)$ and is given by $X \sim N(0, 1)$.

Similarly $\int_{-\infty}^{\infty} f_2(x, y)\, dy$ is the marginal p.d.f. of X,

where $(X, Y) \sim BVN(0, 0, 1, 1, -\rho)$ and is given by $X \sim N(0, 1)$.

$$\therefore \quad f_X(x) = \frac{1}{2}\left[\frac{1}{\sqrt{2\pi}} e^{-x^2/2} + \frac{1}{\sqrt{2\pi}} e^{-x^2/2}\right] = \frac{1}{\sqrt{2\pi}} e^{-x^2/2}; -\infty < x < \infty \qquad \text{... }(i)$$

$\Rightarrow$ $X \sim N(0, 1)$ *i.e.*, the marginal distribution of X in (20·3c) is normal.

Similarly, we can show that the marginal p.d.f. of Y in (20·3c) is given by:

$$f_Y(y) = \frac{1}{2}\left[\frac{1}{\sqrt{2\pi}} e^{-y^2/2} + \frac{1}{\sqrt{2\pi}} e^{-y^2/2}\right] = \frac{1}{\sqrt{2\pi}} e^{-y^2/2}; -\infty < y < \infty \qquad \text{... }(ii)$$

$\Rightarrow$ $Y \sim N(0, 1)$.

Hence, *if the marginal distributions of X and Y are normal (Gaussion), it does not necessarily imply that the joint distribution of (X, Y) is bivariate normal.*

We further note that for the joint p.d.f. (20·3c), using (*i*) and (*ii*) we have

$$E(X) = 0, \quad \sigma_X^2 = 1 \quad \text{and } E(Y) = 0, \quad \sigma_Y^2 = 1.$$

$$\therefore \quad \text{Cov}(X, Y) = \frac{E(XY) - E(X)E(Y)}{\sigma_X \sigma_Y} = E(XY)$$

$$= \frac{1}{2}\left[\int_{-\infty}^{\infty}\int_{-\infty}^{\infty} xy\, f_1(x, y)\, dx\, dy + \int_{-\infty}^{\infty}\int_{-\infty}^{\infty} xy\, f_2(x, y)\, dx\, dy\right]$$

$$= \frac{1}{2}\left[\rho + (-\rho)\right] = 0$$

because, for $f_1(x, y), (X, Y) \sim BVN(0, 0, 1, 1, \rho)$

and for $f_2(x, y), (X, Y) \sim BVN(0, 0, 1, 1, -\rho)$.

$$\therefore \quad \text{Corr.}(X, Y) = \frac{\text{cov}(X, Y)}{\sigma_X \sigma_Y} = 0$$

However, we have ; [From (*i*) and (*ii*)]

$$f_X(x) \cdot f_Y(y) = \frac{1}{2\pi} \exp\left\{-\frac{1}{2}(x^2 + y^2)\right\} \neq f(x, y) \Rightarrow X \text{ and } Y \text{ are not independent.}$$

The above example illustrates that *we may have a joint density (non-Gaussian) of rv's (X, Y) in which the marginal p.d.f.'s of X and Y are normal and ρ (X, Y) = 0 and yet X and Y are not independent.*

20·1·3. Conditional Distributions

Conditional distribution of X for fixed Y is given by :

$$f_{X \mid Y}(x \mid y) = \frac{f_{X,Y}(x, y)}{f_Y(Y)}$$

$$= \frac{1}{\sigma_1\sqrt{2\pi}\sqrt{(1-\rho^2)}} \exp\left[-\frac{1}{2(1-\rho^2)}\left\{\left(\frac{x-\mu_1}{\sigma_1}\right)^2 - 2\rho\left(\frac{x-\mu_1}{\sigma_1}\right)\left(\frac{y-\mu_2}{\sigma_2}\right) + \left(\frac{y-\mu_2}{\sigma_2}\right)^2\left(1-(1-\rho^2)\right)\right\}\right]$$

$$= \frac{1}{\sigma_1\sqrt{2\pi}\sqrt{1-\rho^2}} \times \exp\left[-\frac{1}{2(1-\rho^2)\sigma_1^2}\left\{(x-\mu_1) - \rho\frac{\sigma_1}{\sigma_2}(y-\mu_2)\right\}^2\right]$$

$$= \frac{1}{\sigma_1\sqrt{2\pi}\sqrt{1-\rho^2}} \times \exp\left[-\frac{1}{2(1-\rho^2)\sigma_1^2}\left\{x - \left(\mu_1 + \rho\frac{\sigma_1}{\sigma_2}(y-\mu_2)\right)\right\}^2\right],$$

which is the probability function of a univariate normal distribution with mean and variance given by:

$$E(X \mid Y = y) = \mu_1 + \rho\frac{\sigma_1}{\sigma_2}(y - \mu_2) \quad \text{and} \quad V(X \mid Y = y) = \sigma_1^2(1-\rho^2)$$

Hence, the conditional distribution of X for fixed Y is also normal, given by:

$$(X \mid Y = y) \sim N\left[\mu_1 + \rho\frac{\sigma_1}{\sigma_2}(y - \mu_2), \sigma_1^2(1-\rho^2)\right] \qquad \text{... (20·3 d)}$$

Similarly, the conditional distribution of random variable Y for a fixed X is

$$f_{Y|X}(y \mid x) = \frac{f_{X,Y}(x,y)}{f_X(x)}$$

$$= \frac{1}{\sigma_2\sqrt{2\pi}\sqrt{1-\rho^2}} \exp\left[-\frac{1}{2(1-\rho^2)\sigma_2^2}\left\{(y-\mu_2)-\rho\frac{\sigma_2}{\sigma_1}(x-\mu_1)\right\}^2\right], -\infty < y < \infty$$

Thus, the conditional distribution of Y for fixed X is also normal, given by:

$$(Y \mid X = x) : N\left[\mu_2 + \rho\frac{\sigma_2}{\sigma_1}(x-\mu_1), \sigma_2^2(1-\rho^2)\right] \qquad \text{... (20.3e)}$$

It is apparent from the above results that *the array means are collinear, i.e., the regression equations are linear (involving linear functions of the independent variables) and the array variances are constant (i.e., free from independent variable).*

We express this by saying that *the regression equations of Y on X, and X on Y are linear and homoscedastic.*

For $\rho = 0$, the conditional variance $V(Y \mid X)$ is equal to the marginal variance σ_2^2 and the conditional mean $E(Y \mid X)$ is equal to the marginal mean μ_2 and the two variables become independent, which is also apparent from joint probability function. In between the two extremes when $\rho = \pm 1$, the correlation coefficient ρ provides a measures of degree of association or inter-dependence between the two variables.

Example 20·1. *Show that for the bivariate normal distribution :*

$$dP = const.\ \exp\left[-\frac{1}{2(1-\rho^2)}(x^2 - 2\rho xy + y^2)\right] dx\, dy,$$

(*i*) *M.G.F. is* $M(t_1, t_2) = \exp\left[\frac{1}{2}(t_1^2 + 2\rho t_1 t_2 + t_2^2)\right]$

(*ii*) *Moments obey the recurrence relation :*

$$\mu_{rs} = (r+s-1)\rho\mu_{r-1,s-1} + (r-1)(s-1)(1-\rho^2)\mu_{r-2,s-2}$$

Hence or otherwise, show that

$$\mu_{rs} = 0, \text{ if } r+s \text{ is odd}, \quad \mu_{31} = 3\rho, \quad \mu_{22} = 1 + 2\rho^2$$

Solution:

(*i*) From the given probability function, we see that

$$\mu_1 = 0 = \mu_2 \quad \text{and} \quad \sigma_1^2 = 1 = \sigma_2^2$$

$$\therefore \quad M = M(t_1, t_2) = \exp\left[\frac{1}{2}(t_1^2 + 2\rho t_1 t_2 + t_2^2)\right]$$

(*ii*)
$$\frac{\partial M}{\partial t_1} = M(t_1 + \rho t_2) \quad \text{and} \quad \frac{\partial M}{\partial t_2} = M(t_2 + \rho t_1)$$

and
$$\frac{\partial^2 M}{\partial t_1 \partial t_2} = \frac{\partial}{\partial t_1}\left(\frac{\partial M}{\partial t_2}\right) = \frac{\partial}{\partial t_1}[M(t_2 + \rho t_1)]$$

$$= M\rho + (t_2 + \rho t_1)(t_1 + \rho t_2) M$$

$$\therefore \quad \frac{\partial^2 M}{\partial t_1 \partial t_2} - \rho t_1 \frac{\partial M}{\partial t_1} - \rho t_2 \frac{\partial M}{\partial t_2}$$

$$= M\rho + (t_2 + \rho t_1)(t_1 + \rho t_2) M - \rho t_1 (t_1 + \rho t_2) M - \rho t_2 (t_2 + \rho t_1) M$$

$$= M . \left(t_1 t_2 + \rho - \rho^2 t_1 t_2\right) \quad \text{(On simplification)}$$

$$= M\rho + \left(1 - \rho^2\right) M t_1 t_2$$

$$\therefore \quad \frac{\partial^2 M}{\partial t_1 \partial t_2} = \rho t_1 \frac{\partial M}{\partial t_1} + \rho t_2 \frac{\partial M}{\partial t_2} + M\rho + M\left(1 - \rho^2\right) t_1 t_2 \qquad \ldots(*)$$

But
$$M = \exp\left[\frac{1}{2}\left(t_1^2 + 2\rho t_1 t_2 + t_2^2\right)\right] = \sum_{r=0}^{\infty} \sum_{s=0}^{\infty} \mu_{rs} . \frac{t_1^r t_2^s}{r!\, s!}$$

$\therefore$ (*) gives

$$\sum_{r=1}^{\infty} \sum_{s=1}^{\infty} \mu_{rs} . \frac{\left(t_1^{r-1} t_2^{s-1}\right)}{(r-1)!\,(s-1)!}$$

$$= \left[\rho \sum_{r=1}^{\infty} \sum_{s=0}^{\infty} r\mu_{rs} . \frac{t_1^r t_2^s}{r!\, s!} + \rho \sum_{r=0}^{\infty} \sum_{s=1}^{\infty} s\mu_{rs} . \frac{t_1^r t_2^s}{r!\, s!}\right.$$

$$\left. + \rho \sum_{r=0}^{\infty} \sum_{s=0}^{\infty} \mu_{rs} . \frac{t_1^r t_2^s}{r!\, s!} + \left(1 - \rho^2\right) \sum_{r=0}^{\infty} \sum_{s=0}^{\infty} \mu_{rs} . \frac{t_1^{r+1} t_2^{s+1}}{r!\, s!}\right]$$

Equating the coefficients of $\dfrac{t_1^{r-1}}{(r-1)!} . \dfrac{t_2^{s-1}}{(s-1)!}$ on both sides, we get

$$\mu_{rs} = \left[\rho (r-1) \mu_{r-1,\, s-1} + \rho (s-1) \mu_{r-1,\, s-1} + \rho^2 \mu_{r-1,\, s-1} + \left(1 - \rho^2\right)(r-1)(s-1) \mu_{r-2,\, s-2}\right]$$

$$\Rightarrow \quad \mu_{rs} = (r + s - 1) \rho \mu_{r-1,\, s-1} + (r-1)(s-1)\left(1 - \rho^2\right) \mu_{r-2,\, s-2}$$

In particular

$$\mu_{31} = 3\rho\, \mu_{20} + 0 = 3\rho \sigma_1^2 = 3\rho \qquad \left(Q\ \sigma_1^2 = 1\right)$$

$$\mu_{22} = 3\rho\, \mu_{11} + \left(1 - \rho^2\right) \mu_{00} = 3\rho^2 + \left(1 - \rho^2\right) . 1 \qquad (Q\ \mu_{11} = \rho \sigma_1 \sigma_2 = \rho)$$

$$= \left(1 + 2\rho\right)$$

Also
$$\mu_{03} = \mu_{30} = 0$$

$$\mu_{12} = 2\rho\, \mu_{01} + 0 = 0 \qquad (Q\ \mu_{01} = \mu_{10} = 0)$$

$$\mu_{23} = 4\rho\, \mu_{12} + 1.2\left(1 - \rho^2\right) \mu_{01} = 0$$

Similarly, we will get: $\mu_{21} = 0, \quad \mu_{32} = 0$

If $r + s$ is odd, so is $(r-1)+(s-1), (r-2)+(s-2)$, and so on.

And since, $\mu_{30} = \mu_{03} = 0, \quad \mu_{12} = 0 = \mu_{21}, \quad \mu_{23} = 0 = \mu_{32}$,......., we finally get,

$$\mu_{rs} = 0, \text{ if } r+s \text{ is odd.}$$

Example 20·2: *Show that if X_1 and X_2 are standard normal variate with correlation coefficient ρ between them, then the correlation coefficient between X_1^2 and X_2^2 is given by ρ^2.*

Solution: Since X_1 and X_2, are two standard normal variates, we have

$$E(X_1) = E(X_2) = 0 \text{ and } V(X_1) = E(X_1^2) = 1 = V(X_2) = E(X_2^2)$$

$$\therefore \quad M_{X_1,X_2}(t_1,t_2) = \exp\left[\frac{1}{2}\left(t_1^2 + 2\rho t_1 t_2 + t_2^2\right)\right] \qquad \text{[c.f. (20·2}a\text{)]}$$

$$\rho\left(X_1^2, X_2^2\right) = \frac{E\left(X_1^2\, X_2^2\right) - E\left(X_1^2\right)E\left(X_2^2\right)}{\sqrt{\left[E\left(X_1^4\right) - \left\{E\left(X_1^2\right)\right\}^2\right]}\sqrt{\left[E\left(X_2^4\right) - \left\{E\left(X_2^2\right)\right\}^2\right]}}$$

where $$E\left(X_1^2\, X_2^2\right) = \text{Coefficient of } \frac{t_1^2}{2!}\cdot\frac{t_2^2}{2!} \text{ in } M(t_1,t_2) = \left(2\rho^2+1\right)$$

$$E\left(X \quad\right) = \text{Coefficient of } \frac{t_1^4}{4!} \text{ in } M(t_1,t_2) = 3$$

$$E\left(X_2^4\right) = \text{Coefficient of } \frac{t_2^4}{4!} \text{ in } M(t_1,t_2) = 3$$

$$\therefore \quad \rho\left(X_1^2, X_2^2\right) = \frac{2\rho^2+1-1}{\sqrt{(3-1)}\sqrt{(3-1)}} = \rho^2$$

Example 20·3: *The variables X and Y with zero means and standard deviations σ_1 and σ_2 are normally correlated with correlation coefficient ρ. Show that U and V defined as:*

$$U = \frac{X}{\sigma_1} + \frac{Y}{\sigma_2} \quad \text{and} \quad V = \frac{X}{\sigma_1} - \frac{Y}{\sigma_2}$$

are independent normal variates with variances 2(1 + ρ) and 2(1 − ρ) *respectively.*

Solution: Since $(X, Y) \sim BVN\,(0, 0, \sigma_1^2, \sigma_2^2, \rho)$, its joint p.d.f. is given by:

$$f(x,y) = \frac{1}{2\pi\sigma_1\sigma_2\sqrt{(1-\rho^2)}}\exp\left[\frac{-1}{2\left(1-\rho^2\right)}\left\{\frac{x^2}{\sigma_1^2} - \frac{2\rho xy}{\sigma_1\sigma_2} + \frac{y^2}{\sigma_2^2}\right\}\right]; -\infty < (x,y) < \infty$$

Also $$u = \frac{x}{\sigma_1} + \frac{y}{\sigma_2}, \quad v = \frac{x}{\sigma_1} - \frac{y}{\sigma_2},$$

$$\therefore \quad \frac{1}{2}(u+v) = \frac{x}{\sigma_1} \text{ and } \frac{1}{2}(u-v) = \frac{y}{\sigma_2} \Rightarrow x = \frac{\sigma_1}{2}(u+v) \text{ and } y = \frac{\sigma_2}{2}(u-v)$$

Jacobian of transformation J is given by:

$$J = \frac{\partial(x,y)}{\partial(u,v)} = \begin{vmatrix} \frac{1}{2}\sigma_1 & \frac{1}{2}\sigma_1 \\ \frac{1}{2}\sigma_2 & -\frac{1}{2}\sigma_2 \end{vmatrix} = -\frac{\sigma_1\sigma_2}{2}$$

Also $$\frac{x^2}{\sigma_1^2} + \frac{y^2}{\sigma_2^2} = \frac{1}{4}\left[(u+v)^2 + (u-v)^2\right] = \frac{1}{2}\left[u^2 + v^2\right]$$

The joint p.d.f. $g(u,v)$ of U and V is given by:

$$g(u,v) = \frac{1}{2\pi\sigma_1\sigma_2\sqrt{(1-\rho^2)}} \exp\left[-\frac{1}{2(1-\rho^2)}\left\{\frac{1}{2}(u^2+v^2) - 2\rho\left(\frac{u^2-v^2}{4}\right)\right\}\right]\frac{\sigma_1\sigma_2}{2}$$

$$= \frac{1}{2\pi.\,2\sqrt{(1-\rho^2)}} \exp\left[-\frac{1}{4(1-\rho^2)}\left\{(1-\rho)u^2 + (1+\rho)v^2\right\}\right] du\,dv$$

$$= \frac{1}{2\pi\sqrt{2(1-\rho)}\,\sqrt{2(1+\rho)}} \exp\left[-\frac{u^2}{4(1+\rho)} - \frac{v^2}{4(1-\rho)}\right] du\,dv$$

$$= \left[\frac{1}{\sqrt{2\pi}\sqrt{2(1+\rho)}} \exp\left\{-\frac{u^2}{2(1+\rho)2}\right\}\right] \times \left[\frac{1}{\sqrt{2\pi}\sqrt{2(1-\rho)}} \exp\left\{-\frac{v^2}{2(1-\rho)2}\right\}\right]$$

$$= \left[f_1(u)\,du\right]\left[f_2(v)\,dv\right], \text{ (say)}$$

where $$f_1(u) = \frac{1}{\sqrt{2\pi}\sqrt{2(1+\rho)}} \cdot \exp\left\{-\frac{u^2}{2\times2(1+\rho)}\right\}$$

and $$f_2(v) = \frac{1}{\sqrt{2\pi}\sqrt{2(1-\rho)}} \cdot \exp\left\{-\frac{v^2}{2\times2(1-\rho)}\right\}$$

(*) $\Rightarrow$ U and V are Independently distributed,

$U \sim N[0, 2(1+\rho)]$ and $V \sim N[0, 2(1-\rho)]$

Aliter: *Find joint m.g.f. of U and V viz.,*

$$M(t_1, t_2) = E\left(e^{t_1U + t_2V}\right) = E\left[\exp[\{X(t_1+t_2)/\sigma_1\} + \{Y(t_1-t_2)/\sigma_2\}]\right]$$

and use $$E\left(e^{t_1X + t_2Y}\right) = \exp\left[\left(t_1^2\sigma_1^2 + t_2^2\sigma_2^2 + 2\rho t_1t_2\sigma_1\sigma_2\right)/2\right]$$

Example 20·4. *If X and Y are standard normal variates with co-efficient of correlation ρ, show that*

(i) *Regression of Y and X is linear.*

(ii) *X + Y and X − Y are independently distributed.*

(iii) $Q = \dfrac{X^2 - 2\rho XY + Y^2}{(1-\rho^2)}$ *is distributed like a chi-square, i.e., as that of the sum of the squares of standard normal variates.*

Solution:

(*i*) *c.f.* equation (20·3*e*) § 20·1·3.

(*ii*) $(X, Y) \sim BVN\ (0, 0, 1, 1, \rho)$

$$\therefore \quad f(x, y) = \frac{1}{2\pi\sqrt{1-\rho^2}} \exp\left[-\frac{1}{2(1-\rho^2)}(x^2 - 2\rho xy + y^2)\right]$$

Let $u = x + y$ and $v = x - y \quad \Rightarrow \quad x = \dfrac{u+v}{2}, \quad y = \dfrac{u-v}{2}$

$$\therefore \quad J = \frac{\partial(x,y)}{\partial(u,v)} = \begin{vmatrix} \dfrac{\partial x}{\partial u} & \dfrac{\partial x}{\partial v} \\ \dfrac{\partial y}{\partial u} & \dfrac{\partial y}{\partial v} \end{vmatrix} = \begin{vmatrix} \dfrac{1}{2} & \dfrac{1}{2} \\ \dfrac{1}{2} & -\dfrac{1}{2} \end{vmatrix} = -\frac{1}{2}$$

The joint pd.f. $g\ (u\ , v)$ of U and V is given by:

$$g\,(u, v) = C \exp\left[-\frac{1}{2(1-\rho^2).\,4}\left\{2(u^2+v^2) - 2\rho(u^2 - v^2)\right\}\right],$$

where $\quad C = \dfrac{1}{4\pi\sqrt{1-\rho^2}}$

$$\therefore \quad g\,(u, v) = C \exp\left[-\frac{1}{4(1-\rho^2)}\left\{(1-\rho)u^2 + (1+\rho)v^2\right\}\right]$$

$$= \left[C_1 \exp\left\{-\frac{u^2}{4(1+\rho)}\right\}\right] \cdot \left[C_2 \exp\left\{-\frac{v^2}{4(1-\rho)}\right\}\right]$$

$$= [g_1\,(u)]\,[g_2\,(v)],\ \text{(say)}.$$

Hence, $\quad U$ and V are independently distributed.

(*iii*) $M_Q(t) = \displaystyle\int_{-\infty}^{\infty}\int_{-\infty}^{\infty} e^{tQ} f\,(x,y)\,dx\,dy$

$$= \frac{1}{2\pi\sqrt{(1-\rho^2)}} \int_{-\infty}^{\infty}\int_{-\infty}^{\infty} \exp\,(tQ) \times \exp\left[-\frac{1}{2(1-\rho^2)}\left\{x^2 - 2\rho xy + y^2\right\}\right] dx\,dy$$

$$= \frac{1}{2\pi\sqrt{(1-\rho^2)}} \int_{-\infty}^{\infty}\int_{-\infty}^{\infty} \exp\left(tQ - \frac{Q}{2}\right) dx\,dy$$

$$= \frac{1}{2\pi\sqrt{(1-\rho^2)}} \int_{-\infty}^{\infty} \int_{-\infty}^{\infty} \exp\left[-\frac{Q}{2}(1-2t)\right] dx\, dy$$

Put $\sqrt{(1-2t)}\, x = u$ and $\sqrt{(1-2t)}\, y = v$ $\Rightarrow$ $dx = \frac{du}{\sqrt{(1-2t)}}$ and $dy = \frac{dv}{\sqrt{(1-2t)}}$

Also $Q = \frac{1}{(1-\rho^2)}\left[x^2 - 2\rho xy + y^2\right] = \frac{1}{(1-\rho^2)}\left[\frac{u^2 - 2\rho uv + v^2}{1-2t}\right]$

$$M_Q(t) = \frac{1}{2\pi\sqrt{(1-\rho^2)}\,(1-2t)} \times \int_{-\infty}^{\infty} \int_{-\infty}^{\infty} \exp\left[-\frac{1}{2(1-\rho^2)}\left(u^2 - 2\rho uv + v^2\right)\right] du\, dv$$

$$= \frac{1}{(1-2t)} \cdot 1 = (1-2t)^{-1}$$

which is the m.g.f. of chi-square (χ^2) variate with n (= 2) degrees of freedom.

(For chi-square distribution, see Chapter 15)

Example 20·5. *Let X and Y be independent standard normal variates. Obtain the m.g.f. of XY.*

Solution: We have, by def.,

$$M_{XY}(t) = E\left(e^{tXY}\right) = \int_{-\infty}^{\infty} \int_{-\infty}^{\infty} \left(e^{txy}\right) . f(x,y)\, dxdy$$

Since X and Y are independent standard normal variates, their joint p.d.f. $f(x,y)$ is given by:

$$f(x,y) = f_1(x).\, f_2(y) = \frac{1}{2\pi} e^{-x^2/2}\, e^{-y^2/2}; \quad -\infty < (x,y) < \infty$$

$$\therefore M_{XY}(t) = \frac{1}{2\pi} \int_{-\infty}^{\infty} \int_{-\infty}^{\infty} \exp\left[-\frac{1}{2}\left(x^2 - 2txy + y^2\right)\right] dx\, dy$$

$$= \frac{1}{2\pi} \int_{-\infty}^{\infty} \int_{-\infty}^{\infty} \exp\left[-\frac{1}{2(1-t^2)} \times \left\{\frac{x^2}{1/(1-t^2)} - \frac{2t.x.y}{\left(1/\sqrt{1-t^2}\right)\left(1/\sqrt{1-t^2}\right)} + \frac{y^2}{1/(1-t^2)}\right\}\right] dxdy \quad \ldots(*)$$

If $(U, V) \sim BVN\left(0, 0, \sigma_1{}^2, \sigma_2{}^2, \rho,\right)$, then we have

$$\int_{-\infty}^{\infty} \int_{-\infty}^{\infty} \frac{1}{2\pi\sigma_1\sigma_2\sqrt{1-\rho^2}} . e^{-\frac{1}{2(1-\rho^2)}\left\{\frac{x^2}{\sigma_1{}^2} - \frac{2\rho xy}{\sigma_1\sigma_2} + \frac{y^2}{\sigma_2{}^2}\right\}} dxdy = 1$$

$$\Rightarrow \int_{-\infty}^{\infty} \int_{-\infty}^{\infty} e^{-\frac{1}{2(1-\rho^2)}\left\{\frac{x^2}{\sigma_1^2} - \frac{2\rho xy}{\sigma_1\sigma_2} + \frac{y^2}{\sigma_2{}^2}\right\}} dxdy = 2\pi\sigma_1\sigma_2\sqrt{1-\rho^2} \quad \ldots(**)$$

Comparing (*) and (**) with $\sigma_1^2 = \sigma_2^2 = \dfrac{1}{(1-t^2)}$ and $\rho = t$, we get

$$M_{XY}(t) = \frac{1}{2\pi} 2\pi \frac{1}{\sqrt{1-t^2}} \cdot \frac{1}{\sqrt{1-t^2}} \sqrt{1-t^2} \Rightarrow M_{XY}(t) = (1-t^2)^{-1/2}; -1 < t < 1.$$

Example 20·6: *Let X and Y have bivariate normal distribution with parameters:*

$$\mu_X = 5, \mu_Y = 10, \ \sigma_X^2 = 1, \ \ \sigma_Y^2 = 25 \ \text{and} \ \text{Corr}(X,Y) = \rho.$$

(*a*) If $\rho > 0$, find ρ when $P(4 < Y < 16 \mid X = 5) = 0{\cdot}954$

(*b*) If $\rho = 0$, find $P(X + Y) \le 16$.

Solution: Since $X, Y : BVN(\mu_X, \mu_Y, \sigma_X^2, \sigma_Y^2, \rho^2)$, the conditional distribution of Y given $X = x$ is also normal.

$$(Y \mid X = x) : N\left[\mu = \mu_Y + \frac{\rho\sigma_Y}{\sigma_X}(x - \mu_X), \ \sigma^2 = \sigma_Y^2(1-\rho^2)\right]. \text{ Therefore}$$

$$(Y \mid X = 5) : N\left[\mu = 10 + \rho \times \frac{5}{1}(5-5), \ \sigma^2 = 25(1-\rho^2)\right] = N\left[\mu = 10, \sigma^2 = 25(1-\rho^2)\right]$$

We want ρ so that $P(4 < Y < 16 \mid X = 5) = 0{\cdot}954$

where $$Z = \frac{Y - \mu}{\sigma} = \frac{Y - 10}{5\sqrt{(1-\rho^2)}} : N(0, 1)$$

$$\Rightarrow \quad P\left(\frac{4-10}{\sigma} < Z < \frac{16-10}{\sigma}\right) = 0{\cdot}954 \ \Rightarrow \ P\left(\frac{-6}{\sigma} < Z < \frac{6}{\sigma}\right) = 0{\cdot}954 \quad (*)$$

But we know that if $Z : N(0, 1)$, then $P(-2 < Z < 2) = 0{\cdot}954$ ... (**)

Comparising (*) and (**), we get

$$\frac{6}{\sigma} = 2 \ \Rightarrow \sigma = 3 \ \Rightarrow \ \sigma^2 = 9 = 25(1-\rho^2)$$

$$\therefore \quad 1-\rho^2 = \frac{9}{25} \ \Rightarrow \ \rho^2 = \frac{16}{25} \ \Rightarrow \ \rho = \frac{4}{5} = 0{\cdot}8 \qquad (\because \rho > 0)$$

(*b*) Since (X, Y) have bivariate normal distribution, $\rho = 0 \Rightarrow X$ and Y are independent *r.v's*,

and $X : N(\mu_X = 5, \ \sigma_X^2 = 1)$ and $Y : N(\mu_Y = 10, \ \sigma_Y^2 = 25)$

Thus $(X + Y) : N(\mu = \mu_X + \mu_Y, \sigma^2 = \sigma_X^2 + \sigma_Y^2) = N(15, 26)$

$$\therefore \quad P(X + Y \le 16) = P\left(Z \le \frac{16-15}{\sqrt{26}}\right), \text{ where } Z = \frac{(X+Y) - \mu}{\sigma} : N(0,1)$$

$$\Rightarrow \quad P(X + Y \le 16) = P\left(Z \le \frac{1}{\sqrt{26}}\right) = \Phi(1/\sqrt{26}),$$

where $\Phi(z) = P(Z \le z)$, is the distribution function of standard normal variate.

Remark. $P(X+Y \le 16) = P\left(Z \le \frac{1}{5\cdot099}\right) = P(Z \le 0\cdot196)$

$$= 0\cdot5 + P(0 \le Z \le 0\cdot196) = 0\cdot5 + 0\cdot0793 \text{ (approx.)}$$

$$= 0\cdot5793.$$

Example 20·7. *If* $(X, Y) \sim BVN(0, 0, 1, 1, \rho)$, then find:

(a) $E(Max(X,Y))$, (b) $E(Min(X,Y))$, (c) $E[Max(X,Y) + Min(X,Y)]$

Solution.

$$(X,Y) \sim BVN(0,0,1,1,\rho) \Rightarrow X \sim N(0,1) \text{ and } Y \sim N(0,1) \quad \text{... (i)}$$

$$\Rightarrow \left.\begin{array}{l} E(X) = E(Y) = 0;\ \text{Var}(X) = \text{Var}(Y) = 1;\ \rho(X,Y) = \rho; \\ \text{Cov}(X,Y) = \rho\sigma_X\sigma_Y = \rho \end{array}\right\} \quad \text{...(ii)}$$

$$\left.\begin{array}{l} E(X+Y) = E(X) + E(Y) = 0;\ E(X-Y) = E(X) - E(Y) = 0 \\ \text{Var}(X+Y) = \text{Var}(X) + \text{Var}(Y) + 2\,\text{Cov}(X,Y) = 1+1+2\rho = 2(1+\rho) \\ \text{Var}(X-Y) = \text{Var}(X) + \text{Var}(Y) - 2\,\text{Cov}(X,Y) = 1+1-2\rho = 2(1-\rho) \end{array}\right\} \quad \text{...(iii)}$$

Also $X+Y \sim N\left[0, \sigma_1^2 = 2(1+\rho)\right]$; $X-Y \sim N\left[0, \sigma_2^2 = 2(1-\rho)\right]$ ... (iv)

Let $$Z = (X-Y) \sim N(0, \sigma_2^2)$$

$\Rightarrow$ $E|Z| = E|X-Y| =$ Mean Deviation of $N(0, \sigma_2^2)$ distribution

$$= \sqrt{\frac{2}{\pi}} \cdot \sigma_2 = \sqrt{\frac{2}{\pi}} \cdot \sqrt{2(1-\rho)} = 2 \cdot \sqrt{\frac{1-\rho}{\pi}} \quad \text{... (v)}$$

$$[\text{Max}(X,Y)] = \frac{1}{2}\left[(X+Y) + |X-Y|\right]; \quad [\text{Min}(X,Y)] = \frac{1}{2}\left[(X+Y) - |X-Y|\right] \quad \text{... (vi)}$$

(a) $E[\text{Max}(X,Y)] = \frac{1}{2} E[(X+Y) + |X-Y|]$

$$= \frac{1}{2}\left[E(X+Y) + E|X-Y|\right] = \frac{1}{2}\left[0 + 2\sqrt{\frac{1-\rho}{\pi}}\right] = \sqrt{\frac{1-\rho}{\pi}} \quad \text{[From (iii) and (iv)]}$$

(b) $E[\text{Min}(X,Y)] = \frac{1}{2}\left[E(X+Y) - E|X-Y|\right] = \frac{1}{2}\left[0 - 2\sqrt{\frac{1-\rho}{\pi}}\right] = -\sqrt{\frac{1-\rho}{\pi}}$

(c) $E[\text{Max}(X,Y) + \text{Min}(X,Y)] = E[\text{Max}(X,Y)] + E[\text{Min}(X,Y)]$

$$= \sqrt{\frac{1-\rho}{\pi}} - \sqrt{\frac{1-\rho}{\pi}} = 0 \quad \text{[From (a) and (b)]}$$

20·2. MULTIVARIATE NORMAL (MVN) DISTRIBUTION

The Multivariate normal distribution is among the most important of multivariate distributions, particularly in statistical inference and study of Gaussian Processes such as Brownian motion. The distribution arises naturally from linear transformations of independent normal variables.

Multivariate normal distribution is the generalisation of univariate normal distribution to two or more variables. It is the distribution for random vectors of correlated variables, each of which has a univariate normal distribution.

20·2·1 Multivariate Normal Distribution (p.d.f.)

The p.d.f. of multivariate normal distribution is of the form:

$$f(x) = f(x_1, x_2, \ldots, x_p) = k\exp\left[-\frac{1}{2}(x-b)' A(x-b)\right] \;;\; -\infty < x_i < \infty \;;\; i = 1, 2, \cdots, p \quad \ldots (20.4)$$

where, $x = \begin{pmatrix} x_1 \\ x_2 \\ \vdots \\ x_p \end{pmatrix}$, is a random vector,

$b = \begin{pmatrix} b_1 \\ b_2 \\ \vdots \\ x_p \end{pmatrix}$, is a real vector,

$A = \left(a_{ij}\right)_{pxp} = \begin{pmatrix} a_{11} & a_{12} & \cdots & a_{1p} \\ a_{21} & a_{22} & \cdots & a_{2p} \\ \vdots & \vdots & \cdots & \vdots \\ a_{p1} & a_{p2} & \cdots & a_{pp} \end{pmatrix}$, is any positive definite matrix, and $k > 0$, is a constant.

Since A_{pxp} is positive definite matrix, there exists a non-singular matrix $C_{p\times p}$ such that:

$$C'AC = I_p = \begin{pmatrix} 1 & 0 & \cdots & 0 \\ 0 & 1 & \cdots & 0 \\ \vdots & \vdots & & \vdots \\ 0 & 0 & \cdots & 1 \end{pmatrix}_{pxp}, \text{ where } C' = \text{Transpose of } C. \quad \ldots (20{\cdot}4a)$$

Writing $|C|$ = Determinant of matrix C, and noting that $|C'| = |C|$, on taking determinant of both sides in (20·4*a*), we get:

$$|C'|\cdot|A|\cdot|C| = |I_p| = 1 \quad \Rightarrow \quad |C|^2 \cdot |A| = 1 \quad \Rightarrow \quad |C| = \frac{1}{|A|^{1/2}} \quad \ldots (20{\cdot}4b)$$

Let us transform $(x_1, x_2, \ldots, x_p)$ to $(y_1, y_2, \ldots, y_p)$ by the non-singular linear transformation :

$$x - b = C_{pp} \cdot y_{p\times 1} \quad \text{where} \quad y' = (y_1, y_2, \cdots, y_p) \quad \ldots (20.4c)$$

$$\Rightarrow \quad \begin{pmatrix} x_1 - b_1 \\ x_2 - b_2 \\ \vdots \\ x_p - b_p \end{pmatrix} = \begin{pmatrix} c_{11} & c_{12} & \cdots & c_{1p} \\ c_{21} & c_{22} & \cdots & c_{2p} \\ \vdots & \vdots & & \vdots \\ c_{p1} & c_{p2} & \cdots & c_{pp} \end{pmatrix} \begin{pmatrix} y_1 \\ y_2 \\ \vdots \\ y_p \end{pmatrix}$$

$$\Rightarrow \quad \begin{cases} x_1 - b_1 = c_{11}y_1 + c_{12}y_2 + \cdots + c_{1p}y_p \\ x_2 - b_2 = c_{21}y_1 + c_{22}y_2 + \cdots + c_{2p}y_p \\ \quad\vdots \qquad\qquad \vdots \qquad\qquad \vdots \\ x_p - b_p = c_{p1}y_1 + c_{p2}y_2 + \cdots + c_{pp}y_p \end{cases}$$

Jacobian of transformation (20·4c) is given by:

$$J = \frac{\partial(x_1, x_2, \cdots, x_p)}{\partial(y_1, y_2, \cdots, y_p)} = \begin{vmatrix} \frac{\partial x_1}{\partial y_1} & \frac{\partial x_1}{\partial y_2} & \cdots & \frac{\partial x_1}{\partial y_p} \\ \frac{\partial x_2}{\partial y_1} & \frac{\partial x_2}{\partial y_2} & \cdots & \frac{\partial x_2}{\partial y_p} \\ \vdots & \vdots & & \vdots \\ \frac{\partial x_p}{\partial y_1} & \frac{\partial x_p}{\partial y_2} & \cdots & \frac{\partial x_p}{\partial y_p} \end{vmatrix}$$

$$= \begin{vmatrix} c_{11} & c_{12} & \cdots & c_{1p} \\ c_{21} & c_{22} & \cdots & c_{2p} \\ \vdots & \vdots & & \vdots \\ c_{p1} & c_{p2} & \cdots & c_{pp} \end{vmatrix} = |C| = \frac{1}{|A|^{1/2}} \quad \text{... (20.4d)}$$

Also,
$$(\boldsymbol{x} - \boldsymbol{b})' A(\boldsymbol{x} - \boldsymbol{b}) = (C\boldsymbol{y})' A (C\boldsymbol{y}) = \boldsymbol{y}'(C'AC)\boldsymbol{y}$$

$$= \boldsymbol{y}'(I_p)\boldsymbol{y} = \boldsymbol{y}' \cdot \boldsymbol{y} = \sum_{i=1}^{p} y_i^2 \qquad \text{[From (20.4a)]} \quad \text{... (20·4e)}$$

The joint p.d.f. $g(\cdot)$ of $(y_1, y_2, ..., y_p)$ is given by:

$$g(\boldsymbol{y}) = g(y_1, y_2, \cdots, y_p) = f(\boldsymbol{x}) \cdot \text{mod}(J) \quad \text{.... (20·4f)}$$

$$= \frac{k}{|A|^{1/2}} \exp\left[-\frac{1}{2}\sum_{i=1}^{p} y_i^2\right]; \ -\infty < y_i < \infty, \ i = 1, 2, \cdots, p$$

$$\Rightarrow \qquad y_i \underset{\sim}{\text{i.i.d}} N(0, 1) \ ; \ i = 1, 2, \cdots, p \quad \text{... (20·4g)}$$

i.e., $y_1, y_2, ..., y_p$ are *i.i.d.* $N(0, 1)$ variates.

The constant k is determined from the fact that total probability is 1; *i.e.*,

$$\int_{-\infty}^{\infty}\int_{-\infty}^{\infty} \cdots \int_{-\infty}^{\infty} g(y_1, y_2, \cdots, y_p) \, dy_1 \, dy_2 \cdots dy_p = 1$$

$$\Rightarrow \qquad \frac{k}{|A|^{1/2}} \prod_{i=1}^{p} \left[\int_{-\infty}^{\infty} \exp\left(-\frac{1}{2}y_i^2\right) dy_i\right] = 1$$

$$\Rightarrow \qquad \frac{k}{|A|^{1/2}} \cdot \prod_{i=1}^{p} \left(\sqrt{2\pi}\right) = 1 \quad \Rightarrow \quad k = \frac{|A|^{1/2}}{(2\pi)^{p/2}}$$

Substituting in (20·4a), the p.d.f. of multivariate normal distribution is of the form:

$$f(\boldsymbol{x}) = f(x_1, x_2, \cdots, x_p) = \frac{|A|^{1/2}}{(\sqrt{2\pi})^p} \cdot \exp\left[-\frac{1}{2}(\boldsymbol{x} - \boldsymbol{b})' A (\boldsymbol{x} - \boldsymbol{b})\right]; \ -\infty < x_i < \infty \quad \text{...(20.4h)}$$

Significance of Vector *b* and Matrix *A* in (20·4*h*)

We have proved in (20·4*h*) that when we transform x's in (20·4) to y's by the transformation (20·4*c*), then :

y_i ; $(i = 1, 2, \ldots, p)$ are *i.i.d.* $N(0,1)$ variates

$$\Rightarrow \quad E(y) = E\begin{pmatrix} y_1 \\ y_2 \\ \vdots \\ y_p \end{pmatrix} = \begin{pmatrix} E(y_1) \\ E(y_2) \\ \vdots \\ E(y_p) \end{pmatrix} = \begin{pmatrix} 0 \\ 0 \\ \vdots \\ 0 \end{pmatrix} = \mathbf{0} \quad \ldots (20 \cdot 4h)$$

$$Var(y) = Var\begin{pmatrix} y_1 \\ y_1 \\ \vdots \\ y_p \end{pmatrix} = \begin{pmatrix} Var(y_1) & Cov(y_1, y_2) & \cdots & Cov(y_1, y_p) \\ Cov(y_2, y_1) & Var(y_2) & \cdots & Cov(y_2, y_p) \\ \vdots & \vdots & & \vdots \\ Cov(y_p, y_1) & Cov(y_p, y_2) & \cdots & Var(y_p) \end{pmatrix}$$

$$= \begin{pmatrix} 1 & 0 & 0 & \cdots & 0 \\ 0 & 1 & 0 & \cdots & 0 \\ 0 & 0 & 1 & \cdots & 0 \\ \vdots & \vdots & \vdots & & \vdots \\ 0 & 0 & 0 & & 1 \end{pmatrix} = I_p \qquad \left[\because y_i's \text{ are } i.i.d. N(0, 1)\right] \quad \ldots (20 \cdot 4i)$$

Also, $\quad x - b = Cy \quad \Rightarrow \quad E(x) - b = C \cdot E(y) = 0.$ [Form (20·4*h*)]

$$\Rightarrow \quad b = E(x) = \begin{pmatrix} E(x_1) \\ E(x_2) \\ \vdots \\ E(x_p) \end{pmatrix} = \begin{pmatrix} \mu_1 \\ \mu_2 \\ \vdots \\ \mu_p \end{pmatrix} = \mu_{p \times 1}\,(say), \quad \ldots (20 \cdot 4j)$$

where, $\quad E(x_i) = \mu_i \; ; \quad i = 1, 2, \ldots, p.$

$$Var(x - b) = Var(Cy) = C\,Var(y)C' \qquad [\because Var(Ax) = A \cdot Var(x) \cdot A']$$

$$\Rightarrow \quad Var(x) = C\,Var(y)C' = CI_pC' = CC' \qquad \text{[Form (20·4}i\text{)]} \quad \ldots (20 \cdot 4k)$$

The variance – covariance matrix of x as given by is:

$$Var\,(x) = \begin{pmatrix} \sigma_1^2 & \sigma_{12} & \sigma_{13} & \cdots & \sigma_{1p} \\ \sigma_{21} & \sigma_2^2 & \sigma_{23} & \cdots & \sigma_{2p} \\ \vdots & \vdots & \vdots & & \vdots \\ \sigma_{p1} & \sigma_{p2} & \sigma_{p3} & \cdots & \sigma_p{}^2 \end{pmatrix} = \sum, (say) \quad \ldots (20 \cdot 4l)$$

where, $\quad \sigma_i^2 = Var(x_i) \quad \text{and} \quad \sigma_{ij} = Cov(x_i, x_j), i \neq j = 1, 2, \ldots, p$

From (20·4*k*) and (20·4*l*), we get:

$$Var(x) = \Sigma = CC' \quad \ldots (20 \cdot 4m)$$

But, $\quad C'AC = I_p$ [From (20·4*a*)]

Pre-multiplying by $(C')^{-1}$ and post-multiplying by C^{-1} both sides, we get

$$(C')^{-1}C'AC(C^{-1}) = (C')^{-1} \cdot I_p \cdot C^{-1} = (C')^{-1} \cdot C^{-1}$$

$$\Rightarrow \quad A = (CC')^{-1} \quad \Rightarrow \quad A^{-1} = CC' \quad \ldots (20 \cdot 4n)$$

Substituting in (20·4*m*), we get:

$$Var(x) = \Sigma = A^{-1} \quad \Rightarrow \quad A = (\Sigma)^{-1} \quad \ldots (20 \cdot 4o)$$

Taking determinant on both sides, we get

$$|A| = \left|\Sigma^{-1}\right| = |\Sigma|^{-1} = \frac{1}{|\Sigma|} \quad \ldots (20 \cdot 5p)$$

where, $|\Sigma|$ is the determinant of matrix Σ. Substituting from (20·4*j*), (20·4*o*) and (20·4*p*) in (20·4*h*), we obtain the following result.

Multivariate Normal Distribution

The p.d.f. of the *p*-component random vector $\boldsymbol{x}$ where $\boldsymbol{x}' = (x_1, x_2, \cdots x_p)_{1\times p}$ having multivariate normal distribution with mean vector $\boldsymbol{\mu}$ and the variance-covariance matrix $|\Sigma|$ is given by:

$$f(\boldsymbol{x}) = f(x_1, x_2, \cdots, x_p) = \frac{1}{(\sqrt{2\pi})^p |\Sigma|^{1/2}} \cdot \exp\left[-\frac{1}{2}(\boldsymbol{x}-\boldsymbol{\mu})' \Sigma^{-1} (\boldsymbol{x}-\boldsymbol{\mu})\right] ; -\infty < \boldsymbol{x}; < \infty; \quad \ldots (20·4q)$$

where, $\quad \boldsymbol{\mu} = E(\boldsymbol{x})$ and $\Sigma = \text{Var}(\boldsymbol{x})$ are given in (20·4*j*) and (20·4*l*).

We write: $\quad X \sim N_p(\boldsymbol{\mu}, \Sigma)$.

Remark: The matrix:
$$(\boldsymbol{x}-\boldsymbol{\mu})' \Sigma^{-1} (\boldsymbol{x}-\boldsymbol{\mu})$$
in the non-degenerate MVND in equation (20·4*q*) is known as the "*Square of Mahalanobis Distance*", which represents the distance of the point $\boldsymbol{x}$ from the mean $\boldsymbol{\mu}$.

Note that in case when $p = 1$, MVND reduces to univariate normal distribution $N(\mu, \sigma^2)$, say, and the "*Mahalanobis Distance*" reduces to *standard score* $[(x - \mu)/\sigma]$.

Deductions 1: Univariate Normal Distribution

For univariate normal distribution, we replace the vectors $\boldsymbol{x}$ and $\boldsymbol{\mu}$ by single values x and $E(x) = \mu$, and the variance - covarnce matrix Σ by single value $\text{Var}(x) = \sigma^2$ and take $p = 1$. Thus (20·4*q*) reduces to:

$$f(x) = \frac{1}{\sqrt{2\pi}\cdot\sigma} \exp\left[-\frac{1}{2}\left(\frac{x-\mu}{\sigma}\right)^2\right] ; \quad -\infty < x < \infty \quad \ldots (20·4r)$$

$$\left[\Sigma = \sigma^2 \Rightarrow |\Sigma| = \sigma^2 \text{ and } \Sigma^{-1} = \frac{1}{\sigma^2}\right]$$

which is the p.d.f. of univariate normal distribution and we write $X \sim N(\mu, \sigma^2)$.

2. Bivariate Normal Distribution. In (20·4*q*) Let us take : $p = 2$ and

$$\boldsymbol{x} = \begin{pmatrix} x_1 \\ x_2 \end{pmatrix} ; \quad \boldsymbol{\mu} = \begin{pmatrix} \mu_1 \\ \mu_2 \end{pmatrix}, \quad \sum = Var\, \boldsymbol{x} = \begin{pmatrix} \sigma_1^2 & \rho\sigma_1\sigma_2 \\ \rho\sigma_1\sigma_2 & \sigma_2^2 \end{pmatrix}$$

where $\rho = \text{corr.}(x_1, x_2)$.

$$\therefore \quad |\Sigma| = \sigma_1^2 \sigma_2^2 (1 - \rho^2),$$

$$\sum^{-1} = \frac{1}{\sigma_1^2\sigma_2^2(1-\rho^2)} \cdot \begin{pmatrix} \sigma_2^2 & -\rho\sigma_1\sigma_2 \\ -\rho\sigma_1\sigma_2 & \sigma_1^2 \end{pmatrix} \quad \left[\because \begin{pmatrix} a & b \\ c & d \end{pmatrix}^{-1} = \frac{1}{ad - bc}\begin{pmatrix} d & -b \\ -c & a \end{pmatrix}\right]$$

and $\quad (\boldsymbol{x}-\boldsymbol{\mu})' \sum^{-1} (\boldsymbol{x}-\boldsymbol{\mu})$

$$= (x_1 - \mu_1 \quad x_2 - \mu_2) \cdot \frac{1}{\sigma_1^2\sigma_2^2(1-\rho^2)} \begin{pmatrix} \sigma_2^2 & -\rho\sigma_1\sigma_2 \\ -\rho\sigma_1\sigma_2 & \sigma_1^2 \end{pmatrix} \cdot \begin{pmatrix} x_1 - \mu_1 \\ x_2 - \mu_2 \end{pmatrix}$$

$$= \frac{1}{1-\rho^2}\left[\left(\frac{x_1-\mu_1}{\sigma_1}\right)^2 - 2\rho\left(\frac{x_1-\mu_1}{\sigma_1}\right)\left(\frac{x_2-\mu_2}{\sigma_2}\right) + \left(\frac{x_2-\mu_2}{\sigma_2}\right)^2\right]$$

$$\left[\because (x \quad y)\begin{pmatrix} a & b \\ c & d \end{pmatrix}\begin{pmatrix} x \\ y \end{pmatrix} = ax^2 + (b+c)xy + dy^2\right]$$

Substituting in (20·4q), we get:

$$f(x_1, x_2) = \frac{1}{2\pi\sigma_1\sigma_2\sqrt{1-\rho^2}}$$

$$\times \exp\left[-\frac{1}{2(1-\rho^2)}\left\{\left(\frac{x_1-\mu_1}{\sigma_1}\right)^2 - 2\rho\left(\frac{x_1-\mu_1}{\sigma_1}\right)\left(\frac{x_2-\mu_2}{\sigma_2}\right) + \left(\frac{x_2-\mu_2}{\sigma_2}\right)^2\right\}\right] (-\infty < (x_1, x_2) < \infty) \qquad \text{... (20·4s)}$$

which is the p.d.f. of bivariate normal distribution: BVN $(\mu_1, \mu_2, \sigma_1^2, \sigma_2^2, \rho)$.

20·2·2 Moment Generating Function of $N_p(\mu, \Sigma)$.

$$M_x(t) = E(e^{t'x}) \quad \text{where } t = \begin{pmatrix} t_1 \\ t_2 \\ \vdots \\ t_p \end{pmatrix} \text{ and } x = \begin{pmatrix} x_1 \\ x_2 \\ \vdots \\ x_p \end{pmatrix}$$

Transform the variables x to y by non-singular linear transformation

$$x - \mu = Cy \quad \Rightarrow \quad x = \mu + Cy \qquad \text{... (*)}$$

$$\therefore \quad M_x(t) = E\left[e^{t'x}\right] = E\left[e^{t'(\mu + Cy)}\right] = e^{t'\mu} \cdot E\left(e^{t'Cy}\right)$$

$$= e^{t'\mu} E\left(e^{(C't)'y}\right) = e^{t'\mu} \cdot E\left(e^{u'y}\right) \ ; \ u = C't \qquad \text{...(**)}$$

We have proved in (20·4f), that if we transform x to y by transformation (*), then $(y_1, y_2, \ldots, y_p)$ are *i.i.d.* $N(0,1)$.

$\therefore$ The m.g.f. of y is given by:

$$M_y(u) = E\left(e^{u'y}\right) \ ; \ u' = (u_1, u_2, \ldots, u_p) \qquad \text{... (*)}$$

$$= E\left[e^{u_1 y_1} \cdot e^{u_2 y_2} \ldots e^{u_p y_p}\right] = E\left[\prod_{i=1}^{p} e^{u_i y_i}\right]$$

$$= \prod_{i=1}^{p} E\left(e^{u_i y_i}\right) ; \quad \left[\because y_i\text{'s are independent}\right]$$

$$= \prod_{i=1}^{p} \left[M_{y_i}(u_i)\right] = \prod_{i=1}^{p} e^{\frac{1}{2}u_i^2} ; \quad [\because y_i \sim N(0,1)]$$

$$\Rightarrow \quad M_y(u) = \exp\left[\frac{1}{2}\sum_{i=1}^{p} u_i^2\right] = \exp\left[\frac{1}{2}u'u\right] \qquad \text{... (20.4t)}$$

$$\therefore \quad M_x(t) = \exp(t'\mu) \cdot M_y(u) \qquad \text{[From (*)]}$$

$$= \exp(t'\mu) \cdot \exp\left(\frac{1}{2}u'u\right) \qquad \text{[From (20.4t)]}$$

$$= \exp\left[t'\mu + \frac{1}{2} (C't)' (C't) \right]$$

$$= \exp\left[t'\mu + \frac{1}{2} t' (CC') t \right]$$

$$= \exp\left[t'\mu + \frac{1}{2} t' \Sigma t \right] \qquad [\because \Sigma = CC' ;\ \text{From } (20\cdot 4m)]$$

Hence, the m.g.f. of MVND - $N_p(\mu, \Sigma)$ is

$$M_x(t) = \exp\left[t'\mu + \frac{1}{2} t'\Sigma t \right] \qquad \ldots (20.4u)$$

Deduction. The m.g.f. of Bivariate Normal Distribution

In particular, for Bivariate Normal Distribution, taking:

$$x = \begin{pmatrix} x_1 \\ x_2 \end{pmatrix} \ ; \ t = \begin{pmatrix} t_1 \\ t_2 \end{pmatrix} \ ; \ \Sigma = \begin{pmatrix} \sigma_1^2 & \rho\sigma_1\sigma_2 \\ \rho\sigma_1\sigma_2 & \sigma_2^2 \end{pmatrix} \text{ and } |\Sigma| = \sigma_1^2\sigma_2^2(1-\rho^2)$$

$$t'\mu = (t_1 \quad t_2)\begin{pmatrix} \mu_1 \\ \mu_2 \end{pmatrix} = t_1\mu_1 + t_2\mu_2$$

$$\text{and } t'\Sigma t = (t_1 \quad t_2)\begin{pmatrix} \sigma_1^2 & \rho\sigma_1\sigma_2 \\ \rho\sigma_1\sigma_2 & \rho_2^2 \end{pmatrix}\begin{pmatrix} t_1 \\ t_2 \end{pmatrix} = t_1^2\sigma_1^2 + 2\rho\sigma_1\sigma_2 t_1 t_2 + t_2^2\sigma_2^2$$

$$\left[\because (x \quad y)\begin{pmatrix} a & b \\ c & d \end{pmatrix}\begin{pmatrix} x \\ y \end{pmatrix} = ax^2 + (b+c)xy + dy^2 \right]$$

Substituting in (20·4u)

$$M_{x_1, x_2}(t_1, t_2) = \exp\left[t_1\mu_1 + t_2\mu_2 + \frac{1}{2}\left(t_1^2\sigma_1^2 + 2\rho\sigma_1\sigma_2 t_1 t_2 + t_2^2\sigma_2^2\right)\right]$$

which is the m.g.f. of Bivariate Normal Distribution.

Characteristic Function. Changing t to it, $(i = \sqrt{-1})$, in (20·4u), the characteristic function of multivariate normal distribution $N_p(\mu, \Sigma)$ is given by:

$$\phi_x(t) = \exp\left[(it)'\mu + \frac{1}{2}(it)' \Sigma (it) \right]$$

$$= \exp\left[it'\mu - \frac{1}{2} t'\Sigma t \right] \qquad \ldots (20\cdot 4v)$$

20·2·3 Marginal Distributions

Let $x \sim N_p(\mu, \Sigma)$.

Let us partition the $(p \times 1)$ vectors x, μ and t as follows:

$$x = \begin{pmatrix} \underset{k\times 1}{x^{(1)}} \\ \underset{(p-k)\times 1}{x^{(2)}} \end{pmatrix} \qquad \mu = \begin{pmatrix} \underset{k\times 1}{\mu^{(1)}} \\ \underset{(p-k)\times 1}{\mu^{(2)}} \end{pmatrix}, \qquad t = \begin{pmatrix} \underset{k\times 1}{t^{(1)}} \\ \underset{(p-k)\times 1}{t^{(2)}} \end{pmatrix}$$

where, $$x^{(1)} = \begin{pmatrix} x_1 \\ x_2 \\ \vdots \\ x_k \end{pmatrix}_{k\times 1} \quad \text{and} \quad x^{(2)} = \begin{pmatrix} x_{k+1} \\ x_{k+1} \\ \vdots \\ x_p \end{pmatrix}_{(p-k)\times 1}$$

And similarly for $\mu^{(1)}, \mu^{(2)}, t^{(1)}$ and $t^{(2)}$. Let the variance-covariance matrix Σ of x be accordingly partitioned as follows:

$$\Sigma = \begin{pmatrix} \underset{(k\times k)}{\Sigma_{11}} & \underset{k\times(p-k)}{\Sigma_{12}} \\ \underset{(p-k)\times k}{\Sigma_{21}} & \underset{(p-k)\times(p-k)}{\Sigma_{22}} \end{pmatrix}$$

We want the marginal distributions of $x^{(1)}$ and $x^{(2)}$.

$$M_x(t) = E\left[\exp(t'x)\right] = E\left[\exp\left\{\begin{pmatrix} t'^{(1)} & t'^{(2)} \end{pmatrix}\begin{pmatrix} x^{(1)} \\ x^{(2)} \end{pmatrix}\right\}\right]$$

$$= E\left[\exp\left(t'^{(1)} \cdot x^{(1)} + t'^{(2)} \cdot x^{(2)}\right)\right] \qquad \ldots (20.4v)$$

$$= \exp\left(t'\mu + \frac{1}{2} t' \Sigma t\right)\Big] \qquad \text{[From (20·4}u\text{)]}$$

$$= \exp\left[\begin{pmatrix} t'^{(1)} & t'^{(2)} \end{pmatrix}\begin{pmatrix} \mu^{(1)} \\ \mu^{(1)} \end{pmatrix}\right.$$

$$\left. + \frac{1}{2}\begin{pmatrix} t'^{(1)} & t'^{(2)} \end{pmatrix}\begin{pmatrix} \Sigma_{11} & \Sigma_{12} \\ \Sigma_{21} & \Sigma_{22} \end{pmatrix}\begin{pmatrix} t^{(1)} \\ t^{(2)} \end{pmatrix}\right] \qquad \ldots(*)$$

The marginal m.g.f. of $x^{(1)}$ is given by:

$$M_{x^{(1)}}\left(t^{(1)}\right) = E\left[\exp\left\{t'^{(1)}x^{(1)}\right\}\right] = \left[M_x(t)\right]_{t^{(2)}=0}$$

$$= \exp\left[\begin{pmatrix} t'^{(1)} & O \end{pmatrix}\right]\begin{pmatrix} \mu^{(1)} \\ \mu^{(2)} \end{pmatrix} \cdot + \frac{1}{2}\begin{pmatrix} t'^{(1)} & O \end{pmatrix}\begin{pmatrix} \Sigma_{11} & \Sigma_{12} \\ \Sigma_{21} & \Sigma_{22} \end{pmatrix}\begin{pmatrix} t^{(1)} \\ O \end{pmatrix} \qquad \text{[From *]}$$

$$= \exp\left[t'^{(1)}\mu^{(1)} + \frac{1}{2}\begin{pmatrix} t'^{(1)}\Sigma_{11} & t'^{(1)}\Sigma_{12} \end{pmatrix}\begin{pmatrix} t^{(1)} \\ O \end{pmatrix}\right]$$

$$= \exp\left[t'^{(1)}\mu^{(1)} + \frac{1}{2}\left(t'^{(1)}\Sigma_{11}t^{(1)}\right)\right] \qquad \ldots (20{\cdot}4w)$$

which is the m.g.f. of $N_k\left(\mu^{(1)}, \Sigma_{11}\right)$.

Hence, by uniqueness theorem of m.g.f., $x^{(1)} \sim N_k\left(\mu^{(1)}, \Sigma_{11}\right)$,

Similarly, by taking $t^{(1)} = O$ in (21·4v) and (*), we shall get

$$M_{x^{(2)}}\left(t^{(2)}\right) = \exp\left[t'^{(2)}\mu^{(2)} + \frac{1}{2}t'^{(2)}\Sigma_{22}\,t^{(2)}\right]$$

$$\Rightarrow \qquad x^{(2)} \sim N_{p-\kappa}\left(\mu^{(2)}, \Sigma_{22}\right) \qquad \ldots (20·4x)$$

Hence, we obtain the following result.

If $x \sim N_p(\mu, \Sigma)$, *then the marginal distribution of any subset of the components of* x *is again a multivariate normal distribution with means and variance-covariance matrix obtained by taking proper components of* μ *and* Σ.

Note: In order to obtain the marginal distribution over a subset of multivariate normal random variables, we need only to drop the irrelevant variables (*i.e.*, the variables we want to marginalise out) from the mean vector and the variance-covariance matrix.

20·2·4 Conditional Distribution

If $X \sim N_p(\mu, \Sigma)$, and

$$X = \begin{pmatrix} X^{(1)}_{k\times 1} \\ X^{(2)}_{(p-k)\times 1} \end{pmatrix} \quad ; \quad \mu = \begin{pmatrix} \mu^{(1)}_{k\times 1} \\ \mu^{(2)}_{(p-k)\times 1} \end{pmatrix}$$

and

$$\Sigma = \begin{pmatrix} \underset{(k\times k)}{\Sigma_{11}} & \underset{k\times(p-k)}{\Sigma_{12}} \\ \underset{(p-k)\times k}{\Sigma_{21}} & \underset{(p-k)\times(p-k)}{\Sigma_{22}} \end{pmatrix}$$

Then, p.d.f. of $X^{(1)}$ conditional on $X^{(2)} = x^{(2)}$ is a multivariate normal distribution *viz.*,

$$\left(X^{(1)} \mid X^{(2)} = x^{(2)}\right) \sim N(\bar{\mu}, \bar{\Sigma})$$

where,
$$\left.\begin{aligned} \bar{\mu} &= \mu^{(1)} + \Sigma_{12}\Sigma_{22}^{-1}\left(x^{(2)} - \mu^{(2)}\right) \\ \bar{\Sigma} &= \Sigma_{11} - \Sigma_{12}\Sigma_{22}^{-1}\Sigma_{21} \end{aligned}\right\} \qquad \ldots(*)$$
and

Proof. To prove the above result, we need a result from Matrix Algebra, stated below in the form of a theorem.

Theorem. *If* $\Sigma = \begin{pmatrix} \Sigma_{11} & \Sigma_{12} \\ \Sigma_{21} & \Sigma_{22} \end{pmatrix}$, then $\Sigma^{-1} = \begin{pmatrix} A & B \\ C & D \end{pmatrix}$

where
$$A = \left(\Sigma_{11} - \Sigma_{12}\Sigma_{22}^{-1}\Sigma_{21}\right)^{-1} = \left(\bar{\Sigma}\right)^{-1}$$

$$B = -A\Sigma_{12}\Sigma_{22}^{-1} = -\left(\bar{\Sigma}\right)^{-1}\Sigma_{12}\Sigma_{22}^{-1}$$

$$C = -\Sigma_{22}^{-1}\Sigma_{21}A = -\Sigma_{22}^{-1}\Sigma_{21}\left(\bar{\Sigma}\right)^{-1}$$

$$D = +\Sigma_{22}^{-1} - \Sigma_{22}^{-1}\Sigma_{21}B = \Sigma_{22}^{-1} + \Sigma_{22}^{-1}\Sigma_{21}\left(\bar{\Sigma}\right)^{-1}\Sigma_{12}\Sigma_{22}^{-1}$$

Also,
$$|\Sigma| = |\Sigma_{22}|\left|\Sigma_{11} - \Sigma_{12}\Sigma_{22}^{-1}\Sigma_{21}\right| = |\Sigma_{22}||\bar{\Sigma}| \qquad \ldots(**)$$

The conditional p.d.f. of $X^{(1)}$ given $X^{(2)} = x^{(2)}$ is given by:

$$f\left(x^{(1)} \mid X^{(2)} = x^{(2)}\right) = \frac{f(x^{(1)}, x^{(2)})}{f_2(x^{(2)})} = \frac{f(x)}{f_2(x^{(2)})}$$

$$= \frac{\dfrac{1}{(2\pi)^{p/2}|\Sigma|^{1/2}} \cdot \exp\left[-\dfrac{1}{2}(x-\mu)'\Sigma^{-1}(x-\mu)\right]}{\dfrac{1}{(2\pi)^{(p-k)/2}|\Sigma_{22}|^{1/2}} \cdot \exp\left[-\dfrac{1}{2}\left(x^{(2)}-\mu^{(2)}\right)'\Sigma_{22}^{-1}\left(x^{(2)}-\mu^{(2)}\right)\right]}$$

$$\left[\because x^{(2)} \sim N_{p-k}\left(\mu^{(2)}, \Sigma_{22}\right)\right]$$

$$= \frac{1}{(2\pi)^{k/2}} \cdot \frac{1}{|\bar{\Sigma}|^{1/2}} \times \exp\left[-\frac{1}{2}\left\{(x-\mu)'\Sigma^{-1}(x-\mu) - \left(x^{(2)}-\mu^{(2)}\right)'\Sigma_{22}^{-1}\left(x^{(2)}-\mu^{(2)}\right)\right\}\right]$$

$$\left[\because \frac{|\Sigma_{22}|}{|\Sigma|} = \frac{1}{|\bar{\Sigma}|} \quad ; \quad \text{[From (**) on page 20·25]}\right]$$

$$= P \cdot \exp\left(-\frac{1}{2}Q\right), \text{ (say)},$$

where, $$P = \frac{1}{(2\pi)^{k/2}|\bar{\Sigma}|^{1/2}}$$

and
$$Q = \left(\left(x^{(1)}-\mu^{(1)}\right)' \quad \left(x^{(2)}-\mu^{(2)}\right)'\right)$$
$$\times \begin{pmatrix} (\bar{\Sigma})^{-1} & -(\bar{\Sigma})^{-1}\Sigma_{12}\Sigma_{22}^{-1} \\ -\Sigma_{22}^{-1}\Sigma_{21}(\bar{\Sigma})^{-1} & \Sigma_{22}^{-1}+\Sigma_{22}^{-1}\Sigma_{21}(\bar{\Sigma})^{-1}\Sigma_{12}\Sigma_{22}^{-1} \end{pmatrix} \times \begin{pmatrix} x^{(1)}-\mu^{(1)} \\ x^{(2)}-\mu^{(2)} \end{pmatrix}$$
$$- \left[\left(x^{(2)}-\mu^{(2)}\right)'\Sigma_{22}^{-1}\left(x^{(2)}-\mu^{(2)}\right)\right]$$

$$= \left(x^{(1)}-\mu^{(1)}\right)'\left(\bar{\Sigma}\right)^{-1}\left(x^{(1)}-\mu^{(1)}\right)$$
$$- \left(x^{(1)}-\mu^{(1)}\right)'\left(\bar{\Sigma}\right)^{-1}\Sigma_{12}\Sigma_{22}^{-1}\left(x^{(2)}-\mu^{(2)}\right)$$
$$- \left(x^{(2)}-\mu^{(2)}\right)'\Sigma_{22}^{-1}\Sigma_{21}\left(\bar{\Sigma}\right)^{-1}\left(x^{(1)}-\mu^{(1)}\right)$$
$$+ \left(x^{(2)}-\mu^{(2)}\right)'\left(\Sigma_{22}^{-1}+\Sigma_{22}^{-1}\Sigma_{21}\left(\bar{\Sigma}\right)^{-1}\Sigma_{12}\Sigma_{22}^{-1}\right)\left(x^{(2)}-\mu^{(2)}\right)$$
$$- \left(x^{(2)}-\mu^{(2)}\right)' \cdot \Sigma_{22}^{-1} \cdot \left(x^{(2)}-\mu^{(2)}\right)$$

$$= \left(x^{(1)}-\mu^{(1)}\right)'\left(\bar{\Sigma}\right)^{-1}\left(x^{(1)}-\mu^{(1)}\right)$$
$$- \left(x^{(1)}-\mu^{(1)}\right)'\left(\bar{\Sigma}\right)^{-1}\Sigma_{12} \cdot \Sigma_{22}^{-1} \cdot \left(x^{(2}-\mu^{(2)}\right)$$
$$- \left(x^{(2)}-\mu^{(2)}\right)'\Sigma_{22}^{-1}\Sigma_{21}\left(\bar{\Sigma}\right)^{-1}\left(x^{(1)}-\mu^{(1)}\right)$$
$$+ \left(x^{(2)}-\mu^{(2)}\right)' \cdot \left(\Sigma_{22}^{-1}\Sigma_{21}\left(\bar{\Sigma}\right)^{-1}\Sigma_{12}\Sigma_{22}^{-1}\right)\left(x^{(2)}-\mu^{(2)}\right) \qquad \ldots(***)$$

We have to prove that: $\left(X^{(1)} \mid X^{(2)} = x^{(2)}\right) \sim N_k\left(\bar{\mu}, \bar{\Sigma}\right)$

i.e., its p.d.f. is:

$$f\left(x^{(1)} \mid X^{(2)} = x^{(2)}\right) = \frac{1}{(2\pi)^{k/2} \mid \bar{\Sigma} \mid^{1/2}} \cdot \exp\left[-\frac{1}{2}\left(x^{(1)} - \bar{\mu}\right)'\left(\bar{\Sigma}\right)^{-1}\left(x^{(1)} - \bar{\mu}\right)\right]$$

where $\bar{\mu}$ and $\bar{\Sigma}$ are given in (*) on page 20·25. Now we have to prove that:

$$\left(x^{(1)} - \bar{\mu}\right)'\left(\bar{\Sigma}\right)^{-1}\left(x^{(1)} - \bar{\mu}\right) = Q$$

$$\begin{aligned}
\left(x^{(1)} - \bar{\mu}\right)'\left(\bar{\Sigma}\right)^{-1}\left(x^{(1)} - \bar{\mu}\right) &= \left[x^{(1)} - \mu^{(1)} - \Sigma_{12}\Sigma_{22}^{-1}\left(x^{(2)} - \mu^{(2)}\right)\right]' \\
&\quad \times \left(\bar{\Sigma}\right)^{-1} \cdot \left[x^{(1)} - \mu^{(1)} - \Sigma_{12}\Sigma_{22}^{-1}\left(x^{(2)} - \mu^{(2)}\right)\right] \\
&= \left[\left(x^{(1)} - \mu^{(1)}\right)' - \left(x^{(2)} - \mu^{(2)}\right)'\Sigma_{22}^{-1}\Sigma_{21}\right] \\
&\quad \times \left(\bar{\Sigma}\right)^{-1} \cdot \left[\left(x^{(1)} - \mu^{(1)}\right) - \Sigma_{12}\Sigma_{22}^{-1}\left(x^{(2)} - \mu^{(2)}\right)\right]
\end{aligned}$$

$$\left[\because (ABC)' = C'B'A' \;;\; \left(\Sigma_{22}^{-1}\right)' = \left(\Sigma_{22}^{-1}\right),\; \Sigma_{12}' = \Sigma_{21}\right]$$

$$\begin{aligned}
&= \left(x^{(1)} - \mu^{(1)}\right)'\left(\bar{\Sigma}\right)^{-1}\left(x^{(1)} - \mu^{(1)}\right) \\
&\quad - \left(x^{(1)} - \mu^{(1)}\right)'\left(\bar{\Sigma}\right)^{-1}\Sigma_{12}\Sigma_{22}^{-1}\left(x^{(2)} - \mu^{(2)}\right) \\
&\quad - \left(x^{(2)} - \mu^{(2)}\right)'\Sigma_{22}^{-1}\Sigma_{21}\left(\bar{\Sigma}\right)^{-1}\left(x^{(1)} - \mu^{(1)}\right) \\
&\quad + \left(x^{(2)} - \mu^{(2)}\right)'\Sigma_{22}^{-1}\Sigma_{21}\left(\bar{\Sigma}\right)^{-1}\Sigma_{12}\Sigma_{22}^{-1}\left(x^{(2)} - \mu^{(2)}\right) \\
&= Q \qquad \text{[From (***), on page 20·26]}
\end{aligned}$$

which completes the proof.

Remarks :

1. Here, Σ_{22}^{-1} is the generalised inverse of Σ_{22}.
2. The random vectors

$$X^{(2)} \text{ and } Y^{(1)} = X^{(1)} - \Sigma_{12}\Sigma_{22}^{-1}X^{(2)}$$

are independent.

3. The Matrix $\Sigma_{12}\Sigma_{22}^{-1}$ is known as the matrix of *regression coefficients.*

20·2·5 Standard Multivariate Normal Distribution

The adjective 'standard' implies that the mean of the distribution is equal to zero **(0)** and its variance-covariance matrix is equal to unity matrix.

Thus, taking $\mu_{p\times1} = O_{p\times1}$ and $\Sigma_{pp} = I_p \;\Rightarrow\; |\Sigma| = 1$ in (20·4*q*), the p.d.f. of *standard multivariate normal distribution* becomes:

$$f(x) = \frac{1}{(2\pi)^{p/2} \cdot |\Sigma|^{1/2}} \exp\left[-\frac{1}{2} x' \Sigma^{-1} x\right]$$

$$= \frac{1}{(2\pi)^{p/2} \cdot 1} \exp\left[-\frac{1}{2}\left(x' I_p x\right)\right]$$

$$= \frac{1}{(2\pi)^{p/2}} \exp\left[-\frac{1}{2} x' x\right] ; \quad -\infty < (x_1, x_2, \ldots, x_p) < \infty \qquad \text{.... (*)}$$

Relation to Univariate Normal Distribution

$$f(x) = \frac{1}{(2\pi)^{p/2}} \exp\left[-\frac{1}{2}\sum_{i=1}^{p} x_i^2\right] = \prod_{i=1}^{p}\left[\frac{1}{\sqrt{2\pi}} \cdot \exp\left(-\frac{1}{2}x_i^2\right)\right]$$

$\Rightarrow$

$$\left.\begin{aligned} f(x_1, x_2, \cdots, x_p) &= f(x_1)\, f(x_2) \cdots f(x_p), \\ f(x_i) &= \frac{1}{\sqrt{2\pi}} \cdot \exp\left(-\frac{1}{2}x_i^2\right) ; -\infty < x_i < \infty \end{aligned}\right\} \qquad \text{... (**)}$$

where,

is the p.d.f. of *standard normal distribution. The p-components* $(x_1, x_2, \ldots, x_p)$ *of multivariate standard normal distribution are mutually independent standard normal variates.*

M.G.F.

$$M_x(t) = E\left(\exp(t'x)\right) = E\left[\exp\left(t_1 x_1 + t_2 x_2 + \ldots + t_p x_p\right)\right]$$

$$\Rightarrow \quad M_x(t) = E\left(\prod_{i=1}^{p} \exp(t_i . x_i)\right) = \prod_{i=1}^{p} E\left\{\exp(t_i . x_i)\right\} \quad \left[\because x_i's \text{ are independent}\right]$$

$$= \prod_{i=1}^{p} M_{x_i}(t_i) = \prod_{i=1}^{p} \exp\left(\frac{1}{2}t_i^2\right) ; \qquad [\because x_i \sim N(0,1)]$$

$$\Rightarrow \quad M_x(t) = \exp\left[\frac{1}{2}\sum_{i=1}^{p} t_i^2\right] = \exp\left(\frac{1}{2}t' t\right) \qquad \text{... (***)}$$

Characteristic Function.

Changing t to it in (***), the characteristic function of standard multivariate normal distribution is:

$$\phi_x(t) = \exp\frac{1}{2}\left[(it)'(it)\right] = \exp\left[\frac{1}{2} i^2\, t' t\right] = \exp\left[-\frac{1}{2} t' t\right] \qquad \text{... (****)}$$

Entropy. If $X \sim N_p(\mu, \Sigma)$, *its differential entropy is given by:*

$$h(f) = -\int_{-\infty}^{\infty}\int_{-\infty}^{\infty} \cdots \int_{-\infty}^{\infty} f(x) \cdot \log f(x) dx$$

$$= \frac{1}{2}\log\left[(2\pi e)^p |\Sigma|\right]$$

$$= \frac{p}{2}\left[\log(2\pi) + 1\right] + \frac{1}{2}\log|\Sigma|,$$

where, $|\Sigma|$ is the determinant of matrix Σ.

20·2·6 Characterisation of Multi-Variate Normal Distribution

A random vector $\mathbf{X}_{p\times 1}$ is said to have a p-variate normal distribution if every linear combination of its components

$$Y = a_1X_1 + a_2X_2 + \cdots + a_pX_p = a'X$$

where $\quad \boldsymbol{a} = \begin{pmatrix} a_1 \\ a_2 \\ \vdots \\ a_p \end{pmatrix}_{p\times 1} ; \ \mathbf{X} = \begin{pmatrix} X_1 \\ X_2 \\ \vdots \\ X_p \end{pmatrix}_{p\times 1}$, is normally distributed.

i.e., $\quad Y = \boldsymbol{a}'\mathbf{X},$ has univariate normal distribution.

Example 20·8 *If* $\mathbf{X}_{p\times 1} \sim N_p(\boldsymbol{\mu}, \Sigma)$, *then prove that:*

$$Z = (\mathbf{X} - \boldsymbol{\mu})'\Sigma^{-1}(\mathbf{X} - \boldsymbol{\mu}) \sim \chi^2_{(p)}$$

i.e., $(\mathbf{X} - \boldsymbol{\mu})'\,\Sigma^{-1}(\mathbf{X} - \boldsymbol{\mu})$ *has chi-square distribution with p degrees of freedom.*

Solution. The p.d.f. of multivariate normal distribution $N_p(\boldsymbol{\mu}, \Sigma)$ is :

$$f(x) = f(x_1, x_2, \cdots, x_p) = \frac{1}{(2\pi)^{p/2}|\Sigma|^{1/2}} \cdot \exp\left[-\frac{1}{2}(x-\boldsymbol{\mu})'\Sigma^{-1}(x-\boldsymbol{\mu})\right]$$

$$; \ -\infty < (x_1, x_2, \cdots, x_p) < \infty. \qquad \ldots (*)$$

$\Sigma_{p\times p}$ is positive definite matrix $\quad \Rightarrow \quad \Sigma^{-1}_{p\times p}$ is also positive definite matrix.

Hence, there exists a non-singular matrix $C_{p\times p}$, so that

$$C'\Sigma^{-1}C = I_p \ \Rightarrow \left|C'\right|\left|\Sigma^{-1}\right|\left|C\right| = \left|I_p\right| \Rightarrow \left|C\right|^2\left|\Sigma^{-1}\right| = 1 \ \Rightarrow \ \left|C\right| = \left|\Sigma\right|^{1/2} \qquad \ldots (**)$$

where, $|C|$ is determinant of C ; $\ |C| = |C'|$.

Let us transform the variables $(x_1, x_2, \ldots, x_p)$ to variables $(y_1, y_2, \ldots, y_p)$ by non-singular linear transformation:

$$\mathbf{X} - \boldsymbol{\mu} = C\,Y \ \Rightarrow \ x - \boldsymbol{\mu} = Cy, \quad \text{where} \quad y' = (y_1 \ y_2 \cdots \ y_p)_{1\times p}$$

The Jacobian of transformation is:

$$J = \frac{\partial(x_1, x_2, \cdots, x_n)}{\partial(y_1, y_1, \cdots, y_n)} = |C| = |\Sigma|^{1/2} \qquad \text{[From (**)]}$$

The p.d.f. $g(.)$ of $(y_1, y_2, \ldots, y_p)$ is given by $g(y) = f(x)\cdot|J|$

$$g(y) = \frac{1}{(2\pi)^{p/2}|\Sigma|^{1/2}} \cdot \exp\left[-\frac{1}{2}(Cy)'\,\Sigma^{-1}\,(Cy)\right] \times \left[\ \text{mod}|\Sigma|^{1/2}\right]$$

$$= \frac{1}{(2\pi)^{p/2}} \exp\left[-\frac{1}{2}y'\left(C'\,\Sigma^{-1}\,C\right)y\right]$$

$$= \frac{1}{(2\pi)^{p/2}} \exp\left(-\frac{1}{2}y'y\right) = \frac{1}{(2\pi)^{p/2}} \exp\left[-\frac{1}{2}\sum_{i=1}^{p} y_i^{\,2}\right], \quad \left[\because C'\Sigma^{-1}C = I_p\right]$$

$$\Rightarrow \qquad Y_1, Y_2, \cdots, Y_p \ \text{are } \textit{i.i.d.}, \ N(0, 1) \qquad \ldots (***)$$

We have:

$$Z = (X - \mu)' \Sigma^{-1} \cdot (X - \mu) = (CY)' \Sigma^{-1} (CY) = Y' (C' \Sigma^{-1} C) Y$$

$$= Y' I_p Y = Y' Y = \sum_{i=1}^{p} y_i^2 \qquad (\because C' \Sigma^{-1} C = I_p)$$

Since, $Y_i' s$; $(i = 1, 2, \ldots p)$ are *i.i.d.* $N(0, 1)$, Z, being the sum of squares of 'p' *i.i.d.* standard normal variates is a chi-square variate with 'p' degrees of freedom.

$$\therefore \qquad Z = (X - \mu)' \Sigma^{-1} (X - \mu) \sim \chi_p^2$$

Example. *If* $X_{p \times 1} \sim N_p (\mu, \Sigma)$, *show that*

$$DX \sim N_p \left(D\mu, D \Sigma D' \right),$$

where $D_{p \times p}$ *is any non-singular matrix.*

Solution. $\qquad X \sim N_p(\mu, \Sigma)$,

$$\therefore \; M_X(t) = E\left[\exp \left(t' X \right) \right] = \exp \left[t' \mu + \frac{1}{2} t' \Sigma \, t \right] \qquad \ldots (*)$$

The m.g.f. of $Z = DX$ is given by:

$$M_Z(t) = E \left(\exp \left(t' Z \right) \right) = E \left(\exp \left(t' DX \right) \right) = E \left(\exp \left(t_1' \, X \right) \right), \text{ (say)},$$

where, $$t_1' = t' D \;\Rightarrow\; t_1 = (t' D)' = D' t$$

$$\Rightarrow \qquad M_Z(t) = \exp \left[t_1' \, \mu + \frac{1}{2} t_1' \Sigma \, t_1 \right] \qquad [\text{Using} (*)]$$

$$= \exp \left[t' D \, \mu + \frac{1}{2} t' \, (D \Sigma D') \, t \right]$$

which is the m.g.f. of $N_p \left(D\mu, D \Sigma D' \right)$

Hence, by uniqueness theorem of m.g.f, $Z = DX \sim N_p \left(D\mu, D \Sigma D' \right)$.

DISCUSSION AND REVIEW QUESTIONS

1. *It* X and Y follow bivariate normal distribution with parameters: $(\mu_X, \mu_Y, \sigma_X, \sigma_Y, \rho)$
   (*i*) Find the marginal distributions of X and Y.
   (*ii*) Show that the conditional mean of X is dependent on the given Y, but the conditional variance is independent of it.
2. State and prove the necessary and sufficient condition for a random vector (X, Y) to possess a bivariate normal distribution.
3. If $(X, Y) : \; BVN \left(\mu_1, \mu_2, \sigma_1^2, \sigma_2^2, \rho \right)$ then prove that the marginal distributions of X and Y are also normal.

   Give an example to show that the converse is not true *i.e.,* if the marginal distributions of X and Y are normal, it does not necessarily imply that joint distribution of (X, Y) is bivariate normal.

4. Fill in the blanks:

If $(X, Y) \sim BVN\ (0, 0, 1, 1, \rho)$, then :

(*i*) $M_{X,Y}\ (t_1, t_2) = ...$

(*ii*) $\rho\ (X^2, Y^2) = ...$

(*iii*) μ_{rs} = Recurrence Relation for moments.

(*iv*) Regression of Y on X is ...

(*v*) $(X + Y)$ and $(X - Y)$ are ...

(*vi*) $\dfrac{X^2 - 2\rho XY + Y^2}{(1 - \rho^2)}$ has ... distribution.

ASSORTED REVIEW PROBLEMS FOR SELF ASSESSMENT

20·1.

(*a*) The marks X and Y scored by candidates in an examination in two subjects Mathemathics and Statistics are known to follow a bivariate normal distribution. The mean of X is 52 and its standard deviation is 15, while Y has mean 48 and standard deviation 13. Also the coefficient of correlation between X and Y is 0·6.

Write down the joint distribution of X and Y. If 100 marks in the aggregate are needed for a pass in the examination, show how to calculate the proportion of candidates who pass the examination?

(*b*) A manufacturer of electric bulbs, in his desire for putting only good bulbs for sale, rejects all bulbs for which a certain quality characteristic X of the filament is less than 65 units Assume that the quality characteristic X and the life Y, of the bulb in hours are jointly normally distributed with parameters given below:

	X	Y
Mean	80	100
Standard deviation	10	10
Correlation cofficient ρ (X, Y) = 0·60		

Find:

(*i*) the proportion of bulbs produced that will burn for less than 1,000 hours,

(*ii*) the proportion of bulbs produced that will be put for sale,

(*iii*) the average life of bulbs put for sale.

(*c*) Suppose that the joint distribution of marks (per cent) obtained by Economics students in M.A. first year and second year, X and Y, respectively, have a bivariate normal distribution with $\mu_X = 50$, $\mu_Y = 50$, $\sigma_X = 4$, $\sigma_Y = 2$ and $\rho = 0{\cdot}8$ Find $P(Y > X)$, the probability that a student fares better in the second year than in the first year.

20·2

(*a*) Determine the parameters of the bivariate normal distribution :

$$f(x, y) = k \exp\left[-\frac{8}{27}\left\{(x-7)^2 - 2(x-7)(y+5) + 4(y+5)^2\right\}\right].$$

Also find the value of k.

(*b*) for the bivariate normal distribution:

$$(X, Y) \sim BVN\left(1, 2, 4^2; 5^2, \frac{12}{13}\right), \text{ find } (i)\ P(X > 2), \quad (ii)\ P(X > 2 \mid Y = 2).$$

(*c*) If the exponent of e in a bivariate normal density is :

$$-\frac{1}{102}\left[(x+2)^2 - 2{\cdot}8\,(x+2)(y-1) + 4\,(y-1)^2\right], \text{ Obtain}$$

(*i*) the means and variances of X and Y and the correlation between X and Y.
(*ii*) the mean and variance of the conditional distribution of Y given X.

(*d*) The bivariate random variable (X_1, X_2) have a bivariate normal distribution with means 60 and 75 and standard deviations 6 and 12 with a correlation coefficient of 0·55. Find the following probabilities:

(*i*) $P(65 \le X_1 \le 75)$ (*ii*) $P(71 \le X_2 \le 80 \mid X_1 = 55)$ and (*iii*) $P(|X_1 - X_2| \ge 15)$

20·3. For the bivariate normal distribution :

$$dF = k \exp\left[-\frac{2}{3}\left(x^2 - xy + y^2 - 3x + 3y + 3\right)\right] dx\, dy,$$

obtain (*i*) marginal distribution of Y and (*ii*) the conditional distribution of Y given X.

Also obtain the characteristic function of the above bivariate normal distribution and hence the convariance between X and Y.

20·4. For a bivariate normal distribution :

$$f_{X,Y}(x,y) = \frac{1}{2\pi\sqrt{(1-\rho^2)}} \cdot \exp\left\{-\frac{1}{2(1-\rho^2)}\left(x^2 - 2\rho xy + y^2\right)\right\}, -\infty < (x,y) < \infty$$

Find

(*i*) marginal distributions of X and Y,
(*ii*) conditional distribution of Y given X,
(*iii*) distribution of $\dfrac{1}{(1-\rho^2)}\left[x^2 - 2\rho xy + y^2\right]$ and
(*iv*) show that, in general, X and Y are stochastically dependent and will be independent if and only if $\rho = 0$.

20·5. Let the joint *p.d.f.* of X and Y be :

$$f(x,y) = \frac{1}{2\pi\sigma_1\sigma_2\sqrt{(1-\rho^2)}} \cdot \exp\left\{-\frac{1}{2(1-\rho^2)}\left[\frac{(x-\mu_1)^2}{\sigma_1^2} - 2\rho\frac{(x-\mu_1)}{\sigma_1}\cdot\frac{(y-\mu_2)}{\sigma_2} + \frac{(y-\mu_2)^2}{\sigma_2^2}\right]\right\}$$

where $-\infty < x < \infty, -\infty < y < \infty, 1 < \rho < 1$.

(*i*) Find the marginal distribution of X.
(*ii*) Find the conditional distribution of Y given $X = x$.
(*iii*) Show that the regression of Y on X is linear and homoscedastic.
(*iv*) Find $P(3 < Y < 8 \mid X = 7)$, given that $\mu_1 = 3, \mu_2 = 1, \sigma_1^2 = 16, \sigma_2^2 = 25, \rho = 0{\cdot}6$.

(*v*) Find the probability of the simultaneous materialization of the inequalities,

$$X > E(X) \text{ and } Y > E(Y)$$

20·6.

(*a*) Let the random variables X and Y be assumed to have a joint bivariate normal distribution with, $\mu_1 = \mu_2 = 0, \sigma_1 = 4, \sigma_2 = 3$ and $\rho(X, Y) = 0{\cdot}8$.

(*i*) Write down the joint density function of X and Y.

(*ii*) Write down the regression of Y on X.

(*iii*) Obtain the joint density of X + Y and X – Y.

(*b*) For the distribution of random variables X and Y given by:

$$dF = k \exp\left[-\frac{1}{2(1-\rho^2)}\left(x^2 - 2\rho xy + y^2\right)\right] dx\, dy\,; -\infty < x < \infty, -\infty < y < \infty,$$

Find:

(*i*) the constant *k*,

(*ii*) the distributions of X and Y.

(*iii*) the distributions of X for given Y and of Y for given X.

(*iv*) the curves of regression of Y on X and of X on Y, and

(*v*) the distributions of X – Y and X – Y.

20·7. Let (X,Y) be a bivariate normal *r.v.* with $E(X) = E(Y) = 0$, Var (X) = Var (Y) = 1 and Cov (X,Y) = ρ. Show that the *r.v.* Z = Y/X has Cauchy distribution.

20·8.

(*a*) If $X, Y \sim N\left(\mu_X, \mu_Y, \sigma_X^2, \sigma_Y^2, \rho\right)$, prove that: $P(X > \mu_X \cap Y > \mu_Y) = \frac{1}{4} + \frac{\sin^{-1}\rho}{2\pi}$

(*b*) If $(X, Y) \sim BVN(0, 0, 1, 1, \rho,)$, then prove that

(*i*) $P(X > 0 \cap Y > 0) = \frac{1}{4} + \frac{\sin^{-1}\rho}{2\pi}$

(*ii*) $P(XY > 0) = \frac{1}{2} + \frac{\sin^{-1}\rho}{\pi}$

(*iii*) $P(XY < 0) = \frac{1}{\pi}\cos^{-1}(\rho)$

Hints:

(*i*) $p = P(X > 0 \cap Y > 0)$

$$= \frac{1}{2\pi\sqrt{1-\rho^2}} \int_0^\infty \int_0^\infty \exp\left[-\frac{1}{2(1-\rho^2)}\left\{x^2 - 2\rho xy + y^2\right\}\right] dx\, dy$$

Put $x = r\cos\theta,\ y = r\sin\theta \quad \Rightarrow |J| = r\,; 0 < r < \infty, 0 \le \theta \le \pi/2$

$$\therefore \quad p = \frac{1}{2\pi\sqrt{1-\rho^2}} \int_0^\infty \int_0^{\pi/2} \exp\left[-\frac{r^2}{2(1-\rho^2)}(1 - \rho\sin 2\theta)\right] r\, dr\, d\theta$$

Now integrate first w.r.t. *r* and then w.r.t. θ.

20·9. Let X_1 and X_2 be two independent normally distributed variables with zero means and unit variances. Let Y_1 and Y_2 be the linear functions of X_1 and X_2 defined by

$$Y_1 = m_1 + l_{11} X_1 + l_{12} X_2, \qquad Y_2 = m_2 + l_{21} X_1 + l_{22} X_2$$

Show that Y_1 and Y_2 are normally distributed with means m_1 and m_2, variances $\mu_{20} = l_{11}^2 + l_{12}^2$, $\mu_{02} = l_{21}^2 + l_{22}^2$, and covariance $\mu_{11} = l_{11} l_{21} + l_{12} l_{22}$.

20·10. Let f and g be the *p.d.f.'s* with corresponding distribution functions F and G. Also let

$$h(x,y) = f(x)\, g(y) \left[1 + \alpha \{2F(x) - 1\}\{2G(y) - 1\}\right],$$

where $|\alpha| \le 1$, is a constant and h is a bivariate p.d.f. with marginal p.d.f.'s f and g. Further, let f and g be p.d.f.'s of N (0,1) distribution. Then, prove that:

(*i*) $\text{Cov}(X, Y) = \alpha / \pi$.

(*ii*) $Z = (X + Y)$ is not normally distributed, except in the trivial case $\alpha = 0$.

20·11. If $(X, Y) \sim BVN\left(\mu_1, \mu_2, \sigma_1^2, \sigma_2^2, \rho\right)$, compute the correlation coefficient between e^X and e^Y.

Hints: Let $U = \exp(X)$; $V = \exp(Y)$

$$\mu'_{rs} = E\left[U^r V^s\right] = E\left[e^{rX+sY}\right]$$

$$= \exp\left[r\mu_1 + s\mu_2 + \frac{1}{2}\left(r^2\sigma_1^2 + s^2\sigma_2^2 + 2\rho rs\right)\right]$$

$$E(U) = \mu'_{10}\ ; \quad E\left(U^2\right) = \mu'_{20}\ ; \quad E(UV) = \mu'_{11} \text{ and so on.}$$

Ans: $$\rho(U,V) = \frac{\exp(\rho\sigma_1\sigma_2) - 1}{\left[\left\{\exp\left(\sigma_1^2\right) - 1\right\}\left\{\exp\left(\sigma_2^2\right) - 1\right\}\right]^{1/2}}.$$

20·12.

(*a*) If $(X, Y) \sim BVN(0, 0, 1, 1, \rho)$, find $E\left[\max(X, Y)\right]$.

(*b*) $(X, Y) \sim BVN\left(0, 0, \sigma_1^2, \sigma_2^2, \rho\right)$, show that rth cumulant of XY is given by:

$$\kappa_r = \frac{1}{2}(r-1)!\,\sigma_1^r \sigma_2^r\left[(\rho+1)^r + (\rho-1)^r\right].$$ Deduce that: $E\left(X^2 Y^2\right) = \sigma_1^2 \sigma_2^2 \left(1 + 2\rho^2\right)$

20·13. The joint density of r.v's (X, Y) is given by:

$$f(x,y) = \frac{1}{2\pi}\cdot \exp\left[-\left(x^2 + y^2\right)/2\right] \times \left[1 + xy \exp\left\{-\left(x^2 + y^2 - 2\right)/2\right\}\right];\ -\infty < (x,y) < \infty$$

(*i*) Verify that $f(x, y)$ is a p.d.f.

(*ii*) Show that the marginal distribution of each of X and Y is normal, $N(0,1)$.

(*iii*) Are X and Y independent?

Ans: (*iii*) No.

20·14. Show by means of an example that the normality of conditional p.d.f.'s does not imply that the bivariate density is normal.

Hints: Consider:

$$f(x,y) = c.\exp\left[-\left(1+x^2\right)\left(1+y^2\right)\right];\ -\infty < (x,y) < \infty$$

Then: $(Y \mid X = x) \sim N\left(0, \dfrac{1}{2\left(1+x^2\right)}\right)$ and $(X \mid Y = y) \sim N\left(0, \dfrac{1}{2\left(1+y^2\right)}\right)$

20·15. For a bivariate normal *r.v.* (X,Y), does the conditional *p.d.f.* of (X, Y) given $X+Y=c$, (constant) exist? If so find it. If not, why not?

Ans: No, Since $P(X+Y=c) = 0$

20·16. If X and Y are bivariate normal variates with standard deviations unity and with correlation coefficient ρ, show that the regression of X^2 (Y^2) on Y^2 (X^2) is strictly linear. Also show that the regression of X (Y) on Y^2 (X^2) is not linear.

20·17. Let

$$f(x,y) = \frac{1}{2}\left[\begin{array}{l}\dfrac{1}{2\pi\sqrt{1-\rho^2}}\exp\left\{-\dfrac{1}{2\left(1-\rho^2\right)}\left(x^2-2\rho xy+y^2\right)\right\} \\ +\dfrac{1}{2\pi\sqrt{1-\rho^2}}\exp\left\{-\dfrac{1}{2\left(1-\rho^2\right)}\left(x^2+2\rho xy+y^2\right)\right\}\end{array}\right];\ -\infty < x < \infty,\ -\infty < y < \infty$$

then show that:

(*i*) $f(x,y)$ is a joint p.d.f. such that both marginal densities are normal but $f(x, y)$ is not bivariate normal.

(*ii*) X and Y have zero correlation but X and Y are not independent.

20·18. Let X, Y be normally correlated variates with zero means and variances ${\sigma_1}^2$, ${\sigma_2}^2$ and if

$$W = \frac{X}{\sigma_1},\quad Z = \frac{1}{\sqrt{\left(1-\rho^2\right)}}\left(\frac{Y}{\sigma_2} - \frac{\rho X}{\sigma_1}\right),$$

show that $\dfrac{\partial(w,z)}{\partial(x,y)} = \dfrac{1}{\sigma_1\sigma_2\sqrt{\left(1-\rho^2\right)}}$ and $W^2 + Z^2 = \dfrac{1}{\left(1-\rho^2\right)}\left[\dfrac{X^2}{{\sigma_1}^2} - \dfrac{2\rho XY}{\sigma_1\sigma_2} + \dfrac{Y^2}{{\sigma_2}^2}\right]$

Deduce that the joint probability differential of W and Z is $dP = \dfrac{1}{2\pi}.\exp\left[-\dfrac{1}{2}\left(w^2+z^2\right)\right]dwdz$

and hence, that W, Z are independent normal variates with zero means and unit S.D.'s

Hence, or otherwise obtain the *m.g.f.* of the bivariate normal distribution.

20·19. From a standard bivariate normal population, a random sample of n obsevations $(X_i, Y_i), (i = 1, 2, ..., n)$ is drawn. Show that the distribution of $Z_1 = \frac{1}{n}\sum_{i=1}^{n} X_i^2$ and $Z_2 = \frac{1}{n}\sum_{i=1}^{n} Y_i^2$ has the moment generating function: Constant. $\left[\left(1 - \frac{2t_1}{n}\right)\left(1 - \frac{2t_2}{n}\right) - \frac{4\rho^2 t_1 t_2}{n^2}\right]^{-n/2}$.

Hint :

$$M_{Z_1, Z_2}(t_1, t_2) = \left[E \exp\left(\frac{t_1 x^2}{n} + \frac{t_2 y^2}{n}\right)\right]^n \qquad ...(*)$$

$$E\left[\exp\left(\frac{t_1 x^2}{n} + \frac{t_2 y^2}{n}\right)\right] = \frac{1}{2\pi\sqrt{c}}\int_{-\infty}^{\infty}\int_{-\infty}^{\infty} \exp\left[x^2\left(\frac{t_1}{n} - \frac{1}{2c}\right) + \frac{\rho xy}{c} + y^2\left(\frac{t_2}{n} - \frac{1}{2c}\right)\right] dx\, dy\ ;$$

where $c = (1 - \rho^2)$

Now Use : $\int_{-\infty}^{\infty}\int_{-\infty}^{\infty} \exp\left[-(ax^2 + 2hxy + y^2)\, dxdy = \frac{\pi}{\sqrt{ab - h^2}}\right.$ and simplify.

20·20. If $f = f(x, y)$ is the p.d.f. of $BVN(0, 0, 1, 1, \rho)$ distribution, verity that: $\frac{\partial f}{\partial \rho} = \frac{\partial^2 f}{\partial x \partial y}$.

Further, if two new random variables U and V are defined by the relation:

$$U = P(0 \le Z \le x) \text{ and } V = P(0 \le Z \le y), \text{ where } Z - N(0,1),$$

prove that the marginal distributions of both U and V are uniform in the interval $\left(-\frac{1}{2}, \frac{1}{2}\right)$ and their common variance is $\frac{1}{12}$.

Hence, prove that $R = Corr.(U, V)$, satisfies the relation: $\rho = 2\sin(\pi R / 6)$.

20·21. If $(X, Y) \sim BVN(\mu_X, \mu_Y, \sigma_X^2, \sigma_Y^2, \rho)$, then prove that $a + bX + cY, (b \ne 0, c \ne 0)$ is distributed as: $N\left(a + b\mu_X + c\mu_Y, (b^2\sigma_X^2 + c^2\sigma_Y^2 + 2bc\rho\sigma_X\sigma_Y\right)$

20·22. Let X_1, X_2, X_3 be a random sample of size $n = 3$ from $N(0, 1)$ distribution.

(*a*) Show that $Y_1 = X_1 + \delta X_3$, $Y_2 = X_2 + \delta X_3$ has a bivariate normal distribution.

(*b*) Find the value of δ so that $\rho(Y_1, Y_2) = \frac{1}{2}$.

(*c*) What additional transformation involving Y_1 and Y_2 would produce a bivariate normal distribution with means μ_1 and μ_2, variances σ_1^2 and σ_2^2, and the same correlation coefficient ρ.

Ans:

(*a*) $\rho(Y_1, Y_2) = \frac{\delta^2}{1 + \delta^2} = \frac{1}{2} \quad \Rightarrow \quad \delta = \pm 1$

(*b*) $U_1 = \mu_1 + \frac{\sigma_1 Y_1}{\sqrt{1 + \delta^2}}, \quad U_2 = \mu_2 + \frac{\sigma_2 Y_2}{\sqrt{1 + \delta^2}}$

20·23 .Write down, but do not derive, the moment generating function for a pair of random variables which have a bivariate normal distribution with both means equal to zero.

The independent random variables X, Y, Z, are each normally distributed with means 0 and variance 1. If $U = X + Y + Z$ and $V = X - Y + 2Z$, show that U and V have bivariate normal distribution. Find the correlation of U and V and the expectation of U when V is equal to 1.

Ans: $\rho(U,V) = 2\sqrt{2}/3.$ $\quad E[U|V=1] = \frac{2}{3}$

20·24. If $(X, Y) \sim B.V.N.\left(0, 0, \sigma_1^2, \sigma_2^2, \rho\right)$; $(\lambda = \sigma_2/\sigma_1)$, prove that

$$p = P\left[|X| \le |Y|\right] = \frac{1}{\pi}\tan^{-1}\left[\frac{2\lambda\sqrt{1-\rho^2}}{1-\lambda^2}\right].$$

Hence, show that, irrespective of the value of ρ in its permissible range of variation, $p >$ or $< (1/2)$ according as $\sigma_1 <$ or $> \sigma_2$.

Hint: $\left[U = (X/\sigma_1),\ V = (Y/\sigma_2)\right] \sim B.V.N\ (0, 0, 1, 1, \rho)$,

then, $$Z = \frac{U}{V} \sim C\left(\rho, \sqrt{1-\rho^2}\right) \quad \ldots (*)$$

$$p = P\left[|X| \le |Y|\right] = P\left[|\sigma_1 U| \le |\sigma_2 V|\right] = \left[P\left|\frac{U}{V}\right| \le \lambda\right] = P\left[|Z| \le \lambda\right]$$

Now, use (*).

20·25. If X and Y are independent normal variates with zero means and variances σ_1^2 and σ_2^2 respectively, the points of inflexion of the curve of intersection of the normal correlation surface by planes through the z-axis, lies on the elliptical cylinder

$$\frac{}{\sigma_1^2} + \frac{}{\sigma_2^2} \quad 1.$$

Hint: Normal correlation surface is: $z = f(x, y)$ ... (*)

and the equation of the plane through z – axis is:

$y = mx$, where m is constant. ... (**)

Now, find the points of inflexion of the curve of intersection of (*) and (**).

20·26. Let $X_1, X_2, \ldots, X_n$ be $i.i.d.\ N(\mu, \sigma^2)$ variates. Find the joint p.d.f. of $U = \sum_{i=1}^{n} a_i X_i$ and $V = \sum_{i=1}^{n} b_i X_i$. Hence or otherwise show that necessary and sufficient condition that U and V are independent is that $\sum_i a_i b_i = 0$

Hints: Find joint *m.g.f.* of (U, V).

$$(U, V) \sim BVN\left[\mu_1, \mu_2, \sigma_1^2, \sigma_1^2, \rho\right], \text{ where}$$

$$\mu_1 = \left(\sum_{i=1}^{n} a_i\right)\mu, \quad \mu_2 = \left(\sum_{i=1}^{n} b_i\right)\mu\ ; \quad \sigma_1^2 = \left(\sum_{i=1}^{n} a_i^2\right)\sigma^2$$

$$\sigma^2 = \left(\sum {}^2\right)\sigma^2 \text{ and } \rho = \frac{\sum a_i b_i}{\left[\sum_{i=1}^{n} a_i^2 \sum_{i=1}^{n} b_i^2\right]^{1/2}}$$

Since (U,V) is BVN ; U and V will be independent if and only if $\rho = 0$.

20·27. Let U, V, W be *i.i.d.* $N(0,1)$ variates and T be $U[0, 1]$ variate. Define

$$X = U\sqrt{T} + V\sqrt{1-T} \quad ; \quad Y = W\sqrt{T} + V\sqrt{1-T}$$

Show that $X \sim N(0, 1)$, $Y \sim N(0,1)$ but (X,Y) is not bivariate normal.

20.28. Let $(X, Y) \sim BVN\left(0, 0, \sigma_x^2, \sigma_y^2, \rho\right)$ Show that the area of the ellipse

$$\frac{x^2}{\sigma_x^2} - \frac{2\rho xy}{\sigma_x \sigma_y} + \frac{y^2}{\sigma_y^2} = \lambda^2,$$

of constant probability density is $\pi \lambda^2 \sigma_x \sigma_y / \sqrt{1-\rho^2}$

Hence, show that the area of the strip between the ellipses corresponding to the parameter value λ and $d\lambda$ is $\left[2\pi \lambda \sigma_x \sigma_y / \sqrt{1-\rho^2}\right] d\lambda$. Deduce that the probability that a pair of values (X, Y) chosen at random will be represented by a point inside the strip is

$$\exp\left[-\frac{\lambda^2}{2\left(1-\rho^2\right)}\right] \cdot \frac{\lambda\, d\lambda}{\left(1-\rho^2\right)}.$$

Hence, show that for a given value of $d\lambda$, the probability that (X, Y) will fall in the strip λ to $\lambda + d\lambda$ is maximum when $\lambda^2 = 1-\rho^2$.

Also prove that

(*i*) The probability that the point will fall inside the ellipse λ is : $p = 1 - \exp\left[-\frac{\lambda^2}{2\left(1-\rho^2\right)}\right]$

(*ii*) The probability (X, Y) lies outside the ellips λ is: $\exp\left[-\frac{\lambda^2}{2\left(1-\rho^2\right)}\right]$.

20·29. If $(X, Y) \sim BVN\left(\mu_X, \mu_Y, \sigma_X^2, \sigma_Y^2, \rho\right)$, then prove that:

$$r\left[X, F_Y(Y)\right] = \rho \cdot \left(\frac{3}{\pi}\right)^{1/2}$$

20·30. If $(X,Y) \sim BVN(0, 0, 1, 1, \rho)$, prove that the correlatiom coefficient between $F_X(X)$ and $F_Y(Y)$ is :

$$r\left[F_X(X),\ F_Y(Y)\right] = \frac{6}{\pi}\sin^{-1}(\rho / 2)$$

where $F_X(X)$ and $F_Y(Y)$ are distributions function of X and Y.

CHAPTER

21 Some Additional Topics

LEARNING OBJECTIVES

Upon completion of this Chapter, you should be able to:

- Discuss the genesis and derivation of the Pearsonian System of Frequency Curves.
- Explain the technique and the use of the Variate Transformations in stabilising the variance of the sampling distribution of a statistic.
- Understand the technique of the Curve Fitting by Orthogonal Polynomials and discuss its utility over the usual method of Curve Fitting by the Principle of Least Squares.
- Define Hermite Polynomials and Gram-Charlier Polynomials, and discuss their important Properties.
- Understand the use of Gram-Charlier Type-A Series to approximate the p.d.f. of the random variable from the given statistical moments.
- Know about the Edgeworth Type – A Series for approximating the p.d.f. of a random variable in terms of its cumulants.
- Know about the Gram-Charlier Type-B Series for approximating the p.m.f. of a discrete random variable through the use of Gram-Charlier's Polynomials.

CHAPTER OUTLINES

21·1· PEARSONIAN DISTRIBUTIONS AND THEIR GENESIS

Given a set of observations from a population, the first question that arises in our mind is about the nature of the parent population. A vague idea is provided by the frequency polygon (or frequency curve) but the information is totally inadequate and unreliable, because the sample observations may not cover the entire range of the parent distribution. Moreover, an unusually high frequency in one class, arising out of sheer chance, may completely distort the shape of the frequency curve.

Consequently, to determine the frequency curve, we resort to the technique of curve fitting to the given data. The failure of normal distribution to fit many distributions which are observed in practice for continuous variable necessitated the development of generalised system of frequency curves. Since a trial and error approach is clearly undesirable, an elastic system of frequency curves must be evolved, which should incorporate, if not all, at least the most common of the distributions. Pearsonian system of frequency curves is one of the most important approaches in this direction, in which we decide about the shape of the curve on the basis of a 'criterion κ' calculated from the sample observations.

Karl Pearson's first memoir dealing with generalised frequency curves appeared in 1895. In this paper and the subsequent two papers published in 1901 and 1916. Karl Pearson developed a set of frequency curves which could be obtained by assigning values to the parameters in a certain first order differential equation.

Genesis of Pearson's Frequency Curves : Experience tells that most of the frequency distributions possess the following obvious and common characteristic:

"They rise from a low frequency to a maximum frequency and then again fall to the low frequency as the variable X increases. This suggests a unimodal frequency curve $y = f(x)$. Accordingly, Karl Pearson proposed the following differential equation for the frequency curve

$$y = f(x), \qquad \frac{dy}{dx} = \frac{y(x-a)}{F(x)} \qquad \text{.... (21·1)}$$

where $F(x)$ is an arbitrary function of x not vanishing at $x = a$, the mode of the distribution. Expanding $F(x)$ by Maclaurin's theorem, we get: $F(x) = b_0 + b_1 x + b_2 x^2 + \ldots$ and retaining only the first three terms, we get the differential equation of the Pearsonian system of frequency curves as:

$$\frac{dy}{dx} = \frac{y(x-a)}{b_0 + b_1 x + b_2 x^2} \quad \Rightarrow \quad \frac{df(x)}{dx} = f'(x) = \frac{(x-a)f(x)}{b_0 + b_1 x + b_2 x^2} \qquad \text{... (21·1 a)}$$

where a, b_0, b_1 and b_2 are the constants to be calculated from the given data.

Remark. Equation (21·1a) can also be obtained as a limiting case of Hyper-geometric distribution (c.f. Advance Statistics Vol. II by Kendall).

21·1·1. Determination of the Constants of the Equation (21·1a) in Terms of Moments

Multiplying both sides of (21·1a) by x^n for integral $n > 0$ and integrating over the entire range of the variable X, say, (α β), we get:

$$\int_\alpha^\beta x^n \left(b_0 + b_1 x + b_2 x^2\right) f'(x)\, dx = \int_\alpha^\beta x^n (x-a) f(x)\, dx \qquad \text{... (21·1b)}$$

$$\Rightarrow \quad \left| x^n \left(b_0 + b_1 x + b_2 x^2\right) f(x) \right|_\alpha^\beta - \int_\alpha^\beta \left[n b_0 x^{n-1} + (n+1) b_1 x^n + (n+2) b_2 x^{n+1} \right] f(x) dx$$

$$= \int_\alpha^\beta x^{n+1} f(x)\, dx - a \int_\alpha^\beta x^n f(x)\, dx$$

Assuming high order contact at the extremities so that

$$\left| x^n f(x) \right|_{\alpha}^{\beta} = 0, \qquad i.e., \qquad x^r f(x) \to 0 \text{ as } x \to \alpha \text{ or } \beta, \qquad \ldots (*)$$

we get $\quad -\left[nb_0 \mu_{n-1} + (n+1)\, b_1 \mu_n + (n+2)\, b_2 \mu_{n+1} \right] = \mu_{n+1} - a\mu_n$

(assuming that X is measured from mean and this we can do without any loss of generality). Thus the recurrence relation between the moments becomes

$$nb_0 \mu_{n-1} + \left[(n+1)\, b_1 - a \right] \mu_n + \left[(n+2)\, b_2 + 1 \right] \mu_{n+1} = 0, \quad n = 1, 2, 3 \ldots \qquad \ldots (**)$$

Integrating (21·1 a) w.r.t. x within th limits (α, β) and using (*), we get

$$(b_1 - a) + (2b_2 + 1) = 0$$

Putting $n = 1, 2$ and 3 in (**) and solving these equations and (***) with the help of determinants and using $\mu_0 = 1$, $\mu_1 = 0$, we get

$$\left.\begin{aligned} b_0 &= -\frac{\mu_2 \left(4\mu_2 \mu_4 - 3\mu_3^2 \right)}{2\left(5\mu_2 \mu_4 - 9\mu_2^3 - 6\mu_3^2 \right)} = -\frac{\sigma^2 (4\beta_2 - 3\beta_1)}{2(5\beta_2 - 6\beta_1 - 9)} \\ a = b_1 &= -\frac{\mu_3 \left(\mu_4 + 3\mu_2^2 \right)}{2\left(5\mu_2 \mu_4 - 9\mu_2^3 - 6\mu_3^2 \right)} = -\frac{\sigma \sqrt{\beta_1} (\beta_2 + 3)}{2(5\beta_2 - 6\beta_1 - 9)} \\ b_2 &= \frac{\left(2\mu_2\mu_4 - 3\mu_3^2 - 6\mu_2^3 \right)}{2\left(5\mu_2 \mu_4 - 9\mu_2^3 - 6\mu_3^2 \right)} = \frac{(2\beta_2 - 3\beta_1 - 6)}{2(5\beta_2 - 6\beta_1 - 9)} \end{aligned}\right\} \qquad \ldots (21·2)$$

where $\mu_2 = \sigma^2$, $\beta_1 = \mu_3^2 / \mu_2^3$ and $\beta_2 = \mu_4 / \mu_2^2$.

Thus the Pearson's system (21·1 a) is completely specified by the first four moments.

21·1·2. Pearson's Measure of Skewness

$$\text{Skewness} = \frac{\text{Mean} - \text{Mode}}{\text{Standard Deviation}} = \frac{0 - a}{\sqrt{\mu_2}} = \frac{\sqrt{\beta_1}\,(\beta_2 + 3)}{2\,(5\beta_2 - 6\beta_1 - 9)} \qquad \ldots (21·3)$$

21·1·3· "Criterion κ". Equation (21·1 a) can be re-written as :

$$\frac{df}{f} = \frac{(x - a)\, dx}{b_0 + b_1 x + b_2 x^2}, \quad f = f(x)$$

Integrating, we get

$$\log (f / c) = \int \frac{(x - a)\, dx}{b_0 + b_1 x + b_2 x^2}\, dx = I \; (say),$$

where c is the constant of integration.

$$\therefore \qquad f(x) = c \exp [I] \qquad \ldots (21·4)$$

Thus, f depends on I, which further depends on the roots of the equation

$$b_0 + b_1 x + b_2 x^2 = 0 \qquad \ldots (21·5)$$

We have:

$$b_0 + b_1 x + b_2 x^2 = b_2\left[x^2 + \frac{b_1}{b_2}\,x + \frac{b_0}{b_2}\right]$$

$$= b_2\left[x - \frac{-b_1 + \sqrt{b_1^2 - 4\,b_0 b_2}}{2b_2}\right] \times \left[x - \frac{-b_1 - \sqrt{b_1^2 - 4\,b_0 b_2}}{2b_2}\right]$$

$$= b_2\left[x + \frac{b_1}{2b_2} - \frac{\sqrt{b_1^2 - 4\,b_0 b_2}}{\sqrt{4b_2^2}}\right] \times \left[x + \frac{b_1}{2b_2} + \frac{\sqrt{b_1^2 - 4\,b_0 b_2}}{\sqrt{4b_2^2}}\right]$$

$$= b_2\left[x + \frac{b_1}{2b_2} - \sqrt{\frac{b_0}{b_2}(\kappa - 1)}\right] \times \left[x + \frac{b_1}{2b_2} + \sqrt{\frac{b_0}{b_2}(\kappa - 1)}\right],$$

where,
$$\kappa = \frac{b_1^2}{4\,b_0 b_2}, \qquad \text{... (21·6)}$$

determines the criterion for obtaining the form of the frequency curve.

A brief description of various Pearsonian curves for different values of κ is given below:

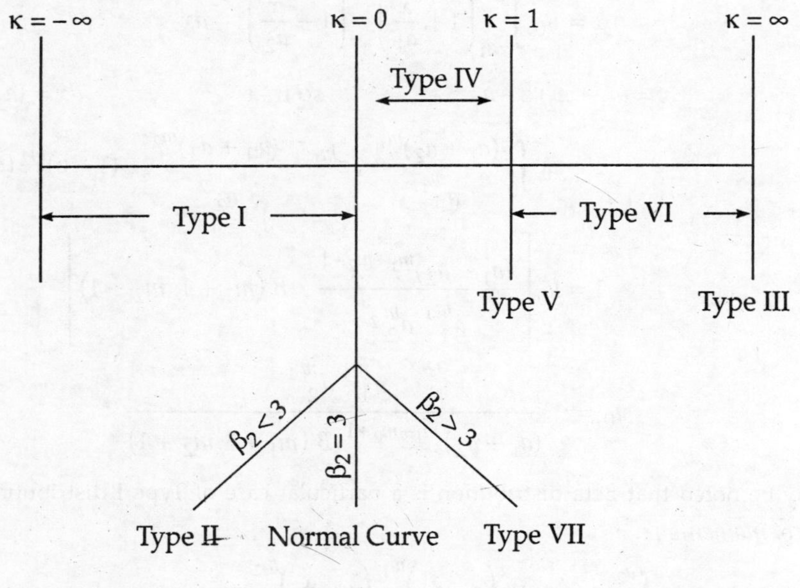

21·1·4· Pearson's Main Type 1

This curve is obtained where the roots of the quadratic equation (21·5) are real and of opposite sign, i.e., when $\kappa < 0$.

Shifting the origin to the mode $x = a$, the equation (21·1*a*) becomes

$$\frac{df}{dx} = \frac{x\,f}{B_0 + B_1 x + B_2 x^2} = \frac{x\,f}{B_2(x + \alpha)\,(x - \beta)} \qquad \text{... (21·6}a\text{)}$$

where $\quad B_0 = b_0,\ B_1 = b_1$ and $B_2 = b_2$

$$\Rightarrow \quad \frac{d}{dx}(\log f) = \frac{1}{B_2(\alpha+\beta)}\left[\frac{\alpha}{x+\alpha}+\frac{\beta}{x-\beta}\right] \qquad \ldots (21{\cdot}6b)$$

Integrating both side w.r.t. x, we get

$$\log f = \log (x+\alpha)^{\alpha/B_2(\alpha+\beta)} + \log (x-\beta)^{\beta/B_2(\alpha+\beta)} + \log C$$

$$\therefore \quad f = C\,(x+\alpha)^{\alpha/B_2(\alpha+\beta)}\,(x-\beta)^{\beta/B_2(\alpha+\beta)}$$

$$\Rightarrow \quad f = y_0\left(1+\frac{x}{\alpha}\right)^{\alpha/B_2(\alpha+\beta)}\left(1-\frac{x}{\beta}\right)^{\beta/B_2(\alpha+\beta)} \quad ; -\alpha \le x \le \beta \qquad \ldots(*)$$

Let $\alpha = a_1$, $\beta = a_2$, $m_1 = \dfrac{\alpha}{B_2(\alpha+\beta)}$ and $m_2 = \dfrac{\beta}{B_2(\alpha+\beta)}$, so that

$$\frac{m_1}{a_1} = \frac{m_2}{a_2} = \frac{1}{B_2(a_1+a_2)},$$ then (*) may be written as

$$f(x) = y_0\left(1+\frac{x}{a_1}\right)^{m_1}\left(1-\frac{x}{a_2}\right)^{m_2}, \quad -a_1 \le x \le a_2$$

which is a standard form of the Type 1.

Determination of y_0 :

$$1 = y_0 \int_{-a_1}^{a_2}\left(1+\frac{x}{a_1}\right)^{m_1}\left(1-\frac{x}{a_2}\right)^{m_2} dx$$

Put $x = (a_1+a_2)\,z - a_1$ so that $dx = (a_1+a_2)\,dz$

$$\Rightarrow \quad 1 = y_0 \int_0^1 \frac{(a_1+a_2)^{m_1}}{a_1^{m_1}}\cdot z^{m_1}\cdot\frac{(a_1+a_2)^{m_2}}{a_2^{m_2}}\cdot(1-z)^{m_2}\,(a_1+a_2)dz$$

$$\Rightarrow \quad 1 = y_0\left[\frac{(a_1+a_2)^{m_1+m_2+1}}{a_1^{m_1}\,a_2^{m_2}}\cdot B(m_1+1, m_2+1)\right]$$

$$\therefore \quad y_0 = \frac{a_1^{m_1}\,a_2^{m_2}}{(a_1+a_2)^{m_1+m_2+1}.B(m_1+1, m_2+1)}$$

Remark. It may be noted that Beta distribution is a particular case of Type I distribution.

Determination of moments :

$$\mu_n' = y_0 \int_{-a_1}^{a_2}(x+a_1)^n\left(1+\frac{x}{a_1}\right)^{m_1}\left(1-\frac{x}{a_2}\right)^{m_2} dx$$

$$= y_0\,\frac{(a_1+a_2)^{m_1+m_2+n+1}}{a_1^{m_1}\,a_2^{m_2}}\,B(n+m_1+1, m_2+1), \text{ where } x = (a_1+a_2)z - a_1.$$

$$= \frac{(a_1+a_2)^n}{B(m_1+1, m_2+1)}.\,B(n+m_1+1, m_2+1) \qquad \text{[On simplification]}$$

21·1·5. Pearson's Type IV

This curve is obtained when the roots of (21·5) are imaginary or when

$$B_1^2 < 4\,B_0 B_2\,,\ \textit{i.e.,}\ 0 < \kappa < 1$$

$$\frac{1}{f}\cdot\frac{df}{dx} = \frac{x}{B_0 + B_1 x + B_2 x^2} \qquad \text{(Origin at mode)}$$

$$\Rightarrow \quad \frac{d}{dx}(\log f) = \frac{x}{B_2\left[\left(x + \dfrac{B_1}{2B_2}\right)^2 + \left(\dfrac{B_0}{B_2} - \dfrac{B_1^2}{4B_2^2}\right)\right]}$$

$$= \frac{(x+\gamma) - \gamma}{B_2\left[(x+\gamma)^2 + \delta^2\right]} = \frac{2(x+\gamma)}{2B_2\left[(x+\gamma)^2+\delta^2\right]} - \frac{2\gamma}{2B_2\left[(x+\gamma)^2+\delta^2\right]}$$

$$\Rightarrow \quad \log f = \frac{1}{2B_2}\log\left[(x+\gamma)^2 + \delta^2\right] - \frac{\gamma}{B_2}\cdot\frac{1}{\delta}\tan^{-1}\left(\frac{x+\gamma}{\delta}\right) + \log C$$

$$\therefore \quad f = C\left[(x+\gamma)^2 + \delta^2\right]^{\frac{1}{2B_2}} \exp\left\{-\frac{\gamma}{B_2\delta}\tan^{-1}\left(\frac{x+\gamma}{\delta}\right)\right\}$$

Put $\dfrac{x+\gamma}{\delta} = \dfrac{x}{a}$ and $\dfrac{\gamma}{B_2\,\delta} = v$

Hence, $$f(x) = y_0\left(1 + \frac{x^2}{a^2}\right)^{-m} \exp\left[-v\tan^{-1}(x/a)\right];\ -\infty < x < \infty, (m, v) > 0 \qquad \text{... (21·8)}$$

which is standard form of Type IV with origin at $\left(-\dfrac{B_1}{2B_2}, 0\right)$. The curve is skew and has unlimited range in both the directions.

Determination of y_0:

$$1 = y_0\int_{-\infty}^{\infty}\left(1 + \frac{x^2}{a^2}\right)^{-m}\exp\left[-v\tan^{-1}(x/a)\right]dx$$

$$= a\,y_0\int_{-\pi/2}^{\pi/2}\cos^{2m-2}\theta\cdot e^{-v\theta}\,d\theta = a\,y_o\,.F\,(2m-2, v) \qquad [\text{Put } x = a\tan\theta]$$

$$\therefore \quad y_0 = \frac{1}{a.F\,(2m-2, v)}$$

Hence, $$f(x) = \frac{1}{a.F(2m-2, v)}\left(1 + \frac{x^2}{a^2}\right)^{-m}\exp\left[-v\tan^{-1}(x/a)\right];\ -\infty < x < \infty$$

21·1·6. Pearson's Type VI

This curve is obtained when the roots of (21·5) are real and are of the same sign. This is obtained when B_0, B_2 are of the same sign or, in other words, when $\kappa > 1$.

Let $-\alpha_1$ and $-\alpha_2$ be the roots of the quadratic equation (21·5) then

$$\frac{d}{dx}(\log f) = \frac{x}{B_0 + B_1 x + B_2 x^2} = \frac{x}{B_2(x+\alpha_1)(x+\alpha_2)}$$

$$= \frac{\alpha_1}{B_2(\alpha_1 - \alpha_2)} \cdot \frac{1}{(x+\alpha_1)} - \frac{\alpha_2}{B_2(\alpha_1-\alpha_2)} \cdot \frac{1}{(x+\alpha_2)}$$

$$\therefore \quad \log f = \frac{\alpha_1}{B_2(\alpha_1 - \alpha_2)} \log(x+\alpha_1) - \frac{\alpha_2}{B_2(\alpha_1-\alpha_2)} \log(x+\alpha_2) + \log C$$

$$\Rightarrow \quad f = C(x+\alpha_1)^{\alpha_1/B_2(\alpha_1-\alpha_2)} \cdot (x+\alpha_2)^{-\alpha_2/B_2(\alpha_1-\alpha_2)}$$

Hence, the probability density function is :

$$f(x) = y_0 \left(1 + \frac{x}{a_1}\right)^{m_1} \left(1 + \frac{x}{a_2}\right)^{-m_2} \qquad \ldots (21·9)$$

where $\alpha_1 = a_1, \quad \alpha_2 = a_2$ and $\frac{m_1}{a_1} = -\frac{m_2}{a_2}$; $a_1, a_2 > 0$

This equation can also be written (on shifting the origin to $-\alpha_2$ or $-a_2$) as

$$f(x) = y_0 (x-a)^{q_2} x^{-q_1} \; ; \; a \le x < \infty,$$

where $q_1 = \frac{\alpha_2}{B_2(\alpha_1 - \alpha_2)}$, $q_2 = \frac{\alpha_1}{B_2(\alpha_1-\alpha_2)}$, $a = -\alpha_1$; $y_0 = \frac{a^{q_1 - q_2 + 1}}{\beta(q_2 + 1, q_1 - q_2 - 1)}$

Remark. The curve is bell shaped if $q_2 > 0$ and J - shaped if $q_2 < 0$.

Determination of y_0 :

$$1 = y_0 \int_a^\infty (x-a)^{q_2} x^{-q_1} \, dx$$

$$= y_0 \int_0^1 \left(\frac{a}{1-z}\right)^{-q_1} \left(\frac{a}{1-z} - a\right)^{q_2} \frac{a}{(1-z)^2} dz$$

$$= y_0 \, a^{q_2 - q_1 + 1} \int_0^1 z^{q_2} \frac{1}{(1-z)^{q_2 - q_1 + 2}} dz$$

$$y_0 = \frac{a^{q_1 - q_2 - 1}}{B(q_2 + 1, q_1 - q_2 - 1)}$$

Hence, $$f(x) = \frac{a^{q_1 - q_2 - 1}}{B(q_2+1, q_1 - q_2 - 1)} x^{-q_1}(x-a)^{q_2}, \; a \le x < \infty .$$

The above discussion covers almost the whole range of κ but in limiting cases we get simple cases. The following are more important of the transition curves when one of main type changes into another.

21·1·7· Pearson's Type III

This is a transition curve and is obtained where $B_2 = 0, B_1 \neq 0$ *or* $\kappa \to \pm \infty$.

$$\frac{d}{dx}(\log f) = \frac{x}{B_0 + B_1 x}, \qquad \text{(origin is at mode).}$$

$$= \frac{B_1 x + B_0 - B_0}{B_1(B_0 + B_1 x)} = \frac{1}{B_1} - \frac{B_0}{B_1(B_0 + B_1 x)}$$

$$\Rightarrow \qquad \log f = \frac{x}{B_1} - \frac{B_0}{B_1^2} \log (B_0 + B_1 x) + \text{constant}$$

$$\Rightarrow \qquad f = \text{constant}.\ e^{x/B_1}(B_0 + B_1 x)^{-B_0/B_1^2}$$

$$\Rightarrow \qquad f(x) = y_0 \left(1 + \frac{x}{a}\right)^p e^{-px/a};\ a \le x < \infty \qquad \text{... (21·10)}$$

where $\dfrac{B_0}{B_1} = a$ and $\dfrac{B_0}{B_1^2} = -p$.

This gives the Type III curve with origin at mode. The curve is usally bell shaped but becomes J-shaped when $\beta_1 > 4$.

Remark. The distribution can be transformed into the gamma form by using the transformation $y = \dfrac{p}{a}(x + a)$ when the curve reduces to

$$f(y) = \frac{1}{\Gamma(p+1)} e^{-y} y^p,\ 0 \le y < \infty.$$

21·1·8. Pearson's Type V

This transition type is obtained when the roots of (21·5) *are equal, i.e., when* $B_1^2 = 4 B_0 B_2$ *or* $\kappa = 1$.

$$\frac{d}{dx}(\log f) = \frac{x}{B_2\left[\left(x + \dfrac{B_1}{2B_2}\right)^2\right]} = \frac{2\left[x + \dfrac{B_1}{2B_2}\right] - \dfrac{B_1}{B_2}}{2B_2\left[\left(x + \dfrac{B_1}{2B_2}\right)^2\right]}$$

$$\therefore \qquad \log f = \frac{1}{2B_2} \log\left(x + \frac{B_1}{2B_2}\right)^2 + \frac{B_1}{2B_2^2} \frac{1}{\left[x + \dfrac{B_1}{2B_2}\right]} + \text{constant.}$$

$$\Rightarrow \qquad f = \text{constant}.\left(x + \frac{B_1}{2B_2}\right)^{1/B_2}.\exp\left[\frac{B_1}{2B_2^2}\left(x + \frac{B_1}{2B_2}\right)^{-1}\right]$$

$$\therefore \qquad f(x) = y_0 X^{-p} e^{-q/X},\ 0 \le X < \infty, \qquad \text{... (21·11)}$$

where $X = \left(x + \dfrac{B_1}{2B_2}\right)$, $\dfrac{B_1}{2B_2^2} = -q$ and $\dfrac{1}{B_2} = -p$.

21·1·9. Pearson's Type II

This curve is obtained when $B_1 = 0$ and B_0, B_2 are of oppsitie sign, i.e., $\kappa = 0$. The equation to the curve is.

$$f(x) = y_0 \left[1 - \frac{x^2}{a^2}\right]^m, \; -\infty < x < \infty \qquad \text{... (21·12)}$$

where $m = \frac{1}{2B_2} > 0,\; a^2 = -\frac{B_0}{B_2}$ with origin at mean (mode).

The curve is symmetrical bell shaped if $m > 0$, U – shaped if $m < 0$, and is rectangular curve if $m = 0$.

21·1·10· Pearson's Type VII

This curve is obtained when $B_1 = 0$ and B_0, B_2 are of the same sign, i.e., $\kappa = 0$ and $B_0 B_2 > 0$. The equation to the curve is

$$f(x) = y_0 \left[1 + \frac{x^2}{a^2}\right]^{-m}, \; -\infty < x < \infty \qquad \text{... (21·13)}$$

where $a^2 = \frac{B_0}{B_2}$ and $m = -\frac{1}{2B_2}$, with origin being at the mean (mode).

This curve is usally bell shaped, symmetrical and of unlimited range in both the directions.

21·1·11. Zero Type (Normal curve)

When $B_1 = B_2 = 0 \Rightarrow b_1 = b_2 = 0$ [from (21·6 b)] , (21·2) implies that $\beta_1 = 0$ and $\beta_2 = 3$ and we have, from (21·6 a) :

$$\frac{d}{dx}(\log f) = \frac{x}{B_0} \qquad \Rightarrow \qquad \log f = \frac{x^2}{2B_0} + \log C,$$

where C is the constant of integration.

$$\therefore \qquad f = C \exp\left(x^2 / 2B_0\right) = C \exp\left(-x^2 / 2\sigma^2\right), \; -\infty < x < \infty \qquad \text{... (21·14)}$$

where $B_0 = -\sigma^2$ and the origin is at mean. This is the normal distribution with mean zero and variance σ^2.

21·1·12 Pearson's Type VIII

When $B_0 = 0,\; B_1 > 0$,

$$f(x) = \frac{1-m}{a}\left(1 + \frac{x}{a}\right)^{-m}, \; -a \le x \le 0 \qquad \text{... (21·15)}$$

21·1·13 Type IX

When $B_0 = 0$, $B_1 < 0$ and $\kappa < 0$,

$$f(x) = \frac{1+m}{a}\left(1 + \frac{x}{a}\right)^{m}, \; -a \le x \le 0 \qquad \text{... (21·16)}$$

21·1·14 Type X

When $B_0 = 0$ and $B_2 = 0$, $\quad f(x) = \dfrac{1}{\sigma} \exp\left(-\dfrac{x}{\sigma}\right), 0 \le x < \infty, \sigma > 0$

This is the p.d.f. of simple exponential distribution with parameter $\sigma > 0$.

21·1·15 Type XI

When $B_0 = B_1 = 0$, and $\kappa > 1$; $\quad f(x) = b^{m-1}(m-1)x^{m-1}, \; b \le x < \infty$

21·1·16 Type XII

When $(5\beta_2 - 6\beta_1 - 9) = 0$, $\kappa < 0$,

$$f(x) = \left(\frac{a_1}{a_2}\right)^m \frac{1}{(a_1 + a_2)\, B(1+m, 1-m)} \cdot \frac{\left(1 + \dfrac{x}{a_1}\right)^m}{\left(1 - \dfrac{x}{a_2}\right)^m}, \; a_1 \le x \le a_2$$

Example 21·1. *Show that for a Pearson distribution :* $\dfrac{df}{f} = \dfrac{(a+x)dx}{b_0 + b_1 x + b_2 x^2}$, *the charactristic function* ϕ *obeys the relation :*

$$b_2 \theta \frac{d^2\varphi}{d\theta^2} + (1 + 2b_2 + b_1\theta)\frac{d\varphi}{d\theta} + (a + b_1 + b_0\theta)\,\phi = 0, \text{ where } \theta = it.$$

Deduce the recurrence relation for moments.

Show also that the cumulant generating function ψ *obeys the relation:*

$$b_2\,\theta\left\{\frac{d^2\psi}{d\theta^2} + \left(\frac{d\psi}{d\theta}\right)^2\right\} + (1 + 2b_2 + b_1\theta)\frac{d\psi}{d\theta} + (a + b_1 + b_0\theta) = 0.$$

Hence show that the cumulants obey the recurrence relation :

$$\left[1 + (r+2)\,b_2\right]\kappa_{r+1} + rb_1\,\kappa_r$$

$$+ r b_2 \left\{\binom{r-1}{1}\kappa_2\kappa_{r-1} + \binom{r-1}{2}\kappa_3\,\kappa_{r-2} + \ldots + \binom{r-1}{j}\kappa_{j+1}\kappa_{r-j} + \ldots + \binom{r-1}{r-2}\kappa_{r-1}\kappa_2\right\} = 0$$

Solution: $\left(b_0 + b_1 x + b_2 x^2\right)\dfrac{df}{dx} = (a+x) f \;\Rightarrow\; e^{\theta x}\left(b_0 + b_1 x + b_2 x^2\right)\dfrac{df}{dx} = e^{\theta x}(a+x) f; \quad (\theta = \text{it}).$

Integrating w.r.t. x, for the total range of x, assuming that integrals vanish at either limit, we get

$$\int_{-\infty}^{\infty} e^{\theta x}\left(b_0 + b_1 x + b_2 x^2\right)\frac{df}{dx}\,dx = \int_{-\infty}^{\infty} e^{\theta x}(a+x) f\,dx$$

$$\Rightarrow \quad e^{\theta x}\left(b_0 + b_1 x + b_2 x^2\right) f\Big]_{-\infty}^{\infty} - \int_{-\infty}^{\infty}\left\{\theta e^{\theta x}\left(b_0 + b_1 x + b_2 x^2\right) + e^{\theta x}(b_1 + 2b_2 x)\right\}$$

$$= \int_{-\infty}^{\infty} e^{\theta x}(a+x) f\,dx$$

$$\Rightarrow \quad 0-\theta\left[b_0\,\varphi+b_1\frac{d\varphi}{d\theta}+b_2\frac{d^2\varphi}{d\theta^2}\right]-\left[b_1\,\varphi+2b_2\frac{d\varphi}{d\theta}\right]=\left(a\varphi+\frac{d\varphi}{d\theta}\right)^*$$

$$\Rightarrow \quad b_2\theta\,\frac{d^2\varphi}{d\theta^2}+(1+2b_2+b_1\,\theta)\,\frac{d\varphi}{d\,\theta}+(a+b_1+b_0\theta)\,\varphi=0 \qquad \text{(On Simplification)} \qquad \ldots (i)$$

Differentiating n times w.r.t. θ, using Leibnitz Theorem, we get

$$b_2\left[\theta\,\frac{d^{n+2}\varphi}{d\theta^{n+2}}+n.\,\frac{d^{n+1}\varphi}{d\theta^{n+1}}.1\right]+\left[\frac{d^{n+1}\varphi}{d\theta^{n+1}}(1+2b_2+b_1\theta)+\frac{d^n\varphi}{d\theta^n}.nb_1\right]$$

$$+\left[\frac{d^n\varphi}{d\theta^n}(a+b_1+b_0\,\theta)+\left\{\frac{d^{n-1}\varphi}{d\theta^{n-1}}.nb_0\right\}\right]=0 \qquad \ldots (ii)$$

Putting $\theta=0$ and using the relation $\left[\dfrac{d^n\varphi}{d\theta^n}\right]_{\theta=0}=\mu_n'$, we get

$$nb_2\,\mu'_{n+1}+(2b_2+1)\,\mu'_{n+1}+nb_1\mu_n'+(b_1+a)\,\mu_n'+nb_0\,\mu_{n-1}'=0$$

Shifting the origin to the mean, we get

$$\left[(n+2)b_2+1\right]\mu_{\,n+1}+\left[(n+1)\,b_1+a\,\right]\mu_{\,n}+nb_0\,\mu_{n-1}=0 \qquad \ldots (iii)$$

Cumulant generating function is given by :

$$\psi=\log\phi \quad\Rightarrow\quad \varphi=e^{\,\psi}, \qquad \frac{d\varphi}{d\theta}=e^{\psi}\frac{d\psi}{d\theta} \quad \text{and} \quad \frac{d^2\varphi}{d\theta^2}=e^{\psi}\left[\frac{d^2\psi}{d\theta^2}+\left(\frac{d\psi}{d\theta}\right)^2\right]$$

Substituting these values in (i) and on simplification, we get

$$b_2\,\theta\left[\frac{d^2\psi}{d\theta^2}+\left(\frac{d\psi}{d\theta}\right)^2\right]+(1+2b_2+b_1\theta)\,\frac{d\psi}{d\theta}+(a+b_1+b_0\theta)=0 \qquad \ldots (iv)$$

Differentiating (iv) r times w.r.t θ using Leibnitz Theorem, we get

$$b_2\,\theta\left[\frac{d^{r+2}\psi}{d\theta^{r+2}}+\frac{d^r}{d\theta^r}\left(\frac{d\psi}{d\theta}\right)^2\right]+\binom{r}{1}b_2\left[\frac{d^{r+1}\psi}{d\theta^{r+1}}+\frac{d^{r-1}}{d\theta^{r-1}}\left(\frac{d\psi}{d\theta}\right)^2\right]$$

$$+(1+2b_2+b_1\theta)\,\frac{d^{r+1}\psi}{d\theta^{r+1}}+\binom{r}{1}b_1.\frac{d^r\psi}{d\theta^r}=0$$

$*\quad \because \quad \phi=E\left(e^{itx}\right)=\int_{-\infty}^{\infty}(e^{\theta x})\,f\,dx=\int_{-\infty}^{\infty}e^{itx}\,f\,dx, \qquad [\because\ it=\theta]$

$$\frac{d\varphi}{d\theta}=\int_{-\infty}^{\infty}x\,e^{\theta x}\,f\,dx \quad\text{and}\quad \frac{d^2\varphi}{d\theta^2}=\int_{-\infty}^{\infty}x^2e^{\theta x}\,f\,dx,$$

assuming differentiation is valid under integral sign.

$$\Rightarrow \quad b_2\theta\left[\frac{d^{r+2}\psi}{d\theta^{r+2}}+\left\{\frac{d^{r-1}}{d\theta^{r-1}}\left(2\frac{d\psi}{d\theta}\cdot\frac{d^2\psi}{d\theta^2}\right)\right\}\right]+rb_2\left[\frac{d^{r+1}\psi}{d\theta^{r+1}}+\frac{d^{r-2}}{d\theta^{r-2}}\left\{2\frac{d\psi}{d\theta}\cdot\frac{d^2\psi}{d\theta^2}\right\}\right]$$

$$+\ (1+2b_2+b_1\theta)\frac{d^{r+1}\psi}{d\theta^{r+1}}+rb_1\frac{d^r\psi}{d\theta^r}=0$$

Putting $\theta = 0$ and using the relation : $\left[\dfrac{d^n\psi}{d\theta^n}\right]_{\theta=0} = \kappa_n$, we get

$$\{1+(r+2)\,b_2\}\,\kappa_{r+1}+\binom{r}{1}\,b_1\,\kappa_r$$

$$+rb_2\left\{\binom{r-1}{1}\kappa_2.\kappa_{r-1}+\binom{r-1}{2}\kappa_3\kappa_{r-2}+\ \ldots\ +\binom{r-1}{r-2}\kappa_{r-1}\kappa_2\right\}=0$$

(On simplification)

21·2· VARIATE TRANSFORMATIONS

Let T be any statistic which is asymptotically normally distributed with mean θ and variance $\psi(\theta)$, where $\psi(\theta)$ is some function of the parameter θ *i.e.*, $T \sim N\{\theta, \psi(\theta)\}$. Let us transform T by function g as $g(T)$ where g is a function which possesses first order derivative which is continuous and $g'(\theta) \neq 0$, where (′) denotes differentiation w.r. to the parameter θ. Then $g(T)$ is normally distributed about mean $g(\theta)$ and variance $[g'(\theta)]^2\,\Psi(\theta)$, *i.e.*,

$$g(T) \sim N\left[g(\theta), \{g'(\theta)\}^2\,\psi(\theta)\right], \qquad \ldots (21\cdot18)$$

asymptotically, provided $g'(\theta) \neq 0$, is continuous in the neighbourhood of θ.

In general Var $(T) = \psi(\theta)$ will be dependent on the parameter θ. We are interested in obtaining a function g such that the asymptotic variance of the transformed statistic $g(T)$ is independent of θ, *i.e.*, it should be constant. In other words, we want g such that

$$\text{Var}\left[g(T)\right] = \{g'(\theta)\}^2 . \psi(\theta) = \text{constant} = c^2 \text{ say}, \quad ie., \quad |g'(\theta)| = \frac{c}{\sqrt{\psi(\theta)}} \qquad \ldots (21\cdot18\,a)$$

Integrating both sides w.r. to θ we get:

$$g(\theta) = \int \frac{c}{\sqrt{\psi(\theta)}}\,d\theta \qquad \ldots (21\cdot19)$$

21·2·1· Uses of Variate Transformations

As discussed above, when we transform statistic T to *a* function $g(T)$, the distribution of $g(T)$ is approximately normal and its asymptotic variance is independent of the population parameter θ.

Hence the use of statistic $g(T)$ gives better results and confidence intervals than the original statistic T. The commonly used transformations are :

(1) Square Root Transformation (2) Sine Inverse or $\sin^{-1}$ Transformation

(3) Logarithmic Transformation (4) Fisher's Z-Transformation

In the following sections we shall discuss these transformations briefly.

21·2·2 Square Root Transformation

Square root transformation is a transformation for the Poisson variate. If a variance X follows Poisson distribution with parameter λ (assuming to be large), then we know that asymptotic distribution of X is normal ($\lambda \to \infty$) with $E(X) = \lambda$ and Var $(X) = \lambda = \psi(\lambda)$, in the above notations.

Then (21·19) gives the function

$$g(\lambda) = \int \frac{c}{\sqrt{\lambda}}\, d\lambda = 2c\sqrt{\lambda}$$

We select c in such a way that $\quad 2c = 1, \quad$ *i.e.,* $\; c = \frac{1}{2} \;$ so that $g(\lambda) = \sqrt{\lambda}$. ... (*)

Hence, the transformed variable is $g(X) = \sqrt{X}$. Using (21·18), the transformed variable

$\sqrt{X}$ has Mean $= g(\lambda) = \sqrt{\lambda}$. and $\quad \text{Var}(\sqrt{X}) = [g'(\lambda)]^2 \cdot \psi(\lambda) = \left(\frac{1}{2\sqrt{\lambda}}\right)^2 . \lambda = \frac{1}{4}$. . (**)

or alternatively $\quad \text{Var}\left(\sqrt{X}\right) = c^2 = (1/2)^2 = 1/4.$ [From (21-18 *a*) and (*)]

Hence, $\quad \sqrt{X} \sim N\left(\sqrt{\lambda}, 1/4\right)$, asymptotically. ... (21·19*a*)

Ancombi has suggested the transformation $\sqrt{X + b}$, where b is a constant suitably chosen.

21·2·3. Sine Inverse or sin $^{-1}$ Transformation

Sine-inverse is the transformation for stabilizing the variance of a binomial variate. If p is the observed proportion of successes in a series of n independent trials with constant probability P of success for each trial then we know that the asymptotic distribution of p is normal (as $n \to \infty$) with

$$E(p) = P \quad \text{and Var}(p) = PQ/n,\; Q = 1 - P,$$

i.e., $$p \sim N\left(P, \frac{PQ}{n}\right), \text{ as } n \to \infty.$$

In the usual notations: $\quad \psi(P) = \text{Var}(p) = \frac{PQ}{n} = \frac{P(1-P)}{n}$

Using (21·19) , the transforming function is

$$g(P) = \int \frac{c}{\sqrt{\psi(P)}}\, dP = c\sqrt{n} \int \frac{dP}{\sqrt{P(1-P)}}$$

$$= 2c\sqrt{n}\,\sin^{-1}\left(\sqrt{P}\right) \qquad \left[\because \frac{d}{dP}\sin^{-1}\left(\sqrt{P}\right) = \frac{1}{2\sqrt{P}\,\sqrt{1-P}}\right] \quad ... (i)$$

Choosing the constant c so that: $\quad 2c\sqrt{n} = 1 \quad \Rightarrow \quad c = \frac{1}{2\sqrt{n}}$ ... (*ii*)

We get : $\quad g(P) = \sin^{-1}\left(\sqrt{P}\right)$ [From (*i*)]

Hence, the transformed statistic is $g(p) = \sin^{-1}.\left(\sqrt{p}\right)$ Using (21·18), $g(p)$ has

$$\text{Mean} = g(P) = \sin^{-1}\sqrt{P}$$

and $\quad \text{Var}\left(\sin \sqrt{p}\right) = [g'(P)] \cdot \psi(P) = \left[\dfrac{1}{2\sqrt{P}\sqrt{1 \quad P}}\right] \times \dfrac{P(1 \quad P)}{n} = \dfrac{1}{n}$

or $\quad \text{Var}\left(\sin^{-1} \sqrt{p}\right) = c^2 = \left(\dfrac{1}{\sqrt{\ }}\right) = \dfrac{1}{\ }$ [From (21.18 *a*) and (*ii*)]

Hence, $\quad \sin^{-1}\sqrt{p} \sim N\left(\sin^{-1}\sqrt{P}, \dfrac{1}{4n}\right)$, asymptotically ... (21·20)

If r is the observed number of successes in n trials so that so that $p = r/n$, then Ancombi has suggested that instead of $\sin^{-1}\left(\sqrt{p}\right) = \sin^{-1}\sqrt{r/n}$, the transformation should be

$$\sin^{-1}\sqrt{\frac{r + (3/8)}{n + (3/8)}} \qquad ...(21·21)$$

21·2·4· Logarithmic Transformation

Log transformation is the transformation for stabilizing the variance of the distribtuion of sample variance. If s^2 is the sample variance in a sample of size n from normal population with variance σ^2, then the sampling distribution of s^2 is asymptotically normal (as $n \to \infty$) with

$$E(s^2) = \sigma^2 \qquad \text{and} \qquad \text{var}\left(s^2\right) = \frac{2\sigma^4}{n} \quad \text{(for large } n\text{)},$$

[*c.f.* Remark to Theorem 15·4]. In the usual notations we have $\quad \psi\left(\sigma^2\right) = 2\sigma^4/n$

Using (21·19), the transforming function is: $g\left(\sigma^2\right) = \displaystyle\int \frac{c\sqrt{n}}{\sqrt{2}\,\sigma^2}\, d\sigma^2 = \frac{c\sqrt{n}}{\sqrt{2}} \log \sigma^2.$

We select c in such a way that

$$\frac{c\sqrt{n}}{\sqrt{2}} = 1 \quad \Rightarrow \quad c = \frac{\sqrt{2}}{\sqrt{n}}, \quad \text{so that} \quad g\left(\sigma^2\right) = \log_e \sigma^2. \qquad ... (*)$$

Hence, the transformation for the statistic s^2 is $g(s^2) = \log s^2$ and using (21·19) the transformed statistic is normally distributed with mean $g(\sigma^2) = \log \sigma^2$ and

$$\text{Var}\left[g\left(s^2\right)\right] = \left\{g'\left(\sigma^2\right)\right\}^2 \cdot \psi\left(\sigma^2\right) = \left(\frac{1}{\sigma^2}\right)^2 \cdot \frac{2\sigma^4}{n} = \frac{2}{n}$$

or $\quad \text{Var}\left[g\left(s^2\right)\right] = c^2 = \left(\dfrac{\sqrt{2}}{\sqrt{n}}\right)^2 = \dfrac{2}{n}$ [From (21·18 *a*) and (*)]

Hence, $\quad \log_e s^2 \sim N\left(\log_e \sigma^2, \dfrac{2}{n}\right)$, for large. ... (21·22)

21·2·5. Fisher's Z-Transformation

This transformation is suggested for stabilizing the variance of sampling distribution of correlation coefficient (*c.f.* chapter 15). If r is the sample correlation coefficient in sampling from a correlated bivariate normal population with correlation coefficient ρ, then the asymptotic distribution of r, as $n \to \infty$ is normal with:

$$E(r) = \rho \quad \text{and} \quad \text{Var}(r) = \frac{(1-\rho^2)^2}{n} = \psi(\rho), \text{ for large } n.$$

Using (21·19), we get

$$g(\rho) = \int \frac{\sqrt{n}\,c}{1-\rho^2}\,d\rho = \frac{\sqrt{n}\,c}{2} \log_e \left(\frac{1+\rho}{1-\rho}\right).$$

We select c in such a way that

$$\sqrt{n}\,c = 1 \Rightarrow \qquad c = \frac{1}{\sqrt{n}}, \qquad \text{so that} \qquad g(\rho) = \frac{1}{2} \log_e \left(\frac{1+\rho}{1-\rho}\right). \qquad (**)$$

Hence, using (21·19) the transformed statistic is:

$g(r) = \frac{1}{2} \log_e \left(\frac{1+r}{1-r}\right)$', which is denoted by Z, is normally distributed with mean

$$g(\rho) = \frac{1}{2} \log_e \left(\frac{1+\rho}{1-\rho}\right) \quad \text{and} \quad Var[g(r)] = c^2 = \frac{1}{n} \qquad [\text{From } (**)]$$

or
$$\text{Var}[g(r)] = [g'(\rho)]^2\, \psi(\rho) = \left(\frac{1}{1-\rho^2}\right)^2 \cdot \frac{(1-\rho^2)^2}{n} = \frac{1}{n}, \text{ for large } n.$$

Hence,
$$Z = \frac{1}{2} \log_e \left[\frac{1+r}{1-r}\right] \sim N\left[\frac{1}{2} \log_e \left(\frac{1+\rho}{1-\rho}\right), \frac{1}{n}\right], \quad \text{for large } n. \qquad \text{... (21·23)}$$

Prof. *R.A.* Fisher proved that the transformed statistic $Z = g(r)$ is normally distributed even if n is small and that for exact samples (small n),

$$Z \sim N\left[\frac{1}{2} \log_e \left(\frac{1+\rho}{1-\rho}\right), \frac{1}{n-3}\right]$$

21·3· CURVE FITTING BY ORTHOGONAL POLYNOMIALS

Suppose that the polynomial of pth degree of Y on X is

$$Y = a_0 + a_1 X + a_2 X^2 + \ldots + a_p X^p \qquad \text{... (21·24)}$$

The normal equations for determining the constants a_i' s are obtained by the principle of least squares by minimising the residual or error sum of squares :

$$E = \Sigma \left(y - a_0 - a_1 x - a_2 x^2 - \ldots - a_p x^p\right)^2, \qquad \text{... (21·24a)}$$

summation being extended over the given set of observations. The normal equations are:

$$\frac{\partial E}{\partial a_j} = 0, (j = 0, 1, 2, \ldots, p)$$

i.e.,
$$\sum x^j \left(y - a_0 - a_1 x - a_2 x^2 - \ldots - a_p x_p\right) = 0; \; [j = 0, 1, 2, \ldots p] \qquad \text{... (21·25)}$$

Assume that X and Y are measured from their means (and this we can do without any loss of generality) so that:

$$\mu_r = \mu'_r = E(X^r) = \frac{1}{N} \Sigma x^r$$

and write $$\mu_{j1} = \frac{1}{N} \Sigma x^j y$$

where N is number of observations taken on each of the variables X and Y.

Hence, (21·25) gives

$$\mu_{j1} - a_0 \mu_j - a_1 \mu_{j+1} - a_2 \mu_{j+2} - \dots - a_p \mu_{j+p} = 0; \; j = 0, 1, 2, \dots, p$$

$$\Rightarrow \quad a_0 \mu_j + a_1 \mu_{j+1} + a_2 \mu_{j+2} + \dots + a_p \mu_{j+p} = \mu_{j1}; \; j = 0, 1, 2, \dots, p \qquad \dots (21 \cdot 26)$$

Putting $j = 0, 1, 2, \dots, p$, in (21·26) we get respectively

$$\left.\begin{array}{l} a_0\mu_0 + a_1\mu_1 + a_2\mu_2 + \dots + a_p\mu_p = \mu_{01} \\ a_0\mu_1 + a_1\mu_2 + a_2\mu_3 + \dots + a_p\mu_{p+1} = \mu_{11} \\ \vdots \qquad \vdots \qquad \vdots \qquad \vdots \\ a_0\mu_p + a_1\mu_{p+1} + a_2\mu_{p+2} + \dots + a_p\mu_{2p} = \mu_{p1} \end{array}\right\} \qquad \dots (21 \cdot 27)$$

Solving (21·27) for $a_0, a_1, \dots, a_p$ in terms of the moments μ_j's and μ_{j1}'s, $j = 0, 1, 2, \dots, \dots p$ and substituting in (21·24) we get the required curve of best fit.

Let $$\Delta^{(p)} = \begin{vmatrix} \mu_0 & \mu_1 & \mu_2 & \cdots & \mu_p \\ \mu_1 & \mu_2 & \mu_3 & \cdots & \mu_{p+1} \\ \vdots & \vdots & \vdots & & \vdots \\ \mu_p & \mu_{p+1} & \mu_{p+2} & \cdots & \mu_{2p} \end{vmatrix} \qquad \dots (21 \cdot 28)$$

and $\Delta_j^{(p)}$ be the determinant obtained on replacing $(j+1)$th column of $\Delta^{(p)}$ by the column

$$\begin{bmatrix} 01 \\ 11 \\ \\ \end{bmatrix}; \qquad \text{then} \qquad \frac{(\ )}{(\ )} \qquad \dots (21 \cdot 29)$$

The required curve of best fit is the eliminant of a_i's in (21·24) and (21·27).

and is given by $$\begin{vmatrix} Y & 1 & X & X^2 & \cdots & X^p \\ \mu_{01} & \mu_0 & \mu_1 & \mu_2 & \cdots & \mu_p \\ \mu_{11} & \mu_1 & \mu_2 & \mu_3 & \cdots & \mu_{p+1} \\ \vdots & \vdots & \vdots & & & \\ \mu_{p1} & \mu_p & \mu_{p+1} & \mu_{p+2} & \cdots & \mu_{2p} \end{vmatrix} = 0 \qquad .. (21 \cdot 29a)$$

The use of equation (21·29*a*) is subject to one serious drawback. If we have a set of data and apart from inspection if there is no guide regarding the order of the polynomial to be fitted, the only way left to us is to try curves of order 1, 2, 3, ... until we reach the point where further terms do not improve the fit. Every time we add a new term, the a_j's given by (21·29) change and accordingly the determinantal arithmetic has to be done afresh. For example, if we want to fit a polynomial curve of third or higher degree to the same data then we cannot use the coefficients which we computed while fitting a second degree parabola. To overcome this drawback Prof. R. A. Fisher suggested a method which involved the fitting of *Orthogonal Polynomials* by the principle of least squares, so that each term is independent of the other, *i.e.*, each of the coefficients in the polynomial is independent of the other so that each of them can be calculated independently. In this method, the coefficients computed earlier remain the same and we have to compute the coefficient only for the added term.

21·3·1· Orthogonal Polynomials (Definition.)

Two polynomials $P_1(x)$ and $P_2(x)$ are said to be orthogonal to each other if

$$\Sigma P_1(x)\, P_2(x) = 0, \qquad \text{... (21·30)}$$

where summation is taken over a specified set of values of x. If x were a continuous variable in the range from a to b, the condition for orthogonality gives :

$$\int_a^b P_1(x) P_2(x)\, dx = 0 \qquad \text{... (21·30a)}$$

For example, if we take :

$$P_0 = 1,\ \ P_1(x) = x - 4,\ \ P_2(x) = x^2 - 8x + 12,\ \ P_3(x) = x^3 - 12x^2 + 41x - 36 \qquad \text{(21·30b)}$$

then these are orthogonal to each other for a set of integral values of x from 1 to 7 as explained in the following table. Other examples of orthogonal polynomials are Hermite polynomials [§ 21·4], Gram Charlier's polynomials [§ 21·7], Legender's polynomials, etc.

x	P_0P_1	P_0P_2	P_0P_3	P_1P_2	P_1P_3	P_2P_3
1	– 3	5	– 6	– 15	18	– 30
2	– 2	0	6	0	– 12	0
3	– 1	– 3	6	3	– 6	– 18
4	0	– 4	0	0	0	0
5	1	– 3	– 6	– 3	– 6	18
6	2	0	– 6	0	– 12	0
7	3	5	6	15	18	30
Total	0	0	0	0	0	0

21·3·2. Fitting of Orthogonal Polynomials

The pth degree polynomial (21·24) can be rewritten as:

$$Y = b_0 P_0 + b_1 P_1 + b_2 P_2 + ... + b_p P_p \qquad \text{...(21·31)}$$

where P's are polynomials in x, P_j being a polynomial of degree j, (j = 0, 1, 2, ..., p). We shall determine P's so that they satisfy the condition of orthogonality, viz.,

$$\Sigma P_j P_k = \sum_x P_j(x) P_k(x) = 0\ ; j \neq k, \qquad \text{(21·32)}$$

the summation being extended over the observed values of x. The normal equations for estimating the constants b_j's in (21·31) are obtained on minimising

$$E = \Sigma \left(y - b_0P_0 - b_1P_1 - ... - b_jP_j - \ ... \ - b_pP_p\right)^2 \qquad \text{... (21·33)}$$

and are given by : $\dfrac{\partial E}{\partial b_j} = 0 \quad \Rightarrow \quad \Sigma P_j\left(y - b_0P_0 - b_1P_1 - ... - b_pP_p\right) = 0;\ j = 0, 1, 2, ..., p$

Simplifying and using (21·32), we get

$$\Sigma P_j \cdot y - b_j \Sigma P_j^2 = 0. \quad \Rightarrow \quad b_j = \frac{\Sigma y P_j}{\Sigma P_j^2};\ \ j = 0, 1, 2, ..., p. \qquad \text{... (21·34)}$$

Thus, b_j is determined by P_j. If having fitted a curve of order p, we wish to go a step further by adding a term $b_{p+1} P_{p+1}$, the coefficients already obtained in (21·34) remain unaltered.

Moreover, the use of orthogonal polynomials will give us a very convenient method of determining, step by step, the goodness of fit of the polynomial curve. For pth degree polynomial (21·31), the error sum of squares is [*c.f.* (21·33)]

$$E = \Sigma\left(y - b_0P_0 - b_1P_1 - \ldots - b_pP_p\right)^2$$

$$= \Sigma y^2 + b_0^2\,\Sigma P_0^2 + b_1^2\,\Sigma P_1^2 + \ldots + b_p^2\,\Sigma P_p^2 - 2b_0\,\Sigma y P_0 - 2b_1\Sigma y P_1 - \ldots - 2b_p\,\Sigma y P_p,$$

other terms vanish because of orthogonality conditions (21·32). Using (21·34), we finally obtain

$$E = \Sigma y^2 - b_0^2\,\Sigma P_0^2 - b_1^2\,\Sigma P_1^2 - \ldots - b_p^2\,\Sigma P_p^2 \qquad \ldots (21\cdot35)$$

Thus, the effect of adding any term b_jP_j is to reduce the error (residual) sum of squares E by $b_j^2\,\Sigma P_j^2$ and we may examine the effect of this term on E separately. If we find that the addition of any term $b_p\,P_p$ does not reduce E significantly, we may conclude that it is not desired (as far as the representation of the given data by a polynomial curve is concerned).

21·3·3. Finding The Orthogonal Polynomial P_p, in Fitting a Polynomial of Degree p

Let P_p, the polynomial of degree p in x be given by :

$$P_p = \sum_{j=0}^{p} c_{pj}x^j \qquad \ldots (21\cdot36)$$

This contains $(p+1)$ unknown constants $c_{p0}, c_{p1}, \ldots, c_{pp}$. Hence, in all the polynomials in (21·31) up to and including those of pth order, there are :

$$1+2+3+\ldots+(p+1) = \frac{(p+1)\,(p+2)}{2}$$

unknown constants. The orthogonality conditions $\quad \Sigma P_iP_j = 0, i \neq j = 0, 1, 2, \ldots, p,$

provide $\binom{p+1}{2} = \frac{(p+1)p}{2}$ conditions on the c's so that there are

$$\frac{(p+1)(p+2)}{2} - \frac{(p+1)p}{2} = p+1,$$

constants which can be assigned arbitrarily. We will take one for each polynomial $P_j\,(j=0,1,2,\ldots p)$ and assign it such that the coefficient of x^j in P_j is unity *i.e.,*

$$c_{jj} = 1,\ j = 0, 1, 2, \ldots, p, \qquad \ldots (21\cdot37)$$

In particular $\quad c_{00} = P_0 = 1.$

The orthogonality conditions give:

$$\sum_x P_pP_j = 0,\ j < p\ (j=0,1,2,\ldots,p-1). \qquad (21\cdot38)$$

$j = 0$, gives $\sum P_pP_0 = 0 \Rightarrow \sum P_p = 0, \qquad (\because P_0 = 1)\ldots (*)$

$j = 1$, gives $\sum P_pP_1 = 0 \Rightarrow \sum P_p\,(x+k) = 0$ or $\sum P_p\,.\,x + k\sum P_p = 0$

$\Rightarrow \sum xP_p = 0 \qquad$ [Using (*)] $\qquad \ldots (**)$

$j = 2$, gives $\sum P_p\,P_2 = 0 \Rightarrow \sum P_p\,(x^2 + k_1\,x + k_2) = 0$

$\Rightarrow \sum x^2.\ P_p = 0 \qquad$ [Using (*) and (**)]

Similarly proceeding, we shall get , in general,

$$\sum_x P_px^r = 0,\ r = 0,1,2,\ldots,p-1. \qquad \ldots (21\cdot39)$$

$$\Rightarrow \qquad \sum_x \left[\left(\sum_{j=0}^{p} c_{pj} \, . x^j \right) x^r \right] = 0 \text{ or } \sum_{j=0}^{p} \left(c_{pj} \sum_x x^{j+r} \right) = 0$$

Dividing both sides by N, the number observations on each of the variables X and Y, we get :

$$\sum_{j=0}^{p} c_{pj} \, \mu_{j+r} = 0; \; r = 0, 1, 2, ..., (p-1) \qquad ... (21·40)$$

where x is assumed to be measured from mean. Putting $r = 0, 1, 2, ..., (p-1)$ in, (21·40), we get respectively

$$\begin{matrix} c_{p0}\,\mu_0 & + & c_{p1}\,\mu_1 & + ... + & c_{pj}\,\mu_j & + ... + & c_{pp-1}\,\mu_{p-1} & + & c_{pp}\,\mu_p & = 0 \\ c_{p0}\,\mu_1 & + & c_{p1}\,\mu_2 & + ... + & c_{pj}\,\mu_{j+1} & + ... + & c_{pp-1}\,\mu_p & + & c_{pp}\,\mu_{p+1} & = 0 \\ \vdots & & \vdots & & \vdots & & \vdots & & \vdots & \\ c_{p0}\,\mu_{p-1} & + & c_{p1}\,\mu_p & + ... + & c_{pj}\,\mu_{j+p-1} & + ... + & c_{pp-1}\,\mu_{2p-2} & + & c_{pp}\,\mu_{2p-1} & = 0 \end{matrix}$$

Noting that $C_{pp} = 1$, solving the above equations for c's, we get

$$c_{pj} = \frac{\begin{vmatrix} \mu_0 & \mu_1 & \cdots & -\mu_p & \cdots & \mu_{p-1} \\ \mu_1 & \mu_2 & \cdots & -\mu_{p+1} & \cdots & \mu_p \\ \vdots & \vdots & & \vdots & & \vdots \\ \mu_{p-1} & \mu_p & \cdots & -\mu_{2p-1} & \cdots & \mu_{2p-2} \end{vmatrix}}{\begin{vmatrix} \mu_0 & \mu_1 & \cdots & \mu_j & \cdots & \mu_{p-1} \\ \mu_1 & \mu_2 & \cdots & \mu_{j+1} & \cdots & \mu_p \\ \vdots & \vdots & & \vdots & & \vdots \\ \mu_{p-1} & \mu_p & \cdots & \mu_{j+p-1} & \cdots & \mu_{2p-2} \end{vmatrix}} = \frac{\Delta^{(p)}_{pj}}{\Delta^{(p-1)}}, \qquad ...(21·41)$$

where $\Delta^{(p)}$ has been defined in (21·28) and $\Delta^{(p)}{}_{pj}$ is the minor of the element in the last row and (j + 1)th column in $\Delta^{(p)}$. Substituting this value of c_{pj} in (21·36), we get

$$P_p = \sum_{j=0}^{p} c_{pj}\, x^j = \sum_{j=0}^{p} \frac{\Delta^{(p)}_{pj}}{\Delta^{(p-1)}} . \, x^j = \frac{1}{\Delta^{(p-1)}} \sum_{j=0}^{p} \Delta^{(p)}{}_{pj} . x^j$$

$$= \frac{1}{\Delta^{(p-1)}} \left[\Delta^{(p)}{}_{p0} + x\Delta^{(p)}{}_{p1} + x^2 \Delta^{(p)}{}_{p2} + ... + x^p . \Delta^{(p)}{}_{pp} \right]$$

$$P_p = \frac{1}{\Delta^{(p-1)}} \begin{vmatrix} \mu_0 & \mu_1 & \mu_2 & \cdots & \mu_p \\ \mu_1 & \mu_2 & \mu_3 & \cdots & \mu_{p+1} \\ \vdots & \vdots & \vdots & \vdots & \vdots \\ \mu_{p-1} & \mu_p & \mu_{p+1} & \cdots & \mu_{2p-1} \\ 1 & x & x^2 & \cdots & x^p \end{vmatrix} \qquad ... (21·42)$$

In particular if $\mu_0 = 1, \mu_1 = 0$ and $\mu_2 = 1$, *i.e.*, if x is a standardised variate then the orthogonal polynomials are given by

$$P_0 = 1; \qquad P_1(x) = \frac{\begin{vmatrix} \mu_0 & \mu_1 \\ 1 & x \end{vmatrix}}{\mu_0} = x \qquad ... (21·42\,a)$$

$$P_2(x) = \frac{\begin{vmatrix} \mu_0 & \mu_1 & \mu_2 \\ \mu_1 & \mu_2 & \mu_3 \\ 1 & x & x^2 \end{vmatrix}}{\begin{vmatrix} \mu_0 & \mu_1 \\ \mu_1 & \mu_2 \end{vmatrix}} = x^2 - \mu_3 x - 1 \quad [\because \mu_0 = 1, \mu_1 = 0, \mu_2 = 1] \qquad (21·42\,b)$$

$$P_3(x) = \begin{vmatrix} \mu_0 & \mu_1 & \mu_2 & \mu_3 \\ \mu_1 & \mu_2 & \mu_3 & \mu_4 \\ \mu_2 & \mu_3 & \mu_4 & \mu_5 \\ 1 & x & x^2 & x^3 \end{vmatrix} \div \begin{vmatrix} \mu_0 & \mu_1 & \mu_2 \\ \mu_1 & \mu_2 & \mu_3 \\ \mu_2 & \mu_3 & \mu_4 \end{vmatrix} \qquad \ldots (21·42c)$$

and so on.

If we further assume that x is a standard normal variate so that $\mu_3 = \mu_5 = \ldots = \mu_{2r+1} = 0$, then the above orthogonal polynomials are called Hermite polynomials and are given by

$$P_0 = 1\,;\; P_1(x) = x\,;\; P_2(x) = x^2 - 1;\; P_3(x) = x^3 - 3x;\; P_4(x) = x^4 - 6x^2 + 3; \qquad \ldots (21·42d)$$

and so on, where x is a continuous *r.v.* taking values from $-\infty$ to ∞.

Hermite Polynomials defined in (21·42*d*) are orthogonal w.r.t. the weight function

$$\alpha(x) = \frac{1}{\sqrt{2\pi}} \exp -\left(\frac{x^2}{2}\right);\; -\infty < x < \infty \quad i.e., \quad \int_{-\infty}^{\infty} P_i(x)\, P_j(x)\, \alpha(x)\, dx = 0;\, i \neq j \qquad \ldots (21·42e)$$

where $P_1(x), P_2(x), P_3(x), P_4(x)$ are defined in (21·42*d*).

Note. For detailed discussion on Hermite polynomials, see § (21·4)

21·3·4. Determination of the Coefficients b_j's in (21·31).

From (21·34), we get

$$b_p = \sum y P_p \Big/ \sum {P_p}^2 \qquad (21·43)$$

$$\sum P_p^2 = \sum P_p \cdot P_p = \sum_x \{P_p [c_{p0} + c_{p1}x + c_{p2}x^2 + \ldots + c_{pp}x^p]\} = \sum_x P_p \cdot x^p,$$

on using (21·39) and the fact that $c_{pp} = 1$.

$$\therefore \sum P_p^2 = \sum_x \left\{\left(\sum_{j=0}^{p} c_{pj}\, x^j\right) x^p\right\} = \sum_{j=0}^{p} \left(c_{pj} \sum_x x^{p+j}\right)$$

$$= N \sum_{j=0}^{p} c_{pj}\, \mu_{p+j} = N \sum_{j=0}^{p} \frac{\Delta^{(p)}_{pj}}{\Delta^{(p-1)}} . \mu_{p+j} \qquad \text{[From (21·41)]}$$

$$= \frac{N}{\Delta^{(p-1)}} \begin{vmatrix} \mu_0 & \mu_1 & \cdots & \mu_p \\ \mu_1 & \mu_2 & \cdots & \mu_{p+1} \\ \vdots & \vdots & & \vdots \\ \mu_{p-1} & \mu_p & \cdots & \mu_{2p-1} \\ \mu_p & \mu_{p+1} & \cdots & \mu_{2p} \end{vmatrix} = \frac{N\Delta^{(p)}}{\Delta^{(p-1)}} \qquad \text{[From (21·28)]} \qquad \ldots (21·44)$$

[Proceeding exactly as we obtained (21·42)]

Similarly, $\sum y P_p = N \sum_{j=0}^{p} \frac{\Delta^{(p)}{}_{pj}}{\Delta^{(p-1)}} \cdot \mu_{j1}$

$$= \frac{N}{\Delta^{(p-1)}} \begin{vmatrix} \mu_0 & \mu_1 & \cdots & \mu_p \\ \mu_1 & \mu_2 & \cdots & \mu_{p+1} \\ \vdots & \vdots & & \vdots \\ \mu_{p-1} & \mu_p & \cdots & \mu_{2p-1} \\ \mu_{01} & \mu_{11} & \cdots & \mu_{p1} \end{vmatrix} = \frac{N\Delta_j^{(p)}}{\Delta^{(p-1)}} \qquad \text{...(21·44}a\text{)}$$

where $\Delta^{(p)}$ and $\Delta^{(p)}{}_j$ are defined in (21·28.) Substituting in (21·43) we get,

$$b_p = \frac{\Delta_j^{(p)}}{\Delta^{(p)}} \qquad \text{... (21·44}b\text{)}$$

If the variable x takes the integral values 1, 2, ..., N, then the first seven of these orthogonal polynomials P_j's; $j = 0, 1, 2, 3, ..., 6$ are given by :

$$P_0(x) = 1, \quad P_1(x) = \lambda_1 . \xi, \quad P_2(x) = \lambda_2 \left\{\xi^2 - \frac{N^2-1}{12}\right\}, \quad P_3(x) = \lambda_3 \left\{\xi^3 - \frac{3N^2-7}{20}\xi\right\};$$

$$P_4(x) = \lambda_4 \left\{\xi^4 - \frac{3N^2-13}{14}\xi^2 + \frac{3}{560}\left(N^2-1\right)\left(N^2-9\right)\right\}$$

$$P_5(x) = \lambda_5 \left\{\xi^5 - \frac{5}{18}\left(N^2-7\right)\xi^3 + \frac{1}{1,008}\left(15N^4 - 230N^2 + 407\right)\xi\right\}$$

$$P_6(x) = \lambda_6 \left\{\xi^6 - \frac{5}{44}\left(3N^2-31\right)\xi^4 + \frac{1}{176}\left(5N^4 - 110N^2 + 329\right)\xi^2 - \frac{5}{14,784}\left(N^2-1\right)\left(N^2-9\right)\left(N^2-25\right)\right\} \qquad \text{... (21·44 }c\text{)}$$

and so on, where $\xi = x - \bar{x}$ so that $\Sigma\, \xi = 0$ and λ_i' s are arbitrary constants.

If $y = b_0 + b_1 P_1(x) + b_2 P_2(x) + ... + b_p P_p(x)$, is the orthogonal polynomial fitted to the given data then, using (21·34), we get

$$b_0 = \frac{\sum y P_0}{\sum P_0^2} = \frac{\sum y}{N}\ ; \ (\because P_0 = 1)\ ; \qquad b_i = \frac{\sum_x y P_i}{\sum_x P_i^2}\ ; \ (i = 1, 2, ... p) \qquad \text{... (21·45)}$$

The origin of P_i's is so chosen that $\Sigma\, P_i = 0$.

If N, the number of observations is odd, then we take : $\quad \xi_i = \dfrac{x_i - A}{h}$

and if N is even then we take : $\quad \xi = \dfrac{x_i - A_1}{(h/2)}$

where $\quad h$ = length of the interval (for values of x)

$\quad A$ = middle value (item) of the data

and $\quad A_1$ = Arithmetic mean of two middle values of the data.

The values of P_i's and λ_i's are obtained from 'Statistical Tables' by R.A. Fisher for the values of N from 3 to 75. In these tables the orthogonal polynomials P_i's are denoted by φ_i's. We reproduce below these tables for $N = 3$ to $N = 6$.

TABLE OF ORTHOGONAL POLYNOMIALS

	$N = 3$		$N = 4$			$N = 5$			
	φ_1	φ_2	φ_1	φ_2	φ_3	φ_1	φ_2	φ_3	φ_4
	− 1	1	− 3	1	− 1	− 2	2	− 1	1
	0	− 2	− 1	− 1	+ 3	− 1	− 1	2	−4
	1	1	1	− 1	− 3	0	− 2	–	6
			3	1	1	1	− 1	−2	−4
						2	2	1	1
$\sum_x \varphi_i^2$	2	6	20	4	20	10	14	10	70
λ_i	1	3	2	1	3	1	1	$\frac{5}{6}$	$\frac{35}{22}$

	$N = 6$				
	φ_1	φ_2	φ_3	φ_4	φ_5
	− 5	5	−5	1	− 1
	−3	−1	7	−3	5
	−1	−4	4	2	−10
	1	−4	−4	+2	10
	3	−1	−7	−3	− 5
	5	5	5	1	1
$\sum_x \varphi_i^2$	70	84	180	28	252
λ_i	2	$\frac{3}{2}$	$\frac{5}{3}$	$\frac{7}{12}$	$\frac{21}{10}$

Example 21.2 *Fit a straight line $y = a + bx$... (*) to the following data by using orthogonal polynomials.*

x :	0	1	2	3	4
y :	1	1·8	3·3	4·5	6·3

Solution. Here $N = 5$. Let us transform to the variable $\xi = \dfrac{x-2}{1} = x - 2$ so that $\Sigma \xi = 0$

Let the orthogonal polynomial form of straight line (*) be

$$y = b_0 + b_1 P_1(x) = b_0 + b_1 \phi_1(x) \qquad \text{... (**)}$$

x	$\xi = x - 2$	y	ϕ_1	$y\,\phi_1$
0	-2	1	-2	-2
1	-1	1·8	-1	$-1{\cdot}8$
2	0	3·3	0	0
3	1	4·5	1	4·5
4	2	6·3	2	12·6
Total		16·9	0	13·3

The values of ϕ_1 are noted from the tables for $N = 5$, the given number of points (x, y). From tables we also find $\sum \phi_1{}^2 = 10,\ \lambda_1 = 1.$

Now using (21·45), $b_0 = \dfrac{\Sigma y}{N} = \dfrac{16{\cdot}9}{5} = 3{\cdot}38\,;$ $\qquad b_1 = \dfrac{\Sigma y\,\phi_1}{\Sigma\,\phi_1{}^2} = \dfrac{13{\cdot}3}{10} = 1{\cdot}33$

$$\phi_1(x) = \lambda_1\,\xi = 1.\,(x-2) = x - 2 \qquad \text{[From (21·44}c\text{)]}$$

Substituting in (**), the required straight line is

$$y = 3{\cdot}38 + 1{\cdot}33\,(x-2) \quad \Rightarrow \quad y = 1{\cdot}33x + 0{\cdot}72$$

Example 21·3. *Fit a second degree parabola to the following data, using the method of orthogonal polynomials:*

x :	0·5	1·0	1·5	2·0	2·5	320
y :	72	110	158	214	290	380

Solution. Let the second degree parabola be : $\qquad y = a + bx + cx^2 \qquad$ ... (*)

and its orthogonal polynomial transform be : $\qquad y = b_0 + b_1\,\phi_1(x) + b_2\,\phi_2(x) \qquad$ ... (**)

Here we have, $N = 6$. Let us transform to

$$\xi = \frac{x - \frac{1}{2}(1{\cdot}5 + 2{\cdot}0)}{\frac{1}{2}(0{\cdot}5)} = 4\,(x - 1{\cdot}75) = 4x - 7, \text{ so that } \Sigma\,\xi = 0.$$

From Fisher's tables we note the values of ϕ_1 and ϕ_2 (as given in the following table) and also

$$\sum \phi_1^2 = 70, \quad \sum \phi_2^2 = 84; \quad \lambda_1 = 2, \quad \lambda_2 = 3/2$$

x	$\xi = 4x - 7$	y	ϕ_1	ϕ_2	$y\,\phi_1$	$y\,\phi_2$
0·5	-5	72	-5	5	-360	360
1·0	-3	110	-3	-1	-330	-110
1·5	-1	158	-1	-4	-158	-632
2·0	1	214	1	-4	214	-856
2·5	3	290	3	-1	870	-290
3·0	5	380	5	5	1900	1900
Total		1224	0	0	2136	372

$$b_0 = \frac{\sum y}{N} = \frac{1,224}{6} = 204\,; \qquad b_1 = \frac{\sum y\phi_1}{\sum \phi_1^2} = \frac{2,136}{70} = 30{\cdot}51$$

$$b_2 = \frac{\sum y\phi_2}{\sum \phi_2^2} = \frac{372}{84} = 4{\cdot}43; \qquad \phi_1(x) = \lambda_1\,\xi = 2\,[4x-7] = 8x-14$$

$$\phi_2(x) = \lambda_2\left(\xi^2 - \frac{N^2-1}{12}\right) = \frac{3}{2}\left\{(4x-7)^2 - \frac{36-1}{12}\right\}$$

$$= \frac{3}{2}\left(16x^2 + 49 - 56x - \frac{35}{12}\right) = 24x^2 - 84x + 69{\cdot}125$$

Substituting in (**), $y = 204 + 30{\cdot}51\,(8x-14) + 4{\cdot}43\,(24x^2 - 84x + 69{\cdot}125)$

$\Rightarrow$ $y = 106{\cdot}32x^2 - 128{\cdot}04x + 83{\cdot}08$

which is the required second degree parabola of best fit.

21·4. HERMITE POLYNOMIALS

In mathematics Hermite polynomials are a classical orthogonal polynomial sequence. These polynomials are used :

(i) In Probability/Statistics such as in

(a) Gram Charlier's and Edgewoth's form of Type A series representation of a frequency function,

(ii) In 'Numerical Analysis' as Gaussian Quadrature,

(iii) In Combinatorics and,

(iv) In Physics (in harmonic oscillator in Quantum Mechanics).

Hermite polynomials were defined by Laplace (1810) and studied in details by Chebyshev (1859). They were named later after Charles Hermite (24 Dec. 1822 to 14 Jan. 1901), a French mathematician, who wrote on the polynomials in 1864.

Definition. **Hermite Polynomials** or Chebyshev's Hermite polynomial $H_n(x)$ of degree n is defined by the equation : $H_n(x)\,\alpha(x) = (-D)^n \cdot \alpha(x)$ ... (21·46)

where, $$\alpha(x) = \frac{1}{\sqrt{2\pi}}e^{-x^2/2} \quad \text{and} \quad D^n \equiv \left(\frac{d}{dx}\right)^n \qquad \text{... (21·46 } a)$$

$\Rightarrow$ $$H_n(x) = \frac{1}{\alpha(x)} \cdot (-D)^n\,\alpha(x) \qquad \text{... (21·47)}$$

or $$H_n(x) = e^{x^2/2} \cdot (-D)^n e^{-x^2/2}\ ; \quad n = 0, 1, 2, \cdots \qquad \text{... (21·47 } a)$$

By convention, we take $H_0(x) = 1$.

Generating Function for $H_n(x)$.

$$\alpha(x-t) = \frac{1}{\sqrt{2\pi}}e^{-(x-t)^2/2} \qquad \text{[From (21·46 } a)]$$

$$= \frac{1}{\sqrt{2\pi}} \cdot e^{-x^2/2} \cdot e^{tx - \frac{t^2}{2}}$$

$$\Rightarrow \qquad \alpha(x-t) = \alpha(x)\cdot e^{tx-\frac{1}{2}t^2}$$

$$\Rightarrow \qquad \alpha(x)\cdot e^{tx-\frac{1}{2}t^2} = \alpha(x-t) = \sum_{n=0}^{\infty} \frac{(-t)^n}{n!} D^n \cdot \alpha(x)$$

[By Taylor's Series expansion]

where, $$D \equiv \frac{d}{dx} \quad \text{and} \quad D^n \equiv \frac{d^n}{dx^n}.$$

$$\alpha(x)\cdot e^{tx-\frac{1}{2}t^2} = \sum_{n=0}^{\infty} \frac{t^n}{n!}(-D)^n \alpha(x)$$

$$\Rightarrow \qquad = \sum_{n=0}^{\infty} \frac{t^n}{n!} H_n(x)\cdot \alpha(x)$$ [From (21·46)]

$$\Rightarrow \qquad e^{tx-\frac{1}{2}t^2} = \sum_{n=0}^{\infty} \frac{t^n}{n!} H_n(x) \qquad \text{...(21·48)}$$

$$\Rightarrow \qquad H_n(x) = \text{Coefficient of } (t^n/n!) \text{ in } e^{tx-\frac{1}{2}t^2}$$

$\therefore$ Generating Function of Hermite Polynomials $= e^{tx-\frac{1}{2}t^2}$ ... (21·49)

Explicit Expression for $H_r(x)$. From (21·48) we get :

$$\sum_{r=0}^{\infty} \frac{t^r}{r!} H_r(x) = e^{tx} \times e^{-\frac{1}{2}t^2} = \sum_{i=0}^{\infty} \frac{(tx)^i}{i!} \times \sum_{j=0}^{\infty} \frac{(-t^2/2)^j}{j!}$$

$$\Rightarrow \qquad 1 + tH_1(x) + \frac{t^2}{2!}H_2(x) + \frac{t^3}{3!}H_3(x) + \cdots + \frac{t^r}{r!}\cdot H_r(x) + \cdots$$

$$\equiv \left[1 + tx + \frac{(tx)^2}{2!} + \frac{(tx)^3}{3!} + \cdots + \frac{(tx)^{r-2j}}{(r-2j)!} + \cdots\right]$$

$$\times \left[1 - \frac{t^2}{2} + \frac{(t^2/2)^2}{2!} - \frac{(t^2/2)^3}{3!} + \cdots + \frac{(-t^2/2)^j}{j!} + \cdots\right] \qquad \text{...(21·50)}$$

Comparing coefficients of like powers of t on both sides of (21·50), we get :

$$H_0(x) \equiv 1\ ; \qquad H_1(x) = \text{Coefficient of } t = x$$

Coefficient of t^2 : $\frac{1}{2}H_2(x) = -\frac{1}{2} + \frac{x^2}{2!}$ $\Rightarrow$ $H_2(x) = x^2 - 1$

Coefficient of t^3 : $\frac{1}{3!}H_3(x) = -\frac{x}{2} + \frac{x^3}{3!}$ $\Rightarrow$ $H_3(x) = x^3 - 3x$

Coefficient of t^4 : $\frac{1}{4!}H_4(x) = \frac{1}{8} - \frac{x^2}{4} + \frac{x^4}{4!}$ $\Rightarrow$ $H_4(x) = x^4 - 6x^2 + 3$

Similarly, we can obtain the higher degree polynomials.

Expression for $H_r(x)$: Comparing coefficient of t^r on both sides of (21·50), we get:

$$\frac{H_r(x)}{r!} = \sum_{j=0}^{r/2}\left[\frac{x^{r-2j}}{(r-2j)!} \times \frac{\left(-\frac{1}{2}\right)^j}{j!}\right] \quad ; \quad r - 2j \geq 0 \Rightarrow j \leq \frac{r}{2}.$$

$$\Rightarrow \qquad H_r(x) = r! \sum_{j=0}^{r/2}\left[\frac{x^{r-2j}\left(-\frac{1}{2}\right)^j}{(r-2j)!\,j!}\right]$$

$$\Rightarrow \qquad H_r(x) = x^r + r!\frac{x^{r-2}}{(r-2)!}\left(-\frac{1}{2}\right) + r!\frac{x^{r-4}}{(r-4)!\,2!}\left(-\frac{1}{2}\right)^2 + \ldots$$

$$\Rightarrow \qquad H_r(x) = x^r - \frac{1}{2}r^{(2)} \cdot x^{r-2} + \frac{r^{(4)}x^{r-4}}{2^2 \cdot 2!} - \frac{r^{(6)}x^{r-6}}{2^3 \cdot 3!} + \cdots \; ; \; r = 0, 1, 2, 3, \qquad \ldots (21·51)$$

where $\quad r^{(k)} = r(r-1)(r-2)\cdots(r-k+1)\cdot$

(21·51) gives the general expression for Hermite polynomial of rth degree; $r = 0, 1, 2, \ldots$.

Putting $r = 0, 1, 2, 3, \ldots$; we can deduce the expressions for $H_0(x)$, $H_1(x)$, $H_2(x)$, $H_3(x)$, $H_4(x)$, as obtained earlier.

The first 11 Hermite polynomials are summarised below:

$$\left.\begin{aligned}
H_0(x) &= 1\\
H_1(x) &= x\\
H_2(x) &= x^2 - 1\\
H_3(x) &= x^3 - 3x\\
H_4(x) &= x^4 - 6x^2 + 3\\
H_5(x) &= x^5 - 10x^3 + 15x\\
H_6(x) &= x^6 - 15x^4 + 45x^2 - 15\\
H_7(x) &= x^7 - 21x^5 + 105x^3 - 105x\\
H_8(x) &= x^8 - 28x^6 + 210x^4 - 420x^2 + 105\\
H_9(x) &= x^9 - 36x^7 + 378x^5 - 1260x^3 + 945x\\
H_{10}(x) &= x^{10} - 45x^8 + 630x^6 - 3150x^4 + 4725x^2 - 945
\end{aligned}\right\} \qquad \ldots (21·52)$$

Note that:

(*i*) $H_n(x)$ is a polynomial of nth degree in x ; $n = 0, 1, 2, \ldots$.

(*ii*) The leading coefficient (*i.e.*, coefficient of x^n in $H_n(x)$) is 1.

(*iii*) Polynomials of even degrees (*viz.*, $n = 2, 4, 6, \ldots$) contain only even degree terms in x and polynomials of odd degrees (*viz.*, $n = 1, 3, 5, \ldots$) contain only odd degree terms in x. Also, the terms in each polynomial alternate in sign.

21·4·1 Properties of Hermite Polynomials

Property 1. From (21·52), we get :

$$H_r(-x) = (-1)^r H_r(x) \; ; \; r = 0, 1, 2, 3, \cdots \qquad \ldots (21·53)$$

$\Rightarrow$ *$H_r(x)$ is an even function of x when r is even and $H_r(x)$ is an odd function of x when r is odd.*

Property 2. Hermite polynomials satisfy the following differential equations:

(*i*) $H_r'(x) = r\, H_{r-1}(x)$ (*ii*) $H_r^{(j)}(x) = r^{(j)} H_{r-j}(x)$ (*iii*) $H_r''(x) - x H_r'(x) + r\, H_r(x) = 0$

where dash (') denotes differentiation w.r.t. x and $H_r^{(j)}(x) = \dfrac{d^j}{dx^j}(H_r(x))$.

Proof. We have [From (21.40)]:

$$e^{tx - \frac{1}{2}t^2} = \sum_{r=0}^{\infty} \frac{t^r}{r!} H_r(x)$$

Differentiating w.r.t. 't', we get

$$(x-t)\, e^{tx - \frac{1}{2}t^2} = \sum_{r=1}^{\infty} \frac{t^{r-1}}{(r-1)!} H_r(x) \qquad \text{... (*)}$$

$$\Rightarrow \quad (x-t) \sum_{r=0}^{\infty} \frac{t^r}{r!} H_r(x) = \sum_{r=1}^{\infty} \frac{t^{r-1}}{(r-1)!} H_r(x) \qquad \text{... (**)}$$

Comparing coefficient of t^r on both sides of (**), we get

$$\frac{x}{r!} H_r(x) - \frac{H_{r-1}(x)}{(r-1)!} = \frac{H_{r+1}(x)}{r!}$$

$$\Rightarrow \quad xH_r(x) - r\, H_{r-1}(x) = H_{r+1}(x)$$

or

$$xH_{r-1}(x) - (r-1)\, H_{r-2}(x) = H_r(x) \qquad \text{... (***)}$$

[Changing r to $(r-1)$ in above equation]

Differentiating (*) 'j' times w.r.t. x, we get

$$t^j e^{tx - \frac{1}{2}t^2} = \sum_{r=0}^{\infty} \frac{t^r}{r!} H_r^{(j)}(x)$$

$$\Rightarrow \quad t^j \sum_{r=0}^{\infty} \frac{t^r}{r!} H_r(x) = \sum_{r=0}^{\infty} \frac{t^r}{r!} H_r^{(j)}(x)$$

Comparing coefficient of t^r on both sides of the above equation, we get

$$\frac{H_{r-j}(x)}{(r-j)!} = \frac{H_r^{(j)}(x)}{r!} \quad \Rightarrow \quad H_r^{(j)}(x) = \frac{r!}{(r-j)!} H_{r-j}(x)$$

$$\therefore \quad H_r^{(j)}(x) = r^{(j)} H_{r-j}(x) \qquad \text{... (21·54)}$$

which established the Property 2 (*ii*).

Taking $j = 1$ and 2 in (21·54), we get respectively:

$$H_r'(x) = r\, H_{r-1}(x) \quad ; \quad H_r''(x) = r(r-1) H_{r-2}(x) \qquad \text{... (21·55)}$$

which proves the Property 2 (*i*).

Substituting the values of $H_{r-1}(x)$ and $H_{r-2}(x)$ from (21·55) in (***), we get

$$\frac{x \cdot H_r'(x)}{r} - \frac{H_r''(x)}{r} = H_r(x)$$

$$\Rightarrow \quad H_r''(x) - x\, H_r'(x) + rH_r(x) = 0 \qquad \text{... (21·56)}$$

which proves the Property 2(*iii*).

Property 3. Orthogonal Property of Hermite Polynomials. Hermite Polynomials are a set of orthogonal polynomials over the domain $(-\infty, \infty)$ with the weighting function $\alpha(x)$, *i.e.*,

$$\int_{-\infty}^{\infty} H_m(x) H_n(x)\, \alpha(x)\, dx = \begin{cases} 0 & ; \quad m \neq n \\ n! & ; \quad m = n \end{cases} \qquad \text{... (21·57)}$$

$$\Rightarrow \quad \int_{-\infty}^{\infty} H_m(x) H_n(x)\, e^{-x^2/2}\, dx = \begin{cases} 0 & ; \quad m \neq n \\ n!\ \sqrt{2\pi} & ; \quad m = n \end{cases} \qquad \text{... (21·57a)}$$

Proof. Let

$$I = \int_{-\infty}^{\infty} H_m(x)\, H_n(x)\, \alpha(x)\, dx \quad ; \quad m < n. \qquad \text{... (21·46)}$$

$$= \int_{-\infty}^{\infty} H_m(x) \cdot (-D)^n \cdot \alpha(x)\, dx \qquad \text{[from (21·46)]}$$

Integrating by parts, we get

$$I = (-1)^n \left| D^{n-1}\alpha(x) \cdot H_m(x) \right|_{-\infty}^{\infty} - (-1)^n \int_{-\infty}^{\infty} D^{n-1}\alpha(x) . H_m'(x) dx$$

$$= (-1)\left[\left\{(-D)^{n-1}\alpha(x)\right\} H_m(x) \Big|_{-\infty}^{\infty} - (-1)^n \int_{-\infty}^{\infty} \left\{D^{n-1}\alpha(x)\right\} . m H_{m-1}(x) dx\right.$$

$$= -\left[H_m(x) H_{n-1}(x) \alpha(x) \right]_{-\infty}^{\infty} + (-1)^2 . m \int_{-\infty}^{\infty} H_{m-1}(x) . H_{n-1}(x) . \alpha(x)\, dx \qquad \text{... (21·58)}$$

$H_m(x)\ H_n\ (x)$ is a polynomial of degree $(m + n - 1)$ in x and since $\alpha(x) = \dfrac{1}{\sqrt{2\pi}} e^{-x^2/2}$, dominates the function,

$$\left| x^N \cdot e^{-x^2/2} \right|_{x=-\infty}^{\infty} = 0, \text{ however large } N, \text{ may be.}$$

The first term in (21·58) $\rightarrow$ 0. ... (21·58a)

$$\therefore \quad I = m \int_{-\infty}^{\infty} H_{m-1}(x) . (-D)^{n-1} \alpha(x) dx$$

By repeated application of this result, we get

$$I = m(m-1) \int_{-\infty}^{\infty} H_{m-2}(x)\, (-D)^{n-2}\, \alpha\, (x)\, dx$$

$$\vdots \qquad \vdots \qquad \vdots$$

$$= m! \int_{-\infty}^{\infty} H_0(x) (-D)^{n-m}\, \alpha(x) dx$$

$$= m! \int_{-\infty}^{\infty} (-D)^{n-m}\, \alpha(x) dx \ ; \quad m < n. \qquad [\because H_0(x) = 1] \qquad \text{... (21·58b)}$$

$$\therefore \quad I = m!(-1)^{n-m} \cdot \left| D^{n-m-1} \alpha(x) \right|_{x=-\infty}^{\infty}$$

$$= m!(-1) \left| (-D)^{n-m-1} \alpha(x) \right|_{-\infty}^{\infty}$$

$$= m!(-1) \left| H_{n-m-1}(x) \; \alpha(x) \right|_{-\infty}^{\infty} \qquad \text{... (21·59)}$$

$$= 0 \;\; ; \;\; m \neq n.$$

[By same argument as in (21·58*a*)]

If $m = n$, then [From (21·58*b*)]

$$I = n! \int_{-\infty}^{\infty} \alpha(x) dx = n! \times 1 = n! \qquad \text{... (21·60)}$$

because $\alpha(x)$ is the p.d.f. of standard normal distribution.

This completes the proof of (21·57) or (21·57*a*).

21·4·2. Physicists' Hermite Polynomials

The definition of Hermite polynomials in (21·46) is conventionally preferred by probabilists or statisticians because $\alpha(x)$ is the p.d.f. of standard normal distribution.

However, the *physicists' Hermite polynomials* are defined by the equation: (*Rodrigues' formula*).

$$h_n(x) = e^{x^2} \left[(-D)^n e^{-x^2} \right]; \; D \equiv \frac{d}{dx} \; ; n = 0, 1, 2, ... \qquad \text{... (21·61)}$$

Thus, Physicists Hermite Polynomials in (21·61) and Probabilists Hermite polynomials are not exactly identical. However, each one is a rescaling of the other and they are connected by the following relation.

$$\left. \begin{aligned} h_n(x) &= 2^{n/2} H_n\left(\sqrt{2}\, x\right) \\ \text{and} \qquad H_n(x) &= 2^{-n/2} \, h_n\left(\frac{x}{\sqrt{2}}\right) \end{aligned} \right\} \qquad \text{... (21·62)}$$

The *generating function for* $h_n(x)$ is given by :

$$e^{2tx - t^2} = \sum_{n=0}^{\infty} \frac{h_n(x) \, . \, t^n}{n!}; \; n = 0, 1, 2, \;\; \cdots, \qquad \text{... (21·63)}$$

where t is a dummy variable.

Also, $$h_n(-x) = (-1)^n \, h_n(x) \;\; ; \;\; n = 0, 1, 2, 3, \cdots \qquad \text{... (21·64)}$$

i.e., $h_n(x)$ *is an even function of x if n is even. and* $h_n(x)$ *is an odd function of x if n is odd.*

The recurrence relation for $h_n(x)$ is given by:

$$h_{n+1}(x) - 2x \, h_n(x) + 2n \, h_{n-1}(x) = 0 \qquad \text{... (21·65)}$$

$h_n(x)$ satisfies the following differential equations:

$$h_n'(x) = 2n \, h_{n-1}(x) \;\; ; \;\; \text{where } h_n'(x) = \frac{d}{dx}\left[h_n(x) \right] \qquad \text{... (21·66)}$$

and $$h_n''(x) - 2x \, h_n'(x) \;\; + \;\; 2n \, h_n(x) = 0 \qquad \text{... (21·67)}$$

Note. *The hermite differential equation* (which arises in the treatment of the harmonic oscillator in quantum mechanics), is given by :

$$y'' - 2xy' + 2ny = 0 \qquad \text{... (*)}$$

where, n is a real number.

If n takes non-negative integer values in (*) *i.e.*, $n = 0, 1, 2, 3, \ldots$, then the solutions of (*) are referred to as the physicists hermite polynomials.

Orthogonality. Physicists Hermite polynomials are orthogonal on the interval $(-\infty, \infty)$ with the weighting function $\exp(-x^2)$ *i.e.*,

$$\int_{-\infty}^{\infty} e^{-x^2} h_m(x)\, h_n(x) dx = \begin{cases} 0 & ; \ m \neq n \\ 2^n \cdot n! \sqrt{\pi} & ; \ m = n \end{cases} \qquad \text{... (21·68)}$$

The first five Physicists' Hermite polynomials are :

$$h_0(x) = 1 \ ; \qquad h_1(x) = 2x \qquad h_2(x) = 4x^2 - 2$$

$$h_3(x) = 8x^3 - 12x \ ; \quad h_4(x) = 16x^4 - 48x^2 + 12$$

Note that the leading coefficient *i.e.*, the coefficient of x^n in $h_n(x)$ is 2^n, as compared to 1 in Probabilists' Hermite polynomial.

21·4·3. Expected Value of Hermite Polynomials.

If X is a random variable following normal distribution with mean μ and unit variance, then:

$$E[H_n(X)] = \mu^n \qquad \text{... (21·69)}$$

where, $H_n(x)$ is the Hermite polynomial of degree n.

Proof. $X \sim N(\mu, 1)$, its p.d.f. is : $f(x) = \dfrac{1}{\sqrt{2\pi}} e^{-(x-\mu)^2/2} \ ; \ -\infty < x < \infty$

$$E\left[H_n(X)\right] = \int_{-\infty}^{\infty} H_n(x) \cdot f(x) dx = \frac{1}{\sqrt{2\pi}} \int_{-\infty}^{\infty} H_n(x) \cdot e^{-(x-\mu)^2/2} dx$$

$$= \int_{-\infty}^{\infty} \left[\frac{1}{\alpha(x)} (-D)^n \alpha(x)\right] \cdot \frac{1}{\sqrt{2\pi}} e^{-(x-\mu)^2/2} dx$$

where $\qquad \alpha(x) = \dfrac{1}{\sqrt{2\pi}} e^{-x^2/2}$ and $H_n(x)\, \alpha(x) = (-D)^n \alpha(x)$.

$$\therefore \qquad E\left[H_n(X)\right] = \int_{-\infty}^{\infty} \left[(-D)^n \alpha(x)\right] \cdot \frac{\alpha(x-\mu)}{\alpha(x)} dx$$

$$= \int_{-\infty}^{\infty} \left[(-D)^n \alpha(x)\right] \cdot e^{\mu x - \frac{1}{2}\mu^2} dx$$

$$= \int_{-\infty}^{\infty} \left[(-D)^n \alpha(x)\right] \cdot \left(\sum_{m=0}^{\infty} H_m(x) \frac{\mu^m}{m!}\right) dx$$

$\left[\because \; e^{\mu x - \frac{\mu^2}{2}}\right.$ is the generating function of Hermite polynomials, with dummy variable t replaced by μ. $\left.\right]$

$$\Rightarrow \quad E[H_n(x)] = \int_{-\infty}^{\infty} \left[\sum_{m=0}^{\infty}\left(H_m(x)\cdot\frac{\mu^m}{m!}\right)\right] H_n(x)\,\alpha(x)\,dx$$

$$= \frac{\mu^n}{n!}\int_{-\infty}^{\infty} H_n^2(x)\,\alpha(x)\,dx = \frac{\mu^n}{n!}(n!) = \mu^n$$

[Expanding, integrating term by term, and using the orthogonal property of Hermite polynomials.]

21·5· GRAM-CHARLIER TYPE-A SERIES

It is often desirable to approximate the p.d.f. of a random variable from the given statistical moments. The *Gram-Charlier Type-A series* (named in honour of *Jargan Pedersen Gram and Carl Charlier*), is one well known method of such representations.

Suppose that the frequency function can be expanded formally in a series of derivatives of $\alpha(x)$. We then have:

$$f(x) = \sum_{j=0}^{\infty} C_j(-D)^j\,\alpha(x) \qquad \ldots (21\cdot70)$$

where,
$$\alpha(x) = \frac{1}{\sqrt{2\pi}}e^{-x^2/2}\ ;\ D^j \equiv \frac{d^j}{dx^j} \qquad \ldots (21\cdot70a)$$

$$\Rightarrow \quad f(x) = \sum_{j=0}^{\infty} C_j H_j(x)\,\alpha(x) \qquad \ldots (21\cdot71)$$

where $H_j(x)$ is the Hermite polynomial of degree j in x.

Multiplying both sides of (21·71) by $H_r(x)$ and integrating w.r.t x from $-\infty$ to ∞, we get

$$\int_{-\infty}^{\infty} f(x)H_r(x)\,dx = \int_{-\infty}^{\infty}\left[\left(\sum_{j=0}^{\infty} C_j\,H_j(x)\,\alpha(x)\right)\cdot H_r(x)\right]dx$$

$$= C_r\int_{-\infty}^{\infty} H_r^2(x)\cdot\alpha(x)\cdot dx = C_r\cdot r!$$

(By othogonal property of Hermite polynomials)

$$\Rightarrow \quad C_r = \frac{1}{r!}\int_{-\infty}^{\infty} f(x)H_r(x)\,dx \qquad \ldots (21\cdot72)$$

$$\Rightarrow \quad C_r = \frac{1}{r!}\int_{-\infty}^{\infty}\left[f(x)\left\{x^r - \frac{r^{(2)}x^{r-2}}{2\cdot 1!} + \frac{r^{(4)}x^{r-4}}{2^2\cdot 2!} - \frac{r^{(6)}x^{r-6}}{2^3\cdot 3!} + \cdots\right\}\right]dx$$

$$= \frac{1}{r!}\left[\mu_r' - \frac{r^{(2)}}{2\cdot 1!}\mu_{r-2}' + \frac{r^{(4)}}{2^2\cdot 2!}\mu_{r-4}' - \cdots\right] \qquad \ldots (21\cdot73)$$

where, $\mu_r' = \int_{-\infty}^{\infty} x^r f(x)\,dx,$ is the rth moment about origin.

If x is measured from mean, then

$\mu_r' = \mu_r = r^{th}$ moment about mean, and in this case : $[E(X) = 0]$

$$C_r = \frac{1}{r!}\left[\mu_r - \frac{1}{2}r^{(2)}\mu_{r-2} + \frac{r^{(4)}}{2^2 \cdot 2!}\mu_{r-4} - \cdots\right] \qquad \text{... (21·74)}$$

In particular, we get:

$$C_0 = 1 \;\; ; \;\; C_1 = 0 \qquad [\because \mu_0 = 1 \text{ and } \mu_1 = 0] \qquad \text{... (21·75)}$$

$$C_2 = \frac{1}{2!}\left[\mu_2 - \frac{1}{2}\cdot 2^{(2)}\mu_0\right] = \frac{1}{2}(\mu_2 - 1) \qquad \text{... (21·75}a\text{)}$$

$$C_3 = \frac{1}{6}\mu_3 \; (\because \mu_1 = 0) \;\; ; \;\; C_4 = \frac{1}{24}(\mu_4 - 6\mu_2 + 3) \qquad \text{... (21·75}b\text{)}$$

Substituting the values of C_r's in (21·71), we get

$$f(x) = \alpha(x)\left[C_0 H_0(x) + C_1 H_1(x) + C_2 H_2(x) + C_3 H_3(x) + \cdots\right]$$

$$\Rightarrow \quad f(x) = \alpha(x)\left[1 + \frac{1}{2}(\mu_2 - 1)H_2(x) + \frac{1}{6}\mu_3 H_3(x) + \frac{1}{24}(\mu_4 - 6\mu_2 + 3)H_4(x) + \cdots\right] \qquad \text{... (21·76)}$$

Further, if X is a standard variable so that $E(X) = 0$ and $\text{Var}(X) = \mu_2 = 1$, then we get [From (21·76)] :

$$f(x) = \alpha(x)\left[1 + \frac{1}{6}\mu_3 H_3(x) + \frac{1}{24}(\mu_4 - 3)H_4(x) + \cdots\right] \qquad \text{... (21·77)}$$

In this case, $(E(X) = 0, \mu_2 = 1)$, we have:

$$\gamma_1 = \frac{\mu_3}{\mu_2^{3/2}} = \mu_3 \text{ and } \gamma_2 = \beta_2 - 3 = \frac{\mu_4}{\mu_2^2} - 3 = \mu_4 - 3$$

$$\therefore \quad f(x) = \alpha(x)\left[1 + \frac{1}{6}\gamma_1 H_3(x) + \frac{1}{24}\gamma_2 H_4(x) + \cdots\right] \qquad \text{... (21·78)}$$

Hence, if X is a standard variable, then the coefficients of second and third term in Gram-Charlier Type-A series are measures of skewness and kurtfosis respectively.

21·6. EDGEWORTH TYPE – A SERIES

Edgeworth series, (named in honour of Francis Ysidro Edgeworth), is a series that approximates a probability distribution in terms of cumulants.

Edgeworth gives the following form of Type–A Series:

$$f(x) = \left[\exp\left\{-\kappa_3 \cdot \frac{D^3}{3!} + \kappa_4 \cdot \frac{D^4}{4!} + \cdots\right\}\right]\alpha(x), \qquad \text{... (21·79)}$$

where, $$D = \frac{d}{dx} \text{ and } \alpha(x) = \frac{1}{\sqrt{2\pi}}e^{-x^2/2} \text{ and } \kappa_r = r^{th} \text{ cumulant.} \qquad \text{... (21·80)}$$

The Fourier Transform (F.T.) of $\alpha(x)$ is given by :

$$F.T.[\alpha(x)] = \int_{-\infty}^{\infty} e^{itx}\alpha(x)dx = \text{c.f. of S.N.V.}$$

$$= e^{-t^2/2} = \sqrt{2\pi}\cdot\alpha(t) \qquad \text{[from (21.80)]}$$

$$\Rightarrow \quad \int_{-\infty}^{\infty} e^{itx}\alpha(x)dx = \sqrt{2\pi}.\alpha(t) \qquad \text{... (21·81)}$$

Differentiating both sides of (21·81) w.r.t. 't' we get

$$\sqrt{2\pi}\cdot D^r[\alpha(t)] = \int_{-\infty}^{\infty} (ix)^r\cdot e^{itx}\alpha(x)dx \;;\; \left(D \equiv \frac{d}{dt}\right)$$

$$\Rightarrow \quad \sqrt{2\pi}(-i)^r D^r[\alpha(t)] = \int_{-\infty}^{\infty} x^r\alpha(x)e^{itx}dx \quad ; \quad (\because -i^2 = 1)$$

$$\Rightarrow \quad \sqrt{2\pi}\, i^r\left[(-D)^r\alpha(t)\right] = \int_{-\infty}^{\infty} e^{itx}\, x^r\alpha(x)dx$$

$$\Rightarrow \quad \sqrt{2\pi}\, i^r[H_r(t)\,\alpha(t)] = F.T.\left[x^r\,\alpha(x)\right] \qquad \text{... (21·82)}$$

Hence, by Inversion Theorem of Fourier Transforms, we get:

$$x^r\alpha(x) = \frac{1}{2\pi}\int_{-\infty}^{\infty} e^{-itx}.\sqrt{2\pi}\, i^r\, H_r(t)\,\alpha(t)dt$$

Interchanging t and x, we get:

$$t^r\alpha(t) = \frac{(i^r)}{\sqrt{2\pi}}\int_{-\infty}^{\infty} e^{-itx}H_r(x)\,\alpha(x)dx$$

$$\Rightarrow \quad \sqrt{2\pi}(-i)^r t^r.\alpha(t) = \int_{-\infty}^{\infty} e^{-itx}H_r(x)\,\alpha(x)dx$$

Changing t to $-t$, we get:

$$\int_{-\infty}^{\infty} e^{itx}H_r(x)\,\alpha(x)dx = \sqrt{2\pi}\,(it)^r\alpha(t)$$

$$\Rightarrow \quad \text{F.T.}\left[H_r(x)\,\alpha(x)\right] = \sqrt{2\pi}\,(it)^r\cdot\alpha(t) \qquad \text{... (21·83)}$$

$$\text{F.T.}\left[e^{\kappa_r D^r}.\alpha(x)\right] = \text{F.T.}\left[\sum_{j=0}^{\infty}\frac{(\kappa_r D^r)^j}{j!}\cdot\alpha(x)\right] \;;\; D \equiv \frac{d}{dx}$$

$$= \text{F.T.}\left[\sum_{j=0}^{\infty}\frac{\kappa_r^j D^{rj}}{j!}\,\alpha(x)\right] = \text{F.T.}\left[\sum_{j=0}^{\infty}(-1)^{rj}\frac{\kappa_r^j}{j!}(-D)^{rj}\,\alpha(x)\right]$$

$$= \text{F.T.}\left[\sum_{j=0}^{\infty}(-1)^{rj}\frac{\kappa_r^j}{j!}\cdot H_{rj}(x)\,\alpha(x)\right] = \sum_{j=0}^{\infty}\left[(-1)^{rj}\frac{\kappa_r^j}{j!}\text{F.T.}\big(H_{rj}(x)\,\alpha(x)\big)\right]$$

$$= \sum_{j=0}^{\infty}\left[(-1)^{rj}\frac{\kappa_r^j}{j!}\sqrt{2\pi}\,.\,(it)^{rj}\,\alpha(t)\right] \qquad \text{... [From (21·83)]}$$

$$= \sqrt{2\pi}\cdot\alpha(t)\sum_{j=0}^{\infty}\left[\frac{\kappa_r^j}{j!}(-it)^{rj}\right] = \sqrt{2\pi}\cdot\alpha(t)\sum_{j=0}^{\infty}\left[\frac{\{\kappa_r\cdot(-it)^r\}^j}{j!}\right]$$

$$\Rightarrow \qquad \text{F.T.}\left[\exp\left(\kappa_r\cdot D^r\right).\alpha(x)\right] = \sqrt{2\pi}\cdot\alpha(t)\ \exp\left[(-it)^r.\,\kappa_r\right] \qquad \text{... (21·84)}$$

Repeated application of (21·84) gives:

$$\text{F.T.}\left[\left\{\exp\left(-\frac{\kappa\cdot\ -a}{1!}\cdot D+\frac{\kappa\ \ -b}{2!}\cdot D\ \ -\frac{\kappa_3}{3!}D\ \ +\frac{\kappa}{4!}D\ \ -\ldots\right)\right\}\alpha(x)\right]$$

$$= \sqrt{2\pi}\cdot\alpha(t)\left[\exp\left\{\frac{\kappa_1-a}{1!}(it)+\frac{\kappa_2-b}{2!}(-t)^2+\frac{\kappa_3}{3!}(it)^3+\cdots\right\}\right] \qquad \text{... (21·85)}$$

Let

$$g(x) = \frac{1}{\sqrt{2\pi}\cdot\sigma}\exp\left[-\frac{(x-\mu)^2}{2\sigma^2}\right]\ ;\ -\infty < x < \infty$$

i.e., $g(x)$ be p.d.f. of $N(\mu, \sigma^2)$ distribution.

$$\text{F.T.}[g(x)] = \int_{-\infty}^{\infty} e^{itx}g(x)dx = e^{it\mu - t^2\sigma^2/2}$$

[c.f. of $N(\mu, \sigma^2)$ distribution]

$$\Rightarrow \qquad \text{F.T. of}\left[g(x)\right] = e^{it\mu}\,.\,e^{-t^2\sigma^2/2} = e^{it\mu}.\sqrt{2\pi}\,\alpha(t\sigma) \qquad \text{... (21·86)}$$

Let us consider the function:

$$f(x) = \exp\left\{-\frac{\kappa_1-a}{1!}D+\frac{\kappa_2-b}{2!}D^2-\frac{\kappa_3}{3!}D^3+\cdots\right\}\cdot g(x) \qquad \text{... (21·87)}$$

Then,

$$F.T.\left[f(x)\right] = \sqrt{2\pi}\,e^{it\mu}\cdot\alpha(t\sigma)\times\left[\exp\left\{\frac{\kappa_1-a}{1!}(it)+\frac{\kappa_2-b}{2!}(it)^2+\frac{\kappa_3}{3!}(it)^3+\cdots\right\}\right] \qquad \text{... (21.87a)}$$

[From (21·85) and (21·86)]

Suppose that $f(x)$ represents a probability function, then its cumulant generating function $\kappa(t)$ is given by:

$$\kappa(t) = \log\,(c.f.\,of\;\; f(x)) = \log\left[\text{F.T.}(f(x))\right]$$

$$= i\mu t - \frac{t^2\sigma^2}{2!} + \frac{\kappa_1 - a}{1!}\,(it) + \frac{\kappa_2 - b}{2!}\,(it)^2 + \frac{\kappa_3}{3!}\,(it)^3 + \ldots$$

$$\left[\because\; \sqrt{2\pi}\, e^{it\mu}\,.\;\alpha(t\sigma) = \sqrt{2\pi}\cdot e^{it\mu}\,.\,\frac{e^{-t^2\sigma^2/2}}{\sqrt{2\pi}} = e^{it\mu - t^2\sigma^2/2}\right]$$

$\Rightarrow$ $\kappa(t)$ = Cumulant generating function of $f(x)$

$$= (\kappa_1 - a + \mu)\,(it) + \left(\frac{\kappa_2 - b + \sigma^2}{2!}\right)(it)^2 + \frac{\kappa_3}{3!}\,(it)^3 + \frac{\kappa_4}{4!}\,(it)^4 + \cdots \qquad \ldots (21\cdot 88)$$

Hence, the first four cumulants are given by:

$$\kappa_1 - a + \mu, \quad \kappa_2 - b + \sigma^2, \quad \kappa_3 \;\text{ and }\; \kappa_4 .$$

In particular if $a = \mu$, $b = \sigma^2$, then the cumulants of the distribution represented by $f(x)$ in (21·87) are $\kappa_1, \kappa_2, \kappa_3$ and κ_4.

If these are in fact the cumulants of a probability distribution, then $f(x)$ in (21·87) must be equal to its p.d.f. provided that:

(*i*) The series converges to a probability function and

(*ii*) It is uniquely determined by its cumulants/moments.

If the variable x is a standardised variable so that κ_1 = Mean $= \mu = 0$ and $\kappa_2 = \mu_2 = \sigma^2 = 1$, and $g(x) = \alpha(x)$, then, the probability function $f(x)$ in (21·87) becomes:

$$f(x) = \left[\exp\left(-\frac{\kappa_3}{3!}D^3 + \frac{\kappa_4}{4!}D^4 - \cdots\right)\right]\alpha(x) \qquad \ldots (21\cdot 89)$$

21·7. GRAM CHARLIER'S POLYNOMIALS

Charlier Polynomials, (also called *Poisson Charlier Polynomials*), are a family of orthogonal polynomials introduced by Carl Charlier.

Gram-Charlier polynomials $G_r = G_r\,(m, x)$ are defined by the equation.

$$G_r(m, x)\,Y(m,x) = \frac{d^r}{dm^r}Y(m\,,\,x) \quad \text{or} \quad G_r = \frac{1}{Y}D^r Y \qquad \ldots (21\cdot 90)$$

where,
$$Y = Y(m,x) = \frac{e^{-m}m^x}{x!} \;\; ; \;\; D \equiv \frac{d}{dm} \qquad \ldots (21\cdot 91)$$

is the Poisson probability function, for non-negative integer values of x.

$G_r = G_r\,(m, x)$, is called the *Gram-Charlier polynomial of* r^{th} *degree.*

Generating Function (G.F.) of $G_r\,(m, x)$

$$\sum_{r=0}^{\infty}\left[G_r(m,x)\cdot\frac{t^r}{r!}\right] = \sum_{r=0}^{\infty}\left(\frac{1}{Y}D^r(Y)\right)\frac{t^r}{r!}$$

$$= \frac{1}{Y}\sum_{r=0}^{\infty}\frac{t^r}{r!}\;D^r(Y) \;\; ; \;\; D \equiv \frac{d}{dm} \qquad \ldots (21\cdot 92)$$

$$Y(m+t) = \sum_{r=0}^{\infty} \frac{t^r}{r!} D^r(Y) \quad \text{(By Taylor's series expansion)}$$

Substituting in (21·92),

$$\sum_{r=0}^{\infty} \left[G_r(m,x) \frac{t^r}{r!} \right] = \frac{Y(m+t)}{Y(m)} = \frac{e^{-(m+t)}(m+t)^x \cdot x!}{x!\ e^{-m} \cdot m^x} = e^{-t}\left(1+\frac{t}{m}\right)^x$$

$\therefore$ $$\text{G.F. of } G_r(m,x) = e^{-t}\left(1+\frac{t}{m}\right)^x \quad \text{... (21·93)}$$

Explicit Expression for G_r (m, x). From (21·93), we get:

$$\sum_{r=0}^{\infty} \left[G_r(x) \cdot \frac{t^r}{r!} \right] = \left[\sum_{j=0}^{\infty} \frac{(-t)^j}{j!} \right] \times \left[\sum_{i=0}^{x} \binom{x}{i} \left(\frac{t}{m}\right)^i \right]$$

Since both the series on the R.H.S., are convergent, we can express the product of two series as a double-series.

$\therefore$ $$\sum_{r=0}^{\infty} \left[G_r(x) \cdot \frac{t^r}{r!} \right] = \sum_{i=0}^{x} \sum_{j=0}^{\infty} \left[\frac{(-t)^j}{j!} \binom{x}{i} \left(\frac{t}{m}\right)^i \right]$$

Comparing coefficient of t^r on both sides, we get:

$$\frac{G_r(x)}{r!} = \sum_{j=0}^{r} \left[\binom{x}{r-j} \frac{(-1)^j}{j!\ m^{r-j}} \right] \quad ; \quad (r-j \geq 0 \ \Rightarrow \ j \leq r)$$

$\Rightarrow$ $$G_r(x) = \frac{r!}{m^r} \sum_{j=0}^{r} \left[\frac{(-m)^j}{j!} \binom{x}{r-j} \right] \quad ; \quad r = 0, 1, 2, \cdots \quad \text{... (21·94)}$$

which gives the explicit expression for $G_r(m, x)$.

Putting $r = 0, 1, 2, \ldots$ in (21·94), we get respectively:

$$G_0(x) = 1$$

$$G_1(x) = \frac{1}{m} \sum_{j=0}^{1} \frac{(-m)^j}{j!} \binom{x}{1-j} = \frac{1}{m}\left[\binom{x}{1} - m\binom{x}{0}\right] = \frac{1}{m}(x-m)$$

$$G_2(x) = \frac{2!}{m^2} \sum_{j=0}^{2} \left[\frac{(-m)^j}{j!} \binom{x}{2-j} \right] = \frac{2}{m^2}\left[\binom{x}{2} - m\binom{x}{1} + \frac{m^2}{2!}\binom{x}{0}\right]$$

$$= \frac{2}{m^2}\left[\frac{x(x-1)}{2} - mx + \frac{m^2}{2}\right] = \frac{1}{m^2}\left[x^2 - (1+2m)x + m^2\right]$$

Similarly proceeding, we will get:

$$G_3(x) = \frac{1}{m^3}\left[m^3 - \left(2 + 3m + 3m^2\right)x + 3(1+m)x^2 - x^3\right]$$

Hence, the first four polynomials are given by:

$$\left.\begin{aligned} &G_0(x) = 1, \quad G_1(x) = \frac{1}{m}(x - m), \quad G_2(x) = \frac{1}{m^2}\left[x^2 - (1 + 2m)x + m^2\right] \\ &G_3(x) = \frac{1}{m^3}\left[m^3 - 3(1 + m)x^2 + (2 + 3m + 3m^2)\, x + -x^3\right] \end{aligned}\right\} \quad \text{... (21·95)}$$

21·7·1. Properties of Gram-Charlier Polynomials

Property 1. $\nabla \equiv -D \quad \Rightarrow \quad \nabla^r \equiv (-D)^r$ ... (21·96)

where ∇ is the backward difference operator, defined by:

$$\nabla f(x) = f(x) - f(x - 1) \quad \text{... (21·96 a)}$$

and D is the differential operator : $D \equiv \dfrac{d}{dm}$.

Note that ∇ Operates on x and D operates on m.

Proof.

$$Y = Y(m, x) = \frac{e^{-m} \cdot m^x}{.x!}$$

$$\Rightarrow \quad D\,(Y) = \frac{d}{dm}(Y) = \frac{1}{x!}\left[-e^{-m} \cdot m^x + e^{-m} \cdot x m^{x-1}\right]$$

$$\Rightarrow \quad D(Y) = -\frac{e^{-m} m^x}{x!}\left[1 - \frac{x}{m}\right] \quad \text{... (*)}$$

$$\nabla Y = Y(x) - Y(x - 1) = \frac{e^{-m} m^x}{x!} - \frac{e^{-m}\, m^{x-1}}{(x - 1)\,!}$$

$$\Rightarrow \quad D(Y) = -\frac{e^{-m} m^x}{x!}\left[1 - \frac{x}{m}\right] \quad \text{... (**)}$$

From (*) and (**), we get : $\nabla \equiv -D \quad \Rightarrow \quad \nabla^r = (-D)^r \ ; \quad D^r \equiv \dfrac{d^r}{dm^r}$

Property 2. Orthogonal Property. *Gram-Charlier polynomials* $G_r = G_r(m, x)$ *are orthogonal with respect to the weighting function* $Y = Y\,(m, x)$ *defined in* (21·91) , *i.e.,*

$$\sum_x \left[G_r(m, x) \cdot G_s(m,\ x) \cdot Y(m, x)\right] = \begin{cases} 0 \ ; & r \neq s \\ \dfrac{r!}{m^r} \ ; & r = s \end{cases} \quad \text{... (21·97)}$$

Property 3. Recurrence Relation. $G_r(m, x)$ satisfies the following recurrence relation :

$$G_{r+1}(m, x) = \frac{x}{m} G_r(m, x - 1) - G_r(m, x)$$

21·7·2. Gram Charlier Type B-Series.

In this series, the *p.m.f.* $f(x)$ of a discrete random variable X is represented by the series:

$$f(x) = \sum_{j=0}^{\infty} C_j \ \nabla^j\, Y(x) \quad \text{... (21·98)}$$

where $Y = Y(x) = Y\,(m, x)$ is the Poisson probability function with parameter m and the backward difference ∇ is defined as:

$$\nabla Y(x) = \nabla Y(m, x) = Y(x) - Y(x - 1) \quad \text{... (21·98 a)}$$

$$\therefore \quad f(x) = \sum_{j=0}^{\infty} C_j (-D)^j Y(m, x) \; ; \; D \equiv \frac{d}{dm}. \qquad \text{[From (21·96)]}$$

$$\Rightarrow \quad f(x) = \sum_{j=0}^{\infty} C_j (-1)^j \; G_j(m, x)\, Y(m, x) \qquad \text{... (21·99)}$$

where $\quad G_r(m,x)\; Y(m,x) = \frac{}{dm} Y(m, x) \quad \Rightarrow \quad G_r = -D\, Y$ [From (21·90)]

The problem is to find the constants C_j's.

Multiplying both sides of (21·99) by $G_r\,(m, x)$ and summing over x from 0 to ∞, we get:

$$\sum_{x=0}^{\infty} G_r(m,x)\, f(x) = \sum_{x=0}^{\infty}\sum_{j=0}^{\infty} C_j(-1)^j G_j(m, x) G_r(m, x)\, Y(m, x)$$

$$= C_r \sum_{x=0}^{\infty} (-1)^r G_r^2(m,x)\cdot Y(m,x) = C_r \cdot \frac{(-1)^r \cdot r!}{m^r}$$

[All other terms are zero because of the orthogonal property of $G_r\,(m, x)$].

$$\Rightarrow \quad C_r = (-1)^r \cdot \frac{m^r}{r!} \sum_{x=0}^{\infty} G_r(m, x)\, f(x) \qquad \text{... (21·100)}$$

$$\therefore \quad C_r = (-1)^r \cdot \frac{m^r}{r!} \sum_{x=0}^{\infty} \left\{ \left[\frac{r!}{m^r} \sum_{j=0}^{r} \frac{(-m)^j}{j!} \binom{x}{r-j} \right] f(x) \right\} \qquad \text{[From (21·94)]}$$

$$= (-1)^r \sum_{x=0}^{\infty} \left[\left\{ \binom{x}{r} - m \binom{x}{r-1} + \frac{m^2}{2!} \binom{x}{r-2} - \cdots + \frac{(-m)^r}{r!} \binom{x}{0} \right\} f(x) \right]$$

$$\Rightarrow \quad C_r = (-1)^r \left[\frac{\mu'_{(r)}}{r!} - \frac{m}{(r-1)!} \mu'_{(r-1)} + \frac{m^2}{2!} \frac{\mu'_{(r-2)}}{(r-2)!} - \cdots + \frac{(-m)^{r-1}}{(r-1)!} \frac{\mu'_{(1)}}{1} + \frac{(-m)^r}{r!} \cdot 1 \right] \qquad \text{... (21·100}a\text{)}$$

where, $\mu'_{(r)} = r^{th}$ factiorial moment of X about origin.

In particular, taking $r = 0, 1, 2, \ldots$ in (21·100), we get, on using (21·95), respectively:

$$C_0 = \sum_{x=0}^{\infty} G_0(x) f(x) = \sum_{x=0}^{\infty} 1 \cdot f(x) = 1 \qquad \text{... (21·101)}$$

$$C_1 = (-m) \sum_{x=0}^{\infty} G_1(x) f(x) = -m \sum_{x=0}^{\infty} \left[\frac{1}{m}(x - m) f(x) \right]$$

$$= -\left[\sum_x x f(x) - m \sum_x f(x) \right] = -\left[\mu'_1 - m \right] \qquad \text{... (21·101}a\text{)}$$

$$C_2 = (-1)^2 \frac{m^2}{2!} \sum_x G_2(x) f(x) = \frac{m^2}{2!} \sum_x \left[\frac{1}{m^2} \left(x^2 - (1 + 2m)x + m^2 \right) f(x) \right]$$

$$= \frac{1}{2}\left[\mu_2' - (1+2m)\,\mu_1' + m^2\right] \qquad \text{... (21·101}b\text{)}$$

where, $\mu_r' = \sum_{x=0}^{\infty} x^r f(x)$, is the rth moment of X about origin. ... (21·101c)

To determine the constants $C_0, C_1, C_2, \ldots$ in terms of the moments of the distribution, we consider the representation by the first three terms of the series and restrict our treatment to a distribution of equally distant ordinates at non-negative integral values of x. Thus, we get:

$$f(x) = \sum_{j=0}^{\infty} C_j (-1)^j G_j(m,x)\, Y(m,x) \qquad \text{[From (21·99)]}$$

$$= C_0\, G_0(m,x)\; Y(m,x) + C_1 G_1(m,x)\cdot Y(m,x) + C_2\, G_2(m,x)\cdot Y(m,x)$$

$$= Y(m,x)\left[1 + (m-\mu_1')\,G_1(m,x) + \frac{1}{2}\left\{\mu_2' - (1+2m)\,\mu_1' + m^2\right\}G_2(m,x)\right] \qquad \text{... (21·102)}$$

If 'm' is taken to denote the first raw moment of X about origin *i.e.*, if we take:

$$\mu_1' = \sum_{x=0}^{\infty} x f(x) = m,$$

Then, we get :

$$f(x) = Y(m,x)\left[1 + \frac{1}{2}(\mu_2 - m)\,G_2(m,x)\right].$$

$$\left[\begin{aligned} \because\ \mu_2' - (1+2m)\,\mu_1' + m^2 &= \mu_2' - m - m^2 \qquad (\because \mu_1' = m) \\ &= \left(\mu_2' - \mu_1'^2\right) - m = \mu_2 - m \end{aligned}\right]$$

Obviously, we can find more terms, if necessary.

Appendix 1 – Bibliography

(Suggested Further Readings)

Brunk, H.D. *An Introduction to Mathematical Statistics 3/e*, John Wiley, New York.

Chou, Ya-Lun. *Statistical Analysis, Holt Rinehart and Winston*, New York.

Cramer Harold. *Mathematical Methods of Statistics (1946), Princeton Univ. Press, N.J.*

David, D.N., *Order Statistics.*

Feller William. *An Introduction to Probability Theory and its Applications: Vol. I, 3/e* (1968), Vol. II (1966), John Wiley, New York.

Fisher, R.A. *Statistical Methods for Research Workers, Oliver and Boyd, Ltd.*, London.

Fisher, R.A. and Yates, F. *Statistical Tables for Biological, Agricultural and Medical Research, 3rd Edn. Haftner Publlishing Company*, New York, 1948.

Freund, John. E. *Mathematical Statistics with Applications 7/e (Reprint 2004), Pearson Education Inc.* (India).

Freund J.E., Milter, I., and Johnson, R.A., *Probability and Statistics for Engineers Prentice Hall Inc.*, 1990.

Gibbons, J.D., *Non- Parametric Interference, Mc Graw Hill*, New York, 1971.

Gupta S.C., Gupta Sanjeev and Gupta Vikas., *Problems and Solutions in Mathematical Statistics.* Sultan Chand and Sons, Darya Ganj, Delhi.

Hoel., P., *Introduction to Mathematical Statistics., 3rd edn.*, New York, John Wiley and Sons, 1962.

Hoel, P., Port S.C., and Stone C.J., *Introduction to Probability Theory, Boston.*, Houghton Mifflin Company, 1971.

Hogg, R.V. and Craig A.T. *Introduction to Mathematical Statistics, 5/e (Reprint 2002), Pearson Education Inc.* (India).

Johnson N.L. and Kotz Samuel. *Distributions in Statistics.*

(i) Discrete Distributions (1972),

(ii) Continuous Univariate Distributions, Vol. 1 (1973), Vol. 2 (1973), Houghton Mifflin Company, Boston.

Keeping E.S. *Introduction to Statistical Inference, Priceton, N.J., D. Van Nostrand* Co., Inc, 1962.

Kenny, J.F., and Keeping E.S. *Mathematics of Statistics, Part II, D. Van Nostrand Co., Ltd.*

Kendal, M. and Stuart, A. *The Advanced Theory of Statistics, Vol. 1, 5/e (1986), and Vol. 2, (4/e)* (1979); G. Griffin & Co., London, U.K.

Lehman, E.L., *Testing Statistical Hyoithesis; 2nd edn., New York, John Wiley and Sons* Inc. 1986.

Lehman. E.L. *Theory of Point Estimation; John Wiley and Sons,* New York, 1983.

Lindgren, B.W. *Statistical Theory, 3/e (1976),* McMillian Publishing Co., N.Y.

Loeve, M. *Probability Theory, 3rd edn., Van Nostrand, Princeton,* 1963.

Meyer, Paul L. *Introductory Probability and Statistical Applications, 2/e (1979), Addison Wesley,* N.Y.

Mood, A.M., Graybill F.A. and Boes, D.C. *Introduction to Theory of Statistics, (International Student Edition) McGraw Hill, Koga Kisha,* Tokyo, 1974.

Mukhopadhyay Parimal. *Mathematical Statistics, New Central Book Agency,* Culcutta.

Parzen, E. *Modern Probability Theory and its Applications; John Wiley,* N.Y.

Rao, C.R. *Advanced Statistical Methods in Biometric Research., N.Y., John Wiley and Sons,* Inc., 1952.

Rao, C.R. *Linear Statistical Inference and its Applications; John Wiley N.Y.,* 1965.

Rahman, N.A. *Exercises in Probability and Statistics, 2/e (1983), Griffin,* (London).

Rohatgi, V.K. *An Introduction to Probability Theory and Mathematical Statistics , (1976) , John Wiley,* N.Y.

Rohatgi, V.K. *Statistical Inference (1984), John Wiley and Sons,* N.Y.

Ross, Sheldon M. *A First Course in Probability, 6/e (Reprint 2002), Pearson Inc.* (India).

Ross, Sheldon M. *Introduction to Probability Models, 8/e, (Reprint 2013), Elsevier Inc.* (New Delhi).

Savage, L.J. *The Foundations of Statistical Inference (1962),* (Methuen , London).

Serffing , R.J. *Approximation Theorems of Mathematical Statistics (1980), John Wiley,* N.Y.

Upspensky, J.V. *Introduction to Mathematical Probability (1973), McGraw Hill,* N.Y.

Wald, A. *Sequential Analysis., John Wiley* New York, 1947.

Weatherburn, C.F. *Mathematical Statistics,* Cambridge University Press.

Wilks, S.S. *Mathematical Statistics (1962), John Wiley,* N.Y.

Yule and Kendall. *An Introduction to the Theory of Statistics,* Charles Griffin and Co., London.

Appendix 2 – Numerical Tables

TABLE I : LOGARITHMS

	0	1	2	3	4	5	6	7	8	9	1	2	3	4	5	6	7	8	9
10	.0000	0043	0086	O120	O170	0212	0253	0294	0334	0374	4	8	12	17	21	25	29	33	37
11	.0414	0453	0492	0531	0569	0607	0645	0682	0719	0755	4	8	11	15	19	23	26	30	34
12	.0792	0828	0864	0899	0934	0969	1004	1038	1072	ll06	3	7	10	14	17	21	24	28	31
13	.1139	1173	1206	1239	1271	1303	1335	1367	1399	1430	3	6	10	13	16	19	23	26	29
14	.1461	1492	1523	1553	1584	1614	1644	1673	1703	1732	3	6	9	12	15	18	21	24	27
15	.1761	1790	1818	1847	1875	1903	1931	1959	1987	2014	3	6	8	11	14	17	20	22	25
16	.2041	2068	2095	2122	2148	2175	2201	2227	2253	2279	3	5	8	11	13	16	18	21	24
17	.2304	2330	2355	2380	2405	2430	2455	2480	2504	2529	2	5	7	10	12	15	17	20	22
18	.2553	2577	2601	2625	2648	2672	2695	2718	2742	2765	2	5	7	9	12	14	16	19	21
19	.2788	2810	2833	2856	2878	2900	2923	2945	2967	2989	2	4	7	9	11	13	16	18	20
20	.3010	3032	3054	3075	3096	3118	3139	3160	3181	3201	2	4	6	8	11	13	15	17	19
21	.3222	3243	3263	3284	3304	3324	3345	3365	3385	3404	2	4	6	8	10	12	14	16	18
22	.3424	3444	3464	3483	3502	3522	3541	3562	3579	3598	2	4	6	8	10	12	14	15	17
23	.3617	3636	3655	3674	3692	3711	3729	3747	3766	3784	2	4	6	7	9	ll	13	15	17
24	.3802	3820	3838	3856	3874	3892	3909	3927	3945	3962	2	4	5	7	9	11	12	14	16
25	.3979	3997	4014	4031	4048	4065	4082	4099	4116	4133	2	3	5	7	9	10	12	14	15
26	.4150	4166	4183	4200	4216	4232	4249	4265	4281	4298	2	3	5	7	8	10	11	13	15
27	.4314	4330	4346	4362	4378	4393	4409	4425	4440	4456	2	3	5	6	8	9	11	13	14
28	.4472	4487	4502	4518	4533	4548	4564	4579	4594	4609	2	3	5	6	8	9	11	12	14
29	.4624	4639	4654	4669	4683	4698	4713	4728	4742	4757	1	3	4	6	7	9	10	12	13
30	.4771	4786	4800	4814	4829	4843	4857	4871	4886	4900	1	3	4	6	7	9	10	11	13
31	.4914	4928	4942	4955	4969	4983	4997	50ll	5024	5038	1	3	4	6	7	8	10	11	12
32	.5051	5065	5079	5092	5105	5119	5132	5145	5159	5172	1	3	4	5	7	8	9	11	12
33	.5185	5198	5211	5224	5237	5250	5263	5276	5289	5302	1	3	4	5	6	8	9	10	12
34	.5315	5315	5340	5353	5366	5378	5391	5403	5416	5428	1	3	4	5	6	8	9	1O	11
35	.5441	5453	5465	5478	5490	5502	5514	5527	5539	5551	1	2	4	5	6	7	9	10	11
36	.5563	5575	5587	5599	5611	5623	5635	5647	5658	5670	1	2	4	5	6	7	8	10	11
37	.5682	5694	5705	5717	5729	5740	5752	5763	5775	5786	1	2	3	5	6	7	8	9	10
38	.5798	5809	5821	5832	5843	5855	5866	5877	5888	5899	1	2	3	5	6	7	8	9	10
39	.5911	5922	5933	5944	5955	5966	5977	5988	5999	6010	1	2	3	4	5	7	8	9	10
40	.6021	6031	6042	6053	6065	6075	6085	6096.	6107	6117	1	2	3	4	5	6	8	9	10
41	.6128	6138	6149	6160	6170	6180	6191	6201	6212	6222	1	2	3	4	5	6	7	8	9
42	.6232	6243	6253	6263	6274	6284	6294	6304	6314	6325	1	2	3	4	5	6	7	8	9
43	.6335	6345	6355	6365	6375	6385	6395	6405	6415	6425	1	2	3	4	5	6	7	8	9
44	.6435	6444	6454	6465	6474	6484	6493	6503	6513	6522	1	2	3	4	5	6	7	8	9
45	.6532	6542	6551	6561	6571	6580	6590	6599	6609	6618	1	2	3	4	5	6	7	8	9
46	.6628	6637	6646	6656	6665	6675	6684	6693	6702	6712	1	2	3	4	5	6	7	7	8
47	.6721	6730	6739	6749	6758	6767	6776	6785	6794	6803	1	2	3	4	5	5	6	7	8
48	.6812	6821	6830	6839	6848	6857	6866	6875	6884	6893	1	2	3	4	4	5	6	7	8
49	.6902	6911	6920	6928	6937	6946	6955	6964	6972	6981	1	2	3	4	4	5	6	7	8
50	.6990	6993	7007	7016	7024	7033	7042	7050	7059	7067	1	2	3	3	4	5	6	7	8
51	.7076	7Q84	7093	7101	7110	7118	7126	7135	7143	7152	1	2	3	3	4	5	6	7	8
52	.7160	7168	7177	7185	7193	7202	7210	7218	7226	7235	1	2	2	3	4	5	6	7	7
53	.7243	7251	7259	7267	7275	7284	7292	7300	7308	7316	1	2	2	3	4	5	6	6	7

TABLE I : LOGARITHMS

	0	1	2	3	4	5	6	7	8	9	1	2	3	4	5	6	7	8	9
54	.7324	7332	7340	7348	7356	7364	7372	7380	7388	7396	1	2	2	3	4	5	6	6	7
55	.7404	7412	7419	7427	7435	7443	7451	7459	7466	7474	1	2	1	3	4	5	5	6	7
56	.7482	7490	7497	7505	7513	7520	7528	7536	7543	7551	1	2	2	3	4	5	5	6	7
57	.7559	7566	7574	7582	7589	7597	7604	7612	7619	7627	1	2	2	3	4	5	5	6	7
58	.7634	7642	7649	7657	7664	7672	7679	7686	7694	7701	1	1	2	3	4	4	5	6	7
59	.7709	7716	7723	7731	7738	7745	7752	7760	7767	7774	1	1	2	3	4	4	5	6	7
60	.7782	7789	7796	7803	7810	7818	7825	7832	7839	7846	1	1	2	3	4	4	5	6	6
61	.7853	7860	7868	7875	7882	7889	7896	7903	7910	7917	1	1	2	3	4	4	5	6	6
62	.7924	7931	7938	7945	7952	7959	7966	7973	7980	7987	1	1	2	3	3	4	5	6	6
63	.7993	8000	8007	8014	8021	8028	8035	8041	8048	8055	1	1	2	3	3	4	5	5	6
64	.8062	8069	8075	8082	8989	8096	8102	8109	8116	8122	1	1	2	3	3	4	5	5	6
65	.8129	8136	8142	8149	8156	8162	8169	8176	8182	8189	1	1	2	3	3	4	5	5	6
66	.8195	8202	8209	8215	8222	8228	8235	8241	8248	8254	1	1	2	3	3	4	5	5	6
67	.8261	8267	8274	8280	8287	8293	8299	8306	8312	8319	1	1	2	3	3	4	5	5	6
68	.8325	8331	8338	8344	8351	8357	8363	8370	8376	8382	1	1	2	3	3	4	4	5	6
69	.8388	8395	8401	8407	8414	8420	8426	8432	8439	8445	1	1	2	2	3	4	4	5	6
70	.8451	8457	8463	8470	8476	8482	8488	8494	8500	8506	1	1	2	2	3	4	4	5	6
71	.8513	8519	8525	8531	8537	8543	8549	8555	8561	8567	1	1	2	2	3	4	4	5	5
72	.8573	8579	8585	8591	8597	8603	8609	8615	8621	8627	1	1	2	2	3	4	4	5	5
73	.8633	8639	8645	8651	8657	8663	8669	8675	8681	8686	1	1	2	2	3	4	4	5	5
74	.8692	8698	8704	8710	8716	8722	8727	8733	8739	8745	1	1	2	2	3	4	4	5	5
75	.8751	8756	8762	8768	8774	8779	8785	8791	8797	8802	1	1	2	2	3	3	4	5	5
76	.8808	8814	8820	8825	8831	8837	8842	8848	8854	8859	1	1	2	2	3	3	4	5	5
77	.8865	8871	8876	8882	8887	8893	8899	8904	8910	8915	1	1	2	2	3	3	4	4	5
78	.8921	8927	8932	8938	8943	8949	8954	8960	8965	8971	1	1	1	2	2	3	3	4	4
79	.8976	8982	8987	8993	8998	9004	9009	9015	9020	9025	1	1	2	2	3	3	4	4	5
80	.9031	9036	9042	9047	9053	9058	6063	9069	9074	9079	1	1	2	2	3	3	4	4	5
81	.9085	9090	9096	9101	9106	9112	9117	9122	9128	9133	1	1	2	2	3	3	4	4	5
82	.9138	9143	9149	9154	9159	9165	9170	9175	9180	9186	1	1	2	2	3	3	4	4	5
83	.9191	9196	9201	9206	9212	9217	9222	9227	9232	9238	1	1	2	2	3	3	4	4	5
84	.9243	9248	9253	9258	9263	9269	9274	9279	9284	9289	1	1	2	2	3	3	4	4	5
85	.9294	9299	9304	9309	9315	9320	9325	9330	9335	9340	1	1	2	2	3	3	4	4	5
86	.9345	9350	9355	9360	9365	9370	9375	9380	9385	9390	1	1	2	2	3	3	4	4	5
87	.9395	9400	9405	9410	9415	9420	9425	9430	9435	9440	0	1	1	2	2	3	3	4	4
88	.9445	9450	9455	9460	9465	9469	9474	9479	9484	9489	0	1	1	2	2	3	3	4	4
89	.9494	9499	9504	9509	9513	9518	9523	9528	9533	9538	0	1	1	2	2	3	3	4	4
90	.9542	9547	9552	9557	9562	9566	9571	9576	9581	9586	0	1	1	2	2	3	3	4	4
91	.9590	9595	9600	9605	9609	9614	9619	9624	9628	9933	0	1	1	2	2	3	3	4	4
92	.9638	9643	9647	9652	9657	9661	9666	9671	9675	9680	0	1	1	2	2	3	3	4	4
93	.9685	9689	9694	9699	9703	9708	9713	9717	9722	9727	0	1	1	2	2	3	3	4	4
94	.9731	9736	9741	9745	9750	9754	9759	9763	9768	9773	0	1	1	2	2	3	3	4	4
95	.9777	9782	9786	9791	9795	9800	9805	9809	9814	9818	0	1	1	2	2	3	3	4	4
96	.9823	9827	9832	9836	9841	9845	9850	9854	9859	9863	0	1	1	2	2	3	3	4	4
97	.9868	9872	9877	9881	9886	9890	9894	9899	9903	9908	0	1	1	2	2	3	3	4	4
98	.9912	9917	9921	9926	9930	9934	9939	9943	9948	9952	0	1	1	2	2	3	3	4	4
99	.9956	9961	9965	9969	9974	9978	9983	9987	9991	9996	0	1	1	2	2	3	3	3	4

TABLE II : ANTILOGARITHMS

	0	1	2	3	4	5	6	7	8	9	1	2	3	4	5	6	7	8	9
.00	1000	1002	1005	1007	1009	1012	1014	1016	1019	1021	0	0	1	1	1	1	2	2	2
.01	1023	1026	1028	1030	1033	1035	1038	1040	1042	1045	0	0	1	1	1	1	2	2	2
.02	1047	1050	1052	1054	1057	1059	1062	1064	1067	1069	0	0	1	1	1	1	2	2	2
.03	1072	1074	1076	1079	1081	1084	1086	1089	1091	1094	0	0	1	1	1	1	2	2	2
.04	1096	1099	1102	1104	1107	1109	1112	1114	1117	1119	0	1	1	1	1	2	2	2	2
.05	1122	1125	1!27	1130	1132	1135	1138	1140	1143	1146	0	1	1	1	1	2	2.	2	2
.06	1148	1151	1153	1156	1159	1161	1164	1167	1169	1172	0	1	1	1	1	2	2	2	2
.07	1175	1178	1180	1183	1186	1189	1191	1194	1197	1199	0	1	1	1	1	2	2	2	2
.08	1202	1205	1208	1211	1213	1216	1219	1222	1225	1227	0	1	1	1	1	2	2	2	3
.09	1230	1233	1236	1239	1242	1245	1247	1250	1253	1256	0	1	1	1.	1	2	2	2	3
.10	1259	1262	1265	1268	1271	1274	1276	1279	1282	1285	0	1	1	1	1	2	2	2	3
.11	1288	1291	1294	1297	1300	1303	1306	1309	1312	1315	0	1	1	1	2	2	2	2	3
.12	1318	1321	1324	1327	1330	1334	1337	1340	1343	1346	0	1	1	1	2	2	2	2	3
.13	1349	1352	1355	1358	1361	1365	1368	1371	1374	1377	0	1	1	1	2	2	2	3	3
.14	1380	1384	1387	1390	1393	1396	1400	1403	1406	1409	0	1	1	1	2	2	2	3	3
.15	1413	1416	1419	1422	1426	1429	1432	14,35	1439	1442	0	1	1	1	2	2	2	3	3
.16	1445	1449	1452	1455	1459	1462	1466	1469	1472	1476	0	1	1	1	2	2	2	3'	3
.17	1479	1483	1486	1489	1493	1496	1500	1503	1507	1510	0	1	1	1	2	2	2	3	3
.18	1514	1517	1521	1524	1528	1531	1535	1538	1542	1545	0	1	1	1	2	2	2	3	3
.19	1549	1552	1556	1560	1563	1567	1570	1574	1578	1581	0	1	1	1	2	2	3	3	3
.20	1585	1589	1592	1596	1600	1603	1607	1611	1614	1618	0	1	1	1	2	2	3	3	3
.21	1622	1626	1629	1633	1637	1641	1644	1648	1652	1656	0	1	1	2	2	2	3	3	3
.22	1660	1663	1667	1671	1675	1679	1683	1687	1690	1694	0	1	1	2	2	2	3	3	3
.23	1698	1702	1706	1710	1714	1718	1722	1726	1730	1734	0	1	1	2	2	2	3	3	4
.24	1738	1742	1746	1750	1754	1758	1762	1766	1770	1774	0	1	1	2	2	2	3	3	4
.25	1778	1782	1786	1791	1795	1799	1803	1807	1811	1816	0	1	1	2	2	2	3	3	4
.26	1820	1824	1828	1832	1837	1841	1845	1849	1854	1858	0	1	1	2	2	3	3	3	4
.27	1862	1866	1871	1875	1879	1884	1888	1892	1897	1901	0	1	1	2	2	3	3	3	4
.28	1905	1910	1914	1919	1923	1928	1932	1936	1941	1945	0	1	1	2	2	3	3	4	4
.29	1950	1954	1959	1963	1968	1972	1977	1982	1986	1991	0	1	1	2	2	3	3	4	4
.30	1995	2000	2004	2009	2014	2018	2023	2028	2032	2037	0	1	1	2	2	3	3	4	4
.31	2042	2046	2051	2056	2061	2065	2070	2075	2080	2084	0	1	1	2	2	3	3	4	4
.32	2089	2094	2099	2104	2109	2113	2118	2123	2128	2133	0	1	1	2	2	3	3	4	4
.33	2138	2143	2148	2153	2158	2163	2168	2173	2178	2183	0	1	1	2	2	3	3	4	4
.34	2188	2193	2198	2203	2208	2213	2218	2223	2228	2234	1	1	2	2	3	3	4	4	5
.35	2239	2244	2249	2254	2259	2265	2270	2275	2280	2286	1	1	2	2	3	3	4	4	5
.36	2291	2296	2301	2307	2312	2317	2323	2328	2333	2339	1	1	2	2	3	3	4	4	5
.37	2344	2350	2355	2360	2366	2371	2377	2382	2388	2393	1	1	2	2	3	3	4	4	5
.38	2399	2404	2410	2415	2421	2427	2432	2438	2443	2449	1	1	2	2	3	3	4	4	5
.39	2455	2460	2466	2472	2477	2483	2489	2495	2500	2506	1	1	2	2	3	3	4	5	5
.40	2513	2518	2523	2529	2535	2541	2547	2553	2559	2564	1	1	2	2	3	4	4	5	5
.41	2570	2576	2582	2588	2594	2600	2606	2612	2618	2624	1	1	2	2	3	4	4	5	5
.42	2630	2636	2642	2649	2655	2661	2667	2673	2679	2685	1	1	2	2	3	4	4	5	6
.43	2692	2698	2704	2710	2716	2723	2729	2735	2742	2748	1	1	2	3	3	4	4	5	6
.44	2754	2761	2767	2773	2780	2786	2793	2799	2805	2812	1	1	2	3	3	4	4	5	6
.45	2818	2825	2831	2838	2844	2851	2858	2864	2871	2877	1	1	2	3	3	4	5	5	6
.46	2884	2891	2897	2904	2911	2917	2924	2931	2938	2944	1	1	2	3	3	4	5	5	6
.47	2951	2958	2965	2972	2979	2985	2992	2999	3006	3013	1	1	2	3	3	4	5	5	6
.48	3020	3027	3034	3041	3048	3055	3062	3069	3076	3083	1	1	2	3	4	4	5	6	6
.49	3090	3097	3105	3112	3119	3126	3133	3141	3148	3155	1	1	2	3	4	4	5	6	6

TABLE II : ANTILOGARITHMS

	0	1	2	3	4	5	6	7	8	9	1	2	3	4	5	6	7	8	9
.50	3162	3170	3177	3184	3192	3199	3206	3214	3221	3228	1	1	2	3	4	4	5	6	7
.51	3236	3243	3251	3258	3266	3273	3281	3289	3296	3304	1	2	2	3	4	5	5	6	7
.52	3311	3319	3327	3334	3342	3350	3357	3365	3373	3381	1	2	2	3	4	5	5	6	7
.53	3388	3396	3404	3412	3420	3428	3436	3443	3451	3459	1	2	2	3	4	5	6	6	7
.54	3467	3475	3483	3491	3499	3508	3516	3524	3532	3540	1	2	2	3	4	5	6	6	7
.55	3548	3556	3565	3573	3581	3589	3597	3606	3614	3622	1	2	2	3	4	5	6	7	7
.56	3631	3639	3648	3656	3664	3673	3681	3690	3698	3707	1	2	3	3	4	5	6	7	8
.57	3715	3724	3733	3741	3750	3758	3767	3776	3784	3793	1	2	3	3	4	5	6	7	8
.58	3802	3811	3819	3828	3837	3846	3855	3864	3873	3882	1	2	3	4	4	5	6	7	8
.59	3890	3899	3908	3917	3926	3936	3945	3954	3963	3972	1	2	3	4	5	5	6	7	8
.60	3981	3990	3999	4009	4018	4027	4036	4046	4055	4064	1	2	3	4	5	6	6	7	8
.61	4074	4083	4093	4102	4111	4121	4130	4140	4150	4159	1	2	3	4	5	6	7	8	9
.62	4169	4178	4188	4198	4207	4217	4227	4236	4246	4256	1	2	3	4	5	6	7	8	9
.63	4266	4276	4285	4295	4305	4315	4325	4335	4345	4355	1	2	3	4	5	6	7	8	9
.64	4365	4375	4385	4395	4406	4416	4426	4436	4446	4457	1	2	3	4	5	6	7	8	9
.65	4467	4477	4487	4498	4508	4519	4529	4539	4550	4560	1	2	3	4	5	6	7	8	9
.66	4571	4581	4592	4603	4613	4624	4634	4645	4656	4667	1	2	3	4	5	6	7	9	10
.67	4677	4688	4699	4710	4721	4732	4742	4753	4764	4775	1	2	3	4	5	7	8	9	10
.68	4786	4797	4808	4819	4831	4842	4853	4864	4875	4887	1	2	3	4	6	7	8	9	10
.69	4898	4909	4920	4932	4943	4955	4966	4977	4989	5000	1	2	3	5	6	7	8	9	10
.70	5012	5023	5035	5047	5058	5070	5082	5093	5105	5117	1	2	4	5	6	7	8	9	11
.71	5129	5140	5152	5164	5176	5188	5200	5212	5224	5236	1	2	4	5	6	7	8	10	11
.72	5248	5260	5272	5284	5297	5309	5321	5333	5346	5358	1	2	4	5	6	7	9	10	11
.73	5370	5383	5395	5408	5420	5433	5445	5458	5470	5483	1	3	4	5	6	8	9	10	11
.74	5495	5508	5521	5534	5546	5559	5572	5585	5598	5610	1	3	4	5	6	8	9	10	12
.75	5623	5636	5649	5662	5675	5689	5702	5715	5728	5741	1	3	4	5	7	8	9	10	12
.76	5754	5768	5781	5794	5808	5821	5834	5848	5861	5875	1	3	4	5	7	8	9	11	12
.77	5888	5902	5916	5929	5943	5957	5970	5984	5998	6012	1	3	4	5	7	8	10	11	12
.78	6026	6039	6053	6067	6081	6095	6109	6124	6138	6152	1	3	4	6	7	8	10	11	13
.79	6166	6180	6194	6209	6223	6237	6252	6266	6281	6295	1	3	4	6	7	9	10	11	13
.80	6310	6324	6339	6353	6368	6383	6397	6412	6427	6442	1	3	4	6	7	9	10	12	13
.81	6457	6471	6486	6501	6516	6531	6546	6561	6577	6592	2	3	5	6	8	9	11	12	14
.82	6607	6622	6637	6653	6668	6683	6699	6714	6730	6745	2	3	5	6	8	9	11	12	14
.83	6761	6776	6792	6808	6823	6839	6855	6871	6887	6902	2	3	5	6	8	9	11	13	14
.84	6918	6934	6950	6966	6982	6998	7015	7031	7047	7063	2	3	5	6	8	10	11	13	15
.85	7079	7096	7112	7129	7145	7161	7178	7194	7211	7228	2	3	5	7	8	10	12	13	15
.86	7244	7261	7278	7295	7311	7328	7345	7362	7379	7396	2	3	5	7	8	10	12	13	15
.87	7413	7430	7447	7464	7482	7499	7516	7534	7551	7568	2	3	5	7	9	10	12	14	16
.88	7586	7706	7621	7638	7656	7674	7691	7709	7727	7745	2	4	5	7	9	11	12	14	16
.89	7762	7780	7798	7816	7834	7852	7870	7889	7907	7925	2	4	5	7	9	11	13	14	16
90	7943	7962	7980	7998	8017	8035	8054	8072	8091	8110	2	4	6	7	9	11	13	15	17
.91	8128	8147	8166	8185	8204	8222	8241	8260	8279	8299	2	4	6	8	9	11	13	15	17
.92	8318	8337	8356	8375	8395	8414	8433	8453	8472	8492	2	4	6	8	10	12	14	15	17
.93	8511	8531	8551	8570	8590	8610	8630	8650	8670	8690	2	4	6	8	10	12	14	16	18
.94	8710	8730	8750	8770	8790	8810	8831	8851	8872	8892	2	4	6	8	10	12	14	16	18
.95	8913	8933	8954	8974	8995	9016	9036	9057	9078	9099	2	4	6	8	10	12	15	17	19
.96	9120	9141	9162	9183	9204	9226	9247	9268	9290	9311	2	4	6	8	11	13	15	17	19
.97	9333	9354	9376	9397	9419	9441	9462	9484	9506	9528	2	4	7	9	11	13	15	17	20
.98	9550	9572	9594	9616	9638	9661	9683	9705	9227	9750	2	4	7	9	11	13	16	18	20
.99	9772	9795	9817	9840	9863	9886	9908	9931	9954	9977	2	5	7	9	11	14	16	18	20

TABLE III : POWERS, ROOTS AND RECIPROCALS

n	n^2	n^3	$\sqrt{n}$	$\sqrt[3]{n}$	$\sqrt{10n}$	$\sqrt[3]{10n}$	$\sqrt{100}$	$\frac{1}{n}$
1	1	1	1	1	3.162	2.!54	4.642	1
2	4	8	1.414	1.260	4.772	2.714	5.848	.5000
3	9	27	1.732	1.442	5.477	3.107	6.694	.3333
4	16	64	2	1.587	6.325	3.420	7.368	.2500
5	25	125	2.236	1.710	7.671	3.684	7.937	.2000
6	36	216	2.449	1.817	7.745	3.915	8.434	.1667
7	49	343	2.646	1.913	8.361	4.121	8.879	.1429
8	64	512	2.828	2.000	8.944	4.309	9.283	.1250
9	81	729	3.000	2.080	9.487	4.481	9.655	.1111
10	100	1000	3.162	2.154	10.0	4.642	10.000	.1000
11	121	1331	3.317	2.224	10.488	4.791	10.323	.09091
12	144	1728	3.464	2.289	10.954	4.932	10.627	.08333
13	169	2197	3.606	2.351	11.402	5.066	10.914	.07692
14	196	2744	3.742	2.410	11.832	5.192	11.187	.07143
15	225	3375	3.873	2.466	12.247	5.313	11.447	.06667
16	256	4096	4.000	2.520	12.649	5.429	11.696	.06250
17	289	4913	4.123	2.571	13.038	5.540	11.935	.05882
18	324	5832	4.243	2.621	13.416	5.646	12.164	.05556
19	361	6859	4.359	2.568	13.784	5.749	12.386	.05263
20	400	8000	4.472	2.714	14.142	5.848	12.599	.05000
21	441	9261	4.583	2.759	14.491	5.944	12.806	.04762
22	484	10648	4.690	2.802	14.832	6.037	13.006	.04545
23	529	12167	4.796	2.844	15.166	6.127	13.200	.04348
24	576	13824	4.899	2.884	15.492	6.214	13.389	.04167
25	625	15625	5.000	2.924	15.811	6.300	13.572	.04000
26	676	17576	5.099	2.962	16.125	6.383	13.751	.03846
27	729	19683	5.196	3.000	16.432	6.463	13.925	.03704
28	784	21952	5.292	3.037	16.733	6.542	14.095	.03571
29	841	24389	5.385	3.072	17.029	6.619	14.260	.03448
30	900	27000	5.477	3.107	17.321	6.694	14.422	.03333
31	961	29791	5.568	3.141	17.607	6.768	14.581	.03226
32	1024	32768	5.657	3.175	17.889	6.840	17.736	.03125
33	1089	35937	5.745	3.208	18.166	6.910	14.888	.03030
34	1156	39304	5.831	3.240	18.439	6.980	15.037	.02941
35	1225	42875	5.916	3.271	18.708	7.047	15.183	.02857
36	1296	46656	6.000	3.302	18.974	7.114	15.326	.02778
37	1369	50653	6.083	3.332	19.235	7.179	15.467	.02703
38	1444	54872	6.164	3.362	19.494	7.243	15.605	.02632
39	1521	59319	6.245	3.391	19.748	7.306	15.741	.02504
40	1600	64000	6.325	3.420	20.000	7.368	15.874	.02500
41	1681	68921	6.403	3.448	20.248	7.429	16.005	.02439
42	1764	74088	6.481	3.476	20.494	7.489	16.134	.02381
43	1849	79507	6.557	3.503	20.736	7.548	16.261	.02326
44	!936	85184	6.633	3.530	20.976	7.606	16.386	.02273
45	2025	91125	6.708	3.557	21.213	7.663	16.510	.02222
46	2116	97336	6.782	3.583	21.448	7.719	16.631	.02174
47	2209	103823	6.856	3.609	21.679	7.775	16.751	.02128
48	2304	110592	6.928	3.634	21.909	7.830	16.869	.02083
49	2401	117649	7.000	3.659	22.136	7.884	16.985	.02041
50	2500	125000	7.071	3.684	22.361	7.937	17.100	.02000

TABLE III : POWERS, ROOTS AND RECIPROCALS

n	n^2	n^3	$\sqrt{n}$	$\sqrt[3]{n}$	$\sqrt{10n}$	$\sqrt[3]{10n}$	$\sqrt[3]{100n}$	$\frac{1}{n}$
51	2601	132651	7.141	3.708	22.583	7.990	17.213	.01961
52	2704	140608	7.211	3.733	22.804	8.041	17.325	.01923
53	2809	148877	7.280	3.756	23.022	8.093	17.435	.01887
54	2916	157464	7.348	3.780	23.238	8.143	17.544	.01852
55	3025	166375	7.416	3.803	23.452	8.193	17.652	.01818
56	3136	175616	7.483	3.832	23.664	8.243	17.758	.01786
57	3249	185193	7.550	3.849	23.875	8.291	17.863	.01754
58	3364	195112	7.616	3.871	24.083	8.340	17.967	.01724
59	3481	205379	7.681	3.893	24.290	8.387	18.070	.01695
60	3600	216000	7.746	3.915	24.495	8.334	18.171	.01667
61	3721	226981	7.810	3.936	24.698	8.481	18.272	.01639
62	3844	238328	7.874	3.958	24.900	8.527	18.371	.01613
63	3969	250047	7.937	3.979	25.100	8.573	18.469	.01587
64	4096	262144	8.000	4.000	25.298	8.618	18.566	.01562
65	4225	274625	8.062	4.021	25.495	8.662	18.663	.01538
66	4356	287496	8.124	4.041	25.690	8.707	18.758	.01515
67	4489	300763	8.185	4.062	25.884	8.750	18.852	.01493
68	4624	314432	8.246	4.082	26.077	8.794	18.945	.01471
69	4761	328509	8.307	4.102	26.268	8.837	19.038	.01449
70	4900	343000	8.367	4.121	26.458	8.879	19.129	.01429
71	5041	357911	8.426	4.141	26.646	8.921	19.220	.01408
72	5184	373248	8.485	4.160	26.833	8.963	19.310	.01389
73	5329	389017	8.544	4.179	27.019	9.004	19.399	.01370
74	5476	405224	8.602	4.198	27.203	9.045	19.487	.01351
75	5625	421875	8.660	4.217	27.386	9.086	19.574	.01333
76	5776	438976	8.718	4.236	27.568	9.126	19.661	.01316
77	5929	456533	8.775	4.254	27.740	9.166	19.747	.01299
78	6084	474552	8.832	4.273	27.928	9.205	19.832	.01282
79	6241	493039	8.888	4.291	28.107	9.244	19.916	.01266
80	6400	512000	$.944	4.309	28.284	9.283	20.000	.01250
81	6561	531441	9.000	4.327	28.460	9.322	20.083	.01235
82	6724	551368	9.055	4.344	28.636	9.360	20.165	.01220
83	6889	571787	9.110	4.362	28.810	9.398	20.247	.01205
84	7056	592704	9.165	4.380	28.983	9.435	20.328	.01190
85	7225	614125	9.220	4.397	29.155	9.473	20.408	.01176
86	7396	636056	9.274	4.414	29.326	9.510	20.488	.01163
87	7569	658503	9.327	4.431	29.496	9.546	20.507	.01149
88	7744	681472	9.381	4.448	29.665	9.583	20.646	.01136
89	7921	704969	7.434	4.465	29.833	9.619	20.224	.01124
90	8100	729000	9.487	4.487	30.000	9.655	20.801	.01111
91	8281	753571	9.539	4.498	30.166	9.691	20.878	.01099
92	8464	775688	9.592	4.514	30.332	9.726	20.954	.01087
93	8649	804357	9.644,	4.531	30.496	9.761	21.029	.01075
94	8836	830584	9.695	4.547	30.659	9.796	21.105	.01064
95	9025	857375	9.747;	4.563	30.822	9.830	21.179	.01053
96	9216	884736	9.798	4.579	30.984	9.865	21.253	.01042
97	9409	912673	9.849	4.595	31.145	9.899	21.327	.01031
98	9604	941192	9.899	4.610	31.305	9.933	21.400	.01020
99	9801	970299	9.900	4.626	31.464	9.697	21.472	.01010
100	10000	1000000	10.000	4.642	31.623	10.000	21.544	.01000

TABLE IV : BINOMIAL COEFFICIENTS

n	$\binom{n}{0}$	$\binom{n}{1}$	$\binom{n}{2}$	$\binom{n}{3}$	$\binom{n}{4}$	$\binom{n}{5}$	$\binom{n}{6}$	$\binom{n}{7}$	$\binom{n}{8}$	$\binom{n}{9}$	$\binom{n}{10}$
1	1	1									
2	1	2	1								
3	1	3	3								
4	1	4	6	4							
5	1	5	10	10	5	1					
6	1	6	15	20	15	6	1				
7	1	7	21	35	35	21	7	1			
8	1	8	28	56	70	56	28	8	1		
9	1	9	36	84	126	126	84	36	9	1	
10	1	10	45	120	210	252	210	120	45	10	
11	1	11	55	165	330	462	462	330	165	55	11
12	1	12	66	220	495	792	924	792	495	220	66
13	1	13	78	286	715	1287	1716	1716	1287	715	286
14	1	14	91	364	1001	2002	3003	3432	3003	2002	1001
15	1	15	105	455	1365	3003	5005	6435	6435	5005	3003
16	1	16	120	560	1820	4368	8008	11440	12870	11440	8008
17	1	17	136	680	2380	6188	12376	19448	24310	24310	19448
18	1	18	153	816	3060	8568	18564	31824	43758	48620	43758
19	1	19	171	969	3876	11628	27132	50388	75582	92378	92378
20	1	20	190	1140	4845	15504	38760	77520	125970	167960	184756

TABLE V: VALUES OF e^{-m} or exp $(-m)$

$(0 < m < 1)$

m	0	1	2	3	4	5	6	7	8	9
0.0	1.0000	.9900	.9802	.9704	.9608	.9512	.9418	.9324	.9231	.9139
0.1	.9084	.8958	.8869	.8781	.8694	.8607	.8521	.8437	.8353	.8270
0.2	.8187	.8106	.8025	.7945	.7866	.7788	.7711	.7634	.7558	.7483
0.3	.7408	.7334	.7261	.7189	.7118	.7047	.6977	.6907	.6839	.6771
0.4	.6703	.6636	.6570	.6505	.6440	.6376	.6313	.6250	.6188	.6126
0.5	.6065	.6005	.5945	.5886	.5827	.5770	.5712	.5655	.5599	.5543
0.6	.5488	.5434	.5379	.5326	.5273	.5220	.5169	.5117	.5066	.5016
0.7	.4966	.4916	.4868	.4819	.4771	.4724	.4677	.4630	.4584	.4538
0.8	.4493	.4449	.4404	.4360	.4317	.4274	.4232	.4190	.4184	.4107
0.9	.4066	.4025	.3985	.3946	.3906	.3867	.3829	.3791	.3753	.3716

$(m = 1, 2, 3, \ldots, 10)$

m	1	2	3	4	5	6	7	8	9	10
e^{-m}	.36788	.13534	.04979	.01832	.00638	.002479	.00091	.000335	.000153	.000045

Note: To obtain values of e^{-m} for other values of m, use the laws of exponents.

Example: $e^{-2.35} = (e^{-2.00})\,(e^{-0.35}) = (.13534) \times (.7047) = 0.095374$

TABLE VI : AREAS UNDER STANDARD NORMAL PROBABILITY CURVE

Normal Probability curve is given by : $f(x) = \dfrac{1}{\sigma\sqrt{2\pi}} \exp\left\{-\dfrac{1}{2}\left(\dfrac{x-\mu}{\sigma}\right)^2\right\}; -\infty < x < \infty$

and standard normal probability curve is given by:

$$\phi(z) = \frac{1}{\sqrt{2\pi}} \exp\left(-\frac{1}{2}z^2\right), -\infty < z < \infty$$

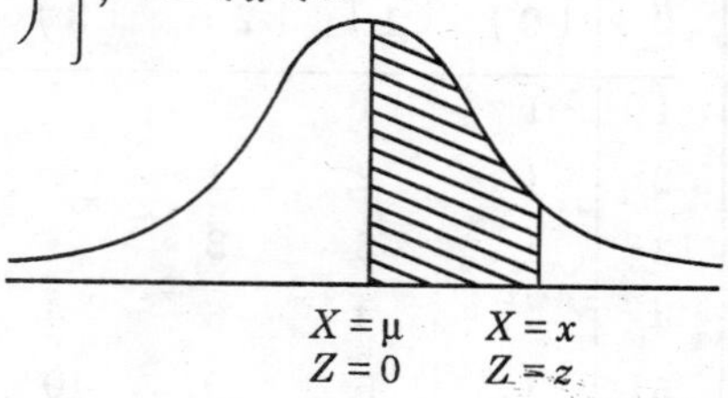

where $Z = \dfrac{X - E(X)}{\sigma_X} = \dfrac{X - \mu}{\sigma} \sim N(0, 1)$

The following table gives the shaded area in the diagram, *viz.* $P(0 < Z < z)$, $z > 0$, for different values of z.

AREAS UNDER STANDARD NORMAL PROBABILITY CURVE

z \ z	0.00	0.01	0.02	0.03	0.04	0.05	0.06	0.07	0.08	0.09
0.0	0.0000	0.0040	0.0080	0.0120	0.0160	0.0199	0.239	0.0279	0.0319	0.0359
0.1	0.0398	0.0438	0.0478	0.0517	0.0557	0.0596	0.0636	0.0675	0.0714	0.0753
0.2	0.0793	0.0832	0.0871	0.0910	0.0948	0.0987	0.1026	0.1064	0.1103	0.1141
0.3	0.1179	0.1217	0.1255	0.1293	0.1331	0.1368	0.1406	0.1443	0.1480	0.1517
0.4	0.1554	0.1591	0.1628	0.1664	0.1700	0.1736	0.1772	0.1808	0.1844	0.1879
0.5	0.1915	0.1950	0.1985	0.2019	0.2054	0.2088	0.2123	0.2157	0.2190	0.2224
0.6	0.2257	0.2291	0.2324	0.2357	0.2389	0.2422	0.2454	0.2486	0.2517	0.2549
0.7	0.2580	0.2611	0.2642	0.2673	0.2704	0.2734	0.2764	0.2794	0.2823	0.2852
0.8	0.2881	0.2910	0.2939	0.2967	0.2995	0.3023	0.3051	0.3078	0.3106	0.3133
0.9	0.3159	0.3186	0.3212	0.3238	0.3264	0.3289	0.3315	0.3340	0.3365	0.3389
1.0	0.3413	0.3438	0.3461	0.3485	0.3508	0.3531	0.3554	0.3577	0.3599	0.3621
1.1	0.3643	0.3665	0.3686	0.3708	0.3729	0.3749	0.3770	0.3790	0.3810	0.3830
1.2	0.3849	0.3869	0.3888	0.3907	0.3925	0.3944	0.3962	0.3980	0.3997	0.4015
1.3	0.4032	0.4049	0.4066	0.4082	0.4099	0.4115	0.4131	0.4147	0.4162	0.4177
1.4	0.4192	0.4207	0.4222	0.4236	0.4251	0.4265	0.4279	0.4292	0.4306	0.4319
1.5	0.4332	0.4345	0.4357	0.4370	0.4382	0.4394	0.4406	0.4418	0.4429	0.4441
1.6	0.4452	0.4463	0.4474	0.4484	0.4495	0.4505	0.4515	0.4525	0.4535	0.4545
1.7	0.4554	0.4564	0.4573	0.4582	0.4591	0.4599	0.4608	0.4616	0.4625	0.4633
1.8	0.4641	0.4649	0.4656	0.4664	0.4671	0.4678	0.4686	0.4693	0.4699	0.4706
1.9	0.4713	0.4719	0.4726	0.4732	0.4738	0.4744	0.4750	0.4756	0.4761	0.4767
2.0	0.4772	0.4778	0.4783	0.4788	0.4793	0.4798	0.4803	0.4808	0.4812	0.4817
2.1	0.4821	0.4826	0.4830	0.4834	0.4838	0.4842	0.4846	0.4850	0.4854	0.4857
2.2	0.4861	0.4864	0.4868	0.4871	0.4875	0.4878	0.4881	0.4884	0.4887	0.4890
2.3	0.4893	0.4896	0.4898	0.4901	0.4904	0.4906	0.4909	0.4911	0.4913	0.4916
2.4	0.4918	0.4920	0.4922	0.4925	0.4927	0.4929	0.4931	0.4932	0.4934	0.4936
2.5	0.4938	0.4940	0.4941	0.4943	0.4945	0.4946	0.4948	0.4949	0.4951	0.4952
2.6	0.4953	0.4955	0.4956	0.4957	0.4959	0.4960	0.4961	0.4962	0.4963	0.4964
2.7	0.4965	0.4966	0.4967	0.4968	0.4969	0.4970	0.4971	0.4972	0.4973	0.4974
2.8	0.4974	0.4975	0.4976	0.4977	0.4977	0.4978	0.4979	0.4979	0.4980	0.4981
2.9	0.4981	0.4982	0.4982	0.4983	0.4984	0.4984	0.4985	0.4985	0.4986	0.4986
3.0	0.4987	0.4987	0.4987	0.4988	0.4988	0.4989	0.4989	0.4989	0.4990	0.4990
3.1	0.4990	0.4991	0.4991	0.4991	0.4992	0.4992	0.4992	0.4992	0.4993	0.4993
3.2	0.4993	0.4993	0.4994	0.4994	0.4994	0.4994	0.4994	0.4995	0.4995	0.4995
3.3	0.4995	0.4995	0.4995	0.4996	0.4996	0.4996	0.4996	0.4996	0.4996	0.4997
3.4	0.4997	0.4997	0.4997	0.4997	0.4997	0.4997	0.4997	0.4997	0.4997	0.4998
3.5	0.4998	0.4998	0.4998	0.4998	0.4998	0.4998	0.4998	0.4998	0.4998	0.4998
3.6	0.4998	0.4998	0.4999	0.4999	0.4999	0.4999	0.4999	0.4999	0.4999	0.4999
3.7	0.4999	0.4999	0.4999	0.4999	0.4999	0.4999	0.4999	0.4999	0.4999	0.4999
3.8	0.4999	0.4999	0.4999	0.4999	0.4999	0.4999	0.4999	0.4999	0.4999	0.4999
3.9	0.5000	0.5000	0.5000	0.5000	0.5000	0.5000	0.5000	0.5000	0.5000	0.5000

TABLE VI A : ORDINATES OF THE NORMAL PROBABILITY CURVE

The following table gives the ordinates of the standard normal probability curve, *i.e.*, it gives the value of

$$\phi(z) = \frac{1}{\sqrt{2\pi}} \exp\left(-\frac{1}{2}z^2\right), \; -\infty < z < \infty$$

for different values of z. where

$$Z = \frac{X - E(X)}{\sigma_X} = \frac{X - \mu}{\sigma} \sim N(0, 1)$$

Obviously, $\phi(-z) = \phi(z)$

↓Z→	.00	.01	.02	.03	.04	.05	.06	.07	.08	.09
0.0	.3989	.3989	.3989	.3988	.3986	.3984	.3982	.3980	.3977	.3973
0.1	.3970	.3965	.3961	.3956	.3951	.3945	.3939	.3932	.3925	.3918
0.2	.3910	.3902	.3894	.3885	.3876	.3867	.3857	.3847	.3836	.3825
0.3	.3814	.3802	.3790	.8778	.3765	.3752	.3739	.3725	.3712	.3538
0.4	.3683	.3668	.3653	.3637	.3621	.3605	.3589	.3572	.3555	.3697
0.5	.3521	.3503	.3485	.3467	.3448	.3429	.3410	.3391	.3372	.3352
0.6	.3332	.3312	.3292	.3271	.3251	.3230	.3209	.3187	.3166	.3144
0.7	.3123	.3101	.3079	.3056	.3034	.3011	.2989	.2966	.2943	.2920
0.8	.2897	.2874	.2850	.2827	.2803	.2780	.2756	.2732	.2709	.2685
0.9	.2661	.2637	.2613	.2589	.2565	.2541	.2516	.2492	.2468	.2444
1.0	.2420	.2396	.2371	.2347	.2323	.2299	.2275	.2251	.2227	.2203
1.1	.2179	.2155	.2131	.2107	.2083	.2059	.2036	.2012	.1989	.1965
1.2	.1942	.1 919	.1895	.1 872	.1849	.1 826	.1804	.1781	.1758	.1736
1.3	.1714	.1691	.1669	.1647	.1626	.1 604	.1582	.1561	.1539	.1518
1.4	.1497	.1476	.1456	.1435	.1415	. 1394	.1374	.1354	.1334	.1315
1.5	.1295	.1276	.1257	.1238	.1219	.1200	.1182	.1163	.1145	.1127
1.6	.1109	.1092	.1074	.1057	.1040	.1023	.1006	.0989	.0973	.0957
1.7	.0940	.0925	.0909	.0893	.0878	.0863	.0848	.0833	.0818	.0804
1.8	.0790	.0775	.0761	.0748	.0734	.0721	.0707	.0694	.0681	.0669
1.9	.0656	.0644	.0632	.0620	.0608	.0596	.0584	.0573	.0562	.0551
2.0	.0540	.0529	.0519	.0508	.0498	.0488	.0478	.0468	.0459	.0449
2.1	.0440	.0431	.0422	.0413	.0404	.0396	.0387	.0379	.0371	.0363
2.2	.0355	.0347	.0339	.0332	.0325	.0317	.0310	.0303	.0297	.0290
2.3	.0283	.0277	.0270	.0264	.0258	.0252	.0246	.0241	.0235	.0229
2.4	.0224	.0219	.0213	.0208	.0203	.0198	.0194	.0189	.0184	.0180
2.5	.0175	.0171	.0167	.0163	.0158	.0154	.0151	.0147	.0143	.0139
2.6	.0136	.0132	.0129	.0126	.0122	.0119	.0116	.0113	0110	.0107
2.7	.0104	.0101	.0099	.0096	.0093	.0091	.0088	.0086	.0084	.0081
2.8	.0079	.0077	.0075	.0073	.0071	.0069	.0067	.0065	.0063	.0061
2.9	.0060	.0058	.0056	.0055	.0053	.0051	.0050	.0048	.0047	.0046
3.0	.0044	.0043	.0042	.0040	.0039	.0038	.0037	.0036	.0035	.0034
3.1	.0033	.0032	.0031	.0030	.0029	.0028	.0027	.0026	.0025	.0025
3.2	.0024	.0023	.0022	.0022	.0021	.0020	.0020	.0019	.0018	.0018
3.3	.0017	.0017	.0016	.0016	.0015	.0015	.0014	.0014	.0013	.0009
3.4	.0012	,0012	.0012	.0011	.0011	.0010	.0010	.0010	.0009	.0013
3.5	.0009	.0008	.0008	.0008	.0008	.0007	.0007	.0007	.0007	.0006
3.6	.0006	.0006	.0006	.0005	.0005	.0005	.0005	.0005	.0005	.0003
3.7	.0004	.0004	.0004	.0004	.0004	.0004	.0003	.0003	.0003	.0004
3.8	.0003	.0003	.0003	.0003	.0003	.0002	.0002	.0002	.0002	.0002
3.9	.0002	.0002	.0002	.0002	.0002	.0002	.0002	.0002	.0001	.0001

TABLE VII : CRITICAL VALUES $\chi^2(\alpha)$ OF CHI-SQUARE DISTRIBUTION (RIGHT TAIL AREAS) FOR GIVEN PROBABILITY α

where

$$P\left[\chi^2 > \chi^2_\nu(\alpha)\right] = \alpha$$

and ν is degrees of freedom ($d.f$).

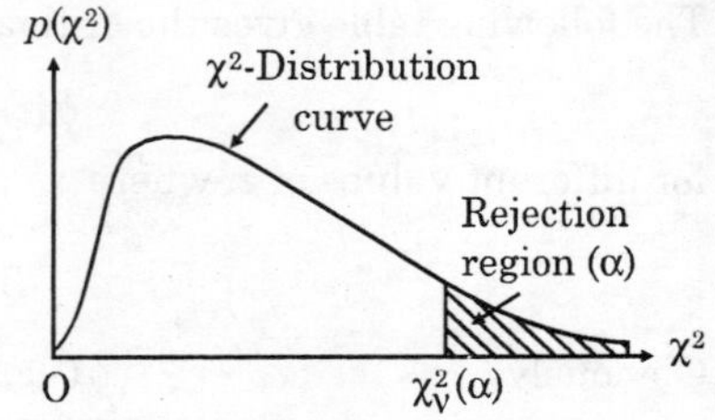

Degrees of	*Probability* (α)							
Freedom (ν)	.99	.975	.95	.90	.10	.05	.025	.01
1	0.000	0.001	0.004	0.016	2.706	3.841	5.024	6.635
2	0.020	0.051	0.103	0.211	4.605	5.991	7.378	9.210
3	0.115	0.216	0.352	0.584	6.251	7.815	9.348	11.345
4	0.297	0.484	0.711	1.064	7.779	9.488	11.143	13.277
5	0.554	0.831	1.145	1.610	9.236	11.070	12.833	15.086
6	0.872	1.237	1.634	2.204	10.645	12.592	14.449	16.812
7	1.239	1.690	2.167	2.833	12.017	14.067	16.013	18.475
8	1.646	2.180	2.733	3.490	13.362	15.507	17.535	20.090
9	2.088	2.700	3.325	4.168	14.684	16.919	19.023	21.666
10	2.558	3.247	3.940	4.865	15.987	18.307	20.483	23 .209
11	3.053	3.816	4.575	5.578	17.275	19.675	21.920	24.725
12	3.571	4.404	5.226	6.304	18.549	21.026	23.337	26.217
13	4.107	5.009	5.892	7.042	19.812	22.362	24.736	27.688
14	4.660	5.629	6.571	7.790	21.064	23.685	26.119	29.141
15	5.229	6.262	7.261	8.547	22.307	24.996	27.488	30.578
16	5.812	6.908	7.962	9.312	23.542	26.296	28.845	32.000
17	6.408	7.564	8.672	10.085	24.769	27.587	30.191	33.409
18	7.015	8.231	8.231	10.865	25.989	28.869	31.526	34.805
19	7.633	8.907	10.117	11.651	27.204	30.144	23.85,2	36.191
20	8.260	9.591	10.851	12.443	28.412	31.410	34.170	37.566
21	8.897	10.283	11.591	13.240	29.615	23.671	35.479	38.932
22	9.542	10.982	12.338	14.042	30.813	33.924	36.781	40.289
23	10.196	11.689	13.091	14.848	32.007	35.172	38.076	41.638
24	10.856	12.401	13.848	15.659	33.196	36.415	39. 364	42.980
25	11.524	13.120	14.611	16.473	34.382	37.652	40.646	44.314
26	12.198	13.844	15.379	17.292	35.563	38.885	41.923	45.642
27	12.879	14.573	16.151	18.114	36.741	40.113	43.194	46.963
28	13.565	15.308	16.928	18.939	37.916	41.337	44.461	48.278
29	14.256	16.047	17.708	19.768	39.087	42.557	45.722	49.588
30	14.953	16.791	18.493	20.599	40.256	43.773	46.979	50.892

For large values of ν, the quantity $\sqrt{2\chi^2} - \sqrt{2\nu - 1}$ may be used as a standard normal variable.

TABLE VIII : CRITICAL VALUES OF STUDENT'S *t*-DISTRIBUTION

	LEVEL OF SIGNIFICANCE FOR ONE-TAILED TEST						
d.f.	.25	.10	.05	0.025	.01	.005	.0005
	LEVEL OF SIGNIFICANCE FOR TWO-TAILED TEST						
ν	.50	.20	.10	.05	.02	.01	.001
1	1.000	3.078	6.314	12.706	31.821	63 .657	636.619
2	.816	1.886	2.920	4.303	6.965	9.925	31.599
3	.765	1.638	2.353	3.182	4.541	5.841	12.924
4	.741	1.533	2.132	2.776	3.747	4.604	8.610
5	.727	1.476	2.015	2.571	3.365	4.032	6.869
6	.718	1.440	1.943	2.447	3.143	3.707	5.959
7	.711	1.415	1.895	2.365	2.998	3.499	5.408
8	.706	1.397	1.860	2.306	2.896	3.355	5.041
9	.703	1.383	1.833	2.262	2.821	3.250	4.781
10	.700	1.372	1.812	2.228	2.764	3.169	4.587
11	.697	1.363	1.796	2.201	2.718	3.106	4.437
12	.695	1.356	1.782	2.179	2.681	3.055	4.318
13	.694	1.350	1.771	2.160	2.650	3.012	4.221
14	.692	1.345	1.761	2.145	2.624	2.977	4.140
15	.691	1.341	1.753	2.131	2.602	2.947	4.073
16	.690	1.337	1.746	2.120	2.583	2.921	4.015
17	.689	1.333	1.740	2.110	2.567	2.898	3.965
18	.688	1.330	1.734	2.101	2.552	2.878	3.922
19	.688	1.328	1.729	2.093	2.539	2.861	3.883
20	.687	1.325	1.725	2.086	2.528	2.845	3.850
21	.686	1.323	1.721	2.080	2.518	2.831	3.819
22	.686	1.321	1.717	2.074	2.508	2.819	3.792
23	.685	1.319	1.714	2.069	2.500	2.807	3.768
24	.865	1.318	1.711	2.064	2.492	2.797	3.745
25	.684	1.316	1.708	2.060	2.485	2.787	3.725
26	.684	1.315	1.706	2.056	2.479	2.779	3.707
27	.684	1.314	1.703	2.052	2.473	2.771	3.690
28	.683	1.313	1.701	2.048	2.467	2.763	3.674
29	.683	1.311	1.699	2.045	2.462	2.756	3.659
30	.683	1.310	1.697	2.042	2.457	2.750	3.646
40	.681	1.303	1.684	2.021	2.423	2.704	3.551
60	.679	1.296	1.671	2.000	2.390	2.660	3.460
120	.677	1.289	1.658	1.980	2.358	2.617	3.373
∞	.674	1.282	1.645	1.960	2.326	2.576	3.291

TABLE IX : CRITICAL VALUES OF THE VARIANCE-RATIO F-DISTRIBUTIONAL (RIGHT TAIL AREAS) — 5 PER CENT POINTS

υ_2 \ υ_1	1	2	3	4	5	6	7	8	9	10	12	15	20	24	30	40
1	161	200	216	225	230	234	237	239	241	242	244	246	248	249	250	251
2	18.5	19.0	19.2	19.2	19.3	19.3	19.4	19.4	19.4	19.4	19.4	19.4	19.4	19.5	19.5	19.5
3	10.1	9.55	9.28	9.12	9.01	8.94	8.89	8.85	8.81	8.79	8.74	8.70	8.66	8.64	8.62	8.59
4	7.71	6.94	6.59	6.39	6.26	6.16	6.09	6.04	6.00	5.96	5.91	5.86	5.80	5.77	5.75	5.72
5	6.61	5.79	5.41	5.19	5.05	4.95	4.88	4.82	4.77	4.74	4.68	4.62	4.56	4.53	4.50	4.46
6	5.99	5.14	4.76	4.53	4.39	4.28	4.21	4.15	4.10	4.06	4.00	3.94	3.87	3.84	3.81	3.77
7	5.59	4.74	4.35	4.12	3.97	3.87	3.79	3.73	3.68	3.64	3.57	3.51	3.44	3.41	3.38	3.34
8	5.32	4.46	4.07	3.84	3.69	3.58	3.50	3.44	3.39	3.35	3.28	3.22	3.15	3.12	3.08	3.04
9	5.12	4.26	3.86	3.63	3.48	3.37	3.29	3.23	3.18	3.14	3.07	3.01	2.94	2.90	2.86	2.83
10	4.96	4.10	3.71	3.48	3.33	3.22	3.14	3.07	3.02	2.98	2.91	2.85	2.77	2.74	2.70	2.66
11	4.84	3.98	3.59	3.36	3.20	3.09	3.01	2.95	2.90	2.85	2.79	2.72	2.65	2.61	2.57	2.53
12	4.75	3.89	3.49	3.26	3.1	3.00	2.91	2.85	2.80	2.75	2.69	2.62	2.54	2.51	2.47	2.43
13	4.67	3.81	3.41	3.18	3.03	2.92	2.83	2.77	2.71	2.67	2.60	2.53	2.46	2.42	2.38	2.34
14	4.60	3.74	3.34	3.11	2.96	2.85	2.76	2.70	2.65	2.60	2.53	2.46	2.39	2.35	2.31	2.27
15	4.54	3.68	3.29	3.06	2.90	2.79	2.71	2.64	2.59	2.54	2.48	2.40	2.33	2.29	2.25	2.20
16	4.49	3.63	3.24	3.01	2.85	2.74	2.66	2.59	2.54	2.49	2.42	2.35	2.28	2.24	2.19	2.15
17	4.45	3.59	3.20	2.96	2.81	2.70	2.61	2.55	2.49	2.45	2.38	2.31	2.23	2.19	2.15	2.10
18	4.41	3.55	3.16	2.93	2.77	2.66	2.58	2.51	2.46	2.41	2.34	2.27	2.19	2.15	2.11	2.06
19	4.38	3.52	3.13	2.90	2.74	2.63	2.54	2.48	2.42	2.38	2.31	2.23	2.16	2.11	2.07	2.03
20	4.35	3.49	3.10	2.87	2.71	2.60	2.51	2.45	2.39	2.35	2.28	2.20	2.12	2.08	2.04	1.99
21	4.32	3.47	3.07	2.84	2.68	2.57	2.49	2.42	2.37	2.32	2.25	2.18	2.10	2.05	2.01	1.96
22	4.30	3.44	3.05	2.82	2.66	2.55	2.46	2.40	2.34	2.30	2.23	2.15	2.07	2.03	1.98	1.94
23	4.28	3.42	3.03	2.80	2.64	2.53	2.44	2.37	2.32	2.27	2.20	2.13	2.05	2.01	1.96	1.91
24	4.26	3.40	3.01	2.78	2.62	2.51	2.42	2.36	2.30	2.25	2.18	2.11	2.03	1.98	1.94	1.89
25	4.24	3.39	2.99	2.76	2.60	2.49	2.40	2.34	2.28	2.24	2.16	2.09	2.01	1.96	1.92	1.87
30	4.17	3.32	2.92	2.69	2.53	2.42	2.33	2.27	2.21	2.16	2.09	2.01	1.93	1.89	1.84	1.79
40	4.08	3.23	2.84	2.61	2.45	2.34	2.25	2.18	2.12	2.08	2.00	1.92	1.84	1.79	1.74	1.69
60	4.00	3.15	2.76	2.53	2.37	2.25	2.17	2.10	2.04	1.99	1.92	1.84	1.75	1.70	1.65	1.59
120	3.92	3.07	2.68	2.45	2.29	2.18	2.09	2.02	1.96	1.91	1.83	1.75	1.66	1.61	1.55	1.50
∞	3.84	3.00	2.60	2.37	2.21	2.10	2.01	1.94	1.88	1.83	1.75	1.67	1.57	1.52	1.46	1.39

υ_1 : Degrees of freedom for the numerator.

υ_2 : Degrees of freedom for the denominator.

TABLE IX-A : CRITICAL VALUES OF THE VARIANCE-RATIO *F*-DISTRIBUTION (RIGHT TAIL AREAS)- 1 PER CENT POINT

v_2 \ v_1	1	2	3	4	5	6	7	8	9	10	12	15	20	24	30	40
1	4052	5000	5403	5625	5764	5859	5928	5981	6022	6056	6106	6157	6209	6235	6261	6287
2	98.5	99.0	99.2	99.2	99.3	99.3	99.4	99.4	99.4	99.4	99.4	99.4	99.4	99.5	99.5	99.5
3	34.1	30.8	29.5	28.7	28.2	27.9	27.7	27.5	27.3	27.2	27.1	26.9	26.7	26.6	26.5	26.4
4	21.2	18.0	16.7	16.0	15.5	15 .2	15.0	14.8	14.7	14.5	14.4	14.2	14.0	13.9	13.8	13.7
5	16.3	13.3	12.1	11.4	11.0	10.7	10.5	10.3	10.2	10.1	9.89	9.72	9.55	9.47	9.38	9.29
6	13.7	10.9	9.78	9.15	8.75	8.47	8.26	8.10	7.98	7.87	7.72	7.56	7.40	7.31	7.23	7.14
7	12.2	9.55	8.45	7.85	7.46	7.19	6.99	6.84	6.72	6.62	6.47	6.31	6.16	6.07	5.99	5.91
8	11.3	8.65	7.59	7.01	6.63	6.37	6.18	6.03	5.91	5.81	5.67	5.52	5.36	5.28	5.20	5.12
9	10.6	8.02	6.99	6.42	6.06	5.80	5.61	5.47	5.35	5.26	5. 11	4.96	4.81	4.73	4.65	4.57
10	10.0	7.56	6.55	5.99	5.64	5.39	5.20	5.06	4.94	4.85	4.71	4.56	4.41	4.33	4.25	4.17
11	9.65	7.21	6.22	5.67	5.32	5.07	4.89	4.74	4.63	4.54	4.40	4.25	4.10	4.02	3.94	3.86
12.	9.33	6.93	5.95	5.41	5.06	4.82	4.64	4.50	4.39	4.30	4.16	4.01	3.86	3.78	3.70	3.62
13	9.07	6.70	5.74	5.21	4.86	4.62	4.44	4.30	4.19	4.10	3.96	3.82	3.66	3.59	3.51	3.43
14	8.86	6.51	5.56	5.04	4.69	4.46	4.28	4.14	4.03	3.94	3.80	3.66	3.51	3.43	3.35	3.27
15	8.68	6.36	5.42	4.89	4.56	4.32	4.14	4.00	3.89	3.80	3.67	3.52	3.37	3.29	3.21	3.13
16	8.53	6.23	5.29	4.77	4.44	4.20	4.03	3.89	3.78	3.69	3.55	3.41	3.26	3.18	3.10	3.02
17	8.40	6.11	5.18	4.67	4.34	4.10	3.93	3.79	3.68	3.59	3.46	3.31	3.16	3.08	3.00	2.92
18	8.29	6.01	5.09	4.58	4.25	4.01	3.84	3.71	3.60	3.51	3.37	3.23	3.08	3.00	2.92	2.84
19	8.18	5.93	5.01	4.50	4.17	3.94	3.77	3.63	3.52	3.43	3.30	3.15	3.00	2.92	2.84	2.76
20	8.10	5.85	4.94	4.43	4.10	3.87	3.70	3.56	3.46	3.37	3.23	3.09	2.94	2.86	2.78	2.69
21	8.02	5.78	4.87	4.37	4.04	3.81	3.64	3.51	3.40	3.31	3.17	3.03	2.88	2.80	2.72	2.64
22	7.95	5.72	4.82	4.31	3.99	3.76	3.59	3.45	3.35	3.26	3.12	2.98	2.83	2.75	2.67	2.58
23	7.88	5.66	4.76	4.26	3.94	3.71	3.54	3.41	3.30	3.21	3.07	2.93	2.78	2.70	2.62	2.54
24	7.82	5.61	4.72	4.22	3.90	3.67	3.50	3.36	3.26	3.17	3.03	2.89	2.74	2.66	2.58	2.49
25	7.77	5.57	4.68	4.18	3.85	3.63	3.46	3.32	3.22	3.13	2.99	2.85	2.70	2.62	2.54	2.45
30	7.56	5.39	4.51	4.02	3:70	3.47	3.30	3.17	3.07	2.98	2.84	2.70	2.55	2.47	2.39	2.30
40	7.31	5.18	4.31	3.83	3.51	3.29	3.12	2.99	2.89	2.80	2.66	2.52	2.37	2.29	2.20	2.11
60	7.08	4.98	4.13	3.65	3.34	3.12	2.95	2.82	2.72	2.63	2.50	2.35	2.20	2.12	2.03	1.94
120	6.85	4.79	3.95	3.48	3.17	2.96	2.79	2.66	2.56	2.47	2.34	2.19	2.03	1.95	1.86	1.76
∞	6.63	4.61	3.78	3.32	3.02	2.80	2.64	2.51	2.41	2.32	2.18	2.04	1.88	1.79	1.70	1.59

v_1 : Degrees of freedom for the numerator.
v_2 : Degrees of freedom for the denominator.

TABLE IX-B : CRITICAL VALUES OF THE VARIANCE RATIO *F*-DISTRIBUTION (RIGHT-TAIL AREAS) 2.5 PERCENT POINTS

v_2 \ v_1	1	2	3	4	5	6	7	8	9	10	12	15	20	24	30	40
1	674.79	799.50	864.16	899.58	921.85	937.11	948.22	956.66	963.28	968.63	976.71	984.87	993.10	997.25	1001.4	1005.6
2	38.51	39.00	39.17	39.25	39.30	39.33	39.33	39.37	39.39	39.40	39.42	39.43	39.45	39.46	39.47	39.47
3	17.44	16.04	15.44	15.01	14.88	14.73	14.62	14.54	14.47	14.42	14.34	14.25	14.17	14.12	14.08	14.04
4	12.22	10.65	9.98	9.60	9.36	9.20	9.07	8.98	8.91	8.84	8.75	8.66	8.56	8.51	8.46	8.41
5	10.01	8.43	7.76	7.39	7.15	6.98	6.85	6.76	6.68	6.62	6.52	6.43	6.33	6.28	6.23	6.18
6	8.81	7.26	6.60	6.23	5.99	5.82	5.70	5.60	5.52	5.46	5.37	5.27	5.17	5.12	5.07	5.01
7	8.07	6.54	5.89	5.52	5.29	5.12	4.99	4.90	4.82	4.76	4.67	4.57	4.47	4.42	4.36	4.31
8	7.57	6.06	5.42	5.05	4.82	4.65	4.53	4.43	4.36	4.30	4.20	4.10	4.00	3.95	3.89	3.84
9	7.21	5.71	5.08	4.72	4.48	4.32	4.20	4.10	4.03	3.96	3.87	3.77	3.67	3.61	3.56	3.51
10	6.94	5.46	4.83	4.47	4.24	4.07	3.95	3.85	3.78	3.72	3.62	3.52	3.42	3.37	3.31	3.26
11	6.72	5.26	4.63	4.28	4.04	3.88	3.76	3.66	3.59	3.53	3.43	3.33	3.23	3.17	3.12	3.06
12	6.55	5.10	4.47	4.12	3.89	3.73	3.61	3.51	3.44	3.37	3.28	3.18	3.07	3.02	2.96	2.91
13	6.41	4.97	4.35	3.99	3.77	3.60	3.48	3.39	3.31	3.25	3.15	3.05	2.95	2.89	2.84	2.78
14	6.30	4.86	4.24	3.89	3.66	3.50	3.38	3.29	3.21	3.15	3.05	2.95	2.84	2.79	2.73	2.67
15	6.20	4.77	4.15	3.80	3.58	3.41	3.29	3.20	3.12	3.06	2.96	2.86	2.76	2.70	2.64	~.59
16	6.12	4.69	4.08	3.73	3.50	3.34	3.22	3.12	3.05	2.99	2.89	2.79	2.68	2.63	2.57	2.S1
17	6.04	4.62	4.01	3.66	3.44	3.28	3.16	3.06	2.98	2.92	2.82	2.72	2.62	2.56	2.50	2.44
18	5.98	4.56	3.95	3.61	3.38	3.22	3.10	3.01	2.93	2.87	2.77	2.67	2.56	2.50	2.44	2.38
19	5.92	4.51	3.90	3.56	3.33	3.17	3.05	2.96	2.88	2.82	2.72	2.62	2.51	2.45	2.39	2.33
20	5.87	4.46	3.86	3.51	3.29	3.13	3.01	2.91	2.84	2.77	2.68	2.57	2.46	2.41	2.35	2.29
21	5.83	4.42	3.82	3.48	3.25	3.09	2.97	2.87	2.80	2.73	2.64	2.53	2.42	2.37	2.31	2.25
22	5.79	4.38	3.78	3.44	3.22	3.05	2.94	2.84	2.76	2.70	2.60	2.50	2.39	2.33	2.27	2.21
23	5.75	4.35	3.75	3.41	3.18	3.02	2.90	2.81	2.73	2.67	2.57	2.47	2.36	2.30	2.24	2.18
24	5.72	4.32	3.72	3.38	3.15	2.99	2.87	2.78	2.70	2.64	2.54	2.44	2.33	2.27	2.21	2.15
25	5.69	4.29	3.69	3.35	3.13	2.97	2.85	2.75	2.68	2.61	2.51	2.41	2.30	2.24	2.18	2.12
26	5.66	4.27	3.67	3.33	3.10	2.94	2.82	2.73	2.65	2.59	2.49	2.39	2.28	2.22	2.16	2.09
27	5.63	4.24	3.65	3.31	3.08	2.92	2.80	2.71	2.63	2.57	2.47	2.36	2.25	2.19	2.13	2.07
28	5.61	4.22	3.63	3.29	3.06	2.90	2.78	2.69	2.61	2.55	2.45	2.34	2.23	2.17	2.11	2.05
29	5.59	4.20	3.61	3.27	3.04	2.88	2.76	2.67	2.59	2.53	2.43	2.32	2.21	2.15	2.09	2.03
30	5.57	4.18	3.59	3.25	3.03	2.87	2.75	2.65	2.57	2.51	2.41	2.31	2.20	2.14	2.07	2.01
40	5.42	4.05	3.46	3.13	2.90	2.74	2.62	2.53	2.45	2.39	2.29	2.18	2.07	2.01	1.94	1.88
60	5.29	3.93	3.34	3.01	2.79	2.63	2.51	2.41	2.33	2.27	2.17	2.06	1.94	1.88	1.82	1.74
120	5.15	3.80	3.23	2.89	2.67	2.52	2.39	2.30	2.22	2.16	2.05	1.95	1.82	1.76	1.69	1.61
∞	5.00	3.67	3.12	2.79	2.57	2.41	2.29	2.19	2.11	2.05	1.94	1.85	1.71	1.64	1.57	1.48

v_1 : Degrees of freedom for the numerator.
v_2 : Degrees of freedom for the denominator.

TABLE X : CONVERSION OF PEARSON'S *r* INTO CORRESPONDING FISHER'S *Z* STATISTIC

$$Z = \frac{1}{2}\log_e\left(\frac{1+r}{1-r}\right) = 1.1513\log_{10}\left(\frac{1+r}{1-r}\right)$$

r	*Z*	*r*	*Z*	*r*	*Z*	*r*	*Z*	*r*	*Z*	*r*	*Z*
.25	.26	.40	.42	.55	.62	.70	.87	.85	1.26	.950	1.83
.26	.27	.41	.44	.56	.63	.71	.89	.86	1.29	.955	1.89
.27	.28	.42	.45	.57	.65	.72	.91	.87	1.33	.960	1.95
.28	.29	.43	.46	.58	.66	.73	.93	.88	1.38	.965	2.01
.29	.30	.44	.47	.59	.68	.74	.95	.89	1.42	.970	2.09
.30	.31	.45	.48	.60	.69	.75	.97	.90	1.47	.975	2.18
.31	.32	.46	.50	.61	.71	.76	1.00	.905	1.50	.980	2.30
.32	.33	.47	.51	.62	.73	.77	1.02	.910	1.53	.985	2.44
.33	.34	.48	.52	.63	.74	.78	1.05	.915	1.56	.990	2.65
.34	.35	.49	.54	.64	.76	.79	1.07	.920	1.59	.995	2.99
.35	.37	.50	.55	.65	.78	.80	1.10	.925	1.62		
.36	.38	.51	.56	.66	.79	.81	1.13	.930	1.66		
.37	.39	.52	.58	.67	.81	.82	1.16	.935	1.70		
.38	.40	.53	.59	.68	.83	.83	1.19	.940	1.74		
.39	.41	.54	.60	.69	.85	.84	1.22	.945	1.78		

* *r*'s under 0.25 may be taken as equivalent to *Z*'s

Index